FAO yearbook annuaire anuario

FAO Fisheries Series
No. 63
FAO Statistics Series
No. 173

Collection FAO:
Pêches N° 63
Collection FAO:
Statistiques N° 173

Colección FAO:
Pesca N° 63
Colección FAO:
Estadística N° 173

Fishery statistics

Capture production

Statistiques des pêches

Captures

Estadísticas de pesca

Capturas

Vol. 92/1

2001

FOOD AND AGRICULTURE ORGANIZATION OF THE UNITED NATIONS
Rome, 2003

ORGANISATION DES NATIONS UNIES POUR L'ALIMENTATION ET L'AGRICULTURE
Rome, 2003

ORGANIZACIÓN DE LAS NACIONES UNIDAS PARA LA AGRICULTURA Y LA ALIMENTACIÓN
Roma, 2003

Prepared by the Fishery Information, Data and Statistics Unit of the Fisheries Department, FAO, on the basis of information available as of 31 December 2002.

Information from this publication may be quoted if reference is made to the source.

FAO provides time series of fishery statistical data in computer readable form to users of the *FAO Yearbook of Fishery Statistics: Capture Production.* The time series are given as annual data for countries or area/species groups/major fishing areas for a period of years starting with the year 1950 up to the latest year published in the Yearbook.

Enquiries about the technical content of this publication should be addressed to:

The Senior Fishery Statistician
Fishery Information, Data and Statistics Unit
Fisheries Department
FAO
Viale delle Terme di Caracalla
00100 Rome,Italy

E-mail: FIDI-Inquiries@fao.org

Requests for copies of this publication should be sent to the sales agents listed at the back, or to:

Sales and Marketing Group
FAO
Viale delle Terme di Caracalla
00100 Rome, Italy

Information on world fisheries and aquaculture can be obtained from the Internet site: **www.fao.org/fi**

Préparé par l'Unité de l'information, des données et des statistiques sur les pêches, Département des pêches, FAO, sur la base des renseignements disponibles au 31 décembre 2002.

Il est possible de reproduire les articles et renseignements contenus dans cette publication sous réserve d'en indiquer la source.

La FAO met en vente des séries de données statistiques publiées dans l'*Annuaire statistique des pêches: Captures.* Les séries chronologiques donnent des chiffres annuels par pays ou zones/groupes d'espèces/zones de pêche principales, qui portent sur une période allant de l'année 1950 à la dernière année publiée dans l'Annuaire.

Adresser toutes demandes d'ordre technique concernant cette publication au:

Statisticien principal des pêches
Unité de l'information, des données et des statistiques sur les pêches
Département des pêches
FAO
Viale delle Terme di Caracalla
00100 Rome, Italie

Mél.:FIDI-Inquiries@fao.org

Adresser les commandes de la publication soit aux dépositaires (liste en dernière page), soit au:

Groupe des ventes et de la commercialisation
FAO
Viale delle Terme di Caracalla
00100 Rome, Italie

Des renseignements sur la pêche et l'aquaculture mondiales peuvent être obtenus en consultant le site Internet : **www.fao.org/fi**

Preparado por la Dependencia de Información, Datos y Estadísticas de Pesca, Departamento de Pesca, FAO, teniendo en cuenta los datos disponibles hasta el 31 de diciembre de 2002.

Las informaciones contenidas en esta publicación se pueden reproducir si se cita su origen.

La FAO ofrece series de datos estadísticos para su lectura en computadoras a las personas que utilizan el *Anuario estadístico de pesca: Capturas.* Las series cronológicas se facilitan como datos anuales por países o áreas/grupos de especies/principales áreas de pesca para el período que inicia con el año 1950 y termina con el último año publicado en el Anuario.

La correspondencia concerniente al contenido técnico de esta publicación debe dirigirse al:

Estadístico Superior de Pesca
Dependencia de Información, Datos y Estadísticas de Pesca
Departamento de Pesca
FAO
Viale delle Terme di Caracalla
00100 Roma, Italia

Correo electrónico: FIDI-Inquiries@fao.org

Los pedidos de esta publicación deben ser dirigidos a los agentes de ventas que figuran en la última página, o al:

Grupo de Ventas y Comercialización
FAO
Viale delle Terme di Caracalla
00100 Roma, Italia

Se puede obtener información sobre la pesca y la acuicultura mundiales en la página de Internet: **www.fao.org/fi**

FAO Fishery Information, Data and Statistics Unit/Unité de l'information, des données et des statistiques sur les pêches/Dependencia de Información, Datos y Estadísticas de Pesca.
Capture production 2001.
Captures 2001.
Capturas 2001.
FAO yearbook. Fishery statistics: Capture production/Annuaire FAO. Statistiques des pêches: Captures/Anuario/FAO. Estadísticas de pesca: Capturas.
Vol. 92/1. Rome/Roma, FAO. 2003. 627p.

Designations of countries or areas	**Désignations des pays ou zones**	**Denominaciones de los países o zonas**
The designations employed and the presentation of material in this publication do not imply the expression of any opinion whatsoever on the part of the Food and Agriculture Organization of the United Nations concerning the legal or development status of any country, territory, city or area or of its authorities, or concerning the delimitation of its frontiers or boundaries.	Les appellations employées dans cette publication et la présentation des données qui y figurent n'impliquent de la part de l'Organisation des Nations Unies pour l'alimentation et l'agriculture aucune prise de position quant au statut juridique ou au stade de développement des pays, territoires, villes ou zones, ou de leurs autorités, ni quant au tracé de leurs frontières ou limites.	Las denominaciones empleadas en esta publicación y la forma en que aparecen presentados los datos que contiene no implican, de parte de la Organización de las Naciones Unidas para la Agricultura y la Alimentación, juicio alguno sobre la condición jurídica o el nivel de desarrollo de países, territorios, ciudades o zonas, o de sus autoridades, ni respecto de la delimitación de sus fronteras o límites.

ISBN 92-5-004953-6

Table of Contents / Table des matières / Tabla de materias

	Standard symbols	Signes conventionnels	Símbolos convencionales
...	Data not available; unobtainable; data not separately available but included in another category	Données non disponibles; données que l'on n'a pas pu obtenir; données non disponibles séparément, mais comprises dans une autre catégorie	No hay datos; no se han podido obtener datos; datos no disponibles por separado pero incluidos en otra partida estadística
—	None; magnitude known to be nil or zero	Néant; quantité que l'on sait égale à zéro	Ninguna; cantidad que se sabe es nula o cero
0	More than zero but less than half the unit used	Quantité supérieure à zéro, mais inférieure à la moitié de l'unité utilisée	Más de cero pero inferior a la mitad de la unidad empleada
F	FAO estimate from available sources of information or calculation based on specific assumptions	Estimations de la FAO d'après les sources d'informations disponibles ou calculée sur la base de suppositions spécifiques	Estimación de la FAO partiendo de fuentes de información disponibles o calculada sobre la base de suposiciones específicas
mt	metric tons = tonnes	tonnes métriques = tonnes	toneladas métricas = toneladas
kg	kilograms	kilogrammes	kilogramos
no	number	nombre	número
nei	not elsewhere included	**nca** non compris ailleurs	**nep** no especificado en otra partida
...A	FAO English name of the species item not ascertainable	Le nom anglais utilisé par la FAO pour la catégorie d'espèces n'est pas vérifiable	Se desconoce el nombre utilizado por la FAO en inglés para la partida de especies
...B	FAO French name of the species item not ascertainable	Le nom français utilisé par la FAO pour la catégorie d'espèces n'est pas vérifiable	Se desconoce el nombre utilizado por la FAO en francés para la partida de especies
...C	FAO Spanish name of the species item not ascertainable	Le nom espagnol utilisé par la FAO pour la catégorie d'espèces n'est pas vérifiable	Se desconoce el nombre utilizado por la FAO en español para la partida de especies
S	Summation of catches	Somme des captures	Suma de las capturas

Introduction

1. This Volume of the **Yearbook of Fishery Statistics** presents the annual statistics, for a varying series of recent years ending in 2001, on a worldwide basis, on nominal catches of fish, crustaceans, molluscs and other aquatic animals, residues and plants, taken for all purposes (commercial, industrial, recreational and subsistence) by all types and classes of fishing units (fishermen, vessels, gear, etc.) operating both in inland, fresh and brackish water areas, and in inshore, offshore and highseas fishing areas. Beginning with Volume 82 statistics for mariculture, aquaculture and other kinds of fish farming, are excluded from all national, regional and global totals. **Beginning with Volume 90/1, the names and species composition of some groups of the FAO International Standard Statistical Classification of Aquatic Animals and Plants (ISSCAAP) have been revised. See paragraph 3 of the NOTES ON SPECIES ITEMS.**

2. Despite the importance of recreational fishing regarding some species and for certain countries, figures include recreational catches only where available.

3. The annual period used is the calendar year (1 January-31 December), with the exception of data on catches in the waters around Antarctica for which the split-year (1 July-30 June) is used. Split-year data are shown under the calendar year in which the split-year ends.

4. Catches are expressed in metric tons, except those for whales, seals and crocodiles which are given in numbers and corals, pearls and sponges which are given in kilograms. See paragraph 11 of the NOTES ON COUNTRIES OR AREAS.

5. Beginning with Volume 48 data on aquatic mammals (tables B-61, B-62 and B-63 expressed in numbers and B-64) and data on aquatic plants (A-6, B-91, B-92, B-93 and B-94) are excluded from all national, regional and global totals. For practice prior to Volume 48, see notes in respective Yearbook. Beginning with Volume 60, data on corals, expressed in kilograms given in table B-82 are also excluded from all national, regional and global totals. Beginning with Volume 62 data on pearls, expressed in kilograms in table B-81 and sponges, expressed in kilograms in table B-83 are likewise excluded. Beginning with Volume 66 data on crocodiles expressed in numbers in table B-73 are excluded from all national, regional and global totals.

6. FAO has completed the separation of the aquaculture time series from the capture production time series dating back to 1950. This separation was based on national reporting, when available, and other sources of historical information. The entire global time series for aquaculture production can be downloaded using the FishStat Plus software available at: http://www.fao.org/fi/statist/FISOFT/FISHPLUS.asp

7. Where necessary the data for 1950-2000 published in the preceding volumes of the Yearbook of Fishery Statistics have been revised. Where figures in this volume differ from those previously published, the amended data represent the most recent version. Some statistics provided to FAO by national offices, in particular those for 2001, are provisional and may be amended in future volumes, and in other FAO publications.

Introduction

Dans ce volume de l'**Annuaire statistique des pêches** figurent, pour diverses périodes récentes prenant fin en 2001, les statistiques établies sur une base mondiale des captures nominales de poissons, crustacés, mollusques et autres animaux aquatiques, résidus et plantes aquatiques, effectuées à toutes fins (commerciale, industrielles, récréatives et de subsistance) par tous les types et catégories d'unités de pêche (pêcheurs, bateaux, engins, etc.) opérant tant dans les eaux continentales, douces et saumâtres, que dans les zones de pêche du littoral, du large et de la haute mer. A partir du Volume 82, les statistiques ayant trait à la mariculture, à l'aquaculture et à d'autres types de pisciculture, sont exclues de toutes les données au niveau national, régional et global. **A partir de Volume 90/1, les dénominations et la composition par espèces de quelques groupes de la Classification statistique internationale type des animaux et des plantes aquatiques (CSITAPA) de la FAO, ont subi des révisions. Voir le paragraphe 3 des NOTES SUR LES CATEGORIES D'ESPECES.**

Malgré la grande incidence de la pêche récréative sur certaines espèces et pour certains pays, les chiffres indiqués comprennent les captures de la pêche récréative, seulement lorsque de telles données sont disponibles.

La période annuelle utilisée est l'année civile (1er janvier - 31 décembre), sauf pour les données relatives aux captures effectuées dans les eaux environnant l'Antarctique pour lesquelles on utilise l'année fractionnée (1er juillet – 30 juin). Les captures relatives aux années fractionnées figurent sous l'année civile durant laquelle se termine l'année fractionnée.

Les captures débarquées sont exprimées en tonnes métriques, à l'exception des données relatives aux baleines, phoques et crocodiles qui sont indiquées numériquement et les données sur les coraux, les perles et les éponges qui sont indiquées en kilogrammes. Voir le paragraphe 11 des NOTES SUR LES PAYS OU ZONES.

A partir du volume 48, les données sur les mammifères aquatiques (tableaux B-61, B-62 et B-63 exprimées en nombres et B-64) et les données sur les plantes aquatiques (A-6, B-91, B-92, B-93 et B-94) sont exclues de tous les totaux au niveau national, régional et global. Pour la méthode suivie avant le volume 48, se reporter aux notes de l'Annuaire en question. A partir du volume 60, les données sur les coraux exprimées en kilogrammes indiquées au tableau B-82 sont également exclues de tous les totaux au niveau national, régional et global. A partir du volume 62, les données sur les perles exprimées en kilogrammes indiquées au tableau B-81, et sur les éponges exprimées en kilogrammes indiquées au tableau B-83 sont, de même, exclues. A partir du volume 66, les données sur les crocodiles exprimées en nombre au tableau B-73 sont exclues de tous les totaux au niveau national, régional et global.

La FAO a complété la séparation des séries de production en aquaculture et capture à partir de 1950. Cette séparation a été basée sur des reportages nationaux, si disponibles, et d' autres sources d' information historique. La série mondiale complète pour la production d'aquaculture peut être téléchargée en utilisant le logiciel FishStat Plus disponible à: http://www.fao.org/fi/statist/FISOFT/FISHPLUS.asp

Le cas échéant, les données relatives aux années 1950-2000 publiées dans les volumes précédents de l'Annuaire statistique des pêches ont été révisées. Lorsque les chiffres indiqués dans le présent volume diffèrent de ceux déjà publiés, les données modifiées représentent la version la plus récente. Certaines statistiques communiquées à la FAO par les services nationaux sont provisoires, notamment pour 2001, et pourront être modifiées dans les futurs volumes ainsi que dans d'autres publications de la FAO.

Introducción

En el presente volumen del **Anuario Estadístico de Pesca** se presentan las estadísticas mundiales, para diversas series de los últimos años que terminan en 2001, de las capturas nominales de peces, crustáceos, moluscos y demás animales, residuos y plantas acuáticos, hechas con cualquier fin (comercial, industrial, recreativo y de subsistencia), por unidades de pesca de todos los tipos y categorías (pescadores, barcos, artes, etc.) en aguas continentales, dulces y salobres y en áreas de pesca de bajura, media altura o altura. A partir del volumen 82 las estadísticas correspondientes a maricultura, acuicultura y otros tipos de psicultura, están excluidas de las cantidades a nivel nacional, regional y global. **A partir de volumen 90/1, las denominaciones y la composición por especies de algunos de los grupos de especies de la Clasificación Estadística Internacional Uniforme de los Animales y Plantas Acuáticos (CEIUAPA) de la FAO fueron revisados. Véase el párrafo 3 en las NOTAS SOBRE LAS PARTIDAS DE ESPECIES.**

A pesar de la importancia que la pesca recreativa tiene para algunas especies de pescado y para ciertos países, las cifras incluyen las capturas de la pesca recreativa solamente cuando se hallan disponibles.

El período anual utilizado es el año civil (1 de enero - 31 de diciembre), excepto en el caso de las capturas hechas alrededor de la Antártida, para las que se utiliza el año emergente (1 de julio - 30 de junio). Los datos correspondientes a los años emergentes se incluyen en el año civil en que termina el año emergente.

Las capturas se expresan en toneladas métricas, excepto en el caso de ballenas, focas y cocodrilos que se dan en número y en el caso de corales, perlas y esponjas que se dan en kilogramos. Véase el párrafo 11 en las NOTAS SOBRE LOS PAISES O AREAS.

A partir del volumen 48 los datos relativos a los mamíferos acuáticos (cuadros B-61, B-62 y B-63 expresados en números y B-64) y los datos relativos a las plantas acuáticas (A-6, B-91, B-92, B-93 y B-94) están excluidos de las cantidades totales a nivel nacional, regional y global. Con respecto a la práctica seguida antes del volumen 48, véanse las notas en el Anuario respectivo. A partir del volumen 60 los datos relativos a los corales expresados en kilogramos que aparecen en el cuadro B-82 están también todos excluidos de las cantidades totales a nivel nacional, regional y global. A partir del volumen 62 los datos relativos a las perlas expresados en kilogramos que aparecen en el cuadro B-81 y a las esponjas expresados en kilogramos que aparecen en el cuadro B-83 están igualmente excluidos. A partir del volumen 66 los datos relativos a los cocodrilos, expresados en números, en el cuadro B-73 están excluidos de las cantidades totales a nivel nacional, regional y global.

La FAO ha completado la separación de las series cronológicas de acuicultura y de producción de captura a partir de 1950. Esta separación se basó en los informes nacionales, cuándo eran disponibles, y en otras fuentes de información histórica. La serie cronológica mundial completa para la producción de acuicultura puede ser descargada utilizando el software de estadística FishStat Plus disponible en: http://www.fao.org/fi/statist/FISOFT/FISHPLUS.asp

Siempre que ha sido necesario se han revisado los datos correspondientes a 1950-2000 publicados en volúmenes anteriores del Anuario Estadístico de Pesca. Cuando las cifras que aparecen en este volumen difieren de las publicadas anteriormente, los nuevos datos representan la última versión disponible. Algunas estadísticas facilitadas a la FAO por las oficinas nacionales, en particular las correspondientes a 2001, son provisionales y podrán modificarse en volúmenes futuros y en otras publicaciones de la FAO.

Introduction

8. National focal points for fishery statistics, in particular those of countries fishing in more than one major fishing area, report their annual catches to various fishery commissions, as well as to FAO. To eliminate duplication in requests to these national offices, FAO cooperates with regional fishery bodies, particularly through the Coordinating Working Party on Fishery Statistics (CWP), to standardize reporting forms, procedures, definitions, classifications and other related documentation. This system reduces discrepancies between the figures appearing in the Yearbook of Fishery Statistics and those published in the bulletins issued by the commissions. Some discrepancies may still exist, but effort is constantly being made to eliminate them.

9. For the time being, the flag of the vessel is used to assign its nationality unless the wording of chartering and joint operation contracts indicates otherwise.

10. To facilitate use of the Yearbook, a series of notes and lists are included.

11. As usual, government officers and staff of international organizations have made possible the timely publication of this Yearbook by their prompt attention to our requests, and the care they devoted to checking material submitted to them. FAO expresses its thanks to them and welcomes the support of national and international organizations, and of interested individuals, in improving the scope and accuracy of the Yearbook.

12. Great care is taken by FAO in ensuring as far as possible the quality of the data presented in this Yearbook, supplementing data reported by countries with information from other sources, where available, including regional fishery bodies, field projects and independent surveys, specialist literature and fishery-independent sources. However, the accuracy and reliability of world aggregations of fishery statistics ultimately depends upon the quality of national data sources, collection methods, periodicity of updating and reporting. Fishery data quality is known to be very uneven among countries. Although improvements to data quality are made by FAO on a continuing basis, it is clear that many more can be made. Any input from data users in this regard will be most welcome.

Introduction

Les centres nationaux des statistiques des pêches, notamment ceux des pays exploitant plus d'une principale zone de pêche, déclarent leurs captures annuelles aux diverses commissions des pêches, de même qu'à la FAO. Pour éviter que les demandes adressées à ces services nationaux ne fassent double emploi, la FAO coopère avec les organes régionaux des pêches, en particulier par l'entremise du Groupe de travail chargé de coordonner les statistiques des pêches (CWP), pour normaliser les formules de déclaration, les procédures, les définitions, les classifications et toute la documentation connexe. Ce système permet de réduire les divergences entre les chiffres figurant dans l'Annuaire statistique des pêches et ceux des bulletins publiés par les commissions. Certaines de ces divergences demeurent, mais on s'efforce toujours de les éliminer.

Pour le moment, le pavillon du navire est utilisé pour déterminer sa nationalité, à moins que le libellé des contrats d'affrètement et d'opérations conjointes ne l'indique autrement.

Pour faciliter l'emploi de l'Annuaire, on a introduit une série de notes et de listes.

Comme de coutume, c'est grâce à la rapidité et au soin avec lesquels les fonctionnaires des services publics et le personnel des organisations internationales ont répondu à nos demandes et vérifié les matériaux qui leur étaient soumis, que cet Annuaire a pu être publié en temps voulu. La FAO tient à les remercier et se félicite de l'aide que lui apportent les organisations tant nationales qu'internationales et les personnes intéressées, afin d'améliorer l'étendue et la précision de l'Annuaire.

La FAO essaie autant que possible, avec le plus grand soin, d'assurer la qualité des données présentées dans cet Annuaire, en complétant les données communiquées par les pays par des informations provenant d'autres sources telles que: organes régionaux des pêches, projets de terrain, enquêtes indépendantes, documents rédigés par des spécialistes et sources extérieures au secteur des pêches. Néanmoins, la précision et la fiabilité des agrégats mondiaux des statistiques des pêches dépendent en dernière analyse de la qualité des sources nationales de données, des méthodes de collecte, de la périodicité de leur mise à jour et de leur présentation. La qualité des données halieutiques peut être fort inégale d'un pays à l'autre. Même si la FAO s'engage à améliorer constamment la qualité des données, il ne fait pas de doute que nombre d'autres améliorations peuvent être apportées. Toute contribution des utilisateurs des données à cet égard sera la bienvenue.

Introducción

Los centros nacionales de estadísticas pesqueras, en particular los de aquellos países que pescan en más de una de las áreas principales de pesca, comunican sus datos sobre capturas anuales a las diversas comisiones de pesca y a la FAO. Para eliminar duplicaciones en la solicitud de datos a esas oficinas nacionales, la FAO colabora con los órganos regionales de pesca, particularmente mediante el Grupo Coordinador de Trabajo sobre Estadísticas de Pesca (CWP), para uniformar los formularios de comunicación de datos, y los procedimientos, definiciones, clasificaciones, etc. Esto permite además reducir las discrepancias entre las cifras que aparacen en el Anuario Estadístico de Pesca y las publicadas en los boletines preparados por las diversas comisiones. Pueden subsistir aún algunas discrepancias, pero siempre se intenta eliminarlas.

Por el momento, se entiende que el pabellón del barco determina su nacionalidad, a menos que en los contratos de fletes o de operación conjunta se indique otra cosa.

Para facilitar la consulta del Anuario se incluyen una serie de notas y listas.

Igual que en ocasiones anteriores, la pronta publicación de este Anuario ha sido posible gracias a la colaboración de funcionarios estatales y de organizaciones internacionales que han respondido con prontitud a nuestras peticiones y han controlado con atención el material que se les ha presentado. La FAO desea manifestar a todos ellos su reconocimiento y agradece ya desde ahora la ayuda que organizaciones nacionales e internacionales o personas privadas le presten para mejorar el alcance y exactitud de este Anuario.

La FAO presta gran atención a asegurar en la medida de lo posible la buena calidad de los datos que se presentan en este Anuario, complementando los datos comunicados por los países con información de otras procedencias, cuando se dispone, tales como los órganos regionales de pesca, los projectos de campo y encuestas independientes, bibliografía especializada y fuentes que no dependen de la pesca. Sin embargo, la exactitud y fiabilidad del acopio mundial de estadísticas de pesca se basan fundamentalmente en la calidad de las fuentes de datos nacionales, en los métodos de recopilación de los mismos, en la periodicidad de la actualización y la comunicación. Como se sabe, la calidad de los datos pesqueros nacionales puede variar mucho de un país a otro. Aunque la FAO está mejorando constantemente la calidad de los datos, es evidente que se puede mejorar todavía mucho. Cualquier aportación de los usuarios de los datos a este respecto será muy bien recibida.

Notes and lists

Notes et listes

Notas y listas

General Notes

1. This volume of the **Yearbook of Fishery Statistics** presents, for the most recent series of calendar years and split-years, annual statistics on NOMINAL CATCHES of freshwater, brackishwater and marine species of fish, crustaceans, molluscs and other aquatic animals and plants, killed, caught, trapped or collected for all commercial, industrial, recreational and subsistence purposes.

2. In view of the importance of recreational fishing regarding some stocks and for certain countries, and the difficulty of distinguishing in many cases between recreational and subsistence fishing (and in accordance with the recommendation of the 16th Session of the Coordinating Working Party on Fishery Statistics - CWP, Madrid, Spain, 20-25 March 1995), data should cover recreational fisheries.

3. The NOMINAL CATCHES concept refers to the landings converted to a *live weight* basis; the closely related concept LANDINGS refers to the quantities on a *landed weight* basis. In many fisheries the landed quantities (LANDINGS) are identical to the quantities caught (NOMINAL CATCHES).

4. There are many instances where the catches on board fishing vessels or factory ships are gutted, eviscerated, filleted, salted, dried, etc., or reduced to meals, oil, etc. The data on the LANDINGS of such species and products require conversion by accurate yield rates (conversion factors) to establish the live weight equivalents (nominal catches) at the time of their capture. See diagram on the following pages.

5. Many national statistical publications do not use the terms "landings" and "catches" with their precise meanings as described above. In such publications the term "catches" is used sometimes to refer to quantities on a landed weight basis, i.e. "landings" which might consist of gutted, eviscerated and filleted fish, as well as of meals and oils. However, only where the "primary production" (this phrase "primary production" is used in its economic and not in its biological sense) is landed whole is it correct to describe such landed quantities as "catches".

6. In some national statistics the following terms are in common use to refer to NOMINAL CATCHES, i.e. landings on a live weight basis:

   a) landings on a round, fresh basis;

   b) landings on a round, whole basis;

   c) landings on an ex-water weight basis.

7. For the "primary production" data on seaweeds, pearls, shells, corals, sponges, etc., it is preferable to use the term PRODUCTION as being more appropriate than NOMINAL CATCH. The term CATCH is also more suitable than NOMINAL CATCH in the case of whales and seals, where the "primary production" data for these species are not expressed in weight units but in numbers.

8. Data concerning the nominal catch of fish included within group 36 (tunas, bonitos and billfishes) are generally reviewed in collaboration with the regional agency concerned with tuna statistics (i.e. ICCAT, IOTC, SPC and IATTC). Due to differences in the date by which these agencies require data to be submitted, figures for the most recent year are often subject to significant revision.

Notes générales

Dans ce volume de l'**Annuaire statistique des pêches** figurent, pour les plus récentes séries d'années civiles et d'années fractionnées, les statistiques annuelles des CAPTURES NOMINALES de poissons, crustacés, mollusques et autres animaux et plantes aquatiques d'eau douce, d'eau saumâtre et d'eau marine, tués, capturés, piégés ou ramassés à des fins commerciales, industrielles, récréative et de subsistance.

Etant donné que la pêche sportive a une grande incidence sur certains stocks et pour certains pays, qu'il est souvent difficile d'établir une distinction entre celle-ci et la pêche de subsistance (et en conformité avec la recommandation de la seizième session du Groupe de travail chargé de coordonner les statistiques de pêches - CWP, Madrid, Espagne, 20-25 mars 1995), les données doivent concerner la pêche sportive.

L'expression CAPTURES NOMINALES désigne l'équivalent en *poids vif* des quantités débarquées; l'expression étroitement apparentée QUANTITES DEBARQUEES désigne le *poids mis à terre*. Dans de nombreuses pêcheries, les mises à terre (QUANTITES DEBARQUEES) sont identiques aux quantités capturées (CAPTURES NOMINALES).

Dans de nombreux cas, les captures sont vidées, éviscérées, filetées, salées, séchées, etc., ou réduites, en farine, en huile, etc., à bord des bateaux de pêche ou de navires-usines. Les données sur les QUANTITES DEBARQUEES de la sorte, à savoir après traitement, doivent être converties à l'aide de taux de rendement précis (coefficients de conversion) afin de déterminer les équivalents en poids vif (captures nominales) au moment de la prise. Voir le diagramme aux pages suivantes.

De nombreuses publications statistiques nationales n'emploient pas les expressions «quantités débarquées» et «captures» avec le sens précis indiqué ci-dessus. Le terme «captures» y sert parfois à désigner les quantités sur la base du poids mis à terre, c.-à-d. les «quantités débarquées» qui peuvent être des poissons vidés, éviscérés et filetés, ainsi que des farines et des huiles. Toutefois, c'est seulement lorsque la «production primaire» (au sens économique et non au sens biologique) est débarquée à l'état entier qu'il est correct de parler de «captures» pour ces mises à terre.

Dans certaines statistiques nationales, on utilise couramment les expressions suivantes pour désigner les CAPTURES NOMINALES, c.-à.-d. l'équivalent en poids vif des quantités débarquées:

a) quantités débarquées sur la base du poisson entier, frais;

b) quantités débarquées sur la base du poids du poisson entier;

c) quantités débarquées sur la base du poids du poisson à sa sortie de l'eau.

Pour les données sur la «production primaire» des algues, perles, coquillages, coraux, éponges, etc., il est préférable d'utiliser le terme PRODUCTION, qui est plus approprié que l'expression CAPTURES NOMINALES. Par ailleurs, le terme CAPTURES convient mieux que l'expression CAPTURES NOMINALES dans les cas des baleines et des phoques où la «production primaire» ne s'exprime pas en unités pondérales mais en nombres.

Les données concernant les captures nominales de poissons appartenant au groupe 36 (thonidés, bonites et marlins) sont en général révisées en collaboration avec l'organisation régionale chargée des statistiques des thonidés (CICTA, CTOI, CPS et CITT). Comme les dates pour lesquelles ces organisations demandent les données ne correspondent pas toujours, les chiffres pour l'année la plus récente sont souvent sujets à des révisions significatives.

Notas generales

En el presente volumen del **Anuario Estadístico de Pesca** se presentan, para las más recientes series de años civiles y emergentes, estadísticas anuales de las CAPTURAS NOMINALES de especies de aguas dulces, salobres y marinas, de peces, crustáceos, moluscos y otros animales y plantas acuáticos, matados, capturados, entrampados o cobrados para todo fin de carácter comercial, industrial, pesca de recreo o de subsistencia.

Dada la importancia que la pesca de recreo tiene para algunas poblaciones de pescado y para ciertos países, y la dificultad de distinguir en muchos casos entre dicha pesca y la de subsistencia (y en conformidad con la recomendación de la decimosexta sesión del Grupo Coordinador de Trabajo sobre Estadísticas de Pesca - CWP, Madrid, España, 20-25 de marzo de 1995), los datos deberán incluir la pesca de recreo.

El concepto de CAPTURAS NOMINALES se refiere a los desembarques expresados en su *peso en vivo*; el concepto DESEMBARQUES íntimamente relacionado, se refiere al *peso descargado*. En muchas pesquerías las cantidades desembarcadas (DESEMBARQUES), son idénticas a las cantidades capturadas (CAPTURAS NOMINALES).

En muchos casos, las capturas a bordo de las embarcaciones de pesca o de los buques factoría se destripan, evisceran, filetean, salan, secan, etc., o se reducen en harina, en aceite, etc. Los datos sobre los DESEMBARQUES de estas especies y productos hay que convertirlos, mediante índices precisos de rendimiento (factores de conversión) para fijar su equivalente en peso en vivo (capturas nominales) en el momento de la captura. Véase el diagrama en las páginas siguientes.

Muchas publicaciones estadísticas nacionales no utilizan los términos «desembarques» y «capturas» con la significación precisa que se describe anteriormente. En tales publicaciones, el término «captura» se emplea algunas veces para referirse a cantidades basadas en el peso desembarcado, es decir «desembarques» que pueden consistir en pescado destripado, eviscerado y fileteado, así como en harinas y aceites. Sin embargo, sólo cuando la «producción primaria» (esta frase «producción primaria» se utiliza en su sentido económico y no en el biológico) se desembarca entera es correcto describir estas cantidades desembarcadas como «capturas».

En algunas estadísticas nacionales son de uso común las siguientes expresiones para referirse a las CAPTURAS NOMINALES, es decir, los desembarques expresados según su peso en vivo:

a) desembarques de pescado fresco, entero;

b) desembarques de pescado entero;

c) desembarques según el peso del pescado al sacarlo del agua.

Para los datos de «producción primaria» de algas marinas, perlas, mariscos, corales, esponjas, etc., es preferible usar el término PRODUCCION, ya que es más apropiado que la expresión CAPTURA NOMINAL. El término CAPTURA también es más adecuado que la expresión CAPTURA NOMINAL en el caso de ballenas y focas donde los datos de «producción primaria» para estas especies no se expresan en unidades de peso sino en números.

Los datos relativos a la captura nominal de pescado incluido en el grupo 36 (atún, bonitos y agujas) se revisan generalmente en colaboración con los organismos regionales que se encargan de las estadísticas atuneras (CICAA, CAOI, SCP y CIAT). Debido a las diferentes fechas en que esos organismos necesitan recibir los datos, es frecuente que las cifras correspondientes a los años más recientes se sometan a una considerable revisión.

CATCH CONCEPTS: DIAGRAMMATIC PRESENTATION

FISH ENCOUNTERING FISHING GEAR

LIVE ESCAPEMENT

The total weight of fish which encountered the fishing gear but escaped alive

GROSS REMOVAL

The total live weight of fish caught, or killed, during fishing operation

PRE-CATCH LOSSES

The total live weight of fish which die as a result of fishing operation and which are lost and not caught, including losses caused through gear lost during fishing

GROSS CATCH

The total live weight of fish caught

DISCARDED CATCH - DEAD

The total live weight of undersized, unsaleable or otherwise undesirable whole fish discarded at the time of capture or shortly afterwards

DISCARDED CATCH - LIVE

The total live weight of undersized, unsaleable, or otherwise undesirable whole fish discarded at the time of capture or shortly afterwards

RETAINED CATCH

The total live weight of fish retained

LOSSES DUE TO DRESSING HANDLING AND PROCESSING

- Dumped viscera, heads and other parts
- Loss of fluid content

UTILIZATION AND LOSSES PRIOR TO LANDING

- Consumption by crew
- Use for bait
- Spoilage and subsequent dumping
- Losses in handling at sea and when landing

UNRECORDED REJECTED OR DUMPED LANDINGS

- Unrecorded landings dumped at sea
- Black market landings
- Unrecorded quantities landed for home consumption, etc

GAINS PRIOR TO LANDINGS

Gain of fluid content: addition of liquids or solids during shipboard processing

LANDINGS

The net weight of the quantities landed as recorded at the time of the landing of:
- whole or eviscerated fish, fillets, livers, roes, etc
- fresh, iced, chilled or frozen, cured or canned products, etc
- fish meals, liver oils, body oils, etc
- other edible or inedible fishery products, etc

NOMINAL CATCHES = (LANDINGS + LOSSES DUE TO DRESSING, HANDLING AND PROCESSING - GAINS PRIOR TO LANDINGS) • CONVERSION FACTORS

NOMINAL CATCH

The live weight equivalent of the landings i.e.:
- *landings on a round, fresh basis;*
- *landings on a round, whole basis;*
- *landings on an ex-water weight basis*

CONCEPTS RELATIFS AUX CAPTURES: DIAGRAMME EXPLICATIF

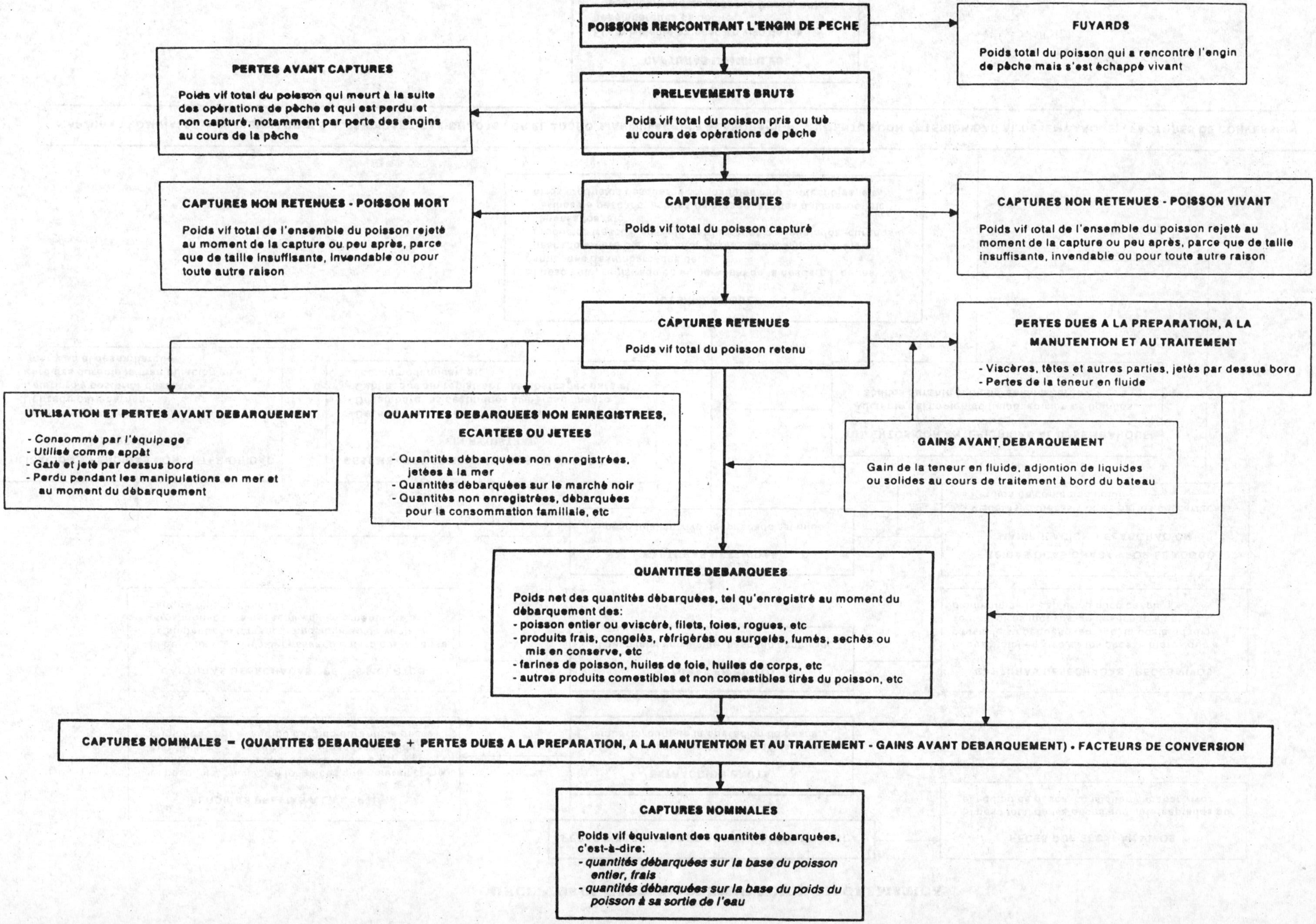

POISSONS RENCONTRANT L'ENGIN DE PECHE

FUYARDS

Poids total du poisson qui a rencontré l'engin de pêche mais s'est échappé vivant

PRELEVEMENTS BRUTS

Poids vif total du poisson pris ou tué au cours des opérations de pêche

PERTES AVANT CAPTURES

Poids vif total du poisson qui meurt à la suite des opérations de pêche et qui est perdu et non capturé, notamment par perte des engins au cours de la pêche

CAPTURES BRUTES

Poids vif total du poisson capturé

CAPTURES NON RETENUES - POISSON MORT

Poids vif total de l'ensemble du poisson rejeté au moment de la capture ou peu après, parce que de taille insuffisante, invendable ou pour toute autre raison

CAPTURES NON RETENUES - POISSON VIVANT

Poids vif total de l'ensemble du poisson rejeté au moment de la capture ou peu après, parce que de taille insuffisante, invendable ou pour toute autre raison

CAPTURES RETENUES

Poids vif total du poisson retenu

PERTES DUES A LA PREPARATION, A LA MANUTENTION ET AU TRAITEMENT

- Viscères, têtes et autres parties, jetés par dessus bord
- Pertes de la teneur en fluide

UTILISATION ET PERTES AVANT DEBARQUEMENT

- Consommé par l'équipage
- Utilisé comme appât
- Gaté et jeté par dessus bord
- Perdu pendant les manipulations en mer et au moment du débarquement

QUANTITES DEBARQUEES NON ENREGISTREES, ECARTEES OU JETEES

- Quantités débarquées non enregistrées, jetées à la mer
- Quantités débarquées sur le marché noir
- Quantités non enregistrées, débarquées pour la consommation familiale, etc

GAINS AVANT DEBARQUEMENT

Gain de la teneur en fluide, adjontion de liquides ou solides au cours de traitement à bord du bateau

QUANTITES DEBARQUEES

Poids net des quantités débarquées, tel qu'enregistré au moment du débarquement des:
- poisson entier ou eviscéré, filets, foies, rogues, etc
- produits frais, congelés, réfrigérés ou surgelés, fumés, sechés ou mis en conserve, etc
- farines de poisson, huiles de foie, huiles de corps, etc
- autres produits comestibles et non comestibles tirés du poisson, etc

CAPTURES NOMINALES = (QUANTITES DEBARQUEES + PERTES DUES A LA PREPARATION, A LA MANUTENTION ET AU TRAITEMENT - GAINS AVANT DEBARQUEMENT) · FACTEURS DE CONVERSION

CAPTURES NOMINALES

Poids vif équivalent des quantités débarquées, c'est-à-dire:
- *quantités débarquées sur la base du poisson entier, frais*
- *quantités débarquées sur la base du poids du poisson à sa sortie de l'eau*

CONCEPTOS DE CAPTURA: PRESENTACION DIAGRAMATICA

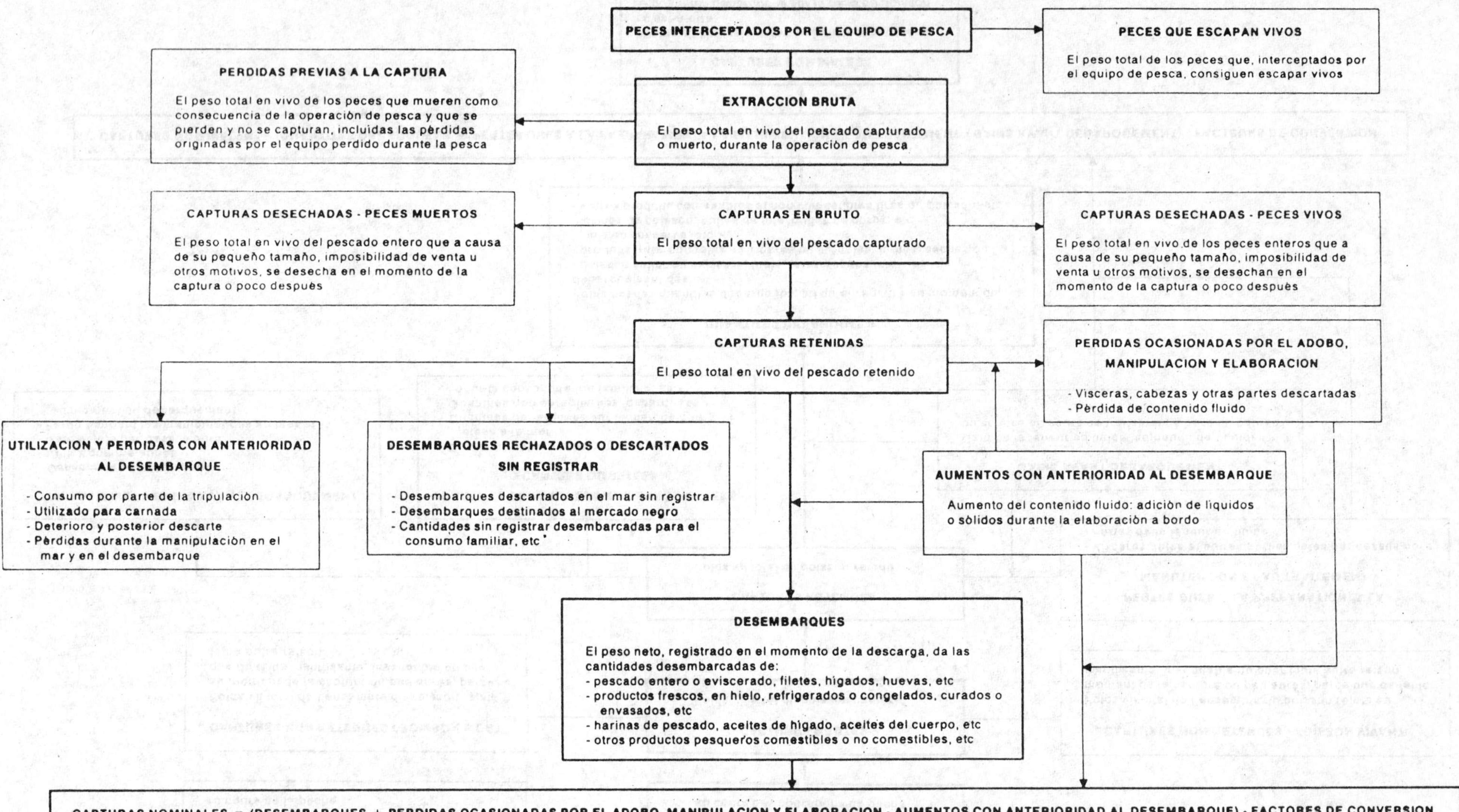

Notes on species items

1. Data available world-wide on nominal catches of aquatic animals and plants, taken in inland and marine waters, for all kinds of commercial, industrial, recreational and subsistence purposes, are at present broken down at either the species, genus, family or higher taxonomic levels into 1291 statistical categories called *species items.*

2. These 1291 *species items* are arranged by FAO within the 50 *groups of species* constituting the nine *divisions* of the FAO "International Standard Statistical Classification of Aquatic Animals and Plants" (ISSCAAP). For these *groups of species* and *divisions* see the ISSCAAP list on the following page. The two-digit group codes shown in this list are reflected in the table numbers in Section B.

3. Following a recommendation of the 19th Session of the Coordinating Working Party on Fishery Statistics - CWP (Nouméa, New Caledonia, 10-13 July 2001) the names and composition of former groups 33, 34 and 37 of the FAO International Standard Statistical Classification of Aquatic Animals and Plants (ISSCAAP) were revised. The species items of the former group 33 "Redfishes, basses, congers" were classified as coastal or demersal fishes and accordingly assigned to the new groups 33 "Miscellaneous coastal fishes" and 34 "Miscellaneous demersal fishes". The species formerly included in group 34 "Jacks, mullets, sauries" were moved to group 37, which was renamed "Miscellaneous pelagic fishes".

4. Each *species item* is identified by means of the following descriptors (subject to constant review and improvement):

a) FAO English name;

b) FAO French name;

c) FAO Spanish name;

d) scientific name (at the species, genus, family or higher taxonomic levels);

e) taxonomic code; and

f) inter-agency 3-alpha code.

5. The taxonomic code descriptors are taken from FAO's **Aquatic Sciences and Fisheries Information System**. The following illustrates the meaning assigned to the different digits of this code:

Main grouping	Order or other taxonomic level	Family	Genus	Species
1,	75	(04)	003,	01

6. In all tables, except A-1(e) (where selected species items of significant commercial importance are listed in the order of 2001 catch size), the species items are arranged in order of their taxonomic codes within their respective ISSCAAP groups.

7. Several countries still report their catches by large groups of species. In these circumstances the catch data presented by individual species items are likely to be underestimated. Therefore, when examining the statistics for a particular species, it should be noted that an unknown proportion of the catches for that species might have been reported by the national office under the generic, family or order name of the species, or even more roughly as, for example, miscellaneous fishes. Consequently, **species items totals frequently underestimate the real catch of the individual species**.

Notes sur les catégories d'espèces

Les statistiques disponibles sur une base mondiale des captures nominales d'animaux et de plantes aquatiques effectuées dans les eaux continentales et dans les eaux marines à toutes fins commerciales, industrielles, récréatives et de subsistance sont, à l'heure actuelle, ventilées au niveau de l'espèce, du genre, de la famille ou des niveaux taxonomiques supérieurs dans 1291 catégories statistiques qui sont appelées des *catégories d'espèces.*

Ces 1291 *catégories d'espèces* sont rassemblées par la FAO en cinquante *groupes d'espèces* constituant les neuf *divisions* de la 'Classification statistique internationale type des animaux et des plantes aquatiques' (CSITAPA) de la FAO. Pour ces *groupes d'espèces* et ces *divisions*, on se reportera à la liste de la CSITAPA figurant à la page suivante. Les codes de groupe à deux chiffres indiqués sur la liste se retrouvent dans les numéros des tableaux de la section B.

Sur la base d'une recommandation de la 19ème session du Groupe de travail chargé de coordonner les statistiques des pêches - CWP (Nouméa, Nouvelle-Calédonie, 10-13 juillet 2001), la dénomination et la composition des anciens groupes 33, 34 et 37 de la Classification statistique internationale type des animaux et des plantes aquatiques (CSITAPA) de la FAO ont été révisées. Les catégories d'espèces de l'ancien groupe 33 "Rascasses, perches de mer, congres" ont été classifiées comme espèces côtières ou démersales et réparties sur cette base entre les nouveaux groupes 33 "Poissons côtiers divers" et 34 "Poissons démersaux divers". Les espèces auparavant incluses dans le groupe 34 "Chinchards, mulets, balaous" ont été classées dans le groupe 37 rebaptisé "Poissons pélagiques divers".

Chaque *catégorie d'espèces* est identifiée au moyen des descripteurs suivants (sous réserve de révision et d'amélioration constantes):

a) nom anglais utilisé par la FAO;

b) nom français utilisé par la FAO;

c) nom espagnol utilisé par la FAO;

d) nom scientifique (de l'espèce, du genre, de la famille ou des niveaux taxonomiques supérieurs);

e) code taxonomique; et

f) code interinstitutions alpha-3.

Les descripteurs de code taxonomique proviennent du **Système d'information sur les sciences aquatiques et la pêche** de la FAO. On trouvera ci-après la signification des divers chiffres constituant ce code taxonomique.

Groupe principal	Ordre ou autre niveau taxonomique	Famille	Genre	Espèce
1,	75	(04)	003,	01

Dans tous les tableaux, à l'exception de A-1(e) (où certaines catégories d'espèces ayant une importance commerciale notable sont énumérées selon le chiffre des captures de 2001), les catégories d'espèces sont disposées selon l'ordre du code taxonomique dans leurs groupes CSITAPA, respectifs.

Plusieurs pays continuent à communiquer leurs captures par grands groupes d'espèces. Dans ces circonstances les données des captures par catégories d'espèces individuelles sont donc probablement sous-estimées. Par conséquent, lorsqu'on examine les statistiques d'une espèce donnée, il convient de noter qu'une proportion inconnue des captures de cette espèce peut avoir été consignée par l'organisme national sous le nom du genre, de la famille ou de l'ordre de l'espèce, ou même, plus approximativement dans la catégorie, par exemple, poissons divers. C'est pourquoi **les totaux correspondant aux espèces données sous-estiment souvent la capture réelle de chacune des espèces**.

Notas sobre las partidas de especies

Los datos procedentes de todo el mundo sobre las capturas nominales de animales y plantas acuáticos capturados en aguas continentales y marinas con fines comerciales, industriales, de recreo o de subsistencia se desglosan en la actualidad en unas 1291 categorías estadísticas, denominadas *partidas de especies,* que corresponden a los niveles taxonómicos de especie, género, familia o superiores.

Utilizando esas 1291 *partidas de especies*, la FAO desglosa los cincuenta *grupos de especies* que constituyen las nueve *divisiones* de la Clasificación Estadística Internacional Uniforme de los Animales y Plantas Acuáticos (CEIUAPA) de la FAO. Estos *grupos de especies* y las nueve *divisiones* indicadas pueden verse en la lista de la CEIUAPA que aparece en la página siguiente. Los códigos de dos dígitos que en esa lista identifican a cada uno de los grupos se utilizan luego en los cuadros de la Sección B.

Sobre la base de una recomendación de la 19a sesión del Grupo Coordinador de Trabajo sobre Estadísticas de Pesca - CWP (Noumea, Nueva Caledonia, 10-13 de julio 2001) fue revisada la denominación y la composición de los antiguos grupos 33, 34, y 37 de la Clasificación Estadística Internacional Uniforme de los Animales y Plantas Acuáticos (CEIUAPA) de la FAO. Las partidas de especies del antiguo grupo 33 "Gallinetas, lubinas, congrios" fueron clasificadas como especies costeras o demersales y, de acuerdo con esto, distribuidas entre los nuevos grupos: 33 "Peces costeros diversos" y 34 "Peces demersales diversos". Las partidas de especies anteriormente comprendidas en el grupo 34 "Jureles, lisas, papardas" se incluyeron en el grupo 37, que pasó a denominarse "Peces pelágicos diversos".

Cada una de las *partidas de especies*, se identifica con los siguientes descriptores (que se revisan y mejoran continuamente):

a) nombre inglés utilizado por la FAO;

b) nombre francés utilizado por la FAO;

c) nombre español utilizado por la FAO;

d) nombre científico (de la especie, género, familia o niveles taxonómicos superiores);

e) código taxonómico; e

f) código interinstitucional alfa-3.

Los descriptores del código taxonómico se han tomado del **Sistema de información de las ciencias acuáticas y la pesca** de la FAO. El cuadro siguiente ilustra el significado de los distintos dígitos de dicho código:

Grupo principal	Orden u otro nivel taxonómico	Familia	Género	Especie
1,	75	(04)	003,	01

En todos los cuadros, excepto en el cuadro A-1(e) (en el cual algunas especies de gran importancia comercial aparecen ordenadas según el volumen de sus capturas en 2001), las partidas de especies aparecen por orden de su código taxonómico dentro de los respectivos grupos de la CEIUAPA.

Varios países siguen informando de sus capturas por grandes grupos de especies. En estos casos, es probable que aparezcan subestimados los datos sobre la captura de las distintas categorías de especies. Por consiguiente, al examinar las estadísticas de una especie particular, debe tenerse en cuenta que una proporción desconocida de las capturas de esa especie puede haber sido ya comunicada por la oficina nacional con el nombre genérico, el de la familia o el del orden de la especie, o incluso más aproximativamente como, por ejemplo, peces diversos. Por consiguiente, **los totales de las categorías de especies subestiman frecuentemente la captura efectiva de la especie individual**.

Code Code Código	ISSCAAP DIVISION Group of species	CSITAPA DIVISION Groupe d'espèces	CEIUAPA DIVISION Grupo de especies
1	**Freshwater fishes**	**Poissons d'eau douce**	**Peces de agua dulce**
11	Carps, barbels and other cyprinids	Carpes, barbeaux et autres cyprinidés	Carpas, barbos y otros ciprínidos
12	Tilapias and other cichlids	Tilapias et autres cichlidés	Tilapias y otros cíclidos
13	Miscellaneous freshwater fishes	Poissons d'eau douce divers	Peces de agua dulce diversos
2	**Diadromous fishes**	**Poissons diadromes**	**Peces diádromos**
21	Sturgeons, paddlefishes	Esturgeons, spatules	Esturiones, sollos
22	River eels	Anguilles	Anguilas
23	Salmons, trouts, smelts	Saumons, truites, éperlans	Salmones, truchas, eperlanos
24	Shads	Aloses	Sábalos
25	Miscellaneous diadromous fishes	Poissons diadromes divers	Peces diádromos diversos
3	**Marine fishes**	**Poissons marins**	**Peces marinos**
31	Flounders, halibuts, soles	Flets, flétans, soles	Platijas, halibuts, lenguados
32	Cods, hakes, haddocks	Morues, merlus, églefins	Bacalaos, merluzas, eglefinos
33	Miscellaneous coastal fishes	Poissons côtiers divers	Peces costeros diversos
34	Miscellaneous demersal fishes	Poissons démersaux divers	Peces demersales diversos
35	Herrings, sardines, anchovies	Harengs, sardines, anchois	Arenques, sardinas, anchoas
36	Tunas, bonitos, billfishes	Thons, pélamides, marlins	Atunes, bonitos, agujas
37	Miscellaneous pelagic fishes	Poissons pélagiques divers	Peces pelágicos diversos
38	Sharks, rays, chimaeras	Squales, raies, chimères	Tiburones, rayas, quimeras
39	Marine fishes not identified	Poissons marins non identifiés	Peces marinos no identificados
4	**Crustaceans**	**Crustacés**	**Crustáceos**
41	Freshwater crustaceans	Crustacés d'eau douce	Crustáceos de agua dulce
42	Crabs, sea-spiders	Crabes, araignées de mer	Cangrejos, centollas
43	Lobsters, spiny-rock lobsters	Homards, langoustes	Bogavantes, langostas
44	King crabs, squat-lobsters	Crabes royaux, galatées	Cangrejos reales, galateidos
45	Shrimps, prawns	Crevettes	Gambas, camarones
46	Krill, planktonic crustaceans	Krill, crustacés planctoniques	Krill, crustáceos planctónicos
47	Miscellaneous marine crustaceans	Crustacés marins divers	Crustáceos marinos diversos
5	**Molluscs**	**Mollusques**	**Moluscos**
51	Freshwater molluscs	Mollusques d'eau douce	Moluscos de agua dulce
52	Abalones, winkles, conchs	Ormeaux, bigorneaux, strombes	Orejas de mar, bígaros, estrombos
53	Oysters	Huîtres	Ostras
54	Mussels	Moules	Mejillones
55	Scallops, pectens	Coquilles St-Jacques	Vieiras
56	Clams, cockles, arkshells	Clams, coques, arches	Almejas, berberechos, arcas
57	Squids, cuttlefishes, octopuses	Encornets, seiches, poulpes	Calamares, jibias, pulpos
58	Miscellaneous marine molluscs	Mollusques marins divers	Moluscos marinos diversos
* 6	**Whales, seals and other aquatic mammals**	**Baleines, phoques et autres mammifères aquatiques**	**Ballenas, focas y otros mamíferos acuáticos**
* 61	Blue-whales, fin-whales	Baleines bleues, rorquals communs	Ballenas azules, rorcuales
* 62	Sperm-whales, pilot-whales	Cachalots, globicéphales	Cachalotes, calderones
* 63	Eared seals, hair seals, walruses	Otaries, phoques, morses	Lobos marinos, focas, morsas
* 64	Miscellaneous aquatic mammals	Mammifères aquatiques divers	Mamíferos acuáticos diversos
7	**Miscellaneous aquatic animals**	**Animaux aquatiques divers**	**Animales acuáticos diversos**
71	Frogs and other amphibians	Grenouilles et autres amphibies	Ranas y otros anfibios
72	Turtles	Tortues	Tortugas
* 73	Crocodiles and alligators	Crocodiles et alligators	Cocodrilos y aligatores
74	Sea-squirts and other tunicates	Ascidiens et autres tuniciers	Ascidias y otros tunicados
75	Horseshoe crabs and other arachnoids	Limules et autres arachnoïdés	Límulos y otros arácnidos
76	Sea-urchins and other echinoderms	Oursins et autres échinodermes	Erizos de mar y otros equinodermos
77	Miscellaneous aquatic invertebrates	Invertébrés aquatiques divers	Invertebrados acuáticos diversos
* 8	**Miscellaneous aquatic animal products**	**Produits divers d'animaux aquatiques**	**Diversos productos de animales acuáticos**
* 81	Pearls, mother-of-pearl, shells	Perles, nacres, coquilles	Perlas, madreperlas, conchas
* 82	Corals	Coraux	Corales
* 83	Sponges	Eponges	Esponjas
* 9	**Aquatic plants**	**Plantes aquatiques**	**Plantas acuáticas**
* 91	Brown seaweeds	Algues brunes	Algas pardas
* 92	Red seaweeds	Algues rouges	Algas rojas
* 93	Green seaweeds	Algues vertes	Algas verdes
* 94	Miscellaneous aquatic plants	Plantes aquatiques diverses	Diversas plantas acuáticas

ISSCAAP = International Standard Statistical Classification of Aquatic Animals and Plants

CSITAPA = Classification statistique internationale type des animaux et des plantes aquatiques

CEIUAPA = Clasificación Estadística Internacional Uniforme de los Animales y Plantas Acuáticos

* See paragraph 5 of the INTRODUCTION

* Voir le paragraphe 5 de l'INTRODUCTION

* Véase al párrafo 5 en la INTRODUCCION

Systematic list of aquatic organisms / Liste systématique des organismes aquatiques / Lista sistemática de los organismos acuáticos

1 PISCES

1,02 PETROMYZONTIFORMES

1,02(01) PETROMYZONTIDAE
1,02(01)XXX,XX *Petromyzontidae*
1,02(01)001,01 *Petromyzon marinus*
1,02(01)002,01 *Lampetra fluviatilis*

1,05 HEXANCHIFORMES

1,05(02) HEXANCHIDAE
1,05(02)005,02 *Notorynchus cepedianus*

1,06 LAMNIFORMES

1,06(01) CETORHINIDAE
1,06(01)003,01 *Cetorhinus maximus*

1,06(02) ODONTASPIDIDAE
1,06(02)005,01 *Carcharias taurus*

1,06(06) ALOPIIDAE
1,06(06)006,XX *Alopias spp*
1,06(06)006,01 *Alopias vulpinus*
1,06(06)006,03 *Alopias superciliosus*

1,06(08) LAMNIDAE
1,06(08)002,XX *Isurus spp*
1,06(08)002,01 *Isurus oxyrinchus*
1,06(08)002,03 *Isurus paucus*
1,06(08)003,01 *Lamna nasus*
1,06(08)007,01 *Carcharodon carcharias*

1,07 ORECTOLOBIFORMES

1,07(03) GINGLYMOSTOMATIDAE
1,07(03)009,01 *Ginglymostoma cirratum*

1,08 CARCHARHINIFORMES

1,08(01) SCYLIORHINIDAE
1,08(01)003,XX *Scyliorhinus spp*
1,08(01)003,01 *Scyliorhinus canicula*
1,08(01)003,02 *Scyliorhinus stellaris*

1,08(02) CARCHARHINIDAE
1,08(02)XXX,XX *Carcharhinidae*
1,08(02)004,01 *Prionace glauca*
1,08(02)010,01 *Carcharhinus plumbeus*
1,08(02)010,03 *Carcharhinus limbatus*
1,08(02)010,16 *Carcharhinus obscurus*
1,08(02)010,17 *Carcharhinus falciformis*
1,08(02)010,20 *Carcharhinus brachyurus*
1,08(02)017,03 *Galeocerdo cuvier*

1,08(03) SPHYRNIDAE
1,08(03)XXX,XX *Sphyrnidae*
1,08(03)005,01 *Sphyrna zygaena*
1,08(03)005,06 *Sphyrna lewini*

1,08(04) TRIAKIDAE
1,08(04)007,XX *Mustelus spp*
1,08(04)007,03 *Mustelus canis*
1,08(04)007,07 *Mustelus henlei*
1,08(04)007,09 *Mustelus lenticulatus*
1,08(04)007,12 *Mustelus schmitti*
1,08(04)007,13 *Mustelus mustelus*
1,08(04)011,03 *Galeorhinus galeus*

1,09 SQUALIFORMES

1,09(01) SQUALIDAE
1,09(01)XXX,XX *Squalidae*
1,09(01)XXX,XX *Squalidae, Scyliorhinidae*
1,09(01)002,01 *Somniosus microcephalus*
1,09(01)002,03 *Somniosus pacificus*
1,09(01)007,04 *Squalus acanthias*
1,09(01)008,01 *Centrophorus granulosus*
1,09(01)008,03 *Centrophorus squamosus*
1,09(01)010,XX *Etmopterus spp*
1,09(01)014,01 *Deania calcea*
1,09(01)016,01 *Centroscymnus coelolepis*
1,09(01)016,02 *Centroscymnus crepidater*
1,09(01)018,01 *Dalatias licha*
1,09(01)019,01 *Centroscyllium fabricii*

1,09(03) SQUATINIDAE
1,09(03)XXX,XX *Squatinidae*
1,09(03)004,01 *Squatina squatina*
1,09(03)004,04 *Squatina argentina*

1,09(05) OXYNOTIDAE
1,09(05)006,01 *Oxynotus centrina*

1,10 RAJIFORMES

1,10(01) RHINOBATIDAE
1,10(01)XXX,XX *Rhinobatidae*
1,10(01)005,09 *Rhinobatos percellens*
1,10(01)005,10 *Rhinobatos planiceps*

1,10(02) PRISTIDAE
1,10(02)XXX,XX *Pristidae*

1,10(04) RAJIDAE
1,10(04)XXX,XX *Rajidae*
1,10(04)001,XX *Raja spp*
1,10(04)001,01 *Raja batis*
1,10(04)001,02 *Raja clavata*
1,10(04)001,03 *Raja radiata*
1,10(04)001,04 *Raja montagui*
1,10(04)001,06 *Raja circularis*
1,10(04)001,07 *Raja fullonica*
1,10(04)001,09 *Raja microocellata*
1,10(04)001,10 *Raja naevus*
1,10(04)001,11 *Raja oxyrinchus*
1,10(04)001,32 *Raja georgiana*
1,10(04)002,XX *Bathyraja spp*
1,10(04)002,01 *Bathyraja eatonii*

1,10(05) DASYATIDAE
1,10(05)003,XX *Dasyatis spp*
1,10(05)003,01 *Dasyatis akajei*
1,10(05)003,26 *Dasyatis pastinaca*

1,10(07) MYLIOBATIDAE
1,10(07)XXX,XX *Myliobatidae*

1,10(08) MOBULIDAE
1,10(08)XXX,XX *Mobulidae*

1,11 TORPEDINIFORMES

1,11(01) TORPEDINIDAE
1,11(01)002,XX *Torpedo spp*

1,12 CHIMAERIFORMES

1,12(01) CHIMAERIDAE
1,12(01)003,01 *Chimaera monstrosa*
1,12(01)004,XX *Hydrolagus spp*
1,12(01)004,11 *Hydrolagus novaezealandiae*

1,12(02) RHINOCHIMAERIDAE
1,12(02)001,01 *Rhinochimaera atlantica*

1,12(03) CALLORHINCHIDAE
1,12(03)001,XX *Callorhinchus spp*
1,12(03)001,01 *Callorhinchus milii*
1,12(03)001,03 *Callorhinchus capensis*

1,16 POLYPTERIFORMES

1,16(02) PROTOPTERIDAE
1,16(02)002,XX *Protopterus spp*

1,17 ACIPENSERIFORMES

1,17(01) ACIPENSERIDAE
1,17(01)XXX,XX *Acipenseridae*
1,17(01)001,02 *Acipenser gueldenstaedtii*
1,17(01)001,04 *Acipenser ruthenus*
1,17(01)001,05 *Acipenser stellatus*
1,17(01)001,09 *Acipenser transmontanus*
1,17(01)001,12 *Acipenser baerii*
1,17(01)001,15 *Acipenser medirostris*
1,17(01)005,01 *Huso huso*

1,17(02) POLYODONTIDAE
1,17(02)002,01 *Polyodon spathula*

1,21 CLUPEIFORMES

1,21(05) CLUPEIDAE
1,21(05)XXX,XX *Stolothrissa, Limnothrissa spp*
1,21(05)001,05 *Clupea harengus*
1,21(05)001,07 *Clupea pallasii*
1,21(05)011,XX *Alosa spp*

Systematic list of aquatic organisms / Liste systématique des organismes aquatiques / Lista sistemática de los organismos acuáticos

1,21(05)011,XX *Alosa alosa, A.fallax*
1,21(05)011,02 *Alosa pontica*
1,21(05)011,03 *Alosa sapidissima*
1,21(05)011,04 *Alosa alosa*
1,21(05)011,05 *Alosa fallax*
1,21(05)011,06 *Alosa pseudoharengus*
1,21(05)011,07 *Alosa aestivalis*
1,21(05)011,08 *Alosa mediocris*
1,21(05)012,XX *Sardinella spp*
1,21(05)012,03 *Sardinella gibbosa*
1,21(05)012,04 *Sardinella longiceps*
1,21(05)012,10 *Sardinella aurita*
1,21(05)012,17 *Sardinella maderensis*
1,21(05)012,22 *Sardinella zunasi*
1,21(05)012,23 *Sardinella lemuru*
1,21(05)012,24 *Sardinella brasiliensis*
1,21(05)013,01 *Sardinops melanostictus*
1,21(05)013,02 *Sardinops caeruleus*
1,21(05)013,03 *Sardinops sagax*
1,21(05)013,05 *Sardinops ocellatus*
1,21(05)013,09 *Sardinops neopilchardus*
1,21(05)014,XX *Caspialosa spp*
1,21(05)018,01 *Dorosoma cepedianum*
1,21(05)023,01 *Anodontostoma chacunda*
1,21(05)024,01 *Brevoortia aurea*
1,21(05)024,02 *Brevoortia pectinata*
1,21(05)024,03 *Brevoortia tyrannus*
1,21(05)024,04 *Brevoortia patronus*
1,21(05)029,01 *Dussumieria acuta*
1,21(05)029,02 *Dussumieria elopsoides*
1,21(05)030,02 *Ethmalosa fimbriata*
1,21(05)031,01 *Etrumeus teres*
1,21(05)031,04 *Etrumeus whiteheadi*
1,21(05)033,XX *Harengula spp*
1,21(05)034,05 *Hilsa kelee*
1,21(05)038,01 *Tenualosa ilisha*
1,21(05)038,04 *Tenualosa toli*
1,21(05)042,01 *Opisthonema libertate*
1,21(05)042,02 *Opisthonema oglinum*
1,21(05)049,01 *Spratelloides gracilis*
1,21(05)058,03 *Clupanodon thrissa*
1,21(05)059,01 *Clupeonella cultriventris*
1,21(05)060,01 *Konosirus punctatus*
1,21(05)064,01 *Sardina pilchardus*
1,21(05)066,01 *Sprattus sprattus*
1,21(05)066,02 *Sprattus fuegensis*
1,21(05)067,02 *Ethmidium maculatum*
1,21(05)072,01 *Herklotsichthys quadrimaculat.*
1,21(05)078,01 *Strangomera bentincki*

1,21(06) ENGRAULIDAE
1,21(06)XXX,XX *Engraulidae*
1,21(06)002,01 *Engraulis encrasicolus*
1,21(06)002,02 *Engraulis japonicus*
1,21(06)002,06 *Engraulis anchoita*
1,21(06)002,07 *Engraulis mordax*
1,21(06)002,08 *Engraulis ringens*
1,21(06)002,12 *Engraulis capensis*
1,21(06)015,01 *Cetengraulis edentulus*
1,21(06)015,03 *Cetengraulis mysticetus*
1,21(06)020,11 *Anchoa hepsetus*
1,21(06)038,04 *Lycengraulis grossidens*
1,21(06)050,XX *Stolephorus spp*

1,21(11) CHIROCENTRIDAE
1,21(11)002,XX *Chirocentrus spp*
1,21(11)002,01 *Chirocentrus dorab*

1,21(12) PRISTIGASTERIDAE
1,21(12)001,03 *Ilisha elongata*
1,21(12)001,12 *Ilisha africana*
1,21(12)003,03 *Pellona ditchela*

1,22 GONORYNCHIFORMES

1,22(02) CHANIDAE
1,22(02)001,01 *Chanos chanos*

1,23 SALMONIFORMES

1,23(01) SALMONIDAE
1,23(01)004,XX *Salmo spp*
1,23(01)004,01 *Salmo salar*
1,23(01)004,02 *Salmo trutta*
1,23(01)009,XX *Oncorhynchus spp*
1,23(01)009,02 *Oncorhynchus gorbuscha*
1,23(01)009,03 *Oncorhynchus keta*
1,23(01)009,05 *Oncorhynchus masou*
1,23(01)009,06 *Oncorhynchus nerka*
1,23(01)009,07 *Oncorhynchus tshawytscha*
1,23(01)009,08 *Oncorhynchus kisutch*
1,23(01)009,09 *Oncorhynchus mykiss*
1,23(01)010,XX *Salvelinus spp*
1,23(01)010,02 *Salvelinus fontinalis*
1,23(01)010,05 *Salvelinus alpinus*
1,23(01)010,07 *Salvelinus namaycush*
1,23(01)020,02 *Hucho hucho*

1,23(02) THYMALLIDAE
1,23(02)005,01 *Thymallus thymallus*

1,23(03) PLECOGLOSSIDAE
1,23(03)016,01 *Plecoglossus altivelis*

1,23(04) OSMERIDAE
1,23(04)XXX,XX *Osmerus spp, Hypomesus spp*
1,23(04)002,01 *Mallotus villosus*
1,23(04)003,01 *Osmerus eperlanus*
1,23(04)003,03 *Osmerus mordax*
1,23(04)008,01 *Hypomesus olidus*
1,23(04)008,02 *Hypomesus pretiosus*
1,23(04)012,01 *Thaleichthys pacificus*

1,23(05) ARGENTINIDAE
1,23(05)015,XX *Argentina spp*
1,23(05)029,01 *Glossanodon semifasciatus*

1,23(12) COREGONIDAE
1,23(12)001,XX *Coregonus spp*
1,23(12)001,01 *Coregonus albula*
1,23(12)001,02 *Coregonus lavaretus*
1,23(12)001,03 *Coregonus oxyrinchus*
1,23(12)001,12 *Coregonus clupeaformis*
1,23(12)001,13 *Coregonus artedi*

1,23(14) ALEPOCEPHALIDAE
1,23(14)007,01 *Alepocephalus bairdii*

1,24 ESOCIFORMES

1,24(03) ESOCIDAE
1,24(03)001,01 *Esox lucius*

1,25 STOMIIFORMES

1,25(02) STERNOPTYCHIDAE
1,25(02)030,01 *Maurolicus muelleri*

1,28 OSTEOGLOSSIFORMES

1,28(01) OSTEOGLOSSIDAE
1,28(01)001,01 *Arapaima gigas*
1,28(01)004,XX *Heterotis spp*

1,28(02) NOTOPTERIDAE
1,28(02)003,XX *Notopterus spp*

1,28(06) MORMYRIDAE
1,28(06)013,XX *Mormyrus spp*

1,29 ELOPIFORMES

1,29(01) ELOPIDAE
1,29(01)003,02 *Elops saurus*
1,29(01)003,04 *Elops lacerta*

1,29(02) MEGALOPIDAE
1,29(02)004,01 *Megalops atlanticus*
1,29(02)004,02 *Megalops cyprinoides*

1,30 ALBULIFORMES

1,30(01) ALBULIDAE
1,30(01)005,01 *Albula vulpes*
1,30(01)006,01 *Pterothrissus belloci*

1,31 AULOPIFORMES

1,31(12) CHLOROPHTHALMIDAE
1,31(12)XXX,XX *Chlorophthalmidae*

1,31(16) SYNODONTIDAE
1,31(16)XXX,XX *Synodontidae*
1,31(16)001,02 *Harpadon nehereus*
1,31(16)068,01 *Saurida tumbil*
1,31(16)068,04 *Saurida undosquamis*

1,32 MYCTOPHIFORMES

1,32(08) MYCTOPHIDAE
1,32(08)XXX,XX *Myctophidae*
1,32(08)017,01 *Lampanyctodes hectoris*

Systematic list of aquatic organisms / Liste systématique des organismes aquatiques / Lista sistemática de los organismos acuáticos

1,38 CHARACIFORMES

1,38(01) CHARACIDAE
1,38(01)XXX,XX *Characidae*
1,38(01)046,05 *Colossoma macropomum*
1,38(01)076,01 *Piaractus brachypomus*

1,38(12) CURIMATIDAE
1,38(12)001,XX *Prochilodus spp*
1,38(12)001,20 *Prochilodus reticulatus*

1,40 CYPRINIFORMES

1,40(01) CATOSTOMIDAE
1,40(01)XXX,XX *Catostomidae*
1,40(01)011,XX *Ictiobus spp*

1,40(02) CYPRINIDAE
1,40(02)XXX,XX *Cyprinidae*
1,40(02)001,XX *Abramis spp*
1,40(02)001,02 *Abramis brama*
1,40(02)002,01 *Cyprinus carpio*
1,40(02)007,01 *Tinca tinca*
1,40(02)012,01 *Alburnus alburnus*
1,40(02)013,01 *Barbus barbus*
1,40(02)014,01 *Chondrostoma nasus*
1,40(02)016,01 *Carassius carassius*
1,40(02)016,02 *Carassius auratus*
1,40(02)018,XX *Rutilus spp*
1,40(02)018,01 *Rutilus rutilus*
1,40(02)018,02 *Rutilus frisii*
1,40(02)019,01 *Scardinius erythrophthalmus*
1,40(02)020,XX *Leuciscus spp*
1,40(02)020,01 *Leuciscus idus*
1,40(02)020,05 *Leuciscus leuciscus*
1,40(02)020,08 *Leuciscus cephalus*
1,40(02)024,XX *Labeo spp*
1,40(02)025,02 *Cirrhinus molitorella*
1,40(02)035,01 *Ctenopharyngodon idellus*
1,40(02)043,01 *Hypophthalmichthys molitrix*
1,40(02)043,02 *Hypophthalmichthys nobilis*
1,40(02)070,01 *Rastrineobola argentea*
1,40(02)097,01 *Vimba vimba*
1,40(02)100,01 *Pelecus cultratus*
1,40(02)115,01 *Aspius aspius*
1,40(02)132,01 *Leptobarbus hoeveni*
1,40(02)144,01 *Mylopharyngodon piceus*
1,40(02)151,01 *Megalobrama amblycephala*
1,40(02)161,XX *Puntius spp*
1,40(02)161,02 *Puntius javanicus*

1,41 SILURIFORMES

1,41(02) ARIIDAE
1,41(02)XXX,XX *Ariidae*
1,41(02)042,08 *Galeichthys feliceps*

1,41(06) PLOTOSIDAE
1,41(06)064,XX *Plotosus spp*

1,41(07) SILURIDAE
1,41(07)031,01 *Silurus glanis*
1,41(07)050,XX *Kryptopterus spp*

1,41(08) BAGRIDAE
1,41(08)078,XX *Chrysichthys spp*
1,41(08)078,02 *Chrysichthys nigrodigitatus*
1,41(08)111,XX *Bagrus spp*

1,41(10) ICTALURIDAE
1,41(10)002,XX *Ictalurus spp*
1,41(10)004,07 *Ameiurus nebulosus*

1,41(18) CLARIIDAE
1,41(18)030,XX *Clarias spp*
1,41(18)030,01 *Clarias batrachus*
1,41(18)030,03 *Clarias gariepinus*
1,41(18)030,07 *Clarias anguillaris*

1,41(30) PANGASIIDAE
1,41(30)002,XX *Pangasius spp*

1,41(32) MOCHOKIDAE
1,41(32)008,XX *Synodontis spp*

1,43 ANGUILLIFORMES

1,43(02) ANGUILLIDAE
1,43(02)002,XX *Anguilla spp*
1,43(02)002,01 *Anguilla anguilla*
1,43(02)002,04 *Anguilla japonica*
1,43(02)002,06 *Anguilla rostrata*
1,43(02)002,07 *Anguilla australis*

1,43(06) MURAENIDAE
1,43(06)XXX,XX *Muraenidae*

1,43(09) MURAENESOCIDAE
1,43(09)011,XX *Muraenesox spp*
1,43(09)011,02 *Muraenesox cinereus*

1,43(13) CONGRIDAE
1,43(13)XXX,XX *Congridae*
1,43(13)001,01 *Conger conger*
1,43(13)001,02 *Conger orbignyanus*
1,43(13)001,04 *Conger oceanicus*
1,43(13)001,05 *Conger myriaster*

1,47 BELONIFORMES

1,47(01) BELONIDAE
1,47(01)XXX,XX *Belonidae*
1,47(01)001,01 *Belone belone*
1,47(01)013,XX *Tylosurus spp*

1,47(02) SCOMBERESOCIDAE
1,47(02)002,01 *Scomberesox saurus*
1,47(02)007,01 *Cololabis saira*

1,47(03) HEMIRAMPHIDAE
1,47(03)003,11 *Hyporhamphus sajori*
1,47(03)004,XX *Hemiramphus spp*
1,47(03)004,06 *Hemiramphus brasiliensis*

1,47(04) EXOCOETIDAE
1,47(04)XXX,XX *Exocoetidae*
1,47(04)010,02 *Cypselurus agoo*

1,48 GADIFORMES

1,48(01) MURAENOLEPIDIDAE
1,48(01)001,XX *Muraenolepis spp*
1,48(01)001,01 *Muraenolepis microps*

1,48(02) MORIDAE
1,48(02)XXX,XX *Moridae*
1,48(02)010,01 *Mora moro*
1,48(02)014,01 *Pseudophycis bachus*
1,48(02)030,01 *Antimora rostrata*
1,48(02)040,01 *Salilota australis*

1,48(03) BREGMACEROTIDAE
1,48(03)018,01 *Bregmaceros mcclellandi*

1,48(04) GADIDAE
1,48(04)001,01 *Brosme brosme*
1,48(04)002,02 *Gadus morhua*
1,48(04)002,11 *Gadus macrocephalus*
1,48(04)002,12 *Gadus ogac*
1,48(04)003,01 *Lota lota*
1,48(04)005,01 *Molva molva*
1,48(04)005,02 *Molva dypterygia*
1,48(04)006,01 *Phycis blennoides*
1,48(04)008,01 *Urophycis brasiliensis*
1,48(04)008,02 *Urophycis chuss*
1,48(04)008,03 *Urophycis tenuis*
1,48(04)010,01 *Melanogrammus aeglefinus*
1,48(04)012,01 *Eleginus navaga*
1,48(04)012,02 *Eleginus gracilis*
1,48(04)013,01 *Microgadus proximus*
1,48(04)013,02 *Microgadus tomcod*
1,48(04)015,01 *Pollachius virens*
1,48(04)015,02 *Pollachius pollachius*
1,48(04)016,01 *Theragra chalcogramma*
1,48(04)019,01 *Boreogadus saida*
1,48(04)028,XX *Gaidropsarus spp*
1,48(04)032,01 *Trisopterus esmarkii*
1,48(04)032,02 *Trisopterus minutus*
1,48(04)032,03 *Trisopterus luscus*
1,48(04)033,01 *Micromesistius poutassou*
1,48(04)033,02 *Micromesistius australis*
1,48(04)034,01 *Merlangius merlangus*

1,48(05) MERLUCCIIDAE
1,48(05)004,XX *Merluccius spp*
1,48(05)004,XX *Merluccius capensis,M.paradox.*
1,48(05)004,01 *Merluccius merluccius*
1,48(05)004,02 *Merluccius senegalensis*
1,48(05)004,03 *Merluccius australis*
1,48(05)004,04 *Merluccius bilinearis*
1,48(05)004,05 *Merluccius gayi*

1,48(05)004,06 *Merluccius hubbsi*
1,48(05)004,07 *Merluccius productus*
1,48(05)004,08 *Merluccius polli*
1,48(05)004,09 *Merluccius capensis*
1,48(05)004,12 *Merluccius albidus*
1,48(05)017,XX *Macruronus spp*
1,48(05)017,01 *Macruronus magellanicus*
1,48(05)017,02 *Macruronus novaezelandiae*

1,48(06) MACROURIDAE
1,48(06)XXX,XX *Macrouridae*
1,48(06)001,XX *Macrourus spp*
1,48(06)001,03 *Macrourus berglax*
1,48(06)001,04 *Macrourus whitsoni*
1,48(06)004,01 *Coryphaenoides rupestris*
1,48(06)030,01 *Lepidorhynchus denticulatus*

1,50 GASTEROSTEIFORMES

1,50(01) GASTEROSTEIDAE
1,50(01)001,01 *Gasterosteus aculeatus*

1,50(05) HYPOPTYCHIDAE
1,50(05)009,01 *Hypoptychus dybowskii*

1,51 SYNGNATHIFORMES

1,51(03) MACRORAMPHOSIDAE
1,51(03)004,01 *Macroramphosus scolopax*

1,52 LAMPRIFORMES

1,52(01) LAMPRIDAE
1,52(01)001,02 *Lampris guttatus*

1,52(04) TRACHIPTERIDAE
1,52(04)002,XX *Trachipterus spp*

1,52(05) REGALECIDAE
1,52(05)001,01 *Regalecus glesne*

1,58 OPHIDIIFORMES

1,58(02) OPHIDIIDAE
1,58(02)XXX,XX *Ophidiidae*
1,58(02)001,XX *Genypterus spp*
1,58(02)001,01 *Genypterus blacodes*
1,58(02)001,02 *Genypterus chilensis*
1,58(02)001,03 *Genypterus maculatus*
1,58(02)001,05 *Genypterus capensis*
1,58(02)005,02 *Brotula barbata*

1,61 BERYCIFORMES

1,61(02) BERYCIDAE
1,61(02)003,XX *Beryx spp*
1,61(02)012,01 *Centroberyx affinis*

1,61(05) TRACHICHTHYIDAE
1,61(05)XXX,XX *Trachichthyidae*
1,61(05)002,02 *Hoplostethus atlanticus*

1,61(11) HOLOCENTRIDAE
1,61(11)XXX,XX *Holocentridae*

1,62 ZEIFORMES

1,62(01) ZEIDAE
1,62(01)XXX,XX *Zeidae*
1,62(01)001,01 *Zeus faber*
1,62(01)004,01 *Zenopsis conchifer*
1,62(01)004,02 *Zenopsis nebulosus*

1,62(03) CAPROIDAE
1,62(03)XXX,XX *Caproidae*

1,62(04) OREOSOMATIDAE
1,62(04)XXX,XX *Oreosomatidae*

1,63 ATHERINIFORMES

1,63(02) ATHERINIDAE
1,63(02)XXX,XX *Atherinidae*
1,63(02)003,01 *Atherina boyeri*
1,63(02)013,02 *Menidia menidia*

1,65 MUGILIFORMES

1,65(01) MUGILIDAE
1,65(01)XXX,XX *Mugilidae*
1,65(01)001,02 *Mugil cephalus*
1,65(01)001,12 *Mugil liza*
1,65(01)001,28 *Mugil soiuy*
1,65(01)010,01 *Joturus pichardi*
1,65(01)012,06 *Liza saliens*

1,70 PERCOIDEI

1,70(00) CAESIONIDAE
1,70(00)112,XX *Caesio spp*

1,70(01) CENTROPOMIDAE
1,70(01)025,XX *Centropomus spp*
1,70(01)025,10 *Centropomus undecimalis*
1,70(01)167,XX *Lates spp*
1,70(01)167,01 *Lates calcarifer*
1,70(01)167,07 *Lates niloticus*

1,70(02) SERRANIDAE
1,70(02)XXX,XX *Serranidae*
1,70(02)040,XX *Mycteroperca spp*
1,70(02)040,01 *Mycteroperca bonaci*
1,70(02)040,04 *Mycteroperca microlepis*
1,70(02)040,06 *Mycteroperca phenax*
1,70(02)040,09 *Mycteroperca venenosa*
1,70(02)040,10 *Mycteroperca xenarcha*
1,70(02)042,XX *Epinephelus spp*
1,70(02)042,01 *Epinephelus marginatus*
1,70(02)042,02 *Epinephelus aeneus*
1,70(02)042,19 *Epinephelus tauvina*
1,70(02)042,20 *Epinephelus striatus*
1,70(02)042,22 *Epinephelus analogus*
1,70(02)042,24 *Epinephelus flavolimbatus*
1,70(02)042,25 *Epinephelus guttatus*
1,70(02)042,28 *Epinephelus morio*
1,70(02)042,30 *Epinephelus niveatus*
1,70(02)042,34 *Epinephelus goreensis*
1,70(02)042,36 *Epinephelus nigritus*
1,70(02)072,01 *Lepidoperca pulchella*
1,70(02)081,02 *Centropristis striata*
1,70(02)259,01 *Acanthistius brasilianus*
1,70(02)405,05 *Paralabrax humeralis*

1,70(04) TERAPONTIDAE
1,70(04)089,XX *Terapon spp*

1,70(05) POLYPRIONIDAE
1,70(05)058,01 *Polyprion americanus*
1,70(05)058,02 *Polyprion oxygeneios*
1,70(05)220,02 *Stereolepis gigas*

1,70(06) MORONIDAE
1,70(06)006,02 *Morone saxatilis*
1,70(06)006,03 *Morone americana*
1,70(06)345,XX *Dicentrarchus spp*
1,70(06)345,01 *Dicentrarchus punctatus*
1,70(06)345,03 *Dicentrarchus labrax*

1,70(08) PERCICHTHYIDAE
1,70(08)297,01 *Lateolabrax japonicus*

1,70(10) CENTRARCHIDAE
1,70(10)014,02 *Micropterus salmoides*

1,70(11) PRIACANTHIDAE
1,70(11)026,XX *Priacanthus spp*
1,70(11)026,05 *Priacanthus macracanthus*

1,70(12) APOGONIDAE
1,70(12)XXX,XX *Apogonidae*

1,70(13) ACROPOMATIDAE
1,70(13)007,01 *Synagrops japonicus*

1,70(14) PERCIDAE
1,70(14)002,01 *Perca fluviatilis*
1,70(14)002,02 *Perca flavescens*
1,70(14)015,XX *Stizostedion spp*
1,70(14)015,01 *Stizostedion vitreum*
1,70(14)015,02 *Stizostedion canadense*
1,70(14)015,03 *Stizostedion lucioperca*
1,70(14)059,01 *Gymnocephalus cernuus*

1,70(15) SILLAGINIDAE
1,70(15)XXX,XX *Sillaginidae*

1,70(16) BRANCHIOSTEGIDAE
1,70(16)XXX,XX *Branchiostegidae*
1,70(16)001,03 *Caulolatilus princeps*
1,70(16)001,04 *Caulolatilus chrysops*
1,70(16)400,02 *Lopholatilus chamaeleonticeps*

Systematic list of aquatic organisms

Liste systématique des organismes aquatiques

Lista sistemática de los organismos acuáticos

Code	Family	Species
1,70(19)	LACTARIIDAE	
1,70(19)165,02		*Lactarius lactarius*
1,70(20)	POMATOMIDAE	
1,70(20)213,01		*Pomatomus saltatrix*
1,70(22)	RACHYCENTRIDAE	
1,70(22)221,01		*Rachycentron canadum*
1,70(23)	CARANGIDAE	
1,70(23)XXX,XX		*Carangidae*
1,70(23)004,XX		*Trachurus spp*
1,70(23)004,01		*Trachurus trachurus*
1,70(23)004,02		*Trachurus picturatus*
1,70(23)004,03		*Trachurus japonicus*
1,70(23)004,05		*Trachurus murphyi*
1,70(23)004,06		*Trachurus symmetricus*
1,70(23)004,08		*Trachurus mediterraneus*
1,70(23)004,11		*Trachurus lathami*
1,70(23)004,13		*Trachurus capensis*
1,70(23)004,14		*Trachurus trecae*
1,70(23)004,15		*Trachurus declivis*
1,70(23)011,27		*Pseudocaranx dentex*
1,70(23)043,XX		*Decapterus spp*
1,70(23)043,07		*Decapterus maruadsi*
1,70(23)043,08		*Decapterus russelli*
1,70(23)044,XX		*Caranx spp*
1,70(23)044,26		*Caranx crysos*
1,70(23)044,29		*Caranx hippos*
1,70(23)044,31		*Caranx ruber*
1,70(23)044,42		*Caranx rhonchus*
1,70(23)046,04		*Selene setapinnis*
1,70(23)046,05		*Selene dorsalis*
1,70(23)047,XX		*Trachinotus spp*
1,70(23)047,01		*Trachinotus blochii*
1,70(23)047,03		*Trachinotus carolinus*
1,70(23)048,XX		*Seriola spp*
1,70(23)048,01		*Seriola dumerili*
1,70(23)048,02		*Seriola quinqueradiata*
1,70(23)048,06		*Seriola lalandi*
1,70(23)072,02		*Lichia amia*
1,70(23)090,03		*Alectis alexandrinus*
1,70(23)099,01		*Parastromateus niger*
1,70(23)134,01		*Elagatis bipinnulata*
1,70(23)151,01		*Gnathanodon speciosus*
1,70(23)179,01		*Megalaspis cordyla*
1,70(23)231,XX		*Scomberoides spp*
1,70(23)268,01		*Chloroscombrus chrysurus*
1,70(23)268,02		*Chloroscombrus orqueta*
1,70(23)283,01		*Parona signata*
1,70(23)291,01		*Selar crumenophthalmus*
1,70(23)422,01		*Selaroides leptolepis*
1,70(23)425,01		*Seriolina nigrofasciata*
1,70(26)	MENIDAE	
1,70(26)327,01		*Mene maculata*
1,70(27)	BRAMIDAE	
1,70(27)XXX,XX		*Bramidae*
1,70(27)003,01		*Brama brama*
1,70(28)	CORYPHAENIDAE	
1,70(28)071,01		*Coryphaena hippurus*
1,70(29)	ARRIPIDAE	
1,70(29)051,01		*Arripis georgianus*
1,70(29)051,02		*Arripis trutta*
1,70(30)	EMMELICHTHYIDAE	
1,70(30)XXX,XX		*Emmelichthyidae*
1,70(30)010,01		*Emmelichthys nitidus*
1,70(32)	LUTJANIDAE	
1,70(32)XXX,XX		*Lutjanidae*
1,70(32)027,XX		*Lutjanus spp*
1,70(32)027,02		*Lutjanus argentimaculatus*
1,70(32)027,20		*Lutjanus argentiventris*
1,70(32)027,22		*Lutjanus purpureus*
1,70(32)027,25		*Lutjanus campechanus*
1,70(32)027,33		*Lutjanus synagris*
1,70(32)028,01		*Ocyurus chrysurus*
1,70(32)225,01		*Rhomboplites aurorubens*
1,70(33)	NEMIPTERIDAE	
1,70(33)XXX,XX		*Nemipteridae*
1,70(33)184,XX		*Nemipterus spp*
1,70(33)184,05		*Nemipterus virgatus*
1,70(33)230,XX		*Scolopsis spp*
1,70(34)	LOBOTIDAE	
1,70(34)029,01		*Lobotes surinamensis*
1,70(35)	LEIOGNATHIDAE	
1,70(35)XXX,XX		*Leiognathidae*
1,70(35)169,XX		*Leiognathus spp*
1,70(36)	HAEMULIDAE	
1,70(36)XXX,XX		*Haemulidae (=Pomadasyidae)*
1,70(36)032,01		*Orthopristis chrysoptera*
1,70(36)163,01		*Isacia conceptionis*
1,70(36)207,05		*Plectorhinchus mediterraneus*
1,70(36)209,04		*Pomadasys argenteus*
1,70(36)209,18		*Pomadasys incisus*
1,70(36)209,19		*Pomadasys jubelini*
1,70(36)263,03		*Brachydeuterus auritus*
1,70(36)395,01		*Conodon nobilis*
1,70(37)	SCIAENIDAE	
1,70(37)XXX,XX		*Sciaenidae*
1,70(37)005,01		*Totoaba macdonaldi*
1,70(37)009,02		*Sciaena gilberti*
1,70(37)009,21		*Sciaena umbra*
1,70(37)016,XX		*Cynoscion spp*
1,70(37)016,04		*Cynoscion analis*
1,70(37)016,09		*Cynoscion nebulosus*
1,70(37)016,14		*Cynoscion regalis*
1,70(37)016,17		*Cynoscion striatus*
1,70(37)038,XX		*Micropogonias spp*
1,70(37)038,02		*Micropogonias furnieri*
1,70(37)038,04		*Micropogonias undulatus*
1,70(37)039,XX		*Menticirrhus spp*
1,70(37)039,07		*Menticirrhus saxatilis*
1,70(37)039,08		*Menticirrhus littoralis*
1,70(37)070,01		*Umbrina cirrosa*
1,70(37)070,03		*Umbrina canosai*
1,70(37)070,09		*Umbrina canariensis*
1,70(37)096,01		*Macrodon ancylodon*
1,70(37)106,01		*Argyrosomus regius*
1,70(37)106,06		*Argyrosomus hololepidotus*
1,70(37)108,01		*Atractoscion aequidens*
1,70(37)108,03		*Atractoscion nobilis*
1,70(37)147,01		*Genyonemus lineatus*
1,70(37)166,07		*Pteroscion peli*
1,70(37)186,03		*Otolithes ruber*
1,70(37)194,02		*Paralonchurus peruanus*
1,70(37)210,02		*Pogonias cromis*
1,70(37)298,05		*Nibea mitsukurii*
1,70(37)303,04		*Larimichthys croceus*
1,70(37)303,05		*Larimichthys polyactis*
1,70(37)363,01		*Leiostomus xanthurus*
1,70(37)411,01		*Sciaenops ocellatus*
1,70(37)457,XX		*Pseudotolithus spp*
1,70(37)457,01		*Pseudotolithus brachygnathus*
1,70(37)457,02		*Pseudotolithus senegalensis*
1,70(37)457,05		*Pseudotolithus elongatus*
1,70(37)553,01		*Atrobucca nibe*
1,70(37)561,01		*Pennahia argentata*
1,70(38)	LETHRINIDAE	
1,70(38)XXX,XX		*Lethrinidae*
1,70(38)155,XX		*Gymnocranius spp*
1,70(38)172,04		*Lethrinus atlanticus*
1,70(39)	SPARIDAE	
1,70(39)XXX,XX		*Sparidae*
1,70(39)008,XX		*Pagellus spp*
1,70(39)008,01		*Pagellus bogaraveo*
1,70(39)008,02		*Pagellus erythrinus*
1,70(39)008,03		*Pagellus acarne*
1,70(39)008,07		*Pagellus bellottii*
1,70(39)033,XX		*Diplodus spp*
1,70(39)033,01		*Diplodus argenteus*
1,70(39)033,03		*Diplodus sargus*
1,70(39)034,XX		*Calamus spp*
1,70(39)060,XX		*Dentex spp*
1,70(39)060,02		*Dentex macrophthalmus*
1,70(39)060,04		*Dentex canariensis*
1,70(39)060,06		*Dentex dentex*
1,70(39)060,10		*Dentex angolensis*
1,70(39)060,11		*Dentex congoensis*
1,70(39)063,02		*Spondyliosoma cantharus*
1,70(39)076,01		*Oblada melanura*
1,70(39)102,01		*Archosargus probatocephalus*
1,70(39)105,01		*Argyrops spinifer*
1,70(39)107,01		*Argyrozona argyrozona*
1,70(39)118,02		*Cheimerius nufar*
1,70(39)191,XX		*Pagrus spp*
1,70(39)191,01		*Pagrus caeruleostictus*
1,70(39)191,03		*Pagrus pagrus*
1,70(39)191,15		*Pagrus auratus*
1,70(39)203,01		*Petrus rupestris*
1,70(39)220,01		*Pterogymnus laniarius*

Systematic list of aquatic organisms | Liste systématique des organismes aquatiques | Lista sistemática de los organismos acuáticos

1,70(39)224,01 *Rhabdosargus globiceps*
1,70(39)235,08 *Sparus aurata*
1,70(39)261,01 *Boops boops*
1,70(39)264,XX *Chrysoblephus spp*
1,70(39)277,XX *Lithognathus spp*
1,70(39)277,01 *Lithognathus lithognathus*
1,70(39)277,02 *Lithognathus mormyrus*
1,70(39)293,01 *Sarpa salpa*
1,70(39)330,04 *Acanthopagrus schlegeli*
1,70(39)330,05 *Acanthopagrus latus*
1,70(39)353,01 *Stenotomus chrysops*

1,70(40) CENTRACANTHIDAE
1,70(40)075,XX *Spicara spp*
1,70(40)075,01 *Spicara maena*

1,70(41) MULLIDAE
1,70(41)XXX,XX *Mullidae*
1,70(41)007,XX *Mullus spp*
1,70(41)007,01 *Mullus surmuletus*
1,70(41)007,02 *Mullus barbatus*
1,70(41)251,XX *Upeneus spp*
1,70(41)499,03 *Pseudupeneus prayensis*

1,70(42) ECHENEIDAE
1,70(42)XXX,XX *Echeneidae*

1,70(46) GERREIDAE
1,70(46)XXX,XX *Gerreidae*
1,70(46)036,XX *Gerres spp*

1,70(47) KYPHOSIDAE
1,70(47)003,02 *Girella tricuspidata*
1,70(47)003,03 *Girella nigricans*
1,70(47)035,XX *Kyphosus spp*
1,70(47)035,01 *Kyphosus cinerascens*

1,70(50) DREPANIDAE
1,70(50)132,01 *Drepane punctata*
1,70(50)132,02 *Drepane africana*

1,70(57) PENTACEROTIDAE
1,70(57)004,02 *Paristiopterus labiosus*
1,70(57)007,01 *Pseudopentaceros richardsoni*

1,70(59) CICHLIDAE
1,70(59)XXX,XX *Cichlidae*
1,70(59)051,XX *Oreochromis (=Tilapia) spp*
1,70(59)051,01 *Oreochromis mossambicus*
1,70(59)051,02 *Oreochromis niloticus*
1,70(59)051,03 *Oreochromis aureus*
1,70(59)052,01 *Sarotherodon galilaeus*
1,70(59)053,06 *Cichlasoma managuense*
1,70(59)054,04 *Cichla ocellaris*
1,70(59)313,XX *Astronotus spp*
1,70(59)316,XX *Haplochromis spp*

1,70(63) LABRIDAE
1,70(63)XXX,XX *Labridae*
1,70(63)005,01 *Labrus bergylta*
1,70(63)243,01 *Tautoga onitis*
1,70(63)244,02 *Tautogolabrus adspersus*
1,70(63)331,02 *Semicossyphus pulcher*

1,70(65) SCARIDAE
1,70(65)XXX,XX *Scaridae*
1,70(65)055,06 *Sparisoma cretense*

1,70(66) POMACANTHIDAE
1,70(66)XXX,XX *Pomacanthidae*

1,70(70) CHEILODACTYLIDAE
1,70(70)270,02 *Cheilodactylus bergi*
1,70(70)270,03 *Cheilodactylus variegatus*
1,70(70)305,XX *Nemadactylus spp*
1,70(70)305,04 *Nemadactylus macropterus*

1,70(71) LATRIDAE
1,70(71)XXX,XX *Latridae*

1,70(77) POLYNEMIDAE
1,70(77)XXX,XX *Polynemidae*
1,70(77)001,09 *Polydactylus quadrifilis*
1,70(77)002,01 *Eleutheronema tetradactylum*
1,70(77)003,02 *Galeoides decadactylus*
1,70(77)004,01 *Pentanemus quinquarius*

1,70(92) NOTOTHENIIDAE
1,70(92)XXX,XX *Nototheniidae*
1,70(92)007,02 *Eleginops maclovinus*
1,70(92)015,XX *Dissostichus spp*
1,70(92)015,01 *Dissostichus mawsoni*
1,70(92)015,02 *Dissostichus eleginoides*
1,70(92)019,01 *Notothenia rossii*
1,70(92)019,03 *Notothenia gibberifrons*
1,70(92)019,04 *Notothenia neglecta*
1,70(92)019,06 *Notothenia squamifrons*
1,70(92)430,02 *Nototheniops nudifrons*
1,70(92)440,01 *Patagonotothen brevicauda*
1,70(92)440,02 *Patagonotothen ramsayi*
1,70(92)448,XX *Trematomus spp*
1,70(92)535,01 *Pleuragramma antarcticum*

1,70(94) CHANNICHTHYIDAE
1,70(94)XXX,XX *Channichthyidae*
1,70(94)416,01 *Chaenocephalus aceratus*
1,70(94)417,01 *Champsocephalus gunnari*
1,70(94)418,01 *Pseudochaenichthys georgianus*
1,70(94)466,01 *Chionodraco rastrospinosus*
1,70(94)470,01 *Channichthys rhinoceratus*
1,70(94)480,01 *Chaenodraco wilsoni*

1,70(95) AMBASSIDAE
1,70(95)XXX,XX *Ambassidae*

1,70(96) EPIGONIDAE
1,70(96)373,XX *Epigonus spp*
1,70(96)373,01 *Epigonus telescopus*

1,71 ZOARCOIDEI

1,71(02) ANARHICHADIDAE
1,71(02)001,XX *Anarhichas spp*
1,71(02)001,01 *Anarhichas lupus*
1,71(02)001,03 *Anarhichas minor*

1,71(15) ZOARCIDAE
1,71(15)004,01 *Zoarces viviparus*
1,71(15)012,02 *Macrozoarces americanus*
1,71(15)024,XX *Lycodes spp*

1,72 TRACHINOIDEI

1,72(04) AMMODYTIDAE
1,72(04)002,XX *Ammodytes spp*
1,72(04)002,04 *Ammodytes personatus*

1,72(08) PERCOPHIDAE
1,72(08)202,01 *Percophis brasilianus*

1,72(09) PINGUIPEDIDAE
1,72(09)006,02 *Pseudopercis semifasciata*
1,72(09)196,02 *Parapercis colias*

1,72(12) TRACHINIDAE
1,72(12)010,02 *Trachinus draco*

1,72(13) URANOSCOPIDAE
1,72(13)352,02 *Uranoscopus scaber*
1,72(13)484,01 *Kathetostoma giganteum*

1,72(72) TRICHODONTIDAE
1,72(72)103,01 *Arctoscopus japonicus*

1,73 GOBIOIDEI

1,73(20) ELEOTRIDAE
1,73(20)XXX,XX *Eleotridae*

1,73(21) GOBIIDAE
1,73(21)XXX,XX *Gobiidae*
1,73(21)XXX,XX *Gobiidae*
1,73(21)003,XX *Gobius spp*
1,73(21)003,03 *Gobius niger*

1,74 ACANTHUROIDEI

1,74(02) ACANTHURIDAE
1,74(02)XXX,XX *Acanthuridae*

1,74(05) EPHIPPIDAE
1,74(05)XXX,XX *Ephippidae*
1,74(05)206,XX *Platax spp*

1,74(06) SCATOPHAGIDAE
1,74(06)330,XX *Scatophagus spp*

1,74(07) SIGANIDAE
1,74(07)001,XX *Siganus spp*

1,75 SCOMBROIDEI

Systematic list of aquatic organisms	Liste systématique des organismes aquatiques	Lista sistemática de los organismos acuáticos

1,75(01) SCOMBRIDAE
1,75(01)XXX,XX *Scombridae*
1,75(01)XXX,XX *Thunnini*
1,75(01)001,01 *Sarda sarda*
1,75(01)001,02 *Sarda orientalis*
1,75(01)001,04 *Sarda chiliensis*
1,75(01)002,XX *Scomber spp*
1,75(01)002,01 *Scomber japonicus*
1,75(01)002,05 *Scomber scombrus*
1,75(01)002,07 *Scomber australasicus*
1,75(01)006,01 *Orcynopsis unicolor*
1,75(01)010,01 *Acanthocybium solandri*
1,75(01)014,XX *Rastrelliger spp*
1,75(01)014,01 *Rastrelliger brachysoma*
1,75(01)014,03 *Rastrelliger kanagurta*
1,75(01)015,XX *Scomberomorus spp*
1,75(01)015,03 *Scomberomorus commerson*
1,75(01)015,04 *Scomberomorus guttatus*
1,75(01)015,05 *Scomberomorus lineolatus*
1,75(01)015,06 *Scomberomorus cavalla*
1,75(01)015,07 *Scomberomorus maculatus*
1,75(01)015,08 *Scomberomorus regalis*
1,75(01)015,09 *Scomberomorus sierra*
1,75(01)015,11 *Scomberomorus tritor*
1,75(01)015,12 *Scomberomorus niphonius*
1,75(01)015,15 *Scomberomorus brasiliensis*
1,75(01)023,XX *Auxis thazard, A.rochei*
1,75(01)023,01 *Auxis thazard*
1,75(01)024,01 *Euthynnus alletteratus*
1,75(01)024,04 *Euthynnus lineatus*
1,75(01)024,06 *Euthynnus affinis*
1,75(01)025,01 *Katsuwonus pelamis*
1,75(01)026,01 *Thunnus thynnus*
1,75(01)026,02 *Thunnus orientalis*
1,75(01)026,03 *Thunnus tonggol*
1,75(01)026,04 *Thunnus atlanticus*
1,75(01)026,05 *Thunnus alalunga*
1,75(01)026,08 *Thunnus maccoyii*
1,75(01)026,10 *Thunnus albacares*
1,75(01)026,12 *Thunnus obesus*

1,75(03) ISTIOPHORIDAE
1,75(03)XXX,XX *Istiophoridae*
1,75(03)004,02 *Istiophorus platypterus*
1,75(03)004,04 *Istiophorus albicans*
1,75(03)005,02 *Makaira mazara*
1,75(03)005,05 *Makaira nigricans*
1,75(03)005,07 *Makaira indica*
1,75(03)009,03 *Tetrapturus audax*
1,75(03)009,04 *Tetrapturus albidus*
1,75(03)009,05 *Tetrapturus angustirostris*
1,75(03)009,06 *Tetrapturus pfluegeri*

1,75(04) XIPHIIDAE
1,75(04)003,01 *Xiphias gladius*

1,75(05) GEMPYLIDAE
1,75(05)001,01 *Thyrsites atun*
1,75(05)002,01 *Thyrsitops lepidopoides*
1,75(05)005,01 *Lepidocybium flavobrunneum*
1,75(05)007,01 *Ruvettus pretiosus*
1,75(05)009,01 *Rexea solandri*

1,75(06) TRICHIURIDAE
1,75(06)XXX,XX *Trichiuridae*
1,75(06)003,02 *Trichiurus lepturus*
1,75(06)006,01 *Lepidopus caudatus*
1,75(06)012,01 *Aphanopus carbo*

1,76 STROMATEOIDEI, ANABANTOIDEI

1,76(03) STROMATEIDAE
1,76(03)XXX,XX *Stromateidae*
1,76(03)004,01 *Stromateus fiatola*
1,76(03)009,XX *Pampus spp*
1,76(03)009,01 *Pampus argenteus*
1,76(03)011,XX *Peprilus spp*
1,76(03)011,01 *Peprilus alepidotus*
1,76(03)011,04 *Peprilus triacanthus*
1,76(03)011,05 *Peprilus simillimus*

1,76(04) ARIOMMATIDAE
1,76(04)012,01 *Ariomma indica*

1,76(05) ANABANTIDAE
1,76(05)002,01 *Anabas testudineus*

1,76(08) CENTROLOPHIDAE
1,76(08)XXX,XX *Centrolophidae*
1,76(08)010,XX *Seriolella spp*
1,76(08)010,01 *Seriolella porosa*
1,76(08)010,02 *Seriolella brama*
1,76(08)010,04 *Seriolella punctata*
1,76(08)010,05 *Seriolella caerulea*
1,76(08)015,02 *Hyperoglyphe antarctica*
1,76(08)015,03 *Hyperoglyphe bythites*
1,76(08)020,01 *Psenopsis anomala*

1,76(09) OSPHRONEMIDAE
1,76(09)007,01 *Osphronemus goramy*

1,76(10) BELONTIIDAE
1,76(10)013,02 *Trichogaster pectoralis*

1,76(11) HELOSTOMATIDAE
1,76(11)006,01 *Helostoma temminckii*

1,77 OTHER PERCIFORMES

1,77(10) SPHYRAENIDAE
1,77(10)001,XX *Sphyraena spp*

1,77(19) CHANNIDAE
1,77(19)001,XX *Channa spp*
1,77(19)001,01 *Channa argus*
1,77(19)001,03 *Channa striata*
1,77(19)001,04 *Channa micropeltes*

1,78 SCORPAENIFORMES

1,78(01) SCORPAENIDAE
1,78(01)XXX,XX *Scorpaenidae*
1,78(01)001,XX *Sebastes spp*
1,78(01)001,01 *Sebastes marinus*
1,78(01)001,03 *Sebastes entomelas*
1,78(01)001,04 *Sebastes flavidus*
1,78(01)001,09 *Sebastes alutus*
1,78(01)001,10 *Sebastes paucispinis*
1,78(01)001,12 *Sebastes mentella*
1,78(01)001,13 *Sebastes capensis*
1,78(01)001,15 *Sebastes pinniger*
1,78(01)001,16 *Sebastes goodei*
1,78(01)001,32 *Sebastes melanops*
1,78(01)017,03 *Helicolenus dactylopterus*
1,78(01)079,01 *Sebastolobus alascanus*

1,78(02) TRIGLIDAE
1,78(02)XXX,XX *Triglidae*
1,78(02)003,01 *Chelidonichthys kumu*
1,78(02)003,02 *Chelidonichthys lucerna*
1,78(02)003,03 *Chelidonichthys cuculus*
1,78(02)003,07 *Chelidonichthys capensis*
1,78(02)020,XX *Prionotus spp*
1,78(02)025,01 *Pterygotrigla polyommata*
1,78(02)025,02 *Pterygotrigla picta*
1,78(02)070,01 *Eutrigla gurnardus*

1,78(07) HEXAGRAMMIDAE
1,78(07)005,01 *Ophiodon elongatus*
1,78(07)014,02 *Pleurogrammus azonus*

1,78(08) ANOPLOPOMATIDAE
1,78(08)004,01 *Anoplopoma fimbria*

1,78(09) PLATYCEPHALIDAE
1,78(09)XXX,XX *Platycephalidae*
1,78(09)018,01 *Platycephalus indicus*

1,78(13) COTTIDAE
1,78(13)XXX,XX *Cottidae*
1,78(13)012,XX *Myoxocephalus spp*
1,78(13)030,01 *Scorpaenichthys marmoratus*

1,78(16) NORMANICHTHYIDAE
1,78(16)022,05 *Normanichthys crockeri*

1,78(20) CYCLOPTERIDAE
1,78(20)003,01 *Cyclopterus lumpus*

1,83 PLEURONECTIFORMES

1,83(01) BOTHIDAE
1,83(01)XXX,XX *Bothidae*

1,83(02) PLEURONECTIDAE
1,83(02)002,01 *Hippoglossus hippoglossus*
1,83(02)002,02 *Hippoglossus stenolepis*
1,83(02)004,05 *Pleuronectes platessa*
1,83(02)004,08 *Pleuronectes quadrituberculat.*
1,83(02)004,15 *Pleuronectes vetulus*
1,83(02)005,01 *Reinhardtius hippoglossoides*
1,83(02)008,01 *Atheresthes stomias*

Systematic list of aquatic organisms	Liste systématique des organismes aquatiques	Lista sistemática de los organismos acuáticos

1,83(02)008,02 *Atheresthes evermanni*
1,83(02)010,01 *Eopsetta jordani*
1,83(02)011,02 *Glyptocephalus cynoglossus*
1,83(02)011,03 *Glyptocephalus zachirus*
1,83(02)014,01 *Hippoglossoides platessoides*
1,83(02)014,04 *Hippoglossoides elassodon*
1,83(02)024,02 *Limanda aspera*
1,83(02)024,04 *Limanda ferruginea*
1,83(02)024,05 *Limanda limanda*
1,83(02)043,01 *Lepidopsetta bilineata*
1,83(02)045,03 *Microstomus pacificus*
1,83(02)045,04 *Microstomus kitt*
1,83(02)048,02 *Platichthys flesus*
1,83(02)049,01 *Psettichthys melanostictus*
1,83(02)050,01 *Pseudopleuronectes herzenst.*
1,83(02)050,03 *Pseudopleuronectes americanus*
1,83(02)053,XX *Rhombosolea spp*
1,83(02)059,02 *Pleuronichthys decurrens*

1,83(03) SOLEIDAE
1,83(03)XXX,XX *Soleidae*
1,83(03)007,01 *Solea solea*
1,83(03)007,04 *Solea lascaris*
1,83(03)017,01 *Dicologlossa cuneata*
1,83(03)057,XX *Austroglossus spp*
1,83(03)057,01 *Austroglossus microlepis*
1,83(03)057,02 *Austroglossus pectoralis*
1,83(03)081,XX *Microchirus spp*

1,83(04) CYNOGLOSSIDAE
1,83(04)XXX,XX *Cynoglossidae*

1,83(05) SCOPHTHALMIDAE
1,83(05)XXX,XX *Scophthalmidae*
1,83(05)003,XX *Lepidorhombus spp*
1,83(05)003,01 *Lepidorhombus whiffiagonis*
1,83(05)064,01 *Scophthalmus rhombus*
1,83(05)064,02 *Scophthalmus aquosus*
1,83(05)092,01 *Psetta maxima*

1,83(07) PSETTODIDAE
1,83(07)001,01 *Psettodes erumei*
1,83(07)001,02 *Psettodes belcheri*

1,83(08) PARALICHTHYIDAE
1,83(08)009,02 *Citharichthys sordidus*
1,83(08)046,XX *Paralichthys spp*
1,83(08)046,01 *Paralichthys olivaceus*
1,83(08)046,03 *Paralichthys californicus*
1,83(08)046,06 *Paralichthys dentatus*

1,90 TETRAODONTIFORMES

1,90(01) OSTRACIIDAE
1,90(01)XXX,XX *Ostraciidae*

1,90(02) TETRAODONTIDAE
1,90(02)XXX,XX *Tetraodontidae*
1,90(02)006,XX *Sphoeroides spp*
1,90(02)006,12 *Sphoeroides maculatus*
1,90(02)011,01 *Takifugu vermicularis*

1,90(08) MOLIDAE
1,90(08)002,01 *Mola mola*

1,90(09) MONACANTHIDAE
1,90(09)004,XX *Cantherhines(=Navodon) spp*
1,90(09)010,01 *Stephanolepis cirrhifer*
1,90(09)018,01 *Parika scaber*

1,90(10) BALISTIDAE
1,90(10)XXX,XX *Balistidae*
1,90(10)002,01 *Balistes carolinensis*

1,93 BATRACHOIDIFORMES

1,93(01) BATRACHOIDIDAE
1,93(01)XXX,XX *Batrachoididae*

1,95 LOPHIIFORMES

1,95(01) LOPHIIDAE
1,95(01)XXX,XX *Lophiidae*
1,95(01)001,01 *Lophius piscatorius*
1,95(01)001,02 *Lophius budegassa*
1,95(01)001,03 *Lophius americanus*
1,95(01)001,05 *Lophius vaillanti*
1,95(01)001,06 *Lophius vomerinus*

1,99 PISCES MISCELLANEA

2 CRUSTACEA

2,02 ANOSTRACA

2,02(02) ARTEMIIDAE
2,02(02)001,01 *Artemia salina*

2,13 THORACICA

2,13(01) SCALPELLIDAE
2,13(01)005,01 *Mitella pollicipes*

2,13(02) LEPADIDAE
2,13(02)007,XX *Lepas spp*

2,13(03) BALANIDAE
2,13(03)015,01 *Megabalanus psittacus*

2,25 STOMATOPODA

2,25(01) SQUILLIDAE
2,25(01)XXX,XX *Squillidae*
2,25(01)001,02 *Squilla mantis*

2,26 EUPHAUSIACEA

2,26(01) EUPHAUSIIDAE
2,26(01)005,01 *Euphausia superba*
2,26(01)011,01 *Meganyctiphanes norvegica*

2,28 NATANTIA

2,28(01) PENAEIDAE
2,28(01)XXX,XX *Xiphopenaeus,Trachypenaeus spp*
2,28(01)001,XX *Penaeus spp*
2,28(01)001,01 *Penaeus aztecus*
2,28(01)001,03 *Penaeus merguiensis*
2,28(01)001,04 *Penaeus californiensis*
2,28(01)001,05 *Penaeus duorarum*
2,28(01)001,09 *Penaeus japonicus*
2,28(01)001,10 *Penaeus stylirostris*
2,28(01)001,11 *Penaeus vannamei*
2,28(01)001,12 *Penaeus monodon*
2,28(01)001,16 *Penaeus chinensis*
2,28(01)001,17 *Penaeus kerathurus*
2,28(01)001,19 *Penaeus brasiliensis*
2,28(01)001,20 *Penaeus semisulcatus*
2,28(01)001,22 *Penaeus setiferus*
2,28(01)001,23 *Penaeus brevirostris*
2,28(01)001,25 *Penaeus indicus*
2,28(01)001,28 *Penaeus latisulcatus*
2,28(01)001,29 *Penaeus occidentalis*
2,28(01)001,30 *Penaeus penicillatus*
2,28(01)001,31 *Penaeus notialis*
2,28(01)001,32 *Penaeus paulensis*
2,28(01)016,XX *Metapenaeus spp*
2,28(01)016,01 *Metapenaeus monoceros*
2,28(01)016,06 *Metapenaeus endeavouri*
2,28(01)016,07 *Metapenaeus joyneri*
2,28(01)017,01 *Parapenaeus longirostris*
2,28(01)019,XX *Parapenaeopsis spp*
2,28(01)019,02 *Parapenaeopsis atlantica*
2,28(01)022,01 *Xiphopenaeus kroyeri*
2,28(01)022,02 *Xiphopenaeus riveti*
2,28(01)043,02 *Trachypenaeus curvirostris*
2,28(01)067,01 *Artemesia longinaris*

2,28(02) ARISTAEIDAE
2,28(02)XXX,XX *Aristeidae*
2,28(02)021,01 *Plesiopenaeus edwardsianus*
2,28(02)031,01 *Aristeus antennatus*
2,28(02)031,02 *Aristeus varidens*

2,28(04) PANDALIDAE
2,28(04)XXX,XX *Pandalus spp, Pandalopsis spp*
2,28(04)002,XX *Pandalus spp*
2,28(04)002,01 *Pandalus hypsinotus*
2,28(04)002,03 *Pandalus borealis*
2,28(04)002,05 *Pandalus montagui*
2,28(04)002,07 *Pandalus goniurus*
2,28(04)002,08 *Pandalus kessleri*
2,28(04)005,01 *Heterocarpus reedi*
2,28(04)005,08 *Heterocarpus vicarius*
2,28(04)037,02 *Pandalopsis japonica*

2,28(07) SERGESTIDAE
2,28(07)XXX,XX *Sergestidae*
2,28(07)009,03 *Acetes japonicus*

Systematic list of aquatic organisms | Liste systématique des organismes aquatiques | Lista sistemática de los organismos acuáticos

2,28(12) PALAEMONIDAE
- 2,28(12)XXX,XX *Palaemonidae*
- 2,28(12)XXX,XX *Palaemonidae*
- 2,28(12)003,02 *Nematopalaemon schmitti*
- 2,28(12)018,10 *Palaemon serratus*
- 2,28(12)023,XX *Macrobrachium spp*
- 2,28(12)023,07 *Macrobrachium rosenbergii*

2,28(23) CRANGONIDAE
- 2,28(23)XXX,XX *Crangonidae*
- 2,28(23)003,03 *Crangon crangon*
- 2,28(23)006,XX *Sclerocrangon spp*

2,28(28) SICYONIIDAE
- 2,28(28)028,01 *Sicyonia brevirostris*
- 2,28(28)028,06 *Sicyonia ingentis*

2,28(29) SOLENOCERIDAE
- 2,28(29)006,01 *Pleoticus muelleri*
- 2,28(29)006,03 *Pleoticus robustus*
- 2,28(29)072,01 *Solenocera agassizii*
- 2,28(29)073,XX *Haliporoides spp*
- 2,28(29)073,01 *Haliporoides triarthrus*
- 2,28(29)073,02 *Haliporoides sibogae*
- 2,28(29)073,03 *Haliporoides diomedeae*

2,29 REPTANTIA

2,29(01) PALINURIDAE
- 2,29(01)XXX,XX *Palinuridae*
- 2,29(01)001,XX *Panulirus spp*
- 2,29(01)001,01 *Panulirus longipes*
- 2,29(01)001,08 *Panulirus argus*
- 2,29(01)001,15 *Panulirus cygnus*
- 2,29(01)001,16 *Panulirus gracilis*
- 2,29(01)002,01 *Jasus lalandii*
- 2,29(01)002,02 *Jasus frontalis*
- 2,29(01)002,03 *Jasus verreauxi*
- 2,29(01)002,05 *Jasus tristani*
- 2,29(01)002,06 *Jasus edwardsii*
- 2,29(01)002,08 *Jasus paulensis*
- 2,29(01)008,XX *Palinurus spp*
- 2,29(01)008,02 *Palinurus mauritanicus*
- 2,29(01)008,04 *Palinurus elephas*
- 2,29(01)008,05 *Palinurus delagoae*
- 2,29(01)008,06 *Palinurus gilchristi*

2,29(03) ASTACIDAE
- 2,29(03)027,02 *Astacus astacus*
- 2,29(03)076,01 *Pacifastacus leniusculus*
- 2,29(03)139,01 *Austropotamobius pallipes*

2,29(04) CAMBARIDAE
- 2,29(04)001,01 *Procambarus clarkii*

2,29(05) PARASTACIDAE
- 2,29(05)XXX,XX *Parastacidae*
- 2,29(05)158,01 *Euastacus armatus*

2,29(15) SCYLLARIDAE
- 2,29(15)XXX,XX *Scyllaridae*
- 2,29(15)001,04 *Ibacus ciliatus*
- 2,29(15)005,01 *Thenus orientalis*

2,29(42) NEPHROPIDAE
- 2,29(42)005,02 *Metanephrops mozambicus*
- 2,29(42)005,03 *Metanephrops andamanicus*
- 2,29(42)005,04 *Metanephrops challengeri*
- 2,29(42)006,02 *Nephrops norvegicus*
- 2,29(42)007,01 *Homarus americanus*
- 2,29(42)007,18 *Homarus gammarus*

2,29(49) UPOGEBIIDAE
- 2,29(49)001,03 *Upogebia pugettensis*

2,29(59) CALLIANASSIDAE
- 2,29(59)001,XX *Callianassa spp*

2,30 ANOMURA

2,30(19) GALATHEIDAE
- 2,30(19)XXX,XX *Galatheidae*
- 2,30(19)001,01 *Pleuroncodes planipes*
- 2,30(19)001,02 *Pleuroncodes monodon*
- 2,30(19)003,01 *Cervimunida johni*

2,30(20) LITHODIDAE
- 2,30(20)XXX,XX *Lithodidae*
- 2,30(20)018,XX *Paralithodes spp*
- 2,30(20)018,01 *Paralithodes camtschaticus*
- 2,30(20)018,02 *Paralithodes platypus*
- 2,30(20)018,03 *Paralithodes brevipes*
- 2,30(20)070,01 *Lithodes antarcticus*
- 2,30(20)070,05 *Lithodes aequispina*
- 2,30(20)123,01 *Paralomis granulosa*
- 2,30(20)123,03 *Paralomis spinosissima*
- 2,30(20)123,04 *Paralomis formosa*

2,31 BRACHYURA

2,31(09) CANCRIDAE
- 2,31(09)010,03 *Cancer irroratus*
- 2,31(09)010,04 *Cancer magister*
- 2,31(09)010,06 *Cancer pagurus*
- 2,31(09)010,07 *Cancer borealis*
- 2,31(09)010,08 *Cancer productus*

2,31(10) XANTHIDAE
- 2,31(10)050,01 *Menippe mercenaria*

2,31(11) PORTUNIDAE
- 2,31(11)004,XX *Portunus spp*
- 2,31(11)004,01 *Portunus pelagicus*
- 2,31(11)004,04 *Portunus trituberculatus*
- 2,31(11)012,02 *Callinectes sapidus*
- 2,31(11)012,05 *Callinectes danae*
- 2,31(11)090,01 *Carcinus maenas*
- 2,31(11)090,02 *Carcinus aestuarii*
- 2,31(11)140,01 *Scylla serrata*

2,31(21) MAJIDAE
- 2,31(21)001,02 *Mithrax armatus*
- 2,31(21)005,01 *Maja squinado*
- 2,31(21)145,XX *Chionoecetes spp*
- 2,31(21)145,01 *Chionoecetes opilio*

2,31(23) ATELECYCLIDAE
- 2,31(23)001,01 *Erimacrus isenbeckii*

2,31(43) GERYONIDAE
- 2,31(43)088,XX *Geryon spp*
- 2,31(43)088,01 *Geryon quinquedens*

2,99 CRUSTACEA MISCELLANEA

3 MOLLUSCA

3,07 GASTROPODA

3,07(01) LITTORINIDAE
- 3,07(01)001,XX *Littorina spp*
- 3,07(01)001,01 *Littorina littorea*

3,07(02) MURICIDAE
- 3,07(02)002,XX *Murex spp*
- 3,07(02)018,XX *Rapana spp*
- 3,07(02)023,01 *Concholepas concholepas*

3,07(03) HALIOTIDAE
- 3,07(03)001,XX *Haliotis spp*
- 3,07(03)001,09 *Haliotis gigantea*
- 3,07(03)001,11 *Haliotis midae*
- 3,07(03)001,13 *Haliotis rubra*
- 3,07(03)001,14 *Haliotis tuberculata*

3,07(04) TROCHIDAE
- 3,07(04)006,XX *Ex Trochus spp*

3,07(05) TURBINIDAE
- 3,07(05)002,XX *Ex Turbo spp*
- 3,07(05)002,02 *Turbo cornutus*

3,07(06) STROMBIDAE
- 3,07(06)002,XX *Strombus spp*

3,07(08) BUCCINIDAE
- 3,07(08)001,01 *Buccinum undatum*

3,07(09) MELONGENIDAE
- 3,07(09)003,XX *Busycon spp*

3,07(42) VOLUTIDAE
- 3,07(42)004,XX *Cymbium spp*
- 3,07(42)007,01 *Zidona dufresnei*

3,16 BIVALVIA

3,16(04) ARCIDAE
- 3,16(04)001,XX *Arca spp*
- 3,16(04)005,08 *Scapharca subcrenata*

Systematic list of aquatic organisms	Liste systématique des organismes aquatiques	Lista sistemática de los organismos acuáticos

3,16(04)071,XX *Anadara spp*
3,16(04)071,01 *Anadara granosa*

3,16(05) UNIONIDAE
3,16(05)XXX,XX *Ex Unionidae*

3,16(06) PTERIIDAE
3,16(06)006,XX *Ex Pinctada spp*

3,16(07) OSTREIDAE
3,16(07)002,XX *Ostrea spp*
3,16(07)002,03 *Ostrea chilensis*
3,16(07)002,05 *Ostrea edulis*
3,16(07)002,06 *Ostrea lurida*
3,16(07)002,20 *Ostrea lutaria*
3,16(07)008,XX *Crassostrea spp*
3,16(07)008,01 *Crassostrea gigas*
3,16(07)008,02 *Crassostrea rhizophorae*
3,16(07)008,03 *Crassostrea virginica*
3,16(07)008,11 *Crassostrea iredalei*

3,16(08) PECTINIDAE
3,16(08)XXX,XX *Pectinidae*
3,16(08)001,05 *Aequipecten opercularis*
3,16(08)003,09 *Pecten maximus*
3,16(08)003,11 *Pecten jacobaeus*
3,16(08)003,13 *Pecten novaezelandiae*
3,16(08)012,01 *Zygochlamis delicatula*
3,16(08)012,02 *Zygochlamis patagonica*
3,16(08)014,04 *Placopecten magellanicus*
3,16(08)030,01 *Argopecten gibbus*
3,16(08)030,02 *Argopecten irradians*
3,16(08)030,03 *Argopecten purpuratus*
3,16(08)030,04 *Argopecten ventricosus*
3,16(08)036,03 *Chlamys islandica*
3,16(08)066,01 *Patinopecten caurinus*
3,16(08)066,07 *Patinopecten yessoensis*

3,16(09) ARCTICIDAE
3,16(09)045,01 *Arctica islandica*

3,16(10) MYTILIDAE
3,16(10)XXX,XX *Mytilidae*
3,16(10)001,01 *Mytilus coruscus*
3,16(10)001,03 *Mytilus chilensis*
3,16(10)001,05 *Mytilus edulis*
3,16(10)001,08 *Mytilus platensis*
3,16(10)001,12 *Mytilus galloprovincialis*
3,16(10)001,17 *Mytilus planulatus*
3,16(10)026,01 *Choromytilus chorus*
3,16(10)028,XX *Modiolus spp*
3,16(10)032,01 *Perna perna*
3,16(10)032,02 *Perna viridis*
3,16(10)032,03 *Perna canaliculus*
3,16(10)038,01 *Aulacomya ater*

3,16(11) VENERIDAE
3,16(11)XXX,XX *Veneridae*
3,16(11)001,05 *Chamelea gallina*
3,16(11)003,01 *Venerupis pullastra*
3,16(11)007,04 *Chione stutchburyi*
3,16(11)017,XX *Meretrix spp*
3,16(11)017,01 *Meretrix lusoria*
3,16(11)020,01 *Ruditapes decussatus*
3,16(11)020,02 *Ruditapes philippinarum*
3,16(11)024,01 *Tawera gayi*
3,16(11)025,02 *Tivela mactroides*
3,16(11)037,02 *Saxidomus giganteus*
3,16(11)041,XX *Paphia spp*
3,16(11)055,02 *Protothaca staminea*
3,16(11)055,03 *Protothaca thaca*
3,16(11)075,01 *Mercenaria mercenaria*

3,16(12) MACTRIDAE
3,16(12)001,02 *Pseudocardium sybillae*
3,16(12)005,XX *Tresus spp*
3,16(12)020,01 *Spisula solidissima*
3,16(12)020,02 *Spisula polynyma*
3,16(12)020,05 *Spisula solida*
3,16(12)040,XX *Mulinia spp*

3,16(15) DONACIDAE
3,16(15)002,XX *Donax spp*

3,16(16) SOLENIDAE
3,16(16)003,XX *Solen spp*
3,16(16)005,02 *Ensis directus*
3,16(16)007,01 *Siliqua patula*

3,16(17) MYIDAE
3,16(17)006,01 *Mya arenaria*

3,16(18) HIATELLIDAE
3,16(18)089,01 *Panopea abrupta*

3,16(21) CORBICULIDAE
3,16(21)025,02 *Corbicula japonica*

3,16(23) CARDIIDAE
3,16(23)XXX,XX *Cardiidae*
3,16(23)002,03 *Cerastoderma edule*

3,16(24) MESODESMATIDAE
3,16(24)002,01 *Paphies australis*
3,16(24)039,01 *Mesodesma donacium*

3,16(39) SEMELIDAE
3,16(39)001,03 *Semele solida*

3,21 CEPHALOPODA

3,21(02) SEPIIDAE
3,21(02)XXX,XX *Sepiidae, Sepiolidae*
3,21(02)002,02 *Sepia officinalis*

3,21(04) LOLIGINIDAE
3,21(04)001,XX *Loligo spp*
3,21(04)001,02 *Loligo gahi*
3,21(04)001,05 *Loligo pealei*
3,21(04)001,12 *Loligo reynaudi*

3,21(05) OMMASTREPHIDAE
3,21(05)XXX,XX *Loliginidae, Ommastrephidae*
3,21(05)003,01 *Ommastrephes bartrami*
3,21(05)010,01 *Illex illecebrosus*
3,21(05)010,02 *Illex coindetii*
3,21(05)010,03 *Illex argentinus*
3,21(05)023,01 *Dosidicus gigas*
3,21(05)058,01 *Todarodes sagittatus*
3,21(05)058,03 *Todarodes pacificus*
3,21(05)059,01 *Nototodarus sloani*
3,21(05)060,01 *Martialia hyadesi*

3,21(09) OCTOPODIDAE
3,21(09)XXX,XX *Octopodidae*
3,21(09)005,07 *Octopus vulgaris*
3,21(09)024,XX *Eledone spp*

3,99 MOLLUSCA MISCELLANEA

4 MAMMALIA

4,06 PINNIPEDIA

4,06(01) OTARIIDAE
4,06(01)001,01 *Eumetopias jubatus*
4,06(01)002,01 *Callorhinus ursinus*
4,06(01)006,01 *Arctocephalus australis*
4,06(01)006,03 *Arctocephalus pusillus*
4,06(01)016,01 *Otaria byronia*

4,06(02) ODOBENIDAE
4,06(02)004,01 *Odobenus rosmarus*

4,06(03) PHOCIDAE
4,06(03)005,01 *Phoca groenlandica*
4,06(03)005,02 *Phoca vitulina*
4,06(03)005,03 *Phoca hispida*
4,06(03)005,04 *Phoca fasciata*
4,06(03)005,05 *Phoca caspica*
4,06(03)005,06 *Phoca sibirica*
4,06(03)005,07 *Phoca largha*
4,06(03)008,01 *Erignathus barbatus*
4,06(03)010,01 *Cystophora cristata*
4,06(03)011,01 *Halichoerus grypus*

4,14 SIRENIA

4,14(02) TRICHECHIDAE
4,14(02)003,02 *Trichechus inunguis*

4,22 ODONTOCETI

4,22(02) ZIPHIIDAE
4,22(02)005,01 *Hyperoodon ampullatus*
4,22(02)019,02 *Berardius bairdii*

4,22(03) PHYSETERIDAE
4,22(03)001,01 *Physeter catodon*

4,22(04) DELPHINIDAE
4,22(04)004,02 *Globicephala melas*
4,22(04)004,03 *Globicephala macrorhynchus*
4,22(04)006,01 *Delphinus delphis*
4,22(04)022,01 *Orcinus orca*
4,22(04)029,01 *Tursiops truncatus*

4,22(05) PHOCOENIDAE
4,22(05)002,01 *Phocoena phocoena*

4,22(06) MONODONTIDAE
4,22(06)014,01 *Delphinapterus leucas*
4,22(06)018,01 *Monodon monoceros*

4,23 MYSTICETI

4,23(02) BALAENOPTERIDAE
4,23(02)001,01 *Balaenoptera acutorostrata*
4,23(02)001,02 *Balaenoptera edeni*
4,23(02)001,03 *Balaenoptera borealis*
4,23(02)001,04 *Balaenoptera musculus*
4,23(02)001,06 *Balaenoptera physalus*
4,23(02)003,01 *Megaptera novaeangliae*

4,23(04) ESCHRICHTIIDAE
4,23(04)001,01 *Eschrichtius robustus*

4,99 MAMMALIA MISCELLANEA

5 AMPHIBIA, REPTILIA

5,12 ANURA

5,12(01) RANIDAE
5,12(01)001,XX *Rana spp*

5,31 TESTUDINES

5,31(06) EMYDIDAE
5,31(06)021,XX *Malaclemys spp*

5,31(07) CHELONIIDAE
5,31(07)005,02 *Chelonia mydas*
5,31(07)017,01 *Eretmochelys imbricata*
5,31(07)018,01 *Caretta caretta*

5,31(11) TRIONYCHIDAE
5,31(11)024,01 *Trionyx sinensis*

5,36 CROCODILIA

5,36(01) CROCODYLIDAE
5,36(01)XXX,XX *Crocodylidae*
5,36(01)001,03 *Caiman crocodilus*
5,36(01)002,01 *Alligator mississippiensis*
5,36(01)003,01 *Crocodylus porosus*
5,36(01)003,02 *Crocodylus siamensis*
5,36(01)003,03 *Crocodylus johnstoni*
5,36(01)003,04 *Crocodylus niloticus*
5,36(01)003,05 *Crocodylus novaeguineae*
5,36(01)003,06 *Crocodylus rhombifer*
5,36(01)003,07 *Crocodylus moreletii*
5,36(01)003,08 *Crocodylus acutus*
5,36(01)008,01 *Paleosuchus palpebrosus*
5,36(01)008,02 *Paleosuchus trigonatus*

6 INVERTEBRATA AQUATICA

6,15 DEMOSPONGIAE

6,15(01) SPONGIDAE
6,15(01)XXX,XX *Spongidae*

6,18 SCYPHOZOA

6,18(41) RHIZOSTOMIDAE
6,18(41)007,XX *Rhopilema spp*

6,19 ANTHOZOA

6,19(01) CORALLIIDAE
6,19(01)003,01 *Corallium rubrum*
6,19(01)003,02 *Corallium japonicum*
6,19(01)003,03 *Corallium elatius*
6,19(01)003,04 *Corallium konojoi*
6,19(01)003,05 *Corallium secundum*
6,19(01)003,07 *Corallium sp. nov.*

6,49 POLYCHAETA

6,56 XIPHOSURA

6,56(01) LIMULIDAE
6,56(01)001,02 *Limulus polyphemus*

6,89 ECHINODERMATA

6,91 ASTEROIDEA

6,91(05) ASTERIIDAE
6,91(05)001,01 *Asterias rubens*

6,93 ECHINOIDEA

6,93(02) STRONGYLOCENTROTIDAE
6,93(02)004,XX *Strongylocentrotus spp*

6,93(04) ECHINIDAE
6,93(04)007,01 *Paracentrotus lividus*
6,93(04)014,01 *Echinus esculentus*
6,93(04)017,01 *Loxechinus albus*

6,94 HOLOTHURIOIDEA

6,94(14) STICHOPODIDAE
6,94(14)004,01 *Stichopus japonicus*

6,96 ASCIDIACEA

6,96(09) PYURIDAE
6,96(09)005,01 *Pyura stolonifera*
6,96(09)005,02 *Pyura chilensis*
6,96(09)037,01 *Microcosmus sulcatus*

6,99 INVERTEBRATA AQUATICA MISCELL.

7 PLANTAE AQUATICAE

7,11 CYANOPHYCEAE

7,11(01) OSCILLATORIACEAE
7,11(01)001,XX *Spirulina spp*

7,41 CHLOROPHYCEAE

7,41(08) ULVACEAE
7,41(08)002,04 *Ulva pertusa*

7,71 PHAEOPHYCEAE

7,71(02) LAMINARIACEAE
7,71(02)002,01 *Laminaria digitata*
7,71(02)002,02 *Laminaria japonica*
7,71(02)002,04 *Laminaria hyperborea*

7,71(04) ALARIACEAE
7,71(04)003,01 *Undaria pinnatifida*

7,71(05) LESSONIACEAE
7,71(05)001,XX *Lessonia spp*
7,71(05)001,01 *Lessonia nigrescens*
7,71(05)001,02 *Lessonia trabeculata*
7,71(05)002,XX *Macrocystis spp*

7,71(06) FUCACEAE
7,71(06)006,01 *Ascophyllum nodosum*

7,71(08) DURVILLAEACEAE
7,71(08)001,01 *Durvillaea antartica*

7,71(09) CYSTOSEIRACEAE
7,71(09)001,01 *Cystoseira barbata*

7,87 RHODOPHYCEAE

7,87(02) PHYLLOPHORACEAE
7,87(02)001,XX *Phyllophora spp*
7,87(02)002,04 *Gymnogongrus furcellatus*

7,87(12) GRACILARIACEAE
7,87(12)004,XX *Gracilaria spp*

7,87(16) GIGARTINACEAE
7,87(16)001,04 *Chondrus crispus*
7,87(16)002,XX *Gigartina spp*
7,87(16)002,03 *Gigartina skottsbergii*

Systematic list of aquatic organisms	Liste systématique des organismes aquatiques	Lista sistemática de los organismos acuáticos

7,87(16)003,XX		*Iridaea spp*
7,87(20)	BANGIACEAE	
7,87(20)002,XX		*Porphyra spp*
7,87(20)002,08		*Porphyra tenera*
7,87(22)	RHODOMELACEAE	
7,87(22)002,02		*Digenea simplex*
7,87(26)	GELIDIACEAE	
7,87(26)002,XX		*Gelidium spp*
7,87(27)	KALLYMENIACEAE	
7,87(27)001,01		*Callophyllis variegata*
7,92	**ANGIOSPERMAE**	
7,92(02)	ZOSTERACEAE	
7,92(02)001,01		*Zostera marina*
7,92(04)	CYPERACEAE	
7,92(04)001,XX		*Scirpus spp*
7,99	**PLANTAE AQUATICAE MISCELLANEA**	

See paragraph 4 of the NOTES ON SPECIES ITEMS

Voir le paragraphe 4 des NOTES SUR LES CATEGORIES D'ESPECES

Véase al párrafo 4 en las NOTAS SOBRE LA PARTIDAS DE ESPECIES

Notes on major fishing areas

1. The 26 *major fishing areas*, internationally established for fishery statistical purposes, consist of:

   a) seven major inland fishing areas, covering the inland waters of the continents;

   b) nineteen major marine fishing areas, covering the waters of the Atlantic, Indian, Pacific and Southern Oceans with their adjacent seas.

2. The list of these 26 areas appears on the preceding page facing the world map on which the boundaries of the various major fishing areas are shown together with their identifying two-digit codes. These two-digit area codes are reflected in the table numbers in Section C.

3. The major fishing areas, inland and marine, are identified in all tables by two kinds of descriptors: their FAO names and/or two-digit codes. Space restrictions require the use in column headings of the word FISHING AREA when referring to *major fishing areas.*

4. Breakdown of catch statistics by subareas, divisions, subdivisions, etc., within these major fishing areas, are not given in this volume. These details appear in statistical bulletins issued regularly by various regional fishery organizations, councils, commissions, committees, etc., e.g. ICES, NAFO, ICCAT, GFCM, CECAF, etc.

5. Changes have been made in some of the boundaries between major fishing areas. The catch statistics for all areas and all years in this volume have been adjusted in accordance with the boundaries as now shown in the world map.

   The fishing area 07 ("Former USSR area – Inland waters") referred to the area that was formerly the Union of Soviet Socialist Republics. Since 1988, information for each new independent Republic is shown separately. The new independent Republics are: Armenia, Azerbaijan, Georgia, Kazakhstan, Kyrgyzstan, Tajikistan, Turkmenistan, Uzbekistan (statistics now assigned to the fishing area "Asia – Inland waters") and Belarus, Estonia, Latvia, Lithuania, Republic of Moldova, Russian Federation, Ukraine (assigned to the fishing area "Europe – Inland waters").

   Since 1981, fishing area 41 includes the waters lying between 55° and 60° South latitude and between 50° and 60° West longitude which during the 1970-80 period were included in the fishing area 48.

   Division of catches between fishing areas 57 and 71 was estimated following boundary change in the Strait of Malacca in 1986.

   Beginning with the 1989 data, catches have been adjusted in accordance with the boundary change between fishing areas 87 and 81; and between 87 and 77.

   Boundary between fishing areas 51 and 57 was modified starting with the 1999 data. As a consequence data for Sri Lanka have been moved from area 51 to area 57.

   Boundary between fishing areas 57 and 71 in the Australian-Indonesian region was modified starting with the 2001 data (see paragraphs on Indonesia and Timor-Leste of the NOTES ON INDIVIDUAL COUNTRIES OR AREAS).

6. Freshwater species are usually recorded as caught in inland waters. Small quantities of several freshwater species are caught regularly in parts of some seas with low salinities; these catches are included in the statistics of the appropriate marine area. Similarly, the catches of diadromous (anadromous and catadromous) species are shown in either the marine or inland area where caught.

Notes sur les principales zones de pêche

Les 26 *principales zones de pêche* établies au plan international à des fins statistiques, comprennent:

a) sept principales zones de pêche continentales, couvrant les eaux continentales des continents;

b) dix-neuf principales zones de pêche maritimes, couvrant les eaux des océans Atlantique, Indien, Pacifique et Austral et leurs mers limitrophes.

La liste de ces 26 zones figure à la page précédente en regard de la carte mondiale sur laquelle on a porté les limites des principales zones de pêche avec les codes à deux chiffres qui les identifient. Ces codes à deux chiffres se retrouvent dans les numéros des tableaux figurant à la Section C.

Les principales zones de pêche, tant continentales que maritimes, sont identifiées dans tous les tableaux par deux types de descripteurs: leur nom FAO et/ou leur code à deux chiffres. Faute d'espace, il a fallu utiliser le terme ZONE DE PÊCHE dans les têtes de colonne pour indiquer les *principales zones de pêche.*

Les ventilations des statistiques des captures par sous-zones, divisions, sous-divisions, etc., au sein de ces principales zones de pêche ne sont pas données dans le présent volume. Ces détails figurent dans les bulletins statistiques qui sont publiés périodiquement par les divers organismes régionaux de pêche, conseils, commissions ou comités des pêches, par exemple, CIEM, NAFO, CICTA, CGPM, COPACE, etc.

Des modifications ont été apportées à certaines des limites tracées des principales zones de pêche. Les statistiques des captures relatives à toutes les zones de pêche et à toutes les années incluses dans le présent volume ont été ajustées de manière à correspondre aux limites qui sont maintenant indiquées sur la carte mondiale.

La zone de pêche 07 (Zone de l'ex-URSS – Eaux continentales) correspondait au territoire de l'ancienne Union des Républiques socialistes soviétiques. Depuis 1988 les renseignements sont donnés séparément pour chaque république indépendante, à savoir: Arménie, Azerbaïdjan, Géorgie, Kazakhstan, Kirghizistan, Tadjikistan, Turkménistan, Ouzbékistan (statistiques sont attribuées à la zone de pêche Asie - Eaux continentales) et Bélarus, Estonie, Lettonie, Lituanie, République de Moldova, Fédération de Russie, Ukraine (statistiques attribuées à la zone de pêche "Europe - Eaux continentales").

Depuis 1981 la zone de pêche 41 comprend les eaux situées entre 55° et 60° de latitude sud et entre 50° et 60° de longitude ouest qui durant la période 1970-1980 étaient incluses dans la zone de pêche 48.

La répartition des captures entre les zones de pêche 57 et 71 était estimée après changement apporté en 1986 à la limite relative au détroit de Malacca.

À partir des données de 1989 les captures ont été ajustées de manière à correspondre à la modification apportée aux limites entre les zones de pêche 87 et 81, et entre les zones 87 et 77.

La limite entre les zones de pêche 51 et 57 a été modifiée à partir des données de 1999. En conséquence les données pour le Sri Lanka ont été portées de la zone de pêche 51 à la zone de pêche 57.

La limite entre les zones de pêche 57 et 71 dans la région australien-indonésienne a été modifiée à partir des données de 2001 (voir les paragraphes sur Indonésie et Timor-Leste des NOTES SUR DIVERS PAYS OU ZONES).

Les poissons d'eau douce sont d'ordinaire enregistrés comme captures faites dans les eaux continentales. De petites quantités de poissons d'eau douce de plusieurs espèces sont couramment capturées dans certaines mers dont la salinité est faible; ces captures figurent dans les statistiques relatives à la zone maritime appropriée. De même, les captures de poissons d'espèces diadromes (anadromes et catadromes) apparaissent dans les statistiques ayant trait soit aux zones maritimes, soit aux zones continentales où elles ont été effectuées.

Notas sobre las áreas principales de pesca

Las 26 *áreas principales de pesca* establecidas internacionalmente con fines estadísticos se dividen en:

a) siete áreas principales de pesca continental, que cubren las aguas continentales de los continentes;

b) diecinueve áreas principales de pesca marítimas, que cubren las aguas de los océanos Atlántico, Indico, Pacífico y Austral así como las de sus mares adyacentes.

La lista de estas 26 áreas de pesca puede verse en la página opuesta a la carta mundial, en la que se indican los límites de las distintas áreas principales de pesca con los códigos de dos dígitos correspondientes a cada área de pesca. Estos dos dígitos se utilizan luego en los números de los cuadros de la Sección C.

Las áreas principales de pesca, continentales y marinas, se identifican en todos los cuadros con dos tipos de descriptores: su nombre FAO y/o código de dos dígitos. Por razones de espacio, en los encabezamientos de las columnas se utiliza el término ÁREA DE PESCA para designar las *áreas principales de pesca.*

En el presente volumen no se desglosan las estadísticas de captura por subáreas, divisiones, subdivisiones, etc., dentro de las áreas principales de pesca. Esos desgloses pueden verse en los boletines estadísticos publicados regularmente por las distintas organizaciones regionales de pesca, consejos, comisiones, comités, etc., como el CIEM, la NAFO, la CICAA, el CGPM, el CPACO, etc.

Los límites de algunas de las áreas principales de pesca han sido modificados en los años pasados. Las estadísticas de captura de todas las áreas y años que figuran en este volumen se han ajustado en consecuencia, teniendo en cuenta los límites que ahora aparecen en la carta mundial.

La área de pesca 07 (Área de la ex URSS – Aguas continentales) correspondía a la antigua Unión de Repúblicas Socialistas Soviéticas. A partir del 1988 la información de cada una de las nuevas repúblicas independientes se expone por separado. Las nuevas repúblicas independientes de la ex URSS son: Armenia, Azerbaiyán, Georgia, Kazajstán, Kirguistán, Tayikistán, Turkmenistán, Uzbekistán (datos han sido atribuidos a la área de pesca Asia - Aguas continentales) y Belarús, Estonia, Letonia, Lituania, República de Moldova, Federación de Rusia, Ucrania (datos atribuidos a la área de pesca "Europa – Aguas continentales").

Desde 1981 la área de pesca 41 incluye las aguas comprendidas entre 55° y 60° de latitud Sur y entre 50° y 60° de longitud Oeste que durante el período 1970-80 eran incluidas en la área de pesca 48.

La distribución de las capturas entre las áreas de pesca 57 y 71 fue una estimación debida a la modificación hecha en 1986 al límite relativo al estrecho de Malaca.

A partir de datos de 1989 las capturas se han ajustado teniendo en cuenta las modificaciones hechas a los límites entre las áreas de pesca 87 y 81, y entre las áreas 87 y 77.

El límite entre las áreas de pesca 51 y 57 ha sido modificado a partir de datos de 1999. Consecuentemente se cambiaron los datos por Sri Lanka desde la área de pesca 51 a la 57.

El límite entre las áreas de pesca 57 y 71 en la región australiana-indonesia ha sido modificado a partir de datos de 2001 (véase los párrafos sobre Indonesia y Timor-Leste en las NOTAS SOBRE LOS DISTINTOS PAISES O ÁREAS).

Las especies de agua dulce se consignan de ordinario considerando que se han capturado en aguas continentales. Pequeñas cantidades de varias especies de agua dulce se capturan regularmente en algunas zonas marítimas de poca salinidad; dichas capturas se incluyen en las estadísticas correspondientes al área marítima en cuestión. De igual forma, las capturas de especies diadromas (anadromas y catadromas) se incluyen en el área marítima o continental en que se han capturadas.

List of major fishing areas

Liste des principales zones de pêche

Lista de las áreas principales de pesca

Code Code Código	Major fishing areas	Principales zones de pêche	Areas principales de pesca	Km²	%
	INLAND WATERS	EAUX CONTINENTALES	AGUAS CONTINENTALES	...	...
01	Africa - Inland waters	Afrique - Eaux continentales	Africa - Aguas continentales	...	...
02	America, North - Inland waters	Amérique du Nord - Eaux continentales	América del Norte - Aguas continentales	...	...
03	America, South - Inland waters	Amérique du Sud - Eaux continentales	América del Sur - Aguas continentales	...	...
04	Asia -Inland waters	Asie - Eaux continentales	Asia - Aguas continentales	...	...
05	Europe - Inland waters	Europe - Eaux continentales	Europa - Aguas continentales	...	...
06	Oceania - Inland waters	Océanie - Eaux continentales	Oceanía - Aguas continentales	...	...
07	(Former USSR area – Inland waters)	(Zone de l'ex-URSS – Eaux continentales)	(Área de la ex URSS – Aguas continentales)	...	...
08	Antarctica - Inland waters	Antarctique - Eaux continentales	Antártida - Aguas continentales	...	...
	MARINE AREAS	ZONES MARITIMES	AREAS MARITIMAS	360 900 000	100.0
	Atlantic Ocean and adjacent seas	*Océan Atlantique et mers limitrophes*	*Océano Atlántico y mares adyacentes*		
18	Arctic Sea	Mer Arctique	Mar Artico	9 300 000	2.6
21	Atlantic, Northwest	Atlantique, nord-ouest	Atlántico, noroeste	6 300 000	1.7
27	Atlantic, Northeast	Atlantique, nord-est	Atlántico, nordeste	14 400 000	4.0
31	Atlantic, Western Central	Atlantique, centre-ouest	Atlántico, centro-occidental	14 500 000	4.0
34	Atlantic, Eastern Central	Atlantique, centre-est	Atlántico, centro-oriental	14 100 000	3.9
37	Mediterranean and Black Sea	Méditerranée et mer Noire	Mediterráneo y mar Negro	3 000 000	0.8
41	Atlantic, Southwest	Atlantique, sud-ouest	Atlántico, sudoccidental	17 500 000	4.8
47	Atlantic, Southeast	Atlantique, sud-est	Atlántico, sudoriental	18 300 000	5.1
	Indian Ocean	*Océan Indien*	*Océano Indico*		
51	Indian Ocean, Western	Océan Indien, ouest	Océano Indico, occidental	29 300 000	8.1
57	Indian Ocean, Eastern	Océan Indien, est	Océano Indico, oriental	31 100 000	8.6
	Pacific Ocean	*Océan Pacifique*	*Océano Pacífico*		
61	Pacific, Northwest	Pacifique, nord-ouest	Pacífico, noroeste	21 500 000	6.0
67	Pacific, Northeast	Pacifique, nord-est	Pacífico, nordeste	7 600 000	2.1
71	Pacific, Western Central	Pacifique, centre-ouest	Pacífico, centro-occidental	33 300 000	9.2
77	Pacific, Eastern Central	Pacifique, centre-est	Pacífico, centro-oriental	48 100 000	13.3
81	Pacific, Southwest	Pacifique, sud-ouest	Pacífico, sudoccidental	27 700 000	7.7
87	Pacific, Southeast	Pacifique, sud-est	Pacífico, sudoriental	30 800 000	8.5
	Southern Ocean	*Océan Austral*	*Océano Austral*		
‣ 48	Atlantic, Antarctic	Atlantique, Antarctique	Atlántico, Antártico	11 800 000	3.3
‣ 58	Indian Ocean, Antarctic	Océan Indien, Antarctique	Océano Índico, Antártico	12 700 000	3.5
‣ 88	Pacific, Antarctic	Pacifique, Antarctique	Pacífico, Antártico	9 600 000	2.7

‣ See paragraph 3 of the INTRODUCTION

See paragraph 5 of the NOTES ON MAJOR FISHING AREAS for the changes of boundaries between some major fishing areas.

‣ Voir le paragraphe 3 de l'INTRODUCTION

Voir le paragraphe 5 de NOTES SUR LES PRINCIPALES ZONES DE PECHE par les modifications apportées à certaines des limites tracées des principales zones de pêche.

‣ Véase el párrafo 3 en la INTRODUCCION

Véase el párrafo 5 en las NOTAS SOBRE LOS DISTINTOS PAÍSES O ÁREAS por las modificaciones de los límites de algunas de las áreas principales de pesca.

List of countries or areas — Liste de pays ou zones — Lista de países o áreas

A	B	C	D	E	F	G	H	I	J
Afghanistan	AFG	AF	004	4	2	32	Afghanistan	Afghanistan	Afganistán
Albania	ALB	AL	008	5	1	51	Albania	Albanie	Albania
Algeria	DZA	DZ	012	1	2	11	Algeria	Algérie	Argelia
Amer Samoa	ASM	AS	016	6	2	60	American Samoa	Samoa américaines	Samoa Americana
Andorra	AND	AD	020	5	1	43	Andorra	Andorre	Andorra
Angola	AGO	AO	024	1	2	13	Angola	Angola	Angola
Anguilla	AIA	AI	660	2	2	23	Anguilla	Anguilla	Anguila
Antigua Barb	ATG	AG	028	2	2	23	Antigua and Barbuda	Antigua-et-Barbuda	Antigua y Barbuda
Argentina	ARG	AR	032	3	2	25	Argentina	Argentine	Argentina
Armenia	ARM	AM	051	4	1	54	Armenia	Arménie	Armenia
Aruba	ABW	AW	533	2	2	23	Aruba	Aruba	Aruba
Australia	AUS	AU	036	6	1	31	Australia	Australie	Australia
Austria	AUT	AT	040	5	1	41	Austria	Autriche	Austria
Azerbaijan	AZE	AZ	031	4	1	54	Azerbaijan	Azerbaïdjan	Azerbaiyán
Bahamas	BHS	BS	044	2	2	23	Bahamas	Bahamas	Bahamas
Bahrain	BHR	BH	048	4	2	32	Bahrain	Bahreïn	Bahrein
Bangladesh	BGD	BD	050	4	2	41	Bangladesh	Bangladesh	Bangladesh
Barbados	BRB	BB	052	2	2	23	Barbados	Barbade	Barbados
Belarus	BLR	BY	112	5	1	53	Belarus	Bélarus	Belarús
Belgium	BEL	BE	056	5	1	41	Belgium	Belgique	Bélgica
Belize	BLZ	BZ	084	2	2	22	Belize	Belize	Belice
Benin	BEN	BJ	204	1	2	12	Benin	Bénin	Benin
Bermuda	BMU	BM	060	2	2	21	Bermuda	Bermudes	Bermudas
Bhutan	BTN	BT	064	4	2	41	Bhutan	Bhoutan	Bhután
Bolivia	BOL	BO	068	3	2	25	Bolivia	Bolivie	Bolivia
Bosnia Herzg	BIH	BA	070	5	1	51	Bosnia and Herzegovina	Bosnie-Herzégovine	Bosnia y Herzegovina
Botswana	BWA	BW	072	1	2	15	Botswana	Botswana	Botswana
Brazil	BRA	BR	076	3	2	25	Brazil	Brésil	Brasil
Br Ind Oc Tr	IOT	IO	086	1	2	14	British Indian Ocean Ter	Terr. brit. océan Indien	Ter. brit. océano Indico
Br Virgin Is	VGB	VG	092	2	2	23	British Virgin Islands	Iles Vierges britanniq.	Islas Vírgenes Britán․
Brunei Darsm	BRN	BN	096	4	2	42	Brunei Darussalam	Brunéi Darussalam	Brunei Darussalam
Bulgaria	BGR	BG	100	5	1	51	Bulgaria	Bulgarie	Bulgaria
Burkina Faso	BFA	BF	854	1	2	12	Burkina Faso	Burkina Faso	Burkina Faso
Burundi	BDI	BI	108	1	2	14	Burundi	Burundi	Burundi
Cambodia	KHM	KH	116	4	2	42	Cambodia	Cambodge	Camboya
Cameroon	CMR	CM	120	1	2	13	Cameroon	Cameroun	Camerún
Canada	CAN	CA	124	2	1	10	Canada	Canada	Canadá
Cape Verde	CPV	CV	132	1	2	12	Cape Verde	Cap-Vert	Cabo Verde
Cayman Is	CYM	KY	136	2	2	23	Cayman Islands	Iles Caïmanes	Islas Caimán
Cent Afr Rep	CAF	CF	140	1	2	13	Central African Republic	Rép. centrafricaine	República Centroafricana
Chad	TCD	TD	148	1	2	13	Chad	Tchad	Chad
Channel Is	...	..	830	5	1	42	Channel Islands	Iles Anglo-Normandes	Islas Anglonormandas
Chile	CHL	CL	152	3	2	24	Chile	Chili	Chile
China	CHN	CN	156	4	2	51	China	Chine	China
China,H.Kong	HKG	HK	344	4	2	51	China, Hong Kong SAR	Chine, RAS de Hong-Kong	China, RAE de Hong Kong
China, Macao	MAC	MO	446	4	2	51	China, Macao SAR	Chine, RAS de Macao	China, RAE de Macao
China,Taiwan	TWN	TW	158	4	2	51	Taiwan Province of China	Prov. chinoise de Taïwan	Prov. china de Taiwán
Christmas Is	CXR	CX	162	6	2	60	Christmas Island	Ile Christmas	Isla Christmas
Cocos Is	CCK	CC	166	6	2	60	Cocos (Keeling) Islands	Iles des Cocos (Keeling)	Islas Cocos (Keeling)
Colombia	COL	CO	170	3	2	25	Colombia	Colombie	Colombia
Comoros	COM	KM	174	1	2	14	Comoros	Comores	Comoras
Congo Dem R	COD	CD	180	1	2	13	Congo, Dem. Rep. of the	Rép. dém. du Congo	Rep. Dem. del Congo
Congo Rep	COG	CG	178	1	2	13	Congo, Republic of	République du Congo	República del Congo
Cook Is	COK	CK	184	6	2	60	Cook Islands	Iles Cook	Islas Cook
Costa Rica	CRI	CR	188	2	2	22	Costa Rica	Costa Rica	Costa Rica
Côte dIvoire	CIV	CI	384	1	2	12	Côte d'Ivoire	Côte d'Ivoire	Côte d'Ivoire
Croatia	HRV	HR	191	5	1	51	Croatia	Croatie	Croacia
Cuba	CUB	CU	192	2	2	23	Cuba	Cuba	Cuba
Cyprus	CYP	CY	196	4	2	32	Cyprus	Chypre	Chipre
Czech Rep	CZE	CZ	203	5	1	51	Czech Republic	République tchèque	República Checa
Czechoslovak	CSK	CS	200	5	1	51	Czechoslovakia	Tchécoslovaquie	Checoslovaquia
Denmark	DNK	DK	208	5	1	41	Denmark	Danemark	Dinamarca
Djibouti	DJI	DJ	262	1	2	14	Djibouti	Djibouti	Djibouti
Dominica	DMA	DM	212	2	2	23	Dominica	Dominique	Dominica
Dominican Rp	DOM	DO	214	2	2	23	Dominican Republic	République dominicaine	República Dominicana
Ecuador	ECU	EC	218	3	2	24	Ecuador	Equateur	Ecuador
Egypt	EGY	EG	818	1	2	31	Egypt	Egypte	Egipto
El Salvador	SLV	SV	222	2	2	22	El Salvador	El Salvador	El Salvador
Eq Guinea	GNQ	GQ	226	1	2	13	Equatorial Guinea	Guinée équatoriale	Guinea Ecuatorial
Eritrea	ERI	ER	232	1	2	14	Eritrea	Erythrée	Eritrea
Estonia	EST	EE	233	5	1	53	Estonia	Estonie	Estonia
Ethiopia	ETH	ET	231	1	2	14	Ethiopia	Ethiopie	Etiopía
Faeroe Is	FRO	FO	234	5	1	42	Faeroe Islands	Iles Féroé	Islas Feroe
Falkland Is	FLK	FK	238	3	2	25	Falkland Is.(Malvinas)	Iles Falkland(Malvinas)	Islas Malvinas(Falkland)
Fiji Islands	FJI	FJ	242	6	2	60	Fiji Islands	Iles Fidji	Islas Fiji
Finland	FIN	FI	246	5	1	41	Finland	Finlande	Finlandia
France	FRA	FR	250	5	1	41	France	France	Francia
Fr Guiana	GUF	GF	254	3	2	25	French Guiana	Guyane française	Guayana Francesa
Fr Polynesia	PYF	PF	258	6	2	60	French Polynesia	Polynésie française	Polinesia Francesa
Fr South Tr	ATF	TF	260	1	2	15	French Southern Terr	Terres australes fr.	Tierras Australes Fr.
Gabon	GAB	GA	266	1	2	13	Gabon	Gabon	Gabón

List of countries or areas   Liste de pays ou zones   Lista de países o áreas

A	B	C	D	E	F	G	H	I	J
Gambia	GMB	GM	270	1	2	12	Gambia	Gambie	Gambia
Gaza Strip	PSE	PS	274	4	2	32	Gaza Strip(Palestine)	Bande de Gaza(Palestine)	Faja de Gaza(Palestine)
Georgia	GEO	GE	268	4	1	54	Georgia	Géorgie	Georgia
Germany	DEU	DE	276	5	1	41	Germany	Allemagne	Alemania
Ghana	GHA	GH	288	1	2	12	Ghana	Ghana	Ghana
Gibraltar	GIB	GI	292	5	1	43	Gibraltar	Gibraltar	Gibraltar
Greece	GRC	GR	300	5	1	41	Greece	Grèce	Grecia
Greenland	GRL	GL	304	2	2	21	Greenland	Groenland	Groenlandia
Grenada	GRD	GD	308	2	2	23	Grenada	Grenade	Granada
Guadeloupe	GLP	GP	312	2	2	23	Guadeloupe	Guadeloupe	Guadalupe
Guam	GUM	GU	316	6	2	60	Guam	Guam	Guam
Guatemala	GTM	GT	320	2	2	22	Guatemala	Guatemala	Guatemala
Guinea	GIN	GN	324	1	2	12	Guinea	Guinée	Guinea
GuineaBissau	GNB	GW	624	1	2	12	Guinea-Bissau	Guinée-Bissau	Guinea-Bissau
Guyana	GUY	GY	328	3	2	25	Guyana	Guyana	Guyana
Haiti	HTI	HT	332	2	2	23	Haiti	Haïti	Haití
Honduras	HND	HN	340	2	2	22	Honduras	Honduras	Honduras
Hungary	HUN	HU	348	5	1	51	Hungary	Hongrie	Hungría
Iceland	ISL	IS	352	5	1	42	Iceland	Islande	Islandia
India	IND	IN	356	4	2	41	India	Inde	India
Indonesia	IDN	ID	360	4	2	42	Indonesia	Indonésie	Indonesia
Iran	IRN	IR	364	4	2	32	Iran (Islamic Rep. of)	Iran (Rép. islamique d')	Irán (Rep. Islámica del)
Iraq	IRQ	IQ	368	4	2	32	Iraq	Iraq	Iraq
Ireland	IRL	IE	372	5	1	41	Ireland	Irlande	Irlanda
Isle of Man	IMY	IM	833	5	1	42	Isle of Man	Ile de Man	Isla de Man
Israel	ISR	IL	376	4	1	20	Israel	Israël	Israel
Italy	ITA	IT	380	5	1	41	Italy	Italie	Italia
Jamaica	JAM	JM	388	2	2	23	Jamaica	Jamaïque	Jamaica
Japan	JPN	JP	392	4	1	20	Japan	Japon	Japón
Jordan	JOR	JO	400	4	2	32	Jordan	Jordanie	Jordania
Kazakhstan	KAZ	KZ	398	4	1	54	Kazakhstan	Kazakhstan	Kazajstán
Kenya	KEN	KE	404	1	2	14	Kenya	Kenya	Kenya
Kiribati	KIR	KI	296	6	2	60	Kiribati	Kiribati	Kiribati
Korea D P Rp	PRK	KP	408	4	2	42	Korea, Dem. People's Rep	Rép. pop. dém. de Corée	Rep. Pop. Dem. de Corea
Korea Rep	KOR	KR	410	4	2	42	Korea, Republic of	République de Corée	República de Corea
Kuwait	KWT	KW	414	4	2	32	Kuwait	Koweït	Kuwait
Kyrgyzstan	KGZ	KG	417	4	1	54	Kyrgyzstan	Kirghizistan	Kirguistán
Laos	LAO	LA	418	4	2	42	Lao People's Dem. Rep.	Rép. dém. pop. lao	Rep. Dem. Pop. Lao
Latvia	LVA	LV	428	5	1	53	Latvia	Lettonie	Letonia
Lebanon	LBN	LB	422	4	2	32	Lebanon	Liban	Líbano
Lesotho	LSO	LS	426	1	2	15	Lesotho	Lesotho	Lesotho
Liberia	LBR	LR	430	1	2	12	Liberia	Libéria	Liberia
Libya	LBY	LY	434	1	2	31	Libyan Arab Jamahiriya	Jamahiriya arabe libyen.	Jamahiriya Arabe Libia
Liechtensten	LIE	LI	438	5	1	43	Liechtenstein	Liechtenstein	Liechtenstein
Lithuania	LTU	LT	440	5	1	53	Lithuania	Lituanie	Lituania
Luxembourg	LUX	LU	442	5	1	41	Luxembourg	Luxembourg	Luxemburgo
Macedonia	MKD	MK	807	5	1	51	Macedonia, Fmr Yug Rp of	Ex-Rép.youg.de Macédoine	Ex Rep.Yug. de Macedonia
Madagascar	MDG	MG	450	1	2	14	Madagascar	Madagascar	Madagascar
Malawi	MWI	MW	454	1	2	14	Malawi	Malawi	Malawi
Malaysia	MYS	MY	458	4	2	42	Malaysia	Malaisie	Malasia
Maldives	MDV	MV	462	4	2	41	Maldives	Maldives	Maldivas
Mali	MLI	ML	466	1	2	12	Mali	Mali	Malí
Malta	MLT	MT	470	5	1	43	Malta	Malte	Malta
Marshall Is	MHL	MH	584	6	2	60	Marshall Islands	Iles Marshall	Islas Marshall
Martinique	MTQ	MQ	474	2	2	23	Martinique	Martinique	Martinica
Mauritania	MRT	MR	478	1	2	12	Mauritania	Mauritanie	Mauritania
Mauritius	MUS	MU	480	1	2	14	Mauritius	Maurice	Mauricio
Mayotte	MYT	YT	175	1	2	14	Mayotte	Mayotte	Mayotte
Mexico	MEX	MX	484	2	2	22	Mexico	Mexique	México
Micronesia	FSM	FM	583	6	2	60	Micronesia,Fed.States of	Micronésie(Etats féd.de)	Micronesia(Estados Fed.)
Moldova Rep	MDA	MD	498	5	1	53	Moldova, Republic of	République de Moldova	República de Moldova
Monaco	MCO	MC	492	5	1	43	Monaco	Monaco	Mónaco
Mongolia	MNG	MN	496	4	2	42	Mongolia	Mongolie	Mongolia
Montserrat	MSR	MS	500	2	2	23	Montserrat	Montserrat	Montserrat
Morocco	MAR	MA	504	1	2	11	Morocco	Maroc	Marruecos
Mozambique	MOZ	MZ	508	1	2	14	Mozambique	Mozambique	Mozambique
Myanmar	MMR	MM	104	4	2	42	Myanmar	Myanmar	Myanmar
Namibia	NAM	NA	516	1	2	15	Namibia	Namibie	Namibia
Nauru	NRU	NR	520	6	2	60	Nauru	Nauru	Nauru
Nepal	NPL	NP	524	4	2	41	Nepal	Népal	Nepal
Netherlands	NLD	NL	528	5	1	41	Netherlands	Pays-Bas	Países Bajos
NethAntilles	ANT	AN	530	2	2	23	Netherlands Antilles	Antilles néerlandaises	Antillas Neerlandesas
NewCaledonia	NCL	NC	540	6	2	60	New Caledonia	Nouvelle-Calédonie	Nueva Caledonia
New Zealand	NZL	NZ	554	6	1	31	New Zealand	Nouvelle-Zélande	Nueva Zelandia
Nicaragua	NIC	NI	558	2	2	22	Nicaragua	Nicaragua	Nicaragua
Niger	NER	NE	562	1	2	12	Niger	Niger	Níger
Nigeria	NGA	NG	566	1	2	12	Nigeria	Nigéria	Nigeria
Niue	NIU	NU	570	6	2	60	Niue	Nioué	Niue
Norfolk Is	NFK	NF	574	6	2	60	Norfolk Island	Ile Norfolk	Isla Norfolk

List of countries or areas | Liste de pays ou zones | Lista de países o áreas

A	B	C	D	E	F	G	H	I	J
N Marianas	MNP	MP	580	6	2	60	Northern Mariana Is.	Iles Mariannes septentr.	Islas Marianas Septent.
Norway	NOR	NO	578	5	1	42	Norway	Norvège	Noruega
Oman	OMN	OM	512	4	2	32	Oman	Oman	Omán
Pakistan	PAK	PK	586	4	2	41	Pakistan	Pakistan	Pakistán
Palau	PLW	PW	585	6	2	60	Palau	Palaos	Palau
Panama	PAN	PA	591	2	2	22	Panama	Panama	Panamá
Papua N Guin	PNG	PG	598	6	2	60	Papua New Guinea	Papouasie-Nlle-Guinée	Papua Nueva Guinea
Paraguay	PRY	PY	600	3	2	25	Paraguay	Paraguay	Paraguay
Peru	PER	PE	604	3	2	24	Peru	Pérou	Perú
Philippines	PHL	PH	608	4	2	42	Philippines	Philippines	Filipinas
Pitcairn Is	PCN	PN	612	6	2	60	Pitcairn Islands	Iles Pitcairn	Islas Pitcairn
Poland	POL	PL	616	5	1	51	Poland	Pologne	Polonia
Portugal	PRT	PT	620	5	1	41	Portugal	Portugal	Portugal
Puerto Rico	PRI	PR	630	2	2	23	Puerto Rico	Porto Rico	Puerto Rico
Qatar	QAT	QA	634	4	2	32	Qatar	Qatar	Qatar
Réunion	REU	RE	638	1	2	14	Réunion	Réunion	Reunión
Romania	ROU	RO	642	5	1	51	Romania	Roumanie	Rumania
Russian Fed	RUS	RU	643	5	1	53	Russian Federation	Fédération de Russie	Federación de Rusia
Rwanda	RWA	RW	646	1	2	14	Rwanda	Rwanda	Rwanda
St Helena	SHN	SH	654	1	2	13	Saint Helena	Sainte-Hélène	Santa Elena
St Kitts Nev	KNA	KN	659	2	2	23	Saint Kitts and Nevis	Saint-Kitts-et-Nevis	Saint Kitts y Nevis
St Lucia	LCA	LC	662	2	2	23	Saint Lucia	Sainte-Lucie	Santa Lucía
St Pier Mq	SPM	PM	666	2	2	21	St. Pierre and Miquelon	Saint-Pierre-et-Miquelon	San Pedro y Miquelón
St Vincent	VCT	VC	670	2	2	23	Saint Vincent/Grenadines	Saint-Vincent/Grenadines	San Vicente/Grenadinas
Samoa	WSM	WS	882	6	2	60	Samoa	Samoa	Samoa
Sao Tome Prn	STP	ST	678	1	2	13	Sao Tome and Principe	Sao Tomé-et-Principe	Santo Tomé y Príncipe
Saudi Arabia	SAU	SA	682	4	2	32	Saudi Arabia	Arabie saoudite	Arabia Saudita
Senegal	SEN	SN	686	1	2	12	Senegal	Sénégal	Senegal
Seychelles	SYC	SC	690	1	2	14	Seychelles	Seychelles	Seychelles
Sierra Leone	SLE	SL	694	1	2	12	Sierra Leone	Sierra Leone	Sierra Leona
Singapore	SGP	SG	702	4	2	42	Singapore	Singapour	Singapur
Slovakia	SVK	SK	703	5	1	51	Slovakia	Slovaquie	Eslovaquia
Slovenia	SVN	SI	705	5	1	51	Slovenia	Slovénie	Eslovenia
Solomon Is	SLB	SB	090	6	2	60	Solomon Islands	Iles Salomon	Islas Salomón
Somalia	SOM	SO	706	1	2	14	Somalia	Somalie	Somalia
South Africa	ZAF	ZA	710	1	1	32	South Africa	Afrique du Sud	Sudáfrica
Spain	ESP	ES	724	5	1	41	Spain	Espagne	España
Sri Lanka	LKA	LK	144	4	2	41	Sri Lanka	Sri Lanka	Sri Lanka
Sudan	SDN	SD	736	1	2	31	Sudan	Soudan	Sudán
Suriname	SUR	SR	740	3	2	25	Suriname	Suriname	Suriname
Svalbard Is	SJM	SJ	744	5	1	42	Svalbard and Jan Mayen	Iles Svalbard/Jan Mayen	Islas Svalbard/Jan Mayen
Swaziland	SWZ	SZ	748	1	2	15	Swaziland	Swaziland	Swazilandia
Sweden	SWE	SE	752	5	1	41	Sweden	Suède	Suecia
Switzerland	CHE	CH	756	5	1	43	Switzerland	Suisse	Suiza
Syria	SYR	SY	760	4	2	32	Syrian Arab Republic	Rép. arabe syrienne	República Arabe Siria
Tajikistan	TJK	TJ	762	4	1	54	Tajikistan	Tadjikistan	Tayikistán
Tanzania	TZA	TZ	834	1	2	14	Tanzania, United Rep. of	Rép.-Unie de Tanzanie	Rep. Unida de Tanzanía
Thailand	THA	TH	764	4	2	42	Thailand	Thaïlande	Tailandia
Timor-Leste	TLS	TL	626	4	2	42	Timor-Leste	Timor-Leste	Timor-Leste
Togo	TGO	TG	768	1	2	12	Togo	Togo	Togo
Tokelau	TKL	TK	772	6	2	60	Tokelau	Tokélaou	Tokelau
Tonga	TON	TO	776	6	2	60	Tonga	Tonga	Tonga
Trinidad Tob	TTO	TT	780	2	2	23	Trinidad and Tobago	Trinité-et-Tobago	Trinidad y Tabago
Tunisia	TUN	TN	788	1	2	11	Tunisia	Tunisie	Túnez
Turkey	TUR	TR	792	4	2	32	Turkey	Turquie	Turquía
Turkmenistan	TKM	TM	795	4	1	54	Turkmenistan	Turkménistan	Turkmenistán
Turks Caicos	TCA	TC	796	2	2	23	Turks and Caicos Is.	Iles Turques et Caïques	Islas Turcas y Caicos
Tuvalu	TUV	TV	798	6	2	60	Tuvalu	Tuvalu	Tuvalu
Uganda	UGA	UG	800	1	2	14	Uganda	Ouganda	Uganda
Ukraine	UKR	UA	804	5	1	53	Ukraine	Ukraine	Ucrania
Untd Arab Em	ARE	AE	784	4	2	32	United Arab Emirates	Emirats arabes unis	Emiratos Arabes Unidos
UK	GBR	GB	826	5	1	41	United Kingdom	Royaume-Uni	Reino Unido
USA	USA	US	840	2	1	10	United States of America	Etats-Unis d'Amérique	EstadosUnidos de América
US Minor Is	UMI	UM	581	6	2	60	US Minor Outlying Is.	Iles Mineures EloignésEU	Is Menores PeriféricasEU
US Virgin Is	VIR	VI	850	2	2	23	US Virgin Islands	Iles Vierges américaines	Islas Vírgenes de los EU
Uruguay	URY	UY	858	3	2	25	Uruguay	Uruguay	Uruguay
Uzbekistan	UZB	UZ	860	4	1	54	Uzbekistan	Ouzbékistan	Uzbekistán
Vanuatu	VUT	VU	548	6	2	60	Vanuatu	Vanuatu	Vanuatu
Venezuela	VEN	VE	862	3	2	25	Venezuela	Venezuela	Venezuela
Viet Nam	VNM	VN	704	4	2	42	Viet Nam	Viet Nam	Viet Nam
Wallis Fut I	WLF	WF	876	6	2	60	Wallis and Futuna Is.	Iles Wallis-et-Futuna	Islas Wallis y Futuna
Westn Sahara	ESH	EH	732	1	2	11	Western Sahara	Sahara occidental	Sahara Occidental
Yemen	YEM	YE	887	4	2	32	Yemen	Yémen	Yemen
Yugoslavia	YUG	YU	891	5	1	51	Yugoslavia, Fed. Rep. of	Rép. féd. de Yougoslavie	Rep. Fed. de Yugoslavia
Zambia	ZMB	ZM	894	1	2	14	Zambia	Zambie	Zambia
Zimbabwe	ZWE	ZW	716	1	2	14	Zimbabwe	Zimbabwe	Zimbabwe

List of countries or areas / Liste de pays ou zones / Lista de países o áreas

A	B	C	D	E	F	G	H	I	J
Other nei	...	...	896	9	1	80	Other nei	Autres nca	Otros nep

Column A = FAO multilingual country or area code (maximum twelve characters) used for statistical purposes.

Column B = Alpha-3 country or area code: ISO (International Organization for Standardization).

Column C = Alpha-2 country or area code: ISO (International Organization for Standardization).

Column D = Three-digit UN numerical country or area code as published in the "STANDARD COUNTRY OR AREA CODES FOR STATISTICAL USE" Statistical Papers, Series M., No. 49, Rev. 4; United Nations, New York, 2002.

Column E = Continents:
- 1 = Africa
- 2 = America, North
- 3 = America, South
- 4 = Asia
- 5 = Europe
- 6 = Oceania
- 9 = Other nei

Column F = Economic classes.

Column G = Regions and sub-regions.

Column H = Country or area names in English (maximum twenty-four characters).

Column I = Country or area names in French (maximum twenty-four characters).

Column J = Country or area names in Spanish (maximum twenty-four characters).

Colonne A = Noms de pays ou zone en code multilingue FAO (n'excédant pas douze lettres) utilisés à des fins statistiques.

Colonne B = Code Alpha-3 des pays ou zones: ISO (Organisation internationale de normalisation).

Colonne C = Code Alpha-2 des pays ou zones: ISO (Organisation internationale de normalisation).

Colonne D = Code numérique à trois chiffres pour les pays ou zones des Nations Unies, tel que publié dans "STANDARD COUNTRY OR AREA CODES FOR STATISTICAL USE" (Code type pour les pays ou les zones des Nations Unies à des fins statistiques), Statistical Papers, Series M., No. 49, Rev. 4; United Nations, New York, 2002.

Colonne E = Continents:
- 1 = Afrique
- 2 = Amérique du Nord
- 3 = Amérique du Sud
- 4 = Asie
- 5 = Europe
- 6 = Océanie
- 9 = Autres nca

Colonne F = Catégories économiques.

Colonne G = Régions et sous-régions.

Colonne H = Noms de pays ou zone en anglais (n'excédant pas vingt-quatre espaces typographiques).

Colonne I = Noms de pays ou zone en français (n'excédant pas vingt-quatre espaces typographiques).

Colonne J = Noms de pays ou zone en espagnol (n'excédant pas vingt-quatre espaces typographiques).

Columna A = Nombres multilingües de los países o áreas utilizados por la FAO para fines estadísticos (máximo, doce espacios).

Columna B = Código Alfa-3 de los países o áreas: ISO (Organización Internacional de Normalización).

Columna C = Código Alfa-2 de los países o áreas: ISO (Organización Internacional de Normalización).

Columna D = Código numérico de tres dígitos atribuido al país o área en cuestión por las Naciones Unidas, tal como aparece en la publicación "STANDARD COUNTRY OR AREA CODES FOR STATISTICAL USE" (Código estandar por los países o áreas de las Naciones Unidas para fines estadísticos), Statistical Papers, Series M., No. 49. Rev. 4; United Nations, New York, 2002.

Columna E = Continentes:
- 1 = Africa
- 2 = América del Norte
- 3 = América del Sur
- 4 = Asia
- 5 = Europa
- 6 = Oceanía
- 9 = Otros nep

Columna F = Clases económicas.

Columna G = Regiones y subregiones.

Columna H = Nombres de los países o áreas en inglés (máximo, veinte y cuatro espacios).

Columna I = Nombres de los países o áreas en francés (máximo, veinte y cuatro espacios).

Columna J = Nombres de los países o áreas en español (máximo, veinte y cuatro espacios).

Notes on countries or areas

1. The designations employed and the presentation of material in this publication do not imply the expression of any opinion whatsoever on the part of the Food and Agriculture Organization of the United Nations concerning the legal status of any country, territory, city or area or of its authorities, or concerning the delimitation of its frontiers or boundaries.

2. The term *country or area* as used in the stubs and the column headings of the tables also covers territories, cities, land areas, as well as provinces, districts, enclaves, exclaves and other parts of territories or combinations of countries or areas such as economic or customs unions.

3. In all tables a country or area entry is designated by a multilingual code of not more than 12 characters. This code is keyed in the LIST OF COUNTRIES OR AREAS to the other descriptors for each country or area entry - the names in English, French and Spanish, the ISO's 3-alpha and 2-alpha codes, the UN's three-digit numerical code, etc.

4. Country or area names and designations are subject to nationally announced changes. Name changes announced after 31 December 2001 have not necessarily been incorporated in this volume but will be reflected in future ones.

5. The flag of the vessel performing the essential part of the operation catching the fish should be considered the paramount indication of the nationality assigned to the catch data.

   The flag State of the vessel performing the essential part of the fishing operation should be responsible for the provision of catch and landing data.

   Where a foreign flag vessel is fishing in the waters under the national jurisdiction of another State, the flag State of the vessel should have at all times the responsibility to provide relevant catch and landing data. The only exceptions to this shall be:

   (a) where the vessel undertakes fishing under a charter agreement or arrangement to augment the local fishing fleet, and the vessel has become for all practical purposes a local fishing vessel of the host country;

   (b) where the vessel undertakes fishing pursuant to a joint venture or similar arrangement in waters under the national jurisdiction of another State and the vessel is operating for all practical purposes as a local vessel, or its operation has become, or is intended to become, an integral part of the economy of the host country.

   In any situation where there is uncertainty as to the application of these criteria, any agreement, charter, joint venture or other similar arrangement should contain a provision setting out clearly the responsibility for reporting catch and landing data, which should be reported to the flag State and, where relevant, to any coastal State in whose waters fishing operations are to take place or competent subregional, regional or global fisheries organization or arrangement.

6. National data cover all quantities caught by fishing craft flying the flag of the reporting country and landed not only in the domestic harbours of the reporting country but also in foreign harbours. National catch excludes quantities caught by foreign fishing craft and landed in domestic ports.

Notes sur les pays ou zones

Les appellations employées dans cette publication et la présentation de données qui y figurent n'impliquent de la part de l'Organisation des Nations Unies pour l'alimentation et l'agriculture aucune prise de position quant au statut juridique des pays, territoires, villes ou zones, ou de leurs autorités, ni quant au tracé de leurs frontières ou limites.

Le terme *pays ou zone* employé dans les talons et les têtes de colonne des tableaux, doit s'entendre également des territoires, villes, zones terrestres, ainsi que des provinces, districts, enclaves, parties détachées d'un Etat, autres parties de territoires ou regroupements de pays ou zones, tels qu'unions économiques ou douanières.

Dans tous les tableaux, les entrées des pays ou zones sont désignées par un code multilingue n'excédant pas douze espaces typographiques. Dans la LISTE DES PAYS OU ZONES ce code est associé aux autres descripteurs, c'est-à-dire aux noms en anglais, français et espagnol, aux codes ISO alpha-3 et alpha-2, au code numérique à trois chiffres des Nations Unies, etc.

La liste des appellations des pays ou zones est sujette de temps à autre à des modifications annoncées par les pays ou zones. Les modifications de noms survenues après le 31 décembre 2001 n'ont pas toujours été incorporées dans le présent volume, mais il en sera tenu compte dans les futurs volumes.

Le pavillon du navire effectuant la partie essentielle de l'opération de pêche devrait être considéré comme le facteur déterminant du pays ou zone auquel sont attribuées les données des captures.

L'État du pavillon du navire effectuant la partie essentielle de l'opération de pêche devrait fournir les données sur les captures et les débarquements.

Lorsqu'un navire battant pavillon étranger pêche dans des eaux placées sous la juridiction nationale d'un autre Etat, l'État du pavillon du navire devrait fournir à tout moment les données pertinentes sur les captures et les débarquements. Les seules exceptions à ce principe sont les suivantes:

(a) lorsque le navire pêche au titre d'un contrat ou d'un arrangement d'affrètement visant à compléter la flottille de pêche locale et que le navire est devenu à toutes fins utiles un navire de pêche local du pays d'accueil;

(b) lorsque le navire pêche dans le cadre d'une opération conjointe ou d'arrangements similaires dans des eaux placées sous la juridiction nationale d'un autre Etat et que le navire opère à toutes fins utiles en tant que navire local ou que ses activités font, ou sont appelées à faire, partie intégrante de l'économie du pays d'accueil.

Dans toute situation où il existe une incertitude quant à l'application de ces critères, l'accord, le contrat, l'opération conjointe ou tout autre arrangement similaire devrait contenir une disposition fixant clairement les responsabilités en matière de notification des données sur les captures et les débarquements. Celles-ci devraient être notifiées à l'État du pavillon et, le cas échéant, à tout État côtier dans les eaux duquel les opérations de pêche ont lieu ou à tout organisme ou accord de pêche sous-régional, régional ou mondial compétent.

Sont comprises dans les données nationales toutes les quantités capturées par les bateaux de pêche battant le pavillon du pays déclarant et débarquées non seulement dans les ports du pays déclarant, mais aussi dans des ports étrangers. Ne sont pas comprises dans les captures nationales, les quantités capturées par des bateaux étrangers et débarquées dans des ports nationaux.

Notas sobre los países o áreas

Las denominaciones empleadas en esta publicación y la forma en que aparecen presentados los datos que contiene no implican, da parte de la Organización de las Naciones Unidas para la Agricultura y la Alimentación, juicio alguno sobre la condición jurídica de países, territorios, ciudades o zonas, o de sus autoridades, ni respecto de la delimitación de sus fronteras o límites.

En los encabezamientos de las columnas o renglones de los cuadros, el término *país o área* puede referirse también a territorios, ciudades, zonas terrestres, provincias, distritos, enclaves, exclaves, otras partes de territorio o grupos de países o áreas, como uniones económicas o aduaneras.

En todos los cuadros, los países o áreas se indican con un código multilingue que no utiliza más de doce espacios. En la LISTA DE PAÍSES O ÁREAS dicho código aparece acompañado de otros descriptores - los nombres en inglés, francés y español, los códigos ISO alfa-3 y alfa-2, y el código numérico de tres dígitos de las Naciones Unidas, etc.

Los países o áreas anuncian en algunas ocasiones cambios de sus nombres y denominaciones. En el presente volumen no se han incorporado todos los cambios de nombres anunciados después del 31 de diciembre de 2001, pero se tendrán presentes en volúmenes futuros.

El pabellón de la embarcación que efectúa la majoría de las operaciones de pesca debería considerarse como indicación decisiva para establecer a qué país o área hay que asignar los datos de captura.

El Estado del pabellón del barco que efectúa la parte principal de la operación pesquera debería ser el responsable de facilitar datos sobre capturas y desembarques.

Cuando un barco de pabellón extranjero faene en aguas sometidas a la jurisdicción nacional de otro Estado, el Estado del pabellón de ese barco debería tener en todo momento la responsabilidad de facilitar datos pertinentes sobre capturas y desembarques. Las únicas excepciones a esto serán:

(a) cuando el barco faene con arreglo a un acuerdo de fletamento o como medida para aumentar la flota pesquera local, y el barco se ha convertido a todos los efectos prácticos en un barco pesquero local del país hospedante;

(b) cuando el barco emprenda las operaciones pesqueras en virtud de un acuerdo de empresa mixta o análogo en aguas sometidas a la jurisdicción nacional de otro Estado y faene a todos los efectos prácticos como un barco local, o sus operaciones se hayan convertido, o se tenga intención de que se conviertan, en parte integrante de la economía del país hospedante.

En cualquier situación en que haya incertidumbre respecto de la aplicación de estos criterios, cualquier acuerdo, fletamento, empresa mixta u otro acuerdo similar debería contener una disposición por la que se establezca claramente la responsabilidad de la presentación de datos sobre capturas y desembarques que deberían notificarse al Estado del pabellón y, según proceda, a cualquier Estado ribereño en cuyas aguas se vayan a realizar operaciones pesqueras o a cualquier organización o acuerdo pesquero subregional, regional o mundial competente.

Los datos nacionales incluyen todas las cantidades capturadas por embarcaciones pesqueras que enarbolan el pabellón del país que comunica los datos y desembarcadas en los puertos del país en cuestión o en puertos extranjeros. Las capturas nacionales no incluyen las cantidades capturadas por embarcaciones pesqueras extranjeras y desembarcadas en puertos del país al que se refieren los datos.

Notes on countries or areas

7. Final data have been provided by many national offices; others submitted provisional figures only. Whenever national offices failed to report their annual catch statistics in time for publication, FAO, in the absence of other information, has repeated the data previously reported by the country (R designated figures), and in cases where statistical information was available, FAO has estimated the quantities to reflect more realistic catch data (F designated figures).

8. Where national figures in this volume differ from those previously published by FAO, the most recently published data represent the latest revisions.

9. The statistics in national publications may sometimes differ from those published in this Yearbook. Landings data, reported on a landed weight bases and for a "landing" year (e.g. catch of December of a year, may be landed in January of the following year), are at times called 'catches' in national publications. The nominal catch data presented in this Yearbook are the live weight equivalent of the landed quantities caught during the annual period covered.

10. The NOTES ON INDIVIDUAL COUNTRIES OR AREAS list the few national exceptions to the standards described above and also to those standards specified in other NOTES in this Yearbook.

11. Users are reminded that, although data are shown to the nearest metric ton, they are not necessarily of this degree of accuracy.

12. Countries included in the low-income food-deficit countries (LIFDCs) grouping are those classified (i) by the World Bank as low-income in terms of GNP per capita, and (ii) by FAO as having a trade deficit for food in terms of calorie content. Countries which have formally objected to being included in the grouping are not included.

Notes sur les pays ou zones

De nombreux services nationaux ont communiqué des données définitives; d'autres n'ont fourni que des chiffres préliminaires. Chaque fois que les services nationaux n'ont pu déclarer leurs statistiques de captures annuelles en temps utile pour la publication, la FAO, en l'absence d'autres informations, a repris les données précédemment communiquées par le pays (chiffres désignés par R) et, dans les cas où elle disposait de statistiques, la FAO a procédé à des estimations pour arriver à des chiffres de captures plus proches de la réalité (chiffres désignés par F).

Lorsque les chiffres nationaux indiqués dans le présent volume diffèrent de ceux déjà publiés par la FAO, les données publiées le plus récemment représentent les dernières révisions.

Les statistiques figurant dans les publications nationales peuvent parfois différer de celles publiées dans le présent Annuaire. Les statistiques des quantités débarquées, déclarées sur la base du poids de poisson débarqué et pour une année de 'mises à terre' (par exemple, les captures relatives au mois de décembre d'une année peuvent être mises à terre au mois de janvier de l'année suivante), sont parfois appelées «captures» dans les publications nationales. Les statistiques relatives aux captures nominales présentées dans cet Annuaire représentent l'équivalent en poids vif des quantités débarquées qui ont été capturées pendant la période annuelle correspondante.

Les quelques exceptions nationales à ces normes ainsi qu'aux normes indiquées dans les autres NOTES figurant dans le présent Annuaire sont indiquées dans les NOTES SUR DIVERS PAYS OU ZONES.

Il est rappelé aux utilisateurs que les chiffres sont arrondis à la tonne la plus proche, mais qu'ils n'ont pas nécessairement un tel degré de précision.

Les pays inclus dans le groupe pays à faible revenu et à déficit vivrier (PFRDV) sont ceux classifiés (I) par la Banque mondiale comme ayant un revenu par habitant bas en termes de PIB, et (ii) par la FAO comme ayant un déficit commercial alimentaire en termes de contenu en calories. Les pays qui se sont formellement opposés à leur inclusion dans ce groupe n'y figurent pas.

Notas sobre los países o áreas

Muchas oficinas nacionales han facilitado datos difinitivos; otras, en cambio, sólo han presentado cifras provisionales. Cuando las oficinas nacionales no han comunicado a tiempo sus estadísticas nacionales de captura, la FAO, a falta de otra información, ha repetido los datos comunicados previamente por el país (cifras designadas R) y cuando había información estadística, la FAO ha estimado las cantidades para que reflejaran más objetivamente los datos de captura (cifras designadas F).

Cuando las cifras correspondientes a un país o área incluidas en este volumen difieren de las publicadas anteriormente por la FAO, los últimos datos publicados representan la última revisión de dichas cifras.

En algunas ocasiones, las estadísticas que aparecen en publicaciones nacionales pueden diferir de las publicadas en este Anuario. Los datos sobre los desembarques, comunicados en peso desembarcado y en el año en que se han descargado (por ejemplo, las capturas hechas en un año en diciembre pueden descargarse en enero del año siguiente), se denominan a veces «capturas» en las publicaciones nacionales. Los datos de capturas nominales presentados en este Anuario representan el equivalente en peso en vivo de las cantidades desembarcadas, capturadas en el período anual correspondiente.

En las NOTAS SOBRE LOS DISTINTOS PAÍSES O ÁREAS pueden verse algunas excepciones nacionales a estas normas; véanse también las que aparecen en otras NOTAS de este Anuario.

Se recuerda a los usarios que, si bien los datos están redondeados a la tonelada más próxima, no tienen necesariamente ese grado de precisión.

Los países incluidos en el grupo países de bajos ingresos y con déficit de alimentos (PBIDA) son aquellos clasificados (i) por el Banco Mundial como teniendo una renta per cápita baja en términos de PNL, y (ii) por la FAO como teniendo un déficit comercial alimentario en términos de contenido en calorías. Los países que se han formalmente opuesto a su inclusión en este grupo no están incluidos.

A - Summaries

A - Résumés

A - Resúmenes

A-1 (a)

Fish, crustaceans, molluscs, etc — World capture production
Poissons, crustacés, mollusques, etc — Captures mondiales
Peces, crustáceos, moluscos, etc — Capturas mundiales

	Division, fishing area, continent / Division, zone de pêche, continent / División, área de pesca, continente	1995 mt	1996 mt	1997 mt	1998 mt	1999 mt	2000 mt	2001 mt
	World total / Total mondial / Total mundial	92 301 906	93 750 402	94 215 594	87 593 312	93 601 896	95 439 820	92 356 034
	Capture production in inland waters / Captures dans les eaux continentales / Capturas en aguas continentales	***7 263 859***	***7 427 541***	***7 562 466***	***8 043 505***	***8 513 361***	***8 788 721***	***8 692 758***
	By ISSCAAP divisions / Par divisions de la CSITAPA / Por divisiones de la CEIUAPA							
1	Freshwater fishes / Poissons d'eau douce / Peces de agua dulce	5 802 072	5 874 183	5 948 790	6 206 016	6 798 441	6 860 730	6 904 104
2	Diadromous fishes / Poissons diadromes / Peces diádromos	447 398	475 658	451 061	510 954	565 267	645 285	401 354
3	Marine fishes / Poissons marins / Peces marinos	57 744	61 971	72 278	67 884	67 820	78 267	86 096
4	Crustaceans / Crustacés / Crustáceos	352 323	448 371	575 112	672 500	524 516	604 051	669 923
5	Molluscs / Mollusques / Moluscos	597 863	562 516	508 781	581 126	552 829	595 755	626 064
6	Whales, seals and other aquatic mammals / Baleines, phoques et autres mammifères aquatiques / Ballenas, focas y otros mamíferos acuáticos	.	.	.	.	.	.	.
7	Miscellaneous aquatic animals / Animaux aquatiques divers / Animales acuáticos diversos	6 459	4 842	6 444	5 025	4 488	4 633	5 217
8	Miscellaneous aquatic animal products / Produits divers d'animaux aquatiques / Diversos productos de animales acuáticos	.	.	.	.	.	.	.
9	Aquatic plants / Plantes aquatiques / Plantas acuáticas	.	.	.	.	.	.	.
	By inland fishing areas/continents / Par zones de pêche continentales/continents / Por áreas de pesca continentales/continentes							
01	Africa - Inland waters / Afrique - Eaux continentales / Africa - Aguas continentales	1 959 418	1 846 647	1 909 656	1 972 873	1 995 628	2 059 555	2 052 001
02	America, North - Inland waters / Amérique du Nord - Eaux continentales / América del Norte - Aguas continentales	219 457	215 209	211 519	196 571	186 505	190 849	174 734
03	America, South - Inland waters / Amérique du Sud - Eaux continentales / América del Sur - Aguas continentales	367 402	344 197	332 798	334 638	347 938	345 276	341 631
04	Asia - Inland waters / Asie - Eaux continentales / Asia - Aguas continentales	4 302 719	4 597 415	4 702 743	5 099 470	5 509 887	5 738 765	5 754 476
05	Europe - Inland waters / Europe - Eaux continentales / Europa - Aguas continentales	394 922	404 689	385 052	418 104	450 743	431 944	347 421
06	Oceania - Inland waters / Océanie - Eaux continentales / Oceanía - Aguas continentales	19 941	19 384	20 698	21 849	22 660	22 332	22 495
	Capture production in marine fishing areas / Captures dans les zones de pêche maritimes / Capturas en áreas de pesca marítimas	***85 038 047***	***86 322 861***	***86 653 128***	***79 549 807***	***85 088 535***	***86 651 099***	***83 663 276***
	By ISSCAAP divisions / Par divisions de la CSITAPA / Por divisiones de la CEIUAPA							
1	Freshwater fishes / Poissons d'eau douce / Peces de agua dulce	31 626	34 119	33 825	32 222	32 793	32 030	33 275
2	Diadromous fishes / Poissons diadromes / Peces diádromos	1 381 159	1 274 530	1 190 307	1 175 230	1 223 469	1 118 706	1 224 426
3	Marine fishes / Poissons marins / Peces marinos	71 947 066	73 390 093	72 620 458	65 782 875	70 309 369	71 731 125	69 099 983
4	Crustaceans / Crustacés / Crustáceos	4 770 704	4 984 553	5 204 032	5 576 850	5 754 135	5 897 327	5 780 729
5	Molluscs / Mollusques / Moluscos	6 380 892	6 145 524	6 947 674	6 340 417	7 100 390	7 270 970	6 943 009
6	Whales, seals and other aquatic mammals / Baleines, phoques et autres mammifères aquatiques / Ballenas, focas y otros mamíferos acuáticos	.	.	.	.	.	.	.
7	Miscellaneous aquatic animals / Animaux aquatiques divers / Animales acuáticos diversos	526 600	494 042	656 832	642 213	668 379	600 941	581 854
8	Miscellaneous aquatic animal products / Produits divers d'animaux aquatiques / Diversos productos de animales acuáticos	.	.	.	.	.	.	.
9	Aquatic plants / Plantes aquatiques / Plantas acuáticas	.	.	.	.	.	.	.

A-1 (a)

Fish, crustaceans, molluscs, etc — World capture production
Poissons, crustacés, mollusques, etc — Captures mondiales
Peces, crustáceos, moluscos, etc — Capturas mundiales

Division, fishing area, continent Division, zone de pêche, continent División, área de pesca, continente	1995 mt	1996 mt	1997 mt	1998 mt	1999 mt	2000 mt	2001 mt
By marine fishing areas	*Par zones de pêche maritimes*					*Por áreas de pesca marítimas*	
21 Atlantic, Northwest Atlantique, nord-ouest Atlántico, noroeste	2 001 037	2 064 442	2 050 053	1 965 303	2 035 191	2 080 779	2 238 371
27 Atlantic, Northeast Atlantique, nord-est Atlántico, nordeste	11 022 702	11 064 598	11 761 184	10 984 475	10 513 335	11 024 004	11 164 413
31 Atlantic, Western Central Atlantique, centre-ouest Atlántico, centro-occidental	1 767 597	1 693 548	1 804 094	1 773 670	1 763 871	1 819 345	1 689 685
34 Atlantic, Eastern Central Atlantique, centre-est Atlántico, centro-oriental	3 395 675	3 573 113	3 630 738	3 788 220	3 684 078	3 637 192	3 817 448
37 Mediterranean and Black Sea Méditerranée et mer Noire Mediterráneo y Mar Negro	1 700 787	1 527 116	1 439 482	1 395 558	1 524 723	1 492 302	1 535 345
41 Atlantic, Southwest Atlantique, sud-ouest Atlántico, sudoccidental	2 336 972	2 480 181	2 756 887	2 379 440	2 622 540	2 375 434	2 265 714
47 Atlantic, Southeast Atlantique, sud-est Atlántico, sudoriental	1 590 334	1 325 597	1 355 409	1 542 066	1 529 255	1 635 484	1 652 992
48 Atlantic, Antarctic Atlantique, Antarctique Atlántico, Antártico	120 837	106 143	84 972	84 236	107 925	113 156	102 561
51 Indian Ocean, Western Océan Indien, ouest Océano Indico, occidental	3 807 617	3 884 762	3 978 510	3 792 545	4 037 596	3 997 956	3 948 676
57 Indian Ocean, Eastern Océan Indien, est Océano Indico, oriental	4 161 740	4 141 894	4 394 809	4 703 942	4 590 510	4 830 059	4 767 860
58 Indian Ocean, Antarctic Océan Indien, Antarctique Océano Indico, Antártico	10 912	4 954	8 234	8 058	13 146	9 523	10 638
61 Pacific, Northwest Pacifique, nord-ouest Pacífico, noroeste	21 751 722	23 482 610	24 606 259	24 753 317	24 112 564	23 180 042	22 532 921
67 Pacific, Northeast Pacifique, nord-est Pacífico, nordeste	3 031 197	2 893 342	2 846 092	2 792 459	2 551 689	2 477 803	2 759 090
71 Pacific, Western Central Pacifique, centre-ouest Pacífico, centro-occidental	8 902 133	8 730 401	8 966 335	9 282 908	9 597 552	9 740 281	9 939 821
77 Pacific, Eastern Central Pacifique, centre-est Pacífico, centro-oriental	1 555 851	1 619 983	1 705 176	1 407 631	1 445 221	1 713 080	1 830 592
81 Pacific, Southwest Pacifique, sud-ouest Pacífico, sudoccidental	813 727	661 821	837 649	861 701	782 603	713 373	750 967
87 Pacific, Southeast Pacifique, sud-est Pacífico, sudoriental	17 067 207	17 068 356	14 427 245	8 034 222	14 176 392	15 810 412	12 655 449
88 Pacific, Antarctic Pacifique, Antarctique Pacífico, Antártico	-	-	0	56	344	874	733
By continents	*Par continents*					*Por continentes*	
1 Africa Afrique Africa	3 891 574	3 732 137	4 031 629	4 134 865	4 329 760	4 556 957	4 838 229
2 America, North Amérique du Nord América del Norte	7 764 597	7 675 975	7 796 976	7 319 461	7 388 195	7 632 280	7 902 479
3 America, South Amérique du Sud América del Sur	19 257 130	19 450 538	16 922 450	10 378 356	16 178 654	17 655 524	14 537 766
4 Asia Asie Asia	36 049 797	37 355 814	39 139 366	39 699 531	40 207 954	39 804 598	39 507 304
5 Europe Europe Europa	16 777 041	17 087 064	17 526 567	16 681 361	15 623 101	15 737 884	15 615 152
6 Oceania Océanie Oceanía	988 708	832 114	1 064 687	1 147 965	1 145 466	1 048 079	1 085 814
9 Other nei Autres nca Otros nep	309 200	189 219	171 453	188 268	215 405	215 777	176 532

* See paragraph 5 of the INTRODUCTION.

* Voir le paragraphe 5 de l'INTRODUCTION.

* Véase el párrafo 5 en la INTRODUCCION.

A-1 (b)

Fish, crustaceans, molluscs, etc — Capture production by groupings of major fishing areas
Poissons, crustacés, molluques, etc — Captures par groupes de principales zones de pêche
Peces, crustáceos, moluscos, etc — Capturas por grupos de áreas principales de pesca

Group of major fishing areas Groupe de principales zones de pêche Grupo de áreas principales de pesca		1995 mt	1996 mt	1997 mt	1998 mt	1999 mt	2000 mt	2001 mt
World total Total mondial Total mundial		**92 301 906**	**93 750 402**	**94 215 594**	**87 593 312**	**93 601 896**	**95 439 820**	**92 356 034**
Inland waters Eaux continentales Aguas continentales		***7 263 859***	***7 427 541***	***7 562 466***	***8 043 505***	***8 513 361***	***8 788 721***	***8 692 758***
Marine areas Zones maritimes Areas marítimas		***85 038 047***	***86 322 861***	***86 653 128***	***79 549 807***	***85 088 535***	***86 651 099***	***83 663 276***
Atlantic Ocean Ocean Atlantique Oceano Atlantico	18+21+27+31+34+37+41+47	23 815 104	23 728 595	24 797 847	23 828 732	23 672 993	24 064 540	24 363 968
Indian Ocean Ocean Indien Ocean Indico	51+57	7 969 357	8 026 656	8 373 319	8 496 487	8 628 106	8 828 015	8 716 536
Pacific Ocean Ocean Pacifique Oceano Pacifico	61+67+71+77+81+87	53 121 837	54 456 513	53 388 756	47 132 238	52 666 021	53 634 991	50 468 840
Southern Ocean Océan Austral Océano Austral	48+58+88	131 749	111 097	93 206	92 350	121 415	123 553	113 932
Atlantic, Northern Atlantique du Nord Atlantico del Norte	18+21+27	13 023 739	13 129 040	13 811 237	12 949 778	12 548 526	13 104 783	13 402 784
Atlantic, Central Atlantique central Atlantico central	31+34+37	6 864 059	6 793 777	6 874 314	6 957 448	6 972 672	6 948 839	7 042 478
Atlantic, Southern Atlantique du Sud Atlantico del Sur	41+47+48	4 048 143	3 911 921	4 197 268	4 005 742	4 259 720	4 124 074	4 021 267
Indian Ocean Ocean Indien Oceano Indico	51+57+58	7 980 269	8 031 610	8 381 553	8 504 545	8 641 252	8 837 538	8 727 174
Pacific, Northern Pacifique du Nord Pacifico del Norte	61+67	24 782 919	26 375 952	27 452 351	27 545 776	26 664 253	25 657 845	25 292 011
Pacific, Central Pacifique central Pacifico central	71+77	10 457 984	10 350 384	10 671 511	10 690 539	11 042 773	11 453 361	11 770 413
Pacific, Southern Pacifique du Sud Pacifico del Sur	81+87+88	17 880 934	17 730 177	15 264 894	8 895 979	14 959 339	16 524 659	13 407 149
Northern regions Regions septentrionales Regiones del Norte	18+21+27+61+67	37 806 658	39 504 992	41 263 588	40 495 554	39 212 779	38 762 628	38 694 795
Central regions Regions centrales Regiones centrales	31+34+37+51+57+71+77	25 291 400	25 170 817	25 919 144	26 144 474	26 643 551	27 230 215	27 529 427
Southern regions Regions meridionales Regiones del Sur	41+47+48+58+81+87+88	21 939 989	21 647 052	19 470 396	12 909 779	19 232 205	20 658 256	17 439 054

A-1 (c)

Fish, crustaceans, molluscs, etc — **Capture production by principal producers in 2001**
Poissons, crustacés, mollusques, etc — **Captures par principaux producteurs en 2001**
Peces, crustáceos, moluscos, etc — **Capturas por productores principales en 2001**

Country or area Pays ou zone País o área	1992 mt	1993 mt	1994 mt	1995 mt	1996 mt	1997 mt	1998 mt	1999 mt	2000 mt	2001 mt
China (1)	8 322 552	9 351 437	10 866 836	12 562 706	14 182 107	15 722 344	17 229 927	17 240 032	16 987 325	16 529 389
Peru	7 502 192	9 004 777	11 999 217	8 937 342	9 515 048	7 869 871	4 338 437	8 428 601	10 658 620	7 986 103
USA	5 190 717	5 523 216	5 535 324	5 224 566	5 001 483	4 983 440	4 708 980	4 749 646	4 745 321	4 944 406
Japan	7 730 580	7 254 943	6 617 118	5 966 456	5 933 659	5 926 113	5 299 399	5 194 186	4 971 412	4 719 152
Indonesia	2 891 318	3 077 540	3 305 815	3 511 145	3 553 276	3 791 240	3 964 897	3 986 919	4 069 691	4 203 830
Chile	6 432 324	5 949 565	7 720 578	7 433 902	6 690 942	5 810 764	3 265 383	5 050 528	4 300 160	3 797 143
India	2 844 102	3 062 712	3 257 607	3 265 240	3 447 954	3 523 448	3 373 492	3 472 150	3 742 296	3 762 600
Russian Fed	5 509 994	4 370 015	3 705 082	4 311 809	4 675 738	4 661 853	4 454 759	4 141 158	3 973 535	3 628 323
Thailand	2 875 456	2 927 689	3 015 196	3 031 074	3 013 961	2 902 898	2 930 354	2 952 008	2 911 173	2 881 316
Norway	2 430 823	2 415 131	2 366 119	2 524 111	2 648 457	2 863 059	2 861 223	2 627 534	2 703 415	2 687 303
Korea Rep	2 321 048	2 257 192	2 357 891	2 319 915	2 413 713	2 204 047	2 026 934	2 119 668	1 823 175	1 988 002
Iceland	1 574 682	1 715 581	1 556 962	1 612 548	2 060 168	2 205 944	1 681 951	1 736 267	1 982 522	1 980 715
Philippines	1 885 041	1 834 323	1 845 335	1 860 701	1 783 601	1 805 806	1 833 380	1 872 818	1 893 017	1 945 217
Denmark	1 953 828	1 614 289	1 873 316	1 999 033	1 681 517	1 826 852	1 557 335	1 405 005	1 534 089	1 510 439
Viet Nam	868 107	932 143	1 025 909	1 084 939	1 223 644	1 276 325	1 293 954	1 386 300	1 450 590 F	1 491 123 F
Mexico	1 157 573	1 102 932	1 191 646	1 329 340	1 464 188	1 489 112	1 179 860	1 205 603	1 315 581	1 398 592
Malaysia	1 025 289	1 049 321	1 067 650	1 112 375	1 130 372	1 172 922	1 153 719	1 251 768	1 289 245	1 234 733
Myanmar	731 544	739 702	746 241	751 232	601 788	780 295	830 117	919 410	1 069 726	1 166 868
Spain	1 077 764 F	1 085 959 F	1 094 524 F	1 181 286 F	1 174 635	1 204 913	1 263 275	1 169 462	1 045 488	1 084 820
Morocco	550 937	624 511	754 327	848 951	642 886	791 906	710 436	745 431	896 620	1 083 276
Canada	1 292 050	1 132 862	1 022 447	849 411	904 862	972 294	1 013 977	1 027 511	1 010 489	1 049 508
China,Taiwan	1 063 947	1 133 676	967 189	1 010 022	967 483	1 038 048	1 091 768	1 099 715	1 093 889	1 005 199
Bangladesh	709 332	764 824	770 790	792 389	814 787	829 426	839 141	959 215	1 004 264	1 000 000 F
Argentina	703 425	931 303	951 387	1 170 124	1 291 355	1 400 162	1 164 829	1 078 313	913 622	923 322
Brazil	741 320 F	717 090 F	740 100 F	706 708	715 482	744 585	706 789	703 941	766 846	770 000 F
South Africa	692 339	562 374	523 449	575 547	440 428	515 095	559 049	588 144	643 238	755 345
UK	813 082	860 229	878 018	909 928	865 145	886 261	923 085	840 898	747 358	741 106
Pakistan	540 354	608 344	537 277	525 849	537 489	589 795	596 980	654 530	614 069	607 020
France	585 093	617 479	622 372	610 808	557 510	567 422	542 560	586 487	623 755	606 194
Ecuador	226 817	286 633	330 248	505 395	702 974	548 988	310 022	497 872	592 547	586 570
New Zealand	460 822	426 668	449 660	555 559	423 635	610 254	639 238	598 475	548 559	561 110
Namibia	655 110	790 332	649 199	569 833	517 828	513 216	612 343	580 084	589 905	547 492
Turkey	447 498	548 758	589 803	633 970	527 826	459 153	487 200	573 824	503 348	527 730
Faeroe Is	248 430	246 294	237 708	287 796	304 587	329 145	363 816	358 133	454 530	524 837
Netherlands	432 970	461 756	420 053	438 092	410 798	451 799	536 626	514 611	495 804	518 162
Nigeria	301 327	238 409	267 059	349 482	337 993	387 923	463 024	455 628	441 377	452 146
Ghana	423 425	372 619	335 437	352 844	477 173	447 088	442 641	492 776	452 070	445 287
Egypt	258 700	272 700	283 900	310 790	320 230	342 759	362 741	380 504	384 314	428 651
Venezuela	330 964	395 861	438 218	501 045	496 287	470 255	503 504	399 799	357 115	417 947
Senegal	370 242	382 212	352 421	354 617	411 759	457 366	403 872	412 125	402 047	405 409
Cambodia	102 600	101 000	95 000	102 999	94 710	102 800	107 900	269 156	284 368	397 200 F
Ireland	249 386	278 774	294 165	384 632	333 030	292 673	324 274	280 957	281 806	356 309
Ukraine	458 939	305 334	267 180	378 650	417 119	373 005	462 308	407 853	392 724	351 260
Iran	305 706	322 006	308 101	341 383	351 725	342 287	367 212	387 200	384 000	336 450
Tanzania	331 235	331 267	288 399	359 800	323 921	356 960	348 000	310 509	332 779	335 900
Sweden	307 545	341 897	386 814	404 572	370 881	357 406	410 886	351 254	338 534	311 816
Italy	396 457	397 533	398 730	396 791	365 899	343 693	306 096	282 790	302 149	310 397
Sri Lanka	202 668	215 400	218 400	229 171	228 945	235 099	263 330	276 080	297 410	279 640
Angola	113 625	126 200	132 413	122 781	137 815	146 304	163 149	175 799	238 351	252 518
Panama	172 250	184 822	185 147	202 993	150 764	162 349	202 687	120 848	222 631	235 000 F
Poland	475 697	404 422	437 950	426 235	341 299	353 661	238 336	235 111	218 354	225 916
Uganda	264 900	219 814	213 129	208 789	195 088	218 026	220 628	226 097	219 356	220 726
Germany	216 890	253 010	230 161	239 890	236 411	259 352	266 622	238 925	205 689	211 282
Congo Dem R	187 840	196 789	155 897	158 627	163 010	162 211	178 041	208 448	208 448 F	208 448 F
Korea D P Rp	900 000 F	870 398 F	371 961	327 083	253 125	236 462	220 000 F	210 000 F	200 850 F	200 000 F
55 countries 55 pays 55 países	***80 858 856***	***82 192 067***	***87 252 875***	***87 163 276***	***88 828 510***	***89 316 474***	***82 600 486***	***88 438 106***	***90 129 596***	***87 129 247***
Other countries Autres pays Otros países	***4 943 079***	***4 780 951***	***4 750 962***	***5 138 630***	***4 921 892***	***4 899 120***	***4 992 826***	***5 163 790***	***5 310 224***	***5 226 787***
World total Total mondial Total mundial	**85 801 935**	**86 973 018**	**92 003 837**	**92 301 906**	**93 750 402**	**94 215 594**	**87 593 312**	**93 601 896**	**95 439 820**	**92 356 034**

These countries or areas are those with capture production of 200 000 metric tons or more in 2001.

(1) See note on China in NOTES ON INDIVIDUAL COUNTRIES OR AREAS.

Ces pays ou zones sont ceux dont les captures ont été de 200 000 tonnes métriques ou plus en 2001.

(1) Voir la note sur la Chine dans les NOTES SUR DIVERS PAYS OU ZONES.

Estos países o áreas son referentes a los que totalizan unas capturas de 200 000 toneladas métricas o más en 2001.

(1) Véase la nota sobre China en las NOTAS SOBRE LOS DISTINTOS PAISES O AREAS.

A-1 (d)

Fish, crustaceans, molluscs, etc — **Capture production by groups of species**
Poissons, crustacés, molluques, etc — **Captures par groupes d'espèces**
Peces, crustáceos, moluscos, etc — **Capturas por grupos de especies**

Species group Groupe d'espèces Grupo de especies	1995 mt	1996 mt	1997 mt	1998 mt	1999 mt	2000 mt	2001 mt
11 Carps, barbels and other cyprinids Carpes, barbeaux et autres cyprinidés Carpas, barbos y otros ciprínidos	780 522	663 445	608 078	604 993	613 612	569 212	547 623
12 Tilapias and other cichlids Tilapias et autres cichlidés Tilapias y otros cíclidos	594 623	560 516	561 796	597 782	637 294	686 877	682 956
13 Miscellaneous freshwater fishes Poissons d'eau douce divers Peces de agua dulce diversos	4 458 553	4 684 341	4 812 741	5 035 463	5 580 328	5 636 671	5 706 800
21 Sturgeons, paddlefishes Esturgeons, spatules Esturiones, sollos	5 906	4 624	4 407	3 777	2 849	2 603	2 010
22 River eels Anguilles Anguilas	14 837	17 117	14 747	12 296	12 489	16 558	15 355
23 Salmons, trouts, smelts Saumons, truites, éperlans Salmones, truchas, eperlanos	1 153 628	1 029 998	921 952	889 982	913 373	805 298	891 058
24 Shads Aloses Sábalos	588 771	634 257	634 292	699 649	782 579	855 571	629 146
25 Miscellaneous diadromous fishes Poissons diadromes divers Peces diádromos diversos	65 415	64 192	65 970	80 480	77 446	83 961	88 211
31 Flounders, halibuts, soles Flets, flétans, soles Platijas, halibuts, lenguados	917 610	941 866	1 027 219	935 176	955 892	1 007 098	945 235
32 Cods, hakes, haddocks Morues, merlus, églefins Bacalaos, merluzas, eglefinos	10 740 688	10 784 133	10 371 936	10 333 230	9 326 770	8 654 744	9 223 810
33 Miscellaneous coastal fishes Poissons côtiers divers Peces costeros diversos	6 008 888	6 222 784	6 863 390	6 739 692	6 691 022	6 648 761	6 692 983
34 Miscellaneous demersal fishes Poissons démersaux divers Peces demersales diversos	2 701 501	2 689 794	2 738 859	2 969 462	2 947 202	3 018 968	3 027 620
35 Herrings, sardines, anchovies Harengs, sardines, anchois Arenques, sardinas, anchoas	22 005 770	22 387 417	21 733 290	16 666 477	22 635 103	24 896 523	20 460 641
36 Tunas, bonitos, billfishes Thons, pélamides, marlins Atunes, bonitos, agujas	4 885 121	4 847 666	5 190 889	5 808 852	5 975 503	5 865 522	5 821 240
37 Miscellaneous pelagic fishes Poissons pélagiques divers Peces pelágicos diversos	13 935 678	14 332 358	13 789 294	11 213 305	10 683 986	10 664 082	12 218 558
38 Sharks, rays, chimaeras Squales, raies, chimères Tiburones, rayas, quimeras	763 306	815 086	829 957	820 211	834 629	856 716	824 772
39 Marine fishes not identified Poissons marins non identifiés Peces marinos no identificados	10 046 248	10 430 960	10 147 902	10 364 354	10 327 082	10 196 978	9 971 220
41 Freshwater crustaceans Crustacés d'eau douce Crustáceos de agua dulce	315 761	410 590	527 699	645 856	498 530	567 933	630 533
42 Crabs, sea-spiders Crabes, araignées de mer Cangrejos, centollas	951 645	999 335	990 278	1 074 977	1 058 754	1 087 976	1 096 371
43 Lobsters, spiny-rock lobsters Homards, langoustes Bogavantes, langostas	220 151	209 590	233 317	216 165	229 426	227 043	225 136
44 King crabs, squat-lobsters Crabes royaux, galatées Cangrejos reales, galateidos	76 666	75 977	70 685	79 223	77 639	68 004	44 969
45 Shrimps, prawns Crevettes Gambas, camarones	2 437 418	2 554 183	2 628 466	2 749 225	3 021 115	3 074 771	2 950 834

A-1 (d)

Fish, crustaceans, molluscs, etc — Capture production by groups of species
Poissons, crustacés, molluques, etc — Captures par groupes d'espèces
Peces, crustáceos, moluscos, etc — Capturas por grupos de especies

Species group Groupe d'espèces Grupo de especies	1995 mt	1996 mt	1997 mt	1998 mt	1999 mt	2000 mt	2001 mt
46 Krill, planktonic crustaceans Krill, crustacés planctoniques Krill, crustáceos planctónicos	118 714	101 714	82 508	81 290	103 318	104 263	98 245
47 Miscellaneous marine crustaceans Crustacés marins divers Crustáceos marinos diversos	1 002 672	1 081 535	1 246 191	1 402 614	1 289 869	1 371 388	1 404 564
51 Freshwater molluscs Mollusques d'eau douce Moluscos de agua dulce	596 567	561 187	507 773	580 047	552 452	595 286	625 628
52 Abalones, winkles, conchs Ormeaux, bigorneaux, strombes Orejas de mar, bígaros, estrombos	109 322	116 290	124 433	99 163	105 826	113 117	120 958
53 Oysters Huîtres Ostras	192 661	185 516	182 405	159 105	156 541	276 905	199 015
54 Mussels Moules Mejillones	244 397	203 439	239 939	237 747	219 183	241 471	257 315
55 Scallops, pectens Coquilles St-Jacques Vieiras	537 324	535 168	532 893	554 767	612 702	661 171	702 525
56 Clams, cockles, arkshells Clams, coques, arches Almejas, berberechos, arcas	960 684	919 151	815 300	838 560	841 586	818 239	808 945
57 Squids, cuttlefishes, octopuses Encornets, seiches, poulpes Calamares, jibias, pulpos	2 938 392	3 149 764	3 456 670	2 857 620	3 597 670	3 655 150	3 346 828
58 Miscellaneous marine molluscs Mollusques marins divers Moluscos marinos diversos	1 399 408	1 037 525	1 597 042	1 594 534	1 567 259	1 505 386	1 507 859
71 Frogs and other amphibians Grenouilles et autres amphibies Ranas y otros anfibios	3 627	2 957	3 622	3 009	1 807	2 315	2 883
72 Turtles Tortues Tortugas	842	1 104	977	1 187	1 243	998	1 188
74 Sea-squirts and other tunicates Ascidiens et autres tuniciers Ascidias y otros tunicados	9 064	21 331	5 976	3 443	3 905	3 858	2 427
75 Horseshoe crabs and other arachnoids Limules et autres arachnoïdés Límulos y otros arácnidos	926	1 598	2 607	3 252	2 397	1 696	1 299
76 Sea-urchins and other echinoderms Oursins et autres échinodermes Erizos de mar y otros equinodermos	136 827	127 725	118 915	109 308	120 550	121 356	105 876
77 Miscellaneous aquatic invertebrates Invertébrés aquatiques divers Invertebrados acuáticos diversos	381 773	344 169	531 179	527 039	542 965	475 351	473 398
World total Total mondial Total mundial	**92 301 906**	**93 750 402**	**94 215 594**	**87 593 312**	**93 601 896**	**95 439 820**	**92 356 034**

A-1 (e)

Fish, crustaceans, molluscs, etc — Capture production by principal species in 2001
Poissons, crustacés, mollusques, etc — Captures par principales espèces en 2001
Peces, crustáceos, moluscos, etc — Capturas por especies principales en 2001

3-alpha code Code alpha-3 Código alfa-3	English name Nom anglais Nombre inglés	Scientific name Nom scientifique Nombre científico	1997 mt	1998 mt	1999 mt	2000 mt	2001 mt
VET	Anchoveta(=Peruvian anchovy)	*Engraulis ringens*	7 685 098	1 729 064	8 723 265	11 276 357	7 213 077
ALK	Alaska pollock(=Walleye poll.)	*Theragra chalcogramma*	4 486 510	4 049 223	3 270 286	2 930 230	3 136 465
CJM	Chilean jack mackerel	*Trachurus murphyi*	3 597 117	2 025 758	1 423 447	1 540 494	2 508 834
HER	Atlantic herring	*Clupea harengus*	2 533 909	2 421 462	2 411 408	2 380 683	1 952 975
JAN	Japanese anchovy	*Engraulis japonicus*	1 666 503	2 093 888	1 820 259	1 725 685	1 836 502
SKJ	Skipjack tuna	*Katsuwonus pelamis*	1 608 811	1 888 792	1 988 826	1 955 430	1 836 438
WHB	Blue whiting(=Poutassou)	*Micromesistius poutassou*	712 279	1 185 003	1 319 002	1 472 105	1 823 305
MAS	Chub mackerel	*Scomber japonicus*	2 427 389	1 924 691	1 950 177	1 474 776	1 798 704
CAP	Capelin	*Mallotus villosus*	1 603 338	982 312	904 045	1 488 629	1 670 906
LHT	Largehead hairtail	*Trichiurus lepturus*	1 206 353	1 436 303	1 416 408	1 477 722	1 471 657
YFT	Yellowfin tuna	*Thunnus albacares*	1 087 363	1 071 243	1 079 917	1 035 107	1 202 312
PIL	European pilchard(=Sardine)	*Sardina pilchardus*	998 734	949 776	907 591	947 147	1 126 832
COD	Atlantic cod	*Gadus morhua*	1 375 079	1 211 067	1 094 072	941 269	943 661
SQA	Argentine shortfin squid	*Illex argentinus*	980 300	693 542	1 144 998	930 781	743 024
MAC	Atlantic mackerel	*Scomber scombrus*	555 621	666 965	618 014	690 063	710 411
CPI	California pilchard	*Sardinops caeruleus*	498 653	381 898	405 402	546 079	685 497
ANE	European anchovy	*Engraulis encrasicolus*	502 390	506 844	630 845	626 371	660 552
SPR	European sprat	*Sprattus sprattus*	700 239	696 228	684 189	658 484	647 417
AKS	Akiami paste shrimp	*Acetes japonicus*	495 680	587 376	598 602	639 219	577 497
SQJ	Japanese flying squid	*Todarodes pacificus*	603 367	378 605	497 887	570 427	528 523
MHG	Gulf menhaden	*Brevoortia patronus*	597 565	497 461	694 242	591 434	528 500
NPH	Japanese Spanish mackerel	*Scomberomorus niphonius*	365 585	551 780	595 103	539 094	522 756
IOS	Indian oil sardine	*Sardinella longiceps*	289 142	247 065	208 898	402 936	437 328
KAW	Kawakawa	*Euthynnus affinis*	396 333	415 693	439 527	458 957	423 490
HEP	Pacific herring	*Clupea pallasii*	437 897	508 627	471 307	457 197	406 938
SAP	Pacific saury	*Cololabis saira*	388 643	180 973	187 898	306 069	376 173
BET	Bigeye tuna	*Thunnus obesus*	380 958	392 586	400 394	410 595	372 110
PIN	Pink(=Humpback)salmon	*Oncorhynchus gorbuscha*	318 717	371 552	386 928	285 338	360 973
HMC	Cape horse mackerel	*Trachurus capensis*	407 482	481 500	400 279	423 607	354 046
GAZ	Gazami crab	*Portunus trituberculatus*	252 502	283 971	284 851	351 051	350 168
PRA	Northern prawn	*Pandalus borealis*	280 316	317 693	338 393	370 434	345 681
JAP	Japanese pilchard	*Sardinops melanostictus*	417 939	295 788	515 477	305 767	339 377
PCO	Pacific cod	*Gadus macrocephalus*	443 895	411 371	402 245	370 912	330 884
CRY	Yellow croaker	*Larimichthys polyactis*	236 960	278 807	323 564	425 047	330 455
POK	Saithe(=Pollock)	*Pollachius virens*	319 307	329 793	340 351	313 580	325 750
CKI	Araucanian herring	*Strangomera bentincki*	441 154	317 564	782 142	722 522	324 617
CHU	Chum(=Keta=Dog)salmon	*Oncorhynchus keta*	347 560	311 965	281 260	276 355	307 662
HKP	Argentine hake	*Merluccius hubbsi*	648 301	527 229	372 167	242 727	302 798
SAA	Round sardinella	*Sardinella aurita*	461 627	509 079	411 645	360 708	300 445
GRM	Patagonian grenadier	*Macruronus magellanicus*	123 678	473 633	447 063	233 570	296 294
JSC	Yesso scallop	*Patinopecten yessoensis*	266 957	294 211	305 510	310 104	293 268
THG	Golden threadfin bream	*Nemipterus virgatus*	263 399	266 383	250 591	296 319	290 920
ANC	Southern African anchovy	*Engraulis capensis*	62 640	110 296	180 954	267 986	289 323
NIP	Nile perch	*Lates niloticus*	329 244	339 183	306 282	302 905	282 245
ATK	Atka mackerel	*Pleurogrammus azonus*	308 894	344 350	265 205	267 848	271 488
MHA	Atlantic menhaden	*Brevoortia tyrannus*	322 239	276 230	208 000	207 122	261 407
DPC	Daggertooth pike conger	*Muraenesox cinereus*	201 111	258 172	253 978	238 364	260 108
HOM	Atlantic horse mackerel	*Trachurus trachurus*	455 109	350 298	321 195	231 059	258 618
GRN	Blue grenadier	*Macruronus novaezelandiae*	285 580	324 750	281 105	283 688	257 782
SCA	American sea scallop	*Placopecten magellanicus*	102 028	99 432	131 962	196 993	254 196
GIT	Giant tiger prawn	*Penaeus monodon*	178 139	240 301	264 218	251 076	249 366
TLN	Nile tilapia	*Oreochromis niloticus*	187 928	229 684	227 342	245 951	248 081
TRV	Southern rough shrimp	*Trachypenaeus curvirostris*	180 255	179 544	403 027	312 968	247 940
PHA	South Pacific hake	*Merluccius gayi*	265 840	162 683	141 053	193 754	246 656
JJM	Japanese jack mackerel	*Trachurus japonicus*	350 619	340 565	227 290	271 501	235 874
HAD	Haddock	*Melanogrammus aeglefinus*	334 105	283 689	249 451	212 807	228 821
GIS	Jumbo flying squid	*Dosidicus gigas*	162 504	27 466	134 773	182 399	223 784
ALB	Albacore	*Thunnus alalunga*	217 621	228 353	243 609	204 753	221 473
HIL	Hilsa shad	*Tenualosa ilisha*	216 065	207 269	215 991	220 372	220 507
COM	Narrow-barred Spanish mackerel	*Scomberomorus commerson*	176 096	184 797	190 489	208 567	213 244
PIA	Southern African pilchard	*Sardinops ocellatus*	144 681	196 581	175 969	161 448	200 100
VEP	Pacific anchoveta	*Cetengraulis mysticetus*	195 630	180 705	70 358	120 943	198 636
RAG	Indian mackerel	*Rastrelliger kanagurta*	303 946	294 606	296 904	196 837	183 372
SAG	Goldstripe sardinella	*Sardinella gibbosa*	156 914	174 691	162 710	172 219	176 610
OYA	American cupped oyster	*Crassostrea virginica*	144 707	135 222	132 207	249 306	175 042
NHA	North Pacific hake	*Merluccius productus*	227 705	227 754	217 000	206 720	172 051
BOA	Bonga shad	*Ethmalosa fimbriata*	177 345	173 641	181 546	160 659	169 543
CLB	Atlantic surf clam	*Spisula solidissima*	141 421	131 700	142 370	165 765	166 310
MUS	Blue mussel	*Mytilus edulis*	136 147	137 268	130 020	147 972	163 911
SCD	Blue swimming crab	*Portunus pelagicus*	138 733	138 123	147 161	154 647	162 387
RUS	Indian scad	*Decapterus russelli*	150 027	145 747	162 437	171 546	160 899
POS	Southern blue whiting	*Micromesistius australis*	166 205	201 363	185 310	152 830	155 780
BUC	Bombay-duck	*Harpadon nehereus*	226 491	194 188	194 307	143 250	153 204
CLQ	Ocean quahog	*Arctica islandica*	168 150	157 282	147 933	124 131	150 260
74 species 74 espèces 74 especies			***50 224 769***	***41 820 697***	***48 320 631***	***50 389 767***	***47 854 240***
Other species Autres espèces Otras especies			***43 990 825***	***45 772 615***	***45 281 265***	***45 050 053***	***44 501 794***
World total Total mondial Total mundial			**94 215 594**	**87 593 312**	**93 601 896**	**95 439 820**	**92 356 034**

These selected species are those with capture production of 150 000 metric tons or more in 2001.

Ces espèces sont celles dont les captures ont été de 150 000 tonnes métriques ou plus en 2001.

Estas especies se refieren a las que totalizan unas capturas de 150 000 toneladas métricas o más en 2001.

A-2

Fish, crustaceans, molluscs, etc	Capture production by countries or areas	All fishing areas
Poissons, crustacés, mollusques, etc	Captures par pays ou zones	Toutes les zones de pêche
Peces, crustáceos, moluscos, etc	Capturas por países o áreas	Todas las áreas de pesca

Country or area Pays ou zone País o área	1992 mt	1993 mt	1994 mt	1995 mt	1996 mt	1997 mt	1998 mt	1999 mt	2000 mt	2001 mt
Afghanistan	1 200 F	1 200 F	1 300 F	1 300 F	1 300 F	1 250 F	1 200 F	1 200 F	1 000 F	800 F
Albania	2 696	2 500 F	2 100 F	1 379	2 125	1 013	2 683	2 745	3 320	3 310
Algeria	95 270	101 895	135 402	105 872	81 989	91 580	92 346	102 396	100 000 F	100 000 F
Amer Samoa	45	27	111	152	210	431	586	518	830	3 663
Andorra	0	0	0	0	0	0	0	0	0	0
Angola	113 625	126 200	132 413	122 781	137 815	146 304	163 149	175 799	238 351	252 518
Anguilla	386	330	333	150 F	200 F	250 F	250 F	250 F	250 F	250 F
Antigua Barb	1 712	642	696	1 311	1 209	1 437	1 415	1 361	1 481	1 583
Argentina	703 425	931 303	951 387	1 170 124	1 291 355	1 400 162	1 164 829	1 078 313	913 622	923 322
Armenia	1 885	1 850	1 033	821	580	580	698	1 144	1 133	866
Aruba	300	260	260	140	160	205	182	175	163	163
Australia	232 867	228 729	202 484	205 464	201 310	196 831	203 334	214 954	194 631	192 682
Austria	479	420	388	404	450	465	451	432	439	362
Azerbaijan	30 339	21 733	18 901	10 545	6 702	5 161	4 760	20 866	18 797	10 893
Bahamas	9 846	10 073	10 311	9 944	10 197	10 439	10 124	10 473	11 070	9 290
Bahrain	7 983	8 958	7 628	9 389	12 940	10 050	9 849	10 620	11 718	11 230
Bangladesh	709 332	764 824	770 790	792 389	814 787	829 426	839 141	959 215	1 004 264	1 000 000 F
Barbados	3 574	3 214	2 818	3 581	3 512	2 809	3 644	3 250	3 100	2 676
Belarus	1 507	2 993	786	715	821	499	457	514	553	943
Belgium	37 120	36 089	34 254	35 599	30 823	30 499	30 835	29 876	29 800	30 209
Belize	2 414	3 528	7 670	9 657	12 038	24 710	26 309	44 587	60 448	14 370
Benin	32 488	39 221	39 923	44 379	42 175	43 785	42 139	40 436	32 324	38 415
Bermuda	432	404	394	449	465	461	466	453	286	315
Bhutan	315 F	320 F	310 F	310 F	300 F	300 F	300 F	300 F	300 F	300 F
Bolivia	4 905	5 518	5 353	5 692	5 988	6 038	6 055	6 052	6 106	5 940
Bosnia Herzg	2 000 F	2 500 F	2 500 F	2 500 F	2 500 F	2 500 F	2 500 F	2 500 F	2 500 F	2 500 F
Botswana	800 F	600 F	400 F	200 F	81	160	191	157	166	118
Brazil	741 320 F	717 090 F	740 100 F	706 708	715 482	744 585	706 789	703 941	766 846	770 000 F
Br Ind Oc Tr	0	0	0	0	0	0	0	0	0	0
Br Virgin Is	453 F	343	470	532	506	105	116	115	43	50 F
Brunei Darsm	1 692	1 728	4 445	4 719	7 405	4 521	5 049	3 186	2 487	1 492
Bulgaria	23 986	13 688	6 405 F	8 191	8 854	11 237	18 946	10 556	6 998	6 530
Burkina Faso	7 500	7 000	8 000	8 000	8 000	8 000	8 335	7 600	8 500	8 500 F
Burundi	24 073	17 000 F	22 000 F	21 101	3 041	20 296	13 426	9 199	17 315	8 964
Cambodia	102 600	101 000	95 000	102 999	94 710	102 800	107 900	269 156	284 368	397 200 F
Cameroon	71 975	65 257	79 000 F	94 131 F	98 400 F	102 000 F	106 800 F	110 000 F	112 109	111 031
Canada	1 292 050	1 132 862	1 022 447	849 411	904 862	972 294	1 013 977	1 027 511	1 010 489	1 049 508
Cape Verde	6 573	7 000	8 256	8 495	9 155	9 705	9 424	10 360	10 586	9 653
Cayman Is	825	445	125	125	110	125	125	125	125	125
Cent Afr Rep	13 000	13 250 F	13 500 F	13 750 F	14 000 F	14 250 F	14 500 F	15 000	15 000 F	15 000 F
Chad	80 000	87 300	80 000	90 000	100 000	85 000	84 000	84 000 F	84 000 F	84 000 F
Channel Is	2 835	2 854 F	2 783	2 949 F	4 346	4 238	4 117	3 601	3 589	3 927
Chile	6 432 324	5 949 565	7 720 578	7 433 902	6 690 942	5 810 764	3 265 383	5 050 528	4 300 160	3 797 143
China	8 322 552	9 351 437	10 866 836	12 562 706	14 182 107	15 722 344	17 229 927	17 240 032	16 987 325	16 529 389
China,H.Kong	220 181	217 544	211 010	194 999	183 856	186 000	180 000	127 780	157 012	173 972
China, Macao	2 668	1 898	1 890	1 604	1 418	1 500 F	1 500 F	1 500 F	1 500 F	1 500 F
China,Taiwan	1 063 947	1 133 676	967 189	1 010 022	967 483	1 038 048	1 091 768	1 099 715	1 093 889	1 005 199
Christmas Is	0	0	0	0	0	0	0	0	0	0
Cocos Is	0	0	0	0	0	0	0	0	0	0
Colombia	141 170	123 497	97 799	120 699	130 829	147 918	132 908	117 995	129 644	125 000 F
Comoros	11 825	11 645	12 976	13 000 F	12 700 F	12 500 F	12 500 F	12 000	13 200	12 180
Congo Dem R	187 840	196 789	155 897	158 627	163 010	162 211	178 041	208 448	208 448 F	208 448 F
Congo Rep	39 992	46 748	42 664	45 776	45 473	38 082	42 955 F	43 696 F	44 000 F	42 000 F
Cook Is	967	985 F	932 F	1 090	900 F	820 F	770 F	750 F	720 F	700 F
Costa Rica	16 009	16 170	17 132	17 437	24 327	26 669	24 757	28 218	35 398	34 733
Côte dIvoire	87 039	76 972	73 978	70 189	69 168	64 169	69 572	74 365	75 772	73 556
Croatia	26 728	26 626	17 477	16 265	18 233	17 035	21 938	18 900	20 718	18 090
Cuba	100 130	84 752	76 790	80 059	85 603	84 911	67 076	67 381	56 146 F	56 000 F
Cyprus	9 336	10 016	9 427	9 320	12 526	24 819	19 295	39 638	67 482	75 803
Czech Rep	...	3 185	3 955	3 929	3 524	3 321	3 952	4 190	4 654	4 646
Czechoslovak	4 350 F	-	-	-	-	-	-	-	-	-
Denmark	1 953 828	1 614 289	1 873 316	1 999 033	1 681 517	1 826 852	1 557 335	1 405 005	1 534 089	1 510 439
Djibouti	275 F	300 F	320 F	350 F	350 F	350 F	350 F	350 F	350 F	350 F
Dominica	711	794	882	950	1 030	1 079	1 212	1 200 F	1 200 F	1 150 F
Dominican Rp	12 840	12 857	22 832	17 874	12 894	14 860	10 283	8 433	11 029	13 217
Ecuador	226 817	286 633	330 248	505 395	702 974	548 988	310 022	497 872	592 547	586 570
Egypt	258 700	272 700	283 900	310 790	320 230	342 759	362 741	380 504	384 314	428 651
El Salvador	12 076	12 451	13 559	15 071	14 434	11 897	12 657	16 524	9 590	17 747
Eq Guinea	3 600 F	3 507	5 069	2 306	5 040	6 090	6 005	7 001	3 634	3 500 F
Eritrea	...	475	2 706	3 559	3 252	1 038	1 629	6 891	12 612	8 820
Estonia	131 347	147 185	123 680	132 028	108 446	123 613	118 714	111 793	113 146	105 167
Ethiopia	4 585	4 175	5 285	6 325	8 770	10 370	14 000	15 858	15 681	15 390
Faeroe Is	248 430	246 294	237 708	287 796	304 587	329 145	363 816	358 133	454 530	524 837
Falkland Is	1 855	1 974	5 914	27 190	31 540	17 113	43 616	39 164	62 928	59 824
Fiji Islands	24 605	27 817	28 964 F	28 836	25 029	27 755	28 158	36 713	40 000 F	42 972
Finland	151 753	156 294	164 269	167 484	179 077	180 185	171 681	160 560	156 501	150 096
France	585 093	617 479	622 372	610 808	557 510	567 422	542 560	586 487	623 755	606 194
Fr Guiana	7 617	6 931	7 819	8 089	7 377 F	6 602	6 709	6 271 F	5 237 F	5 194 F
Fr Polynesia	3 425	8 075	8 175	9 020	9 910	11 670	12 473	12 336	13 899	15 404
Fr South Tr	464	460	524	519 F	437 F	375 F	388 F	425 F	272 F	263 F
Gabon	24 000 F	31 789 F	31 015 F	40 437 F	46 113	43 584	53 609	51 143	47 470	40 457
Gambia	18 045	21 308	22 781	23 752	31 601	32 254	29 002	30 000	29 016	34 527
Gaza Strip	...	...	...	1 229	2 493	3 791	3 625	3 600 F	3 600 F	3 000 F
Georgia	38 100 F	18 240 F	7 413 F	3 560 F	2 453	2 583	3 001	1 413	2 200	1 830
Germany	216 890	253 010	230 161	239 890	236 411	259 352	266 622	238 925	205 689	211 282
Ghana	423 425	372 619	335 437	352 844	477 173	447 088	442 641	492 776	452 070	445 287
Gibraltar	0	0	0	0	0	0	0	0	0	0
Greece	152 608	159 098	181 122	151 788	149 435	157 088	108 580	118 771	99 280	94 388
Greenland	113 267	116 650	117 417	128 890	116 018	120 596	128 542	160 253	159 711	158 485
Grenada	2 052	2 103	1 599	1 497	1 577	1 530	1 837	1 658	1 701	2 247

A-2

Fish, crustaceans, molluscs, etc	Capture production by countries or areas	All fishing areas
Poissons, crustacés, mollusques, etc	Captures par pays ou zones	Toutes les zones de pêche
Peces, crustáceos, moluscos, etc	Capturas por países o áreas	Todas las áreas de pesca

Country or area Pays ou zone País o área	1992 mt	1993 mt	1994 mt	1995 mt	1996 mt	1997 mt	1998 mt	1999 mt	2000 mt	2001 mt
Guadeloupe	8 540	8 600	8 800	9 500	9 570	10 480	9 084	9 114	10 100	10 100 F
Guam	345	431	435	185	121	158	253	223	275	278
Guatemala	6 578	7 582	7 211	8 253	7 653	6 896	10 847	21 018	37 978	10 100 F
Guinea	55 000 F	60 600 F	63 800 F	67 860	63 360	62 441	69 764	87 314	91 513	90 000 F
GuineaBissau	5 200 F	5 350 F	6 000 F	6 328	7 000 F	7 250 F	6 000 F	5 000 F	5 000 F	5 000 F
Guyana	41 252	44 123	46 367	47 900 F	48 583	53 998	52 840	53 844	48 887	53 405
Haiti	5 000 F	5 150 F	5 500 F	5 517 F	5 245 F	5 301 F	5 259 F	5 000 F	5 000 F	5 000 F
Honduras	16 486	19 577	9 279	19 337	13 528	18 357	7 393 F	12 754	11 684	7 451
Hungary	8 678	7 886	8 307	7 314	7 606	7 406	7 265	7 514	7 101	6 638
Iceland	1 574 682	1 715 581	1 556 962	1 612 548	2 060 168	2 205 944	1 681 951	1 736 267	1 982 522	1 980 715
India	2 844 102	3 062 712	3 257 607	3 265 240	3 447 954	3 523 448	3 373 492	3 472 150	3 742 296	3 762 600
Indonesia	2 891 318	3 077 540	3 305 815	3 511 145	3 553 276	3 791 240	3 964 897	3 986 919	4 069 691	4 203 830
Iran	305 706	322 006	308 101	341 383	351 725	342 287	367 212	387 200	384 000	336 450
Iraq	18 073	19 941	25 147	28 208	30 737	31 302	22 574	22 423	20 767	20 800 F
Ireland	249 386	278 774	294 165	384 632	333 030	292 673	324 274	280 957	281 806	356 309
Isle of Man	4 481	4 850	3 571	3 734	3 537	4 289	2 214	2 609	3 552	3 112
Israel	5 460	5 125	4 061	4 941	5 229	5 204	6 300	5 884	5 818	5 000 F
Italy	396 457	397 533	398 730	396 791	365 899	343 693	306 096	282 790	302 149	310 397
Jamaica	9 350 F	10 000 F	10 327 F	10 367 F	12 504 F	8 198	6 560	8 508	5 676	5 700 F
Japan	7 730 580	7 254 943	6 617 118	5 966 456	5 933 659	5 926 113	5 299 399	5 194 186	4 971 412	4 719 152
Jordan	380	395	410	425	440	450	470	510	550	520
Kazakhstan	65 124	56 700	46 433	48 402	44 273	31 826	25 000 F	36 170	36 620	30 654
Kenya	162 210	182 052	202 577	192 706	180 988	161 054	172 592	205 287	215 106	164 151
Kiribati	28 211	27 072	27 753	30 306	31 871	30 052	35 304	52 741	25 563	32 375
Korea D P Rp	900 000 F	870 398 F	371 961	327 083	253 125	236 462	220 000 F	210 000 F	200 850 F	200 000 F
Korea Rep	2 321 048	2 257 192	2 357 891	2 319 915	2 413 713	2 204 047	2 026 934	2 119 668	1 823 175	1 988 002
Kuwait	7 871	8 466	7 752	8 616	8 255	7 826	7 799	6 271	6 000 F	5 846
Kyrgyzstan	201	127	131	185	160 F	120 F	80 F	48	52	57
Laos	19 240	19 500 F	23 800	27 370	23 000 F	18 857	19 642	30 041	29 250	30 000 F
Latvia	156 737	141 973	138 176	149 194	142 644	105 682	102 331	125 389	136 403	125 433
Lebanon	1 720 F	2 020	2 225	4 085	4 135	3 655	3 520	3 560	3 666	3 670
Lesotho	16 F	22 F	22 F	26 F	28 F	30 F	30 F	30	32	24
Liberia	8 891	7 782	7 721	8 829	8 308	8 580	10 830	15 472	11 726	11 286
Libya	28 820 F	31 085 F	33 500 F	34 400 F	32 976 F	31 877 F	32 911 F	32 850 F	33 387 F	33 239 F
Liechtensten	0	0	0	0	0	0	0	0	0	0
Lithuania	188 467	117 171	49 162	57 368	88 514	44 002	66 578	72 962	78 987	151 931
Luxembourg	0	0	0	0	0	0	0	0	0	0
Macedonia	195	164	196	208	78	130	131	135	208	128
Madagascar	104 405	114 261	116 431	115 653	114 475	116 391	126 395	129 630	132 093	135 583
Malawi	69 261	67 951	58 579	53 664	63 569	56 340	41 111	45 392	43 000 F	40 619
Malaysia	1 025 289	1 049 321	1 067 650	1 112 375	1 130 372	1 172 922	1 153 719	1 251 768	1 289 245	1 234 733
Maldives	90 399	100 851	118 673	119 521	121 137	122 387	132 566	134 962	136 420	125 814
Mali	68 467	64 300	62 850	132 900	111 910	99 550	98 000	98 536	109 870	100 000 F
Malta	565	823	2 351	4 621	9 183	1 019	1 143	1 224	1 059	882
Marshall Is	326 F	471 F	393 F	381 F	2 772	400 F	500 F	500 F	8 155	37 098
Martinique	4 538	5 853	5 800 F	5 300	3 500 F	5 500	5 500	6 000	6 310	6 200
Mauritania	66 054 F	59 452 F	51 746 F	53 147 F	60 324 F	57 756 F	61 660 F	76 026 F	80 849 F	83 596 F
Mauritius	18 861	20 579	18 145	16 395	11 869	14 025	12 093	12 205	9 615	10 694
Mayotte	1 100 F	500 F	600 F	700 F	1 000	1 300 F	2 000 F	2 000	5 500	5 500
Mexico	1 157 573	1 102 932	1 191 646	1 329 340	1 464 188	1 489 112	1 179 860	1 205 603	1 315 581	1 398 592
Micronesia	16 547 F	17 888 F	22 926 F	7 629 F	9 069 F	10 127 F	16 581 F	12 838 F	23 715 F	18 062 F
Moldova Rep	410	630	708	709	603	569	491	309	344	387
Monaco	3 F	3 F	3 F	3 F	3 F	3 F	3 F	3 F	3 F	3 F
Mongolia	120 F	165	184	158	221	180	311	524	425	117
Montserrat	23	58	62	48	38	45	46	50 F	50 F	50 F
Morocco	550 937	624 511	754 327	848 951	642 886	791 906	710 436	745 431	896 620	1 083 276
Mozambique	31 608	30 195	27 456	26 833	34 915	39 703	36 677	33 989	39 065	32 512
Myanmar	731 544	739 702	746 241	751 232	601 788	780 295	830 117	919 410	1 069 726	1 166 868
Namibia	655 110	790 332	649 199	569 833	517 828	513 216	612 343	580 084	589 905	547 492
Nauru	377	500	500	450 F	400 F	400 F	400 F	400 F	400 F	400 F
Nepal	5 895	7 418	7 340	11 230	11 230	11 230	12 000	12 752	16 700	16 700
Netherlands	432 970	461 756	420 053	438 092	410 798	451 799	536 626	514 611	495 804	518 162
NethAntilles	1 150 F	1 200 F	1 100 F	1 020 F	1 000 F	950 F	950 F	950 F	20 494 F	950 F
NewCaledonia	3 184	2 670	3 185	2 644	2 998	2 438	3 105	3 152	3 386	3 337 F
New Zealand	460 822	426 668	449 660	555 559	423 635	610 254	639 238	598 475	548 559	561 110
Nicaragua	6 718	8 169	10 034	10 995	15 442	16 176	19 892	23 909	28 008	22 799
Niger	2 456	2 162	2 516	3 616	4 156	6 328	7 013	11 000	16 250	20 800
Nigeria	301 327	238 409	267 059	349 482	337 993	387 923	463 024	455 628	441 377	452 146
Niue	120	120 F	150 F	150 F	200 F	200 F	200 F	200 F	200 F	200 F
Norfolk Is	0	0	0	0	0	0	0	0	0	0
N Marianas	140	136	176	192	225	250	235	193	189	197
Norway	2 430 823	2 415 131	2 366 119	2 524 111	2 648 457	2 863 059	2 861 223	2 627 534	2 703 415	2 687 303
Oman	109 218	105 772	118 572	139 861	121 618	118 995	106 171	108 809	120 421	126 531
Pakistan	540 354	608 344	537 277	525 849	537 489	589 795	596 980	654 530	614 069	607 020
Palau	1 918	2 066	1 820	1 956	1 990	1 751	1 777	1 800 F	2 000 F	2 000 F
Panama	172 250	184 822	185 147	202 993	150 764	162 349	202 687	120 848	222 631	235 000 F
Papua N Guin	26 241 F	25 741 F	26 219 F	39 077 F	37 823 F	46 185 F	78 029 F	65 431 F	96 069 F	122 419 F
Paraguay	17 925	19 000 F	20 000 F	21 000 F	22 000	28 000	25 000 F	25 000 F	25 000 F	25 000 F
Peru	7 502 192	9 004 777	11 999 217	8 937 342	9 515 048	7 869 871	4 338 437	8 428 601	10 658 620	7 986 103
Philippines	1 885 041	1 834 323	1 845 335	1 860 701	1 783 601	1 805 806	1 833 380	1 872 818	1 893 017	1 945 217
Pitcairn Is	8	8 F	8 F	8 F	8 F	8 F	8 F	8 F	8 F	8 F
Poland	475 697	404 422	437 950	426 235	341 299	353 661	238 336	235 111	218 354	225 916
Portugal	293 170	287 802	263 065	260 253	259 846	221 879	224 234	208 459	187 570	191 214
Puerto Rico	1 812	1 877	2 275	3 173	2 701	3 187	3 006	3 020	4 154	3 794
Qatar	7 845	6 994	5 086	4 271	4 739	5 032	5 279	4 207	7 142	8 606
Réunion	1 103	1 679	2 531	2 500	3 607	4 288	4 579	4 043	4 079	3 635
Romania	70 761	13 819	22 251	49 275	18 259	8 446	9 061	7 843	7 372	7 637
Russian Fed	5 509 994	4 370 015	3 705 082	4 311 809	4 675 738	4 661 853	4 454 759	4 141 158	3 973 535	3 628 323
Rwanda	3 644	3 500 F	3 400 F	3 300 F	2 952	4 428	6 641	6 433	6 726	6 828
St Helena	651	726	702	915	819	897	1 060	632	718	866

A-2

Fish, crustaceans, molluscs, etc	Capture production by countries or areas	All fishing areas
Poissons, crustacés, mollusques, etc	Captures par pays ou zones	Toutes les zones de pêche
Peces, crustáceos, moluscos, etc	Capturas por países o áreas	Todas las áreas de pesca

Country or area Pays ou zone País o área	1992 mt	1993 mt	1994 mt	1995 mt	1996 mt	1997 mt	1998 mt	1999 mt	2000 mt	2001 mt
St Kitts Nev	300 F	250 F	212	192	352	272	533	555	492	591
St Lucia	1 073	1 336	1 252	1 188	1 274	1 308	1 589	1 718	1 855	1 983
St Pier Mq	16 443	282	294	317	747	3 571	6 108	5 892	6 485	3 802
St Vincent	2 157	1 945	3 840	1 005	921 F	6 093	33 973	17 759	27 694	45 778
Samoa	1 296	673	1 108	2 519	2 727	7 041	7 547	10 204	13 004	12 966 F
Sao Tome Prn	2 094	2 334	3 391	3 565	3 980	3 338	3 477	3 756	3 500 F	3 500 F
Saudi Arabia	45 910	48 021	54 612	45 609	47 698	49 314	51 206	46 618	49 650	49 167
Senegal	370 242	382 212	352 421	354 617	411 759	457 366	403 872	412 125	402 047	405 409
Seychelles	6 663	5 178	4 469	4 008	4 707	14 043	23 886	34 297	32 359	47 550
Sierra Leone	68 510	66 288	62 439	64 870	67 304 F	72 628	63 065	59 407	74 730	75 210
Singapore	9 201	9 304	11 301	10 102	9 943	9 250	7 733	6 489	5 371	3 342
Slovakia	...	1 185	1 627	1 950	1 414	1 364	1 361	1 396	1 368	1 531
Slovenia	3 905	2 284	2 346	2 167	2 367	2 367	2 228	2 027	1 856	1 827
Solomon Is	45 601 F	42 970 F	46 904 F	62 019 F	50 990 F	64 005 F	61 332 F	58 428 F	24 788 F	30 075 F
Somalia	25 750 F	27 750 F	29 850 F	27 950 F	26 050 F	24 150 F	22 250 F	20 250 F	20 200 F	20 000 F
South Africa	692 339	562 374	523 449	575 547	440 428	515 095	559 049	588 144	643 238	755 345
Spain	1 077 764 F	1 085 959 F	1 094 524 F	1 181 286 F	1 174 635	1 204 913	1 263 275	1 169 462	1 045 488	1 084 820
Sri Lanka	202 668	215 400	218 400	229 171	228 945	235 099	263 330	276 080	297 410	279 640
Sudan	35 000	40 000	44 000	44 000	45 000	47 000	49 500	49 500	53 000 F	58 000
Suriname	10 930	9 500	14 465	13 000 F	13 000 F	14 000 F	16 195	16 200	17 500 F	18 915
Svalbard Is	0	0	0	0	0	0	0	0	0	0
Swaziland	60 F	68 F	65 F	60 F	60 F	65 F	70 F	70 F	70 F	70 F
Sweden	307 545	341 897	386 814	404 572	370 881	357 406	410 886	351 254	338 534	311 816
Switzerland	2 715	1 822	1 481	1 588	1 841	1 859	1 809	1 840	1 659	1 715
Syria	4 538	4 554	5 520	5 782	5 773	6 131	7 097	7 938	6 572	8 291
Tajikistan	149	253	127	100	40 F	75 F	100 F	80 F	78 F	137
Tanzania	331 235	331 267	288 399	359 800	323 921	356 960	348 000	310 509	332 779	335 900
Thailand	2 875 456	2 927 689	3 015 196	3 031 074	3 013 961	2 902 898	2 930 354	2 952 008	2 911 173	2 881 316
Timor-Leste	...	...	...	...	...	...	...	408 F	362	356
Togo	10 749	16 964	13 052	12 201	15 098	14 290	16 655	22 924	22 277	23 163
Tokelau	191	200 F	200 F	200 F	200 F	200 F	200 F	200 F	200 F	200 F
Tonga	2 237	2 375	2 499	2 597	2 915	2 871	4 076	4 221	3 760	4 673
Trinidad Tob	13 000 F	9 637	14 483	12 000 F	9 435	11 283	9 175	8 826	9 786	11 408
Tunisia	86 397	82 914	85 610	83 355	83 734	87 012	88 075	93 186	95 550	98 482
Turkey	447 498	548 758	589 803	633 970	527 826	459 153	487 200	573 824	503 348	527 730
Turkmenistan	31 777	16 080	15 140	9 740	9 014	8 179	7 014	9 058	12 228	12 749
Turks Caicos	1 191	1 458	1 419 F	1 395 F	1 297 F	1 250 F	1 318	1 300 F	1 300 F	1 300 F
Tuvalu	499	1 460	561	399	400 F	500 F	500 F	500 F	500 F	500 F
Uganda	264 900	219 814	213 129	208 789	195 088	218 026	220 628	226 097	219 356	220 726
Ukraine	458 939	305 334	267 180	378 650	417 119	373 005	462 308	407 853	392 724	351 260
Untd Arab Em	95 046	99 600	108 600	105 884	107 000	114 358	114 739	117 607	110 056	110 000 F
UK	813 082	860 229	878 018	909 928	865 145	886 261	923 085	840 898	747 358	741 105
USA	5 190 717	5 523 216	5 535 324	5 224 566	5 001 483	4 983 440	4 708 980	4 749 646	4 745 321	4 944 405
US Minor Is	0	0	0	0	0	0	0	0	0	0
US Virgin Is	750 F	650 F	550 F	470 F	400 F	350 F	300 F	263	300 F	300 F
Uruguay	125 751	118 815	120 732	126 446	123 330	136 954	140 707	103 012	116 588	105 034
Uzbekistan	9 346	4 358	3 095	3 611	1 494	3 075	2 799	2 871	3 387	4 070
Vanuatu	44 791	40 515 F	47 190 F	57 646 F	45 615 F	70 862 F	74 908 F	93 041 F	69 260 F	26 690 F
Venezuela	330 964	395 861	438 218	501 045	496 287	470 255	503 504	399 799	357 115	417 947
Viet Nam	868 107	932 143	1 025 909	1 084 939	1 223 644	1 276 325	1 293 954	1 386 300	1 450 590 F	1 491 123 F
Wallis Fut I	143	150	193	170	180	176	300	300 F	300 F	300 F
Westn Sahara	0	0	0	0	0	0	0	0	0	0
Yemen	79 547	82 356	81 885	107 970	104 955	115 600	127 620	124 385	114 751	142 198
Yugoslavia	5 334	4 074	4 172	4 167	4 030	3 873	2 610 F	1 251	1 096	1 088
Zambia	67 864	65 768	70 057	70 546	66 332	65 923	69 938	67 327	66 671	65 000 F
Zimbabwe	21 601	21 230	20 219	16 463	16 387	18 156	16 371	12 410	13 114	13 000 F
Other nei	117 926	135 877	137 788	309 200	189 219	171 453	188 268	215 405	215 777	176 532
World total Total mondial Total mundial	**85 801 935**	**86 973 018**	**92 003 837**	**92 301 906**	**93 750 402**	**94 215 594**	**87 593 312**	**93 601 896**	**95 439 820**	**92 356 034**
World excl. China Monde excl. Chine Mundo excl. China	**77 479 383**	**77 621 581**	**81 137 001**	**79 739 200**	**79 568 295**	**78 493 250**	**70 363 385**	**76 361 864**	**78 452 495**	**75 826 645**

A-3

Fish, crustaceans, molluscs, etc
Poissons, crustacés, mollusques, etc
Peces, crustáceos, moluscos, etc

Capture production by countries or areas
Captures par pays ou zones
Capturas por países o áreas

Inland waters
Eaux continentales
Aguas continentales

Country or area Pays ou zone País o área	1992 mt	1993 mt	1994 mt	1995 mt	1996 mt	1997 mt	1998 mt	1999 mt	2000 mt	2001 mt
Afghanistan	1 200 F	1 200 F	1 300 F	1 300 F	1 300 F	1 250 F	1 200 F	1 200 F	1 000 F	800 F
Albania	916	850 F	700 F	252	357	180	823	814	955	1 466
Algeria	0	0	0	0	0	0	2	0	0	0
Amer Samoa	0	0	0	0	0	0	0	0	0	0
Andorra	0	0	0	0	0	0	0	0	0	0
Angola	7 000	7 000	7 000	6 000	6 000	6 000	6 000	6 000 F	6 000 F	6 000 F
Anguilla	0	0	0	0	0	0	0	0	0	0
Antigua Barb	0	0	0	0	0	0	0	0	0	0
Argentina	11 227	11 800	12 785	17 191	19 189	22 735	23 197	27 558	30 418	23 860
Armenia	1 885	1 850	1 033	821	580	580	698	1 144	1 133	866
Aruba	0	0	0	0	0	0	0	-	-	-
Australia	3 757	3 211	584	1 724	1 766	1 790	1 867	2 328	1 984 F	1 939 F
Austria	479	420	388	404	450	465	451	432	439	362
Azerbaijan	30 339	21 733	18 901	10 545	6 702	5 161	4 760	20 866	18 797	10 893
Bahamas	0	0	0	0	0	0	0	0	0	0
Bahrain	0	0	0	0	0	0	0	0	-	-
Bangladesh	429 205	452 109	517 746	527 739	535 617	534 285	538 689	649 418	670 465	670 000 F
Barbados	0	0	0	0	0	0	0	0	0	-
Belarus	1 507	2 993	786	715	821	499	457	514	553	943
Belgium	511	511	511	511	511	511	511	536	511	511
Belize	1	1	1	0	0	0	0	0	0	0
Benin	26 566	32 805	32 707	37 449	34 193	32 871	31 778	31 894	26 400	30 000
Bermuda	0	0	0	0	0	0	0	0	0	0
Bhutan	315 F	320 F	310 F	310 F	300 F	300 F	300 F	300 F	300 F	300 F
Bolivia	4 905	5 518	5 353	5 692	5 988	6 038	6 055	6 052	6 106	5 940
Bosnia Herzg	2 000 F	2 500 F	2 500 F	2 500 F	2 500 F	2 500 F	2 500 F	2 500 F	2 500 F	2 500 F
Botswana	800 F	600 F	400 F	200 F	81	160	191	157	166	118
Brazil	182 540 F	186 990 F	191 485 F	193 042	193 309	178 871	174 190	185 471	199 159	200 000 F
Br Ind Oc Tr	0	0	0	0	0	0	0	0	0	0
Br Virgin Is	0	0	0	0	0	0	0	0	0	0
Brunei Darsm	25	25	4	7	15	17	35	26	23	16
Bulgaria	1 611	1 675	995 F	762	1 127	1 881	2 336	2 475	861	1 650
Burkina Faso	7 500	7 000	8 000	8 000	8 000	8 000	8 335	7 600	8 500	8 500 F
Burundi	24 073	17 000 F	22 000 F	21 101	3 041	20 296	13 426	9 199	17 315	8 964
Cambodia	68 900	67 900	65 000	72 499	63 510	73 000	75 700	231 000	245 600	360 000 F
Cameroon	22 000	23 000	27 000 F	30 000 F	35 000 F	40 000 F	45 000 F	50 000 F	55 000	52 500 F
Canada	42 633	36 327	36 333	38 756	38 295	38 798	40 744	40 587	40 667	35 120
Cape Verde	0	0	0	0	0	0	0	0	0	0
Cayman Is	0	0	0	0	0	0	0	0	0	0
Cent Afr Rep	13 000	13 250 F	13 500 F	13 750 F	14 000 F	14 250 F	14 500 F	15 000	15 000 F	15 000 F
Chad	80 000	87 300	80 000	90 000	100 000	85 000	84 000	84 000 F	84 000 F	84 000 F
Channel Is	0	0	0	0	0	0	0	0	0	0
Chile	14	7	5	...	...	...	4	...	...	...
China	998 028	1 182 390	1 327 785	1 607 385	1 762 860	1 886 967	2 280 244	2 285 364	2 233 230	2 149 932
China,H.Kong	0	0	0	0	0	0	0	0	0	0
China, Macao	0	0	0	0	0	0	0	0	0	0
China,Taiwan	1 566	1 216	1 059	695	407	403	449	561	549	591
Christmas Is	0	0	0	0	0	0	0	0	0	0
Cocos Is	0	0	0	0	0	0	0	0	0	0
Colombia	33 759	30 538	34 983	23 524	23 061	20 610	21 673	28 788	24 854	25 000 F
Comoros	0	0	0	0	0	0	0	0	0	0
Congo Dem R	184 040	192 589	152 117	154 751	159 037	158 367	174 087	204 503	204 503 F	204 503 F
Congo Rep	21 049	27 850	24 752	26 811	25 873	18 987	25 455	25 455 F	26 000 F	24 500 F
Cook Is	0	0	0	10	0	0	0	0	0	0
Costa Rica	406	710	840	900	1 090	840	1 000 F	1 000 F	1 000 F	1 000 F
Côte dIvoire	15 404	13 477	15 604	11 335	11 562	12 032	12 501	10 656	10 502	10 530 F
Croatia	198	284	340	364	434	408	10 F	10	17	34
Cuba	13 820	10 178	9 823	8 893	10 324	8 525	5 954	4 624	4 600 F	4 600 F
Cyprus	...	...	5	65	64	70	70	70	78	70 F
Czech Rep	...	3 185	3 955	3 929	3 524	3 321	3 952	4 190	4 654	4 646
Czechoslovak	4 350 F	-	-	-	-	-	-	-	-	-
Denmark	653	337	243	264	196	232	349	206	183	77
Djibouti	0	0	0	0	0	0	0	0	0	0
Dominica	0	0	0	0	0	0	0	0	0	0
Dominican Rp	1 024	2 037	3 774	2 106	288	1 067	1 095	598	187	1 158
Ecuador	332	372	300	300	300	400	400	400	400	400
Egypt	180 400	186 700	199 300	228 930	230 660	243 609	250 181	225 300	253 470	295 422
El Salvador	5 136	4 461	3 819	4 324	2 968	2 808	2 443	2 653	2 831	1 692
Eq Guinea	370 F	600	700	450	900	850	970	1 101 F	1 076	1 000 F
Eritrea	...	0	0	0	0	0	0	0	0	0
Estonia	3 509	2 411	1 909	2 366	2 361	2 439	3 878	3 108	3 190	2 461
Ethiopia	4 485	4 175	5 285	6 325	8 770	10 370	14 000	15 858	15 681	15 390
Faeroe Is	0	0	0	0	0	0	0	0	0	0
Falkland Is	-	-	-	-	1	1	1	1	1	1
Fiji Islands	4 084	2 907	2 964 F	3 586	3 034	4 325	5 111	5 622	5 700 F	5 921
Finland	51 407	51 522	47 895	48 436	47 618	47 618	36 813	36 813	34 840	34 820
France	4 350	4 400	4 450	4 500	4 540	4 540	4 500	2 134 F	2 131 F	2 130 F
Fr Guiana	0	0	0	0	0	0	0	0	0	0
Fr Polynesia	0	0	0	0	0	53	53	53	53	53
Fr South Tr	0	0	0	0	0	0	0	0	0	0
Gabon	2 000 F	3 500 F	4 500 F	7 648 F	9 433	9 441	9 000	10 000	10 417	9 850
Gambia	2 500	2 400	2 400	2 500	2 500	2 500	2 500	2 500 F	2 500 F	2 500 F
Gaza Strip	...	...	...	0	0	0	0	0	0	0
Georgia	190	549	16	90	6	1	4	17	22	8
Germany	8 770	10 837	10 908	22 987	22 987	22 916	22 916	22 868	22 868	22 818
Ghana	56 000	52 000	54 700	60 000	73 580	70 000	74 500	74 500	74 500	74 500
Greece	2 370	2 960	3 452	3 606	2 903	2 601	2 818	3 280	3 433	3 181
Greenland	0	0	0	0	0	0	0	0	0	0
Grenada	0	0	0	0	0	0	0	0	0	0
Guadeloupe	0	0	0	0	0	0	0	0	0	0

A-3

Fish, crustaceans, molluscs, etc — Capture production by countries or areas — Inland waters
Poissons, crustacés, mollusques, etc — Captures par pays ou zones — Eaux continentales
Peces, crustáceos, moluscos, etc — Capturas por países o áreas — Aguas continentales

Country or area / Pays ou zone / País o área	1992 mt	1993 mt	1994 mt	1995 mt	1996 mt	1997 mt	1998 mt	1999 mt	2000 mt	2001 mt
Guam	0	0	0	0	0	-	6	-	-	-
Guatemala	3 702	4 228	3 776	4 025	4 000	5 121	6 523	6 976	7 301	7 300 F
Guinea	4 000 F	4 600	3 800	3 100	2 780	3 600	4 000	4 000	4 000	4 000 F
GuineaBissau	200 F	250 F	250 F	250 F	250 F	250 F	200 F	200 F	200 F	200 F
Guyana	800	800	800	700 F	800	625	625	603	800	800
Haiti	500 F	600 F	500 F	500 F	500 F	500 F	500 F	500 F	500 F	500 F
Honduras	85	86	92	127	98	126	119	102	61	111
Hungary	8 678	7 886	8 307	7 314	7 606	7 406	7 265	7 514	7 101	6 638
Iceland	886	907	698	739	608	404	416	370	176	160
India	373 287	575 905	552 874	608 378	633 425	641 775	692 439	696 083	955 620	974 710
Indonesia	300 896	308 648	336 083	329 790	335 696	304 258	288 666	327 627	305 212	306 560
Iran	61 737	75 021	89 157	88 800	109 286	103 795	140 263	143 400	123 500	73 645
Iraq	17 530	17 808	20 926	22 955	19 049	20 519	9 111	9 330	8 378	8 400 F
Ireland	2 991	2 929	3 604	3 761	3 806	3 804	3 976	865	961	798
Isle of Man	0	0	0	0	0	0	0	0	0	0
Israel	2 214	1 813	1 478	1 214	1 845	1 476	2 164	2 145	1 852	1 600 F
Italy	9 075	9 515	9 921	10 035	6 764	6 690	4 667	5 436	4 565	5 527
Jamaica	450 F	450 F	450 F	450 F	450 F	450 F	450 F	450 F	450 F	450 F
Japan	96 739	91 032	92 219	91 455	93 501	85 612	78 822	71 270	70 612	61 354
Jordan	350	350	350	350	350	350	350	350	400	350
Kazakhstan	65 124	56 700	46 433	48 402	44 273	31 826	25 000 F	36 170	36 620	30 654
Kenya	155 644	176 435	198 805	187 241	174 692	154 955	165 992	198 653	210 343	156 763
Kiribati	0	0	0	0	0	0	0	0	0	0
Korea D P Rp	35 000 F	35 000 F	20 000	20 000	20 000	20 000	20 000 F	20 000 F	20 000 F	20 000 F
Korea Rep	25 483	12 263	10 492	9 646	8 034	6 934	6 845	6 316	7 141	5 971
Kuwait	0	0	0	0	0	0	0	0	0	0
Kyrgyzstan	201	127	131	185	160 F	120 F	80 F	48	52	57
Laos	19 240	19 500 F	23 800	27 370	23 000 F	18 857	19 642	30 041	29 250	30 000 F
Latvia	551	553	495	514	536	544	501	610	612	581
Lebanon	20 F	20	20	20	20	20	20	20	20	20
Lesotho	16 F	22 F	22 F	26 F	28 F	30 F	30 F	30	32	24
Liberia	4 000	4 000	4 000	4 000	4 000	4 000	4 000	4 000	4 000	4 000
Libya	0	0	0	0	0	0	0	0	0	0
Liechtenstein	0	0	0	0	0	0	0	0	0	0
Lithuania	1 522	1 146	1 187	1 260	1 295	1 713	1 737	1 715	1 911	1 854
Luxembourg	0	0	0	0	0	0	0	0	0	0
Macedonia	195	164	196	208	78	130	131	135	208	128
Madagascar	27 500	30 000	30 000	30 000	30 000	30 000	30 000	30 000	30 000	30 000
Malawi	69 261	67 951	58 579	53 664	63 569	56 340	41 111	45 392	43 000 F	40 619
Malaysia	1 773	1 971	2 064	3 939	3 683	3 949	4 626	3 366	3 549	3 446
Maldives	0	0	0	0	0	0	0	0	0	0
Mali	68 467	64 300	62 850	132 900	111 910	99 550	98 000	98 536	109 870	100 000 F
Malta	0	0	0	0	0	0	0	0	0	0
Marshall Is	0	0	0	0	0	0	0	0	0	0
Martinique	0	3	0	0	0	0	0	0	0	0
Mauritania	5 000 F	5 000 F	5 000 F	5 000 F	5 000 F	5 000 F	5 000 F	5 000 F	5 000 F	5 000 F
Mauritius	0	3	0	0	0	0	0	0	0	0
Mexico	108 707	111 178	111 125	122 020	122 501	113 552	100 335	91 462	106 817	92 154
Micronesia	5 F	4 F	4 F	5 F	5 F	5 F	5 F	5 F	5 F	5 F
Moldova Rep	410	630	708	709	603	569	491	309	344	387
Mongolia	120 F	165	184	158	221	180	311	524	425	117
Montserrat	0	0	0	0	0	0	0	0	0	0
Morocco	1 794	1 617	1 750	1 500	1 500	2 100	1 703	2 163	1 608	983
Mozambique	3 800	4 689	4 925	5 093	7 510	11 668	8 994	10 243	13 088	8 076
Myanmar	141 281	142 065	146 365	148 347	146 494	149 069	149 279	159 746	189 708	235 376
Namibia	1 102	1 200	1 200	1 200	1 200	1 500	1 500	1 500	1 500	1 500
Nepal	5 895	7 418	7 340	11 230	11 230	11 230	12 000	12 752	16 700	16 700
Netherlands	2 287	1 601	2 446	4 107	2 157	2 293	1 547	2 303	2 280 F	2 200 F
NethAntilles	0	0	0	0	0	0	0	0	0	0
NewCaledonia	0	0	0	0	0	0	0	0	0	0
New Zealand	1 252	1 160	1 100	1 115	1 079	1 025	1 307	1 152	1 089	1 076
Nicaragua	348	547	824	538	1 142	1 293	1 256	1 120	1 076	1 051
Niger	2 456	2 162	2 516	3 616	4 156	6 328	7 013	11 000	16 250	20 800
Nigeria	93 281	95 627	103 800	117 903	89 521	93 644	139 020	139 393	132 315	154 175
Niue	0	0	0	0	0	0	0	0	0	0
Norfolk Is	0	0	0	0	0	0	0	0	0	0
N Marianas	0	0	0	1	0	0	0	0	0	0
Norway	580	435	432	413	338	439	507	514	578	570 F
Oman	0	0	0	0	0	0	0	0	0	0
Pakistan	109 087	109 185	118 703	121 405	142 092	167 530	163 524	179 865	176 468	180 100
Palau	0	0	0	0	0	0	0	0	0	0
Panama	80	28	285	130	80	91	23	20	20	20 F
Papua N Guin	13 500 F	13 500 F	13 500 F	13 500 F	13 500 F	13 500 F	13 500 F	13 500 F	13 500 F	13 500 F
Paraguay	17 925	19 000 F	20 000 F	21 000 F	22 000	28 000	25 000 F	25 000 F	25 000 F	25 000 F
Peru	32 734	38 290	48 837	50 789	28 890	32 221	35 327	36 223	32 297	35 653
Philippines	229 673	210 775	204 325	186 006	177 355	159 353	146 004	146 234	151 753	135 845
Pitcairn Is	0	0	0	0	0	0	0	0	0	0
Poland	20 950	31 391	27 500	24 889	22 037	13 832	13 236	13 875	17 543	17 789
Portugal	3	3	4	2	0	0	0	0	0	1
Puerto Rico	0	0	0	0	0	0	0	0	0	0
Qatar	0	0	0	0	0	0	0	0	0	0
Réunion	0	0	0	0	0	0	0	0	0	0
Romania	9 890	8 562	10 598	9 048	6 145	4 574	4 630	5 336	4 896	5 206
Russian Fed	275 125	216 866	217 950	212 874	233 272	227 091	271 311	307 823	292 368	206 430
Rwanda	3 644	3 500 F	3 400 F	3 300 F	2 952	4 428	6 641	6 433	6 726	6 828
St Helena	0	0	0	0	0	-	-	-	-	-
St Kitts Nev	0	0	0	0	0	0	-	0	0	0
St Lucia	0	0	0	0	0	0	0	0	0	0
St Pier Mq	0	0	0	0	0	0	0	0	0	0
St Vincent	0	0	0	0	2	1	0	0	0	0

A-3

Fish, crustaceans, molluscs, etc | Capture production by countries or areas | Inland waters
Poissons, crustacés, mollusques, etc | Captures par pays ou zones | Eaux continentales
Peces, crustáceos, moluscos, etc | Capturas por países o áreas | Aguas continentales

Country or area Pays ou zone País o área	1992 mt	1993 mt	1994 mt	1995 mt	1996 mt	1997 mt	1998 mt	1999 mt	2000 mt	2001 mt
Samoa	0	0	0	0	0	0	0	0	1	1 F
Sao Tome Prn	0	0	0	0	0	0	0	0	0	0
Saudi Arabia	0	0	0	0	0	0	0	0	0	0
Senegal	24 750	27 650	30 000 F	31 000 F	23 000 F	31 000 F	21 000 F	34 000 F	22 450	20 000 F
Seychelles	0	0	0	0	0	0	0	0	0	0
Sierra Leone	14 000	14 000	15 000	15 000	14 500 F	14 500	14 190	14 480	14 000	14 000
Singapore	24	25	23	0	0	0	0	0	0	0
Slovakia	...	1 185	1 627	1 950	1 414	1 364	1 361	1 396	1 368	1 531
Slovenia	293	297	339	316	289	302	269	243	226	206
Solomon Is	0	0	0	0	0	0	0	0	0	0
Somalia	250 F	250 F	250 F	250 F	250 F	250 F	250 F	250 F	200 F	200 F
South Africa	832	832	800	800	850	850	900	900 F	900 F	900 F
Spain	8 796	9 215	6 284	8 869	8 710 F	8 710 F	8 710 F	8 710 F	8 710 F	8 710 F
Sri Lanka	17 500	15 000	9 500	15 000	22 250	27 250	29 900	31 450	36 700	29 870
Sudan	33 000	37 500	40 000	40 000	40 500	42 000	44 000	44 000	48 000 F	53 000
Suriname	561	187	138	140 F	150 F	200 F	200 F	200 F	200 F	200
Swaziland	60 F	68 F	65 F	60 F	60 F	65 F	70 F	70 F	70 F	70 F
Sweden	2 308	2 273	2 254	1 934	1 810	2 011	1 559	1 478	1 459	1 234
Switzerland	2 715	1 822	1 481	1 588	1 841	1 859	1 809	1 840	1 659	1 715
Syria	2 584	2 535	3 570	3 832	3 103	3 557	4 347	5 338	3 991	5 969
Tajikistan	149	253	127	100	40 F	75 F	100 F	80 F	78 F	137
Tanzania	275 150	294 582	247 614	317 029	262 276	306 750	300 000	260 020	280 000	283 000
Thailand	135 239	175 140	196 397	186 665	205 903	203 671	200 715	206 540	201 100	209 977
Timor-Leste	...	...	...	...	...	...	...	0	0	0
Togo	5 500	6 000	5 000	4 998	5 000	5 000	5 000	5 000	5 000	5 000
Tokelau	0	0	0	0	0	0	0	0	0	0
Tonga	1	1	0	0	0	0	0	0	0	0
Trinidad Tob	0	0	0	0	0	0	0	0	0	0
Tunisia	0	400	243	440	706	1 010	896	808	832	860
Turkey	40 495	44 801	45 067	47 976	49 600	50 460	54 500	50 190	42 824	43 323
Turkmenistan	31 777	16 080	15 140	9 740	9 014	8 179	7 014	9 058	12 228	12 749
Turks Caicos	0	0	0	0	0	0	0	0	0	0
Tuvalu	0	0	0	0	0	0	0	0	0	0
Uganda	264 900	219 814	213 129	208 789	195 088	218 026	220 628	226 097	219 356	220 726
Ukraine	24 801	13 189	14 786	6 847	9 468	6 215	4 898	4 728	4 429	4 343
Untd Arab Em	0	0	0	0	0	0	0	0	0	0
UK	2 035	1 909	2 191	2 146	1 930	1 491	4 569	4 835	2 743	3 142
USA	48 639	54 377	42 648	36 688	33 471	38 347	36 129	36 413	25 339	29 578
US Virgin Is	0	0	0	0	0	0	0	0	0	-
Uruguay	323	621	966	849	598	2 216	1 931	2 423	2 302	451
Uzbekistan	9 346	4 358	3 095	3 611	1 494	3 075	2 799	2 871	3 387	4 070
Vanuatu	0	0	0	0	0	0	0	0	0	0
Venezuela	20 114	28 251	35 412	54 175	49 911	40 881	46 035	35 219	23 739	24 326
Viet Nam	138 154	146 839	79 587	94 689	164 936	177 589	138 800	169 107	170 000 F	170 000 F
Wallis Fut I	0	0	0	0	0	0	0	0	0	0
Yugoslavia	5 111	3 797	3 912	3 803	3 653	3 500	2 200 F	828	672	672 F
Zambia	67 864	65 768	70 057	70 546	66 332	65 923	69 938	67 327	66 671	65 000 F
Zimbabwe	21 601	21 230	20 219	16 463	16 387	18 156	16 371	12 410	13 114	13 000 F
World total Total mondial Total mundial	**6 203 027**	**6 590 313**	**6 711 306**	**7 263 859**	**7 427 541**	**7 562 466**	**8 043 505**	**8 513 361**	**8 788 721**	**8 692 758**
World excl. China Monde excl. Chine Mundo excl. China	**5 204 999**	**5 407 923**	**5 383 521**	**5 656 474**	**5 664 681**	**5 675 499**	**5 763 261**	**6 227 997**	**6 555 491**	**6 542 826**

A-4

Fish, crustaceans, molluscs, etc | Capture production by countries or areas | Marine fishing areas
Poissons, crustacés, mollusques, etc | Captures par pays ou zones | Zones de pêche maritimes
Peces, crustáceos, moluscos, etc | Capturas por países o áreas | Areas de pesca marítimas

Country or area Pays ou zone País o área	1992 mt	1993 mt	1994 mt	1995 mt	1996 mt	1997 mt	1998 mt	1999 mt	2000 mt	2001 mt
Albania	1 780	1 650 F	1 400 F	1 127	1 768	833	1 860	1 931	2 365	1 844
Algeria	95 270	101 895	135 402	105 872	81 989	91 580	92 344	102 396	100 000 F	100 000 F
Amer Samoa	45	27	111	152	210	431	586	518	830	3 663
Angola	106 625	119 200	125 413	116 781	131 815	140 304	157 149	169 799	232 351	246 518
Anguilla	386	330	333	150 F	200 F	250 F	250 F	250 F	250 F	250 F
Antigua Barb	1 712	642	696	1 311	1 209	1 437	1 415	1 361	1 481	1 583
Argentina	692 198	919 503	938 602	1 152 933	1 272 166	1 377 427	1 141 632	1 050 755	883 204	899 462
Aruba	300	260	260	140	160	205	182	175	163	163
Australia	229 110 F	225 518 F	201 900 F	203 740 F	199 544 F	195 041	201 467	212 626	192 647	190 743
Bahamas	9 846	10 073	10 311	9 944	10 197	10 439	10 124	10 473	11 070	9 290
Bahrain	7 983	8 958	7 628	9 389	12 940	10 050	9 849	10 620	11 718	11 230
Bangladesh	280 127	312 715	253 044	264 650	279 170	295 141	300 452	309 797	333 799	330 000 F
Barbados	3 574	3 214	2 818	3 581	3 512	2 809	3 644	3 250	3 100	2 676
Belgium	36 609	35 578	33 743	35 088	30 312	29 988	30 324	29 340	29 289	29 698
Belize	2 413	3 527	7 669	9 657	12 038	24 710	26 309	44 587	60 448	14 370
Benin	5 922	6 416	7 216	6 930	7 982	10 914	10 361	8 542	5 924	8 415
Bermuda	432	404	394	449	465	461	466	453	286	315
Bosnia Herzg	0	0	0	0	0	0	0	0	0	0
Brazil	558 780 F	530 100 F	548 615 F	513 666	522 173	565 714	532 599	518 470	567 687	570 000 F
Br Ind Oc Tr	0	0	0	0	0	0	0	0	0	0
Br Virgin Is	453 F	343	470	532	506	105	116	115	43	50 F
Brunei Darsm	1 667	1 703	4 441	4 712	7 390	4 504	5 014	3 160	2 464	1 476
Bulgaria	22 375	12 013	5 410 F	7 429	7 727	9 356	16 610	8 081	6 137	4 880
Cambodia	33 700	33 100	30 000	30 500	31 200	29 800	32 200	38 156	38 768	37 200 F
Cameroon	49 975	42 257	52 000 F	64 131	63 400	62 000 F	61 800 F	60 000 F	57 109	58 531
Canada	1 249 417	1 096 535	986 114	810 655	866 567	933 496	973 233	986 924	969 822	1 014 388
Cape Verde	6 573	7 000	8 256	8 495	9 155	9 705	9 424	10 360	10 586	9 653
Cayman Is	825	445	125	125	110	125	125	125	125	125
Channel Is	2 835	2 854 F	2 783	2 949 F	4 346	4 238	4 117	3 601	3 589	3 927
Chile	6 432 310	5 949 558	7 720 573	7 433 902	6 690 942	5 810 764	3 265 379	5 050 528	4 300 160	3 797 143
China	7 324 524	8 169 047	9 539 051	10 955 321	12 419 247	13 835 377	14 949 683	14 954 668	14 754 095	14 379 457
China,H.Kong	220 181	217 544	211 010	194 999	183 856	186 000	180 000	127 780	157 012	173 972
China, Macao	2 668	1 898	1 890	1 604	1 418	1 500 F	1 500 F	1 500 F	1 500 F	1 500 F
China,Taiwan	1 062 381 F	1 132 460 F	966 130 F	1 009 327 F	967 076	1 037 645	1 091 319	1 099 154	1 093 340	1 004 608
Christmas Is	0	0	0	0	0	0	0	0	0	0
Cocos Is	0	0	0	0	0	0	0	0	0	0
Colombia	107 411	92 959	62 816	97 175	107 768	127 308	111 235	89 207	104 790	100 000 F
Comoros	11 825	11 645	12 976	13 000 F	12 700 F	12 500 F	12 500 F	12 000	13 200	12 180
Congo Dem R	3 800	4 200	3 780	3 876	3 973	3 844	3 954	3 945	3 945 F	3 945 F
Congo Rep	18 943	18 898	17 912	18 965	19 600	19 095	17 500 F	18 241 F	18 000 F	17 500 F
Cook Is	967	985 F	932 F	1 080	900 F	820 F	770 F	750 F	720 F	700 F
Costa Rica	15 603	15 460	16 292	16 537	23 237	25 829	23 757	27 218	34 398	33 733
Côte dIvoire	71 635	63 495	58 374	58 854	57 606	52 137	57 071	63 709	65 270	63 026
Croatia	26 530	26 342	17 137	15 901	17 799	16 627	21 928	18 890	20 701	18 056
Cuba	86 310	74 574	66 967	71 166	75 279	76 386	61 122	62 757	51 546 F	51 400 F
Cyprus	9 336	10 016	9 422	9 255	12 462	24 749	19 225	39 568	67 404	75 733
Denmark	1 953 175	1 613 952	1 873 073	1 998 769	1 681 321	1 826 620	1 556 986	1 404 799	1 533 906	1 510 362
Djibouti	275 F	300 F	320 F	350 F	350 F	350 F	350 F	350 F	350 F	350 F
Dominica	711	794	882	950	1 030	1 079	1 212	1 200 F	1 200 F	1 150 F
Dominican Rp	11 816	10 820	19 058	15 768	12 606	13 793	9 188	7 835	10 842	12 059
Ecuador	226 485	286 261	329 948	505 095	702 674	548 588	309 622	497 472	592 147	586 170
Egypt	78 300	86 000	84 600	81 860	89 570	99 150	112 560	155 204	130 844	133 229
El Salvador	6 940	7 990	9 740	10 747	11 466	9 089	10 214	13 871	6 759	16 055
Eq Guinea	3 230 F	2 907	4 369	1 856	4 140	5 240	5 035	5 900 F	2 558	2 500 F
Eritrea	...	475	2 706	3 559	3 252	1 038	1 629	6 891	12 612	8 820
Estonia	127 838	144 774	121 771	129 662	106 085	121 174	114 836	108 685	109 956	102 706
Ethiopia	100	-	-	-	-	-	-	-	-	-
Faeroe Is	248 430	246 294	237 708	287 796	304 587	329 145	363 816	358 133	454 530	524 837
Falkland Is	1 855	1 974	5 914	27 190	31 539	17 112	43 615	39 163	62 927	59 823
Fiji Islands	20 521	24 910	26 000 F	25 250	21 995	23 430	23 047	31 091	34 300 F	37 051
Finland	100 346	104 772	116 374	119 048	131 459	132 567	134 868	123 747	121 661	115 276
France	580 743	613 079	617 922	606 308	552 970	562 882	538 060	584 353	621 624	604 064
Fr Guiana	7 617	6 931	7 819	8 089	7 377 F	6 602	6 709	6 271 F	5 237 F	5 194 F
Fr Polynesia	3 425	8 075	8 175	9 020	9 910	11 617	12 420	12 283	13 846	15 351
Fr South Tr	464	460	524	519 F	437 F	375 F	388 F	425 F	272 F	263 F
Gabon	22 000 F	28 289	26 515	32 789	36 680	34 143	44 609	41 143	37 053	30 607
Gambia	15 545	18 908	20 381	21 252	29 101	29 754	26 502	27 500	26 516	32 027
Gaza Strip	...	...	...	1 229	2 493	3 791	3 625	3 600 F	3 600 F	3 000 F
Georgia	37 910 F	17 691 F	7 397 F	3 470 F	2 447	2 582	2 997	1 396	2 178	1 822
Germany	208 120	242 173	219 253	216 903	213 424	236 436	243 706	216 057	182 821	188 464
Ghana	367 425	320 619	280 737	292 844	403 593	377 088	368 141	418 276	377 570	370 787
Gibraltar	0	0	0	0	0	0	0	0	0	0
Greece	150 238	156 138	177 670	148 182	146 532	154 487	105 762	115 491	95 847	91 207
Greenland	113 267	116 650	117 417	128 890	116 018	120 596	128 542	160 253	159 711	158 485
Grenada	2 052	2 103	1 599	1 497	1 577	1 530	1 837	1 658	1 701	2 247
Guadeloupe	8 540	8 600	8 800	9 500	9 570	10 480	9 084	9 114	10 100	10 100 F
Guam	345	431	435	185	121	158	247	223	275	278
Guatemala	2 876	3 354	3 435	4 228	3 653	1 775	4 324	14 042	30 677	2 800 F
Guinea	51 000 F	56 000 F	60 000 F	64 760	60 580	58 841	65 764	83 314	87 513	86 000 F
GuineaBissau	5 000 F	5 100 F	5 750 F	6 078 F	6 750 F	7 000 F	5 800 F	4 800 F	4 800 F	4 800 F
Guyana	40 452	43 323	45 567	47 200 F	47 783	53 373	52 215	53 241	48 087	52 605
Haiti	4 500 F	4 550 F	5 000 F	5 017 F	4 745 F	4 801 F	4 759 F	4 500 F	4 500 F	4 500 F
Honduras	16 401	19 491	9 187	19 210	13 430	18 231	7 274 F	12 652	11 623	7 340
Iceland	1 573 796	1 714 674	1 556 264	1 611 809	2 059 560	2 205 540	1 681 535	1 735 897	1 982 346	1 980 555
India	2 470 815	2 486 807	2 704 733	2 656 862	2 814 529	2 881 673	2 681 053	2 776 067	2 786 676	2 787 890
Indonesia	2 590 422	2 768 892	2 969 732	3 181 355	3 217 580	3 486 982	3 676 231	3 659 292	3 764 479	3 897 270
Iran	243 969	246 985	218 944	252 583	242 439	238 492	226 949	243 800	260 500	262 805
Iraq	543	2 133	4 221	5 253	11 688	10 783	13 463	13 093	12 389	12 400 F
Ireland	246 395	275 845	290 561	380 871	329 224	288 869	320 298	280 092	280 845	355 511
Isle of Man	4 481	4 850	3 571	3 734	3 537	4 289	2 214	2 609	3 552	3 112

A-4

Fish, crustaceans, molluscs, etc	Capture production by countries or areas	Marine fishing areas
Poissons, crustacés, mollusques, etc	Captures par pays ou zones	Zones de pêche maritimes
Peces, crustáceos, moluscos, etc	Capturas por países o áreas	Areas de pesca marítimas

Country or area Pays ou zone País o área	1992 mt	1993 mt	1994 mt	1995 mt	1996 mt	1997 mt	1998 mt	1999 mt	2000 mt	2001 mt
Israel	3 246	3 312	2 583	3 727	3 384	3 728	4 136	3 739	3 966	3 400 F
Italy	387 382	388 018	388 809	386 756	359 135	337 003	301 429	277 354	297 584	304 870
Jamaica	8 900 F	9 550 F	9 877 F	9 917 F	12 054 F	7 748	6 110	8 058	5 226	5 250 F
Japan	7 633 841	7 163 911	6 524 899	5 875 001	5 840 158	5 840 501	5 220 577	5 122 916	4 900 800	4 657 798
Jordan	30	45	60	75	90	100	120	160	150	170
Kenya	6 566	5 617	3 772	5 465	6 296	6 099	6 600	6 634	4 763	7 388
Kiribati	28 211	27 072	27 753	30 306	31 871	30 052	35 304	52 741	25 563	32 375
Korea D P Rp	865 000 F	835 398 F	351 961	307 083	233 125	216 462	200 000 F	190 000 F	180 850 F	180 000 F
Korea Rep	2 295 565	2 244 929	2 347 399	2 310 269	2 405 679	2 197 113	2 020 089	2 113 352	1 816 034	1 982 031
Kuwait	7 871	8 466	7 752	8 616	8 255	7 826	7 799	6 271	6 000 F	5 846
Latvia	156 186	141 420	137 681	148 680	142 108	105 138	101 830	124 779	135 791	124 852
Lebanon	1 700 F	2 000	2 205	4 065	4 115	3 635	3 500	3 540	3 646	3 650
Liberia	4 891	3 782	3 721	4 829	4 308	4 580	6 830	11 472	7 726	7 286
Libya	28 820 F	31 085 F	33 500 F	34 400 F	32 976 F	31 877 F	32 911 F	32 850 F	33 387 F	33 239 F
Lithuania	186 945	116 025	47 975	56 108	87 219	42 289	64 841	71 247	77 076	150 077
Madagascar	76 905	84 261	86 431	85 653	84 475	86 391	96 395	99 630	102 093	105 583
Malaysia	1 023 516	1 047 350	1 065 586	1 108 436	1 126 689	1 168 973	1 149 093	1 248 402	1 285 696	1 231 287
Maldives	90 399	100 851	118 673	119 521	121 137	122 387	132 566	134 962	136 420	125 814
Malta	565	823	2 351	4 621	9 183	1 019	1 143	1 224	1 059	882
Marshall Is	326 F	471 F	393 F	381 F	2 772	400 F	500 F	500 F	8 155	37 098
Martinique	4 538	5 850	5 800 F	5 300	3 500 F	5 500	5 500	6 000	6 310	6 200
Mauritania	61 054 F	54 452 F	46 746 F	48 147 F	55 324 F	52 756 F	56 660 F	71 026 F	75 849 F	78 596 F
Mauritius	18 861	20 576	18 145	16 395	11 869	14 025	12 093	12 205	9 615	10 694
Mayotte	1 100 F	500 F	600 F	700 F	1 000	1 300 F	2 000 F	2 000	5 500	5 500
Mexico	1 048 866	991 754	1 080 521	1 207 320	1 341 687	1 375 560	1 079 525	1 114 141	1 208 764	1 306 438
Micronesia	16 542 F	17 884 F	22 922 F	7 624 F	9 064 F	10 122 F	16 576 F	12 833 F	23 710 F	18 057 F
Monaco	3 F	3 F	3 F	3 F	3 F	3 F	3 F	3 F	3 F	3 F
Montserrat	23	58	62	48	38	45	46	50 F	50 F	50 F
Morocco	549 143	622 894	752 577	847 451	641 386	789 806	708 733	743 268	895 012	1 082 293
Mozambique	27 808	25 506	22 531	21 740	27 405	28 035	27 683	23 746	25 977	24 436
Myanmar	590 263	597 637	599 876	602 885	455 294	631 226	680 838	759 664	880 018	931 492
Namibia	654 008	789 132	647 999	568 633	516 628	511 716	610 843	578 584	588 405	545 992
Nauru	377	500	500	450 F	400 F	400 F	400 F	400 F	400 F	400 F
Netherlands	430 683	460 155	417 607	433 985	408 641	449 506	535 079	512 308	493 524	515 962
NethAntilles	1 150 F	1 200 F	1 100 F	1 020 F	1 000 F	950 F	950 F	950 F	20 494 F	950 F
NewCaledonia	3 184	2 670	3 185	2 644	2 998	2 438	3 105	3 152	3 386	3 337 F
New Zealand	459 570	425 508	448 560	554 444	422 556	609 229	637 931	597 323	547 470	560 034
Nicaragua	6 370	7 622	9 210	10 457	14 300	14 883	18 636	22 789	26 932	21 748
Nigeria	208 046	142 782	163 259	231 579	248 472	294 279	324 004	316 235	309 062	297 971
Niue	120	120 F	150 F	150 F	200 F	200 F	200 F	200 F	200 F	200 F
Norfolk Is	0	0	0	0	0	0	0	0	0	0
N Marianas	140	136	176	191	225	250	235	193	189	197
Norway	2 430 243	2 414 696	2 365 687	2 523 698	2 648 119	2 862 620	2 860 716	2 627 020	2 702 837	2 686 733
Oman	109 218	105 772	118 572	139 861	121 618	118 995	106 171	108 809	120 421	126 531
Pakistan	431 267	499 159	418 574	404 444	395 397	422 265	433 456	474 665	437 601	426 920
Palau	1 918	2 066	1 820	1 956	1 990	1 751	1 777	1 800 F	2 000 F	2 000 F
Panama	172 170	184 794	184 862	202 863	150 684	162 258	202 664	120 828	222 611	234 980 F
Papua N Guin	12 741 F	12 241 F	12 719 F	25 577 F	24 323 F	32 685 F	64 529 F	51 931 F	82 569 F	108 919 F
Peru	7 469 458	8 966 487	11 950 380	8 886 553	9 486 158	7 837 650	4 303 110	8 392 378	10 626 323	7 950 450
Philippines	1 655 368	1 623 548	1 641 010	1 674 695	1 606 246	1 646 453	1 687 376	1 726 584	1 741 264	1 809 372
Pitcairn Is	8	8 F	8 F	8 F	8 F	8 F	8 F	8 F	8 F	8 F
Poland	454 747	373 031	410 450	401 346	319 262	339 829	225 100	221 236	200 811	208 127
Portugal	293 167	287 799	263 061	260 251	259 846	221 879	224 234	208 459	187 570	191 213
Puerto Rico	1 812	1 877	2 275	3 173	2 701	3 187	3 006	3 020	4 154	3 794
Qatar	7 845	6 994	5 086	4 271	4 739	5 032	5 279	4 207	7 142	8 606
Réunion	1 103	1 679	2 531	2 500	3 607	4 288	4 579	4 043	4 079	3 635
Romania	60 871	5 257	11 653	40 227	12 114	3 872	4 431	2 507	2 476	2 431
Russian Fed	5 234 869	4 153 149	3 487 132	4 098 935	4 442 466	4 434 762	4 183 448	3 833 335	3 681 167	3 421 893
St Helena	651	726	702	915	819	897	1 060	632	718	866
St Kitts Nev	300 F	250 F	212	192	352	272	533	555	492	591
St Lucia	1 073	1 336	1 252	1 188	1 274	1 308	1 589	1 718	1 855	1 983
St Pier Mq	16 443	282	294	317	747	3 571	6 108	5 892	6 485	3 802
St Vincent	2 157	1 945	3 840	1 005	919 F	6 092	33 973	17 759	27 694	45 778
Samoa	1 296	673	1 108	2 519	2 727	7 041	7 547	10 204	13 003	12 965 F
Sao Tome Prn	2 094	2 334	3 391	3 565	3 980	3 338	3 477	3 756	3 500 F	3 500 F
Saudi Arabia	45 910	48 021	54 612	45 609	47 698	49 314	51 206	46 618	49 650	49 167
Senegal	345 492	354 562	322 421	323 617	388 759	426 366	382 872	378 125	379 597	385 409
Seychelles	6 663	5 178	4 469	4 008	4 707	14 043	23 886	34 297	32 359	47 550
Sierra Leone	54 510	52 288	47 439	49 870	52 804 F	58 128	48 875	44 927	60 730	61 210
Singapore	9 177	9 279	11 278	10 102	9 943	9 250	7 733	6 489	5 371	3 342
Slovenia	3 612	1 987	2 007	1 851	2 078	2 065	1 959	1 784	1 630	1 621
Solomon Is	45 601 F	42 970 F	46 904 F	62 019 F	50 990 F	64 005 F	61 332 F	58 428 F	24 788 F	30 075 F
Somalia	25 500 F	27 500 F	29 600	27 700 F	25 800 F	23 900 F	22 000 F	20 000 F	20 000 F	19 800 F
South Africa	691 507	561 542	522 649	574 747	439 578	514 245	558 149	587 244 F	642 338 F	754 445
Spain	1 068 968 F	1 076 744 F	1 088 240 F	1 172 417 F	1 165 925	1 196 203	1 254 565	1 160 752	1 036 778	1 076 110
Sri Lanka	185 168	200 400	208 900	214 171	206 695	207 849	233 430	244 630	260 710	249 770
Sudan	2 000	2 500	4 000	4 000	4 500	5 000	5 500	5 500	5 000 F	5 000
Suriname	10 369	9 313	14 327	12 860 F	12 850 F	13 800 F	15 995	16 000	17 300 F	18 715
Svalbard Is	0	0	0	0	0	0	0	0	0	0
Sweden	305 237	339 624	384 560	402 638	369 071	355 395	409 327	349 776	337 075	310 582
Syria	1 954	2 019	1 950	1 950	2 670	2 574	2 750	2 600	2 581	2 322
Tanzania	56 085	36 685	40 785	42 771	61 645	50 210	48 000	50 489	52 779	52 900
Thailand	2 740 217	2 752 549	2 818 799	2 844 409	2 808 058	2 699 227	2 729 639	2 745 468	2 710 073	2 671 339
Timor-Leste	...	...	...	...	...	...	...	408 F	362	356
Togo	5 249	10 964	8 052	7 203	10 098	9 290	11 655	17 924	17 277	18 163
Tokelau	191	200 F	200 F	200 F	200 F	200 F	200 F	200 F	200 F	200 F
Tonga	2 236	2 374	2 499	2 597	2 915	2 871	4 076	4 221	3 760	4 673
Trinidad Tob	13 000 F	9 637	14 483	12 000 F	9 435	11 283	9 175	8 826	9 786	11 408
Tunisia	86 397	82 514	85 367	82 915	83 028	86 002	87 179	92 378	94 718	97 622
Turkey	407 003	503 957	544 736	585 994	478 226	408 693	432 700	523 634	460 524	484 407

A-4

Fish, crustaceans, molluscs, etc	Capture production by countries or areas	Marine fishing areas
Poissons, crustacés, mollusques, etc	Captures par pays ou zones	Zones de pêche maritimes
Peces, crustáceos, moluscos, etc	Capturas por países o áreas	Areas de pesca marítimas

Country or area Pays ou zone País o área	1992 mt	1993 mt	1994 mt	1995 mt	1996 mt	1997 mt	1998 mt	1999 mt	2000 mt	2001 mt
Turks Caicos	1 191	1 458	1 419 F	1 395 F	1 297 F	1 250 F	1 318	1 300 F	1 300 F	1 300 F
Tuvalu	499	1 460	561	399	400 F	500 F	500 F	500 F	500 F	500 F
Ukraine	434 138	292 145	252 394	371 803	407 651	366 790	457 410	403 125	388 295	346 917
Untd Arab Em	95 046	99 600	108 600	105 884	107 000	114 358	114 739	117 607	110 056	110 000 F
UK	811 047	858 320	875 827	907 782	863 215	884 770	918 516	836 063	744 615	737 964
USA	5 142 078	5 468 839	5 492 676	5 187 878	4 968 012	4 945 093	4 672 851	4 713 233	4 719 982	4 914 828
US Minor Is	0	0	0	0	0	0	0	0	0	0
US Virgin Is	750 F	650 F	550 F	470 F	400 F	350 F	300 F	263	300 F	300 F
Uruguay	125 428	118 194	119 766	125 597	122 732	134 738	138 776	100 589	114 286	104 583
Vanuatu	44 791	40 515 F	47 190 F	57 646 F	45 615 F	70 862 F	74 908 F	93 041 F	69 260 F	26 690 F
Venezuela	310 850	367 610	402 806	446 870	446 376	429 374	457 469	364 580	333 376	393 621
Viet Nam	729 953	785 304	946 322	990 250	1 058 708	1 098 736	1 155 154	1 217 193	1 280 590	1 321 123
Wallis Fut I	143	150	193	170	180	176	300	300 F	300 F	300 F
Westn Sahara	0	0	0	0	0	0	0	0	0	0
Yemen	79 547	82 356	81 885	107 970	104 955	115 600	127 620	124 385	114 751	142 198
Yugoslavia	223	277	260	364	377	373	410	423	424	416
Other nei	117 926	135 877	137 788	309 200	189 219	171 453	188 268	215 405	215 777	176 532
World total Total mondial Total mundial	**79 598 908**	**80 382 705**	**85 292 531**	**85 038 047**	**86 322 861**	**86 653 128**	**79 549 807**	**85 088 535**	**86 651 099**	**83 663 276**
World excl. China Monde excl. Chine Mundo excl. China	**72 274 384**	**72 213 658**	**75 753 480**	**74 082 726**	**73 903 614**	**72 817 751**	**64 600 124**	**70 133 867**	**71 897 004**	**69 283 819**

A-5

Fish, crustaceans, molluscs, etc — Capture production by low-income food-deficit countries (LIFDCs)
Poissons, crustacés, molluques, etc — Captures par pays à faible revenu et à déficit vivrier (PFRDV)
Peces, crustáceos, moluscos, etc — Capturas por países de bajos ingresos y con déficit de alimentos (PBIDA)

Country or area Pays ou zone País o área	Population in 2001** Population en 2001** Población en 2001** '000	1995 mt	1996 mt	1997 mt	1998 mt	1999 mt	2000 mt	2001 mt
Africa Afrique Africa	*703 294*	*4 399 848*	*4 336 809*	*4 605 677*	*4 608 837*	*4 811 416*	*5 041 139*	*5 233 589*
Angola	*13 527*	122 781	137 815	146 304	163 149	175 799	238 351	252 518
Benin	*6 446*	44 379	42 175	43 785	42 139	40 436	32 324	38 415
Burkina Faso	*11 856*	8 000	8 000	8 000	8 335	7 600	8 500	8 500 F
Burundi	*6 502*	21 101	3 041	20 296	13 426	9 199	17 315	8 964
Cameroon	*15 203*	94 131 F	98 400 F	102 000 F	106 800 F	110 000 F	112 109	111 031
Cape Verde	*437*	8 495	9 155	9 705	9 424	10 360	10 586	9 653
Cent Afr Rep	*3 782*	13 750 F	14 000 F	14 250 F	14 500 F	15 000	15 000 F	15 000 F
Chad	*8 135*	90 000	100 000	85 000	84 000	84 000 F	84 000 F	84 000 F
Comoros	*727*	13 000 F	12 700 F	12 500 F	12 500 F	12 000	13 200	12 180
Congo Dem R	*52 522*	158 627	163 010	162 211	178 041	208 448	208 448 F	208 448 F
Congo Rep	*3 110*	45 776	45 473	38 082	42 955 F	43 696 F	44 000 F	42 000 F
Côte dIvoire	*16 349*	70 189	69 168	64 169	69 572	74 365	75 772	73 556
Djibouti	*644*	350 F	350 F	350 F	350 F	350 F	350 F	350 F
Egypt	*69 080*	310 790	320 230	342 759	362 741	380 504	384 314	428 651
Eq Guinea	*470*	2 306	5 040	6 090	6 005	7 001	3 634	3 500 F
Eritrea	*3 816*	3 559	3 252	1 038	1 629	6 891	12 612	8 820
Ethiopia	*64 459*	6 325	8 770	10 370	14 000	15 858	15 681	15 390
Gambia	*1 337*	23 752	31 601	32 254	29 002	30 000	29 016	34 527
Ghana	*19 734*	352 844	477 173	447 088	442 641	492 776	452 070	445 287
Guinea	*8 274*	67 860	63 360	62 441	69 764	87 314	91 513	90 000 F
GuineaBissau	*1 227*	6 328	7 000 F	7 250 F	6 000 F	5 000 F	5 000 F	5 000 F
Kenya	*31 293*	192 706	180 988	161 054	172 592	205 287	215 106	164 151
Lesotho	*2 057*	26 F	28 F	30 F	30 F	30	32	24
Liberia	*3 108*	8 829	8 308	8 580	10 830	15 472	11 726	11 286
Madagascar	*16 437*	115 653	114 475	116 391	126 395	129 630	132 093	135 583
Malawi	*11 572*	53 664	63 569	56 340	41 111	45 392	43 000 F	40 619
Mali	*11 677*	132 900	111 910	99 550	98 000	98 536	109 870	100 000 F
Mauritania	*2 747*	53 147 F	60 324 F	57 756 F	61 660 F	76 026 F	80 849 F	83 596 F
Morocco	*30 430*	848 951	642 886	791 906	710 436	745 431	896 620	1 083 276
Mozambique	*18 644*	26 833	34 915	39 703	36 677	33 989	39 065	32 512
Niger	*11 227*	3 616	4 156	6 328	7 013	11 000	16 250	20 800
Nigeria	*116 929*	349 482	337 993	387 923	463 024	455 628	441 377	452 146
Rwanda	*7 949*	3 300 F	2 952	4 428	6 641	6 433	6 726	6 828
Sao Tome Prn	*140*	3 565	3 980	3 338	3 477	3 756	3 500 F	3 500 F
Senegal	*9 662*	354 617	411 759	457 366	403 872	412 125	402 047	405 409
Sierra Leone	*4 587*	64 870	67 304 F	72 628	63 065	59 407	74 730	75 210
Somalia	*9 157*	27 950 F	26 050 F	24 150 F	22 250 F	20 250 F	20 200 F	20 000 F
Sudan	*31 809*	44 000	45 000	47 000	49 500	49 500	53 000 F	58 000
Swaziland	*938*	60 F	60 F	65 F	70 F	70 F	70 F	70 F
Tanzania	*35 965*	359 800	323 921	356 960	348 000	310 509	332 779	335 900
Togo	*4 657*	12 201	15 098	14 290	16 655	22 924	22 277	23 163
Uganda	*24 023*	208 789	195 088	218 026	220 628	226 097	219 356	220 726
Zambia	*10 649*	70 546	66 332	65 923	69 938	67 327	66 671	65 000 F
America, North Amérique du Nord América del Norte	*42 977*	*124 161*	*127 471*	*131 641*	*110 467*	*130 062*	*138 816*	*101 350*
Cuba	*11 237*	80 059	85 603	84 911	67 076	67 381	56 146 F	56 000 F
Guatemala	*11 687*	8 253	7 653	6 896	10 847	21 018	37 978	10 100 F
Haiti	*8 270*	5 517 F	5 245 F	5 301 F	5 259 F	5 000 F	5 000 F	5 000 F
Honduras	*6 575*	19 337	13 528	18 357	7 393 F	12 754	11 684	7 451
Nicaragua	*5 208*	10 995	15 442	16 176	19 892	23 909	28 008	22 799
America, South Amérique du Sud América del Sur	*21 396*	*511 087*	*708 962*	*555 026*	*316 077*	*503 924*	*598 653*	*592 510*
Bolivia	*8 516*	5 692	5 988	6 038	6 055	6 052	6 106	5 940
Ecuador	*12 880*	505 395	702 974	548 988	310 022	497 872	592 547	586 570
Asia Asie Asia	*3 070 519*	*23 479 486*	*25 184 353*	*27 132 128*	*28 748 235*	*29 288 482*	*29 436 583*	*29 279 718*
Afghanistan	*22 474*	1 300 F	1 300 F	1 250 F	1 200 F	1 200 F	1 000 F	800 F
Armenia	*3 788*	821	580	580	698	1 144	1 133	866
Azerbaijan	*8 096*	10 545	6 702	5 161	4 760	20 866	18 797	10 893
Bangladesh	*140 369*	792 389	814 787	829 426	839 141	959 215	1 004 264	1 000 000 F
Bhutan	*2 141*	310 F	300 F	300 F	300 F	300 F	300 F	300 F
Cambodia	*13 441*	102 999	94 710	102 800	107 900	269 156	284 368	397 200 F
China	*1 262 609*	12 562 706	14 182 107	15 722 344	17 229 927	17 240 032	16 987 325	16 529 389
Georgia	*5 239*	3 560 F	2 453	2 583	3 001	1 413	2 200	1 830
India	*1 025 096*	3 265 240	3 447 954	3 523 448	3 373 492	3 472 150	3 742 296	3 762 600
Indonesia	*214 840*	3 511 145	3 553 276	3 791 240	3 964 897	3 986 919	4 069 691	4 203 830
Korea D P Rp	*22 428*	327 083	253 125	236 462	220 000 F	210 000 F	200 850 F	200 000 F
Kyrgyzstan	*4 986*	185	160 F	120 F	80 F	48	52	57
Laos	*5 403*	27 370	23 000 F	18 857	19 642	30 041	29 250	30 000 F
Maldives	*300*	119 521	121 137	122 387	132 566	134 962	136 420	125 814
Mongolia	*2 559*	158	221	180	311	524	425	117
Nepal	*23 593*	11 230	11 230	11 230	12 000	12 752	16 700	16 700
Pakistan	*144 971*	525 849	537 489	589 795	596 980	654 530	614 069	607 020
Philippines	*77 131*	1 860 701	1 783 601	1 805 806	1 833 380	1 872 818	1 893 017	1 945 217
Sri Lanka	*19 104*	229 171	228 945	235 099	263 330	276 080	297 410	279 640
Syria	*16 610*	5 782	5 773	6 131	7 097	7 938	6 572	8 291
Tajikistan	*6 135*	100	40 F	75 F	100 F	80 F	78 F	137
Turkmenistan	*4 835*	9 740	9 014	8 179	7 014	9 058	12 228	12 749
Uzbekistan	*25 257*	3 611	1 494	3 075	2 799	2 871	3 387	4 070
Yemen	*19 114*	107 970	104 955	115 600	127 620	124 385	114 751	142 198

A-5

Fish, crustaceans, molluscs, etc — **Capture production by low-income food-deficit countries (LIFDCs)**
Poissons, crustacés, molluques, etc — **Captures par pays à faible revenu et à déficit vivrier (PFRDV)**
Peces, crustáceos, moluscos, etc — **Capturas por países de bajos ingresos y con déficit de alimentos (PBIDA)**

Country or area Pays ou zone País o área	Population in 2001** Population en 2001** Población en 2001** '000	1995 mt	1996 mt	1997 mt	1998 mt	1999 mt	2000 mt	2001 mt
Europe Europe Europa	*9 256*	*4 087*	*4 703*	*3 643*	*5 314*	*5 380*	*6 028*	*5 938*
Albania	*3 145*	1 379	2 125	1 013	2 683	2 745	3 320	3 310
Bosnia Herzg	*4 067*	2 500 F	2 500 F	2 500 F	2 500 F	2 500 F	2 500 F	2 500 F
Macedonia	*2 044*	208	78	130	131	135	208	128
Oceania Océanie Oceanía	*5 838*	*191 966*	*169 426*	*218 645*	*257 620*	*280 345*	*229 184*	*225 025*
Kiribati	*84*	30 306	31 871	30 052	35 304	52 741	25 563	32 375
Papua N Guin	*4 920*	39 077 F	37 823 F	46 185 F	78 029 F	65 431 F	96 069 F	122 419 F
Samoa	*159*	2 519	2 727	7 041	7 547	10 204	13 004	12 966 F
Solomon Is	*463*	62 019 F	50 990 F	64 005 F	61 332 F	58 428 F	24 788 F	30 075 F
Tuvalu	*10*	399	400 F	500 F	500 F	500 F	500 F	500 F
Vanuatu	*202*	57 646 F	45 615 F	70 862 F	74 908 F	93 041 F	69 260 F	26 690 F
LIFDCs PFRDV PBIDA	***3 853 280***	***28 710 635***	***30 531 724***	***32 646 760***	***34 046 550***	***35 019 609***	***35 450 403***	***35 438 130***
Other countries Autres pays Otros países	***2 280 858***	***63 591 271***	***63 218 678***	***61 568 834***	***53 546 762***	***58 582 287***	***59 989 417***	***56 917 904***
World total Total mondial Total mundial	***6 134 138***	**92 301 906**	**93 750 402**	**94 215 594**	**87 593 312**	**93 601 896**	**95 439 820**	**92 356 034**

** Population (mid-2001) estimated by the Population Division of the United Nations, New York.

** Les données relatives à la population (mi-2001) ont été estimées par la Division de la Population des Nations Unies, New York.

** Los datos relativos a la población (mitad-2001) están estimados por la División de Población de las Naciones Unidas, Nueva York.

A-6 Seaweeds and other aquatic plants — Capture production by countries or areas — All fishing areas
Algues et autres plantes aquatiques — Captures par pays ou zones — Toutes les zones de pêche
Algas y otras plantas acuáticas — Capturas por países o áreas — Todas las áreas de pesca

Country or area Pays ou zone País o área	1992 mt	1993 mt	1994 mt	1995 mt	1996 mt	1997 mt	1998 mt	1999 mt	2000 mt	2001 mt
Argentina	1 767	2 418	3 139	2 403	2 000 F	900 F	500 F	100 F	3	3 F
Australia	16 108	18 275	17 519	22 316	18 570	21 152	20 811	20 774	13 650	13 547
Canada	28 824	21 560	30 214	26 671	24 332	34 696	27 472	30 660	34 507	14 932
Chile	78 759	107 109	116 755	250 038	216 815	178 839	197 495	230 203	247 376	234 253
China	100 000	124 250	151 000	150 000	151 630	184 270	170 520	215 550	204 290	266 870
China,H.Kong	2	1	0	0	0	0	0	0	0	0
China,Taiwan	1 688	206	212	96	106	124	87	123	125	91
Cook Is	30 F	40 F	40 F	40 F	45 F	50 F	50 F	50 F	50 F	50 F
Estonia	0	0	411	548	163	2 444	2 880	1 319	201	325
Fiji Islands	41	47	50 F	50 F	40 F	50 F	60	67	65 F	60 F
France	81 893	60 333	79 527	75 550	84 372	75 502	67 032	70 996	68 232	66 283
Iceland	8 529	11 303	13 982	11 841	14 411	19 505	17 770	18 147	17 501	20 367
India	85 000 F	87 500 F	90 000 F	92 500 F	95 000 F	97 500 F	100 000 F	100 000 F	100 000 F	100 000 F
Indonesia	101 762	118 395	110 438	111 575	161 544	125 979	47 515	21 559	42 712	42 760
Ireland	30 690	29 923	33 417	35 456	33 360	35 386	36 100 F	36 100 F	36 100 F	36 100 F
Italy	...	...	600	500	1 850	1 950	2 000	2 000 F	2 000 F	2 000 F
Japan	209 441	167 604	137 722	151 042	154 078	149 910	116 892	120 877	119 030	122 145
Korea Rep	27 542	22 096	26 899	22 373	22 916	23 234	12 599	12 895	13 023	14 933
Latvia	10	1	0	0	0	0	0	0	0	0
Madagascar	285	423	702	787	787	1 000	2 510	1 933	5 792	5 045
Mexico	62 276	60 600	36 777	49 334	34 598	42 067	12 456	32 090	33 555	46 927
Morocco	7 783	7 108	5 357	7 858	7 625	8 094	7 049	5 920	6 080	10 015
Namibia	1 272	226	175	799	936	851	897	660	829	829 F
New Zealand	0	0	0	0	10	10	4	1	0	13
Norway	189 294	169 606	185 065	185 033	173 160	191 681	179 762	178 542	192 426	175 210
Peru	593	243	170	413	307	155	1 650	1 733	1 312	5 505
Philippines	1 049	1 144	1 062	919	884	494	417	433	413	447
Portugal	4 000 F	3 000 F	2 686	2 816	1 703	1 743	968	1 949	1 224	1 198
Russian Fed	28 356	6 753	7 661	10 546	18 667	26 441	30 279	25 715	53 653	27 541
Senegal	50 F	50 F	50 F	50 F	50 F	50 F	100 F	150 F	200 F	223
South Africa	9 264	7 960	4 376	6 252	8 500 F	11 500 F	14 400 F	17 300 F	20 511	32 138
Spain	14 153 F	13 939 F	14 291 F	14 392 F	14 871 F	14 885 F	13 940	13 818	14 214	14 011
Tanzania	2 500 F	3 600	3 500 F	3 500 F	3 500 F	3 610	4 000	4 500	5 000	5 000 F
Timor-Leste	...	...	...	...	...	...	...	1	1	1
Ukraine	1 361	424	4 238	3 172	1 190	331	314	160	26	...
UK	5 627	4 769	5 683	7 400	8 400	8 100	...	-	-	-
USA	83 773	85 108	116 841	73 999	59 295	70 729	25 689	79 236	42 058	37 165
World total Total mondial Total mundial	**1 183 722**	**1 136 014**	**1 200 559**	**1 320 269**	**1 315 715**	**1 333 232**	**1 114 218**	**1 245 561**	**1 276 159**	**1 295 987**

The data in this A-6 table and in tables B-91, B-92, B-93 and B-94 refer to the annual production of seaweeds and other aquatic plants (expressed in metric tons and on a wet-weight basis) through harvesting of wild stocks.

See paragraph 5 of the INTRODUCTION.

Les données présentées dans les tableaux A-6 et B-91, B-92, B-93 et B-94 se rapportent à la production annuelle d'algues marines et autres plantes aquatiques (exprimée en tonnes et en poids vert) fournie par la récolte des espèces sauvages.

Voir le paragraphe 5 de l'INTRODUCTION.

Los datos de este cuadro A-6 y de los cuadros B-91, B-92, B-93 y B-94 se refieren a la producción anual de algas marinas y otras plantas acuáticas (expresada en toneladas métricas y peso en húmedo) obtenidas mediante la recolección de plantas silvestres.

Véase el párrafo 5 en la INTRODUCCION.

B - Capture production: by species groups

B - Captures: par groupes d'espèces

B - Captures: por grupos de especies

B-00 (a)

Fish, crustaceans, molluscs, etc — Capture production by species groups and fishing areas
Poissons, crustacés, mollusques, etc — Captures par groupes d'espèces et zones de pêche
Peces, crustáceos, moluscos, etc — Capturas por grupos de especies y áreas de pesca

Species group Groupe d'espèces Grupo de especies	Fishing area Zone de pêche Area de pesca	1995 mt	1996 mt	1997 mt	1998 mt	1999 mt	2000 mt	2001 mt
11 Carps, barbels and other cyprinids	01	82 453	81 926	86 704	78 960	96 396	97 962	98 007
Carpes, barbeaux et autres cyprinidés	02	27 654	28 105	24 924	21 241	20 971	22 094	21 943
Carpas, barbos y otros ciprínidos	03	663	502	213	265	307	1 745	360
	04	549 997	429 416	376 916	377 296	382 408	334 543	309 825
	05	113 394	117 076	111 636	120 201	106 254	103 830	108 456
	06	0	0	28	2	3	4	6
	27	5 362	5 834	6 674	6 311	6 574	8 172	8 399
	37	938	476	882	617	643	854	575
	61	61	110	101	100	56	8	52
	Group total	780 522	663 445	608 078	604 993	613 612	569 212	547 623
12 Tilapias and other cichlids	01	374 857	366 526	361 520	394 271	423 730	467 477	475 290
Tilapias et autres cichlidés	02	90 324	86 191	84 500	72 068	69 538	76 849	68 642
Tilapias y otros cíclidos	03	18 599	16 173	15 460	15 370	14 242	16 226	16 283
	04	106 051	86 676	95 384	112 104	125 424	121 673	117 829
	06	2 316	2 347	2 350	2 604	2 600	2 590	2 588
	34	2 476	2 603	2 582	1 365	1 760	2 062	2 324
	Group total	594 623	560 516	561 796	597 782	637 294	686 877	682 956
13 Miscellaneous freshwater fishes	01	1 464 558	1 366 584	1 425 683	1 461 868	1 438 772	1 447 646	1 426 331
Poissons d'eau douce divers	02	53 373	51 168	49 460	50 912	51 409	57 866	48 114
Peces de agua dulce diversos	03	344 763	322 890	312 455	315 347	328 254	323 038	320 821
	04	2 442 316	2 791 205	2 883 899	3 071 956	3 633 592	3 663 275	3 779 712
	05	121 451	118 041	108 305	102 393	95 419	114 807	100 894
	06	9 303	9 357	9 353	9 158	9 122	9 105	9 003
	27	13 781	14 971	17 105	17 928	17 062	11 206	11 466
	31	7 705	7 303	3 656	2 870	4 130	6 772	6 937
	37	1 303	2 822	2 825	3 031	2 568	2 956	3 522
	Group total	4 458 553	4 684 341	4 812 741	5 035 463	5 580 328	5 636 671	5 706 800
21 Sturgeons, paddlefishes	02	433	526	615	525	420	281	6
Esturgeons, spatules	04	2 139	2 179	1 733	1 531	1 320	1 288	1 139
Esturiones, sollos	05	2 347	1 248	1 331	1 222	856	704	633
	21	157	86	91	69	9	3	4
	27	0	0	0	1	2	-	-
	37	826	584	637	429	242	85	33
	67	4	1	0	0	0	242	195
	Group total	5 906	4 624	4 407	3 777	2 849	2 603	2 010
22 River eels	01	898	637	985	801	959	2 164	2 052
Anguilles	02	610	730	796	739	770	942	423
Anguilas	04	2 467	5 796	2 089	2 064	2 255	5 580	5 790
	05	3 568	3 827	3 194	3 133	3 092	3 101	2 554
	06	1 134	931	810	1 286	1 209	1 151	1 244
	21	631	551	455	496	353	395	220
	27	3 195	2 959	4 327	2 290	2 622	2 247	2 098
	31	43	35	19	9	2	1	...
	34	0	0	27	4	0	-	-
	37	849	919	1 254	914	682	464	504
	57	259	208	203	160	129	133	157
	81	1 183	524	588	400	416	380	313
	Group total	14 837	17 117	14 747	12 296	12 489	16 558	15 355
23 Salmons, trouts, smelts	01	196	11	16	24	29	31	67
Saumons, truites, éperlans	02	30 964	32 374	27 972	29 200	25 267	22 515	22 669
Salmones, truchas, eperlanos	03	685	463	1 208	733	527	566	658
	04	37 527	38 676	35 498	33 261	28 219	29 420	24 738
	05	67 327	64 252	69 832	66 252	83 397	82 877	80 978
	21	1 410	1 301	1 458	1 374	428	1 070	405
	27	11 696	12 027	10 663	11 199	10 262	14 507	9 324
	37	-	18	22	26	-	-	0
	61	442 427	455 779	471 418	428 221	380 303	350 771	401 629
	67	558 396	423 012	301 117	318 745	382 944	300 917	349 383
	77	2 998	2 084	2 747	944	1 996	2 623	1 206
	81	2	1	1	3	1	1	1
	Group total	1 153 628	1 029 998	921 952	889 982	913 373	805 298	891 058
24 Shads	01	366	290	253	359	270	284	258
Aloses	02	2 088	3 187	3 941	3 825	5 199	3 070	2 661
Sábalos	03	425	642	455	567	523	407	300
	04	204 042	219 321	209 777	238 029	249 306	365 220	203 098
	05	83 078	96 744	86 878	120 291	156 813	120 766	47 425
	06	480	480	480	480	480	480	480
	21	7 159	6 415	6 887	8 281	7 367	5 919	4 007
	27	133	122	113	58	59	140	103
	31	273	343	246	280	117	288	171
	34	6 918	11 585	8 810	8 113	4 608	7 489	5 633
	37	10 522	3 862	3 439	6 391	13 550	14 141	30 727

B-00 (a)

Fish, crustaceans, molluscs, etc — Capture production by species groups and fishing areas
Poissons, crustacés, mollusques, etc — Captures par groupes d'espèces et zones de pêche
Peces, crustáceos, moluscos, etc — Capturas por grupos de especies y áreas de pesca

Species group / Groupe d'espèces / Grupo de especies	Fishing area / Zone de pêche / Area de pesca	1995 mt	1996 mt	1997 mt	1998 mt	1999 mt	2000 mt	2001 mt
	51	4 226	5 547	5 265	4 607	4 548	4 736	6 965
	57	178 744	197 740	192 141	180 806	190 310	194 226	186 598
	61	78 769	75 549	106 577	116 818	137 932	126 545	127 786
	67	158	15	34	62	-	105	178
	71	11 390	12 415	8 996	10 682	11 497	11 755	12 756
	77	-	-	0	-	-	-	-
	Group total	588 771	634 257	634 292	699 649	782 579	855 571	629 146
25 Miscellaneous diadromous fishes	02	815	995	1 169	966	1 046	1 221	1 226
Poissons diadromes divers	04	3 997	369	34	3 000	83	163	201
Peces diádromos diversos	05	559	569	467	1 042	1 243	1 354	1 104
	06	1 253	1 411	1 528	1 624	1 984	1 700	1 650
	21	1 655	2 144	2 785	3 032	3 000	3 123	2 934
	27	18	67	135	101	40	571	658
	31	7	9	12	14	4	12	15
	37	-	-	-	-	-	-	8
	51	376	395	334	335	341	125	134
	57	12 274	13 991	13 792	23 062	15 687	15 923	17 360
	61	1	18	9	26	47	32	81
	71	44 460	44 224	45 704	47 278	53 971	59 730	62 840
	77	0	0	1	0	0	7	...
	Group total	65 415	64 192	65 970	80 480	77 446	83 961	88 211
31 Flounders, halibuts, soles	01	...	...	547	381	763	2 490	2 134
Flets, flétans, soles	21	77 559	80 472	88 100	92 236	113 568	118 531	122 421
Platijas, halibuts, lenguados	27	338 917	318 365	310 781	298 034	317 559	300 443	303 030
	31	2 696	2 527	1 546	1 891	1 673	1 765	1 695
	34	34 934	35 323	38 371	46 234	41 604	50 540	49 780
	37	16 027	13 340	11 718	11 059	10 469	11 134	10 853
	41	11 870	10 356	11 976	10 902	8 701	8 681	7 524
	47	2 830	1 996	2 671	2 960	3 069	2 928	4 021
	51	21 718	20 730	25 299	20 676	15 878	27 872	14 432
	57	20 915	30 824	33 443	28 204	26 965	27 146	25 741
	61	155 341	181 394	186 917	197 870	210 735	225 682	203 918
	67	202 798	217 113	273 510	198 371	174 192	203 808	173 733
	71	16 353	14 922	24 987	16 791	20 281	16 545	17 501
	77	8 477	9 229	8 937	4 853	6 470	6 229	4 756
	81	5 385	4 514	8 005	4 278	3 554	2 939	3 257
	87	1 790	761	411	436	411	365	439
	Group total	917 610	941 866	1 027 219	935 176	955 892	1 007 098	945 235
32 Cods, hakes, haddocks	02	22	23	24	25	25	25	...
Morues, merlus, églefins	05	193	98	163	177	22	11	55
Bacalaos, merluzas, eglefinos	21	108 617	119 254	127 256	140 007	143 120	137 455	148 393
	27	3 160 141	3 192 725	3 133 827	3 256 629	3 289 075	3 331 770	3 607 968
	31	0	...	4	2	2	6	3
	34	30 346	31 447	25 650	25 150	25 207	19 473	25 314
	37	94 006	90 708	72 072	75 371	64 641	73 338	66 333
	41	829 684	860 118	815 951	779 466	618 103	506 425	534 286
	47	277 212	287 275	261 732	307 178	312 014	301 401	325 120
	48	11	24	17	22	12	5	3
	51	1 147	1 251	1 348	1 456	747	1 472	2 436
	57	3 400	2 716	2 864	5 890	7 566	9 653	7 623
	58	-	-	1	50	52	343	266
	61	3 632 526	3 469 239	3 436 455	2 975 625	2 456 168	1 951 990	1 877 369
	67	1 815 668	1 725 540	1 749 564	1 764 162	1 511 756	1 631 326	1 830 568
	71	34	24	21	33	765	29	33
	77	89	138	1 273	806	379	718	146
	81	276 918	251 354	351 951	435 214	389 941	349 651	338 911
	87	510 674	752 199	391 763	565 958	507 147	339 571	458 922
	88	-	-	-	9	28	82	61
	Group total	10 740 688	10 784 133	10 371 936	10 333 230	9 326 770	8 654 744	9 223 810
33 Miscellaneous coastal fishes	01	19 921	15 875	18 986	20 939	20 750	24 658	28 359
Poissons côtiers divers	02	470	368	549	573	470	577	506
Peces costeros diversos	04	28 359	37 088	41 467	35 911	35 721	39 120	40 978
	05	31	52	109	219	451	360	489
	06	1 850	1 850	1 850	1 850	1 850	1 850	1 850
	21	15 573	20 492	20 055	19 826	28 628	19 522	20 343
	27	1 155 876	880 758	1 267 537	1 065 549	789 791	768 539	938 408
	31	189 681	164 649	157 164	159 994	130 246	154 467	143 807
	34	239 654	263 259	291 496	267 169	292 111	277 528	282 701
	37	174 592	183 355	170 147	173 388	176 952	172 695	159 825
	41	243 275	226 233	234 505	213 688	207 297	236 136	233 662
	47	24 271	23 740	19 664	52 202	36 793	47 795	61 670
	48	5	-	-	0	10	6	4
	51	724 827	821 096	856 133	777 597	848 376	805 353	796 320
	57	455 354	488 328	485 143	511 999	511 995	497 472	497 936
	58	-	15	5	3	14	-	0
	61	1 479 446	1 856 071	2 033 717	2 223 641	2 245 815	2 343 594	2 193 022
	67	72 259	92 432	62 763	54 050	54 531	47 878	59 881
	71	1 027 055	1 026 184	1 076 151	1 030 523	1 159 461	1 083 291	1 112 637

B-00 (a)

Fish, crustaceans, molluscs, etc — Capture production by species groups and fishing areas
Poissons, crustacés, mollusques, etc — Captures par groupes d'espèces et zones de pêche
Peces, crustáceos, moluscos, etc — Capturas por grupos de especies y áreas de pesca

Species group / Groupe d'espèces / Grupo de especies	Fishing area / Zone de pêche / Area de pesca	1995 mt	1996 mt	1997 mt	1998 mt	1999 mt	2000 mt	2001 mt
	77	37 114	43 004	46 049	44 181	57 328	46 293	44 880
	81	37 223	25 463	26 477	23 120	27 367	26 153	25 066
	87	82 052	52 472	53 423	63 270	65 065	55 474	50 638
	88	-	-	-	0	0	0	1
	Group total	6 008 888	6 222 784	6 863 390	6 739 692	6 691 022	6 648 761	6 692 983
34 Miscellaneous demersal fishes	04	849	490	447	270	494	402	625
Poissons démersaux divers	21	63 380	56 720	59 857	66 917	66 707	73 940	81 218
Peces demersales diversos	27	450 650	435 011	418 833	425 131	393 489	419 979	422 804
	31	10 724	12 416	14 408	19 486	17 505	15 196	22 095
	34	54 211	94 585	75 913	76 738	49 631	38 243	37 292
	37	18 196	21 004	23 071	18 757	17 683	17 558	19 894
	41	75 912	58 259	59 688	74 886	66 381	54 868	56 136
	47	55 094	58 742	65 833	74 252	53 312	51 583	55 866
	48	3 273	3 822	2 389	3 266	4 570	8 874	4 316
	51	67 602	91 138	189 740	118 295	166 920	168 828	162 545
	57	70 170	68 457	72 515	94 099	94 505	86 016	114 831
	58	9 585	4 925	8 208	7 982	13 063	9 079	10 294
	61	1 443 897	1 469 582	1 394 197	1 655 751	1 637 181	1 696 006	1 712 536
	67	102 446	107 235	111 003	88 790	97 110	86 495	79 323
	71	47 741	43 313	47 776	48 100	55 290	53 504	53 033
	77	10 159	9 816	9 547	7 942	5 446	4 517	4 418
	81	177 518	128 843	160 195	158 749	155 139	153 216	144 355
	87	40 094	25 436	25 239	30 009	52 479	79 913	45 377
	88	...	-	0	42	297	751	662
	Group total	2 701 501	2 689 794	2 738 859	2 969 462	2 947 202	3 018 968	3 027 620
35 Herrings, sardines, anchovies	01	1 963	1 900	1 850	1 800	1 798	1 500	1 700
Harengs, sardines, anchois	04	1 086	1 481	2 497	1 517	1 143	917	917
Arenques, sardinas, anchoas	21	609 107	575 240	606 757	521 264	472 439	460 612	543 766
	27	2 881 853	2 875 382	3 089 362	3 013 492	2 995 072	2 890 306	2 432 932
	31	665 064	679 660	758 862	734 996	865 288	711 720	639 070
	34	1 518 897	1 753 281	1 828 230	1 869 052	1 691 950	1 752 143	1 868 923
	37	824 466	674 739	642 385	625 771	781 261	718 037	767 875
	41	109 509	139 838	161 501	115 964	59 628	58 850	75 529
	47	496 577	233 659	328 352	419 566	495 703	582 167	604 524
	51	308 233	373 255	433 033	415 773	407 553	546 667	581 721
	57	444 653	435 811	497 823	490 762	393 684	411 637	406 070
	61	1 967 502	1 990 976	2 556 096	2 932 317	2 856 880	2 534 514	2 614 151
	67	75 577	71 607	84 451	73 503	68 758	74 750	87 058
	71	1 006 676	1 014 952	1 070 572	1 051 737	999 286	1 017 746	1 030 263
	77	523 437	563 272	621 114	544 447	485 986	705 994	857 221
	81	654	525	829	956	908	1 303	1 431
	87	10 570 516	11 001 839	9 049 576	3 853 560	10 057 766	12 427 660	7 947 490
	Group total	22 005 770	22 387 417	21 733 290	16 666 477	22 635 103	24 896 523	20 460 641
36 Tunas, bonitos, billfishes	21	9 312	10 713	7 657	7 844	10 656	8 394	8 735
Thons, pélamides, marlins	27	55 026	47 546	43 940	40 330	40 814	40 293	28 760
Atunes, bonitos, agujas	31	80 752	93 947	83 900	103 556	80 011	89 621	106 341
	34	358 023	354 648	325 041	343 064	356 019	300 656	326 035
	37	75 459	80 668	75 120	81 206	74 790	71 577	71 176
	41	55 722	73 088	72 141	63 456	64 590	67 042	64 878
	47	59 468	48 720	38 640	51 053	52 576	51 338	48 543
	51	862 346	798 132	811 442	794 035	903 097	927 100	870 908
	57	347 226	391 680	411 012	518 788	483 296	488 499	406 893
	61	555 308	567 298	748 063	924 890	961 585	917 339	850 012
	67	6 694	10 063	15 888	11 654	4 710	11 220	12 162
	71	1 761 877	1 717 100	1 775 537	2 117 045	2 066 217	2 117 445	2 160 123
	77	394 902	372 775	416 717	422 246	393 369	378 100	452 131
	81	31 799	37 555	36 171	42 007	29 628	35 374	36 389
	87	231 207	243 733	329 620	287 678	454 145	361 524	378 154
	Group total	4 885 121	4 847 666	5 190 889	5 808 852	5 975 503	5 865 522	5 821 240
37 Miscellaneous pelagic fishes	01	1 241	1 326	1 229	1 150	1 180	3 182	5 154
Poissons pélagiques divers	02	...	...	...	372	487	392	558
Peces pelágicos diversos	03	850	850	850	850	850	850	760
	04	909	562	1 600	1 500	1 455	1 583	1 685
	05	-	8	110	350	361	350	326
	21	32 967	77 943	66 326	77 936	58 871	49 971	65 508
	27	2 126 591	2 547 671	2 630 715	2 000 933	1 857 940	2 413 135	2 637 042
	31	39 198	35 367	36 538	38 063	32 601	30 731	30 009
	34	442 341	439 833	444 064	513 196	433 700	573 643	446 307
	37	127 593	116 031	110 585	100 465	97 217	103 880	114 782
	41	48 163	40 265	42 416	39 028	31 276	41 721	33 802
	47	593 098	554 646	507 629	534 140	488 612	501 729	414 020
	51	429 274	555 851	430 288	437 632	451 621	289 167	276 474
	57	562 919	543 343	537 462	584 058	560 031	521 921	535 505
	61	2 727 467	3 267 951	3 162 978	2 674 518	2 462 646	2 469 381	2 667 858
	67	148	388	206	1 593	251	302	537
	71	1 466 658	1 394 097	1 477 317	1 513 492	1 616 073	1 650 036	1 696 951
	77	36 178	28 341	48 042	51 052	85 841	85 372	34 086
	81	63 933	57 373	63 536	62 766	77 222	55 728	57 310

B-00 (a)

Fish, crustaceans, molluscs, etc — **Capture production by species groups and fishing areas**
Poissons, crustacés, mollusques, etc — **Captures par groupes d'espèces et zones de pêche**
Peces, crustáceos, moluscos, etc — **Capturas por grupos de especies y áreas de pesca**

Species group Groupe d'espèces Grupo de especies	Fishing area Zone de pêche Area de pesca	1995 mt	1996 mt	1997 mt	1998 mt	1999 mt	2000 mt	2001 mt
	87	5 236 150	4 670 512	4 227 403	2 580 211	2 425 751	1 871 008	3 199 884
	Group total	13 935 678	14 332 358	13 789 294	11 213 305	10 683 986	10 664 082	12 218 558
38 Sharks, rays, chimaeras	21	46 806	56 131	48 219	51 296	49 186	49 360	39 716
Squales, raies, chimères	27	86 729	74 119	106 720	94 115	98 564	104 582	107 084
Tiburones, rayas, quimeras	31	32 509	31 629	33 133	29 426	25 717	23 072	21 440
	34	28 155	35 435	67 933	52 264	51 106	50 923	50 310
	37	19 222	16 047	16 334	14 886	12 157	13 037	11 473
	41	51 050	56 595	58 696	57 896	61 308	60 184	62 780
	47	4 321	3 058	7 781	6 531	8 334	10 688	15 221
	48	90	40	30	14	15	4	13
	51	118 048	177 712	117 951	120 758	115 589	126 542	117 656
	57	116 109	117 206	116 166	131 197	126 253	121 361	117 915
	58	-	1	7	23	15	98	78
	61	33 508	31 201	37 365	34 383	57 887	56 079	35 970
	67	5 406	8 700	6 496	5 142	9 340	8 628	8 313
	71	150 402	138 974	140 048	143 300	141 931	144 330	158 986
	77	37 787	38 188	35 128	38 346	39 191	43 727	36 932
	81	21 012	17 075	24 720	18 773	22 584	20 112	22 391
	87	12 152	12 975	13 230	21 856	15 433	23 948	18 487
	88	-	-	-	5	19	41	7
	Group total	763 306	815 086	829 957	820 211	834 629	856 716	824 772
39 Marine fishes not identified	21	17 618	14 896	14 378	5 869	7 492	8 029	4 483
Poissons marins non identifiés	27	102 557	90 392	74 377	81 245	48 856	72 385	37 427
Peces marinos no identificados	31	233 163	223 781	259 985	220 341	174 182	159 083	167 559
	34	418 352	287 958	295 819	283 454	396 161	271 779	431 215
	37	112 010	100 777	83 640	87 647	63 217	75 454	63 933
	41	172 960	158 570	165 118	157 405	181 165	194 956	201 983
	47	54 442	90 578	106 531	60 388	56 941	57 768	90 626
	48	10	-	-	7	-	-	-
	51	816 105	600 508	649 586	617 389	648 207	628 354	646 074
	57	1 490 614	1 350 902	1 499 417	1 682 392	1 625 684	1 874 633	1 893 936
	58	61	-	1	0	0	0	0
	61	4 058 374	4 864 097	4 442 968	4 372 457	4 368 442	4 055 185	3 641 860
	67	64 914	106 820	106 601	97 853	98 191	31 003	74 345
	71	2 144 552	2 206 263	2 122 217	2 120 946	2 261 522	2 305 847	2 354 058
	77	269 721	213 172	181 727	159 041	114 636	126 141	130 511
	81	21 989	49 339	48 209	27 498	10 734	12 611	42 338
	87	68 806	72 907	97 328	390 422	271 652	323 750	190 872
	Group total	10 046 248	10 430 960	10 147 902	10 364 354	10 327 082	10 196 978	9 971 220
41 Freshwater crustaceans	01	9 360	8 042	8 423	8 995	7 801	9 140	9 398
Crustacés d'eau douce	02	11 753	10 524	15 099	14 230	9 694	3 843	7 902
Crustáceos de agua dulce	03	1 412	2 677	2 157	1 506	3 235	2 437	2 440
	04	289 226	386 235	498 702	617 997	474 588	548 383	605 912
	05	2 974	2 774	2 989	2 783	2 800	3 758	4 507
	06	1 036	338	329	345	412	372	374
	Group total	315 761	410 590	527 699	645 856	498 530	567 933	630 533
42 Crabs, sea-spiders	04	329	363	267	44	47	77	150
Crabes, araignées de mer	21	134 345	142 599	150 626	147 582	172 414	166 103	173 491
Cangrejos, centollas	27	48 058	42 033	58 913	53 020	51 036	53 654	56 441
	31	54 670	56 663	68 561	67 358	52 565	60 456	48 998
	34	773	2 331	4 114	6 558	5 586	5 231	6 689
	37	1 268	2 195	2 104	2 520	4 491	2 891	2 579
	41	15 107	13 845	16 668	15 285	16 091	18 000	15 305
	47	7 163	5 329	3 415	5 639	4 855	6 874	5 575
	51	28 533	7 652	9 814	11 083	11 839	20 952	21 593
	57	24 091	23 284	24 528	31 378	31 429	46 272	46 238
	61	407 664	488 600	431 585	442 492	437 240	500 701	500 522
	67	63 545	65 038	73 923	131 368	102 383	34 839	33 594
	71	134 464	126 806	127 556	144 422	142 022	148 807	163 208
	77	12 523	15 131	12 455	8 663	7 818	13 382	12 640
	81	10 019	1 094	1 120	1 138	429	385	390
	87	9 093	6 372	4 629	6 427	18 509	9 352	8 958
	Group total	951 645	999 335	990 278	1 074 977	1 058 754	1 087 976	1 096 371
43 Lobsters, spiny-rock lobsters	21	70 631	71 866	78 146	77 155	83 105	83 062	83 803
Homards, langoustes	27	59 874	52 107	58 122	55 760	59 818	50 364	55 364
Bogavantes, langostas	31	29 278	29 472	29 172	27 191	30 660	33 444	27 702
	34	2 143	2 174	9 378	4 010	2 821	2 886	2 982
	37	7 552	7 455	7 203	4 927	4 623	4 068	4 224
	41	10 817	8 026	7 502	6 002	6 334	6 469	6 400
	47	3 499	3 032	3 117	3 348	2 931	2 690	3 452
	51	3 562	3 452	3 315	3 634	3 187	2 918	2 804
	57	17 160	16 338	16 286	16 600	20 985	22 904	19 895
	61	2 398	2 221	1 724	1 794	1 842	3 316	3 531
	67	...	...	...	...	25	27	18
	71	6 182	6 731	10 099	8 512	6 347	7 001	7 668

B-00 (a)

Fish, crustaceans, molluscs, etc — Capture production by species groups and fishing areas
Poissons, crustacés, mollusques, etc — Captures par groupes d'espèces et zones de pêche
Peces, crustáceos, moluscos, etc — Capturas por grupos de especies y áreas de pesca

Species group Groupe d'espèces Grupo de especies	Fishing area Zone de pêche Area de pesca	1995 mt	1996 mt	1997 mt	1998 mt	1999 mt	2000 mt	2001 mt
	77	1 959	2 621	2 734	2 506	2 306	3 572	3 396
	81	4 846	3 953	6 418	3 980	3 873	3 976	3 763
	87	250	142	101	746	569	346	134
	Group total	220 151	209 590	233 317	216 165	229 426	227 043	225 136
44 King crabs, squat-lobsters	27	124	175	177	313	471	468	814
Crabes royaux, galatées	41	381	201	414	457	271	369	218
Cangrejos reales, galateidos	48	-	497	-	-	-	3	14
	58	-	-	-	0	-	3	0
	61	55 246	48 254	39 019	41 217	45 646	36 020	23 244
	67	6 656	9 526	8 177	10 941	7 675	6 848	7 281
	77	356	164	-	-	-	1	2
	87	13 903	17 160	22 898	26 295	23 576	24 292	13 396
	Group total	76 666	75 977	70 685	79 223	77 639	68 004	44 969
45 Shrimps, prawns	01	3 117	3 000	2 850	2 700	2 574	2 100	2 400
Crevettes	04	33 116	34 418	44 296	23 900	23 365	33 941	36 840
Gambas, camarones	21	170 713	200 651	174 549	207 259	243 053	269 324	261 408
	27	162 378	169 575	191 183	176 695	173 335	177 013	145 079
	31	177 954	155 882	164 564	174 996	169 575	203 495	209 786
	34	54 931	57 810	53 206	83 505	70 361	59 356	57 739
	37	27 350	25 384	30 279	27 525	29 407	34 531	29 191
	41	47 953	45 681	47 343	58 911	48 673	76 603	119 616
	47	4 289	5 915	5 569	16 230	5 333	10 421	13 329
	51	292 066	297 846	272 896	336 556	322 351	311 254	295 312
	57	212 724	217 698	231 576	182 547	234 840	247 010	238 721
	61	706 720	776 922	834 742	932 180	1 150 109	1 119 120	991 000
	67	22 148	24 422	24 331	11 471	18 086	20 381	22 952
	71	416 888	435 421	440 526	414 267	438 590	435 591	457 761
	77	72 906	72 867	73 384	66 791	67 417	58 662	52 429
	81	2 154	1 824	1 949	1 684	2 410	2 647	2 593
	87	30 011	28 867	35 223	32 008	21 636	13 322	14 678
	Group total	2 437 418	2 554 183	2 628 466	2 749 225	3 021 115	3 074 771	2 950 834
46 Krill, planktonic crustaceans	27	-	-	-	88	-	-	36
Krill, crustacés planctoniques	41	-	-	-	74	-	4	-
Krill, crustáceos planctónicos	47	-	-	-	254	-	-	-
	48	117 448	101 708	82 508	80 874	103 318	104 259	98 209
	58	1 266	6	-	-	-	-	-
	Group total	118 714	101 714	82 508	81 290	103 318	104 263	98 245
47 Miscellaneous marine crustaceans	21	1	2	-	1	5	6	1
Crustacés marins divers	27	2 239	1 621	1 783	1 389	1 276	1 116	1 673
Crustáceos marinos diversos	31	440	363	373	422	200	347	186
	34	1 668	1 034	1 317	1 743	875	249	562
	37	10 363	9 804	8 827	7 873	9 927	10 200	9 882
	41	860	313	849	2 161	6 293	227	1 375
	47	5	2	-	-	-	-	1 062
	51	12 078	15 600	12 367	12 797	11 460	11 251	9 416
	57	35 664	40 568	45 940	52 272	54 445	36 561	36 623
	61	934 355	1 007 693	1 165 004	1 314 628	1 198 281	1 303 443	1 333 362
	67	...	...	...	...	...	49	63
	71	2 963	2 751	8 275	7 359	4 498	5 203	6 943
	77	832	748	714	1 179	1 252	1 307	1 369
	81	523	156	157	92	670	802	537
	87	681	880	585	698	687	627	1 510
	Group total	1 002 672	1 081 535	1 246 191	1 402 614	1 289 869	1 371 388	1 404 564
51 Freshwater molluscs	01	488	530	610	625	606	921	851
Mollusques d'eau douce	02	198	241	122	185	240	565	2
Moluscos de agua dulce	04	593 312	557 746	503 071	574 737	546 606	588 720	619 401
	06	2 569	2 670	3 970	4 500	5 000	5 080	5 300
	61	-	-	-	-	-	-	74
	Group total	596 567	561 187	507 773	580 047	552 452	595 286	625 628
52 Abalones, winkles, conchs	21	3 044	4 513	3 668	3 149	5 721	4 567	6 339
Ormeaux, bigorneaux, strombes	27	18 538	29 391	33 733	18 337	22 331	33 270	33 900
Orejas de mar, bígaros, estrombos	31	15 627	13 785	18 903	15 433	16 300	18 919	18 651
	34	8 440	8 235	6 598	9 849	7 290	6 637	9 659
	37	4 166	4 518	5 682	4 829	5 545	5 716	4 675
	41	574	558	1 322	1 010	683	1 621	1 406
	47	615	735	330	524	481	490	5 265
	51	43	43	40	40	29	45	51
	57	4 896	5 081	4 905	4 920	5 297	5 204	5 301
	61	27 340	23 352	24 084	25 540	22 257	20 249	19 385
	67	4 257	6 369	...	...	...	-	4
	71	483	448	183	348	282	241	250
	77	5 558	4 309	3 465	2 374	2 108	2 673	3 182

B-00 (a)

Fish, crustaceans, molluscs, etc — **Capture production by species groups and fishing areas**
Poissons, crustacés, mollusques, etc — **Captures par groupes d'espèces et zones de pêche**
Peces, crustáceos, moluscos, etc — **Capturas por grupos de especies y áreas de pesca**

Species group / Groupe d'espèces / Grupo de especies	Fishing area / Zone de pêche / Area de pesca	1995 mt	1996 mt	1997 mt	1998 mt	1999 mt	2000 mt	2001 mt
	81	1 592	1 364	1 515	1 627	1 170	1 265	1 064
	87	14 149	13 589	20 005	11 183	16 332	12 220	11 826
	Group total	109 322	116 290	124 433	99 163	105 826	113 117	120 958
53 Oysters / Huîtres / Ostras	21	32 882	39 569	49 099	33 329	47 106	26 475	18 571
	27	1 487	1 400	1 852	1 434	881	683	899
	31	130 876	113 081	99 310	106 852	89 056	226 333	162 117
	34	223	100	109	89	125	101	151
	37	3 005	2 246	1 863	1 204	952	319	157
	41	726	873	828	744	1 547	884	870
	51	14	32	24	28	34	32	29
	57	-	-	-	-	-	350	380
	61	18 269	18 259	17 210	9 905	11 610	15 943	10 092
	67	...	4 989	6 187	11	6	14	20
	71	1 655	1 887	1 830	2 118	2 120	2 010	2 203
	77	3 510	3 071	4 079	3 383	3 028	3 745	3 317
	81	9	4	4	2	65	2	2
	87	5	5	10	6	11	14	207
	Group total	192 661	185 516	182 405	159 105	156 541	276 905	199 015
54 Mussels / Moules / Mejillones	04	136	4	-	-	-	-	-
	21	15 723	17 585	17 228	15 051	15 651	21 098	18 353
	27	132 575	101 062	118 949	122 237	114 376	128 707	146 696
	31	155	223	295	3 802	451	316	1 081
	34	8	10	6	7	...	401	102
	37	40 764	39 046	53 310	38 899	44 453	46 143	46 092
	41	2 905	1 234	1 548	1 606	1 381	1 215	1 292
	57	243	71	-	-	-	-	-
	61	6 655	3 028	7 309	7 934	8 265	6 952	4 955
	67	1 317	455	...	...	1	675	448
	71	20 413	20 216	18 017	17 454	6 554	6 462	6 422
	77	290	583	2 206	2 230	832	643	90
	81	195	454	1	665	2 978	4 468	2 271
	87	23 018	19 468	21 070	27 862	24 241	24 391	29 513
	Group total	244 397	203 439	239 939	237 747	219 183	241 471	257 315
55 Scallops, pectens / Coquilles St-Jacques / Vieiras	21	135 702	122 704	116 169	108 136	135 115	201 288	256 929
	27	62 165	55 638	71 668	73 299	73 406	74 764	81 361
	31	12 313	684	581	834	223	256	253
	37	23	93	137	65	80	588	154
	41	10 592	36 952	39 817	28 441	42 700	37 404	42 598
	57	8 159	6 872	6 093	5 543	7 622	8 301	4 024
	61	279 382	276 406	266 957	294 211	305 510	310 104	293 268
	67	1 950	2 372	2	3 228	2 642	2 012	1 052
	71	6 986	7 276	3 488	6 392	5 403	5 093	6 086
	77	1 256	17 290	2 320	2 726	1 864	6 287	3 100
	81	14 318	5 209	19 050	4 684	6 281	2 912	7 015
	87	4 478	3 672	6 611	27 208	31 856	12 162	6 685
	Group total	537 324	535 168	532 893	554 767	612 702	661 171	702 525
56 Clams, cockles, arkshells / Clams, coques, arches / Almejas, berberechos, arcas	04	835	844	956	914	304	303	260
	21	389 779	384 556	345 924	315 694	325 053	329 462	347 200
	27	86 742	58 500	57 086	116 751	94 266	65 922	49 247
	31	59 707	39 693	50 135	38 568	46 339	55 625	56 306
	34	178	54	69	139	148	120	111
	37	50 534	52 633	39 346	36 511	43 249	47 403	44 929
	41	0	126	196	664	2 443	279	287
	47	0	0	-	-	-	-	-
	57	46 187	61 448	63 188	34 444	42 944	45 342	42 685
	61	185 430	181 775	144 085	156 515	145 198	132 789	119 870
	67	14 999	7 111	5 415	8 262	6 777	6 517	6 658
	71	65 252	72 085	57 595	74 531	83 453	84 621	83 064
	77	8 950	7 270	6 792	5 770	6 913	7 143	5 711
	81	1 791	1 274	1 120	1 984	1 577	1 920	1 950
	87	50 300	51 782	43 393	47 813	42 922	40 793	50 667
	Group total	960 684	919 151	815 300	838 560	841 586	818 239	808 945
57 Squids, cuttlefishes, octopuses / Encornets, seiches, poulpes / Calamares, jibias, pulpos	21	35 583	38 908	48 733	44 725	26 767	26 402	18 409
	27	54 033	53 371	50 982	49 948	53 067	58 808	44 474
	31	20 691	31 119	20 751	24 135	20 113	24 384	23 237
	34	191 176	190 620	150 352	183 729	250 294	216 421	212 920
	37	62 545	59 266	60 139	54 281	50 692	53 846	53 773
	41	645 234	747 875	1 016 557	749 986	1 192 738	1 001 022	802 928
	47	7 433	8 170	4 145	7 801	8 301	7 506	4 654
	48	-	52	28	53	-	-	2
	51	110 641	94 868	148 579	112 432	121 305	119 020	140 124
	57	88 363	104 648	109 444	115 730	111 791	124 084	115 986
	58	-	-	1	-	-	-	-
	61	1 006 539	1 165 463	1 154 074	1 033 618	1 117 839	1 221 972	1 137 512
	67	333	685	2 787	2 293	3 132	748	2 997

B-00 (a)

Fish, crustaceans, molluscs, etc — **Capture production by species groups and fishing areas**
Poissons, crustacés, mollusques, etc — **Captures par groupes d'espèces et zones de pêche**
Peces, crustáceos, moluscos, etc — **Capturas por grupos de especies y áreas de pesca**

Species group Groupe d'espèces Grupo de especies	Fishing area Zone de pêche Area de pesca	1995 mt	1996 mt	1997 mt	1998 mt	1999 mt	2000 mt	2001 mt
	71	355 772	343 912	360 034	366 123	364 643	435 288	390 472
	77	111 316	201 487	213 136	30 406	150 585	202 531	167 068
	81	139 581	73 299	84 833	71 006	43 625	35 499	57 273
	87	109 152	36 021	32 095	11 354	82 778	127 619	174 999
	Group total	2 938 392	3 149 764	3 456 670	2 857 620	3 597 670	3 655 150	3 346 828
58 Miscellaneous marine molluscs	04	325	481	52	165	73	166	250
Mollusques marins divers	21	2 031	2 394	600	168	1 920	1 469	1 384
Moluscos marinos diversos	27	1 021	865	1 015	1 294	648	498	424
	31	3 836	673	1 872	2 921	6 706	2 716	1 913
	34	1 806	746	1 489	12 599	2 258	1 189	384
	37	16 650	18 120	15 480	15 120	13 389	10 245	15 404
	41	3 678	1 175	1 851	1 408	4 937	2 474	2 839
	47	17	-	-	-	-	-	-
	51	3 363	15 996	7 363	5 015	3 747	5 034	2 452
	57	401	381	263	844	8 396	8 307	8 517
	61	1 302 416	938 870	1 502 215	1 488 978	1 457 426	1 410 260	1 405 909
	67	2 155	1 873	4 593	2 879	1 850	1 904	1 649
	71	59 506	53 322	54 772	59 749	59 051	53 484	60 027
	77	1 380	1 307	1 857	1 462	1 742	3 816	2 951
	81	266	342	173	230	1 386	1 311	1 483
	87	557	980	3 447	1 702	3 730	2 513	2 273
	Group total	1 399 408	1 037 525	1 597 042	1 594 534	1 567 259	1 505 386	1 507 859
71 Frogs and other amphibians	02	564	422	2 034	1 201	337	325	31
Grenouilles et autres amphibies	03	5	0	0	0	0	7	9
Ranas y otros anfibios	04	3 058	2 535	1 550	1 767	1 435	1 957	2 843
	05	...	...	38	41	35	26	-
	Group total	3 627	2 957	3 622	3 009	1 807	2 315	2 883
72 Turtles	02	6	18	17	26	15	53	51
Tortues	03	0	0	0	0	0	0	0
Tortugas	04	-	36	48	27	1	3	-
	31	209	228	86	141	31	32	36
	34	22	42	164	189	463	122	315
	41	0	0	0	0	0	0	0
	57	291	353	248	337	273	372	381
	71	300	417	403	452	448	384	393
	77	0	0	0	3	1	21	...
	87	14	10	11	12	11	11	12
	Group total	842	1 104	977	1 187	1 243	998	1 188
74 Sea-squirts and other tunicates	37	...	28	22	22	30	30	76
Ascidiens et autres tuniciers	51	-	-	-	-	-	95	-
Ascidias y otros tunicados	58	-	7	-	-	-	-	-
	61	5 767	16 747	2 780	891	1 171	1 443	1 053
	87	3 297	4 549	3 174	2 530	2 704	2 290	1 298
	Group total	9 064	21 331	5 976	3 443	3 905	3 858	2 427
75 Horseshoe crabs and other arachnoids	21	926	1 598	2 606	3 182	2 238	1 557	1 071
Limules et autres arachnoïdés	31	-	-	1	70	159	139	228
Erizos de mar y otros equinodermos								
	Group total	926	1 598	2 607	3 252	2 397	1 696	1 299
76 Sea-urchins and other echinoderms	21	17 726	15 139	12 201	13 126	14 298	13 440	8 778
Oursins et autres échinodermes	27	942	911	612	564	643	462	478
Erizos de mar y otros equinodermos	31	15	10	15	15	15	10	49
	37	71	69	55	60	76	204	104
	51	3 415	3 658	3 693	2 407	767	1 139	1 230
	57	327	419	610	833	859	1 048	1 090
	61	28 633	27 020	28 009	25 353	24 156	24 161	22 753
	67	9 369	7 576	9 044	8 081	7 329	7 115	6 678
	71	7 847	8 372	7 648	6 825	6 099	6 484	6 220
	77	12 807	12 244	10 401	6 221	8 682	9 328	8 600
	81	817	282	627	845	645	718	864
	87	54 858	52 025	46 000	44 978	56 981	57 247	49 030
	88	-	-	-	-	-	-	2
	Group total	136 827	127 725	118 915	109 308	120 550	121 356	105 876
77 Miscellaneous aquatic invertebrates	02	183	337	297	483	617	231	...
Invertébrés aquatiques divers	04	2 643	1 494	2 460	1 480	2 048	2 031	2 283
Invertebrados acuáticos diversos	21	0	-	223	299	921	202	481
	27	1	0	-	-	-	-	24
	31	11	6	2	4	0	139	-
	37	487	909	904	1 764	1 737	908	2 592
	47	-	-	-	-	-	106	44

B-00 (a)

Fish, crustaceans, molluscs, etc — **Capture production by species groups and fishing areas**
Poissons, crustacés, mollusques, etc — **Captures par groupes d'espèces et zones de pêche**
Peces, crustáceos, moluscos, etc — **Capturas por grupos de especies y áreas de pesca**

Species group / Groupe d'espèces / Grupo de especies	Fishing area / Zone de pêche / Area de pesca	1995 mt	1996 mt	1997 mt	1998 mt	1999 mt	2000 mt	2001 mt
	48	-	-	-	-	-	5	-
	57	20 597	23 527	29 747	7 077	35 524	35 684	37 454
	58	-	-	11	-	2	-	-
	61	210 281	278 735	410 601	441 444	410 337	346 443	340 147
	71	146 224	38 289	86 583	74 429	91 748	89 354	89 923
	77	1 346	872	351	59	31	248	450
	Group total	381 773	344 169	531 179	527 039	542 965	475 351	473 398
World total / Total mondial / Total mundial		**92 301 906**	**93 750 402**	**94 215 594**	**87 593 312**	**93 601 896**	**95 439 820**	**92 356 034**

B-00 (b)

Fish, crustaceans, molluscs, etc — Capture production by species items
Poissons, crustacés, mollusques, etc — Captures par catégories d'espèces
Peces, crustáceos, moluscos, etc — Capturas por partidas de especies

3-alpha code Code alpha-3 Código alfa-3	Scientific name Nom scientifique Nombre científico	Species group Groupe d'espèces Grupo de especies	1995 mt	1996 mt	1997 mt	1998 mt	1999 mt	2000 mt	2001 mt
BUF	*Ictiobus spp*	11	480	793	991	959	697	1 280	1 569
CTM	*Catostomidae*	11	662	542	550	464	465	231	187
FBM	*Abramis brama*	11	60 979	59 110	50 388	51 460	46 470	47 222	51 279
FBR	*Abramis spp*	11	2 376	2 426	1 507	2 427	2 680	2 414	2 594
FCP	*Cyprinus carpio*	11	82 816	79 592	69 865	75 779	70 751	65 463	60 842
FTE	*Tinca tinca*	11	1 504	1 157	1 879	1 211	1 419	2 336	2 473
ALR	*Alburnus alburnus*	11	153	168	130	240	268	247	546
PTB	*Barbus barbus*	11	161	186	239	215	176	118	158
HON	*Chondrostoma nasus*	11	31	24	26	28	63	67	59
FCC	*Carassius carassius*	11	9 026	8 584	7 691	7 096	6 539	6 406	6 321
CGO	*Carassius auratus*	11	3 562	3 454	2 048	2 243	2 831	2 677	2 582
FRO	*Rutilus rutilus*	11	5 897	7 807	7 197	6 698	6 717	6 929	7 360
RFR	*Rutilus frisii*	11	8 435	9 210	2 320	6 624	6 905	10 120	7 199
FRX	*Rutilus spp*	11	31 465	32 707	25 069	23 876	18 093	21 550	21 948
SRE	*Scardinius erythrophthalmus*	11	29	36	43	123	145	97	56
FID	*Leuciscus idus*	11	2 911	3 612	3 660	3 799	3 095	2 943	2 811
FIE	*Leuciscus leuciscus*	11	228	220	255	305	180	104	91
LUH	*Leuciscus cephalus*	11	...	...	92	166	167	70	35
LEW	*Leuciscus spp*	11	63	47	39	32	38	33	31
RHI	*Labeo spp*	11	605	3 456	2 362	3 375	3 840	3 672	3 609
MUC	*Cirrhinus molitorella*	11	5	3	4	7	7	6	6
FCG	*Ctenopharyngodon idellus*	11	15 149	20 206	21 615	9 911	5 881	16 047	19 164
SVC	*Hypophthalmichthys molitrix*	11	23 968	24 480	22 527	15 069	18 047	11 271	6 863
BIC	*Hypophthalmichthys nobilis*	11	2 372	2 356	2 036	3 008	4 251	4 826	778
ENA	*Rastrineobola argentea*	11	56 827	49 670	40 315	42 336	48 816	49 618	41 384
VIV	*Vimba vimba*	11	310	270	390	331	361	279	269
FSC	*Pelecus cultratus*	11	569	498	403	1 780	1 826	1 235	1 590
ASU	*Aspius aspius*	11	588	580	454	385	453	598	664
FCH	*Leptobarbus hoeveni*	11	5 454	6 892	5 836	3 241	4 608	3 149	3 260
BKC	*Mylopharyngodon piceus*	11	20	15	16	22	27	34	36
WUB	*Megalobrama amblycephala*	11	0	0	0	-	-	-	-
FJB	*Puntius javanicus*	11	18 102	19 601	19 469	20 189	17 939	17 124	17 080
FAB	*Puntius spp*	11	34 812	39 004	37 272	56 480	56 774	52 745	54 220
FCY	*Cyprinidae*	11	410 963	286 739	281 390	265 114	283 083	238 301	230 559
TLM	*Oreochromis mossambicus*	12	15 603	19 253	20 025	20 175	23 482	21 429	21 860
TLN	*Oreochromis niloticus*	12	242 322	207 777	187 928	229 684	227 342	245 951	248 081
OEA	*Oreochromis aureus*	12	8 455	9 704	7 654	5 210	4 630	4 567	4 560
TLP	*Oreochromis (=Tilapia) spp*	12	276 383	281 916	302 170	299 248	322 149	353 564	347 460
SAR	*Sarotherodon galilaeus*	12	316	350	462	391	405	262	230
CHL	*Cichlasoma managuense*	12	608	566	367	344	351	324	410
CLO	*Cichla ocellaris*	12	120	65	35	12	7	4	4
AST	*Astronotus spp*	12	320	308	177	141	251	188	183
CIM	*Haplochromis spp*	12	16 782	10 472	13 053	13 077	10 731	11 699	9 695
CIX	*Cichlidae*	12	33 714	30 105	29 925	29 500	47 946	48 889	50 473
FLU	*Protopterus spp*	13	8 931	7 653	12 313	11 091	11 428	8 827	9 911
DAG	*Stolothrissa, Limnothrissa spp*	13	95 570	70 277	100 636	87 869	78 245	81 992	75 114
FPI	*Esox lucius*	13	24 060	25 807	23 981	24 020	23 288	25 289	24 459
ARP	*Arapaima gigas*	13	420	457	465	210	338	273	204
HTT	*Heterotis spp*	13	6 902	4 451	5 716	8 385	9 872	9 514	10 127
FKN	*Notopterus spp*	13	5 173	3 891	3 456	3 743	3 878	4 046	4 072
MZU	*Mormyrus spp*	13	...	...	...	...	...	7 691	7 000
CSM	*Colossoma macropomum*	13	11 379	4 991	3 600	3 014	3 921	5 928	5 904
CSD	*Piaractus brachypomus*	13	...	397	...	166	648	324	487
CHA	*Characidae*	13	141 924	144 295	128 789	140 913	125 605	121 684	114 090
PLR	*Prochilodus reticulatus*	13	5 088	7 448	2 007	6 117	9 768	10 942	11 220
PRL	*Prochilodus spp*	13	20 430	18 740	10 927	15 670	25 320	27 313	20 743
SOM	*Silurus glanis*	13	7 487	8 374	9 326	11 362	9 266	9 349	8 561
CAG	*Kryptopterus spp*	13	15 283	17 560	15 938	13 943	15 927	12 404	13 490
CSR	*Chrysichthys nigrodigitatus*	13	4 000	8 505	5 814	15 538	10 679	7 311	12 413
CST	*Chrysichthys spp*	13	8 609	7 596	6 513	7 100	5 592	7 800	7 397
CAN	*Bagrus spp*	13	14 437	9 865	12 458	13 386	16 239	11 495	11 574
CAF	*Ictalurus spp*	13	9 974	9 745	11 822	9 524	13 710	11 567	10 801
ITE	*Ameiurus nebulosus*	13	1	1	5	12	14	10	8
CLZ	*Clarias gariepinus*	13	33 855	20 601	24 703	21 956	27 220	36 520	39 867
CLN	*Clarias anguillaris*	13	19 608	20 370	22 433	21 618	21 511	31 506	39 535
CTO	*Clarias spp*	13	54 508	36 382	36 769	54 160	48 456	56 634	54 516
PGZ	*Pangasius spp*	13	1 000	541	522	917	1 061	1 300	1 358
CSY	*Synodontis spp*	13	13 852	9 267	11 560	13 384	12 856	12 571	13 737
FSI	*Siluroidei*	13	97 035	95 574	100 229	167 437	178 915	219 691	183 368
FBU	*Lota lota*	13	3 027	3 334	3 354	3 437	3 107	3 070	2 944
NIP	*Lates niloticus*	13	367 504	312 017	329 244	339 183	306 282	302 905	282 245
PEX	*Lates spp*	13	2 888	1 205	2 378	3 925	1 523	3 070	1 657
MPS	*Micropterus salmoides*	13	1 212	1 014	1 058	684	784	903	693
FPE	*Perca fluviatilis*	13	24 497	25 594	27 095	24 357	23 066	21 336	21 464
FPY	*Perca flavescens*	13	2 528	2 151	2 695	2 639	2 421	2 414	2 605
STV	*Stizostedion vitreum*	13	7 849	7 736	7 755	8 212	8 784	9 164	6 959
SZC	*Stizostedion canadense*	13	4	1	...	0	...	...	...
FPP	*Stizostedion lucioperca*	13	22 489	26 259	18 022	18 708	17 166	15 936	15 394
ACC	*Gymnocephalus cernuus*	13	19	18	10	9	56	99	64
FGB	*Eleotridae*	13	3 364	2 961	3 140	2 928	3 093	2 835	2 820
FGX	*Gobiidae*	13	4 607	4 670	5 350	4 863	4 961	5 370	5 196
FPC	*Anabas testudineus*	13	6 651	3 905	3 754	4 637	6 340	6 700	6 999
FGS	*Trichogaster pectoralis*	13	25 090	30 793	21 728	22 422	23 776	21 575	21 990
FGO	*Helostoma temminckii*	13	19 166	12 614	18 376	16 598	23 320	17 927	18 320
CNA	*Channa argus*	13	-	169	120	...	28	...	...
FSS	*Channa striata*	13	27 828	30 966	28 646	21 520	23 784	26 886	28 098
FIS	*Channa micropeltes*	13	9 021	11 615	10 117	8 253	8 787	7 771	7 060
FSN	*Channa spp*	13	126 493	128 341	136 916	188 042	88 915	93 248	93 767

B-00 (b)
Fish, crustaceans, molluscs, etc — Capture production by species items
Poissons, crustacés, mollusques, etc — Captures par catégories d'espèces
Peces, crustáceos, moluscos, etc — Capturas por partidas de especies

3-alpha code Code alpha-3 Código alfa-3	Scientific name Nom scientifique Nombre científico	Species group Groupe d'espèces Grupo de especies	1995 mt	1996 mt	1997 mt	1998 mt	1999 mt	2000 mt	2001 mt
FRF	*Osteichthyes*	13	3 204 790	3 546 190	3 643 001	3 713 511	4 380 378	4 373 481	4 508 569
APG	*Acipenser gueldenstaedtii*	21	139	132	140	125	50	42	26
APR	*Acipenser ruthenus*	21	37	36	15	10	37	15	12
APE	*Acipenser stellatus*	21	9	18	54	19	28	28	24
APN	*Acipenser transmontanus*	21	-	-	-	-	-	206	185
AAM	*Acipenser medirostris*	21	-	-	-	-	-	36	10
HUH	*Huso huso*	21	25	30	45	51	45	51	28
STU	*Acipenseridae*	21	5 696	4 408	4 153	3 572	2 689	2 225	1 725
ELE	*Anguilla anguilla*	22	8 900	8 684	10 187	7 442	7 454	8 152	7 330
ELJ	*Anguilla japonica*	22	1 023	1 014	916	904	830	765	677
ELA	*Anguilla rostrata*	22	1 284	1 316	1 270	1 244	1 125	1 338	643
ELU	*Anguilla australis*	22	739	621	523	453	425	433	457
ELX	*Anguilla spp*	22	2 891	5 482	1 851	2 253	2 655	5 870	6 248
SAL	*Salmo salar*	23	6 894	6 495	5 579	4 744	4 608	4 690	4 665
TRS	*Salmo trutta*	23	5 151	4 963	4 677	4 896	3 917	3 450	3 844
TRO	*Salmo spp*	23	2 738	3 183	3 186	2 879	2 990	3 029	2 997
PIN	*Oncorhynchus gorbuscha*	23	394 735	294 915	318 717	371 552	386 928	285 338	360 973
CHU	*Oncorhynchus keta*	23	424 595	411 395	347 560	311 965	281 260	276 355	307 662
CHE	*Oncorhynchus masou*	23	2 290	2 552	1 808	2 577	1 989	1 814	1 610
SOC	*Oncorhynchus nerka*	23	189 612	188 584	132 075	78 972	130 128	124 782	108 618
CHI	*Oncorhynchus tshawytscha*	23	13 801	10 509	13 000	9 840	8 735	8 437	8 771
COH	*Oncorhynchus kisutch*	23	28 922	28 201	13 191	19 386	15 449	18 035	20 006
TRR	*Oncorhynchus mykiss*	23	4 351	4 391	3 370	6 031	5 761	3 614	3 758
ORC	*Oncorhynchus spp*	23	-	-	-	-	-	50	204
SVF	*Salvelinus fontinalis*	23	4	3	4	4	4	5	7
ACH	*Salvelinus alpinus*	23	125	96	130	125	67	76	55
LAT	*Salvelinus namaycush*	23	1 037	975	1 109	1 054	985	1 129	1 130
CHR	*Salvelinus spp*	23	136	92	109	129	139	165	146
HUC	*Hucho hucho*	23	1	1	1	1	1	1	1
TLV	*Thymallus thymallus*	23	53	52	43	38	41	37	39
PCA	*Plecoglossus altivelis*	23	13 702	12 732	12 624	11 386	11 380	11 172	11 148
SME	*Osmerus eperlanus*	23	7 374	5 387	4 921	5 752	5 639	9 917	4 454
SMR	*Osmerus mordax*	23	3 106	2 429	2 862	3 334	2 024	1 977	1 242
PSM	*Hypomesus olidus*	23	5 517	3 981	5 964	6 471	5 691	3 254	4 241
SUS	*Hypomesus pretiosus*	23	1	1	-	-	-	-	
EUL	*Thaleichthys pacificus*	23	226	34	27	6	8	13	143
SMX	*Osmerus spp, Hypomesus spp*	23	3 319	2 127	5 262	3 921	3 567	5 008	4 410
FVE	*Coregonus albula*	23	5 709	7 675	7 648	6 177	6 048	6 138	6 081
PLN	*Coregonus lavaretus*	23	6 893	5 742	5 639	5 495	5 313	5 125	4 815
HOU	*Coregonus oxyrinchus*	23	25	63	30	61	35	31	9
WHL	*Coregonus clupeaformis*	23	14 506	14 393	14 218	14 277	13 683	14 227	14 108
CIS	*Coregonus artedi*	23	1 417	1 145	1 223	1 048	804	836	870
WHF	*Coregonus spp*	23	10 189	12 770	13 097	13 181	12 976	12 583	12 183
SLX	*Salmonoidei*	23	7 199	5 112	3 878	4 680	3 203	4 010	2 868
SHC	*Alosa pontica*	24	712	925	983	831	117	228	308
SHA	*Alosa sapidissima*	24	1 326	1 869	1 527	2 048	1 259	1 280	1 644
ASD	*Alosa alosa*	24	57	59	56	2	...	38	45
TSD	*Alosa fallax*	24	1	2	1	...	...	12	16
ALE	*Alosa pseudoharengus*	24	7 410	6 731	7 517	9 062	8 355	7 135	3 502
SHH	*Alosa mediocris*	24	36	88	75	47	62	51	90
SHZ	*Alosa spp*	24	2 620	2 318	1 915	2 822	2 703	2 172	3 025
SHD	*Alosa alosa, A.fallax*	24	39	24	17	22	18	23	30
ASP	*Caspialosa spp*	24	42 545	58 957	62 746	88 366	99 714	79 274	45 408
SHG	*Dorosoma cepedianum*	24	906	1 272	1 989	1 291	3 007	977	1 781
CHG	*Anodontostoma chacunda*	24	6 493	6 752	4 546	3 850	5 488	5 156	5 226
HIX	*Hilsa kelee*	24	92 707	100 591	95 677	105 281	89 354	221 899	97 337
HIL	*Tenualosa ilisha*	24	215 209	227 308	216 065	207 269	215 991	220 372	220 507
TOL	*Tenualosa toli*	24	1 876	2 131	3 011	3 714	3 505	2 645	2 860
DAS	*Clupanodon thrissa*	24	7 961	4 933	13 846	11 360	9 538	6 387	9 194
CLA	*Clupeonella cultriventris*	24	117 415	118 857	106 151	136 557	197 860	164 322	97 406
DOD	*Konosirus punctatus*	24	23 707	18 647	14 850	20 787	17 770	12 335	17 210
ANR	*Lycengraulis grossidens*	24	80	90	100	100	120	120	100
EIL	*Ilisha elongata*	24	47 101	51 969	77 881	84 671	110 624	107 823	101 382
ILI	*Ilisha africana*	24	6 918	11 585	8 797	8 073	4 608	7 467	5 571
PEO	*Pellona ditchela*	24	11 136	16 077	14 176	10 763	10 014	13 544	13 817
DCX	*Clupeoidei*	24	2 516	3 072	2 366	2 733	2 472	2 311	2 667
LAU	*Petromyzon marinus*	25	17	34	20	17	38	36	32
LAR	*Lampetra fluviatilis*	25	96	141	87	95	129	163	118
LAS	*Petromyzontidae*	25	53	76	33	39	73	45	54
MIL	*Chanos chanos*	25	7 994	4 180	839	3 312	562	2 733	668
GTA	*Gasterosteus aculeatus*	25	424	385	462	992	1 047	1 681	1 566
GIP	*Lates calcarifer*	25	54 367	56 228	60 563	72 013	71 551	74 947	81 598
STB	*Morone saxatilis*	25	1 662	2 153	2 802	3 046	3 004	3 135	2 949
PEW	*Morone americana*	25	802	995	1 164	966	1 042	1 221	1 226
LEF	*Bothidae*	31	169	62	29	39	140	278	274
HAL	*Hippoglossus hippoglossus*	31	3 712	3 771	4 000	3 339	3 493	3 573	4 330
HAP	*Hippoglossus stenolepis*	31	26 299	28 110	39 291	41 609	43 620	40 848	40 161
PLE	*Pleuronectes platessa*	31	136 501	118 029	122 045	104 322	113 992	113 760	111 975
RFE	*Pleuronectes vetulus*	31	...	...	...	425	1 100	1 140	1 395
GHL	*Reinhardtius hippoglossoides*	31	96 111	101 604	92 657	85 295	114 824	111 261	112 787
ARF	*Atheresthes stomias*	31	5 167	10 681	6 320	8 280	12 083	18 774	14 401
KAF	*Atheresthes evermanni*	31	7 039	9 739	8 476	9 567	10 743	23 473	19 049
EOJ	*Eopsetta jordani*	31	...	...	1 937	1 464	1 480	1 870	1 822
WIT	*Glyptocephalus cynoglossus*	31	14 949	15 036	17 825	19 224	18 979	18 805	20 052
GLZ	*Glyptocephalus zachirus*	31	...	...	...	289	544	541	571

B-00 (b)

Fish, crustaceans, molluscs, etc — **Capture production by species items**
Poissons, crustacés, mollusques, etc — **Captures par catégories d'espèces**
Peces, crustáceos, moluscos, etc — **Capturas por partidas de especies**

3-alpha code Code alpha-3 Código alfa-3	Scientific name Nom scientifique Nombre científico	Species group Groupe d'espèces Grupo de especies	1995 mt	1996 mt	1997 mt	1998 mt	1999 mt	2000 mt	2001 mt
PLA	*Hippoglossoides platessoides*	31	15 210	16 025	17 223	19 223	21 624	16 883	18 486
FTS	*Hippoglossoides elassodon*	31	9 073	11 392	12 867	18 886	14 318	16 266	16 092
YES	*Limanda aspera*	31	96 765	101 354	149 302	80 500	56 830	69 971	54 918
YEL	*Limanda ferruginea*	31	3 430	3 840	5 266	9 604	13 960	20 971	24 273
DAB	*Limanda limanda*	31	16 141	17 967	17 509	22 134	20 917	16 550	17 172
ROS	*Lepidopsetta bilineata*	31	24 969	26 202	32 752	15 199	17 192	27 517	24 213
MIP	*Microstomus pacificus*	31	...	...	12 342	9 992	10 558	9 412	7 441
LEM	*Microstomus kitt*	31	11 062	12 078	12 283	14 378	14 203	13 974	15 461
FLE	*Platichthys flesus*	31	11 215	10 667	11 421	15 333	11 899	13 961	15 344
YSE	*Psettichthys melanostictus*	31	...	...	...	34	605	79	129
YFL	*Pseudopleuronectes herzenst.*	31	13 683	18 066	18 079	20 135	19 569	15 423	14 503
FLW	*Pseudopleuronectes americanus*	31	6 806	8 155	8 313	6 904	6 417	8 185	9 187
FSA	*Rhombosolea spp*	31	169	251	278	28	57	1	37
NYD	*Pleuronichthys decurrens*	31	-	-	-	-	-	1	5
SOL	*Solea solea*	31	55 677	43 338	36 682	44 378	44 078	48 617	48 439
SOS	*Solea lascaris*	31	369	246	288	305	222	297	339
CET	*Dicologlossa cuneata*	31	1 000	885	656	488	713	805	700
SOW	*Austroglossus microlepis*	31	463	409	517	393	463	571	589
SOE	*Austroglossus pectoralis*	31	769	909	837	859	768	800	844
SOA	*Austroglossus spp*	31	1 365	480	1 197	773	926	965	650
THS	*Microchirus spp*	31	48	50	57	77	71	71	80
SOX	*Soleidae*	31	9 182	12 606	13 552	16 050	9 960	6 225	5 194
TOX	*Cynoglossidae*	31	27 487	28 387	29 018	28 598	30 502	35 490	34 596
MEG	*Lepidorhombus whiffiagonis*	31	15 935	15 866	15 243	14 020	12 668	13 461	12 915
LEZ	*Lepidorhombus spp*	31	6 275	6 405	7 006	7 446	6 969	8 361	8 450
BLL	*Scophthalmus rhombus*	31	2 982	3 221	2 764	2 548	2 535	2 983	3 068
FLD	*Scophthalmus aquosus*	31	759	46	48	521	166	268	177
TUR	*Psetta maxima*	31	10 941	8 852	7 445	7 527	7 591	9 180	9 086
SCF	*Scophthalmidae*	31	1 923	1 377	964	528	478	643	622
HAI	*Psettodes erumei*	31	11 088	15 571	23 309	13 626	15 267	14 259	16 795
SOT	*Psettodes belcheri*	31	0	0	2	0	-	1	2
IYO	*Citharichthys sordidus*	31	-	-	-	-	-	150	-
BAH	*Paralichthys olivaceus*	31	9 472	10 628	9 953	9 617	8 877	9 179	8 436
YSF	*Paralichthys californicus*	31	...	...	...	523	620	390	393
FLS	*Paralichthys dentatus*	31	6 943	5 425	4 775	6 348	5 795	6 063	6 554
BAX	*Paralichthys spp*	31	12 269	11 428	11 605	9 780	7 472	7 249	5 680
FLX	*Pleuronectiformes*	31	244 193	262 708	271 086	264 568	270 604	277 553	237 548
MOY	*Muraenolepis microps*	32	-	-	-	-	4	5	-
MRL	*Muraenolepis spp*	32	-	-	-	0	1	2	3
RIB	*Mora moro*	32	1 192	694	1 410	1 324	1 122	1 358	1 211
NEC	*Pseudophycis bachus*	32	16 159	10 596	11 087	16 495	12 555	5 365	4 530
ANT	*Antimora rostrata*	32	1	18	2	29	30	31	26
SAO	*Salilota australis*	32	11 772	8 071	6 741	11 806	15 888	14 913	6 169
MOR	*Moridae*	32	3 758	-	567	11 190	31 362	39 378	32 742
UNC	*Bregmaceros mcclellandi*	32	1 232	1 287	1 351	2 587	2 095	1 629	2 467
USK	*Brosme brosme*	32	32 192	30 296	24 745	30 986	34 748	32 528	28 524
COD	*Gadus morhua*	32	1 270 501	1 340 757	1 375 079	1 211 067	1 094 072	941 269	943 661
PCO	*Gadus macrocephalus*	32	430 196	429 475	443 895	411 371	402 245	370 912	330 884
GRC	*Gadus ogac*	32	2 528	2 121	1 729	1 697	1 903	931	1 133
LIN	*Molva molva*	32	54 463	54 619	50 988	60 826	53 949	43 320	36 988
BLI	*Molva dypterygia*	32	9 742	9 788	9 970	12 768	15 699	16 146	19 347
GFB	*Phycis blennoides*	32	2 478	5 056	8 339	7 807	6 535	5 865	5 680
HKU	*Urophycis brasiliensis*	32	3 703	3 152	2 570	6 081	2 842	3 496	2 564
HKR	*Urophycis chuss*	32	2 254	2 907	2 850	2 803	3 227	3 365	4 008
HKW	*Urophycis tenuis*	32	10 810	8 303	6 830	5 959	6 378	8 158	8 324
HAD	*Melanogrammus aeglefinus*	32	317 794	361 970	334 105	283 689	249 451	212 807	228 821
COW	*Eleginus navaga*	32	1 014	659	1 152	1 185	480	674	1 166
SAF	*Eleginus gracilis*	32	25 566	21 110	27 803	40 426	47 032	35 763	33 753
MGX	*Microgadus proximus*	32	...	...	...	3	1	0	0
TOM	*Microgadus tomcod*	32	131	140	90	119	42	55	57
POK	*Pollachius virens*	32	371 709	355 383	319 307	329 793	340 351	313 580	325 750
POL	*Pollachius pollachius*	32	14 180	12 988	12 164	10 952	10 379	10 952	11 937
ALK	*Theragra chalcogramma*	32	4 809 011	4 548 585	4 486 510	4 049 223	3 270 286	2 930 230	3 136 465
POC	*Boreogadus saida*	32	26 430	20 788	6 826	3 592	22 005	40 743	39 445
ROL	*Gaidropsarus spp*	32	...	...	...	...	24	43	20
NOP	*Trisopterus esmarkii*	32	389 800	275 667	211 394	97 478	112 613	204 845	93 405
POD	*Trisopterus minutus*	32	586	603	428	428	637	890	755
BIB	*Trisopterus luscus*	32	16 722	14 881	14 213	12 762	13 300	18 751	18 203
WHB	*Micromesistius poutassou*	32	544 310	632 471	712 279	1 185 003	1 319 002	1 472 105	1 823 305
POS	*Micromesistius australis*	32	191 599	156 154	166 205	201 363	185 310	152 830	155 780
WHG	*Merlangius merlangus*	32	107 753	98 609	88 163	75 081	76 115	75 154	59 923
HKE	*Merluccius merluccius*	32	119 906	92 332	76 958	66 088	70 285	72 890	58 141
HKM	*Merluccius senegalensis*	32	14 779	16 485	16 430	17 396	18 187	13 515	17 923
HKN	*Merluccius australis*	32	39 463	37 022	39 047	43 200	47 560	52 478	48 642
HKS	*Merluccius bilinearis*	32	32 237	42 428	33 565	31 892	27 567	24 237	33 000
PHA	*Merluccius gayi*	32	256 585	323 470	265 840	162 683	141 053	193 754	246 656
HKP	*Merluccius hubbsi*	32	636 385	681 999	648 301	527 229	372 167	242 727	302 798
NHA	*Merluccius productus*	32	177 117	195 406	227 705	227 754	217 000	206 720	172 051
HKB	*Merluccius polli*	32	0	39	0	38	147	44	139
HKK	*Merluccius capensis*	32	26	832	777	906	2 060	658	1 863
HOF	*Merluccius albidus*	32	...	32	...	5	12	5	2
HKX	*Merluccius spp*	32	133	2 075	1 126	2 064	1 674	1 940	1 822
HKC	*Merluccius capensis,M.paradox.*	32	277 186	286 443	260 955	306 269	309 954	300 743	323 257
GRM	*Macruronus magellanicus*	32	243 072	434 797	123 678	473 633	447 063	233 570	296 294
GRN	*Macruronus novaezelandiae*	32	211 984	201 017	285 580	324 750	281 105	283 688	257 782
RHG	*Macrourus berglax*	32	1 384	1 080	4 545	7 170	8 158	8 511	1 912
WGR	*Macrourus whitsoni*	32	-	-	-	-	1	5	49
GRV	*Macrourus spp*	32	754	925	1 057	3 041	2 651	10 841	261

B-00 (b)

Fish, crustaceans, molluscs, etc — Capture production by species items
Poissons, crustacés, mollusques, etc — Captures par catégories d'espèces
Peces, crustáceos, moluscos, etc — Capturas por partidas de especies

3-alpha code Code alpha-3 Código alfa-3	Scientific name Nom scientifique Nombre científico	Species group Groupe d'espèces Grupo de especies	1995 mt	1996 mt	1997 mt	1998 mt	1999 mt	2000 mt	2001 mt
RNG	*Coryphaenoides rupestris*	32	15 708	14 196	19 906	23 260	18 213	30 914	53 669
LDE	*Lepidorhynchus denticulatus*	32	745	670	2 361	4 627	3 678	3 833	4 783
RTX	*Macrouridae*	32	1 131	1 889	6 241	4 921	4 056	3 498	5 631
GAD	*Gadiformes*	32	42 507	43 848	29 072	20 411	22 496	12 080	40 089
LAD	*Elops saurus*	33	177	143	746	980	1 978	141	480
CEC	*Elops lacerta*	33	1 474	379	735	1 546	1 822	755	558
TAR	*Megalops atlanticus*	33	1 504	515	763	604	331	349	344
TAI	*Megalops cyprinoides*	33	1 112	1 096	1 158	1 058	1 433	1 284	1 342
BOF	*Albula vulpes*	33	163	383	1 558	1 674	370	249	526
BUC	*Harpadon nehereus*	33	170 852	196 431	226 491	194 188	194 307	143 250	153 204
LIG	*Saurida tumbil*	33	15 951	14 751	10 874	10 931	10 791	11 203	8 342
LIB	*Saurida undosquamis*	33	320	333	321	286	103	846	810
LIX	*Synodontidae*	33	136 237	124 080	135 924	139 055	137 330	169 604	167 704
CAX	*Ariidae*	33	275 938	289 022	311 731	297 711	323 201	318 795	336 210
CAE	*Plotosus spp*	33	2 633	2 644	2 210	2 860	2 380	2 844	2 606
MUI	*Muraenidae*	33	837	710	268	165	335	358	354
KLA	*Hypoptychus dybowskii*	33	9 677	6 613	8 832	4 801	4 806	-	462
HCZ	*Holocentridae*	33	92	121	92	101	90	88	107
MUF	*Mugil cephalus*	33	32 936	32 311	30 701	24 283	27 320	29 306	26 724
MUB	*Mugil liza*	33	3 249	3 287	1 550	2 876	2 855	3 186	2 537
MYZ	*Mugil soiuy*	33	775	1 039	2 718	3 674	5 364	5 575	2 810
MUA	*Joturus pichardi*	33	614	659	339	374	404	497	504
LZS	*Liza saliens*	33	1	3	2	2	2	0	10
MUL	*Mugilidae*	33	306 813	301 698	344 064	370 630	381 301	383 047	399 649
FUS	*Caesio spp*	33	58 208	48 183	57 080	51 423	54 251	48 235	49 329
SNO	*Centropomus undecimalis*	33	5 908	5 132	4 867	5 012	5 119	5 034	5 903
ROB	*Centropomus spp*	33	3 650	4 058	3 792	4 925	7 830	8 349	8 577
MAB	*Mycteroperca bonaci*	33	...	...	...	...	...	9	7
MKM	*Mycteroperca microlepis*	33	...	...	...	...	...	15	10
MKH	*Mycteroperca phenax*	33	...	...	...	1	14	27	17
MKV	*Mycteroperca venenosa*	33	...	...	...	...	...	3	1
GBS	*Mycteroperca xenarcha*	33	207	180	125	155	82	80	210
GPB	*Mycteroperca spp*	33	1 117	423	584	518	628	536	530
GPD	*Epinephelus marginatus*	33	2 597	1 715	1 563	1 105	420	317	504
GPW	*Epinephelus aeneus*	33	743	682	603	561	633	683	655
GPN	*Epinephelus striatus*	33	464	459	596	554	429	276	331
GPS	*Epinephelus analogus*	33	22	43	48	32	39	30	28
EEL	*Epinephelus flavolimbatus*	33	...	...	15	18	...	73	36
EEU	*Epinephelus guttatus*	33	81	60	42	52	72	67	88
GPR	*Epinephelus morio*	33	1 846	625	528	309	806	1 373	1 357
EFV	*Epinephelus niveatus*	33	...	...	...	...	...	6	4
EEG	*Epinephelus goreensis*	33	102	196	63	76	290	277	62
ELG	*Epinephelus nigritus*	33	37	16	23	23	-	45	44
GPX	*Epinephelus spp*	33	119 411	125 479	137 330	150 940	159 024	168 688	169 521
LDP	*Lepidoperca pulchella*	33	...	...	...	193	32	...	46
BSB	*Centropristis striata*	33	1 297	1 978	1 590	1 531	1 717	1 516	1 667
BSZ	*Acanthistius brasilianus*	33	10 987	10 799	9 164	5 258	5 914	4 204	4 701
BAP	*Paralabrax humeralis*	33	5 837	4 954	2 789	2 554	3 278	4 373	2 011
BSX	*Serranidae*	33	49 217	48 123	48 157	53 679	55 088	66 510	70 171
THO	*Terapon spp*	33	-	-	-	1	2	0	1
TEJ	*Stereolepis gigas*	33	0	1	1	3	2	2	3
SPU	*Dicentrarchus punctatus*	33	399	1 273	798	963	782	379	460
BSS	*Dicentrarchus labrax*	33	9 451	8 263	7 051	6 686	7 648	8 513	9 118
BSE	*Dicentrarchus spp*	33	5 204	5 243	5 788	7 230	5 505	7 157	5 953
BAJ	*Lateolabrax japonicus*	33	8 831	9 512	10 558	12 180	12 167	10 538	12 256
BIR	*Priacanthus macracanthus*	33	5 486	7 386	4 824	3 670	2 756	3 986	3 079
BIG	*Priacanthus spp*	33	79 967	91 272	87 311	89 071	90 814	91 486	92 267
APO	*Apogonidae*	33	11	211	180	60	365	60	82
WHS	*Sillaginidae*	33	21 690	19 649	21 190	20 535	24 014	22 704	22 520
MOO	*Mene maculata*	33	5 066	5 540	12 842	12 539	14 132	13 214	14 580
RUF	*Arripis georgianus*	33	1 063	1 302	1 287	1 008	1 066	1 143	1 005
ASA	*Arripis trutta*	33	10 345	8 165	7 550	6 847	8 306	8 361	7 453
RES	*Lutjanus argentimaculatus*	33	14 326	13 488	13 242	9 354	16 129	14 442	13 944
HUS	*Lutjanus argentiventris*	33	3 930	4 960	3 163	3 490	3 076	3 472	3 388
SNC	*Lutjanus purpureus*	33	7 413	6 400	7 446	7 100	10 731	8 226	8 329
SNR	*Lutjanus campechanus*	33	6 126	6 346	6 264	5 166	5 412	4 841	4 869
SNL	*Lutjanus synagris*	33	3 642	3 682	3 807	4 250	3 384	3 164	3 413
SNA	*Lutjanus spp*	33	64 576	73 384	84 765	86 541	92 149	83 655	85 657
SNY	*Ocyurus chrysurus*	33	6 942	7 332	7 436	6 145	6 967	6 462	6 614
RPU	*Rhomboplites aurorubens*	33	...	...	678	480	731	911	1 057
SNX	*Lutjanidae*	33	80 915	74 743	79 690	85 403	86 073	87 372	82 428
THG	*Nemipterus virgatus*	33	228 238	242 261	263 399	266 383	250 591	296 319	290 920
THB	*Nemipterus spp*	33	214 116	205 683	198 987	221 903	225 200	218 848	219 664
MOB	*Scolopsis spp*	33	1 342	1 342	1 294	1 473	1 090	1 754	1 695
THD	*Nemipteridae*	33	3 438	2 750	4 703	3 469	4 066	2 917	4 506
POY	*Leiognathus spp*	33	2 404	2 690	2 500	3 211	3 180	2 573	2 395
PON	*Leiognathidae*	33	189 471	198 581	218 807	186 246	215 875	186 865	189 987
PIG	*Orthopristis chrysoptera*	33	1	-	1	1	-	0	0
GRP	*Isacia conceptionis*	33	1 507	2 048	2 034	2 133	2 947	3 293	3 321
GBR	*Plectorhinchus mediterraneus*	33	1 079	1 869	904	459	890	297	271
GRL	*Pomadasys argenteus*	33	1 175	1 644	1 156	2 847	2 235	1 972	1 593
BGR	*Pomadasys incisus*	33	1	1	5	-	-	0	0
BUR	*Pomadasys jubelini*	33	1 101	1 439	1 101	950	951	922	724
GRB	*Brachydeuterus auritus*	33	23 063	20 033	29 274	18 853	22 385	21 939	21 167
BRG	*Conodon nobilis*	33	339	172	118	114	84	39	40
GRX	*Haemulidae (=Pomadasyidae)*	33	67 169	75 707	74 663	74 580	74 208	77 709	83 562
IAG	*Sciaena gilberti*	33	5 592	9 099	3 565	6 154	6 442	4 744	4 328
CBM	*Sciaena umbra*	33	369	430	222	240	186	128	171

B-00 (b)

Fish, crustaceans, molluscs, etc — **Capture production by species items**
Poissons, crustacés, mollusques, etc — **Captures par catégories d'espèces**
Peces, crustáceos, moluscos, etc — **Capturas por partidas de especies**

3-alpha code Code alpha-3 Código alfa-3	Scientific name Nom scientifique Nombre científico	Species group Groupe d'espèces Grupo de especies	1995 mt	1996 mt	1997 mt	1998 mt	1999 mt	2000 mt	2001 mt
WEP	*Cynoscion analis*	33	9 406	7 887	5 797	11 187	8 929	6 339	4 438
SWF	*Cynoscion nebulosus*	33	3 950	3 696	3 875	6 865	6 944	6 486	1 937
STG	*Cynoscion regalis*	33	3 095	3 261	3 317	3 821	3 146	2 438	2 273
WKS	*Cynoscion striatus*	33	32 635	31 641	39 318	32 394	19 588	22 874	22 193
WKX	*Cynoscion spp*	33	49 575	43 883	40 593	44 830	51 244	58 817	59 153
CKM	*Micropogonias furnieri*	33	88 569	78 698	76 857	64 502	45 117	61 469	67 149
CKA	*Micropogonias undulatus*	33	7 013	9 049	12 434	11 522	12 180	12 138	13 018
CRX	*Micropogonias spp*	33	4 352	6 062	5 852	6 709	5 391	4 830	4 797
KGF	*Menticirrhus saxatilis*	33	34	49	21	17	13	28	39
KGG	*Menticirrhus littoralis*	33	968	1 070	1 443	2 061	1 822	961	690
KIX	*Menticirrhus spp*	33	1 330	1 144	1 303	1 402	1 062	1 352	1 336
COB	*Umbrina cirrosa*	33	1 226	796	545	429	469	382	450
CKY	*Umbrina canosai*	33	14 537	13 018	6 658	5 284	7 660	9 221	9 116
UCA	*Umbrina canariensis*	33	6	11	10	10	839	9	2 213
WKK	*Macrodon ancylodon*	33	5 345	7 266	7 625	9 229	5 595	6 642	6 886
MGR	*Argyrosomus regius*	33	2 944	3 435	3 905	3 605	4 923	4 095	4 007
KOB	*Argyrosomus hololepidotus*	33	1 501	1 497	1 126	1 297	1 442	1 272	4 098
AWE	*Atractoscion aequidens*	33	397	381	519	632	444	380	383
WEW	*Atractoscion nobilis*	33	33	46	28	71	112	101	124
KIC	*Genyonemus lineatus*	33	256	242	167	64	74	105	137
DRS	*Pteroscion peli*	33	7 820	1 879	1 291	1 190	1 260	1 126	1 159
LKR	*Otolithes ruber*	33	-	-	-	13	10	...	...
PDR	*Paralonchurus peruanus*	33	5 543	4 263	2 737	4 363	6 063	5 729	4 167
BDM	*Pogonias cromis*	33	850	505	1 755	3 297	1 181	2 876	2 864
HOC	*Nibea mitsukurii*	33	2 164	1 940	1 177	1 285	1 566	1 999	2 156
LYC	*Larimichthys croceus*	33	89 630	102 297	71 515	71 229	66 076	123 274	76 715
CRY	*Larimichthys polyactis*	33	181 553	280 036	236 960	278 807	323 564	425 047	330 455
SPT	*Leiostomus xanthurus*	33	3 521	2 554	3 073	3 359	2 599	3 141	3 091
RDM	*Sciaenops ocellatus*	33	2	1	2	3	6	6	3
CKL	*Pseudotolithus brachygnathus*	33	813	715	618	354	318	811	1 083
PSS	*Pseudotolithus senegalensis*	33	2 936	2 727	2 781	4 659	4 837	3 949	4 800
PSE	*Pseudotolithus elongatus*	33	13 095	12 662	11 497	9 461	6 855	13 405	9 761
CKW	*Pseudotolithus spp*	33	20 826	29 349	23 762	32 890	32 669	31 520	30 821
CRL	*Atrobucca nibe*	33	197	297	288	237	223	450	644
CRV	*Pennahia argentata*	33	7 369	8 599	6 085	5 089	4 940	4 180	3 796
CDX	*Sciaenidae*	33	660 001	527 546	565 757	526 451	613 004	504 585	475 049
LBR	*Gymnocranius spp*	33	215	228	192	168	128	145	160
LTN	*Lethrinus atlanticus*	33	137	231	334	380	...	...	...
EMP	*Lethrinidae*	33	81 015	81 553	80 751	80 925	86 181	95 664	98 126
SBR	*Pagellus bogaraveo*	33	3 386	2 739	5 078	3 453	2 684	2 022	1 323
PAC	*Pagellus erythrinus*	33	3 788	4 716	5 745	5 077	4 933	5 354	5 216
SBA	*Pagellus acarne*	33	1 520	1 305	1 462	1 680	1 368	1 607	1 420
PAR	*Pagellus bellottii*	33	4 627	7 621	8 047	10 278	13 734	2 980	3 997
PAX	*Pagellus spp*	33	13 308	11 018	9 680	8 151	10 791	10 188	9 120
DIG	*Diplodus argenteus*	33	8	39	84	15	5	2	...
SWA	*Diplodus sargus*	33	584	463	681	468	695	759	580
SRG	*Diplodus spp*	33	7 251	8 010	7 376	6 597	9 026	8 965	8 527
PRG	*Calamus spp*	33	1 266	1 052	1 541	553	1 266	512	520
DEL	*Dentex macrophthalmus*	33	1 470	2 442	2 733	3 580	5 479	4 581	2 266
DEN	*Dentex canariensis*	33	-	-	-	-	-	20	
DEC	*Dentex dentex*	33	4 396	3 293	1 715	1 662	1 451	1 388	1 119
DEA	*Dentex angolensis*	33	591	763	490	1 416	1 767	504	838
DNC	*Dentex congoensis*	33	43	102	119	392	1 272	350	571
DEX	*Dentex spp*	33	15 866	8 054	6 400	13 863	13 599	17 596	19 668
BRB	*Spondyliosoma cantharus*	33	4 070	5 147	5 148	4 640	5 128	5 487	5 427
SBS	*Oblada melanura*	33	1 098	1 040	881	862	781	980	660
SPH	*Archosargus probatocephalus*	33	2 305	2 024	2 292	1 939	1 653	2 105	1 720
KBR	*Argyrops spinifer*	33	2 711	2 784	3 013	2 992	3 015	3 803	4 026
SLF	*Argyrozona argyrozona*	33	729	883	780	505	541	500	287
SLD	*Cheimerius nufar*	33	66	73	62	0	29	25	25
RPG	*Pagrus pagrus*	33	6 321	6 978	6 771	6 691	9 299	7 348	5 248
GSU	*Pagrus auratus*	33	27 515	29 114	29 618	29 014	31 465	31 307	29 670
SBP	*Pagrus spp*	33	4 651	5 346	6 522	6 439	4 026	6 504	10 368
RER	*Petrus rupestris*	33	56	28	35	...	22	10	7
PGA	*Pterogymnus laniarius*	33	999	1 200	1 323	1 042	1 349	900	1 058
WSN	*Rhabdosargus globiceps*	33	168	237	165	296	335	300	169
SBG	*Sparus aurata*	33	6 361	5 843	5 564	5 806	7 479	8 301	9 552
BOG	*Boops boops*	33	30 767	30 408	27 054	29 456	28 219	28 467	25 389
RSX	*Chrysoblephus spp*	33	267	268	224	154	74	70	78
SNW	*Lithognathus lithognathus*	33	11	7	8	9	2	...	...
SSB	*Lithognathus mormyrus*	33	1 276	968	1 228	2 509	1 999	2 230	3 047
STW	*Lithognathus spp*	33	195	6	136	2	116	56	20
SLM	*Sarpa salpa*	33	1 940	1 941	2 421	2 237	2 304	2 056	2 128
MLM	*Acanthopagrus schlegeli*	33	197	207	277	404	604	717	878
YWF	*Acanthopagrus latus*	33	273	234	249	281	464	350	271
SCP	*Stenotomus chrysops*	33	2 845	2 952	2 196	1 893	1 676	1 206	1 845
SBX	*Sparidae*	33	140 516	133 241	148 793	144 442	163 506	192 810	207 972
BPI	*Spicara maena*	33	218	285	505	395	419	255	381
PIC	*Spicara spp*	33	10 177	12 319	12 271	13 210	9 367	8 227	8 953
MUR	*Mullus surmuletus*	33	15 205	15 167	14 233	13 177	12 352	14 609	14 008
MUT	*Mullus barbatus*	33	5 115	5 331	4 531	4 878	5 290	4 170	5 405
MUX	*Mullus spp*	33	19 363	26 576	18 518	17 103	22 209	23 167	21 045
GOX	*Upeneus spp*	33	48 117	49 504	52 579	51 738	55 362	51 657	63 206
GOA	*Pseudupeneus prayensis*	33	1 025	1 719	1 808	1 396	1 146	833	3 371
MUM	*Mullidae*	33	27 131	24 836	17 003	14 991	15 724	16 001	17 859
MOJ	*Gerres spp*	33	9 972	9 499	8 189	7 641	8 118	8 552	9 322
GDJ	*Gerreidae*	33	16 783	16 597	9 857	7 432	6 219	5 459	5 633
GIY	*Girella tricuspidata*	33	83	561	588	576	579	631	558
GIQ	*Girella nigricans*	33	-	-	-	-	-	3	2

B-00 (b)

Fish, crustaceans, molluscs, etc — **Capture production by species items**
Poissons, crustacés, mollusques, etc — **Captures par catégories d'espèces**
Peces, crustáceos, moluscos, etc — **Capturas por partidas de especies**

3-alpha code Code alpha-3 Código alfa-3	Scientific name Nom scientifique Nombre científico	Species group Groupe d'espèces Grupo de especies	1995 mt	1996 mt	1997 mt	1998 mt	1999 mt	2000 mt	2001 mt
KYC	*Kyphosus cinerascens*	33	-	-	8	-	-	-	-
KYP	*Kyphosus spp*	33	...	...	...	...	...	...	8
SPS	*Drepane punctata*	33	699	939	757	612	976	792	806
SIC	*Drepane africana*	33	1 706	2 774	3 790	2 268	1 545	2 856	2 790
USB	*Labrus bergylta*	33	-	-	-	-	2	1	-
TAU	*Tautoga onitis*	33	148	119	116	116	95	111	138
CUN	*Tautogolabrus adspersus*	33	0	1	1	3	4	3	9
YFH	*Semicossyphus pulcher*	33	...	...	...	118	58	78	68
WRA	*Labridae*	33	23 464	21 983	22 662	19 486	18 746	18 744	19 455
PRR	*Sparisoma cretense*	33	-	-	-	-	56	89	162
PWT	*Scaridae*	33	1 195	1 051	1 381	1 296	1 894	1 096	1 294
ANW	*Pomacanthidae*	33	-	0	0	7	17	1	3
TGA	*Polydactylus quadrifilis*	33	3 598	7 359	10 705	10 023	10 770	12 237	12 447
FOT	*Eleutheronema tetradactylum*	33	4 281	3 673	4 614	4 217	2 414	6 980	1 770
GAL	*Galeoides decadactylus*	33	12 903	17 768	18 385	12 890	10 764	13 948	17 101
PET	*Pentanemus quinquarius*	33	3 734	3 005	3 356	4 430	4 162	3 612	3 947
THF	*Polynemidae*	33	54 624	335 621	391 071	571 370	618 862	550 273	534 342
BLP	*Eleginops maclovinus*	33	344	439	194	1 301	2 220	1 919	170
NOR	*Notothenia rossii*	33	2	-	1	-	1	0	0
NOG	*Notothenia gibberifrons*	33	1	-	-	-	5	1	2
NON	*Notothenia neglecta*	33	-	-	-	-	0	-	2
NOS	*Notothenia squamifrons*	33	-	15	4	3	15	5	0
NOD	*Nototheniops nudifrons*	33	-	-	-	-	-	0	0
TRT	*Trematomus spp*	33	-	-	-	-	0	-	0
ANS	*Pleuragramma antarcticum*	33	-	-	-	-	-	-	0
NOX	*Nototheniidae*	33	36	16	34	3	229	3	16
AIB	*Ambassidae*	33	3 554	3 075	2 594	2 186	1 733	1 575	1 664
PRC	*Percoidei*	33	79 778	155 971	110 833	95 569	136 617	117 051	104 423
ELP	*Zoarces viviparus*	33	152	147	97	54	43	28	37
OPT	*Macrozoarces americanus*	33	24	41	15	17	18	19	18
PAS	*Ammodytes personatus*	33	108 124	115 766	108 666	90 688	82 918	66 129	92 967
SAN	*Ammodytes spp*	33	1 133 234	858 452	1 242 837	1 039 873	762 248	740 852	912 436
FLA	*Percophis brasilianus*	33	9 320	9 231	11 887	10 191	7 236	7 862	7 640
UPR	*Pseudopercis semifasciata*	33	3 023	3 438	2 483	2 222	2 679	1 905	1 887
NEB	*Parapercis colias*	33	11 262	2 115	2 227	2 313	2 286	2 130	2 441
WEG	*Trachinus draco*	33	278	265	337	395	308	331	374
GBN	*Gobius niger*	33	-		-	-	-	1	1
GPA	*Gobiidae*	33	27 282	27 350	27 646	51 521	41 788	43 596	39 012
SUR	*Acanthuridae*	33	5 260	4 952	7 672	8 236	6 294	5 864	6 504
BAT	*Platax spp*	33	1 600	1 584	2 975	2 793	2 877	2 598	2 607
SPA	*Ephippidae*	33	47	80	61	87	309	359	345
SCT	*Scatophagus spp*	33	4 782	4 204	2 152	2 496	2 541	2 698	2 800
SPI	*Siganus spp*	33	29 527	29 700	27 320	29 119	30 229	32 996	34 917
CLI	*Ophiodon elongatus*	33	5 484	4 754	3 589	2 559	3 154	3 214	2 698
ATK	*Pleurogrammus azonus*	33	269 693	301 393	308 894	344 350	265 205	267 848	271 488
FLI	*Platycephalus indicus*	33	2 520	2 900	4 196	2 857	2 248	2 310	1 699
FLH	*Platycephalidae*	33	4 136	4 389	4 576	3 305	4 772	4 801	4 137
SMQ	*Scorpaenichthys marmoratus*	33	...	...	...	168	173	148	119
SWU	*Cottidae*	33	1	-	2	5	3	64	75
NRC	*Normanichthys crockeri*	33	2 692	3 924	20 426	236	4 843	853	223
BXF	*Ostraciidae*	33	...	...	1	1	55	68	118
PUF	*Sphoeroides maculatus*	33	19	11	18	17	37	36	35
PUP	*Takifugu vermicularis*	33	10 178	9 708	7 471	3 897	4 787	-	-
PUX	*Tetraodontidae*	33	9 857	8 512	7 510	8 536	9 666	14 008	11 638
FLF	*Cantherhines(=Navodon) spp*	33	124 038	211 059	297 227	236 189	241 209	225 403	204 270
FIL	*Stephanolepis cirrhifer*	33	1 755	1 772	16 318	9 364	2 999	-	-
PKB	*Parika scaber*	33	329	442	1 095	312	738	1 279	1 142
TRG	*Balistes carolinensis*	33	59	38	53	59	36	50	66
TRI	*Balistidae*	33	3 782	4 070	8 531	6 047	5 919	3 544	2 617
TFD	*Batrachoididae*	33	...	...	1	6	74	112	102
ARG	*Argentina spp*	34	20 463	24 477	27 431	46 509	31 629	28 630	49 036
DES	*Glossanodon semifasciatus*	34	7 705	8 134	7 431	7 142	6 312	5 970	5 414
ALC	*Alepocephalus bairdii*	34	1	0	0	0	0	12	616
LAN	*Lampanyctodes hectoris*	34	0	33	243	6 553	0	...	...
LXX	*Myctophidae*	34	2 002	0	0	0	5	67	335
DPC	*Muraenesox cinereus*	34	170 522	189 864	201 111	258 172	253 978	238 364	260 108
PCX	*Muraenesox spp*	34	17 366	19 286	20 507	19 365	24 172	19 427	19 868
COE	*Conger conger*	34	16 940	17 800	17 290	17 360	15 936	14 830	14 238
COS	*Conger orbignyanus*	34	203	89	138	189	180	122	100
COA	*Conger oceanicus*	34	32	29	17	48	43	49	40
ELS	*Conger myriaster*	34	19 667	17 314	19 136	11 913	10 160	8 304	7 676
COX	*Congridae*	34	17 302	15 724	14 864	13 328	12 089	11 561	11 205
SNS	*Macroramphosus scolopax*	34	20	-	89	443	-	-	44
CUS	*Genypterus blacodes*	34	49 355	43 643	55 223	59 402	55 118	49 057	51 685
CUC	*Genypterus chilensis*	34	1 082	982	745	584	415	608	730
CUB	*Genypterus maculatus*	34	1 193	1 343	1 661	2 753	1 943	3 542	3 889
KCP	*Genypterus capensis*	34	6 651	6 807	5 955	6 203	7 820	7 922	11 462
CEX	*Genypterus spp*	34	1 631	1 121	439	425	196	614	552
BRD	*Brotula barbata*	34	504	811	1 842	2 213	3 862	1 720	1 926
OPH	*Ophidiidae*	34	304	294	355	489	557	518	528
ALF	*Beryx spp*	34	6 190	8 109	8 026	6 516	5 638	9 469	9 520
CXF	*Centroberyx affinis*	34	89	1 407	1 551	1 984	1 689	1 185	977
ORY	*Hoplostethus atlanticus*	34	48 546	48 988	44 545	42 420	36 924	27 252	25 258
TRC	*Trachichthyidae*	34	-	833	1 052	40	807	1 489	2 346
JOD	*Zeus faber*	34	5 746	6 161	5 615	6 517	7 698	8 624	8 344
JOS	*Zenopsis conchifer*	34	37	36	12	55	22	28	62
ZNE	*Zenopsis nebulosus*	34	361	352	1 356	1 831	1 035	502	529
ZEX	*Zeidae*	34	338	288	641	754	763	778	685
BOR	*Caproidae*	34	-	-	-	5	-	-	7

B-00 (b)

Fish, crustaceans, molluscs, etc — **Capture production by species items**
Poissons, crustacés, mollusques, etc — **Captures par catégories d'espèces**
Peces, crustáceos, moluscos, etc — **Capturas por partidas de especies**

3-alpha code Code alpha-3 Código alfa-3	Scientific name Nom scientifique Nombre científico	Species group Groupe d'espèces Grupo de especies	1995 mt	1996 mt	1997 mt	1998 mt	1999 mt	2000 mt	2001 mt
ORD	*Oreosomatidae*	34	21 839	18 799	22 038	21 102	22 690	22 960	24 413
WRF	*Polyprion americanus*	34	863	914	736	584	558	617	475
WHA	*Polyprion oxygeneios*	34	1 569	1 178	1 687	1 597	1 555	1 504	1 589
ULP	*Caulolatilus princeps*	34	...	...	...	535	627	1 077	984
CKZ	*Caulolatilus chrysops*	34	-	-	-	-	0	6	806
TIL	*Lopholatilus chamaeleonticeps*	34	1 284	1 463	1 890	1 419	616	688	285
TIS	*Branchiostegidae*	34	9 978	10 188	9 185	8 907	8 831	9 481	10 527
EMM	*Emmelichthys nitidus*	34	2 488	2 061	1 826	2 188	2 959	2 875	1 918
EMT	*Emmelichthyidae*	34	1 261	624	438	653	452	582	526
LOB	*Lobotes surinamensis*	34	-	-	-	-	-	1	1
SWH	*Paristiopterus labiosus*	34	21	19	27	75	6	9	3
EDR	*Pseudopentaceros richardsoni*	34	106	305	53	156	121	127	19
CTA	*Cheilodactylus bergi*	34	13 443	3 142	4 893	11 508	3 260	1 431	1 318
HAW	*Cheilodactylus variegatus*	34	93	283	462	140	288	380	306
TAK	*Nemadactylus macropterus*	34	5 149	4 366	5 441	5 239	5 589	5 739	6 129
MOW	*Nemadactylus spp*	34	1 420	1 277	1 620	1 287	1 412	1 240	1 285
TRU	*Latridae*	34	685	747	776	601	735	651	587
TOA	*Dissostichus mawsoni*	34	-	-	-	42	296	751	626
TOP	*Dissostichus eleginoides*	34	44 047	31 746	28 035	33 712	38 908	37 435	36 347
TOT	*Dissostichus spp*	34	-	-	-	-	-	-	572
NOT	*Patagonotothen brevicauda*	34	1	-	-	13	3	0	-
SSI	*Chaenocephalus aceratus*	34	-	-	-	-	1	0	1
ANI	*Champsocephalus gunnari*	34	3 946	5	217	73	339	4 195	1 890
SGI	*Pseudochaenichthys georgianus*	34	-	-	-	-	3	0	6
KIF	*Chionodraco rastrospinosus*	34	-	-	1	-	1	-	1
LIC	*Channichthys rhinoceratus*	34	-	1	7	6	3	2	1
WIC	*Chaenodraco wilsoni*	34	-	-	-	-	0	-	11
ICX	*Channichthyidae*	34	-	-	-	0	0	0	2
EPI	*Epigonus telescopus*	34	3 783	3 085	4 515	2 894	2 873	4 343	2 132
CDL	*Epigonus spp*	34	232	513	1 727	5 284	2 999	5 792	4 648
CAA	*Anarhichas lupus*	34	20 395	27 004	30 981	37 624	39 593	39 749	39 046
CAS	*Anarhichas minor*	34	700	1 109	1 180	1 599	1 545	1 896	3 257
CAT	*Anarhichas spp*	34	11 472	8 695	15 046	18 437	8 618	8 242	15 019
ELZ	*Lycodes spp*	34	45	18	-	2	1	28	48
UUC	*Uranoscopus scaber*	34	...	...	...	...	15	50	46
STZ	*Kathetostoma giganteum*	34	9 597	2 123	3 991	2 196	3 370	3 638	4 233
JAS	*Arctoscopus japonicus*	34	7 571	9 220	8 403	8 285	9 064	8 223	10 039
SNK	*Thyrsites atun*	34	38 456	32 829	39 627	45 914	41 693	40 234	41 421
WSM	*Thyrsitops lepidopoides*	34	66	1 232	309	285	136	-	-
LEC	*Lepidocybium flavobrunneum*	34	0	0	1	55	57	125	134
OIL	*Ruvettus pretiosus*	34	3 767	2 687	2 661	3 110	2 739	2 697	3 805
GEM	*Rexea solandri*	34	7 026	4 344	2 253	1 899	1 360	1 478	1 282
LHT	*Trichiurus lepturus*	34	1 244 673	1 283 139	1 206 353	1 436 303	1 416 408	1 477 722	1 471 657
SFS	*Lepidopus caudatus*	34	15 951	17 397	13 945	16 503	11 991	4 730	8 989
BSF	*Aphanopus carbo*	34	10 765	13 658	9 930	9 678	9 398	12 387	14 834
CUT	*Trichiuridae*	34	114 272	119 436	219 561	150 705	190 272	206 611	211 859
DRI	*Ariomma indica*	34	287	120	79	49	35	40	45
SEO	*Seriolella porosa*	34	...	...	...	...	...	3 542	3 990
SEM	*Seriolella brama*	34	1 081	1 760	4 108	3 101	3 881	4 259	4 101
SEP	*Seriolella punctata*	34	13 377	4 926	11 253	10 993	9 029	11 218	11 268
SEU	*Seriolella caerulea*	34	2 348	1 467	2 432	2 296	2 366	2 407	1 962
BSP	*Seriolella spp*	34	13 638	10 382	8 922	6 890	8 619	8 982	11 392
BWA	*Hyperoglyphe antarctica*	34	2 719	2 432	2 974	2 630	2 755	2 793	2 954
HGY	*Hyperoglyphe bythites*	34	...	...	...	...	2	7	5
BUP	*Psenopsis anomala*	34	14 191	13 290	11 777	13 734	10 871	10 721	10 942
CEN	*Centrolophidae*	34	654	459	654	597	1 495	987	719
REG	*Sebastes marinus*	34	12 073	11 034	11 413	7 968	7 136	13 642	13 598
WRO	*Sebastes entomelas*	34	...	...	7 751	4 897	4 243	3 604	2 609
YRO	*Sebastes flavidus*	34	...	...	2 744	3 602	3 468	3 170	2 077
OPP	*Sebastes alutus*	34	25 808	31 523	28 224	27 264	29 717	25 631	25 001
SBC	*Sebastes paucispinis*	34	...	...	724	596	197	27	33
REB	*Sebastes mentella*	34	7 753	4 842	4 457	4 607	3 574	6 745	9 046
REC	*Sebastes capensis*	34	1 353	1 759	1 630	1 095	1 214	1 389	1 673
SPG	*Sebastes pinniger*	34	...	...	1 262	1 313	772	60	49
SGO	*Sebastes goodei*	34	...	...	1 850	1 273	918	445	618
RMG	*Sebastes melanops*	34	...	...	...	...	...	109	148
RED	*Sebastes spp*	34	305 384	273 169	238 690	247 172	244 207	251 182	240 456
BRF	*Helicolenus dactylopterus*	34	7 495	3 807	6 082	5 721	6 406	5 731	2 275
SJU	*Sebastolobus alascanus*	34	...	...	...	...	...	331	268
SCO	*Scorpaenidae*	34	73 891	74 466	69 824	50 960	51 339	48 832	46 456
KUG	*Chelidonichthys kumu*	34	11 997	2 395	3 061	2 972	2 242	2 742	3 810
GUU	*Chelidonichthys lucerna*	34	-	5	4	5	22	1 184	15
GUR	*Chelidonichthys cuculus*	34	200	214	153	188	319	475	2 424
GUC	*Chelidonichthys capensis*	34	559	497	642	839	578	686	609
SRA	*Prionotus spp*	34	1 033	812	550	889	1 187	1 863	1 839
BEG	*Pterygotrigla polyommata*	34	121	58	136	92	94	119	148
JGU	*Pterygotrigla picta*	34	107	87	71	87	74	55	65
GUG	*Eutrigla gurnardus*	34	1 241	1 514	1 319	1 216	1 497	1 682	1 446
GUX	*Triglidae*	34	16 111	17 906	15 825	15 774	40 365	70 335	21 383
SAB	*Anoplopoma fimbria*	34	34 444	30 764	28 006	24 232	26 471	25 351	22 948
LUM	*Cyclopterus lumpus*	34	10 417	11 340	15 551	7 138	9 823	7 081	9 494
MON	*Lophius piscatorius*	34	61 482	69 104	72 930	60 156	54 680	55 405	54 777
ANK	*Lophius budegassa*	34	900	1 021	735	1 156	654	383	520
ANG	*Lophius americanus*	34	26 871	25 826	29 580	27 826	26 320	22 075	25 178
MVA	*Lophius vaillanti*	34	...	...	...	...	...	7	5
MVO	*Lophius vomerinus*	34	16 526	15 433	17 222	24 729	21 783	21 402	22 007
ANF	*Lophiidae*	34	8 539	12 014	5 444	4 512	4 200	6 867	9 821
DPX	*Perciformes*	34	37 687	47 836	53 690	57 070	54 273	45 062	49 189
HER	*Clupea harengus*	35	2 352 857	2 328 688	2 533 909	2 421 462	2 411 408	2 380 683	1 952 975

B-00 (b) Fish, crustaceans, molluscs, etc — Capture production by species items
Poissons, crustacés, mollusques, etc — Captures par catégories d'espèces
Peces, crustáceos, moluscos, etc — Capturas por partidas de especies

3-alpha code Code alpha-3 Código alfa-3	Scientific name Nom scientifique Nombre científico	Species group Groupe d'espèces Grupo de especies	1995 mt	1996 mt	1997 mt	1998 mt	1999 mt	2000 mt	2001 mt
HEP	*Clupea pallasii*	35	211 676	257 870	437 897	508 627	471 307	457 197	406 938
SAG	*Sardinella gibbosa*	35	161 096	157 104	156 914	174 691	162 710	172 219	176 610
IOS	*Sardinella longiceps*	35	197 601	223 355	289 142	247 065	208 898	402 936	437 328
SAA	*Sardinella aurita*	35	405 458	554 692	461 627	509 079	411 645	360 708	300 445
SAE	*Sardinella maderensis*	35	80 281	125 701	139 827	131 970	129 066	140 943	128 945
JSS	*Sardinella zunasi*	35	18 345	10 663	5 593	1 973	6 674	4 603	766
SAM	*Sardinella lemuru*	35	98 905	88 590	138 636	153 965	89 286	88 744	94 280
BSR	*Sardinella brasiliensis*	35	60 212	97 093	117 642	82 283	25 518	17 053	35 000
SIX	*Sardinella spp*	35	756 581	913 128	1 014 443	1 127 220	1 069 589	1 043 481	995 623
JAP	*Sardinops melanostictus*	35	733 427	430 837	417 939	295 788	515 477	305 767	339 377
CPI	*Sardinops caeruleus*	35	363 464	450 596	498 653	381 898	405 402	546 079	685 497
CHP	*Sardinops sagax*	35	1 503 131	1 493 936	722 807	937 269	442 790	338 131	135 712
PIA	*Sardinops ocellatus*	35	158 002	106 381	144 681	196 581	175 969	161 448	200 100
SRP	*Sardinops neopilchardus*	35	209	169	385	519	894	1 253	1 399
MHS	*Brevoortia aurea*	35	11 133	6 294	2 396	2 936	2 202	1 123	1 100
MHP	*Brevoortia pectinata*	35	424	511	1 112	519	291	337	103
MHA	*Brevoortia tyrannus*	35	365 736	304 665	322 239	276 230	208 000	207 122	261 407
MHG	*Brevoortia patronus*	35	472 039	491 612	597 565	497 461	694 242	591 434	528 500
RAS	*Dussumieria acuta*	35	39 904	41 347	38 045	42 960	48 265	36 754	37 644
RAL	*Dussumieria elopsoides*	35	154	48	19	17	10	15	17
BOA	*Ethmalosa fimbriata*	35	140 432	142 558	177 345	173 641	181 546	160 659	169 543
RRH	*Etrumeus teres*	35	54 189	85 563	58 362	58 569	36 837	29 407	32 861
WRR	*Etrumeus whiteheadi*	35	78 792	67 773	97 279	57 669	59 032	38 877	56 762
SAS	*Harengula spp*	35	1 117	839	1 023	708	946	923	911
THP	*Opisthonema libertate*	35	62 135	73 558	69 411	90 002	60 999	84 051	91 166
THA	*Opisthonema oglinum*	35	14 407	11 082	18 912	22 674	26 418	27 257	19 062
SRH	*Spratelloides gracilis*	35	977	657	785	896	561	669	505
PIL	*Sardina pilchardus*	35	1 208 681	996 334	998 734	949 776	907 591	947 147	1 126 832
SPR	*Sprattus sprattus*	35	601 998	671 616	700 239	696 228	684 189	658 484	647 417
FAS	*Sprattus fuegensis*	35	0	0	0	29	17	97	4
MES	*Ethmidium maculatum*	35	5 487	9 792	13 241	40 845	29 436	23 991	14 016
HES	*Herklotsichthys quadrimaculat.*	35	185	140	120	22	20	20	...
CKI	*Strangomera bentincki*	35	126 715	446 669	441 154	317 564	782 142	722 522	324 617
ANE	*Engraulis encrasicolus*	35	618 646	527 610	502 390	506 844	630 845	626 371	660 552
JAN	*Engraulis japonicus*	35	972 008	1 254 487	1 666 503	2 093 888	1 820 259	1 725 685	1 836 502
ANA	*Engraulis anchoita*	35	24 498	21 023	25 211	13 417	13 025	12 163	11 602
NPA	*Engraulis mordax*	35	27 150	14 103	7 925	2 335	11 137	19 460	19 677
VET	*Engraulis ringens*	35	8 644 576	8 863 714	7 685 098	1 729 064	8 723 265	11 276 357	7 213 077
ANC	*Engraulis capensis*	35	218 331	41 792	62 640	110 296	180 954	267 986	289 323
AVA	*Cetengraulis edentulus*	35	41	8	0	119	0	0	2
VEP	*Cetengraulis mysticetus*	35	161 984	113 020	195 630	180 705	70 358	120 943	198 636
ENP	*Anchoa hepsetus*	35	...	...	1	16	8	0	0
STO	*Stolephorus spp*	35	260 917	267 262	296 356	279 854	274 846	278 834	280 342
ANX	*Engraulidae*	35	444 764	328 256	288 793	964 051	235 545	231 949	362 983
DOB	*Chirocentrus dorab*	35	18 279	11 873	12 392	14 432	17 326	17 629	17 365
DOS	*Chirocentrus spp*	35	51 021	46 855	56 304	45 340	50 265	49 883	53 514
CLU	*Clupeoidei*	35	277 805	307 553	315 971	326 980	357 893	317 129	313 604
BON	*Sarda sarda*	36	20 976	26 390	30 999	47 445	37 629	28 960	30 704
BIP	*Sarda orientalis*	36	788	370	498	162	134	95	287
BEP	*Sarda chiliensis*	36	35 175	23 927	18 929	7 239	3 187	972	1 471
BOP	*Orcynopsis unicolor*	36	640	2 136	476	224	861	1 199	1 053
WAH	*Acanthocybium solandri*	36	1 924	2 289	2 931	2 627	3 183	2 698	3 131
COM	*Scomberomorus commerson*	36	165 451	166 537	176 096	184 797	190 489	208 567	213 244
GUT	*Scomberomorus guttatus*	36	44 436	41 883	41 581	50 829	55 241	61 297	76 701
STS	*Scomberomorus lineolatus*	36	87	96	901	107	135	147	381
KGM	*Scomberomorus cavalla*	36	11 495	14 154	15 551	12 396	14 326	11 625	11 454
SSM	*Scomberomorus maculatus*	36	9 680	12 594	9 084	8 840	9 744	7 367	7 236
CER	*Scomberomorus regalis*	36	429	307	481	441	230	190	147
SIE	*Scomberomorus sierra*	36	6 646	7 154	6 813	6 651	8 657	9 086	7 640
MAW	*Scomberomorus tritor*	36	2 051	1 296	1 377	1 456	703	834	963
NPH	*Scomberomorus niphonius*	36	259 301	301 356	365 585	551 780	595 103	539 094	522 756
BRS	*Scomberomorus brasiliensis*	36	7 849	8 435	8 065	7 923	5 674	6 643	5 849
KGX	*Scomberomorus spp*	36	44 259	41 659	42 974	43 728	43 714	41 329	39 053
FRI	*Auxis thazard*	36	717	1 134	791	487	486	505	872
FRZ	*Auxis thazard, A.rochei*	36	179 026	179 248	209 273	193 979	251 742	215 065	207 364
LTA	*Euthynnus alletteratus*	36	7 429	9 948	11 495	11 406	10 350	11 222	8 453
BKJ	*Euthynnus lineatus*	36	261	544	134	611	140	290	1 810
KAW	*Euthynnus affinis*	36	332 806	370 568	396 333	415 693	439 527	458 957	423 490
SKJ	*Katsuwonus pelamis*	36	1 648 300	1 576 637	1 608 811	1 888 792	1 988 826	1 955 430	1 836 438
BFT	*Thunnus thynnus*	36	49 180	52 581	48 710	41 416	35 172	35 959	35 682
PBF	*Thunnus orientalis*	36	7 251	16 098	10 954	7 593	16 845	16 185	9 143
LOT	*Thunnus tonggol*	36	121 653	104 316	99 962	102 847	116 784	137 114	121 482
BLF	*Thunnus atlanticus*	36	3 534	3 572	2 884	2 898	2 990	2 505	4 385
ALB	*Thunnus alalunga*	36	191 526	194 327	217 621	228 353	243 609	204 753	221 473
SBF	*Thunnus maccoyii*	36	13 192	14 987	13 462	16 426	17 606	14 173	15 543
YFT	*Thunnus albacares*	36	991 299	951 616	1 087 363	1 071 243	1 079 917	1 035 107	1 202 312
BET	*Thunnus obesus*	36	368 156	357 353	380 958	392 586	400 394	410 595	372 110
TUN	*Thunnini*	36	76	297	1 330	112	165	162	175
SFA	*Istiophorus platypterus*	36	12 189	13 144	15 557	16 858	16 339	16 521	15 641
SAI	*Istiophorus albicans*	36	1 967	2 263	1 819	1 814	1 901	2 361	1 433
BLZ	*Makaira mazara*	36	24 960	22 975	21 206	25 255	23 958	24 887	22 673
BUM	*Makaira nigricans*	36	3 550	4 156	3 604	3 637	3 694	3 076	2 515
BLM	*Makaira indica*	36	2 135	2 062	1 771	2 794	2 509	2 328	2 266
MLS	*Tetrapturus audax*	36	15 086	12 965	12 021	14 294	10 211	9 253	7 286
WHM	*Tetrapturus albidus*	36	1 438	1 164	1 026	1 021	957	1 175	398
SSP	*Tetrapturus angustirostris*	36	8	44	20	5	8	5	0
SPF	*Tetrapturus pfluegeri*	36	82	61	69	90	133	136	60

B-00 (b)

Fish, crustaceans, molluscs, etc — **Capture production by species items**
Poissons, crustacés, mollusques, etc — **Captures par catégories d'espèces**
Peces, crustáceos, moluscos, etc — **Capturas por partidas de especies**

3-alpha code Code alpha-3 Código alfa-3	Scientific name Nom scientifique Nombre científico	Species group Groupe d'espèces Grupo de especies	1995 mt	1996 mt	1997 mt	1998 mt	1999 mt	2000 mt	2001 mt
BIL	*Istiophoridae*	36	25 156	29 625	29 793	37 603	38 201	29 365	26 531
SWO	*Xiphias gladius*	36	98 076	88 641	100 905	106 252	97 320	106 418	92 152
TUX	*Scombroidei*	36	174 881	186 757	190 676	298 142	206 709	251 872	267 483
CAP	*Mallotus villosus*	37	748 800	1 527 422	1 603 338	982 312	904 045	1 488 629	1 670 906
GAR	*Belone belone*	37	2 997	3 027	3 453	3 203	2 771	2 147	2 507
NED	*Tylosurus spp*	37	41 264	36 501	40 184	39 549	40 129	43 204	45 303
BEN	*Belonidae*	37	816	1 228	431	397	481	727	742
SAU	*Scomberesox saurus*	37	534	484	814	1 368	1 154	481	5 171
SAP	*Cololabis saira*	37	350 287	276 111	388 643	180 973	187 898	306 069	376 173
HAJ	*Hyporhamphus sajori*	37	1 531	990	1 193	1 160	913	956	613
BAL	*Hemiramphus brasiliensis*	37	631	494	442	1 500	864	1 036	1 021
HAX	*Hemiramphus spp*	37	15 263	11 885	10 305	9 468	9 644	12 007	10 726
JFL	*Cypselurus agoo*	37	7 881	8 501	7 486	8 933	6 738	9 615	8 286
FLY	*Exocoetidae*	37	73 806	42 392	55 480	69 731	363 026	104 722	71 999
LAG	*Lampris guttatus*	37	223	172	126	369	542	466	538
TRP	*Trachipterus spp*	37	127	49	60	74	94	87	128
REL	*Regalecus glesne*	37	5	10	64	60	34	20	1
ATB	*Atherina boyeri*	37	15	39	123	375	381	434	423
SSA	*Menidia menidia*	37	236	170	259	255	583	319	661
SIL	*Atherinidae*	37	14 646	16 672	17 646	14 315	22 512	24 229	23 769
TRF	*Lactarius lactarius*	37	7 533	7 345	8 709	10 321	7 743	6 736	7 994
BLU	*Pomatomus saltatrix*	37	22 436	23 265	19 105	17 326	13 950	16 318	26 632
CBA	*Rachycentron canadum*	37	5 732	5 072	5 583	5 549	6 634	8 557	8 997
HOM	*Trachurus trachurus*	37	559 867	474 625	455 109	350 298	321 195	231 059	258 618
JJM	*Trachurus japonicus*	37	330 104	349 166	350 619	340 565	227 290	271 501	235 874
CJM	*Trachurus murphyi*	37	4 955 186	4 378 843	3 597 117	2 025 758	1 423 447	1 540 494	2 508 834
PJM	*Trachurus symmetricus*	37	1 875	2 176	1 160	1 793	1 126	1 316	3 839
HMM	*Trachurus mediterraneus*	37	20 162	21 002	17 065	15 122	12 898	19 111	19 308
RSC	*Trachurus lathami*	37	732	976	601	328	495	107	40
HMC	*Trachurus capensis*	37	506 180	470 372	407 482	481 500	400 279	423 607	354 046
HMZ	*Trachurus trecae*	37	83 450	86 118	94 493	57 816	92 476	70 651	49 307
HMG	*Trachurus declivis*	37	10 592	15 441	10 656	9 379	15 545	12 239	7 634
JAX	*Trachurus spp*	37	322 589	247 950	236 646	321 010	314 148	367 263	334 697
TRZ	*Pseudocaranx dentex*	37	3 910	3 833	3 813	4 166	4 671	3 984	3 495
RSA	*Decapterus maruadsi*	37	75 735	62 378	69 764	71 168	67 689	40 803	47 460
RUS	*Decapterus russelli*	37	133 544	145 320	150 027	145 747	162 437	171 546	160 899
SDX	*Decapterus spp*	37	1 049 724	1 132 890	1 042 486	1 079 873	1 032 812	1 034 891	1 107 187
RUB	*Caranx crysos*	37	1 074	816	923	851	823	830	845
CVJ	*Caranx hippos*	37	4 610	3 059	2 529	3 432	4 591	2 292	3 564
CXR	*Caranx ruber*	37	-	-	-	-	3	5	7
HMY	*Caranx rhonchus*	37	3 483	3 301	3 344	2 864	2 710	3 935	281
TRE	*Caranx spp*	37	118 560	168 746	169 096	169 726	184 139	160 941	165 076
MOA	*Selene setapinnis*	37	5 012	3 445	4 024	3 850	3 043	2 362	4 813
LUK	*Selene dorsalis*	37	1 073	1 015	832	755	772	1 236	1 936
POO	*Trachinotus blochii*	37	-	-	-	4	0	31	-
POM	*Trachinotus carolinus*	37	54	44	297	321	207	242	181
POX	*Trachinotus spp*	37	11 167	9 238	5 218	5 366	6 499	2 372	1 789
AMB	*Seriola dumerili*	37	713	1 057	724	1 315	1 847	2 004	1 886
AMJ	*Seriola quinqueradiata*	37	7 564	246	...	...	...	...	...
YTC	*Seriola lalandi*	37	1 437	1 042	961	1 517	885	971	954
AMX	*Seriola spp*	37	69 595	60 974	63 417	80 195	69 133	97 491	103 559
LEE	*Lichia amia*	37	4 915	2 795	3 577	3 793	4 788	1 777	1 222
ALA	*Alectis alexandrinus*	37	124	160	742	1 023	563	502	862
POB	*Parastromateus niger*	37	45 018	44 676	48 464	44 524	47 590	48 135	49 578
RRU	*Elagatis bipinnulata*	37	13 450	12 402	13 147	18 940	15 874	14 923	16 233
GLT	*Gnathanodon speciosus*	37	434	441	471	536	489	1 125	1 144
HAS	*Megalaspis cordyla*	37	64 260	59 552	70 059	77 525	80 113	79 349	78 778
QUE	*Scomberoides spp*	37	24 686	25 067	22 364	25 282	27 630	24 588	24 358
BUA	*Chloroscombrus chrysurus*	37	5 597	4 534	6 186	12 485	12 250	9 810	13 501
HSO	*Chloroscombrus orqueta*	37	17 999	1 706	952	565	1 409	...	1 008
PAO	*Parona signata*	37	1 869	2 364	2 236	2 209	2 070	1 850	1 478
BIS	*Selar crumenophthalmus*	37	46 476	72 630	80 881	93 486	99 057	102 246	110 034
TRY	*Selaroides leptolepis*	37	26 737	31 031	35 139	35 852	44 537	46 888	42 488
RNJ	*Seriolina nigrofasciata*	37	6 930	7 269	7 096	6 042	6 045	5 962	5 926
CGX	*Carangidae*	37	289 958	293 721	296 228	293 154	301 845	282 199	278 915
POA	*Brama brama*	37	11 730	12 631	10 439	10 061	10 192	10 037	17 727
BRZ	*Bramidae*	37	-	-	-	-	-	2	5
DOL	*Coryphaena hippurus*	37	37 446	24 498	30 304	44 144	31 231	38 086	42 261
ECN	*Echeneidae*	37	-	-	-	3	1	12	...
MAS	*Scomber japonicus*	37	1 575 108	2 177 781	2 427 389	1 924 691	1 950 177	1 474 776	1 798 704
MAC	*Scomber scombrus*	37	794 335	559 259	555 621	666 965	618 014	690 063	710 411
MAA	*Scomber australasicus*	37	7 967	2 994	8 777	7 260	15 874	12 108	11 801
MAZ	*Scomber spp*	37	10 621	13 169	11 878	10 879	8 964	8 572	9 148
RAB	*Rastrelliger brachysoma*	37	26 200	25 224	22 978	23 350	25 713	26 771	28 289
RAG	*Rastrelliger kanagurta*	37	341 201	399 938	303 946	294 606	296 904	196 837	183 372
RAX	*Rastrelliger spp*	37	479 891	425 521	427 233	458 361	477 519	467 476	472 636
MAX	*Scombridae*	37	44 975	51 342	60 706	63 435	49 331	44 438	53 866
BLB	*Stromateus fiatola*	37	466	240	323	292	636	320	4 432
SIP	*Pampus argenteus*	37	38 753	37 452	31 369	30 252	32 246	32 508	36 074
XPO	*Pampus spp*	37	209 031	220 364	242 547	303 024	337 919	338 848	352 493
HVF	*Peprilus alepidotus*	37	25	14	15	13	16	23	18
BTG	*Peprilus spp*	37	1 881	1 889	568	1 175	1 004	809	742
BUX	*Stromateidae*	37	83 171	61 130	61 324	59 923	59 896	45 523	45 549
BAR	*Sphyraena spp*	37	79 121	83 739	84 885	97 486	109 083	92 030	96 733
MOX	*Mola mola*	37	...	...	...	...	...	...	2
PPX	*Perciformes*	37	72 548	70 923	69 603	80 012	93 710	75 446	90 051
NTC	*Notorynchus cepedianus*	38	...	...	...	2	3	4	5
BSK	*Cetorhinus maximus*	38	123	1 984	1 169	192	210	389	287

B-00 (b)

Fish, crustaceans, molluscs, etc — Capture production by species items
Poissons, crustacés, mollusques, etc — Captures par catégories d'espèces
Peces, crustáceos, moluscos, etc — Capturas por partidas de especies

3-alpha code Code alpha-3 Código alfa-3	Scientific name Nom scientifique Nombre científico	Species group Groupe d'espèces Grupo de especies	1995 mt	1996 mt	1997 mt	1998 mt	1999 mt	2000 mt	2001 mt
CCT	*Carcharias taurus*	38	-	-	-	-	-	1	-
ALV	*Alopias vulpinus*	38	29	20	67	393	495	650	614
BTH	*Alopias superciliosus*	38	...	...	149	125	5	5	2
THR	*Alopias spp*	38	...	...	34	55	66	...	...
SMA	*Isurus oxyrinchus*	38	514	455	2 108	2 286	1 159	2 132	2 280
LMA	*Isurus paucus*	38	0	-	1	1	-	4	3
MAK	*Isurus spp*	38	12	...	92	38	...	116	47
POR	*Lamna nasus*	38	2 130	1 496	1 917	2 221	2 690	2 865	2 134
WSH	*Carcharodon carcharias*	38	-	-	-	-	-	2	0
GNC	*Ginglymostoma cirratum*	38	214	-	-	-	-	407	89
SYC	*Scyliorhinus canicula*	38	5 363	5 144	5 613	5 740	5 818	6 182	7 072
SYT	*Scyliorhinus stellaris*	38	243	306	378	258	274	274	264
SCL	*Scyliorhinus spp*	38	56	54	78	51	275	525	508
BSH	*Prionace glauca*	38	437	1 281	6 811	3 496	4 615	8 401	10 087
CCP	*Carcharhinus plumbeus*	38	1	-	-	-	-	41	24
CCL	*Carcharhinus limbatus*	38	...	3	9	10	11	601	521
DUS	*Carcharhinus obscurus*	38	0	-	7	0	...	80	0
FAL	*Carcharhinus falciformis*	38	21 418	21 502	15 281	20 875	20 880	16 309	14 700
BRO	*Carcharhinus brachyurus*	38	...	...	...	15	14	25	38
TIG	*Galeocerdo cuvier*	38	-	-	-	-	-	-	1
RSK	*Carcharhinidae*	38	51 002	52 477	44 502	47 760	41 816	38 195	35 607
SPZ	*Sphyrna zygaena*	38	12	10	223	109	19	35	27
SPL	*Sphyrna lewini*	38	12	37	180	10	40	48	50
SPY	*Sphyrnidae*	38	69	...	998	1 028	147	1 457	1 451
CTI	*Mustelus canis*	38	0	...	...	...	...	334	321
CTK	*Mustelus henlei*	38	...	...	...	3	5	3	4
MTL	*Mustelus lenticulatus*	38	2 787	1 350	3 464	1 707	1 662	1 643	1 563
SDP	*Mustelus schmitti*	38	11 343	10 456	10 130	13 422	12 274	8 156	9 933
SMD	*Mustelus mustelus*	38	-	-	-	-	-	15	-
SDV	*Mustelus spp*	38	17 322	15 206	11 533	15 368	10 536	12 437	12 837
GAG	*Galeorhinus galeus*	38	4 189	3 594	3 478	3 654	4 259	3 826	3 733
GSK	*Somniosus microcephalus*	38	55	61	73	87	51	45	58
SON	*Somniosus pacificus*	38	-	-	-	-	1	-	-
DGS	*Squalus acanthias*	38	26 046	23 622	23 980	20 382	25 569	31 730	28 886
GUP	*Centrophorus granulosus*	38	...	...	...	...	73	54	93
GUQ	*Centrophorus squamosus*	38	51	53	58	133	452	506	538
SHL	*Etmopterus spp*	38	3	0	2	...	573	...	4
DCA	*Deania calcea*	38	...	...	...	36	17	46	333
CYO	*Centroscymnus coelolepis*	38	60	336	280	232	717	1 154	2 319
CYP	*Centroscymnus crepidater*	38	-	-	-	3	-	-	-
SCK	*Dalatias licha*	38	303	175	352	434	373	628	564
CFB	*Centroscyllium fabricii*	38	1	4	0	-	-	271	271
DGX	*Squalidae*	38	31 549	37 123	30 262	29 031	22 393	9 520	8 759
DGH	*Squalidae, Scyliorhinidae*	38	2 117	2 022	2 083	2 011	2 113	3 032	2 700
AGN	*Squatina squatina*	38	35	18	34	44	25	20	22
SUG	*Squatina argentina*	38	3 802	4 281	4 410	4 311	3 368	3 123	3 339
ASK	*Squatinidae*	38	435	2 041	271	272	363	497	547
OXY	*Oxynotus centrina*	38	...	...	...	...	81	33	63
SHX	*Squaliformes*	38	1 983	3 603	1 392	987	1 733	722	886
GUD	*Rhinobatos percellens*	38	162	404	...	...	...	...	...
GUF	*Rhinobatos planiceps*	38	121	460	333	344	95	2 624	1 060
GTF	*Rhinobatidae*	38	1 288	1 535	1 550	1 882	1 955	4 230	3 810
SAW	*Pristidae*	38	23	...	48	...	41	42	...
RJB	*Raja batis*	38	530	508	441	411	558	794	817
RJC	*Raja clavata*	38	1 749	1 773	1 588	1 367	1 366	1 275	1 296
RJR	*Raja radiata*	38	1 749	1 493	1 431	1 252	996	1 076	1 211
RJM	*Raja montagui*	38	925	977	1 163	1 179	1 260	1 341	1 563
RJI	*Raja circularis*	38	431	438	438	410	435	369	330
RJF	*Raja fullonica*	38	75	65	55	50	86	65	105
RJE	*Raja microocellata*	38	-	-	-	1	11	-	-
RJN	*Raja naevus*	38	3 762	4 077	4 721	4 015	3 638	3 064	2 885
RJO	*Raja oxyrinchus*	38	359	346	311	327	194	140	89
SRR	*Raja georgiana*	38	-	-	-	...	11	36	7
SKA	*Raja spp*	38	40 420	47 768	55 330	58 826	63 094	64 879	60 472
BEA	*Bathyraja eatonii*	38	-	-	-	...	1	5	0
BHY	*Bathyraja spp*	38	-	-	-	-	1	-	-
RAJ	*Rajidae*	38	-	-	-	...	6	-	-
WST	*Dasyatis akajei*	38	3 985	4 029	3 959	4 329	4 407	5 388	4 312
JDP	*Dasyatis pastinaca*	38	...	...	...	-	-	4	11
STI	*Dasyatis spp*	38	-	1	2	5	6	10	7
EAG	*Myliobatidae*	38	2	0	1	1	15	12	16
MAN	*Mobulidae*	38	...	...	...	342	802	931	106
SRX	*Rajiformes*	38	153 475	160 642	171 419	166 754	178 598	176 613	175 242
TOE	*Torpedo spp*	38	20	16	18	19	34	32	43
CMO	*Chimaera monstrosa*	38	106	21	15	30	12	14	123
CYV	*Hydrolagus novaezealandiae*	38	1 593	1 614	2 064	1 956	1 975	1 819	1 572
HYD	*Hydrolagus spp*	38	...	...	0	36	491	1 548	3 019
RCT	*Rhinochimaera atlantica*	38	-	-	-	-	-	-	2
CHB	*Callorhinchus milii*	38	769	595	913	951	1 260	1 228	1 189
CHM	*Callorhinchus capensis*	38	386	366	484	482	356	380	405
ELF	*Callorhinchus spp*	38	1 841	2 265	2 151	3 186	2 609	1 981	1 125
HOL	*Chimaeriformes*	38	7	50	5	5	21	40	76
SKH	*Selachimorpha(Pleurotremata)*	38	22 679	7 694	36 309	25 125	26 390	33 412	33 220
SKX	*Elasmobranchii*	38	343 133	389 235	373 789	370 144	378 730	396 821	377 075
GRO	*Osteichthyes*	39	22 394	20 101	10 220	11 240	7 665	10 257	4 708
PEL	*Osteichthyes*	39	18 840	25 717	8 375	15 791	4 483	2 979	325
FIN	*Osteichthyes*	39	58 855	41 461	49 272	55 357	40 669	62 311	33 694
MZZ	*Osteichthyes*	39	9 946 159	10 343 681	10 080 035	10 281 966	10 274 265	10 121 431	9 932 493

B-00 (b)

Fish, crustaceans, molluscs, etc — **Capture production by species items**
Poissons, crustacés, mollusques, etc — **Captures par catégories d'espèces**
Peces, crustáceos, moluscos, etc — **Capturas por partidas de especies**

3-alpha code Code alpha-3 Código alfa-3	Scientific name Nom scientifique Nombre científico	Species group Groupe d'espèces Grupo de especies	1995 mt	1996 mt	1997 mt	1998 mt	1999 mt	2000 mt	2001 mt
PRF	*Macrobrachium rosenbergii*	41	5 837	7 755	6 769	5 436	6 072	5 225	5 539
PPF	*Macrobrachium spp*	41	5 826	7 093	5 872	4 875	7 528	6 020	5 670
PPZ	*Palaemonidae*	41	14 491	13 076	12 324	10 645	11 150	12 226	12 469
AAS	*Astacus astacus*	41	...	...	10	10	10	...	...
PCL	*Pacifastacus leniusculus*	41	-	-	-	-	10	81	80
AUP	*Austropotamobius pallipes*	41	0	0	0	0	0	0	0
RCW	*Procambarus clarkii*	41	2 773	2 513	2 524	2 519	2 521	2 522	2 503
AYA	*Euastacus armatus*	41	-	-	-	-	66	24	22
CJF	*Parastacidae*	41	20	6	6	14	6	0	1
AYS	*Astacidae, Cambaridae*	41	7 991	6 971	12 130	11 909	7 084	2 214	6 507
FCX	*Crustacea*	41	278 823	373 176	488 064	610 448	464 083	539 621	597 742
CRK	*Cancer irroratus*	42	6 259	4 341	7 530	7 696	7 033	9 516	8 294
DUN	*Cancer magister*	42	26 282	34 503	21 251	18 487	19 024	19 941	22 210
CRE	*Cancer pagurus*	42	33 382	29 185	38 011	40 635	39 341	43 039	44 143
CRJ	*Cancer borealis*	42	332	334	745	1 255	1 549	1 114	1 245
ROC	*Cancer productus*	42	...	...	...	574	359	494	537
STC	*Menippe mercenaria*	42	2 569	3 675	3 355	3 347	3 194	4 023	3 931
SCD	*Portunus pelagicus*	42	115 576	139 950	138 733	138 123	147 161	154 647	162 387
GAZ	*Portunus trituberculatus*	42	265 295	303 170	252 502	283 971	284 851	351 051	350 168
CRS	*Portunus spp*	42	5 098	2 761	4 790	4 262	8 993	9 348	9 525
CRB	*Callinectes sapidus*	42	103 599	115 559	121 643	115 996	105 238	92 645	79 797
CRZ	*Callinectes danae*	42	2 062	2 020	2 600	3 014	1 626	1 597	1 580
CRG	*Carcinus maenas*	42	1 113	837	951	995	1 066	889	1 144
CMR	*Carcinus aestuarii*	42	16	65	44	66	44	30	37
MUD	*Scylla serrata*	42	20 260	17 073	14 012	13 579	16 293	16 359	16 835
MXT	*Mithrax armatus*	42	-	-	-	114	110	54	35
SCR	*Maja squinado*	42	6 572	5 705	6 193	5 919	5 828	6 376	7 518
CRQ	*Chionoecetes opilio*	42	66 372	66 831	74 341	77 517	98 633	104 252	110 044
PCR	*Chionoecetes spp*	42	45 748	57 001	85 914	139 755	110 115	60 190	55 999
HBZ	*Erimacrus isenbeckii*	42	...	789	612	409	440	198	162
CRR	*Geryon quinquedens*	42	1 304	1 978	3 528	2 743	3 382	8 391	6 720
GER	*Geryon spp*	42	7 587	7 370	4 884	5 734	5 829	7 490	5 563
CRA	*Brachyura*	42	242 219	206 188	208 639	210 786	198 645	196 332	208 497
LOJ	*Panulirus longipes*	43	1 154	1 106	1 082	1 098	1 166	1 716	1 924
SLC	*Panulirus argus*	43	40 095	37 498	36 674	33 188	36 989	39 908	34 098
LOA	*Panulirus cygnus*	43	10 886	9 902	9 896	10 400	17 720	19 380	16 017
NUG	*Panulirus gracilis*	43	235	207	221	791	678	805	765
SLV	*Panulirus spp*	43	9 479	9 490	11 300	10 064	10 296	11 280	12 237
LBC	*Jasus lalandii*	43	2 180	1 767	1 879	2 076	2 097	2 058	1 974
LOF	*Jasus frontalis*	43	29	36	32	21	22	17	21
LOG	*Jasus verreauxi*	43	4 706	5 017	5 046	4 739	2 049	2 476	2 221
LBT	*Jasus tristani*	43	344	327	321	376	336	316	425
LOR	*Jasus edwardsii*	43	3 568	3 121	5 009	2 707	2 818	2 789	2 551
JSP	*Jasus paulensis*	43	439	357	295	308	345	192	183
PSL	*Palinurus mauritanicus*	43	3	1	0	25	11	9	5
SLO	*Palinurus elephas*	43	323	440	477	271	230	200	258
SLN	*Palinurus delagoae*	43	13	10	10	6	7	8	10
SLS	*Palinurus gilchristi*	43	966	918	892	864	429	305	1 053
CRW	*Palinurus spp*	43	2 236	1 967	9 143	3 984	1 876	2 586	2 768
VLO	*Palinuridae*	43	248	332	233	239	204	228	199
IBC	*Ibacus ciliatus*	43	1 224	1 115	642	696	676	1 600	1 607
THQ	*Thenus orientalis*	43	335	299	177	533	37	36	36
LOS	*Scyllaridae*	43	3 036	3 814	3 753	4 100	1 888	1 847	1 873
NEM	*Metanephrops mozambicus*	43	179	132	156	192	152	180	141
MEC	*Metanephrops challengeri*	43	1 078	670	1 093	989	925	1 034	1 093
NEP	*Nephrops norvegicus*	43	63 774	56 588	61 596	57 379	61 770	51 872	56 365
LBA	*Homarus americanus*	43	70 631	71 866	78 146	77 155	83 105	83 062	83 803
LBE	*Homarus gammarus*	43	2 981	2 590	3 219	2 933	3 285	2 527	2 781
UOP	*Upogebia pugettensis*	43	...	...	...	...	...	10	4
CZP	*Callianassa spp*	43	-	-	-	-	25	17	14
LOX	*Reptantia*	43	9	20	2 025	1 031	290	585	710
LQL	*Pleuroncodes planipes*	44	356	164	-	-	-	-	-
PQG	*Pleuroncodes monodon*	44	4 938	7 726	8 939	12 602	12 710	11 129	1 754
CZJ	*Cervimunida johni*	44	5 743	6 402	10 322	9 426	7 273	5 069	2 178
LOQ	*Galatheidae*	44	92	105	106	81	130	352	84
KCD	*Paralithodes camtschaticus*	44	55 018	34 370	23 333	32 718	37 413	28 924	16 794
KCI	*Paralithodes platypus*	44	...	8 762	10 268	4 508	5 455	5 233	4 500
KCY	*Paralithodes brevipes*	44	...	204	418	194	256	347	254
KCS	*Paralithodes spp*	44	6 916	9 848	8 331	11 073	7 792	6 938	7 464
KCR	*Lithodes antarcticus*	44	2 286	1 959	2 573	3 222	2 425	3 004	2 995
KAQ	*Lithodes aequispina*	44	...	4 666	4 917	3 897	2 746	1 797	2 245
PAG	*Paralomis granulosa*	44	1 317	1 274	1 478	1 502	1 439	5 205	6 687
KCV	*Paralomis spinosissima*	44	-	497	-	-	-	0	3
KCF	*Paralomis formosa*	44	-	-	-	-	-	3	11
KCX	*Lithodidae*	44	-	-	-	0	-	3	0
ABS	*Penaeus aztecus*	45	57 130	55 369	47 836	50 722	61 206	63 817	68 869
PBA	*Penaeus merguiensis*	45	71 155	72 746	71 809	79 674	82 053	82 956	90 502
YPS	*Penaeus californiensis*	45	356	345	186	321	250	158	242
APS	*Penaeus duorarum*	45	11 121	15 510	10 875	12 452	8 868	8 529	10 084
KUP	*Penaeus japonicus*	45	7 027	5 606	6 995	5 113	6 096	10 193	5 963
PNV	*Penaeus vannamei*	45	7 258	5 735	5 046	4 277	2 182	1 619	2 710
GIT	*Penaeus monodon*	45	207 097	177 054	178 139	240 301	264 218	251 076	249 366
FLP	*Penaeus chinensis*	45	44 449	56 534	71 317	79 595	70 725	85 545	97 583
TGS	*Penaeus kerathurus*	45	4 880	4 542	6 078	5 498	5 966	7 700	4 952
PNB	*Penaeus brasiliensis*	45	6 565	8 743	10 758	7 796	9 092	10 728	10 600

B-00 (b)

Fish, crustaceans, molluscs, etc — **Capture production by species items**
Poissons, crustacés, mollusques, etc — **Captures par catégories d'espèces**
Peces, crustáceos, moluscos, etc — **Capturas por partidas de especies**

3-alpha code Code alpha-3 Código alfa-3	Scientific name Nom scientifique Nombre científico	Species group Groupe d'espèces Grupo de especies	1995 mt	1996 mt	1997 mt	1998 mt	1999 mt	2000 mt	2001 mt
TIP	*Penaeus semisulcatus*	45	2 620	3 469	3 517	3 019	2 821	2 782	2 765
PST	*Penaeus setiferus*	45	39 959	28 808	32 841	39 799	44 633	52 593	40 724
CSP	*Penaeus brevirostris*	45	3 676	2 136	2 776	2 090	2 462	2 604	2 729
WKP	*Penaeus latisulcatus*	45	3 072	3 350	3 629	3 386	4 145	4 099	4 055
WWP	*Penaeus occidentalis*	45	686	1 158	1 505	1 261	2 706	2 020	1 259
REP	*Penaeus penicillatus*	45	3 564	5 020	2 473	647	316	308	312
SOP	*Penaeus notialis*	45	21 484	18 371	22 233	19 172	32 721	25 250	23 659
PPS	*Penaeus paulensis*	45	0	-	177	13	12	56	23
PEN	*Penaeus spp*	45	292 552	274 833	294 475	252 852	224 500	231 954	230 383
MPN	*Metapenaeus monoceros*	45	-	-	-	-	-	-	757
ENS	*Metapenaeus endeavouri*	45	2 182	2 400	2 339	2 930	2 691	2 299	2 173
SHI	*Metapenaeus joyneri*	45	2 168	2 211	1 976	3 651	3 633	2 621	2 385
MET	*Metapenaeus spp*	45	51 536	48 322	57 307	63 831	56 911	61 805	66 790
DPS	*Parapenaeus longirostris*	45	15 829	16 198	19 262	22 789	19 076	22 903	25 459
NPP	*Parapenaeopsis spp*	45	12 919	14 047	16 722	14 689	12 889	11 945	11 576
BOB	*Xiphopenaeus kroyeri*	45	18 802	29 074	36 721	30 358	28 612	36 730	44 151
TIT	*Xiphopenaeus riveti*	45	1 037	2 683	2 830	1 970	1 752	2 000	2 341
TRV	*Trachypenaeus curvirostris*	45	154 623	167 723	180 255	179 544	403 027	312 968	247 940
ASH	*Artemesia longinaris*	45	250	263	166	146	37	37	...
BOS	*Xiphopenaeus,Trachypenaeus spp*	45	9 898	12 156	7 354	9 881	6 457	4 720	3 755
SSH	*Plesiopenaeus edwardsianus*	45	400	352	479	337	605	54	39
ARA	*Aristeus antennatus*	45	1 333	1 428	1 359	1 672	1 526	1 957	2 541
ARV	*Aristeus varidens*	45	1 570	1 578	1 323	2 808	1 827	3 024	3 405
ARI	*Aristeidae*	45	2 551	2 258	2 406	1 231	2 128	4 463	1 833
PYX	*Pandalus hypsinotus*	45	...	...	467	388	288	275	359
PRA	*Pandalus borealis*	45	275 601	287 830	280 316	317 693	338 393	370 434	345 681
AES	*Pandalus montagui*	45	...	...	...	...	...	697	609
DUJ	*Pandalus goniurus*	45	...	...	-	1 199	330	1 200	247
DUK	*Pandalus kessleri*	45	...	...	123	55	97	75	94
PAN	*Pandalus spp*	45	26 813	28 015	28 649	30 463	30 649	35 640	31 308
CHS	*Heterocarpus reedi*	45	10 620	10 535	10 239	7 301	7 951	5 448	4 863
HUV	*Heterocarpus vicarius*	45	-	-	-	40	37	20	131
NDJ	*Pandalopsis japonica*	45	...	...	12	86	35	11	36
PSH	*Pandalus spp, Pandalopsis spp*	45	15 130	16 737	20 419	6 890	14 651	16 530	19 049
AKS	*Acetes japonicus*	45	406 495	460 871	495 680	587 376	598 602	639 219	577 497
SHS	*Sergestidae*	45	60 377	61 611	48 290	43 427	43 402	38 125	36 505
NLC	*Nematopalaemon schmitti*	45	-	-	-	-	-	1 464	1 382
CPR	*Palaemon serratus*	45	629	583	611	509	534	530	542
PAL	*Palaemonidae*	45	313	400	373	510	589	4 110	281
CSH	*Crangon crangon*	45	30 761	32 444	37 922	30 683	37 133	33 461	32 148
CVL	*Sclerocrangon spp*	45	...	...	38	59	82	20	45
CRN	*Crangonidae*	45	-	-	31	41	-	42	45
RSH	*Sicyonia brevirostris*	45	3 848	10 549	1 789	4 409	1 826	3 254	2 909
YII	*Sicyonia ingentis*	45	...	...	...	185	630	756	165
LAA	*Pleoticus muelleri*	45	6 705	9 874	6 479	23 203	15 928	36 769	78 077
RRS	*Pleoticus robustus*	45	252	198	209	195	286	391	305
SOK	*Solenocera agassizii*	45	...	...	...	...	...	...	686
KNS	*Haliporoides triarthrus*	45	2 036	1 771	1 510	1 882	1 611	1 766	1 738
HJD	*Haliporoides diomedeae*	45	5	15	32	29	135	169	309
DCP	*Natantia*	45	538 084	592 728	592 113	548 777	561 783	572 852	553 898
KRI	*Euphausia superba*	46	118 714	101 714	82 508	81 202	103 318	104 263	98 209
NKR	*Meganyctiphanes norvegica*	46	-	-	-	88	-	-	36
AMS	*Artemia salina*	47	...	...	...	513	691	561	538
GOO	*Lepas spp*	47	0	2	-	-	-	-	-
MBZ	*Megabalanus psittacus*	47	681	879	579	683	620	620	685
MTS	*Squilla mantis*	47	4 613	5 446	4 507	3 680	6 023	6 331	6 606
SQY	*Squillidae*	47	982	695	2 133	2 219	2 068	2 141	2 561
SVX	*Stomatopoda*	47	1	181	176	459	871	860	850
CRU	*Crustacea*	47	996 395	1 074 332	1 238 796	1 395 060	1 279 596	1 360 875	1 393 324
CMJ	*Corbicula japonica*	51	27 597	27 314	22 209	20 609	21 015	19 295	17 369
MOF	*Mollusca*	51	568 970	533 873	485 564	559 438	531 437	575 991	608 259
PEE	*Littorina littorea*	52	301	2 968	3 299	2 753	3 140	2 311	2 781
PER	*Littorina spp*	52	2 534	1 989	3 153	2 311	1 716	1 378	1 506
MUE	*Murex spp*	52	791	1 304	1 264	2 584	1 290	1 580	2 132
RPN	*Rapana spp*	52	4 123	4 349	5 494	4 671	4 419	4 713	3 753
SNE	*Concholepas concholepas*	52	4 031	5 269	7 520	3 394	4 583	2 524	1 372
ABG	*Haliotis gigantea*	52	1 980	1 941	2 218	2 269	2 109	2 146	1 982
ABP	*Haliotis midae*	52	615	735	330	524	481	490	5 265
ABR	*Haliotis rubra*	52	5 208	5 425	5 240	5 247	5 297	5 204	5 301
HLT	*Haliotis tuberculata*	52	49	62	75	36	37	64	62
ABX	*Haliotis spp*	52	3 486	2 909	2 649	2 470	2 144	2 241	1 955
TOS	*Turbo cornutus*	52	18 864	17 480	19 010	21 748	18 397	17 120	16 515
CON	*Strombus spp*	52	17 946	15 447	20 140	15 896	16 830	20 331	20 753
WHE	*Buccinum undatum*	52	15 852	24 723	27 659	13 768	17 794	29 818	30 146
WHX	*Busycon spp*	52	4 337	5 487	4 379	3 643	6 181	4 853	6 078
CXY	*Cymbium spp*	52	7 454	6 648	5 166	4 678	5 737	4 989	5 422
ZDF	*Zidona dufresnei*	52	574	558	1 322	1 010	683	1 621	1 406
GAS	*Gastropoda*	52	21 177	18 996	15 515	12 161	14 988	11 734	14 529
OCH	*Ostrea chilensis*	53	-	-	5	1	6	9	202
OYF	*Ostrea edulis*	53	4 340	3 456	3 598	2 565	1 776	880	959
OFO	*Ostrea lurida*	53	...	...	...	...	6	7	8
DRY	*Ostrea lutaria*	53	1	2	2	2	3	2	2
OYX	*Ostrea spp*	53	1 803	1 356	1 908	1 276	2 489	1 828	2 239
OYG	*Crassostrea gigas*	53	20 141	25 160	25 692	12 101	12 272	17 983	11 240
OYM	*Crassostrea rhizophorae*	53	5 224	4 110	3 702	4 895	3 906	3 467	5 632

B-00 (b)

Fish, crustaceans, molluscs, etc — Capture production by species items
Poissons, crustacés, mollusques, etc — Captures par catégories d'espèces
Peces, crustáceos, moluscos, etc — Capturas por partidas de especies

3-alpha code Code alpha-3 Código alfa-3	Scientific name Nom scientifique Nombre científico	Species group Groupe d'espèces Grupo de especies	1995 mt	1996 mt	1997 mt	1998 mt	1999 mt	2000 mt	2001 mt
OYA	*Crassostrea virginica*	53	158 534	148 540	144 707	135 222	132 207	249 306	175 042
CSI	*Crassostrea iredalei*	53	324	291	152	89	95	79	83
OYC	*Crassostrea spp*	53	2 294	2 601	2 639	2 954	3 781	3 344	3 608
MUK	*Mytilus coruscus*	54	2 942	2 191	3 211	1 469	1 414	1 133	1 085
MYC	*Mytilus chilensis*	54	5 129	5 715	4 724	4 900	4 344	5 237	6 758
MUS	*Mytilus edulis*	54	146 205	117 550	136 147	137 268	130 020	147 972	163 911
MSR	*Mytilus platensis*	54	687	370	354	375	474	412	492
MSM	*Mytilus galloprovincialis*	54	40 764	39 046	53 310	38 899	44 453	46 143	46 092
MYA	*Mytilus planulatus*	54	247	75	1	1	1	1	1
CHC	*Choromytilus chorus*	54	307	323	266	127	155	217	166
MOD	*Modiolus spp*	54	7	20	30	20	7	2	2
MSL	*Perna perna*	54	155	223	295	3 802	451	316	1 081
MSV	*Perna viridis*	54	20 549	20 220	18 017	17 454	6 554	6 462	6 422
MSC	*Aulacomya ater*	54	17 580	13 428	16 078	22 831	19 738	18 933	22 584
MSX	*Mytilidae*	54	9 825	4 278	7 506	10 601	11 572	14 643	8 721
OSC	*Aequipecten opercularis*	55	8 040	7 183	11 443	14 486	15 746	14 891	20 464
SCE	*Pecten maximus*	55	23 471	31 521	35 657	35 543	36 385	37 710	30 590
SJA	*Pecten jacobaeus*	55	23	52	95	50	68	570	150
SCZ	*Pecten novaezelandiae*	55	14 160	5 080	18 848	4 592	6 152	2 912	6 792
ZYE	*Zygochlamis delicatula*	55	135	124	201	91	128	0	222
ZYP	*Zygochlamis patagonica*	55	10 592	36 952	39 817	28 441	42 700	37 404	42 598
SCA	*Placopecten magellanicus*	55	121 230	109 382	102 028	99 432	131 962	196 993	254 196
SCC	*Argopecten gibbus*	55	10 003	...	...	...	...	...	...
SCB	*Argopecten irradians*	55	1 593	230	452	690	216	154	30
SCQ	*Argopecten purpuratus*	55	3 113	2 095	4 013	23 546	30 141	11 830	6 544
SCH	*Argopecten ventricosus*	55	1 256	17 290	2 320	2 726	1 864	6 287	3 100
SC	*Chlamys islandica*	55	23 570	22 752	24 673	18 946	12 018	13 471	9 469
SCG	*Patinopecten caurinus*	55	1 950	2 372	2	3 228	2 642	2 012	1 052
ISC	*Patinopecten yessoensis*	55	279 382	276 406	266 957	294 211	305 510	310 104	293 268
SCX	*Pectinidae*	55	38 806	23 729	26 387	28 785	27 170	26 833	34 050
ARK	*Arca spp*	56	37 398	35 504	44 151	33 209	42 559	48 202	46 928
MCL	*Scapharca subcrenata*	56	15 426	16 328	14 133	10 120	10 413	7 308	4 899
BLC	*Anadara granosa*	56	1 415	995	493	12 555	8 260	5 917	2 745
BLS	*Anadara spp*	56	45 180	47 233	43 872	34 395	36 904	37 448	35 477
CLQ	*Arctica islandica*	56	185 881	180 114	168 150	157 282	147 933	124 131	150 260
SVE	*Chamelea gallina*	56	45 820	50 501	38 794	36 798	44 605	49 822	48 213
CTS	*Venerupis pullastra*	56	1 305	2 826	4 058	5 166	4 969	2 388	2 420
KNU	*Chione stutchburyi*	56	1 220	815	541	1 325	1 396	1 789	1 748
HCJ	*Meretrix lusoria*	56	5 031	4 394	4 133	6 147	4 110	2 973	2 289
HCX	*Meretrix spp*	56	18 023	13 481	14 027	17 146	14 767	14 177	14 480
CTG	*Ruditapes decussatus*	56	2 019	1 469	1 286	1 943	1 899	1 487	1 317
CLJ	*Ruditapes philippinarum*	56	64 523	56 095	56 514	51 392	57 051	56 540	51 026
TWG	*Tawera gayi*	56	...	...	...	...	...	1	291
TVM	*Tivela mactroides*	56	...	126	196	664	1 684	273	270
BCL	*Saxidomus giganteus*	56	1 728	1 431	1 293	1 726	1 019	1 113	1 159
NCL	*Paphia spp*	56	31 207	53 131	35 968	49 866	70 160	69 153	68 803
PTS	*Protothaca staminea*	56	-	-	-	-	-	-	102
TCL	*Protothaca thaca*	56	17 162	20 016	12 475	24 254	16 429	16 303	26 483
CLH	*Mercenaria mercenaria*	56	16 888	22 964	6 287	1 234	2 536	15 150	11 793
CLV	*Veneridae*	56	7 761	6 394	5 293	4 968	5 915	6 844	5 939
HCL	*Pseudocardium sybillae*	56	17 056	14 791	14 720	11 104	16 428	13 033	12 323
TQZ	*Tresus spp*	56	-	-	-	-	-	1	2
CLB	*Spisula solidissima*	56	156 056	155 584	141 421	131 700	142 370	165 765	166 310
CLT	*Spisula polynyma*	56	24 017	25 612	27 365	26 008	26 722	22 985	20 273
ULO	*Spisula solida*	56	...	...	...	...	765	1 335	1 440
MUN	*Mulinia spp*	56	1 852	999	2 757	2 549	1 536	1 491	1 699
DON	*Donax spp*	56	197	0	0	0	490	495	628
RAZ	*Solen spp*	56	2 670	1 873	581	632	701	468	608
CLR	*Ensis directus*	56	-	-	14	49	64	99	36
RAP	*Siliqua patula*	56	-	-	-	-	-	14	61
CLS	*Mya arenaria*	56	8 654	5 824	6 777	8 131	7 793	8 248	10 069
GEC	*Panopea abrupta*	56	2 056	1 768	3 719	4 027	4 139	3 821	3 894
COC	*Cerastoderma edule*	56	63 185	37 089	38 132	86 750	70 400	47 266	23 764
COZ	*Cardiidae*	56	428	2 583	1 609	1 078	130	460	2 069
AFQ	*Paphies australis*	56	605	702	1 295	1 496	976	1 085	1 249
CLM	*Mesodesma donacium*	56	8 113	7 204	7 831	7 042	1 728	1 259	1 396
TUW	*Semele solida*	56	2 523	4 418	2 199	1 900	2 071	4 212	3 054
CLX	*Bivalvia*	56	175 285	146 887	115 216	105 904	92 664	85 183	83 428
CTC	*Sepia officinalis*	57	10 352	11 408	13 758	12 449	14 850	12 724	13 987
CTL	*Sepiidae, Sepiolidae*	57	411 367	369 998	455 667	429 064	438 264	484 725	519 313
SQP	*Loligo gahi*	57	85 174	68 504	21 711	51 693	42 505	67 016	57 730
SQL	*Loligo pealei*	57	18 926	12 490	16 161	18 879	18 749	16 942	14 211
CHO	*Loligo reynaudi*	57	7 047	7 549	3 696	6 670	7 169	6 000	3 373
SQC	*Loligo spp*	57	216 817	216 651	227 601	218 114	199 843	214 991	213 605
OFJ	*Ommastrephes bartrami*	57	...	...	49 870	54 951	36 076	47 368	23 870
SQI	*Illex illecebrosus*	57	18 550	28 971	34 837	26 589	9 883	11 222	5 698
SQM	*Illex coindetii*	57	571	395	411	216	338	402	250
SQA	*Illex argentinus*	57	520 938	656 481	980 300	693 542	1 144 998	930 781	743 024
GIS	*Dosidicus gigas*	57	136 288	142 186	162 504	27 466	134 773	182 399	223 784
SQE	*Todarodes sagittatus*	57	5 171	5 875	5 571	6 639	4 890	4 909	4 261
SQJ	*Todarodes pacificus*	57	513 413	715 908	603 367	378 605	497 887	570 427	528 523
TSQ	*Nototodarus sloani*	57	94 098	53 699	64 602	55 570	31 358	25 603	45 119
SQS	*Martialia hyadesi*	57	23 986	3 845	8 376	55	27	686	4
SQU	*Loliginidae, Ommastrephidae*	57	360 839	363 683	308 900	344 475	453 581	481 972	379 177
OCC	*Octopus vulgaris*	57	66 764	72 936	60 313	72 465	56 055	50 478	53 103
OCM	*Eledone spp*	57	2 500	1 939	2 258	1 758	1 433	1 990	2 357

B-00 (b)

Fish, crustaceans, molluscs, etc — **Capture production by species items**
Poissons, crustacés, mollusques, etc — **Captures par catégories d'espèces**
Peces, crustáceos, moluscos, etc — **Capturas por partidas de especies**

3-alpha code Code alpha-3 Código alfa-3	Scientific name Nom scientifique Nombre científico	Species group Groupe d'espèces Grupo de especies	1995 mt	1996 mt	1997 mt	1998 mt	1999 mt	2000 mt	2001 mt
OCT	*Octopodidae*	57	235 021	230 773	210 845	245 385	295 013	259 722	261 750
CEP	*Cephalopoda*	57	210 570	186 473	225 922	213 035	209 978	284 793	253 689
MOL	*Mollusca*	58	1 399 408	1 037 525	1 597 042	1 594 534	1 567 259	1 505 386	1 507 859
FRG	*Rana spp*	71	3 627	2 957	3 622	3 009	1 807	2 315	2 883
TTG	*Malaclemys spp*	72	0	-	-	0	-	0	-
TUG	*Chelonia mydas*	72	59	64	28	28	15	15	18
TTH	*Eretmochelys imbricata*	72	20	23	19	18	12	12	12
TTL	*Caretta caretta*	72	11	10	7	7	5	5	5
TUL	*Testudinata*	72	6	54	65	53	16	56	51
TTX	*Testudinata*	72	746	953	858	1 081	1 195	910	1 102
SSE	*Pyura chilensis*	74	3 297	4 549	3 174	2 530	2 704	2 290	1 298
SSG	*Microcosmus sulcatus*	74	...	28	22	22	30	30	76
SSX	*Ascidiacea*	74	5 767	16 754	2 780	891	1 171	1 538	1 053
HSC	*Limulus polyphemus*	75	926	1 598	2 607	3 252	2 397	1 696	1 299
ECH	*Echinodermata*	76	7 433	7 054	6 521	4 040	5 702	6 936	7 009
STF	*Asteroidea*	76	9	4	0	13	3	6	11
URC	*Strongylocentrotus spp*	76	53 354	45 583	45 724	39 344	39 656	36 744	33 097
URM	*Paracentrotus lividus*	76	78	63	48	59	84	198	101
URS	*Echinus esculentus*	76	933	425	25	1	13	1	5
UCH	*Loxechinus albus*	76	54 609	51 437	45 560	44 843	55 654	54 096	46 794
CUJ	*Stichopus japonicus*	76	8 494	9 205	9 377	8 391	7 866	8 376	8 129
CUX	*Holothurioidea*	76	11 917	13 954	11 660	12 617	11 572	14 999	10 730
JEL	*Rhopilema spp*	77	338 761	326 850	516 748	513 399	529 403	458 991	458 545
WOR	*Polychaeta*	77	59	100	347	408	422	257	931
INV	*Invertebrata*	77	42 953	17 219	14 084	13 232	13 140	16 103	13 922
World total Total mondial Total mundial			**92 301 906**	**93 750 402**	**94 215 594**	**87 593 312**	**93 601 896**	**95 439 820**	**92 356 034**

B-11

Carps, barbels and other cyprinids — **Capture production by species, fishing areas and countries or areas**
Carpes, barbeaux et autres cyprinidés — **Captures par espèces, zones de pêche et pays ou zones**
Carpas, barbos y otros ciprínidos — **Capturas por especies, áreas de pesca y países o áreas**

Species, Fishing area Espèce, Zone de pêche Especie, Area de pesca	1992 mt	1993 mt	1994 mt	1995 mt	1996 mt	1997 mt	1998 mt	1999 mt	2000 mt	2001 mt
Buffalofishes nei	**Poissons-taureaux nca**		**...C**		***Ictiobus spp***			**1,40(01)011,XX**		**BUF**
02 USA	1 049	982	896	480	793	991	959	697	1 280	1 569
02 Fishing area total	*1 049*	*982*	*896*	*480*	*793*	*991*	*959*	*697*	*1 280*	*1 569*
Species total	*1 049*	*982*	*896*	*480*	*793*	*991*	*959*	*697*	*1 280*	*1 569*
Suckers nei	**Cyprins sucets nca**		**Chupadores nep**		***Catostomidae***			**1,40(01)XXX,XX**		**CTM**
02 USA	663	872	292	662	542	550	464	465	231	187
02 Fishing area total	*663*	*872*	*292*	*662*	*542*	*550*	*464*	*465*	*231*	*187*
Species total	*663*	*872*	*292*	*662*	*542*	*550*	*464*	*465*	*231*	*187*
Freshwater bream	**Brème d'eau douce**		**Brema común**		***Abramis brama***			**1,40(02)001,02**		**FBM**
04 Azerbaijan	518	412	309	219	402	346	314	52	55	127
Georgia	16	18	11	27	3	1	2	1	3	1
Kazakhstan	20 872	20 420	19 302	20 520	18 770 F	11 000 F	9 800 F	11 000 F	12 000 F	12 630
Kyrgyzstan	-	20	19	11	9 F	7 F	5 F	3 F	3 F	4
Tajikistan	91	123	82	58	...	...	...	...	...	37
Turkey	...	...	...	...	...	...	...	259	200	151
Turkmenistan	329	170	158	137	75	100	142	147	153	126
Uzbekistan	726	598	579	474	220 F	289	387	353	335	540
04 Fishing area total	*22 552*	*21 761*	*20 460*	*21 446*	*19 479 F*	*11 743 F*	*10 650 F*	*11 815 F*	*12 749 F*	*13 616*
05 Belarus	360	464	201	145	244	182	130	98	27	198
Belgium	60	60	60	60	60	60	60	60	60	60
Bulgaria	...	...	70 F	74	91	90	82	71	25	18
Czech Rep	...	202	265	286	247	232	253	297	261	247
Czechoslovak	200 F	-	-	-	-	-	-	-	-	-
Denmark	369	130	90	85	64	93	141	79	47	4
Estonia	...	318	152	155	167	222	240	181	194	318
Finland	3 451	3 449	3 266	3 239	2 334	2 334	1 590	1 590	1 660	1 660
Latvia	193	233	145	156	151	172	135	158	155	152
Lithuania	377	276	324	420	397	448	454	466	467	470
Netherlands	316	229	30	...	75	65	399	355	350 F	350 F
Poland	2 500	2 400	2 400	1 043	3 053	1 498	1 013	1 291	1 883	1 859
Romania	711	468	1 090	1 604	827	328	951	1 052	936	800
Russian Fed	29 089	27 441	28 154	29 142	29 050	29 507	31 715	25 360	23 794	27 264
Slovenia	21	9	18	16	10	11	11	10	11	10
Sweden	8	7	3	13	24	16	20	12	-	-
Switzerland	122	70	59	22	24	20	16	16	13	9
Ukraine	3 203	2 033	1 922	908	986	810	721	832	873	942
UK	1	1	1	2	1	9	8	8	2	0
05 Fishing area total	*40 981 F*	*37 790*	*38 250 F*	*37 370*	*37 805*	*36 097*	*37 939*	*31 936*	*30 758 F*	*34 361 F*
27 Denmark	1	4	3	2	2	1	3	1	1	0
Estonia	4	6	6	9	8	8	7	13	10	10
Finland	1 227	1 216	1 010	986	809	941	1 264	1 255	886	987
Latvia	18	33	70	81	91	69	56	77	63	94
Lithuania	-	9	-	-	-	-	-	-	-	-
Poland	522	453	469	462	569	869	1 044	1 030	1 062	717
Russian Fed	-	-	-	-	-	-	-	-	1 351	1 382
Sweden	8	2	3	1	1	2	2	8	6	4
27 Fishing area total	*1 780*	*1 723*	*1 561*	*1 541*	*1 480*	*1 890*	*2 376*	*2 384*	*3 379*	*3 194*
37 Russian Fed	441	1 360	387	618	324	636	474	327	328	108
Ukraine	4	9	9	4	22	22	21	8	8	-
37 Fishing area total	*445*	*1 369*	*396*	*622*	*346*	*658*	*495*	*335*	*336*	*108*
Species total	*65 758 F*	*62 643*	*60 667 F*	*60 979*	*59 110 F*	*50 388 F*	*51 460 F*	*46 470 F*	*47 222 F*	*51 279 F*
Freshwater breams nei	**Brèmes d'eau douce nca**		**Bremas nep**		***Abramis spp***			**1,40(02)001,XX**		**FBR**
04 Iran	20	17	29	5	3	7	20	9	20	10
04 Fishing area total	*20*	*17*	*29*	*5*	*3*	*7*	*20*	*9*	*20*	*10*
05 Lithuania	-	-	5	4	7	10	12	11	13	24
Poland	50	50	50	267	373	...	...	235	490	400
Romania	-	-	-	-	-	-	3	1	-	-
Russian Fed	2 065	2 246	1 236	1 720	1 664	1 046	2 058	1 951	1 593	1 751
Slovakia	...	79	102	141	111	102	99	98	94	95
Slovenia	-	-	5	5	5	5	5	4	0	1
Ukraine	816	55	484	53	101	62	86	122	80	137
05 Fishing area total	*2 931*	*2 430*	*1 882*	*2 190*	*2 261*	*1 225*	*2 263*	*2 422*	*2 270*	*2 408*
27 Germany	101	53	196	181	162	275	144	249	124	174
Lithuania	-	-	-	-	-	-	-	-	-	2
27 Fishing area total	*101*	*53*	*196*	*181*	*162*	*275*	*144*	*249*	*124*	*176*
Species total	*3 052*	*2 500*	*2 107*	*2 376*	*2 426*	*1 507*	*2 427*	*2 680*	*2 414*	*2 594*
Common carp	**Carpe commune**		**Carpa**		***Cyprinus carpio***			**1,40(02)002,01**		**FCP**
01 Ethiopia	...	...	...	74	94	27	62	71	75	74
Kenya	189	279	305	360	334	216	48	52	47	49
Lesotho	13 F	15 F	15 F	16 F	16 F	18 F	18 F	18 F	18 F	8
01 Fishing area total	*202 F*	*294 F*	*320 F*	*450 F*	*444 F*	*261 F*	*128 F*	*141 F*	*140 F*	*131*

B-11 Carps, barbels and other cyprinids — Capture production by species, fishing areas and countries or areas
Carpes, barbeaux et autres cyprinidés — Captures par espèces, zones de pêche et pays ou zones
Carpas, barbos y otros ciprínidos — Capturas por especies, áreas de pesca y países o áreas

Species, Fishing area Espèce, Zone de pêche Especie, Area de pesca	1992 mt	1993 mt	1994 mt	1995 mt	1996 mt	1997 mt	1998 mt	1999 mt	2000 mt	2001 mt
02 Canada	494	222	649	619	687	543	354	741	516	506
Dominican Rp	233	1 579	2 881	597	62	109	27	180	...	397
Mexico	20 393	18 171	16 187	22 677	21 237	16 787	14 345	10 945	15 300	14 700
USA	780	1 106	824	852	1 034	1 076	1 072	1 103	724	704
02 Fishing area total	*21 900*	*21 078*	*20 541*	*24 745*	*23 020*	*18 515*	*15 798*	*12 969*	*16 540*	*16 307*
03 Chile	-	-	-	-	-	-	4	-	-	-
Venezuela	0	0	4	544	218	1	3	5	1 390	0
03 Fishing area total	*0*	*0*	*4*	*544*	*218*	*1*	*7*	*5*	*1 390*	*0*
04 Armenia	9	1	0	1	91	32	14	12	9	7
Azerbaijan	1 188	1 446	803	370	84	49	87	92	93	51
China,Taiwan	85	104	54	44	26	19	22	45	49	50
Georgia	91	499	0	49	0	0	2	11	12	5
Indonesia	5 613	5 287	5 490	5 613	7 081	6 644	7 082	7 127	6 826	6 770
Iraq	4 677	7 703	5 394	5 691	5 538	4 163	2 336	...	...	...
Israel	111	108	147	89	35	38	150	165	189	160 F
Japan	6 178	5 338	4 968	4 896	4 771	4 607	4 477	4 259	4 079	3 558
Kazakhstan	168	431	537	477	440 F	320 F	230 F	500 F	650 F	710
Korea Rep	-	-	1 469	1 684	1 979	842	874	438	...	...
Kyrgyzstan	-	-	-	36	31 F	23 F	15 F	9 F	10 F	11
Nepal	8	25	-	-	-	-	-	-	-	-
Tajikistan	...	...	...	...	...	...	...	48	59	24
Thailand	6 614	8 692	8 200	10 144	7 420	7 418	11 508	13 689	7 000	7 310
Turkey	15 545	16 035	15 900	17 081	15 631	16 000	20 000	17 797	14 137	12 265
Turkmenistan	93	342	774	484	115	154	140	150	144	93
Uzbekistan	1 246	771	605	864	193 F	843	804	826	617	906
04 Fishing area total	*41 626*	*46 782*	*44 341*	*47 523*	*43 435 F*	*41 152 F*	*47 741 F*	*45 168 F*	*33 874 F*	*31 920 F*
05 Albania	14	20 F	30 F	34	45	38	230	216	230	300
Belarus	118	59	18	39	39	12	8	5	17	15
Belgium	30	30	30	30	30	30	30	30	30	30
Bulgaria	-	-	-	19	16	281	251	302	143	880
Croatia	60	72	97	96	143	126	1 F	1	3	1
Czech Rep	...	2 284	2 832	2 919	2 522	2 312	2 899	3 006	3 558	3 560
Czechoslovak	2 830 F	-	-	-	-	-	-	-	-	-
Denmark	2	1	1	2	1	1	1	0	0	0
Germany	66	66	386	386	386	386	386	386	386	386
Greece	...	...	...	...	208	256	279	247	220	198
Hungary	3 699	3 103	3 265	2 856	2 717	2 255	3 373	3 279	3 212	2 470
Latvia	-	-	5	5	3	3	6	5	3	5
Lithuania	-	52	41	27	31	14	13	12	16	16
Macedonia	11	9	13	32	10	9	...	...	25	6
Moldova Rep	334	515	469	472	408	349	280	178	192	212
Netherlands	1	0	-	-	-	-	-	-	-	-
Poland	-	-	-	-	77	82	78	37	45	50
Portugal	0	0	0	0	0	0	0	0	0	0
Romania	678	562	560	577	441	173	147	310	458	566
Russian Fed	4 403	1 491	2 637	752	3 933	2 713	3 212	3 507	4 007	2 698
Slovakia	...	612	779	1 063	778	746	778	822	854	967
Slovenia	64	96	97	89	86	90	94	78	71	75
Switzerland	-	-	-	1	1	2	1	1	1	1
Ukraine	215	149	232	153	598	27	33	41	44	39
05 Fishing area total	*12 525 F*	*9 121 F*	*11 492 F*	*9 552*	*12 473*	*9 905*	*12 100 F*	*12 463*	*13 515*	*12 475*
06 New Zealand	-	-	-	0	0	28	2	3	4	6
06 Fishing area total	-	-	-	*0*	*0*	*28*	*2*	*3*	*4*	*6*
27 Germany	-	-	-	-	-	-	-	1	-	1
27 Fishing area total	-	-	-	-	-	-	-	*1*	-	*1*
37 Russian Fed	8	29	5	2	2	3	3	1	-	1
Ukraine	-	-	-	-	-	-	-	-	-	1
37 Fishing area total	*8*	*29*	*5*	*2*	*2*	*3*	*3*	*1*	-	*2*
Species total	*76 261 F*	*77 304 F*	*76 703 F*	*82 816 F*	*79 592 F*	*69 865 F*	*75 779 F*	*70 751 F*	*65 463 F*	*60 842 F*
Tench	**Tanche**		**Tenca**			***Tinca tinca***			**1,40(02)007,01**	**FTE**
04 Azerbaijan	18	5	0	0	2	0	0	0	0	-
Turkey	...	...	...	...	...	...	...	...	690	778
04 Fishing area total	*18*	*5*	*0*	*0*	*2*	*0*	*0*	*0*	*690*	*778*
05 Belarus	14	23	11	13	8	4	3	90	41	7
Belgium	15	15	15	15	15	15	15	15	15	15
Croatia	-	1	1	2	3	0	-	-	-	-
Czech Rep	...	36	29	30	23	21	29	30	27	24
Czechoslovak	30 F	-	-	-	-	-	-	-	-	-
Denmark	0	0	0	0	0	0	0	0	1	0
Germany	34	33	36	36	36	36	36	36	36	36
Hungary	9	-	-	-	-	5	-	-	-	-
Latvia	3	5	16	15	15	24	22	41	29	35
Lithuania	-	6	9	10	9	12	9	11	13	15
Poland	100	100	100	70	97	101	91	102	160	113
Romania	-	-	-	-	-	-	2	7	4	31
Russian Fed	1 398	879	1 501	1 300	936	1 647	990	1 071	1 307	1 409
Slovakia	...	5	5	8	8	9	8	8	8	5
Slovenia	3	2	1	2	1	2	2	3	2	1

B-11 Carps, barbels and other cyprinids / Carpes, barbeaux et autres cyprinidés / Carpas, barbos y otros ciprínidos

Capture production by species, fishing areas and countries or areas / Captures par espèces, zones de pêche et pays ou zones / Capturas por especies, áreas de pesca y países o áreas

Species, Fishing area Espèce, Zone de pêche Especie, Area de pesca	1992 mt	1993 mt	1994 mt	1995 mt	1996 mt	1997 mt	1998 mt	1999 mt	2000 mt	2001 mt
Sweden	0	0	0	0	0	0	1	1	-	-
Switzerland	5	3	3	3	4	3	3	3	3	3
Ukraine	22	...	11	...	...	...	...	...	...	...
05 Fishing area total	*1 633 F*	*1 108*	*1 738*	*1 504*	*1 155*	*1 879*	*1 211*	*1 418*	*1 646*	*1 694*
27 Denmark	0	0	0	0	0	-	0	0	-	0
Germany	-	-	-	-	-	-	-	1	-	1
27 Fishing area total	*0*	*0*	*0*	*0*	*0*	-	*0*	*1*	-	*1*
Species total	*1 651 F*	*1 113*	*1 738*	*1 504*	*1 157*	*1 879*	*1 211*	*1 419*	*2 336*	*2 473*
Bleak	**Ablette**		**Alburno**		***Alburnus alburnus***			**1,40(02)012,01**		**ALR**
05 Albania	...	...	...	151	162	68	149	160	190	478
Bulgaria	-	-	-	2	6	62	67	92	24	4
Lithuania	-	-	-	-	-	-	-	-	3	4
Romania	-	-	-	-	-	-	24	16	30	60
Ukraine	92	...	12	...	...	...	...	...	...	...
05 Fishing area total	*92*	...	*12*	*153*	*168*	*130*	*240*	*268*	*247*	*546*
Species total	*92*	...	*12*	*153*	*168*	*130*	*240*	*268*	*247*	*546*
Barbel	**Barbeau fluviatile**		**Barbo común**		***Barbus barbus***			**1,40(02)013,01**		**PTB**
05 Bulgaria	...	...	...	92	120	142	113	93	43	42
Czechoslovak	40 F	-	-	-	-	-	-	-	-	-
Hungary	...	39	66	32	37	64	46	50	30	52
Romania	-	-	-	-	-	-	32	11	17	34
Slovakia	...	16	20	24	19	21	15	10	17	19
Slovenia	12	16	15	13	10	12	9	12	11	11
05 Fishing area total	*52 F*	*71*	*101*	*161*	*186*	*239*	*215*	*176*	*118*	*158*
Species total	*52 F*	*71*	*101*	*161*	*186*	*239*	*215*	*176*	*118*	*158*
Common nase	**Nase commun**		**Condrostoma común**		***Chondrostoma nasus***			**1,40(02)014,01**		**HON**
05 Bulgaria	...	...	...	...	...	3	3	4	8	14
Slovakia	...	33	32	31	24	23	25	24	24	20
Slovenia	-	-	-	-	-	-	-	35	35	25
05 Fishing area total	...	*33*	*32*	*31*	*24*	*26*	*28*	*63*	*67*	*59*
Species total	...	*33*	*32*	*31*	*24*	*26*	*28*	*63*	*67*	*59*
Crucian carp	**Carassin(=Cyprin)**		**Carpín**		***Carassius carassius***			**1,40(02)016,01**		**FCC**
01 Ethiopia	-	-	-	47	61	191	88	101	110	108
01 Fishing area total	-	-	-	*47*	*61*	*191*	*88*	*101*	*110*	*108*
04 Armenia	65	25	10	2	28	23	42	26	38	32
Azerbaijan	...	...	...	...	...	...	...	6	4	17
China,Taiwan	62	35	23	14	5	5	5	28	35	37
Japan	5 321	4 921	4 402	4 286	4 205	4 008	3 881	3 493	3 423	2 948
Kazakhstan	5 800	4 500	4 000	4 317	3 950 F	2 840 F	2 060 F	1 900 F	1 800 F	1 707
Turkmenistan	99	41	75	63	85	1	225	233	228	154
Uzbekistan	...	...	...	...	...	...	...	...	...	331
04 Fishing area total	*11 347*	*9 522*	*8 510*	*8 682*	*8 273 F*	*6 877 F*	*6 213 F*	*5 686 F*	*5 528 F*	*5 226*
05 Albania	11	10 F	9 F	...	...	...	77	62	65	326
Belarus	188	384	60	64	69	35	106	138	154	188
Bulgaria	...	...	...	...	...	328	392	360	179	138
Latvia	-	-	5	7	7	11	10	14	15	24
Moldova Rep	33	61	171	164	128	166	159	104	132	127
Poland	30	30	30	60	46	49	47	46	92	84
Portugal	2	3	4	2	0	0	0	0	0	0
Romania	-	-	-	-	-	33	2	23	128	97
Slovakia	...	2	-	-	-	-	2	1	0	...
Slovenia	-	-	-	-	-	-	0	1	0	0
05 Fishing area total	*264*	*490 F*	*279 F*	*297*	*250*	*622*	*795*	*749*	*765*	*984*
27 Denmark	0	0	0	0	0	0	-	0	0	0
Poland	-	-	-	-	-	1	0	3	3	3
27 Fishing area total	*0*	*0*	*0*	*0*	*0*	*1*	*0*	*3*	*3*	*3*
Species total	*11 611*	*10 012 F*	*8 789 F*	*9 026*	*8 584 F*	*7 691 F*	*7 096 F*	*6 539 F*	*6 406 F*	*6 321*
Goldfish	**Poisson rouge(=Cyprin doré)**		**Pez rojo**		***Carassius auratus***			**1,40(02)016,02**		**CGO**
02 USA	...	...	...	5	5	10	10	7	9	10
02 Fishing area total	...	...	...	*5*	*5*	*10*	*10*	*7*	*9*	*10*
04 Kyrgyzstan	2	1	3	2	2 F	1 F	1 F	1 F	1 F	2
Uzbekistan	...	...	...	361	170 F	340	300	300	298	110
04 Fishing area total	*2*	*1*	*3*	*363*	*172 F*	*341 F*	*301 F*	*301 F*	*299 F*	*112*
05 Czech Rep	...	30	49	39	49	40	40	43	35	37
Greece	...	...	...	...	530	296	235	548	448	415
Lithuania	13	26	41	41	30	30	38	32	45	38
Romania	2 368	2 135	2 130	2 320	1 954	817	1 113	1 199	1 212	1 149
Russian Fed	97	219	92	-	-	-	-	-	-	-

B-11

Carps, barbels and other cyprinids — **Capture production by species, fishing areas and countries or areas**
Carpes, barbeaux et autres cyprinidés — **Captures par espèces, zones de pêche et pays ou zones**
Carpas, barbos y otros ciprínidos — **Capturas por especies, áreas de pesca y países o áreas**

Species, Fishing area Espèce, Zone de pêche Especie, Area de pesca	1992 mt	1993 mt	1994 mt	1995 mt	1996 mt	1997 mt	1998 mt	1999 mt	2000 mt	2001 mt
Slovakia	...	48	60	90	76	53	54	62	0	73
Slovenia	-	-	-	-	-	-	0	1	1	1
Ukraine	1 403	737	783	704	638	461	452	638	628	747
05 Fishing area total	*3 881*	*3 195*	*3 155*	*3 194*	*3 277*	*1 697*	*1 932*	*2 523*	*2 369*	*2 460*
Species total	*3 883*	*3 196*	*3 158*	*3 562*	*3 454 F*	*2 048 F*	*2 243 F*	*2 831 F*	*2 677 F*	*2 582*
Roach	**Gardon**		**Rutilo**		***Rutilus rutilus***				**1,40(02)018,01**	**FRO**
04 Azerbaijan	78	63	53	19	74	89	62	81	8	64
Turkmenistan	110	111	110	88	71	69	48	39	41	1
Uzbekistan	870	247	88	200	90 F	379	392	613	1 035	1 300
04 Fishing area total	*1 058*	*421*	*251*	*307*	*235 F*	*537*	*502*	*733*	*1 084*	*1 365*
05 Belgium	160	160	160	160	160	160	160	150	160	160
Bulgaria	...	...	...	...	...	...	10	40	-	-
Denmark	43	84	24	44	30	27	64	28	35	1
Estonia	-	50	70	92	209	150	128	167	234	231
Latvia	100	59	42	47	36	52	45	53	60	49
Lithuania	569	353	318	314	431	593	645	647	635	643
Poland	2 000	2 000	2 000	1 050	2 185	792	593	653	947	837
Slovenia	5	2	4	2	2	2	2	2	3	4
Switzerland	409	350	312	169	146	164	168	154	136	137
Ukraine	8 195	4 467	4 513	717	798	795	1 140	666	449	519
UK	-	-	-	-	-	4	5	5	-	-
05 Fishing area total	*11 481*	*7 525*	*7 443*	*2 595*	*3 997*	*2 739*	*2 960*	*2 565*	*2 659*	*2 581*
27 Denmark	20	36	32	26	13	11	22	11	8	7
Estonia	148	212	188	240	293	342	321	157	244	272
Finland	1 992	1 983	1 385	1 439	2 172	2 271	1 465	1 465	1 412	1 459
Germany	-	-	430	337	176	242	347	462	269	320
Latvia	18	15	16	7	10	12	11	14	10	11
Poland	765	675	645	937	910	1 042	1 069	1 309	1 241	1 338
Sweden	0	0	0	1	0	1	-	-	-	-
27 Fishing area total	*2 943*	*2 921*	*2 696*	*2 987*	*3 574*	*3 921*	*3 235*	*3 418*	*3 184*	*3 407*
37 Ukraine	4	...	6	8	1	0	1	1	2	7
37 Fishing area total	*4*	*...*	*6*	*8*	*1*	*0*	*1*	*1*	*2*	*7*
Species total	*15 486*	*10 867*	*10 396*	*5 897*	*7 807 F*	*7 197*	*6 698*	*6 717*	*6 929*	*7 360*
Kutum	**...B**		**...C**		***Rutilus frisii***				**1,40(02)018,02**	**RFR**
04 Iran	12 000	12 727	9 277	8 435	9 210	2 320	6 624	6 905	10 120	7 199
04 Fishing area total	*12 000*	*12 727*	*9 277*	*8 435*	*9 210*	*2 320*	*6 624*	*6 905*	*10 120*	*7 199*
Species total	*12 000*	*12 727*	*9 277*	*8 435*	*9 210*	*2 320*	*6 624*	*6 905*	*10 120*	*7 199*
Roaches nei	**Gardons nca**		**Rutilos nep**		***Rutilus spp***				**1,40(02)018,XX**	**FRX**
04 Iran	120	714	1 366	1 178	878	203	607	626	1 515	1 316
Kazakhstan	3 409	3 702	2 251	2 650	2 420 F	1 740 F	1 260 F	1 000 F	850 F	579
04 Fishing area total	*3 529*	*4 416*	*3 617*	*3 828*	*3 298 F*	*1 943 F*	*1 867 F*	*1 626 F*	*2 365 F*	*1 895*
05 Albania	8	6 F	4 F	-	-	-	-	-	-	-
Belarus	82	417	62	83	95	42	24	19	9	33
Finland	5 721	5 793	6 850	6 915	6 775	6 775	5 362	5 362	3 999	3 999
Greece	...	...	...	...	164	81	252	309	345	324
Netherlands	11	10	20	54	100	123	107	100	100 F	100 F
Romania	709	670	1 295	693	301	125	292	234	174	278
Russian Fed	28 443	25 557	22 180	19 630	21 888	15 864	15 950	10 355	13 973	14 545
Slovenia	9	10	10	8	6	8	9	10	11	7
Sweden	2	2	2	4	0	1	-	-	-	-
05 Fishing area total	*34 985*	*32 465 F*	*30 423 F*	*27 387*	*29 329*	*23 019*	*21 996*	*16 389*	*18 611 F*	*19 286 F*
27 Russian Fed	-	-	-	-	-	-	-	-	501	653
27 Fishing area total	*-*	*-*	*-*	*-*	*-*	*-*	*-*	*-*	*501*	*653*
37 Russian Fed	76	102	325	250	80	107	13	78	73	114
37 Fishing area total	*76*	*102*	*325*	*250*	*80*	*107*	*13*	*78*	*73*	*114*
Species total	*38 590*	*36 983 F*	*34 365 F*	*31 465*	*32 707 F*	*25 069 F*	*23 876 F*	*18 093 F*	*21 550 F*	*21 948 F*
Rudd	**Rotengle**		**Escardinio**		***Scardinius erythrophthalmus***				**1,40(02)019,01**	**SRE**
05 Bulgaria	...	...	...	...	...	...	71	90	39	3
Denmark	-	-	-	-	-	-	-	0	2	-
Lithuania	...	11	16	17	14	18	14	13	17	20
Slovenia	-	-	6	7	7	6	4	4	1	0
Ukraine	44	21	15	5	15	19	34	38	38	33
05 Fishing area total	*44*	*32*	*37*	*29*	*36*	*43*	*123*	*145*	*97*	*56*
Species total	*44*	*32*	*37*	*29*	*36*	*43*	*123*	*145*	*97*	*56*
Orfe(=Ide)	**Ide mélanote**		**Cachuelo**		***Leuciscus idus***				**1,40(02)020,01**	**FID**
05 Albania	10	8 F	5 F	-	-	-	-	-	-	-
Belarus	31	45	22	18	19	11	14	11	5	18
Bulgaria	-	-	-	7	8	10	12	14	-	0

B-11 Carps, barbels and other cyprinids / Carpes, barbeaux et autres cyprinidés / Carpas, barbos y otros ciprínidos

Capture production by species, fishing areas and countries or areas / Captures par espèces, zones de pêche et pays ou zones / Capturas por especies, áreas de pesca y países o áreas

Species, Fishing area Espèce, Zone de pêche Especie, Area de pesca	1992 mt	1993 mt	1994 mt	1995 mt	1996 mt	1997 mt	1998 mt	1999 mt	2000 mt	2001 mt
Czech Rep	...	34	42	31	27	0	0	0	0	0
Czechoslovak	90 F	-	-	-	-	-	-	-	-	-
Estonia	-	3	1	1	1	2	1	2	4	4
Finland	311	311	442	442	229	229	158	158	181	181
Latvia	7	7	1	2	-	1	1	1	-	1
Lithuania	-	1	1	3	1	1	-	-	1	1
Poland	20	20	20	0	0	-	-	-	9	7
Romania	-	-	-	-	-	1	5	1	-	-
Russian Fed	2 263	2 560	2 093	1 977	2 914	3 037	3 280	2 604	2 378	2 257
Slovenia	18	22	25	19	17	17	16	14	15	14
Ukraine	21	...	6	...	...	...	...	...	...	...
05 Fishing area total	*2 771 F*	*3 011 F*	*2 658 F*	*2 500*	*3 216*	*3 309*	*3 487*	*2 805*	*2 593*	*2 483*
27 Estonia	98	187	165	98	131	88	69	50	61	36
Finland	359	356	302	309	263	262	242	238	287	290
Latvia	7	5	7	4	2	1	1	2	2	2
27 Fishing area total	*464*	*548*	*474*	*411*	*396*	*351*	*312*	*290*	*350*	*328*
Species total	*3 235 F*	*3 559 F*	*3 132 F*	*2 911*	*3 612*	*3 660*	*3 799*	*3 095*	*2 943*	*2 811*
Common dace	**Vandoise**		**Leucisco**			*Leuciscus leuciscus*			**1,40(02)020,05**	**FIE**
04 Turkey	221	242	215	223	215	250	300	176	104	91
04 Fishing area total	*221*	*242*	*215*	*223*	*215*	*250*	*300*	*176*	*104*	*91*
05 Slovenia	-	-	5	5	5	5	5	4	0	0
05 Fishing area total	-	-	*5*	*5*	*5*	*5*	*5*	*4*	*0*	*0*
Species total	*221*	*242*	*220*	*228*	*220*	*255*	*305*	*180*	*104*	*91*
Chub	**...B**		**...C**			*Leuciscus cephalus*			**1,40(02)020,08**	**LUH**
05 Bulgaria	...	...	...	...	...	92	121	150	43	4
Lithuania	-	-	-	-	-	-	-	-	3	3
Romania	-	-	-	-	-	-	45	17	24	28
05 Fishing area total	...	...	...	...	...	*92*	*166*	*167*	*70*	*35*
Species total	...	...	...	...	...	*92*	*166*	*167*	*70*	*35*
Chubs nei	**...B**		**...C**			*Leuciscus spp*			**1,40(02)020,XX**	**LEW**
05 Slovakia	...	46	81	63	47	39	32	38	33	31
05 Fishing area total	...	*46*	*81*	*63*	*47*	*39*	*32*	*38*	*33*	*31*
Species total	...	*46*	*81*	*63*	*47*	*39*	*32*	*38*	*33*	*31*
Rhinofishes nei	**Labéos nca**		**Labeos nep**			*Labeo spp*			**1,40(02)024,XX**	**RHI**
01 Ethiopia	...	...	...	...	1 994	2 007	3 168	3 621	3 451	3 387
Kenya	212	175	143	605	1 462	355	207	219	221	222
01 Fishing area total	*212*	*175*	*143*	*605*	*3 456*	*2 362*	*3 375*	*3 840*	*3 672*	*3 609*
Species total	*212*	*175*	*143*	*605*	*3 456*	*2 362*	*3 375*	*3 840*	*3 672*	*3 609*
Mud carp	**Carpe de vase**		**Carpa de fango**			*Cirrhinus molitorella*			**1,40(02)025,02**	**MUC**
04 China,Taiwan	11	6	6	5	3	4	7	7	6	6
04 Fishing area total	*11*	*6*	*6*	*5*	*3*	*4*	*7*	*7*	*6*	*6*
Species total	*11*	*6*	*6*	*5*	*3*	*4*	*7*	*7*	*6*	*6*
Grass carp(=White amur)	**Carpe herbivore(=chinoise)**		**Carpa china**			*Ctenopharyngodon idellus*			**1,40(02)035,01**	**FCG**
01 Egypt	...	...	...	10 000	15 343	16 553	4 233	707	12 826	17 918
01 Fishing area total	...	...	...	*10 000*	*15 343*	*16 553*	*4 233*	*707*	*12 826*	*17 918*
02 USA	8	8	8	19	10	15	13	11	17	31
02 Fishing area total	*8*	*8*	*8*	*19*	*10*	*15*	*13*	*11*	*17*	*31*
04 China,Taiwan	168	179	109	90	79	83	81	95	104	132
Iran	3 455	3 935	3 727	4 392	4 205	4 517	5 039	4 600	2 500	580
Kazakhstan	6	1	-	2	2 F	1 F	1 F	1 F	-	-
Kyrgyzstan	3	-	-	-	-	-	-	-	-	-
Nepal	2	6	-	-	-	-	-	-	-	-
Uzbekistan	93	500	13	19	10 F	30	92	7	17	13
04 Fishing area total	*3 727*	*4 621*	*3 849*	*4 503*	*4 296 F*	*4 631 F*	*5 213 F*	*4 703 F*	*2 621*	*725*
05 Albania	1	1 F	1 F	0	1	0	3	5	45	10
Bulgaria	-	-	-	3	8	8	17	20	12	18
Czech Rep	...	47	44	51	47	49	53	70	60	60
Czechoslovak	50 F	-	-	-	-	-	-	-	-	-
Germany	-	-	10	10	10	10	10	5	5	5
Hungary	358	343	381	366	346	305	301	318	356	309
Poland	2	2	0	3	4	11	4	2	4	4
Romania	544	397	360	170	124	12	46	21	83	73
Slovakia	...	14	19	23	15	21	16	17	15	9
Slovenia	-	-	-	-	-	-	2	2	3	2
Ukraine	2	-	3	1	2	-	-	-	0	-
05 Fishing area total	*957 F*	*804 F*	*818 F*	*627*	*557*	*416*	*452*	*460*	*583*	*490*

B-11

Carps, barbels and other cyprinids — **Capture production by species, fishing areas and countries or areas**
Carpes, barbeaux et autres cyprinidés — **Captures par espèces, zones de pêche et pays ou zones**
Carpas, barbos y otros ciprínidos — **Capturas por especies, áreas de pesca y países o áreas**

Species, Fishing area Espèce, Zone de pêche Especie, Area de pesca	1992 mt	1993 mt	1994 mt	1995 mt	1996 mt	1997 mt	1998 mt	1999 mt	2000 mt	2001 mt
Species total	*4 692 F*	*5 433 F*	*4 675 F*	*15 149*	*20 206 F*	*21 615 F*	*9 911 F*	*5 881 F*	*16 047*	*19 164*
Silver carp	**Carpe argentée**		**Carpa plateada**			***Hypophthalmichthys molitrix***		**1,40(02)043,01**		**SVC**
04 China,Taiwan	72	59	49	37	29	35	34	11	9	11
Iran	4 242	4 225	8 246	16 965	17 510	15 825	11 629	14 400	8 750	3 680
Israel	103	103	181	40	40	17	11	9	32	30 F
Kazakhstan	387	362	272	267	240 F	170 F	120 F	70 F	30 F	-
Kyrgyzstan	-	-	-	28	24 F	18 F	12 F	7 F	8 F	19
Nepal	9	42	-	-	-	-	-	-	-	-
Uzbekistan	2 754	916	402	1 003	481 F	893	249	322	586	544
04 Fishing area total	*7 567*	*5 707*	*9 150*	*18 340*	*18 324 F*	*16 958 F*	*12 055 F*	*14 819 F*	*9 415 F*	*4 284 F*
05 Albania	133	100 F	70 F	11	52	33	104	130	140	101
Bulgaria	...	...	250 F	415	488	471	553	488	42	403
Germany	-	-	76	76	76	76	76	39	39	39
Hungary	846	1 354	1 136	521	862	1 483	731	676	365	997
Poland	6	6	0	198	211	180	106	136	185	120
Romania	1 982	1 779	1 708	1 950	1 272	1 940	428	1 308	634	644
Slovakia	...	2	-	-	-	-	4	5	7	8
Slovenia	-	-	6	7	7	6	4	4	-	0
Ukraine	4 544	2 896	3 083	2 448	3 187	1 380	1 008	442	444	267
05 Fishing area total	*7 511*	*6 137 F*	*6 329 F*	*5 626*	*6 155*	*5 569*	*3 014*	*3 228*	*1 856*	*2 579*
37 Ukraine	-	-	-	2	1	-	-	-	-	-
37 Fishing area total	-	-	-	*2*	*1*	-	-	-	-	-
Species total	*15 078*	*11 844 F*	*15 479 F*	*23 968*	*24 480 F*	*22 527 F*	*15 069 F*	*18 047 F*	*11 271 F*	*6 863 F*
Bighead carp	**Carpe à grosse tête**		**Carpa cabezona**			***Hypophthalmichthys nobilis***		**1,40(02)043,02**		**BIC**
04 China,Taiwan	310	244	147	105	67	72	99	241	204	202
Iran	666	819	866	1 414	1 751	1 565	2 522	3 600	4 250	245
Nepal	5	25	-	-	-	-	-	-	-	-
04 Fishing area total	*981*	*1 088*	*1 013*	*1 519*	*1 818*	*1 637*	*2 621*	*3 841*	*4 454*	*447*
05 Croatia	-	-	-	1	-	-	-	-	-	-
Czech Rep	...	4	3	13	3	5	6	8	10	12
Germany	-	-	12	12	12	12	12	6	6	6
Hungary	513	291	45	127	74	83	-	-	-	-
Romania	1 148	1 047	1 103	700	449	299	369	396	356	313
05 Fishing area total	*1 661*	*1 342*	*1 163*	*853*	*538*	*399*	*387*	*410*	*372*	*331*
Species total	*2 642*	*2 430*	*2 176*	*2 372*	*2 356*	*2 036*	*3 008*	*4 251*	*4 826*	*778*
Silver cyprinid	**...B**		**...C**			***Rastrineobola argentea***		**1,40(02)070,01**		**ENA**
01 Kenya	35 414	42 505	69 134	56 827	49 670	40 315	42 336	48 816	49 618	41 384
01 Fishing area total	*35 414*	*42 505*	*69 134*	*56 827*	*49 670*	*40 315*	*42 336*	*48 816*	*49 618*	*41 384*
Species total	*35 414*	*42 505*	*69 134*	*56 827*	*49 670*	*40 315*	*42 336*	*48 816*	*49 618*	*41 384*
Vimba bream	**...B**		**...C**			***Vimba vimba***		**1,40(02)097,01**		**VIV**
05 Bulgaria	...	...	...	...	...	84	15	50	4	1
Estonia	-	0	0	0	1	0	0	0	0	0
Latvia	-	-	-	1	8	14	13	13	19	20
Lithuania	-	-	1	3	2	3	3	11	48	40
Poland	...	...	...	...	...	...	...	...	5	2
Romania	-	-	-	-	-	26	10	31	3	2
Slovakia	...	28	30	40	20	25	27	14	11	10
Slovenia	-	-	-	-	-	-	1	1	2	2
Ukraine	41	29	16	24	25	10	18	13	13	20
05 Fishing area total	*41*	*57*	*47*	*68*	*56*	*162*	*87*	*133*	*105*	*97*
27 Estonia	70	118	104	188	164	185	165	122	101	83
Latvia	35	49	39	54	50	43	79	106	73	89
27 Fishing area total	*105*	*167*	*143*	*242*	*214*	*228*	*244*	*228*	*174*	*172*
Species total	*146*	*224*	*190*	*310*	*270*	*390*	*331*	*361*	*279*	*269*
Sichel	**...B**		**Peleco**			***Pelecus cultratus***		**1,40(02)100,01**		**FSC**
04 Kazakhstan	-	-	-	27	25 F	20 F	15 F	10 F	10 F	8
Tajikistan	0	3	1	0	...	...	...	...	...	8
Turkmenistan	78	24	20	16	13	1	0	0	0	-
04 Fishing area total	*78*	*27*	*21*	*43*	*38 F*	*21 F*	*15 F*	*10 F*	*10 F*	*16*
05 Lithuania	-	-	-	-	-	-	4	3	8	12
Russian Fed	505	721	871	435	412	256	1 645	1 571	550	1 022
Ukraine	377	...	108	37	2	13	11	14	7	7
05 Fishing area total	*882*	*721*	*979*	*472*	*414*	*269*	*1 660*	*1 588*	*565*	*1 041*
27 Russian Fed	-	-	-	-	-	-	-	-	384	348
27 Fishing area total	-	-	-	-	-	-	-	-	*384*	*348*
37 Russian Fed	104	88	53	54	39	86	79	215	267	178
Ukraine	-	-	-	-	7	27	26	13	9	7

B-11 Carps, barbels and other cyprinids — Capture production by species, fishing areas and countries or areas
Carpes, barbeaux et autres cyprinidés — Captures par espèces, zones de pêche et pays ou zones
Carpas, barbos y otros ciprínidos — Capturas por especies, áreas de pesca y países o áreas

Species, Fishing area Espèce, Zone de pêche Especie, Area de pesca	1992 mt	1993 mt	1994 mt	1995 mt	1996 mt	1997 mt	1998 mt	1999 mt	2000 mt	2001 mt
37 Fishing area total	*104*	*88*	*53*	*54*	*46*	*113*	*105*	*228*	*276*	*185*
Species total	*1 064*	*836*	*1 053*	*569*	*498 F*	*403 F*	*1 780 F*	*1 826 F*	*1 235 F*	*1 590*
Asp	**Aspe**		**Aspio**		***Aspius aspius***			**1,40(02)115,01**		**ASU**
04 Azerbaijan	63	31	14	5	9	6	4	2	1	5
Kazakhstan	565	313	314	415	380 F	270 F	195 F	250 F	350 F	476
Tajikistan	0	11	8	6	...	...	...	...	...	6
Turkmenistan	6	8	7	6	10	6	12	8	11	16
Uzbekistan	...	...	...	44	20 F	38	32	37	56	22
04 Fishing area total	*634*	*363*	*343*	*476*	*419 F*	*320 F*	*243 F*	*297 F*	*418 F*	*525*
05 Bulgaria	-	-	-	3	4	5	7	8	9	7
Czech Rep	...	-	-	-	10	15	16	16	13	17
Czechoslovak	30 F	-	-	-	-	-	-	-	-	-
Hungary	...	39	52	20	22	44	38	42	38	21
Lithuania	-	0	5	3	4	6	9	6	6	5
Poland	...	...	...	...	...	...	...	...	6	5
Romania	-	-	-	-	-	-	11	7	17	4
Russian Fed	141	51	...	55	84	36	45	53	51	56
Slovakia	...	8	16	13	12	9	9	9	8	8
Slovenia	-	1	-	-	-	-	0	-	0	0
Ukraine	32	20	16	18	25	19	7	15	26	7
05 Fishing area total	*203 F*	*119*	*89*	*112*	*161*	*134*	*142*	*156*	*174*	*130*
27 Russian Fed	-	-	-	-	-	-	-	-	6	9
27 Fishing area total	-	-	-	-	-	-	-	-	*6*	*9*
Species total	*837 F*	*482*	*432*	*588*	*580 F*	*454 F*	*385 F*	*453 F*	*598 F*	*664*
Hoven's carp	**Barbus d'Hoven**		**Barbo de Hoven**		***Leptobarbus hoeveni***			**1,40(02)132,01**		**FCH**
04 Indonesia	3 702	4 606	5 376	5 454	6 892	5 836	3 241	4 608	3 149	3 260
04 Fishing area total	*3 702*	*4 606*	*5 376*	*5 454*	*6 892*	*5 836*	*3 241*	*4 608*	*3 149*	*3 260*
Species total	*3 702*	*4 606*	*5 376*	*5 454*	*6 892*	*5 836*	*3 241*	*4 608*	*3 149*	*3 260*
Black carp	**Carpe noire**		**Carpa negra**		***Mylopharyngodon piceus***			**1,40(02)144,01**		**BKC**
04 China,Taiwan	34	22	20	20	15	16	22	27	34	36
04 Fishing area total	*34*	*22*	*20*	*20*	*15*	*16*	*22*	*27*	*34*	*36*
Species total	*34*	*22*	*20*	*20*	*15*	*16*	*22*	*27*	*34*	*36*
Wuchang bream	**Carpe de Wuchang**		**Carpa de Wuchang**		***Megalobrama amblycephala***			**1,40(02)151,01**		**WUB**
05 Albania	1	1 F	1 F	0	0	0	-	-	-	-
05 Fishing area total	*1*	*1 F*	*1 F*	*0*	*0*	*0*	-	-	-	-
Species total	*1*	*1 F*	*1 F*	*0*	*0*	*0*	-	-	-	-
Java barb	**Barbeau de Java**		**Barbo de Java**		***Puntius javanicus***			**1,40(02)161,02**		**FJB**
04 Indonesia	16 082	15 027	19 084	18 102	19 601	19 469	20 189	17 939	17 124	17 080
04 Fishing area total	*16 082*	*15 027*	*19 084*	*18 102*	*19 601*	*19 469*	*20 189*	*17 939*	*17 124*	*17 080*
Species total	*16 082*	*15 027*	*19 084*	*18 102*	*19 601*	*19 469*	*20 189*	*17 939*	*17 124*	*17 080*
Asian barbs nei	**Barbeaux d'Asie nca**		**Barbos de Asia nep**		***Puntius spp***			**1,40(02)161,XX**		**FAB**
04 Indonesia	10 489	10 523	12 113	12 344	13 254	11 976	12 131	11 263	11 745	11 420
Thailand	22 361	23 061	25 567	22 468	25 750	25 296	44 349	45 511	41 000	42 800
04 Fishing area total	*32 850*	*33 584*	*37 680*	*34 812*	*39 004*	*37 272*	*56 480*	*56 774*	*52 745*	*54 220*
Species total	*32 850*	*33 584*	*37 680*	*34 812*	*39 004*	*37 272*	*56 480*	*56 774*	*52 745*	*54 220*
Cyprinids nei	**Cyprinidés nca**		**Ciprínidos nep**		***Cyprinidae***			**1,40(02)XXX,XX**		**FCY**
01 Egypt	...	928	2 154	1 059	1 261	1 386	2 117	4 515	1 232	5 040
Ethiopia	-	-	-	362	387	639	768	878	860	843
Kenya	145	136	127	118	0	0	0	0	0	0
Madagascar	6 000	5 100	4 313	4 123	4 000	4 000	4 000	4 000	4 000	4 000
Malawi	3 142	6 775	7 265	538	505	522	299	8 302	7 500 F	6 291
Morocco	230	500	1 000	900	800	800	900	1 200	1 000	600
Nigeria	2 054	3 293	3 021	4 476	4 460	6 199	5 006	6 296	4 823	5 901
Uganda	2 000	1 149	2 872	2 948	1 539	13 476	15 710	17 600	12 181	12 182
01 Fishing area total	*13 571*	*17 881*	*20 752*	*14 524*	*12 952*	*27 022*	*28 800*	*42 791*	*31 596 F*	*34 857*
02 Mexico	2 886	3 086	3 255	1 739	3 732	4 838	3 982	6 805	3 988	3 806
USA	6	3	10	4	3	5	15	17	29	33
02 Fishing area total	*2 892*	*3 089*	*3 265*	*1 743*	*3 735*	*4 843*	*3 997*	*6 822*	*4 017*	*3 839*
03 Brazil	100 F	100 F	100 F	119	284	212	258	302	355	360 F
03 Fishing area total	*100 F*	*100 F*	*100 F*	*119*	*284*	*212*	*258*	*302*	*355*	*360 F*
04 Armenia	231	312	46	42	33	7	37	21	19	15
China,H.Kong	0	0	0	0	0	0	0	0	0	0
India	195 074	258 502	314 939	339 828	220 994	194 608	167 186	177 394	152 350	148 622

B-11 Carps, barbels and other cyprinids — Capture production by species, fishing areas and countries or areas
Carpes, barbeaux et autres cyprinidés — Captures par espèces, zones de pêche et pays ou zones
Carpas, barbos y otros ciprínidos — Capturas por especies, áreas de pesca y países o áreas

Species, Fishing area Espèce, Zone de pêche Especie, Area de pesca		1992 mt	1993 mt	1994 mt	1995 mt	1996 mt	1997 mt	1998 mt	1999 mt	2000 mt	2001 mt
	Iran	9 783	7 748	8 001	7 802	12 364	12 870	19 573	10 800	10 000	2 680
	Iraq	7 455	4 836	10 213	9 981	7 694	4 782	1 703	5 723	4 013	4 000
	Israel	1 346	1 015	352	483	1 167	641	1 235	1 128	1 112	960 F
	Korea Rep	4 219	6 262	4 626	3 311	1 010	2 221	1 977	2 294	...	...
	Kyrgyzstan	111	49	30	39	35 F	26 F	18 F	11 F	12 F	8
	Laos	2 500 F	2 500 F	3 200 F	4 000 F	3 500 F	2 800 F	3 000 F	4 500 F	4 400 F	4 500 F
	Lebanon	10 F	10	10	10	10	10	10	10	10	10
	Philippines	11 444	7 117	5 568	8 880	6 497	5 717	6 453	4 677	5 032	5 562
	Tajikistan	-	-	-	-	-	-	-	-	-	31
	Turkey	1 031	1 303	1 499	1 535	1 380	1 900	1 800	406	699	626
	Uzbekistan	1 253	350	585	0	-	-	-	-	87	-
04	*Fishing area total*	*234 457 F*	*290 004 F*	*349 069 F*	*375 911 F*	*254 684 F*	*225 582 F*	*202 992 F*	*206 964 F*	*177 734 F*	*167 014 F*
05	Belarus	164	221	122	105	92	61	47	27	19	77
	Belgium	50	50	50	50	50	50	50	50	50	50
	Bulgaria	...	...	...	...	...	1	251	250	102	2
	Czechoslovak	355 F	-	-	-	-	-	-	-	-	-
	Finland	98	193	277	277	15	15	-	-	-	-
	Germany	1 280	1 156	1 653	1 653	1 653	1 653	1 653	1 653	1 653	1 653
	Greece	...	...	...	...	176	107	107	74	90	80
	Hungary	1 623	1 342	1 557	1 744	1 731	1 730	1 510	1 666	1 710	2 155
	Italy	2 503	2 450	2 628	2 540	1 146	2 378	1 155	1 900	725	1 821
	Latvia	-	-	30	36	34	42	26	27	-	-
	Lithuania	9	5	1	3	1	2	-	-	-	-
	Poland	0	0	0	0	0	395	330	-	2	2
	Romania	300	274	337	153	112	16	396	126	39	298
	Russian Fed	15 750	16 754	9 306	11 887	9 852	17 067	23 353	20 331	19 906	18 008
	Slovakia	...	51	71	64	8	44	37	38	53	12
	Slovenia	65	60	71	59	54	40	42	-	-	-
	Switzerland	49	4	4	34	42	20	10	6	8	13
	Ukraine	4	-	-	-	-	-	-	-	-	-
05	*Fishing area total*	*22 250 F*	*22 560*	*16 107*	*18 605*	*14 966*	*23 621*	*28 967*	*26 148*	*24 357*	*24 171*
27	Finland	-	-	-	-	8	8	-	-	-	-
	Lithuania	-	-	-	-	-	-	-	-	-	19
	Russian Fed	-	-	-	-	-	-	-	-	67	88
	Sweden	0	0	1	0	0	0	-	-	-	-
27	*Fishing area total*	*0*	*0*	*1*	*0*	*8*	*8*	-	-	*67*	*107*
37	Russian Fed	-	5	-	-	-	1	-	-	167	159
37	*Fishing area total*	-	*5*	-	-	-	*1*	-	-	*167*	*159*
61	Russian Fed	-	-	55	61	110	101	100	56	8	52
61	*Fishing area total*	...	...	*55*	*61*	*110*	*101*	*100*	*56*	*8*	*52*
Species total		*273 270 F*	*333 639 F*	*389 349 F*	*410 963 F*	*286 739 F*	*281 390 F*	*265 114 F*	*283 083 F*	*238 301 F*	*230 559 F*
Group total		***619 683***	***673 996***	***756 801***	***780 522***	***663 445***	***608 078***	***604 993***	***613 612***	***569 212***	***547 623***

B-12

Tilapias and other cichlids — Capture production by species, fishing areas and countries or areas
Tilapias et autres cichlidés — Captures par espèces, zones de pêche et pays ou zones
Tilapias y otros cíclidos — Capturas por especies, áreas de pesca y países o áreas

Species, Fishing area Espèce, Zone de pêche Especie, Area de pesca	1992 mt	1993 mt	1994 mt	1995 mt	1996 mt	1997 mt	1998 mt	1999 mt	2000 mt	2001 mt
Mozambique tilapia	**Tilapia du Mozambique**		**Tilapia del Mozambique**		***Oreochromis mossambicus***			**1,70(59)051,01**		**TLM**
04 Indonesia	9 749	15 849	15 337	13 293	16 943	17 715	17 865	21 172	19 119	19 550
04 Fishing area total	*9 749*	*15 849*	*15 337*	*13 293*	*16 943*	*17 715*	*17 865*	*21 172*	*19 119*	*19 550*
06 Papua N Guin	2 310 F	2 310 F	2 310 F	2 310	2 310 F	2 310 F	2 310 F	2 310 F	2 310 F	2 310 F
06 Fishing area total	*2 310 F*	*2 310 F*	*2 310 F*	*2 310*	*2 310 F*	*2 310 F*	*2 310 F*	*2 310 F*	*2 310 F*	*2 310 F*
Species total	*12 059 F*	*18 159 F*	*17 647 F*	*15 603*	*19 253 F*	*20 025 F*	*20 175 F*	*23 482 F*	*21 429 F*	*21 860 F*
Nile tilapia	**Tilapia du Nil**		**Tilapia del Nilo**		***Oreochromis niloticus***			**1,70(59)051,02**		**TLN**
01 Burundi	40	30 F	50 F	50	50	50	50	50	120	120
Egypt	93 495	97 679	95 819	122 207	125 307	130 992	128 446	112 811	131 276	145 291
Kenya	16 300	12 196	11 319	11 827	10 765	13 953	14 652	17 524	19 347	7 292
Mali	20 540	19 290	18 855	39 842	30 450	1 719	26 675	26 821	32 961	30 000 F
Niger	2	...	...	...	...	...	...	...	...	...
Rwanda	...	...	...	...	...	...	2 233	2 640	2 646	2 650
Sudan	...	...	...	10 000	10 500	11 000	16 000	16 000	18 000 F	20 000
01 Fishing area total	*130 377*	*129 195 F*	*126 043 F*	*183 926*	*177 072*	*157 714*	*188 056*	*175 846*	*204 350 F*	*205 353 F*
02 El Salvador	3 863	2 801	2 323	2 494	1 265	1 297	1 017	1 216	1 171	560
Jamaica	150 F	150 F	150 F	150 F	150 F	150 F	150 F	150 F	150 F	150 F
02 Fishing area total	*4 013 F*	*2 951 F*	*2 473 F*	*2 644 F*	*1 415 F*	*1 447 F*	*1 167 F*	*1 366 F*	*1 321 F*	*710 F*
04 Thailand	41 365	53 903	63 400	55 746	29 253	28 727	40 173	49 840	40 000	41 740
04 Fishing area total	*41 365*	*53 903*	*63 400*	*55 746*	*29 253*	*28 727*	*40 173*	*49 840*	*40 000*	*41 740*
06 Fiji Islands	12	7	8 F	6	37	40	288	290	280 F	278
06 Fishing area total	*12*	*7*	*8 F*	*6*	*37*	*40*	*288*	*290*	*280 F*	*278*
Species total	*175 767 F*	*186 056 F*	*191 924 F*	*242 322 F*	*207 777 F*	*187 928 F*	*229 684 F*	*227 342 F*	*245 951 F*	*248 081 F*
Blue tilapia	**...B**		**...C**		***Oreochromis aureus***			**1,70(59)051,03**		**OEA**
02 Cuba	13 699	9 348	9 453	8 386	9 672	7 553	5 122	4 564	4 500 F	4 500 F
02 Fishing area total	*13 699*	*9 348*	*9 453*	*8 386*	*9 672*	*7 553*	*5 122*	*4 564*	*4 500 F*	*4 500 F*
04 Israel	188	160	228	69	32	101	88	66	67	60 F
04 Fishing area total	*188*	*160*	*228*	*69*	*32*	*101*	*88*	*66*	*67*	*60 F*
Species total	*13 887*	*9 508*	*9 681*	*8 455*	*9 704*	*7 654*	*5 210*	*4 630*	*4 567 F*	*4 560 F*
Tilapias nei	**Tilapias nca**		**Tilapias nep**		***Oreochromis (=Tilapia) spp***			**1,70(59)051,XX**		**TLP**
01 Benin	7 390 F	10 200 F	10 235	11 713	11 500 F	11 500 F	11 500 F	11 648	9 600 F	10 900 F
Botswana	400 F	300 F	200 F	100 F	48	80	88	93	92	88
Ethiopia	4 485	4 175	5 285	3 703	5 066	6 076	7 952	9 088	7 000	6 870
Gabon	1 250 F	1 500 F	1 500 F	2 600 F	3 022	3 630	3 500	3 800	3 800	3 122
Gambia	1 100	1 000	1 000	1 050	1 050	1 050	1 050	1 050 F	1 050 F	1 050 F
Kenya	8 395	7 886	6 354	7 568	7 635	19 329	19 732	18 203	19 853	19 121
Malawi	13 357	10 219	6 635	3 863	5 080	4 472	5 104	6 808	6 200 F	5 154
Mauritius	0	3	0	0	0	0	0	0	0	0
Nigeria	11 046	8 003	7 874	9 060	10 074	12 614	16 300	19 662	13 402	18 332
Senegal	...	...	...	...	...	...	...	...	9 133	8 100 F
Tanzania	41 650	20 213	19 610	25 869	35 160	25 100	24 000	38 000	40 000	45 000
Togo	4 500	5 000	3 500	3 500	3 500	3 500	3 500	3 500	3 500	3 500
Uganda	109 000	74 280	80 176	83 223	75 027	81 379	78 500	84 540	96 468	96 172
Zimbabwe	760	370	440	300	320	523	412	420	830	800 F
01 Fishing area total	*203 333 F*	*143 149 F*	*142 809 F*	*152 549 F*	*157 482 F*	*169 253 F*	*171 638 F*	*196 812 F*	*210 928 F*	*218 209 F*
02 Dominican Rp	255	328	659	1 188	100	207	77	320	...	529
Mexico	72 364	75 835	75 024	74 646	74 354	74 814	65 178	59 343	68 772	60 336
Nicaragua	...	...	...	...	...	...	...	750	680	660
Panama	-	-	-	10	15	56	11	13	16	16 F
USA	2 409	1 732	1 098	2 717	4	0	0	2 651	1 040	1 277
02 Fishing area total	*75 028*	*77 895*	*76 781*	*78 561*	*74 473*	*75 077*	*65 266*	*63 077*	*70 508*	*62 818 F*
03 Brazil	9 115 F	8 400 F	7 700 F	6 234	7 300	6 927	7 444	6 616	7 893	7 900 F
03 Fishing area total	*9 115 F*	*8 400 F*	*7 700 F*	*6 234*	*7 300*	*6 927*	*7 444*	*6 616*	*7 893*	*7 900 F*
04 China,Taiwan	632	427	607	319	145	146	152	86	79	98
Philippines	19 458	19 071	17 649	21 244	17 663	20 935	23 477	25 278	28 874	28 881
Sri Lanka	17 500	15 000	9 500	15 000	22 250	27 250	29 900	28 520	33 220	27 230
04 Fishing area total	*37 590*	*34 498*	*27 756*	*36 563*	*40 058*	*48 331*	*53 529*	*53 884*	*62 173*	*56 209*
06 Guam	-	-	-	-	-	-	6	-	-	-
06 Fishing area total	-	-	-	-	-	-	*6*	-	-	-
34 Gabon	...	...	...	...	598	520	19	378	421	301
Gambia	18	75	5	94	198	85	23	20	12	27
Senegal	1 600	1 024	2 803	2 382	1 807	1 977	1 323	1 362	1 629	1 996
34 Fishing area total	*1 618*	*1 099*	*2 808*	*2 476*	*2 603*	*2 582*	*1 365*	*1 760*	*2 062*	*2 324*
Species total	*326 684 F*	*265 041 F*	*257 854 F*	*276 383 F*	*281 916 F*	*302 170 F*	*299 248 F*	*322 149 F*	*353 564 F*	*347 460 F*
Mango tilapia	**...B**		**...C**		***Sarotherodon galilaeus***			**1,70(59)052,01**		**SAR**
04 Israel	248	215	362	316	350	462	391	405	262	230 F

B-12 Tilapias and other cichlids / Tilapias et autres cichlidés / Tilapias y otros cíclidos

Capture production by species, fishing areas and countries or areas / Captures par espèces, zones de pêche et pays ou zones / Capturas por especies, áreas de pesca y países o áreas

Species, Fishing area Espèce, Zone de pêche Especie, Area de pesca	1992 mt	1993 mt	1994 mt	1995 mt	1996 mt	1997 mt	1998 mt	1999 mt	2000 mt	2001 mt
04 Fishing area total	*248*	*215*	*362*	*316*	*350*	*462*	*391*	*405*	*262*	*230 F*
Species total	*248*	*215*	*362*	*316*	*350*	*462*	*391*	*405*	*262*	*230 F*
Jaguar guapote	**...B**		**Guapote tigre**			***Cichlasoma managuense***			**1,70(59)053,06**	**CHL**
02 El Salvador	394	464	382	608	566	367	344	351	324	410
02 Fishing area total	*394*	*464*	*382*	*608*	*566*	*367*	*344*	*351*	*324*	*410*
Species total	*394*	*464*	*382*	*608*	*566*	*367*	*344*	*351*	*324*	*410*
Peacock cichlid	**...B**		**Sargento**			***Cichla ocellaris***			**1,70(59)054,04**	**CLO**
02 Panama	80	28	285	120	65	35	12	7	4	4 F
02 Fishing area total	*80*	*28*	*285*	*120*	*65*	*35*	*12*	*7*	*4*	*4 F*
Species total	*80*	*28*	*285*	*120*	*65*	*35*	*12*	*7*	*4*	*4 F*
Velvety cichlids	**...B**		**Acarahuazu**			***Astronotus spp***			**1,70(59)313,XX**	**AST**
03 Peru	...	...	...	320	308	177	141	251	188	183
03 Fishing area total	*...*	*...*	*...*	*320*	*308*	*177*	*141*	*251*	*188*	*183*
Species total	*...*	*...*	*...*	*320*	*308*	*177*	*141*	*251*	*188*	*183*
Mouthbrooding cichlids	**...B**		**...C**			***Haplochromis spp***			**1,70(59)316,XX**	**CIM**
01 Kenya	3 018	3 506	4 196	4 822	3 914	2 453	2 577	2 731	2 699	1 195
Tanzania	13 500	9 122	13 204	11 960	6 558	10 600	10 500	8 000	9 000	8 500
01 Fishing area total	*16 518*	*12 628*	*17 400*	*16 782*	*10 472*	*13 053*	*13 077*	*10 731*	*11 699*	*9 695*
Species total	*16 518*	*12 628*	*17 400*	*16 782*	*10 472*	*13 053*	*13 077*	*10 731*	*11 699*	*9 695*
Cichlids nei	**Cichlidés nca**		**Cíclidos nep**			***Cichlidae***			**1,70(59)XXX,XX**	**CIX**
01 Madagascar	18 000	21 100	21 500	21 600	21 500	21 500	21 500	21 500	21 500	21 500
Malawi	...	...	...	...	...	...	...	18 841	19 000 F	20 533
01 Fishing area total	*18 000*	*21 100*	*21 500*	*21 600*	*21 500*	*21 500*	*21 500*	*40 341*	*40 500 F*	*42 033*
02 Guatemala	-	-	2	5	0	21	157	173	192	200 F
02 Fishing area total	*-*	*-*	*2*	*5*	*0*	*21*	*157*	*173*	*192*	*200 F*
03 Brazil	14 720 F	15 100 F	15 450 F	12 045	8 565	8 356	7 785	7 375	8 145	8 200 F
03 Fishing area total	*14 720 F*	*15 100 F*	*15 450 F*	*12 045*	*8 565*	*8 356*	*7 785*	*7 375*	*8 145*	*8 200 F*
04 Israel	95	97	90	64	40	48	58	57	52	40 F
04 Fishing area total	*95*	*97*	*90*	*64*	*40*	*48*	*58*	*57*	*52*	*40 F*
Species total	*32 815 F*	*36 297 F*	*37 042 F*	*33 714*	*30 105*	*29 925*	*29 500*	*47 946*	*48 889 F*	*50 473 F*
Group total	***578 452***	***528 396***	***532 577***	***594 623***	***560 516***	***561 796***	***597 782***	***637 294***	***686 877***	***682 956***

B-13

Miscellaneous freshwater fishes — **Capture production by species, fishing areas and countries or areas**
Poissons d'eau douce divers — **Captures par espèces, zones de pêche et pays ou zones**
Peces de agua dulce diversos — **Capturas por especies, áreas de pesca y países o áreas**

Species, Fishing area Espèce, Zone de pêche Especie, Area de pesca	1992 mt	1993 mt	1994 mt	1995 mt	1996 mt	1997 mt	1998 mt	1999 mt	2000 mt	2001 mt
African lungfishes	**Protoptères d'Afrique**		**Protopteros africanos**		***Protopterus spp***				**1,16(02)002,XX**	**FLU**
01 Benin	32 F	100 F	127	121	110 F	105 F	100 F	100 F	100 F	110 F
Kenya	1 544	146	398	408	164	1 704	1 717	1 600	1 608	1 822
Nigeria	587	798	436	678	1 284	3 950	1 679	536	1 364	2 183
Uganda	4 500	3 949	7 868	7 724	6 095	6 554	7 595	9 192	5 755	5 796
01 Fishing area total	*6 663 F*	*4 993 F*	*8 829*	*8 931*	*7 653 F*	*12 313 F*	*11 091 F*	*11 428 F*	*8 827 F*	*9 911 F*
Species total	*6 663 F*	*4 993 F*	*8 829*	*8 931*	*7 653 F*	*12 313 F*	*11 091 F*	*11 428 F*	*8 827 F*	*9 911 F*
Dagaas	**Dagaas**		**Dagaas**		***Stolothrissa, Limnothrissa spp***				**1,21(05)XXX,XX**	**DAG**
01 Burundi	19 990	14 100 F	18 200 F	18 163	1 508	17 868	8 646	7 030	10 839	6 138
Mozambique	-	689	925	3 093	5 574	9 921	7 313	9 052	11 813	7 076
Tanzania	36 000	44 582	51 467	50 360	40 179	48 000	46 800	42 000	40 000	43 000
Zambia	8 658	9 722	8 910	8 674	7 593	7 813	9 822	8 955	8 863	8 500 F
Zimbabwe	18 941	19 958	19 232	15 280	15 423	17 034	15 288	11 208	10 477	10 400 F
01 Fishing area total	*83 589*	*89 051 F*	*98 734 F*	*95 570*	*70 277*	*100 636*	*87 869*	*78 245*	*81 992*	*75 114 F*
Species total	*83 589*	*89 051 F*	*98 734 F*	*95 570*	*70 277*	*100 636*	*87 869*	*78 245*	*81 992*	*75 114 F*
Northern pike	**Brochet du Nord**		**Lucio**		***Esox lucius***				**1,24(03)001,01**	**FPI**
01 Algeria	-	-	-	-	-	-	2	...	...	...
01 Fishing area total	-	-	-	-	-	-	*2*	...	...	...
02 Canada	3 266	3 424	1 787	2 496	2 406	2 334	2 568	2 615	2 808	2 464
02 Fishing area total	*3 266*	*3 424*	*1 787*	*2 496*	*2 406*	*2 334*	*2 568*	*2 615*	*2 808*	*2 464*
04 Azerbaijan	85	82	27	15	7	23	18	21	28	25
Kazakhstan	1 044	733	564	637	580 F	420 F	300 F	550 F	700 F	819
Turkey	277	304	406	453	225	350	200	276	224	192
Uzbekistan	10	282	122	36	20 F	7	10	60	23	24
04 Fishing area total	*1 416*	*1 401*	*1 119*	*1 141*	*832 F*	*800 F*	*528 F*	*907 F*	*975 F*	*1 060*
05 Belarus	60	468	42	29	27	11	9	12	162	171
Belgium	20	20	20	20	20	20	20	20	20	20
Bulgaria	-	-	-	0	0	19	41	74	14	2
Croatia	18	23	22	30	30	30	1 F	1	1	1
Czech Rep	...	133	173	82	163	159	168	183	180	176
Czechoslovak	265 F	-	-	-	-	-	-	-	-	-
Denmark	17	7	8	6	5	7	11	8	5	8
Estonia	...	103	47	42	97	82	114	133	156	180
Finland	13 320	13 334	9 704	9 681	9 763	9 763	7 621	7 621	7 887	7 887
Germany	476	480	205	205	205	205	205	205	205	205
Greece	...	...	...	...	12	9	5	8	10	9
Hungary	168	165	190	37	46	203	158	241	280	191
Ireland	1 500	1 493	2 000	2 000	2 000	2 000	2 000	...	...	...
Latvia	25	30	41	55	56	47	55	72	72	73
Lithuania	18	62	78	78	66	71	56	62	71	68
Moldova Rep	25	39	45	56	53	39	38	25	19	36
Netherlands	1	1	-	-	-	-	-	-	-	-
Norway	...	...	...	...	...	13	7	...	...	...
Poland	350	350	350	245	1 076	280	226	262	363	302
Portugal	0	0	0	0	0	0	-	-	0	0
Romania	27	8	10	11	8	4	40	47	47	95
Russian Fed	6 062	5 338	4 367	4 938	5 688	4 646	5 567	6 118	8 843	8 427
Slovakia	...	89	193	134	103	85	68	69	76	73
Slovenia	9	9	12	11	11	12	9	10	11	9
Sweden	205	168	158	154	156	149	139	177	175	145
Switzerland	49	35	38	60	36	34	33	34	48	48
Ukraine	266	10	136	8	26	23	15	18	15	14
UK	7	5	7	4	5	2	4	4	6	5
Yugoslavia	...	...	...	...	...	...	...	22	...	...
05 Fishing area total	*22 888 F*	*22 370*	*17 846*	*17 886*	*19 652*	*17 913*	*16 610 F*	*15 426*	*18 666*	*18 145*
27 Denmark	7	6	6	5	6	7	5	3	4	3
Estonia	37	65	37	30	42	23	17	19	21	19
Finland	4 196	4 202	2 175	2 150	2 582	2 614	4 028	4 042	2 562	2 541
Germany	-	-	135	112	83	101	99	131	117	104
Poland	4	1	0	0	0	-	-	-	-	9
Russian Fed	-	-	-	-	-	-	-	-	21	17
Sweden	326	279	272	240	204	189	163	145	115	97
27 Fishing area total	*4 570*	*4 553*	*2 625*	*2 537*	*2 917*	*2 934*	*4 312*	*4 340*	*2 840*	*2 790*
Species total	*32 140 F*	*31 748*	*23 377*	*24 060*	*25 807 F*	*23 981 F*	*24 020 F*	*23 288 F*	*25 289 F*	*24 459*
Arapaima	**...B**		**Paiche**		***Arapaima gigas***				**1,28(01)001,01**	**ARP**
03 Peru	...	...	...	420	457	465	210	338	273	204
03 Fishing area total	...	...	...	*420*	*457*	*465*	*210*	*338*	*273*	*204*
Species total	...	...	...	*420*	*457*	*465*	*210*	*338*	*273*	*204*
Bonytongues nei	**...B**		**...C**		***Heterotis spp***				**1,28(01)004,XX**	**HTT**
01 Benin	504 F	400 F	349	742	700 F	650 F	600 F	600 F	500 F	550 F
Mali	685	643	629	1 328	453	329	397	399	1 099	1 000 F
Nigeria	4 533	4 375	2 897	4 832	3 298	4 737	7 388	8 873	7 915	8 577

B-13

Miscellaneous freshwater fishes — Capture production by species, fishing areas and countries or areas
Poissons d'eau douce divers — Captures par espèces, zones de pêche et pays ou zones
Peces de agua dulce diversos — Capturas por especies, áreas de pesca y países o áreas

Species, Fishing area Espèce, Zone de pêche Especie, Area de pesca	1992 mt	1993 mt	1994 mt	1995 mt	1996 mt	1997 mt	1998 mt	1999 mt	2000 mt	2001 mt
01 Fishing area total	*5 722 F*	*5 418 F*	*3 875*	*6 902*	*4 451 F*	*5 716 F*	*8 385 F*	*9 872 F*	*9 514 F*	*10 127 F*
Species total	*5 722 F*	*5 418 F*	*3 875*	*6 902*	*4 451 F*	*5 716 F*	*8 385 F*	*9 872 F*	*9 514 F*	*10 127 F*
Knifefishes	**...B**		**...C**		***Notopterus spp***				**1,28(02)003,XX**	**FKN**
01 Nigeria	633	...	703	383	10	1	144	33	39	22
01 Fishing area total	*633*	*...*	*703*	*383*	*10*	*1*	*144*	*33*	*39*	*22*
04 Indonesia	6 342	4 256	3 866	4 790	3 881	3 455	3 599	3 845	4 007	4 050
04 Fishing area total	*6 342*	*4 256*	*3 866*	*4 790*	*3 881*	*3 455*	*3 599*	*3 845*	*4 007*	*4 050*
Species total	*6 975*	*4 256*	*4 569*	*5 173*	*3 891*	*3 456*	*3 743*	*3 878*	*4 046*	*4 072*
Bottlenose fishes nei	**...B**		**...C**		***Mormyrus spp***				**1,28(06)013,XX**	**MZU**
01 Mali	...	...	...	...	...	...	...	...	7 691	7 000 F
01 Fishing area total	*...*	*...*	*...*	*...*	*...*	*...*	*...*	*...*	*7 691*	*7 000 F*
Species total	*...*	*...*	*...*	*...*	*...*	*...*	*...*	*...*	*7 691*	*7 000 F*
Cachama	**...B**		**Cachama**		***Colossoma macropomum***				**1,38(01)046,05**	**CSM**
03 Brazil	...	...	...	11 379	4 298	2 838	2 760	2 905	4 965	5 000 F
Peru	...	...	...	...	693	762	254	1 016	963	904
03 Fishing area total	*...*	*...*	*...*	*11 379*	*4 991*	*3 600*	*3 014*	*3 921*	*5 928*	*5 904 F*
Species total	*...*	*...*	*...*	*11 379*	*4 991*	*3 600*	*3 014*	*3 921*	*5 928*	*5 904 F*
Pirapatinga	**...B**		**Cachama blanca**		***Piaractus brachypomus***				**1,38(01)076,01**	**CSD**
03 Peru	...	...	...	...	397	...	166	648	324	487
03 Fishing area total	*...*	*...*	*...*	*...*	*397*	*...*	*166*	*648*	*324*	*487*
Species total	*...*	*...*	*...*	*...*	*397*	*...*	*166*	*648*	*324*	*487*
Characins nei	**Characinidés nca**		**Carácidos nep**		***Characidae***				**1,38(01)XXX,XX**	**CHA**
01 Mali	3 423	3 215	3 142	6 650	9 504	6 969	8 316	8 362	5 493	5 000 F
Nigeria	337	4 309	7 169	8 802	7 141	7 646	19 835	11 883	12 500	10 274
Uganda	6 500	10 033	10 000	10 400	12 160	11 718	11 744	13 165	10 331	10 331
01 Fishing area total	*10 260*	*17 557*	*20 311*	*25 852*	*28 805*	*26 333*	*39 895*	*33 410*	*28 324*	*25 605 F*
03 Argentina	7 000	8 000	8 700 F	11 700 F	13 000 F	15 500 F	16 000 F	2 800 F	3 000 F	2 300 F
Brazil	85 000 F	83 000 F	81 000 F	81 433	80 598	65 273	60 983	65 149	67 297	67 500 F
Colombia	24 870	12 834	8 832	5 698	5 504	3 730	6 427	8 667	6 600 F	6 600 F
Paraguay	6 500	7 000 F	7 350 F	7 700 F	8 000 F	10 000 F	9 000 F	9 000 F	9 000 F	9 000 F
Uruguay	275	439	872	763	338	1 924	1 687	1 762	1 690	107
Venezuela	3 811	4 878	5 827	8 778	8 050	6 029	6 921	4 817	5 773	2 978
03 Fishing area total	*127 456 F*	*116 151 F*	*112 581 F*	*116 072 F*	*115 490 F*	*102 456 F*	*101 018 F*	*92 195 F*	*93 360 F*	*88 485 F*
Species total	*137 716 F*	*133 708 F*	*132 892 F*	*141 924 F*	*144 295 F*	*128 789 F*	*140 913 F*	*125 605 F*	*121 684 F*	*114 090 F*
Netted prochilod	**Prochilode réticulé**		**Boquichico reticulado**		***Prochilodus reticulatus***				**1,38(12)001,20**	**PLR**
03 Peru	...	...	...	5 088	7 448	2 007	6 117	9 768	10 942	11 220
03 Fishing area total	*...*	*...*	*...*	*5 088*	*7 448*	*2 007*	*6 117*	*9 768*	*10 942*	*11 220*
Species total	*...*	*...*	*...*	*5 088*	*7 448*	*2 007*	*6 117*	*9 768*	*10 942*	*11 220*
Prochilods nei	**Prochilodes nca**		**Sábalos sudamericanos nep**		***Prochilodus spp***				**1,38(12)001,XX**	**PRL**
03 Argentina	...	...	...	...	...	...	...	16 000 F	17 700 F	14 000 F
Venezuela	5 023	9 767	16 907	20 430	18 740	10 927	15 670	9 320	9 613	6 743
03 Fishing area total	*5 023*	*9 767*	*16 907*	*20 430*	*18 740*	*10 927*	*15 670*	*25 320 F*	*27 313 F*	*20 743 F*
Species total	*5 023*	*9 767*	*16 907*	*20 430*	*18 740*	*10 927*	*15 670*	*25 320 F*	*27 313 F*	*20 743 F*
Wels(=Som)catfish	**Silure glane**		**Siluro**		***Silurus glanis***				**1,41(07)031,01**	**SOM**
04 Azerbaijan	78	63	19	6	11	9	8	8	9	8
Georgia	4	1	0	2	-	-	-	-	-	-
Kazakhstan	2 155	1 855	1 496	1 513	1 380 F	990 F	700 F	750 F	750 F	780
Tajikistan	0	15	7	12	...	...	...	...	...	10
Turkey	302	723	857	896	705	1 000	1 000	958	1 019	813
Turkmenistan	82	42	54	33	42	7	8	9	7	3
Uzbekistan	10	13	4	20	10 F	9	14	16	8	16
04 Fishing area total	*2 631*	*2 712*	*2 437*	*2 482*	*2 148 F*	*2 015 F*	*1 730 F*	*1 741 F*	*1 793 F*	*1 630*
05 Bulgaria	21	18	20 F	30	27	34	44	106	59	25
Croatia	6	10	20	19	24	20	1 F	1	1	1
Czech Rep	...	49	38	42	36	44	49	52	53	57
Czechoslovak	30 F	-	-	-	-	-	-	-	-	-
Germany	50	55	12	12	12	12	12	12	12	12
Greece	...	...	...	...	11	13	20	14	15	18
Hungary	133	126	153	205	201	121	113	145	104	120
Latvia	...	...	1	1	1	2	1	2	-	-
Lithuania	-	-	0	0	0	-	-	-	0	0
Poland	6	6	0	0	0	0	1	1	7	4

B-13 Miscellaneous freshwater fishes / Poissons d'eau douce divers / Peces de agua dulce diversos

Capture production by species, fishing areas and countries or areas / Captures par espèces, zones de pêche et pays ou zones / Capturas por especies, áreas de pesca y países o áreas

Species, Fishing area Espèce, Zone de pêche Especie, Area de pesca	1992 mt	1993 mt	1994 mt	1995 mt	1996 mt	1997 mt	1998 mt	1999 mt	2000 mt	2001 mt
Romania	49	66	113	72	22	9	58	87	73	116
Russian Fed	5 820	3 923	3 915	4 590	5 864	7 026	9 305	7 031	7 199	6 540
Slovakia	...	14	20	26	21	21	20	20	22	28
Slovenia	4	5	6	6	4	7	7	5	7	6
Switzerland	0	1	1	1	1	1	0	1	1	1
Ukraine	27	0	59	1	2	1	1	1	3	3
Yugoslavia	...	...	...	...	...	...	...	47	...	...
05 Fishing area total	*6 146 F*	*4 273*	*4 358 F*	*5 005*	*6 226*	*7 311*	*9 632 F*	*7 525*	*7 556*	*6 931*
Species total	*8 777 F*	*6 985*	*6 795 F*	*7 487*	*8 374 F*	*9 326 F*	*11 362 F*	*9 266 F*	*9 349 F*	*8 561*
Glass catfishes	...B		...C			*Kryptopterus spp*			1,41(07)050,XX	CAG
04 Indonesia	16 185	16 466	15 602	15 283	17 560	15 938	13 943	15 927	12 404	13 490
04 Fishing area total	*16 185*	*16 466*	*15 602*	*15 283*	*17 560*	*15 938*	*13 943*	*15 927*	*12 404*	*13 490*
Species total	*16 185*	*16 466*	*15 602*	*15 283*	*17 560*	*15 938*	*13 943*	*15 927*	*12 404*	*13 490*
Bagrid catfish	...B		...C			*Chrysichthys nigrodigitatus*			1,41(08)078,02	CSR
01 Nigeria	-	-	2 000	4 000	8 505	5 814	15 538	10 679	7 311	12 413
01 Fishing area total	-	-	*2 000*	*4 000*	*8 505*	*5 814*	*15 538*	*10 679*	*7 311*	*12 413*
Species total	-	-	*2 000*	*4 000*	*8 505*	*5 814*	*15 538*	*10 679*	*7 311*	*12 413*
Black catfishes nei	...B		...C			*Chrysichthys spp*			1,41(08)078,XX	CST
01 Benin	987 F	1 100 F	1 143	1 289	1 300 F	1 400 F	1 500 F	1 542	1 300 F	1 500 F
Gabon	...	1 000 F	1 500 F	2 000 F	2 124	1 131	1 950	380	887	797
Mali	2 739	2 572	2 514	5 320	4 172	3 982	3 650	3 670	4 395	4 000 F
Senegal	...	...	...	...	...	...	...	...	1 218	1 100 F
01 Fishing area total	*3 726 F*	*4 672 F*	*5 157 F*	*8 609 F*	*7 596 F*	*6 513 F*	*7 100 F*	*5 592*	*7 800 F*	*7 397 F*
Species total	*3 726 F*	*4 672 F*	*5 157 F*	*8 609 F*	*7 596 F*	*6 513 F*	*7 100 F*	*5 592*	*7 800 F*	*7 397 F*
Naked catfishes	...B		...C			*Bagrus spp*			1,41(08)111,XX	CAN
01 Ethiopia	...	...	...	...	...	...	94	107	120	119
Kenya	78	34	37	127	157	158	107	147	161	178
Nigeria	3 461	8 903	5 944	6 531	3 594	4 726	5 463	5 421	4 539	4 902
Tanzania	4 500	1 954	1 801	2 900	1 136	2 800	2 500	2 000	2 300	2 000
Uganda	4 200	10 110	4 681	4 879	4 978	4 774	5 222	8 564	4 375	4 375
01 Fishing area total	*12 239*	*21 001*	*12 463*	*14 437*	*9 865*	*12 458*	*13 386*	*16 239*	*11 495*	*11 574*
Species total	*12 239*	*21 001*	*12 463*	*14 437*	*9 865*	*12 458*	*13 386*	*16 239*	*11 495*	*11 574*
Catfishes nei	Barbottes nca		Bagres nep			*Ictalurus spp*			1,41(10)002,XX	CAF
02 Canada	54	39	43	37	37	29	27	35	19	19
El Salvador	228	205	178	250	240	230	213	208	142	169
Mexico	3 469	3 815	3 165	4 423	5 358	4 691	4 027	4 232	3 845	3 135
USA	4 906	5 463	5 262	5 264	4 110	6 872	5 257	9 235	7 561	7 478
02 Fishing area total	*8 657*	*9 522*	*8 648*	*9 974*	*9 745*	*11 822*	*9 524*	*13 710*	*11 567*	*10 801*
Species total	*8 657*	*9 522*	*8 648*	*9 974*	*9 745*	*11 822*	*9 524*	*13 710*	*11 567*	*10 801*
Brown bullhead	...B		...C			*Ameiurus nebulosus*			1,41(10)004,07	ITE
06 New Zealand	0	0	0	1	1	5	12	14	10	8
06 Fishing area total	*0*	*0*	*0*	*1*	*1*	*5*	*12*	*14*	*10*	*8*
Species total	*0*	*0*	*0*	*1*	*1*	*5*	*12*	*14*	*10*	*8*
North African catfish	Poisson-chat nord-africain		Pez-gato			*Clarias gariepinus*			1,41(18)030,03	CLZ
01 Ethiopia	...	...	...	643	812	1 200	1 677	1 917	4 000	3 926
Lesotho	-	2 F	2 F	2 F	2 F	2 F	2 F	2 F	2 F	2
Mali	17 117	16 075	15 712	33 210	17 129	14 933	15 009	15 091	27 468	25 000 F
Nigeria	-	-	-	-	2 658	8 568	5 268	9 994	4 474	10 419
01 Fishing area total	*17 117*	*16 077 F*	*15 714 F*	*33 855 F*	*20 601 F*	*24 703 F*	*21 956 F*	*27 004 F*	*35 944 F*	*39 347 F*
04 Turkey	556	759	...	...	...	...	...	216	576	520
04 Fishing area total	*556*	*759*	...	...	...	...	...	*216*	*576*	*520*
Species total	*17 673*	*16 836 F*	*15 714 F*	*33 855 F*	*20 601 F*	*24 703 F*	*21 956 F*	*27 220 F*	*36 520 F*	*39 867 F*
Mudfish	...B		...C			*Clarias anguillaris*			1,41(18)030,07	CLN
01 Egypt	19 610	23 680	19 895	19 608	20 370	22 433	21 618	21 511	31 506	39 535
01 Fishing area total	*19 610*	*23 680*	*19 895*	*19 608*	*20 370*	*22 433*	*21 618*	*21 511*	*31 506*	*39 535*
Species total	*19 610*	*23 680*	*19 895*	*19 608*	*20 370*	*22 433*	*21 618*	*21 511*	*31 506*	*39 535*
Torpedo-shaped catfishes nei	...B		...C			*Clarias spp*			1,41(18)030,XX	CTO
01 Benin	1 701 F	1 600 F	1 399	2 325	2 100 F	1 900 F	1 700 F	1 537	1 300 F	1 500 F
Kenya	589	254	581	574	339	1 724	1 532	1 570	1 592	1 611
Malawi	...	...	...	...	4 556	...	3 935	8 471	7 600 F	6 367
Nigeria	13 985	13 292	13 585	20 745	8 480	10 202	17 220	10 963	14 012	10 925

B-13 Miscellaneous freshwater fishes / Poissons d'eau douce divers / Peces de agua dulce diversos

Capture production by species, fishing areas and countries or areas / Captures par espèces, zones de pêche et pays ou zones / Capturas por especies, áreas de pesca y países o áreas

Species, Fishing area Espèce, Zone de pêche Especie, Area de pesca	1992 mt	1993 mt	1994 mt	1995 mt	1996 mt	1997 mt	1998 mt	1999 mt	2000 mt	2001 mt
Tanzania	12 000	4 531	1 613	7 920	2 107	7 750	7 500	2 500	2 000	2 500
Uganda	4 800	3 616	2 380	2 284	1 916	2 130	2 170	2 833	2 987	2 987
01 Fishing area total	*33 075 F*	*23 293 F*	*19 558*	*33 848*	*19 498 F*	*23 706 F*	*34 057 F*	*27 874*	*29 491 F*	*25 890 F*
04 China,Taiwan	3	-	-	-	-	-	6	-	-	-
Indonesia	5 962	7 603	7 485	9 911	8 388	7 257	7 535	6 413	5 343	5 790
Philippines	4 646	6 574	2 100	2 669	2 696	2 396	1 628	2 058	2 200	2 366
Thailand	6 725	8 137	7 100	8 080	5 800	3 410	10 934	12 111	19 600	20 470
04 Fishing area total	*17 336*	*22 314*	*16 685*	*20 660*	*16 884*	*13 063*	*20 103*	*20 582*	*27 143*	*28 626*
Species total	*50 411 F*	*45 607 F*	*36 243*	*54 508*	*36 382 F*	*36 769 F*	*54 160 F*	*48 456*	*56 634 F*	*54 516 F*
Pangas catfishes nei	**...B**		**...C**			***Pangasius spp***			**1,41(30)002,XX**	**PGZ**
04 Thailand	817	1 110	1 311	1 000	541	522	917	1 061	1 300	1 358
04 Fishing area total	*817*	*1 110*	*1 311*	*1 000*	*541*	*522*	*917*	*1 061*	*1 300*	*1 358*
Species total	*817*	*1 110*	*1 311*	*1 000*	*541*	*522*	*917*	*1 061*	*1 300*	*1 358*
Upsidedown catfishes	**...B**		**...C**			***Synodontis spp***			**1,41(32)008,XX**	**CSY**
01 Benin	243 F	130 F	21	125	120 F	110 F	110 F	110 F	100 F	110 F
Egypt	...	...	...	...	...	...	...	...	2 678	3 937
Kenya	37	25	37	28	24	408	418	506	443	48
Mali	2 054	1 929	1 886	3 990	4 657	5 973	4 075	4 097	3 296	3 000 F
Nigeria	6 844	3 781	10 909	9 709	4 466	5 069	8 781	8 143	6 054	6 642
01 Fishing area total	*9 178 F*	*5 865 F*	*12 853*	*13 852*	*9 267 F*	*11 560 F*	*13 384 F*	*12 856 F*	*12 571 F*	*13 737 F*
Species total	*9 178 F*	*5 865 F*	*12 853*	*13 852*	*9 267 F*	*11 560 F*	*13 384 F*	*12 856 F*	*12 571 F*	*13 737 F*
Freshwater siluroids nei	**Silurides d'eau douce nca**		**Siluroideos de agua dulce nep**			***Siluroidei***			**1,41(XX)XXX,XX**	**FSI**
01 Togo	500	500	500	500	500	500	500	500	500	500
Uganda	1 400	1 365	1 500	1 600	120	868	887	1 000	...	...
01 Fishing area total	*1 900*	*1 865*	*2 000*	*2 100*	*620*	*1 368*	*1 387*	*1 500*	*500*	*500*
03 Argentina	1 781	1 800	1 900 F	2 500 F	2 800 F	3 000 F	3 000 F	3 500 F	3 500 F	3 000 F
Brazil	42 853 F	43 900 F	44 900 F	40 048	47 302	52 601	49 866	57 014	58 474	58 600 F
Colombia	1 062	473	6 414	8 718	1 392	2 648	9 215	10 544	10 454 F	10 600 F
Paraguay	8 000	8 500 F	9 000 F	9 500 F	10 000 F	13 000 F	12 000 F	12 000 F	12 000 F	12 000 F
Uruguay	36	32	21	36	72	252	203	319	293	81
Venezuela	9 275	11 216	9 890	19 497	18 143	15 248	14 833	7 470	6 226	8 967
03 Fishing area total	*63 007 F*	*65 921 F*	*72 125 F*	*80 299 F*	*79 709 F*	*86 749 F*	*89 117 F*	*90 847 F*	*90 947 F*	*93 248 F*
04 India	...	...	...	...	...	...	63 724	71 152	114 099	76 809
Indonesia	11 658	11 941	13 451	13 215	13 508	10 117	12 466	13 854	13 167	11 810
Iran	1 000	670	28	5	22	6	32	24	25	1
Iraq	795	500	1 338	1 416	1 436	1 738	711	1 402	953	1 000
Korea Rep	...	...	-	-	279	251	...	136	...	...
04 Fishing area total	*13 453*	*13 111*	*14 817*	*14 636*	*15 245*	*12 112*	*76 933*	*86 568*	*128 244*	*89 620*
Species total	*78 360 F*	*80 897 F*	*88 942 F*	*97 035 F*	*95 574 F*	*100 229 F*	*167 437 F*	*178 915 F*	*219 691 F*	*183 368 F*
Burbot	**Lotte de rivière**		**Lota**			***Lota lota***			**1,48(04)003,01**	**FBU**
02 USA	16	15	16	15	14	9	28	13	20	9
02 Fishing area total	*16*	*15*	*16*	*15*	*14*	*9*	*28*	*13*	*20*	*9*
05 Belarus	11	28	16	14	17	13	18	20	7	29
Bulgaria	-	-	-	-	-	-	-	-	1	2
Czech Rep	...	2	-	-	-	-	-	-	-	-
Czechoslovak	0	-	-	-	-	-	-	-	-	-
Denmark	1	0	0	0	0	0	0	0	0	-
Estonia	-	16	8	29	33	27	18	52	38	37
Finland	1 763	1 765	1 384	1 384	1 227	1 227	679	679	831	831
Germany	-	-	2	2	2	2	2	2	2	2
Lithuania	14	4	3	5	5	14	10	13	13	9
Norway	...	...	...	...	...	1	1	...	...	...
Poland	0	0	0	0	0	-	-	-	0	-
Russian Fed	2 106	1 746	881	993	1 531	1 533	2 121	1 763	1 765	1 628
Slovakia	...	1	2	3	1	3	4	4	2	4
Sweden	77	83	112	93	66	56	61	59	-	-
Switzerland	10	12	18	11	6	9	6	7	8	8
05 Fishing area total	*3 982*	*3 657*	*2 426*	*2 534*	*2 888*	*2 885*	*2 920*	*2 599*	*2 667*	*2 550*
27 Estonia	1	1	3	2	3	4	3	1	2	1
Finland	719	724	517	465	420	436	468	475	345	337
Germany	-	-	-	-	-	1	1	1	1	2
Poland	-	-	-	-	-	9	10	12	18	26
Russian Fed	-	-	-	-	-	-	-	-	13	16
Sweden	12	9	12	11	9	10	7	6	4	3
27 Fishing area total	*732*	*734*	*532*	*478*	*432*	*460*	*489*	*495*	*383*	*385*
Species total	*4 730*	*4 406*	*2 974*	*3 027*	*3 334*	*3 354*	*3 437*	*3 107*	*3 070*	*2 944*
Nile perch	**Perche du Nil**		**Perca del Nilo**			***Lates niloticus***			**1,70(01)167,07**	**NIP**
01 Egypt	...	...	...	2 495	2 730	2 551	3 255	2 572	3 278	5 328
Ethiopia	...	...	...	908	270	230	191	75	65	63

B-13

Miscellaneous freshwater fishes — **Capture production by species, fishing areas and countries or areas**
Poissons d'eau douce divers — **Captures par espèces, zones de pêche et pays ou zones**
Peces de agua dulce diversos — **Capturas por especies, áreas de pesca y países o áreas**

Species, Fishing area Espèce, Zone de pêche Especie, Area de pesca	1992 mt	1993 mt	1994 mt	1995 mt	1996 mt	1997 mt	1998 mt	1999 mt	2000 mt	2001 mt
Kenya	77 599	100 037	104 102	102 546	97 145	73 555	76 663	103 014	109 815	78 534
Mali	4 108	3 858	3 771	7 980	5 712	4 978	5 024	5 052	6 592	6 000 F
Nigeria	7 939	9 062	6 118	4 993	3 746	4 224	5 250	6 366	4 447	6 139
Senegal	...	...	...	...	...	...	...	...	1 451	1 300 F
Tanzania	100 000	156 401	123 557	155 860	121 161	152 000	150 000	100 000	90 000	96 000
Uganda	130 000	95 005	101 208	92 722	81 253	91 706	98 800	89 203	87 257	88 881
01 Fishing area total	*319 646*	*364 363*	*338 756*	*367 504*	*312 017*	*329 244*	*339 183*	*306 282*	*302 905*	*282 245 F*
Species total	*319 646*	*364 363*	*338 756*	*367 504*	*312 017*	*329 244*	*339 183*	*306 282*	*302 905*	*282 245 F*
Freshwater perches nei	**Perches d'eau douce nca**		**Percas de agua dulce nep**		***Lates spp***				**1,70(01)167,XX**	**PEX**
01 Burundi	2 897	2 000 F	2 650 F	2 888	1 205	2 378	3 925	1 523	3 070	1 657
01 Fishing area total	*2 897*	*2 000 F*	*2 650 F*	*2 888*	*1 205*	*2 378*	*3 925*	*1 523*	*3 070*	*1 657*
Species total	*2 897*	*2 000 F*	*2 650 F*	*2 888*	*1 205*	*2 378*	*3 925*	*1 523*	*3 070*	*1 657*
Largemouth black bass	**...B**		**Perca atruchada**		***Micropterus salmoides***				**1,70(10)014,02**	**MPS**
02 Mexico	...	...	996	1 212	1 014	1 058	684	784	903	693
02 Fishing area total	*...*	*...*	*996*	*1 212*	*1 014*	*1 058*	*684*	*784*	*903*	*693*
Species total	*...*	*...*	*996*	*1 212*	*1 014*	*1 058*	*684*	*784*	*903*	*693*
European perch	**Perche européenne**		**Perca**		***Perca fluviatilis***				**1,70(14)002,01**	**FPE**
04 Kazakhstan	1 043	817	728	736	670 F	480 F	350 F	450 F	500 F	547
04 Fishing area total	*1 043*	*817*	*728*	*736*	*670 F*	*480 F*	*350 F*	*450 F*	*500 F*	*547*
05 Belarus	110	215	98	93	101	75	49	38	34	86
Belgium	25	25	25	25	25	25	25	15	25	25
Bulgaria	...	...	...	...	...	63	81	102	26	6
Czech Rep	...	39	38	31	33	34	36	37	34	34
Czechoslovak	20 F	-	-	-	-	-	-	-	-	-
Denmark	45	37	32	27	21	16	35	35	24	14
Estonia	-	525	672	623	634	887	815	680	561	300
Finland	13 774	13 754	13 043	13 028	12 634	12 634	9 470	9 470	9 599	9 599
Germany	293	277	215	215	215	215	215	215	215	215
Greece	...	...	...	...	24	15	19	24	15	23
Ireland	200	200	...	...	...	...	200	...	...	...
Latvia	15	16	20	27	22	29	34	34	34	34
Lithuania	60	92	115	114	83	114	104	116	115	114
Netherlands	360	340	136	219	376	336	155	177	170 F	150 F
Norway	...	...	...	...	...	9	3	...	...	...
Poland	120	120	120	224	601	179	163	186	297	245
Romania	-	-	-	39	38	8	40	60	28	42
Russian Fed	4 564	2 221	1 976	2 234	2 628	3 527	4 459	3 455	3 726	3 824
Slovakia	...	18	30	32	18	11	13	13	13	14
Slovenia	1	1	2	2	1	1	1	0	0	1
Sweden	192	220	211	216	173	278	251	171	150	135
Switzerland	960	460	396	373	475	364	422	491	395	262
Ukraine	268	125	288	319	126	80	107	79	155	74
05 Fishing area total	*21 007 F*	*18 685*	*17 417*	*17 841*	*18 228*	*18 900*	*16 697*	*15 398*	*15 616 F*	*15 197 F*
27 Denmark	125	75	72	63	42	67	65	78	76	60
Estonia	803	940	551	384	396	315	237	296	280	386
Finland	6 116	6 143	3 670	3 848	5 013	5 226	5 251	5 224	3 775	3 796
Germany	1 114	536	672	733	549	567	607	499	286	326
Ireland	-	-	-	-	-	-	-	-	73	-
Latvia	32	54	34	37	34	27	21	53	24	49
Lithuania	-	-	-	-	-	-	-	-	-	1
Poland	580	404	744	728	568	1 405	1 009	938	625	874
Russian Fed	-	-	-	-	-	-	-	-	-	161
Sweden	158	112	145	127	94	108	120	130	81	66
27 Fishing area total	*8 928*	*8 264*	*5 888*	*5 920*	*6 696*	*7 715*	*7 310*	*7 218*	*5 220*	*5 719*
37 Russian Fed	-	-	-	-	-	-	-	-	-	1
37 Fishing area total	*-*	*-*	*-*	*-*	*-*	*-*	*-*	*-*	*-*	*1*
Species total	*30 978 F*	*27 766*	*24 033*	*24 497*	*25 594 F*	*27 095 F*	*24 357 F*	*23 066 F*	*21 336 F*	*21 464 F*
American yellow perch	**Perchaude**		**Perca canadiense**		***Perca flavescens***				**1,70(14)002,02**	**FPY**
02 Canada	...	...	1 439	1 282	1 440	2 073	2 083	1 884	1 847	1 965
USA	1 566	1 412	1 387	1 246	711	622	556	537	567	640
02 Fishing area total	*1 566*	*1 412*	*2 826*	*2 528*	*2 151*	*2 695*	*2 639*	*2 421*	*2 414*	*2 605*
Species total	*1 566*	*1 412*	*2 826*	*2 528*	*2 151*	*2 695*	*2 639*	*2 421*	*2 414*	*2 605*
Walleye	**Sandre américain**		**Lucioperca americana**		***Stizostedion vitreum***				**1,70(14)015,01**	**STV**
02 Canada	7 708	8 160	7 543	7 832	7 718	7 751	8 206	8 782	9 160	6 949
USA	10	8	22	17	18	4	6	2	4	10
02 Fishing area total	*7 718*	*8 168*	*7 565*	*7 849*	*7 736*	*7 755*	*8 212*	*8 784*	*9 164*	*6 959*
Species total	*7 718*	*8 168*	*7 565*	*7 849*	*7 736*	*7 755*	*8 212*	*8 784*	*9 164*	*6 959*

B-13 Miscellaneous freshwater fishes / Poissons d'eau douce divers / Peces de agua dulce diversos

Capture production by species, fishing areas and countries or areas / Captures par espèces, zones de pêche et pays ou zones / Capturas por especies, áreas de pesca y países o áreas

Species, Fishing area Espèce, Zone de pêche Especie, Area de pesca	1992 mt	1993 mt	1994 mt	1995 mt	1996 mt	1997 mt	1998 mt	1999 mt	2000 mt	2001 mt
Sauger	**Sandre canadien**		**Lucioperca canadiense**		***Stizostedion canadense***				**1,70(14)015,02**	**SZC**
02 USA	3	3	1	4	1	-	0	-	-	-
02 Fishing area total	*3*	*3*	*1*	*4*	*1*	...	*0*	...	...	...
Species total	*3*	*3*	*1*	*4*	*1*	...	*0*	...	...	...
Pike-perch	**Sandre**		**Lucioperca**		***Stizostedion lucioperca***				**1,70(14)015,03**	**FPF**
04 Azerbaijan	45	33	27	19	22	11	7	5	5	19
Georgia	3	-	-	-	-	-	-	-	-	-
Iran	100	16	95	10	6	4	105	19	20	26
Kazakhstan	5 156	5 983	5 436	6 089	5 570 F	4 000 F	2 900 F	2 500 F	2 000 F	1 628
Kyrgyzstan	21	15	17	7	6 F	5 F	3 F	2 F	2 F	1
Tajikistan	52	87	22	20	...	...	...	...	...	21
Turkey	1 897	5 683	5 181	5 877	8 042	1 500	3 000	1 906	1 633	1 644
Turkmenistan	128	69	78	47	30	28	109	87	96	42
Uzbekistan	766	379	161	282	130 F	117	175	118	127	136
04 Fishing area total	*8 168*	*12 265*	*11 017*	*12 351*	*13 806 F*	*5 665 F*	*6 299 F*	*4 637 F*	*3 883 F*	*3 517*
05 Albania	34	25 F	20 F	5	10	4	-	-	-	45
Belarus	49	67	8	9	6	3	3	10	17	24
Belgium	15	15	15	15	15	15	15	10	15	15
Bulgaria	34	29	30 F	22	26	45	54	68	27	7
Croatia	6	13	12	12	12	14	1 F	1	1	1
Czech Rep	...	101	164	86	130	157	125	130	134	139
Czechoslovak	150 F	-	-	-	-	-	-	-	-	-
Denmark	29	12	10	13	11	39	38	13	19	13
Estonia	...	554	460	393	393	280	750	655	652	484
Finland	1 441	1 454	978	994	1 128	1 128	985	985	759	759
Germany	580	582	251	251	251	251	251	251	251	251
Hungary	213	212	243	226	224	199	156	169	200	196
Latvia	-	-	17	22	19	20	21	33	35	38
Lithuania	110	73	55	44	56	65	51	60	78	87
Moldova Rep	3	5	4	6	5	4	1	2	1	12
Netherlands	42	44	185	79	100	89	61	104	100 F	100 F
Norway	...	...	...	...	...	5	3	...	...	...
Poland	230	230	230	208	289	150	128	136	203	162
Romania	182	361	859	353	94	30	78	154	155	92
Russian Fed	7 110	4 975	3 212	3 271	3 994	4 082	3 161	3 644	3 863	3 427
Slovakia	...	52	89	94	76	70	65	64	56	62
Slovenia	5	1	5	5	5	5	5	4	5	4
Sweden	353	356	354	298	279	311	287	307	291	241
Switzerland	4	6	10	13	10	6	7	6	5	9
Ukraine	690	409	291	159	210	152	163	94	58	93
UK	1	1	3	1	1	0	0	2	1	1
05 Fishing area total	*11 281 F*	*9 577 F*	*7 505 F*	*6 579*	*7 344*	*7 124*	*6 409 F*	*6 902*	*6 926 F*	*6 262 F*
27 Denmark	5	1	0	1	1	12	15	2	9	4
Estonia	169	458	167	264	333	180	141	116	25	33
Finland	1 090	1 192	1 294	1 352	1 348	1 502	2 256	2 203	1 065	1 027
Germany	1 020	719	327	319	295	309	240	273	257	243
Latvia	79	48	33	35	54	20	17	25	13	21
Lithuania	-	1	-	-	-	-	-	-	-	28
Netherlands	-	-	-	-	-	-	-	-	1	4
Poland	302	242	266	232	223	330	253	401	343	316
Russian Fed	-	-	-	-	-	-	-	-	422	409
Sweden	57	89	71	53	33	55	47	39	36	26
27 Fishing area total	*2 722*	*2 750*	*2 158*	*2 256*	*2 287*	*2 408*	*2 969*	*3 059*	*2 171*	*2 111*
37 Russian Fed	843	546	866	950	2 221	2 022	2 229	1 676	1 843	1 875
Ukraine	61	100	217	353	601	803	802	892	1 113	1 629
37 Fishing area total	*904*	*646*	*1 083*	*1 303*	*2 822*	*2 825*	*3 031*	*2 568*	*2 956*	*3 504*
Species total	*23 075 F*	*25 238 F*	*21 763 F*	*22 489*	*26 259 F*	*18 022 F*	*18 708 F*	*17 166 F*	*15 936 F*	*15 394 F*
Walleyes nei	**Sandres nca**		**Luciopercas nep**		***Stizostedion spp***				**1,70(14)015,XX**	**STF**
05 Ukraine	6	...	3	...	...	...	-	-	-	-
05 Fishing area total	*6*	...	*3*	...	...	...	-	-	-	-
Species total	*6*	...	*3*	...	...	...	-	-	-	-
Ruffe	**Grémille**		**Acerina**		***Gymnocephalus cernuus***				**1,70(14)059,01**	**AC**
05 Denmark	7	2	5	15	13	5	9	3	2	-
Latvia	-	-	5	4	5	5	-	1	-	-
Lithuania	-	-	-	-	-	-	-	52	97	64
05 Fishing area total	*7*	*2*	*10*	*19*	*18*	*10*	*9*	*56*	*99*	*64*
Species total	*7*	*2*	*10*	*19*	*18*	*10*	*9*	*56*	*99*	*64*
Gudgeons, sleepers nei	**Gudgeons, dormeurs nca**		**Durmientes nep**		***Eleotridae***				**1,73(20)XXX,XX**	**FGI**
04 Indonesia	1 483	1 080	1 207	1 514	1 111	1 290	1 078	1 243	985	970
04 Fishing area total	*1 483*	*1 080*	*1 207*	*1 514*	*1 111*	*1 290*	*1 078*	*1 243*	*985*	*970*
06 Papua N Guin	1 850 F	1 850 F	1 850 F	1 850	1 850 F	1 850 F	1 850 F	1 850 F	1 850 F	1 850 F
06 Fishing area total	*1 850 F*	*1 850 F*	*1 850 F*	*1 850*	*1 850 F*	*1 850 F*	*1 850 F*	*1 850 F*	*1 850 F*	*1 850 F*
Species total	*3 333 F*	*2 930 F*	*3 057 F*	*3 364*	*2 961 F*	*3 140 F*	*2 928 F*	*3 093 F*	*2 835 F*	*2 820 F*

B-13 Miscellaneous freshwater fishes / Poissons d'eau douce divers / Peces de agua dulce diversos

Capture production by species, fishing areas and countries or areas / Captures par espèces, zones de pêche et pays ou zones / Capturas por especies, áreas de pesca y países o áreas

Species, Fishing area Espèce, Zone de pêche Especie, Area de pesca	1992 mt	1993 mt	1994 mt	1995 mt	1996 mt	1997 mt	1998 mt	1999 mt	2000 mt	2001 mt
Freshwater gobies nei	**Gobies d'eau douce nca**		**Góbidos de agua dulce nep**		***Gobiidae***			**1,73(21)XXX,XX**		**FGX**
01 Benin	483 F	1 100 F	1 302	914	900 F	850 F	800 F	800 F	700 F	800 F
01 Fishing area total	*483 F*	*1 100 F*	*1 302*	*914*	*900 F*	*850 F*	*800 F*	*800 F*	*700 F*	*800 F*
04 Philippines	7 257	9 160	4 466	3 431	3 585	4 300	3 803	4 027	4 563	4 280
Turkey	826	189	230	262	185	200	200	118	107	116
04 Fishing area total	*8 083*	*9 349*	*4 696*	*3 693*	*3 770*	*4 500*	*4 003*	*4 145*	*4 670*	*4 396*
05 Ukraine	1	-	...	...	...	...	60	16	...	...
05 Fishing area total	*1*	-	...	...	...	...	*60*	*16*	...	...
Species total	*8 567 F*	*10 449 F*	*5 998*	*4 607*	*4 670 F*	*5 350 F*	*4 863 F*	*4 961 F*	*5 370 F*	*5 196 F*
Climbing perch	**Anabas**		**Perca trepadora**		***Anabas testudineus***			**1,76(05)002,01**		**FPC**
04 Thailand	5 774	8 132	3 956	6 651	3 905	3 754	4 637	6 340	6 700	6 999
04 Fishing area total	*5 774*	*8 132*	*3 956*	*6 651*	*3 905*	*3 754*	*4 637*	*6 340*	*6 700*	*6 999*
Species total	*5 774*	*8 132*	*3 956*	*6 651*	*3 905*	*3 754*	*4 637*	*6 340*	*6 700*	*6 999*
Snakeskin gourami	**Gourami peau de serpent**		**Gurami piel de serpiente**		***Trichogaster pectoralis***			**1,76(10)013,02**		**FGS**
04 Indonesia	24 670	26 911	23 587	24 904	30 408	21 375	20 936	23 265	20 875	21 260
Thailand	542	751	200	186	385	353	1 486	511	700	730
04 Fishing area total	*25 212*	*27 662*	*23 787*	*25 090*	*30 793*	*21 728*	*22 422*	*23 776*	*21 575*	*21 990*
Species total	*25 212*	*27 662*	*23 787*	*25 090*	*30 793*	*21 728*	*22 422*	*23 776*	*21 575*	*21 990*
Kissing gourami	**Gourami embrasseur**		**Gurami besador**		***Helostoma temminckii***			**1,76(11)006,01**		**FGO**
04 Indonesia	18 500	18 954	16 675	19 166	12 614	18 376	16 598	23 320	17 927	18 320
04 Fishing area total	*18 500*	*18 954*	*16 675*	*19 166*	*12 614*	*18 376*	*16 598*	*23 320*	*17 927*	*18 320*
Species total	*18 500*	*18 954*	*16 675*	*19 166*	*12 614*	*18 376*	*16 598*	*23 320*	*17 927*	*18 320*
Snakehead	**Poisson tête de serpent**		**Cabeza de serpiente**		***Channa argus***			**1,77(19)001,01**		**CNA**
04 Korea Rep	-	-	-	-	169	120	...	28	...	...
04 Fishing area total	-	-	-	-	*169*	*120*	...	*28*	...	...
Species total	-	-	-	-	*169*	*120*	...	*28*	...	...
Striped snakehead	**Tête de serpent strié**		**Cabeza de serpiente cabrio**		***Channa striata***			**1,77(19)001,03**		**FSS**
04 Philippines	7 219	13 104	5 619	6 018	5 457	4 547	4 856	5 789	6 386	6 698
Thailand	13 986	18 591	21 400	21 810	25 509	24 099	16 664	17 995	20 500	21 400
04 Fishing area total	*21 205*	*31 695*	*27 019*	*27 828*	*30 966*	*28 646*	*21 520*	*23 784*	*26 886*	*28 098*
Species total	*21 205*	*31 695*	*27 019*	*27 828*	*30 966*	*28 646*	*21 520*	*23 784*	*26 886*	*28 098*
Indonesian snakehead	**Tête de serpent d'Indonésie**		**Cabeza de serpiente rojo**		***Channa micropeltes***			**1,77(19)001,04**		**FIS**
04 Indonesia	7 910	7 903	13 236	9 021	11 615	10 117	8 253	8 787	7 771	7 060
04 Fishing area total	*7 910*	*7 903*	*13 236*	*9 021*	*11 615*	*10 117*	*8 253*	*8 787*	*7 771*	*7 060*
Species total	*7 910*	*7 903*	*13 236*	*9 021*	*11 615*	*10 117*	*8 253*	*8 787*	*7 771*	*7 060*
Snakeheads(=Murrels) nei	**Poissons tête de serpent nca**		**Cabezas de serpiente nep**		***Channa spp***			**1,77(19)001,XX**		**FSN**
01 Nigeria	1 235	1 073	185	368	921	86	2 589	2 038	2 990	2 951
01 Fishing area total	*1 235*	*1 073*	*185*	*368*	*921*	*86*	*2 589*	*2 038*	*2 990*	*2 951*
04 India	65 273	75 127	90 762	93 933	94 944	105 496	159 065	50 349	58 669	58 856
Indonesia	27 052	31 407	31 634	31 940	32 356	31 326	26 108	36 309	31 381	31 820
Kazakhstan	4	12	9	10	10 F	7 F	5 F	10 F	10 F	12
Turkmenistan	0	0	0	0	0	1	0	0	0	1
Uzbekistan	-	-	-	242	110 F	...	275	209	198	127
04 Fishing area total	*92 329*	*106 546*	*122 405*	*126 125*	*127 420 F*	*136 830 F*	*185 453 F*	*86 877 F*	*90 258 F*	*90 816*
Species total	*93 564*	*107 619*	*122 590*	*126 493*	*128 341 F*	*136 916 F*	*188 042 F*	*88 915 F*	*93 248 F*	*93 767*
Freshwater fishes nei	**Poissons d'eau douce nca**		**Peces de agua dulce nep**		***Osteichthyes***			**1,99(XX)XXX,XX**		**FRF**
01 Algeria	0	0	0	0	0	0	0	0	0	0
Angola	7 000	7 000	7 000	6 000	6 000	6 000	6 000	6 000 F	6 000 F	6 000 F
Benin	6 413 F	7 595 F	7 464	9 552	6 563 F	5 096 F	3 838 F	3 733 F	3 000 F	3 440 F
Botswana	400 F	300 F	200 F	100 F	33	80	103	64	74	30
Br Ind Oc Tr	0	0	0	0	0	0	0	0	0	0
Burkina Faso	7 500	7 000	8 000	8 000	8 000	8 000	8 335	7 600	8 500	8 500 F
Burundi	1 146	870 F	1 100 F	...	278	...	805	596	3 286	1 049
Cameroon	22 000	23 000	27 000 F	30 000 F	35 000 F	40 000 F	45 000 F	50 000 F	55 000	52 500 F
Cape Verde	0	0	0	0	0	0	0	0	0	0
Cent Afr Rep	13 000	13 250 F	13 500 F	13 750 F	14 000 F	14 250 F	14 500 F	15 000 F	15 000 F	15 000 F
Chad	80 000	87 300	80 000	90 000	100 000	85 000	84 000	84 000 F	84 000 F	84 000 F
Comoros	0	0	0	0	0	0	0	0	0	0
Congo Dem R	184 040	192 589	152 117	154 751	159 037	158 367	174 087	204 503	204 503 F	204 503 F
Congo Rep	21 049	27 850	24 752	26 811	25 873	18 987	25 455	25 455 F	26 000 F	24 500 F

B-13 Miscellaneous freshwater fishes / Poissons d'eau douce divers / Peces de agua dulce diversos

Capture production by species, fishing areas and countries or areas / Captures par espèces, zones de pêche et pays ou zones / Capturas por especies, áreas de pesca y países o áreas

Species, Fishing area Espèce, Zone de pêche Especie, Area de pesca	1992 mt	1993 mt	1994 mt	1995 mt	1996 mt	1997 mt	1998 mt	1999 mt	2000 mt	2001 mt
Côte dIvoire	13 904	12 477	14 604	10 835	11 162 F	11 732 F	12 301 F	10 556 F	10 475	10 500
Djibouti	0	0	0	0	0	0	0	0	0	0
Egypt	50 845	45 041	62 506	47 525	45 498	45 989	64 964	58 697	34 196	37 282
Eq Guinea	370 F	600	700	450	900	850	970	1 101 F	1 076	1 000 F
Eritrea	...	0	0	0	0	0	0	0	0	0
Ethiopia	...	...	...	588	86	-	-	-	-	-
Fr South Tr	0	0	0	0	0	0	0	0	0	0
Gabon	750 F	1 000 F	1 500 F	3 018	4 251	4 636	3 500	5 814	5 726	5 922
Gambia	1 400	1 400	1 400	1 450	1 450	1 450	1 450	1 450 F	1 450 F	1 450 F
Ghana	56 000	52 000	54 700	60 000	73 580	70 000	74 500	74 500	74 500	74 500
Guinea	4 000 F	4 600	3 800	3 100	2 780	3 600	4 000	4 000	4 000	4 000 F
GuineaBissau	200 F	250 F	250 F	250 F	250 F	250 F	200 F	200 F	200 F	200 F
Kenya	12 001	9 103	2 015	1 215	3 059	745	5 960	4 221	4 886	5 237
Lesotho	3 F	5 F	5 F	8 F	10 F	10 F	10 F	10 F	12 F	14
Liberia	4 000	4 000	4 000	4 000	4 000	4 000	4 000	4 000	4 000	4 000
Libya	0	0	0	0	0	0	0	0	0	0
Madagascar	3 500	3 800	4 187	4 277	4 500	4 500	4 500	4 500	4 500	4 500
Malawi	52 762	50 957	44 679	49 263	53 428	51 346	31 773	2 970	2 700 F	2 274
Mali	17 801	16 718	16 341	34 580	39 833	60 667	34 854	35 044	20 875	19 000 F
Mauritania	5 000 F	5 000 F	5 000 F	5 000 F	5 000 F	5 000 F	5 000 F	5 000 F	5 000 F	5 000 F
Morocco	1 500	1 000	600	500	600	900	500	700	500	300
Mozambique	3 800	4 000	4 000	2 000	1 936	1 747	1 681	1 191	1 275	1 000
Namibia	1 102	1 200	1 200	1 200	1 200	1 500	1 500	1 500	1 500	1 500
Niger	2 454	2 162	2 516	3 616	4 156	6 328	7 013	11 000	16 250	20 800
Nigeria	40 627	38 738	42 959	43 326	30 884	19 808	28 559	38 506	48 445	54 495
Réunion	0	0	0	0	0	0	0	0	0	0
Rwanda	3 644	3 500 F	3 400 F	3 300 F	2 952	4 428	4 408	3 793	4 080	4 178
St Helena	0	0	0	0	0	-	-	-	-	-
Sao Tome Prn	0	0	0	0	0	0	0	0	0	0
Senegal	24 750	27 650	30 000 F	31 000 F	23 000 F	31 000 F	21 000 F	34 000 F	10 648	9 500 F
Seychelles	0	0	0	0	0	0	0	0	0	0
Sierra Leone	14 000	14 000	15 000	15 000	14 500 F	14 500	14 190	14 480	14 000	14 000
Somalia	250 F	250 F	250 F	250 F	250 F	250 F	250 F	250 F	200 F	200 F
South Africa	832	832	800	800	850	850	900	900 F	900 F	900 F
Sudan	33 000	37 500	40 000	30 000	30 000	31 000	28 000	28 000	30 000 F	33 000
Swaziland	60 F	68 F	65 F	60 F	60 F	65 F	70 F	70 F	70 F	70 F
Tanzania	67 500	57 779	36 362	62 160	55 975	60 500	58 700	67 520	96 700	86 000
Togo	500	500	1 000	998	1 000	1 000	1 000	1 000	1 000	1 000
Tunisia	0	400	243	440	706	1 010	896	808	832	860
Uganda	2 500	20 307	2 444	3 009	12 000	5 421	-	-	2	2
Zambia	59 206	56 046	61 147	61 872	58 739	58 110	60 116	58 372	57 808	56 500 F
Zimbabwe	1 900	902	547	883	644	599	671	782	1 807	1 800 F
01 Fishing area total	*822 709 F*	*840 539 F*	*779 353 F*	*824 937 F*	*844 023 F*	*839 571 F*	*839 559 F*	*871 886 F*	*864 976 F*	*860 506 F*
02 Anguilla	0	0	0	0	0	0	0	0	0	0
Antigua Barb	0	0	0	0	0	0	0	0	0	0
Aruba	0	0	0	0	0	0	0	-	-	-
Bahamas	0	0	0	0	0	0	0	0	0	0
Barbados	0	0	0	0	0	0	0	0	0	-
Belize	1	1	1	0	0	0	0	0	0	0
Bermuda	0	0	0	0	0	0	0	0	0	0
Br Virgin Is	0	0	0	0	0	0	0	0	0	0
Canada	13 422	5 809	6 707	8 445	9 649	8 089	8 914	9 317	10 756	8 070
Cayman Is	0	0	0	0	0	0	0	0	0	0
Costa Rica	406	710	840	900	1 090	840	1 000 F	1 000 F	1 000 F	1 000 F
Cuba	48	486	65	161	230	...	...	4	40 F	40 F
Dominica	0	0	0	0	0	0	0	0	0	0
Dominican Rp	536	130	234	315	105	751	990	96	179	183
El Salvador	627	776	932	967	891	912	863	866	1 177	535
Greenland	0	0	0	0	0	0	0	0	0	0
Grenada	0	0	0	0	0	0	0	0	0	0
Guadeloupe	0	0	0	0	0	0	0	0	0	0
Guatemala	3 702	4 228	3 774	4 020	4 000	5 100	6 366	6 803	7 109	7 100 F
Haiti	500 F	600 F	500 F	500 F	500 F	500 F	500 F	500 F	500 F	500 F
Honduras	85	86	92	127	98	126	119	102	61	111
Jamaica	300 F	300 F	300 F	300 F	300 F	300 F	300 F	300 F	300 F	300 F
Martinique	0	0	0	0	0	0	0	0	0	0
Mexico	5 919	4 850	8 166	10 724	9 766	5 185	6 482	3 284	8 716	5 574
Montserrat	0	0	0	0	0	0	0	0	0	0
NethAntilles	0	0	0	0	0	0	0	0	0	0
Nicaragua	348	547	824	538	1 142	1 293	1 256	363	396	388
Puerto Rico	0	0	0	0	0	0	0	0	0	0
St Kitts Nev	0	0	0	0	0	0	0	0	0	0
St Lucia	0	0	0	0	0	0	0	0	0	0
St Pier Mq	0	0	0	0	0	0	0	0	0	0
St Vincent	0	0	0	0	2	1	0	0	0	0
Trinidad Tob	0	0	0	0	0	0	0	0	0	0
Turks Caicos	0	0	0	0	0	0	0	0	0	0
USA	1 432	1 934	1 568	2 298	328	690	467	447	756	782
US Virgin Is	0	0	0	0	0	0	0	0	0	-
02 Fishing area total	*27 326 F*	*20 457 F*	*24 003 F*	*29 295 F*	*28 101 F*	*23 787 F*	*27 257 F*	*23 082 F*	*30 990 F*	*24 583 F*
03 Argentina	1 996	1 754	1 854 F	2 908 F	3 297 F	4 135 F	4 097 F	5 138 F	6 098 F	4 460 F
Bolivia	3 036	4 742	4 452	4 726	4 800	4 850	4 865	4 860	4 911	4 900
Brazil	24 169 F	29 760 F	35 410 F	40 372	42 285	40 507	43 588	42 875	49 593	50 000 F
Colombia	7 827	17 231	19 737	9 108	16 165	14 232	6 031	9 577	7 800 F	7 800 F
Ecuador	332	372	300	300	300	400	400	400	400	400
Fr Guiana	0	0	0	0	0	0	0	0	0	0
Guyana	800	800	800	700 F	800	625	625	603	800	800

B-13 Miscellaneous freshwater fishes / Poissons d'eau douce divers / Peces de agua dulce diversos

Capture production by species, fishing areas and countries or areas / Captures par espèces, zones de pêche et pays ou zones / Capturas por especies, áreas de pesca y países o áreas

Species, Fishing area Espèce, Zone de pêche Especie, Area de pesca	1992 mt	1993 mt	1994 mt	1995 mt	1996 mt	1997 mt	1998 mt	1999 mt	2000 mt	2001 mt
Paraguay	3 425	3 500 F	3 650 F	3 800 F	4 000 F	5 000 F	4 000 F	4 000 F	4 000 F	4 000 F
Peru	32 734	37 987	48 837	44 452	19 524	27 941	28 047	24 018	19 387	22 464
Suriname	561	187	138	140 F	150 F	200 F	200 F	200 F	200 F	200
Uruguay	12	150	73	45	188	40	41	342	312	254
Venezuela	1 933	2 185	2 532	4 524	4 149	8 321	8 141	13 204	450	5 252
03 Fishing area total	*76 825 F*	*98 668 F*	*117 783 F*	*111 075 F*	*95 658 F*	*106 251 F*	*100 035 F*	*105 217 F*	*93 951 F*	*100 530 F*
04 Afghanistan	1 200 F	1 200 F	1 300 F	1 300 F	1 300 F	1 250 F	1 200 F	1 200 F	1 000 F	800 F
Armenia	0	0	0	0	0	0	0	0	0	-
Azerbaijan	2	1	4	25	16	21	13	8	-	-
Bahrain	0	0	0	0	0	0	0	0	-	-
Bangladesh	331 838	354 459	445 676	443 319	445 377	451 055	457 055	575 609	591 300	590 000 F
Bhutan	315 F	320 F	310 F	310 F	300 F	300 F	300 F	300 F	300 F	300 F
Brunei Darsm	10	14	1	2	1	0	0	0	0	0
Cambodia	68 881	67 880	64 960	72 420	63 440	72 900	75 600	230 700	245 300	359 600 F
China	650 385	738 015	794 129	894 855	994 971	1 032 861	1 218 152	1 394 610	1 222 955	1 033 302
China, Macao	0	0	0	0	0	0	0	0	0	0
China,Taiwan	75	35	11	21	16	11	18	18	17	19
Cyprus	...	...	5	65	64	70	70	70	78	70 F
Gaza Strip	...	...	...	0	0	0	0	0	0	0
Georgia	0	0	0	0	0	0	0	5	7	2
India	73 642	195 021	88 484	92 964	224 990	245 595	225 100	330 448	416 490	591 550
Indonesia	113 110	110 071	125 078	122 559	115 327	103 812	96 597	110 347	108 387	107 180
Iran	3 767	7 005	851	157	1 845	1 131	2 529	1 637	1 575	5 970
Iraq	4 603	4 769	3 981	5 867	4 381	9 836	4 361	2 205	3 412	3 400 F
Japan	16 898	14 515	13 643	12 943	13 344	13 163	12 073	11 020	10 341	9 610
Jordan	350	350	350	350	350	350	350	350	400	350
Kazakhstan	116	64	8	6	269 F	358 F	364 F	10 809 F	13 715 F	10 526
Korea D P Rp	35 000 F	35 000 F	20 000	20 000	20 000	20 000	20 000 F	20 000 F	20 000 F	20 000 F
Korea Rep	19 023	4 049	2 891	3 309	3 418	2 633	2 755	1 681	6 402	5 254
Kuwait	0	0	0	0	0	0	0	0	0	0
Laos	16 740 F	17 000 F	20 600 F	23 370 F	19 500 F	16 057 F	16 642 F	25 541 F	24 850 F	25 500 F
Lebanon	10 F	10	10	10	10	10	10	10	10	10
Malaysia	1 773	1 971	2 064	3 939	3 683	3 949	4 626	3 366	3 549	3 446
Maldives	0	0	0	0	0	0	0	0	0	0
Mongolia	120 F	165	184	158	221	180	311	524	425	117
Myanmar	141 281	142 065	146 365	148 347	146 494	149 069	149 279	159 746	189 708	235 376
Nepal	5 871	7 320	7 340	11 230	11 230	11 230	12 000	12 752	16 700	16 700
Oman	0	0	0	0	0	0	0	0	0	0
Pakistan	109 087	109 185	118 703	121 405	142 092	167 530	163 524	179 865	176 468	180 100
Philippines	15 851	9 131	7 632	9 425	7 143	10 413	6 632	8 309	9 800	9 384
Qatar	0	0	0	0	0	0	0	0	0	0
Saudi Arabia	0	0	0	0	0	0	0	0	0	0
Singapore	24	25	23	0	0	0	0	0	0	0
Sri Lanka	...	...	...	...	...	...	...	2 930	3 480	2 640
Syria	2 584	2 535	3 570	3 832	3 103	3 557	4 347	5 338	3 991	5 969
Tajikistan	6	14	7	4	40 F	75 F	100 F	32 F	19 F	
Thailand	37 054	52 763	64 587	60 272	105 726	108 551	70 011	59 376	64 300	67 170
Timor-Leste	...	...	...	...	...	...	...	...	0	0
Turkey	2 475	1 882	3 139	3 424	4 621	1 800	1 700	2 434	1 697	3 290
Turkmenistan	22	3	9	6	27	2	2	1	2	9
Untd Arab Em	0	0	0	0	0	0	0	0	0	0
Uzbekistan	1 618	302	536	66	40 F	130	69	10	-	1
Viet Nam	137 154	145 839	79 087	94 189	163 936	176 589	137 800	168 107	169 000 F	169 000 F
04 Fishing area total	*1 790 885 F*	*2 022 978 F*	*2 015 538 F*	*2 150 149 F*	*2 497 275 F*	*2 604 488 F*	*2 683 590 F*	*3 319 358 F*	*3 305 678 F*	*3 456 645 F*
05 Albania	481	500 F	440 F	-	-	-	-	-	-	-
Andorra	0	0	0	0	0	0	0	0	0	0
Austria	479	420	388	404	450	465	451	432	439	362
Belarus	22	2	-	-	-	-	-	-	-	-
Bosnia Herzg	2 000 F	2 500 F	2 500 F	2 500 F	2 500 F	2 500 F	2 500 F	2 500 F	2 500 F	2 500 F
Bulgaria	1 521	1 576	590 F	39	182	0	1	-	-	26
Channel Is	0	0	0	0	0	0	0	0	0	0
Croatia	106	161	184	202	217	213	5 F	5	8	12
Czech Rep	...	89	153	192	102	121	134	168	151	131
Denmark	0	0	0	0	3	-	-	0	0	-
Estonia	3 509	247	212	218	123	160	134	168	170	209
Faeroe Is	0	0	0	0	0	0	0	0	0	0
Finland	834	696	640	517	476	476	444	444	393	393
France	4 340	4 400	4 450	4 500	4 500	4 500	4 460	2 000 F	2 000 F	2 000 F
Germany	3 880	6 103	7 000	19 079	19 079	19 008	19 008	19 008	19 008	19 008
Greece	2 370	2 960	3 452	3 606	1 730	1 624	1 410	1 592	1 803	1 663
Hungary	695	595	703	733	733	776	648	714	718	89
Iceland	0	0	0	0	0	0	0	0	0	0
Ireland	0	0	0	0	0	0	0	...	...	...
Isle of Man	0	0	0	0	0	0	0	0	0	0
Italy	2 952	3 250	3 518	3 900	3 647	2 895	2 346	2 316	2 819	2 643
Latvia	83	89	9	8	9	9	9	13	41	40
Liechtensten	0	0	0	0	0	0	0	0	0	0
Lithuania	19	2	17	50	45	90	88	10	6	2
Luxembourg	0	0	0	0	0	0	0	0	0	0
Macedonia	6	70	46	146	31	68	107	113	52	7
Malta	0	0	0	0	0	0	0	0	0	0
Moldova Rep	15	10	19	11	9	11	13	-	-	-
Netherlands	17	22	228	410	350	362	176	154	150 F	150 F
Poland	14 138	24 769	20 898	20 871	13 345	9 500	10 000	10 200	12 120	13 035
Portugal	1	0	0	0	0	0	0	0	0	1
Romania	326	369	313	27	67	5	-	-	169	188
Russian Fed	1 140	8 261	18 195	5 952	8 095	3 618	1 784	2 742	15 857	4 451

B-13 Miscellaneous freshwater fishes / Poissons d'eau douce divers / Peces de agua dulce diversos

Capture production by species, fishing areas and countries or areas / Captures par espèces, zones de pêche et pays ou zones / Capturas por especies, áreas de pesca y países o áreas

Species, Fishing area Espèce, Zone de pêche Especie, Area de pesca		1992 mt	1993 mt	1994 mt	1995 mt	1996 mt	1997 mt	1998 mt	1999 mt	2000 mt	2001 mt
	Slovenia	8	5	0	0	0	17	0	3	-	-
	Spain	4 856	4 439	3 981	3 944	4 000 F	4 000 F	4 000 F	4 000 F	4 000 F	4 000 F
	Sweden	458	413	243	206	126	229	134	154	197	155
	Switzerland	1	0	0	8	0	0	3	0	4	1
	Ukraine	996	994	485	261	213	15	1	2	...	7
	Yugoslavia	5 111	3 797	3 912	3 803	3 653	3 500	2 200 F	759	672	672 F
05	*Fishing area total*	*50 364 F*	*66 739 F*	*72 576 F*	*71 587 F*	*63 685 F*	*54 162 F*	*50 056 F*	*47 497 F*	*63 277 F*	*51 745 F*
06	Amer Samoa	0	0	0	0	0	0	0	0	0	0
	Australia	3 337	2 834	122	301	252	194	200	154	134	135
	Christmas Is	0	0	0	0	0	0	0	0	0	0
	Cocos Is	0	0	0	0	0	0	0	0	0	0
	Cook Is	0	0	0	0	0	0	0	0	0	0
	Fr Polynesia	0	0	0	0	0	50	50	50	50	50
	Guam	0	0	0	0	0	-	-	-	-	-
	Kiribati	0	0	0	0	0	0	0	0	0	0
	Marshall Is	0	0	0	0	0	0	0	0	0	0
	Micronesia	5 F	4 F	4 F	5 F	5 F	5 F	5 F	5 F	5 F	5 F
	NewCaledonia	0	0	0	0	0	0	0	0	0	0
	New Zealand	500	550	500	500	600	600	400	400	400	300
	Niue	0	0	0	0	0	0	0	0	0	0
	Norfolk Is	0	0	0	0	0	0	0	0	0	0
	N Marianas	0	0	0	1	0	0	0	0	0	0
	Palau	0	0	0	0	0	0	0	0	0	0
	Papua N Guin	6 650 F	6 652 F	6 644 F	6 645 F	6 649 F	6 649 F	6 641 F	6 649 F	6 655 F	6 654 F
	Pitcairn Is	0	0	0	0	0	0	0	0	0	0
	Samoa	0	0	0	0	0	0	0	0	1	1 F
	Solomon Is	0	0	0	0	0	0	0	0	0	0
	Tokelau	0	0	0	0	0	0	0	0	0	0
	Tonga	1	1	0	0	0	0	0	0	0	0
	Tuvalu	0	0	0	0	0	0	0	0	0	0
	Vanuatu	0	0	0	0	0	0	0	0	0	0
	Wallis Fut I	0	0	0	0	0	0	0	0	0	0
06	*Fishing area total*	*10 493 F*	*10 041 F*	*7 270 F*	*7 452 F*	*7 506 F*	*7 498 F*	*7 296 F*	*7 258 F*	*7 245 F*	*7 145 F*
27	Denmark	0	-	-	-	0	-	-	-	-	-
	Finland	356	425	412	486	481	452	229	292	179	209
	Germany	-	-	-	-	-	3	1	2	3	1
	Latvia	2	-	-	1	11	4	4	7	22	10
	Poland	451	884	608	384	500	574	327	217	388	241
	Russian Fed	-	1 544	2 516	1 719	1 642	2 554	2 287	1 429	-	-
	Sweden	10	1	0	0	5	1	0	3	-	-
27	*Fishing area total*	*819*	*2 854*	*3 536*	*2 590*	*2 639*	*3 588*	*2 848*	*1 950*	*592*	*461*
31	Venezuela	5 022	5 101	5 404	7 705	7 303	3 656	2 870	4 130	6 772	6 937
31	*Fishing area total*	*5 022*	*5 101*	*5 404*	*7 705*	*7 303*	*3 656*	*2 870*	*4 130*	*6 772*	*6 937*
37	Russian Fed	-	-	-	-	-	-	-	-	-	17
37	*Fishing area total*	*-*	*-*	*-*	*-*	*-*	*-*	*-*	*-*	*-*	*17*
Species total		*2 784 443 F*	*3 067 377 F*	*3 025 463 F*	*3 204 790 F*	*3 546 190 F*	*3 643 001 F*	*3 713 511 F*	*4 380 378 F*	*4 373 481 F*	*4 508 569 F*
Group total		***3 862 595***	***4 227 661***	***4 158 164***	***4 458 553***	***4 684 341***	***4 812 741***	***5 035 463***	***5 580 328***	***5 636 671***	***5 706 800***

B-21 Sturgeons, paddlefishes / Esturgeons, spatules / Esturiones, sollos

Capture production by species, fishing areas and countries or areas / Captures par espèces, zones de pêche et pays ou zones / Capturas por especies, áreas de pesca y países o áreas

Species, Fishing area Espèce, Zone de pêche Especie, Area de pesca	1992 mt	1993 mt	1994 mt	1995 mt	1996 mt	1997 mt	1998 mt	1999 mt	2000 mt	2001 mt
Danube sturgeon(=Osetr)	**Esturgeon du Danube**		**Esturión del Danube**		***Acipenser gueldenstaedtii***			**1,17(01)001,02**		**APG**
05 Bulgaria	-	-	-	4	2	4	5	4	1	1
Romania	-	-	-	-	-	-	5	10	19	17
Ukraine	...	...	...	...	1	2	1	...	2	-
05 Fishing area total	...	...	...	*4*	*3*	*6*	*11*	*14*	*22*	*18*
37 Bulgaria	-	-	-	1	0	2	2	2	-	-
Ukraine	...	...	...	134	129	132	112	34	20	8
37 Fishing area total	...	...	...	*135*	*129*	*134*	*114*	*36*	*20*	*8*
Species total	...	...	...	*139*	*132*	*140*	*125*	*50*	*42*	*26*
Sterlet sturgeon	**Sterlet**		**Esterlete**		***Acipenser ruthenus***			**1,17(01)001,04**		**APR**
05 Bulgaria	-	-	-	0	1	1	1	2	2	1
Hungary	...	14	15	36	34	14	9	35	12	11
Slovakia	...	0	0	1	1	0	0	0	1	-
05 Fishing area total	...	*14*	*15*	*37*	*36*	*15*	*10*	*37*	*15*	*12*
37 Bulgaria	-	-	-	-	-	-	-	-	0	-
37 Fishing area total	-	-	-	-	-	-	-	-	*0*	-
Species total	...	*14*	*15*	*37*	*36*	*15*	*10*	*37*	*15*	*12*
Starry sturgeon	**Esturgeon étoilé**		**Esturión estrellado**		***Acipenser stellatus***			**1,17(01)001,05**		**APE**
05 Bulgaria	-	-	-	0	0	0	4	6	1	1
Romania	-	-	-	-	-	5	2	11	22	20
05 Fishing area total	-	-	-	*0*	*0*	*5*	*6*	*17*	*23*	*21*
37 Ukraine	...	...	...	9	18	49	13	11	5	3
37 Fishing area total	...	...	...	*9*	*18*	*49*	*13*	*11*	*5*	*3*
Species total	...	...	...	*9*	*18*	*54*	*19*	*28*	*28*	*24*
White sturgeon	**Esturgeon blanc**		**Esturión blanco**		***Acipenser transmontanus***			**1,17(01)001,09**		**APN**
67 USA	-	-	-	-	-	-	-	-	206	185
67 Fishing area total	-	-	-	-	-	-	-	-	*206*	*185*
Species total	-	-	-	-	-	-	-	-	*206*	*185*
Green sturgeon	**...B**		**...C**		***Acipenser medirostris***			**1,17(01)001,15**		**AAM**
67 USA	-	-	-	-	-	-	-	-	36	10
67 Fishing area total	-	-	-	-	-	-	-	-	*36*	*10*
Species total	-	-	-	-	-	-	-	-	*36*	*10*
Beluga	**Béluga**		**Esturión beluga**		***Huso huso***			**1,17(01)005,01**		**HUH**
05 Bulgaria	...	...	...	21	24	31	31	27	18	7
Romania	-	-	-	-	-	2	7	7	32	20
Slovenia	-	-	-	0	1	1	1	1	0	1
05 Fishing area total	...	...	...	*21*	*25*	*34*	*39*	*35*	*50*	*28*
37 Bulgaria	-	-	-	4	5	11	12	10	1	0
37 Fishing area total	-	-	-	*4*	*5*	*11*	*12*	*10*	*1*	*0*
Species total	...	...	...	*25*	*30*	*45*	*51*	*45*	*51*	*28*
Sturgeons nei	**Esturgeons nca**		**Esturiones nep**		***Acipenseridae***			**1,17(01)XXX,XX**		**STU**
02 Canada	252	267	281	286	212	330	349	282	281	6
USA	220	217	147	147	314	285	176	138	0	-
02 Fishing area total	*472*	*484*	*428*	*433*	*526*	*615*	*525*	*420*	*281*	*6*
04 Azerbaijan	98	84	92	76	69	63	61	69	70	76
Iran	2 692	1 710	1 700	1 500	1 600	1 300	1 200	1 000	1 000	870
Kazakhstan	1 705	1 109	635	563	510 F	370 F	270 F	240 F	215 F	190
Turkmenistan	...	...	...	...	...	...	...	11	3	3
04 Fishing area total	*4 495*	*2 903*	*2 427*	*2 139*	*2 179 F*	*1 733 F*	*1 531 F*	*1 320 F*	*1 288 F*	*1 139*
05 Bulgaria	12	10	5 F	-	-	-	-	-	-	-
Germany	-	-	-	-	0	0	0	0	0	0
Macedonia	4	8	0	0	0	2	6	-	-	-
Romania	2	2	2	9	5	5	1	3	-	1
Russian Fed	7 921	3 819	3 048	2 276	1 179	1 264	1 149	750	594	553
Ukraine	-	2	-	-	-	-	-	-	-	-
05 Fishing area total	*7 939*	*3 841*	*3 055 F*	*2 285*	*1 184*	*1 271*	*1 156*	*753*	*594*	*554*
21 Canada	14	181	100	157	86	91	69	9	3	4
USA	...	...	15	0	-	-	-	-	-	-
21 Fishing area total	*14*	*181*	*115*	*157*	*86*	*91*	*69*	*9*	*3*	*4*
27 Denmark	-	-	-	-	-	-	0	-	-	-
France	0	0	0	0	0	0	0	1	-	-
Portugal	0	0	0	0	0	-	-	-	-	-

B-21 Sturgeons, paddlefishes / Esturgeons, spatules / Esturiones, sollos

Capture production by species, fishing areas and countries or areas / Captures par espèces, zones de pêche et pays ou zones / Capturas por especies, áreas de pesca y países o áreas

Species, Fishing area Espèce, Zone de pêche Especie, Area de pesca	1992 mt	1993 mt	1994 mt	1995 mt	1996 mt	1997 mt	1998 mt	1999 mt	2000 mt	2001 mt
Spain	-	-	-	-	-	-	1	1	-	-
27 Fishing area total	*0*	*0*	*0*	*0*	*0*	*0*	*1*	*2*	*-*	*-*
37 Georgia	...	...	...	...	...	...	...	3	4	3
Romania	13	17	6	5	2	2	6	1	1	1
Russian Fed	878	904	1 012	673	430	441	284	181	54	18
Ukraine	133	284	227	...	...	...	-	-	-	-
37 Fishing area total	*1 024*	*1 205*	*1 245*	*678*	*432*	*443*	*290*	*185*	*59*	*22*
67 Canada	5	3	2	4	1	0	0	0	-	-
67 Fishing area total	*5*	*3*	*2*	*4*	*1*	*0*	*0*	*0*	*-*	*-*
Species total	*13 949*	*8 617*	*7 272 F*	*5 696*	*4 408 F*	*4 153 F*	*3 572 F*	*2 689 F*	*2 225 F*	*1 725*
Group total	***13 949***	***8 631***	***7 287***	***5 906***	***4 624***	***4 407***	***3 777***	***2 849***	***2 603***	***2 010***

B-22 River eels / Anguilles / Anguilas

Capture production by species, fishing areas and countries or areas
Captures par espèces, zones de pêche et pays ou zones
Capturas por especies, áreas de pesca y países o áreas

Species, Fishing area Espèce, Zone de pêche Especie, Area de pesca	1992 mt	1993 mt	1994 mt	1995 mt	1996 mt	1997 mt	1998 mt	1999 mt	2000 mt	2001 mt
European eel	**Anguille d'Europe**		**Anguila europea**		***Anguilla anguilla***				**1,43(02)002,01**	**ELE**
01 Egypt	...	...	...	798	537	585	501	709	2 064	1 979
Morocco	50	100	150	100	100	400	300	250	100	73
01 Fishing area total	*50*	*100*	*150*	*898*	*637*	*985*	*801*	*959*	*2 164*	*2 052*
04 Turkey	245	261	329	390	342	400	300	99	176	122
04 Fishing area total	*245*	*261*	*329*	*390*	*342*	*400*	*300*	*99*	*176*	*122*
05 Albania	188	150 F	100 F	39	50	21	58	63	70	98
Belarus	22	19	26	15	20	15	18	16	14	25
Belgium	30	30	30	30	30	30	30	30	30	30
Czech Rep	...	31	32	31	28	27	28	28	24	29
Czechoslovak	40 F	-	-	-	-	-	-	-	-	-
Denmark	108	57	60	62	34	39	40	30	20	36
Estonia	-	49	44	32	35	38	22	32	40	40
Finland	0	0	0	0	21	21	...	...	...	...
France	10	-	-	-	40	40	40	134	131	130 F
Germany	776	774	550	550	550	550	550	550	550	500
Greece	...	...	...	...	10	10	9	11	10	8
Hungary	421	263	501	411	579	124	182	179	76	27
Ireland	234	260	300	400	400	400	400	250	250	110
Italy	920	815	550	270	346	326	269	283	329	217
Latvia	18	18	38	26	25	27	25	15	13	17
Lithuania	12	10	12	10	12	11	17	18	11	12
Netherlands	22	375	310	393	300	285	322	332	330 F	300 F
Norway	...	...	...	...	...	30	22	28	...	...
Poland	800	800	800	390	373	256	223	248	257	263
Romania	-	-	-	-	-	1	1	0	26	-
Russian Fed	-	16	8	8	10	13	24	4	6	3
Slovakia	...	7	20	13	7	8	8	8	4	6
Spain	0	0	0	0	0	0	0	0	0	0
Sweden	132	129	171	127	97	142	112	140	113	118
Switzerland	7	4	5	5	3	2	3	3	2	2
UK	749	714	833	756	857	778	730	690	795	583
05 Fishing area total	*4 489 F*	*4 521 F*	*4 390 F*	*3 568*	*3 827*	*3 194*	*3 133*	*3 092*	*3 101 F*	*2 554 F*
27 Denmark	1 341	1 024	1 140	842	700	757	560	686	600	635
Estonia	8	10	10	6	20	18	22	28	27	27
Finland	0	0	0	0	1	1	...	...	...	...
France	213	275	346	300	237	1 435	102	78	203	180
Germany	250	253	35	35	146	196	167	197	136	138
Ireland	200	200	200	200	150	150	250	250	265	...
Latvia	1	-	1	2	1	2	2	2	2	2
Netherlands	66	43	49	39	36	30	23	40	21	34
Norway	373	340	472	454	352	467	341	447	281	304
Poland	295	316	290	237	266	233	231	226	172	163
Portugal	52	-	-	-	-	-	-	30	29	37
Russian Fed	53	19	25	33	36	34	25	19	40	53
Spain	37	22	30	23	31	39	23	18	23	51
Sweden	1 048	1 015	1 127	972	945	931	533	594	447	462
UK	33	38	40	52	38	34	11	7	1	12
27 Fishing area total	*3 970*	*3 555*	*3 765*	*3 195*	*2 959*	*4 327*	*2 290*	*2 622*	*2 247*	*2 098*
34 Gambia	-	-	-	-	-	26	0	0	-	-
Morocco	1	4	0	0	0	1	4	-	-	-
34 Fishing area total	*1*	*4*	*0*	*0*	*0*	*27*	*4*	*0*	-	-
37 Algeria	-	-	-	-	-	-	10	...	...	...
Croatia	7	5	5	7	6	7	-	-	-	-
France	941	589	261	20	126	307	307	77	65	105
Greece	15	17	23	31	21	21	34	16	24	24
Italy	742	492	436	616	537	684	413	362	220	229
Spain	60 F	55 F	50 F	45 F	37	33	-	21	47	11
Tunisia	224	373	390	130	192	202	150	206	108	135
37 Fishing area total	*1 989 F*	*1 531 F*	*1 165 F*	*849 F*	*919*	*1 254*	*914*	*682*	*464*	*504*
Species total	*10 744 F*	*9 972 F*	*9 799 F*	*8 900 F*	*8 684*	*10 187*	*7 442*	*7 454*	*8 152 F*	*7 330 F*
Japanese eel	**Anguille du Japon**		**Anguila japonesa**		***Anguilla japonica***				**1,43(02)002,04**	**ELJ**
04 Japan	1 092	970	949	899	901	860	860	817	765	677
Korea Rep	111	96	93	124	113	56	44	13	...	...
04 Fishing area total	*1 203*	*1 066*	*1 042*	*1 023*	*1 014*	*916*	*904*	*830*	*765*	*677*
Species total	*1 203*	*1 066*	*1 042*	*1 023*	*1 014*	*916*	*904*	*830*	*765*	*677*
American eel	**Anguille d'Amérique**		**Anguila americana**		***Anguilla rostrata***				**1,43(02)002,06**	**ELA**
02 Canada	314	278	272	319	289	311	278	278	286	29
Cuba	0	-	-	-	-	-	-	-	-	-
Dominican Rp	...	...	...	6	...	...	1	2	7	1
USA	615	523	724	285	441	485	460	490	649	393
02 Fishing area total	*929*	*801*	*996*	*610*	*730*	*796*	*739*	*770*	*942*	*423*
21 Canada	753	848	752	631	551	455	496	353	395	220
21 Fishing area total	*753*	*848*	*752*	*631*	*551*	*455*	*496*	*353*	*395*	*220*
31 Mexico	-	-	-	43	35	19	9	2	1	...

B-22 River eels / Anguilles / Anguilas

Capture production by species, fishing areas and countries or areas
Captures par espèces, zones de pêche et pays ou zones
Capturas por especies, áreas de pesca y países o áreas

Species, Fishing area Espèce, Zone de pêche Especie, Area de pesca	1992 mt	1993 mt	1994 mt	1995 mt	1996 mt	1997 mt	1998 mt	1999 mt	2000 mt	2001 mt
31 Fishing area total	-	-	-	*43*	*35*	*19*	*9*	*2*	*1*	...
Species total	*1 682*	*1 649*	*1 748*	*1 284*	*1 316*	*1 270*	*1 244*	*1 125*	*1 338*	*643*
Short-finned eel	**Anguille d'Australie**		**Anguila australiana**		***Anguilla australis***			**1,43(02)002,07**		**ELU**
06 Australia	284	239	385	480	413	320	293	296	300 F	300 F
06 Fishing area total	*284*	*239*	*385*	*480*	*413*	*320*	*293*	*296*	*300 F*	*300 F*
57 Australia	252	349	365	259	208	203	160	129	133	157
57 Fishing area total	*252*	*349*	*365*	*259*	*208*	*203*	*160*	*129*	*133*	*157*
Species total	*536*	*588*	*750*	*739*	*621*	*523*	*453*	*425*	*433 F*	*457 F*
River eels nei	**Anguilles nca**		**Anguilas nep**		***Anguilla spp***			**1,43(02)002,XX**		**ELX**
04 Indonesia	1 137	791	3 056	926	4 340	657	787	1 212	4 446	4 790
Philippines	498	194	118	128	100	116	73	114	193	201
04 Fishing area total	*1 635*	*985*	*3 174*	*1 054*	*4 440*	*773*	*860*	*1 326*	*4 639*	*4 991*
06 Australia	135	138	77	40	40	98	100	178	176	182
New Zealand	752	610	600	614	478	392	893	735	675	762
06 Fishing area total	*887*	*748*	*677*	*654*	*518*	*490*	*993*	*913*	*851*	*944*
81 New Zealand	913	865	1 362	1 183	524	588	400	416	380	313
81 Fishing area total	*913*	*865*	*1 362*	*1 183*	*524*	*588*	*400*	*416*	*380*	*313*
Species total	*3 435*	*2 598*	*5 213*	*2 891*	*5 482*	*1 851*	*2 253*	*2 655*	*5 870*	*6 248*
Group total	***17 600***	***15 873***	***18 552***	***14 837***	***17 117***	***14 747***	***12 296***	***12 489***	***16 558***	***15 355***

B-23 Salmons, trouts, smelts / Saumons, truites, éperlans / Salmones, truchas, eperlanos

Capture production by species, fishing areas and countries or areas / Captures par espèces, zones de pêche et pays ou zones / Capturas por especies, áreas de pesca y países o áreas

Species, Fishing area Espèce, Zone de pêche Especie, Area de pesca	1992 mt	1993 mt	1994 mt	1995 mt	1996 mt	1997 mt	1998 mt	1999 mt	2000 mt	2001 mt
Atlantic salmon	**Saumon de l'Atlantique**		**Salmón del Atlántico**		***Salmo salar***			**1,23(01)004,01**		**SAL**
05 Denmark	1	0	1	0	0	0	1	0	1	0
Estonia	-	1	0	0	-	-	-	-	0	0
Finland	325	317	385	383	523	523	222	222	221	221
Iceland	636	657	448	489	358	154	166	120	85	88
Ireland	72	43	29	86	131	105	101	515	611	688
Latvia	-	-	-	-	-	-	-	-	1	3
Norway	520	351	351	325	267	236	331	327	423	420 F
Poland	-	-	-	-	-	-	-	-	20	5
Russian Fed	-	73	87	68	56	32	73	55	63	52
Spain	11	8	11	9	10 F	10 F	10 F	10 F	10 F	10 F
Sweden	78	86	59	51	44	47	44	34	36	56
UK	614	674	790	823	568	457	419	403	447	482
05 Fishing area total	*2 257*	*2 210*	*2 161*	*2 234*	*1 957 F*	*1 564 F*	*1 367 F*	*1 686 F*	*1 918 F*	*2 025 F*
21 Canada	283	164	136	103	82	77	7	1	-	-
Denmark	-	-	-	-	-	-	-	-	-	0
Greenland	236	0	0	68	82	43	...	...	24	42
St Pier Mq	0	2	3	0	1	1	1	1	1	1
21 Fishing area total	*519*	*166*	*139*	*171*	*165*	*121*	*8*	*2*	*25*	*43*
27 Denmark	656	583	745	560	528	493	486	389	412	434
Estonia	31	31	5	9	9	10	7	14	21	14
Faeroe Is	32	78	12	3	-	-	5	-	0	0
Finland	2 172	1 895	1 184	1 258	1 191	1 267	798	690	743	596
France	-	-	42	0	0	0	0	2	10	12
Germany	56	55	13	13	27	35	42	30	45	39
Greenland	6	-	-	2	0	1	-	-	-	-
Iceland	460	496	308	289	239	48	36	22	2	-
Ireland	631	541	804	790	687	570	624	511	-	-
Isle of Man	0	0	0	0	-	-	-	-	-	-
Latvia	268	243	130	139	151	169	125	166	150	135
Lithuania	20	15	5	2	10	4	5	6	6	4
Netherlands	1	0	1	1	2	1	1	1	-	0
Norway	520	571	649	520	526	402	422	500	631	705
Poland	462	191	184	133	125	110	114	118	125	156
Portugal	-	1	-	-	-	-	-	0	0	-
Russian Fed	115	114	120	124	117	116	91	72	91	109
Sweden	992	971	729	646	761	668	613	399	478	356
UK	-	-	-	-	-	-	-	-	33	37
27 Fishing area total	*6 422*	*5 785*	*4 931*	*4 489*	*4 373*	*3 894*	*3 369*	*2 920*	*2 747*	*2 597*
Species total	*9 198*	*8 161*	*7 231*	*6 894*	*6 495 F*	*5 579 F*	*4 744 F*	*4 608 F*	*4 690 F*	*4 665 F*
Sea trout	**Truite de mer**		**Trucha marina**		***Salmo trutta***			**1,23(01)004,02**		**TRS**
03 Falkland Is	-	-	-	-	1	1	1	1	1	1
03 Fishing area total	*-*	*-*	*-*	*-*	*1*	*1*	*1*	*1*	*1*	*1*
04 Turkey	558	479	554	594	395	200	200	263	277	364
04 Fishing area total	*558*	*479*	*554*	*594*	*395*	*200*	*200*	*263*	*277*	*364*
05 Belgium	100	100	100	100	100	100	100	150	100	100
Bulgaria	-	-	-	-	-	8	11	11	4	1
Czechoslovak	150 F	-	-	-	-	-	-	-	-	-
Denmark	16	3	8	6	3	3	4	3	4	0
Ireland	880	880	1 200	1 200	1 200	1 200	1 200	100	100	...
Lithuania	-	-	-	-	-	-	-	-	1	1
Romania	29	33	40	39	39	28	10	63	18	48
Russian Fed	-	31	-	-	-	-	-	-	1	3
Slovakia	...	35	32	46	34	40	42	41	37	38
Slovenia	11	15	14	14	14	13	11	10	9	7
Spain	2 371	2 389	2 018	2 163	2 200 F	2 200 F	2 200 F	2 200 F	2 200 F	2 200 F
Switzerland	-	-	-	10	8	13	12	11	13	13
UK	275	214	256	260	198	141	572	489	142	560
05 Fishing area total	*3 832 F*	*3 700*	*3 668*	*3 838*	*3 796 F*	*3 746 F*	*4 162 F*	*3 078 F*	*2 629 F*	*2 971 F*
27 Denmark	45	68	50	80	74	49	55	98	66	52
Estonia	9	15	8	6	15	11	8	10	13	13
Finland	1 270	1 256	603	615	672	661	460	441	437	416
France	400	109	890	0	0	0	0	0	-	1
Ireland	-	-	-	-	-	-	-	10	-	-
Isle of Man	0	0	0	0	-	-	-	-	-	-
Latvia	5	19	18	14	10	7	7	10	14	11
Lithuania	-	-	-	3	-	2	3	4	4	2
Portugal	0	0	0	1	0	-	-	1	1	1
Russian Fed	-	-	-	-	-	-	-	1	-	1
UK	0	0	-	-	-	-	0	0	8	11
27 Fishing area total	*1 729*	*1 467*	*1 569*	*719*	*771*	*730*	*533*	*575*	*543*	*508*
Species total	*6 119 F*	*5 646*	*5 791*	*5 151*	*4 963 F*	*4 677 F*	*4 896 F*	*3 917 F*	*3 450 F*	*3 844 F*
Trouts nei	**Truites nca**		**Truchas nep**		***Salmo spp***			**1,23(01)004,XX**		**TRO**
04 Armenia	50	5	7	0	0	6	0	163	186	180
Azerbaijan	-	-	-	-	-	-	-	-	-	5
Japan	1 244	1 189	1 106	1 080	1 222	1 207	1 187	1 122	1 136	1 053
04 Fishing area total	*1 294*	*1 194*	*1 113*	*1 080*	*1 222*	*1 213*	*1 187*	*1 285*	*1 322*	*1 238*

B-23

Salmons, trouts, smelts — **Capture production by species, fishing areas and countries or areas**
Saumons, truites, éperlans — **Captures par espèces, zones de pêche et pays ou zones**
Salmones, truchas, eperlanos — **Capturas por especies, áreas de pesca y países o áreas**

Species, Fishing area Espèce, Zone de pêche Especie, Area de pesca	1992 mt	1993 mt	1994 mt	1995 mt	1996 mt	1997 mt	1998 mt	1999 mt	2000 mt	2001 mt
05 Czech Rep	...	51	52	54	53	57	70	64	55	56
Finland	959	952	1 025	1 022	1 286	1 286	879	879	610	610
Iceland	250	250	250	250	250	250	250	250	91	72
Macedonia	174	77	137	30	37	51	18	22	131	115
Netherlands	0	0	0	0	-	-	-	-	-	-
Poland	40	-	-	-	40	-	-	-	139	48
Slovenia	3	2	2	2	3	2	2	2	2	1
Sweden	8	14	12	14	18	22	26	22	24	24
UK	-	-	-	-	-	-	-	-	-	254
05 Fishing area total	*1 434*	*1 346*	*1 478*	*1 372*	*1 687*	*1 668*	*1 245*	*1 239*	*1 052*	*1 180*
21 Canada	30	53	34	20	4	1	0	0	0	-
21 Fishing area total	*30*	*53*	*34*	*20*	*4*	*1*	*0*	*0*	*0*	-
27 Germany	-	-	0	1	7	8	6	9	12	11
Poland	-	272	222	187	150	200	329	385	579	529
Sweden	174	177	123	78	113	96	112	72	64	39
27 Fishing area total	*174*	*449*	*345*	*266*	*270*	*304*	*447*	*466*	*655*	*579*
Species total	*2 932*	*3 042*	*2 970*	*2 738*	*3 183*	*3 186*	*2 879*	*2 990*	*3 029*	*2 997*
Pink(=Humpback)salmon	**Saumon rose**		**Salmón rosado**		***Oncorhynchus gorbuscha***			**1,23(01)009,02**		**PIN**
02 USA	13	209	51	319	3	31	0	1	-	-
02 Fishing area total	*13*	*209*	*51*	*319*	*3*	*31*	*0*	*1*	-	-
04 Japan	1 482	729	3 598	1 511	3 134	947	2 091	927	1 947	376
04 Fishing area total	*1 482*	*729*	*3 598*	*1 511*	*3 134*	*947*	*2 091*	*927*	*1 947*	*376*
05 Russian Fed	7 465	1 190	7 420	8 862	8 804	18 597	14 057	28 582	24 277	17 961
05 Fishing area total	*7 465*	*1 190*	*7 420*	*8 862*	*8 804*	*18 597*	*14 057*	*28 582*	*24 277*	*17 961*
27 Russian Fed	-	-	-	-	-	-	-	39	10	184
27 Fishing area total	-	-	-	-	-	-	-	*39*	*10*	*184*
61 Japan	20 420	24 116	28 476	23 526	29 461	14 897	23 246	15 975	24 655	9 389
Russian Fed	79 414	104 855	117 533	139 369	104 377	169 070	177 382	158 560	132 851	149 421
61 Fishing area total	*99 834*	*128 971*	*146 009*	*162 895*	*133 838*	*183 967*	*200 628*	*174 535*	*157 506*	*158 810*
67 Canada	14 913	16 046	3 383	19 767	8 597	12 241	3 920	9 529	7 158	10 575
USA	92 403	155 423	165 604	201 381	140 539	102 934	150 856	173 315	94 440	173 067
67 Fishing area total	*107 316*	*171 469*	*168 987*	*221 148*	*149 136*	*115 175*	*154 776*	*182 844*	*101 598*	*183 642*
77 USA	-	-	-	-	-	0	-	-	-	0
77 Fishing area total	-	-	-	-	-	*0*	-	-	-	*0*
Species total	*216 110*	*302 568*	*326 065*	*394 735*	*294 915*	*318 717*	*371 552*	*386 928*	*285 338*	*360 973*
Chum(=Keta=Dog)salmon	**Saumon chien**		**Keta**		***Oncorhynchus keta***			**1,23(01)009,03**		**CHU**
02 USA	2 708	769	1 717	518	1 567	738	981	325	59	4
02 Fishing area total	*2 708*	*769*	*1 717*	*518*	*1 567*	*738*	*981*	*325*	*59*	*4*
04 Japan	8 623	13 608	17 460	17 736	19 026	18 346	16 069	11 684	12 326	9 599
04 Fishing area total	*8 623*	*13 608*	*17 460*	*17 736*	*19 026*	*18 346*	*16 069*	*11 684*	*12 326*	*9 599*
05 Russian Fed	7 231	6 599	10 115	10 049	6 670	6 999	7 309	8 137	14 546	14 127
05 Fishing area total	*7 231*	*6 599*	*10 115*	*10 049*	*6 670*	*6 999*	*7 309*	*8 137*	*14 546*	*14 127*
61 Japan	149 540	195 892	205 647	249 982	280 855	250 837	190 553	171 182	151 123	207 760
Russian Fed	14 060	12 610	14 603	12 632	16 413	15 899	18 737	20 025	21 944	17 940
61 Fishing area total	*163 600*	*208 502*	*220 250*	*262 614*	*297 268*	*266 736*	*209 290*	*191 207*	*173 067*	*225 700*
67 Canada	17 964	17 273	20 323	12 115	6 524	8 685	19 903	4 937	2 783	5 549
USA	38 301	40 245	58 900	121 563	80 340	46 056	58 413	64 970	73 574	52 683
67 Fishing area total	*56 265*	*57 518*	*79 223*	*133 678*	*86 864*	*54 741*	*78 316*	*69 907*	*76 357*	*58 232*
Species total	*238 427*	*286 996*	*328 765*	*424 595*	*411 395*	*347 560*	*311 965*	*281 260*	*276 355*	*307 662*
Masu(=Cherry) salmon	**Saumon du Japon**		**Salmón japonés**		***Oncorhynchus masou***			**1,23(01)009,05**		**CHE**
04 Japan	802	796	852	842	830	810	836	849	856	826
04 Fishing area total	*802*	*796*	*852*	*842*	*830*	*810*	*836*	*849*	*856*	*826*
05 Russian Fed	8	3	3	13	41	4	3	7	3	4
05 Fishing area total	*8*	*3*	*3*	*13*	*41*	*4*	*3*	*7*	*3*	*4*
61 Japan	2 360	1 543	1 694	1 431	1 677	990	1 734	1 130	955	779
Russian Fed	-	-	4	4	4	4	4	3	-	1
61 Fishing area total	*2 360*	*1 543*	*1 698*	*1 435*	*1 681*	*994*	*1 738*	*1 133*	*955*	*780*
Species total	*3 170*	*2 342*	*2 553*	*2 290*	*2 552*	*1 808*	*2 577*	*1 989*	*1 814*	*1 610*
Sockeye(=Red)salmon	**Saumon rouge**		**Salmón rojo**		***Oncorhynchus nerka***			**1,23(01)009,06**		**SOC**
02 USA	318	171	176	0	115	78	190	56	13	0
02 Fishing area total	*318*	*171*	*176*	*0*	*115*	*78*	*190*	*56*	*13*	*0*

B-23 Salmons, trouts, smelts / Saumons, truites, éperlans / Salmones, truchas, eperlanos

Capture production by species, fishing areas and countries or areas / Captures par espèces, zones de pêche et pays ou zones / Capturas por especies, áreas de pesca y países o áreas

Species, Fishing area Espèce, Zone de pêche Especie, Area de pesca	1992 mt	1993 mt	1994 mt	1995 mt	1996 mt	1997 mt	1998 mt	1999 mt	2000 mt	2001 mt
04 Japan	90	80	83	79	72	88	118	90	52	22
04 Fishing area total	*90*	*80*	*83*	*79*	*72*	*88*	*118*	*90*	*52*	*22*
05 Russian Fed	9 308	7 651	5 150	8 229	9 261	3 479	3 947	7 097	6 872	12 106
05 Fishing area total	*9 308*	*7 651*	*5 150*	*8 229*	*9 261*	*3 479*	*3 947*	*7 097*	*6 872*	*12 106*
61 Japan	5 761	7 725	3 704	6 155	5 651	9 158	2 650	2 660	2 095	2 718
Russian Fed	6 408	5 490	5 658	5 998	13 630	6 698	8 820	7 792	12 676	10 369
61 Fishing area total	*12 169*	*13 215*	*9 362*	*12 153*	*19 281*	*15 856*	*11 470*	*10 452*	*14 771*	*13 087*
67 Canada	20 938	42 529	30 828	10 533	15 525	25 353	5 041	1 653	8 665	6 231
USA	156 886	178 967	138 087	158 618	144 330	87 221	58 206	110 780	94 409	77 172
67 Fishing area total	*177 824*	*221 496*	*168 915*	*169 151*	*159 855*	*112 574*	*63 247*	*112 433*	*103 074*	*83 403*
Species total	*199 709*	*242 613*	*183 686*	*189 612*	*188 584*	*132 075*	*78 972*	*130 128*	*124 782*	*108 618*
Chinook(=Spring=King)salmon	**Saumon royal**		**Salmón real**			***Oncorhynchus tshawytscha***			**1,23(01)009,07**	**CHI**
02 USA	2 108	1 549	1 907	2 146	2 111	1 621	1 367	1 344	390	236
02 Fishing area total	*2 108*	*1 549*	*1 907*	*2 146*	*2 111*	*1 621*	*1 367*	*1 344*	*390*	*236*
05 Russian Fed	1 013	1 063	721	738	401	445	340	483	264	163
05 Fishing area total	*1 013*	*1 063*	*721*	*738*	*401*	*445*	*340*	*483*	*264*	*163*
61 Japan	661	615	364	195	250	825	534	270	147	111
Russian Fed	248	221	380	137	120	191	216	310	215	336
61 Fishing area total	*909*	*836*	*744*	*332*	*370*	*1 016*	*750*	*580*	*362*	*447*
67 Canada	5 336	4 817	3 573	1 510	455	1 662	1 386	742	507	636
USA	4 999	5 619	5 271	6 080	5 093	5 518	5 057	3 594	4 300	6 095
67 Fishing area total	*10 335*	*10 436*	*8 844*	*7 590*	*5 548*	*7 180*	*6 443*	*4 336*	*4 807*	*6 731*
77 USA	728	1 152	1 407	2 993	2 078	2 737	937	1 991	2 613	1 193
77 Fishing area total	*728*	*1 152*	*1 407*	*2 993*	*2 078*	*2 737*	*937*	*1 991*	*2 613*	*1 193*
81 New Zealand	-	-	-	2	1	1	3	1	1	1
81 Fishing area total	-	-	-	*2*	*1*	*1*	*3*	*1*	*1*	*1*
Species total	*15 093*	*15 036*	*13 623*	*13 801*	*10 509*	*13 000*	*9 840*	*8 735*	*8 437*	*8 771*
Coho(=Silver)salmon	**Saumon argenté**		**Salmón plateado**			***Oncorhynchus kisutch***			**1,23(01)009,08**	**COH**
02 USA	2 753	2 314	2 996	441	4 021	696	855	282	802	659
02 Fishing area total	*2 753*	*2 314*	*2 996*	*441*	*4 021*	*696*	*855*	*282*	*802*	*659*
05 Russian Fed	3 546	1 315	1 761	1 213	1 577	898	1 449	1 054	1 419	1 090
05 Fishing area total	*3 546*	*1 315*	*1 761*	*1 213*	*1 577*	*898*	*1 449*	*1 054*	*1 419*	*1 090*
61 Japan	567	207	-	270	701	575	746	508	376	502
Russian Fed	1 038	1 013	504	266	399	412	870	614	859	944
61 Fishing area total	*1 605*	*1 220*	*504*	*536*	*1 100*	*987*	*1 616*	*1 122*	*1 235*	*1 446*
67 Canada	7 328	4 316	7 713	4 866	3 871	751	16	14	31	46
USA	23 179	16 259	32 614	21 861	17 626	9 856	15 450	12 977	14 544	16 764
67 Fishing area total	*30 507*	*20 575*	*40 327*	*26 727*	*21 497*	*10 607*	*15 466*	*12 991*	*14 575*	*16 810*
77 USA	5	0	0	5	6	3	0	-	4	1
77 Fishing area total	*5*	*0*	*0*	*5*	*6*	*3*	*0*	-	*4*	*1*
87 Chile	17	27	14	-	-	-	-	-	-	-
87 Fishing area total	*17*	*27*	*14*	-	-	-	-	-	-	-
Species total	*38 433*	*25 451*	*45 602*	*28 922*	*28 201*	*13 191*	*19 386*	*15 449*	*18 035*	*20 006*
Rainbow trout	**Truite arc-en-ciel**		**Trucha arco iris**			***Oncorhynchus mykiss***			**1,23(01)009,09**	**TRR**
01 Morocco	12	15	-	-	-	-	-	-	-	-
01 Fishing area total	*12*	*15*	-	-	-	-	-	-	-	-
02 Mexico	301	297	300	1 349	1 653	102	95	91	232	223
USA	1 041	268	174	161	168	137	358	82	145	221
02 Fishing area total	*1 342*	*565*	*474*	*1 510*	*1 821*	*239*	*453*	*173*	*377*	*444*
03 Argentina	400	186	261	3	2	0	0	0	-	-
Bolivia	245	141	116	116	338	338	340	342	345	280
Brazil	10 F	10 F	10 F	0	0	0	0	0	0	0
Peru	...	303	...	509	63	869	392	184	220	191
Venezuela	-	50	92	57	59	...	...	...	...	186
03 Fishing area total	*655 F*	*690 F*	*479 F*	*685*	*462*	*1 207*	*732*	*526*	*565*	*657*
04 China,Taiwan	1	-	-	-	-	-	-	-	-	-
Georgia	76	31	5	12	3	...	...	...	...	...
Japan	563	568	567	597	728	672	618	562	536	484
Korea Rep	11	12	15	13	...	...	...	...	...	...
Uzbekistan	...	-	0	-	-	-	-	-	-	-
04 Fishing area total	*651*	*611*	*587*	*622*	*731*	*672*	*618*	*562*	*536*	*484*
05 Bulgaria	-	-	-	-	-	12	10	10	5	17
Croatia	2	4	4	2	5	5	1 F	1	3	18

B-23 Salmons, trouts, smelts — Capture production by species, fishing areas and countries or areas
Saumons, truites, éperlans — Captures par espèces, zones de pêche et pays ou zones
Salmones, truchas, eperlanos — Capturas por especies, áreas de pesca y países o áreas

Species, Fishing area Espèce, Zone de pêche Especie, Area de pesca	1992 mt	1993 mt	1994 mt	1995 mt	1996 mt	1997 mt	1998 mt	1999 mt	2000 mt	2001 mt
Czech Rep	...	30	24	23	28	29	30	38	39	48
Czechoslovak	40 F	-	-	-	-	-	-	-	-	-
Denmark	6	0	1	1	1	1	-	6	-	0
Finland	954	954	782	782	813	813	975	975	660	660
Ireland	105	53	75	75	75	99	75	...	...	...
Poland	...	...	...	...	...	...	...	...	3	1
Romania	...	...	...	...	...	3	25	25	32	65
Slovakia	...	11	10	15	16	16	17	16	19	30
Slovenia	45	33	28	36	32	33	23	19	21	21
Switzerland	-	-	-	40	0	0	0	0	0	0
UK	388	300	300	300	300	100	2 831	3 224	1 267	1 174
05 Fishing area total	*1 540 F*	*1 385*	*1 224*	*1 274*	*1 270*	*1 111*	*3 987 F*	*4 314*	*2 049*	*2 034*
27 Finland	320	307	225	252	104	105	210	171	77	128
Poland	-	-	-	-	-	35	27	14	9	11
Sweden	13	9	-	4	-	-	-	0	-	-
27 Fishing area total	*333*	*316*	*225*	*256*	*104*	*140*	*237*	*185*	*86*	*139*
67 Canada	18	8	5	4	3	1	4	1	1	-
67 Fishing area total	*18*	*8*	*5*	*4*	*3*	*1*	*4*	*1*	*1*	-
Species total	*4 551 F*	*3 590 F*	*2 994 F*	*4 351*	*4 391*	*3 370*	*6 031 F*	*5 761*	*3 614*	*3 758*
Pacific salmons nei	**Saumons du Pacifique nca**		**Salmones del Pacífico nep**			***Oncorhynchus spp***			**1,23(01)009,XX**	**ORC**
67 USA	-	-	-	-	-	-	-	-	50	204
67 Fishing area total	-	-	-	-	-	-	-	-	*50*	*204*
Species total	-	-	-	-	-	-	-	-	*50*	*204*
Brook trout	**Saumon de fontaine**		**Trucha de arroyo**			***Salvelinus fontinalis***			**1,23(01)010,02**	**SVF**
05 Bulgaria	-	-	-	-	-	1	1	1	1	3
Czech Rep	...	4	1	3	2	2	2	3	3	3
Czechoslovak	0	-	-	-	-	-	-	-	-	-
Slovakia	...	0	0	1	0	1	1	0	0	1
Slovenia	-	-	-	-	1	0	0	0	1	0
05 Fishing area total	*0*	*4*	*1*	*4*	*3*	*4*	*4*	*4*	*5*	*7*
Species total	*0*	*4*	*1*	*4*	*3*	*4*	*4*	*4*	*5*	*7*
Arctic char	**Omble-chevalier**		**Trucha alpina**			***Salvelinus alpinus***			**1,23(01)010,05**	**ACH**
05 Slovenia	1	1	1	1	0	1	0	0	1	1
Sweden	69	57	65	47	25	24	28	21	24	18
Switzerland	31	27	22	22	28	27	21	22	22	16
05 Fishing area total	*101*	*85*	*88*	*70*	*53*	*52*	*49*	*43*	*47*	*35*
21 Greenland	46	75	22	55	43	78	76	24	29	20
21 Fishing area total	*46*	*75*	*22*	*55*	*43*	*78*	*76*	*24*	*29*	*20*
Species total	*147*	*160*	*110*	*125*	*96*	*130*	*125*	*67*	*76*	*55*
Lake trout(=Char)	**Touladi (=Omble du Canada)**		**Trucha lacustre**			***Salvelinus namaycush***			**1,23(01)010,07**	**LAT**
02 Canada	481	575	690	631	689	617	554	491	556	679
USA	313	324	298	406	286	492	500	494	573	451
02 Fishing area total	*794*	*899*	*988*	*1 037*	*975*	*1 109*	*1 054*	*985*	*1 129*	*1 130*
Species total	*794*	*899*	*988*	*1 037*	*975*	*1 109*	*1 054*	*985*	*1 129*	*1 130*
Chars nei	**Ombles nca**		**Salvelinos nep**			***Salvelinus spp***			**1,23(01)010,XX**	**CHR**
05 Norway	60	84	81	88	71	67	74	85	102	100 F
UK	-	-	-	-	-	-	-	-	-	1
05 Fishing area total	*60*	*84*	*81*	*88*	*71*	*67*	*74*	*85*	*102*	*101 F*
21 Canada	4	48	60	32	11	31	35	46	46	34
21 Fishing area total	*4*	*48*	*60*	*32*	*11*	*31*	*35*	*46*	*46*	*34*
27 Norway	-	13	16	16	10	11	20	8	17	11
Sweden	1	0	1	-	-	-	-	-	-	-
27 Fishing area total	*1*	*13*	*17*	*16*	*10*	*11*	*20*	*8*	*17*	*11*
Species total	*65*	*145*	*158*	*136*	*92*	*109*	*129*	*139*	*165*	*146 F*
Huchen	**...B**		**...C**			***Hucho hucho***			**1,23(01)020,02**	**HUC**
05 Slovakia	...	0	0	1	1	1	1	1	1	1
05 Fishing area total	...	*0*	*0*	*1*	*1*	*1*	*1*	*1*	*1*	*1*
Species total	...	*0*	*0*	*1*	*1*	*1*	*1*	*1*	*1*	*1*
Grayling	**Ombre commun**		**Tímalo**			***Thymallus thymallus***			**1,23(02)005,01**	**TLV**
05 Belgium	5	5	5	5	5	5	5	5	5	5
Czech Rep	...	16	16	16	19	15	13	16	16	15
Czechoslovak	30 F	-	-	-	-	-	-	-	-	-
Norway	...	...	...	...	...	1	...	...	...	...

B-23 Salmons, trouts, smelts — Capture production by species, fishing areas and countries or areas
Saumons, truites, éperlans — Captures par espèces, zones de pêche et pays ou zones
Salmones, truchas, eperlanos — Capturas por especies, áreas de pesca y países o áreas

Species, Fishing area Espèce, Zone de pêche Especie, Area de pesca	1992 mt	1993 mt	1994 mt	1995 mt	1996 mt	1997 mt	1998 mt	1999 mt	2000 mt	2001 mt
Romania	-	-		-		-	-	2	-	
Slovakia	...	14	16	24	18	16	16	14	13	17
Slovenia	9	7	6	7	7	6	4	4	3	2
Sweden	...	...	1	1	3	0	0	0	-	-
UK	...	...	1	-	-	-	-	-	0	0
05 Fishing area total	*44 F*	*42*	*45*	*53*	*52*	*43*	*38*	*41*	*37*	*39*
Species total	*44 F*	*42*	*45*	*53*	*52*	*43*	*38*	*41*	*37*	*39*
Ayu sweetfish	**Ayu**		**Ayu**		***Plecoglossus altivelis***				**1,23(03)016,01**	**PCA**
04 China,Taiwan	6	8	-	2	-	-	-	-	-	-
Japan	17 677	14 242	14 272	13 700	12 732	12 619	11 386	11 380	11 172	11 148
Korea Rep	-	-	-	-	-	5	-	-	-	-
04 Fishing area total	*17 683*	*14 250*	*14 272*	*13 702*	*12 732*	*12 624*	*11 386*	*11 380*	*11 172*	*11 148*
Species total	*17 683*	*14 250*	*14 272*	*13 702*	*12 732*	*12 624*	*11 386*	*11 380*	*11 172*	*11 148*
European smelt	**Eperlan européen**		**Eperlano europeo**		***Osmerus eperlanus***				**1,23(04)003,01**	**SME**
05 Denmark	-	-	-	-	7	0	1	-	0	0
Estonia	...	502	224	710	478	401	1 421	947	1 104	623
Finland	628	603	864	1 100	854	854	450	450	366	366
Latvia	...	...	3	3	2	4	18	8	-	1
Lithuania	301	136	131	105	81	190	200	147	214	178
Netherlands	1 517	580	1 537	2 952	856	1 033	327	1 081	1 080 F	1 050 F
Poland	-	-	-	-	-	38	2	33	2	1
Sweden	7	7	18	11	9	10	8	9	-	-
05 Fishing area total	*2 453*	*1 828*	*2 777*	*4 881*	*2 287*	*2 530*	*2 427*	*2 675*	*2 766 F*	*2 219 F*
27 Denmark	91	71	70	76	46	34	18	20	29	25
Estonia	92	3	3	21	6	14	10	61	90	139
Finland	1 052	1 427	822	1 137	1 369	1 044	759	880	413	533
France	69	60	71	84	95	106	69	75	103	96
Germany	20	46	46	12	29	87	32	46	4	6
Latvia	505	331	2	351	384	331	200	172	261	127
Lithuania	-	-	-	-	-	-	134	218	-	182
Netherlands	-	-	-	-	-	-	-	16	74	111
Poland	-	-	-	-	-	-	51	179	17	19
Russian Fed	3 711	1 351	1 150	805	1 022	760	835	409	844	976
UK	7	1	57	7	149	15	1 217	888	5 316	21
27 Fishing area total	*5 547*	*3 290*	*2 221*	*2 493*	*3 100*	*2 391*	*3 325*	*2 964*	*7 151*	*2 235*
Species total	*8 000*	*5 118*	*4 998*	*7 374*	*5 387*	*4 921*	*5 752*	*5 639*	*9 917 F*	*4 454 F*
Rainbow smelt	**Eperlan arc-en-ciel**		**Eperlano arco iris**		***Osmerus mordax***				**1,23(04)003,03**	**SMR**
02 USA	1 597	1 666	1 183	968	711	522	321	328	398	209
02 Fishing area total	*1 597*	*1 666*	*1 183*	*968*	*711*	*522*	*321*	*328*	*398*	*209*
05 Russian Fed	2 990	1 143	338	1 006	640	1 113	1 758	1 340	608	719
05 Fishing area total	*2 990*	*1 143*	*338*	*1 006*	*640*	*1 113*	*1 758*	*1 340*	*608*	*719*
21 Canada	1 087	1 036	1 396	1 132	1 078	1 227	1 255	356	970	308
USA	27	10	9	0	-	0	0	-	0	0
21 Fishing area total	*1 114*	*1 046*	*1 405*	*1 132*	*1 078*	*1 227*	*1 255*	*356*	*970*	*308*
27 Russian Fed	-	-	-	-	-	-	-	-	1	6
27 Fishing area total	*-*	*-*	*-*	*-*	*-*	*-*	*-*	*-*	*1*	*6*
Species total	*5 701*	*3 855*	*2 926*	*3 106*	*2 429*	*2 862*	*3 334*	*2 024*	*1 977*	*1 242*
Pond smelt	**Eperlan à petite bouche**		**Eperlano de estanque**		***Hypomesus olidus***				**1,23(04)008,01**	**PSM**
02 Canada	5 769	7 975	4 826	5 517	3 981	5 964	6 471	5 691	3 254	4 241
02 Fishing area total	*5 769*	*7 975*	*4 826*	*5 517*	*3 981*	*5 964*	*6 471*	*5 691*	*3 254*	*4 241*
Species total	*5 769*	*7 975*	*4 826*	*5 517*	*3 981*	*5 964*	*6 471*	*5 691*	*3 254*	*4 241*
Surf smelt	**Eperlan du Pacifique**		**Eperlano del Pacífico**		***Hypomesus pretiosus***				**1,23(04)008,02**	**SUS**
67 Canada	2	2	3	1	1	-	-	-	-	-
67 Fishing area total	*2*	*2*	*3*	*1*	*1*	*-*	*-*	*-*	*-*	*-*
Species total	*2*	*2*	*3*	*1*	*1*	*-*	*-*	*-*	*-*	*-*
Eulachon	**Eulakane**		**...C**		***Thaleichthys pacificus***				**1,23(04)012,01**	**EUL**
02 USA	1 668	233	20	200	4	27	3	8	-	-
02 Fishing area total	*1 668*	*233*	*20*	*200*	*4*	*27*	*3*	*8*	*-*	*-*
67 Canada	20	9	6	26	30	0	0	-	-	1
USA	-	-	-	-	-	0	3	-	13	142
67 Fishing area total	*20*	*9*	*6*	*26*	*30*	*0*	*3*	*-*	*13*	*143*
Species total	*1 688*	*242*	*26*	*226*	*34*	*27*	*6*	*8*	*13*	*143*

B-23 Salmons, trouts, smelts — Capture production by species, fishing areas and countries or areas
Saumons, truites, éperlans — Captures par espèces, zones de pêche et pays ou zones
Salmones, truchas, eperlanos — Capturas por especies, áreas de pesca y países o áreas

Species, Fishing area Espèce, Zone de pêche Especie, Area de pesca	1992 mt	1993 mt	1994 mt	1995 mt	1996 mt	1997 mt	1998 mt	1999 mt	2000 mt	2001 mt
Smelts nei	**Eperlans nca**		**Eperlanos nep**			***Osmerus spp, Hypomesus spp***			**1,23(04)XXX,XX**	**SMX**
05 Russian Fed	1 719	3 750	2 701	3 118	1 927	4 095	3 102	2 740	3 825	3 568
05 Fishing area total	*1 719*	*3 750*	*2 701*	*3 118*	*1 927*	*4 095*	*3 102*	*2 740*	*3 825*	*3 568*
61 Russian Fed	880	316	39	130	122	321	322	390	735	612
61 Fishing area total	*880*	*316*	*39*	*130*	*122*	*321*	*322*	*390*	*735*	*612*
67 USA	35	76	218	71	78	839	490	432	442	218
67 Fishing area total	*35*	*76*	*218*	*71*	*78*	*839*	*490*	*432*	*442*	*218*
77 USA	-	-	-	-	-	7	7	5	6	12
77 Fishing area total	*-*	*-*	*-*	*-*	*-*	*7*	*7*	*5*	*6*	*12*
Species total	*2 634*	*4 142*	*2 958*	*3 319*	*2 127*	*5 262*	*3 921*	*3 567*	*5 008*	*4 410*
Vendace	**Corégone blanc**		**Coregono blanco**			***Coregonus albula***			**1,23(12)001,01**	**FVE**
05 Estonia	-	-	-	45	127	153	159	47	1	-
Finland	3 495	3 609	3 555	4 174	5 845	5 845	4 990	4 990	4 837	4 837
Latvia	-	-	-	-	-	-	-	-	-	1
Norway	...	...	...	...	...	10	4	10	6	5 F
Poland	550	500	500	232	234	297	217	286	275	227
Sweden	489	528	593	492	604	522	257	182	261	123
05 Fishing area total	*4 534*	*4 637*	*4 648*	*4 943*	*6 810*	*6 827*	*5 627*	*5 515*	*5 380*	*5 193 F*
27 Finland	93	102	105	135	143	130	125	135	185	187
Latvia	6	4	5	5	5	5	6	7	4	4
Sweden	1 080	1 104	965	626	717	686	419	391	569	697
27 Fishing area total	*1 179*	*1 210*	*1 075*	*766*	*865*	*821*	*550*	*533*	*758*	*888*
Species total	*5 713*	*5 847*	*5 723*	*5 709*	*7 675*	*7 648*	*6 177*	*6 048*	*6 138*	*6 081 F*
European whitefish	**Corégone lavaret**		**Lavareto**			***Coregonus lavaretus***			**1,23(12)001,02**	**PLN**
05 Bulgaria	-	-	-	-	-	0	0	0	-	-
Czech Rep	...	3	-	-	2	2	1	1	1	1
Czechoslovak	0	-	-	-	-	-	-	-	-	-
Denmark	9	4	3	3	3	1	3	1	1	1
Finland	3 607	3 612	4 216	4 014	3 042	3 042	2 531	2 531	2 421	2 421
Norway	...	...	...	...	...	57	52	54	47	45 F
Poland	2	2	2	28	33	24	14	21	29	13
Sweden	226	199	247	202	180	195	185	181	171	182
05 Fishing area total	*3 844*	*3 820*	*4 468*	*4 247*	*3 260*	*3 321*	*2 786*	*2 789*	*2 670*	*2 663 F*
27 Denmark	24	28	36	35	27	30	26	24	6	76
Estonia	10	8	10	6	21	20	20	28	33	33
Finland	2 880	2 783	2 083	2 140	2 084	1 961	2 351	2 172	2 120	1 826
Germany	-	-	-	-	-	5	8	21	47	63
Sweden	307	354	571	465	350	302	304	279	249	154
27 Fishing area total	*3 221*	*3 173*	*2 700*	*2 646*	*2 482*	*2 318*	*2 709*	*2 524*	*2 455*	*2 152*
Species total	*7 065*	*6 993*	*7 168*	*6 893*	*5 742*	*5 639*	*5 495*	*5 313*	*5 125*	*4 815 F*
Houting	**Bondelle**		**Coregono picudo**			***Coregonus oxyrinchus***			**1,23(12)001,03**	**HOU**
05 Bulgaria	-	-	-	-	-	0	0	0	-	-
Denmark	-	-	-	-	-	-	1	-	22	0
Estonia	...	18	14	25	63	30	60	35	9	9
05 Fishing area total	*...*	*18*	*14*	*25*	*63*	*30*	*61*	*35*	*31*	*9*
Species total	*...*	*18*	*14*	*25*	*63*	*30*	*61*	*35*	*31*	*9*
Lake(=Common)whitefish	**Corégone de lac**		**Coregono de lago**			***Coregonus clupeaformis***			**1,23(12)001,12**	**WHL**
02 Canada	9 038	7 730	9 480	9 196	9 121	8 376	8 599	8 330	9 028	9 624
USA	4 603	5 162	5 110	5 310	5 272	5 842	5 678	5 353	5 199	4 484
02 Fishing area total	*13 641*	*12 892*	*14 590*	*14 506*	*14 393*	*14 218*	*14 277*	*13 683*	*14 227*	*14 108*
Species total	*13 641*	*12 892*	*14 590*	*14 506*	*14 393*	*14 218*	*14 277*	*13 683*	*14 227*	*14 108*
Lake cisco	**Cisco de lac**		**Coregono de artedi**			***Coregonus artedi***			**1,23(12)001,13**	**CIS**
02 Canada	699	608	1 301	1 101	919	1 008	764	565	581	567
USA	279	388	345	316	226	215	284	239	255	303
02 Fishing area total	*978*	*996*	*1 646*	*1 417*	*1 145*	*1 223*	*1 048*	*804*	*836*	*870*
Species total	*978*	*996*	*1 646*	*1 417*	*1 145*	*1 223*	*1 048*	*804*	*836*	*870*
Whitefishes nei	**Corégones nca**		**Coregonos nep**			***Coregonus spp***			**1,23(12)001,XX**	**WHF**
02 USA	1 909	1 839	2 436	2 246	1 415	1 506	2 178	1 586	1 030	768
02 Fishing area total	*1 909*	*1 839*	*2 436*	*2 246*	*1 415*	*1 506*	*2 178*	*1 586*	*1 030*	*768*
04 Armenia	1 530	1 507	970	776	428	512	605	922	881	632
Kazakhstan	42	32	27	29	27 F	20 F	15 F	20 F	30 F	35
Kyrgyzstan	64	42	62	62	53 F	40 F	26 F	15 F	16 F	12
04 Fishing area total	*1 636*	*1 581*	*1 059*	*867*	*508 F*	*572 F*	*646 F*	*957 F*	*927 F*	*679*
05 Belarus	5	4	1	3	4	1	2	17	18	31
Belgium	1	1	1	1	1	1	1	1	1	1

B-23 Salmons, trouts, smelts / Saumons, truites, éperlans / Salmones, truchas, eperlanos

Capture production by species, fishing areas and countries or areas / Captures par espèces, zones de pêche et pays ou zones / Capturas por especies, áreas de pesca y países o áreas

Species, Fishing area Espèce, Zone de pêche Especie, Area de pesca	1992 mt	1993 mt	1994 mt	1995 mt	1996 mt	1997 mt	1998 mt	1999 mt	2000 mt	2001 mt
Germany	982	956	459	459	459	459	459	459	459	459
Latvia	-	-	1	3	2	1	1	-	-	-
Lithuania	18	20	13	8	18	20	10	11	7	7
Russian Fed	10 442	9 705	5 408	5 781	9 307	9 350	8 783	8 867	9 132	9 053
Switzerland	1 068	850	613	816	1 050	1 182	1 098	1 074	980	1 174
05 Fishing area total	*12 516*	*11 536*	*6 496*	*7 071*	*10 841*	*11 014*	*10 354*	*10 429*	*10 597*	*10 725*
27 Isle of Man	1	0	0	0	-	-	-	-	-	-
Lithuania	-	-	-	-	-	-	-	-	-	3
Russian Fed	-	4	6	5	6	5	3	4	29	8
27 Fishing area total	*1*	*4*	*6*	*5*	*6*	*5*	*3*	*4*	*29*	*11*
Species total	*16 062*	*14 960*	*9 997*	*10 189*	*12 770 F*	*13 097 F*	*13 181 F*	*12 976 F*	*12 583 F*	*12 183*
Salmonoids nei	**Salmonoidés nca**			**Salmonoideos nep**		***Salmonoidei***			**1,23(XX)XXX,XX**	**SLX**
01 Kenya	88	112	17	196	11	16	24	29	31	67
01 Fishing area total	*88*	*112*	*17*	*196*	*11*	*16*	*24*	*29*	*31*	*67*
02 Canada	-	-	-	139	112	-	2	1	-	-
02 Fishing area total	-	-	-	*139*	*112*	-	*2*	*1*	-	-
04 Iran	130	677	242	433	8	7	8	3	5	2
Korea Rep	63	109	90	61	18	19	102	219	...	...
04 Fishing area total	*193*	*786*	*332*	*494*	*26*	*26*	*110*	*222*	*5*	*2*
05 Albania	17	16 F	13 F	11	35	15	102	104	110	56
Czechoslovak	0	-	-	-	-	-	-	-	-	-
Finland	364	364	293	293	426	426	323	323	282	282
Germany	353	354	29	29	29	29	29	29	29	29
Greece	...	...	...	...	1	6	38	11	50	42
Italy	2 700	3 000	3 225	3 325	1 625	1 091	897	937	692	846
Russian Fed	621	465	288	340	664	657	716	619	616	783
05 Fishing area total	*4 055*	*4 199 F*	*3 848 F*	*3 998*	*2 780*	*2 224*	*2 105*	*2 023*	*1 779*	*2 038*
27 Finland	...	...	39	39	46	46	6	6	6	0
France	16	-	-	-	-	-	-	30	16	1
Latvia	-	5	-	1	-	-	-	-	-	-
Spain	-	-	-	-	0	2	-	8	33	13
Sweden	14	19	9	0	0	1	0	0	-	-
27 Fishing area total	*30*	*24*	*48*	*40*	*46*	*49*	*6*	*44*	*55*	*14*
37 Spain	-	-	-	-	18	22	26	-	-	0
37 Fishing area total	-	-	-	-	*18*	*22*	*26*	-	-	*0*
61 Korea Rep	1 530	1 335	1 215	1 455	1 445	633	1 020	57	28	19
Russian Fed	1 381	1 110	995	877	674	908	1 387	827	2 112	728
61 Fishing area total	*2 911*	*2 445*	*2 210*	*2 332*	*2 119*	*1 541*	*2 407*	*884*	*2 140*	*747*
Species total	*7 277*	*7 566 F*	*6 455 F*	*7 199*	*5 112*	*3 878*	*4 680*	*3 203*	*4 010*	*2 868*
Group total	***827 005***	***981 551***	***996 184***	***1 153 628***	***1 029 998***	***921 952***	***889 982***	***913 373***	***805 298***	***891 058***

B-24 Shads / Aloses / Sábalos

Capture production by species, fishing areas and countries or areas
Captures par espèces, zones de pêche et pays ou zones
Capturas por especies, áreas de pesca y países o áreas

Species, Fishing area Espèce, Zone de pêche Especie, Area de pesca	1992 mt	1993 mt	1994 mt	1995 mt	1996 mt	1997 mt	1998 mt	1999 mt	2000 mt	2001 mt
Pontic shad	**Alose de la mer Noire**		**Sábalo del Mar Negro**		***Alosa pontica***				**1,21(05)011,02**	**SHC**
05 Bulgaria	23	42	30 F	31	124	86	87	32	29	16
Romania	833	391	678	331	392	485	378	16	101	125
Russian Fed	-	-	2	1	1	1	-	-	-	-
Ukraine	321	220	364	223	277	329	213	21	83	146
05 Fishing area total	*1 177*	*653*	*1 074 F*	*586*	*794*	*901*	*678*	*69*	*213*	*287*
37 Bulgaria	20	14	15 F	112	109	79	84	41	10	16
Romania	13	-	-	-	8	2	68	4	5	3
Russian Fed	17	11	1	1	8	1	1	-	-	-
Ukraine	64	61	19	13	6	-	-	3	0	2
37 Fishing area total	*114*	*86*	*35 F*	*126*	*131*	*82*	*153*	*48*	*15*	*21*
Species total	*1 291*	*739*	*1 109 F*	*712*	*925*	*983*	*831*	*117*	*228*	*308*
American shad	**Alose savoureuse**		**Sábalo americano**		***Alosa sapidissima***				**1,21(05)011,03**	**SHA**
02 Canada	21	20	22	23	22	22	23	23	23	1
USA	554	673	282	308	815	523	935	557	485	768
02 Fishing area total	*575*	*693*	*304*	*331*	*837*	*545*	*958*	*580*	*508*	*769*
21 Canada	80	112	80	120	116	118	41	41	56	21
Cuba	3	-	-	-	-	-	-	-	-	-
USA	786	758	457	444	558	584	707	521	323	505
21 Fishing area total	*869*	*870*	*537*	*564*	*674*	*702*	*748*	*562*	*379*	*526*
31 USA	337	265	294	273	343	246	280	117	288	171
31 Fishing area total	*337*	*265*	*294*	*273*	*343*	*246*	*280*	*117*	*288*	*171*
67 USA	16	8	34	158	15	34	62	-	105	178
67 Fishing area total	*16*	*8*	*34*	*158*	*15*	*34*	*62*	-	*105*	*178*
77 USA	-	-	-	-	-	0	-	-	-	-
77 Fishing area total	-	-	-	-	-	*0*	-	-	-	-
Species total	*1 797*	*1 836*	*1 169*	*1 326*	*1 869*	*1 527*	*2 048*	*1 259*	*1 280*	*1 644*
Allis shad	**Alose vraie (=Grande alose)**		**Sábalo común**		***Alosa alosa***				**1,21(05)011,04**	**ASD**
27 France	7	29	69	57	59	56	2	...	38	45
27 Fishing area total	*7*	*29*	*69*	*57*	*59*	*56*	*2*	...	*38*	*45*
Species total	*7*	*29*	*69*	*57*	*59*	*56*	*2*	...	*38*	*45*
Twaite shad	**Alose feinte**		**Saboga(=Alosa)**		***Alosa fallax***				**1,21(05)011,05**	**TSD**
27 France	...	4	3	1	2	1	...	...	11	11
Netherlands	-	-	-	-	-	-	-	-	1	5
27 Fishing area total	...	*4*	*3*	*1*	*2*	*1*	...	...	*12*	*16*
Species total	...	*4*	*3*	*1*	*2*	*1*	...	...	*12*	*16*
Alewife	**Gaspareau**		**Pinchagua**		***Alosa pseudoharengus***				**1,21(05)011,06**	**ALE**
02 Canada	1 089	1 200	1 272	811	990	1 327	1 527	1 527	1 527	...
USA	513	67	0	4	-	5	2	23	7	21
02 Fishing area total	*1 602*	*1 267*	*1 272*	*815*	*990*	*1 332*	*1 529*	*1 550*	*1 534*	*21*
21 Canada	6 801	6 775	5 825	6 212	5 299	5 664	6 941	6 146	5 256	2 790
Cuba	20	26	-	-	-	-	-	-	-	-
USA	1 452	765	117	383	442	521	592	659	284	691
21 Fishing area total	*8 273*	*7 566*	*5 942*	*6 595*	*5 741*	*6 185*	*7 533*	*6 805*	*5 540*	*3 481*
27 Spain	...	...	...	...	...	...	...	...	61	-
27 Fishing area total	...	...	...	...	...	...	...	...	*61*	-
31 USA	362	24	321	0	0	0	0	-	0	0
31 Fishing area total	*362*	*24*	*321*	*0*	*0*	*0*	*0*	-	*0*	*0*
Species total	*10 237*	*8 857*	*7 535*	*7 410*	*6 731*	*7 517*	*9 062*	*8 355*	*7 135*	*3 502*
Hickory shad	**Alose américaine**		**...C**		***Alosa mediocris***				**1,21(05)011,08**	**SHH**
02 USA	13	40	31	36	88	75	47	62	51	90
02 Fishing area total	*13*	*40*	*31*	*36*	*88*	*75*	*47*	*62*	*51*	*90*
Species total	*13*	*40*	*31*	*36*	*88*	*75*	*47*	*62*	*51*	*90*
Allis and twaite shads	**Aloses vraie et feinte**		**Sábalo común y saboga**		***Alosa alosa, A.fallax***				**1,21(05)011,XX**	**SHD**
05 UK	0	0	-	-	-	-	-	-	0	0
05 Fishing area total	*0*	*0*	-	-	-	-	-	-	*0*	*0*
27 Portugal	36	40	40	39	18	17	21	17	20	22
UK	0	0	0	0	6	0	1	1	3	8
27 Fishing area total	*36*	*40*	*40*	*39*	*24*	*17*	*22*	*18*	*23*	*30*
Species total	*36*	*40*	*40*	*39*	*24*	*17*	*22*	*18*	*23*	*30*

B-24 Shads / Aloses / Sábalos

Capture production by species, fishing areas and countries or areas
Captures par espèces, zones de pêche et pays ou zones
Capturas por especies, áreas de pesca y países o áreas

Species, Fishing area Espèce, Zone de pêche Especie, Area de pesca	1992 mt	1993 mt	1994 mt	1995 mt	1996 mt	1997 mt	1998 mt	1999 mt	2000 mt	2001 mt
Shads nei	**Aloses nca**		**Sábalos nep**		***Alosa spp***				**1,21(05)011,XX**	**SHZ**
01 Morocco	2	2	0	0	0	0	1	-	-	-
01 Fishing area total	*2*	*2*	*0*	*0*	*0*	*0*	*1*	*-*	*-*	*-*
05 Albania	3	3 F	2 F	1	2	1	-	-	-	2
Greece	...	...	...	...	2	5	4	6	3	2
Romania	-	-	-	-	-	-	3	-	1	-
Switzerland	...	...	...	...	7	12	6	11	20	18
Ukraine	-	19	32	27	22	46	22	5	0	-
05 Fishing area total	*3*	*22 F*	*34 F*	*28*	*33*	*64*	*35*	*22*	*24*	*22*
27 France	8	7	3	1	2	3	4	41	6	12
27 Fishing area total	*8*	*7*	*3*	*1*	*2*	*3*	*4*	*41*	*6*	*12*
34 Greece	-	-	-	-	-	13	36	-	-	-
Morocco	7	4	2	-	-	-	4	-	22	62
34 Fishing area total	*7*	*4*	*2*	*-*	*-*	*13*	*40*	*-*	*22*	*62*
37 France	-	-	-	-	-	-	-	8	19	2
Greece	333	409	706	856	1 008	1 266	1 744	1 892	1 280	2 215
Morocco	-	1	0	7	7	10	0	-	25	0
Romania	85	189	92	106	101	43	111	60	76	22
Tunisia	2	0	-	0	0	0	-	-	-	-
Turkey	2 609	5 101	4 331	1 590	1 166	505	880	680	720	690
Ukraine	38	27	...	32	1	11	7	-	-	-
37 Fishing area total	*3 067*	*5 727*	*5 129*	*2 591*	*2 283*	*1 835*	*2 742*	*2 640*	*2 120*	*2 929*
Species total	*3 087*	*5 762 F*	*5 168 F*	*2 620*	*2 318*	*1 915*	*2 822*	*2 703*	*2 172*	*3 025*
Caspian shads	**Aloses de la mer Caspienne**		**Sábalos del Mar Caspio**		***Caspialosa spp***				**1,21(05)014,XX**	**ASP**
04 Azerbaijan	...	150	209	67	75	42	82	60	1	52
Iran	21 527	28 730	51 000	41 000	57 000	60 400	85 000	95 000	78 000	45 180
Kazakhstan	6	3	7	0	0	0	0	0	0	0
04 Fishing area total	*21 533*	*28 883*	*51 216*	*41 067*	*57 075*	*60 442*	*85 082*	*95 060*	*78 001*	*45 232*
05 Russian Fed	250	1 510	1 328	1 478	1 882	2 304	3 284	4 654	1 273	176
05 Fishing area total	*250*	*1 510*	*1 328*	*1 478*	*1 882*	*2 304*	*3 284*	*4 654*	*1 273*	*176*
Species total	*21 783*	*30 393*	*52 544*	*42 545*	*58 957*	*62 746*	*88 366*	*99 714*	*79 274*	*45 408*
American gizzard shad	**Alose noyer**		**Sábalo molleja**		***Dorosoma cepedianum***				**1,21(05)018,01**	**SHG**
02 USA	532	774	1 124	906	1 272	1 989	1 291	3 007	977	1 781
02 Fishing area total	*532*	*774*	*1 124*	*906*	*1 272*	*1 989*	*1 291*	*3 007*	*977*	*1 781*
Species total	*532*	*774*	*1 124*	*906*	*1 272*	*1 989*	*1 291*	*3 007*	*977*	*1 781*
Chacunda gizzard shad	**Alose chaconde**		**Sábalo chacunda**		***Anodontostoma chacunda***				**1,21(05)023,01**	**CHG**
57 Malaysia	2 435	2 633	2 595	2 338	2 167	1 916	1 474	3 009	2 645	2 769
57 Fishing area total	*2 435*	*2 633*	*2 595*	*2 338*	*2 167*	*1 916*	*1 474*	*3 009*	*2 645*	*2 769*
71 Malaysia	396	412	1 557	1 200	1 824	1 370	1 185	1 364	1 371	1 268
Philippines	801	325	288	2 955	2 761	1 260	1 191	1 115	1 140	1 189
71 Fishing area total	*1 197*	*737*	*1 845*	*4 155*	*4 585*	*2 630*	*2 376*	*2 479*	*2 511*	*2 457*
Species total	*3 632*	*3 370*	*4 440*	*6 493*	*6 752*	*4 546*	*3 850*	*5 488*	*5 156*	*5 226*
Kelee shad	**Alose palli**		**Sábalo chandano**		***Hilsa kelee***				**1,21(05)034,05**	**HIX**
04 India	39 298	47 255	36 478	49 441	48 302	44 519	53 729	44 810	174 399	54 554
04 Fishing area total	*39 298*	*47 255*	*36 478*	*49 441*	*48 302*	*44 519*	*53 729*	*44 810*	*174 399*	*54 554*
51 India	3 543	3 682	3 072	2 552	3 837	3 634	3 077	3 076	3 896	6 458
51 Fishing area total	*3 543*	*3 682*	*3 072*	*2 552*	*3 837*	*3 634*	*3 077*	*3 076*	*3 896*	*6 458*
57 India	36 939	40 857	36 311	40 714	48 452	47 524	48 475	41 468	43 604	36 325
57 Fishing area total	*36 939*	*40 857*	*36 311*	*40 714*	*48 452*	*47 524*	*48 475*	*41 468*	*43 604*	*36 325*
Species total	*79 780*	*91 794*	*75 861*	*92 707*	*100 591*	*95 677*	*105 281*	*89 354*	*221 899*	*97 337*
Hilsa shad	**Alose hilsa**		**Sábalo hilsa**		***Tenualosa ilisha***				**1,21(05)038,01**	**HIL**
04 Bangladesh	96 596	96 950	71 370	84 420	90 240	83 230	81 634	73 809	79 165	80 000 F
04 Fishing area total	*96 596*	*96 950*	*71 370*	*84 420*	*90 240*	*83 230*	*81 634*	*73 809*	*79 165*	*80 000 F*
51 Kuwait	519	500	928	1 198	1 148	1 034	919	970	650 F	337
Pakistan	823	796	658	476	562	597	611	502	190	170
51 Fishing area total	*1 342*	*1 296*	*1 586*	*1 674*	*1 710*	*1 631*	*1 530*	*1 472*	*840 F*	*507*
57 Bangladesh	125 574	130 295	121 161	129 115	135 358	131 204	124 105	140 710	140 367	140 000 F
57 Fishing area total	*125 574*	*130 295*	*121 161*	*129 115*	*135 358*	*131 204*	*124 105*	*140 710*	*140 367*	*140 000 F*
Species total	*223 512*	*228 541*	*194 117*	*215 209*	*227 308*	*216 065*	*207 269*	*215 991*	*220 372 F*	*220 507 F*

B-24 Shads / Aloses / Sábalos

Capture production by species, fishing areas and countries or areas
Captures par espèces, zones de pêche et pays ou zones
Capturas por especies, áreas de pesca y países o áreas

Species, Fishing area Espèce, Zone de pêche Especie, Area de pesca	1992 mt	1993 mt	1994 mt	1995 mt	1996 mt	1997 mt	1998 mt	1999 mt	2000 mt	2001 mt
Toli shad	**Alose toli**		**Sábalo toli**		*Tenualosa toli*				**1,21(05)038,04**	**TOL**
57 Indonesia	315	519	626	456	429	476	617	630	528	570
57 Fishing area total	*315*	*519*	*626*	*456*	*429*	*476*	*617*	*630*	*528*	*570*
71 Indonesia	3 540	4 136	2 077	1 420	1 702	2 535	3 097	2 875	2 117	2 310
71 Fishing area total	*3 540*	*4 136*	*2 077*	*1 420*	*1 702*	*2 535*	*3 097*	*2 875*	*2 117*	*2 310*
Species total	*3 855*	*4 655*	*2 703*	*1 876*	*2 131*	*3 011*	*3 714*	*3 505*	*2 645*	*2 880*
Chinese gizzard shad	**Alose à museau court**		**Alosa chata**		*Clupanodon thrissa*				**1,21(05)058,03**	**DAS**
61 China,Taiwan	-	4	54	30	25	10	11	27	21	74
Korea Rep	6 806	6 760	7 370	7 931	4 908	13 836	11 349	9 511	6 366	9 120
61 Fishing area total	*6 806*	*6 764*	*7 424*	*7 961*	*4 933*	*13 846*	*11 360*	*9 538*	*6 387*	*9 194*
Species total	*6 806*	*6 764*	*7 424*	*7 961*	*4 933*	*13 846*	*11 360*	*9 538*	*6 387*	*9 194*
Azov sea sprat	**Clupeonelle**		**Espadín del Mar d'Azov**		*Clupeonella cultriventris*				**1,21(05)059,01**	**CLA**
04 Azerbaijan	28 100	19 323	17 254	9 651	5 828	4 420	4 043	20 460	18 520	10 389
Kazakhstan	22 611	16 329	10 815	10 113	9 000 F	8 800	6 400	6 100	3 000	-
Turkmenistan	30 830	15 270	13 855	8 860	8 546	7 800	6 324	8 370	11 540	12 300
04 Fishing area total	*81 541*	*50 922*	*41 924*	*28 624*	*23 374 F*	*21 020*	*16 767*	*34 930*	*33 060*	*22 689*
05 Russian Fed	110 985	73 470	78 447	80 205	91 821	81 638	115 502	150 587	117 914	46 036
Ukraine	3 214	1 003	1 924	781	2 214	1 971	792	1 481	1 342	904
05 Fishing area total	*114 199*	*74 473*	*80 371*	*80 986*	*94 035*	*83 609*	*116 294*	*152 068*	*119 256*	*46 940*
37 Romania	16	-	47	42	4	2	52	4	5	11
Russian Fed	960	306	665	820	369	273	912	2 347	4 897	9 681
Ukraine	3 761	1 503	4 227	6 943	1 075	1 247	2 532	8 511	7 104	18 085
37 Fishing area total	*4 737*	*1 809*	*4 939*	*7 805*	*1 448*	*1 522*	*3 496*	*10 862*	*12 006*	*27 777*
Species total	*200 477*	*127 204*	*127 234*	*117 415*	*118 857 F*	*106 151*	*136 557*	*197 860*	*164 322*	*97 406*
Dotted gizzard shad	**Alose tachetée**		**Alosa manchada**		*Konosirus punctatus*				**1,21(05)060,01**	**DOD**
61 Japan	...	...	...	23 707	18 647	14 850	20 787	17 770	12 335	17 210
61 Fishing area total	*...*	*...*	*...*	*23 707*	*18 647*	*14 850*	*20 787*	*17 770*	*12 335*	*17 210*
Species total	*...*	*...*	*...*	*23 707*	*18 647*	*14 850*	*20 787*	*17 770*	*12 335*	*17 210*
Atlantic sabretooth anchovy	**Anchois goulard**		**Anchoa dentona**		*Lycengraulis grossidens*				**1,21(06)038,04**	**ANR**
03 Argentina	50	60	70 F	80 F	90 F	100 F	100 F	120 F	120 F	100 F
03 Fishing area total	*50*	*60*	*70 F*	*80 F*	*90 F*	*100 F*	*100 F*	*120 F*	*120 F*	*100 F*
Species total	*50*	*60*	*70 F*	*80 F*	*90 F*	*100 F*	*100 F*	*120 F*	*120 F*	*100 F*
Elongate ilisha	**Alose gracile**		**Sardineta grácil**		*Ilisha elongata*				**1,21(12)001,03**	**EIL**
61 China	30 014	28 624	32 540	46 635	51 339	77 532	84 290	110 359	107 689	101 342
Korea Rep	282	342	299	466	630	349	381	265	134	40
61 Fishing area total	*30 296*	*28 966*	*32 839*	*47 101*	*51 969*	*77 881*	*84 671*	*110 624*	*107 823*	*101 382*
Species total	*30 296*	*28 966*	*32 839*	*47 101*	*51 969*	*77 881*	*84 671*	*110 624*	*107 823*	*101 382*
West African ilisha	**Alose rasoir**		**Sardineta africana**		*Ilisha africana*				**1,21(12)001,12**	**ILI**
34 Benin	446 F	483 F	544 F	528 F	602 F	822 F	781 F	671	450 F	923
Côte dIvoire	357	350	802	288	491	534	452	314	152	50
Ghana	4 951	2 995	2 810	3 051	7 343	4 305	3 632	3 262	3 341	3 600
Liberia	104	44	36	6	124	26	63	242	110	198
Portugal	-	-	-	-	-	-	-	1	-	-
Sierra Leone	3 065	3 027	3 017	3 013	3 010 F	3 009	3 110	5	2 996	4
Togo	38	36	152	32	15	101	35	113	418	796
34 Fishing area total	*8 961 F*	*6 935 F*	*7 361 F*	*6 918 F*	*11 585 F*	*8 797 F*	*8 073 F*	*4 608*	*7 467 F*	*5 571*
Species total	*8 961 F*	*6 935 F*	*7 361 F*	*6 918 F*	*11 585 F*	*8 797 F*	*8 073 F*	*4 608*	*7 467 F*	*5 571*
Indian pellona	**Alose-écaille indienne**		**Sardineta índica**		*Pellona ditchela*				**1,21(12)003,03**	**PEO**
57 Malaysia	5 522	6 225	7 260	6 091	11 227	10 919	6 113	4 448	7 069	6 884
57 Fishing area total	*5 522*	*6 225*	*7 260*	*6 091*	*11 227*	*10 919*	*6 113*	*4 448*	*7 069*	*6 884*
71 Malaysia	2 648	2 187	3 526	3 868	4 058	2 455	3 808	4 740	5 636	5 780
Philippines	1 439	1 443	1 451	1 177	792	802	842	826	839	1 153
71 Fishing area total	*4 087*	*3 630*	*4 977*	*5 045*	*4 850*	*3 257*	*4 650*	*5 566*	*6 475*	*6 933*
Species total	*9 609*	*9 855*	*12 237*	*11 136*	*16 077*	*14 176*	*10 763*	*10 014*	*13 544*	*13 817*
Diadromous clupeoids nei	**Clupéoidés diadromes nca**		**Clupeoideos diádromos nep**		*Clupeoidei*				**1,21(XX)XXX,XX**	**DCX**
01 Benin	183 F	160 F	137	260	250 F	230 F	210 F	210 F	200 F	230 F
Egypt	...	...	...	106	40	23	148	60	84	28
01 Fishing area total	*183 F*	*160 F*	*137*	*366*	*290 F*	*253 F*	*358 F*	*270 F*	*284 F*	*258 F*
03 Venezuela	72	155	160	345	552	355	467	403	287	200
03 Fishing area total	*72*	*155*	*160*	*345*	*552*	*355*	*467*	*403*	*287*	*200*

B-24 Shads / Aloses / Sábalos

Capture production by species, fishing areas and countries or areas
Captures par espèces, zones de pêche et pays ou zones
Capturas por especies, áreas de pesca y países o áreas

Species, Fishing area Espèce, Zone de pêche Especie, Area de pesca	1992 mt	1993 mt	1994 mt	1995 mt	1996 mt	1997 mt	1998 mt	1999 mt	2000 mt	2001 mt
04 Iran	35	893	920	490	330	566	817	697	595	623
04 Fishing area total	*35*	*893*	*920*	*490*	*330*	*566*	*817*	*697*	*595*	*623*
06 Papua N Guin	480 F	480 F	480 F	480 F	480 F	480 F	480 F	480 F	480 F	480 F
06 Fishing area total	*480 F*	*480 F*	*480 F*	*480 F*	*480 F*	*480 F*	*480 F*	*480 F*	*480 F*	*480 F*
27 Portugal	-	43	38	35	35	36	30	-	-	-
27 Fishing area total	*-*	*43*	*38*	*35*	*35*	*36*	*30*	*-*	*-*	*-*
57 Malaysia	-	45	17	30	107	102	22	45	13	50
57 Fishing area total	*-*	*45*	*17*	*30*	*107*	*102*	*22*	*45*	*13*	*50*
71 Malaysia	396	370	412	770	1 278	574	559	577	652	1 056
71 Fishing area total	*396*	*370*	*412*	*770*	*1 278*	*574*	*559*	*577*	*652*	*1 056*
Species total	*1 166 F*	*2 146 F*	*2 164 F*	*2 516 F*	*3 072 F*	*2 366 F*	*2 733 F*	*2 472 F*	*2 311 F*	*2 667 F*
Group total	***606 927***	***558 764***	***535 242***	***588 771***	***634 257***	***634 292***	***699 649***	***782 579***	***855 571***	***629 146***

B-25

Miscellaneous diadromous fishes — **Capture production by species, fishing areas and countries or areas**
Poissons diadromes divers — **Captures par espèces, zones de pêche et pays ou zones**
Peces diádromos diversos — **Capturas por especies, áreas de pesca y países o áreas**

Species, Fishing area Espèce, Zone de pêche Especie, Area de pesca	1992 mt	1993 mt	1994 mt	1995 mt	1996 mt	1997 mt	1998 mt	1999 mt	2000 mt	2001 mt
Sea lamprey	**Lamproie marine**		**Lamprea marina**		***Petromyzon marinus***				**1,02(01)001,01**	**LAU**
27 Estonia	11	0	-	0	18	3	4	7	8	3
France	15	42	13	13	13	15	11	27	22	23
Portugal	12	12	10	4	3	2	2	4	6	6
27 Fishing area total	*38*	*54*	*23*	*17*	*34*	*20*	*17*	*38*	*36*	*32*
Species total	*38*	*54*	*23*	*17*	*34*	*20*	*17*	*38*	*36*	*32*
River lamprey	**Lamproie de rivière**		**Lamprea de río**		***Lampetra fluviatilis***				**1,02(01)002,01**	**LAR**
05 Estonia	-	25	5	1	0	7	16	9	26	25
Latvia	107	96	113	95	140	80	79	120	135	88
Lithuania	2	1	-	-	1	-	-	-	-	3
UK	-	-	-	-	-	-	-	-	2	2
05 Fishing area total	*109*	*122*	*118*	*96*	*141*	*87*	*95*	*129*	*163*	*118*
Species total	*109*	*122*	*118*	*96*	*141*	*87*	*95*	*129*	*163*	*118*
Lampreys nei	**Lamproies nca**		**Lampreas nep**		***Petromyzontidae***				**1,02(01)XXX,XX**	**LAS**
02 USA	4	8	6	13	-	-	0	4	0	-
02 Fishing area total	*4*	*8*	*6*	*13*	-	-	*0*	*4*	*0*	-
05 Poland	6	6	0	0	0	-	-	-	-	-
Russian Fed	59	109	40	40	76	31	37	67	24	21
05 Fishing area total	*65*	*115*	*40*	*40*	*76*	*31*	*37*	*67*	*24*	*21*
27 Poland	-	-	-	-	-	2	2	2	6	5
Russian Fed	-	-	-	-	-	-	-	-	15	28
27 Fishing area total	-	-	-	-	-	*2*	*2*	*2*	*21*	*33*
Species total	*69*	*123*	*46*	*53*	*76*	*33*	*39*	*73*	*45*	*54*
Milkfish	**Chano**		**Chano**		***Chanos chanos***				**1,22(02)001,01**	**MIL**
04 Philippines	2 295	2 323	4 652	3 997	369	34	3 000	83	163	201
04 Fishing area total	*2 295*	*2 323*	*4 652*	*3 997*	*369*	*34*	*3 000*	*83*	*163*	*201*
31 Mexico	0	0	0	0	0	0	0	0	0	...
31 Fishing area total	*0*	*0*	*0*	*0*	*0*	*0*	*0*	*0*	*0*	...
51 Eritrea	...	...	0	2	0	2	3	1	3	1
Saudi Arabia	-	-	10	137	130	70	82	81	64	73
Untd Arab Em	30	40	35	50	51	53	54	55	58	60 F
51 Fishing area total	*30*	*40*	*45*	*189*	*181*	*125*	*139*	*137*	*125*	*134 F*
61 China,Taiwan	-	-	20	1	4	-	-	-	-	1
61 Fishing area total	-	-	*20*	*1*	*4*	-	-	-	-	*1*
71 Fiji Islands	18	19	19 F	20	34	26	21	30	30 F	32
Kiribati	280	280	285	285	290	290	...	80	2 175	58
Philippines	1 118	1 095	495	3 502	3 302	363	152	232	233	242
71 Fishing area total	*1 416*	*1 394*	*799 F*	*3 807*	*3 626*	*679*	*173*	*342*	*2 438 F*	*332*
77 Mexico	0	0	0	0	0	1	0	0	7	...
77 Fishing area total	*0*	*0*	*0*	*0*	*0*	*1*	*0*	*0*	*7*	...
Species total	*3 741*	*3 757*	*5 516 F*	*7 994*	*4 180*	*839*	*3 312*	*562*	*2 733 F*	*668 F*
Three-spined stickleback	**Epinoche à trois épines**		**Espinoso**		***Gasterosteus aculeatus***				**1,50(01)001,01**	**GTA**
05 Belarus	271	577	99	85	80	34	26	13	29	41
Latvia	-	-	3	1	1	1	-	-	-	-
Lithuania	-	16	-	-	-	-	-	13	22	18
Russian Fed	577	564	330	337	271	314	884	1 021	1 116	906
05 Fishing area total	*848*	*1 157*	*432*	*423*	*352*	*349*	*910*	*1 047*	*1 167*	*965*
27 Denmark	-	-	-	-	-	4	-	-	-	-
Latvia	61	14	7	1	33	109	82	-	-	-
Lithuania	6	-	-	-	-	-	-	-	-	-
Russian Fed	-	-	-	-	-	-	-	-	514	593
27 Fishing area total	*67*	*14*	*7*	*1*	*33*	*113*	*82*	-	*514*	*593*
37 Russian Fed	-	-	-	-	-	-	-	-	-	8
37 Fishing area total	-	-	-	-	-	-	-	-	-	*8*
Species total	*915*	*1 171*	*439*	*424*	*385*	*462*	*992*	*1 047*	*1 681*	*1 566*
Barramundi(=Giant seaperch)	**Perche barramundi**		**Perca gigante**		***Lates calcarifer***				**1,70(01)167,01**	**GIP**
06 Australia	-	...	...	903	1 061	1 178	1 274	1 634	1 350 F	1 300 F
Papua N Guin	350 F	350 F	350 F	350 F	350 F	350 F	350 F	350 F	350 F	350 F
06 Fishing area total	*350 F*	*350 F*	*350 F*	*1 253 F*	*1 411 F*	*1 528 F*	*1 624 F*	*1 984 F*	*1 700 F*	*1 650 F*
51 Pakistan	237	210	193	187	214	209	196	204	-	-
51 Fishing area total	*237*	*210*	*193*	*187*	*214*	*209*	*196*	*204*	-	-
57 Australia	60	45	43	38	46	-	-	-	-	-

B-25

Miscellaneous diadromous fishes — Capture production by species, fishing areas and countries or areas
Poissons diadromes divers — Captures par espèces, zones de pêche et pays ou zones
Peces diádromos diversos — Capturas por especies, áreas de pesca y países o áreas

Species, Fishing area Espèce, Zone de pêche Especie, Area de pesca	1992 mt	1993 mt	1994 mt	1995 mt	1996 mt	1997 mt	1998 mt	1999 mt	2000 mt	2001 mt
Indonesia	7 947	11 722	10 325	12 134	13 735	13 614	21 154	15 602	15 752	17 240
Malaysia	61	73	64	56	42	47	901	55	141	91
Thailand	-	-	2 688	46	168	131	1 007	30	30	29
57 Fishing area total	*8 068*	*11 840*	*13 120*	*12 274*	*13 991*	*13 792*	*23 062*	*15 687*	*15 923*	*17 360*
61 China,Taiwan	95	2	2	0	14	9	26	47	32	80
61 Fishing area total	*95*	*2*	*2*	*0*	*14*	*9*	*26*	*47*	*32*	*80*
71 Australia	928	966	984	1 002	1 226	1 178	1 306	1 690	1 886	2 088
Indonesia	19 530	25 079	28 121	35 493	34 577	42 328	44 039	49 571	53 036	58 060
Malaysia	1 670	1 529	1 604	1 263	1 280	827	979	1 342	1 560	1 427
Papua N Guin	38	25	19	9	6	43	45	158	73	149
Philippines	1 188	2 960	1 920	2 821	3 049	553	655	784	642	678
Singapore	31	33	64	51	39	58	39	29	41	52
Thailand	-	-	411	14	421	38	42	55	54	54
71 Fishing area total	*23 385*	*30 592*	*33 123*	*40 653*	*40 598*	*45 025*	*47 105*	*53 629*	*57 292*	*62 508*
Species total	*32 135 F*	*42 994 F*	*46 788 F*	*54 367 F*	*56 228 F*	*60 563 F*	*72 013 F*	*71 551 F*	*74 947 F*	*81 598 F*
Striped bass	**Bar d'Amérique**		**Lubina estriada**		***Morone saxatilis***				**1,70(06)006,02**	**STB**
02 USA	-	-	-	-	-	5	-	-	-	-
02 Fishing area total	-	-	-	-	-	*5*	-	-	-	-
21 Canada	-	3	-	18	15	0	-	-	-	-
USA	572	759	639	1 637	2 129	2 785	3 032	3 000	3 123	2 934
21 Fishing area total	*572*	*762*	*639*	*1 655*	*2 144*	*2 785*	*3 032*	*3 000*	*3 123*	*2 934*
31 USA	7	4	109	7	9	12	14	4	12	15
31 Fishing area total	*7*	*4*	*109*	*7*	*9*	*12*	*14*	*4*	*12*	*15*
Species total	*579*	*766*	*748*	*1 662*	*2 153*	*2 802*	*3 046*	*3 004*	*3 135*	*2 949*
White perch	**Bar blanc d'Amérique**		**Lubina blanca**		***Morone americana***				**1,70(06)006,03**	**PEW**
02 USA	818	935	747	802	995	1 164	966	1 042	1 221	1 226
02 Fishing area total	*818*	*935*	*747*	*802*	*995*	*1 164*	*966*	*1 042*	*1 221*	*1 226*
Species total	*818*	*935*	*747*	*802*	*995*	*1 164*	*966*	*1 042*	*1 221*	*1 226*
Group total	***38 404***	***49 922***	***54 425***	***65 415***	***64 192***	***65 970***	***80 480***	***77 446***	***83 961***	***88 211***

B-31

Flounders, halibuts, soles — Capture production by species, fishing areas and countries or areas
Flets, flétans, soles — Captures par espèces, zones de pêche et pays ou zones
Platijas, halibuts, lenguados — Capturas por especies, áreas de pesca y países o áreas

Species, Fishing area Espèce, Zone de pêche Especie, Area de pesca	1992 mt	1993 mt	1994 mt	1995 mt	1996 mt	1997 mt	1998 mt	1999 mt	2000 mt	2001 mt
Lefteye flounders nei	**Arnoglosses, rombous nca**		**Rodaballos, rombos, etc. nep**		*Bothidae*			**1,83(01)XXX,XX**		**LEF**
27 Portugal	...	...	...	...	...	...	...	89	117	107
27 Fishing area total	*...*	*...*	*...*	*...*	*...*	*...*	*...*	*89*	*117*	*107*
31 Korea Rep	-	-	-	-	-	-	-	-	2	-
31 Fishing area total	-	-	-	-	-	-	-	-	*2*	-
34 Cameroon	4	1	0	0	0	0	0	0	0	-
Korea Rep	27	-	7	65	2	-	-	-	15	-
Portugal	-	-	-	18	22	2	-	-	2	12
34 Fishing area total	*31*	*1*	*7*	*83*	*24*	*2*	*0*	*0*	*17*	*12*
41 Korea Rep	95	194	-	-	-	-	-	-	-	20
41 Fishing area total	*95*	*194*	-	-	-	-	-	-	-	*20*
51 Eritrea	...	...	-	-	-	0	1	23	119	125
Korea Rep	-	-	-	-	-	-	8	4	10	-
51 Fishing area total	*...*	*...*	-	-	-	*0*	*9*	*27*	*129*	*125*
71 Korea Rep	11	22	115	86	38	27	30	15	8	10
71 Fishing area total	*11*	*22*	*115*	*86*	*38*	*27*	*30*	*15*	*8*	*10*
77 Korea Rep	-	-	-	-	-	-	-	9	-	-
77 Fishing area total	-	-	-	-	-	-	-	*9*	-	-
87 Korea Rep	-	-	-	-	-	-	-	-	5	-
87 Fishing area total	-	-	-	-	-	-	-	-	*5*	-
Species total	*137*	*217*	*122*	*169*	*62*	*29*	*39*	*140*	*278*	*274*
Atlantic halibut	**Flétan de l'Atlantique**		**Fletán del Atlántico**		*Hippoglossus hippoglossus*			**1,83(02)002,01**		**HAL**
21 Canada	1 581	1 444	1 264	895	1 097	1 362	1 300	1 188	1 219	1 647
Cuba	9	14	1	1	-	0	0	0	-	-
Estonia	-	651	-	-	-	-	-	-	-	0
Faeroe Is	-	-	-	-	8	-	7	-	-	-
Germany	14	0	0	-	-	-	-	0	-	-
Greenland	51	47	40	23	35	22	22	45	9	0
Norway	0	-	-	-	0	1	0	-	-	-
Poland	-	-	-	-	-	-	-	-	-	488
Portugal	79	53	45	17	11	17	31	51	29	44
Russian Fed	-	-	-	-	-	-	-	6	5	2
St Pier Mq	36	-	-	-	-	1	1	1	0	1
Spain	-	1	67	68	58	46	65	98	96	138
UK	-	-	-	-	-	-	1	-	-	-
USA	22	19	21	16	13	14	8	12	11	11
21 Fishing area total	*1 792*	*2 229*	*1 438*	*1 020*	*1 222*	*1 463*	*1 435*	*1 401*	*1 369*	*2 331*
27 Belgium	3	6	2	15	5	4	3	2	2	2
Denmark	54	59	57	61	71	70	62	51	53	53
Faeroe Is	871	722	1 090	654	451	479	384	432	318	205
France	62	26	27	27	34	16	16	21	29	44
Germany	374	138	139	30	43	23	28	42	23	20
Greenland	34	346	241	37	16	4	12	1	10	19
Iceland	1 184	1 486	1 427	888	837	677	501	567	493	589
Ireland	1	1	6	6	8	4	9	11	1	17
Netherlands	1	1	3	5	3	5	4	2	1	-
Norway	659	589	754	551	678	878	672	696	1 039	868
Portugal	0	0	0	0	3	-	0	0	1	1
Spain	-	-	-	-	-	-	4	5	16	14
Sweden	26	25	17	13	8	10	8	8	10	8
UK	216	169	186	405	392	367	201	254	208	159
27 Fishing area total	*3 485*	*3 568*	*3 949*	*2 692*	*2 549*	*2 537*	*1 904*	*2 092*	*2 204*	*1 999*
Species total	*5 277*	*5 797*	*5 387*	*3 712*	*3 771*	*4 000*	*3 339*	*3 493*	*3 573*	*4 330*
Pacific halibut	**Flétan du Pacifique**		**Fletán del Pacífico**		*Hippoglossus stenolepis*			**1,83(02)002,02**		**HAP**
67 Canada	4 577	6 379	5 862	5 745	5 430	7 448	7 714	7 103	6 095	4 766
Poland	-	-	-	-	-	-	-	-	-	4
USA	31 085	29 294	27 028	20 551	22 676	31 841	33 892	36 515	34 753	35 391
67 Fishing area total	*35 662*	*35 673*	*32 890*	*26 296*	*28 106*	*39 289*	*41 606*	*43 618*	*40 848*	*40 161*
77 USA	-	-	-	3	4	2	3	2	-	0
77 Fishing area total	-	-	-	*3*	*4*	*2*	*3*	*2*	-	*0*
Species total	*35 662*	*35 673*	*32 890*	*26 299*	*28 110*	*39 291*	*41 609*	*43 620*	*40 848*	*40 161*
European plaice	**Plie d'Europe**		**Solla europea**		*Pleuronectes platessa*			**1,83(02)004,05**		**PLE**
27 Belgium	14 823	12 775	10 398	9 290	7 684	7 469	7 369	8 186	9 148	8 487
Channel Is	8	8 F	3	3 F	26	19	17	23	18	13
Denmark	31 931	27 159	27 954	24 092	22 571	24 621	18 927	23 123	23 902	26 849
Faeroe Is	417	319	957	425	443	506	443	325	259	250
France	5 026	4 582	4 667	3 982	3 843	4 326	4 667	5 642	4 703	4 247
Germany	6 910	7 038	5 832	6 533	4 935	4 304	3 050	3 462	4 501	4 842
Greenland	40	-	-	-	-	-	2	-	-	-
Iceland	10 500	12 516	11 851	10 649	11 070	10 557	7 111	7 064	5 218	4 905

B-31

Flounders, halibuts, soles — **Capture production by species, fishing areas and countries or areas**
Flets, flétans, soles — **Captures par espèces, zones de pêche et pays ou zones**
Platijas, halibuts, lenguados — **Capturas por especies, áreas de pesca y países o áreas**

Species, Fishing area Espèce, Zone de pêche Especie, Area de pesca	1992 mt	1993 mt	1994 mt	1995 mt	1996 mt	1997 mt	1998 mt	1999 mt	2000 mt	2001 mt
Ireland	2 549	1 661	1 470	1 590	1 679	1 699	1 731	1 424	932	841
Isle of Man	15	13	14	20	16	11	14	5	6	1
Latvia	-	-	8	-	-	-	-	-	-	-
Lithuania	9	120	262	194	330	624	736	571	618	1 137
Netherlands	51 124	48 555	50 289	44 262	35 539	34 272	30 592	37 543	35 079	33 835
Norway	1 510	1 399	1 105	1 166	1 731	2 857	1 872	1 816	1 943	2 773
Portugal	197	218	143	147	137	89	115	95	124	145
Russian Fed	2 886	4 984	4 404	4 431	2 161	3 531	3 729	3 911	3 114	1 250
Spain	17	2	1	12	14	3	6	3	39	41
Sweden	666	475	565	511	531	558	431	405	449	416
UK	37 390	36 278	32 802	29 194	25 319	26 599	23 510	20 394	23 701	21 936
27 Fishing area total	*166 018*	*158 102 F*	*152 725*	*136 501 F*	*118 029*	*122 045*	*104 322*	*113 992*	*113 754*	*111 968*
37 France	-	-	-	-	-	-	-	0	6	7
37 Fishing area total	-	-	-	-	-	-	-	*0*	*6*	*7*
Species total	*166 018*	*158 102 F*	*152 725*	*136 501 F*	*118 029*	*122 045*	*104 322*	*113 992*	*113 760*	*111 975*
English sole	**Carlottin anglais**		**Soya inglesa**		***Pleuronectes vetulus***			**1,83(02)004,15**		**RFE**
67 USA	...	...	...	...	...	...	198	874	963	1 198
67 Fishing area total	...	...	...	...	...	...	*198*	*874*	*963*	*1 198*
77 USA	...	...	...	...	...	...	227	226	177	197
77 Fishing area total	...	...	...	...	...	...	*227*	*226*	*177*	*197*
Species total	...	...	...	...	...	...	*425*	*1 100*	*1 140*	*1 395*
Greenland halibut	**Flétan noir**		**Fletán negro**		***Reinhardtius hippoglossoides***			**1,83(02)005,01**		**GHL**
21 Canada	12 802	9 019	6 791	7 752	10 700	13 153	12 269	12 442	16 331	13 814
Cuba	-	1	-	-	-	-	-	-	-	-
Denmark	-	-	-	-	0	0	-	-	-	0
Estonia	-	631	-	-	-	-	-	-	181	1 100
Faeroe Is	1 354	1 552	850	13	870	579	272	748	1 257	1 000 F
Germany	42	46	217	-	474	445	350	415	444	537
Greenland	12 174	13 417	17 311	17 872	19 152	23 146	21 908	34 481	23 219	19 111
Iceland	-	1	2	-	-	-	-	-	507	14
Japan	6 325	-	-	-	-	-	-	-	-	2 814
Korea Rep	49	5	-	-	-	-	-	-	-	-
Latvia	-	83	-	-	-	-	-	-	215	291
Lithuania	-	-	-	-	-	-	-	-	21	392
Norway	2 432	2 739	3 121	2 378	1 786	1 817	1 339	1 335	1 431	1 419
Portugal	10 547	8 811	5 967	1 936	3 316	3 343	3 242	3 995	4 688	5 026
Russian Fed	9 386	4 234	3 763	677	565	-	2 433	3 669	4 128	4 687
St Pier Mq	203	-	-	-	-	439	1 431	1 132	2	7
Spain	34 520	35 640	40 772	9 135	7 311	7 945	7 238	9 022	9 537	11 571
UK	9	-	49	-	-	-	-	-	-	-
21 Fishing area total	*89 843*	*76 179*	*78 843*	*39 763*	*44 174*	*50 867*	*50 482*	*67 239*	*61 961*	*61 783 F*
27 Denmark	3	3	5	1	3	2	1	1	1	1
Faeroe Is	2 159	4 469	6 361	4 481	5 371	4 545	3 515	3 363	4 327	3 128
France	538	177	172	542	603	647	282	268	158	230
Germany	401	438	985	848	3 450	3 401	3 520	3 117	3 298	2 865
Greenland	450	297	870	545	1 286	1 129	943	4 970	1 756	1 573
Iceland	31 955	34 043	27 793	27 408	22 125	18 631	10 751	11 187	14 553	16 628
Ireland	-	-	5	2	2	2	21	78	22	71
Lithuania	-	-	-	-	-	-	-	-	-	3
Norway	8 518	12 184	10 330	11 695	15 287	10 526	10 609	18 369	11 597	13 733
Poland	-	-	-	-	-	12	31	8	4	4
Portugal	0	43	36	84	79	50	99	49	37	40
Russian Fed	723	1 235	283	804	2 000	1 075	2 711	3 961	4 751	4 881
Spain	23	-	1	1 188	214	255	246	589	864	1 766
UK	139	863	682	2 890	2 358	1 515	1 777	1 611	1 746	1 690
27 Fishing area total	*44 909*	*53 752*	*47 523*	*50 488*	*52 778*	*41 790*	*34 506*	*47 571*	*43 114*	*46 613*
67 USA	1 635	8 461	7 115	5 860	4 652	...	307	14	6 186	4 391
67 Fishing area total	*1 635*	*8 461*	*7 115*	*5 860*	*4 652*	...	*307*	*14*	*6 186*	*4 391*
Species total	*136 387*	*138 392*	*133 481*	*96 111*	*101 604*	*92 657*	*85 295*	*114 824*	*111 261*	*112 787 F*
Arrow-tooth flounder	**Faux flétan du Pacifique**		**Halibut del Pacífico**		***Atheresthes stomias***			**1,83(02)008,01**		**ARF**
67 Canada	144	108	129	75	24	-	33	44	39	59
USA	27 361	13 589	4 335	5 091	10 656	6 319	8 246	12 037	18 735	14 342
67 Fishing area total	*27 505*	*13 697*	*4 464*	*5 166*	*10 680*	*6 319*	*8 279*	*12 081*	*18 774*	*14 401*
77 USA	-	-	3	1	1	1	1	2	0	0
77 Fishing area total	-	-	*3*	*1*	*1*	*1*	*1*	*2*	*0*	*0*
Species total	*27 505*	*13 697*	*4 467*	*5 167*	*10 681*	*6 320*	*8 280*	*12 083*	*18 774*	*14 401*
Kamchatka flounder	**Faux flétan du Japon**		**Halibut japonés**		***Atheresthes evermanni***			**1,83(02)008,02**		**KAF**
61 Russian Fed	5 868	1 276	6 944	7 039	9 739	8 476	9 567	10 743	23 473	19 049
61 Fishing area total	*5 868*	*1 276*	*6 944*	*7 039*	*9 739*	*8 476*	*9 567*	*10 743*	*23 473*	*19 049*
Species total	*5 868*	*1 276*	*6 944*	*7 039*	*9 739*	*8 476*	*9 567*	*10 743*	*23 473*	*19 049*

B-31 Flounders, halibuts, soles / Flets, flétans, soles / Platijas, halibuts, lenguados

Capture production by species, fishing areas and countries or areas / Captures par espèces, zones de pêche et pays ou zones / Capturas por especies, áreas de pesca y países o áreas

Species, Fishing area Espèce, Zone de pêche Especie, Area de pesca	1992 mt	1993 mt	1994 mt	1995 mt	1996 mt	1997 mt	1998 mt	1999 mt	2000 mt	2001 mt
Petrale sole	**Charlottin pétrale**		**Limanda petrale**			***Eopsetta jordani***			**1,83(02)010,01**	**EOJ**
67 USA	...	...	...	...	...	1 454	1 197	1 221	1 632	1 554
67 Fishing area total	...	...	...	...	...	*1 454*	*1 197*	*1 221*	*1 632*	*1 554*
77 USA	...	...	...	...	...	483	267	259	238	268
77 Fishing area total	...	...	...	...	...	*483*	*267*	*259*	*238*	*268*
Species total	...	...	...	...	...	*1 937*	*1 464*	*1 480*	*1 870*	*1 822*
Witch flounder	**Plie cynoglosse**		**Mendo**			***Glyptocephalus cynoglossus***			**1,83(02)011,02**	**WIT**
21 Canada	10 037	7 510	1 610	1 244	1 600	1 675	2 094	1 806	2 168	2 128
Cuba	2	11	-	-	-	2	-	-	-	-
Estonia	-	-	-	-	-	-	-	-	5	3
Faeroe Is	-	-	-	6	11	-	-	4	-	-
Germany	5	-	-	-	-	-	-	-	-	-
Japan	50	-	-	-	-	-	-	-	-	3
Korea Rep	44	-	-	-	-	-	-	-	-	-
Lithuania	-	-	-	-	-	-	-	-	-	3
Portugal	851	292	573	389	240	348	381	508	228	579
Russian Fed	-	-	-	-	-	-	52	110	114	65
St Pier Mq	62	-	-	-	-	8	57	35	7	120
Spain	859	69	295	1 086	1 636	1 344	1 349	1 180	1 482	1 355
UK	3	-	-	-	-	-	-	-	-	-
USA	2 235	2 595	2 693	2 197	2 082	1 775	1 855	2 123	2 439	3 020
21 Fishing area total	*14 148*	*10 477*	*5 171*	*4 922*	*5 569*	*5 152*	*5 788*	*5 766*	*6 443*	*7 276*
27 Denmark	1 069	835	864	897	999	1 563	1 906	2 116	2 338	2 049
Faeroe Is	-	2	5	2	2	3	3	...	...	
France	559	570	650	734	719	816	602	484	587	582
Germany	5	5	5	10	7	9	13	8	13	8
Iceland	2 564	1 594	1 771	1 755	1 486	1 272	947	1 408	1 098	1 132
Ireland	284	396	371	601	615	605	657	713	555	915
Isle of Man	0	0	0	2	0	1	0	0	0	-
Netherlands	7	13	14	7	0	1	4	9	7	1
Norway	180	152	118	100	80	86	140	135	97	88
Portugal	37	38	25	33	30	32	22	27	26	20
Spain	-	-	2 028	2 339	1 751	4 183	4 805	4 148	2 961	2 870
Sweden	498	400	312	357	299	355	448	501	578	576
UK	2 381	2 359	2 638	3 187	3 479	3 748	3 889	3 661	4 102	4 535
27 Fishing area total	*7 584*	*6 362*	*8 798*	*10 027*	*9 467*	*12 673*	*13 436*	*13 213*	*12 362*	*12 776*
Species total	*21 732*	*16 839*	*13 969*	*14 949*	*15 036*	*17 825*	*19 224*	*18 979*	*18 805*	*20 052*
Rex sole	**Plie cynoglosse royale**		**...C**			***Glyptocephalus zachirus***			**1,83(02)011,03**	**GLZ**
67 USA	...	...	...	...	...	...	121	391	413	473
67 Fishing area total	...	...	...	...	...	...	*121*	*391*	*413*	*473*
77 USA	...	...	...	...	...	...	168	153	128	98
77 Fishing area total	...	...	...	...	...	...	*168*	*153*	*128*	*98*
Species total	...	...	...	...	...	...	*289*	*544*	*541*	*571*
Amer. plaice(=Long rough dab)	**Balai (=Plie canadienne)**		**Platija americana**			***Hippoglossoides platessoides***			**1,83(02)014,01**	**PLA**
21 Canada	19 481	10 900	3 093	3 095	2 525	3 583	3 546	4 359	3 904	4 466
Cuba	49	78	13	26	-	22	51	19	0	...
Estonia	-	-	-	-	-	-	-	-	94	54
Faeroe Is	1	244	-	-	8	-	-	3	2	...
Germany	11	-	-	-	0	-	-	-	-	-
Greenland	3	0	0	0	0	0	5	3	0	4
Japan	73	-	-	-	-	-	-	-	-	6
Korea Rep	528	16	-	-	-	-	-	-	-	-
Lithuania	-	-	-	-	-	-	-	-	-	3
Poland	-	-	-	-	-	-	-	-	-	1
Portugal	453	323	344	171	291	389	357	719	399	633
Russian Fed	80	15	-	-	-	-	-	151	368	257
St Pier Mq	52	-	0	0	0	23	27	24	41	112
Spain	802	443	394	554	660	1 002	1 141	1 461	1 438	1 158
UK	44	-	-	-	-	-	-	-	-	-
USA	6 652	5 808	5 062	4 612	4 397	3 937	3 662	3 134	4 213	4 425
21 Fishing area total	*28 229*	*17 827*	*8 906*	*8 458*	*7 881*	*8 956*	*8 789*	*9 873*	*10 459*	*11 119*
27 Denmark	23	39	2	1	0	5	8	31	5	31
France	-	-	17	13	13	14	11	14	10	11
Greenland	-	-	28	-	-	-	-	-	-	-
Iceland	1 468	1 342	2 734	5 418	7 027	6 468	3 329	3 833	3 176	3 473
Ireland	-	-	-	-	-	-	-	-	99	-
Norway	4	-	-	-	-	119	24	15	0	15
Portugal	0	29	27	48	85	57	288	70	111	71
Russian Fed	-	-	2 920	1 157	778	1 482	6 569	7 727	2 839	3 497
Spain	-	24	13	115	218	116	205	61	184	254
Sweden	24	0	0	0	22	6	0	-	-	-
UK	2	0	0	-	1	0	0	0	0	15
27 Fishing area total	*1 521*	*1 434*	*5 741*	*6 752*	*8 144*	*8 267*	*10 434*	*11 751*	*6 424*	*7 367*
Species total	*29 750*	*19 261*	*14 647*	*15 210*	*16 025*	*17 223*	*19 223*	*21 624*	*16 883*	*18 486*

B-31 Flounders, halibuts, soles / Flets, flétans, soles / Platijas, halibuts, lenguados

Capture production by species, fishing areas and countries or areas / Captures par espèces, zones de pêche et pays ou zones / Capturas por especies, áreas de pesca y países o áreas

Species, Fishing area Espèce, Zone de pêche Especie, Area de pesca	1992 mt	1993 mt	1994 mt	1995 mt	1996 mt	1997 mt	1998 mt	1999 mt	2000 mt	2001 mt
Flathead sole	**Balai du Japon**		**Platija japonesa**		***Hippoglossoides elassodon***			**1,83(02)014,04**		**FTS**
67 USA	2 044	5 865	4 835	9 073	11 392	12 867	18 886	14 318	16 266	16 092
67 Fishing area total	*2 044*	*5 865*	*4 835*	*9 073*	*11 392*	*12 867*	*18 886*	*14 318*	*16 266*	*16 092*
Species total	*2 044*	*5 865*	*4 835*	*9 073*	*11 392*	*12 867*	*18 886*	*14 318*	*16 266*	*16 092*
Yellowfin sole	**Limande du Japon**		**Limanda japonesa**		***Limanda aspera***			**1,83(02)024,02**		**YES**
67 USA	123 660	105 793	107 674	96 765	101 354	149 302	80 500	56 830	69 971	54 918
67 Fishing area total	*123 660*	*105 793*	*107 674*	*96 765*	*101 354*	*149 302*	*80 500*	*56 830*	*69 971*	*54 918*
Species total	*123 660*	*105 793*	*107 674*	*96 765*	*101 354*	*149 302*	*80 500*	*56 830*	*69 971*	*54 918*
Yellowtail flounder	**Limande à queue jaune**		**Limanda**		***Limanda ferruginea***			**1,83(02)024,04**		**YEL**
21 Canada	8 656	8 650	2 537	1 483	1 178	1 728	5 242	8 222	12 790	15 648
Cuba	-	0	0	-	-	-	-	-	-	-
Estonia	-	-	-	-	-	-	-	-	53	47
Honduras	-	20	-	-	-	-	-	-	-	-
Korea Rep	3 826	2	-	-	-	-	-	-	-	-
Lithuania	-	-	-	-	-	-	-	-	-	1
Portugal	-	20	-	-	-	-	85	426	153	351
Russian Fed	-	-	-	-	-	-	-	96	212	148
St Pier Mq	36	-	-	-	-	18	59	33	60	152
Spain	122	14	315	65	259	656	562	752	775	622
USA	5 631	3 607	3 094	1 882	2 403	2 864	3 656	4 431	6 928	7 304
21 Fishing area total	*18 271*	*12 313*	*5 946*	*3 430*	*3 840*	*5 266*	*9 604*	*13 960*	*20 971*	*24 273*
Species total	*18 271*	*12 313*	*5 946*	*3 430*	*3 840*	*5 266*	*9 604*	*13 960*	*20 971*	*24 273*
Common dab	**Limande**		**Lenguadina**		***Limanda limanda***			**1,83(02)024,05**		**DAB**
27 Belgium	602	666	569	557	690	789	961	980	865	850
Denmark	3 676	3 957	5 067	4 484	3 952	3 211	2 646	2 514	2 113	2 300
Faeroe Is	5	69	7	27	37	39	39	29	...	...
France	1 514	1 581	1 258	1 062	1 120	1 446	1 576	1 194	1 106	950
Germany	627	1 102	1 944	2 074	1 880	1 384	1 129	1 104	1 124	1 074
Iceland	3 045	4 222	5 159	5 558	7 954	7 891	5 061	3 981	3 015	4 373
Ireland	229	80	90	95	76	113	109	66	39	34
Isle of Man	0	0	0	0	0	0	-	-	-	-
Netherlands	...	...	...	...	...	...	7 983	8 656	6 544	5 969
Norway	-	-	-	-	-	-	-	-	-	54
Spain	-	-	-	-	-	0	130	129	29	24
Sweden	46	38	38	59	37	46	33	16	10	14
UK	1 476	1 360	1 706	2 225	2 221	2 590	2 467	2 248	1 705	1 530
27 Fishing area total	*11 220*	*13 075*	*15 838*	*16 141*	*17 967*	*17 509*	*22 134*	*20 917*	*16 550*	*17 172*
Species total	*11 220*	*13 075*	*15 838*	*16 141*	*17 967*	*17 509*	*22 134*	*20 917*	*16 550*	*17 172*
Rock sole	**Fausse limande du Pacifique**		**Lenguado del Pacífico**		***Lepidopsetta bilineata***			**1,83(02)043,01**		**ROS**
67 USA	40 601	69 745	23 448	24 962	26 198	32 743	15 189	17 185	27 510	24 206
67 Fishing area total	*40 601*	*69 745*	*23 448*	*24 962*	*26 198*	*32 743*	*15 189*	*17 185*	*27 510*	*24 206*
77 USA	4	6	4	7	4	9	10	7	7	7
77 Fishing area total	*4*	*6*	*4*	*7*	*4*	*9*	*10*	*7*	*7*	*7*
Species total	*40 605*	*69 751*	*23 452*	*24 969*	*26 202*	*32 752*	*15 199*	*17 192*	*27 517*	*24 213*
Dover sole	**Limande sole du Pacifique**		**Soya escurridiza**		***Microstomus pacificus***			**1,83(02)045,03**		**MIP**
67 USA	...	...	...	...	...	8 496	8 174	8 384	7 546	6 038
67 Fishing area total	*...*	*...*	*...*	*...*	*...*	*8 496*	*8 174*	*8 384*	*7 546*	*6 038*
77 USA	...	...	...	...	...	3 846	1 818	2 174	1 866	1 403
77 Fishing area total	*...*	*...*	*...*	*...*	*...*	*3 846*	*1 818*	*2 174*	*1 866*	*1 403*
Species total	*...*	*...*	*...*	*...*	*...*	*12 342*	*9 992*	*10 558*	*9 412*	*7 441*
Lemon sole	**Limande sole**		**Mendo limón**		***Microstomus kitt***			**1,83(02)045,04**		**LEM**
27 Belgium	776	710	709	1 006	1 094	975	1 256	1 020	1 057	1 076
Channel Is	0	0	0	0	0	0	1	1	3	1
Denmark	2 034	2 008	1 291	1 208	1 108	1 172	1 591	1 812	2 037	1 820
Faeroe Is	261	201	276	265	236	332	464	433	389	694
France	1 455	1 898	2 005	1 944	2 396	1 782	1 522	1 349	1 308	1 366
Germany	30	42	29	71	67	78	151	68	74	77
Iceland	915	697	692	741	984	1 135	1 432	1 860	1 438	1 371
Ireland	505	531	390	724	581	667	527	531	468	440
Isle of Man	7	2	3	2	4	0	4	3	3	1
Netherlands	...	...	...	...	...	...	839	681	492	456
Norway	30	31	33	31	47	63	59	59	60	53
Spain	-	-	-	-	-	235	1 197	1 282	2 207	4 040
Sweden	125	147	127	96	117	121	105	94	71	61
UK	5 825	5 443	5 117	4 974	5 444	5 723	5 230	5 010	4 367	4 005
27 Fishing area total	*11 963*	*11 710*	*10 672*	*11 062*	*12 078*	*12 283*	*14 378*	*14 203*	*13 974*	*15 461*

B-31 Flounders, halibuts, soles — Capture production by species, fishing areas and countries or areas
Flets, flétans, soles — Captures par espèces, zones de pêche et pays ou zones
Platijas, halibuts, lenguados — Capturas por especies, áreas de pesca y países o áreas

Species, Fishing area Espèce, Zone de pêche Especie, Area de pesca	1992 mt	1993 mt	1994 mt	1995 mt	1996 mt	1997 mt	1998 mt	1999 mt	2000 mt	2001 mt
Species total	*11 963*	*11 710*	*10 672*	*11 062*	*12 078*	*12 283*	*14 378*	*14 203*	*13 974*	*15 461*
European flounder	**Flet d'Europe**		**Platija europea**		***Platichthys flesus***				**1,83(02)048,02**	**FLE**
27 Belgium	170	213	223	348	278	146	307	363	322	316
Denmark	2 799	4 031	3 585	4 235	5 469	5 378	4 725	3 528	4 604	6 066
Estonia	164	165	162	102	297	333	355	416	419	482
Finland	1 072	1 092	564	575	715	702	555	558	449	500
France	149	127	169	206	236	212	152	198	219	242
Germany	2 056	1 630	6 018	4 488	1 637	2 449	2 159	2 347	2 782	2 349
Ireland	17	17	19	24	13	13	13	13	12	18
Latvia	691	501	329	362	294	367	364	509	418	613
Netherlands	31	35	...	...	...	...	4 942	3 159	2 658	2 621
Norway	-	-	-	-	-	-	-	-	-	3
Portugal	0	0	0	0	0	-	-	-	-	-
Russian Fed	-	-	-	-	-	-	-	-	1 392	1 351
Spain	-	-	-	-	74	319	873	323	88	139
Sweden	269	210	275	459	1 262	1 073	526	274	341	467
UK	310	276	383	388	357	379	293	149	201	148
27 Fishing area total	*7 728*	*8 297*	*11 727*	*11 187*	*10 632*	*11 371*	*15 264*	*11 837*	*13 905*	*15 315*
37 Albania	...	...	...	0	3	9	42	41	45	15
France	29	12	6	3	2	11	11	6	2	1
Slovenia	-	-	-	-	-	-	-	-	1	3
Ukraine	51	66	68	25	30	30	16	15	8	10
37 Fishing area total	*80*	*78*	*74*	*28*	*35*	*50*	*69*	*62*	*56*	*29*
Species total	*7 808*	*8 375*	*11 801*	*11 215*	*10 667*	*11 421*	*15 333*	*11 899*	*13 961*	*15 344*
Pacific sand sole	**Plie à points noirs**		**...C**		***Psettichthys melanostictus***				**1,83(02)049,01**	**YSE**
67 USA	...	...	...	...	...	...	15	582	44	85
67 Fishing area total	...	...	...	...	...	...	*15*	*582*	*44*	*85*
77 USA	...	...	...	...	...	...	19	23	35	44
77 Fishing area total	...	...	...	...	...	...	*19*	*23*	*35*	*44*
Species total	...	...	...	...	...	...	*34*	*605*	*79*	*129*
Yellow striped flounder	**Limande-plie du Japon**		**Acedía del Japón**		***Pseudopleuronectes herzenst.***				**1,83(02)050,01**	**YFL**
61 Korea Rep	14 631	13 505	13 343	13 683	18 066	18 079	20 135	19 569	15 423	14 503
61 Fishing area total	*14 631*	*13 505*	*13 343*	*13 683*	*18 066*	*18 079*	*20 135*	*19 569*	*15 423*	*14 503*
Species total	*14 631*	*13 505*	*13 343*	*13 683*	*18 066*	*18 079*	*20 135*	*19 569*	*15 423*	*14 503*
Winter flounder	**Limande-plie rouge**		**Solla roja**		***Pseudopleuronectes americanus***				**1,83(02)050,03**	**FLW**
21 Canada	3 499	2 667	3 448	2 804	2 468	2 548	1 804	1 763	2 342	2 257
Estonia	-	-	-	-	-	-	-	-	25	-
USA	6 414	5 267	3 592	4 002	5 687	5 765	5 100	4 654	5 818	6 930
21 Fishing area total	*9 913*	*7 934*	*7 040*	*6 806*	*8 155*	*8 313*	*6 904*	*6 417*	*8 185*	*9 187*
31 USA	-	-	-	-	-	-	0	-	-	-
31 Fishing area total	-	-	-	-	-	-	*0*	-	-	-
Species total	*9 913*	*7 934*	*7 040*	*6 806*	*8 155*	*8 313*	*6 904*	*6 417*	*8 185*	*9 187*
Sand flounders	**Rhombosoles**		**Sollas de arena**		***Rhombosolea spp***				**1,83(02)053,XX**	**FSA**
57 Australia	6	5	49	34	26	26	26	29	1	...
57 Fishing area total	*6*	*5*	*49*	*34*	*26*	*26*	*26*	*29*	*1*	...
81 Australia	34	34	32	32	1	1	2	28	...	...
New Zealand	...	...	...	103	224	251	...	...	...	37
81 Fishing area total	*34*	*34*	*32*	*135*	*225*	*252*	*2*	*28*	...	*37*
Species total	*40*	*39*	*81*	*169*	*251*	*278*	*28*	*57*	*1*	*37*
Curlfin sole	**...B**		**...C**		***Pleuronichthys decurrens***				**1,83(02)059,02**	**NYD**
67 USA	-	-	-	-	-	-	-	-	1	5
67 Fishing area total	-	-	-	-	-	-	-	-	*1*	*5*
Species total	-	-	-	-	-	-	-	-	*1*	*5*
Common sole	**Sole commune**		**Lenguado común**		***Solea solea***				**1,83(03)007,01**	**SOL**
01 Egypt	...	...	...	...	...	547	381	763	2 490	2 134
01 Fishing area total	...	...	...	...	...	*547*	*381*	*763*	*2 490*	*2 134*
27 Belgium	4 509	5 231	5 703	5 457	5 150	4 514	4 102	4 492	4 479	4 975
Channel Is	2	2 F	2	2 F	9	26	21	21	26	25
Denmark	2 722	3 091	3 073	3 039	2 083	1 478	1 050	1 433	1 804	1 324
France	9 127	9 401	10 050	8 687	7 179	6 838	6 554	8 255	8 255	7 644
Germany	1 881	1 387	1 749	1 569	685	513	786	1 462	1 291	959
Ireland	547	483	573	561	463	483	526	492	383	375
Isle of Man	7	4	5	12	4	5	3	1	1	1
Netherlands	18 681	22 014	22 925	20 927	15 563	10 370	15 308	16 329	15 343	13 737

B-31

Flounders, halibuts, soles	Capture production by species, fishing areas and countries or areas
Flets, flétans, soles	Captures par espèces, zones de pêche et pays ou zones
Platijas, halibuts, lenguados	Capturas por especies, áreas de pesca y países o áreas

Species, Fishing area Espèce, Zone de pêche Especie, Area de pesca	1992 mt	1993 mt	1994 mt	1995 mt	1996 mt	1997 mt	1998 mt	1999 mt	2000 mt	2001 mt
Portugal	1 019	118	237	235	167	151	113	121	152	189
Spain	190	103	52	39	74	178	228	248	279	224
Sweden	107	138	94	89	61	52	41	43	30	20
UK	3 794	3 433	3 589	3 530	3 022	2 671	2 561	2 808	2 443	2 720
27 Fishing area total	*42 586*	*45 405 F*	*48 052*	*44 147 F*	*34 460*	*27 279*	*31 293*	*35 705*	*34 486*	*32 193*
34 Congo Rep	814 F	812 F	326	230	200 F	200 F	100 F	100 F	...	50 F
France	-	1	1	0	0	0	-	-	-	-
Greece	259	246	92	194	148	87	125	110	66	82
Italy	10 111	8 384	8 429	299	193	864	731	410	291	749
Liberia	169	131	112	158	48	150	149	217	129	206
Morocco	1 106	1 152	1 342	1 451	1 340	1 633	1 520	2 594	5 925	3 775
Portugal	4	169	183	0	0	0	0	...	0	12
Spain	1 022	...	...	...	-	5	5 035	-	99	4 297
34 Fishing area total	*13 485 F*	*10 895 F*	*10 485*	*2 332*	*1 929 F*	*2 939 F*	*7 660 F*	*3 431 F*	*6 510*	*9 171 F*
37 Albania	...	...	...	25	27	21	35	31	41	14
Algeria	320 F	350 F	400 F	313	387	333	285	234	240 F	240 F
Egypt	360	814	710	473	751	762	653	965	1 012	1 041
France	340	330	187	189	159	443	519	147	192	184
Greece	1 849	2 068	1 742	1 259	1 087	1 082	620	509	621	480
Israel	10	9	100	100	...	...	...	...	...	...
Italy	5 302	4 002	4 097	5 766	3 404	2 221	1 907	1 842	1 874	2 217
Malta	0	0	0	0	0	0	0	0	0	0
Morocco	51	63	65	94	64	42	20	20	26	13
Romania	-	-	3	-	-	-	4	5	6	9
Slovenia	2	9	7	2	1	1	1	1	2	3
Spain	700 F	650 F	650 F	650 F	500	370	520	-	715	284
Tunisia	751	572	493	327	569	642	480	425	402	456
37 Fishing area total	*9 685 F*	*8 867 F*	*8 454 F*	*9 198 F*	*6 949*	*5 917*	*5 044*	*4 179*	*5 131 F*	*4 941 F*
Species total	*65 756 F*	*65 167 F*	*66 991 F*	*55 677 F*	*43 338 F*	*36 682 F*	*44 378 F*	*44 078 F*	*48 617 F*	*48 439 F*
Sand sole	**Sole-pole**		**Lenguado de arena**		***Solea lascaris***			**1,83(03)007,04**		**SOS**
27 Channel Is	1	1 F	1	1 F	2	3	3	1	1	1
France	106	140	169	160	127	150	137	118	131	153
Ireland	45	14	7	25	13	12	15	1	2	1
Portugal	49	57	51	155	77	95	118	90	116	142
UK	12	16	31	28	27	28	32	12	47	42
27 Fishing area total	*213*	*228 F*	*259*	*369 F*	*246*	*288*	*305*	*222*	*297*	*339*
Species total	*213*	*228 F*	*259*	*369 F*	*246*	*288*	*305*	*222*	*297*	*339*
Wedge sole	**Céteau**		**Acedía**		***Dicologlossa cuneata***			**1,83(03)017,01**		**CET**
27 France	551	713	847	941	798	647	488	602	686	579
Portugal	...	...	...	...	...	...	...	102	109	112
UK	-	-	-	-	-	1	0	0	-	-
27 Fishing area total	*551*	*713*	*847*	*941*	*798*	*648*	*488*	*704*	*795*	*691*
34 Portugal	-	-	39	59	87	8	0	9	10	9
34 Fishing area total	*-*	*-*	*39*	*59*	*87*	*8*	*0*	*9*	*10*	*9*
Species total	*551*	*713*	*886*	*1 000*	*885*	*656*	*488*	*713*	*805*	*700*
West coast sole	**Sole australe occidentale**		**Lenguado austral oeste**		***Austroglossus microlepis***			**1,83(03)057,01**		**SOW**
47 Namibia	75	529	661	462	339	514	393	463	571	589
South Africa	134	45	-	1	16	3	...	...	...	...
Spain	1	-	-	-	54	-	-	-	-	-
47 Fishing area total	*210*	*574*	*661*	*463*	*409*	*517*	*393*	*463*	*571*	*589*
Species total	*210*	*574*	*661*	*463*	*409*	*517*	*393*	*463*	*571*	*589*
Mud sole	**Sole de vase**		**Lenguado de fango**		***Austroglossus pectoralis***			**1,83(03)057,02**		**SOE**
47 South Africa	700	764	942	769	909	837	859	768	800 F	844
47 Fishing area total	*700*	*764*	*942*	*769*	*909*	*837*	*859*	*768*	*800 F*	*844*
Species total	*700*	*764*	*942*	*769*	*909*	*837*	*859*	*768*	*800 F*	*844*
Southeast Atlantic soles nei	**Soles de l'Atl.sud-est nca**		**Lenguados del Atl.sudeste nep**		***Austroglossus spp***			**1,83(03)057,XX**		**SOA**
47 Korea Rep	130	122	362	1 365	480	1 197	773	926	965	650
47 Fishing area total	*130*	*122*	*362*	*1 365*	*480*	*1 197*	*773*	*926*	*965*	*650*
Species total	*130*	*122*	*362*	*1 365*	*480*	*1 197*	*773*	*926*	*965*	*650*
Thickback soles	**Soles-perdix**		**Golletas**		***Microchirus spp***			**1,83(03)081,XX**		**THS**
27 France	63	65	46	48	50	57	77	71	71	80
27 Fishing area total	*63*	*65*	*46*	*48*	*50*	*57*	*77*	*71*	*71*	*80*
Species total	*63*	*65*	*46*	*48*	*50*	*57*	*77*	*71*	*71*	*80*
Soles nei	**Soles nca**		**Lenguados nep**		***Soleidae***			**1,83(03)XXX,XX**		**SOX**
27 France	-	-	-	-	-	-	-	13	51	14

B-31

Flounders, halibuts, soles — **Capture production by species, fishing areas and countries or areas**
Flets, flétans, soles — **Captures par espèces, zones de pêche et pays ou zones**
Platijas, halibuts, lenguados — **Capturas por especies, áreas de pesca y países o áreas**

Species, Fishing area Espèce, Zone de pêche Especie, Area de pesca	1992 mt	1993 mt	1994 mt	1995 mt	1996 mt	1997 mt	1998 mt	1999 mt	2000 mt	2001 mt
Norway	...	...	...	...	...	...	...	...	...	95
Portugal	-	-	-	-	-	-	-	797	1 033	831
27 Fishing area total	*...*	*...*	*...*	*...*	*...*	*...*	*...*	*810*	*1 084*	*940*
34 Ghana	28	-	-	-	-	-	-	-	-	-
Korea Rep	67	263	-	29	102	-	-	256	1 016	-
Mauritania	600 F	600 F	1 000 F	680 F	760 F	1 120 F	830 F	449 F	444 F	...
Nigeria	2 050	3 711	4 511	3 807	4 640	6 084	7 633	6 583	3 301	3 171
Portugal	6	70	11	164	248	124	60	57	91	73
Spain	1 232	2 500 F	3 500 F	4 500 F	6 846	6 208	7 519	1 781	227	979
Togo	1	5	5	2	10	16	8	6	48	19
34 Fishing area total	*3 984 F*	*7 149 F*	*9 027 F*	*9 182 F*	*12 606 F*	*13 552 F*	*16 050 F*	*9 132 F*	*5 127 F*	*4 242*
37 France	-	-	-	-	-	-	-	18	14	12
37 Fishing area total	-	-	-	-	-	-	-	*18*	*14*	*12*
Species total	*3 984 F*	*7 149 F*	*9 027 F*	*9 182 F*	*12 606 F*	*13 552 F*	*16 050 F*	*9 960 F*	*6 225 F*	*5 194*
Tonguefishes	**Cynoglossidés**		**Cinoglósidos**		***Cynoglossidae***			**1,83(04)XXX,XX**		**TOX**
31 Korea Rep	-	-	-	-	-	-	-	59	39	20
31 Fishing area total	-	-	-	-	-	-	-	*59*	*39*	*20*
34 Cameroon	701	653	700 F	753	748	700 F	594	455	488	439
Congo Dem R	80 F	90 F	80 F	82 F	84 F	81 F	83 F	80 F	80 F	80 F
Congo Rep	...	...	316	300 F	270 F	240 F	200 F	180 F	158	150 F
Côte dIvoire	316	255	211	208	177	139	217	217	211	197
Gabon	220 F	242	178	209	310	411	492	386	620	538
Gambia	95	188	211	859	541	307	441	450	725	2 262
Ghana	195	247	231	407	295	339	347	284	394	227
Korea Rep	137	451	943	1 239	1 413	1 736	1 128	1 322	669	1 956
Nigeria	3 062	684	650	1 771	1 281	1 311	1 781	3 812	7 057	6 655
Sierra Leone	356	352	351	377	380 F	585	206	714	2 326	500
34 Fishing area total	*5 162 F*	*3 162 F*	*3 871 F*	*6 205 F*	*5 499 F*	*5 849 F*	*5 489 F*	*7 900 F*	*12 728 F*	*13 004 F*
41 Italy	1 348	1 118	1 124	36	23	-	-	-	-	-
Korea Rep	-	-	-	-	-	-	-	19	48	-
41 Fishing area total	*1 348*	*1 118*	*1 124*	*36*	*23*	-	-	*19*	*48*	-
47 Italy	-	-	-	4	3	-	-	-	-	-
South Africa	-	-	-	-	-	-	-	-	-	10
47 Fishing area total	-	-	-	*4*	*3*	-	-	-	-	*10*
51 Italy	2 022	1 677	1 686	61	40	-	-	-	-	-
Korea Rep	1	-	3	32	41	130	2	24	13	1
Pakistan	2 041	2 024	1 963	1 982	2 205	2 390	2 149	2 037	2 124	1 915
South Africa	-	-	-	-	-	3	7	5 F	0	7
51 Fishing area total	*4 064*	*3 701*	*3 652*	*2 075*	*2 286*	*2 523*	*2 158*	*2 066 F*	*2 137*	*1 923*
57 Korea Rep	-	-	-	-	-	-	-	7	-	-
Malaysia	2 103	2 403	2 644	2 430	2 123	2 602	2 390	2 298	2 120	1 551
Thailand	3 199	3 203	3 164	4 383	8 162	8 313	9 292	9 011	8 888	8 833
57 Fishing area total	*5 302*	*5 606*	*5 808*	*6 813*	*10 285*	*10 915*	*11 682*	*11 316*	*11 008*	*10 384*
61 China,H.Kong	929	1 008	1 583	1 326	1 160	1 261	1 177	800 F	1 000 F	1 100 F
Korea Rep	3 050	2 691	2 498	1 944	1 644	1 289	654	836	1 148	1 013
61 Fishing area total	*3 979*	*3 699*	*4 081*	*3 270*	*2 804*	*2 550*	*1 831*	*1 636 F*	*2 148 F*	*2 113 F*
71 Korea Rep	1	-	2	2	21	37	83	49	58	79
Malaysia	457	404	673	664	617	427	550	1 008	956	866
Thailand	4 516	4 781	7 319	8 418	6 849	6 717	6 805	6 322	6 235	6 197
71 Fishing area total	*4 974*	*5 185*	*7 994*	*9 084*	*7 487*	*7 181*	*7 438*	*7 379*	*7 249*	*7 142*
77 Korea Rep	-	-	-	-	-	-	-	105	95	-
77 Fishing area total	-	-	-	-	-	-	-	*105*	*95*	-
81 Korea Rep	-	-	-	-	-	-	-	9	-	-
81 Fishing area total	-	-	-	-	-	-	-	*9*	-	-
87 Korea Rep	-	-	-	-	-	-	-	13	38	-
87 Fishing area total	-	-	-	-	-	-	-	*13*	*38*	-
Species total	*24 829 F*	*22 471 F*	*26 530 F*	*27 487 F*	*28 387 F*	*29 018 F*	*28 598 F*	*30 502 F*	*35 490 F*	*34 596 F*
Megrim	**Cardine franche**		**Gallo del Norte**		***Lepidorhombus whiffiagonis***			**1,83(05)003,01**		**MEG**
27 Belgium	83	63	129	225	208	188	142	143	132	87
Channel Is	-	-	1	1 F	-	-	-	-	-	1
Denmark	6	10	1	2	7	6	26	21	30	55
France	4 771	4 165	3 718	4 591	4 157	3 877	3 491	3 421	4 015	4 000
Germany	3	4	1	2	1	2	3	1	3	1
Iceland	246	213	301	405	419	281	221	124	97	96
Ireland	3 147	2 999	3 053	3 839	3 507	3 063	3 383	3 162	3 364	3 713
Isle of Man	-	-	-	-	-	3	2	...	...	-
Netherlands	27	31	29	26	11	23	31	28	20	11
Portugal	219	19	13	53	58	26	47	54	47	22
Sweden	0	0	0	1	-	-	-	-	-	-
UK	5 186	5 566	6 039	6 563	7 212	7 479	6 520	5 606	5 513	4 672
27 Fishing area total	*13 688*	*13 070*	*13 285*	*15 708 F*	*15 580*	*14 948*	*13 866*	*12 560*	*13 221*	*12 658*

B-31

Flounders, halibuts, soles — **Capture production by species, fishing areas and countries or areas**
Flets, flétans, soles — **Captures par espèces, zones de pêche et pays ou zones**
Platijas, halibuts, lenguados — **Capturas por especies, áreas de pesca y países o áreas**

Species, Fishing area Espèce, Zone de pêche Especie, Area de pesca	1992 mt	1993 mt	1994 mt	1995 mt	1996 mt	1997 mt	1998 mt	1999 mt	2000 mt	2001 mt
34 Portugal	0	12	0	-	-	-	-	-	-	-
Spain	-	-	-	-	-	-	36	-	38	52
34 Fishing area total	*0*	*12*	*0*	*-*	*-*	*-*	*36*	*-*	*38*	*52*
37 Albania	...	...	...	1	1	0	-	-	-	1
France	203	183	103	51	110	110	110	108	139	125
Spain	150 F	160 F	170 F	175 F	175	185	8	-	63	79
37 Fishing area total	*353 F*	*343 F*	*273 F*	*227 F*	*286*	*295*	*118*	*108*	*202*	*205*
Species total	*14 041 F*	*13 425 F*	*13 558 F*	*15 935 F*	*15 866*	*15 243*	*14 020*	*12 668*	*13 461*	*12 915*
Megrims nei	**Cardines nca**		**Gallos nep**		***Lepidorhombus spp***				**1,83(05)003,XX**	**LEZ**
27 Germany	...	...	...	...	...	...	...	...	...	289
Spain	5 842	5 771	5 968	6 275	6 405	7 006	7 446	6 969	8 361	8 161
27 Fishing area total	*5 842*	*5 771*	*5 968*	*6 275*	*6 405*	*7 006*	*7 446*	*6 969*	*8 361*	*8 450*
Species total	*5 842*	*5 771*	*5 968*	*6 275*	*6 405*	*7 006*	*7 446*	*6 969*	*8 361*	*8 450*
Brill	**Barbue**		**Rémol**		***Scophthalmus rhombus***				**1,83(05)064,01**	**BLL**
27 Belgium	361	425	379	378	422	349	341	365	423	491
Channel Is	7	7 F	18	17 F	10	10	10	6	17	17
Denmark	235	337	363	281	220	148	189	244	237	163
France	480	509	450	452	417	357	351	373	486	464
Germany	84	73	95	72	47	48	60	54	80	65
Ireland	216	135	113	128	126	181	141	126	119	95
Isle of Man	1	1	1	1	1	0	0	1	1	0
Netherlands	1 196	1 647	1 235	943	736	598	811	809	1 005	1 093
Norway	29	24	23	20	21	26	26	30	27	26
Portugal	53	64	49	57	48	39	33	39	46	57
Spain	-	-	-	-	460	409	38	36	29	23
Sweden	15	16	18	15	7	11	13	18	17	13
UK	537	575	608	618	688	568	513	410	470	535
27 Fishing area total	*3 214*	*3 813 F*	*3 352*	*2 982 F*	*3 203*	*2 744*	*2 526*	*2 511*	*2 957*	*3 042*
37 France	...	...	...	...	...	...	...	24	26	26
Spain	-	-	-	-	18	20	22	-	-	-
37 Fishing area total	*...*	*...*	*...*	*...*	*18*	*20*	*22*	*24*	*26*	*26*
Species total	*3 214*	*3 813 F*	*3 352*	*2 982 F*	*3 221*	*2 764*	*2 548*	*2 535*	*2 983*	*3 068*
Windowpane flounder	**Turbot de sable**		**Rodaballo aranero**		***Scophthalmus aquosus***				**1,83(05)064,02**	**FLD**
21 USA	2 099	1 603	526	759	46	48	521	166	268	177
21 Fishing area total	*2 099*	*1 603*	*526*	*759*	*46*	*48*	*521*	*166*	*268*	*177*
Species total	*2 099*	*1 603*	*526*	*759*	*46*	*48*	*521*	*166*	*268*	*177*
Turbot	**Turbot**		**Rodaballo**		***Psetta maxima***				**1,83(05)092,01**	**TUR**
27 Belgium	456	480	480	499	382	337	327	368	464	506
Channel Is	2	2 F	6	6 F	5	5	3	4	6	9
Denmark	1 642	1 531	1 572	1 396	1 117	908	770	727	809	872
Faeroe Is	-	320	-	-	2	-	-	-	-	-
Finland	...	...	...	...	...	...	...	...	6	4
France	927	1 043	1 728	817	770	515	498	540	632	624
Germany	300	385	384	399	256	330	267	309	454	364
Iceland	0	0	0	1	0	0	0	0	0	0
Ireland	247	223	194	233	232	257	234	261	236	185
Isle of Man	1	1	0	0	1	1	0	0	-	0
Latvia	-	-	-	49	42	46	36	54	16	6
Lithuania	-	-	-	-	-	-	62	58	23	18
Netherlands	3 495	2 938	2 724	2 476	1 780	1 866	1 700	1 812	2 287	2 277
Norway	73	66	62	53	54	57	45	48	68	94
Portugal	66	65	54	57	40	28	27	34	63	83
Russian Fed	-	-	-	-	-	-	-	-	53	69
Spain	222	271	225	239	264	320	218	241	113	108
Sweden	104	114	113	195	296	294	188	159	106	64
UK	1 243	1 531	1 490	1 281	1 270	1 148	974	851	877	1 001
27 Fishing area total	*8 778*	*8 970 F*	*9 032*	*7 701 F*	*6 511*	*6 112*	*5 349*	*5 466*	*6 213*	*6 284*
37 Bulgaria	...	...	...	60	62	60	64	54	55	57
France	62	89	77	5	40	131	131	13	18	15
Greece	152	182	115	102	60	60	47	65	63	77
Romania	-	6	6	4	6	1	-	2	2	13
Spain	19 F	18 F	18 F	18 F	18	19	13	11	11	14
Tunisia	0	1	2	0	0	0	-	-	-	-
Turkey	437	1 636	2 159	2 955	2 035	980	1 860	1 870	2 700	2 455
Ukraine	254	167	139	96	120	82	63	110	118	171
37 Fishing area total	*924 F*	*2 099 F*	*2 516 F*	*3 240 F*	*2 341*	*1 333*	*2 178*	*2 125*	*2 967*	*2 802*
Species total	*9 702 F*	*11 069 F*	*11 548 F*	*10 941 F*	*8 852*	*7 445*	*7 527*	*7 591*	*9 180*	*9 086*
Turbots nei	**Turbots nca**		**Rodaballos nep**		***Scophthalmidae***				**1,83(05)XXX,XX**	**SCF**
37 Italy	1 766	1 288	1 213	1 923	1 377	964	528	478	643	622
37 Fishing area total	*1 766*	*1 288*	*1 213*	*1 923*	*1 377*	*964*	*528*	*478*	*643*	*622*

B-31 Flounders, halibuts, soles / Flets, flétans, soles / Platijas, halibuts, lenguados

Capture production by species, fishing areas and countries or areas / Captures par espèces, zones de pêche et pays ou zones / Capturas por especies, áreas de pesca y países o áreas

Species, Fishing area Espèce, Zone de pêche Especie, Area de pesca	1992 mt	1993 mt	1994 mt	1995 mt	1996 mt	1997 mt	1998 mt	1999 mt	2000 mt	2001 mt
Species total	*1 766*	*1 288*	*1 213*	*1 923*	*1 377*	*964*	*528*	*478*	*643*	*622*
Indian halibut	**Turbot épineux-indien**		**Lenguado espinudo-indio**		*Psettodes erumei*			**1,83(07)001,01**		**HAI**
51 Eritrea	...	...	9	4	8	1	0	-	-	-
Tanzania	100	40	27	43	148	50	45	50	52	75
Yemen	368	89	422	974	724	760 F	900 F	800 F	750 F	950 F
51 Fishing area total	*468*	*129*	*458*	*1 021*	*880*	*811 F*	*945 F*	*850 F*	*802 F*	*1 025 F*
57 Indonesia	2 092	2 819	3 881	3 352	4 937	6 102	6 167	6 954	7 081	8 560
Thailand	262	914	2 689	2 741	5 926	6 110	1 612	518	511	508
57 Fishing area total	*2 354*	*3 733*	*6 570*	*6 093*	*10 863*	*12 212*	*7 779*	*7 472*	*7 592*	*9 068*
71 Indonesia	2 319	3 131	1 979	2 317	2 460	8 973	3 772	5 117	4 062	4 910
Thailand	1 071	968	1 216	1 657	1 368	1 313	1 130	1 828	1 803	1 792
71 Fishing area total	*3 390*	*4 099*	*3 195*	*3 974*	*3 828*	*10 286*	*4 902*	*6 945*	*5 865*	*6 702*
Species total	*6 212*	*7 961*	*10 223*	*11 088*	*15 571*	*23 309 F*	*13 626 F*	*15 267 F*	*14 259 F*	*16 795 F*
Spottail spiny turbot	**Turbot épineux tacheté**		**Perro**		*Psettodes belcheri*			**1,83(07)001,02**		**SOT**
34 Togo	-	1	0	0	0	2	0	-	1	2
34 Fishing area total	-	*1*	*0*	*0*	*0*	*2*	*0*	-	*1*	*2*
Species total	-	*1*	*0*	*0*	*0*	*2*	*0*	-	*1*	*2*
Pacific sanddab	**...B**		**...C**		*Citharichthys sordidus*			**1,83(08)009,02**		**IYO**
67 USA	-	-	-	-	-	-	-	-	150	-
67 Fishing area total	-	-	-	-	-	-	-	-	*150*	-
Species total	-	-	-	-	-	-	-	-	*150*	-
Bastard halibut	**Cardeau hirame**		**Falso halibut del Japón**		*Paralichthys olivaceus*			**1,83(08)046,01**		**BAH**
61 Japan	6 817	6 464	6 667	7 558	8 311	8 361	7 615	7 198	7 572	6 729
Korea Rep	2 180	2 458	2 035	1 914	2 317	1 592	2 002	1 679	1 607	1 707
61 Fishing area total	*8 997*	*8 922*	*8 702*	*9 472*	*10 628*	*9 953*	*9 617*	*8 877*	*9 179*	*8 436*
Species total	*8 997*	*8 922*	*8 702*	*9 472*	*10 628*	*9 953*	*9 617*	*8 877*	*9 179*	*8 436*
California flounder	**Cardeau californien**		**Lenguado de California**		*Paralichthys californicus*			**1,83(08)046,03**		**YSF**
67 USA	...	...	...	...	...	...	13	3	0	4
67 Fishing area total	...	...	...	...	...	...	*13*	*3*	*0*	*4*
77 USA	...	...	...	...	...	...	510	617	390	389
77 Fishing area total	...	...	...	...	...	...	*510*	*617*	*390*	*389*
Species total	...	...	...	...	...	...	*523*	*620*	*390*	*393*
Summer flounder	**Cardeau d'été**		**Falso halibut del Canadá**		*Paralichthys dentatus*			**1,83(08)046,06**		**FLS**
21 Cuba	0	0	-	-	-	-	-	-	-	-
USA	7 622	5 771	4 961	6 943	5 425	4 775	5 548	5 166	5 344	5 350
21 Fishing area total	*7 622*	*5 771*	*4 961*	*6 943*	*5 425*	*4 775*	*5 548*	*5 166*	*5 344*	*5 350*
31 USA	...	...	...	...	...	...	800	629	719	1 204
31 Fishing area total	...	...	...	...	...	...	*800*	*629*	*719*	*1 204*
Species total	*7 622*	*5 771*	*4 961*	*6 943*	*5 425*	*4 775*	*6 348*	*5 795*	*6 063*	*6 554*
Bastard halibuts nei	**Cardeaux nca**		**Falsos halibuts nep**		*Paralichthys spp*			**1,83(08)046,XX**		**BAX**
31 USA	1 311	1 327	3 918	1 926	2 192	1 059	533	391	503	-
31 Fishing area total	*1 311*	*1 327*	*3 918*	*1 926*	*2 192*	*1 059*	*533*	*391*	*503*	-
41 Argentina	8 591	9 557	7 512	10 213	8 753	10 044	8 751	6 668	6 490	5 373
Uruguay	124	99	113	130	483	502	496	413	256	307
41 Fishing area total	*8 715*	*9 656*	*7 625*	*10 343*	*9 236*	*10 546*	*9 247*	*7 081*	*6 746*	*5 680*
Species total	*10 026*	*10 983*	*11 543*	*12 269*	*11 428*	*11 605*	*9 780*	*7 472*	*7 249*	*5 680*
Flatfishes nei	**Poissons plats nca**		**Peces planos nep**		*Pleuronectiformes*			**1,83(XX)XXX,XX**		**FLX**
21 Canada	7 012	6 268	4 498	2 214	1 519	1 283	1 025	984	896	617
Cuba	16	19	3	13	-	-	-	1	0	...
Estonia	-	-	-	-	-	-	-	-	1	-
Faeroe Is	-	36	-	14	2	-	-	-	-	-
Germany	10	-	-	-	-	-	-	-	-	-
Japan	22	4 425	2 601	3 168	2 576	1 907	2 131	2 541	2 621	303
Korea Rep	-	13	-	-	-	-	-	-	-	-
USA	78	58	70	49	63	70	9	54	13	5
21 Fishing area total	*7 138*	*10 819*	*7 172*	*5 458*	*4 160*	*3 260*	*3 165*	*3 580*	*3 531*	*925*
27 Faeroe Is	-	1 634	-	327	-	757	1 059	292	...	...
France	26	5	14	14	14	17	24	15	3 067	1 851
Germany	512	179	1 600	-	404	508	279	293	370	-

B-31 Flounders, halibuts, soles / Flets, flétans, soles / Platijas, halibuts, lenguados

Capture production by species, fishing areas and countries or areas / Captures par espèces, zones de pêche et pays ou zones / Capturas por especies, áreas de pesca y países o áreas

Species, Fishing area Espèce, Zone de pêche Especie, Area de pesca	1992 mt	1993 mt	1994 mt	1995 mt	1996 mt	1997 mt	1998 mt	1999 mt	2000 mt	2001 mt
Iceland	40	30	6	10	11	13	3	5	-	2
Ireland	6	35	-	68	57	74	184	37	172	15
Japan	434	283	399	140	-	-	-	-	-	-
Norway	782	650	555	337	376	477	389	475	346	153
Poland	3 905	5 101	4 900	8 964	8 836	6 168	5 835	5 779	5 601	6 725
Portugal	2 525	1 646	1 504	1 652	1 291	1 082	964	8	10	5
Spain	6 651	5 800	4 070	4 251	8 307	13 955	11 569	9 972	830	653
Sweden	-	0	0	1	0	1	0	0	-	-
UK	-	-	-	132	172	172	-	-	158	171
27 Fishing area total	*14 881*	*15 363*	*13 048*	*15 896*	*19 468*	*23 224*	*20 306*	*16 876*	*10 554*	*9 575*
31 Korea Rep	-	-	-	-	-	-	-	16	7	-
Mexico	103	83	100	166	182	201	225	197	113	175
USA	643	702	728	604	153	286	333	330	371	296
Venezuela	...	...	...	...	...	...	...	51	11	...
31 Fishing area total	*746*	*785*	*828*	*770*	*335*	*487*	*558*	*594*	*502*	*471*
34 Benin	13 F	14 F	16 F	23 F	30 F	37 F	45 F	52	30 F	14
Cuba	0	0	0	-	-	-	-	-	-	-
Eq Guinea	320 F	290 F	450 F	190 F	420 F	530 F	325	380 F	40 F	40 F
Gabon	...	12	31	24	75	93	65	84	242	...
Guinea	200 F	240 F	250 F	350	254	256	179	148	1 032	1 000 F
GuineaBissau	110 F	112 F	92	87	64	70 F	60 F	50 F	50 F	50 F
Italy	1 618	1 570	1 570	-	-	261	244	96	107	117
Korea Rep	90	-	12	-	73	27	-	294	166	-
Lithuania	5	146	-	-	-	-	-	-	-	-
Mauritania	480 F	430 F	490 F	570 F	580 F	700 F	600 F	1 751 F	1 756 F	2 200 F
Morocco	5 175	3 266	3 348	3 709	3 710	2 230	4 896	5 748	5 000	3 869
Nigeria	37	0	0	-	-	-	-	-	-	-
Portugal	14	6	28	10	13	1	0	3	1	1
Senegal	7 417	7 224	11 857	10 510	8 113	8 002	7 132	7 335	8 113	9 059
Spain	-	800 F	1 200 F	1 600 F	1 846	3 812	3 453	5 191	9 572	6 938
34 Fishing area total	*15 479 F*	*14 110 F*	*19 344 F*	*17 073 F*	*15 178 F*	*16 019 F*	*16 999 F*	*21 132 F*	*26 109 F*	*23 288 F*
37 Croatia	143	217	144	124	130	133	150	65	113	111
France	-	-	-	-	-	-	-	5	8	4
Gaza Strip	...	...	...	...	7	15	25	25 F	25 F	20 F
Georgia	-	-	-	-	-	-	-	5	9	11
Lebanon	5 F	5	5	5	5	5	10	5	11	15
Morocco	16	22	17	37	40	57	19	25	38	29
Russian Fed	77	72	96	44	43	44	34	36	31	25
Spain	-	-	-	-	100	467	674	683	655	528
Tunisia	118	114	177	96	52	110	178	217	190	205
Turkey	817	1 779	1 549	1 092	1 947	2 300	2 000	2 400	1 000	1 250
Yugoslavia	2	11	8	13	10	8	10	9	9	11
37 Fishing area total	*1 178 F*	*2 220*	*1 996*	*1 411*	*2 334*	*3 139*	*3 100*	*3 475 F*	*2 089 F*	*2 209 F*
41 Brazil	2 510 F	2 500 F	2 500 F	1 491	1 091	1 430	1 655	1 590	1 844	1 820 F
Italy	216	209	209	-	-	-	-	-	-	-
Korea Rep	-	-	5	-	6	-	-	7	34	-
Spain	-	-	-	-	-	-	-	4	9	4
41 Fishing area total	*2 726 F*	*2 709 F*	*2 714 F*	*1 491*	*1 097*	*1 430*	*1 655*	*1 601*	*1 887*	*1 824 F*
47 Angola	467	784	428	97	195	120	928	912	592	1 928
Korea Rep	1 096	1 900	2 065	132	-	-	-	-	-	-
Russian Fed	-	-	-	-	-	-	7	-	-	-
47 Fishing area total	*1 563*	*2 684*	*2 493*	*229*	*195*	*120*	*935*	*912*	*592*	*1 928*
51 India	24 572	20 724	27 043	18 533	17 456	21 696	17 460	12 840	24 709	11 271
Italy	324	314	314	-	-	-	-	-	-	-
Korea Rep	2	-	-	38	50	194	14	11	11	3
Saudi Arabia	30	30	42	51	58	75	90	84	84	85
51 Fishing area total	*24 928*	*21 068*	*27 399*	*18 622*	*17 564*	*21 965*	*17 564*	*12 935*	*24 804*	*11 359*
57 Australia	-	-	1	2	2	-	-	-	-	-
India	4 932	4 764	4 084	6 148	6 741	7 556	6 202	5 184	5 362	2 344
Indonesia	592	982	1 036	1 509	2 369	2 216	2 138	2 586	2 841	3 460
Malaysia	284	332	281	316	538	518	377	378	342	485
57 Fishing area total	*5 808*	*6 078*	*5 402*	*7 975*	*9 650*	*10 290*	*8 717*	*8 148*	*8 545*	*6 289*
61 China,Taiwan	273	121	122	145	345	346	142	146	137	198
Japan	75 314	76 905	70 544	71 986	80 418	76 671	72 813	68 750	68 445	60 713
Korea D P Rp	...	...	2 318	2 953	1 966	6 972	4 000 F	4 000 F	3 800 F	3 800 F
Korea Rep	7	-	-	25	248	102	83	-	-	-
Russian Fed	87 136	66 025	38 550	46 768	57 180	63 768	79 682	97 014	103 077	95 106
61 Fishing area total	*162 730*	*143 051*	*111 534*	*121 877*	*140 157*	*147 859*	*156 720 F*	*169 910 F*	*175 459 F*	*159 817 F*
67 Canada	7 944	9 917	8 501	7 963	5 222	4 300	5 000	5 651	5 877	5 459
USA	59 714	44 674	23 767	26 713	29 509	18 740	18 886	13 040	7 627	4 748
67 Fishing area total	*67 658*	*54 591*	*32 268*	*34 676*	*34 731*	*23 040*	*23 886*	*18 691*	*13 504*	*10 207*
71 Indonesia	458	916	605	705	1 045	5 191	1 582	2 488	1 395	1 700
Korea Rep	1	-	-	-	-	-	-	84	-	-
Malaysia	1 343	1 389	1 581	1 699	1 695	1 675	2 180	2 648	1 299	1 227
Philippines	1 517	1 186	1 072	805	829	627	659	722	729	720
71 Fishing area total	*3 319*	*3 491*	*3 258*	*3 209*	*3 569*	*7 493*	*4 421*	*5 942*	*3 423*	*3 647*
77 Mexico	2 202	1 973	1 192	1 980	2 342	2 540	1 165	2 071	2 568	1 664
USA	7 317	5 343	4 912	6 486	6 878	2 056	665	822	725	686

B-31

Flounders, halibuts, soles — Capture production by species, fishing areas and countries or areas
Flets, flétans, soles — Captures par espèces, zones de pêche et pays ou zones
Platijas, halibuts, lenguados — Capturas por especies, áreas de pesca y países o áreas

Species, Fishing area Espèce, Zone de pêche Especie, Area de pesca	1992 mt	1993 mt	1994 mt	1995 mt	1996 mt	1997 mt	1998 mt	1999 mt	2000 mt	2001 mt
77 Fishing area total	*9 519*	*7 316*	*6 104*	*8 466*	*9 220*	*4 596*	*1 830*	*2 893*	*3 293*	*2 350*
81 Australia	5	5	3	3	5	6	6	12	...	...
New Zealand	3 590	5 751	4 873	5 247	4 284	7 747	4 270	3 505	2 939	3 220
81 Fishing area total	*3 595*	*5 756*	*4 876*	*5 250*	*4 289*	*7 753*	*4 276*	*3 517*	*2 939*	*3 220*
87 Chile	557	726	1 191	220	203	154	75	84	95	76
Colombia	52	15	9	11	30	45	131	48	50 F	50 F
Korea Rep	-	-	-	-	-	-	-	3	-	-
Peru	2 076	1 195	732	1 559	528	212	230	263	177	313
87 Fishing area total	*2 685*	*1 936*	*1 932*	*1 790*	*761*	*411*	*436*	*398*	*322 F*	*439 F*
Species total	*323 953 F*	*291 977 F*	*240 368 F*	*244 193 F*	*262 708 F*	*271 086 F*	*264 568 F*	*270 604 F*	*277 553 F*	*237 548 F*
Group total	***1 158 401***	***1 097 451***	***982 980***	***917 610***	***941 866***	***1 027 219***	***935 176***	***955 892***	***1 007 098***	***945 235***

B-32 Cods, hakes, haddocks / Morues, merlus, églefins / Bacalaos, merluzas, eglefinos

Capture production by species, fishing areas and countries or areas / Captures par espèces, zones de pêche et pays ou zones / Capturas por especies, áreas de pesca y países o áreas

Species, Fishing area Espèce, Zone de pêche Especie, Area de pesca	1992 mt	1993 mt	1994 mt	1995 mt	1996 mt	1997 mt	1998 mt	1999 mt	2000 mt	2001 mt
Smalleye moray cod	**Gadomurène petit oeil**		**Gadimorena ojichica**		***Muraenolepis microps***			**1,48(01)001,01**		**MOY**
88 New Zealand	-	-	-	-	-	-	-	4	5	-
88 Fishing area total	-	-	-	-	-	-	-	*4*	*5*	-
Species total	-	-	-	-	-	-	-	*4*	*5*	-
Moray cods nei	**Gadomurènes nca**		**Gadimorenas nep**		***Muraenolepis spp***			**1,48(01)001,XX**		**MRL**
88 New Zealand	-	-	-	-	-	-	0	1	2	3
88 Fishing area total	-	-	-	-	-	-	*0*	*1*	*2*	*3*
Species total	-	-	-	-	-	-	*0*	*1*	*2*	*3*
Common mora	**Moro commun**		**Mollera moranella**		***Mora moro***			**1,48(02)010,01**		**RIB**
81 New Zealand	...	...	...	1 192	694	1 410	1 324	1 122	1 355	1 209
Ukraine	-	-	-	-	-	-	-	-	3	2
81 Fishing area total	...	...	...	*1 192*	*694*	*1 410*	*1 324*	*1 122*	*1 358*	*1 211*
Species total	...	...	...	*1 192*	*694*	*1 410*	*1 324*	*1 122*	*1 358*	*1 211*
Red codling	**Morue rouge**		**Brotolilla**		***Pseudophycis bachus***			**1,48(02)014,01**		**NEC**
81 Japan	377	467	313	105	24	14	15	27	70	26
New Zealand	9 372	12 217	10 198	15 916	10 572	11 073	16 429	12 528	5 232	4 454
Russian Fed	-	34	130	138	-	-	-	-	-	-
Ukraine	11	98	...	...	...	...	51	...	63	50
81 Fishing area total	*9 760*	*12 816*	*10 641*	*16 159*	*10 596*	*11 087*	*16 495*	*12 555*	*5 365*	*4 530*
Species total	*9 760*	*12 816*	*10 641*	*16 159*	*10 596*	*11 087*	*16 495*	*12 555*	*5 365*	*4 530*
Blue antimora	**Antimora bleu**		**Mollera azul**		***Antimora rostrata***			**1,48(02)030,01**		**ANT**
21 Spain	-	-	-	-	16	-	26	24	21	-
21 Fishing area total	-	-	-	-	*16*	-	*26*	*24*	*21*	-
27 Iceland	-	-	-	-	2	-	-	-	-	-
27 Fishing area total	-	-	-	-	*2*	-	-	-	-	-
48 Argentina	-	-	-	1	-	-	-	-	-	-
Korea Rep	-	-	-	-	0	1	0	-	-	-
South Africa	-	-	-	-	-	-	1	-	-	-
UK	-	-	-	-	-	1	0	-	-	-
48 Fishing area total	-	-	-	*1*	*0*	*2*	*1*	-	-	-
58 South Africa	-	-	-	-	-	-	2	6	10	22
58 Fishing area total	-	-	-	-	-	-	*2*	*6*	*10*	*22*
88 New Zealand	-	-	-	-	-	-	0	0	0	3
South Africa	-	-	-	-	-	-	-	-	-	1
88 Fishing area total	-	-	-	-	-	-	*0*	*0*	*0*	*4*
Species total	-	-	-	*1*	*18*	*2*	*29*	*30*	*31*	*26*
Tadpole codling	**More têtard**		**Bacalao criollo**		***Salilota australis***			**1,48(02)040,01**		**SAO**
41 Argentina	...	...	...	2 241	1 742	2 632	3 604	6 607	8 433	1 834
Australia	-	-	-	-	-	-	85	60	-	-
Belize	-	-	-	-	-	-	-	28	237	42
Falkland Is	32	69	100	1 530	2 033	817	1 491	2 692	1 886	1 371
France	-	-	16	21	31	25	11	5	29	-
Honduras	394	374	107	106	189	-	-	-	-	-
Namibia	-	-	-	-	-	20	99	128	-	-
Panama	658	101	8	90	93	-	-	-	-	-
Poland	1	-	-	-	-	-	-	-	-	-
Portugal	36	21	-	-	-	-	-	-	12	-
St Vincent	-	-	-	-	-	-	-	-	-	14
Seychelles	-	-	-	-	-	56	-	-	-	-
Spain	4 168	4 084	3 061	5 942	3 484	2 505	6 140	5 935	3 914	2 250
UK	-	140	6	16	18	39	24	188	30	17
41 Fishing area total	*5 289*	*4 789*	*3 298*	*9 946*	*7 590*	*6 094*	*11 454*	*15 643*	*14 541*	*5 528*
87 Chile	4 578	2 908	1 800	1 826	481	647	352	245	372	641
87 Fishing area total	*4 578*	*2 908*	*1 800*	*1 826*	*481*	*647*	*352*	*245*	*372*	*641*
Species total	*9 867*	*7 697*	*5 098*	*11 772*	*8 071*	*6 741*	*11 806*	*15 888*	*14 913*	*6 169*
Moras nei	**Mores nca**		**Moras nep**		***Moridae***			**1,48(02)XXX,XX**		**MOR**
27 France	-	-	-	-	-	-	75	67	60	73
Norway	-	-	-	-	-	-	-	-	-	277
UK	-	-	-	-	-	415	0	-	2	-
27 Fishing area total	-	-	-	-	-	*415*	*75*	*67*	*62*	*350*
61 Russian Fed	8 040	8 081	13 811	3 758		152	11 115	31 295	39 316	32 392
61 Fishing area total	*8 040*	*8 081*	*13 811*	*3 758*	-	*152*	*11 115*	*31 295*	*39 316*	*32 392*
Species total	*8 040*	*8 081*	*13 811*	*3 758*	-	*567*	*11 190*	*31 362*	*39 378*	*32 742*

B-32 Cods, hakes, haddocks / Morues, merlus, églefins / Bacalaos, merluzas, eglefinos

Capture production by species, fishing areas and countries or areas / Captures par espèces, zones de pêche et pays ou zones / Capturas por especies, áreas de pesca y países o áreas

Species, Fishing area Espèce, Zone de pêche Especie, Area de pesca	1992 mt	1993 mt	1994 mt	1995 mt	1996 mt	1997 mt	1998 mt	1999 mt	2000 mt	2001 mt
Unicorn cod	**Bregmacère de l'océan Indien**		**Bregmacero**			***Bregmaceros mcclellandi***			**1,48(03)018,01**	**UNC**
51 India	2 300	1 271	753	770	939	971	1 113	458	1 123	2 095
Mauritius	330	302	309	301	312	367	336	285	347	340
51 Fishing area total	*2 630*	*1 573*	*1 062*	*1 071*	*1 251*	*1 338*	*1 449*	*743*	*1 470*	*2 435*
57 India	222	113	152	161	36	13	1 138	1 352	159	32
57 Fishing area total	*222*	*113*	*152*	*161*	*36*	*13*	*1 138*	*1 352*	*159*	*32*
Species total	*2 852*	*1 686*	*1 214*	*1 232*	*1 287*	*1 351*	*2 587*	*2 095*	*1 629*	*2 467*
Tusk(=Cusk)	**Brosme**		**Brosmio**			***Brosme brosme***			**1,48(04)001,01**	**USK**
21 Canada	5 077	2 957	1 692	2 010	1 405	1 801	1 641	1 065	1 097	1 498
Cuba	17	10	-	-	-	-	-	-	-	-
Norway	0	-	-	-	-	-	-	-	-	-
USA	1 581	1 428	1 081	772	468	443	354	230	188	180
21 Fishing area total	*6 675*	*4 395*	*2 773*	*2 782*	*1 873*	*2 244*	*1 995*	*1 295*	*1 285*	*1 678*
27 Denmark	199	134	92	89	130	146	105	177	225	274
Faeroe Is	5 018	3 313	4 678	4 490	2 562	2 593	2 389	2 714	2 585	2 922
France	650	525	465	433	439	439	453	428	335	282
Germany	46	39	42	29	59	26	19	16	13	10
Greenland	-	1	-	20	-	-	-	-	-	-
Iceland	6 440	4 747	4 612	5 245	5 226	4 847	4 118	5 796	4 741	3 425
Ireland	13	67	52	76	64	45	43	43	113	122
Norway	26 143	26 792	20 375	18 682	19 483	13 797	21 032	23 274	21 912	18 778
Portugal	0	0	0	0	0	-	-	-	-	-
Russian Fed	-	-	-	-	-	-	-	-	46	83
Spain	-	-	-	-	62	98	106	150	249	72
Sweden	16	12	12	5	7	3	3	3	8	6
UK	327	294	334	341	391	507	723	852	1 016	872
27 Fishing area total	*38 852*	*35 924*	*30 662*	*29 410*	*28 423*	*22 501*	*28 991*	*33 453*	*31 243*	*26 846*
Species total	*45 527*	*40 319*	*33 435*	*32 192*	*30 296*	*24 745*	*30 986*	*34 748*	*32 528*	*28 524*
Atlantic cod	**Morue de l'Atlantique**		**Bacalao del Atlántico**			***Gadus morhua***			**1,48(04)002,02**	**COD**
21 Canada	187 526	76 556	22 719	12 438	15 541	29 899	37 741	55 410	46 046	40 325
Cuba	92	15	2	1	2	0	1	-	-	-
Estonia	-	-	-	-	-	-	-	-	6	44
Faeroe Is	773	3 049	2 250	1 016	701	-	-	-	-	-
Germany	293	-	-	-	16	-	-	-	-	-
Greenland	5 727	1 925	2 117	1 710	944	904	326	2	764	1 680
Iceland	-	-	93	-	-	-	-	-	-	-
Japan	3	-	-	-	-	-	-	-	-	-
Korea Rep	258	10	-	-	-	-	-	-	-	-
Latvia	31	-	-	-	-	-	-	-	-	-
Norway	2	0	1	-	-	-	-	-	-	-
Portugal	5 986	3 657	2 636	1 669	1 318	1 546	549	327	191	357
Russian Fed	96	298	-	-	-	-	5	26	140	254
St Pier Mq	14 397	103	86	60	44	1 547	3 123	3 171	4 682	2 350
Spain	9 964	6 055	3 735	563	181	2	-	3	5	-
UK	79	-	-	-	129	23	-	-	-	-
USA	27 798	22 908	17 533	13 440	14 253	12 982	11 119	9 727	11 367	15 064
21 Fishing area total	*253 025*	*114 576*	*51 172*	*30 897*	*33 129*	*46 903*	*52 864*	*68 666*	*63 201*	*60 074*
27 Belgium	4 248	4 347	3 611	5 938	4 491	5 677	6 893	4 540	3 693	3 207
Channel Is	2	2 F	3	3 F	6	14	19	21	11	6
Denmark	64 067	47 924	55 221	78 332	90 741	80 491	69 025	70 547	57 018	46 185
Estonia	1 369	70	905	1 049	1 392	1 174	1 070	1 059	514	755
Faeroe Is	19 308	24 731	33 882	44 332	59 803	57 921	39 689	33 725	32 601	38 706
Finland	489	230	529	1 861	3 139	1 543	1 037	1 572	1 824	1 723
France	16 039	19 037	16 607	17 669	21 500	25 123	20 640	14 431	11 886	11 336
Germany	24 954	18 602	22 055	31 892	37 613	26 491	23 075	21 990	18 414	19 222
Greenland	4 706	5 620	6 955	7 493	6 542	6 777	5 272	4 115	2 234	3 934
Iceland	266 684	260 544	214 656	202 900	204 058	208 636	242 968	260 643	238 324	240 002
Ireland	4 837	3 635	4 963	5 650	7 992	5 702	5 294	3 860	2 923	2 647
Isle of Man	96	57	27	22	27	19	34	9	11	1
Japan	7	2	-	-	-	-	-	-	-	-
Latvia	1 250	1 333	2 379	6 471	8 741	6 187	7 778	6 914	6 280	6 298
Lithuania	2 141	574	1 886	3 629	5 520	4 694	3 296	4 371	4 721	3 852
Netherlands	11 143	10 220	6 512	11 189	9 307	11 838	14 724	9 075	6 000	3 656
Norway	219 094	275 238	373 577	365 333	358 395	401 277	321 428	256 555	220 120	208 856
Poland	13 314	8 909	14 426	25 001	35 968	34 295	27 705	28 056	23 340	23 310
Portugal	1 134	1 788	5 644	5 654	6 765	7 533	5 493	3 885	3 587	4 027
Russian Fed	183 325	250 567	319 679	297 770	309 391	316 147	248 714	215 590	170 878	188 630
Spain	6 217	8 800	14 929	15 580	16 175	17 226	14 392	10 156	8 765	20 283
Sweden	22 361	17 971	30 986	33 186	41 827	34 797	22 475	22 597	23 174	24 111
UK	65 207	67 289	68 855	78 650	78 235	74 614	77 182	51 695	41 750	32 840
27 Fishing area total	*931 992*	*1 027 490 F*	*1 198 287*	*1 239 604 F*	*1 307 628*	*1 328 176*	*1 158 203*	*1 025 406*	*878 068*	*883 587*
Species total	*1 185 017*	*1 142 066 F*	*1 249 459*	*1 270 501 F*	*1 340 757*	*1 375 079*	*1 211 067*	*1 094 072*	*941 269*	*943 661*
Pacific cod	**Morue du Pacifique**		**Bacalao del Pacífico**			***Gadus macrocephalus***			**1,48(04)002,11**	**PCO**
61 Japan	75 690	62 286	65 778	56 561	57 576	58 477	57 243	55 292	51 052	43 550
Korea Rep	3 966	11 626	3 702	2 476	2 740	3 984	6 249	6 509	10 098	13 110

B-32 Cods, hakes, haddocks / Morues, merlus, églefins / Bacalaos, merluzas, eglefinos

Capture production by species, fishing areas and countries or areas / Captures par espèces, zones de pêche et pays ou zones / Capturas por especies, áreas de pesca y países o áreas

Species, Fishing area Espèce, Zone de pêche Especie, Area de pesca	1992 mt	1993 mt	1994 mt	1995 mt	1996 mt	1997 mt	1998 mt	1999 mt	2000 mt	2001 mt
Russian Fed	154 297	95 823	81 445	100 730	93 711	79 927	94 282	101 929	68 415	59 783
61 Fishing area total	*233 953*	*169 735*	*150 925*	*159 767*	*154 027*	*142 388*	*157 774*	*163 730*	*129 565*	*116 443*
67 Canada	10 125	8 123	3 550	2 172	700	1 537	1 400	836	712	474
Japan	-	200	220	-	-	-	-	-	-	-
Russian Fed	-	-	-	-	179	-	-	-	-	-
USA	249 717	218 996	208 785	268 257	274 569	299 970	252 197	237 679	240 635	213 967
67 Fishing area total	*259 842*	*227 319*	*212 555*	*270 429*	*275 448*	*301 507*	*253 597*	*238 515*	*241 347*	*214 441*
Species total	*493 795*	*397 054*	*363 480*	*430 196*	*429 475*	*443 895*	*411 371*	*402 245*	*370 912*	*330 884*
Greenland cod	**Morue ogac**		**Bacalao de Groenlandia**		***Gadus ogac***			**1,48(04)002,12**		**GRC**
21 Canada	0	0	0	0	-	-	-	-	-	-
Greenland	1 778	1 892	1 854	2 525	2 120	1 728	1 695	1 899	931	1 128
21 Fishing area total	*1 778*	*1 892*	*1 854*	*2 525*	*2 120*	*1 728*	*1 695*	*1 899*	*931*	*1 128*
27 Greenland	-	-	4	3	1	1	2	4	-	5
27 Fishing area total	-	-	*4*	*3*	*1*	*1*	*2*	*4*	-	*5*
Species total	*1 778*	*1 892*	*1 858*	*2 528*	*2 121*	*1 729*	*1 697*	*1 903*	*931*	*1 133*
Ling	**Lingue**		**Maruca**		***Molva molva***			**1,48(04)005,01**		**LIN**
27 Belgium	180	168	171	181	153	124	124	103	120	88
Channel Is	16	16 F	28	26 F	20	37	25	19	13	3
Denmark	1 199	1 237	950	790	868	969	823	837	741	910
Faeroe Is	2 361	2 047	2 799	3 686	3 132	4 056	3 547	2 998	2 358	2 558
France	7 526	5 158	5 590	5 644	5 738	5 463	5 510	5 112	3 099	2 987
Germany	223	973	1 041	877	1 409	965	308	247	215	110
Iceland	4 556	4 333	4 049	3 729	3 670	3 634	3 603	3 976	3 223	2 864
Ireland	452	880	1 158	1 542	1 379	1 305	1 272	1 138	1 089	1 463
Isle of Man	1	2	2	1	3	2	1	1	1	0
Netherlands	-	-	-	-	-	-	-	-	5	4
Norway	19 369	18 351	17 907	18 172	18 931	15 295	22 719	19 217	16 899	13 562
Poland	-	-	-	-	-	-	-	-	-	19
Portugal	7	-	-	-	-	-	-	-	-	-
Russian Fed	-	-	-	-	-	-	-	-	8	2
Spain	3 187	3 233	4 724	5 026	5 301	6 153	9 256	8 907	6 259	4 276
Sweden	128	144	120	94	73	61	44	44	46	47
UK	7 919	10 747	11 752	14 695	13 942	12 924	13 594	11 350	9 244	8 095
27 Fishing area total	*47 124*	*47 289 F*	*50 291*	*54 463 F*	*54 619*	*50 988*	*60 826*	*53 949*	*43 320*	*36 988*
Species total	*47 124*	*47 289 F*	*50 291*	*54 463 F*	*54 619*	*50 988*	*60 826*	*53 949*	*43 320*	*36 988*
Blue ling	**Lingue bleue**		**Maruca azul**		***Molva dypterygia***			**1,48(04)005,02**		**BLI**
21 Faeroe Is	-	1	3	-	-	-	-	-	-	-
21 Fishing area total	-	*1*	*3*	-	-	-	-	-	-	-
27 Denmark	30	18	15	16	8	14	4	7	15	27
Estonia	-	-	-	-	-	-	-	-	-	85
Faeroe Is	4 326	3 116	1 781	2 398	1 624	1 172	1 274	2 136	1 756	2 454
France	4 229	4 772	3 278	3 590	4 127	4 736	5 955	4 794	5 571	3 666
Germany	63	298	176	202	119	11	16	15	110	26
Greenland	5	3	-	2	-	-	-	-	-	-
Iceland	2 584	5 317	1 842	1 636	1 284	1 320	1 208	2 321	1 623	765
Ireland	-	3	74	14	-	1	22	43	91	827
Lithuania	-	-	-	-	-	-	-	-	-	16
Norway	2 140	1 602	973	734	530	497	420	544	834	1 020
Portugal	31	33	42	29	25	21	14	10	14	9
Spain	-	-	-	-	298	1 165	1 404	1 710	3 113	4 472
Sweden	1	1	0	0	0	2	0	-	-	-
UK	131	323	258	1 121	1 772	1 030	2 450	4 119	3 019	5 980
27 Fishing area total	*13 540*	*15 486*	*8 439*	*9 742*	*9 787*	*9 969*	*12 767*	*15 699*	*16 146*	*19 347*
37 Spain	-	-	-	-	1	1	1	-	-	-
37 Fishing area total	-	-	-	-	*1*	*1*	*1*	-	-	-
Species total	*13 540*	*15 487*	*8 442*	*9 742*	*9 788*	*9 970*	*12 768*	*15 699*	*16 146*	*19 347*
Greater forkbeard	**Phycis de fond**		**Brótola de fango**		***Phycis blennoides***			**1,48(04)006,01**		**GFB**
27 France	684	637	524	499	562	605	476	526	729	743
Germany	-	-	-	-	-	-	-	1	8	12
Iceland	-	-	1	0	-	-	-	-	-	-
Ireland	5	60	111	163	154	228	318	379	399	679
Norway	-	-	-	-	-	-	-	-	-	1 340
Portugal	82	116	135	79	47	31	44	52	98	89
Russian Fed	-	-	-	-	-	-	-	-	2	11
Spain	-	-	-	-	1 601	5 065	4 505	3 504	2 369	1 175
UK	77	345	423	1 044	2 022	2 054	1 887	1 495	1 563	1 204
27 Fishing area total	*848*	*1 158*	*1 194*	*1 785*	*4 386*	*7 983*	*7 230*	*5 957*	*5 168*	*5 253*
34 Morocco	357	279	407	397	400	100	230	325	269	350
Portugal	-	-	6	50	66	15	1	2	0	3
Russian Fed	-	1	-	-	-	-	-	-	-	-
Spain	2	-	-	-	-	-	-	42	-	-
34 Fishing area total	*359*	*280*	*413*	*447*	*466*	*115*	*231*	*369*	*269*	*353*

B-32 Cods, hakes, haddocks / Morues, merlus, églefins / Bacalaos, merluzas, eglefinos

Capture production by species, fishing areas and countries or areas
Captures par espèces, zones de pêche et pays ou zones
Capturas por especies, áreas de pesca y países o áreas

Species, Fishing area Espèce, Zone de pêche Especie, Area de pesca	1992 mt	1993 mt	1994 mt	1995 mt	1996 mt	1997 mt	1998 mt	1999 mt	2000 mt	2001 mt
37 Algeria	...	...	...	...	...	...	...	4	4 F	4 F
France	...	...	...	...	...	...	...	2	4	3
Malta	-	-	-	-	-	2	3	4	5	0
Morocco	69	48	24	31	40	45	27	20	365	26
Spain	350 F	300 F	250 F	200 F	140	144	166	129	-	6
Tunisia	127	-	-	-	-	-	-	-	-	-
Turkey	1	13	7	15	24	50	150	50	50	35
37 Fishing area total	*547 F*	*361 F*	*281 F*	*246 F*	*204*	*241*	*346*	*209*	*428 F*	*74 F*
Species total	*1 754 F*	*1 799 F*	*1 888 F*	*2 478 F*	*5 056*	*8 339*	*7 807*	*6 535*	*5 865 F*	*5 680 F*
Brazilian codling	**Phycis brésilien**		**Brótola brasileña**			***Urophycis brasiliensis***			**1,48(04)008,01**	**HKU**
41 Argentina	2 122	2 095	1 440	416	618	182	3 329	754	1 036	...
Brazil	1 000 F	1 000 F	1 000 F	2 924	2 311	2 116	2 408	1 807	2 225	2 200 F
Portugal	-	-	-	-	-	-	-	-	-	3
Russian Fed	5	-	-	-	-	-	-	-	-	-
Uruguay	261	357	280	363	223	272	344	281	235	361
41 Fishing area total	*3 388 F*	*3 452 F*	*2 720 F*	*3 703*	*3 152*	*2 570*	*6 081*	*2 842*	*3 496*	*2 564 F*
Species total	*3 388 F*	*3 452 F*	*2 720 F*	*3 703*	*3 152*	*2 570*	*6 081*	*2 842*	*3 496*	*2 564 F*
Red hake	**Merluche écureuil**		**Locha roja**			***Urophycis chuss***			**1,48(04)008,02**	**HKR**
21 Canada	338	173	100	135	372	248	120	156	81	130
Cuba	244	171	72	170	430	259	118	86	0	...
Portugal	467	366	267	230	125	56	18	77	42	273
Russian Fed	114	-	-	-	-	-	4	2	120	118
Spain	-	-	54	112	893	958	1 200	1 350	1 551	1 755
USA	2 157	1 643	1 701	1 607	1 087	1 329	1 343	1 556	1 571	1 732
21 Fishing area total	*3 320*	*2 353*	*2 194*	*2 254*	*2 907*	*2 850*	*2 803*	*3 227*	*3 365*	*4 008*
Species total	*3 320*	*2 353*	*2 194*	*2 254*	*2 907*	*2 850*	*2 803*	*3 227*	*3 365*	*4 008*
White hake	**Merluche blanche**		**Locha blanca**			***Urophycis tenuis***			**1,48(04)008,03**	**HKW**
21 Canada	12 975	10 209	7 348	6 492	4 827	4 309	3 102	3 318	4 246	4 140
Cuba	1	0	37	-	-	-	-	-	-	-
Estonia	-	-	-	-	-	-	-	-	3	2
St Pier Mq	36	-	-	-	-	0	1	9	122	10
Spain	-	-	28	39	187	304	491	426	802	689
USA	8 993	7 460	4 695	4 279	3 289	2 217	2 365	2 624	2 984	3 482
21 Fishing area total	*22 005*	*17 669*	*12 108*	*10 810*	*8 303*	*6 830*	*5 959*	*6 377*	*8 157*	*8 323*
31 USA	...	...	...	...	...	...	...	1	1	1
31 Fishing area total	...	...	...	...	...	...	...	*1*	*1*	*1*
Species total	*22 005*	*17 669*	*12 108*	*10 810*	*8 303*	*6 830*	*5 959*	*6 378*	*8 158*	*8 324*
Haddock	**Eglefin**		**Eglefino**			***Melanogrammus aeglefinus***			**1,48(04)010,01**	**HAD**
21 Canada	21 975	12 986	6 955	7 933	10 311	9 776	11 771	10 550	12 683	15 594
Cuba	83	68	12	32	40	27	12	4	-	-
Faeroe Is	-	-	1	-	-	-	-	-	-	-
Japan	2	-	-	-	-	-	-	-	-	-
Portugal	165	10	10	2	-	39	6	10	13	23
Russian Fed	50	27	-	-	-	-	1	-	2	33
St Pier Mq	159	-	-	-	-	9	27	16	10	78
Spain	-	-	-	-	-	4	0	-	-	-
USA	2 318	879	328	398	570	1 504	2 836	3 146	4 002	5 826
21 Fishing area total	*24 752*	*13 970*	*7 306*	*8 365*	*10 921*	*11 359*	*14 653*	*13 726*	*16 710*	*21 554*
27 Belgium	820	835	706	648	394	746	976	569	512	840
Channel Is	...	...	...	...	44	0	0	-	-	-
Denmark	5 292	5 204	4 768	4 479	5 050	5 227	5 786	3 130	2 707	4 001
Faeroe Is	6 856	5 143	7 759	8 494	13 896	21 409	22 598	19 697	16 212	16 061
France	3 128	4 597	4 187	4 167	6 027	8 060	4 983	3 582	4 379	5 970
Germany	1 044	1 522	4 259	3 978	2 718	2 441	1 712	1 039	1 225	1 368
Greenland	1 031	880	770	1 351	1 524	1 876	762	...	176	547
Iceland	46 098	46 932	58 426	60 125	56 223	43 256	40 712	44 729	41 698	39 825
Ireland	2 594	3 118	2 860	3 417	4 462	6 234	6 572	4 898	5 812	5 404
Isle of Man	13	18	24	27	38	9	13	7	19	1
Netherlands	151	192	95	146	111	494	289	115	121	295
Norway	39 937	43 931	73 791	79 834	97 115	106 155	79 008	53 243	45 920	51 648
Poland	-	-	-	-	18	35	27	24	16	96
Portugal	-	583	755	605	208	168	49	38	131	105
Russian Fed	19 707	35 071	51 822	54 516	73 857	41 228	20 559	30 978	24 892	34 937
Spain	89	76	22	62	718	501	541	780	669	2 217
Sweden	2 033	1 344	959	1 265	1 226	1 519	1 013	895	964	1 087
UK	54 270	87 625	93 698	86 315	87 420	83 388	83 436	72 001	50 644	42 865
27 Fishing area total	*183 063*	*237 071*	*304 901*	*309 429*	*351 049*	*322 746*	*269 036*	*235 725*	*196 097*	*207 267*
Species total	*207 815*	*251 041*	*312 207*	*317 794*	*361 970*	*334 105*	*283 689*	*249 451*	*212 807*	*228 821*
Navaga(=Wachna cod)	**Morue arctique**		**Bacalao navaga**			***Eleginus navaga***			**1,48(04)012,01**	**COW**
05 Russian Fed	-	-	112	193	98	163	177	22	11	55
05 Fishing area total	-	-	*112*	*193*	*98*	*163*	*177*	*22*	*11*	*55*

B-32 Cods, hakes, haddocks / Morues, merlus, églefins / Bacalaos, merluzas, eglefinos

Capture production by species, fishing areas and countries or areas / Captures par espèces, zones de pêche et pays ou zones / Capturas por especies, áreas de pesca y países o áreas

Species, Fishing area Espèce, Zone de pêche Especie, Area de pesca	1992 mt	1993 mt	1994 mt	1995 mt	1996 mt	1997 mt	1998 mt	1999 mt	2000 mt	2001 mt
27 Russian Fed	3 261	624	1 106	821	561	989	1 008	458	663	1 111
27 Fishing area total	*3 261*	*624*	*1 106*	*821*	*561*	*989*	*1 008*	*458*	*663*	*1 111*
Species total	*3 261*	*624*	*1 218*	*1 014*	*659*	*1 152*	*1 185*	*480*	*674*	*1 166*
Saffron cod	**Morne boréale**		**Bacalao del Artico**		***Eleginus gracilis***			**1,48(04)012,02**		**SAF**
61 Russian Fed	50 624	43 079	13 213	25 566	21 110	27 803	40 426	47 032	35 763	33 753
61 Fishing area total	*50 624*	*43 079*	*13 213*	*25 566*	*21 110*	*27 803*	*40 426*	*47 032*	*35 763*	*33 753*
Species total	*50 624*	*43 079*	*13 213*	*25 566*	*21 110*	*27 803*	*40 426*	*47 032*	*35 763*	*33 753*
Pacific tomcod	**Poulamon du Pacifique**		**...C**		***Microgadus proximus***			**1,48(04)013,01**		**MGX**
67 USA	...	...	...	...	...	...	3	1	-	0
67 Fishing area total	...	...	...	...	...	...	*3*	*1*	-	*0*
77 USA	...	...	...	...	...	...	0	-	0	-
77 Fishing area total	...	...	...	...	...	...	*0*	-	*0*	-
Species total	...	...	...	...	...	...	*3*	*1*	*0*	*0*
Atlantic tomcod	**Poulamon atlantique**		**Microgado**		***Microgadus tomcod***			**1,48(04)013,02**		**TOM**
02 Canada	26	20	21	22	23	24	25	25	25	...
02 Fishing area total	*26*	*20*	*21*	*22*	*23*	*24*	*25*	*25*	*25*	...
21 Canada	20	74	40	109	117	66	94	17	30	57
21 Fishing area total	*20*	*74*	*40*	*109*	*117*	*66*	*94*	*17*	*30*	*57*
Species total	*46*	*94*	*61*	*131*	*140*	*90*	*119*	*42*	*55*	*57*
Saithe(=Pollock)	**Lieu noir**		**Carbonero(=Colín)**		***Pollachius virens***			**1,48(04)015,01**		**POK**
21 Canada	34 068	21 583	15 584	10 222	9 739	12 604	15 092	8 552	6 528	7 190
Cuba	966	605	13	61	124	57	8	6	-	-
Japan	61	-	-	-	-	-	-	-	-	-
Portugal	29	41	13	-	-	-	-	-	-	-
Russian Fed	869	112	-	-	-	-	1	-	-	-
St Pier Mq	91	-	-	-	-	14	13	6	38	13
Spain	-	-	-	-	-	6	-	-	-	-
USA	7 186	5 670	3 737	3 244	2 962	4 251	5 583	4 595	4 043	4 109
21 Fishing area total	*43 270*	*28 011*	*19 347*	*13 527*	*12 825*	*16 932*	*20 697*	*13 159*	*10 609*	*11 312*
27 Belgium	269	223	169	236	161	264	256	208	126	30
Channel Is	1	1 F	0	0	2	4	0	2	-	-
Denmark	4 834	4 314	4 331	4 395	4 708	4 517	3 973	4 501	3 536	3 592
Estonia	-	-	-	-	-	16	-	-	-	-
Faeroe Is	40 195	37 189	35 218	31 979	20 398	22 598	26 751	34 423	35 997	45 792
France	21 777	33 472	29 317	19 882	19 598	17 802	18 218	24 638	26 941	28 533
Germany	17 407	18 762	12 405	13 393	15 197	15 993	13 562	13 307	12 385	13 320
Greenland	734	78	15	53	165	318	437	...	601	1 526
Iceland	77 832	69 985	63 333	47 466	39 297	36 548	30 532	30 729	32 947	31 941
Ireland	1 998	2 329	2 355	2 929	2 579	1 841	1 687	1 704	2 848	2 048
Isle of Man	8	5	4	11	11	9	7	2	1	0
Japan	16	4	-	-	-	-	-	-	-	-
Netherlands	179	78	17	9	19	42	8	7	11	19
Norway	168 161	188 364	188 869	218 853	221 638	183 451	194 452	198 387	169 653	169 506
Poland	1 238	937	151	592	365	822	813	862	747	727
Portugal	0	1	0	5	24	13	49	37	64	86
Romania	-	-	31	-	-	-	-	-	-	-
Russian Fed	976	9 509	1 223	1 148	1 177	1 802	3 836	3 932	4 564	4 953
Spain	6	-	1	13	33	77	397	82	158	152
Sweden	3 302	4 955	5 366	1 998	1 773	1 649	1 857	1 929	1 468	1 628
UK	15 783	15 278	14 678	15 220	15 413	14 609	12 261	12 442	10 924	10 585
27 Fishing area total	*354 716*	*385 484 F*	*357 483*	*358 182*	*342 558*	*302 375*	*309 096*	*327 192*	*302 971*	*314 438*
Species total	*397 986*	*413 495 F*	*376 830*	*371 709*	*355 383*	*319 307*	*329 793*	*340 351*	*313 580*	*325 750*
Pollack	**Lieu jaune**		**Abadejo**		***Pollachius pollachius***			**1,48(04)015,02**		**POL**
27 Belgium	120	109	144	158	115	119	113	108	116	137
Channel Is	24	24 F	24	22 F	27	35	52	75	97	57
Denmark	2 030	2 295	1 301	1 036	1 049	637	564	480	490	358
Faeroe Is	-	5	2	2	-	1	-	-	-	-
France	4 628	3 837	4 554	3 843	3 881	3 626	3 359	2 935	3 775	3 649
Germany	60	167	-	87	102	117	43	63	39	41
Ireland	1 038	1 149	947	1 190	1 288	1 052	946	1 049	24	1 382
Isle of Man	4	1	1	15	16	11	11	2	1	-
Netherlands	7	18	14	17	19	15	7	5	5	1
Norway	2 517	3 274	2 473	3 071	2 318	2 230	2 247	2 928	3 385	2 888
Portugal	2	1	3	2	2	2	1	1	15	41
Spain	238	189	193	216	185	213	218	175	175	436
Sweden	461	659	350	510	355	261	180	160	124	108
UK	3 378	3 966	3 531	4 011	3 631	3 845	3 211	2 398	2 706	2 839
27 Fishing area total	*14 507*	*15 694 F*	*13 537*	*14 180 F*	*12 988*	*12 164*	*10 952*	*10 379*	*10 952*	*11 937*
Species total	*14 507*	*15 694 F*	*13 537*	*14 180 F*	*12 988*	*12 164*	*10 952*	*10 379*	*10 952*	*11 937*

B-32

Cods, hakes, haddocks — Capture production by species, fishing areas and countries or areas
Morues, merlus, églefins — Captures par espèces, zones de pêche et pays ou zones
Bacalaos, merluzas, eglefinos — Capturas por especies, áreas de pesca y países o áreas

Species, Fishing area Espèce, Zone de pêche Especie, Area de pesca	1992 mt	1993 mt	1994 mt	1995 mt	1996 mt	1997 mt	1998 mt	1999 mt	2000 mt	2001 mt
Alaska pollock(=Walleye poll.)	**Lieu de l'Alaska**		**Colín de Alaska**			***Theragra chalcogramma***			**1,48(04)016,01**	**ALK**
61 China	131 727	135 000	130 174	189 459	166 900	258 478	141 433	64 520	60 338	39 665
China,Taiwan	1	2	9	37	12	7	9	9	9	-
Japan	498 756	382 293	379 338	338 507	331 163	338 785	315 987	382 385	300 001	241 881
Korea D P Rp	...	...	75 065	120 219	15 369	66 578	60 000 F	55 000 F	52 000 F	52 000 F
Korea Rep	254 381	181 288	296 932	334 921	224 430	223 065	236 278	146 165	86 143	197 396
Poland	297 732	235 208	269 979	249 257	116 257	125 413	81 889	65 508	33 192	16 590
Russian Fed	2 340 700	2 114 456	1 746 629	2 208 410	2 439 651	2 252 742	1 930 650	1 500 450	1 215 065	1 145 016
61 Fishing area total	*3 523 297*	*3 048 247*	*2 898 126*	*3 440 810*	*3 293 782*	*3 265 068*	*2 766 246 F*	*2 214 037 F*	*1 746 748 F*	*1 692 548 F*
67 Canada	3 249	8 121	4 706	3 295	2 150	1 800	800	1 233	1 044	1 747
China	40 000	40 000	40 000	60 000	60 000	80 000	50 000	...	...	...
Japan	-	15	13	-	-	-	-	-	-	-
Korea Rep	81 041	44 810	14 642	10 967	2 480	-	-	-	-	-
Russian Fed	-	-	-	-	329	-	-	-	-	-
USA	1 339 077	1 477 815	1 417 278	1 293 939	1 189 844	1 139 642	1 232 177	1 055 016	1 182 438	1 442 170
67 Fishing area total	*1 463 367*	*1 570 761*	*1 476 639*	*1 368 201*	*1 254 803*	*1 221 442*	*1 282 977*	*1 056 249*	*1 183 482*	*1 443 917*
Species total	*4 986 664*	*4 619 008*	*4 374 765*	*4 809 011*	*4 548 585*	*4 486 510*	*4 049 223 F*	*3 270 286 F*	*2 930 230 F*	*3 136 465 F*
Polar cod	**Morue polaire**		**Bacalao polar**			***Boreogadus saida***			**1,48(04)019,01**	**POC**
21 Greenland	2	0	1	-	4	-	-	-	-	-
21 Fishing area total	*2*	*0*	*1*	-	*4*	-	-	-	-	-
27 Germany	-	-	58	-	-	-	-	-	-	-
Russian Fed	20 768	50 638	6 126	24 030	20 784	6 826	3 592	22 005	40 730	39 445
27 Fishing area total	*20 768*	*50 638*	*6 184*	*24 030*	*20 784*	*6 826*	*3 592*	*22 005*	*40 730*	*39 445*
61 Russian Fed	-	35	-	2 400	-	-	-	-	13	-
61 Fishing area total	-	*35*	-	*2 400*	-	-	-	-	*13*	-
Species total	*20 770*	*50 673*	*6 185*	*26 430*	*20 788*	*6 826*	*3 592*	*22 005*	*40 743*	*39 445*
Rocklings nei	**Motelles nca**		**Barbadas nep**			***Gaidropsarus spp***			**1,48(04)028,XX**	**ROL**
27 France	-	-	-	-	-	-	-	4	24	1
27 Fishing area total	-	-	-	-	-	-	-	*4*	*24*	*1*
37 France	...	...	...	...	...	...	...	20	19	19
37 Fishing area total	...	...	...	...	...	...	...	*20*	*19*	*19*
Species total	...	...	...	...	...	...	...	*24*	*43*	*20*
Norway pout	**Tacaud norvégien**		**Faneca noruega**			***Trisopterus esmarkii***			**1,48(04)032,01**	**NOP**
27 Denmark	266 933	190 071	166 928	262 515	162 943	153 047	63 678	57 441	150 040	62 913
Faeroe Is	22 416	31 471	32 358	8 960	9 133	11 215	6 222	4 045	1 754	2 429
Germany	-	3	-	38	-	-	-	-	2	-
Iceland	-	-	-	0	0	-	-	-	-	160
Ireland	0	0	0	0	-	-	-	-	1	-
Netherlands	110	-	24	138	13	85	3	1	3	-
Norway	163 460	102 766	91 694	118 081	103 126	47 032	27 575	51 124	52 912	27 123
Sweden	5	-	-	68	237	2	-	-	133	780
UK	2	7	1	0	215	13	-	2	0	0
27 Fishing area total	*452 926*	*324 318*	*291 005*	*389 800*	*275 667*	*211 394*	*97 478*	*112 613*	*204 845*	*93 405*
Species total	*452 926*	*324 318*	*291 005*	*389 800*	*275 667*	*211 394*	*97 478*	*112 613*	*204 845*	*93 405*
Poor cod	**Capelan de Méditerranée**		**Capellán**			***Trisopterus minutus***			**1,48(04)032,02**	**POD**
37 France	1 047	1 084	711	586	603	428	428	637	888	754
Slovenia	-	-	-	-	-	-	-	-	2	1
37 Fishing area total	*1 047*	*1 084*	*711*	*586*	*603*	*428*	*428*	*637*	*890*	*755*
Species total	*1 047*	*1 084*	*711*	*586*	*603*	*428*	*428*	*637*	*890*	*755*
Pouting(=Bib)	**Tacaud commun**		**Faneca**			***Trisopterus luscus***			**1,48(04)032,03**	**BIB**
27 Belgium	489	451	400	305	377	336	323	364	468	561
Channel Is	0	0	0	0	1	1	0	1	5	6
Denmark	1	1	2	1	1	2	2	1	1	3
France	7 869	6 727	6 539	5 678	6 119	6 283	6 108	6 333	7 619	7 293
Ireland	6	3	5	5	2	12	1	21	10	28
Netherlands	-	-	-	-	-	-	-	-	612	645
Portugal	4 300	3 863	2 981	3 070	2 491	2 051	2 254	2 792	3 299	4 511
Spain	4 207	3 714	4 112	4 656	2 515	3 254	1 875	2 088	2 689	2 587
UK	1 394	1 286	1 191	919	962	813	554	458	885	899
27 Fishing area total	*18 266*	*16 045*	*15 230*	*14 634*	*12 468*	*12 752*	*11 117*	*12 058*	*15 588*	*16 533*
34 Morocco	1 907	1 926	1 795	2 083	2 100	1 073	1 243	573	1 216	1 013
Portugal	15	11	2	0	0	0	-	-	-	-
34 Fishing area total	*1 922*	*1 937*	*1 797*	*2 083*	*2 100*	*1 073*	*1 243*	*573*	*1 216*	*1 013*
37 Morocco	3	1	2	5	5	5	8	21	1 001	4
Spain	-	-	-	-	308	383	394	648	946	653
37 Fishing area total	*3*	*1*	*2*	*5*	*313*	*388*	*402*	*669*	*1 947*	*657*

B-32

Cods, hakes, haddocks — Capture production by species, fishing areas and countries or areas
Morues, merlus, églefins — Captures par espèces, zones de pêche et pays ou zones
Bacalaos, merluzas, eglefinos — Capturas por especies, áreas de pesca y países o áreas

Species, Fishing area Espèce, Zone de pêche Especie, Area de pesca	1992 mt	1993 mt	1994 mt	1995 mt	1996 mt	1997 mt	1998 mt	1999 mt	2000 mt	2001 mt
Species total	*20 191*	*17 983*	*17 029*	*16 722*	*14 881*	*14 213*	*12 762*	*13 300*	*18 751*	*18 203*
Blue whiting(=Poutassou)	**Merlan bleu**		**Bacaladilla**			**Micromesistius poutassou**			**1,48(04)033,01**	**WHB**
27 Channel Is	-	-	-	-	-	1	1	1	-	-
Denmark	43 966	69 378	22 835	46 182	52 699	33 486	69 305	79 810	62 074	65 058
Estonia	6 156	1 077	4 342	13 715	10 982	5 678	6 321	0	-	-
Faeroe Is	12 731	14 984	24 404	25 936	21 483	28 773	71 217	105 106	152 687	259 761
France	...	...	...	6	6 442	12 446	7 992	6 343	16 042	19 054
Germany	1 320	100	5 919	6 314	6 865	4 722	17 970	3 170	12 654	19 059
Greenland	3	0	-	-	-	-	-	-	-	-
Iceland	-	-	-	369	513	10 480	68 514	160 424	259 157	365 101
Ireland	781	0	3	222	1 709	25 987	45 538	33 687	26 067	29 910
Japan	906	2 002	3 248	1 127	-	-	-	-	-	-
Latvia	10 742	10 328	2 582	-	-	-	-	-	-	-
Lithuania	13 809	2 418	-	400	651	-	-	1 231	-	-
Netherlands	11 037	18 481	21 076	22 685	16 407	24 132	27 693	32 889	43 145	63 625
Norway	154 555	199 981	226 235	261 362	356 054	348 268	570 665	534 570	553 478	573 686
Portugal	2 163	1 222	1 987	2 346	3 565	2 448	1 900	2 676	2 169	1 763
Russian Fed	159 365	137 796	123 258	93 824	87 310	118 656	130 042	182 637	241 905	315 586
Spain	31 965	34 256	30 506	33 397	30 262	37 900	30 549	30 926	28 000	28 822
Sweden	2 058	37 265	3 705	13 000	4 038	4 568	6 034	15 511	3 362	2 058
UK	7 472	2 294	4 470	5 495	14 326	33 701	98 936	106 491	45 048	51 889
27 Fishing area total	*459 029*	*531 582*	*474 570*	*526 380*	*613 306*	*691 246*	*1 152 677*	*1 295 472*	*1 445 788*	*1 795 372*
34 Greece	-	-	-	1	1	-	-	-	-	-
Portugal	10	-	-	-	-	-		-	-	-
Spain	-	-	-	-	310	68	48	-	-	-
34 Fishing area total	*10*	-	-	*1*	*311*	*68*	*48*	-	-	-
37 Albania	...	...	...	0	2	0	-	-	-	-
France	13	48	26	...	21	21	21	29	34	24
Greece	1 449	1 858	2 281	1 944	1 226	1 558	846	630	566	471
Italy	3 185	2 752	1 998	1 769	1 546	1 300	1 449	1 451	1 261	1 167
Spain	4 400 F	4 450 F	4 500 F	4 500 F	4 541	3 086	2 762	4 445	6 276	5 461
Turkey	7 168	9 734	10 598	9 716	11 518	15 000	27 200	16 975	18 180	20 810
37 Fishing area total	*16 215 F*	*18 842 F*	*19 403 F*	*17 929 F*	*18 854*	*20 965*	*32 278*	*23 530*	*26 317*	*27 933*
Species total	*475 254 F*	*550 424 F*	*493 973 F*	*544 310 F*	*632 471*	*712 279*	*1 185 003*	*1 319 002*	*1 472 105*	*1 823 305*
Southern blue whiting	**Merlan bleu austral**		**Polaca austral**			**Micromesistius australis**			**1,48(04)033,02**	**POS**
41 Argentina	85 549	109 829	86 084	104 208	85 040	79 945	71 643	55 097	61 313	54 005
Australia	-	-	-	-	-	-	23	165	-	-
Belize	-	-	-	-	-	-	-	-	257	206
Bulgaria	9 377	3 138	-	-	-	-	-	-	-	-
Chile	-	-	-	-	-	3 744	14 215	5 036	2 726	6 709
Falkland Is	1	74	279	1 616	1 083	727	1 977	2 127	2 704	4 581
France	-	-	23	0	67	-	-	-	-	-
Honduras	43	19	7	-	3	-	-	-	-	-
Japan	...	8 797	22 047	12 872	12 493	16 340	16 935	18 028	14 121	8 918
Latvia	-	298	-	-	2	-	-	-	-	-
Lithuania	9	-	-	-	-	-	-	-	-	-
Namibia	-	-	-	-	-	83	282	29	-	-
Panama	24	-	1	-	-	-	-	-	-	-
Poland	14 932	8 351	10 553	8 923	3 402	-	-	-	-	-
Portugal	1	-	124	-	-	-	-	-	1	-
Russian Fed	9 671	306	-	-	-	-	-	-	-	30
Spain	9 862	4 553	5 259	10 711	2 471	1 591	3 435	3 128	3 346	5 243
Ukraine	3 129	-	-	-	-	-	-	-	-	-
UK	-	2	4	30	108	20	48	85	22	30
Uruguay	-	-	-	-	-	-	-	-	9	0
41 Fishing area total	*132 598*	*135 367*	*124 381*	*138 360*	*104 669*	*102 450*	*108 558*	*83 695*	*84 499*	*79 722*
81 Japan	34 148	20 455	11 068	17 333	19 197	20 173	23 374	22 827	17 096	23 956
New Zealand	35 906	9 589	4 620	11 357	2 753	10 234	35 059	39 012	23 000	29 789
Poland	245	-	-	-	-	-	-	-	-	-
Russian Fed	18 337	2 254	1 950	22	377	610	-	-	-	-
Ukraine	3 533	145	934	3 610	3 713	3 607	7 730	8 306	3 502	267
81 Fishing area total	*92 169*	*32 443*	*18 572*	*32 322*	*26 040*	*34 624*	*66 163*	*70 145*	*43 598*	*54 012*
87 Chile	5 149	27 607	4 664	20 917	25 445	29 131	26 642	31 470	24 733	22 046
87 Fishing area total	*5 149*	*27 607*	*4 664*	*20 917*	*25 445*	*29 131*	*26 642*	*31 470*	*24 733*	*22 046*
Species total	*229 916*	*195 417*	*147 617*	*191 599*	*156 154*	*166 205*	*201 363*	*185 310*	*152 830*	*155 780*
Whiting	**Merlan**		**Plegonero**			**Merlangius merlangus**			**1,48(04)034,01**	**WHG**
27 Belgium	1 460	1 319	1 490	1 250	1 281	989	856	1 072	826	732
Channel Is	0	0	1	1 F	1	0	3	2	3	3
Denmark	2 469	2 126	1 410	789	391	196	144	175	326	326
Faeroe Is	656	601	815	966	1 042	1 018	1 724	1 756	1 593	1 289
France	23 811	28 531	32 355	27 821	21 315	21 590	20 074	21 570	19 026	19 414
Germany	1 399	1 216	1 075	1 186	711	276	191	371	754	680
Iceland	510	230	315	560	430	443	531	931	1 349	1 179
Ireland	6 082	7 118	8 736	11 262	10 340	9 394	7 762	7 643	6 505	6 621
Isle of Man	44	55	44	41	28	24	33	5	2	1
Netherlands	5 390	4 799	3 863	3 640	3 411	2 554	1 981	1 806	1 899	2 619

B-32

Cods, hakes, haddocks — Capture production by species, fishing areas and countries or areas
Morues, merlus, églefins — Captures par espèces, zones de pêche et pays ou zones
Bacalaos, merluzas, eglefinos — Capturas por especies, áreas de pesca y países o áreas

Species, Fishing area Espèce, Zone de pêche Especie, Area de pesca		1992 mt	1993 mt	1994 mt	1995 mt	1996 mt	1997 mt	1998 mt	1999 mt	2000 mt	2001 mt
	Norway	452	355	287	334	210	140	116	143	145	237
	Poland	-	-	-	-	-	-	1	-	-	-
	Portugal	210	234	306	169	184	139	115	76	77	38
	Spain	2	88	136	6	44	72	187	233	353	299
	Sweden	1 045	871	473	670	374	101	90	128	177	153
	UK	44 657	46 056	42 202	40 383	36 788	35 189	27 243	25 561	23 458	15 287
27	*Fishing area total*	*88 187*	*93 599*	*93 508*	*89 078 F*	*76 550*	*72 125*	*61 051*	*61 472*	*56 493*	*48 878*
37	Bulgaria	-	-	-	-	-	-	-	-	9	8
	France	...	...	...	...	...	...	...	2	2	2
	Georgia	70	172	187	146	223	58	53	41	...	32
	Romania	1 357	599	432	327	372	441	640	272	275	306
	Russian Fed	-	16	125	91	11	10	119	184	341	642
	Slovenia	-	-	-	-	-	-	13	16	14	37
	Turkey	20 197	20 487	16 615	18 094	21 450	15 500	13 150	14 110	18 000	10 000
	Ukraine	...	5	64	17	3	29	55	18	20	18
37	*Fishing area total*	*21 624*	*21 279*	*17 423*	*18 675*	*22 059*	*16 038*	*14 030*	*14 643*	*18 661*	*11 045*
Species total		*109 811*	*114 878*	*110 931*	*107 753 F*	*98 609*	*88 163*	*75 081*	*76 115*	*75 154*	*59 923*
European hake		**Merlu européen**		**Merluza europea**		***Merluccius merluccius***			**1,48(05)004,01**		**HKE**
27	Belgium	183	116	105	76	42	54	76	92	117	124
	Channel Is	0	0	0	0	0	0	0	0	-	-
	Denmark	3 371	3 179	2 128	1 487	868	670	591	846	811	1 043
	Faeroe Is	-	6	4	14	1	6	5	5	...	...
	France	15 570	12 953	14 758	14 530	8 899	8 592	5 165	7 378	10 122	8 007
	Germany	51	120	84	110	83	76	69	68	46	73
	Ireland	2 513	2 539	2 175	2 186	1 741	2 270	1 971	2 090	2 037	1 124
	Isle of Man	6	7	25	23	18	28	30	3	3	1
	Netherlands	181	144	75	78	111	62	75	98	43	72
	Norway	872	887	589	783	938	981	825	609	696	635
	Portugal	4 620	3 273	2 685	3 109	2 636	2 198	2 346	3 103	3 058	3 030
	Spain	23 757	21 015	20 001	27 168	18 076	20 161	20 057	22 931	24 856	12 137
	Sweden	201	200	170	69	45	33	26	27	34	63
	UK	7 577	7 380	6 120	5 614	5 380	5 716	5 168	5 537	4 519	2 775
27	*Fishing area total*	*58 902*	*51 819*	*48 919*	*55 247*	*38 838*	*40 847*	*36 404*	*42 787*	*46 342*	*29 084*
34	Greece	34	46	39	85	70	9	20	1	9	-
	Italy	-	-	-	371	-	-	-	-	-	483
	Morocco	5 799	3 735	5 090	6 038	3 086	2 473	1 161	2 313	2 331	3 812
	Portugal	126	1 041	687	357	986	380	217	114	3	2
	Spain	6 615	6 300 F	6 000 F	5 800 F	5 536	3 293	1 741	926	-	7
34	*Fishing area total*	*12 574*	*11 122 F*	*11 816 F*	*12 651 F*	*9 678*	*6 155*	*3 139*	*3 354*	*2 343*	*4 304*
37	Albania	454	400 F	300 F	227	293	185	340	341	330	380
	Algeria	900 F	1 000 F	1 200 F	1 115	1 154	1 012	1 310	1 681	1 650 F	1 650 F
	Croatia	1 057	1 589	1 230	1 129	929	828	935	650	583	557
	Cyprus	4	14	13	7	3	4	2	5	6	8
	France	4 275	4 006	1 955	1 775	1 423	1 423	1 423	1 782	1 513	2 022
	Greece	4 141	5 043	6 390	5 369	4 579	4 248	3 032	3 127	2 960	2 753
	Israel	54	61	120	120	131	86	134	60	62	50 F
	Italy	28 433	34 001	36 334	37 680	30 707	17 971	13 166	9 754	9 219	8 765
	Malta	1	1	1	1	2	4	5	6	6	0
	Morocco	106	66	77	95	84	74	96	37	461	197
	Slovenia	1	2	1	4	4	4	2	1	0	2
	Spain	4 100 F	4 000 F	3 800 F	3 600 F	3 400	2 850	5 402	5 497	6 327	7 015
	Syria	130 F	134 F	129 F	128 F	250	300	125	110	87	52
	Tunisia	710	818	840	744	685	922	537	1 069	974	1 284
	Turkey	71	583	3	1	150	25	15	5	10	-
	Yugoslavia	0	1	1	13	22	20	21	19	17	18
37	*Fishing area total*	*44 437 F*	*51 719 F*	*52 394 F*	*52 008 F*	*43 816*	*29 956*	*26 545*	*24 144*	*24 205 F*	*24 753 F*
Species total		*115 913 F*	*114 660 F*	*113 129 F*	*119 906 F*	*92 332*	*76 958*	*66 088*	*70 285*	*72 890 F*	*58 141 F*
Senegalese hake		**Merlu du Sénégal**		**Merluza del Senegal**		***Merluccius senegalensis***			**1,48(05)004,02**		**HKM**
34	Bulgaria	-	-	-	-	-	-	10	-	-	-
	Latvia	190	-	43	8	68	27	16	320	280	125
	Lithuania	511	-	-	...	...	...	180	307	180	42
	Poland	-	-	-	-	64	-	-	-	-	87
	Portugal	135	28	6	38	223	102	42	17	-	-
	Russian Fed	405	-	1	132	1 112	1 081	1 171	1 230	604	207
	Senegal	18	33	8	1	7	162	22	335	113	98
	Spain	13 474	13 800 F	14 200 F	14 600 F	15 001	14 478	13 594	15 678	11 211	16 716
	Ukraine	24	11	177	...	10	580	2 361	300	1 127	648
34	*Fishing area total*	*14 757*	*13 872 F*	*14 435 F*	*14 779 F*	*16 485*	*16 430*	*17 396*	*18 187*	*13 515*	*17 923*
Species total		*14 757*	*13 872 F*	*14 435 F*	*14 779 F*	*16 485*	*16 430*	*17 396*	*18 187*	*13 515*	*17 923*
Southern hake		**Merlu austral**		**Merluza sureña**		***Merluccius australis***			**1,48(05)004,03**		**HKN**
41	Argentina	4 094	3 026	1 650	3 899	4 115	3 011	3 125	3 471	7 035	4 648
	Chile	17	5	3	-	-	-	-	-	-	1
	Korea Rep	203	...	...	...	...	...	-	-	-	-
41	*Fishing area total*	*4 314*	*3 031*	*1 653*	*3 899*	*4 115*	*3 011*	*3 125*	*3 471*	*7 035*	*4 649*
81	Korea Rep	180	19	229	...	454	1 178	1 976	2 512	2 142	2 224
	New Zealand	5 189	6 192	2 870	9 707	8 317	9 692	15 047	15 499	12 799	12 956

B-32

Cods, hakes, haddocks — **Capture production by species, fishing areas and countries or areas**
Morues, merlus, églefins — **Captures par espèces, zones de pêche et pays ou zones**
Bacalaos, merluzas, eglefinos — **Capturas por especies, áreas de pesca y países o áreas**

Species, Fishing area Espèce, Zone de pêche Especie, Area de pesca	1992 mt	1993 mt	1994 mt	1995 mt	1996 mt	1997 mt	1998 mt	1999 mt	2000 mt	2001 mt
Norway	645	542	338	57	210	117	16	-	-	-
Poland	69	-	-	-	-	-	-	-	-	-
Russian Fed	-	-	278	414	27	202	-	-	-	-
Ukraine	84	353	167	774	111	181	578	1 422	1 100	8
81 Fishing area total	*6 167*	*7 106*	*3 882*	*10 952*	*9 119*	*11 370*	*17 617*	*19 433*	*16 041*	*15 188*
87 Chile	38 110	20 112	23 169	24 612	23 788	24 666	22 458	24 656	29 402	28 805
87 Fishing area total	*38 110*	*20 112*	*23 169*	*24 612*	*23 788*	*24 666*	*22 458*	*24 656*	*29 402*	*28 805*
Species total	*48 591*	*30 249*	*28 704*	*39 463*	*37 022*	*39 047*	*43 200*	*47 560*	*52 478*	*48 642*
Silver hake	**Merlu argenté**		**Merluza norteamericana**		***Merluccius bilinearis***			**1,48(05)004,04**		**HKS**
21 Canada	133	13	79	283	3 485	5 333	10 489	9 676	10 378	17 929
Cuba	16 724	22 714	8 055	16 785	22 279	12 697	6 281	3 852	-	-
Germany	0	0	0	0	-	-	-	-	-	-
Japan	523	-	-	-	-	-	-	-	-	-
Russian Fed	11 764	7 052	-	-	639	-	163	-	1 679	2 055
Spain	-	-	-	-	-	-	-	-	4	9
USA	16 281	16 111	16 040	15 169	16 025	15 535	14 959	14 039	12 176	13 007
21 Fishing area total	*45 425*	*45 890*	*24 174*	*32 237*	*42 428*	*33 565*	*31 892*	*27 567*	*24 237*	*33 000*
31 USA	-	-	2	0	-	-	-	-	-	-
31 Fishing area total	-	-	*2*	*0*	-	-	-	-	-	-
Species total	*45 425*	*45 890*	*24 176*	*32 237*	*42 428*	*33 565*	*31 892*	*27 567*	*24 237*	*33 000*
South Pacific hake	**Merlu du Pacifique sud**		**Merluza común**		***Merluccius gayi***			**1,48(05)004,05**		**PHA**
87 Chile	62 644	64 262	68 107	75 403	88 555	87 620	80 151	103 789	110 143	121 200
Colombia	...	...	25	...	...	267	165	143	250 F	391
Peru	30 410	88 700	135 705	181 182	234 915	177 953	82 367	37 121	83 361	125 065
87 Fishing area total	*93 054*	*152 962*	*203 837*	*256 585*	*323 470*	*265 840*	*162 683*	*141 053*	*193 754 F*	*246 656*
Species total	*93 054*	*152 962*	*203 837*	*256 585*	*323 470*	*265 840*	*162 683*	*141 053*	*193 754 F*	*246 656*
Argentine hake	**Merlu d'Argentine**		**Merluza argentina**		***Merluccius hubbsi***			**1,48(05)004,06**		**HKP**
41 Argentina	368 998	422 195	435 788	574 317	597 557	584 048	458 433	311 953	191 440	248 804
Australia	-	-	-	-	-	-	3	10	-	-
Belize	-	-	-	-	-	-	-	35	63	4
Brazil	4 500 F	3 000 F	1 500 F	255	0	-	-	128	226	220 F
Bulgaria	12	-	-	-	-	-	-	-	-	-
Falkland Is	30	88	43	194	383	267	959	1 031	1 000	562
France	-	-	20	4	17	4	3	3	0	-
Honduras	89	223	53	78	19	-	-	-	-	-
Italy	-	-	-	44	-	-	-	-	-	-
Japan	3	68	29	72	83	53	30	27	59	3
Korea Rep	92	...	...	...	...	...	-	-	-	-
Latvia	-	62	-	-	-	-	-	-	-	-
Namibia	-	-	-	-	-	12	15	37	-	-
Panama	121	70	18	15	14	-	-	-	-	-
Poland	24	32	-	-	-	-	-	35	-	-
Portugal	23	6	1 143	2 371	4 253	603	310	-	3	-
Russian Fed	252	44	-	-	-	-	-	-	-	197
St Vincent	-	-	-	-	-	-	-	-	-	5
Seychelles	-	-	-	-	-	27	-	-	-	-
Spain	3 091	1 693	967	1 161	21 697	14 816	18 298	26 810	22 196	25 302
Ukraine	34	-	-	-	1	-	-	-	-	-
UK	-	62	10	0	38	104	67	53	30	83
Uruguay	74 509	69 910	56 981	57 874	57 937	48 367	49 111	32 045	27 710	27 618
41 Fishing area total	*451 778 F*	*497 453 F*	*496 552 F*	*636 385*	*681 999*	*648 301*	*527 229*	*372 167*	*242 727*	*302 798 F*
Species total	*451 778 F*	*497 453 F*	*496 552 F*	*636 385*	*681 999*	*648 301*	*527 229*	*372 167*	*242 727*	*302 798 F*
North Pacific hake	**Merlu du Pacifique nord**		**Merluza del Pacífico norte**		***Merluccius productus***			**1,48(05)004,07**		**NHA**
67 Poland	-	-	-	-	-	-	-	-	977	
USA	56 131	140 697	252 742	177 038	195 289	226 615	227 502	216 889	205 350	172 050
67 Fishing area total	*56 131*	*140 697*	*252 742*	*177 038*	*195 289*	*226 615*	*227 502*	*216 889*	*206 327*	*172 050*
77 Mexico	100	106	17	78	116	1 089	250	111	392	...
USA	0	1	3	1	1	1	2	-	1	1
77 Fishing area total	*100*	*107*	*20*	*79*	*117*	*1 090*	*252*	*111*	*393*	*1*
Species total	*56 231*	*140 804*	*252 762*	*177 117*	*195 406*	*227 705*	*227 754*	*217 000*	*206 720*	*172 051*
Benguela hake	**Merlu d'Afrique tropicale**		**Merluza de Benguela**		***Merluccius polli***			**1,48(05)004,08**		**HKB**
34 Belize	-	-	-	-	-	-	-	54	...	...
Cameroon	-	3	-	-	-	-	-	-	-	-
Ghana	0	0	0	0	1	0	34	3	0	-
Korea Rep	28	-	-	-	-	-	4	-	-	-
Liberia	-	-	-	-	38	-	-	-	-	-
Nigeria	-	-	-	-	-	-	-	64	-	-
Panama	-	14	-	-	-	-	-	-	-	-
St Vincent	0	-	-	-	-	-	-	-	-	-
Sierra Leone	0	1	-	-	-	0	-	-	-	-
Spain	-	-	-	-	-	-	-	18	-	-

B-32 Cods, hakes, haddocks / Morues, merlus, églefins / Bacalaos, merluzas, eglefinos

Capture production by species, fishing areas and countries or areas / Captures par espèces, zones de pêche et pays ou zones / Capturas por especies, áreas de pesca y países o áreas

Species, Fishing area Espèce, Zone de pêche Especie, Area de pesca	1992 mt	1993 mt	1994 mt	1995 mt	1996 mt	1997 mt	1998 mt	1999 mt	2000 mt	2001 mt
Other nei	5	-	-	-	-	-	-	8	44	139
34 Fishing area total	*33*	*18*	*0*	*0*	*39*	*0*	*38*	*147*	*44*	*139*
Species total	*33*	*18*	*0*	*0*	*39*	*0*	*38*	*147*	*44*	*139*
Shallow-water Cape hake	**Merlu côtier du Cap**		**Merluza del Cabo**		***Merluccius capensis***				**1,48(05)004,09**	**HKK**
47 Angola	219	92	24	26	832	777	906	2 060	658	1 863
47 Fishing area total	*219*	*92*	*24*	*26*	*832*	*777*	*906*	*2 060*	*658*	*1 863*
Species total	*219*	*92*	*24*	*26*	*832*	*777*	*906*	*2 060*	*658*	*1 863*
Offshore silver hake	**Merlu argenté du large**		**Merluza blanca de altura**		***Merluccius albidus***				**1,48(05)004,12**	**HOF**
21 USA	...	...	115	...	32	...	5	12	5	2
21 Fishing area total	...	...	*115*	...	*32*	...	*5*	*12*	*5*	*2*
Species total	...	...	*115*	...	*32*	...	*5*	*12*	*5*	*2*
Cape hakes	**Merlus du Cap**		**Merluzas del Cabo**		***Merluccius capensis,M.paradox.***				**1,48(05)004,XX**	**HKC**
47 Iceland	-	-	-	-	-	352	206	-	-	-
Italy	-	-	-	5	-	-	-	-	-	-
Japan	483	-	33	55	-	-	-	-	-	-
Korea Rep	-	44	4	-	-	-	-	-	-	-
Namibia	87 588	108 102	112 229	130 374	129 462	117 683	154 422	166 562	162 803	173 461
Poland	-	-	-	-	3	-	-	-	-	-
Portugal	-	-	-	-	-	-	-	-	-	1
Russian Fed	7	976	321	10	98	252	321	67	41	49
South Africa	135 805	108 336	136 917	137 742	155 155	141 076	151 317	141 165	135 000 F	146 393
Spain	20 474	18 000 F	12 500 F	9 000 F	1 724	1 592	3	2 154	2 877	3 349
Ukraine	24	11	20	-	-	-	-	6	22	4
Other nei	-	-	2	-	1	-	-	-	-	-
47 Fishing area total	*244 381*	*235 469 F*	*262 026 F*	*277 186 F*	*286 443*	*260 955*	*306 269*	*309 954*	*300 743 F*	*323 257*
Species total	*244 381*	*235 469 F*	*262 026 F*	*277 186 F*	*286 443*	*260 955*	*306 269*	*309 954*	*300 743 F*	*323 257*
Hakes nei	**Merlus nca**		**Merluzas nep**		***Merluccius spp***				**1,48(05)004,XX**	**HKX**
34 Belize	-	-	-	-	-	30	29	60	165	8
China	21	0	-	10	-	-	0	-	-	-
Cyprus	-	-	-	-	...	...	40	164	189	110
France	-	-	-	-	-	...	44	5	-	-
Germany	-	-	-	16	-	-	-	-	-	-
Guinea	-	-	-	-	-	-	-	-	-	3
Honduras	0	0	-	2	-	-	-	-	-	-
Korea Rep	-	-	-	-	-	-	-	-	-	15
Mauritania	30 F	90 F	100 F	40 F	150 F	110 F	818 F	940	1 558	1 270
Panama	-	-	-	-	40	29	-	-	-	-
Portugal	0	0	0	49	1 515	914	1 027	474	-	11
St Vincent	-	-	-	-	-	10	106	31	28	51
34 Fishing area total	*51 F*	*90 F*	*100 F*	*117 F*	*1 705 F*	*1 093 F*	*2 064 F*	*1 674*	*1 940*	*1 468*
41 Korea Rep	22	119	50	16	370	33	-	-	-	-
Lithuania	-	-	8	-	-	-	-	-	-	-
Portugal	-	-	-	-	-	-	-	-	-	354
41 Fishing area total	*22*	*119*	*58*	*16*	*370*	*33*	-	-	-	*354*
47 Panama	9	-	-	-	-	-	-	-	-	-
47 Fishing area total	*9*	-	-	-	-	-	-	-	-	-
Species total	*82 F*	*209 F*	*158 F*	*133 F*	*2 075 F*	*1 126 F*	*2 064 F*	*1 674*	*1 940*	*1 822*
Patagonian grenadier	**Grenadier patagonien**		**Merluza de cola**		***Macruronus magellanicus***				**1,48(05)017,01**	**GRM**
41 Argentina	7 747	39 373	17 251	22 796	44 065	41 835	96 157	117 571	123 480	111 349
Australia	-	-	-	-	-	-	31	377	-	-
Belize	-	-	-	-	-	-	-	84	1 720	374
Brazil	...	...	...	30	18	20	0	0	0	0
Bulgaria	37	-	-	-	-	-	-	-	-	-
Chile	-	-	-	-	-	-	361	181	23	1 308
Estonia	-	-	-	-	113	-	-	-	-	108
Falkland Is	44	58	98	864	2 569	1 829	4 246	5 109	3 404	5 452
France	-	-	114	15	29	-	-	2	0	-
Honduras	940	251	797	1 053	92	-	-	-	-	-
Japan	14	409	18	568	544	644	844	400	1 889	866
Korea Rep	-	64	-	76	33	-	31	282	1 076	1 553
Latvia	-	-	-	32	15	-	-	-	-	-
Lithuania	-	-	2	-	-	-	-	-	-	-
Namibia	-	-	-	-	-	98	205	308	-	-
Panama	1 683	214	191	223	339	-	-	-	-	-
Poland	1 321	258	33	-	146	-	-	86	-	73
Portugal	48	4	-	-	-	-	-	-	32	-
Russian Fed	908	128	-	-	-	-	-	-	-	173
Seychelles	-	-	-	-	-	35	5	-	-	-
Spain	5 080	5 608	6 736	10 601	7 733	7 422	16 104	11 132	9 794	13 599
Ukraine	874	-	-	-	-	-	-	-	-	-
UK	-	169	3	80	86	166	2	347	42	30
Uruguay	...	...	...	...	...	150	1 824	1 461	800	635

B-32 Cods, hakes, haddocks / Morues, merlus, églefins / Bacalaos, merluzas, eglefinos

Capture production by species, fishing areas and countries or areas / Captures par espèces, zones de pêche et pays ou zones / Capturas por especies, áreas de pesca y países o áreas

Species, Fishing area Espèce, Zone de pêche Especie, Area de pesca	1992 mt	1993 mt	1994 mt	1995 mt	1996 mt	1997 mt	1998 mt	1999 mt	2000 mt	2001 mt
41 Fishing area total	*18 696*	*46 536*	*25 243*	*36 338*	*55 782*	*52 199*	*119 810*	*137 340*	*142 260*	*135 520*
87 Chile	214 324	82 580	81 310	206 734	379 015	71 479	353 823	309 723	91 310	160 774
87 Fishing area total	*214 324*	*82 580*	*81 310*	*206 734*	*379 015*	*71 479*	*353 823*	*309 723*	*91 310*	*160 774*
Species total	*233 020*	*129 116*	*106 553*	*243 072*	*434 797*	*123 678*	*473 633*	*447 063*	*233 570*	*296 294*
Blue grenadier	**Grenadier bleu**		**Cola de rata azul**		***Macruronus novaezelandiae***				**1,48(05)017,02**	**GRN**
57 Australia	3 526	3 039	3 048	3 239	2 680	2 851	4 752	6 209	9 493	7 561
57 Fishing area total	*3 526*	*3 039*	*3 048*	*3 239*	*2 680*	*2 851*	*4 752*	*6 209*	*9 493*	*7 561*
81 Japan	92 936	65 190	57 346	31 463	26 031	25 349	26 802	17 269	12 139	15 734
Korea Rep	423	373	376	340	1 725	6 091	4 883	5 834	8 694	9 484
New Zealand	143 394	131 086	151 718	152 161	145 308	229 890	267 616	236 652	234 029	223 703
Norway	10 357	20 236	13 321	6 100	6 614	5 576	4 633	-	-	-
Poland	596	-	-	-	-	-	-	-	-	-
Russian Fed	24 438	14 654	20 409	7 907	3 905	3 191	-	-	-	-
Ukraine	4 507	12 654	18 269	10 774	14 754	12 632	16 064	15 141	19 333	1 300
81 Fishing area total	*276 651*	*244 193*	*261 439*	*208 745*	*198 337*	*282 729*	*319 998*	*274 896*	*274 195*	*250 221*
Species total	*280 177*	*247 232*	*264 487*	*211 984*	*201 017*	*285 580*	*324 750*	*281 105*	*283 688*	*257 782*
Roughhead grenadier	**Grenadier berglax**		**Grenadero berglax**		***Macrourus berglax***				**1,48(06)001,03**	**RHG**
21 Estonia	-	-	-	-	-	-	-	-	1	-
Lithuania	-	-	-	-	-	-	-	-	1	28
Norway	-	-	-	-	-	-	-	-	-	1
Portugal	2 004	1 993	2 223	1 377	787	762	1 090	1 299	395	610
Spain	-	-	-	-	257	3 740	6 050	5 705	8 097	1 103
21 Fishing area total	*2 004*	*1 993*	*2 223*	*1 377*	*1 044*	*4 502*	*7 140*	*7 004*	*8 494*	*1 742*
27 France	-	-	-	1	21	25	29	116	4	4
Iceland	-	-	28	6	15	4	1	-	5	3
Norway	-	-	-	-	-	-	-	-	-	147
UK	-	-	-	-	-	14	-	1 038	8	16
27 Fishing area total	*-*	*-*	*28*	*7*	*36*	*43*	*30*	*1 154*	*17*	*170*
Species total	*2 004*	*1 993*	*2 251*	*1 384*	*1 080*	*4 545*	*7 170*	*8 158*	*8 511*	*1 912*
Whitson's grenadier	**...B**		**...C**		***Macrourus whitsoni***				**1,48(06)001,04**	**WGR**
48 UK	-	-	-	-	-	-	-	-	-	1
48 Fishing area total	*-*	*-*	*-*	*-*	*-*	*-*	*-*	*-*	*-*	*1*
88 New Zealand	-	-	-	-	-	-	-	1	5	48
88 Fishing area total	*-*	*-*	*-*	*-*	*-*	*-*	*-*	*1*	*5*	*48*
Species total	*-*	*-*	*-*	*-*	*-*	*-*	*-*	*1*	*5*	*49*
Grenadiers nei	**Grenadiers nca**		**Granaderos nep**		***Macrourus spp***				**1,48(06)001,XX**	**GRV**
41 Argentina	16	16	16	664	821	1 041	2 972	2 580	10 503	...
Poland	-	3	-	-	-	-	-	13	-	-
Portugal	-	-	-	80	80	-	-	-	-	1
Russian Fed	3 622	-	-	-	-	-	-	-	-	-
Uruguay	-	-	-	-	-	-	-	-	-	8
41 Fishing area total	*3 638*	*19*	*16*	*744*	*901*	*1 041*	*2 972*	*2 593*	*10 503*	*9*
48 Argentina	-	-	-	8	8	-	-	-	-	-
Chile	-	-	-	0	12	-	0	-	-	-
Korea Rep	-	-	-	-	4	9	3	-	-	-
Russian Fed	-	-	0	2	-	-	-	-	-	0
South Africa	-	-	-	-	-	-	13	1	-	-
Spain	-	-	-	-	-	1	1	-	-	-
UK	-	-	-	-	-	5	4	8	5	2
Uruguay	-	-	-	-	-	-	0	3	-	-
48 Fishing area total	*-*	*-*	*0*	*10*	*24*	*15*	*21*	*12*	*5*	*2*
58 Australia	0	-	0	-	-	1	-	1	3	-
France	-	-	-	-	-	-	25	-	173	103
South Africa	-	-	-	-	-	-	23	45	157	141
58 Fishing area total	*0*	*-*	*0*	*-*	*-*	*1*	*48*	*46*	*333*	*244*
81 Ukraine	2	34	121	-	-	-	-	-	-	-
81 Fishing area total	*2*	*34*	*121*	*-*	*-*	*-*	*-*	*-*	*-*	*-*
88 South Africa	-	-	-	-	-	-	-	-	-	6
88 Fishing area total	*-*	*-*	*-*	*-*	*-*	*-*	*-*	*-*	*-*	*6*
Species total	*3 640*	*53*	*137*	*754*	*925*	*1 057*	*3 041*	*2 651*	*10 841*	*261*
Roundnose grenadier	**Grenadier de roche**		**Granadero de roca**		***Coryphaenoides rupestris***				**1,48(06)004,01**	**RNG**
21 Canada	568	1 298	543	695	236	75	2	-	1	2
Estonia	-	-	-	-	-	-	-	-	20	21
Faeroe Is	9	23	-	2	1	1	-	10	23	...
Germany	34	2	12	-	5	4	5	-	-	5
Greenland	19	28	21	46	100	146	19	33	17	22

B-32 Cods, hakes, haddocks / Morues, merlus, églefins / Bacalaos, merluzas, eglefinos

Capture production by species, fishing areas and countries or areas / Captures par espèces, zones de pêche et pays ou zones / Capturas por especies, áreas de pesca y países o áreas

Species, Fishing area / Espèce, Zone de pêche / Especie, Area de pesca		1992 mt	1993 mt	1994 mt	1995 mt	1996 mt	1997 mt	1998 mt	1999 mt	2000 mt	2001 mt
	Japan	147	222	134	106	35	40	37	40	27	146
	Norway	-	-	-	-	-	-	-	-	-	3
	Poland	-	-	-	-	-	-	-	-	-	1
	Russian Fed	34	17	14	191	59	-	96	53	284	186
	St Pier Mq	-	-	-	-	-	4	-	-	-	-
	Spain	4 970	2 054	1 720	2 646	3 096	-	-	-	-	5 126
21	*Fishing area total*	*5 781*	*3 644*	*2 444*	*3 686*	*3 532*	*270*	*159*	*136*	*372*	*5 512*
27	Denmark	2 856	1 591	1 911	2 227	1 174	2 124	4 429	2 521	1 981	2 229
	Estonia	-	-	-	-	-	-	-	-	-	680
	Faeroe Is	746	641	743	764	233	198	84	98	30	91
	France	9 109	8 813	8 241	8 414	7 650	7 389	6 820	7 851	9 968	8 494
	Germany	168	61	48	22	9	34	116	101	42	18
	Greenland	1	18	5	14	19	26	3	1	11	5
	Iceland	210	280	210	398	216	207	120	146	70	57
	Ireland	...	144	6	59	1	4	-	1	45	-
	Latvia	1 684	2 176	675	-	-	-	-	-	-	-
	Lithuania	-	-	-	-	-	-	-	-	-	137
	Norway	-	-	-	-	-	-	-	-	-	75
	Poland	-	-	-	-	-	5 867	6 769	546	-	178
	Portugal	0	0	0	0	0	-	-	-	-	-
	Russian Fed	295	473	-	-	208	1 041	825	588	2 343	1 806
	Spain	-	-	-	32	970	2 476	3 935	6 224	15 465	33 099
	Sweden	755	-	42	1	0	42	0	-	-	258
	UK	143	1	12	91	184	228	-	-	587	1 030
27	*Fishing area total*	*15 967*	*14 198*	*11 893*	*12 022*	*10 664*	*19 636*	*23 101*	*18 077*	*30 542*	*48 157*
Species total		*21 748*	*17 842*	*14 337*	*15 708*	*14 196*	*19 906*	*23 260*	*18 213*	*30 914*	*53 669*
Thorntooth grenadier		**...B**		**...C**			***Lepidorhynchus denticulatus***			**1,48(06)030,01**	**LDE**
81	New Zealand	...	...	...	745	670	2 361	4 627	3 678	3 833	4 783
81	*Fishing area total*	...	...	...	*745*	*670*	*2 361*	*4 627*	*3 678*	*3 833*	*4 783*
Species total		...	...	...	*745*	*670*	*2 361*	*4 627*	*3 678*	*3 833*	*4 783*
Grenadiers, rattails nei		**...B**		**Granaderos, colas de ratón nep**			***Macrouridae***			**1,48(06)XXX,XX**	**RTX**
61	Russian Fed	2 681	764	121	222	310	1 044	62	52	579	2 232
61	*Fishing area total*	*2 681*	*764*	*121*	*222*	*310*	*1 044*	*62*	*52*	*579*	*2 232*
67	USA	...	...	...	...	...	...	83	102	170	160
67	*Fishing area total*	...	...	...	...	...	...	*83*	*102*	*170*	*160*
77	Korea Rep	-	-	-	-	-	-	-	-	140	-
	USA	...	...	...	...	...	...	417	210	145	145
77	*Fishing area total*	...	...	...	...	...	...	*417*	*210*	*285*	*145*
81	New Zealand	...	...	...	909	1 579	5 197	4 350	3 670	2 394	3 094
	Russian Fed	8	-	-	-	-	-	-	-	-	-
81	*Fishing area total*	*8*	...	...	*909*	*1 579*	*5 197*	*4 350*	*3 670*	*2 394*	*3 094*
87	Lithuania	10	-	-	-	-	-	-	-	-	-
87	*Fishing area total*	*10*	-	-	-	-	-	-	-	-	-
88	New Zealand	...	...	...	-	-	-	9	22	70	-
88	*Fishing area total*	...	...	...	-	-	-	*9*	*22*	*70*	-
Species total		*2 699*	*764*	*121*	*1 131*	*1 889*	*6 241*	*4 921*	*4 056*	*3 498*	*5 631*
Gadiformes nei		**Gadiformes nca**		**Gadiformes nep**			***Gadiformes***			**1,48(XX)XXX,XX**	**GAD**
21	Japan	4	-	1	-	2	2	-	-	13	-
	Norway	3	14	-	-	-	5	7	3	7	-
	Portugal	-	-	-	-	-	-	-	-	2	-
	USA	69	138	98	48	1	0	18	8	16	3
21	*Fishing area total*	*76*	*152*	*99*	*48*	*3*	*7*	*25*	*11*	*38*	*3*
27	Channel Is	-	-	-	-	-	-	-	1	-	-
	Faeroe Is	1 083	372	192	229	449	899	57	348	...	...
	France	-	11	11	-	-	406	825	10	2 435	3 093
	Germany	-	463	-	-	-	-	-	-	-	-
	Ireland	-	-	44	55	105	190	-	279	26	87
	Japan	-	4	5	12	-	-	-	-	-	-
	Norway	1 182	307	502	263	217	257	742	435	861	-
	Portugal	728	663	658	599	586	653	668	545	500	483
	Spain	20 369	21 046	19 892	30 166	31 053	18 246	10 701	13 526	2 832	25 971
	Sweden	0	0	1	0	0	-	-	-	57	160
27	*Fishing area total*	*23 362*	*22 866*	*21 305*	*31 324*	*32 410*	*20 651*	*12 993*	*15 144*	*6 711*	*29 794*
31	USA	-	-	1	0	-	4	2	1	5	2
31	*Fishing area total*	-	-	*1*	*0*	-	*4*	*2*	*1*	*5*	*2*
34	Eq Guinea	360 F	320 F	480 F	200 F	450 F	570 F	603	710 F	80 F	80 F
	Korea Rep	-	-	-	-	-	59	277	89	-	-
	Portugal	109	404	182	68	138	71	58	51	12	9
	Spain	-	-	-	-	75	16	53	53	54	25
34	*Fishing area total*	*469 F*	*724 F*	*662 F*	*268 F*	*663 F*	*716 F*	*991*	*903 F*	*146 F*	*114 F*

B-32 Cods, hakes, haddocks / Morues, merlus, églefins / Bacalaos, merluzas, eglefinos

Capture production by species, fishing areas and countries or areas / Captures par espèces, zones de pêche et pays ou zones / Capturas por especies, áreas de pesca y países o áreas

Species, Fishing area Espèce, Zone de pêche Especie, Area de pesca	1992 mt	1993 mt	1994 mt	1995 mt	1996 mt	1997 mt	1998 mt	1999 mt	2000 mt	2001 mt
37 France	3	31	22	27	24	36	36	0	1	-
Lebanon	25 F	20	20	30	30	25	30	30	32	30
Spain	3 000 F	3 500 F	4 000 F	4 500 F	4 804	3 994	1 275	759	838	1 067
37 Fishing area total	*3 028 F*	*3 551 F*	*4 042 F*	*4 557 F*	*4 858*	*4 055*	*1 341*	*789*	*871*	*1 097*
41 Argentina	13	13	13	-	-	-	-	-	-	-
Japan	9	31	42	-	148	165	132	115	108	-
Korea Rep	110	282	-	293	1 392	87	105	237	1 256	538
Spain	-	-	-	-	-	-	-	-	-	2 601
UK	-	-	-	-	-	-	-	-	-	3
41 Fishing area total	*132*	*326*	*55*	*293*	*1 540*	*252*	*237*	*352*	*1 364*	*3 142*
47 Lithuania	-	-	3	-	-	-	-	-	-	-
Norway	-	-	-	-	-	-	3	-	-	-
47 Fishing area total	-	-	*3*	-	-	-	*3*	-	-	-
51 Italy	-	-	-	76	-	-	-	-	-	-
South Africa	-	3	-	-	-	10	7	4 F	2	1
51 Fishing area total	-	*3*	-	*76*	-	*10*	*7*	*4 F*	*2*	*1*
57 Australia	-	-	1	-	-	-	-	5	1	0
Korea Rep	-	-	-	-	-	-	-	-	-	30
57 Fishing area total	-	-	*1*	-	-	-	-	*5*	*1*	*30*
61 Korea Rep	-	-	-	-	-	-	-	17	-	-
Russian Fed	-	-	8	3	10	-	2	5	6	1
61 Fishing area total	-	-	*8*	*3*	*10*	-	*2*	*22*	*6*	*1*
71 Australia	55	32	55	34	24	21	33	21	22	33
Korea Rep	-	-	-	-	-	-	-	744	7	-
71 Fishing area total	*55*	*32*	*55*	*34*	*24*	*21*	*33*	*765*	*29*	*33*
77 Korea Rep	-	-	-	-	-	-	-	26	-	-
Mexico	323	415	50	10	21	183	137	32	40	...
77 Fishing area total	*323*	*415*	*50*	*10*	*21*	*183*	*137*	*58*	*40*	...
81 Japan	5 378	7 200	3 080	5 584	3 981	3 019	3 966	3 300	2 398	2 356
Korea Rep	-	1 275	222	237	302	132	452	1 130	455	2 511
New Zealand	...	...	...	73	36	22	34	11	14	8
Norway	820	-	-	-	-	-	188	-	-	-
Russian Fed	280	705	12	-	-	-	-	1	-	-
Ukraine	-	-	-	-	-	-	-	-	-	997
81 Fishing area total	*6 478*	*9 180*	*3 314*	*5 894*	*4 319*	*3 173*	*4 640*	*4 442*	*2 867*	*5 872*
Species total	*33 923 F*	*37 249 F*	*29 595 F*	*42 507 F*	*43 848 F*	*29 072 F*	*20 411*	*22 496 F*	*12 080 F*	*40 089 F*
Group total	***10 466 260***	***9 963 399***	***9 729 320***	***10 740 688***	***10 784 133***	***10 371 936***	***10 333 230***	***9 326 770***	***8 654 744***	***9 223 810***

B-33

Miscellaneous coastal fishes — Capture production by species, fishing areas and countries or areas
Poissons côtiers divers — Captures par espèces, zones de pêche et pays ou zones
Peces costeros diversos — Capturas por especies, áreas de pesca y países o áreas

Species, Fishing area Espèce, Zone de pêche Especie, Area de pesca	1992 mt	1993 mt	1994 mt	1995 mt	1996 mt	1997 mt	1998 mt	1999 mt	2000 mt	2001 mt
Ladyfish	**Guinée-machète**		**Malacho**			*Elops saurus*			**1,29(01)003,02**	**LAD**
21 USA	-	-	-	-	-	1	1	1 963	1	0
21 Fishing area total	-	-	-	-	-	*1*	*1*	*1 963*	*1*	*0*
31 Colombia	0	21	53	30	143	20	4	11	20 F	20 F
USA	-	-	-	-	-	725	975	4	120	460
31 Fishing area total	*0*	*21*	*53*	*30*	*143*	*745*	*979*	*15*	*140 F*	*480 F*
41 Brazil	80 F	80 F	80 F	147	-	-	-	-	-	-
41 Fishing area total	*80 F*	*80 F*	*80 F*	*147*	-	-	-	-	-	-
Species total	*80 F*	*101 F*	*133 F*	*177*	*143*	*746*	*980*	*1 978*	*141 F*	*480 F*
West African ladyfish	**...B**		**...C**			*Elops lacerta*			**1,29(01)003,04**	**CEC**
34 Benin	...	...	...	...	...	...	...	...	...	15
Gambia	...	...	...	...	...	...	...	...	...	12
Nigeria	579	579	544	1 474	379	735	1 546	1 822	755	531
34 Fishing area total	*579*	*579*	*544*	*1 474*	*379*	*735*	*1 546*	*1 822*	*755*	*558*
Species total	*579*	*579*	*544*	*1 474*	*379*	*735*	*1 546*	*1 822*	*755*	*558*
Tarpon	**Tarpon argenté**		**Tarpón**			*Megalops atlanticus*			**1,29(02)004,01**	**TAR**
31 Colombia	6	21	12	13	135	2	2	-	-	-
Dominican Rp	308	78	40	268	31	27	47	12	14	14
Mexico	-	-	-	2	1	14	4	4	4	...
31 Fishing area total	*314*	*99*	*52*	*283*	*167*	*43*	*53*	*16*	*18*	*14*
41 Brazil	1 410 F	1 410 F	1 410 F	1 221	348	720	551	315	331	330 F
41 Fishing area total	*1 410 F*	*1 410 F*	*1 410 F*	*1 221*	*348*	*720*	*551*	*315*	*331*	*330 F*
Species total	*1 724 F*	*1 509 F*	*1 462 F*	*1 504*	*515*	*763*	*604*	*331*	*349*	*344 F*
Indo-Pacific tarpon	**Tarpon indo-pacifique**		**Tarpón Indo-Pacífico**			*Megalops cyprinoides*			**1,29(02)004,02**	**TAI**
57 Malaysia	1	7	8	7	6	4	13	29	31	12
57 Fishing area total	*1*	*7*	*8*	*7*	*6*	*4*	*13*	*29*	*31*	*12*
71 Malaysia	158	92	90	126	93	510	325	197	102	92
Philippines	118	129	222	979	997	644	720	1 207	1 151	1 238
71 Fishing area total	*276*	*221*	*312*	*1 105*	*1 090*	*1 154*	*1 045*	*1 404*	*1 253*	*1 330*
Species total	*277*	*228*	*320*	*1 112*	*1 096*	*1 158*	*1 058*	*1 433*	*1 284*	*1 342*
Bonefish	**Banane de mer**		**Macabí**			*Albula vulpes*			**1,30(01)005,01**	**BOF**
31 Dominican Rp	129	36	65	143	151	16	34	125	10	11
31 Fishing area total	*129*	*36*	*65*	*143*	*151*	*16*	*34*	*125*	*10*	*11*
34 Côte dIvoire	17	37	20	20 F	43	1 430	1 617	29	21	13
Gabon	...	...	...	...	25	21	2	6	10	...
Liberia	-	-	-	-	-	-	21	104	27	6
Russian Fed	-	-	-	-	10	-	-	-	-	-
Sierra Leone	...	...	...	...	...	91	...	106	181	496
34 Fishing area total	*17*	*37*	*20*	*20 F*	*78*	*1 542*	*1 640*	*245*	*239*	*515*
47 Russian Fed	-	-	125	-	154	-	-	-	-	-
47 Fishing area total	-	-	*125*	-	*154*	-	-	-	-	-
Species total	*146*	*73*	*210*	*163 F*	*383*	*1 558*	*1 674*	*370*	*249*	*526*
Bombay-duck	**Scopelidé**		**Bumalo**			*Harpadon nehereus*			**1,31(16)001,02**	**BUC**
04 India	-	-	1 380	448	-	-	-	-	-	488
04 Fishing area total	-	-	*1 380*	*448*	-	-	-	-	-	*488*
51 India	153 738	142 074	126 396	133 786	159 362	187 887	144 774	146 591	133 156	142 944
Pakistan	187	170	121	98	101	95	91	72	65	55
51 Fishing area total	*153 925*	*142 244*	*126 517*	*133 884*	*159 463*	*187 982*	*144 865*	*146 663*	*133 221*	*142 999*
57 India	23 923	6 595	11 884	25 485	25 750	25 229	35 141	35 229	1 041	1 067
Indonesia	231	293	168	318	1 382	2 413	2 682	2 808	539	520
Malaysia	...	...	161	-	-	-	-	-	-	-
57 Fishing area total	*24 154*	*6 888*	*12 213*	*25 803*	*27 132*	*27 642*	*37 823*	*38 037*	*1 580*	*1 587*
71 Indonesia	12 309	12 205	12 643	10 717	9 836	10 867	11 500	9 607	8 449	8 130
Malaysia	...	...	647	-	-	-	-	-	-	-
71 Fishing area total	*12 309*	*12 205*	*13 290*	*10 717*	*9 836*	*10 867*	*11 500*	*9 607*	*8 449*	*8 130*
Species total	*190 388*	*161 337*	*153 400*	*170 852*	*196 431*	*226 491*	*194 188*	*194 307*	*143 250*	*153 204*
Greater lizardfish	**Anoli tumbil**		**Lagarto tumbil**			*Saurida tumbil*			**1,31(16)068,01**	**LIG**
51 Pakistan	82	96	67	43	45	28	22	-	-	-
Qatar	0	0	1	0	1	0	0	0	0	0
51 Fishing area total	*82*	*96*	*68*	*43*	*46*	*28*	*22*	*0*	*0*	*0*

B-33 Miscellaneous coastal fishes / Poissons côtiers divers / Peces costeros diversos

Capture production by species, fishing areas and countries or areas / Captures par espèces, zones de pêche et pays ou zones / Capturas por especies, áreas de pesca y países o áreas

Species, Fishing area Espèce, Zone de pêche Especie, Area de pesca	1992 mt	1993 mt	1994 mt	1995 mt	1996 mt	1997 mt	1998 mt	1999 mt	2000 mt	2001 mt
61 China,Taiwan	6 924	6 801	4 423	6 744	6 561	3 208	3 049	3 075	3 612	2 248
Japan	10 388	9 392	9 315	9 164	8 144	7 638	7 860	7 716	7 591	6 094
61 Fishing area total	*17 312*	*16 193*	*13 738*	*15 908*	*14 705*	*10 846*	*10 909*	*10 791*	*11 203*	*8 342*
Species total	*17 394*	*16 289*	*13 806*	*15 951*	*14 751*	*10 874*	*10 931*	*10 791*	*11 203*	*8 342*
Brushtooth lizardfish	**Anoli à grandes écailles**		**Lagarto escamoso**		***Saurida undosquamis***			**1,31(16)068,04**		**LIB**
37 Israel	227	214	80	80	124	61	37	48	21	20 F
37 Fishing area total	*227*	*214*	*80*	*80*	*124*	*61*	*37*	*48*	*21*	*20 F*
51 Kuwait	18	23	22	31	47	53	34	20	30 F	32
51 Fishing area total	*18*	*23*	*22*	*31*	*47*	*53*	*34*	*20*	*30 F*	*32*
61 Korea Rep	146	131	296	209	162	207	215	35	795	758
61 Fishing area total	*146*	*131*	*296*	*209*	*162*	*207*	*215*	*35*	*795*	*758*
Species total	*391*	*368*	*398*	*320*	*333*	*321*	*286*	*103*	*846 F*	*810 F*
Lizardfishes nei	**Anolis nca**		**Lagartos nep**		***Synodontidae***			**1,31(16)XXX,XX**		**LIX**
37 Egypt	822	980	1 134	993	1 307	1 075	941	1 162	1 265	1 065
Gaza Strip	...	...	...	...	16	32	32	30 F	30 F	30 F
Turkey	85	65	397	142	161	150	165	190	250	55
37 Fishing area total	*907*	*1 045*	*1 531*	*1 135*	*1 484*	*1 257*	*1 138*	*1 382 F*	*1 545 F*	*1 150 F*
51 Egypt	5 820	6 073	4 845	4 696	4 151	5 117	7 994	7 213	10 543	7 913
Eritrea	...	...	651	166	3	5	0	1 905	3 177	2 574
India	13 835	16 883	18 062	21 436	16 520	11 252	12 012	11 539	4 029	2 671
Saudi Arabia	72	76	195	195	172	188	215	214	169	199
51 Fishing area total	*19 727*	*23 032*	*23 753*	*26 493*	*20 846*	*16 562*	*20 221*	*20 871*	*17 918*	*13 357*
57 India	2 383	2 302	2 466	2 553	3 150	2 276	753	1 375	35 544	35 075
Indonesia	1 118	1 477	1 211	1 043	825	1 098	1 051	1 244	996	1 250
Malaysia	3 702	4 521	5 358	5 835	7 010	9 218	8 155	7 468	9 492	9 997
Thailand	6 472	11 068	13 302	11 955	10 680	8 918	41 178	12 774	12 595	12 522
57 Fishing area total	*13 675*	*19 368*	*22 337*	*21 386*	*21 665*	*21 510*	*51 137*	*22 861*	*58 627*	*58 844*
61 China,H.Kong	20 010	11 352	11 614	10 888	9 034	8 070	6 619	4 700 F	5 800 F	6 400 F
61 Fishing area total	*20 010*	*11 352*	*11 614*	*10 888*	*9 034*	*8 070*	*6 619*	*4 700 F*	*5 800 F*	*6 400 F*
71 Indonesia	4 560	5 299	5 035	6 019	6 401	14 060	10 947	11 700	12 387	15 480
Malaysia	2 084	2 437	2 849	3 018	5 039	5 272	6 269	7 582	8 075	7 627
Philippines	9 228	6 700	7 887	8 630	8 435	6 671	7 345	7 649	5 539	5 499
Singapore	371	286	306	186	172	125	90	51	28	6
Thailand	31 840	42 486	35 593	58 482	51 004	62 397	35 289	60 534	59 685	59 341
71 Fishing area total	*48 083*	*57 208*	*51 670*	*76 335*	*71 051*	*88 525*	*59 940*	*87 516*	*85 714*	*87 953*
Species total	*102 402*	*112 005*	*110 905*	*136 237*	*124 080*	*135 924*	*139 055*	*137 330 F*	*169 604 F*	*167 704 F*
Sea catfishes nei	**Mâchoirons nca**		**Bagres marinos nep**		***Ariidae***			**1,41(02)XXX,XX**		**CAX**
04 Philippines	2 369	3 646	2 495	2 170	2 964	2 701	3 207	3 785	4 046	3 868
04 Fishing area total	*2 369*	*3 646*	*2 495*	*2 170*	*2 964*	*2 701*	*3 207*	*3 785*	*4 046*	*3 868*
06 Papua N Guin	1 850 F	1 850 F	1 850 F	1 850 F	1 850 F	1 850 F	1 850 F	1 850 F	1 850 F	1 850 F
06 Fishing area total	*1 850 F*	*1 850 F*	*1 850 F*	*1 850 F*	*1 850 F*	*1 850 F*	*1 850 F*	*1 850 F*	*1 850 F*	*1 850 F*
31 Colombia	299	30	37	44	24	20	17	9	20 F	20 F
Mexico	4 742	4 129	4 365	5 038	5 085	5 902	6 270	5 333	6 008	6 308
USA	-	-	-	-	-	-	-	-	0	3
Venezuela	7 892	11 227	12 436	21 548	17 041	8 963	10 098	11 100	14 279	11 929
31 Fishing area total	*12 933*	*15 386*	*16 838*	*26 630*	*22 150*	*14 885*	*16 385*	*16 442*	*20 307 F*	*18 260 F*
34 Benin	19 F	21 F	23 F	22 F	20 F	17 F	15 F	12	8 F	2
Cameroon	313	334	600 F	976	886	640 F	370 F	520 F	455	410
Congo Rep	656 F	654 F	269	230 F	200 F	170 F	130 F	110 F	73	70 F
Côte dIvoire	19	21	18	19	16	21	25	16	20	13
Gabon	200 F	1 335	256	415	2 780	1 565	1 290	1 351	921	1 140
Gambia	434	357	302	846	158	1 234	517	540	749	950
Ghana	0	2	0	0	2	24	6	0	1	0
Guinea	2 530 F	3 030 F	3 110 F	4 381	3 589	2 462	2 610	2 378	4 593	4 500 F
GuineaBissau	...	...	175	211	195	200 F	170 F	140 F	140 F	140 F
Liberia	114	59	63	7	77	-	4	31	21	210
Lithuania	-	5	-	-	-	-	-	-	-	-
Mauritania	...	...	...	...	...	...	...	750 F	750 F	750 F
Nigeria	4 977	4	6 756	12 570	12 676	14 062	10 862	17 014	14 885	16 537
Romania	23	-	-	-	-	-	-	-	-	-
Russian Fed	682	25	-	-	-	-	-	2	2	-
Senegal	5 433	4 743	5 874	3 912	4 457	10 162	10 094	6 006	7 432	9 279
Sierra Leone	869	859	856	858	860 F	973	52	62	2 092	748
Ukraine	-	-	-	-	-	-	-	-	1	-
34 Fishing area total	*16 269 F*	*11 449 F*	*18 302 F*	*24 447 F*	*25 916 F*	*31 530 F*	*26 145 F*	*28 932 F*	*32 143 F*	*34 749 F*
41 Argentina	-	-	21	16	6	13	18	5	5	...
Brazil	13 760 F	13 730 F	13 760 F	14 226	16 946	17 736	20 489	25 224	29 473	29 200 F
Uruguay	102	111	169	119	18	15	24	78	51	42
41 Fishing area total	*13 862 F*	*13 841 F*	*13 950 F*	*14 361*	*16 970*	*17 764*	*20 531*	*25 307*	*29 529*	*29 242 F*

B-33 Miscellaneous coastal fishes / Poissons côtiers divers / Peces costeros diversos

Capture production by species, fishing areas and countries or areas / Captures par espèces, zones de pêche et pays ou zones / Capturas por especies, áreas de pesca y países o áreas

Species, Fishing area Espèce, Zone de pêche Especie, Area de pesca	1992 mt	1993 mt	1994 mt	1995 mt	1996 mt	1997 mt	1998 mt	1999 mt	2000 mt	2001 mt
47 Angola	0	35	17	8	1	43	90	69	407	2 424
Namibia	8	-	7	13	16	31	42	32	52	11
South Africa	0	1	1	0	2	2	0	1	-	-
47 Fishing area total	*8*	*36*	*25*	*21*	*19*	*76*	*132*	*102*	*459*	*2 435*
51 Eritrea	...	...	52	60	9	0	149	205	851	436
India	26 138	30 763	32 803	31 848	27 249	31 392	36 082	38 432	31 467	45 850
Oman	395	376	140	499	372	679	1 024	1 161	1 306	681
Pakistan	27 648	37 840	42 112	45 444	49 428	54 437	55 934	51 665	39 168	38 215
Qatar	0	0	0	0	0	0	0	0	0	0
Saudi Arabia	201	212	226	303	302	366	315	309	564	534
Tanzania	950	647	456	780	913	850	810	850	880	850
Untd Arab Em	122	129	139	139	140	150	150	154	763	760 F
Yemen	1 364	1 411	1 043	1 697	1 700	1 780 F	2 000 F	1 900 F	1 750 F	2 200 F
51 Fishing area total	*56 818*	*71 378*	*76 971*	*80 770*	*80 113*	*89 654 F*	*96 464 F*	*94 676 F*	*76 749 F*	*89 526 F*
57 Australia	3	6	14	14	218	-	-	-	-	-
India	47 711	67 422	53 074	42 453	45 876	46 758	36 809	47 664	49 350	49 616
Indonesia	6 423	7 082	6 847	7 154	7 698	6 235	7 994	7 992	8 653	9 000
Malaysia	3 393	4 285	4 719	4 543	5 111	4 232	4 088	4 368	4 797	5 017
Thailand	91	952	60	58	2 004	1 924	4 207	4 250	4 192	4 166
57 Fishing area total	*57 621*	*79 747*	*64 714*	*54 222*	*60 907*	*59 149*	*53 098*	*64 274*	*66 992*	*67 799*
61 China,Taiwan	471	427	302	274	122	134	208	257	725	435
61 Fishing area total	*471*	*427*	*302*	*274*	*122*	*134*	*208*	*257*	*725*	*435*
71 Australia	17	41	8	17	17	18	-	-	-	-
Indonesia	38 818	45 052	49 575	51 541	54 714	72 343	58 304	61 654	61 613	64 110
Malaysia	9 530	10 251	8 995	9 279	10 080	9 367	11 094	10 973	10 008	9 059
Philippines	5 441	8 194	5 284	7 990	7 533	6 324	4 590	4 527	4 350	4 554
Singapore	329	163	452	359	333	385	358	241	141	76
Thailand	706	6 878	839	565	3 680	3 780	3 334	8 348	8 234	8 183
71 Fishing area total	*54 841*	*70 579*	*65 153*	*69 751*	*76 357*	*92 217*	*77 680*	*85 743*	*84 346*	*85 982*
77 El Salvador	162	122	107	94	107	62	114	101	172	192
Mexico	1 048	1 023	1 079	1 234	1 101	1 338	1 755	1 603	1 377	1 772
77 Fishing area total	*1 210*	*1 145*	*1 186*	*1 328*	*1 208*	*1 400*	*1 869*	*1 704*	*1 549*	*1 964*
81 Australia	1	1	0	0	21	21	21	32	...	...
Russian Fed	-	1	-	-	-	-	-	-	-	-
81 Fishing area total	*1*	*2*	*0*	*0*	*21*	*21*	*21*	*32*	*...*	*...*
87 Colombia	178	119	223	114	425	350	121	97	100 F	100 F
87 Fishing area total	*178*	*119*	*223*	*114*	*425*	*350*	*121*	*97*	*100 F*	*100 F*
Species total	*218 431 F*	*269 605 F*	*262 009 F*	*275 938 F*	*289 022 F*	*311 731 F*	*297 711 F*	*323 201 F*	*318 795 F*	*336 210 F*
Eeltail catfishes	**Balibots**		**Patunas**		***Plotosus spp***				**1,41(06)064,XX**	**CA**
57 Malaysia	700	916	851	1 046	807	596	634	813	1 370	1 157
Thailand	-	-	60	58	74	116	108	75	74	74
57 Fishing area total	*700*	*916*	*911*	*1 104*	*881*	*712*	*742*	*888*	*1 444*	*1 231*
71 Malaysia	388	1 217	1 649	964	996	851	1 025	936	852	830
Thailand	-	-	839	565	767	647	1 093	556	548	545
71 Fishing area total	*388*	*1 217*	*2 488*	*1 529*	*1 763*	*1 498*	*2 118*	*1 492*	*1 400*	*1 375*
Species total	*1 088*	*2 133*	*3 399*	*2 633*	*2 644*	*2 210*	*2 860*	*2 380*	*2 844*	*2 606*
Morays	**Murènes**		**Morenas**		***Muraenidae***				**1,43(06)XXX,XX**	**MU**
27 Portugal	...	...	...	...	...	...	...	193	164	142
27 Fishing area total	*...*	*...*	*...*	*...*	*...*	*...*	*...*	*193*	*164*	*142*
34 Portugal	...	...	...	...	...	...	...	...	5	3
Senegal	111	166	910	837	710	268	165	142	189	209
34 Fishing area total	*111*	*166*	*910*	*837*	*710*	*268*	*165*	*142*	*194*	*212*
Species total	*111*	*166*	*910*	*837*	*710*	*268*	*165*	*335*	*358*	*354*
Korean sandlance	**...B**		**...C**		***Hypoptychus dybowskii***				**1,50(05)009,01**	**KL**
61 Korea Rep	8 005	9 632	9 466	9 677	6 613	8 832	4 801	4 806	-	462
61 Fishing area total	*8 005*	*9 632*	*9 466*	*9 677*	*6 613*	*8 832*	*4 801*	*4 806*	*-*	*462*
Species total	*8 005*	*9 632*	*9 466*	*9 677*	*6 613*	*8 832*	*4 801*	*4 806*	*-*	*462*
Squirrelfishes nei	**Marignons nca**		**Candiles nep**		***Holocentridae***				**1,61(11)XXX,XX**	**HC**
31 Antigua Barb	...	...	...	...	...	...	...	...	...	29
Dominican Rp	79	80	144	81	70	35	56	16	24	22
Grenada	1	0	1	1	1	1	3	2	3	3
Puerto Rico	...	...	...	...	...	...	...	9	13	12
St Kitts Nev	...	...	...	3	9	5	8	9	1	1
31 Fishing area total	*80*	*80*	*145*	*85*	*80*	*41*	*67*	*36*	*41*	*67*
51 Saudi Arabia	...	...	...	7	40	50	34	53	46	40
51 Fishing area total	*...*	*...*	*...*	*7*	*40*	*50*	*34*	*53*	*46*	*40*

B-33 Miscellaneous coastal fishes / Poissons côtiers divers / Peces costeros diversos

Capture production by species, fishing areas and countries or areas / Captures par espèces, zones de pêche et pays ou zones / Capturas por especies, áreas de pesca y países o áreas

Species, Fishing area Espèce, Zone de pêche Especie, Area de pesca	1992 mt	1993 mt	1994 mt	1995 mt	1996 mt	1997 mt	1998 mt	1999 mt	2000 mt	2001 mt
77 Amer Samoa	...	...	0	0	1	1	0	1	1	0
77 Fishing area total	*...*	*...*	*0*	*0*	*1*	*1*	*0*	*1*	*1*	*0*
Species total	*80*	*80*	*145*	*92*	*121*	*92*	*101*	*90*	*88*	*107*
Flathead grey mullet	**Mulet à grosse tête**		**Pardete**		***Mugil cephalus***			**1,65(01)001,02**		**MUF**
01 Egypt	6 049	6 567	5 441	...	...	...	...	...	...	...
01 Fishing area total	*6 049*	*6 567*	*5 441*	*...*	*...*	*...*	*...*	*...*	*...*	*...*
04 Israel	123	115	118	153	181	169	231	315	138	120 F
04 Fishing area total	*123*	*115*	*118*	*153*	*181*	*169*	*231*	*315*	*138*	*120 F*
05 Greece	-	-	-	-	27	54	67	58	51	46
05 Fishing area total	*-*	*-*	*-*	*-*	*27*	*54*	*67*	*58*	*51*	*46*
31 Mexico	6 573	6 295	6 016	7 636	7 882	7 128	4 299	4 872	5 190	4 838
Venezuela	8 244	9 103	10 678	9 363	8 839	6 465	6 284	5 151	5 467	5 008
31 Fishing area total	*14 817*	*15 398*	*16 694*	*16 999*	*16 721*	*13 593*	*10 583*	*10 023*	*10 657*	*9 846*
34 Greece	4	12	9	1	-	1	3	-	-	-
34 Fishing area total	*4*	*12*	*9*	*1*	*-*	*1*	*3*	*-*	*-*	*-*
37 Bulgaria	...	...	...	23	26	28	11	14	15	47
Croatia	144	134	116	123	105	105	195	120	68	14
Cyprus	0	2	2	2	5	5	2	6	7	7
Egypt	620	479	381	...	...	...	...	...	...	...
Greece	2 392	3 217	2 689	2 816	3 797	3 022	1 690	1 965	1 697	1 357
Israel	169	150 F	150 F	99	258	381	283	284	172	150 F
Slovenia	15	34	11	3	12	2	27	13	22	49
Tunisia	3 529	2 765	2 223	2 062	1 345	1 309	1 525	238	2 048	229
Yugoslavia	14	18	19	26	25	24	26	25	26	25
37 Fishing area total	*6 883*	*6 799 F*	*5 591 F*	*5 154*	*5 573*	*4 876*	*3 759*	*2 665*	*4 055*	*1 878 F*
51 Kuwait	58	57	53	96	39	65	69	89	60 F	34
51 Fishing area total	*58*	*57*	*53*	*96*	*39*	*65*	*69*	*89*	*60 F*	*34*
61 China,Taiwan	2 372	1 460	932	1 638	1 302	2 446	606	863	1 890	1 559
Japan	4 710	4 569	4 585	4 579	4 268	3 933	4 003	3 629	3 725	3 457
Korea Rep	5 179	5 899	4 909	4 316	4 200	5 564	4 962	9 678	8 730	9 784
61 Fishing area total	*12 261*	*11 928*	*10 426*	*10 533*	*9 770*	*11 943*	*9 571*	*14 170*	*14 345*	*14 800*
Species total	*40 195*	*40 876 F*	*38 332 F*	*32 936*	*32 311*	*30 701*	*24 283*	*27 320*	*29 306 F*	*26 724 F*
Lebranche mullet	**Mulet lebranche**		**Lebranche**		***Mugil liza***			**1,65(01)001,12**		**MUB**
31 Colombia	780	43	16	21	39	-	2	-	-	-
Venezuela	4 070	4 402	3 644	3 228	3 248	1 550	2 874	2 855	3 186	2 537
31 Fishing area total	*4 850*	*4 445*	*3 660*	*3 249*	*3 287*	*1 550*	*2 876*	*2 855*	*3 186*	*2 537*
Species total	*4 850*	*4 445*	*3 660*	*3 249*	*3 287*	*1 550*	*2 876*	*2 855*	*3 186*	*2 537*
So-iuy mullet	**Mulet so-iuy**		**Lisa so-iuy**		***Mugil soiuy***			**1,65(01)001,28**		**MYZ**
05 Ukraine	-	-	-	-	-	-	13	190	169	382
05 Fishing area total	*-*	*-*	*-*	*-*	*-*	*-*	*13*	*190*	*169*	*382*
37 Ukraine	36	73	334	775	1 039	2 718	3 661	5 174	5 406	2 428
37 Fishing area total	*36*	*73*	*334*	*775*	*1 039*	*2 718*	*3 661*	*5 174*	*5 406*	*2 428*
Species total	*36*	*73*	*334*	*775*	*1 039*	*2 718*	*3 674*	*5 364*	*5 575*	*2 810*
Bobo mullet	**Mulet bobo**		**Bobo**		***Joturus pichardi***			**1,65(01)010,01**		**MUA**
31 Mexico	766	665	245	520	572	283	323	322	451	400
31 Fishing area total	*766*	*665*	*245*	*520*	*572*	*283*	*323*	*322*	*451*	*400*
77 Mexico	184	142	37	94	87	56	51	82	46	104
77 Fishing area total	*184*	*142*	*37*	*94*	*87*	*56*	*51*	*82*	*46*	*104*
Species total	*950*	*807*	*282*	*614*	*659*	*339*	*374*	*404*	*497*	*504*
Leaping mullet	**Mulet sauteur**		**Galúa**		***Liza saliens***			**1,65(01)012,06**		**LZS**
37 Bulgaria	5	6	6 F	1	3	2	2	2	0	10
37 Fishing area total	*5*	*6*	*6 F*	*1*	*3*	*2*	*2*	*2*	*0*	*10*
Species total	*5*	*6*	*6 F*	*1*	*3*	*2*	*2*	*2*	*0*	*10*
Mullets nei	**Mulets nca**		**Lizas nep**		***Mugilidae***			**1,65(01)XXX,XX**		**MUL**
01 Benin	298 F	600 F	763	637	1 000 F	1 500 F	2 000 F	2 198	1 800 F	2 050 F
Egypt	6 972	5 840	6 116	15 543	11 836	14 538	15 356	17 041	16 697	20 688
01 Fishing area total	*7 270 F*	*6 440 F*	*6 879*	*16 180*	*12 836 F*	*16 038 F*	*17 356 F*	*19 239*	*18 497 F*	*22 738 F*
04 Azerbaijan	66	40	90	73	103	82	61	2	3	55
India	-	-	2 855	3 319	9 852	9 424	4 205	3 833	11 195	12 533
Iran	2 200	5 135	2 809	5 014	2 554	3 074	4 558	4 080	5 125	5 263
Japan	981	943	973	832	850	1 053	887	1 000	1 140	1 084

B-33 Miscellaneous coastal fishes / Poissons côtiers divers / Peces costeros diversos

Capture production by species, fishing areas and countries or areas / Captures par espèces, zones de pêche et pays ou zones / Capturas por especies, áreas de pesca y países o áreas

Species, Fishing area Espèce, Zone de pêche Especie, Area de pesca	1992 mt	1993 mt	1994 mt	1995 mt	1996 mt	1997 mt	1998 mt	1999 mt	2000 mt	2001 mt
Kazakhstan	35	34	32	31	30 F	20 F	15 F	10 F	10 F	7
Korea Rep	-	-	-	-	-	-	-	133	-	-
Philippines	539	628	2 006	2 209	2 440	1 675	613	853	768	667
Turkey	13 717	13 584	13 699	13 767	14 207	22 000	21 200	20 752	16 352	16 558
Turkmenistan	...	...	...	...	...	10	4	3	3	1
04 Fishing area total	*17 538*	*20 364*	*22 464*	*25 245*	*30 036 F*	*37 338 F*	*31 543 F*	*30 666 F*	*34 596 F*	*36 168*
05 Albania	-	-	-	-	-	-	100	74	105	50
Russian Fed	-	-	27	31	25	55	39	129	35	11
05 Fishing area total	-	-	*27*	*31*	*25*	*55*	*139*	*203*	*140*	*61*
21 USA	227	276	3	298	2 645	553	370	2 323	546	412
21 Fishing area total	*227*	*276*	*3*	*298*	*2 645*	*553*	*370*	*2 323*	*546*	*412*
27 Channel Is	5	5 F	9	8 F	23	14	11	13	12	9
Denmark	10	16	22	31	22	29	24	26	22	29
France	989	1 041	885	938	685	787	794	850	819	974
Germany	-	-	-	-	1	3	6	21	13	23
Ireland	6	4	5	22	40	33	15	29	11	3
Netherlands	...	...	...	4	-	0	-	17	36	184
Portugal	457	381	352	398	297	303	336	324	336	376
Spain	56	188	171	130	145	90	178	210	117	75
UK	115	89	166	203	63	126	89	115	67	65
27 Fishing area total	*1 638*	*1 724 F*	*1 610*	*1 734 F*	*1 276*	*1 385*	*1 453*	*1 605*	*1 433*	*1 738*
31 Colombia	402	5	34	10	83	34	27	16	30 F	30 F
Cuba	144	158	124	108	93	159	122	21	20 F	20 F
Dominican Rp	275	108	324	320	43	59	41	22	21	28
Mexico	3 402	5 298	6 401	6 663	6 023	8 014	7 897	8 311	9 057	7 096
Puerto Rico	...	...	...	...	...	...	...	39	44	41
USA	9 962	14 037	13 849	9 790	5 073	8 352	7 759	2 219	5 607	6 148
31 Fishing area total	*14 185*	*19 606*	*20 732*	*16 891*	*11 315*	*16 618*	*15 846*	*10 628*	*14 779 F*	*13 363 F*
34 Benin	...	...	...	...	...	...	...	3	2 F	5
Gabon	...	...	...	51	271	254	233	382	365	1 061
Gambia	22	23	18	279	475	278	66	70	123	69
Georgia	25 F	10 F	-	-	-	-	-	-	-	-
Guinea	1 040 F	1 240 F	1 280 F	1 800	1 901	1 244	1 534	1 600	1 894	1 850 F
GuineaBissau	2 750 F	2 800 F	2 870 F	2 930 F	3 050 F	3 200 F	2 650 F	2 200 F	2 200 F	2 200 F
Korea Rep	10	-	-	-	-	-	-	-	-	-
Latvia	-	-	-	58	21	1	20	44	19	...
Liberia	-	4	3	-	-	22	18	63	85	...
Lithuania	31	-	-	-	-	-	-	-	-	-
Mauritania	...	...	...	...	...	...	...	2 000 F	2 000 F	2 000 F
Morocco	1 208	1 819	787	1 609	1 600	2 570	2 865	3 730	2 743	4 338
Nigeria	6 048	1 580	3 333	3 421	1 364	7 073	8 450	10 226	8 535	8 728
Portugal	3	4	0	-	-	-	-	-	-	-
Romania	48	10	-	-	-	-	-	-	-	-
Russian Fed	129	19	6	27	135	52	26	102	126	151
Senegal	5 980	3 428	3 415	1 947	2 109	2 780	1 690	2 512	4 226	2 546
Sierra Leone	608	590	599	598	600 F	597	150	233	595	...
Togo	3	4	4	0	0	0	1	-	3	0
Ukraine	126	...	...	...	...	...	...	78	94	27
34 Fishing area total	*18 031 F*	*11 531 F*	*12 315 F*	*12 720 F*	*11 526 F*	*18 071 F*	*17 703 F*	*23 243 F*	*23 010 F*	*22 975 F*
37 Albania	196	150 F	100 F	52	105	42	36	66	45	130
Egypt	987	1 152	2 160	2 600	2 896	2 012	2 304	1 995	1 396	2 085
France	2 353	2 104	589	206	401	505	505	608	453	511
Gaza Strip	...	...	...	...	32	19	35	35 F	35 F	30 F
Georgia	-	-	-	-	-	-	-	9	19	28
Italy	4 753	5 012	4 991	5 655	5 172	5 281	5 344	4 799	4 095	5 023
Lebanon	125 F	170	200	400	400	300	300	300	...	400
Malta	0	0	0	0	0	0	0	0	0	0
Morocco	8	7	44	27	30	58	42	20	15	9
Russian Fed	2	55	224	250	575	1 163	1 730	2 380	2 499	1 421
Spain	-	-	-	-	344	366	482	-	-	-
Tunisia	700	516	594	375	932	1 472	2 272	1 383	2 306	3 016
Turkey	13 688	11 873	14 943	17 710	23 308	20 500	24 150	26 000	27 000	22 000
Ukraine	0	0	0	4	3	0	5	12	27	51
37 Fishing area total	*22 812 F*	*21 039 F*	*23 845 F*	*27 279*	*34 198*	*31 718*	*37 205*	*37 607 F*	*37 890 F*	*34 704 F*
41 Argentina	8	8	8	18	10	10	10	10	5	-
Brazil	14 970 F	14 940 F	14 970 F	13 324	7 722	10 262	10 006	10 886	11 987	11 900 F
Uruguay	92	156	206	100	346	214	272	57	194	219
41 Fishing area total	*15 070 F*	*15 104 F*	*15 184 F*	*13 442*	*8 078*	*10 486*	*10 288*	*10 953*	*12 186*	*12 119 F*
47 Angola	0	0	0	0	0	0	-	-	-	-
Namibia	-	-	-	112	...	...	...	...	161	117
South Africa	914	1 253	1 106	1 147	1 086	972	908	757	400 F	8
47 Fishing area total	*914*	*1 253*	*1 106*	*1 259*	*1 086*	*972*	*908*	*757*	*561 F*	*125*
51 Bahrain	92	270	86	201	99	60	73	10	59	39
Djibouti	0	0	0	0	0	0	0	0	0	0
Egypt	-	-	-	31	67	59	352	1 811	2 365	3 760
Eritrea	...	...	...	...	3	4	2	4	3	3
India	3 807	3 761	5 616	5 255	5 747	7 623	5 760	6 104	6 799	6 004
Kenya	117	116	109	127	153	120	107	146	181	199
Kuwait	340	359	356	579	540	628	1 068	965	700 F	422
Mauritius	131	121	125	115	121	103	111	120	76	92

B-33 Miscellaneous coastal fishes / Poissons côtiers divers / Peces costeros diversos

Capture production by species, fishing areas and countries or areas / Captures par espèces, zones de pêche et pays ou zones / Capturas por especies, áreas de pesca y países o áreas

Species, Fishing area Espèce, Zone de pêche Especie, Area de pesca	1992 mt	1993 mt	1994 mt	1995 mt	1996 mt	1997 mt	1998 mt	1999 mt	2000 mt	2001 mt
Oman	90	100	1 207	98	79	139	176	158	123	543
Pakistan	10 176	22 485	19 039	17 280	17 631	18 935	17 580	12 336	9 618	9 723
Qatar	6	12	8	8	8	7	9	5	4	4
Saudi Arabia	145	112	29	400	320	369	510	317	326	397
Tanzania	500	436	234	470	618	550	300	350	400	250
Untd Arab Em	330	334	460	846	1 360	1 453	1 458	1 494	86	90 F
Yemen	0	0	0	391	380	400 F	500 F	400 F	400 F	500 F
51 Fishing area total	*15 734*	*28 106*	*27 269*	*25 801*	*27 126*	*30 450 F*	*28 006 F*	*24 220 F*	*21 140 F*	*22 026 F*
57 Australia	754	776	773	751	607	674	679	361	306	265
India	27 287	11 043	9 518	6 791	8 666	9 824	9 492	9 935	10 843	11 343
Indonesia	5 843	7 305	7 876	7 365	10 158	9 556	8 720	8 791	9 046	9 090
Malaysia	1 391	1 481	1 594	3 124	3 388	2 420	3 279	2 705	3 905	2 669
Thailand	831	737	1 125	793	1 135	1 322	1 644	2 798	2 760	2 743
57 Fishing area total	*36 106*	*21 342*	*20 886*	*18 824*	*23 954*	*23 796*	*23 814*	*24 590*	*26 860*	*26 110*
61 China	...	...	...	71 426	67 423	85 967	99 023	113 454	107 243	119 246
Russian Fed	-	-	2	17	-	-	1	7	33	94
61 Fishing area total	*...*	*...*	*2*	*71 443*	*67 423*	*85 967*	*99 024*	*113 461*	*107 276*	*119 340*
71 Australia	2 349	1 448	1 486	1 350	1 627	1 630	1 484	2 579	1 828	2 456
Fiji Islands	1 289	2 482	2 530 F	1 795	759	860	1 195	3 067	2 800 F	2 915
Indonesia	21 125	21 637	23 099	24 563	25 292	25 922	26 862	26 646	27 031	27 150
Kiribati	450	440	440	450	450	500	149	994	611	1 300
Malaysia	1 837	1 617	2 839	2 087	1 942	2 130	2 348	2 157	1 444	1 334
NewCaledonia	140	90	48	33	20	61	64	63	75	75 F
Philippines	18 719	13 724	17 272	14 113	12 969	13 364	13 591	14 807	14 183	14 950
Singapore	51	34	21	173	54	55	34	32	42	27
Thailand	3 771	3 273	3 961	3 678	3 952	4 075	3 905	2 700	2 663	2 647
71 Fishing area total	*49 731*	*44 745*	*51 696 F*	*48 242*	*47 065*	*48 597*	*49 632*	*53 045*	*50 677 F*	*52 854 F*
77 Cook Is	5 F	5 F	5 F	5 F	5 F	5 F	5 F	5 F	5 F	5 F
Mexico	3 786	4 107	3 832	3 952	3 764	3 467	2 987	3 449	3 129	3 166
USA	8	5	3	3	5	2	1	-	-	-
77 Fishing area total	*3 799 F*	*4 117 F*	*3 840 F*	*3 960 F*	*3 774 F*	*3 474 F*	*2 993 F*	*3 454 F*	*3 134 F*	*3 171 F*
81 Australia	3 403	3 671	5 683	5 674	4 258	4 306	4 451	3 456	3 098	3 384
New Zealand	935	837	809	886	897	872	720	846	748	910
81 Fishing area total	*4 338*	*4 508*	*6 492*	*6 560*	*5 155*	*5 178*	*5 171*	*4 302*	*3 846*	*4 294*
87 Chile	619	496	335	278	244	78	68	134	132	93
Colombia	70	28	43	25	20	26	36	28	30 F	28
Peru	23 333	14 711	16 964	16 601	13 916	13 264	29 075	20 843	26 314	27 330
87 Fishing area total	*24 022*	*15 235*	*17 342*	*16 904*	*14 180*	*13 368*	*29 179*	*21 005*	*26 476 F*	*27 451*
Species total	*231 415 F*	*215 390 F*	*231 692 F*	*306 813 F*	*301 698 F*	*344 064 F*	*370 630 F*	*381 301 F*	*383 047 F*	*399 649 F*
Fusiliers	**Fusiliers**		**Fusileros**		***Caesio spp***			**1,70(00)112,XX**		**FUS**
51 Jordan	...	...	...	...	...	...	15	18	17	15
51 Fishing area total	*...*	*...*	*...*	*...*	*...*	*...*	*15*	*18*	*17*	*15*
57 Indonesia	3 925	4 794	4 975	5 333	4 163	5 306	4 857	5 568	5 088	5 170
Malaysia	51	5	23	57	35	23	5	1	19	14
57 Fishing area total	*3 976*	*4 799*	*4 998*	*5 390*	*4 198*	*5 329*	*4 862*	*5 569*	*5 107*	*5 184*
71 Indonesia	17 213	17 204	23 325	37 360	28 551	33 052	29 285	32 376	28 624	29 070
Malaysia	1 751	1 877	1 547	1 543	1 646	1 936	1 312	1 323	961	973
Philippines	13 504	15 476	15 292	13 914	13 788	16 754	15 931	14 952	13 516	14 086
Singapore	104	22	1	1	-	9	18	13	10	1
71 Fishing area total	*32 572*	*34 579*	*40 165*	*52 818*	*43 985*	*51 751*	*46 546*	*48 664*	*43 111*	*44 130*
Species total	*36 548*	*39 378*	*45 163*	*58 208*	*48 183*	*57 080*	*51 423*	*54 251*	*48 235*	*49 329*
Common snook	**Crossie blanc**		**Róbalo blanco**		***Centropomus undecimalis***			**1,70(01)025,10**		**SNO**
31 Mexico	1 995	2 139	2 090	2 885	2 955	3 307	2 990	3 415	3 184	4 000
Venezuela	2 593	4 783	4 569	3 023	2 177	1 560	2 022	1 704	1 850	1 903
31 Fishing area total	*4 588*	*6 922*	*6 659*	*5 908*	*5 132*	*4 867*	*5 012*	*5 119*	*5 034*	*5 903*
Species total	*4 588*	*6 922*	*6 659*	*5 908*	*5 132*	*4 867*	*5 012*	*5 119*	*5 034*	*5 903*
Snooks(=Robalos) nei	**Crossies nca**		**Róbalos nep**		***Centropomus spp***			**1,70(01)025,XX**		**ROB**
31 Colombia	387	86	13	22	111	5	0	-	...	...
Dominican Rp	63	52	94	35	20	24	64	12	32	17
Grenada	1	0	0	0	0	0	0	1	0	0
Mexico	844	836	852	1 124	1 313	1 140	1 017	1 125	979	1 360
Nicaragua	...	600	...	...	...	...	...	2 000	2 060	1 960
Puerto Rico	...	...	...	...	...	...	...	32	33	8
31 Fishing area total	*1 295*	*1 574*	*959*	*1 181*	*1 444*	*1 169*	*1 081*	*3 170*	*3 104*	*3 345*
41 Brazil	2 310 F	2 310 F	2 320 F	1 820	1 686	1 866	2 996	3 602	4 366	4 300 F
41 Fishing area total	*2 310 F*	*2 310 F*	*2 320 F*	*1 820*	*1 686*	*1 866*	*2 996*	*3 602*	*4 366*	*4 300 F*
77 Mexico	481	757	1 026	563	587	684	752	996	809	862
77 Fishing area total	*481*	*757*	*1 026*	*563*	*587*	*684*	*752*	*996*	*809*	*862*
87 Colombia	97	64	63	86	341	73	96	62	70 F	70 F

B-33

Miscellaneous coastal fishes — **Capture production by species, fishing areas and countries or areas**
Poissons côtiers divers — **Captures par espèces, zones de pêche et pays ou zones**
Peces costeros diversos — **Capturas por especies, áreas de pesca y países o áreas**

Species, Fishing area Espèce, Zone de pêche Especie, Area de pesca	1992 mt	1993 mt	1994 mt	1995 mt	1996 mt	1997 mt	1998 mt	1999 mt	2000 mt	2001 mt
87 Fishing area total	*97*	*64*	*63*	*86*	*341*	*73*	*96*	*62*	*70 F*	*70 F*
Species total	*4 183 F*	*4 705 F*	*4 368 F*	*3 650*	*4 058*	*3 792*	*4 925*	*7 830*	*8 349 F*	*8 577 F*
Black grouper	**Badèche bonaci**		**Cuna bonací**		***Mycteroperca bonaci***				**1,70(02)040,01**	**MAB**
31 USA	...	...	...	...	...	...	...	...	9	7
31 Fishing area total	...	...	...	...	...	...	...	...	*9*	*7*
Species total	...	...	...	...	...	...	...	...	*9*	*7*
Gag	**Badèche baillou**		**Cuna aguají**		***Mycteroperca microlepis***				**1,70(02)040,04**	**MKM**
31 USA	...	...	...	...	...	...	...	...	15	10
31 Fishing area total	...	...	...	...	...	...	...	...	*15*	*10*
Species total	...	...	...	...	...	...	...	...	*15*	*10*
Scamp	**Badèche galopin**		**Cuna garopa**		***Mycteroperca phenax***				**1,70(02)040,06**	**MKH**
31 USA	...	...	...	...	...	...	1	14	27	17
31 Fishing area total	...	...	...	...	...	...	*1*	*14*	*27*	*17*
Species total	...	...	...	...	...	...	*1*	*14*	*27*	*17*
Yellowfin grouper	**Badèche de roche**		**Cuna de piedra**		***Mycteroperca venenosa***				**1,70(02)040,09**	**MKV**
31 USA	...	...	...	...	...	...	...	...	3	1
31 Fishing area total	...	...	...	...	...	...	...	...	*3*	*1*
Species total	...	...	...	...	...	...	...	...	*3*	*1*
Broomtail grouper	**Badèche balai**		**Garropa jaspeada**		***Mycteroperca xenarcha***				**1,70(02)040,10**	**GBS**
87 Colombia	...	1 447	428	207	180	125	155	82	80 F	210
87 Fishing area total	...	*1 447*	*428*	*207*	*180*	*125*	*155*	*82*	*80 F*	*210*
Species total	...	*1 447*	*428*	*207*	*180*	*125*	*155*	*82*	*80 F*	*210*
Brazilian groupers nei	**Badèches nca**		**Cunas nep**		***Mycteroperca spp***				**1,70(02)040,XX**	**GPB**
41 Brazil	1 110 F	1 110 F	1 110 F	1 117	423	584	518	628	536	530 F
41 Fishing area total	*1 110 F*	*1 110 F*	*1 110 F*	*1 117*	*423*	*584*	*518*	*628*	*536*	*530 F*
Species total	*1 110 F*	*1 110 F*	*1 110 F*	*1 117*	*423*	*584*	*518*	*628*	*536*	*530 F*
Dusky grouper	**Mérou noir**		**Mero moreno**		***Epinephelus marginatus***				**1,70(02)042,01**	**GPD**
27 France	-	-	-	-	-	-	-	1	-	-
Portugal	33	28	36	21	120	11	24	-	-	-
Spain	180	229	267	305	207	187	243	95	24	51
UK	-	-	-	-	-	12	1	-	-	-
27 Fishing area total	*213*	*257*	*303*	*326*	*327*	*210*	*268*	*96*	*24*	*51*
34 Greece	3	4	4	3	2	3	1	2	1	2
Italy	2 025	1 718	1 718	-	-	-	-	-	-	-
Portugal	0	0	0	2	3	2	3	8	2	2
34 Fishing area total	*2 028*	*1 722*	*1 722*	*5*	*5*	*5*	*4*	*10*	*3*	*4*
37 Greece	136	114	108	192	110	91	56	62	89	97
Italy	2 367	2 163	1 814	1 454	558	640	124	89	97	252
Spain	-	-	-	-	15	17	13	28	19	20
Turkey	452	513	808	620	700	600	640	135	85	80
37 Fishing area total	*2 955*	*2 790*	*2 730*	*2 266*	*1 383*	*1 348*	*833*	*314*	*290*	*449*
Species total	*5 196*	*4 769*	*4 755*	*2 597*	*1 715*	*1 563*	*1 105*	*420*	*317*	*504*
White grouper	**Mérou blanc**		**Cherna de ley**		***Epinephelus aeneus***				**1,70(02)042,02**	**GPW**
34 Côte dIvoire	3	3	1	1 F	1 F	1 F	1 F	...	2	3
Greece	355	396	343	293	306	125	123	148	137	161
Mauritania	240 F	250 F	330 F	420 F	340 F	450 F	390 F	450 F	450 F	450 F
Morocco	50	94	37	21	25	23	43	29	46	33
Portugal	-	-	-	3	5	0	-	-	-	1
34 Fishing area total	*648 F*	*743 F*	*711 F*	*738 F*	*677 F*	*599 F*	*557 F*	*627 F*	*635 F*	*648 F*
37 Morocco	15	11	12	5	5	4	4	6	48	7
37 Fishing area total	*15*	*11*	*12*	*5*	*5*	*4*	*4*	*6*	*48*	*7*
Species total	*663 F*	*754 F*	*723 F*	*743 F*	*682 F*	*603 F*	*561 F*	*633 F*	*683 F*	*655 F*
Nassau grouper	**Mérou rayé**		**Cherna criolla**		***Epinephelus striatus***				**1,70(02)042,20**	**GPN**
31 Bahamas	...	...	665	358	331	514	511	381	226	281
Colombia	120	48	65	25	11	5	3	0	...	...
Cuba	97	70	83	81	117	77	40	48	50 F	50 F
31 Fishing area total	*217*	*118*	*813*	*464*	*459*	*596*	*554*	*429*	*276 F*	*331 F*
Species total	*217*	*118*	*813*	*464*	*459*	*596*	*554*	*429*	*276 F*	*331 F*

B-33 Miscellaneous coastal fishes / Poissons côtiers divers / Peces costeros diversos

Capture production by species, fishing areas and countries or areas
Captures par espèces, zones de pêche et pays ou zones
Capturas por especies, áreas de pesca y países o áreas

Species, Fishing area Espèce, Zone de pêche Especie, Area de pesca	1992 mt	1993 mt	1994 mt	1995 mt	1996 mt	1997 mt	1998 mt	1999 mt	2000 mt	2001 mt
Spotted grouper	**Mérou cabrilla**		**Mero moteado**		***Epinephelus analogus***			**1,70(02)042,22**		**GPS**
87 Colombia	43	16	28	22	43	48	32	39	30 F	28
87 Fishing area total	*43*	*16*	*28*	*22*	*43*	*48*	*32*	*39*	*30 F*	*28*
Species total	*43*	*16*	*28*	*22*	*43*	*48*	*32*	*39*	*30 F*	*28*
Yellowedge grouper	**Mérou aile jaune**		**Mero aleta amarilla**		***Epinephelus flavolimbatus***			**1,70(02)042,24**		**EEL**
31 USA	...	...	...	...	...	15	18	...	73	36
31 Fishing area total	...	...	...	...	...	*15*	*18*	...	*73*	*36*
Species total	...	...	...	...	...	*15*	*18*	...	*73*	*36*
Red hind	**Mérou couronné**		**Mero colorado**		***Epinephelus guttatus***			**1,70(02)042,25**		**EEU**
31 Grenada	45	42	56	81	60	42	52	72	67	88
31 Fishing area total	*45*	*42*	*56*	*81*	*60*	*42*	*52*	*72*	*67*	*88*
Species total	*45*	*42*	*56*	*81*	*60*	*42*	*52*	*72*	*67*	*88*
Red grouper	**Mérou rouge**		**Mero americano**		***Epinephelus morio***			**1,70(02)042,28**		**GPR**
31 Cuba	429	341	296	211	276	198	86	119	120 F	120 F
Dominican Rp	252	280	504	763	29	92	11	21	33	37
31 Fishing area total	*681*	*621*	*800*	*974*	*305*	*290*	*97*	*140*	*153 F*	*157 F*
41 Brazil	1 810 F	1 810 F	1 820 F	872	320	238	212	666	1 220	1 200 F
41 Fishing area total	*1 810 F*	*1 810 F*	*1 820 F*	*872*	*320*	*238*	*212*	*666*	*1 220*	*1 200 F*
Species total	*2 491 F*	*2 431 F*	*2 620 F*	*1 846*	*625*	*528*	*309*	*806*	*1 373 F*	*1 357 F*
Snowy grouper	**Mérou neige**		**Cherna pintada**		***Epinephelus niveatus***			**1,70(02)042,30**		**EFV**
31 USA	...	...	...	...	...	...	...	...	6	4
31 Fishing area total	...	...	...	...	...	...	...	...	*6*	*4*
Species total	...	...	...	...	...	...	...	...	*6*	*4*
Dungat grouper	**Mérou de Gorée**		**Mero de Gorea**		***Epinephelus goreensis***			**1,70(02)042,34**		**EEG**
34 Sierra Leone	-	-	120	102	196	63	76	290	277	62
34 Fishing area total	-	-	*120*	*102*	*196*	*63*	*76*	*290*	*277*	*62*
Species total	-	-	*120*	*102*	*196*	*63*	*76*	*290*	*277*	*62*
Warsaw grouper	**Mérou Varsovie**		**Mero negro**		***Epinephelus nigritus***			**1,70(02)042,36**		**ELG**
31 USA	...	...	54	37	16	23	23	-	45	44
31 Fishing area total	...	...	*54*	*37*	*16*	*23*	*23*	-	*45*	*44*
Species total	...	...	*54*	*37*	*16*	*23*	*23*	-	*45*	*44*
Groupers nei	**Mérous nca**		**Meros nep**		***Epinephelus spp***			**1,70(02)042,XX**		**GPX**
31 Aruba	...	...	...	20	20	25	22	25	18	15
Bahamas	539	618	514	513	475	246	227	193	132	177
Bermuda	44	36	45	42	42	42	43	48	27	45
Br Virgin Is	...	...	10	12	69	1	3	4	1	1 F
Cuba	256	159	77	85	130	89	89	44	40 F	40 F
Dominican Rp	505	809	1 456	2 316	455	338	379	662	873	942
Mexico	14 018	13 827	13 526	15 193	12 389	11 783	11 692	13 031	13 289	10 572
Puerto Rico	...	...	...	...	...	...	...	89	104	111
St Kitts Nev	...	...	...	8	18	10	11	11	7	4
USA	4 968	6 376	9 582	4 701	4 387	4 583	4 399	1 038	5 665	5 984
US Virgin Is	...	...	...	...	...	...	...	21	25 F	25 F
Venezuela	1 680	1 772	1 930	2 040	2 185	2 235	1 990	1 591	1 408	811
31 Fishing area total	*22 010*	*23 597*	*27 140*	*24 930*	*20 170*	*19 352*	*18 855*	*16 757*	*21 589 F*	*18 727 F*
34 Benin	19 F	21 F	23 F	22 F	25 F	25 F	23 F	20	14 F	16
Cameroon	3	1	1 F	2	0	0	1	0	1	1
Congo Rep	3 F	3 F	0	5	5 F	4 F	3 F	3 F	2	2 F
Gabon	160 F	88	114	264	250	94	190	136	105	40
Gambia	33	42	54	118	62	53	30	30	49	63
Ghana	225	385	169	306	426	478	1 361	181	94	138
Liberia	12	9	6	22	-	5	110	71	25	44
Mauritania	110 F	120 F	190 F	240 F	180 F	240 F	220 F	300 F	300 F	300 F
Portugal	2	0	85	116	203	36	11	9	2	5
Sao Tome Prn	24 F	29 F	42 F	44 F	33	31	40 F	45 F	40 F	40 F
Spain	-	-	-	-	119	32	21	28	42	36
Togo	21	29	33	41	14	19	215	120	16	12
34 Fishing area total	*612 F*	*727 F*	*717 F*	*1 180 F*	*1 317 F*	*1 017 F*	*2 225 F*	*943 F*	*690 F*	*697 F*
37 Albania	...	...	...	0	2	1	-	-	-	-
Cyprus	33	26	47	25	22	22	29	23	27	24
Egypt	842	614	604	721	1 010	778	905	1 943	1 621	1 123
Gaza Strip	...	...	...	...	46	45	44	45 F	45 F	40 F
Greece	65	32	51	76	54	91	65	85	63	142

B-33 Miscellaneous coastal fishes / Poissons côtiers divers / Peces costeros diversos

Capture production by species, fishing areas and countries or areas
Captures par espèces, zones de pêche et pays ou zones
Capturas por especies, áreas de pesca y países o áreas

Species, Fishing area Espèce, Zone de pêche Especie, Area de pesca	1992 mt	1993 mt	1994 mt	1995 mt	1996 mt	1997 mt	1998 mt	1999 mt	2000 mt	2001 mt
Israel	356	336	250	298	406	384	394	381	263	220 F
Libya	3 450 F	3 700 F	4 000 F	4 100 F	4 000 F	4 000 F	4 000 F	4 000 F	4 000 F	4 000 F
Malta	0	1	0	15	19	27	15	0	15	15
Tunisia	669	381	302	236	265	416	334	382	103	539
37 Fishing area total	*5 415 F*	*5 090 F*	*5 254 F*	*5 471 F*	*5 824 F*	*5 764 F*	*5 786 F*	*6 859 F*	*6 137 F*	*6 103 F*
41 Brazil	2 310 F	2 300 F	2 300 F	2 286	2 341	2 170	2 315	1 710	2 670	2 630 F
Italy	270	229	229	-	-	-	-	-	-	-
41 Fishing area total	*2 580 F*	*2 529 F*	*2 529 F*	*2 286*	*2 341*	*2 170*	*2 315*	*1 710*	*2 670*	*2 630 F*
47 Korea Rep	2	37	23	-	-	-	-	-	-	-
St Helena	28	20	37	32	48	41	59	33	17	18
47 Fishing area total	*30*	*57*	*60*	*32*	*48*	*41*	*59*	*33*	*17*	*18*
51 Bahrain	530	469	498	459	532	300	331	525	670	794
Djibouti	84 F	92 F	98 F	100 F	100 F	100 F	100 F	100 F	100 F	100 F
Egypt	58	334	247	387	647	689	722	1 215	3 126	3 576
Eritrea	...	...	...	...	1	56	50	...	48	...
Italy	405	344	344	-	-	-	-	-	-	-
Korea Rep	598	300	632	209	389	174	119	820	331	227
Kuwait	532	431	413	341	287	241	264	237	250 F	268
Oman	2 789	2 075	3 735	4 031	3 409	3 366	5 345	4 829	5 013	3 799
Pakistan	4 863	6 255	7 617	8 600	9 793	10 474	13 991	17 355	16 012	15 928
Qatar	1 200	1 145	950	728	768	736	804	913	1 215	1 820
Tanzania	700	278	335	290	652	400	350	500	450	500
51 Fishing area total	*11 759 F*	*11 723 F*	*14 869 F*	*15 145 F*	*16 578 F*	*16 536 F*	*22 076 F*	*26 494 F*	*27 215 F*	*27 012 F*
57 Indonesia	8 132	10 995	10 491	12 647	13 866	15 351	16 724	14 726	14 798	15 710
Korea Rep	-	-	-	-	-	-	-	234	22	32
Malaysia	2 391	2 166	1 671	1 513	1 533	1 654	1 487	1 347	1 491	1 156
57 Fishing area total	*10 523*	*13 161*	*12 162*	*14 160*	*15 399*	*17 005*	*18 211*	*16 307*	*16 311*	*16 898*
61 China	21 598	22 000	22 469	22 999	23 241	30 201	36 098	40 245	41 513	44 975
China,Taiwan	2 002	2 906	3 019	1 771	2 457	2 387	1 414	1 265	1 202	1 610
61 Fishing area total	*23 600*	*24 906*	*25 488*	*24 770*	*25 698*	*32 588*	*37 512*	*41 510*	*42 715*	*46 585*
71 Fiji Islands	1 034	1 500	1 529 F	854	1 017	1 060	1 160	1 600	1 450 F	1 544
Guam	...	...	3	1	0	1	0	1	1	1
Indonesia	13 635	19 020	29 430	21 357	24 420	26 813	27 042	28 746	33 624	35 690
Korea Rep	-	1	-	472	-	-	1	-	-	-
Malaysia	6 772	5 710	5 538	5 523	6 741	7 470	9 114	10 882	10 683	9 320
N Marianas	...	...	2	1	3	5	3	2	2	3
Palau	35	14	8	9	15	1	3	10 F	13 F	13 F
Singapore	51	31	129	109	120	94	72	56	43	38
71 Fishing area total	*21 527*	*26 276*	*36 639 F*	*28 326*	*32 316*	*35 444*	*37 395*	*41 297 F*	*45 816 F*	*46 609 F*
77 Amer Samoa	...	...	2	2	3	2	2	1	1	1
Cook Is	130 F	132 F	125 F	123 F	110 F	100 F	90 F	80 F	60 F	60 F
Mexico	735	1 928	1 414	2 588	5 254	7 126	6 158	6 800	5 306	4 037
77 Fishing area total	*865 F*	*2 060 F*	*1 541 F*	*2 713 F*	*5 367 F*	*7 228 F*	*6 250 F*	*6 881 F*	*5 367 F*	*4 098 F*
87 Colombia	269	1 447	428	207	180	125	155	82	70 F	57
Peru	...	16	110	191	241	60	101	151	91	87
87 Fishing area total	*269*	*1 463*	*538*	*398*	*421*	*185*	*256*	*233*	*161 F*	*144*
Species total	*99 190 F*	*111 589 F*	*126 937 F*	*119 411 F*	*125 479 F*	*137 330 F*	*150 940 F*	*159 024 F*	*168 688 F*	*169 521 F*
Orange perch	**...B**		**...C**			***Lepidoperca pulchella***			**1,70(02)072,01**	**LDP**
81 New Zealand	...	...	...	...	...	...	193	32	...	46
81 Fishing area total	...	...	...	...	...	...	*193*	*32*	...	*46*
Species total	...	...	...	...	...	...	*193*	*32*	...	*46*
Black seabass	**Fanfre noir**		**Serrano estriado**			***Centropristis striata***			**1,70(02)081,02**	**BSB**
21 USA	1 424	1 453	1 339	904	1 683	1 215	1 190	1 412	1 189	1 276
21 Fishing area total	*1 424*	*1 453*	*1 339*	*904*	*1 683*	*1 215*	*1 190*	*1 412*	*1 189*	*1 276*
31 USA	506	425	693	393	295	375	341	305	327	391
31 Fishing area total	*506*	*425*	*693*	*393*	*295*	*375*	*341*	*305*	*327*	*391*
Species total	*1 930*	*1 878*	*2 032*	*1 297*	*1 978*	*1 590*	*1 531*	*1 717*	*1 516*	*1 667*
Argentine seabass	**Serran argentin**		**Mero sureño**			***Acanthistius brasilianus***			**1,70(02)259,01**	**BSZ**
41 Argentina	6 384	9 195	7 756	10 926	10 714	9 113	5 216	5 895	4 116	4 692
Uruguay	5	9	6	61	85	51	42	19	88	9
41 Fishing area total	*6 389*	*9 204*	*7 762*	*10 987*	*10 799*	*9 164*	*5 258*	*5 914*	*4 204*	*4 701*
Species total	*6 389*	*9 204*	*7 762*	*10 987*	*10 799*	*9 164*	*5 258*	*5 914*	*4 204*	*4 701*
Peruvian rock seabass	**Serran cabrilla**		**Cabrilla loca**			***Paralabrax humeralis***			**1,70(02)405,05**	**BAP**
87 Peru	4 895	3 647	3 104	5 837	4 954	2 789	2 554	3 278	4 373	2 011
87 Fishing area total	*4 895*	*3 647*	*3 104*	*5 837*	*4 954*	*2 789*	*2 554*	*3 278*	*4 373*	*2 011*
Species total	*4 895*	*3 647*	*3 104*	*5 837*	*4 954*	*2 789*	*2 554*	*3 278*	*4 373*	*2 011*

B-33

Miscellaneous coastal fishes — Capture production by species, fishing areas and countries or areas
Poissons côtiers divers — Captures par espèces, zones de pêche et pays ou zones
Peces costeros diversos — Capturas por especies, áreas de pesca y países o áreas

Species, Fishing area Espèce, Zone de pêche Especie, Area de pesca	1992 mt	1993 mt	1994 mt	1995 mt	1996 mt	1997 mt	1998 mt	1999 mt	2000 mt	2001 mt
Groupers, seabasses nei	**Serranidés nca**		**Meros, chernas, nep**		*Serranidae*			**1,70(02)XXX,XX**		**BSX**
21 USA	...	...	...	...	...	...	23	4 702	18	18
21 Fishing area total	...	...	...	...	...	...	*23*	*4 702*	*18*	*18*
27 Germany	-	-	-	-	-	-	-	-	-	1
Netherlands	-	-	-	-	-	-	17	14	-	-
Portugal	730	579	488	372	437	273	329	325	283	171
Spain	-	-	-	-	142	432	442	444	593	421
27 Fishing area total	*730*	*579*	*488*	*372*	*579*	*705*	*788*	*783*	*876*	*593*
31 Antigua Barb	...	...	...	...	...	...	...	...	...	217
Colombia	23	23	39	67	86	17	37	26	30 F	30 F
Costa Rica	...	...	...	...	...	...	...	1	3	0
Grenada	9	7	4	4	10	11	16	20	22	38
Nicaragua	...	...	...	...	...	...	...	440	460	435
USA	...	...	...	...	...	...	...	...	55	16
Venezuela	101	133	264	279	265	322	384	328	140	123
31 Fishing area total	*133*	*163*	*307*	*350*	*361*	*350*	*437*	*815*	*710 F*	*859 F*
34 Greece	-	0	1	1	-	3	-	-	-	192
Korea Rep	2	-	-	-	-	177	356	413	536	351
Nigeria	2 726	3 113	3 468	3 102	1 285	2 001	1 741	2 486	2 117	2 487
Portugal	137	270	180	11	24	10	7	11	9	8
Senegal	3 159	4 243	5 269	4 175	4 411	4 317	3 488	3 162	3 067	3 685
Spain	-	-	-	-	58	12	156	176	66	82
34 Fishing area total	*6 024*	*7 626*	*8 918*	*7 289*	*5 778*	*6 520*	*5 748*	*6 248*	*5 795*	*6 805*
37 Algeria	...	...	...	...	...	...	...	6	6 F	6 F
Cyprus	5	4	6	13	16	15	11	17	3	5
Greece	287	322	756	428	410	616	330	336	286	271
Lebanon	100 F	125	150	250	250	150	250	250	230	240
Malta	0	0	0	1	1	1	1	2	2	2
Spain	-	-	-	-	14	23	30	71	4	121
Tunisia	146	49	121	75	84	130	3 161	748	98	98
Turkey	269	210	280	339	790	1 000	700	350	400	410
37 Fishing area total	*807 F*	*710*	*1 313*	*1 106*	*1 565*	*1 935*	*4 483*	*1 780*	*1 029 F*	*1 153 F*
41 Portugal	-	-	-	-	-	15	2	-	-	1
41 Fishing area total	-	-	-	-	-	*15*	*2*	-	-	*1*
47 Angola	5	5	66	235	1 002	460	1 326	347	2 341	4 214
South Africa	-	-	-	-	-	-	-	-	-	15
Spain	-	-	-	-	-	-	91	-	-	-
47 Fishing area total	*5*	*5*	*66*	*235*	*1 002*	*460*	*1 417*	*347*	*2 341*	*4 229*
51 Eritrea	...	...	14	83	140	39	67	257	378	270
Mauritius	912	1 026	1 035	1 022	905	933	931	826	863	938
Réunion	...	...	...	...	...	...	...	47	20	11
Saudi Arabia	5 731	5 752	5 707	3 514	4 207	4 403	5 053	5 430	5 402	5 273
Seychelles	286	249	133	71	66	124	45	143	56	107
South Africa	43	-	33	25	30	-	-	-	...	...
Spain	-	-	-	-	-	-	-	2	-	-
Untd Arab Em	5 562	5 997	6 330	6 696	6 767	7 232	7 256	7 437	24 045	24 050 F
Yemen	2 038	2 487	2 400	2 260	1 743	1 820 F	2 100 F	2 000 F	1 800 F	2 300 F
51 Fishing area total	*14 572*	*15 511*	*15 652*	*13 671*	*13 858*	*14 551 F*	*15 452 F*	*16 142 F*	*32 564 F*	*32 949 F*
57 Australia	83	95	323	377	657	1	4	60	74	82
Thailand	-	-	2 600	3 778	3 502	3 295	4 290	2 585	2 550	2 534
57 Fishing area total	*83*	*95*	*2 923*	*4 155*	*4 159*	*3 296*	*4 294*	*2 645*	*2 624*	*2 616*
61 China,H.Kong	2 383	2 109	1 699	1 392	1 349	1 318	1 240	900 F	1 100 F	1 200 F
Korea Rep	639	782	440	557	436	583	765	551	947	640
61 Fishing area total	*3 022*	*2 891*	*2 139*	*1 949*	*1 785*	*1 901*	*2 005*	*1 451 F*	*2 047 F*	*1 840 F*
71 Australia	55	32	55	34	33	22	34	0	3	0
Philippines	14 557	14 358	13 654	14 226	12 776	12 197	13 160	13 675	12 492	13 285
Thailand	-	-	5 679	5 431	5 847	5 649	5 020	5 465	5 390	5 357
71 Fishing area total	*14 612*	*14 390*	*19 388*	*19 691*	*18 656*	*17 868*	*18 214*	*19 140*	*17 885*	*18 642*
77 Costa Rica	747	438	328	281	177	387	644	395	201	103
Nicaragua	...	...	...	...	...	...	...	600	380	335
77 Fishing area total	*747*	*438*	*328*	*281*	*177*	*387*	*644*	*995*	*581*	*438*
81 Australia	46	55	58	85	170	132	141	11	22	10
Japan	51	51	25	10	3	13	21	10	11	13
81 Fishing area total	*97*	*106*	*83*	*95*	*173*	*145*	*162*	*21*	*33*	*23*
87 Chile	11	17	39	23	30	24	10	19	7	5
87 Fishing area total	*11*	*17*	*39*	*23*	*30*	*24*	*10*	*19*	*7*	*5*
Species total	*40 843 F*	*42 531*	*51 644*	*49 217*	*48 123*	*48 157 F*	*53 679 F*	*55 088 F*	*66 510 F*	*70 171 F*
Therapon pearch	**...B**		**...C**		*Terapon spp*			**1,70(04)089,XX**		**THO**
51 Saudi Arabia	-	-	-	-	-	-	1	2	0	1
51 Fishing area total	-	-	-	-	-	-	*1*	*2*	*0*	*1*
Species total	-	-	-	-	-	-	*1*	*2*	*0*	*1*

B-33 Miscellaneous coastal fishes / Poissons côtiers divers / Peces costeros diversos

Capture production by species, fishing areas and countries or areas
Captures par espèces, zones de pêche et pays ou zones
Capturas por especies, áreas de pesca y países o áreas

Species, Fishing area Espèce, Zone de pêche Especie, Area de pesca	1992 mt	1993 mt	1994 mt	1995 mt	1996 mt	1997 mt	1998 mt	1999 mt	2000 mt	2001 mt
Giant seabass	**Bar gigantesque**		**Lubina gigante**		***Stereolepis gigas***			**1,70(05)220,02**		**TEJ**
77 USA	40	72	0	0	1	1	3	2	2	3
77 Fishing area total	*40*	*72*	*0*	*0*	*1*	*1*	*3*	*2*	*2*	*3*
Species total	*40*	*72*	*0*	*0*	*1*	*1*	*3*	*2*	*2*	*3*
Spotted seabass	**Bar tacheté**		**Baila**		***Dicentrarchus punctatus***			**1,70(06)345,01**		**SPU**
27 France	49	38	55	76	68	42	75	68	71	79
27 Fishing area total	*49*	*38*	*55*	*76*	*68*	*42*	*75*	*68*	*71*	*79*
34 Mauritania	130 F	120 F	160 F	140 F	160 F	130 F	190 F	100 F	45 F	...
Senegal	...	...	...	9	619	472	527	170	67	125
34 Fishing area total	*130 F*	*120 F*	*160 F*	*149 F*	*779 F*	*602 F*	*717 F*	*270 F*	*112 F*	*125*
37 Egypt	98	100	604	73	424	119	135	203	186	238
37 Fishing area total	*98*	*100*	*604*	*73*	*424*	*119*	*135*	*203*	*186*	*238*
51 Egypt	...	...	...	101	2	35	36	241	10	18
51 Fishing area total	*...*	*...*	*...*	*101*	*2*	*35*	*36*	*241*	*10*	*18*
Species total	*277 F*	*258 F*	*819 F*	*399 F*	*1 273 F*	*798 F*	*963 F*	*782 F*	*379 F*	*460*
European seabass	**Bar européen**		**Lubina**		***Dicentrarchus labrax***			**1,70(06)345,03**		**BSS**
27 Channel Is	36	37 F	83	77 F	56	74	79	107	129	80
Denmark	0	1	0	1	1	1	2	1	5	2
France	2 483	2 317	2 305	2 351	3 163	2 845	2 626	3 224	3 881	3 957
Ireland	0	0	0	0	-	-	-	-	-	-
Netherlands	-	-	-	-	8	1	49	32	60	79
Portugal	53	71	107	68	57	40	38	37	49	43
Spain	362	365	457	446	534	474	457	383	473	326
UK	157	249	549	722	582	572	501	687	406	457
27 Fishing area total	*3 091*	*3 040 F*	*3 501*	*3 665 F*	*4 401*	*4 007*	*3 752*	*4 471*	*5 003*	*4 944*
34 Greece	213	169	80	137	108	19	7	26	9	20
34 Fishing area total	*213*	*169*	*80*	*137*	*108*	*19*	*7*	*26*	*9*	*20*
37 Albania	...	...	...	14	32	14	30	30	50	70
Croatia	...	...	...	...	...	...	31	20	22	13
Egypt	-	419	...	429	727	453	559	662	626	800
France	582	984	199	179	167	167	167	279	271	251
Greece	408	412	413	392	347	361	251	263	336	280
Italy	2 116	2 127	2 528	4 633	2 481	2 030	1 889	1 881	2 195	2 735
Malta	-	50	0	0	0	0	0	15	0	0
Slovenia	-	-	-	2	-	-	-	1	1	5
37 Fishing area total	*3 106*	*3 992*	*3 140*	*5 649*	*3 754*	*3 025*	*2 927*	*3 151*	*3 501*	*4 154*
Species total	*6 410*	*7 201 F*	*6 721*	*9 451 F*	*8 263*	*7 051*	*6 686*	*7 648*	*8 513*	*9 118*
Seabasses nei	**Bars nca**		**Lubinas nep**		***Dicentrarchus spp***			**1,70(06)345,XX**		**BSE**
01 Egypt	1 470	2 413	1 272	2 706	2 397	2 233	2 738	815	4 075	3 789
01 Fishing area total	*1 470*	*2 413*	*1 272*	*2 706*	*2 397*	*2 233*	*2 738*	*815*	*4 075*	*3 789*
27 Portugal	-	-	-	-	-	5	6	336	405	367
27 Fishing area total	*-*	*-*	*-*	*-*	*-*	*5*	*6*	*336*	*405*	*367*
34 Portugal	-	-	-	-	-	-	-	-	-	11
34 Fishing area total	*-*	*-*	*-*	*-*	*-*	*-*	*-*	*-*	*-*	*11*
37 Cyprus	...	...	...	...	...	...	...	...	...	3
Spain	-	-	-	-	-	-	-	158	197	259
Tunisia	527	316	316	379	430	96	529	540	573	317
Turkey	1 188	1 160	1 884	2 116	2 411	3 450	3 950	3 650	1 900	1 200
Yugoslavia	1	2	3	3	5	4	7	6	7	7
37 Fishing area total	*1 716*	*1 478*	*2 203*	*2 498*	*2 846*	*3 550*	*4 486*	*4 354*	*2 677*	*1 786*
Species total	*3 186*	*3 891*	*3 475*	*5 204*	*5 243*	*5 788*	*7 230*	*5 505*	*7 157*	*5 953*
Japanese seabass	**Bar du Japon**		**Serránido japonés**		***Lateolabrax japonicus***			**1,70(08)297,01**		**BAJ**
61 Japan	6 719	6 906	7 302	7 713	8 334	9 057	9 223	9 234	9 337	10 690
Korea Rep	1 214	1 728	1 217	1 118	1 178	1 501	2 957	2 933	1 201	1 566
61 Fishing area total	*7 933*	*8 634*	*8 519*	*8 831*	*9 512*	*10 558*	*12 180*	*12 167*	*10 538*	*12 256*
Species total	*7 933*	*8 634*	*8 519*	*8 831*	*9 512*	*10 558*	*12 180*	*12 167*	*10 538*	*12 256*
Red bigeye	**Beauclaire Pacifique**		**Catalufa Pacífico**		***Priacanthus macracanthus***			**1,70(11)026,05**		**BIR**
61 China,Taiwan	3 909	5 373	3 314	5 486	7 386	4 824	3 670	2 756	3 986	3 079
61 Fishing area total	*3 909*	*5 373*	*3 314*	*5 486*	*7 386*	*4 824*	*3 670*	*2 756*	*3 986*	*3 079*
Species total	*3 909*	*5 373*	*3 314*	*5 486*	*7 386*	*4 824*	*3 670*	*2 756*	*3 986*	*3 079*

B-33 Miscellaneous coastal fishes / Poissons côtiers divers / Peces costeros diversos

Capture production by species, fishing areas and countries or areas / Captures par espèces, zones de pêche et pays ou zones / Capturas por especies, áreas de pesca y países o áreas

Species, Fishing area Espèce, Zone de pêche Especie, Area de pesca	1992 mt	1993 mt	1994 mt	1995 mt	1996 mt	1997 mt	1998 mt	1999 mt	2000 mt	2001 mt
Bigeyes nei	**Beauclaires nca**		**Catalufas nep**		***Priacanthus spp***			**1,70(11)026,XX**		**BIG**
34 Côte dIvoire	9	12	15	15 F	20 F	20 F	25 F	29	12	1
34 Fishing area total	*9*	*12*	*15*	*15 F*	*20 F*	*20 F*	*25 F*	*29*	*12*	*1*
41 Brazil	...	...	...	70	59	55	46	60	67	70 F
41 Fishing area total	*...*	*...*	*...*	*70*	*59*	*55*	*46*	*60*	*67*	*70 F*
47 St Helena	2	3	8	3	1	1	3	7	2	1
47 Fishing area total	*2*	*3*	*8*	*3*	*1*	*1*	*3*	*7*	*2*	*1*
51 Eritrea	...	...	0	5	0	-	-	0	-	-
Saudi Arabia	-	-	-	-	-	-	2	1	0	1
51 Fishing area total	*...*	*...*	*0*	*5*	*0*	*...*	*2*	*1*	*0*	*1*
57 Indonesia	1 044	1 139	1 253	1 052	1 200	1 221	1 360	1 605	1 583	1 810
Thailand	8 399	10 129	7 614	11 456	14 711	15 150	15 361	11 608	11 449	11 379
57 Fishing area total	*9 443*	*11 268*	*8 867*	*12 508*	*15 911*	*16 371*	*16 721*	*13 213*	*13 032*	*13 189*
61 China,H.Kong	27 833	7 528	6 034	6 781	5 391	4 962	4 149	3 000 F	3 600 F	4 000 F
61 Fishing area total	*27 833*	*7 528*	*6 034*	*6 781*	*5 391*	*4 962*	*4 149*	*3 000 F*	*3 600 F*	*4 000 F*
71 Indonesia	2 516	2 165	2 258	2 862	2 479	3 227	3 252	3 377	4 620	5 280
Thailand	36 221	49 710	44 680	57 723	67 411	62 675	64 873	71 127	70 153	69 725
71 Fishing area total	*38 737*	*51 875*	*46 938*	*60 585*	*69 890*	*65 902*	*68 125*	*74 504*	*74 773*	*75 005*
Species total	*76 024*	*70 686*	*61 862*	*79 967 F*	*91 272 F*	*87 311 F*	*89 071 F*	*90 814 F*	*91 486 F*	*92 267 F*
Cardinalfishes, etc. nei	**Apogonidés nca**		**Peces cardenal, etc. nep**		***Apogonidae***			**1,70(12)XXX,XX**		**APO**
27 France	-	-	-	-	-	-	-	294	-	-
Germany	-	-	-	-	-	-	-	-	-	10
Ireland	-	-	-	-	-	-	-	-	-	5
27 Fishing area total	*-*	*-*	*-*	*-*	*-*	*-*	*-*	*294*	*-*	*15*
34 Portugal	-	-	-	11	11	0	-	-	0	-
34 Fishing area total	*-*	*-*	*-*	*11*	*11*	*0*	*-*	*-*	*0*	*-*
71 Fiji Islands	133	210	214 F	...	200	180	60	71	60 F	67
71 Fishing area total	*133*	*210*	*214 F*	*...*	*200*	*180*	*60*	*71*	*60 F*	*67*
Species total	*133*	*210*	*214 F*	*11*	*211*	*180*	*60*	*365*	*60 F*	*82*
Sillago-whitings	**Sillaginidés**		**Sillagínidos**		***Sillaginidae***			**1,70(15)XXX,XX**		**WHS**
51 Pakistan	317	321	365	423	289	266	218	201	194	204
51 Fishing area total	*317*	*321*	*365*	*423*	*289*	*266*	*218*	*201*	*194*	*204*
57 Australia	377	352	294	303	899	1 105	1 219	924	1 122	1 083
Malaysia	1 167	1 178	1 191	1 042	899	1 301	738	930	856	1 275
Thailand	1 228	1 444	1 593	2 781	2 532	3 144	3 345	6 240	6 155	6 117
57 Fishing area total	*2 772*	*2 974*	*3 078*	*4 126*	*4 330*	*5 550*	*5 302*	*8 094*	*8 133*	*8 475*
61 China,Taiwan	1 599	1 342	552	407	547	401	346	190	188	132
Korea Rep	120	109	157	199	248	199	48	109	100	10
61 Fishing area total	*1 719*	*1 451*	*709*	*606*	*795*	*600*	*394*	*299*	*288*	*142*
71 Australia	557	1 411	1 835	2 569	2 321	2 140	1 828	1 233	833	525
Malaysia	746	798	873	906	996	633	1 071	1 084	956	782
Philippines	8 654	8 294	8 405	7 417	6 800	8 662	8 918	9 529	9 472	9 481
Singapore	103	80	39	32	48	45	41	24	10	14
Thailand	2 016	1 956	2 490	3 785	2 901	2 594	2 156	1 746	1 722	1 712
71 Fishing area total	*12 076*	*12 539*	*13 642*	*14 709*	*13 066*	*14 074*	*14 014*	*13 616*	*12 993*	*12 514*
81 Australia	1 684	1 580	2 002	1 826	1 169	700	607	1 804	1 096	1 185
81 Fishing area total	*1 684*	*1 580*	*2 002*	*1 826*	*1 169*	*700*	*607*	*1 804*	*1 096*	*1 185*
Species total	*18 568*	*18 865*	*19 796*	*21 690*	*19 649*	*21 190*	*20 535*	*24 014*	*22 704*	*22 520*
Moonfish	**Luneur**		**Lunero**		***Mene maculata***			**1,70(26)327,01**		**MOO**
61 China,Taiwan	6 581	6 147	1 823	859	725	902	1 010	1 747	1 496	1 582
61 Fishing area total	*6 581*	*6 147*	*1 823*	*859*	*725*	*902*	*1 010*	*1 747*	*1 496*	*1 582*
71 Philippines	7 253	4 656	6 175	4 201	4 810	11 933	11 516	12 372	11 703	12 991
Singapore	42	47	15	6	5	7	13	13	15	7
71 Fishing area total	*7 295*	*4 703*	*6 190*	*4 207*	*4 815*	*11 940*	*11 529*	*12 385*	*11 718*	*12 998*
Species total	*13 876*	*10 850*	*8 013*	*5 066*	*5 540*	*12 842*	*12 539*	*14 132*	*13 214*	*14 580*
Ruff	**Saumon rude**		**Salmón áspero**		***Arripis georgianus***			**1,70(29)051,01**		**RUF**
57 Australia	2 500 F	1 500 F	1 095	1 063	1 302	1 287	1 008	1 066	1 143	1 005
57 Fishing area total	*2 500 F*	*1 500 F*	*1 095*	*1 063*	*1 302*	*1 287*	*1 008*	*1 066*	*1 143*	*1 005*
Species total	*2 500 F*	*1 500 F*	*1 095*	*1 063*	*1 302*	*1 287*	*1 008*	*1 066*	*1 143*	*1 005*
Australian salmon	**Saumon australien**		**Salmón de Australia**		***Arripis trutta***			**1,70(29)051,02**		**ASA**
57 Australia	1 699	2 946	2 720	4 840	3 507	3 457	3 623	3 354	4 087	3 799
57 Fishing area total	*1 699*	*2 946*	*2 720*	*4 840*	*3 507*	*3 457*	*3 623*	*3 354*	*4 087*	*3 799*

B-33

Miscellaneous coastal fishes — **Capture production by species, fishing areas and countries or areas**
Poissons côtiers divers — **Captures par espèces, zones de pêche et pays ou zones**
Peces costeros diversos — **Capturas por especies, áreas de pesca y países o áreas**

Species, Fishing area Espèce, Zone de pêche Especie, Area de pesca	1992 mt	1993 mt	1994 mt	1995 mt	1996 mt	1997 mt	1998 mt	1999 mt	2000 mt	2001 mt
81 Australia	787	786	489	1 086	1 162	1 316	296	172	449	720
New Zealand	6 100	6 512	4 548	4 419	3 496	2 777	2 928	4 780	3 825	2 934
81 Fishing area total	*6 887*	*7 298*	*5 037*	*5 505*	*4 658*	*4 093*	*3 224*	*4 952*	*4 274*	*3 654*
Species total	*8 586*	*10 244*	*7 757*	*10 345*	*8 165*	*7 550*	*6 847*	*8 306*	*8 361*	*7 453*
Mangrove red snapper	**Vivaneau des mangroves**		**Pargo de manglar**		***Lutjanus argentimaculatus***				**1,70(32)027,02**	**RES**
51 Pakistan	1 222	2 178	2 524	3 145	2 002	2 394	3 192	3 195	3 003	2 900
51 Fishing area total	*1 222*	*2 178*	*2 524*	*3 145*	*2 002*	*2 394*	*3 192*	*3 195*	*3 003*	*2 900*
57 Malaysia	1 267	1 110	901	659	662	658	1 227	615	789	608
57 Fishing area total	*1 267*	*1 110*	*901*	*659*	*662*	*658*	*1 227*	*615*	*789*	*608*
71 Malaysia	12 678	10 163	8 629	10 522	10 824	10 190	4 935	12 319	10 650	10 436
71 Fishing area total	*12 678*	*10 163*	*8 629*	*10 522*	*10 824*	*10 190*	*4 935*	*12 319*	*10 650*	*10 436*
Species total	*15 167*	*13 451*	*12 054*	*14 326*	*13 488*	*13 242*	*9 354*	*16 129*	*14 442*	*13 944*
Yellow snapper	**Vivaneau jaune**		**Pargo amarillo (=Huachinango)**		***Lutjanus argentiventris***				**1,70(32)027,20**	**HUS**
77 Mexico	4 734	4 357	4 081	3 810	4 917	3 123	3 390	2 994	3 392	3 388
77 Fishing area total	*4 734*	*4 357*	*4 081*	*3 810*	*4 917*	*3 123*	*3 390*	*2 994*	*3 392*	*3 388*
87 Colombia	178	103	63	120	43	40	100	82	80 F	...
87 Fishing area total	*178*	*103*	*63*	*120*	*43*	*40*	*100*	*82*	*80 F*	*...*
Species total	*4 912*	*4 460*	*4 144*	*3 930*	*4 960*	*3 163*	*3 490*	*3 076*	*3 472 F*	*3 388*
Southern red snapper	**Vivaneau rouge**		**Pargo colorado**		***Lutjanus purpureus***				**1,70(32)027,22**	**SNC**
31 Colombia	104	195	235	304	17	...	290	172	250 F	250 F
Cuba	604	476	552	809	963	865	716	698	700 F	700 F
Dominican Rp	413	430	774	484	316	496	157	71	126	355
Guyana	...	...	...	...	...	...	...	...	570	524
31 Fishing area total	*1 121*	*1 101*	*1 561*	*1 597*	*1 296*	*1 361*	*1 163*	*941*	*1 646 F*	*1 829 F*
41 Brazil	3 620 F	3 610 F	3 620 F	5 816	5 104	6 085	5 937	9 790	6 580	6 500 F
41 Fishing area total	*3 620 F*	*3 610 F*	*3 620 F*	*5 816*	*5 104*	*6 085*	*5 937*	*9 790*	*6 580*	*6 500 F*
Species total	*4 741 F*	*4 711 F*	*5 181 F*	*7 413*	*6 400*	*7 446*	*7 100*	*10 731*	*8 226 F*	*8 329 F*
Northern red snapper	**Vivaneau campèche**		**Pargo del Golfo**		***Lutjanus campechanus***				**1,70(32)027,25**	**SNR**
31 Mexico	6 741	7 198	4 903	4 714	4 555	4 219	3 392	3 445	2 726	2 717
USA	1 510	1 513	1 520	1 412	1 791	2 045	1 774	1 967	2 115	2 152
31 Fishing area total	*8 251*	*8 711*	*6 423*	*6 126*	*6 346*	*6 264*	*5 166*	*5 412*	*4 841*	*4 869*
Species total	*8 251*	*8 711*	*6 423*	*6 126*	*6 346*	*6 264*	*5 166*	*5 412*	*4 841*	*4 869*
Lane snapper	**Vivaneau gazou**		**Pargo biajaiba**		***Lutjanus synagris***				**1,70(32)027,33**	**SNL**
31 Colombia	548	224	58	225	236	35	12	11	20 F	20 F
Cuba	1 445	1 201	1 788	1 943	1 848	2 472	2 609	1 712	1 700 F	1 700 F
Mexico	2 463	2 745	1 875	968	950	641	591	630	430	693
31 Fishing area total	*4 456*	*4 170*	*3 721*	*3 136*	*3 034*	*3 148*	*3 212*	*2 353*	*2 150 F*	*2 413 F*
41 Brazil	...	...	...	506	648	659	1 038	1 031	1 014	1 000 F
41 Fishing area total	*...*	*...*	*...*	*506*	*648*	*659*	*1 038*	*1 031*	*1 014*	*1 000 F*
Species total	*4 456*	*4 170*	*3 721*	*3 642*	*3 682*	*3 807*	*4 250*	*3 384*	*3 164 F*	*3 413 F*
Snappers nei	**Vivaneaux nca**		**Pargos tropicales nep**		***Lutjanus spp***				**1,70(32)027,XX**	**SNA**
31 Bahamas	228	329	258	297	341	751	781	866	721	777
Br Virgin Is	...	...	53	20	6	1	4	3	1	1 F
Nicaragua	...	100	...	...	...	...	...	540	760	910
St Kitts Nev	...	...	...	4	9	5	8	15	21	10
St Lucia	...	...	25	56	69	31	34	45	68	82
31 Fishing area total	*228*	*429*	*336*	*377*	*425*	*788*	*827*	*1 469*	*1 571*	*1 780 F*
34 Benin	50 F	54 F	60 F	58 F	66 F	90 F	86 F	92	60 F	32
Cameroon	2	1	300 F	510	511	400 F	320 F	337 F	337	293
Congo Rep	102 F	102 F	61 F	50 F	40 F	30 F	20 F	10 F	1	1 F
Gabon	180 F	192	117	463	611	480	907	941	795	765
Gambia	56	61	67	127	27	150	86	90	90	126
Ghana	635	895	716	626	328	181	294	163	491	447
Korea Rep	-	-	-	-	-	-	555	474	524	601
Liberia	1	24	13	1	9	17	27	339	132	201
Nigeria	4 205	2 222	2 902	2 999	3 779	6 473	8 685	8 345	7 235	6 619
Russian Fed	120	-	-	-	-	-	-	-	-	-
Senegal	383	417	488	368	236	296	237	221	366	499
Sierra Leone	...	...	...	3	3 F	155	...	1 728	294	381
Togo	75	71	65	87	203	42	433	129	24	16
Ukraine	3	-	-	-	-	-	-	-	-	-
34 Fishing area total	*5 812 F*	*4 039 F*	*4 789 F*	*5 292 F*	*5 813 F*	*8 314 F*	*11 650 F*	*12 869 F*	*10 349 F*	*9 981 F*
51 Bahrain	183	97	75	117	133	207	100	294	157	103

B-33 Miscellaneous coastal fishes / Poissons côtiers divers / Peces costeros diversos

Capture production by species, fishing areas and countries or areas / Captures par espèces, zones de pêche et pays ou zones / Capturas por especies, áreas de pesca y países o áreas

Species, Fishing area Espèce, Zone de pêche Especie, Area de pesca	1992 mt	1993 mt	1994 mt	1995 mt	1996 mt	1997 mt	1998 mt	1999 mt	2000 mt	2001 mt
Eritrea	...	...	43	205	281	219	294	365	149	280
Kuwait	93	95	134	117	84	78	64	51	40 F	27
Qatar	212	178	151	111	157	157	143	73	181	230
Réunion	...	...	...	...	...	...	...	112	38	45
Seychelles	517	454	515	358	381	242	424	489	213	602
51 Fishing area total	*1 005*	*824*	*918*	*908*	*1 036*	*903*	*1 025*	*1 384*	*778 F*	*1 287*
57 Indonesia	10 528	12 577	10 867	10 972	13 352	11 738	12 886	14 121	13 651	13 810
Malaysia	181	286	234	138	127	149	241	115	239	356
57 Fishing area total	*10 709*	*12 863*	*11 101*	*11 110*	*13 479*	*11 887*	*13 127*	*14 236*	*13 890*	*14 166*
61 Russian Fed	-	-	15	-	-	-	-	-	-	-
61 Fishing area total	-	-	*15*	-	-	-	-	-	-	-
71 Fiji Islands	1 473	1 650	1 682 F	1 117	1 347	1 315	1 155	1 710	1 600 F	1 728
Indonesia	38 772	43 276	47 471	41 855	46 986	57 847	53 394	52 371	48 655	49 210
Malaysia	2 721	2 665	2 598	2 725	3 120	3 061	4 245	4 551	4 031	3 976
NewCaledonia	81	48	56	38	45	36	43	22	24	24 F
Singapore	156	135	368	310	321	289	238	154	122	80
71 Fishing area total	*43 203*	*47 774*	*52 175 F*	*46 045*	*51 819*	*62 548*	*59 075*	*58 808*	*54 432 F*	*55 018 F*
77 Cook Is	25 F	25 F	24 F	24 F	20 F	20 F	20 F	20 F	20 F	20 F
Nicaragua	...	1 600	...	...	...	...	...	2 100	1 620	2 410
USA	65	83	191	242	215	187	157	175	191	138
77 Fishing area total	*90 F*	*1 708 F*	*215 F*	*266 F*	*235 F*	*207 F*	*177 F*	*2 295 F*	*1 831 F*	*2 568 F*
87 Colombia	446	387	646	278	460	43	452	515	600 F	802
Peru	119	87	23	300	117	75	208	573	204	55
87 Fishing area total	*565*	*474*	*669*	*578*	*577*	*118*	*660*	*1 088*	*804 F*	*857*
Species total	*61 612 F*	*68 111 F*	*70 218 F*	*64 576 F*	*73 384 F*	*84 765 F*	*86 541 F*	*92 149 F*	*83 655 F*	*85 657 F*
Yellowtail snapper	**Vivaneau queue jaune**		**Rabirrubia**		***Ocyurus chrysurus***			**1,70(32)028,01**		**SNY**
31 Br Virgin Is	...	...	...	...	...	5	9	9	0	0
Cuba	745	539	592	592	1 176	727	457	409	400 F	400 F
Dominican Rp	267	273	671	248	793	529	190	234	249	356
Mexico	1 132	910	1 184	825	858	840	1 900	1 554	1 357	1 600
Venezuela	659	678	684	511	338	335	272	220	291	158
31 Fishing area total	*2 803*	*2 400*	*3 131*	*2 176*	*3 165*	*2 436*	*2 828*	*2 426*	*2 297 F*	*2 514 F*
41 Brazil	2 810 F	2 800 F	2 800 F	4 766	4 167	5 000	3 317	4 541	4 165	4 100 F
41 Fishing area total	*2 810 F*	*2 800 F*	*2 800 F*	*4 766*	*4 167*	*5 000*	*3 317*	*4 541*	*4 165*	*4 100 F*
Species total	*5 613 F*	*5 200 F*	*5 931 F*	*6 942*	*7 332*	*7 436*	*6 145*	*6 967*	*6 462 F*	*6 614 F*
Vermilion snapper	**Vivaneau ti-yeux**		**Pargo cunaro**		***Rhomboplites aurorubens***			**1,70(32)225,01**		**RPU**
31 USA	...	...	...	...	...	678	480	731	911	1 057
31 Fishing area total	...	...	...	...	...	*678*	*480*	*731*	*911*	*1 057*
Species total	...	...	...	...	...	*678*	*480*	*731*	*911*	*1 057*
Snappers, jobfishes nei	**Lutianidés nca**		**Lutjánidos nep**		***Lutjanidae***			**1,70(32)XXX,XX**		**SNX**
31 Antigua Barb	...	...	...	...	...	...	...	...	...	284
Aruba	...	...	...	50	60	60	50	50	45	45
Barbados	35	19	26	41	40	26	32	24	25	20
Bermuda	41	41	39	37	38	29	29	28	23	32
Colombia	94	24	36	75	55	35	0	-	-	-
Cuba	369	301	364	416	601	403	509	95	100 F	100 F
Dominican Rp	100	84	151	84	130	235	875	49	753	231
Grenada	39	40	17	22	32	25	28	36	48	56
Mexico	922	850	335	920	827	1 495	3 675	3 929	3 533	1 543
Puerto Rico	...	...	...	...	...	...	...	566	762	744
USA	2 762	3 611	18 304	2 599	2 329	1 876	1 594	1 672	1 476	1 529
US Virgin Is	...	...	...	...	...	...	...	57	65 F	65 F
Venezuela	4 989	8 094	6 028	7 745	8 205	9 273	8 652	3 511	7 621	4 986
31 Fishing area total	*9 351*	*13 064*	*25 300*	*11 989*	*12 317*	*13 457*	*15 444*	*10 017*	*14 451 F*	*9 635 F*
41 Brazil	...	...	...	4 550	3 690	4 488	4 390	4 549	3 859	3 800 F
41 Fishing area total	...	...	...	*4 550*	*3 690*	*4 488*	*4 390*	*4 549*	*3 859*	*3 800 F*
47 South Africa	-	-	-	-	-	-	-	-	-	1
47 Fishing area total	-	-	-	-	-	-	-	-	-	*1*
51 Djibouti	60 F	65 F	69 F	80 F	80 F	80 F	80 F	80 F	80 F	80 F
Egypt	...	...	4 875	5 053	4 044	5 165	8 784	4 265	8 830	5 014
Eritrea	...	...	2	0	43	59	69	95	392	29
Kenya	155	129	116	112	147	144	106	151	120	177
Korea Rep	-	-	-	-	-	-	14	-	-	-
Mauritius	...	...	...	...	...	...	...	...	...	2 184
Oman	554	387	254	689	426	657	474	597	669	380
Saudi Arabia	2 712	3 019	3 054	1 662	1 704	2 148	2 302	2 022	1 647	2 092
Seychelles	637	904	601	572	466	464	602	825	552	703
South Africa	-	-	-	-	2	-	-	-	...	...
Untd Arab Em	2 842	3 082	3 241	3 438	3 474	3 713	3 725	3 818	1 718	1 720 F
Yemen	1 815	3 575	2 660	2 006	1 460	1 530 F	1 700 F	1 700 F	1 500 F	1 900 F
51 Fishing area total	*8 775 F*	*11 161 F*	*14 872 F*	*13 612 F*	*11 846 F*	*13 960 F*	*17 856 F*	*13 553 F*	*15 508 F*	*14 279 F*

B-33 Miscellaneous coastal fishes / Poissons côtiers divers / Peces costeros diversos

Capture production by species, fishing areas and countries or areas / Captures par espèces, zones de pêche et pays ou zones / Capturas por especies, áreas de pesca y países o áreas

Species, Fishing area Espèce, Zone de pêche Especie, Area de pesca	1992 mt	1993 mt	1994 mt	1995 mt	1996 mt	1997 mt	1998 mt	1999 mt	2000 mt	2001 mt
57 Australia	266	363	2 977	533	660	952	1 019	1 834	1 450	1 547
Malaysia	488	474	426	320	189	128	81	414	147	196
Thailand	1 768	5 798	6 476	5 532	5 473	4 359	3 188	4 280	4 222	4 196
57 Fishing area total	*2 522*	*6 635*	*9 879*	*6 385*	*6 322*	*5 439*	*4 288*	*6 528*	*5 819*	*5 939*
61 China,H.Kong	1 008	1 037	436	406	430	439	304	250 F	300 F	330 F
China,Taiwan	5 799	3 889	3 607	1 595	2 334	1 408	808	467	352	385
61 Fishing area total	*6 807*	*4 926*	*4 043*	*2 001*	*2 764*	*1 847*	*1 112*	*717 F*	*652 F*	*715 F*
71 Australia	1	16	647	496	460	734	771	722	585	814
Guam	...	...	20	3	1	2	3	10	6	5
Kiribati	1 990	1 930	1 910	1 930	1 950	1 940	1 047	1 505	2 141	2 794
Malaysia	3 023	2 512	1 258	3 416	5 149	4 566	6 004	7 373	5 082	4 575
N Marianas	...	...	2	9	10	15	7	16	5	11
Palau	62	48	20	40	33	8	25	25 F	27 F	27 F
Philippines	10 732	19 100	16 353	16 332	15 005	18 378	15 442	19 192	19 154	19 603
Singapore	265	68	62	52	61	66	50	31	32	16
Thailand	7 887	13 569	11 008	12 679	9 180	8 469	11 559	8 961	8 838	8 784
71 Fishing area total	*23 960*	*37 243*	*31 280*	*34 957*	*31 849*	*34 178*	*34 908*	*37 835 F*	*35 870 F*	*36 629 F*
77 Amer Samoa	...	...	10	7	8	7	4	5	5	9
Costa Rica	547	701	664	550	352	244	478	526	417	347
El Salvador	291	358	418	334	197	117	252	230	282	481
Fr Polynesia	...	...	...	...	...	100	81	109	127	112
Mexico	2 328	1 899	1 645	1 260	1 463	1 154	1 178	1 116	1 463	1 581
Panama	2 341	4 597	4 525	5 170	3 857	4 659	5 358	10 433	8 480	8 500 F
USA	283	292	53	100	78	40	54	154	148	125
77 Fishing area total	*5 790*	*7 847*	*7 315*	*7 421*	*5 955*	*6 321*	*7 405*	*12 573*	*10 922*	*11 155 F*
81 Australia	...	...	...	...	...	...	...	301	291	275
81 Fishing area total	...	...	...	...	...	...	...	*301*	*291*	*275*
Species total	*57 205 F*	*80 876 F*	*92 689 F*	*80 915 F*	*74 743 F*	*79 690 F*	*85 403 F*	*86 073 F*	*87 372 F*	*82 428 F*
Golden threadfin bream	**Cohana doré**		**Baga dorada**			***Nemipterus virgatus***			**1,70(33)184,05**	**THG**
61 China	...	...	...	224 574	238 000	258 998	262 702	246 601	291 495	287 384
China,Taiwan	8 933	9 841	4 142	3 664	4 261	4 401	3 681	3 990	4 824	3 536
61 Fishing area total	*8 933*	*9 841*	*4 142*	*228 238*	*242 261*	*263 399*	*266 383*	*250 591*	*296 319*	*290 920*
Species total	*8 933*	*9 841*	*4 142*	*228 238*	*242 261*	*263 399*	*266 383*	*250 591*	*296 319*	*290 920*
Threadfin breams nei	**Cohanas nca**		**Bagas nep**			***Nemipterus spp***			**1,70(33)184,XX**	**THB**
51 Egypt	...	...	...	...	...	...	...	...	2 081	4 974
Eritrea	...	...	928	206	4	15	0	1 674	1 757	1 115
Kuwait	8	6	4	6	10	17	49	19	20 F	24
Pakistan	368	526	752	952	825	884	3 192	7 166	8 940	8 466
Saudi Arabia	-	-	21	221	131	87	66	49	144	153
Tanzania	400	191	123	206	937	250	195	500	400	400
Untd Arab Em	313	371	348	392	396	423	424	435	432	430 F
51 Fishing area total	*1 089*	*1 094*	*2 176*	*1 983*	*2 303*	*1 676*	*3 926*	*9 843*	*13 774 F*	*15 562 F*
57 Indonesia	4 300	6 938	6 613	6 051	6 495	6 378	6 754	7 271	7 578	7 810
Malaysia	9 240	10 886	9 271	13 291	10 457	9 369	10 917	10 187	9 113	6 891
Thailand	13 722	17 424	19 260	22 148	24 843	24 773	36 912	23 088	22 772	22 633
57 Fishing area total	*27 262*	*35 248*	*35 144*	*41 490*	*41 795*	*40 520*	*54 583*	*40 546*	*39 463*	*37 334*
61 China,H.Kong	37 322	27 921	26 113	23 772	19 568	20 024	19 449	14 000 F	17 000 F	18 800 F
61 Fishing area total	*37 322*	*27 921*	*26 113*	*23 772*	*19 568*	*20 024*	*19 449*	*14 000 F*	*17 000 F*	*18 800 F*
71 Indonesia	15 948	17 582	18 665	21 409	25 098	22 962	24 183	31 926	26 640	27 460
Malaysia	19 759	19 875	19 992	18 032	19 077	20 733	29 410	29 507	23 397	22 019
Philippines	31 196	40 074	34 177	35 538	32 884	29 839	30 511	29 301	29 487	29 870
Singapore	128	123	305	255	209	239	158	128	96	48
Thailand	51 655	57 903	55 850	71 637	64 749	62 994	59 683	69 949	68 991	68 571
71 Fishing area total	*118 686*	*135 557*	*128 989*	*146 871*	*142 017*	*136 767*	*143 945*	*160 811*	*148 611*	*147 968*
Species total	*184 359*	*199 820*	*192 422*	*214 116*	*205 683*	*198 987*	*221 903*	*225 200 F*	*218 848 F*	*219 664 F*
Monocle breams	**Mamilas**		**Besugatos**			***Scolopsis spp***			**1,70(33)230,XX**	**MOB**
57 Malaysia	21	3	0	2	1	1	2	96	86	88
Thailand	-	1	0	0	0	0	0	0	0	0
57 Fishing area total	*21*	*4*	*0*	*2*	*1*	*1*	*2*	*96*	*86*	*88*
71 Malaysia	1 007	952	1 110	1 263	1 243	949	827	940	1 615	1 554
Thailand	78	72	...	77	98	344	644	54	53	53
71 Fishing area total	*1 085*	*1 024*	*1 110*	*1 340*	*1 341*	*1 293*	*1 471*	*994*	*1 668*	*1 607*
Species total	*1 106*	*1 028*	*1 110*	*1 342*	*1 342*	*1 294*	*1 473*	*1 090*	*1 754*	*1 695*
Threadfin and dwarf breams nei	**Cohanas, mamilas nca**		**Bagas, besugatos nep**			***Nemipteridae***			**1,70(33)XXX,XX**	**THD**
51 Oman	988	639	415	260	233	2 068	466	1 166	317	1 206
Qatar	23	11	5	5	13	15	3	0	0	0
South Africa	-	-	-	3	-	-	-	...	...	...
Yemen	4 279	3 592	3 718	3 170	2 504	2 620 F	3 000 F	2 900 F	2 600 F	3 300 F

B-33 Miscellaneous coastal fishes / Poissons côtiers divers / Peces costeros diversos

Capture production by species, fishing areas and countries or areas
Captures par espèces, zones de pêche et pays ou zones
Capturas por especies, áreas de pesca y países o áreas

Species, Fishing area Espèce, Zone de pêche Especie, Area de pesca	1992 mt	1993 mt	1994 mt	1995 mt	1996 mt	1997 mt	1998 mt	1999 mt	2000 mt	2001 mt
51 Fishing area total	*5 290*	*4 242*	*4 138*	*3 438*	*2 750*	*4 703 F*	*3 469 F*	*4 066 F*	*2 917 F*	*4 506 F*
Species total	*5 290*	*4 242*	*4 138*	*3 438*	*2 750*	*4 703 F*	*3 469 F*	*4 066 F*	*2 917 F*	*4 506 F*
Ponyfishes(=Slipmouths)	**Sapsap**		**Motambos**		***Leiognathus spp***				**1,70(35)169,XX**	**POY**
57 Malaysia	84	164	38	34	101	78	86	84	161	71
57 Fishing area total	*84*	*164*	*38*	*34*	*101*	*78*	*86*	*84*	*161*	*71*
71 Fiji Islands	120	127	129 F	37	76	75	69	84	80 F	87
Malaysia	1 822	1 754	1 895	2 250	2 438	2 284	3 004	2 965	2 300	2 214
Singapore	67	57	141	83	75	63	52	47	32	23
71 Fishing area total	*2 009*	*1 938*	*2 165 F*	*2 370*	*2 589*	*2 422*	*3 125*	*3 096*	*2 412 F*	*2 324*
Species total	*2 093*	*2 102*	*2 203 F*	*2 404*	*2 690*	*2 500*	*3 211*	*3 180*	*2 573 F*	*2 395*
Ponyfishes(=Slipmouths) nei	**Sapsap nca**		**Motambos nep**		***Leiognathidae***				**1,70(35)XXX,XX**	**PON**
51 Eritrea	...	...	-	-	-	-	-	19	38	137
India	7 865	10 919	9 348	7 192	10 343	12 341	9 115	8 934	4 551	6 951
Untd Arab Em	810	852	1 109	723	731	781	784	804	780	780 F
51 Fishing area total	*8 675*	*11 771*	*10 457*	*7 915*	*11 074*	*13 122*	*9 899*	*9 757*	*5 369*	*7 868 F*
57 India	41 454	44 490	51 638	56 202	58 239	55 028	36 953	46 528	44 729	44 839
Indonesia	10 766	12 281	12 610	12 555	15 496	17 249	20 801	22 608	17 631	17 850
57 Fishing area total	*52 220*	*56 771*	*64 248*	*68 757*	*73 735*	*72 277*	*57 754*	*69 136*	*62 360*	*62 689*
71 Indonesia	34 771	40 519	44 852	53 665	55 905	72 154	58 731	68 611	51 881	52 540
Philippines	60 005	60 046	59 547	59 134	57 867	61 254	59 862	68 371	67 255	66 890
71 Fishing area total	*94 776*	*100 565*	*104 399*	*112 799*	*113 772*	*133 408*	*118 593*	*136 982*	*119 136*	*119 430*
Species total	*155 671*	*169 107*	*179 104*	*189 471*	*198 581*	*218 807*	*186 246*	*215 875*	*186 865*	*189 987 F*
Pigfish	**Goret mule**		**Corocoro burro**		***Orthopristis chrysoptera***				**1,70(36)032,01**	**PIG**
21 USA	2	-	-	1	-	1	1	-	0	0
21 Fishing area total	*2*	-	-	*1*	-	*1*	*1*	-	*0*	*0*
Species total	*2*	-	-	*1*	-	*1*	*1*	-	*0*	*0*
Cabinza grunt	**Cagna cabinza**		**Cabinza**		***Isacia conceptionis***				**1,70(36)163,01**	**GRP**
87 Chile	338	161	240	165	93	142	54	156	42	28
Peru	1 985	987	505	1 342	1 955	1 892	2 079	2 791	3 251	3 293
87 Fishing area total	*2 323*	*1 148*	*745*	*1 507*	*2 048*	*2 034*	*2 133*	*2 947*	*3 293*	*3 321*
Species total	*2 323*	*1 148*	*745*	*1 507*	*2 048*	*2 034*	*2 133*	*2 947*	*3 293*	*3 321*
Rubberlip grunt	**Diagramme gris**		**Burro chiclero**		***Plectorhinchus mediterraneus***				**1,70(36)207,05**	**GBR**
27 Portugal	...	...	...	...	...	...	...	251	77	12
27 Fishing area total	...	...	...	...	...	...	...	*251*	*77*	*12*
34 Gambia	54	43	39	102	156	101	160	170	107	124
Mauritania	20 F	20 F	30 F	20 F	20 F	20 F	10 F	...	9 F	...
Portugal	1 791	1 246	707	712	1 436	544	115	16	10	-
Romania	9	-	-	-	-	-	-	-	-	-
Spain	216	225 F	235 F	245 F	253	234	170	453	94	133
34 Fishing area total	*2 090 F*	*1 534 F*	*1 011 F*	*1 079 F*	*1 865 F*	*899 F*	*455 F*	*639*	*220 F*	*257*
37 Spain	-	-	-	-	4	5	4	-	-	2
37 Fishing area total	-	-	-	-	*4*	*5*	*4*	-	-	*2*
Species total	*2 090 F*	*1 534 F*	*1 011 F*	*1 079 F*	*1 869 F*	*904 F*	*459 F*	*890*	*297 F*	*271*
Silver grunt	**Grondeur argenté**		**Corocoro plateado**		***Pomadasys argenteus***				**1,70(36)209,04**	**GRL**
51 Eritrea	...	...	...	...	0	0	3	...	472	...
51 Fishing area total	...	...	...	...	*0*	*0*	*3*	...	*472*	...
57 Malaysia	456	501	423	323	625	539	666	548	581	507
57 Fishing area total	*456*	*501*	*423*	*323*	*625*	*539*	*666*	*548*	*581*	*507*
71 Malaysia	693	756	1 239	852	1 019	617	2 178	1 687	919	1 086
71 Fishing area total	*693*	*756*	*1 239*	*852*	*1 019*	*617*	*2 178*	*1 687*	*919*	*1 086*
Species total	*1 149*	*1 257*	*1 662*	*1 175*	*1 644*	*1 156*	*2 847*	*2 235*	*1 972*	*1 593*
Bastard grunt	**Grondeur métis**		**Ronco mestizo**		***Pomadasys incisus***				**1,70(36)209,18**	**BGR**
34 Portugal	-	-	-	1	1	0	-	-	0	0
Romania	92	-	-	-	-	-	-	-	-	-
Spain	-	-	-	-	0	5	-	-	-	-
34 Fishing area total	*92*	-	-	*1*	*1*	*5*	-	-	*0*	*0*
Species total	*92*	-	-	*1*	*1*	*5*	-	-	*0*	*0*

B-33 Miscellaneous coastal fishes / Poissons côtiers divers / Peces costeros diversos

Capture production by species, fishing areas and countries or areas / Captures par espèces, zones de pêche et pays ou zones / Capturas por especies, áreas de pesca y países o áreas

Species, Fishing area Espèce, Zone de pêche Especie, Area de pesca	1992 mt	1993 mt	1994 mt	1995 mt	1996 mt	1997 mt	1998 mt	1999 mt	2000 mt	2001 mt
Sompat grunt	**Grondeur sompat**		**Ronco sompat**		***Pomadasys jubelini***			**1,70(36)209,19**		**BU**
34 Cameroon	...	...	...	534	533	400 F	337 F	345 F	304	48
Congo Rep	...	...	6	5 F	5 F	5 F	5 F	5 F	...	4 F
Côte dIvoire	240	220	233	235 F	247	143	227	236	231	148
Gambia	83	142	120	307	498	350	220	230	276	423
GuineaBissau	...	...	47	20	153	160 F	130	100 F	100 F	100 F
Togo	4	4	2	0	3	43	31	35	11	1
34 Fishing area total	*327*	*366*	*408*	*1 101 F*	*1 439 F*	*1 101 F*	*950 F*	*951 F*	*922 F*	*724 F*
Species total	*327*	*366*	*408*	*1 101 F*	*1 439 F*	*1 101 F*	*950 F*	*951 F*	*922 F*	*724 F*
Bigeye grunt	**Lippu pelon**		**Burro ojón**		***Brachydeuterus auritus***			**1,70(36)263,03**		**GR**
34 Congo Dem R	380 F	420 F	380 F	388 F	398 F	385 F	396 F	400 F	400 F	400 F
Congo Rep	481 F	480 F	122	126	125 F	115 F	100 F	100 F	95	90 F
Côte dIvoire	5 453	4 292	4 950	4 848	4 632	4 732	1 617	3 964	5 175	1 689
Gabon	150 F	150	150	200	...	...	...	...	...	...
Ghana	11 024	11 940	18 216	14 807	13 552	19 816	12 059	12 724	10 032	13 845
GuineaBissau	6 F	7 F	9 F	12 F	14 F	15 F	10 F	10 F	10 F	10 F
Lithuania	-	2 652	461	-	-	-	-	-	-	-
Nigeria	42	...	...	...	556	2 537	2 887	2 141	3 345	275
Romania	1 104	-	-	-	-	-	-	-	-	-
Russian Fed	1 180	1 727	393	1 282	-	-	36	-	-	-
Senegal	986	1 687	1 367	1 255	610	1 445	1 481	2 414	1 949	2 701
Sierra Leone	...	...	...	...	...	...	...	...	...	132
Togo	22	49	44	21	0	1	5	465	742	844
34 Fishing area total	*20 828 F*	*23 404 F*	*26 092 F*	*22 939 F*	*19 887 F*	*29 046 F*	*18 591 F*	*22 218 F*	*21 748 F*	*19 986 F*
47 Angola	-	17	75	124	146	228	262	167	191	1 176
Russian Fed	-	-	-	-	-	-	-	-	-	5
47 Fishing area total	-	*17*	*75*	*124*	*146*	*228*	*262*	*167*	*191*	*1 181*
Species total	*20 828 F*	*23 421 F*	*26 167 F*	*23 063 F*	*20 033 F*	*29 274 F*	*18 853 F*	*22 385 F*	*21 939 F*	*21 167 F*
Barred grunt	**Cagna rayée**		**Ronco canario**		***Conodon nobilis***			**1,70(36)395,01**		**BR**
41 Brazil	1 000 F	700 F	500 F	339	172	118	114	84	39	40 F
41 Fishing area total	*1 000 F*	*700 F*	*500 F*	*339*	*172*	*118*	*114*	*84*	*39*	*40 F*
Species total	*1 000 F*	*700 F*	*500 F*	*339*	*172*	*118*	*114*	*84*	*39*	*40 F*
Grunts, sweetlips nei	**Grondeurs, diagrammes nca**		**Burros, roncos nep**		***Haemulidae (=Pomadasyidae)***			**1,70(36)XXX,XX**		**GR**
27 Ireland	-	-	-	-	-	-	-	-	-	5
Portugal	649	94	61	28	28	45	55	44	17	9
Spain	-	351	52	304	476	205	261	100	198	163
27 Fishing area total	*649*	*445*	*113*	*332*	*504*	*250*	*316*	*144*	*215*	*177*
31 Antigua Barb	...	...	...	...	...	...	...	...	...	167
Bahamas	20	13	9	8	14	67	90	66	62	67
Colombia	781	75	100	115	224	12	6	1	5 F	5 F
Cuba	1 627	1 247	1 991	2 128	1 723	1 451	1 207	1 303	1 300 F	1 300 F
Dominican Rp	392	398	716	278	249	325	205	423	348	385
Grenada	2	1	0	2	1	1	1	3	1	1
Mexico	6 000	5 224	5 512	5 032	5 901	9 645	6 076	5 039	3 594	1 161
Puerto Rico	...	...	...	...	...	...	...	76	97	104
St Kitts Nev	...	...	...	1	10	1	3	3	1	1
USA	385	484	536	387	293	243	292	316	403	429
Venezuela	5 305	5 693	5 387	6 310	5 466	6 336	7 685	4 108	4 191	9 132
31 Fishing area total	*14 512*	*13 135*	*14 251*	*14 261*	*13 881*	*18 081*	*15 565*	*11 338*	*10 002 F*	*12 752 F*
34 Benin	169 F	183 F	206 F	200 F	228 F	311 F	296 F	269	180 F	222
Cameroon	21	17	...	-	-	-	-	-	-	-
Congo Rep	19 F	19 F	6	15	12 F	9 F	6 F	3 F	-	-
Gabon	620 F	531	508	897	780	853	1 728	1 062	819	1 050
Ghana	449	659	575	523	1 191	477	411	476	150	847
Guinea	200 F	240 F	240 F	341	193	198	185	1 598	430	420 F
Korea Rep	1 351	3 363	-	5 620	4 969	-	-	-	-	538
Latvia	-	-	-	-	-	2	-	-	-	-
Liberia	16	13	180	196	105	99	85	216	102	180
Lithuania	861	150	4	-	-	-	-	-	-	-
Mauritania	10 F	10 F	10 F	10 F	10 F	10 F	10 F	...	2 F	...
Morocco	2 105	1 588	1 538	1 665	1 670	2 590	3 049	2 228	2 715	3 158
Nigeria	3 028	4 194	5 235	4 152	2 585	2 611	5 000	4 561	5 152	4 804
Portugal	212	994	90	83	158	42	14	3	0	-
Russian Fed	107	23	6	9	-	-	-	9	1	-
Sao Tome Prn	1 F	1 F	1 F	1 F	1	0	0	0	0	0
Senegal	4 426	3 495	5 418	2 121	11 187	11 283	7 828	10 902	7 403	8 048
Sierra Leone	793	784	781	800	800 F	1 086	433	413	1 357	1 975
Ukraine	66	25	-	-	-	-	-	-	132	6
34 Fishing area total	*14 454 F*	*16 289 F*	*14 798 F*	*16 633 F*	*23 889 F*	*19 571 F*	*19 045 F*	*21 740 F*	*18 443 F*	*21 248 F*
41 Brazil	...	...	...	1 812	2 081	1 936	2 000	2 382	3 024	3 000 F
41 Fishing area total	...	...	...	*1 812*	*2 081*	*1 936*	*2 000*	*2 382*	*3 024*	*3 000 F*
47 Angola	-	67	78	149	1 031	557	438	412	799	1 688
Japan	5	-	-	-	-	-	-	-	-	-
Korea Rep	-	5	-	-	-	-	-	-	-	-
Portugal	-	-	-	-	-	-	-	-	18	18
Russian Fed	-	-	-	-	-	-	2	-	-	8

B-33 Miscellaneous coastal fishes / Poissons côtiers divers / Peces costeros diversos

Capture production by species, fishing areas and countries or areas / Captures par espèces, zones de pêche et pays ou zones / Capturas por especies, áreas de pesca y países o áreas

Species, Fishing area Espèce, Zone de pêche Especie, Area de pesca	1992 mt	1993 mt	1994 mt	1995 mt	1996 mt	1997 mt	1998 mt	1999 mt	2000 mt	2001 mt
South Africa	-	-	-	-	-	-	-	-	-	1
47 Fishing area total	*5*	*72*	*78*	*149*	*1 031*	*557*	*440*	*412*	*817*	*1 715*
51 Bahrain	174	177	231	282	253	223	355	325	297	236
Eritrea	...	...	61	469	367	13	45	594	536	435
Kenya	61	65	52	67	65	72	55	78	63	85
Korea Rep	-	-	-	-	563	-	130	20	25	2
Kuwait	124	104	145	195	180	163	174	245	210 F	191
Oman	376	478	526	1 122	803	1 111	713	627	1 747	541
Pakistan	4 491	5 417	4 849	5 537	5 268	6 010	6 221	8 147	9 961	9 752
Qatar	576	731	622	606	653	542	583	433	789	900
Saudi Arabia	254	351	489	814	1 063	1 014	779	846	1 350	1 206
South Africa	-	-	5	2	-	6	11	5 F	1	3
Untd Arab Em	1 236	1 387	1 380	1 768	1 787	1 910	1 916	1 964	4 196	4 200 F
Yemen	569	754	1 147	1 813	1 318	1 380 F	1 600 F	1 500 F	1 350 F	1 700 F
51 Fishing area total	*7 861*	*9 464*	*9 507*	*12 675*	*12 320*	*12 444 F*	*12 582 F*	*14 784 F*	*20 525 F*	*19 251 F*
57 Indonesia	2 875	2 836	2 766	3 717	4 339	4 052	4 253	4 544	4 738	4 930
Malaysia	68	139	50	38	20	27	21	9	39	81
57 Fishing area total	*2 943*	*2 975*	*2 816*	*3 755*	*4 359*	*4 079*	*4 274*	*4 553*	*4 777*	*5 011*
61 Japan	...	...	...	5 555	5 303	4 801	5 703	5 282	5 136	4 995
Russian Fed	-	-	32	-	-	-	-	3	-	-
61 Fishing area total	*...*	*...*	*32*	*5 555*	*5 303*	*4 801*	*5 703*	*5 285*	*5 136*	*4 995*
71 Indonesia	8 666	9 669	9 525	9 492	10 515	10 605	10 950	10 804	12 630	13 160
Malaysia	911	915	730	978	1 093	1 047	1 057	1 548	1 468	1 336
Singapore	54	64	39	33	53	45	27	25	18	18
71 Fishing area total	*9 631*	*10 648*	*10 294*	*10 503*	*11 661*	*11 697*	*12 034*	*12 377*	*14 116*	*14 514*
77 Mexico	1 135	1 108	1 169	604	484	868	1 276	375	471	480
77 Fishing area total	*1 135*	*1 108*	*1 169*	*604*	*484*	*868*	*1 276*	*375*	*471*	*480*
87 Ecuador	...	...	...	...	...	...	1 091	500	...	...
Peru	195	416	185	890	194	379	254	318	183	419
87 Fishing area total	*195*	*416*	*185*	*890*	*194*	*379*	*1 345*	*818*	*183*	*419*
Species total	*51 385 F*	*54 552 F*	*53 243 F*	*67 169 F*	*75 707 F*	*74 663 F*	*74 580 F*	*74 208 F*	*77 709 F*	*83 562 F*
Corvina	**...B**		**...C**		***Sciaena gilberti***				**1,70(37)009,02**	**IAG**
87 Chile	1 995	2 150	1 868	1 239	1 179	1 350	1 069	747	1 052	1 033
Peru	4 109	4 098	4 275	4 353	7 920	2 215	5 085	5 695	3 692	3 295
87 Fishing area total	*6 104*	*6 248*	*6 143*	*5 592*	*9 099*	*3 565*	*6 154*	*6 442*	*4 744*	*4 328*
Species total	*6 104*	*6 248*	*6 143*	*5 592*	*9 099*	*3 565*	*6 154*	*6 442*	*4 744*	*4 328*
Brown meagre	**Corb commun**		**Corvallo**		***Sciaena umbra***				**1,70(37)009,21**	**CBM**
34 Korea Rep	-	-	-	234	208	...	...	...	...	...
34 Fishing area total	*-*	*-*	*-*	*234*	*208*	*...*	*...*	*...*	*...*	*...*
37 Albania	...	...	...	2	6	2	-	-	-	10
Cyprus	2	2	2	0	2	5	3	7	4	3
Tunisia	159	116	162	90	164	180	192	114	104	138
Turkey	116	233	143	43	50	35	45	65	20	20
37 Fishing area total	*277*	*351*	*307*	*135*	*222*	*222*	*240*	*186*	*128*	*171*
Species total	*277*	*351*	*307*	*369*	*430*	*222*	*240*	*186*	*128*	*171*
Peruvian weakfish	**Acoupa du Pérou**		**Corvinata ayanque**		***Cynoscion analis***				**1,70(37)016,04**	**WEP**
87 Chile	12	21	52	249	11	12	18	19	14	14
Colombia	208	193	336	255	401	284	374	352	330	317
Peru	2 850	9 676	5 248	8 902	7 475	5 501	10 795	8 558	5 995	4 107
87 Fishing area total	*3 070*	*9 890*	*5 636*	*9 406*	*7 887*	*5 797*	*11 187*	*8 929*	*6 339*	*4 438*
Species total	*3 070*	*9 890*	*5 636*	*9 406*	*7 887*	*5 797*	*11 187*	*8 929*	*6 339*	*4 438*
Spotted weakfish	**Acoupa pintade**		**Corvinata pintada**		***Cynoscion nebulosus***				**1,70(37)016,09**	**SWF**
21 USA	138	151	19	206	86	78	79	202	119	36
21 Fishing area total	*138*	*151*	*19*	*206*	*86*	*78*	*79*	*202*	*119*	*36*
31 Mexico	1 855	1 716	1 253	3 107	3 213	3 448	6 600	6 549	6 227	1 785
USA	936	910	1 043	637	397	349	186	193	140	116
31 Fishing area total	*2 791*	*2 626*	*2 296*	*3 744*	*3 610*	*3 797*	*6 786*	*6 742*	*6 367*	*1 901*
Species total	*2 929*	*2 777*	*2 315*	*3 950*	*3 696*	*3 875*	*6 865*	*6 944*	*6 486*	*1 937*
Squeteague(=Gray weakfish)	**Acoupa royal**		**Corvinata real**		***Cynoscion regalis***				**1,70(37)016,14**	**STG**
21 USA	2 073	1 977	1 093	2 578	2 729	2 697	3 212	2 736	2 329	2 119
21 Fishing area total	*2 073*	*1 977*	*1 093*	*2 578*	*2 729*	*2 697*	*3 212*	*2 736*	*2 329*	*2 119*
31 USA	1 314	1 213	1 674	517	532	620	609	410	109	154
31 Fishing area total	*1 314*	*1 213*	*1 674*	*517*	*532*	*620*	*609*	*410*	*109*	*154*
Species total	*3 387*	*3 190*	*2 767*	*3 095*	*3 261*	*3 317*	*3 821*	*3 146*	*2 438*	*2 273*

B-33 Miscellaneous coastal fishes / Poissons côtiers divers / Peces costeros diversos

Capture production by species, fishing areas and countries or areas / Captures par espèces, zones de pêche et pays ou zones / Capturas por especies, áreas de pesca y países o áreas

Species, Fishing area Espèce, Zone de pêche Especie, Area de pesca	1992 mt	1993 mt	1994 mt	1995 mt	1996 mt	1997 mt	1998 mt	1999 mt	2000 mt	2001 mt
Striped weakfish	**Acoupa rayé**		**Corvinata pescadilla**			***Cynoscion striatus***			**1,70(37)016,17**	**WI**
41 Argentina	10 102	6 239	15 661	19 218	18 987	24 132	17 108	11 107	9 434	11 303
Uruguay	8 786	6 962	10 323	13 417	12 654	15 186	15 286	8 481	13 440	10 890
41 Fishing area total	*18 888*	*13 201*	*25 984*	*32 635*	*31 641*	*39 318*	*32 394*	*19 588*	*22 874*	*22 193*
Species total	*18 888*	*13 201*	*25 984*	*32 635*	*31 641*	*39 318*	*32 394*	*19 588*	*22 874*	*22 193*
Weakfishes nei	**Acoupas nca**		**Corvinatas nep**			***Cynoscion spp***			**1,70(37)016,XX**	**WI**
21 USA	-	-	-	-	-	14	-	11	-	-
21 Fishing area total	-	-	-	-	-	*14*	-	*11*	-	-
31 Dominican Rp	10	26	47	83	56	14	14	7	10	10
USA	127	140	190	91	76	60	56	84	74	53
Venezuela	14 086	18 541	19 078	20 045	13 879	10 503	13 481	6 303	12 716	10 734
31 Fishing area total	*14 223*	*18 707*	*19 315*	*20 219*	*14 011*	*10 577*	*13 551*	*6 394*	*12 800*	*10 797*
41 Brazil	33 000 F	30 000 F	28 000 F	27 075	26 470	25 960	26 677	39 332	43 523	43 000
41 Fishing area total	*33 000 F*	*30 000 F*	*28 000 F*	*27 075*	*26 470*	*25 960*	*26 677*	*39 332*	*43 523*	*43 000*
77 Mexico	2 707	2 643	2 788	2 281	3 402	4 042	4 602	5 507	2 494	5 356
77 Fishing area total	*2 707*	*2 643*	*2 788*	*2 281*	*3 402*	*4 042*	*4 602*	*5 507*	*2 494*	*5 356*
Species total	*49 930 F*	*51 350 F*	*50 103 F*	*49 575*	*43 883*	*40 593*	*44 830*	*51 244*	*58 817*	*59 153*
Whitemouth croaker	**Tambour rayé**		**Corvinón rayado**			***Micropogonias furnieri***			**1,70(37)038,02**	**CI**
31 Venezuela	5 387	6 697	6 058	7 043	6 065	3 694	4 871	1 900	3 262	6 848
31 Fishing area total	*5 387*	*6 697*	*6 058*	*7 043*	*6 065*	*3 694*	*4 871*	*1 900*	*3 262*	*6 848*
41 Argentina	10 619	12 709	18 261	29 989	23 514	26 108	9 451	6 641	5 264	4 479
Brazil	28 000 F	25 000 F	23 000 F	22 024	23 374	23 311	27 927	21 926	28 797	28 500
Uruguay	28 271	25 804	29 012	29 513	25 745	23 744	22 253	14 650	24 146	27 322
41 Fishing area total	*66 890 F*	*63 513 F*	*70 273 F*	*81 526*	*72 633*	*73 163*	*59 631*	*43 217*	*58 207*	*60 301*
Species total	*72 277 F*	*70 210 F*	*76 331 F*	*88 569*	*78 698*	*76 857*	*64 502*	*45 117*	*61 469*	*67 149*
Atlantic croaker	**Tambour brésilien**		**Corvinón brasileño**			***Micropogonias undulatus***			**1,70(37)038,04**	**CI**
21 USA	1 223	3 270	2 787	6 462	8 250	11 038	10 438	11 451	11 632	12 410
21 Fishing area total	*1 223*	*3 270*	*2 787*	*6 462*	*8 250*	*11 038*	*10 438*	*11 451*	*11 632*	*12 410*
31 USA	760	687	2 205	551	799	1 396	1 084	729	506	608
31 Fishing area total	*760*	*687*	*2 205*	*551*	*799*	*1 396*	*1 084*	*729*	*506*	*608*
Species total	*1 983*	*3 957*	*4 992*	*7 013*	*9 049*	*12 434*	*11 522*	*12 180*	*12 138*	*13 018*
Croakers nei	**Tambours nca**		**Corvinones nep**			***Micropogonias spp***			**1,70(37)038,XX**	**CI**
77 Mexico	3 755	3 666	3 867	3 432	5 036	4 966	5 476	4 602	3 705	3 728
77 Fishing area total	*3 755*	*3 666*	*3 867*	*3 432*	*5 036*	*4 966*	*5 476*	*4 602*	*3 705*	*3 728*
87 Chile	109	104	113	111	128	101	9	0	0	7
Colombia	200	168	192	105	165	179	156	173	170 F	237
Peru	217	1 369	602	704	733	606	1 068	616	955	825
87 Fishing area total	*526*	*1 641*	*907*	*920*	*1 026*	*886*	*1 233*	*789*	*1 125 F*	*1 069*
Species total	*4 281*	*5 307*	*4 774*	*4 352*	*6 062*	*5 852*	*6 709*	*5 391*	*4 830 F*	*4 797*
Northern kingfish	**Bourrugue renard**		**Lambe zorro**			***Menticirrhus saxatilis***			**1,70(37)039,07**	**K**
21 USA	34	41	34	34	49	21	17	13	28	39
21 Fishing area total	*34*	*41*	*34*	*34*	*49*	*21*	*17*	*13*	*28*	*39*
Species total	*34*	*41*	*34*	*34*	*49*	*21*	*17*	*13*	*28*	*39*
Gulf kingcroaker	**Bourrugue du Golfe**		**Lambe verrugato**			***Menticirrhus littoralis***			**1,70(37)039,08**	**K**
31 Mexico	454	582	869	948	1 070	1 220	1 824	1 576	606	484
St Lucia	...	...	6	20	...	...	...	...	...	...
USA	-	-	-	-	-	223	237	246	355	206
31 Fishing area total	*454*	*582*	*875*	*968*	*1 070*	*1 443*	*2 061*	*1 822*	*961*	*690*
Species total	*454*	*582*	*875*	*968*	*1 070*	*1 443*	*2 061*	*1 822*	*961*	*690*
Kingcroakers nei	**Bourrugues nca**		**Lambes nep**			***Menticirrhus spp***			**1,70(37)039,XX**	**K**
41 Brazil	...	...	...	1 330	1 144	1 303	1 395	1 058	1 348	1 330
Uruguay	-	-	-	-	-	-	7	4	4	6
41 Fishing area total	...	...	...	*1 330*	*1 144*	*1 303*	*1 402*	*1 062*	*1 352*	*1 336*
Species total	...	...	...	*1 330*	*1 144*	*1 303*	*1 402*	*1 062*	*1 352*	*1 336*
Shi drum	**Ombrine côtière**		**Verrugato fusco**			***Umbrina cirrosa***			**1,70(37)070,01**	**C**
34 Greece	136	97	29	63	86	52	43	83	93	57
Togo	0	8	0	0	0	0	0	-	-	[illegible]

B-33 Miscellaneous coastal fishes / Poissons côtiers divers / Peces costeros diversos

Capture production by species, fishing areas and countries or areas
Captures par espèces, zones de pêche et pays ou zones
Capturas por especies, áreas de pesca y países o áreas

Species, Fishing area Espèce, Zone de pêche Especie, Area de pesca	1992 mt	1993 mt	1994 mt	1995 mt	1996 mt	1997 mt	1998 mt	1999 mt	2000 mt	2001 mt
34 Fishing area total	*136*	*105*	*29*	*63*	*86*	*52*	*43*	*83*	*93*	*57*
37 Greece	51	23	87	87	74	26	66	53	58	46
Italy	652	555	762	956	495	351	138	138	158	259
Tunisia	29	32	30	29	25	51	32	40	28	33
Turkey	913	748	786	91	116	65	150	155	45	55
37 Fishing area total	*1 645*	*1 358*	*1 665*	*1 163*	*710*	*493*	*386*	*386*	*289*	*393*
Species total	*1 781*	*1 463*	*1 694*	*1 226*	*796*	*545*	*429*	*469*	*382*	*450*
Argentine croaker	**Ombrine d'Argentine**		**Verrugato pargo**		***Umbrina canosai***			**1,70(37)070,03**		**CKY**
41 Argentina	305	84	884	3 561	6 825	2 753	1 956	1 410	421	...
Brazil	14 000 F	11 000 F	9 000 F	9 268	4 983	3 358	2 216	4 849	7 729	7 700 F
Uruguay	1 210	903	1 576	1 708	1 210	547	1 112	1 401	1 071	1 416
41 Fishing area total	*15 515 F*	*11 987 F*	*11 460 F*	*14 537*	*13 018*	*6 658*	*5 284*	*7 660*	*9 221*	*9 116 F*
Species total	*15 515 F*	*11 987 F*	*11 460 F*	*14 537*	*13 018*	*6 658*	*5 284*	*7 660*	*9 221*	*9 116 F*
Canary drum (=Baardman)	**Ombrine bronze**		**Verrugato de Canarias**		***Umbrina canariensis***			**1,70(37)070,09**		**UCA**
27 France	2	1	1	1	1	-	-	3	7	7
Spain	-	-	-	-	-	-	-	-	-	4
27 Fishing area total	*2*	*1*	*1*	*1*	*1*	*-*	*-*	*3*	*7*	*11*
34 Mauritania	10 F	5 F	5 F	5 F	10 F	10 F	10 F	6 F	2 F	...
34 Fishing area total	*10 F*	*5 F*	*5 F*	*5 F*	*10 F*	*10 F*	*10 F*	*6 F*	*2 F*	*...*
47 Angola	...	...	...	...	...	...	...	830	...	2 202
47 Fishing area total	*...*	*...*	*...*	*...*	*...*	*...*	*...*	*830*	*...*	*2 202*
Species total	*12 F*	*6 F*	*6 F*	*6 F*	*11 F*	*10 F*	*10 F*	*839 F*	*9 F*	*2 213*
King weakfish	**Acoupa chasseur**		**Pescadilla real**		***Macrodon ancylodon***			**1,70(37)096,01**		**WKK**
41 Argentina	-	-	-	149	380	224	167	515	44	...
Brazil	...	...	...	3 677	5 143	6 006	6 708	4 278	5 475	5 400 F
Uruguay	163	402	1 082	1 519	1 743	1 395	2 354	802	1 123	1 486
41 Fishing area total	*163*	*402*	*1 082*	*5 345*	*7 266*	*7 625*	*9 229*	*5 595*	*6 642*	*6 886 F*
Species total	*163*	*402*	*1 082*	*5 345*	*7 266*	*7 625*	*9 229*	*5 595*	*6 642*	*6 886 F*
Meagre	**Maigre commun**		**Corvina**		***Argyrosomus regius***			**1,70(37)106,01**		**MGR**
01 Egypt	40	76	136	127	168	128	106	92	201	51
01 Fishing area total	*40*	*76*	*136*	*127*	*168*	*128*	*106*	*92*	*201*	*51*
27 France	154	134	100	40	250	409	457	349	189	162
Portugal	-	-	-	-	-	0	3	3	4	6
Russian Fed	-	-	-	-	-	-	-	-	5	-
27 Fishing area total	*154*	*134*	*100*	*40*	*250*	*409*	*460*	*352*	*198*	*168*
34 Côte dIvoire	10	9	15	15 F	10 F	10 F	5 F	3	1	1
Gambia	18	...	27	50	125	72	8	10	22	33
GuineaBissau	...	...	290	222	482	500 F	430 F	350 F	350 F	350 F
Mauritania	350 F	390 F	450 F	500 F	480 F	580 F	510 F	600 F	600 F	600 F
Morocco	1 469	926	1 381	904	868	1 540	1 263	2 473	1 755	1 523
Portugal	0	0	-	-	0	-	-	-	-	24
Spain	194	170 F	150 F	130 F	-	-	-	-	-	-
34 Fishing area total	*2 041 F*	*1 495 F*	*2 313 F*	*1 821 F*	*1 965 F*	*2 702 F*	*2 216 F*	*3 436 F*	*2 728 F*	*2 531 F*
37 Egypt	654	545	547	541	908	615	768	882	575	987
Gaza Strip	...	...	...	...	39	7	4	5 F	5 F	5 F
Israel	50	47	80	80	33	2	2	9	288	200 F
Malta	1	0	-	-	-	-	-	-	-	-
Morocco	0	0	3	0	1	2	0	-	30	15
Turkey	327	353	295	290	71	40	30	65	70	50
37 Fishing area total	*1 032*	*945*	*925*	*911*	*1 052*	*666*	*804*	*961 F*	*968 F*	*1 257 F*
51 Egypt	-	-	158	45	0	0	19	82	-	-
51 Fishing area total	*-*	*-*	*158*	*45*	*0*	*0*	*19*	*82*	*-*	*-*
Species total	*3 267 F*	*2 650 F*	*3 632 F*	*2 944 F*	*3 435 F*	*3 905 F*	*3 605 F*	*4 923 F*	*4 095 F*	*4 007 F*
Southern meagre(=Mulloway)	**Maigre du Sud**		**Corvina del Sur**		***Argyrosomus hololepidotus***			**1,70(37)106,06**		**KOB**
47 Angola	...	...	...	...	...	...	...	259	...	2 540
Namibia	234	664	595	496	464	184	424	273	409	596
Portugal	-	-	-	-	-	-	-	-	-	1
South Africa	993	1 052	765	772	829	748	690	671	600 F	597
47 Fishing area total	*1 227*	*1 716*	*1 360*	*1 268*	*1 293*	*932*	*1 114*	*1 203*	*1 009 F*	*3 734*
51 South Africa	134	1	15	22	21	17	...	...	19	20
51 Fishing area total	*134*	*1*	*15*	*22*	*21*	*17*	*...*	*...*	*19*	*20*
57 Australia	24	22	26	26	44	44	44	96	73	138
57 Fishing area total	*24*	*22*	*26*	*26*	*44*	*44*	*44*	*96*	*73*	*138*
71 Australia	-	-	54	45	44	45	48	49	93	142

B-33

Miscellaneous coastal fishes — Capture production by species, fishing areas and countries or areas
Poissons côtiers divers — Captures par espèces, zones de pêche et pays ou zones
Peces costeros diversos — Capturas por especies, áreas de pesca y países o áreas

Species, Fishing area Espèce, Zone de pêche Especie, Area de pesca	1992 mt	1993 mt	1994 mt	1995 mt	1996 mt	1997 mt	1998 mt	1999 mt	2000 mt	2001 mt
71 Fishing area total	*-*	*-*	*54*	*45*	*44*	*45*	*48*	*49*	*93*	*142*
81 Australia	147	148	140	140	95	88	91	94	78	64
81 Fishing area total	*147*	*148*	*140*	*140*	*95*	*88*	*91*	*94*	*78*	*64*
Species total	*1 532*	*1 887*	*1 595*	*1 501*	*1 497*	*1 126*	*1 297*	*1 442*	*1 272 F*	*4 098*
Geelbek croaker	**Téraglin**		**Corvinata prieta**		***Atractoscion aequidens***				**1,70(37)108,01**	**AWE**
47 South Africa	498	579	444	353	346	474	605	415	380 F	383
47 Fishing area total	*498*	*579*	*444*	*353*	*346*	*474*	*605*	*415*	*380 F*	*383*
51 South Africa	22	-	11	17	15	18	...	...	...	...
51 Fishing area total	*22*	*-*	*11*	*17*	*15*	*18*	*...*	*...*	*...*	*...*
81 Australia	48	47	27	27	20	27	27	29	...	...
81 Fishing area total	*48*	*47*	*27*	*27*	*20*	*27*	*27*	*29*	*...*	*...*
Species total	*568*	*626*	*482*	*397*	*381*	*519*	*632*	*444*	*380 F*	*383*
White weakfish	**Acoupa blanc**		**Corvinata blanca**		***Atractoscion nobilis***				**1,70(37)108,03**	**WEV**
77 USA	54	44	35	33	46	28	71	112	101	124
77 Fishing area total	*54*	*44*	*35*	*33*	*46*	*28*	*71*	*112*	*101*	*124*
Species total	*54*	*44*	*35*	*33*	*46*	*28*	*71*	*112*	*101*	*124*
White croaker	**Courbine blanche**		**Roncador blanco**		***Genyonemus lineatus***				**1,70(37)147,01**	**KIC**
77 USA	320	296	205	256	242	167	64	74	105	137
77 Fishing area total	*320*	*296*	*205*	*256*	*242*	*167*	*64*	*74*	*105*	*137*
Species total	*320*	*296*	*205*	*256*	*242*	*167*	*64*	*74*	*105*	*137*
Boe drum	**Courbine pélin**		**Bombache boe**		***Pteroscion peli***				**1,70(37)166,07**	**DRS**
34 Benin	...	...	...	...	...	...	...	61	40 F	115
Congo Rep	944 F	942 F	566 F	637	630 F	620 F	600 F	600 F	602	600 F
Portugal	-	-	5	350	582	136	101	18	2	0
Senegal	1 123	1 227	1 802	6 833	667	527	489	581	472	444
Sierra Leone	...	...	...	...	...	8	0	...	10	-
34 Fishing area total	*2 067 F*	*2 169 F*	*2 373 F*	*7 820*	*1 879 F*	*1 291 F*	*1 190 F*	*1 260 F*	*1 126 F*	*1 159 F*
Species total	*2 067 F*	*2 169 F*	*2 373 F*	*7 820*	*1 879 F*	*1 291 F*	*1 190 F*	*1 260 F*	*1 126 F*	*1 159 F*
Tigertooth croaker	**Grande verrue tigre**		**Bombache tigre mayor**		***Otolithes ruber***				**1,70(37)186,03**	**LKR**
51 South Africa	-	-	-	-	-	-	13	10 F	...	...
51 Fishing area total	*-*	*-*	*-*	*-*	*-*	*-*	*13*	*10 F*	*...*	*...*
Species total	*-*	*-*	*-*	*-*	*-*	*-*	*13*	*10 F*	*...*	*...*
Peruvian banded croaker	**Bourrugue coco**		**Lambe coco**		***Paralonchurus peruanus***				**1,70(37)194,02**	**PDR**
87 Peru	6 078	7 550	3 788	5 543	4 263	2 737	4 363	6 063	5 729	4 167
87 Fishing area total	*6 078*	*7 550*	*3 788*	*5 543*	*4 263*	*2 737*	*4 363*	*6 063*	*5 729*	*4 167*
Species total	*6 078*	*7 550*	*3 788*	*5 543*	*4 263*	*2 737*	*4 363*	*6 063*	*5 729*	*4 167*
Black drum	**Grand tambour**		**Corvinón negro**		***Pogonias cromis***				**1,70(37)210,02**	**BDM**
21 USA	75	65	102	32	39	72	41	88	52	39
21 Fishing area total	*75*	*65*	*102*	*32*	*39*	*72*	*41*	*88*	*52*	*39*
31 Mexico	274	250	204	166	134	362	392	179	207	...
USA	-	-	-	-	-	1 158	1 926	631	2 260	2 489
31 Fishing area total	*274*	*250*	*204*	*166*	*134*	*1 520*	*2 318*	*810*	*2 467*	*2 489*
41 Argentina	53	96	11	40	159	81	260	182	13	...
Brazil	770 F	770 F	770 F	38	1	-	-	-	-	-
Uruguay	281	273	220	574	172	82	678	100	344	336
41 Fishing area total	*1 104 F*	*1 139 F*	*1 001 F*	*652*	*332*	*163*	*938*	*282*	*357*	*336*
77 Mexico	6	6	1	0	0	0	0	1	0	...
77 Fishing area total	*6*	*6*	*1*	*0*	*0*	*0*	*0*	*1*	*0*	*...*
Species total	*1 459 F*	*1 460 F*	*1 308 F*	*850*	*505*	*1 755*	*3 297*	*1 181*	*2 876*	*2 864*
Honnibe croaker	**Tambour honnibe**		**Corvina honnibe**		***Nibea mitsukurii***				**1,70(37)298,05**	**HOC**
61 Korea Rep	2 625	1 920	2 364	2 164	1 940	1 177	1 285	1 566	1 999	2 156
61 Fishing area total	*2 625*	*1 920*	*2 364*	*2 164*	*1 940*	*1 177*	*1 285*	*1 566*	*1 999*	*2 156*
Species total	*2 625*	*1 920*	*2 364*	*2 164*	*1 940*	*1 177*	*1 285*	*1 566*	*1 999*	*2 156*
Large yellow croaker	**Tambour à gros yeux**		**Corvina japonesa**		***Larimichthys croceus***				**1,70(37)303,04**	**LYC**
61 China	36 437	34 820	69 181	67 031	80 072	69 950	70 935	65 806	122 920	76 289
Korea Rep	24 122	20 680	26 613	22 599	22 225	1 565	294	270	354	426

B-33 Miscellaneous coastal fishes / Poissons côtiers divers / Peces costeros diversos

Capture production by species, fishing areas and countries or areas / Captures par espèces, zones de pêche et pays ou zones / Capturas por especies, áreas de pesca y países o áreas

Species, Fishing area Espèce, Zone de pêche Especie, Area de pesca	1992 mt	1993 mt	1994 mt	1995 mt	1996 mt	1997 mt	1998 mt	1999 mt	2000 mt	2001 mt
61 Fishing area total	*60 559*	*55 500*	*95 794*	*89 630*	*102 297*	*71 515*	*71 229*	*66 076*	*123 274*	*76 715*
Species total	*60 559*	*55 500*	*95 794*	*89 630*	*102 297*	*71 515*	*71 229*	*66 076*	*123 274*	*76 715*
Yellow croaker	**...B**		**Verrugato de Manchuria**		***Larimichthys polyactis***			**1,70(37)303,05**		**CRY**
61 China	63 047	78 311	102 976	153 048	253 482	212 631	262 786	308 907	404 637	321 614
China,Taiwan	807	1 021	2 815	2 786	2 859	2 228	1 010	1 167	780	903
Japan	141	93	225	-	-	-	-	-	-	-
Korea Rep	39 672	31 119	37 488	25 719	23 695	22 101	15 011	13 490	19 630	7 938
61 Fishing area total	*103 667*	*110 544*	*143 504*	*181 553*	*280 036*	*236 960*	*278 807*	*323 564*	*425 047*	*330 455*
Species total	*103 667*	*110 544*	*143 504*	*181 553*	*280 036*	*236 960*	*278 807*	*323 564*	*425 047*	*330 455*
Spot croaker	**Tambour croca**		**Verrugato croca**		***Leiostomus xanthurus***			**1,70(37)363,01**		**SPT**
21 USA	1 713	1 837	1 940	1 982	1 738	1 975	2 354	1 715	2 159	1 877
21 Fishing area total	*1 713*	*1 837*	*1 940*	*1 982*	*1 738*	*1 975*	*2 354*	*1 715*	*2 159*	*1 877*
31 USA	1 410	1 526	2 042	1 539	816	1 098	1 005	884	982	1 214
31 Fishing area total	*1 410*	*1 526*	*2 042*	*1 539*	*816*	*1 098*	*1 005*	*884*	*982*	*1 214*
Species total	*3 123*	*3 363*	*3 982*	*3 521*	*2 554*	*3 073*	*3 359*	*2 599*	*3 141*	*3 091*
Red drum	**Tambour rouge**		**Corvinón ocelado**		***Sciaenops ocellatus***			**1,70(37)411,01**		**RDM**
21 USA	2	4	5	2	1	2	3	6	6	3
21 Fishing area total	*2*	*4*	*5*	*2*	*1*	*2*	*3*	*6*	*6*	*3*
Species total	*2*	*4*	*5*	*2*	*1*	*2*	*3*	*6*	*6*	*3*
Law croaker	**Otolithe gabo**		**Corvina reina**		***Pseudotolithus brachygnathus***			**1,70(37)457,01**		**CKL**
34 Gambia	161	357	216	449	355	225	264	270	454	856
Sierra Leone	...	357	356	364	360 F	393	90	48	357	227
34 Fishing area total	*161*	*714*	*572*	*813*	*715 F*	*618*	*354*	*318*	*811*	*1 083*
Species total	*161*	*714*	*572*	*813*	*715 F*	*618*	*354*	*318*	*811*	*1 083*
Cassava croaker	**Otolithe sénégalais**		**Corvina casava**		***Pseudotolithus senegalensis***			**1,70(37)457,02**		**PSS**
34 Cameroon	...	...	...	2 686	2 540	2 600 F	2 900 F	3 200 F	1 972	2 679
Côte dIvoire	28	30	...	...	5 F	10 F	20 F	30	39	39
Gambia	24	28	37	230	182	160	109	120	58	400
Liberia	...	79	40	20	...	11	...	...	...	...
Senegal	...	...	...	...	...	...	1 630	1 487	1 880	1 682
34 Fishing area total	*52*	*137*	*77*	*2 936*	*2 727 F*	*2 781 F*	*4 659 F*	*4 837 F*	*3 949*	*4 800*
Species total	*52*	*137*	*77*	*2 936*	*2 727 F*	*2 781 F*	*4 659 F*	*4 837 F*	*3 949*	*4 800*
Bobo croaker	**Otolithe bobo**		**Corvina bobo**		***Pseudotolithus elongatus***			**1,70(37)457,05**		**PSE**
34 Cameroon	4 202	1 005	3 500 F	4 436	4 371	4 400 F	4 400 F	1 900 F	3 400	3 218
Congo Rep	5 F	5 F	3 F	3 F	3 F	3 F	3 F	3 F	...	2 F
Côte dIvoire	8	4	4	0	5	3	4	5	...	4
Gabon	530 F	176	904	986	946	769	1 584	2 079	1 784	1 555
Gambia	116	92	50	328	242	214	163	170	138	120
Guinea	2 130 F	2 540 F	2 610 F	3 685	3 386	2 751	2 781	2 142	4 015	4 000 F
Liberia	...	100	40	7	109	27	...	...	...	...
Senegal	...	...	...	...	...	...	137	146	33	114
Sierra Leone	4 633	3 637	3 625	3 650	3 600 F	3 330	389	410	4 035	748
34 Fishing area total	*11 624 F*	*7 559 F*	*10 736 F*	*13 095 F*	*12 662 F*	*11 497 F*	*9 461 F*	*6 855 F*	*13 405*	*9 761 F*
Species total	*11 624 F*	*7 559 F*	*10 736 F*	*13 095 F*	*12 662 F*	*11 497 F*	*9 461 F*	*6 855 F*	*13 405*	*9 761 F*
West African croakers nei	**Otolithes nca**		**Corvinas africanas nep**		***Pseudotolithus spp***			**1,70(37)457,XX**		**CKW**
34 Cameroon	5	-	-	-	-	-	-	-	-	-
Congo Rep	2 230 F	2 225 F	864	581	500 F	450 F	400 F	350 F	298	300 F
Côte dIvoire	861	604	807	653	551	466	545	378	356	407
Gabon	2 320 F	3 365	1 570	2 095	4 020	3 068	4 653	3 367	2 344	...
Gambia	232	769	202	414	473	412	327	340	629	504
Georgia	27 F	20 F	10 F	-	-	-	-	-	-	-
Ghana	2 340	1 301	1 104	698	1 128	1 995	962	937	739	1 070
Guinea	2 380 F	2 840 F	2 920 F	4 115	2 856	2 206	2 570	1 796	3 423	3 400 F
Latvia	3	-	-	-	-	-	-	-	-	-
Liberia	1 016	800	646	1 008	364	510	433	1 025	327	210
Lithuania	2	-	-	-	-	-	-	-	-	-
Mauritania	10 F	10 F	10 F	10 F	10 F	10 F	96 F	...	9 F	...
Nigeria	17 523	8 114	12 013	9 264	14 544	11 762	15 035	14 591	12 604	15 084
Portugal	5	92	254	115	157	36	15	49	635	25
Russian Fed	-	22	15	146	-	15	32	13	13	18
Sierra Leone	...	587	585	677	680 F	1 210	687	869	2 014	1 011
Togo	23	61	51	15	22	16	11	6	37	43
Ukraine	218	2	...	...	...	...	28	12	87	5
34 Fishing area total	*29 195 F*	*20 812 F*	*21 051 F*	*19 791 F*	*25 305 F*	*22 156 F*	*25 794 F*	*23 733 F*	*23 515 F*	*22 077 F*
47 Angola	1 044	1 745	2 962	1 035	4 044	1 606	7 095	8 936	8 005	8 725
Russian Fed	-	2	-	-	-	-	1	-	-	19

B-33

Miscellaneous coastal fishes — Capture production by species, fishing areas and countries or areas
Poissons côtiers divers — Captures par espèces, zones de pêche et pays ou zones
Peces costeros diversos — Capturas por especies, áreas de pesca y países o áreas

Species, Fishing area Espèce, Zone de pêche Especie, Area de pesca	1992 mt	1993 mt	1994 mt	1995 mt	1996 mt	1997 mt	1998 mt	1999 mt	2000 mt	2001 mt
47 Fishing area total	*1 044*	*1 747*	*2 962*	*1 035*	*4 044*	*1 606*	*7 096*	*8 936*	*8 005*	*8 744*
Species total	*30 239 F*	*22 559 F*	*24 013 F*	*20 826 F*	*29 349 F*	*23 762 F*	*32 890 F*	*32 669 F*	*31 520 F*	*30 821 F*
Blackmouth croaker	**Maigre noire**		**Corvina negra**		***Atrobucca nibe***				**1,70(37)553,01**	**CRL**
61 China,Taiwan	2 559	2 481	495	197	297	288	237	223	450	644
61 Fishing area total	*2 559*	*2 481*	*495*	*197*	*297*	*288*	*237*	*223*	*450*	*644*
Species total	*2 559*	*2 481*	*495*	*197*	*297*	*288*	*237*	*223*	*450*	*644*
Silver croaker	**Maigre argenté**		**Corvina plateada**		***Pennahia argentata***				**1,70(37)561,01**	**CRW**
61 China,Taiwan	6 227	7 289	8 467	7 369	8 599	6 085	5 089	4 940	4 180	3 796
61 Fishing area total	*6 227*	*7 289*	*8 467*	*7 369*	*8 599*	*6 085*	*5 089*	*4 940*	*4 180*	*3 796*
Species total	*6 227*	*7 289*	*8 467*	*7 369*	*8 599*	*6 085*	*5 089*	*4 940*	*4 180*	*3 796*
Croakers, drums nei	**Sciaenidés nca**		**Esciénidos nep**		***Sciaenidae***				**1,70(37)XXX,XX**	**CDX**
04 India	-	-	201	174	3 271	748	667	686	213	213
04 Fishing area total	*-*	*-*	*201*	*174*	*3 271*	*748*	*667*	*686*	*213*	*213*
27 Portugal	...	...	...	...	...	...	...	228	132	168
27 Fishing area total	*...*	*...*	*...*	*...*	*...*	*...*	*...*	*228*	*132*	*168*
31 Korea Rep	-	-	-	-	-	-	-	818	1 037	137
Mexico	0	0	0	0	1	2	183	179	91	-
31 Fishing area total	*0*	*0*	*0*	*0*	*1*	*2*	*183*	*997*	*1 128*	*137*
34 Bahamas	-	-	-	25	-	-	-	-	-	-
Belize	-	17	3	67	-	1	65	299	580	100
Cameroon	6 046	2 300	3 000 F	3 679	3 539	3 500 F	3 500 F	3 500 F	3 260	2 399
China	393	748	113	1 065	530	490	1 568	1 027	2 301	2 061
Congo Rep	27 F	27 F	16 F	15 F	10 F	10 F	5 F	5 F	0	0
Gambia	0	0	-	-	-	-	-	-	-	-
GuineaBissau	990 F	1 010 F	900 F	1 000 F	1 020 F	1 040 F	850 F	650 F	650 F	650 F
Honduras	62	184	32	192	91	130	19	73	19	62
Italy	1 618	1 372	1 372	-	-	-	-	-	-	-
Korea D P Rp	-	276	-	-	-	-	-	-	224	-
Korea Rep	1 686	4 187	7 479	5 119	4 793	12 986	9 924	13 036	13 614	14 750
Liberia	6	7	4	87	-	-	-	-	-	-
Mauritania	200 F	200 F	200 F	200 F	200 F	200 F	200 F	300 F	300 F	300 F
Nigeria	1 698	1 453	1 543	4 605	2 165	6 335	4 056	7 846	5 826	5 900
Panama	60	75	-	54	4	1	0	-	-	-
Portugal	607	805	310	60	110	52	30	11	4	6
Russian Fed	-	68	15	127	58	-	-	-	-	-
St Vincent	8	14	-	33	-	-	-	-	-	-
Sao Tome Prn	52 F	66 F	96 F	101 F	71	30	40 F	45 F	40 F	40 F
Senegal	3 035	4 649	5 363	2 115	8 047	7 620	5 572	3 295	3 199	3 700
Spain	-	-	-	-	117	-	166	219	10	9
Other nei	4	5	-	-	-	-	-	-	-	12
34 Fishing area total	*16 492 F*	*17 463 F*	*20 446 F*	*18 544 F*	*20 755 F*	*32 395 F*	*25 995 F*	*30 306 F*	*30 027 F*	*29 989 F*
37 Korea Rep	-	-	-	22	-	-	-	-	-	-
Spain	-	-	-	-	-	-	18	29	164	32
37 Fishing area total	*-*	*-*	*-*	*22*	*-*	*-*	*18*	*29*	*164*	*32*
41 Brazil	...	...	...	69	150	52	57	36	34	30 F
Italy	216	183	183	-	-	-	-	-	-	-
Korea Rep	109	203	80	20	-	-	60	710	654	270
41 Fishing area total	*325*	*386*	*263*	*89*	*150*	*52*	*117*	*746*	*688*	*300 F*
47 Angola	129	45	33	5	-	12	70	91	29	594
China	86	-	-	-	-	-	-	-	64	-
Honduras	-	2	-	0	-	-	-	-	-	-
Korea Rep	362	1 464	2 937	4 853	4 004	4 625	3 049	2 802	2 202	1 965
Panama	1 597	-	-	-	-	-	-	-	-	-
Russian Fed	-	-	-	-	-	-	-	-	-	59
South Africa	-	-	-	-	-	-	-	-	-	5
Spain	-	-	-	-	-	-	-	27	-	17
47 Fishing area total	*2 174*	*1 511*	*2 970*	*4 858*	*4 004*	*4 637*	*3 119*	*2 920*	*2 295*	*2 640*
51 India	238 172	247 198	277 827	242 660	250 080	270 961	233 160	280 556	235 744	214 665
Italy	324	274	274	-	-	-	-	-	-	-
Korea Rep	672	633	176	381	381	2 385	751	1 127	951	361
Kuwait	436	798	1 089	1 572	974	1 127	1 211	1 385	1 100 F	853
Oman	2 023	3 444	3 859	4 070	3 709	5 468	2 218	2 121	1 926	1 873
Pakistan	16 606	19 740	22 808	25 201	19 934	20 428	19 625	24 665	21 976	21 725
Portugal	-	-	-	-	-	-	-	-	2	1
South Africa	-	-	1	9	8	7	...	...	1	0
51 Fishing area total	*258 233*	*272 087*	*306 034*	*273 893*	*275 086*	*300 376*	*256 965*	*309 854*	*261 700 F*	*239 478*
57 India	25 041	28 554	30 639	44 743	46 207	41 505	41 413	36 510	32 116	31 180
Indonesia	8 041	8 031	9 713	10 698	9 802	8 505	9 817	11 827	12 095	12 590
Korea Rep	-	-	-	92	-	-	-	273	152	202
Malaysia	10 260	10 631	10 089	10 313	12 221	11 763	11 484	10 956	12 216	10 507
Thailand	14 581	15 773	15 396	19 750	24 643	24 913	29 750	20 975	20 688	20 562
57 Fishing area total	*57 923*	*62 989*	*65 837*	*85 596*	*92 873*	*86 686*	*92 464*	*80 541*	*77 267*	*75 041*

B-33 Miscellaneous coastal fishes / Poissons côtiers divers / Peces costeros diversos

Capture production by species, fishing areas and countries or areas / Captures par espèces, zones de pêche et pays ou zones / Capturas por especies, áreas de pesca y países o áreas

Species, Fishing area Espèce, Zone de pêche Especie, Area de pesca	1992 mt	1993 mt	1994 mt	1995 mt	1996 mt	1997 mt	1998 mt	1999 mt	2000 mt	2001 mt
61 China	...	...	...	107 176	-	-	-	-	-	-
China,H.Kong	11 551	11 367	8 725	5 547	3 776	4 881	3 992	2 800 F	3 500 F	3 900 F
China,Taiwan	2 244	1 483	3 488	3 269	3 070	2 497	1 422	1 580	1 708	3 841
Japan	10 907	7 518	8 310	9 008	7 062	5 998	5 430	4 850	4 791	4 362
Korea Rep	62 457	71 872	70 259	70 394	57 274	67 804	68 218	81 690	40 314	28 486
61 Fishing area total	*87 159*	*92 240*	*90 782*	*195 394*	*71 182*	*81 180*	*79 062*	*90 920 F*	*50 313 F*	*40 589 F*
71 Indonesia	26 691	28 329	27 687	29 100	35 431	36 332	40 297	45 164	40 159	41 790
Korea Rep	3 812	8 087	10 885	8 282	5 881	6 180	5 606	5 401	4 789	4 072
Malaysia	5 936	6 683	7 684	7 079	8 129	8 605	10 986	11 232	11 223	18 253
Singapore	389	280	162	136	123	180	160	114	68	45
Thailand	3 956	4 760	3 933	4 195	5 118	5 048	3 896	15 616	15 402	15 308
71 Fishing area total	*40 784*	*48 139*	*50 351*	*48 792*	*54 682*	*56 345*	*60 945*	*77 527*	*71 641*	*79 468*
77 El Salvador	236	289	407	233	313	337	252	334	279	575
Korea Rep	2 090	1 935	809	793	1 320	1 301	1 404	7 325	3 030	441
Mexico	437	426	450	326	710	690	763	721	511	-
Panama	445	557	548	346	662	666	1 138	1 533	4 352	4 300 F
77 Fishing area total	*3 208*	*3 207*	*2 214*	*1 698*	*3 005*	*2 994*	*3 557*	*9 913*	*8 172*	*5 316 F*
81 Korea Rep	-	-	-	-	-	-	9	305	-	-
81 Fishing area total	-	-	-	-	-	-	*9*	*305*	-	-
87 Chile	21	38	37	31	27	25	50	39	25	15
Colombia	47	22	8	-	24	1	0	-	-	-
Ecuador	1 977	2 532	10 427	30 910	2 486	316	3 300	7 320	...	1 558
Korea Rep	-	-	-	-	-	-	-	673	820	105
87 Fishing area total	*2 045*	*2 592*	*10 472*	*30 941*	*2 537*	*342*	*3 350*	*8 032*	*845*	*1 678*
Species total	*468 343 F*	*500 614 F*	*549 570 F*	*660 001 F*	*527 546 F*	*565 757 F*	*526 451 F*	*613 004 F*	*504 585 F*	*475 049 F*
Largeeye breams	**...B**		**...C**		***Gymnocranius spp***			**1,70(38)155,XX**		**LBR**
61 China,H.Kong	822	270	211	196	208	170	143	100 F	120 F	130 F
61 Fishing area total	*822*	*270*	*211*	*196*	*208*	*170*	*143*	*100 F*	*120 F*	*130 F*
71 Fiji Islands	125	13	13 F	19	20	22	25	28	25 F	30
71 Fishing area total	*125*	*13*	*13 F*	*19*	*20*	*22*	*25*	*28*	*25 F*	*30*
Species total	*947*	*283*	*224 F*	*215*	*228*	*192*	*168*	*128 F*	*145 F*	*160 F*
Atlantic emperor	**Empereur atlantique**		**Emperador atlántico**		***Lethrinus atlanticus***			**1,70(38)172,04**		**LTN**
34 Guinea	...	...	...	137	231	334	380	...	...	...
34 Fishing area total	...	...	...	*137*	*231*	*334*	*380*	...	...	...
Species total	...	...	...	*137*	*231*	*334*	*380*	...	...	...
Emperors(=Scavengers) nei	**Empereurs nca**		**Emperadores nep**		***Lethrinidae***			**1,70(38)XXX,XX**		**EMP**
47 South Africa	-	-	-	-	-	-	-	-	-	3
47 Fishing area total	-	-	-	-	-	-	-	-	-	*3*
51 Bahrain	1 736	1 603	1 194	1 486	1 506	892	944	1 227	1 403	1 377
Egypt	...	...	...	429	425	574	577	1 040	1 147	2 696
Eritrea	...	...	217	1 010	767	90	104	371	443	346
Jordan	...	...	...	...	...	...	2	2	1	2
Kenya	477	441	353	396	433	361	412	358	334	466
Korea Rep	-	-	-	-	-	-	93	160	97	459
Kuwait	21	36	17	15	15	20	38	35	50 F	78
Mauritius	5 320	6 296	6 312	6 216	5 137	5 018	4 981	4 598	4 698	4 008
Oman	2 877	4 358	3 521	6 013	4 087	5 767	6 630	6 954	7 664	6 526
Pakistan	1 010	1 466	1 660	1 643	1 549	1 911	2 334	3 323	5 173	5 044
Qatar	1 811	1 271	922	722	1 031	1 172	1 326	798	1 442	1 820
Saudi Arabia	7 715	8 281	7 524	6 598	7 314	6 904	6 796	7 233	7 078	7 448
Seychelles	559	355	285	294	295	220	280	285	423	483
South Africa	1	-	-	-	-	-	-	-	-	-
Tanzania	11 500	4 308	4 566	6 490	7 304	7 350	7 025	7 500	8 000	7 850
Untd Arab Em	8 960	9 272	10 515	11 336	11 455	12 242	12 283	12 590	19 647	19 650 F
Yemen	3 275	5 478	4 390	3 214	2 437	2 550 F	2 900 F	2 800 F	2 500 F	3 200 F
51 Fishing area total	*45 262*	*43 165*	*41 476*	*45 862*	*43 755*	*45 071 F*	*46 725 F*	*49 274 F*	*60 100 F*	*61 453 F*
57 Indonesia	2 614	3 227	4 228	4 950	5 046	5 921	6 791	7 060	6 648	6 460
57 Fishing area total	*2 614*	*3 227*	*4 228*	*4 950*	*5 046*	*5 921*	*6 791*	*7 060*	*6 648*	*6 460*
71 Fiji Islands	1 550	3 097	2 956 F	1 359	1 230	1 780	1 731	2 990	2 700 F	2 883
Guam	...	...	8	1	1	1	1	1	1	3
Indonesia	17 553	17 654	19 330	26 852	29 513	25 987	24 915	25 497	24 009	23 320
Kiribati	2 020	1 950	1 930	1 950	1 970	1 960	675	1 299	2 137	3 930
Korea Rep	-	-	-	-	-	-	-	5	1	
N Marianas	...	...	1	2	5	13	50	4	4	8
Palau	36	31	29	37	28	12	37	50 F	60 F	60 F
71 Fishing area total	*21 159*	*22 732*	*24 254 F*	*30 201*	*32 747*	*29 753*	*27 409*	*29 846 F*	*28 912 F*	*30 204 F*
77 Amer Samoa	...	...	4	2	5	2	0	1	4	6
77 Fishing area total	...	...	*4*	*2*	*5*	*2*	*0*	*1*	*4*	*6*
81 New Zealand	-	-	-	-	-	4	-	-	-	-

B-33 Miscellaneous coastal fishes / Poissons côtiers divers / Peces costeros diversos

Capture production by species, fishing areas and countries or areas / Captures par espèces, zones de pêche et pays ou zones / Capturas por especies, áreas de pesca y países o áreas

Species, Fishing area Espèce, Zone de pêche Especie, Area de pesca	1992 mt	1993 mt	1994 mt	1995 mt	1996 mt	1997 mt	1998 mt	1999 mt	2000 mt	2001 mt
81 Fishing area total	*-*	*-*	*-*	*-*	*-*	*4*	*-*	*-*	*-*	*-*
Species total	*69 035*	*69 124*	*69 962 F*	*81 015*	*81 553*	*80 751 F*	*80 925 F*	*86 181 F*	*95 664 F*	*98 126 F*
Blackspot(=red) seabream	**Dorade rose**		**Besugo**			***Pagellus bogaraveo***			**1,70(39)008,01**	**SBR**
27 France	37	16	21	8	15	19	20	31	22	13
Ireland	16	7	0	3	8	8	6	1	...	11
Latvia	-	-	75	-	-	-	-	-	-	-
Netherlands	-	-	-	-	-	-	-	-	-	2
Portugal	1 203	1 030	1 159	1 364	1 305	1 242	1 370	1 373	1 019	1 116
Spain	733	1 567	1 754	1 206	739	3 075	1 395	947	699	94
UK	254	475	484	766	481	606	501	149	159	37
27 Fishing area total	*2 243*	*3 095*	*3 493*	*3 347*	*2 548*	*4 950*	*3 292*	*2 501*	*1 899*	*1 273*
34 Netherlands	-	-	-	-	38	-	-	28	71	-
Portugal	-	-	-	27	141	108	141	117	14	12
34 Fishing area total	*-*	*-*	*-*	*27*	*179*	*108*	*141*	*145*	*85*	*12*
37 France	43	...	42	12	12	20	20	38	38	38
37 Fishing area total	*43*	*...*	*42*	*12*	*12*	*20*	*20*	*38*	*38*	*38*
Species total	*2 286*	*3 095*	*3 535*	*3 386*	*2 739*	*5 078*	*3 453*	*2 684*	*2 022*	*1 323*
Common pandora	**Pageot commun**		**Breca**			***Pagellus erythrinus***			**1,70(39)008,02**	**PAC**
27 France	4	17	1	1	2	0	1	2	3	3
Portugal	183	201	115	87	102	152	132	104	151	128
27 Fishing area total	*187*	*218*	*116*	*88*	*104*	*152*	*133*	*106*	*154*	*131*
34 Portugal	107	69	17	7	20	6	2	1	0	-
Spain	-	-	-	-	-	34	139	204	-	-
34 Fishing area total	*107*	*69*	*17*	*7*	*20*	*40*	*141*	*205*	*0*	*-*
37 Algeria	...	...	...	1 204	1 633	2 661	1 503	1 145	1 100 F	1 100 F
Cyprus	46	36	32	32	32	25	19	37	42	34
France	353	281	118	12	76	77	77	165	123	94
Israel	36	34	150	150	100	65	47	67	517	400 F
Malta	1	2	1	3	5	6	6	6	5	2
Slovenia	-	-	-	-	-	-	1	1	3	7
Spain	-	-	-	-	-	213	183	73	305	229
Tunisia	3 173	2 820	1 981	2 292	2 746	2 506	2 967	3 128	3 105	3 219
37 Fishing area total	*3 609*	*3 173*	*2 282*	*3 693*	*4 592*	*5 553*	*4 803*	*4 622*	*5 200 F*	*5 085 F*
Species total	*3 903*	*3 460*	*2 415*	*3 788*	*4 716*	*5 745*	*5 077*	*4 933*	*5 354 F*	*5 216 F*
Axillary seabream	**Pageot acarne**		**Aligote**			***Pagellus acarne***			**1,70(39)008,03**	**SBA**
27 France	27	20	32	23	17	8	5	32	25	12
Portugal	29	1 208	977	1 044	1 180	966	878	996	1 297	1 201
27 Fishing area total	*56*	*1 228*	*1 009*	*1 067*	*1 197*	*974*	*883*	*1 028*	*1 322*	*1 213*
34 Portugal	616	554	32	8	18	4	5	1	1	1
Spain	-	-	-	-	-	7	122	-	-	-
34 Fishing area total	*616*	*554*	*32*	*8*	*18*	*11*	*127*	*1*	*1*	*1*
37 Algeria	...	...	...	332	...	...	362	...	...	...
Cyprus	34	29	32	11	22	25	23	50	25	20
France	208	104	80	95	63	31	31	286	257	186
Malta	...	...	...	7	5	4	3	3	2	-
Spain	-	-	-	-	-	417	251	-	-	-
37 Fishing area total	*242*	*133*	*112*	*445*	*90*	*477*	*670*	*339*	*284*	*206*
Species total	*914*	*1 915*	*1 153*	*1 520*	*1 305*	*1 462*	*1 680*	*1 368*	*1 607*	*1 420*
Red pandora	**Pageot à tache rouge**		**Breca chata**			***Pagellus bellottii***			**1,70(39)008,07**	**PAR**
34 Congo Rep	34 F	34 F	2	7	5 F	4 F	3 F	2 F	...	2 F
Ghana	8 724	7 915	5 635	4 505	7 541	7 933	10 029	13 265	2 916	3 206
Portugal	15	12	-	-	1	-	-	-	-	-
Romania	1 036	-	-	-	-	-	-	-	-	-
Spain	-	-	-	-	-	43	44	42	-	-
Togo	60	110	93	115	74	67	202	237	64	45
34 Fishing area total	*9 869 F*	*8 071 F*	*5 730*	*4 627*	*7 621 F*	*8 047 F*	*10 278 F*	*13 546 F*	*2 980*	*3 253 F*
47 Angola	...	...	...	...	...	...	...	188	...	744
47 Fishing area total	*...*	*...*	*...*	*...*	*...*	*...*	*...*	*188*	*...*	*744*
Species total	*9 869 F*	*8 071 F*	*5 730*	*4 627*	*7 621 F*	*8 047 F*	*10 278 F*	*13 734 F*	*2 980*	*3 997 F*
Pandoras nei	**Pageots nca**		**Brecas nep**			***Pagellus spp***			**1,70(39)008,XX**	**PAX**
34 Benin	11 F	12 F	14 F	14 F	16 F	22 F	20 F	15	10 F	...
Côte dIvoire	541	346	399	361	328	346	1 093	1 137	913	697
Gabon	...	...	68	538	200	191	301	127	376	274
Gambia	-	-	-	123	76	9	12	20	12	...
Greece	1 015	1 215	906	751	615	449	384	483	620	602
Portugal	61	42	42	-	-	-	-	-	-	-
Sao Tome Prn	64 F	79 F	115 F	121 F	88	86	100 F	110 F	100 F	100 F

B-33 Miscellaneous coastal fishes / Poissons côtiers divers / Peces costeros diversos

Capture production by species, fishing areas and countries or areas / Captures par espèces, zones de pêche et pays ou zones / Capturas por especies, áreas de pesca y países o áreas

Species, Fishing area Espèce, Zone de pêche Especie, Area de pesca	1992 mt	1993 mt	1994 mt	1995 mt	1996 mt	1997 mt	1998 mt	1999 mt	2000 mt	2001 mt
Senegal	5 422	3 890	5 051	4 679	4 380	5 086	3 719	4 319	4 728	3 932
Spain	208	600 F	1 000 F	1 500 F	2 327	47	26	764	57	101
Togo	94	48	37	208	91	65	365	1 055	120	75
34 Fishing area total	*7 416 F*	*6 232 F*	*7 632 F*	*8 295 F*	*8 121 F*	*6 301 F*	*6 020 F*	*8 030 F*	*6 936 F*	*5 781 F*
37 Albania	...	...	...	12	27	25	33	35	34	15
Greece	751	897	1 257	1 187	828	1 095	455	517	430	335
Italy	1 350	1 237	1 181	2 937	1 152	1 445	836	751	1 171	949
Spain	1 000 F	900 F	850 F	800 F	756	716	682	1 353	1 552	1 955
Syria	78 F	80 F	77 F	77 F	134	98	125	100	65	85
37 Fishing area total	*3 179 F*	*3 114 F*	*3 365 F*	*5 013 F*	*2 897*	*3 379*	*2 131*	*2 756*	*3 252*	*3 339*
51 Oman	532	419	...	...	...	...	...	...	...	...
Spain	-	-	-	-	-	-	-	5	-	-
51 Fishing area total	*532*	*419*	*...*	*...*	*...*	*...*	*...*	*5*	*...*	*...*
Species total	*11 127 F*	*9 765 F*	*10 997 F*	*13 308 F*	*11 018 F*	*9 680 F*	*8 151 F*	*10 791 F*	*10 188 F*	*9 120 F*
South American silver porgy	**...B**		**...C**		*Diplodus argenteus*			**1,70(39)033,01**		**DIG**
41 Argentina	-	-	6	8	39	84	15	5	2	...
41 Fishing area total	*-*	*-*	*6*	*8*	*39*	*84*	*15*	*5*	*2*	*...*
Species total	*-*	*-*	*6*	*8*	*39*	*84*	*15*	*5*	*2*	*...*
White seabream	**Sar commun**		**Sargo**		*Diplodus sargus*			**1,70(39)033,03**		**SWA**
27 France	43	54	52	45	54	41	35	55	64	51
27 Fishing area total	*43*	*54*	*52*	*45*	*54*	*41*	*35*	*55*	*64*	*51*
37 Greece	517	659	629	539	409	509	327	503	581	347
Spain	-	-	-	-	-	131	106	137	114	182
37 Fishing area total	*517*	*659*	*629*	*539*	*409*	*640*	*433*	*640*	*695*	*529*
Species total	*560*	*713*	*681*	*584*	*463*	*681*	*468*	*695*	*759*	*580*
Sargo breams nei	**Sars, sparaillons nca**		**Sargos, raspallones nep**		*Diplodus spp*			**1,70(39)033,XX**		**SRG**
27 Portugal	...	...	...	...	...	...	...	1 050	977	793
27 Fishing area total	*...*	*...*	*...*	*...*	*...*	*...*	*...*	*1 050*	*977*	*793*
34 Greece	1	4	12	31	19	7	22	18	117	88
Latvia	-	14	5	18	19	20	13	81	90	176
Mauritania	20 F	30 F	30 F	20 F	20 F	20 F	70 F	99	145	277
Morocco	1 641	1 063	1 194	837	840	1 002	975	1 062	1 217	1 089
Portugal	294	260	40	34	65	18	4	1	2	2
Romania	118	-	-	-	-	-	-	-	-	-
Senegal	-	-	-	-	-	4	221	194	364	453
Spain	-	-	-	-	151	117	111	77	74	98
34 Fishing area total	*2 074 F*	*1 371 F*	*1 281 F*	*940 F*	*1 114 F*	*1 188 F*	*1 416 F*	*1 532*	*2 009*	*2 183*
37 Bulgaria	-	-	-	-	-	-	-	-	-	90
Egypt	113	113	0	346	590	382	390	841	812	793
France	183	166	99	60	80	109	109	82	79	83
Gaza Strip	...	...	...	...	4	17	24	20 F	20 F	20 F
Italy	1 546	1 361	1 722	1 123	1 069	706	382	340	321	462
Malta	1	0	0	0	2	0	4	4	2	2
Morocco	4	5	15	92	100	47	7	11	240	199
Slovenia	-	-	-	-	-	-	-	-	1	2
Tunisia	413	407	345	2 292	2 706	3 177	2 055	3 036	3 084	3 241
Turkey	1 817	2 420	2 887	2 398	2 345	1 750	2 210	2 110	1 420	655
37 Fishing area total	*4 077*	*4 472*	*5 068*	*6 311*	*6 896*	*6 188*	*5 181*	*6 444 F*	*5 979 F*	*5 547 F*
47 Portugal	-	-	-	-	-	-	-	-	-	4
47 Fishing area total	*-*	*-*	*-*	*-*	*-*	*-*	*-*	*-*	*-*	*4*
Species total	*6 151 F*	*5 843 F*	*6 349 F*	*7 251 F*	*8 010 F*	*7 376 F*	*6 597 F*	*9 026 F*	*8 965 F*	*8 527 F*
Porgies	**Daubenets**		**Plumas**		*Calamus spp*			**1,70(39)034,XX**		**PRG**
31 Cuba	363	286	373	378	385	333	270	259	260 F	260 F
Dominican Rp	426	400	720	631	428	1 112	17	430	17	10
Mexico	183	160	168	257	239	96	266	577	235	250
31 Fishing area total	*972*	*846*	*1 261*	*1 266*	*1 052*	*1 541*	*553*	*1 266*	*512 F*	*520 F*
Species total	*972*	*846*	*1 261*	*1 266*	*1 052*	*1 541*	*553*	*1 266*	*512 F*	*520 F*
Large-eye dentex	**Denté à gros yeux**		**Cachucho**		*Dentex macrophthalmus*			**1,70(39)060,02**		**DEL**
27 Portugal	32	34	25	12	8	26	2	2	0	0
27 Fishing area total	*32*	*34*	*25*	*12*	*8*	*26*	*2*	*2*	*0*	*0*
34 Georgia	46 F	30 F	10 F	-	-	-	-	-	-	-
Greece	10	8	15	1	2	16	9	19	20	12
Latvia	-	-	-	-	-	24	57	91	190	71
Liberia	23	12	8	...	9	39	...	...	...	...
Lithuania	-	-	80	-	-	-	-	-	-	-
Portugal	364	1 330	90	8	63	48	76	64	-	0
Romania	6	-	-	-	-	-	-	-	-	-

B-33 Miscellaneous coastal fishes / Poissons côtiers divers / Peces costeros diversos

Capture production by species, fishing areas and countries or areas
Captures par espèces, zones de pêche et pays ou zones
Capturas por especies, áreas de pesca y países o áreas

Species, Fishing area Espèce, Zone de pêche Especie, Area de pesca	1992 mt	1993 mt	1994 mt	1995 mt	1996 mt	1997 mt	1998 mt	1999 mt	2000 mt	2001 mt
Russian Fed	751	142	172	329	528	445	1 506	2 217	592	374
Senegal	...	...	...	...	447	...	451	531	684	499
Togo	0	2	0	0	0	0	1	-	-	-
Ukraine	485	9	12	...	52	...	207	199	468	130
34 Fishing area total	*1 685 F*	*1 533 F*	*387 F*	*338*	*1 101*	*572*	*2 307*	*3 121*	*1 954*	*1 086*
37 Greece	282	221	550	710	651	1 010	535	497	378	318
37 Fishing area total	*282*	*221*	*550*	*710*	*651*	*1 010*	*535*	*497*	*378*	*318*
47 Estonia	-	-	-	181	-	-	-	-	-	-
Russian Fed	35	113	40	229	682	1 121	736	1 786	1 793	807
Ukraine	13	...	...	...	-	4	-	73	456	55
47 Fishing area total	*48*	*113*	*40*	*410*	*682*	*1 125*	*736*	*1 859*	*2 249*	*862*
Species total	*2 047 F*	*1 901 F*	*1 002 F*	*1 470*	*2 442*	*2 733*	*3 580*	*5 479*	*4 581*	*2 266*
Canary dentex	**Denté à tache rouge**		**Chacarona de Canarias**		***Dentex canariensis***				**1,70(39)060,04**	**DEN**
34 Netherlands	-	-	-	-	-	-	-	-	20	-
34 Fishing area total	-	-	-	-	-	-	-	-	*20*	-
Species total	-	-	-	-	-	-	-	-	*20*	-
Common dentex	**Denté commun**		**Dentón**		***Dentex dentex***				**1,70(39)060,06**	**DEC**
27 Portugal	15	19	10	13	11	18	36	24	13	16
27 Fishing area total	*15*	*19*	*10*	*13*	*11*	*18*	*36*	*24*	*13*	*16*
34 Greece	936	780	383	623	550	228	387	254	134	317
Italy	4 046	3 430	3 430	-	-	-	-	-	-	-
Latvia	-	-	101	694	436	8	...	...	...	19
Portugal	27	29	8	5	9	3	2	1	0	-
34 Fishing area total	*5 009*	*4 239*	*3 922*	*1 322*	*995*	*239*	*389*	*255*	*134*	*336*
37 Albania	...	...	...	...	...	...	...	...	...	26
Croatia	45	49	47	53	56	76	70	50	55	10
Cyprus	11	18	47	23	38	18	18	22	18	19
France	10	8	8	6	10	9	9	5	2	1
Greece	230	227	398	291	194	220	135	220	230	163
Italy	4 398	2 250	2 279	2 270	1 253	389	190	205	309	201
Malta	0	0	1	0	0	1	0	1	1	1
Spain	35 F	40 F	50 F	55 F	60	62	61	149	265	56
Tunisia	357	214	250	167	227	346	375	291	250	219
Turkey	217	229	218	191	442	330	370	220	100	60
Yugoslavia	2	4	4	5	7	7	9	9	11	11
37 Fishing area total	*5 305 F*	*3 039 F*	*3 302 F*	*3 061 F*	*2 287*	*1 458*	*1 237*	*1 172*	*1 241*	*767*
Species total	*10 329 F*	*7 297 F*	*7 234 F*	*4 396 F*	*3 293*	*1 715*	*1 662*	*1 451*	*1 388*	*1 119*
Angolan dentex	**Denté angolais**		**Dentón angoleño**		***Dentex angolensis***				**1,70(39)060,10**	**DEA**
34 Ghana	284	428	183	591	489	490	1 416	1 767	504	838
34 Fishing area total	*284*	*428*	*183*	*591*	*489*	*490*	*1 416*	*1 767*	*504*	*838*
47 Korea Rep	-	-	-	-	274	-	-	-	-	-
47 Fishing area total	-	-	-	-	*274*	-	-	-	-	-
Species total	*284*	*428*	*183*	*591*	*763*	*490*	*1 416*	*1 767*	*504*	*838*
Congo dentex	**Denté congolais**		**Dentón congolés**		***Dentex congoensis***				**1,70(39)060,11**	**DNC**
34 Ghana	151	364	112	43	102	119	392	1 272	350	571
34 Fishing area total	*151*	*364*	*112*	*43*	*102*	*119*	*392*	*1 272*	*350*	*571*
Species total	*151*	*364*	*112*	*43*	*102*	*119*	*392*	*1 272*	*350*	*571*
Dentex nei	**Dentés nca**		**Dentones, samas, etc. nep**		***Dentex spp***				**1,70(39)060,XX**	**DEX**
27 Lithuania	-	-	-	8	-	-	-	-	-	-
Portugal	9	14	2	1	2	2	3	3	1	1
Spain	39	60	54	22	35	39	23	31	386	182
27 Fishing area total	*48*	*74*	*56*	*31*	*37*	*41*	*26*	*34*	*387*	*183*
34 Benin	70 F	76 F	86 F	83 F	95 F	100 F	90 F	83	60 F	20
Congo Rep	422 F	420 F	252 F	220 F	180 F	140 F	100 F	60 F	22	20 F
Estonia	49	91	-	-	-	-	-	-	-	-
Gabon	930 F	756	477	776	820	522	1 047	423	371	498
Ghana	840	1 192	813	1 372	1 584	1 084	1 278	1 874	892	675
Korea Rep	2 093	155	-	4	4	-	-	-	-	-
Liberia	82	72	84	313	327	346	974	936	588	671
Portugal	474	404	7	10	13	0	60	25	5	1
Senegal	3 129	2 731	1 906	4 318	1 608	2 301	1 132	788	1 033	833
Spain	422	400 F	350 F	350 F	300	299	471	618	473	687
34 Fishing area total	*8 511 F*	*6 297 F*	*3 975 F*	*7 446 F*	*4 931 F*	*4 792 F*	*5 152 F*	*4 807 F*	*3 444 F*	*3 405 F*
41 Italy	539	457	457	-	-	-	-	-	-	-
41 Fishing area total	*539*	*457*	*457*	-	-	-	-	-	-	-
47 Angola	941	2 490	2 883	7 283	2 624	1 567	8 683	8 758	13 765	16 080

B-33 Miscellaneous coastal fishes / Poissons côtiers divers / Peces costeros diversos

Capture production by species, fishing areas and countries or areas / Captures par espèces, zones de pêche et pays ou zones / Capturas por especies, áreas de pesca y países o áreas

Species, Fishing area Espèce, Zone de pêche Especie, Area de pesca	1992 mt	1993 mt	1994 mt	1995 mt	1996 mt	1997 mt	1998 mt	1999 mt	2000 mt	2001 mt
Other nei	-	-	124	1 106	462	-	-	-	-	-
47 Fishing area total	*941*	*2 490*	*3 007*	*8 389*	*3 086*	*1 567*	*8 683*	*8 758*	*13 765*	*16 080*
51 Italy	809	686	686	-	-	-	-	-	-	-
Ukraine	-	-	-	-	-	-	2	-	-	-
51 Fishing area total	*809*	*686*	*686*	-	-	-	*2*	-	-	-
Species total	*10 848 F*	*10 004 F*	*8 181 F*	*15 866 F*	*8 054 F*	*6 400 F*	*13 863 F*	*13 599 F*	*17 596 F*	*19 668 F*
Black seabream	**Dorade grise**		**Chopa**		***Spondyliosoma cantharus***			**1,70(39)063,02**		**BRB**
27 France	2 395	2 857	2 666	2 219	2 802	2 884	3 115	3 030	3 011	2 784
Portugal	...	...	...	...	...	...	...	164	177	158
UK	-	-	-	-	-	-	5	260	240	311
27 Fishing area total	*2 395*	*2 857*	*2 666*	*2 219*	*2 802*	*2 884*	*3 120*	*3 454*	*3 428*	*3 253*
34 Benin	9 F	10 F	20 F	20 F	30 F	45 F	48 F	56	40 F	93
Greece	5	9	13	25	13	2	1	5	18	6
Liberia	-	-	-	-	-	-	-	-	-	10
Morocco	284	143	151	103	167	121	109	139	220	265
Portugal	25	33	12	13	23	5	4	1	0	0
Senegal	1 951	1 744	1 628	862	1 196	1 251	597	922	1 067	1 035
Spain	252	270 F	300 F	320 F	340	351	318	-	-	-
Togo	3	8	15	10	6	5	2	6	25	12
34 Fishing area total	*2 529 F*	*2 217 F*	*2 139 F*	*1 353 F*	*1 775 F*	*1 780 F*	*1 079 F*	*1 129*	*1 370 F*	*1 421*
37 Croatia	31	31	36	52	51	56	60	23	19	5
France	9	1	0	0	1	1	1	1	18	15
Greece	352	394	541	319	378	316	218	325	158	132
Morocco	37	31	9	53	62	64	17	7	215	35
Spain	-	-	-	-	-	9	7	7	18	23
Turkey	132	346	1 208	72	73	35	40	50	45	45
Yugoslavia	0	1	1	2	5	3	5	3	4	4
37 Fishing area total	*561*	*804*	*1 795*	*498*	*570*	*484*	*348*	*416*	*477*	*259*
47 Angola	...	...	...	...	...	...	93	129	212	494
47 Fishing area total	...	...	...	...	...	...	*93*	*129*	*212*	*494*
Species total	*5 485 F*	*5 878 F*	*6 600 F*	*4 070 F*	*5 147 F*	*5 148 F*	*4 640 F*	*5 128*	*5 487 F*	*5 427*
Saddled seabream	**Oblade**		**Oblada**		***Oblada melanura***			**1,70(39)076,01**		**SBS**
01 Egypt	...	...	...	...	...	33	67	47	158	70
01 Fishing area total	...	...	...	...	...	*33*	*67*	*47*	*158*	*70*
34 Greece	-	-	5	29	19	-	-	-	-	-
Portugal	-	-	-	-	1	0	1	-	0	0
34 Fishing area total	-	-	*5*	*29*	*20*	*0*	*1*	-	*0*	*0*
37 Croatia	196	183	175	180	183	169	185	130	120	55
Cyprus	6	3	3	10	4	4	3	7	8	11
France	6	5	4	4	13	14	14	10	14	8
Greece	1 417	1 206	1 072	651	588	399	332	239	443	287
Malta	2	0	0	1	1	2	1	2	2	1
Spain	-	-	-	-	-	30	44	78	66	81
Tunisia	40	17	22	24	43	36	29	146	84	51
Turkey	286	208	330	196	184	190	180	115	80	90
Yugoslavia	2	1	2	3	4	4	6	7	5	6
37 Fishing area total	*1 955*	*1 623*	*1 608*	*1 069*	*1 020*	*848*	*794*	*734*	*822*	*590*
Species total	*1 955*	*1 623*	*1 613*	*1 098*	*1 040*	*881*	*862*	*781*	*980*	*660*
Sheepshead	**Rondeau mouton**		**Sargo chopa**		***Archosargus probatocephalus***			**1,70(39)102,01**		**SPH**
02 USA	514	495	611	470	368	549	573	470	577	506
02 Fishing area total	*514*	*495*	*611*	*470*	*368*	*549*	*573*	*470*	*577*	*506*
21 USA	15	13	0	29	137	31	20	87	27	19
21 Fishing area total	*15*	*13*	*0*	*29*	*137*	*31*	*20*	*87*	*27*	*19*
31 USA	1 973	2 167	2 130	1 806	1 519	1 712	1 346	1 096	1 501	1 195
31 Fishing area total	*1 973*	*2 167*	*2 130*	*1 806*	*1 519*	*1 712*	*1 346*	*1 096*	*1 501*	*1 195*
Species total	*2 502*	*2 675*	*2 741*	*2 305*	*2 024*	*2 292*	*1 939*	*1 653*	*2 105*	*1 720*
King soldier bream	**Spare royal**		**Sargo real**		***Argyrops spinifer***			**1,70(39)105,01**		**KBR**
51 Qatar	216	174	110	85	130	177	146	98	199	426
Untd Arab Em	2 370	2 470	2 703	2 626	2 654	2 836	2 846	2 917	3 604	3 600 F
51 Fishing area total	*2 586*	*2 644*	*2 813*	*2 711*	*2 784*	*3 013*	*2 992*	*3 015*	*3 803*	*4 026 F*
Species total	*2 586*	*2 644*	*2 813*	*2 711*	*2 784*	*3 013*	*2 992*	*3 015*	*3 803*	*4 026 F*
Carpenter seabream	**Denté charpentier**		**Dentón carpintero**		***Argyrozona argyrozona***			**1,70(39)107,01**		**SLF**
47 South Africa	7	692	16	729	883	780	505	541	500 F	287
47 Fishing area total	*7*	*692*	*16*	*729*	*883*	*780*	*505*	*541*	*500 F*	*287*
Species total	*7*	*692*	*16*	*729*	*883*	*780*	*505*	*541*	*500 F*	*287*

B-33 Miscellaneous coastal fishes / Poissons côtiers divers / Peces costeros diversos

Capture production by species, fishing areas and countries or areas / Captures par espèces, zones de pêche et pays ou zones / Capturas por especies, áreas de pesca y países o áreas

Species, Fishing area Espèce, Zone de pêche Especie, Area de pesca	1992 mt	1993 mt	1994 mt	1995 mt	1996 mt	1997 mt	1998 mt	1999 mt	2000 mt	2001 mt
Santer seabream	**Denté nufar**		**Dentón nufar**		***Cheimerius nufar***				**1,70(39)118,02**	**SLD**
47 South Africa	50	81	48	41	40	34	0	29	25 F	25
47 Fishing area total	*50*	*81*	*48*	*41*	*40*	*34*	*0*	*29*	*25 F*	*25*
51 South Africa	45	-	32	25	33	28	...	...	...	...
51 Fishing area total	*45*	*-*	*32*	*25*	*33*	*28*	*...*	*...*	*...*	*...*
Species total	*95*	*81*	*80*	*66*	*73*	*62*	*0*	*29*	*25 F*	*25*
Red porgy	**Pagre rouge**		**Pargo**		***Pagrus pagrus***				**1,70(39)191,03**	**RPG**
27 France	1	2	2	2	3	1	1	2	1	1
Portugal	97	90	83	91	106	162	317	540	441	243
Spain	-	-	-	-	22	302	720	243	86	100
27 Fishing area total	*98*	*92*	*85*	*93*	*131*	*465*	*1 038*	*785*	*528*	*344*
34 Greece	1 563	1 184	840	1 011	909	527	569	587	572	820
Portugal	-	-	-	27	65	112	318	54	35	26
34 Fishing area total	*1 563*	*1 184*	*840*	*1 038*	*974*	*639*	*887*	*641*	*607*	*846*
37 Cyprus	43	71	103	45	35	27	25	31	23	21
Egypt	775	1 080	1 508	1 021	1 825	1 425	1 230	2 984	1 847	575
Gaza Strip	...	...	...	...	35	47	35	35 F	35 F	30 F
Greece	517	422	487	416	371	406	284	300	341	319
Malta	6	3	5	6	8	9	8	6	6	4
Spain	110 F	115 F	115 F	120 F	120	123	121	114	136	81
Tunisia	305	237	230	272	254	327	362	388	460	416
Turkey	1 956	2 217	1 624	2 012	750	560	675	480	540	320
37 Fishing area total	*3 712 F*	*4 145 F*	*4 072 F*	*3 892 F*	*3 398*	*2 924*	*2 740*	*4 338 F*	*3 388 F*	*1 766 F*
41 Argentina	2 640	1 216	1 248	1 203	1 590	1 159	574	2 159	1 301	805
Brazil	130 F	130 F	130 F	83	884	1 572	1 448	1 362	1 497	1 480 F
Italy	270	229	229	-	-	-	-	-	-	-
Lithuania	-	-	3	-	-	-	-	-	-	-
Uruguay	58	21	19	12	1	12	4	14	27	7
41 Fishing area total	*3 098 F*	*1 596 F*	*1 629 F*	*1 298*	*2 475*	*2 743*	*2 026*	*3 535*	*2 825*	*2 292 F*
Species total	*8 471 F*	*7 017 F*	*6 626 F*	*6 321 F*	*6 978*	*6 771*	*6 691*	*9 299 F*	*7 348 F*	*5 248 F*
Silver seabream	**Dorade**		**Dorada del Pacífico**		***Pagrus auratus***				**1,70(39)191,15**	**GSU**
57 Australia	1 776	2 321	2 611	3 025	3 293	3 213	2 487	2 588	2 326	...
57 Fishing area total	*1 776*	*2 321*	*2 611*	*3 025*	*3 293*	*3 213*	*2 487*	*2 588*	*2 326*	*...*
61 China,H.Kong	22	...	...	...	...	...	...	...	...	...
China,Taiwan	1 573	1 585	1 270	1 580	1 809	2 445	2 073	3 333	4 726	6 722
Japan	14 243	14 159	14 442	15 007	16 468	15 611	15 375	15 731	15 041	14 633
Korea Rep	716	979	456	552	762	1 149	1 657	924	986	913
61 Fishing area total	*16 554*	*16 723*	*16 168*	*17 139*	*19 039*	*19 205*	*19 105*	*19 988*	*20 753*	*22 268*
71 Australia	634	570	1 181	783	644	649	872	395	311	394
71 Fishing area total	*634*	*570*	*1 181*	*783*	*644*	*649*	*872*	*395*	*311*	*394*
81 Australia	590	588	523	403	324	318	272	1 618	1 065	799
Japan	40	71	9	4	-	-	-	-	-	-
New Zealand	7 734	7 151	6 701	6 161	5 814	6 233	6 278	6 876	6 852	6 209
81 Fishing area total	*8 364*	*7 810*	*7 233*	*6 568*	*6 138*	*6 551*	*6 550*	*8 494*	*7 917*	*7 008*
Species total	*27 328*	*27 424*	*27 193*	*27 515*	*29 114*	*29 618*	*29 014*	*31 465*	*31 307*	*29 670*
Pargo breams nei	**Dorades nca**		**Pargos nep**		***Pagrus spp***				**1,70(39)191,XX**	**SBP**
34 Ghana	1 200	1 862	834	1 546	1 448	1 347	1 682	...	...	2 189
Greece	3	1	4	-	1	5	14	-	-	-
Korea Rep	-	35	-	30	10	-	-	-	-	-
Morocco	65	67	13	16	20	19	100	92	124	156
Portugal	332	862	437	124	196	31	11	4	6	23
Senegal	2 498	2 261	2 394	1 775	2 570	2 763	1 881	1 931	2 206	2 488
Sierra Leone	250	221	220	239	240 F	933	713	752	1 106	...
Spain	6	10 F	40 F	70 F	100	106	223	313	177	141
34 Fishing area total	*4 354*	*5 319 F*	*3 942 F*	*3 800 F*	*4 585 F*	*5 204*	*4 624*	*3 092*	*3 619*	*4 997*
37 Algeria	...	...	...	71	81	58	255	254	250 F	250 F
Greece	336	371	720	697	530	460	357	435	386	326
Morocco	2	1	2	5	5	4	36	0	300	112
Syria	39 F	40 F	39 F	39 F	50	80	90	62	74	77
37 Fishing area total	*377 F*	*412 F*	*761 F*	*812 F*	*666*	*602*	*738*	*751*	*1 010 F*	*765 F*
47 Angola	151	334	246	39	95	716	1 077	183	1 840	4 484
Portugal	-	-	-	-	-	-	-	-	35	122
47 Fishing area total	*151*	*334*	*246*	*39*	*95*	*716*	*1 077*	*183*	*1 875*	*4 606*
Species total	*4 882 F*	*6 065 F*	*4 949 F*	*4 651 F*	*5 346 F*	*6 522*	*6 439*	*4 026*	*6 504 F*	*10 368 F*
Red steenbras	**Denté du Cap**		**Dentón del Cabo**		***Petrus rupestris***				**1,70(39)203,01**	**RER**
47 South Africa	-	73	37	56	28	35	...	22	10 F	7

B-33 Miscellaneous coastal fishes / Poissons côtiers divers / Peces costeros diversos

Capture production by species, fishing areas and countries or areas / Captures par espèces, zones de pêche et pays ou zones / Capturas por especies, áreas de pesca y países o áreas

Species, Fishing area Espèce, Zone de pêche Especie, Area de pesca	1992 mt	1993 mt	1994 mt	1995 mt	1996 mt	1997 mt	1998 mt	1999 mt	2000 mt	2001 mt
47 Fishing area total	-	*73*	*37*	*56*	*28*	*35*	...	*22*	*10 F*	*7*
Species total	-	*73*	*37*	*56*	*28*	*35*	...	*22*	*10 F*	*7*
Panga seabream	**Spare panga**		**Panga**		***Pterogymnus laniarius***			**1,70(39)220,01**		**PGA**
47 Japan	634	-	-	-	-	-	-	-	-	-
Namibia	-	-	246	291	470	416	199	383	...	...
South Africa	849	1 000	690	708	730	907	843	966	900 F	1 058
47 Fishing area total	*1 483*	*1 000*	*936*	*999*	*1 200*	*1 323*	*1 042*	*1 349*	*900 F*	*1 058*
Species total	*1 483*	*1 000*	*936*	*999*	*1 200*	*1 323*	*1 042*	*1 349*	*900 F*	*1 058*
White stumpnose	**Sargue australe**		**Pargo ñato**		***Rhabdosargus globiceps***			**1,70(39)224,01**		**WSN**
47 South Africa	134	139	143	168	237	165	296	335	300 F	169
47 Fishing area total	*134*	*139*	*143*	*168*	*237*	*165*	*296*	*335*	*300 F*	*169*
Species total	*134*	*139*	*143*	*168*	*237*	*165*	*296*	*335*	*300 F*	*169*
Gilthead seabream	**Dorade royale**		**Dorada**		***Sparus aurata***			**1,70(39)235,08**		**SBG**
01 Egypt	780	1 197	899	908	474	554	672	557	1 727	1 711
01 Fishing area total	*780*	*1 197*	*899*	*908*	*474*	*554*	*672*	*557*	*1 727*	*1 711*
27 Channel Is	...	...	...	...	...	...	...	15	-	-
France	163	133	147	126	155	115	107	152	206	162
Portugal	114	141	164	200	209	188	173	151	183	213
Spain	304	273	377	405	511	363	258	118	100	152
27 Fishing area total	*581*	*547*	*688*	*731*	*875*	*666*	*538*	*436*	*489*	*527*
34 Italy	2 024	1 718	1 718	-	-	-	-	-	-	-
Morocco	0	0	0	0	0	0	4	0	6	25
Portugal	-	-	-	2	4	1	0	0	-	-
Spain	-	-	-	-	-	15	24	735	593	1 263
34 Fishing area total	*2 024*	*1 718*	*1 718*	*2*	*4*	*16*	*28*	*735*	*599*	*1 288*
37 Albania	...	...	...	17	27	11	20	20	23	90
Croatia	9	9	18	17	13	44	84	27	25	11
Cyprus	-	-	-	1	1	0	0	0	-	37
Egypt	335	378	475	451	754	533	553	1 398	751	601
France	237	387	79	103	132	106	106	226	170	207
Greece	172	310	445	201	199	138	125	142	248	176
Italy	1 772	1 944	3 086	2 179	1 743	1 859	1 717	1 754	1 939	2 675
Morocco	...	...	...	...	...	...	...	...	200	0
Slovenia	-	-	-	2	-	-	-	1	1	4
Spain	250 F	230 F	210 F	190 F	170	168	226	103	536	749
Tunisia	385	330	309	125	107	265	333	409	757	399
Turkey	1 984	1 593	1 334	1 432	1 340	1 200	1 400	1 665	830	1 070
Yugoslavia	0	1	1	2	4	4	4	6	6	7
37 Fishing area total	*5 144 F*	*5 182 F*	*5 957 F*	*4 720 F*	*4 490*	*4 328*	*4 568*	*5 751*	*5 486*	*6 026*
Species total	*8 529 F*	*8 644 F*	*9 262 F*	*6 361 F*	*5 843*	*5 564*	*5 806*	*7 479*	*8 301*	*9 552*
Bogue	**Bogue**		**Boga**		***Boops boops***			**1,70(39)261,01**		**BOG**
27 France	39	83	85	61	17	10	26	20	76	71
Portugal	904	628	500	428	380	375	316	354	642	929
Spain	463	521	399	644	406	639	1 036	2 985	1 043	933
27 Fishing area total	*1 406*	*1 232*	*984*	*1 133*	*803*	*1 024*	*1 378*	*3 359*	*1 761*	*1 933*
34 Greece	0	0	0	3	2	-	-	-	-	-
Lithuania	-	11	-	-	-	-	-	-	-	-
Morocco	393	324	393	504	500	510	250	60	2 024	785
Portugal	-	-	-	23	37	45	42	72	28	29
Romania	70	-	-	-	-	-	-	-	-	-
Sao Tome Prn	...	...	...	...	...	17	20 F	25 F	20 F	20 F
Senegal	32	66	35	139	10	20	38	7	17	14
Spain	-	-	-	-	-	6	13	98	37	30
Togo	33	27	11	3	40	18	109	198	71	99
34 Fishing area total	*528*	*428*	*439*	*672*	*589*	*616*	*472 F*	*460 F*	*2 197 F*	*977 F*
37 Albania	196	150 F	100 F	52	104	65	220	220	220	120
Algeria	...	...	...	2 282	1 425	1 631	3 234	3 452	3 400 F	3 400 F
Croatia	548	501	396	329	303	342	335	140	147	109
Cyprus	252	278	178	290	285	230	233	259	354	216
Egypt	2 177	3 118	2 638	2 173	2 609	2 499	1 956	...	1 450	1 222
France	476	486	197	37	172	193	193	197	223	258
Gaza Strip	...	...	...	...	9	81	162	160 F	160 F	130 F
Greece	9 438	12 398	14 592	6 842	6 742	5 973	4 228	4 658	4 096	3 674
Israel	130	123	50	102	34	34	47	73	91	80 F
Italy	5 186	5 543	4 457	5 659	5 281	4 178	4 074	3 105	3 541	3 537
Libya	2 150 F	2 300 F	2 500 F	2 550 F	2 500 F	2 500 F	2 500 F	2 500 F	2 500 F	2 500 F
Malta	46	43	19	19	17	16	15	12	21	27
Morocco	1 195	1 766	1 799	2 877	2 900	2 000	2 951	3 276	2 864	2 471
Slovenia	-	-	-	-	-	-	2	1	3	2
Spain	660 F	740 F	820 F	900 F	950	997	870	815	863	856
Tunisia	1 522	1 529	1 491	1 636	1 928	2 205	2 466	3 890	3 052	2 852
Turkey	3 633	3 154	4 236	3 196	3 736	2 450	4 100	1 620	1 500	1 000

B-33 Miscellaneous coastal fishes / Poissons côtiers divers / Peces costeros diversos

Capture production by species, fishing areas and countries or areas / Captures par espèces, zones de pêche et pays ou zones / Capturas por especies, áreas de pesca y países o áreas

Species, Fishing area Espèce, Zone de pêche Especie, Area de pesca	1992 mt	1993 mt	1994 mt	1995 mt	1996 mt	1997 mt	1998 mt	1999 mt	2000 mt	2001 mt
Yugoslavia	13	17	18	18	21	20	20	22	24	25
37 Fishing area total	*27 622 F*	*32 146 F*	*33 491 F*	*28 962 F*	*29 016 F*	*25 414 F*	*27 606 F*	*24 400 F*	*24 509 F*	*22 479 F*
Species total	*29 556 F*	*33 806 F*	*34 914 F*	*30 767 F*	*30 408 F*	*27 054 F*	*29 456 F*	*28 219 F*	*28 467 F*	*25 389 F*
Daggerhead breams nei	**Spares australs nca**		**Sargos australes nep**		***Chrysoblephus spp***				**1,70(39)264,XX**	**RSX**
47 South Africa	227	171	231	197	176	139	154	74	70 F	78
47 Fishing area total	*227*	*171*	*231*	*197*	*176*	*139*	*154*	*74*	*70 F*	*78*
51 South Africa	153	-	109	70	92	85	...	...	...	0
51 Fishing area total	*153*	*-*	*109*	*70*	*92*	*85*	*...*	*...*	*...*	*0*
Species total	*380*	*171*	*340*	*267*	*268*	*224*	*154*	*74*	*70 F*	*78*
White steenbras	**Marbré du Cap**		**Herrera del Cabo**		***Lithognathus lithognathus***				**1,70(39)277,01**	**SNW**
47 South Africa	11	4	1	11	7	8	9	2	...	...
47 Fishing area total	*11*	*4*	*1*	*11*	*7*	*8*	*9*	*2*	*...*	*...*
Species total	*11*	*4*	*1*	*11*	*7*	*8*	*9*	*2*	*...*	*...*
Sand steenbras	**Marbré**		**Herrera**		***Lithognathus mormyrus***				**1,70(39)277,02**	**SSB**
27 France	30	54	47	78	36	41	30	46	59	36
Portugal	...	...	...	...	...	...	...	178	158	109
27 Fishing area total	*30*	*54*	*47*	*78*	*36*	*41*	*30*	*224*	*217*	*145*
34 Senegal	-	-	-	-	-	-	223	45	8	77
34 Fishing area total	*-*	*-*	*-*	*-*	*-*	*-*	*223*	*45*	*8*	*77*
37 France	54	45	24	2	24	80	80	66	59	61
Slovenia	-	-	-	-	-	-	-	-	1	2
Spain	-	-	-	-	-	270	323	255	290	246
Tunisia	...	...	...	917	694	661	781	683	656	781
37 Fishing area total	*54*	*45*	*24*	*919*	*718*	*1 011*	*1 184*	*1 004*	*1 006*	*1 090*
47 Angola	...	642	580	279	214	176	1 072	726	999	1 735
47 Fishing area total	*...*	*642*	*580*	*279*	*214*	*176*	*1 072*	*726*	*999*	*1 735*
Species total	*84*	*741*	*651*	*1 276*	*968*	*1 228*	*2 509*	*1 999*	*2 230*	*3 047*
Steenbrasses nei	**Marbrés nca**		**Herreras nep**		***Lithognathus spp***				**1,70(39)277,XX**	**STW**
47 Namibia	0	0	7	195	6	136	2	116	56	20
47 Fishing area total	*0*	*0*	*7*	*195*	*6*	*136*	*2*	*116*	*56*	*20*
Species total	*0*	*0*	*7*	*195*	*6*	*136*	*2*	*116*	*56*	*20*
Salema	**Saupe**		**Salema**		***Sarpa salpa***				**1,70(39)293,01**	**SLM**
27 France	3	2	4	3	-	5	2	4	8	10
Portugal	...	...	...	...	...	...	...	336	246	320
27 Fishing area total	*3*	*2*	*4*	*3*	*...*	*5*	*2*	*340*	*254*	*330*
37 Croatia	164	166	142	149	127	132	125	42	24	28
Cyprus	1	1	0	4	5	3	4	1	4	5
Egypt	40	0	0	0	0	0	-	-	-	-
France	291	110	36	15	30	77	77	60	58	60
Greece	733	656	1 160	592	687	485	369	404	408	346
Malta	0	0	0	0	0	0	1	0	0	0
Spain	-	-	-	-	-	242	199	167	147	164
Tunisia	1 764	974	972	916	750	1 127	1 151	1 124	951	1 024
Turkey	1 089	1 479	1 765	256	331	340	300	155	200	160
Yugoslavia	4	1	4	4	7	7	9	10	10	10
37 Fishing area total	*4 086*	*3 387*	*4 079*	*1 936*	*1 937*	*2 413*	*2 235*	*1 963*	*1 802*	*1 797*
47 Russian Fed	-	-	-	-	-	-	-	-	-	1
South Africa	-	-	1	1	4	3	0	1	...	...
47 Fishing area total	*-*	*-*	*1*	*1*	*4*	*3*	*0*	*1*	*...*	*1*
Species total	*4 089*	*3 389*	*4 084*	*1 940*	*1 941*	*2 421*	*2 237*	*2 304*	*2 056*	*2 128*
Blackhead seabream	**Pagre tête noire**		**...C**		***Acanthopagrus schlegeli***				**1,70(39)330,04**	**MLM**
61 China,Taiwan	323	178	153	197	207	277	404	604	717	878
61 Fishing area total	*323*	*178*	*153*	*197*	*207*	*277*	*404*	*604*	*717*	*878*
Species total	*323*	*178*	*153*	*197*	*207*	*277*	*404*	*604*	*717*	*878*
Yellowfin seabream	**Pagre à nageoires jaunes**		**Sargo aleta amarilla**		***Acanthopagrus latus***				**1,70(39)330,05**	**YWF**
51 Kuwait	158	167	210	273	234	249	280	464	350 F	271
51 Fishing area total	*158*	*167*	*210*	*273*	*234*	*249*	*280*	*464*	*350 F*	*271*
Species total	*158*	*167*	*210*	*273*	*234*	*249*	*280*	*464*	*350 F*	*271*

B-33 Miscellaneous coastal fishes / Poissons côtiers divers / Peces costeros diversos

Capture production by species, fishing areas and countries or areas / Captures par espèces, zones de pêche et pays ou zones / Capturas por especies, áreas de pesca y países o áreas

Species, Fishing area Espèce, Zone de pêche Especie, Area de pesca	1992 mt	1993 mt	1994 mt	1995 mt	1996 mt	1997 mt	1998 mt	1999 mt	2000 mt	2001 mt
Scup	**Spare doré**		**Sargo de América del Norte**		***Stenotomus chrysops***				**1,70(39)353,01**	**SCP**
21 USA	5 874	4 090	4 027	2 845	2 952	2 196	1 893	1 676	1 206	1 845
21 Fishing area total	*5 874*	*4 090*	*4 027*	*2 845*	*2 952*	*2 196*	*1 893*	*1 676*	*1 206*	*1 845*
Species total	*5 874*	*4 090*	*4 027*	*2 845*	*2 952*	*2 196*	*1 893*	*1 676*	*1 206*	*1 845*
Porgies, seabreams nei	**Dentés, spares nca**		**Dentones, sargos nep**		***Sparidae***				**1,70(39)XXX,XX**	**SBX**
27 Channel Is	2	2 F	14	13 F	9	48	126	132	105	108
France	-	-	-	-	-	-	-	2	-	3
Portugal	2 990	2 406	2 379	2 317	1 955	1 949	1 834	-	21	3
Spain	1 952	1 173	535	1 272	582	1 857	2 006	1 648	1 940	1 950
27 Fishing area total	*4 944*	*3 581 F*	*2 928*	*3 602 F*	*2 546*	*3 854*	*3 966*	*1 782*	*2 066*	*2 064*
31 Antigua Barb	...	...	...	...	...	...	...	...	...	9
Cuba	138	76	95	104	91	97	126	72	70 F	70 F
Korea Rep	-	-	-	-	-	-	309	18	4	-
Puerto Rico	...	...	...	...	...	...	...	22	24	25
USA	483	583	644	431	176	398	336	93	168	219
Venezuela	...	...	...	...	...	...	...	...	6	...
31 Fishing area total	*621*	*659*	*739*	*535*	*267*	*495*	*771*	*205*	*272 F*	*323 F*
34 Belize	-	54	7	88	-	183	1 152	2 815	2 162	30
Bulgaria	-	-	-	-	-	-	35	-	-	-
Cameroon	1	14	0	0	2	2 F	5 F	5 F	4	34
Chile	-	-	3	-	-	-	-	-	-	-
China	417	1 028	76	1 057	904	3 011	2 442	2 783	2 544	3 638
Côte dIvoire	0	0	1	1	1 F	1 F	1 F	1	3	...
Cuba	-	-	-	-	263	-	-	-	-	-
Cyprus	-	-	-	-	...	...	38	80	93	191
France	-	-	-	-	-	...	77	53	-	-
Gambia	520	763	146	446	129		...	-	-	-
Georgia	323 F	170 F	60 F	-	-	-	-	-	-	-
Germany	-	-	-	23	-	-	-	-	-	-
Ghana	239	443	165	157	220	129	316	3 986	1 238	532
Greece	0	0	0	0	-	-	-	-	-	-
Guinea	2 720 F	3 250 F	3 340 F	4 709	3 814	3 649	3 019	3 265	1 838	1 800 F
GuineaBissau	7 F	10 F	10 F	28	14 F	15 F	10 F	10 F	10 F	10 F
Honduras	991	2 170	83	1 768	736	565	233	705	120	438
Italy	10 112	8 573	8 573	-	-	-	-	-	-	-
Korea D P Rp	-	49	-	-	-	-	-	-	273	-
Korea Rep	523	352	581	287	408	933	1 924	637	224	444
Latvia	1 279	-	-	-	-	-	48	80	53	-
Liberia	30	41	18	42	-	-	-	-	-	-
Lithuania	1 444	7	38	...	...	...	192	157	155	32
Mauritania	270 F	360 F	490 F	350 F	490 F	520 F	870 F	900 F	900 F	900 F
Morocco	8 137	8 447	8 195	8 396	9 588	10 326	7 415	9 320	10 169	9 568
Nigeria	205	297	148	395	-	-	-	-	-	-
Norway	-	-	61	-	-	-	-	-	-	-
Panama	1 072	498	-	396	402	46	319	66	-	-
Poland	-	-	-	-	7	5	-	-	-	-
Portugal	36	40	27	298	522	95	25	6	0	0
Russian Fed	794	201	187	362	408	615	1 265	1 492	954	860
St Vincent	202	267	-	16	30	10	59	39	5	30
Sao Tome Prn	26 F	33 F	48 F	50 F	37	49	60 F	70 F	60 F	60 F
Senegal	-	-	-	-	-	-	-	-	-	295
Sierra Leone	1 772	2 836	37	885	2 565	1 967	522	500	...	1 189
Spain	122	400 F	600 F	800 F	754	100	226	145	117	163
Ukraine	1 354	100	197	155	153	1 245	757	1 259	1 966	363
Vanuatu	2	95	0	0	0	-	0	-	-	-
Other nei	57	-	-	-	-	-	-	63	49	341
34 Fishing area total	*32 655 F*	*30 498 F*	*23 091 F*	*20 709 F*	*21 447 F*	*23 466 F*	*21 010 F*	*28 437 F*	*22 937 F*	*20 918 F*
37 Albania	82	70 F	50 F	3	9	1	-	-	-	1
Algeria	5 000 F	5 500 F	6 000 F	-	-	-	-	7	10 F	10 F
Cyprus	64	76	130	61	65	83	64	74	63	46
Israel	205	193	100	100	100	...	...	...	...	...
Lebanon	200 F	250	300	450	450	350	400	450	450	400
Libya	3 450 F	3 700 F	4 000 F	4 100 F	4 000 F	4 000 F	4 000 F	4 000 F	4 000 F	4 000 F
Morocco	835	885	1 131	1 336	1 400	54	36	830	851	1 156
Spain	900 F	1 100 F	1 300 F	1 500 F	1 808	813	853	444	230	232
Tunisia	4 313	3 262	2 865	2 693	2 381	2 714	1 841	2 973	3 084	2 825
Turkey	514	495	792	334	412	220	180	240	110	135
37 Fishing area total	*15 563 F*	*15 531 F*	*16 668 F*	*10 577 F*	*10 625 F*	*8 235 F*	*7 374 F*	*9 018 F*	*8 798 F*	*8 805 F*
41 Italy	1 348	1 143	1 143	-	-	-	-	-	-	-
Korea Rep	...	...	148	-	-	-	-	2	-	-
Spain	-	-	-	-	-	-	-	-	1 054	1 221
41 Fishing area total	*1 348*	*1 143*	*1 291*	-	-	-	-	*2*	*1 054*	*1 221*
47 Angola	1 344	-	-	-	-	-	-	-	-	-
China	202	-	-	-	-	-	-	-	124	-
Honduras	-	27	-	5	-	-	-	-	-	-
Japan	10	-	-	-	121	-	-	-	-	-
Korea Rep	1 658	1 227	2 409	665	-	-	-	250	198	-
Namibia	343	717	973	2 178	2 279	2 608	1 364	4 434	8 890	6 100
Panama	2 940	-	-	-	-	-	-	-	-	-
Portugal	-	-	-	-	-	-	-	-	-	0
Russian Fed	16	1	-	-	-	-	76	-	407	8
South Africa	485	606	836	274	305	311	365	274	250 F	147

B-33

Miscellaneous coastal fishes — **Capture production by species, fishing areas and countries or areas**
Poissons côtiers divers — **Captures par espèces, zones de pêche et pays ou zones**
Peces costeros diversos — **Capturas por especies, áreas de pesca y países o áreas**

Species, Fishing area Espèce, Zone de pêche Especie, Area de pesca	1992 mt	1993 mt	1994 mt	1995 mt	1996 mt	1997 mt	1998 mt	1999 mt	2000 mt	2001 mt
47 Fishing area total	*6 998*	*2 578*	*4 218*	*3 122*	*2 705*	*2 919*	*1 805*	*4 958*	*9 869 F*	*6 255*
51 Bahrain	487	294	283	439	496	493	757	582	591	401
Djibouti	31 F	34 F	36 F	40 F	40 F	40 F	40 F	40 F	40 F	40 F
Egypt	...	...	...	4 012	2 428	2 175	1 282	182	4 731	3 545
Eritrea	...	...	12	22	12	0	0	15	48	11
Italy	2 427	2 059	2 059	-	-	-	-	-	-	-
Korea Rep	4 832	3 785	2 492	2 633	2 453	1 926	1 479	3 503	1 988	1 218
Oman	4 520	4 503	5 770	5 090	4 692	4 322	4 016	6 098	4 419	3 976
Pakistan	3 284	3 939	3 866	3 358	3 097	3 058	1 255	4 220	4 510	4 411
Qatar	224	213	183	124	126	221	188	182	248	288
Saudi Arabia	845	1 064	1 910	1 469	1 722	2 481	2 822	2 771	2 488	2 646
South Africa	3	-	26	6	19	15	1	5 F	...	...
51 Fishing area total	*16 653 F*	*15 891 F*	*16 637 F*	*17 193 F*	*15 085 F*	*14 731 F*	*11 840 F*	*17 598 F*	*19 063 F*	*16 536 F*
57 Australia	260	320	643	552	278	113	159	325	318	289
Korea Rep	-	-	-	3	-	3	-	666	1 674	677
57 Fishing area total	*260*	*320*	*643*	*555*	*278*	*116*	*159*	*991*	*1 992*	*966*
61 China	57 903	57 565	56 785	58 576	56 174	71 043	74 905	78 149	105 785	125 692
China,H.Kong	698	486	932	1 215	925	883	852	600 F	700 F	780 F
China,Taiwan	5 821	6 306	6 515	5 626	5 401	6 334	6 440	7 312	6 288	9 953
Japan	10 692	11 433	10 994	11 496	11 484	11 256	11 284	10 675	9 066	9 545
Korea Rep	8 546	6 341	6 062	2 290	2 161	2 637	2 424	1 748	2 056	1 931
61 Fishing area total	*83 660*	*82 131*	*81 288*	*79 203*	*76 145*	*92 153*	*95 905*	*98 484 F*	*123 895 F*	*147 901 F*
71 Australia	199	136	127	133	160	169	157	156	185	198
Korea Rep	1 969	2 519	3 990	3 196	2 048	434	94	62	42	70
71 Fishing area total	*2 168*	*2 655*	*4 117*	*3 329*	*2 208*	*603*	*251*	*218*	*227*	*268*
77 Korea Rep	2	-	-	-	-	-	-	-	2	-
77 Fishing area total	*2*	-	-	-	-	-	-	-	*2*	-
81 Australia	599	575	727	722	502	498	439	378	318	288
Japan	1	2	1	6	-	-	-	-	-	-
Korea Rep	271	754	1 197	961	939	1 715	922	1 422	2 309	2 427
Russian Fed	144	-	9	2	494	8	-	-	-	-
Ukraine	-	-	-	-	-	-	-	5	-	-
81 Fishing area total	*1 015*	*1 331*	*1 934*	*1 691*	*1 935*	*2 221*	*1 361*	*1 805*	*2 627*	*2 715*
87 Korea Rep	-	-	-	-	-	-	-	8	8	-
87 Fishing area total	-	-	-	-	-	-	-	*8*	*8*	-
Species total	*165 887 F*	*156 318 F*	*153 554 F*	*140 516 F*	*133 241 F*	*148 793 F*	*144 442 F*	*163 506 F*	*192 810 F*	*207 972 F*
Blotched picarel	**Mendole**		**Chucla**		***Spicara maena***				**1,70(40)075,01**	**BPI**
37 Tunisia	508	135	203	218	285	505	395	419	255	381
37 Fishing area total	*508*	*135*	*203*	*218*	*285*	*505*	*395*	*419*	*255*	*381*
Species total	*508*	*135*	*203*	*218*	*285*	*505*	*395*	*419*	*255*	*381*
Picarels nei	**Mendoles, picarels nca**		**Chuclas, carameles nep**		***Spicara spp***				**1,70(40)075,XX**	**PIC**
27 Portugal	...	...	...	...	...	...	...	43	22	25
27 Fishing area total	...	...	...	...	...	...	...	*43*	*22*	*25*
34 Greece	19	6	-	-	-	17	8	-	-	-
34 Fishing area total	*19*	*6*	-	-	-	*17*	*8*	-	-	-
37 Albania	127	100 F	50 F	...	11	0	7	7	10	5
Croatia	664	510	315	312	290	340	255	245	231	189
Cyprus	760	715	513	711	764	650	709	546	533	671
France	213	99	31	...	8	9	9	10	9	5
Greece	6 678	7 432	14 118	6 354	8 307	8 354	4 541	4 663	4 029	4 031
Italy	1 182	971	810	1 138	953	647	545	547	385	313
Lebanon	20 F	50	50	50	100	100	100	100	100	95
Malta	4	3	3	3	7	7	7	8	9	6
Slovenia	-	-	-	-	-	-	3	3	4	5
Spain	-	-	-	-	1	-	-	494	712	759
Tunisia	401	406	345	382	335	485	3 258	1 005	646	572
Turkey	6 612	5 306	5 378	1 210	1 525	1 650	3 700	1 680	1 500	2 250
Ukraine	-	-	-	-	-	-	-	-	3	-
Yugoslavia	7	7	8	8	11	12	13	16	21	27
37 Fishing area total	*16 668 F*	*15 599 F*	*21 621 F*	*10 168*	*12 312*	*12 254*	*13 147*	*9 324*	*8 192*	*8 928*
47 Angola	52	407	113	9	7	0	55	...	13	-
47 Fishing area total	*52*	*407*	*113*	*9*	*7*	*0*	*55*	...	*13*	-
Species total	*16 739 F*	*16 012 F*	*21 734 F*	*10 177*	*12 319*	*12 271*	*13 210*	*9 367*	*8 227*	*8 953*
Red mullet	**Rouget de roche**		**Salmonete de roca**		***Mullus surmuletus***				**1,70(41)007,01**	**MUF**
27 Channel Is	2	2 F	5	5 F	4	11	17	22	23	19
Denmark	0	0	0	0	1	1	1	3	2	5
France	1 567	1 414	1 171	2 886	2 936	1 861	3 666	2 592	4 054	2 969
Ireland	-	-	-	-	-	-	38	-	-	-
Isle of Man	-	-	-	-	-	-	-	-	-	4
Netherlands	-	-	-	-	1	0	-	-	235	560

B-33 Miscellaneous coastal fishes — Capture production by species, fishing areas and countries or areas
Poissons côtiers divers — Captures par espèces, zones de pêche et pays ou zones
Peces costeros diversos — Capturas por especies, áreas de pesca y países o áreas

pecies, Fishing area spèce, Zone de pêche specie, Area de pesca	1992 mt	1993 mt	1994 mt	1995 mt	1996 mt	1997 mt	1998 mt	1999 mt	2000 mt	2001 mt
Portugal	67	69	109	64	52	69	66	180	155	191
Spain	-	-	-	-	292	300	132	136	278	352
UK	110	218	172	228	192	173	180	147	220	310
27 Fishing area total	*1 746*	*1 703 F*	*1 457*	*3 183 F*	*3 478*	*2 415*	*4 100*	*3 080*	*4 967*	*4 410*
37 Croatia	359	479	395	420	311	280	285	200	277	480
Cyprus	213	184	189	215	240	228	176	184	159	132
Greece	2 452	2 101	3 320	2 535	2 004	1 944	1 095	1 274	1 368	1 238
Libya	3 450 F	3 700 F	4 000 F	4 100 F	4 000 F	4 000 F	4 000 F	4 000 F	4 000 F	4 000 F
Slovenia	-	-	-	-	-	-	1	1	1	4
Tunisia	2 008	1 448	1 506	1 140	1 161	2 406	1 459	1 504	1 527	2 163
Turkey	3 769	4 728	3 704	3 602	3 962	2 950	2 050	2 100	2 300	1 570
Yugoslavia	4	3	5	10	11	10	11	9	10	11
37 Fishing area total	*12 255 F*	*12 643 F*	*13 119 F*	*12 022 F*	*11 689 F*	*11 818 F*	*9 077 F*	*9 272 F*	*9 642 F*	*9 598 F*
pecies total	*14 001 F*	*14 346 F*	*14 576 F*	*15 205 F*	*15 167 F*	*14 233 F*	*13 177 F*	*12 352 F*	*14 609 F*	*14 008 F*
triped mullet	**Rouget de vase**		**Salmonete de fango**		***Mullus barbatus***				**1,70(41)007,02**	**MUT**
37 Bulgaria	-	-	-	-	-	-	-	-	5	26
Cyprus	125	113	121	88	108	119	115	145	103	91
Tunisia	1 058	1 168	1 032	1 108	1 287	1 394	1 228	1 250	1 600	2 632
Turkey	5 911	4 774	4 447	3 906	3 936	3 000	3 500	3 865	2 450	2 455
Ukraine	5	12	10	13	...	18	35	30	12	201
37 Fishing area total	*7 099*	*6 067*	*5 610*	*5 115*	*5 331*	*4 531*	*4 878*	*5 290*	*4 170*	*5 405*
pecies total	*7 099*	*6 067*	*5 610*	*5 115*	*5 331*	*4 531*	*4 878*	*5 290*	*4 170*	*5 405*
urmullets(=Red mullets) nei	**Rougets nca**		**Salmonetes nep**		***Mullus spp***				**1,70(41)007,XX**	**MUX**
27 Germany	-	-	-	-	-	-	-	-	13	10
27 Fishing area total	-	-	-	-	-	-	-	-	*13*	*10*
34 Gabon	...	...	177	224	100	573	76	57	448	311
Italy	2 023	1 959	2 004	-	-	-	-	-	-	-
Liberia	0	2	3	-	-	46	-	-	-	-
Mauritania	30 F	30 F	50 F	30 F	30 F	30 F	141 F	150 F	150 F	150 F
Morocco	372	346	319	431	266	528	385	426	884	1 062
Nigeria	2 198	...	...	...	4 575	...	...	...	-	-
Portugal	-	13	4	0	3	0	1	0	0	-
Senegal	845	998	1 714	352	731	896	1 087	1 223	1 110	1 908
Spain	-	-	-	-	276	47	56	26	15	28
34 Fishing area total	*5 468 F*	*3 348 F*	*4 271 F*	*1 037 F*	*5 981 F*	*2 120 F*	*1 746 F*	*1 882 F*	*2 607 F*	*3 459 F*
37 Albania	537	400 F	200 F	76	64	61	143	145	140	170
Algeria	1 300 F	1 400 F	1 800 F	1 576	1 950	1 320	1 480	1 420	1 400 F	1 400 F
Egypt	999	1 516	2 098	1 512	2 314	2 533	1 681	2 727	1 611	1 717
France	454	267	186	189	149	149	149	266	270	206
Gaza Strip	...	...	...	...	56	57	85	85 F	85 F	70 F
Georgia	-	-	-	-	-	14	11	8	3	22
Greece	2 631	2 808	3 016	2 550	2 495	2 426	1 744	1 735	1 810	1 541
Israel	363	300 F	250 F	162	227	185	276	350	340	300 F
Italy	10 220	10 448	9 921	9 441	11 325	7 499	7 491	8 771	9 044	7 121
Lebanon	100 F	150	200	200	200	150	200	200	250	200
Malta	1	2	3	4	7	7	8	12	7	5
Morocco	354	374	254	343	416	374	421	396	605	362
Romania	-	-	5	6	1	3	3	1	2	3
Russian Fed	37	2	25	324	76	68	119	92	127	119
Spain	1 700 F	1 400 F	1 100 F	800 F	478	480	491	2 643	2 477	2 485
Syria	117 F	121 F	117 F	116 F	232	250	116	130	122	125
37 Fishing area total	*18 813 F*	*19 188 F*	*19 175 F*	*17 299 F*	*19 990*	*15 576*	*14 418*	*18 981 F*	*18 293 F*	*15 846 F*
41 Argentina	133	20	26	63	75	73	99	229	498	...
Brazil	260 F	260 F	260 F	964	530	749	840	1 117	1 756	1 730 F
Italy	270	261	267	-	-	-	-	-	-	-
41 Fishing area total	*663 F*	*541 F*	*553 F*	*1 027*	*605*	*822*	*939*	*1 346*	*2 254*	*1 730 F*
pecies total	*24 944 F*	*23 077 F*	*23 999 F*	*19 363 F*	*26 576 F*	*18 518 F*	*17 103 F*	*22 209 F*	*23 167 F*	*21 045 F*
oatfishes	**Rougets-souris**		**Salmonetes**		***Upeneus spp***				**1,70(41)251,XX**	**GOX**
51 Bahrain	43	29	85	36	139	65	16	414	82	192
Eritrea	...	...	61	7	2	1	1	21	19	9
India	26 616	9 620	12 213	11 316	10 300	6 148	7 061	7 186	8 457	14 748
Mauritius	605	403	416	457	512	479	436	509	541	556
Saudi Arabia	10	10	52	65	83	73	107	123	72	122
51 Fishing area total	*27 274*	*10 062*	*12 827*	*11 881*	*11 036*	*6 766*	*7 621*	*8 253*	*9 171*	*15 627*
57 India	8 654	8 151	7 768	7 553	7 829	9 722	10 199	10 706	2 079	1 988
Indonesia	4 062	8 548	8 526	8 586	10 758	13 357	15 752	15 757	16 228	17 510
Malaysia	2 523	1 958	2 085	1 407	966	1 227	343	2 059	5 021	6 488
57 Fishing area total	*15 239*	*18 657*	*18 379*	*17 546*	*19 553*	*24 306*	*26 294*	*28 522*	*23 328*	*25 986*
61 China,H.Kong	1 568	1 322	706	593	718	477	393	300 F	350 F	400 F
China,Taiwan	505	469	602	399	508	440	335	478	477	300
61 Fishing area total	*2 073*	*1 791*	*1 308*	*992*	*1 226*	*917*	*728*	*778 F*	*827 F*	*700 F*
71 Fiji Islands	172	117	119 F	9	108	190	155	160	140 F	157
Indonesia	7 270	8 582	8 244	9 026	9 966	10 846	9 455	10 495	11 720	12 640

B-33 Miscellaneous coastal fishes / Poissons côtiers divers / Peces costeros diversos

Capture production by species, fishing areas and countries or areas / Captures par espèces, zones de pêche et pays ou zones / Capturas por especies, áreas de pesca y países o áreas

Species, Fishing area Espèce, Zone de pêche Especie, Area de pesca	1992 mt	1993 mt	1994 mt	1995 mt	1996 mt	1997 mt	1998 mt	1999 mt	2000 mt	2001 mt
Kiribati	440	430	430	440	450	450	149	582	639	1 609
Malaysia	2 364	3 429	5 104	8 194	7 141	9 075	7 307	6 547	5 824	6 472
Singapore	265	216	35	29	24	29	29	25	8	15
71 Fishing area total	*10 511*	*12 774*	*13 932 F*	*17 698*	*17 689*	*20 590*	*17 095*	*17 809*	*18 331 F*	*20 893*
Species total	*55 097*	*43 284*	*46 446 F*	*48 117*	*49 504*	*52 579*	*51 738*	*55 362 F*	*51 657 F*	*63 206*
West African goatfish	**Rouget du Sénégal**		**Salmonete barbudo**			***Pseudupeneus prayensis***			**1,70(41)499,03**	**GC**
34 Côte dIvoire	84	47	53	53 F	68	44	38	61	44	40
Ghana	247	163	190	65	586	1 035	553	247	39	285
Greece	771	1 306	1 024	907	1 065	692	805	838	715	1 081
Senegal	-	-	-	-	-	-	-	-	-	1 846
Sierra Leone	...	...	...	...	...	37	...	...	35	119
Togo	1	3	5	0	0	0	0	-	-	-
34 Fishing area total	*1 103*	*1 519*	*1 272*	*1 025 F*	*1 719*	*1 808*	*1 396*	*1 146*	*833*	*3 371*
Species total	*1 103*	*1 519*	*1 272*	*1 025 F*	*1 719*	*1 808*	*1 396*	*1 146*	*833*	*3 371*
Goatfishes, red mullets nei	**Rougets, etc. nca**		**Salmonetes, etc. nep**			***Mullidae***			**1,70(41)XXX,XX**	**MU**
31 Dominican Rp	190	325	585	376	265	178	168	90	171	161
Grenada	0	0	1	1	1	1	0	0	0	0
Puerto Rico	...	...	...	...	...	...	...	17	17	15
St Kitts Nev	...	...	...	3	10	1	2	1	1	1
USA	-	-	-	-	-	-	-	-	-	36
31 Fishing area total	*190*	*325*	*586*	*380*	*276*	*180*	*170*	*108*	*189*	*213*
34 Korea Rep	52	12	-	18		-	-	-	-	-
34 Fishing area total	*52*	*12*	-	*18*	-	-	-	-	-	-
47 Angola	-	11	26	0	-	0	-	-	-	-
Korea Rep	-	42	-	-	-	-	-	-	-	-
47 Fishing area total	-	*53*	*26*	*0*	-	*0*	-	-	-	-
51 Egypt	668	689	777	1 077	716	744	439	876	914	2 590
Italy	405	392	401	-	-	-	-	-	-	-
Qatar	16	14	10	15	16	14	14	20	1	0
Untd Arab Em	134	140	152	130	154	172	185	192	178	180
51 Fishing area total	*1 223*	*1 235*	*1 340*	*1 222*	*886*	*930*	*638*	*1 088*	*1 093*	*2 770*
61 Korea Rep	-	58	161	-	-	-	-	-	-	-
61 Fishing area total	-	*58*	*161*	-	-	-	-	-	-	-
71 N Marianas	...	...	0	1	12	9	1	1	1	1
Philippines	22 278	25 919	22 091	25 510	23 662	15 884	14 182	14 527	14 718	14 875
71 Fishing area total	*22 278*	*25 919*	*22 091*	*25 511*	*23 674*	*15 893*	*14 183*	*14 528*	*14 719*	*14 876*
Species total	*23 743*	*27 602*	*24 204*	*27 131*	*24 836*	*17 003*	*14 991*	*15 724*	*16 001*	*17 859*
Mojarras(=Silver-biddies) nei	**Blanches nca**		**Mojarras nep**			***Gerres spp***			**1,70(46)036,XX**	**MO**
31 USA	-	-	-	-	-	-	-	-	159	180
31 Fishing area total	-	-	-	-	-	-	-	-	*159*	*180*
51 Bahrain	210	64	233	160	219	350	347	261	334	372
Qatar	152	115	53	53	50	56	92	77	78	82
Saudi Arabia	4	5	164	337	360	438	445	520	529	510
Untd Arab Em	1 051	1 150	1 383	1 222	1 235	1 320	1 324	1 351	1 651	1 650
51 Fishing area total	*1 417*	*1 334*	*1 833*	*1 772*	*1 864*	*2 164*	*2 208*	*2 209*	*2 592*	*2 614*
71 Fiji Islands	24	12	12 F	9	6	10	13	11	10 F	11
Kiribati	1 770	1 710	1 690	1 700	1 720	1 730	566	890	674	1 455
Philippines	4 602	7 266	7 911	6 491	5 909	4 285	4 854	5 008	5 117	5 062
71 Fishing area total	*6 396*	*8 988*	*9 613 F*	*8 200*	*7 635*	*6 025*	*5 433*	*5 909*	*5 801 F*	*6 528*
Species total	*7 813*	*10 322*	*11 446 F*	*9 972*	*9 499*	*8 189*	*7 641*	*8 118*	*8 552 F*	*9 322*
Mojarras, etc. nei	**Blanches, etc. nca**		**Mojarras, etc. nep**			***Gerreidae***			**1,70(46)XXX,XX**	**GD**
31 Colombia	4	71	90	38	125	6	23	0	5 F	5
Cuba	2 242	1 637	2 144	2 221	1 719	1 346	1 199	1 013	1 000 F	1 000
Dominican Rp	20	23	41	60	45	12	20	9	14	15
Mexico	7 796	7 554	7 549	7 379	7 867	4 547	4 245	3 069	2 701	2 800
Puerto Rico	...	...	...	...	...	...	...	14	15	13
Venezuela	0	0	0	0	0	0	0	0	0	0
31 Fishing area total	*10 062*	*9 285*	*9 824*	*9 698*	*9 756*	*5 911*	*5 487*	*4 105*	*3 735 F*	*3 833*
77 Mexico	4 947	4 792	4 786	7 085	6 841	3 946	1 945	2 114	1 724	1 800
77 Fishing area total	*4 947*	*4 792*	*4 786*	*7 085*	*6 841*	*3 946*	*1 945*	*2 114*	*1 724*	*1 800*
Species total	*15 009*	*14 077*	*14 610*	*16 783*	*16 597*	*9 857*	*7 432*	*6 219*	*5 459 F*	*5 633*
Parore	**...B**		**...C**			***Girella tricuspidata***			**1,70(47)003,02**	**GI**
81 Australia	...	...	...	...	501	507	496	503	537	487
New Zealand	-	-	-	83	60	81	80	76	94	71
81 Fishing area total	...	...	...	*83*	*561*	*588*	*576*	*579*	*631*	*558*

B-33 Miscellaneous coastal fishes / Poissons côtiers divers / Peces costeros diversos

Capture production by species, fishing areas and countries or areas / Captures par espèces, zones de pêche et pays ou zones / Capturas por especies, áreas de pesca y países o áreas

pecies, Fishing area spèce, Zone de pêche specie, Area de pesca	1992 mt	1993 mt	1994 mt	1995 mt	1996 mt	1997 mt	1998 mt	1999 mt	2000 mt	2001 mt
pecies total	...	...	...	*83*	*561*	*588*	*576*	*579*	*631*	*558*
paleye	**...B**		**...C**		***Girella nigricans***				**1,70(47)003,03**	**GIQ**
77 USA	-	-	-	-	-	-	-	-	3	2
77 Fishing area total	-	-	-	-	-	-	-	-	*3*	*2*
pecies total	-	-	-	-	-	-	-	-	*3*	*2*
lue sea chub	**Calicagère bleue**		**Chopa azul**		***Kyphosus cinerascens***				**1,70(47)035,01**	**KYC**
51 Saudi Arabia	-	-	-	-	-	8	-	-	-	-
51 Fishing area total	-	-	-	-	-	*8*	-	-	-	-
pecies total	-	-	-	-	-	*8*	-	-	-	-
ea chubs nei	**Calicagères nca**		**Chopas nep**		***Kyphosus spp***				**1,70(47)035,XX**	**KYP**
31 Antigua Barb	...	...	...	...	...	...	...	...	...	8
31 Fishing area total	...	...	...	...	...	...	...	...	...	*8*
pecies total	...	...	...	...	...	...	...	...	...	*8*
potted sicklefish	**Forgeron tacheté**		**Catemo manchado**		***Drepane punctata***				**1,70(50)132,01**	**SPS**
57 Malaysia	87	67	33	17	51	23	30	58	53	26
57 Fishing area total	*87*	*67*	*33*	*17*	*51*	*23*	*30*	*58*	*53*	*26*
71 Malaysia	388	525	516	619	824	560	486	785	625	663
Philippines	191	121	369	63	64	174	96	133	114	117
71 Fishing area total	*579*	*646*	*885*	*682*	*888*	*734*	*582*	*918*	*739*	*780*
Species total	*666*	*713*	*918*	*699*	*939*	*757*	*612*	*976*	*792*	*806*
frican sicklefish	**Forgeron ailé**		**Catemo africano**		***Drepane africana***				**1,70(50)132,02**	**SIC**
34 Benin	14 F	16 F	15 F	15 F	15 F	18 F	14 F	12	8 F	6
Cameroon	96	39	50 F	111	72	75 F	78	94	73	70
Congo Rep	3 F	3 F	7	15	15 F	15 F	15 F	20 F	24	20 F
Côte dIvoire	30	19	26	35	23	18	28	21	35	20
Gabon	210 F	269	189	330	270	222	281	290	86	71
Gambia	43	44	42	118	151	104	23	30	60	85
Ghana	18	20	7	34	6	46	4	24	2	8
Guinea	...	...	...	89	324	149	...	...	...	...
GuineaBissau	...	...	24	71	172	180 F	150 F	120 F	120 F	120 F
Liberia	2	8	12	30	15	-	70	256	94	104
Nigeria	303	251	440	597	954	2 047	603	...	1 316	1 429
Sao Tome Prn	...	...	...	...	...	15	20 F	25 F	20 F	20 F
Senegal	607	530	605	248	741	606	325	439	356	421
Sierra Leone	...	...	...	12	10 F	293	657	214	662	416
Togo	5	7	4	1	6	2	0	0	-	-
34 Fishing area total	*1 331 F*	*1 206 F*	*1 421 F*	*1 706 F*	*2 774 F*	*3 790 F*	*2 268 F*	*1 545 F*	*2 856 F*	*2 790 F*
Species total	*1 331 F*	*1 206 F*	*1 421 F*	*1 706 F*	*2 774 F*	*3 790 F*	*2 268 F*	*1 545 F*	*2 856 F*	*2 790 F*
Ballan wrasse	**Vieille commune**		**Maragota**		***Labrus bergylta***				**1,70(63)005,01**	**USB**
27 Channel Is	-	-	-	-	-	-	-	2	1	-
27 Fishing area total	-	-	-	-	-	-	-	*2*	*1*	-
Species total	-	-	-	-	-	-	-	*2*	*1*	-
Tautog	**Tautogue noir**		**...C**		***Tautoga onitis***				**1,70(63)243,01**	**TAU**
21 USA	458	325	206	148	119	116	116	95	111	138
21 Fishing area total	*458*	*325*	*206*	*148*	*119*	*116*	*116*	*95*	*111*	*138*
Species total	*458*	*325*	*206*	*148*	*119*	*116*	*116*	*95*	*111*	*138*
Cunner	**Tanche-tautogue**		**...C**		***Tautogolabrus adspersus***				**1,70(63)244,02**	**CUN**
21 USA	2	0	1	0	1	1	3	4	3	9
21 Fishing area total	*2*	*0*	*1*	*0*	*1*	*1*	*3*	*4*	*3*	*9*
Species total	*2*	*0*	*1*	*0*	*1*	*1*	*3*	*4*	*3*	*9*
California sheephead	**Labre californien**		**Vieja de California**		***Semicossyphus pulcher***				**1,70(63)331,02**	**YFH**
77 USA	...	...	...	...	...	...	118	58	78	68
77 Fishing area total	...	...	...	...	...	...	*118*	*58*	*78*	*68*
Species total	...	...	...	...	...	...	*118*	*58*	*78*	*68*
Wrasses, hogfishes, etc. nei	**Pourceaux, donzelles, etc. nca**		**Lábridos(=Tordos,maragotas)nep**		***Labridae***				**1,70(63)XXX,XX**	**WRA**
27 France	...	...	...	...	...	...	...	222	250	250
Norway	-	-	-	-	-	-	-	-	-	6
Portugal	...	...	...	...	...	...	...	44	36	39
UK	27	21	22	30	17	16	20	20	22	25

B-33 Miscellaneous coastal fishes / Poissons côtiers divers / Peces costeros diversos

Capture production by species, fishing areas and countries or areas / Captures par espèces, zones de pêche et pays ou zones / Capturas por especies, áreas de pesca y países o áreas

Species, Fishing area Espèce, Zone de pêche Especie, Area de pesca	1992 mt	1993 mt	1994 mt	1995 mt	1996 mt	1997 mt	1998 mt	1999 mt	2000 mt	2001 mt
27 Fishing area total	*27*	*21*	*22*	*30*	*17*	*16*	*20*	*286*	*308*	*320*
31 Dominican Rp	745	741	1 334	821	529	1 456	52	52	69	22
Puerto Rico	...	...	...	...	...	...	...	30	47	46
31 Fishing area total	*745*	*741*	*1 334*	*821*	*529*	*1 456*	*52*	*82*	*116*	*68*
34 Portugal	-	-	-	-	-	-	-	-	1	1
34 Fishing area total	-	-	-	-	-	-	-	-	*1*	*1*
51 Saudi Arabia	...	...	...	165	108	105	66	155	91	79
Tanzania	4 000	2 040	2 583	2 810	3 725	3 200	3 000	3 000	3 200	3 500
51 Fishing area total	*4 000*	*2 040*	*2 583*	*2 975*	*3 833*	*3 305*	*3 066*	*3 155*	*3 291*	*3 579*
57 Australia	46	75	105	121	77	77	77	98	137	86
Malaysia	0	3	10	8	7	3	3	2	0	-
57 Fishing area total	*46*	*78*	*115*	*129*	*84*	*80*	*80*	*100*	*137*	*86*
71 Australia	0	0	0	0	0	0	0	-	0	1
Fiji Islands	562	383	390 F	218	204	375	180	399	360 F	400
Guam	-	-	-	-	-	-	1	2	0	1
Malaysia	311	78	59	192	219	2 299	2 694	2 400	2 139	1 948
Philippines	9 638	16 780	21 017	19 099	17 097	15 107	13 391	12 318	12 384	13 045
71 Fishing area total	*10 511*	*17 241*	*21 466 F*	*19 509*	*17 520*	*17 781*	*16 266*	*15 119*	*14 883 F*	*15 395*
81 Australia	-	-	-	-	-	-	-	1	4	4
New Zealand	-	-	-	0	0	24	2	3	4	2
81 Fishing area total	-	-	-	*0*	*0*	*24*	*2*	*4*	*8*	*6*
Species total	*15 329*	*20 121*	*25 520 F*	*23 464*	*21 983*	*22 662*	*19 486*	*18 746*	*18 744 F*	*19 455*
Parrotfish	**Perroquet vieillard**		**Loro viejo**			***Sparisoma cretense***			**1,70(65)055,06**	**PR**
27 Portugal	-	-	-	-	-	-	-	56	87	161
27 Fishing area total	-	-	-	-	-	-	-	*56*	*87*	*161*
34 Portugal	-	-	-	-	-	-	-	-	2	1
34 Fishing area total	-	-	-	-	-	-	-	-	*2*	*1*
Species total	-	-	-	-	-	-	-	*56*	*89*	*162*
Parrotfishes nei	**Perroquets nca**		**Loros nep**			***Scaridae***			**1,70(65)XXX,XX**	**PW**
31 Antigua Barb	...	...	...	...	...	...	...	...	...	173
Dominican Rp	239	195	351	149	80	94	109	7	18	19
Grenada	1	0	0	0	0	1	1	42	18	41
Puerto Rico	...	...	...	...	...	...	...	52	63	66
St Kitts Nev	...	...	...	7	19	5	8	8	2	8
31 Fishing area total	*240*	*195*	*351*	*156*	*99*	*100*	*118*	*109*	*101*	*307*
34 Gambia	...	...	...	...	...	...	...	...	13	1
34 Fishing area total	...	...	...	...	...	...	...	...	*13*	*1*
37 Cyprus	32	44	45	29	32	37	78	58	33	42
37 Fishing area total	*32*	*44*	*45*	*29*	*32*	*37*	*78*	*58*	*33*	*42*
51 Bahrain	34	34	26	63	56	46	63	66	32	21
Egypt	...	...	...	...	...	227	283	918	...	...
Eritrea	...	...	2	15	9	0	0	3	2	2
Saudi Arabia	183	242	176	837	777	906	702	655	756	758
51 Fishing area total	*217*	*276*	*204*	*915*	*842*	*1 179*	*1 048*	*1 642*	*790*	*781*
61 China,Taiwan	23	13	447	8	11	17	12	5	29	27
61 Fishing area total	*23*	*13*	*447*	*8*	*11*	*17*	*12*	*5*	*29*	*27*
71 Guam	...	...	1	0	1	1	4	1	0	0
N Marianas	...	...	5	1	3	7	1	2	4	13
Palau	62	79	55	86	55	37	25	70 F	120 F	120 F
71 Fishing area total	*62*	*79*	*61*	*87*	*59*	*45*	*30*	*73 F*	*124 F*	*133 F*
77 Amer Samoa	...	...	1	0	8	3	10	7	6	3
77 Fishing area total	...	...	*1*	*0*	*8*	*3*	*10*	*7*	*6*	*3*
Species total	*574*	*607*	*1 109*	*1 195*	*1 051*	*1 381*	*1 296*	*1 894 F*	*1 096 F*	*1 294 F*
Angelfishes nei	**Demoiselles nca**		**Angeles nep**			***Pomacanthidae***			**1,70(66)XXX,XX**	**AN**
51 Eritrea	...	...	-	-	0	0	0	3	0	0
Saudi Arabia	-	-	-	-	-	-	7	14	1	3
51 Fishing area total	...	...	-	-	*0*	*0*	*7*	*17*	*1*	*3*
Species total	...	...	-	-	*0*	*0*	*7*	*17*	*1*	*3*
Giant African threadfin	**Gros capitaine**		**Barbudo gigante africano**			***Polydactylus quadrifilis***			**1,70(77)001,09**	**TG**
34 Cameroon	69	39	50 F	76	54	40 F	12	34	43	170
Gambia	67	154	72	94	179	110	83	90	169	141
GuineaBissau	...	...	76	22	55	60 F	50 F	40 F	40 F	40 F
Nigeria	7 984	1 569	2 821	2 755	6 415	9 815	9 803	10 544	11 287	11 645
Sierra Leone	...	638	636	635	640 F	679	75	61	690	445

B-33 Miscellaneous coastal fishes / Poissons côtiers divers / Peces costeros diversos

Capture production by species, fishing areas and countries or areas / Captures par espèces, zones de pêche et pays ou zones / Capturas por especies, áreas de pesca y países o áreas

Species, Fishing area Espèce, Zone de pêche Especie, Area de pesca	1992 mt	1993 mt	1994 mt	1995 mt	1996 mt	1997 mt	1998 mt	1999 mt	2000 mt	2001 mt
Togo	0	1	15	16	16	1	0	1	8	6
34 Fishing area total	*8 120*	*2 401*	*3 670 F*	*3 598*	*7 359 F*	*10 705 F*	*10 023 F*	*10 770 F*	*12 237 F*	*12 447 F*
Species total	*8 120*	*2 401*	*3 670 F*	*3 598*	*7 359 F*	*10 705 F*	*10 023 F*	*10 770 F*	*12 237 F*	*12 447 F*
Fourfinger threadfin	**Barbure mamalí**		**Barbudo mamalí**		***Eleutheronema tetradactylum***			**1,70(77)002,01**		**FOT**
51 Pakistan	512	753	653	812	516	1 783	969	...	63	55
51 Fishing area total	*512*	*753*	*653*	*812*	*516*	*1 783*	*969*	*...*	*63*	*55*
61 China,Taiwan	10 543	10 310	4 603	3 469	3 157	2 831	3 248	2 414	6 917	1 715
61 Fishing area total	*10 543*	*10 310*	*4 603*	*3 469*	*3 157*	*2 831*	*3 248*	*2 414*	*6 917*	*1 715*
Species total	*11 055*	*11 063*	*5 256*	*4 281*	*3 673*	*4 614*	*4 217*	*2 414*	*6 980*	*1 770*
Lesser African threadfin	**Petit capitaine**		**Barbudo de diez barbas**		***Galeoides decadactylus***			**1,70(77)003,02**		**GAL**
34 Cameroon	542	350	600 F	949	810	500 F	13	150	477	462
Congo Rep	380 F	379 F	418	350 F	300 F	250 F	190 F	150 F	63	60 F
Côte dIvoire	380	279	408	374	268	173	278	196	224	168
Gabon	1 590 F	1 598	1 045	1 876	3 805	3 174	3 484	2 808	2 516	3 201
Gambia	71	99	41	88	140	146	57	60	87	116
Ghana	1 826	2 120	3 247	1 969	3 146	1 477	774	586	1 947	2 892
Guinea	250 F	300 F	300 F	431	338	204	43	88	123	120 F
GuineaBissau	...	...	45	101	108	110 F	100 F	90 F	90 F	90 F
Liberia	15	13	11	26	-	-	-	-	-	-
Nigeria	1 217	819	578	983	3 194	5 974	2 343	2 884	4 279	5 131
Senegal	3 595	915	1 456	5 357	5 255	5 453	4 849	3 016	2 844	3 849
Sierra Leone	...	328	327	392	390 F	921	758	716	1 149	914
Togo	27	28	24	7	14	3	1	20	149	98
34 Fishing area total	*9 893 F*	*7 228 F*	*8 500 F*	*12 903 F*	*17 768 F*	*18 385 F*	*12 890 F*	*10 764 F*	*13 948 F*	*17 101 F*
Species total	*9 893 F*	*7 228 F*	*8 500 F*	*12 903 F*	*17 768 F*	*18 385 F*	*12 890 F*	*10 764 F*	*13 948 F*	*17 101 F*
Royal threadfin	**Capitaine royal**		**Barbudo real**		***Pentanemus quinquarius***			**1,70(77)004,01**		**PET**
34 Cameroon	2 100	1 777	1 500 F	610	584	1 000 F	2 368	1 778	1 363	3 276
Congo Rep	330 F	329 F	200 F	200 F	200 F	200 F	180 F	190 F	189	180 F
Gabon	...	...	...	...	...	...	0	292	207	389
Guinea	590 F	700 F	720 F	1 020	435	233	...	...	...	...
Liberia	63	50	25	134	36	155	138	118	92	102
Sierra Leone	1 799	1 778	1 773	1 770	1 750 F	1 768	1 744	1 784	1 761	...
34 Fishing area total	*4 882 F*	*4 634 F*	*4 218 F*	*3 734 F*	*3 005 F*	*3 356 F*	*4 430 F*	*4 162 F*	*3 612*	*3 947 F*
Species total	*4 882 F*	*4 634 F*	*4 218 F*	*3 734 F*	*3 005 F*	*3 356 F*	*4 430 F*	*4 162 F*	*3 612*	*3 947 F*
Threadfins, tasselfishes nei	**Barbures, capitaines nca**		**Barbudos nep**		***Polynemidae***			**1,70(77)XXX,XX**		**THF**
04 India	-	-	-	-	467	416	243	229	66	66
04 Fishing area total	*-*	*-*	*-*	*-*	*467*	*416*	*243*	*229*	*66*	*66*
34 Benin	805 F	700 F	650 F	550 F	500 F	500 F	450 F	358	250 F	324
Korea Rep	468	195	-	-	-	-	141	-	-	-
Nigeria	677	1 638	1 845	2 156	2 930	3 184	-	-	-	-
Sao Tome Prn	54 F	67 F	97 F	102 F	75	75	100 F	110 F	100 F	100 F
Sierra Leone	...	...	...	2	2 F	-	-	-	-	-
34 Fishing area total	*2 004 F*	*2 600 F*	*2 592 F*	*2 810 F*	*3 507 F*	*3 759 F*	*691 F*	*468 F*	*350 F*	*424 F*
47 Angola	19	136	309	259	370	265	520	1 378	872	1 834
Korea Rep	10	327	272	25	-	-	-	-	-	-
Spain	-	-	-	-	-	2	-	-	-	-
47 Fishing area total	*29*	*463*	*581*	*284*	*370*	*267*	*520*	*1 378*	*872*	*1 834*
51 India	20 493	7 186	5 417	3 837	1 584	1 535	1 544	2 780	2 243	2 817
Korea Rep	-	-	-	-	239	-	456	-	-	-
Réunion	...	...	...	...	...	...	...	49	5	10
51 Fishing area total	*20 493*	*7 186*	*5 417*	*3 837*	*1 823*	*1 535*	*2 000*	*2 829*	*2 248*	*2 827*
57 Australia	61	193	42	255	333	-	-	-	-	-
India	5 093	5 610	5 707	5 810	5 001	5 144	5 977	7 049	3 467	3 299
Indonesia	4 532	5 336	4 746	6 445	6 528	6 983	7 543	7 035	7 363	7 790
Malaysia	1 129	1 272	1 037	939	1 270	1 032	1 507	928	1 624	1 951
Thailand	754	788	3 811	1 194	444	38	109	28	28	27
57 Fishing area total	*11 569*	*13 199*	*15 343*	*14 643*	*13 576*	*13 197*	*15 136*	*15 040*	*12 482*	*13 067*
61 China	...	...	...	...	283 784	340 302	517 528	565 764	496 566	476 690
61 Fishing area total	*...*	*...*	*...*	*...*	*283 784*	*340 302*	*517 528*	*565 764*	*496 566*	*476 690*
71 Australia	28	28	475	385	573	216	272	914	1 275	1 455
Indonesia	15 346	16 683	18 050	25 834	24 183	24 726	26 677	26 300	30 919	32 730
Korea Rep	-	-	-	-	-	-	644	-	-	-
Malaysia	2 861	2 810	3 687	3 268	3 828	2 669	3 436	2 983	2 691	2 377
Philippines	1 172	1 690	2 616	2 548	2 383	2 933	3 017	2 524	2 371	2 455
Singapore	0	0	14	5	5	6	7	13	25	11
Thailand	1 134	2 365	2 132	1 010	1 106	1 041	1 196	404	398	396
71 Fishing area total	*20 541*	*23 576*	*26 974*	*33 050*	*32 078*	*31 591*	*35 249*	*33 138*	*37 679*	*39 424*
87 Colombia	-	-	-	-	16	4	3	16	10 F	10 F
87 Fishing area total	*-*	*-*	*-*	*-*	*16*	*4*	*3*	*16*	*10 F*	*10 F*

B-33 Miscellaneous coastal fishes / Poissons côtiers divers / Peces costeros diversos

Capture production by species, fishing areas and countries or areas / Captures par espèces, zones de pêche et pays ou zones / Capturas por especies, áreas de pesca y países o áreas

Species, Fishing area Espèce, Zone de pêche Especie, Area de pesca	1992 mt	1993 mt	1994 mt	1995 mt	1996 mt	1997 mt	1998 mt	1999 mt	2000 mt	2001 mt
Species total	*54 636 F*	*47 024 F*	*50 907 F*	*54 624 F*	*335 621 F*	*391 071 F*	*571 370 F*	*618 862 F*	*550 273 F*	*534 342 F*
Patagonian blennie	**Guite de Patagonie**		**Róbalo patagónico**		***Eleginops maclovinus***			**1,70(92)007,02**		**BLP**
41 Argentina	329	156	624	66	148	57	1 194	2 023	1 745	...
Falkland Is	3	4	4	4	4	4	4	4	10	61
41 Fishing area total	*332*	*160*	*628*	*70*	*152*	*61*	*1 198*	*2 027*	*1 755*	*61*
87 Chile	820	782	615	274	287	133	103	179	164	109
Estonia	-	-	-	-	-	-	-	14	-	-
87 Fishing area total	*820*	*782*	*615*	*274*	*287*	*133*	*103*	*193*	*164*	*109*
Species total	*1 152*	*942*	*1 243*	*344*	*439*	*194*	*1 301*	*2 220*	*1 919*	*170*
Marbled rockcod	**Bocasse marbrée**		**Trama jaspeada**		***Notothenia rossii***			**1,70(92)019,02**		**NOR**
48 Argentina	-	-	1	2	-	-	-	-	-	-
UK	1	-	1	-	-	-	-	-	0	-
USA	-	-	-	-	-	-	-	0	-	0
48 Fishing area total	*1*	-	*2*	*2*	-	-	-	*0*	*0*	*0*
58 France	-	-	0	-	-	1	-	1	-	-
Ukraine	-	2	-	-	-	-	-	-	-	-
58 Fishing area total	-	*2*	*0*	-	-	*1*	-	*1*	-	-
Species total	*1*	*2*	*2*	*2*	-	*1*	-	*1*	*0*	*0*
Humped rockcod	**Bocasse bossue**		**Trama jorobada**		***Notothenia gibberifrons***			**1,70(92)019,03**		**NOG**
48 Argentina	-	-	1	1	-	-	-	-	-	-
France	-	-	-	-	-	-	-	-	-	0
UK	4	-	3	-	-	-	-	-	1	0
USA	-	-	-	-	-	-	-	5	-	2
48 Fishing area total	*4*	-	*4*	*1*	-	-	-	*5*	*1*	*2*
Species total	*4*	-	*4*	*1*	-	-	-	*5*	*1*	*2*
Yellowbelly rockcod	**Bocasse jaune**		**Trama amarilla**		***Notothenia neglecta***			**1,70(92)019,04**		**NON**
48 USA	-	-	-	-	-	-	-	0	-	2
48 Fishing area total	-	-	-	-	-	-	-	*0*	-	*2*
Species total	-	-	-	-	-	-	-	*0*	-	*2*
Grey rockcod	**Bocasse grise**		**Trama gris**		***Notothenia squamifrons***			**1,70(92)019,06**		**NOS**
48 UK	0	-	0	-	-	-	-	-	5	0
USA	-	-	-	-	-	-	-	5	-	0
48 Fishing area total	*0*	-	*0*	-	-	-	-	*5*	*5*	*0*
58 Australia	2	-	0	-	-	4	3	10	-	0
France	-	-	-	-	15	-	-	-	-	-
Russian Fed	3	-	-	-	-	-	-	-	-	-
Ukraine	1	-	-	-	0	-	-	-	-	-
58 Fishing area total	*6*	-	*0*	-	*15*	*4*	*3*	*10*	-	*0*
Species total	*6*	-	*0*	-	*15*	*4*	*3*	*15*	*5*	*0*
Yellowfin notie	**Bocassette dégarnie**		**Doradillo pobre**		***Nototheniops nudifrons***			**1,70(92)430,02**		**NOD**
48 UK	0	-	-	-	-	-	-	-	0	0
USA	-	-	-	-	-	-	-	-	-	0
48 Fishing area total	*0*	-	-	-	-	-	-	-	*0*	*0*
Species total	*0*	-	-	-	-	-	-	-	*0*	*0*
Antarctic rockcods nei	**Bocassons nca**		**Austrobacalaos nep**		***Trematomus spp***			**1,70(92)448,XX**		**TRT**
48 USA	-	-	-	-	-	-	-	0	-	0
48 Fishing area total	-	-	-	-	-	-	-	*0*	-	*0*
Species total	-	-	-	-	-	-	-	*0*	-	*0*
Antarctic silverfish	**Calandre antarctique**		**Diablillo antártico**		***Pleuragramma antarcticum***			**1,70(92)535,01**		**ANS**
48 USA	-	-	-	-	-	-	-	-	-	0
48 Fishing area total	-	-	-	-	-	-	-	-	-	*0*
Species total	-	-	-	-	-	-	-	-	-	*0*
Antarctic rockcods, noties nei	**Bocasses, bocassons nca**		**Tramas, doradillos nep**		***Nototheniidae***			**1,70(92)XXX,XX**		**NOX**
41 Poland	-	-	-	-	-	-	-	207	-	-
41 Fishing area total	-	-	-	-	-	-	-	*207*	-	-
48 Argentina	-	-	0	2	-	-	-	-	-	-
Spain	-	-	-	-	-	-	0	-	-	-
UK	-	-	-	-	-	-	-	-	-	0

B-33 Miscellaneous coastal fishes / Poissons côtiers divers / Peces costeros diversos

Capture production by species, fishing areas and countries or areas / Captures par espèces, zones de pêche et pays ou zones / Capturas por especies, áreas de pesca y países o áreas

Species, Fishing area Espèce, Zone de pêche Especie, Area de pesca	1992 mt	1993 mt	1994 mt	1995 mt	1996 mt	1997 mt	1998 mt	1999 mt	2000 mt	2001 mt
USA	-	-	-	-	-	-	-	0	-	0
48 Fishing area total	*-*	*-*	*0*	*2*	*-*	*-*	*0*	*0*	*-*	*0*
58 Australia	-	-	-	-	-	-	-	3	-	0
58 Fishing area total	*-*	*-*	*-*	*-*	*-*	*-*	*-*	*3*	*-*	*0*
81 New Zealand	...	...	...	34	16	34	3	19	3	15
81 Fishing area total	*...*	*...*	*...*	*34*	*16*	*34*	*3*	*19*	*3*	*15*
88 New Zealand	-	-	-	-	-	-	0	0	0	1
88 Fishing area total	*-*	*-*	*-*	*-*	*-*	*-*	*0*	*0*	*0*	*1*
Species total	*...*	*...*	*0*	*36*	*16*	*34*	*3*	*229*	*3*	*16*
Glassfishes	**...B**		**...C**		*Ambassidae*			**1,70(95)XXX,XX**		**AIB**
71 Philippines	6 076	4 831	3 969	3 554	3 075	2 594	2 186	1 733	1 575	1 664
71 Fishing area total	*6 076*	*4 831*	*3 969*	*3 554*	*3 075*	*2 594*	*2 186*	*1 733*	*1 575*	*1 664*
Species total	*6 076*	*4 831*	*3 969*	*3 554*	*3 075*	*2 594*	*2 186*	*1 733*	*1 575*	*1 664*
Percoids nei	**Percoides nca**		**Percoideos nep**		*Percoidei*			**1,70(XX)XXX,XX**		**PRC**
41 Brazil	1 500 F	1 000 F	800 F	720	664	692	1 838	947	1 465	1 450 F
41 Fishing area total	*1 500 F*	*1 000 F*	*800 F*	*720*	*664*	*692*	*1 838*	*947*	*1 465*	*1 450 F*
51 India	47 809	10 563	8 922	11 273	89 717	58 969	52 230	53 600	58 297	40 053
Mauritius	196	196	201	186	183	158	153	165	181	152
Tanzania	700	312	269	340	333	350	200	250	300	200
51 Fishing area total	*48 705*	*11 071*	*9 392*	*11 799*	*90 233*	*59 477*	*52 583*	*54 015*	*58 778*	*40 405*
57 Australia	9	0	7	6	351	-	-	-	-	-
India	33 470	33 640	32 090	26 710	26 162	28 741	10 268	36 350	36 264	40 301
Malaysia	-	-	744	-	-	-	-	-	-	-
57 Fishing area total	*33 479*	*33 640*	*32 841*	*26 716*	*26 513*	*28 741*	*10 268*	*36 350*	*36 264*	*40 301*
71 Australia	0	0	0	0	0	0	0	-	-	-
Kiribati	1 630	1 580	1 560	1 570	1 590	1 580	10 713	24 588	632	1 069
Malaysia	-	-	431	-	-	-	-	-	-	-
Philippines	29 800	41 228	38 425	38 843	36 825	20 212	20 036	20 717	19 912	21 198
71 Fishing area total	*31 430*	*42 808*	*40 416*	*40 413*	*38 415*	*21 792*	*30 749*	*45 305*	*20 544*	*22 267*
81 Australia	146	145	130	130	146	131	131	-	-	-
81 Fishing area total	*146*	*145*	*130*	*130*	*146*	*131*	*131*	*-*	*-*	*-*
Species total	*115 260 F*	*88 664 F*	*83 579 F*	*79 778*	*155 971*	*110 833*	*95 569*	*136 617*	*117 051*	*104 423 F*
Eelpout	**Loquette d'Europe**		**Viruela**		*Zoarces viviparus*			**1,71(15)004,01**		**ELP**
27 Denmark	14	9	9	2	3	7	2	7	3	5
Estonia	5	9	1	2	2	8	9	2	1	1
Germany	18	7	1	5	3	2	2	2	1	5
Latvia	20	79	164	143	139	80	41	32	23	26
27 Fishing area total	*57*	*104*	*175*	*152*	*147*	*97*	*54*	*43*	*28*	*37*
Species total	*57*	*104*	*175*	*152*	*147*	*97*	*54*	*43*	*28*	*37*
Ocean pout	**Loquette d'Amérique**		**...C**		*Macrozoarces americanus*			**1,71(15)012,02**		**OPT**
21 USA	462	226	196	24	41	15	17	18	19	18
21 Fishing area total	*462*	*226*	*196*	*24*	*41*	*15*	*17*	*18*	*19*	*18*
Species total	*462*	*226*	*196*	*24*	*41*	*15*	*17*	*18*	*19*	*18*
Pacific sandlance	**Lançon du Pacifique**		**Lanzón del Pacífico**		*Ammodytes personatus*			**1,72(04)002,04**		**PAS**
61 Japan	124 273	106 568	109 344	108 124	115 766	108 666	90 688	82 918	49 819	88 164
Korea Rep	-	-	-	-	-	-	-	-	16 293	4 803
Russian Fed	-	-	-	-	-	-	-	-	17	-
61 Fishing area total	*124 273*	*106 568*	*109 344*	*108 124*	*115 766*	*108 666*	*90 688*	*82 918*	*66 129*	*92 967*
Species total	*124 273*	*106 568*	*109 344*	*108 124*	*115 766*	*108 666*	*90 688*	*82 918*	*66 129*	*92 967*
Sandeels(=Sandlances) nei	**Lançons nca**		**Lanzones nep**		*Ammodytes spp*			**1,72(04)002,XX**		**SAN**
21 Canada	-	3	-	-	2	1	0	-	2	-
USA	1	7	14	1	-	1	1	1	0	0
21 Fishing area total	*1*	*10*	*14*	*1*	*2*	*2*	*1*	*1*	*2*	*0*
27 Channel Is	1	1 F	-	-	2	63	61	41	41	41
Denmark	952 794	631 549	839 825	844 512	669 035	840 774	646 905	528 551	567 350	666 295
Faeroe Is	9 139	2 934	10 288	7 485	5 023	11 221	11 071	7 487	8 513	6 030
France	78	131	89	114	95	159	70	92	88	50
Iceland	-	-	-	-	-	-	-	-	-	8
Ireland	-	-	-	-	-	-	-	389	-	-
Norway	92 822	101 519	168 155	263 490	160 702	350 672	343 625	187 589	119 015	187 459
Portugal	0	9	17	64	41	18	9	13	29	40
Spain	-	-	-	-	542	492	183	390	811	584
Sweden	568	-	20	40	-	1	8 585	23 225	28 165	50 559

B-33 Miscellaneous coastal fishes / Poissons côtiers divers / Peces costeros diversos

Capture production by species, fishing areas and countries or areas / Captures par espèces, zones de pêche et pays ou zones / Capturas por especies, áreas de pesca y países o áreas

Species, Fishing area Espèce, Zone de pêche Especie, Area de pesca	1992 mt	1993 mt	1994 mt	1995 mt	1996 mt	1997 mt	1998 mt	1999 mt	2000 mt	2001 mt
UK	10 762	9 121	19 123	17 528	22 936	39 271	29 080	14 102	16 530	1 264
27 Fishing area total	*1 066 164*	*745 264 F*	*1 037 517*	*1 133 233*	*858 376*	*1 242 671*	*1 039 589*	*761 879*	*740 542*	*912 330*
37 France	...	...	...	...	...	...	...	1	2	1
Spain	-	-	-	-	74	164	283	367	306	105
37 Fishing area total	...	...	...	...	*74*	*164*	*283*	*368*	*308*	*106*
Species total	*1 066 165*	*745 274 F*	*1 037 531*	*1 133 234*	*858 452*	*1 242 837*	*1 039 873*	*762 248*	*740 852*	*912 436*
Brazilian flathead	**Platête brésilien**		**Pez palo**			***Percophis brasilianus***			**1,72(08)202,01**	**FLA**
41 Argentina	3 822	5 607	6 187	8 680	8 771	11 475	9 677	6 526	7 255	7 040
Brazil	...	...	...	640	460	412	514	709	607	600 F
Uruguay	-	-	-	-	-	-	-	1	-	0
41 Fishing area total	*3 822*	*5 607*	*6 187*	*9 320*	*9 231*	*11 887*	*10 191*	*7 236*	*7 862*	*7 640 F*
Species total	*3 822*	*5 607*	*6 187*	*9 320*	*9 231*	*11 887*	*10 191*	*7 236*	*7 862*	*7 640 F*
Argentinian sandperch	**...B**		**Salmón de mar**			***Pseudopercis semifasciata***			**1,72(09)006,02**	**UPR**
41 Argentina	2 509	3 589	2 907	3 023	3 438	2 483	2 222	2 679	1 905	1 887
41 Fishing area total	*2 509*	*3 589*	*2 907*	*3 023*	*3 438*	*2 483*	*2 222*	*2 679*	*1 905*	*1 887*
Species total	*2 509*	*3 589*	*2 907*	*3 023*	*3 438*	*2 483*	*2 222*	*2 679*	*1 905*	*1 887*
New Zealand blue cod	**...B**		**...C**			***Parapercis colias***			**1,72(09)196,02**	**NEB**
81 New Zealand	1 505	1 986	1 910	11 262	2 115	2 227	2 313	2 286	2 130	2 441
81 Fishing area total	*1 505*	*1 986*	*1 910*	*11 262*	*2 115*	*2 227*	*2 313*	*2 286*	*2 130*	*2 441*
Species total	*1 505*	*1 986*	*1 910*	*11 262*	*2 115*	*2 227*	*2 313*	*2 286*	*2 130*	*2 441*
Greater weever	**Grande vive**		**Escorpión**			***Trachinus draco***			**1,72(12)010,02**	**WEG**
27 Denmark	207	120	94	76	45	50	36	52	39	48
France	129	131	149	169	127	132	147	213	218	203
Netherlands	-	-	-	-	-	-	-	-	-	6
Portugal	0	0	0	0	0	0	2	7	12	11
Spain	-	-	-	-	-	-	-	-	2	-
Sweden	25	27	21	25	10	1	3	12	7	26
27 Fishing area total	*361*	*278*	*264*	*270*	*182*	*183*	*188*	*284*	*278*	*294*
37 France	4	4	3	2	2	37	37	5	6	3
Malta	-	-	-	1	2	2	0	3	0	0
Spain	-	-	-	-	60	69	67	-	28	57
Tunisia	-	24	8	5	19	46	103	16	19	20
37 Fishing area total	*4*	*28*	*11*	*8*	*83*	*154*	*207*	*24*	*53*	*80*
Species total	*365*	*306*	*275*	*278*	*265*	*337*	*395*	*308*	*331*	*374*
Black goby	**Gobie noir**		**Chaparrudo**			***Gobius niger***			**1,73(21)003,03**	**GBN**
37 Slovenia	-	-	-	-	-	-	-	-	1	1
37 Fishing area total	-	-	-	-	-	-	-	-	*1*	*1*
Species total	-	-	-	-	-	-	-	-	*1*	*1*
Gobies nei	**Gobies nca**		**Góbidos nep**			***Gobiidae***			**1,73(21)XXX,XX**	**GPA**
27 Portugal	0	0	0	0	0	0	-	0	-	-
27 Fishing area total	*0*	*0*	*0*	*0*	*0*	*0*	-	*0*	-	-
37 Albania	...	...	...	6	10	4	-	-	-	5
Bulgaria	20	10	11 F	580	477	424	381	437	145	142
France	3	3	3	0	4	2	0	4	3	1
Georgia	...	...	...	...	...	...	...	2	3	5
Italy	1 763	1 187	1 583	1 452	1 311	1 085	991	800	712	665
Malta	3	2	0	0	0	0	0	0	0	-
Romania	2	3	22	13	8	2	6	30	42	24
Russian Fed	1	2	-	-	2	1	3	5	12	25
Spain	-	-	-	-	-	-	-	478	311	378
Turkey	808	1 024	1 248	233	390	305	250	325	300	335
Ukraine	214	302	401	201	72	96	286	601	825	1 510
37 Fishing area total	*2 814*	*2 533*	*3 268 F*	*2 485*	*2 274*	*1 919*	*1 917*	*2 682*	*2 353*	*3 090*
47 Namibia	86	172	19	5	552	287	20 998	16	3	...
47 Fishing area total	*86*	*172*	*19*	*5*	*552*	*287*	*20 998*	*16*	*3*	...
61 Korea Rep	2 626	2 788	3 570	1 941	1 598	1 722	1 299	897	788	703
Russian Fed	38 499	27 479	12 549	14 071	14 522	16 176	20 551	30 172	32 733	27 321
61 Fishing area total	*41 125*	*30 267*	*16 119*	*16 012*	*16 120*	*17 898*	*21 850*	*31 069*	*33 521*	*28 024*
71 Philippines	5 141	4 090	7 856	8 780	8 404	7 542	6 756	8 021	7 719	7 898
71 Fishing area total	*5 141*	*4 090*	*7 856*	*8 780*	*8 404*	*7 542*	*6 756*	*8 021*	*7 719*	*7 898*
Species total	*49 166*	*37 062*	*27 262 F*	*27 282*	*27 350*	*27 646*	*51 521*	*41 788*	*43 596*	*39 012*

B-33 Miscellaneous coastal fishes / Poissons côtiers divers / Peces costeros diversos

Capture production by species, fishing areas and countries or areas / Captures par espèces, zones de pêche et pays ou zones / Capturas por especies, áreas de pesca y países o áreas

Species, Fishing area Espèce, Zone de pêche Especie, Area de pesca	1992 mt	1993 mt	1994 mt	1995 mt	1996 mt	1997 mt	1998 mt	1999 mt	2000 mt	2001 mt
Surgeonfishes nei	**Chirurgiens nca**		**Navajones nep**		***Acanthuridae***				**1,74(02)XXX,XX**	**SUR**
31 Antigua Barb	...	...	...	...	...	...	...	...	...	158
Grenada	0	0	0	0	0	0	0	1	1	2
St Kitts Nev	...	...	...	5	11	4	9	6	1	1
31 Fishing area total	*0*	*0*	*0*	*5*	*11*	*4*	*9*	*7*	*2*	*161*
34 Nigeria	-	-	31	...	...	...	...	...	-	-
34 Fishing area total	-	-	*31*	...	...	...	...	...	-	-
51 Saudi Arabia	-	-	68	113	238	173	253	205	149	343
51 Fishing area total	-	-	*68*	*113*	*238*	*173*	*253*	*205*	*149*	*343*
71 Fiji Islands	175	346	353 F	278	125	105	180	206	180 F	196
Guam	...	...	5	0	1	...	3	0	-	0
N Marianas	...	...	...	1	3	-	0	3	3	10
Palau	34	52	57	111	92	23	8	60 F	128 F	128 F
Philippines	2 798	5 172	6 058	4 752	4 474	7 367	7 770	5 797	5 392	5 663
71 Fishing area total	*3 007*	*5 570*	*6 473 F*	*5 142*	*4 695*	*7 495*	*7 961*	*6 066 F*	*5 703 F*	*5 997 F*
77 Amer Samoa	...	...	...	...	8	-	13	16	10	3
77 Fishing area total	...	...	...	...	*8*	-	*13*	*16*	*10*	*3*
Species total	*3 007*	*5 570*	*6 572 F*	*5 260*	*4 952*	*7 672*	*8 236*	*6 294 F*	*5 864 F*	*6 504 F*
Batfishes	**Poules d'eau**		**Dalapuganos**		***Platax spp***				**1,74(05)206,XX**	**BAT**
51 Eritrea	...	...	-	-	-	0	1	7	1	0
51 Fishing area total	...	...	-	-	-	*0*	*1*	*7*	*1*	*0*
71 Philippines	1 644	1 506	1 928	1 600	1 584	2 975	2 792	2 870	2 597	2 607
71 Fishing area total	*1 644*	*1 506*	*1 928*	*1 600*	*1 584*	*2 975*	*2 792*	*2 870*	*2 597*	*2 607*
Species total	*1 644*	*1 506*	*1 928*	*1 600*	*1 584*	*2 975*	*2 793*	*2 877*	*2 598*	*2 607*
Spadefishes nei	**Chèvres, disques nca**		**Pagualas nep**		***Ephippidae***				**1,74(05)XXX,XX**	**SPA**
21 USA	-	2	2	5	6	-	14	21	14	20
21 Fishing area total	-	*2*	*2*	*5*	*6*	-	*14*	*21*	*14*	*20*
31 USA	-	-	-	-	-	-	-	-	15	1
31 Fishing area total	-	-	-	-	-	-	-	-	*15*	*1*
34 Benin	1 F	1 F	1 F	2 F	3 F	5 F	4 F	4	3 F	2
Togo	0	2	0	-	-	1	1	0	1	-
34 Fishing area total	*1 F*	*3 F*	*1 F*	*2 F*	*3 F*	*6 F*	*5 F*	*4*	*4 F*	*2*
41 Brazil	...	...	...	40	71	55	51	283	325	320 F
41 Fishing area total	...	...	...	*40*	*71*	*55*	*51*	*283*	*325*	*320 F*
51 Saudi Arabia	-	-	-	-	-	-	17	1	1	2
51 Fishing area total	-	-	-	-	-	-	*17*	*1*	*1*	*2*
Species total	*1 F*	*5 F*	*3 F*	*47 F*	*80 F*	*61 F*	*87 F*	*309*	*359 F*	*345 F*
Scats	**Pavillons**		**Pingos**		***Scatophagus spp***				**1,74(06)330,XX**	**SCT**
04 Philippines	51	32	18	169	169	95	20	40	61	55
04 Fishing area total	*51*	*32*	*18*	*169*	*169*	*95*	*20*	*40*	*61*	*55*
71 Philippines	1 710	2 767	3 228	4 613	4 035	2 057	2 476	2 501	2 637	2 745
71 Fishing area total	*1 710*	*2 767*	*3 228*	*4 613*	*4 035*	*2 057*	*2 476*	*2 501*	*2 637*	*2 745*
Species total	*1 761*	*2 799*	*3 246*	*4 782*	*4 204*	*2 152*	*2 496*	*2 541*	*2 698*	*2 800*
Spinefeet(=Rabbitfishes) nei	**Sigans nca**		**Síganos nep**		***Siganus spp***				**1,74(07)001,XX**	**SPI**
37 Cyprus	11	19	3	1	11	7	36	19	12	31
Egypt	...	...	...	246	750	372	379	481	624	904
Gaza Strip	...	...	...	...	2	11	10	10 F	10 F	10 F
Israel	90	85	50	50	...	...	...	...	...	...
37 Fishing area total	*101*	*104*	*53*	*297*	*763*	*390*	*425*	*510 F*	*646 F*	*945 F*
51 Bahrain	1 146	1 242	1 251	1 543	2 185	1 612	1 523	1 241	2 114	1 899
Egypt	...	...	...	185	128	129	175	199	42	1 402
Eritrea	...	...	-	-	0	0	1	1	1	0
Jordan	...	...	...	...	...	...	10	12	10	11
Kenya	496	440	365	387	404	347	356	304	299	403
Mauritius	544	438	455	465	514	430	494	461	448	450
Oman	110	116	69	142	59	259	122	97	131	363
Qatar	346	281	194	161	225	237	285	240	387	451
Saudi Arabia	1 385	1 388	1 887	2 341	2 779	2 173	1 761	1 691	1 832	1 822
Tanzania	4 500	2 319	2 527	3 470	3 816	4 000	3 550	3 500	3 525	3 000
Untd Arab Em	422	503	472	535	541	578	580	595	1 825	1 820 F
51 Fishing area total	*8 949*	*6 727*	*7 220*	*9 229*	*10 651*	*9 765*	*8 857*	*8 341*	*10 614*	*11 621 F*
57 Malaysia	173	134	104	86	81	88	237	27	109	87
57 Fishing area total	*173*	*134*	*104*	*86*	*81*	*88*	*237*	*27*	*109*	*87*
71 Fiji Islands	103	76	77 F	62	73	80	93	100	90 F	96
Guam	...	...	6	0	0	...	1	2	0	0

B-33

Miscellaneous coastal fishes — Capture production by species, fishing areas and countries or areas
Poissons côtiers divers — Captures par espèces, zones de pêche et pays ou zones
Peces costeros diversos — Capturas por especies, áreas de pesca y países o áreas

Species, Fishing area Espèce, Zone de pêche Especie, Area de pesca	1992 mt	1993 mt	1994 mt	1995 mt	1996 mt	1997 mt	1998 mt	1999 mt	2000 mt	2001 mt
Malaysia	1 503	1 354	1 090	1 058	1 071	1 247	1 227	1 181	1 284	1 407
N Marianas	...	...	...	0	...	...	2	3	5	4
Palau	43	32	26	29	35	9	7	60 F	135 F	135 F
Philippines	14 006	24 293	19 175	18 766	17 012	15 720	18 246	19 977	20 108	20 614
Singapore	60	48	20	-	14	21	24	28	5	8
71 Fishing area total	*15 715*	*25 803*	*20 394 F*	*19 915*	*18 205*	*17 077*	*19 600*	*21 351 F*	*21 627 F*	*22 264 F*
Species total	*24 938*	*32 768*	*27 771 F*	*29 527*	*29 700*	*27 320*	*29 119*	*30 229 F*	*32 996 F*	*34 917 F*
Lingcod	**Morue-lingue**		**Bacalao largo(=Lorcha)**		***Ophiodon elongatus***			**1,78(07)005,01**		**CLI**
67 Canada	4 334	5 234	4 568	3 793	2 500	1 700	1 900	2 523	3 042	2 511
USA	1 909	1 681	1 851	1 310	1 902	1 525	564	538	144	153
67 Fishing area total	*6 243*	*6 915*	*6 419*	*5 103*	*4 402*	*3 225*	*2 464*	*3 061*	*3 186*	*2 664*
77 USA	492	497	441	381	352	364	95	93	28	34
77 Fishing area total	*492*	*497*	*441*	*381*	*352*	*364*	*95*	*93*	*28*	*34*
Species total	*6 735*	*7 412*	*6 860*	*5 484*	*4 754*	*3 589*	*2 559*	*3 154*	*3 214*	*2 698*
Atka mackerel	**Maquereau de atka**		**Lorcha de atka**		***Pleurogrammus azonus***			**1,78(07)014,02**		**ATK**
61 Japan	97 564	135 529	152 503	176 603	181 513	206 763	240 971	169 481	165 118	161 160
Korea D P Rp	...	...	3 477	3 480	6 487	3 535	3 000 F	3 000 F	2 800 F	2 800 F
Korea Rep	2 942	5 846	3 552	3 342	4 103	2 983	7 911	1 005	2 554	1 261
Russian Fed	22 635	16 945	18 283	19 112	21 260	36 075	40 889	40 283	52 784	49 171
61 Fishing area total	*123 141*	*158 320*	*177 815*	*202 537*	*213 363*	*249 356*	*292 771 F*	*213 769 F*	*223 256 F*	*214 392 F*
67 USA	60 228	70 499	62 501	67 156	88 030	59 538	51 579	51 436	44 592	57 096
67 Fishing area total	*60 228*	*70 499*	*62 501*	*67 156*	*88 030*	*59 538*	*51 579*	*51 436*	*44 592*	*57 096*
Species total	*183 369*	*228 819*	*240 316*	*269 693*	*301 393*	*308 894*	*344 350 F*	*265 205 F*	*267 848 F*	*271 488 F*
Bartail flathead	**Platycéphale indien**		**Chato índico**		***Platycephalus indicus***			**1,78(09)018,01**		**FLI**
61 Korea Rep	3 329	3 489	3 036	2 520	2 900	4 196	2 857	2 248	2 310	1 699
61 Fishing area total	*3 329*	*3 489*	*3 036*	*2 520*	*2 900*	*4 196*	*2 857*	*2 248*	*2 310*	*1 699*
Species total	*3 329*	*3 489*	*3 036*	*2 520*	*2 900*	*4 196*	*2 857*	*2 248*	*2 310*	*1 699*
Flatheads nei	**Platycéphalidés nca**		**Platicefálidos nep**		***Platycephalidae***			**1,78(09)XXX,XX**		**FLH**
51 Eritrea	...	...	18	14	0	0	0	21	5	11
Saudi Arabia	-	-	-	-	-	-	18	7	5	6
51 Fishing area total	*...*	*...*	*18*	*14*	*0*	*0*	*18*	*28*	*10*	*17*
57 Australia	657	541	460	1 360	1 640	1 477	989	3 102	2 779	2 423
57 Fishing area total	*657*	*541*	*460*	*1 360*	*1 640*	*1 477*	*989*	*3 102*	*2 779*	*2 423*
71 Australia	65	...	...	...	...	...	...	72	72	57
71 Fishing area total	*65*	*...*	*...*	*...*	*...*	*...*	*...*	*72*	*72*	*57*
81 Australia	3 299	2 629	2 647	2 760	2 748	3 098	2 296	1 564	1 933	1 621
New Zealand	-	-	-	2	1	1	2	6	7	19
81 Fishing area total	*3 299*	*2 629*	*2 647*	*2 762*	*2 749*	*3 099*	*2 298*	*1 570*	*1 940*	*1 640*
Species total	*4 021*	*3 170*	*3 125*	*4 136*	*4 389*	*4 576*	*3 305*	*4 772*	*4 801*	*4 137*
Cabezon	**...B**		**...C**		***Scorpaenichthys marmoratus***			**1,78(13)030,01**		**SMQ**
67 USA	...	...	...	...	...	...	7	33	37	50
67 Fishing area total	*...*	*...*	*...*	*...*	*...*	*...*	*7*	*33*	*37*	*50*
77 USA	...	...	...	...	...	...	161	140	111	69
77 Fishing area total	*...*	*...*	*...*	*...*	*...*	*...*	*161*	*140*	*111*	*69*
Species total	*...*	*...*	*...*	*...*	*...*	*...*	*168*	*173*	*148*	*119*
Sculpins nei	**Chabots nca**		**Cótidos nep**		***Cottidae***			**1,78(13)XXX,XX**		**SWU**
21 USA	0	0	-	1	-	2	5	2	1	1
21 Fishing area total	*0*	*0*	*-*	*1*	*-*	*2*	*5*	*2*	*1*	*1*
67 USA	-	-	-	-	-	-	-	1	63	71
67 Fishing area total	*-*	*-*	*-*	*-*	*-*	*-*	*-*	*1*	*63*	*71*
77 USA	-	-	-	-	-	-	-	-	-	3
77 Fishing area total	*-*	*-*	*-*	*-*	*-*	*-*	*-*	*-*	*-*	*3*
Species total	*0*	*0*	*-*	*1*	*-*	*2*	*5*	*3*	*64*	*75*
...A	**...B**		**Camotillo**		***Normanichthys crockeri***			**1,78(16)022,05**		**NRC**
21 Faeroe Is	-	-	-	2	3	-	-	-	-	-
21 Fishing area total	*-*	*-*	*-*	*2*	*3*	*-*	*-*	*-*	*-*	*-*
87 Chile	12 587	0	25	2 690	3 921	20 426	236	4 843	853	223
87 Fishing area total	*12 587*	*0*	*25*	*2 690*	*3 921*	*20 426*	*236*	*4 843*	*853*	*223*

B-33 Miscellaneous coastal fishes / Poissons côtiers divers / Peces costeros diversos

Capture production by species, fishing areas and countries or areas
Captures par espèces, zones de pêche et pays ou zones
Capturas por especies, áreas de pesca y países o áreas

Species, Fishing area Espèce, Zone de pêche Especie, Area de pesca	1992 mt	1993 mt	1994 mt	1995 mt	1996 mt	1997 mt	1998 mt	1999 mt	2000 mt	2001 mt
Species total	*12 587*	*0*	*25*	*2 692*	*3 924*	*20 426*	*236*	*4 843*	*853*	*223*
Boxfishes nei	**Coffres nca**		**Toritos nep**		***Ostraciidae***				**1,90(01)XXX,XX**	**BXF**
31 Antigua Barb	...	...	...	...	...	...	...	...	...	66
Br Virgin Is	...	...	...	...	...	1	1	1	0	0
Puerto Rico	...	...	...	...	...	...	...	54	68	52
31 Fishing area total	...	...	...	...	...	*1*	*1*	*55*	*68*	*118*
Species total	...	...	...	...	...	*1*	*1*	*55*	*68*	*118*
Northern puffer	**Compère bigaré**		**Tamboril norteño**		***Sphoeroides maculatus***				**1,90(02)006,12**	**PUF**
21 USA	205	57	32	19	11	18	17	37	36	35
21 Fishing area total	*205*	*57*	*32*	*19*	*11*	*18*	*17*	*37*	*36*	*35*
Species total	*205*	*57*	*32*	*19*	*11*	*18*	*17*	*37*	*36*	*35*
Purple puffer	**Compère rouge**		**Tamboril rojo**		***Takifugu vermicularis***				**1,90(02)011,01**	**PUP**
61 Korea Rep	9 459	5 683	4 191	10 178	9 708	7 471	3 897	4 787	-	-
61 Fishing area total	*9 459*	*5 683*	*4 191*	*10 178*	*9 708*	*7 471*	*3 897*	*4 787*	-	-
Species total	*9 459*	*5 683*	*4 191*	*10 178*	*9 708*	*7 471*	*3 897*	*4 787*	-	-
Puffers nei	**Compères nca**		**Tamboriles nep**		***Tetraodontidae***				**1,90(02)XXX,XX**	**PUX**
34 Benin	...	...	...	...	...	...	...	...	...	2
Gambia	...	...	16	125	28	100	-	-	4	33
34 Fishing area total	...	...	*16*	*125*	*28*	*100*	...	...	*4*	*35*
41 Brazil	...	...	...	468	18	88	23	16	35	30 F
41 Fishing area total	...	...	...	*468*	*18*	*88*	*23*	*16*	*35*	*30 F*
51 Saudi Arabia	-	-	-	-	-	-	-	0	0	18
51 Fishing area total	-	-	-	-	-	-	-	*0*	*0*	*18*
57 Australia	294	160	68	62	158	122	115	3	2	...
57 Fishing area total	*294*	*160*	*68*	*62*	*158*	*122*	*115*	*3*	*2*	...
61 Japan	...	...	...	8 991	8 238	7 103	8 329	9 647	10 989	7 820
Korea Rep	-	-	-	-	-	-	-	-	2 978	3 735
61 Fishing area total	...	...	...	*8 991*	*8 238*	*7 103*	*8 329*	*9 647*	*13 967*	*11 555*
81 Australia	164	163	195	211	70	97	69	-	-	-
81 Fishing area total	*164*	*163*	*195*	*211*	*70*	*97*	*69*	-	-	-
Species total	*458*	*323*	*279*	*9 857*	*8 512*	*7 510*	*8 536*	*9 666*	*14 008*	*11 638 F*
Filefishes	**Bourses**		**Cachúas**		***Cantherhines(=Navodon) spp***				**1,90(09)004,XX**	**FLF**
61 China	157 965	95 500	196 321	122 358	210 188	296 781	235 603	240 214	221 683	201 733
China,Taiwan	1 439	1 689	803	1 680	871	446	586	995	829	959
Korea Rep	-	-	-	-	-	-	-	-	2 891	1 578
61 Fishing area total	*159 404*	*97 189*	*197 124*	*124 038*	*211 059*	*297 227*	*236 189*	*241 209*	*225 403*	*204 270*
Species total	*159 404*	*97 189*	*197 124*	*124 038*	*211 059*	*297 227*	*236 189*	*241 209*	*225 403*	*204 270*
Threadsail filefish	**...B**		**...C**		***Stephanolepis cirrhifer***				**1,90(09)010,01**	**FIL**
61 Korea Rep	34 872	11 365	4 382	1 755	1 772	16 318	9 364	2 999	-	-
61 Fishing area total	*34 872*	*11 365*	*4 382*	*1 755*	*1 772*	*16 318*	*9 364*	*2 999*	-	-
Species total	*34 872*	*11 365*	*4 382*	*1 755*	*1 772*	*16 318*	*9 364*	*2 999*	-	-
Velvet leatherjacket	**...B**		**...C**		***Parika scaber***				**1,90(09)018,01**	**PKB**
81 New Zealand	337	290	351	329	442	1 095	312	738	1 279	1 142
81 Fishing area total	*337*	*290*	*351*	*329*	*442*	*1 095*	*312*	*738*	*1 279*	*1 142*
Species total	*337*	*290*	*351*	*329*	*442*	*1 095*	*312*	*738*	*1 279*	*1 142*
Grey triggerfish	**Baliste cabri**		**Pejepuerco blanco**		***Balistes carolinensis***				**1,90(10)002,01**	**TRG**
27 France	-	-	-	-	-	1	1	2	-	4
27 Fishing area total	-	-	-	-	-	*1*	*1*	*2*	-	*4*
37 Tunisia	-	34	23	59	38	52	58	34	50	62
37 Fishing area total	-	*34*	*23*	*59*	*38*	*52*	*58*	*34*	*50*	*62*
Species total	-	*34*	*23*	*59*	*38*	*53*	*59*	*36*	*50*	*66*
Triggerfishes, durgons nei	**Balistes nca**		**Peces-ballesta nep**		***Balistidae***				**1,90(10)XXX,XX**	**TRI**
21 USA	-	-	-	-	-	6	5	61	3	2
21 Fishing area total	-	-	-	-	-	*6*	*5*	*61*	*3*	*2*
27 Portugal	-	-	-	-	-	-	-	42	38	21
27 Fishing area total	-	-	-	-	-	-	-	*42*	*38*	*21*

B-33

Miscellaneous coastal fishes	Capture production by species, fishing areas and countries or areas
Poissons côtiers divers	Captures par espèces, zones de pêche et pays ou zones
Peces costeros diversos	Capturas por especies, áreas de pesca y países o áreas

Species, Fishing area Espèce, Zone de pêche Especie, Area de pesca	1992 mt	1993 mt	1994 mt	1995 mt	1996 mt	1997 mt	1998 mt	1999 mt	2000 mt	2001 mt
31 Antigua Barb	...	...	...	...	...	...	...	...	...	18
Dominican Rp	447	450	810	607	31	26	67	7	10	21
Grenada	0	0	0	0	0	0	0	0	0	0
Mexico	419	365	385	492	446	449	401	348	73	...
Puerto Rico	...	...	...	...	...	...	...	32	34	41
St Kitts Nev	...	...	...	3	7	2	5	2	2	2
USA	309	403	374	355	326	74	244	135	165	178
US Virgin Is	...	...	...	...	...	...	...	31	35 F	35 F
31 Fishing area total	*1 175*	*1 218*	*1 569*	*1 457*	*810*	*551*	*717*	*555*	*319 F*	*295 F*
34 Gambia	-	-	-	3	51	9	2	2	0	3
Ghana	198	9	11	2	17	-	1	-	2	2
Korea Rep	2	2	-	10	15	-	-	16	-	-
Liberia	-	-	-	-	-	5	2	-	-	-
Portugal	2	4	3	4	9	4	1	2	0	0
Sao Tome Prn	...	...	...	...	...	34	40 F	45 F	40 F	40 F
Senegal	-	-	-	-	-	-	10	7	16	52
Sierra Leone	0	0	0	1	1 F	3	195	...	5	-
Togo	0	5	2	0	0	3	4	0	32	0
34 Fishing area total	*202*	*20*	*16*	*20*	*93 F*	*58*	*255 F*	*72 F*	*95 F*	*97 F*
51 Eritrea	...	...	9	1	0	0	0	3	1	0
Korea Rep	-	-	-	-	291	-	-	-	-	-
Saudi Arabia	-	-	-	-	-	26	8	8	5	7
51 Fishing area total	*...*	*...*	*9*	*1*	*291*	*26*	*8*	*11*	*6*	*7*
57 Malaysia	1 164	1 571	498	540	708	543	250	248	115	193
57 Fishing area total	*1 164*	*1 571*	*498*	*540*	*708*	*543*	*250*	*248*	*115*	*193*
71 Fiji Islands	12	35	36 F	6	4	8	10	14	10 F	...
Malaysia	895	742	956	852	898	1 398	1 543	2 578	1 313	2 002
71 Fishing area total	*907*	*777*	*992 F*	*858*	*902*	*1 406*	*1 553*	*2 592*	*1 323 F*	*2 002*
77 Mexico	1 310	1 279	1 349	906	1 266	5 787	3 259	2 338	1 645	...
77 Fishing area total	*1 310*	*1 279*	*1 349*	*906*	*1 266*	*5 787*	*3 259*	*2 338*	*1 645*	*...*
81 Russian Fed	-	-	-	-	-	154	-	-	-	-
81 Fishing area total	*-*	*-*	*-*	*-*	*-*	*154*	*-*	*-*	*-*	*-*
Species total	*4 758*	*4 865*	*4 433 F*	*3 782*	*4 070 F*	*8 531*	*6 047 F*	*5 919 F*	*3 544 F*	*2 617 F*
Toadfishes, etc. nei	**Crapauds, etc. nca**		**Sapos, etc. nep**		***Batrachoididae***				**1,93(01)XXX,XX**	**TFD**
21 USA	...	...	...	...	...	1	6	4	21	27
21 Fishing area total	*...*	*...*	*...*	*...*	*...*	*1*	*6*	*4*	*21*	*27*
27 Portugal	-	-	-	-	-	-	-	70	91	75
27 Fishing area total	*-*	*-*	*-*	*-*	*-*	*-*	*-*	*70*	*91*	*75*
Species total	*...*	*...*	*...*	*...*	*...*	*1*	*6*	*74*	*112*	*102*
Group total	***5 146 488***	***4 811 359***	***5 350 172***	***6 008 888***	***6 222 784***	***6 863 390***	***6 739 692***	***6 691 022***	***6 648 761***	***6 692 983***

B-34

Miscellaneous demersal fishes — Capture production by species, fishing areas and countries or areas
Poissons démersaux divers — Captures par espèces, zones de pêche et pays ou zones
Peces demersales diversos — Capturas por especies, áreas de pesca y países o áreas

Species, Fishing area Espèce, Zone de pêche Especie, Area de pesca	1992 mt	1993 mt	1994 mt	1995 mt	1996 mt	1997 mt	1998 mt	1999 mt	2000 mt	2001 mt
Argentines	**Argentines**		**Argentinas**		***Argentina spp***				**1,23(05)015,XX**	**ARG**
21 Canada	-	428	-	229	259	591	51	12	8	17
Cuba	56	134	15	113	-	553	4	5	-	-
Japan	-	1	-	-	-	-	-	-	-	-
Russian Fed	2	-	-	-	-	-	-	-	5	-
21 Fishing area total	*58*	*563*	*15*	*342*	*259*	*1 144*	*55*	*17*	*13*	*17*
27 Denmark	3 565	2 353	13 741	1 069	1 446	1 455	748	1 420	1 039	907
Faeroe Is	1 439	1 069	960	7 131	9 496	8 433	17 167	8 186	6 388	9 952
France	1	0	0	-	-	-	-	114	55	41
Germany	1	-	43	357	1 394	1 498	633	24	483	189
Iceland	657	1 255	613	492	808	3 367	13 387	5 495	4 595	2 478
Ireland	320	0	-	6	295	1 089	405	396	4 709	7 505
Netherlands	5 287	1 466	6 256	4 136	3 953	4 696	4 964	8 033	3 636	3 659
Norway	8 931	8 480	6 116	6 419	6 817	5 167	8 654	7 823	6 107	14 668
Portugal	0	0	0	-	-	-	-	-	-	-
Russian Fed	4	-	-	-	-	-	-	-	1 214	496
Spain	-	-	-	-	-	-	-	-	34	34
Sweden	...	...	...	...	...	541	428	0	273	1 010
UK	568	465	2 009	485	-	-	-	28	-	7 955
27 Fishing area total	*20 773*	*15 088*	*29 738*	*20 095*	*24 209*	*26 246*	*46 386*	*31 519*	*28 533*	*48 894*
37 France	-	-	-	-	-	-	-	7	4	4
Spain	-	-	-	-	-	-	-	23	38	65
37 Fishing area total	-	-	-	-	-	-	-	*30*	*42*	*69*
81 Japan	-	-	-	-	1	-	-	-	-	-
New Zealand	...	...	...	26	8	41	68	63	42	56
81 Fishing area total	...	...	...	*26*	*9*	*41*	*68*	*63*	*42*	*56*
Species total	*20 831*	*15 651*	*29 753*	*20 463*	*24 477*	*27 431*	*46 509*	*31 629*	*28 630*	*49 036*
Deepsea smelt	**Argentina du Pacifique**		**Argentina del Pacífico**		***Glossanodon semifasciatus***				**1,23(05)029,01**	**DES**
61 Japan	7 468	7 776	8 978	7 705	8 134	7 431	7 142	6 312	5 970	5 414
61 Fishing area total	*7 468*	*7 776*	*8 978*	*7 705*	*8 134*	*7 431*	*7 142*	*6 312*	*5 970*	*5 414*
Species total	*7 468*	*7 776*	*8 978*	*7 705*	*8 134*	*7 431*	*7 142*	*6 312*	*5 970*	*5 414*
Baird's slickhead	**Alépocéphale de Baird**		**...C**		***Alepocephalus bairdii***				**1,23(14)007,01**	**ALC**
27 Estonia	-	-	-	-	-	-	-	-	-	153
Faeroe Is	-	2	-	-	-	-	-	-	-	-
Germany	-	-	-	-	-	-	-	-	12	1
Iceland	10	3	1	1	0	0	0	0	0	2
Lithuania	-	-	-	-	-	-	-	-	-	460
27 Fishing area total	*10*	*5*	*1*	*1*	*0*	*0*	*0*	*0*	*12*	*616*
Species total	*10*	*5*	*1*	*1*	*0*	*0*	*0*	*0*	*12*	*616*
Hector's lanternfish	**Lanternule de Hector**		**Linternillas de Hector**		***Lampanyctodes hectoris***				**1,32(08)017,01**	**LAN**
47 South Africa	656	1 177	871	0	33	243	6 553	0	...	...
47 Fishing area total	*656*	*1 177*	*871*	*0*	*33*	*243*	*6 553*	*0*	...	...
Species total	*656*	*1 177*	*871*	*0*	*33*	*243*	*6 553*	*0*	...	...
Lanternfishes nei	**Lanternules nca**		**Peces linterna nep**		***Myctophidae***				**1,32(08)XXX,XX**	**LXX**
48 Russian Fed	47 013	-	114	-	-	-	-	5	67	-
Ukraine	4 902	-	-	-	-	-	-	-	-	-
48 Fishing area total	*51 915*	-	*114*	-	-	-	-	*5*	*67*	-
51 Iran	2 000	2 000	2 000	2 000	0	0	0	0	0	335
Ukraine	-	-	-	2	-	-	-	-	-	-
51 Fishing area total	*2 000*	*2 000*	*2 000*	*2 002*	*0*	*0*	*0*	*0*	*0*	*335*
Species total	*53 915*	*2 000*	*2 114*	*2 002*	*0*	*0*	*0*	*5*	*67*	*335*
Daggertooth pike conger	**Murénésoce-dague**		**Morenocio dentón**		***Muraenesox cinereus***				**1,43(09)011,02**	**DPC**
57 Malaysia	1 505	1 635	1 158	1 146	2 099	3 707	3 812	2 390	2 160	2 083
Thailand	1 508	1 034	789	1 321	694	734	465	1 178	1 162	1 155
57 Fishing area total	*3 013*	*2 669*	*1 947*	*2 467*	*2 793*	*4 441*	*4 277*	*3 568*	*3 322*	*3 238*
61 China	20 305	50 000	142 092	154 867	177 470	184 843	239 874	234 314	220 497	243 888
China,Taiwan	5 349	4 567	5 095	3 548	2 846	3 246	5 733	9 001	5 874	5 084
Japan	4 899	3 478	3 820	3 055	1 989	2 060	2 081	2 298	2 400	2 738
Korea Rep	2 616	3 760	2 243	1 604	1 411	2 518	1 506	190	1 862	1 080
61 Fishing area total	*33 169*	*61 805*	*153 250*	*163 074*	*183 716*	*192 667*	*249 194*	*245 803*	*230 633*	*252 790*
71 Malaysia	1 765	1 328	2 259	2 249	2 434	3 137	3 523	3 440	3 258	2 936
Thailand	1 043	2 155	3 112	2 732	921	866	1 178	1 167	1 151	1 144
71 Fishing area total	*2 808*	*3 483*	*5 371*	*4 981*	*3 355*	*4 003*	*4 701*	*4 607*	*4 409*	*4 080*
Species total	*38 990*	*67 957*	*160 568*	*170 522*	*189 864*	*201 111*	*258 172*	*253 978*	*238 364*	*260 108*

B-34 Miscellaneous demersal fishes / Poissons démersaux divers / Peces demersales diversos

Capture production by species, fishing areas and countries or areas / Captures par espèces, zones de pêche et pays ou zones / Capturas por especies, áreas de pesca y países o áreas

Species, Fishing area Espèce, Zone de pêche Especie, Area de pesca	1992 mt	1993 mt	1994 mt	1995 mt	1996 mt	1997 mt	1998 mt	1999 mt	2000 mt	2001 mt
Pike-congers nei	**Murénésoces nca**		**Morenocios nep**			***Muraenesox spp***			**1,43(09)011,XX**	**PCX**
04 India	-	-	55	849	490	447	270	494	402	625
04 Fishing area total	-	-	*55*	*849*	*490*	*447*	*270*	*494*	*402*	*625*
51 India	5 198	14 884	6 317	6 230	6 514	7 327	5 820	5 414	5 236	4 831
Pakistan	8 335	9 484	5 725	4 692	4 904	5 637	5 627	8 377	5 937	5 834
South Africa	-	-	-	-	-	1	9	5 F	4	5
51 Fishing area total	*13 533*	*24 368*	*12 042*	*10 922*	*11 418*	*12 965*	*11 456*	*13 796 F*	*11 177*	*10 670*
57 India	3 174	4 326	4 111	2 466	4 097	4 192	5 189	8 182	5 748	6 273
57 Fishing area total	*3 174*	*4 326*	*4 111*	*2 466*	*4 097*	*4 192*	*5 189*	*8 182*	*5 748*	*6 273*
61 China,H.Kong	6 428	3 436	3 354	3 129	3 281	2 903	2 450	1 700 F	2 100 F	2 300 F
61 Fishing area total	*6 428*	*3 436*	*3 354*	*3 129*	*3 281*	*2 903*	*2 450*	*1 700 F*	*2 100 F*	*2 300 F*
Species total	*23 135*	*32 130*	*19 562*	*17 366*	*19 286*	*20 507*	*19 365*	*24 172 F*	*19 427 F*	*19 868 F*
European conger	**Congre d'Europe**		**Congrio común**			***Conger conger***			**1,43(13)001,01**	**COE**
27 Belgium	53	52	54	58	75	73	79	53	49	56
Channel Is	220	222 F	81	75 F	57	17	32	30	24	25
France	5 075	5 787	3 960	4 288	4 262	3 715	4 179	4 118	4 872	4 651
Ireland	36	77	86	144	142	202	374	295	14	253
Isle of Man	0	0	0	0	0	0	...	-	-	-
Netherlands	-	-	-	-	-	-	-	-	-	1
Norway	0	0	0	0	0	0	0	1	0	0
Portugal	2 255	2 772	3 628	3 247	2 865	2 541	2 577	2 254	1 986	1 888
Spain	3 124	3 341	3 544	3 542	4 130	4 424	4 299	4 029	2 993	2 537
UK	779	808	1 023	1 024	980	957	1 044	1 081	1 157	1 255
27 Fishing area total	*11 542*	*13 059 F*	*12 376*	*12 378 F*	*12 511*	*11 929*	*12 584*	*11 861*	*11 095*	*10 666*
34 Morocco	1 849	1 446	1 963	1 922	2 000	1 630	1 386	1 350	1 152	1 092
Portugal	635	807	344	192	490	326	284	219	16	8
Sierra Leone	...	...	...	...	...	...	348	...	...	...
Spain	1	0	0	0	179	75	137	71	45	66
Togo	-	-	-	-	-	0	0	1	0	0
34 Fishing area total	*2 485*	*2 253*	*2 307*	*2 114*	*2 669*	*2 031*	*2 155*	*1 641*	*1 213*	*1 166*
37 Albania	...	...	...	0	1	2	-	-	-	-
Croatia	148	197	143	145	122	114	130	49	38	25
France	815	706	497	358	395	395	395	494	598	574
Greece	...	...	1 419	1 036	1 160	1 922	1 293	1 211	1 019	1 062
Lebanon	5 F	5	5	5	5	5	10	5	8	10
Malta	3	5	4	2	2	4	3	2	2	3
Morocco	1	3	5	5	5	5	0	1	169	2
Slovenia	-	-	-	-	-	-	-	-	0	1
Spain	770 F	810 F	850 F	880 F	910	850	757	650	630	708
Tunisia	-	-	11	6	0	14	14	5	42	7
Yugoslavia	2	6	6	11	20	19	19	17	16	14
37 Fishing area total	*1 744 F*	*1 732 F*	*2 940 F*	*2 448 F*	*2 620*	*3 330*	*2 621*	*2 434*	*2 522*	*2 406*
Species total	*15 771 F*	*17 044 F*	*17 623 F*	*16 940 F*	*17 800*	*17 290*	*17 360*	*15 936*	*14 830*	*14 238*
Argentine conger	**Congre argentin**		**Congrio argentino**			***Conger orbignyanus***			**1,43(13)001,02**	**COS**
41 Argentina	16	13	27	21	22	28	81	18	14	...
Brazil	560 F	560 F	560 F	182	67	110	108	162	108	100 F
41 Fishing area total	*576 F*	*573 F*	*587 F*	*203*	*89*	*138*	*189*	*180*	*122*	*100 F*
Species total	*576 F*	*573 F*	*587 F*	*203*	*89*	*138*	*189*	*180*	*122*	*100 F*
American conger	**Congre d'Amérique**		**Congrio americano**			***Conger oceanicus***			**1,43(13)001,04**	**COA**
21 USA	113	80	87	32	29	17	48	43	49	40
21 Fishing area total	*113*	*80*	*87*	*32*	*29*	*17*	*48*	*43*	*49*	*40*
Species total	*113*	*80*	*87*	*32*	*29*	*17*	*48*	*43*	*49*	*40*
Whitespotted conger	**Congre du Pacifique nord-ouest**		**Congrio del Pacífico**			***Conger myriaster***			**1,43(13)001,05**	**ELS**
61 Korea Rep	24 163	29 882	21 703	19 667	17 314	19 136	11 913	10 160	8 304	7 676
61 Fishing area total	*24 163*	*29 882*	*21 703*	*19 667*	*17 314*	*19 136*	*11 913*	*10 160*	*8 304*	*7 676*
Species total	*24 163*	*29 882*	*21 703*	*19 667*	*17 314*	*19 136*	*11 913*	*10 160*	*8 304*	*7 676*
Conger eels, etc. nei	**Congres, etc. nca**		**Congrios, etc. nep**			***Congridae***			**1,43(13)XXX,XX**	**COX**
34 Cameroon	13	14	10 F	5	6	6 F	5 F	5 F	8	7
Congo Dem R	70 F	80 F	70 F	71 F	73 F	71 F	73 F	70 F	70 F	70 F
Congo Rep	494 F	493 F	296 F	250 F	210 F	170 F	120 F	90 F	52	50 F
Côte dIvoire	33	29	19	20 F	22	26	32	12	12	17
Liberia	87	42	81	70	41	117	85	128	49	76
Nigeria	...	...	...	...	...	...	301	...	0	-
Senegal	40	11	28	127	83	226	241	227	236	112
34 Fishing area total	*737 F*	*669 F*	*504 F*	*543 F*	*435 F*	*616 F*	*857 F*	*532 F*	*427 F*	*332 F*
37 Turkey	572	1 266	1 413	304	416	310	300	680	200	340
37 Fishing area total	*572*	*1 266*	*1 413*	*304*	*416*	*310*	*300*	*680*	*200*	*340*

B-34 Miscellaneous demersal fishes / Poissons démersaux divers / Peces demersales diversos

Capture production by species, fishing areas and countries or areas / Captures par espèces, zones de pêche et pays ou zones / Capturas por especies, áreas de pesca y países o áreas

Species, Fishing area Espèce, Zone de pêche Especie, Area de pesca	1992 mt	1993 mt	1994 mt	1995 mt	1996 mt	1997 mt	1998 mt	1999 mt	2000 mt	2001 mt
41 Korea Rep	-	-	-	-	-	-	-	-	20	-
Russian Fed	31	-	-	-	-	-	-	-	-	-
41 Fishing area total	*31*	-	-	-	-	-	-	-	*20*	-
47 Portugal	-	-	-	-	-	-	-	-	-	1
St Helena	5	7	5	5	1	3	3	3	3	5
47 Fishing area total	*5*	*7*	*5*	*5*	*1*	*3*	*3*	*3*	*3*	*6*
51 Korea Rep	-	-	-	-	-	6	64	19	53	-
51 Fishing area total	-	-	-	-	-	*6*	*64*	*19*	*53*	-
61 Japan	...	...	...	12 978	12 007	11 706	9 444	8 168	8 364	7 999
61 Fishing area total	...	...	...	*12 978*	*12 007*	*11 706*	*9 444*	*8 168*	*8 364*	*7 999*
71 Korea Rep	210	37	525	318	69	57	-	8	49	78
Philippines	2 682	2 131	2 384	3 061	2 687	2 053	2 540	2 459	2 349	2 344
71 Fishing area total	*2 892*	*2 168*	*2 909*	*3 379*	*2 756*	*2 110*	*2 540*	*2 467*	*2 398*	*2 422*
81 Korea Rep	-	-	-	-	-	-	-	132	-	-
New Zealand	...	...	...	93	109	113	97	88	96	106
81 Fishing area total	...	...	...	*93*	*109*	*113*	*97*	*220*	*96*	*106*
87 Korea Rep	-	-	-	-	-	-	23		-	-
87 Fishing area total	-	-	-	-	-	-	*23*		-	-
Species total	*4 237 F*	*4 110 F*	*4 831 F*	*17 302 F*	*15 724 F*	*14 864 F*	*13 328 F*	*12 089 F*	*11 561 F*	*11 205 F*
Longspine snipefish	**Bécasse de mer**		**Trompetero**		***Macroramphosus scolopax***				**1,51(03)004,01**	**SNS**
27 Portugal	-	2	0	0	-	-	-	-	-	-
27 Fishing area total	-	*2*	*0*	*0*	-	-	-	-	-	-
34 Lithuania	-	3 169	-	-	-	-	-	-	-	-
Russian Fed	329	-	-	20	-	89	443	-	-	44
34 Fishing area total	*329*	*3 169*	-	*20*	-	*89*	*443*	-	-	*44*
Species total	*329*	*3 171*	*0*	*20*	-	*89*	*443*	-	-	*44*
Pink cusk-eel	**Abadèche rosé**		**Congribadejo rosado**		***Genypterus blacodes***				**1,58(02)001,01**	**CUS**
41 Argentina	22 994	23 788	20 338	23 265	21 933	21 917	25 086	21 503	15 019	19 591
Australia	-	-	-	-	-	-	2	10	-	-
Belize	-	-	-	-	-	-	-	15	87	8
Chile	-	-	2	-	-	-	-	-	-	-
Falkland Is	4	8	22	116	297	154	253	451	304	347
France	-	-	3	3	2	1	0	0	-	-
Honduras	106	196	39	61	59	-	-	-	-	-
Korea Rep	186	25	9	-	516	30	32	23	-	327
Namibia	-	-	-	-	-	5	24	45	-	-
Panama	246	56	5	33	46	-	-	-	-	-
Portugal	6	1	-	-	-	-	-	-	13	89
Seychelles	-	-	-	-	-	10	-	-	-	-
Spain	869	771	576	1 299	706	779	1 800	1 901	1 392	1 408
UK	-	36	1	5	6	11	7	32	7	9
Uruguay	1 105	1 645	436	105	43	41	86	206	368	756
41 Fishing area total	*25 516*	*26 526*	*21 431*	*24 887*	*23 608*	*22 948*	*27 290*	*24 186*	*17 190*	*22 535*
57 Australia	0	0	0	0	1 397	1 923	1 830	1 881	2 039	1 696
57 Fishing area total	*0*	*0*	*0*	*0*	*1 397*	*1 923*	*1 830*	*1 881*	*2 039*	*1 696*
81 Australia	80	78	73	85	...	...	...	...	...	...
Korea Rep	-	410	814	549	404	1 348	1 210	1 871	1 684	1 277
New Zealand	16 115	15 629	15 120	18 396	12 454	22 594	22 215	21 424	21 617	18 620
Poland	48	-	-	-	-	-	-	-	-	-
Ukraine	61	7	...	...	...	...	21	35	258	35
81 Fishing area total	*16 304*	*16 124*	*16 007*	*19 030*	*12 858*	*23 942*	*23 446*	*23 330*	*23 559*	*19 932*
87 Chile	6 483	4 643	4 624	5 438	5 780	6 410	6 836	5 721	6 269	7 522
87 Fishing area total	*6 483*	*4 643*	*4 624*	*5 438*	*5 780*	*6 410*	*6 836*	*5 721*	*6 269*	*7 522*
Species total	*48 303*	*47 293*	*42 062*	*49 355*	*43 643*	*55 223*	*59 402*	*55 118*	*49 057*	*51 685*
Red cusk-eel	**Abadèche rouge**		**Congribadejo colorado**		***Genypterus chilensis***				**1,58(02)001,02**	**CUC**
87 Chile	1 203	1 411	1 712	1 082	982	745	584	415	608	730
87 Fishing area total	*1 203*	*1 411*	*1 712*	*1 082*	*982*	*745*	*584*	*415*	*608*	*730*
Species total	*1 203*	*1 411*	*1 712*	*1 082*	*982*	*745*	*584*	*415*	*608*	*730*
Black cusk-eel	**Abadèche noir**		**Congribadejo negro**		***Genypterus maculatus***				**1,58(02)001,03**	**CUB**
87 Chile	1 908	2 023	1 125	1 193	1 343	1 661	2 753	1 943	3 542	3 889
87 Fishing area total	*1 908*	*2 023*	*1 125*	*1 193*	*1 343*	*1 661*	*2 753*	*1 943*	*3 542*	*3 889*
Species total	*1 908*	*2 023*	*1 125*	*1 193*	*1 343*	*1 661*	*2 753*	*1 943*	*3 542*	*3 889*
Kingklip	**Abadèche du Cap**		**Congribadejo(=Rosada)del Cabo**		***Genypterus capensis***				**1,58(02)001,05**	**KCP**
47 Iceland	-	-	-	-	-	31	2	-	-	-

B-34 Miscellaneous demersal fishes / Poissons démersaux divers / Peces demersales diversos

Capture production by species, fishing areas and countries or areas / Captures par espèces, zones de pêche et pays ou zones / Capturas por especies, áreas de pesca y países o áreas

Species, Fishing area Espèce, Zone de pêche Especie, Area de pesca	1992 mt	1993 mt	1994 mt	1995 mt	1996 mt	1997 mt	1998 mt	1999 mt	2000 mt	2001 mt
Korea Rep	18	13	69	-	-	-	-	-	-	-
Namibia	295	747	1 646	3 853	3 667	2 506	2 820	3 706	3 922	6 609
Russian Fed	-	-	-	-	-	-	-	-	-	5
South Africa	2 225	2 601	2 679	2 798	3 140	3 418	3 381	4 114	4 000 F	4 848
Spain	32	-	-	-	-	-	-	-	-	-
47 Fishing area total	*2 570*	*3 361*	*4 394*	*6 651*	*6 807*	*5 955*	*6 203*	*7 820*	*7 922 F*	*11 462*
Species total	*2 570*	*3 361*	*4 394*	*6 651*	*6 807*	*5 955*	*6 203*	*7 820*	*7 922 F*	*11 462*
Cusk-eels nei	**Abadèches nca**		**Congribadejos nep**		***Genypterus spp***				**1,58(02)001,XX**	**CE**
41 Korea Rep	-	-	-	-	-	-	-	-	57	-
41 Fishing area total	-	-	-	-	-	-	-	-	*57*	-
87 Peru	539	640	639	1 631	1 121	439	425	196	557	552
87 Fishing area total	*539*	*640*	*639*	*1 631*	*1 121*	*439*	*425*	*196*	*557*	*552*
Species total	*539*	*640*	*639*	*1 631*	*1 121*	*439*	*425*	*196*	*614*	*552*
Bearded brotula	**Brotule barbée**		**Brótula de barbas**		***Brotula barbata***				**1,58(02)005,02**	**BR**
31 USA	...	...	...	...	...	...	5	1	1	0
31 Fishing area total	...	...	...	...	...	...	*5*	*1*	*1*	*0*
34 Congo Rep	23 F	23 F	14 F	20 F	25 F	30 F	30 F	40 F	46	40 F
Côte dIvoire	450	333	154	298	192	158	342	513	280	156
Korea Rep	-	-	-	-	-	-	-	-	16	-
Liberia	279	108	216	46	4	10	66	48	52	...
Senegal	1 679	1 609	847	140	590	1 644	1 770	3 260	1 325	1 669
Spain	-	-	-	-	-	-	-	-	-	61
34 Fishing area total	*2 431 F*	*2 073 F*	*1 231 F*	*504 F*	*811 F*	*1 842 F*	*2 208 F*	*3 861 F*	*1 719*	*1 926 F*
Species total	*2 431 F*	*2 073 F*	*1 231 F*	*504 F*	*811 F*	*1 842 F*	*2 213 F*	*3 862 F*	*1 720*	*1 926 F*
Cusk-eels, brotulas nei	**Abadèches, brotules nca**		**Brótulas, congribadejos nep**		***Ophidiidae***				**1,58(02)XXX,XX**	**OP**
31 Cuba	232	187	204	198	118	97	37	13	10 F	10 F
31 Fishing area total	*232*	*187*	*204*	*198*	*118*	*97*	*37*	*13*	*10 F*	*10 F*
41 Brazil	...	...	...	106	176	258	452	544	507	500 F
Russian Fed	-	-	-	-	-	-	-	-	-	18
41 Fishing area total	...	...	...	*106*	*176*	*258*	*452*	*544*	*507*	*518 F*
77 Korea Rep	-	-	-	-	-	-	-	-	1	-
77 Fishing area total	-	-	-	-	-	-	-	-	*1*	-
Species total	*232*	*187*	*204*	*304*	*294*	*355*	*489*	*557*	*518 F*	*528 F*
Alfonsinos nei	**Béryx nca**		**Alfonsinos nep**		***Beryx spp***				**1,61(02)003,XX**	**ALF**
21 Russian Fed	-	-	-	541	141	-	-	-	-	-
21 Fishing area total	-	-	-	*541*	*141*	-	-	-	-	-
27 Faeroe Is	-	-	1	3	-	5	-	-	-	-
France	4	3	5	0	0	3	27	75	40	52
Iceland	-	-	-	-	0	-	-	-	-	-
Portugal	-	-	-	-	-	-	-	87	87	60
Russian Fed	-	-	-	-	-	-	-	-	5	-
UK	-	-	-	-	-	4	-	-	7	16
27 Fishing area total	*4*	*3*	*6*	*3*	*0*	*12*	*27*	*162*	*139*	*128*
31 Iceland	-	-	-	-	7	-	-	-	-	-
Russian Fed	-	-	-	278	15	-	-	-	-	-
31 Fishing area total	-	-	-	*278*	*22*	-	-	-	-	-
34 Latvia	3	7	-	-	-	-	-	-	-	-
Lithuania	41	-	-	-	-	-	-	-	-	-
Norway	-	-	-	-	-	-	-	-	71	-
Portugal	55	60	74	33	126	58	51	42	1	-
Russian Fed	-	20	-	-	-	-	10	21	12	6
Spain	-	-	-	-	24	47	18	-	-	243
34 Fishing area total	*99*	*87*	*74*	*33*	*150*	*105*	*79*	*63*	*84*	*249*
37 France	-	-	-	-	-	-	-	-	8	-
37 Fishing area total	-	-	-	-	-	-	-	-	*8*	-
41 Chile	-	-	-	-	-	-	144	-	-	-
Korea Rep	-	-	-	-	-	-	-	-	4	-
41 Fishing area total	-	-	-	-	-	-	*144*	-	*4*	-
47 Iceland	-	-	-	-	-	466	126	-	-	-
Namibia	-	-	-	909	1 805	369	147	123	59	300
Norway	...	...	...	-	-	836	1 066	-	242	-
Poland	-	-	-	-	-	1 964	-	-	-	-
Portugal	-	-	-	-	-	-	-	3	1	7
Russian Fed	4	-	-	-	-	48	69	-	-	1
South Africa	-	-	-	-	-	-	-	-	-	10
Spain	-	-	-	-	-	186	402	-	-	-
Ukraine	-	172	-	-	747	392	-	-	-	-

B-34 Miscellaneous demersal fishes / Poissons démersaux divers / Peces demersales diversos

Capture production by species, fishing areas and countries or areas / Captures par espèces, zones de pêche et pays ou zones / Capturas por especies, áreas de pesca y países o áreas

Species, Fishing area Espèce, Zone de pêche Especie, Area de pesca	1992 mt	1993 mt	1994 mt	1995 mt	1996 mt	1997 mt	1998 mt	1999 mt	2000 mt	2001 mt
47 *Fishing area total*	*4*	*172*	...	*909*	*2 552*	*4 261*	*1 810*	*126*	*302*	*318*
51 Norway	-	-	-	-	-	-	-	-	11	-
Russian Fed	-	-	-	-	-	-	-	-	-	210
Spain	-	-	-	-	-	-	-	-	79	4
Ukraine	314	462	1 534	2 249	3 079	1 031	859	1 964	1 578	371
51 *Fishing area total*	*314*	*462*	*1 534*	*2 249*	*3 079*	*1 031*	*859*	*1 964*	*1 668*	*585*
61 Russian Fed	-	-	17	-	6	-	4	38	18	14
61 *Fishing area total*	-	-	*17*	-	*6*	-	*4*	*38*	*18*	*14*
71 NewCaledonia	0	0	0	0	0	0	0	0	0	0
71 *Fishing area total*	*0*	*0*	*0*	*0*	*0*	*0*	*0*	*0*	*0*	*0*
81 Korea Rep	-	-	-	-	-	-	77	-	-	-
New Zealand	1 711	1 713	2 595	2 177	2 159	2 617	3 516	2 579	2 880	3 044
81 *Fishing area total*	*1 711*	*1 713*	*2 595*	*2 177*	*2 159*	*2 617*	*3 593*	*2 579*	*2 880*	*3 044*
87 Chile	-	-	-	-	-	-	-	706	4 366	5 182
87 *Fishing area total*	-	-	-	-	-	-	-	*706*	*4 366*	*5 182*
Species total	*2 132*	*2 437*	*4 226*	*6 190*	*8 109*	*8 026*	*6 516*	*5 638*	*9 469*	*9 520*
Redfish	**...B**		**...C**			***Centroberyx affinis***			**1,61(02)012,01**	**CXF**
81 Australia	...	...	...	...	1 284	1 357	1 812	1 555	1 009	775
New Zealand	...	...	...	89	123	194	172	134	176	202
81 *Fishing area total*	...	...	...	*89*	*1 407*	*1 551*	*1 984*	*1 689*	*1 185*	*977*
Species total	...	...	...	*89*	*1 407*	*1 551*	*1 984*	*1 689*	*1 185*	*977*
Orange roughy	**Hoplostète orange**		**Reloj anaranjado**			***Hoplostethus atlanticus***			**1,61(05)002,02**	**ORY**
27 Faeroe Is	-	60	259	732	950	854	747	349	155	1
France	4 050	2 159	1 939	998	1 067	1 012	1 110	1 330	1 048	1 254
Iceland	382	717	158	64	43	79	28	14	68	18
Ireland	-	-	-	-	-	-	-	-	-	2 759
Portugal	-	-	-	-	-	-	-	117	157	161
Russian Fed	-	-	-	-	-	-	-	-	14	-
Spain	-	-	-	-	22	26	26	38	20	15
UK	-	-	-	-	0	0	0	12	2	35
27 *Fishing area total*	*4 432*	*2 936*	*2 356*	*1 794*	*2 082*	*1 971*	*1 911*	*1 860*	*1 464*	*4 243*
47 Namibia	-	-	30	6 377	13 379	18 516	10 945	3 473	1 542	857
Norway	-	-	-	-	-	22	12	-	-	-
47 *Fishing area total*	-	-	*30*	*6 377*	*13 379*	*18 538*	*10 957*	*3 473*	*1 542*	*857*
51 China	-	-	-	-	-	-	-	-	623	710
Norway	-	-	-	-	-	-	-	-	642	-
Spain	-	-	-	-	-	-	-	-	-	1
51 *Fishing area total*	-	-	-	-	-	-	-	-	*1 265*	*711*
57 Australia	627	432	668	227	357	350	4 857	7 553	4 974	5 145
57 *Fishing area total*	*627*	*432*	*668*	*227*	*357*	*350*	*4 857*	*7 553*	*4 974*	*5 145*
81 Australia	18 187	12 050	9 977	7 070	4 526	3 129	3 207	28	26	16
Korea Rep	-	-	-	-	-	-	-	230	-	47
New Zealand	36 568	29 681	31 718	33 077	28 639	20 545	21 485	23 780	17 879	14 044
Norway	2	1 602	665	1	5	12	3	-	-	-
Ukraine	-	-	-	-	-	-	-	-	102	195
81 *Fishing area total*	*54 757*	*43 333*	*42 360*	*40 148*	*33 170*	*23 686*	*24 695*	*24 038*	*18 007*	*14 302*
Species total	*59 816*	*46 701*	*45 414*	*48 546*	*48 988*	*44 545*	*42 420*	*36 924*	*27 252*	*25 258*
Slimeheads nei	**Poissons-montres nca**		**Relojes nep**			***Trachichthyidae***			**1,61(05)XXX,XX**	**TRC**
27 Portugal	-	-	-	-	-	-	-	-	-	235
Spain	-	-	-	-	833	1 052	33	25	3	6
27 *Fishing area total*	-	-	-	-	*833*	*1 052*	*33*	*25*	*3*	*241*
34 Portugal	-	-	-	-	-	-	-	-	-	235
34 *Fishing area total*	-	-	-	-	-	-	-	-	-	*235*
81 New Zealand	-	-	-	-	-	-	7	3	4	2
81 *Fishing area total*	-	-	-	-	-	-	*7*	*3*	*4*	*2*
87 Chile	-	-	-	-	-	-	-	779	1 482	1 868
87 *Fishing area total*	-	-	-	-	-	-	-	*779*	*1 482*	*1 868*
Species total	-	-	-	-	*833*	*1 052*	*40*	*807*	*1 489*	*2 346*
John dory	**Saint Pierre**		**Pez de San Pedro**			***Zeus faber***			**1,62(01)001,01**	**JOD**
27 Channel Is	0	0	1	1 F	0	0	1	1	2	1
France	668	658	703	715	692	670	767	900	1 268	1 355
Ireland	86	95	81	147	125	112	98	145	174	169
Portugal	311	371	300	160	154	173	288	324	431	457
Spain	-	-	-	-	225	467	599	595	507	899
UK	125	138	262	287	220	159	136	181	296	267

B-34 Miscellaneous demersal fishes / Poissons démersaux divers / Peces demersales diversos

Capture production by species, fishing areas and countries or areas / Captures par espèces, zones de pêche et pays ou zones / Capturas por especies, áreas de pesca y países o áreas

Species, Fishing area Espèce, Zone de pêche Especie, Area de pesca	1992 mt	1993 mt	1994 mt	1995 mt	1996 mt	1997 mt	1998 mt	1999 mt	2000 mt	2001 mt
27 Fishing area total	*1 190*	*1 262*	*1 347*	*1 310 F*	*1 416*	*1 581*	*1 889*	*2 146*	*2 678*	*3 148*
34 Greece	6	2	34	9	40	20	15	6	11	28
Mauritania	10 F	5 F	10 F	10 F	5 F	10 F	20 F	...	...	...
Morocco	520	529	478	558	459	587	564	621	940	510
Portugal	108	91	14	6	12	6	29	42	0	0
Russian Fed	-	-	-	-	-	-	-	-	2	-
Senegal	65	63	794	661	1 157	142	53	139	282	161
Spain	9	15 F	20 F	25 F	33	14	1	-	-	-
Ukraine	-	-	-	-	-	-	99	5	59	9
34 Fishing area total	*718 F*	*705 F*	*1 350 F*	*1 269 F*	*1 706 F*	*779 F*	*781 F*	*813*	*1 294*	*708*
37 Albania	...	...	...	2	0	0	-	-	-	-
France	60	103	18	5	20	20	20	6	7	9
Greece	277	221	366	413	447	289	259	195	185	268
Malta	0	0	0	0	0	1	1	2	1	0
Morocco	7	4	5	3	2	3	3	1	2	2
Spain	-	-	-	-	26	29	21	20	34	50
Tunisia	16	20	25	19	28	32	107	59	56	43
Turkey	32	33	44	35	73	50	120	135	100	130
37 Fishing area total	*392*	*381*	*458*	*477*	*596*	*424*	*531*	*418*	*385*	*502*
47 Angola	27	299	770	72	315	307	1 339	1 668	1 582	1 303
Iceland	-	-	-	-	-	7	-	-	-	-
Namibia	3	1	1	3	0	5	25	14	4	138
Russian Fed	2	-	-	-	-	-	-	-	-	-
South Africa	1 103	1 098	1 069	1 070	1 156	1 274	964	1 022	1 000 F	1 177
Spain	42	-	-	-	-	-	-	-	-	-
Ukraine	...	...	...	-	...	...	...	2	-	-
47 Fishing area total	*1 177*	*1 398*	*1 840*	*1 145*	*1 471*	*1 593*	*2 328*	*2 706*	*2 586 F*	*2 618*
51 South Africa	-	3	-	-	-	3	4	5 F	6	6
51 Fishing area total	-	*3*	-	-	-	*3*	*4*	*5 F*	*6*	*6*
57 Australia	3	2	1	0	3	0	1	21	26	22
Korea Rep	-	-	-	-	-	-	-	-	535	216
57 Fishing area total	*3*	*2*	*1*	*0*	*3*	*0*	*1*	*21*	*561*	*238*
81 Australia	333	309	403	401	119	118	102	154	170	155
Japan	247	239	98	38	16	19	17	8	2	46
Korea Rep	105	-	218	265	105	298	36	540	101	9
New Zealand	776	777	825	841	729	800	828	882	841	914
81 Fishing area total	*1 461*	*1 325*	*1 544*	*1 545*	*969*	*1 235*	*983*	*1 584*	*1 114*	*1 124*
87 Japan	-	-	-	-	-	-	-	5	-	-
87 Fishing area total	-	-	-	-	-	-	-	*5*	-	-
Species total	*4 941 F*	*5 076 F*	*6 540 F*	*5 746 F*	*6 161 F*	*5 615 F*	*6 517 F*	*7 698 F*	*8 624 F*	*8 344*
Silvery John dory	**Saint Pierre argenté**		**San Pedro plateado**		***Zenopsis conchifer***				**1,62(01)004,01**	**JOS**
21 USA	5	2	10	34	27	6	49	19	28	62
21 Fishing area total	*5*	*2*	*10*	*34*	*27*	*6*	*49*	*19*	*28*	*62*
34 Portugal	-	-	-	3	9	6	6	3	-	0
34 Fishing area total	-	-	-	*3*	*9*	*6*	*6*	*3*	-	*0*
Species total	*5*	*2*	*10*	*37*	*36*	*12*	*55*	*22*	*28*	*62*
Mirror dory	**...B**		**...C**		***Zenopsis nebulosus***				**1,62(01)004,02**	**ZNE**
57 Australia	11	81	0	0	4	9	37	1 033	500	528
57 Fishing area total	*11*	*81*	*0*	*0*	*4*	*9*	*37*	*1 033*	*500*	*528*
81 Australia	695	467	453	361	348	1 347	1 794	2	2	1
81 Fishing area total	*695*	*467*	*453*	*361*	*348*	*1 347*	*1 794*	*2*	*2*	*1*
Species total	*706*	*548*	*453*	*361*	*352*	*1 356*	*1 831*	*1 035*	*502*	*529*
Dories nei	**Saint Pierres nca**		**Peces de San Pedro nep**		***Zeidae***				**1,62(01)XXX,XX**	**ZEX**
81 New Zealand	-	-	-	338	288	641	754	763	778	685
81 Fishing area total	-	-	-	*338*	*288*	*641*	*754*	*763*	*778*	*685*
Species total	-	-	-	*338*	*288*	*641*	*754*	*763*	*778*	*685*
Boarfishes nei	**Sangliers nca**		**Ochavos nep**		***Caproidae***				**1,62(03)XXX,XX**	**BOR**
47 Russian Fed	-	-	-	-	-	-	5	-	-	-
47 Fishing area total	-	-	-	-	-	-	*5*	-	-	-
51 Spain	-	-	-	-	-	-	-	-	-	7
51 Fishing area total	-	-	-	-	-	-	-	-	-	*7*
Species total	-	-	-	-	-	-	*5*	-	-	*7*
Oreo dories nei	**Oréos nca**		**Oreós nep**		***Oreosomatidae***				**1,62(04)XXX,XX**	**ORD**
47 Namibia	-	-	-	6	17	188	...	42	10	54

B-34 Miscellaneous demersal fishes / Poissons démersaux divers / Peces demersales diversos

Capture production by species, fishing areas and countries or areas / Captures par espèces, zones de pêche et pays ou zones / Capturas por especies, áreas de pesca y países o áreas

Species, Fishing area Espèce, Zone de pêche Especie, Area de pesca	1992 mt	1993 mt	1994 mt	1995 mt	1996 mt	1997 mt	1998 mt	1999 mt	2000 mt	2001 mt
Norway	-	-	-	-	-	-	6	-	-	-
47 Fishing area total	*-*	*-*	*-*	*6*	*17*	*188*	*6*	*42*	*10*	*54*
51 China	-	-	-	-	-	-	-	-	-	180
Norway	-	-	-	-	-	-	-	-	175	-
51 Fishing area total	*-*	*-*	*-*	*-*	*-*	*-*	*-*	*-*	*175*	*180*
61 Russian Fed	-	-	2	-	-	-	-	2	-	14
61 Fishing area total	*-*	*-*	*2*	*-*	*-*	*-*	*-*	*2*	*-*	*14*
81 New Zealand	19 544	23 216	22 602	21 833	18 776	21 850	21 095	22 646	22 775	24 165
Norway	3	1	11	-	1	-	1	-	-	-
Russian Fed	51	-	18	-	5	-	-	-	-	-
Ukraine	4	-	-	-	-	-	-	-	-	-
81 Fishing area total	*19 602*	*23 217*	*22 631*	*21 833*	*18 782*	*21 850*	*21 096*	*22 646*	*22 775*	*24 165*
Species total	*19 602*	*23 217*	*22 633*	*21 839*	*18 799*	*22 038*	*21 102*	*22 690*	*22 960*	*24 413*
Wreckfish	**Cernier commun**		**Cherna**			***Polyprion americanus***			**1,70(05)058,01**	**WRF**
27 Channel Is	-	-	-	-	-	2	1	0	-	-
France	20	33	4	2	4	10	13	20	30	22
Ireland	-	-	-	-	-	-	5	-	1	1
Portugal	542	647	831	619	460	347	304	275	338	306
Spain	10	21	48	42	115	265	124	152	72	84
UK	0	1	1	2	8	0	0	0	0	1
27 Fishing area total	*572*	*702*	*884*	*665*	*587*	*624*	*447*	*447*	*441*	*414*
31 USA	-	-	0	112	82	14	6	1	-	-
31 Fishing area total	*-*	*-*	*0*	*112*	*82*	*14*	*6*	*1*	*-*	*-*
34 Greece	0	0	0	0	-	-	-	-	-	-
Portugal	92	161	205	78	184	63	43	26	2	4
Spain	-	-	-	-	49	12	31	27	2	9
34 Fishing area total	*92*	*161*	*205*	*78*	*233*	*75*	*74*	*53*	*4*	*13*
37 Albania	...	...	...	0	1	0	-	-	-	-
France	...	...	...	...	...	...	...	22	22	-
Malta	23	29	16	8	9	14	8	8	8	8
Spain	-	-	-	-	2	3	7	7	5	4
37 Fishing area total	*23*	*29*	*16*	*8*	*12*	*17*	*15*	*37*	*35*	*12*
41 Argentina	...	...	...	...	...	...	...	...	129	...
Spain	-	-	-	-	-	-	-	-	-	35
41 Fishing area total	*...*	*...*	*...*	*...*	*...*	*...*	*...*	*...*	*129*	*35*
47 Portugal	-	-	-	-	-	6	42	20	8	-
47 Fishing area total	*-*	*-*	*-*	*-*	*-*	*6*	*42*	*20*	*8*	*-*
51 Spain	-	-	-	-	-	-	-	-	-	1
51 Fishing area total	*-*	*-*	*-*	*-*	*-*	*-*	*-*	*-*	*-*	*1*
Species total	*687*	*892*	*1 105*	*863*	*914*	*736*	*584*	*558*	*617*	*475*
Hapuku wreckfish	**Cernier de Nouvelle Zélande**		**Cherna hapuku**			***Polyprion oxygeneios***			**1,70(05)058,02**	**WHA**
81 New Zealand	1 263	1 439	1 448	1 536	1 155	1 657	1 571	1 547	1 497	1 579
81 Fishing area total	*1 263*	*1 439*	*1 448*	*1 536*	*1 155*	*1 657*	*1 571*	*1 547*	*1 497*	*1 579*
87 Chile	51	42	37	33	23	30	26	8	7	10
87 Fishing area total	*51*	*42*	*37*	*33*	*23*	*30*	*26*	*8*	*7*	*10*
Species total	*1 314*	*1 481*	*1 485*	*1 569*	*1 178*	*1 687*	*1 597*	*1 555*	*1 504*	*1 589*
Ocean whitefish	**Tile fin**		**Blanquillo fino**			***Caulolatilus princeps***			**1,70(16)001,03**	**ULP**
77 Mexico	...	...	...	...	...	...	535	622	1 073	979
USA	-	-	-	-	-	-	-	5	4	5
77 Fishing area total	*...*	*...*	*...*	*...*	*...*	*...*	*535*	*627*	*1 077*	*984*
Species total	*...*	*...*	*...*	*...*	*...*	*...*	*535*	*627*	*1 077*	*984*
Atlantic goldeneye tilefish	**Tile oeil doré**		**Blanquillo ojo amarillo**			***Caulolatilus chrysops***			**1,70(16)001,04**	**CKZ**
21 USA	-	-	-	-	-	-	-	0	2	790
21 Fishing area total	*-*	*-*	*-*	*-*	*-*	*-*	*-*	*0*	*2*	*790*
31 USA	-	-	-	-	-	-	-	0	4	16
31 Fishing area total	*-*	*-*	*-*	*-*	*-*	*-*	*-*	*0*	*4*	*16*
Species total	*-*	*-*	*-*	*-*	*-*	*-*	*-*	*0*	*6*	*806*
Great Northern tilefish	**Tile chameau**		**Blanquillo camello**			***Lopholatilus chamaeleonticeps***			**1,70(16)400,02**	**TIL**
21 USA	1 683	1 873	783	673	1 349	1 493	1 339	528	515	117
21 Fishing area total	*1 683*	*1 873*	*783*	*673*	*1 349*	*1 493*	*1 339*	*528*	*515*	*117*
31 USA	745	800	735	611	114	397	80	88	173	168
31 Fishing area total	*745*	*800*	*735*	*611*	*114*	*397*	*80*	*88*	*173*	*168*

B-34 Miscellaneous demersal fishes / Poissons démersaux divers / Peces demersales diversos

Capture production by species, fishing areas and countries or areas / Captures par espèces, zones de pêche et pays ou zones / Capturas por especies, áreas de pesca y países o áreas

Species, Fishing area Espèce, Zone de pêche Especie, Area de pesca	1992 mt	1993 mt	1994 mt	1995 mt	1996 mt	1997 mt	1998 mt	1999 mt	2000 mt	2001 mt
Species total	*2 428*	*2 673*	*1 518*	*1 284*	*1 463*	*1 890*	*1 419*	*616*	*688*	*285*
Tilefishes nei	**Tiles nca**		**Blanquillos, paletas nep**		***Branchiostegidae***			**1,70(16)XXX,XX**		**TIS**
31 Mexico	...	...	...	...	...	...	53	68	45	28
USA	-	-	-	-	-	28	294	318	483	314
31 Fishing area total	...	...	...	...	...	*28*	*347*	*386*	*528*	*342*
34 Côte dIvoire	30	22	14	14 F	20 F	30 F	40 F	45	44	15
34 Fishing area total	*30*	*22*	*14*	*14 F*	*20 F*	*30 F*	*40 F*	*45*	*44*	*15*
41 Brazil	340 F	340 F	340 F	812	1 098	1 000	786	524	547	540 F
41 Fishing area total	*340 F*	*340 F*	*340 F*	*812*	*1 098*	*1 000*	*786*	*524*	*547*	*540 F*
61 China,H.Kong	2 311	1 642	2 548	2 583	3 186	4 187	4 879	3 500 F	4 300 F	4 750 F
China,Taiwan	2 025	1 879	1 299	579	1 227	626	372	496	448	512
Japan	...	...	...	4 194	3 648	2 994	2 284	1 949	1 678	1 781
Korea Rep	-	-	-	-	-	-	-	1 651	1 664	1 049
61 Fishing area total	*4 336*	*3 521*	*3 847*	*7 356*	*8 061*	*7 807*	*7 535*	*7 596 F*	*8 090 F*	*8 092 F*
87 Chile	576	383	195	252	117	28	80	134	155	53
Peru	433	736	738	1 544	892	292	119	146	117	1 485
87 Fishing area total	*1 009*	*1 119*	*933*	*1 796*	*1 009*	*320*	*199*	*280*	*272*	*1 538*
Species total	*5 715 F*	*5 002 F*	*5 134 F*	*9 978 F*	*10 188 F*	*9 185 F*	*8 907 F*	*8 831 F*	*9 481 F*	*10 527 F*
Cape bonnetmouth	**Andorrève du Cap**		**Andorrero del Cabo**		***Emmelichthys nitidus***			**1,70(30)010,01**		**EMM**
47 Russian Fed	-	-	-	-	-	70	7	-	-	-
South Africa	418	434	121	96	216	121	117	113	50 F	37
47 Fishing area total	*418*	*434*	*121*	*96*	*216*	*191*	*124*	*113*	*50 F*	*37*
81 New Zealand	...	...	...	2 392	1 845	1 635	2 064	2 846	2 825	1 881
81 Fishing area total	...	...	...	*2 392*	*1 845*	*1 635*	*2 064*	*2 846*	*2 825*	*1 881*
Species total	*418*	*434*	*121*	*2 488*	*2 061*	*1 826*	*2 188*	*2 959*	*2 875 F*	*1 918*
Bonnetmouths, rubyfishes nei	**Andorrèves, poissons rubis nca**		**Andorreros, peces rubí nep**		***Emmelichthyidae***			**1,70(30)XXX,XX**		**EMT**
47 Russian Fed	-	-	-	-	-	-	-	-	-	6
47 Fishing area total	-	-	-	-	-	-	-	-	-	*6*
51 Ukraine	468	551	227	144	28	7	275	181	-	86
51 Fishing area total	*468*	*551*	*227*	*144*	*28*	*7*	*275*	*181*	-	*86*
81 New Zealand	...	...	...	616	596	431	378	271	582	434
Russian Fed	-	29	-	-	-	-	-	-	-	-
Ukraine	-	-	-	501	-	-	-	-	-	-
81 Fishing area total	...	*29*	...	*1 117*	*596*	*431*	*378*	*271*	*582*	*434*
Species total	*468*	*580*	*227*	*1 261*	*624*	*438*	*653*	*452*	*582*	*526*
Tripletail	**...B**		**...C**		***Lobotes surinamensis***			**1,70(34)029,01**		**LOB**
31 USA	-	-	-	-	-	-	-	-	1	1
31 Fishing area total	-	-	-	-	-	-	-	-	*1*	*1*
Species total	-	-	-	-	-	-	-	-	*1*	*1*
Giant boarfish	**...B**		**...C**		***Paristiopterus labiosus***			**1,70(57)004,02**		**SWH**
81 New Zealand	-	-	-	21	19	27	75	6	9	3
81 Fishing area total	-	-	-	*21*	*19*	*27*	*75*	*6*	*9*	*3*
Species total	-	-	-	*21*	*19*	*27*	*75*	*6*	*9*	*3*
Pelagic armourhead	**Tête casquée pélagique**		**...C**		***Pseudopentaceros richardsoni***			**1,70(57)007,01**		**EDR**
47 Ukraine	-	435	-	49	281	18	-	-	-	-
47 Fishing area total	-	*435*	-	*49*	*281*	*18*	-	-	-	-
51 China	-	-	-	-	-	-	-	-	44	-
Ukraine	45	...	40	54	17	33	78	108	77	12
51 Fishing area total	*45*	...	*40*	*54*	*17*	*33*	*78*	*108*	*121*	*12*
81 New Zealand	-	-	-	3	7	2	78	13	6	7
81 Fishing area total	-	-	-	*3*	*7*	*2*	*78*	*13*	*6*	*7*
Species total	*45*	*435*	*40*	*106*	*305*	*53*	*156*	*121*	*127*	*19*
Castaneta	**Castanette pontude**		**Castañeta**		***Cheilodactylus bergi***			**1,70(70)270,02**		**CTA**
41 Argentina	55	1 346	20 383	10 409	204	744	1 827	155	80	49
Uruguay	1	75	4 289	3 034	2 938	4 149	9 681	3 105	1 351	1 269
41 Fishing area total	*56*	*1 421*	*24 672*	*13 443*	*3 142*	*4 893*	*11 508*	*3 260*	*1 431*	*1 318*
Species total	*56*	*1 421*	*24 672*	*13 443*	*3 142*	*4 893*	*11 508*	*3 260*	*1 431*	*1 318*

B-34 Miscellaneous demersal fishes / Poissons démersaux divers / Peces demersales diversos

Capture production by species, fishing areas and countries or areas
Captures par espèces, zones de pêche et pays ou zones
Capturas por especies, áreas de pesca y países o áreas

Species, Fishing area Espèce, Zone de pêche Especie, Area de pesca	1992 mt	1993 mt	1994 mt	1995 mt	1996 mt	1997 mt	1998 mt	1999 mt	2000 mt	2001 mt
Peruvian morwong	**Castanette pintadille**		**Pintadilla**			***Cheilodactylus variegatus***			**1,70(70)270,03**	**HAW**
87 Chile	-	-	-	-	-	51	50	52	45	46
Peru	141	111	95	93	283	411	90	236	335	260
87 Fishing area total	*141*	*111*	*95*	*93*	*283*	*462*	*140*	*288*	*380*	*306*
Species total	*141*	*111*	*95*	*93*	*283*	*462*	*140*	*288*	*380*	*306*
Tarakihi	**...B**		**...C**			***Nemadactylus macropterus***			**1,70(70)305,04**	**TAK**
81 Japan	191	223	207	41	-	-	-	-	-	-
New Zealand	5 261	4 847	4 771	5 108	4 366	5 441	5 239	5 589	5 739	6 129
81 Fishing area total	*5 452*	*5 070*	*4 978*	*5 149*	*4 366*	*5 441*	*5 239*	*5 589*	*5 739*	*6 129*
Species total	*5 452*	*5 070*	*4 978*	*5 149*	*4 366*	*5 441*	*5 239*	*5 589*	*5 739*	*6 129*
Morwongs	**...B**		**...C**			***Nemadactylus spp***			**1,70(70)305,XX**	**MOW**
57 Australia	57	48	74	75	77	94	95	638	630	775
57 Fishing area total	*57*	*48*	*74*	*75*	*77*	*94*	*95*	*638*	*630*	*775*
81 Australia	1 206	1 146	1 199	1 225	1 108	1 419	1 083	696	511	418
New Zealand	119	127	94	120	92	107	109	78	99	92
81 Fishing area total	*1 325*	*1 273*	*1 293*	*1 345*	*1 200*	*1 526*	*1 192*	*774*	*610*	*510*
Species total	*1 382*	*1 321*	*1 367*	*1 420*	*1 277*	*1 620*	*1 287*	*1 412*	*1 240*	*1 285*
Trumpeters nei	**...B**		**Tromperos nep**		***Latridae***				**1,70(71)XXX,XX**	**TRU**
57 Australia	0	1	0	0	0	0	0	153	166	69
57 Fishing area total	*0*	*1*	*0*	*0*	*0*	*0*	*0*	*153*	*166*	*69*
81 Australia	10	7	16	16	20	15	34	8	9	13
New Zealand	393	548	512	669	727	761	567	574	476	505
81 Fishing area total	*403*	*555*	*528*	*685*	*747*	*776*	*601*	*582*	*485*	*518*
Species total	*403*	*556*	*528*	*685*	*747*	*776*	*601*	*735*	*651*	*587*
Antarctic toothfish	**Légine antarctique**		**Austromerluza antártica**			***Dissostichus mawsoni***			**1,70(92)015,01**	**TOA**
48 Chile	-	-	-	-	-	-	1	-	-	-
USA	-	-	-	-	-	-	-	0	-	0
48 Fishing area total	-	-	-	-	-	-	*1*	*0*	-	*0*
88 New Zealand	-	-	-	-	-	-	41	296	751	582
South Africa	-	-	-	-	-	-	-	-	-	21
Uruguay	-	-	-	-	-	-	-	-	-	23
88 Fishing area total	-	-	-	-	-	-	*41*	*296*	*751*	*626*
Species total	-	-	-	-	-	-	*42*	*296*	*751*	*626*
Patagonian toothfish	**Légine australe**		**Austromerluza negra**			***Dissostichus eleginoides***			**1,70(92)015,02**	**TOP**
41 Argentina	525	3 651	10 840	19 180	14 811	8 793	9 950	7 692	7 771	6 408
Australia	-	-	-	-	-	-	15	24	-	-
Belize	-	-	-	-	-	-	-	16	27	11
Chile	-	-	-	-	-	-	-	-	-	831
Falkland Is	1	0	18	34	50	178	570	1 113	927	1 460
France	-	-	0	1	3	0	3	4	0	-
Honduras	22	30	5	26	7	-	-	-	-	-
Korea Rep	-	-	-	-	-	514	1 051	933	1 292	686
Namibia	-	-	-	-	-	2	21	28	-	-
Panama	22	7	3	9	-	-	-	-	-	-
Portugal	3	1	-	-	-	-	-	-	3	-
Russian Fed	93	-	-	-	-	-	-	-	-	1
Seychelles	-	-	-	-	-	1	-	-	-	
Spain	626	259	147	191	79	109	355	572	538	277
UK	-	2	1	1	1	2	18	30	6	3
Uruguay	-	-	-	-	-	-	1 345	888	558	336
41 Fishing area total	*1 292*	*3 950*	*11 014*	*19 442*	*14 951*	*9 599*	*13 328*	*11 300*	*11 122*	*10 013*
47 Chile	-	-	-	-	-	-	-	-	-	5
Uruguay	-	-	-	-	-	-	-	-	320	-
47 Fishing area total	-	-	-	-	-	-	-	-	*320*	*5*
48 Argentina	-	-	0	816	101	-	-	10	-	-
Bulgaria	115	223	70	179	-	-	-	-	-	-
Chile	2 920	2 125	151	1 876	3 065	1 275	1 478	1 668	1 609	531
Japan	-	-	-	-	-	-	76	-	-	-
Korea Rep	-	-	135	381	366	425	170	255	380	467
Russian Fed	307	283	152	10	103	-	-	-	-	89
South Africa	-	-	-	-	-	-	487	449	324	227
Spain	-	-	-	-	-	291	197	154	264	487
Ukraine	407	458		-	-				128	99
UK	1	-	1	-	-	398	589	1 238	1 221	988
USA	-	-	-	-	187	-	-	-	-	
Uruguay	-	-	-	-	-	-	262	517	767	460
48 Fishing area total	*3 750*	*3 089*	*509*	*3 262*	*3 822*	*2 389*	*3 259*	*4 291*	*4 693*	*3 348*

B-34 Miscellaneous demersal fishes / Poissons démersaux divers / Peces demersales diversos

Capture production by species, fishing areas and countries or areas / Captures par espèces, zones de pêche et pays ou zones / Capturas por especies, áreas de pesca y países o áreas

Species, Fishing area Espèce, Zone de pêche Especie, Area de pesca	1992 mt	1993 mt	1994 mt	1995 mt	1996 mt	1997 mt	1998 mt	1999 mt	2000 mt	2001 mt
51 Uruguay	-	-	-	-	-	-	-		1 628	7 002
51 Fishing area total	-	-	-	-	-	-	-	-	*1 628*	*7 002*
58 Australia	0	-	0	-	-	860	2 417	5 451	2 579	1 765
France	1 589	826	4 197	4 089	3 652	3 675	3 832	6 277	5 503	6 634
Japan	-	-	-	-	264	335	-	-	-	-
Russian Fed	1 258	-	-	-	-	-	-	-	-	-
South Africa	-	-	-	-	-	2 106	663	499	914	789
Ukraine	5 903	1 874	942	1 560	1 003	1 007	997	760	-	65
Uruguay	-	-	-	-	-	-	-	-	-	99
58 Fishing area total	*8 750*	*2 700*	*5 139*	*5 649*	*4 919*	*7 983*	*7 909*	*12 987*	*8 996*	*9 352*
81 New Zealand	...	...	...	...	1 061	5	43	1	0	14
81 Fishing area total	...	...	...	...	*1 061*	*5*	*43*	*1*	*0*	*14*
87 Chile	26 918	20 997	20 902	15 694	6 993	8 059	9 172	10 328	10 676	6 568
Spain	-	-	-	-	-	-	-	-	-	11
87 Fishing area total	*26 918*	*20 997*	*20 902*	*15 694*	*6 993*	*8 059*	*9 172*	*10 328*	*10 676*	*6 579*
88 New Zealand	...	...	...	...	-	0	1	1	0	30
South Africa	-	-	-	-	-	-	-	-	-	4
Uruguay	-	-	-	-	-	-	-	-	-	0
88 Fishing area total	...	...	...	...	-	*0*	*1*	*1*	*0*	*34*
Species total	*40 710*	*30 736*	*37 564*	*44 047*	*31 746*	*28 035*	*33 712*	*38 908*	*37 435*	*36 347*
Antarctic toothfishes nei	**Légines antarctiques nca**		**Austromerluzas nep**		***Dissostichus spp***				**1,70(92)015,XX**	**TOT**
51 Korea Rep	-	-	-	-	-	-	-	-	-	122
51 Fishing area total	-	-	-	-	-	-	-	-	-	*122*
57 Korea Rep	-	-	-	-	-	-	-	-	-	450
57 Fishing area total	-	-	-	-	-	-	-	-	-	*450*
Species total	-	-	-	-	-	-	-	-	-	*572*
Patagonian rockcod	**Bocasse de Patagonie**		**Trama patagónica**		***Patagonotothen brevicauda***				**1,70(92)440,01**	**NOT**
41 UK	-	-	-	-	-	-	13	-	-	-
41 Fishing area total	-	-	-	-	-	-	*13*	-	-	-
48 Argentina	-	-	0	1	-	-	-	-		-
Russian Fed	-	-	-	-	-	-	-	3	0	-
UK	-	-	1	-	-	-	-	-	0	-
48 Fishing area total	-	-	*1*	*1*	-	-	-	*3*	*0*	-
Species total	-	-	*1*	*1*	-	-	*13*	*3*	*0*	-
Longtail Southern cod	**Notothénia queue longue**		**Nototenia coluda**		***Patagonotothen ramsayi***				**1,70(92)440,02**	**PAT**
41 Russian Fed	49	89	115	-	-	-	-	-	-	-
Ukraine	104	-	-	-	-	-	-	-	-	-
41 Fishing area total	*153*	*89*	*115*	-	-	-	-	-	-	-
Species total	*153*	*89*	*115*	-	-	-	-	-	-	-
Blackfin icefish	**Grande-gueule antarctique**		**Draco antártico**		***Chaenocephalus aceratus***				**1,70(94)416,01**	**SSI**
48 UK	2	-	2	-	-	-	-	-	0	-
USA	-	-	-	-	-	-	-	1	-	1
48 Fishing area total	*2*	-	*2*	-	-	-	-	*1*	*0*	*1*
Species total	*2*	-	*2*	-	-	-	-	*1*	*0*	*1*
Mackerel icefish	**Poisson des glaces antarctique**		**Draco rayado**		***Champsocephalus gunnari***				**1,70(94)417,01**	**ANI**
48 Argentina	-	-	10	10	-	-	-	-	-	-
Chile	-	-	-	-	-	-	6	-	715	365
France	-	-	-	-	-	-	-	-	-	386
Russian Fed	-	-	-	-	-	-	-	265	3 395	-
UK	5	-	3	-	-	-	-	-	4	208
USA	-	-	-	-	-	-	-	1	-	1
48 Fishing area total	*5*	-	*13*	*10*	-	-	*6*	*266*	*4 114*	*960*
58 Australia	2	-	3	-	-	217	67	73	81	930
France	0	-	12	84	5	0	-	-	-	-
Russian Fed	13	-	-	-	-	-	-	-	-	-
Ukraine	44	-	-	3 852	-	-	-	-	-	-
58 Fishing area total	*59*	-	*15*	*3 936*	*5*	*217*	*67*	*73*	*81*	*930*
Species total	*64*	-	*28*	*3 946*	*5*	*217*	*73*	*339*	*4 195*	*1 890*
South Georgia icefish	**Crocodile de Géorgie**		**Draco cocodrilo**		***Pseudochaenichthys georgianus***				**1,70(94)418,01**	**SGI**
48 France	-	-	-	-	-	-	-	-	-	0
UK	2	-	1	-	-	-	-	-	0	6
USA	-	-	-	-	-	-	-	3	-	0
48 Fishing area total	*2*	-	*1*	-	-	-	-	*3*	*0*	*6*

B-34 Miscellaneous demersal fishes / Poissons démersaux divers / Peces demersales diversos

Capture production by species, fishing areas and countries or areas / Captures par espèces, zones de pêche et pays ou zones / Capturas por especies, áreas de pesca y países o áreas

Species, Fishing area Espèce, Zone de pêche Especie, Area de pesca	1992 mt	1993 mt	1994 mt	1995 mt	1996 mt	1997 mt	1998 mt	1999 mt	2000 mt	2001 mt
Species total	*2*	-	*1*	-		-	-	*3*	*0*	*6*
Ocellated icefish	**Grande-gueule ocellée**		**Draco ocelado**		***Chionodraco rastrospinosus***			**1,70(94)466,01**		**KIF**
48 USA	-	-	-	-	-	-	-	1	-	1
48 Fishing area total	-	-	-	-	-	-	-	*1*	-	*1*
58 Australia	-	-	-	-	-	1	-	-	-	-
58 Fishing area total	-	-	-	-	-	*1*	-	-	-	-
Species total	-	-	-	-	-	*1*	-	*1*	-	*1*
Unicorn icefish	**Grande-gueule à long nez**		**Draco rinoceronte**		***Channichthys rhinoceratus***			**1,70(94)470,01**		**LIC**
58 Australia	0	-	1	-	-	2	5	2	2	1
France	-	-	-	-	1	5	1	1	-	-
58 Fishing area total	*0*	-	*1*	-	*1*	*7*	*6*	*3*	*2*	*1*
Species total	*0*	-	*1*	-	*1*	*7*	*6*	*3*	*2*	*1*
Spiny icefish	**Grande-gueule épineuse**		**Draco espinudo**		***Chaenodraco wilsoni***			**1,70(94)480,01**		**WIC**
48 Latvia	-	71	-	-	-	-	-	-	-	-
USA	-	-	-	-	-	-	-	0	-	0
48 Fishing area total	-	*71*	-	-	-	-	-	*0*	-	*0*
58 Australia	-	-	-	-	-	-	-	-	-	11
58 Fishing area total	-	-	-	-	-	-	-	-	-	*11*
Species total	-	*71*	-	-	-	-	-	*0*	-	*11*
Icefishes nei	**Poissons des glaces nca**		**Dracos nep**		***Channichthyidae***			**1,70(94)XXX,XX**		**ICX**
48 France	-	-	-	-	-	-	-	-	-	0
USA	-	-	-	-	-	-	-	0	-	0
48 Fishing area total	-	-	-	-	-	-	-	*0*	-	*0*
88 New Zealand	-	-	-	-	-	-	0	0	0	2
88 Fishing area total	-	-	-	-	-	-	*0*	*0*	*0*	*2*
Species total	-	-	-	-	-	-	*0*	*0*	*0*	*2*
Black cardinal fish	**Poisson cardinal**		**Boca negra(Pez del diablo)**		***Epigonus telescopus***			**1,70(96)373,01**		**EPI**
27 Faeroe Is	-	41	45	38	31	129	94	4	...	...
France	...	26	231	95	52	52	232	...	197	153
Germany	-	-	-	-	-	-	-	-	50	-
UK	-	-	-	-	-	-	-	-	1	22
27 Fishing area total	*...*	*67*	*276*	*133*	*83*	*181*	*326*	*4*	*248*	*175*
81 New Zealand	1 803	2 049	4 291	3 650	3 002	4 334	2 568	2 869	4 095	1 957
81 Fishing area total	*1 803*	*2 049*	*4 291*	*3 650*	*3 002*	*4 334*	*2 568*	*2 869*	*4 095*	*1 957*
Species total	*1 803*	*2 116*	*4 567*	*3 783*	*3 085*	*4 515*	*2 894*	*2 873*	*4 343*	*2 132*
Cardinal fishes nei	**Poissons-cardinaux nca**		**Peces cardenal nep**		***Epigonus spp***			**1,70(96)373,XX**		**CDL**
87 Chile	579	862	137	232	513	1 727	5 284	2 999	5 792	4 648
87 Fishing area total	*579*	*862*	*137*	*232*	*513*	*1 727*	*5 284*	*2 999*	*5 792*	*4 648*
Species total	*579*	*862*	*137*	*232*	*513*	*1 727*	*5 284*	*2 999*	*5 792*	*4 648*
Atlantic wolffish	**Loup atlantique**		**Perro del Norte**		***Anarhichas lupus***			**1,71(02)001,01**		**CAA**
21 Spain	-	-	8	116	7	23	7	-	2	7
21 Fishing area total	-	-	*8*	*116*	*7*	*23*	*7*	-	*2*	*7*
27 Belgium	319	180	161	206	99	125	208	201	290	175
Denmark	955	569	392	248	195	220	273	298	294	223
Faeroe Is	313	176	132	141	146	196	264	291	...	...
France	2	8	3	4	3	1	2	14	9	7
Germany	285	145	139	176	94	44	88	67	86	66
Greenland	4	4	28	6	15	6	42	7	12	16
Iceland	16 024	12 945	12 766	12 574	14 638	11 685	11 844	13 769	15 043	17 953
Ireland	9	30	45	42	39	22	39	35	66	27
Japan	9	7	5	5	-	-	-	-	-	-
Latvia	1	-	-	-	-	-	-	-	-	-
Norway	...	...	...	...	...	...	...	...	...	1 111
Poland	-	-	-	-	-	19	40	6	18	8
Russian Fed	2 947	4 300	3 944	6 280	10 523	18 247	23 730	23 756	22 756	19 087
Spain	0	134	140	174	-	20	37	20	66	98
Sweden	295	262	173	146	117	174	157	163	154	95
UK	605	571	454	277	1 128	199	893	928	917	140
27 Fishing area total	*21 768*	*19 331*	*18 382*	*20 279*	*26 997*	*30 958*	*37 617*	*39 555*	*39 711*	*39 006*
61 Russian Fed	-	-	-	-	-	-	-	38	36	33
61 Fishing area total	-	-	-	-	-	-	-	*38*	*36*	*33*
Species total	*21 768*	*19 331*	*18 390*	*20 395*	*27 004*	*30 981*	*37 624*	*39 593*	*39 749*	*39 046*

B-34 Miscellaneous demersal fishes / Poissons démersaux divers / Peces demersales diversos

Capture production by species, fishing areas and countries or areas / Captures par espèces, zones de pêche et pays ou zones / Capturas por especies, áreas de pesca y países o áreas

Species, Fishing area Espèce, Zone de pêche Especie, Area de pesca	1992 mt	1993 mt	1994 mt	1995 mt	1996 mt	1997 mt	1998 mt	1999 mt	2000 mt	2001 mt
Spotted wolffish	**Loup tacheté**		**Perro pintado**		***Anarhichas minor***				**1,71(02)001,03**	**CAS**
27 Iceland	851	1 244	916	700	1 109	1 180	1 599	1 545	1 896	2 126
Norway	...	...	...	...	...	...	...	...	...	1 111
Portugal	...	...	...	...	...	...	...	...	...	20
27 Fishing area total	*851*	*1 244*	*916*	*700*	*1 109*	*1 180*	*1 599*	*1 545*	*1 896*	*3 257*
Species total	*851*	*1 244*	*916*	*700*	*1 109*	*1 180*	*1 599*	*1 545*	*1 896*	*3 257*
Wolffishes(=Catfishes) nei	**Loups nca**		**Perritos del Norte nep**		***Anarhichas spp***				**1,71(02)001,XX**	**CAT**
21 Canada	1 289	1 073	485	303	422	856	526	694	678	578
Cuba	1	0	-	-	-	-	-	-	-	-
Estonia	-	-	-	-	-	-	-	-	6	5
Faeroe Is	-	5	-	1	4	-	-	-	-	-
Germany	1	-	-	-	-	-	-	-	-	-
Greenland	190	156	101	50	47	67	30	26	47	58
Iceland	-	-	2	-	-	-	-	-	-	-
Japan	2	-	1	33	20	17	26	21	15	53
Norway	6	16	-	0	-	-	-	-	-	1
Portugal	1 697	2 302	3 219	1 358	123	185	141	549	61	141
Russian Fed	58	14	-	57	-	-	38	-	7	23
St Pier Mq	27	-	-	-	-	3	3	2	0	1
Spain	-	-	184	195	695	535	473	435	518	785
UK	-	-	-	-	0	-	-	-	-	-
USA	464	506	479	464	363	309	296	258	200	250
21 Fishing area total	*3 735*	*4 072*	*4 471*	*2 461*	*1 674*	*1 972*	*1 533*	*1 985*	*1 532*	*1 895*
27 Faeroe Is	32	15	49	37	27	64	60	118	154	155
Netherlands	101	71	58	50	6	16	36	21	10	2
Norway	3 651	3 143	6 505	7 589	6 819	12 769	16 332	6 398	6 378	12 204
Portugal	6	530	594	502	169	225	476	96	168	163
UK	1 400	1 242	1 007	833	...	...	...	...	...	600
27 Fishing area total	*5 190*	*5 001*	*8 213*	*9 011*	*7 021*	*13 074*	*16 904*	*6 633*	*6 710*	*13 124*
Species total	*8 925*	*9 073*	*12 684*	*11 472*	*8 695*	*15 046*	*18 437*	*8 618*	*8 242*	*15 019*
Eelpouts	**Loquettes**		**Viruelas**		***Lycodes spp***				**1,71(15)024,XX**	**ELZ**
27 Russian Fed	-	-	-	-	-	-	-	-	-	1
27 Fishing area total	-	-	-	-	-	-	-	-	-	*1*
61 Russian Fed	-	380	5	45	18	-	2	1	28	47
61 Fishing area total	-	*380*	*5*	*45*	*18*	-	*2*	*1*	*28*	*47*
Species total	-	*380*	*5*	*45*	*18*	-	*2*	*1*	*28*	*48*
Stargazer	**Uranoscope**		**Rata**		***Uranoscopus scaber***				**1,72(13)352,02**	**UUC**
27 Portugal	...	...	...	...	...	...	...	15	50	46
27 Fishing area total	...	...	...	...	...	...	...	*15*	*50*	*46*
Species total	...	...	...	...	...	...	...	*15*	*50*	*46*
Giant stargazer	**Uranoscope géant**		**Miracielo gigante**		***Kathetostoma giganteum***				**1,72(13)484,01**	**STZ**
81 New Zealand	2 801	2 856	99	9 597	2 122	3 990	2 195	3 370	3 638	4 233
Norway	11	10	10	-	1	1	1	-	-	-
81 Fishing area total	*2 812*	*2 866*	*109*	*9 597*	*2 123*	*3 991*	*2 196*	*3 370*	*3 638*	*4 233*
Species total	*2 812*	*2 866*	*109*	*9 597*	*2 123*	*3 991*	*2 196*	*3 370*	*3 638*	*4 233*
Japanese sandfish	**...B**		**...C**		***Arctoscopus japonicus***				**1,72(72)103,01**	**JAS**
61 Japan	5 552	5 796	5 877	5 506	6 719	6 209	6 795	6 615	6 652	8 753
Korea Rep	4 202	3 731	1 466	2 065	2 501	2 194	1 490	2 449	1 571	1 286
61 Fishing area total	*9 754*	*9 527*	*7 343*	*7 571*	*9 220*	*8 403*	*8 285*	*9 064*	*8 223*	*10 039*
Species total	*9 754*	*9 527*	*7 343*	*7 571*	*9 220*	*8 403*	*8 285*	*9 064*	*8 223*	*10 039*
Snoek	**Escolier**		**Sierra**		***Thyrsites atun***				**1,75(05)001,01**	**SNK**
34 Russian Fed	-	-	26	-	-	-	-	-	-	-
34 Fishing area total	-	-	*26*	-	-	-	-	-	-	-
47 Iceland	-	-	-	-	-	1	-	-	-	-
Lithuania	430	-	-	-	-	-	-	-	-	-
Namibia	816	939	683	856	622	895	701	1 212	966	1 699
Russian Fed	-	5	103	-	15	11	-	-	-	-
South Africa	18 417	15 248	13 180	15 299	12 400	11 126	14 135	11 188	10 200 F	10 328
Spain	80	-	-	-	-	-	-	-	-	-
Ukraine	-	-	-	-	-	-	-	2	-	-
47 Fishing area total	*19 743*	*16 192*	*13 966*	*16 155*	*13 037*	*12 033*	*14 836*	*12 402*	*11 166 F*	*12 027*
57 Australia	371	370 F	370 F	400 F	400 F	300	200	88	120	154
57 Fishing area total	*371*	*370 F*	*370 F*	*400 F*	*400 F*	*300*	*200*	*88*	*120*	*154*
81 Australia	500 F	400 F	400 F	450 F	450 F	...	...	14	20	29

B-34 Miscellaneous demersal fishes / Poissons démersaux divers / Peces demersales diversos

Capture production by species, fishing areas and countries or areas / Captures par espèces, zones de pêche et pays ou zones / Capturas por especies, áreas de pesca y países o áreas

Species, Fishing area Espèce, Zone de pêche Especie, Area de pesca	1992 mt	1993 mt	1994 mt	1995 mt	1996 mt	1997 mt	1998 mt	1999 mt	2000 mt	2001 mt
Japan	2 354	2 502	1 582	588	28	12	3	23	59	189
New Zealand	16 197	22 348	14 624	18 428	15 849	22 047	25 972	20 642	21 905	25 222
Russian Fed	-	1 963	91	-	-	-	-	-	-	-
Ukraine	931	...	...	1 748	2 244	3 898	3 881	7 920	6 113	2 970
81 Fishing area total	*19 982 F*	*27 213 F*	*16 697 F*	*21 214 F*	*18 571 F*	*25 957*	*29 856*	*28 599*	*28 097*	*28 410*
87 Chile	582	427	572	687	821	1 337	1 022	604	851	830
87 Fishing area total	*582*	*427*	*572*	*687*	*821*	*1 337*	*1 022*	*604*	*851*	*830*
Species total	*40 678 F*	*44 202 F*	*31 631 F*	*38 456 F*	*32 829 F*	*39 627*	*45 914*	*41 693*	*40 234 F*	*41 421*
White snake mackerel	**Escolier blanc**		**Escolar sierra**			***Thyrsitops lepidopoides***			**1,75(05)002,01**	**WSM**
41 Argentina	32	32	32	66	1 232	309	285	136	-	-
41 Fishing area total	*32*	*32*	*32*	*66*	*1 232*	*309*	*285*	*136*	-	-
Species total	*32*	*32*	*32*	*66*	*1 232*	*309*	*285*	*136*	-	-
Escolar	**Escolier noir**		**Escolar negro**			***Lepidocybium flavobrunneum***			**1,75(05)005,01**	**LEC**
21 USA	...	...	...	...	...	...	1	1	2	4
21 Fishing area total	...	...	...	...	...	...	*1*	*1*	*2*	*4*
31 USA	...	...	...	...	...	...	51	33	70	40
31 Fishing area total	...	...	...	...	...	...	*51*	*33*	*70*	*40*
77 USA	...	...	...	...	...	...	2	1	6	3
77 Fishing area total	...	...	...	...	...	...	*2*	*1*	*6*	*3*
81 New Zealand	-	-	-	0	0	1	1	22	47	87
81 Fishing area total	-	-	-	*0*	*0*	*1*	*1*	*22*	*47*	*87*
Species total	...	...	...	*0*	*0*	*1*	*55*	*57*	*125*	*134*
Oilfish	**Rouvet**		**Escolar clavo**			***Ruvettus pretiosus***			**1,75(05)007,01**	**OIL**
27 Portugal	-	-	-	-	-	-	-	14	9	9
27 Fishing area total	-	-	-	-	-	-	-	*14*	*9*	*9*
31 USA	...	...	...	...	...	...	10	11	38	30
31 Fishing area total	...	...	...	...	...	...	*10*	*11*	*38*	*30*
34 Portugal	-	-	-	-	-	-	-	-	5	2
34 Fishing area total	-	-	-	-	-	-	-	-	*5*	*2*
47 Norway	-	-	-	-	-	5	-	-	-	-
47 Fishing area total	-	-	-	-	-	*5*	-	-	-	-
61 China,Taiwan	4 351	2 341	2 094	3 729	2 634	2 622	3 043	2 661	2 584	3 678
61 Fishing area total	*4 351*	*2 341*	*2 094*	*3 729*	*2 634*	*2 622*	*3 043*	*2 661*	*2 584*	*3 678*
81 New Zealand	-	-	-	38	53	34	57	53	61	86
81 Fishing area total	-	-	-	*38*	*53*	*34*	*57*	*53*	*61*	*86*
Species total	*4 351*	*2 341*	*2 094*	*3 767*	*2 687*	*2 661*	*3 110*	*2 739*	*2 697*	*3 805*
Silver gemfish	**Escolier tifiati**		**Escolar plateado**			***Rexea solandri***			**1,75(05)009,01**	**GEM**
57 Australia	3 000 F	3 000 F	3 000 F	3 000 F	2 000 F	...	...	4	2	...
57 Fishing area total	*3 000 F*	*3 000 F*	*3 000 F*	*3 000 F*	*2 000 F*	...	...	*4*	*2*	...
81 Australia	3 000 F	3 000 F	2 000 F	2 000 F	1 000 F	339	598	454	447	455
New Zealand	2 570	2 861	2 365	2 026	1 344	1 914	1 301	902	1 029	827
81 Fishing area total	*5 570 F*	*5 861 F*	*4 365 F*	*4 026 F*	*2 344 F*	*2 253*	*1 899*	*1 356*	*1 476*	*1 282*
Species total	*8 570 F*	*8 861 F*	*7 365 F*	*7 026 F*	*4 344 F*	*2 253*	*1 899*	*1 360*	*1 478*	*1 282*
Largehead hairtail	**Poisson-sabre commun**		**Pez sable**			***Trichiurus lepturus***			**1,75(06)003,02**	**LHT**
21 USA	...	...	...	...	...	...	0	4	5	1
21 Fishing area total	...	...	...	...	...	...	*0*	*4*	*5*	*1*
27 Ireland	-	-	-	-	-	-	-	-	-	776
Portugal	-	-	-	-	-	-	-	3	0	0
27 Fishing area total	-	-	-	-	-	-	-	*3*	*0*	*776*
31 Russian Fed	-	-	-	-	5	-	-	-	-	-
USA	5	4	20	32	1	10	5	2	36	6
Venezuela	3 907	3 560	3 944	4 933	4 609	5 050	5 408	4 017	3 716	8 793
31 Fishing area total	*3 912*	*3 564*	*3 964*	*4 965*	*4 615*	*5 060*	*5 413*	*4 019*	*3 752*	*8 799*
34 Belize	-	-	-	-	-	300	317	644	265	7
Benin	100 F	150 F	200 F	250 F	300 F	350 F	400 F	543	370 F	584
Bulgaria	-	-	-	-	-	-	1 383	-	-	-
Cameroon	5	5	2 F	1	4	10 F	16	1	6	59
Côte dIvoire	188	221	229	231 F	180	265	492	518	321	200
Cyprus	-	-	-	-	...	...	5 061	130	230	3 949
France	-	-	-	-	-	...	3 400	509	-	-
Georgia	68 F	40 F	10 F	-	-	-	-	-	-	-

B-34

Miscellaneous demersal fishes — Capture production by species, fishing areas and countries or areas
Poissons démersaux divers — Captures par espèces, zones de pêche et pays ou zones
Peces demersales diversos — Capturas por especies, áreas de pesca y países o áreas

Species, Fishing area Espèce, Zone de pêche Especie, Area de pesca	1992 mt	1993 mt	1994 mt	1995 mt	1996 mt	1997 mt	1998 mt	1999 mt	2000 mt	2001 mt
Ghana	4 341	1 445	1 140	1 823	2 543	2 866	2 047	1 267	1 664	1 849
Guinea	...	...	...	324	401	409	400	400	418	400 F
GuineaBissau	3 F	5 F	27	17	17	20 F	15 F	10 F	10 F	10 F
Latvia	4 106	250	16	-	12	-	1 232	1 502	544	13
Liberia	7	6	2	12	7	10	9	33	34	169
Lithuania	14 237	-	-	...	...	...	9 708	13	32	69
Mauritania	80 F	100 F	60 F	100 F	160 F	110 F	-	...	...	...
Morocco	1 706	3 475	3 519	5 450	5 500	8 340	8 064	6 595	6 176	5 124
Netherlands	-	-	-	-	33	-	103	401	115	697
Portugal	-	0	0	0	0	0	0	-	-	-
Romania	1 153	34	-	-	-	-	-	-	-	-
Russian Fed	22 438	17 441	19 564	23 709	51 773	33 368	7 303	6 777	3 297	1 146
St Vincent	-	-	-	-	-	130	1 393	35	24	700
Senegal	286	195	463	345	432	1 114	948	838	781	1 083
Sierra Leone	...	...	...	...	...	...	204	75	35	48
Spain	-	-	-	-	-	-	-	-	-	3
Togo	1	11	2	0	0	1	0	6	2	4
Ukraine	-	-	-	-	-	-	2 490	1 314	1 423	-
Other nei	-	-	-	-	-	-	-	147	948	523
34 Fishing area total	*48 719 F*	*23 378 F*	*25 234 F*	*32 262 F*	*61 362 F*	*47 293 F*	*44 985 F*	*21 758 F*	*16 695 F*	*16 637 F*
37 Egypt	...	...	...	564	914	679	774	711	809	1 107
Morocco	-	-	83	1	5	19	0	-	0	0
37 Fishing area total	*...*	*...*	*83*	*565*	*919*	*698*	*774*	*711*	*809*	*1 107*
41 Brazil	1 110 F	1 110 F	1 110 F	1 197	736	938	1 405	1 230	1 665	1 650 F
41 Fishing area total	*1 110 F*	*1 110 F*	*1 110 F*	*1 197*	*736*	*938*	*1 405*	*1 230*	*1 665*	*1 650 F*
47 Angola	...	...	...	...	...	...	...	28	-	-
Namibia	-	-	-	13	...	...	...	...	346	691
Russian Fed	-	1	2	-	-	97	119	440	1 549	1 180
47 Fishing area total	*...*	*1*	*2*	*13*	*...*	*97*	*119*	*468*	*1 895*	*1 871*
51 Egypt	-	-	-	-	-	-	-	12	2	-
Pakistan	4 755	3 474	6 320	6 093	9 073	11 583	12 337	31 623	28 754	27 355
51 Fishing area total	*4 755*	*3 474*	*6 320*	*6 093*	*9 073*	*11 583*	*12 337*	*31 635*	*28 756*	*27 355*
57 Malaysia	6 243	4 123	2 713	2 653	2 524	3 372	14 038	5 995	1 957	1 681
Thailand	1 243	2 972	5 278	10 015	11 712	13 033	16 223	9 509	9 379	9 322
57 Fishing area total	*7 486*	*7 095*	*7 991*	*12 668*	*14 236*	*16 405*	*30 261*	*15 504*	*11 336*	*11 003*
61 China	622 243	635 315	878 144	1 039 684	1 071 914	1 014 598	1 223 360	1 222 454	1 285 469	1 282 698
China,Taiwan	17 822	17 227	19 313	16 210	11 830	8 955	7 991	9 375	7 271	8 834
Japan	31 539	31 712	31 577	28 207	26 644	20 932	22 268	26 200	22 947	16 615
Korea Rep	87 325	58 035	101 052	94 596	74 461	67 170	74 851	64 445	81 050	79 898
61 Fishing area total	*758 929*	*742 289*	*1 030 086*	*1 178 697*	*1 184 849*	*1 111 655*	*1 328 470*	*1 322 474*	*1 396 737*	*1 388 045*
71 Malaysia	2 983	3 313	3 443	3 594	3 844	7 901	9 913	12 014	9 617	8 030
Singapore	162	119	192	137	144	170	140	127	83	50
Thailand	1 741	2 333	2 812	4 482	3 361	4 553	2 486	6 461	6 372	6 333
71 Fishing area total	*4 886*	*5 765*	*6 447*	*8 213*	*7 349*	*12 624*	*12 539*	*18 602*	*16 072*	*14 413*
Species total	*829 797 F*	*786 676 F*	*1 081 237 F*	*1 244 673 F*	*1 283 139 F*	*1 206 353 F*	*1 436 303 F*	*1 416 408 F*	*1 477 722 F*	*1 471 657 F*
Silver scabbardfish	**Sabre argenté**		**Pez cinto**			***Lepidopus caudatus***		**1,75(06)006,01**		**SFS**
27 France	-	-	-	-	-	-	-	30	8	16
Germany	-	2	-	-	-	-	-	-	4	-
Latvia	1 905	1 458	-	8	-	-	-	-	-	-
Portugal	1 518	2 662	1 429	6 479	2 057	2 853	2 155	319	66	82
Russian Fed	110	19	-	-	-	-	-	-	-	2
Spain	-	-	-	-	9	651	1 377	1 584	14	256
UK	-	-	-	-	-	-	-	-	12	5
27 Fishing area total	*3 533*	*4 141*	*1 429*	*6 487*	*2 066*	*3 504*	*3 532*	*1 933*	*104*	*361*
34 Germany	-	-	-	-	-	-	-	64	-	-
Portugal	6 800	6 367	7 931	2 571	8 939	4 731	3 357	2 966	0	4
Spain	-	-	-	-	636	-	-	-	-	-
34 Fishing area total	*6 800*	*6 367*	*7 931*	*2 571*	*9 575*	*4 731*	*3 357*	*3 030*	*0*	*4*
37 Albania	...	...	...	7	0	0	18	19	18	0
France	-	-	-	-	41	11	11	225	8	15
Spain	-	-	-	-	1 552	1 347	1 627	1 163	270	3 031
Tunisia	140	263	173	-	26	7	57	423	411	375
37 Fishing area total	*140*	*263*	*173*	*7*	*1 619*	*1 365*	*1 713*	*1 830*	*707*	*3 421*
47 South Africa	14 227	12 196	4 697	4 931	3 196	2 001	4 557	2 560	2 300 F	2 316
Spain	-	-	-	-	-	2	-	-	-	-
47 Fishing area total	*14 227*	*12 196*	*4 697*	*4 931*	*3 196*	*2 003*	*4 557*	*2 560*	*2 300 F*	*2 316*
81 New Zealand	2 295	1 709	2 476	1 955	941	2 342	3 344	2 638	1 536	2 876
Norway	0	-	-	-	-	-	-	-	-	-
Russian Fed	-	6	-	-	-	-	-	-	-	-
Ukraine	-	-	-	-	-	-	-	-	83	11
81 Fishing area total	*2 295*	*1 715*	*2 476*	*1 955*	*941*	*2 342*	*3 344*	*2 638*	*1 619*	*2 887*
Species total	*26 995*	*24 682*	*16 706*	*15 951*	*17 397*	*13 945*	*16 503*	*11 991*	*4 730 F*	*8 989*

B-34 Miscellaneous demersal fishes / Poissons démersaux divers / Peces demersales diversos

Capture production by species, fishing areas and countries or areas / Captures par espèces, zones de pêche et pays ou zones / Capturas por especies, áreas de pesca y países o áreas

Species, Fishing area Espèce, Zone de pêche Especie, Area de pesca	1992 mt	1993 mt	1994 mt	1995 mt	1996 mt	1997 mt	1998 mt	1999 mt	2000 mt	2001 mt
Black scabbardfish	**Sabre noir**		**Sable negro**		***Aphanopus carbo***				**1,75(06)012,01**	**BSF**
27 Estonia	-	-	-	-	-	-	-	-	-	224
Faeroe Is	375	1 315	893	550	256	126	89	45	116	412
France	3 618	3 421	2 512	2 448	2 868	2 118	1 710	1 833	3 707	5 070
Germany	-	149	94	3	2	-	-	-	-	-
Iceland	-	-	1	0	0	1	0	9	18	8
Ireland	-	8	-	-	0	1	-	1	12	299
Lithuania	-	-	-	-	-	-	-	-	-	3
Netherlands	-	-	-	-	-	-	-	11	7	-
Portugal	4 424	4 521	3 428	4 275	3 686	3 553	3 153	2 776	2 867	2 745
Spain	-	-	-	-	41	106	127	117	1 029	1 323
UK	-	-	2	20	40	2	159	201	428	742
27 Fishing area total	*8 417*	*9 414*	*6 930*	*7 296*	*6 893*	*5 907*	*5 238*	*4 993*	*8 184*	*10 826*
31 Iceland	-	-	-	-	17	-	-	-	-	-
31 Fishing area total	-	-	-	-	*17*	-	-	-	-	-
34 Portugal	2 495	3 467	3 133	3 469	6 748	4 023	4 430	4 405	4 203	4 008
Spain	-	-	-	-	-	-	10	-	-	-
34 Fishing area total	*2 495*	*3 467*	*3 133*	*3 469*	*6 748*	*4 023*	*4 440*	*4 405*	*4 203*	*4 008*
Species total	*10 912*	*12 881*	*10 063*	*10 765*	*13 658*	*9 930*	*9 678*	*9 398*	*12 387*	*14 834*
Hairtails, scabbardfishes nei	**Poissons-sabres, sabres nca**		**Peces sable, cintos nep**		***Trichiuridae***				**1,75(06)XXX,XX**	**CUT**
27 Spain	-	-	-	-	-	-	-	-	-	13
27 Fishing area total	-	-	-	-	-	-	-	-	-	*13*
31 Korea Rep	-	-	-	-	-	-	-	24	15	6
Mexico	...	...	...	...	...	...	6 349	6 723	7 263	5 085
31 Fishing area total	...	...	...	...	...	...	*6 349*	*6 747*	*7 278*	*5 091*
34 Congo Rep	67 F	67 F	187	150 F	120 F	90 F	60 F	30 F	-	-
Gambia	...	...	...	0	10	12	5	5	-	28
Korea Rep	36	106	646	255	349	901	706	963	904	1 001
Nigeria	5 422	2 557	3 806	5 175	4 826	6 399	7 750	5 366	754	1 156
34 Fishing area total	*5 525 F*	*2 730 F*	*4 639*	*5 580 F*	*5 305 F*	*7 402 F*	*8 521 F*	*6 364 F*	*1 658*	*2 185*
41 Korea Rep	-	103	203	-	-	-	51	1 205	1 888	155
41 Fishing area total	-	*103*	*203*	-	-	-	*51*	*1 205*	*1 888*	*155*
47 Korea Rep	208	52	224	-	-	5	-	-	-	-
47 Fishing area total	*208*	*52*	*224*	-	-	*5*	-	-	-	-
51 India	32 048	25 568	44 583	25 409	38 771	129 662	58 595	92 334	100 570	93 970
Korea Rep	5 274	3 709	3 350	4 301	5 595	7 918	7 035	7 782	5 147	1 663
Oman	3 906	2 971	3 883	8 163	8 132	10 384	4 767	1 776	4 367	2 617
South Africa	-	-	-	-	-	6	0	...	...	...
Untd Arab Em	53	50	58	51	52	65	72	78	80	80 F
51 Fishing area total	*41 281*	*32 298*	*51 874*	*37 924*	*52 550*	*148 035*	*70 469*	*101 970*	*110 164*	*98 330 F*
57 India	12 501	11 730	20 884	25 173	15 170	16 456	14 801	22 872	20 026	50 504
Indonesia	8 622	6 736	9 222	11 030	13 049	15 135	19 959	18 101	16 793	18 110
Korea Rep	-	-	-	17	-	29	-	1 085	1 374	90
57 Fishing area total	*21 123*	*18 466*	*30 106*	*36 220*	*28 219*	*31 620*	*34 760*	*42 058*	*38 193*	*68 704*
61 China,H.Kong	4 673	3 578	2 951	3 470	3 326	2 522	1 690	1 200 F	1 500 F	1 650 F
61 Fishing area total	*4 673*	*3 578*	*2 951*	*3 470*	*3 326*	*2 522*	*1 690*	*1 200 F*	*1 500 F*	*1 650 F*
71 Indonesia	11 428	12 478	12 229	15 559	15 322	17 846	17 195	18 557	21 284	22 960
Korea Rep	1 215	1 972	5 215	2 987	1 769	1 531	1 248	694	700	361
Philippines	16 730	15 307	16 175	12 522	12 266	9 412	9 877	10 363	8 641	8 797
71 Fishing area total	*29 373*	*29 757*	*33 619*	*31 068*	*29 357*	*28 789*	*28 320*	*29 614*	*30 625*	*32 118*
77 Korea Rep	8	-	-	10	1	-	-	606	538	-
Mexico	...	...	...	...	...	...	-	1	2	4
77 Fishing area total	*8*	...	...	*10*	*1*	...	-	*607*	*540*	*4*
81 Korea Rep	-	-	-	-	678	1 188	545	422	1 099	227
81 Fishing area total	-	-	-	-	*678*	*1 188*	*545*	*422*	*1 099*	*227*
87 Ecuador	...	...	...	...	...	...	...	...	13 196	3 382
Korea Rep	-	-	-	-	-	-	-	85	470	-
87 Fishing area total	...	...	...	...	...	...	...	*85*	*13 666*	*3 382*
Species total	*102 191 F*	*86 984 F*	*123 616*	*114 272 F*	*119 436 F*	*219 561 F*	*150 705 F*	*190 272 F*	*206 611 F*	*211 859 F*
Indian driftfish	**Ariomme indienne**		**Arioma índica**		***Ariomma indica***				**1,76(04)012,01**	**DRI**
61 China,H.Kong	394	133	289	287	120	79	49	35 F	40 F	45 F
61 Fishing area total	*394*	*133*	*289*	*287*	*120*	*79*	*49*	*35 F*	*40 F*	*45 F*
Species total	*394*	*133*	*289*	*287*	*120*	*79*	*49*	*35 F*	*40 F*	*45 F*
Choicy ruff	**Sériolelle argentine**		**Cojinoba savorín**		***Seriolella porosa***				**1,76(08)010,01**	**SEO**
41 Argentina	...	...	...	...	...	...	...	...	3 542	3 990
41 Fishing area total	...	...	...	...	...	...	...	...	*3 542*	*3 990*

B-34 Miscellaneous demersal fishes / Poissons démersaux divers / Peces demersales diversos

Capture production by species, fishing areas and countries or areas / Captures par espèces, zones de pêche et pays ou zones / Capturas por especies, áreas de pesca y países o áreas

Species, Fishing area Espèce, Zone de pêche Especie, Area de pesca	1992 mt	1993 mt	1994 mt	1995 mt	1996 mt	1997 mt	1998 mt	1999 mt	2000 mt	2001 mt
Species total	*...*	*...*	*...*	*...*	*...*	*...*	*...*	*...*	*3 542*	*3 990*
Common warehou	**...B**		**...C**			***Seriolella brama***			**1,76(08)010,02**	**SEM**
81 New Zealand	2 570	2 705	1 312	1 081	1 760	4 108	3 101	3 881	4 259	4 101
81 Fishing area total	*2 570*	*2 705*	*1 312*	*1 081*	*1 760*	*4 108*	*3 101*	*3 881*	*4 259*	*4 101*
Species total	*2 570*	*2 705*	*1 312*	*1 081*	*1 760*	*4 108*	*3 101*	*3 881*	*4 259*	*4 101*
Silver warehou	**...B**		**...C**			***Seriolella punctata***			**1,76(08)010,04**	**SEP**
81 New Zealand	3 120	4 518	5 113	13 377	4 926	11 253	10 993	9 029	11 218	11 268
81 Fishing area total	*3 120*	*4 518*	*5 113*	*13 377*	*4 926*	*11 253*	*10 993*	*9 029*	*11 218*	*11 268*
Species total	*3 120*	*4 518*	*5 113*	*13 377*	*4 926*	*11 253*	*10 993*	*9 029*	*11 218*	*11 268*
White warehou	**...B**		**...C**			***Seriolella caerulea***			**1,76(08)010,05**	**SEU**
81 New Zealand	1 124	819	1 072	2 348	1 467	2 432	2 296	2 366	2 407	1 962
81 Fishing area total	*1 124*	*819*	*1 072*	*2 348*	*1 467*	*2 432*	*2 296*	*2 366*	*2 407*	*1 962*
Species total	*1 124*	*819*	*1 072*	*2 348*	*1 467*	*2 432*	*2 296*	*2 366*	*2 407*	*1 962*
South Pacific breams nei	**Sériolelles nca**		**Cojinobas nep**			***Seriolella spp***			**1,76(08)010,XX**	**BSP**
57 Australia	2 458	2 242	2 603	2 558	3 406	4 020	3 356	3 327	3 449	4 190
57 Fishing area total	*2 458*	*2 242*	*2 603*	*2 558*	*3 406*	*4 020*	*3 356*	*3 327*	*3 449*	*4 190*
81 Norway	26	85	63	-	121	70	4	-	-	-
Russian Fed	-	922	721	150	185	352	34	28	-	-
Ukraine	905	24	4	...	484	1 183	320	328	1 558	666
81 Fishing area total	*931*	*1 031*	*788*	*150*	*790*	*1 605*	*358*	*356*	*1 558*	*666*
87 Chile	2 659	4 944	4 038	3 232	2 482	2 909	2 671	3 347	2 502	3 336
Colombia	...	...	...	...	...	...	...	...	...	8
Peru	11 211	2 795	8 892	7 698	3 704	388	505	1 589	1 473	3 192
87 Fishing area total	*13 870*	*7 739*	*12 930*	*10 930*	*6 186*	*3 297*	*3 176*	*4 936*	*3 975*	*6 536*
Species total	*17 259*	*11 012*	*16 321*	*13 638*	*10 382*	*8 922*	*6 890*	*8 619*	*8 982*	*11 392*
Bluenose warehou	**Rouffe antarctique**		**Rufo antártico**			***Hyperoglyphe antarctica***			**1,76(08)015,02**	**BWA**
81 New Zealand	2 262	2 387	42	2 719	2 432	2 974	2 630	2 755	2 793	2 954
81 Fishing area total	*2 262*	*2 387*	*42*	*2 719*	*2 432*	*2 974*	*2 630*	*2 755*	*2 793*	*2 954*
Species total	*2 262*	*2 387*	*42*	*2 719*	*2 432*	*2 974*	*2 630*	*2 755*	*2 793*	*2 954*
Black driftfish	**...B**		**...C**			***Hyperoglyphe bythites***			**1,76(08)015,03**	**HGY**
31 USA	...	...	...	...	...	...	...	2	7	5
31 Fishing area total	*...*	*...*	*...*	*...*	*...*	*...*	*...*	*2*	*7*	*5*
Species total	*...*	*...*	*...*	*...*	*...*	*...*	*...*	*2*	*7*	*5*
Pacific rudderfish	**Stromaté du Japon**		**Pámpano del Pacífico**			***Psenopsis anomala***			**1,76(08)020,01**	**BUP**
61 China,H.Kong	5 576	3 259	1 820	1 494	940	1 460	1 160	800 F	1 000 F	1 100 F
China,Taiwan	5 940	6 338	6 711	9 212	8 834	6 001	8 258	5 075	4 506	5 646
Japan	3 458	3 203	4 470	3 485	3 516	4 316	4 316	4 996	5 215	4 196
61 Fishing area total	*14 974*	*12 800*	*13 001*	*14 191*	*13 290*	*11 777*	*13 734*	*10 871 F*	*10 721 F*	*10 942 F*
Species total	*14 974*	*12 800*	*13 001*	*14 191*	*13 290*	*11 777*	*13 734*	*10 871 F*	*10 721 F*	*10 942 F*
Ruffs, barrelfishes nei	**Centrolophes nca**		**Rufos, romerillos nep**			***Centrolophidae***			**1,76(08)XXX,XX**	**CEN**
47 Ukraine	...	...	...	...	36	-	-	-	-	-
47 Fishing area total	*...*	*...*	*...*	*...*	*36*	*-*	*-*	*-*	*-*	*-*
51 Spain	-	-	-	-	-	-	-	-	36	-
Ukraine	828	301	732	485	254	440	395	753	360	299
51 Fishing area total	*828*	*301*	*732*	*485*	*254*	*440*	*395*	*753*	*396*	*299*
71 NewCaledonia	0	0	0	0	0	0	0	0	0	0
71 Fishing area total	*0*	*0*	*0*	*0*	*0*	*0*	*0*	*0*	*0*	*0*
81 Korea Rep	-	-	-	-	-	-	-	493	317	209
New Zealand	121	154	201	169	169	214	202	249	274	211
81 Fishing area total	*121*	*154*	*201*	*169*	*169*	*214*	*202*	*742*	*591*	*420*
Species total	*949*	*455*	*933*	*654*	*459*	*654*	*597*	*1 495*	*987*	*719*
Golden redfish	**Sébaste doré**		**Gallineta dorada**			***Sebastes marinus***			**1,78(01)001,01**	**REG**
27 Denmark	21	2	7	1	4	4	7	5	8	11
Faeroe Is	19 532	17 402	12 161	12 071	11 028	11 402	7 951	7 129	13 324	13 572
Greenland	...	...	...	...	...	...	...	...	41	13
Portugal	0	0	0	1	2	7	10	2	269	2
27 Fishing area total	*19 553*	*17 404*	*12 168*	*12 073*	*11 034*	*11 413*	*7 968*	*7 136*	*13 642*	*13 598*

B-34

Miscellaneous demersal fishes / Poissons démersaux divers / Peces demersales diversos

Capture production by species, fishing areas and countries or areas / Captures par espèces, zones de pêche et pays ou zones / Capturas por especies, áreas de pesca y países o áreas

Species, Fishing area Espèce, Zone de pêche Especie, Area de pesca	1992 mt	1993 mt	1994 mt	1995 mt	1996 mt	1997 mt	1998 mt	1999 mt	2000 mt	2001 mt
Species total	*19 553*	*17 404*	*12 168*	*12 073*	*11 034*	*11 413*	*7 968*	*7 136*	*13 642*	*13 598*
Widow rockfish	**Sébaste rocote**		**Rocote**			***Sebastes entomelas***			**1,78(01)001,03**	**WRO**
67 USA	...	...	...	...	...	6 886	4 357	3 841	3 575	2 481
67 Fishing area total	...	...	...	...	...	*6 886*	*4 357*	*3 841*	*3 575*	*2 481*
77 USA	...	...	...	...	...	865	540	402	29	128
77 Fishing area total	...	...	...	...	...	*865*	*540*	*402*	*29*	*128*
Species total	...	...	...	...	...	*7 751*	*4 897*	*4 243*	*3 604*	*2 609*
Yellowtail rockfish	**Sébaste à queue jaune**		**Chancharro cola amarilla**			***Sebastes flavidus***			**1,78(01)001,04**	**YRO**
67 USA	...	...	...	...	...	2 512	3 349	3 427	3 138	2 073
67 Fishing area total	...	...	...	...	...	*2 512*	*3 349*	*3 427*	*3 138*	*2 073*
77 USA	...	...	...	...	...	232	253	41	32	4
77 Fishing area total	...	...	...	...	...	*232*	*253*	*41*	*32*	*4*
Species total	...	...	...	...	...	*2 744*	*3 602*	*3 468*	*3 170*	*2 077*
Pacific ocean perch	**Sébaste du Pacifique**		**Gallineta del Pacífico**			***Sebastes alutus***			**1,78(01)001,09**	**OPP**
61 Japan	1 621	2 760	1 520	1 449	1 730	767	896	766	458	668
Korea Rep	4	-	-	-	-	-	-	-	-	-
Russian Fed	560	1 153	3 447	3 539	2 615	2 095	1 544	864	1 017	793
61 Fishing area total	*2 185*	*3 913*	*4 967*	*4 988*	*4 345*	*2 862*	*2 440*	*1 630*	*1 475*	*1 461*
67 Canada	3 990	4 609	5 649	5 207	6 174	5 782	6 184	5 849	6 177	5 832
Poland	-	-	-	-	-	-	-	-	21	-
Russian Fed	-	-	-	-	-	-	195	1 614	6	-
USA	20 630	20 649	14 000	15 613	21 004	19 580	18 445	20 608	17 940	17 696
67 Fishing area total	*24 620*	*25 258*	*19 649*	*20 820*	*27 178*	*25 362*	*24 824*	*28 071*	*24 144*	*23 528*
77 Estonia	-	-	-	-	-	-	-	8	-	-
USA	1	-	-	-	-	-	-	8	12	12
77 Fishing area total	*1*	-	-	-	-	-	-	*16*	*12*	*12*
Species total	*26 806*	*29 171*	*24 616*	*25 808*	*31 523*	*28 224*	*27 264*	*29 717*	*25 631*	*25 001*
Bocaccio rockfish	**Sébaste bocace**		**Chancharro bocacio**			***Sebastes paucispinis***			**1,78(01)001,10**	**SBC**
67 USA	...	...	...	...	...	448	491	135	3	0
67 Fishing area total	...	...	...	...	...	*448*	*491*	*135*	*3*	*0*
77 USA	...	...	...	...	...	276	105	62	24	33
77 Fishing area total	...	...	...	...	...	*276*	*105*	*62*	*24*	*33*
Species total	...	...	...	...	...	*724*	*596*	*197*	*27*	*33*
Beaked redfish	**Sébaste du Nord**		**Gallineta nórdica**			***Sebastes mentella***			**1,78(01)001,12**	**REB**
21 Denmark	-	-	-	-	-	-	-	-	-	0
Greenland	...	...	...	...	...	...	...	...	671	124
21 Fishing area total	...	...	...	...	...	...	...	...	*671*	*124*
27 Bulgaria	628	3 163	-	-	-	-	-	-	-	-
Denmark	16	18	35	14	17	19	20	48	35	89
Greenland	...	...	...	...	...	...	...	...	3 509	3 258
Spain	-	-	-	4 554	4 307	4 438	4 587	3 526	2 530	5 575
Ukraine	-	2 782	5 561	3 185	518	-	-	-	-	-
27 Fishing area total	*644*	*5 963*	*5 596*	*7 753*	*4 842*	*4 457*	*4 607*	*3 574*	*6 074*	*8 922*
Species total	*644*	*5 963*	*5 596*	*7 753*	*4 842*	*4 457*	*4 607*	*3 574*	*6 745*	*9 046*
Cape redfish	**Sébaste du Cap**		**Gallineta del Cabo**			***Sebastes capensis***			**1,78(01)001,13**	**REC**
47 Namibia	28	21	183	271	576	673	343	448	639	891
South Africa	1 229	1 476	723	1 082	1 183	957	752	766	750 F	782
47 Fishing area total	*1 257*	*1 497*	*906*	*1 353*	*1 759*	*1 630*	*1 095*	*1 214*	*1 389 F*	*1 673*
Species total	*1 257*	*1 497*	*906*	*1 353*	*1 759*	*1 630*	*1 095*	*1 214*	*1 389 F*	*1 673*
Canary rockfish	**Sébaste citron**		**Chancharro flioma**			***Sebastes pinniger***			**1,78(01)001,15**	**SPG**
67 USA	...	...	...	...	...	1 133	1 222	736	57	45
67 Fishing area total	...	...	...	...	...	*1 133*	*1 222*	*736*	*57*	*45*
77 USA	...	...	...	...	...	129	91	36	3	4
77 Fishing area total	...	...	...	...	...	*129*	*91*	*36*	*3*	*4*
Species total	...	...	...	...	...	*1 262*	*1 313*	*772*	*60*	*49*
Chilipepper rockfish	**Sébaste piment**		**Chancharro pimienta**			***Sebastes goodei***			**1,78(01)001,16**	**SGO**
67 USA	...	...	...	...	...	38	11	10	31	9
67 Fishing area total	...	...	...	...	...	*38*	*11*	*10*	*31*	*9*

B-34

Miscellaneous demersal fishes — **Capture production by species, fishing areas and countries or areas**
Poissons démersaux divers — **Captures par espèces, zones de pêche et pays ou zones**
Peces demersales diversos — **Capturas por especies, áreas de pesca y países o áreas**

Species, Fishing area Espèce, Zone de pêche Especie, Area de pesca	1992 mt	1993 mt	1994 mt	1995 mt	1996 mt	1997 mt	1998 mt	1999 mt	2000 mt	2001 mt
77 USA	...	...	...	...	...	1 812	1 262	908	414	609
77 Fishing area total	...	...	...	...	...	*1 812*	*1 262*	*908*	*414*	*609*
Species total	...	...	...	...	...	*1 850*	*1 273*	*918*	*445*	*618*
Black rockfish	**...B**		**...C**			***Sebastes melanops***			**1,78(01)001,32**	**RMG**
67 USA	...	...	...	...	...	...	...	...	109	148
67 Fishing area total	...	...	...	...	...	...	...	...	*109*	*148*
Species total	...	...	...	...	...	...	...	...	*109*	*148*
Atlantic redfishes nei	**Sébastes de l'Atlantique nca**			**Gallinetas del Atlántico nep**		***Sebastes spp***			**1,78(01)001,XX**	**RED**
21 Canada	99 365	84 787	50 774	17 954	21 571	18 751	25 763	19 682	19 852	19 710
Cuba	6 453	2 880	23	70	-	88	13	39	0	...
Estonia	33	4 115	135	863	-	-	-	-	841	186
Faeroe Is	19	61	12	15	1	-	-	-	-	-
Germany	4 898	295	74	-	-	-	-	154	4 476	817
Greenland	333	907	1 063	922	864	969	929	779	721	305
Iceland	-	-	10	751	-	-	-	-	1	-
Japan	1 620	1 161	570	868	1 169	565	891	460	140	187
Korea Rep	15 215	3 687	-	-	-	-	-	-	-	-
Latvia	7 441	8 502	149	304	-	-	-	-	13	11
Lithuania	3 868	3 904	37	-	-	-	-	-	430	4 396
Norway	114	105	8	3	-	-	1	-	113	-
Poland	-	-	-	-	-	-	-	-	-	4
Portugal	6 584	9 831	8 609	3 291	2 153	1 125	2 368	6 081	5 675	5 625
Russian Fed	12 547	14 560	5 060	5 395	86	375	15	546	9 799	14 152
St Pier Mq	1 208	-	-	-	-	430	654	423	196	129
Spain	948	136	870	628	560	1 388	2 686	5 775	4 694	3 709
UK	31	-	-	-	-	-	1	-	-	-
USA	847	800	439	436	322	251	320	353	322	363
21 Fishing area total	*161 524*	*135 731*	*67 833*	*31 500*	*26 726*	*23 942*	*33 641*	*34 292*	*47 273*	*49 594*
27 Belgium	111	124	54	16	19	16	2	3	5	6
Estonia	1 810	6 365	17 875	16 854	7 091	3 720	3 968	2 108	7 811	599
France	3 534	2 449	2 293	2 520	2 244	2 567	1 604	1 254	943	1 076
Germany	10 193	34 462	30 404	20 472	22 512	21 096	20 342	18 263	9 802	11 973
Greenland	1 585	918	428	4 799	280	193	1 476	4 437	139	94
Iceland	107 725	116 325	142 101	117 999	120 751	111 652	116 132	110 345	116 301	92 527
Ireland	5	5	15	18	15	48	71	171	186	433
Japan	935	1 077	1 818	1 148	416	31	31	-	-	-
Latvia	780	6 803	13 205	5 003	1 084	-	-	-	-	-
Lithuania	6 656	7 899	7 404	22 893	10 649	-	1 769	3 884	6 257	15 786
Netherlands	1	1	8	29	41	53	20	16	19	8
Norway	38 287	32 974	28 934	23 279	28 679	22 687	28 559	30 856	25 540	28 657
Poland	-	-	-	-	-	777	12	6	2	5
Portugal	976	1 040	2 873	6 063	2 904	4 211	4 260	4 370	4 024	2 694
Russian Fed	8 679	18 478	30 629	51 377	47 574	41 613	31 200	29 069	29 338	33 186
Spain	16	65	34	67	408	4 577	2 532	2 460	1 047	943
Sweden	-	-	4	1	0	-	0	1	-	-
UK	954	1 384	743	1 346	1 776	1 507	1 553	2 672	2 495	2 875
27 Fishing area total	*182 247*	*230 369*	*278 822*	*273 884*	*246 443*	*214 748*	*213 531*	*209 915*	*203 909*	*190 862*
Species total	*343 771*	*366 100*	*346 655*	*305 384*	*273 169*	*238 690*	*247 172*	*244 207*	*251 182*	*240 456*
Blackbelly rosefish	**Sébaste chèvre**		**Gallineta**			***Helicolenus dactylopterus***			**1,78(01)017,03**	**BRF**
21 USA	-	-	-	-	-	-	-	-	3	0
21 Fishing area total	-	-	-	-	-	-	-	-	*3*	*0*
27 France	234	174	130	91	104	115	140	129	123	125
Portugal	...	...	...	...	...	...	...	334	436	313
27 Fishing area total	*234*	*174*	*130*	*91*	*104*	*115*	*140*	*463*	*559*	*438*
41 Argentina	228	580	628	4 670	1 499	2 563	1 192	3 354	3 050	...
Spain	-	-	-	-	-	-	-	8	8	6
Uruguay	2 499	1 994	4 898	2 734	2 204	3 404	4 389	2 581	2 111	1 830
41 Fishing area total	*2 727*	*2 574*	*5 526*	*7 404*	*3 703*	*5 967*	*5 581*	*5 943*	*5 169*	*1 836*
51 Spain	-	-	-	-	-	-	-	-	-	1
51 Fishing area total	-	-	-	-	-	-	-	-	-	*1*
Species total	*2 961*	*2 748*	*5 656*	*7 495*	*3 807*	*6 082*	*5 721*	*6 406*	*5 731*	*2 275*
Shortspine thornyhead	**...B**		**...C**			***Sebastolobus alascanus***			**1,78(01)079,01**	**SJU**
67 USA	...	...	...	...	...	...	...	...	331	268
67 Fishing area total	...	...	...	...	...	...	...	...	*331*	*268*
Species total	...	...	...	...	...	...	...	...	*331*	*268*
Scorpionfishes nei	**Rascasses, etc. nca**			**Rascacios, gallinetas nep**		***Scorpaenidae***			**1,78(01)XXX,XX**	**SCO**
27 France	90	50	38	22	31	63	1	8	16	18
Portugal	185	969	1 070	905	730	658	682	294	256	363
Spain	777	717	1 188	736	886	1 546	822	1 004	1 227	653
UK	-	-	1	0	15	...	116	253	185	188

B-34 Miscellaneous demersal fishes / Poissons démersaux divers / Peces demersales diversos

Capture production by species, fishing areas and countries or areas / Captures par espèces, zones de pêche et pays ou zones / Capturas por especies, áreas de pesca y países o áreas

Species, Fishing area Espèce, Zone de pêche Especie, Area de pesca	1992 mt	1993 mt	1994 mt	1995 mt	1996 mt	1997 mt	1998 mt	1999 mt	2000 mt	2001 mt
27 Fishing area total	*1 052*	*1 736*	*2 297*	*1 663*	*1 662*	*2 267*	*1 621*	*1 559*	*1 684*	*1 222*
34 Greece	49	169	24	28	64	35	36	42	53	42
Italy	-	-	-	-	-	-	-	-	-	14
Lithuania	529	-	-	-	-	-	-	-	-	-
Mauritania	5 F	5 F	10 F	5 F	5 F	5 F	0	1 F	0	...
Morocco	242	227	201	212	220	201	195	257	517	382
Nigeria	774	791	942	1 893	1 032	1 278	2 071	1 149	3 170	2 568
Portugal	203	129	127	-	-	-	-	-	15	9
Senegal	212	275	657	539	581	491	343	267	471	471
Spain	13	-	-	-	-	-	-	-	-	-
34 Fishing area total	*2 027 F*	*1 596 F*	*1 961 F*	*2 677 F*	*1 902 F*	*2 010 F*	*2 645*	*1 716 F*	*4 226*	*3 486*
37 Croatia	184	287	183	172	137	133	185	46	40	39
Cyprus	10	9	10	11	16	19	10	3	8	11
France	157	119	77	19	25	80	80	30	27	24
Greece	704	900	1 538	1 039	777	748	482	592	587	684
Lebanon	20 F	50	50	100	100	100	100	150	150	100
Malta	2	5	4	6	8	3	8	12	11	0
Morocco	23	10	6	7	10	25	4	6	56	10
Russian Fed	-	-	-	-	-	-	-	-	1	2
Spain	-	-	-	-	236	287	307	239	419	332
Tunisia	560	420	428	311	339	472	755	442	400	444
Turkey	436	547	434	539	585	435	515	315	360	640
Yugoslavia	2	1	3	6	7	7	7	7	8	7
37 Fishing area total	*2 098 F*	*2 348*	*2 733*	*2 210*	*2 240*	*2 309*	*2 453*	*1 842*	*2 067*	*2 293*
41 Italy	270	229	229	-	-	-	-	-	-	-
Korea Rep	-	48	-	-	293	99	-	-	-	4
41 Fishing area total	*270*	*277*	*229*	-	*293*	*99*	-	-	-	*4*
47 Iceland	-	-	-	-	-	5	-	-	-	-
Portugal	-	-	-	-	-	-	3	4	2	-
Spain	-	-	-	-	-	48	18	-	-	-
47 Fishing area total	-	-	-	-	-	*53*	*21*	*4*	*2*	-
51 Italy	405	344	344	-	-	-	-	-	-	-
South Africa	-	2	-	-	-	3	1	1 F	2	2
51 Fishing area total	*405*	*346*	*344*	-	-	*3*	*1*	*1 F*	*2*	*2*
61 Japan	3 793	3 594	3 059	2 888	2 317	2 082	1 638	1 314	1 244	1 132
Korea Rep	6 287	6 556	5 823	5 901	4 647	5 414	4 602	4 665	4 879	5 152
Russian Fed	-	-	-	-	-	-	-	-	45	54
61 Fishing area total	*10 080*	*10 150*	*8 882*	*8 789*	*6 964*	*7 496*	*6 240*	*5 979*	*6 168*	*6 338*
67 Canada	21 792	20 034	18 100	16 462	16 626	14 118	15 356	18 086	17 188	16 351
USA	46 554	47 268	34 262	32 514	35 305	34 251	15 772	17 465	13 662	12 436
67 Fishing area total	*68 346*	*67 302*	*52 362*	*48 976*	*51 931*	*48 369*	*31 128*	*35 551*	*30 850*	*28 787*
71 Korea Rep	154	-	-	-	2	-	-	-	-	-
71 Fishing area total	*154*	-	-	-	*2*	-	-	-	-	-
77 Korea Rep	-	-	-	-	-	-	4	-	-	-
USA	10 177	7 464	7 335	8 363	7 678	4 482	4 326	1 614	1 291	1 679
77 Fishing area total	*10 177*	*7 464*	*7 335*	*8 363*	*7 678*	*4 482*	*4 330*	*1 614*	*1 291*	*1 679*
81 Australia	...	...	...	...	316	349	371	356	363	373
Korea Rep	4	-	-	-	128	212	99	371	219	198
New Zealand	...	...	...	986	1 133	1 643	1 843	2 088	1 819	1 897
Russian Fed	1	2	-	-	5	-	-	-	-	-
81 Fishing area total	*5*	*2*	*...*	*986*	*1 582*	*2 204*	*2 313*	*2 815*	*2 401*	*2 468*
87 Chile	570	330	287	227	212	532	208	258	141	177
87 Fishing area total	*570*	*330*	*287*	*227*	*212*	*532*	*208*	*258*	*141*	*177*
Species total	*95 184 F*	*91 551 F*	*76 430 F*	*73 891 F*	*74 466 F*	*69 824 F*	*50 960*	*51 339 F*	*48 832*	*46 456*
Bluefin gurnard	**Grondin aile bleue**		**Testolín de aleta azul**		***Chelidonichthys kumu***			**1,78(02)003,01**		**KUG**
57 Australia	0	1	1	1	0	1	1	0	-	-
57 Fishing area total	*0*	*1*	*1*	*1*	*0*	*1*	*1*	*0*	-	-
61 Japan	984	1 038	1 132	-	-	-	-	-	-	-
Korea Rep	59	141	70	40	75	86	146	43	79	140
61 Fishing area total	*1 043*	*1 179*	*1 202*	*40*	*75*	*86*	*146*	*43*	*79*	*140*
71 Australia	-	-	-	-	14	0	0	0	-	-
71 Fishing area total	-	-	-	-	*14*	*0*	*0*	*0*	-	-
81 Australia	342	363	147	123	46	349	371	...	...	...
Japan	15	23	24	4	-	-	-	-	-	-
New Zealand	3 456	3 562	3 041	11 829	2 260	2 625	2 454	2 199	2 663	3 670
81 Fishing area total	*3 813*	*3 948*	*3 212*	*11 956*	*2 306*	*2 974*	*2 825*	*2 199*	*2 663*	*3 670*
Species total	*4 856*	*5 128*	*4 415*	*11 997*	*2 395*	*3 061*	*2 972*	*2 242*	*2 742*	*3 810*
Tub gurnard	**Grondin perlon**		**Begel**		***Chelidonichthys lucerna***			**1,78(02)003,02**		**GUU**
27 Denmark	-	-	-	-	5	4	5	19	15	12

B-34 Miscellaneous demersal fishes / Poissons démersaux divers / Peces demersales diversos

Capture production by species, fishing areas and countries or areas / Captures par espèces, zones de pêche et pays ou zones / Capturas por especies, áreas de pesca y países o áreas

Species, Fishing area Espèce, Zone de pêche Especie, Area de pesca	1992 mt	1993 mt	1994 mt	1995 mt	1996 mt	1997 mt	1998 mt	1999 mt	2000 mt	2001 mt
Netherlands	-	-	-	-	-	-	-	-	1 164	-
Portugal	-	-	-	-	-	-	-	3	5	3
27 Fishing area total	*-*	*-*	*-*	*-*	*5*	*4*	*5*	*22*	*1 184*	*15*
Species total	*-*	*-*	*-*	*-*	*5*	*4*	*5*	*22*	*1 184*	*15*
Red gurnard	**Grondin rouge**		**Arete**		***Chelidonichthys cuculus***				**1,78(02)003,03**	**GUR**
27 Belgium	140	113	119	116	145	128	157	262	418	493
Channel Is	7	7 F	3	3 F	10	8	6	10	7	-
Ireland	-	-	-	-	-	-	25	47	...	...
Isle of Man	0	0	-	-	-	-	-	-	-	-
Netherlands	-	-	-	-	-	-	-	-	46	1 724
UK	50	69	54	81	59	17	...	...	4	207
27 Fishing area total	*197*	*189 F*	*176*	*200 F*	*214*	*153*	*188*	*319*	*475*	*2 424*
Species total	*197*	*189 F*	*176*	*200 F*	*214*	*153*	*188*	*319*	*475*	*2 424*
Cape gurnard	**Grondin du Cap**		**Rubio del Cabo**		***Chelidonichthys capensis***				**1,78(02)003,07**	**GUC**
47 Namibia	10	20	28	57	51	214	216	71	236	144
South Africa	676	1 003	576	502	446	428	623	507	450 F	465
47 Fishing area total	*686*	*1 023*	*604*	*559*	*497*	*642*	*839*	*578*	*686 F*	*609*
Species total	*686*	*1 023*	*604*	*559*	*497*	*642*	*839*	*578*	*686 F*	*609*
Atlantic searobins	**Grondins**		**Rubios americanos**		***Prionotus spp***				**1,78(02)020,XX**	**SRA**
21 USA	34	17	49	111	20	10	31	38	24	39
21 Fishing area total	*34*	*17*	*49*	*111*	*20*	*10*	*31*	*38*	*24*	*39*
41 Brazil	930 F	930 F	930 F	922	792	540	858	1 149	1 839	1 800 F
41 Fishing area total	*930 F*	*930 F*	*930 F*	*922*	*792*	*540*	*858*	*1 149*	*1 839*	*1 800 F*
Species total	*964 F*	*947 F*	*979 F*	*1 033*	*812*	*550*	*889*	*1 187*	*1 863*	*1 839 F*
Latchet(=Sharpbeak gurnard)	**Grondin pointu**		**Cabete picudo**		***Pterygotrigla polyommata***				**1,78(02)025,01**	**BEG**
57 Australia	22	43	26	0	0	60	25	55	66	78
57 Fishing area total	*22*	*43*	*26*	*0*	*0*	*60*	*25*	*55*	*66*	*78*
81 Australia	120	133	127	121	58	76	67	39	53	70
81 Fishing area total	*120*	*133*	*127*	*121*	*58*	*76*	*67*	*39*	*53*	*70*
Species total	*142*	*176*	*153*	*121*	*58*	*136*	*92*	*94*	*119*	*148*
Spotted gurnard	**...B**		**...C**		***Pterygotrigla picta***				**1,78(02)025,02**	**JGU**
81 New Zealand	-	-	-	107	87	71	87	74	55	65
81 Fishing area total	*-*	*-*	*-*	*107*	*87*	*71*	*87*	*74*	*55*	*65*
Species total	*-*	*-*	*-*	*107*	*87*	*71*	*87*	*74*	*55*	*65*
Grey gurnard	**Grondin gris**		**Borracho**		***Eutrigla gurnardus***				**1,78(02)070,01**	**GUG**
27 Belgium	116	125	79	59	119	61	47	49	44	33
Denmark	8 137	840	121	84	70	36	56	85	96	289
Iceland	-	-	-	-	1	0	0	0	0	-
Ireland	-	-	-	-	-	-	38	71	...	...
Netherlands	-	-	-	-	-	-	-	-	459	295
Sweden	10	9	12	7	4	5	8	133	5	5
UK	10	25	24	21	56	59	...	...	...	46
27 Fishing area total	*8 273*	*999*	*236*	*171*	*250*	*161*	*149*	*338*	*604*	*668*
37 Egypt	2 015	1 099	1 077	1 070	1 264	1 158	1 067	1 159	1 078	778
37 Fishing area total	*2 015*	*1 099*	*1 077*	*1 070*	*1 264*	*1 158*	*1 067*	*1 159*	*1 078*	*778*
Species total	*10 288*	*2 098*	*1 313*	*1 241*	*1 514*	*1 319*	*1 216*	*1 497*	*1 682*	*1 446*
Gurnards, searobins nei	**Grondins, cavillones nca**		**Cabetes, rubios nep**		***Triglidae***				**1,78(02)XXX,XX**	**GUX**
27 Belgium	236	209	270	255	235	245	200	160	161	177
Channel Is	...	...	...	...	...	10	31	22	10	41
Faeroe Is	-	-	-	-	-	-	-	1	...	...
France	6 023	6 148	5 999	5 902	5 845	5 633	5 394	5 644	5 830	6 197
Germany	23	140	123	133	94	136	141	187	180	150
Ireland	47	26	67	85	77	82	-	-	79	97
Isle of Man	6	0	2	5	2	1	1	1	...	1
Portugal	915	816	560	616	627	611	500	496	650	616
Russian Fed	-	-	-	-	-	-	-	2 426	26 081	3 155
Spain	-	-	-	-	1 079	1 264	620	681	755	506
UK	653	651	600	599	583	484	629	631	953	1 131
27 Fishing area total	*7 903*	*7 990*	*7 621*	*7 595*	*8 542*	*8 466*	*7 516*	*10 249*	*34 699*	*12 071*
34 Benin	...	...	...	...	...	...	...	11	7 F	-
Ghana	7	140	14	8	0	76	...	53	0	37
Greece	0	4	18	9	8	4	3	-	-	-
Italy	2 024	1 718	1 718	-	-	11	13	13	31	129
Mauritania	10 F	10 F	10 F	10 F	10 F	0	70 F	...	...	...

B-34 Miscellaneous demersal fishes / Poissons démersaux divers / Peces demersales diversos

Capture production by species, fishing areas and countries or areas
Captures par espèces, zones de pêche et pays ou zones
Capturas por especies, áreas de pesca y países o áreas

Species, Fishing area Espèce, Zone de pêche Especie, Area de pesca	1992 mt	1993 mt	1994 mt	1995 mt	1996 mt	1997 mt	1998 mt	1999 mt	2000 mt	2001 mt
Morocco	1 540	1 765	1 798	2 002	2 269	2 211	2 827	3 140	4 858	4 160
Portugal	32	20	11	5	14	4	4	7	0	-
Russian Fed	-	175	-	-	-	-	628	-	-	-
Senegal	...	...	...	...	...	...	31	81	41	94
Spain	-	-	-	-	28	-	10	9	-	-
34 Fishing area total	*3 613 F*	*3 832 F*	*3 569 F*	*2 034 F*	*2 329 F*	*2 306*	*3 586 F*	*3 314*	*4 937 F*	*4 420*
37 Albania	...	...	...	2	15	0	-	-	-	34
France	325	235	148	326	50	50	50	207	388	339
Greece	1 394	1 829	421	321	368	376	234	229	207	169
Italy	5 060	4 269	4 304	4 116	3 908	3 462	3 280	2 655	2 137	2 135
Malta	-	-	-	1	2	0	2	4	2	1
Morocco	37	36	24	9	17	5	7	12	16	9
Slovenia	-	-	-	-	-	-	-	-	1	1
Spain	-	-	-	-	249	261	238	175	359	346
Syria	130 F	134 F	129 F	128 F	0	60	46	35	32	32
Tunisia	83	61	66	79	107	87	115	141	168	198
Turkey	3 287	2 253	1 921	1 500	2 319	750	700	710	260	200
37 Fishing area total	*10 316 F*	*8 817 F*	*7 013 F*	*6 482 F*	*7 035*	*5 051*	*4 672*	*4 168*	*3 570*	*3 464*
47 Angola	0	0	0	0	0	0	-	-	-	-
Japan	17	-	-	-	-	-	-	-	-	-
47 Fishing area total	*17*	*0*	*0*	*0*	*0*	*0*	*-*	*-*	*-*	*-*
51 South Africa	-	-	-	-	-	2	-	...	0	0
51 Fishing area total	*-*	*-*	*-*	*-*	*-*	*2*	*-*	*...*	*0*	*0*
87 Ecuador	...	...	...	...	...	...	...	22 634	27 129	1 428
87 Fishing area total	*...*	*...*	*...*	*...*	*...*	*...*	*...*	*22 634*	*27 129*	*1 428*
Species total	*21 849 F*	*20 639 F*	*18 203 F*	*16 111 F*	*17 906 F*	*15 825*	*15 774 F*	*40 365*	*70 335 F*	*21 383*
Sablefish	**Morue charbonnière**		**Bacalao negro**		***Anoplopoma fimbria***			**1,78(08)004,01**		**SAB**
61 Russian Fed	-	-	-	8	502	-	-	-	6	6
61 Fishing area total	*-*	*-*	*-*	*8*	*502*	*-*	*-*	*-*	*6*	*6*
67 Canada	5 372	5 310	5 201	4 542	3 069	4 000	4 500	4 583	2 811	2 967
USA	32 166	33 796	31 173	28 108	25 057	22 255	18 908	20 756	21 446	19 017
67 Fishing area total	*37 538*	*39 106*	*36 374*	*32 650*	*28 126*	*26 255*	*23 408*	*25 339*	*24 257*	*21 984*
77 USA	2 058	1 343	1 186	1 786	2 136	1 751	824	1 132	1 088	958
77 Fishing area total	*2 058*	*1 343*	*1 186*	*1 786*	*2 136*	*1 751*	*824*	*1 132*	*1 088*	*958*
Species total	*39 596*	*40 449*	*37 560*	*34 444*	*30 764*	*28 006*	*24 232*	*26 471*	*25 351*	*22 948*
Lumpfish(=Lumpsucker)	**Lompe**		**Liebre de mar**		***Cyclopterus lumpus***			**1,78(20)003,01**		**LUM**
21 Canada	-	18	-	25	18	175	8	1	8	-
Greenland	115	246	579	448	426	1 157	2 142	3 056	1 211	3 216
St Pier Mq	38	126	158	226	218	363	249	422	536	146
21 Fishing area total	*153*	*390*	*737*	*699*	*662*	*1 695*	*2 399*	*3 479*	*1 755*	*3 362*
27 Denmark	1 272	924	1 443	1 069	1 973	1 553	188	835	425	422
Germany	-	2	2	1	2	2	0	4	1	3
Greenland	-	-	-	-	-	-	1	1	-	-
Iceland	5 715	3 318	5 324	4 563	4 201	6 520	3 165	3 373	2 458	412
Norway	3 792	4 613	5 618	4 015	4 355	5 652	1 365	2 059	2 374	5 184
Russian Fed	-	-	-	-	-	-	-	-	20	28
Sweden	141	75	109	70	147	129	20	72	47	83
UK	-	-	-	-	0	0	0	0	1	-
27 Fishing area total	*10 920*	*8 932*	*12 496*	*9 718*	*10 678*	*13 856*	*4 739*	*6 344*	*5 326*	*6 132*
Species total	*11 073*	*9 322*	*13 233*	*10 417*	*11 340*	*15 551*	*7 138*	*9 823*	*7 081*	*9 494*
Angler(=Monk)	**Baudroie commune**		**Rape**		***Lophius piscatorius***			**1,95(01)001,01**		**MON**
27 Belgium	823	1 024	1 638	1 772	1 369	1 113	961	818	1 047	1 354
Channel Is	1	1 F	0	0	2	3	3	2	3	5
Denmark	2 349	1 863	1 974	1 350	1 835	2 040	1 873	1 858	1 724	1 917
Faeroe Is	1 210	993	1 065	1 433	1 610	1 765	1 885	2 578	2 216	2 006
France	13 724	13 557	15 407	18 168	18 476	17 145	14 053	10 566	12 498	13 049
Germany	73	452	385	893	542	1 137	1 446	847	568	364
Iceland	743	685	641	552	669	787	850	977	1 570	1 353
Ireland	3 399	2 362	2 535	2 929	3 348	3 880	4 251	4 298	3 839	3 112
Isle of Man	19	16	22	27	34	27	28	9	5	2
Netherlands	409	559	567	362	227	319	259	169	170	168
Norway	1 328	4 454	2 721	1 731	2 071	1 447	2 646	3 239	4 357	4 974
Portugal	1 899	225	107	104	224	323	179	1 471	880	617
Russian Fed	-	-	-	-	-	-	-	-	-	1
Spain	3 991	3 336	3 274	3 518	3 534	5 161	6 301	6 986	5 607	5 245
Sweden	68	96	78	55	38	44	44	48	107	92
UK	17 680	18 894	19 999	24 456	31 451	29 783	21 395	16 989	15 955	16 249
27 Fishing area total	*47 716*	*48 517 F*	*50 413*	*57 350*	*65 430*	*64 974*	*56 174*	*50 855*	*50 546*	*50 508*
37 Albania	...	...	...	0	46	0	42	48	44	-
France	475	264	128	101	225	225	225	272	557	494
Greece	729	732	1 314	1 508	942	912	757	739	882	694
Italy	2 572	2 149	2 888	2 409	2 345	6 672	2 845	1 705	1 269	1 244

B-34 Miscellaneous demersal fishes / Poissons démersaux divers / Peces demersales diversos

Capture production by species, fishing areas and countries or areas / Captures par espèces, zones de pêche et pays ou zones / Capturas por especies, áreas de pesca y países o áreas

Species, Fishing area Espèce, Zone de pêche Especie, Area de pesca	1992 mt	1993 mt	1994 mt	1995 mt	1996 mt	1997 mt	1998 mt	1999 mt	2000 mt	2001 mt
Malta	0	0	0	0	0	1	0	2	0	-
Morocco	191	143	92	107	110	113	83	66	785	310
Spain	-	-	-	-	-	-	-	949	1 283	1 492
Tunisia	-	32	21	7	6	33	30	44	39	35
37 Fishing area total	*3 967*	*3 320*	*4 443*	*4 132*	*3 674*	*7 956*	*3 982*	*3 825*	*4 859*	*4 269*
Species total	*51 683*	*51 837 F*	*54 856*	*61 482*	*69 104*	*72 930*	*60 156*	*54 680*	*55 405*	*54 777*
Blackbellied angler	**Baudroie rousse**		**Rape negro**		***Lophius budegassa***			**1,95(01)001,02**		**ANK**
34 Greece	-	-	-	-	-	4	7	4	27	-
Mauritania	10 F	10 F	10 F	20 F	20 F	20 F	20 F	20	45	23
Morocco	223	269	350	348	350	320	579	498	85	126
Portugal	113	41	58	82	127	55	26	16	18	2
Spain	176	250 F	350 F	450 F	524	336	524	116	208	369
34 Fishing area total	*522 F*	*570 F*	*768 F*	*900 F*	*1 021 F*	*735 F*	*1 156 F*	*654*	*383*	*520*
Species total	*522 F*	*570 F*	*768 F*	*900 F*	*1 021 F*	*735 F*	*1 156 F*	*654*	*383*	*520*
American angler	**Baudroie d'Amérique**		**Rape americano**		***Lophius americanus***			**1,95(01)001,03**		**ANG**
21 Canada	1 684	1 543	2 340	1 722	1 623	2 073	1 466	1 274	1 265	1 900
Cuba	18	52	2	5	-	-	-	-	-	-
Portugal	37	8	-	2	-	-	-	-	-	-
Russian Fed	13	-	-	-	-	-	-	-	-	-
St Pier Mq	6	-	-	-	-	0	1	0	0	1
Spain	-	-	-	-	-	1	2	0	3	9
USA	16 308	18 765	19 138	25 142	24 203	27 481	26 345	25 027	20 798	23 256
21 Fishing area total	*18 066*	*20 368*	*21 480*	*26 871*	*25 826*	*29 555*	*27 814*	*26 301*	*22 066*	*25 166*
31 USA	-	-	-	-	-	25	12	19	9	12
31 Fishing area total	-	-	-	-	-	*25*	*12*	*19*	*9*	*12*
Species total	*18 066*	*20 368*	*21 480*	*26 871*	*25 826*	*29 580*	*27 826*	*26 320*	*22 075*	*25 178*
Shortspine African angler	**Baudroie africaine**		**Rape africano**		***Lophius vaillanti***			**1,95(01)001,05**		**MVA**
34 Congo Rep	...	...	...	...	...	...	...	...	7	5 F
34 Fishing area total	...	...	...	...	...	...	...	...	*7*	*5 F*
Species total	...	...	...	...	...	...	...	...	*7*	*5 F*
Devil anglerfish	**Baudroie diable**		**Rape diablo**		***Lophius vomerinus***			**1,95(01)001,06**		**MVO**
47 Iceland	-	-	-	-	-	5	2	-	-	-
Namibia	8 078	9 226	12 158	10 130	9 236	10 259	16 701	14 802	14 358	12 401
South Africa	4 858	4 249	4 089	6 246	6 139	6 958	7 903	6 949	7 000 F	9 554
Spain	1 030	600 F	200 F	150 F	58	-	123	32	44	52
47 Fishing area total	*13 966*	*14 075 F*	*16 447 F*	*16 526 F*	*15 433*	*17 222*	*24 729*	*21 783*	*21 402 F*	*22 007*
Species total	*13 966*	*14 075 F*	*16 447 F*	*16 526 F*	*15 433*	*17 222*	*24 729*	*21 783*	*21 402 F*	*22 007*
Anglerfishes nei	**Baudroies, etc. nca**		**Rapes, etc. nep**		***Lophiidae***			**1,95(01)XXX,XX**		**ANF**
41 Brazil	...	...	...	366	294	399	542	794	1 937	1 900 F
Spain	-	-	-	-	-	-	-	-	-	3
UK	-	-	-	-	-	-	-	-	-	2 105
41 Fishing area total	...	...	...	*366*	*294*	*399*	*542*	*794*	*1 937*	*4 008 F*
61 Korea Rep	7 872	6 312	5 064	8 173	11 720	5 045	3 970	3 406	4 930	5 813
61 Fishing area total	*7 872*	*6 312*	*5 064*	*8 173*	*11 720*	*5 045*	*3 970*	*3 406*	*4 930*	*5 813*
Species total	*7 872*	*6 312*	*5 064*	*8 539*	*12 014*	*5 444*	*4 512*	*4 200*	*6 867*	*9 821 F*
Demersal percomorphs nei	**Percomorphes démersaux nca**		**Percomorfos demersales nep**		***Perciformes***			**1,99(XX)XXX,XX**		**DPX**
31 Colombia	2	14	27	11	32	16	4	-	20 F	20 F
Mexico	990	472	634	198	169	190	180	149	171	3 046
Puerto Rico	436	809	421	741	1 059	1 249	1 375	-	-	-
Trinidad Tob	1 300 F	1 462	2 011	2 400 F	2 656	2 572	2 001	1 918	1 812	2 828
Venezuela	2 052	3 071	3 703	1 210	3 532	4 760	3 616	4 118	1 322	1 687
31 Fishing area total	*4 780 F*	*5 828*	*6 796*	*4 560 F*	*7 448*	*8 787*	*7 176*	*6 185*	*3 325 F*	*7 581 F*
34 Cape Verde	...	...	...	...	...	1 450	1 150	1 079	1 314	1 307
Eq Guinea	230 F	210 F	320 F	140 F	310 F	390 F	255	300 F	30 F	30 F
Ghana	300	0	0	0	0	0	0	-	-	-
34 Fishing area total	*530 F*	*210 F*	*320 F*	*140 F*	*310 F*	*1 840 F*	*1 405*	*1 379 F*	*1 344 F*	*1 337 F*
37 France	498	205	2	1	3	3	3	19	43	26
Spain	-	-	-	-	-	-	-	-	841	758
Syria	495 F	511 F	494 F	492 F	606	450	626	530	392	449
37 Fishing area total	*993 F*	*716 F*	*496 F*	*493 F*	*609*	*453*	*629*	*549*	*1 276*	*1 233*
41 Argentina	2 259	936	1 270	1 763	2 420	2 714	2 537	3 540	-	-
Brazil	7 840 F	7 820 F	7 830 F	5 217	5 697	9 825	9 828	12 050	7 644	7 600 F
Japan	-	17	-	62	25	4	89	6	4	-
Uruguay	14	14	18	22	3	57	-	334	51	34
41 Fishing area total	*10 113 F*	*8 787 F*	*9 118 F*	*7 064*	*8 145*	*12 600*	*12 454*	*15 930*	*7 699*	*7 634 F*

B-34 Miscellaneous demersal fishes / Poissons démersaux divers / Peces demersales diversos

Capture production by species, fishing areas and countries or areas / Captures par espèces, zones de pêche et pays ou zones / Capturas por especies, áreas de pesca y países o áreas

Species, Fishing area Espèce, Zone de pêche Especie, Area de pesca	1992 mt	1993 mt	1994 mt	1995 mt	1996 mt	1997 mt	1998 mt	1999 mt	2000 mt	2001 mt
47 Japan	75	260	408	319	27	1 147	25	-	-	-
47 Fishing area total	*75*	*260*	*408*	*319*	*27*	*1 147*	*25*	*...*	*...*	*...*
51 Egypt	5 562	5 600	-	-	-	-	-	-	-	-
Japan	-	-	-	-	-	-	-	-	-	4 416
Kenya	680	407	418	211	1 247	1 188	1 196	1 366	1 224	2 060
Oman	4 245	4 818	6 232	5 175	7 529	4 554	6 938	8 760	5 735	2 421
Seychelles	469	364	208	293	344	163	145	247	372	244
Yemen	3 766	2 441	...	2 050	5 599	9 727 F	14 078 F	6 115 F	6 086 F	7 700 F
51 Fishing area total	*14 722*	*13 630*	*6 858*	*7 729*	*14 719*	*15 632 F*	*22 357 F*	*16 488 F*	*13 417 F*	*16 841 F*
57 Australia	2 300 F	2 300 F	2 000 F	3 000 F	2 500 F	-	-	-	-	-
Sri Lanka	9 870	15 277	10 585	7 088	8 968	9 100	9 210	10 440	14 910	12 290
57 Fishing area total	*12 170 F*	*17 577 F*	*12 585 F*	*10 088 F*	*11 468 F*	*9 100*	*9 210*	*10 440*	*14 910*	*12 290*
71 Australia	96	96	100 F	100 F	100 F	-	-	-	-	-
Fiji Islands	320	210	214 F	...	380	250	-	-	-	-
71 Fishing area total	*416*	*306*	*314 F*	*100 F*	*480 F*	*250*	*-*	*-*	*-*	*-*
77 Mexico	-	-	2	0	1	0	0	0	0	0
77 Fishing area total	*-*	*-*	*2*	*0*	*1*	*0*	*0*	*0*	*0*	*0*
81 Australia	2 000 F	1 500 F	2 000 F	1 000 F	-	-	-	-	-	-
Japan	11 281	2 736	4 358	5 136	4 459	3 661	3 653	3 008	2 891	2 073
81 Fishing area total	*13 281 F*	*4 236 F*	*6 358 F*	*6 136 F*	*4 459*	*3 661*	*3 653*	*3 008*	*2 891*	*2 073*
87 Colombia	150	128	148	1 058	170	220	161	211	200 F	200 F
Ecuador	1 223	1 168	-	-	-	-	-	-	-	-
Japan	-	-	-	-	-	-	-	83	-	-
87 Fishing area total	*1 373*	*1 296*	*148*	*1 058*	*170*	*220*	*161*	*294*	*200 F*	*200 F*
Species total	*58 453 F*	*52 846 F*	*43 403 F*	*37 687 F*	*47 836 F*	*53 690 F*	*57 070 F*	*54 273 F*	*45 062 F*	*49 189 F*
Group total	***2 237 081***	***2 154 697***	***2 521 962***	***2 701 501***	***2 689 794***	***2 738 859***	***2 969 462***	***2 947 202***	***3 018 968***	***3 027 620***

B-35 Herrings, sardines, anchovies / Harengs, sardines, anchois / Arenques, sardinas, anchoas

Capture production by species, fishing areas and countries or areas / Captures par espèces, zones de pêche et pays ou zones / Capturas por especies, áreas de pesca y países o áreas

Species, Fishing area Espèce, Zone de pêche Especie, Area de pesca	1992 mt	1993 mt	1994 mt	1995 mt	1996 mt	1997 mt	1998 mt	1999 mt	2000 mt	2001 mt
Atlantic herring	**Hareng de l'Atlantique**		**Arenque del Atlántico**		***Clupea harengus***				**1,21(05)001,05**	**HE**
21 Canada	215 750	197 597	206 777	193 690	188 843	186 550	190 861	202 046	203 261	199 295
Cuba	129	50	1	6	213	253	134	121	-	-
Japan	56	-	-	-	-	-	-	-	-	-
Latvia	-	289	-	-	-	-	-	-	-	-
Lithuania	85	-	-	-	-	-	-	-	-	-
Russian Fed	173	2	-	-	-	-	5	-	4	1 664
St Pier Mq	-	-	-	-	-	-	-	2	0	0
USA	59 924	56 392	48 667	76 980	104 002	97 706	81 813	79 597	74 037	106 561
21 Fishing area total	*276 117*	*254 330*	*255 445*	*270 676*	*293 058*	*284 509*	*272 813*	*281 766*	*277 302*	*307 520*
27 Belgium	242	58	145	12	2	1	1	1	1	11
Denmark	155 557	169 477	177 529	191 415	153 009	125 300	139 710	137 578	153 899	141 263
Estonia	29 556	32 982	34 493	43 866	45 296	52 435	42 721	44 038	41 735	41 738
Faeroe Is	11 865	4 592	5 500	66 023	57 853	65 949	70 214	56 476	65 270	35 172
Finland	74 051	79 234	98 958	95 897	94 548	91 544	86 350	83 042	81 648	82 867
France	12 622	15 053	24 461	33 240	11 742	21 222	23 398	25 531	25 398	27 577
Germany	66 785	68 594	57 077	55 916	42 153	42 749	47 028	50 857	47 048	50 680
Iceland	123 323	116 617	151 229	284 473	265 413	291 117	277 461	298 435	287 663	178 950
Ireland	52 538	51 468	51 006	46 643	71 953	57 155	58 248	45 334	42 114	40 640
Isle of Man	856	775	716	615	693	821	0	1	...	35
Japan	-	1	0	-	-	-	-	-	-	-
Latvia	25 845	21 949	22 676	24 972	27 523	29 330	24 417	27 163	26 768	26 652
Lithuania	5 768	3 775	4 988	7 058	4 257	3 330	2 368	1 313	1 198	1 639
Netherlands	85 714	89 188	86 007	99 447	77 605	65 448	77 090	78 741	75 221	66 357
Norway	227 472	352 240	549 621	686 705	763 073	923 165	831 844	829 008	799 731	581 161
Poland	52 864	50 833	49 111	45 676	31 246	28 939	21 873	19 229	24 516	37 611
Portugal	0	0	0	0	2	0	0	1	0	2
Romania	-	-	5 565	6 588	1 794	-	-	-	-	-
Russian Fed	43 770	59 093	98 287	121 561	134 380	181 179	139 561	170 601	174 200	125 756
Spain	-	-	-	-	-	-	-	-	232	232
Sweden	195 295	165 158	153 109	157 503	132 153	166 311	201 738	157 541	174 081	125 749
UK	106 209	106 443	104 002	114 571	120 935	103 405	104 627	104 752	82 658	81 363
27 Fishing area total	*1 270 332*	*1 387 530*	*1 674 480*	*2 082 181*	*2 035 630*	*2 249 400*	*2 148 649*	*2 129 642*	*2 103 381*	*1 645 455*
Species total	*1 546 449*	*1 641 860*	*1 929 925*	*2 352 857*	*2 328 688*	*2 533 909*	*2 421 462*	*2 411 408*	*2 380 683*	*1 952 975*
Pacific herring	**Hareng du Pacifique**		**Arenque del Pacífico**		***Clupea pallasii***				**1,21(05)001,07**	**HE**
61 China	2 209	994	1 330	2 325	1 665	15 780	21 796	17 936	15 258	51 950
Japan	2 854	1 482	2 146	3 873	2 021	1 926	2 531	2 579	2 260	2 385
Korea Rep	6 079	3 983	3 865	8 622	5 525	13 214	13 340	21 446	15 203	8 491
Russian Fed	109 279	115 148	85 218	116 787	171 810	313 397	395 595	359 194	361 241	278 511
61 Fishing area total	*120 421*	*121 607*	*92 559*	*131 607*	*181 021*	*344 317*	*433 262*	*401 155*	*393 962*	*341 337*
67 Canada	34 867	41 321	40 909	26 780	22 221	31 500	33 500	28 847	29 290	24 189
USA	66 920	44 492	48 211	48 666	49 300	52 892	39 878	39 038	31 092	38 891
67 Fishing area total	*101 787*	*85 813*	*89 120*	*75 446*	*71 521*	*84 392*	*73 378*	*67 885*	*60 382*	*63 080*
77 USA	5 226	3 849	2 972	4 623	5 328	9 188	1 987	2 267	2 853	2 521
77 Fishing area total	*5 226*	*3 849*	*2 972*	*4 623*	*5 328*	*9 188*	*1 987*	*2 267*	*2 853*	*2 521*
Species total	*227 434*	*211 269*	*184 651*	*211 676*	*257 870*	*437 897*	*508 627*	*471 307*	*457 197*	*406 938*
Goldstripe sardinella	**Sardinelle dorée**		**Sardinela dorada**		***Sardinella gibbosa***				**1,21(05)012,03**	**SA**
57 Indonesia	21 805	28 703	32 992	26 450	29 836	26 335	31 715	34 785	37 358	38 310
57 Fishing area total	*21 805*	*28 703*	*32 992*	*26 450*	*29 836*	*26 335*	*31 715*	*34 785*	*37 358*	*38 310*
71 Indonesia	117 547	123 857	133 460	134 646	127 268	130 579	142 976	127 925	134 861	138 300
71 Fishing area total	*117 547*	*123 857*	*133 460*	*134 646*	*127 268*	*130 579*	*142 976*	*127 925*	*134 861*	*138 300*
Species total	*139 352*	*152 560*	*166 452*	*161 096*	*157 104*	*156 914*	*174 691*	*162 710*	*172 219*	*176 610*
Indian oil sardine	**Sardinelle indienne**		**Sardinela aceitera**		***Sardinella longiceps***				**1,21(05)012,04**	**IO**
51 India	131 974	72 676	30 136	33 351	65 370	120 214	110 288	122 254	277 842	287 628
Oman	...	29 204	30 267	33 053	26 741	16 765	20 650	21 710	40 044	58 960
Pakistan	74 553	92 704	65 050	55 177	52 290	51 930	44 079	30 629	31 167	31 201
Untd Arab Em	7 100	2 953	6 085	3 455	3 491	4 077	4 085	4 144	...	...
Yemen	...	...	...	...	4 120	4 310 F	4 900 F	4 700 F	4 300 F	5 500 F
51 Fishing area total	*213 627*	*197 537*	*131 538*	*125 036*	*152 012*	*197 296 F*	*184 002 F*	*183 437 F*	*353 353 F*	*383 289 F*
57 India	57 041	64 765	66 179	72 565	71 343	91 846	63 063	25 461	49 583	54 039
57 Fishing area total	*57 041*	*64 765*	*66 179*	*72 565*	*71 343*	*91 846*	*63 063*	*25 461*	*49 583*	*54 039*
Species total	*270 668*	*262 302*	*197 717*	*197 601*	*223 355*	*289 142 F*	*247 065 F*	*208 898 F*	*402 936 F*	*437 328 F*
Round sardinella	**Allache**		**Alacha**		***Sardinella aurita***				**1,21(05)012,10**	**SA**
21 USA	-	-	-	-	-	-	-	340	-	-
21 Fishing area total	-	-	-	-	-	-	-	*340*	-	-
31 Cuba	...	...	...	...	...	5	93	-	-	-
Mexico	741	782	3 119	1 643	1 073	2 028	4 253	1 384	1 424	373
USA	805	954	1 053	191	571	512	489	196	614	623
Venezuela	80 079	85 751	112 540	153 037	153 782	140 571	186 060	126 468	73 534	71 168
31 Fishing area total	*81 625*	*87 487*	*116 712*	*154 871*	*155 426*	*143 116*	*190 895*	*128 048*	*75 572*	*72 164*

B-35

Herrings, sardines, anchovies — **Capture production by species, fishing areas and countries or areas**
Harengs, sardines, anchois — **Captures par espèces, zones de pêche et pays ou zones**
Arenques, sardinas, anchoas — **Capturas por especies, áreas de pesca y países o áreas**

Species, Fishing area Espèce, Zone de pêche Especie, Area de pesca	1992 mt	1993 mt	1994 mt	1995 mt	1996 mt	1997 mt	1998 mt	1999 mt	2000 mt	2001 mt
34 Côte dIvoire	24 403	14 492	11 819	12 098	18 491	10 002	10 884	12 206	19 358	18 864
Georgia	3 750 F	2 170 F	800 F	-	-	-	-	-	-	-
Ghana	125 814	92 735	69 895	67 835	118 408	49 394	55 965	57 170	102 043	67 321
Lithuania	35 262	22 810	7 393	...	...	...	10 575	8 680	6 324	4 167
Romania	6 857	292	-	-	-	-	-	-	-	-
Russian Fed	65 582	31 930	17 216	53 303	116 705	101 457	106 318	109 445	42 400	22 557
Senegal	145 518	142 197	128 138	116 766	142 457	152 713	129 429	93 512	112 970	112 120
Spain	-	-	-	-	2 224	3 210	3 411	17	41	49
Togo	74	342	361	335	739	1 300	1 045	1 921	1 912	3 123
34 Fishing area total	*407 260 F*	*306 968 F*	*235 622 F*	*250 337*	*399 024*	*318 076*	*317 627*	*282 951*	*285 048*	*228 201*
37 Israel	498	470	250	250	242	435	557	306	88	80 F
37 Fishing area total	*498*	*470*	*250*	*250*	*242*	*435*	*557*	*306*	*88*	*80 F*
Species total	*489 383 F*	*394 925 F*	*352 584 F*	*405 458*	*554 692*	*461 627*	*509 079*	*411 645*	*360 708*	*300 445 F*
Madeiran sardinella	**Grande allache**		**Machuelo**		***Sardinella maderensis***			**1,21(05)012,17**		**SAE**
34 Côte dIvoire	1 350	2 827	4 492	2 273	1 500 F	1 000 F	500 F	175	426	716
Ghana	14 410	16 984	12 467	13 211	13 619	14 183	15 468	12 105	14 970	15 906
Nigeria	...	...	...	...	2 230	9 387	12 226	11 041	10 158	11 931
Portugal	0	0	0	0	0	1	5	11	2	3
Romania	4 161	53	-	-	-	-	-	-	-	-
Senegal	57 566	73 935	51 405	64 711	108 186	114 824	103 363	105 120	114 749	99 944
Togo	160	150	162	86	166	432	408	614	638	445
34 Fishing area total	*77 647*	*93 949*	*68 526*	*80 281*	*125 701 F*	*139 827 F*	*131 970 F*	*129 066*	*140 943*	*128 945*
Species total	*77 647*	*93 949*	*68 526*	*80 281*	*125 701 F*	*139 827 F*	*131 970 F*	*129 066*	*140 943*	*128 945*
Japanese sardinella	**Sardinelle japonaise**		**Sardinela del Japón**		***Sardinella zunasi***			**1,21(05)012,22**		**JSS**
61 Korea Rep	3 597	24 383	23 974	18 345	10 663	5 593	1 973	6 674	4 603	766
61 Fishing area total	*3 597*	*24 383*	*23 974*	*18 345*	*10 663*	*5 593*	*1 973*	*6 674*	*4 603*	*766*
Species total	*3 597*	*24 383*	*23 974*	*18 345*	*10 663*	*5 593*	*1 973*	*6 674*	*4 603*	*766*
Bali sardinella	**Sardinelle de Bali**		**Sardinela de Balí**		***Sardinella lemuru***			**1,21(05)012,23**		**SAM**
57 Indonesia	52 261	69 363	63 010	40 363	29 647	72 666	92 294	34 143	31 237	33 190
57 Fishing area total	*52 261*	*69 363*	*63 010*	*40 363*	*29 647*	*72 666*	*92 294*	*34 143*	*31 237*	*33 190*
71 Indonesia	84 761	52 676	65 192	58 542	58 943	65 970	61 671	55 143	57 507	61 090
71 Fishing area total	*84 761*	*52 676*	*65 192*	*58 542*	*58 943*	*65 970*	*61 671*	*55 143*	*57 507*	*61 090*
Species total	*137 022*	*122 039*	*128 202*	*98 905*	*88 590*	*138 636*	*153 965*	*89 286*	*88 744*	*94 280*
Brazilian sardinella	**Sardinelle de Brésil**		**Sardinela del Brasil**		***Sardinella brasiliensis***			**1,21(05)012,24**		**BSR**
31 Grenada	4	24	13	0	1	0	0	0	0	0
31 Fishing area total	*4*	*24*	*13*	*0*	*1*	*0*	*0*	*0*	*0*	*0*
41 Brazil	64 842	49 991	84 635	60 212	97 092	117 642	82 283	25 518	17 053	35 000
41 Fishing area total	*64 842*	*49 991*	*84 635*	*60 212*	*97 092*	*117 642*	*82 283*	*25 518*	*17 053*	*35 000*
Species total	*64 846*	*50 015*	*84 648*	*60 212*	*97 093*	*117 642*	*82 283*	*25 518*	*17 053*	*35 000*
Sardinellas nei	**Sardinelles nca**		**Sardinelas nep**		***Sardinella spp***			**1,21(05)012,XX**		**SIX**
27 Spain	520	6	1	30	-	-	3	1 003	-	1
27 Fishing area total	*520*	*6*	*1*	*30*	*-*	*-*	*3*	*1 003*	*-*	*1*
34 Belize	-	-	-	-	-	2 400	2 389	5 445	5 535	1 271
Benin	1 223 F	1 250 F	1 200 F	1 150 F	1 100 F	1 200 F	1 100 F	1 072	750 F	1 434
Bulgaria	-	-	-	-	-	-	3 216	-	-	-
Cameroon	16 008	16 006	18 600 F	24 002	24 002	23 500 F	23 000 F	23 500 F	21 611	21 645
Congo Dem R	1 520 F	1 680 F	1 500 F	1 551 F	1 590 F	1 538 F	1 582 F	1 595 F	1 595 F	1 595 F
Congo Rep	9 492 F	9 469 F	10 914	11 600 F	12 000 F	11 600 F	11 000 F	11 200 F	12 000 F	11 500 F
Cyprus	-	-	-	-	-	...	2 172	4 492	15 043	8 418
Estonia	...	4 786	1 076	-	252	415	3 171	2 474	-	-
France	-	-	-	-	-	...	7 228	624	5 517	1 220
Gabon	...	...	...	1 174	1 897	878	746	128	1 414	1 083
Gambia	4	3	0	1	11	86	64	70	10	81
Germany	-	-	-	-	10 115	23 866	17 448	24 150	-	1 683
Ghana	1 435	1 333	7 545	14 074	20 070	35 985	33 264	19 664	18 358	8 345
Guinea	3 050 F	3 650 F	3 760 F	5 281	6 107	4 176	6 292	8 909	13 288	13 000 F
Ireland	-	-	-	-	-	-	-	-	-	52 980
Latvia	-	-	7 083	17 032	24 209	6 497	6 064	15 031	7 886	7 306
Liberia	410	330	222	876	199	485	620	1 112	887	1 358
Mauritania	2 660 F	2 750 F	1 340 F	3 020 F	5 440 F	4 870 F	1 060 F	4 000 F	4 000 F	4 000 F
Morocco	1 823	251	12	19	20	18	15	81	77	45
Netherlands	-	-	-	-	41 488	86 635	110 091	115 753	122 783	134 490
Nigeria	49 207	26 264	25 825	76 585	102 230	97 286	95 495	83 593	80 892	61 927
Poland	-	-	-	-	7 166	2 553	-	-	-	3 463
St Vincent	-	-	-	-	-	2 000	10 038	7 962	3 369	6 971
Sierra Leone	7 796	7 705	7 681	7 669	7 670 F	7 661	7 450	7 850	7 628	9 845
Ukraine	29 900	11 192	5 862	21 578	83 333	97 009	161 821	74 729	38 120	10 613
Other nei	-	-	-	-	-	-	-	5 812	2 306	2 353
34 Fishing area total	*124 528 F*	*86 669 F*	*92 620 F*	*185 612 F*	*348 899 F*	*410 658 F*	*505 326 F*	*419 246 F*	*363 069 F*	*366 626 F*

B-35 Herrings, sardines, anchovies / Harengs, sardines, anchois / Arenques, sardinas, anchoas

Capture production by species, fishing areas and countries or areas / Captures par espèces, zones de pêche et pays ou zones / Capturas por especies, áreas de pesca y países o áreas

Species, Fishing area Espèce, Zone de pêche Especie, Area de pesca	1992 mt	1993 mt	1994 mt	1995 mt	1996 mt	1997 mt	1998 mt	1999 mt	2000 mt	2001 mt
37 Algeria	9 000 F	10 000 F	13 000 F	17 887	8 150	15 741	9 992	11 393	11 000 F	11 000 F
Croatia	310	435	296	243	163	344	170	25	35	37
Egypt	7 072	10 634	8 147	8 935	8 267	10 621	23 920	39 563	19 689	20 037
Gaza Strip	...	...	...	...	978	2 483	1 780	1 750 F	1 750 F	1 450 F
Libya	6 000 F	6 400 F	7 000 F	7 100 F	7 000 F	7 000 F	7 000 F	7 000 F	7 000 F	7 000 F
Morocco	971	912	1 176	3 005	2 880	2 168	1 230	661	726	3 256
Spain	3 600 F	4 400 F	5 200 F	6 000 F	6 879	7 156	7 993	8 782	7 579	13 555
Syria	339 F	350 F	338 F	336 F	550	300	292	338	251	197
Tunisia	8 832	8 856	7 569	8 591	7 738	8 084	6 327	7 491	11 800	12 942
Yugoslavia	14	7	6	10	11	11	13	17	13	16
37 Fishing area total	*36 138 F*	*41 994 F*	*42 732 F*	*52 107 F*	*42 616 F*	*53 908 F*	*58 717 F*	*77 020 F*	*59 843 F*	*69 490 F*
47 Angola	14 742	12 357	19 731	34 211	17 484	21 014	45 483	57 579	108 211	57 896
Korea Rep	-	42	-	-	-	-	-	-	-	-
Lithuania	-	4 324	1 088	-	-	-	-	-	-	-
Panama	2	-	-	-	-	-	-	-	-	-
Russian Fed	-	-	15 046	7 241	229	2 739	9 537	22 169	5 645	443
Ukraine	-	818	124	-	-	-	-	-	-	-
47 Fishing area total	*14 744*	*17 541*	*35 989*	*41 452*	*17 713*	*23 753*	*55 020*	*79 748*	*113 856*	*58 339*
51 Comoros	1 000	1 000	1 000	1 000 F	1 000 F	1 000 F	1 000 F	950 F	1 050 F	1 000
Egypt	1 613	3 763	2 973	2 822	6 833	5 639	4 973	5 383	5 705	4 343
Eritrea	...	...	...	...	...	...	...	...	2	-
Qatar	0	0	0	0	0	-	0	0	0	0
Tanzania	5 000	5 472	8 563	3 750	14 323	5 000	4 450	14 000	15 000	15 500
Untd Arab Em	8 889	7 573	10 121	8 840	8 933	9 200	9 236	9 510	6 140	6 140 F
51 Fishing area total	*16 502*	*17 808*	*22 657*	*16 412 F*	*31 089 F*	*20 839 F*	*19 659 F*	*29 843 F*	*27 897 F*	*26 983 F*
57 Thailand	20 893	38 440	29 455	54 849	53 086	51 084	56 813	54 321	53 577	53 250
57 Fishing area total	*20 893*	*38 440*	*29 455*	*54 849*	*53 086*	*51 084*	*56 813*	*54 321*	*53 577*	*53 250*
61 China,H.Kong	68	31	19	24	30	12	8	5 F	5 F	6 F
61 Fishing area total	*68*	*31*	*19*	*24*	*30*	*12*	*8*	*5 F*	*5 F*	*6 F*
71 Fiji Islands	73	120	122 F	240	120	140	30	47	35 F	...
Philippines	195 879	243 610	253 286	264 675	257 804	302 341	302 599	279 864	298 466	294 968
Thailand	142 634	113 860	125 179	141 180	161 771	151 708	129 045	128 492	126 733	125 960
71 Fishing area total	*338 586*	*357 590*	*378 587 F*	*406 095*	*419 695*	*454 189*	*431 674*	*408 403*	*425 234 F*	*420 928*
Species total	*551 979 F*	*560 079 F*	*602 060 F*	*756 581 F*	*913 128 F*	*1 014 443 F*	*1 127 220 F*	*1 069 589 F*	*1 043 481 F*	*995 623 F*
Japanese pilchard	**Pilchard du Japon**		**Sardina japonesa**			***Sardinops melanostictus***			**1,21(05)013,01**	**JAP**
61 China	52 986	46 846	68 453	58 434	92 918	124 844	121 120	147 125	153 944	160 825
Japan	2 223 766	1 713 687	1 188 848	661 391	319 354	284 054	167 073	351 207	149 616	178 423
Korea D P Rp	...	...	19 701	63	5	0	...	...	...	...
Korea Rep	46 511	31 285	36 709	13 539	18 560	9 041	7 595	17 142	2 207	129
Russian Fed	165 270	4 314	28	-	-	-	-	3	-	-
61 Fishing area total	*2 488 533*	*1 796 132*	*1 313 739*	*733 427*	*430 837*	*417 939*	*295 788*	*515 477*	*305 767*	*339 377*
Species total	*2 488 533*	*1 796 132*	*1 313 739*	*733 427*	*430 837*	*417 939*	*295 788*	*515 477*	*305 767*	*339 377*
California pilchard	**Pilchard de Californie**		**Sardina monterrey**			***Sardinops caeruleus***			**1,21(05)013,02**	**CPI**
67 USA	-	-	-	1	-	-	22	775	14 368	23 908
67 Fishing area total	*-*	*-*	*-*	*1*	*-*	*-*	*22*	*775*	*14 368*	*23 908*
77 Mexico	279 293	272 630	287 620	320 999	418 093	455 837	339 317	345 458	478 191	609 777
USA	...	...	13 094	42 464	32 503	42 816	42 559	59 169	53 520	51 812
77 Fishing area total	*279 293*	*272 630*	*300 714*	*363 463*	*450 596*	*498 653*	*381 876*	*404 627*	*531 711*	*661 589*
Species total	*279 293*	*272 630*	*300 714*	*363 464*	*450 596*	*498 653*	*381 898*	*405 402*	*546 079*	*685 497*
South American pilchard	**Pilchard sudaméricain**		**Sardina sudamericana**			***Sardinops sagax***			**1,21(05)013,03**	**CHP**
87 Chile	808 838	481 119	194 499	161 557	81 043	40 473	27 966	246 045	60 189	33 271
Ecuador	5 010	23 652	212	75 916	356 480	57 191	1 012	8 821	51 648	42 143
Peru	2 243 225	1 461 759	1 551 833	1 265 658	1 056 413	625 143	908 291	187 924	226 294	60 298
87 Fishing area total	*3 057 073*	*1 966 530*	*1 746 544*	*1 503 131*	*1 493 936*	*722 807*	*937 269*	*442 790*	*338 131*	*135 712*
Species total	*3 057 073*	*1 966 530*	*1 746 544*	*1 503 131*	*1 493 936*	*722 807*	*937 269*	*442 790*	*338 131*	*135 712*
Southern African pilchard	**Pilchard de l'Afrique australe**		**Sardina de Africa austral**			***Sardinops ocellatus***			**1,21(05)013,05**	**PIA**
47 Namibia	80 784	114 812	116 429	42 797	1 171	27 685	68 562	44 653	25 388	7 940
South Africa	53 436	50 702	93 438	115 205	105 210	116 995	128 019	131 316	136 060	192 160
Ukraine	-	1 261	34	-	-	-	-	-	-	-
47 Fishing area total	*134 220*	*166 775*	*209 901*	*158 002*	*106 381*	*144 680*	*196 581*	*175 969*	*161 448*	*200 100*
51 South Africa	-	-	-	-	-	1	...	...	0	0
51 Fishing area total	*-*	*-*	*-*	*-*	*-*	*1*	*...*	*...*	*0*	*0*
Species total	*134 220*	*166 775*	*209 901*	*158 002*	*106 381*	*144 681*	*196 581*	*175 969*	*161 448*	*200 100*
Australian pilchard	**...B**		**...C**			***Sardinops neopilchardus***			**1,21(05)013,09**	**SRP**
81 New Zealand	...	...	...	209	169	385	519	894	1 253	1 399
81 Fishing area total	*...*	*...*	*...*	*209*	*169*	*385*	*519*	*894*	*1 253*	*1 399*

B-35 Herrings, sardines, anchovies / Harengs, sardines, anchois / Arenques, sardinas, anchoas

Capture production by species, fishing areas and countries or areas / Captures par espèces, zones de pêche et pays ou zones / Capturas por especies, áreas de pesca y países o áreas

Species, Fishing area Espèce, Zone de pêche Especie, Area de pesca	1992 mt	1993 mt	1994 mt	1995 mt	1996 mt	1997 mt	1998 mt	1999 mt	2000 mt	2001 mt
Species total	...	...	...	*209*	*169*	*385*	*519*	*894*	*1 253*	*1 399*
Brazilian menhaden	**Menhaden du Brésil**		**Savela**			***Brevoortia aurea***			**1,21(05)024,01**	**MHS**
41 Brazil	3 420 F	3 400 F	3 410 F	11 133	6 294	2 396	2 936	2 202	1 123	1 100 F
41 Fishing area total	*3 420 F*	*3 400 F*	*3 410 F*	*11 133*	*6 294*	*2 396*	*2 936*	*2 202*	*1 123*	*1 100 F*
Species total	*3 420 F*	*3 400 F*	*3 410 F*	*11 133*	*6 294*	*2 396*	*2 936*	*2 202*	*1 123*	*1 100 F*
Argentine menhaden	**Menhaden d'Argentine**		**Lacha**			***Brevoortia pectinata***			**1,21(05)024,02**	**MHP**
41 Argentina	357	137	573	294	427	893	104	205	271	...
Uruguay	49	116	483	130	84	219	415	86	66	103
41 Fishing area total	*406*	*253*	*1 056*	*424*	*511*	*1 112*	*519*	*291*	*337*	*103*
Species total	*406*	*253*	*1 056*	*424*	*511*	*1 112*	*519*	*291*	*337*	*103*
Atlantic menhaden	**Menhaden tyran**		**Lacha tirana**			***Brevoortia tyrannus***			**1,21(05)024,03**	**MHA**
21 USA	313 109	327 241	249 047	338 422	280 496	322 239	248 451	189 185	183 310	236 246
21 Fishing area total	*313 109*	*327 241*	*249 047*	*338 422*	*280 496*	*322 239*	*248 451*	*189 185*	*183 310*	*236 246*
31 USA	27 437	20 565	37 455	27 314	24 169	...	27 779	18 815	23 812	25 161
31 Fishing area total	*27 437*	*20 565*	*37 455*	*27 314*	*24 169*	...	*27 779*	*18 815*	*23 812*	*25 161*
Species total	*340 546*	*347 806*	*286 502*	*365 736*	*304 665*	*322 239*	*276 230*	*208 000*	*207 122*	*261 407*
Gulf menhaden	**Menhaden écailleux**		**Lacha escamuda**			***Brevoortia patronus***			**1,21(05)024,04**	**MHG**
31 USA	432 848	551 822	767 448	472 039	491 612	597 565	497 461	694 242	591 434	528 500
31 Fishing area total	*432 848*	*551 822*	*767 448*	*472 039*	*491 612*	*597 565*	*497 461*	*694 242*	*591 434*	*528 500*
Species total	*432 848*	*551 822*	*767 448*	*472 039*	*491 612*	*597 565*	*497 461*	*694 242*	*591 434*	*528 500*
Rainbow sardine	**Sardine arc-en-ciel**		**Sardina arco iris**			***Dussumieria acuta***			**1,21(05)029,01**	**RAS**
57 Indonesia	2 380	3 021	4 659	4 604	5 013	4 909	4 729	6 072	5 081	5 190
57 Fishing area total	*2 380*	*3 021*	*4 659*	*4 604*	*5 013*	*4 909*	*4 729*	*6 072*	*5 081*	*5 190*
71 Indonesia	11 380	13 657	14 356	16 045	16 893	17 920	17 592	18 113	18 680	19 080
Philippines	22 519	9 271	21 079	19 255	19 441	15 216	20 639	24 080	12 993	13 374
71 Fishing area total	*33 899*	*22 928*	*35 435*	*35 300*	*36 334*	*33 136*	*38 231*	*42 193*	*31 673*	*32 454*
Species total	*36 279*	*25 949*	*40 094*	*39 904*	*41 347*	*38 045*	*42 960*	*48 265*	*36 754*	*37 644*
Slender rainbow sardine	**Sardine arc-en-ciel gracile**		**Sardina arco iris grácil**			***Dussumieria elopsoides***			**1,21(05)029,02**	**RAL**
61 China,H.Kong	373	64	64	154	48	19	17	10 F	15 F	17 F
61 Fishing area total	*373*	*64*	*64*	*154*	*48*	*19*	*17*	*10 F*	*15 F*	*17 F*
Species total	*373*	*64*	*64*	*154*	*48*	*19*	*17*	*10 F*	*15 F*	*17 F*
Bonga shad	**Ethmalose d'Afrique**		**Sábalo africano**			***Ethmalosa fimbriata***			**1,21(05)030,02**	**BOA**
01 Benin	1 990 F	2 100 F	2 095	1 963	1 900 F	1 850 F	1 800 F	1 798	1 500 F	1 700 F
01 Fishing area total	*1 990 F*	*2 100 F*	*2 095*	*1 963*	*1 900 F*	*1 850 F*	*1 800 F*	*1 798*	*1 500 F*	*1 700 F*
34 Benin	...	...	...	...	...	...	...	8	5 F	8
Cameroon	16 000	16 000	18 600 F	24 000	24 000	23 500 F	23 000 F	23 500 F	21 609	21 645
Congo Rep	37 F	37 F	22 F	20 F	20 F	15 F	10 F	10 F	...	10 F
Côte dIvoire	12 000	9 500	10 000	10 000 F	9 500 F	9 500 F	11 000 F	11 006 F	11 009 F	10 512 F
Gabon	9 500 F	10 000 F	10 000	11 787	13 046	14 695	19 284	17 408	14 788	12 733
Gambia	12 019	14 053	16 897	13 897	22 648	21 523	21 952	22 750	20 508	18 516
Ghana	2 408	1 138	573	1 073	1 197	9 762	158	766	963	282
Guinea	22 000 F	23 000 F	25 000 F	23 596	26 051	29 529	27 852	33 780	29 015	28 500 F
Liberia	34	14	26	9	70	17	33	123	37	123
Lithuania	-	224	-	-		-	-	-	-	-
Mauritania	...	...	...	...	...	...	...	...	2 F	...
Nigeria	38 187	24 501	25 645	15 072	4 643	28 000	30 216	18 529	17 570	19 049
Senegal	12 164	12 229	13 503	17 277	17 783	17 147	16 496	29 468	22 032	31 675
Sierra Leone	22 096	21 838	21 771	21 738	21 700 F	21 807	21 840	22 400	21 621	24 790
34 Fishing area total	*146 445 F*	*132 534 F*	*142 037 F*	*138 469 F*	*140 658 F*	*175 495 F*	*171 841 F*	*179 748 F*	*159 159 F*	*167 843 F*
Species total	*148 435 F*	*134 634 F*	*144 132 F*	*140 432 F*	*142 558 F*	*177 345 F*	*173 641 F*	*181 546 F*	*160 659 F*	*169 543 F*
Red-eye round herring	**Shadine ronde**		**Sardineta canalera**			***Etrumeus teres***			**1,21(05)031,01**	**RRH**
37 Egypt	-	-	-	-	-	-	-	1 403	...	...
37 Fishing area total	-	-	-	-	-	-	-	*1 403*	...	...
51 Egypt	-	-	-	-	-	-	-	2 135	...	...
51 Fishing area total	-	-	-	-	-	-	-	*2 135*	...	...
61 China,Taiwan	1 410	2 144	1 615	1 653	1 462	2 152	1 248	893	1 118	1 306
Japan	61 208	60 095	68 374	47 590	49 752	55 043	48 441	28 712	23 833	31 525
61 Fishing area total	*62 618*	*62 239*	*69 989*	*49 243*	*51 214*	*57 195*	*49 689*	*29 605*	*24 951*	*32 831*
77 USA	-	-	-	-	-	72	7	58	42	2

B-35 Herrings, sardines, anchovies / Harengs, sardines, anchois / Arenques, sardinas, anchoas

Capture production by species, fishing areas and countries or areas / Captures par espèces, zones de pêche et pays ou zones / Capturas por especies, áreas de pesca y países o áreas

Species, Fishing area Espèce, Zone de pêche Especie, Area de pesca	1992 mt	1993 mt	1994 mt	1995 mt	1996 mt	1997 mt	1998 mt	1999 mt	2000 mt	2001 mt
77 Fishing area total	*-*	*-*	*-*	*-*	*-*	*72*	*7*	*58*	*42*	*2*
87 Ecuador	8 185	23 484	30 748	4 946	34 349	1 095	8 873	3 636	4 414	28
87 Fishing area total	*8 185*	*23 484*	*30 748*	*4 946*	*34 349*	*1 095*	*8 873*	*3 636*	*4 414*	*28*
Species total	*70 803*	*85 723*	*100 737*	*54 189*	*85 563*	*58 362*	*58 569*	*36 837*	*29 407*	*32 861*
Whitehead's round herring	**Shadine de Angola**		**Sardina angoleña**		***Etrumeus whiteheadi***				**1,21(05)031,04**	**WRR**
47 Namibia	1 663	5 090	1 378	1 934	20 656	5 070	5 193	176	1 127	1 432
South Africa	47 341	56 329	54 147	76 858	47 117	92 209	52 476	58 856	37 750	55 330
47 Fishing area total	*49 004*	*61 419*	*55 525*	*78 792*	*67 773*	*97 279*	*57 669*	*59 032*	*38 877*	*56 762*
Species total	*49 004*	*61 419*	*55 525*	*78 792*	*67 773*	*97 279*	*57 669*	*59 032*	*38 877*	*56 762*
Scaled sardines	**Harengules**		**Sardinetas**		***Harengula spp***				**1,21(05)033,XX**	**SAS**
31 Cuba	692	943	1 112	1 045	707	947	649	766	770 F	770 F
Dominican Rp	126	80	44	72	111	64	55	18	27	21
Grenada	0	3	2	0	1	0	2	0	0	0
31 Fishing area total	*818*	*1 026*	*1 158*	*1 117*	*819*	*1 011*	*706*	*784*	*797 F*	*791 F*
41 Brazil	-	-	-	-	20	12	2	162	126	120 F
41 Fishing area total	*-*	*-*	*-*	*-*	*20*	*12*	*2*	*162*	*126*	*120 F*
Species total	*818*	*1 026*	*1 158*	*1 117*	*839*	*1 023*	*708*	*946*	*923 F*	*911 F*
Pacific thread herring	**Chardin du Pacifique**		**Machuelo hebra pinchagua**		***Opisthonema libertate***				**1,21(05)042,01**	**THP**
77 Panama	32 746	29 752	40 663	21 224	32 517	26 266	49 472	38 746	63 532	71 280 F
77 Fishing area total	*32 746*	*29 752*	*40 663*	*21 224*	*32 517*	*26 266*	*49 472*	*38 746*	*63 532*	*71 280 F*
87 Ecuador	29 229	20 314	69 892	40 911	41 041	43 145	40 530	22 253	20 519	19 886
87 Fishing area total	*29 229*	*20 314*	*69 892*	*40 911*	*41 041*	*43 145*	*40 530*	*22 253*	*20 519*	*19 886*
Species total	*61 975*	*50 066*	*110 555*	*62 135*	*73 558*	*69 411*	*90 002*	*60 999*	*84 051*	*91 166 F*
Atlantic thread herring	**Chardin fil**		**Machuelo hebra atlántico**		***Opisthonema oglinum***				**1,21(05)042,02**	**THA**
21 USA	-	-	-	9	1 686	9	-	1 148	0	-
21 Fishing area total	*-*	*-*	*-*	*9*	*1 686*	*9*	*-*	*1 148*	*0*	*-*
31 Cuba	1 572	886	1 484	2 005	2 361	1 900	1 286	1 756	1 750 F	1 750 F
Dominican Rp	239	121	118	369	127	188	111	99	130	106
Grenada	0	0	0	0	0	0	0	0	0	0
USA	3 688	5 360	2 583	5 056	2 852	7 539	2 576	422	3 056	1 256
Venezuela	455	206	269	307	294	5 564	10 619	14 789	9 866	3 650
31 Fishing area total	*5 954*	*6 573*	*4 454*	*7 737*	*5 634*	*15 191*	*14 592*	*17 066*	*14 802 F*	*6 762 F*
41 Brazil	...	...	...	6 661	3 762	3 712	8 082	8 204	12 455	12 300 F
41 Fishing area total	*...*	*...*	*...*	*6 661*	*3 762*	*3 712*	*8 082*	*8 204*	*12 455*	*12 300 F*
Species total	*5 954*	*6 573*	*4 454*	*14 407*	*11 082*	*18 912*	*22 674*	*26 418*	*27 257 F*	*19 062 F*
Silver-stripe round herring	**Hareng gracile**		**Arenquillo de banda**		***Spratelloides gracilis***				**1,21(05)049,01**	**SRH**
61 China,Taiwan	741	624	1 127	827	517	650	886	521	639	505
61 Fishing area total	*741*	*624*	*1 127*	*827*	*517*	*650*	*886*	*521*	*639*	*505*
71 Fiji Islands	66	80	82 F	150	140	135	10	40	30 F	...
71 Fishing area total	*66*	*80*	*82 F*	*150*	*140*	*135*	*10*	*40*	*30 F*	*...*
Species total	*807*	*704*	*1 209 F*	*977*	*657*	*785*	*896*	*561*	*669 F*	*505*
European pilchard(=Sardine)	**Sardine commune**		**Sardina europea**		***Sardina pilchardus***				**1,21(05)064,01**	**PIL**
27 Denmark	37 265	53 394	39 349	36 196	13 704	1 740	17 337	17 676	7 893	1 399
France	10 637	9 962	9 904	11 197	10 378	14 321	11 907	27 156	13 042	13 536
Germany	4	-	2	35	-	13	166	143	307	463
Ireland	-	-	-	-	-	-	-	3 195	2 592	7 855
Netherlands	42	109	20	116	93	518	2 709	5 698	6 825	807
Portugal	83 316	90 405	94 468	87 710	86 853	81 475	82 985	71 956	66 283	71 907
Spain	46 932	53 662	44 329	39 590	38 822	33 652	35 068	31 594	22 460	28 846
Sweden	-	-	-	-	-	-	-	-	-	1 031
UK	4 494	4 917	2 081	7 133	7 304	7 400	6 873	4 815	4 358	10 427
27 Fishing area total	*182 690*	*212 449*	*190 153*	*181 977*	*157 154*	*139 119*	*157 045*	*162 233*	*123 760*	*136 271*
34 Belize	-	-	-	-	-	2 126	2 229	595	903	127
Bulgaria	2 062	-	-	-	-	-	48	-	-	-
China	-	-	0	-	-	-	-	-	-	2
Cyprus	-	-	-	-	...	...	-	12	559	29
Estonia	6 782	1 152	1 231	163	274	480	-	-	-	-
France	-	-	-	-	-	...	30	24	205	-
Georgia	12 830 F	6 900 F	2 550 F	900 F	-	-	-	-	-	-
Germany	-	-	-	-	50	3 282	4 615	1 303	-	37
Honduras	-	-	-	1	-	-	-	-	-	-
Latvia	8 011	7 878	1 387	-	474	-	7	23	633	54
Lithuania	6 773	7 202	...	...	...	...	20	6	292	22
Morocco	298 352	364 710	476 947	556 152	376 925	487 496	426 474	415 188	522 504	751 812

B-35 Herrings, sardines, anchovies / Harengs, sardines, anchois / Arenques, sardinas, anchoas

Capture production by species, fishing areas and countries or areas / Captures par espèces, zones de pêche et pays ou zones / Capturas por especies, áreas de pesca y países o áreas

Species, Fishing area Espèce, Zone de pêche Especie, Area de pesca	1992 mt	1993 mt	1994 mt	1995 mt	1996 mt	1997 mt	1998 mt	1999 mt	2000 mt	2001 mt
Netherlands	-	-	-	-	1 149	5 970	6 289	1 926	11 037	10 979
Norway	-	-	-	-	-	-	3 421	-	-	-
Panama	2	-	-	-	-	-	-	-	-	-
Poland	-	-	-	-	2 439	1 269	-	-	-	-
Portugal	31	7	3	1	2	2	7	16	36	40
Romania	8 806	19	-	-	-	-	-	-	-	-
Russian Fed	144 627	67 523	53 845	47 526	56 168	24 864	5 100	5 504	11 200	1 829
St Vincent	-	-	-	-	-	46	26	96	340	41
Sao Tome Prn	14 F	17 F	25 F	26 F	20	...	20 F	30 F	30 F	30 F
Spain	96 830	105 000 F	115 000 F	125 000 F	132 912	114 103	127 523	59 029	9 615	1 118
Ukraine	139 065	50 827	32 400	49 354	44 221	10 126	12 828	49 136	40 902	27 939
Other nei	0	-	-	-	-	-	-	226	1 876	541
34 Fishing area total	*724 185 F*	*611 235 F*	*683 388 F*	*779 123 F*	*614 634*	*649 764*	*588 637 F*	*533 114 F*	*600 132 F*	*794 600 F*
37 Albania	34	50 F	100 F	235	196	28	28	40	45	123
Algeria	60 000 F	63 000 F	87 000 F	58 989	49 906	49 142	49 295	56 724	55 000 F	55 000 F
Croatia	15 194	12 773	7 105	6 377	9 199	6 996	12 500	10 500	11 226	9 097
Cyprus	-	-	-	-	-	-	-	-	5	1
France	12 421	12 243	10 589	13 452	6 174	6 652	6 652	11 681	15 962	15 539
Greece	20 691	20 702	20 273	20 413	18 896	20 561	17 734	15 214	16 026	14 395
Italy	26 113	34 013	29 679	36 825	42 129	38 174	36 387	28 876	25 805	23 980
Morocco	23 890	17 478	15 371	14 762	16 437	10 325	9 325	14 544	17 281	11 411
Romania	-	-	-	-	2	-	-	-	-	-
Slovenia	3 432	1 748	1 907	1 719	1 982	1 973	1 788	1 614	1 415	1 219
Spain	47 500 F	48 000 F	48 500 F	49 000 F	50 222	44 688	37 829	37 608	48 953	41 180
Tunisia	10 775	9 577	13 140	11 940	10 389	10 767	8 907	13 394	15 001	13 988
Turkey	29 765	32 911	26 399	33 812	18 972	20 500	23 600	22 000	16 500	10 000
Yugoslavia	74	68	58	57	42	45	49	49	36	28
37 Fishing area total	*249 889 F*	*252 563 F*	*260 121 F*	*247 581 F*	*224 546*	*209 851*	*204 094*	*212 244*	*223 255 F*	*195 961 F*
Species total	*1 156 764 F*	*1 076 247 F*	*1 133 662 F*	*1 208 681 F*	*996 334*	*998 734*	*949 776 F*	*907 591 F*	*947 147 F*	*1 126 832 F*
European sprat	**Sprat**		**Espadín**		***Sprattus sprattus***				**1,21(05)066,01**	**SPR**
27 Belgium	6	1	1	2	3	7	4	2	2	2
Denmark	99 343	136 871	240 233	258 179	226 135	284 489	270 439	282 299	276 878	256 517
Estonia	4 139	5 763	9 079	13 051	22 493	39 693	32 165	36 407	41 394	40 777
Faeroe Is	170	-	1 827	598	100	-	753	1 719	...	...
Finland	892	205	497	4 104	14 351	19 851	27 014	18 886	23 242	15 850
France	98	37	293	36	597	68	0	83	97	9
Germany	608	8 278	374	231	161	427	4 551	183	22	791
Ireland	508	2 352	232	799	4 214	2 085	1 578	5 826	6 032	455
Latvia	17 388	12 553	20 132	24 383	34 211	49 314	44 858	42 834	46 186	42 769
Lithuania	3 279	2 779	2 789	4 799	10 165	6 018	4 460	3 117	1 682	3 135
Netherlands	229	326	609	402	293	806	54	264	307	136
Norway	32 961	47 038	44 078	40 969	59 115	7 051	35 166	22 214	6 425	12 465
Poland	30 127	33 700	44 556	46 182	77 472	105 298	59 090	71 706	84 324	85 757
Portugal	-	-	-	1	0	-	-	-	-	-
Russian Fed	9 112	10 745	15 904	14 934	18 287	22 194	21 078	31 627	30 369	31 959
Spain	10	877	317	-	5	17	-	6	17	17
Sweden	59 349	96 920	170 900	165 604	168 582	126 361	149 664	112 452	91 164	88 562
UK	10 217	7 352	9 338	5 906	7 172	8 317	7 021	15 168	8 336	5 091
27 Fishing area total	*268 436*	*365 797*	*561 159*	*580 180*	*643 356*	*671 996*	*657 895*	*644 793*	*616 477*	*584 292*
37 Bulgaria	2 353	2 174	2 200 F	2 874	3 535	3 646	3 275	3 595	1 737	695
France	0	0	0	-	-	-	-	-	-	-
Georgia	830	232	308	292	185	85	24	45	...	30
Greece	352	148	169	178	262	279	216	110	266	474
Romania	2 074	2 439	2 203	1 982	2 014	3 318	3 293	1 933	1 803	1 792
Russian Fed	3 221	694	1 013	1 263	1 537	706	1 243	4 473	5 543	11 122
Slovenia	26	37	14	11	7	1	-	2	3	8
Ukraine	11 492	9 154	12 615	15 218	20 720	20 208	30 282	29 238	32 655	49 004
37 Fishing area total	*20 348*	*14 878*	*18 522 F*	*21 818*	*28 260*	*28 243*	*38 333*	*39 396*	*42 007*	*63 125*
Species total	*288 784*	*380 675*	*579 681 F*	*601 998*	*671 616*	*700 239*	*696 228*	*684 189*	*658 484*	*647 417*
Falkland sprat	**Sprat des îles Falkland**		**Espadín de las Malvinas**		***Sprattus fuegensis***				**1,21(05)066,02**	**FAS**
41 Falkland Is	0	0	0	0	0	0	29	17	97	4
41 Fishing area total	*0*	*0*	*0*	*0*	*0*	*0*	*29*	*17*	*97*	*4*
Species total	*0*	*0*	*0*	*0*	*0*	*0*	*29*	*17*	*97*	*4*
Pacific menhaden	**Menhaden du Pacifique**		**Machete**		***Ethmidium maculatum***				**1,21(05)067,02**	**MES**
87 Chile	1 883	1 553	1 992	2 347	4 023	6 106	1 534	3 588	4 977	4 931
Peru	6 018	5 860	4 348	3 140	5 769	7 135	39 311	25 848	19 014	9 085
87 Fishing area total	*7 901*	*7 413*	*6 340*	*5 487*	*9 792*	*13 241*	*40 845*	*29 436*	*23 991*	*14 016*
Species total	*7 901*	*7 413*	*6 340*	*5 487*	*9 792*	*13 241*	*40 845*	*29 436*	*23 991*	*14 016*
Bluestripe herring	**Hareng à bande bleue**		**Arenque banda azul**		***Herklotsichthys quadrimaculat.***				**1,21(05)072,01**	**HES**
71 Fiji Islands	71	65	66 F	185	140	120	22	20	20 F	...
71 Fishing area total	*71*	*65*	*66 F*	*185*	*140*	*120*	*22*	*20*	*20 F*	*...*
Species total	*71*	*65*	*66 F*	*185*	*140*	*120*	*22*	*20*	*20 F*	*...*

B-35 Herrings, sardines, anchovies / Harengs, sardines, anchois / Arenques, sardinas, anchoas

Capture production by species, fishing areas and countries or areas / Captures par espèces, zones de pêche et pays ou zones / Capturas por especies, áreas de pesca y países o áreas

Species, Fishing area Espèce, Zone de pêche Especie, Area de pesca	1992 mt	1993 mt	1994 mt	1995 mt	1996 mt	1997 mt	1998 mt	1999 mt	2000 mt	2001 mt
Araucanian herring	**Hareng araucian**		**Sardina araucana**			***Strangomera bentincki***			**1,21(05)078,01**	**CKI**
87 Chile	452 012	244 125	341 250	126 715	446 669	441 154	317 564	782 142	722 522	324 617
87 Fishing area total	*452 012*	*244 125*	*341 250*	*126 715*	*446 669*	*441 154*	*317 564*	*782 142*	*722 522*	*324 617*
Species total	*452 012*	*244 125*	*341 250*	*126 715*	*446 669*	*441 154*	*317 564*	*782 142*	*722 522*	*324 617*
European anchovy	**Anchois**		**Boquerón**			***Engraulis encrasicolus***			**1,21(06)002,01**	**ANE**
27 Denmark	-	-	-	759	-	-	-	-	-	0
France	13 993	20 120	17 247	10 296	15 506	10 995	18 945	15 301	17 233	16 339
Germany	-	-	-	-	0	-	16	-	-	-
Latvia	6	3	-	-	-	-	-	-	-	-
Netherlands	-	24	2	20	6	1	16	3	-	3
Portugal	138	23	244	2 530	2 775	633	1 657	1 408	310	855
Russian Fed	-	-	-	-	-	-	-	-	998	3 215
Spain	13 452	10 229	12 378	23 407	10 693	10 662	13 611	20 460	19 555	25 094
UK	...	...	...	...	...	...	79	3	0	274
27 Fishing area total	*27 589*	*30 399*	*29 871*	*37 012*	*28 980*	*22 291*	*34 324*	*37 175*	*38 096*	*45 780*
34 Belize	-	-	-	-	-	300	278	2 118	6 686	2 245
Benin	92 F	100 F	125 F	200 F	250 F	350 F	400 F	478	330 F	852
Bulgaria	-	-	-	-	-	-	848	-	-	-
Cyprus	-	-	-	-	...	...	35	11 879	14 705	13 710
France	-	-	-	-	-	...	69	1	-	-
Georgia	2 720 F	1 470 F	550 F	100 F	-	-	-	-	-	-
Germany	-	-	-	-	-	-	-	-	-	42
Ghana	85 384	81 350	60 519	65 497	98 341	82 724	44 644	32 107	83 501	68 175
Greece	8	4	6	0	-	-	2	-	-	-
Latvia	22	-	-	-	-	-	1 978	4 876	10 142	8 623
Lithuania	2	-	-	...	...	...	3 612	13 774	16 137	8 441
Morocco	16 654	10 353	7 516	10 489	12 042	24 272	40 442	34 280	21 708	47 130
Portugal	0	0	0	-	-	-	-	-	-	-
Romania	255	-		-	-	-	-	-	-	-
Russian Fed	2 343	850	-	416	549	18 298	44 893	31 139	27 850	12 344
St Vincent	-	-	-	-	-	1 100	7 182	3 702	4 841	5 653
Senegal	353	147	8 197	73	34	31	307	1 209	964	1 015
Sierra Leone	185	183	182	182	180 F	182	150	43	181	...
Spain	-	-	-	-	3 084	-	164	140	18	60
Togo	3 551	7 830	4 573	4 779	7 072	4 758	6 325	6 678	7 164	6 660
Ukraine	7	...	...	...	...	...	-	-	-	-
Other nei	-	-	-	-	-	-	-	3 560	4 419	3 815
34 Fishing area total	*111 576 F*	*102 287 F*	*81 668 F*	*81 736 F*	*121 552 F*	*132 015 F*	*151 329 F*	*145 984*	*198 646 F*	*178 765*
37 Albania	...	...	...	0	2	0	-	-	-	4
Algeria	2 500 F	2 700 F	3 500 F	1 913	1 330	1 855	3 511	3 141	3 000 F	3 000 F
Bulgaria	...	...	...	35	23	44	48	36	64	102
Croatia	715	555	298	359	220	545	990	3 000	3 735	2 850
France	5 214	6 533	5 444	5 489	3 993	5 228	6 463	9 375	9 246	7 360
Georgia	6 871	1 656	857	1 301	1 232	2 288	2 346	1 264	2 080	1 652
Greece	11 512	14 600	17 333	13 876	15 073	14 583	17 097	16 456	9 863	10 770
Italy	18 224	21 219	30 840	42 746	40 541	53 439	44 429	39 783	50 728	53 047
Morocco	404	93	379	691	405	683	512	5 933	388	318
Romania	85	374	197	189	138	45	146	155	204	186
Russian Fed	7 294	2 137	4 600	10 071	2 954	3 283	2 465	2 268	5 292	7 766
Slovenia	31	20	18	29	24	33	51	75	96	97
Spain	22 400 F	20 600 F	18 800 F	17 000 F	16 046	15 558	11 103	10 470	8 536	11 908
Tunisia	106	20	106	119	19	55	114	199	2	269
Turkey	174 626	227 130	294 418	387 574	290 680	241 000	228 000	350 000	280 000	320 000
Ukraine	12 304	12 858	15 987	18 505	4 398	9 444	3 914	5 527	16 390	16 668
Yugoslavia	0	0	0	1	0	1	2	4	5	10
37 Fishing area total	*262 286 F*	*310 495 F*	*392 777 F*	*499 898 F*	*377 078*	*348 084*	*321 191*	*447 686*	*389 629 F*	*436 007 F*
Species total	*401 451 F*	*443 181 F*	*504 316 F*	*618 646 F*	*527 610 F*	*502 390 F*	*506 844 F*	*630 845*	*626 371 F*	*660 552 F*
Japanese anchovy	**Anchois japonais**		**Anchoíta japonesa**			***Engraulis japonicus***			**1,21(06)002,02**	**JAN**
61 China	192 720	557 237	438 955	489 066	671 376	1 201 964	1 373 328	1 096 916	1 142 884	1 260 712
China,Taiwan	693	415	239	305	466	515	425	650	589	695
Japan	300 892	194 511	188 034	251 958	345 517	233 113	470 616	484 230	381 020	301 168
Korea Rep	168 235	249 209	193 398	230 679	237 128	230 911	249 519	238 463	201 192	273 927
61 Fishing area total	*662 540*	*1 001 372*	*820 626*	*972 008*	*1 254 487*	*1 666 503*	*2 093 888*	*1 820 259*	*1 725 685*	*1 836 502*
Species total	*662 540*	*1 001 372*	*820 626*	*972 008*	*1 254 487*	*1 666 503*	*2 093 888*	*1 820 259*	*1 725 685*	*1 836 502*
Argentine anchovy	**Anchois d'Argentine**		**Anchoíta**			***Engraulis anchoita***			**1,21(06)002,06**	**ANA**
41 Argentina	19 289	19 149	19 458	24 457	21 001	25 198	13 350	9 832	12 157	11 284
Uruguay	66	29	25	41	22	13	67	3 193	6	318
41 Fishing area total	*19 355*	*19 178*	*19 483*	*24 498*	*21 023*	*25 211*	*13 417*	*13 025*	*12 163*	*11 602*
Species total	*19 355*	*19 178*	*19 483*	*24 498*	*21 023*	*25 211*	*13 417*	*13 025*	*12 163*	*11 602*
Californian anchovy	**Anchois de Californie**		**Anchoa de California**			***Engraulis mordax***			**1,21(06)002,07**	**NPA**
67 USA	43	44	78	130	86	59	103	98	0	70
67 Fishing area total	*43*	*44*	*78*	*130*	*86*	*59*	*103*	*98*	*0*	*70*
77 Mexico	3 406	343	195	24 071	9 598	2 147	782	5 814	7 973	418
USA	6 162	4 395	3 702	2 949	4 419	5 719	1 450	5 225	11 487	19 189

B-35 Herrings, sardines, anchovies / Harengs, sardines, anchois / Arenques, sardinas, anchoas

Capture production by species, fishing areas and countries or areas / Captures par espèces, zones de pêche et pays ou zones / Capturas por especies, áreas de pesca y países o áreas

Species, Fishing area Espèce, Zone de pêche Especie, Area de pesca	1992 mt	1993 mt	1994 mt	1995 mt	1996 mt	1997 mt	1998 mt	1999 mt	2000 mt	2001 mt
77 Fishing area total	*9 568*	*4 738*	*3 897*	*27 020*	*14 017*	*7 866*	*2 232*	*11 039*	*19 460*	*19 607*
Species total	*9 611*	*4 782*	*3 975*	*27 150*	*14 103*	*7 925*	*2 335*	*11 137*	*19 460*	*19 677*
Anchoveta(=Peruvian anchovy)	**Anchois du Pérou**		**Anchoveta**			***Engraulis ringens***			**1,21(06)002,08**	**VET**
87 Chile	1 287 303	1 472 929	2 720 388	2 086 468	1 400 567	1 757 499	522 742	1 983 040	1 700 640	852 789
Ecuador	-	-	-	-	-	-	-	-	-	2 071
Peru	4 869 966	7 009 534	9 800 223	6 558 108	7 463 147	5 927 599	1 206 322	6 740 225	9 575 717	6 358 217
87 Fishing area total	*6 157 269*	*8 482 463*	*12 520 611*	*8 644 576*	*8 863 714*	*7 685 098*	*1 729 064*	*8 723 265*	*11 276 357*	*7 213 077*
Species total	*6 157 269*	*8 482 463*	*12 520 611*	*8 644 576*	*8 863 714*	*7 685 098*	*1 729 064*	*8 723 265*	*11 276 357*	*7 213 077*
Southern African anchovy	**Anchois de l'Afrique australe**		**Anchoa de Africa austral**			***Engraulis capensis***			**1,21(06)002,12**	**ANC**
47 Namibia	38 821	63 074	25 121	48 023	1 080	2 545	2 748	412	146	2 133
South Africa	347 312	235 606	155 554	170 308	40 712	60 095	107 548	180 542	267 840	287 190
47 Fishing area total	*386 133*	*298 680*	*180 675*	*218 331*	*41 792*	*62 640*	*110 296*	*180 954*	*267 986*	*289 323*
Species total	*386 133*	*298 680*	*180 675*	*218 331*	*41 792*	*62 640*	*110 296*	*180 954*	*267 986*	*289 323*
Atlantic anchoveta	**Anchois queue jaune**		**Anchoveta rabo amarillo**			***Cetengraulis edentulus***			**1,21(06)015,01**	**AVA**
31 Venezuela	10	33	35	41	8	0	119	0	0	2
31 Fishing area total	*10*	*33*	*35*	*41*	*8*	*0*	*119*	*0*	*0*	*2*
Species total	*10*	*33*	*35*	*41*	*8*	*0*	*119*	*0*	*0*	*2*
Pacific anchoveta	**Anchois chuchueco**		**Anchoveta chuchueco**			***Cetengraulis mysticetus***			**1,21(06)015,03**	**VEP**
77 Panama	69 607	89 220	72 111	106 743	60 322	77 726	107 730	27 356	86 681	100 000 F
77 Fishing area total	*69 607*	*89 220*	*72 111*	*106 743*	*60 322*	*77 726*	*107 730*	*27 356*	*86 681*	*100 000 F*
87 Colombia	20 186	24 240	19 453	31 823	26 344	28 747	28 501	15 781	20 500 F	25 099
Ecuador	46 162	57 742	27 164	23 418	26 354	89 157	44 474	27 221	13 762	73 537
87 Fishing area total	*66 348*	*81 982*	*46 617*	*55 241*	*52 698*	*117 904*	*72 975*	*43 002*	*34 262 F*	*98 636*
Species total	*135 955*	*171 202*	*118 728*	*161 984*	*113 020*	*195 630*	*180 705*	*70 358*	*120 943 F*	*198 636 F*
Broad-striped anchovy	**Anchois rayé**		**Anchoa legítima**			***Anchoa hepsetus***			**1,21(06)020,11**	**ENP**
31 Grenada	...	...	...	...	...	1	16	8	0	0
31 Fishing area total	...	...	...	...	...	*1*	*16*	*8*	*0*	*0*
Species total	...	...	...	...	...	*1*	*16*	*8*	*0*	*0*
Stolephorus anchovies	**Anchois Stolephorus**		**Boquerones**			***Stolephorus spp***			**1,21(06)050,XX**	**STO**
51 Untd Arab Em	7 786	8 968	8 793	9 533	9 633	10 295	10 329	10 587	2 729	2 730 F
51 Fishing area total	*7 786*	*8 968*	*8 793*	*9 533*	*9 633*	*10 295*	*10 329*	*10 587*	*2 729*	*2 730 F*
57 Indonesia	39 263	50 449	55 155	60 881	60 885	67 686	70 535	66 358	63 933	65 340
Malaysia	28 260	12 802	11 936	13 703	13 764	13 481	14 908	11 422	11 152	7 881
57 Fishing area total	*67 523*	*63 251*	*67 091*	*74 584*	*74 649*	*81 167*	*85 443*	*77 780*	*75 085*	*73 221*
61 China,H.Kong	57	34	14	89	31	20	17	10 F	15 F	17 F
61 Fishing area total	*57*	*34*	*14*	*89*	*31*	*20*	*17*	*10 F*	*15 F*	*17 F*
71 Indonesia	94 647	92 337	95 413	96 335	100 896	115 905	96 273	96 759	110 011	112 420
Malaysia	10 010	11 983	10 427	8 860	10 597	10 291	10 743	11 623	11 364	9 842
Philippines	84 652	81 354	67 507	71 516	71 456	78 678	77 049	78 087	79 630	82 112
71 Fishing area total	*189 309*	*185 674*	*173 347*	*176 711*	*182 949*	*204 874*	*184 065*	*186 469*	*201 005*	*204 374*
Species total	*264 675*	*257 927*	*249 245*	*260 917*	*267 262*	*296 356*	*279 854*	*274 846 F*	*278 834 F*	*280 342 F*
Anchovies, etc. nei	**Anchois, etc. nca**		**Anchoas, etc. nep**			***Engraulidae***			**1,21(06)XXX,XX**	**ANX**
31 Mexico	1 179	1 027	1 083	1 564	1 464	903	1 762	1 613	1 163	1 328
31 Fishing area total	*1 179*	*1 027*	*1 083*	*1 564*	*1 464*	*903*	*1 762*	*1 613*	*1 163*	*1 328*
41 Brazil	3 420 F	3 400 F	3 410 F	3 806	7 672	7 102	2 654	2 682	4 039	4 000 F
41 Fishing area total	*3 420 F*	*3 400 F*	*3 410 F*	*3 806*	*7 672*	*7 102*	*2 654*	*2 682*	*4 039*	*4 000 F*
51 Comoros	1 000	870	870	900 F	900 F	900 F	900 F	850 F	950 F	850
India	73 288	23 637	45 989	48 104	67 492	68 198	68 487	54 308	54 987	64 249
Oman	1 983	1 000	1 272	2 073	1 017	1 189	941	485	5 126	970
Pakistan	14 544	29 260	19 098	17 564	14 091	16 113	13 165	15 154	15 191	15 001
51 Fishing area total	*90 815*	*54 767*	*67 229*	*68 641 F*	*83 500 F*	*86 400 F*	*83 493 F*	*70 797 F*	*76 254 F*	*81 070*
57 Australia	85	178	735	470	775	-	-	-	-	-
India	13 923	14 323	13 167	12 875	13 202	12 320	12 719	14 435	13 621	7 268
Thailand	36 596	41 584	66 630	44 892	39 547	40 112	35 771	31 295	30 867	30 678
57 Fishing area total	*50 604*	*56 085*	*80 532*	*58 237*	*53 524*	*52 432*	*48 490*	*45 730*	*44 488*	*37 946*
61 Russian Fed	-	-	-	19	22	5	23	34	69	114
61 Fishing area total	-	-	-	*19*	*22*	*5*	*23*	*34*	*69*	*114*
71 Thailand	123 288	123 751	102 729	123 095	122 423	117 229	121 443	103 445	102 029	101 406
71 Fishing area total	*123 288*	*123 751*	*102 729*	*123 095*	*122 423*	*117 229*	*121 443*	*103 445*	*102 029*	*101 406*

B-35 Herrings, sardines, anchovies / Harengs, sardines, anchois / Arenques, sardinas, anchoas

Capture production by species, fishing areas and countries or areas / Captures par espèces, zones de pêche et pays ou zones / Capturas por especies, áreas de pesca y países o áreas

Species, Fishing area Espèce, Zone de pêche Especie, Area de pesca	1992 mt	1993 mt	1994 mt	1995 mt	1996 mt	1997 mt	1998 mt	1999 mt	2000 mt	2001 mt
81 Australia	12	12	13	13	12	19	19	2	39	21
81 Fishing area total	*12*	*12*	*13*	*13*	*12*	*19*	*19*	*2*	*39*	*21*
87 Peru	-	63 420	39 844	189 389	59 639	24 703	706 167	11 242	3 868	137 098
87 Fishing area total	*-*	*63 420*	*39 844*	*189 389*	*59 639*	*24 703*	*706 167*	*11 242*	*3 868*	*137 098*
Species total	*269 318 F*	*302 462 F*	*294 840 F*	*444 764 F*	*328 256 F*	*288 793 F*	*964 051 F*	*235 545 F*	*231 949 F*	*362 983 F*
Dorab wolf-herring	**Chirocentre dorab**		**Arencón dorab**		***Chirocentrus dorab***			**1,21(11)002,01**		**DOB**
51 Pakistan	865	1 070	1 204	2 289	1 580	1 931	2 051	2 266	2 775	2 604
Qatar	0	0	0	0	0	-	0	0	0	0
Saudi Arabia	-	-	-	-	-	-	8	7	10	9
51 Fishing area total	*865*	*1 070*	*1 204*	*2 289*	*1 580*	*1 931*	*2 059*	*2 273*	*2 785*	*2 613*
57 Thailand	2 214	3 124	3 794	6 090	7 551	7 744	7 648	8 451	8 335	8 284
57 Fishing area total	*2 214*	*3 124*	*3 794*	*6 090*	*7 551*	*7 744*	*7 648*	*8 451*	*8 335*	*8 284*
61 China,Taiwan	12	4	1	3	1	2	1	4	1	-
61 Fishing area total	*12*	*4*	*1*	*3*	*1*	*2*	*1*	*4*	*1*	*-*
71 Thailand	5 550	6 016	7 352	9 897	2 741	2 715	4 724	6 598	6 508	6 468
71 Fishing area total	*5 550*	*6 016*	*7 352*	*9 897*	*2 741*	*2 715*	*4 724*	*6 598*	*6 508*	*6 468*
Species total	*8 641*	*10 214*	*12 351*	*18 279*	*11 873*	*12 392*	*14 432*	*17 326*	*17 629*	*17 365*
Wolf-herrings nei	**Chirocentres nca**		**Arencones nep**		***Chirocentrus spp***			**1,21(11)002,XX**		**DOS**
51 India	9 544	8 325	8 942	8 351	5 802	11 921	10 201	4 663	6 813	10 038
Untd Arab Em	74	81	81	71	72	77	78	80	75	70 F
51 Fishing area total	*9 618*	*8 406*	*9 023*	*8 422*	*5 874*	*11 998*	*10 279*	*4 743*	*6 888*	*10 108 F*
57 India	12 166	14 050	13 033	12 121	13 708	13 232	4 926	12 848	9 261	8 688
Indonesia	3 898	4 374	5 342	6 615	5 701	5 027	6 487	6 610	7 089	7 370
Malaysia	2 850	2 449	1 333	1 111	1 375	1 762	1 448	1 426	1 393	1 382
57 Fishing area total	*18 914*	*20 873*	*19 708*	*19 847*	*20 784*	*20 021*	*12 861*	*20 884*	*17 743*	*17 440*
71 Indonesia	13 300	13 968	14 739	19 598	17 358	21 500	19 111	20 903	22 065	22 940
Malaysia	1 893	2 049	2 858	2 938	2 638	2 461	2 695	3 307	2 791	2 607
Philippines	742	1 178	193	135	114	245	317	377	354	389
Singapore	53	35	96	81	87	79	77	51	42	30
71 Fishing area total	*15 988*	*17 230*	*17 886*	*22 752*	*20 197*	*24 285*	*22 200*	*24 638*	*25 252*	*25 966*
Species total	*44 520*	*46 509*	*46 617*	*51 021*	*46 855*	*56 304*	*45 340*	*50 265*	*49 883*	*53 514 F*
Clupeoids nei	**Clupéoidés nca**		**Clupeoideos nep**		***Clupeoidei***			**1,21(XX)XXX,XX**		**CLU**
04 India	-	-	1 183	1 086	1 481	2 497	1 517	1 143	917	917
04 Fishing area total	*-*	*-*	*1 183*	*1 086*	*1 481*	*2 497*	*1 517*	*1 143*	*917*	*917*
27 France	76	1	-	-	-	-	-	765	165	1 086
Lithuania	-	-	-	35	2 400	-	-	-	-	-
Portugal	-	-	-	1	-	-	-	-	-	-
Spain	15 768	9 839	7 593	433	7 858	6 539	15 572	19 461	8 427	20 047
UK	2	3	5	4	4	17	4	-	-	-
27 Fishing area total	*15 846*	*9 843*	*7 598*	*473*	*10 262*	*6 556*	*15 576*	*20 226*	*8 592*	*21 133*
31 Br Virgin Is	...	...	...	...	...	5	5	5	0	0
Colombia	-	-	-	-	179	126	196	61	150 F	150 F
Korea Rep	-	-	-	-	-	-	-	1	-	-
Martinique	50 F	60 F	60 F	50 F	50 F	100 F	500 F	3 500	3 700	4 000
Mexico	70	79	74	251	222	629	332	266	188	6
Puerto Rico	14	27	18	20	18	19	14	20	20	17
Trinidad Tob	500 F	506	56	60 F	58	196	619	859	82	189
31 Fishing area total	*634 F*	*672 F*	*208 F*	*381 F*	*527 F*	*1 075 F*	*1 666 F*	*4 712*	*4 140 F*	*4 362 F*
34 Benin	...	...	...	...	...	...	...	13	10 F	-
Congo Rep	...	...	1 903	1 684	1 200 F	900 F	600 F	300 F	1	10 F
Eq Guinea	170 F	160 F	240 F	100 F	220 F	280 F	850	1 000 F	2 000	1 900 F
Gambia	...	...	...	53	27	8	12	10	-	-
Ghana	1 443	620	312	236	477	1 060	162	132	293	1 824
Korea Rep	-	-	-	25	-	-	-	-	-	-
Liberia	229	121	115	75	189	147	383	386	318	208
Mauritania	-	-	-	-	-	-	-	-	-	1 F
Portugal	1	0	-	-	-	-	-	-	-	-
Sierra Leone	722	705	711	713	700 F	...	315	-	6	-
Spain	-	-	-	-	-	-	-	-	2 518	-
Togo	272	528	676	453	0	0	0	0	0	0
34 Fishing area total	*2 837 F*	*2 134 F*	*3 957 F*	*3 339 F*	*2 813 F*	*2 395 F*	*2 322 F*	*1 841 F*	*5 146 F*	*3 943 F*
37 Algeria	2 700 F	2 900 F	3 500 F	2 012	1 182	1 261	2 279	2 634	2 500 F	2 500 F
France	-	-	-	-	-	-	-	72	15	17
Lebanon	500 F	500	500	800	800	600	600	500	700	500
Malta	0	0	0	0	0	0	0	0	-	-
Spain	-	-	-	-	15	3	-	-	-	195
37 Fishing area total	*3 200 F*	*3 400 F*	*4 000 F*	*2 812*	*1 997*	*1 864*	*2 879*	*3 206*	*3 215 F*	*3 212 F*
41 Argentina	1	0	0	-	-	-	-	-	-	-

B-35 Herrings, sardines, anchovies / Harengs, sardines, anchois / Arenques, sardinas, anchoas

Capture production by species, fishing areas and countries or areas
Captures par espèces, zones de pêche et pays ou zones
Capturas por especies, áreas de pesca y países o áreas

Species, Fishing area Espèce, Zone de pêche Especie, Area de pesca	1992 mt	1993 mt	1994 mt	1995 mt	1996 mt	1997 mt	1998 mt	1999 mt	2000 mt	2001 mt
Brazil	...	...	...	2 775	3 464	4 314	6 042	7 527	11 457	11 300 F
41 Fishing area total	*1*	*0*	*0*	*2 775*	*3 464*	*4 314*	*6 042*	*7 527*	*11 457*	*11 300 F*
51 India	65 423	28 410	40 006	38 098	51 568	66 868	70 487	63 435	36 678	32 823
Iran	25 000	20 000	5 000	8 000	10 000	10 000	9 708	13 030	14 955	17 453
Kenya	358	166	162	112	217	189	155	167	119	164
Mauritius	0	0	0	0	0	-	-	-	-	-
Pakistan	42 141	43 475	18 111	31 426	27 576	26 650	25 487	26 934	24 810	24 306
Réunion	3	3	5	5	5	6	7	10	12	4
Saudi Arabia	346	369	391	259	201	560	108	162	187	178
51 Fishing area total	*133 271*	*92 423*	*63 675*	*77 900*	*89 567*	*104 273*	*105 952*	*103 738*	*76 761*	*74 928*
57 Australia	9 768	11 020	11 484	12 820	12 697	13 028	7 801	4 377	2 304	1 646
India	11 273	7 686	14 483	20 943	25 072	24 625	25 906	25 727	27 205	26 049
Malaysia	5 521	5 023	3 811	3 516	4 388	4 766	3 199	4 583	3 391	4 275
Sri Lanka	35 097	37 379	38 870	49 785	48 221	47 200	50 800	51 370	56 250	53 230
57 Fishing area total	*61 659*	*61 108*	*68 648*	*87 064*	*90 378*	*89 619*	*87 706*	*86 057*	*89 150*	*85 200*
61 China,Taiwan	7 036	6 042	6 243	6 348	3 873	4 222	4 338	3 721	4 221	4 393
Japan	63 381	59 674	59 401	55 408	58 232	59 619	52 427	79 405	74 581	58 286
61 Fishing area total	*70 417*	*65 716*	*65 644*	*61 756*	*62 105*	*63 841*	*56 765*	*83 126*	*78 802*	*62 679*
71 Kiribati	3 420	3 300	3 270	3 300	3 340	3 320	613	2 638	2 725	2 079
Malaysia	18 064	21 913	30 870	35 477	40 137	33 344	43 116	40 934	30 222	36 475
Philippines	238	405	613	286	266	325	663	634	602	650
Singapore	315	265	353	240	379	351	329	206	78	73
71 Fishing area total	*22 037*	*25 883*	*35 106*	*39 303*	*44 122*	*37 340*	*44 721*	*44 412*	*33 627*	*39 277*
77 Costa Rica	931	323	365	311	438	1 175	906	1 788	1 628	2 207
Mexico	46	41	45	53	54	168	237	105	87	15
77 Fishing area total	*977*	*364*	*410*	*364*	*492*	*1 343*	*1 143*	*1 893*	*1 715*	*2 222*
81 Australia	465	464	432	432	333	417	417	...	...	...
New Zealand	85	230	286	...	11	8	1	12	11	11
Ukraine	-	2	-	-	-	-	-	-	-	-
81 Fishing area total	*550*	*696*	*718*	*432*	*344*	*425*	*418*	*12*	*11*	*11*
87 Colombia	0	89	2 502	120	1	429	273	0	100 F	100 F
Ghana	-	-	-	-	-	-	-	-	3 496	4 320
87 Fishing area total	*0*	*89*	*2 502*	*120*	*1*	*429*	*273*	*0*	*3 596 F*	*4 420 F*
Species total	*311 429 F*	*262 328 F*	*253 649 F*	*277 805 F*	*307 553 F*	*315 971 F*	*326 980 F*	*357 893 F*	*317 129 F*	*313 604 F*
Group total	***21 195 603***	***21 993 743***	***25 912 131***	***22 005 770***	***22 387 417***	***21 733 290***	***16 666 477***	***22 635 103***	***24 896 523***	***20 460 641***

B-36 Tunas, bonitos, billfishes / Thons, pélamides, marlins / Atunes, bonitos, agujas

Capture production by species, fishing areas and countries or areas / Captures par espèces, zones de pêche et pays ou zones / Capturas por especies, áreas de pesca y países o áreas

Species, Fishing area Espèce, Zone de pêche Especie, Area de pesca	1992 mt	1993 mt	1994 mt	1995 mt	1996 mt	1997 mt	1998 mt	1999 mt	2000 mt	2001 mt
Atlantic bonito	**Bonite à dos rayé**		**Bonito del Atlántico**		*Sarda sarda*			**1,75(01)001,01**		**BON**
21 USA	226	93	126	102	153	134	74	73	44	38
21 Fishing area total	*226*	*93*	*126*	*102*	*153*	*134*	*74*	*73*	*44*	*38*
27 France	-	52	-	-	-	-	-	24	32	42
Portugal	132	120	25	77	82	48	97	98	161	47
27 Fishing area total	*132*	*172*	*25*	*77*	*82*	*48*	*97*	*122*	*193*	*89*
31 Br Virgin Is	...	...	8	0	6	8	6	9	0	0
Grenada	-	-	-	-	24	6	14	16	7	10
Martinique	770 F	1 000	990	990 F	610	610 F	610 F	610 F	610 F	530 F
Mexico	657	779	674	1 143	1 279	2 040	2 194	2 314	1 721	1 506
St Lucia	3	4	1	1	1	0	0	0	0	0
Trinidad Tob	-	17	703	169	266	220	30	117	117	56
USA	278	294	718	110	5	27	10	10	4	10
Venezuela	1 443	1 541	1 646	1 503	1 348	1 294	1 647	1 597	1 376	1 815
31 Fishing area total	*3 151 F*	*3 635*	*4 740*	*3 916 F*	*3 539*	*4 205 F*	*4 511 F*	*4 673 F*	*3 835 F*	*3 927 F*
34 Germany	-	-	-	-	714	417	42	143	-	51
Greece	-	0	0	0	-	-	-	-	-	-
Latvia	4	-	3	19	301	887	318	510	416	396
Lithuania	10	-	-	-	-	-	-	-	-	-
Morocco	1 068	1 246	584	699	894	1 259	1 557	1 390	2 163	1 700
Netherlands	-	-	-	-	1 694	1 625	2 171	966	1 507	1 791
NethAntilles	-	-	-	-	-	-	-	-	2	...
Poland	-	-	-	-	225	39	-	-	-	521
Portugal	1	25	31	1	1	1	1	-	-	-
Romania	84	-	-	-	-	-	-	-	-	-
Russian Fed	135	6	-	6	175	1 937	4 960	2 156	837	538
Senegal	215	402	600	354	570	564	1 723	349	179	120
Sierra Leone	6	...	...	0	0	0	4	...	11	245
Spain	39	5	3	2	2	1	-	12	12	17
Togo	107	311	254	145	197	338	294	426	423	663
Ukraine	25	...	...	...	342	2 786	1 918	1 114	399	231
34 Fishing area total	*1 694*	*1 995*	*1 475*	*1 226*	*5 115*	*9 854*	*12 988*	*7 066*	*5 949*	*6 273*
37 Albania	...	...	...	1	2	0	12	30	25	30
Algeria	315	471	418	506	277	357	511	475	405	350
Bulgaria	12	8	...	25	33	16	51	20	35	49
Croatia	128	230	70	182	159	171	158	120	120	54
Cyprus	-	-	-	-	-	-	-	-	14	13
Egypt	518	640	648	697	985	725	724	1 442	1 128	1 072
France	5	6	-	-	-	-	-	-	-	28
Greece	2 690	2 690	1 581	2 116	1 752	1 559	945	2 135	1 914	1 550
Italy	1 288	1 662	1 828	1 512	2 233	4 580	2 121	1 614	1 116	1 006
Libya	71	70	...	...	...	...	...	...	...	...
Malta	0	0	0	0	2	7	2	2	1	1
Morocco	31	25	93	37	67	45	39	120	115	5
Spain	228	200	344	632	690	628	333	433	342	343
Tunisia	643	792	305	413	560	611	855	1 350	1 528	1 183
Turkey	8 863	19 548	10 093	8 944	10 284	7 810	24 000	17 900	12 000	13 460
Yugoslavia	1	3	2	6	10	12	12	14	17	17
Other nei	311	300	300	300	300	75	-	-	-	-
37 Fishing area total	*15 104*	*26 645*	*15 682*	*15 371*	*17 354*	*16 596*	*29 763*	*25 655*	*18 760*	*19 161*
41 Argentina	1 559	434	4	138	108	130	12	38	19	0
Brazil	86	142	142	137	-	-	-	-	-	-
Uruguay	0	-	-	-	-	-	-	-	1	23
41 Fishing area total	*1 645*	*576*	*146*	*275*	*108*	*130*	*12*	*38*	*20*	*23*
47 Angola	4	49	20	9	39	32	-	2	118	1 157
Russian Fed	-	-	-	-	-	-	-	-	41	36
South Africa	0	0	0	0	0	-	-	-	-	-
47 Fishing area total	*4*	*49*	*20*	*9*	*39*	*32*	*-*	*2*	*159*	*1 193*
Species total	*21 956 F*	*33 165*	*22 214*	*20 976 F*	*26 390*	*30 999 F*	*47 445 F*	*37 629 F*	*28 960 F*	*30 704 F*
Striped bonito	**Bonite oriental**		**Bonito mono**		*Sarda orientalis*			**1,75(01)001,02**		**BIP**
51 Oman	298	224	155	788	370	498	162	134	95	287
51 Fishing area total	*298*	*224*	*155*	*788*	*370*	*498*	*162*	*134*	*95*	*287*
Species total	*298*	*224*	*155*	*788*	*370*	*498*	*162*	*134*	*95*	*287*
Eastern Pacific bonito	**Bonite du Pacifique oriental**		**Bonito del Pacífico oriental**		*Sarda chiliensis*			**1,75(01)001,04**		**BEP**
77 Mexico	189	472	6 333	6 718	399	875	423	1 775	429	146
USA	1 060	390	431	71	449	290	1 094	87	44	6
77 Fishing area total	*1 249*	*862*	*6 764*	*6 789*	*848*	*1 165*	*1 517*	*1 862*	*473*	*152*
87 Chile	233	288	172	52	14	28	584	368	55	19
Colombia	...	...	...	3	6	5	8	9	10 F	13
Peru	35 023	36 976	31 125	28 331	23 059	17 731	5 130	948	434	1 287
87 Fishing area total	*35 256*	*37 264*	*31 297*	*28 386*	*23 079*	*17 764*	*5 722*	*1 325*	*499 F*	*1 319*
Species total	*36 505*	*38 126*	*38 061*	*35 175*	*23 927*	*18 929*	*7 239*	*3 187*	*972 F*	*1 471*

B-36 Tunas, bonitos, billfishes / Thons, pélamides, marlins / Atunes, bonitos, agujas

Capture production by species, fishing areas and countries or areas / Captures par espèces, zones de pêche et pays ou zones / Capturas por especies, áreas de pesca y países o áreas

Species, Fishing area Espèce, Zone de pêche Especie, Area de pesca	1992 mt	1993 mt	1994 mt	1995 mt	1996 mt	1997 mt	1998 mt	1999 mt	2000 mt	2001 mt
Plain bonito	**Palomette**		**Tasarte**			***Orcynopsis unicolor***			**1,75(01)006,01**	**BOP**
34 Benin	1	1 F	1 F	1 F	1 F	3	1	1	-	-
Morocco	423	348	598	524	2 003	246	28	626	1 048	830
Senegal	-	-	-	-	-	-	65	17	6	69
34 Fishing area total	*424*	*349 F*	*599 F*	*525 F*	*2 004 F*	*249*	*94*	*644*	*1 054*	*899*
37 Algeria	135	198	153	92	119	224	128	216	135	145
Libya	40	40 F	...	...	...	...	...	...	...	...
Morocco	1	14	23	23	13	3	2	1	10	9
37 Fishing area total	*176*	*252 F*	*176*	*115*	*132*	*227*	*130*	*217*	*145*	*154*
Species total	*600*	*601 F*	*775 F*	*640 F*	*2 136 F*	*476*	*224*	*861*	*1 199*	*1 053*
Wahoo	**Thazard-bâtard**		**Peto**			***Acanthocybium solandri***			**1,75(01)010,01**	**WAH**
21 USA	...	...	...	...	...	1	1	2	2	1
21 Fishing area total	...	...	...	...	...	*1*	*1*	*2*	*2*	*1*
31 Aruba	50	80	125	40	50	65	70	60	60	60
Barbados	51	91	82	42	35	52	52	41	41	24
Bermuda	80	58	50	93	99	105	108	104	61	56
Dominica	59	59	59	58	58	58	58	50	46	46
Dominican Rp	13	7	...	...	...	325	112	31	42	37
Grenada	59	55	46	49	56	56	59	82	51	71
NethAntilles	260 F	270	250	230	230 F	230 F	230 F	230 F	230 F	230 F
Puerto Rico	...	...	...	...	...	...	...	...	...	6
St Lucia	150	141	98	80	221	224	223	310	243	213
St Vincent	33	41	28	16	23	10	65	52	46	311
Trinidad Tob	1	-	-	-	0	1	1	1	2	1
USA	...	...	...	...	...	1	64	29	79	56
Venezuela	331	513	538	445	479	498	349	448	150	297
31 Fishing area total	*1 087 F*	*1 315*	*1 276*	*1 053*	*1 251 F*	*1 625 F*	*1 391 F*	*1 438 F*	*1 051 F*	*1 408 F*
34 Cape Verde	350	326	361	408	503	603	429	587	487	578
Sao Tome Prn	-	-	-	-	80	52	52 F	52 F	52 F	52 F
Spain	32	22	20	15	25	25	29	28	32	38
34 Fishing area total	*382*	*348*	*381*	*423*	*608*	*680*	*510 F*	*667 F*	*571 F*	*668 F*
41 Brazil	71	33	26	1	16	58	41	-	-	-
41 Fishing area total	*71*	*33*	*26*	*1*	*16*	*58*	*41*	-	-	-
47 St Helena	17	35	26	25	23	19	10	15	15	22
47 Fishing area total	*17*	*35*	*26*	*25*	*23*	*19*	*10*	*15*	*15*	*22*
51 India	10	0	0	5	14	13	14	12	13	...
Réunion	9	67	67	67	80	66	59	57	50	4
Seychelles	3	-	-	10	-	-	3	4	6	-
Spain	-	-	-	-	-	-	-	4	-	-
51 Fishing area total	*22*	*67*	*67*	*82*	*94*	*79*	*76*	*77*	*69*	*4*
57 Australia	-	-	-	-	1	4	4	8	6	9
India	-	1	0	0	9	8	9	17	18	...
Sri Lanka	133	433	268	129	128	156	196	488	545	520
57 Fishing area total	*133*	*434*	*268*	*129*	*138*	*168*	*209*	*513*	*569*	*529*
61 China,Taiwan	0	-	-	-	-	-	-	-	-	-
61 Fishing area total	*0*	-	-	-	-	-	-	-	-	-
71 Fiji Islands	329	92	94 F	179	130	145	148	160	150 F	167
Guam	28	36	48	24	19	20	30	19	20	23
N Marianas	6	1	2	3	5	4	2	4	2	2
71 Fishing area total	*363*	*129*	*144 F*	*206*	*154*	*169*	*180*	*183*	*172 F*	*192*
77 Amer Samoa	1	1	4	5	5	7	12	17	20	47
Fr Polynesia	...	...	...	...	...	119	188	269	229	259
USA	...	...	...	...	...	-	9	2	0	1
77 Fishing area total	*1*	*1*	*4*	*5*	*5*	*126*	*209*	*288*	*249*	*307*
81 New Zealand	-	-	-	-	-	6	-	-	-	-
81 Fishing area total	-	-	-	-	-	*6*	-	-	-	-
Species total	*2 076 F*	*2 362*	*2 192 F*	*1 924*	*2 289 F*	*2 931 F*	*2 627 F*	*3 183 F*	*2 698 F*	*3 131 F*
Narrow-barred Spanish mackerel	**Thazard rayé indo-pacifique**		**Carite estriado Indo-Pacífico**			***Scomberomorus commerson***			**1,75(01)015,03**	**COM**
51 Bahrain	114	77	69	109	158	47	85	44	66	109
Egypt	191	191	1 242	1 297	8 880	8 503	9 933	7 442	14 879	16 352
Eritrea	...	...	...	49	191	200	210	250	217	280
India	20 830	16 903	20 746	24 085	15 235	14 460	19 921	17 123	18 759	27 650
Iran	3 328	2 869	3 300	11 067	3 560	4 290	4 034	4 609	7 075	2 474
Israel	36	-	-	-	-	-	-	-	-	-
Kenya	75	46	103	74	93	69	139	122	94	136
Kuwait	33	36	59	68	85	73	124	127	130 F	135
Madagascar	8 000	10 000	10 000	10 000	10 000	10 000	12 000	12 000	12 000	12 000
Oman	3 616	3 143	3 764	6 185	5 243	5 944	3 145	3 390	2 559	2 785
Pakistan	12 133	12 252	7 157	8 618	10 108	12 009	12 232	11 734	9 366	8 405
Qatar	766	636	406	255	307	411	552	496	768	1 019
Saudi Arabia	9 872	10 374	10 851	6 342	5 276	5 511	6 722	6 032	6 057	5 292
South Africa	22	54	15	4	13	...	...	...	...	...

B-36 Tunas, bonitos, billfishes / Thons, pélamides, marlins / Atunes, bonitos, agujas

Capture production by species, fishing areas and countries or areas / Captures par espèces, zones de pêche et pays ou zones / Capturas por especies, áreas de pesca y países o áreas

Species, Fishing area Espèce, Zone de pêche Especie, Area de pesca	1992 mt	1993 mt	1994 mt	1995 mt	1996 mt	1997 mt	1998 mt	1999 mt	2000 mt	2001 mt
Sudan	...	...	...	...	...	...	19	24	19	34
Untd Arab Em	5 789	5 800	6 475	6 584	6 653	7 110	7 133	7 311	6 644	6 640 F
Yemen	2 551	3 092	3 255	3 047	3 521	3 680	3 580	3 580 F	3 580 F	3 580 F
Other nei	-	-	-	3 060	3 060	3 060	-	-	-	-
51 Fishing area total	*67 356*	*65 473*	*67 442*	*80 844*	*72 383*	*75 367*	*79 829*	*74 284 F*	*82 213 F*	*86 891 F*
57 Australia	325	445	513	444	471	622	540	336	307	490
India	3 748	3 116	3 878	4 502	9 378	8 900	12 260	23 195	25 411	17 808
Indonesia	11 573	14 731	13 481	14 090	16 349	14 286	20 137	16 215	18 986	20 080
Sri Lanka	119	163	203	199	817	999	1 246	856	766	2 180
57 Fishing area total	*15 765*	*18 455*	*18 075*	*19 235*	*27 015*	*24 807*	*34 183*	*40 602*	*45 470*	*40 558*
61 China,Taiwan	4 665	3 316	2 845	3 211	2 541	2 701	2 417	2 674	3 551	4 153
61 Fishing area total	*4 665*	*3 316*	*2 845*	*3 211*	*2 541*	*2 701*	*2 417*	*2 674*	*3 551*	*4 153*
71 Australia	644	779	796	785	680	830	1 054	-	-	-
Fiji Islands	1 230	783	828 F	1 424	1 247	1 025	1 455	2 296	2 000 F	2 120
Indonesia	44 609	43 375	46 833	49 342	52 105	60 106	55 064	61 496	66 444	70 270
Philippines	9 072	12 962	9 234	10 593	10 557	11 237	10 772	9 137	8 889	9 252
71 Fishing area total	*55 555*	*57 899*	*57 691 F*	*62 144*	*64 589*	*73 198*	*68 345*	*72 929*	*77 333 F*	*81 642*
81 Australia	47	47	17	17	9	23	23	-	-	-
81 Fishing area total	*47*	*47*	*17*	*17*	*9*	*23*	*23*	-	-	-
Species total	*143 388*	*145 190*	*146 070 F*	*165 451*	*166 537*	*176 096*	*184 797*	*190 489 F*	*208 567 F*	*213 244 F*
Indo-Pacific king mackerel	**Thazard ponctué indo-pacifique**		**Carite del Indo-Pacífico**		***Scomberomorus guttatus***			**1,75(01)015,04**		**GUT**
51 India	10 330	14 163	11 367	11 936	7 838	8 205	13 965	12 003	13 150	43 112
Iran	2 218	1 636	1 650	5 418	4 340	4 129	3 883	3 476	4 100	6 071
Kuwait	92	126	60	156	172	206	166	211	210 F	204
Saudi Arabia	-	-	-	-	-	114	300	303	303	303
South Africa	-	-	-	-	-	20	46	16	22	49
51 Fishing area total	*12 640*	*15 925*	*13 077*	*17 510*	*12 350*	*12 674*	*18 360*	*16 009*	*17 785 F*	*49 739*
57 India	7 164	7 467	4 988	5 238	4 824	5 050	8 595	16 260	17 814	707
Indonesia	1 872	8 880	5 723	10 601	13 784	13 152	11 821	11 217	11 668	11 860
Sri Lanka	0	6	52	1	0	-	-	-	-	-
57 Fishing area total	*9 036*	*16 353*	*10 763*	*15 840*	*18 608*	*18 202*	*20 416*	*27 477*	*29 482*	*12 567*
61 China,Taiwan	674	586	795	1 814	1 611	1 607	1 128	1 298	1 249	1 395
61 Fishing area total	*674*	*586*	*795*	*1 814*	*1 611*	*1 607*	*1 128*	*1 298*	*1 249*	*1 395*
71 Indonesia	8 384	9 368	8 603	9 272	9 314	9 098	10 925	10 457	12 781	13 000
71 Fishing area total	*8 384*	*9 368*	*8 603*	*9 272*	*9 314*	*9 098*	*10 925*	*10 457*	*12 781*	*13 000*
Species total	*30 734*	*42 232*	*33 238*	*44 436*	*41 883*	*41 581*	*50 829*	*55 241*	*61 297 F*	*76 701*
Streaked seerfish	**Thazard cirrus**		**Carite rayado**		***Scomberomorus lineolatus***			**1,75(01)015,05**		**STS**
51 India	14	74	31	59	59	558	67	58	64	...
51 Fishing area total	*14*	*74*	*31*	*59*	*59*	*558*	*67*	*58*	*64*	...
57 India	1 013	5	15	28	37	343	40	77	83	381
Sri Lanka	3	2	0	0	0	-	-	-	-	-
57 Fishing area total	*1 016*	*7*	*15*	*28*	*37*	*343*	*40*	*77*	*83*	*381*
Species total	*1 030*	*81*	*46*	*87*	*96*	*901*	*107*	*135*	*147*	*381*
King mackerel	**Thazard barré**		**Carite lucio**		***Scomberomorus cavalla***			**1,75(01)015,06**		**KGM**
21 USA	320	322	1	280	1 625	385	341	1 171	315	232
21 Fishing area total	*320*	*322*	*1*	*280*	*1 625*	*385*	*341*	*1 171*	*315*	*232*
31 Dominica	-	-	-	-	-	-	-	36	35	35
Dominican Rp	782	791	1 330	2 042	1 648	589	288	...	271	261
Grenada	-	-	-	-	2	4	28	14	9	4
Guyana	-	-	-	-	-	270	440	398	214	239
Mexico	3 014	3 289	3 097	3 050	4 377	5 370	4 598	5 002	4 576	5 199
Trinidad Tob	1 044	1 192	...	471	1 029	875	746	447	432	410
USA	1 714	2 136	2 012	1 769	444	2 130	2 020	1 239	1 931	1 962
Venezuela	1 307	800	2 484	2 555	2 139	3 530	340	2 424	1 498	1 861
31 Fishing area total	*7 861*	*8 208*	*8 923*	*9 887*	*9 639*	*12 768*	*8 460*	*9 560*	*8 966*	*9 971*
41 Brazil	979	1 380	1 365	1 328	2 890	2 398	3 595	3 595	2 344	1 251
41 Fishing area total	*979*	*1 380*	*1 365*	*1 328*	*2 890*	*2 398*	*3 595*	*3 595*	*2 344*	*1 251*
Species total	*9 160*	*9 910*	*10 289*	*11 495*	*14 154*	*15 551*	*12 396*	*14 326*	*11 625*	*11 454*
Atlantic Spanish mackerel	**Thazard atlantique**		**Carite atlántico**		***Scomberomorus maculatus***			**1,75(01)015,07**		**SSM**
21 USA	420	361	228	96	1 411	364	215	694	368	363
21 Fishing area total	*420*	*361*	*228*	*96*	*1 411*	*364*	*215*	*694*	*368*	*363*
31 Mexico	9 181	10 066	8 300	7 673	11 049	7 389	7 381	8 382	5 717	5 320
USA	1 261	1 997	2 487	1 911	134	1 331	1 244	668	1 282	1 553
31 Fishing area total	*10 442*	*12 063*	*10 787*	*9 584*	*11 183*	*8 720*	*8 625*	*9 050*	*6 999*	*6 873*
Species total	*10 862*	*12 424*	*11 015*	*9 680*	*12 594*	*9 084*	*8 840*	*9 744*	*7 367*	*7 236*

B-36 Tunas, bonitos, billfishes / Thons, pélamides, marlins / Atunes, bonitos, agujas

Capture production by species, fishing areas and countries or areas / Captures par espèces, zones de pêche et pays ou zones / Capturas por especies, áreas de pesca y países o áreas

Species, Fishing area Espèce, Zone de pêche Especie, Area de pesca	1992 mt	1993 mt	1994 mt	1995 mt	1996 mt	1997 mt	1998 mt	1999 mt	2000 mt	2001 mt
Cero	**Thazard franc**		**Carite chinigua**		***Scomberomorus regalis***				**1,75(01)015,08**	**CER**
31 Dominican Rp	79	50	90	29	57	231	191	230	190	147
Martinique	310 F	400	400 F	400 F	250	250 F	250 F	...	...	...
St Vincent	1	0	0	0	-	-	-	-	-	-
31 Fishing area total	*390 F*	*450*	*490 F*	*429 F*	*307*	*481 F*	*441 F*	*230*	*190*	*147*
Species total	*390 F*	*450*	*490 F*	*429 F*	*307*	*481 F*	*441 F*	*230*	*190*	*147*
Pacific sierra	**Thazard sierra (Pacifique)**		**Carite sierra**		***Scomberomorus sierra***				**1,75(01)015,09**	**SIE**
77 Japan	1	-	-	-	-	-	-	-	-	-
Mexico	4 599	5 756	5 625	5 137	5 742	5 405	3 896	5 265	6 261	5 959
Panama	282	576	567	463	489	665	982	159	1 395	1 000 F
77 Fishing area total	*4 882*	*6 332*	*6 192*	*5 600*	*6 231*	*6 070*	*4 878*	*5 424*	*7 656*	*6 959 F*
87 Colombia	285	444	694	360	484	645	912	521	500 F	500 F
Peru	772	924	352	686	439	98	861	2 712	930	181
87 Fishing area total	*1 057*	*1 368*	*1 046*	*1 046*	*923*	*743*	*1 773*	*3 233*	*1 430 F*	*681 F*
Species total	*5 939*	*7 700*	*7 238*	*6 646*	*7 154*	*6 813*	*6 651*	*8 657*	*9 086 F*	*7 640 F*
West African Spanish mackerel	**Thazard blanc**		**Carite lusitánico**		***Scomberomorus tritor***				**1,75(01)015,11**	**MAW**
34 Benin	202	214	194	188	188 F	362	511	205	205 F	203
Ghana	899	466	-	-	-	-	-	-	-	-
Korea Rep	41	-	-	-	-	-	-	-	-	-
Lithuania	4	-	-	-	-	-	-	-	-	-
Mauritania	...	...	...	...	...	...	...	...	12 F	...
Russian Fed	-	19	-	-	44	-	14	19	7	4
Sao Tome Prn	-	-	-	-	8	-	-	-	-	-
Senegal	1 075	1 060	766	1 863	1 056	1 015	931	479	589	756
Ukraine	90	-	-	-	-	-	-	-	21	...
34 Fishing area total	*2 311*	*1 759*	*960*	*2 051*	*1 296 F*	*1 377*	*1 456*	*703*	*834 F*	*963*
Species total	*2 311*	*1 759*	*960*	*2 051*	*1 296 F*	*1 377*	*1 456*	*703*	*834 F*	*963*
Japanese Spanish mackerel	**Thazard oriental**		**Carite oriental**		***Scomberomorus niphonius***				**1,75(01)015,12**	**NPH**
61 China	146 756	145 480	202 811	226 520	283 784	340 302	517 528	565 764	496 566	476 690
China,Taiwan	11 407	9 130	11 217	8 971	7 546	11 761	8 579	4 516	5 955	11 497
Japan	5 159	4 200	5 059	6 381	3 607	2 349	2 864	5 321	10 932	9 056
Korea Rep	8 230	13 951	8 673	17 429	6 419	11 173	22 809	19 502	25 641	25 513
61 Fishing area total	*171 552*	*172 761*	*227 760*	*259 301*	*301 356*	*365 585*	*551 780*	*595 103*	*539 094*	*522 756*
Species total	*171 552*	*172 761*	*227 760*	*259 301*	*301 356*	*365 585*	*551 780*	*595 103*	*539 094*	*522 756*
Serra Spanish mackerel	**Thazard serra**		**Serra**		***Scomberomorus brasiliensis***				**1,75(01)015,15**	**BRS**
31 Grenada	0	0	0	0	0	0	1	1	1	0
Guyana	-	-	-	-	211	571	625	1 143	308	329
Trinidad Tob	2 153	2 130	2 130	1 816	1 568	1 699	2 130	1 328	1 722	2 207
Venezuela	2 772	5 077	3 882	4 725	3 609	3 670	3 651	1 686	3 624	3 062
31 Fishing area total	*4 925*	*7 207*	*6 012*	*6 541*	*5 388*	*5 940*	*6 407*	*4 158*	*5 655*	*5 598*
41 Brazil	1 149	842	1 149	1 308	3 047	2 125	1 516	1 516	988	251
41 Fishing area total	*1 149*	*842*	*1 149*	*1 308*	*3 047*	*2 125*	*1 516*	*1 516*	*988*	*251*
Species total	*6 074*	*8 049*	*7 161*	*7 849*	*8 435*	*8 065*	*7 923*	*5 674*	*6 643*	*5 849*
Seerfishes nei	**Thazards nca**		**Carites nep**		***Scomberomorus spp***				**1,75(01)015,XX**	**KGX**
31 Barbados	51	55	36	42	49	...	...	...	...	...
Br Virgin Is	...	...	...	...	...	6	6	5	0	0
Colombia	107	79	217	180	539	22	30	55	50 F	50 F
Cuba	611	310	409	548	613	466	236	247	247 F	247 F
Nicaragua	...	...	...	...	...	...	...	240	250	260
Puerto Rico	53	84	86	134	106	119	109	123	145	124
St Lucia	150	141	98	80	51	4	0	...	60	7
St Vincent	-	-	-	-	1	1	1	1	138	0
31 Fishing area total	*972*	*669*	*846*	*984*	*1 359*	*618*	*382*	*671*	*890 F*	*688 F*
34 Gabon	-	-	140	145	79	-	85	-	-	-
Korea Rep	-	-	-	-	14	-	-	-	8	-
34 Fishing area total	*-*	*-*	*140*	*145*	*93*	*-*	*85*	*-*	*8*	*-*
51 Comoros	230	230	271	271	270	260	260	250	270 F	250
Djibouti	52	57	61	67	65	61	60	60 F	60 F	60 F
India	50	51	733	670	753	2 064	-	-	-	-
Korea Rep	-	-	-	-	38	-	-	-	-	17
South Africa	-	-	-	-	-	4	5	3	4	4
Tanzania	1 500	594	544	680	766	750	750	650	400	500
Untd Arab Em	1 390	1 400	1 566	1 267	1 594	2 500	2 886	3 117	2 876	2 870 F
Yemen	500	538	538	500	500	520	510	510 F	510 F	510 F
Other nei	-	-	-	431	431	431	-	-	-	-
51 Fishing area total	*3 722*	*2 870*	*3 713*	*3 886*	*4 417*	*6 590*	*4 471*	*4 590 F*	*4 120 F*	*4 211 F*
57 Australia	117	111	126	89	131	171	-	-	22	55

B-36 Tunas, bonitos, billfishes / Thons, pélamides, marlins / Atunes, bonitos, agujas

Capture production by species, fishing areas and countries or areas / Captures par espèces, zones de pêche et pays ou zones / Capturas por especies, áreas de pesca y países o áreas

	Species, Fishing area Espèce, Zone de pêche Especie, Area de pesca	1992 mt	1993 mt	1994 mt	1995 mt	1996 mt	1997 mt	1998 mt	1999 mt	2000 mt	2001 mt
	Bangladesh	30	40	50	50	40	50	60	60	60	60 F
	India	19	18	203	181	463	1 271	-	-	-	-
	Korea Rep	-	-	-	-	-	1	-	-	-	-
	Malaysia	5 662	4 394	4 375	3 043	3 330	3 652	4 857	3 473	4 090	3 323
	Sri Lanka	-	-	-	-	-	-	-	168	20	-
	Thailand	3 911	3 498	6 473	7 780	5 882	5 625	6 750	5 707	5 752	4 224
57	*Fishing area total*	*9 739*	*8 061*	*11 227*	*11 143*	*9 846*	*10 770*	*11 667*	*9 408*	*9 944*	*7 662 F*
61	China,H.Kong	4 252	3 093	2 665	2 790	2 036	1 711	1 723	1 200 F	1 500 F	1 650 F
	China,Taiwan	1 880	1 795	1 495	1 085	1 103	1 143	1 478	2 883	2 260	1 433
61	*Fishing area total*	*6 132*	*4 888*	*4 160*	*3 875*	*3 139*	*2 854*	*3 201*	*4 083 F*	*3 760 F*	*3 083 F*
71	Australia	386	248	398	339	495	-	-	-	-	-
	Japan	69	10	40	28	7	-	-	-	-	-
	Korea Rep	95	78	43	25	9	-	38	5	-	1
	Malaysia	22 112	16 424	11 225	11 858	11 070	10 082	11 421	13 774	11 233	11 647
	NewCaledonia	35	20	11	12	19	3	16	41	4	4 F
	Singapore	47	98	90	76	76	71	70	79	78	46
	Thailand	8 414	11 085	9 904	10 660	9 360	8 875	9 480	9 826	9 691	9 632
71	*Fishing area total*	*31 158*	*27 963*	*21 711*	*22 998*	*21 036*	*19 031*	*21 025*	*23 725*	*21 006*	*21 330 F*
81	Australia	497	493	0	0	0	-	-	-	-	-
	Korea Rep	1 303	2 574	1 146	1 228	1 769	3 111	2 897	1 237	1 601	2 079
81	*Fishing area total*	*1 800*	*3 067*	*1 146*	*1 228*	*1 769*	*3 111*	*2 897*	*1 237*	*1 601*	*2 079*
	Species total	*53 523*	*47 518*	*42 943*	*44 259*	*41 659*	*42 974*	*43 728*	*43 714 F*	*41 329 F*	*39 053 F*
	Frigate tuna	**Auxide**		**Melva**			***Auxis thazard***			**1,75(01)023,01**	**FRI**
34	NethAntilles	-	-	-	-	-	-	-	-	215	...
	Panama	57	118	341	327	240	91	-	-	-	-
	Other nei	61	150	409	390	894	700	487	486	290	872
34	*Fishing area total*	*118*	*268*	*750*	*717*	*1 134*	*791*	*487*	*486*	*505*	*872*
	Species total	*118*	*268*	*750*	*717*	*1 134*	*791*	*487*	*486*	*505*	*872*
	Frigate and bullet tunas	**Auxide et bonitou**		**Melva y melvera**			***Auxis thazard, A.rochei***			**1,75(01)023,XX**	**FRZ**
21	USA	0	0	0	0	-	-	1	17	9	3
21	*Fishing area total*	*0*	*0*	*0*	*0*	-	-	*1*	*17*	*9*	*3*
27	Portugal	0	0	0	-	-	-	28	263	494	208
	Spain	157	57	43	15	2	-	2	1	17	4
27	*Fishing area total*	*157*	*57*	*43*	*15*	*2*	-	*30*	*264*	*511*	*212*
31	Grenada	-	-	-	-	0	1	0	0	0	1
	Spain	14	5	0	-	-	-	-	-	-	-
	Trinidad Tob	...	17	...	56	199	368	127	138	138	...
	Venezuela	327	881	2 597	2 161	3 053	2 813	1 926	1 524	1 410	1 342
31	*Fishing area total*	*341*	*903*	*2 597*	*2 217*	*3 252*	*3 182*	*2 053*	*1 662*	*1 548*	*1 343*
34	Cape Verde	82	115	86	13	6	22	191	154	81	171
	Côte dIvoire	...	5 174	...	...	5 269	4 458	4 502	5 772	6 768	...
	France	121	63	105	126	161	147	146	0	91	-
	Korea Rep	-	-	25	7	-	-	-	-	-	-
	Morocco	332	274	122	645	543	2 614	2 137	494	582	418
	Portugal	0	0	0	0	-	1	31	5	9	28
	Russian Fed	627	220	505	459	46	500	2 433	460	408	1 028
	Sao Tome Prn	-	-	-	-	79	323	...	...	...	...
	Spain	57	300	254	371	945	581	568	22	-	709
	Ukraine	-	-	-	-	-	-	-	36	48	...
34	*Fishing area total*	*1 219*	*6 146*	*1 097*	*1 621*	*7 049*	*8 646*	*10 008*	*6 943*	*7 987*	*2 354*
37	Algeria	270	348	306	230	237	179	299	173	225	230
	Croatia	21	52	22	28	26	16	12	0	-	-
	France	4	0	0	1	-	-	-	-	-	-
	Greece	1 400	1 400	1 400	1 400	1 426	1 426	...	...	196	125
	Italy	305	379	531	1 435	229	499	254	439	215	375
	Malta	11	10	1	2	3	6	6	3	1	1
	Morocco	1 644	170	1 726	621	1 673	562	1 140	682	763	256
	Spain	1 210	648	1 124	1 472	2 296	604	487	669	1 024	522
	Tunisia	35	20	13	14	13	26	87	38	7	2 292
	Yugoslavia	1	0	0	2	6	6	6	7	8	9
37	*Fishing area total*	*4 901*	*3 027*	*5 123*	*5 205*	*5 909*	*3 324*	*2 291*	*2 011*	*2 439*	*3 810*
41	Brazil	291	608	906	558	527	215	162	166	106	98
41	*Fishing area total*	*291*	*608*	*906*	*558*	*527*	*215*	*162*	*166*	*106*	*98*
47	Angola	0	4	6	21	29	12	31	2	38	206
	Russian Fed	-	-	-	-	-	-	-	-	12	25
47	*Fishing area total*	*0*	*4*	*6*	*21*	*29*	*12*	*31*	*2*	*50*	*231*
51	France	-	-	-	-	-	-	-	-	-	15
	India	7 892	3 968	12 440	5 906	8 906	8 462	6 152	12 722	13 101	7 031
	Iran	300	436	200	4 438	755	544	509	590	785	562
	Maldives	3 389	5 455	4 018	3 938	6 484	2 489	4 218	3 401	3 991	3 982
	Oman	210	354	391	786	613	846	611	583	488	638
	Pakistan	2	2	36	36	49	54	56	59	42	150
	Spain	-	-	-	-	1 227	208	-	-	315	201

B-36

Tunas, bonitos, billfishes — **Capture production by species, fishing areas and countries or areas**
Thons, pélamides, marlins — **Captures par espèces, zones de pêche et pays ou zones**
Atunes, bonitos, agujas — **Capturas por especies, áreas de pesca y países o áreas**

Species, Fishing area Espèce, Zone de pêche Especie, Area de pesca	1992 mt	1993 mt	1994 mt	1995 mt	1996 mt	1997 mt	1998 mt	1999 mt	2000 mt	2001 mt
Untd Arab Em	650	651	737	572	578	618	620	636	376	380 F
Yemen	23	25	25	20	20	20	20	20 F	20 F	20 F
Other nei	28	13	10	100	100	100	-	18	367	110
51 Fishing area total	*12 494*	*10 904*	*17 857*	*15 796*	*18 732*	*13 341*	*12 186*	*18 029 F*	*19 485 F*	*13 089 F*
57 India	3	8	23	11	2 213	2 102	1 528	4 461	5 810	...
Japan	-	-	-	-	-	-	1	-	-	-
Sri Lanka	3 026	4 011	3 226	4 006	5 334	6 521	8 133	3 515	4 583	8 240
Thailand	1 360	2 770	2 300	4 237	2 989	2 723	3 071	2 683	2 645	...
57 Fishing area total	*4 389*	*6 789*	*5 549*	*8 254*	*10 536*	*11 346*	*12 733*	*10 659*	*13 038*	*8 240*
61 China,Taiwan	1 964	2 261	2 098	2 881	2 482	4 280	4 634	4 558	3 073	2 089
Japan	25 968	26 233	24 078	27 376	20 339	32 495	21 597	29 514	27 139	37 234
61 Fishing area total	*27 932*	*28 494*	*26 176*	*30 257*	*22 821*	*36 775*	*26 231*	*34 072*	*30 212*	*39 323*
71 China,Taiwan	0	0	0	0	0	0	0	0	0	0
Japan	887	1 663	239	10	35	79	15	3	-	13
Philippines	125 655	110 266	109 866	88 426	88 969	108 494	106 433	111 301	112 227	115 905
Thailand	33 700	26 800	26 900	19 250	18 850	17 000	17 600	17 700	17 800	17 000
71 Fishing area total	*160 242*	*138 729*	*137 005*	*107 686*	*107 854*	*125 573*	*124 048*	*129 004*	*130 027*	*132 918*
77 Japan	-	125	-	-	-	-	-	-	-	-
77 Fishing area total	-	*125*	-	-	-	-	-	-	-	-
81 New Zealand	-	-	-	0	0	2	4	-	5	5
81 Fishing area total	-	-	-	*0*	*0*	*2*	*4*	-	*5*	*5*
87 Ecuador	...	...	9 051	7 396	2 537	6 857	4 201	48 913	9 648	5 738
87 Fishing area total	...	...	*9 051*	*7 396*	*2 537*	*6 857*	*4 201*	*48 913*	*9 648*	*5 738*
Species total	*211 966*	*195 786*	*205 410*	*179 026*	*179 248*	*209 273*	*193 979*	*251 742 F*	*215 065 F*	*207 364 F*

Little tunny(=Atl.black skipj)	**Thonine commune**		**Bacoreta**			***Euthynnus alletteratus***			**1,75(01)024,01**	**LTA**
21 USA	96	116	49	136	81	199	120	405	110	121
21 Fishing area total	*96*	*116*	*49*	*136*	*81*	*199*	*120*	*405*	*110*	*121*
27 Portugal	73	45	72	72	218	320	171	14	-	-
Spain	1	-	-	-	-	-	11	1	2	-
27 Fishing area total	*74*	*45*	*72*	*72*	*218*	*320*	*182*	*15*	*2*	-
31 Bermuda	11	5	6	6	7	6	5	4	2	4
Cuba	33	13	15	27	23	17	9	-	-	-
Puerto Rico	-	...	...	...	...	...	...	...	...	14
St Lucia	-	-	-	-	-	2	2	2	-	1
St Vincent	0	1	0	0	-	-	-	-	-	-
USA	469	241	110	14	9	252	180	109	110	236
Venezuela	1 409	1 889	2 115	1 627	1 840	2 064	2 815	2 389	2 040	1 948
31 Fishing area total	*1 922*	*2 149*	*2 246*	*1 674*	*1 879*	*2 341*	*3 011*	*2 504*	*2 152*	*2 203*
34 Benin	49	53	60	58	58 F	196	83	69	69 F	69
Cape Verde	148	17	23	72	63	86	110	776	491	178
Côte dIvoire	142	2 314	251	253	2 337	1 880	1 818	2 352	2 789	...
France	13	8	54	59	22	215	21	86	21	-
Gabon	-	-	-	-	182	-	18	159	301	213
Ghana	11 608	359	994	513	113	2 025	359	306	707	730
Morocco	370	44	43	230	588	195	189	67	101	87
Panama	-	65	-	-	-	-	-	-	-	-
Portugal	0	0	0	-	-	0	-	-	50	-
Russian Fed	306	265	189	96	49	-	88	-	-	-
Sao Tome Prn	-	-	-	-	40	159	...	...	...	...
Senegal	1 521	1 496	1 628	1 133	1 066	1 662	1 604	460	1 146	1 613
Spain	-	-	-	10	55	27	99	5	-	322
Other nei	-	72	-	53	-	-	3	2	3	-
34 Fishing area total	*14 157*	*4 693*	*3 242*	*2 477*	*4 573 F*	*6 445*	*4 392*	*4 282*	*5 678 F*	*3 212*
37 Algeria	585	495	459	552	554	448	384	562	494	407
Croatia	3	2	15	7	9	9	16	0	-	-
Cyprus	21	11	23	10	19	30	10	16	14	13
Gaza Strip	...	...	...	...	90	59	61	60 F	60 F	50 F
Israel	126	119	119	215	119	103	73	90	113	100 F
Libya	-	-	-	-	-	45	52	-	5	4
Malta	1	8	8	8	3	3	0	0	0	5
Morocco	0	0	0	1	0	1	14	8	...	-
Spain	-	-	-	15	18	9	15	-	8	70
Syria	156	161	156	155	270	350	417	390	370	370
Tunisia	664	242	204	696	824	333	1 113	752	1 453	1 036
Yugoslavia	10	28	21	35	22	18	20	18	16	16
Other nei	200	200	200	200	200	200	200	200	-	-
37 Fishing area total	*1 766*	*1 266*	*1 205*	*1 894*	*2 128*	*1 608*	*2 375*	*2 096 F*	*2 533 F*	*2 071 F*
41 Brazil	935	985	1 225	1 059	834	507	920	930	615	615
41 Fishing area total	*935*	*985*	*1 225*	*1 059*	*834*	*507*	*920*	*930*	*615*	*615*
47 Angola	14	175	121	117	235	75	406	118	132	231
47 Fishing area total	*14*	*175*	*121*	*117*	*235*	*75*	*406*	*118*	*132*	*231*
Species total	*18 964*	*9 429*	*8 160*	*7 429*	*9 948 F*	*11 495*	*11 406*	*10 350 F*	*11 222 F*	*8 453 F*

B-36

Tunas, bonitos, billfishes — Capture production by species, fishing areas and countries or areas
Thons, pélamides, marlins — Captures par espèces, zones de pêche et pays ou zones
Atunes, bonitos, agujas — Capturas por especies, áreas de pesca y países o áreas

Species, Fishing area / Espèce, Zone de pêche / Especie, Area de pesca	1992 mt	1993 mt	1994 mt	1995 mt	1996 mt	1997 mt	1998 mt	1999 mt	2000 mt	2001 mt
Black skipjack	Thonine noire		Barrilete negro			*Euthynnus lineatus*			1,75(01)024,04	BKJ
77 Belize	-	-	-	-	40	-	-	-	-	-
Ecuador	0	0	50	-	270	-	-	-	-	150
Panama	20	-	-	-	-	-	10	-	10	-
USA	-	63	80	101	84	44	231	90	0	0
Venezuela	-	10	-	-	50	40	10	-	10	-
77 Fishing area total	*20*	*73*	*130*	*101*	*444*	*84*	*251*	*90*	*20*	*150*
87 Belize	-	-	-	-	-	-	20	-	-	-
Ecuador	50	80	90	160	100	50	260	10	270	1 650
Panama	-	-	-	-	-	-	-	-	-	10
Vanuatu	-	-	-	-	-	-	10	-	-	-
Venezuela	-	-	-	-	-	-	70	40	-	-
87 Fishing area total	*50*	*80*	*90*	*160*	*100*	*50*	*360*	*50*	*270*	*1 660*
Species total	*70*	*153*	*220*	*261*	*544*	*134*	*611*	*140*	*290*	*1 810*
Kawakawa	Thonine orientale		Bacoreta oriental			*Euthynnus affinis*			1,75(01)024,06	KAW
51 Comoros	200	180	180	180	170	160	160	150	170 F	160
Egypt	76	128	43	138	318	755	841	326	344	209
Eritrea	...	...	...	...	4	6	...	...	0	36
India	22 758	17 989	14 683	17 559	11 837	18 763	12 376	16 757	17 255	851
Iran	722	518	2 100	3 911	5 665	8 451	7 947	10 858	13 500	12 474
Maldives	2 450	3 569	2 656	2 694	3 789	2 089	3 624	1 692	1 898	2 149
Oman	1 439	701	1 113	2 064	2 335	2 388	1 731	1 522	1 550	1 961
Pakistan	1 812	1 933	1 716	1 449	2 351	2 571	2 684	2 715	2 340	1 755
Réunion	-	27	24	28	26	24	28	22	21	13
Saudi Arabia	-	-	-	121	162	304	332	256	264	260
Seychelles	262	163	170	125	93	97	41	158	81	52
South Africa	1	1		1	-	1	2	2	0	0
Untd Arab Em	2 395	1 400	2 725	2 394	2 418	2 500	2 512	2 605	868	870 F
Yemen	1 615	504	1 164	1 226	1 183	1 240	1 210	1 210 F	1 210 F	1 210 F
Other nei	-	-	-	140	140	140	-	-	-	-
51 Fishing area total	*33 730*	*27 113*	*26 574*	*32 030*	*30 491*	*39 489*	*33 488*	*38 273 F*	*39 501 F*	*22 000 F*
57 Australia	0	0	0	0	0	1	-	-	-	0
India	633	1 208	1 022	1 222	2 941	4 662	3 075	4 884	6 363	5 132
Indonesia	53 224	57 890	71 243	67 513	72 522	69 599	86 208	87 295	82 091	81 720
Malaysia	5 571	3 524	1 964	2 494	4 134	4 297	6 283	5 616	7 127	8 978
Sri Lanka	1 055	412	1 546	2 086	2 262	2 765	3 449	2 167	2 167	2 130
Thailand	8 380	17 003	14 146	25 993	18 337	16 700	18 841	16 452	16 223	2 926
57 Fishing area total	*68 863*	*80 037*	*89 921*	*99 308*	*100 196*	*98 024*	*117 856*	*116 414*	*113 971*	*100 886*
61 China,Taiwan	1 400 F	1 000 F	800 F	500 F	0	0	0	0	0	0
61 Fishing area total	*1 400 F*	*1 000 F*	*800 F*	*500 F*	*0*	*0*	*0*	*0*	*0*	*0*
71 China,Taiwan	3 147 F	2 266 F	2 525 F	2 954 F	2 616	2 577	2 307	2 065	1 977	2 136
Indonesia	102 437	103 060	115 245	116 887	135 985	142 912	150 465	148 816	168 431	167 680
Malaysia	26 809	30 520	22 881	24 948	29 810	44 201	43 126	51 665	51 005	47 133
Papua N Guin	0	0	0	0	0	-	-	-	-	-
Philippines	31 943	21 714	29 669	27 308	24 345	26 573	24 424	25 406	27 963	27 890
Thailand	51 187	40 602	40 927	28 871	47 125	42 557	44 027	56 888	56 109	55 765
71 Fishing area total	*215 523 F*	*198 162 F*	*211 247 F*	*200 968 F*	*239 881*	*258 820*	*264 349*	*284 840*	*305 485*	*300 604*
Species total	*319 516 F*	*306 312 F*	*328 542 F*	*332 806 F*	*370 568*	*396 333*	*415 693*	*439 527 F*	*458 957 F*	*423 490 F*
Skipjack tuna	Listao		Listado			*Katsuwonus pelamis*			1,75(01)025,01	SKJ
21 Canada	-	-	-	-	-	0	-	-	-	-
USA	880	54	30	53	35	9	24	42	2	4
21 Fishing area total	*880*	*54*	*30*	*53*	*35*	*9*	*24*	*42*	*2*	*4*
27 Panama	-	-	-	-	-	-	-	-	126	...
Portugal	2 609	2 315	3 384	613	6 271	3 596	3 691	1 465	1 040	1 611
Spain	-	-	-	-	-	-	1	0	-	-
27 Fishing area total	*2 609*	*2 315*	*3 384*	*613*	*6 271*	*3 596*	*3 692*	*1 465*	*1 166*	*1 611*
31 Barbados	5	6	6	6	5	5	10	3	3	1
Bermuda	0	0	0	0	0	0	0	0	0	1
China,Taiwan	-	-	-	-	-	-	0	0	-	-
Colombia	-	2 074	789	1 583	...	...	...	...	...	...
Cuba	1 627	752	1 151	881	1 000	1 282	1 302	750	750 F	750 F
Dominica	41	24	43	33	33	33	33	85	86	...
Dominican Rp	135	143	257	146	123	231	158	72	23	32
Grenada	17	14	11	12	11	15	23	23	23	15
Mexico	8	1	0	0	2	4	6	51	25	91
NethAntilles	40 F	45	40	35	30	30 F	30 F	30 F	30 F	30 F
Panama	-	-	-	-	-	-	-	-	328	...
Puerto Rico	...	...	...	...	...	...	...	...	...	26
St Lucia	39	53	86	72	38	100	263	153	216	151
St Vincent	20	66	56	53	37	42	57	37	68	97
Spain	1 120	397	-	-	-	-	-	1	1	1 959
Trinidad Tob	-	-	-	3	...	0	-	-	-	-
USA	0	0	0	0	-	0	0	-	-	0
Venezuela	7 834	11 172	6 697	2 387	3 574	3 834	4 114	2 981	3 003	6 870
31 Fishing area total	*10 886 F*	*14 747*	*9 136*	*5 211*	*4 853*	*5 576 F*	*5 996 F*	*4 186 F*	*4 556 F*	*10 023 F*

B-36 Tunas, bonitos, billfishes / Thons, pélamides, marlins / Atunes, bonitos, agujas

Capture production by species, fishing areas and countries or areas / Captures par espèces, zones de pêche et pays ou zones / Capturas por especies, áreas de pesca y países o áreas

Species, Fishing area / Espèce, Zone de pêche / Especie, Area de pesca		1992 mt	1993 mt	1994 mt	1995 mt	1996 mt	1997 mt	1998 mt	1999 mt	2000 mt	2001 mt
34	Benin	2	2 F	2 F	2 F	2 F	7	3	2	2 F	...
	Cape Verde	864	860	1 007	1 314	470	591	684	962	789	794
	China	-	-	-	-	-	-	4	-	-	-
	China,Taiwan	-	-	-	1	5	24	42	3	28	2
	Congo Rep	9	10	7	7	6	6	6	6	6	...
	Cuba	11	272	117	5	-	-	-	-	-	-
	France	21 890	33 735	32 779	25 188	23 107	17 023	18 382	20 344	18 183	16 594
	Gabon	-	1	11	51	26	...	59	76	21	101
	Germany	-	-	-	-	3	-	-	-	-	-
	Ghana	18 967	20 225	21 258	18 607	19 602	27 667	34 150	43 460	29 950	43 340
	Japan	1 401	-	-	-	-	-	-	-	-	-
	Korea Rep	-	-	-	-	-	-	-	7	7	-
	Morocco	559	310	248	4 981	675	4 509	2 481	848	1 198	268
	NethAntilles	-	-	-	-	-	-	-	-	11 074	...
	Panama	8 719	13 027	12 978	14 853	5 855	1 300	572	1 308	1 563	...
	Portugal	4 862	3 336	4 136	4 357	2 000	797	849	345	266	496
	Romania	73	-	-	-	-	-	-	-	-	-
	Russian Fed	1 110	540	1 471	1 466	381	1 146	2 086	1 426	374	-
	Sao Tome Prn	25 F	15 F	...	...	...	7	-	-	-	-
	Senegal	260	53	193	293	265	430	1 836	1 422	1 009	3 768
	Spain	53 319	63 660	50 538	51 594	38 538	38 513	36 007	44 519	37 226	30 954
	Other nei	6 273	14 426	15 233	14 388	21 856	13 634	11 815	15 661	3 728	13 147
34	*Fishing area total*	*118 344 F*	*150 472 F*	*139 978 F*	*137 107 F*	*112 791 F*	*105 654*	*108 976*	*130 389*	*105 424 F*	*109 464*
37	Algeria	-	-	-	-	-	-	171	43	89	77
	Morocco	0	2	0	43	9	4	5	10	1	-
37	*Fishing area total*	*0*	*2*	*0*	*43*	*9*	*4*	*176*	*53*	*90*	*77*
41	Argentina	123	50	1	0	1	-	-	-	-	-
	Brazil	18 535	17 771	20 588	16 560	22 528	26 564	23 789	23 188	25 164	24 146
	China,Taiwan	26	9	6	0	2	12	1	0	16	9
	Panama	-	-	-	-	-	-	-	-	640	...
41	*Fishing area total*	*18 684*	*17 830*	*20 595*	*16 560*	*22 531*	*26 576*	*23 790*	*23 188*	*25 820*	*24 155*
47	Angola	41	13	7	3	15	52	2	32	14	687
	China,Taiwan	3	2	1	4	8	38	32	1	62	49
	Japan	-	-	-	-	-	-	-	-	-	1
	Namibia	-	-	5	27	1	1	...	...	...	8
	Panama	-	-	-	-	-	-	-	-	598	...
	Portugal	6	0	8	26	26	6	4	...	1	61
	St Helena	16	65	55	115	86	294	298	13	64	205
	South Africa	7	6	4	4	1	6	2	1	-	1
47	*Fishing area total*	*73*	*86*	*80*	*179*	*137*	*397*	*338*	*47*	*739*	*1 012*
51	China,Taiwan	29	84	10	61	48	48	5	4	17	64
	Comoros	1 847	1 847	2 185	2 185	2 150	2 070	2 070	2 000	2 200 F	2 000
	France	45 048	48 192	58 430	48 652	40 003	30 425	25 423	41 956	39 862	32 023
	India	4 762	4 940	8 959	6 386	5 488	4 884	831	5 707	5 878	21 789
	Iran	4 291	4 353	7 400	1 133	3 239	5 970	6 514	16 583	20 091	26 058
	Italy	-	-	-	-	-	1 733	2 316	3 416	...	1 681
	Japan	30 886	29 413	10 067	556	398	2 264	62	523	448	245
	Kenya	70	72	150	116	108	114	98	109	86	183
	Korea Rep	1	-	-	-	7	-	-	-	-	-
	Maldives	58 577	58 741	69 410	70 372	66 502	69 015	78 410	92 888	79 683	88 044
	Mauritius	6 061	6 902	5 166	3 848	1 898	3 055	1 685	2 361	305	8
	Oman	-	-	372	775	408	730	227	320	293	1
	Pakistan	8 200	8 950	8 134	7 089	4 140	4 480	4 372	4 505	4 308	3 968
	Réunion	59	71	64	105	91	77	92	89	84	79
	Seychelles	643	-	-	-	-	4 695	9 108	15 846	11 604	26 147
	South Africa	1	3	1	1	-	1	1	2	1	0
	Spain	46 756	51 372	61 626	69 587	66 115	62 728	53 966	74 025	77 099	68 414
	Yemen	13	14	14	15	88	90	90	90 F	90 F	90 F
	Other nei	20 818	25 397	32 712	43 776	34 193	34 472	40 398	49 549	61 826	24 846
51	*Fishing area total*	*228 062*	*240 351*	*264 700*	*254 657*	*224 876*	*226 851*	*225 668*	*309 973 F*	*303 875 F*	*295 640 F*
57	Australia	334	29	1 204	466	212	905	2 247	4 986	2 857	2 079
	China,Taiwan	47	136	4	45	11	11	2	0	12	13
	France	-	-	-	-	54	851	4 916	669	73	-
	India	97	48	295	191	1 362	1 212	206	-	108	...
	Indonesia	25 742	30 623	31 979	30 732	42 012	52 730	45 768	47 144	36 237	38 810
	Iran	-	-	-	-	2	3	157	-	-	-
	Italy	-	-	-	-	-	21	708	-	...	-
	Japan	120	28	-	15 431	6 632	4 452	4 739	3 973	1 890	1 465
	Korea Rep	-	-	-	-	1	-	-	-	-	-
	Seychelles	-	-	-	-	-	245	1 597	-	-	-
	Spain	-	-	-	-	161	185	4 680	260	88	...
	Sri Lanka	24 146	24 832	21 548	18 288	22 754	27 815	34 691	51 940	51 940	45 640
	Thailand	-	-	-	-	-	-	-	-	1 110	-
	Other nei	11	36	8	12	110	46	1 054	18	43	...
57	*Fishing area total*	*50 497*	*55 732*	*55 038*	*65 165*	*73 311*	*88 476*	*100 765*	*108 990*	*94 358*	*88 007*
61	China,Taiwan	3 196	3 205	1 599	3 502	2 413	2 670	616	1 864	2 893	16 159
	Japan	108 103	181 405	123 799	120 380	103 190	185 322	205 303	136 316	174 287	140 171
	Russian Fed	-	-	-	-	-	-	-	-	-	33
61	*Fishing area total*	*111 299*	*184 610*	*125 398*	*123 882*	*105 603*	*187 992*	*205 919*	*138 180*	*177 180*	*156 363*
67	Japan	405	-	-	-	-	-	-	-	-	-
	USA	-	-	-	-	-	-	-	-	0	1
67	*Fishing area total*	*405*	-	-	-	-	-	-	-	*0*	*1*

B-36 Tunas, bonitos, billfishes / Thons, pélamides, marlins / Atunes, bonitos, agujas

Capture production by species, fishing areas and countries or areas
Captures par espèces, zones de pêche et pays ou zones
Capturas por especies, áreas de pesca y países o áreas

Species, Fishing area Espèce, Zone de pêche Especie, Area de pesca	1992 mt	1993 mt	1994 mt	1995 mt	1996 mt	1997 mt	1998 mt	1999 mt	2000 mt	2001 mt
71 Australia	-	-	-	-	-	4	2	1	1	1
China	-	-	-	-	-	-	-	-	1 050	2 750
China,Taiwan	81 586	113 190	136 358	155 167	170 949	116 862	196 181	161 321	194 499	182 531
Fiji Islands	3 705	3 178	3 379	4 319	3 124	987	459	507	343	431
Guam	54	68	94	23	17	24	28	19	61	60
Indonesia	126 296	116 668	125 684	128 935	140 137	134 476	181 300	197 703	200 038	214 240
Japan	175 539	159 572	164 970	169 215	176 008	113 275	174 788	145 073	164 031	133 980
Kiribati	248	184	1 016	2 520	4 111	2 855	5 544	4 493	3 701	3 286
Korea Rep	113 954	73 989	145 541	137 848	128 434	114 042	140 685	106 068	132 966	134 766
Marshall Is	-	-	-	-	-	-	-	-	6 715	33 468
Micronesia	11 657	11 692	17 531	4 216	6 745	5 501	11 314	6 972	15 843	10 310
NewCaledonia	-	-	1	2	0	1	1	0	0	0
N Marianas	30	35	39	60	75	64	61	48	56	61
Palau	61	100	100	100	100	100	100	100	100	100
Papua N Guin	-	-	987	9 811	9 512	11 270	37 214	29 949	52 289	64 355
Philippines	83 179	68 065	84 560	110 111	110 004	110 097	116 673	108 778	113 011	112 238
Singapore	0	0	6	5	5	47	12	23	2	10
Solomon Is	24 219	20 080	26 661	40 136	26 485	36 311	38 662	35 613	8 368	12 530
Tuvalu	6	292	310	259	260 F	300 F	300 F	300 F	300 F	300 F
USA	112 370	142 457	134 394	125 504	112 467	92 503	105 200	144 148	92 780	86 441
Vanuatu	-	-	730	5 577	8 080	16 730	28 982	36 321	33 075	...
71 Fishing area total	*732 904*	*709 570*	*842 361*	*893 808*	*896 513 F*	*755 449 F*	*1 037 506 F*	*977 437 F*	*1 019 229 F*	*991 858 F*
77 Amer Samoa	32	11	68	80	32	16	8	20	14	56
Belize	-	-	20	70	720	500	290	880	310	...
China	-	-	-	-	-	-	-	-	-	1 130
China,Taiwan	0	5	81	45	17	40	164	710	9	5
Colombia	...	...	...	...	...	...	...	...	160	270
Costa Rica	...	...	...	...	80	...	...	...	...	...
Cyprus	30	10	-	710	700	840	-	-	-	-
Ecuador	110	650	570	-	-	-	-	-	-	-
El Salvador	28	...	...	...	...	...	750	320	...	4 476
Fr Polynesia	1 122	1 125	1 004	1 400	1 400	1 126	1 560	1 386	1 189	1 557
Guatemala	-	-	-	-	-	-	-	3 360	5 390	...
Honduras	-	-	-	-	540	1 120	630	590	50	...
Japan	6 477	4 200	2 462	3 368	266	4 032	109	179	33	299
Korea Rep	1 340	4	-	-	1 446	1 885	2 705	-	2	138
Mexico	8 779	13 291	7 045	30 694	16 728	24 286	16 450	18 356	14 385	7 520
Nicaragua	-	-	-	-	-	-	-	250	400	...
Panama	140	210	230	410	1 230	820	940	2 210	1 190	1 320
Spain	-	90	50	520	720	1 170	14 060	23 420	13 837	9 821
Tonga	3	4	3	3	2	4	7	3	2	12
USA	57 351	15 728	18 666	30 732	13 224	17 958	17 679	7 150	2 092	2 583
Vanuatu	130	230	280	3 340	3 900	6 000	5 300	3 260	1 060	1 000
Venezuela	1 010	750	510	3 530	2 800	5 990	2 940	2 970	2 410	930
Other nei	-	-	-	40	-	-	10	-	-	9 380
77 Fishing area total	*76 552*	*36 308*	*30 989*	*74 942*	*43 805*	*65 787*	*63 602*	*65 064*	*42 533*	*40 497*
81 Australia	6 958	4 313	3 514	4 297	2 767	4 689	849	429	1 862	1 090
China,Taiwan	0	285	98	12	36	84	270	-	-	7
Japan	13	15	3	12	151	2 189	431	1 227	723	512
New Zealand	995	963	3 212	1 428	3 631	4 792	8 156	5 688	9 699	3 691
81 Fishing area total	*7 966*	*5 576*	*6 827*	*5 749*	*6 585*	*11 754*	*9 706*	*7 344*	*12 284*	*5 300*
87 Belize	-	10	2 020	5 090	4 420	5 760	4 150	3 450	5 570	...
Chile	-	-	-	-	-	53	47	9	0	57
Colombia	5 182	12 635	3 807	7 391	15 676	24 341	3 526	27 861	6 000	2 250
Costa Rica	...	...	...	120	540	...	...	...	...	...
Cuba	0	-	-	-	-	-	-	-	-	-
Cyprus	3 390	3 290	3 850	3 490	2 510	4 360	300	-	-	-
Ecuador	28 621	23 210	15 431	31 599	37 468	67 400	67 453	126 992	105 146	68 217
El Salvador	...	...	...	...	...	...	...	2 810	...	...
Guatemala	-	-	-	-	-	-	-	3 390	7 480	...
Honduras	-	-	-	-	240	1 990	-	3 690	1 960	...
Japan	1	26	7	4	26	8	30	26	2	7
Korea Rep	-	-	-	-	-	-	-	3 705	4 040	2 665
Liberia	-	-	-	-	410	-	-	-	-	-
Mexico	507	152	505	1 066	1 600	881	1 241	753	440	610
Nicaragua	-	-	-	-	-	-	-	-	30	...
Panama	3 200	670	1 690	3 540	2 210	3 170	30	2 870	11 290	5 330
Peru	481	500	109	151	85	823	9 373	802	711	81
St Vincent	-	-	1 200	-	-	-	-	-	-	-
Spain	2 020	4 730	1 780	4 940	3 070	8 480	6 070	17 460	8 960	12 950
USA	2 017	-	-	-	2 822	1 684	764	-	2 555	9
Vanuatu	12 530	10 160	9 190	11 130	7 620	8 130	5 830	17 570	9 980	6 980
Venezuela	7 720	5 410	4 910	1 080	620	1 290	3 170	11 080	2 710	1 250
Other nei	440	850	640	730	-	2 320	650	-	1 300	12 020
87 Fishing area total	*66 109*	*61 643*	*45 139*	*70 331*	*79 317*	*130 690*	*102 634*	*222 468*	*168 174*	*112 426*
Species total	*1 425 270 F*	*1 479 296 F*	*1 543 655 F*	*1 648 300 F*	*1 576 637 F*	*1 608 811 F*	*1 888 792 F*	*1 988 826 F*	*1 955 430 F*	*1 836 438 F*
Atlantic bluefin tuna	**Thon rouge de l'Atlantique**		**Atún rojo del Atlántico**		***Thunnus thynnus***			**1,75(01)026,01**		**BFT**
21 Canada	400	330	412	578	599	507	596	452	550	512
Japan	1 156	1 170	752	657	913	594	790	1 121	939	1 056
St Pier Mq	-	-	-	-	-	-	-	1	0	0
USA	913	981	1 019	877	720	1 006	1 041	1 032	1 033	1 175
21 Fishing area total	*2 469*	*2 481*	*2 183*	*2 112*	*2 232*	*2 107*	*2 427*	*2 606*	*2 522*	*2 743*

B-36 Tunas, bonitos, billfishes — Capture production by species, fishing areas and countries or areas
Thons, pélamides, marlins — Captures par espèces, zones de pêche et pays ou zones
Atunes, bonitos, agujas — Capturas por especies, áreas de pesca y países o áreas

Species, Fishing area Espèce, Zone de pêche Especie, Area de pesca		1992 mt	1993 mt	1994 mt	1995 mt	1996 mt	1997 mt	1998 mt	1999 mt	2000 mt	2001 mt
27	China,Taiwan	-	-	-	6	9	9	-	146	99	100
	Denmark	0	37	-	0	0	-	1	-	-	-
	Faeroe Is	-	-	-	-	-	-	67	104	118	...
	France	894	1 099	336	725	563	269	613	588	542	630
	Germany	-	-	-	1	-	-	-	-	-	-
	Iceland	-	-	-	-	-	1	2	33	29	-
	Ireland	-	-	-	-	-	14	21	52	24	10
	Japan	2 194	1 084	1 212	3 190	2 464	2 026	2 148	1 988	1 842	1 361
	Norway	0	0	-	-	-	-	-	5	0	-
	Portugal	34	23	20	23	139	372	158	408	440	402
	Spain	2 338	5 047	3 081	3 815	6 017	5 842	3 761	3 328	3 448	1 497
	Sweden	0	0	0	0	-	-	-	-	-	-
	UK	-	-	-	1	0	1	1	12	0	-
	Other nei	144	223	68	189	71	208	-	-	-	-
27	*Fishing area total*	*5 604*	*7 513*	*4 717*	*7 950*	*9 263*	*8 742*	*6 772*	*6 664*	*6 542*	*4 000*
31	Bermuda	-	-	-	-	1	2	2	1	1	1
	China,Taiwan	-	-	-	0	0	-	456	0	-	-
	Japan	36	2	-	-	34	-	-	-	1	-
	Mexico	-	-	-	5	14	7	14	16	35	10
	St Lucia	14	2	43	9	3	-	-	-	-	-
	USA	62	64	45	29	21	18	18	10	88	57
	Other nei	17	-	-	-	2	-	-	429	270	49
31	*Fishing area total*	*129*	*68*	*88*	*43*	*75*	*27*	*490*	*456*	*395*	*117*
34	China	-	-	-	-	-	-	85	103	80	68
	China,Taiwan	-	6	11	2	91	97	-	46	277	101
	Greece	8	4	-	3	4	1	1	-	-	-
	Guinea	-	-	330	...	...	...	...	...	...	...
	Japan	367	264	486	590	486	709	911	342	216	211
	Korea Rep	-	438	642	85	76	83	-	-	-	-
	Libya	312	-	-	-	576	477	511	450	487	...
	Morocco	562	415	720	678	1 035	2 068	1 866	1 591	2 228	2 497
	Panama	-	-	1	19	550	255	-	13	-	-
	Portugal	4	2	220	12	60	340	165	3	1	2
	Sierra Leone	...	...	...	...	...	...	...	...	93	118
	Spain	29	31	56	4	157	-	39	32	26	2 099
	Other nei	-	-	-	-	-	-	66	-	-	-
34	*Fishing area total*	*1 282*	*1 160*	*2 466*	*1 393*	*3 035*	*4 030*	*3 644*	*2 580*	*3 408*	*5 096*
37	Algeria	1 104	1 097	1 560	156	156	157	1 947	2 142	2 330	2 012
	China	-	-	97	137	93	49	-	-	-	-
	China,Taiwan	-	328	713	493	372	398	-	58	31	196
	Croatia	1 076	1 058	1 410	1 220	1 360	1 105	906	970	930	903
	Cyprus	10	14	10	10	10	10	21	31	61	90
	France	7 376	6 995	11 843	9 604	9 127	8 201	7 100	6 153	6 780	6 119
	Greece	447	439	886	1 004	874	1 217	286	248	622	361
	Israel	-	-	-	-	14	-	-	-	-	-
	Italy	5 005	5 328	6 882	7 062	10 006	9 548	4 059	3 279	3 845	4 377
	Japan	186	607	521	734	664	167	410	371	142	187
	Korea Rep	-	-	-	458	591	410	-	-	-	-
	Libya	425	635	1 422	1 540	812	552	820	745	1 063	1 940
	Malta	80	251	572	587	399	393	407	447	376	219
	Morocco	205	79	1 092	1 035	586	535	564	636	695	511
	Panama	484	467	1 499	1 498	2 850	236	-	-	-	-
	Portugal	320	183	428	446	274	37	54	76	61	64
	Spain	2 165	2 018	2 741	4 607	2 588	2 205	2 000	2 003	2 772	2 148
	Tunisia	1 195	2 132	2 503	1 897	2 393	2 200	1 745	2 352	2 184	2 493
	Turkey	2 817	3 084	3 466	4 220	4 616	5 093	5 899	1 200	1 070	2 100
	Yugoslavia	0	0	0	2	4	4	6	7	4	5
	Other nei	1 398	-	1 200	850	171	1 167	1 855	2 135	126	-
37	*Fishing area total*	*24 293*	*24 715*	*38 845*	*37 560*	*37 960*	*33 684*	*28 079*	*22 853*	*23 092*	*23 725*
41	Argentina	0		-	-	-	-	-		-	-
	Brazil	0	0	0	0	0	0	0	13	0	0
	UK	-	-	-	-	-	0	1	-	-	-
	Uruguay	0	1	0	2	-	-	-	-	-	1
41	*Fishing area total*	*0*	*1*	*0*	*2*	*0*	*0*	*1*	*13*	*0*	*1*
47	Japan	-	14	-	-	-	-	3	-	-	-
	Korea Rep	-	-	42	120	16	120	-	-	-	-
47	*Fishing area total*	-	*14*	*42*	*120*	*16*	*120*	*3*	-	-	-
Species total		*33 777*	*35 952*	*48 341*	*49 180*	*52 581*	*48 710*	*41 416*	*35 172*	*35 959*	*35 682*
Pacific bluefin tuna		**Thon bleu du Pacifique**		**Atún aleta azul del Pacífico**		***Thunnus orientalis***				**1,75(01)026,02**	**PBF**
61	China,Taiwan	1 045 F	1 074 F	539 F	31	157	505	695	2	1 014	580
	Japan	4 588	4 390	7 624	5 844	6 651	6 469	3 633	11 145	10 044	5 676
61	*Fishing area total*	*5 633 F*	*5 464 F*	*8 163 F*	*5 875*	*6 808*	*6 974*	*4 328*	*11 147*	*11 058*	*6 256*
67	Japan	-	-	1	-	-	-	-	-	-	-
	USA	1	4	12	1	2	3	8	7	3	0
67	*Fishing area total*	*1*	*4*	*13*	*1*	*2*	*3*	*8*	*7*	*3*	*0*
71	China,Taiwan	20 F	30 F	20 F	283	799	1 308	1 215	3 087	1 676	1 504
	Japan	341	494	4	344	2	6	18	17	79	150
	Korea Rep	-	-	23	-	-	-	-	-	-	-

B-36

Tunas, bonitos, billfishes — **Capture production by species, fishing areas and countries or areas**
Thons, pélamides, marlins — **Captures par espèces, zones de pêche et pays ou zones**
Atunes, bonitos, agujas — **Capturas por especies, áreas de pesca y países o áreas**

Species, Fishing area Espèce, Zone de pêche Especie, Area de pesca	1992 mt	1993 mt	1994 mt	1995 mt	1996 mt	1997 mt	1998 mt	1999 mt	2000 mt	2001 mt
71 Fishing area total	*361 F*	*524 F*	*47 F*	*627*	*801*	*1 314*	*1 233*	*3 104*	*1 755*	*1 654*
77 China,Taiwan	0	0	0	0	0	0	-	-	-	-
Japan	28	18	14	18	12	2	9	11	2	2
Korea Rep	81	43	-	-	-	-	-	-	-	-
Mexico	23	20	60	83	3 700	370	34	2 370	3 030	860
USA	1 066	552	919	642	4 767	2 269	1 954	170	313	317
77 Fishing area total	*1 198*	*633*	*993*	*743*	*8 479*	*2 641*	*1 997*	*2 551*	*3 345*	*1 179*
81 Japan	12	4	15	3	3	8	7	14	3	4
New Zealand	-	-	-	2	5	12	20	21	21	50
81 Fishing area total	*12*	*4*	*15*	*5*	*8*	*20*	*27*	*35*	*24*	*54*
87 Japan	-	-	-	-	-	2	-	1	-	-
87 Fishing area total	-	-	-	-	-	*2*	-	*1*	-	-
Species total	*7 205 F*	*6 629 F*	*9 231 F*	*7 251*	*16 098*	*10 954*	*7 593*	*16 845*	*16 185*	*9 143*
Longtail tuna	**Thon mignon**		**Atún tongol**			***Thunnus tonggol***			**1,75(01)026,03**	**LOT**
51 China,Taiwan	17	0	0	4 544	3 437	2 611	-	-	-	-
Eritrea	...	...	...	...	...	6	22	...	0	-
India	2 487	4 324	4 917	6 985	3 415	4 263	3 805	2 275	2 342	258
Iran	9 758	8 150	12 100	27 188	17 147	20 112	19 694	23 465	41 407	34 896
Oman	5 065	6 705	5 088	3 967	5 316	5 020	4 379	4 798	5 318	6 011
Pakistan	2 770	2 510	5 807	5 006	4 121	5 360	5 220	5 600	5 315	6 935
Saudi Arabia	-	-	-	234	115	101	181	136	143	180
Untd Arab Em	3 959	4 000	4 466	5 715	5 775	3 671	3 678	3 739	1 725	1 720 F
Yemen	1 324	1 707	2 291	2 204	1 887	1 970	1 920	1 920 F	1 920 F	1 920 F
Other nei	-	-	-	351	351	351	-	-	85	-
51 Fishing area total	*25 380*	*27 396*	*34 669*	*56 194*	*41 564*	*43 465*	*38 899*	*41 933 F*	*58 255 F*	*51 920 F*
57 Australia	13	0	2	1	0	0	0	31	8	58
China,Taiwan	20	0	0	1 600	1 298	986	-	-	-	-
India	0	0	36	51	848	1 059	945	293	381	283
Malaysia	2 614	1 930	972	1 233	2 045	2 125	3 106	2 824	3 586	-
Sri Lanka	0	12	57	0	0	-	-	-	-	-
Thailand	2 169	20 950	32 762	18 709	22 036	20 035	18 626	16 676	15 264	5 685
57 Fishing area total	*4 816*	*22 892*	*33 829*	*21 594*	*26 227*	*24 205*	*22 677*	*19 824*	*19 239*	*6 026*
61 China,Taiwan	16 318	19 000	15 150	595	514	391	303	6 434	10 087	14 321
61 Fishing area total	*16 318*	*19 000*	*15 150*	*595*	*514*	*391*	*303*	*6 434*	*10 087*	*14 321*
71 Australia	26	7	7	16	10	-	-	-	-	0
China,Taiwan	1 714	2 249	1 844	4 010	3 304	2 510	6 163	2 772	4 342	4 300
Papua N Guin	0	0	0	0	0	-	-	-	-	-
Thailand	72 277	39 396	32 006	38 824	32 347	29 127	34 805	45 818	45 191	44 915
71 Fishing area total	*74 017*	*41 652*	*33 857*	*42 850*	*35 661*	*31 637*	*40 968*	*48 590*	*49 533*	*49 215*
77 China,Taiwan	-	-	-	388	327	249	-	-	-	-
77 Fishing area total	-	-	-	*388*	*327*	*249*	-	-	-	-
81 Australia	2	2	3	3	3	-	-	3	-	-
China,Taiwan	0	0	0	29	20	15	-	-	-	-
81 Fishing area total	*2*	*2*	*3*	*32*	*23*	*15*	-	*3*	-	-
Species total	*120 533*	*110 942*	*117 508*	*121 653*	*104 316*	*99 962*	*102 847*	*116 784 F*	*137 114 F*	*121 482 F*
Blackfin tuna	**Thon à nageoires noires**		**Atún aleta negra**			***Thunnus atlanticus***			**1,75(01)026,04**	**BLF**
21 USA	1	1	0	2	28	5	1	20	16	2
21 Fishing area total	*1*	*1*	*0*	*2*	*28*	*5*	*1*	*20*	*16*	*2*
31 Bermuda	6	5	7	4	5	4	6	6	5	4
Cuba	196	54	223	156	287	301	226	309	309 F	309 F
Dominica	14	15	19	30	...	...	...	79	83	83
Dominican Rp	110	133	239	892	518	323	89	73	114	517
Grenada	83	144	189	123	164	126	233	94	164	222
Guadeloupe	470	440	440	480	500	500	500	500	500	500
Martinique	700 F	700 F	890 F	890 F	540 F	540 F	540 F	540 F	540 F	470 F
NethAntilles	60 F	65	60	50	45	45 F	45 F	45 F	45 F	45 F
Puerto Rico	...	...	...	...	...	...	...	...	...	17
St Lucia	13	16	82	47	35	40	100	41	45	108
St Vincent	7	53	19	20	18	22	17	15	23	24
Spain	307	46	-	-	-	-	-	-	-	-
USA	127	126	107	63	25	62	52	21	34	33
Venezuela	2 148	1 224	21	624	758	498	1 034	1 192	589	1 902
31 Fishing area total	*4 241 F*	*3 021 F*	*2 296 F*	*3 379 F*	*2 895 F*	*2 461 F*	*2 842 F*	*2 915 F*	*2 451 F*	*4 234 F*
41 Brazil	49	22	37	153	649	418	55	55	38	149
41 Fishing area total	*49*	*22*	*37*	*153*	*649*	*418*	*55*	*55*	*38*	*149*
Species total	*4 291 F*	*3 044 F*	*2 333 F*	*3 534 F*	*3 572 F*	*2 884 F*	*2 898 F*	*2 990 F*	*2 505 F*	*4 385 F*
Albacore	**Germon**		**Atún blanco**			***Thunnus alalunga***			**1,75(01)026,05**	**ALB**
21 Canada	1	9	32	11	24	31	24	39	122	51
China,Taiwan	0	25	358	181	70	17	8	102	290	465
Japan	331	337	365	113	185	186	148	175	222	492

B-36 Tunas, bonitos, billfishes / Thons, pélamides, marlins / Atunes, bonitos, agujas

Capture production by species, fishing areas and countries or areas / Captures par espèces, zones de pêche et pays ou zones / Capturas por especies, áreas de pesca y países o áreas

Species, Fishing area Espèce, Zone de pêche Especie, Area de pesca	1992 mt	1993 mt	1994 mt	1995 mt	1996 mt	1997 mt	1998 mt	1999 mt	2000 mt	2001 mt
USA	191	242	284	318	88	148	187	174	108	121
21 Fishing area total	*523*	*613*	*1 039*	*623*	*367*	*382*	*367*	*490*	*742*	*1 129*
27 China,Taiwan	0	477	1 515	1 030	3	3	0	384	941	423
France	6 924	6 293	5 934	5 304	4 694	4 618	3 711	7 189	6 019	6 344
Ireland	451	1 946	2 534	918	874	1 913	3 750	4 858	3 464	2 085
Japan	72	60	70	197	104	46	89	71	146	117
Panama	382	210	363	289	369	58	58	-	-	-
Portugal	1 225	3 161	918	6 267	834	182	49	246	264	411
Spain	17 874	17 775	16 837	19 538	15 580	16 245	12 964	13 355	15 726	9 177
UK	59	499	613	196	49	33	117	343	15	2
27 Fishing area total	*26 987*	*30 421*	*28 784*	*33 739*	*22 507*	*23 098*	*20 738*	*26 446*	*26 575*	*18 559*
31 Barbados	-	-	-	-	-	1	1	1	...	2
Bermuda	-	-	-	-	-	1	...	2	2	2
China,Taiwan	3 164	2 069	2 660	3 057	3 349	2 856	1 241	5 191	5 196	9 240
Grenada	-	0	0	2	1	6	7	6	12	21
Japan	22	8	14	9	35	23	33	140	125	281
Philippines	-	-	-	-	-	-	-	4	0	-
St Lucia	1	1	0	1	1	0	0	0	1	3
St Vincent	-	2	0	0	-	-	-	1	2 820	5 662
Spain	3	2	1	2	0	4	8	37	-	-
Trinidad Tob	247	639	...	...	...	2	1	1	2	11
USA	7	9	21	10	5	15	3	3	5	8
Venezuela	205	246	282	279	315	49	107	91	1 374	349
31 Fishing area total	*3 649*	*2 976*	*2 978*	*3 360*	*3 706*	*2 957*	*1 401*	*5 477*	*9 537*	*15 579*
34 Belize	-	-	-	2	-	-	-	8	2	...
China	-	-	14	8	20	-	-	21	16	57
China,Taiwan	1 000	2 028	1 891	1 741	576	492	346	76	657	849
Cuba	5	3	-	0	-	-	-	-	-	-
Japan	42	93	58	73	149	193	131	51	194	275
Korea Rep	-	-	-	-	-	5	-	-	-	-
Liberia	...	...	...	...	41	-	-	-	-	-
Panama	129	168	213	12	22	-	3	14	-	-
Philippines	-	-	-	-	-	-	5	4	0	-
Portugal	413	224	56	203	800	213	42	78	14	765
Sierra Leone	...	...	...	...	...	...	...	...	...	91
Spain	2 115	1 252	842	918	804	1 139	494	2 128	240	...
Other nei	281	159	133	110	180	50	50	50	-	-
34 Fishing area total	*3 985*	*3 927*	*3 207*	*3 067*	*2 592*	*2 092*	*1 071*	*2 430*	*1 123*	*2 037*
37 Cyprus	-	-	-	-	-	-	-	-	6	-
France	11	64	23	3	0	5	5	-	-	-
Greece	500	1	1	0	952	741	1 152	2 005	1 786	1 840
Italy	1 464	1 275	1 107	1 109	1 769	1 414	1 414	2 561	3 630	2 826
Malta	-	-	-	-	-	1	1	1	4	...
Spain	227	290	218	475	404	380	126	284	152	77
Other nei	-	500	-	-	-	-	-	-	-	-
37 Fishing area total	*2 202*	*2 130*	*1 349*	*1 587*	*3 125*	*2 541*	*2 698*	*4 851*	*5 578*	*4 743*
41 Argentina	306	0	2	0	0	120	-	-	-	-
Brazil	2 710	3 613	1 227	923	819	652	3 418	1 872	4 414	6 862
China,Taiwan	24 162	13 345	12 429	7 408	15 088	12 866	8 087	10 221	4 040	4 644
Japan	177	119	233	39	36	88	41	27	86	64
Spain	126	135	149	196	123	162	12	-	-	-
Uruguay	31	28	16	49	75	56	110	69	90	40
41 Fishing area total	*27 512*	*17 240*	*14 056*	*8 615*	*16 141*	*13 944*	*11 668*	*12 189*	*8 630*	*11 610*
47 China	-	-	-	-	-	-	-	39	89	26
China,Taiwan	7 741	11 339	10 035	8 910	3 825	3 262	9 525	7 187	6 779	3 685
France	449	564	129	82	190	38	40	13	23	16
Japan	1 052	645	720	609	615	692	867	800	786	456
Namibia	2 241	3 524	3 075	1 861	1 521	1 199	1 422	1 162	2 418	3 419
Portugal	184	483	1 185	655	494	256	124	232	486	41
St Helena	28	38	5	82	47	18	1	1	58	12
South Africa	6 360	6 881	6 931	5 214	5 634	6 708	8 412	5 101	2 072	7 236
Spain	-	-	-	-	-	-	-	871	282	...
Other nei	122	68	55	63	41	13	218	-	723	-
47 Fishing area total	*18 177*	*23 542*	*22 135*	*17 476*	*12 367*	*12 186*	*20 609*	*15 406*	*13 716*	*14 891*
51 China	-	-	-	-	-	-	-	88	3	2
China,Taiwan	4 694	13 484	11 847	11 330	12 074	10 843	18 196	18 191	8 455	4 795
France	1 402	310	292	350	391	539	460	154	350	647
Iran	-	-	-	-	10	16	9	-	-	-
Italy	-	-	-	-	-	12	58	-	...	56
Japan	820	444	840	625	1 028	1 589	2 123	1 141	1 046	1 590
Korea Rep	5	4	9	3	14	102	118	26	95	31
Mauritius	2	2	2	2	2	7	15	12	...	18
Philippines	-	-	-	-	-	-	2	180	101	68
Réunion	55	120	175	163	347	306	318	357	579	648
Seychelles	-	-	-	-	-	-	183	65	214	210
South Africa	-	2	1	2	-	-	7	1	26	21
Spain	1 456	904	1 773	561	826	1 031	267	275	532	572
Other nei	886	2 125	-	-	-	4 751	8 398	-	-	92
51 Fishing area total	*9 320*	*17 395*	*14 939*	*13 036*	*14 692*	*19 196*	*30 154*	*20 490*	*11 401*	*8 750*
57 Australia	49	103	127	27	4	23	26	31	28	137
China	-	-	-	-	-	-	-	101	-	20

B-36 Tunas, bonitos, billfishes / Thons, pélamides, marlins / Atunes, bonitos, agujas

Capture production by species, fishing areas and countries or areas / Captures par espèces, zones de pêche et pays ou zones / Capturas por especies, áreas de pesca y países o áreas

Species, Fishing area Espèce, Zone de pêche Especie, Area de pesca	1992 mt	1993 mt	1994 mt	1995 mt	1996 mt	1997 mt	1998 mt	1999 mt	2000 mt	2001 mt
China,Taiwan	9 846	2 835	2 560	2 879	4 856	4 361	3 376	4 323	1 308	290
Japan	244	509	647	1 166	1 138	1 238	690	905	1 088	1 172
Korea Rep	-	-	4	3	-	-	4	1	-	-
Philippines	-	-	-	-	-	-	499	10	0	-
Seychelles	-	-	-	-	-	-	-	9	33	8
Spain	-	-	-	-	-	-	6	-	...	...
Thailand	-	-	-	-	-	-	-	-	12	-
Other nei	1 962	626	1 592	2 240	2 711	1 097	2 658	1 655	1 810	...
57 Fishing area total	*12 101*	*4 073*	*4 930*	*6 315*	*8 709*	*6 719*	*7 259*	*7 035*	*4 279*	*1 627*
61 China	-	-	-	-	-	-	-	-	-	528
China,Taiwan	6 467	241	77	4 269	2 662	3 518	651	1 306	1 310	713
Japan	28 336	33 011	40 098	33 857	48 364	66 487	54 206	85 662	45 852	49 507
61 Fishing area total	*34 803*	*33 252*	*40 175*	*38 126*	*51 026*	*70 005*	*54 857*	*86 968*	*47 162*	*50 748*
67 Canada	512	551	589	792	457	80	137	308	2 535	3 061
China,Taiwan	-	-	-	26	0	0	0	55	1 165	310
Japan	394	-	-	-	-	-	-	-	-	-
USA	3 848	4 822	8 288	5 853	9 581	14 508	11 380	4 312	7 482	8 762
67 Fishing area total	*4 754*	*5 373*	*8 877*	*6 671*	*10 038*	*14 588*	*11 517*	*4 675*	*11 182*	*12 133*
71 Australia	...	...	...	...	...	...	...	179	147	173
China	0	1	8	5	8	2	1	3 473	2 056	2 232
China,Taiwan	2 084	4 455	11 943	8 891	5 810	6 298	5 095	3 017	5 784	6 958
Fiji Islands	243	463	842	702	1 446	1 842	2 121	2 279	6 065	7 971
Japan	8 405	12 173	19 144	17 490	2 124	5 027	6 481	3 527	7 926	8 678
Korea Rep	15	26	-	20	114	666	690	147	20	332
Micronesia	-	-	-	-	-	1	-	2	3	4
NewCaledonia	692	755	840	332	414	277	860	690	895	1 020
Papua N Guin	-	-	-	6	38	101	40	85	102	49
Solomon Is	...	...	...	24	100	109	370	136	224	54
USA	-	-	2	0	-	-	5 110	2 628	-	-
Vanuatu	310 F	240 F	180 F	109	192	95	10	-	-	-
71 Fishing area total	*11 749 F*	*18 113 F*	*32 959 F*	*27 579*	*10 246*	*14 418*	*20 778*	*16 163*	*23 222*	*27 471*
77 Amer Samoa	...	0	1	27	86	309	446	338	624	3 253
China	-	-	-	-	-	-	-	-	1 239	1 624
China,Taiwan	8 705	5 003	6 093	3 867	4 454	4 721	5 409	8 575	6 837	6 729
Cook Is	...	...	21	32	14	10 F	10 F	10 F	10 F	10 F
Fr Polynesia	267	959	913	1 100	1 750	2 717	3 235	2 642	3 580	4 432
Japan	6 526	6 450	4 839	5 098	4 401	4 717	5 187	4 506	2 372	3 285
Korea Rep	202	75	96	48	588	1 171	3 177	954	569	1 557
Mexico	10	11	6	5	21	53	8	32	159	40
Samoa	922	213	641	1 883	1 775	4 108	4 742	4 027	4 067	4 820
Tonga	199	231	343	379	431	493	616	801	862	1 268
USA	1 424	1 707	3 579	3 504	5 772	4 236	3 593	6 976	5 909	3 814
77 Fishing area total	*18 255*	*14 649*	*16 532*	*15 943*	*19 292*	*22 535 F*	*26 423 F*	*28 861 F*	*26 228 F*	*30 832 F*
81 Australia	204	226	351	401	468	317	418	225	212	222
Canada	235	235	235	235	136	149	167	253	351	206
China,Taiwan	34 902	14 624	4 006	2 875	5 479	5 797	7 849	5 619	7 452	5 491
Japan	2 144	2 759	2 167	2 307	2 267	2 876	3 287	1 565	1 052	2 415
Korea Rep	-	-	-	-	-	-	9	-	-	-
New Zealand	3 794	3 613	6 352	6 423	7 150	3 220	6 525	3 903	4 500	5 353
USA	2 918	1 123	605	2 170	3 223	-	-	-	-	5 942
81 Fishing area total	*44 197*	*22 580*	*13 716*	*14 411*	*18 723*	*12 359*	*18 255*	*11 565*	*13 567*	*19 629*
87 Chile	18	19	22	15	21	-	-	-	3	5
China,Taiwan	156	422	57	3	162	171	85	55	1 454	866
Japan	681	1 658	1 695	960	613	430	473	457	354	864
Korea Rep	-	-	-	-	-	-	-	51	-	-
87 Fishing area total	*855*	*2 099*	*1 774*	*978*	*796*	*601*	*558*	*563*	*1 811*	*1 735*
Species total	*219 069 F*	*198 383 F*	*207 450 F*	*191 526*	*194 327*	*217 621 F*	*228 353 F*	*243 609 F*	*204 753 F*	*221 473 F*
Southern bluefin tuna	**Thon rouge du Sud**		**Atún rojo del Sur**			***Thunnus maccoyii***			**1,75(01)026,08**	**SBF**
41 China,Taiwan	204	119	6	7	18	5	6	3	3	-
Japan	2	-	1	-	-	-	-	-	-	-
41 Fishing area total	*206*	*119*	*7*	*7*	*18*	*5*	*6*	*3*	*3*	-
47 China,Taiwan	203	119	186	161	139	42	228	158	283	76
Japan	2 480	3 757	1 642	2 290	2 103	409	1 840	1 908	1 986	2 421
Korea Rep	-	-	-	-	-	-	-	28	62	19
South Africa	-	-	-	-	-	-	1	-	2	0
47 Fishing area total	*2 683*	*3 876*	*1 828*	*2 451*	*2 242*	*451*	*2 069*	*2 094*	*2 333*	*2 516*
51 China,Taiwan	47	360	555	1 098	1 195	488	1 049	805	1 400	1 473
Japan	1 061	543	715	257	848	2 139	1 133	855	370	1 232
Korea Rep	15	-	98	216	314	1 056	1 415	463	784	363
South Africa	-	-	-	-	-	-	-	-	2	-
Other nei	84	40	74	237	243	294	342	372	24	-
51 Fishing area total	*1 207*	*943*	*1 442*	*1 808*	*2 600*	*3 977*	*3 939*	*2 495*	*2 580*	*3 068*
57 Australia	2 184	2 152	2 499	2 998	4 732	4 912	4 320	5 444	4 629	5 463
China,Taiwan	35	261	106	208	258	105	142	781	192	106
Japan	1 516	843	1 806	2 392	2 743	1 824	3 733	3 219	2 667	1 907
Korea Rep	-	-	-	99	597	172	147	773	112	347
Sri Lanka	-	-	-	-	68	83	104	121	120	...

B-36 Tunas, bonitos, billfishes — Capture production by species, fishing areas and countries or areas
Thons, pélamides, marlins — Captures par espèces, zones de pêche et pays ou zones
Atunes, bonitos, agujas — Capturas por especies, áreas de pesca y países o áreas

Species, Fishing area Espèce, Zone de pêche Especie, Area de pesca	1992 mt	1993 mt	1994 mt	1995 mt	1996 mt	1997 mt	1998 mt	1999 mt	2000 mt	2001 mt
Other nei	63	29	14	45	52	39	134	111	7	...
57 Fishing area total	*3 798*	*3 285*	*4 425*	*5 742*	*8 450*	*7 135*	*8 580*	*10 449*	*7 727*	*7 823*
71 Australia	...	...	...	...	...	...	...	0	3	-
71 Fishing area total	...	...	...	...	...	...	...	*0*	*3*	-
81 Australia	2 974	2 743	2 198	2 270	623	1 028	471	355	74	98
China,Taiwan	-	-	-	-	0	0	14	4	4	
Japan	1 787	2 157	1 307	716	973	728	1 010	1 746	1 069	1 680
New Zealand	33	50	52	181	81	138	337	460	380	358
81 Fishing area total	*4 794*	*4 950*	*3 557*	*3 167*	*1 677*	*1 894*	*1 832*	*2 565*	*1 527*	*2 136*
87 Japan	-	-	1	17	-	-	-	-		
87 Fishing area total	-	-	*1*	*17*	-	-	-	-		
Species total	*12 688*	*13 173*	*11 260*	*13 192*	*14 987*	*13 462*	*16 426*	*17 606*	*14 173*	*15 543*
Yellowfin tuna	**Albacore**		**Rabil**		***Thunnus albacares***			**1,75(01)026,10**		**YFT**
21 Canada	25	72	52	174	155	100	57	22	105	125
China,Taiwan	0	0	17	3	1	1	0	22	40	143
Japan	544	68	277	73	19	42	190	80	20	12
Spain	0	-	0	-	-	-	23	4	46	...
USA	1 959	641	598	1 637	1 144	854	546	755	888	740
21 Fishing area total	*2 528*	*781*	*944*	*1 887*	*1 319*	*997*	*816*	*883*	*1 099*	*1 020*
27 China,Taiwan	0	0	18	1	0	0	6	62	40	11
Faeroe Is	-	-	-	-	-	-	-	-	1	...
Ireland	-	-	-	-	-	-	-	-	-	3
Japan	31	4	5	34	28	4	2	-	-	2
Panama	-	-	-	-	-	-	-	-	10	...
Portugal	13	3	6	19	11	5	7	11	1	2
Spain	29	5	5	18	19	17	22	17	14	...
27 Fishing area total	*73*	*12*	*34*	*72*	*58*	*26*	*37*	*90*	*66*	*18*
31 Barbados	179	161	156	255	160	149	150	155	155	142
Bermuda	42	58	44	44	67	55	53	59	31	37
Br Virgin Is	...	...	...	...	...	3	2	3	1	1 F
China,Taiwan	700	1 162 F	361	158	567	381	437	198	456	995
Colombia	95	2 404	3 418	7 172	238	46	46	46	50 F	50 F
Cuba	11	1	14	54	40	6	14	34	34 F	34 F
Dominica	23	30	31	9	...	...	...	80	78	78
Dominican Rp	-	-	-	-	-	-	-	-	272	263
Grenada	340	490	385	410	523	411	484	430	403	759
Japan	287	196	141	151	484	415	503	883	1 090	860
Korea Rep	-	-	-	-	11	-	-	-	-	-
Mexico	742	855	1 093	1 518	814	1 089	1 135	1 920	1 446	1 084
NethAntilles	160 F	170	155	140	130	130 F	130 F	130 F	130 F	130 F
Panama	-	-	-	-	-	-	-	1	332	...
Philippines	-	-	-	-	-	-	7	103	78	2
Puerto Rico	...	...	...	...	...	...	...	...	...	24
St Lucia	58	92	130	144	110	109	276	123	134	145
St Vincent	22	65	16	43	37	35	48	38	1 989	1 365
Spain	1 290	810	0	0	-	-	-	-	-	672
Trinidad Tob	4	4	120	79	183	223	213	163	112	122
USA	4 474	2 821	2 561	1 757	1 783	2 407	1 722	1 168	2 043	1 443
Venezuela	13 773	16 663	24 789	9 714	13 772	14 671	13 995	11 187	10 549	18 652
Other nei	3 836	2 671	4 404	4 202	5 962	6 100	8 339	7 409	5 269	2 874
31 Fishing area total	*26 036 F*	*28 653 F*	*37 818*	*25 850*	*24 881*	*26 230 F*	*27 554 F*	*24 130 F*	*24 652 F*	*29 732 F*
34 Benin	1	1 F	1 F	1 F	1 F	3	1	1	1 F	1
Cape Verde	1 426	1 536	1 727	1 781	1 448	1 721	1 418	1 663	1 851	1 684
China	-	139	156	200	124	84	71	1 535	1 652	586
China,Taiwan	100	750 F	2 338	2 040	2 835	1 903	2 161	1 667	1 512	2 072
Congo Rep	18	17	14	13	12	12	12	12	12	12
Côte dIvoire	-	-	-	-	...	2	...	...	...	...
Cuba	653	541	238	212	257	269	-	-	-	-
France	33 964	36 064	35 468	29 567	33 819	29 966	30 739	31 246	29 789	32 210
Gabon	...	12	88	218	225	225	295	225	162	270
Gambia	15	...	...	14	...	...	1	1	5	1
Georgia	22 F	10 F	-	-	-	-	-	-	-	-
Ghana	9 331	13 283	9 984	9 268	12 160	16 504	17 807	28 328	17 010	30 642
Japan	1 836	1 934	2 293	2 879	3 283	1 785	2 307	1 484	1 352	1 136
Korea Rep	159	176	340	266	221	60	-	-	-	-
Latvia	54	16	-	55	151	223	97	25	36	72
Liberia	...	...	...	...	...	185	310	369	227	166
Libya	-	-	-	-	-	-	-	-	-	208
NethAntilles	-	-	-	-	-	-	-	-	5 626	...
Panama	10 624	10 972	12 066	13 442	7 713	4 293	2 111	1 317	1 099	...
Philippines	-	-	-	-	-	-	126	173	86	0
Portugal	47	40	10	49	18	22	47	23	9	2
Russian Fed	1 862	2 160	1 503	2 936	2 696	4 275	4 931	4 359	737	-
Sao Tome Prn	229 F	140 F	...	...	1	4	4 F	4 F	4 F	-
Senegal	40	6	83	108	68	152	222	358	218	1 118
Spain	49 851	40 393	40 591	38 249	34 848	24 513	31 297	19 910	24 651	30 937
Other nei	10 921	9 875	8 544	8 970	12 749	12 781	7 875	9 797	4 873	13 163
34 Fishing area total	*121 153 F*	*118 065 F*	*115 444 F*	*110 268 F*	*112 629 F*	*98 982*	*101 832 F*	*102 497 F*	*90 912 F*	*114 280*
41 Argentina	1	-	-	-	-	-	-	-	-	-

B-36 Tunas, bonitos, billfishes / Thons, pélamides, marlins / Atunes, bonitos, agujas

Capture production by species, fishing areas and countries or areas / Captures par espèces, zones de pêche et pays ou zones / Capturas por especies, áreas de pesca y países o áreas

	Species, Fishing area Espèce, Zone de pêche Especie, Area de pesca	1992 mt	1993 mt	1994 mt	1995 mt	1996 mt	1997 mt	1998 mt	1999 mt	2000 mt	2001 mt
	Brazil	4 228	5 131	4 169	4 021	2 767	2 705	2 514	4 127	6 145	6 239
	China	-	-	-	-	-	-	628	655	22	470
	China,Taiwan	891	1 400 F	1 676	784	2 190	1 470	1 517	1 156	663	642
	Japan	557	451	401	261	241	244	147	140	282	192
	Korea Rep	-	-	75	105	25	1	-	-	-	-
	Panama	-	-	-	-	-	-	-	4	203	...
	Philippines	-	-	-	-	-	-	29	3	-	10
	Spain	24	179	7	4	36	34	23	26	125	...
	Uruguay	74	20	59	53	171	53	88	52	54	98
41	*Fishing area total*	*5 775*	*7 181 F*	*6 387*	*5 228*	*5 430*	*4 507*	*4 946*	*6 163*	*7 494*	*7 651*
47	Angola	441	211	137	216	78	70	115	170	35	...
	China,Taiwan	57	404 F	2 104	1 713	1 060	711	1 206	1 306	1 174	1 140
	Japan	2 629	2 976	3 643	3 753	2 738	1 624	3 999	3 221	2 708	1 117
	Korea Rep	60	4	21	82	77	179	65	94	143	21
	Namibia	-	-	33	19	3	69	3	147	59	165
	Panama	-	-	-	-	-	-	-	-	70	...
	Portugal	135	85	110	163	259	149	213	143	185	0
	St Helena	166	171	150	181	151	109	181	116	136	70
	Seychelles	-	-	-	-	-	-	-	-	6	-
	South Africa	69	266	486	183	157	116	229	318	353	316
	Spain	22	5	16	11	12	20	18	20	16	...
47	*Fishing area total*	*3 579*	*4 122 F*	*6 700*	*6 321*	*4 535*	*3 047*	*6 029*	*5 535*	*4 885*	*2 829*
51	China	-	-	-	-	-	-	-	129	306	484
	China,Taiwan	19 388	69 531	24 616	19 329	25 311	16 714	23 082	15 663	18 974	18 064
	Comoros	4 742	4 742	5 609	5 609	5 520	5 310	5 310	5 200	5 600 F	5 200
	France	45 282	39 539	35 819	39 635	35 564	30 308	18 983	30 009	37 675	31 377
	India	17	218	169	177	5 888	3 307	2 772	1 547	1 596	7 324
	Iran	14 565	21 636	27 162	27 175	30 233	21 248	21 333	26 871	15 743	20 153
	Italy	-	-	-	-	-	1 156	1 850	2 626	...	1 332
	Japan	14 342	14 631	9 272	4 790	9 359	11 435	13 736	9 727	10 038	10 767
	Korea Rep	3 861	4 681	3 608	2 426	3 426	3 607	2 218	718	1 738	1 240
	Maldives	8 309	9 603	13 126	12 504	12 440	18 619	17 164	15 079	15 706	15 247
	Mauritius	2 285	2 537	1 858	1 725	713	1 095	1 443	742	226	125
	Oman	13 419	11 366	20 707	28 477	20 718	15 905	14 897	7 377	8 377	7 945
	Pakistan	23 394	30 817	4 604	5 140	5 250	3 838	3 795	8 884	4 946	5 921
	Philippines	-	-	-	-	-	-	609	473	316	295
	Portugal	-	-	-	-	-	-	-	-	10	22
	Réunion	388	410	492	402	628	636	609	534	656	584
	Seychelles	225	-	-	5	67	2 152	5 855	9 955	11 737	13 092
	South Africa	13	24	6	26	-	28	106	149	199	96
	Spain	37 782	47 713	43 159	65 143	59 390	60 838	32 728	51 796	52 112	47 936
	Tanzania	-	-	-	-	-	-	-	350	700	800
	Uruguay	-	-	-	-	-	-	-	-	-	14
	Yemen	748	804	804	800	800	840	820	820 F	820 F	820 F
	Other nei	25 373	43 244	32 371	42 781	41 734	36 903	32 601	42 089	48 816	18 112
51	*Fishing area total*	*214 133*	*301 496*	*223 382*	*256 144*	*257 041*	*233 939*	*199 911*	*230 738 F*	*236 291 F*	*206 950 F*
57	Australia	14	91	647	263	107	305	275	478	395	1 016
	China	-	-	-	138	494	750	402	2 206	2 055	1 288
	China,Taiwan	1 754	6 287	4 655	7 690	8 708	12 391	6 441	8 240	5 014	4 636
	France	-	-	-	-	14	920	3 398	791	19	-
	India	23	10	16	17	1 439	799	846	526	562	...
	Iran	-	-	-	-	-	2	197	-	-	-
	Italy	-	-	-	-	-	184	449	-	...	-
	Japan	636	399	743	6 160	5 022	3 281	3 270	4 251	3 851	2 821
	Korea Rep	224	-	14	18	17	35	47	190	73	161
	Philippines	-	-	-	-	-	-	14	146	24	29
	Seychelles	-	-	-	-	-	726	1 597	11	42	66
	Spain	-	-	-	-	41	148	5 860	123	67	...
	Sri Lanka	10 446	11 616	11 939	8 696	12 889	15 756	19 651	27 538	22 091	27 910
	Thailand	-	-	-	-	-	-	-	-	478	...
	Timor-Leste	...	...	...	...	...	...	...	1	3	...
	Other nei	1 100	157	622	-	1 538	719	3 715	814	1 653	...
57	*Fishing area total*	*14 197*	*18 560*	*18 636*	*22 982*	*30 269*	*36 016*	*46 162*	*45 315*	*36 327*	*37 927*
61	China	-	-	-	-	-	-	-	-	-	1 079
	China,Taiwan	48 763	38 614 F	866	4 091	6 269	3 310	5 139	2 975	5 085	1 568
	Japan	16 660	25 151	13 132	14 382	9 860	11 283	11 040	16 437	16 688	12 698
	Korea Rep	-	-	-	-	-	-	6	30	-	-
61	*Fishing area total*	*65 423*	*63 765 F*	*13 998*	*18 473*	*16 129*	*14 593*	*16 185*	*19 442*	*21 773*	*15 345*
67	Japan	3	-	-	-	-	-	-	-	-	-
	USA	1	2	-	0	-	0	0	-	1	13
67	*Fishing area total*	*4*	*2*	-	*0*	-	*0*	*0*	-	*1*	*13*
71	Australia	...	...	...	...	...	627	593	1 226	498	1 031
	China	1 315	2 754	4 823	5 837	2 757	1 419	1 435	2 237	2 207	1 015
	China,Taiwan	19 130	39 000 F	58 914	48 801	35 703	64 215	82 004	63 464	66 006	78 374
	Fiji Islands	597	947	1 368	1 507	1 540	1 016	869	725	2 467	2 126
	Guam	33	42	29	23	15	17	25	15	18	10
	Japan	74 217	76 876	63 999	60 469	34 663	67 411	50 694	50 819	49 975	52 710
	Kiribati	303	108	273	1 025	651	2 223	2 076	1 423	1 209	1 220
	Korea Rep	68 633	54 224	51 985	38 545	20 163	43 885	56 876	31 954	31 127	40 428
	Marshall Is	3	69	27	18	-	-	-	-	905	2 993
	Micronesia	3 453	4 623	3 985	2 062	891	2 845	3 212	3 403	5 394	5 316
	NewCaledonia	373	433	437	839	554	466	185	373	250	570
	N Marianas	9	5	6	9	17	11	5	11	7	7

B-36 Tunas, bonitos, billfishes / Thons, pélamides, marlins / Atunes, bonitos, agujas

Capture production by species, fishing areas and countries or areas / Captures par espèces, zones de pêche et pays ou zones / Capturas por especies, áreas de pesca y países o áreas

Species, Fishing area Espèce, Zone de pêche Especie, Area de pesca		1992 mt	1993 mt	1994 mt	1995 mt	1996 mt	1997 mt	1998 mt	1999 mt	2000 mt	2001 mt
	Palau	14	...	...	...	...	...	...	...	...	...
	Papua N Guin	-	8	374	2 722	971	6 968	11 726	7 610	14 446	25 812
	Philippines	45 026	38 083	63 179	60 957	61 280	67 342	79 215	90 353	90 328	96 450
	Solomon Is	5 630	7 193	6 671	8 114	11 003	9 588	8 114	8 843	3 196	4 490
	Tuvalu	2	292	110	13	15 F	20 F	20 F	20 F	20 F	20 F
	USA	42 043	33 793	59 821	28 334	29 849	47 384	44 300	37 975	27 154	27 217
	Vanuatu	130 F	140 F	85	1 090	982	8 827	9 603	10 337	2 415	...
71	*Fishing area total*	*260 911 F*	*258 590 F*	*316 086*	*260 365*	*201 054 F*	*324 264 F*	*350 952 F*	*310 788 F*	*297 622 F*	*339 789 F*
77	Amer Samoa	-	1	1	2	12	22	42	64	86	183
	Belize	-	-	190	10	670	590	380	90	190	...
	China	-	-	-	-	-	-	-	-	580	1 994
	China,Taiwan	3 440	8 090 F	1 832	380	94	108	161	192	126	1 750
	Colombia	...	...	...	...	...	...	...	...	7 470	7 260
	Cook Is	...	...	12	23	7	5 F	5 F	5 F	5 F	5 F
	Costa Rica	...	...	...	10	50	...	...	...	...	...
	Cyprus	840	940	180	100	600	920	-	-	-	-
	Ecuador	1 880	380	400	-	-	-	-	-	-	-
	El Salvador	...	...	...	...	...	...	920	260	...	2 165
	Fr Polynesia	398	602	436	820	811	860	843	1 225	1 762	1 514
	Guatemala	-	-	-	-	-	-	-	1 050	2 340	...
	Honduras	-	-	-	-	40	40	-	40	-	-
	Japan	12 809	15 771	19 035	14 501	8 974	12 714	9 089	7 003	12 752	10 392
	Korea Rep	10 873	8 250	8 709	10 436	11 327	13 455	14 091	8 812	13 640	15 208
	Mexico	116 190	100 200	113 037	93 384	108 606	127 497	107 521	109 966	95 829	127 000
	Nicaragua	-	-	-	-	-	-	-	3 060	3 320	...
	Panama	1 770	3 640	1 370	370	3 150	3 660	3 230	3 310	910	4 980
	Samoa	50	81	73	216	573	1 327	801	681	1 120	470
	Spain	-	640	110	290	690	530	3 900	6 966	5 325	3 007
	Tonga	19	64	46	59	88	100	125	163	175	259
	USA	25 577	10 429	19 474	12 866	12 574	12 685	14 882	3 725	3 901	5 188
	Vanuatu	20 150	14 040	12 680	16 130	7 250	14 750	13 370	11 590	4 310	4 670
	Venezuela	36 230	31 940	26 350	41 500	55 750	43 320	42 350	46 430	50 700	68 630
	Other nei	-	-	50	220	-	-	-	-	1 530	12 240
77	*Fishing area total*	*230 226*	*195 068 F*	*203 985*	*191 317*	*211 266*	*232 583 F*	*211 710 F*	*204 632 F*	*206 071 F*	*266 915 F*
81	Australia	978	878	1 229	1 259	1 743	1 109	1 243	958	802	896
	China,Taiwan	346	1 000 F	764	138	138	158	329	57	173	174
	Japan	1 374	1 200	1 615	2 615	2 564	1 912	975	345	155	567
	Korea Rep	-	7	-	-	-	-	24	202	545	419
	New Zealand	18	17	46	138	181	118	127	153	107	137
81	*Fishing area total*	*2 716*	*3 102 F*	*3 654*	*4 150*	*4 626*	*3 297*	*2 698*	*1 715*	*1 782*	*2 193*
87	Belize	-	1 100	850	880	1 540	1 280	3 110	2 400	1 630	...
	Chile	14	82	118	43	32	57	78	48	77	66
	China,Taiwan	-	-	17	0	15	17	1	2	16	5
	Colombia	37 889	35 589	18 890	34 771	18 283	42 344	14 501	29 328	8 080	17 610
	Cyprus	2 130	2 590	520	850	2 290	1 230	100	-	-	-
	Ecuador	11 452	17 910	23 173	15 921	19 314	19 603	31 052	50 200	36 955	57 563
	El Salvador	...	...	...	...	...	...	...	2 990	...	...
	Guatemala	-	-	-	-	-	-	-	610	2 310	...
	Honduras	-	-	-	-	-	190	870	1 600	620	...
	Japan	3 723	3 752	5 077	3 154	2 900	2 006	2 478	2 044	4 457	5 192
	Korea Rep	6	178	73	73	-	22	27	1 255	2 195	1 480
	Liberia	-	-	-	-	180	-	-	-	-	-
	Mexico	5 252	4 911	7 244	12 256	18 395	11 700	9 106	9 998	5 065	7 410
	Nicaragua	-	-	-	-	-	-	-	-	1 630	...
	Panama	2 000	2 220	2 170	1 690	490	1 920	290	3 210	5 300	8 000
	Peru	582	3 573	269	914	953	908	12 747	2 784	2 548	4 175
	St Vincent	-	-	950	-	-	-	-	-	-	-
	Spain	1 010	1 740	1 120	1 740	3 270	2 390	1 840	3 670	3 520	8 430
	USA	1 486	-	-	-	1 837	895	461	-	289	9
	Vanuatu	8 290	11 900	15 200	8 400	4 950	7 690	5 270	6 550	9 890	6 050
	Venezuela	9 660	12 770	16 210	7 170	7 930	16 010	20 260	11 300	19 100	41 080
	Other nei	130	990	1 040	380	-	620	220	-	2 450	20 580
87	*Fishing area total*	*83 624*	*99 305*	*92 921*	*88 242*	*82 379*	*108 882*	*102 411*	*127 989*	*106 132*	*177 650*
Species total		*1 030 378 F*	*1 098 702 F*	*1 039 989 F*	*991 299 F*	*951 616 F*	*1 087 363 F*	*1 071 243 F*	*1 079 917 F*	*1 035 107 F*	*1 202 312 F*
Bigeye tuna		**Thon obèse(=Patudo)**		**Patudo**			***Thunnus obesus***			**1,75(01)026,12**	**BET**
21	Canada	67	125	111	148	144	166	129	263	327	241
	China,Taiwan	0	0	13	0	0	0	0	0	12	2
	Japan	1 624	961	856	162	356	405	498	944	290	342
	USA	458	683	799	846	410	440	506	659	328	453
21	*Fishing area total*	*2 149*	*1 769*	*1 779*	*1 156*	*910*	*1 011*	*1 133*	*1 866*	*957*	*1 038*
27	China,Taiwan	0	0	21	15	4	3	38	17	27	20
	Faeroe Is	-	-	-	-	-	-	-	11	8	...
	France	-	-	-	-	-	-	-	28	15	28
	Iceland	-	-	-	-	-	-	-	-	5	-
	Ireland	-	-	-	-	-	-	-	-	-	8
	Japan	489	149	108	300	486	379	368	94	269	339
	Portugal	2 582	4 180	1 965	4 964	1 771	2 590	4 246	2 045	968	426
	Spain	12	11	12	16	77	52	304	362	562	559
	Other nei	-	-	-	-	-	4	-	-	-	-
27	*Fishing area total*	*3 083*	*4 340*	*2 106*	*5 295*	*2 338*	*3 028*	*4 956*	*2 557*	*1 854*	*1 380*
31	Barbados	-	-	-	-	-	24	17	18	18	6

B-36 Tunas, bonitos, billfishes / Thons, pélamides, marlins / Atunes, bonitos, agujas

Capture production by species, fishing areas and countries or areas
Captures par espèces, zones de pêche et pays ou zones
Capturas por especies, áreas de pesca y países o áreas

Species, Fishing area Espèce, Zone de pêche Especie, Area de pesca	1992 mt	1993 mt	1994 mt	1995 mt	1996 mt	1997 mt	1998 mt	1999 mt	2000 mt	2001 mt
China,Taiwan	1 726	3 060 F	269	3	308	236	166	34	205	114
Grenada	14	11	10	10	-	1	-	-	-	-
Japan	346	423	407	407	496	600	1 196	2 216	2 134	2 719
Korea Rep	-	-	-	-	33	-	-	-	-	-
Mexico	-	-	-	-	-	6	8	6	2	2
Panama	-	-	-	-	-	-	-	-	49	...
Philippines	-	-	-	-	-	-	21	442	260	34
Puerto Rico	-	-	-	-	-	54	-	-	-	-
St Lucia	1	0	0	0	0	0	0	0	-	1
St Vincent	1	3	0	0	4	2	2	1	1 216	506
Spain	1	5	2	2	5	32	35	-	-	-
Trinidad Tob	-	3	29	27	37	36	24	19	5	11
USA	95	56	143	45	19	60	65	65	72	67
Venezuela	270	809	457	235	189	274	222	140	226	708
31 Fishing area total	*2 454*	*4 370 F*	*1 317*	*729*	*1 091*	*1 325*	*1 756*	*2 941*	*4 187*	*4 168*
34 Benin	7	8	9	9 F	9 F	30	13	11	...	...
Cape Verde	105	85	209	66	16	10	1	1	2	-
China	-	70	428	476	520	427	656	2 520	393	2 897
China,Taiwan	1 000	1 769 F	7 260	7 062	11 360	8 703	6 844	5 539	9 314	8 279
Congo Rep	12	14	9	9	8	8	8	8	8	8
Cuba	56	36	7	7	5		-	-	-	-
France	6 888	12 719	12 263	8 363	9 171	5 980	5 624	5 501	5 934	4 948
Gabon	-	1	87	10	-	-	-	184	150	121
Ghana	2 866	3 577	4 738	5 517	5 805	7 431	13 252	11 460	5 586	14 095
Japan	8 726	15 693	13 866	17 009	16 982	10 360	14 109	10 021	13 101	8 625
Korea Rep	831	363	321	368	721	361	-	-	-	-
Liberia	42	65	53	57	105	340	108	112	201	175
Libya	508	1 085	500	400	400	400	400	400	400	31
Morocco	-	-	-	-	-	-	-	700	770	857
NethAntilles	-	-	-	-	-	-	-	-	2 627	...
Panama	10 002	10 410	13 087	9 927	4 777	2 098	1 252	318	492	...
Philippines	-	-	-	-	-	-	1 038	1 627	715	29
Portugal	2 944	1 206	881	4 423	3 723	2 767	1 956	1 107	384	491
Russian Fed	-	-	-	-	13	38	4	8	91	-
Sao Tome Prn	-	-	-	-	-	5	-	-	-	-
Senegal	5	4	126	177	135	218	791	2 007	860	2 169
Sierra Leone	...	...	...	...	...	...	...	...	6	2
Spain	14 309	16 632	21 943	17 673	15 160	12 245	6 469	13 261	10 091	9 364
Togo	2	86	23	6	33	17	6	66	32	26
Other nei	8 856	9 299	14 036	15 894	19 674	22 406	30 174	33 128	16 788	15 158
34 Fishing area total	*57 159*	*73 122 F*	*89 846*	*87 453 F*	*88 617 F*	*73 844*	*82 705*	*87 979*	*67 945*	*67 275*
41 Argentina	0	-	-	-	-	-	-	-	-	-
Brazil	790	1 256	596	1 935	1 707	1 237	644	2 024	2 768	2 659
China	-	-	-	-	-	-	847	-	-	-
China,Taiwan	1 273	4 440 F	4 485	4 134	8 017	6 142	3 437	2 711	3 356	3 732
Japan	6 169	4 266	4 190	1 913	1 729	1 795	796	753	1 255	1 133
Korea Rep	-	-	65	53	109	4	-	-	-	-
Panama	-	-	-	-	-	-	-	-	448	...
Philippines	-	-	-	-		-	95	44	-	314
Spain	167	-	9	13	11	123	183	29	243	...
Uruguay	56	48	37	80	124	69	59	27	29	47
41 Fishing area total	*8 455*	*10 010 F*	*9 382*	*8 128*	*11 697*	*9 370*	*6 061*	*5 588*	*8 099*	*7 885*
47 China	-	-	-	-	-	-	-	4 827	6 170	4 313
China,Taiwan	750	2 612 F	5 426	6 808	5 426	4 158	5 829	8 536	9 126	8 895
Japan	13 358	20 293	20 767	16 708	13 652	10 982	9 486	8 145	6 894	4 926
Korea Rep	35	14	-	-	164	212	163	124	70	4
Namibia	-	-	751	352	63	45	16	423	589	640
Panama	-	28	147	-	-	-	-	-	7	...
Portugal	270	230	253	275	316	80	132	161	146	689
St Helena	10	6	6	10	10	12	17	6	8	4
South Africa	43	88	76	27	7	10	41	41	225	167
Spain	167	134	130	145	140	61	123	29	243	...
47 Fishing area total	*14 633*	*23 405 F*	*27 556*	*24 325*	*19 778*	*15 560*	*15 807*	*22 292*	*23 478*	*19 638*
51 China	-	-	-	-	-	-	-	69	877	889
China,Taiwan	13 560	20 340	16 119	27 210	21 358	24 371	35 002	35 156	42 712	48 908
Comoros	27	27	32	32	30	30	30	30	30 F	30
France	3 882	5 016	5 367	7 280	6 896	7 648	3 854	8 324	6 657	5 090
India	-	839	1 042	1 042	-	-	-	-	-	-
Iran	-	-	-	-	153	261	310	592	347	-
Italy	-	-	-	-	-	450	296	848	...	57
Japan	4 878	5 820	9 464	6 560	7 115	8 403	10 293	6 042	5 516	4 953
Korea Rep	4 382	7 146	8 179	6 106	10 737	10 129	3 154	608	3 091	1 145
Maldives	388	505	506	473	630	540	606	604	472	-
Mauritius	873	664	729	570	271	546	260	250	37	5
Philippines	-	-	-	-	-	-	1 465	1 177	1 354	876
Réunion	-	6	7	15	98	91	112	213	167	64
Seychelles	11	-	-	5	75	858	1 494	3 127	2 164	3 049
South Africa	2	-	-	-	-	-	8	13	29	25
Spain	3 629	5 414	5 950	12 233	11 339	15 863	8 447	16 024	10 803	8 029
Uruguay	-	-	-	-	-	-	-	-	-	16
Other nei	5 962	10 504	7 042	13 077	14 482	16 061	20 019	21 148	21 606	3 091
51 Fishing area total	*37 594*	*56 281*	*54 437*	*74 603*	*73 184*	*85 251*	*85 350*	*94 225*	*95 862 F*	*76 227*
57 Australia	125	475	146	84	25	57	187	476	468	553
China	-	-	-	140	466	1 652	2 165	2 113	1 822	2 105

B-36 Tunas, bonitos, billfishes — Capture production by species, fishing areas and countries or areas
Thons, pélamides, marlins — Captures par espèces, zones de pêche et pays ou zones
Atunes, bonitos, agujas — Capturas por especies, áreas de pesca y países o áreas

Species, Fishing area Espèce, Zone de pêche Especie, Area de pesca	1992 mt	1993 mt	1994 mt	1995 mt	1996 mt	1997 mt	1998 mt	1999 mt	2000 mt	2001 mt
China,Taiwan	2 806	5 865	7 871	9 079	13 063	17 230	8 526	6 353	10 933	4 780
France	-	-	-	-	12	176	2 535	192	16	-
India	-	27	34	34	-	-	4	-	-	-
Iran	-	-	-	-	-	1	95	-	-	-
Italy	-	-	-	-	-	7	316	-	...	-
Japan	1 725	4 126	7 022	12 643	9 029	7 903	7 247	8 832	7 800	8 137
Korea Rep	154	-	60	48	48	77	33	737	129	256
Philippines	-	-	-	-	-	-	86	714	107	135
Seychelles	-	-	-	-	-	78	591	27	107	116
Spain	-	-	-	-	35	46	2 833	68	13	...
Sri Lanka	169	1 009	1 012	2 108	491	600	749	462	348	470
Thailand	-	-	-	-	-	-	-	-	280	...
Other nei	975	2 291	2 886	2 975	4 899	2 939	5 115	5 250	5 887	...
57 Fishing area total	*5 954*	*13 793*	*19 031*	*27 111*	*28 068*	*30 766*	*30 482*	*25 224*	*27 910*	*16 552*
61 China	-	-	-	-	-	-	-	-	-	2 003
China,Taiwan	5 721	1 000	160	52	12	21	2 440	1 060	1 268	316
Japan	11 007	12 686	10 841	8 130	6 598	9 653	7 703	9 023	8 110	6 254
Korea Rep	-	-	-	-	-	-	2	90	-	-
61 Fishing area total	*16 728*	*13 686*	*11 001*	*8 182*	*6 610*	*9 674*	*10 145*	*10 173*	*9 378*	*8 573*
67 Japan	1	-	-	-	-	-	1	-	-	-
USA	-	-	-	-	-	-	-	-	-	1
67 Fishing area total	*1*	-	-	-	-	-	*1*	-	-	*1*
71 Australia	...	...	...	...	...	419	845	646	317	628
China	1 400	3 665	7 846	4 744	3 261	2 243	1 836	1 805	1 981	831
China,Taiwan	1 686	501	4 305	5 260	4 748	13 523	12 974	15 996	4 698	5 914
Fiji Islands	187	204	249	378	593	409	460	462	687	662
Japan	27 304	18 529	18 061	15 623	9 529	20 291	19 815	17 468	17 344	25 629
Kiribati	-	-	26	66	69	130	99	157	63	113
Korea Rep	3 764	2 235	1 980	1 483	1 120	2 624	2 313	3 131	3 147	2 370
Marshall Is	3	67	26	13	-	-	-	-	35	137
Micronesia	362	473	360	211	183	430	705	1 011	1 025	882
NewCaledonia	27	106	78	103	233	234	498	553	517	128
Papua N Guin	-	-	20	161	50	1 060	1 461	998	1 471	4 981
Solomon Is	709	733	593	1 391	1 109	1 434	1 232	1 070	577	576
USA	-	3	1	2	864	504	-	3 289	1 863	-
Vanuatu	...	...	5	151	151	907	583	999	140	-
71 Fishing area total	*35 442*	*26 516*	*33 550*	*29 586*	*21 910*	*44 208*	*42 821*	*47 585*	*33 865*	*42 851*
77 Amer Samoa	...	0	0	1	4	4	10	9	21	74
Belize	-	-	-	120	-	20	10	0	60	...
China	-	-	-	-	-	-	-	-	750	3 074
China,Taiwan	6 298	1 540	1 042	463	93	177	465	1 032	678	169
Cook Is	...	...	8	16	4	5 F	5 F	5 F	5 F	5 F
Costa Rica	...	...	...	...	120	...	...	...	...	...
Cyprus	-	-	-	20	430	160	-	-	-	-
Ecuador	0	20	690	-	-	-	-	-	-	-
El Salvador	-	-	-	-	-	-	10	-	-	2 059
Fr Polynesia	57	163	165	184	186	310	403	278	712	746
Guatemala	-	-	-	-	-	-	-	760	2 270	...
Honduras	-	-	-	-	1 000	850	140	-	20	...
Japan	68 519	57 640	51 112	40 822	30 385	30 533	37 124	25 234	27 870	32 366
Korea Rep	15 985	14 633	19 557	18 599	15 486	17 597	27 275	21 236	22 704	26 868
Mexico	270	80	40	299	495	102	5	91	6	90
Nicaragua	-	-	-	-	-	-	-	30	10	...
Panama	-	10	-	60	200	40	10	790	-	470
Samoa	...	3	14	40	27	63	334	283	177	185
Spain	-	-	-	170	500	250	2 960	7 347	8 254	2 884
Tonga	5	34	19	23	60	69	86	112	120	191
USA	1 776	4 385	2 147	8 506	4 574	5 255	7 063	2 896	2 986	5 378
Vanuatu	10	-	130	1 990	2 240	1 490	1 920	1 240	220	540
Venezuela	-	-	10	-	-	80	-	-	0	-
Other nei	-	-	-	-	-	-	300	-	-	2 710
77 Fishing area total	*92 920*	*78 508*	*74 934*	*71 313*	*55 804*	*57 005 F*	*78 120 F*	*61 343 F*	*66 863 F*	*77 809 F*
81 Australia	37	40	120	155	293	563	1 024	257	355	367
China,Taiwan	562	200	465	31	59	112	528	325	139	107
Japan	685	586	484	587	903	668	1 203	875	468	1 108
Korea Rep	-	15	-	-	-	-	43	497	938	692
New Zealand	39	74	69	60	86	140	388	420	421	480
81 Fishing area total	*1 323*	*915*	*1 138*	*833*	*1 341*	*1 483*	*3 186*	*2 374*	*2 321*	*2 754*
87 Belize	-	-	2 360	1 150	2 600	4 700	1 150	320	500	...
Chile	-	-	8	15	16	6	29	6	20	5
China,Taiwan	-	-	42	0	50	94	1	2	16	8
Colombia	...	...	...	...	7 270	3 090	560	1 420	1 030	150
Costa Rica	-	-	-	150	720	-	-	-	-	-
Cyprus	270	480	2 110	1 580	750	1 290	30	-	-	-
Ecuador	1 412	1 806	3 276	10 193	17 892	26 148	17 909	22 278	29 398	23 440
El Salvador	-	-	-	-	-	-	-	230	-	-
Guatemala	-	-	-	-	-	-	-	820	8 420	...
Honduras	-	-	-	-	230	380	-	420	-	-
Japan	8 723	5 605	7 176	5 752	4 331	3 608	5 732	3 026	5 016	7 716
Korea Rep	51	434	442	162	-	187	59	135	-	-
Liberia	-	-	-	-	310	-	-	-	-	-
Panama	30	-	-	570	870	1 280	-	-	3 930	1 250
St Vincent	-	-	600	-	-	-	-	-	-	-

B-36 Tunas, bonitos, billfishes / Thons, pélamides, marlins / Atunes, bonitos, agujas

Capture production by species, fishing areas and countries or areas / Captures par espèces, zones de pêche et pays ou zones / Capturas por especies, áreas de pesca y países o áreas

Species, Fishing area Espèce, Zone de pêche Especie, Area de pesca	1992 mt	1993 mt	1994 mt	1995 mt	1996 mt	1997 mt	1998 mt	1999 mt	2000 mt	2001 mt
Spain	430	420	850	2 050	2 040	2 950	2 370	4 720	12 870	4 580
USA	-	-	-	-	656	400	343	-	466	-
Vanuatu	1 010	1 360	6 620	7 250	7 840	3 710	1 640	2 860	5 850	3 250
Venezuela	130	180	450	470	430	170	240	10	210	-
Other nei	-	70	190	100	-	420	-	-	150	5 560
87 Fishing area total	*12 056*	*10 355*	*24 124*	*29 442*	*46 005*	*48 433*	*30 063*	*36 247*	*67 876*	*45 959*
Species total	*289 951*	*317 070 F*	*350 201*	*368 156 F*	*357 353 F*	*380 958 F*	*392 586 F*	*400 394 F*	*410 595 F*	*372 110 F*
Tunas nei	**Thonidés nca**		**Atunes nep**		***Thunnini***				**1,75(01)XXX,XX**	**TUN**
21 USA	176	117	57	45	115	18	15	29	35	13
21 Fishing area total	*176*	*117*	*57*	*45*	*115*	*18*	*15*	*29*	*35*	*13*
31 USA	103	116	29	23	6	33	17	25	33	47
31 Fishing area total	*103*	*116*	*29*	*23*	*6*	*33*	*17*	*25*	*33*	*47*
37 Japan	-	-	-	1	-	1	-	-	-	-
37 Fishing area total	-	-	-	*1*	-	*1*	-	-	-	-
51 Jordan	...	...	...	...	...	...	70	96	90	110
51 Fishing area total	...	...	...	...	...	...	*70*	*96*	*90*	*110*
67 USA	0	1	0	0	-	1 263	-	1	-	0
67 Fishing area total	*0*	*1*	*0*	*0*	-	*1 263*	-	*1*	-	*0*
71 USA	-	-	-	-	168	-	-	-	-	-
71 Fishing area total	-	-	-	-	*168*	-	-	-	-	-
77 USA	10	9	27	7	8	15	10	14	4	5
77 Fishing area total	*10*	*9*	*27*	*7*	*8*	*15*	*10*	*14*	*4*	*5*
Species total	*289*	*243*	*113*	*76*	*297*	*1 330*	*112*	*165*	*162*	*175*
Indo-Pacific sailfish	**Voilier indo-pacifique**		**Pez vela del Indo-Pacífico**		***Istiophorus platypterus***				**1,75(03)004,02**	**SFA**
51 China,Taiwan	0	626	513	432	142	91	133	344	706	1 602
Comoros	212	212	250	250	250	240	240	200	250 F	200
Eritrea	...	...	-	-	-	1	0	0	1	-
India	-	-	-	-	13	12	-	-	8	...
Iran	170	740	1 085	3 619	2 306	1 928	1 813	3 193	3 080	3 160
Japan	10	7	19	29	26	125	124	148	117	98
Korea Rep	6	-	-	-	3	5	-	-	-	-
Oman	715	540	357	664	581	1 261	591	399	448	218
Pakistan	998	731	855	910	980	41	45	46	...	998
Réunion	-	6	8	7	10	11	17	18	30	17
Saudi Arabia	-	-	-	-	-	1	2	1	1	8
Seychelles	7	-	-	-	1	8	2	17	-	7
South Africa	-	1	-	-	-	0	0	0	0	0
Spain	-	-	-	-	-	3	9	7	1	1
51 Fishing area total	*2 118*	*2 863*	*3 087*	*5 911*	*4 312*	*3 727*	*2 976*	*4 373*	*4 642 F*	*6 309*
57 Australia	-	-	-	-	-	-	3	1	-	-
China,Taiwan	0	202	315	99	29	18	11	47	278	202
Honduras	-	-	-	-	-	-	-	-	-	205
India	-	-	-	-	-	-	14	9	1	...
Japan	5	1	3	28	7	15	23	42	57	36
Sri Lanka	1 264	1 872	4 638	3 335	5 360	6 552	8 172	6 979	8 878	7 680
57 Fishing area total	*1 269*	*2 075*	*4 956*	*3 462*	*5 396*	*6 585*	*8 223*	*7 078*	*9 214*	*8 123*
61 China,Taiwan	4 530 F	3 257 F	1 758	1 057	1 784	3 046	762	3 062	1 539	201
Japan	576	285	546	621	859	382	1 059	779	541	382
61 Fishing area total	*5 106 F*	*3 542 F*	*2 304*	*1 678*	*2 643*	*3 428*	*1 821*	*3 841*	*2 080*	*583*
67 Japan	1	-	-	-	-	-	-	-	-	-
67 Fishing area total	*1*	-	-	-	-	-	-	-	-	-
71 China,Taiwan	355 F	320 F	200	438	302	186	3 027	153	4	63
Guam	-	-	-	1	0	0	1	0	1	1
Japan	274	232	305	215	11	132	76	18	56	40
Korea Rep	-	-	-	-	4	-	-	9	2	-
71 Fishing area total	*629 F*	*552 F*	*505*	*654*	*317*	*318*	*3 104*	*180*	*63*	*104*
77 Amer Samoa	...	...	1	3	2	3	2	3	1	1
China,Taiwan	-	-	-	7	39	7	41	4	0	112
Japan	274	207	205	226	175	140	234	358	326	295
Korea Rep	403	589	1 200	224	243	1 297	389	210	132	33
77 Fishing area total	*677*	*796*	*1 406*	*460*	*459*	*1 447*	*666*	*575*	*459*	*441*
81 China,Taiwan	-	-	-	0	0	2	1	0	0	36
Japan	66	33	29	17	9	6	13	5	1	10
Korea Rep	-	-	-	-	-	-	-	277	51	-
81 Fishing area total	*66*	*33*	*29*	*17*	*9*	*8*	*14*	*282*	*52*	*46*
87 Japan	28	13	11	7	8	44	54	10	11	35
87 Fishing area total	*28*	*13*	*11*	*7*	*8*	*44*	*54*	*10*	*11*	*35*
Species total	*9 894 F*	*9 874 F*	*12 298*	*12 189*	*13 144*	*15 557*	*16 858*	*16 339*	*16 521 F*	*15 641*

B-36 Tunas, bonitos, billfishes / Thons, pélamides, marlins / Atunes, bonitos, agujas

Capture production by species, fishing areas and countries or areas / Captures par espèces, zones de pêche et pays ou zones / Capturas por especies, áreas de pesca y países o áreas

Species, Fishing area Espèce, Zone de pêche Especie, Area de pesca	1992 mt	1993 mt	1994 mt	1995 mt	1996 mt	1997 mt	1998 mt	1999 mt	2000 mt	2001 mt
Atlantic sailfish	**Voilier de l'Atlantique**		**Pez vela del Atlántico**		***Istiophorus albicans***				**1,75(03)004,04**	**SAI**
21 Japan	1	-	-	-	-	-	6	-	-	-
21 Fishing area total	*1*	-	-	-	-	-	*6*	-	-	-
31 Aruba	5	10	10	10	10	...	...	...	...	...
Barbados	42	50	46	74	25	71	58	44	44	...
China,Taiwan	10 F	223	101	38	37	17	1	8	0	-
Cuba	70	42	46	37	37	41	28	199	199 F	199 F
Dominican Rp	98	50	90	40	25	49	89	56	81	81
Grenada	177	141	151	119	56	83	151	148	164	186
Japan	-	-	-	-	1	7	2	9	6	2
Mexico	-	2	19	6	10	7	21	33	37	36
NethAntilles	10 F	15	15 F	15 F	15 F	15 F	15 F	15 F	15 F	15 F
St Vincent	4	4	4	2	1	3	-	1	-	2
Spain	-	-	-	-	-	7	-	-	-	-
Trinidad Tob	4	56	101	101	104	10	...	4	3	7
Venezuela	71	206	162	103	165	185	258	179	93	126
Other nei	-	31	30	30	30	30	-	-	-	-
31 Fishing area total	*491 F*	*830*	*775 F*	*575 F*	*516 F*	*525 F*	*623 F*	*696 F*	*642 F*	*654 F*
34 Benin	21	20	20 F	20	19	6	4	5	5 F	12
China	-	-	3	3	3	3	5	9	4	5
China,Taiwan	-	420	233	155	65	29	38	34	57	-
Côte dIvoire	69	40	54	66	91	65	35	80	45	47
Cuba	200	77	83	72	533	-	-	-	-	-
France	150	182	160	128	97	110	138	131	98	-
Ghana	297	693	450	353	303	196	351	305	275	...
Japan	6	28	39	49	48	19	47	21	37	11
Portugal	1	2	1	0	0	-	-	37	16	3
Sao Tome Prn	...	...	...	...	...	139	...	...	...	...
Senegal	260	41	172	204	206	509	192	79	447	266
Spain	0	2	0	1	2	6	2	-	1	...
Other nei	-	11	15	10	10	10	-	-	-	-
34 Fishing area total	*1 004*	*1 516*	*1 230 F*	*1 061*	*1 377*	*1 092*	*812*	*701*	*985 F*	*344*
41 Brazil	351	243	128	245	310	137	184	356	598	412
China	-	-	3	3	3	3	3	9	4	3
China,Taiwan	23 F	-	-	-	-	-	3	44	16	-
Japan	22	20	19	4	4	17	6	14	13	4
Spain	7	5	3	36	3	8	20	6	14	...
41 Fishing area total	*403 F*	*268*	*153*	*288*	*320*	*165*	*216*	*429*	*645*	*419*
47 China,Taiwan	-	-	-	-	-	-	111	13	62	-
Japan	14	28	34	31	27	17	30	27	17	16
Portugal	-	-	-	-	-	-	-	16	2	0
Spain	2	28	7	12	23	20	16	19	8	...
47 Fishing area total	*16*	*56*	*41*	*43*	*50*	*37*	*157*	*75*	*89*	*16*
Species total	*1 915 F*	*2 670*	*2 199 F*	*1 967 F*	*2 263 F*	*1 819 F*	*1 814 F*	*1 901 F*	*2 361 F*	*1 433 F*
Indo-Pacific blue marlin	**Makaire bleu indo-pacifique**		**Aguja azul del Indo-Pacífico**		***Makaira mazara***				**1,75(03)005,02**	**BLZ**
51 China,Taiwan	2 523	3 715	208	1 727	1 522	1 522	2 560	3 208	3 843	4 085
Japan	244	239	500	259	464	916	1 110	578	695	372
Korea Rep	32	-	3	7	1	75	101	10	79	16
Philippines	-	-	-	-	-	-	63	78	46	40
Seychelles	-	-	-	-	-	-	1	13	19	10
Spain	-	-	-	-	-	-	-	16	1	3
Other nei	517	655	299	596	838	825	1 279	938	1 047	-
51 Fishing area total	*3 316*	*4 609*	*1 010*	*2 589*	*2 825*	*3 338*	*5 114*	*4 841*	*5 730*	*4 526*
57 Australia	0	1	33	4	0	2	-	-	-	-
China,Taiwan	680	1 098	54	435	421	421	240	492	1 046	282
Japan	49	54	97	171	112	215	213	358	304	142
Korea Rep	-	-	-	-	-	-	2	6	-	-
Philippines	-	-	-	-	-	-	6	36	22	1
Seychelles	-	-	-	-	-	-	-	2	9	1
Sri Lanka	3 481	2 031	6 589	37	0	-	-	-	-	-
Other nei	137	170	78	152	248	233	548	462	515	...
57 Fishing area total	*4 347*	*3 354*	*6 851*	*799*	*781*	*871*	*1 009*	*1 356*	*1 896*	*426*
61 China,Taiwan	833	3 995	3 255	4 210	7 025	4 399	3 323	2 458	2 122	2 016
Japan	1 576	2 476	2 041	1 975	2 200	1 087	1 582	1 422	1 184	1 152
61 Fishing area total	*2 409*	*6 471*	*5 296*	*6 185*	*9 225*	*5 486*	*4 905*	*3 880*	*3 306*	*3 168*
71 China,Taiwan	5 430	345	2 379	4 018	4 462	4 399	6 385	7 153	7 117	6 442
Guam	25	32	66	31	15	24	13	14	30	15
Japan	2 751	3 029	3 023	2 671	1 066	1 312	1 506	1 522	1 813	1 969
Korea Rep	-	-	-	-	17	88	112	25	15	164
N Marianas	...	...	1	3	3	4	2	2	2	1
Philippines	1 172	679	991	1 178	1 191	1 096	1 470	2 208	2 229	2 304
71 Fishing area total	*9 378*	*4 085*	*6 460*	*7 901*	*6 754*	*6 923*	*9 488*	*10 924*	*11 206*	*10 895*
77 Amer Samoa	...	...	6	11	14	18	16	13	17	5
China,Taiwan	204	242	1 668	128	70	38	47	105	91	871
Japan	6 283	6 383	8 126	6 786	3 009	3 969	3 458	2 432	2 335	2 407
Korea Rep	110	84	5	1	6	223	567	256	207	170
77 Fishing area total	*6 597*	*6 709*	*9 805*	*6 926*	*3 099*	*4 248*	*4 088*	*2 806*	*2 650*	*3 453*

B-36

Tunas, bonitos, billfishes — Capture production by species, fishing areas and countries or areas
Thons, pélamides, marlins — Captures par espèces, zones de pêche et pays ou zones
Atunes, bonitos, agujas — Capturas por especies, áreas de pesca y países o áreas

Species, Fishing area Espèce, Zone de pêche Especie, Area de pesca	1992 mt	1993 mt	1994 mt	1995 mt	1996 mt	1997 mt	1998 mt	1999 mt	2000 mt	2001 mt
81 China,Taiwan	4	15	248	1	4	2	1	2	16	26
Japan	57	62	49	66	31	54	83	26	10	15
Korea Rep	-	-	-	-	-	-	23	-	-	-
81 Fishing area total	*61*	*77*	*297*	*67*	*35*	*56*	*107*	*28*	*26*	*41*
87 China,Taiwan	28	37	19	0	3	1	1	0	1	1
Japan	495	301	697	493	253	271	543	120	72	163
Korea Rep	-	-	-	-	-	12	-	3	-	-
87 Fishing area total	*523*	*338*	*716*	*493*	*256*	*284*	*544*	*123*	*73*	*164*
Species total	*26 631*	*25 643*	*30 435*	*24 960*	*22 975*	*21 206*	*25 255*	*23 958*	*24 887*	*22 673*
Atlantic blue marlin	**Makaire bleu de l'Atlantique**		**Aguja azul del Atlántico**		***Makaira nigricans***			**1,75(03)005,05**		**BUM**
21 Japan	9	1	7	1	-	1	12	-	2	-
21 Fishing area total	*9*	*1*	*7*	*1*	-	*1*	*12*	-	*2*	-
27 China,Taiwan	-	-	-	-	-	-	1	-	-	-
Japan	1	-	4	3	1	-	1	-	-	-
Spain	6	7	5	-	20	3	3	3	8	...
27 Fishing area total	*7*	*7*	*9*	*3*	*21*	*3*	*5*	*3*	*8*	...
31 Barbados	18	21	19	31	25	30	25	19	19	...
Bermuda	19	11	15	15	15	3	5	1	2	2
China,Taiwan	59	157	21	10	7	6	4	7	0	-
Cuba	135	69	39	85	43	53	12	30	30 F	30 F
Dominican Rp	-	-	-	-	-	-	-	-	23	23
Grenada	30	33	52	50	26	47	60	100	87	103
Japan	60	70	31	3	105	101	84	153	175	39
NethAntilles	40 F	40 F	40 F	40 F	40 F	40 F	40 F	40 F	40 F	40 F
Panama	-	-	-	-	-	-	-	-	3	...
Philippines	-	-	-	-	-	-	-	-	38	-
St Vincent	1	2	2	2	0	1	-	-	-	-
Spain	1	-	1	0	2	1	3	-	17	...
Trinidad Tob	4	3	27	46	21	81	70	33	55	17
Venezuela	66	74	122	106	137	130	205	220	28	72
31 Fishing area total	*433 F*	*480 F*	*369 F*	*388 F*	*421 F*	*493 F*	*508 F*	*603 F*	*517 F*	*326 F*
34 Benin	6	6 F	5	5 F	5 F	5	5	5	5 F	...
China	-	-	41	48	41	51	79	133	9	31
China,Taiwan	253	980	180	234	334	268	256	189	27	249
Côte dIvoire	79	139	212	177	157	222	182	275	206	196
France	116	146	133	126	96	82	80	83	79	-
Ghana	123	236	441	472	422	491	447	624	639	...
Japan	216	514	671	716	864	485	529	331	338	131
Korea Rep	-	-	-	-	-	7	-	-	-	-
Liberia	...	...	...	...	114	122	59	37	187	131
Philippines	-	-	-	-	-	-	5	38	-	-
Portugal	2	15	11	10	7	3	47	8	17	18
Sao Tome Prn	-	-	-	-	-	35	-	-	-	-
Spain	9	7	16	27	84	57	44	77	38	...
Ukraine	5	-	-	-	-	-	-	-	-	-
Other nei	-	57	100	100	100	100	-	-	-	-
34 Fishing area total	*809*	*2 100 F*	*1 810*	*1 915 F*	*2 224 F*	*1 928*	*1 733*	*1 800*	*1 545 F*	*756*
41 Brazil	125	147	81	180	331	193	486	509	467	780
China,Taiwan	400	584	272	124	222	178	138	107	0	106
Japan	272	194	275	118	101	127	77	53	51	46
Panama	-	-	-	-	-	-	-	-	38	...
Spain	14	11	5	11	4	15	38	32	78	...
Uruguay	-	-	-	-	-	-	23	-	-	-
41 Fishing area total	*811*	*936*	*633*	*433*	*658*	*513*	*762*	*701*	*634*	*932*
47 China	-	-	21	25	21	27	41	68	15	61
China,Taiwan	20	183	127	99	97	78	179	183	62	323
Cuba	69	-	-	-	-	-	-	-	-	-
Japan	299	434	556	586	614	461	395	303	293	115
Philippines	-	-	-	-	-	-	2	33	-	-
Other nei	-	117	100	100	100	100	-	-	-	-
47 Fishing area total	*388*	*734*	*804*	*810*	*832*	*666*	*617*	*587*	*370*	*499*
51 Uruguay	-	-	-	-	-	-	-	-	-	2
51 Fishing area total	-	-	-	-	-	-	-	-	-	*2*
Species total	*2 457 F*	*4 258 F*	*3 632 F*	*3 550 F*	*4 156 F*	*3 604 F*	*3 637 F*	*3 694 F*	*3 076 F*	*2 515 F*
Black marlin	**Makaire noir**		**Aguja negra**		***Makaira indica***			**1,75(03)005,07**		**BLM**
34 Belize	-	-	-	-	-	-	-	10	4	...
China,Taiwan	-	3	2	21	14	3	45	99	27	-
Korea Rep	-	-	-	-	-	3	-	-	-	-
34 Fishing area total	-	*3*	*2*	*21*	*14*	*6*	*45*	*109*	*31*	...
47 China,Taiwan	-	-	4	10	3	1	97	22	62	618
47 Fishing area total	-	-	*4*	*10*	*3*	*1*	*97*	*22*	*62*	*618*
51 China,Taiwan	553	977	41	376	247	247	226	171	512	94
Iran	-	-	-	-	3	3	5	1	1	-

B-36 Tunas, bonitos, billfishes / Thons, pélamides, marlins / Atunes, bonitos, agujas

Capture production by species, fishing areas and countries or areas / Captures par espèces, zones de pêche et pays ou zones / Capturas por especies, áreas de pesca y países o áreas

Area	Species, Fishing area Espèce, Zone de pêche Especie, Area de pesca	1992 mt	1993 mt	1994 mt	1995 mt	1996 mt	1997 mt	1998 mt	1999 mt	2000 mt	2001 mt
	Japan	51	38	47	35	-	-	-	-	-	-
	Korea Rep	2	-	-	21	8	40	20	2	25	10
	Seychelles	-	-	-	-	-	-	1	1	2	1
	Spain	-	-	14	1	0	-	-	1	0	0
	Other nei	171	13	95	168	145	79	151	90	85	-
51	*Fishing area total*	*777*	*1 028*	*197*	*601*	*403*	*369*	*403*	*266*	*625*	*105*
57	Australia	-	-	-	-	-	4	-	-	-	-
	China,Taiwan	33	326	11	194	121	121	74	59	296	-
	Japan	27	28	20	54	-	-	-	-	-	-
	Korea Rep	-	-	-	-	-	-	-	11	-	13
	Seychelles	-	-	-	-	-	-	-	0	1	0
	Sri Lanka	877	2 612	46	0	39	48	59	69	68	...
	Other nei	12	15	17	30	69	35	74	86	82	...
57	*Fishing area total*	*949*	*2 981*	*94*	*278*	*229*	*208*	*207*	*225*	*447*	*13*
61	China,Taiwan	1 087	340	1 138	483	830	659	285	681	332	193
61	*Fishing area total*	*1 087*	*340*	*1 138*	*483*	*830*	*659*	*285*	*681*	*332*	*193*
71	China,Taiwan	1 921	548	864	255	253	324	1 332	389	28	59
	Korea Rep	-	-	-	76	9	10	-	7	16	115
71	*Fishing area total*	*1 921*	*548*	*864*	*331*	*262*	*334*	*1 332*	*396*	*44*	*174*
77	China,Taiwan	-	2	4	1	0	0	2	11	29	-
	Korea Rep	615	1 375	483	395	300	182	403	793	735	1 153
	Tonga	26	28	22	15	20	10	14	6	13	10
77	*Fishing area total*	*641*	*1 405*	*509*	*411*	*320*	*192*	*419*	*810*	*777*	*1 163*
81	China,Taiwan	-	3	10	-	-	1	2	0	9	-
81	*Fishing area total*	-	*3*	*10*	-	-	*1*	*2*	*0*	*9*	-
87	China,Taiwan	-	-	-	-	1	1	0	0	1	-
	Korea Rep	-	-	-	-	-	-	4	-	-	-
87	*Fishing area total*	-	-	-	-	*1*	*1*	*4*	*0*	*1*	-
Species total		*5 375*	*6 308*	*2 818*	*2 135*	*2 062*	*1 771*	*2 794*	*2 509*	*2 328*	*2 266*
Striped marlin		**Marlin rayé**		**Marlín rayado**			***Tetrapturus audax***			**1,75(03)009,03**	**MLS**
51	China,Taiwan	1 212	2 482	1 318	2 720	2 138	2 138	1 793	1 148	1 601	1 602
	Japan	99	77	136	132	173	192	175	141	169	63
	Korea Rep	1	3	2	38	-	65	43	-	20	2
	Philippines	-	-	-	-	-	-	103	35	21	11
	Réunion	...	...	...	...	...	...	...	...	120	117
	Seychelles	-	-	-	-	-	-	-	4	10	6
	Spain	-	-	-	-	-	-	-	1	0	0
	Other nei	323	1 012	620	1 043	1 023	469	601	520	552	-
51	*Fishing area total*	*1 635*	*3 574*	*2 076*	*3 933*	*3 334*	*2 864*	*2 715*	*1 849*	*2 493*	*1 801*
57	Australia	0	3	9	7	3	17	12	62	3	...
	China,Taiwan	270	368	510	917	828	828	567	591	613	202
	Japan	95	32	58	92	105	119	95	134	144	71
	Korea Rep	-	-	-	-	-	-	-	1	-	1
	Philippines	-	-	-	-	-	-	18	19	11	2
	Seychelles	-	-	-	-	-	-	-	2	6	1
	Sri Lanka	1	8	-	-	0	0	0	-	-	-
	Other nei	77	182	128	215	459	286	513	332	353	...
57	*Fishing area total*	*443*	*593*	*705*	*1 231*	*1 395*	*1 250*	*1 205*	*1 141*	*1 130*	*277*
61	China,Taiwan	465	227	322	173	148	160	108	505	408	1
	Japan	3 619	4 484	4 347	4 297	3 712	2 308	4 220	3 588	2 973	2 875
61	*Fishing area total*	*4 084*	*4 711*	*4 669*	*4 470*	*3 860*	*2 468*	*4 328*	*4 093*	*3 381*	*2 876*
71	China,Taiwan	1 324	209	1 001	199	188	311	1 404	278	107	64
	Japan	676	682	653	450	218	749	237	145	155	186
	Korea Rep	4	11	-	-	51	20	11	13	17	3
71	*Fishing area total*	*2 004*	*902*	*1 654*	*649*	*457*	*1 080*	*1 652*	*436*	*279*	*253*
77	China,Taiwan	132	126	104	84	57	51	164	74	74	112
	Japan	2 670	2 643	2 266	2 878	1 594	1 992	1 419	1 419	809	656
	Korea Rep	59	651	590	374	400	1 087	739	571	600	343
77	*Fishing area total*	*2 861*	*3 420*	*2 960*	*3 336*	*2 051*	*3 130*	*2 322*	*2 064*	*1 483*	*1 111*
81	China,Taiwan	272	186	171	33	56	50	80	98	189	36
	Japan	408	573	637	813	728	449	914	330	133	266
	New Zealand	0	0	0	0	0	1	-	-	-	-
81	*Fishing area total*	*680*	*759*	*808*	*846*	*784*	*500*	*994*	*428*	*322*	*302*
87	China,Taiwan	0	39	1	0	1	1	-	2	3	-
	Japan	1 060	935	967	621	1 083	728	1 076	194	162	666
	Korea Rep	-	-	-	-	-	-	2	4	-	-
87	*Fishing area total*	*1 060*	*974*	*968*	*621*	*1 084*	*729*	*1 078*	*200*	*165*	*666*
Species total		*12 767*	*14 933*	*13 840*	*15 086*	*12 965*	*12 021*	*14 294*	*10 211*	*9 253*	*7 286*
Atlantic white marlin		**Makaire blanc de l'Atlantique**		**Aguja blanca del Atlántico**			***Tetrapturus albidus***			**1,75(03)009,04**	**WHM**
21	China,Taiwan	-	-	4	3	0	0	-	-	-	-
	Japan	7	1	1	-	-	-	2	-	-	-

B-36 Tunas, bonitos, billfishes / Thons, pélamides, marlins / Atunes, bonitos, agujas

Capture production by species, fishing areas and countries or areas / Captures par espèces, zones de pêche et pays ou zones / Capturas por especies, áreas de pesca y países o áreas

Species, Fishing area Espèce, Zone de pêche Especie, Area de pesca	1992 mt	1993 mt	1994 mt	1995 mt	1996 mt	1997 mt	1998 mt	1999 mt	2000 mt	2001 mt
21 Fishing area total	*7*	*1*	*5*	*3*	*0*	*0*	*2*	-	-	-
27 China,Taiwan	-	1	8	13	0	0	-	2	-	1
Japan	-	-	-	-	-	1	-	-	-	-
27 Fishing area total	-	*1*	*8*	*13*	*0*	*1*	-	*2*	-	*1*
31 Barbados	24	29	26	43	15	41	33	25	25	...
Bermuda	1	1	1	1	1	1	1	1	0	1
China,Taiwan	11	75	42	133	38	29	45	22	3	11
Costa Rica	...	...	...	...	...	...	...	3	14	-
Cuba	5	0	-	-	-	-	-	-	-	-
Grenada	-	-	-	-	-	-	-	-	1	15
Japan	30	6	6	6	29	6	4	5	15	7
St Vincent	0	1	0	0	0	-	-	-	-	-
Spain	1	1	1	0	7	4	0	43	45	...
Venezuela	187	226	148	171	164	90	80	61	13	72
31 Fishing area total	*259*	*339*	*224*	*354*	*254*	*171*	*163*	*160*	*116*	*106*
34 Belize	-	-	-	-	-	1	-	1	0	...
China	-	-	6	7	6	7	10	20	1	7
China,Taiwan	33	350	293	128	141	110	124	175	321	-
Côte dIvoire	-	-	-	-	1	2	1	5	1	2
Cuba	5	0	-	-	-	-	-	-	-	-
France	10	12	11	9	7	7	9	8	7	-
Ghana	14	22	1	2	1	3	7	6	8	21
Japan	20	51	36	30	58	19	31	23	50	11
Philippines	-	-	-	-	-	-	0	4	-	-
Sao Tome Prn	-	-	-	-	-	45	-	-	-	-
Spain	17	14	24	9	68	66	64	45	72	...
Other nei	-	46	50	50	50	50	-	-	-	-
34 Fishing area total	*99*	*495*	*421*	*235*	*332*	*310*	*246*	*287*	*460*	*41*
37 Spain	-	-	-	1	-	1	1	-	1	...
37 Fishing area total	-	-	-	*1*	-	*1*	*1*	-	*1*	...
41 Argentina	0	-	-	-	-	-	-	-	-	-
Brazil	211	301	91	105	75	105	217	158	106	172
China,Taiwan	240	395	366	336	311	242	115	101	118	6
Japan	36	23	24	5	5	8	3	6	6	14
Spain	1	-	0	14	5	2	4	28	61	...
Uruguay	3	0	6	1	2	50	22	-	-	0
41 Fishing area total	*491*	*719*	*487*	*461*	*398*	*407*	*361*	*293*	*291*	*192*
47 China	-	-	3	4	3	4	5	10	1	13
China,Taiwan	1	134	186	295	77	60	223	164	281	17
Japan	33	33	32	18	23	21	19	16	18	28
Philippines	-	-	-	-	-	-	1	8	-	-
Spain	3	8	-	4	27	1	-	17	7	...
Other nei	-	68	50	50	50	50	-	-	-	-
47 Fishing area total	*37*	*243*	*271*	*371*	*180*	*136*	*248*	*215*	*307*	*58*
Species total	*893*	*1 798*	*1 416*	*1 438*	*1 164*	*1 026*	*1 021*	*957*	*1 175*	*398*

Shortbill spearfish — **Makaire à rostre court** — **Marlín trompa corta** — ***Tetrapturus angustirostris*** — **1,75(03)009,05** — **SSP**

Species, Fishing area	1992	1993	1994	1995	1996	1997	1998	1999	2000	2001
51 Réunion	-	-	-	-	2	1	3	5	5	-
Spain	-	-	-	-	-	1	2	2	0	0
51 Fishing area total	-	-	-	-	*2*	*2*	*5*	*7*	*5*	*0*
57 Australia	-	-	-	-	-	-	-	1	-	0
57 Fishing area total	-	-	-	-	-	-	-	*1*	-	*0*
81 New Zealand	-	-	-	8	42	18	-	-	-	-
81 Fishing area total	-	-	-	*8*	*42*	*18*	-	-	-	-
Species total	-	-	-	*8*	*44*	*20*	*5*	*8*	*5*	*0*

Longbill spearfish — **Makaire bécune** — **Aguja picuda** — ***Tetrapturus pfluegeri*** — **1,75(03)009,06** — **SPF**

Species, Fishing area	1992	1993	1994	1995	1996	1997	1998	1999	2000	2001
31 Trinidad Tob	-	62	-	-	-	-	-	-	-	-
Venezuela	-	-	0	-	1	0	1	0	-	4
31 Fishing area total	-	*62*	*0*	-	*1*	*0*	*1*	*0*	-	*4*
34 France	92	112	98	78	59	68	86	81	60	-
Spain	0	0	0	3	1	-	-	2	3	...
34 Fishing area total	*92*	*112*	*98*	*81*	*60*	*68*	*86*	*83*	*63*	...
41 Brazil	-	-	-	-	-	-	-	-	12	56
Spain	0	2	-	1	-	0	-	22	50	...
41 Fishing area total	*0*	*2*	-	*1*	-	*0*	-	*22*	*62*	*56*
47 China	-	-	-	-	-	-	2	-	-	-
Spain	0	8	-	-	-	1	1	28	11	...
47 Fishing area total	*0*	*8*	-	-	-	*1*	*3*	*28*	*11*	...
Species total	*92*	*184*	*98*	*82*	*61*	*69*	*90*	*133*	*136*	*60*

B-36 Tunas, bonitos, billfishes / Thons, pélamides, marlins / Atunes, bonitos, agujas

Capture production by species, fishing areas and countries or areas / Captures par espèces, zones de pêche et pays ou zones / Capturas por especies, áreas de pesca y países o áreas

Species, Fishing area Espèce, Zone de pêche Especie, Area de pesca		1992 mt	1993 mt	1994 mt	1995 mt	1996 mt	1997 mt	1998 mt	1999 mt	2000 mt	2001 mt
Marlins,sailfishes,etc. nei		**Makaires,marlins,voiliers nca**		**Agujas,marlines,peces vela nep**		***Istiophoridae***			**1,75(03)XXX,XX**		**BIL**
31	Barbados	-	-	-	-	-	-	-	-	-	85
	Bermuda	-	5	-	-	-	3	1	-	-	-
	China,Taiwan	-	-	-	2	15	3	-	-	-	-
	Korea Rep	-	-	-	-	17	-	-	-	-	-
	Mexico	-	2	8	22	73	77	164	104	133	97
	St Lucia	-	-	-	-	-	4	1	-	14	5
	St Vincent	-	-	-	-	1	0	2	1	343	339
	Spain	-	-	0	3	6	11	29	-	-	-
	Trinidad Tob	-	-	-	-	-	-	-	-	-	2
	Venezuela	-	-	-	-	-	-	-	-	-	8
31	*Fishing area total*	-	*7*	*8*	*27*	*112*	*98*	*197*	*105*	*490*	*536*
34	China	-	-	-	-	18	-	-	-	-	-
	France	-	-	-	-	-	-	-	-	10	-
	Gabon	0	4	5	0	523	7	-	-	-	1
	Korea Rep	13	-	-	-	72	36	-	-	-	-
	Liberia	38	27	112	120	145	71	781	513	683	163
	Portugal	-	0	-	-	0	-	15	13	74	444
	Spain	-	-	1	10	10	17	28	-	-	-
	Togo	...	...	...	...	...	32	...	110	77	205
34	*Fishing area total*	*51*	*31*	*118*	*130*	*768*	*163*	*824*	*636*	*844*	*813*
37	Korea Rep	-	-	-	-	-	28	-	-	-	-
	Malta	-	-	-	1	1	1	-	-	-	-
	Portugal	-	-	-	-	-	-	-	-	1	25
37	*Fishing area total*	-	-	-	*1*	*1*	*29*	-	-	*1*	*25*
41	Brazil	-	-	-	-	-	-	-	-	18	2
	China,Taiwan	-	3	17	78	9	2	24	0	0	8
	Korea Rep	-	-	92	62	24	1	-	-	-	-
	Spain	19	26	42	-	-	27	38	-	-	-
	UK	-	-	-	-	-	-	82	-	-	-
41	*Fishing area total*	*19*	*29*	*151*	*140*	*33*	*30*	*144*	*0*	*18*	*10*
47	France	-	-	-	-	-	-	-	-	56	-
	Korea Rep	27	-	-	-	12	20	-	-	-	-
	Spain	7	24	4	-	1	40	32	-	-	-
47	*Fishing area total*	*34*	*24*	*4*	-	*13*	*60*	*32*	-	*56*	-
51	China	-	-	-	-	-	-	-	-	142	121
	China,Taiwan	390	-	-	-	-	-	-	-	-	57
	Comoros	115	115	136	136	130	120	120	100	130 F	120
	India	1 162	1 274	1 577	1 301	1 864	2 128	1 557	1 188	1 303	407
	Kenya	-	-	-	-	73	53	38	82	80	78
	Korea Rep	978	1 548	2 003	1 248	2 128	945	219	4	174	74
	Mauritius	264	194	227	196	190	639	295	287	287	2
	Pakistan	2 400	2 245	2 932	2 684	2 834	2 198	2 264	2 340	2 215	1 177
	Portugal	-	-	-	-	-	-	5	-	-	2
	Réunion	35	61	72	87	120	110	136	109	3	1
	Seychelles	-	-	2	2	80	32	6	8	4	3
	South Africa	1	1	-	-	-	2	3	5	4	4
	Tanzania	300	531	760	580	656	700	800	780	800	850
	Untd Arab Em	199	193	193	232	239	230	233	233	250	250 F
	Other nei	139	235	41	-	127	313	285	195	218	-
51	*Fishing area total*	*5 983*	*6 397*	*7 943*	*6 466*	*8 441*	*7 470*	*5 961*	*5 331*	*5 610 F*	*3 146 F*
57	China	-	-	-	-	-	-	-	287	344	258
	China,Taiwan	240	-	-	-	-	-	-	-	-	700
	Honduras	-	-	-	-	-	-	-	-	-	182
	India	366	585	300	242	2 076	2 374	1 824	3 014	3 279	801
	Korea Rep	58	-	-	25	9	22	15	14	1	4
	Malaysia	0	0	3	3	-	5	0	0	0	1
	Seychelles	-	-	-	-	-	-	-	1	2	0
	Sri Lanka	48	252	347	5 196	5 675	6 937	8 653	7 253	5 728	6 340
	Thailand	-	-	-	-	-	-	-	-	16	...
	Other nei	88	60	38	36	17	35	80	105	117	...
57	*Fishing area total*	*800*	*897*	*688*	*5 502*	*7 777*	*9 373*	*10 572*	*10 674*	*9 487*	*8 286*
61	China	-	-	-	-	-	-	-	-	-	146
	Korea Rep	-	-	-	-	-	-	2	3	-	-
61	*Fishing area total*	-	-	-	-	-	-	*2*	*3*	-	*146*
67	Japan	32	-	-	-	-	-	-	-	-	-
67	*Fishing area total*	*32*	-	-	-	-	-	-	-	-	-
71	Australia	...	...	...	...	199	1 084	1 929	2 185	2 604	2 573
	China	-	-	-	-	-	-	-	114	469	171
	China,Taiwan	-	-	-	-	-	-	-	-	-	42
	Korea Rep	983	502	103	87	254	230	361	59	40	100
	Malaysia	598	469	230	450	274	357	324	2 046	161	154
	Palau	0	1	4	1	2	...	...	4 F	5 F	5 F
	Papua N Guin	...	...	0	0	16	6	87	230	418	368
	Philippines	3 418	4 575	2 423	4 648	4 317	4 799	5 338	5 082	4 969	5 247
	Solomon Is	50 F	50 F	50 F	50 F	50 F	50 F	50 F	50 F	50 F	50 F
	Vanuatu	...	...	...	209	98	97	99	14	...	-
71	*Fishing area total*	*5 049 F*	*5 597 F*	*2 810 F*	*5 445 F*	*5 210 F*	*6 623 F*	*8 188 F*	*9 784 F*	*8 716 F*	*8 710 F*
77	China	-	-	-	-	-	-	-	-	45	103
	Cook Is	...	...	26	40	30 F	30 F	30 F	30 F	20 F	20 F

B-36 Tunas, bonitos, billfishes / Thons, pélamides, marlins / Atunes, bonitos, agujas

Capture production by species, fishing areas and countries or areas
Captures par espèces, zones de pêche et pays ou zones
Capturas por especies, áreas de pesca y países o áreas

Species, Fishing area Espèce, Zone de pêche Especie, Area de pesca	1992 mt	1993 mt	1994 mt	1995 mt	1996 mt	1997 mt	1998 mt	1999 mt	2000 mt	2001 mt
Costa Rica	380	388	1 226	1 764	1 892	1 876	1 647	2 143	2 035	2 206
Fr Polynesia	...	...	465	598	587	598	518	703	566	551
Korea Rep	2 520	2 723	2 466	3 309	2 840	1 952	3 240	1 333	1 112	1 479
Mexico	315	235	87	204	553	1 391	2 375	283	242	250
USA	1 202	1 324	1 048	1 530	1 368	100	134	214	123	250
77 Fishing area total	*4 417*	*4 670*	*5 318*	*7 445*	*7 270 F*	*5 947 F*	*7 944 F*	*4 706 F*	*4 143 F*	*4 859 F*
81 Korea Rep	-	-	-	-	-	-	12	-	-	-
81 Fishing area total	-	-	-	-	-	-	*12*	-	-	-
87 Ecuador	-	-	-	-	-	-	3 727	6 962	...	...
Korea Rep	2	22	-	-	-	-	-	-	-	-
87 Fishing area total	*2*	*22*	-	-	-	-	*3 727*	*6 962*	...	...
Species total	*16 387 F*	*17 674 F*	*17 040 F*	*25 156 F*	*29 625 F*	*29 793 F*	*37 603 F*	*38 201 F*	*29 365 F*	*26 531 F*
Swordfish	**Espadon**		**Pez espada**			***Xiphias gladius***			**1,75(04)003,01**	**SWC**
21 Canada	1 547	2 230	1 675	1 568	739	1 089	1 115	1 118	968	1 079
China,Taiwan	-	-	-	1	-	-	-	1	3	-
Japan	199	138	75	27	37	37	81	128	34	1
USA	1 617	1 531	1 248	1 215	1 661	918	1 093	1 111	1 163	948
21 Fishing area total	*3 363*	*3 899*	*2 998*	*2 811*	*2 437*	*2 044*	*2 289*	*2 358*	*2 168*	*2 028*
27 China,Taiwan	-	-	-	38	-	-	2	19	10	1
Faeroe Is	-	-	-	-	-	-	-	5	4	...
France	75	95	46	84	97	164	110	104	122	101
Iceland	-	-	-	-	-	-	-	2	2	-
Ireland	-	-	-	-	15	15	132	81	36	14
Japan	85	47	37	60	111	42	56	46	18	1
Portugal	495	1 931	1 542	1 576	1 652	864	741	764	660	665
Spain	5 387	4 834	4 561	5 001	4 654	3 827	2 517	2 076	2 422	2 047
UK	...	2	3	1	5	11	0	2	1	-
27 Fishing area total	*6 042*	*6 909*	*6 189*	*6 760*	*6 534*	*4 923*	*3 558*	*3 099*	*3 275*	*2 829*
31 Barbados	-	-	-	-	33	16	16	12	13	19
Bermuda	-	-	-	1	1	5	5	3	3	2
Br Virgin Is	...	...	58	19	54	6	5	5	2	2 F
China,Taiwan	100 F	90	70	19	59	41	14	15	50	10
Costa Rica	...	...	...	...	...	...	...	1	2	-
Cuba	25	8	10	15	5	9	10	5	5 F	5 F
Grenada	2	7	0	1	4	47	33	42	84	73
Japan	35	28	33	27	69	50	78	104	41	8
Mexico	-	6	14	13	11	14	28	24	37	27
Panama	-	-	-	-	-	-	-	17	-	-
Philippines	-	-	-	-	-	-	-	-	-	1
St Lucia	-	-	1	0	-	-	-	-	-	-
St Vincent	3	23	0	4	3	1	0	1	0	22
Spain	1 283	1 761	1 619	1 948	886	1 309	1 562	1 917	2 161	2 000
Trinidad Tob	562	11	180	150	158	110	130	138	41	75
USA	1 378	1 445	1 184	1 270	486	1 123	1 122	1 068	1 281	816
Venezuela	103	73	69	54	85	20	37	30	30	21
Other nei	35	111	-	-	-	-	-	-	-	-
31 Fishing area total	*3 526 F*	*3 563*	*3 238*	*3 521*	*1 854*	*2 751*	*3 040*	*3 382*	*3 750 F*	*3 081 F*
34 Belize	-	-	-	1	-	-	-	17	8	...
Benin	26	28	25	24	24 F	10	0	3	...	41
Cape Verde	0	-	-	-	-	-	-	-	-	-
China	-	73	86	104	132	40	38	304	22	102
China,Taiwan	32 F	18 F	618	1 356	1 737	1 206	649	606	1 666	1 986
Côte dIvoire	13	14	20	19	26	18	25	26	20	19
Cuba	248	200	492	849	62	-	-	-	-	-
Ghana	69	121	51	103	140	44	106	121	117	...
Greece	-	3	6	2	1	-	-	-	-	-
Japan	1 277	2 724	1 870	1 986	2 116	878	1 483	955	562	262
Korea Rep	17	-	-	-	18	33	-	-	-	-
Liberia	7	14	26	28	112	543	21	39	42	34
Lithuania	-	-	794	-	-	-	-	-	-	-
Morocco	69	39	36	79	462	267	191	119	114	523
Nigeria	3	0	857	-	9	-	-	-	-	-
Portugal	47	30	57	41	51	39	11	10	13	6
Sao Tome Prn	-	-	-	-	-	14	14 F	14 F	-	-
Senegal	-	-	-	-	-	-	169	127	39	35
Sierra Leone	...	...	...	...	...	...	...	...	2	2
Spain	2	3	5	4	7	4	-	-	12	...
Togo	5	8	14	14	64	2	23	37	7	17
34 Fishing area total	*1 815 F*	*3 275 F*	*4 957*	*4 610*	*4 961 F*	*3 098*	*2 730 F*	*2 378 F*	*2 624*	*3 027*
37 Albania	...	...	...	0	13	0	-	-	-	2
Algeria	395	562	600	807	807	807	825	709	816	1 081
China,Taiwan	-	-	-	-	1	1	-	-	-	1
Croatia	...	...	...	...	...	...	10	20	...	...
Cyprus	73	116	159	89	40	51	61	92	82	135
France	-	-	-	-	-	-	-	-	-	12
Greece	1 456	1 568	2 520	974	1 237	750	1 650	1 520	1 960	1 730
Italy	7 595	6 330	7 765	6 725	5 286	6 104	6 104	6 312	7 515	6 388
Japan	2	6	4	6	6	3	7	4	1	1
Libya	-	-	-	-	-	-	11	...	8	6
Malta	71	76	42	58	58	83	116	167	160	89

B-36 Tunas, bonitos, billfishes — Capture production by species, fishing areas and countries or areas
Thons, pélamides, marlins — Captures par espèces, zones de pêche et pays ou zones
Atunes, bonitos, agujas — Capturas por especies, áreas de pesca y países o áreas

Species, Fishing area Espèce, Zone de pêche Especie, Area de pesca	1992 mt	1993 mt	1994 mt	1995 mt	1996 mt	1997 mt	1998 mt	1999 mt	2000 mt	2001 mt
Morocco	2 692	2 589	2 654	1 696	2 734	4 900	3 228	3 238	2 708	3 026
Portugal	-	-	-	-	-	-	-	-	13	115
Spain	822	1 358	1 503	1 379	1 186	1 264	1 443	905	1 436	1 475
Tunisia	178	354	298	378	352	346	414	468	483	567
Turkey	136	292	533	306	320	350	450	230	373	360
Other nei	1 292	-	-	-	-	-	-	-	-	-
37 Fishing area total	*14 712*	*13 251*	*16 078*	*12 418*	*12 040*	*14 659*	*14 319*	*13 665*	*15 555*	*14 988*
41 Argentina	88	14	24	0	-	-	-	-	-	5
Brazil	2 609	2 013	1 571	1 975	1 892	4 100	3 847	4 721	4 697	4 082
China	-	-	-	-	-	-	328	-	-	-
China,Taiwan	1 071 F	541	1 354	970	1 023	710	366	212	696	430
Japan	907	587	657	174	170	149	71	93	143	166
Panama	-	-	-	-	-	-	-	87	-	-
Philippines	-	-	-	-	-	-	-	-	-	6
Portugal	-	-	-	153	389	81	221	3	153	126
Spain	2 948	4 545	5 186	7 339	3 671	4 475	3 119	3 150	4 496	4 100
Uruguay	210	260	165	499	644	760	889	661	713	636
41 Fishing area total	*7 833 F*	*7 960*	*8 957*	*11 110*	*7 789*	*10 275*	*8 841*	*8 927*	*10 898*	*9 551*
47 Angola	0	0	0	0	0	0	0	0	0	...
China	-	-	-	-	-	-	-	534	344	200
China,Taiwan	200 F	100 F	540	981	573	398	402	602	824	747
France	-	-	-	-	-	-	-	-	4	-
Japan	1 883	2 458	3 173	1 876	1 670	877	1 043	768	604	353
Namibia	-	-	-	-	-	-	-	730	469	751
Norway	-	-	-	-	-	1	-	-	-	-
Panama	-	-	-	-	-	-	-	18	-	-
Portugal	1	-	-	227	-	360	184	381	299	330
South Africa	9	4	1	4	1	1	169	76	230	269
Spain	2 703	2 429	2 751	3 951	5 951	3 986	2 713	2 608	1 892	1 748
47 Fishing area total	*4 796 F*	*4 991 F*	*6 465*	*7 039*	*8 195*	*5 623*	*4 511*	*5 717*	*4 666*	*4 398*
51 China	-	-	-	-	-	-	-	8	79	93
China,Taiwan	3 773 F	3 960 F	3 631	16 312	7 493	13 369	13 778	12 901	10 914	11 528
India	-	-	-	-	60	56	-	-	-	-
Iran	-	-	-	-	9	9	13	2	1	-
Japan	1 023	628	999	559	927	993	1 406	676	689	650
Korea Rep	60	20	17	74	51	196	147	8	63	18
Mauritius	...	...	...	...	...	...	...	...	...	34
Mozambique	-	-	-	312	358	524	1 039	447	...	...
Philippines	-	-	-	-	-	-	260	236	140	134
Portugal	-	-	-	-	-	-	161	285	262	754
Réunion	65	286	734	769	1 336	1 586	2 080	1 930	1 744	1 514
Seychelles	-	-	-	22	141	317	216	323	374	57
South Africa	-	-	-	-	-	-	236	48	20	229
Spain	-	207	694	19	29	508	1 425	2 013	983	1 870
Uruguay	-	-	-	-	-	-	-	-	-	80
Other nei	1 791	3 520	2 689	5 356	6 709	4 463	5 965	5 199	6 097	-
51 Fishing area total	*6 712 F*	*8 621 F*	*8 764*	*23 423*	*17 113*	*22 021*	*26 726*	*24 076*	*21 366*	*16 961*
57 Australia	32	189	115	62	22	43	337	1 360	1 798	2 900
China	-	-	-	71	238	255	117	262	294	170
China,Taiwan	679	520	545	1 949	2 127	3 794	3 051	1 826	3 213	2 620
Honduras	-	-	-	-	-	-	-	-	-	165
India	-	-	-	-	2	2	25	15	-	-
Japan	122	317	360	684	758	598	538	639	576	497
Korea Rep	-	-	-	2	-	8	2	21	-	19
Philippines	-	-	-	-	-	-	12	89	53	15
Seychelles	-	-	-	-	-	-	-	10	43	263
Sri Lanka	872	4 662	2 407	2 558	2 591	3 167	3 950	2 132	5 545	2 860
Thailand	-	-	-	-	-	-	-	-	19	...
Other nei	246	285	103	406	2 073	1 706	1 781	1 815	2 242	...
57 Fishing area total	*1 951*	*5 973*	*3 530*	*5 732*	*7 811*	*9 573*	*9 813*	*8 169*	*13 783*	*9 509*
61 China	-	-	-	-	-	-	-	-	-	104
China,Taiwan	500 F	1 180	785	508	489	763	760	577	2 620	1 710
Japan	7 840	8 789	8 392	7 376	5 493	4 991	7 259	7 040	7 242	5 506
61 Fishing area total	*8 340 F*	*9 969*	*9 177*	*7 884*	*5 982*	*5 754*	*8 019*	*7 617*	*9 862*	*7 320*
67 Japan	25	-	5	-	-	-	-	-	-	-
USA	405	313	90	22	23	34	128	27	34	14
67 Fishing area total	*430*	*313*	*95*	*22*	*23*	*34*	*128*	*27*	*34*	*14*
71 China	-	-	-	-	-	-	-	396	72	95
China,Taiwan	80 F	94 F	1 259	777	712	1 477	1 741	2 177	101	17
Japan	837	692	789	645	406	1 000	398	369	427	589
Korea Rep	9	27	-	9	5	513	19	3	24	100
71 Fishing area total	*926 F*	*813 F*	*2 048*	*1 431*	*1 123*	*2 990*	*2 158*	*2 945*	*624*	*801*
77 Amer Samoa	-	-	-	-	-	-	2	0	1	1
China	-	-	-	-	-	-	-	-	68	319
China,Taiwan	180 F	210 F	134	93	21	22	48	162	118	592
Cook Is	...	...	26	44	40 F	40 F	30 F	30 F	30 F	30 F
Costa Rica	161	35	17	29	433	1 072	419	99	407	653
Fr Polynesia	...	65	71	62	84	56	58	66	47	79
Japan	3 712	2 709	2 101	1 979	2 114	1 678	2 620	1 608	1 952	3 029
Korea Rep	78	82	15	13	55	198	401	439	578	894
Mexico	1 160	806	567	424	428	2 351	3 575	1 112	2 179	2 300

B-36 Tunas, bonitos, billfishes — Capture production by species, fishing areas and countries or areas

Thons, pélamides, marlins — Captures par espèces, zones de pêche et pays ou zones

Atunes, bonitos, agujas — Capturas por especies, áreas de pesca y países o áreas

Species, Fishing area Espèce, Zone de pêche Especie, Area de pesca	1992 mt	1993 mt	1994 mt	1995 mt	1996 mt	1997 mt	1998 mt	1999 mt	2000 mt	2001 mt
Tonga	4	5	8	10	7	6	8	5	53	8
USA	5 543	6 668	4 400	3 409	3 672	4 088	4 503	5 071	5 598	2 490
77 Fishing area total	*10 838 F*	*10 580 F*	*7 339*	*6 063*	*6 854 F*	*9 511 F*	*11 664 F*	*8 592 F*	*11 031 F*	*10 395 F*
81 China,Taiwan	40 F	51 F	223	26	60	61	128	117	106	17
Japan	2 178	1 278	1 419	1 096	1 387	1 029	1 164	751	537	488
New Zealand	-	-	89	93	152	170	564	1 004	975	1 029
81 Fishing area total	*2 218 F*	*1 329 F*	*1 731*	*1 215*	*1 599*	*1 260*	*1 856*	*1 872*	*1 618*	*1 534*
87 Chile	6 379	4 712	3 801	2 594	3 145	4 040	4 492	2 925	2 973	3 262
China,Taiwan	-	-	6	-	20	21	1	1	9	2
Colombia	-	-	-	-	-	-	6	-	-	1
Ecuador	350	33	0	222	...	...	...	...	...	...
Japan	1 027	605	824	523	388	270	694	404	355	1 267
Korea Rep	-	-	-	-	-	19	4	3	-	-
Peru	21	19	5	-	1	-	57	42	20	356
Spain	2 435	928	576	698	772	2 039	1 346	1 121	1 807	828
87 Fishing area total	*10 212*	*6 297*	*5 212*	*4 037*	*4 326*	*6 389*	*6 600*	*4 496*	*5 164*	*5 716*
Species total	*83 714 F*	*87 743 F*	*86 778*	*98 076*	*88 641 F*	*100 905 F*	*106 252 F*	*97 320 F*	*106 418 F*	*92 152 F*
Tuna-like fishes nei	**Poissons type thon nca**		**Peces parec.a los atunes nep**		***Scombroidei***				**1,75(XX)XXX,XX**	**TUX**
21 Canada	3	8	-	5	-	-	0	-	-	-
Russian Fed	-	-	-	-	-	-	-	-	3	-
21 Fishing area total	*3*	*8*	-	*5*	-	-	*0*	-	*3*	-
27 Portugal	0	0	25	417	252	155	263	87	101	61
Spain	13	-	-	-	-	-	-	-	-	-
27 Fishing area total	*13*	*0*	*25*	*417*	*252*	*155*	*263*	*87*	*101*	*61*
31 Antigua Barb	-	2	...	...	...	...	...	...	...	28
Barbados	112	167	162	255	68	-	-	-	11	-
Br Virgin Is	...	...	20	15	19	-	-	-	-	-
Colombia	3 831	228	112	265	14 647	188	23 259	380	5 000 F	5 000 F
Costa Rica	1	3	0	0	0	0	0	6	26	3
Dominican Rp	125	86	155	138	85	624	196	394	249	-
Jamaica	...	...	...	...	239	275	...	...	78	86
Panama	-	-	-	-	-	-	-	2	2	...
Puerto Rico	53	62	64	82	88	72	121	99	111	17
St Kitts Nev	...	...	...	1	3	7	16	24	24	24
St Lucia	45	71	15	10	8	1	3	3	1	-
Seychelles	-	-	-	-	-	-	-	-	127	-
Trinidad Tob	3 361	-	-	25	134	206	92	81	1 380	405
Venezuela	111	-	862	-	-	-	-	-	-	13
Other nei	106	146	150	216	194	-	-	-	-	-
31 Fishing area total	*7 745*	*765*	*1 540*	*1 007*	*15 485*	*1 373*	*23 687*	*989*	*7 009 F*	*5 576 F*
34 Belize	-	-	-	-	-	-	-	107	143	115
Bulgaria	-	-	-	-	-	-	225	-	-	-
Cameroon	6	6	1	1	0	0	0	0	1	0
Cape Verde	-	-	-	-	-	245	-	-	-	-
China	-	41	68	76	80	-	-	150	6	187
Côte dIvoire	11 150	-	-	-	-	-	-	-	-	-
Cuba	-	3	30	3	21	-	-	-	-	-
Cyprus	-	-	-	-	...	...	39	72	173	262
Eq Guinea	360	390	380	340	216	570 F	392	460 F	50 F	50 F
Estonia	1	-	-	-	-	-	-	-	-	-
Greece	10	7	4	-	-	-	-	-	-	3
GuineaBissau	3 F	5 F	5 F	6 F	6 F	6 F	5 F	4 F	4 F	4 F
Italy	-	-	-	-	-	1	-	-	-	-
Korea Rep	5	24	15	39	120	8	-	-	-	-
Latvia	-	-	-	-	-	-	147	27	-	-
Liberia	-	-	-	-	-	-	3	2	14	8
Lithuania	160	73	-	...	...	...	467	110	80	153
Mauritania	377	746	54	263	2 479	2 170	1 304	...	...	...
Morocco	-	-	-	-	-	-	-	82	-	-
Nigeria	90	157	109	119	200	55	73	7	51	5
Philippines	-	-	-	-	-	-	15	5	-	-
Portugal	-	-	-	-	-	-	-	-	-	16
Russian Fed	300	-	-	-	-	-	-	-	-	-
St Vincent	-	-	-	-	-	50	387	167	...	...
Sao Tome Prn	300 F	183 F	-	-	-	9	-	-	-	-
Senegal	537	3	3	52	-	-	-	-	-	6
Sierra Leone	608	601	599	598	254	2 618	4 980	2 166	623	6 639
Spain	-	-	-	-	-	-	-	-	789	-
Ukraine	1	3	4	...	...	...	303	-	28	213
Other nei	-	-	-	-	-	-	-	-	744	-
34 Fishing area total	*13 908 F*	*2 242 F*	*1 272 F*	*1 497 F*	*3 376 F*	*5 732 F*	*8 340 F*	*3 359 F*	*2 706 F*	*7 661 F*
37 Cyprus	-	-	-	-	-	-	-	-	7	-
Egypt	-	-	-	530	1 071	594	576	2 004	1 676	778
Gaza Strip	...	...	...	...	50	102	92	100 F	100 F	80 F
Greece	510	510	116	116	116	145	300	...	195	125
Korea Rep	-	-	-	2	-	248	-	-	-	-
Lebanon	150 F	175	200	500	500	700	400	400	500	450
Malta	0	8	0	0	0	0	0	-	-	-
Morocco	-	-	-	-	-	-	-	71	-	-
Spain	-	-	-	-	8	-	-	-	-	-

B-36 Tunas, bonitos, billfishes — Capture production by species, fishing areas and countries or areas
Thons, pélamides, marlins — Captures par espèces, zones de pêche et pays ou zones
Atunes, bonitos, agujas — Capturas por especies, áreas de pesca y países o áreas

Species, Fishing area Espèce, Zone de pêche Especie, Area de pesca	1992 mt	1993 mt	1994 mt	1995 mt	1996 mt	1997 mt	1998 mt	1999 mt	2000 mt	2001 mt
Tunisia	20	309	105	115	215	657	6	814	905	989
Other nei	-	-	-	-	50	-	-	-	-	-
37 Fishing area total	*680 F*	*1 002*	*421*	*1 263*	*2 010*	*2 446*	*1 374*	*3 389 F*	*3 383 F*	*2 422 F*
41 Argentina	1	2	0	0	0	-	-	-	-	-
Brazil	173	287	140	58	-	446	258	570	151	5
Korea Rep	-	-	3	9	-	-	-	-	-	-
Panama	-	-	-	-	-	-	-	68	13	...
Portugal	-	-	-	-	-	2	7	-	78	4
Uruguay	0	0	0	0	2	50	94	136	95	20
41 Fishing area total	*174*	*289*	*143*	*67*	*2*	*498*	*359*	*774*	*337*	*29*
47 Cambodia	-	-	-	-	-	-	-	56	-	-
China	-	-	-	-	-	90	-	265	229	346
Honduras	-	-	-	13	10	15	5	20	-	-
Korea Rep	-	-	9	66	36	105	62	23	10	3
Panama	-	-	-	-	-	-	-	7	-	-
Philippines	-	-	-	-	-	-	-	7	-	-
Portugal	-	-	-	72	0	7	19	43	31	42
Russian Fed	514	-	-	-	-	-	-	-	-	-
47 Fishing area total	*514*	*...*	*9*	*151*	*46*	*217*	*86*	*421*	*270*	*391*
51 China	-	-	-	-	-	-	-	-	179	125
Comoros	434	434	513	513	510	490	490	450	520 F	450
Djibouti	12	13	14	15	15	14	15	15 F	15 F	15 F
Eritrea	...	...	7	1	30	44	111	64	42	0
India	8 201	7 588	3 130	4 951	21	60	4 931	-	-	-
Iran	-	-	-	-	4	5	9	-	-	9 152
Italy	-	-	-	-	-	-	-	-	-	8
Japan	58	650	-	-	-	-	-	-	-	-
Korea Rep	464	796	584	577	1 036	1 545	705	182	529	294
Maldives	336	627	388	438	625	489	470	426	451	647
Mauritius	129	344	141	189	70	199	44	681	726	745
Mozambique	7 338	7 000	3 914	3 347	2 461	3 602	7 140	2 635	5 081	3 096
Oman	178	477	588	240	184	366	124	99	521	188
Pakistan	500	635	-	-	1 990	1 500	1 592	4 610	4 240	2 877
Portugal	-	-	-	-	-	-	7	8	11	61
Qatar	0	0	0	0	0	-	0	0	0	0
Réunion	-	-	-	-	38	-	-	-	3	79
Russian Fed	17 165	10 893	19 379	-	-	-	-	-	-	-
Saudi Arabia	343	713	923	690	804	861	773	797	783	854
Seychelles	19	10	29	7	1	6	12	28	1 252	10
South Africa	36	44	3	11	-	8	72	25	12	7
Spain	-	41	138	1	2	27	60	130	28	-
Tanzania	800	538	1 002	670	757	850	650	500	250	250
Ukraine	2	-	-	-	-	-	-	-	-	-
Uruguay	-	-	-	-	-	-	-	-	-	14
Yemen	294	316	316	300	300	310	300	300 F	300 F	300 F
Other nei	151	1 452	45	36	1	-	-	-	-	-
51 Fishing area total	*36 460*	*32 571*	*31 114*	*11 986*	*8 849*	*10 376*	*17 505*	*10 950 F*	*14 943 F*	*19 172 F*
57 Australia	36	19	40	24	16	9	116	34	8	12
China	-	-	-	96	299	307	396	712	309	168
India	618	316	147	247	1	2	1 225	-	1	...
Indonesia	23 045	20 218	18 927	27 002	26 519	25 843	72 950	31 845	49 807	50 840
Japan	-	-	-	-	-	-	24	-	-	236
Korea Rep	58	-	-	-	46	14	19	62	-	29
Seychelles	-	-	-	-	-	-	-	-	-	99
Sri Lanka	0	14	2	2	0	0	0	11	17	90
Timor-Leste	...	...	...	...	...	...	...	1	3	...
Other nei	9	15	6	5	0	-	-	-	-	-
57 Fishing area total	*23 766*	*20 582*	*19 122*	*27 376*	*26 881*	*26 175*	*74 730*	*32 665*	*50 145*	*51 474*
61 China	...	...	...	23 477	15 556	14 203	13 691	19 397	29 820	2 002
China,H.Kong	15	0	0	18	18	1	4	0	0	0
Japan	6 830	5 900	10 352	15 206	8 234	13 798	11 942	9 595	10 790	8 096
Korea Rep	-	459	848	1 816	2 792	3 115	3 393	2 893	3 245	3 294
Russian Fed	-	-	-	-	-	-	6	11	19	18
61 Fishing area total	*6 845*	*6 359*	*11 200*	*40 517*	*26 600*	*31 117*	*29 036*	*31 896*	*43 874*	*13 410*
67 Japan	3	-	-	-	-	-	-	-	-	-
67 Fishing area total	*3*	*...*	*...*	*...*	*...*	*...*	*...*	*...*	*...*	*...*
71 Australia	0	0	0	0	0	-	-	-	-	-
Guam	0	0	1	1	1	2	1	1	0	2
Indonesia	67 406	56 432	70 403	74 686	89 031	90 371	95 172	104 629	113 434	115 790
Japan	3 715	3 498	2 929	3 185	370	574	165	328	36	621
Korea Rep	221	263	56	27	99	186	135	593	517	256
NewCaledonia	70	101	300	255	244	169	285	236	294	310
N Marianas	...	...	3	4	7	7	8	8	7	3
Palau	-	26	23	21	27	7	18	30 F	51 F	51 F
Philippines	4 256	4 627	3 641	4 202	4 002	5 554	4 789	3 902	3 621	3 809
Russian Fed	2 150	18 937	5 840	4 981	-	-	-	-	-	-
Singapore	0	0	0	0	0	-	-	-	-	-
Tuvalu	1	219	21	15	15 F	20 F	20 F	20 F	20 F	20 F
Viet Nam	...	...	...	...	...	3 200 F	7 400 F	7 000 F	6 500 F	15 800 F
71 Fishing area total	*77 819*	*84 103*	*83 217*	*87 377*	*93 796 F*	*100 090 F*	*107 993 F*	*116 747 F*	*124 480 F*	*136 662 F*
77 Amer Samoa	3	3	1	1	3	0	0	0	0	1

B-36 Tunas, bonitos, billfishes / Thons, pélamides, marlins / Atunes, bonitos, agujas

Capture production by species, fishing areas and countries or areas / Captures par espèces, zones de pêche et pays ou zones / Capturas por especies, áreas de pesca y países o áreas

Species, Fishing area Espèce, Zone de pêche Especie, Area de pesca	1992 mt	1993 mt	1994 mt	1995 mt	1996 mt	1997 mt	1998 mt	1999 mt	2000 mt	2001 mt
Belize	-	-	30	-	-	-	-	-	-	-
Costa Rica	639	292	613	275	2 572	990	1 213	1 057	1 110	1 160
Cyprus	-	10	-	-	-	-	-	-	-	-
Ecuador	0	-	-	-	-	-	-	-	-	-
Fr Polynesia	...	...	...	...	...	...	...	2	0	0
Japan	88	137	108	60	877	12	-	-	-	-
Korea Rep	2 368	1 854	1 573	2 769	2 740	2 972	4 159	2 574	2 484	3 408
Panama	-	10	-	-	20	-	-	-	-	-
Samoa	60	-	-	-	-	-	-	-	479	470 F
Spain	10	10	-	-	-	-	1 040	20	-	10
Tonga	25	28	18	8	1	8	14	34	42	85
USA	2	1	-	-	-	-	-	-	-	-
Vanuatu	30	-	-	-	-	-	-	-	-	750
Other nei	-	-	-	-	-	-	-	-	-	20
77 Fishing area total	*3 225*	*2 345*	*2 343*	*3 113*	*6 213*	*3 982*	*6 426*	*3 687*	*4 115*	*5 904 F*
81 Australia	...	...	...	...	289	281	328	-	-	-
Japan	-	-	-	-	-	-	-	1	-	-
Korea Rep	-	4	-	-	-	-	-	78	167	228
New Zealand	-	-	-	41	36	83	66	101	69	88
Russian Fed	5	-	1	13	-	-	-	-	-	-
Ukraine	11	-	-	-	-	-	-	-	-	-
81 Fishing area total	*16*	*4*	*1*	*54*	*325*	*364*	*394*	*180*	*236*	*316*
87 Colombia	21 130	1 336	3	2	2 921	8 111	26 149	1 501	220	90
Ecuador	990	220	380	-	-	-	-	-	-	23 255
Korea Rep	3	145	71	49	-	-	-	64	-	-
Panama	-	-	-	-	-	-	-	-	30	-
Russian Fed	27	-	-	-	-	-	-	-	-	-
Spain	-	-	-	-	1	-	1 790	-	-	-
Vanuatu	-	10	-	-	-	40	-	-	20	1 050
Venezuela	-	-	-	-	-	-	10	-	-	-
Other nei	-	-	10	-	-	-	-	-	-	10
87 Fishing area total	*22 150*	*1 711*	*464*	*51*	*2 922*	*8 151*	*27 949*	*1 565*	*270*	*24 405*
Species total	*193 321 F*	*151 981 F*	*150 871 F*	*174 881 F*	*186 757 F*	*190 676 F*	*298 142 F*	*206 709 F*	*251 872 F*	*267 483 F*
Group total	***4 543 929***	***4 621 000***	***4 745 195***	***4 885 121***	***4 847 666***	***5 190 889***	***5 808 852***	***5 975 503***	***5 865 522***	***5 821 240***

B-37 Miscellaneous pelagic fishes / Poissons pélagiques divers / Peces pelágicos diversos

Capture production by species, fishing areas and countries or areas / Captures par espèces, zones de pêche et pays ou zones / Capturas por especies, áreas de pesca y países o áreas

Species, Fishing area Espèce, Zone de pêche Especie, Area de pesca	1992 mt	1993 mt	1994 mt	1995 mt	1996 mt	1997 mt	1998 mt	1999 mt	2000 mt	2001 mt
Capelin	**Capelan**		**Capelán**		***Mallotus villosus***			**1,23(04)002,01**		**CAP**
21 Canada	30 995	48 647	2 249	293	32 628	21 945	38 249	23 495	21 352	19 747
Cuba	71	3	0	-	-	-	-	-	-	-
Greenland	118	110	158	68	83	41	21	34	22	3
St Pier Mq	3	1	2	1	0	1	-	2	0	1
21 Fishing area total	*31 187*	*48 761*	*2 409*	*362*	*32 711*	*21 987*	*38 270*	*23 531*	*21 374*	*19 751*
27 Denmark	2 115	0	294	-	60 898	48 524	40 349	3 837	20 807	17 588
Estonia	31	0	-	-	-	-	-	-	-	-
Faeroe Is	41 724	40 980	12 310	3 306	39 777	43 466	41 966	24 275	59 855	32 110
France	-	-	-	-	-	-	-	-	1	1
Germany	-	-	-	-	-	-	5 001	-	-	-
Greenland	6	11 064	1 953	1 797	7 099	12 121	16 914	24 261	24 623	18 638
Iceland	797 743	940 947	753 466	715 551	1 179 051	1 319 191	750 065	703 694	892 405	918 417
Ireland	-	-	-	-	-	0	1	-	-	-
Norway	810 457	530 401	113 393	27 740	207 706	157 889	88 226	91 813	374 580	482 835
Russian Fed	424 781	169 858	522	-	-	-	-	32 485	94 693	180 098
Ukraine	6 273	-	-	-	-	-	-	-	-	-
UK	-	-	-	-	-	-	1 115	79	-	-
27 Fishing area total	*2 083 130*	*1 693 250*	*881 938*	*748 394*	*1 494 531*	*1 581 191*	*943 637*	*880 444*	*1 466 964*	*1 649 687*
61 Russian Fed	823	675	57	44	180	160	405	70	291	1 468
61 Fishing area total	*823*	*675*	*57*	*44*	*180*	*160*	*405*	*70*	*291*	*1 468*
Species total	*2 115 140*	*1 742 686*	*884 404*	*748 800*	*1 527 422*	*1 603 338*	*982 312*	*904 045*	*1 488 629*	*1 670 906*
Garfish	**Orphie**		**Aguja**		***Belone belone***			**1,47(01)001,01**		**GAR**
27 Channel Is	0	0	0	0	0	1	2	0	2	2
Denmark	641	755	641	622	708	994	648	571	714	558
Estonia	15	40	119	193	405	400	167	122	135	111
France	110	71	53	38	59	70	43	55	59	53
Germany	103	100	140	152	168	130	58	125	82	73
Latvia	-	-	-	-	-	-	-	-	-	11
Netherlands	-	-	-	-	-	-	-	-	-	2
Norway	1	1	2	3	1	2	1	1	0	0
Portugal	1	52	95	72	41	35	43	54	55	57
Spain	179	72	176	295	68	148	651	582	165	152
Sweden	44	47	35	33	14	12	27	30	7	9
UK	1	0	0	0	0	0	1	4	3	2
27 Fishing area total	*1 095*	*1 138*	*1 261*	*1 408*	*1 464*	*1 792*	*1 641*	*1 544*	*1 222*	*1 030*
34 Russian Fed	-	-	-	-	-	-	-	13	82	-
34 Fishing area total	-	-	-	-	-	-	-	*13*	*82*	-
37 Bulgaria	-	-	-	-	2	2	4	4	9	16
Croatia	84	144	88	81	63	85	60	10	5	11
France	-	-	-	-	-	-	-	-	1	2
Greece	225	371	491	323	331	253	188	83	184	126
Italy	354	364	304	554	243	216	238	209	134	139
Romania	-	-	2	-	-	-	-	-	-	-
Russian Fed	-	-	-	-	-	-	-	2	1	1
Slovenia	-	-	-	-	-	-	-	-	1	1
Spain	-	-	-	-	387	361	347	18	105	228
Tunisia	308	209	114	48	141	273	272	382	98	307
Turkey	2 384	1 150	1 564	581	395	470	450	500	300	640
Ukraine	1	-	1	1	0	0	1	1	0	-
Yugoslavia	1	0	0	1	1	1	2	5	5	6
37 Fishing area total	*3 357*	*2 238*	*2 564*	*1 589*	*1 563*	*1 661*	*1 562*	*1 214*	*843*	*1 477*
Species total	*4 452*	*3 376*	*3 825*	*2 997*	*3 027*	*3 453*	*3 203*	*2 771*	*2 147*	*2 507*
Needlefishes nei	**Aiguilles nca**		**Maraos nep**		***Tylosurus spp***			**1,47(01)013,XX**		**NED**
51 Egypt	...	...	...	123	48	32	17	28	11	28
Oman	...	...	779	508	201	252	116	209	131	130
Qatar	61	54	31	19	14	25	12	20	1	2
Saudi Arabia	-	-	461	324	96	111	131	140	107	126
51 Fishing area total	*61*	*54*	*1 271*	*974*	*359*	*420*	*276*	*397*	*250*	*286*
57 Indonesia	5 661	6 589	5 723	6 081	6 584	8 335	8 335	7 848	9 154	9 660
57 Fishing area total	*5 661*	*6 589*	*5 723*	*6 081*	*6 584*	*8 335*	*8 335*	*7 848*	*9 154*	*9 660*
71 Indonesia	23 246	17 583	19 726	24 038	20 104	21 462	21 406	21 561	23 716	25 030
Philippines	10 926	11 730	11 386	10 171	9 454	9 967	9 532	10 323	10 084	10 327
71 Fishing area total	*34 172*	*29 313*	*31 112*	*34 209*	*29 558*	*31 429*	*30 938*	*31 884*	*33 800*	*35 357*
Species total	*39 894*	*35 956*	*38 106*	*41 264*	*36 501*	*40 184*	*39 549*	*40 129*	*43 204*	*45 303*
Needlefishes, etc. nei	**Aiguilles, orphies nca**		**Agujones, maraos nep**		***Belonidae***			**1,47(01)XXX,XX**		**BEN**
31 Dominican Rp	68	81	46	146	39	5	4	7	15	45
Grenada	1	2	1	1	0	2	3	0	0	0
St Kitts Nev	...	...	...	12	27	26	60	58	66	41
31 Fishing area total	*69*	*83*	*47*	*159*	*66*	*33*	*67*	*65*	*81*	*86*
34 Benin	59 F	55 F	50 F	45 F	40 F	35 F	30 F	27	20 F	38
Côte dIvoire	1	0	1	1 F	1 F	1 F	1 F	...	...	3

B-37 Miscellaneous pelagic fishes / Poissons pélagiques divers / Peces pelágicos diversos

Capture production by species, fishing areas and countries or areas / Captures par espèces, zones de pêche et pays ou zones / Capturas por especies, áreas de pesca y países o áreas

Species, Fishing area Espèce, Zone de pêche Especie, Area de pesca	1992 mt	1993 mt	1994 mt	1995 mt	1996 mt	1997 mt	1998 mt	1999 mt	2000 mt	2001 mt
Liberia	-	-	-	-	-	-	54	90	30	31
Lithuania	8 472	-	-	-	-	-	-	-	-	-
Nigeria	576	1 708	1 570	590	1 082	272	99	166	535	431
Portugal	-	-	-	1	1	-	19	-	-	-
Senegal	...	...	...	...	...	...	5	37	8	0
Togo	2	60	13	18	37	90	122	96	53	153
34 Fishing area total	*9 110 F*	*1 823 F*	*1 634 F*	*655 F*	*1 161 F*	*398 F*	*330 F*	*416*	*646 F*	*656*
81 Australia	2	2	2	2	1	0	0	-	-	-
81 Fishing area total	*2*	*2*	*2*	*2*	*1*	*0*	*0*	-	-	-
Species total	*9 181 F*	*1 908 F*	*1 683 F*	*816 F*	*1 228 F*	*431 F*	*397 F*	*481*	*727 F*	*742*
Atlantic saury	**Balaou atlantique**		**Paparda del Atlántico**		***Scomberesox saurus***			**1,47(02)002,01**		**SAU**
27 Portugal	9	0	1	0	0	0	54	0	1	1
Spain	44	1	619	526	483	639	907	574	184	992
27 Fishing area total	*53*	*1*	*620*	*526*	*483*	*639*	*961*	*574*	*185*	*993*
87 Chile	-	-	37	8	1	175	407	580	296	4 178
87 Fishing area total	-	-	*37*	*8*	*1*	*175*	*407*	*580*	*296*	*4 178*
Species total	*53*	*1*	*657*	*534*	*484*	*814*	*1 368*	*1 154*	*481*	*5 171*
Pacific saury	**Balaou du Japon**		**Paparda del Pacífico**		***Cololabis saira***			**1,47(02)007,01**		**SAP**
61 China,Taiwan	34 235	36 435	12 550	13 772	8 236	21 887	12 794	12 541	27 868	39 764
Japan	265 884	277 461	261 587	273 510	229 227	290 812	144 983	141 011	216 471	269 797
Korea Rep	31 400	38 435	34 810	37 865	28 368	68 853	18 531	28 784	43 280	25 782
Russian Fed	50 172	48 145	26 385	25 140	10 280	7 091	4 665	4 808	17 390	40 407
61 Fishing area total	*381 691*	*400 476*	*335 332*	*350 287*	*276 111*	*388 643*	*180 973*	*187 144*	*305 009*	*375 750*
67 Korea Rep	1 408	2 454	272	-	-	-	-	-	-	-
67 Fishing area total	*1 408*	*2 454*	*272*	-	-	-	-	-	-	-
77 Korea Rep	1 345	-	-	-	-	-	-	754	1 060	423
77 Fishing area total	*1 345*	-	-	-	-	-	-	*754*	*1 060*	*423*
Species total	*384 444*	*402 930*	*335 604*	*350 287*	*276 111*	*388 643*	*180 973*	*187 898*	*306 069*	*376 173*
Japanese halfbeak	**Demi-bec du Japon**		**Agujeta del Japón**		***Hyporhamphus sajori***			**1,47(03)003,11**		**HAJ**
61 Korea Rep	1 629	2 005	1 480	1 531	990	1 193	1 160	913	956	613
61 Fishing area total	*1 629*	*2 005*	*1 480*	*1 531*	*990*	*1 193*	*1 160*	*913*	*956*	*613*
Species total	*1 629*	*2 005*	*1 480*	*1 531*	*990*	*1 193*	*1 160*	*913*	*956*	*613*
Ballyhoo halfbeak	**Demi-bec brésilien**		**Agujeta brasileña**		***Hemiramphus brasiliensis***			**1,47(03)004,06**		**BAL**
31 Puerto Rico	...	...	...	...	...	...	...	32	47	41
31 Fishing area total	...	...	...	...	...	...	...	*32*	*47*	*41*
41 Brazil	500 F	500 F	500 F	631	494	442	1 500	832	989	980 F
41 Fishing area total	*500 F*	*500 F*	*500 F*	*631*	*494*	*442*	*1 500*	*832*	*989*	*980 F*
Species total	*500 F*	*500 F*	*500 F*	*631*	*494*	*442*	*1 500*	*864*	*1 036*	*1 021 F*
Halfbeaks nei	**Demi-becs nca**		**Agujetas nep**		***Hemiramphus spp***			**1,47(03)004,XX**		**HAX**
21 USA	-	-	-	-	-	-	-	258	-	-
21 Fishing area total	-	-	-	-	-	-	-	*258*	-	-
31 Grenada	5	32	2	1	1	2	2	0	1	1
USA	398	523	500	528	398	293	441	92	414	354
31 Fishing area total	*403*	*555*	*502*	*529*	*399*	*295*	*443*	*92*	*415*	*355*
34 Liberia	-	-	-	-	-	-	77	85	96	97
Sao Tome Prn	211 F	262 F	381 F	401 F	293	126	150 F	160 F	150 F	150 F
Senegal	-	-	-	-	-	-	70	30	14	14
Togo	0	0	3	1	2	4	102	3	4	18
34 Fishing area total	*211 F*	*262 F*	*384 F*	*402 F*	*295*	*130*	*399 F*	*278 F*	*264 F*	*279 F*
51 India	913	582	747	3 828	1 057	1 172	1 312	1 928	1 650	1 474
Tanzania	1 800	1 213	1 066	1 640	1 483	1 850	1 175	1 200	1 250	1 300
Untd Arab Em	30	32	35	29	36	42	58	61	55	50 F
51 Fishing area total	*2 743*	*1 827*	*1 848*	*5 497*	*2 576*	*3 064*	*2 545*	*3 189*	*2 955*	*2 824 F*
57 Australia	239	158	163	174	138	565	580	0	0	-
India	5 128	2 876	2 870	4 846	5 061	4 402	3 360	3 461	5 993	4 696
57 Fishing area total	*5 367*	*3 034*	*3 033*	*5 020*	*5 199*	*4 967*	*3 940*	*3 461*	*5 993*	*4 696*
71 Australia	-	-	-	-	-	-	-	-	3	-
Fiji Islands	138	99	101 F	49	70	62	49	61	50 F	56
Philippines	3 045	2 719	2 843	3 555	3 231	1 655	1 949	2 281	2 309	2 503
71 Fishing area total	*3 183*	*2 818*	*2 944 F*	*3 604*	*3 301*	*1 717*	*1 998*	*2 342*	*2 362 F*	*2 559*
81 Australia	280	280	193	193	99	115	115	0	0	-
New Zealand	28	14	27	18	16	17	28	24	18	13
81 Fishing area total	*308*	*294*	*220*	*211*	*115*	*132*	*143*	*24*	*18*	*13*

B-37 Miscellaneous pelagic fishes / Poissons pélagiques divers / Peces pelágicos diversos

Capture production by species, fishing areas and countries or areas / Captures par espèces, zones de pêche et pays ou zones / Capturas por especies, áreas de pesca y países o áreas

Species, Fishing area Espèce, Zone de pêche Especie, Area de pesca	1992 mt	1993 mt	1994 mt	1995 mt	1996 mt	1997 mt	1998 mt	1999 mt	2000 mt	2001 mt
Species total	*12 215 F*	*8 790 F*	*8 931 F*	*15 263 F*	*11 885*	*10 305*	*9 468 F*	*9 644 F*	*12 007 F*	*10 726 F*
Japanese flyingfish	**Poisson-volant du Japon**		**Volador japonés**		***Cypselurus agoo***				**1,47(04)010,02**	**JFL**
61 Japan	7 192	8 039	8 606	7 881	8 501	7 486	8 933	6 738	9 615	8 286
61 Fishing area total	*7 192*	*8 039*	*8 606*	*7 881*	*8 501*	*7 486*	*8 933*	*6 738*	*9 615*	*8 286*
Species total	*7 192*	*8 039*	*8 606*	*7 881*	*8 501*	*7 486*	*8 933*	*6 738*	*9 615*	*8 286*
Flyingfishes nei	**Exocets nca**		**Voladores nep**		***Exocoetidae***				**1,47(04)XXX,XX**	**FLY**
31 Barbados	1 461	1 987	1 640	1 766	2 042	1 566	2 680	2 075	1 916	1 673
Grenada	74	27	15	6	12	0	5	1	14	10
Martinique	0	0	0	0	0	0	0	0	0	0
St Kitts Nev	...	...	...	21	54	23	38	22	24	53
St Lucia	...	...	48	50	40	34	112	67	99	323
31 Fishing area total	*1 535*	*2 014*	*1 703*	*1 843*	*2 148*	*1 623*	*2 835*	*2 165*	*2 053*	*2 059*
34 Benin	28 F	30 F	33 F	32 F	28 F	25 F	20 F	15	10 F	96
Liberia	...	...	...	...	69	8	10	-	38	19
Nigeria	37	7	7	7	17	...	87	69	8	55
Sao Tome Prn	184 F	229 F	333 F	350 F	256	939	1 000 F	1 100 F	1 000 F	1 000 F
Togo	60	94	115	109	204	287	610	211	130	186
34 Fishing area total	*309 F*	*360 F*	*488 F*	*498 F*	*574 F*	*1 259 F*	*1 727 F*	*1 395 F*	*1 186 F*	*1 356 F*
41 Brazil	650 F	650 F	650 F	1 036	743	1 082	1 084	760	388	380 F
41 Fishing area total	*650 F*	*650 F*	*650 F*	*1 036*	*743*	*1 082*	*1 084*	*760*	*388*	*380 F*
51 Bahrain	20	39	91	85	92	85	145	299	52	112
India	104	42	64	82	41	99	57	102	114	47
51 Fishing area total	*124*	*81*	*155*	*167*	*133*	*184*	*202*	*401*	*166*	*159*
57 India	1 149	1 895	2 772	2 263	1 845	1 995	555	634	2 597	2 399
Indonesia	2 368	2 562	2 417	2 992	4 613	5 205	7 562	5 585	4 627	4 870
57 Fishing area total	*3 517*	*4 457*	*5 189*	*5 255*	*6 458*	*7 200*	*8 117*	*6 219*	*7 224*	*7 269*
61 China,Taiwan	492	657	699	572	662	2 497	1 077	618	287	366
61 Fishing area total	*492*	*657*	*699*	*572*	*662*	*2 497*	*1 077*	*618*	*287*	*366*
71 Indonesia	8 979	9 155	8 373	10 188	11 921	10 921	11 399	12 237	15 353	16 170
Japan	2	2	-	-	-	-	-	-	-	-
Kiribati	1 820	1 760	1 740	1 760	1 780	1 770	2 525	2 547	836	1 594
Philippines	54 112	13 799	14 019	16 826	17 300	27 801	36 134	38 280	36 050	37 498
71 Fishing area total	*64 913*	*24 716*	*24 132*	*28 774*	*31 001*	*40 492*	*50 058*	*53 064*	*52 239*	*55 262*
77 Cook Is	39 F	40 F	38 F	38 F	30 F	30 F	30 F	30 F	30 F	30 F
Fr Polynesia	54	25	26	...	...	55	92	...	88	82
77 Fishing area total	*93 F*	*65 F*	*64 F*	*38 F*	*30 F*	*85 F*	*122 F*	*30 F*	*118 F*	*112 F*
81 New Zealand	-	-	-	4	4	2	3	1	2	1
81 Fishing area total	-	-	-	*4*	*4*	*2*	*3*	*1*	*2*	*1*
87 Peru	13 594	9 210	926	35 619	639	1 056	4 506	298 373	41 059	5 035
87 Fishing area total	*13 594*	*9 210*	*926*	*35 619*	*639*	*1 056*	*4 506*	*298 373*	*41 059*	*5 035*
Species total	*85 227 F*	*42 210 F*	*34 006 F*	*73 806 F*	*42 392 F*	*55 480 F*	*69 731 F*	*363 026 F*	*104 722 F*	*71 999 F*
Opah	**Opah**		**Opa**		***Lampris guttatus***				**1,52(01)001,02**	**LAG**
21 USA	...	...	...	...	...	...	1	1	2	1
21 Fishing area total	...	...	...	...	...	...	*1*	*1*	*2*	*1*
27 Faeroe Is	-	-	-	-	-	-	-	1	...	...
27 Fishing area total	-	-	-	-	-	-	-	*1*	...	...
31 USA	...	...	...	...	...	...	-	-	-	1
31 Fishing area total	...	...	...	...	...	...	-	-	-	*1*
77 Amer Samoa	-	-	-	-	-	-	-	1	1	1
Fr Polynesia	...	...	77	93	96	...	...	137	124	148
USA	...	...	...	...	...	...	114	67	56	47
77 Fishing area total	...	...	*77*	*93*	*96*	...	*114*	*205*	*181*	*196*
81 New Zealand	-	-	-	130	76	126	254	335	283	340
81 Fishing area total	-	-	-	*130*	*76*	*126*	*254*	*335*	*283*	*340*
Species total	...	...	*77*	*223*	*172*	*126*	*369*	*542*	*466*	*538*
Dealfishes	**...B**		**...C**		***Trachipterus spp***				**1,52(04)002,XX**	**TRP**
27 Portugal	...	...	...	...	...	...	...	29	20	25
27 Fishing area total	...	...	...	...	...	...	...	*29*	*20*	*25*
81 New Zealand	-	-	-	127	49	60	74	65	67	103
81 Fishing area total	-	-	-	*127*	*49*	*60*	*74*	*65*	*67*	*103*
Species total	...	...	...	*127*	*49*	*60*	*74*	*94*	*87*	*128*

B-37 Miscellaneous pelagic fishes / Poissons pélagiques divers / Peces pelágicos diversos

Capture production by species, fishing areas and countries or areas / Captures par espèces, zones de pêche et pays ou zones / Capturas por especies, áreas de pesca y países o áreas

Species, Fishing area Espèce, Zone de pêche Especie, Area de pesca	1992 mt	1993 mt	1994 mt	1995 mt	1996 mt	1997 mt	1998 mt	1999 mt	2000 mt	2001 mt
King of herrings	**Roi des harengs**		**Rey de los arenques**		*Regalecus glesne*				**1,52(05)001,01**	**REL**
81 New Zealand	-	-	-	5	10	64	60	34	20	1
81 Fishing area total	-	-	-	*5*	*10*	*64*	*60*	*34*	*20*	*1*
Species total	-	-	-	*5*	*10*	*64*	*60*	*34*	*20*	*1*
Big-scale sand smelt	**Joël**		**Pejerrey mediterráneo**		*Atherina boyeri*				**1,63(02)003,01**	**ATB**
05 Greece	-	-	-	-	8	110	350	350	350	326
Russian Fed	-	-	-	-	-	-	-	11	-	-
05 Fishing area total	-	-	-	-	*8*	*110*	*350*	*361*	*350*	*326*
37 Bulgaria	-	-	-	-	-	-	0	1	21	2
Russian Fed	10	9	5	15	31	13	25	19	63	95
37 Fishing area total	*10*	*9*	*5*	*15*	*31*	*13*	*25*	*20*	*84*	*97*
Species total	*10*	*9*	*5*	*15*	*39*	*123*	*375*	*381*	*434*	*423*
Atlantic silverside	**Capucette**		**Pejerrey del Atlántico**		*Menidia menidia*				**1,63(02)013,02**	**SSA**
21 Canada	20	83	60	223	151	238	231	558	304	646
USA	18	39	32	13	19	21	24	25	15	15
21 Fishing area total	*38*	*122*	*92*	*236*	*170*	*259*	*255*	*583*	*319*	*661*
Species total	*38*	*122*	*92*	*236*	*170*	*259*	*255*	*583*	*319*	*661*
Silversides(=Sand smelts) nei	**Athérinidés nca**		**Pejerreyes nep**		*Atherinidae*				**1,63(02)XXX,XX**	**SIL**
01 Egypt	-	927	1 375	1 241	1 326	1 229	1 150	1 180	3 182	5 154
01 Fishing area total	-	*927*	*1 375*	*1 241*	*1 326*	*1 229*	*1 150*	*1 180*	*3 182*	*5 154*
02 Mexico	...	...	...	...	...	...	372	487	392	558
02 Fishing area total	...	...	...	...	...	...	*372*	*487*	*392*	*558*
03 Bolivia	1 624	635	785	850	850	850	850	850	850	760
03 Fishing area total	*1 624*	*635*	*785*	*850*	*850*	*850*	*850*	*850*	*850*	*760*
04 Turkey	987	953	899	909	562	1 600	1 500	1 455	1 583	1 685
04 Fishing area total	*987*	*953*	*899*	*909*	*562*	*1 600*	*1 500*	*1 455*	*1 583*	*1 685*
27 France	98	98	78	76	68	72	65	78	54	52
Portugal	17	11	10	1 002	3	6	22	6	3	0
Spain	-	-	-	-	262	281	212	213	192	192
UK	-	-	-	-	-	-	-	-	-	1
27 Fishing area total	*115*	*109*	*88*	*1 078*	*333*	*359*	*299*	*297*	*249*	*245*
37 Albania	20	20 F	15 F	8	20	8	11	15	20	10
Croatia	176	224	176	175	148	171	260	50	44	20
Egypt	1 993	2 027	3 740	2 596	5 316	4 066	4 558	3 017	1 732	3 490
France	309	154	80	3	31	55	55	57	29	44
Italy	1 818	1 762	1 595	1 530	1 112	1 101	883	851	725	736
Lebanon	-	-	-	25	25	50	50	50	50	50
Romania	15	-	4	3	3	10	73	33	42	29
Slovenia	-	-	-	-	-	-	-	-	1	2
Spain	-	-	-	-	35	32	30	31	62	20
Tunisia	...	...	254	594	141	116	338	6	57	309
Turkey	3 545	7 682	3 834	1 171	412	640	800	1 300	500	575
Ukraine	1 037	937	439	195	326	396	632	388	653	332
Yugoslavia	23	6	3	6	6	7	9	9	13	14
37 Fishing area total	*8 936*	*12 812 F*	*10 140 F*	*6 306*	*7 575*	*6 652*	*7 699*	*5 807*	*3 928*	*5 631*
41 Argentina	823	571	489	411	583	520	48	618	14	...
Brazil	170 F	170 F	170 F	40	52	62	52	17	39	40 F
Falkland Is	1	1	2	2	2	2	2	2	4	1
41 Fishing area total	*994 F*	*742 F*	*661 F*	*453*	*637*	*584*	*102*	*637*	*57*	*41 F*
51 Egypt	-	-	-	-	-	-	-	78	-	-
51 Fishing area total	-	-	-	-	-	-	-	*78*	-	-
71 Fiji Islands	54	31	32 F	30	31	25	80	91	80 F	86
Philippines	1 990	2 867	855	864	866	669	596	618	543	544
71 Fishing area total	*2 044*	*2 898*	*887 F*	*894*	*897*	*694*	*676*	*709*	*623 F*	*630*
77 Mexico	...	...	...	...	...	...	1 062	906	793	704
77 Fishing area total	...	...	...	...	...	...	*1 062*	*906*	*793*	*704*
87 Chile	750	341	701	558	690	494	560	3 414	1 357	833
Peru	2 033	1 395	2 207	2 357	3 802	5 184	45	6 692	11 215	7 528
87 Fishing area total	*2 783*	*1 736*	*2 908*	*2 915*	*4 492*	*5 678*	*605*	*10 106*	*12 572*	*8 361*
Species total	*17 483 F*	*20 812 F*	*17 743 F*	*14 646*	*16 672*	*17 646*	*14 315*	*22 512*	*24 229 F*	*23 769 F*
False trevally	**Péliau chanos**		**Pagapa**		*Lactarius lactarius*				**1,70(19)165,02**	**TRF**
51 India	4 238	3 169	3 934	4 517	6 401	7 056	7 995	5 503	4 623	5 992
Pakistan	2	3	2	3	2	4	5	-	-	-
51 Fishing area total	*4 240*	*3 172*	*3 936*	*4 520*	*6 403*	*7 060*	*8 000*	*5 503*	*4 623*	*5 992*

B-37 Miscellaneous pelagic fishes / Poissons pélagiques divers / Peces pelágicos diversos

Capture production by species, fishing areas and countries or areas / Captures par espèces, zones de pêche et pays ou zones / Capturas por especies, áreas de pesca y países o áreas

Species, Fishing area Espèce, Zone de pêche Especie, Area de pesca	1992 mt	1993 mt	1994 mt	1995 mt	1996 mt	1997 mt	1998 mt	1999 mt	2000 mt	2001 mt
57 India	2 492	1 778	2 376	2 570	529	886	1 533	1 324	1 438	1 291
Malaysia	207	1 202	112	25	3	211	212	92	18	2
57 Fishing area total	*2 699*	*2 980*	*2 488*	*2 595*	*532*	*1 097*	*1 745*	*1 416*	*1 456*	*1 293*
71 Malaysia	921	580	531	393	384	350	363	589	404	437
Philippines	71	54	28	25	26	202	213	235	253	272
71 Fishing area total	*992*	*634*	*559*	*418*	*410*	*552*	*576*	*824*	*657*	*709*
Species total	*7 931*	*6 786*	*6 983*	*7 533*	*7 345*	*8 709*	*10 321*	*7 743*	*6 736*	*7 994*
Bluefish	**Tassergal**		**Anjova**		***Pomatomus saltatrix***			**1,70(20)213,01**		**BLU**
21 USA	4 289	3 892	3 081	3 367	4 137	3 900	3 446	3 145	3 499	3 825
21 Fishing area total	*4 289*	*3 892*	*3 081*	*3 367*	*4 137*	*3 900*	*3 446*	*3 145*	*3 499*	*3 825*
27 Portugal	199	123	72	57	44	40	25	21	20	15
Spain	386	601	651	674	1 833	1 874	979	787	39	38
27 Fishing area total	*585*	*724*	*723*	*731*	*1 877*	*1 914*	*1 004*	*808*	*59*	*53*
31 USA	970	905	1 376	434	107	322	318	214	162	168
Venezuela	975	987	1 130	1 024	651	825	581	542	950	1 158
31 Fishing area total	*1 945*	*1 892*	*2 506*	*1 458*	*758*	*1 147*	*899*	*756*	*1 112*	*1 326*
34 Benin	1 176 F	1 150 F	1 100 F	1 050 F	1 000 F	950 F	900 F	875	600 F	697
Gambia	...	...	20	23	31	8	75	80	35	70
Georgia	23 F	10 F	-	-	-	-	-	-	-	-
Greece	0	0	0	-	-	-	-	-	-	-
GuineaBissau	1 F	2 F	2 F	2 F	3 F	3 F	3 F	3 F	3 F	3 F
Latvia	49	17	-	7	155	116	31	116	144	17
Lithuania	-	41	32	-	-	-	-	-	-	-
Mauritania	...	...	...	...	...	...	...	...	1 F	...
Morocco	19	3	9	11	10	54	161	47	56	193
Portugal	2	21	14	13	18	8	12	1	0	0
Romania	84	13	-	-	-	-	-	-	-	-
Russian Fed	159	150	5	74	96	406	283	536	226	157
Senegal	1 099	2 142	1 614	461	691	2 075	3 118	327	277	1 149
Spain	-	-	-	-	1 801	-	-	-	-	-
Ukraine	57	-	-	-	-	209	13	238	97	29
34 Fishing area total	*2 669 F*	*3 549 F*	*2 796 F*	*1 641 F*	*3 805 F*	*3 829 F*	*4 596 F*	*2 223 F*	*1 439 F*	*2 315 F*
37 Bulgaria	4	8	8 F	12	10	12	10	8	18	2
Egypt	1	0	37	174	307	147	56	153	326	468
Gaza Strip	...	...	...	...	36	8	31	30 F	30 F	30 F
Greece	439	457	377	346	128	111	144	259	265	511
Morocco	14	17	4	1	5	20	2	-	45	5
Romania	-	-	2	-	-	-	12	3	4	10
Spain	-	-	-	-	251	240	190	171	222	168
Tunisia	3 153	2 049	1 270	653	1 317	1 426	787	748	1 096	1 037
Turkey	9 697	16 442	8 078	5 456	4 117	3 050	3 350	2 995	4 250	13 060
Ukraine	-	-	1	0	-	-	-	-	-	-
37 Fishing area total	*13 308*	*18 973*	*9 777 F*	*6 642*	*6 171*	*5 014*	*4 582*	*4 367 F*	*6 256 F*	*15 291 F*
41 Argentina	1 105	1 331	202	565	342	416	15	286	416	...
Brazil	5 520 F	5 510 F	5 520 F	7 588	5 866	2 616	2 504	2 064	3 314	3 300 F
Uruguay	9	13	63	56	11	11	84	18	48	93
41 Fishing area total	*6 634 F*	*6 854 F*	*5 785 F*	*8 209*	*6 219*	*3 043*	*2 603*	*2 368*	*3 778*	*3 393 F*
47 Angola	-	-	-	-	-	-	-	-	-	167
South Africa	42	68	34	57	38	15	4	33	20 F	3
47 Fishing area total	*42*	*68*	*34*	*57*	*38*	*15*	*4*	*33*	*20 F*	*170*
57 Australia	48	45	49	54	56	-	-	-	-	-
57 Fishing area total	*48*	*45*	*49*	*54*	*56*	-	-	-	-	-
71 Australia	121	156	98	181	103	179	128	204	155	259
71 Fishing area total	*121*	*156*	*98*	*181*	*103*	*179*	*128*	*204*	*155*	*259*
81 Australia	100	98	96	96	101	64	64	46	...	...
81 Fishing area total	*100*	*98*	*96*	*96*	*101*	*64*	*64*	*46*	...	...
Species total	*29 741 F*	*36 251 F*	*24 945 F*	*22 436 F*	*23 265 F*	*19 105 F*	*17 326 F*	*13 950 F*	*16 318 F*	*26 632 F*
Cobia	**Mafou**		**Cobia**		***Rachycentron canadum***			**1,70(22)221,01**		**CBA**
21 USA	6	4	3	18	142	6	13	72	16	13
21 Fishing area total	*6*	*4*	*3*	*18*	*142*	*6*	*13*	*72*	*16*	*13*
31 Mexico	475	363	215	347	347	630	588	565	303	1 405
USA	157	177	182	152	45	127	129	65	96	84
31 Fishing area total	*632*	*540*	*397*	*499*	*392*	*757*	*717*	*630*	*399*	*1 489*
34 Gambia	-	-	-	33	0	6	0	0	-	-
34 Fishing area total	-	-	-	*33*	*0*	*6*	*0*	*0*	-	-
41 Brazil	...	...	...	256	498	367	622	1 818	1 580	1 550 F
41 Fishing area total	...	...	...	*256*	*498*	*367*	*622*	*1 818*	*1 580*	*1 550 F*
51 Bahrain	22	30	17	32	38	19	6	9	9	20
Eritrea	...	...	2	10	38	2	6	8	31	31

B-37 Miscellaneous pelagic fishes / Poissons pélagiques divers / Peces pelágicos diversos

Capture production by species, fishing areas and countries or areas / Captures par espèces, zones de pêche et pays ou zones / Capturas por especies, áreas de pesca y países o áreas

Species, Fishing area Espèce, Zone de pêche Especie, Area de pesca	1992 mt	1993 mt	1994 mt	1995 mt	1996 mt	1997 mt	1998 mt	1999 mt	2000 mt	2001 mt
Kuwait	...	...	...	...	...	...	...	...	...	38
Oman	112	202	104	234	103	180	115	124	100	54
Pakistan	1 281	1 459	1 541	2 306	1 574	1 449	1 254	1 136	2 896	2 797
Qatar	55	62	54	47	56	52	56	44	56	95
Saudi Arabia	52	48	71	124	155	155	130	137	138	167
Untd Arab Em	39	50	47	50	52	56	57	58	632	630 F
51 Fishing area total	*1 561*	*1 851*	*1 836*	*2 803*	*2 016*	*1 913*	*1 624*	*1 516*	*3 862*	*3 832 F*
57 Malaysia	164	131	109	62	76	111	87	111	136	150
57 Fishing area total	*164*	*131*	*109*	*62*	*76*	*111*	*87*	*111*	*136*	*150*
61 China,Taiwan	23	843	978	779	692	987	815	655	1 014	486
61 Fishing area total	*23*	*843*	*978*	*779*	*692*	*987*	*815*	*655*	*1 014*	*486*
71 Malaysia	447	476	374	279	292	274	436	645	504	407
Philippines	1 076	1 714	1 344	1 003	964	1 162	1 235	1 187	1 046	1 070
71 Fishing area total	*1 523*	*2 190*	*1 718*	*1 282*	*1 256*	*1 436*	*1 671*	*1 832*	*1 550*	*1 477*
Species total	*3 909*	*5 559*	*5 041*	*5 732*	*5 072*	*5 583*	*5 549*	*6 634*	*8 557*	*8 997 F*
Atlantic horse mackerel	**Chinchard d'Europe**		**Jurel**			***Trachurus trachurus***			**1,70(23)004,01**	**HOM**
27 Belgium	34	74	58	51	28	19	19	21	19	20
Bulgaria	-	226	-	-	-	-	-	-	-	-
Channel Is	0	0	0	0	5	7	11	10	8	8
Denmark	49 663	49 426	53 616	56 167	63 929	63 430	32 597	32 047	25 083	23 631
Estonia	293	-	55	-	80	203	34	0	-	-
Faeroe Is	10 877	10 531	505	950	863	1 005	216	3 643	2 014	180
France	7 800	7 986	6 152	16 000	25 396	26 862	28 103	27 087	21 628	19 852
Germany	21 722	29 381	17 277	20 407	21 815	37 584	33 750	23 803	16 778	12 464
Iceland	0	0	-	0	-	-	-	-	-	-
Ireland	50 538	68 556	85 804	178 355	127 876	75 002	74 253	58 201	55 438	63 497
Latvia	457	2 803	-	6	-	-	-	-	-	-
Lithuania	-	-	-	232	7 400	-	-	421	5	344
Netherlands	122 553	146 423	101 845	113 828	135 965	122 683	103 248	84 891	65 994	84 011
Norway	107 555	128 338	94 648	96 132	15 556	46 491	13 366	46 657	2 084	7 988
Portugal	25 033	25 307	19 040	17 701	14 054	18 720	21 364	15 533	15 471	15 305
Romania	-	-	-	-	360	-	-	-	-	-
Russian Fed	839	1 197	996	1 709	804	554	345	121	86	16
Sweden	0	2	207	447	166	1 761	3 418	2 004	1 162	119
Ukraine	-	264	74	-	-	-	-	-	-	-
UK	16 385	16 456	32 143	48 203	49 927	51 909	32 832	21 025	17 100	19 581
27 Fishing area total	*413 749*	*486 970*	*412 420*	*550 188*	*464 224*	*446 230*	*343 556*	*315 464*	*222 870*	*247 016*
34 Germany	-	-	-	-	213	493	626	574	-	-
Portugal	621	207	14	2	11	19	40	2	-	-
Romania	2 123	-	-	-	-	-	-	-	-	-
34 Fishing area total	*2 744*	*207*	*14*	*2*	*224*	*512*	*666*	*576*	-	-
37 Greece	481	582	1 628	1 860	2 484	2 956	1 360	942	774	747
Israel	112	200 F	300 F	340	65	226	172	178	175	160 F
Slovenia	38	41	12	7	9	8	4	5	4	4
Syria	39 F	40 F	39 F	39 F	60	77	40	30	36	56
Turkey	8 913	27 321	20 019	7 431	7 559	5 100	4 500	4 000	7 200	10 635
37 Fishing area total	*9 583 F*	*28 184 F*	*21 998 F*	*9 677 F*	*10 177*	*8 367*	*6 076*	*5 155*	*8 189*	*11 602 F*
Species total	*426 076 F*	*515 361 F*	*434 432 F*	*559 867 F*	*474 625*	*455 109*	*350 298*	*321 195*	*231 059*	*258 618 F*
Japanese jack mackerel	**Chinchard du Japon**		**Jurel japonés**			***Trachurus japonicus***			**1,70(23)004,03**	**JJM**
61 China,Taiwan	90	193	358	4 841	4 218	4 711	7 121	2 661	6 003	3 903
Japan	223 412	311 949	326 130	312 994	330 406	323 142	311 311	211 077	245 988	214 434
Korea Rep	27 715	38 095	38 433	12 269	14 542	22 766	22 132	13 552	19 510	17 537
61 Fishing area total	*251 217*	*350 237*	*364 921*	*330 104*	*349 166*	*350 619*	*340 564*	*227 290*	*271 501*	*235 874*
71 Korea Rep	-	3	38	-	-	-	1	-	-	-
71 Fishing area total	-	*3*	*38*	-	-	-	*1*	-	-	-
Species total	*251 217*	*350 240*	*364 959*	*330 104*	*349 166*	*350 619*	*340 565*	*227 290*	*271 501*	*235 874*
Chilean jack mackerel	**Chinchard du Chili**		**Jurel chileno**			***Trachurus murphyi***			**1,70(23)004,05**	**CJM**
87 Chile	3 212 060	3 236 244	4 041 447	4 404 193	3 883 326	2 917 064	1 612 912	1 219 689	1 234 299	1 649 933
Cuba	3 196	-	-	-	-	-	-	-	-	-
Ecuador	22 818	9 946	23 723	174 393	56 781	30 302	25 900	19 072	7 144	134 011
Estonia	376	-	-	-	-	-	-	-	-	-
Ghana	-	-	-	-	-	-	-	-	2 472	1 157
Japan	-	-	-	-	-	-	-	7	-	-
Latvia	2 298	-	-	-	-	-	-	-	-	-
Lithuania	7 842	-	-	-	-	-	-	-	-	-
Peru	96 660	130 681	196 771	376 600	438 736	649 751	386 946	184 679	296 579	723 733
Russian Fed	31 357	-	-	-	-	-	-	-	-	-
87 Fishing area total	*3 376 607*	*3 376 871*	*4 261 941*	*4 955 186*	*4 378 843*	*3 597 117*	*2 025 758*	*1 423 447*	*1 540 494*	*2 508 834*
Species total	*3 376 607*	*3 376 871*	*4 261 941*	*4 955 186*	*4 378 843*	*3 597 117*	*2 025 758*	*1 423 447*	*1 540 494*	*2 508 834*
Pacific jack mackerel	**Chinchard gros yeux**		**Chicharro ojotón**			***Trachurus symmetricus***			**1,70(23)004,06**	**PJM**
67 Russian Fed	-	-	-	-	-	-	230	10	-	-

B-37 Miscellaneous pelagic fishes / Poissons pélagiques divers / Peces pelágicos diversos

Capture production by species, fishing areas and countries or areas / Captures par espèces, zones de pêche et pays ou zones / Capturas por especies, áreas de pesca y países o áreas

Species, Fishing area Espèce, Zone de pêche Especie, Area de pesca	1992 mt	1993 mt	1994 mt	1995 mt	1996 mt	1997 mt	1998 mt	1999 mt	2000 mt	2001 mt
USA	1	277	200	147	-	2	731	166	181	215
67 Fishing area total	*1*	*277*	*200*	*147*	*-*	*2*	*961*	*176*	*181*	*215*
77 USA	1 189	1 504	2 697	1 728	2 176	1 158	832	950	1 135	3 624
77 Fishing area total	*1 189*	*1 504*	*2 697*	*1 728*	*2 176*	*1 158*	*832*	*950*	*1 135*	*3 624*
Species total	*1 190*	*1 781*	*2 897*	*1 875*	*2 176*	*1 160*	*1 793*	*1 126*	*1 316*	*3 839*
Mediterranean horse mackerel	**Chinchard à queue jaune**		**Jurel mediterráneo**		***Trachurus mediterraneus***			**1,70(23)004,08**		**HMM**
37 Bulgaria	82	79	80 F	70	68	36	40	30	111	130
Croatia	589	787	566	453	361	336	200	90	75	192
Greece	8 578	8 263	11 323	8 331	8 039	7 169	4 350	3 534	3 902	3 408
Malta	7	6	5	8	5	4	2	4	0	0
Romania	22	30	35	23	13	1	15	3	8	17
Turkey	20 421	8 027	11 742	11 260	12 500	9 500	10 500	9 220	15 000	15 545
Ukraine	0	0	1	2	...	5	-	-	1	1
Yugoslavia	3	7	10	15	16	14	15	17	14	15
37 Fishing area total	*29 702*	*17 199*	*23 762 F*	*20 162*	*21 002*	*17 065*	*15 122*	*12 898*	*19 111*	*19 308*
Species total	*29 702*	*17 199*	*23 762 F*	*20 162*	*21 002*	*17 065*	*15 122*	*12 898*	*19 111*	*19 308*
Rough scad	**Chinchard frappeur**		**Chicharro garretón**		***Trachurus lathami***			**1,70(23)004,11**		**RSC**
41 Argentina	227	207	214	196	587	288	247	470	67	...
Brazil	...	...	...	536	389	313	81	25	40	40 F
Latvia	49	-	-	-	-	-	-	-	-	-
41 Fishing area total	*276*	*207*	*214*	*732*	*976*	*601*	*328*	*495*	*107*	*40 F*
Species total	*276*	*207*	*214*	*732*	*976*	*601*	*328*	*495*	*107*	*40 F*
Cape horse mackerel	**Chinchard du Cap**		**Jurel del Cabo**		***Trachurus capensis***			**1,70(23)004,13**		**HMC**
47 Estonia	33 127	31 447	31 372	28 655	-	-	-	-	-	-
Georgia	5 350 F	2 000 F	1 000 F	-	-	-	-	-	-	-
Japan	1 951	-	27	18	-	-	-	-	-	-
Korea Rep	-	189	575	-	-	-	-	-	-	-
Lithuania	19 546	16 460	1 726	-	-	-	-	-	-	-
Namibia	427 373	474 611	364 801	310 836	321 322	301 847	311 836	322 075	344 314	309 381
Poland	-	-	-	3 058	1 700	-	-	-	-	3 098
Russian Fed	95 575	136 576	140 525	116 695	78 429	69 248	104 935	55 642	50 456	28 215
South Africa	33 640	35 429	20 031	10 262	31 995	31 206	46 384	17 970	15 000 F	9 659
Spain	198	-	-	-	-	2	-	-	-	-
Ukraine	66 319	88 706	50 719	18 060	27 121	5 179	18 345	4 592	13 837	3 693
Other nei	-	-	2 877	18 596	9 805	-	-	-	-	-
47 Fishing area total	*683 079 F*	*785 418 F*	*613 653 F*	*506 180*	*470 372*	*407 482*	*481 500*	*400 279*	*423 607 F*	*354 046*
Species total	*683 079 F*	*785 418 F*	*613 653 F*	*506 180*	*470 372*	*407 482*	*481 500*	*400 279*	*423 607 F*	*354 046*
Cunene horse mackerel	**Chinchard du Cunène**		**Jurel de Cunene**		***Trachurus trecae***			**1,70(23)004,14**		**HMZ**
34 Belize	-	1	-	-	-	-	-	-	-	-
China	10	0	-	-	-	-	-	-	-	-
Ghana	762	1 893	2 741	4 215	7 714	6 962	11 690	9 964	572	1 540
Honduras	-	3	-	2	-	-	-	-	-	-
Liberia	3	3	3	12	-	-	-	-	-	-
Romania	14 583	441	-	-	-	-	-	-	-	-
Sierra Leone	14	8	0	4	11	44	40	5	-	434
Togo	92	107	224	149	163	107	82	815	449	501
Vanuatu	-	13	-	-	-	-	-	-	-	-
Other nei	17	-	-	-	-	-	-	-	-	-
34 Fishing area total	*15 481*	*2 469*	*2 968*	*4 382*	*7 888*	*7 113*	*11 812*	*10 784*	*1 021*	*2 475*
47 Angola	31 475	43 970	29 459	25 308	19 764	35 797	39 739	47 719	53 243	40 898
Russian Fed	72 866	57 980	70 408	53 760	58 466	51 583	6 265	33 973	16 387	5 934
47 Fishing area total	*104 341*	*101 950*	*99 867*	*79 068*	*78 230*	*87 380*	*46 004*	*81 692*	*69 630*	*46 832*
Species total	*119 822*	*104 419*	*102 835*	*83 450*	*86 118*	*94 493*	*57 816*	*92 476*	*70 651*	*49 307*
Greenback horse mackerel	**Chinchard dos vert**		**Jurel verde**		***Trachurus declivis***			**1,70(23)004,15**		**HMG**
81 Australia	...	...	...	...	68	30	18	16	26	57
Russian Fed	2 892	4 260	1 804	1 602	2 280	886	52	223	-	-
Ukraine	2 878	7 937	4 192	8 990	13 093	9 740	9 309	15 306	12 213	7 577
81 Fishing area total	*5 770*	*12 197*	*5 996*	*10 592*	*15 441*	*10 656*	*9 379*	*15 545*	*12 239*	*7 634*
Species total	*5 770*	*12 197*	*5 996*	*10 592*	*15 441*	*10 656*	*9 379*	*15 545*	*12 239*	*7 634*
Jack and horse mackerels nei	**Chinchards noirs nca**		**Jureles nep**		***Trachurus spp***			**1,70(23)004,XX**		**JAX**
27 Spain	28 075	29 828	29 895	31 863	32 283	39 846	36 894	38 160	36 989	39 824
27 Fishing area total	*28 075*	*29 828*	*29 895*	*31 863*	*32 283*	*39 846*	*36 894*	*38 160*	*36 989*	*39 824*
34 Belize	-	4	0	-	-	1 600	1 626	4 788	7 619	3 492
Bulgaria	-	-	-	-	-	-	1 669	-	-	-
China	1	4	2	-	-	-	1	-	0	3
Congo Dem R	1 370 F	1 510 F	1 370 F	1 397 F	1 432 F	1 386 F	1 426 F	1 400 F	1 400 F	1 400 F
Congo Rep	368 F	367 F	26	64	60 F	50 F	40 F	30 F	23	20 F
Cyprus	-	-	-	-	...	...	5 968	17 892	27 326	40 612

B-37

Miscellaneous pelagic fishes / **Poissons pélagiques divers** / **Peces pelágicos diversos**

Capture production by species, fishing areas and countries or areas
Captures par espèces, zones de pêche et pays ou zones
Capturas por especies, áreas de pesca y países o áreas

Species, Fishing area Espèce, Zone de pêche Especie, Area de pesca	1992 mt	1993 mt	1994 mt	1995 mt	1996 mt	1997 mt	1998 mt	1999 mt	2000 mt	2001 mt
Estonia	12 305	32 036	12 364	1 989	1 903	2 080	1 550	-	-	-
France	-	-	-	-	-	...	4 623	1 316	162	-
Gambia	93	128	104	336	312	133	119	130	175	246
Georgia	2 500 F	1 350 F	500 F	-	-	-	-	-	-	-
Germany	-	-	-	-	-	-	-	-	-	708
Ghana	-	-	-	5 289	3 201	10 512	4 892	1 904	2 183	2 109
Greece	-	-	-	-	-	-	-	-	1	-
Guinea	2 910 F	3 470 F	3 570 F	4 781	3 576	1 546	7 109	508	6 084	6 000 F
Honduras	28	-	-	1	-	-	-	-	0	-
Korea Rep	83	-	-	12	12	1	-	-	-	-
Latvia	33 947	34 799	34 764	38 824	14 818	4 881	8 710	14 284	22 591	17 617
Lithuania	25 913	16 580	11 716	...	...	...	11 902	20 657	25 464	15 226
Mauritania	1 420 F	1 360 F	970 F	1 510 F	1 220 F	1 060 F	940 F	...	...	58
Morocco	17 987	24 881	23 717	23 130	10 660	8 694	7 320	10 938	19 825	9 930
Netherlands	-	-	-	-	1 938	3 245	3 163	2 847	9 053	14 476
Panama	-	3	-	-	-	-	-	-	-	-
Poland	-	-	-	-	3 583	281	-	-	-	1 449
Portugal	1 170	551	299	206	599	764	660	495	562	386
Russian Fed	84 857	53 644	40 908	86 123	73 446	56 008	84 734	71 260	70 947	56 229
St Vincent	-	-	-	-	-	1 300	9 765	3 546	8 275	16 369
Senegal	1 352	1 710	1 093	1 027	521	1 078	1 061	2 108	1 020	1 877
Spain	-	-	-	-	345	124	52	57	133	215
Ukraine	28 823	43 242	41 246	56 236	42 471	38 311	64 222	53 322	66 266	37 233
Other nei	-	-	-	-	-	-	-	3 375	8 994	10 439
34 Fishing area total	*215 127 F*	*215 639 F*	*172 649 F*	*220 925 F*	*160 097 F*	*133 054 F*	*221 552 F*	*210 857 F*	*278 103 F*	*236 094 F*
37 Albania	...	...	...	50	68	18	85	92	90	21
Algeria	4 500 F	5 000 F	7 000 F	6 552	3 644	5 491	4 827	6 212	6 000 F	6 000 F
Egypt	346	354	...	...	...	...	...	...	...	...
France	603	519	272	311	422	2 148	2 148	571	753	1 038
Gaza Strip	...	...	...	...	90	100	115	115 F	115 F	100 F
Georgia	-	-	-	-	-	18	13	...	35	7
Italy	6 014	6 214	6 164	7 458	6 790	5 168	6 314	4 315	3 428	3 927
Libya	2 550 F	2 750 F	3 000 F	3 050	3 000 F	3 000 F	3 000 F	3 000 F	3 000 F	3 000 F
Morocco	3 631	3 916	5 073	7 344	4 811	3 818	2 496	1 849	2 270	2 237
Russian Fed	-	-	1	1	-	-	2	2	2	6
Spain	-	-	-	-	-	-	-	5 533	7 346	6 957
Tunisia	2 503	2 382	3 704	4 116	4 755	6 326	3 899	5 635	4 917	4 204
37 Fishing area total	*20 147 F*	*21 135 F*	*25 214 F*	*28 882*	*23 580 F*	*26 087 F*	*22 899 F*	*27 324 F*	*27 956 F*	*27 497 F*
47 Korea Rep	1 872	1 000	20	-	33	45	-	-	-	-
Russian Fed	19 871	24 836	-	-	-	-	-	-	-	-
47 Fishing area total	*21 743*	*25 836*	*20*	*...*	*33*	*45*	*...*	*...*	*...*	*...*
51 Korea Rep	-	-	-	-	-	-	-	-	-	6
Russian Fed	-	-	65	-	-	-	-	-	-	-
Ukraine	...	127	...	3	...	...	15	11	-	7
Yemen	921	990	990	4 412	1 380	1 440 F	1 600 F	1 600 F	1 400 F	1 800 F
51 Fishing area total	*921*	*1 117*	*1 055*	*4 415*	*1 380*	*1 440 F*	*1 615 F*	*1 611 F*	*1 400 F*	*1 813 F*
81 Japan	16 627	14 044	10 034	4 111	335	29	8	11	12	655
Korea Rep	1 315	1 260	359	474	1 157	2 087	1 983	2 182	259	307
New Zealand	25 087	34 838	34 002	31 919	29 085	34 057	36 059	34 003	22 544	28 507
Norway	3	3	-	-	0	1	-	-	-	-
81 Fishing area total	*43 032*	*50 145*	*44 395*	*36 504*	*30 577*	*36 174*	*38 050*	*36 196*	*22 815*	*29 469*
Species total	*329 045 F*	*343 700 F*	*273 228 F*	*322 589 F*	*247 950 F*	*236 646 F*	*321 010 F*	*314 148 F*	*367 263 F*	*334 697 F*
White trevally	**Carangue dentue**		**Jurel dentón**		***Pseudocaranx dentex***				**1,70(23)011,27**	**TRZ**
81 Australia	...	...	...	...	964	872	558	479	381	379
New Zealand	2 900	3 393	3 406	3 910	2 869	2 941	3 608	4 192	3 603	3 116
81 Fishing area total	*2 900*	*3 393*	*3 406*	*3 910*	*3 833*	*3 813*	*4 166*	*4 671*	*3 984*	*3 495*
Species total	*2 900*	*3 393*	*3 406*	*3 910*	*3 833*	*3 813*	*4 166*	*4 671*	*3 984*	*3 495*
Japanese scad	**Comète japonaise**		**Macarela japonesa**		***Decapterus maruadsi***				**1,70(23)043,07**	**RSA**
61 China,Taiwan	1 937	1 987	3 356	3 626	5 059	19 667	12 090	20 532	4 387	6 224
Japan	62 101	49 900	47 810	72 109	57 319	50 097	59 078	47 157	36 416	41 236
61 Fishing area total	*64 038*	*51 887*	*51 166*	*75 735*	*62 378*	*69 764*	*71 168*	*67 689*	*40 803*	*47 460*
71 Japan	-	13	-	-	-	-	-	-	-	-
71 Fishing area total	*-*	*13*	*-*	*-*	*-*	*-*	*-*	*-*	*-*	*-*
Species total	*64 038*	*51 900*	*51 166*	*75 735*	*62 378*	*69 764*	*71 168*	*67 689*	*40 803*	*47 460*
Indian scad	**Comète indienne**		**Macarela índica**		***Decapterus russelli***				**1,70(23)043,08**	**RUS**
57 Malaysia	6 969	10 507	9 979	10 392	8 748	6 953	13 189	8 423	5 117	5 811
Thailand	8 434	8 984	35 994	22 981	32 557	30 658	28 270	28 113	27 728	27 559
57 Fishing area total	*15 403*	*19 491*	*45 973*	*33 373*	*41 305*	*37 611*	*41 459*	*36 536*	*32 845*	*33 370*
61 China,Taiwan	380	927	678	299	382	687	6 158	7 703	3 927	598
61 Fishing area total	*380*	*927*	*678*	*299*	*382*	*687*	*6 158*	*7 703*	*3 927*	*598*
71 Malaysia	37 676	54 215	51 662	45 231	50 985	64 231	40 237	61 737	79 086	71 583
Thailand	42 525	46 186	38 394	54 641	52 648	47 498	57 893	56 461	55 688	55 348
71 Fishing area total	*80 201*	*100 401*	*90 056*	*99 872*	*103 633*	*111 729*	*98 130*	*118 198*	*134 774*	*126 931*

B-37 Miscellaneous pelagic fishes — Capture production by species, fishing areas and countries or areas
Poissons pélagiques divers — Captures par espèces, zones de pêche et pays ou zones
Peces pelágicos diversos — Capturas por especies, áreas de pesca y países o áreas

Species, Fishing area Espèce, Zone de pêche Especie, Area de pesca	1992 mt	1993 mt	1994 mt	1995 mt	1996 mt	1997 mt	1998 mt	1999 mt	2000 mt	2001 mt
Species total	*95 984*	*120 819*	*136 707*	*133 544*	*145 320*	*150 027*	*145 747*	*162 437*	*171 546*	*160 899*
Scads nei	**Comètes nca**		**Macarelas nep**		*Decapterus spp*			**1,70(23)043,XX**		**SDX**
31 Grenada	103	104	64	94	82	61	101	59	38	52
31 Fishing area total	*103*	*104*	*64*	*94*	*82*	*61*	*101*	*59*	*38*	*52*
34 Gabon	170 F	216	216	106	33	20	18	76	21	...
Ghana	993	1 235	963	1 760	1 462	1 654	2 989	3	263	2 466
Poland	-	-	-	-	54	-	-	-	-	-
Sao Tome Prn	...	...	...	...	...	43	55 F	60 F	60 F	60 F
Senegal	3 362	2 524	1 958	1 973	2 265	2 428	3 249	2 401	4 963	4 209
Sierra Leone	...	...	...	...	...	112	39	277	141	...
34 Fishing area total	*4 525 F*	*3 975*	*3 137*	*3 839*	*3 814*	*4 257*	*6 350 F*	*2 817 F*	*5 448 F*	*6 735 F*
47 Ukraine	-	327	-	181	...	...	-	-	-	-
47 Fishing area total	-	*327*	-	*181*	...	...	-	-	-	-
51 Jordan	...	...	...	...	...	...	20	25	25	26
Pakistan	1 200	1 675	1 875	1 920	1 010	1 225	3 505	4 661	4 600	4 355
Qatar	0	0	0	0	0	0	0	0	0	0
Untd Arab Em	1 596	1 600	1 783	1 790	1 809	1 933	1 939	1 987	580	580 F
51 Fishing area total	*2 796*	*3 275*	*3 658*	*3 710*	*2 819*	*3 158*	*5 464*	*6 673*	*5 205*	*4 961 F*
57 Indonesia	34 045	35 314	30 807	33 763	36 299	33 653	39 497	41 602	42 051	42 290
57 Fishing area total	*34 045*	*35 314*	*30 807*	*33 763*	*36 299*	*33 653*	*39 497*	*41 602*	*42 051*	*42 290*
61 China	392 021	260 758	430 860	515 298	607 686	505 991	532 986	502 590	502 289	544 728
China,H.Kong	5 895	3 874	4 600	5 254	5 368	4 143	2 962	2 100 F	2 600 F	2 900 F
China,Taiwan	15 529	16 863	15 041	9 168	32 586	12 743	3 172	2 836	2 491	2 670
61 Fishing area total	*413 445*	*281 495*	*450 501*	*529 720*	*645 640*	*522 877*	*539 120*	*507 526 F*	*507 380 F*	*550 298 F*
71 Guam	-	-	-	0	0	3	2	5	4	5
Indonesia	161 664	169 037	189 086	213 542	214 990	243 271	238 096	219 536	213 324	214 560
N Marianas	...	...	1	5	2	4	0	5	10	13
Philippines	297 902	274 029	235 973	264 472	228 757	234 849	250 809	254 178	260 999	287 810
Singapore	157	138	249	209	227	212	222	156	163	106
71 Fishing area total	*459 723*	*443 204*	*425 309*	*478 228*	*443 976*	*478 339*	*489 129*	*473 880*	*474 500*	*502 494*
77 El Salvador	247	138	94	156	237	130	149	226	187	270
77 Fishing area total	*247*	*138*	*94*	*156*	*237*	*130*	*149*	*226*	*187*	*270*
81 New Zealand	...	...	...	33	23	11	63	29	82	87
81 Fishing area total	...	...	...	*33*	*23*	*11*	*63*	*29*	*82*	*87*
Species total	*914 884 F*	*767 832*	*913 570*	*1 049 724*	*1 132 890*	*1 042 486*	*1 079 873 F*	*1 032 812 F*	*1 034 891 F*	*1 107 187 F*
Blue runner	**Carangue coubali**		**Cojinúa negra**		*Caranx crysos*			**1,70(23)044,26**		**RUB**
21 USA	-	-	-	-	-	-	-	49	1	0
21 Fishing area total	-	-	-	-	-	-	-	*49*	*1*	*0*
31 Dominican Rp	286	300	540	205	194	256	132	53	74	68
USA	599	820	552	531	119	152	276	131	130	157
31 Fishing area total	*885*	*1 120*	*1 092*	*736*	*313*	*408*	*408*	*184*	*204*	*225*
41 Brazil	...	...	...	338	503	515	443	590	625	620 F
41 Fishing area total	...	...	...	*338*	*503*	*515*	*443*	*590*	*625*	*620 F*
Species total	*885*	*1 120*	*1 092*	*1 074*	*816*	*923*	*851*	*823*	*830*	*845 F*
Crevalle jack	**Carangue crevalle**		**Jurel común**		*Caranx hippos*			**1,70(23)044,29**		**CVJ**
21 USA	-	-	-	-	-	-	4	232	2	1
21 Fishing area total	-	-	-	-	-	-	*4*	*232*	*2*	*1*
31 USA	-	-	-	146	136	264	393	91	316	304
31 Fishing area total	-	-	-	*146*	*136*	*264*	*393*	*91*	*316*	*304*
34 Ghana	5 321	4 399	5 603	4 422	2 884	2 207	2 891	3 906	79	2 525
Sao Tome Prn	...	...	...	...	...	42	55 F	60 F	60 F	60 F
34 Fishing area total	*5 321*	*4 399*	*5 603*	*4 422*	*2 884*	*2 249*	*2 946 F*	*3 966 F*	*139 F*	*2 585 F*
47 Angola	0	15	43	42	39	16	89	302	1 835	674
47 Fishing area total	*0*	*15*	*43*	*42*	*39*	*16*	*89*	*302*	*1 835*	*674*
Species total	*5 321*	*4 414*	*5 646*	*4 610*	*3 059*	*2 529*	*3 432 F*	*4 591 F*	*2 292 F*	*3 564 F*
Bar jack	**Carangue comade**		**Cojinúa carbonera**		*Caranx ruber*			**1,70(23)044,31**		**CXR**
31 USA	-	-	-	-	-	-	-	3	5	7
31 Fishing area total	-	-	-	-	-	-	-	*3*	*5*	*7*
Species total	-	-	-	-	-	-	-	*3*	*5*	*7*
False scad	**Comète coussut**		**Macarela real**		*Caranx rhonchus*			**1,70(23)044,42**		**HMY**
34 Ghana	2 472	2 213	4 040	3 483	3 301	3 337	2 753	2 275	3 800	147

B-37 Miscellaneous pelagic fishes / Poissons pélagiques divers / Peces pelágicos diversos

Capture production by species, fishing areas and countries or areas / Captures par espèces, zones de pêche et pays ou zones / Capturas por especies, áreas de pesca y países o áreas

Species, Fishing area Espèce, Zone de pêche Especie, Area de pesca	1992 mt	1993 mt	1994 mt	1995 mt	1996 mt	1997 mt	1998 mt	1999 mt	2000 mt	2001 mt
Liberia	-	-	-	-	-	7	111	435	135	134
Romania	12 506	367	-	-	-	-	-	-	-	-
34 Fishing area total	*14 978*	*2 580*	*4 040*	*3 483*	*3 301*	*3 344*	*2 864*	*2 710*	*3 935*	*281*
Species total	*14 978*	*2 580*	*4 040*	*3 483*	*3 301*	*3 344*	*2 864*	*2 710*	*3 935*	*281*
Jacks, crevalles nei	**Chinchards, carangues nca**		**Jureles, pámpanos nep**		***Caranx spp***			**1,70(23)044,XX**		**TRE**
31 Br Virgin Is	...	...	...	...	...	13	15	14	1	1 F
Colombia	137	217	496	510	1 111	350	239	117	260 F	260 F
Cuba	323	268	427	344	348	234	211	167	170 F	170 F
Dominican Rp	254	270	486	144	424	453	75	23	43	43
Mexico	1 580	2 587	2 766	5 437	5 911	8 398	8 735	6 365	5 779	6 002
Trinidad Tob	400 F	501	328	400 F	504	562	203	189	202	210
Venezuela	5 247	5 136	4 271	3 439	3 732	2 823	2 898	2 642	3 462	2 115
31 Fishing area total	*7 941 F*	*8 979*	*8 774*	*10 274 F*	*12 030*	*12 833*	*12 376*	*9 517*	*9 917 F*	*8 801 F*
34 Congo Rep	0	0	16	15 F	12 F	9 F	5 F	3 F	0	0
Côte dIvoire	81	415	377	128	69	51	282	290	144	119
Estonia	-	-	-	-	-	-	-	1 622	-	-
Gabon	...	...	31	47	594	404	91	4	29	31
Gambia	55	73	88	73	174	124	147	160	137	288
Ghana	108	31	14	57	13	11	1	-	1	-
GuineaBissau	45 F	46 F	50 F	100 F	100 F	100 F	80 F	70 F	70 F	70 F
Latvia	286	66	-	-	-	11	-	-	-	-
Liberia	170	97	96	76	62	30	178	394	235	271
Lithuania	-	209	-	-	-	-	-	-	-	-
Nigeria	2 049	3 290	2 414	636	591	2 525	414	1 099	825	317
Portugal	32	102	188	44	134	114	95	32	1	1
Russian Fed	220	126	13	53	15	24	-	32	2	-
Sao Tome Prn	...	...	...	...	...	129	150 F	160 F	150 F	150 F
Sierra Leone	475	469	468	467	470 F	474	18	7	587	327
Ukraine	76	47	...	...	...	...	-	-	-	-
34 Fishing area total	*3 597 F*	*4 971 F*	*3 755 F*	*1 696 F*	*2 234 F*	*4 006 F*	*1 461 F*	*3 873 F*	*2 181 F*	*1 574 F*
37 Egypt	-	-	-	433	716	441	402	326	525	1 351
37 Fishing area total	-	-	-	*433*	*716*	*441*	*402*	*326*	*525*	*1 351*
41 Brazil	4 520 F	4 510 F	4 520 F	3 656	3 905	5 838	5 635	4 164	6 950	6 900 F
41 Fishing area total	*4 520 F*	*4 510 F*	*4 520 F*	*3 656*	*3 905*	*5 838*	*5 635*	*4 164*	*6 950*	*6 900 F*
47 Russian Fed	-	-	-	-	-	-	-	-	-	1
47 Fishing area total	-	-	-	-	-	-	-	-	-	*1*
51 Egypt	4 841	14 361	11 823	...	...	...	...	...	...	
India	70 976	53 501	48 378	5 263	50 436	33 042	52 315	51 741	22 808	31 418
Korea Rep	81	-	5	-	46	-	11	-	-	-
Oman	647	1 914	1 773	3 392	2 927	3 896	2 957	2 552	1 806	1 218
Pakistan	2 852	3 933	4 003	4 631	3 972	5 391	6 523	8 407	9 111	8 928
Qatar	416	361	201	175	222	247	308	289	409	388
Ukraine	3	-	-	-	-	-	-	2	-	-
Untd Arab Em	5 974	6 000	6 682	6 833	6 905	7 379	7 403	7 588	3 107	3 100 F
Yemen	509	174	480	431	413	430 F	500 F	500 F	400 F	500 F
51 Fishing area total	*86 299*	*80 244*	*73 345*	*20 725*	*64 921*	*50 385 F*	*70 017 F*	*71 079 F*	*37 641 F*	*45 552 F*
57 India	13 923	11 117	11 731	16 176	17 443	17 791	14 595	13 758	16 763	15 422
Indonesia	8 093	8 273	6 097	8 162	7 848	8 186	14 941	9 508	10 398	10 980
Malaysia	4 239	4 859	6 465	7 847	6 524	5 319	834	5 020	6 117	5 807
57 Fishing area total	*26 255*	*24 249*	*24 293*	*32 185*	*31 815*	*31 296*	*30 370*	*28 286*	*33 278*	*32 209*
61 China,H.Kong	286	50	72	106	66	33	20	10 F	15 F	17 F
China,Taiwan	10 237	9 643	8 270	6 595	7 065	11 433	8 228	4 882	6 355	3 273
61 Fishing area total	*10 523*	*9 693*	*8 342*	*6 701*	*7 131*	*11 466*	*8 248*	*4 892 F*	*6 370 F*	*3 290 F*
71 Fiji Islands	1 347	1 795	1 829 F	644	384	695	647	730	650 F	709
Indonesia	19 120	18 632	19 989	20 863	22 197	23 911	24 502	24 712	25 923	27 390
Kiribati	530	510	500	500	500	510	2 530	1 910	1 108	2 858
Malaysia	19 291	24 347	19 893	18 683	19 845	24 561	10 719	31 624	33 497	31 318
Singapore	345	331	352	297	312	313	234	175	139	66
71 Fishing area total	*40 633*	*45 615*	*42 563 F*	*40 987*	*43 238*	*49 990*	*38 632*	*59 151*	*61 317 F*	*62 341*
77 Cook Is	54 F	55 F	52 F	51 F	40 F	40 F	40 F	40 F	40 F	40 F
Mexico	3 164	983	1 146	1 723	2 049	2 281	2 198	1 931	2 487	2 868
USA	-	-	-	-	-	-	-	-	0	-
77 Fishing area total	*3 218 F*	*1 038 F*	*1 198 F*	*1 774 F*	*2 089 F*	*2 321 F*	*2 238 F*	*1 971 F*	*2 527 F*	*2 908 F*
87 Colombia	151	194	181	129	538	504	278	273	200 F	138
Peru	...	...	18	-	129	16	69	607	35	11
87 Fishing area total	*151*	*194*	*199*	*129*	*667*	*520*	*347*	*880*	*235 F*	*149*
Species total	*183 137 F*	*179 493 F*	*166 989 F*	*118 560 F*	*168 746 F*	*169 096 F*	*169 726 F*	*184 139 F*	*160 941 F*	*165 076 F*
Atlantic moonfish	**Musso atlantique**		**Jorobado lamparosa**		***Selene setapinnis***			**1,70(23)046,04**		**MOA**
31 Grenada	0	0	0	0	0	0	0	0	0	0
Venezuela	2 266	2 467	2 603	2 544	1 766	2 110	2 338	1 529	976	3 443
31 Fishing area total	*2 266*	*2 467*	*2 603*	*2 544*	*1 766*	*2 110*	*2 338*	*1 529*	*976*	*3 443*
41 Brazil	...	...	...	2 468	1 679	1 914	1 512	1 514	1 386	1 370 F

B-37 Miscellaneous pelagic fishes / Poissons pélagiques divers / Peces pelágicos diversos

Capture production by species, fishing areas and countries or areas / Captures par espèces, zones de pêche et pays ou zones / Capturas por especies, áreas de pesca y países o áreas

Species, Fishing area Espèce, Zone de pêche Especie, Area de pesca	1992 mt	1993 mt	1994 mt	1995 mt	1996 mt	1997 mt	1998 mt	1999 mt	2000 mt	2001 mt
41 Fishing area total	*...*	*...*	*...*	*2 468*	*1 679*	*1 914*	*1 512*	*1 514*	*1 386*	*1 370 F*
Species total	*2 266*	*2 467*	*2 603*	*5 012*	*3 445*	*4 024*	*3 850*	*3 043*	*2 362*	*4 813 F*
African moonfish	**Musso africain**		**Jorobado africano**		***Selene dorsalis***			**1,70(23)046,05**		**LUK**
34 Benin	21 F	22 F	25 F	24 F	27 F	37 F	35 F	57	40 F	121
Congo Rep	75 F	75 F	45 F	50 F	55 F	60 F	60 F	70 F	72	70 F
Georgia	17 F	10 F	-	-	-	-	-	-	-	-
Ghana	1 202	882	501	976	907	712	381	469	738	1 100
Latvia	204	-	-	-	-	-	-	-	-	-
Russian Fed	171	194	176	20	15	12	1	-	-	-
Senegal	-	-	-	-	-	9	278	176	251	326
Sierra Leone	0	0	0	1	1 F	...	...	...	135	319
Togo	4	4	3	2	10	2	0	-	0	-
Ukraine	15	91	...	...	...	...	...	...	...	...
34 Fishing area total	*1 709 F*	*1 278 F*	*750 F*	*1 073 F*	*1 015 F*	*832 F*	*755 F*	*772 F*	*1 236 F*	*1 936 F*
Species total	*1 709 F*	*1 278 F*	*750 F*	*1 073 F*	*1 015 F*	*832 F*	*755 F*	*772 F*	*1 236 F*	*1 936 F*
Snubnose pompano	**Pompaneau lune**		**Pámpano lunero**		***Trachinotus blochii***			**1,70(23)047,01**		**POO**
51 Saudi Arabia	-	-	-	-	-	-	4	0	31	-
51 Fishing area total	*-*	*-*	*-*	*-*	*-*	*-*	*4*	*0*	*31*	*-*
Species total	*-*	*-*	*-*	*-*	*-*	*-*	*4*	*0*	*31*	*-*
Florida pompano	**Pompaneau sole**		**Pámpano amarillo**		***Trachinotus carolinus***			**1,70(23)047,03**		**POM**
21 USA	0	0	1	0	44	1	2	139	0	0
21 Fishing area total	*0*	*0*	*1*	*0*	*44*	*1*	*2*	*139*	*0*	*0*
31 USA	-	-	-	54	-	296	319	68	242	181
31 Fishing area total	*-*	*-*	*-*	*54*	*-*	*296*	*319*	*68*	*242*	*181*
Species total	*0*	*0*	*1*	*54*	*44*	*297*	*321*	*207*	*242*	*181*
Pompanos nei	**Pompaneaux nca**		**Pámpanos(=Palometas) nep**		***Trachinotus spp***			**1,70(23)047,XX**		**POX**
31 Dominican Rp	101	58	104	76	57	19	47	5	10	7
Mexico	582	643	602	599	735	429	556	542	410	333
Venezuela	128	140	122	244	307	285	435	146	132	117
31 Fishing area total	*811*	*841*	*828*	*919*	*1 099*	*733*	*1 038*	*693*	*552*	*457*
34 Estonia	-	10	-	-	-	-	-	-	-	-
Gambia	-	-	-	8	20	7	1	2	1	22
Portugal	-	-	-	0	0	1	2	3	-	-
Senegal	...	98	55	64	132	83	28	207	27	158
Sierra Leone	...	...	...	3	3 F	63	279	7	-	-
Spain	-	-	-	-	-	-	-	4	-	7
34 Fishing area total	*...*	*108*	*55*	*75*	*155 F*	*154*	*310*	*223*	*28*	*187*
41 Brazil	...	...	...	424	534	166	172	144	286	280 F
41 Fishing area total	*...*	*...*	*...*	*424*	*534*	*166*	*172*	*144*	*286*	*280 F*
51 India	49 874	53 087	3 649	6 765	5 706	3 697	2 533	2 329	12	11
51 Fishing area total	*49 874*	*53 087*	*3 649*	*6 765*	*5 706*	*3 697*	*2 533*	*2 329*	*12*	*11*
57 India	782	0	1 043	2 132	1 078	-	205	-	-	-
57 Fishing area total	*782*	*0*	*1 043*	*2 132*	*1 078*	*-*	*205*	*-*	*-*	*-*
77 Mexico	275	303	298	286	206	200	329	309	274	222
77 Fishing area total	*275*	*303*	*298*	*286*	*206*	*200*	*329*	*309*	*274*	*222*
87 Peru	41	895	580	566	460	268	779	2 801	1 220	632
87 Fishing area total	*41*	*895*	*580*	*566*	*460*	*268*	*779*	*2 801*	*1 220*	*632*
Species total	*51 783*	*55 234*	*6 453*	*11 167*	*9 238 F*	*5 218*	*5 366*	*6 499*	*2 372*	*1 789 F*
Greater amberjack	**Sériole couronnée**		**Pez de limón**		***Seriola dumerili***			**1,70(23)048,01**		**AMB**
21 USA	...	...	...	...	...	...	...	368	-	-
21 Fishing area total	*...*	*...*	*...*	*...*	*...*	*...*	*...*	*368*	*-*	*-*
27 Portugal	...	...	...	...	...	...	...	10	21	20
27 Fishing area total	*...*	*...*	*...*	*...*	*...*	*...*	*...*	*10*	*21*	*20*
31 USA	...	...	...	...	...	...	560	149	538	452
31 Fishing area total	*...*	*...*	*...*	*...*	*...*	*...*	*560*	*149*	*538*	*452*
37 Albania	...	...	...	0	2	1	-	-	-	2
Algeria	...	...	...	87	32	19	48	75	70 F	70 F
Croatia	16	30	55	78	80	64	62	30	25	19
Cyprus	13	12	38	16	28	21	8	17	25	14
Greece	199	114	199	322	697	336	252	205	176	396
Israel	...	34	30	79	33	80	185	98	307	250 F
Malta	2	0	0	4	9	6	6	6	3	2
Spain	-	-	-	-	-	-	-	704	375	430
Syria	52 F	54 F	52 F	52 F	90	75	108	88	52	96

B-37 Miscellaneous pelagic fishes / Poissons pélagiques divers / Peces pelágicos diversos

Capture production by species, fishing areas and countries or areas
Captures par espèces, zones de pêche et pays ou zones
Capturas por especies, áreas de pesca y países o áreas

Species, Fishing area Espèce, Zone de pêche Especie, Area de pesca	1992 mt	1993 mt	1994 mt	1995 mt	1996 mt	1997 mt	1998 mt	1999 mt	2000 mt	2001 mt
Tunisia	122	72	27	59	75	113	75	88	400	122
Yugoslavia	5	12	8	16	11	9	11	9	12	13
37 Fishing area total	*409 F*	*328 F*	*409 F*	*713 F*	*1 057*	*724*	*755*	*1 320*	*1 445 F*	*1 414 F*
Species total	*409 F*	*328 F*	*409 F*	*713 F*	*1 057*	*724*	*1 315*	*1 847*	*2 004 F*	*1 886 F*
Japanese amberjack	**Sériole du Japon**		**Medregal del Japón**		***Seriola quinqueradiata***			**1,70(23)048,02**		**AMJ**
61 Japan	-	-	-	7 564	246	...	...	...	...	...
61 Fishing area total	-	-	-	*7 564*	*246*	...	...	...	...	...
Species total	-	-	-	*7 564*	*246*	...	...	...	...	...
Yellowtail amberjack	**Sériole chicard**		**Medregal rabo amarillo**		***Seriola lalandi***			**1,70(23)048,06**		**YTC**
41 Argentina	60	24	13	9	16	17	7	9	10	...
Brazil	...	...	...	653	538	466	880	544	611	600 F
41 Fishing area total	*60*	*24*	*13*	*662*	*554*	*483*	*887*	*553*	*621*	*600 F*
47 Russian Fed	-	-	-	6	-	-	-	-	-	-
South Africa	566	858	771	769	488	477	519	302	300 F	315
47 Fishing area total	*566*	*858*	*771*	*775*	*488*	*477*	*519*	*302*	*300 F*	*315*
51 South Africa	-	-	-	-	-	1	...	...	...	...
51 Fishing area total	-	-	-	-	-	*1*	...	...	...	...
77 USA	-	-	-	-	-	-	111	30	50	39
77 Fishing area total	-	-	-	-	-	-	*111*	*30*	*50*	*39*
Species total	*626*	*882*	*784*	*1 437*	*1 042*	*961*	*1 517*	*885*	*971 F*	*954 F*
Amberjacks nei	**Sérioles nca**		**Medregales nep**		***Seriola spp***			**1,70(23)048,XX**		**AMX**
21 USA	...	...	...	...	...	...	4	5	8	3
21 Fishing area total	...	...	...	...	...	...	*4*	*5*	*8*	*3*
31 Dominican Rp	87	90	62	59	50	46	50	12	16	18
Mexico	1 127	981	1 035	1 148	1 109	1 590	1 574	1 615	1 805	...
USA	1 711	1 631	1 995	1 294	1 123	829	243	184	370	383
Venezuela	430	286	380	288	338	355	336	450	610	399
31 Fishing area total	*3 355*	*2 988*	*3 472*	*2 789*	*2 620*	*2 820*	*2 203*	*2 261*	*2 801*	*800*
34 Portugal	-	-	-	10	33	45	101	32	8	7
Russian Fed	-	11	-	-	82	26	-	4	-	-
Senegal	-	-	-	-	-	-	62	60	64	165
Spain	-	-	-	-	114	142	161	71	88	116
Ukraine	-	-	-	-	-	-	-	15	...	...
34 Fishing area total	-	*11*	-	*10*	*229*	*213*	*324*	*182*	*160*	*288*
41 Brazil	...	...	...	856	825	1 048	119	104	128	120 F
41 Fishing area total	...	...	...	*856*	*825*	*1 048*	*119*	*104*	*128*	*120 F*
51 Kenya	72	44	42	89	79	63	60	78	71	92
Korea Rep	13	-	-	-	-	21	5	1	-	-
Tanzania	400	198	143	215	51	250	275	100	75	60
51 Fishing area total	*485*	*242*	*185*	*304*	*130*	*334*	*340*	*179*	*146*	*152*
57 Australia	5	7	102	105	100	10	10	1	2	2
57 Fishing area total	*5*	*7*	*102*	*105*	*100*	*10*	*10*	*1*	*2*	*2*
61 Japan	55 325	43 243	53 801	54 101	50 333	47 211	45 484	54 918	77 461	66 925
Korea Rep	2 520	2 893	3 710	3 745	4 093	6 064	9 620	8 653	4 814	6 475
61 Fishing area total	*57 845*	*46 136*	*57 511*	*57 846*	*54 426*	*53 275*	*55 104*	*63 571*	*82 275*	*73 400*
67 Japan	47	-	-	-	-	-	-	-	-	-
67 Fishing area total	*47*	-	-	-	-	-	-	-	-	-
71 Australia	-	-	-	-	-	-	-	-	-	0
Japan	6	5	1	-	-	-	-	-	-	-
Korea Rep	-	-	48	72	34	34	-	-	-	-
71 Fishing area total	*6*	*5*	*49*	*72*	*34*	*34*	-	-	-	*0*
77 Japan	49	-	-	-	-	-	-	-	-	-
Mexico	9	5	6	11	7	12	6	13	27	...
Panama	808	864	850	341	470	333	469	159	418	400 F
77 Fishing area total	*866*	*869*	*856*	*352*	*477*	*345*	*475*	*172*	*445*	*400 F*
81 Australia	401	401	347	348	194	84	76	1	1	0
New Zealand	532	489	281	315	381	349	327	317	296	278
Russian Fed	-	-	-	-	-	206	-	209	-	-
81 Fishing area total	*933*	*890*	*628*	*663*	*575*	*639*	*403*	*527*	*297*	*278*
87 Colombia	...	...	17	...	...	51	109	47	70 F	91
Peru	3 992	3 084	3 325	6 598	1 558	4 648	21 104	2 084	11 159	28 025
87 Fishing area total	*3 992*	*3 084*	*3 342*	*6 598*	*1 558*	*4 699*	*21 213*	*2 131*	*11 229 F*	*28 116*
Species total	*67 534*	*54 232*	*66 145*	*69 595*	*60 974*	*63 417*	*80 195*	*69 133*	*97 491 F*	*103 559 F*

B-37 Miscellaneous pelagic fishes / Poissons pélagiques divers / Peces pelágicos diversos

Capture production by species, fishing areas and countries or areas / Captures par espèces, zones de pêche et pays ou zones / Capturas por especies, áreas de pesca y países o áreas

Species, Fishing area Espèce, Zone de pêche Especie, Area de pesca	1992 mt	1993 mt	1994 mt	1995 mt	1996 mt	1997 mt	1998 mt	1999 mt	2000 mt	2001 mt
Leerfish	**Liche**		**Palometón**			***Lichia amia***			**1,70(23)072,02**	**LEE**
27 Lithuania	-	-	-	1	-	-	-	-	-	-
Portugal	1	0	0	0	0	0	-	0	0	0
Spain	593	2	0	4	-	5	2	1	1	10
27 Fishing area total	*594*	*2*	*0*	*5*	*0*	*5*	*2*	*1*	*1*	*10*
34 Côte dIvoire	-	-	-	-	-	-	-	20	16	3
Italy	1 213	1 029	1 029	-	-	-	-	-	-	-
Latvia	41	27	-	18	152	236	127	172	274	96
Liberia	-	-	-	-	2	0	-	-	-	-
Lithuania	30	72	11	-	-	-	-	-	-	-
Portugal	1	0	0	0	0	0	-	-	-	-
Romania	74	-	-	-	-	-	-	-	-	-
Russian Fed	43	24	40	14	84	393	736	622	200	83
Sao Tome Prn	18 F	22 F	32 F	34 F	25	36	45 F	50 F	50 F	50 F
Senegal	1 828	501	920	2 450	228	257	540	489	490	341
Spain	-	-	-	-	5	-	-	-	-	-
Ukraine	142	31	2	...	...	428	-	266	109	31
34 Fishing area total	*3 390 F*	*1 706 F*	*2 034 F*	*2 516 F*	*496*	*1 350*	*1 448 F*	*1 619 F*	*1 139 F*	*604 F*
37 Italy	744	776	629	948	643	400	197	249	185	183
Spain	-	-	-	-	19	27	36	27	42	35
Tunisia	176	134	100	101	93	145	160	112	90	134
Turkey	563	923	950	1 345	1 544	1 650	1 950	2 780	320	255
37 Fishing area total	*1 483*	*1 833*	*1 679*	*2 394*	*2 299*	*2 222*	*2 343*	*3 168*	*637*	*607*
47 Russian Fed	-	-	-	-	-	-	-	-	-	1
47 Fishing area total	-	-	-	-	-	-	-	-	-	*1*
Species total	*5 467 F*	*3 541 F*	*3 713 F*	*4 915 F*	*2 795*	*3 577*	*3 793 F*	*4 788 F*	*1 777 F*	*1 222 F*
Alexandria pompano	**Cordonnier bossu**		**Jurel de Alejandría**			***Alectis alexandrinus***			**1,70(23)090,03**	**ALA**
34 Senegal	...	...	324	124	160	742	1 023	563	502	862
34 Fishing area total	...	...	*324*	*124*	*160*	*742*	*1 023*	*563*	*502*	*862*
Species total	...	...	*324*	*124*	*160*	*742*	*1 023*	*563*	*502*	*862*
Black pomfret	**Castagnoline noire**		**Palometa negra**			***Parastromateus niger***			**1,70(23)099,01**	**POB**
51 Pakistan	1 295	1 961	2 199	3 066	2 221	2 322	2 109	2 917	2 027	1 975
51 Fishing area total	*1 295*	*1 961*	*2 199*	*3 066*	*2 221*	*2 322*	*2 109*	*2 917*	*2 027*	*1 975*
57 Indonesia	4 004	4 296	3 295	4 809	5 637	5 638	7 043	7 041	7 276	7 660
Malaysia	...	...	2 317	2 709	2 430	2 785	2 816	3 856	3 115	2 612
Thailand	1 275	1 181	1 957	2 080	2 900	2 981	1 013	1 219	1 202	1 195
57 Fishing area total	*5 279*	*5 477*	*7 569*	*9 598*	*10 967*	*11 404*	*10 872*	*12 116*	*11 593*	*11 467*
61 China,Taiwan	5 174	3 832	4 668	5 444	3 427	2 534	1 744	1 307	1 671	2 129
61 Fishing area total	*5 174*	*3 832*	*4 668*	*5 444*	*3 427*	*2 534*	*1 744*	*1 307*	*1 671*	*2 129*
71 Indonesia	16 375	19 556	19 568	21 225	22 358	27 000	25 708	24 819	26 817	28 240
Malaysia	...	...	2 517	2 102	2 220	1 932	2 186	1 771	1 431	1 199
Thailand	1 979	1 895	2 545	3 583	3 483	3 272	1 905	4 660	4 596	4 568
71 Fishing area total	*18 354*	*21 451*	*24 630*	*26 910*	*28 061*	*32 204*	*29 799*	*31 250*	*32 844*	*34 007*
Species total	*30 102*	*32 721*	*39 066*	*45 018*	*44 676*	*48 464*	*44 524*	*47 590*	*48 135*	*49 578*
Rainbow runner	**Comète saumon**		**Macarela salmón**			***Elagatis bipinnulata***			**1,70(23)134,01**	**RRU**
31 Grenada	11	19	6	32	14	20	20	36	23	15
31 Fishing area total	*11*	*19*	*6*	*32*	*14*	*20*	*20*	*36*	*23*	*15*
34 Togo	-	-	-	-	-	1	-	27	4	1
34 Fishing area total	-	-	-	-	-	*1*	-	*27*	*4*	*1*
51 Saudi Arabia	-	-	-	-	-	415	98	132	5	52
51 Fishing area total	...	...	...	...	...	*415*	*98*	*132*	*5*	*52*
57 Indonesia	1 799	1 677	2 206	1 669	1 477	1 476	6 497	1 883	1 275	1 480
Malaysia	232	262	479	470	262	97	68	27	446	75
57 Fishing area total	*2 031*	*1 939*	*2 685*	*2 139*	*1 739*	*1 573*	*6 565*	*1 910*	*1 721*	*1 555*
71 Guam	-	-	-	-	-	-	1	2	1	2
Indonesia	4 770	5 156	5 071	6 046	6 016	5 579	7 636	8 431	8 708	10 110
Malaysia	477	369	359	267	128	1 484	322	836	119	120
Philippines	2 442	11 608	4 503	4 966	4 505	4 075	4 298	4 500	4 342	4 378
71 Fishing area total	*7 689*	*17 133*	*9 933*	*11 279*	*10 649*	*11 138*	*12 257*	*13 769*	*13 170*	*14 610*
Species total	*9 731*	*19 091*	*12 624*	*13 450*	*12 402*	*13 147*	*18 940*	*15 874*	*14 923*	*16 233*
Golden trevally	**Carangue royale**		**Jurel dorado**			***Gnathanodon speciosus***			**1,70(23)151,01**	**GLT**
51 Qatar	223	224	123	98	101	108	172	116	185	204
Untd Arab Em	595	600	673	336	340	363	364	373	940	940 F
51 Fishing area total	*818*	*824*	*796*	*434*	*441*	*471*	*536*	*489*	*1 125*	*1 144 F*
Species total	*818*	*824*	*796*	*434*	*441*	*471*	*536*	*489*	*1 125*	*1 144 F*

B-37 Miscellaneous pelagic fishes / Poissons pélagiques divers / Peces pelágicos diversos

Capture production by species, fishing areas and countries or areas / Captures par espèces, zones de pêche et pays ou zones / Capturas por especies, áreas de pesca y países o áreas

Species, Fishing area Espèce, Zone de pêche Especie, Area de pesca	1992 mt	1993 mt	1994 mt	1995 mt	1996 mt	1997 mt	1998 mt	1999 mt	2000 mt	2001 mt
Torpedo scad	**Comète torpille**		**Macarela torpedo**		***Megalaspis cordyla***				**1,70(23)179,01**	**HAS**
51 Pakistan	4 525	5 863	3 369	6 511	3 000	2 100	1 100	1 450	2 017	1 825
Untd Arab Em	952	960	1 079	944	954	1 019	1 022	1 048	1 100	1 100 F
51 Fishing area total	*5 477*	*6 823*	*4 448*	*7 455*	*3 954*	*3 119*	*2 122*	*2 498*	*3 117*	*2 925 F*
57 Indonesia	10 182	10 498	7 395	7 438	7 182	5 720	5 628	6 764	7 250	7 860
Malaysia	4 459	4 724	9 332	12 896	8 324	7 849	10 377	5 872	3 402	2 929
Thailand	4 517	4 716	14 143	7 347	14 457	16 038	16 207	14 743	14 541	14 452
57 Fishing area total	*19 158*	*19 938*	*30 870*	*27 681*	*29 963*	*29 607*	*32 212*	*27 379*	*25 193*	*25 241*
61 China,Taiwan	221	454	205	65	233	6 847	52	113	327	198
61 Fishing area total	*221*	*454*	*205*	*65*	*233*	*6 847*	*52*	*113*	*327*	*198*
71 Indonesia	7 643	7 923	6 592	6 665	7 733	9 217	11 933	12 693	13 235	14 360
Malaysia	7 955	10 585	7 474	6 574	6 588	5 813	8 408	14 023	13 930	11 023
Philippines	10 316	8 344	7 988	6 097	5 864	11 429	14 817	16 033	16 274	17 802
Thailand	18 067	18 581	20 809	9 723	5 217	4 027	7 981	7 374	7 273	7 229
71 Fishing area total	*43 981*	*45 433*	*42 863*	*29 059*	*25 402*	*30 486*	*43 139*	*50 123*	*50 712*	*50 414*
Species total	*68 837*	*72 648*	*78 386*	*64 260*	*59 552*	*70 059*	*77 525*	*80 113*	*79 349*	*78 778 F*
Queenfishes	**Sauteurs**		**Jureles saltadores**		***Scomberoides spp***				**1,70(23)231,XX**	**QUE**
51 Eritrea	...	...	...	...	8	4	...	44	243	...
Oman	...	281	518	1 019	823	703	528	693	408	528
Qatar	90	72	45	49	57	56	80	42	57	78
Saudi Arabia	...	...	154	312	349	385	572	456	572	543
Untd Arab Em	1 700	1 700	1 931	1 981	2 002	2 140	2 147	2 201	609	610 F
51 Fishing area total	*1 790*	*2 053*	*2 648*	*3 361*	*3 239*	*3 288*	*3 327*	*3 436*	*1 889*	*1 759 F*
57 Indonesia	3 232	2 250	2 884	3 722	4 678	2 263	2 304	2 588	2 328	2 300
Malaysia	449	579	454	219	389	256	341	380	395	357
57 Fishing area total	*3 681*	*2 829*	*3 338*	*3 941*	*5 067*	*2 519*	*2 645*	*2 968*	*2 723*	*2 657*
71 Indonesia	9 233	10 743	9 662	11 185	10 945	11 310	13 568	13 391	12 224	12 090
Malaysia	1 763	1 730	2 701	3 093	2 851	2 616	2 912	3 431	3 087	3 170
Philippines	3 648	4 184	4 182	3 106	2 965	2 631	2 830	4 404	4 665	4 682
71 Fishing area total	*14 644*	*16 657*	*16 545*	*17 384*	*16 761*	*16 557*	*19 310*	*21 226*	*19 976*	*19 942*
Species total	*20 115*	*21 539*	*22 531*	*24 686*	*25 067*	*22 364*	*25 282*	*27 630*	*24 588*	*24 358 F*
Atlantic bumper	**Sapater**		**Casabe**		***Chloroscombrus chrysurus***				**1,70(23)268,01**	**BUA**
34 Benin	-	-	-	-	-	-	-	-	-	853
Côte dIvoire	838	1 103	1 178	1 188	1 116	1 302	1 107	1 120	1 374	2 364
Ghana	5 153	6 742	4 644	2 482	2 466	3 847	7 264	7 611	6 861	6 418
Lithuania	396	2 220	-	-	-	-	-	-	-	-
Russian Fed	139	3 274	702	59	-	-	49	-	-	-
Senegal	827	-	-	-	-	-	-	-	-	2 045
Sierra Leone	...	...	...	0	0	-	-	-	-	227
Ukraine	-	-	-	-	-	-	-	-	2	44
34 Fishing area total	*7 353*	*13 339*	*6 524*	*3 729*	*3 582*	*5 149*	*8 420*	*8 731*	*8 237*	*11 951*
41 Brazil	...	...	...	1 868	952	1 035	4 065	3 519	1 573	1 550 F
41 Fishing area total	*...*	*...*	*...*	*1 868*	*952*	*1 035*	*4 065*	*3 519*	*1 573*	*1 550 F*
47 Russian Fed	-	-	-	-	-	2	-	-	-	-
47 Fishing area total	-	-	-	-	-	*2*	-	-	-	-
Species total	*7 353*	*13 339*	*6 524*	*5 597*	*4 534*	*6 186*	*12 485*	*12 250*	*9 810*	*13 501 F*
Pacific bumper	**Sapater du Pacifique**		**Casabe orqueta**		***Chloroscombrus orqueta***				**1,70(23)268,02**	**HSO**
87 Ecuador	...	...	8 346	17 999	1 706	952	565	1 409	...	1 008
87 Fishing area total	*...*	*...*	*8 346*	*17 999*	*1 706*	*952*	*565*	*1 409*	*...*	*1 008*
Species total	*...*	*...*	*8 346*	*17 999*	*1 706*	*952*	*565*	*1 409*	*...*	*1 008*
Parona leatherjacket	**Sauteur parone**		**Parona**		***Parona signata***				**1,70(23)283,01**	**PAO**
41 Argentina	677	648	2 128	1 478	1 925	1 805	1 744	1 710	1 479	860
Uruguay	377	491	664	391	439	431	465	360	371	618
41 Fishing area total	*1 054*	*1 139*	*2 792*	*1 869*	*2 364*	*2 236*	*2 209*	*2 070*	*1 850*	*1 478*
Species total	*1 054*	*1 139*	*2 792*	*1 869*	*2 364*	*2 236*	*2 209*	*2 070*	*1 850*	*1 478*
Bigeye scad	**Sélar coulisou**		**Chicharro ojón**		***Selar crumenophthalmus***				**1,70(23)291,01**	**BIS**
31 Grenada	324	341	191	48	100	181	53	72	137	97
St Kitts Nev	...	...	...	-	-	16	20	36	35	28
Venezuela	2 091	2 801	2 531	2 836	2 353	2 425	2 653	3 765	1 704	1 774
31 Fishing area total	*2 415*	*3 142*	*2 722*	*2 884*	*2 453*	*2 622*	*2 726*	*3 873*	*1 876*	*1 899*
57 Thailand	-	-	-	-	1 984	1 904	3 830	3 442	3 395	3 374
57 Fishing area total	-	-	-	-	*1 984*	*1 904*	*3 830*	*3 442*	*3 395*	*3 374*
71 Philippines	37 766	33 200	49 937	43 592	43 660	54 167	61 999	65 776	71 365	79 307
Thailand	-	-	-	-	24 533	22 188	24 931	25 966	25 610	25 454

B-37

Miscellaneous pelagic fishes — **Capture production by species, fishing areas and countries or areas**
Poissons pélagiques divers — **Captures par espèces, zones de pêche et pays ou zones**
Peces pelágicos diversos — **Capturas por especies, áreas de pesca y países o áreas**

Species, Fishing area Espèce, Zone de pêche Especie, Area de pesca	1992 mt	1993 mt	1994 mt	1995 mt	1996 mt	1997 mt	1998 mt	1999 mt	2000 mt	2001 mt
71 Fishing area total	*37 766*	*33 200*	*49 937*	*43 592*	*68 193*	*76 355*	*86 930*	*91 742*	*96 975*	*104 761*
Species total	*40 181*	*36 342*	*52 659*	*46 476*	*72 630*	*80 881*	*93 486*	*99 057*	*102 246*	*110 034*
Yellowstripe scad	**Sélar à bande dorée**		**Chicharro banda dorada**		***Selaroides leptolepis***				**1,70(23)422,01**	**TRY**
51 Untd Arab Em	355	360	403	2 878	2 908	3 108	3 118	3 196	2 635	2 630 F
51 Fishing area total	*355*	*360*	*403*	*2 878*	*2 908*	*3 108*	*3 118*	*3 196*	*2 635*	*2 630 F*
57 Malaysia	2 708	2 946	3 684	2 903	3 656	5 505	3 643	4 871	3 437	3 693
57 Fishing area total	*2 708*	*2 946*	*3 684*	*2 903*	*3 656*	*5 505*	*3 643*	*4 871*	*3 437*	*3 693*
71 Malaysia	22 427	26 510	23 870	20 956	24 467	26 526	29 091	36 470	40 816	36 165
71 Fishing area total	*22 427*	*26 510*	*23 870*	*20 956*	*24 467*	*26 526*	*29 091*	*36 470*	*40 816*	*36 165*
Species total	*25 490*	*29 816*	*27 957*	*26 737*	*31 031*	*35 139*	*35 852*	*44 537*	*46 888*	*42 488 F*
Blackbanded trevally	**Sériole amourez**		**...C**		***Seriolina nigrofasciata***				**1,70(23)425,01**	**RNJ**
57 Thailand	-	-	1 373	2 882	3 666	3 564	2 954	2 700	2 663	2 647
57 Fishing area total	-	-	*1 373*	*2 882*	*3 666*	*3 564*	*2 954*	*2 700*	*2 663*	*2 647*
71 Thailand	-	-	3 934	4 048	3 603	3 532	3 088	3 345	3 299	3 279
71 Fishing area total	-	-	*3 934*	*4 048*	*3 603*	*3 532*	*3 088*	*3 345*	*3 299*	*3 279*
Species total	-	-	*5 307*	*6 930*	*7 269*	*7 096*	*6 042*	*6 045*	*5 962*	*5 926*
Carangids nei	**Carangidés nca**		**Carángidos nep**		***Carangidae***				**1,70(23)XXX,XX**	**CGX**
27 Portugal	...	...	...	...	...	...	...	43	28	28
27 Fishing area total	...	...	...	...	...	...	...	*43*	*28*	*28*
31 Antigua Barb	...	...	...	...	...	...	...	...	...	33
Bahamas	79	113	106	72	91	103	92	79	81	101
Barbados	31	28	15	24	28	19	14	6	28	11
Bermuda	56	60	59	86	70	50	48	51	30	41
Dominican Rp	65	18	32	80	30	10	27	6	22	28
Grenada	8	17	8	7	10	17	10	16	12	14
Mexico	236	205	217	164	201	454	364	342	111	-
Puerto Rico	...	...	...	...	...	...	...	50	70	66
Venezuela	37	41	32	25	6	23	31	11	26	5
31 Fishing area total	*512*	*482*	*469*	*458*	*436*	*676*	*586*	*561*	*380*	*299*
34 Benin	369 F	400 F	500 F	600 F	700 F	800 F	900 F	1 027	700 F	475
Cameroon	30	5	5 F	6	5	5 F	2	2	4	5
Côte dIvoire	2 109	1 282	626	200	169	1 400	1 267	1 025	509	500 F
Greece	-	-	-	-	-	-	11	-	4	1
Guinea	...	...	...	253	311	257	392	326	764	750 F
Latvia	-	-	-	-	-	36	-	-	-	-
Mauritania	200 F	200 F	200 F	100 F	100 F	100 F	100 F	100 F	112 F	...
Nigeria	278	9 850	5 468	5 199	2 923	1 035	700	4 451	3 521	833
Portugal	2	34	7	-	-	-	-	2	1	2
Senegal	6 988	3 373	4 636	4 772	3 299	4 806	7 491	4 456	5 725	2 616
Sierra Leone	...	...	...	...	...	52	158	-	39	-
Togo	163	326	659	332	273	260	291	1 613	2 433	2 482
34 Fishing area total	*10 139 F*	*15 470 F*	*12 101 F*	*11 462 F*	*7 780 F*	*8 751 F*	*11 312 F*	*13 002 F*	*13 812 F*	*7 664 F*
37 Israel	108	102	30	171	...	...	...	...	...	...
Lebanon	125 F	150	150	450	450	350	400	350	450	400
Malta	49	72	29	13	7	4	13	23	28	8
Spain	-	-	-	-	-	-	-	39	34	393
37 Fishing area total	*282 F*	*324*	*209*	*634*	*457*	*354*	*413*	*412*	*512*	*801*
41 Brazil	7 630 F	7 610 F	7 630 F	-	-	-	-	-	-	-
Italy	162	137	137	-	-	-	-	-	-	-
Latvia	-	61	-	-	-	-	-	-	-	-
41 Fishing area total	*7 792 F*	*7 808 F*	*7 767 F*	-	-	-	-	-	-	-
47 Russian Fed	-	-	-	7	40	227	81	-	38	805
St Helena	3	3	2	2	2	3	2	3	1	1
47 Fishing area total	*3*	*3*	*2*	*9*	*42*	*230*	*83*	*3*	*39*	*806*
51 Bahrain	361	515	382	495	668	498	359	524	414	450
Comoros	500	500	500	500 F	490 F	470 F	470 F	450 F	500 F	500
Djibouti	14 F	15 F	16 F	20 F	20 F	20 F	20 F	20 F	20 F	20 F
Eritrea	...	...	48	697	818	13	184	386	2 026	1 172
India	25 782	6 180	9 922	13 341	28 831	28 764	21 426	22 589	5 679	5 909
Italy	243	206	206	-	-	-	-	-	-	-
Kenya	141	68	73	76	82	111	86	101	85	119
Kuwait	66	92	113	111	97	138	74	73	150 F	242
Mauritius	152	165	167	165	43	58	46	33	53	51
Oman	2 612	2 113	2 123	3 607	2 708	3 216	2 323	2 556	2 485	1 302
Pakistan	9 628	13 111	13 760	16 495	15 957	19 002	18 689	17 779	16 545	15 988
Qatar	443	452	250	291	308	308	326	287	420	552
Réunion	...	...	...	...	...	...	...	107	130	131
Saudi Arabia	4 444	4 920	5 730	4 547	5 524	5 125	4 795	5 534	6 612	5 943
Seychelles	1 925	1 492	1 066	1 284	1 568	1 562	1 008	1 474	1 764	1 285
Tanzania	2 000	1 216	1 016	1 380	3 669	1 800	2 000	2 000	2 500	2 000
Untd Arab Em	5 696	5 700	6 371	5 800	5 861	6 205	6 810	7 020	3 107	3 100 F

B-37

Miscellaneous pelagic fishes — **Capture production by species, fishing areas and countries or areas**
Poissons pélagiques divers — **Captures par espèces, zones de pêche et pays ou zones**
Peces pelágicos diversos — **Capturas por especies, áreas de pesca y países o áreas**

Species, Fishing area Espèce, Zone de pêche Especie, Area de pesca	1992 mt	1993 mt	1994 mt	1995 mt	1996 mt	1997 mt	1998 mt	1999 mt	2000 mt	2001 mt
51 Fishing area total	*54 007 F*	*36 745 F*	*41 743 F*	*48 809 F*	*66 644 F*	*67 290 F*	*58 616 F*	*60 933 F*	*42 490 F*	*38 764 F*
57 India	5 102	7 249	9 301	9 148	5 394	3 988	4 845	7 327	5 450	7 182
Indonesia	19 572	24 273	24 549	28 076	27 313	27 764	29 932	29 830	30 059	30 920
Sri Lanka	8 948	10 878	8 000	6 910	6 088	6 900	8 500	8 680	10 450	9 950
Thailand	3 197	14 304	13 180	8 226	8 663	8 391	10 395	8 582	8 465	8 413
57 Fishing area total	*36 819*	*56 704*	*55 030*	*52 360*	*47 458*	*47 043*	*53 672*	*54 419*	*54 424*	*56 465*
71 Guam	...	...	5	1	1	1	1	2	1	1
Indonesia	80 900	81 673	89 381	88 693	88 781	97 740	98 527	98 965	99 854	102 730
N Marianas	...	...	1	0	1	2	2	2	1	3
Palau	13	16	10	22	28	7	19	30 F	56 F	56 F
Philippines	47 066	44 624	47 539	39 682	37 456	32 175	31 472	35 204	34 713	35 889
Singapore	117	89	9	22	37	33	36	30	21	12
Thailand	42 531	42 224	55 616	47 456	44 365	41 356	35 599	35 668	35 180	34 965
71 Fishing area total	*170 627*	*168 626*	*192 561*	*175 876*	*170 669*	*171 314*	*165 656*	*169 901 F*	*169 826 F*	*173 656 F*
77 Amer Samoa	...	...	2	1	1	2	1	1	2	1
Mexico	175	171	180	61	73	318	1 628	735	379	-
77 Fishing area total	*175*	*171*	*182*	*62*	*74*	*320*	*1 629*	*736*	*381*	*1*
87 Chile	294	251	143	89	96	205	348	203	307	363
Ecuador	465	497	1 328	199	65	45	839	1 632	...	68
87 Fishing area total	*759*	*748*	*1 471*	*288*	*161*	*250*	*1 187*	*1 835*	*307*	*431*
Species total	*281 115 F*	*287 081 F*	*311 535 F*	*289 958 F*	*293 721 F*	*296 228 F*	*293 154 F*	*301 845 F*	*282 199 F*	*278 915 F*
Atlantic pomfret	**Grande castagnole**		**Japuta**			***Brama brama***			**1,70(27)003,01**	**POA**
27 Denmark	0	0	0	-	-	-	-	-	-	0
France	30	40	23	10	1	0	2	2	2	-
Ireland	-	-	-	-	-	-	-	-	1	184
Portugal	232	589	37	18	34	91	9	9	5	7
Spain	3 819	5 191	5 478	5 592	5 313	3 334	2 821	2 233	559	244
27 Fishing area total	*4 081*	*5 820*	*5 538*	*5 620*	*5 348*	*3 425*	*2 832*	*2 244*	*567*	*435*
34 Portugal	101	45	14	-	-	-	-	-	3	3
Russian Fed	-	6	-	2	-	26	-	103	21	-
Ukraine	-	-	-	-	-	-	-	-	9	-
34 Fishing area total	*101*	*51*	*14*	*2*	*-*	*26*	*-*	*103*	*33*	*3*
37 France	...	...	...	...	...	...	...	5	6	5
Spain	-	-	-	-	26	157	87	113	27	104
37 Fishing area total	*...*	*...*	*...*	*...*	*26*	*157*	*87*	*118*	*33*	*109*
47 Namibia	73	62	39	20	19	9	15	99	493	684
Russian Fed	16	4	4	-	103	30	-	9	-	-
South Africa	868	1 717	2 019	1 709	1 070	381	307	324	350 F	437
47 Fishing area total	*957*	*1 783*	*2 062*	*1 729*	*1 192*	*420*	*322*	*432*	*843 F*	*1 121*
81 New Zealand	...	...	...	449	480	413	488	465	401	903
81 Fishing area total	*...*	*...*	*...*	*449*	*480*	*413*	*488*	*465*	*401*	*903*
87 Chile	-	-	1 186	3 930	5 585	5 998	6 332	6 830	8 160	15 156
87 Fishing area total	*-*	*-*	*1 186*	*3 930*	*5 585*	*5 998*	*6 332*	*6 830*	*8 160*	*15 156*
Species total	*5 139*	*7 654*	*8 800*	*11 730*	*12 631*	*10 439*	*10 061*	*10 192*	*10 037 F*	*17 727*
Pomfrets, ocean breams nei	**Castagnoles nca**		**Japutas nep**			***Bramidae***			**1,70(27)XXX,XX**	**BRZ**
77 Amer Samoa	-	-	-	-	-	-	-	-	-	1
USA	-	-	-	-	-	-	-	-	-	4
77 Fishing area total	*-*	*-*	*-*	*-*	*-*	*-*	*-*	*-*	*-*	*5*
81 Russian Fed	-	2	-	-	-	-	-	-	-	-
Ukraine	2	-	-	-	-	-	-	-	2	-
81 Fishing area total	*2*	*2*	*-*	*-*	*-*	*-*	*-*	*-*	*2*	*-*
Species total	*2*	*2*	*-*	*-*	*-*	*-*	*-*	*-*	*2*	*5*
Common dolphinfish	**Coryphène commune**		**Lampuga**			***Coryphaena hippurus***			**1,70(28)071,01**	**DOL**
27 Portugal	-	-	-	-	-	-	-	1	1	3
27 Fishing area total	*-*	*-*	*-*	*-*	*-*	*-*	*-*	*1*	*1*	*3*
31 Antigua Barb	...	...	...	...	...	...	...	...	...	4
Barbados	1 470	513	499	758	849	721	482	745	728	574
Br Virgin Is	...	...	5	2	6	3	2	3	1	1 F
Dominican Rp	200	225	405	89	237	113	151	175	255	232
Grenada	157	103	124	182	130	171	153	162	167	221
Guadeloupe	640 F	650 F	650 F	700 F	730	800 F	670 F	670 F	700 F	700 F
Martinique	330 F	400 F	400 F	350 F	250 F	350 F	320 F	300 F	250 F	220 F
Mexico	39	49	175	88	244	871	1 078	376	308	...
Puerto Rico	...	...	...	...	...	...	...	83	111	74
St Kitts Nev	...	...	...	3	13	20	34	13	29	29
St Lucia	...	...	144	200	351	455	264	588	552	427
USA	506	496	620	980	739	763	322	555	498	411
Venezuela	208	244	274	447	...	...	...	290	141	...
31 Fishing area total	*3 550 F*	*2 680 F*	*3 296 F*	*3 799 F*	*3 549 F*	*4 267 F*	*3 476 F*	*3 960 F*	*3 740 F*	*2 893 F*

B-37

Miscellaneous pelagic fishes — Capture production by species, fishing areas and countries or areas
Poissons pélagiques divers — Captures par espèces, zones de pêche et pays ou zones
Peces pelágicos diversos — Capturas por especies, áreas de pesca y países o áreas

Species, Fishing area Espèce, Zone de pêche Especie, Area de pesca	1992 mt	1993 mt	1994 mt	1995 mt	1996 mt	1997 mt	1998 mt	1999 mt	2000 mt	2001 mt
34 Benin	...	...	...	...	...	...	...	4	3 F	48
Liberia	...	...	...	...	...	31	20	48	45	22
Portugal	-	-	-	1	1	0	1	-	0	1
Senegal	-	-	-	-	-	-	103	115	150	258
Togo	3	3	2	1	3	3	0	10	36	148
34 Fishing area total	*3*	*3*	*2*	*2*	*4*	*34*	*124*	*177*	*234 F*	*477*
37 Malta	188	174	334	334	307	295	363	349	234	303
Spain	-	-	-	-	-	-	-	92	137	70
Tunisia	179	340	366	434	399	538	1 088	919	667	784
37 Fishing area total	*367*	*514*	*700*	*768*	*706*	*833*	*1 451*	*1 360*	*1 038*	*1 157*
41 Brazil	1 710 F	1 710 F	1 720 F	3 186	2 500	4 028	4 117	2 848	4 359	4 300 F
41 Fishing area total	*1 710 F*	*1 710 F*	*1 720 F*	*3 186*	*2 500*	*4 028*	*4 117*	*2 848*	*4 359*	*4 300 F*
47 South Africa	-	-	-	-	-	-	-	-	-	1
47 Fishing area total	-	-	-	-	-	-	-	-	-	*1*
51 Oman	...	11	...	...	...	...	...	-	-	-
Pakistan	1 577	1 875	2 054	2 570	1 841	1 658	1 892	3 109	1 954	1 869
Réunion	...	...	...	...	...	...	...	221	194	141
South Africa	-	-	-	-	-	1	-	-	-	-
Untd Arab Em	42	64	46	55	56	60	61	63	55	50 F
51 Fishing area total	*1 619*	*1 950*	*2 100*	*2 625*	*1 897*	*1 719*	*1 953*	*3 393*	*2 203*	*2 060 F*
57 Japan	-	-	-	-	-	-	3	-	-	-
57 Fishing area total	-	-	-	-	-	-	*3*	-	-	-
61 China,Taiwan	13 200	10 385	6 881	12 203	3 209	8 089	17 157	8 560	5 558	7 427
Japan	14 765	9 187	7 586	9 777	8 164	10 218	14 909	9 269	8 953	9 123
61 Fishing area total	*27 965*	*19 572*	*14 467*	*21 980*	*11 373*	*18 307*	*32 066*	*17 829*	*14 511*	*16 550*
67 Japan	66	-	-	-	-	-	-	-	-	-
67 Fishing area total	*66*	-	-	-	-	-	-	-	-	-
71 Guam	143	180	90	72	35	43	79	40	34	53
Japan	3 878	2 243	525	428	357	10	32	11	6	19
N Marianas	...	...	7	11	16	17	9	6	3	6
Palau	1	0	1	1	0	0	0	0	0	-
Philippines	898	1 491	1 799	4 253	3 544	359	189	223	211	226
71 Fishing area total	*4 920*	*3 914*	*2 422*	*4 765*	*3 952*	*429*	*309*	*280*	*254*	*304*
77 Amer Samoa	...	...	5	5	5	17	10	13	15	16
Costa Rica	...	...	...	...	...	...	...	...	8 370	11 221
Fr Polynesia	45	467	256	178	257	427	437	429	446	651
Japan	92	-	1	7	6	1	-	-	-	-
Mexico	162	230	9	26	70	113	115	104	201	...
Nicaragua	...	...	...	...	...	...	...	710	2 540	2 470
USA	...	...	...	...	...	5	3	17	43	7
77 Fishing area total	*299*	*697*	*271*	*216*	*338*	*563*	*565*	*1 273*	*11 615*	*14 365*
81 New Zealand	-	-	-	1	0	0	0	1	0	15
81 Fishing area total	-	-	-	*1*	*0*	*0*	*0*	*1*	*0*	*15*
87 Chile	173	374	206	104	179	124	80	109	131	136
87 Fishing area total	*173*	*374*	*206*	*104*	*179*	*124*	*80*	*109*	*131*	*136*
Species total	*40 672 F*	*31 414 F*	*25 184 F*	*37 446 F*	*24 498 F*	*30 304 F*	*44 144 F*	*31 231 F*	*38 086 F*	*42 261 F*
Suckerfishes, remoras nei	**Rémoras nca**		**Remoras, pegas nep**		***Echeneidae***			**1,70(42)XXX,XX**		**ECN**
34 Liberia	-	-	-	-	-	-	3	1	12	...
34 Fishing area total	-	-	-	-	-	-	*3*	*1*	*12*	...
Species total	-	-	-	-	-	-	*3*	*1*	*12*	...
Chub mackerel	**Maquereau espagnol**		**Estornino**		***Scomber japonicus***			**1,75(01)002,01**		**MAS**
27 France	26	122	412	374	126	29	103	104	109	110
Portugal	9 139	7 606	4 810	4 351	5 385	6 116	7 218	14 061	10 703	4 492
27 Fishing area total	*9 165*	*7 728*	*5 222*	*4 725*	*5 511*	*6 145*	*7 321*	*14 165*	*10 812*	*4 602*
31 Mexico	7	9	9	2	2	1	0	0	3	0
USA	0	2	19	0	-	6	18	10	7	13
Venezuela	285	431	401	377	416	549	753	631	399	514
31 Fishing area total	*292*	*442*	*429*	*379*	*418*	*556*	*771*	*641*	*409*	*527*
34 Belize	-	-	-	-	-	1 100	1 051	1 522	4 850	995
Benin	344 F	330 F	320 F	300 F	280 F	300 F	250 F	205	140 F	314
Bulgaria	120	-	-	-	-	-	486	-	-	-
Cape Verde	...	...	...	...	...	...	182	...	...	...
China	-	-	4	-	-	2	-	8	59	36
Côte dIvoire	340	230	8	679	313	1 355	1 340	2 494	259	110
Cyprus	-	-	-	-	...	...	2 758	1 906	6 207	5 500
Eq Guinea	270 F	240 F	360 F	150 F	330 F	420 F	400	470 F	50 F	50 F
Estonia	1 107	3 390	692	2 991	4 486	1 980	7 215	3 416	-	-
France	-	-	-	-	-	...	1 491	213	334	-
Georgia	1 315 F	700 F	260 F	-	-	-	-	-	-	-

B-37 Miscellaneous pelagic fishes / Poissons pélagiques divers / Peces pelágicos diversos

Capture production by species, fishing areas and countries or areas / Captures par espèces, zones de pêche et pays ou zones / Capturas por especies, áreas de pesca y países o áreas

Species, Fishing area Espèce, Zone de pêche Especie, Area de pesca	1992 mt	1993 mt	1994 mt	1995 mt	1996 mt	1997 mt	1998 mt	1999 mt	2000 mt	2001 mt
Germany	-	-	-	-	87	645	315	334	-	20
Ghana	11 982	4 286	9 769	12 473	15 590	19 883	30 160	15 482	28 465	14 338
Greece	3	0	1	1	-	13	-	-	-	-
Guinea	...	...	...	...	2 043	1 006	849	...	...	...
Korea Rep	-	-	-	-	-	-	-	61	37	26
Latvia	3 953	2 612	4 056	3 494	3 765	1 931	2 562	3 123	7 151	9 227
Liberia	52	37	12	134	24	8	45	218	238	34
Lithuania	4 204	1 887	2 351	...	...	...	5 420	2 105	3 871	2 798
Mauritania	200 F	180 F	230 F	-	110 F	80 F	150 F	...	...	...
Morocco	14 252	16 970	38 215	28 350	16 154	28 018	11 261	15 786	62 160	25 679
Netherlands	-	-	-	-	857	3 202	1 836	1 561	12 005	12 708
Panama	-	-	-	201	-	-	397	-	-	-
Poland	-	-	-	-	4 086	480	-	-	-	1 666
Portugal	787	1 232	1 082	859	2 273	1 665	553	897	890	446
Romania	3 003	67	-	-	-	-	-	-	-	-
Russian Fed	23 490	14 093	29 583	52 551	72 874	61 309	65 958	47 215	43 654	29 619
St Vincent	-	-	-	-	-	350	2 644	988	2 951	4 475
Senegal	603	1 394	1 149	1 748	947	1 670	2 117	988	2 656	2 710
Spain	-	-	-	-	894	99	352	302	3 697	1 138
Togo	82	129	97	158	112	252	183	454	308	204
Ukraine	14 268	7 361	15 072	65 926	98 944	120 272	72 959	42 283	44 961	19 988
Vanuatu	-	54	-	-	-	-	-	-	-	-
Other nei	-	-	-	-	-	-	-	781	4 002	3 338
34 Fishing area total	*80 375 F*	*55 192 F*	*103 261 F*	*170 015 F*	*224 169 F*	*246 040 F*	*212 934 F*	*142 812 F*	*228 945 F*	*135 419 F*
37 Cyprus	...	...	...	...	...	...	...	...	...	3
France	-	-	-	-	58	69	69	45	27	77
Gaza Strip	...	...	...	...	130	120	337	340 F	340 F	280 F
Greece	5 146	10 996	14 308	6 468	6 520	5 627	2 173	1 850	2 286	1 829
Israel	108	80 F	50 F	12	71	47	54	44	8	10 F
Malta	10	10	10	13	23	29	40	19	34	32
Morocco	1 663	1 341	980	1 756	834	757	360	1 597	1 221	342
Slovenia	2	1	4	3	10	4	17	3	4	3
Tunisia	4 910	3 263	2 333	2 562	1 439	4 118	3 233	3 515	2 200	3 727
Turkey	14 762	15 908	16 748	17 410	10 444	10 850	10 120	10 200	9 000	4 500
37 Fishing area total	*26 601*	*31 599 F*	*34 433 F*	*28 224*	*19 529*	*21 621*	*16 403*	*17 613 F*	*15 120 F*	*10 803 F*
41 Argentina	4 913	6 689	10 297	13 442	11 195	10 468	3 224	7 012	10 122	3 360
Brazil	5 120 F	5 100 F	5 110 F	8 049	5 670	8 306	9 422	1 595	6 377	6 300 F
Estonia	-	24	-	-	-	-	-	-	-	-
Korea Rep	-	44	-	-	-	-	-	-	-	-
Uruguay	5	3	5	5	4	5	1	5	-	0
41 Fishing area total	*10 038 F*	*11 860 F*	*15 412 F*	*21 496*	*16 869*	*18 779*	*12 647*	*8 612*	*16 499*	*9 660 F*
47 Iceland	-	-	-	-	-	28	-	-	-	-
Japan	56	-	-	-	-	-	-	-	-	-
Korea Rep	3	69	759	9	-	31	-	-	-	-
Lithuania	-	-	6	-	-	-	-	-	-	-
Namibia	...	...	...	...	...	...	...	...	1 641	6 329
Portugal	-	-	-	-	-	-	-	4	-	-
Russian Fed	20	8	40	80	66	3 439	2 369	1 150	2 182	2 090
St Helena	9	7	8	7	7	12	2	8	4	11
South Africa	4 263	4 721	4 508	4 677	4 139	8 039	3 201	1 765	1 600 F	1 620
Spain	-	-	-	-	-	1	26	-	-	-
Ukraine	-	-	-	284	...	...	-	-	-	-
47 Fishing area total	*4 351*	*4 805*	*5 321*	*5 057*	*4 212*	*11 550*	*5 598*	*2 927*	*5 427 F*	*10 050*
51 Egypt	185	453	3 041	1 926	2 042	2 392	810	378	3 561	2 747
Korea Rep	-	-	32	-	-	-	-	4	-	-
South Africa	-	1	-	-	11	5	3	3 F	...	-
Ukraine	940	-	-	-	-	-	-	-	-	-
51 Fishing area total	*1 125*	*454*	*3 073*	*1 926*	*2 053*	*2 397*	*813*	*385 F*	*3 561*	*2 747*
61 China	243 143	272 604	336 091	372 038	374 400	408 933	385 183	402 540	350 750	381 561
China,Taiwan	42 728	41 908	54 914	54 155	53 906	47 279	35 527	45 339	28 635	24 298
Japan	266 609	664 298	633 062	469 447	760 430	848 967	511 238	381 865	346 220	375 273
Korea Rep	116 422	174 684	210 442	200 481	415 003	160 448	172 925	177 540	145 908	203 717
Russian Fed	-	-	-	-	-	-	-	-	-	1
61 Fishing area total	*668 902*	*1 153 494*	*1 234 509*	*1 096 121*	*1 603 739*	*1 465 627*	*1 104 873*	*1 007 284*	*871 513*	*984 850*
67 Russian Fed	-	-	-	-	-	-	41	-	-	-
USA	785	30	33	1	388	204	591	75	121	322
67 Fishing area total	*785*	*30*	*33*	*1*	*388*	*204*	*632*	*75*	*121*	*322*
71 Philippines	3 344	4 842	6 605	3 307	2 991	2 025	1 507	1 485	1 422	1 520
71 Fishing area total	*3 344*	*4 842*	*6 605*	*3 307*	*2 991*	*2 025*	*1 507*	*1 485*	*1 422*	*1 520*
77 Korea Rep	-	-	-	-	-	-	-	4	-	-
Mexico	19 839	20 608	12 199	22 832	12 959	24 245	22 990	69 375	45 202	3 813
USA	18 197	10 873	10 007	8 606	9 589	18 186	19 814	8 639	21 221	6 925
77 Fishing area total	*38 036*	*31 481*	*22 206*	*31 438*	*22 548*	*42 431*	*42 804*	*78 018*	*66 423*	*10 738*
87 Chile	72 364	96 023	27 171	110 210	146 649	211 649	71 769	120 123	95 789	365 031
Ecuador	26 322	45 322	38 991	57 950	79 484	192 182	44 716	28 307	84 324	85 378
Ghana	-	-	-	-	-	-	-	-	1 148	855
Japan	-	-	-	-	-	-	-	1	-	-
Latvia	4	-	-	-	-	-	-	-	-	-
Lithuania	36	-	-	-	-	-	-	-	-	-
Peru	17 939	29 504	44 115	44 259	49 221	206 183	401 903	527 729	73 263	176 202

B-37 Miscellaneous pelagic fishes / Poissons pélagiques divers / Peces pelágicos diversos

Capture production by species, fishing areas and countries or areas / Captures par espèces, zones de pêche et pays ou zones / Capturas por especies, áreas de pesca y países o áreas

Species, Fishing area Espèce, Zone de pêche Especie, Area de pesca	1992 mt	1993 mt	1994 mt	1995 mt	1996 mt	1997 mt	1998 mt	1999 mt	2000 mt	2001 mt
Russian Fed	970	-	-	-	-	-	-	-	-	-
87 Fishing area total	*117 635*	*170 849*	*110 277*	*212 419*	*275 354*	*610 014*	*518 388*	*676 160*	*254 524*	*627 466*
Species total	*960 649 F*	*1 472 776 F*	*1 540 781 F*	*1 575 108 F*	*2 177 781 F*	*2 427 389 F*	*1 924 691 F*	*1 950 177 F*	*1 474 776 F*	*1 798 704 F*
Atlantic mackerel	**Maquereau commun**		**Caballa del Atlántico**		***Scomber scombrus***			**1,75(01)002,05**		**MAC**
21 Canada	25 974	26 474	20 612	18 300	21 027	21 305	19 557	16 317	17 637	24 488
Cuba	334	666	44	61	78	115	7	13	-	-
Japan	11	-	-	-	-	-	-	-	-	-
Lithuania	704	-	-	-	-	-	-	-	-	-
Russian Fed	1 202	22	-	-	-	-	1	-	-	3
St Pier Mq	4	4	3	1	1	1	3	1	26	7
USA	12 629	4 691	8 959	8 495	16 022	15 395	14 429	12 041	5 649	12 339
21 Fishing area total	*40 858*	*31 857*	*29 618*	*26 857*	*37 128*	*36 816*	*33 997*	*28 372*	*23 312*	*36 837*
27 Belgium	104	193	353	108	64	106	125	178	151	98
Bulgaria	1 997	1 883	-	-	-	-	-	-	-	-
Channel Is	1	1 F	1	1 F	9	9	23	18	16	14
Denmark	37 429	42 056	46 839	36 758	26 238	24 054	27 415	29 705	31 642	31 370
Estonia	616	1 100	3 302	2 286	3 741	6 324	7 356	3 595	2 673	218
Faeroe Is	14 898	13 978	21 570	34 924	19 530	8 401	10 654	11 334	21 022	24 005
France	17 694	20 011	25 787	22 807	13 167	14 368	18 764	17 400	20 897	20 958
Germany	26 502	28 734	26 492	24 417	16 229	15 864	21 490	19 960	22 980	25 325
Iceland	-	-	0	0	92	927	357	144	0	1
Ireland	89 620	94 979	86 274	78 534	49 966	53 094	67 310	59 609	74 871	76 586
Isle of Man	1	0	1	1	0	0	0	4	0	8
Japan	-	4	-	-	-	-	-	-	-	-
Latvia	640	3 424	1 508	534	233	-	-	-	-	-
Lithuania	581	1 000	707	6 236	7 334	-	2 823	4 936	2 085	1 949
Netherlands	38 913	42 532	44 335	35 787	24 246	23 702	30 163	27 816	32 403	33 109
Norway	206 963	223 871	260 121	202 209	136 699	137 256	158 340	161 046	174 173	180 603
Poland	-	-	-	-	-	22	-	-	-	-
Portugal	3 576	2 015	2 149	3 073	3 009	2 083	2 898	2 035	2 254	3 121
Romania	-	-	2 903	30 844	7 265	-	-	-	-	-
Russian Fed	47 091	46 694	29 509	46 249	43 046	53 732	67 837	51 348	50 772	41 568
Spain	5 146	10 452	11 452	10 595	13 748	20 301	25 541	24 026	25 384	24 382
Sweden	5 025	3 610	7 515	6 268	5 387	4 390	5 161	5 003	4 500	5 098
Ukraine	2 991	4 514	7 193	-	-	-	-	-	-	-
UK	234 583	257 736	238 457	218 417	144 964	149 448	179 711	166 658	193 638	198 953
27 Fishing area total	*734 371*	*798 787 F*	*816 468*	*760 048 F*	*514 967*	*514 081*	*625 968*	*584 815*	*659 461*	*667 366*
34 Greece	0	0	0	6	6	13	-	-	-	-
34 Fishing area total	*0*	*0*	*0*	*6*	*6*	*13*	...	...	...	...
37 Algeria	382 F	379 F	539 F	1 243	590	750	1 055	1 053	1 000 F	1 000 F
France	1 149	1 326	1 313	789	741	2 199	2 199	856	1 743	1 342
Greece	2 183	1 551	1 360	680	880	292	141	200	443	155
Slovenia	5	1	-	-	5	5	10	13	14	16
Spain	4 600 F	4 500 F	4 400 F	4 300 F	4 234	955	2 671	1 610	2 806	2 958
Syria	117 F	121 F	117 F	116 F	116	210	274	245	384	187
Turkey	1 151	2 311	1 105	296	592	300	650	850	900	550
37 Fishing area total	*9 587 F*	*10 189 F*	*8 834 F*	*7 424 F*	*7 158*	*4 711*	*7 000*	*4 827*	*7 290 F*	*6 208 F*
Species total	*784 816 F*	*840 833 F*	*854 920 F*	*794 335 F*	*559 259*	*555 621*	*666 965*	*618 014*	*690 063 F*	*710 411 F*
Blue mackerel	**Maquereau tacheté**		**Caballa pintoja**		***Scomber australasicus***			**1,75(01)002,07**		**MAA**
81 Japan	2 477	380	292	358	-	-	-	-	-	-
Korea Rep	-	-	-	-	1	-	5	-	-	-
New Zealand	10 172	9 554	5 467	7 534	2 837	8 768	7 041	12 417	10 431	9 761
Russian Fed	-	326	204	75	-	-	-	-	-	-
Ukraine	213	94	133	...	156	9	214	3 457	1 677	2 040
81 Fishing area total	*12 862*	*10 354*	*6 096*	*7 967*	*2 994*	*8 777*	*7 260*	*15 874*	*12 108*	*11 801*
Species total	*12 862*	*10 354*	*6 096*	*7 967*	*2 994*	*8 777*	*7 260*	*15 874*	*12 108*	*11 801*
Scomber mackerels nei	**Maquereaux scomber nca**		**Caballas scomber nep**		***Scomber spp***			**1,75(01)002,XX**		**MAZ**
34 Spain	-	-	-	-	1 489	-	-	23	-	-
34 Fishing area total	-	-	-	-	*1 489*	-	-	*23*	-	-
37 Albania	...	...	...	0	10	5	4	4	4	11
Croatia	1 224	1 541	580	615	650	998	585	175	33	91
Italy	5 139	7 997	5 862	6 949	8 012	7 866	7 277	5 748	5 522	6 033
Libya	2 550 F	2 750 F	3 000 F	3 050 F	3 000 F	3 000 F	3 000 F	3 000 F	3 000 F	3 000 F
Yugoslavia	2	5	5	7	8	9	13	14	13	13
37 Fishing area total	*8 915 F*	*12 293 F*	*9 447 F*	*10 621 F*	*11 680 F*	*11 878 F*	*10 879 F*	*8 941 F*	*8 572 F*	*9 148 F*
Species total	*8 915 F*	*12 293 F*	*9 447 F*	*10 621 F*	*13 169 F*	*11 878 F*	*10 879 F*	*8 964 F*	*8 572 F*	*9 148 F*
Short mackerel	**Maquereau trapu**		**Caballa rechoncha**		***Rastrelliger brachysoma***			**1,75(01)014,01**		**RAB**
71 Philippines	23 703	26 133	27 517	26 200	25 224	22 978	23 350	25 713	26 771	28 289
71 Fishing area total	*23 703*	*26 133*	*27 517*	*26 200*	*25 224*	*22 978*	*23 350*	*25 713*	*26 771*	*28 289*
Species total	*23 703*	*26 133*	*27 517*	*26 200*	*25 224*	*22 978*	*23 350*	*25 713*	*26 771*	*28 289*

B-37

Miscellaneous pelagic fishes — **Capture production by species, fishing areas and countries or areas**
Poissons pélagiques divers — **Captures par espèces, zones de pêche et pays ou zones**
Peces pelágicos diversos — **Capturas por especies, áreas de pesca y países o áreas**

Species, Fishing area Espèce, Zone de pêche Especie, Area de pesca	1992 mt	1993 mt	1994 mt	1995 mt	1996 mt	1997 mt	1998 mt	1999 mt	2000 mt	2001 mt
Indian mackerel	**Maquereau des Indes**		**Caballa de la India**		*Rastrelliger kanagurta*				**1,75(01)014,03**	**RAG**
51 Egypt	...	...	...	1 363	1 151	1 914	652	1 004	430	1 442
Eritrea	...	...	40	75	58	2	0	20	197	...
India	78 834	113 309	180 680	168 757	267 692	157 541	142 669	138 086	62 026	29 938
Oman	964	1 955	1 282	2 122	1 712	2 207	1 994	2 024	2 426	3 223
Saudi Arabia	1 658	1 741	3 240	3 069	1 549	1 990	2 072	1 979	1 525	1 803
Seychelles	308	304	586	322	425	298	25	101	295	182
Tanzania	3 000	2 542	3 248	3 490	4 618	4 000	4 050	4 500	4 500	5 000
Untd Arab Em	4 728	4 750	5 382	4 309	4 354	4 653	4 669	4 786	4 775	4 770 F
Yemen	6 567	6 000	6 500	6 958	752	790	900 F	900 F	900 F	1 150 F
51 Fishing area total	*96 059*	*130 601*	*200 958*	*190 465*	*282 311*	*173 395*	*157 031 F*	*153 400 F*	*77 074 F*	*47 508 F*
57 India	24 542	34 418	25 642	28 476	25 834	32 831	41 750	41 292	16 795	28 968
Thailand	8 547	13 743	13 695	25 118	23 801	23 400	24 289	20 973	20 686	20 560
57 Fishing area total	*33 089*	*48 161*	*39 337*	*53 594*	*49 635*	*56 231*	*66 039*	*62 265*	*37 481*	*49 528*
71 Fiji Islands	211	716	730 F	452	400	312	224	721	650 F	722
Philippines	63 152	57 945	59 505	51 352	46 264	54 732	51 919	53 606	55 088	59 232
Thailand	31 577	35 986	50 898	45 338	21 328	19 276	19 393	26 912	26 544	26 382
71 Fishing area total	*94 940*	*94 647*	*111 133 F*	*97 142*	*67 992*	*74 320*	*71 536*	*81 239*	*82 282 F*	*86 336*
Species total	*224 088*	*273 409*	*351 428 F*	*341 201*	*399 938*	*303 946*	*294 606 F*	*296 904 F*	*196 837 F*	*183 372 F*
Indian mackerels nei	**Maquereaux(Indo-pacifiq.) nca**		**Caballas Indo-Pacífico nep**		*Rastrelliger spp*				**1,75(01)014,XX**	**RAX**
51 Comoros	250	220	220	250 F	250 F	240 F	240 F	230 F	250 F	220
Seychelles	210	161	260	205	159	116	111	219	174	83
51 Fishing area total	*460*	*381*	*480*	*455 F*	*409 F*	*356 F*	*351 F*	*449 F*	*424 F*	*303*
57 Indonesia	54 866	62 123	64 351	65 365	67 318	70 536	72 027	67 919	75 083	76 850
Malaysia	46 066	36 104	63 771	101 001	73 780	66 586	79 446	84 665	65 733	65 971
Thailand	32 953	66 985	65 499	46 945	48 061	46 999	43 927	38 935	38 402	38 168
57 Fishing area total	*133 885*	*165 212*	*193 621*	*213 311*	*189 159*	*184 121*	*195 400*	*191 519*	*179 218*	*180 989*
71 Indonesia	122 226	111 823	130 531	128 525	121 592	130 868	132 736	133 547	131 954	135 070
Malaysia	31 184	31 871	29 875	25 169	21 584	20 215	22 626	26 700	32 322	33 498
Singapore	110	101	210	151	12	51	165	129	97	68
Thailand	96 598	76 997	82 021	112 280	92 765	91 622	107 083	125 175	123 461	122 708
71 Fishing area total	*250 118*	*220 792*	*242 637*	*266 125*	*235 953*	*242 756*	*262 610*	*285 551*	*287 834*	*291 344*
Species total	*384 463*	*386 385*	*436 738*	*479 891 F*	*425 521 F*	*427 233 F*	*458 361 F*	*477 519 F*	*467 476 F*	*472 636*
Mackerels nei	**Maquereaux nca**		**Caballas nep**		*Scombridae*				**1,75(01)XXX,XX**	**MAX**
27 France	-	9	-	-	-	120	25	13	1	2 642
Portugal	12	47	72	106	236	340	246	-	-	-
Spain	13 399	11 898	17 133	21 899	26 414	34 628	36 547	19 200	13 609	22 949
27 Fishing area total	*13 411*	*11 954*	*17 205*	*22 005*	*26 650*	*35 088*	*36 818*	*19 213*	*13 610*	*25 591*
31 Guadeloupe	1 270 F	1 280 F	1 290 F	1 400 F	1 550	1 700 F	1 430 F	1 430 F	1 600 F	1 600 F
31 Fishing area total	*1 270 F*	*1 280 F*	*1 290 F*	*1 400 F*	*1 550*	*1 700 F*	*1 430 F*	*1 430 F*	*1 600 F*	*1 600 F*
34 Gabon	...	...	81	145	64	65	114	158	304	201
GuineaBissau	8 F	10 F	10 F	12 F	14 F	15 F	14 F	10 F	10 F	10 F
Nigeria	1 059	2 871	2 833	2 254	3 745	3 326	3 640	2 919	2 710	3 759
Portugal	2	2	0	1	1	0	1	-	-	-
34 Fishing area total	*1 069 F*	*2 883 F*	*2 924 F*	*2 412 F*	*3 824 F*	*3 406 F*	*3 769 F*	*3 087 F*	*3 024 F*	*3 970 F*
37 France	-	-	-	-	-	-	-	-	3	7
Lebanon	150 F	150	150	500	500	300	300	300	350	350
37 Fishing area total	*150 F*	*150*	*150*	*500*	*500*	*300*	*300*	*300*	*353*	*357*
41 Portugal	-	-	-	-	-	-	-	21	0	-
41 Fishing area total	-	-	-	-	-	-	-	*21*	*0*	-
47 Iceland	-	-	-	-	-	11	-	-	-	-
47 Fishing area total	-	-	-	-	-	*11*	-	-	-	-
51 Mauritius	13	12	12	11	12	10	11	11	13	13
Portugal	-	-	-	-	-	-	-	14	-	-
Saudi Arabia	...	...	...	...	...	...	12	16	11	10
South Africa	-	-	-	-	-	-	-	-	0	0
51 Fishing area total	*13*	*12*	*12*	*11*	*12*	*10*	*23*	*41*	*24*	*23*
57 Australia	444	400	462	471	502	2	...	1 478	785	525
Sri Lanka	13 557	10 854	16 450	17 642	17 700	20 000	20 900	21 350	21 480	18 760
57 Fishing area total	*14 001*	*11 254*	*16 912*	*18 113*	*18 202*	*20 002*	*20 900*	*22 828*	*22 265*	*19 285*
71 Australia	0	0	0	0	85	87	91	1 792	2 500	2 432
NewCaledonia	26	65	104	126	119	102	104	161	161	161 F
71 Fishing area total	*26*	*65*	*104*	*126*	*204*	*189*	*195*	*1 953*	*2 661*	*2 593 F*
81 Australia	41	71	414	408	400	...	...	458	901	447
81 Fishing area total	*41*	*71*	*414*	*408*	*400*	...	...	*458*	*901*	*447*
87 Cuba	1	0	-	-	-	-	-	-	-	-
87 Fishing area total	*1*	*0*	-	-	-	-	-	-	-	-
Species total	*29 982 F*	*27 669 F*	*39 011 F*	*44 975 F*	*51 342 F*	*60 706 F*	*63 435 F*	*49 331 F*	*44 438 F*	*53 866 F*

B-37 Miscellaneous pelagic fishes / Poissons pélagiques divers / Peces pelágicos diversos

Capture production by species, fishing areas and countries or areas / Captures par espèces, zones de pêche et pays ou zones / Capturas por especies, áreas de pesca y países o áreas

Species, Fishing area Espèce, Zone de pêche Especie, Area de pesca	1992 mt	1993 mt	1994 mt	1995 mt	1996 mt	1997 mt	1998 mt	1999 mt	2000 mt	2001 mt
Blue butterfish	**Fiatole**		**Palometa fiatola**		***Stromateus fiatola***			**1,76(03)004,01**		**BLB**
34 Congo Rep	...	...	...	...	...	...	...	...	1	1 F
Côte dIvoire	12	3	6	6 F	10 F	15 F	20 F	22	45	27
Estonia	-	262	-	-	-	-	-	-	-	-
Korea Rep	122	77	-	-	-	-	-	-	-	-
Latvia	15	-	-	-	-	-	-	-	-	-
Liberia	334	299	307	382	143	236	248	573	181	183
Nigeria	112	0	0	-	-	-	-	-	-	4 106
Portugal	-	-	-	63	87	71	3	-	1	-
Russian Fed	-	-	-	15	-	-	-	4	-	-
Senegal	-	-	-	-	-	-	-	20	43	57
Spain	...	...	...	...	...	...	...	...	20	58
Ukraine	4	-	-	-	-	-	-	-	1	-
34 Fishing area total	*599*	*641*	*313*	*466 F*	*240 F*	*322 F*	*271 F*	*619*	*292*	*4 432 F*
47 Angola	0	0	0	0	0	-	-	-	-	-
Portugal	-	-	-	-	-	1	21	17	28	-
47 Fishing area total	*0*	*0*	*0*	*0*	*0*	*1*	*21*	*17*	*28*	-
Species total	*599*	*641*	*313*	*466 F*	*240 F*	*323 F*	*292 F*	*636*	*320*	*4 432 F*
Silver pomfret	**Aileron argenté**		**Palometón platero**		***Pampus argenteus***			**1,76(03)009,01**		**SIP**
51 Kuwait	1 103	1 094	1 112	1 101	862	560	501	259	200 F	133
Saudi Arabia	-	-	5	41	31	16	14	30	43	25
51 Fishing area total	*1 103*	*1 094*	*1 117*	*1 142*	*893*	*576*	*515*	*289*	*243 F*	*158*
57 Indonesia	4 434	4 547	4 446	4 985	4 849	5 100	6 304	6 236	5 919	6 310
Malaysia	...	...	2 501	2 924	2 623	3 006	3 040	1 487	1 201	1 400
Thailand	372	434	1 036	1 371	1 100	1 096	225	299	295	293
57 Fishing area total	*4 806*	*4 981*	*7 983*	*9 280*	*8 572*	*9 202*	*9 569*	*8 022*	*7 415*	*8 003*
61 China,Taiwan	4 476	4 258	4 049	6 978	5 113	2 703	1 621	2 477	2 852	4 245
Korea Rep	16 588	5 131	4 527	4 806	4 806	919	1 029	3 374	-	-
61 Fishing area total	*21 064*	*9 389*	*8 576*	*11 784*	*9 919*	*3 622*	*2 650*	*5 851*	*2 852*	*4 245*
71 Indonesia	11 783	12 945	12 261	13 763	15 083	15 361	15 004	15 104	19 573	20 860
Malaysia	...	...	2 717	2 269	2 397	2 087	2 361	2 884	2 330	2 714
Thailand	1 281	1 223	1 515	515	588	521	153	96	95	94
71 Fishing area total	*13 064*	*14 168*	*16 493*	*16 547*	*18 068*	*17 969*	*17 518*	*18 084*	*21 998*	*23 668*
Species total	*40 037*	*29 632*	*34 169*	*38 753*	*37 452*	*31 369*	*30 252*	*32 246*	*32 508 F*	*36 074*
Silver pomfrets nei	**Ailerons nca**		**Palometones nep**		***Pampus spp***			**1,76(03)009,XX**		**XPO**
61 China	73 208	116 553	138 335	209 031	220 364	242 547	303 024	337 919	338 848	352 493
61 Fishing area total	*73 208*	*116 553*	*138 335*	*209 031*	*220 364*	*242 547*	*303 024*	*337 919*	*338 848*	*352 493*
Species total	*73 208*	*116 553*	*138 335*	*209 031*	*220 364*	*242 547*	*303 024*	*337 919*	*338 848*	*352 493*
North Atlantic harvestfish	**Stromaté lune**		**Palometa mono**		***Peprilus alepidotus***			**1,76(03)011,01**		**HVF**
31 USA	23	36	13	25	14	15	13	16	23	18
31 Fishing area total	*23*	*36*	*13*	*25*	*14*	*15*	*13*	*16*	*23*	*18*
Species total	*23*	*36*	*13*	*25*	*14*	*15*	*13*	*16*	*23*	*18*
Gulf butterfishes, etc. nei	**Stromatés du Golfe, etc. nca**		**Pámpanos del Golfo, etc. nep**		***Peprilus spp***			**1,76(03)011,XX**		**BTG**
31 Venezuela	1 158	2 856	1 801	1 881	1 889	568	1 175	1 004	809	742
31 Fishing area total	*1 158*	*2 856*	*1 801*	*1 881*	*1 889*	*568*	*1 175*	*1 004*	*809*	*742*
Species total	*1 158*	*2 856*	*1 801*	*1 881*	*1 889*	*568*	*1 175*	*1 004*	*809*	*742*
Butterfishes, pomfrets nei	**Stromatés, ailerons nca**		**Pámpanos, palometónes nep**		***Stromateidae***			**1,76(03)XXX,XX**		**BUX**
21 USA	2 927	4 607	3 641	2 127	3 611	3 357	1 944	2 116	1 438	4 416
21 Fishing area total	*2 927*	*4 607*	*3 641*	*2 127*	*3 611*	*3 357*	*1 944*	*2 116*	*1 438*	*4 416*
27 Portugal	-	-	-	-	-	-	-	94	35	74
Spain	-	-	-	-	-	-	-	-	-	7
27 Fishing area total	-	-	-	-	-	-	-	*94*	*35*	*81*
31 USA	524	142	993	789	782	65	633	644	681	539
31 Fishing area total	*524*	*142*	*993*	*789*	*782*	*65*	*633*	*644*	*681*	*539*
41 Korea Rep	-	45	-	-	-	-	-	-	-	-
Portugal	-	-	-	-	-	-	5	-	-	20
41 Fishing area total	-	*45*	-	-	-	-	*5*	-	-	*20*
51 India	24 134	22 416	30 925	34 723	19 529	18 036	15 200	14 897	10 417	8 093
Kuwait	156	153	191	289	230	186	111	170	-	-
Pakistan	2 552	3 103	2 985	4 156	2 799	3 786	4 089	4 605	3 945	3 454
51 Fishing area total	*26 842*	*25 672*	*34 101*	*39 168*	*22 558*	*22 008*	*19 400*	*19 672*	*14 362*	*11 547*
57 India	7 399	22 395	31 998	23 255	17 967	18 695	18 119	16 288	14 974	15 663
Malaysia	4 884	4 631	228	268	240	275	278	165	133	109

B-37 Miscellaneous pelagic fishes / Poissons pélagiques divers / Peces pelágicos diversos

Capture production by species, fishing areas and countries or areas / Captures par espèces, zones de pêche et pays ou zones / Capturas por especies, áreas de pesca y países o áreas

Species, Fishing area Espèce, Zone de pêche Especie, Area de pesca	1992 mt	1993 mt	1994 mt	1995 mt	1996 mt	1997 mt	1998 mt	1999 mt	2000 mt	2001 mt
Thailand	-	-	-	-	-	-	-	-	16	
57 Fishing area total	*12 283*	*27 026*	*32 226*	*23 523*	*18 207*	*18 970*	*18 397*	*16 453*	*15 123*	*15 772*
61 China,H.Kong	3 409	2 321	2 331	2 428	2 114	1 973	2 172	1 500 F	1 900 F	2 100 F
China,Taiwan	2 816	1 894	378	226	246	336	231	280	131	146
Korea Rep	8 870	8 101	9 849	10 896	9 484	10 770	13 210	15 235	7 838	6 819
61 Fishing area total	*15 095*	*12 316*	*12 558*	*13 550*	*11 844*	*13 079*	*15 613*	*17 015 F*	*9 869 F*	*9 065 F*
71 Malaysia	5 177	5 498	248	208	220	192	217	405	328	268
Philippines	7 508	2 434	2 044	2 382	2 281	1 001	1 366	1 547	1 522	1 564
Singapore	53	82	123	103	100	94	103	94	75	48
71 Fishing area total	*12 738*	*8 014*	*2 415*	*2 693*	*2 601*	*1 287*	*1 686*	*2 046*	*1 925*	*1 880*
77 USA	-	-	-	-	-	-	1	2	3	7
77 Fishing area total	-	-	-	-	-	-	*1*	*2*	*3*	*7*
81 Japan	5 348	4 232	2 769	1 314	1 518	2 126	2 194	1 837	2 077	2 211
Korea Rep	-	-	-	-	-	-	39	-	-	-
81 Fishing area total	*5 348*	*4 232*	*2 769*	*1 314*	*1 518*	*2 126*	*2 233*	*1 837*	*2 077*	*2 211*
87 Chile	5	4	7	7	9	6	11	12	10	11
Ecuador	517	2 748	20 545	...	...	426	...	...	...	...
Japan	-	-	-	-	-	-	-	5	-	-
87 Fishing area total	*522*	*2 752*	*20 552*	*7*	*9*	*432*	*11*	*17*	*10*	*11*
Species total	*76 279*	*84 806*	*109 255*	*83 171*	*61 130*	*61 324*	*59 923*	*59 896 F*	*45 523 F*	*45 549 F*
Barracudas nei	**Bécunes nca**		**Picudas nep**			***Sphyraena spp***			**1,77(10)001,XX**	**BAR**
27 Portugal	...	...	...	...	...	...	...	33	41	33
27 Fishing area total	...	...	...	...	...	...	...	*33*	*41*	*33*
31 Antigua Barb	...	...	...	...	...	...	...	...	...	6
Dominican Rp	101	15	17	14	3	-	8	4	5	7
Grenada	38	37	28	24	40	30	41	35	57	43
Mexico	437	320	460	781	654	1 102	899	987	445	...
Puerto Rico	...	...	...	...	...	...	...	16	21	13
USA	...	...	...	...	...	0	1	-	53	55
Venezuela	666	1 039	998	923	899	998	1 123	881	731	1 165
31 Fishing area total	*1 242*	*1 411*	*1 503*	*1 742*	*1 596*	*2 130*	*2 072*	*1 923*	*1 312*	*1 289*
34 Benin	277 F	300 F	338 F	328 F	330 F	400 F	350 F	297	200 F	262
Cameroon	30	16	20 F	27	37	30 F	5	28	23	293
Congo Rep	0	0	0	0	0	0	0	0	1	1 F
Côte dIvoire	152	406	285	287 F	366	151	195	184	160	75
Gabon	160 F	215	184	420	1 055	883	1 238	975	976	1 636
Gambia	104	202	105	170	355	144	116	120	284	631
Ghana	1 753	1 970	1 095	1 372	1 763	1 684	872	2 591	759	1 156
Guinea	...	...	...	452	309	431	400	300	226	220 F
Korea Rep	230	54	-	-	9	-	-	-	-	-
Liberia	93	42	31	202	35	-	67	343	174	196
Lithuania	-	11	-	-	-	-	-	-	-	-
Nigeria	4 048	4 941	5 039	3 238	3 556	5 212	7 671	10 490	10 799	10 947
Portugal	11	8	7	14	20	9	10	10	2	9
Romania	59	-	-	-	-	-	-	-	-	-
Sao Tome Prn	16 F	20 F	29 F	30 F	22	56	70 F	80 F	70 F	70 F
Senegal	974	1 072	1 717	1 117	1 530	2 339	1 456	1 474	1 680	1 519
Sierra Leone	688	680	678	681	680 F	772	237	124	1 095	2 757
Spain	-	-	-	-	24	31	19	28	22	37
Togo	39	45	59	33	96	38	25	144	226	54
34 Fishing area total	*8 634 F*	*9 982 F*	*9 587 F*	*8 371 F*	*10 187 F*	*12 180 F*	*12 731 F*	*17 188 F*	*16 697 F*	*19 863 F*
37 Algeria	200 F	100 F	100 F	...	...	...	...	6	6 F	6 F
Egypt	729	730	...	1 731	1 098	1 721	1 764	1 159	1 191	1 085
Gaza Strip	...	...	...	...	49	101	90	90 F	90 F	80 F
Israel	223	210	50	84	76	104	105	160	77	70 F
Lebanon	75 F	100	125	200	200	200	200	200	200	250
Spain	-	-	-	-	-	-	-	29	103	130
Syria	52 F	54 F	52 F	52 F	82	90	88	130	76	77
Tunisia	70	52	52	36	49	69	100	103	95	96
Turkey	4 580	1 594	1 507	506	250	200	120	170	150	130
37 Fishing area total	*5 929 F*	*2 840 F*	*1 886 F*	*2 609 F*	*1 804*	*2 485*	*2 467*	*2 047 F*	*1 988 F*	*1 924 F*
41 Brazil	180 F	180 F	180 F	20	11	12	828	217	527	520 F
41 Fishing area total	*180 F*	*180 F*	*180 F*	*20*	*11*	*12*	*828*	*217*	*527*	*520 F*
47 Angola	...	...	...	...	...	...	...	119	-	-
Russian Fed	-	-	-	-	-	-	-	79	-	1
47 Fishing area total	...	...	...	...	...	...	...	*198*	-	*1*
51 Bahrain	73	89	66	107	159	82	156	6	8	7
Djibouti	13 F	14 F	15 F	18 F	20 F	20 F	20 F	20 F	20 F	20 F
Egypt	-	-	-	12	3	1	-	3	1 199	485
Eritrea	...	...	37	109	185	21	57	150	684	497
India	4 224	3 036	2 893	4 692	4 202	3 588	9 903	9 926	350	382
Kenya	53	57	55	65	54	54	71	108	83	99
Oman	1 645	1 580	1 713	2 244	1 948	2 023	1 789	2 347	1 431	1 781
Pakistan	1 578	2 485	2 923	2 324	2 878	2 683	2 664	3 520	3 981	3 889
Qatar	79	18	52	48	54	59	21	2	62	86

B-37 Miscellaneous pelagic fishes / Poissons pélagiques divers / Peces pelágicos diversos

Capture production by species, fishing areas and countries or areas / Captures par espèces, zones de pêche et pays ou zones / Capturas por especies, áreas de pesca y países o áreas

pecies, Fishing area / spèce, Zone de pêche / specie, Area de pesca	1992 mt	1993 mt	1994 mt	1995 mt	1996 mt	1997 mt	1998 mt	1999 mt	2000 mt	2001 mt
Saudi Arabia	1 913	1 990	2 249	1 065	1 251	1 246	1 489	1 267	1 563	1 562
Seychelles	110	286	152	163	190	183	150	323	217	253
Tanzania	100	23	30	26	29	30	15	20	25	25
Untd Arab Em	2 310	2 444	2 810	2 094	2 116	2 261	2 269	2 326	543	540 F
Yemen	2 670	3 279	2 716	2 356	1 813	1 900 F	2 100 F	2 100 F	1 900 F	2 400 F
51 Fishing area total	*14 768 F*	*15 301 F*	*15 711 F*	*15 323 F*	*14 902 F*	*14 151 F*	*20 704 F*	*22 118 F*	*12 066 F*	*12 026 F*
57 India	4 057	3 335	4 760	4 378	4 671	4 239	5 062	5 010	3 812	4 482
Indonesia	3 639	3 886	4 439	5 135	5 624	5 732	7 004	6 480	7 442	7 780
Malaysia	1 334	1 036	815	730	844	1 012	709	872	735	554
Thailand	5 669	5 523	5 487	7 726	10 427	10 554	10 817	11 297	11 142	11 074
57 Fishing area total	*14 699*	*13 780*	*15 501*	*17 969*	*21 566*	*21 537*	*23 592*	*23 659*	*23 131*	*23 890*
61 China,Taiwan	929	927	866	429	547	761	771	519	362	439
61 Fishing area total	*929*	*927*	*866*	*429*	*547*	*761*	*771*	*519*	*362*	*439*
71 Fiji Islands	1 097	3 247	2 883 F	1 735	567	1 562	1 626	2 979	2 700 F	2 809
Guam	1	1	2	0	1	1	1	1	2	4
Indonesia	22 209	23 474	19 641	9 093	10 928	12 474	13 436	14 098	12 116	12 660
Kiribati	2 200	2 130	2 110	2 130	2 160	450	987	855	420	732
Malaysia	2 579	2 856	3 148	3 966	4 089	5 642	6 445	7 231	6 712	6 057
Philippines	12 200	11 936	11 039	10 654	10 003	6 707	7 722	8 976	7 778	8 363
Singapore	183	153	232	195	170	134	149	105	79	86
Thailand	2 261	2 389	4 011	4 335	3 962	3 691	3 208	5 563	5 487	5 453
71 Fishing area total	*42 730*	*46 186*	*43 066 F*	*32 108*	*31 880*	*30 661*	*33 574*	*39 808*	*35 294 F*	*36 164*
77 Amer Samoa	...	...	2	2	2	4	1	0	0	0
Mexico	205	195	37	33	68	485	561	164	104	...
USA	...	...	...	...	...	...	59	95	76	72
77 Fishing area total	*205*	*195*	*39*	*35*	*70*	*489*	*621*	*259*	*180*	*72*
81 Korea Rep	-	-	-	-	-	80	1	349	432	512
Russian Fed	-	911	1 597	515	1 176	399	125	765	-	
81 Fishing area total	-	*911*	*1 597*	*515*	*1 176*	*479*	*126*	*1 114*	*432*	*512*
ecies total	*89 316 F*	*91 713 F*	*89 936 F*	*79 121 F*	*83 739 F*	*84 885 F*	*97 486 F*	*109 083 F*	*92 030 F*	*96 733 F*
cean sunfish	**Poisson lune**		**Pez luna**			***Mola mola***			**1,90(08)002,01**	**MOX**
47 Namibia	...	...	...	...	...	...	...	...	...	2
47 Fishing area total	...	...	...	...	...	...	...	...	...	*2*
ecies total	...	...	...	...	...	...	...	...	...	*2*
lagic percomorphs nei	**Percomorphes pélagiques nca**		**Percomorfos pelágicos nep**		***Perciformes***				**1,99(XX)XXX,XX**	**PPX**
31 Puerto Rico	96	123	79	186	98	97	83	-	-	-
Venezuela	696	403	520	3 579	759	442	381	219	182	109
31 Fishing area total	*792*	*526*	*599*	*3 765*	*857*	*539*	*464*	*219*	*182*	*109*
34 Cape Verde	...	...	...	...	...	4 414	4 899	4 463	4 823	4 288
Eq Guinea	170 F	160 F	240 F	100 F	220 F	280 F	170	200 F	20 F	10 F
Spain	-	-	-	-	-	-	-	-	1	2
34 Fishing area total	*170 F*	*160 F*	*240 F*	*100 F*	*220 F*	*4 694 F*	*5 069*	*4 663 F*	*4 844 F*	*4 300 F*
41 Brazil	290 F	290 F	290 F	-	-	-	-	-	-	-
Uruguay	8	15	2	3	2	243	150	10	18	0
41 Fishing area total	*298 F*	*305 F*	*292 F*	*3*	*2*	*243*	*150*	*10*	*18*	*0*
47 Cuba	-	-	-	-	-	-	-	2 427	...	-
Japan	1	-	-	-	-	-	-	-	-	-
47 Fishing area total	*1*	-	-	-	-	-	-	*2 427*	...	-
51 Kenya	319	290	114	286	358	351	611	528	378	1 003
Oman	39 796	8 447	7 788	1 269	2 045	2 949	6 055	10 790	5 353	3 391
Yemen	36 134	35 938	39 227	60 721	62 563	60 707 F	67 630 F	74 000 F	63 900 F	80 877 F
51 Fishing area total	*76 249*	*44 675*	*47 129*	*62 276*	*64 966*	*64 007 F*	*74 296 F*	*85 318 F*	*69 631 F*	*85 271 F*
57 Australia	3 200 F	3 700 F	3 500 F	5 000 F	4 000 F	-	-	-	-	-
57 Fishing area total	*3 200 F*	*3 700 F*	*3 500 F*	*5 000 F*	*4 000 F*	-	-	-	-	-
71 Australia	15 F	15 F	20 F	20 F	20 F	-	-	-	-	-
71 Fishing area total	*15 F*	*15 F*	*20 F*	*20 F*	*20 F*	-				
81 Australia	3 000 F	2 500 F	2 000 F	1 000 F	-	-	-	-	-	-
Japan	24	12	10	2	-	-	-	-	-	-
81 Fishing area total	*3 024 F*	*2 512 F*	*2 010 F*	*1 002 F*	-	-	-	-	-	-
87 Peru	...	464	152	382	858	120	33	1 073	771	371
87 Fishing area total	...	*464*	*152*	*382*	*858*	*120*	*33*	*1 073*	*771*	*371*
ecies total	*83 749 F*	*52 357 F*	*53 942 F*	*72 548 F*	*70 923 F*	*69 603 F*	*80 012 F*	*93 710 F*	*75 446 F*	*90 051 F*
oup total	***12 678 200***	***13 056 902***	***13 116 612***	***13 935 678***	***14 332 358***	***13 789 294***	***11 213 305***	***10 683 986***	***10 664 082***	***12 218 558***

B-38

Sharks, rays, chimaeras — **Capture production by species, fishing areas and countries or areas**
Squales, raies, chimères — **Captures par espèces, zones de pêche et pays ou zones**
Tiburones, rayas, quimeras — **Capturas por especies, áreas de pesca y países o áreas**

Species, Fishing area Espèce, Zone de pêche Especie, Area de pesca	1992 mt	1993 mt	1994 mt	1995 mt	1996 mt	1997 mt	1998 mt	1999 mt	2000 mt	2001 mt
Broadnose sevengill shark	**Platnez**		**Cañabota gata**			***Notorynchus cepedianus***			**1,05(02)005,02**	**NT**
81 New Zealand	...	...	...	...	...	...	2	3	4	5
81 Fishing area total	...	...	...	...	...	...	*2*	*3*	*4*	*5*
Species total	...	...	...	...	...	...	*2*	*3*	*4*	*5*
Basking shark	**Pèlerin**		**Peregrino**			***Cetorhinus maximus***			**1,06(01)003,01**	**BS**
27 France	-		-	0	0	1	0	3	-	-
Norway	3 658	2 910	1 762	108	1 979	1 159	137	77	293	200
Portugal	0	0	1	1	1	1	-	1	1	3
27 Fishing area total	*3 658*	*2 910*	*1 763*	*109*	*1 980*	*1 161*	*137*	*81*	*294*	*203*
37 Spain	-	-	-	-	2	6	6	-	-	-
37 Fishing area total	-	-	-	-	*2*	*6*	*6*	-	-	-
81 New Zealand	...	...	...	14	2	2	49	129	95	84
81 Fishing area total	...	...	...	*14*	*2*	*2*	*49*	*129*	*95*	*84*
Species total	*3 658*	*2 910*	*1 763*	*123*	*1 984*	*1 169*	*192*	*210*	*389*	*287*
Sand tiger shark	**Requin taureau**		**Toro bacota**			***Carcharias taurus***			**1,06(02)005,01**	**CC**
21 USA	-	-	2	-	-	-	-	-	1	-
21 Fishing area total	-	-	*2*	-	-	-	-	-	*1*	-
Species total	-	-	*2*	-	-	-	-	-	*1*	-
Thresher	**Renard**		**Zorro**			***Alopias vulpinus***			**1,06(06)006,01**	**AL**
21 USA	105	14	23	1	-	-	-	-	8	11
21 Fishing area total	*105*	*14*	*23*	*1*	-	-	-	-	*8*	*11*
27 France	14	14	11	13	7	13	7	21	116	113
Portugal	...	...	...	...	...	...	...	13	20	27
27 Fishing area total	*14*	*14*	*11*	*13*	*7*	*13*	*7*	*34*	*136*	*140*
34 Liberia	...	...	...	...	...	...	...	151	146	...
Portugal	...	...	...	...	...	...	...	...	...	7
Spain	...	...	...	...	...	30	45	...	...	...
34 Fishing area total	...	...	...	...	...	*30*	*45*	*151*	*146*	*7*
37 France	...	...	...	...	...	...	...	14	12	19
Portugal	...	...	...	...	...	...	...	...	...	2
37 Fishing area total	...	...	...	...	...	...	...	*14*	*12*	*21*
41 Portugal	...	...	...	...	...	...	...	...	...	3
41 Fishing area total	...	...	...	...	...	...	...	...	...	*3*
67 USA	...	...	...	...	...	...	17	2	48	76
67 Fishing area total	...	...	...	...	...	...	*17*	*2*	*48*	*76*
77 USA	...	...	...	...	...	...	303	262	249	299
77 Fishing area total	...	...	...	...	...	...	*303*	*262*	*249*	*299*
81 New Zealand	...	...	...	15	13	24	21	32	51	57
81 Fishing area total	...	...	...	*15*	*13*	*24*	*21*	*32*	*51*	*57*
Species total	*119*	*28*	*34*	*29*	*20*	*67*	*393*	*495*	*650*	*614*
Bigeye thresher	**Renard à gros yeux**		**Zorro ojón**			***Alopias superciliosus***			**1,06(06)006,03**	**BT**
31 Spain	-	-	-	-	-	1	...	...	...	...
31 Fishing area total	-	-	-	-	-	*1*	...	...	...	...
34 Spain	...	...	...	...	...	148	114	...	...	...
34 Fishing area total	...	...	...	...	...	*148*	*114*	...	...	...
77 USA	...	...	...	...	...	...	11	5	5	2
77 Fishing area total	...	...	...	...	...	...	*11*	*5*	*5*	*2*
Species total	...	...	...	...	...	*149*	*125*	*5*	*5*	*2*
Thresher sharks nei	**Renards de mer nca**		**Zorros nep**			***Alopias spp***			**1,06(06)006,XX**	**TH**
34 Spain	...	...	...	...	...	34	55	66	...	...
34 Fishing area total	...	...	...	...	...	*34*	*55*	*66*	...	...
Species total	...	...	...	...	...	*34*	*55*	*66*	...	...
Shortfin mako	**Taupe bleue**		**Marrajo dientuso**			***Isurus oxyrinchus***			**1,06(08)002,01**	**SM**
21 Portugal	...	...	...	...	...	...	...	...	10	-
USA	59	71	115	5	-	-	-	-	19	19
21 Fishing area total	*59*	*71*	*115*	*5*	...	...	...	...	*29*	*19*
27 Portugal	...	...	...	...	...	...	...	160	183	186
UK	-	-	-	-	-	-	0	2	3	2

B-38 Sharks, rays, chimaeras / Squales, raies, chimères / Tiburones, rayas, quimeras

Capture production by species, fishing areas and countries or areas / Captures par espèces, zones de pêche et pays ou zones / Capturas por especies, áreas de pesca y países o áreas

Species, Fishing area Espèce, Zone de pêche Especie, Area de pesca	1992 mt	1993 mt	1994 mt	1995 mt	1996 mt	1997 mt	1998 mt	1999 mt	2000 mt	2001 mt
27 Fishing area total	...	...	...	...	...	...	*0*	*162*	*186*	*188*
31 Portugal	-	-	-	-	-	-	0	-	-	-
Spain	-	-	-	-	-	73	33	...	...	...
USA	-	-	-	-	-	-	-	-	5	5
31 Fishing area total	-	-	-	-	-	*73*	*33*	...	*5*	*5*
34 Benin	...	...	...	...	...	...	...	4	3 F	1
China	-	-	-	-	-	...	...	...	3	...
Philippines	-	-	-	-	-	-	-	3	-	-
Portugal	...	...	...	...	...	...	...	...	42	42
34 Fishing area total	...	...	...	...	...	...	...	*7*	*48 F*	*43*
37 Portugal	-	-	-	-	-	-	-	-	1	6
37 Fishing area total	-	-	-	-	-	-	-	-	*1*	*6*
41 Brazil	100 F	100 F	...	...	83	190	...	100	120	120 F
Portugal	...	...	...	...	...	...	...	...	15	26
41 Fishing area total	*100 F*	*100 F*	...	...	*83*	*190*	...	*100*	*135*	*146 F*
47 China	-	-	-	-	-	-	-	-	150	...
Portugal	-	-	-	1	...	...	4	3	350	158
South Africa	...	...	...	...	...	...	...	...	...	2
Spain	-	-	-	-	-	587	304	335	264	228
47 Fishing area total	...	...	...	*1*	...	*587*	*308*	*338*	*764*	*388*
51 Portugal	-	-	-	-	-	-	18	...	58	95
51 Fishing area total	-	-	-	-	-	-	*18*	...	*58*	*95*
67 USA	...	...	...	...	...	...	3	-	0	0
67 Fishing area total	...	...	...	...	...	...	*3*	-	*0*	*0*
77 Estonia	-	-	-	-	-	-	-	2	-	-
Fr Polynesia	...	...	...	...	...	...	...	...	27	53
USA	...	...	...	...	...	...	93	61	79	46
77 Fishing area total	...	...	...	...	...	...	*93*	*63*	*106*	*99*
81 New Zealand	...	...	...	33	52	40	74	110	208	327
81 Fishing area total	...	...	...	*33*	*52*	*40*	*74*	*110*	*208*	*327*
87 Chile	702	581	450	475	320	888	830	379	592	964
Spain	-	-	-	-	-	330	927	...	...	...
87 Fishing area total	*702*	*581*	*450*	*475*	*320*	*1 218*	*1 757*	*379*	*592*	*964*
Species total	*861 F*	*752 F*	*565*	*514*	*455*	*2 108*	*2 286*	*1 159*	*2 132 F*	*2 280 F*
Longfin mako	**Petite taupe**		**Marrajo carite**		***Isurus paucus***				**1,06(08)002,03**	**LMA**
21 USA	12	0	5	0	-	-	-	-	1	0
21 Fishing area total	*12*	*0*	*5*	*0*	-	-	-	-	*1*	*0*
31 USA	-	-	-	-	-	1	1	-	3	3
31 Fishing area total	-	-	-	-	-	*1*	*1*	-	*3*	*3*
Species total	*12*	*0*	*5*	*0*	-	*1*	*1*	-	*4*	*3*
Mako sharks	**Taupes**		**Marrajos**		***Isurus spp***				**1,06(08)002,XX**	**MAK**
21 USA	-	-	-	-	-	-	-	-	-	47
21 Fishing area total	-	-	-	-	-	-	-	-	-	*47*
34 Côte dIvoire	13	7	17	12	...	92	38	...	...	...
Liberia	...	...	...	...	...	...	...	...	116	...
34 Fishing area total	*13*	*7*	*17*	*12*	...	*92*	*38*	...	*116*	...
Species total	*13*	*7*	*17*	*12*	...	*92*	*38*	...	*116*	*47*
Porbeagle	**Requin-taupe commun**		**Marrajo sardinero**		***Lamna nasus***				**1,06(08)003,01**	**POR**
21 Canada	-	-	-	1 388	1 029	1 343	1 006	958	904	499
Faeroe Is	559	250	-	-	-	-	-	-	-	-
St Pier Mq	-	-	-	7	40	13	20	0	23	2
USA	13	39	64	0	-	-	-	-	3	1
21 Fishing area total	*572*	*289*	*64*	*1 395*	*1 069*	*1 356*	*1 026*	*958*	*930*	*502*
27 Channel Is	...	...	...	...	...	...	1	0	2	2
Denmark	80	91	94	86	71	69	85	107	73	76
Faeroe Is	20	76	48	44	7	9	7	10	13	8
France	496	633	820	565	267	315	219	318	410	368
Germany	-	-	22	-	-	-	-	0	17	1
Iceland	1	3	4	6	5	3	4	2	2	3
Ireland	-	-	-	-	-	-	-	8	1	6
Norway	42	24	25	27	28	17	28	33	22	17
Portugal	0	0	0	0	0	0	0	0	6	2
Spain	-	-	-	-	31	124	679	1 001	1 184	1 007
Sweden	4	3	2	2	1	1	1	1	1	1
UK	-	-	-	-	-	-	-	6	6	10
27 Fishing area total	*643*	*830*	*1 015*	*730*	*410*	*538*	*1 024*	*1 486*	*1 737*	*1 501*

B-38 Sharks, rays, chimaeras / Squales, raies, chimères / Tiburones, rayas, quimeras

Capture production by species, fishing areas and countries or areas / Captures par espèces, zones de pêche et pays ou zones / Capturas por especies, áreas de pesca y países o áreas

Species, Fishing area Espèce, Zone de pêche Especie, Area de pesca	1992 mt	1993 mt	1994 mt	1995 mt	1996 mt	1997 mt	1998 mt	1999 mt	2000 mt	2001 mt
34 Portugal	...	...	...	...	...	...	...	...	10	2
34 Fishing area total	*...*	*...*	*...*	*...*	*...*	*...*	*...*	*...*	*10*	*2*
37 Malta	0	0	0	0	1	0	0	0	0	0
Spain	-	-	-	-	-	-	-	-	-	1
37 Fishing area total	*0*	*0*	*0*	*0*	*1*	*0*	*0*	*0*	*0*	*1*
51 Spain	-	-	-	-	-	-	-	-	-	1
51 Fishing area total	-	-	-	-	-	-	-	-	-	*1*
58 Australia	-	-	-	-	-	2	-	-	-	-
58 Fishing area total	-	-	-	-	-	*2*	-	-	-	-
81 New Zealand	...	...	...	5	16	21	164	246	188	127
81 Fishing area total	*...*	*...*	*...*	*5*	*16*	*21*	*164*	*246*	*188*	*127*
87 Spain	-	-	-	-	-	...	7	...	...	...
87 Fishing area total	-	-	-	-	-	*...*	*7*	*...*	*...*	*...*
Species total	*1 215*	*1 119*	*1 079*	*2 130*	*1 496*	*1 917*	*2 221*	*2 690*	*2 865*	*2 134*
Great white shark	**Grand requin blanc**		**Jaquetón blanco**			***Carcharodon carcharias***			**1,06(08)007,01**	**WS**
21 USA	-	-	-	-	-	-	-	-	1	0
21 Fishing area total	-	-	-	-	-	-	-	-	*1*	*0*
77 USA	-	-	-	-	-	-	-	-	1	0
77 Fishing area total	-	-	-	-	-	-	-	-	*1*	*0*
Species total	-	-	-	-	-	-	-	-	*2*	*0*
Nurse shark	**Requin-nourrice**		**Gata nodriza**			***Ginglymostoma cirratum***			**1,07(03)009,01**	**GN**
21 USA	-	-	0	214	-	-	-	-	0	-
21 Fishing area total	-	-	*0*	*214*	-	-	-	-	*0*	-
31 Dominican Rp	-	-	-	-	-	-	-	-	407	89
31 Fishing area total	-	-	-	-	-	-	-	-	*407*	*89*
Species total	-	-	*0*	*214*	-	-	-	-	*407*	*89*
Small-spotted catshark	**Petite roussette**		**Pintarroja**			***Scyliorhinus canicula***			**1,08(01)003,01**	**SY**
27 France	5 247	5 019	4 975	5 022	4 871	5 338	5 480	5 707	5 914	6 253
Ireland	...	...	...	...	...	...	...	...	...	633
UK	...	...	...	341	273	275	260	111	238	155
27 Fishing area total	*5 247*	*5 019*	*4 975*	*5 363*	*5 144*	*5 613*	*5 740*	*5 818*	*6 152*	*7 041*
37 France	...	...	...	...	...	...	...	...	30	31
37 Fishing area total	*...*	*...*	*...*	*...*	*...*	*...*	*...*	*...*	*30*	*31*
Species total	*5 247*	*5 019*	*4 975*	*5 363*	*5 144*	*5 613*	*5 740*	*5 818*	*6 182*	*7 072*
Nursehound	**Grande roussette**		**Alitán**			***Scyliorhinus stellaris***			**1,08(01)003,02**	**SY**
27 France	623	370	196	174	197	292	181	135	159	180
UK	...	...	...	69	109	86	77	139	115	84
27 Fishing area total	*623*	*370*	*196*	*243*	*306*	*378*	*258*	*274*	*274*	*264*
Species total	*623*	*370*	*196*	*243*	*306*	*378*	*258*	*274*	*274*	*264*
Catsharks, nursehounds nei	**Roussettes nca**		**Alitanes, pintarrojas nep**			***Scyliorhinus spp***			**1,08(01)003,XX**	**SC**
27 France	1	1	7	8	18	6	51	13	68	7
27 Fishing area total	*1*	*1*	*7*	*8*	*18*	*6*	*51*	*13*	*68*	*7*
37 Spain	-	-	-	-	-	-	-	213	331	379
Tunisia	87	78	84	48	36	72	...	49	126	122
37 Fishing area total	*87*	*78*	*84*	*48*	*36*	*72*	*...*	*262*	*457*	*501*
Species total	*88*	*79*	*91*	*56*	*54*	*78*	*51*	*275*	*525*	*508*
Blue shark	**Peau bleue**		**Tiburón azul**			***Prionace glauca***			**1,08(02)004,01**	**BS**
21 Portugal	...	...	...	...	...	...	...	...	169	-
21 Fishing area total	*...*	*...*	*...*	*...*	*...*	*...*	*...*	*...*	*169*	-
27 Channel Is	...	...	...	...	...	...	1	0	-	-
Denmark	1	0	-	-	3	1	1	1	2	1
France	276	322	350	266	278	213	163	230	395	205
Ireland	-	-	-	-	-	-	-	67	23	66
Portugal	...	...	...	...	...	...	...	887	21	1 006
UK	-	-	-	-	-	-	-	-	12	9
27 Fishing area total	*277*	*322*	*350*	*266*	*281*	*214*	*165*	*1 185*	*453*	*1 287*
31 Portugal	-	-	-	-	-	-	17	-	-	-
Spain	-	-	-	-	-	1 700	418	...	...	...
31 Fishing area total	-	-	-	-	-	*1 700*	*435*	*...*	*...*	*...*

B-38

Sharks, rays, chimaeras — **Capture production by species, fishing areas and countries or areas**
Squales, raies, chimères — **Captures par espèces, zones de pêche et pays ou zones**
Tiburones, rayas, quimeras — **Capturas por especies, áreas de pesca y países o áreas**

Species, Fishing area Espèce, Zone de pêche Especie, Area de pesca	1992 mt	1993 mt	1994 mt	1995 mt	1996 mt	1997 mt	1998 mt	1999 mt	2000 mt	2001 mt
34 Liberia	...	...	...	...	...	...	...	76	70	...
Portugal	...	...	...	...	...	...	...	...	351	557
34 Fishing area total	...	...	...	...	...	...	...	*76*	*421*	*557*
37 France	...	...	...	...	...	...	...	3	1	2
Portugal	-	-	-	-	-	-	-	-	3	40
37 Fishing area total	...	...	...	...	...	...	...	*3*	*4*	*42*
41 Brazil	150 F	800 F	...	...	743	1 103	...	500	580	570 F
Portugal	...	...	...	...	...	...	...	...	56	988
41 Fishing area total	*150 F*	*800 F*	...	...	*743*	*1 103*	...	*500*	*636*	*1 558 F*
47 Portugal	-	-	-	21	...	...	8	18	1 908	949
South Africa	...	...	...	...	...	...	...	...	...	1
Spain	-	-	-	-	-	3 560	1 463	2 233	2 803	2 784
47 Fishing area total	...	...	...	*21*	...	*3 560*	*1 471*	*2 251*	*4 711*	*3 734*
51 Portugal	-	-	-	-	-	-	60	...	575	1 123
51 Fishing area total	-	-	-	-	-	-	*60*	...	*575*	*1 123*
67 USA	-	-	-	-	-	-	-	-	1	2
67 Fishing area total	-	-	-	-	-	-	-	-	*1*	*2*
77 USA	...	...	...	...	...	...	1	-	0	0
77 Fishing area total	...	...	...	...	...	...	*1*	-	*0*	*0*
81 New Zealand	...	...	...	111	246	120	540	593	1 169	1 328
81 Fishing area total	...	...	...	*111*	*246*	*120*	*540*	*593*	*1 169*	*1 328*
87 Chile	175	237	33	39	11	114	10	7	262	445
Spain	-	-	-	-	-	...	814	...	...	11
87 Fishing area total	*175*	*237*	*33*	*39*	*11*	*114*	*824*	*7*	*262*	*456*
Species total	*602 F*	*1 359 F*	*383*	*437*	*1 281*	*6 811*	*3 496*	*4 615*	*8 401*	*10 087 F*
Sandbar shark	**Requin gris**		**Tiburón trozo**		*Carcharhinus plumbeus*				**1,08(02)010,01**	**CCP**
21 USA	55	31	24	1	-	-	-	-	41	24
21 Fishing area total	*55*	*31*	*24*	*1*	-	-	-	-	*41*	*24*
Species total	*55*	*31*	*24*	*1*	-	-	-	-	*41*	*24*
Blacktip shark	**Requin bordé**		**Tiburón macuira**		*Carcharhinus limbatus*				**1,08(02)010,03**	**CCL**
21 USA	-	-	7	-	-	-	-	-	21	1
21 Fishing area total	-	-	*7*	-	-	-	-	-	*21*	*1*
31 Trinidad Tob	...	...	...	...	3	8	10	11	...	...
USA	-	-	-	-	-	1	0	-	580	520
31 Fishing area total	...	...	...	...	*3*	*9*	*10*	*11*	*580*	*520*
Species total	...	...	*7*	...	*3*	*9*	*10*	*11*	*601*	*521*
Dusky shark	**Requin de sable**		**Tiburón arenero**		*Carcharhinus obscurus*				**1,08(02)010,16**	**DUS**
21 USA	69	23	20	0	-	-	-	-	80	0
21 Fishing area total	*69*	*23*	*20*	*0*	-	-	-	-	*80*	*0*
51 South Africa	-	-	-	-	-	7	0	...	...	...
51 Fishing area total	-	-	-	-	-	*7*	*0*	...	...	...
Species total	*69*	*23*	*20*	*0*	-	*7*	*0*	...	*80*	*0*
Silky shark	**Requin soyeux**		**Tiburón jaquetón**		*Carcharhinus falciformis*				**1,08(02)010,17**	**FAL**
34 Côte dIvoire	19	4	13	18	...	2	...	...	...	...
Liberia	...	...	...	...	...	...	...	110	99	...
34 Fishing area total	*19*	*4*	*13*	*18*	...	*2*	...	*110*	*99*	...
41 Brazil	...	...	...	...	502	279	...	70	80	80 F
41 Fishing area total	...	...	...	...	*502*	*279*	...	*70*	*80*	*80 F*
57 Sri Lanka	13 700	21 800	25 400	21 400	21 000	15 000	20 875	20 700	16 130	14 620
57 Fishing area total	*13 700*	*21 800*	*25 400*	*21 400*	*21 000*	*15 000*	*20 875*	*20 700*	*16 130*	*14 620*
Species total	*13 719*	*21 804*	*25 413*	*21 418*	*21 502*	*15 281*	*20 875*	*20 880*	*16 309*	*14 700 F*
Copper shark	**Requin cuivre**		**Tiburón cobrizo**		*Carcharhinus brachyurus*				**1,08(02)010,20**	**BRO**
81 New Zealand	...	...	...	...	...	...	15	14	25	38
81 Fishing area total	...	...	...	...	...	...	*15*	*14*	*25*	*38*
Species total	...	...	...	...	...	...	*15*	*14*	*25*	*38*
Tiger shark	**Requin tigre commun**		**Tintorera tigre**		*Galeocerdo cuvier*				**1,08(02)017,03**	**TIG**
21 USA	-	-	-	-	-	-	-	-	-	1
21 Fishing area total	-	-	-	-	-	-	-	-	-	*1*

B-38

Sharks, rays, chimaeras — Capture production by species, fishing areas and countries or areas
Squales, raies, chimères — Captures par espèces, zones de pêche et pays ou zones
Tiburones, rayas, quimeras — Capturas por especies, áreas de pesca y países o áreas

Species, Fishing area Espèce, Zone de pêche Especie, Area de pesca	1992 mt	1993 mt	1994 mt	1995 mt	1996 mt	1997 mt	1998 mt	1999 mt	2000 mt	2001 mt
Species total	-	-	-	-	-	-	-	-	-	*1*
Requiem sharks nei	**Requins nca**		**Cazones picudos,tintoreras nep**			***Carcharhinidae***			**1,08(02)XXX,XX**	**RSK**
31 Mexico	6 564	6 426	5 544	4 741	5 166	3 785	3 991	3 281	3 108	3 196
Nicaragua	...	...	...	...	...	...	...	40	40	175
Venezuela	6 103	6 101	6 656	7 468	6 979	6 000	4 597	2 973	3 343	2 536
31 Fishing area total	*12 667*	*12 527*	*12 200*	*12 209*	*12 145*	*9 785*	*8 588*	*6 294*	*6 491*	*5 907*
51 Eritrea	...	...	13	6	15	13	17	38	130	109
Mozambique	-	-	-	165	21	...	...	...	...	...
Pakistan	27 773	28 780	30 226	32 288	34 447	31 179	35 357	32 535	28 245	26 524
Qatar	0	0	0	0	0	-	0	0	0	0
Spain	-	-	-	-	-	43	810	-	9	8
51 Fishing area total	*27 773*	*28 780*	*30 239*	*32 459*	*34 483*	*31 235*	*36 184*	*32 573*	*28 384*	*26 641*
77 Mexico	6 169	6 764	5 985	6 334	5 849	3 482	2 988	2 789	3 210	2 859
Nicaragua	...	...	...	...	...	...	...	160	110	200
77 Fishing area total	*6 169*	*6 764*	*5 985*	*6 334*	*5 849*	*3 482*	*2 988*	*2 949*	*3 320*	*3 059*
Species total	*46 609*	*48 071*	*48 424*	*51 002*	*52 477*	*44 502*	*47 760*	*41 816*	*38 195*	*35 607*
Smooth hammerhead	**Requin-marteau commun**		**Cornuda cruz(=Pez martillo)**			***Sphyrna zygaena***			**1,08(03)005,01**	**SPZ**
27 Portugal	...	...	...	...	...	...	...	8	8	4
27 Fishing area total	...	...	...	...	...	...	...	*8*	*8*	*4*
34 Portugal	...	...	...	...	...	...	...	...	7	1
34 Fishing area total	...	...	...	...	...	...	...	...	*7*	*1*
41 Portugal	...	...	...	...	...	...	...	...	3	-
41 Fishing area total	...	...	...	...	...	...	...	...	*3*	-
47 Portugal	...	...	...	...	...	...	...	...	4	5
Spain	-	-	-	-	-	220	103	-	-	-
47 Fishing area total	...	...	...	...	...	*220*	*103*	...	*4*	*5*
81 New Zealand	...	...	...	12	10	3	6	11	13	17
81 Fishing area total	...	...	...	*12*	*10*	*3*	*6*	*11*	*13*	*17*
Species total	...	...	...	*12*	*10*	*223*	*109*	*19*	*35*	*27*
Scalloped hammerhead	**Requin-marteau halicorne**		**Cornuda común**			***Sphyrna lewini***			**1,08(03)005,06**	**SPL**
34 GuineaBissau	...	...	2	12	12	10 F	10 F	10 F	10 F	10 F
34 Fishing area total	...	...	*2*	*12*	*12*	*10 F*	*10 F*	*10 F*	*10 F*	*10 F*
41 Brazil	...	100 F	...	...	25	170	...	30	38	40 F
41 Fishing area total	...	*100 F*	...	...	*25*	*170*	...	*30*	*38*	*40 F*
Species total	...	*100 F*	*2*	*12*	*37*	*180 F*	*10 F*	*40 F*	*48 F*	*50 F*
Hammerhead sharks, etc. nei	**Requins marteau, etc. nca**		**Cornudas, etc. nep**			***Sphyrnidae***			**1,08(03)XXX,XX**	**SPY**
31 Spain	-	-	-	-	-	3	2	...	...	...
31 Fishing area total	-	-	-	-	-	*3*	*2*	...	...	...
34 Côte dIvoire	69	55	66	69	...	190	125	...	...	...
Liberia	...	...	...	...	...	...	...	127	152	...
Senegal	-	-	-	-	-	-	156	20	1 305	1 451
Spain	...	...	...	...	...	805	739	...	...	...
34 Fishing area total	*69*	*55*	*66*	*69*	...	*995*	*1 020*	*147*	*1 457*	*1 451*
77 USA	...	...	...	...	...	...	1	-	-	-
77 Fishing area total	...	...	...	...	...	...	*1*	-	-	-
87 Spain	-	-	-	-	-	...	5	...	...	...
87 Fishing area total	-	-	-	-	-	...	*5*	...	...	...
Species total	*69*	*55*	*66*	*69*	...	*998*	*1 028*	*147*	*1 457*	*1 451*
Dusky smooth-hound	**Emissole douce**		**Boca dulce**			***Mustelus canis***			**1,08(04)007,03**	**CTI**
21 USA	...	...	280	0	...	...	...	...	334	321
21 Fishing area total	...	...	*280*	*0*	...	...	...	...	*334*	*321*
Species total	...	...	*280*	*0*	...	...	...	...	*334*	*321*
Brown smooth-hound	**Emissole brune**		**Musola parda**			***Mustelus henlei***			**1,08(04)007,07**	**CTK**
77 USA	...	...	...	...	...	...	3	5	3	4
77 Fishing area total	...	...	...	...	...	...	*3*	*5*	*3*	*4*
Species total	...	...	...	...	...	...	*3*	*5*	*3*	*4*
Spotted estuary smooth-hound	**Emissole grivelée**		**Musola manchada**			***Mustelus lenticulatus***			**1,08(04)007,09**	**MTL**
81 New Zealand	1 555	1 701	1 806	2 787	1 350	3 464	1 707	1 662	1 643	1 563
81 Fishing area total	*1 555*	*1 701*	*1 806*	*2 787*	*1 350*	*3 464*	*1 707*	*1 662*	*1 643*	*1 563*

B-38

Sharks, rays, chimaeras — **Capture production by species, fishing areas and countries or areas**
Squales, raies, chimères — **Captures par espèces, zones de pêche et pays ou zones**
Tiburones, rayas, quimeras — **Capturas por especies, áreas de pesca y países o áreas**

Species, Fishing area Espèce, Zone de pêche Especie, Area de pesca	1992 mt	1993 mt	1994 mt	1995 mt	1996 mt	1997 mt	1998 mt	1999 mt	2000 mt	2001 mt
Species total	*1 555*	*1 701*	*1 806*	*2 787*	*1 350*	*3 464*	*1 707*	*1 662*	*1 643*	*1 563*
Narrownose smooth-hound	**Emissole gatuso**		**Gatuso**		***Mustelus schmitti***			**1,08(04)007,12**		**SDP**
41 Argentina	10 094	11 070	11 450	11 057	10 252	9 956	11 266	9 062	7 119	8 780
Uruguay	...	329	319	286	204	174	2 156	3 212	1 037	1 153
41 Fishing area total	*10 094*	*11 399*	*11 769*	*11 343*	*10 456*	*10 130*	*13 422*	*12 274*	*8 156*	*9 933*
Species total	*10 094*	*11 399*	*11 769*	*11 343*	*10 456*	*10 130*	*13 422*	*12 274*	*8 156*	*9 933*
Smooth-hound	**Emissole lisse**		**Musola**		***Mustelus mustelus***			**1,08(04)007,13**		**SMD**
27 UK	-	-	-	-	-	-	-	-	15	-
27 Fishing area total	-	-	-	-	-	-	-	-	*15*	-
Species total	-	-	-	-	-	-	-	-	*15*	-
Smooth-hounds nei	**Emissoles nca**		**Tollos nep**		***Mustelus spp***			**1,08(04)007,XX**		**SDV**
27 France	274	298	352	414	578	624	749	824	1 050	1 249
Portugal	...	...	...	...	...	...	...	76	41	43
27 Fishing area total	*274*	*298*	*352*	*414*	*578*	*624*	*749*	*900*	*1 091*	*1 292*
31 Colombia	286	307	102	46	253	27	45	3	30 F	30 F
31 Fishing area total	*286*	*307*	*102*	*46*	*253*	*27*	*45*	*3*	*30 F*	*30 F*
34 Greece	131	75	64	102	87	24	43	23	39	21
Portugal	129	86	66	97	187	27	14	5	-	-
Senegal	...	...	...	...	...	...	...	...	464	1 908
Spain	-	-	-	-	118	-	-	-	-	-
34 Fishing area total	*260*	*161*	*130*	*199*	*392*	*51*	*57*	*28*	*503*	*1 929*
37 Albania	...	...	...	20	12	3	12	12	32	5
France	31	0	0	0	0	-	-	0	-	-
Gaza Strip	...	...	...	...	24	11	9	10 F	10 F	10 F
Greece	227	267	377	360	353	493	316	553	578	351
Italy	5 778	4 675	9 999	5 942	2 659	621	636	440	462	369
Slovenia	-	-	-	-	-	-	-	-	2	4
Spain	-	-	-	-	-	-	-	21	15	19
Syria	39 F	40 F	39 F	39 F	50	-	-	-	-	-
Tunisia	427	187	142	128	640	132	826	997	121	192
Turkey	2 404	1 436	2 880	1 783	2 158	1 720	1 450	1 625	2 880	1 000
37 Fishing area total	*8 906 F*	*6 605 F*	*13 437 F*	*8 272 F*	*5 896*	*2 980*	*3 249*	*3 658 F*	*4 100 F*	*1 950 F*
57 Australia	...	...	...	3 911	3 878	4 169	2 858	2 462	2 198	2 579
57 Fishing area total	...	...	...	*3 911*	*3 878*	*4 169*	*2 858*	*2 462*	*2 198*	*2 579*
87 Chile	481	398	588	193	225	108	56	208	143	128
Colombia	459	316	365	162	754	408	316	385	330 F	281
Peru	8 578	8 747	3 431	4 125	3 230	3 166	8 038	2 892	4 042	4 648
87 Fishing area total	*9 518*	*9 461*	*4 384*	*4 480*	*4 209*	*3 682*	*8 410*	*3 485*	*4 515 F*	*5 057*
Species total	*19 244 F*	*16 832 F*	*18 405 F*	*17 322 F*	*15 206*	*11 533*	*15 368*	*10 536 F*	*12 437 F*	*12 837 F*
Tope shark	**Requin-hâ**		**Cazón**		***Galeorhinus galeus***			**1,08(04)011,03**		**GAG**
27 Denmark	-	-	-	-	2	2	3	4	7	4
France	279	291	301	317	403	454	369	386	450	469
Ireland	...	...	...	...	...	...	...	...	...	4
UK	68	62	71	63	53	55	55	74	110	82
27 Fishing area total	*347*	*353*	*372*	*380*	*458*	*511*	*427*	*464*	*567*	*559*
34 Portugal	...	...	...	...	...	...	...	...	2	1
34 Fishing area total	...	...	...	...	...	...	...	...	*2*	*1*
41 Argentina	58	230	75	104	92	103	92	89	109	...
Spain	-	-	-	-	-	-	-	-	-	37
41 Fishing area total	*58*	*230*	*75*	*104*	*92*	*103*	*92*	*89*	*109*	*37*
67 USA	...	...	...	...	...	...	1	-	3	1
67 Fishing area total	...	...	...	...	...	...	*1*	-	*3*	*1*
77 USA	...	...	...	...	...	...	51	73	45	44
77 Fishing area total	...	...	...	...	...	...	*51*	*73*	*45*	*44*
81 New Zealand	2 472	2 577	2 618	3 705	3 044	2 864	3 083	3 633	3 100	3 091
81 Fishing area total	*2 472*	*2 577*	*2 618*	*3 705*	*3 044*	*2 864*	*3 083*	*3 633*	*3 100*	*3 091*
Species total	*2 877*	*3 160*	*3 065*	*4 189*	*3 594*	*3 478*	*3 654*	*4 259*	*3 826*	*3 733*
Greenland shark	**Laimargue du Groenland**		**Tollo de Groenlandia**		***Somniosus microcephalus***			**1,09(01)002,01**		**GSK**
27 Iceland	68	41	42	44	61	73	87	51	45	57
Portugal	-	9	1	11	0	0	-	0	0	1
27 Fishing area total	*68*	*50*	*43*	*55*	*61*	*73*	*87*	*51*	*45*	*58*
Species total	*68*	*50*	*43*	*55*	*61*	*73*	*87*	*51*	*45*	*58*

B-38 Sharks, rays, chimaeras / Squales, raies, chimères / Tiburones, rayas, quimeras

Capture production by species, fishing areas and countries or areas / Captures par espèces, zones de pêche et pays ou zones / Capturas por especies, áreas de pesca y países o áreas

Species, Fishing area Espèce, Zone de pêche Especie, Area de pesca	1992 mt	1993 mt	1994 mt	1995 mt	1996 mt	1997 mt	1998 mt	1999 mt	2000 mt	2001 mt
Pacific sleeper shark	**Laimargue dormeur**		**Tollo negro dormilón**			***Somniosus pacificus***			**1,09(01)002,03**	**SON**
58 Australia	-	-	-	-	-	-	-	1	-	-
58 Fishing area total	-	-	-	-	-	-	-	*1*	-	-
Species total	-	-	-	-	-	-	-	*1*	-	-
Picked dogfish	**Aiguillat commun**		**Mielga**			***Squalus acanthias***			**1,09(01)007,04**	**DGS**
21 Canada	868	1 271	1 690	957	431	446	1 081	2 456	2 828	3 760
Cuba	11	-	-	-	-	6	-	-	-	-
St Pier Mq	0	0	0	0	0	0	-	-	0	0
Spain	-	-	-	-	63	0	0	-	-	1
USA	...	...	...	128	...	...	...	...	7 873	2 234
21 Fishing area total	*879*	*1 271*	*1 690*	*1 085*	*494*	*452*	*1 081*	*2 456*	*10 701*	*5 995*
27 Belgium	56	47	21	14	16	15	17	10	11	13
Denmark	800	486	211	146	142	196	126	131	146	156
Faeroe Is	...	...	...	308	51	212	356	484	354	613
France	2 406	1 911	1 661	1 349	1 719	1 708	1 410	1 192	1 097	1 333
Germany	56	8	-	-	-	-	-	45	188	303
Iceland	181	109	97	166	157	106	78	57	109	136
Ireland	1 383	3 424	3 624	2 435	2 095	1 407	1 259	962	880	1 301
Netherlands	-	-	-	-	-	-	-	-	28	39
Norway	7 114	6 945	4 546	3 939	2 749	1 567	1 293	1 461	1 643	1 424
Portugal	2	5	7	5	2	2	2	21	2	3
Spain	-	-	-	-	-	0	27	94	372	363
Sweden	230	188	95	104	154	197	140	114	124	238
UK	13 812	10 032	8 072	10 815	9 423	8 691	8 926	7 527	7 138	7 306
27 Fishing area total	*26 040*	*23 155*	*18 334*	*19 281*	*16 508*	*14 101*	*13 634*	*12 098*	*12 092*	*13 228*
37 Bulgaria	14	12	12 F	80	64	40	28	25	102	126
France	29	21	19	...	7	7	7	5	3	2
Malta	28	33	29	24	28	28	23	18	19	17
Romania	53	6	3	7	-	-	-	-	-	-
Slovenia	8	4	2	4	0	0	1	1	-	-
Spain	...	...	...	...		...	...	...	9	8
Ukraine	595	409	148	67	44	20	38	94	71	134
37 Fishing area total	*727*	*485*	*213 F*	*182*	*143*	*95*	*97*	*143*	*204*	*287*
67 Canada	2 356	830	1 776	2 744	4 000	2 100	2 500	5 897	4 696	4 543
USA	...	...	...	...	...	...	1	542	667	638
67 Fishing area total	*2 356*	*830*	*1 776*	*2 744*	*4 000*	*2 100*	*2 501*	*6 439*	*5 363*	*5 181*
77 USA	1	3	0	1	-	0	5	24	8	3
77 Fishing area total	*1*	*3*	*0*	*1*	-	*0*	*5*	*24*	*8*	*3*
81 New Zealand	2 592	5 429	3 601	2 753	2 477	7 232	3 064	4 409	3 362	4 192
81 Fishing area total	*2 592*	*5 429*	*3 601*	*2 753*	*2 477*	*7 232*	*3 064*	*4 409*	*3 362*	*4 192*
Species total	*32 595*	*31 173*	*25 614 F*	*26 046*	*23 622*	*23 980*	*20 382*	*25 569*	*31 730*	*28 886*
Gulper shark	**Squale-chagrin commun**		**Quelvacho**			***Centrophorus granulosus***			**1,09(01)008,01**	**GUP**
27 Portugal	...	...	...	...	...	...	...	73	54	93
27 Fishing area total	...	...	...	...	...	...	...	*73*	*54*	*93*
Species total	...	...	...	...	...	...	...	*73*	*54*	*93*
Leafscale gulper shark	**Squale-chagrin de l'Atlantique**		**Quelvacho negro**			***Centrophorus squamosus***			**1,09(01)008,03**	**GUQ**
27 Faeroe Is	-	-	3	51	53	58	129	11	...	...
Norway	-	-	-	-	-	-	-	-	-	1
Portugal	...	...	...	...	...	...	...	440	478	510
27 Fishing area total	...	...	*3*	*51*	*53*	*58*	*129*	*451*	*478*	*511*
34 Portugal	...	...	...	...	...	...	...	...	28	27
34 Fishing area total	...	...	...	...	...	...	...	...	*28*	*27*
81 New Zealand	...	...	...	...	...	...	4	1	0	0
81 Fishing area total	...	...	...	...	...	...	*4*	*1*	*0*	*0*
Species total	...	...	*3*	*51*	*53*	*58*	*133*	*452*	*506*	*538*
Lanternsharks nei	**Sagres nca**		**Tollos lucero nep**			***Etmopterus spp***			**1,09(01)010,XX**	**SHL**
27 Spain	...	...	...	...	...	...	...	573	...	...
27 Fishing area total	...	...	...	...	...	...	...	*573*	...	...
81 New Zealand	...	...	...	3	0	2	-	-	-	4
81 Fishing area total	...	...	...	*3*	*0*	*2*	-	-	-	*4*
Species total	...	...	...	*3*	*0*	*2*	...	*573*	...	*4*
Birdbeak dogfish	**Squale savate**		**Tollo pajarito**			***Deania calcea***			**1,09(01)014,01**	**DCA**
27 Ireland	...	...	...	...	...	...	...	...	...	216
Portugal	...	...	...	...	...	...	...	...	18	50
UK	-	-	-	-	-	-	-	-	-	1

B-38

Sharks, rays, chimaeras — **Capture production by species, fishing areas and countries or areas**
Squales, raies, chimères — **Captures par espèces, zones de pêche et pays ou zones**
Tiburones, rayas, quimeras — **Capturas por especies, áreas de pesca y países o áreas**

Species, Fishing area Espèce, Zone de pêche Especie, Area de pesca	1992 mt	1993 mt	1994 mt	1995 mt	1996 mt	1997 mt	1998 mt	1999 mt	2000 mt	2001 mt
27 Fishing area total	*...*	*...*	*...*	*...*	*...*	*...*	*...*	*...*	*18*	*267*
81 New Zealand	...	...	...	...	...	...	36	17	28	66
81 Fishing area total	*...*	*...*	*...*	*...*	*...*	*...*	*36*	*17*	*28*	*66*
Species total	*...*	*...*	*...*	*...*	*...*	*...*	*36*	*17*	*46*	*333*
Portuguese dogfish	**Pailona commun**		**Pailona**		***Centroscymnus coelolepis***			**1,09(01)016,01**		**CYO**
27 Faeroe Is	-	22	0	60	282	228	80	35	...	...
Iceland	1	1	0	-	-	-	5	0	0	...
Norway	-	-	-	-	-	-	-	-	-	13
Portugal	...	...	...	...	...	...	...	607	633	620
UK	-	-	-	-	54	52	147	75	514	1 663
27 Fishing area total	*1*	*23*	*0*	*60*	*336*	*280*	*232*	*717*	*1 147*	*2 296*
37 Portugal	-	-	-	-	-	-	-	-	7	23
37 Fishing area total	*-*	*-*	*-*	*-*	*-*	*-*	*-*	*-*	*7*	*23*
Species total	*1*	*23*	*0*	*60*	*336*	*280*	*232*	*717*	*1 154*	*2 319*
Longnose velvet dogfish	**Pailona à long nez**		**Sapata negra**		***Centroscymnus crepidater***			**1,09(01)016,02**		**CYP**
27 UK	-	-	-	-	-	-	3	-	-	-
27 Fishing area total	*-*	*-*	*-*	*-*	*-*	*-*	*3*	*-*	*-*	*-*
Species total	*-*	*-*	*-*	*-*	*-*	*-*	*3*	*-*	*-*	*-*
Kitefin shark	**Squale liche**		**Carocho**		***Dalatias licha***			**1,09(01)018,01**		**SCK**
27 Portugal	...	...	...	...	...	...	...	45	311	189
27 Fishing area total	*...*	*...*	*...*	*...*	*...*	*...*	*...*	*45*	*311*	*189*
81 New Zealand	...	...	...	303	175	352	434	328	317	375
81 Fishing area total	*...*	*...*	*...*	*303*	*175*	*352*	*434*	*328*	*317*	*375*
Species total	*...*	*...*	*...*	*303*	*175*	*352*	*434*	*373*	*628*	*564*
Black dogfish	**Aiguillat noir**		**Tollo negro merga**		***Centroscyllium fabricii***			**1,09(01)019,01**		**CFB**
27 France	-	-	-	-	-	-	-	-	269	271
Iceland	1	0	-	1	4	0	-	-	2	-
27 Fishing area total	*1*	*0*	*-*	*1*	*4*	*0*	*-*	*-*	*271*	*271*
Species total	*1*	*0*	*-*	*1*	*4*	*0*	*-*	*-*	*271*	*271*
Dogfishes and hounds nei	**Squales et émissoles nca**		**Galludos, tollos y musolas nep**		***Squalidae, Scyliorhinidae***			**1,09(01)XXX,XX**		**DGH**
27 Belgium	391	288	379	377	415	430	365	345	390	396
Channel Is	14	14 F	3	3 F	48	30	22	15	33	51
Portugal	2 621	1 815	1 570	1 594	1 341	1 376	1 266	754	803	810
UK	366	638	487	143	218	247	358	999	1 806	1 443
27 Fishing area total	*3 392*	*2 755 F*	*2 439*	*2 117 F*	*2 022*	*2 083*	*2 011*	*2 113*	*3 032*	*2 700*
Species total	*3 392*	*2 755 F*	*2 439*	*2 117 F*	*2 022*	*2 083*	*2 011*	*2 113*	*3 032*	*2 700*
Dogfish sharks nei	**Squales nca**		**Galludos, tollos, nep**		***Squalidae***			**1,09(01)XXX,XX**		**DGX**
21 Cuba	30	29	-	-	-	-	-	-	-	-
Germany	-	-	-	-	0	-	-	-	-	-
Greenland	-	11	34	47	134	...	...	...	-	-
Russian Fed	27	-	-	-	-	-	-	-	-	-
Spain	-	-	-	-	138	211	608	550	493	692
UK	-	-	-	-	-	-	4	-	-	-
USA	17 168	20 804	14 171	22 577	27 447	20 086	21 379	15 858	1 753	287
21 Fishing area total	*17 225*	*20 844*	*14 205*	*22 624*	*27 719*	*20 297*	*21 991*	*16 408*	*2 246*	*979*
27 Faeroe Is	106	141	225	12	7	8	8	3	...	...
France	3 259	3 610	3 421	3 125	3 135	2 811	2 288	3 077	4 224	4 194
Germany	-	-	-	-	19	12	16	235	271	433
Greenland	5	3	0	20	2	6	...	...	-	-
Ireland	...	...	...	1 676	1 170	917	1 144	683	441	30
Isle of Man	66	60	54	24	25	25	12	19	11	3
Lithuania	-	-	-	-	-	-	-	-	-	14
Norway	-	-	-	-	-	-	-	-	-	313
Portugal	4	1 210	889	1 137	977	999	905	-	-	-
Spain	-	-	-	-	534	1 226	627	321	419	365
UK	-	-	-	-	-	-	-	-	-	477
27 Fishing area total	*3 440*	*5 024*	*4 589*	*5 994*	*5 869*	*6 004*	*5 000*	*4 338*	*5 366*	*5 829*
31 USA	1	0	4 478	26	138	310	334	222	104	8
31 Fishing area total	*1*	*0*	*4 478*	*26*	*138*	*310*	*334*	*222*	*104*	*8*
34 Greece	3	15	3	2	5	2	1	0	0	-
34 Fishing area total	*3*	*15*	*3*	*2*	*5*	*2*	*1*	*0*	*0*	*-*
37 Albania	...	...	...	1	64	13	-	-	-	10
Croatia	300	535	317	315	260	239	105	53	50	74
France	-	-	-	-	-	-	-	-	12	17

B-38 Sharks, rays, chimaeras / Squales, raies, chimères / Tiburones, rayas, quimeras

Capture production by species, fishing areas and countries or areas / Captures par espèces, zones de pêche et pays ou zones / Capturas por especies, áreas de pesca y países o áreas

Species, Fishing area Espèce, Zone de pêche Especie, Area de pesca	1992 mt	1993 mt	1994 mt	1995 mt	1996 mt	1997 mt	1998 mt	1999 mt	2000 mt	2001 mt
Greece	170	124	205	266	285	241	289	258	270	224
Malta	5	7	10	5	4	5	3	1	2	3
Spain	-	-	-	-	4	5	-	68	11	20
Tunisia	1 183	860	674	596	19	806	44	25	680	883
Yugoslavia	5	2	3	7	10	10	8	9	9	7
37 Fishing area total	*1 663*	*1 528*	*1 209*	*1 190*	*646*	*1 319*	*449*	*414*	*1 034*	*1 238*
67 USA	2 030	1 957	2 324	1 300	2 053	625	555	1	-	-
67 Fishing area total	*2 030*	*1 957*	*2 324*	*1 300*	*2 053*	*625*	*555*	*1*	*-*	*-*
81 New Zealand	...	...	...	413	693	1 705	701	1 010	770	705
81 Fishing area total	*...*	*...*	*...*	*413*	*693*	*1 705*	*701*	*1 010*	*770*	*705*
Species total	*24 362*	*29 368*	*26 808*	*31 549*	*37 123*	*30 262*	*29 031*	*22 393*	*9 520*	*8 759*
Angelshark	**Ange de mer commun**		**Angelote**		***Squatina squatina***				**1,09(03)004,01**	**AGN**
37 Tunisia	10	53	18	35	18	34	44	25	20	22
37 Fishing area total	*10*	*53*	*18*	*35*	*18*	*34*	*44*	*25*	*20*	*22*
Species total	*10*	*53*	*18*	*35*	*18*	*34*	*44*	*25*	*20*	*22*
Argentine angelshark	**Ange de mer argentin**		**Angelote argentino**		***Squatina argentina***				**1,09(03)004,04**	**SUG**
41 Argentina	3 569	3 975	3 622	3 802	4 281	4 410	4 311	3 368	3 123	3 339
41 Fishing area total	*3 569*	*3 975*	*3 622*	*3 802*	*4 281*	*4 410*	*4 311*	*3 368*	*3 123*	*3 339*
Species total	*3 569*	*3 975*	*3 622*	*3 802*	*4 281*	*4 410*	*4 311*	*3 368*	*3 123*	*3 339*
Angelsharks, sand devils nei	**Anges de mer nca**		**Angelotes, peces ángel nep**		***Squatinidae***				**1,09(03)XXX,XX**	**ASK**
27 France	1	2	2	2	1	0	0	1	1	1
UK	0	0	0	0	0	47	-	-	-	-
27 Fishing area total	*1*	*2*	*2*	*2*	*1*	*47*	*0*	*1*	*1*	*1*
37 Albania	...	...	...	0	53	20	31	30	30	16
Malta	0	0	0	0	0	0	0	0	0	0
Turkey	13	13	15	31	42	15	140	70	60	20
37 Fishing area total	*13*	*13*	*15*	*31*	*95*	*35*	*171*	*100*	*90*	*36*
41 Brazil	...	...	...	113	1 587	...	...	...	...	...
41 Fishing area total	*...*	*...*	*...*	*113*	*1 587*	*...*	*...*	*...*	*...*	*...*
87 Peru	93	228	159	289	358	189	101	262	406	510
87 Fishing area total	*93*	*228*	*159*	*289*	*358*	*189*	*101*	*262*	*406*	*510*
Species total	*107*	*243*	*176*	*435*	*2 041*	*271*	*272*	*363*	*497*	*547*
Angular roughshark	**Centrine commune**		**Cerdo marino**		***Oxynotus centrina***				**1,09(05)006,01**	**OXY**
27 Portugal	...	...	...	...	...	...	...	81	33	63
27 Fishing area total	*...*	*...*	*...*	*...*	*...*	*...*	*...*	*81*	*33*	*63*
Species total	*...*	*...*	*...*	*...*	*...*	*...*	*...*	*81*	*33*	*63*
Dogfish sharks, etc. nei	**Squaliformes nca**		**Squaliformes nep**		***Squaliformes***				**1,09(XX)XXX,XX**	**SHX**
21 Canada	1 004	1 122	1 000	293	106	169	115	143	120	92
USA	1 100	1 773	1 026	1 470	3 317	1 083	772	1 530	590	784
21 Fishing area total	*2 104*	*2 895*	*2 026*	*1 763*	*3 423*	*1 252*	*887*	*1 673*	*710*	*876*
34 Congo Rep	202 F	202 F	260	220 F	180 F	140 F	100 F	60 F	12	10 F
Togo	-	4	0	-	-	-	-	-	0	0
34 Fishing area total	*202 F*	*206 F*	*260*	*220 F*	*180 F*	*140 F*	*100 F*	*60 F*	*12*	*10 F*
Species total	*2 306 F*	*3 101 F*	*2 286*	*1 983 F*	*3 603 F*	*1 392 F*	*987 F*	*1 733 F*	*722*	*886 F*
Chola guitarfish	**Poisson-guitare chola**		**Guitarra chola**		***Rhinobatos percellens***				**1,10(01)005,09**	**GUD**
41 Brazil	1 110 F	1 110 F	1 110 F	162	404	...	...	...	...	...
41 Fishing area total	*1 110 F*	*1 110 F*	*1 110 F*	*162*	*404*	*...*	*...*	*...*	*...*	*...*
Species total	*1 110 F*	*1 110 F*	*1 110 F*	*162*	*404*	*...*	*...*	*...*	*...*	*...*
Pacific guitarfish	**Poisson-guitare du Pacifique**		**Guitarra del Pacífico**		***Rhinobatos planiceps***				**1,10(01)005,10**	**GUF**
87 Peru	42	89	...	121	460	333	344	95	2 624	1 060
87 Fishing area total	*42*	*89*	*...*	*121*	*460*	*333*	*344*	*95*	*2 624*	*1 060*
Species total	*42*	*89*	*...*	*121*	*460*	*333*	*344*	*95*	*2 624*	*1 060*
Guitarfishes, etc. nei	**Guitares, etc. nca**		**Guitarras, etc. nep**		***Rhinobatidae***				**1,10(01)XXX,XX**	**GTF**
34 Liberia	...	...	...	...	...	...	54	175	16	...
Senegal	...	...	...	...	...	...	171	59	1 930	1 772
34 Fishing area total	*...*	*...*	*...*	*...*	*...*	*...*	*225*	*234*	*1 946*	*1 772*
37 Albania	...	...	...	0	1	0	-	-	-	-
Gaza Strip	...	...	...	...	...	6	6	5 F	5 F	5 F

B-38 Sharks, rays, chimaeras — Capture production by species, fishing areas and countries or areas
Squales, raies, chimères — Captures par espèces, zones de pêche et pays ou zones
Tiburones, rayas, quimeras — Capturas por especies, áreas de pesca y países o áreas

Species, Fishing area Espèce, Zone de pêche Especie, Area de pesca	1992 mt	1993 mt	1994 mt	1995 mt	1996 mt	1997 mt	1998 mt	1999 mt	2000 mt	2001 mt
Greece	3	20	117	79	112	63	87	73	94	89
37 Fishing area total	*3*	*20*	*117*	*79*	*113*	*69*	*93*	*78 F*	*99 F*	*94 F*
51 Eritrea	...	...	3	1	0	0	-	-	0	-
Pakistan	1 438	1 500	1 442	1 208	1 422	1 481	1 564	1 643	2 185	1 944
51 Fishing area total	*1 438*	*1 500*	*1 445*	*1 209*	*1 422*	*1 481*	*1 564*	*1 643*	*2 185*	*1 944*
Species total	*1 441*	*1 520*	*1 562*	*1 288*	*1 535*	*1 550*	*1 882*	*1 955 F*	*4 230 F*	*3 810 F*
Sawfishes	**Poissons-scies**		**Peces sierra**		***Pristidae***				**1,10(02)XXX,XX**	**SAW**
34 Liberia	...	...	...	...	...	48	...	39	42	...
Senegal	...	...	...	...	...	...	...	2	-	...
34 Fishing area total	...	...	...	...	...	*48*	...	*41*	*42*	...
41 Brazil	690 F	690 F	690 F	...	...	...	...	...	...	...
41 Fishing area total	*690 F*	*690 F*	*690 F*	...	...	...	...	...	...	...
51 Pakistan	2	32	28	23	-	-	-	-	-	-
51 Fishing area total	*2*	*32*	*28*	*23*	-	-	-	-	-	-
Species total	*692 F*	*722 F*	*718 F*	*23*	...	*48*	...	*41*	*42*	...
Blue skate	**Pocheteau gris**		**Noriega**		***Raja batis***				**1,10(04)001,01**	**RJB**
27 Denmark	-	-	-	-	32	9	7	11	47	0
France	266	245	239	285	295	314	296	467	653	667
Iceland	363	274	299	245	181	118	108	80	94	85
Norway	...	...	...	...	...	...	...	...	...	65
27 Fishing area total	*629*	*519*	*538*	*530*	*508*	*441*	*411*	*558*	*794*	*817*
Species total	*629*	*519*	*538*	*530*	*508*	*441*	*411*	*558*	*794*	*817*
Thornback ray	**Raie bouclée**		**Raya de clavos**		***Raja clavata***				**1,10(04)001,02**	**RJC**
27 France	2 255	1 745	1 577	1 749	1 756	1 579	1 343	1 290	1 222	1 214
27 Fishing area total	*2 255*	*1 745*	*1 577*	*1 749*	*1 756*	*1 579*	*1 343*	*1 290*	*1 222*	*1 214*
37 France	...	...	...	...	...	...	...	45	29	17
Ukraine	...	...	...	...	17	9	24	31	24	65
37 Fishing area total	...	...	...	...	*17*	*9*	*24*	*76*	*53*	*82*
Species total	*2 255*	*1 745*	*1 577*	*1 749*	*1 773*	*1 588*	*1 367*	*1 366*	*1 275*	*1 296*
Starry ray	**Raie radiée**		**Raya radiante**		***Raja radiata***				**1,10(04)001,03**	**RJR**
27 Iceland	317	295	1 206	1 749	1 493	1 431	1 252	996	1 076	1 211
27 Fishing area total	*317*	*295*	*1 206*	*1 749*	*1 493*	*1 431*	*1 252*	*996*	*1 076*	*1 211*
Species total	*317*	*295*	*1 206*	*1 749*	*1 493*	*1 431*	*1 252*	*996*	*1 076*	*1 211*
Spotted ray	**Raie douce**		**Raya pintada**		***Raja montagui***				**1,10(04)001,04**	**RJM**
27 France	1 172	1 120	953	925	977	1 163	1 179	1 260	1 341	1 563
27 Fishing area total	*1 172*	*1 120*	*953*	*925*	*977*	*1 163*	*1 179*	*1 260*	*1 341*	*1 563*
Species total	*1 172*	*1 120*	*953*	*925*	*977*	*1 163*	*1 179*	*1 260*	*1 341*	*1 563*
Sandy ray	**Raie circulaire**		**Raya falsa vela**		***Raja circularis***				**1,10(04)001,06**	**RJI**
27 France	430	356	397	431	438	438	410	435	369	330
27 Fishing area total	*430*	*356*	*397*	*431*	*438*	*438*	*410*	*435*	*369*	*330*
Species total	*430*	*356*	*397*	*431*	*438*	*438*	*410*	*435*	*369*	*330*
Shagreen ray	**Raie chardon**		**Raya cardadora**		***Raja fullonica***				**1,10(04)001,07**	**RJF**
27 France	88	71	56	51	46	39	38	65	38	68
Iceland	-	2	12	24	19	16	12	21	27	37
27 Fishing area total	*88*	*73*	*68*	*75*	*65*	*55*	*50*	*86*	*65*	*105*
Species total	*88*	*73*	*68*	*75*	*65*	*55*	*50*	*86*	*65*	*105*
Small-eyed ray	**Raie mêlée**		**Raya colorada**		***Raja microocellata***				**1,10(04)001,09**	**RJE**
27 France	-	-	-	-	-	-	1	11	-	-
27 Fishing area total	-	-	-	-	-	-	*1*	*11*	-	-
Species total	-	-	-	-	-	-	*1*	*11*	-	-
Cuckoo ray	**Raie fleurie**		**Raya santiguesa**		***Raja naevus***				**1,10(04)001,10**	**RJN**
27 France	3 676	3 050	3 365	3 762	4 077	4 721	4 015	3 638	3 064	2 885
27 Fishing area total	*3 676*	*3 050*	*3 365*	*3 762*	*4 077*	*4 721*	*4 015*	*3 638*	*3 064*	*2 885*
Species total	*3 676*	*3 050*	*3 365*	*3 762*	*4 077*	*4 721*	*4 015*	*3 638*	*3 064*	*2 885*

B-38

Sharks, rays, chimaeras — **Capture production by species, fishing areas and countries or areas**
Squales, raies, chimères — **Captures par espèces, zones de pêche et pays ou zones**
Tiburones, rayas, quimeras — **Capturas por especies, áreas de pesca y países o áreas**

Species, Fishing area Espèce, Zone de pêche Especie, Area de pesca	1992 mt	1993 mt	1994 mt	1995 mt	1996 mt	1997 mt	1998 mt	1999 mt	2000 mt	2001 mt
Longnosed skate	**Pocheteau noir**		**Raya picuda**			***Raja oxyrinchus***			**1,10(04)001,11**	**RJO**
27 France	393	387	347	359	346	311	327	194	140	89
27 Fishing area total	*393*	*387*	*347*	*359*	*346*	*311*	*327*	*194*	*140*	*89*
Species total	*393*	*387*	*347*	*359*	*346*	*311*	*327*	*194*	*140*	*89*
Antarctic starry skate	**Raie étoilée antarctique**		**Raya estrellada antártica**			***Raja georgiana***			**1,10(04)001,32**	**SRR**
48 UK	-	-	-	-	-	-	-	-	0	0
48 Fishing area total	-	-	-	-	-	-	-	-	*0*	*0*
88 New Zealand	-	-	-	-	-	-	...	11	36	7
88 Fishing area total	-	-	-	-	-	-	...	*11*	*36*	*7*
Species total	-	-	-	-	-	-	...	*11*	*36*	*7*
Raja rays nei	**Pocheteaux et raies raja nca**		**Rayas raja nep**			***Raja spp***			**1,10(04)001,XX**	**SKA**
21 Canada	495	319	6 362	6 263	3 900	4 373	3 044	3 119	2 066	2 625
Cuba	0	10	-	5	-	0	1	-	-	-
Estonia	-	-	-	-	-	-	-	-	240	1 023
Faeroe Is	-	-	-	-	-	-	-	3	12	...
Germany	-	-	2	-	0	-	-	-	-	-
Greenland	0	0	5	-	-	-	-	-	-	-
Korea Rep	1 045	5	-	-	-	-	-	-	-	-
Lithuania	-	-	-	-	-	-	-	-	-	4
Norway	1	7	-	-	3	-	-	-	-	-
Poland	-	-	-	-	-	-	-	-	-	2
Portugal	7 019	7 605	6 239	2 060	794	904	1 104	2 168	671	880
Russian Fed	199	14	-	6	7	-	3	160	3 578	2 575
St Pier Mq	46	12	4	4	3	3	9	4	21	38
Spain	209	2 126	5 485	4 511	4 578	9 329	8 106	9 390	14 075	10 547
UK	3	-	-	-	-	-	5	-	-	-
USA	12 473	8 103	8 846	6 454	13 891	10 142	13 932	12 619	13 335	13 122
21 Fishing area total	*21 490*	*18 201*	*26 943*	*19 303*	*23 176*	*24 751*	*26 204*	*27 463*	*33 998*	*30 816*
27 Belgium	1 385	1 430	1 307	1 275	1 363	1 259	1 232	1 351	1 231	1 527
Channel Is	144	145 F	156	144 F	180	33	223	262	181	241
Denmark	47	35	64	57	44	40	20	46	87	122
Estonia	-	-	-	-	-	-	-	-	-	56
Faeroe Is	259	208	174	230	169	187	151	180	113	90
France	3 276	3 438	2 967	2 752	2 902	3 196	2 870	3 408	3 111	3 195
Germany	3	20	57	35	65	74	81	102	130	27
Ireland	2 270	1 755	1 524	2 098	2 212	2 715	2 120	2 283	2 096	2 140
Isle of Man	15	7	6	9	10	6	6	3	5	1
Lithuania	1 289	-	-	-	-	-	-	-	-	-
Netherlands	-	-	-	-	-	-	550	480	631	748
Norway	988	1 112	1 060	951	795	591	752	791	778	725
Portugal	1 595	1 696	1 467	1 599	1 665	1 711	1 703	1 560	1 654	1 683
Russian Fed	580	502	626	-	-	476	929	815	907	339
Spain	1 884	1 659	3 825	1 313	3 972	10 026	11 912	14 997	11 158	9 198
Sweden	29	31	35	17	9	8	2	3	3	12
UK	8 043	7 538	7 781	8 373	9 157	8 088	7 532	6 233	6 457	6 392
27 Fishing area total	*21 807*	*19 576 F*	*21 049*	*18 853 F*	*22 543*	*28 410*	*30 083*	*32 514*	*28 542*	*26 496*
37 Greece	181	528	1 380	1 120	1 002	900	715	718	746	579
37 Fishing area total	*181*	*528*	*1 380*	*1 120*	*1 002*	*900*	*715*	*718*	*746*	*579*
47 Angola	696	860	568	215	21	16	750	1 399	593	1 430
Iceland	-	-	-	-	-	14	-	-	-	-
South Africa	1 292	1 110	1 130	929	1 026	1 239	1 074	1 000	1 000 F	1 151
47 Fishing area total	*1 988*	*1 970*	*1 698*	*1 144*	*1 047*	*1 269*	*1 824*	*2 399*	*1 593 F*	*2 581*
Species total	*45 466*	*40 275 F*	*51 070*	*40 420 F*	*47 768*	*55 330*	*58 826*	*63 094*	*64 879 F*	*60 472*
Eaton's skate	**Raie d'Eaton**		**Raya de Eaton**			***Bathyraja eatonii***			**1,10(04)002,01**	**BEA**
88 New Zealand	-	-	-	-	-	-	...	1	5	0
88 Fishing area total	-	-	-	-	-	-	...	*1*	*5*	*0*
Species total	-	-	-	-	-	-	...	*1*	*5*	*0*
Bathyraja rays nei	**Raies bathyraja nca**		**Rayas bathyraja nep**			***Bathyraja spp***			**1,10(04)002,XX**	**BHY**
88 New Zealand	-	-	-	-	-	-	-	1	-	-
88 Fishing area total	-	-	-	-	-	-	-	*1*	-	-
Species total	-	-	-	-	-	-	-	*1*	-	-
Rays and skates nei	**Rajidés nca**		**Rayidos nep**			***Rajidae***			**1,10(04)XXX,XX**	**RAJ**
88 New Zealand	-	-	-	-	-	-	...	6	-	-
88 Fishing area total	-	-	-	-	-	-	...	*6*	-	-
Species total	-	-	-	-	-	-	...	*6*	-	-
Whip stingray	**Pastenague du Pacifique**		**Raya látigo del Pacífico**			***Dasyatis akajei***			**1,10(05)003,01**	**WST**
61 Japan	4 585	4 247	4 040	3 985	4 029	3 959	4 329	4 407	5 388	4 312
61 Fishing area total	*4 585*	*4 247*	*4 040*	*3 985*	*4 029*	*3 959*	*4 329*	*4 407*	*5 388*	*4 312*

B-38 Sharks, rays, chimaeras / Squales, raies, chimères / Tiburones, rayas, quimeras

Capture production by species, fishing areas and countries or areas
Captures par espèces, zones de pêche et pays ou zones
Capturas por especies, áreas de pesca y países o áreas

Species, Fishing area Espèce, Zone de pêche Especie, Area de pesca	1992 mt	1993 mt	1994 mt	1995 mt	1996 mt	1997 mt	1998 mt	1999 mt	2000 mt	2001 mt
Species total	*4 585*	*4 247*	*4 040*	*3 985*	*4 029*	*3 959*	*4 329*	*4 407*	*5 388*	*4 312*
Common stingray	**Pastenague commune**		**Raya látigo común**		***Dasyatis pastinaca***				**1,10(05)003,26**	**JDP**
37 Ukraine	...	...	...	...	...	...	-	-	4	11
37 Fishing area total	...	...	...	...	...	...	-	-	*4*	*11*
Species total	...	...	...	...	...	...	-	-	*4*	*11*
Stingrays nei	**Pastenagues nca**		**Pastinacas nep**		***Dasyatis spp***				**1,10(05)003,XX**	**STI**
27 France	3	1	2	-	1	2	5	6	10	7
27 Fishing area total	*3*	*1*	*2*	-	*1*	*2*	*5*	*6*	*10*	*7*
Species total	*3*	*1*	*2*	-	*1*	*2*	*5*	*6*	*10*	*7*
Eagle rays	**Aigles de mer**		**Aguilas de mar**		***Myliobatidae***				**1,10(07)XXX,XX**	**EAG**
27 France	5	3	2	2	0	0	0	2	2	2
Portugal	...	...	...	...	...	...	...	11	8	9
27 Fishing area total	*5*	*3*	*2*	*2*	*0*	*0*	*0*	*13*	*10*	*11*
81 New Zealand	...	...	...	0	0	1	1	2	2	5
81 Fishing area total	...	...	...	*0*	*0*	*1*	*1*	*2*	*2*	*5*
Species total	*5*	*3*	*2*	*2*	*0*	*1*	*1*	*15*	*12*	*16*
Mantas	**Mantes**		**Mantas**		***Mobulidae***				**1,10(08)XXX,XX**	**MAN**
34 Liberia	...	...	...	...	...	...	342	802	931	106
34 Fishing area total	...	...	...	...	...	...	*342*	*802*	*931*	*106*
Species total	...	...	...	...	...	...	*342*	*802*	*931*	*106*
Rays, stingrays, mantas nei	**Raies, pastenagues, mantes nca**		**Rayas, pastinacas, mantas nep**		***Rajiformes***				**1,10(XX)XXX,XX**	**SRX**
31 Cuba	0	12	0	-	955	1 359	1 335	1 352	1 350 F	1 350 F
Dominican Rp	46	10	18	90	39	96	62	134	111	123
Korea Rep	240	-	-	-	-	-	175	42	97	3
Martinique	4 F	5 F	5 F	5 F	3 F	5 F	5 F	5 F	5 F	5 F
Mexico	4 585	3 992	4 212	5 165	4 782	5 496	6 197	4 201	2 873	2 618
Venezuela	1 867	1 748	1 994	2 450	1 812	1 896	2 111	2 287	2 148	2 182
31 Fishing area total	*6 742 F*	*5 767 F*	*6 229 F*	*7 710 F*	*7 591 F*	*8 852 F*	*9 885 F*	*8 021 F*	*6 584 F*	*6 281 F*
34 Benin	46 F	50 F	56 F	54 F	62 F	70 F	60 F	47	30 F	38
Cameroon	77	70	100 F	160	186	170 F	161	211	150	130
Congo Rep	396 F	395 F	185	160 F	135 F	110 F	85 F	60 F	33	30 F
Côte dIvoire	...	...	...	...	218	146	202	227	241	168
Gabon	...	0	3	33	172	173	90	197	141	88
Ghana	231	272	296	338	261	185	172	869	231	814
Italy	1 618	1 372	1 372	26	6	-	-	-	-	12
Korea Rep	67	369	635	565	576	653	57	249	402	334
Liberia	29	50	56	33	12	38	50	119	103	29
Mauritania	20 F	20 F	20 F	10 F	10 F	20 F	180 F	350 F	350 F	350 F
Morocco	1 107	892	1 262	1 643	1 650	1 350	1 187	1 348	2 139	1 487
Nigeria	2 446	1 126	3 102	3 670	4 103	5 004	7 382	7 622	5 753	7 352
Portugal	112	57	24	57	126	239	155	74	0	-
Senegal	1 844	2 205	2 886	3 308	2 962	4 515	4 051	3 332	1 585	1 667
Sierra Leone	1 424	1 408	1 403	1 401	1 400 F	1 404	15	10	18	...
Spain	24	-	-	-	203	-	18	10	0	-
Togo	7	18	9	5	11	2	0	2	16	5
34 Fishing area total	*9 448 F*	*8 304 F*	*11 409 F*	*11 463 F*	*12 093 F*	*14 079 F*	*13 865 F*	*14 727 F*	*11 192 F*	*12 504 F*
37 Albania	10	10 F	15 F	67	23	24	86	78	85	14
Algeria	...	...	...	124	272	120	450	207	200 F	200 F
Croatia	170	276	224	190	141	119	120	68	57	42
France	172	135	86	15	75	75	75	46	70	64
Gaza Strip	...	...	...	...	29	16	23	20 F	20 F	20 F
Italy	3 086	3 011	2 358	4 552	2 301	5 325	2 807	1 117	507	543
Malta	7	7	5	5	7	8	5	6	7	0
Morocco	31	40	37	27	30	30	19	115	...	23
Russian Fed	54	20	24	14	22	16	18	40	13	30
Spain	-	-	-	-	248	288	289	501	536	375
Tunisia	1 534	614	551	460	489	803	836	922	974	1 113
Turkey	1 557	1 124	1 238	337	524	340	385	420	1 100	555
Ukraine	318	3	4	15	...	1	-	-	-	-
Yugoslavia	6	9	8	14	12	12	12	12	11	11
37 Fishing area total	*6 945*	*5 249 F*	*4 550 F*	*5 820*	*4 173*	*7 177*	*5 125*	*3 552 F*	*3 580 F*	*2 990 F*
41 Argentina	761	910	5 701	7 190	12 478	12 119	14 855	12 116	13 265	15 179
Australia	-	-	-	-	-	-	3	23	-	-
Belize	-	-	-	-	-	-	-	519	48	201
Brazil	3 700 F	3 500 F	3 000 F	3 948	3 104	3 010	4 673	4 277	4 867	4 800 F
Chile	-	-	16	-	-	-	4	4	-	-
Falkland Is	32	98	63	117	184	204	216	314	353	417
France	-	-	20	5	9	3	1	0	0	-
Honduras	1 148	1 948	876	615	460	-	-	-	-	-
Italy	216	183	183	3	1	-	-	-	-	-

B-38 Sharks, rays, chimaeras / Squales, raies, chimères / Tiburones, rayas, quimeras

Capture production by species, fishing areas and countries or areas / Captures par espèces, zones de pêche et pays ou zones / Capturas por especies, áreas de pesca y países o áreas

Species, Fishing area Espèce, Zone de pêche Especie, Area de pesca	1992 mt	1993 mt	1994 mt	1995 mt	1996 mt	1997 mt	1998 mt	1999 mt	2000 mt	2001 mt
Korea Rep	1 640	1 696	2 631	5 506	5 243	5 118	1 503	5 539	7 246	6 382
Namibia	-	-	-	-	-	3	14	12	-	-
Panama	1 257	611	372	85	170	-	-	-	-	-
Portugal	13	11	28	80	135	18	34	-	0	82
Russian Fed	-	-	-	-	-	-	-	-	-	203
Seychelles	-	-	-	-	-	4	1	-	-	-
UK	-	29	3	12	8	17	13	40	17	32
Uruguay	...	128	1 032	1 469	2 614	2 342	398	1 575	1 004	989
41 Fishing area total	*8 767 F*	*9 114 F*	*13 925 F*	*19 030*	*24 406*	*22 838*	*21 715*	*24 419*	*26 800*	*28 285 F*
47 Korea Rep	29	475	364	540	227	283	435	343	375	323
Namibia	4	1	0	62	133	191	80	377	966	1 204
Russian Fed	-	-	-	-	-	-	8	-	-	-
Spain	30	-	-	-	-	-	-	-	-	-
47 Fishing area total	*63*	*476*	*364*	*602*	*360*	*474*	*523*	*720*	*1 341*	*1 527*
48 Argentina	-	-	-	28	0	-	-	-	-	-
Chile	-	-	-	20	21	-	0	-	-	-
Korea Rep	-	-	11	42	19	24	5	4	1	7
Russian Fed	1	-	0	0	-	-	-	-	-	0
South Africa	-	-	-	-	-	-	7	1	-	-
Spain	-	-	-	-	-	2	1	-	-	-
UK	1	-	-	-	-	4	1	9	3	6
USA	-	-	-	-	-	-	-	0	-	0
Uruguay	-	-	-	-	-	-	0	1	-	-
48 Fishing area total	*2*	*-*	*11*	*90*	*40*	*30*	*14*	*15*	*4*	*13*
51 Italy	324	274	274	5	1	-	-	-	-	-
Korea Rep	291	315	629	27	106	114	329	162	14	23
Mauritius	2	2	2	2	2	2	2	2	2	2
Oman	297	288	188	538	372	359	189	289	240	198
Pakistan	16 532	16 093	18 481	16 445	15 563	15 769	17 576	20 780	20 740	20 801
Saudi Arabia	-	-	-	-	-	-	4	8	4	6
Tanzania	2 500	2 511	2 474	3 200	4 006	3 500	3 350	3 500	3 600	3 500
Yemen	-	-	-	156	...	100 F	100 F	100 F	100 F	130 F
51 Fishing area total	*19 946*	*19 483*	*22 048*	*20 373*	*20 050*	*19 844 F*	*21 550 F*	*24 841 F*	*24 700 F*	*24 660 F*
57 Australia	-	-	17	22	55	55	55	8	10	18
Indonesia	9 056	12 841	12 606	10 194	9 472	10 188	16 961	12 453	11 682	12 360
Korea Rep	-	-	-	7	-	-	-	18	8	-
Malaysia	5 430	5 608	4 645	4 211	4 541	5 766	3 915	4 833	5 389	5 176
Thailand	1 908	2 099	2 370	3 520	5 075	5 238	4 185	4 688	4 624	4 596
57 Fishing area total	*16 394*	*20 548*	*19 638*	*17 954*	*19 143*	*21 247*	*25 116*	*22 000*	*21 713*	*22 150*
58 Australia	1	-	0	-	-	3	1	4	-	-
France	0	-	2	-	-	-	21	7	94	69
South Africa	-	-	-	-	-	-	1	2	4	9
Ukraine	-	-	-	-	1	-	-	-	-	-
58 Fishing area total	*1*	*-*	*2*	*-*	*1*	*3*	*23*	*13*	*98*	*78*
61 China,Taiwan	464	673	533	647	2 457	1 367	246	235	351	851
Korea Rep	5 767	9 918	6 914	6 235	6 787	6 638	4 375	3 948	2 565	211
Russian Fed	-	-	-	6	6	9	8	304	1 427	1 602
61 Fishing area total	*6 231*	*10 591*	*7 447*	*6 888*	*9 250*	*8 014*	*4 629*	*4 487*	*4 343*	*2 664*
67 Canada	264	249	570	982	1 293	1 584	900	1 406	1 661	1 618
Korea Rep	-	-	13	-	-	-	-	-	-	-
USA	14 713	139	17	370	1 344	2 173	1 145	1 411	1 549	1 399
67 Fishing area total	*14 977*	*388*	*600*	*1 352*	*2 637*	*3 757*	*2 045*	*2 817*	*3 210*	*3 017*
71 Australia	-	-	-	-	-	-	-	-	-	0
Indonesia	20 497	22 845	24 000	24 623	26 985	33 136	31 330	32 874	33 578	35 510
Korea Rep	237	356	3 123	3 007	1 422	1 289	1 121	2 155	1 401	1 526
Malaysia	8 101	8 996	9 355	11 496	11 387	11 516	12 189	12 200	11 184	11 356
Philippines	5 488	6 230	4 906	4 980	4 756	2 125	2 174	2 299	2 248	2 405
Singapore	388	311	411	320	327	308	336	250	261	187
Thailand	3 458	3 702	7 772	6 448	4 903	5 115	4 104	7 591	7 487	7 441
71 Fishing area total	*38 169*	*42 440*	*49 567*	*50 874*	*49 780*	*53 489*	*51 254*	*57 369*	*56 159*	*58 425*
77 Korea Rep	102	25	92	24	167	239	162	1 069	596	106
Mexico	4 017	3 922	4 137	4 411	5 150	4 578	4 757	4 875	4 944	4 405
USA	66	29	31	60	210	315	195	205	167	101
77 Fishing area total	*4 185*	*3 976*	*4 260*	*4 495*	*5 527*	*5 132*	*5 114*	*6 149*	*5 707*	*4 612*
81 Australia	22	29	75	68	70	72	67	43	41	37
Korea Rep	98	188	340	177	135	47	21	197	82	168
New Zealand	1 427	2 788	2 598	2 116	1 582	2 227	2 313	2 821	2 634	2 784
81 Fishing area total	*1 547*	*3 005*	*3 013*	*2 361*	*1 787*	*2 346*	*2 401*	*3 061*	*2 757*	*2 989*
87 Chile	1 239	1 971	2 899	2 622	2 675	2 958	2 011	3 365	4 151	2 974
Colombia	-	-	-	-	3	2	2	1	1 F	1
Korea Rep	-	668	120	-	-	-	-	252	247	38
Peru	2 771	3 632	1 658	1 841	1 126	1 177	1 477	2 789	4 026	2 034
87 Fishing area total	*4 010*	*6 271*	*4 677*	*4 463*	*3 804*	*4 137*	*3 490*	*6 407*	*8 425 F*	*5 047*
88 New Zealand	-	-	-	-	-	-	5	-	0	-
88 Fishing area total	*-*	*-*	*-*	*-*	*-*	*-*	*5*	*-*	*0*	*-*
Species total	*137 427 F*	*135 612 F*	*147 740 F*	*153 475 F*	*160 642 F*	*171 419 F*	*166 754 F*	*178 598 F*	*176 613 F*	*175 242 F*

B-38 Sharks, rays, chimaeras / Squales, raies, chimères / Tiburones, rayas, quimeras

Capture production by species, fishing areas and countries or areas / Captures par espèces, zones de pêche et pays ou zones / Capturas por especies, áreas de pesca y países o áreas

Species, Fishing area Espèce, Zone de pêche Especie, Area de pesca	1992 mt	1993 mt	1994 mt	1995 mt	1996 mt	1997 mt	1998 mt	1999 mt	2000 mt	2001 mt
Torpedo rays	**Torpillés**		**Tremolinas**			*Torpedo spp*			**1,11(01)002,XX**	**TOE**
27 France	15	21	21	20	16	18	19	34	32	43
27 Fishing area total	*15*	*21*	*21*	*20*	*16*	*18*	*19*	*34*	*32*	*43*
Species total	*15*	*21*	*21*	*20*	*16*	*18*	*19*	*34*	*32*	*43*
Rabbit fish	**Chimère commune**		**Quimera**			*Chimaera monstrosa*			**1,12(01)003,01**	**CMO**
27 Denmark	-	-	-	-	-	-	-	-	-	1
Iceland	106	3	60	106	21	15	29	11	5	1
Ireland	...	...	...	...	...	...	...	...	5	15
Norway	-	-	-	-	-	-	-	-	-	70
UK	-	-	-	-	-	0	1	1	4	36
27 Fishing area total	*106*	*3*	*60*	*106*	*21*	*15*	*30*	*12*	*14*	*123*
Species total	*106*	*3*	*60*	*106*	*21*	*15*	*30*	*12*	*14*	*123*
Dark ghost shark	**...B**		**...C**			*Hydrolagus novaezealandiae*			**1,12(01)004,11**	**CYV**
81 New Zealand	959	1 089	1 455	1 593	1 614	2 064	1 956	1 975	1 819	1 572
81 Fishing area total	*959*	*1 089*	*1 455*	*1 593*	*1 614*	*2 064*	*1 956*	*1 975*	*1 819*	*1 572*
Species total	*959*	*1 089*	*1 455*	*1 593*	*1 614*	*2 064*	*1 956*	*1 975*	*1 819*	*1 572*
Ratfishes nei	**Chimères nca**		**Quimeras nep**			*Hydrolagus spp*			**1,12(01)004,XX**	**HYD**
27 France	-	-	-	-	-	-	-	38	573	822
Ireland	...	...	...	...	...	...	...	...	...	5
Norway	-	-	-	-	-	-	-	-	-	6
UK	-	-	-	-	-	-	-	-	-	2
27 Fishing area total	*...*	*...*	*...*	*...*	*...*	*...*	*...*	*38*	*573*	*835*
81 New Zealand	...	...	...	...	...	0	36	453	975	2 184
81 Fishing area total	*...*	*...*	*...*	*...*	*...*	*0*	*36*	*453*	*975*	*2 184*
Species total	*...*	*...*	*...*	*...*	*...*	*0*	*36*	*491*	*1 548*	*3 019*
Straightnose rabbitfish	**Chimère à nez mou**		**Narigón sierra**			*Rhinochimaera atlantica*			**1,12(02)001,01**	**RCT**
27 UK	-	-	-	-	-	-	-	-	-	2
27 Fishing area total	*-*	*-*	*-*	*-*	*-*	*-*	*-*	*-*	*-*	*2*
Species total	*-*	*-*	*-*	*-*	*-*	*-*	*-*	*-*	*-*	*2*
Ghost shark	**Masca laboureur**		**...C**			*Callorhinchus milii*			**1,12(03)001,01**	**CHB**
81 New Zealand	612	587	639	769	595	913	951	1 260	1 228	1 189
81 Fishing area total	*612*	*587*	*639*	*769*	*595*	*913*	*951*	*1 260*	*1 228*	*1 189*
Species total	*612*	*587*	*639*	*769*	*595*	*913*	*951*	*1 260*	*1 228*	*1 189*
Cape elephantfish	**Masca du Cap**		**Pejegallo del Cabo**			*Callorhinchus capensis*			**1,12(03)001,03**	**CHM**
47 South Africa	542	983	262	386	366	484	482	356	380 F	405
47 Fishing area total	*542*	*983*	*262*	*386*	*366*	*484*	*482*	*356*	*380 F*	*405*
Species total	*542*	*983*	*262*	*386*	*366*	*484*	*482*	*356*	*380 F*	*405*
Elephantfishes nei	**Mascas nca**		**Pejegallos nep**			*Callorhinchus spp*			**1,12(03)001,XX**	**ELF**
41 Argentina	479	704	1 096	921	815	1 329	1 770	1 977	1 378	...
41 Fishing area total	*479*	*704*	*1 096*	*921*	*815*	*1 329*	*1 770*	*1 977*	*1 378*	*...*
87 Chile	4 729	2 516	1 570	920	1 450	822	1 416	632	603	1 125
87 Fishing area total	*4 729*	*2 516*	*1 570*	*920*	*1 450*	*822*	*1 416*	*632*	*603*	*1 125*
Species total	*5 208*	*3 220*	*2 666*	*1 841*	*2 265*	*2 151*	*3 186*	*2 609*	*1 981*	*1 125*
Chimaeras, etc. nei	**Chimères, etc. nca**		**Quimeras, etc. nep**			*Chimaeriformes*			**1,12(XX)XXX,XX**	**HOL**
27 Iceland	0	2	0	2	1	0	-	-	-	-
27 Fishing area total	*0*	*2*	*0*	*2*	*1*	*0*	*-*	*-*	*-*	*-*
81 New Zealand	...	...	...	5	49	5	5	21	40	76
81 Fishing area total	*...*	*...*	*...*	*5*	*49*	*5*	*5*	*21*	*40*	*76*
Species total	*0*	*2*	*0*	*7*	*50*	*5*	*5*	*21*	*40*	*76*
Various sharks nei	**Requins divers nca**		**Escualos diversos nep**			*Selachimorpha(Pleurotremata)*			**1,99(XX)XXX,XX**	**SKH**
27 Belgium	23	22	19	20	19	18	11	14	15	18
Channel Is	42	43 F	32	30 F	2	3	3	7	1	-
Denmark	5	5	3	4	-	-	-	-	-	-
Germany	2	133	440	292	309	139	110	-	-	-
Ireland	-	17	16	40	23	32	-	-	233	455
Netherlands	-	-	-	-	-	-	-	-	-	3
Norway	0	-	0	0	0	1	0	13	119	72

B-38 Sharks, rays, chimaeras / Squales, raies, chimères / Tiburones, rayas, quimeras

Capture production by species, fishing areas and countries or areas
Captures par espèces, zones de pêche et pays ou zones
Capturas por especies, áreas de pesca y países o áreas

Species, Fishing area Espèce, Zone de pêche Especie, Area de pesca	1992 mt	1993 mt	1994 mt	1995 mt	1996 mt	1997 mt	1998 mt	1999 mt	2000 mt	2001 mt
Portugal	618	1 047	969	1 853	1 659	1 874	1 882	352	297	217
Spain	6 551	6 914	10 998	18 101	3 642	30 377	20 450	23 662	31 793	30 815
UK	1 119	1 393	1 944	2 339	2 040	3 865	2 669	2 342	954	1 640
27 Fishing area total	*8 360*	*9 574 F*	*14 421*	*22 679 F*	*7 694*	*36 309*	*25 125*	*26 390*	*33 412*	*33 220*
Species total	*8 360*	*9 574 F*	*14 421*	*22 679 F*	*7 694*	*36 309*	*25 125*	*26 390*	*33 412*	*33 220*
Sharks, rays, skates, etc. nei	**Requins, raies, etc. nca**			**Tiburones, rayas, etc. nep**		***Elasmobranchii***			**1,99(XX)XXX,XX**	**SKX**
21 China,Taiwan	-	-	-	18	12	12	-	2	6	8
Japan	345	553	450	397	238	99	107	123	83	116
Portugal	-	-	-	-	-	-	-	103	1	-
21 Fishing area total	*345*	*553*	*450*	*415*	*250*	*111*	*107*	*228*	*90*	*124*
27 China,Taiwan	-	-	-	22	15	15	-	10	14	8
France	18	-	-	2	-	-	-	-	-	-
Japan	107	174	168	376	132	108	211	72	35	82
Poland	-	-	-	-	-	-	-	-	-	11
Portugal	1 047	...	...	...	...	...	...	41	40	40
Sweden	1	0	0	0	0	0	0	0	0	-
27 Fishing area total	*1 173*	*174*	*168*	*400*	*147*	*123*	*211*	*123*	*89*	*141*
31 Antigua Barb	...	...	...	...	...	...	...	...	...	8
Bahamas	-	37	0	0	5	3	2	1	0	0
Barbados	24	18	22	24	25	14	12	10	14	10
Belize	...	...	...	...	...	1	0	2	6	...
Bermuda	12	14	10	17	13	9	12	24	10	5
Br Virgin Is	...	...	...	...	...	1	1	1	0	0
China,Taiwan	38 F	33 F	23 F	67	47	47	18	4	100	195
Costa Rica	7	32	11	27	11	1	92	64	144	108
Cuba	1 314	881	1 383	1 365	1 409	1 932	1 737	1 495	1 500 F	1 500 F
Grenada	7	12	4	14	4	9	18	24	29	29
Guyana	...	...	...	...	765	1 892	...	2 175	...	...
Japan	7	13	30	17	9	17	16	63	42	75
Korea Rep	-	-	-	-	-	-	-	2	-	-
Martinique	100 F	120 F	120 F	100 F	70 F	90 F	80 F	70 F	50 F	40 F
Mexico	7 359	6 674	6 696	6 860	7 417	4 994	4 617	4 743	4 370	3 986
Portugal	-		-	-		-		22		
Puerto Rico	...	...	...	...	...	...	...	28	35	32
St Lucia	12	...	6	6	11	3	8	6	5	5
St Vincent	...	...	...	...	2	...	...	3	...	2
Trinidad Tob	531	440	488	550 F	621	545	635	701	755	756
USA	4 630	3 582	5 134	3 471	1 090	2 814	2 845	1 728	1 808	1 846
31 Fishing area total	*14 041 F*	*11 856 F*	*13 927 F*	*12 518 F*	*11 499 F*	*12 372 F*	*10 093 F*	*11 166 F*	*8 868 F*	*8 597 F*
34 Benin	181 F	160 F	140 F	120 F	100 F	100 F	80 F	59	40 F	87
Cameroon	157	92	80 F	59	48	50 F	55	86	67	146
Cape Verde	...	...	1	1	...	...	...	...	...	...
China	-	-	-	-	-	2	5	31	-	-
China,Taiwan	38 F	33 F	23 F	388	270	270	45	82	411	116
Côte dIvoire	278	269	160	159 F	70	71	42	38	521	19
Cuba	1 482	1 915	2 008	1 691	1 051	-	-	-	-	-
Eq Guinea	370 F	330 F	500 F	220 F	490 F	620 F	779	910 F	100 F	100 F
Gabon	...	0	2	22	1 267	626	1 933	1 338	659	375
Gambia	194	316	480	498	415	3 223	606	630	720	3 982
Ghana	914	1 981	1 171	1 115	1 106	709	1 764	3 998	1 670	2 092
Guinea	...	...	...	726	506	505	700	800	969	950 F
Italy	2 023	1 715	1 715	-	-	-	-	-	-	-
Japan	521	1 412	977	925	729	464	709	228	270	265
Korea Rep	-	-	34	-	2	11	1	-	2	-
Liberia	23	100	309	358	207	386	210	-	-	512
Mauritania	160 F	50 F	60 F	80 F	10 F	10 F	350 F	500 F	500 F	500 F
Morocco	1 146	1 423	1 108	1 615	1 600	1 210	2 234	1 985	3 460	2 181
Nigeria	6 466	4 723	5 951	2 801	4 285	3 817	6 587	7 751	7 485	7 274
Panama	-	-	-	-	-	-	-	202	...	-
Portugal	236	170	93	859	2 366	1 127	849	737	141	21
Russian Fed	-	-	-	-	-	-	101	11	-	-
Sao Tome Prn	178 F	221 F	321 F	337 F	247	130	175 F	190 F	180 F	180 F
Senegal	2 159	1 791	3 347	4 169	3 803	4 470	4 887	4 808	5 473	3 260
Sierra Leone	...	...	...	2	2 F	1	68	41	1 672	164
Spain	3	-	-	-	3 977	34 443	14 145	9 992	9 481	9 536
Togo	4	22	4	15	202	57	67	230	132	130
34 Fishing area total	*16 533 F*	*16 723 F*	*18 484 F*	*16 160 F*	*22 753 F*	*52 302 F*	*36 392 F*	*34 647 F*	*33 953 F*	*31 890 F*
37 Algeria	751	1 127	1 200 F	1 000	965	415	867	854	850 F	850 F
Cyprus	24	30	19	21	14	17	10	12	14	28
Egypt	1 152	1 000	1 226	1 172	1 120	1 629	1 211	1 383	1 197	2 143
France	...	...	...	...	28	...	...	43	-	-
Georgia	14	131	45	31	71	1	550	18	21	27
Israel	68	60 F	50 F	48	330	49	59	58	...	40 F
Japan	1	3	5	8	3	1	-	1	-	-
Lebanon	50 F	50	50	50	50	50	50	50	60	55
Malta	5	1	1	4	3	2	11	4	13	0
Morocco	46	31	44	21	25	45	9	19	...	11
Portugal	-	-	-	-	-	-	-	-	3	1
Russian Fed	15	5	11	90	19	9	6	9	12	27
Spain	-	-	-	-	1 277	1 420	2 140	663	426	377
37 Fishing area total	*2 126 F*	*2 438 F*	*2 651 F*	*2 445*	*3 905*	*3 638*	*4 913*	*3 114*	*2 596 F*	*3 559 F*

B-38

Sharks, rays, chimaeras — Capture production by species, fishing areas and countries or areas
Squales, raies, chimères — Captures par espèces, zones de pêche et pays ou zones
Tiburones, rayas, quimeras — Capturas por especies, áreas de pesca y países o áreas

Species, Fishing area Espèce, Zone de pêche Especie, Area de pesca		1992 mt	1993 mt	1994 mt	1995 mt	1996 mt	1997 mt	1998 mt	1999 mt	2000 mt	2001 mt
41	Argentina	3 954	2 044	1 707	2 230	2 251	1 070	1 220	905	719	740
	Brazil	14 750 F	12 000 F	11 000 F	10 658	8 446	10 189	12 596	13 576	15 900	15 700 F
	China,Taiwan	75 F	65 F	46 F	650	450	452	222	357	263	142
	Italy	270	229	229	-	-	-	-	-	-	-
	Japan	185	185	581	65	69	81	35	34	25	46
	Korea Rep	6	-	-	-	-	-	-	-	-	-
	Portugal	-	-	-	-	-	114	103	55	11	41
	Spain	1 223	838	469	395	225	3 855	1 935	1 653	1 817	1 818
	UK	-	-	-	-	-	16	31	-	-	4
	Uruguay	1 198	803	949	1 577	1 760	2 367	444	1 901	991	868
41	*Fishing area total*	*21 661 F*	*16 164 F*	*14 981 F*	*15 575*	*13 201*	*18 144*	*16 586*	*18 481*	*19 726*	*19 359 F*
47	Angola	7	29	35	755	379	90	376	...	157	3 354
	China,Taiwan	75 F	65 F	46 F	268	186	186	341	639	255	568
	Honduras	...	...	...	...	...	10	4	...	...	...
	Japan	658	1 140	1 295	676	398	369	565	603	324	205
	Korea Rep	5	-	-	-	-	-	-	-	-	-
	Namibia	-	-	4	7	5	4	6	1	769	1 875
	Poland	-	-	1	-	-	-	-	-	-	-
	Portugal	-	-	-	12	...	...	...	488	20	239
	St Helena	...	...	...	...	...	...	...	...	...	6
	South Africa	651	699	630	389	317	433	500	430	350 F	314
	Spain	22	35 F	50 F	60 F	-	95	28	109	20	20
47	*Fishing area total*	*1 418 F*	*1 968 F*	*2 061 F*	*2 167 F*	*1 285*	*1 187*	*1 820*	*2 270*	*1 895 F*	*6 581*
51	China	-	-	-	-	-	-	-	-	4	...
	China,Taiwan	468 F	405 F	286 F	331	360	360	261	392	802	1 826
	Comoros	58	58	...	...	...	...	...	...	...	...
	Egypt	32	89	69	137	122	180	135	182	244	263
	Eritrea	...	...	...	...	...	6	7	6	...	...
	India	29 794	32 169	32 514	36 642	93 268	37 588	33 418	34 088	37 060	34 036
	Iran	...	...	...	...	...	...	1	...	...	...
	Italy	405	343	343	-	-	-	-	-	-	-
	Japan	321	196	502	282	620	430	442	276	185	158
	Kenya	173	152	166	176	191	140	134	131	115	175
	Korea Rep	16	-	30	82	76	28	64	33	36	49
	Maldives	6 921	9 168	11 212	11 245	11 856	10 643	10 887	6 883	13 523	11 935
	Mauritius	18	16	17	15	17	58	9	9	25	12
	Oman	5 248	4 540	3 503	6 566	5 870	6 342	4 805	4 020	3 651	3 632
	Philippines	-	-	-	-	-	-	11	11	7	29
	Portugal	-	-	-	-	-	-	262	487	354	728
	Réunion	...	36	33	37	46	89	111	81	78	60
	Saudi Arabia	40	42	125	467	398	543	697	497	653	651
	Seychelles	93	82	117	116	84	57	102	68	150	95
	South Africa	135	141	187	129	10	230	199	240	229	112
	Spain	-	-	-	-	-	90	509	-	5 357	-
	Sudan	...	...	...	...	...	...	45	56	44	79
	Tanzania	2 000	962	1 389	1 310	1 594	1 500	1 325	1 375	1 400	1 500
	Untd Arab Em	1 581	1 600	1 802	1 553	1 902	1 832	1 881	1 945	1 530	1 530 F
	Uruguay	-	-	-	-	-	-	-	-	-	22
	Yemen	6 067	6 537	6 455	4 480	4 878	5 000 F	5 800 F	5 600 F	5 000 F	6 300 F
	Other nei	-	232	158	416	465	268	277	152	193	...
51	*Fishing area total*	*53 370 F*	*56 768 F*	*58 908 F*	*63 984*	*121 757*	*65 384 F*	*61 382 F*	*56 532 F*	*70 640 F*	*63 192 F*
57	Australia	7 603	8 483	6 775	2 695	2 635	5 939	4 881	5 897	5 111	5 755
	China	-	-	-	-	-	-	-	187	95	...
	China,Taiwan	234 F	206 F	150 F	799	758	758	108	11	240	345
	Honduras	-	-	-	-	-	-	-	-	-	85
	India	29 936	44 435	51 175	40 436	38 892	34 403	41 286	42 714	38 997	38 532
	Indonesia	16 033	15 983	16 949	17 905	18 048	16 811	22 786	18 694	19 572	20 490
	Japan	164	244	185	554	437	436	227	378	323	271
	Korea Rep	12	-	-	-	-	-	-	-	3	-
	Malaysia	769	694	709	824	810	762	964	685	1 271	1 105
	Philippines	-	-	-	-	-	-	22	3	2	-
	Seychelles	-	-	-	-	-	-	-	0	1	2
	Sri Lanka	4 606	7 311	8 475	7 077	6 954	11 920	7 625	8 660	11 884	8 240
	Thailand	632	577	879	2 501	4 615	4 623	4 431	3 816	3 764	3 741
	Other nei	-	30	20	53	36	98	18	46	57	...
57	*Fishing area total*	*59 989 F*	*77 963 F*	*85 317 F*	*72 844*	*73 185*	*75 750*	*82 348*	*81 091*	*81 320*	*78 566*
58	Australia	-	-	-	-	-	2	-	-	-	-
	South Africa	-	-	-	-	-	-	-	1	-	-
58	*Fishing area total*	-	-	-	-	-	*2*	-	*1*	-	-
61	China,H.Kong	817	848	688	485	456	420	382	300 F	330 F	370 F
	China,Taiwan	9 470 F	8 200 F	5 750 F	4 646	4 910	4 913	3 171	24 718	22 759	8 230
	Japan	22 743	23 330	18 864	16 633	12 556	19 494	21 369	23 516	22 997	19 905
	Korea Rep	1 696	5 253	1 512	871	...	565	503	449	262	389
	Russian Fed	-	-	-	-	-	-	-	10	-	100
61	*Fishing area total*	*34 726 F*	*37 631 F*	*26 814 F*	*22 635*	*17 922*	*25 392*	*25 425*	*48 993 F*	*46 348 F*	*28 994 F*
67	Japan	39	-	14	-	-	-	-	-	-	-
	USA	1 013	63	14	10	10	14	20	81	3	36
67	*Fishing area total*	*1 052*	*63*	*28*	*10*	*10*	*14*	*20*	*81*	*3*	*36*
71	Australia	495	529	1 006	938	1 040	1 243	1 215	3 995	1 535	1 773
	China	-	-	-	-	-	-	-	160	-	-
	China,Taiwan	53 650 F	46 400 F	32 600 F	36 224	31 687	31 703	35 559	16 431	20 650	29 961
	Guam	...	...	5	0	0	0	-	0	0	0

B-38

Sharks, rays, chimaeras — **Capture production by species, fishing areas and countries or areas**
Squales, raies, chimères — **Captures par espèces, zones de pêche et pays ou zones**
Tiburones, rayas, quimeras — **Capturas por especies, áreas de pesca y países o áreas**

Species, Fishing area Espèce, Zone de pêche Especie, Area de pesca	1992 mt	1993 mt	1994 mt	1995 mt	1996 mt	1997 mt	1998 mt	1999 mt	2000 mt	2001 mt
Indonesia	34 573	35 469	39 221	45 376	39 891	35 863	39 711	44 372	48 794	51 100
Japan	811	780	1 124	480	246	374	227	297	203	155
Kiribati	1 890	1 830	1 800	1 820	1 840	1 830	2 381	3 012	1 581	1 273
Korea Rep	-	-	1	2	78	49	6	43	101	8
Malaysia	6 471	5 600	6 180	7 613	7 269	6 721	6 875	7 407	6 677	7 558
Philippines	3 497	4 698	4 175	4 079	3 839	1 690	2 086	2 174	2 071	2 223
Singapore	262	241	124	104	94	93	80	59	43	32
Solomon Is	40 F	60 F	140 F	80 F	50 F	4 000 F	600 F	310 F	300 F	300 F
Thailand	1 578	1 934	2 208	2 812	3 160	2 993	3 306	6 302	6 216	6 178
71 Fishing area total	*103 267 F*	*97 541 F*	*88 584 F*	*99 528 F*	*89 194 F*	*86 559 F*	*92 046 F*	*84 562 F*	*88 171 F*	*100 561 F*
77 Amer Samoa	...	...	2	0	-	4	-	-	-	-
China,Taiwan	-	-	-	4	6	6	54	52	72	105
Cook Is	31 F	32 F	30 F	30 F	20 F	20 F	20 F	20 F	20 F	20 F
Costa Rica	2 206	2 550	2 855	2 914	3 486	5 548	7 632	7 833	12 757	9 551
El Salvador	620	287	980	759	347	1 186	266	176	364	759
Fr Polynesia	...	...	420	365	387	367	347	427	582	705
Guatemala	103	225	225	207	81	146	237	203	151	150 F
Japan	5 322	3 591	3 645	5 215	2 982	2 400	3 173	1 829	1 163	956
Korea Rep	289	216	-	38	119	646	1 173	841	852	703
Mexico	14 573	15 825	16 348	15 959	16 841	13 330	13 982	15 350	16 755	15 654
Samoa	...	...	...	...	...	...	...	...	20	20 F
USA	585	1 443	1 183	1 466	2 543	2 861	2 892	2 930	1 547	187
77 Fishing area total	*23 729 F*	*24 169 F*	*25 688 F*	*26 957 F*	*26 812 F*	*26 514 F*	*29 776 F*	*29 661 F*	*34 283 F*	*28 810 F*
81 Australia	675	887	1 326	1 324	1 040	1 152	1 382	659	795	808
Japan	1 625	1 675	1 022	862	901	639	1 088	842	412	727
Korea Rep	710	858	1 396	815	636	191	380	1 051	1 105	861
New Zealand	...	...	...	3 129	2 375	1 580	673	1 062	6	-
Ukraine	5	...	...	...	...	...	...	...	...	1
81 Fishing area total	*3 015*	*3 420*	*3 744*	*6 130*	*4 952*	*3 562*	*3 523*	*3 614*	*2 318*	*2 397*
87 Japan	1 032	996	1 415	671	857	526	1 167	441	438	648
Korea Rep	-	-	-	-	-	5	-	-	-	-
Peru	2 087	1 212	548	694	1 506	1 915	4 335	2 951	4 307	3 618
Spain	-	-	-	-	-	289	-	774	1 776	2
87 Fishing area total	*3 119*	*2 208*	*1 963*	*1 365*	*2 363*	*2 735*	*5 502*	*4 166*	*6 521*	*4 268*
Species total	*339 564 F*	*349 639 F*	*343 768 F*	*343 133 F*	*389 235 F*	*373 789 F*	*370 144 F*	*378 730 F*	*396 821 F*	*377 075 F*
Group total	***728 237***	***741 802***	***757 387***	***763 306***	***815 086***	***829 957***	***820 211***	***834 629***	***856 716***	***824 772***

B-39

Marine fishes not identified / **Poissons marins non identifiés** / **Peces marinos no identificados**

Capture production by species, fishing areas and countries or areas
Captures par espèces, zones de pêche et pays ou zones
Capturas por especies, áreas de pesca y países o áreas

Species, Fishing area Espèce, Zone de pêche Especie, Area de pesca	1992 mt	1993 mt	1994 mt	1995 mt	1996 mt	1997 mt	1998 mt	1999 mt	2000 mt	2001 mt
Groundfishes nei	**Poissons de fond nca**		**Peces de fondo nep**		***Osteichthyes***			**1,99(XX)XXX,XX**		**GRO**
21 Canada	883	829	1 176	872	597	475	365	770	1 003	831
Cuba	7	9	4	19	415	164	2	5	-	-
Denmark	349	-	-	-	-	-	-	-	-	-
Faeroe Is	326	94	-	-	-	-	-	-	-	-
Greenland	3	432	639	623	691	1 273	588	-	-	-
Japan	48	7	28	130	57	32	35	31	11	6
Latvia	-	-	-	43	-	-	-	-	-	-
UK	-	-	-	-	-	-	283	-	-	-
USA	2	0	126	0	-	-	-	-	-	0
21 Fishing area total	*1 618*	*1 371*	*1 973*	*1 687*	*1 760*	*1 944*	*1 273*	*806*	*1 014*	*837*
27 Belgium	797	1 081	1 158	1 027	1 290	1 389	979	774	643	501
Denmark	2	2	1	0	0	0	0	-	-	-
Faeroe Is	-	-	-	-	-	474	133	-	-	-
France	220	211	409	277	228	-	-	-	-	-
Iceland	725	244	512	337	237	179	2	81	-	45
Ireland	61	90	11	0	65	-	-	-	2	1 520
Japan	5	11	3	-	-	-	-	-	-	-
Poland	12	7	29	165	36	39	41	-	-	-
Portugal	1 031	-	-	-	-	-	-	-	-	-
Spain	13 757	12 504	16 589	13 806	12 354	2 506	5 411	3 506	7 687	828
UK	3 796	4 706	5 262	5 095	4 131	3 689	3 401	2 498	911	977
27 Fishing area total	*20 406*	*18 856*	*23 974*	*20 707*	*18 341*	*8 276*	*9 967*	*6 859*	*9 243*	*3 871*
Species total	*22 024*	*20 227*	*25 947*	*22 394*	*20 101*	*10 220*	*11 240*	*7 665*	*10 257*	*4 708*
Pelagic fishes nei	**Poissons pélagiques nca**		**Peces pelágicos nep**		***Osteichthyes***			**1,99(XX)XXX,XX**		**PEL**
21 Canada	-	-	-	-	36	128	25	31	29	208
Cuba	-	-	-	-	-	-	19	13	-	-
Japan	-	1	-	-	-	-	-	-	-	-
Russian Fed	-	-	-	102	8	-	-	-	-	-
21 Fishing area total	-	*1*	-	*102*	*44*	*128*	*44*	*44*	*29*	*208*
27 Belgium	4	10	20	7	4	2	1	1	2	4
Ireland	-	-	-	-	-	-	-	-	1	13
Poland	23	12	50	281	60	66	61	-	-	-
Portugal	221	-	-	-	-	-	-	-	-	-
Spain	14 406	30 034	11 375	18 450	25 609	8 169	15 685	4 426	2 947	100
UK	947	182	165	-	-	10	-	12	-	-
27 Fishing area total	*15 601*	*30 238*	*11 610*	*18 738*	*25 673*	*8 247*	*15 747*	*4 439*	*2 950*	*117*
Species total	*15 601*	*30 239*	*11 610*	*18 840*	*25 717*	*8 375*	*15 791*	*4 483*	*2 979*	*325*
Finfishes nei	**Poissons téléostéens nca**		**Peces de escama nep**		***Osteichthyes***			**1,99(XX)XXX,XX**		**FIN**
21 Canada	437	140	2 316	125	1	-	-	1	-	-
Germany	65	-	-	-	-	-	-	-	-	-
Greenland	-	-	-	-			-	-	769	589
Lithuania	-	-	-	-	-	-	-	-	-	1
Norway	-	-	-	-	-	-	-	-	3	-
Poland	-	-	-	-	-	-	-	-	-	1
Portugal	325	238	12	14	22	114	40	122	164	41
Russian Fed	448	11	-	17	72	-	26	57	63	162
Spain	2 783	601	-	72	56	20	30	35	68	74
USA	3 584	5 802	5 588	15 595	12 935	12 153	4 431	6 398	5 907	2 566
21 Fishing area total	*7 642*	*6 792*	*7 916*	*15 823*	*13 086*	*12 287*	*4 527*	*6 613*	*6 974*	*3 434*
27 Denmark	1 224	1 094	400	545	118	103	116	78	142	107
Estonia	195	78	60	29	26	23	28	32	41	91
France	5 357	5 688	5 199	5 167	3 771	3 732	2 344	2 179	8 759	12 661
Germany	1 580	-	248	316	270	291	164	150	66	37
Isle of Man	-	-	1	0	-	-	0	-	-	-
Lithuania	-	107	45	66	188	152	41	40	204	2
Norway	835	796	165	128	158	322	2 791	3 389	3 198	1 909
Poland	73	85	43	53	23	34	23	-	17	306
Portugal	16 702	15 684	10 467	12 809	9 330	9 293	10 727	4 962	2 350	1 531
Russian Fed	11 187	1 935	1 295	1 071	729	520	577	788	22 596	536
Spain	11 445	11 781	10 362	10 031	10 665	19 512	33 681	22 056	17 424	12 572
Sweden	2 573	755	741	12 650	3 065	3 003	338	382	540	508
Ukraine	328	316	557	167	32	-	-	-	-	-
27 Fishing area total	*51 499*	*38 319*	*29 583*	*43 032*	*28 375*	*36 985*	*50 830*	*34 056*	*55 337*	*30 260*
Species total	*59 141*	*45 111*	*37 499*	*58 855*	*41 461*	*49 272*	*55 357*	*40 669*	*62 311*	*33 694*
Marine fishes nei	**Poissons marins nca**		**Peces marinos nep**		***Osteichthyes***			**1,99(XX)XXX,XX**		**MZZ**
21 Japan	14	18	29	6	6	1	7	29	12	4
St Pier Mq	0	-	-	-	-	18	18	-	-	-
21 Fishing area total	*14*	*18*	*29*	*6*	*6*	*19*	*25*	*29*	*12*	*4*
27 Channel Is	16	16 F	-	-	22	203	27	40	-	-
Faeroe Is	...	...	...	...	1 201	1 362	177	18	4 053	2 911
Greenland	19	144	244	80	168	189	1 187	...	200	264
Japan	2	2	-	3	3	1	1	2	5	4
Latvia	-	-	660	464	-	-	4	-	-	-
Netherlands	16 156	16 687	20 267	19 457	16 596	19 114	3 305	3 442	597	-

B-39 Marine fishes not identified / Poissons marins non identifiés / Peces marinos no identificados

Capture production by species, fishing areas and countries or areas / Captures par espèces, zones de pêche et pays ou zones / Capturas por especies, áreas de pesca y países o áreas

Species, Fishing area Espèce, Zone de pêche Especie, Area de pesca	1992 mt	1993 mt	1994 mt	1995 mt	1996 mt	1997 mt	1998 mt	1999 mt	2000 mt	2001 mt
Romania	-	-	94	76	13	-	-	-	-	-
Svalbard Is	0	0	0	0	0	0	0	0	0	0
27 Fishing area total	*16 193*	*16 849 F*	*21 265*	*20 080*	*18 003*	*20 869*	*4 701*	*3 502*	*4 855*	*3 179*
31 Anguilla	271	232	234	105 F	140 F	180 F	180 F	180 F	180 F	180 F
Antigua Barb	1 323	450	487	1 116	1 045	1 242	1 013	1 041	1 164	66
Aruba	245	170	125	20	20	55	40	40	40	43
Bahamas	444	544	428	381	385	264	156	139	110	133
Barbados	71	69	83	220	113	74	62	72	60	109
Belize	317	217	145	166	400	121	133	202	119	66
Bermuda	112	94	101	93	96	108	113	80	55	57
Br Virgin Is	385 F	291	192	389	265	39	42	41	27	34 F
Cayman Is	125	125	125	125	110	125	125	125	125	125
China,Taiwan	100 F	110 F	80 F	71	89	53	48	184	29	32
Colombia	1 306	1 524	2 015	2 879	4 384	3 079	1 701	1 061	5 752 F	5 595 F
Costa Rica	33	133	111	197	73	163	141	367	563	583
Cuba	23 420	16 282	20 748	18 589	15 443	24 015	20 732	19 121	19 150 F	19 050 F
Dominica	574	666	730	820	939	988	1 121	870 F	872 F	908 F
Dominican Rp	430	188	7	-	2 650	2 292	994	1 998	2 382	4 391
Fr Guiana	3 630	3 500	3 578	3 634	3 000 F	2 500	2 500	2 500 F	2 500 F	2 500 F
Grenada	466	347	147	135	173	130	158	29	5	10
Guadeloupe	5 540 F	5 590 F	5 765	6 270 F	6 220	6 860 F	5 800 F	5 800 F	6 600 F	6 600 F
Guatemala	100	92	120	270	270	195	213	196	203	200 F
Guyana	37 097	37 151	38 122	37 000 F	33 971	34 841	38 124	37 534	27 666	24 662
Haiti	3 300 F	3 200 F	3 730 F	3 600 F	4 000 F	4 000 F	4 000 F	3 800 F	3 800 F	3 800 F
Honduras	537	564	760	878	529	590	160	349	607	465
Jamaica	7 200 F	7 300 F	7 000 F	7 300 F	8 500 F	5 305	4 161	6 283	4 586	4 616 F
Japan	-	-	1	-	-	2	5	8	8	5
Korea Rep	-	-	-	-	-	-	-	50	1 023	1
Martinique	2 137	3 036 F	2 810 F	2 390 F	1 647 F	3 430 F	3 060 F	810 F	45 F	35 F
Mexico	147 185	136 458	142 846	77 964	81 517	108 324	91 863	70 679	59 872	60 815
Montserrat	23	58	62	48	38	45	46	50 F	50 F	50 F
NethAntilles	575 F	590 F	535 F	505 F	505 F	455 F	455 F	455 F	455 F	455 F
Nicaragua	720	392	1 301	1 367	2 279	2 291	4 088	741	631	605
Puerto Rico	691	195	979	983	569	691	50	264	298	174
Russian Fed	-	-	-	82	-	-	-	-	-	-
St Kitts Nev	200 F	200 F	172	80	96	97	142	222	175	278
St Lucia	559	790	442	385	308	264	243	325	352	435
St Vincent	1 289	1 218	933	774	737	821	1 152	874	675	653
Suriname	10 108	9 250	14 091	11 855 F	10 400 F	11 777 F	11 845	9 800	10 500 F	11 300
Trinidad Tob	1 339 F	1 298	7 364	4 947 F	1 609	2 818	1 424	1 928	2 074	3 151
Turks Caicos	300	307	310 F	300 F	300 F	300 F	300	300 F	300 F	300 F
USA	20 997	22 352	26 959	14 266	7 694	4 527	7 320	5 545	5 427	3 635
US Virgin Is	650 F	560 F	470 F	400 F	340 F	300 F	255 F	119	139 F	139 F
Venezuela	24 551	27 223	27 691	32 559	32 927	36 624	16 376	-	464	11 303
31 Fishing area total	*298 350 F*	*282 766 F*	*311 799 F*	*233 163 F*	*223 781 F*	*259 985 F*	*220 341 F*	*174 182 F*	*159 083 F*	*167 559 F*
34 Belize	-	80	7	27	-	372	1 611	3 126	6 828	810
Benin	27 F	455 F	1 110 F	889 F	1 785 F	3 628 F	3 294 F	1 677	1 150 F	370
Bulgaria	-	-	-	-	-	-	269	-	-	-
Cameroon	3 000	3 000	3 740 F	-	-	-	-	-	954	907
Cape Verde	3 482	3 980	4 774	4 780	6 612	538	333	640	719	633
Chile	-	-	10	-	-	-	-	-	-	-
China	3 075	2 325	653	4 389	4 295	3 689	2 713	4 536	6 604	10 803
China,Taiwan	100 F	110 F	80 F	71	89	53	73	374	197	23
Congo Dem R	380 F	420 F	380 F	387 F	396 F	383 F	394 F	400 F	400 F	400 F
Congo Rep	939 F	935 F	556 F	1 383 F	2 256 F	3 100 F	2 850 F	4 009 F	3 606 F	3 765 F
Côte dIvoire	9 138	17 748	20 135	23 132	10 290 F	11 378 F	16 101 F	18 239 F	12 053	25 100
Cuba	8	0	0	0	172	-	-	-	-	-
Cyprus	-	-	-	-	2 632	13 640	276	691	647	690
Eq Guinea	350 F	277 F	649 F	96 F	789 F	695 F	700	820 F	117 F	169 F
Estonia	19 298	15 162	2 200	31	148	-	469	-	0	-
France	8 812	10 000	7 210	12 669	13 319	13 220	1 868	24	1	1
Gabon	4 100 F	8 538 F	9 142	8 256	278	1 325	1 154	3 949	2 896	...
Gambia	17	186	37	27	18	33	-	-	-	6
Georgia	1 086 F	600 F	250 F	-	-	-	-	-	-	-
Germany	-	-	-	-	40	45	53	110	-	132
Ghana	18 840	23 620	19 552	19 946	25 128	27 896	43 399	115 933	22 235	40 993
Greece	4 327	4 038	2 579	2 198	1 986	1 104	1 998	1 223	1 644	1 298
Guinea	10 665 F	11 290 F	12 539 F	8 001	3 695	7 089	7 631	23 580	18 646	18 043 F
GuineaBissau	-	-	5	17	-	-	-	-	-	-
Honduras	1 595	1 710	125	1 624	1 638	1 600	481	801	752	620
Italy	-	-	-	3 752	1 460	2 241	1 631	1 302	955	2 239
Japan	19	60	62	50	51	22	37	31	98	46
Korea D P Rp	-	28	-	-	-	-	-	-	154	-
Korea Rep	949	1 377	10 584	4 335	4 856	3 501	2 393	4 403	5 398	5 329
Latvia	19 441	17 530	18 231	22 475	19 130	3 118	1 103	3 247	1 616	759
Liberia	1 306	710	598	134	362	157	539	797	281	640
Lithuania	1 401	-	-	9 572	33 330	25 680	3 495	1 099	899	80 150
Malta	-	-	1 236	3 465	8 197	-	-	-	-	-
Marshall Is	-	-	-	-	2 403	-	-	-	-	-
Mauritania	27 382 F	18 591 F	13 917 F	15 404 F	18 595 F	19 711 F	22 179 F	37 674 F	39 997 F	42 659 F
Morocco	30 931	38 141	40 527	40 909	39 138	84 415	63 098	53 179	5 000	13 220
Netherlands	-	-	-	-	318	292	822	455	604	1 781
Nigeria	23 682	12 260	20 292	39 520	32 995	21 516	40 295	33 770	51 541	49 205
Panama	444	257	-	34	6	49	50	166	199	-
Poland	-	-	2 220	-	2 142	641	-	-	-	5 999
Portugal	2 361	577	631	327	734	479	401	256	3	58
Romania	699	9	-	-	-	-	-	-	-	-

B-39 Marine fishes not identified / Poissons marins non identifiés / Peces marinos no identificados

Capture production by species, fishing areas and countries or areas / Captures par espèces, zones de pêche et pays ou zones / Capturas por especies, áreas de pesca y países o áreas

Species, Fishing area Espèce, Zone de pêche Especie, Area de pesca	1992 mt	1993 mt	1994 mt	1995 mt	1996 mt	1997 mt	1998 mt	1999 mt	2000 mt	2001 mt
Russian Fed	15 673	3 661	840	6 275	6 815	6 140	6 304	-	6 379	2 242
St Vincent	151	92	-	9	-	148	1 008	161	536	644
Sao Tome Prn	679 F	927 F	1 837 F	1 933 F	2 579	683	1 167 F	1 284 F	1 234 F	1 238 F
Senegal	31 141	38 369	10 778	18 088	13 522	13 782	7 590	6 679	12 146	14 708
Sierra Leone	0	0	0	0	0	0	0	4	2 732	3 608
Spain	17 884	14 701 F	19 043 F	19 522 F	10 033	15 237	18 797	11 070	7 185	9 756
Togo	209	341	226	-	176	823	594	2 059	1 424	1 027
Ukraine	10 058	5 567	4 034	20 488	15 528	7 167	26 274	57 653	53 274	90 636
Vanuatu	29	16	7	11	9	27	10	-	-	-
Westn Sahara	0	0	0	0	0	0	0	0	0	0
Other nei	22 417	3 070	6 110	124 126	13	202	-	740	675	508
34 Fishing area total	*296 095 F*	*260 758 F*	*236 906 F*	*418 352 F*	*287 958 F*	*295 819 F*	*283 454 F*	*396 161 F*	*271 779 F*	*431 215 F*
37 Albania	48	250 F	450 F	105	327	108	370	375	789	313
Algeria	1 873 F	1 913 F	2 737 F	3 890	3 964	3 799	2 363	3 403	3 700 F	3 892 F
Bosnia Herzg	0	0	0	0	0	0	0	0	0	0
Bulgaria	14	7	8 F	70	49	51	107	-	-	2
Croatia	1 111	1 094	771	738	619	613	590	619	1 080	863
Cyprus	516	405	339	322	425	363	359	253	274	238
Egypt	12 219	6 357	7 335	3 746	2 632	5 602	6 865	365	4 109	3 637
France	1 502	1 337	984	350	715	700	700	907	750	1 748
Gaza Strip	...	...	...	1 057	559	140	194	200 F	200 F	160 F
Georgia	-	-	-	-	25	-	-	1	4	5
Gibraltar	0	0	0	0	0	0	0	0	0	0
Greece	12 501	11 635	12 518	11 197	9 100	11 329	7 085	8 383	10 951	10 055
Israel	-	203 F	84 F	727	496	912	1 273	1 129	1 143	1 000 F
Italy	24 848	28 595	27 503	34 458	31 259	26 017	26 142	22 053	23 177	24 946
Korea Rep	-	-	-	2	110	-	-	-	-	-
Libya	3 864 F	3 955 F	4 078 F	4 410 F	3 688 F	2 903 F	3 617 F	3 755 F	3 924 F	3 550 F
Malta	0	4	3	1	2	1	12	16	29	82
Monaco	3 F	3 F	3 F	3 F	3 F	3 F	3 F	3 F	3 F	3 F
Morocco	527	998	2 279	2 885	3 135	4 266	4 679	1 649	2 600	1 241
Portugal	-	-	-	-	-	-	-	-	7	12
Romania	-	244	1	12	10	2	2	1	1	5
Russian Fed	6	21	4 450	9	2	32	1	1	727	-
Slovenia	23	35	15	12	8	7	5	4	-	-
Spain	26 885 F	28 147 F	28 704 F	27 190 F	22 864	20 406	18 105	8 741	8 889	6 200
Syria	80 F	85 F	81 F	91 F	90	150	328	342	580	462
Tunisia	5 058	16 251	22 469	18 828	18 813	5 112	12 965	9 617	5 125	4 269
Turkey	708	1 596	1 412	1 829	1 855	1 095	1 861	1 375	7 365	1 230
Ukraine	957	211	64	57	8	4	1	-	-	20
Yugoslavia	14	24	31	21	19	25	20	25	27	-
37 Fishing area total	*92 757 F*	*103 370 F*	*116 319 F*	*112 010 F*	*100 777 F*	*83 640 F*	*87 647 F*	*63 217 F*	*75 454 F*	*63 933 F*
41 Argentina	-	1 919	4 661	10 945	4 960	7 855	6 702	5 998	7 115	21 767
Australia	-	-	-	-	-	-	234	389	-	-
Belize	-	-	-	-	-	-	-	7	222	43
Brazil	190 877 F	195 450 F	191 029 F	145 188	143 419	146 661	138 925	169 938	176 695	165 661 F
Chile	-	-	-	-	-	-	31	5	-	-
Falkland Is	22	78	17	50	370	181	1 033	1 217	1 250	774
France	-	-	8	15	0	0	15	-	-	-
Honduras	18	57	2	4	12	-	-	-	-	-
Italy	-	-	-	447	174	-	-	-	-	-
Japan	15	57	30	43	44	5	21	15	53	21
Korea Rep	17 776	16 171	12 881	11 043	6 136	5 873	4 020	1 176	5 724	10 621
Latvia	594	-	413	76	68	-	-	-	-	-
Lithuania	2		-	-	-	-	-	-	-	-
Namibia	-	-	-	-	-	4	16	96	-	-
Panama	8	-	-	4	43	-	-	-	-	-
Poland	-	-	72	-	-	-	-	8	-	-
Portugal	-	0	5	5	7	2	6	1	8	7
Russian Fed	1 753	37	-	-		-	-	-	-	14
Seychelles	-	-	-	-	-	6	7	-	-	-
Spain	1 053	550	219	1 037	-	-	-	1 379	207	370
Ukraine	793	-	-	-	-	-	-	-	-	-
UK	-	5	0	20	48	21	60	0	13	19
Uruguay	3 044	2 851	3 072	4 083	3 289	4 510	6 335	936	3 669	2 686
41 Fishing area total	*215 955 F*	*217 175 F*	*212 409 F*	*172 960*	*158 570*	*165 118*	*157 405*	*181 165*	*194 956*	*201 983 F*
47 Angola	53 026	52 634	64 541	44 922	80 633	74 578	43 038	32 185	31 674	80 740
China	106	-	-	-	-	-	-	-	5	-
Honduras	-	65	-	20	-	-	-	-	-	-
Iceland	-	-	-	-	-	-	4	-	-	-
Italy	-	-	-	50	20	-	-	-	-	-
Japan	276	956	527	896	922	388	825	763	676	441
Korea Rep	1 382	499	746	1 244	1 091	2 550	3 473	3 074	2 599	2 041
Lithuania	185	-	-	-	-	-	-	-	-	-
Namibia	2 532	3 411	3 143	3 846	5 727	15 552	7 868	9 016	11 801	1 502
Panama	3 142	-	-	-	-	-	-	-	-	-
Poland	-	-	-	120	33	-	-	-	-	2
Portugal	-	-	-	-	-	5	3	32	54	74
Russian Fed	111	48	98	260	301	147	3 742	8 138	3 732	146
St Helena	6	9	79	82	82	39	59	68	69	70
South Africa	1 708	770	2 184	1 766	1 549	13 108	1 075	3 586	2 865 F	4 635
Spain	97	82 F	75 F	51 F	-	56	301	79	512	444
Ukraine	-	4 690	2 459	1 185	220	108	-	-	3 781	531
47 Fishing area total	*62 571*	*63 164 F*	*73 852 F*	*54 442 F*	*90 578*	*106 531*	*60 388*	*56 941*	*57 768 F*	*90 626*
48 Argentina	-	-	-	10	-	-	-	-	-	-

B-39

Marine fishes not identified
Poissons marins non identifiés
Peces marinos no identificados

Capture production by species, fishing areas and countries or areas
Captures par espèces, zones de pêche et pays ou zones
Capturas por especies, áreas de pesca y países o áreas

Species, Fishing area Espèce, Zone de pêche Especie, Area de pesca	1992 mt	1993 mt	1994 mt	1995 mt	1996 mt	1997 mt	1998 mt	1999 mt	2000 mt	2001 mt
Chile	-	-	-	-	-	-	7	-	-	-
Russian Fed	-	-	0	-	-	-	-	-	-	-
UK	1	-	-	-	-	-	-	-	-	-
48 Fishing area total	*1*	-	*0*	*10*	-	-	*7*	-	-	-
51 Bahrain	851	764	785	1 043	1 185	1 029	852	938	859	1 077
Br Ind Oc Tr	0	0	0	0	0	0	0	0	0	0
China	-	-	-	-	-	-	-	-	330	169
China,Taiwan	3 000 F	3 190 F	2 240 F	2 312	1 953	1 720	267	440	2 081	905
Comoros	1 200	1 200	1 200	1 159 F	1 010 F	1 190 F	1 190 F	1 120 F	1 260 F	1 180
Djibouti	9 F	10 F	11 F	10 F	10 F	15 F	15 F	15 F	15 F	15 F
Egypt	19 091	13 219	12 049	7 193	10 224	15 538	11 673	33 315	8 231	7 907
Eritrea	...	475	449	314	202	215	211	246	8	0
Ethiopia	100	-	-	-	-	-	-	-	-	-
Fr South Tr	63	56	75	75 F	75 F	75 F	75 F	80 F	80 F	80 F
India	235 026	413 427	404 127	492 372	256 427	307 120	264 758	288 658	293 193	335 781
Iran	173 578	175 209	148 553	149 090	157 397	143 212	139 107	130 225	122 620	115 396
Iraq	543	2 133	4 221	5 253	11 688	10 783	13 463	13 093	12 389	12 400 F
Israel	62	80	110	150	225	171	137	98	-	-
Italy	-	-	-	769	299	-	-	-	-	-
Japan	168	272	169	189	194	310	426	270	269	195
Jordan	30	45	60	75	90	100	3	7	7	6
Kenya	2 351	2 611	728	1 842	1 187	1 006	1 675	1 774	474	667
Korea Rep	6 864	5 536	2 350	3 048	1 979	4 057	4 878	2 269	4 952	1 426
Kuwait	495	1 579	753	729	893	922	1 055	231	550 F	580
Madagascar	57 121	62 929	60 112	61 544	59 965	61 596	69 652	74 417	75 107	78 601
Maldives	9 880	13 081	17 181	17 620	18 526	17 914	16 788	13 450	19 618	3 571
Mauritius	618	586	594	550	589	525	502	515	441	568
Mayotte	1 100 F	500 F	600 F	700 F	1 000	1 300 F	2 000 F	2 000	5 500	5 500
Mozambique	9 354	6 108	7 437	6 209	13 058	10 374	6 990	8 363	7 816	8 766
Norway	-	-	-	-	-	-	-	-	42	-
Oman	4 685	2 980	1 422	2	101	355	432	1	384	1 273
Pakistan	54 764	31 359	40 725	14 742	16 311	21 042	35 352	39 473	36 451	35 450
Portugal	-	-	-	-	-	-	-	10	-	4
Qatar	820	839	627	593	357	332	38	10	587	79
Réunion	486	582	845	810	775	1 279	1 000	67	127	98
Russian Fed	10	-	-	-	-	-	-	-	-	11
Saudi Arabia	4 578	4 325	3 074	1 480	1 205	820	768	587	574	804
Seychelles	339	273	295	123	221	108	122	416	245	228
Somalia	24 500 F	26 600 F	28 700 F	26 800 F	24 900 F	23 000 F	21 100 F	19 100 F	19 100 F	18 900 F
South Africa	123	36	62	7	85	104	21	100 F	5	6
Spain	-	-	-	-	-	-	-	21	661	69
Sudan	2 000	2 500	4 000	4 000	4 500	5 000	5 436	5 420	4 937 F	4 887
Tanzania	10 000	7 112	6 108	6 721	6 334	8 300	8 395	2 125	2 000	2 000
Ukraine	29	50	100	33	102	59	1	334	-	35
Untd Arab Em	6 909	14 874	9 908	8 548	7 441	10 015	9 007	9 019	7 441	7 440 F
51 Fishing area total	*630 747 F*	*794 540 F*	*759 670 F*	*816 105 F*	*600 508 F*	*649 586 F*	*617 389 F*	*648 207 F*	*628 354 F*	*646 074 F*
57 Australia	21 247 F	20 820 F	5 031 F	13 431 F	11 499 F	11 243	10 581	12 157	7 354	7 679
Bangladesh	136 885	161 860	110 314	115 122	119 484	135 860	145 260	137 285	161 977	158 640 F
China,Taiwan	100 F	110 F	80 F	34	29	25	11	9	96	132
Christmas Is	0	0	0	0	0	0	0	0	0	0
Cocos Is	0	0	0	0	0	0	0	0	0	0
India	147 262	119 554	169 430	168 837	199 275	196 114	235 105	184 787	268 175	224 004
Indonesia	64 336	69 076	65 154	59 673	57 340	46 353	68 638	56 135	54 577	54 760
Japan	325	209	578	1 062	1 093	1 140	1 498	1 360	1 198	921
Korea Rep	11	-	-	7	-	142	-	428	950	1 443
Malaysia	160 844	166 280	163 315	174 506	177 896	174 448	206 661	183 297	205 880	197 436
Myanmar	580 263	585 637	584 876	581 073	435 868	607 814	656 406	731 664	849 018	900 492
Sri Lanka	45 702	35 090	41 998	55 428	36 404	23 398	25 084	16 106	26 090	26 740
Thailand	386 986	425 370	300 258	321 441	312 014	302 880	333 148	302 056	298 968	321 339
Timor-Leste	...	...	...	...	...	...	...	400 F	350	350
57 Fishing area total	*1 543 961 F*	*1 584 006 F*	*1 441 034 F*	*1 490 614 F*	*1 350 902 F*	*1 499 417*	*1 682 392*	*1 625 684 F*	*1 874 633*	*1 893 936 F*
58 Australia	0	-	0	-	-	1	-	0	0	-
South Africa	-	-	-	-	-	-	0	-	-	0
Ukraine	10	14	-	61	-	-	-	-	-	-
58 Fishing area total	*10*	*14*	*0*	*61*	-	*1*	*0*	*0*	*0*	*0*
61 China	2 933 682	3 094 592	3 237 857	3 321 368	4 110 802	3 745 598	3 703 180	3 767 795	3 368 652	2 972 032
China,H.Kong	53 901	103 926	104 611	98 215	101 269	101 062	108 206	76 460 F	94 272 F	104 550 F
China, Macao	1 495	1 200	1 259	1 151	970	1 020 F	1 020 F	1 020 F	1 020 F	1 020 F
China,Taiwan	100 797 F	108 251 F	75 219 F	77 082	48 677	60 098	71 179	72 750	79 064	74 853
Japan	472 759	374 123	319 991	312 769	287 326	286 882	264 415	256 730	271 352	247 560
Korea D P Rp	843 900 F	810 900 F	220 011	155 973	166 325	109 770	107 900 F	102 900 F	97 500 F	97 500 F
Korea Rep	97 046	90 327	91 418	91 708	148 719	136 916	115 298	90 683	77 224	101 327
Poland	-	-	-	108	9	1	-	-	25	-
Russian Fed	4 190	3 780	5 941	-	-	1 621	1 259	104	66 076	43 018
61 Fishing area total	*4 507 770 F*	*4 587 099 F*	*4 056 307 F*	*4 058 374*	*4 864 097*	*4 442 968 F*	*4 372 457 F*	*4 368 442 F*	*4 055 185 F*	*3 641 860 F*
67 Canada	102 141	67 490	121 769	62 475	105 185	102 100	96 130	97 253	26 979	69 582
Japan	147	485	430	-	-	-	-	-	-	-
Russian Fed	-	-	-	-	32	-	-	-	-	-
USA	11 498	28 572	1 832	2 439	1 603	4 501	1 723	938	4 024	4 763
67 Fishing area total	*113 786*	*96 547*	*124 031*	*64 914*	*106 820*	*106 601*	*97 853*	*98 191*	*31 003*	*74 345*
71 Australia	4 914 F	4 866 F	3 827 F	4 500 F	5 326 F	197	422	5 354	3 738	5 179
Brunei Darsm	1 231	1 330	3 131	4 389	6 818	4 405	4 930	3 081	2 356	1 186
Cambodia	23 825	24 835	22 400	22 500	23 000	21 970	23 740	28 100	26 606	26 600 F

B-39 Marine fishes not identified / Poissons marins non identifiés / Peces marinos no identificados

Capture production by species, fishing areas and countries or areas
Captures par espèces, zones de pêche et pays ou zones
Capturas por especies, áreas de pesca y países o áreas

Species, Fishing area Espèce, Zone de pêche Especie, Area de pesca	1992 mt	1993 mt	1994 mt	1995 mt	1996 mt	1997 mt	1998 mt	1999 mt	2000 mt	2001 mt
China	103	1 377	1 738	926	380	110	174	818	1 047	425
China,Taiwan	-	-	-	141	119	105	3 493	66	2	-
Fiji Islands	1 087	632	724	1 414	1 800	1 035	1 358	1 218	1 200 F	1 113
Guam	61	72	52	4	13	18	49	85	92	89
Indonesia	230 940	295 620	333 283	321 392	376 137	366 727	387 908	414 441	454 389	455 900
Japan	2 208	1 743	1 606	1 237	1 273	156	396	427	355	285
Kiribati	2 500	2 420	2 323	2 350	2 400	2 380	3 875	4 216	2 396	3 198
Korea Rep	10 489	10 038	16 691	13 965	11 628	7 965	5 920	3 420	4 321	3 177
Malaysia	148 016	166 241	195 030	193 020	174 941	157 279	176 145	190 892	205 348	215 813
Marshall Is	320 F	335 F	340 F	350 F	369	400 F	500 F	500 F	500 F	500 F
Micronesia	1 040 F	1 070 F	1 010 F	1 100 F	1 200 F	1 300 F	1 300 F	1 400 F	1 400 F	1 500 F
Nauru	377	500	500	450 F	400 F	400 F	400 F	400 F	400 F	400 F
NewCaledonia	585	229	471	363	415	385	458	415	509	514 F
N Marianas	94	94	104	79	61	88	80	75	75	49
Palau	1 490	1 590	1 439	1 457	1 532	1 503	1 489	1 300 F	1 230 F	1 230 F
Papua N Guin	9 500 F	9 500 F	9 500 F	10 000 F	10 500 F	10 500 F	10 000 F	10 000 F	10 000 F	10 000 F
Philippines	8 029	9 039	14 645	19 052	37 385	10 176	13 104	12 783	12 043	13 428
Singapore	2 892	3 992	4 231	4 465	4 308	3 933	2 812	2 740	2 339	1 325
Solomon Is	14 200 F	14 500 F	12 500 F	12 000 F	12 000 F	12 000 F	12 000 F	12 000 F	12 000 F	12 000 F
Thailand	819 913	818 746	863 460	805 867	724 548	685 437	619 329	643 247	634 442	613 570
Tuvalu	490	657	120	112	110 F	160 F	160 F	160 F	160 F	160 F
Vanuatu	1 123 F	1 200 F	1 160 F	1 200 F	1 200 F	1 300 F	1 300 F	1 400 F	1 400 F	1 500 F
Viet Nam	620 427	669 832	719 661	722 055	808 226	832 118	849 310	922 690	927 205	984 623
Wallis Fut I	143	150	187	164	174	170	294	294 F	294 F	294 F
71 Fishing area total	*1 905 997 F*	*2 040 608 F*	*2 210 133 F*	*2 144 552 F*	*2 206 263 F*	*2 122 217 F*	*2 120 946 F*	*2 261 522 F*	*2 305 847 F*	*2 354 058 F*
77 Amer Samoa	9	11	1	3	10	9	5	6	0	0
China	-	-	-	-	-	-	-	-	-	303
China,Taiwan	100 F	110 F	80 F	32	27	24	124	0	0	16
Cook Is	377 F	384 F	268 F	360 F	335 F	300 F	300 F	300 F	300 F	280 F
Costa Rica	6 692	6 280	5 141	5 224	7 543	10 564	7 727	9 612	3 978	3 544
El Salvador	1 640	1 501	2 246	2 553	2 267	1 791	1 595	1 842	1 925	1 678
Fr Polynesia	1 469	4 665	4 339	4 200	4 332	4 820	4 585	4 538	4 274	4 362
Guatemala	736	458	514	375	375	305	488	537	310	460 F
Honduras	542	448	593	503	638	1 157	928	1 026	226	720
Japan	2 717	1 712	1 160	1 232	1 268	729	972	1 230	785	861
Korea Rep	430	651	561	327	77	65	30	573	711	744
Mexico	147 620	120 349	174 176	238 947	180 690	131 611	113 369	66 776	83 729	87 087
Nicaragua	1 753	494	1 891	3 894	3 580	4 595	5 442	926	757	1 270
Niue	120	120 F	150 F	150 F	200 F	200 F	200 F	200 F	200 F	200 F
Panama	5 136	4 069	6 340	6 731	5 920	17 521	15 310	15 458	18 064	18 000 F
Samoa	241	345	350	350	322	1 513	1 640	5 163	5 289	5 200 F
Tokelau	191	200 F	200 F	200 F	200 F	200 F	200 F	200 F	200 F	200 F
Tonga	1 870	1 900	1 950	2 000	2 100	2 000	2 936	2 890	2 305	2 548
USA	18 632	16 453	2 425	2 640	3 288	4 323	3 190	3 359	3 088	3 038
US Minor Is	0	0	0	0	0	0	0	0	0	0
77 Fishing area total	*190 275 F*	*160 150 F*	*202 385 F*	*269 721 F*	*213 172 F*	*181 727 F*	*159 041 F*	*114 636 F*	*126 141 F*	*130 511 F*
81 Australia	8 984 F	10 281 F	9 109 F	7 797 F	12 829 F	13 997	19 258	4 299	5 736	4 755
China,Taiwan	2 000 F	2 130 F	1 500 F	0	0	-	-	0	112	7
Japan	5 003	4 742	3 003	220	226	2 271	269	278	245	403
Korea Rep	7 522	6 472	5 964	7 952	8 420	4 361	3 962	3 478	3 297	2 880
New Zealand	7 765	9 401	8 696	5 598	26 527	26 266	65	113	122	119
Norfolk Is	0	0	0	0	0	0	0	0	0	0
Norway	924	1 494	882	208	235	155	187	-	-	-
Pitcairn Is	8	8 F	8 F	8 F	8 F	8 F	8 F	8 F	8 F	8 F
Poland	1	-	-	-	-	-	-	-	-	-
Russian Fed	7 942	2 003	3 905	175	289	210	57	754	-	-
Ukraine	477	1 412	1 557	31	805	941	3 692	1 804	3 091	34 166
81 Fishing area total	*40 626 F*	*37 943 F*	*34 624 F*	*21 989 F*	*49 339 F*	*48 209 F*	*27 498 F*	*10 734 F*	*12 611 F*	*42 338 F*
87 Chile	4 394	3 238	2 119	2 088	2 253	1 014	897	2 464	6 414	5 861
Colombia	2 926	940	1 947	2 204	4 229	4 633	3 443	2 032	38 029 F	23 523 F
Cuba	-	-	-	-	-	-	-	4 234	...	...
Ecuador	23 054	39 674	34 544	5 530	20 426	8 296	9 510	97 237	186 519	38 255
Ghana	-	-	-	-	-	-	-	-	13	-
Japan	465	291	420	205	211	182	600	470	386	771
Korea Rep	-	-	-	-	-	-	188	360	484	12
Latvia	721	-	-	-	-	-	-	-	-	-
Peru	21 754	20 146	15 686	58 779	45 788	83 203	375 784	164 855	91 905	122 447
Spain	-	-	-	-	-	-	-	-	-	3
87 Fishing area total	*53 314*	*64 289*	*54 716*	*68 806*	*72 907*	*97 328*	*390 422*	*271 652*	*323 750 F*	*190 872 F*
Species total	*9 968 422 F*	*10 309 296 F*	*9 855 479 F*	*9 946 159 F*	*10 343 681 F*	*10 080 035 F*	*10 281 966 F*	*10 274 265 F*	*10 121 431 F*	*9 932 493 F*
Group total	***10 065 188***	***10 404 873***	***9 930 535***	***10 046 248***	***10 430 960***	***10 147 902***	***10 364 354***	***10 327 082***	***10 196 978***	***9 971 220***

B-41 Freshwater crustaceans / Crustacés d'eau douce / Crustáceos de agua dulce

Capture production by species, fishing areas and countries or areas / Captures par espèces, zones de pêche et pays ou zones / Capturas por especies, áreas de pesca y países o áreas

Species, Fishing area Espèce, Zone de pêche Especie, Area de pesca	1992 mt	1993 mt	1994 mt	1995 mt	1996 mt	1997 mt	1998 mt	1999 mt	2000 mt	2001 mt
Giant river prawn	**Bouquet géant**		**Langostino de río**			***Macrobrachium rosenbergii***		**2,28(12)023,07**		**PRF**
02 Martinique	-	3	-	-	-	-	-	-	-	-
02 Fishing area total	-	3	-	-	-	-	-	-	-	-
04 Brunei Darsm	15	11	3	5	14	17	35	26	23	16
Indonesia	6 163	4 927	5 393	5 524	6 127	5 208	5 362	5 937	5 199	5 520
Thailand	1	-	676	308	1 614	1 541	36	106	-	-
04 Fishing area total	6 179	4 938	6 072	5 837	7 755	6 766	5 433	6 069	5 222	5 536
06 Fr Polynesia	-	-	0	0	0	3	3	3	3	3
06 Fishing area total	-	-	0	0	0	3	3	3	3	3
Species total	6 179	4 941	6 072	5 837	7 755	6 769	5 436	6 072	5 225	5 539
River prawns nei	**Bouquets d'eau douce nca**		**Camarones de agua dulce nep**			***Macrobrachium spp***		**2,28(12)023,XX**		**PPF**
01 Benin	113 F	120 F	139	188	150 F	130 F	120 F	120 F	100 F	110 F
01 Fishing area total	113 F	120 F	139	188	150 F	130 F	120 F	120 F	100 F	110 F
02 Mexico	2 244	4 456	3 226	4 202	4 261	3 580	3 244	4 161	3 478	3 112
Nicaragua	-	-	-	-	-	-	-	7	0	3
USA	445	20	19	19	-	-	-	-	0	-
02 Fishing area total	2 689	4 476	3 245	4 221	4 261	3 580	3 244	4 168	3 478	3 115
03 Brazil	6 560 F	6 700 F	6 900 F	1 412	2 677	2 157	1 506	3 235	2 437	2 440 F
03 Fishing area total	6 560 F	6 700 F	6 900 F	1 412	2 677	2 157	1 506	3 235	2 437	2 440 F
06 Papua N Guin	5 F	5 F	5 F	5 F	5 F	5 F	5 F	5 F	5 F	5 F
06 Fishing area total	5 F	5 F	5 F	5 F	5 F	5 F	5 F	5 F	5 F	5 F
Species total	9 367 F	11 301 F	10 289 F	5 826 F	7 093 F	5 872 F	4 875 F	7 528 F	6 020 F	5 670 F
Freshwater prawns, shrimps nei	**Crevettes d'eau douce nca**		**Gambas,camaron.(agua dulce)nep**			***Palaemonidae***		**2,28(12)XXX,XX**		**PPZ**
01 Egypt	1 139	2 352	3 687	3 169	2 286	2 387	2 886	1 388	2 411	2 450
Morocco	0	0	0	0	0	0	2	5	3	4
01 Fishing area total	1 139	2 352	3 687	3 169	2 286	2 387	2 888	1 393	2 414	2 454
03 Chile	14	7	5	-	-	-	-	-	-	-
03 Fishing area total	14	7	5	-	-	-	-	-	-	-
04 Indonesia	3 625	3 336	3 738	4 024	3 354	3 242	3 871	3 900	4 116	4 350
Japan	3 280	3 013	3 140	2 717	2 222	2 413	1 872	1 942	1 676	1 158
Philippines	8 756	3 351	3 938	4 581	5 214	4 282	2 014	3 915	4 020	4 507
04 Fishing area total	15 661	9 700	10 816	11 322	10 790	9 937	7 757	9 757	9 812	10 015
Species total	16 814	12 059	14 508	14 491	13 076	12 324	10 645	11 150	12 226	12 469
Noble crayfish	**Ecrevisse à pieds rouges**		**Cangrejo de río de patas rojas**			***Astacus astacus***		**2,29(03)027,02**		**AAS**
05 Norway	...	...	...	...	...	10	10	10	...	...
05 Fishing area total	...	...	...	...	...	10	10	10	...	...
Species total	...	...	...	...	...	10	10	10	...	...
Signal crayfish	**Ecrevisse signal**		**...C**			***Pacifastacus leniusculus***		**2,29(03)076,01**		**PCL**
05 UK	-	-	-	-	-	-	-	10	81	80
05 Fishing area total	-	-	-	-	-	-	-	10	81	80
Species total	-	-	-	-	-	-	-	10	81	80
White-clawed crayfish	**Ecrevisse à pattes blanches**		**Cangrejo a pinzas blancas**			***Austropotamobius pallipes***		**2,29(03)139,01**		**AUP**
05 Spain	0	0	0	0	0	0	0	0	0	0
05 Fishing area total	0	0	0	0	0	0	0	0	0	0
Species total	0	0	0	0	0	0	0	0	0	0
Red swamp crawfish	**Ecrevisse rouge de marais**		**Cangrejo de las marismas**			***Procambarus clarkii***		**2,29(04)001,01**		**RCW**
01 Kenya	35	41	40	20	13	24	19	21	22	3
01 Fishing area total	35	41	40	20	13	24	19	21	22	3
05 Spain	1 558	2 379	274	2 753	2 500 F	2 500 F	2 500 F	2 500 F	2 500 F	2 500 F
05 Fishing area total	1 558	2 379	274	2 753	2 500 F	2 500 F	2 500 F	2 500 F	2 500 F	2 500 F
Species total	1 593	2 420	314	2 773	2 513 F	2 524 F	2 519 F	2 521 F	2 522 F	2 503 F
Australian crayfish	**Ecrevisse d'Australie**		**Cangrejo de río de Australia**			***Euastacus armatus***		**2,29(05)158,01**		**AYA**
06 Australia	1	-	-	-	-	-	-	66	24	22
06 Fishing area total	1	-	-	-	-	-	-	66	24	22
Species total	1	-	-	-	-	-	-	66	24	22

B-41

Freshwater crustaceans — **Capture production by species, fishing areas and countries or areas**
Crustacés d'eau douce — **Captures par espèces, zones de pêche et pays ou zones**
Crustáceos de agua dulce — **Capturas por especies, áreas de pesca y países o áreas**

Species, Fishing area Espèce, Zone de pêche Especie, Area de pesca	1992 mt	1993 mt	1994 mt	1995 mt	1996 mt	1997 mt	1998 mt	1999 mt	2000 mt	2001 mt
Oceanian crayfishes nei	**...B**		**...C**		***Parastacidae***				**2,29(05)XXX,XX**	**CJF**
06 Cook Is	...	...	...	10	...	...	...	...	...	...
Papua N Guin	5 F	3	11	10	6	6	14	6	0	1
06 Fishing area total	*5 F*	*3*	*11*	*20*	*6*	*6*	*14*	*6*	*0*	*1*
Species total	*5 F*	*3*	*11*	*20*	*6*	*6*	*14*	*6*	*0*	*1*
Euro-American crayfishes nei	**Ecrevisses euro-américain. nca**		**Cangrejos de río nep**		***Astacidae, Cambaridae***				**2,29(XX)XXX,XX**	**AYS**
02 Mexico	...	...	157	170	198	101	99	163	108	17
USA	12 175	22 105	11 130	7 073	5 689	10 496	10 082	5 322	217	4 676
02 Fishing area total	*12 175*	*22 105*	*11 287*	*7 243*	*5 887*	*10 597*	*10 181*	*5 485*	*325*	*4 693*
04 Turkey	324	404	524	551	850	1 100	1 500	1 372	1 681	1 634
04 Fishing area total	*324*	*404*	*524*	*551*	*850*	*1 100*	*1 500*	*1 372*	*1 681*	*1 634*
05 Bulgaria	-	-	-	-	-	-	-	-	-	1
Denmark	-	0	0	0	0	0	0	0	0	0
Estonia	-	-	0	...	0	...	...	...	1	1
Finland	362	362	191	191	227	227	134	134	134	114
Greece	...	...	...	...	...	15	23	28	23	27
Lithuania	-	0	1	1	1	1	0	1	1	0
Romania	-	-	-	-	-	181	65	56	32	-
Sweden	4	4	5	5	6	9	6	8	17	37
05 Fishing area total	*366*	*366*	*197*	*197*	*234*	*433*	*228*	*227*	*208*	*180*
Species total	*12 865*	*22 875*	*12 008*	*7 991*	*6 971*	*12 130*	*11 909*	*7 084*	*2 214*	*6 507*
Freshwater crustaceans nei	**Crustacés d'eau douce nca**		**Crustáceos de agua dulce nep**		***Crustacea***				**2,99(XX)XXX,XX**	**FCX**
01 Benin	3 218 F	3 800 F	3 794	4 503	4 600 F	4 700 F	4 800 F	4 924	4 100 F	4 600 F
Côte dIvoire	1 500	1 000	1 000	500	400 F	300 F	200 F	100 F	27	30 F
Egypt	...	...	...	950	557	838	918	1 237	2 473	2 192
Gabon	...	...	...	30	36	44	50	6	4	9
01 Fishing area total	*4 718 F*	*4 800 F*	*4 794*	*5 983*	*5 593 F*	*5 882 F*	*5 968 F*	*6 267 F*	*6 604 F*	*6 831 F*
02 Cuba	-	292	253	284	350	922	802	30	30 F	30 F
Dominican Rp	-	-	-	-	21	-	-	-	1	48
El Salvador	24	177	3	5	5	0	3	11	9	16
02 Fishing area total	*24*	*469*	*256*	*289*	*376*	*922*	*805*	*41*	*40 F*	*94 F*
03 Brazil	9 F	10 F	10 F	0	0	0	0	0	0	0
03 Fishing area total	*9 F*	*10 F*	*10 F*	*0*	*0*	*0*	*0*	*0*	*0*	*0*
04 Cambodia	19	20	40	79	70	100	100	300	300	400 F
China	123 818	154 891	203 372	270 536	364 441	479 645	601 966	455 761	530 026	586 985
India	-	-	33	26	803	71	114	85	112	77
Indonesia	244	164	242	86	223	49	93	132	132	150
Korea Rep	76	138	181	216	248	-	2	111	93	48
Philippines	-	1 232	732	73	55	34	32	1	5	67
Viet Nam	1 000 F	1 000 F	500 F	500 F	1 000 F	1 000 F	1 000 F	1 000 F	1 000 F	1 000 F
04 Fishing area total	*125 157 F*	*157 445 F*	*205 100 F*	*271 516 F*	*366 840 F*	*480 899 F*	*603 307 F*	*457 390 F*	*531 668 F*	*588 727 F*
05 Albania	15	10 F	5 F	0	0	-	-	-	-	-
Germany	0	1	12	12	12	12	12	12	12	12
Russian Fed	34	7	-	12	28	34	33	41	957	1 733
Ukraine	1	...	0	...	...	...	...	...	0	2
05 Fishing area total	*50*	*18 F*	*17 F*	*24*	*40*	*46*	*45*	*53*	*969*	*1 747*
06 Fiji Islands	260	102	104 F	1 011	327	315	323	332	340 F	343
06 Fishing area total	*260*	*102*	*104 F*	*1 011*	*327*	*315*	*323*	*332*	*340 F*	*343*
Species total	*130 218 F*	*162 844 F*	*210 281 F*	*278 823 F*	*373 176 F*	*488 064 F*	*610 448 F*	*464 083 F*	*539 621 F*	*597 742 F*
Group total	***177 042***	***216 443***	***253 483***	***315 761***	***410 590***	***527 699***	***645 856***	***498 530***	***567 933***	***630 533***

B-42

Crabs, sea-spiders — Capture production by species, fishing areas and countries or areas
Crabes, araignées de mer — Captures par espèces, zones de pêche et pays ou zones
Cangrejos, centollas — Capturas por especies, áreas de pesca y países o áreas

Species, Fishing area Espèce, Zone de pêche Especie, Area de pesca	1992 mt	1993 mt	1994 mt	1995 mt	1996 mt	1997 mt	1998 mt	1999 mt	2000 mt	2001 mt
Atlantic rock crab	**Tourteau poïnclos**		**Jaiba de roca amarilla**			***Cancer irroratus***			**2,31(09)010,03**	**CRK**
21 Canada	969	2 134	3 743	5 415	4 286	6 410	6 338	5 732	7 672	7 968
USA	743	880	718	844	55	1 120	1 358	1 301	1 844	326
21 Fishing area total	*1 712*	*3 014*	*4 461*	*6 259*	*4 341*	*7 530*	*7 696*	*7 033*	*9 516*	*8 294*
Species total	*1 712*	*3 014*	*4 461*	*6 259*	*4 341*	*7 530*	*7 696*	*7 033*	*9 516*	*8 294*
Dungeness crab	**Dormeur du Pacifique**		**Buey del Pacífico**			***Cancer magister***			**2,31(09)010,04**	**DUN**
67 Canada	3 334	6 225	5 995	4 586	5 025	3 923	2 968	2 944	2 832	5 685
USA	18 149	24 594	19 202	20 116	28 581	15 735	14 034	15 346	16 337	15 685
67 Fishing area total	*21 483*	*30 819*	*25 197*	*24 702*	*33 606*	*19 658*	*17 002*	*18 290*	*19 169*	*21 370*
77 USA	759	559	1 666	1 580	897	1 593	1 485	734	772	840
77 Fishing area total	*759*	*559*	*1 666*	*1 580*	*897*	*1 593*	*1 485*	*734*	*772*	*840*
Species total	*22 242*	*31 378*	*26 863*	*26 282*	*34 503*	*21 251*	*18 487*	*19 024*	*19 941*	*22 210*
Edible crab	**Tourteau**		**Buey de mar**			***Cancer pagurus***			**2,31(09)010,06**	**CRE**
27 Belgium	195	274	244	268	175	222	180	81	106	108
Channel Is	1 620	1 633 F	1 459	1 349 F	2 043	2 268	2 214	1 655	1 440	1 560
Denmark	15	15	17	17	15	14	21	25	33	53
France	5 859	6 495	6 395	6 185	5 927	6 554	5 598	6 503	6 549	6 604
Germany	8	8	-	-	-	38	44	57	64	44
Ireland	3 699	4 218	6 374	7 049	5 649	7 572	7 392	7 772	9 865	9 738
Isle of Man	16	137	52	282	94	478	274	231	142	170
Netherlands	-	-	-	-	-	-	-	-	146	300
Norway	1 316	1 642	1 781	1 807	1 889	2 204	2 984	2 837	2 897	3 476
Portugal	22	11	6	5	4	15	13	21	14	13
Spain	114	127	157	149	120	51	23	49	48	35
Sweden	97	95	161	105	87	79	93	122	128	133
UK	13 392	11 289	14 241	16 166	13 182	18 516	21 799	19 988	21 607	21 909
27 Fishing area total	*26 353*	*25 944 F*	*30 887*	*33 382 F*	*29 185*	*38 011*	*40 635*	*39 341*	*43 039*	*44 143*
Species total	*26 353*	*25 944 F*	*30 887*	*33 382 F*	*29 185*	*38 011*	*40 635*	*39 341*	*43 039*	*44 143*
Jonah crab	**Tourteau jona**		**Jaiba de roca jonás**			***Cancer borealis***			**2,31(09)010,07**	**CRJ**
21 USA	1 195	1 099	1 120	332	334	745	1 255	1 549	1 114	1 245
21 Fishing area total	*1 195*	*1 099*	*1 120*	*332*	*334*	*745*	*1 255*	*1 549*	*1 114*	*1 245*
Species total	*1 195*	*1 099*	*1 120*	*332*	*334*	*745*	*1 255*	*1 549*	*1 114*	*1 245*
Pacific rock crab	**Tourteau du Pacifique**		**Jaiba del Pacífico**			***Cancer productus***			**2,31(09)010,08**	**ROC**
67 USA	...	...	...	...	...	...	3	5	4	5
67 Fishing area total	...	...	...	...	...	...	*3*	*5*	*4*	*5*
77 USA	...	...	...	...	...	...	571	354	490	532
77 Fishing area total	...	...	...	...	...	...	*571*	*354*	*490*	*532*
Species total	...	...	...	...	...	...	*574*	*359*	*494*	*537*
Black stone crab	**Crabe caillou noir**		**Cangrejo de piedra negro**			***Menippe mercenaria***			**2,31(10)050,01**	**STC**
31 Bahamas	16	40	47	40	25	42	39	50	46	47
Belize	14	14	20	28	53	210	25	29	16	8
Cuba	7	3	2	5	28	10	35	883	880 F	880 F
Mexico	65	60	58	120	353	610	681	355	120	...
USA	1 961	2 483	3 177	2 376	3 216	2 483	2 567	1 877	2 961	2 996
31 Fishing area total	*2 063*	*2 600*	*3 304*	*2 569*	*3 675*	*3 355*	*3 347*	*3 194*	*4 023 F*	*3 931 F*
Species total	*2 063*	*2 600*	*3 304*	*2 569*	*3 675*	*3 355*	*3 347*	*3 194*	*4 023 F*	*3 931 F*
Blue swimming crab	**Etrille bleue**		**Jaiba azul**			***Portunus pelagicus***			**2,31(11)004,01**	**SCD**
37 Egypt	110	318	657	...	...	...	...	...	...	...
37 Fishing area total	*110*	*318*	*657*	...	...	...	...	...	...	...
57 Australia	270	308	644	607	522	740	865	717	810	1 004
Indonesia	1 179	768	1 542	1 975	1 830	2 207	3 025	3 594	3 301	4 120
Thailand	4 470	6 118	4 930	5 781	5 696	5 173	9 397	7 386	7 285	7 240
57 Fishing area total	*5 919*	*7 194*	*7 116*	*8 363*	*8 048*	*8 120*	*13 287*	*11 697*	*11 396*	*12 364*
61 China	13 714	15 000	20 000	25 168	51 288	45 749	48 190	52 577	56 092	59 416
China,Taiwan	2 166	1 579	1 690	2 808	1 391	1 141	966	511	1 966	1 703
61 Fishing area total	*15 880*	*16 579*	*21 690*	*27 976*	*52 679*	*46 890*	*49 156*	*53 088*	*58 058*	*61 119*
71 Australia	1 802	2 573	4 047	4 590	4 816	4 507	4 390	3 732	4 708	5 294
Indonesia	5 287	6 470	6 787	8 909	9 728	13 181	9 345	10 682	10 752	13 410
Philippines	41 287	26 151	26 128	29 536	27 660	30 358	23 919	34 076	36 303	36 973
Thailand	31 784	33 641	35 157	35 414	36 219	34 916	37 281	33 864	33 400	33 197
71 Fishing area total	*80 160*	*68 835*	*72 119*	*78 449*	*78 423*	*82 962*	*74 935*	*82 354*	*85 163*	*88 874*
81 Australia	686	627	662	788	800	761	745	22	30	30
81 Fishing area total	*686*	*627*	*662*	*788*	*800*	*761*	*745*	*22*	*30*	*30*

B-42 Crabs, sea-spiders / Crabes, araignées de mer / Cangrejos, centollas

Capture production by species, fishing areas and countries or areas / Captures par espèces, zones de pêche et pays ou zones / Capturas por especies, áreas de pesca y países o áreas

Species, Fishing area Espèce, Zone de pêche Especie, Area de pesca	1992 mt	1993 mt	1994 mt	1995 mt	1996 mt	1997 mt	1998 mt	1999 mt	2000 mt	2001 mt
Species total	*102 755*	*93 553*	*102 244*	*115 576*	*139 950*	*138 733*	*138 123*	*147 161*	*154 647*	*162 387*
Gazami crab	**Crabe gazami**		**Jaiba gazami**		***Portunus trituberculatus***			**2,31(11)004,04**		**GAZ**
61 China	155 548	117 264	272 102	243 485	283 394	237 960	266 630	270 280	335 078	333 556
Japan	3 270	2 958	3 564	4 159	4 022	3 112	3 528	2 752	3 131	3 596
Korea Rep	17 317	10 419	21 483	17 651	15 754	11 430	13 813	11 819	12 842	13 016
61 Fishing area total	*176 135*	*130 641*	*297 149*	*265 295*	*303 170*	*252 502*	*283 971*	*284 851*	*351 051*	*350 168*
Species total	*176 135*	*130 641*	*297 149*	*265 295*	*303 170*	*252 502*	*283 971*	*284 851*	*351 051*	*350 168*
Portunus swimcrabs nei	**Etrilles nca**		**Jaibas, nécoras nep**		***Portunus spp***			**2,31(11)004,XX**		**CRS**
27 France	326	336	361	254	225	261	155	163	242	267
Ireland	...	...	...	...	...	...	314	309	...	214
Portugal	36	94	30	32	29	16	19	0	59	67
Spain	207	198	186	169	0	6	20	18		-
UK	2 442	2 515	2 886	3 835	1 459	2 929	2 513	2 282	1 672	2 418
27 Fishing area total	*3 011*	*3 143*	*3 463*	*4 290*	*1 713*	*3 212*	*3 021*	*2 772*	*1 973*	*2 966*
31 Trinidad Tob	-	-	-	-	-	-	0	1	1	1
Venezuela	502	650	197	1	1	289	224	4 041	4 994	4 001
31 Fishing area total	*502*	*650*	*197*	*1*	*1*	*289*	*224*	*4 042*	*4 995*	*4 002*
51 Bahrain	914	690	821	807	1 047	1 289	1 017	2 179	2 380	2 556
Eritrea	...	...	-		-		-	-	-	1
51 Fishing area total	*914*	*690*	*821*	*807*	*1 047*	*1 289*	*1 017*	*2 179*	*2 380*	*2 557*
Species total	*4 427*	*4 483*	*4 481*	*5 098*	*2 761*	*4 790*	*4 262*	*8 993*	*9 348*	*9 525*
Blue crab	**Crabe bleu**		**Cangrejo azul**		***Callinectes sapidus***			**2,31(11)012,02**		**CRB**
21 USA	43 212	73 171	39 760	57 740	67 337	62 784	56 834	60 392	42 408	39 511
21 Fishing area total	*43 212*	*73 171*	*39 760*	*57 740*	*67 337*	*62 784*	*56 834*	*60 392*	*42 408*	*39 511*
31 Cuba	1 138	513	840	745	904	704	1 065	567	570 F	570 F
Mexico	9 372	11 076	11 026	10 657	13 736	14 412	12 920	12 714	8 602	7 291
Nicaragua	-	-	-	-	-	-	131	106	71	69
USA	45 074	40 870	56 889	34 457	33 582	43 743	45 046	31 459	40 994	32 356
31 Fishing area total	*55 584*	*52 459*	*68 755*	*45 859*	*48 222*	*58 859*	*59 162*	*44 846*	*50 237 F*	*40 286 F*
Species total	*98 796*	*125 630*	*108 515*	*103 599*	*115 559*	*121 643*	*115 996*	*105 238*	*92 645 F*	*79 797 F*
Dana swimcrab	**Crabe lénée**		**Cangrejo sirí**		***Callinectes danae***			**2,31(11)012,05**		**CRZ**
41 Brazil	3 300 F	2 800 F	2 500 F	2 062	2 020	2 600	3 014	1 626	1 597	1 580 F
41 Fishing area total	*3 300 F*	*2 800 F*	*2 500 F*	*2 062*	*2 020*	*2 600*	*3 014*	*1 626*	*1 597*	*1 580 F*
Species total	*3 300 F*	*2 800 F*	*2 500 F*	*2 062*	*2 020*	*2 600*	*3 014*	*1 626*	*1 597*	*1 580 F*
Green crab	**Crabe vert**		**Cangrejo verde**		***Carcinus maenas***			**2,31(11)090,01**		**CRG**
21 USA	2	0	0	0	-	54	86	14	16	67
21 Fishing area total	*2*	*0*	*0*	*0*	-	*54*	*86*	*14*	*16*	*67*
27 France	296	327	342	292	350	359	204	465	558	541
Ireland	...	...	...	...	...	...	79	169	...	68
Portugal	297	739	382	351	200	125	156	77	111	125
Spain	-	-	-	-	1	2	8	8	10	12
UK	481	512	502	470	286	411	462	333	194	331
27 Fishing area total	*1 074*	*1 578*	*1 226*	*1 113*	*837*	*897*	*909*	*1 052*	*873*	*1 077*
Species total	*1 076*	*1 578*	*1 226*	*1 113*	*837*	*951*	*995*	*1 066*	*889*	*1 144*
Mediterranean shore crab	**Crabe vert de la Méditerranée**		**Cangrejo verde mediterráneo**		***Carcinus aestuarii***			**2,31(11)090,02**		**CMR**
37 France	10	7	28	...	16	16	16	7	14	9
Greece	166	82	59	16	49	28	50	37	16	17
Tunisia	-	-	-	-	-	-	-	-	-	11
37 Fishing area total	*176*	*89*	*87*	*16*	*65*	*44*	*66*	*44*	*30*	*37*
Species total	*176*	*89*	*87*	*16*	*65*	*44*	*66*	*44*	*30*	*37*
Indo-Pacific swamp crab	**Crabe de palétuviers**		**Cangrejo de manglares**		***Scylla serrata***			**2,31(11)140,01**		**MUD**
04 China,Taiwan	-	-	-	-	-	-	2	1	10	0
Philippines	156	321	145	329	363	267	42	46	67	150
04 Fishing area total	*156*	*321*	*145*	*329*	*363*	*267*	*44*	*47*	*77*	*150*
51 Mauritius	-	-	-	22	23	19	23	21	25	24
51 Fishing area total	-	-	-	*22*	*23*	*19*	*23*	*21*	*25*	*24*
57 Indonesia	2 378	2 225	3 018	2 351	2 597	2 714	3 296	3 638	3 490	3 660
Thailand	2 296	1 816	3 653	3 508	2 527	2 421	1 884	1 973	1 946	1 934
57 Fishing area total	*4 674*	*4 041*	*6 671*	*5 859*	*5 124*	*5 135*	*5 180*	*5 611*	*5 436*	*5 594*
61 China,Taiwan	158	84	169	1 339	935	180	213	268	289	230
61 Fishing area total	*158*	*84*	*169*	*1 339*	*935*	*180*	*213*	*268*	*289*	*230*

B-42 Crabs, sea-spiders / Crabes, araignées de mer / Cangrejos, centollas

Capture production by species, fishing areas and countries or areas
Captures par espèces, zones de pêche et pays ou zones
Capturas por especies, áreas de pesca y países o áreas

Species, Fishing area Espèce, Zone de pêche Especie, Area de pesca	1992 mt	1993 mt	1994 mt	1995 mt	1996 mt	1997 mt	1998 mt	1999 mt	2000 mt	2001 mt
71 Fiji Islands	435	198	202 F	234	208	290	270	281	250 F	268
Indonesia	3 351	8 284	3 423	5 629	4 745	5 584	4 865	5 069	5 284	5 530
Marshall Is	0	0	0	0	0	0	0	0	0	0
Micronesia	3 F	4 F	4 F	5 F	5 F	5 F	5 F	5 F	5 F	5 F
Palau	13	14	13	14	15	15	15	30 F	48 F	48 F
Papua N Guin	20 F	20 F	27 F	28 F	25 F	24 F	23 F	22 F	24 F	24 F
Philippines	7 632	6 080	5 490	4 506	3 895	866	1 082	1 165	1 180	1 262
Singapore	32	38	33	27	19	15	9	9	28	9
Thailand	2 434	2 972	2 541	2 268	1 716	1 610	1 848	3 763	3 711	3 689
71 Fishing area total	*13 920 F*	*17 610 F*	*11 733 F*	*12 711 F*	*10 628 F*	*8 409 F*	*8 117 F*	*10 344 F*	*10 530 F*	*10 835 F*
77 Fr Polynesia	-	-	-	-	-	2	2	2	2	2
77 Fishing area total	-	-	-	-	-	*2*	*2*	*2*	*2*	*2*
Species total	*18 908 F*	*22 056 F*	*18 718 F*	*20 260 F*	*17 073 F*	*14 012 F*	*13 579 F*	*16 293 F*	*16 359 F*	*16 835 F*
Harbour spidercrab	**Araignée portuaire**		**Araña porteña**		***Mithrax armatus***				**2,31(21)001,02**	**MXT**
31 Nicaragua	-	-	-	-	-	-	114	110	54	35
31 Fishing area total	-	-	-	-	-	-	*114*	*110*	*54*	*35*
Species total	-	-	-	-	-	-	*114*	*110*	*54*	*35*
Spinous spider crab	**Araignée européenne**		**Centolla europea**		***Maja squinado***				**2,31(21)005,01**	**SCR**
27 Channel Is	485	489 F	573	530 F	868	445	460	553	522	428
Denmark	2	1	1	1	0	1	1	1	1	2
France	3 765	3 611	3 631	3 467	3 169	3 433	3 124	3 380	4 422	5 439
Ireland	106	130	264	153	192	153	185	299	163	264
Portugal	42	69	50	49	40	47	58	60	59	51
Spain	102	154	137	151	181	206	193	209	198	108
UK	994	812	1 352	2 178	1 224	1 849	1 843	1 310	990	1 199
27 Fishing area total	*5 496*	*5 266 F*	*6 008*	*6 529 F*	*5 674*	*6 134*	*5 864*	*5 812*	*6 355*	*7 491*
37 Croatia	33	37	39	43	25	54	50	15	14	21
France	1	7	11	...	2	2	2	0	6	1
Spain	-	-	-	-	4	3	3	1	1	5
Yugoslavia	0	0	0	0	0	0	0	0	0	0
37 Fishing area total	*34*	*44*	*50*	*43*	*31*	*59*	*55*	*16*	*21*	*27*
Species total	*5 530*	*5 310 F*	*6 058*	*6 572 F*	*5 705*	*6 193*	*5 919*	*5 828*	*6 376*	*7 518*
Queen crab	**Crabe des neiges**		**Cangrejo de las nieves**		***Chionoecetes opilio***				**2,31(21)145,01**	**CRQ**
21 Canada	37 079	48 099	60 400	65 372	65 825	71 416	75 216	95 148	93 505	95 299
Greenland	3	1	72	998	817	2 557	1 947	2 896	10 236	14 247
St Pier Mq	0	1	2	2	189	368	354	589	511	498
21 Fishing area total	*37 082*	*48 101*	*60 474*	*66 372*	*66 831*	*74 341*	*77 517*	*98 633*	*104 252*	*110 044*
Species total	*37 082*	*48 101*	*60 474*	*66 372*	*66 831*	*74 341*	*77 517*	*98 633*	*104 252*	*110 044*
Tanner crabs nei	**Crabes des neiges du Pac. nca**		**Cangrejos de las nieves (Pac.)**		***Chionoecetes spp***				**2,31(21)145,XX**	**PCR**
61 Japan	6 595	5 612	6 919	9 090	3 447	4 870	4 677	4 892	5 640	5 355
Korea Rep	-	-	-	-	-	-	-	-	17 037	13 974
Russian Fed	...	...	...	...	22 769	27 112	20 848	21 234	21 848	24 493
61 Fishing area total	*6 595*	*5 612*	*6 919*	*9 090*	*26 216*	*31 982*	*25 525*	*26 126*	*44 525*	*43 822*
67 USA	158 777	116 000	72 382	36 658	30 785	53 932	114 230	83 989	15 665	12 177
67 Fishing area total	*158 777*	*116 000*	*72 382*	*36 658*	*30 785*	*53 932*	*114 230*	*83 989*	*15 665*	*12 177*
Species total	*165 372*	*121 612*	*79 301*	*45 748*	*57 001*	*85 914*	*139 755*	*110 115*	*60 190*	*55 999*
Hair crab	**Crabe velu**		**Cangrejo peludo**		***Erimacrus isenbeckii***				**2,31(23)001,01**	**HBZ**
61 Russian Fed	...	...	...	...	789	612	409	440	198	162
61 Fishing area total	...	...	...	...	*789*	*612*	*409*	*440*	*198*	*162*
Species total	...	...	...	...	*789*	*612*	*409*	*440*	*198*	*162*
Red crab	**Gériocrabe rouge**		**Geriocangrejo rojo**		***Geryon quinquedens***				**2,31(43)088,01**	**CRR**
21 USA	1 061	1 440	...	521	465	96	-	-	3 130	4 004
21 Fishing area total	*1 061*	*1 440*	...	*521*	*465*	*96*	-	-	*3 130*	*4 004*
31 USA	-	-	-	-	-	-	-	-	2	-
31 Fishing area total	-	-	-	-	-	-	-	-	*2*	-
34 UK	-	-	-	-	-	-	-	-	-	31
34 Fishing area total	-	-	-	-	-	-	-	-	-	*31*
41 UK	-	-	-	-	-	-	61	-	-	596
Uruguay	...	254	852	783	1 513	3 432	2 682	3 382	5 259	2 089
41 Fishing area total	...	*254*	*852*	*783*	*1 513*	*3 432*	*2 743*	*3 382*	*5 259*	*2 685*
Species total	*1 061*	*1 694*	*852*	*1 304*	*1 978*	*3 528*	*2 743*	*3 382*	*8 391*	*6 720*

B-42

Crabs, sea-spiders — Capture production by species, fishing areas and countries or areas
Crabes, araignées de mer — Captures par espèces, zones de pêche et pays ou zones
Cangrejos, centollas — Capturas por especies, áreas de pesca y países o áreas

Species, Fishing area Espèce, Zone de pêche Especie, Area de pesca	1992 mt	1993 mt	1994 mt	1995 mt	1996 mt	1997 mt	1998 mt	1999 mt	2000 mt	2001 mt
Geryons nei	**Géryons nca**		**Geriones nep**		***Geryon spp***			**2,31(43)088,XX**		**GER**
27 UK	-	-	-	10	1 477	325	587	1 015	763	382
27 Fishing area total	*-*	*-*	*-*	*10*	*1 477*	*325*	*587*	*1 015*	*763*	*382*
34 Mauritania	...	...	...	...	...	...	...	...	6	14
Spain	453	-	-	-	-	-	-	-	-	-
34 Fishing area total	*453*	*...*	*...*	*...*	*...*	*...*	*...*	*...*	*6*	*14*
47 Japan	7 005	6 780	5 249	4 855	3 250	1 937	1 562	1 730	2 980	1 650
Namibia	2 790	3 190	3 598	2 008	1 709	1 478	2 283	2 074	2 699	2 667
South Africa	0	0	-	-	-	-	-	-	-	-
Spain	-	300 F	300 F	300 F	370	-	300	135	156	168
47 Fishing area total	*9 795*	*10 270 F*	*9 147 F*	*7 163 F*	*5 329*	*3 415*	*4 145*	*3 939*	*5 835*	*4 485*
51 Mozambique	418	406	345	414	564	1 144	911	795	832	629
South Africa	187	138	-	-	-	-	91	80 F	54	53
51 Fishing area total	*605*	*544*	*345*	*414*	*564*	*1 144*	*1 002*	*875 F*	*886*	*682*
Species total	*10 853*	*10 814 F*	*9 492 F*	*7 587 F*	*7 370*	*4 884*	*5 734*	*5 829 F*	*7 490*	*5 563*
Marine crabs nei	**Crabes de mer nca**		**Cangrejos de mar nep**		***Brachyura***			**2,31(XX)XXX,XX**		**CRA**
21 Canada	0	34	806	2 489	1 588	1 948	2 267	1 835	3 052	4 306
Spain	-	-	-	-	11	-	-	-	-	-
USA	297	228	295	632	1 692	3 128	1 927	2 958	2 615	6 020
21 Fishing area total	*297*	*262*	*1 101*	*3 121*	*3 291*	*5 076*	*4 194*	*4 793*	*5 667*	*10 326*
27 Channel Is	-	-	-	-	-	-	2	2	2	-
Denmark	226	236	238	188	171	193	220	193	205	189
Faeroe Is	-	-	-	1	3	1	4	1	6	1
France	11	14	17	14	160	5	5	-	13	10
Germany	-	4	-	-	13	6	4	-	-	-
Iceland	0	0	0	0	0	0	-	-	-	1
Ireland	424	518	355	487	312	272	-	1	268	-
Norway	-	-	-	-	-	-	-	-	-	2
Portugal	-	-	-	-	-	-	-	19	39	45
Spain	1 319	1 119	178	575	1 849	9 652	1 712	738	101	74
UK	695	1 008	1 827	1 469	639	205	57	90	17	60
27 Fishing area total	*2 675*	*2 899*	*2 615*	*2 734*	*3 147*	*10 334*	*2 004*	*1 044*	*651*	*382*
31 Colombia	-	-	65	181	21	131	32	...	100 F	100 F
Cuba	283	0	0	0	-	-	-	-	-	-
Dominican Rp	96	74	92	75	32	14	37	7	14	52
Haiti	...	...	...	17	5	71	59	50 F	50 F	50 F
Honduras	-	10	9	20	47	10	5	2	20	...
Jamaica	...	...	...	34	11	9	9	9	8	8 F
Korea Rep	-	-	-	-	5	-	-	-	-	2
Puerto Rico	...	...	...	...	...	...	...	2	2	6
Suriname	0	4	5	5 F	6 F	10 F	10 F	10 F	20	25
USA	196	160	137	358	47	473	179	97	392	346
Venezuela	2 243	3 940	6 072	5 551	4 591	5 340	4 180	196	539	155
31 Fishing area total	*2 818*	*4 188*	*6 380*	*6 241 F*	*4 765 F*	*6 058 F*	*4 511 F*	*373 F*	*1 145 F*	*744 F*
34 Benin	33 F	36 F	40 F	40 F	40 F	50 F	45 F	32	20 F	9
Cameroon	42	40	40 F	32	18	20 F	20 F	20 F	27	0
Congo Rep	1 F	1 F	0	5 F	10 F	5 F	20 F	30 F	44	10 F
Côte dIvoire	0	0	0	0	0	0	0	1	2	2 F
Gabon	10	26	34	65	120	145	158	283	289	142
Gambia	...	...	...	...	...	...	...	...	6	2
Ghana	75	195	462	218	399	576	271	145	74	155
Greece	141	610	58	71	60	84	90	74	59	54
GuineaBissau	22 F	23 F	23 F	27 F	30 F	30 F	25 F	20 F	20 F	20 F
Italy	-	-	-	-	-	-	-	-	8	-
Korea Rep	-	-	5	-	-	1	1	-	-	-
Liberia	13	22	10	28	-	35	38	32	27	122
Lithuania	1	0	-	-	-	-	-	-	-	-
Morocco	46	28	56	46	50	286	236	400	370	364
Nigeria	306	92	28	48	57	593	3 384	4 018	3 211	4 374
Portugal	-	-	-	21	33	61	36	17	79	193
Senegal	25	29	520	158	217	727	356	186	343	389
Sierra Leone	100	100	100	14	15 F	140	83	107	142	269
Spain	-	-	-	-	1 282	1 361	1 794	218	471	516
Togo	0	0	0	0	0	0	1	3	33	23
34 Fishing area total	*815 F*	*1 202 F*	*1 376 F*	*773 F*	*2 331 F*	*4 114 F*	*6 558 F*	*5 586 F*	*5 225 F*	*6 644 F*
37 Egypt	949	848	684	1 026	981	1 063	1 249	3 524	1 725	1 160
Gaza Strip	...	...	...	66	72	25	65	65 F	65 F	50 F
Morocco	2	1	7	0	0	0	0	-	-	1
Spain	-	-	-	-	741	595	842	663	863	1 031
Turkey	16	92	83	117	305	318	243	179	187	273
37 Fishing area total	*967*	*941*	*774*	*1 209*	*2 099*	*2 001*	*2 399*	*4 431 F*	*2 840 F*	*2 515 F*
41 Argentina	57	54	168	345	390	403	326	170	9	-
Brazil	13 560 F	13 530 F	13 560 F	11 917	9 922	10 233	9 202	10 913	11 135	11 000 F
Russian Fed	-	12	17	-	-	-	-	-	-	-
UK	-	-	-	-	-	-	-	-	-	40
41 Fishing area total	*13 617 F*	*13 596 F*	*13 745 F*	*12 262*	*10 312*	*10 636*	*9 528*	*11 083*	*11 144*	*11 040 F*
47 Angola	...	...	...	...	...	...	...	602	800	836
Spain	-	-	-	-	-	-	1 494	314	239	254

B-42 Crabs, sea-spiders — Capture production by species, fishing areas and countries or areas
Crabes, araignées de mer — Captures par espèces, zones de pêche et pays ou zones
Cangrejos, centollas — Capturas por especies, áreas de pesca y países o áreas

Species, Fishing area Espèce, Zone de pêche Especie, Area de pesca		1992 mt	1993 mt	1994 mt	1995 mt	1996 mt	1997 mt	1998 mt	1999 mt	2000 mt	2001 mt
47	*Fishing area total*	...	...	...	...	...	...	*1 494*	*916*	*1 039*	*1 090*
51	Egypt	-	-	-	10 049	0	4	149	1 177	2 141	744
	India	1 235	5 908	25 244	13 859	521	642	369	394	6 193	8 009
	Kenya	58	69	59	70	112	100	117	135	166	134
	Korea Rep	-	-	-	-	-	10	-	-	167	5
	Madagascar	849	1 085	1 300	1 300	1 000	1 000	1 500	868	1 030	1 347
	Pakistan	437	480	650	877	3 200	3 989	5 680	5 109	5 187	5 099
	Portugal	-	-	-	-	-	-	-	-	2	5
	Qatar	28	52	30	39	47	47	69	39	26	21
	Réunion	3	4	5	5	5	6	7	5	7	11
	Saudi Arabia	169	256	1 126	1 035	1 060	1 371	1 062	959	1 021	1 002
	Seychelles	20	14	8	9	17	20	24	11	11	26
	South Africa	7	-	-	-	-	113	4	5 F	...	2
	Spain	-	-	-	-	-	-	-	-	-	188
	Untd Arab Em	47	50	58	47	56	60	60	62	1 710	1 700 F
	Yemen	...	...	...	...	...	...	...	...	...	37
51	*Fishing area total*	*2 853*	*7 918*	*28 480*	*27 290*	*6 018*	*7 362*	*9 041*	*8 764 F*	*17 661*	*18 330 F*
57	India	3 172	12 157	11 627	2 846	1 026	1 595	1 420	1 299	17 457	17 035
	Korea Rep	-	-	-	-	-	-	-	10	25	-
	Malaysia	3 892	3 703	3 372	3 781	3 406	3 793	5 679	5 864	5 105	4 434
	Thailand	2 839	2 803	2 475	3 242	5 680	5 885	5 812	6 947	6 852	6 810
	Timor-Leste	...	...	...	...	...	...	...	1 F	1	1
57	*Fishing area total*	*9 903*	*18 663*	*17 474*	*9 869*	*10 112*	*11 273*	*12 911*	*14 121 F*	*29 440*	*28 280*
61	China,H.Kong	2 601	1 224	1 896	1 601	1 408	1 105	1 108	800 F	1 000 F	1 100 F
	China,Taiwan	2 976	2 155	3 124	4 747	4 830	4 630	4 553	5 433	5 898	7 373
	Japan	40 021	40 148	40 295	38 812	37 265	34 894	33 678	30 859	30 310	27 353
	Korea Rep	30 000	40 620	56 547	58 804	61 308	58 741	43 879	35 375	9 372	9 001
	Russian Fed	-	-	-	-	-	49	-	-	-	194
61	*Fishing area total*	*75 598*	*84 147*	*101 862*	*103 964*	*104 811*	*99 419*	*83 218*	*72 467 F*	*46 580 F*	*45 021 F*
67	USA	2 395	892	1 453	2 185	647	333	133	99	1	42
67	*Fishing area total*	*2 395*	*892*	*1 453*	*2 185*	*647*	*333*	*133*	*99*	*1*	*42*
71	Cambodia	2 270	2 300	2 200	2 300	2 350	2 240	2 420	2 850	3 593	3 600 F
	Fiji Islands	267	57	58 F	288	200	280	255	300	270 F	290
	Korea Rep	-	-	-	-	200	103	119	54	-	23
	Malaysia	4 414	5 424	6 210	6 416	5 596	5 938	8 564	8 566	7 534	7 864
	NewCaledonia	47	28	7	8	9	29	20	58	22	22 F
	Palau	22	28	7	2	2	2	4	10 F	16 F	16 F
	Singapore	221	196	235	196	176	203	264	175	189	203
	Thailand	584	1 452	2 194	2 093	921	989	1 723	1 510	1 489	1 480
	Viet Nam	11 300 F	10 800 F	28 000 F	32 000 F	28 300 F	26 400 F	48 000 F	35 800 F	40 000 F	50 000 F
	Wallis Fut I	...	...	1	1	1	1	1	1 F	1 F	1 F
71	*Fishing area total*	*19 125 F*	*20 285 F*	*38 912 F*	*43 304 F*	*37 755 F*	*36 185 F*	*61 370 F*	*49 324 F*	*53 114 F*	*63 499 F*
77	Cook Is	5 F	5 F	5 F	5 F	5 F	5 F	5 F	5 F	5 F	5 F
	Costa Rica	...	...	...	1	3	50	9	8	4	3
	Honduras	51	23	47	50	70	100	30	4	40	...
	Mexico	2 125	2 706	4 955	10 395	13 602	10 073	6 503	6 670	12 036	11 204
	USA	725	576	507	492	554	632	58	41	33	54
77	*Fishing area total*	*2 906 F*	*3 310 F*	*5 514 F*	*10 943 F*	*14 234 F*	*10 860 F*	*6 605 F*	*6 728 F*	*12 118 F*	*11 266 F*
81	New Zealand	293	326	492	9 231	294	359	393	407	355	360
81	*Fishing area total*	*293*	*326*	*492*	*9 231*	*294*	*359*	*393*	*407*	*355*	*360*
87	Chile	3 616	4 949	5 085	5 733	3 989	3 541	5 063	6 501	6 758	6 770
	Colombia	-	-	33	57	28	35	12	11	20 F	20 F
	Ecuador	530	683	750	750	750	750	600	600	600	600
	Peru	1 265	1 027	1 383	2 553	1 605	303	752	11 397	1 974	1 568
87	*Fishing area total*	*5 411*	*6 659*	*7 251*	*9 093*	*6 372*	*4 629*	*6 427*	*18 509*	*9 352 F*	*8 958 F*
Species total		*139 673 F*	*165 288 F*	*227 429 F*	*242 219 F*	*206 188 F*	*208 639 F*	*210 786 F*	*198 645 F*	*196 332 F*	*208 497 F*
Group total		***818 709***	***797 684***	***985 161***	***951 645***	***999 335***	***990 278***	***1 074 977***	***1 058 754***	***1 087 976***	***1 096 371***

B-43 Lobsters, spiny-rock lobsters / Homards, langoustes / Bogavantes, langostas

Capture production by species, fishing areas and countries or areas / Captures par espèces, zones de pêche et pays ou zones / Capturas por especies, áreas de pesca y países o áreas

Species, Fishing area Espèce, Zone de pêche Especie, Area de pesca	1992 mt	1993 mt	1994 mt	1995 mt	1996 mt	1997 mt	1998 mt	1999 mt	2000 mt	2001 mt
Longlegged spiny lobster	**Langouste diablotin**		**Langosta duende**		***Panulirus longipes***			**2,29(01)001,01**		**LOJ**
61 China,Taiwan	244	23	21	18	14	14	14	12	11	23
Japan	1 194	1 238	1 118	1 136	1 092	1 068	1 084	1 154	1 244	1 486
Korea Rep	-	-	-	-	-	-	-	-	461	415
61 Fishing area total	*1 438*	*1 261*	*1 139*	*1 154*	*1 106*	*1 082*	*1 098*	*1 166*	*1 716*	*1 924*
Species total	*1 438*	*1 261*	*1 139*	*1 154*	*1 106*	*1 082*	*1 098*	*1 166*	*1 716*	*1 924*
Caribbean spiny lobster	**Langouste blanche**		**Langosta común del Caribe**		***Panulirus argus***			**2,29(01)001,08**		**SLC**
31 Anguilla	97	90	90	40 F	50 F	60 F	60 F	60 F	60 F	60 F
Antigua Barb	188	120	140	149	125	160	357	274	275	272
Bahamas	8 156	7 848	7 589	7 750	7 938	7 798	7 553	8 225	9 023	7 042
Belize	463	388	490	608	448	534	468	552	503	394
Bermuda	8	16	17	10	10	38	30	36	29	21
Br Virgin Is	68 F	52	92	32	27	5	6	4	3	3 F
Colombia	524	218	97	449	185	108	319	175	250 F	250 F
Costa Rica	176	29	32	93	196	196	40	163	271	39
Cuba	9 340	8 501	9 694	9 405	9 375	8 996	9 417	9 879	9 850 F	9 850 F
Dominican Rp	532	537	967	619	420	1 061	863	828	1 286	1 209
Grenada	22	26	30	57	23	14	31	72	47	32
Haiti	750 F	830 F	780 F	900 F	190 F	200 F	200 F	200 F	200 F	200 F
Honduras	1 353	1 045	1 080	1 520	470	1 006	306	570	670	850
Jamaica	200 F	250 F	300 F	350 F	394	271	170	330	517	500 F
Martinique	117	113	110 F	110 F	70 F	110 F	120 F	150 F	200	190
Mexico	621	881	945	896	756	844	613	645	747	782
Nicaragua	2 378	2 191	2 805	2 260	4 357	4 012	3 729	5 141	6 327	3 909
Puerto Rico	...	...	...	...	...	...	...	209	212	190
St Kitts Nev	...	...	19	12	17	12	29	34	27	35
Trinidad Tob	-	-	-	-	-	-	2	0	1	0
Turks Caicos	452	413	410 F	400 F	350 F	300 F	230	230 F	230 F	230 F
USA	1 792	2 548	3 420	2 934	3 373	2 783	2 343	2 749	2 571	1 527
US Virgin Is	69 F	64 F	59 F	55 F	50 F	45 F	40 F	34	35 F	35 F
Venezuela	371	940	763	629	648	619	260	95	105	78
31 Fishing area total	*27 677 F*	*27 100 F*	*29 929 F*	*29 278 F*	*29 472 F*	*29 172 F*	*27 186 F*	*30 655 F*	*33 439 F*	*27 698 F*
41 Brazil	9 127	9 100 F	9 120 F	10 817	8 026	7 502	6 002	6 334	6 469	6 400 F
41 Fishing area total	*9 127*	*9 100 F*	*9 120 F*	*10 817*	*8 026*	*7 502*	*6 002*	*6 334*	*6 469*	*6 400 F*
Species total	*36 804 F*	*36 200 F*	*39 049 F*	*40 095 F*	*37 498 F*	*36 674 F*	*33 188 F*	*36 989 F*	*39 908 F*	*34 098 F*
Australian spiny lobster	**Langouste d'Australie**		**Langosta de Australia**		***Panulirus cygnus***			**2,29(01)001,15**		**LOA**
57 Australia	12 194	12 366	11 046	10 886	9 902	9 896	10 400	17 720	19 380	16 017
57 Fishing area total	*12 194*	*12 366*	*11 046*	*10 886*	*9 902*	*9 896*	*10 400*	*17 720*	*19 380*	*16 017*
Species total	*12 194*	*12 366*	*11 046*	*10 886*	*9 902*	*9 896*	*10 400*	*17 720*	*19 380*	*16 017*
Green spiny lobster	**Langouste verte**		**Langosta barbona**		***Panulirus gracilis***			**2,29(01)001,16**		**NUG**
77 Nicaragua	9	9	17	14	101	152	66	131	476	652
77 Fishing area total	*9*	*9*	*17*	*14*	*101*	*152*	*66*	*131*	*476*	*652*
87 Colombia	13	9	6	3	4	7	6	1	1 F	1 F
Ecuador	82	33	50	50	50	50	50	50	50	50
Peru	4	14	52	168	52	12	669	496	278	62
87 Fishing area total	*99*	*56*	*108*	*221*	*106*	*69*	*725*	*547*	*329 F*	*113 F*
Species total	*108*	*65*	*125*	*235*	*207*	*221*	*791*	*678*	*805 F*	*765 F*
Tropical spiny lobsters nei	**Langoustes tropicales nca**		**Langostas tropicales nep**		***Panulirus spp***			**2,29(01)001,XX**		**SLV**
31 Bermuda	-	-	-	-	-	-	5	5	5	4
31 Fishing area total	-	-	-	-	-	-	*5*	*5*	*5*	*4*
34 Cameroon	1	0	1	0	0	0	1	1	0	0
Cape Verde	36	26	28	20	12	8	9	12	10	7
Ghana	218	369	510	230	134	203	65	...	39	342
Sierra Leone	20	10	10	2	4 F	57	29	51	56	30
Togo	1	1	1	0	0	0	0	2	2	0
34 Fishing area total	*276*	*406*	*550*	*252*	*150 F*	*268*	*104*	*66*	*107*	*379*
41 Italy	-	2	2	-	-	-	-	-	-	-
41 Fishing area total	*...*	*2*	*2*	*...*	*...*	*...*	*...*	*...*	*...*	*...*
47 St Helena	0	0	0	0	0	-	-	-	-	-
47 Fishing area total	*0*	*0*	*0*	*0*	*0*	-	-	-	-	-
51 Djibouti	0	0	0	0	0	0	0	0	0	0
Iran	120	45	35	30	49	76	65	35	25	20
Italy	-	3	3	-	-	-	-	-	-	-
Kenya	52	47	44	119	177	136	39	52	52	76
Madagascar	554	358	390	390	390	390	341	338	329	359
Maldives	...	...	...	...	...	...	...	...	7	13
Mauritius	15	13	13	12	14	17	17	17	17	19
Oman	546	701	623	608	397	263	336	180	402	379
Pakistan	502	507	669	615	724	765	782	1 077	807	756
Russian Fed	3	-	-	-	-	-	-	-	-	-

B-43 Lobsters, spiny-rock lobsters — Capture production by species, fishing areas and countries or areas
Homards, langoustes — Captures par espèces, zones de pêche et pays ou zones
Bogavantes, langostas — Capturas por especies, áreas de pesca y países o áreas

Species, Fishing area Espèce, Zone de pêche Especie, Area de pesca	1992 mt	1993 mt	1994 mt	1995 mt	1996 mt	1997 mt	1998 mt	1999 mt	2000 mt	2001 mt
Saudi Arabia	6	8	23	21	13	18	13	19	8	14
Seychelles	0	0	0	2	1	-	0	7	14	5
Somalia	500 F	400 F	400 F	400 F	400 F	400 F	400 F	400 F	400 F	400 F
Ukraine	-	-	-	-	-	-	1	-	-	-
Yemen	1 204	1 021	475	328	323	482	828	334	178	202
51 Fishing area total	*3 502 F*	*3 103 F*	*2 675 F*	*2 525 F*	*2 488 F*	*2 547 F*	*2 822 F*	*2 459 F*	*2 239 F*	*2 243 F*
57 Australia	0	0	0	1	4	0	0	-	-	-
Indonesia	1 126	734	1 420	1 398	1 242	1 310	1 040	1 285	1 393	1 630
Malaysia	24	27	33	20	14	15	12	15	38	104
Timor-Leste	...	...	...	...	...	...	...	2 F	2	2
57 Fishing area total	*1 150*	*761*	*1 453*	*1 419*	*1 260*	*1 325*	*1 052*	*1 302 F*	*1 433*	*1 736*
61 China,H.Kong	75	60	50	20	0	0	0	0	0	0
61 Fishing area total	*75*	*60*	*50*	*20*	*0*	*0*	*0*	*0*	*0*	*0*
71 Australia	147	174	185	182	201	233	551	520	432	358
Fiji Islands	394	96	98 F	184	105	130	211	220	200 F	224
Guam	-	-	-	0	0	0	1	1	2	1
Indonesia	1 272	474	601	1 454	1 222	2 711	1 354	1 959	2 203	2 590
Malaysia	969	1 091	1 170	682	800	821	1 025	1 079	1 065	1 508
Marshall Is	0	0	0	0	0	0	0	0	-	-
Micronesia	10 F	10 F	15 F	15 F	20 F	20 F	20 F	20 F	20 F	20 F
NewCaledonia	8	6	26	16	17	14	17	17	13	13 F
N Marianas	1	1	2	1	2	0	2	1	2	2
Palau	22	25	18	16	16	16	16	10 F	6 F	6 F
Papua N Guin	100 F	104	182	206	165	205	217	203	197	131
Philippines	655	896	877	559	522	421	214	249	250	269
Singapore	33	54	2	1	-	5	11	8	8	7
Wallis Fut I	...	...	2	2	2	2	2	2 F	2 F	2 F
71 Fishing area total	*3 611 F*	*2 931 F*	*3 178 F*	*3 318 F*	*3 072 F*	*4 578 F*	*3 641 F*	*4 289 F*	*4 400 F*	*5 131 F*
77 Amer Samoa	0	0	0	-	1	1	2	1	1	1
Costa Rica	3	1	4	5	7	7	3	4	14	17
El Salvador	7	-	-	-	-	-	-	-	3	2
Fr Polynesia	3	3	3	20	20	40	51	50	51	58
Guatemala	1	2	4	3	5	7	16	18	2	5 F
Honduras	3	1	1	2	2	9	2	1	-	-
Mexico	1 408	1 136	1 294	1 421	1 799	1 708	1 599	1 328	2 052	1 727
Panama	323	280	279	197	291	309	417	486	612	610 F
USA	418	208	256	297	395	501	350	287	361	324
77 Fishing area total	*2 166*	*1 631*	*1 841*	*1 945*	*2 520*	*2 582*	*2 440*	*2 175*	*3 096*	*2 744 F*
Species total	*10 780 F*	*8 894 F*	*9 749 F*	*9 479 F*	*9 490 F*	*11 300 F*	*10 064 F*	*10 296 F*	*11 280 F*	*12 237 F*
Cape rock lobster	**Langouste du Cap**		**Langosta del Cabo**		***Jasus lalandii***				**2,29(01)002,01**	**LBI**
47 Namibia	133	136	134	224	251	199	350	304	365	363
South Africa	2 077	2 176	2 192	1 956	1 516	1 680	1 726	1 793	1 693	1 611
47 Fishing area total	*2 210*	*2 312*	*2 326*	*2 180*	*1 767*	*1 879*	*2 076*	*2 097*	*2 058*	*1 974*
Species total	*2 210*	*2 312*	*2 326*	*2 180*	*1 767*	*1 879*	*2 076*	*2 097*	*2 058*	*1 974*
Juan Fernandez rock lobster	**Langouste Juan Fernandez**		**Langosta de Juan Fernández**		***Jasus frontalis***				**2,29(01)002,02**	**LOI**
87 Chile	12	30	24	29	36	32	21	22	17	21
87 Fishing area total	*12*	*30*	*24*	*29*	*36*	*32*	*21*	*22*	*17*	*21*
Species total	*12*	*30*	*24*	*29*	*36*	*32*	*21*	*22*	*17*	*21*
Green rock lobster	**Langouste d'Océanie**		**Langosta de Oceanía**		***Jasus verreauxi***				**2,29(01)002,03**	**LOI**
57 Australia	5 520	5 164	4 635	4 507	4 856	4 888	4 615	1 926	2 055	2 106
57 Fishing area total	*5 520*	*5 164*	*4 635*	*4 507*	*4 856*	*4 888*	*4 615*	*1 926*	*2 055*	*2 106*
71 Australia	...	...	...	...	...	...	...	...	269	...
71 Fishing area total	*...*	*...*	*...*	*...*	*...*	*...*	*...*	*...*	*269*	*...*
81 Australia	98	100	143	151	97	104	108	110	117	105
New Zealand	15	8	5	48	64	54	16	13	35	10
81 Fishing area total	*113*	*108*	*148*	*199*	*161*	*158*	*124*	*123*	*152*	*115*
Species total	*5 633*	*5 272*	*4 783*	*4 706*	*5 017*	*5 046*	*4 739*	*2 049*	*2 476*	*2 221*
Tristan da Cunha rock lobster	**Langouste de Tristan da Cunha**		**Langosta de Tristán da Cunha**		***Jasus tristani***				**2,29(01)002,05**	**LBT**
47 St Helena	361	362	321	344	327	321	376	336	316	425
47 Fishing area total	*361*	*362*	*321*	*344*	*327*	*321*	*376*	*336*	*316*	*425*
Species total	*361*	*362*	*321*	*344*	*327*	*321*	*376*	*336*	*316*	*425*
Red rock lobster	**...B**		**Langosta roja**		***Jasus edwardsii***				**2,29(01)002,06**	**LOI**
81 New Zealand	2 755	3 039	2 730	3 568	3 121	5 009	2 707	2 818	2 789	2 551
81 Fishing area total	*2 755*	*3 039*	*2 730*	*3 568*	*3 121*	*5 009*	*2 707*	*2 818*	*2 789*	*2 551*
Species total	*2 755*	*3 039*	*2 730*	*3 568*	*3 121*	*5 009*	*2 707*	*2 818*	*2 789*	*2 551*

B-43

Lobsters, spiny-rock lobsters — Capture production by species, fishing areas and countries or areas
Homards, langoustes — Captures par espèces, zones de pêche et pays ou zones
Bogavantes, langostas — Capturas por especies, áreas de pesca y países o áreas

pecies, Fishing area spèce, Zone de pêche specie, Area de pesca	1992 mt	1993 mt	1994 mt	1995 mt	1996 mt	1997 mt	1998 mt	1999 mt	2000 mt	2001 mt
t.Paul rock lobster	**Langouste de St.Paul**		**Langosta de St.Paul**		***Jasus paulensis***			**2,29(01)002,08**		**JSP**
51 Fr South Tr	395	403	443	439	357	295	308	345	192	183
51 Fishing area total	*395*	*403*	*443*	*439*	*357*	*295*	*308*	*345*	*192*	*183*
pecies total	*395*	*403*	*443*	*439*	*357*	*295*	*308*	*345*	*192*	*183*
ink spiny lobster	**Langouste rose**		**Langosta mora**		***Palinurus mauritanicus***			**2,29(01)008,02**		**PSL**
27 France	33	18	11	3	1	0	25	11	9	5
27 Fishing area total	*33*	*18*	*11*	*3*	*1*	*0*	*25*	*11*	*9*	*5*
37 France	2	0	0	0	0	0	0	0	-	-
37 Fishing area total	*2*	*0*	*0*	*0*	*0*	*0*	*0*	*0*	*-*	*-*
pecies total	*35*	*18*	*11*	*3*	*1*	*0*	*25*	*11*	*9*	*5*
ommon spiny lobster	**Langouste rouge**		**Langosta común**		***Palinurus elephas***			**2,29(01)008,04**		**SLO**
27 France	173	137	128	112	104	90	44	58	59	65
27 Fishing area total	*173*	*137*	*128*	*112*	*104*	*90*	*44*	*58*	*59*	*65*
37 Albania	...	...	...	...	...	...	...	...	...	1
Cyprus	2	1	3	2	5	2	4	2	4	5
France	31	44	33	8	9	9	9	3	3	3
Italy	300	211	198	197	312	331	174	161	123	166
Turkey	10	21	2	4	10	45	40	6	11	18
37 Fishing area total	*343*	*277*	*236*	*211*	*336*	*387*	*227*	*172*	*141*	*193*
pecies total	*516*	*414*	*364*	*323*	*440*	*477*	*271*	*230*	*200*	*258*
atal spiny lobster	**Langouste du Natal**		**Langosta del Natal**		***Palinurus delagoae***			**2,29(01)008,05**		**SLN**
51 South Africa	31	33	10	13	10	10	6	7 F	8	10
51 Fishing area total	*31*	*33*	*10*	*13*	*10*	*10*	*6*	*7 F*	*8*	*10*
pecies total	*31*	*33*	*10*	*13*	*10*	*10*	*6*	*7 F*	*8*	*10*
outhern spiny lobster	**Langouste du Sud**		**Langosta del Sur**		***Palinurus gilchristi***			**2,29(01)008,06**		**SLS**
47 South Africa	1 009	985	1 021	966	918	892	864	429	305	1 053
47 Fishing area total	*1 009*	*985*	*1 021*	*966*	*918*	*892*	*864*	*429*	*305*	*1 053*
pecies total	*1 009*	*985*	*1 021*	*966*	*918*	*892*	*864*	*429*	*305*	*1 053*
alinurid spiny lobsters nei	**Langoustes palinurus nca**		**Langostas palinurus nep**		***Palinurus spp***			**2,29(01)008,XX**		**CRW**
27 Channel Is	1	1 F	0	0	1	0	1	1	1	3
France	-	-	-	-	-	-	-	1	-	-
Ireland	166	108	...	84	62	48	46	35	41	35
Portugal	54	27	12	13	23	15	27	20	12	7
Spain	6	4	20	7	11	22	17	26	4	21
UK	72	50	59	67	7	42	19	15	15	15
27 Fishing area total	*299*	*190 F*	*91*	*171*	*104*	*127*	*110*	*98*	*73*	*81*
34 Benin	5 F	4 F	4 F	3 F	3 F	5 F	4 F	3	2 F	2
Cape Verde	70	50	40	40	25	17	18	23	19	13
Congo Rep	34 F	34 F	1	3 F	4 F	3 F	7 F	9 F	10	4 F
Côte dIvoire	2	0	0	0	2 F	4 F	5 F	6	10	2
Cyprus	-	-	-	-	-	-	-	-	0	1
Gabon	...	23	16	42	103	33	85	50	32	57
Gambia	22	125	44	26	84	37	86	80	130	75
Italy	-	14	14	-	-	-	-	-	-	-
Liberia	...	4	6	10	-	8	26	4	41	36
Mauritania	40 F	20 F	10 F	30 F	50 F	110 F	100 F	80 F	82 F	83 F
Morocco	10	3	5	13	20	58	31	38	32	394
Nigeria	1 233	1 049	1 751	1 496	1 133	8 315	3 071	1 245	1 939	1 699
Portugal	299	66	154	84	116	30	-	-	8	5
Senegal	96	103	787	121	130	196	144	39	37	53
Spain	-	-	-	-	9	...	...	13	3	46
34 Fishing area total	*1 811 F*	*1 495 F*	*2 832 F*	*1 868 F*	*1 679 F*	*8 816 F*	*3 577 F*	*1 590 F*	*2 345 F*	*2 470 F*
37 Algeria	20 F	25 F	30 F	22	9	44	141	70	70 F	70 F
Morocco	6	3	3	1	6	3	2	0	10	62
Spain	130 F	130 F	130 F	130 F	129	82	86	57	39	44
Tunisia	62	62	48	44	40	71	68	61	49	41
37 Fishing area total	*218 F*	*220 F*	*211 F*	*197 F*	*184*	*200*	*297*	*188*	*168 F*	*217 F*
ecies total	*2 328 F*	*1 905 F*	*3 134 F*	*2 236 F*	*1 967 F*	*9 143 F*	*3 984 F*	*1 876 F*	*2 586 F*	*2 768 F*
iny lobsters nei	**Langoustes diverses nca**		**Langostas diversas nep**		***Palinuridae***			**2,29(01)XXX,XX**		**VLO**
51 Eritrea	...	...	...	...	1	1	2	1	0	0
Mozambique	278	313	307	248	331	232	237	203	228	199
51 Fishing area total	*278*	*313*	*307*	*248*	*332*	*233*	*239*	*204*	*228*	*199*
ecies total	*278*	*313*	*307*	*248*	*332*	*233*	*239*	*204*	*228*	*199*

B-43 Lobsters, spiny-rock lobsters — Capture production by species, fishing areas and countries or areas
Homards, langoustes — Captures par espèces, zones de pêche et pays ou zones
Bogavantes, langostas — Capturas por especies, áreas de pesca y países o áreas

Species, Fishing area Espèce, Zone de pêche Especie, Area de pesca	1992 mt	1993 mt	1994 mt	1995 mt	1996 mt	1997 mt	1998 mt	1999 mt	2000 mt	2001 mt
Japanese fan lobster	**Cigale japonaise**		**Cigarra japonesa**		***Ibacus ciliatus***				**2,29(15)001,04**	**IBC**
61 China,Taiwan	514	456	236	1 224	1 115	642	696	676	1 600	1 607
61 Fishing area total	*514*	*456*	*236*	*1 224*	*1 115*	*642*	*696*	*676*	*1 600*	*1 607*
Species total	*514*	*456*	*236*	*1 224*	*1 115*	*642*	*696*	*676*	*1 600*	*1 607*
Flathead lobster	**Cigale raquette**		**Cigarra chata**		***Thenus orientalis***				**2,29(15)005,01**	**TH**
57 Thailand	134	199	225	335	299	177	533	37	36	36
57 Fishing area total	*134*	*199*	*225*	*335*	*299*	*177*	*533*	*37*	*36*	*36*
Species total	*134*	*199*	*225*	*335*	*299*	*177*	*533*	*37*	*36*	*36*
Slipper lobsters nei	**Cigales nca**		**Cigarros nep**		***Scyllaridae***				**2,29(15)XXX,XX**	**LO**
34 Senegal	-	-	-	-	-	-	1	2	4	6
34 Fishing area total	*-*	*-*	*-*	*-*	*-*	*-*	*1*	*2*	*4*	*6*
47 St Helena	0	0	0	0	0	0	1	1	1	0
47 Fishing area total	*0*	*0*	*0*	*0*	*0*	*0*	*1*	*1*	*1*	*0*
51 Bahrain	107	209	149	150	125	56	51	6	2	2
Qatar	35	33	23	8	8	16	15	13	5	20
South Africa	-	-	-	-	-	2	1	1 F	2	4
51 Fishing area total	*142*	*242*	*172*	*158*	*133*	*74*	*67*	*20 F*	*9*	*26*
57 Australia	0	0	18	13	21	0	0	0	0	-
57 Fishing area total	*0*	*0*	*18*	*13*	*21*	*0*	*0*	*0*	*0*	*-*
71 Australia	506	603	901	773	600	717	789	...	...	...
Philippines	160	342	938	350	334	8	65	89	90	105
Singapore	44	50	16	11	9	11	12	9	6	7
Thailand	931	1 236	861	1 730	2 716	2 785	3 005	1 760	1 736	1 725
71 Fishing area total	*1 641*	*2 231*	*2 716*	*2 864*	*3 659*	*3 521*	*3 871*	*1 858*	*1 832*	*1 837*
81 Australia	102	114	28	1	1	158	156	7	0	0
New Zealand	-	-	-	0	0	0	4	0	1	4
81 Fishing area total	*102*	*114*	*28*	*1*	*1*	*158*	*160*	*7*	*1*	*4*
Species total	*1 885*	*2 587*	*2 934*	*3 036*	*3 814*	*3 753*	*4 100*	*1 888 F*	*1 847*	*1 873*
Mozambique lobster	**Langoustine du Mozambique**		**Cigala del Mozambique**		***Metanephrops mozambicus***				**2,29(42)005,02**	**NE**
51 Mozambique	156	443	261	179	132	156	147	92	105	69
South Africa	70	83	-	-	-	-	45	60 F	75	72
51 Fishing area total	*226*	*526*	*261*	*179*	*132*	*156*	*192*	*152 F*	*180*	*141*
Species total	*226*	*526*	*261*	*179*	*132*	*156*	*192*	*152 F*	*180*	*141*
New Zealand lobster	**Langoustine de N.lle Zélande**		**Cigala de Nueva Zelandia**		***Metanephrops challengeri***				**2,29(42)005,04**	**ME**
81 New Zealand	1 035	926	1 067	1 078	670	1 093	989	925	1 034	1 093
81 Fishing area total	*1 035*	*926*	*1 067*	*1 078*	*670*	*1 093*	*989*	*925*	*1 034*	*1 093*
Species total	*1 035*	*926*	*1 067*	*1 078*	*670*	*1 093*	*989*	*925*	*1 034*	*1 093*
Norway lobster	**Langoustine**		**Cigala**		***Nephrops norvegicus***				**2,29(42)006,02**	**NE**
27 Belgium	384	418	306	413	188	316	240	349	254	284
Denmark	2 635	2 942	3 178	3 608	4 176	4 282	4 982	5 455	5 083	4 813
Faeroe Is	...	55	105	91	66	40	57	80	73	51
France	10 177	9 844	9 271	9 782	8 623	7 125	6 611	5 843	6 639	6 992
Germany	2	16	23	17	77	70	70	110	86	141
Iceland	2 230	2 381	2 238	1 027	1 623	1 215	1 411	1 389	1 230	1 420
Ireland	4 481	4 765	5 312	7 241	2 769	7 020	6 950	8 492	2 945	7 074
Isle of Man	14	32	14	29	20	24	17	10	3	2
Netherlands	133	130	158	253	423	627	694	662	572	853
Norway	230	211	234	166	188	187	293	383	346	281
Portugal	230	237	223	282	185	162	175	216	211	282
Spain	1 611	1 621	1 500	1 486	959	1 256	898	994	844	907
Sweden	780	864	764	914	1 105	1 130	1 317	1 263	1 232	1 067
UK	25 224	28 803	30 462	31 893	29 218	31 713	29 212	31 312	28 422	28 536
27 Fishing area total	*48 131*	*52 319*	*53 788*	*57 202*	*49 620*	*55 167*	*52 927*	*56 558*	*47 940*	*52 703*
34 Greece	0	5	1	-	1	-	10	-	2	-
Morocco	3	2	1	3	6	3	4	2	2	13
Portugal	2	0	0	0	-	-	12	42	78	88
Spain	-	-	-	-	303	261	271	1 100	328	3
34 Fishing area total	*5*	*7*	*2*	*3*	*310*	*264*	*297*	*1 144*	*410*	*104*
37 Albania	...	...	...	0	3	0	-	-	-	10
Algeria	120 F	130 F	150 F	129	100	89	138	96	100 F	100
Croatia	470	593	657	524	486	473	500	240	250	274
France	...	...	...	...	...	...	...	0	-	-
Greece	709	799	1 087	1 104	486	351	445	243	266	242
Italy	5 858	4 864	5 263	4 312	5 101	4 834	2 582	3 033	2 485	2 287
Morocco	1	0	1	0	1	1	1	1	3	2
Spain	530 F	515 F	500 F	485 F	473	411	478	448	407	630
Tunisia	-	1	15	2	3	1	4	2	4	4
Yugoslavia	1	10	5	13	5	5	7	5	7	9

B-43 Lobsters, spiny-rock lobsters — Capture production by species, fishing areas and countries or areas
Homards, langoustes — Captures par espèces, zones de pêche et pays ou zones
Bogavantes, langostas — Capturas por especies, áreas de pesca y países o áreas

Species, Fishing area Espèce, Zone de pêche Especie, Area de pesca	1992 mt	1993 mt	1994 mt	1995 mt	1996 mt	1997 mt	1998 mt	1999 mt	2000 mt	2001 mt
37 Fishing area total	*7 689 F*	*6 912 F*	*7 678 F*	*6 569 F*	*6 658*	*6 165*	*4 155*	*4 068*	*3 522 F*	*3 558 F*
Species total	*55 825 F*	*59 238 F*	*61 468 F*	*63 774 F*	*56 588*	*61 596*	*57 379*	*61 770*	*51 872 F*	*56 365 F*
American lobster	**Homard américain**		**Bogavante americano**			***Homarus americanus***			**2,29(42)007,01**	**LBA**
21 Canada	41 827	40 917	41 537	40 508	39 369	40 079	41 030	43 428	45 331	51 412
St Pier Mq	0	1	0	1	1	1	-	1	1	2
USA	25 307	25 634	30 126	30 122	32 496	38 066	36 125	39 676	37 730	32 389
21 Fishing area total	*67 134*	*66 552*	*71 663*	*70 631*	*71 866*	*78 146*	*77 155*	*83 105*	*83 062*	*83 803*
Species total	*67 134*	*66 552*	*71 663*	*70 631*	*71 866*	*78 146*	*77 155*	*83 105*	*83 062*	*83 803*
European lobster	**Homard européen**		**Bogavante**			***Homarus gammarus***			**2,29(42)007,18**	**LBE**
27 Belgium	-	-	-	3	-	-	-	1	1	6
Channel Is	128	129 F	123	114 F	260	221	214	212	153	178
Denmark	21	22	26	27	44	39	18	11	11	11
France	202	202	223	266	267	327	219	308	332	329
Germany	0	2	2	9	17	0	0	0	-	0
Ireland	552	470	824	564	567	513	611	597	533	781
Isle of Man	1	3	12	14	0	26	25	14	8	12
Netherlands	-	-	-	-	-	-	-	13	12	33
Norway	28	28	30	34	30	35	45	59	52	40
Portugal	5	4	3	2	3	2	2	1	2	2
Spain	14	22	17	18	21	14	12	11	7	8
Sweden	22	18	26	29	26	27	26	25	20	18
UK	1 093	1 045	1 162	1 306	1 043	1 534	1 482	1 819	1 141	1 086
27 Fishing area total	*2 066*	*1 945 F*	*2 448*	*2 386 F*	*2 278*	*2 738*	*2 654*	*3 071*	*2 272*	*2 504*
34 Greece	9	176	11	13	14	17	17	12	9	12
Morocco	3	1	6	7	19	12	14	7	11	11
Portugal	-	-	-	-	-	1	-	-	-	-
Spain	-	-	-	-	2	-	-	-	-	-
34 Fishing area total	*12*	*177*	*17*	*20*	*35*	*30*	*31*	*19*	*20*	*23*
37 Algeria	0	0	0	0	0	0	0	-	-	-
Croatia	32	23	43	30	31	43	40	18	18	13
France	2	3	1	0	0	0	0	0	-	-
Greece	134	118	320	510	198	357	138	158	192	221
Morocco	0	0	3	1	1	2	1	1	-	-
Spain	-	-	-	-	7	3	3	2	3	3
Tunisia	1	2	0	0	1	1	-	1	1	0
Turkey	11	8	18	33	34	40	60	10	15	10
Yugoslavia	1	0	1	1	5	5	6	5	6	7
37 Fishing area total	*181*	*154*	*386*	*575*	*277*	*451*	*248*	*195*	*235*	*254*
Species total	*2 259*	*2 276 F*	*2 851*	*2 981 F*	*2 590*	*3 219*	*2 933*	*3 285*	*2 527*	*2 781*
Blue mud shrimp	**...B**		**...C**			***Upogebia pugettensis***			**2,29(49)001,03**	**UOP**
67 USA	...	...	...	...	...	...	...	...	10	4
67 Fishing area total	...	...	...	...	...	...	...	...	*10*	*4*
Species total	...	...	...	...	...	...	...	...	*10*	*4*
Ghost shrimps	**...B**		**...C**			***Callianassa spp***			**2,29(59)001,XX**	**CZP**
67 USA	-	-	-	-	-	-	-	25	17	14
67 Fishing area total	-	-	-	-	-	-	-	*25*	*17*	*14*
Species total	-	-	-	-	-	-	-	*25*	*17*	*14*
Lobsters nei	**Langoustes, homards nca**		**Langostas nep**			***Reptantia***			**2,29(XX)XXX,XX**	**LOX**
27 Portugal	-	-	-	-	-	-	-	22	11	6
27 Fishing area total	-	-	-	-	-	-	-	*22*	*11*	*6*
37 France	-	-	-	-	-	-	-	0	2	2
37 Fishing area total	-	-	-	-	-	-	-	*0*	*2*	*2*
47 Portugal	-	-	-	9	9	25	17	67	10	-
Spain	-	-	-	-	11	-	14	1	-	-
47 Fishing area total	-	-	-	*9*	*20*	*25*	*31*	*68*	*10*	-
51 Portugal	-	-	-	-	-	-	-	-	62	2
51 Fishing area total	-	-	-	-	-	-	-	-	*62*	*2*
71 Viet Nam	...	...	...	...	...	2 000 F	1 000 F	200 F	500 F	700 F
71 Fishing area total	...	...	...	...	...	*2 000 F*	*1 000 F*	*200 F*	*500 F*	*700 F*
Species total	...	...	...	*9*	*20*	*2 025 F*	*1 031 F*	*290 F*	*585 F*	*710 F*
Group total	***205 899***	***206 632***	***217 287***	***220 151***	***209 590***	***233 317***	***216 165***	***229 426***	***227 043***	***225 136***

B-44 King crabs, squat-lobsters / Crabes royaux, galatées / Cangrejos reales, galateidos

Capture production by species, fishing areas and countries or areas / Captures par espèces, zones de pêche et pays ou zones / Capturas por especies, áreas de pesca y países o áreas

Species, Fishing area Espèce, Zone de pêche Especie, Area de pesca	1992 mt	1993 mt	1994 mt	1995 mt	1996 mt	1997 mt	1998 mt	1999 mt	2000 mt	2001 mt
Pelagic red crab	**Galatée pélagique**		**Langostino pelágico**		***Pleuroncodes planipes***			**2,30(19)001,01**		**LQL**
77 El Salvador	0	-	299	356	164	-	-	-	-	-
77 Fishing area total	*0*	*-*	*299*	*356*	*164*	*-*	*-*	*-*	*-*	*-*
Species total	*0*	*-*	*299*	*356*	*164*	*-*	*-*	*-*	*-*	*-*
Carrot squat lobster	**Galatée orange**		**Langostino colorado**		***Pleuroncodes monodon***			**2,30(19)001,02**		**PQ**
87 Chile	4 002	3 334	2 422	4 938	7 726	8 939	12 602	12 710	11 129	1 754
87 Fishing area total	*4 002*	*3 334*	*2 422*	*4 938*	*7 726*	*8 939*	*12 602*	*12 710*	*11 129*	*1 754*
Species total	*4 002*	*3 334*	*2 422*	*4 938*	*7 726*	*8 939*	*12 602*	*12 710*	*11 129*	*1 754*
Blue squat lobster	**Galatée bleue**		**Langostino amarillo**		***Cervimunida johni***			**2,30(19)003,01**		**CZJ**
87 Chile	3 736	2 224	4 842	5 743	6 402	10 322	9 426	7 273	5 069	2 178
87 Fishing area total	*3 736*	*2 224*	*4 842*	*5 743*	*6 402*	*10 322*	*9 426*	*7 273*	*5 069*	*2 178*
Species total	*3 736*	*2 224*	*4 842*	*5 743*	*6 402*	*10 322*	*9 426*	*7 273*	*5 069*	*2 178*
Craylets, squat lobsters	**Galatées**		**Camaroncillos,langostinos,etc.**		***Galatheidae***			**2,30(19)XXX,XX**		**LO**
27 France	127	104	89	85	90	94	81	102	98	84
UK	10	10	15	7	15	12	-	28	-	-
27 Fishing area total	*137*	*114*	*104*	*92*	*105*	*106*	*81*	*130*	*98*	*84*
87 Chile	-	-	-	-	-	-	-	-	254	-
87 Fishing area total	*-*	*-*	*-*	*-*	*-*	*-*	*-*	*-*	*254*	*-*
Species total	*137*	*114*	*104*	*92*	*105*	*106*	*81*	*130*	*352*	*84*
Red king crab	**Crabe royal du Kamtchatka**		**Cangrejo real rojo**		***Paralithodes camtschaticus***			**2,30(20)018,01**		**KC**
27 Norway	...	...	32	32	70	71	124	202	203	434
Russian Fed	-	-	-	-	-	-	71	-	78	252
UK	-	-	-	-	-	-	37	139	89	44
27 Fishing area total	*...*	*...*	*32*	*32*	*70*	*71*	*232*	*341*	*370*	*730*
61 Russian Fed	38 830	35 330	38 068	54 986	34 300	23 262	32 486	37 072	28 554	16 064
61 Fishing area total	*38 830*	*35 330*	*38 068*	*54 986*	*34 300*	*23 262*	*32 486*	*37 072*	*28 554*	*16 064*
Species total	*38 830*	*35 330*	*38 100*	*55 018*	*34 370*	*23 333*	*32 718*	*37 413*	*28 924*	*16 794*
Blue king crab	**Crabe royal bleu**		**Cangrejo real azul**		***Paralithodes platypus***			**2,30(20)018,02**		**KCI**
61 Russian Fed	...	...	...	...	8 762	10 268	4 508	5 455	5 233	4 500
61 Fishing area total	*...*	*...*	*...*	*...*	*8 762*	*10 268*	*4 508*	*5 455*	*5 233*	*4 500*
Species total	*...*	*...*	*...*	*...*	*8 762*	*10 268*	*4 508*	*5 455*	*5 233*	*4 500*
Brown king crab	**Crabe royal brun**		**Cangrejo real marrón**		***Paralithodes brevipes***			**2,30(20)018,03**		**KC**
61 Russian Fed	...	...	...	...	204	418	194	256	347	254
61 Fishing area total	*...*	*...*	*...*	*...*	*204*	*418*	*194*	*256*	*347*	*254*
Species total	*...*	*...*	*...*	*...*	*204*	*418*	*194*	*256*	*347*	*254*
King crabs	**Crabes royaux**		**Cangrejos rusos**		***Paralithodes spp***			**2,30(20)018,XX**		**KCS**
61 Japan	2 058	313	472	260	322	154	132	117	89	181
Korea Rep	11	9	-	-	-	-	-	-	-	-
61 Fishing area total	*2 069*	*322*	*472*	*260*	*322*	*154*	*132*	*117*	*89*	*181*
67 USA	8 644	11 218	5 425	6 656	9 526	8 177	10 941	7 675	6 848	7 281
67 Fishing area total	*8 644*	*11 218*	*5 425*	*6 656*	*9 526*	*8 177*	*10 941*	*7 675*	*6 848*	*7 281*
77 USA	-	-	-	-	-	-	-	-	1	2
77 Fishing area total	*-*	*-*	*-*	*-*	*-*	*-*	*-*	*-*	*1*	*2*
Species total	*10 713*	*11 540*	*5 897*	*6 916*	*9 848*	*8 331*	*11 073*	*7 792*	*6 938*	*7 464*
Southern king crab	**Crabe royal patagonien**		**Centolla patagónica**		***Lithodes antarcticus***			**2,30(20)070,01**		**KCF**
41 Argentina	143	158	295	380	200	413	456	270	102	58
Uruguay	1	2	0	0	0	0	0	0	0	0
41 Fishing area total	*144*	*160*	*295*	*380*	*200*	*413*	*456*	*270*	*102*	*58*
87 Chile	1 571	1 980	1 673	1 906	1 759	2 160	2 766	2 155	2 902	2 937
87 Fishing area total	*1 571*	*1 980*	*1 673*	*1 906*	*1 759*	*2 160*	*2 766*	*2 155*	*2 902*	*2 937*
Species total	*1 715*	*2 140*	*1 968*	*2 286*	*1 959*	*2 573*	*3 222*	*2 425*	*3 004*	*2 995*
Golden king crab	**Crabe royal doré**		**Centolla dorada**		***Lithodes aequispina***			**2,30(20)070,05**		**KAC**
61 Russian Fed	...	...	...	...	4 666	4 917	3 897	2 746	1 797	2 245
61 Fishing area total	*...*	*...*	*...*	*...*	*4 666*	*4 917*	*3 897*	*2 746*	*1 797*	*2 245*
Species total	*...*	*...*	*...*	*...*	*4 666*	*4 917*	*3 897*	*2 746*	*1 797*	*2 245*

B-44

King crabs, squat-lobsters — **Capture production by species, fishing areas and countries or areas**
Crabes royaux, galatées — **Captures par espèces, zones de pêche et pays ou zones**
Cangrejos reales, galateidos — **Capturas por especies, áreas de pesca y países o áreas**

Species, Fishing area Espèce, Zone de pêche Especie, Area de pesca	1992 mt	1993 mt	1994 mt	1995 mt	1996 mt	1997 mt	1998 mt	1999 mt	2000 mt	2001 mt
Softshell red crab	**Crabe royal hérisson**		**Centollón**			***Paralomis granulosa***			**2,30(20)123,01**	**PAG**
41 Argentina	...	...	...	...	...	...	...	...	266	155
Falkland Is	0	1	1	1	1	1	1	1	1	5
41 Fishing area total	*0*	*1*	*1*	*1*	*1*	*1*	*1*	*1*	*267*	*160*
87 Chile	1 326	955	2 221	1 316	1 273	1 477	1 501	1 438	4 938	6 527
87 Fishing area total	*1 326*	*955*	*2 221*	*1 316*	*1 273*	*1 477*	*1 501*	*1 438*	*4 938*	*6 527*
Species total	*1 326*	*956*	*2 222*	*1 317*	*1 274*	*1 478*	*1 502*	*1 439*	*5 205*	*6 687*
Antarctic stone crab	**Crabe royal de l'Antartique**		**Centolla antártica**			***Paralomis spinosissima***			**2,30(20)123,03**	**KCV**
48 UK	-	-	-	-	-	-	-	-	0	3
USA	-	299	-	-	497	-	-	-	-	-
48 Fishing area total	-	*299*	-	-	*497*	-	-	-	*0*	*3*
Species total	-	*299*	-	-	*497*	-	-	-	*0*	*3*
Globose king crab	**Crabe royal sphérique**		**Centolla redonda**			***Paralomis formosa***			**2,30(20)123,04**	**KCF**
48 UK	-	-	-	-	-	-	-	-	3	11
48 Fishing area total	-	-	-	-	-	-	-	-	*3*	*11*
Species total	-	-	-	-	-	-	-	-	*3*	*11*
King crabs, stone crabs nei	**Crabes royaux, etc. nca**		**Centollas, centollones nep**			***Lithodidae***			**2,30(20)XXX,XX**	**KCX**
58 South Africa	-	-	-	-	-	-	0	-	3	0
58 Fishing area total	-	-	-	-	-	-	*0*	-	*3*	*0*
Species total	-	-	-	-	-	-	*0*	-	*3*	*0*
Group total	***60 459***	***55 937***	***55 854***	***76 666***	***75 977***	***70 685***	***79 223***	***77 639***	***68 004***	***44 969***

B-45 Shrimps, prawns / Crevettes / Gambas, camarones

Capture production by species, fishing areas and countries or areas
Captures par espèces, zones de pêche et pays ou zones
Capturas por especies, áreas de pesca y países o áreas

Species, Fishing area Espèce, Zone de pêche Especie, Area de pesca	1992 mt	1993 mt	1994 mt	1995 mt	1996 mt	1997 mt	1998 mt	1999 mt	2000 mt	2001 mt
Northern brown shrimp	**Crevette royale grise**		**Camarón café norteño**		***Penaeus aztecus***			**2,28(01)001,01**		**ABS**
21 USA	312	187	-	19	666	3 377	-	679	1 104	458
21 Fishing area total	*312*	*187*	*-*	*19*	*666*	*3 377*	*-*	*679*	*1 104*	*458*
31 Costa Rica	0	0	0	4	8	0	0	0	0	-
USA	53 822	52 851	51 574	57 107	54 695	44 459	50 722	60 527	62 713	68 411
31 Fishing area total	*53 822*	*52 851*	*51 574*	*57 111*	*54 703*	*44 459*	*50 722*	*60 527*	*62 713*	*68 411*
Species total	*54 134*	*53 038*	*51 574*	*57 130*	*55 369*	*47 836*	*50 722*	*61 206*	*63 817*	*68 869*
Banana prawn	**Crevette banane**		**Langostino banana**		***Penaeus merguiensis***			**2,28(01)001,03**		**PBA**
57 Australia	...	...	...	...	...	...	...	517	329	281
Indonesia	6 964	6 953	9 092	9 852	10 318	10 073	10 382	9 569	10 087	10 650
Thailand	3 867	4 478	3 283	6 100	6 512	5 464	4 850	4 574	4 511	4 484
57 Fishing area total	*10 831*	*11 431*	*12 375*	*15 952*	*16 830*	*15 537*	*15 232*	*14 660*	*14 927*	*15 415*
71 Australia	2 508	4 058	2 433	4 490	4 347	4 546	3 711	3 608	2 222	6 286
Indonesia	40 762	36 972	38 145	40 625	43 596	43 851	51 810	54 610	56 557	59 720
Papua N Guin	800 F	813	628	858	820	676	1 233	949	1 136	1 017
Solomon Is	2 F	10 F	3 F	5 F	20 F	30 F	40 F	20 F	20 F	20 F
Thailand	7 336	8 130	9 108	9 225	7 133	7 169	7 648	8 206	8 094	8 044
71 Fishing area total	*51 408 F*	*49 983 F*	*50 317 F*	*55 203 F*	*55 916 F*	*56 272 F*	*64 442 F*	*67 393 F*	*68 029 F*	*75 087 F*
Species total	*62 239 F*	*61 414 F*	*62 692 F*	*71 155 F*	*72 746 F*	*71 809 F*	*79 674 F*	*82 053 F*	*82 956 F*	*90 502 F*
Yellowleg shrimp	**Crevette pattes jaunes**		**Camarón patiamarillo**		***Penaeus californiensis***			**2,28(01)001,04**		**YPS**
77 Costa Rica	18	12	10	8	6	2	12	12	6	2
El Salvador	110	-	-	-	-	-	-	-	-	-
Guatemala	231	137	393	275	266	124	259	218	132	200 F
77 Fishing area total	*359*	*149*	*403*	*283*	*272*	*126*	*271*	*230*	*138*	*202 F*
87 Ecuador	-	-	-	73	73	60	50	20	20	40
87 Fishing area total	*-*	*-*	*-*	*73*	*73*	*60*	*50*	*20*	*20*	*40*
Species total	*359*	*149*	*403*	*356*	*345*	*186*	*321*	*250*	*158*	*242 F*
Northern pink shrimp	**Crevette rose du Nord**		**Camarón rosado norteño**		***Penaeus duorarum***			**2,28(01)001,05**		**APS**
21 USA	124	13	-	-	10 753	114	-	5 062	16	1
21 Fishing area total	*124*	*13*	*-*	*-*	*10 753*	*114*	*-*	*5 062*	*16*	*1*
31 Cuba	2 262	2 738	2 229	1 851	1 710	2 001	1 734	2 943	2 940 F	2 940 F
USA	5 351	7 477	5 757	9 270	3 047	8 760	10 718	863	5 573	7 143
31 Fishing area total	*7 613*	*10 215*	*7 986*	*11 121*	*4 757*	*10 761*	*12 452*	*3 806*	*8 513 F*	*10 083 F*
Species total	*7 737*	*10 228*	*7 986*	*11 121*	*15 510*	*10 875*	*12 452*	*8 868*	*8 529 F*	*10 084 F*
Kuruma prawn	**Crevette kuruma**		**Langostino japonés**		***Penaeus japonicus***			**2,28(01)001,09**		**KUP**
37 Israel	-	-	-	-	-	84	...	...	...	...
37 Fishing area total	*-*	*-*	*-*	*-*	*-*	*84*	*...*	*...*	*...*	*...*
61 China,Taiwan	2 982	2 379	2 178	1 960	1 561	2 665	1 906	4 071	8 168	4 179
Japan	2 498	2 263	2 685	2 668	2 262	2 144	2 069	1 523	1 447	1 271
Korea Rep	2 311	2 531	2 265	2 399	1 783	2 102	1 138	502	578	513
61 Fishing area total	*7 791*	*7 173*	*7 128*	*7 027*	*5 606*	*6 911*	*5 113*	*6 096*	*10 193*	*5 963*
Species total	*7 791*	*7 173*	*7 128*	*7 027*	*5 606*	*6 995*	*5 113*	*6 096*	*10 193*	*5 963*
Whiteleg shrimp	**Crevette pattes blanches**		**Camarón patiblanco**		***Penaeus vannamei***			**2,28(01)001,11**		**PNV**
77 El Salvador	...	830	1 175	1 258	1 238	1 046	1 277	1 082	519	460
77 Fishing area total	*...*	*830*	*1 175*	*1 258*	*1 238*	*1 046*	*1 277*	*1 082*	*519*	*460*
87 Ecuador	13 900	14 048	10 000	6 000	4 497	4 000	3 000	1 100	1 100	2 250
87 Fishing area total	*13 900*	*14 048*	*10 000*	*6 000*	*4 497*	*4 000*	*3 000*	*1 100*	*1 100*	*2 250*
Species total	*13 900*	*14 878*	*11 175*	*7 258*	*5 735*	*5 046*	*4 277*	*2 182*	*1 619*	*2 710*
Giant tiger prawn	**Crevette géante tigrée**		**Langostino jumbo**		***Penaeus monodon***			**2,28(01)001,12**		**GIT**
51 India	109 703	133 063	184 476	140 028	112 342	106 622	167 904	168 942	145 857	136 443
Pakistan	124	157	138	132	141	140	122	138	139	140
51 Fishing area total	*109 827*	*133 220*	*184 614*	*140 160*	*112 483*	*106 762*	*168 026*	*169 080*	*145 996*	*136 583*
57 India	20 867	41 611	35 159	36 261	38 251	39 489	36 661	54 761	58 731	58 913
Indonesia	4 496	4 905	6 169	8 189	4 825	8 489	8 158	8 080	8 047	9 510
Thailand	396	536	153	799	1 993	1 837	871	1 209	1 192	1 185
57 Fishing area total	*25 759*	*47 052*	*41 481*	*45 249*	*45 069*	*49 815*	*45 690*	*64 050*	*67 970*	*69 608*
61 China,Taiwan	250	38	28	67	49	45	38	92	57	91
61 Fishing area total	*250*	*38*	*28*	*67*	*49*	*45*	*38*	*92*	*57*	*91*
71 Australia	4 906	3 477	3 339	4 368	3 841	3 055	3 686	3 477	2 726	2 697
Indonesia	11 153	11 211	10 791	16 312	14 571	17 440	21 889	26 143	32 940	38 920
Papua N Guin	150 F	144	123	137	109	99	209	164	126	117

B-45 Shrimps, prawns / Crevettes / Gambas, camarones

Capture production by species, fishing areas and countries or areas
Captures par espèces, zones de pêche et pays ou zones
Capturas por especies, áreas de pesca y países o áreas

Species, Fishing area Espèce, Zone de pêche Especie, Area de pesca	1992 mt	1993 mt	1994 mt	1995 mt	1996 mt	1997 mt	1998 mt	1999 mt	2000 mt	2001 mt
Philippines	...	...	1 216	431	360	292	422	169	232	328
Thailand	...	...	488	373	572	631	341	1 043	1 029	1 022
71 Fishing area total	*16 209 F*	*14 832*	*15 957*	*21 621*	*19 453*	*21 517*	*26 547*	*30 996*	*37 053*	*43 084*
Species total	*152 045 F*	*195 142*	*242 080*	*207 097*	*177 054*	*178 139*	*240 301*	*264 218*	*251 076*	*249 366*
Fleshy prawn	**Crevette charnue**		**Langostino carnoso**		***Penaeus chinensis***			**2,28(01)001,16**		**FLP**
61 China	37 852	16 582	45 770	43 043	55 292	69 406	78 350	69 911	84 334	97 002
Korea Rep	1 128	897	1 363	1 406	1 242	1 911	1 245	814	1 211	581
61 Fishing area total	*38 980*	*17 479*	*47 133*	*44 449*	*56 534*	*71 317*	*79 595*	*70 725*	*85 545*	*97 583*
Species total	*38 980*	*17 479*	*47 133*	*44 449*	*56 534*	*71 317*	*79 595*	*70 725*	*85 545*	*97 583*
Caramote prawn	**Caramote**		**Langostino**		***Penaeus kerathurus***			**2,28(01)001,17**		**TGS**
34 Greece	160	597	181	130	96	100	40	13	8	1
Italy	239	-	-	193	-	-	-	-	-	-
Spain	14	25 F	35 F	50 F	63	-	-	6	-	-
34 Fishing area total	*413*	*622 F*	*216 F*	*373 F*	*159*	*100*	*40*	*19*	*8*	*1*
37 Albania	...	...	...	30	8	4	18	18	20	23
France	...	...	...	...	...	...	...	...	4	6
Greece	395	473	494	737	1 165	1 554	1 459	1 103	1 440	1 502
Spain	150 F	150 F	150 F	150 F	150	53	196	97	63	64
Tunisia	1 556	1 657	2 287	3 587	3 057	4 367	3 785	4 729	6 165	3 356
37 Fishing area total	*2 101 F*	*2 280 F*	*2 931 F*	*4 504 F*	*4 380*	*5 978*	*5 458*	*5 947*	*7 692*	*4 951*
47 Italy	-	-	-	3	3	-	-	-	-	-
47 Fishing area total	*-*	*-*	*-*	*3*	*3*	*-*	*-*	*-*	*-*	*-*
Species total	*2 514 F*	*2 902 F*	*3 147 F*	*4 880 F*	*4 542*	*6 078*	*5 498*	*5 966*	*7 700*	*4 952*
Redspotted shrimp	**Crevette royale rose**		**Camarón rosado con manchas**		***Penaeus brasiliensis***			**2,28(01)001,19**		**PNB**
41 Brazil	6 163	6 150 F	6 160 F	6 565	8 743	10 758	7 796	9 092	10 728	10 600 F
41 Fishing area total	*6 163*	*6 150 F*	*6 160 F*	*6 565*	*8 743*	*10 758*	*7 796*	*9 092*	*10 728*	*10 600 F*
Species total	*6 163*	*6 150 F*	*6 160 F*	*6 565*	*8 743*	*10 758*	*7 796*	*9 092*	*10 728*	*10 600 F*
Green tiger prawn	**Crevette tigrée verte**		**Langostino tigre verde**		***Penaeus semisulcatus***			**2,28(01)001,20**		**TIP**
51 Qatar	3	0	1	0	0	-	0	0	0	0
51 Fishing area total	*3*	*0*	*1*	*0*	*0*	*-*	*0*	*0*	*0*	*0*
57 Thailand	441	673	760	1 909	2 427	2 449	2 180	2 212	2 182	2 168
57 Fishing area total	*441*	*673*	*760*	*1 909*	*2 427*	*2 449*	*2 180*	*2 212*	*2 182*	*2 168*
71 Thailand	325	323	642	711	1 042	1 068	839	609	600	597
71 Fishing area total	*325*	*323*	*642*	*711*	*1 042*	*1 068*	*839*	*609*	*600*	*597*
Species total	*769*	*996*	*1 403*	*2 620*	*3 469*	*3 517*	*3 019*	*2 821*	*2 782*	*2 765*
Northern white shrimp	**Crevette ligubam du Nord**		**Camarón blanco norteño**		***Penaeus setiferus***			**2,28(01)001,22**		**PST**
21 USA	94	96	-	-	1 347	913	-	619	313	28
21 Fishing area total	*94*	*96*	*-*	*-*	*1 347*	*913*	*-*	*619*	*313*	*28*
31 USA	39 814	32 264	36 838	39 959	27 461	31 928	39 799	44 014	52 280	40 696
31 Fishing area total	*39 814*	*32 264*	*36 838*	*39 959*	*27 461*	*31 928*	*39 799*	*44 014*	*52 280*	*40 696*
Species total	*39 908*	*32 360*	*36 838*	*39 959*	*28 808*	*32 841*	*39 799*	*44 633*	*52 593*	*40 724*
Crystal shrimp	**Crevette cristal**		**Camarón cristal**		***Penaeus brevirostris***			**2,28(01)001,23**		**CSP**
77 Costa Rica	1 140	974	1 028	526	408	392	276	548	350	456
El Salvador	426	615	955	860	397	318	483	177	80	153
Guatemala	90	72	84	58	33	18	21	19	16	20 F
Panama	2 300	2 150	1 990	2 232	1 298	2 048	1 310	1 718	2 158	2 100 F
77 Fishing area total	*3 956*	*3 811*	*4 057*	*3 676*	*2 136*	*2 776*	*2 090*	*2 462*	*2 604*	*2 729 F*
Species total	*3 956*	*3 811*	*4 057*	*3 676*	*2 136*	*2 776*	*2 090*	*2 462*	*2 604*	*2 729 F*
Western king prawn	**Crevette royale occidentale**		**Langostino marfil**		***Penaeus latisulcatus***			**2,28(01)001,28**		**WKP**
57 Thailand	357	739	385	1 475	1 674	1 724	1 501	1 518	1 497	1 488
57 Fishing area total	*357*	*739*	*385*	*1 475*	*1 674*	*1 724*	*1 501*	*1 518*	*1 497*	*1 488*
71 Australia	136	93	88	111	76	67	110	81	91	71
Thailand	1 624	1 399	1 304	1 486	1 600	1 838	1 775	2 546	2 511	2 496
71 Fishing area total	*1 760*	*1 492*	*1 392*	*1 597*	*1 676*	*1 905*	*1 885*	*2 627*	*2 602*	*2 567*
Species total	*2 117*	*2 231*	*1 777*	*3 072*	*3 350*	*3 629*	*3 386*	*4 145*	*4 099*	*4 055*
Western white shrimp	**Crevette royale blanche**		**Camarón blanco del Pacífico**		***Penaeus occidentalis***			**2,28(01)001,29**		**WWP**
87 Colombia	579	549	759	619	1 091	1 445	1 211	2 686	2 000 F	1 219
Ecuador	-	-	-	67	67	60	50	20	20	40
87 Fishing area total	*579*	*549*	*759*	*686*	*1 158*	*1 505*	*1 261*	*2 706*	*2 020 F*	*1 259*

B-45 Shrimps, prawns / Crevettes / Gambas, camarones

Capture production by species, fishing areas and countries or areas / Captures par espèces, zones de pêche et pays ou zones / Capturas por especies, áreas de pesca y países o áreas

Species, Fishing area Espèce, Zone de pêche Especie, Area de pesca	1992 mt	1993 mt	1994 mt	1995 mt	1996 mt	1997 mt	1998 mt	1999 mt	2000 mt	2001 mt
Species total	*579*	*549*	*759*	*686*	*1 158*	*1 505*	*1 261*	*2 706*	*2 020 F*	*1 259*
Redtail prawn	**Crevette queue rouge**		**Camarón rabo colorado**		***Penaeus penicillatus***				**2,28(01)001,30**	**REP**
61 China,Taiwan	2 892	2 538	2 071	3 564	5 020	2 473	647	316	308	312
61 Fishing area total	*2 892*	*2 538*	*2 071*	*3 564*	*5 020*	*2 473*	*647*	*316*	*308*	*312*
Species total	*2 892*	*2 538*	*2 071*	*3 564*	*5 020*	*2 473*	*647*	*316*	*308*	*312*
Southern pink shrimp	**Crevette rose du Sud**		**Camarón rosado sureño**		***Penaeus notialis***				**2,28(01)001,31**	**SOP**
34 Gabon	950 F	512	606	867	950	828	2 272	93	62	...
Gambia	...	...	559	367	339	...	399	400	301	211
Nigeria	11 944	13 755	8 595	14 742	12 073	14 799	12 396	27 341	18 882	18 805
Senegal	3 495	3 597	2 673	4 828	4 302	6 606	4 105	4 887	6 005	4 643
Spain	586	620 F	650 F	680 F	707	-	-	-	-	-
34 Fishing area total	*16 975 F*	*18 484 F*	*13 083 F*	*21 484 F*	*18 371*	*22 233*	*19 172*	*32 721*	*25 250*	*23 659*
Species total	*16 975 F*	*18 484 F*	*13 083 F*	*21 484 F*	*18 371*	*22 233*	*19 172*	*32 721*	*25 250*	*23 659*
Sao Paulo shrimp	**Crevette de Sao Paulo**		**Langostino de Sao Paulo**		***Penaeus paulensis***				**2,28(01)001,32**	**PPS**
41 Uruguay	0	0	0	0	-	177	13	12	56	23
41 Fishing area total	*0*	*0*	*0*	*0*	*-*	*177*	*13*	*12*	*56*	*23*
Species total	*0*	*0*	*0*	*0*	*-*	*177*	*13*	*12*	*56*	*23*
Penaeus shrimps nei	**Crevettes Penaeus nca**		**Langostinos Penaeus nep**		***Penaeus spp***				**2,28(01)001,XX**	**PEN**
01 Benin	3 011 F	3 800 F	3 739	3 117	3 000 F	2 850 F	2 700 F	2 574	2 100 F	2 400 F
01 Fishing area total	*3 011 F*	*3 800 F*	*3 739*	*3 117*	*3 000 F*	*2 850 F*	*2 700 F*	*2 574*	*2 100 F*	*2 400 F*
04 India	...	...	10 421	18 564	14 645	20 558	8 783	7 830	10 873	11 775
04 Fishing area total	*...*	*...*	*10 421*	*18 564*	*14 645*	*20 558*	*8 783*	*7 830*	*10 873*	*11 775*
27 Netherlands	-	-	-	-	-	-	-	1	1	3
Portugal	53	94	176	277	354	497	808	-	-	-
Spain	43	44	50	-	373	368	498	336	99	267
27 Fishing area total	*96*	*138*	*226*	*277*	*727*	*865*	*1 306*	*337*	*100*	*270*
31 Belize	48	28	34	49	38	43	40	35	45	69
Colombia	2 272	1 210	992	391	710	1 745	377	775	3 039 F	3 000 F
Costa Rica	1	2	2	0	0	27	43	61	65	83
Cuba	...	...	...	...	...	...	...	1	1 F	1 F
Dominican Rp	191	40	385	375	47	79	77	49	...	32
Fr Guiana	3 987	3 431	4 241	4 455	4 377	4 102	4 209	3 771	2 737	2 694
Guatemala	-	-	59	120	120	90	115	96	163	150 F
Guyana	147	558	708	400 F	84	79	1 935	1 595	1 132	1 698
Honduras	2 015	2 157	2 253	3 830	1 155	1 880	1 360	864	2 057	1 224
Jamaica	...	...	277	100 F	60	67	70	70	37	40 F
Japan	-	786	630	349	192	41	-	-	-	-
Korea Rep	834	907	750	653	560	143	136	502	-	-
Mexico	22 715	28 058	22 709	23 435	21 450	21 984	23 170	20 155	21 288	21 847
Nicaragua	1 114	1 718	2 632	2 330	3 049	3 148	3 787	3 560	3 630	4 130
Suriname	261	59	231	1 000 F	2 444	2 013	2 094	1 653	2 240 F	2 840
Trinidad Tob	1 550 F	1 296	946	700 F	285	751	648	658	755	856
USA	4 281	4 368	4 232	4 498	3 793	6 675	5 708	2 358	1 800	2 663
Venezuela	6 877	13 875	13 645	10 786	11 735	10 949	6 910	4 607	9 882	12 128
31 Fishing area total	*46 293 F*	*58 493*	*54 726*	*53 471 F*	*50 099*	*53 816*	*50 679*	*40 810*	*48 871 F*	*53 455 F*
34 Belize	-	-	-	-	-	0	549	1 283	1 191	41
Cameroon	489	462	500 F	514	442	450 F	635	326	471	168
Chile	-	-	3	-	-	-	-	-	-	-
China	30	1 758	249	2 372	2 113	2 260	2 504	1 997	1 873	2 087
Côte dIvoire	320	340	275	400	310 F	260 F	300 F	340 F	851	1
Gabon	-	-	-	-	-	556	190	987	1 682	1 892
Guinea	199	25	...	97	196	98	216	396	526	701
Honduras	210	162	20	106	173	193	128	86	79	-
Panama	1	118	-	28	8	39	-	-	-	-
St Vincent	15	-	-	1	-	-	-	-	-	-
Sierra Leone	570	573	169	868	973	2 147	1 690	2 155	1 663	1 280
Togo	0	1	1	1	12	2	2	0	2	0
Vanuatu	134	145	50	137	170	150	70	-	-	-
Other nei	30	-	-	-	-	-	-	19	31	-
34 Fishing area total	*1 998*	*3 584*	*1 267 F*	*4 524*	*4 397 F*	*6 155 F*	*6 284 F*	*7 589 F*	*8 369*	*6 170*
41 Brazil	30 896 F	25 320 F	25 380 F	27 132	15 007	14 506	13 998	12 550	16 509	16 300 F
Italy	317	286	290	23	30	-	-	-	-	-
Korea Rep	709	-	-	-	-	-	-	-	-	-
41 Fishing area total	*31 922 F*	*25 606 F*	*25 670 F*	*27 155*	*15 037*	*14 506*	*13 998*	*12 550*	*16 509*	*16 300 F*
51 Bahrain	755	2 128	1 185	1 662	3 565	2 571	2 530	1 622	2 104	1 359
Eritrea	...	...	15	13	2	0	9	75	519	790
Mozambique	7 997	8 897	7 608	8 615	8 183	9 825	8 559	8 846	9 460	9 479
Pakistan	5 273	6 663	5 883	5 591	5 982	5 975	5 189	5 874	5 920	5 974
Saudi Arabia	3 163	2 658	4 282	5 941	7 423	7 011	7 812	3 612	5 639	4 761
South Africa	6	13	-	-	-	127	178	180 F	168	249
Spain	-	-	-	-	-	-	-	-	219	166

B-45 Shrimps, prawns / Crevettes / Gambas, camarones

Capture production by species, fishing areas and countries or areas
Captures par espèces, zones de pêche et pays ou zones
Capturas por especies, áreas de pesca y países o áreas

Species, Fishing area Espèce, Zone de pêche Especie, Area de pesca	1992 mt	1993 mt	1994 mt	1995 mt	1996 mt	1997 mt	1998 mt	1999 mt	2000 mt	2001 mt
Yemen	89	619	722	984	665	547	904	686	526	1 600
51 Fishing area total	*17 283*	*20 978*	*19 695*	*22 806*	*25 820*	*26 056*	*25 181*	*20 895 F*	*24 555*	*24 378*
57 Australia	5 445	5 221	5 689	6 080	6 226	6 051	6 884	7 516	7 247	6 179
Thailand	11 073	11 010	11 087	10 185	11 807	12 272	9 452	10 965	10 815	10 749
57 Fishing area total	*16 518*	*16 231*	*16 776*	*16 265*	*18 033*	*18 323*	*16 336*	*18 481*	*18 062*	*16 928*
71 Australia	6 705	7 207	6 368	7 440	8 986	8 172	8 303	5 539	6 402	5 962
NewCaledonia	1	-	-	-	-	-	-	0	0	0
Philippines	14 351	14 608	11 567	11 309	10 462	11 508	10 796	11 255	11 063	11 869
Thailand	58 190	59 688	59 604	62 073	62 338	59 253	37 941	34 147	33 679	33 474
71 Fishing area total	*79 247*	*81 503*	*77 539*	*80 822*	*81 786*	*78 933*	*57 040*	*50 941*	*51 144*	*51 305*
77 Costa Rica	662	1 666	924	1 172	1 923	1 588	1 062	1 080	706	634
El Salvador	479	133	192	167	65	96	156	24	8	19
Fr Polynesia	0	...	...	...	...	...	...	...	...	...
Guatemala	634	733	376	557	284	506	1 597	1 103	880	1 000 F
Honduras	8	-	-	8	-	517	302	1	2 091	2 324
Mexico	35 174	40 280	41 134	46 599	44 114	49 083	43 416	46 336	40 309	35 662
Nicaragua	396	518	564	592	934	685	1 077	1 658	801	836
Panama	2 450	2 462	2 458	3 082	2 558	2 014	3 200	2 512	1 952	2 000 F
77 Fishing area total	*39 803*	*45 792*	*45 648*	*52 177*	*49 878*	*54 489*	*50 810*	*52 714*	*46 747*	*42 475 F*
81 Australia	2 273	2 100	2 312	2 153	1 822	1 926	1 683	2 399	2 647	2 588
81 Fishing area total	*2 273*	*2 100*	*2 312*	*2 153*	*1 822*	*1 926*	*1 683*	*2 399*	*2 647*	*2 588*
87 Ecuador	-	-	-	344	344	350	300	125	125	264
Peru	9 237	9 270	9 610	10 877	9 245	15 648	17 752	7 255	1 852	2 075
87 Fishing area total	*9 237*	*9 270*	*9 610*	*11 221*	*9 589*	*15 998*	*18 052*	*7 380*	*1 977*	*2 339*
Species total	*247 681 F*	*267 495 F*	*267 629 F*	*292 552 F*	*274 833 F*	*294 475 F*	*252 852 F*	*224 500 F*	*231 954 F*	*230 383 F*
Speckled shrimp	**Crevette mouchetée**		**Gamba moteada**			***Metapenaeus monoceros***			**2,28(01)016,01**	**MPN**
37 Tunisia	-	-	-	-	-	-	-	-	-	757
37 Fishing area total	-	-	-	-	-	-	-	-	-	*757*
Species total	-	-	-	-	-	-	-	-	-	*757*
Endeavour shrimp	**Crevette devo**		**Camarón devo**			***Metapenaeus endeavouri***			**2,28(01)016,06**	**ENS**
71 Australia	2 043	1 801	1 920	1 945	2 175	2 245	2 885	2 575	2 163	1 999
Philippines	-	-	192	237	225	94	45	116	136	174
71 Fishing area total	*2 043*	*1 801*	*2 112*	*2 182*	*2 400*	*2 339*	*2 930*	*2 691*	*2 299*	*2 173*
Species total	*2 043*	*1 801*	*2 112*	*2 182*	*2 400*	*2 339*	*2 930*	*2 691*	*2 299*	*2 173*
Shiba shrimp	**Crevette siba**		**Camarón siba**			***Metapenaeus joyneri***			**2,28(01)016,07**	**SHI**
61 Korea Rep	6 882	3 834	4 993	2 168	2 211	1 976	3 651	3 633	2 621	2 385
61 Fishing area total	*6 882*	*3 834*	*4 993*	*2 168*	*2 211*	*1 976*	*3 651*	*3 633*	*2 621*	*2 385*
Species total	*6 882*	*3 834*	*4 993*	*2 168*	*2 211*	*1 976*	*3 651*	*3 633*	*2 621*	*2 385*
Metapenaeus shrimps nei	**Crevettes Metapenaeus nca**		**Camarones Metapenaeus nep**			***Metapenaeus spp***			**2,28(01)016,XX**	**MET**
51 Pakistan	8 238	9 468	7 120	6 981	7 602	6 801	6 204	6 791	7 126	7 246
51 Fishing area total	*8 238*	*9 468*	*7 120*	*6 981*	*7 602*	*6 801*	*6 204*	*6 791*	*7 126*	*7 246*
57 Indonesia	4 273	4 175	6 350	9 437	7 573	10 309	9 698	11 419	10 092	11 310
Thailand	1 400	1 386	1 634	3 978	3 587	3 613	2 726	2 559	2 524	2 508
57 Fishing area total	*5 673*	*5 561*	*7 984*	*13 415*	*11 160*	*13 922*	*12 424*	*13 978*	*12 616*	*13 818*
61 China,Taiwan	1 068	794	780	3 767	626	369	299	210	202	241
61 Fishing area total	*1 068*	*794*	*780*	*3 767*	*626*	*369*	*299*	*210*	*202*	*241*
71 Indonesia	11 968	11 639	14 014	13 426	14 715	22 279	31 019	22 428	28 833	32 300
Papua N Guin	140 F	127	138	220	219	159	187	279	328	346
Philippines	4 695	3 497	5 135	6 054	5 431	5 420	6 545	6 419	5 987	6 167
Thailand	6 689	7 209	7 888	7 673	8 569	8 357	7 153	6 806	6 713	6 672
71 Fishing area total	*23 492 F*	*22 472*	*27 175*	*27 373*	*28 934*	*36 215*	*44 904*	*35 932*	*41 861*	*45 485*
Species total	*38 471 F*	*38 295*	*43 059*	*51 536*	*48 322*	*57 307*	*63 831*	*56 911*	*61 805*	*66 790*
Deepwater rose shrimp	**Crevette rose du large**		**Gamba de altura**			***Parapenaeus longirostris***			**2,28(01)017,01**	**DPS**
27 Portugal	-	-	-	-	-	-	-	2 084	1 360	1 080
27 Fishing area total	...	...	...	...	...	...	...	*2 084*	*1 360*	*1 080*
34 Estonia	-	-	-	-	-	-	-	-	3	6
Gabon	...	...	...	5	6	257	56	76	356	55
GuineaBissau	...	...	0	40	124	148	120 F	100 F	100 F	100 F
Liberia	...	...	...	...		8	...	...	...	...
Portugal	-	110	70	53	534	340	646	260	396	424
Russian Fed	118	40	61	-	-	-	-	-	-	-
Senegal	...	...	...	699	954	2 998	3 888	1 167	2 451	2 199
Spain	4 883	4 500 F	3 500 F	2 500 F	2 000	1 146	6 417	4 237	2 301	4 867
34 Fishing area total	*5 001*	*4 650 F*	*3 631 F*	*3 297 F*	*3 618*	*4 897*	*11 127 F*	*5 840 F*	*5 607 F*	*7 651 F*

B-45 Shrimps, prawns / Crevettes / Gambas, camarones

Capture production by species, fishing areas and countries or areas
Captures par espèces, zones de pêche et pays ou zones
Capturas por especies, áreas de pesca y países o áreas

Species, Fishing area Espèce, Zone de pêche Especie, Area de pesca	1992 mt	1993 mt	1994 mt	1995 mt	1996 mt	1997 mt	1998 mt	1999 mt	2000 mt	2001 mt
37 Albania	...	...	...	0	20	8	-	-	-	52
Algeria	700 F	800 F	1 000 F	539	918	1 433	1 639	2 147	2 100 F	2 100 F
Cyprus	4	1	3	3	2	1	1	5	4	2
France	3	2	5	3	4	17	17	1	1	1
Israel	30	28	50	60	50	...	...	...	...	...
Italy	16 367	10 731	11 262	7 998	7 065	7 019	4 410	4 631	7 500	6 980
Spain	1 300 F	1 200 F	1 100 F	1 000 F	881	1 000	1 222	1 463	839	504
Tunisia	1 208	599	354	213	161	641	838	1 014	1 283	1 454
37 Fishing area total	*19 612 F*	*13 361 F*	*13 774 F*	*9 816 F*	*9 101*	*10 119*	*8 127*	*9 261*	*11 727 F*	*11 093 F*
47 Angola	...	796	904	682	1 367	1 175	1 900	1 101	1 600	1 660
Portugal	-	-	-	-	-	-	-	1	-	-
Spain	3 000	2 427	3 625	2 034	2 112	3 071	1 635	789	2 609	3 966
47 Fishing area total	*3 000*	*3 223*	*4 529*	*2 716*	*3 479*	*4 246*	*3 535*	*1 891*	*4 209*	*5 626*
51 Spain	-	-	-	-	-	-	-	-	-	9
51 Fishing area total	-	-	-	-	-	-	-	-	-	*9*
Species total	*27 613 F*	*21 234 F*	*21 934 F*	*15 829 F*	*16 198*	*19 262*	*22 789 F*	*19 076 F*	*22 903 F*	*25 459 F*
Parapenaeopsis shrimps nei	**Crevettes parapenaeopsis nca**		**Camarones parapenaeopsis nep**		***Parapenaeopsis spp***			**2,28(01)019,XX**		**NPP**
51 Pakistan	12 693	18 632	16 023	12 919	14 047	16 722	14 689	12 889	11 945	11 576
51 Fishing area total	*12 693*	*18 632*	*16 023*	*12 919*	*14 047*	*16 722*	*14 689*	*12 889*	*11 945*	*11 576*
Species total	*12 693*	*18 632*	*16 023*	*12 919*	*14 047*	*16 722*	*14 689*	*12 889*	*11 945*	*11 576*
Atlantic seabob	**Crevette seabob(Atlantique)**		**Camarón siete barbas**		***Xiphopenaeus kroyeri***			**2,28(01)022,01**		**BOB**
31 Guyana	3 208	5 614	6 737	9 800 F	12 752	15 720	11 091	10 396	16 733	23 771
Suriname	-	-	-	-	-	-	2 046	4 537	4 540 F	4 550
Trinidad Tob	...	...	...	...	...	...	69	89	94	79
USA	5 661	5 016	3 969	1 724	4 558	5 744	3 397	3 626	3 415	3 951
31 Fishing area total	*8 869*	*10 630*	*10 706*	*11 524 F*	*17 310*	*21 464*	*16 603*	*18 648*	*24 782 F*	*32 351*
41 Brazil	4 860	4 850 F	4 860 F	7 278	11 764	15 257	13 755	9 964	11 948	11 800 F
41 Fishing area total	*4 860*	*4 850 F*	*4 860 F*	*7 278*	*11 764*	*15 257*	*13 755*	*9 964*	*11 948*	*11 800 F*
Species total	*13 729*	*15 480 F*	*15 566 F*	*18 802 F*	*29 074*	*36 721*	*30 358*	*28 612*	*36 730 F*	*44 151 F*
Pacific seabob	**Crevette seabob**		**Camarón botalón**		***Xiphopenaeus riveti***			**2,28(01)022,02**		**TIT**
87 Colombia	686	1 932	1 835	1 037	2 683	2 830	1 970	1 752	2 000 F	2 341
87 Fishing area total	*686*	*1 932*	*1 835*	*1 037*	*2 683*	*2 830*	*1 970*	*1 752*	*2 000 F*	*2 341*
Species total	*686*	*1 932*	*1 835*	*1 037*	*2 683*	*2 830*	*1 970*	*1 752*	*2 000 F*	*2 341*
Southern rough shrimp	**Crevette-archer**		**Camarón fijador arquero**		***Trachypenaeus curvirostris***			**2,28(01)043,02**		**TRV**
61 China	100 664	120 000	167 165	151 746	163 060	174 967	175 618	400 786	312 436	244 202
China,Taiwan	964	815	701	2 877	4 663	5 288	3 926	2 241	532	593
Korea Rep	-	-	-	-	-	-	-	-	-	3 145
61 Fishing area total	*101 628*	*120 815*	*167 866*	*154 623*	*167 723*	*180 255*	*179 544*	*403 027*	*312 968*	*247 940*
Species total	*101 628*	*120 815*	*167 866*	*154 623*	*167 723*	*180 255*	*179 544*	*403 027*	*312 968*	*247 940*
Argentine stiletto shrimp	**Crevette stylet d'Argentine**		**Camarón estilete argentino**		***Artemesia longinaris***			**2,28(01)067,01**		**ASH**
41 Argentina	380	185	215	250	263	166	146	37	37	...
41 Fishing area total	*380*	*185*	*215*	*250*	*263*	*166*	*146*	*37*	*37*	...
Species total	*380*	*185*	*215*	*250*	*263*	*166*	*146*	*37*	*37*	...
Pacific seabobs	**Crevettes seabob(Pacifique)**		**Camaroncillos**		***Xiphopenaeus,Trachypenaeus spp***			**2,28(01)XXX,XX**		**BOS**
77 Costa Rica	382	538	346	544	576	314	586	310	198	228
El Salvador	1 489	2 487	1 827	2 812	5 038	3 079	3 015	1 656	1 472	1 525
Guatemala	970	1 621	1 638	2 352	2 203	355	1 320	1 619	365	400 F
Panama	2 549	3 285	4 924	4 190	4 339	3 606	4 960	2 872	2 685	1 602
77 Fishing area total	*5 390*	*7 931*	*8 735*	*9 898*	*12 156*	*7 354*	*9 881*	*6 457*	*4 720*	*3 755 F*
Species total	*5 390*	*7 931*	*8 735*	*9 898*	*12 156*	*7 354*	*9 881*	*6 457*	*4 720*	*3 755 F*
Scarlet shrimp	**Gambon écarlate**		**Gamba carabinero**		***Plesiopenaeus edwardsianus***			**2,28(02)021,01**		**SSH**
34 Spain	571	500 F	450 F	400 F	352	479	337	505	14	39
34 Fishing area total	*571*	*500 F*	*450 F*	*400 F*	*352*	*479*	*337*	*505*	*14*	*39*
37 Spain	-	-	-	-	-	-	-	100	40	-
37 Fishing area total	-	-	-	-	-	-	-	*100*	*40*	-
Species total	*571*	*500 F*	*450 F*	*400 F*	*352*	*479*	*337*	*605*	*54*	*39*
Blue and red shrimp	**Crevette rouge**		**Gamba rosada**		***Aristeus antennatus***			**2,28(02)031,01**		**ARA**
27 Portugal	...	...	...	...	...	...	...	194	269	182
27 Fishing area total	...	...	...	...	...	...	...	*194*	*269*	*182*
37 Albania	...	...	...	0	3	0	-	-	-	-

B-45 Shrimps, prawns / Crevettes / Gambas, camarones

Capture production by species, fishing areas and countries or areas
Captures par espèces, zones de pêche et pays ou zones
Capturas por especies, áreas de pesca y países o áreas

Species, Fishing area Espèce, Zone de pêche Especie, Area de pesca	1992 mt	1993 mt	1994 mt	1995 mt	1996 mt	1997 mt	1998 mt	1999 mt	2000 mt	2001 mt
Algeria	1 400 F	1 500 F	1 600 F	1 305	1 230	1 186	1 485	880	900 F	900 F
Spain	-	-	-	-	-	-	-	414	772	1 418
Tunisia	334	69	30	28	195	173	187	38	16	41
37 Fishing area total	*1 734 F*	*1 569 F*	*1 630 F*	*1 333*	*1 428*	*1 359*	*1 672*	*1 332*	*1 688 F*	*2 359 F*
Species total	*1 734 F*	*1 569 F*	*1 630 F*	*1 333*	*1 428*	*1 359*	*1 672*	*1 526*	*1 957 F*	*2 541 F*
Striped red shrimp	**Gambon rayé**		**Gamba listada**			***Aristeus varidens***			**2,28(02)031,02**	**ARV**
47 Angola	590	331	364	538	814	543	876	820	1 154	1 200
Spain	928	1 065	1 101	1 032	764	780	1 932	1 007	1 870	2 205
47 Fishing area total	*1 518*	*1 396*	*1 465*	*1 570*	*1 578*	*1 323*	*2 808*	*1 827*	*3 024*	*3 405*
Species total	*1 518*	*1 396*	*1 465*	*1 570*	*1 578*	*1 323*	*2 808*	*1 827*	*3 024*	*3 405*
Aristeid shrimps nei	**Gambons,crevet. aristeidés nca**		**Gambas aristeidos nep**			***Aristeidae***			**2,28(02)XXX,XX**	**ARI**
37 Italy	...	...	...	2 551	2 258	2 406	1 231	2 128	4 463	1 833
37 Fishing area total	*...*	*...*	*...*	*2 551*	*2 258*	*2 406*	*1 231*	*2 128*	*4 463*	*1 833*
Species total	*...*	*...*	*...*	*2 551*	*2 258*	*2 406*	*1 231*	*2 128*	*4 463*	*1 833*
Coonstripe shrimp	**Crevette à front rayé**		**Camarón malacho**			***Pandalus hypsinotus***			**2,28(04)002,01**	**PYX**
61 Russian Fed	...	...	...	...	...	467	388	288	275	359
61 Fishing area total	*...*	*...*	*...*	*...*	*...*	*467*	*388*	*288*	*275*	*359*
Species total	*...*	*...*	*...*	*...*	*...*	*467*	*388*	*288*	*275*	*359*
Northern prawn	**Crevette nordique**		**Camarón norteño**			***Pandalus borealis***			**2,28(04)002,03**	**PRA**
21 Canada	24 189	24 968	28 985	30 213	31 340	48 310	78 867	85 331	100 091	94 328
Cuba	-	-	-	-	-	-	-	119	46	...
Denmark	3 142	897	245	447	403	421	556	235	-	93
Estonia	-	-	1 051	2 379	1 898	3 240	5 533	10 834	12 209	9 917
Faeroe Is	2 028	9 823	6 908	5 815	8 600	7 871	9 815	9 792	7 249	10 000 F
Greenland	79 270	74 124	75 907	77 832	67 441	60 079	66 034	74 563	80 888	81 398
Honduras	-	1 265	-	-	-	-	-	-	-	-
Iceland	-	2 195	2 355	7 481	20 680	7 197	6 572	9 147	8 832	5 065
Japan	1	-	-	-	-	-	-	-	-	-
Latvia	-	-	324	679	1 253	997	1 191	3 080	3 169	3 028
Lithuania	-	-	863	980	1 585	1 785	3 107	3 370	3 595	2 768
Norway	-	7 074	8 625	9 391	5 648	1 886	1 339	2 975	2 742	13 291
Portugal	-	-	-	16	-	170	203	227	289	420
Russian Fed	-	55	350	3 397	4 444	1 090	-	1 103	7 137	5 754
Spain	-	-	187	279	201	423	913	1 029	1 388	885
Ukraine	-	-	-	-	-	-	-	-	-	405
USA	3 377	4 321	3 718	7 416	9 932	7 239	3 926	2 832	2 625	1 309
21 Fishing area total	*112 007*	*124 722*	*129 518*	*146 325*	*153 425*	*140 708*	*178 056*	*204 637*	*230 260*	*228 661 F*
27 Denmark	5 336	4 688	4 016	8 013	8 668	8 131	7 544	4 645	5 721	5 238
Estonia	-	-	-	-	1 322	1 751	1 673	1 614	607	1 318
Faeroe Is	8 461	1 815	2 351	3 475	1 992	2 997	3 170	5 051	5 362	6 175
Germany	-	-	-	-	-	-	-	1 585	-	-
Greenland	2 656	2 337	3 924	4 094	4 545	3 853	3 536	4 615	4 514	4 444
Iceland	46 910	53 881	72 792	76 048	68 953	75 430	56 155	33 811	24 707	25 725
Ireland	...	4	...	...	2 409	-	-	-	1 116	-
Lithuania	-	-	-	-	-	-	233	797	2 781	2 645
Norway	49 098	41 883	29 543	29 859	35 857	40 075	55 802	60 563	63 466	53 045
Portugal	4	1	1	4	1	71	171	835	266	220
Russian Fed	20 944	22 397	7 108	3 564	5 747	1 493	4 895	10 765	19 596	5 846
Spain	-	-	-	-	739	1 006	453	267	906	798
Sweden	2 204	2 300	2 728	2 678	2 176	2 598	2 283	2 297	2 073	2 113
UK	117	545	77	1 541	1 996	417	595	1 815	539	996
27 Fishing area total	*135 730*	*129 851*	*122 540*	*129 276*	*134 405*	*137 822*	*136 510*	*128 660*	*131 654*	*108 563*
61 Russian Fed	-	-	-	-	-	1 786	3 127	5 096	8 520	8 457
61 Fishing area total	*-*	*-*	*-*	*-*	*-*	*1 786*	*3 127*	*5 096*	*8 520*	*8 457*
Species total	*247 737*	*254 573*	*252 058*	*275 601*	*287 830*	*280 316*	*317 693*	*338 393*	*370 434*	*345 681 F*
Aesop shrimp	**Crevette ésope**		**Camarón esópico**			***Pandalus montagui***			**2,28(04)002,05**	**AES**
21 Greenland	...	...	...	...	...	...	...	...	697	609
21 Fishing area total	*...*	*...*	*...*	*...*	*...*	*...*	*...*	*...*	*697*	*609*
Species total	*...*	*...*	*...*	*...*	*...*	*...*	*...*	*...*	*697*	*609*
Humpy shrimp	**Crevette gibbeuse**		**Camarón jiboso**			***Pandalus goniurus***			**2,28(04)002,07**	**DUJ**
61 Russian Fed	...	...	...	...	...	-	1 199	330	1 200	247
61 Fishing area total	*...*	*...*	*...*	*...*	*...*	*-*	*1 199*	*330*	*1 200*	*247*
Species total	*...*	*...*	*...*	*...*	*...*	*-*	*1 199*	*330*	*1 200*	*247*
Hokkai shrimp	**Crevette hokkai**		**Camarón de Hokkai**			***Pandalus kessleri***			**2,28(04)002,08**	**DUK**
61 Russian Fed	...	...	...	...	...	123	55	97	75	94
61 Fishing area total	*...*	*...*	*...*	*...*	*...*	*123*	*55*	*97*	*75*	*94*

B-45 Shrimps, prawns / Crevettes / Gambas, camarones

Capture production by species, fishing areas and countries or areas
Captures par espèces, zones de pêche et pays ou zones
Capturas por especies, áreas de pesca y países o áreas

Species, Fishing area Espèce, Zone de pêche Especie, Area de pesca	1992 mt	1993 mt	1994 mt	1995 mt	1996 mt	1997 mt	1998 mt	1999 mt	2000 mt	2001 mt
Species total	...	...	...	...	...	*123*	*55*	*97*	*75*	*94*
Pandalus shrimps nei	**Crevettes Pandalus nca**		**Camarones Pandalus nep**		***Pandalus spp***				**2,28(04)002,XX**	**PAN**
21 Canada	15 075	17 973	19 676	24 369	24 976	28 601	29 042	30 603	35 180	31 234
21 Fishing area total	*15 075*	*17 973*	*19 676*	*24 369*	*24 976*	*28 601*	*29 042*	*30 603*	*35 180*	*31 234*
27 UK	68	29	37	46	39	48	1 421	46	460	74
27 Fishing area total	*68*	*29*	*37*	*46*	*39*	*48*	*1 421*	*46*	*460*	*74*
61 Russian Fed	2 028	2 462	1 989	2 398	3 000	-	-	-	-	-
61 Fishing area total	*2 028*	*2 462*	*1 989*	*2 398*	*3 000*	-	-	-	-	-
Species total	*17 171*	*20 464*	*21 702*	*26 813*	*28 015*	*28 649*	*30 463*	*30 649*	*35 640*	*31 308*
Chilean nylon shrimp	**Crevette nylon chilienne**		**Camarón nailon**		***Heterocarpus reedi***				**2,28(04)005,01**	**CHS**
87 Chile	8 224	8 237	9 840	10 620	10 535	10 239	7 301	7 951	5 448	4 863
87 Fishing area total	*8 224*	*8 237*	*9 840*	*10 620*	*10 535*	*10 239*	*7 301*	*7 951*	*5 448*	*4 863*
Species total	*8 224*	*8 237*	*9 840*	*10 620*	*10 535*	*10 239*	*7 301*	*7 951*	*5 448*	*4 863*
Northern nylon shrimp	**Crevette nylon nordique**		**Camarón nailon norteño**		***Heterocarpus vicarius***				**2,28(04)005,08**	**HUV**
77 Nicaragua	-	-	-	-	-	-	40	37	20	131
77 Fishing area total	-	-	-	-	-	-	*40*	*37*	*20*	*131*
Species total	-	-	-	-	-	-	*40*	*37*	*20*	*131*
Morotoge shrimp	**Crevette morotoge**		**Camarón morotoje**		***Pandalopsis japonica***				**2,28(04)037,02**	**NDJ**
61 Russian Fed	...	...	...	...	...	12	86	35	11	36
61 Fishing area total	...	...	...	...	...	*12*	*86*	*35*	*11*	*36*
Species total	...	...	...	...	...	*12*	*86*	*35*	*11*	*36*
Pacific shrimps nei	**Crevettes océan Pacifique nca**		**Camarones Océano Pacífico nep**		***Pandalus spp, Pandalopsis spp***				**2,28(04)XXX,XX**	**PSH**
67 USA	37 289	24 058	16 539	13 591	15 030	19 131	6 268	14 015	16 158	18 740
67 Fishing area total	*37 289*	*24 058*	*16 539*	*13 591*	*15 030*	*19 131*	*6 268*	*14 015*	*16 158*	*18 740*
77 USA	184	880	1 287	1 539	1 707	1 288	622	636	372	309
77 Fishing area total	*184*	*880*	*1 287*	*1 539*	*1 707*	*1 288*	*622*	*636*	*372*	*309*
Species total	*37 473*	*24 938*	*17 826*	*15 130*	*16 737*	*20 419*	*6 890*	*14 651*	*16 530*	*19 049*
Akiami paste shrimp	**Chevrette akiami**		**Camaroncillo akiami**		***Acetes japonicus***				**2,28(07)009,03**	**AKS**
61 China	228 726	262 457	326 314	390 000	442 460	480 056	571 383	579 213	625 234	565 792
Korea Rep	29 348	24 324	18 510	16 495	18 411	15 624	15 993	19 389	13 985	11 705
61 Fishing area total	*258 074*	*286 781*	*344 824*	*406 495*	*460 871*	*495 680*	*587 376*	*598 602*	*639 219*	*577 497*
Species total	*258 074*	*286 781*	*344 824*	*406 495*	*460 871*	*495 680*	*587 376*	*598 602*	*639 219*	*577 497*
Sergestid shrimps nei	**Crevettes sergestid nca**		**Camarones sergéstidos nep**		***Sergestidae***				**2,28(07)XXX,XX**	**SHS**
57 Malaysia	...	17 310	14 654	12 799	14 057	13 241	9 684	7 893	9 094	7 026
Thailand	3 232	2 654	2 455	4 194	3 740	2 388	1 057	1 279	1 261	1 254
57 Fishing area total	*3 232*	*19 964*	*17 109*	*16 993*	*17 797*	*15 629*	*10 741*	*9 172*	*10 355*	*8 280*
71 Malaysia	1 149	2 968	5 012	4 855	6 770	1 999	1 652	8 587	1 354	1 341
Philippines	21 351	16 214	15 809	18 997	18 657	15 562	16 719	19 262	20 122	20 629
Thailand	20 751	19 354	21 375	19 532	18 387	15 100	14 315	6 381	6 294	6 255
71 Fishing area total	*43 251*	*38 536*	*42 196*	*43 384*	*43 814*	*32 661*	*32 686*	*34 230*	*27 770*	*28 225*
Species total	*46 483*	*58 500*	*59 305*	*60 377*	*61 611*	*48 290*	*43 427*	*43 402*	*38 125*	*36 505*
Whitebelly prawn	**Bouquet covac**		**Camarón cuac**		***Nematopalaemon schmitti***				**2,28(12)003,02**	**NLC**
31 Guyana	-	-	-	-	-	-	-	-	1 464	1 382
31 Fishing area total	-	-	-	-	-	-	-	-	*1 464*	*1 382*
Species total	-	-	-	-	-	-	-	-	*1 464*	*1 382*
Common prawn	**Bouquet commun**		**Camarón común**		***Palaemon serratus***				**2,28(12)018,10**	**CPR**
27 Denmark	223	248	632	195	144	182	122	141	151	141
France	410	345	562	381	311	288	213	296	307	326
Spain	-	-	-	-	80	91	79	66	47	41
UK	8	16	14	16	8	15	6	22	15	24
27 Fishing area total	*641*	*609*	*1 208*	*592*	*543*	*576*	*420*	*525*	*520*	*532*
37 Algeria	...	...	...	37	40	35	89	9	10 F	10 F
France	1	5	0	0	0	0	0	-	-	-
37 Fishing area total	*1*	*5*	*0*	*37*	*40*	*35*	*89*	*9*	*10 F*	*10 F*
Species total	*642*	*614*	*1 208*	*629*	*583*	*611*	*509*	*534*	*530 F*	*542 F*

B-45 Shrimps, prawns / Crevettes / Gambas, camarones

Capture production by species, fishing areas and countries or areas
Captures par espèces, zones de pêche et pays ou zones
Capturas por especies, áreas de pesca y países o áreas

Species, Fishing area Espèce, Zone de pêche Especie, Area de pesca	1992 mt	1993 mt	1994 mt	1995 mt	1996 mt	1997 mt	1998 mt	1999 mt	2000 mt	2001 mt
Palaemonid shrimps nei	**Crevettes palémonides nca**		**Camarones palemónidos nep**		*Palaemonidae*			**2,28(12)XXX,XX**		**PAL**
27 Ireland	165	269	314	312	399	358	505	551	4 047	268
Portugal	23	-	-	1	1	15	5	38	63	13
27 Fishing area total	*188*	*269*	*314*	*313*	*400*	*373*	*510*	*589*	*4 110*	*281*
Species total	*188*	*269*	*314*	*313*	*400*	*373*	*510*	*589*	*4 110*	*281*
Common shrimp	**Crevette grise**		**Quisquilla**		*Crangon crangon*			**2,28(23)003,03**		**CSH**
27 Belgium	965	981	1 223	1 567	903	743	378	1 053	616	988
Denmark	2 502	1 521	1 743	2 065	2 207	3 250	2 509	2 911	2 322	1 826
France	428	509	541	247	309	237	272	399	472	389
Germany	11 600	13 480	16 769	11 608	15 994	19 890	14 814	17 457	17 423	12 571
Netherlands	9 992	9 672	10 180	13 912	12 067	13 054	11 871	13 772	11 496	14 081
Portugal	1	0	0	1	0	0	-	1	3	0
Spain	-	-	-	-	-	0	-	0	-	1
UK	560	1 004	1 346	1 131	742	598	739	1 453	1 129	2 290
27 Fishing area total	*26 048*	*27 167*	*31 802*	*30 531*	*32 222*	*37 772*	*30 583*	*37 046*	*33 461*	*32 146*
37 France	15	7	0	0	0	0	0	0	-	2
Spain	290 F	270 F	250 F	230 F	222	150	100	87	-	-
37 Fishing area total	*305 F*	*277 F*	*250 F*	*230 F*	*222*	*150*	*100*	*87*	*-*	*2*
Species total	*26 353 F*	*27 444 F*	*32 052 F*	*30 761 F*	*32 444*	*37 922*	*30 683*	*37 133*	*33 461*	*32 148*
Sculptured shrimps nei	**Crevettes sculptées nca**		**Camarones esculpidos nep**		*Sclerocrangon spp*			**2,28(23)006,XX**		**CVL**
61 Russian Fed	...	...	...	...	...	38	59	82	20	45
61 Fishing area total	*...*	*...*	*...*	*...*	*...*	*38*	*59*	*82*	*20*	*45*
Species total	*...*	*...*	*...*	*...*	*...*	*38*	*59*	*82*	*20*	*45*
Crangonid shrimps nei	**Crevettes crangonidés nca**		**Camarones crangónidos nep**		*Crangonidae*			**2,28(23)XXX,XX**		**CRN**
77 USA	-	-	-	-	-	31	41	-	42	45
77 Fishing area total	*-*	*-*	*-*	*-*	*-*	*31*	*41*	*-*	*42*	*45*
Species total	*-*	*-*	*-*	*-*	*-*	*31*	*41*	*-*	*42*	*45*
Rock shrimp	**Boucot ovetgernade**		**Camarón de piedra**		*Sicyonia brevirostris*			**2,28(28)028,01**		**RSH**
21 USA	-	-	-	-	9 470	8	-	481	-	-
21 Fishing area total	*-*	*-*	*-*	*-*	*9 470*	*8*	*-*	*481*	*-*	*-*
31 USA	4 025	3 075	3 993	3 848	1 079	1 781	4 409	1 345	3 254	2 909
31 Fishing area total	*4 025*	*3 075*	*3 993*	*3 848*	*1 079*	*1 781*	*4 409*	*1 345*	*3 254*	*2 909*
Species total	*4 025*	*3 075*	*3 993*	*3 848*	*10 549*	*1 789*	*4 409*	*1 826*	*3 254*	*2 909*
Pacific rock shrimp	**Boucot du Pacifique**		**Camarón de piedra del Pacífico**		*Sicyonia ingentis*			**2,28(28)028,06**		**YII**
77 USA	...	...	...	...	...	...	185	630	756	165
77 Fishing area total	*...*	*...*	*...*	*...*	*...*	*...*	*185*	*630*	*756*	*165*
Species total	*...*	*...*	*...*	*...*	*...*	*...*	*185*	*630*	*756*	*165*
Argentine red shrimp	**Salicoque rouge d'Argentine**		**Camarón langostín argentino**		*Pleoticus muelleri*			**2,28(29)006,01**		**LAA**
41 Argentina	24 397	17 645	15 826	6 705	9 874	6 479	23 203	15 888	36 769	78 077
Uruguay	-	-	-	-	-	-	-	40	0	0
41 Fishing area total	*24 397*	*17 645*	*15 826*	*6 705*	*9 874*	*6 479*	*23 203*	*15 928*	*36 769*	*78 077*
Species total	*24 397*	*17 645*	*15 826*	*6 705*	*9 874*	*6 479*	*23 203*	*15 928*	*36 769*	*78 077*
Royal red shrimp	**Salicoque royale rouge**		**Camarón rojo real**		*Pleoticus robustus*			**2,28(29)006,03**		**RRS**
21 USA	-	-	-	-	14	4	13	78	22	24
21 Fishing area total	*-*	*-*	*-*	*-*	*14*	*4*	*13*	*78*	*22*	*24*
31 USA	203	297	290	252	184	205	182	208	369	281
31 Fishing area total	*203*	*297*	*290*	*252*	*184*	*205*	*182*	*208*	*369*	*281*
Species total	*203*	*297*	*290*	*252*	*198*	*209*	*195*	*286*	*391*	*305*
Kolibri shrimp	**Salicoque colibri**		**Camarón chupaflor**		*Solenocera agassizii*			**2,28(29)072,01**		**SOK**
87 Colombia	...	...	...	...	...	...	...	...	...	686
87 Fishing area total	*...*	*...*	*...*	*...*	*...*	*...*	*...*	*...*	*...*	*686*
Species total	*...*	*...*	*...*	*...*	*...*	*...*	*...*	*...*	*...*	*686*
Knife shrimp	**Salicoque couteau**		**Camarón navaja**		*Haliporoides triarthrus*			**2,28(29)073,01**		**KNS**
51 Mozambique	2 021	2 093	2 512	2 036	1 771	1 510	1 882	1 611	1 766	1 738
51 Fishing area total	*2 021*	*2 093*	*2 512*	*2 036*	*1 771*	*1 510*	*1 882*	*1 611*	*1 766*	*1 738*
Species total	*2 021*	*2 093*	*2 512*	*2 036*	*1 771*	*1 510*	*1 882*	*1 611*	*1 766*	*1 738*

B-45 Shrimps, prawns / Crevettes / Gambas, camarones

Capture production by species, fishing areas and countries or areas
Captures par espèces, zones de pêche et pays ou zones
Capturas por especies, áreas de pesca y países o áreas

Species, Fishing area Espèce, Zone de pêche Especie, Area de pesca	1992 mt	1993 mt	1994 mt	1995 mt	1996 mt	1997 mt	1998 mt	1999 mt	2000 mt	2001 mt
Chilean knife shrimp	**Salicoque couteau du Chili**		**Camarón cuchilla**		***Haliporoides diomedeae***			**2,28(29)073,03**		**HJD**
87 Chile	6	5	20	5	15	32	29	135	169	309
87 Fishing area total	*6*	*5*	*20*	*5*	*15*	*32*	*29*	*135*	*169*	*309*
Species total	*6*	*5*	*20*	*5*	*15*	*32*	*29*	*135*	*169*	*309*
Knife shrimps nei	**Salicoques-couteau nca**		**Camarones navaja nep**		***Haliporoides spp***			**2,28(29)073,XX**		**KNI**
51 South Africa	199	218	-	-	-	-	-	-	-	-
51 Fishing area total	*199*	*218*	-	-	-	-	-	-	-	-
Species total	*199*	*218*	-	-	-	-	-	-	-	-
Natantian decapods nei	**Décapodes natantia nca**		**Decápodos natantia nep**		***Natantia***			**2,28(XX)XXX,XX**		**DCP**
04 China,Taiwan	77	76	33	38	22	12	-	-	-	-
India	...	...	6 083	7 746	13 186	17 396	7 836	7 630	15 835	17 625
Indonesia	6 338	6 323	8 965	6 768	6 565	6 330	7 281	7 905	7 233	7 440
04 Fishing area total	*6 415*	*6 399*	*15 081*	*14 552*	*19 773*	*23 738*	*15 117*	*15 535*	*23 068*	*25 065*
21 Japan	-	-	-	-	-	-	...	...	...	130
Poland	-	-	-	-	-	824	148	894	1 732	263
21 Fishing area total	-	-	-	-	-	*824*	*148*	*894*	*1 732*	*393*
27 France	3	1	3	4	3	0	0	10	19	25
Portugal	45	54	50	196	204	233	382	30	70	84
Spain	323	813	677	1 141	1 030	13 494	5 563	3 814	4 990	1 842
UK	0	0	3	2	2	-	-	-	-	-
27 Fishing area total	*371*	*868*	*733*	*1 343*	*1 239*	*13 727*	*5 945*	*3 854*	*5 079*	*1 951*
31 Colombia	366	252	17	518	139	-	-	-	-	-
Haiti	100 F	120 F	110 F	150 F	150 F	150 F	150 F	150 F	150 F	150 F
Japan	1 162	-	-	-	-	-	-	-	-	-
Korea Rep	-	-	-	-	-	-	-	67	1 099	68
Lithuania	61	-	-	-	-	-	-	-	-	-
31 Fishing area total	*1 689 F*	*372 F*	*127 F*	*668 F*	*289 F*	*150 F*	*150 F*	*217 F*	*1 249 F*	*218 F*
34 Belize	-	-	-	-	-	-	2	6	6	11
Benin	0	0	0	0	0	0	0	31	20 F	3
Cayman Is	700	320	...	...	...	...	...	...	...	...
China	2	-	-	-	-	-	-	-	-	-
Congo Rep	326 F	325 F	23	322	584	320 F	420 F	374	529	400 F
Egypt	-	-	-	-	-	-	-	4	-	-
Ghana	2 236	1 148	1 507	2 228	1 554	1 602	1 448	87	1 446	1 361
Greece	1 312	1 534	932	884	951	746	563	573	553	511
GuineaBissau	1 048 F	1 060 F	1 040 F	1 070 F	1 100 F	1 100 F	900 F	800 F	800 F	800 F
Honduras	1	-	-	-	-	0	-	-	-	-
Italy	2 140	2 149	2 179	-	254	345	456	493	370	686
Korea D P Rp	-	-	-	-	-	-	-	-	52	-
Korea Rep	13	1	13	-	-	4	5	16	4	916
Latvia	-	-	-	-	-	-	-	-	-	19
Liberia	66	205	116	110	28	73	113	302	25	31
Lithuania	12	-	-	-	-	-	-	-	11	-
Mauritania	100 F	110 F	120 F	130 F	160 F	190 F	270 F	222	779	1 378
Morocco	5 626	6 387	8 425	8 947	8 950	7 268	9 860	8 905	11 608	7 259
Nigeria	1 557	2 612	2 207	4 474	3 417	3 456	7 364	2 690	1 564	909
Panama	0	-	-	-	4	2	-	-	-	-
Portugal	0	0	561	688	1 069	446	421	181	164	255
St Vincent	0	-	-	-	-	-	-	-	-	-
Senegal	9	51	-	-	-	-	-	-	-	19
Spain	1 097	2 000 F	4 000 F	6 000 F	12 842	3 790	24 723	9 003	2 160	5 661
Vanuatu	-	-	-	-	-	-	0	-	-	-
Other nei	18	-	-	-	-	-	-	-	17	-
34 Fishing area total	*16 263 F*	*17 902 F*	*21 123 F*	*24 853 F*	*30 913 F*	*19 342 F*	*46 545 F*	*23 687 F*	*20 108 F*	*20 219 F*
37 Bulgaria	0	0	0	1	1	3	2	2	-	-
Egypt	2 656	3 818	3 959	3 971	3 164	4 851	5 071	7 099	4 408	3 668
France	-	-	-	-	-	-	-	5	2	1
Gaza Strip	...	...	...	50	55	161	161	160 F	160 F	130 F
Greece	1 823	1 371	1 319	980	1 327	1 724	1 245	1 239	1 085	844
Israel	117	110	30	200	200	221	225	186	184	170 F
Malta	4	6	4	5	9	16	18	24	23	36
Morocco	136	83	74	196	388	8	910	935	1 044	326
Russian Fed	-	-	-	-	-	2	-	-	1	3
Spain	650 F	900 F	1 200 F	1 500 F	1 710	1 781	1 815	2	3	8
Turkey	3 299	4 275	3 437	1 976	1 100	1 380	1 400	890	2 000	3 000
Ukraine	...	...	...	...	1	1	1	1	1	...
37 Fishing area total	*8 685 F*	*10 563 F*	*10 023 F*	*8 879 F*	*7 955*	*10 148*	*10 848*	*10 543 F*	*8 911 F*	*8 186 F*
41 Korea Rep	-	-	-	-	-	-	-	1 090	556	2 750
Spain	-	-	-	-	-	-	-	-	-	66
41 Fishing area total	-	-	-	-	-	-	-	*1 090*	*556*	*2 816*
47 Angola	598	387	676	-	-	-	-	0	-	-
Panama	0	-	-	-	-	-	-	-	-	-
Poland	-	-	-	-	-	-	543	-	-	-
Spain	-	-	-	-	855	-	9 344	1 615	3 188	4 298
47 Fishing area total	*598*	*387*	*676*	-	*855*	-	*9 887*	*1 615*	*3 188*	*4 298*

B-45 Shrimps, prawns / Crevettes / Gambas, camarones

Capture production by species, fishing areas and countries or areas
Captures par espèces, zones de pêche et pays ou zones
Capturas por especies, áreas de pesca y países o áreas

	Species, Fishing area Espèce, Zone de pêche Especie, Area de pesca	1992 mt	1993 mt	1994 mt	1995 mt	1996 mt	1997 mt	1998 mt	1999 mt	2000 mt	2001 mt
51	Comoros	10	10	10	15 F	20 F	20 F	20 F	20 F	20 F	20
	Egypt	435	614	572	763	644	639	436	1 170	2 655	1 704
	India	99 547	86 919	95 120	85 016	113 259	90 528	97 570	90 957	90 734	86 882
	Iran	6 469	6 985	6 694	6 864	5 837	7 620	5 774	4 570	9 850	6 940
	Italy	476	429	436	40	52	-	-	-	-	-
	Kenya	388	208	379	207	378	491	774	513	458	690
	Korea Rep	406	44	-	-	165	49	29	-	-	279
	Kuwait	3 619	2 810	2 093	1 739	2 358	2 066	1 598	720	1 300 F	1 977
	Lithuania	-	-	11	-	-	-	-	-	-	-
	Madagascar	9 439	8 869	12 279	9 919	10 470	10 755	11 470	10 507	12 127	11 776
	Mauritius	23	22	21	1	0	1	0	1	1	1
	Oman	155	380	451	340	276	376	65	356	432	627
	Portugal	-	-	-	-	-	-	-	-	145	735
	Russian Fed	380	-	-	-	-	-	-	-	-	-
	Spain	-	-	-	-	-	-	-	164	-	-
	Tanzania	2 200	1 829	1 417	2 260	2 664	2 500	2 800	2 100	2 100	2 000
	Yemen	101	65	...	...	...	...	38	7	44	151
51	*Fishing area total*	*123 648*	*109 184*	*119 483*	*107 164 F*	*136 123 F*	*115 045 F*	*120 574 F*	*111 085 F*	*119 866 F*	*113 782*
57	India	20 246	29 167	24 174	20 984	28 736	26 428	15 638	22 077	21 830	23 404
	Indonesia	13 623	13 016	17 726	17 403	14 455	18 150	17 979	19 934	17 333	17 880
	Korea Rep	-	-	-	-	-	-	-	1	-	1
	Malaysia	85 642	55 591	44 007	43 079	45 517	47 599	20 826	41 756	50 237	39 730
	Myanmar	10 000 F	12 000 F	15 000 F	20 000 F	16 000 F	22 000 F	24 000 F	27 000 F	30 000 F	30 000 F
	Timor-Leste	...	...	...	...	...	...	...	1 F	1	1
57	*Fishing area total*	*129 511 F*	*109 774 F*	*100 907 F*	*101 466 F*	*104 708 F*	*114 177 F*	*78 443 F*	*110 769 F*	*119 401 F*	*111 016 F*
61	China,H.Kong	10 065	8 621	7 549	7 155	6 589	6 225	5 285	3 800 F	4 600 F	5 100 F
	China, Macao	716	406	266	181	218	230 F	230 F	230 F	230 F	230 F
	China,Taiwan	24 674	16 572	16 167	24 108	23 748	22 100	15 702	14 943	11 336	11 987
	Japan	40 103	32 756	34 275	31 764	28 641	27 114	25 283	25 630	25 898	24 276
	Korea Rep	24 739	34 739	29 464	18 954	16 086	17 621	24 503	16 861	15 751	8 077
	Russian Fed	-	-	-	-	-	-	-	16	91	80
61	*Fishing area total*	*100 297*	*93 094*	*87 721*	*82 162*	*75 282*	*73 290 F*	*71 003 F*	*61 480 F*	*57 906 F*	*49 750 F*
67	Canada	3 851	4 498	4 502	8 557	9 392	5 200	5 203	4 071	4 223	4 212
67	*Fishing area total*	*3 851*	*4 498*	*4 502*	*8 557*	*9 392*	*5 200*	*5 203*	*4 071*	*4 223*	*4 212*
71	Brunei Darsm	348	299	1 208	272	275	0	0	0	0	0
	Cambodia	4 593	4 500	4 000	4 200	4 300	4 110	4 440	5 250	5 000	5 000 F
	Guam	-	-	-	-	-	-	1	0	0	0
	Indonesia	69 838	66 698	73 426	63 858	74 760	77 640	69 221	83 438	81 547	84 120
	Korea Rep	86	36	49	2	45	94	51	124	138	380
	Malaysia	39 614	29 892	36 872	32 053	33 893	28 606	15 069	32 238	35 291	29 371
	Marshall Is	0	0	0	0	0	0	0	0	-	-
	Micronesia	0	0	0	0	0	0	0	0	0	0
	N Marianas	-	-	-	1	-	-	-	-	-	-
	Papua N Guin	100 F	60	94	84	104	59	50	99	135	117
	Singapore	703	728	908	767	857	706	621	522	422	250
	Viet Nam	48 824	55 170	66 981	82 758	86 166	98 401	93 541	91 500	81 700 F	90 000 F
71	*Fishing area total*	*164 106 F*	*157 383*	*183 538*	*183 995*	*200 400*	*209 616*	*182 994*	*213 171*	*204 233 F*	*209 238 F*
77	Costa Rica	776	1 015	2 356	2 118	1 362	1 022	672	1 004	1 009	508
	Japan	153	-	57	-	-	-	-	-	-	-
	Korea Rep	61	360	88	12	-	-	-	491	74	-
	Panama	263	748	1 697	1 945	4 118	5 252	902	1 674	1 661	1 650 F
77	*Fishing area total*	*1 253*	*2 123*	*4 198*	*4 075*	*5 480*	*6 274*	*1 574*	*3 169*	*2 744*	*2 158 F*
81	Japan	-	3	-	-	-	-	-	-	-	5
	Korea Rep	-	-	-	-	-	-	-	11	-	-
	New Zealand	-	-	-	1	2	23	1	-	-	0
81	*Fishing area total*	*-*	*3*	*-*	*1*	*2*	*23*	*1*	*11*	*-*	*5*
87	Colombia	639	113	282	369	317	559	345	555	570 F	591
	Korea Rep	-	-	-	-	-	-	-	37	18	-
87	*Fishing area total*	*639*	*113*	*282*	*369*	*317*	*559*	*345*	*592*	*588 F*	*591*
	Species total	*557 326 F*	*512 663 F*	*548 394 F*	*538 084 F*	*592 728 F*	*592 113 F*	*548 777 F*	*561 783 F*	*572 852 F*	*553 898 F*
	Group total	***2 104 599***	***2 148 437***	***2 363 142***	***2 437 418***	***2 554 183***	***2 628 466***	***2 749 225***	***3 021 115***	***3 074 771***	***2 950 834***

B-46 Krill, planktonic crustaceans — Capture production by species, fishing areas and countries or areas
Krill, crustacés planctoniques — Captures par espèces, zones de pêche et pays ou zones
Krill, crustáceos planctónicos — Capturas por especies, áreas de pesca y países o áreas

Species, Fishing area Espèce, Zone de pêche Especie, Area de pesca	1992 mt	1993 mt	1994 mt	1995 mt	1996 mt	1997 mt	1998 mt	1999 mt	2000 mt	2001 mt
Antarctic krill	**Krill antarctique**		**Krill antártico**		***Euphausia superba***				**2,26(01)005,01**	**KRI**
41 Japan	-	-	-	-	-	-	-	-	4	-
Poland	-	2 506	-	-	-	-	74	-	-	-
41 Fishing area total	*-*	*2 506*	*-*	*-*	*-*	*-*	*74*	*-*	*4*	*-*
47 Poland	-	-	-	-	-	-	254	-	-	-
47 Fishing area total	*-*	*-*	*-*	*-*	*-*	*-*	*254*	*-*	*-*	*-*
48 Argentina	-	-	-	-	-	-	-	6 524	-	-
Chile	6 066	3 261	3 834	-	-	-	-	-	-	-
Estonia	2 352	-	-	-	-	-	-	-	-	-
Japan	74 275	53 510	61 423	59 037	60 546	58 798	63 233	71 318	67 188	73 523
Korea Rep	519	-	-	-	-	-	1 621	1 228	5 444	5 125
Latvia	-	-	71	-	-	-	-	-	-	-
Panama	-	-	-	141	496	-	-	-	-	-
Poland	8 607	13 406	7 915	9 384	20 610	19 156	15 386	18 554	20 721	14 568
Russian Fed	151 725	4 249	965	-	-	-	-	-	-	-
South Africa	-	-	3	-	-	-	-	-	-	-
Ukraine	61 719	6 083	8 852	48 886	20 056	4 246	-	5 694	985	3 362
UK	-	-	-	-	-	308	634	-	-	-
USA	-	-	-	-	-	-	-	-	-	1 631
Uruguay	-	-	-	-	-	-	-	-	9 921	-
48 Fishing area total	*305 263*	*80 509*	*83 063*	*117 448*	*101 708*	*82 508*	*80 874*	*103 318*	*104 259*	*98 209*
58 India	-	-	-	-	6	-	-	-	-	-
Japan	-	5 762	899	1 266	-	-	-	-	-	-
58 Fishing area total	*-*	*5 762*	*899*	*1 266*	*6*	*-*	*-*	*-*	*-*	*-*
88 Japan	50	-	-	-	-	-	-	-	-	-
88 Fishing area total	*50*	*-*	*-*	*-*	*-*	*-*	*-*	*-*	*-*	*-*
Species total	*305 313*	*88 777*	*83 962*	*118 714*	*101 714*	*82 508*	*81 202*	*103 318*	*104 263*	*98 209*
Norwegian krill	**Krill norvégien**		**Krill de Noruega**		***Meganyctiphanes norvegica***				**2,26(01)011,01**	**NKR**
27 Denmark	-	-	-	-	-	-	88	-	-	36
27 Fishing area total	*-*	*-*	*-*	*-*	*-*	*-*	*88*	*-*	*-*	*36*
Species total	*-*	*-*	*-*	*-*	*-*	*-*	*88*	*-*	*-*	*36*
Group total	***305 313***	***88 777***	***83 962***	***118 714***	***101 714***	***82 508***	***81 290***	***103 318***	***104 263***	***98 245***

B-47

Miscellaneous marine crustaceans	Capture production by species, fishing areas and countries or areas
Crustacés marins divers	Captures par espèces, zones de pêche et pays ou zones
Crustáceos marinos diversos	Capturas por especies, áreas de pesca y países o áreas

Species, Fishing area Espèce, Zone de pêche Especie, Area de pesca	1992 mt	1993 mt	1994 mt	1995 mt	1996 mt	1997 mt	1998 mt	1999 mt	2000 mt	2001 mt
Brine shrimp	**Crevette de salines**		**Artemia**		***Artemia salina***			**2,02(02)001,01**		**AMS**
67 USA	...	...	...	...	...	...	...	...	49	63
67 Fishing area total	*...*	*...*	*...*	*...*	*...*	*...*	*...*	*...*	*49*	*63*
77 USA	...	...	...	...	...	...	513	691	512	475
77 Fishing area total	*...*	*...*	*...*	*...*	*...*	*...*	*513*	*691*	*512*	*475*
Species total	*...*	*...*	*...*	*...*	*...*	*...*	*513*	*691*	*561*	*538*
Goose barnacles	**Balanes**		**Bellotas de mar**		***Lepas spp***			**2,13(02)007,XX**		**GOO**
27 France	11	22	12	0	2	-	-	-	-	-
27 Fishing area total	*11*	*22*	*12*	*0*	*2*	*-*	*-*	*-*	*-*	*-*
Species total	*11*	*22*	*12*	*0*	*2*	*-*	*-*	*-*	*-*	*-*
Giant barnacle	**Balane géante**		**Picoroco gigante**		***Megabalanus psittacus***			**2,13(03)015,01**		**MBZ**
87 Chile	1 488	1 192	809	681	879	579	683	620	620	685
87 Fishing area total	*1 488*	*1 192*	*809*	*681*	*879*	*579*	*683*	*620*	*620*	*685*
Species total	*1 488*	*1 192*	*809*	*681*	*879*	*579*	*683*	*620*	*620*	*685*
Spottail mantis squillid	**Squille ocellée**		**Galera ocelada**		***Squilla mantis***			**2,25(01)001,02**		**MTS**
37 France	13	11	11	2	15	10	10	34	44	33
Italy	4 847	4 037	4 317	4 611	5 431	4 497	3 670	4 767	5 244	5 570
Slovenia	-	-	-	-	-	-	-	-	0	3
Spain	-	-	-	-	-	-	-	1 222	1 043	1 000
37 Fishing area total	*4 860*	*4 048*	*4 328*	*4 613*	*5 446*	*4 507*	*3 680*	*6 023*	*6 331*	*6 606*
Species total	*4 860*	*4 048*	*4 328*	*4 613*	*5 446*	*4 507*	*3 680*	*6 023*	*6 331*	*6 606*
Squillids nei	**Squilles nca**		**Galeras nep**		***Squillidae***			**2,25(01)XXX,XX**		**SQY**
71 Philippines	364	1 005	835	982	695	2 133	2 219	2 068	2 141	2 561
71 Fishing area total	*364*	*1 005*	*835*	*982*	*695*	*2 133*	*2 219*	*2 068*	*2 141*	*2 561*
Species total	*364*	*1 005*	*835*	*982*	*695*	*2 133*	*2 219*	*2 068*	*2 141*	*2 561*
Stomatopods nei	**Stomatopodes nca**		**Estomatopodos nep**		***Stomatopoda***			**2,25(XX)XXX,XX**		**SVX**
21 USA	...	...	0	1	-	-	1	5	6	1
21 Fishing area total	*...*	*...*	*0*	*1*	*-*	*-*	*1*	*5*	*6*	*1*
57 Thailand	-	-	-	-	-	0	1	116	114	114
57 Fishing area total	*-*	*-*	*-*	*-*	*-*	*0*	*1*	*116*	*114*	*114*
71 Thailand	-	-	-	-	181	176	457	750	740	735
71 Fishing area total	*-*	*-*	*-*	*-*	*181*	*176*	*457*	*750*	*740*	*735*
Species total	*...*	*...*	*0*	*1*	*181*	*176*	*459*	*871*	*860*	*850*
Marine crustaceans nei	**Crustacés marins nca**		**Crustáceos marinos nep**		***Crustacea***			**2,99(XX)XXX,XX**		**CRU**
21 Japan	-	-	-	-	2	-	-	-	-	-
21 Fishing area total	*-*	*-*	*-*	*-*	*2*	*-*	*-*	*-*	*-*	*-*
27 Denmark	225	10	6 089	563	117	0	-	-	-	-
France	798	933	864	979	885	1 154	676	700	686	690
Germany	205	-	-	-	-	-	-	-	-	-
Ireland	-	-	-	-	-	-	-	-	49	401
Portugal	196	65	69	95	63	180	123	137	15	25
Spain	757	439	522	600	552	449	590	437	365	545
Sweden	-	-	-	-	-	-	-	2	1	-
UK	0	275	8	2	2	-	-	-	-	12
27 Fishing area total	*2 181*	*1 722*	*7 552*	*2 239*	*1 619*	*1 783*	*1 389*	*1 276*	*1 116*	*1 673*
31 Barbados	0	0	0	0	0	0	0	0	0	-
Cuba	51	170	101	79	22	31	71	3	-	-
Guadeloupe	140 F	150 F	150 F	150 F	140	150	134	134	150	150
Honduras	-	-	-	-	-	9	0	32	172	...
Korea Rep	-	-	-	-	-	-	1	1	-	-
Puerto Rico	123	120	136	199	189	171	184	-	-	-
St Kitts Nev	100 F	50 F	...	...	...	...	...	...	...	...
St Lucia	20	15	14	12	12	12	32	30	25	36
31 Fishing area total	*434 F*	*505 F*	*401 F*	*440 F*	*363*	*373*	*422*	*200*	*347*	*186*
34 Cameroon	-	-	-	-	-	-	-	-	-	26
Eq Guinea	400 F	360 F	550 F	230 F	510 F	650 F	430	500 F	50 F	50 F
Italy	-	-	-	387	172	595	173	126	44	77
Korea Rep	0	-	40	32	6	-	-	-	6	-
Liberia	-	-	12	-	-	-	-	-	-	-
Mauritania	-	-	-	-	-	-	30 F	9	6	20
Morocco	2	4	1	1	2	1	0	-	1	1
Nigeria	0	0	0	9	0	0	0	0	0	-
Portugal	13	1 334	21	2	3	2	-	52	-	3
Sao Tome Prn	3 F	4 F	6 F	6 F	4	5	75	7 F	10 F	10 F

B-47 Miscellaneous marine crustaceans — Capture production by species, fishing areas and countries or areas
Crustacés marins divers — Captures par espèces, zones de pêche et pays ou zones
Crustáceos marinos diversos — Capturas por especies, áreas de pesca y países o áreas

Species, Fishing area Espèce, Zone de pêche Especie, Area de pesca	1992 mt	1993 mt	1994 mt	1995 mt	1996 mt	1997 mt	1998 mt	1999 mt	2000 mt	2001 mt
Spain	2 069	2 000 F	1 500 F	1 000 F	337	64	1 035	181	132	375
Togo	1	2	0	1	0	0	0	0	0	0
Westn Sahara	0	0	0	0	0	0	0	0	0	0
34 Fishing area total	*2 488 F*	*3 704 F*	*2 130 F*	*1 668 F*	*1 034 F*	*1 317 F*	*1 743 F*	*875 F*	*249 F*	*562 F*
37 Algeria	150 F	150 F	250 F	73	67	71	154	83	80 F	80 F
France	22	24	27	...	2	2	2	1	-	-
Italy	2 413	2 143	2 169	3 944	3 428	3 086	3 072	2 168	1 984	1 707
Lebanon	25 F	25	25	25	25	150	50	125	55	55
Morocco	36	32	22	17	27	35	17	-	33	28
Spain	2 200 F	2 000 F	1 800 F	1 600 F	711	891	751	1 456	1 656	1 349
Syria	91 F	94 F	91 F	90 F	90	84	75	70	60	57
Tunisia	0	1	1	1	8	1	72	1	1	0
37 Fishing area total	*4 937 F*	*4 469 F*	*4 385 F*	*5 750 F*	*4 358*	*4 320*	*4 193*	*3 904*	*3 869 F*	*3 276 F*
41 Argentina	-	395	4 200	80	2	288	2	6 027	-	-
Brazil	750 F	750 F	750 F	287	288	310	244	266	227	220 F
Italy	-	-	-	46	20	-	-	-	-	-
Korea Rep	64	2 161	2 292	447	-	251	1 915	-	-	1 155
Uruguay	0	0	0	0	3	-	-	-	-	-
41 Fishing area total	*814 F*	*3 306 F*	*7 242 F*	*860*	*313*	*849*	*2 161*	*6 293*	*227*	*1 375 F*
47 Angola	-	-	-	-	-	-	-	-	-	1 062
Italy	-	-	-	5	2	-	-	-	-	-
47 Fishing area total	-	-	-	*5*	*2*	-	-	-	-	*1 062*
51 India	17 454	3 839	13 441	11 940	15 333	12 130	12 658	11 272	11 047	8 853
Italy	-	-	-	79	35	-	-	-	102	65
Kenya	75	48	72	59	185	223	139	104	101	136
Korea Rep	-	-	-	-	47	14	-	10	-	361
Réunion	...	...	...	...	...	...	...	2	1	1
Spain	-	-	-	-	-	-	-	72	-	-
51 Fishing area total	*17 529*	*3 887*	*13 513*	*12 078*	*15 600*	*12 367*	*12 797*	*11 460*	*11 251*	*9 416*
57 Australia	115	87	277	261	1 017	1 351	1 467	858	963	783
Bangladesh	17 638	20 520	21 519	20 363	24 288	28 027	31 027	31 742	31 395	31 300 F
India	11 361	19 111	16 276	11 745	12 714	13 209	18 603	18 475	3 703	2 651
Indonesia	402	44	407	1 495	47	993	294	172	154	170
Korea Rep	-	-	-	-	-	-	-	2	2	155
Sri Lanka	7 783	7 599	4 900	1 800	2 502	2 360	880	3 080	230	1 450
57 Fishing area total	*37 299*	*47 361*	*43 379*	*35 664*	*40 568*	*45 940*	*52 271*	*54 329*	*36 447*	*36 509 F*
61 China	498 434	754 006	755 775	852 257	914 672	1 086 501	1 231 473	1 131 643	1 213 725	1 264 976
China, Macao	418	271	296	207	200	210 F	210 F	210 F	210 F	210 F
Japan	-	...	...	60 783	58 890	63 028	67 945	49 783	74 072	49 619
Korea D P Rp	21 000 F	24 000 F	26 654	21 091	33 931	15 265	15 000 F	15 000 F	14 300 F	14 300 F
Korea Rep	-	-	136	17	-	-	-	1 645	1 136	4 257
61 Fishing area total	*519 852 F*	*778 277 F*	*782 861*	*934 355*	*1 007 693*	*1 165 004 F*	*1 314 628 F*	*1 198 281 F*	*1 303 443 F*	*1 333 362 F*
71 Australia	1	0	0	0	179	3 966	3 769	964	1 009	1 766
Brunei Darsm	79	39	68	18	281	78	65	44	78	266
Fiji Islands	129	25	26 F	180	85	78	85	91	80 F	87
Indonesia	566	313	2 820	1 098	860	1 590	514	331	774	830
Kiribati	220	210	210	210	220	4	...	...	131	418
Korea Rep	415	864	620	225	-	-	-	-	-	30
NewCaledonia	0	0	0	0	0	0	-	0	0	0
Vanuatu	228 F	250 F	240 F	250 F	250 F	250 F	250 F	250 F	250 F	250 F
71 Fishing area total	*1 638 F*	*1 701 F*	*3 984 F*	*1 981 F*	*1 875 F*	*5 966 F*	*4 683 F*	*1 680 F*	*2 322 F*	*3 647 F*
77 El Salvador	595	405	370	564	498	392	355	294	337	409
Fr Polynesia	10	1	0	0	0	0	0	0	0	-
Guatemala	-	-	6	8	0	8	19	20	14	15 F
Honduras	32	33	47	150	117	199	100 F	15	60	...
Mexico	20	21	10	0	0	2	3	1	1	-
Panama	3	4	3	...	3	3	2	1	1	-
Samoa	20	7	10	10	10	10	10	30	207	200 F
Tonga	85	80	90	100	120	100	177	200	175	270
77 Fishing area total	*765*	*551*	*536*	*832*	*748*	*714*	*666 F*	*561*	*795*	*894 F*
81 Australia	169	167	366	521	156	157	92	670	802	537
Japan	-	3	-	2	-	-	-	-	-	-
81 Fishing area total	*169*	*170*	*366*	*523*	*156*	*157*	*92*	*670*	*802*	*537*
87 Chile	152	26	51	-	1	6	15	67	7	65
Peru	-	-	-	-	-	-	-	-	-	758
Spain	-	-	-	-	-	-	-	-	-	2
87 Fishing area total	*152*	*26*	*51*	-	*1*	*6*	*15*	*67*	*7*	*825*
Species total	*588 258 F*	*845 679 F*	*866 400 F*	*996 395 F*	*1 074 332 F*	*1 238 796 F*	*1 395 060 F*	*1 279 596 F*	*1 360 875 F*	*1 393 324 F*
Group total	***594 981***	***851 946***	***872 384***	***1 002 672***	***1 081 535***	***1 246 191***	***1 402 614***	***1 289 869***	***1 371 388***	***1 404 564***

B-51 Freshwater molluscs — Capture production by species, fishing areas and countries or areas
Mollusques d'eau douce — Captures par espèces, zones de pêche et pays ou zones
Moluscos de agua dulce — Capturas por especies, áreas de pesca y países o áreas

Species, Fishing area Espèce, Zone de pêche Especie, Area de pesca	1992 mt	1993 mt	1994 mt	1995 mt	1996 mt	1997 mt	1998 mt	1999 mt	2000 mt	2001 mt
Japanese corbicula	**Cyrène japonaise**		**Corbicula japonesa**		***Corbicula japonica***			**3,16(21)025,02**		**CMJ**
04 China,Taiwan	30	21	-	-	-	-	-	-	-	-
Japan	29 820	27 134	23 988	26 938	26 714	21 822	19 932	20 009	19 295	17 295
Korea Rep	1 153	755	713	659	600	387	677	1 006	-	-
04 Fishing area total	*31 003*	*27 910*	*24 701*	*27 597*	*27 314*	*22 209*	*20 609*	*21 015*	*19 295*	*17 295*
61 Russian Fed	-	-	-	-	-	-	-	-	-	74
61 Fishing area total	-	-	-	-	-	-	-	-	-	*74*
Species total	*31 003*	*27 910*	*24 701*	*27 597*	*27 314*	*22 209*	*20 609*	*21 015*	*19 295*	*17 369*
Freshwater molluscs nei	**Mollusques d'eau douce nca**		**Moluscos de agua dulce nep**		***Mollusca***			**3,99(XX)XXX,XX**		**MOF**
01 Egypt	-	-	-	488	530	610	625	598	916	845
Morocco	-	-	-	-	-	-	-	8	5	6
01 Fishing area total	-	-	-	*488*	*530*	*610*	*625*	*606*	*921*	*851*
02 El Salvador	-	38	1	-	1	2	3	1	8	2
Mexico	115	120	95	198	240	120	177	239	557	...
USA	25	42	-	-	-	-	5	-	-	-
02 Fishing area total	*140*	*200*	*96*	*198*	*241*	*122*	*185*	*240*	*565*	*2*
04 China	223 825	289 484	330 284	441 994	403 448	374 461	460 126	434 993	480 249	529 645
China,Taiwan	-	-	-	-	-	-	1	2	2	-
Indonesia	227	499	2 201	932	1 343	909	806	597	734	650
Japan	777	705	787	847	1 305	1 361	1 132	900	628	583
Korea Rep	806	832	404	245	200	290	409	255	645	669
Philippines	149 134	134 891	147 076	120 547	122 636	101 841	90 154	87 259	85 575	68 958
Turkey	886	1 250	784	1 150	1 500	2 000	1 500	1 585	1 592	1 601
04 Fishing area total	*375 655*	*427 661*	*481 536*	*565 715*	*530 432*	*480 862*	*554 128*	*525 591*	*569 425*	*602 106*
06 Fiji Islands	3 812	2 798	2 852 F	2 569	2 670	3 970	4 500	5 000	5 080 F	5 300
06 Fishing area total	*3 812*	*2 798*	*2 852 F*	*2 569*	*2 670*	*3 970*	*4 500*	*5 000*	*5 080 F*	*5 300*
Species total	*379 607*	*430 659*	*484 484 F*	*568 970*	*533 873*	*485 564*	*559 438*	*531 437*	*575 991 F*	*608 259*
Group total	***410 610***	***458 569***	***509 185***	***596 567***	***561 187***	***507 773***	***580 047***	***552 452***	***595 286***	***625 628***

B-52 Abalones, winkles, conchs — Capture production by species, fishing areas and countries or areas
Ormeaux, bigorneaux, strombes — Captures par espèces, zones de pêche et pays ou zones
Orejas de mar, bígaros, estrombos — Capturas por especies, áreas de pesca y países o áreas

Species, Fishing area Espèce, Zone de pêche Especie, Area de pesca	1992 mt	1993 mt	1994 mt	1995 mt	1996 mt	1997 mt	1998 mt	1999 mt	2000 mt	2001 mt
Common periwinkle	**Bigorneau**		**Bígaro**		***Littorina littorea***				**3,07(01)001,01**	**PEE**
27 Ireland	2 309	1 770	2 457	301	2 836	3 152	2 636	3 014	2 172	2 781
27 Fishing area total	*2 309*	*1 770*	*2 457*	*301*	*2 836*	*3 152*	*2 636*	*3 014*	*2 172*	*2 781*
37 Spain	-	-	-	-	132	147	117	126	139	-
37 Fishing area total	*-*	*-*	*-*	*-*	*132*	*147*	*117*	*126*	*139*	*-*
Species total	*2 309*	*1 770*	*2 457*	*301*	*2 968*	*3 299*	*2 753*	*3 140*	*2 311*	*2 781*
Periwinkles nei	**Bigorneaux nca**		**Bígaros nep**		***Littorina spp***				**3,07(01)001,XX**	**PER**
21 Canada	1 133	279	230	166	200	274	198	149	98	92
USA	624	384	202	32	19	-	171	223	216	649
21 Fishing area total	*1 757*	*663*	*432*	*198*	*219*	*274*	*369*	*372*	*314*	*741*
27 Denmark	0	0	0	0	4	1	2	-	2	0
France	8	9	3	-	2	-	-	-	-	-
Portugal	0	0	0	0	0	-	-	-	-	-
Spain	17	2	2	5	8	8	15	8	3	5
UK	2 021	1 910	2 263	2 331	1 756	2 870	1 925	1 336	1 059	760
27 Fishing area total	*2 046*	*1 921*	*2 268*	*2 336*	*1 770*	*2 879*	*1 942*	*1 344*	*1 064*	*765*
Species total	*3 803*	*2 584*	*2 700*	*2 534*	*1 989*	*3 153*	*2 311*	*1 716*	*1 378*	*1 506*
Murex	**Rochers**		**Murices**		***Murex spp***				**3,07(02)002,XX**	**MUE**
34 Senegal	...	...	...	748	1 267	1 223	2 543	1 255	1 529	2 080
34 Fishing area total	*...*	*...*	*...*	*748*	*1 267*	*1 223*	*2 543*	*1 255*	*1 529*	*2 080*
37 France	97	88	84	43	37	41	41	35	51	52
37 Fishing area total	*97*	*88*	*84*	*43*	*37*	*41*	*41*	*35*	*51*	*52*
Species total	*97*	*88*	*84*	*791*	*1 304*	*1 264*	*2 584*	*1 290*	*1 580*	*2 132*
Sea snails	**Escargots de mer**		**Caracols de mar**		***Rapana spp***				**3,07(02)018,XX**	**RPN**
37 Bulgaria	...	...	3 000 F	3 120	3 260	4 900	4 300	3 800	3 800	3 353
Georgia	-	-	-	700	711	118	-	-	-	-
Ukraine	14	3	5	303	378	476	371	619	913	400
37 Fishing area total	*14*	*3*	*3 005 F*	*4 123*	*4 349*	*5 494*	*4 671*	*4 419*	*4 713*	*3 753*
Species total	*14*	*3*	*3 005 F*	*4 123*	*4 349*	*5 494*	*4 671*	*4 419*	*4 713*	*3 753*
False abalone	**Rocher loco**		**Loco**		***Concholepas concholepas***				**3,07(02)023,01**	**SNE**
87 Chile	5	8 574	8 111	2 670	2 541	3 154	2 564	2 294	1 274	828
Peru	5 632	2 919	2 557	1 361	2 728	4 366	830	2 289	1 250	544
87 Fishing area total	*5 637*	*11 493*	*10 668*	*4 031*	*5 269*	*7 520*	*3 394*	*4 583*	*2 524*	*1 372*
Species total	*5 637*	*11 493*	*10 668*	*4 031*	*5 269*	*7 520*	*3 394*	*4 583*	*2 524*	*1 372*
Giant abalone	**Ormeau géant**		**Abulón gigante**		***Haliotis gigantea***				**3,07(03)001,09**	**ABG**
61 Japan	2 496	2 353	2 164	1 980	1 941	2 218	2 269	2 109	2 146	1 982
61 Fishing area total	*2 496*	*2 353*	*2 164*	*1 980*	*1 941*	*2 218*	*2 269*	*2 109*	*2 146*	*1 982*
Species total	*2 496*	*2 353*	*2 164*	*1 980*	*1 941*	*2 218*	*2 269*	*2 109*	*2 146*	*1 982*
Perlemoen abalone	**Ormeau de Mida**		**Oreja de mar**		***Haliotis midae***				**3,07(03)001,11**	**ABP**
47 South Africa	738	561	586	615	735	330	524	481	490	5 265
47 Fishing area total	*738*	*561*	*586*	*615*	*735*	*330*	*524*	*481*	*490*	*5 265*
Species total	*738*	*561*	*586*	*615*	*735*	*330*	*524*	*481*	*490*	*5 265*
Blacklip abalone	**Ormeau à lèvres noires**		**Oreja de mar de labios negros**		***Haliotis rubra***				**3,07(03)001,13**	**ABR**
57 Australia	4 744	4 341	4 361	4 896	5 081	4 905	4 920	5 297	5 204	5 301
57 Fishing area total	*4 744*	*4 341*	*4 361*	*4 896*	*5 081*	*4 905*	*4 920*	*5 297*	*5 204*	*5 301*
71 Solomon Is	28	26	0	-	-	-	-	-	-	-
71 Fishing area total	*28*	*26*	*0*	*-*	*-*	*-*	*-*	*-*	*-*	*-*
81 Australia	285	327	312	312	344	335	327	...	...	...
81 Fishing area total	*285*	*327*	*312*	*312*	*344*	*335*	*327*	*...*	*...*	*...*
Species total	*5 057*	*4 694*	*4 673*	*5 208*	*5 425*	*5 240*	*5 247*	*5 297*	*5 204*	*5 301*
Tuberculate abalone	**Ormeau tuberculeux**		**Oreja marina tuberculosa**		***Haliotis tuberculata***				**3,07(03)001,14**	**HLT**
27 Channel Is	...	...	-	-	-	-	-	-	3	2
France	3	11	3	49	62	75	36	37	61	60
27 Fishing area total	*3*	*11*	*3*	*49*	*62*	*75*	*36*	*37*	*64*	*62*
Species total	*3*	*11*	*3*	*49*	*62*	*75*	*36*	*37*	*64*	*62*

B-52 Abalones, winkles, conchs / Ormeaux, bigorneaux, strombes / Orejas de mar, bígaros, estrombos

Capture production by species, fishing areas and countries or areas / Captures par espèces, zones de pêche et pays ou zones / Capturas por especies, áreas de pesca y países o áreas

Species, Fishing area Espèce, Zone de pêche Especie, Area de pesca	1992 mt	1993 mt	1994 mt	1995 mt	1996 mt	1997 mt	1998 mt	1999 mt	2000 mt	2001 mt
Abalones nei	**Ormeaux nca**		**Orejas de mar nep**		***Haliotis spp***			**3,07(03)001,XX**		**ABX**
51 Oman	42	34	36	43	43	40	40	29	45	51
51 Fishing area total	*42*	*34*	*36*	*43*	*43*	*40*	*40*	*29*	*45*	*51*
61 China,Taiwan	15	14	15	14	8	6	2	2	42	1
Korea Rep	320	361	281	260	188	214	71	79	113	104
61 Fishing area total	*335*	*375*	*296*	*274*	*196*	*220*	*73*	*81*	*155*	*105*
67 Canada	30	-	-	-	-	-	-	-	-	-
USA	73	14	45	29	-	-	-	-	-	3
67 Fishing area total	*103*	*14*	*45*	*29*	-	-	-	-	-	*3*
71 Philippines	73	122	240	483	448	183	347	282	241	250
Solomon Is	5	2	-	-	-	-	1	-	-	-
71 Fishing area total	*78*	*124*	*240*	*483*	*448*	*183*	*348*	*282*	*241*	*250*
77 Mexico	3 132	2 180	1 536	1 227	1 075	924	709	574	535	482
USA	252	209	168	150	127	102	-	8	-	-
77 Fishing area total	*3 384*	*2 389*	*1 704*	*1 377*	*1 202*	*1 026*	*709*	*582*	*535*	*482*
81 New Zealand	1 481	1 099	1 080	1 280	1 020	1 180	1 300	1 170	1 265	1 064
81 Fishing area total	*1 481*	*1 099*	*1 080*	*1 280*	*1 020*	*1 180*	*1 300*	*1 170*	*1 265*	*1 064*
Species total	*5 423*	*4 035*	*3 401*	*3 486*	*2 909*	*2 649*	*2 470*	*2 144*	*2 241*	*1 955*
Horned turban	**Troque**		**Peonza cornuda**		***Turbo cornutus***			**3,07(05)002,02**		**TOS**
61 Japan	8 484	9 617	10 393	9 943	10 119	12 132	12 556	11 000	9 839	10 241
Korea Rep	6 282	8 764	9 128	8 921	7 361	6 878	9 192	7 397	7 281	6 274
61 Fishing area total	*14 766*	*18 381*	*19 521*	*18 864*	*17 480*	*19 010*	*21 748*	*18 397*	*17 120*	*16 515*
Species total	*14 766*	*18 381*	*19 521*	*18 864*	*17 480*	*19 010*	*21 748*	*18 397*	*17 120*	*16 515*
Stromboid conchs nei	**Strombes nca**		**Cobos nep**		***Strombus spp***			**3,07(06)002,XX**		**CON**
31 Anguilla	18	8	9	5 F	10 F	10 F	10 F	10 F	10 F	10 F
Antigua Barb	201	70	69	46	39	35	45	46	42	37
Bahamas	358	527	693	494	589	648	670	472	667	661
Belize	1 571	1 137	1 413	1 026	1 105	1 926	1 891	1 051	1 745	1 980
Br Virgin Is	...	...	32	43	54	8	9	8	6	6 F
Cuba	51	90	47	32	717	1 234	487	831	830 F	830 F
Dominican Rp	2 640	2 600	4 680	2 210	1 889	1 594	2 683	1 257	1 778	1 437
Grenada	0	11	1	2	6	1	24	6	0	2
Guadeloupe	470 F	480 F	500 F	500 F	430	470	550	580	550	550
Haiti	350 F	400 F	380 F	350 F	400 F	380 F	350 F	300 F	300 F	300 F
Honduras	722	485	402	410	490	2 987	500 F	44	832	...
Jamaica	1 500	2 000	2 300	2 133	2 850	1 821	1 700	1 366	0	...
Mexico	3 218	4 023	2 670	4 963	2 566	5 218	3 293	7 243	8 295	8 730
NethAntilles	5 F	5 F	5 F	5 F	5 F	5 F	5 F	5 F	5 F	5 F
Nicaragua	-	-	-	-	-	-	162	209	555	956
Puerto Rico	308	375 F	405	758	450	638	1 025	1 025	1 710	1 643
St Kitts Nev	...	...	21	29	49	38	140	91	76	75
St Lucia	8	10	13	15	15	25	28	25	40	41
St Vincent	...	...	32	30	25 F	10	21	7	7	37
Turks Caicos	439	738	699	695	647	650 F	788	770 F	770 F	770 F
US Virgin Is	30 F	25 F	20 F	15 F	10 F	5 F	5 F	1	1 F	1 F
31 Fishing area total	*11 889 F*	*12 984 F*	*14 391 F*	*13 761 F*	*12 346 F*	*17 703 F*	*14 386 F*	*15 347 F*	*18 219 F*	*18 071 F*
77 Mexico	4 866	3 607	4 920	4 163	3 091	2 421	973	1 348	2 097	2 667
Panama	12	3	...	...	...	6	527	125	5	5 F
77 Fishing area total	*4 878*	*3 610*	*4 920*	*4 163*	*3 091*	*2 427*	*1 500*	*1 473*	*2 102*	*2 672 F*
87 Ecuador	20	5	5	22	10	10	10	10	10	10
87 Fishing area total	*20*	*5*	*5*	*22*	*10*	*10*	*10*	*10*	*10*	*10*
Species total	*16 787 F*	*16 599 F*	*19 316 F*	*17 946 F*	*15 447 F*	*20 140 F*	*15 896 F*	*16 830 F*	*20 331 F*	*20 753 F*
Whelk	**Buccin**		**Bocina**		***Buccinum undatum***			**3,07(08)001,01**		**WHE**
21 St Pier Mq	...	...	...	...	...	32	45	...	0	-
21 Fishing area total	...	...	...	...	...	*32*	*45*	...	*0*	-
27 Belgium	269	252	165	164	226	166	117	83	101	119
Channel Is	0	0	0	375 F	550	438	135	8	338	519
Denmark	-	-	-	-	-	-	-	1	-	1
Faeroe Is	-	-	4	-	-	-	-	-	-	-
France	1 726	1 074	1 363	3 679	4 406	12 887	6 266	7 691	12 724	11 030
Iceland	0	0	0	0	520	1 199	13	298	770	678
Ireland	2 082	2 562	4 489	5 952	6 575	3 852	3 667	4 561	4 942	6 364
Isle of Man	-	-	149	148	296	193	...	227	89	2
Netherlands	-	-	-	-	-	-	-	-	121	163
UK	2 481	3 514	3 315	5 534	12 150	8 892	3 525	4 925	10 733	11 270
27 Fishing area total	*6 558*	*7 402*	*9 485*	*15 852 F*	*24 723*	*27 627*	*13 723*	*17 794*	*29 818*	*30 146*
Species total	*6 558*	*7 402*	*9 485*	*15 852 F*	*24 723*	*27 659*	*13 768*	*17 794*	*29 818*	*30 146*
Whelks	**Busycons**		**Busicones**		***Busycon spp***			**3,07(09)003,XX**		**WHX**
21 Canada	33	742	570	935	1 291	1 275	1 062	1 557	1 923	2 041
USA	2 546	3 687	2 237	1 911	3 003	2 087	1 673	3 792	2 330	3 557

B-52

Abalones, winkles, conchs — **Capture production by species, fishing areas and countries or areas**
Ormeaux, bigorneaux, strombes — **Captures par espèces, zones de pêche et pays ou zones**
Orejas de mar, bígaros, estrombos — **Capturas por especies, áreas de pesca y países o áreas**

Species, Fishing area Espèce, Zone de pêche Especie, Area de pesca	1992 mt	1993 mt	1994 mt	1995 mt	1996 mt	1997 mt	1998 mt	1999 mt	2000 mt	2001 mt
21 Fishing area total	*2 579*	*4 429*	*2 807*	*2 846*	*4 294*	*3 362*	*2 735*	*5 349*	*4 253*	*5 598*
31 USA	878	1 181	1 489	1 491	1 193	1 017	908	832	600	480
31 Fishing area total	*878*	*1 181*	*1 489*	*1 491*	*1 193*	*1 017*	*908*	*832*	*600*	*480*
Species total	*3 457*	*5 610*	*4 296*	*4 337*	*5 487*	*4 379*	*3 643*	*6 181*	*4 853*	*6 078*
Volutes nei	**Volutes nca**		**Volutas nep**		***Cymbium spp***				**3,07(42)004,XX**	**CXY**
34 Senegal	...	4 294	4 888	7 454	6 648	5 166	4 678	5 737	4 989	5 422
34 Fishing area total	...	*4 294*	*4 888*	*7 454*	*6 648*	*5 166*	*4 678*	*5 737*	*4 989*	*5 422*
Species total	...	*4 294*	*4 888*	*7 454*	*6 648*	*5 166*	*4 678*	*5 737*	*4 989*	*5 422*
Angulate volute	**Volute angulée**		**Voluta angulosa**		***Zidona dufresnei***				**3,07(42)007,01**	**ZDF**
41 Argentina	128	191	223	574	558	1 322	1 010	683	631	581
Uruguay	...	...	...	...	...	...	...	...	990	825
41 Fishing area total	*128*	*191*	*223*	*574*	*558*	*1 322*	*1 010*	*683*	*1 621*	*1 406*
Species total	*128*	*191*	*223*	*574*	*558*	*1 322*	*1 010*	*683*	*1 621*	*1 406*
Gastropods nei	**Gastropodes nca**		**Gasterópodos nep**		***Gastropoda***				**3,07(XX)XXX,XX**	**GAS**
27 France	-	-	-	-	-	-	-	1	1	-
Portugal	...	...	...	...	...	...	...	141	151	146
27 Fishing area total	...	...	...	...	...	...	...	*142*	*152*	*146*
31 Colombia	...	121	134	329	112	169	135	121	100 F	100 F
Cuba	3	1	14	46	134	14	4	-	-	-
31 Fishing area total	*3*	*122*	*148*	*375*	*246*	*183*	*139*	*121*	*100 F*	*100 F*
34 Gambia	...	...	55	194	227	128	230	250	5	20
Nigeria	...	...	...	...	...	...	2 357	...	...	2 084
Portugal	8	17	23	20	48	41	41	48	49	42
Senegal	4 576	917	682	24	45	40	...	...	65	11
34 Fishing area total	*4 584*	*934*	*760*	*238*	*320*	*209*	*2 628*	*298*	*119*	*2 157*
37 France	-	-	-	-	-	-	-	10	14	19
Spain	-	-	-	-	-	-	-	955	799	851
37 Fishing area total	-	-	-	-	-	-	-	*965*	*813*	*870*
61 China,Taiwan	937	959	816	998	760	1 019	125	0	10	53
Korea Rep	5 852	8 352	8 131	5 224	2 975	1 617	1 325	1 670	818	730
61 Fishing area total	*6 789*	*9 311*	*8 947*	*6 222*	*3 735*	*2 636*	*1 450*	*1 670*	*828*	*783*
67 USA	...	...	3 565	4 228	6 369	...	...	...	-	1
67 Fishing area total	...	...	*3 565*	*4 228*	*6 369*	...	...	...	-	*1*
71 Solomon Is	2	0	1	-	-	-	0	0	-	-
71 Fishing area total	*2*	*0*	*1*	-	-	-	*0*	*0*	-	-
77 Panama	66	9	372	0	5	0	116	17	1	3 F
USA	11	22	22	18	11	12	49	36	35	25
77 Fishing area total	*77*	*31*	*394*	*18*	*16*	*12*	*165*	*53*	*36*	*28 F*
87 Chile	9 990	9 377	7 538	6 403	6 078	5 348	4 649	7 204	6 898	5 429
Colombia	...	...	9	7	17	29	20	10	20 F	20 F
Peru	3 651	2 871	2 504	3 686	2 215	7 098	3 110	4 525	2 768	4 995
87 Fishing area total	*13 641*	*12 248*	*10 051*	*10 096*	*8 310*	*12 475*	*7 779*	*11 739*	*9 686 F*	*10 444 F*
Species total	*25 096*	*22 646*	*23 866*	*21 177*	*18 996*	*15 515*	*12 161*	*14 988*	*11 734 F*	*14 529 F*
Group total	***92 369***	***102 715***	***111 336***	***109 322***	***116 290***	***124 433***	***99 163***	***105 826***	***113 117***	***120 958***

B-53 Oysters / Huîtres / Ostras

Capture production by species, fishing areas and countries or areas
Captures par espèces, zones de pêche et pays ou zones
Capturas por especies, áreas de pesca y países o áreas

Species, Fishing area Espèce, Zone de pêche Especie, Area de pesca	1992 mt	1993 mt	1994 mt	1995 mt	1996 mt	1997 mt	1998 mt	1999 mt	2000 mt	2001 mt
Chilean flat oyster	**Huître plate chilienne**		**Ostra chilena**		*Ostrea chilensis*				**3,16(07)002,03**	**OCH**
87 Chile	11	14	18	-	-	5	1	6	9	202
87 Fishing area total	*11*	*14*	*18*	*-*	*-*	*5*	*1*	*6*	*9*	*202*
Species total	*11*	*14*	*18*	*-*	*-*	*5*	*1*	*6*	*9*	*202*
European flat oyster	**Huître plate européenne**		**Ostra europea**		*Ostrea edulis*				**3,16(07)002,05**	**OYF**
27 Channel Is	4	4 F	-	-	-	-	-	-	-	-
Denmark	0	75	4	12	9	24	6	8	9	23
France	20	20	9	18	6	8	4	10	7	130
Ireland	516	350	1 208	815	415	773	...	...	...	...
Netherlands	1 266	123	-	-	-	-	-	-	-	-
Spain	399	38	17	11	252	388	316	443	152	37
Sweden	9	4	2	2	3	3	2	4	2	1
UK	312	432	521	527	584	553	1 047	407	439	611
27 Fishing area total	*2 526*	*1 046 F*	*1 761*	*1 385*	*1 269*	*1 749*	*1 375*	*872*	*609*	*802*
37 Croatia	...	...	...	...	...	...	35	14	12	24
France	21	80	92	22	43	10	10	2	2	2
Greece	3 771	1 691	1 093	1 096	1 003	344	95	47	105	120
Slovenia	1	5	1	0	0	0	-	-	0	0
Spain	9 F	6 F	3 F	1 F	1	0	0	-	-	-
Tunisia	150	6	-	-	-	-	-	-	1	1
Turkey	2 226	1 222	1 803	1 836	1 140	1 495	1 050	840	150	10
Yugoslavia	0	0	0	0	0	0	0	1	1	0
37 Fishing area total	*6 178 F*	*3 010 F*	*2 992 F*	*2 955 F*	*2 187*	*1 849*	*1 190*	*904*	*271*	*157*
Species total	*8 704 F*	*4 056 F*	*4 753 F*	*4 340 F*	*3 456*	*3 598*	*2 565*	*1 776*	*880*	*959*
Olympia flat oyster	**Huître plate Olympie**		**Ostra Olimpia**		*Ostrea lurida*				**3,16(07)002,06**	**OFO**
67 USA	...	...	...	...	...	...	...	6	7	8
67 Fishing area total	*...*	*...*	*...*	*...*	*...*	*...*	*...*	*6*	*7*	*8*
Species total	*...*	*...*	*...*	*...*	*...*	*...*	*...*	*6*	*7*	*8*
New Zealand dredge oyster	**Huître plate néo-zélandaise**		**Ostra de Nueva Zelandia**		*Ostrea lutaria*				**3,16(07)002,20**	**DRY**
81 New Zealand	1 036	871	584	1	2	2	2	3	2	2
81 Fishing area total	*1 036*	*871*	*584*	*1*	*2*	*2*	*2*	*3*	*2*	*2*
Species total	*1 036*	*871*	*584*	*1*	*2*	*2*	*2*	*3*	*2*	*2*
Flat oysters nei	**Huîtres plates nca**		**Ostras nep**		*Ostrea spp*				**3,16(07)002,XX**	**OYX**
27 Norway	...	...	-	0	-	-	-	-	-	5
UK	0	0	3	2	2	5	-	-	-	-
27 Fishing area total	*0*	*0*	*3*	*2*	*2*	*5*	*-*	*-*	*-*	*5*
77 Mexico	1 667	2 033	615	1 801	1 354	1 903	1 276	2 489	1 828	2 234
77 Fishing area total	*1 667*	*2 033*	*615*	*1 801*	*1 354*	*1 903*	*1 276*	*2 489*	*1 828*	*2 234*
Species total	*1 667*	*2 033*	*618*	*1 803*	*1 356*	*1 908*	*1 276*	*2 489*	*1 828*	*2 239*
Pacific cupped oyster	**Huître creuse du Pacifique**		**Ostión japonés**		*Crassostrea gigas*				**3,16(07)008,01**	**OYG**
27 France	-	-	-	-	4	0	0	2	60	63
Portugal	-	1	-	8	-	-	-	-	-	-
Spain	7	-	-	10	0	-	-	1	9	9
UK	72	99	22	82	125	98	59	6	5	20
27 Fishing area total	*79*	*100*	*22*	*100*	*129*	*98*	*59*	*9*	*74*	*92*
37 France	1 322	36	369	50	50	14	14	48	48	-
Spain	-	-	-	-	9	-	-	-	-	-
37 Fishing area total	*1 322*	*36*	*369*	*50*	*59*	*14*	*14*	*48*	*48*	*-*
61 China,Taiwan	14	8	2	7	-	-	-	-	-	4
Korea Rep	17 526	28 215	20 710	18 262	18 259	17 210	9 905	11 609	15 939	10 056
61 Fishing area total	*17 540*	*28 223*	*20 712*	*18 269*	*18 259*	*17 210*	*9 905*	*11 609*	*15 939*	*10 060*
67 USA	...	...	...	...	4 989	6 187	11	...	...	...
67 Fishing area total	*...*	*...*	*...*	*...*	*4 989*	*6 187*	*11*	*...*	*...*	*...*
77 USA	2 057	2 443	2 583	1 709	1 717	2 176	2 107	539	1 917	1 083
77 Fishing area total	*2 057*	*2 443*	*2 583*	*1 709*	*1 717*	*2 176*	*2 107*	*539*	*1 917*	*1 083*
81 New Zealand	-	-	-	8	2	2	0	62	0	0
81 Fishing area total	*-*	*-*	*-*	*8*	*2*	*2*	*0*	*62*	*0*	*0*
87 Ecuador	9	1	5	5	5	5	5	5	5	5
87 Fishing area total	*9*	*1*	*5*	*5*	*5*	*5*	*5*	*5*	*5*	*5*
Species total	*21 007*	*30 803*	*23 691*	*20 141*	*25 160*	*25 692*	*12 101*	*12 272*	*17 983*	*11 240*
Mangrove cupped oyster	**Huître creuse des Caraïbes**		**Ostión de mangle**		*Crassostrea rhizophorae*				**3,16(07)008,02**	**OYM**
31 Colombia	...	...	20	-	-	-	8	-	-	-

B-53 Oysters / Huîtres / Ostras

Capture production by species, fishing areas and countries or areas
Captures par espèces, zones de pêche et pays ou zones
Capturas por especies, áreas de pesca y países o áreas

Species, Fishing area Espèce, Zone de pêche Especie, Area de pesca	1992 mt	1993 mt	1994 mt	1995 mt	1996 mt	1997 mt	1998 mt	1999 mt	2000 mt	2001 mt
Cuba	729	744	1 248	1 408	1 409	1 991	2 281	1 865	1 870 F	1 870 F
Dominican Rp	31	20	36	41	6	6	12	4	7	8
Venezuela	1 931	2 465	3 144	3 775	2 695	1 705	2 594	2 037	1 590	3 754
31 Fishing area total	*2 691*	*3 229*	*4 448*	*5 224*	*4 110*	*3 702*	*4 895*	*3 906*	*3 467 F*	*5 632 F*
Species total	*2 691*	*3 229*	*4 448*	*5 224*	*4 110*	*3 702*	*4 895*	*3 906*	*3 467 F*	*5 632 F*
American cupped oyster	**Huître creuse américaine**		**Ostión virgínico**			***Crassostrea virginica***			**3,16(07)008,03**	**OYA**
21 Canada	1 739	1 590	2 636	2 312	2 132	1 726	3 124	3 225	4 133	3 009
USA	29 149	21 589	40 442	30 570	37 437	47 373	30 141	43 832	22 307	15 548
21 Fishing area total	*30 888*	*23 179*	*43 078*	*32 882*	*39 569*	*49 099*	*33 265*	*47 057*	*26 440*	*18 557*
31 Mexico	28 891	22 821	33 647	27 609	34 726	38 515	30 715	39 268	48 101	48 570
USA	39 603	62 480	51 747	98 043	74 245	57 093	71 242	45 882	174 765	107 915
31 Fishing area total	*68 494*	*85 301*	*85 394*	*125 652*	*108 971*	*95 608*	*101 957*	*85 150*	*222 866*	*156 485*
Species total	*99 382*	*108 480*	*128 472*	*158 534*	*148 540*	*144 707*	*135 222*	*132 207*	*249 306*	*175 042*
Slipper cupped oyster	**Huître creuse chausson**		**...C**			***Crassostrea iredalei***			**3,16(07)008,11**	**CSI**
71 Philippines	1 041	161	107	324	291	152	89	95	79	83
71 Fishing area total	*1 041*	*161*	*107*	*324*	*291*	*152*	*89*	*95*	*79*	*83*
Species total	*1 041*	*161*	*107*	*324*	*291*	*152*	*89*	*95*	*79*	*83*
Cupped oysters nei	**Huîtres creuses nca**		**Ostiones nep**			***Crassostrea spp***			**3,16(07)008,XX**	**OYC**
21 USA	18 940	6 063	...	...	...	...	64	49	35	14
21 Fishing area total	*18 940*	*6 063*	...	...	...	...	*64*	*49*	*35*	*14*
34 Senegal	...	13	86	223	100	109	89	125	101	151
34 Fishing area total	...	*13*	*86*	*223*	*100*	*109*	*89*	*125*	*101*	*151*
41 Brazil	430 F	430 F	430 F	726	873	828	744	1 547	884	870 F
41 Fishing area total	*430 F*	*430 F*	*430 F*	*726*	*873*	*828*	*744*	*1 547*	*884*	*870 F*
47 South Africa	424	...	...	...	...	...	...	-	-	-
47 Fishing area total	*424*	...	...	...	...	...	...	-	-	-
51 Kenya	13	17	8	14	32	16	9	8	2	1
Mozambique	-	-	-	-	-	8	19	26	30	28
51 Fishing area total	*13*	*17*	*8*	*14*	*32*	*24*	*28*	*34*	*32*	*29*
57 Indonesia	9	44	15	-	-	-	-	-	350	380
57 Fishing area total	*9*	*44*	*15*	-	-	-	-	-	*350*	*380*
61 Russian Fed	-	-	-	-	-	-	-	1	4	32
61 Fishing area total	-	-	-	-	-	-	-	*1*	*4*	*32*
67 USA	...	...	...	...	...	...	...	...	7	12
67 Fishing area total	...	...	...	...	...	...	...	...	*7*	*12*
71 Indonesia	442	102	642	1 331	1 596	1 678	2 029	2 025	1 931	2 120
71 Fishing area total	*442*	*102*	*642*	*1 331*	*1 596*	*1 678*	*2 029*	*2 025*	*1 931*	*2 120*
Species total	*20 258 F*	*6 669 F*	*1 181 F*	*2 294*	*2 601*	*2 639*	*2 954*	*3 781*	*3 344*	*3 608 F*
Group total	***155 797***	***156 316***	***163 872***	***192 661***	***185 516***	***182 405***	***159 105***	***156 541***	***276 905***	***199 015***

B-54 Mussels / Moules / Mejillones

Capture production by species, fishing areas and countries or areas
Captures par espèces, zones de pêche et pays ou zones
Capturas por especies, áreas de pesca y países o áreas

Species, Fishing area Espèce, Zone de pêche Especie, Area de pesca	1992 mt	1993 mt	1994 mt	1995 mt	1996 mt	1997 mt	1998 mt	1999 mt	2000 mt	2001 mt
Korean mussel	**Moule coréenne**		**Mejillón coreano**		***Mytilus coruscus***			**3,16(10)001,01**		**MUK**
61 Korea Rep	6 303	2 271	2 731	2 942	2 191	3 211	1 469	1 414	1 133	1 085
61 Fishing area total	*6 303*	*2 271*	*2 731*	*2 942*	*2 191*	*3 211*	*1 469*	*1 414*	*1 133*	*1 085*
Species total	*6 303*	*2 271*	*2 731*	*2 942*	*2 191*	*3 211*	*1 469*	*1 414*	*1 133*	*1 085*
Chilean mussel	**Moule chilienne**		**Chorito**		***Mytilus chilensis***			**3,16(10)001,03**		**MYC**
41 Falkland Is	1	1	1	1	1	1	1	1	1	0
41 Fishing area total	*1*	*1*	*1*	*1*	*1*	*1*	*1*	*1*	*1*	*0*
87 Chile	7 641	6 892	5 718	5 128	5 714	4 723	4 899	4 343	5 236	6 758
87 Fishing area total	*7 641*	*6 892*	*5 718*	*5 128*	*5 714*	*4 723*	*4 899*	*4 343*	*5 236*	*6 758*
Species total	*7 642*	*6 893*	*5 719*	*5 129*	*5 715*	*4 724*	*4 900*	*4 344*	*5 237*	*6 758*
Blue mussel	**Moule commune**		**Mejillón común**		***Mytilus edulis***			**3,16(10)001,05**		**MUS**
21 Canada	4 359	4 938	6 118	4 505	8 147	9 163	10 273	11 565	15 089	11 829
St Pier Mq	5	9	7	8	4	2	4	0	0	0
USA	21 362	14 931	10 891	11 210	9 434	8 063	4 774	4 086	6 009	6 524
21 Fishing area total	*25 726*	*19 878*	*17 016*	*15 723*	*17 585*	*17 228*	*15 051*	*15 651*	*21 098*	*18 353*
27 Channel Is	2	2 F	-	-	-	-	-	-	-	-
Denmark	136 271	136 677	129 317	107 377	86 002	90 765	108 329	96 215	110 618	122 480
France	1 272	2 050	3 908	8 910	197	6 972	1 355	9 564	8 660	8 015
Ireland	-	4 095	3 033	4 556	1 372	1 963	955	503	-	-
Norway	...	...	0	8	4	0	-	1	10	-
Portugal	45	59	37	45	35	46	24	87	48	74
Spain	-	38	-	-	18	176	84	27	-	33
Sweden	19	30	51	52	-	3	36	0	70	51
UK	7 040	7 751	10 348	9 534	12 337	18 994	11 434	7 972	7 468	14 905
27 Fishing area total	*144 649*	*150 702 F*	*146 694*	*130 482*	*99 965*	*118 919*	*122 217*	*114 369*	*126 874*	*145 558*
Species total	*170 375*	*170 580 F*	*163 710*	*146 205*	*117 550*	*136 147*	*137 268*	*130 020*	*147 972*	*163 911*
River Plata mussel	**Moule de la Plata**		**Mejillón del Plata**		***Mytilus platensis***			**3,16(10)001,08**		**MSR**
41 Argentina	599	648	389	388	164	180	149	332	236	186
Uruguay	314	189	183	299	206	174	226	142	176	306
41 Fishing area total	*913*	*837*	*572*	*687*	*370*	*354*	*375*	*474*	*412*	*492*
Species total	*913*	*837*	*572*	*687*	*370*	*354*	*375*	*474*	*412*	*492*
Mediterranean mussel	**Moule méditerranéenne**		**Mejillón mediterráneo**		***Mytilus galloprovincialis***			**3,16(10)001,12**		**MSM**
37 Albania	...	...	...	...	...	24	-	-	-	...
Bulgaria	-	-	-	-	-	-	-	-	-	7
Croatia	...	...	...	...	...	...	200	60	57	32
France	2 900	490	798	1 912	500	1 078	1 078	23	24	12
Greece	8 206	2 919	2 273	10 806	12 625	24 139	6 389	15 860	469	254
Italy	20 380	28 344	22 702	21 425	22 174	21 430	27 270 F	26 510 F	44 200 F	44 160 F
Morocco	24	1	0	0	0	0	0	14	60	32
Russian Fed	-	-	-	-	-	-	-	4	-	-
Slovenia	-	-	-	1	1	1	-	-	-	0
Spain	-	-	-	-	0	29	0	19	18	24
Turkey	6 757	7 086	8 033	6 042	3 500	6 450	3 880	1 800	1 200	1 500
Ukraine	449	210	226	578	246	159	82	163	115	71
37 Fishing area total	*38 716*	*39 050*	*34 032*	*40 764*	*39 046*	*53 310*	*38 899 F*	*44 453 F*	*46 143 F*	*46 092 F*
Species total	*38 716*	*39 050*	*34 032*	*40 764*	*39 046*	*53 310*	*38 899 F*	*44 453 F*	*46 143 F*	*46 092 F*
Australian mussel	**Moule d'Australie**		**Mejillón de Australia**		***Mytilus planulatus***			**3,16(10)001,17**		**MYA**
57 Australia	2 636	1 000 F	500 F	243	71	-	-	-	-	-
57 Fishing area total	*2 636*	*1 000 F*	*500 F*	*243*	*71*	-	-	-	-	-
81 Australia	1	1	4	4	4	1	1	1	1	1
81 Fishing area total	*1*	*1*	*4*	*4*	*4*	*1*	*1*	*1*	*1*	*1*
Species total	*2 637*	*1 001 F*	*504 F*	*247*	*75*	*1*	*1*	*1*	*1*	*1*
Choro mussel	**Moule choro**		**Choro**		***Choromytilus chorus***			**3,16(10)026,01**		**CHC**
87 Chile	1 339	1 201	927	307	323	266	127	155	217	166
87 Fishing area total	*1 339*	*1 201*	*927*	*307*	*323*	*266*	*127*	*155*	*217*	*166*
Species total	*1 339*	*1 201*	*927*	*307*	*323*	*266*	*127*	*155*	*217*	*166*
Horse mussels nei	**Modioles nca**		**Modiolos nep**		***Modiolus spp***			**3,16(10)028,XX**		**MOD**
27 Norway	-	14	6	7	20	30	20	7	2	2
27 Fishing area total	-	*14*	*6*	*7*	*20*	*30*	*20*	*7*	*2*	*2*
Species total	-	*14*	*6*	*7*	*20*	*30*	*20*	*7*	*2*	*2*

B-54 Mussels / Moules / Mejillones

Capture production by species, fishing areas and countries or areas
Captures par espèces, zones de pêche et pays ou zones
Capturas por especies, áreas de pesca y países o áreas

Species, Fishing area Espèce, Zone de pêche Especie, Area de pesca	1992 mt	1993 mt	1994 mt	1995 mt	1996 mt	1997 mt	1998 mt	1999 mt	2000 mt	2001 mt
South American rock mussel	**Moule de roche sudaméricaine**		**Mejillón de roca sudamericano**		***Perna perna***			**3,16(10)032,01**		**MSL**
31 Venezuela	199	325	366	155	223	295	3 802	451	316	1 081
31 Fishing area total	*199*	*325*	*366*	*155*	*223*	*295*	*3 802*	*451*	*316*	*1 081*
Species total	*199*	*325*	*366*	*155*	*223*	*295*	*3 802*	*451*	*316*	*1 081*
Green mussel	**Moule verte asiatique**		**Mejillón verde**		***Perna viridis***			**3,16(10)032,02**		**MSV**
04 Philippines	-	-	111	136	4	-	-	-	-	-
04 Fishing area total	-	-	*111*	*136*	*4*	-	-	-	-	-
71 Philippines	1 788	1 110	4 797	334	330	47	22	20	17	17
Singapore	131	88	-	-	188	-	-	-	-	-
Thailand	24 007	24 850	25 083	20 079	19 698	17 970	17 432	6 534	6 445	6 405
71 Fishing area total	*25 926*	*26 048*	*29 880*	*20 413*	*20 216*	*18 017*	*17 454*	*6 554*	*6 462*	*6 422*
Species total	*25 926*	*26 048*	*29 991*	*20 549*	*20 220*	*18 017*	*17 454*	*6 554*	*6 462*	*6 422*
Cholga mussel	**Moule cholga**		**Cholga**		***Aulacomya ater***			**3,16(10)038,01**		**MSC**
87 Chile	7 478	7 565	9 640	6 376	7 405	6 409	7 725	5 126	5 563	7 884
Peru	7 791	5 976	7 203	11 204	6 023	9 669	15 106	14 612	13 370	14 700
87 Fishing area total	*15 269*	*13 541*	*16 843*	*17 580*	*13 428*	*16 078*	*22 831*	*19 738*	*18 933*	*22 584*
Species total	*15 269*	*13 541*	*16 843*	*17 580*	*13 428*	*16 078*	*22 831*	*19 738*	*18 933*	*22 584*
Sea mussels nei	**Moules nca**		**Mejillones nep**		***Mytilidae***			**3,16(10)XXX,XX**		**MSX**
27 France	607	1 175	1 988	2 031	1 077	0	0	0	1 792	1 110
Iceland	0	0	0	0	-	-	-	-	-	-
Russian Fed	-	-	-	55	-	-	-	-	39	26
27 Fishing area total	*607*	*1 175*	*1 988*	*2 086*	*1 077*	*0*	*0*	*0*	*1 831*	*1 136*
34 Morocco	0	0	2	8	10	6	7	...	401	102
34 Fishing area total	*0*	*0*	*2*	*8*	*10*	*6*	*7*	...	*401*	*102*
41 Brazil	...	...	...	2 217	863	1 193	1 230	906	802	800 F
41 Fishing area total	...	...	...	*2 217*	*863*	*1 193*	*1 230*	*906*	*802*	*800 F*
61 Korea Rep	5 523	3 334	3 162	3 685	837	4 098	6 456	6 771	5 795	3 828
Russian Fed	76	107	-	28	-	-	9	80	24	42
61 Fishing area total	*5 599*	*3 441*	*3 162*	*3 713*	*837*	*4 098*	*6 465*	*6 851*	*5 819*	*3 870*
67 USA	327	709	689	1 317	455	...	...	1	675	448
67 Fishing area total	*327*	*709*	*689*	*1 317*	*455*	...	...	*1*	*675*	*448*
77 Mexico	675	414	291	118	412	2 036	2 056	737	547	...
USA	38	91	93	172	171	170	174	95	96	90
77 Fishing area total	*713*	*505*	*384*	*290*	*583*	*2 206*	*2 230*	*832*	*643*	*90*
81 New Zealand	...	...	...	191	450	0	664	2 977	4 467	2 270
81 Fishing area total	...	...	...	*191*	*450*	*0*	*664*	*2 977*	*4 467*	*2 270*
87 Ecuador	4	3	3	3	3	3	5	5	5	5
87 Fishing area total	*4*	*3*	*3*	*3*	*3*	*3*	*5*	*5*	*5*	*5*
Species total	*7 250*	*5 833*	*6 228*	*9 825*	*4 278*	*7 506*	*10 601*	*11 572*	*14 643*	*8 721 F*
Group total	***276 569***	***267 594***	***261 629***	***244 397***	***203 439***	***239 939***	***237 747***	***219 183***	***241 471***	***257 315***

B-55 Scallops, pectens / Coquilles St-Jacques / Vieiras

Capture production by species, fishing areas and countries or areas / Captures par espèces, zones de pêche et pays ou zones / Capturas por especies, áreas de pesca y países o áreas

Species, Fishing area Espèce, Zone de pêche Especie, Area de pesca	1992 mt	1993 mt	1994 mt	1995 mt	1996 mt	1997 mt	1998 mt	1999 mt	2000 mt	2001 mt
Queen scallop	**Vanneau**		**Volandeira**		***Aequipecten opercularis***				**3,16(08)001,05**	**QSC**
27 Channel Is	4	4 F	-	-	-	-	-	7	-	-
Faeroe Is	3 500	3 320	3 854	2 781	3 559	3 581	4 751	5 993	3 989	4 053
France	1 245	1 399	2 223	926	311	595	637	2 574	3 475	5 989
Ireland	35	55	27	11	3	7	5	29	3	13
Isle of Man	2 635	3 000	1 455	1 465	1 129	1 630	991	1 255	2 275	1 749
Portugal	0	0	0	0	0	-	-	-	-	-
UK	8 608	7 485	2 979	2 857	2 181	5 630	8 102	5 888	5 149	8 660
27 Fishing area total	*16 027*	*15 263 F*	*10 538*	*8 040*	*7 183*	*11 443*	*14 486*	*15 746*	*14 891*	*20 464*
Species total	*16 027*	*15 263 F*	*10 538*	*8 040*	*7 183*	*11 443*	*14 486*	*15 746*	*14 891*	*20 464*
Great Atlantic scallop	**Coquille St-Jacques atlantique**		**Vieira(=Concha de Santiago)**		***Pecten maximus***				**3,16(08)003,09**	**SCE**
27 Belgium	90	115	163	137	163	208	224	247	292	340
Channel Is	19	19 F	73	65 F	29	109	155	203	295	439
France	14 059	13 348	13 503	12 288	12 169	14 451	12 866	13 707	13 727	16 454
Ireland	1 029	543	918	423	560	633	693	1 497	1 579	1 413
Isle of Man	633	639	931	931	1 064	933	706	794	965	1 115
Netherlands	-	-	-	-	228	188	408	306	249	274
Norway	...	3	100	66	14	39	114	425	570	670
Portugal	-	-	4	-	1	2	0	0	0	1
Spain	313	206	282	185	675	391	299	86	508	84
UK	5 069	5 556	9 020	9 376	16 618	18 703	20 078	19 108	19 507	9 796
27 Fishing area total	*21 212*	*20 429 F*	*24 994*	*23 471 F*	*31 521*	*35 657*	*35 543*	*36 373*	*37 692*	*30 586*
37 France	...	...	...	...	...	...	...	12	18	4
37 Fishing area total	...	...	...	...	...	...	...	*12*	*18*	*4*
Species total	*21 212*	*20 429 F*	*24 994*	*23 471 F*	*31 521*	*35 657*	*35 543*	*36 385*	*37 710*	*30 590*
Great Mediterranean scallop	**Coquille St-Jacques méditerr.**		**Concha de peregrino**		***Pecten jacobaeus***				**3,16(08)003,11**	**SJA**
37 Turkey	0	202	308	23	52	95	50	68	570	150
37 Fishing area total	*0*	*202*	*308*	*23*	*52*	*95*	*50*	*68*	*570*	*150*
Species total	*0*	*202*	*308*	*23*	*52*	*95*	*50*	*68*	*570*	*150*
New Zealand scallop	**Pecten de Nouvelle Zélande**		**Vieira de Nueva Zelandia**		***Pecten novaezelandiae***				**3,16(08)003,13**	**SCZ**
81 New Zealand	10 040	8 928	9 088	14 160	5 080	18 848	4 592	6 152	2 912	6 792
81 Fishing area total	*10 040*	*8 928*	*9 088*	*14 160*	*5 080*	*18 848*	*4 592*	*6 152*	*2 912*	*6 792*
Species total	*10 040*	*8 928*	*9 088*	*14 160*	*5 080*	*18 848*	*4 592*	*6 152*	*2 912*	*6 792*
Delicate scallop	**...B**		**...C**		***Zygochlamis delicatula***				**3,16(08)012,01**	**ZYE**
81 New Zealand	...	...	141	135	124	201	91	128	0	222
81 Fishing area total	...	...	*141*	*135*	*124*	*201*	*91*	*128*	*0*	*222*
Species total	...	...	*141*	*135*	*124*	*201*	*91*	*128*	*0*	*222*
Patagonean scallop	**Peigne patagonien**		**Pecten patagónico**		***Zygochlamis patagonica***				**3,16(08)012,02**	**ZYP**
41 Argentina	159	25	0	10 592	36 952	39 817	28 441	42 700	36 514	38 960
Uruguay	...	...	...	...	...	...	...	...	890	3 638
41 Fishing area total	*159*	*25*	*0*	*10 592*	*36 952*	*39 817*	*28 441*	*42 700*	*37 404*	*42 598*
Species total	*159*	*25*	*0*	*10 592*	*36 952*	*39 817*	*28 441*	*42 700*	*37 404*	*42 598*
American sea scallop	**Pecten d'Amérique**		**Vieira americana**		***Placopecten magellanicus***				**3,16(08)014,04**	**SCA**
21 Canada	85 166	87 219	84 833	58 770	47 807	53 881	56 399	54 756	83 852	89 196
St Pier Mq	18	0	11	4	23	27	3	0	0	2
USA	106 208	56 137	60 616	61 651	61 093	47 966	42 880	77 179	113 026	164 762
21 Fishing area total	*191 392*	*143 356*	*145 460*	*120 425*	*108 923*	*101 874*	*99 282*	*131 935*	*196 878*	*253 960*
31 USA	1 430	565	572	805	459	154	150	27	115	236
31 Fishing area total	*1 430*	*565*	*572*	*805*	*459*	*154*	*150*	*27*	*115*	*236*
Species total	*192 822*	*143 921*	*146 032*	*121 230*	*109 382*	*102 028*	*99 432*	*131 962*	*196 993*	*254 196*
Calico scallop	**Peigne calicot**		**Peine percal**		***Argopecten gibbus***				**3,16(08)030,01**	**SCC**
31 USA	...	...	74 325	10 003	...	...	...	...	...	...
31 Fishing area total	...	...	*74 325*	*10 003*	...	...	...	...	...	...
Species total	...	...	*74 325*	*10 003*	...	...	...	...	...	...
Atlantic bay scallop	**Peigne baie**		**Peine caletero**		***Argopecten irradians***				**3,16(08)030,02**	**SCB**
21 USA	896	1 695	0	88	5	25	6	20	13	13
21 Fishing area total	*896*	*1 695*	*0*	*88*	*5*	*25*	*6*	*20*	*13*	*13*
31 USA	668	975	524	1 505	225	427	684	196	141	17
31 Fishing area total	*668*	*975*	*524*	*1 505*	*225*	*427*	*684*	*196*	*141*	*17*
Species total	*1 564*	*2 670*	*524*	*1 593*	*230*	*452*	*690*	*216*	*154*	*30*

B-55 Scallops, pectens / Coquilles St-Jacques / Vieiras

Capture production by species, fishing areas and countries or areas
Captures par espèces, zones de pêche et pays ou zones
Capturas por especies, áreas de pesca y países o áreas

Species, Fishing area Espèce, Zone de pêche Especie, Area de pesca	1992 mt	1993 mt	1994 mt	1995 mt	1996 mt	1997 mt	1998 mt	1999 mt	2000 mt	2001 mt
Peruvian calico scallop	**Pétoncle éventail**		**Ostión abanico**		***Argopecten purpuratus***				**3,16(08)030,03**	**SCQ**
87 Chile	1	-	5	-	9	4	21	0	20	272
Peru	5 445	2 732	837	3 113	2 086	4 009	23 525	30 141	11 810	6 272
87 Fishing area total	*5 446*	*2 732*	*842*	*3 113*	*2 095*	*4 013*	*23 546*	*30 141*	*11 830*	*6 544*
Species total	*5 446*	*2 732*	*842*	*3 113*	*2 095*	*4 013*	*23 546*	*30 141*	*11 830*	*6 544*
Pacific calico scallop	**Pétoncle volant**		**Peine volador**		***Argopecten ventricosus***				**3,16(08)030,04**	**SCH**
77 Mexico	5 290	5 883	8 570	1 256	17 290	2 320	2 726	1 864	6 287	3 100
77 Fishing area total	*5 290*	*5 883*	*8 570*	*1 256*	*17 290*	*2 320*	*2 726*	*1 864*	*6 287*	*3 100*
Species total	*5 290*	*5 883*	*8 570*	*1 256*	*17 290*	*2 320*	*2 726*	*1 864*	*6 287*	*3 100*
Iceland scallop	**Peigne islandais**		**Peine islándico**		***Chlamys islandica***				**3,16(08)036,03**	**ISC**
21 Canada	6 912	3 746	6 793	9 900	12 180	12 145	6 632	3 160	2 764	1 347
Greenland	1 914	1 566	2 028	5 287	1 373	1 886	2 211	...	1 630	1 593
Iceland	-	-	-	-	2	-	-	-	-	-
St Pier Mq	14	20	8	2	221	239	5	0	3	16
USA	0	0	0	0	-	-	-	-	-	-
21 Fishing area total	*8 840*	*5 332*	*8 829*	*15 189*	*13 776*	*14 270*	*8 848*	*3 160*	*4 397*	*2 956*
27 Iceland	12 429	11 466	8 401	8 381	8 976	10 403	10 098	8 858	9 074	6 499
Norway	-	-	-	-	-	-	-	-	-	14
27 Fishing area total	*12 429*	*11 466*	*8 401*	*8 381*	*8 976*	*10 403*	*10 098*	*8 858*	*9 074*	*6 513*
Species total	*21 269*	*16 798*	*17 230*	*23 570*	*22 752*	*24 673*	*18 946*	*12 018*	*13 471*	*9 469*
Weathervane scallop	**Pecten géant du Pacifique**		**Vieira gigante del Pacífico**		***Patinopecten caurinus***				**3,16(08)066,01**	**SCG**
67 USA	6 884	6 224	4 944	1 950	2 372	2	3 228	2 642	2 012	1 052
67 Fishing area total	*6 884*	*6 224*	*4 944*	*1 950*	*2 372*	*2*	*3 228*	*2 642*	*2 012*	*1 052*
Species total	*6 884*	*6 224*	*4 944*	*1 950*	*2 372*	*2*	*3 228*	*2 642*	*2 012*	*1 052*
Yesso scallop	**Pétoncle du Japon**		**Vieira japonesa**		***Patinopecten yessoensis***				**3,16(08)066,07**	**JSC**
61 China,Taiwan	58	59	58	22	76	63	114	112	90	47
Japan	193 475	223 844	270 890	274 879	271 124	261 164	287 802	299 628	304 286	290 974
Korea Rep	35	94	53	74	122	196	42	6	0	11
Russian Fed	3 002	2 783	3 289	4 407	5 084	5 534	6 253	5 764	5 728	2 236
61 Fishing area total	*196 570*	*226 780*	*274 290*	*279 382*	*276 406*	*266 957*	*294 211*	*305 510*	*310 104*	*293 268*
Species total	*196 570*	*226 780*	*274 290*	*279 382*	*276 406*	*266 957*	*294 211*	*305 510*	*310 104*	*293 268*
Scallops nei	**Peignes nca**		**Peines nep**		***Pectinidae***				**3,16(08)XXX,XX**	**SCX**
27 Channel Is	8	8 F	-	-	-	50	80	-	-	-
Denmark	0	0	0	0	0	0	0	0	0	0
France	220	233	258	213	268	185	149	328	313	430
Norway	6 805	10 252	7 967	7 310	3	16	21	12	14	13
Russian Fed	5 353	4 000	6 211	8 372	7 634	13 878	12 887	11 948	12 717	13 598
Spain	89	41	81	81	53	36	35	141	63	35
UK	2 842	3 771	5 000	6 297	-	-	-	-	-	9 722
27 Fishing area total	*15 317*	*18 305 F*	*19 517*	*22 273*	*7 958*	*14 165*	*13 172*	*12 429*	*13 107*	*23 798*
37 Spain	-	-	-	-	41	42	15	-	-	-
37 Fishing area total	*-*	*-*	*-*	*-*	*41*	*42*	*15*	*-*	*-*	*-*
57 Australia	24 707	27 240	17 735	7 411	6 261	6 087	5 250	7 396	8 300	4 023
Indonesia	11	68	3	730	606	1	290	225	-	-
Thailand	-	-	12	18	5	5	3	1	1	1
57 Fishing area total	*24 718*	*27 308*	*17 750*	*8 159*	*6 872*	*6 093*	*5 543*	*7 622*	*8 301*	*4 024*
71 Australia	4 919	6 384	6 738	5 869	6 054	2 501	4 623	4 177	3 920	4 913
Indonesia	152	195	291	373	536	423	547	614	578	580
Philippines	30	27	107	156	139	61	40	62	53	54
Thailand	-	-	534	588	547	503	1 182	550	542	539
71 Fishing area total	*5 101*	*6 606*	*7 670*	*6 986*	*7 276*	*3 488*	*6 392*	*5 403*	*5 093*	*6 086*
81 Australia	6	6	23	23	5	1	1	1	0	1
81 Fishing area total	*6*	*6*	*23*	*23*	*5*	*1*	*1*	*1*	*0*	*1*
87 Chile	864	1 332	1 225	1 365	1 577	2 598	3 662	1 715	332	141
87 Fishing area total	*864*	*1 332*	*1 225*	*1 365*	*1 577*	*2 598*	*3 662*	*1 715*	*332*	*141*
Species total	*46 006*	*53 557 F*	*46 185*	*38 806*	*23 729*	*26 387*	*28 785*	*27 170*	*26 833*	*34 050*
Group total	***523 289***	***503 412***	***618 011***	***537 324***	***535 168***	***532 893***	***554 767***	***612 702***	***661 171***	***702 525***

B-56 Clams, cockles, arkshells / Clams, coques, arches / Almejas, berberechos, arcas

Capture production by species, fishing areas and countries or areas / Captures par espèces, zones de pêche et pays ou zones / Capturas por especies, áreas de pesca y países o áreas

Species, Fishing area Espèce, Zone de pêche Especie, Area de pesca	1992 mt	1993 mt	1994 mt	1995 mt	1996 mt	1997 mt	1998 mt	1999 mt	2000 mt	2001 mt
Ark clams nei	**Arches nca**		**Arcas nep**		*Arca spp*				**3,16(04)001,XX**	**ARK**
31 Cuba	1 288	1 354	1 553	1 905	1 963	2 989	2 899	2 499	2 500 F	2 500 F
Venezuela	15 793	28 581	31 193	33 987	31 925	39 128	27 981	38 646	44 709	43 675
31 Fishing area total	*17 081*	*29 935*	*32 746*	*35 892*	*33 888*	*42 117*	*30 880*	*41 145*	*47 209 F*	*46 175 F*
61 Korea Rep	914	553	601	473	782	667	1 311	534	318	153
61 Fishing area total	*914*	*553*	*601*	*473*	*782*	*667*	*1 311*	*534*	*318*	*153*
77 Mexico	1 084	837	1 315	1 033	834	1 367	1 018	880	675	600
77 Fishing area total	*1 084*	*837*	*1 315*	*1 033*	*834*	*1 367*	*1 018*	*880*	*675*	*600*
Species total	*19 079*	*31 325*	*34 662*	*37 398*	*35 504*	*44 151*	*33 209*	*42 559*	*48 202 F*	*46 928 F*
Half-crenated ark	**Arche crénelée**		**Arca japonesa**		*Scapharca subcrenata*				**3,16(04)005,08**	**MCL**
61 Japan	14 866	16 689	17 373	15 426	16 328	14 133	10 120	10 413	7 308	4 899
61 Fishing area total	*14 866*	*16 689*	*17 373*	*15 426*	*16 328*	*14 133*	*10 120*	*10 413*	*7 308*	*4 899*
Species total	*14 866*	*16 689*	*17 373*	*15 426*	*16 328*	*14 133*	*10 120*	*10 413*	*7 308*	*4 899*
Blood cockle	**Arche granuleuse**		**Arca del Pacífico occidental**		*Anadara granosa*				**3,16(04)071,01**	**BLC**
57 Thailand	-	-	-	-	-	-	20	7	7	6
57 Fishing area total	-	-	-	-	-	-	*20*	*7*	*7*	*6*
61 Korea Rep	428	3 279	1 245	1 415	995	493	12 114	6 503	4 184	923
Russian Fed	-	-	-	-	-	-	-	-	-	100
61 Fishing area total	*428*	*3 279*	*1 245*	*1 415*	*995*	*493*	*12 114*	*6 503*	*4 184*	*1 023*
71 Thailand	-	-	-	-	-	-	421	1 750	1 726	1 716
71 Fishing area total	-	-	-	-	-	-	*421*	*1 750*	*1 726*	*1 716*
Species total	*428*	*3 279*	*1 245*	*1 415*	*995*	*493*	*12 555*	*8 260*	*5 917*	*2 745*
Anadara clams nei	**Arches anadara nca**		**Arcas anadara nep**		*Anadara spp*				**3,16(04)071,XX**	**BLS**
57 Indonesia	30 279	21 648	25 209	16 712	21 227	22 348	15 443	10 603	9 538	8 960
57 Fishing area total	*30 279*	*21 648*	*25 209*	*16 712*	*21 227*	*22 348*	*15 443*	*10 603*	*9 538*	*8 960*
71 Fiji Islands	765	446	454 F	1 513	1 044	1 950	2 800	2 990	2 750 F	2 884
Indonesia	15 393	20 149	21 738	26 949	24 956	19 571	16 148	23 307	25 157	23 630
Philippines	30	36	62	6	6	3	4	4	3	3
71 Fishing area total	*16 188*	*20 631*	*22 254 F*	*28 468*	*26 006*	*21 524*	*18 952*	*26 301*	*27 910 F*	*26 517*
Species total	*46 467*	*42 279*	*47 463 F*	*45 180*	*47 233*	*43 872*	*34 395*	*36 904*	*37 448 F*	*35 477*
Ocean quahog	**Cyprine d'Islande**		**Almeja de Islandia**		*Arctica islandica*				**3,16(09)045,01**	**CLQ**
21 Canada	-	-	-	124	114	-	-	66	-	164
USA	192 496	197 559	177 874	183 777	173 685	163 799	148 506	144 366	122 547	142 662
21 Fishing area total	*192 496*	*197 559*	*177 874*	*183 901*	*173 799*	*163 799*	*148 506*	*144 432*	*122 547*	*142 826*
27 Iceland	-	-	-	1 980	6 315	4 351	8 776	3 501	1 584	7 434
27 Fishing area total	-	-	-	*1 980*	*6 315*	*4 351*	*8 776*	*3 501*	*1 584*	*7 434*
Species total	*192 496*	*197 559*	*177 874*	*185 881*	*180 114*	*168 150*	*157 282*	*147 933*	*124 131*	*150 260*
Striped venus	**Petite praire**		**Chirla**		*Chamelea gallina*				**3,16(11)001,05**	**SVE**
27 France	827	573	580	651	706	810	715	790	696	627
Portugal	212	48	38	4	7	17	185	129	156	48
Spain	20	16	9	10	1 728	1 728	3 087	3 265	4 390	4 881
27 Fishing area total	*1 059*	*637*	*627*	*665*	*2 441*	*2 555*	*3 987*	*4 184*	*5 242*	*5 556*
37 Albania	76	50 F	20 F	0	0	-	-	-	-	-
Bulgaria	-	-	-	182	-	-	-	-	-	-
Italy	33 853	29 392	19 255	32 609	36 707	28 604	28 830	36 462	34 191	34 916
Slovenia	-	-	-	-	-	-	-	-	0	1
Spain	800 F	700 F	600 F	500 F	428	485	431	374	389	240
Turkey	20 412	30 134	31 869	11 864	10 925	7 150	3 550	3 585	10 000	7 500
37 Fishing area total	*55 141 F*	*60 276 F*	*51 744 F*	*45 155 F*	*48 060*	*36 239*	*32 811*	*40 421*	*44 580*	*42 657*
Species total	*56 200 F*	*60 913 F*	*52 371 F*	*45 820 F*	*50 501*	*38 794*	*36 798*	*44 605*	*49 822*	*48 213*
Pullet carpet shell	**Palourde bleue**		**Almeja babosa**		*Venerupis pullastra*				**3,16(11)003,01**	**CTS**
27 France	1 036	1 036	1 397	638	934	1 493	2 216	2 306	1 311	1 595
Portugal	64	213	573	225	153	419	240	212	87	81
Spain	334	350	432	442	1 734	2 141	2 710	2 310	963	744
27 Fishing area total	*1 434*	*1 599*	*2 402*	*1 305*	*2 821*	*4 053*	*5 166*	*4 828*	*2 361*	*2 420*
37 Spain	-	-	-	-	5	5	-	141	27	-
37 Fishing area total	-	-	-	-	*5*	*5*	-	*141*	*27*	-
Species total	*1 434*	*1 599*	*2 402*	*1 305*	*2 826*	*4 058*	*5 166*	*4 969*	*2 388*	*2 420*

B-56

Clams, cockles, arkshells	Capture production by species, fishing areas and countries or areas
Clams, coques, arches	Captures par espèces, zones de pêche et pays ou zones
Almejas, berberechos, arcas	Capturas por especies, áreas de pesca y países o áreas

Species, Fishing area Espèce, Zone de pêche Especie, Area de pesca	1992 mt	1993 mt	1994 mt	1995 mt	1996 mt	1997 mt	1998 mt	1999 mt	2000 mt	2001 mt
Stutchbury's venus	**...B**		**...C**		***Chione stutchburyi***				**3,16(11)007,04**	**KNU**
81 New Zealand	...	...	...	1 220	815	541	1 325	1 396	1 789	1 748
81 Fishing area total	...	...	...	*1 220*	*815*	*541*	*1 325*	*1 396*	*1 789*	*1 748*
Species total	...	...	...	*1 220*	*815*	*541*	*1 325*	*1 396*	*1 789*	*1 748*
Japanese hard clam	**Cythérée du Japon**		**Mercenaria japonesa**		***Meretrix lusoria***				**3,16(11)017,01**	**HCJ**
61 China,Taiwan	484	419	72	256	135	161	805	526	0	-
Japan	2 581	2 964	2 330	2 060	1 944	1 897	1 870	1 785	1 543	1 245
Korea Rep	2 382	1 407	1 561	2 715	2 315	2 075	3 472	1 799	1 430	1 044
61 Fishing area total	*5 447*	*4 790*	*3 963*	*5 031*	*4 394*	*4 133*	*6 147*	*4 110*	*2 973*	*2 289*
Species total	*5 447*	*4 790*	*3 963*	*5 031*	*4 394*	*4 133*	*6 147*	*4 110*	*2 973*	*2 289*
Hard clams nei	**...B**		**...C**		***Meretrix spp***				**3,16(11)017,XX**	**HCX**
57 Indonesia	3 441	1 392	2 126	3 074	3 693	3 722	6 639	3 420	2 606	2 660
57 Fishing area total	*3 441*	*1 392*	*2 126*	*3 074*	*3 693*	*3 722*	*6 639*	*3 420*	*2 606*	*2 660*
71 Indonesia	481	7 312	10 431	14 949	9 788	10 305	10 507	11 347	11 571	11 820
71 Fishing area total	*481*	*7 312*	*10 431*	*14 949*	*9 788*	*10 305*	*10 507*	*11 347*	*11 571*	*11 820*
Species total	*3 922*	*8 704*	*12 557*	*18 023*	*13 481*	*14 027*	*17 146*	*14 767*	*14 177*	*14 480*
Grooved carpet shell	**Palourde croisée d'Europe**		**Almeja fina**		***Ruditapes decussatus***				**3,16(11)020,01**	**CTG**
27 France	1	14	3	3	100	65	504	469	80	298
Ireland	209	...	...	20	21	23	3	3	3	130
Portugal	15 070	8 028	1 638	169	185	27	33	75	16	21
Spain	448	221	339	448	675	1 036	1 231	1 209	592	270
UK	-	-	-	-	-	-	-	-	-	29
27 Fishing area total	*15 728*	*8 263*	*1 980*	*640*	*981*	*1 151*	*1 771*	*1 756*	*691*	*748*
37 France	179	195	96	36	6	6	6	21	21	6
Spain	-	-	-	-	86	52	109	74	20	19
Tunisia	1 383	1 538	1 036	1 343	396	77	57	48	755	544
37 Fishing area total	*1 562*	*1 733*	*1 132*	*1 379*	*488*	*135*	*172*	*143*	*796*	*569*
Species total	*17 290*	*9 996*	*3 112*	*2 019*	*1 469*	*1 286*	*1 943*	*1 899*	*1 487*	*1 317*
Japanese carpet shell	**Palourde japonaise**		**Almeja japonesa**		***Ruditapes philippinarum***				**3,16(11)020,02**	**CLJ**
61 China,Taiwan	400	26	24	16	-	-	-	-	-	-
Japan	59 038	57 356	46 597	49 466	43 703	39 660	36 807	43 088	35 558	31 022
Korea Rep	13 016	31 202	14 595	15 041	12 392	16 854	14 585	13 963	20 982	20 004
61 Fishing area total	*72 454*	*88 584*	*61 216*	*64 523*	*56 095*	*56 514*	*51 392*	*57 051*	*56 540*	*51 026*
Species total	*72 454*	*88 584*	*61 216*	*64 523*	*56 095*	*56 514*	*51 392*	*57 051*	*56 540*	*51 026*
Gay's little venus	**...B**		**Juliana**		***Tawera gayi***				**3,16(11)024,01**	**TWG**
87 Chile	...	...	...	...	...	...	...	...	1	291
87 Fishing area total	...	...	...	...	...	...	...	...	*1*	*291*
Species total	...	...	...	...	...	...	...	...	*1*	*291*
Triangular tivela	**...B**		**...C**		***Tivela mactroides***				**3,16(11)025,02**	**TVM**
41 Brazil	...	...	...	...	126	196	664	1 684	273	270 F
41 Fishing area total	...	...	...	...	*126*	*196*	*664*	*1 684*	*273*	*270 F*
Species total	...	...	...	...	*126*	*196*	*664*	*1 684*	*273*	*270 F*
Butter clam	**Coque jaune**		**Almeja amarilla**		***Saxidomus giganteus***				**3,16(11)037,02**	**BCL**
67 Canada	1 338	1 352	1 797	1 728	1 431	1 206	1 600	945	949	1 111
USA	...	...	...	...	...	87	126	74	164	48
67 Fishing area total	*1 338*	*1 352*	*1 797*	*1 728*	*1 431*	*1 293*	*1 726*	*1 019*	*1 113*	*1 159*
Species total	*1 338*	*1 352*	*1 797*	*1 728*	*1 431*	*1 293*	*1 726*	*1 019*	*1 113*	*1 159*
Short neck clams nei	**...B**		**...C**		***Paphia spp***				**3,16(11)041,XX**	**NCL**
57 Thailand	4 136	2 738	5 070	10 800	18 329	11 625	6 204	27 366	26 991	26 827
57 Fishing area total	*4 136*	*2 738*	*5 070*	*10 800*	*18 329*	*11 625*	*6 204*	*27 366*	*26 991*	*26 827*
71 Philippines	1 190	517	1 498	30	31	2	1	1	2	2
Thailand	66 439	39 834	27 960	20 060	34 560	24 227	43 457	42 612	42 029	41 772
71 Fishing area total	*67 629*	*40 351*	*29 458*	*20 090*	*34 591*	*24 229*	*43 458*	*42 613*	*42 031*	*41 774*
81 New Zealand	...	...	...	317	211	114	204	181	131	202
81 Fishing area total	...	...	...	*317*	*211*	*114*	*204*	*181*	*131*	*202*
Species total	*71 765*	*43 089*	*34 528*	*31 207*	*53 131*	*35 968*	*49 866*	*70 160*	*69 153*	*68 803*
Pacific littleneck clam	**Palourde commune**		**Almejuela común**		***Protothaca staminea***				**3,16(11)055,02**	**PTS**
67 USA	-	-	-	-	-	-	-	-	-	102
67 Fishing area total	-	-	-	-	-	-	-	-	-	*102*

B-56 Clams, cockles, arkshells — Capture production by species, fishing areas and countries or areas
Clams, coques, arches — Captures par espèces, zones de pêche et pays ou zones
Almejas, berberechos, arcas — Capturas por especies, áreas de pesca y países o áreas

Species, Fishing area Espèce, Zone de pêche Especie, Area de pesca	1992 mt	1993 mt	1994 mt	1995 mt	1996 mt	1997 mt	1998 mt	1999 mt	2000 mt	2001 mt
Species total	-	-	-	-	-	-	-	-	-	*102*
Taca clam	**Palourde taca**		**Taca**		***Protothaca thaca***			**3,16(11)055,03**		**TCL**
87 Chile	34 792	23 068	16 236	17 162	20 016	12 475	24 254	16 429	16 303	26 483
87 Fishing area total	*34 792*	*23 068*	*16 236*	*17 162*	*20 016*	*12 475*	*24 254*	*16 429*	*16 303*	*26 483*
Species total	*34 792*	*23 068*	*16 236*	*17 162*	*20 016*	*12 475*	*24 254*	*16 429*	*16 303*	*26 483*
Northern quahog(=Hard clam)	**Praire**		**Chirla mercenaria**		***Mercenaria mercenaria***			**3,16(11)075,01**		**CLH**
21 Canada	1 555	1 127	1 170	830	1 109	1 656	1 234	2 536	1 252	1 657
USA	24 689	23 026	10 562	16 058	21 855	4 631	...	...	8 217	5 023
21 Fishing area total	*26 244*	*24 153*	*11 732*	*16 888*	*22 964*	*6 287*	*1 234*	*2 536*	*9 469*	*6 680*
27 UK	-	-	-	-	-	0	0	-	175	-
27 Fishing area total	-	-	-	-	-	*0*	*0*	-	*175*	-
31 USA	...	...	...	...	...	...	...	...	5 506	5 113
31 Fishing area total	...	...	...	...	...	...	...	...	*5 506*	*5 113*
Species total	*26 244*	*24 153*	*11 732*	*16 888*	*22 964*	*6 287*	*1 234*	*2 536*	*15 150*	*11 793*
Venus clams nei	**Petites praires nca**		**Almejas(=Veneridos) nep**		***Veneridae***			**3,16(11)XXX,XX**		**CLV**
27 France	-	-	-	-	1	99	-	189	234	289
27 Fishing area total	-	-	-	-	*1*	*99*	-	*189*	*234*	*289*
31 Mexico	1 223	637	805	1 000	814	746	1 001	844	1 403	1 834
Venezuela	472	354	487	517	536	533	295	378	353	388
31 Fishing area total	*1 695*	*991*	*1 292*	*1 517*	*1 350*	*1 279*	*1 296*	*1 222*	*1 756*	*2 222*
37 France	120	130	64	-	-	-	-	18	19	22
37 Fishing area total	*120*	*130*	*64*	-	-	-	-	*18*	*19*	*22*
77 Honduras	3	4	4	5	5	3	0	0	...	...
Mexico	4 944	4 882	5 134	6 239	5 038	3 912	3 672	4 486	4 835	3 406
77 Fishing area total	*4 947*	*4 886*	*5 138*	*6 244*	*5 043*	*3 915*	*3 672*	*4 486*	*4 835*	*3 406*
87 Ecuador	5	0	0	0	0	0	0	0	0	0
87 Fishing area total	*5*	*0*	*0*	*0*	*0*	*0*	*0*	*0*	*0*	*0*
Species total	*6 767*	*6 007*	*6 494*	*7 761*	*6 394*	*5 293*	*4 968*	*5 915*	*6 844*	*5 939*
Imperial surf clam	**Clam**		**Almeja**		***Pseudocardium sybillae***			**3,16(12)001,02**		**HCL**
61 Japan	8 175	7 768	8 003	8 018	8 738	7 541	8 227	8 808	8 883	8 914
Korea Rep	3 416	8 137	11 383	9 038	6 053	7 179	2 877	7 620	4 150	3 409
61 Fishing area total	*11 591*	*15 905*	*19 386*	*17 056*	*14 791*	*14 720*	*11 104*	*16 428*	*13 033*	*12 323*
Species total	*11 591*	*15 905*	*19 386*	*17 056*	*14 791*	*14 720*	*11 104*	*16 428*	*13 033*	*12 323*
Pacific horse clams nei	**...B**		**...C**		***Tresus spp***			**3,16(12)005,XX**		**TQZ**
67 USA	-	-	-	-	-	-	-	-	1	2
67 Fishing area total	-	-	-	-	-	-	-	-	*1*	*2*
Species total	-	-	-	-	-	-	-	-	*1*	*2*
Atlantic surf clam	**Mactre solide**		**Almeja blanca**		***Spisula solidissima***			**3,16(12)020,01**		**CLB**
21 Canada	1 129	885	921	674	1 204	800	444	303	496	602
USA	182 269	179 354	167 600	155 382	154 380	140 621	131 256	142 067	165 269	165 708
21 Fishing area total	*183 398*	*180 239*	*168 521*	*156 056*	*155 584*	*141 421*	*131 700*	*142 370*	*165 765*	*166 310*
Species total	*183 398*	*180 239*	*168 521*	*156 056*	*155 584*	*141 421*	*131 700*	*142 370*	*165 765*	*166 310*
Stimpson's surf clam	**Douceron de Stimpson**		**...C**		***Spisula polynyma***			**3,16(12)020,02**		**CLT**
21 Canada	11 833	19 930	20 735	24 008	25 597	27 322	25 976	26 699	22 979	20 257
USA	113	0	0	9	15	43	32	23	6	16
21 Fishing area total	*11 946*	*19 930*	*20 735*	*24 017*	*25 612*	*27 365*	*26 008*	*26 722*	*22 985*	*20 273*
Species total	*11 946*	*19 930*	*20 735*	*24 017*	*25 612*	*27 365*	*26 008*	*26 722*	*22 985*	*20 273*
Solid surf clam	**Spisule épaisse**		**...C**		***Spisula solida***			**3,16(12)020,05**		**ULO**
27 Denmark	-	-	-	-	-	-	-	-	55	214
France	...	...	...	...	...	...	...	...	127	101
Portugal	-	-	-	-	-	-	-	765	1 153	1 125
27 Fishing area total	...	...	...	...	...	...	...	*765*	*1 335*	*1 440*
Species total	...	...	...	...	...	...	...	*765*	*1 335*	*1 440*
Taquilla clams	**Mactres taquille**		**Taquillas**		***Mulinia spp***			**3,16(12)040,XX**		**MUN**
87 Chile	...	...	2 567	1 852	999	2 757	2 549	1 536	1 491	1 699
87 Fishing area total	...	...	*2 567*	*1 852*	*999*	*2 757*	*2 549*	*1 536*	*1 491*	*1 699*

B-56 Clams, cockles, arkshells / Clams, coques, arches / Almejas, berberechos, arcas

Capture production by species, fishing areas and countries or areas / Captures par espèces, zones de pêche et pays ou zones / Capturas por especies, áreas de pesca y países o áreas

Species, Fishing area Espèce, Zone de pêche Especie, Area de pesca	1992 mt	1993 mt	1994 mt	1995 mt	1996 mt	1997 mt	1998 mt	1999 mt	2000 mt	2001 mt
Species total	...	...	*2 567*	*1 852*	*999*	*2 757*	*2 549*	*1 536*	*1 491*	*1 699*
Donax clams	**Olives de mer**		**Coquinas**		***Donax spp***				**3,16(15)002,XX**	**DON**
27 France	89	124	292	197	-	-	-	34	94	88
Portugal	...	...	...	...	...	...	...	456	401	540
27 Fishing area total	*89*	*124*	*292*	*197*	...	...	...	*490*	*495*	*628*
41 Uruguay	0	0	0	0	0	0	0	0	0	0
41 Fishing area total	*0*	*0*	*0*	*0*	*0*	*0*	*0*	*0*	*0*	*0*
47 South Africa	0	0	0	0	0	-	-	-	-	-
47 Fishing area total	*0*	*0*	*0*	*0*	*0*	-	-	-	-	-
Species total	*89*	*124*	*292*	*197*	*0*	*0*	*0*	*490*	*495*	*628*
Razor clams nei	**Couteaux nca**		**Navajas(=Solénidos) nep**		***Solen spp***				**3,16(16)003,XX**	**RAZ**
27 Ireland	11	15	-	-	-	28	316	407	334	201
Portugal	1 133	1 106	1 225	2 603	1 729	124	15	4	12	214
Spain	5	100	80	21	86	205	166	171	50	129
UK	-	-	41	46	56	220	134	114	67	59
27 Fishing area total	*1 149*	*1 221*	*1 346*	*2 670*	*1 871*	*577*	*631*	*696*	*463*	*603*
37 France	...	...	...	...	...	...	...	5	5	5
Spain	-	-	-	-	2	4	1	-	-	-
37 Fishing area total	...	...	...	...	*2*	*4*	*1*	*5*	*5*	*5*
Species total	*1 149*	*1 221*	*1 346*	*2 670*	*1 873*	*581*	*632*	*701*	*468*	*608*
Atl.jackknife(=Atl.razor clam)	**Couteau de l'Atlantique**		**Navaja del Atlántico**		***Ensis directus***				**3,16(16)005,02**	**CLR**
21 USA	-	6	-	-	-	14	49	64	99	36
21 Fishing area total	-	*6*	-	-	-	*14*	*49*	*64*	*99*	*36*
Species total	-	*6*	-	-	-	*14*	*49*	*64*	*99*	*36*
Pacific razor clam	**Couteau du Pacifique**		**Navaja del Pacífico**		***Siliqua patula***				**3,16(16)007,01**	**RAP**
67 USA	-	-	-	-	-	-	-	-	14	61
67 Fishing area total	-	-	-	-	-	-	-	-	*14*	*61*
Species total	-	-	-	-	-	-	-	-	*14*	*61*
Sand gaper	**Mye des sables**		**Almeja de can**		***Mya arenaria***				**3,16(17)006,01**	**CLS**
21 Canada	3 059	2 152	3 270	3 841	1 469	2 323	2 638	2 680	3 026	3 152
USA	7 885	9 146	5 190	4 813	4 355	4 454	5 493	5 113	5 222	6 794
21 Fishing area total	*10 944*	*11 298*	*8 460*	*8 654*	*5 824*	*6 777*	*8 131*	*7 793*	*8 248*	*9 946*
67 USA	-	-	-	-	-	-	-	-	-	123
67 Fishing area total	-	-	-	-	-	-	-	-	-	*123*
Species total	*10 944*	*11 298*	*8 460*	*8 654*	*5 824*	*6 777*	*8 131*	*7 793*	*8 248*	*10 069*
Pacific geoduck	**Panopée du Pacifique**		**Panopea del Pacífico**		***Panopea abrupta***				**3,16(18)089,01**	**GEC**
67 Canada	2 874	2 455	2 235	2 056	1 768	1 757	1 784	1 688	1 562	1 413
USA	-	-	-	-	-	1 962	2 243	2 451	2 259	2 481
67 Fishing area total	*2 874*	*2 455*	*2 235*	*2 056*	*1 768*	*3 719*	*4 027*	*4 139*	*3 821*	*3 894*
Species total	*2 874*	*2 455*	*2 235*	*2 056*	*1 768*	*3 719*	*4 027*	*4 139*	*3 821*	*3 894*
Common edible cockle	**Coque commune**		**Berberecho común**		***Cerastoderma edule***				**3,16(23)002,03**	**COC**
27 Denmark	2 423	543	31	-	5	2 603	1 993	246	2 089	2 392
France	7 843	4 565	2 152	379	664	731	418	481	81	11
Germany	495	-	-	-	-	-	-	-	-	-
Ireland	25	17	26	20	10	64	296	1	8	6
Netherlands	47 060	43 635	38 350	39 594	6 300	10 923	68 133	50 888	19 633	-
Norway	-	-	-	-	-	-	-	-	-	33
Portugal	1 116	488	531	591	3 522	1 285	1 264	1 409	1 292	683
Spain	903	331	573	627	2 358	2 964	2 472	3 104	3 740	1 486
UK	32 051	21 360	22 330	21 796	24 176	19 493	12 035	14 123	20 306	19 048
27 Fishing area total	*91 916*	*70 939*	*63 993*	*63 007*	*37 035*	*38 063*	*86 611*	*70 252*	*47 149*	*23 659*
34 Senegal	...	...	...	178	54	69	139	147	117	105
34 Fishing area total	...	...	...	*178*	*54*	*69*	*139*	*147*	*117*	*105*
37 Spain	-	-	-	-	-	-	-	1	-	-
37 Fishing area total	-	-	-	-	-	-	-	*1*	-	-
Species total	*91 916*	*70 939*	*63 993*	*63 185*	*37 089*	*38 132*	*86 750*	*70 400*	*47 266*	*23 764*
Cockles nei	**Coques nca**		**Berberechos(=Cárdidos) nep**		***Cardiidae***				**3,16(23)XXX,XX**	**COZ**
61 Korea Rep	3 714	11 226	6 551	428	2 583	1 587	1 078	112	411	2 002
61 Fishing area total	*3 714*	*11 226*	*6 551*	*428*	*2 583*	*1 587*	*1 078*	*112*	*411*	*2 002*

B-56 Clams, cockles, arkshells / Clams, coques, arches / Almejas, berberechos, arcas

Capture production by species, fishing areas and countries or areas / Captures par espèces, zones de pêche et pays ou zones / Capturas por especies, áreas de pesca y países o áreas

Species, Fishing area Espèce, Zone de pêche Especie, Area de pesca	1992 mt	1993 mt	1994 mt	1995 mt	1996 mt	1997 mt	1998 mt	1999 mt	2000 mt	2001 mt
67 USA	-	-	-	-	-	22	-	18	49	67
67 Fishing area total	*-*	*-*	*-*	*-*	*-*	*22*	*-*	*18*	*49*	*67*
Species total	*3 714*	*11 226*	*6 551*	*428*	*2 583*	*1 609*	*1 078*	*130*	*460*	*2 069*
Pipi wedge clam	**...B**		**...C**			***Paphies australis***		**3,16(24)002,01**		**AFQ**
57 Australia	441	304	380	351	454	830	1 041	976	1 085	1 249
57 Fishing area total	*441*	*304*	*380*	*351*	*454*	*830*	*1 041*	*976*	*1 085*	*1 249*
81 Australia	212	315	246	254	248	465	455	...	...	...
81 Fishing area total	*212*	*315*	*246*	*254*	*248*	*465*	*455*	*...*	*...*	*...*
Species total	*653*	*619*	*626*	*605*	*702*	*1 295*	*1 496*	*976*	*1 085*	*1 249*
Macha clam	**Mesodème chilienne**		**Macha**			***Mesodesma donacium***		**3,16(24)039,01**		**CLM**
87 Chile	11 832	8 274	6 415	6 913	6 144	6 770	6 464	1 728	1 249	1 396
Peru	1 483	1 513	1 070	1 200	1 060	1 061	578	-	10	0
87 Fishing area total	*13 315*	*9 787*	*7 485*	*8 113*	*7 204*	*7 831*	*7 042*	*1 728*	*1 259*	*1 396*
Species total	*13 315*	*9 787*	*7 485*	*8 113*	*7 204*	*7 831*	*7 042*	*1 728*	*1 259*	*1 396*
Chilean semele	**Sémèle chilienne**		**Almeja blanca chilena**			***Semele solida***		**3,16(39)001,03**		**TUW**
87 Chile	-	-	4 613	2 523	4 418	2 199	1 900	2 071	4 212	3 054
87 Fishing area total	*-*	*-*	*4 613*	*2 523*	*4 418*	*2 199*	*1 900*	*2 071*	*4 212*	*3 054*
Species total	*-*	*-*	*4 613*	*2 523*	*4 418*	*2 199*	*1 900*	*2 071*	*4 212*	*3 054*
Clams, etc. nei	**Clams, etc. nca**		**Almejas, etc. nep**		***Bivalvia***			**3,16(XX)XXX,XX**		**CLX**
04 Indonesia	1 369	1 775	763	835	844	956	914	304	303	260
04 Fishing area total	*1 369*	*1 775*	*763*	*835*	*844*	*956*	*914*	*304*	*303*	*260*
21 Canada	20	642	0	263	773	259	49	791	345	1 078
USA	0	0	789	0	-	2	17	345	4	51
21 Fishing area total	*20*	*642*	*789*	*263*	*773*	*261*	*66*	*1 136*	*349*	*1 129*
27 Channel Is	5	1 F	76	70 F	-	-	-	-	-	-
Denmark	484	1 651	2 716	3 136	-	-	-	-	-	-
France	3 917	2 370	1 465	2 637	3 746	2 893	5 822	4 273	3 482	4 005
Germany	...	1 301	1 034	5 410	-	-	-	-	-	-
Ireland	-	-	-	-	-	-	-	-	301	126
Portugal	661	3 367	4 656	1 489	1 482	741	1 007	493	395	398
Spain	1 467	1 412	3 251	3 210	1 758	2 459	2 967	2 761	2 000	1 850
UK	255	246	37	326	49	144	13	78	15	91
27 Fishing area total	*6 789*	*10 348 F*	*13 235*	*16 278 F*	*7 035*	*6 237*	*9 809*	*7 605*	*6 193*	*6 470*
31 Colombia	1 007	5	7	1	0	-	0	-	...	...
Martinique	...	...	...	...	...	...	...	...	900	700
USA	7 750	9 999	16 012	22 297	4 455	6 739	6 392	3 972	254	2 096
31 Fishing area total	*8 757*	*10 004*	*16 019*	*22 298*	*4 455*	*6 739*	*6 392*	*3 972*	*1 154*	*2 796*
34 Spain	-	-	-	-	-	-	-	1	3	6
34 Fishing area total	*-*	*-*	*-*	*-*	*-*	*-*	*-*	*1*	*3*	*6*
37 France	-	-	-	-	-	-	-	365	797	691
Morocco	5	10	0	0	0	0	0	-	-	-
Spain	2 500 F	3 000 F	3 500 F	4 000 F	4 078	2 963	3 527	2 155	1 179	985
37 Fishing area total	*2 505 F*	*3 010 F*	*3 500 F*	*4 000 F*	*4 078*	*2 963*	*3 527*	*2 520*	*1 976*	*1 676*
41 Spain	-	-	-	-	-	-	-	759	-	-
Uruguay	16	9	14	...	...	0	0	0	6	17
41 Fishing area total	*16*	*9*	*14*	*...*	*...*	*0*	*0*	*759*	*6*	*17*
57 Malaysia	7 730	18 424	8 670	15 250	17 745	24 663	5 097	572	5 115	2 983
57 Fishing area total	*7 730*	*18 424*	*8 670*	*15 250*	*17 745*	*24 663*	*5 097*	*572*	*5 115*	*2 983*
61 Japan	56 937	57 728	42 515	50 015	52 361	42 987	47 584	35 319	35 529	30 133
Korea Rep	20 917	30 057	33 207	31 063	33 446	8 851	15 665	14 728	12 277	15 705
Russian Fed	-	-	-	-	-	-	-	-	216	317
61 Fishing area total	*77 854*	*87 785*	*75 722*	*81 078*	*85 807*	*51 838*	*63 249*	*50 047*	*48 022*	*46 155*
67 USA	3 380	10 550	10 244	11 215	3 912	381	2 509	1 601	1 519	1 250
67 Fishing area total	*3 380*	*10 550*	*10 244*	*11 215*	*3 912*	*381*	*2 509*	*1 601*	*1 519*	*1 250*
71 Fiji Islands	169	27	28 F	92	54	45	52	50	40 F	43
Malaysia	2 909	3 816	722	732	861	837	904	1 105	1 116	966
Philippines	1 528	1 343	4 282	921	725	375	227	277	222	223
Solomon Is	...	...	...	...	60	280 F	10 F	10 F	5 F	5 F
71 Fishing area total	*4 606*	*5 186*	*5 032 F*	*1 745*	*1 700*	*1 537 F*	*1 193 F*	*1 442 F*	*1 383 F*	*1 237 F*
77 Panama	440	1 043	758	1 669	1 391	1 507	1 072	1 518	1 630	1 700 F
USA	2	16	4	4	2	3	8	29	3	5
77 Fishing area total	*442*	*1 059*	*762*	*1 673*	*1 393*	*1 510*	*1 080*	*1 547*	*1 633*	*1 705 F*
87 Chile	38 405	22 216	21 510	20 075	18 720	17 880	11 900	20 804	16 551	16 775

B-56 Clams, cockles, arkshells / Clams, coques, arches / Almejas, berberechos, arcas

Capture production by species, fishing areas and countries or areas
Captures par espèces, zones de pêche et pays ou zones
Capturas por especies, áreas de pesca y países o áreas

Species, Fishing area Espèce, Zone de pêche Especie, Area de pesca	1992 mt	1993 mt	1994 mt	1995 mt	1996 mt	1997 mt	1998 mt	1999 mt	2000 mt	2001 mt
Colombia	3	5	1	0	4	5	6	6	10 F	10 F
Ecuador	478	2	2	6	10	10	10	10	10	10
Peru	1 862	668	643	569	411	236	152	338	956	949
87 Fishing area total	*40 748*	*22 891*	*22 156*	*20 650*	*19 145*	*18 131*	*12 068*	*21 158*	*17 527 F*	*17 744 F*
Species total	*154 216 F*	*171 683 F*	*156 906 F*	*175 285 F*	*146 887*	*115 216 F*	*105 904 F*	*92 664 F*	*85 183 F*	*83 428 F*
Group total	***1 056 794***	***1 058 818***	***948 741***	***960 684***	***919 151***	***815 300***	***838 560***	***841 586***	***818 239***	***808 945***

B-57 Squids, cuttlefishes, octopuses — Capture production by species, fishing areas and countries or areas
Encornets, seiches, poulpes — Captures par espèces, zones de pêche et pays ou zones
Calamares, jibias, pulpos — Capturas por especies, áreas de pesca y países o áreas

Species, Fishing area Espèce, Zone de pêche Especie, Area de pesca	1992 mt	1993 mt	1994 mt	1995 mt	1996 mt	1997 mt	1998 mt	1999 mt	2000 mt	2001 mt
Common cuttlefish	**Seiche commune**		**Sepia común**			***Sepia officinalis***			**3,21(02)002,02**	**CTC**
27 Belgium	171	206	288	467	386	153	252	222	463	370
Channel Is	4	4 F	2	2 F	12	10	15	22	26	8
Portugal	1 234	1 209	1 125	987	1 636	1 422	1 734	1 162	1 357	1 348
Spain	-	-	-	-	176	336	1 078	1 411	1 895	833
27 Fishing area total	*1 409*	*1 419 F*	*1 415*	*1 456 F*	*2 210*	*1 921*	*3 079*	*2 817*	*3 741*	*2 559*
34 Cyprus	-	-	-	-	-	-	-	-	1	0
Greece	1 604	1 349	1 125	900	654	588	472	391	157	615
Portugal	2	166	360	377	440	218	131	157	75	115
34 Fishing area total	*1 606*	*1 515*	*1 485*	*1 277*	*1 094*	*806*	*603*	*548*	*233*	*730*
37 Albania	...	...	...	39	33	33	51	51	50	22
Algeria	400 F	500 F	600 F	351	567	619	492	312	300 F	300 F
Cyprus	103	174	218	153	111	89	146	140	122	106
France	254	152	130	100	85	85	85	88	106	83
Greece	1 780	1 941	2 853	2 516	1 819	2 381	1 807	2 732	1 609	1 623
Slovenia	12	21	4	10	6	5	18	18	11	72
Spain	-	-	-	-	499	440	483	985	-	879
Tunisia	7 053	6 315	5 121	3 517	4 340	6 479	4 935	6 622	6 002	7 148
Turkey	611	526	717	933	644	900	750	537	550	465
37 Fishing area total	*10 213 F*	*9 629 F*	*9 643 F*	*7 619*	*8 104*	*11 031*	*8 767*	*11 485*	*8 750 F*	*10 698 F*
Species total	*13 228 F*	*12 563 F*	*12 543 F*	*10 352 F*	*11 408*	*13 758*	*12 449*	*14 850*	*12 724 F*	*13 987 F*
Cuttlefish,bobtail squids nei	**Seiches, sépioles nca**		**Sepias,choquitos,globitos nep**			***Sepiidae, Sepiolidae***			**3,21(02)XXX,XX**	**CTL**
27 Denmark	53	5	2	3	2	0	-	26	20	5
France	10 634	13 832	11 932	14 642	14 532	11 982	13 015	14 465	18 939	13 814
Netherlands	-	-	-	-	-	-	-	-	101	162
Spain	1 637	1 267	1 196	1 588	971	1 621	1 446	1 322	256	65
UK	1 157	2 164	2 089	3 631	4 607	2 202	2 760	2 260	3 076	2 705
27 Fishing area total	*13 481*	*17 268*	*15 219*	*19 864*	*20 112*	*15 805*	*17 221*	*18 073*	*22 392*	*16 751*
31 Korea Rep	-	-	-	-	-	-	-	32	-	-
31 Fishing area total	-	-	-	-	-	-	-	*32*	-	-
34 Belize	-	460	80	141	-	406	2 445	4 803	4 862	19
Benin	...	...	...	...	...	...	...	18	12 F	1
Cameroon	0	0	0	2	2	2 F	2 F	2 F	1	1
Chile	-	-	9	-	-	-	-	-	-	-
China	1 276	3 082	500	8 111	6 823	7 485	5 686	5 528	4 930	8 238
Congo Rep	0	0	0	1 F	3 F	2 F	7 F	9 F	10	3 F
Côte dIvoire	-	-	-	-	-	-	-	81	291	140
Gabon	...	27	56	33	157	52	186	92	281	166
Gambia	406	182	62	325	184	137	98	100	422	1 499
Ghana	1 541	1 673	2 396	2 891	2 967	3 355	3 288	4 095	1 805	2 866
Guinea	136	184	31	61	41	95	386	181	124	236
GuineaBissau	...	...	31	17	2	2 F	2 F	2 F	2 F	2 F
Honduras	2 874	2 336	291	2 873	3 108	2 076	889	1 077	687	-
Italy	2 429	2 017	2 017	1 017	790	2 016	1 491	1 360	791	1 239
Korea D P Rp	-	25	-	-	-	-	-	-	61	-
Korea Rep	406	184	228	288	169	112	510	215	29	461
Liberia	13	14	8	-	-	-	12	8	14	29
Mauritania	3 690 F	4 150 F	4 470 F	5 710 F	4 510 F	5 240 F	5.000 F	4 129 F	4 344 F	4 744 F
Morocco	8 837	9 074	14 209	12 131	12 500	16 625	14 536	20 011	32 725	17 544
Nigeria	...	...	...	...	3	...	...	...	...	189
Panama	346	253	-	78	0	53	82	-	-	-
Russian Fed	3	21	30	-	-	-	-	-	-	-
St Vincent	77	84	-	2	-	-	-	-	-	2
Senegal	6 496	6 380	6 945	6 399	5 919	6 864	6 365	5 709	3 953	3 941
Sierra Leone	2 392	835	127	602	1 621	1 626	600	512	294	574
Spain	2 667	2 500 F	2 000 F	1 500 F	702	658	18	3 727	1 262	2 761
Togo	0	32	9	1	12	77	51	5	0	20
Vanuatu	46	22	13	22	33	19	11	-	-	-
Other nei	8	-	-	-	-	-	-	138	154	148
34 Fishing area total	*33 643 F*	*33 535 F*	*33 512 F*	*42 205 F*	*39 546 F*	*46 902 F*	*41 665 F*	*51 802 F*	*57 054 F*	*44 823 F*
37 Croatia	167	203	118	118	102	151	215	145	138	85
Egypt	787	1 067	1 031	1 097	1 365	1 370	1 152	1 449	1 503	1 554
Gaza Strip	...	...	...	41	62	98	144	145 F	145 F	120 F
Italy	7 984	7 682	13 616	11 150	7 974	7 382	6 839	5 720	5 534	6 131
Lebanon	25 F	25	25	25	25	50	25	50	25	25
Malta	0	1	1	0	5	3	3	5	4	0
Morocco	205	215	139	265	178	162	132	157	179	18
Spain	1 100 F	1 000 F	900 F	800 F	746	780	834	446	1 311	263
Yugoslavia	2	6	5	9	10	9	10	10	10	10
37 Fishing area total	*10 270 F*	*10 199 F*	*15 835 F*	*13 505 F*	*10 467*	*10 005*	*9 354*	*8 127 F*	*8 849 F*	*8 206 F*
47 Angola	237	32	-	-	-	-	-	-	-	-
China	47	-	-	-	-	-	-	-	18	-
Honduras	-	36	-	4	-	-	-	-	-	-
Italy	-	-	-	14	11	-	-	-	-	-
Korea Rep	81	44	-	-	-	-	-	-	-	-
Panama	523	-	-	-	-	-	-	-	-	-
Russian Fed	-	-	-	-	-	-	-	-	-	5
Spain	-	-	-	-	113	-	88	33	72	26
47 Fishing area total	*888*	*112*	-	*18*	*124*	-	*88*	*33*	*90*	*31*

B-57 Squids, cuttlefishes, octopuses — Capture production by species, fishing areas and countries or areas
Encornets, seiches, poulpes — Captures par espèces, zones de pêche et pays ou zones
Calamares, jibias, pulpos — Capturas por especies, áreas de pesca y países o áreas

Species, Fishing area Espèce, Zone de pêche Especie, Area de pesca	1992 mt	1993 mt	1994 mt	1995 mt	1996 mt	1997 mt	1998 mt	1999 mt	2000 mt	2001 mt
51 Bahrain	131	138	101	113	285	126	139	48	85	104
Egypt	128	286	486	560	399	414	237	3 525	1 043	1 365
Eritrea	...	...	...	...	51	0	3	11	31	35
Italy	486	403	403	208	162	-	-	-	-	-
Korea Rep	1 232	1 307	1 006	814	1 179	2 730	1 314	1 003	601	840
Oman	2 141	1 948	2 584	2 945	5 036	6 148	4 080	7 478	2 891	3 854
Pakistan	2 507	2 945	3 356	2 455	3 308	4 528	3 225	5 146	5 307	5 256
Saudi Arabia	39	35	353	308	578	598	656	760	593	796
South Africa	10	11	-	-	-	9	5	10 F	16	16
Spain	-	-	-	-	-	-	-	1	-	-
Untd Arab Em	26	20	23	22	25	26	27	28	491	490 F
Yemen	695	906	31	1 457	1 884	8 657	5 092	5 292	8 917	9 330
51 Fishing area total	*7 395*	*7 999*	*8 343*	*8 882*	*12 907*	*23 236*	*14 778*	*23 302 F*	*19 975*	*22 086 F*
57 Australia	12	28	42	50	56	56	56	62	32	48
Indonesia	2 026	2 229	1 845	2 316	4 019	3 221	5 120	5 328	4 848	5 360
Korea Rep	-	-	-	-	-	8	-	399	200	189
Malaysia	7 105	6 283	5 720	7 036	8 014	8 923	13 329	9 570	13 492	9 810
Sri Lanka	200	200	250	300	300	300	300	365	310	290
Thailand	16 960	15 911	14 333	16 561	23 286	23 244	18 493	21 238	20 947	20 819
57 Fishing area total	*26 303*	*24 651*	*22 190*	*26 263*	*35 675*	*35 752*	*37 298*	*36 962*	*39 829*	*36 516*
61 China	69 349	118 536	193 046	213 772	166 319	235 946	221 834	210 775	255 958	310 129
China,H.Kong	1 561	1 209	1 151	1 689	991	841	510	400 F	450 F	500 F
China,Taiwan	11 173	9 536	10 959	12 084	7 191	8 621	7 198	5 019	3 880	4 546
Japan	11 609	9 837	9 926	9 643	9 545	8 052	9 790	9 942	8 241	8 297
Korea Rep	5 814	5 452	3 090	2 567	1 484	2 082	2 563	6 652	1 267	1 443
61 Fishing area total	*99 506*	*144 570*	*218 172*	*239 755*	*185 530*	*255 542*	*241 895*	*232 788 F*	*269 796 F*	*324 915 F*
71 Australia	5	3	3	5	4	6	6	8	9	7
Indonesia	2 997	3 044	3 444	3 623	4 335	4 980	6 353	7 003	7 097	7 850
Korea Rep	292	262	145	22	-	14	70	8	10	7
Malaysia	8 249	9 034	9 382	8 440	10 762	11 668	12 853	12 498	12 822	11 657
NewCaledonia	6	6	1	3	3	1	2	3	1	1 F
Philippines	2 482	2 196	2 188	2 836	2 702	2 803	2 204	2 093	2 016	2 066
Singapore	237	237	233	196	214	235	179	142	134	56
Thailand	48 036	44 456	41 987	45 358	47 239	48 344	44 847	45 009	44 393	44 122
71 Fishing area total	*62 304*	*59 238*	*57 383*	*60 483*	*65 259*	*68 051*	*66 514*	*66 764*	*66 482*	*65 766 F*
81 Australia	334	328	392	392	378	374	251	381	258	219
81 Fishing area total	*334*	*328*	*392*	*392*	*378*	*374*	*251*	*381*	*258*	*219*
Species total	*254 124 F*	*297 900 F*	*371 046 F*	*411 367 F*	*369 998 F*	*455 667 F*	*429 064 F*	*438 264 F*	*484 725 F*	*519 313 F*
Patagonian squid	**Calmar patagon**		**Calamar patagónico**			***Loligo gahi***		**3,21(04)001,02**		**SQP**
41 Australia	-	-	-	-	-	-	3 198	2 486	-	-
Belize	-	-	-	-	-	-	-	-	2	-
Falkland Is	1 683	1 494	5 266	22 310	24 366	12 710	32 029	22 502	50 270	42 909
France	-	-	1 636	7 245	4 394	1 512	4 146	2 309	2 024	-
Honduras	2	1	1	31	8	-	-	-	-	-
Namibia	-	-	-	-	-	74	1	0	-	-
Panama	8	-	0	1	1	-	-	-	-	-
Poland	9 300	3 811	593	-	-	-	19	4 875	-	-
St Vincent	-	-	-	-	-	-	-	-	-	1 795
Seychelles	-	-	-	-	-	1 114	88	-	-	-
Spain	60 843	39 012	46 736	53 671	35 692	3 967	8 933	8 185	9 392	9 011
UK	-	1	1 228	1 916	4 043	2 334	3 279	2 148	5 328	4 015
41 Fishing area total	*71 836*	*44 319*	*55 460*	*85 174*	*68 504*	*21 711*	*51 693*	*42 505*	*67 016*	*57 730*
Species total	*71 836*	*44 319*	*55 460*	*85 174*	*68 504*	*21 711*	*51 693*	*42 505*	*67 016*	*57 730*
Longfin squid	**Calmar totam**		**Calamar pálido**			***Loligo pealei***		**3,21(04)001,05**		**SQL**
21 Japan	1 528	-	-	-	-	-	-	-	-	-
USA	18 134	22 146	22 469	18 887	12 490	16 161	18 879	18 749	16 942	14 211
21 Fishing area total	*19 662*	*22 146*	*22 469*	*18 887*	*12 490*	*16 161*	*18 879*	*18 749*	*16 942*	*14 211*
31 USA	49	54	33	39	-	-	-	-	-	-
31 Fishing area total	*49*	*54*	*33*	*39*	-	-	-	-	-	-
Species total	*19 711*	*22 200*	*22 502*	*18 926*	*12 490*	*16 161*	*18 879*	*18 749*	*16 942*	*14 211*
Cape Hope squid	**Calmar du Cap**		**Calamar del Cabo**			***Loligo reynaudi***		**3,21(04)001,12**		**CHO**
47 South Africa	2 805	6 271	5 814	7 047	7 491	3 696	6 670	7 169	6 000 F	3 373
Spain	-	-	-	-	58	-	-	0	-	-
47 Fishing area total	*2 805*	*6 271*	*5 814*	*7 047*	*7 549*	*3 696*	*6 670*	*7 169*	*6 000 F*	*3 373*
Species total	*2 805*	*6 271*	*5 814*	*7 047*	*7 549*	*3 696*	*6 670*	*7 169*	*6 000 F*	*3 373*
Common squids nei	**Calmars nca**		**Calamares nep**			***Loligo spp***		**3,21(04)001,XX**		**SQC**
27 Belgium	64	84	57	45	21	16	25	23	59	51
Channel Is	1	1 F	2	2 F	1	6	5	11	9	1
Isle of Man	15	15	6	7	3	2	2	2	-	1
Portugal	1 646	879	616	1 267	787	1 514	1 499	701	1 003	1 273
Spain	1 647	1 251	1 280	1 135	1 015	2 520	2 190	1 376	1 371	1 473

B-57 Squids, cuttlefishes, octopuses / Encornets, seiches, poulpes / Calamares, jibias, pulpos

Capture production by species, fishing areas and countries or areas / Captures par espèces, zones de pêche et pays ou zones / Capturas por especies, áreas de pesca y países o áreas

Species, Fishing area Espèce, Zone de pêche Especie, Area de pesca	1992 mt	1993 mt	1994 mt	1995 mt	1996 mt	1997 mt	1998 mt	1999 mt	2000 mt	2001 mt
UK	1 227	1 655	1 947	2 272	3 264	3 004	3 039	2 611	1 757	850
27 Fishing area total	*4 600*	*3 885 F*	*3 908*	*4 728 F*	*5 091*	*7 062*	*6 760*	*4 724*	*4 199*	*3 649*
31 Dominican Rp	67	24	3	37	38	24	84	12	56	51
Mexico	350	572	111	69	112	140	71	91	85	92
Venezuela	946	1 384	1 633	756	771	719	787	102	463	1 261
31 Fishing area total	*1 363*	*1 980*	*1 747*	*862*	*921*	*883*	*942*	*205*	*604*	*1 404*
34 Italy	3 239	2 689	2 689	188	2	11	-	-	-	151
Portugal	17	122	0	2	5	5	0	3	3	1
34 Fishing area total	*3 256*	*2 811*	*2 689*	*190*	*7*	*16*	*0*	*3*	*3*	*152*
37 Albania	...	...	...	7	47	34	93	93	90	65
Algeria	250 F	250 F	300 F	101	53	138	202	240	230 F	230 F
Croatia	263	330	268	273	233	287	290	170	127	162
France	261	220	136	150	135	465	465	241	264	177
Gaza Strip	...	...	...	15	23	30	61	60 F	60 F	50 F
Greece	1 135	1 369	1 101	945	856	623	426	397	560	405
Italy	4 728	4 372	5 282	5 483	5 366	4 141	2 237	1 909	1 890	2 201
Malta	0	0	0	0	2	2	2	2	3	2
Slovenia	4	4	4	2	2	4	3	2	3	8
Spain	800 F	800 F	700 F	700 F	505	462	405	1 002	691	806
Tunisia	201	214	277	240	230	253	288	348	310	327
Turkey	557	397	579	331	364	420	500	360	400	230
Yugoslavia	4	10	7	13	12	13	13	12	13	14
37 Fishing area total	*8 203 F*	*7 966 F*	*8 654 F*	*8 260 F*	*7 828*	*6 872*	*4 985*	*4 836 F*	*4 641 F*	*4 677 F*
41 Argentina	266	957	1 073	914	184	1 999	802	209	268	...
Brazil	1 310 F	1 310 F	1 320 F	1 317	950	1 486	758	1 696	1 187	1 170 F
Italy	432	359	359	22	0	-	-	-	-	-
Japan	5 199	4 118	5 683	5 561	3 187	1 562	2 618	1 847	-	-
Korea Rep	396	-	-	-	-	-	-	-	-	-
Portugal	1 395	1 762	2 296	3 805	3 984	48	0	-	-	-
41 Fishing area total	*8 998 F*	*8 506 F*	*10 731 F*	*11 619*	*8 305*	*5 095*	*4 178*	*3 752*	*1 455*	*1 170 F*
51 China,Taiwan	2 250 F	2 500 F	2 100 F	758 F	-	-	-	-	-	-
Italy	648	538	538	38	0	-	-	-	-	-
Qatar	44	46	34	31	30	37	37	10	22	41
51 Fishing area total	*2 942 F*	*3 084 F*	*2 672 F*	*827 F*	*30*	*37*	*37*	*10*	*22*	*41*
57 China,Taiwan	110 F	100 F	100 F	36 F	-	-	-	-	-	-
Indonesia	5 264	3 778	3 847	9 657	8 001	8 360	10 499	10 205	9 750	10 800
Thailand	13 565	16 295	16 464	18 485	23 229	23 208	24 120	20 522	20 241	20 118
57 Fishing area total	*18 939 F*	*20 173 F*	*20 411 F*	*28 178 F*	*31 230*	*31 568*	*34 619*	*30 727*	*29 991*	*30 918*
61 China,Taiwan	18 610 F	20 500 F	17 000 F	6 143 F	14 853	19 221	20 074	11 083	7 325	5 791
Korea Rep	-	-	-	-	-	-	-	1 629	885	898
61 Fishing area total	*18 610 F*	*20 500 F*	*17 000 F*	*6 143 F*	*14 853*	*19 221*	*20 074*	*12 712*	*8 210*	*6 689*
71 China,Taiwan	40 940 F	45 000 F	37 400 F	13 518 F	9 868	9 181	6 853	4 815	2 248	3 308
Indonesia	13 101	17 136	22 369	17 918	19 168	33 395	21 351	26 502	30 088	33 340
Philippines	39 402	55 789	48 948	56 415	52 458	54 155	48 678	47 115	46 778	47 854
Singapore	226	246	1 000	679	546	470	462	376	348	186
Thailand	51 209	55 867	55 762	59 624	56 006	55 740	68 788	62 613	61 756	61 379
71 Fishing area total	*144 878 F*	*174 038 F*	*165 479 F*	*148 154 F*	*138 046*	*152 941*	*146 132*	*141 421*	*141 218*	*146 067*
87 Ecuador	1 476	85	90	90	90	100	100	100	100	100
Peru	2 621	1 316	1 215	7 766	10 250	3 806	287	1 353	24 548	18 738
87 Fishing area total	*4 097*	*1 401*	*1 305*	*7 856*	*10 340*	*3 906*	*387*	*1 453*	*24 648*	*18 838*
Species total	***215 886 F***	***244 344 F***	***234 596 F***	***216 817 F***	***216 651***	***227 601***	***218 114***	***199 843 F***	***214 991 F***	***213 605 F***
Neon flying squid	**Encornet volant**		**Pota saltadora**			***Ommastrephes bartrami***		**3,21(05)003,01**		**OFJ**
61 Japan	...	...	...	...	...	47 852	53 638	34 551	46 199	21 578
Russian Fed	-	-	-	-	-	-	-	-	405	100
61 Fishing area total	*...*	*...*	*...*	*...*	*...*	*47 852*	*53 638*	*34 551*	*46 604*	*21 678*
67 Japan	-	-	-	-	-	2 008	1 200	1 521	683	2 061
67 Fishing area total	*-*	*-*	*-*	*-*	*-*	*2 008*	*1 200*	*1 521*	*683*	*2 061*
77 Japan	-	-	-	-	-	10	113	4	81	131
77 Fishing area total	*-*	*-*	*-*	*-*	*-*	*10*	*113*	*4*	*81*	*131*
Species total	***...***	***...***	***...***	***...***	***...***	***49 870***	***54 951***	***36 076***	***47 368***	***23 870***
Northern shortfin squid	**Encornet rouge nordique**		**Pota norteña**			***Illex illecebrosus***		**3,21(05)010,01**		**SQI**
21 Canada	1 352	2 802	5 778	999	8 822	15 775	1 944	314	368	65
Cuba	636	2 354	4 036	959	493	2 978	1 085	280	-	-
Honduras	-	13	-	-	-	-	-	-	-	-
Japan	68	-	-	-	1	-	-	-	-	-
Portugal	-	-	-	-	4	-	1	-	-	-
Russian Fed	31	92	-	-	-	-	29	-	12	-
St Pier Mq	2	3	10	0	0	6	-	-	-	-
Spain	-	-	3	-	1	1	3	-	-	-
USA	17 828	18 224	18 419	14 202	17 041	13 630	22 715	7 334	9 010	4 009
21 Fishing area total	*19 917*	*23 488*	*28 246*	*16 160*	*26 362*	*32 390*	*25 777*	*7 928*	*9 390*	*4 074*

B-57

Squids, cuttlefishes, octopuses — **Capture production by species, fishing areas and countries or areas**
Encornets, seiches, poulpes — **Captures par espèces, zones de pêche et pays ou zones**
Calamares, jibias, pulpos — **Capturas por especies, áreas de pesca y países o áreas**

Species, Fishing area Espèce, Zone de pêche Especie, Area de pesca	1992 mt	1993 mt	1994 mt	1995 mt	1996 mt	1997 mt	1998 mt	1999 mt	2000 mt	2001 mt
27 Denmark	-	-	-	-	-	17	16	-	-	-
Faeroe Is	-	-	1	-	-	2	32	23	...	...
Germany	-	0	3	11	-	-	-	-	-	-
Ireland	...	...	...	...	...	...	...	...	...	121
Portugal	777	-	-	-	-	-	-	-	-	-
Spain	3 172	2 089	2 272	2 359	2 600	2 426	762	1 932	1 831	1 503
27 Fishing area total	*3 949*	*2 089*	*2 276*	*2 370*	*2 600*	*2 445*	*810*	*1 955*	*1 831*	*1 624*
31 USA	2	5	148	20	9	2	2	-	1	0
31 Fishing area total	*2*	*5*	*148*	*20*	*9*	*2*	*2*	*-*	*1*	*0*
Species total	*23 868*	*25 582*	*30 670*	*18 550*	*28 971*	*34 837*	*26 589*	*9 883*	*11 222*	*5 698*
Broadtail shortfin squid	**Encornet rouge**		**Pota voladora**			***Illex coindetii***			**3,21(05)010,02**	**SQM**
27 France	833	762	702	571	395	411	216	304	360	219
27 Fishing area total	*833*	*762*	*702*	*571*	*395*	*411*	*216*	*304*	*360*	*219*
37 France	...	...	...	...	...	...	...	34	42	31
37 Fishing area total	*...*	*...*	*...*	*...*	*...*	*...*	*...*	*34*	*42*	*31*
Species total	*833*	*762*	*702*	*571*	*395*	*411*	*216*	*338*	*402*	*250*
Argentine shortfin squid	**Encornet rouge argentin**		**Pota argentina**			***Illex argentinus***			**3,21(05)010,03**	**SQA**
41 Argentina	77 468	193 690	196 893	199 048	292 628	411 994	291 174	342 691	278 988	229 874
Australia	-	-	-	-	-	-	-	167	-	-
Belize	-	-	-	-	-	-	-	3 796	4 066	1 692
Bulgaria	5 503	1 062	-	-	-	-	-	-	-	-
Cambodia	-	-	-	-	-	-	-	-	2 768	1 200
Chile	2 415	841	2	302	-	-	-	-	-	-
China	-	-	-	-	-	-	30 000	61 000	93 130	93 500
China,Taiwan	117 270 F	123 700 F	104 480 F	100 371	101 328	185 775	163 180	264 089	238 334	146 783
Estonia	-	-	-	-	-	-	-	-	-	1 833
Falkland Is	-	-	-	351	196	37	804	2 582	716	1 879
France	-	-	106	23	28	0	0	56	0	-
Honduras	-	556	1 089	679	-	-	-	-	-	-
Japan	98 857	131 707	92 838	75 691	73 896	126 504	77 452	154 129	119 966	70 652
Korea Rep	211 284	128 581	79 130	124 005	144 750	208 160	92 397	271 716	150 149	142 585
Namibia	-	-	-	-	-	3	0	63	-	-
Poland	16 932	6 421	2 390	282	1	-	-	-	970	683
Portugal	24	4	170	353	640	712	1 531	-	-	1 049
Russian Fed	76 733	47 309	23 557	11 762	20 254	884	-	-	3 404	2 578
St Vincent	-	-	-	-	-	-	-	-	-	4
Spain	947	776	3 021	3 889	17 091	25 374	23 829	30 694	26 140	41 318
UK	-	-	-	-	-	-	-	336	6	21
Uruguay	2 390	3 806	2 022	4 182	5 669	20 857	13 175	13 679	12 144	7 373
41 Fishing area total	*609 823 F*	*638 453 F*	*505 698 F*	*520 938*	*656 481*	*980 300*	*693 542*	*1 144 998*	*930 781*	*743 024*
Species total	*609 823 F*	*638 453 F*	*505 698 F*	*520 938*	*656 481*	*980 300*	*693 542*	*1 144 998*	*930 781*	*743 024*
Jumbo flying squid	**Encornet géant**		**Jibia gigante**			***Dosidicus gigas***			**3,21(05)023,01**	**GIS**
77 Japan	828	-	-	-	13 096	19 988	-	348	25 904	211
Mexico	8 549	3 043	1 800	39 657	107 967	120 877	26 611	57 985	56 153	73 741
USA	-	-	-	-	-	3	107	18	1	0
77 Fishing area total	*9 377*	*3 043*	*1 800*	*39 657*	*121 063*	*140 868*	*26 718*	*58 351*	*82 058*	*73 952*
87 Japan	51 187	55 800	84 205	36 515	1 201	3 191	-	2 957	30 921	72 201
Korea Rep	36 101	57 778	66 386	34 440	11 784	2 384	201	18 813	15 625	5 797
Peru	12 695	7 769	42 838	25 676	8 138	16 061	547	54 652	53 795	71 834
87 Fishing area total	*99 983*	*121 347*	*193 429*	*96 631*	*21 123*	*21 636*	*748*	*76 422*	*100 341*	*149 832*
Species total	*109 360*	*124 390*	*195 229*	*136 288*	*142 186*	*162 504*	*27 466*	*134 773*	*182 399*	*223 784*
European flying squid	**Toutenon commun**		**Pota europea**			***Todarodes sagittatus***			**3,21(05)058,01**	**SQE**
27 Iceland	-	-	-	11	3	5	4	3	1	-
Ireland	...	...	...	...	...	...	...	...	...	14
Norway	-	-	-	352	0	190	2	0	0	-
Spain	-	-	-	-	973	2 508	2 137	2 481	1 866	1 588
UK	-	-	-	-	-	18	293	204	186	193
27 Fishing area total	*...*	*...*	*...*	*363*	*976*	*2 721*	*2 436*	*2 688*	*2 053*	*1 795*
37 France	49	11	21	19	87	87	87	-	-	-
Italy	7 800	6 451	5 737	4 789	4 672	2 614	3 995	2 056	2 516	2 346
Malta	-	-	-	-	-	-	2	2	2	0
Spain	-	-	-	-	140	149	119	144	338	120
37 Fishing area total	*7 849*	*6 462*	*5 758*	*4 808*	*4 899*	*2 850*	*4 203*	*2 202*	*2 856*	*2 466*
Species total	*7 849*	*6 462*	*5 758*	*5 171*	*5 875*	*5 571*	*6 639*	*4 890*	*4 909*	*4 261*
Japanese flying squid	**Toutenon japonais**		**Pota japonesa**			***Todarodes pacificus***			**3,21(05)058,03**	**SQJ**
61 China,Taiwan	11 053	10 422 F	10 917	22 243	19 101	12 430	34 840	11 261	6 833	4 716
Japan	394 364	315 934	301 651	290 273	444 189	365 978	180 749	237 346	337 285	298 191
Korea Rep	139 792	222 009	191 857	200 897	252 618	224 959	163 016	249 280	226 309	225 616
61 Fishing area total	*545 209*	*548 365 F*	*504 425*	*513 413*	*715 908*	*603 367*	*378 605*	*497 887*	*570 427*	*528 523*

B-57 Squids, cuttlefishes, octopuses / Encornets, seiches, poulpes / Calamares, jibias, pulpos

Capture production by species, fishing areas and countries or areas / Captures par espèces, zones de pêche et pays ou zones / Capturas por especies, áreas de pesca y países o áreas

Species, Fishing area Espèce, Zone de pêche Especie, Area de pesca	1992 mt	1993 mt	1994 mt	1995 mt	1996 mt	1997 mt	1998 mt	1999 mt	2000 mt	2001 mt
Species total	*545 209*	*548 365 F*	*504 425*	*513 413*	*715 908*	*603 367*	*378 605*	*497 887*	*570 427*	*528 523*
Wellington flying squid	**Encornet minami**		**Pota neozelandesa**		***Nototodarus sloani***				**3,21(05)059,01**	**TSQ**
81 China,Taiwan	5 000 F	6 000 F	7 000 F	8 284	14 747	6 620	3 974	761	-	-
Japan	12 126	8 072	10 180	19 687	11 342	5 182	3 734	1 853	1 853	1 396
New Zealand	44 376	25 530	51 841	59 497	23 474	44 845	42 541	27 282	20 878	35 100
Ukraine	2 932	5 546	10 428	6 630	4 136	7 955	5 321	1 462	2 872	8 623
81 Fishing area total	*64 434 F*	*45 148 F*	*79 449 F*	*94 098*	*53 699*	*64 602*	*55 570*	*31 358*	*25 603*	*45 119*
Species total	*64 434 F*	*45 148 F*	*79 449 F*	*94 098*	*53 699*	*64 602*	*55 570*	*31 358*	*25 603*	*45 119*
Sevenstar flying squid	**Encornet étoile**		**Pota festoneada**		***Martialia hyadesi***				**3,21(05)060,01**	**SQS**
41 Argentina	-	-	-	404	-	-	-	-	653	...
China,Taiwan	1 041 F	1 239 F	392	23 464	3 792	8 348	-	-	-	-
Falkland Is	1	-	-	0	0	0	-	0	0	-
Honduras	-	-	-	118	-	-	-	-	-	-
Japan	-	-	-	-	-	-	2	27	33	2
Poland	-	13	-	-	-		-	-	-	-
Spain	1	0	0	-	1	0	0	0	-	-
41 Fishing area total	*1 043 F*	*1 252 F*	*392*	*23 986*	*3 793*	*8 348*	*2*	*27*	*686*	*2*
48 Korea Rep	-	-	-	-	52	28	53	-	-	2
48 Fishing area total	-	-	-	-	*52*	*28*	*53*	-	-	*2*
Species total	*1 043 F*	*1 252 F*	*392*	*23 986*	*3 845*	*8 376*	*55*	*27*	*686*	*4*
Various squids nei	**Calmars, encornets nca**		**Calamares, jibias, potas nep**		***Loliginidae, Ommastrephidae***				**3,21(05)XXX,XX**	**SQU**
21 USA	1 156	762	1 292	531	56	182	69	90	70	124
21 Fishing area total	*1 156*	*762*	*1 292*	*531*	*56*	*182*	*69*	*90*	*70*	*124*
27 France	5 394	6 881	6 140	6 017	3 852	4 413	4 289	5 946	5 399	4 690
Germany	-	3	-	-	2	4	11	6	5	3
Ireland	260	364	277	298	481	442	610	282	322	242
Korea Rep	1	-	-	-	-	-	-	-	-	-
Lithuania	241	-	-	-	-	-	-	-	-	-
Netherlands	-	-	-	-	-	-	-	-	773	171
Spain	-	-	-	-	-	-	1 459	2 267	1 022	307
Sweden	3	4	0	0	0	1	1	1	-	0
UK	1 085	543	273	666	13	8	8	4	3	815
27 Fishing area total	*6 984*	*7 795*	*6 690*	*6 981*	*4 348*	*4 868*	*6 378*	*8 506*	*7 524*	*6 228*
31 Belize	-	-	-	-	4	-	-	-	-	-
Colombia	247	45	50	27	29	20	8	-	...	...
Cuba	...	...	...	...	...	1	0	-	-	-
Grenada	0	0	0	0	0	0	0	0	0	5
Honduras	3	13	22	28	...	77	1	1	1	...
Korea Rep	2 821	-	-	-	-	-	-	-	1	-
Trinidad Tob	-	-	-	-	-	-	0	2	1	14
USA	-	-	-	-	-	42	110	42	62	74
31 Fishing area total	*3 071*	*58*	*72*	*55*	*33*	*140*	*119*	*45*	*65*	*93*
34 Belize	-	-	-	48	-	21	114	128	31	1
China	4	18	6	57	229	66	200	237	333	290
Cyprus	-	-	-	-	-	-	-	9	1	-
Gabon	...	...	29	3	5	2	0	1	8	10
Gambia	66	126	124	1	...	...	...	...	...	...
Georgia	23 F	10 F	-	-	-	-	-	-	-	-
Greece	231	342	269	55	31	-	1	11	16	11
Guinea	-	-	-	16	1	2	53	27	9	10 F
GuineaBissau	-	-	-	8	-	-	-	-	-	-
Honduras	469	1 003	160	615	212	52	29	23	-	-
Italy	810	673	686	377	307	36	54	101	95	11
Korea Rep	94	80	26	-	-	57	93	115	13	55
Lithuania	4	-	-	...	...	...	233	2	-	-
Mauritania	180 F	250 F	190 F	260 F	230 F	280 F	1 472 F	2 359 F	2 354 F	2 125 F
Morocco	12 917	18 487	12 313	19 872	20 000	8 987	11 350	8 079	15 426	10 622
Panama	66	49	-	3	8	2	1	-	-	-
Portugal	78	7	0	0	0	0	2	0	9	2
Russian Fed	1	-	-	94	53	-	-	-	-	-
St Vincent	45	-	-	-	-	-	-	-	-	4
Senegal	...	...	...	7	49	92	17	140	152	107
Sierra Leone	525	401	114	166	121	90	143	...	...	...
Spain	2 962	5 000 F	6 500 F	8 000 F	8 614	3 770	7 062	11 707	1 637	2 548
Other nei	10	2	-	-	-	-	-	79	91	77
34 Fishing area total	*18 485 F*	*26 448 F*	*20 417 F*	*29 582 F*	*29 860 F*	*13 457 F*	*20 824 F*	*23 018 F*	*20 175 F*	*15 873 F*
37 France	86	86	67	...	43	43	43	-	-	-
Greece	703	567	678	906	812	513	566	675	870	739
Morocco	40	25	122	32	10	2	2	3	6	11
Spain	-	-	-	-	592	547	747	673	790	302
37 Fishing area total	*829*	*678*	*867*	*938*	*1 457*	*1 105*	*1 358*	*1 351*	*1 666*	*1 052*
41 Estonia	6 568	-	-	-	2 425	-	-	-	-	-
Italy	108	90	92	45	37	-	-	-	-	-
Korea Rep	-	-	-	-	-	415	-	-	-	-
Latvia	10 994	4 608	6 280	1 717	3 956	-	-	-	-	-

B-57 Squids, cuttlefishes, octopuses / Encornets, seiches, poulpes / Calamares, jibias, pulpos

Capture production by species, fishing areas and countries or areas / Captures par espèces, zones de pêche et pays ou zones / Capturas por especies, áreas de pesca y países o áreas

Species, Fishing area Espèce, Zone de pêche Especie, Area de pesca	1992 mt	1993 mt	1994 mt	1995 mt	1996 mt	1997 mt	1998 mt	1999 mt	2000 mt	2001 mt
Lithuania	19 251	15 171	3 262	-	3 400	-	-	-	-	-
Ukraine	4 481	4 719	3 202	998	339	-	-	-	-	-
41 Fishing area total	*41 402*	*24 588*	*12 836*	*2 760*	*10 157*	*415*	-	-	-	-
47 Angola	352	228	191	122	42	...	390	463	442	292
Honduras	-	6	-	1	-	-	-	-	-	-
Iceland	-	-	-	-	-	4	0	-	-	-
Italy	-	-	-	8	4	-	-	-	-	-
Japan	15	-	-	-	-	-	-	-	-	-
Korea Rep	3 539	142	317	79	-	176	-	158	601	-
Lithuania	6	-	-	-	-	-	-	-	-	-
Namibia	30	83	22	16	26	34	42	19	28	775
Panama	6	-	-	-	-	-	-	-	-	-
Russian Fed	21	-	-	-	-	-	-	-	-	-
South Africa	300	59	36	17	34	45	29	24	40 F	53
Spain	14	-	-	-	-	-	-	1	-	1
47 Fishing area total	*4 283*	*518*	*566*	*243*	*106*	*259*	*461*	*665*	*1 111 F*	*1 121*
51 Djibouti	0	0	0	0	0	0	0	0	0	0
Eritrea	...	...	13	15	0	0	0	5	38	85
Italy	162	134	137	77	63	-	-	-	-	-
Kenya	56	32	57	345	389	317	30	35	42	78
Korea Rep	316	52	114	465	210	439	757	917	441	612
Mozambique	218	188	106	142	366	602	677	616	538	385
Pakistan	3 225	3 317	3 126	2 832	2 600	4 460	3 300	5 062	4 070	4 024
Portugal	-	-	-	-	-	-	-	-	1	3
South Africa	7	6	-	-	-	2	...	...	...	...
Spain	-	-	-	-	-	-	-	76	-	-
Ukraine	-	-	-	-	-	-	-	3	-	-
51 Fishing area total	*3 984*	*3 729*	*3 553*	*3 876*	*3 628*	*5 820*	*4 764*	*6 714*	*5 130*	*5 187*
57 Australia	685	840	1 315	2 111	930	2 657	1 083	2 745	1 735	3 268
Korea Rep	-	-	-	-	-	1	-	44	13	235
Malaysia	18 515	12 041	11 011	13 990	14 568	18 028	16 873	18 164	28 039	20 128
57 Fishing area total	*19 200*	*12 881*	*12 326*	*16 101*	*15 498*	*20 686*	*17 956*	*20 953*	*29 787*	*23 631*
61 China	...	...	...	...	...	...	117 000	132 000	123 855	81 000
China,H.Kong	11 110	13 613	13 700	10 043	8 559	13 586	8 340	6 000 F	7 300 F	8 100 F
Japan	111 198	61 767	91 145	110 441	121 718	63 116	56 371	56 575	49 607	46 451
Korea D P Rp	...	...	4 635	3 164	8 892	14 192	10 000 F	10 000 F	9 500 F	9 500 F
Korea Rep	35 184	187	12 624	14 928	3 573	1 915	8 471	3 357	-	893
Russian Fed	32 216	24 820	15 750	33 842	33 260	54 913	50 557	54 756	69 835	44 249
61 Fishing area total	*189 708*	*100 387*	*137 854*	*172 418*	*176 002*	*147 722*	*250 739 F*	*262 688 F*	*260 097 F*	*190 193 F*
67 Japan	23 725	-	-	-	-	-	-	-	-	-
Korea Rep	1 569	-	948	-	-	-	311	1 330	-	-
USA	603	193	410	225	488	571	521	221	6	852
67 Fishing area total	*25 897*	*193*	*1 358*	*225*	*488*	*571*	*832*	*1 551*	*6*	*852*
71 Australia	212	211	172	133	167	508	442	512	304	614
Japan	7	6	3	-	-	-	-	-	-	-
Korea Rep	3 313	1 531	2 901	2 948	1 785	2 125	1 423	2 478	700	1 535
Malaysia	15 752	20 581	24 913	17 264	21 702	20 463	21 824	22 119	26 300	25 149
71 Fishing area total	*19 284*	*22 329*	*27 989*	*20 345*	*23 654*	*23 096*	*23 689*	*25 109*	*27 304*	*27 298*
77 China	-	-	-	-	-	-	-	-	-	6 000
Costa Rica	17	14	13	13	20	11	1	2	7	11
El Salvador	14	-	-	-	-	-	-	-	16	9
Guatemala	11	14	16	3	16	21	39	23	231	200 F
Japan	12 981	1 154	-	-	-	-	-	-	-	-
Korea Rep	4 059	-	693	461	205	203	-	316	1 429	-
Panama	237	386	24	1	45	56	19	16	...	...
USA	13 203	32 262	55 108	70 210	78 794	70 915	2 709	90 662	117 711	85 829
77 Fishing area total	*30 522*	*33 830*	*55 854*	*70 688*	*79 080*	*71 206*	*2 768*	*91 019*	*119 394*	*92 049 F*
81 Australia	1 471	1 722	1 454	1 577	787	186	92	384	644	370
Korea Rep	17 798	6 652	13 110	17 436	9 836	13 068	12 278	9 951	8 801	11 380
New Zealand	-	-	-	10	7	17	27	48	74	45
Poland	2	-	-	-	-	-	-	-	-	-
Russian Fed	28 767	15 600	22 098	17 004	8 365	5 809	1 907	1 352	-	-
81 Fishing area total	*48 038*	*23 974*	*36 662*	*36 027*	*18 995*	*19 080*	*14 304*	*11 735*	*9 519*	*11 795*
87 Chile	9 468	7 604	462	55	26	110	179	99	64	3 594
Colombia	360	285	36	14	295	183	35	38	60 F	87
87 Fishing area total	*9 828*	*7 889*	*498*	*69*	*321*	*293*	*214*	*137*	*124 F*	*3 681*
Species total	*422 671 F*	*266 059 F*	*318 834 F*	*360 839 F*	*363 683 F*	*308 900 F*	*344 475 F*	*453 581 F*	*481 972 F*	*379 177 F*
Common octopus	**Pieuvre**		**Pulpo común**			***Octopus vulgaris***		**3,21(09)005,07**		**OCC**
27 Portugal	-	-	-	-	-	-	-	12	624	823
27 Fishing area total	-	-	-	-	-	-	-	*12*	*624*	*823*
31 Dominican Rp	54	21	28	33	36	33	87	39	99	57
Mexico	16 296	15 790	16 817	18 958	28 572	17 776	16 478	19 081	22 463	20 596
31 Fishing area total	*16 350*	*15 811*	*16 845*	*18 991*	*28 608*	*17 809*	*16 565*	*19 120*	*22 562*	*20 653*
34 GuineaBissau	...	...	7	34	1	1 F	1 F	1 F	1 F	1 F
Italy	1 620	1 345	1 345	213	244	612	933	1 845	1 373	1 388

B-57 Squids, cuttlefishes, octopuses / Encornets, seiches, poulpes / Calamares, jibias, pulpos

Capture production by species, fishing areas and countries or areas
Captures par espèces, zones de pêche et pays ou zones
Capturas por especies, áreas de pesca y países o áreas

Species, Fishing area Espèce, Zone de pêche Especie, Area de pesca	1992 mt	1993 mt	1994 mt	1995 mt	1996 mt	1997 mt	1998 mt	1999 mt	2000 mt	2001 mt
Spain	34 432	33 000 F	32 000 F	31 000 F	26 874	22 545	38 235	21 692	11 379	14 913
34 Fishing area total	*36 052*	*34 345 F*	*33 352 F*	*31 247 F*	*27 119*	*23 158 F*	*39 169 F*	*23 538 F*	*12 753 F*	*16 302 F*
37 Albania	...	...	...	66	75	59	93	90	85	24
Croatia	155	231	187	193	180	293	435	...	...	...
France	1 196	1 161	873	1 146	706	439	439	1 493	1 643	1 406
Greece	3 563	2 456	2 886	2 856	2 919	2 952	1 673	1 910	2 193	2 058
Italy	10 300	11 565	9 661	9 757	9 057	8 907	9 478	6 999	7 800	8 648
Lebanon	...	...	...	...	...	...	25	25	25	25
Slovenia	11	25	6	34	2	7	-	-	-	-
Tunisia	7 544	3 350	1 841	1 868	3 460	5 681	3 129	2 349	2 103	1 752
Turkey	579	472	659	602	802	1 000	1 450	510	680	1 400
Yugoslavia	2	6	4	4	8	8	9	9	10	12
37 Fishing area total	*23 350*	*19 266*	*16 117*	*16 526*	*17 209*	*19 346*	*16 731*	*13 385*	*14 539*	*15 325*
Species total	*75 752*	*69 422 F*	*66 314 F*	*66 764 F*	*72 936*	*60 313 F*	*72 465 F*	*56 055 F*	*50 478 F*	*53 103 F*
Horned and musky octopuses	**Elédones communes et musquées**		**Pulpos blancos y almizclados**			***Eledone spp***			**3,21(09)024,XX**	**OCM**
37 Italy	3 632	3 542	3 928	2 500	1 939	1 590	1 506	1 041	1 621	1 661
Tunisia	...	...	...	...	...	668	252	392	369	696
37 Fishing area total	*3 632*	*3 542*	*3 928*	*2 500*	*1 939*	*2 258*	*1 758*	*1 433*	*1 990*	*2 357*
Species total	*3 632*	*3 542*	*3 928*	*2 500*	*1 939*	*2 258*	*1 758*	*1 433*	*1 990*	*2 357*
Octopuses, etc. nei	**Pieuvres, poulpes nca**		**Pulpitos, pulpos nep**			***Octopodidae***			**3,21(09)XXX,XX**	**OCT**
21 USA	-	-	-	5	-	0	0	-	0	0
21 Fishing area total	-	-	-	*5*	-	*0*	*0*	-	*0*	*0*
27 Belgium	44	37	40	62	28	46	49	41	44	29
France	121	226	106	126	67	75	90	246	484	388
Ireland	1	4	6	25	13	7	3	10	10	14
Netherlands	-	-	-	-	-	-	-	-	-	7
Portugal	9 631	7 219	7 479	9 836	11 652	9 181	6 445	9 254	9 072	7 329
Spain	8 670	8 135	5 822	7 318	5 650	6 292	6 350	4 374	6 339	2 895
UK	137	182	313	333	229	148	111	63	135	164
27 Fishing area total	*18 604*	*15 803*	*13 766*	*17 700*	*17 639*	*15 749*	*13 048*	*13 988*	*16 084*	*10 826*
31 Korea Rep	-	-	-	-	-	-	-	157	3	-
Puerto Rico	...	...	...	...	...	...	...	28	39	23
Venezuela	693	901	1 196	724	1 548	1 917	6 507	526	1 110	1 064
31 Fishing area total	*693*	*901*	*1 196*	*724*	*1 548*	*1 917*	*6 507*	*711*	*1 152*	*1 087*
34 Bahamas	-	-	-	4	-	-	-	-	-	-
Belize	-	17	0	86	-	185	785	3 247	1 185	...
Chile	-	-	0	-	-	-	-	-	-	-
China	156	523	9	3 481	2 464	3 193	4 336	7 491	4 578	5 130
Côte dIvoire	-	-	-	-	-	-	-	-	-	88
Cyprus	-	-	-	-	-	-	-	-	0	3
Gabon	-	-	-	-	-	-	5	91	0	0
Gambia	503	277	141	6	...	...	...	...	...	1
Ghana	334	43	73	55	137	67	103	19	4	94
Greece	417	326	201	112	95	75	50	987	56	133
Guinea	-	1	-	114	22	12	183	1 092	96	97
Honduras	1 994	1 994	282	1 657	1 040	468	129	590	54	-
Korea D P Rp	-	20	-	-	-	-	-	-	86	-
Korea Rep	288	276	481	392	162	96	406	2 653	387	32
Latvia	-	-	-	-	-	-	-	-	-	2
Liberia	-	-	-	-	176	2	61	23	16	41
Lithuania	806	-	-	-	-	-	-	-	-	-
Mauritania	22 000 F	23 230 F	21 520 F	18 250 F	18 770 F	14 620 F	18 420 F	12 758 F	13 709 F	13 305 F
Morocco	60 291	63 813	56 239	57 765	58 600	38 089	42 465	84 464	99 279	112 589
Nigeria	11	...	...	...	...	...	...	...	...	...
Norway	-	-	26	-	-	-	-	-	-	-
Panama	227	154	-	82	89	41	102	-	-	-
Portugal	21	27	128	99	107	51	14	91	2	2
Russian Fed	189	9	-	-	-	-	-	-	-	-
St Vincent	278	9	0	-	-	-	-	-	-	-
Senegal	5 072	4 799	8 397	4 191	4 180	2 701	5 530	37 257	6 058	2 985
Sierra Leone	2 744	777	143	381	912	777	328	188	8	11
Vanuatu	3	-	-	-	-	-	0	-	-	-
Other nei	14	22	-	-	-	-	-	196	245	264
34 Fishing area total	*95 348 F*	*96 317 F*	*87 640 F*	*86 675 F*	*86 754 F*	*60 377 F*	*72 917 F*	*151 147 F*	*125 763 F*	*134 777 F*
37 Algeria	...	...	...	382	190	185	543	305	300 F	300 F
Cyprus	258	288	474	300	190	199	228	179	166	173
Greece	708	780	1 166	1 451	809	740	732	794	878	680
Malta	3	3	4	6	11	11	9	11	9	5
Morocco	78	71	25	69	36	98	68	115	112	45
Spain	5 700 F	5 700 F	5 700 F	5 700 F	5 594	4 898	5 134	5 565	7 889	6 634
37 Fishing area total	*6 747 F*	*6 842 F*	*7 369 F*	*7 908 F*	*6 830*	*6 131*	*6 714*	*6 969*	*9 354 F*	*7 837 F*
41 Argentina	12	6	5	34	39	48	17	34	5	-
Brazil	630 F	630 F	630 F	577	459	640	554	906	1 032	1 000 F
Estonia	-	1 314	-	-	-	-	-	-	-	-
Italy	540	448	448	146	123	-	-	-	-	-
Korea Rep	-	-	1	-	14	-	-	516	47	2
41 Fishing area total	*1 182 F*	*2 398 F*	*1 084 F*	*757*	*635*	*688*	*571*	*1 456*	*1 084*	*1 002 F*

B-57 Squids, cuttlefishes, octopuses / Encornets, seiches, poulpes / Calamares, jibias, pulpos

Capture production by species, fishing areas and countries or areas
Captures par espèces, zones de pêche et pays ou zones
Capturas por especies, áreas de pesca y países o áreas

Species, Fishing area Espèce, Zone de pêche Especie, Area de pesca	1992 mt	1993 mt	1994 mt	1995 mt	1996 mt	1997 mt	1998 mt	1999 mt	2000 mt	2001 mt
47 Angola	4	-	-	-	-	-	-	45	-	-
China	4	-	-	-	-	-	-	-	-	-
Honduras	-	57	-	5	-	-	-	-	-	-
Italy	-	-	-	3	3	-	-	-	-	-
Korea Rep	34	-	24	36	-	-	-	-	-	-
Panama	75	-	-	-	-	-	-	-	-	-
Portugal	...	...	...	21	26	101	56	164	144	-
St Helena	-	-	-	27	34	25	48	22	24	16
South Africa	122	149	91	33	46	64	60	87	90 F	93
Spain	-	-	-	-	282	-	418	116	47	20
47 Fishing area total	*239*	*206*	*115*	*125*	*391*	*190*	*582*	*434*	*305 F*	*129*
51 Fr South Tr	6	1	6	5 F	5 F	5 F	5 F	...	...	...
Italy	324	269	269	43	50	-	-	-	-	-
Kenya	49	78	106	460	117	393	155	169	106	154
Korea Rep	-	-	-	-	-	16	-	48	-	11
Mauritius	368	335	343	325	341	306	299	299	303	347
Mozambique	19	53	30	29	49	51	80	104	109	35
Portugal	-	-	-	-	-	-	-	-	2	1
Réunion	...	...	...	...	...	...	...	8	5	3
Saudi Arabia	-	-	-	-	-	1	-	-	0	-
Seychelles	20	67	32	20	32	19	40	78	29	54
South Africa	15	7	-	-	-	10	8	10 F	15	14
Spain	-	-	-	-	-	-	-	10	-	-
Tanzania	600	393	314	490	605	653	690	600	600	650
Yemen	...	...	...	...	...	...	...	...	...	21
51 Fishing area total	*1 401*	*1 203*	*1 100*	*1 372 F*	*1 199 F*	*1 454 F*	*1 277 F*	*1 326 F*	*1 169*	*1 290*
57 Australia	221	218	135	111	74	84	69	121	147	112
Indonesia	288	503	278	288	210	298	309	504	387	600
Malaysia	466	223	304	476	663	925	993	727	774	804
Thailand	5 897	7 027	4 608	5 541	10 702	10 919	14 989	12 880	12 699	12 626
57 Fishing area total	*6 872*	*7 971*	*5 325*	*6 416*	*11 649*	*12 226*	*16 360*	*14 232*	*14 007*	*14 142*
61 China,Taiwan	501	929	909	919	967	615	941	500	287	374
Japan	49 085	51 288	51 459	51 874	50 584	56 593	61 260	57 427	47 374	45 200
Korea Rep	11 203	7 618	7 916	21 700	21 384	22 952	26 432	19 262	19 148	19 873
Russian Fed	6	19	154	317	235	210	34	24	29	64
61 Fishing area total	*60 795*	*59 854*	*60 438*	*74 810*	*73 170*	*80 370*	*88 667*	*77 213*	*66 838*	*65 511*
67 Canada	117	145	89	108	197	208	139	54	56	61
USA	-	-	-	-	-	-	122	6	3	23
67 Fishing area total	*117*	*145*	*89*	*108*	*197*	*208*	*261*	*60*	*59*	*84*
71 Fiji Islands	37	58	59 F	91	48	35	62	61	50 F	57
Guam	-	-	-	0	0	-	1	2	1	2
Indonesia	337	327	416	376	435	971	3 210	1 833	1 598	2 500
Kiribati	2 270	2 200	2 180	2 200	2 230	1 874	650	687	373	69
Korea Rep	8	5	43	14	147	16	336	27	8	47
Malaysia	374	361	351	536	326	345	354	785	777	676
Micronesia	10 F	10 F	15 F	15 F	20 F	20 F	20 F	20 F	20 F	20 F
N Marianas	0	0	0	0	0	0	0	0	-	0
Philippines	3 630	8 915	7 109	9 729	9 025	7 991	5 235	5 813	5 502	6 088
Thailand	14 648	13 681	11 282	10 828	12 721	12 193	16 919	12 120	11 954	11 881
Wallis Fut I	...	...	1	1	1	1	1	1 F	1 F	1 F
71 Fishing area total	*21 314 F*	*25 557 F*	*21 456 F*	*23 790 F*	*24 953 F*	*23 446 F*	*26 788 F*	*21 349 F*	*20 284 F*	*21 341 F*
77 Amer Samoa	-	-	-	-	-	-	0	1	0	0
Cook Is	72 F	73 F	69 F	68 F	60 F	50 F	40 F	30 F	30 F	30 F
Costa Rica	12	15	67	26	27	58	6	76	71	66
Korea Rep	-	-	-	-	-	-	-	8	12	-
Mexico	830	1 196	966	877	1 257	944	755	1 094	883	837
USA	-	-	-	-	-	-	6	2	2	3
77 Fishing area total	*914 F*	*1 284 F*	*1 102 F*	*971 F*	*1 344 F*	*1 052 F*	*807 F*	*1 211 F*	*998 F*	*936 F*
81 Australia	...	...	...	...	...	506	714	...	...	...
Korea Rep	-	-	-	-	-	4	-	3	-	-
New Zealand	161	286	183	9 064	227	267	167	148	119	140
81 Fishing area total	*161*	*286*	*183*	*9 064*	*227*	*777*	*881*	*151*	*119*	*140*
87 Chile	3 286	3 608	3 732	3 796	3 477	4 404	4 877	3 168	1 682	2 008
Ecuador	1	1	0	0	0	0	5	5	5	5
Peru	350	1 245	602	800	760	1 856	5 123	1 593	819	635
87 Fishing area total	*3 637*	*4 854*	*4 334*	*4 596*	*4 237*	*6 260*	*10 005*	*4 766*	*2 506*	*2 648*
Species total	*218 024 F*	*223 621 F*	*205 197 F*	*235 021 F*	*230 773 F*	*210 845 F*	*245 385 F*	*295 013 F*	*259 722 F*	*261 750 F*
Cephalopods nei	**Céphalopodes nca**		**Cefalópodos nep**		***Cephalopoda***			**3,21(XX)XXX,XX**		**CEP**
34 Mauritania	-	-	-	-	-	-	10 F	16	129	134
Spain	-	-	-	-	6 240	5 636	8 541	222	311	129
34 Fishing area total	-	-	-	-	*6 240*	*5 636*	*8 551 F*	*238*	*440*	*263*
37 Algeria	-	-	-	-	-	-	70	29	30 F	30 F
Croatia	552	799	455	431	483	443	265	646	873	885
France	-	-	-	-	-	-	-	55	49	48
Israel	68	64	50	50	50	98	76	120	117	100 F
Spain	-	-	-	-	-	-	-	20	90	61

B-57 Squids, cuttlefishes, octopuses / Encornets, seiches, poulpes / Calamares, jibias, pulpos

Capture production by species, fishing areas and countries or areas
Captures par espèces, zones de pêche et pays ou zones
Capturas por especies, áreas de pesca y países o áreas

Species, Fishing area Espèce, Zone de pêche Especie, Area de pesca	1992 mt	1993 mt	1994 mt	1995 mt	1996 mt	1997 mt	1998 mt	1999 mt	2000 mt	2001 mt
37 Fishing area total	*620*	*863*	*505*	*481*	*533*	*541*	*411*	*870*	*1 159 F*	*1 124 F*
51 India	59 938	62 801	84 061	92 334	74 524	108 412	86 337	84 793	85 939	103 903
Iran	900	2 138	1 550	2 500	1 580	8 620	4 189	4 060	5 685	6 517
Madagascar	180	200	200	350	500	500	550	600	600	600
Somalia	500 F	500 F	500 F	500 F	500 F	500 F	500 F	500 F	500 F	500 F
51 Fishing area total	*61 518 F*	*65 639 F*	*86 311 F*	*95 684 F*	*77 104 F*	*118 032 F*	*91 576 F*	*89 953 F*	*92 724 F*	*111 520 F*
57 India	12 744	7 622	11 048	11 405	10 596	9 212	9 497	8 916	10 469	10 778
Timor-Leste	...	...	...	...	...	...	...	1 F	1	1
57 Fishing area total	*12 744*	*7 622*	*11 048*	*11 405*	*10 596*	*9 212*	*9 497*	*8 917 F*	*10 470*	*10 779*
58 Australia	-	-	-	-	-	1	-	-	-	-
58 Fishing area total	-	-	-	-	-	*1*	-	-	-	-
61 China,Taiwan	-	-	-	-	-	-	-	-	-	3
61 Fishing area total	-	-	-	-	-	-	-	-	-	*3*
71 Viet Nam	32 000 F	33 000 F	87 000 F	103 000 F	92 000 F	92 500 F	103 000 F	110 000 F	180 000 F	130 000 F
71 Fishing area total	*32 000 F*	*33 000 F*	*87 000 F*	*103 000 F*	*92 000 F*	*92 500 F*	*103 000 F*	*110 000 F*	*180 000 F*	*130 000 F*
Species total	*106 882 F*	*107 124 F*	*184 864 F*	*210 570 F*	*186 473 F*	*225 922 F*	*213 035 F*	*209 978 F*	*284 793 F*	*253 689 F*
Group total	***2 766 970***	***2 687 779***	***2 803 421***	***2 938 392***	***3 149 764***	***3 456 670***	***2 857 620***	***3 597 670***	***3 655 150***	***3 346 828***

B-58 Miscellaneous marine molluscs / Mollusques marins divers / Moluscos marinos diversos

Capture production by species, fishing areas and countries or areas / Captures par espèces, zones de pêche et pays ou zones / Capturas por especies, áreas de pesca y países o áreas

Species, Fishing area Espèce, Zone de pêche Especie, Area de pesca	1992 mt	1993 mt	1994 mt	1995 mt	1996 mt	1997 mt	1998 mt	1999 mt	2000 mt	2001 mt
Marine molluscs nei	**Mollusques marins nca**		**Moluscos marinos nep**		***Mollusca***			**3,99(XX)XXX,XX**		**MOL**
04 Indonesia	351	106	197	325	481	52	165	73	166	250
04 Fishing area total	*351*	*106*	*197*	*325*	*481*	*52*	*165*	*73*	*166*	*250*
21 Canada	538	52	705	1 089	663	503	168	644	1 263	1 262
St Pier Mq	-	-	-	-	1	0	0	17	205	115
USA	283	255	2 035	942	1 730	97	0	1 259	1	7
21 Fishing area total	*821*	*307*	*2 740*	*2 031*	*2 394*	*600*	*168*	*1 920*	*1 469*	*1 384*
27 Belgium	35	43	33	20	30	31	26	22	4	10
France	351	-	-	-	-	-	-	-	-	-
Ireland	-	-	-	-	-	-	-	-	66	-
Norway	-	9	3	40	1	14	111	118	60	9
Portugal	625	522	524	640	626	687	797	28	66	118
Spain	271	195	399	319	206	268	360	382	288	286
Sweden	-	-	-	-	-	-	-	-	1	-
UK	110	114	6	2	2	15	-	98	13	1
27 Fishing area total	*1 392*	*883*	*965*	*1 021*	*865*	*1 015*	*1 294*	*648*	*498*	*424*
31 Barbados	0	0	0	0	0	0	0	0	0	-
Colombia	1 231	14	8	8	20	17	3	-	15 F	15 F
Dominican Rp	...	...	...	...	...	...	...	...	48	18
Honduras	-	-	-	-	-	1	...	3	...	...
Korea Rep	732	555	-	-	-	-	-	-	-	-
Puerto Rico	38	82	87	70	124	77	45	9	12	9
USA	355	59 613	1 740	2 660	104	1 177	1 868	886	184	815
US Virgin Is	1 F	1 F	1 F	0	0	0	0	0	0	0
Venezuela	477	993	744	1 098	425	600	1 005	5 808	2 457	1 056
31 Fishing area total	*2 834 F*	*61 258 F*	*2 580 F*	*3 836*	*673*	*1 872*	*2 921*	*6 706*	*2 716 F*	*1 913 F*
34 Cameroon	12	7	0	0	0	0	0	0	0	0
Cape Verde	0	0	0	0	0	0	0	0	0	-
Côte dIvoire	-	-	-	-	-	-	-	-	26	-
Eq Guinea	130 F	120 F	180 F	80 F	180 F	230 F	122	140 F	20 F	20 F
Estonia	-	-	-	-	-	-	-	-	2	-
Greece	1	-	-	-	-	-	-	-	-	-
GuineaBissau	7 F	10 F	10 F	12 F	14 F	15 F	15 F	10 F	10 F	10 F
Italy	-	-	-	1 244	-	187	165	99	12	64
Korea Rep	-	2 970	-	-	-	-	-	-	1	-
Mauritania	-	-	-	-	-	-	20 F	13	1	5
Morocco	359	277	356	172	200	273	1 800	1 750	980	114
Portugal	0	0	0	-	0	0	2	4	2	3
Sao Tome Prn	16 F	19 F	28 F	29 F	21	20	25 F	30 F	30 F	30 F
Senegal	...	3	...	268	247	748	699	212	105	75
Sierra Leone	-	-	-	-	-	-	-	-	-	63
Spain	-	-	-	-	84	16	9 751	-	-	-
Togo	0	0	0	1	0	0	0	0	0	0
Westn Sahara	0	0	0	0	0	0	0	0	0	0
34 Fishing area total	*525 F*	*3 406 F*	*574 F*	*1 806 F*	*746 F*	*1 489 F*	*12 599 F*	*2 258 F*	*1 189 F*	*384 F*
37 Croatia	42	33	32	16	38	313	125	40	27	69
Egypt	...	...	...	140	233	0	198	0	-	4 173
France	47	64	26	-	-	-	-	-	-	-
Greece	2 619	2 382	1 739	2 393	3 076	2 241	1 789	1 736	1 688	2 169
Italy	8 867	9 407	7 864	11 663	11 238	9 563	7 680	7 761	6 270	6 260
Morocco	2	1	3	8	53	2	120	176	68	2
Romania	110	45	-	-	-	-	-	-	-	-
Slovenia	1	-	1	6	5	10	12	8	18	54
Spain	1 000 F	1 000 F	1 200 F	1 200 F	1 011	1 259	1 119	22	6	1
Tunisia	-	-	-	-	-	-	-	-	-	1
Turkey	3 663	3 689	2 632	1 224	2 466	2 092	4 077	3 646	2 168	2 671
Ukraine	-	-	1	0	-	-	-	-	-	3
Yugoslavia	1	0	0	0	0	0	0	0	-	1
37 Fishing area total	*16 352 F*	*16 621 F*	*13 498 F*	*16 650 F*	*18 120*	*15 480*	*15 120*	*13 389*	*10 245*	*15 404*
41 Argentina	1	72	714	13	1	50	-	1 000	-	-
Brazil	3 470 F	3 460 F	3 470 F	3 194	1 085	1 244	98	286	2 096	2 070 F
Italy	-	-	-	148	-	-	-	-	-	-
Korea Rep	-	40	-	-	-	-	-	-	378	769
Uruguay	96	64	170	323	89	557	1 310	3 651	0	-
41 Fishing area total	*3 567 F*	*3 636 F*	*4 354 F*	*3 678*	*1 175*	*1 851*	*1 408*	*4 937*	*2 474*	*2 839 F*
47 Italy	-	-	-	17	-	-	-	-	-	-
47 Fishing area total	-	-	-	*17*	-	-	-	-	-	-
51 Comoros	0	0	0	0	0	0	0	0	0	-
Egypt	-	-	-	-	-	-	-	-	1 718	361
India	3 130	934	1 385	2 409	15 299	4 750	2 718	1 217	815	547
Iran	550	270	115	150	150	1 992	1 583	1 640	1 235	1 144
Italy	-	-	-	254	-	-	-	-	-	-
Korea Rep	-	-	-	19	-	-	-	5	-	-
Madagascar	314	350	350	350	350	350	400	400	400	400
Maldives	30	30	110	143	140	271	314	485	866	-
Mozambique	9	5	11	38	57	-	-	-	-	-
51 Fishing area total	*4 033*	*1 589*	*1 971*	*3 363*	*15 996*	*7 363*	*5 015*	*3 747*	*5 034*	*2 452*
57 Australia	338	721	444	284	358	152	629	7 148	7 205	7 470

B-58 Miscellaneous marine molluscs / Mollusques marins divers / Moluscos marinos diversos

Capture production by species, fishing areas and countries or areas / Captures par espèces, zones de pêche et pays ou zones / Capturas por especies, áreas de pesca y países o áreas

Species, Fishing area Espèce, Zone de pêche Especie, Area de pesca	1992 mt	1993 mt	1994 mt	1995 mt	1996 mt	1997 mt	1998 mt	1999 mt	2000 mt	2001 mt
India	-	-	407	43	0	-	...	1 000	981	922
Indonesia	213	42	28	69	23	111	188	213	81	80
Sri Lanka	0	0	0	0	0	-	-	10	15	20
Thailand	26	13	-	5	-	-	27	25	25	25
57 Fishing area total	*577*	*776*	*879*	*401*	*381*	*263*	*844*	*8 396*	*8 307*	*8 517*
61 China	765 116	942 725	1 103 048	1 294 815	932 374	1 494 902	1 479 065	1 445 303	1 402 625	1 399 810
China,H.Kong	2 168	2 128	2 585	1 849	1 461	1 213	700	500 F	600 F	660 F
China, Macao	39	21	69	65	30	40 F	40 F	40 F	40 F	40 F
China,Taiwan	1 284	883	1 005	932	464	443	663	433	914	118
Korea Rep	16 793	14 692	11 941	633	1 011	837	2 209	564	438	501
Russian Fed	560	800	2 981	4 122	3 530	4 780	6 301	10 586	5 643	4 780
61 Fishing area total	*785 960*	*961 249*	*1 121 629*	*1 302 416*	*938 870*	*1 502 215 F*	*1 488 978 F*	*1 457 426 F*	*1 410 260 F*	*1 405 909 F*
67 Canada	1 233	1 227	1 354	1 567	1 336	2 805	1 835	989	1 272	919
USA	1 881	172	815	588	537	1 788	1 044	861	632	730
67 Fishing area total	*3 114*	*1 399*	*2 169*	*2 155*	*1 873*	*4 593*	*2 879*	*1 850*	*1 904*	*1 649*
71 Australia	27	46	47	86	134	60	187	299	397	296
Brunei Darsm	9	35	34	33	16	21	19	35	30	24
Cambodia	3 012	1 465	1 400	1 500	1 550	1 480	1 600	1 900	801	800 F
Fiji Islands	293	959	977 F	2 570	2 000	3 880	3 200	3 302	3 100 F	3 150
Indonesia	876	809	108	80	644	219	321	331	277	260
Kiribati	4 230	4 100	4 060	4 100	4 150	4 120	571	776	1 947	3 260
Korea Rep	209	-	48	-	20	180	99	114	-	-
NewCaledonia	3	6	7	34	130	95	150	27	6	6 F
Philippines	0	13	68	66	62	-	-	-	-	-
Thailand	226	322	-	-	-	-	99	1 664	1 641	1 631
Vanuatu	597 F	600 F	580 F	600 F	600 F	600 F	600 F	600 F	600 F	600 F
Viet Nam	17 402 F	16 502 F	44 680 F	50 437 F	44 016 F	44 117 F	52 903 F	50 003 F	44 685 F	50 000 F
71 Fishing area total	*26 884 F*	*24 857 F*	*52 009 F*	*59 506 F*	*53 322 F*	*54 772 F*	*59 749 F*	*59 051 F*	*53 484 F*	*60 027 F*
77 Cook Is	204 F	208 F	198 F	196 F	160 F	140 F	120 F	120 F	120 F	120 F
Costa Rica	29	19	66	74	65	109	62	55	42	-
El Salvador	596	825	670	601	598	535	620	1 119	1 115	823
Fr Polynesia	0	0	0	0	0	10	10	10	25	25
Honduras	146	...	...	146	...	7	20	24	465	...
Panama	66	84	...	45	48	71	172	16	76	20
Samoa	3	24	20	20	20	20	20	20	1 644	1 600 F
Tonga	...	...	...	...	...	1	3	7	13	22
USA	351	735	333	298	416	964	435	371	316	341
77 Fishing area total	*1 395 F*	*1 895 F*	*1 287 F*	*1 380 F*	*1 307 F*	*1 857 F*	*1 462 F*	*1 742 F*	*3 816 F*	*2 951 F*
81 Australia	163	163	256	256	326	169	230	1 386	1 309	1 482
New Zealand	1 400	419	1 422	10	16	4	-	-	2	1
81 Fishing area total	*1 563*	*582*	*1 678*	*266*	*342*	*173*	*230*	*1 386*	*1 311*	*1 483*
87 Chile	813	306	103	12	9	69	72	35	28	17
Colombia	26	568	2	77	434	840	84	19	143 F	140 F
Korea Rep	-	-	-	-	-	-	-	-	30	-
Peru	3 066	540	556	468	537	2 538	1 546	3 676	2 312	2 116
87 Fishing area total	*3 905*	*1 414*	*661*	*557*	*980*	*3 447*	*1 702*	*3 730*	*2 513 F*	*2 273 F*
Species total	*853 273 F*	*1 079 978 F*	*1 207 191 F*	*1 399 408 F*	*1 037 525 F*	*1 597 042 F*	*1 594 534 F*	*1 567 259 F*	*1 505 386 F*	*1 507 859 F*
Group total	***853 273***	***1 079 978***	***1 207 191***	***1 399 408***	***1 037 525***	***1 597 042***	***1 594 534***	***1 567 259***	***1 505 386***	***1 507 859***

B-61 Blue-whales, fin-whales / Baleines bleues, rorquals communs / Ballenas azules, rorcuales

Capture production by species, fishing areas and countries or areas / Captures par espèces, zones de pêche et pays ou zones / Capturas por especies, áreas de pesca y países o áreas

Species, Fishing area Espèce, Zone de pêche Especie, Area de pesca	1992 no	1993 no	1994 no	1995 no	1996 no	1997 no	1998 no	1999 no	2000 no	2001 no
Minke whale	**Petit rorqual**		**Rorcual enano**		***Balaenoptera acutorostrata***				**4,23(02)001,01**	**MIW**
98 Japan	288	330	330	330	440	440	438	389	439	440
98 Fishing area total	*288*	*330*	*330*	*330*	*440*	*440*	*438*	*389*	*439*	*440*
99 Australia	-	-	-	-	-	-	1	2	-	-
France	-	1	-	-	-	1	1	-	-	-
Greenland	104	110	105	156	172	157	176	179	154	148
Japan	-	14	37	120	104	127	124	119	69	179
Korea Rep	-	-	-	-	129	78	45	56	79	149
Norway	95	226	273	217	388	503	625	589	487	552
Peru	-	2	-	-	-	-	-	-	-	-
South Africa	-	-	-	-	-	1	-	-	-	-
UK	-	-	-	-	-	1	3	-	3	1
USA	-	-	-	-	12	4	2	-	-	-
99 Fishing area total	*199*	*353*	*415*	*493*	*805*	*872*	*977*	*945*	*792*	*1 029*
Species total	*487*	*683*	*745*	*823*	*1 245*	*1 312*	*1 415*	*1 334*	*1 231*	*1 469*
Bryde's whale	**Rorqual de Bryde**		**Rorcual tropical**		***Balaenoptera edeni***				**4,23(02)001,02**	**BRW**
99 Australia	1	-	-	-	-	-	-	-	1	-
Japan	-	-	-	-	-	-	1	-	43	50
New Zealand	-	-	-	-	-	1	-	-	-	-
St Vincent	-	-	-	-	-	-	-	-	1	-
South Africa	-	-	-	-	-	1	-	-	-	-
Spain	-	-	-	-	-	-	-	-	1	-
99 Fishing area total	*1*	-	-	-	-	*2*	*1*	-	*46*	*50*
Species total	*1*	-	-	-	-	*2*	*1*	-	*46*	*50*
Sei whale	**Rorqual de Rudolphi**		**Rorcual del Norte**		***Balaenoptera borealis***				**4,23(02)001,03**	**SIW**
99 Japan	-	-	-	-	-	-	-	-	-	1
New Zealand	-	1	-	-	-	-	-	-	-	-
99 Fishing area total	-	*1*	-	-	-	-	-	-	-	*1*
Species total	*...*	*1*	*...*	*...*	*...*	*...*	*...*	*...*	*...*	*1*
Blue whale	**Rorqual bleu**		**Ballena azul**		***Balaenoptera musculus***				**4,23(02)001,04**	**BLW**
99 Australia	-	-	-	-	-	-	-	-	-	2
New Zealand	-	-	1	-	-	-	-	-	-	-
USA	-	-	-	-	-	-	1	-	-	-
99 Fishing area total	-	-	*1*	-	-	-	*1*	-	-	*2*
Species total	*...*	*...*	*1*	*...*	*...*	*...*	*1*	*...*	*...*	*2*
Fin whale	**Rorqual commun**		**Rorcual común**		***Balaenoptera physalus***				**4,23(02)001,06**	**FIW**
99 France	2	-	-	1	-	-	-	-	-	-
Greenland	16	13	20	12	19	11	9	7	6	7
Ireland	-	-	-	-	-	-	-	-	1	-
Japan	-	-	-	-	-	1	-	-	-	1
USA	-	-	-	-	1	-	1	-	-	-
99 Fishing area total	*18*	*13*	*20*	*13*	*20*	*12*	*10*	*7*	*7*	*8*
Species total	*18*	*13*	*20*	*13*	*20*	*12*	*10*	*7*	*7*	*8*
Humpback whale	**Baleine à bosse**		**Rorcual jorobado**		***Megaptera novaeangliae***				**4,23(02)003,01**	**HUW**
99 Australia	-	2	-	1	-	-	-	-	-	5
Ecuador	-	-	-	2	-	-	-	-	-	-
Greenland	-	-	1	-	1	-	1	1	2	2
Japan	-	-	-	-	2	1	1	1	1	-
Korea Rep	-	-	-	-	-	-	-	1	-	-
St Vincent	1	2	-	-	1	-	2	2	2	2
USA	-	-	-	2	3	1	5	-	-	-
99 Fishing area total	*1*	*4*	*1*	*5*	*7*	*2*	*9*	*5*	*5*	*9*
Species total	*1*	*4*	*1*	*5*	*7*	*2*	*9*	*5*	*5*	*9*
Gray whale	**Baleine grise**		**Ballena gris**		***Eschrichtius robustus***				**4,23(04)001,01**	**GRW**
61 Russian Fed	-	-	-	-	-	-	-	-	55	64
61 Fishing area total	-	-	-	-	-	-	-	-	*55*	*64*
67 Russian Fed	-	-	-	-	-	-	-	-	11	-
67 Fishing area total	-	-	-	-	-	-	-	-	*11*	-
Species total	-	-	-	-	-	-	-	-	*66*	*64*
Baleen whales nei	**Baleines mysticètes nca**		**Ballenas mysticetas nep**		***Mysticeti***				**4,23(XX)XXX,XX**	**MYS**
99 Australia	1	2	-	3	-	-	-	-	-	3
Brazil	-	-	-	-	-	1	1	-	-	-
Canada	-	-	-	-	1	-	1	-	1	-
France	-	5	-	-	-	-	-	-	-	-
Japan	-	17	6	-	-	-	-	-	-	1

B-61 Blue-whales, fin-whales / Baleines bleues, rorquals communs / Ballenas azules, rorcuales

Capture production by species, fishing areas and countries or areas / Captures par espèces, zones de pêche et pays ou zones / Capturas por especies, áreas de pesca y países o áreas

Species, Fishing area Espèce, Zone de pêche Especie, Area de pesca	1992 no	1993 no	1994 no	1995 no	1996 no	1997 no	1998 no	1999 no	2000 no	2001 no
New Zealand	-	1	2	-	-	-	-	-	-	-
Russian Fed	-	-	44	89	43	79	123	122	114	113
Spain	-	-	-	-	-	-	-	1	-	1
USA	38	42	34	48	41	48	41	43	35	49
99 Fishing area total	*39*	*67*	*86*	*140*	*85*	*128*	*166*	*166*	*150*	*167*
Species total	*39*	*67*	*86*	*140*	*85*	*128*	*166*	*166*	*150*	*167*
Group total	***546***	***768***	***853***	***981***	***1 357***	***1 456***	***1 602***	***1 512***	***1 505***	***1 770***

Fishing area 98: Antarctic, pelagic; split-year data shown under the calendar year in which the split-year ends.

Fishing area 99: Outside the Antarctic; calendar year data.

Data are derived mainly from sources of the International Whaling Commission (IWC).

See paragraph 5 of the INTRODUCTION.

Zone de pêche 98: Antarctique, pélagique; données relatives à des années fractionnées figurant sous l'année civile durant laquelle se termine l'année fractionnée.

Zone de pêche 99: Hors de l'Antarctique; données relatives à l'année civile.

Données provenant principalement des sources de la Commission baleinière internationale.

Voir le paragraphe 5 de l'INTRODUCTION.

Area de pesca 98: Antártico, pelágico; datos correspondientes a los años emergentes que figuran en la columna del año civil en que finaliza el año emergente.

Area de pesca 99: Fuera del Antártico; datos relativos al año civil.

Datos obtenidos por la mayoría desde fuentes de la Comisión Ballenera Internacional.

Véase el párrafo 5 en la INTRODUCCION.

B-62 Sperm-whales, pilot-whales / Cachalots, globicéphales / Cachalotes, calderones

Capture production by species, fishing areas and countries or areas / Captures par espèces, zones de pêche et pays ou zones / Capturas por especies, áreas de pesca y países o áreas

Species, Fishing area Espèce, Zone de pêche Especie, Area de pesca	1992 no	1993 no	1994 no	1995 no	1996 no	1997 no	1998 no	1999 no	2000 no	2001 no
Baird's beaked whale	**Baleine à bec de Baird**		**Zifio de Baird**		***Berardius bairdii***				**4,22(02)019,02**	**BEV**
99 Japan	54	54	54	54	54	54	54	62	62	62
Korea Rep	-	-	-	-	-	1	-	1	-	-
USA	-	-	11	-	-	-	-	-	-	-
99 Fishing area total	*54*	*54*	*65*	*54*	*54*	*55*	*54*	*63*	*62*	*62*
Species total	*54*	*54*	*65*	*54*	*54*	*55*	*54*	*63*	*62*	*62*
Sperm whale	**Cachalot**		**Cachalote**		***Physeter catodon***				**4,22(03)001,01**	**SPV**
99 Brazil	-	-	-	1	-	-	-	-	-	-
Ecuador	-	-	-	4	-	-	-	-	-	-
France	1	6	1	-	-	-	-	-	-	-
Japan	-	-	-	1	-	-	-		5	8
Spain	-	-	-	-	-	-	-	3	5	3
USA	8	22	-	-	1	-	5	-	-	-
99 Fishing area total	*9*	*28*	*1*	*6*	*1*	-	*5*	*3*	*10*	*11*
Species total	*9*	*28*	*1*	*6*	*1*	...	*5*	*3*	*10*	*11*
Long-finned pilot whale	**Globicéphale commun**		**Calderón común**		***Globicephala melas***				**4,22(04)004,02**	**PIW**
99 Brazil	1	6	-	-	-	-	-	-	-	-
Canada	-	14	3	9	6	15	-	-	-	-
Chile	-	-	-	-	-	-	-	1	-	-
Faeroe Is	1 572	-	1 201	228	1 524	1 162	815	608	588	...
France	-	19	2	2	2	3	1	5	1	2
Germany	-	-	-	-	8	-	-	-	-	-
Greenland	-	20	-	-	67	208	365	115	5	43
Ireland	-	-	-	-	4	-	-	-	-	-
Netherlands	5	2	15	-	16	1	-	-	-	-
New Zealand	-	1	-	7	-	-	1	3	-	-
Spain	-	-	2	-	-	-	-	1	-	-
USA	14	31	22	31	12	93	104	370	-	-
99 Fishing area total	*1 592*	*93*	*1 245*	*277*	*1 639*	*1 482*	*1 286*	*1 103*	*594*	*45*
Species total	*1 592*	*93*	*1 245*	*277*	*1 639*	*1 482*	*1 286*	*1 103*	*594*	*45*
Short-finned pilot whale	**Globicéphale tropical**		**Calderón de aletas cortas**		***Globicephala macrorhynchus***				**4,22(04)004,03**	**SHV**
99 Ecuador	-	20	-	-	-	-	-	-	-	-
Japan	360	337	196	239	482	347	229	394	106	87
Korea Rep	-	-	-	-	-	2	-	-	-	-
Peru	-	1	-	-	-	-	-	-	-	-
St Lucia	-	-	-	-	-	-	-	35	-	-
Spain	-	-	-	-	-	-	-	2	-	-
USA	8	81	-	-	-	6	-	-	-	-
Other nei	-	-	-	-	-	5	-	-	1	-
99 Fishing area total	*368*	*439*	*196*	*239*	*482*	*360*	*229*	*431*	*107*	*87*
Species total	*368*	*439*	*196*	*239*	*482*	*360*	*229*	*431*	*107*	*87*
Killer whale	**Orque**		**Orca**		***Orcinus orca***				**4,22(04)022,01**	**KIW**
99 Brazil	-	1	1	-	-	-	-	-	-	-
Japan	-	-	-	-	-	1	-	-	-	-
Korea Rep	-	-	-	-	1	-	-	-	1	-
New Zealand	1	1	-	-	-	-	-	-	-	-
USA	-	1	-	6	-	4	1	4	-	-
99 Fishing area total	*1*	*3*	*1*	*6*	*1*	*5*	*1*	*4*	*1*	-
Species total	*1*	*3*	*1*	*6*	*1*	*5*	*1*	*4*	*1*	-
Harbour porpoise	**Marsouin commun**		**Marsopa común**		***Phocoena phocoena***				**4,22(05)002,01**	**PHR**
99 Canada	1	257	98	-	-	-	-	-	-	-
Denmark	119	4 449	4 449	7 000	-	-	2	-	-	-
Faeroe Is	-	-	-	-	3	-	-	-	-	-
France	-	-	-	-	-	9	1	8	11	12
Germany	9	12	18	8	6	4	5	3	5	8
Greenland	-	-	1 716	1 135	1 824	1 592	2 131	1 830	1 607	1 628
Ireland	11	1 497	-	1	2	3	2	4	-	1
Korea Rep	-	-	-	-	1	-	-	1	-	87
Netherlands	-	4	-	1	-	4	4	-	2	-
Spain	-	-	-	1	-	-	-	1	2	1
Sweden	-	-	25	53	124	8	14	2	3	-
UK	-	740	761	933	752	791	33	19	34	-
USA	922	1 414	2 129	1 537	1 540	1 430	842	475	26	-
99 Fishing area total	*1 062*	*8 373*	*9 196*	*10 669*	*4 252*	*3 841*	*3 034*	*2 343*	*1 690*	*1 737*
Species total	*1 062*	*8 373*	*9 196*	*10 669*	*4 252*	*3 841*	*3 034*	*2 343*	*1 690*	*1 737*
White whale	**Bélouga**		**Beluga**		***Delphinapterus leucas***				**4,22(06)014,01**	**BEL**
61 Russian Fed	10	26	71	20	3	3	27	70	22	7
61 Fishing area total	*10*	*26*	*71*	*20*	*3*	*3*	*27*	*70*	*22*	*7*
67 Russian Fed	-	-	-	-	-	-	-	6	-	-

B-62 Sperm-whales, pilot-whales / Cachalots, globicéphales / Cachalotes, calderones

Capture production by species, fishing areas and countries or areas / Captures par espèces, zones de pêche et pays ou zones / Capturas por especies, áreas de pesca y países o áreas

Species, Fishing area Espèce, Zone de pêche Especie, Area de pesca	1992 no	1993 no	1994 no	1995 no	1996 no	1997 no	1998 no	1999 no	2000 no	2001 no
67 Fishing area total	-	-	-	-	-	-	-	*6*	-	-
99 Canada	618	-	-	-	-	-	-	-	-	375
Greenland	267	475	488	606	542	577	746	493	609	260
USA	325	312	276	229	304	232	327	238	-	-
99 Fishing area total	*1 210*	*787*	*764*	*835*	*846*	*809*	*1 073*	*731*	*609*	*635*
Species total	*1 220*	*813*	*835*	*855*	*849*	*812*	*1 100*	*807*	*631*	*642*
Narwhal	**Narval**		**Narval**		***Monodon monoceros***				**4,22(06)018,01**	**NAR**
99 Canada	325	-	-	-	-	-	-	-	-	559
Greenland	130	741	847	461	738	797	822	912	597	449
99 Fishing area total	*455*	*741*	*847*	*461*	*738*	*797*	*822*	*912*	*597*	*1 008*
Species total	*455*	*741*	*847*	*461*	*738*	*797*	*822*	*912*	*597*	*1 008*
Toothed whales nei	**Baleines odontocètes nca**		**Ballenas odontocetas nep**		***Odontoceti***				**4,22(XX)XXX,XX**	**ODN**
99 Argentina	96	364	117	162	-	-	15	-	445	6
Australia	16	30	30	30	22	31	50	40	22	21
Brazil	194	255	335	223	495	163	132	512	939	23
Canada	-	-	3	1	3	-	-	-	-	-
Chile	-	-	-	-	-	-	1	1	-	-
Ecuador	-	3 741	227	-	-	-	-	-	-	-
Faeroe Is	46	-	266	151	173	350	438	-	255	...
France	472	1 735	56	16	30	312	47	179	212	140
Ireland	-	321	-	-	493	2	29	9	4	1
Japan	14 029	16 126	17 577	14 474	17 173	19 739	13 037	16 764	...	...
Korea Rep	-	-	-	-	90	75	33	46	94	140
Mexico	-	14	-	-	-	1	-	-	6	3
Netherlands	90	14	103	10	32	43	29	-	-	-
New Zealand	22	34	19	21	3	8	15	6	13	19
Peru	3 971	4 154	1 877	252	15	8	48	209	77	24
St Lucia	-	-	-	-	-	-	-	126	-	-
South Africa	-	87	165	95	100	149	45	60	-	37
Spain	-	-	33	3	3	7	6	9	18	11
UK	-	9	64	168	2	9	6	6	12	52
USA	1 162	1 170	1 041	947	1 723	895	467	324	146	5
Other nei	15 108	3 486	4 096	3 274	2 547	3 000	1 877	1 348	1 635	-
99 Fishing area total	*35 206*	*31 540*	*26 009*	*19 827*	*22 904*	*24 792*	*16 275*	*19 639*	*3 878*	*482*
Species total	*35 206*	*31 540*	*26 009*	*19 827*	*22 904*	*24 792*	*16 275*	*19 639*	*3 878*	*482*
Group total	***39 967***	***42 084***	***38 395***	***32 394***	***30 920***	***32 144***	***22 806***	***25 305***	***7 570***	***4 074***

Fishing area 98: Antarctic, pelagic; split-year data shown under the calendar year in which the split-year ends.

Fishing area 99: Outside the Antarctic; calendar year data.

Data are derived mainly from sources of the International Whaling Commission (IWC).

See paragraph 5 of the INTRODUCTION.

Zone de pêche 98: Antarctique, pélagique; données relatives à des années fractionnées figurant sous l'année civile durant laquelle se termine l'année fractionnée.

Zone de pêche 99: Hors de l'Antarctique; données relatives à l'année civile.

Données provenant principalement des sources de la Commission baleinière internationale.

Voir le paragraphe 5 de l'INTRODUCTION.

Area de pesca 98: Antártico, pelágico; datos correspondientes a los años emergentes que figuran en la columna del año civil en que finaliza el año emergente.

Area de pesca 99: Fuera del Antártico; datos relativos al año civil.

Datos obtenidos por la mayoría desde fuentes de la Comisión Ballenera Internacional.

Véase el párrafo 5 en la INTRODUCCION.

B-63 Eared seals, hair seals, walruses / Otaries, phoques, morses / Lobos marinos, focas, morsas

Capture production by species, fishing areas and countries or areas / Captures par espèces, zones de pêche et pays ou zones / Capturas por especies, áreas de pesca y países o áreas

Species, Fishing area Espèce, Zone de pêche Especie, Area de pesca	1992 no	1993 no	1994 no	1995 no	1996 no	1997 no	1998 no	1999 no	2000 no	2001 no
Steller sea lion	**Lion de mer de Steller**		**Lobo marino de Steller**		***Eumetopias jubatus***				**4,06(01)001,01**	**SSL**
67 USA	549	487	416	339	186	164	178	...	164	198
67 Fishing area total	*549*	*487*	*416*	*339*	*186*	*164*	*178*	...	*164*	*198*
Species total	*549*	*487*	*416*	*339*	*186*	*164*	*178*	...	*164*	*198*
Northern fur seal	**Otarie des Pribilofs**		**Lobo fino del Norte**		***Callorhinus ursinus***				**4,06(01)002,01**	**SEN**
61 Russian Fed	7 970	8 738	10 095	8 103	6 487	6 537	7 742	4 500	2 919	1 602
61 Fishing area total	*7 970*	*8 738*	*10 095*	*8 103*	*6 487*	*6 537*	*7 742*	*4 500*	*2 919*	*1 602*
67 USA	1 676	1 837	1 777	1 785	1 823	1 380	1 553	1 193	...	...
67 Fishing area total	*1 676*	*1 837*	*1 777*	*1 785*	*1 823*	*1 380*	*1 553*	*1 193*	...	...
Species total	*9 646*	*10 575*	*11 872*	*9 888*	*8 310*	*7 917*	*9 295*	*5 693*	*2 919*	*1 602*
South American fur seal	**Otarie d'Amérique du Sud**		**Lobo fino austral**		***Arctocephalus australis***				**4,06(01)006,01**	**SEF**
41 Uruguay	102	112	-	-	-	-	-	-	-	-
41 Fishing area total	*102*	*112*	-	-	-	-	-	-	-	-
Species total	*102*	*112*	-	-	-	-	-	-	-	-
South African fur seal	**Otarie du Cap**		**Lobo marino de dos pelos**		***Arctocephalus pusillus***				**4,06(01)006,03**	**SEK**
47 Namibia	23 400	35 730	37 853	20 450	20 814	25 783	29 475	25 161	41 753	41 753 F
47 Fishing area total	*23 400*	*35 730*	*37 853*	*20 450*	*20 814*	*25 783*	*29 475*	*25 161*	*41 753*	*41 753 F*
Species total	*23 400*	*35 730*	*37 853*	*20 450*	*20 814*	*25 783*	*29 475*	*25 161*	*41 753*	*41 753 F*
South American sea lion	**Lion de mer d'Amérique du Sud**		**Lobo común**		***Otaria byronia***				**4,06(01)016,01**	**SEL**
87 Chile	-	-	-	389	96	16	-	-	-	-
87 Fishing area total	-	-	-	*389*	*96*	*16*	-	-	-	-
Species total	-	-	-	*389*	*96*	*16*	-	-	-	-
Walrus	**Morse**		**Morsa**		***Odobenus rosmarus***				**4,06(02)004,01**	**WAL**
21 Greenland	-	-	-	-	305	317	610	311	329	224
21 Fishing area total	-	-	-	-	*305*	*317*	*610*	*311*	*329*	*224*
61 Russian Fed	1 750	856	1 013	1 091	955	731	950	657	534	738
61 Fishing area total	*1 750*	*856*	*1 013*	*1 091*	*955*	*731*	*950*	*657*	*534*	*738*
67 Russian Fed	-	-	-	-	-	-	-	688	210	-
67 Fishing area total	-	-	-	-	-	-	-	*688*	*210*	-
Species total	*1 750*	*856*	*1 013*	*1 091*	*1 260*	*1 048*	*1 560*	*1 656*	*1 073*	*962*
Harp seal	**Phoque du Groenland**		**Foca de Groenlandia**		***Phoca groenlandica***				**4,06(03)005,01**	**SEH**
21 Canada	67 428	25 175	61 176	65 391	242 717	264 204	249 053	228 828	87 084	193 366
Greenland	16 896	6 293	12 323	12 283	70 000	65 000	77 000	89 000	93 500	59 000
21 Fishing area total	*84 324*	*31 468*	*73 499*	*77 674*	*312 717*	*329 204*	*326 053*	*317 828*	*180 584*	*252 366*
27 Greenland	1 465	363	1 036	1 814	4 945	4 663	5 491	6 097	6 347	3 837
Norway	13 321	12 278	17 621	15 048	15 926	7 163	2 716	1 953	18 678	8 192
Russian Fed	32 973	31 500	35 272	29 644	31 528	31 380	13 370	34 850	38 413	39 116
27 Fishing area total	*47 759*	*44 141*	*53 929*	*46 506*	*52 399*	*43 206*	*21 577*	*42 900*	*63 438*	*51 145*
Species total	*132 083*	*75 609*	*127 428*	*124 180*	*365 116*	*372 410*	*347 630*	*360 728*	*244 022*	*303 511*
Harbour seal	**Phoque veau marin**		**Foca común**		***Phoca vitulina***				**4,06(03)005,02**	**SEC**
21 Greenland	-	36	20	26	220	250	190	130	110	50
21 Fishing area total	-	*36*	*20*	*26*	*220*	*250*	*190*	*130*	*110*	*50*
27 Greenland	-	1	1	4	36	45	27	18	14	13
UK	1	2	8	-	-	-	-	-	-	-
27 Fishing area total	*1*	*3*	*9*	*4*	*36*	*45*	*27*	*18*	*14*	*13*
67 USA	2 854	2 736	2 621	2 742	2 741	2 546	2 597	...	2 224	2 031
67 Fishing area total	*2 854*	*2 736*	*2 621*	*2 742*	*2 741*	*2 546*	*2 597*	...	*2 224*	*2 031*
Species total	*2 855*	*2 775*	*2 650*	*2 772*	*2 997*	*2 841*	*2 814*	*148*	*2 348*	*2 094*
Ringed seal	**Phoque annelé ou marbré**		**Foca marbreada**		***Phoca hispida***				**4,06(03)005,03**	**SER**
21 Canada	-	-	-	-	670	-	1 234	742	1 747	2 035
Greenland	34 201	36 444	32 791	20 863	63 000	56 000	57 000	58 000	56 000	40 700
21 Fishing area total	*34 201*	*36 444*	*32 791*	*20 863*	*63 670*	*56 000*	*58 234*	*58 742*	*57 747*	*42 735*
27 Finland	10	16	-	-	-	-	-	-	-	-
Greenland	12 873	8 181	11 370	9 766	27 309	24 387	25 108	25 453	24 365	17 815
Russian Fed	781	530	628	399	155	705	532	599	672	687
27 Fishing area total	*13 664*	*8 727*	*11 998*	*10 165*	*27 464*	*25 092*	*25 640*	*26 052*	*25 037*	*18 502*

B-63 Eared seals, hair seals, walruses — Capture production by species, fishing areas and countries or areas
Otaries, phoques, morses — Captures par espèces, zones de pêche et pays ou zones
Lobos marinos, focas, morsas — Capturas por especies, áreas de pesca y países o áreas

Species, Fishing area Espèce, Zone de pêche Especie, Area de pesca	1992 no	1993 no	1994 no	1995 no	1996 no	1997 no	1998 no	1999 no	2000 no	2001 no
61 Russian Fed	16 455	16 148	9 198	3 142	1 973	1 758	2 443	1 673	2 645	2 030
61 Fishing area total	*16 455*	*16 148*	*9 198*	*3 142*	*1 973*	*1 758*	*2 443*	*1 673*	*2 645*	*2 030*
67 Russian Fed	-	-	-	-	-	-	-	1 641	1 310	-
67 Fishing area total	-	-	-	-	-	-	-	*1 641*	*1 310*	-
Species total	*64 320*	*61 319*	*53 987*	*34 170*	*93 107*	*82 850*	*86 317*	*88 108*	*86 739*	*63 267*
Ribbon seal	**Phoque à rubans**		**Foca fajada**		***Phoca fasciata***			**4,06(03)005,04**		**SLR**
61 Russian Fed	11 915	13 543	3 585	18	24	20	-	8	8	-
61 Fishing area total	*11 915*	*13 543*	*3 585*	*18*	*24*	*20*	-	*8*	*8*	-
Species total	*11 915*	*13 543*	*3 585*	*18*	*24*	*20*	-	*8*	*8*	-
Caspian seal	**Phoque de la Mer Caspienne**		**Foca del Caspio**		***Phoca caspica***			**4,06(03)005,05**		**SAC**
05 Russian Fed	23 200	23 000	11 700	2 110	4 930	4 320	-	-	-	9
05 Fishing area total	*23 200*	*23 000*	*11 700*	*2 110*	*4 930*	*4 320*	-	-	-	*9*
Species total	*23 200*	*23 000*	*11 700*	*2 110*	*4 930*	*4 320*	-	-	-	*9*
Baikal seal	**Phoque du lac Baikal**		**Foca de Baikal**		***Phoca sibirica***			**4,06(03)005,06**		**SBK**
05 Russian Fed	3 487	4 902	2 951	2 272	2 450	2 991	-	193	2 381	2 817
05 Fishing area total	*3 487*	*4 902*	*2 951*	*2 272*	*2 450*	*2 991*	-	*193*	*2 381*	*2 817*
Species total	*3 487*	*4 902*	*2 951*	*2 272*	*2 450*	*2 991*	-	*193*	*2 381*	*2 817*
Larga seal	**Veau marin du Pacifique**		**Foca largha**		***Phoca largha***			**4,06(03)005,07**		**SST**
61 Russian Fed	5 544	4 839	1 778	350	317	167	181	231	331	179
61 Fishing area total	*5 544*	*4 839*	*1 778*	*350*	*317*	*167*	*181*	*231*	*331*	*179*
67 Russian Fed	-	-	-	-	-	-	-	222	-	-
67 Fishing area total	-	-	-	-	-	-	-	*222*	-	-
Species total	*5 544*	*4 839*	*1 778*	*350*	*317*	*167*	*181*	*453*	*331*	*179*
Bearded seal	**Phoque barbu**		**Foca barbuda**		***Erignathus barbatus***			**4,06(03)008,01**		**SEB**
21 Canada	-	-	-	-	45	-	59	50	75	170
Greenland	155	-	-	-	1 630	1 800	1 800	1 790	2 060	1 460
21 Fishing area total	*155*	-	-	-	*1 675*	*1 800*	*1 859*	*1 840*	*2 135*	*1 630*
27 Greenland	83	-	-	-	504	549	554	546	635	448
Russian Fed	34	21	31	15	6	24	30	29	24	26
27 Fishing area total	*117*	*21*	*31*	*15*	*510*	*573*	*584*	*575*	*659*	*474*
61 Russian Fed	2 496	3 255	1 719	1 013	1 114	1 014	890	802	528	340
61 Fishing area total	*2 496*	*3 255*	*1 719*	*1 013*	*1 114*	*1 014*	*890*	*802*	*528*	*340*
67 Russian Fed	-	-	-	-	-	-	-	338	119	-
67 Fishing area total	-	-	-	-	-	-	-	*338*	*119*	-
Species total	*2 768*	*3 276*	*1 750*	*1 028*	*3 299*	*3 387*	*3 333*	*3 555*	*3 441*	*2 444*
Hooded seal	**Phoque à crete**		**Foca capuchina**		***Cystophora cristata***			**4,06(03)010,01**		**SEZ**
21 Canada	119	19	149	857	25 754	7 058	10 020	49	10	257
Greenland	1 155	321	262	13	5 800	4 400	3 700	4 400	3 400	3 200
21 Fishing area total	*1 274*	*340*	*411*	*870*	*31 554*	*11 458*	*13 720*	*4 449*	*3 410*	*3 457*
27 Greenland	2 000	411	1 794	33	4 106	3 100	2 628	3 058	2 434	2 280
Norway	755	384	492	933	811	2 934	6 351	4 446	1 936	3 820
Russian Fed	8 038	-	4 252	-	-	-	-	-	-	-
27 Fishing area total	*10 793*	*795*	*6 538*	*966*	*4 917*	*6 034*	*8 979*	*7 504*	*4 370*	*6 100*
Species total	*12 067*	*1 135*	*6 949*	*1 836*	*36 471*	*17 492*	*22 699*	*11 953*	*7 780*	*9 557*
Grey seal	**Phoque gris**		**Foca de gris**		***Halichoerus grypus***			**4,06(03)011,01**		**SEG**
27 Finland	32	23	-	-	-	-	4	30	62	73
UK	14	35	23	-	-	-	-	-	-	-
27 Fishing area total	*46*	*58*	*23*	-	-	-	*4*	*30*	*62*	*73*
Species total	*46*	*58*	*23*	-	-	-	*4*	*30*	*62*	*73*
Seals nei	**Phoques nca**		**Focas nep**		***Otariidae, Phocidae***			**4,06(XX)XXX,XX**		**SXX**
21 Canada	1 127	1 125	1 798	1 799	190	1 838	55	28	7	39
21 Fishing area total	*1 127*	*1 125*	*1 798*	*1 799*	*190*	*1 838*	*55*	*28*	*7*	*39*
Species total	*1 127*	*1 125*	*1 798*	*1 799*	*190*	*1 838*	*55*	*28*	*7*	*39*
Group total	***294 859***	***239 341***	***265 753***	***202 692***	***539 567***	***523 244***	***503 541***	***497 714***	***393 028***	***428 505***

See paragraph 5 of the INTRODUCTION. Voir le paragraphe 5 de l'INTRODUCTION. Véase el párrafo 5 en la INTRODUCCION.

B-64 Miscellaneous aquatic mammals / Mammifères aquatiques divers / Mamíferos acuáticos diversos

Capture production by species, fishing areas and countries or areas / Captures par espèces, zones de pêche et pays ou zones / Capturas por especies, áreas de pesca y países o áreas

Species, Fishing area Espèce, Zone de pêche Especie, Area de pesca	1992 mt	1993 mt	1994 mt	1995 mt	1996 mt	1997 mt	1998 mt	1999 mt	2000 mt	2001 mt
Aquatic mammals nei	**Mammifères aquatiques nca**		**Mamíferos acuáticos nep**		***Mammalia***			**4,99(XX)XXX,XX**		**MAM**
61 Japan	1 261	1 522	1 605	1 259	1 748	1 883	1 242	1 705	1 767	1 874
61 Fishing area total	*1 261*	*1 522*	*1 605*	*1 259*	*1 748*	*1 883*	*1 242*	*1 705*	*1 767*	*1 874*
Species total	*1 261*	*1 522*	*1 605*	*1 259*	*1 748*	*1 883*	*1 242*	*1 705*	*1 767*	*1 874*
Group total	***1 261***	***1 522***	***1 605***	***1 259***	***1 748***	***1 883***	***1 242***	***1 705***	***1 767***	***1 874***

See paragraph 5 of the INTRODUCTION.

Voir le paragraphe 5 de l'INTRODUCTION.

Véase el párrafo 5 en la INTRODUCCION.

B-71 Frogs and other amphibians / Grenouilles et autres amphibies / Ranas y otros anfibios

Capture production by species, fishing areas and countries or areas
Captures par espèces, zones de pêche et pays ou zones
Capturas por especies, áreas de pesca y países o áreas

Species, Fishing area Espèce, Zone de pêche Especie, Area de pesca	1992 mt	1993 mt	1994 mt	1995 mt	1996 mt	1997 mt	1998 mt	1999 mt	2000 mt	2001 mt
Frogs	**Grenouilles**		**Ranas**		***Rana spp***				**5,12(01)001,XX**	**FRG**
02 Cuba	73	52	52	62	69	46	28	26	30 F	30 F
Mexico	320	312	335	497	351	1 979	1 167	311	295	...
USA	18	20	17	5	2	9	6	-	0	1
02 Fishing area total	*411*	*384*	*404*	*564*	*422*	*2 034*	*1 201*	*337*	*325 F*	*31 F*
03 Uruguay	-	-	-	5	0	0	0	0	7	9
03 Fishing area total	-	-	-	*5*	*0*	*0*	*0*	*0*	*7*	*9*
04 Bangladesh	771	700	700	-	-	-	-	-	-	-
Indonesia	2 666	2 411	2 111	2 194	1 795	1 390	1 667	1 317	1 880	1 970
Turkey	648	750	851	864	740	160	100	118	77	873
04 Fishing area total	*4 085*	*3 861*	*3 662*	*3 058*	*2 535*	*1 550*	*1 767*	*1 435*	*1 957*	*2 843*
05 Romania	2	...	...	...	...	38	41	35	26	-
05 Fishing area total	*2*	...	...	...	...	*38*	*41*	*35*	*26*	-
Species total	*4 498*	*4 245*	*4 066*	*3 627*	*2 957*	*3 622*	*3 009*	*1 807*	*2 315 F*	*2 883 F*
Group total	***4 498***	***4 245***	***4 066***	***3 627***	***2 957***	***3 622***	***3 009***	***1 807***	***2 315***	***2 883***

B-72 Turtles / Tortues / Tortugas

Capture production by species, fishing areas and countries or areas
Captures par espèces, zones de pêche et pays ou zones
Capturas por especies, áreas de pesca y países o áreas

Species, Fishing area Espèce, Zone de pêche Especie, Area de pesca	1992 mt	1993 mt	1994 mt	1995 mt	1996 mt	1997 mt	1998 mt	1999 mt	2000 mt	2001 mt
Diamond back terrapins	**Tortues diamantées**		**Tortugas comestibles**			***Malaclemys spp***			**5,31(06)021,XX**	**TTG**
02 USA	0	0	0	0	-	-	0	-	0	-
02 Fishing area total	*0*	*0*	*0*	*0*	-	-	*0*	-	*0*	-
Species total	*0*	*0*	*0*	*0*	-	-	*0*	-	*0*	-
Green turtle	**Tortue verte**		**Tortuga verde**			***Chelonia mydas***			**5,31(07)005,02**	**TUG**
31 Cuba	258	122	115	46	34	18	20	8	8 F	8 F
Grenada	11	8	4	7	8	6	6	5	5	7
Venezuela	0	0	0	0	0	0	0	-	-	-
31 Fishing area total	*269*	*130*	*119*	*53*	*42*	*24*	*26*	*13*	*13 F*	*15 F*
71 Fiji Islands	22	49	25 F	6	22	4	2	2	2 F	3
71 Fishing area total	*22*	*49*	*25 F*	*6*	*22*	*4*	*2*	*2*	*2 F*	*3*
Species total	*291*	*179*	*144 F*	*59*	*64*	*28*	*28*	*15*	*15 F*	*18 F*
Hawksbill turtle	**Tortue caret**		**Tortuga carey**			***Eretmochelys imbricata***			**5,31(07)017,01**	**TTH**
31 Cuba	193	117	45	20	23	19	18	12	12 F	12 F
Venezuela	0	0	0	0	0	0	0	-	-	-
31 Fishing area total	*193*	*117*	*45*	*20*	*23*	*19*	*18*	*12*	*12 F*	*12 F*
Species total	*193*	*117*	*45*	*20*	*23*	*19*	*18*	*12*	*12 F*	*12 F*
Loggerhead turtle	**Caouane**		**Caguama**			***Caretta caretta***			**5,31(07)018,01**	**TTL**
31 Cuba	64	48	23	11	10	7	7	5	5 F	5 F
Venezuela	0	0	0	0	0	0	0	-	-	-
31 Fishing area total	*64*	*48*	*23*	*11*	*10*	*7*	*7*	*5*	*5 F*	*5 F*
Species total	*64*	*48*	*23*	*11*	*10*	*7*	*7*	*5*	*5 F*	*5 F*
River and lake turtles nei	**Tortues d'eau douce nca**		**Galápagos nep**			***Testudinata***			**5,31(XX)XXX,XX**	**TUL**
02 Cuba	0	0	0	0	3	4	2	-	-	-
USA	35	28	13	6	15	13	24	15	53	51
02 Fishing area total	*35*	*28*	*13*	*6*	*18*	*17*	*26*	*15*	*53*	*51*
03 Brazil	4 F	10 F	5 F	0	0	0	0	0	0	0
Venezuela	0	0	0	0	0	0	0	-	-	-
03 Fishing area total	*4 F*	*10 F*	*5 F*	*0*	*0*	*0*	*0*	*0*	*0*	*0*
04 Indonesia	32	-	56	-	36	48	27	1	3	-
04 Fishing area total	*32*	-	*56*	-	*36*	*48*	*27*	*1*	*3*	-
Species total	*71 F*	*38 F*	*74 F*	*6*	*54*	*65*	*53*	*16*	*56*	*51*
Marine turtles nei	**Tortues de mer nca**		**Tortugas de mar nep**			***Testudinata***			**5,31(XX)XXX,XX**	**TTX**
31 Bahamas	6	4	2	2	3	3	3	1	2	4
Costa Rica	43	...	113	101	149	33	86	0	0	-
Cuba	12	8	4	22	1	0	1	-	-	-
Guadeloupe	10 F	10 F	5 F	0	0	0	0	-	-	-
Martinique	2 F	0	0	0	0	0	0	-	-	-
Puerto Rico	0	0	0	0	0	0	0	-	-	-
Venezuela	0	0	0	0	0	0	0	-	-	-
31 Fishing area total	*73 F*	*22 F*	*124 F*	*125*	*153*	*36*	*90*	*1*	*2*	*4*
34 Benin	...	...	...	...	...	...	...	29	20 F	5
Cape Verde	10	5	0	0	0	0	0	-	-	-
Côte dIvoire	...	...	...	...	...	...	...	...	50	71
Eq Guinea	100 F	50 F	20 F	10 F	5 F	5 F	9	10 F	1 F	1 F
Gabon	...	...	...	12	37	159	180	424	51	238
Liberia	3	4	-	-	-	-	-	-	-	-
34 Fishing area total	*113 F*	*59 F*	*20 F*	*22 F*	*42 F*	*164 F*	*189*	*463 F*	*122 F*	*315 F*
37 Egypt	231	-	-	-	-	-	-	-	-	-
37 Fishing area total	*231*	-	-	-	-	-	-	-	-	-
41 Brazil	3 F	5 F	0	0	0	0	0	0	0	0
41 Fishing area total	*3 F*	*5 F*	*0*	*0*	*0*	*0*	*0*	*0*	*0*	*0*
51 Madagascar	25	20	-	-	-	-	-	-	-	-
Seychelles	...	...	10	-	-	-	-	-	-	-
51 Fishing area total	*25*	*20*	*10*	-	-	-	-	-	-	-
57 Indonesia	82	52	337	291	353	248	337	272	371	380
Timor-Leste	...	...	...	...	...	...	...	1 F	1	1
57 Fishing area total	*82*	*52*	*337*	*291*	*353*	*248*	*337*	*273 F*	*372*	*381*
71 Fiji Islands	4	20	15 F	7	24	7	2	8	6 F	7
Indonesia	231	220	207	284	368	390	444	434	373	380
Micronesia	7 F	2 F	2 F	0	0	0	0	0	0	0
N Marianas	0	0	0	0	0	-	-	-	-	-
Philippines	17	-	-	1	1	-	2	2	1	1
Solomon Is	1	-	-	-	-	-	-	-	-	-

B-72 Turtles / Tortues / Tortugas

Capture production by species, fishing areas and countries or areas
Captures par espèces, zones de pêche et pays ou zones
Capturas por especies, áreas de pesca y países o áreas

Species, Fishing area Espèce, Zone de pêche Especie, Area de pesca	1992 mt	1993 mt	1994 mt	1995 mt	1996 mt	1997 mt	1998 mt	1999 mt	2000 mt	2001 mt
Wallis Fut I	...	...	2	2	2	2	2	2 F	2 F	2 F
71 Fishing area total	*260 F*	*242 F*	*226 F*	*294*	*395*	*399*	*450*	*446 F*	*382 F*	*390 F*
77 Honduras	-	-	-	-	-	-	3 F	1	21	...
Panama	0	0	0	0	0	0	0	0	0	...
77 Fishing area total	*0*	*0*	*0*	*0*	*0*	*0*	*3 F*	*1*	*21*	*...*
87 Ecuador	-	-	-	10	10	10	10	10	10	10
Peru	30	28	6	4	0	1	2	1	1	2
87 Fishing area total	*30*	*28*	*6*	*14*	*10*	*11*	*12*	*11*	*11*	*12*
Species total	*817 F*	*428 F*	*723 F*	*746 F*	*953 F*	*858 F*	*1 081 F*	*1 195 F*	*910 F*	*1 102 F*
Group total	***1 436***	***810***	***1 009***	***842***	***1 104***	***977***	***1 187***	***1 243***	***998***	***1 188***

B-73 Crocodiles and alligators — Capture production by species, fishing areas and countries or areas
Crocodiles et alligators — Captures par espèces, zones de pêche et pays ou zones
Cocodrilos y aligatores — Capturas por especies, áreas de pesca y países o áreas

Species, Fishing area Espèce, Zone de pêche Especie, Area de pesca	1992 no	1993 no	1994 no	1995 no	1996 no	1997 no	1998 no	1999 no	2000 no	2001 no
Spectacled caiman	**Caïman à lunettes**		**Caimán de anteojos**		***Caiman crocodilus***				**5,36(01)001,03**	**CAI**
02 Costa Rica	-	-	-	-	-	-	40	-	-	-
Cuba	-	-	-	-	302	506	5	2	-	-
Honduras	4 000 F	4 000 F	3 000 F	2 000	6 000	...	...	-	-	-
Nicaragua	20 472	9 963	8 919	4 238	10 795	1 590	3 927	250	6 440	...
Panama	-	7 869	2 840	2 005	46	500	3 022	10	10 250	9 926
02 Fishing area total	*24 472 F*	*21 832 F*	*14 759 F*	*8 243*	*17 143*	*2 596*	*6 994*	*262*	*16 690*	*9 926*
03 Bolivia	2 724	...	...	...	...	15 961	1 757	17 500	...	28 170
Brazil	233	7 034	43 633	369	659	7 307	2 092	4 619	8 286	1 253
Colombia	207 696	457 749	536 501	828 533	656 522	452 707	670 389	771 456	832 203	704 313
Guyana	3 459	1 558	685	1 556	2 650	910	...	9 880	9 880	5 917
Paraguay	...	...	5 466	19 793	725	503	4 445	-	9 750	3 792
Venezuela	104 935	78 972	54 038	55 195	29 996	33 528	35 579	24 640	23 655	14 978
03 Fishing area total	*319 047*	*545 313*	*640 323*	*905 446*	*690 552*	*510 916*	*714 262*	*828 095*	*883 774*	*758 423*
04 China,Taiwan	...	21	-	-	-	-	-	-	-	-
04 Fishing area total	*...*	*21*	*-*	*-*	*-*	*-*	*-*	*-*	*-*	*-*
Species total	*343 519 F*	*567 166 F*	*655 082 F*	*913 689*	*707 695*	*513 512*	*721 256*	*828 357*	*900 464*	*768 349*
American alligator	**Alligator américain**		**Caimán americano**		***Alligator mississippiensis***				**5,36(01)002,01**	**AGM**
02 USA	191 449	201 431	185 853	197 368	190 841	223 093	206 620	239 519	248 922	343 110
02 Fishing area total	*191 449*	*201 431*	*185 853*	*197 368*	*190 841*	*223 093*	*206 620*	*239 519*	*248 922*	*343 110*
04 Israel	900 F	1 055	1 815	348	944	210	401	425	233	6
04 Fishing area total	*900 F*	*1 055*	*1 815*	*348*	*944*	*210*	*401*	*425*	*233*	*6*
Species total	*192 349 F*	*202 486*	*187 668*	*197 716*	*191 785*	*223 303*	*207 021*	*239 944*	*249 155*	*343 116*
Estuarine crocodile	**Crocodile d'estuaires**		**Cocodrilo estuarino**		***Crocodylus porosus***				**5,36(01)003,01**	**CDP**
04 Cambodia	3 664	4 816	6 200	14 691	20 200	17 000	40 700	25 380	26 300	25 000 F
Indonesia	1 352	1 064	3 346	...	...	150	3 141	1 087	3 172	...
Malaysia	1 717	2 090	2 522	398	120	120	320	120	559	375
Singapore	962	286	301	1 004	411	296	416	350	481	2 074
Thailand	484	-	1	419	160	440	...	...	...	...
04 Fishing area total	*8 179*	*8 256*	*12 370*	*16 512*	*20 891*	*18 006*	*44 577*	*26 937*	*30 512*	*27 449 F*
06 Australia	4 405	6 886	5 356	7 251	9 054	8 777	9 896	5 048	13 296	11 849
Papua N Guin	5 083	8 415	7 551	12 908	10 597	8 578	12 138	9 388	8 336	...
06 Fishing area total	*9 488*	*15 301*	*12 907*	*20 159*	*19 651*	*17 355*	*22 034*	*14 436*	*21 632*	*11 849*
Species total	*17 667*	*23 557*	*25 277*	*36 671*	*40 542*	*35 361*	*66 611*	*41 373*	*52 144*	*39 298 F*
Siamese crocodile	**Crocodile siamois**		**Cocodrilo del Siam**		***Crocodylus siamensis***				**5,36(01)003,02**	**CDS**
04 Cambodia	-	-	-	-	-	-	-	-	-	30
Thailand	102	19	2 067	4 372	3 186	5 452	...	...	4 945	2 104
04 Fishing area total	*102*	*19*	*2 067*	*4 372*	*3 186*	*5 452*	*...*	*...*	*4 945*	*2 134*
Species total	*102*	*19*	*2 067*	*4 372*	*3 186*	*5 452*	*...*	*...*	*4 945*	*2 134*
Australian crocodile	**Crocodile australien**		**Cocodrilo de Australia**		***Crocodylus johnstoni***				**5,36(01)003,03**	**CRH**
06 Australia	1 870	4 290	2 381	3 132	1 641	194	309	44	10	-
06 Fishing area total	*1 870*	*4 290*	*2 381*	*3 132*	*1 641*	*194*	*309*	*44*	*10*	*-*
Species total	*1 870*	*4 290*	*2 381*	*3 132*	*1 641*	*194*	*309*	*44*	*10*	*-*
Nile crocodile	**Crocodile du Nil**		**Cocodrilo del Nilo**		***Crocodylus niloticus***				**5,36(01)003,04**	**CRI**
01 Botswana	1 324	7 414	587	699	347	338	2	9	11	152
Ethiopia	5	594	2	2 005	...	...	...	991	930	42
Guinea	-	-	-	100	...	...	...	...	...	...
Kenya	2 883	3 721	4 258	2 250	300	1 445	714	3 350	3 460	4 250
Madagascar	1 344	1 909	2 800	2 411	4 589	5 814	6 520	4 302	6 606	9 408
Malawi	266	2 036	1 732	950	636	400	200	199	200	1 256
Mozambique	2 727	3 164	1 042	3 021	523	1 430	810	585	718	477
Namibia	...	...	277	515	210	120	53	115	105	-
South Africa	10 722	18 451	25 416	14 805	2 280	13 322	8 863	27 631	29 733	33 174
Sudan	7 903	-	...	...	...	...	...	...	...	...
Tanzania	459	144	342	915	1 185	630	777	827	1 302	1 589
Uganda	2 500	4 019	4 817	...	...	...	...	...	508	900
Zambia	3 346	8 645	6 059	11 644	2 415	12 228	9 250	19 702	19 740	20 900
Zimbabwe	36 476	54 111	39 271	39 590	38 295	52 829	40 720	63 064	81 898	76 157
01 Fishing area total	*69 955*	*104 208*	*86 603*	*78 905*	*50 780*	*88 556*	*67 909*	*120 775*	*145 211*	*148 305*
03 Brazil	-	-	-	-	-	-	-	720	1 477	50
03 Fishing area total	*-*	*-*	*-*	*-*	*-*	*-*	*-*	*720*	*1 477*	*50*
04 Israel	-	-	-	-	-	-	...	552	1 661	2 289
04 Fishing area total	*-*	*-*	*-*	*-*	*-*	*-*	*...*	*552*	*1 661*	*2 289*
Species total	*69 955*	*104 208*	*86 603*	*78 905*	*50 780*	*88 556*	*67 909*	*122 047*	*148 349*	*150 644*

B-73

Crocodiles and alligators — **Capture production by species, fishing areas and countries or areas**
Crocodiles et alligators — **Captures par espèces, zones de pêche et pays ou zones**
Cocodrilos y aligatores — **Capturas por especies, áreas de pesca y países o áreas**

Species, Fishing area Espèce, Zone de pêche Especie, Area de pesca	1992 no	1993 no	1994 no	1995 no	1996 no	1997 no	1998 no	1999 no	2000 no	2001 no
New Guinea crocodile	**Crocodile de Nouvelle-Guinée**		**Cocodrilo de Nueva Guinea**			***Crocodylus novaeguineae***			**5,36(01)003,05**	**CNG**
04 Indonesia	537	2 263	9 016	...	...	100	8 506	6 574	7 215	...
04 Fishing area total	*537*	*2 263*	*9 016*	*...*	*...*	*100*	*8 506*	*6 574*	*7 215*	*...*
06 Australia	-	-	-	-	-	-	-	139	-	-
Papua N Guin	13 358	19 133	22 596	19 556	14 234	32 911	15 078	15 617	16 018	...
06 Fishing area total	*13 358*	*19 133*	*22 596*	*19 556*	*14 234*	*32 911*	*15 078*	*15 756*	*16 018*	*...*
Species total	*13 895*	*21 396*	*31 612*	*19 556*	*14 234*	*33 011*	*23 584*	*22 330*	*23 233*	*...*
Cuban crocodile	**Crocodile cubain**		**Cocodrilo de Cuba**			***Crocodylus rhombifer***			**5,36(01)003,06**	**CMB**
02 Cuba	-	-	-	99	40	-	3	-	-	-
02 Fishing area total	-	-	-	*99*	*40*	-	*3*	-	-	
Species total	-	-	-	*99*	*40*	-	*3*	-	-	
Morelet's crocodile	**Crocodile de Morelet**		**...C**			***Crocodylus moreletii***			**5,36(01)003,07**	**CME**
02 Mexico	-	-	-	2	20	146	193	2	1 228	3 643
02 Fishing area total	-	-	-	*2*	*20*	*146*	*193*	*2*	*1 228*	*3 643*
Species total	-	-	-	*2*	*20*	*146*	*193*	*2*	*1 228*	*3 643*
American crocodile	**...B**		**...C**			***Crocodylus acutus***			**5,36(01)003,08**	**YUU**
03 Colombia	-	-	-	-	-	-	-	-	-	100
03 Fishing area total	-	-	-	-	-	-	-	-	-	*100*
Species total	-	-	-	-	-	-	-	-	-	*100*
Cuvier's Dwarf caiman	**...B**		**...C**			***Paleosuchus palpebrosus***			**5,36(01)008,01**	**UCB**
03 Guyana	-	-	-	-	-	-	-	409	409	476
03 Fishing area total	-	-	-	-	-	-	-	*409*	*409*	*476*
Species total	-	-	-	-	-	-	-	*409*	*409*	*476*
Smooth-fronted caiman	**...B**		**...C**			***Paleosuchus trigonatus***			**5,36(01)008,02**	**UCI**
03 Guyana	-	-	-	-	-	-	-	270	270	423
03 Fishing area total	-	-	-	-	-	-	-	*270*	*270*	*423*
Species total	-	-	-	-	-	-	-	*270*	*270*	*423*
Group total	***639 357***	***923 122***	***990 690***	***1 254 142***	***1 009 923***	***899 535***	***1 086 886***	***1 254 776***	***1 380 207***	***1 308 183***

The data, derived mainly from sources of the UNEP World Conservation Monitoring Centre, refer to captive breeding, ranching and wild harvest.

See paragraph 5 of the INTRODUCTION.

Les données, provenant principalement des sources du Centre de surveillance continue de la conservation mondial de la nature de PNUE, ont trait à l'élevage en captivité, au 'ranching' et à la récolte sauvage.

Voir le paragraphe 5 de l'INTRODUCTION.

Los datos, obtenidos por la mayoría desde fuentes del Centro de Monitoreo de la Conservación Mundial de PNUMA, se refieren a la cría en cautividad, las grandes explotaciones y las capturas en libertad.

Véase el párrafo 5 en la INTRODUCCION.

B-74 Sea-squirts and other tunicates / Ascidiens et autres tuniciers / Ascidias y otros tunicados

Capture production by species, fishing areas and countries or areas / Captures par espèces, zones de pêche et pays ou zones / Capturas por especies, áreas de pesca y países o áreas

Species, Fishing area Espèce, Zone de pêche Especie, Area de pesca	1992 mt	1993 mt	1994 mt	1995 mt	1996 mt	1997 mt	1998 mt	1999 mt	2000 mt	2001 mt
Red sea squirt	**Violet chilien**		**Piure chileno**		***Pyura chilensis***				**6,96(09)005,02**	**SSE**
87 Chile	3 934	3 992	3 009	3 297	4 549	3 174	2 530	2 704	2 290	1 298
87 Fishing area total	*3 934*	*3 992*	*3 009*	*3 297*	*4 549*	*3 174*	*2 530*	*2 704*	*2 290*	*1 298*
Species total	*3 934*	*3 992*	*3 009*	*3 297*	*4 549*	*3 174*	*2 530*	*2 704*	*2 290*	*1 298*
Grooved sea squirt	**Violet**		**Provecho**		***Microcosmus sulcatus***				**6,96(09)037,01**	**SSG**
37 France	164	147	127	...	28	22	22	30	30	76
37 Fishing area total	*164*	*147*	*127*	*...*	*28*	*22*	*22*	*30*	*30*	*76*
Species total	*164*	*147*	*127*	*...*	*28*	*22*	*22*	*30*	*30*	*76*
Sea squirts nei	**Ascidiens nca**		**Ascidias nep**		***Ascidiacea***				**6,96(XX)XXX,XX**	**SSX**
51 India	-	-	-	-	-	-	-	-	95	-
51 Fishing area total	-	-	-	-	-	-	-	-	*95*	-
58 India	-	-	-	-	7	-	-	-	-	-
58 Fishing area total	-	-	-	-	*7*	-	-	-	-	-
61 Korea Rep	278	402	2 178	5 767	16 747	2 780	891	1 171	1 443	1 053
61 Fishing area total	*278*	*402*	*2 178*	*5 767*	*16 747*	*2 780*	*891*	*1 171*	*1 443*	*1 053*
Species total	*278*	*402*	*2 178*	*5 767*	*16 754*	*2 780*	*891*	*1 171*	*1 538*	*1 053*
Group total	***4 376***	***4 541***	***5 314***	***9 064***	***21 331***	***5 976***	***3 443***	***3 905***	***3 858***	***2 427***

B-75 Horseshoe crabs and other arachnoids / Limules et autres arachnoïdés / Límulos y otros arácnidos

Capture production by species, fishing areas and countries or areas / Captures par espèces, zones de pêche et pays ou zones / Capturas por especies, áreas de pesca y países o áreas

Species, Fishing area Espèce, Zone de pêche Especie, Area de pesca	1992 mt	1993 mt	1994 mt	1995 mt	1996 mt	1997 mt	1998 mt	1999 mt	2000 mt	2001 mt
Horseshoe crab **Limule** **Límulo(=Cangrejo cacerola)** ***Limulus polyphemus*** **6,56(01)001,02** **HSC**										
21 USA	449	810	634	926	1 598	2 606	3 182	2 238	1 557	1 071
21 Fishing area total	*449*	*810*	*634*	*926*	*1 598*	*2 606*	*3 182*	*2 238*	*1 557*	*1 071*
31 USA	-	-	-	-	-	1	70	159	139	228
31 Fishing area total	-	-	-	-	-	*1*	*70*	*159*	*139*	*228*
Species total	*449*	*810*	*634*	*926*	*1 598*	*2 607*	*3 252*	*2 397*	*1 696*	*1 299*
Group total	***449***	***810***	***634***	***926***	***1 598***	***2 607***	***3 252***	***2 397***	***1 696***	***1 299***

B-76 Sea-urchins and other echinoderms — Oursins et autres échinodermes — Erizos de mar y otros equinodermos

Capture production by species, fishing areas and countries or areas — Captures par espèces, zones de pêche et pays ou zones — Capturas por especies, áreas de pesca y países o áreas

Species, Fishing area Espèce, Zone de pêche Especie, Area de pesca	1992 mt	1993 mt	1994 mt	1995 mt	1996 mt	1997 mt	1998 mt	1999 mt	2000 mt	2001 mt
Echinoderms	**Oursins, bèches-de-mer**		**Erizos, cohombros de mar**		***Echinodermata***			**6,89(XX)XXX,XX**		**ECH**
27 Portugal	-	-	-	-	-	-	-	-	-	15
Spain	-	-	-	-	485	586	551	616	306	304
27 Fishing area total	*-*	*-*	*-*	*-*	*485*	*586*	*551*	*616*	*306*	*319*
31 Mexico	7	8	2	0	0	0	0	0	0	0
31 Fishing area total	*7*	*8*	*2*	*0*	*0*	*0*	*0*	*0*	*0*	*0*
37 Spain	-	-	-	-	2	4	9	5	3	2
37 Fishing area total	*-*	*-*	*-*	*-*	*2*	*4*	*9*	*5*	*3*	*2*
61 Korea Rep	2 476	3 944	3 714	3 707	2 802	2 771	1 410	1 182	1 461	1 454
61 Fishing area total	*2 476*	*3 944*	*3 714*	*3 707*	*2 802*	*2 771*	*1 410*	*1 182*	*1 461*	*1 454*
77 Fr Polynesia	0	0	0	0	0	10	10	10	15	15
Mexico	2 430	2 809	3 419	2 791	3 027	2 099	1 138	2 042	2 813	2 252
77 Fishing area total	*2 430*	*2 809*	*3 419*	*2 791*	*3 027*	*2 109*	*1 148*	*2 052*	*2 828*	*2 267*
81 New Zealand	869	848	944	804	277	627	832	643	712	853
81 Fishing area total	*869*	*848*	*944*	*804*	*277*	*627*	*832*	*643*	*712*	*853*
87 Peru	63	13	15	131	461	424	90	1 204	1 626	2 114
87 Fishing area total	*63*	*13*	*15*	*131*	*461*	*424*	*90*	*1 204*	*1 626*	*2 114*
Species total	*5 845*	*7 622*	*8 094*	*7 433*	*7 054*	*6 521*	*4 040*	*5 702*	*6 936*	*7 009*
Starfishes nei	**Astéridés nca**		**Estrellas nep**		***Asteroidea***			**6,91(XX)XXX,XX**		**STF**
21 USA	2	1	9	0	-	-	-	-	0	-
21 Fishing area total	*2*	*1*	*9*	*0*	*-*	*-*	*-*	*-*	*0*	*-*
27 Denmark	-	0	0	0	-	-	-	0	0	0
27 Fishing area total	*-*	*0*	*0*	*0*	*-*	*-*	*-*	*0*	*0*	*0*
77 USA	-	-	-	-	-	-	-	1	-	0
77 Fishing area total	*-*	*-*	*-*	*-*	*-*	*-*	*-*	*1*	*-*	*0*
81 New Zealand	-	-	-	9	4	0	13	2	6	9
81 Fishing area total	*-*	*-*	*-*	*9*	*4*	*0*	*13*	*2*	*6*	*9*
88 New Zealand	-	-	-	-	-	-	-	-	-	2
88 Fishing area total	*-*	*-*	*-*	*-*	*-*	*-*	*-*	*-*	*-*	*2*
Species total	*2*	*1*	*9*	*9*	*4*	*0*	*13*	*3*	*6*	*11*
Sea urchins nei	**Oursins nca**		**Erizos nep**		***Strongylocentrotus spp***			**6,93(02)004,XX**		**URC**
21 Canada	584	1 153	2 318	3 167	3 798	3 679	3 707	3 662	3 125	2 775
St Pier Mq	-	-	-	1	1	-	-	-	0	0
USA	12 039	19 237	16 445	14 558	10 052	8 522	7 013	7 132	6 006	4 499
21 Fishing area total	*12 623*	*20 390*	*18 763*	*17 726*	*13 851*	*12 201*	*10 720*	*10 794*	*9 131*	*7 274*
27 Faeroe Is	-	-	14	-	-	-	-	-	-	-
Russian Fed	-	-	-	-	-	-	-	-	149	151
UK	-	-	-	-	1	0	-	-	-	-
27 Fishing area total	*-*	*-*	*14*	*-*	*1*	*0*	*-*	*-*	*149*	*151*
31 Grenada	1	5	36	0	0	0	0	0	0	-
Martinique	18	16	15 F	15 F	10 F	15 F	15 F	15 F	10	10
31 Fishing area total	*19*	*21*	*51 F*	*15 F*	*10 F*	*15 F*	*15 F*	*15 F*	*10*	*10*
61 China	50	100	150	150	200	200	200	200	200	200
China,Taiwan	65	31	51	63	59	61	39	33	41	50
Japan	13 889	13 713	15 525	13 735	12 996	14 297	13 653	13 530	12 455	11 208
Korea D P Rp	100 F	100 F	100	140	150	150	100 F	100 F	100 F	100 F
Russian Fed	5 917	2 460	2 069	2 344	1 608	1 153	1 560	1 245	1 528	1 612
61 Fishing area total	*20 021 F*	*16 404 F*	*17 895*	*16 432*	*15 013*	*15 861*	*15 552 F*	*15 108 F*	*14 324 F*	*13 170 F*
67 Canada	14 056	7 102	6 161	6 666	5 867	5 542	6 160	5 390	4 887	4 288
USA	5 587	4 048	3 321	1 974	1 218	3 502	1 921	1 711	1 954	2 090
67 Fishing area total	*19 643*	*11 150*	*9 482*	*8 640*	*7 085*	*9 044*	*8 081*	*7 101*	*6 841*	*6 378*
71 Fiji Islands	14	55	56 F	59	40	95	103	100	90 F	96
Philippines	40	74	151	466	452	296	161	143	125	127
71 Fishing area total	*54*	*129*	*207 F*	*525*	*492*	*391*	*264*	*243*	*215 F*	*223*
77 Cook Is	25 F	26 F	25 F	25 F	20 F	20 F	20 F	20 F	20 F	20 F
USA	12 222	9 084	9 401	9 991	9 111	8 192	4 692	6 375	6 054	5 871
77 Fishing area total	*12 247 F*	*9 110 F*	*9 426 F*	*10 016 F*	*9 131 F*	*8 212 F*	*4 712 F*	*6 395 F*	*6 074 F*	*5 891 F*
87 Ecuador	1	0	0	0	0	0	0	0	0	0
87 Fishing area total	*1*	*0*	*0*	*0*	*0*	*0*	*0*	*0*	*0*	*0*
Species total	*64 608 F*	*57 204 F*	*55 838 F*	*53 354 F*	*45 583 F*	*45 724 F*	*39 344 F*	*39 656 F*	*36 744 F*	*33 097 F*
Stony sea urchin	**Oursin-pierre**		**Erizo de mar**		***Paracentrotus lividus***			**6,93(04)007,01**		**URM**
27 France	11	17	17	7	0	1	12	14	6	3

B-76 Sea-urchins and other echinoderms / Oursins et autres échinodermes / Erizos de mar y otros equinodermos

Capture production by species, fishing areas and countries or areas / Captures par espèces, zones de pêche et pays ou zones / Capturas por especies, áreas de pesca y países o áreas

Species, Fishing area Espèce, Zone de pêche Especie, Area de pesca	1992 mt	1993 mt	1994 mt	1995 mt	1996 mt	1997 mt	1998 mt	1999 mt	2000 mt	2001 mt
27 Fishing area total	*11*	*17*	*17*	*7*	*0*	*1*	*12*	*14*	*6*	*3*
37 France	390	240	142	71	63	47	47	70	192	98
37 Fishing area total	*390*	*240*	*142*	*71*	*63*	*47*	*47*	*70*	*192*	*98*
Species total	*401*	*257*	*159*	*78*	*63*	*48*	*59*	*84*	*198*	*101*
European edible sea urchin	**Oursin d'Europe**		**Erizo europeo**		***Echinus esculentus***				**6,93(04)014,01**	**URS**
27 Denmark	-	0	0	0	0	0	0	1	0	0
Iceland	-	713	1 409	923	423	20	-	10	-	-
Ireland	89	26	34	10	2	5	1	2	1	5
27 Fishing area total	*89*	*739*	*1 443*	*933*	*425*	*25*	*1*	*13*	*1*	*5*
Species total	*89*	*739*	*1 443*	*933*	*425*	*25*	*1*	*13*	*1*	*5*
Chilean sea urchin	**Oursin chilien**		**Erizo blanco**		***Loxechinus albus***				**6,93(04)017,01**	**UCH**
87 Chile	29 197	31 300	39 705	54 609	51 437	45 560	44 843	55 654	54 096	46 794
87 Fishing area total	*29 197*	*31 300*	*39 705*	*54 609*	*51 437*	*45 560*	*44 843*	*55 654*	*54 096*	*46 794*
Species total	*29 197*	*31 300*	*39 705*	*54 609*	*51 437*	*45 560*	*44 843*	*55 654*	*54 096*	*46 794*
Japanese sea cucumber	**Bèche-de-mer japonaise**		**Cohombro de mar japonés**		***Stichopus japonicus***				**6,94(14)004,01**	**CUJ**
61 Japan	6 072	5 996	6 106	6 602	7 226	7 160	6 952	6 662	6 957	7 229
Korea Rep	1 583	2 068	2 117	1 892	1 979	2 217	1 439	1 204	1 419	900
61 Fishing area total	*7 655*	*8 064*	*8 223*	*8 494*	*9 205*	*9 377*	*8 391*	*7 866*	*8 376*	*8 129*
Species total	*7 655*	*8 064*	*8 223*	*8 494*	*9 205*	*9 377*	*8 391*	*7 866*	*8 376*	*8 129*
Sea cucumbers nei	**Bèches-de-mer nca**		**Cohombros de mar nep**		***Holothurioidea***				**6,94(XX)XXX,XX**	**CUX**
21 USA	0	0	1 505	0	1 288	-	2 406	3 504	4 309	1 504
21 Fishing area total	*0*	*0*	*1 505*	*0*	*1 288*	-	*2 406*	*3 504*	*4 309*	*1 504*
27 Iceland	-	-	-	2	-	-	-	-	-	-
27 Fishing area total	-	-	-	*2*	-	-	-	-	-	-
31 Mexico	-	-	-	-	-	-	-	-	-	39
31 Fishing area total	-	-	-	-	-	-	-	-	-	*39*
37 Spain	-	-	-	-	4	4	4	1	9	4
37 Fishing area total	-	-	-	-	*4*	*4*	*4*	*1*	*9*	*4*
51 Egypt	-	-	-	-	-	-	-	-	20	139
Kenya	277	14	41	55	15	41	38	15	30	13
Madagascar	423	450	1 800	1 800	1 800	1 800	482	500	500	500
Maldives	119	72	66	94	145	318	85	54	205	226
Mozambique	-	0	0	6	54	7	2	8	12	12
Tanzania	535	980	1 591	1 460	1 644	1 527	1 800	189	372	340
Yemen	48	-	102	-	-	-	-	1	-	-
51 Fishing area total	*1 402*	*1 516*	*3 600*	*3 415*	*3 658*	*3 693*	*2 407*	*767*	*1 139*	*1 230*
57 Indonesia	537	608	548	227	269	338	630	689	903	970
Sri Lanka	65	65	92	100	150	272	203	170	145	120
57 Fishing area total	*602*	*673*	*640*	*327*	*419*	*610*	*833*	*859*	*1 048*	*1 090*
67 USA	481	472	636	729	491	-	-	228	274	300
67 Fishing area total	*481*	*472*	*636*	*729*	*491*	-	-	*228*	*274*	*300*
71 Fiji Islands	447	191	400 F	835	850	790	400	880	800 F	824
Indonesia	1 576	1 756	2 584	2 335	2 174	2 800	2 428	1 928	2 138	2 280
Kiribati	-	-	-	-	-	136	154	89	64	60
NewCaledonia	1 090	777	798	480	776	565	402	493	615	489 F
Palau	6	6	6	6	6	7	7	6 F	...	...
Papua N Guin	1 893	1 440	627	1 335	1 788	1 515	2 037	1 185	1 824	1 453
Philippines	3 679	3 109	1 497	2 062	2 123	1 191	830	849	730	791
Solomon Is	715	316	285	219	113	203	253	376	48	50 F
Vanuatu	39 F	40 F	40 F	50 F	50 F	50 F	50 F	50 F	50 F	50 F
71 Fishing area total	*9 445 F*	*7 635 F*	*6 237 F*	*7 322 F*	*7 880 F*	*7 257 F*	*6 561 F*	*5 856 F*	*6 269 F*	*5 997 F*
77 Mexico	...	...	...	...	...	...	271	234	426	442
Tonga	...	...	...	...	86	80	90	0	0	0
77 Fishing area total	...	...	...	...	*86*	*80*	*361*	*234*	*426*	*442*
81 New Zealand	-	-	-	4	1	0	-	-	-	2
81 Fishing area total	-	-	-	*4*	*1*	*0*	-	-	-	*2*
87 Chile	237	13	4	106	115	1	30	108	1 510	107
Ecuador	152	12	12	12	12	15	15	15	15	15
87 Fishing area total	*389*	*25*	*16*	*118*	*127*	*16*	*45*	*123*	*1 525*	*122*
Species total	*12 319 F*	*10 321 F*	*12 634 F*	*11 917 F*	*13 954 F*	*11 660 F*	*12 617 F*	*11 572 F*	*14 999 F*	*10 730 F*
Group total	***120 116***	***115 508***	***126 105***	***136 827***	***127 725***	***118 915***	***109 308***	***120 550***	***121 356***	***105 876***

B-77 Miscellaneous aquatic invertebrates / Invertébrés aquatiques divers / Invertebrados acuáticos diversos

Capture production by species, fishing areas and countries or areas / Captures par espèces, zones de pêche et pays ou zones / Capturas por especies, áreas de pesca y países o áreas

Species, Fishing area Espèce, Zone de pêche Especie, Area de pesca	1992 mt	1993 mt	1994 mt	1995 mt	1996 mt	1997 mt	1998 mt	1999 mt	2000 mt	2001 mt
Jellyfishes	**Méduses**		**Medusas**		***Rhopilema spp***				**6,18(41)007,XX**	**JEL**
21 USA	-	-	-	-	-	-	-	578	-	-
21 Fishing area total	-	-	-	-	-	-	-	*578*	-	-
31 USA	...	...	...	...	...	...	...	...	137	-
31 Fishing area total	...	...	...	...	...	...	...	...	*137*	-
37 Turkey	564	781	814	487	904	900	1 750	1 203	900	2 000
37 Fishing area total	*564*	*781*	*814*	*487*	*904*	*900*	*1 750*	*1 203*	*900*	*2 000*
47 Namibia	-	-	-	-	-	-	-	-	106	44
47 Fishing area total	-	-	-	-	-	-	-	-	*106*	*44*
48 UK	-	-	-	-	-	-	-	-	5	-
48 Fishing area total	-	-	-	-	-	-	-	-	*5*	-
57 Indonesia	280	675	1 744	1 309	4 018	5 354	-	29 377	28 181	29 450
Malaysia	3 299	2 001	10	205	2 823	2 813	633	4 555	5 926	6 367
Myanmar	...	...	...	1 812	3 426	1 412	432	1 000 F	1 000 F	1 000 F
Thailand	1 012	5 778	7 804	17 060	12 540	19 968	5 595	48	47	47
57 Fishing area total	*4 591*	*8 454*	*9 558*	*20 386*	*22 807*	*29 547*	*6 660*	*34 980 F*	*35 154 F*	*36 864 F*
58 Australia	-	-	-	-	-	10	-	2	-	-
58 Fishing area total	-	-	-	-	-	*10*	-	*2*	-	-
61 China	228 459	132 572	113 354	171 905	265 325	400 483	430 784	402 206	334 869	331 297
Russian Fed	-	-	-	-	-	-	-	-	-	142
61 Fishing area total	*228 459*	*132 572*	*113 354*	*171 905*	*265 325*	*400 483*	*430 784*	*402 206*	*334 869*	*331 439*
71 Indonesia	5 646	25 768	1 059	121 767	2 722	12 365	3 861	3 275	1 335	1 400
Malaysia	12 324	14 292	10 128	7 487	17 079	50 998	11 169	2 627	3 110	3 932
Philippines	75	73	...	61	57	20	10	12	12	12
Thailand	102 179	9 857	78 308	16 668	17 956	22 425	59 165	84 520	83 363	82 854
71 Fishing area total	*120 224*	*49 990*	*89 495*	*145 983*	*37 814*	*85 808*	*74 205*	*90 434*	*87 820*	*88 198*
Species total	*353 838*	*191 797*	*213 221*	*338 761*	*326 850*	*516 748*	*513 399*	*529 403 F*	*458 991 F*	*458 545 F*
Marine worms	**Vers marins**		**Poliquetos**		***Polychaeta***				**6,49(XX)XXX,XX**	**WOR**
02 Mexico	90	98	49	59	58	57	58	48	27	...
02 Fishing area total	*90*	*98*	*49*	*59*	*58*	*57*	*58*	*48*	*27*	...
21 USA	404	400	379	0	-	223	299	343	202	481
21 Fishing area total	*404*	*400*	*379*	*0*	-	*223*	*299*	*343*	*202*	*481*
61 Korea Rep	540	-	-	-	-	-	-	-	-	-
61 Fishing area total	*540*	-	-	-	-	-	-	-	-	-
77 Panama	...	...	...	...	42	67	51	31	28	450 F
77 Fishing area total	...	...	...	...	*42*	*67*	*51*	*31*	*28*	*450 F*
Species total	*1 034*	*498*	*428*	*59*	*100*	*347*	*408*	*422*	*257*	*931 F*
Aquatic invertebrates nei	**Invertébrés aquatiques nca**		**Invertebrados acuáticos nep**		***Invertebrata***				**6,99(XX)XXX,XX**	**INV**
02 Mexico	606	138	170	124	279	240	425	569	204	...
02 Fishing area total	*606*	*138*	*170*	*124*	*279*	*240*	*425*	*569*	*204*	...
04 Indonesia	237	432	140	1 067	49	714	72	830	790	1 350
Japan	1 911	2 281	1 431	1 552	1 445	1 636	1 403	1 216	1 240	933
Korea Rep	21	10	10	24	-	110	5	2	1	-
04 Fishing area total	*2 169*	*2 723*	*1 581*	*2 643*	*1 494*	*2 460*	*1 480*	*2 048*	*2 031*	*2 283*
27 Denmark	0	0	1	1	0	-	-	-	-	-
UK	-	-	-	-	-	-	-	-	-	24
27 Fishing area total	*0*	*0*	*1*	*1*	*0*	-	-	-	-	*24*
31 Mexico	9	5	7	11	6	2	4	0	2	-
31 Fishing area total	*9*	*5*	*7*	*11*	*6*	*2*	*4*	*0*	*2*	-
37 Croatia	-	-	-	-	-	-	4	5	8	592
Spain	-	-	-	-	5	4	10	529	-	-
37 Fishing area total	-	-	-	-	*5*	*4*	*14*	*534*	*8*	*592*
57 Indonesia	196	167	139	211	720	200	417	544	530	590
57 Fishing area total	*196*	*167*	*139*	*211*	*720*	*200*	*417*	*544*	*530*	*590*
58 Australia	-	-	-	-	-	1	-	-	-	-
58 Fishing area total	-	-	-	-	-	*1*	-	-	-	-
61 Japan	124 175	78 567	85 215	37 360	12 537	9 134	9 618	7 130	9 692	7 406
Korea Rep	2 006	913	2 161	1 016	873	984	1 042	1 001	1 882	1 302
61 Fishing area total	*126 181*	*79 480*	*87 376*	*38 376*	*13 410*	*10 118*	*10 660*	*8 131*	*11 574*	*8 708*
71 Indonesia	148	262	271	237	471	771	220	1 275	1 529	1 720
Palau	4	4	4	4	4	4	4	5	5 F	5 F
Thailand	200	-	-	-	-	-	-	34	-	-
71 Fishing area total	*352*	*266*	*275*	*241*	*475*	*775*	*224*	*1 314*	*1 534 F*	*1 725 F*

B-77

Miscellaneous aquatic invertebrates — **Capture production by species, fishing areas and countries or areas**
Invertébrés aquatiques divers — **Captures par espèces, zones de pêche et pays ou zones**
Invertebrados acuáticos diversos — **Capturas por especies, áreas de pesca y países o áreas**

Species, Fishing area Espèce, Zone de pêche Especie, Area de pesca	1992 mt	1993 mt	1994 mt	1995 mt	1996 mt	1997 mt	1998 mt	1999 mt	2000 mt	2001 mt
77 Mexico	0	973	3 326	1 346	830	284	8	0	220	-
77 Fishing area total	*0*	*973*	*3 326*	*1 346*	*830*	*284*	*8*	*0*	*220*	*-*
Species total	*129 513*	*83 752*	*92 875*	*42 953*	*17 219*	*14 084*	*13 232*	*13 140*	*16 103 F*	*13 922 F*
Group total	***484 385***	***276 047***	***306 524***	***381 773***	***344 169***	***531 179***	***527 039***	***542 965***	***475 351***	***473 398***

B-81 Pearls, mother-of-pearl, shells / Perles, nacres, coquilles / Perlas, madreperlas, conchas

Capture production by species, fishing areas and countries or areas / Captures par espèces, zones de pêche et pays ou zones / Capturas por especies, áreas de pesca y países o áreas

Species, Fishing area Espèce, Zone de pêche Especie, Area de pesca	1992 kg	1993 kg	1994 kg	1995 kg	1996 kg	1997 kg	1998 kg	1999 kg	2000 kg	2001 kg
Trochus shells	**Troques**		**Tróquidos**			***Ex Trochus spp***			**3,07(04)006,XX**	**TSH**
51 Eritrea	...	-	-	-	-	-	223 900	106 050	100 000 F	100 000 F
51 Fishing area total	*...*	*-*	*-*	*-*	*-*	*-*	*223 900*	*106 050*	*100 000 F*	*100 000 F*
71 Australia	0	0	0	0	0	0	0	0	0	0
Fiji Islands	88 430	65 899	100 000 F	160 000	160 000 F	150 000 F	115 000	165 000	158 000 F	150 000 F
Marshall Is	100 000 F	100 000 F	120 000 F	120 000 F	120 000 F	110 000 F	100 000 F	100 000 F	100 000 F	100 000 F
Micronesia	172 000	132 000	266 000	150 000 F	150 000 F	150 000 F	150 000 F	150 000 F	150 000 F	150 000 F
NewCaledonia	186 000	223 000	274 000	250 000	197 400	124 700	151 300	98 100	96 400	342 700 F
Palau	229 000	26 636	...	389 090	69 450	...	...	...	210 000	200 000 F
Papua N Guin	282 000	180 946	221 635	320 160	177 987	209 870	281 590	233 319	224 400	345 062
Solomon Is	320 000	394 000	306 000	80 405	30 625	181 670	60 910	261 293	54 431	54 000 F
Vanuatu	150 000	160 000	107 000	100 000 F	100 000 F	100 000 F	100 000 F	100 000 F	100 000 F	100 000 F
Wallis Fut I	17 000	16 000	34 000	30 000 F	30 000 F	30 000 F	25 000 F	25 000 F	25 000 F	25 000 F
71 Fishing area total	*1 544 430 F*	*1 298 481 F*	*1 428 635 F*	*1 599 655 F*	*1 035 462 F*	*1 056 240 F*	*983 800 F*	*1 132 712 F*	*1 118 231 F*	*1 466 762 F*
77 Cook Is	26 000	25 000 F	25 000 F	26 000 F	26 000 F	26 000 F	25 000 F	25 000 F	25 000 F	15 000 F
77 Fishing area total	*26 000*	*25 000 F*	*25 000 F*	*26 000 F*	*26 000 F*	*26 000 F*	*25 000 F*	*25 000 F*	*25 000 F*	*15 000 F*
Species total	*1 570 430 F*	*1 323 481 F*	*1 453 635 F*	*1 625 655 F*	*1 061 462 F*	*1 082 240 F*	*1 232 700 F*	*1 263 762 F*	*1 243 231 F*	*1 581 762 F*
Turban shells nei	**Coquilles turbo nca**		**Turbinas nep**			***Ex Turbo spp***			**3,07(05)002,XX**	**GSH**
71 Malaysia	23 950	20 000	17 800	97 000	90 000 F	90 000 F	80 000 F	80 000 F	80 000 F	80 000 F
71 Fishing area total	*23 950*	*20 000*	*17 800*	*97 000*	*90 000 F*	*90 000 F*	*80 000 F*	*80 000 F*	*80 000 F*	*80 000 F*
Species total	*23 950*	*20 000*	*17 800*	*97 000*	*90 000 F*	*90 000 F*	*80 000 F*	*80 000 F*	*80 000 F*	*80 000 F*
Freshwater mussel shells	**Coquilles des moules eau douce**		**Conchas de mejillón agua dulce**			***Ex Unionidae***			**3,16(05)XXX,XX**	**FSH**
02 USA	2 495	499	1 814	-	19 000	1 183	...	2 716	...	...
02 Fishing area total	*2 495*	*499*	*1 814*	*-*	*19 000*	*1 183*	*...*	*2 716*	*...*	*...*
Species total	*2 495*	*499*	*1 814*	*-*	*19 000*	*1 183*	*...*	*2 716*	*...*	*...*
Pearl oyster shells nei	**Coquilles d'huîtres perl. nca**		**Conchas de ostras perleras nep**			***Ex Pinctada spp***			**3,16(06)006,XX**	**OSH**
04 Japan	237	135	153	165	190	204	214	187	181	...
04 Fishing area total	*237*	*135*	*153*	*165*	*190*	*204*	*214*	*187*	*181*	*...*
51 Sudan	23 000	14 000	13 000	3 000	9 600	10 000	13 000	13 500	10 000 F	7 500
51 Fishing area total	*23 000*	*14 000*	*13 000*	*3 000*	*9 600*	*10 000*	*13 000*	*13 500*	*10 000 F*	*7 500*
57 Australia	200 000 F	220 000 F	230 000 F	240 000 F	240 000 F	250 000 F	250 000 F	250 000 F	250 000 F	250 000 F
57 Fishing area total	*200 000 F*	*220 000 F*	*230 000 F*	*240 000 F*	*240 000 F*	*250 000 F*	*250 000 F*	*250 000 F*	*250 000 F*	*250 000 F*
61 China,Taiwan	-	-	-	-	-	-	-	-	-	1 251
Japan	68 752	72 711	64 892	63 330	56 565	48 307	28 893	24 576	29 905	34 516
61 Fishing area total	*68 752*	*72 711*	*64 892*	*63 330*	*56 565*	*48 307*	*28 893*	*24 576*	*29 905*	*35 767*
71 Fiji Islands	10 900	17 034	17 000 F	20 000 F	20 000 F	20 000 F	10 000	12 020	12 000 F	10 000 F
Indonesia	...	...	...	...	...	...	...	500	68 600	1 000
Papua N Guin	7 000 F	14 739	7 681	10 498	16 460	20 561	35 960	17 051	21 521	1 484
71 Fishing area total	*17 900 F*	*31 773*	*24 681 F*	*30 498 F*	*36 460 F*	*40 561 F*	*45 960*	*29 571*	*102 121 F*	*12 484 F*
77 Fr Polynesia	...	...	...	...	...	752 349	562 213	650 000 F	752 364	820 724
77 Fishing area total	*...*	*...*	*...*	*...*	*...*	*752 349*	*562 213*	*650 000 F*	*752 364*	*820 724*
Species total	*309 889 F*	*338 619 F*	*332 726 F*	*336 993 F*	*342 815 F*	*1 101 421 F*	*900 280 F*	*967 834 F*	*1 144 571 F*	*1 126 475 F*
Marine shells nei	**Coquilles marines nca**		**Conchas marinas nep**			***Ex Mollusca***			**3,99(XX)XXX,XX**	**MSH**
31 Bahamas	10 877	150 294	60 490	35 671	10 807	8 430	39 168	40 000 F	40 000 F	13 000
Mexico	...	...	...	15 055	60 067	80 838	50 977	36 150	43 570	43 000 F
31 Fishing area total	*10 877*	*150 294*	*60 490*	*50 726*	*70 874*	*89 268*	*90 145*	*76 150 F*	*83 570 F*	*56 000 F*
37 Croatia	278 000	274 000	308 000	254 000	296 000	492 000	1 078 000	1 192 000	1 244 000	3 125 000
Russian Fed	192 000	29 000	2 000	...	1 000	440 000	46 000	45 000	182 000	224 000
Yugoslavia	5 000	5 000	3 585	7 520	1 990	2 000 F	2 890	2 500 F	2 000 F	2 000 F
37 Fishing area total	*475 000*	*308 000*	*313 585*	*261 520*	*298 990*	*934 000 F*	*1 126 890*	*1 239 500 F*	*1 428 000 F*	*3 351 000 F*
51 Kenya	232 000	329 000	329 000	178 743	166 351	148 658	170 584	262 386	253 911	260 324
Madagascar	141 000	76 000	187 000	187 000	169 000	156 000	10 000	8 000	74 000	32 000
Tanzania	47 309	183 358	122 077	54 000	95 200	183 172	154 893	250 006	250 000 F	436 743
51 Fishing area total	*420 309*	*588 358*	*638 077*	*419 743*	*430 551*	*487 830*	*335 477*	*520 392*	*577 911 F*	*729 067*
57 Sri Lanka	93 080	121 890	235 670	745 920	723 310	430 840	602 860	652 700	697 600	498 000
57 Fishing area total	*93 080*	*121 890*	*235 670*	*745 920*	*723 310*	*430 840*	*602 860*	*652 700*	*697 600*	*498 000*
71 Philippines	3 130 000	2 855 000	5 512 000	3 842 000	3 824 000	2 801 000	2 389 000	2 358 000	2 373 000	2 433 000
Solomon Is	6 745	3 412	3 286	219 997	54 825	75	15 575	470	35	50 F
71 Fishing area total	*3 136 745*	*2 858 412*	*5 515 286*	*4 061 997*	*3 878 825*	*2 801 075*	*2 404 575*	*2 358 470*	*2 373 035*	*2 433 050 F*
77 Fr Polynesia	82 000	87 000	27 000	40 000 F	55 000 F	68 134	...	90 000 F	105 204	...
Mexico	...	...	...	833 232	481 813	1 321 742	918 563	928 808	420 242	420 000 F
77 Fishing area total	*82 000*	*87 000*	*27 000*	*873 232 F*	*536 813 F*	*1 389 876*	*918 563*	*1 018 808 F*	*525 446*	*420 000 F*
Species total	*4 218 011*	*4 113 954*	*6 790 108*	*6 413 138 F*	*5 939 363 F*	*6 132 889 F*	*5 478 510*	*5 866 020 F*	*5 685 562 F*	*7 487 117 F*
Group total	***6 124 775***	***5 796 553***	***8 596 083***	***8 472 786***	***7 452 640***	***8 407 733***	***7 691 490***	***8 180 332***	***8 153 364***	***10 275 354***

B-81

Pearls, mother-of-pearl, shells
Perles, nacres, coquilles
Perlas, madreperlas, conchas

Capture production by species, fishing areas and countries or areas
Captures par espèces, zones de pêche et pays ou zones
Capturas por especies, áreas de pesca y países o áreas

Species, Fishing area Espèce, Zone de pêche Especie, Area de pesca	1992 kg	1993 kg	1994 kg	1995 kg	1996 kg	1997 kg	1998 kg	1999 kg	2000 kg	2001 kg

Data by Japan for "Pearl oyster shells nei" refer only to production of pearls.

Data of "Pearl oyster shells nei" include aquaculture production.

See paragraph 5 of the INTRODUCTION.

Les données du Japon concernant les "Coquilles d'huîtres perlières nca" se rapportent à la production des perles.

Les données de "Coquilles d'huîtres perlières nca" comprennent les quantités provenant de l'aquaculture.

Voir le paragraphe 5 de l'INTRODUCTION.

Los datos de Japón referidos a "Conchas de ostras perleras nep" se refieren solamente a la producción de perlas.

Los datos correspondientes a "Conchas de ostras perleras nep" incluyen las cantidades procedentes de la acuicultura.

Véase el párrafo 5 en la INTRODUCCION.

B-82 Corals / Coraux / Corales

Capture production by species, fishing areas and countries or areas
Captures par espèces, zones de pêche et pays ou zones
Capturas por especies, áreas de pesca y países o áreas

Species, Fishing area Espèce, Zone de pêche Especie, Area de pesca	1992 kg	1993 kg	1994 kg	1995 kg	1996 kg	1997 kg	1998 kg	1999 kg	2000 kg	2001 kg
Sardinia coral	**Corail Sardaigne**		**Coral Cerdeña**		***Corallium rubrum***			**6,19(01)003,01**		**COL**
27 Spain	-	-	-	-	3 200	3 200	1 000	2 500	3 000	1 600
27 Fishing area total	*-*	*-*	*-*	*-*	*3 200*	*3 200*	*1 000*	*2 500*	*3 000*	*1 600*
34 Morocco	8 500	3 800	2 900	2 600	1 800	1 500	600	...	1 000	600
34 Fishing area total	*8 500*	*3 800*	*2 900*	*2 600*	*1 800*	*1 500*	*600*	*...*	*1 000*	*600*
37 Albania	200	400 F	800 F	1 300	1 200	800	500	900	1 000	1 200
Algeria	7 914	5 971	7 857	6 666	6 800	5 400	3 000	3 000	2 900	5 000
Croatia	3 300	3 500	3 200	1 800	1 700	1 600	1 600	1 500	2 000	1 900
France	2 500	2 500	2 600	2 300	1 900	2 500	4 500	3 800	3 600	3 700
Greece	1 000 F	500	900	1 500	1 200	1 100	1 400	1 750	1 700	1 400
Italy	4 900	4 400	4 800	3 100	2 500	4 100	4 700	3 850	4 350	4 700
Morocco	600	600	300	200	200	100	500	3 100	3 200	3 300
Spain	3 500	4 700	4 400	6 300	5 000	5 200	1 500	4 350	4 300	5 000
Tunisia	1 000	1 200	1 071	1 100	975	1 975	1 196	1 783	2 399	2 005
37 Fishing area total	*24 914 F*	*23 771 F*	*25 928 F*	*24 266*	*21 475*	*22 775*	*18 896*	*24 033*	*25 449*	*28 205*
Species total	*33 414 F*	*27 571 F*	*28 828 F*	*26 866*	*26 475*	*27 475*	*20 496*	*26 533*	*29 449*	*30 405*
Aka coral	**Corail aka**		**Coral aka**		***Corallium japonicum***			**6,19(01)003,02**		**COJ**
61 China,Taiwan	800	-	-	-	600	5	-	650	400	350
Japan	1 043	347	358	565	475	49	492	990	1	1
61 Fishing area total	*1 843*	*347*	*358*	*565*	*1 075*	*54*	*492*	*1 640*	*401*	*351*
Species total	*1 843*	*347*	*358*	*565*	*1 075*	*54*	*492*	*1 640*	*401*	*351*
Momo, boke magai, misu coral	**Corail momo**		**Coral momo**		***Corallium elatius***			**6,19(01)003,03**		**CEL**
61 China,Taiwan	8 000	6 000	5 550	6 000	12 000	240	5 500	4 650	4 100	4 500
Japan	2 326	62	1 512	2 220	815	12	1 169	248	3	2
61 Fishing area total	*10 326*	*6 062*	*7 062*	*8 220*	*12 815*	*252*	*6 669*	*4 898*	*4 103*	*4 502*
Species total	*10 326*	*6 062*	*7 062*	*8 220*	*12 815*	*252*	*6 669*	*4 898*	*4 103*	*4 502*
Shiro, white coral	**Corail blanc**		**Coral blanco**		***Corallium konojoi***			**6,19(01)003,04**		**COK**
61 Japan	42	-	-	145	76	53	111	-	0	-
61 Fishing area total	*42*	*-*	*-*	*145*	*76*	*53*	*111*	*-*	*0*	*-*
Species total	*42*	*-*	*-*	*145*	*76*	*53*	*111*	*-*	*0*	*-*
Midway deep sea coral	**Corail de profondeur de Midway**		**Coral de profundidad de Midway**		***Corallium sp. nov.***			**6,19(01)003,07**		**CDE**
61 Japan	890	-	-	-	-	-	623	-	0	-
61 Fishing area total	*890*	*-*	*-*	*-*	*-*	*-*	*623*	*-*	*0*	*-*
Species total	*890*	*-*	*-*	*-*	*-*	*-*	*623*	*-*	*0*	*-*
Soft corals nei	**Corails mous nca**		**Corales muelles nep**		***Non-Scleractinia***			**6,19(XX)XXX,XX**		**CBL**
31 Mexico	476	350	215	100	582	775	694	574	593	500 F
31 Fishing area total	*476*	*350*	*215*	*100*	*582*	*775*	*694*	*574*	*593*	*500 F*
34 Italy	100 F	-	-	-	-	-	-	-	-	-
34 Fishing area total	*100 F*	*-*	*-*	*-*	*-*	*-*	*-*	*-*	*-*	*-*
77 Mexico	2 074	1 230	556	-	-	-	-	-	-	-
USA	1 000	1 500	2 000	3 000	-	907	14	1 292	...	8 533
77 Fishing area total	*3 074*	*2 730*	*2 556*	*3 000*	*-*	*907*	*14*	*1 292*	*...*	*8 533*
Species total	*3 650 F*	*3 080*	*2 771*	*3 100*	*582*	*1 682*	*708*	*1 866*	*593*	*9 033 F*
Hard corals, madrepores nei	**Madrépores nca**		**Madréporas nep**		***Scleractinia***			**6,19(XX)XXX,XX**		**CSS**
31 Dominican Rp	6 000 F	-	-	-	-	-	-	-	-	-
Haiti	16 000 F	15 000 F	15 000 F	14 000 F	14 000 F	13 000 F	10 000 F	10 000 F	10 000 F	10 000 F
31 Fishing area total	*22 000 F*	*15 000 F*	*15 000 F*	*14 000 F*	*14 000 F*	*13 000 F*	*10 000 F*	*10 000 F*	*10 000 F*	*10 000 F*
57 India	60 000 F	55 000 F	50 000 F	50 000 F	50 000 F	50 000 F	50 000 F	50 000 F	50 000 F	50 000 F
57 Fishing area total	*60 000 F*	*55 000 F*	*50 000 F*	*50 000 F*	*50 000 F*	*50 000 F*	*50 000 F*	*50 000 F*	*50 000 F*	*50 000 F*
61 China,Taiwan	16 000	10 000 F	12 000 F	10 000 F	10 000 F	10 000 F	10 000 F	10 000 F	10 000 F	-
Japan	7 000 F	6 000 F	5 000 F	5 000 F	5 000 F	5 000 F	5 000 F	5 000 F	5 000 F	5 000 F
61 Fishing area total	*23 000 F*	*16 000 F*	*17 000 F*	*15 000 F*	*15 000 F*	*15 000 F*	*15 000 F*	*15 000 F*	*15 000 F*	*5 000 F*
71 Fiji Islands	76 771	120 000	150 000 F	300 000 F	600 000 F	900 000 F	1 279 609	1 000 000	1 000 000 F	1 000 000 F
Indonesia	1 500 000	2 000 000 F	1 800 000 F	1 700 000 F	1 700 000 F	1 500 000 F	1 500 000 F	1 500 000 F	1 200 000 F	1 000 000 F
Malaysia	1 225 810	271 140	872 690	4 743 000	4 500 000 F	4 000 000 F	4 000 000 F	4 000 000 F	4 000 000 F	4 000 000 F
Philippines	600 000 F	600 000 F	550 000 F	550 000 F	550 000 F	500 000 F	-	-	-	-
71 Fishing area total	*3 402 581 F*	*2 991 140 F*	*3 372 690 F*	*7 293 000 F*	*7 350 000 F*	*6 900 000 F*	*6 779 609 F*	*6 500 000 F*	*6 200 000 F*	*6 000 000 F*
Species total	*3 507 581 F*	*3 077 140 F*	*3 454 690 F*	*7 372 000 F*	*7 429 000 F*	*6 978 000 F*	*6 854 609 F*	*6 575 000 F*	*6 275 000 F*	*6 065 000 F*
Group total	***3 557 746***	***3 114 200***	***3 493 709***	***7 410 896***	***7 470 023***	***7 007 516***	***6 883 708***	***6 609 937***	***6 309 546***	***6 109 291***

See paragraph 5 of the INTRODUCTION. Voir le paragraphe 5 de l'INTRODUCTION. Véase el párrafo 5 en la INTRODUCCION.

B-83 Sponges / Eponges / Esponjas

Capture production by species, fishing areas and countries or areas
Captures par espèces, zones de pêche et pays ou zones
Capturas por especies, áreas de pesca y países o áreas

Species, Fishing area Espèce, Zone de pêche Especie, Area de pesca	1992 kg	1993 kg	1994 kg	1995 kg	1996 kg	1997 kg	1998 kg	1999 kg	2000 kg	2001 kg
Sponges	**Eponges**		**Esponjas**		***Spongidae***				**6,15(01)XXX,XX**	**SPO**
31 Bahamas	21 655	46 650	50 526	54 812	62 337	60 162	59 517	86 000	81 000	71 000
Colombia	3 000 F	2 500 F	2 500 F	2 000 F	2 000 F	2 000 F	2 000 F	2 000 F	2 000 F	2 000 F
Cuba	36 800	24 300	40 400	55 200	56 800	81 500	72 100	50 100	50 000 F	50 000 F
USA	141 880	130 740	378 747	338 609	196 157	700 911	1 199 131	2 399 327	303 679	291 326
31 Fishing area total	*203 335 F*	*204 190 F*	*472 173 F*	*450 621 F*	*317 294 F*	*844 573 F*	*1 332 748 F*	*2 537 427 F*	*436 679 F*	*414 326 F*
37 Croatia	500 F	500 F	500 F	1 000 F	2 000 F	3 900	5 400	6 500	3 000	4 000
Egypt	1 800 F	1 700 F	1 600 F	1 500 F	1 400 F	1 200 F	1 000 F	1 000 F	1 000 F	1 000 F
France	14 000 F	12 000 F	10 000 F	8 000 F	7 000 F	5 000 F	5 000 F	4 000 F	3 000 F	2 000 F
Greece	3 800	2 000	2 000	10 000	11 000	10 000	10 000	10 000	10 000	5 000
Italy	4 500 F	4 000 F	4 000 F	3 000 F	3 000 F	2 500 F	2 000 F	1 500 F	1 500 F	1 000 F
Lebanon	11 000 F	12 000	12 000	0	0	0	0	0	0	0
Libya	15 000 F	14 000 F	13 000 F	10 000 F	10 000 F	10 000 F	10 000 F	10 000 F	10 000 F	10 000 F
Spain	900 F	900 F	900 F	800 F	800 F	800 F	800 F	750 F	700 F	700 F
Syria	0	0	0	0	0	0	0	0	0	-
Tunisia	10 000	16 046	33 743	14 803	14 035	27 330	21 400	14 450	15 431	23 092
Turkey	1 000	3 000	2 000	1 000	1 500 F	2 000	1 000	3 000	7 000	3 000
37 Fishing area total	*62 500 F*	*66 146 F*	*79 743 F*	*50 103 F*	*50 735 F*	*62 730 F*	*56 600 F*	*51 200 F*	*51 631 F*	*49 792 F*
61 China,Taiwan	13 000 F	12 000 F	10 000 F	10 000 F	10 000 F	8 000 F	8 000 F	7 000 F	7 000 F	-
Japan	8 000 F	7 000 F	6 000 F	5 000 F	5 000 F	5 000 F	4 000 F	4 000 F	4 000 F	4 000 F
61 Fishing area total	*21 000 F*	*19 000 F*	*16 000 F*	*15 000 F*	*15 000 F*	*13 000 F*	*12 000 F*	*11 000 F*	*11 000 F*	*4 000 F*
71 Philippines	6 000	3 000	1 000	23 000	2 000	16 000	7 000	8 000	8 000	8 000
71 Fishing area total	*6 000*	*3 000*	*1 000*	*23 000*	*2 000*	*16 000*	*7 000*	*8 000*	*8 000*	*8 000*
77 Fr Polynesia	...	...	...	...	...	...	...	...	116	-
77 Fishing area total	*...*	*...*	*...*	*...*	*...*	*...*	*...*	*...*	*116*	*-*
81 New Zealand	-	-	-	-	-	-	4 100	14 000	18 000 F	23 000
81 Fishing area total	*-*	*-*	*-*	*-*	*-*	*-*	*4 100*	*14 000*	*18 000 F*	*23 000*
Species total	*292 835 F*	*292 336 F*	*568 916 F*	*538 724 F*	*385 029 F*	*936 303 F*	*1 412 448 F*	*2 621 627 F*	*525 426 F*	*499 118 F*
Group total	***292 835***	***292 336***	***568 916***	***538 724***	***385 029***	***936 303***	***1 412 448***	***2 621 627***	***525 426***	***499 118***

See paragraph 5 of the INTRODUCTION. Voir le paragraphe 5 de l'INTRODUCTION. Véase el párrafo 5 en la INTRODUCCION.

B-91 Brown seaweeds / Algues brunes / Algas pardas

Capture production by species, fishing areas and countries or areas / Captures par espèces, zones de pêche et pays ou zones / Capturas por especies, áreas de pesca y países o áreas

Species, Fishing area Espèce, Zone de pêche Especie, Area de pesca	1992 mt	1993 mt	1994 mt	1995 mt	1996 mt	1997 mt	1998 mt	1999 mt	2000 mt	2001 mt
Tangle	**Laminaire digitée**		**...C**			***Laminaria digitata***			**7,71(02)002,01**	**LQD**
27 Iceland	...	...	...	...	...	...	4 500	5 239	2 764	4 679
27 Fishing area total	*...*	*...*	*...*	*...*	*...*	*...*	*4 500*	*5 239*	*2 764*	*4 679*
Species total	*...*	*...*	*...*	*...*	*...*	*...*	*4 500*	*5 239*	*2 764*	*4 679*
Japanese kelp	**Laminaire du Japon**		**Laminaria del Japón**			***Laminaria japonica***			**7,71(02)002,02**	**LNJ**
61 Japan	157 318	134 147	103 603	120 957	120 194	122 976	91 752	94 371	93 611	97 261
61 Fishing area total	*157 318*	*134 147*	*103 603*	*120 957*	*120 194*	*122 976*	*91 752*	*94 371*	*93 611*	*97 261*
Species total	*157 318*	*134 147*	*103 603*	*120 957*	*120 194*	*122 976*	*91 752*	*94 371*	*93 611*	*97 261*
North European kelp	**...B**		**...C**			***Laminaria hyperborea***			**7,71(02)002,04**	**LAH**
27 Ireland	2 800	2 700	2 600	2 400	2 300	1 900	2 000 F	2 000 F	2 000 F	2 000 F
27 Fishing area total	*2 800*	*2 700*	*2 600*	*2 400*	*2 300*	*1 900*	*2 000 F*	*2 000 F*	*2 000 F*	*2 000 F*
Species total	*2 800*	*2 700*	*2 600*	*2 400*	*2 300*	*1 900*	*2 000 F*	*2 000 F*	*2 000 F*	*2 000 F*
Wakame	**Wakamé**		**Abeto marino**			***Undaria pinnatifida***			**7,71(04)003,01**	**UDP**
61 Japan	3 685	3 034	3 265	3 148	4 044	2 936	2 839	3 431	3 396	3 034
61 Fishing area total	*3 685*	*3 034*	*3 265*	*3 148*	*4 044*	*2 936*	*2 839*	*3 431*	*3 396*	*3 034*
Species total	*3 685*	*3 034*	*3 265*	*3 148*	*4 044*	*2 936*	*2 839*	*3 431*	*3 396*	*3 034*
Chilean kelp	**...B**		**Chascón**			***Lessonia nigrescens***			**7,71(05)001,01**	**LJX**
87 Chile	49 377	70 565	72 029	123 772	140 770	125 535	136 313	111 766	61 954	87 508
87 Fishing area total	*49 377*	*70 565*	*72 029*	*123 772*	*140 770*	*125 535*	*136 313*	*111 766*	*61 954*	*87 508*
Species total	*49 377*	*70 565*	*72 029*	*123 772*	*140 770*	*125 535*	*136 313*	*111 766*	*61 954*	*87 508*
...A	**...B**		**Huiro palo**			***Lessonia trabeculata***			**7,71(05)001,02**	**LJZ**
87 Chile	...	...	...	...	...	...	...	...	18 107	18 457
87 Fishing area total	*...*	*...*	*...*	*...*	*...*	*...*	*...*	*...*	*18 107*	*18 457*
Species total	*...*	*...*	*...*	*...*	*...*	*...*	*...*	*...*	*18 107*	*18 457*
...A	**...B**		**Chascón nep**			***Lessonia spp***			**7,71(05)001,XX**	**EOZ**
81 New Zealand	...	...	...	...	3	3	1	0	-	3
81 Fishing area total	*...*	*...*	*...*	*...*	*3*	*3*	*1*	*0*	*-*	*3*
Species total	*...*	*...*	*...*	*...*	*3*	*3*	*1*	*0*	*-*	*3*
Kelp nei	**Kelps nca**		**Huiros nep**			***Macrocystis spp***			**7,71(05)002,XX**	**GQC**
41 Argentina	9	57	52	102	80 F	40 F	25 F	...	...	...
41 Fishing area total	*9*	*57*	*52*	*102*	*80 F*	*40 F*	*25 F*	*...*	*...*	*...*
67 USA	-	435	185	191	44	149	107	202	1	2
67 Fishing area total	*-*	*435*	*185*	*191*	*44*	*149*	*107*	*202*	*1*	*2*
77 USA	83 773	84 673	116 656	73 604	58 989	70 017	24 879	78 953	41 998	34 544
77 Fishing area total	*83 773*	*84 673*	*116 656*	*73 604*	*58 989*	*70 017*	*24 879*	*78 953*	*41 998*	*34 544*
81 New Zealand	...	...	...	...	3	5	2	-	0	1
81 Fishing area total	*...*	*...*	*...*	*...*	*3*	*5*	*2*	*-*	*0*	*1*
87 Chile	7 021	4 580	4 325	12 451	18 333	11 903	10 104	11 928	6 084	9 672
87 Fishing area total	*7 021*	*4 580*	*4 325*	*12 451*	*18 333*	*11 903*	*10 104*	*11 928*	*6 084*	*9 672*
Species total	*90 803*	*89 745*	*121 218*	*86 348*	*77 449 F*	*82 114 F*	*35 117 F*	*91 083*	*48 083*	*44 219*
North Atlantic rockweed	**...B**		**...C**			***Ascophyllum nodosum***			**7,71(06)006,01**	**ASN**
21 Canada	26 972	14 110	18 713	15 438	18 958	25 203	21 133	24 157	26 521	9 911
USA	...	...	...	...	...	...	...	...	...	2 394
21 Fishing area total	*26 972*	*14 110*	*18 713*	*15 438*	*18 958*	*25 203*	*21 133*	*24 157*	*26 521*	*12 305*
27 Iceland	...	...	...	...	...	...	13 270	12 908	14 737	15 688
Ireland	27 824	27 152	30 748	32 976	30 980	33 400	34 000 F	34 000 F	34 000 F	34 000 F
27 Fishing area total	*27 824*	*27 152*	*30 748*	*32 976*	*30 980*	*33 400*	*47 270 F*	*46 908 F*	*48 737 F*	*49 688 F*
Species total	*54 796*	*41 262*	*49 461*	*48 414*	*49 938*	*58 603*	*68 403 F*	*71 065 F*	*75 258 F*	*61 993 F*
Bull kelp	**Durvillée antarctique**		**Cochayuyo**			***Durvillaea antartica***			**7,71(08)001,01**	**DVA**
81 New Zealand	-	-	-	-	-	-	-	-	-	4
81 Fishing area total	*-*	*-*	*-*	*-*	*-*	*-*	*-*	*-*	*-*	*4*
87 Chile	1 023	1 749	2 283	464	1 924	1 691	3 934	4 567	2 122	2 098
87 Fishing area total	*1 023*	*1 749*	*2 283*	*464*	*1 924*	*1 691*	*3 934*	*4 567*	*2 122*	*2 098*
Species total	*1 023*	*1 749*	*2 283*	*464*	*1 924*	*1 691*	*3 934*	*4 567*	*2 122*	*2 102*

B-91 Brown seaweeds / Algues brunes / Algas pardas

Capture production by species, fishing areas and countries or areas
Captures par espèces, zones de pêche et pays ou zones
Capturas por especies, áreas de pesca y países o áreas

Species, Fishing area Espèce, Zone de pêche Especie, Area de pesca	1992 mt	1993 mt	1994 mt	1995 mt	1996 mt	1997 mt	1998 mt	1999 mt	2000 mt	2001 mt
Golden Cystoseira	**Cystoseire dorée**		**Cistosira barbuda**			***Cystoseira barbata***			**7,71(09)001,01**	**YQT**
37 Ukraine	-	-	-	-	-	-	4	3	...	...
37 Fishing area total	-	-	-	-	-	-	*4*	*3*	...	...
Species total	-	-	-	-	-	-	*4*	*3*	...	...
Brown seaweeds	**Algues brunes**		**Algas pardas**			***Phaeophyceae***			**7,71(XX)XXX,XX**	**SWB**
21 USA	...	...	...	...	...	...	...	...	...	225
21 Fishing area total	...	...	...	...	...	...	...	...	...	*225*
27 France	74 725	54 728	74 807	71 238	81 293	72 785	64 756	68 397	66 172	64 516
Iceland	8 529	11 303	13 982	11 841	14 411	19 505	-	-	-	-
Norway	189 294	169 606	185 065	185 033	173 160	191 681	179 762	178 542	192 426	175 210
Russian Fed	11 958	3 170	2 732	1 085	4 632	1 627	2 346	4 893	11 583	6 572
Spain	800 F	650 F	500 F	350 F	200 F	100 F	5	28	239	239
UK	5 627	4 769	5 683	7 400	8 400	8 100	...	-	-	-
27 Fishing area total	*290 933 F*	*244 226 F*	*282 769 F*	*276 947 F*	*282 096 F*	*293 798 F*	*246 869*	*251 860*	*270 420*	*246 537*
47 South Africa	2 456	3 768	2 579	3 689	6 000 F	9 000 F	12 000 F	15 000 F	18 216	30 518
47 Fishing area total	*2 456*	*3 768*	*2 579*	*3 689*	*6 000 F*	*9 000 F*	*12 000 F*	*15 000 F*	*18 216*	*30 518*
51 India	16 000 F	16 000 F	16 000 F	16 000 F	16 000 F	16 000 F	16 000 F	16 000 F	16 000 F	16 000 F
51 Fishing area total	*16 000 F*	*16 000 F*	*16 000 F*	*16 000 F*	*16 000 F*	*16 000 F*	*16 000 F*	*16 000 F*	*16 000 F*	*16 000 F*
57 Australia	16 108	18 275	17 519	22 316	18 570	21 152	20 811	20 774	13 650	13 547
57 Fishing area total	*16 108*	*18 275*	*17 519*	*22 316*	*18 570*	*21 152*	*20 811*	*20 774*	*13 650*	*13 547*
61 Japan	...	...	...	8 936	10 834	7 933	7 553	8 326	7 247	7 238
Korea Rep	7 894	13 095	11 366	11 026	16 523	13 131	5 159	8 624	8 158	10 067
Russian Fed	11 657	2 504	2 561	7 526	11 490	20 345	23 639	16 247	32 557	16 406
61 Fishing area total	*19 551*	*15 599*	*13 927*	*27 488*	*38 847*	*41 409*	*36 351*	*33 197*	*47 962*	*33 711*
77 Mexico	53 132	52 343	32 456	44 230	27 663	34 516	6 119	26 470	28 251	38 233
77 Fishing area total	*53 132*	*52 343*	*32 456*	*44 230*	*27 663*	*34 516*	*6 119*	*26 470*	*28 251*	*38 233*
Species total	*398 180 F*	*350 211 F*	*365 250 F*	*390 670 F*	*389 176 F*	*415 875 F*	*338 150 F*	*363 301 F*	*394 499 F*	*378 771 F*
Group total	***757 982***	***693 413***	***719 709***	***776 173***	***785 798***	***811 633***	***683 013***	***746 826***	***701 794***	***700 027***

See paragraph 5 of the INTRODUCTION. Voir le paragraphe 5 de l'INTRODUCTION. Véase el párrafo 5 en la INTRODUCCION.

B-92 Red seaweeds / Algues rouges / Algas rojas

Capture production by species, fishing areas and countries or areas
Captures par espèces, zones de pêche et pays ou zones
Capturas por especies, áreas de pesca y países o áreas

Species, Fishing area Espèce, Zone de pêche Especie, Area de pesca	1992 mt	1993 mt	1994 mt	1995 mt	1996 mt	1997 mt	1998 mt	1999 mt	2000 mt	2001 mt
Phyllophora nei	**Phyllophora nca**		**Phyllophora nep**		***Phyllophora spp***				**7,87(02)001,XX**	**YFQ**
37 Ukraine	...	...	3 974	2 929	820	...	...	...	20	...
37 Fishing area total	*...*	*...*	*3 974*	*2 929*	*820*	*...*	*...*	*...*	*20*	*...*
Species total	*...*	*...*	*3 974*	*2 929*	*820*	*...*	*...*	*...*	*20*	*...*
Gracilaria seaweeds	**Algues gracilaires**		**Gracilarias**		***Gracilaria spp***				**7,87(12)004,XX**	**GLS**
41 Argentina	1 740	2 350	3 060	2 276	1 900 F	850 F	470 F	100 F	3	3 F
41 Fishing area total	*1 740*	*2 350*	*3 060*	*2 276*	*1 900 F*	*850 F*	*470 F*	*100 F*	*3*	*3 F*
47 Namibia	1 272	226	175	799	936	851	897	660	829	829 F
47 Fishing area total	*1 272*	*226*	*175*	*799*	*936*	*851*	*897*	*660*	*829*	*829 F*
61 China,Taiwan	132	8	1	-	-	-	-	-	-	-
61 Fishing area total	*132*	*8*	*1*	*-*	*-*	*-*	*-*	*-*	*-*	*-*
87 Chile	6 586	4 789	3 596	68 436	15 738	4 809	4 601	54 800	103 629	52 431
87 Fishing area total	*6 586*	*4 789*	*3 596*	*68 436*	*15 738*	*4 809*	*4 601*	*54 800*	*103 629*	*52 431*
Species total	*9 730*	*7 373*	*6 832*	*71 511*	*18 574 F*	*6 510 F*	*5 968 F*	*55 560 F*	*104 461*	*53 263 F*
Carragheen (Irish) moss	**Mousse perle**		**...C**		***Chondrus crispus***				**7,87(16)001,04**	**IMS**
21 USA	-	-	-	204	262	563	703	81	59	-
21 Fishing area total	*-*	*-*	*-*	*204*	*262*	*563*	*703*	*81*	*59*	*-*
Species total	*-*	*-*	*-*	*204*	*262*	*563*	*703*	*81*	*59*	*-*
Skottsberg's gigartina	**Gigartine de Skottsberg**		**Gigartina de Skottsberg**		***Gigartina skottsbergii***				**7,87(16)002,03**	**GJK**
87 Chile	...	...	...	...	...	...	...	...	...	22 717
87 Fishing area total	*...*	*...*	*...*	*...*	*...*	*...*	*...*	*...*	*...*	*22 717*
Species total	*...*	*...*	*...*	*...*	*...*	*...*	*...*	*...*	*...*	*22 717*
Gigartina seaweeds	**...B**		**Chicorea de mar**		***Gigartina spp***				**7,87(16)002,XX**	**GIG**
41 Argentina	14	8	24	22	20 F	10 F	5 F	...	...	...
41 Fishing area total	*14*	*8*	*24*	*22*	*20 F*	*10 F*	*5 F*	*...*	*...*	*...*
87 Chile	3 748	4 163	4 572	6 389	6 690	11 745	15 453	23 458	24 778	3 325
87 Fishing area total	*3 748*	*4 163*	*4 572*	*6 389*	*6 690*	*11 745*	*15 453*	*23 458*	*24 778*	*3 325*
Species total	*3 762*	*4 171*	*4 596*	*6 411*	*6 710 F*	*11 755 F*	*15 458 F*	*23 458*	*24 778*	*3 325*
Iridea nei	**Iridea nca**		**Luga-luga**		***Iridaea spp***				**7,87(16)003,XX**	**LGY**
87 Chile	9 325	19 513	28 331	37 376	32 438	22 679	26 181	23 099	30 118	37 606
87 Fishing area total	*9 325*	*19 513*	*28 331*	*37 376*	*32 438*	*22 679*	*26 181*	*23 099*	*30 118*	*37 606*
Species total	*9 325*	*19 513*	*28 331*	*37 376*	*32 438*	*22 679*	*26 181*	*23 099*	*30 118*	*37 606*
Laver (Nori)	**Algue nori**		**Lechuga nori**		***Porphyra tenera***				**7,87(20)002,08**	**PRT**
61 China,Taiwan	23	24	6	3	23	2	3	3	-	-
61 Fishing area total	*23*	*24*	*6*	*3*	*23*	*2*	*3*	*3*	*-*	*-*
Species total	*23*	*24*	*6*	*3*	*23*	*2*	*3*	*3*	*-*	*-*
Nori nei	**Nori nca**		**Luche**		***Porphyra spp***				**7,87(20)002,XX**	**FYS**
87 Chile	73	181	119	6	3	53	45	9	-	-
87 Fishing area total	*73*	*181*	*119*	*6*	*3*	*53*	*45*	*9*	*-*	*-*
Species total	*73*	*181*	*119*	*6*	*3*	*53*	*45*	*9*	*-*	*-*
Gelidium seaweeds	**Algues gélidium**		**Gelidios**		***Gelidium spp***				**7,87(26)002,XX**	**GEL**
61 China,Taiwan	487	19	38	3	4	9	10	33	29	26
61 Fishing area total	*487*	*19*	*38*	*3*	*4*	*9*	*10*	*33*	*29*	*26*
87 Chile	1 606	1 569	1 500	1 144	867	405	762	491	525	402
87 Fishing area total	*1 606*	*1 569*	*1 500*	*1 144*	*867*	*405*	*762*	*491*	*525*	*402*
Species total	*2 093*	*1 588*	*1 538*	*1 147*	*871*	*414*	*772*	*524*	*554*	*428*
Manifold callophyllis	**Callophyllis variable**		**Calofila variable**		***Callophyllis variegata***				**7,87(27)001,01**	**KFV**
87 Chile	-	-	-	-	49	11	73	84	56	5
87 Fishing area total	*-*	*-*	*-*	*-*	*49*	*11*	*73*	*84*	*56*	*5*
Species total	*-*	*-*	*-*	*-*	*49*	*11*	*73*	*84*	*56*	*5*
Red seaweeds	**Algues rouges**		**Algas rojas**		***Rhodophyceae***				**7,87(XX)XXX,XX**	**SWR**
21 Canada	1 852	7 450	11 501	11 233	5 374	9 493	6 339	6 503	7 986	5 021
21 Fishing area total	*1 852*	*7 450*	*11 501*	*11 233*	*5 374*	*9 493*	*6 339*	*6 503*	*7 986*	*5 021*

B-92 Red seaweeds / Algues rouges / Algas rojas

Capture production by species, fishing areas and countries or areas
Captures par espèces, zones de pêche et pays ou zones
Capturas por especies, áreas de pesca y países o áreas

Species, Fishing area Espèce, Zone de pêche Especie, Area de pesca	1992 mt	1993 mt	1994 mt	1995 mt	1996 mt	1997 mt	1998 mt	1999 mt	2000 mt	2001 mt
27 Estonia	0	0	411	548	163	2 444	2 880	1 319	201	325
France	4 864	5 546	4 633	4 239	2 995	2 637	2 164	2 481	1 977	1 653
Ireland	66	71	69	80	80	86	100 F	100 F	100 F	100 F
Latvia	10	1	0	0	0	0	0	0	0	0
Portugal	4 000 F	3 000 F	2 686	2 816	1 703	1 743	968	1 949	1 224	1 198
Russian Fed	457	192	95	40	95	75	85	66	6 525	135
Spain	13 353	13 289	13 791	14 042	14 671	14 785	13 935	13 790	13 975	13 772
27 Fishing area total	*22 750 F*	*22 099 F*	*21 685*	*21 765*	*19 707*	*21 770*	*20 132 F*	*19 705 F*	*24 002 F*	*17 183 F*
34 Morocco	7 783	7 108	5 357	7 858	7 625	8 094	7 049	5 920	6 080	10 015
34 Fishing area total	*7 783*	*7 108*	*5 357*	*7 858*	*7 625*	*8 094*	*7 049*	*5 920*	*6 080*	*10 015*
37 Italy	...	...	600	500	600	700	750	750 F	750 F	750 F
37 Fishing area total	*...*	*...*	*600*	*500*	*600*	*700*	*750*	*750 F*	*750 F*	*750 F*
41 Argentina	4	3	3	3	...	...	...	...	...	...
41 Fishing area total	*4*	*3*	*3*	*3*	*...*	*...*	*...*	*...*	*...*	*...*
47 South Africa	6 808	4 192	1 797	2 563	2 500 F	2 500 F	2 400 F	2 300 F	2 295	1 620
47 Fishing area total	*6 808*	*4 192*	*1 797*	*2 563*	*2 500 F*	*2 500 F*	*2 400 F*	*2 300 F*	*2 295*	*1 620*
51 India	18 000 F	19 000 F	20 000 F	21 000 F	22 000 F	23 000 F	24 000 F	24 000 F	24 000 F	24 000 F
Madagascar	285	423	702	787	787	1 000	2 510	1 933	5 792	5 045
Tanzania	2 500 F	3 600	3 500 F	3 500 F	3 500 F	3 610	4 000	4 500	5 000	5 000 F
51 Fishing area total	*20 785 F*	*23 023 F*	*24 202 F*	*25 287 F*	*26 287 F*	*27 610 F*	*30 510 F*	*30 433 F*	*34 792 F*	*34 045 F*
57 Indonesia	94 258	109 986	96 437	97 483	130 199	96 831	23 641	1 039	417	360
Timor-Leste	...	...	...	...	...	...	...	1	1	1
57 Fishing area total	*94 258*	*109 986*	*96 437*	*97 483*	*130 199*	*96 831*	*23 641*	*1 040*	*418*	*361*
61 Japan	6 859	6 795	5 714	4 204	4 136	3 722	3 489	3 207	2 824	3 218
Korea Rep	14 556	4 730	9 089	8 185	3 971	6 272	3 903	2 904	2 873	3 779
Russian Fed	4 284	887	2 273	1 895	2 250	4 244	4 109	4 509	2 988	4 428
61 Fishing area total	*25 699*	*12 412*	*17 076*	*14 284*	*10 357*	*14 238*	*11 501*	*10 620*	*8 685*	*11 425*
71 Fiji Islands	-	-	3 F	-	0	0	0	2	2 F	2 F
Indonesia	7 504	8 409	14 001	14 092	31 345	29 148	23 874	20 520	42 295	42 400
Philippines	1 049	1 144	1 062	919	884	494	417	433	413	447
71 Fishing area total	*8 553*	*9 553*	*15 066 F*	*15 011*	*32 229*	*29 642*	*24 291*	*20 955*	*42 710 F*	*42 849 F*
77 Mexico	5 283	4 588	4 250	4 977	6 833	7 459	6 324	5 620	5 304	8 694
77 Fishing area total	*5 283*	*4 588*	*4 250*	*4 977*	*6 833*	*7 459*	*6 324*	*5 620*	*5 304*	*8 694*
81 New Zealand	0	0	0	0	4	2	1	1	0	5
81 Fishing area total	*0*	*0*	*0*	*0*	*4*	*2*	*1*	*1*	*0*	*5*
87 Peru	593	243	170	413	307	155	1 650	1 733	1 312	5 505
87 Fishing area total	*593*	*243*	*170*	*413*	*307*	*155*	*1 650*	*1 733*	*1 312*	*5 505*
Species total	*194 368 F*	*200 657 F*	*198 144 F*	*201 377 F*	*242 022 F*	*218 494 F*	*134 588 F*	*105 580 F*	*134 334 F*	*137 473 F*
Group total	***219 374***	***233 507***	***243 540***	***320 964***	***301 772***	***260 481***	***183 791***	***208 398***	***294 380***	***254 817***

See paragraph 5 of the INTRODUCTION. Voir le paragraphe 5 de l'INTRODUCTION. Véase el párrafo 5 en la INTRODUCCION.

B-93 Green seaweeds / Algues vertes / Algas verdes

Capture production by species, fishing areas and countries or areas
Captures par espèces, zones de pêche et pays ou zones
Capturas por especies, áreas de pesca y países o áreas

Species, Fishing area Espèce, Zone de pêche Especie, Area de pesca	1992 mt	1993 mt	1994 mt	1995 mt	1996 mt	1997 mt	1998 mt	1999 mt	2000 mt	2001 mt
Lacy sea lettuce	**...B**		**...C**		***Ulva pertusa***				**7,41(08)002,04**	**UVP**
61 China,Taiwan	106	86	74	66	61	61	58	47	79	46
61 Fishing area total	*106*	*86*	*74*	*66*	*61*	*61*	*58*	*47*	*79*	*46*
Species total	*106*	*86*	*74*	*66*	*61*	*61*	*58*	*47*	*79*	*46*
Green seaweeds	**Algues vertes**		**Algas verdes**		***Chlorophyceae***				**7,41(XX)XXX,XX**	**SWG**
04 Japan	300	160	286	301	252	296	98	83	144	145
04 Fishing area total	*300*	*160*	*286*	*301*	*252*	*296*	*98*	*83*	*144*	*145*
27 France	56	23	45	32	60	80	112	118	83	114
27 Fishing area total	*56*	*23*	*45*	*32*	*60*	*80*	*112*	*118*	*83*	*114*
37 Italy	...	...	...	...	1 250	1 250	1 250	1 250 F	1 250 F	1 250 F
37 Fishing area total	*...*	*...*	*...*	*...*	*1 250*	*1 250*	*1 250*	*1 250 F*	*1 250 F*	*1 250 F*
51 India	51 000 F	52 500 F	54 000 F	55 500 F	57 000 F	58 500 F	60 000 F	60 000 F	60 000 F	60 000 F
51 Fishing area total	*51 000 F*	*52 500 F*	*54 000 F*	*55 500 F*	*57 000 F*	*58 500 F*	*60 000 F*	*60 000 F*	*60 000 F*	*60 000 F*
61 Korea Rep	1 793	962	1 042	1 029	757	582	341	574	284	444
61 Fishing area total	*1 793*	*962*	*1 042*	*1 029*	*757*	*582*	*341*	*574*	*284*	*444*
71 Fiji Islands	41	47	47 F	50 F	40 F	50 F	60	65	63 F	58 F
71 Fishing area total	*41*	*47*	*47 F*	*50 F*	*40 F*	*50 F*	*60*	*65*	*63 F*	*58 F*
Species total	*53 190 F*	*53 692 F*	*55 420 F*	*56 912 F*	*59 359 F*	*60 758 F*	*61 861 F*	*62 090 F*	*61 824 F*	*62 011 F*
Group total	***53 296***	***53 778***	***55 494***	***56 978***	***59 420***	***60 819***	***61 919***	***62 137***	***61 903***	***62 057***

See paragraph 5 of the INTRODUCTION. Voir le paragraphe 5 de l'INTRODUCTION. Véase el párrafo 5 en la INTRODUCCION.

B-94 Miscellaneous aquatic plants / Plantes aquatiques diverses / Diversas plantas acuáticas

Capture production by species, fishing areas and countries or areas / Captures par espèces, zones de pêche et pays ou zones / Capturas por especies, áreas de pesca y países o áreas

Species, Fishing area Espèce, Zone de pêche Especie, Area de pesca	1992 mt	1993 mt	1994 mt	1995 mt	1996 mt	1997 mt	1998 mt	1999 mt	2000 mt	2001 mt
Spirulina nei	**Spirulina nca**		**Spirulina nep**		***Spirulina spp***			**7,11(01)001,XX**		**SIZ**
02 Mexico	3 175	3 435	-	-	-	-	-	-	-	-
02 Fishing area total	*3 175*	*3 435*	-	-	-	-	-	-	-	-
Species total	*3 175*	*3 435*	-	-	-	-	-	-	-	-
Eel-grass	**Grande zostère**		**Gran zostera**		***Zostera marina***			**7,92(02)001,01**		**ZOM**
37 Ukraine	1 361	424	264	243	370	331	310	157	6	...
37 Fishing area total	*1 361*	*424*	*264*	*243*	*370*	*331*	*310*	*157*	*6*	...
Species total	*1 361*	*424*	*264*	*243*	*370*	*331*	*310*	*157*	*6*	...
Tule nei	**Tule nca**		**Tule nep**		***Scirpus spp***			**7,92(04)001,XX**		**CJW**
02 Mexico	686	234	71	127	102	92	13	-	-	-
02 Fishing area total	*686*	*234*	*71*	*127*	*102*	*92*	*13*	-	-	-
Species total	*686*	*234*	*71*	*127*	*102*	*92*	*13*	-	-	-
Aquatic plants nei	**Plantes aquatiques nca**		**Plantas acuáticas nep**		***Plantae aquaticae***			**7,99(XX)XXX,XX**		**APL**
04 Korea Rep	4	-	-	5	11	1	5	-	-	-
04 Fishing area total	*4*	-	-	*5*	*11*	*1*	*5*	-	-	-
27 France	2 248	36	42	41	24	-	-	-	-	-
27 Fishing area total	*2 248*	*36*	*42*	*41*	*24*	...	...	...	...	...
34 Senegal	50 F	50 F	50 F	50 F	50 F	50 F	100 F	150 F	200 F	223
34 Fishing area total	*50 F*	*50 F*	*50 F*	*50 F*	*50 F*	*50 F*	*100 F*	*150 F*	*200 F*	*223*
61 China	100 000	124 250	151 000	150 000	151 630	184 270	170 520	215 550	204 290	266 870
China,H.Kong	2	1	0	0	0	0	0	0	0	0
China,Taiwan	940	69	93	24	18	52	16	40	17	19
Japan	41 279	23 468	24 854	13 496	14 618	12 047	11 161	11 459	11 808	11 249
Korea Rep	3 295	3 309	5 402	2 128	1 654	3 248	3 191	793	1 708	643
Russian Fed	-	-	-	-	200	150	100	-	-	-
61 Fishing area total	*145 516*	*151 097*	*181 349*	*165 648*	*168 120*	*199 767*	*184 988*	*227 842*	*217 823*	*278 781*
77 Cook Is	30 F	40 F	40 F	40 F	45 F	50 F	50 F	50 F	50 F	50 F
77 Fishing area total	*30 F*	*40 F*	*40 F*	*40 F*	*45 F*	*50 F*	*50 F*	*50 F*	*50 F*	*50 F*
87 Chile	-	-	-	-	3	8	29	1	3	32
87 Fishing area total	-	-	-	-	*3*	*8*	*29*	*1*	*3*	*32*
Species total	*147 848 F*	*151 223 F*	*181 481 F*	*165 784 F*	*168 253 F*	*199 876 F*	*185 172 F*	*228 043 F*	*218 076 F*	*279 086 F*
Group total	***153 070***	***155 316***	***181 816***	***166 154***	***168 725***	***200 299***	***185 495***	***228 200***	***218 082***	***279 086***

See paragraph 5 of the INTRODUCTION. Voir le paragraphe 5 de l'INTRODUCTION. Véase el párrafo 5 en la INTRODUCCION.

C - Capture production: by fishing areas

C - Captures: par zones de pêche

C - Capturas: por áreas de pesca

C-00 Fish, crustaceans, molluscs, etc / Poissons, crustacés, mollusques, etc / Peces, crustáceos, moluscos, etc

Capture production by fishing areas and species groups / Captures par zones de pêche et groupes d'espèces / Capturas por áreas de pesca y grupos de especies

Fishing area Zone de pêche Area de pesca	Species group Groupe d'espèces Grupo de especies	1995 mt	1996 mt	1997 mt	1998 mt	1999 mt	2000 mt	2001 mt
01 Africa - Inland waters	11	82 453	81 926	86 704	78 960	96 396	97 962	98 007
Afrique - Eaux continentales	12	374 857	366 526	361 520	394 271	423 730	467 477	475 290
Africa - Aguas continentales	13	1 464 558	1 366 584	1 425 683	1 461 868	1 438 772	1 447 646	1 426 331
	22	898	637	985	801	959	2 164	2 052
	23	196	11	16	24	29	31	67
	24	366	290	253	359	270	284	258
	31	...	...	547	381	763	2 490	2 134
	33	19 921	15 875	18 986	20 939	20 750	24 658	28 359
	35	1 963	1 900	1 850	1 800	1 798	1 500	1 700
	37	1 241	1 326	1 229	1 150	1 180	3 182	5 154
	41	9 360	8 042	8 423	8 995	7 801	9 140	9 398
	45	3 117	3 000	2 850	2 700	2 574	2 100	2 400
	51	488	530	610	625	606	921	851
	Area total	1 959 418	1 846 647	1 909 656	1 972 873	1 995 628	2 059 555	2 052 001
02 America, North - Inland waters	11	27 654	28 105	24 924	21 241	20 971	22 094	21 943
Amérique du Nord - Eaux continentales	12	90 324	86 191	84 500	72 068	69 538	76 849	68 642
América del Norte - Aguas continentales	13	53 373	51 168	49 460	50 912	51 409	57 866	48 114
	21	433	526	615	525	420	281	6
	22	610	730	796	739	770	942	423
	23	30 964	32 374	27 972	29 200	25 267	22 515	22 669
	24	2 088	3 187	3 941	3 825	5 199	3 070	2 661
	25	815	995	1 169	966	1 046	1 221	1 226
	32	22	23	24	25	25	25	...
	33	470	368	549	573	470	577	506
	37	...	...	...	372	487	392	558
	41	11 753	10 524	15 099	14 230	9 694	3 843	7 902
	51	198	241	122	185	240	565	2
	71	564	422	2 034	1 201	337	325	31
	72	6	18	17	26	15	53	51
	77	183	337	297	483	617	231	...
	Area total	219 457	215 209	211 519	196 571	186 505	190 849	174 734
03 America, South - Inland waters	11	663	502	213	265	307	1 745	360
Amérique du Sud - Eaux continentales	12	18 599	16 173	15 460	15 370	14 242	16 226	16 283
América del Sur - Aguas continentales	13	344 763	322 890	312 455	315 347	328 254	323 038	320 821
	23	685	463	1 208	733	527	566	658
	24	425	642	455	567	523	407	300
	37	850	850	850	850	850	850	760
	41	1 412	2 677	2 157	1 506	3 235	2 437	2 440
	71	5	0	0	0	0	7	9
	72	0	0	0	0	0	0	0
	Area total	367 402	344 197	332 798	334 638	347 938	345 276	341 631
04 Asia - Inland waters	11	549 997	429 416	376 916	377 296	382 408	334 543	309 825
Asie - Eaux continentales	12	106 051	86 676	95 384	112 104	125 424	121 673	117 829
Asia - Aguas continentales	13	2 442 316	2 791 205	2 883 899	3 071 956	3 633 592	3 663 275	3 779 712
	21	2 139	2 179	1 733	1 531	1 320	1 288	1 139
	22	2 467	5 796	2 089	2 064	2 255	5 580	5 790
	23	37 527	38 676	35 498	33 261	28 219	29 420	24 738
	24	204 042	219 321	209 777	238 029	249 306	365 220	203 098
	25	3 997	369	34	3 000	83	163	201
	33	28 359	37 088	41 467	35 911	35 721	39 120	40 978
	34	849	490	447	270	494	402	625
	35	1 086	1 481	2 497	1 517	1 143	917	917
	37	909	562	1 600	1 500	1 455	1 583	1 685
	41	289 226	386 235	498 702	617 997	474 588	548 383	605 912
	42	329	363	267	44	47	77	150
	45	33 116	34 418	44 296	23 900	23 365	33 941	36 840
	51	593 312	557 746	503 071	574 737	546 606	588 720	619 401
	54	136	4	-	-	-	-	-
	56	835	844	956	914	304	303	260
	58	325	481	52	165	73	166	250
	71	3 058	2 535	1 550	1 767	1 435	1 957	2 843
	72	-	36	48	27	1	3	-
	77	2 643	1 494	2 460	1 480	2 048	2 031	2 283
	Area total	4 302 719	4 597 415	4 702 743	5 099 470	5 509 887	5 738 765	5 754 476
05 Europe - Inland waters	11	113 394	117 076	111 636	120 201	106 254	103 830	108 456
Europe - Eaux continentales	13	121 451	118 041	108 305	102 393	95 419	114 807	100 894
Europa - Aguas continentales	21	2 347	1 248	1 331	1 222	856	704	633
	22	3 568	3 827	3 194	3 133	3 092	3 101	2 554
	23	67 327	64 252	69 832	66 252	83 397	82 877	80 978
	24	83 078	96 744	86 878	120 291	156 813	120 766	47 425
	25	559	569	467	1 042	1 243	1 354	1 104
	32	193	98	163	177	22	11	55
	33	31	52	109	219	451	360	489
	37	-	8	110	350	361	350	326
	41	2 974	2 774	2 989	2 783	2 800	3 758	4 507

C-00

Fish, crustaceans, molluscs, etc — **Capture production by fishing areas and species groups**
Poissons, crustacés, mollusques, etc — **Captures par zones de pêche et groupes d'espèces**
Peces, crustáceos, moluscos, etc — **Capturas por áreas de pesca y grupos de especies**

Fishing area / Zone de pêche / Area de pesca	Species group / Groupe d'espèces / Grupo de especies	1995 mt	1996 mt	1997 mt	1998 mt	1999 mt	2000 mt	2001 mt
	71	...	...	38	41	35	26	-
	Area total	394 922	404 689	385 052	418 104	450 743	431 944	347 421
06 Oceania - Inland waters	11	0	0	28	2	3	4	6
Océanie - Eaux continentales	12	2 316	2 347	2 350	2 604	2 600	2 590	2 588
Oceanía - Aguas continentales	13	9 303	9 357	9 353	9 158	9 122	9 105	9 003
	22	1 134	931	810	1 286	1 209	1 151	1 244
	24	480	480	480	480	480	480	480
	25	1 253	1 411	1 528	1 624	1 984	1 700	1 650
	33	1 850	1 850	1 850	1 850	1 850	1 850	1 850
	41	1 036	338	329	345	412	372	374
	51	2 569	2 670	3 970	4 500	5 000	5 080	5 300
	Area total	19 941	19 384	20 698	21 849	22 660	22 332	22 495
21 Atlantic, Northwest	21	157	86	91	69	9	3	4
Atlantique, nord-ouest	22	631	551	455	496	353	395	220
Atlántico, noroeste	23	1 410	1 301	1 458	1 374	428	1 070	405
	24	7 159	6 415	6 887	8 281	7 367	5 919	4 007
	25	1 655	2 144	2 785	3 032	3 000	3 123	2 934
	31	77 559	80 472	88 100	92 236	113 568	118 531	122 421
	32	108 617	119 254	127 256	140 007	143 120	137 455	148 393
	33	15 573	20 492	20 055	19 826	28 628	19 522	20 343
	34	63 380	56 720	59 857	66 917	66 707	73 940	81 218
	35	609 107	575 240	606 757	521 264	472 439	460 612	543 766
	36	9 312	10 713	7 657	7 844	10 656	8 394	8 735
	37	32 967	77 943	66 326	77 936	58 871	49 971	65 508
	38	46 806	56 131	48 219	51 296	49 186	49 360	39 716
	39	17 618	14 896	14 378	5 869	7 492	8 029	4 483
	42	134 345	142 599	150 626	147 582	172 414	166 103	173 491
	43	70 631	71 866	78 146	77 155	83 105	83 062	83 803
	45	170 713	200 651	174 549	207 259	243 053	269 324	261 408
	47	1	2	-	1	5	6	1
	52	3 044	4 513	3 668	3 149	5 721	4 567	6 339
	53	32 882	39 569	49 099	33 329	47 106	26 475	18 571
	54	15 723	17 585	17 228	15 051	15 651	21 098	18 353
	55	135 702	122 704	116 169	108 136	135 115	201 288	256 929
	56	389 779	384 556	345 924	315 694	325 053	329 462	347 200
	57	35 583	38 908	48 733	44 725	26 767	26 402	18 409
	58	2 031	2 394	600	168	1 920	1 469	1 384
	75	926	1 598	2 606	3 182	2 238	1 557	1 071
	76	17 726	15 139	12 201	13 126	14 298	13 440	8 778
	77	0	-	223	299	921	202	481
	Area total	2 001 037	2 064 442	2 050 053	1 965 303	2 035 191	2 080 779	2 238 371
27 Atlantic, Northeast	11	5 362	5 834	6 674	6 311	6 574	8 172	8 399
Atlantique, nord-est	13	13 781	14 971	17 105	17 928	17 062	11 206	11 466
Atlántico, nordeste	21	0	0	0	1	2	-	-
	22	3 195	2 959	4 327	2 290	2 622	2 247	2 098
	23	11 696	12 027	10 663	11 199	10 262	14 507	9 324
	24	133	122	113	58	59	140	103
	25	18	67	135	101	40	571	658
	31	338 917	318 365	310 781	298 034	317 559	300 443	303 030
	32	3 160 141	3 192 725	3 133 827	3 256 629	3 289 075	3 331 770	3 607 968
	33	1 155 876	880 758	1 267 537	1 065 549	789 791	768 539	938 408
	34	450 650	435 011	418 833	425 131	393 489	419 979	422 804
	35	2 881 853	2 875 382	3 089 362	3 013 492	2 995 072	2 890 306	2 432 932
	36	55 026	47 546	43 940	40 330	40 814	40 293	28 760
	37	2 126 591	2 547 671	2 630 715	2 000 933	1 857 940	2 413 135	2 637 042
	38	86 729	74 119	106 720	94 115	98 564	104 582	107 084
	39	102 557	90 392	74 377	81 245	48 856	72 385	37 427
	42	48 058	42 033	58 913	53 020	51 036	53 654	56 441
	43	59 874	52 107	58 122	55 760	59 818	50 364	55 364
	44	124	175	177	313	471	468	814
	45	162 378	169 575	191 183	176 695	173 335	177 013	145 079
	46	-	-	-	88	-	-	36
	47	2 239	1 621	1 783	1 389	1 276	1 116	1 673
	52	18 538	29 391	33 733	18 337	22 331	33 270	33 900
	53	1 487	1 400	1 852	1 434	881	683	899
	54	132 575	101 062	118 949	122 237	114 376	128 707	146 696
	55	62 165	55 638	71 668	73 299	73 406	74 764	81 361
	56	86 742	58 500	57 086	116 751	94 266	65 922	49 247
	57	54 033	53 371	50 982	49 948	53 067	58 808	44 474
	58	1 021	865	1 015	1 294	648	498	424
	76	942	911	612	564	643	462	478
	77	1	0	-	-	-	-	24
	Area total	11 022 702	11 064 598	11 761 184	10 984 475	10 513 335	11 024 004	11 164 413
31 Atlantic, Western Central	13	7 705	7 303	3 656	2 870	4 130	6 772	6 937
Atlantique, centre-ouest	22	43	35	19	9	2	1	...

Fishing area Zone de pêche Area de pesca	Species group Groupe d'espèces Grupo de especies	1995 mt	1996 mt	1997 mt	1998 mt	1999 mt	2000 mt	2001 mt
Atlántico, centro-occidental	24	273	343	246	280	117	288	171
	25	7	9	12	14	4	12	15
	31	2 696	2 527	1 546	1 891	1 673	1 765	1 695
	32	0	...	4	2	2	6	3
	33	189 681	164 649	157 164	159 994	130 246	154 467	143 807
	34	10 724	12 416	14 408	19 486	17 505	15 196	22 095
	35	665 064	679 660	758 862	734 996	865 288	711 720	639 070
	36	80 752	93 947	83 900	103 556	80 011	89 621	106 341
	37	39 198	35 367	36 538	38 063	32 601	30 731	30 009
	38	32 509	31 629	33 133	29 426	25 717	23 072	21 440
	39	233 163	223 781	259 985	220 341	174 182	159 083	167 559
	42	54 670	56 663	68 561	67 358	52 565	60 456	48 998
	43	29 278	29 472	29 172	27 191	30 660	33 444	27 702
	45	177 954	155 882	164 564	174 996	169 575	203 495	209 786
	47	440	363	373	422	200	347	186
	52	15 627	13 785	18 903	15 433	16 300	18 919	18 651
	53	130 876	113 081	99 310	106 852	89 056	226 333	162 117
	54	155	223	295	3 802	451	316	1 081
	55	12 313	684	581	834	223	256	253
	56	59 707	39 693	50 135	38 568	46 339	55 625	56 306
	57	20 691	31 119	20 751	24 135	20 113	24 384	23 237
	58	3 836	673	1 872	2 921	6 706	2 716	1 913
	72	209	228	86	141	31	32	36
	75	-	-	1	70	159	139	228
	76	15	10	15	15	15	10	49
	77	11	6	2	4	0	139	-
	Area total	1 767 597	1 693 548	1 804 094	1 773 670	1 763 871	1 819 345	1 689 685
34 Atlantic, Eastern Central Atlantique, centre-est Atlántico, centro-oriental	12	2 476	2 603	2 582	1 365	1 760	2 062	2 324
	22	0	0	27	4	0	-	-
	24	6 918	11 585	8 810	8 113	4 608	7 489	5 633
	31	34 934	35 323	38 371	46 234	41 604	50 540	49 780
	32	30 346	31 447	25 650	25 150	25 207	19 473	25 314
	33	239 654	263 259	291 496	267 169	292 111	277 528	282 701
	34	54 211	94 585	75 913	76 738	49 631	38 243	37 292
	35	1 518 897	1 753 281	1 828 230	1 869 052	1 691 950	1 752 143	1 868 923
	36	358 023	354 648	325 041	343 064	356 019	300 656	326 035
	37	442 341	439 833	444 064	513 196	433 700	573 643	446 307
	38	28 155	35 435	67 933	52 264	51 106	50 923	50 310
	39	418 352	287 958	295 819	283 454	396 161	271 779	431 215
	42	773	2 331	4 114	6 558	5 586	5 231	6 689
	43	2 143	2 174	9 378	4 010	2 821	2 886	2 982
	45	54 931	57 810	53 206	83 505	70 361	59 356	57 739
	47	1 668	1 034	1 317	1 743	875	249	562
	52	8 440	8 235	6 598	9 849	7 290	6 637	9 659
	53	223	100	109	89	125	101	151
	54	8	10	6	7	...	401	102
	56	178	54	69	139	148	120	111
	57	191 176	190 620	150 352	183 729	250 294	216 421	212 920
	58	1 806	746	1 489	12 599	2 258	1 189	384
	72	22	42	164	189	463	122	315
	Area total	3 395 675	3 573 113	3 630 738	3 788 220	3 684 078	3 637 192	3 817 448
37 Mediterranean and Black Sea Méditerranée et mer Noire Mediterráneo y Mar Negro	11	938	476	882	617	643	854	575
	13	1 303	2 822	2 825	3 031	2 568	2 956	3 522
	21	826	584	637	429	242	85	33
	22	849	919	1 254	914	682	464	504
	23	-	18	22	26	-	-	0
	24	10 522	3 862	3 439	6 391	13 550	14 141	30 727
	25	-	-	-	-	-	-	8
	31	16 027	13 340	11 718	11 059	10 469	11 134	10 853
	32	94 006	90 708	72 072	75 371	64 641	73 338	66 333
	33	174 592	183 355	170 147	173 388	176 952	172 695	159 825
	34	18 196	21 004	23 071	18 757	17 683	17 558	19 894
	35	824 466	674 739	642 385	625 771	781 261	718 037	767 875
	36	75 459	80 668	75 120	81 206	74 790	71 577	71 176
	37	127 593	116 031	110 585	100 465	97 217	103 880	114 782
	38	19 222	16 047	16 334	14 886	12 157	13 037	11 473
	39	112 010	100 777	83 640	87 647	63 217	75 454	63 933
	42	1 268	2 195	2 104	2 520	4 491	2 891	2 579
	43	7 552	7 455	7 203	4 927	4 623	4 068	4 224
	45	27 350	25 384	30 279	27 525	29 407	34 531	29 191
	47	10 363	9 804	8 827	7 873	9 927	10 200	9 882
	52	4 166	4 518	5 682	4 829	5 545	5 716	4 675
	53	3 005	2 246	1 863	1 204	952	319	157
	54	40 764	39 046	53 310	38 899	44 453	46 143	46 092
	55	23	93	137	65	80	588	154
	56	50 534	52 633	39 346	36 511	43 249	47 403	44 929
	57	62 545	59 266	60 139	54 281	50 692	53 846	53 773
	58	16 650	18 120	15 480	15 120	13 389	10 245	15 404
	74	...	28	22	22	30	30	76
	76	71	69	55	60	76	204	104
	77	487	909	904	1 764	1 737	908	2 592

C-00

Fish, crustaceans, molluscs, etc — Capture production by fishing areas and species groups
Poissons, crustacés, mollusques, etc — Captures par zones de pêche et groupes d'espèces
Peces, crustáceos, moluscos, etc — Capturas por áreas de pesca y grupos de especies

Fishing area Zone de pêche Area de pesca	Species group Groupe d'espèces Grupo de especies	1995 mt	1996 mt	1997 mt	1998 mt	1999 mt	2000 mt	2001 mt
	Area total	1 700 787	1 527 116	1 439 482	1 395 558	1 524 723	1 492 302	1 535 345
41 Atlantic, Southwest	31	11 870	10 356	11 976	10 902	8 701	8 681	7 524
Atlantique, sud-ouest	32	829 684	860 118	815 951	779 466	618 103	506 425	534 286
Atlántico, sudoccidental	33	243 275	226 233	234 505	213 688	207 297	236 136	233 662
	34	75 912	58 259	59 688	74 886	66 381	54 868	56 136
	35	109 509	139 838	161 501	115 964	59 628	58 850	75 529
	36	55 722	73 088	72 141	63 456	64 590	67 042	64 878
	37	48 163	40 265	42 416	39 028	31 276	41 721	33 802
	38	51 050	56 595	58 696	57 896	61 308	60 184	62 780
	39	172 960	158 570	165 118	157 405	181 165	194 956	201 983
	42	15 107	13 845	16 668	15 285	16 091	18 000	15 305
	43	10 817	8 026	7 502	6 002	6 334	6 469	6 400
	44	381	201	414	457	271	369	218
	45	47 953	45 681	47 343	58 911	48 673	76 603	119 616
	46	-	-	-	74	-	4	-
	47	860	313	849	2 161	6 293	227	1 375
	52	574	558	1 322	1 010	683	1 621	1 406
	53	726	873	828	744	1 547	884	870
	54	2 905	1 234	1 548	1 606	1 381	1 215	1 292
	55	10 592	36 952	39 817	28 441	42 700	37 404	42 598
	56	0	126	196	664	2 443	279	287
	57	645 234	747 875	1 016 557	749 986	1 192 738	1 001 022	802 928
	58	3 678	1 175	1 851	1 408	4 937	2 474	2 839
	72	0	0	0	0	0	0	0
	Area total	2 336 972	2 480 181	2 756 887	2 379 440	2 622 540	2 375 434	2 265 714
47 Atlantic, Southeast	31	2 830	1 996	2 671	2 960	3 069	2 928	4 021
Atlantique, sud-est	32	277 212	287 275	261 732	307 178	312 014	301 401	325 120
Atlántico, sudoriental	33	24 271	23 740	19 664	52 202	36 793	47 795	61 670
	34	55 094	58 742	65 833	74 252	53 312	51 583	55 866
	35	496 577	233 659	328 352	419 566	495 703	582 167	604 524
	36	59 468	48 720	38 640	51 053	52 576	51 338	48 543
	37	593 098	554 646	507 629	534 140	488 612	501 729	414 020
	38	4 321	3 058	7 781	6 531	8 334	10 688	15 221
	39	54 442	90 578	106 531	60 388	56 941	57 768	90 626
	42	7 163	5 329	3 415	5 639	4 855	6 874	5 575
	43	3 499	3 032	3 117	3 348	2 931	2 690	3 452
	45	4 289	5 915	5 569	16 230	5 333	10 421	13 329
	46	-	-	-	254	-	-	-
	47	5	2	-	-	-	-	1 062
	52	615	735	330	524	481	490	5 265
	56	0	0	-	-	-	-	-
	57	7 433	8 170	4 145	7 801	8 301	7 506	4 654
	58	17	-	-	-	-	-	-
	77	-	-	-	-	-	106	44
	Area total	1 590 334	1 325 597	1 355 409	1 542 066	1 529 255	1 635 484	1 652 992
48 Atlantic, Antarctic	32	11	24	17	22	12	5	3
Atlantique, Antarctique	33	5	-	-	0	10	6	4
Atlántico, Antártico	34	3 273	3 822	2 389	3 266	4 570	8 874	4 316
	38	90	40	30	14	15	4	13
	39	10	-	-	7	-	-	-
	44	-	497	-	-	-	3	14
	46	117 448	101 708	82 508	80 874	103 318	104 259	98 209
	57	-	52	28	53	-	-	2
	77	-	-	-	-	-	5	-
	Area total	120 837	106 143	84 972	84 236	107 925	113 156	102 561
51 Indian Ocean, Western	24	4 226	5 547	5 265	4 607	4 548	4 736	6 965
Océan Indien, ouest	25	376	395	334	335	341	125	134
Océano Indico, occidental	31	21 718	20 730	25 299	20 676	15 878	27 872	14 432
	32	1 147	1 251	1 348	1 456	747	1 472	2 436
	33	724 827	821 096	856 133	777 597	848 376	805 353	796 320
	34	67 602	91 138	189 740	118 295	166 920	168 828	162 545
	35	308 233	373 255	433 033	415 773	407 553	546 667	581 721
	36	862 346	798 132	811 442	794 035	903 097	927 100	870 908
	37	429 274	555 851	430 288	437 632	451 621	289 167	276 474
	38	118 048	177 712	117 951	120 758	115 589	126 542	117 656
	39	816 105	600 508	649 586	617 389	648 207	628 354	646 074
	42	28 533	7 652	9 814	11 083	11 839	20 952	21 593
	43	3 562	3 452	3 315	3 634	3 187	2 918	2 804
	45	292 066	297 846	272 896	336 556	322 351	311 254	295 312
	47	12 078	15 600	12 367	12 797	11 460	11 251	9 416
	52	43	43	40	40	29	45	51
	53	14	32	24	28	34	32	29
	57	110 641	94 868	148 579	112 432	121 305	119 020	140 124
	58	3 363	15 996	7 363	5 015	3 747	5 034	2 452
	74	-	-	-	-	-	95	-

C-00

Fish, crustaceans, molluscs, etc — **Capture production by fishing areas and species groups**
Poissons, crustacés, mollusques, etc — **Captures par zones de pêche et groupes d'espèces**
Peces, crustáceos, moluscos, etc — **Capturas por áreas de pesca y grupos de especies**

Fishing area Zone de pêche Area de pesca	Species group Groupe d'espèces Grupo de especies	1995 mt	1996 mt	1997 mt	1998 mt	1999 mt	2000 mt	2001 mt
	76	3 415	3 658	3 693	2 407	767	1 139	1 230
	Area total	3 807 617	3 884 762	3 978 510	3 792 545	4 037 596	3 997 956	3 948 676
57 Indian Ocean, Eastern	22	259	208	203	160	129	133	157
Océan Indien, est	24	178 744	197 740	192 141	180 806	190 310	194 226	186 598
Océano Indico, oriental	25	12 274	13 991	13 792	23 062	15 687	15 923	17 360
	31	20 915	30 824	33 443	28 204	26 965	27 146	25 741
	32	3 400	2 716	2 864	5 890	7 566	9 653	7 623
	33	455 354	488 328	485 143	511 999	511 995	497 472	497 936
	34	70 170	68 457	72 515	94 099	94 505	86 016	114 831
	35	444 653	435 811	497 823	490 762	393 684	411 637	406 070
	36	347 226	391 680	411 012	518 788	483 296	488 499	406 893
	37	562 919	543 343	537 462	584 058	560 031	521 921	535 505
	38	116 109	117 206	116 166	131 197	126 253	121 361	117 915
	39	1 490 614	1 350 902	1 499 417	1 682 392	1 625 684	1 874 633	1 893 936
	42	24 091	23 284	24 528	31 378	31 429	46 272	46 238
	43	17 160	16 338	16 286	16 600	20 985	22 904	19 895
	45	212 724	217 698	231 576	182 547	234 840	247 010	238 721
	47	35 664	40 568	45 940	52 272	54 445	36 561	36 623
	52	4 896	5 081	4 905	4 920	5 297	5 204	5 301
	53	-	-	-	-	-	350	380
	54	243	71	-	-	-	-	-
	55	8 159	6 872	6 093	5 543	7 622	8 301	4 024
	56	46 187	61 448	63 188	34 444	42 944	45 342	42 685
	57	88 363	104 648	109 444	115 730	111 791	124 084	115 986
	58	401	381	263	844	8 396	8 307	8 517
	72	291	353	248	337	273	372	381
	76	327	419	610	833	859	1 048	1 090
	77	20 597	23 527	29 747	7 077	35 524	35 684	37 454
	Area total	4 161 740	4 141 894	4 394 809	4 703 942	4 590 510	4 830 059	4 767 860
58 Indian Ocean, Antarctic	32	-	-	1	50	52	343	266
Océan Indien, Antarctique	33	-	15	5	3	14	-	0
Océano Indico, Antártico	34	9 585	4 925	8 208	7 982	13 063	9 079	10 294
	38	-	1	7	23	15	98	78
	39	61	-	1	0	0	0	0
	44	-	-	-	0	-	3	0
	46	1 266	6	-	-	-	-	-
	57	-	-	1	-	-	-	-
	74	-	7	-	-	-	-	-
	77	-	-	11	-	2	-	-
	Area total	10 912	4 954	8 234	8 058	13 146	9 523	10 638
61 Pacific, Northwest	11	61	110	101	100	56	8	52
Pacifique, nord-ouest	23	442 427	455 779	471 418	428 221	380 303	350 771	401 629
Pacífico, noroeste	24	78 769	75 549	106 577	116 818	137 932	126 545	127 786
	25	1	18	9	26	47	32	81
	31	155 341	181 394	186 917	197 870	210 735	225 682	203 918
	32	3 632 526	3 469 239	3 436 455	2 975 625	2 456 168	1 951 990	1 877 369
	33	1 479 446	1 856 071	2 033 717	2 223 641	2 245 815	2 343 594	2 193 022
	34	1 443 897	1 469 582	1 394 197	1 655 751	1 637 181	1 696 006	1 712 536
	35	1 967 502	1 990 976	2 556 096	2 932 317	2 856 880	2 534 514	2 614 151
	36	555 308	567 298	748 063	924 890	961 585	917 339	850 012
	37	2 727 467	3 267 951	3 162 978	2 674 518	2 462 646	2 469 381	2 667 858
	38	33 508	31 201	37 365	34 383	57 887	56 079	35 970
	39	4 058 374	4 864 097	4 442 968	4 372 457	4 368 442	4 055 185	3 641 860
	42	407 664	488 600	431 585	442 492	437 240	500 701	500 522
	43	2 398	2 221	1 724	1 794	1 842	3 316	3 531
	44	55 246	48 254	39 019	41 217	45 646	36 020	23 244
	45	706 720	776 922	834 742	932 180	1 150 109	1 119 120	991 000
	47	934 355	1 007 693	1 165 004	1 314 628	1 198 281	1 303 443	1 333 362
	51	-	-	-	-	-	-	74
	52	27 340	23 352	24 084	25 540	22 257	20 249	19 385
	53	18 269	18 259	17 210	9 905	11 610	15 943	10 092
	54	6 655	3 028	7 309	7 934	8 265	6 952	4 955
	55	279 382	276 406	266 957	294 211	305 510	310 104	293 268
	56	185 430	181 775	144 085	156 515	145 198	132 789	119 870
	57	1 006 539	1 165 463	1 154 074	1 033 618	1 117 839	1 221 972	1 137 512
	58	1 302 416	938 870	1 502 215	1 488 978	1 457 426	1 410 260	1 405 909
	74	5 767	16 747	2 780	891	1 171	1 443	1 053
	76	28 633	27 020	28 009	25 353	24 156	24 161	22 753
	77	210 281	278 735	410 601	441 444	410 337	346 443	340 147
	Area total	21 751 722	23 482 610	24 606 259	24 753 317	24 112 564	23 180 042	22 532 921
67 Pacific, Northeast	21	4	1	0	0	0	242	195
Pacifique, nord-est	23	558 396	423 012	301 117	318 745	382 944	300 917	349 383
Pacífico, nordeste	24	158	15	34	62	-	105	178
	31	202 798	217 113	273 510	198 371	174 192	203 808	173 733
	32	1 815 668	1 725 540	1 749 564	1 764 162	1 511 756	1 631 326	1 830 568

C-00

Fish, crustaceans, molluscs, etc — **Capture production by fishing areas and species groups**
Poissons, crustacés, mollusques, etc — **Captures par zones de pêche et groupes d'espèces**
Peces, crustáceos, moluscos, etc — **Capturas por áreas de pesca y grupos de especies**

Fishing area Zone de pêche Area de pesca	Species group Groupe d'espèces Grupo de especies	1995 mt	1996 mt	1997 mt	1998 mt	1999 mt	2000 mt	2001 mt
	33	72 259	92 432	62 763	54 050	54 531	47 878	59 881
	34	102 446	107 235	111 003	88 790	97 110	86 495	79 323
	35	75 577	71 607	84 451	73 503	68 758	74 750	87 058
	36	6 694	10 063	15 888	11 654	4 710	11 220	12 162
	37	148	388	206	1 593	251	302	537
	38	5 406	8 700	6 496	5 142	9 340	8 628	8 313
	39	64 914	106 820	106 601	97 853	98 191	31 003	74 345
	42	63 545	65 038	73 923	131 368	102 383	34 839	33 594
	43	...	...	...	...	25	27	18
	44	6 656	9 526	8 177	10 941	7 675	6 848	7 281
	45	22 148	24 422	24 331	11 471	18 086	20 381	22 952
	47	...	...	...	...	...	49	63
	52	4 257	6 369	...	...	...	-	4
	53	...	4 989	6 187	11	6	14	20
	54	1 317	455	...	...	1	675	448
	55	1 950	2 372	2	3 228	2 642	2 012	1 052
	56	14 999	7 111	5 415	8 262	6 777	6 517	6 658
	57	333	685	2 787	2 293	3 132	748	2 997
	58	2 155	1 873	4 593	2 879	1 850	1 904	1 649
	76	9 369	7 576	9 044	8 081	7 329	7 115	6 678
	Area total	3 031 197	2 893 342	2 846 092	2 792 459	2 551 689	2 477 803	2 759 090
71 Pacific, Western Central	24	11 390	12 415	8 996	10 682	11 497	11 755	12 756
Pacifique, centre-ouest	25	44 460	44 224	45 704	47 278	53 971	59 730	62 840
Pacífico, centro-occidental	31	16 353	14 922	24 987	16 791	20 281	16 545	17 501
	32	34	24	21	33	765	29	33
	33	1 027 055	1 026 184	1 076 151	1 030 523	1 159 461	1 083 291	1 112 637
	34	47 741	43 313	47 776	48 100	55 290	53 504	53 033
	35	1 006 676	1 014 952	1 070 572	1 051 737	999 286	1 017 746	1 030 263
	36	1 761 877	1 717 100	1 775 537	2 117 045	2 066 217	2 117 445	2 160 123
	37	1 466 658	1 394 097	1 477 317	1 513 492	1 616 073	1 650 036	1 696 951
	38	150 402	138 974	140 048	143 300	141 931	144 330	158 986
	39	2 144 552	2 206 263	2 122 217	2 120 946	2 261 522	2 305 847	2 354 058
	42	134 464	126 806	127 556	144 422	142 022	148 807	163 208
	43	6 182	6 731	10 099	8 512	6 347	7 001	7 668
	45	416 888	435 421	440 526	414 267	438 590	435 591	457 761
	47	2 963	2 751	8 275	7 359	4 498	5 203	6 943
	52	483	448	183	348	282	241	250
	53	1 655	1 887	1 830	2 118	2 120	2 010	2 203
	54	20 413	20 216	18 017	17 454	6 554	6 462	6 422
	55	6 986	7 276	3 488	6 392	5 403	5 093	6 086
	56	65 252	72 085	57 595	74 531	83 453	84 621	83 064
	57	355 772	343 912	360 034	366 123	364 643	435 288	390 472
	58	59 506	53 322	54 772	59 749	59 051	53 484	60 027
	72	300	417	403	452	448	384	393
	76	7 847	8 372	7 648	6 825	6 099	6 484	6 220
	77	146 224	38 289	86 583	74 429	91 748	89 354	89 923
	Area total	8 902 133	8 730 401	8 966 335	9 282 908	9 597 552	9 740 281	9 939 821
77 Pacific, Eastern Central	23	2 998	2 084	2 747	944	1 996	2 623	1 206
Pacifique, centre-est	24	-	-	0	-	-	-	-
Pacífico, centro-oriental	25	0	0	1	0	0	7	...
	31	8 477	9 229	8 937	4 853	6 470	6 229	4 756
	32	89	138	1 273	806	379	718	146
	33	37 114	43 004	46 049	44 181	57 328	46 293	44 880
	34	10 159	9 816	9 547	7 942	5 446	4 517	4 418
	35	523 437	563 272	621 114	544 447	485 986	705 994	857 221
	36	394 902	372 775	416 717	422 246	393 369	378 100	452 131
	37	36 178	28 341	48 042	51 052	85 841	85 372	34 086
	38	37 787	38 188	35 128	38 346	39 191	43 727	36 932
	39	269 721	213 172	181 727	159 041	114 636	126 141	130 511
	42	12 523	15 131	12 455	8 663	7 818	13 382	12 640
	43	1 959	2 621	2 734	2 506	2 306	3 572	3 396
	44	356	164	-	-	-	1	2
	45	72 906	72 867	73 384	66 791	67 417	58 662	52 429
	47	832	748	714	1 179	1 252	1 307	1 369
	52	5 558	4 309	3 465	2 374	2 108	2 673	3 182
	53	3 510	3 071	4 079	3 383	3 028	3 745	3 317
	54	290	583	2 206	2 230	832	643	90
	55	1 256	17 290	2 320	2 726	1 864	6 287	3 100
	56	8 950	7 270	6 792	5 770	6 913	7 143	5 711
	57	111 316	201 487	213 136	30 406	150 585	202 531	167 068
	58	1 380	1 307	1 857	1 462	1 742	3 816	2 951
	72	0	0	0	3	1	21	...
	76	12 807	12 244	10 401	6 221	8 682	9 328	8 600
	77	1 346	872	351	59	31	248	450
	Area total	1 555 851	1 619 983	1 705 176	1 407 631	1 445 221	1 713 080	1 830 592
81 Pacific, Southwest	22	1 183	524	588	400	416	380	313
Pacifique, sud-ouest	23	2	1	1	3	1	1	1
Pacífico, sudoccidental	31	5 385	4 514	8 005	4 278	3 554	2 939	3 257

C-00

Fish, crustaceans, molluscs, etc — **Capture production by fishing areas and species groups**
Poissons, crustacés, mollusques, etc — **Captures par zones de pêche et groupes d'espèces**
Peces, crustáceos, moluscos, etc — **Capturas por áreas de pesca y grupos de especies**

Fishing area Zone de pêche Area de pesca	Species group Groupe d'espèces Grupo de especies	1995 mt	1996 mt	1997 mt	1998 mt	1999 mt	2000 mt	2001 mt
	32	276 918	251 354	351 951	435 214	389 941	349 651	338 911
	33	37 223	25 463	26 477	23 120	27 367	26 153	25 066
	34	177 518	128 843	160 195	158 749	155 139	153 216	144 355
	35	654	525	829	956	908	1 303	1 431
	36	31 799	37 555	36 171	42 007	29 628	35 374	36 389
	37	63 933	57 373	63 536	62 766	77 222	55 728	57 310
	38	21 012	17 075	24 720	18 773	22 584	20 112	22 391
	39	21 989	49 339	48 209	27 498	10 734	12 611	42 338
	42	10 019	1 094	1 120	1 138	429	385	390
	43	4 846	3 953	6 418	3 980	3 873	3 976	3 763
	45	2 154	1 824	1 949	1 684	2 410	2 647	2 593
	47	523	156	157	92	670	802	537
	52	1 592	1 364	1 515	1 627	1 170	1 265	1 064
	53	9	4	4	2	65	2	2
	54	195	454	1	665	2 978	4 468	2 271
	55	14 318	5 209	19 050	4 684	6 281	2 912	7 015
	56	1 791	1 274	1 120	1 984	1 577	1 920	1 950
	57	139 581	73 299	84 833	71 006	43 625	35 499	57 273
	58	266	342	173	230	1 386	1 311	1 483
	76	817	282	627	845	645	718	864
	Area total	813 727	661 821	837 649	861 701	782 603	713 373	750 967
87 Pacific, Southeast Pacifique, sud-est Pacífico, sudoriental	31	1 790	761	411	436	411	365	439
	32	510 674	752 199	391 763	565 958	507 147	339 571	458 922
	33	82 052	52 472	53 423	63 270	65 065	55 474	50 638
	34	40 094	25 436	25 239	30 009	52 479	79 913	45 377
	35	10 570 516	11 001 839	9 049 576	3 853 560	10 057 766	12 427 660	7 947 490
	36	231 207	243 733	329 620	287 678	454 145	361 524	378 154
	37	5 236 150	4 670 512	4 227 403	2 580 211	2 425 751	1 871 008	3 199 884
	38	12 152	12 975	13 230	21 856	15 433	23 948	18 487
	39	68 806	72 907	97 328	390 422	271 652	323 750	190 872
	42	9 093	6 372	4 629	6 427	18 509	9 352	8 958
	43	250	142	101	746	569	346	134
	44	13 903	17 160	22 898	26 295	23 576	24 292	13 396
	45	30 011	28 867	35 223	32 008	21 636	13 322	14 678
	47	681	880	585	698	687	627	1 510
	52	14 149	13 589	20 005	11 183	16 332	12 220	11 826
	53	5	5	10	6	11	14	207
	54	23 018	19 468	21 070	27 862	24 241	24 391	29 513
	55	4 478	3 672	6 611	27 208	31 856	12 162	6 685
	56	50 300	51 782	43 393	47 813	42 922	40 793	50 667
	57	109 152	36 021	32 095	11 354	82 778	127 619	174 999
	58	557	980	3 447	1 702	3 730	2 513	2 273
	72	14	10	11	12	11	11	12
	74	3 297	4 549	3 174	2 530	2 704	2 290	1 298
	76	54 858	52 025	46 000	44 978	56 981	57 247	49 030
	Area total	17 067 207	17 068 356	14 427 245	8 034 222	14 176 392	15 810 412	12 655 449
88 Pacific, Antarctic Pacifique, Antarctique Pacífico, Antártico	32	-	-	-	9	28	82	61
	33	-	-	-	0	0	0	1
	34	...	-	0	42	297	751	662
	38	-	-	-	5	19	41	7
	76	-	-	-	-	-	-	2
	Area total	...	-	0	56	344	874	733
World total Total mondial Total mundial		***92 301 906***	***93 750 402***	***94 215 594***	***87 593 312***	***93 601 896***	***95 439 820***	***92 356 034***

C-01 (a)

Fish, crustaceans, molluscs, etc — **Capture production by species items** — **Africa - Inland waters**
Poissons, crustacés, mollusques, etc — **Captures par catégories d'espèces** — **Afrique - Eaux continentales**
Peces, crustáceos, moluscos, etc — **Capturas por categorías de especies** — **Africa - Aguas continentales**

English name Nom anglais Nombre inglés	Scientific name Nom scientifique Nombre científico	Species group Groupe d'espèces Grupo de especies	1995 mt	1996 mt	1997 mt	1998 mt	1999 mt	2000 mt	2001 mt
Common carp	*Cyprinus carpio*	11	450	444	261	128	141	140	131
Crucian carp	*Carassius carassius*	11	47	61	191	88	101	110	108
Rhinofishes nei	*Labeo spp*	11	605	3 456	2 362	3 375	3 840	3 672	3 609
Grass carp(=White amur)	*Ctenopharyngodon idellus*	11	10 000	15 343	16 553	4 233	707	12 826	17 918
Silver cyprinid	*Rastrineobola argentea*	11	56 827	49 670	40 315	42 336	48 816	49 618	41 384
Cyprinids nei	*Cyprinidae*	11	14 524	12 952	27 022	28 800	42 791	31 596	34 857
Nile tilapia	*Oreochromis niloticus*	12	183 926	177 072	157 714	188 056	175 846	204 350	205 353
Tilapias nei	*Oreochromis (=Tilapia) spp*	12	152 549	157 482	169 253	171 638	196 812	210 928	218 209
Mouthbrooding cichlids	*Haplochromis spp*	12	16 782	10 472	13 053	13 077	10 731	11 699	9 695
Cichlids nei	*Cichlidae*	12	21 600	21 500	21 500	21 500	40 341	40 500	42 033
African lungfishes	*Protopterus spp*	13	8 931	7 653	12 313	11 091	11 428	8 827	9 911
Dagaas	*Stolothrissa, Limnothrissa spp*	13	95 570	70 277	100 636	87 869	78 245	81 992	75 114
Northern pike	*Esox lucius*	13	-	-	-	2	...	...	...
Bonytongues nei	*Heterotis spp*	13	6 902	4 451	5 716	8 385	9 872	9 514	10 127
Knifefishes	*Notopterus spp*	13	383	10	1	144	33	39	22
Bottlenose fishes nei	*Mormyrus spp*	13	...	...	...	...	...	7 691	7 000
Characins nei	*Characidae*	13	25 852	28 805	26 333	39 895	33 410	28 324	25 605
Bagrid catfish	*Chrysichthys nigrodigitatus*	13	4 000	8 505	5 814	15 538	10 679	7 311	12 413
Black catfishes nei	*Chrysichthys spp*	13	8 609	7 596	6 513	7 100	5 592	7 800	7 397
Naked catfishes	*Bagrus spp*	13	14 437	9 865	12 458	13 386	16 239	11 495	11 574
North African catfish	*Clarias gariepinus*	13	33 855	20 601	24 703	21 956	27 004	35 944	39 347
Mudfish	*Clarias anguillaris*	13	19 608	20 370	22 433	21 618	21 511	31 506	39 535
Torpedo-shaped catfishes nei	*Clarias spp*	13	33 848	19 498	23 706	34 057	27 874	29 491	25 890
Upsidedown catfishes	*Synodontis spp*	13	13 852	9 267	11 560	13 384	12 856	12 571	13 737
Freshwater siluroids nei	*Siluroidei*	13	2 100	620	1 368	1 387	1 500	500	500
Nile perch	*Lates niloticus*	13	367 504	312 017	329 244	339 183	306 282	302 905	282 245
Freshwater perches nei	*Lates spp*	13	2 888	1 205	2 378	3 925	1 523	3 070	1 657
Freshwater gobies nei	*Gobiidae*	13	914	900	850	800	800	700	800
Snakeheads(=Murrels) nei	*Channa spp*	13	368	921	86	2 589	2 038	2 990	2 951
Freshwater fishes nei	*Osteichthyes*	13	824 937	844 023	839 571	839 559	871 886	864 976	860 506
European eel	*Anguilla anguilla*	22	898	637	985	801	959	2 164	2 052
Salmonoids nei	*Salmonoidei*	23	196	11	16	24	29	31	67
Shads nei	*Alosa spp*	24	0	0	0	1	-	-	-
Diadromous clupeoids nei	*Clupeoidei*	24	366	290	253	358	270	284	258
Common sole	*Solea solea*	31	...	...	547	381	763	2 490	2 134
Mullets nei	*Mugilidae*	33	16 180	12 836	16 038	17 356	19 239	18 497	22 738
Seabasses nei	*Dicentrarchus spp*	33	2 706	2 397	2 233	2 738	815	4 075	3 789
Meagre	*Argyrosomus regius*	33	127	168	128	106	92	201	51
Saddled seabream	*Oblada melanura*	33	...	...	33	67	47	158	70
Gilthead seabream	*Sparus aurata*	33	908	474	554	672	557	1 727	1 711
Bonga shad	*Ethmalosa fimbriata*	35	1 963	1 900	1 850	1 800	1 798	1 500	1 700
Silversides(=Sand smelts) nei	*Atherinidae*	37	1 241	1 326	1 229	1 150	1 180	3 182	5 154
River prawns nei	*Macrobrachium spp*	41	188	150	130	120	120	100	110
Freshwater prawns, shrimps nei	*Palaemonidae*	41	3 169	2 286	2 387	2 888	1 393	2 414	2 454
Red swamp crawfish	*Procambarus clarkii*	41	20	13	24	19	21	22	3
Freshwater crustaceans nei	*Crustacea*	41	5 983	5 593	5 882	5 968	6 267	6 604	6 831
Penaeus shrimps nei	*Penaeus spp*	45	3 117	3 000	2 850	2 700	2 574	2 100	2 400
Freshwater molluscs nei	*Mollusca*	51	488	530	610	625	606	921	851
Total			***1 959 418***	***1 846 647***	***1 909 656***	***1 972 873***	***1 995 628***	***2 059 555***	***2 052 001***

C-01 (b)

Fish, crustaceans, molluscs, etc	Capture production by countries or areas	Africa - Inland waters
Poissons, crustacés, mollusques, etc	Captures par pays ou zones	Afrique - Eaux continentales
Peces, crustáceos, moluscos, etc	Capturas por países o áreas	Africa - Aguas continentales

Country or area Pays ou zone País o área	1992 mt	1993 mt	1994 mt	1995 mt	1996 mt	1997 mt	1998 mt	1999 mt	2000 mt	2001 mt
Algeria	0	0	0	0	0	0	2	0	0	0
Angola	7 000	7 000	7 000	6 000	6 000	6 000	6 000	6 000 F	6 000 F	6 000 F
Benin	26 566	32 805	32 707	37 449	34 193	32 871	31 778	31 894	26 400	30 000
Botswana	800 F	600 F	400 F	200 F	81	160	191	157	166	118
Br Ind Oc Tr	0	0	0	0	0	0	0	0	0	0
Burkina Faso	7 500	7 000	8 000	8 000	8 000	8 000	8 335	7 600	8 500	8 500 F
Burundi	24 073	17 000 F	22 000 F	21 101	3 041	20 296	13 426	9 199	17 315	8 964
Cameroon	22 000	23 000	27 000 F	30 000 F	35 000 F	40 000 F	45 000 F	50 000 F	55 000	52 500 F
Cape Verde	0	0	0	0	0	0	0	0	0	0
Cent Afr Rep	13 000	13 250 F	13 500 F	13 750 F	14 000 F	14 250 F	14 500 F	15 000	15 000 F	15 000 F
Chad	80 000	87 300	80 000	90 000	100 000	85 000	84 000	84 000 F	84 000 F	84 000 F
Comoros	0	0	0	0	0	0	0	0	0	0
Congo Dem R	184 040	192 589	152 117	154 751	159 037	158 367	174 087	204 503	204 503 F	204 503 F
Congo Rep	21 049	27 850	24 752	26 811	25 873	18 987	25 455	25 455 F	26 000 F	24 500 F
Côte dIvoire	15 404	13 477	15 604	11 335	11 562	12 032	12 501	10 656	10 502	10 530 F
Djibouti	0	0	0	0	0	0	0	0	0	0
Egypt	180 400	186 700	199 300	228 930	230 660	243 609	250 181	225 300	253 470	295 422
Eq Guinea	370 F	600	700	450	900	850	970	1 101 F	1 076	1 000 F
Eritrea	...	0	0	0	0	0	0	0	0	0
Ethiopia	4 485	4 175	5 285	6 325	8 770	10 370	14 000	15 858	15 681	15 390
Fr South Tr	0	0	0	0	0	0	0	0	0	0
Gabon	2 000 F	3 500 F	4 500 F	7 648 F	9 433	9 441	9 000	10 000	10 417	9 850
Gambia	2 500	2 400	2 400	2 500	2 500	2 500	2 500	2 500 F	2 500 F	2 500 F
Ghana	56 000	52 000	54 700	60 000	73 580	70 000	74 500	74 500	74 500	74 500
Guinea	4 000 F	4 600	3 800	3 100	2 780	3 600	4 000	4 000	4 000	4 000 F
GuineaBissau	200 F	250 F	250 F	250 F	250 F	250 F	200 F	200 F	200 F	200 F
Kenya	155 644	176 435	198 805	187 241	174 692	154 955	165 992	198 653	210 343	156 763
Lesotho	16 F	22 F	22 F	26 F	28 F	30 F	30 F	30	32	24
Liberia	4 000	4 000	4 000	4 000	4 000	4 000	4 000	4 000	4 000	4 000
Libya	0	0	0	0	0	0	0	0	0	0
Madagascar	27 500	30 000	30 000	30 000	30 000	30 000	30 000	30 000	30 000	30 000
Malawi	69 261	67 951	58 579	53 664	63 569	56 340	41 111	45 392	43 000 F	40 619
Mali	68 467	64 300	62 850	132 900	111 910	99 550	98 000	98 536	109 870	100 000 F
Mauritania	5 000 F	5 000 F	5 000 F	5 000 F	5 000 F	5 000 F	5 000 F	5 000 F	5 000 F	5 000 F
Mauritius	0	3	0	0	0	0	0	0	0	0
Morocco	1 794	1 617	1 750	1 500	1 500	2 100	1 703	2 163	1 608	983
Mozambique	3 800	4 689	4 925	5 093	7 510	11 668	8 994	10 243	13 088	8 076
Namibia	1 102	1 200	1 200	1 200	1 200	1 500	1 500	1 500	1 500	1 500
Niger	2 456	2 162	2 516	3 616	4 156	6 328	7 013	11 000	16 250	20 800
Nigeria	93 281	95 627	103 800	117 903	89 521	93 644	139 020	139 393	132 315	154 175
Réunion	0	0	0	0	0	0	0	0	0	0
Rwanda	3 644	3 500 F	3 400 F	3 300 F	2 952	4 428	6 641	6 433	6 726	6 828
St Helena	0	0	0	0	0	-	-	-	-	-
Sao Tome Prn	0	0	0	0	0	0	0	0	0	0
Senegal	24 750	27 650	30 000 F	31 000 F	23 000 F	31 000 F	21 000 F	34 000 F	22 450	20 000 F
Seychelles	0	0	0	0	0	0	0	0	0	0
Sierra Leone	14 000	14 000	15 000	15 000	14 500 F	14 500	14 190	14 480	14 000	14 000
Somalia	250 F	250 F	250 F	250 F	250 F	250 F	250 F	250 F	200 F	200 F
South Africa	832	832	800	800	850	850	900	900 F	900 F	900 F
Sudan	33 000	37 500	40 000	40 000	40 500	42 000	44 000	44 000	48 000 F	53 000
Swaziland	60 F	68 F	65 F	60 F	60 F	65 F	70 F	70 F	70 F	70 F
Tanzania	275 150	294 582	247 614	317 029	262 276	306 750	300 000	260 020	280 000	283 000
Togo	5 500	6 000	5 000	4 998	5 000	5 000	5 000	5 000	5 000	5 000
Tunisia	0	400	243	440	706	1 010	896	808	832	860
Uganda	264 900	219 814	213 129	208 789	195 088	218 026	220 628	226 097	219 356	220 726
Zambia	67 864	65 768	70 057	70 546	66 332	65 923	69 938	67 327	66 671	65 000 F
Zimbabwe	21 601	21 230	20 219	16 463	16 387	18 156	16 371	12 410	13 114	13 000 F
Total	***1 795 259***	***1 820 696***	***1 773 239***	***1 959 418***	***1 846 647***	***1 909 656***	***1 972 873***	***1 995 628***	***2 059 555***	***2 052 001***

C-02 (a)

Fish, crustaceans, molluscs, etc — Capture production by species items — America, North - Inland waters

Poissons, crustacés, mollusques, etc — Captures par catégories d'espèces — Amérique du Nord - Eaux continentales

Peces, crustáceos, moluscos, etc — Capturas por categorías de especies — América del Norte - Aguas continentales

English name Nom anglais Nombre inglés	Scientific name Nom scientifique Nombre científico	Species group Groupe d'espèces Grupo de especies	1995 mt	1996 mt	1997 mt	1998 mt	1999 mt	2000 mt	2001 mt
Buffalofishes nei	*Ictiobus spp*	11	480	793	991	959	697	1 280	1 569
Suckers nei	*Catostomidae*	11	662	542	550	464	465	231	187
Common carp	*Cyprinus carpio*	11	24 745	23 020	18 515	15 798	12 969	16 540	16 307
Goldfish	*Carassius auratus*	11	5	5	10	10	7	9	10
Grass carp(=White amur)	*Ctenopharyngodon idellus*	11	19	10	15	13	11	17	31
Cyprinids nei	*Cyprinidae*	11	1 743	3 735	4 843	3 997	6 822	4 017	3 839
Nile tilapia	*Oreochromis niloticus*	12	2 644	1 415	1 447	1 167	1 366	1 321	710
Blue tilapia	*Oreochromis aureus*	12	8 386	9 672	7 553	5 122	4 564	4 500	4 500
Tilapias nei	*Oreochromis (=Tilapia) spp*	12	78 561	74 473	75 077	65 266	63 077	70 508	62 818
Jaguar guapote	*Cichlasoma managuense*	12	608	566	367	344	351	324	410
Peacock cichlid	*Cichla ocellaris*	12	120	65	35	12	7	4	4
Cichlids nei	*Cichlidae*	12	5	0	21	157	173	192	200
Northern pike	*Esox lucius*	13	2 496	2 406	2 334	2 568	2 615	2 808	2 464
Catfishes nei	*Ictalurus spp*	13	9 974	9 745	11 822	9 524	13 710	11 567	10 801
Burbot	*Lota lota*	13	15	14	9	28	13	20	9
Largemouth black bass	*Micropterus salmoides*	13	1 212	1 014	1 058	684	784	903	693
American yellow perch	*Perca flavescens*	13	2 528	2 151	2 695	2 639	2 421	2 414	2 605
Walleye	*Stizostedion vitreum*	13	7 849	7 736	7 755	8 212	8 784	9 164	6 959
Sauger	*Stizostedion canadense*	13	4	1	...	0	...	...	...
Freshwater fishes nei	*Osteichthyes*	13	29 295	28 101	23 787	27 257	23 082	30 990	24 583
Sturgeons nei	*Acipenseridae*	21	433	526	615	525	420	281	6
American eel	*Anguilla rostrata*	22	610	730	796	739	770	942	423
Pink(=Humpback)salmon	*Oncorhynchus gorbuscha*	23	319	3	31	0	1	-	-
Chum(=Keta=Dog)salmon	*Oncorhynchus keta*	23	518	1 567	738	981	325	59	4
Sockeye(=Red)salmon	*Oncorhynchus nerka*	23	0	115	78	190	56	13	0
Chinook(=Spring=King)salmon	*Oncorhynchus tshawytscha*	23	2 146	2 111	1 621	1 367	1 344	390	236
Coho(=Silver)salmon	*Oncorhynchus kisutch*	23	441	4 021	696	855	282	802	659
Rainbow trout	*Oncorhynchus mykiss*	23	1 510	1 821	239	453	173	377	444
Lake trout(=Char)	*Salvelinus namaycush*	23	1 037	975	1 109	1 054	985	1 129	1 130
Rainbow smelt	*Osmerus mordax*	23	968	711	522	321	328	398	209
Pond smelt	*Hypomesus olidus*	23	5 517	3 981	5 964	6 471	5 691	3 254	4 241
Eulachon	*Thaleichthys pacificus*	23	200	4	27	3	8	-	-
Lake(=Common)whitefish	*Coregonus clupeaformis*	23	14 506	14 393	14 218	14 277	13 683	14 227	14 108
Lake cisco	*Coregonus artedi*	23	1 417	1 145	1 223	1 048	804	836	870
Whitefishes nei	*Coregonus spp*	23	2 246	1 415	1 506	2 178	1 586	1 030	768
Salmonoids nei	*Salmonoidei*	23	139	112	-	2	1	-	-
American shad	*Alosa sapidissima*	24	331	837	545	958	580	508	769
Alewife	*Alosa pseudoharengus*	24	815	990	1 332	1 529	1 550	1 534	21
Hickory shad	*Alosa mediocris*	24	36	88	75	47	62	51	90
American gizzard shad	*Dorosoma cepedianum*	24	906	1 272	1 989	1 291	3 007	977	1 781
Lampreys nei	*Petromyzontidae*	25	13	-	-	0	4	0	-
Striped bass	*Morone saxatilis*	25	-	-	5	-	-	-	-
White perch	*Morone americana*	25	802	995	1 164	966	1 042	1 221	1 226
Atlantic tomcod	*Microgadus tomcod*	32	22	23	24	25	25	25	...
Sheepshead	*Archosargus probatocephalus*	33	470	368	549	573	470	577	506
Silversides(=Sand smelts) nei	*Atherinidae*	37	...	...	...	372	487	392	558
River prawns nei	*Macrobrachium spp*	41	4 221	4 261	3 580	3 244	4 168	3 478	3 115
Euro-American crayfishes nei	*Astacidae, Cambaridae*	41	7 243	5 887	10 597	10 181	5 485	325	4 693
Freshwater crustaceans nei	*Crustacea*	41	289	376	922	805	41	40	94
Freshwater molluscs nei	*Mollusca*	51	198	241	122	185	240	565	2
Frogs	*Rana spp*	71	564	422	2 034	1 201	337	325	31
Diamond back terrapins	*Malaclemys spp*	72	0	-	-	0	-	0	-
River and lake turtles nei	*Testudinata*	72	6	18	17	26	15	53	51
Marine worms	*Polychaeta*	77	59	58	57	58	48	27	...
Aquatic invertebrates nei	*Invertebrata*	77	124	279	240	425	569	204	...
Total			***219 457***	***215 209***	***211 519***	***196 571***	***186 505***	***190 849***	***174 734***

C-02 (b)

Fish, crustaceans, molluscs, etc — Capture production by countries or areas — America, North - Inland waters
Poissons, crustacés, mollusques, etc — Captures par pays ou zones — Amérique du Nord - Eaux continentales
Peces, crustáceos, moluscos, etc — Capturas por países o áreas — América del Norte - Aguas continentales

Country or area Pays ou zone País o área	1992 mt	1993 mt	1994 mt	1995 mt	1996 mt	1997 mt	1998 mt	1999 mt	2000 mt	2001 mt
Anguilla	0	0	0	0	0	0	0	0	0	0
Antigua Barb	0	0	0	0	0	0	0	0	0	0
Aruba	0	0	0	0	0	0	0	-	-	-
Bahamas	0	0	0	0	0	0	0	0	0	0
Barbados	0	0	0	0	0	0	0	0	0	-
Belize	1	1	1	0	0	0	0	0	0	0
Bermuda	0	0	0	0	0	0	0	0	0	0
Br Virgin Is	0	0	0	0	0	0	0	0	0	0
Canada	42 633	36 327	36 333	38 756	38 295	38 798	40 744	40 587	40 667	35 120
Cayman Is	0	0	0	0	0	0	0	0	0	0
Costa Rica	406	710	840	900	1 090	840	1 000 F	1 000 F	1 000 F	1 000 F
Cuba	13 820	10 178	9 823	8 893	10 324	8 525	5 954	4 624	4 600 F	4 600 F
Dominica	0	0	0	0	0	0	0	0	0	0
Dominican Rp	1 024	2 037	3 774	2 106	288	1 067	1 095	598	187	1 158
El Salvador	5 136	4 461	3 819	4 324	2 968	2 808	2 443	2 653	2 831	1 692
Greenland	0	0	0	0	0	0	0	0	0	0
Grenada	0	0	0	0	0	0	0	0	0	0
Guadeloupe	0	0	0	0	0	0	0	0	0	0
Guatemala	3 702	4 228	3 776	4 025	4 000	5 121	6 523	6 976	7 301	7 300 F
Haiti	500 F	600 F	500 F	500 F	500 F	500 F	500 F	500 F	500 F	500 F
Honduras	85	86	92	127	98	126	119	102	61	111
Jamaica	450 F	450 F	450 F	450 F	450 F	450 F	450 F	450 F	450 F	450 F
Martinique	0	3	0	0	0	0	0	0	0	0
Mexico	108 707	111 178	111 125	122 020	122 501	113 552	100 335	91 462	106 817	92 154
Montserrat	0	0	0	0	0	0	0	0	0	0
NethAntilles	0	0	0	0	0	0	0	0	0	0
Nicaragua	348	547	824	538	1 142	1 293	1 256	1 120	1 076	1 051
Panama	80	28	285	130	80	91	23	20	20	20 F
Puerto Rico	0	0	0	0	0	0	0	0	0	0
St Kitts Nev	0	0	0	0	0	0	0	0	0	0
St Lucia	0	0	0	0	0	0	0	0	0	0
St Pier Mq	0	0	0	0	0	0	0	0	0	0
St Vincent	0	0	0	0	2	1	0	0	0	0
Trinidad Tob	0	0	0	0	0	0	0	0	0	0
Turks Caicos	0	0	0	0	0	0	0	0	0	0
USA	48 639	54 377	42 648	36 688	33 471	38 347	36 129	36 413	25 339	29 578
US Virgin Is	0	0	0	0	0	0	0	0	0	-
Total	***225 531***	***225 211***	***214 290***	***219 457***	***215 209***	***211 519***	***196 571***	***186 505***	***190 849***	***174 734***

C-03 (a)

Fish, crustaceans, molluscs, etc — Capture production by species items — America, South - Inland waters
Poissons, crustacés, mollusques, etc — Captures par catégories d'espèces — Amérique du Sud - Eaux continentales
Peces, crustáceos, moluscos, etc — Capturas por categorías de especies — América del Sur - Aguas continentales

English name Nom anglais Nombre inglés	Scientific name Nom scientifique Nombre científico	Species group Groupe d'espèces Grupo de especies	1995 mt	1996 mt	1997 mt	1998 mt	1999 mt	2000 mt	2001 mt
Common carp	*Cyprinus carpio*	11	544	218	1	7	5	1 390	0
Cyprinids nei	*Cyprinidae*	11	119	284	212	258	302	355	360
Tilapias nei	*Oreochromis (=Tilapia) spp*	12	6 234	7 300	6 927	7 444	6 616	7 893	7 900
Velvety cichlids	*Astronotus spp*	12	320	308	177	141	251	188	183
Cichlids nei	*Cichlidae*	12	12 045	8 565	8 356	7 785	7 375	8 145	8 200
Arapaima	*Arapaima gigas*	13	420	457	465	210	338	273	204
Cachama	*Colossoma macropomum*	13	11 379	4 991	3 600	3 014	3 921	5 928	5 904
Pirapatinga	*Piaractus brachypomus*	13	...	397	...	166	648	324	487
Characins nei	*Characidae*	13	116 072	115 490	102 456	101 018	92 195	93 360	88 485
Netted prochilod	*Prochilodus reticulatus*	13	5 088	7 448	2 007	6 117	9 768	10 942	11 220
Prochilods nei	*Prochilodus spp*	13	20 430	18 740	10 927	15 670	25 320	27 313	20 743
Freshwater siluroids nei	*Siluroidei*	13	80 299	79 709	86 749	89 117	90 847	90 947	93 248
Freshwater fishes nei	*Osteichthyes*	13	111 075	95 658	106 251	100 035	105 217	93 951	100 530
Sea trout	*Salmo trutta*	23	-	1	1	1	1	1	1
Rainbow trout	*Oncorhynchus mykiss*	23	685	462	1 207	732	526	565	657
Atlantic sabretooth anchovy	*Lycengraulis grossidens*	24	80	90	100	100	120	120	100
Diadromous clupeoids nei	*Clupeoidei*	24	345	552	355	467	403	287	200
Silversides(=Sand smelts) nei	*Atherinidae*	37	850	850	850	850	850	850	760
River prawns nei	*Macrobrachium spp*	41	1 412	2 677	2 157	1 506	3 235	2 437	2 440
Freshwater crustaceans nei	*Crustacea*	41	0	0	0	0	0	0	0
Frogs	*Rana spp*	71	5	0	0	0	0	7	9
River and lake turtles nei	*Testudinata*	72	0	0	0	0	0	0	0
Total			***367 402***	***344 197***	***332 798***	***334 638***	***347 938***	***345 276***	***341 631***

C-03 (b)

Fish, crustaceans, molluscs, etc — Capture production by countries or areas — America, South - Inland waters
Poissons, crustacés, mollusques, etc — Captures par pays ou zones — Amérique du Sud - Eaux continentales
Peces, crustáceos, moluscos, etc — Capturas por países o áreas — América del Sur - Aguas continentales

Country or area Pays ou zone País o área	1992 mt	1993 mt	1994 mt	1995 mt	1996 mt	1997 mt	1998 mt	1999 mt	2000 mt	2001 mt
Argentina	11 227	11 800	12 785	17 191	19 189	22 735	23 197	27 558	30 418	23 860
Bolivia	4 905	5 518	5 353	5 692	5 988	6 038	6 055	6 052	6 106	5 940
Brazil	182 540 F	186 990 F	191 485 F	193 042	193 309	178 871	174 190	185 471	199 159	200 000 F
Chile	14	7	5	...	...	...	4	...	...	...
Colombia	33 759	30 538	34 983	23 524	23 061	20 610	21 673	28 788	24 854	25 000 F
Ecuador	332	372	300	300	300	400	400	400	400	400
Falkland Is	-	-	-	-	1	1	1	1	1	1
Fr Guiana	0	0	0	0	0	0	0	0	0	0
Guyana	800	800	800	700 F	800	625	625	603	800	800
Paraguay	17 925	19 000 F	20 000 F	21 000 F	22 000	28 000	25 000 F	25 000 F	25 000 F	25 000 F
Peru	32 734	38 290	48 837	50 789	28 890	32 221	35 327	36 223	32 297	35 653
Suriname	561	187	138	140 F	150 F	200 F	200 F	200 F	200 F	200
Uruguay	323	621	966	849	598	2 216	1 931	2 423	2 302	451
Venezuela	20 114	28 251	35 412	54 175	49 911	40 881	46 035	35 219	23 739	24 326
Total	***305 234***	***322 374***	***351 064***	***367 402***	***344 197***	***332 798***	***334 638***	***347 938***	***345 276***	***341 631***

C-04 (a)

Fish, crustaceans, molluscs, etc — Capture production by species items — Asia - Inland waters
Poissons, crustacés, mollusques, etc — Captures par catégories d'espèces — Asie - Eaux continentales
Peces, crustáceos, moluscos, etc — Capturas por categorías de especies — Asia - Aguas continentales

English name Nom anglais Nombre inglés	Scientific name Nom scientifique Nombre científico	Species group Groupe d'espèces Grupo de especies	1995 mt	1996 mt	1997 mt	1998 mt	1999 mt	2000 mt	2001 mt
Freshwater bream	*Abramis brama*	11	21 446	19 479	11 743	10 650	11 815	12 749	13 616
Freshwater breams nei	*Abramis spp*	11	5	3	7	20	9	20	10
Common carp	*Cyprinus carpio*	11	47 523	43 435	41 152	47 741	45 168	33 874	31 920
Tench	*Tinca tinca*	11	0	2	0	0	0	690	778
Crucian carp	*Carassius carassius*	11	8 682	8 273	6 877	6 213	5 686	5 528	5 226
Goldfish	*Carassius auratus*	11	363	172	341	301	301	299	112
Roach	*Rutilus rutilus*	11	307	235	537	502	733	1 084	1 365
Kutum	*Rutilus frisii*	11	8 435	9 210	2 320	6 624	6 905	10 120	7 199
Roaches nei	*Rutilus spp*	11	3 828	3 298	1 943	1 867	1 626	2 365	1 895
Common dace	*Leuciscus leuciscus*	11	223	215	250	300	176	104	91
Mud carp	*Cirrhinus molitorella*	11	5	3	4	7	7	6	6
Grass carp(=White amur)	*Ctenopharyngodon idellus*	11	4 503	4 296	4 631	5 213	4 703	2 621	725
Silver carp	*Hypophthalmichthys molitrix*	11	18 340	18 324	16 958	12 055	14 819	9 415	4 284
Bighead carp	*Hypophthalmichthys nobilis*	11	1 519	1 818	1 637	2 621	3 841	4 454	447
Sichel	*Pelecus cultratus*	11	43	38	21	15	10	10	16
Asp	*Aspius aspius*	11	476	419	320	243	297	418	525
Hoven's carp	*Leptobarbus hoeveni*	11	5 454	6 892	5 836	3 241	4 608	3 149	3 260
Black carp	*Mylopharyngodon piceus*	11	20	15	16	22	27	34	36
Java barb	*Puntius javanicus*	11	18 102	19 601	19 469	20 189	17 939	17 124	17 080
Asian barbs nei	*Puntius spp*	11	34 812	39 004	37 272	56 480	56 774	52 745	54 220
Cyprinids nei	*Cyprinidae*	11	375 911	254 684	225 582	202 992	206 964	177 734	167 014
Mozambique tilapia	*Oreochromis mossambicus*	12	13 293	16 943	17 715	17 865	21 172	19 119	19 550
Nile tilapia	*Oreochromis niloticus*	12	55 746	29 253	28 727	40 173	49 840	40 000	41 740
Blue tilapia	*Oreochromis aureus*	12	69	32	101	88	66	67	60
Tilapias nei	*Oreochromis (=Tilapia) spp*	12	36 563	40 058	48 331	53 529	53 884	62 173	56 209
Mango tilapia	*Sarotherodon galilaeus*	12	316	350	462	391	405	262	230
Cichlids nei	*Cichlidae*	12	64	40	48	58	57	52	40
Northern pike	*Esox lucius*	13	1 141	832	800	528	907	975	1 060
Knifefishes	*Notopterus spp*	13	4 790	3 881	3 455	3 599	3 845	4 007	4 050
Wels(=Som)catfish	*Silurus glanis*	13	2 482	2 148	2 015	1 730	1 741	1 793	1 630
Glass catfishes	*Kryptopterus spp*	13	15 283	17 560	15 938	13 943	15 927	12 404	13 490
North African catfish	*Clarias gariepinus*	13	...	...	...	...	216	576	520
Torpedo-shaped catfishes nei	*Clarias spp*	13	20 660	16 884	13 063	20 103	20 582	27 143	28 626
Pangas catfishes nei	*Pangasius spp*	13	1 000	541	522	917	1 061	1 300	1 358
Freshwater siluroids nei	*Siluroidei*	13	14 636	15 245	12 112	76 933	86 568	128 244	89 620
European perch	*Perca fluviatilis*	13	736	670	480	350	450	500	547
Pike-perch	*Stizostedion lucioperca*	13	12 351	13 806	5 665	6 299	4 637	3 883	3 517
Gudgeons, sleepers nei	*Eleotridae*	13	1 514	1 111	1 290	1 078	1 243	985	970
Freshwater gobies nei	*Gobiidae*	13	3 693	3 770	4 500	4 003	4 145	4 670	4 396
Climbing perch	*Anabas testudineus*	13	6 651	3 905	3 754	4 637	6 340	6 700	6 999
Snakeskin gourami	*Trichogaster pectoralis*	13	25 090	30 793	21 728	22 422	23 776	21 575	21 990
Kissing gourami	*Helostoma temminckii*	13	19 166	12 614	18 376	16 598	23 320	17 927	18 320
Snakehead	*Channa argus*	13	-	169	120	...	28	...	...
Striped snakehead	*Channa striata*	13	27 828	30 966	28 646	21 520	23 784	26 886	28 098
Indonesian snakehead	*Channa micropeltes*	13	9 021	11 615	10 117	8 253	8 787	7 771	7 060
Snakeheads(=Murrels) nei	*Channa spp*	13	126 125	127 420	136 830	185 453	86 877	90 258	90 816
Freshwater fishes nei	*Osteichthyes*	13	2 150 149	2 497 275	2 604 488	2 683 590	3 319 358	3 305 678	3 456 645
Sturgeons nei	*Acipenseridae*	21	2 139	2 179	1 733	1 531	1 320	1 288	1 139
European eel	*Anguilla anguilla*	22	390	342	400	300	99	176	122
Japanese eel	*Anguilla japonica*	22	1 023	1 014	916	904	830	765	677
River eels nei	*Anguilla spp*	22	1 054	4 440	773	860	1 326	4 639	4 991
Sea trout	*Salmo trutta*	23	594	395	200	200	263	277	364
Trouts nei	*Salmo spp*	23	1 080	1 222	1 213	1 187	1 285	1 322	1 238
Pink(=Humpback)salmon	*Oncorhynchus gorbuscha*	23	1 511	3 134	947	2 091	927	1 947	376
Chum(=Keta=Dog)salmon	*Oncorhynchus keta*	23	17 736	19 026	18 346	16 069	11 684	12 326	9 599
Masu(=Cherry) salmon	*Oncorhynchus masou*	23	842	830	810	836	849	856	826
Sockeye(=Red)salmon	*Oncorhynchus nerka*	23	79	72	88	118	90	52	22
Rainbow trout	*Oncorhynchus mykiss*	23	622	731	672	618	562	536	484
Ayu sweetfish	*Plecoglossus altivelis*	23	13 702	12 732	12 624	11 386	11 380	11 172	11 148
Whitefishes nei	*Coregonus spp*	23	867	508	572	646	957	927	679
Salmonoids nei	*Salmonoidei*	23	494	26	26	110	222	5	2
Caspian shads	*Caspialosa spp*	24	41 067	57 075	60 442	85 082	95 060	78 001	45 232
Kelee shad	*Hilsa kelee*	24	49 441	48 302	44 519	53 729	44 810	174 399	54 554
Hilsa shad	*Tenualosa ilisha*	24	84 420	90 240	83 230	81 634	73 809	79 165	80 000
Azov sea sprat	*Clupeonella cultriventris*	24	28 624	23 374	21 020	16 767	34 930	33 060	22 689
Diadromous clupeoids nei	*Clupeoidei*	24	490	330	566	817	697	595	623
Milkfish	*Chanos chanos*	25	3 997	369	34	3 000	83	163	201
Bombay-duck	*Harpadon nehereus*	33	448	-	-	-	-	-	488
Sea catfishes nei	*Ariidae*	33	2 170	2 964	2 701	3 207	3 785	4 046	3 868
Flathead grey mullet	*Mugil cephalus*	33	153	181	169	231	315	138	120
Mullets nei	*Mugilidae*	33	25 245	30 036	37 338	31 543	30 666	34 596	36 168
Croakers, drums nei	*Sciaenidae*	33	174	3 271	748	667	686	213	213
Threadfins, tasselfishes nei	*Polynemidae*	33	-	467	416	243	229	66	66
Scats	*Scatophagus spp*	33	169	169	95	20	40	61	55
Pike-congers nei	*Muraenesox spp*	34	849	490	447	270	494	402	625
Clupeoids nei	*Clupeoidei*	35	1 086	1 481	2 497	1 517	1 143	917	917
Silversides(=Sand smelts) nei	*Atherinidae*	37	909	562	1 600	1 500	1 455	1 583	1 685
Giant river prawn	*Macrobrachium rosenbergii*	41	5 837	7 755	6 766	5 433	6 069	5 222	5 536
Freshwater prawns, shrimps nei	*Palaemonidae*	41	11 322	10 790	9 937	7 757	9 757	9 812	10 015
Euro-American crayfishes nei	*Astacidae, Cambaridae*	41	551	850	1 100	1 500	1 372	1 681	1 634
Freshwater crustaceans nei	*Crustacea*	41	271 516	366 840	480 899	603 307	457 390	531 668	588 727

C-04 (a)

Fish, crustaceans, molluscs, etc — **Capture production by species items** — **Asia - Inland waters**
Poissons, crustacés, mollusques, etc — **Captures par catégories d'espèces** — **Asie - Eaux continentales**
Peces, crustáceos, moluscos, etc — **Capturas por categorías de especies** — **Asia - Aguas continentales**

English name Nom anglais Nombre inglés	Scientific name Nom scientifique Nombre científico	Species group Groupe d'espèces Grupo de especies	1995 mt	1996 mt	1997 mt	1998 mt	1999 mt	2000 mt	2001 mt
Indo-Pacific swamp crab	*Scylla serrata*	42	329	363	267	44	47	77	150
Penaeus shrimps nei	*Penaeus spp*	45	18 564	14 645	20 558	8 783	7 830	10 873	11 775
Natantian decapods nei	*Natantia*	45	14 552	19 773	23 738	15 117	15 535	23 068	25 065
Japanese corbicula	*Corbicula japonica*	51	27 597	27 314	22 209	20 609	21 015	19 295	17 295
Freshwater molluscs nei	*Mollusca*	51	565 715	530 432	480 862	554 128	525 591	569 425	602 106
Green mussel	*Perna viridis*	54	136	4	-	-	-	-	-
Clams, etc. nei	*Bivalvia*	56	835	844	956	914	304	303	260
Marine molluscs nei	*Mollusca*	58	325	481	52	165	73	166	250
Frogs	*Rana spp*	71	3 058	2 535	1 550	1 767	1 435	1 957	2 843
River and lake turtles nei	*Testudinata*	72	-	36	48	27	1	3	-
Aquatic invertebrates nei	*Invertebrata*	77	2 643	1 494	2 460	1 480	2 048	2 031	2 283
Total			***4 302 719***	***4 597 415***	***4 702 743***	***5 099 470***	***5 509 887***	***5 738 765***	***5 754 476***

C-04 (b)

Fish, crustaceans, molluscs, etc — **Capture production by countries or areas** — **Asia - Inland waters**
Poissons, crustacés, mollusques, etc — **Captures par pays ou zones** — **Asie - Eaux continentales**
Peces, crustáceos, moluscos, etc — **Capturas por países o áreas** — **Asia - Aguas continentales**

Country or area Pays ou zone País o área	1992 mt	1993 mt	1994 mt	1995 mt	1996 mt	1997 mt	1998 mt	1999 mt	2000 mt	2001 mt
Afghanistan	1 200 F	1 200 F	1 300 F	1 300 F	1 300 F	1 250 F	1 200 F	1 200 F	1 000 F	800 F
Armenia	1 885	1 850	1 033	821	580	580	698	1 144	1 133	866
Azerbaijan	30 339	21 733	18 901	10 545	6 702	5 161	4 760	20 866	18 797	10 893
Bahrain	0	0	0	0	0	0	0	0	-	-
Bangladesh	429 205	452 109	517 746	527 739	535 617	534 285	538 689	649 418	670 465	670 000 F
Bhutan	315 F	320 F	310 F	310 F	300 F	300 F	300 F	300 F	300 F	300 F
Brunei Darsm	25	25	4	7	15	17	35	26	23	16
Cambodia	68 900	67 900	65 000	72 499	63 510	73 000	75 700	231 000	245 600	360 000 F
China	998 028	1 182 390	1 327 785	1 607 385	1 762 860	1 886 967	2 280 244	2 285 364	2 233 230	2 149 932
China,H.Kong	0	0	0	0	0	0	0	0	0	0
China, Macao	0	0	0	0	0	0	0	0	0	0
China,Taiwan	1 566	1 216	1 059	695	407	403	449	561	549	591
Cyprus	...	...	5	65	64	70	70	70	78	70 F
Gaza Strip	...	...	...	0	0	0	0	0	0	0
Georgia	190	549	16	90	6	1	4	17	22	8
India	373 287	575 905	552 874	608 378	633 425	641 775	692 439	696 083	955 620	974 710
Indonesia	300 896	308 648	336 083	329 790	335 696	304 258	288 666	327 627	305 212	306 560
Iran	61 737	75 021	89 157	88 800	109 286	103 795	140 263	143 400	123 500	73 645
Iraq	17 530	17 808	20 926	22 955	19 049	20 519	9 111	9 330	8 378	8 400 F
Israel	2 214	1 813	1 478	1 214	1 845	1 476	2 164	2 145	1 852	1 600 F
Japan	96 739	91 032	92 219	91 455	93 501	85 612	78 822	71 270	70 612	61 354
Jordan	350	350	350	350	350	350	350	350	400	350
Kazakhstan	65 124	56 700	46 433	48 402	44 273	31 826	25 000 F	36 170	36 620	30 654
Korea D P Rp	35 000 F	35 000 F	20 000	20 000	20 000	20 000	20 000 F	20 000 F	20 000 F	20 000 F
Korea Rep	25 483	12 263	10 492	9 646	8 034	6 934	6 845	6 316	7 141	5 971
Kuwait	0	0	0	0	0	0	0	0	0	0
Kyrgyzstan	201	127	131	185	160 F	120 F	80 F	48	52	57
Laos	19 240	19 500 F	23 800	27 370	23 000 F	18 857	19 642	30 041	29 250	30 000 F
Lebanon	20 F	20	20	20	20	20	20	20	20	20
Malaysia	1 773	1 971	2 064	3 939	3 683	3 949	4 626	3 366	3 549	3 446
Maldives	0	0	0	0	0	0	0	0	0	0
Mongolia	120 F	165	184	158	221	180	311	524	425	117
Myanmar	141 281	142 065	146 365	148 347	146 494	149 069	149 279	159 746	189 708	235 376
Nepal	5 895	7 418	7 340	11 230	11 230	11 230	12 000	12 752	16 700	16 700
Oman	0	0	0	0	0	0	0	0	0	0
Pakistan	109 087	109 185	118 703	121 405	142 092	167 530	163 524	179 865	176 468	180 100
Philippines	229 673	210 775	204 325	186 006	177 355	159 353	146 004	146 234	151 753	135 845
Qatar	0	0	0	0	0	0	0	0	0	0
Saudi Arabia	0	0	0	0	0	0	0	0	0	0
Singapore	24	25	23	0	0	0	0	0	0	0
Sri Lanka	17 500	15 000	9 500	15 000	22 250	27 250	29 900	31 450	36 700	29 870
Syria	2 584	2 535	3 570	3 832	3 103	3 557	4 347	5 338	3 991	5 969
Tajikistan	149	253	127	100	40 F	75 F	100 F	80 F	78 F	137
Thailand	135 239	175 140	196 397	186 665	205 903	203 671	200 715	206 540	201 100	209 977
Timor-Leste	...	...	...	...	...	...	...	0	0	0
Turkey	40 495	44 801	45 067	47 976	49 600	50 460	54 500	50 190	42 824	43 323
Turkmenistan	31 777	16 080	15 140	9 740	9 014	8 179	7 014	9 058	12 228	12 749
Untd Arab Em	0	0	0	0	0	0	0	0	0	0
Uzbekistan	9 346	4 358	3 095	3 611	1 494	3 075	2 799	2 871	3 387	4 070
Viet Nam	138 154	146 839	79 587	94 689	164 936	177 589	138 800	169 107	170 000 F	170 000 F
Total	***3 392 571***	***3 800 089***	***3 958 609***	***4 302 719***	***4 597 415***	***4 702 743***	***5 099 470***	***5 509 887***	***5 738 765***	***5 754 476***

English name Nom anglais Nombre inglés	Scientific name Nom scientifique Nombre científico	Species group Groupe d'espèces Grupo de especies	1995 mt	1996 mt	1997 mt	1998 mt	1999 mt	2000 mt	2001 mt
Freshwater bream	*Abramis brama*	11	37 370	37 805	36 097	37 939	31 936	30 758	34 361
Freshwater breams nei	*Abramis spp*	11	2 190	2 261	1 225	2 263	2 422	2 270	2 408
Common carp	*Cyprinus carpio*	11	9 552	12 473	9 905	12 100	12 463	13 515	12 475
Tench	*Tinca tinca*	11	1 504	1 155	1 879	1 211	1 418	1 646	1 694
Bleak	*Alburnus alburnus*	11	153	168	130	240	268	247	546
Barbel	*Barbus barbus*	11	161	186	239	215	176	118	158
Common nase	*Chondrostoma nasus*	11	31	24	26	28	63	67	59
Crucian carp	*Carassius carassius*	11	297	250	622	795	749	765	984
Goldfish	*Carassius auratus*	11	3 194	3 277	1 697	1 932	2 523	2 369	2 460
Roach	*Rutilus rutilus*	11	2 595	3 997	2 739	2 960	2 565	2 659	2 581
Roaches nei	*Rutilus spp*	11	27 387	29 329	23 019	21 996	16 389	18 611	19 286
Rudd	*Scardinius erythrophthalmus*	11	29	36	43	123	145	97	56
Orfe(=Ide)	*Leuciscus idus*	11	2 500	3 216	3 309	3 487	2 805	2 593	2 483
Common dace	*Leuciscus leuciscus*	11	5	5	5	5	4	0	0
Chub	*Leuciscus cephalus*	11	...	...	92	166	167	70	35
Chubs nei	*Leuciscus spp*	11	63	47	39	32	38	33	31
Grass carp(=White amur)	*Ctenopharyngodon idellus*	11	627	557	416	452	460	583	490
Silver carp	*Hypophthalmichthys molitrix*	11	5 626	6 155	5 569	3 014	3 228	1 856	2 579
Bighead carp	*Hypophthalmichthys nobilis*	11	853	538	399	387	410	372	331
Vimba bream	*Vimba vimba*	11	68	56	162	87	133	105	97
Sichel	*Pelecus cultratus*	11	472	414	269	1 660	1 588	565	1 041
Asp	*Aspius aspius*	11	112	161	134	142	156	174	130
Wuchang bream	*Megalobrama amblycephala*	11	0	0	0	-	-	-	-
Cyprinids nei	*Cyprinidae*	11	18 605	14 966	23 621	28 967	26 148	24 357	24 171
Northern pike	*Esox lucius*	13	17 886	19 652	17 913	16 610	15 426	18 666	18 145
Wels(=Som)catfish	*Silurus glanis*	13	5 005	6 226	7 311	9 632	7 525	7 556	6 931
Burbot	*Lota lota*	13	2 534	2 888	2 885	2 920	2 599	2 667	2 550
European perch	*Perca fluviatilis*	13	17 841	18 228	18 900	16 697	15 398	15 616	15 197
Pike-perch	*Stizostedion lucioperca*	13	6 579	7 344	7 124	6 409	6 902	6 926	6 262
Ruffe	*Gymnocephalus cernuus*	13	19	18	10	9	56	99	64
Freshwater gobies nei	*Gobiidae*	13	...	...	...	60	16	...	...
Freshwater fishes nei	*Osteichthyes*	13	71 587	63 685	54 162	50 056	47 497	63 277	51 745
Danube sturgeon(=Osetr)	*Acipenser gueldenstaedtii*	21	4	3	6	11	14	22	18
Sterlet sturgeon	*Acipenser ruthenus*	21	37	36	15	10	37	15	12
Starry sturgeon	*Acipenser stellatus*	21	0	0	5	6	17	23	21
Beluga	*Huso huso*	21	21	25	34	39	35	50	28
Sturgeons nei	*Acipenseridae*	21	2 285	1 184	1 271	1 156	753	594	554
European eel	*Anguilla anguilla*	22	3 568	3 827	3 194	3 133	3 092	3 101	2 554
Atlantic salmon	*Salmo salar*	23	2 234	1 957	1 564	1 367	1 686	1 918	2 025
Sea trout	*Salmo trutta*	23	3 838	3 796	3 746	4 162	3 078	2 629	2 971
Trouts nei	*Salmo spp*	23	1 372	1 687	1 668	1 245	1 239	1 052	1 180
Pink(=Humpback)salmon	*Oncorhynchus gorbuscha*	23	8 862	8 804	18 597	14 057	28 582	24 277	17 961
Chum(=Keta=Dog)salmon	*Oncorhynchus keta*	23	10 049	6 670	6 999	7 309	8 137	14 546	14 127
Masu(=Cherry) salmon	*Oncorhynchus masou*	23	13	41	4	3	7	3	4
Sockeye(=Red)salmon	*Oncorhynchus nerka*	23	8 229	9 261	3 479	3 947	7 097	6 872	12 106
Chinook(=Spring=King)salmon	*Oncorhynchus tshawytscha*	23	738	401	445	340	483	264	163
Coho(=Silver)salmon	*Oncorhynchus kisutch*	23	1 213	1 577	898	1 449	1 054	1 419	1 090
Rainbow trout	*Oncorhynchus mykiss*	23	1 274	1 270	1 111	3 987	4 314	2 049	2 034
Brook trout	*Salvelinus fontinalis*	23	4	3	4	4	4	5	7
Arctic char	*Salvelinus alpinus*	23	70	53	52	49	43	47	35
Chars nei	*Salvelinus spp*	23	88	71	67	74	85	102	101
Huchen	*Hucho hucho*	23	1	1	1	1	1	1	1
Grayling	*Thymallus thymallus*	23	53	52	43	38	41	37	39
European smelt	*Osmerus eperlanus*	23	4 881	2 287	2 530	2 427	2 675	2 766	2 219
Rainbow smelt	*Osmerus mordax*	23	1 006	640	1 113	1 758	1 340	608	719
Smelts nei	*Osmerus spp, Hypomesus spp*	23	3 118	1 927	4 095	3 102	2 740	3 825	3 568
Vendace	*Coregonus albula*	23	4 943	6 810	6 827	5 627	5 515	5 380	5 193
European whitefish	*Coregonus lavaretus*	23	4 247	3 260	3 321	2 786	2 789	2 670	2 663
Houting	*Coregonus oxyrinchus*	23	25	63	30	61	35	31	9
Whitefishes nei	*Coregonus spp*	23	7 071	10 841	11 014	10 354	10 429	10 597	10 725
Salmonoids nei	*Salmonoidei*	23	3 998	2 780	2 224	2 105	2 023	1 779	2 038
Pontic shad	*Alosa pontica*	24	586	794	901	678	69	213	287
Shads nei	*Alosa spp*	24	28	33	64	35	22	24	22
Allis and twaite shads	*Alosa alosa, A.fallax*	24	-	-	-	-	-	0	0
Caspian shads	*Caspialosa spp*	24	1 478	1 882	2 304	3 284	4 654	1 273	176
Azov sea sprat	*Clupeonella cultriventris*	24	80 986	94 035	83 609	116 294	152 068	119 256	46 940
River lamprey	*Lampetra fluviatilis*	25	96	141	87	95	129	163	118
Lampreys nei	*Petromyzontidae*	25	40	76	31	37	67	24	21
Three-spined stickleback	*Gasterosteus aculeatus*	25	423	352	349	910	1 047	1 167	965
Navaga(=Wachna cod)	*Eleginus navaga*	32	193	98	163	177	22	11	55
Flathead grey mullet	*Mugil cephalus*	33	-	27	54	67	58	51	46
So-iuy mullet	*Mugil soiuy*	33	-	-	-	13	190	169	382
Mullets nei	*Mugilidae*	33	31	25	55	139	203	140	61
Big-scale sand smelt	*Atherina boyeri*	37	-	8	110	350	361	350	326
Noble crayfish	*Astacus astacus*	41	...	...	10	10	10	...	...
Signal crayfish	*Pacifastacus leniusculus*	41	-	-	-	-	10	81	80
White-clawed crayfish	*Austropotamobius pallipes*	41	0	0	0	0	0	0	0
Red swamp crawfish	*Procambarus clarkii*	41	2 753	2 500	2 500	2 500	2 500	2 500	2 500
Euro-American crayfishes nei	*Astacidae, Cambaridae*	41	197	234	433	228	227	208	180
Freshwater crustaceans nei	*Crustacea*	41	24	40	46	45	53	969	1 747
Frogs	*Rana spp*	71	...	...	38	41	35	26	-

C-05 (a)

Fish, crustaceans, molluscs, etc — Capture production by species items — Europe - Inland waters
Poissons, crustacés, mollusques, etc — Captures par catégories d'espèces — Europe - Eaux continentales
Peces, crustáceos, moluscos, etc — Capturas por categorías de especies — Europa - Aguas continentales

English name Nom anglais Nombre inglés	Scientific name Nom scientifique Nombre científico	Species group Groupe d'espèces Grupo de especies	1995 mt	1996 mt	1997 mt	1998 mt	1999 mt	2000 mt	2001 mt
Total			*394 922*	*404 689*	*385 052*	*418 104*	*450 743*	*431 944*	*347 421*

C-05 (b)

Fish, crustaceans, molluscs, etc — Capture production by countries or areas — Europe - Inland waters
Poissons, crustacés, mollusques, etc — Captures par pays ou zones — Europe - Eaux continentales
Peces, crustáceos, moluscos, etc — Capturas por países o áreas — Europa - Aguas continentales

Country or area Pays ou zone País o área	1992 mt	1993 mt	1994 mt	1995 mt	1996 mt	1997 mt	1998 mt	1999 mt	2000 mt	2001 mt
Albania	916	850 F	700 F	252	357	180	823	814	955	1 466
Andorra	0	0	0	0	0	0	0	0	0	0
Austria	479	420	388	404	450	465	451	432	439	362
Belarus	1 507	2 993	786	715	821	499	457	514	553	943
Belgium	511	511	511	511	511	511	511	536	511	511
Bosnia Herzg	2 000 F	2 500 F	2 500 F	2 500 F	2 500 F	2 500 F	2 500 F	2 500 F	2 500 F	2 500 F
Bulgaria	1 611	1 675	995 F	762	1 127	1 881	2 336	2 475	861	1 650
Channel Is	0	0	0	0	0	0	0	0	0	0
Croatia	198	284	340	364	434	408	10 F	10	17	34
Czech Rep	...	3 185	3 955	3 929	3 524	3 321	3 952	4 190	4 654	4 646
Czechoslovak	4 350 F	-	-	-	-	-	-	-	-	-
Denmark	653	337	243	264	196	232	349	206	183	77
Estonia	3 509	2 411	1 909	2 366	2 361	2 439	3 878	3 108	3 190	2 461
Faeroe Is	0	0	0	0	0	0	0	0	0	0
Finland	51 407	51 522	47 895	48 436	47 618	47 618	36 813	36 813	34 840	34 820
France	4 350	4 400	4 450	4 500	4 540	4 540	4 500	2 134 F	2 131 F	2 130 F
Germany	8 770	10 837	10 908	22 987	22 987	22 916	22 916	22 868	22 868	22 818
Greece	2 370	2 960	3 452	3 606	2 903	2 601	2 818	3 280	3 433	3 181
Hungary	8 678	7 886	8 307	7 314	7 606	7 406	7 265	7 514	7 101	6 638
Iceland	886	907	698	739	608	404	416	370	176	160
Ireland	2 991	2 929	3 604	3 761	3 806	3 804	3 976	865	961	798
Isle of Man	0	0	0	0	0	0	0	0	0	0
Italy	9 075	9 515	9 921	10 035	6 764	6 690	4 667	5 436	4 565	5 527
Latvia	551	553	495	514	536	544	501	610	612	581
Liechtensten	0	0	0	0	0	0	0	0	0	0
Lithuania	1 522	1 146	1 187	1 260	1 295	1 713	1 737	1 715	1 911	1 854
Luxembourg	0	0	0	0	0	0	0	0	0	0
Macedonia	195	164	196	208	78	130	131	135	208	128
Malta	0	0	0	0	0	0	0	0	0	0
Moldova Rep	410	630	708	709	603	569	491	309	344	387
Netherlands	2 287	1 601	2 446	4 107	2 157	2 293	1 547	2 303	2 280 F	2 200 F
Norway	580	435	432	413	338	439	507	514	578	570 F
Poland	20 950	31 391	27 500	24 889	22 037	13 832	13 236	13 875	17 543	17 789
Portugal	3	3	4	2	0	0	0	0	0	1
Romania	9 890	8 562	10 598	9 048	6 145	4 574	4 630	5 336	4 896	5 206
Russian Fed	275 125	216 866	217 950	212 874	233 272	227 091	271 311	307 823	292 368	206 430
Slovakia	...	1 185	1 627	1 950	1 414	1 364	1 361	1 396	1 368	1 531
Slovenia	293	297	339	316	289	302	269	243	226	206
Spain	8 796	9 215	6 284	8 869	8 710 F	8 710 F	8 710 F	8 710 F	8 710 F	8 710 F
Sweden	2 308	2 273	2 254	1 934	1 810	2 011	1 559	1 478	1 459	1 234
Switzerland	2 715	1 822	1 481	1 588	1 841	1 859	1 809	1 840	1 659	1 715
Ukraine	24 801	13 189	14 786	6 847	9 468	6 215	4 898	4 728	4 429	4 343
UK	2 035	1 909	2 191	2 146	1 930	1 491	4 569	4 835	2 743	3 142
Yugoslavia	5 111	3 797	3 912	3 803	3 653	3 500	2 200 F	828	672	672 F
Total	*461 833*	*401 160*	*395 952*	*394 922*	*404 689*	*385 052*	*418 104*	*450 743*	*431 944*	*347 421*

C-06 (a)

Fish, crustaceans, molluscs, etc — **Capture production by species items** — **Oceania - Inland waters**
Poissons, crustacés, mollusques, etc — **Captures par catégories d'espèces** — **Océanie - Eaux continentales**
Peces, crustáceos, moluscos, etc — **Capturas por categorías de especies** — **Oceanía - Aguas continentales**

English name Nom anglais Nombre inglés	Scientific name Nom scientifique Nombre científico	Species group Groupe d'espèces Grupo de especies	1995 mt	1996 mt	1997 mt	1998 mt	1999 mt	2000 mt	2001 mt
Common carp	*Cyprinus carpio*	11	0	0	28	2	3	4	6
Mozambique tilapia	*Oreochromis mossambicus*	12	2 310	2 310	2 310	2 310	2 310	2 310	2 310
Nile tilapia	*Oreochromis niloticus*	12	6	37	40	288	290	280	278
Tilapias nei	*Oreochromis (=Tilapia) spp*	12	-	-	-	6	-	-	-
Brown bullhead	*Ameiurus nebulosus*	13	1	1	5	12	14	10	8
Gudgeons, sleepers nei	*Eleotridae*	13	1 850	1 850	1 850	1 850	1 850	1 850	1 850
Freshwater fishes nei	*Osteichthyes*	13	7 452	7 506	7 498	7 296	7 258	7 245	7 145
Short-finned eel	*Anguilla australis*	22	480	413	320	293	296	300	300
River eels nei	*Anguilla spp*	22	654	518	490	993	913	851	944
Diadromous clupeoids nei	*Clupeoidei*	24	480	480	480	480	480	480	480
Barramundi(=Giant seaperch)	*Lates calcarifer*	25	1 253	1 411	1 528	1 624	1 984	1 700	1 650
Sea catfishes nei	*Ariidae*	33	1 850	1 850	1 850	1 850	1 850	1 850	1 850
Giant river prawn	*Macrobrachium rosenbergii*	41	0	0	3	3	3	3	3
River prawns nei	*Macrobrachium spp*	41	5	5	5	5	5	5	5
Australian crayfish	*Euastacus armatus*	41	-	-	-	-	66	24	22
Oceanian crayfishes nei	*Parastacidae*	41	20	6	6	14	6	0	1
Freshwater crustaceans nei	*Crustacea*	41	1 011	327	315	323	332	340	343
Freshwater molluscs nei	*Mollusca*	51	2 569	2 670	3 970	4 500	5 000	5 080	5 300
Total			***19 941***	***19 384***	***20 698***	***21 849***	***22 660***	***22 332***	***22 495***

C-06 (b)

Fish, crustaceans, molluscs, etc — **Capture production by countries or areas** — **Oceania - Inland waters**
Poissons, crustacés, mollusques, etc — **Captures par pays ou zones** — **Océanie - Eaux continentales**
Peces, crustáceos, moluscos, etc — **Capturas por países o áreas** — **Oceanía - Aguas continentales**

Country or area Pays ou zone País o área	1992 mt	1993 mt	1994 mt	1995 mt	1996 mt	1997 mt	1998 mt	1999 mt	2000 mt	2001 mt
Amer Samoa	0	0	0	0	0	0	0	0	0	0
Australia	3 757	3 211	584	1 724	1 766	1 790	1 867	2 328	1 984 F	1 939 F
Christmas Is	0	0	0	0	0	0	0	0	0	0
Cocos Is	0	0	0	0	0	0	0	0	0	0
Cook Is	0	0	0	10	0	0	0	0	0	0
Fiji Islands	4 084	2 907	2 964 F	3 586	3 034	4 325	5 111	5 622	5 700 F	5 921
Fr Polynesia	0	0	0	0	0	53	53	53	53	53
Guam	0	0	0	0	0	-	6	-	-	-
Kiribati	0	0	0	0	0	0	0	0	0	0
Marshall Is	0	0	0	0	0	0	0	0	0	0
Micronesia	5 F	4 F	4 F	5 F	5 F	5 F	5 F	5 F	5 F	5 F
NewCaledonia	0	0	0	0	0	0	0	0	0	0
New Zealand	1 252	1 160	1 100	1 115	1 079	1 025	1 307	1 152	1 089	1 076
Niue	0	0	0	0	0	0	0	0	0	0
Norfolk Is	0	0	0	0	0	0	0	0	0	0
N Marianas	0	0	0	1	0	0	0	0	0	0
Palau	0	0	0	0	0	0	0	0	0	0
Papua N Guin	13 500 F	13 500 F	13 500 F	13 500 F	13 500 F	13 500 F	13 500 F	13 500 F	13 500 F	13 500 F
Pitcairn Is	0	0	0	0	0	0	0	0	0	0
Samoa	0	0	0	0	0	0	0	0	1	1 F
Solomon Is	0	0	0	0	0	0	0	0	0	0
Tokelau	0	0	0	0	0	0	0	0	0	0
Tonga	1	1	0	0	0	0	0	0	0	0
Tuvalu	0	0	0	0	0	0	0	0	0	0
Vanuatu	0	0	0	0	0	0	0	0	0	0
Wallis Fut I	0	0	0	0	0	0	0	0	0	0
Total	***22 599***	***20 783***	***18 152***	***19 941***	***19 384***	***20 698***	***21 849***	***22 660***	***22 332***	***22 495***

C-21 (a)

Fish, crustaceans, molluscs, etc — **Capture production by species items** — **Atlantic, Northwest**
Poissons, crustacés, mollusques, etc — **Captures par catégories d'espèces** — **Atlantique, nord-ouest**
Peces, crustáceos, moluscos, etc — **Capturas por categorías de especies** — **Atlántico, noroeste**

English name Nom anglais Nombre inglés	Scientific name Nom scientifique Nombre científico	Species group Groupe d'espèces Grupo de especies	1995 mt	1996 mt	1997 mt	1998 mt	1999 mt	2000 mt	2001 mt
Sturgeons nei	*Acipenseridae*	21	157	86	91	69	9	3	4
American eel	*Anguilla rostrata*	22	631	551	455	496	353	395	220
Atlantic salmon	*Salmo salar*	23	171	165	121	8	2	25	43
Trouts nei	*Salmo spp*	23	20	4	1	0	0	0	-
Arctic char	*Salvelinus alpinus*	23	55	43	78	76	24	29	20
Chars nei	*Salvelinus spp*	23	32	11	31	35	46	46	34
Rainbow smelt	*Osmerus mordax*	23	1 132	1 078	1 227	1 255	356	970	308
American shad	*Alosa sapidissima*	24	564	674	702	748	562	379	526
Alewife	*Alosa pseudoharengus*	24	6 595	5 741	6 185	7 533	6 805	5 540	3 481
Striped bass	*Morone saxatilis*	25	1 655	2 144	2 785	3 032	3 000	3 123	2 934
Atlantic halibut	*Hippoglossus hippoglossus*	31	1 020	1 222	1 463	1 435	1 401	1 369	2 331
Greenland halibut	*Reinhardtius hippoglossoides*	31	39 763	44 174	50 867	50 482	67 239	61 961	61 783
Witch flounder	*Glyptocephalus cynoglossus*	31	4 922	5 569	5 152	5 788	5 766	6 443	7 276
Amer. plaice(=Long rough dab)	*Hippoglossoides platessoides*	31	8 458	7 881	8 956	8 789	9 873	10 459	11 119
Yellowtail flounder	*Limanda ferruginea*	31	3 430	3 840	5 266	9 604	13 960	20 971	24 273
Winter flounder	*Pseudopleuronectes americanus*	31	6 806	8 155	8 313	6 904	6 417	8 185	9 187
Windowpane flounder	*Scophthalmus aquosus*	31	759	46	48	521	166	268	177
Summer flounder	*Paralichthys dentatus*	31	6 943	5 425	4 775	5 548	5 166	5 344	5 350
Flatfishes nei	*Pleuronectiformes*	31	5 458	4 160	3 260	3 165	3 580	3 531	925
Blue antimora	*Antimora rostrata*	32	-	16	-	26	24	21	-
Tusk(=Cusk)	*Brosme brosme*	32	2 782	1 873	2 244	1 995	1 295	1 285	1 678
Atlantic cod	*Gadus morhua*	32	30 897	33 129	46 903	52 864	68 666	63 201	60 074
Greenland cod	*Gadus ogac*	32	2 525	2 120	1 728	1 695	1 899	931	1 128
Red hake	*Urophycis chuss*	32	2 254	2 907	2 850	2 803	3 227	3 365	4 008
White hake	*Urophycis tenuis*	32	10 810	8 303	6 830	5 959	6 377	8 157	8 323
Haddock	*Melanogrammus aeglefinus*	32	8 365	10 921	11 359	14 653	13 726	16 710	21 554
Atlantic tomcod	*Microgadus tomcod*	32	109	117	66	94	17	30	57
Saithe(=Pollock)	*Pollachius virens*	32	13 527	12 825	16 932	20 697	13 159	10 609	11 312
Polar cod	*Boreogadus saida*	32	-	4	-	-	-	-	-
Silver hake	*Merluccius bilinearis*	32	32 237	42 428	33 565	31 892	27 567	24 237	33 000
Offshore silver hake	*Merluccius albidus*	32	...	32	...	5	12	5	2
Roughhead grenadier	*Macrourus berglax*	32	1 377	1 044	4 502	7 140	7 004	8 494	1 742
Roundnose grenadier	*Coryphaenoides rupestris*	32	3 686	3 532	270	159	136	372	5 512
Gadiformes nei	*Gadiformes*	32	48	3	7	25	11	38	3
Ladyfish	*Elops saurus*	33	-	-	1	1	1 963	1	0
Mullets nei	*Mugilidae*	33	298	2 645	553	370	2 323	546	412
Black seabass	*Centropristis striata*	33	904	1 683	1 215	1 190	1 412	1 189	1 276
Groupers, seabasses nei	*Serranidae*	33	...	...	...	23	4 702	18	18
Pigfish	*Orthopristis chrysoptera*	33	1	-	1	1	-	0	0
Spotted weakfish	*Cynoscion nebulosus*	33	206	86	78	79	202	119	36
Squeteague(=Gray weakfish)	*Cynoscion regalis*	33	2 578	2 729	2 697	3 212	2 736	2 329	2 119
Weakfishes nei	*Cynoscion spp*	33	-	-	14	-	11	-	-
Atlantic croaker	*Micropogonias undulatus*	33	6 462	8 250	11 038	10 438	11 451	11 632	12 410
Northern kingfish	*Menticirrhus saxatilis*	33	34	49	21	17	13	28	39
Black drum	*Pogonias cromis*	33	32	39	72	41	88	52	39
Spot croaker	*Leiostomus xanthurus*	33	1 982	1 738	1 975	2 354	1 715	2 159	1 877
Red drum	*Sciaenops ocellatus*	33	2	1	2	3	6	6	3
Sheepshead	*Archosargus probatocephalus*	33	29	137	31	20	87	27	19
Scup	*Stenotomus chrysops*	33	2 845	2 952	2 196	1 893	1 676	1 206	1 845
Tautog	*Tautoga onitis*	33	148	119	116	116	95	111	138
Cunner	*Tautogolabrus adspersus*	33	0	1	1	3	4	3	9
Ocean pout	*Macrozoarces americanus*	33	24	41	15	17	18	19	18
Sandeels(=Sandlances) nei	*Ammodytes spp*	33	1	2	2	1	1	2	0
Spadefishes nei	*Ephippidae*	33	5	6	-	14	21	14	20
Sculpins nei	*Cottidae*	33	1	-	2	5	2	1	1
...A	*Normanichthys crockeri*	33	2	3	-	-	-	-	-
Northern puffer	*Sphoeroides maculatus*	33	19	11	18	17	37	36	35
Triggerfishes, durgons nei	*Balistidae*	33	-	-	6	5	61	3	2
Toadfishes, etc. nei	*Batrachoididae*	33	...	...	1	6	4	21	27
Argentines	*Argentina spp*	34	342	259	1 144	55	17	13	17
American conger	*Conger oceanicus*	34	32	29	17	48	43	49	40
Alfonsinos nei	*Beryx spp*	34	541	141	-	-	-	-	-
Silvery John dory	*Zenopsis conchifer*	34	34	27	6	49	19	28	62
Atlantic goldeneye tilefish	*Caulolatilus chrysops*	34	-	-	-	-	0	2	790
Great Northern tilefish	*Lopholatilus chamaeleonticeps*	34	673	1 349	1 493	1 339	528	515	117
Atlantic wolffish	*Anarhichas lupus*	34	116	7	23	7	-	2	7
Wolffishes(=Catfishes) nei	*Anarhichas spp*	34	2 461	1 674	1 972	1 533	1 985	1 532	1 895
Escolar	*Lepidocybium flavobrunneum*	34	...	...	...	1	1	2	4
Largehead hairtail	*Trichiurus lepturus*	34	...	...	...	0	4	5	1
Beaked redfish	*Sebastes mentella*	34	...	...	...	...	...	671	124
Atlantic redfishes nei	*Sebastes spp*	34	31 500	26 726	23 942	33 641	34 292	47 273	49 594
Blackbelly rosefish	*Helicolenus dactylopterus*	34	-	-	-	-	-	3	0
Atlantic searobins	*Prionotus spp*	34	111	20	10	31	38	24	39
Lumpfish(=Lumpsucker)	*Cyclopterus lumpus*	34	699	662	1 695	2 399	3 479	1 755	3 362
American angler	*Lophius americanus*	34	26 871	25 826	29 555	27 814	26 301	22 066	25 166
Atlantic herring	*Clupea harengus*	35	270 676	293 058	284 509	272 813	281 766	277 302	307 520
Round sardinella	*Sardinella aurita*	35	-	-	-	-	340	-	-
Atlantic menhaden	*Brevoortia tyrannus*	35	338 422	280 496	322 239	248 451	189 185	183 310	236 246
Atlantic thread herring	*Opisthonema oglinum*	35	9	1 686	9	-	1 148	0	-
Atlantic bonito	*Sarda sarda*	36	102	153	134	74	73	44	38
Wahoo	*Acanthocybium solandri*	36	...	...	1	1	2	2	1
King mackerel	*Scomberomorus cavalla*	36	280	1 625	385	341	1 171	315	232

C-21 (a)

Fish, crustaceans, molluscs, etc — Capture production by species items — Atlantic, Northwest
Poissons, crustacés, mollusques, etc — Captures par catégories d'espèces — Atlantique, nord-ouest
Peces, crustáceos, moluscos, etc — Capturas por categorías de especies — Atlántico, noroeste

English name Nom anglais Nombre inglés	Scientific name Nom scientifique Nombre científico	Species group Groupe d'espèces Grupo de especies	1995 mt	1996 mt	1997 mt	1998 mt	1999 mt	2000 mt	2001 mt
Atlantic Spanish mackerel	*Scomberomorus maculatus*	36	96	1 411	364	215	694	368	363
Frigate and bullet tunas	*Auxis thazard, A.rochei*	36	0	-	-	1	17	9	3
Little tunny(=Atl.black skipj)	*Euthynnus alletteratus*	36	136	81	199	120	405	110	121
Skipjack tuna	*Katsuwonus pelamis*	36	53	35	9	24	42	2	4
Atlantic bluefin tuna	*Thunnus thynnus*	36	2 112	2 232	2 107	2 427	2 606	2 522	2 743
Blackfin tuna	*Thunnus atlanticus*	36	2	28	5	1	20	16	2
Albacore	*Thunnus alalunga*	36	623	367	382	367	490	742	1 129
Yellowfin tuna	*Thunnus albacares*	36	1 887	1 319	997	816	883	1 099	1 020
Bigeye tuna	*Thunnus obesus*	36	1 156	910	1 011	1 133	1 866	957	1 038
Tunas nei	*Thunnini*	36	45	115	18	15	29	35	13
Atlantic sailfish	*Istiophorus albicans*	36	-	-	-	6	-	-	-
Atlantic blue marlin	*Makaira nigricans*	36	1	-	1	12	-	2	-
Atlantic white marlin	*Tetrapturus albidus*	36	3	0	0	2	-	-	-
Swordfish	*Xiphias gladius*	36	2 811	2 437	2 044	2 289	2 358	2 168	2 028
Tuna-like fishes nei	*Scombroidei*	36	5	-	-	0	-	3	-
Capelin	*Mallotus villosus*	37	362	32 711	21 987	38 270	23 531	21 374	19 751
Halfbeaks nei	*Hemiramphus spp*	37	-	-	-	-	258	-	-
Opah	*Lampris guttatus*	37	...	...	...	1	1	2	1
Atlantic silverside	*Menidia menidia*	37	236	170	259	255	583	319	661
Bluefish	*Pomatomus saltatrix*	37	3 367	4 137	3 900	3 446	3 145	3 499	3 825
Cobia	*Rachycentron canadum*	37	18	142	6	13	72	16	13
Blue runner	*Caranx crysos*	37	-	-	-	-	49	1	0
Crevalle jack	*Caranx hippos*	37	-	-	-	4	232	2	1
Florida pompano	*Trachinotus carolinus*	37	0	44	1	2	139	0	0
Greater amberjack	*Seriola dumerili*	37	...	...	...	...	368	-	-
Amberjacks nei	*Seriola spp*	37	...	...	...	4	5	8	3
Atlantic mackerel	*Scomber scombrus*	37	26 857	37 128	36 816	33 997	28 372	23 312	36 837
Butterfishes, pomfrets nei	*Stromateidae*	37	2 127	3 611	3 357	1 944	2 116	1 438	4 416
Sand tiger shark	*Carcharias taurus*	38	-	-	-	-	-	1	-
Thresher	*Alopias vulpinus*	38	1	-	-	-	-	8	11
Shortfin mako	*Isurus oxyrinchus*	38	5		...	...	...	29	19
Longfin mako	*Isurus paucus*	38	0	-	-	-	-	1	0
Mako sharks	*Isurus spp*	38	-	-	-	-	-	-	47
Porbeagle	*Lamna nasus*	38	1 395	1 069	1 356	1 026	958	930	502
Great white shark	*Carcharodon carcharias*	38	-	-	-	-	-	1	0
Nurse shark	*Ginglymostoma cirratum*	38	214	-	-	-	-	0	-
Blue shark	*Prionace glauca*	38	...	...	...	...	...	169	-
Sandbar shark	*Carcharhinus plumbeus*	38	1	-	-	-	-	41	24
Blacktip shark	*Carcharhinus limbatus*	38	-	-	-	-	-	21	1
Dusky shark	*Carcharhinus obscurus*	38	0	-	-	-	-	80	0
Tiger shark	*Galeocerdo cuvier*	38	-	-	-	-	-	-	1
Dusky smooth-hound	*Mustelus canis*	38	0	...	...	...	...	334	321
Picked dogfish	*Squalus acanthias*	38	1 085	494	452	1 081	2 456	10 701	5 995
Dogfish sharks nei	*Squalidae*	38	22 624	27 719	20 297	21 991	16 408	2 246	979
Dogfish sharks, etc. nei	*Squaliformes*	38	1 763	3 423	1 252	887	1 673	710	876
Raja rays nei	*Raja spp*	38	19 303	23 176	24 751	26 204	27 463	33 998	30 816
Sharks, rays, skates, etc. nei	*Elasmobranchii*	38	415	250	111	107	228	90	124
Groundfishes nei	*Osteichthyes*	39	1 687	1 760	1 944	1 273	806	1 014	837
Pelagic fishes nei	*Osteichthyes*	39	102	44	128	44	44	29	208
Finfishes nei	*Osteichthyes*	39	15 823	13 086	12 287	4 527	6 613	6 974	3 434
Marine fishes nei	*Osteichthyes*	39	6	6	19	25	29	12	4
Atlantic rock crab	*Cancer irroratus*	42	6 259	4 341	7 530	7 696	7 033	9 516	8 294
Jonah crab	*Cancer borealis*	42	332	334	745	1 255	1 549	1 114	1 245
Blue crab	*Callinectes sapidus*	42	57 740	67 337	62 784	56 834	60 392	42 408	39 511
Green crab	*Carcinus maenas*	42	0	-	54	86	14	16	67
Queen crab	*Chionoecetes opilio*	42	66 372	66 831	74 341	77 517	98 633	104 252	110 044
Red crab	*Geryon quinquedens*	42	521	465	96	-	-	3 130	4 004
Marine crabs nei	*Brachyura*	42	3 121	3 291	5 076	4 194	4 793	5 667	10 326
American lobster	*Homarus americanus*	43	70 631	71 866	78 146	77 155	83 105	83 062	83 803
Northern brown shrimp	*Penaeus aztecus*	45	19	666	3 377	-	679	1 104	458
Northern pink shrimp	*Penaeus duorarum*	45	-	10 753	114	-	5 062	16	1
Northern white shrimp	*Penaeus setiferus*	45	-	1 347	913	-	619	313	28
Northern prawn	*Pandalus borealis*	45	146 325	153 425	140 708	178 056	204 637	230 260	228 661
Aesop shrimp	*Pandalus montagui*	45	...	...	...	...	...	697	609
Pandalus shrimps nei	*Pandalus spp*	45	24 369	24 976	28 601	29 042	30 603	35 180	31 234
Rock shrimp	*Sicyonia brevirostris*	45	-	9 470	8	-	481	-	-
Royal red shrimp	*Pleoticus robustus*	45	-	14	4	13	78	22	24
Natantian decapods nei	*Natantia*	45	-	-	824	148	894	1 732	393
Stomatopods nei	*Stomatopoda*	47	1	-	-	1	5	6	1
Marine crustaceans nei	*Crustacea*	47	-	2	-	-	-	-	-
Periwinkles nei	*Littorina spp*	52	198	219	274	369	372	314	741
Whelk	*Buccinum undatum*	52	...	...	32	45	...	0	-
Whelks	*Busycon spp*	52	2 846	4 294	3 362	2 735	5 349	4 253	5 598
American cupped oyster	*Crassostrea virginica*	53	32 882	39 569	49 099	33 265	47 057	26 440	18 557
Cupped oysters nei	*Crassostrea spp*	53	...	...	...	64	49	35	14
Blue mussel	*Mytilus edulis*	54	15 723	17 585	17 228	15 051	15 651	21 098	18 353
American sea scallop	*Placopecten magellanicus*	55	120 425	108 923	101 874	99 282	131 935	196 878	253 960
Atlantic bay scallop	*Argopecten irradians*	55	88	5	25	6	20	13	13
Iceland scallop	*Chlamys islandica*	55	15 189	13 776	14 270	8 848	3 160	4 397	2 956
Ocean quahog	*Arctica islandica*	56	183 901	173 799	163 799	148 506	144 432	122 547	142 826

C-21 (a)

Fish, crustaceans, molluscs, etc — Capture production by species items — Atlantic, Northwest
Poissons, crustacés, mollusques, etc — Captures par catégories d'espèces — Atlantique, nord-ouest
Peces, crustáceos, moluscos, etc — Capturas por categorías de especies — Atlántico, noroeste

English name Nom anglais Nombre inglés	Scientific name Nom scientifique Nombre científico	Species group Groupe d'espèces Grupo de especies	1995 mt	1996 mt	1997 mt	1998 mt	1999 mt	2000 mt	2001 mt
Northern quahog(=Hard clam)	*Mercenaria mercenaria*	56	16 888	22 964	6 287	1 234	2 536	9 469	6 680
Atlantic surf clam	*Spisula solidissima*	56	156 056	155 584	141 421	131 700	142 370	165 765	166 310
Stimpson's surf clam	*Spisula polynyma*	56	24 017	25 612	27 365	26 008	26 722	22 985	20 273
Atl.jackknife(=Atl.razor clam)	*Ensis directus*	56	-	-	14	49	64	99	36
Sand gaper	*Mya arenaria*	56	8 654	5 824	6 777	8 131	7 793	8 248	9 946
Clams, etc. nei	*Bivalvia*	56	263	773	261	66	1 136	349	1 129
Longfin squid	*Loligo pealei*	57	18 887	12 490	16 161	18 879	18 749	16 942	14 211
Northern shortfin squid	*Illex illecebrosus*	57	16 160	26 362	32 390	25 777	7 928	9 390	4 074
Various squids nei	*Loliginidae, Ommastrephidae*	57	531	56	182	69	90	70	124
Octopuses, etc. nei	*Octopodidae*	57	5	-	0	0	-	0	0
Marine molluscs nei	*Mollusca*	58	2 031	2 394	600	168	1 920	1 469	1 384
Horseshoe crab	*Limulus polyphemus*	75	926	1 598	2 606	3 182	2 238	1 557	1 071
Starfishes nei	*Asteroidea*	76	0	-	-	-	-	0	-
Sea urchins nei	*Strongylocentrotus spp*	76	17 726	13 851	12 201	10 720	10 794	9 131	7 274
Sea cucumbers nei	*Holothurioidea*	76	0	1 288	-	2 406	3 504	4 309	1 504
Jellyfishes	*Rhopilema spp*	77	-	-	-	-	578	-	-
Marine worms	*Polychaeta*	77	0	-	223	299	343	202	481
Total			***2 001 037***	***2 064 442***	***2 050 053***	***1 965 303***	***2 035 191***	***2 080 779***	***2 238 371***

C-21 (b)

Fish, crustaceans, molluscs, etc — Capture production by countries or areas — Atlantic, Northwest
Poissons, crustacés, mollusques, etc — Captures par pays ou zones — Atlantique, nord-ouest
Peces, crustáceos, moluscos, etc — Capturas por países o áreas — Atlántico, noroeste

Country or area Pays ou zone País o área	1992 mt	1993 mt	1994 mt	1995 mt	1996 mt	1997 mt	1998 mt	1999 mt	2000 mt	2001 mt
Canada	954 158	810 017	676 031	593 299	631 357	685 964	747 053	774 095	826 399	830 296
China,Taiwan	0	25	392	206	83	30	8	127	351	618
Cuba	25 972	29 919	12 333	18 327	24 074	17 221	7 736	4 563	46	...
Denmark	3 491	897	245	447	403	421	556	235	-	93
Estonia	33	5 397	1 186	3 242	1 898	3 240	5 533	10 834	13 685	12 402
Faeroe Is	5 069	15 138	10 024	6 884	10 209	8 451	10 094	10 560	8 543	11 000 F
Germany	5 373	343	305	0	495	449	355	569	4 920	1 359
Greenland	101 982	94 937	101 952	108 574	94 356	94 096	97 953	117 841	121 885	124 149
Honduras	-	1 298	-	-	-	-	-	-	-	-
Iceland	...	2 196	2 462	8 232	20 682	7 197	6 572	9 147	9 340	5 079
Japan	14 774	9 064	6 147	5 741	5 616	3 928	4 961	5 693	4 429	5 671
Korea Rep	20 965	3 738	-	-	-	-	-	-	-	-
Latvia	7 472	8 874	473	1 026	1 253	997	1 191	3 080	3 397	3 330
Lithuania	4 657	3 904	900	980	1 585	1 785	3 107	3 370	4 047	7 596
Norway	2 558	9 955	11 755	11 772	7 437	3 709	2 686	4 313	4 296	14 715
Poland	-	-	-	-	-	824	148	894	1 732	760
Portugal	36 243	35 550	30 157	12 532	9 184	8 998	9 616	16 662	13 180	15 003
Russian Fed	37 093	26 525	9 187	10 383	6 021	1 465	2 872	5 979	27 660	32 138
St Pier Mq	16 443	282	294	317	747	3 571	6 108	5 892	6 485	3 802
Spain	55 177	47 139	54 117	20 069	20 864	27 938	30 973	37 239	45 095	40 235
Ukraine	-	-	-	-	-	-	-	-	-	405
UK	169	-	49	-	129	23	294	-	-	-
USA	1 281 350	1 240 243	1 062 863	1 199 006	1 228 049	1 179 746	1 027 487	1 024 098	985 289	1 129 720
Total	***2 572 979***	***2 345 441***	***1 980 872***	***2 001 037***	***2 064 442***	***2 050 053***	***1 965 303***	***2 035 191***	***2 080 779***	***2 238 371***

English name Nom anglais Nombre inglés	Scientific name Nom scientifique Nombre científico	Species group Groupe d'espèces Grupo de especies	1995 mt	1996 mt	1997 mt	1998 mt	1999 mt	2000 mt	2001 mt
Freshwater bream	*Abramis brama*	11	1 541	1 480	1 890	2 376	2 384	3 379	3 194
Freshwater breams nei	*Abramis spp*	11	181	162	275	144	249	124	176
Common carp	*Cyprinus carpio*	11	-	-	-	-	1	-	1
Tench	*Tinca tinca*	11	0	0	-	0	1	-	1
Crucian carp	*Carassius carassius*	11	0	0	1	0	3	3	3
Roach	*Rutilus rutilus*	11	2 987	3 574	3 921	3 235	3 418	3 184	3 407
Roaches nei	*Rutilus spp*	11	-	-	-	-	-	501	653
Orfe(=Ide)	*Leuciscus idus*	11	411	396	351	312	290	350	328
Vimba bream	*Vimba vimba*	11	242	214	228	244	228	174	172
Sichel	*Pelecus cultratus*	11	-	-	-	-	-	384	348
Asp	*Aspius aspius*	11	-	-	-	-	-	6	9
Cyprinids nei	*Cyprinidae*	11	0	8	8	-	-	67	107
Northern pike	*Esox lucius*	13	2 537	2 917	2 934	4 312	4 340	2 840	2 790
Burbot	*Lota lota*	13	478	432	460	489	495	383	385
European perch	*Perca fluviatilis*	13	5 920	6 696	7 715	7 310	7 218	5 220	5 719
Pike-perch	*Stizostedion lucioperca*	13	2 256	2 287	2 408	2 969	3 059	2 171	2 111
Freshwater fishes nei	*Osteichthyes*	13	2 590	2 639	3 588	2 848	1 950	592	461
Sturgeons nei	*Acipenseridae*	21	0	0	0	1	2	-	-
European eel	*Anguilla anguilla*	22	3 195	2 959	4 327	2 290	2 622	2 247	2 098
Atlantic salmon	*Salmo salar*	23	4 489	4 373	3 894	3 369	2 920	2 747	2 597
Sea trout	*Salmo trutta*	23	719	771	730	533	575	543	508
Trouts nei	*Salmo spp*	23	266	270	304	447	466	655	579
Pink(=Humpback)salmon	*Oncorhynchus gorbuscha*	23	-	-	-	-	39	10	184
Rainbow trout	*Oncorhynchus mykiss*	23	256	104	140	237	185	86	139
Chars nei	*Salvelinus spp*	23	16	10	11	20	8	17	11
European smelt	*Osmerus eperlanus*	23	2 493	3 100	2 391	3 325	2 964	7 151	2 235
Rainbow smelt	*Osmerus mordax*	23	-	-	-	-	-	1	6
Vendace	*Coregonus albula*	23	766	865	821	550	533	758	888
European whitefish	*Coregonus lavaretus*	23	2 646	2 482	2 318	2 709	2 524	2 455	2 152
Whitefishes nei	*Coregonus spp*	23	5	6	5	3	4	29	11
Salmonoids nei	*Salmonoidei*	23	40	46	49	6	44	55	14
Allis shad	*Alosa alosa*	24	57	59	56	2	...	38	45
Twaite shad	*Alosa fallax*	24	1	2	1	...	...	12	16
Alewife	*Alosa pseudoharengus*	24	...	...	...	...	...	61	-
Shads nei	*Alosa spp*	24	1	2	3	4	41	6	12
Allis and twaite shads	*Alosa alosa, A.fallax*	24	39	24	17	22	18	23	30
Diadromous clupeoids nei	*Clupeoidei*	24	35	35	36	30	-	-	-
Sea lamprey	*Petromyzon marinus*	25	17	34	20	17	38	36	32
Lampreys nei	*Petromyzontidae*	25	-	-	2	2	2	21	33
Three-spined stickleback	*Gasterosteus aculeatus*	25	1	33	113	82	-	514	593
Lefteye flounders nei	*Bothidae*	31	...	...	...	...	89	117	107
Atlantic halibut	*Hippoglossus hippoglossus*	31	2 692	2 549	2 537	1 904	2 092	2 204	1 999
European plaice	*Pleuronectes platessa*	31	136 501	118 029	122 045	104 322	113 992	113 754	111 968
Greenland halibut	*Reinhardtius hippoglossoides*	31	50 488	52 778	41 790	34 506	47 571	43 114	46 613
Witch flounder	*Glyptocephalus cynoglossus*	31	10 027	9 467	12 673	13 436	13 213	12 362	12 776
Amer. plaice(=Long rough dab)	*Hippoglossoides platessoides*	31	6 752	8 144	8 267	10 434	11 751	6 424	7 367
Common dab	*Limanda limanda*	31	16 141	17 967	17 509	22 134	20 917	16 550	17 172
Lemon sole	*Microstomus kitt*	31	11 062	12 078	12 283	14 378	14 203	13 974	15 461
European flounder	*Platichthys flesus*	31	11 187	10 632	11 371	15 264	11 837	13 905	15 315
Common sole	*Solea solea*	31	44 147	34 460	27 279	31 293	35 705	34 486	32 193
Sand sole	*Solea lascaris*	31	369	246	288	305	222	297	339
Wedge sole	*Dicologlossa cuneata*	31	941	798	648	488	704	795	691
Thickback soles	*Microchirus spp*	31	48	50	57	77	71	71	80
Soles nei	*Soleidae*	31	...	...	...	...	810	1 084	940
Megrim	*Lepidorhombus whiffiagonis*	31	15 708	15 580	14 948	13 866	12 560	13 221	12 658
Megrims nei	*Lepidorhombus spp*	31	6 275	6 405	7 006	7 446	6 969	8 361	8 450
Brill	*Scophthalmus rhombus*	31	2 982	3 203	2 744	2 526	2 511	2 957	3 042
Turbot	*Psetta maxima*	31	7 701	6 511	6 112	5 349	5 466	6 213	6 284
Flatfishes nei	*Pleuronectiformes*	31	15 896	19 468	23 224	20 306	16 876	10 554	9 575
Blue antimora	*Antimora rostrata*	32	-	2	-	-	-	-	-
Moras nei	*Moridae*	32	-	-	415	75	67	62	350
Tusk(=Cusk)	*Brosme brosme*	32	29 410	28 423	22 501	28 991	33 453	31 243	26 846
Atlantic cod	*Gadus morhua*	32	1 239 604	1 307 628	1 328 176	1 158 203	1 025 406	878 068	883 587
Greenland cod	*Gadus ogac*	32	3	1	1	2	4	-	5
Ling	*Molva molva*	32	54 463	54 619	50 988	60 826	53 949	43 320	36 988
Blue ling	*Molva dypterygia*	32	9 742	9 787	9 969	12 767	15 699	16 146	19 347
Greater forkbeard	*Phycis blennoides*	32	1 785	4 386	7 983	7 230	5 957	5 168	5 253
Haddock	*Melanogrammus aeglefinus*	32	309 429	351 049	322 746	269 036	235 725	196 097	207 267
Navaga(=Wachna cod)	*Eleginus navaga*	32	821	561	989	1 008	458	663	1 111
Saithe(=Pollock)	*Pollachius virens*	32	358 182	342 558	302 375	309 096	327 192	302 971	314 438
Pollack	*Pollachius pollachius*	32	14 180	12 988	12 164	10 952	10 379	10 952	11 937
Polar cod	*Boreogadus saida*	32	24 030	20 784	6 826	3 592	22 005	40 730	39 445
Rocklings nei	*Gaidropsarus spp*	32	-	-	-	-	4	24	1
Norway pout	*Trisopterus esmarkii*	32	389 800	275 667	211 394	97 478	112 613	204 845	93 405
Pouting(=Bib)	*Trisopterus luscus*	32	14 634	12 468	12 752	11 117	12 058	15 588	16 533
Blue whiting(=Poutassou)	*Micromesistius poutassou*	32	526 380	613 306	691 246	1 152 677	1 295 472	1 445 788	1 795 372
Whiting	*Merlangius merlangus*	32	89 078	76 550	72 125	61 051	61 472	56 493	48 878
European hake	*Merluccius merluccius*	32	55 247	38 838	40 847	36 404	42 787	46 342	29 084
Roughhead grenadier	*Macrourus berglax*	32	7	36	43	30	1 154	17	170
Roundnose grenadier	*Coryphaenoides rupestris*	32	12 022	10 664	19 636	23 101	18 077	30 542	48 157
Gadiformes nei	*Gadiformes*	32	31 324	32 410	20 651	12 993	15 144	6 711	29 794
Morays	*Muraenidae*	33	...	...	...	...	193	164	142
Mullets nei	*Mugilidae*	33	1 734	1 276	1 385	1 453	1 605	1 433	1 738

C-27 (a)

Fish, crustaceans, molluscs, etc — Capture production by species items — Atlantic, Northeast
Poissons, crustacés, mollusques, etc — Captures par catégories d'espèces — Atlantique, nord-est
Peces, crustáceos, moluscos, etc — Capturas por categorías de especies — Atlántico, nordeste

English name Nom anglais Nombre inglés	Scientific name Nom scientifique Nombre científico	Species group Groupe d'espèces Grupo de especies	1995 mt	1996 mt	1997 mt	1998 mt	1999 mt	2000 mt	2001 mt
Dusky grouper	*Epinephelus marginatus*	33	326	327	210	268	96	24	51
Groupers, seabasses nei	*Serranidae*	33	372	579	705	788	783	876	593
Spotted seabass	*Dicentrarchus punctatus*	33	76	68	42	75	68	71	79
European seabass	*Dicentrarchus labrax*	33	3 665	4 401	4 007	3 752	4 471	5 003	4 944
Seabasses nei	*Dicentrarchus spp*	33	-	-	5	6	336	405	367
Cardinalfishes, etc. nei	*Apogonidae*	33	-	-	-	-	294	-	15
Rubberlip grunt	*Plectorhinchus mediterraneus*	33	...	...	...	...	251	77	12
Grunts, sweetlips nei	*Haemulidae (=Pomadasyidae)*	33	332	504	250	316	144	215	177
Canary drum (=Baardman)	*Umbrina canariensis*	33	1	1	-	-	3	7	11
Meagre	*Argyrosomus regius*	33	40	250	409	460	352	198	168
Croakers, drums nei	*Sciaenidae*	33	...	...	...	...	228	132	168
Blackspot(=red) seabream	*Pagellus bogaraveo*	33	3 347	2 548	4 950	3 292	2 501	1 899	1 273
Common pandora	*Pagellus erythrinus*	33	88	104	152	133	106	154	131
Axillary seabream	*Pagellus acarne*	33	1 067	1 197	974	883	1 028	1 322	1 213
White seabream	*Diplodus sargus*	33	45	54	41	35	55	64	51
Sargo breams nei	*Diplodus spp*	33	...	...	...	...	1 050	977	793
Large-eye dentex	*Dentex macrophthalmus*	33	12	8	26	2	2	0	0
Common dentex	*Dentex dentex*	33	13	11	18	36	24	13	16
Dentex nei	*Dentex spp*	33	31	37	41	26	34	387	183
Black seabream	*Spondyliosoma cantharus*	33	2 219	2 802	2 884	3 120	3 454	3 428	3 253
Red porgy	*Pagrus pagrus*	33	93	131	465	1 038	785	528	344
Gilthead seabream	*Sparus aurata*	33	731	875	666	538	436	489	527
Bogue	*Boops boops*	33	1 133	803	1 024	1 378	3 359	1 761	1 933
Sand steenbras	*Lithognathus mormyrus*	33	78	36	41	30	224	217	145
Salema	*Sarpa salpa*	33	3	...	5	2	340	254	330
Porgies, seabreams nei	*Sparidae*	33	3 602	2 546	3 854	3 966	1 782	2 066	2 064
Picarels nei	*Spicara spp*	33	...	...	...	...	43	22	25
Red mullet	*Mullus surmuletus*	33	3 183	3 478	2 415	4 100	3 080	4 967	4 410
Surmullets(=Red mullets) nei	*Mullus spp*	33	-	-	-	-	-	13	10
Ballan wrasse	*Labrus bergylta*	33	-	-	-	-	2	1	-
Wrasses, hogfishes, etc. nei	*Labridae*	33	30	17	16	20	286	308	320
Parrotfish	*Sparisoma cretense*	33	-	-	-	-	56	87	161
Eelpout	*Zoarces viviparus*	33	152	147	97	54	43	28	37
Sandeels(=Sandlances) nei	*Ammodytes spp*	33	1 133 233	858 376	1 242 671	1 039 589	761 879	740 542	912 330
Greater weever	*Trachinus draco*	33	270	182	183	188	284	278	294
Gobies nei	*Gobiidae*	33	0	0	0	-	0	-	-
Grey triggerfish	*Balistes carolinensis*	33	-	-	1	1	2	-	4
Triggerfishes, durgons nei	*Balistidae*	33	-	-	-	-	42	38	21
Toadfishes, etc. nei	*Batrachoididae*	33	-	-	-	-	70	91	75
Argentines	*Argentina spp*	34	20 095	24 209	26 246	46 386	31 519	28 533	48 894
Baird's slickhead	*Alepocephalus bairdii*	34	1	0	0	0	0	12	616
European conger	*Conger conger*	34	12 378	12 511	11 929	12 584	11 861	11 095	10 666
Longspine snipefish	*Macroramphosus scolopax*	34	0	-	-	-	-	-	-
Alfonsinos nei	*Beryx spp*	34	3	0	12	27	162	139	128
Orange roughy	*Hoplostethus atlanticus*	34	1 794	2 082	1 971	1 911	1 860	1 464	4 243
Slimeheads nei	*Trachichthyidae*	34	-	833	1 052	33	25	3	241
John dory	*Zeus faber*	34	1 310	1 416	1 581	1 889	2 146	2 678	3 148
Wreckfish	*Polyprion americanus*	34	665	587	624	447	447	441	414
Black cardinal fish	*Epigonus telescopus*	34	133	83	181	326	4	248	175
Atlantic wolffish	*Anarhichas lupus*	34	20 279	26 997	30 958	37 617	39 555	39 711	39 006
Spotted wolffish	*Anarhichas minor*	34	700	1 109	1 180	1 599	1 545	1 896	3 257
Wolffishes(=Catfishes) nei	*Anarhichas spp*	34	9 011	7 021	13 074	16 904	6 633	6 710	13 124
Eelpouts	*Lycodes spp*	34	-	-	-	-	-	-	1
Stargazer	*Uranoscopus scaber*	34	...	...	...	...	15	50	46
Oilfish	*Ruvettus pretiosus*	34	-	-	-	-	14	9	9
Largehead hairtail	*Trichiurus lepturus*	34	-	-	-	-	3	0	776
Silver scabbardfish	*Lepidopus caudatus*	34	6 487	2 066	3 504	3 532	1 933	104	361
Black scabbardfish	*Aphanopus carbo*	34	7 296	6 893	5 907	5 238	4 993	8 184	10 826
Hairtails, scabbardfishes nei	*Trichiuridae*	34	-	-	-	-	-	-	13
Golden redfish	*Sebastes marinus*	34	12 073	11 034	11 413	7 968	7 136	13 642	13 598
Beaked redfish	*Sebastes mentella*	34	7 753	4 842	4 457	4 607	3 574	6 074	8 922
Atlantic redfishes nei	*Sebastes spp*	34	273 884	246 443	214 748	213 531	209 915	203 909	190 862
Blackbelly rosefish	*Helicolenus dactylopterus*	34	91	104	115	140	463	559	438
Scorpionfishes nei	*Scorpaenidae*	34	1 663	1 662	2 267	1 621	1 559	1 684	1 222
Tub gurnard	*Chelidonichthys lucerna*	34	-	5	4	5	22	1 184	15
Red gurnard	*Chelidonichthys cuculus*	34	200	214	153	188	319	475	2 424
Grey gurnard	*Eutrigla gurnardus*	34	171	250	161	149	338	604	668
Gurnards, searobins nei	*Triglidae*	34	7 595	8 542	8 466	7 516	10 249	34 699	12 071
Lumpfish(=Lumpsucker)	*Cyclopterus lumpus*	34	9 718	10 678	13 856	4 739	6 344	5 326	6 132
Angler(=Monk)	*Lophius piscatorius*	34	57 350	65 430	64 974	56 174	50 855	50 546	50 508
Atlantic herring	*Clupea harengus*	35	2 082 181	2 035 630	2 249 400	2 148 649	2 129 642	2 103 381	1 645 455
Sardinellas nei	*Sardinella spp*	35	30	-	-	3	1 003	-	1
European pilchard(=Sardine)	*Sardina pilchardus*	35	181 977	157 154	139 119	157 045	162 233	123 760	136 271
European sprat	*Sprattus sprattus*	35	580 180	643 356	671 996	657 895	644 793	616 477	584 292
European anchovy	*Engraulis encrasicolus*	35	37 012	28 980	22 291	34 324	37 175	38 096	45 780
Clupeoids nei	*Clupeoidei*	35	473	10 262	6 556	15 576	20 226	8 592	21 133
Atlantic bonito	*Sarda sarda*	36	77	82	48	97	122	193	89
Frigate and bullet tunas	*Auxis thazard, A.rochei*	36	15	2	-	30	264	511	212
Little tunny(=Atl.black skipj)	*Euthynnus alletteratus*	36	72	218	320	182	15	2	-
Skipjack tuna	*Katsuwonus pelamis*	36	613	6 271	3 596	3 692	1 465	1 166	1 611
Atlantic bluefin tuna	*Thunnus thynnus*	36	7 950	9 263	8 742	6 772	6 664	6 542	4 000
Albacore	*Thunnus alalunga*	36	33 739	22 507	23 098	20 738	26 446	26 575	18 559
Yellowfin tuna	*Thunnus albacares*	36	72	58	26	37	90	66	18
Bigeye tuna	*Thunnus obesus*	36	5 295	2 338	3 028	4 956	2 557	1 854	1 380
Atlantic blue marlin	*Makaira nigricans*	36	3	21	3	5	3	8	...
Atlantic white marlin	*Tetrapturus albidus*	36	13	0	1	-	2	-	1
Swordfish	*Xiphias gladius*	36	6 760	6 534	4 923	3 558	3 099	3 275	2 829

C-27 (a)

Fish, crustaceans, molluscs, etc — Capture production by species items — Atlantic, Northeast
Poissons, crustacés, mollusques, etc — Captures par catégories d'espèces — Atlantique, nord-est
Peces, crustáceos, moluscos, etc — Capturas por categorías de especies — Atlántico, nordeste

English name Nom anglais Nombre inglés	Scientific name Nom scientifique Nombre científico	Species group Groupe d'espèces Grupo de especies	1995 mt	1996 mt	1997 mt	1998 mt	1999 mt	2000 mt	2001 mt
Tuna-like fishes nei	*Scombroidei*	36	417	252	155	263	87	101	61
Capelin	*Mallotus villosus*	37	748 394	1 494 531	1 581 191	943 637	880 444	1 466 964	1 649 687
Garfish	*Belone belone*	37	1 408	1 464	1 792	1 641	1 544	1 222	1 030
Atlantic saury	*Scomberesox saurus*	37	526	483	639	961	574	185	993
Opah	*Lampris guttatus*	37	-	-	-	-	1	...	...
Dealfishes	*Trachipterus spp*	37	...	...	...	...	29	20	25
Silversides(=Sand smelts) nei	*Atherinidae*	37	1 078	333	359	299	297	249	245
Bluefish	*Pomatomus saltatrix*	37	731	1 877	1 914	1 004	808	59	53
Atlantic horse mackerel	*Trachurus trachurus*	37	550 188	464 224	446 230	343 556	315 464	222 870	247 016
Jack and horse mackerels nei	*Trachurus spp*	37	31 863	32 283	39 846	36 894	38 160	36 989	39 824
Greater amberjack	*Seriola dumerili*	37	...	...	...	...	10	21	20
Leerfish	*Lichia amia*	37	5	0	5	2	1	1	10
Carangids nei	*Carangidae*	37	...	...	...	...	43	28	28
Atlantic pomfret	*Brama brama*	37	5 620	5 348	3 425	2 832	2 244	567	435
Common dolphinfish	*Coryphaena hippurus*	37	-	-	-	-	1	1	3
Chub mackerel	*Scomber japonicus*	37	4 725	5 511	6 145	7 321	14 165	10 812	4 602
Atlantic mackerel	*Scomber scombrus*	37	760 048	514 967	514 081	625 968	584 815	659 461	667 366
Mackerels nei	*Scombridae*	37	22 005	26 650	35 088	36 818	19 213	13 610	25 591
Butterfishes, pomfrets nei	*Stromateidae*	37	-	-	-	-	94	35	81
Barracudas nei	*Sphyraena spp*	37	...	...	...	...	33	41	33
Basking shark	*Cetorhinus maximus*	38	109	1 980	1 161	137	81	294	203
Thresher	*Alopias vulpinus*	38	13	7	13	7	34	136	140
Shortfin mako	*Isurus oxyrinchus*	38	...	...	...	0	162	186	188
Porbeagle	*Lamna nasus*	38	730	410	538	1 024	1 486	1 737	1 501
Small-spotted catshark	*Scyliorhinus canicula*	38	5 363	5 144	5 613	5 740	5 818	6 152	7 041
Nursehound	*Scyliorhinus stellaris*	38	243	306	378	258	274	274	264
Catsharks, nursehounds nei	*Scyliorhinus spp*	38	8	18	6	51	13	68	7
Blue shark	*Prionace glauca*	38	266	281	214	165	1 185	453	1 287
Smooth hammerhead	*Sphyrna zygaena*	38	...	...	...	...	8	8	4
Smooth-hound	*Mustelus mustelus*	38	-	-	-	-	-	15	-
Smooth-hounds nei	*Mustelus spp*	38	414	578	624	749	900	1 091	1 292
Tope shark	*Galeorhinus galeus*	38	380	458	511	427	464	567	559
Greenland shark	*Somniosus microcephalus*	38	55	61	73	87	51	45	58
Picked dogfish	*Squalus acanthias*	38	19 281	16 508	14 101	13 634	12 098	12 092	13 228
Gulper shark	*Centrophorus granulosus*	38	...	...	...	...	73	54	93
Leafscale gulper shark	*Centrophorus squamosus*	38	51	53	58	129	451	478	511
Lanternsharks nei	*Etmopterus spp*	38	...	...	...	...	573	...	...
Birdbeak dogfish	*Deania calcea*	38	...	...	...	...	...	18	267
Portuguese dogfish	*Centroscymnus coelolepis*	38	60	336	280	232	717	1 147	2 296
Longnose velvet dogfish	*Centroscymnus crepidater*	38	-	-	-	3	-	-	-
Kitefin shark	*Dalatias licha*	38	...	...	...	...	45	311	189
Black dogfish	*Centroscyllium fabricii*	38	1	4	0	-	-	271	271
Dogfish sharks nei	*Squalidae*	38	5 994	5 869	6 004	5 000	4 338	5 366	5 829
Dogfishes and hounds nei	*Squalidae, Scyliorhinidae*	38	2 117	2 022	2 083	2 011	2 113	3 032	2 700
Angelsharks, sand devils nei	*Squatinidae*	38	2	1	47	0	1	1	1
Angular roughshark	*Oxynotus centrina*	38	...	...	...	...	81	33	63
Blue skate	*Raja batis*	38	530	508	441	411	558	794	817
Thornback ray	*Raja clavata*	38	1 749	1 756	1 579	1 343	1 290	1 222	1 214
Starry ray	*Raja radiata*	38	1 749	1 493	1 431	1 252	996	1 076	1 211
Spotted ray	*Raja montagui*	38	925	977	1 163	1 179	1 260	1 341	1 563
Sandy ray	*Raja circularis*	38	431	438	438	410	435	369	330
Shagreen ray	*Raja fullonica*	38	75	65	55	50	86	65	105
Small-eyed ray	*Raja microocellata*	38	-	-	-	1	11	-	-
Cuckoo ray	*Raja naevus*	38	3 762	4 077	4 721	4 015	3 638	3 064	2 885
Longnosed skate	*Raja oxyrinchus*	38	359	346	311	327	194	140	89
Raja rays nei	*Raja spp*	38	18 853	22 543	28 410	30 083	32 514	28 542	26 496
Stingrays nei	*Dasyatis spp*	38	-	1	2	5	6	10	7
Eagle rays	*Myliobatidae*	38	2	0	0	0	13	10	11
Torpedo rays	*Torpedo spp*	38	20	16	18	19	34	32	43
Rabbit fish	*Chimaera monstrosa*	38	106	21	15	30	12	14	123
Ratfishes nei	*Hydrolagus spp*	38	...	...	...	...	38	573	835
Straightnose rabbitfish	*Rhinochimaera atlantica*	38	-	-	-	-	-	-	2
Chimaeras, etc. nei	*Chimaeriformes*	38	2	1	0	-	-	-	-
Various sharks nei	*Selachimorpha(Pleurotremata)*	38	22 679	7 694	36 309	25 125	26 390	33 412	33 220
Sharks, rays, skates, etc. nei	*Elasmobranchii*	38	400	147	123	211	123	89	141
Groundfishes nei	*Osteichthyes*	39	20 707	18 341	8 276	9 967	6 859	9 243	3 871
Pelagic fishes nei	*Osteichthyes*	39	18 738	25 673	8 247	15 747	4 439	2 950	117
Finfishes nei	*Osteichthyes*	39	43 032	28 375	36 985	50 830	34 056	55 337	30 260
Marine fishes nei	*Osteichthyes*	39	20 080	18 003	20 869	4 701	3 502	4 855	3 179
Edible crab	*Cancer pagurus*	42	33 382	29 185	38 011	40 635	39 341	43 039	44 143
Portunus swimcrabs nei	*Portunus spp*	42	4 290	1 713	3 212	3 021	2 772	1 973	2 966
Green crab	*Carcinus maenas*	42	1 113	837	897	909	1 052	873	1 077
Spinous spider crab	*Maja squinado*	42	6 529	5 674	6 134	5 864	5 812	6 355	7 491
Geryons nei	*Geryon spp*	42	10	1 477	325	587	1 015	763	382
Marine crabs nei	*Brachyura*	42	2 734	3 147	10 334	2 004	1 044	651	382
Pink spiny lobster	*Palinurus mauritanicus*	43	3	1	0	25	11	9	5
Common spiny lobster	*Palinurus elephas*	43	112	104	90	44	58	59	65
Palinurid spiny lobsters nei	*Palinurus spp*	43	171	104	127	110	98	73	81
Norway lobster	*Nephrops norvegicus*	43	57 202	49 620	55 167	52 927	56 558	47 940	52 703
European lobster	*Homarus gammarus*	43	2 386	2 278	2 738	2 654	3 071	2 272	2 504
Lobsters nei	*Reptantia*	43	-	-	-	-	22	11	6
Craylets, squat lobsters	*Galatheidae*	44	92	105	106	81	130	98	84
Red king crab	*Paralithodes camtschaticus*	44	32	70	71	232	341	370	730
Penaeus shrimps nei	*Penaeus spp*	45	277	727	865	1 306	337	100	270

C-27 (a)

Fish, crustaceans, molluscs, etc — Capture production by species items — Atlantic, Northeast
Poissons, crustacés, mollusques, etc — Captures par catégories d'espèces — Atlantique, nord-est
Peces, crustáceos, moluscos, etc — Capturas por categorías de especies — Atlántico, nordeste

English name Nom anglais Nombre inglés	Scientific name Nom scientifique Nombre científico	Species group Groupe d'espèces Grupo de especies	1995 mt	1996 mt	1997 mt	1998 mt	1999 mt	2000 mt	2001 mt
Deepwater rose shrimp	*Parapenaeus longirostris*	45	...	...	...	...	2 084	1 360	1 080
Blue and red shrimp	*Aristeus antennatus*	45	...	...	...	...	194	269	182
Northern prawn	*Pandalus borealis*	45	129 276	134 405	137 822	136 510	128 660	131 654	108 563
Pandalus shrimps nei	*Pandalus spp*	45	46	39	48	1 421	46	460	74
Common prawn	*Palaemon serratus*	45	592	543	576	420	525	520	532
Palaemonid shrimps nei	*Palaemonidae*	45	313	400	373	510	589	4 110	281
Common shrimp	*Crangon crangon*	45	30 531	32 222	37 772	30 583	37 046	33 461	32 146
Natantian decapods nei	*Natantia*	45	1 343	1 239	13 727	5 945	3 854	5 079	1 951
Norwegian krill	*Meganyctiphanes norvegica*	46	-	-	-	88	-	-	36
Goose barnacles	*Lepas spp*	47	0	2	-	-	-	-	-
Marine crustaceans nei	*Crustacea*	47	2 239	1 619	1 783	1 389	1 276	1 116	1 673
Common periwinkle	*Littorina littorea*	52	301	2 836	3 152	2 636	3 014	2 172	2 781
Periwinkles nei	*Littorina spp*	52	2 336	1 770	2 879	1 942	1 344	1 064	765
Tuberculate abalone	*Haliotis tuberculata*	52	49	62	75	36	37	64	62
Whelk	*Buccinum undatum*	52	15 852	24 723	27 627	13 723	17 794	29 818	30 146
Gastropods nei	*Gastropoda*	52	...	...	...	...	142	152	146
European flat oyster	*Ostrea edulis*	53	1 385	1 269	1 749	1 375	872	609	802
Flat oysters nei	*Ostrea spp*	53	2	2	5	-	-	-	5
Pacific cupped oyster	*Crassostrea gigas*	53	100	129	98	59	9	74	92
Blue mussel	*Mytilus edulis*	54	130 482	99 965	118 919	122 217	114 369	126 874	145 558
Horse mussels nei	*Modiolus spp*	54	7	20	30	20	7	2	2
Sea mussels nei	*Mytilidae*	54	2 086	1 077	0	0	0	1 831	1 136
Queen scallop	*Aequipecten opercularis*	55	8 040	7 183	11 443	14 486	15 746	14 891	20 464
Great Atlantic scallop	*Pecten maximus*	55	23 471	31 521	35 657	35 543	36 373	37 692	30 586
Iceland scallop	*Chlamys islandica*	55	8 381	8 976	10 403	10 098	8 858	9 074	6 513
Scallops nei	*Pectinidae*	55	22 273	7 958	14 165	13 172	12 429	13 107	23 798
Ocean quahog	*Arctica islandica*	56	1 980	6 315	4 351	8 776	3 501	1 584	7 434
Striped venus	*Chamelea gallina*	56	665	2 441	2 555	3 987	4 184	5 242	5 556
Pullet carpet shell	*Venerupis pullastra*	56	1 305	2 821	4 053	5 166	4 828	2 361	2 420
Grooved carpet shell	*Ruditapes decussatus*	56	640	981	1 151	1 771	1 756	691	748
Northern quahog(=Hard clam)	*Mercenaria mercenaria*	56	-	-	0	0	-	175	-
Venus clams nei	*Veneridae*	56	-	1	99	-	189	234	289
Solid surf clam	*Spisula solida*	56	...	...	...	...	765	1 335	1 440
Donax clams	*Donax spp*	56	197	...	...	...	490	495	628
Razor clams nei	*Solen spp*	56	2 670	1 871	577	631	696	463	603
Common edible cockle	*Cerastoderma edule*	56	63 007	37 035	38 063	86 611	70 252	47 149	23 659
Clams, etc. nei	*Bivalvia*	56	16 278	7 035	6 237	9 809	7 605	6 193	6 470
Common cuttlefish	*Sepia officinalis*	57	1 456	2 210	1 921	3 079	2 817	3 741	2 559
Cuttlefish,bobtail squids nei	*Sepiidae, Sepiolidae*	57	19 864	20 112	15 805	17 221	18 073	22 392	16 751
Common squids nei	*Loligo spp*	57	4 728	5 091	7 062	6 760	4 724	4 199	3 649
Northern shortfin squid	*Illex illecebrosus*	57	2 370	2 600	2 445	810	1 955	1 831	1 624
Broadtail shortfin squid	*Illex coindetii*	57	571	395	411	216	304	360	219
European flying squid	*Todarodes sagittatus*	57	363	976	2 721	2 436	2 688	2 053	1 795
Various squids nei	*Loliginidae, Ommastrephidae*	57	6 981	4 348	4 868	6 378	8 506	7 524	6 228
Common octopus	*Octopus vulgaris*	57	-	-	-	-	12	624	823
Octopuses, etc. nei	*Octopodidae*	57	17 700	17 639	15 749	13 048	13 988	16 084	10 826
Marine molluscs nei	*Mollusca*	58	1 021	865	1 015	1 294	648	498	424
Echinoderms	*Echinodermata*	76	-	485	586	551	616	306	319
Starfishes nei	*Asteroidea*	76	0	-	-	-	0	0	0
Sea urchins nei	*Strongylocentrotus spp*	76	-	1	0	-	-	149	151
Stony sea urchin	*Paracentrotus lividus*	76	7	0	1	12	14	6	3
European edible sea urchin	*Echinus esculentus*	76	933	425	25	1	13	1	5
Sea cucumbers nei	*Holothurioidea*	76	2	-	-	-	-	-	-
Aquatic invertebrates nei	*Invertebrata*	77	1	0	-	-	-	-	24
Total			***11 022 702***	***11 064 598***	***11 761 184***	***10 984 475***	***10 513 335***	***11 024 004***	***11 164 413***

C-27 (b)

Fish, crustaceans, molluscs, etc — Capture production by countries or areas — Atlantic, Northeast
Poissons, crustacés, mollusques, etc — Captures par pays ou zones — Atlantique, nord-est
Peces, crustáceos, moluscos, etc — Capturas por países o áreas — Atlántico, nordeste

Country or area Pays ou zone País o área	1992 mt	1993 mt	1994 mt	1995 mt	1996 mt	1997 mt	1998 mt	1999 mt	2000 mt	2001 mt
Belgium	36 609	35 578	33 743	35 088	30 312	29 988	30 324	29 340	29 289	29 698
Bulgaria	2 625	5 272	-	-	-	-	-	-	-	-
Channel Is	2 835	2 854 F	2 783	2 949 F	4 346	4 238	4 117	3 601	3 589	3 927
China,Taiwan	0	478	1 562	1 125	31	30	47	640	1 131	564
Denmark	1 949 684	1 613 055	1 872 828	1 998 322	1 680 918	1 826 199	1 556 430	1 404 564	1 533 906	1 510 269
Estonia	45 840	49 703	71 650	92 410	94 586	112 979	96 898	90 315	96 266	88 357
Faeroe Is	243 361	231 156	227 684	280 912	294 378	320 694	353 722	347 573	445 987	513 837
Finland	100 346	104 772	116 374	119 048	131 459	132 567	134 868	123 747	121 661	115 276
France	360 301	379 989	383 736	384 524	353 922	386 921	363 118	394 383	423 035	429 672
Germany	202 747	241 830	218 948	216 864	201 707	207 239	220 252	188 810	177 901	184 432
Greenland	11 285	21 713	15 465	20 316	21 662	26 500	30 589	42 412	37 826	34 336
Iceland	1 573 796	1 712 478	1 553 802	1 603 577	2 038 854	2 197 419	1 674 623	1 726 750	1 973 006	1 975 476
Ireland	246 395	275 845	290 561	380 871	329 224	288 869	320 298	280 092	280 845	302 531
Isle of Man	4 481	4 850	3 571	3 734	3 537	4 289	2 214	2 609	3 552	3 112
Japan	5 293	4 915	7 082	6 595	3 745	2 638	2 907	2 273	2 315	1 906
Korea Rep	1	-	-	-	-	-	-	-	-	-
Latvia	62 446	64 230	64 755	63 127	73 103	86 123	78 109	78 147	80 329	76 930
Lithuania	33 799	18 697	18 086	45 556	48 904	14 824	15 930	20 967	19 584	31 381
Netherlands	430 683	460 155	417 607	433 985	361 126	348 537	410 604	388 371	336 329	339 040
Norway	2 414 894	2 380 768	2 338 555	2 505 560	2 633 495	2 852 115	2 848 489	2 622 707	2 697 358	2 672 018
Panama	382	210	363	289	369	58	58	-	136	...
Poland	104 937	103 022	116 694	130 214	157 335	187 203	126 787	131 056	143 173	159 164
Portugal	224 037	219 683	202 601	216 641	200 335	188 077	192 964	175 033	159 740	157 247
Romania	-	-	8 593	37 508	9 432	-	-	-	-	-
Russian Fed	970 582	833 147	709 669	736 359	769 774	830 132	727 936	839 509	989 122	1 061 969
Spain	341 104	360 270	346 392	395 247	393 622	476 637	470 576	439 111	392 982	432 988
Svalbard Is	0	0	0	0	0	0	0	0	0	0
Sweden	305 237	339 624	384 560	402 638	369 071	355 395	409 327	349 776	337 075	310 582
Ukraine	9 592	7 876	13 385	3 352	550	-	-	-	-	-
UK	810 861	857 874	874 510	905 702	858 730	881 301	913 288	831 549	737 867	729 701
Other nei	144	223	68	189	71	212	-	-	-	-
Total	***10 494 297***	***10 330 267***	***10 295 627***	***11 022 702***	***11 064 598***	***11 761 184***	***10 984 475***	***10 513 335***	***11 024 004***	***11 164 413***

C-31 (a)

Fish, crustaceans, molluscs, etc — Capture production by species items — Atlantic, Western Central
Poissons, crustacés, mollusques, etc — Captures par catégories d'espèces — Atlantique, centre-ouest
Peces, crustáceos, moluscos, etc — Capturas por categorías de especies — Atlántico, centro-occidental

English name Nom anglais Nombre inglés	Scientific name Nom scientifique Nombre científico	Species group Groupe d'espèces Grupo de especies	1995 mt	1996 mt	1997 mt	1998 mt	1999 mt	2000 mt	2001 mt
Freshwater fishes nei	*Osteichthyes*	13	7 705	7 303	3 656	2 870	4 130	6 772	6 937
American eel	*Anguilla rostrata*	22	43	35	19	9	2	1	...
American shad	*Alosa sapidissima*	24	273	343	246	280	117	288	171
Alewife	*Alosa pseudoharengus*	24	0	0	0	0	-	0	0
Milkfish	*Chanos chanos*	25	0	0	0	0	0	0	...
Striped bass	*Morone saxatilis*	25	7	9	12	14	4	12	15
Lefteye flounders nei	*Bothidae*	31	-	-	-	-	-	2	-
Winter flounder	*Pseudopleuronectes americanus*	31	-	-	-	0	-	-	-
Tonguefishes	*Cynoglossidae*	31	-	-	-	-	59	39	20
Summer flounder	*Paralichthys dentatus*	31	...	...	...	800	629	719	1 204
Bastard halibuts nei	*Paralichthys spp*	31	1 926	2 192	1 059	533	391	503	-
Flatfishes nei	*Pleuronectiformes*	31	770	335	487	558	594	502	471
White hake	*Urophycis tenuis*	32	...	...	...	...	1	1	1
Silver hake	*Merluccius bilinearis*	32	0	-	-	-	-	-	-
Gadiformes nei	*Gadiformes*	32	0	-	4	2	1	5	2
Ladyfish	*Elops saurus*	33	30	143	745	979	15	140	480
Tarpon	*Megalops atlanticus*	33	283	167	43	53	16	18	14
Bonefish	*Albula vulpes*	33	143	151	16	34	125	10	11
Sea catfishes nei	*Ariidae*	33	26 630	22 150	14 885	16 385	16 442	20 307	18 260
Squirrelfishes nei	*Holocentridae*	33	85	80	41	67	36	41	67
Flathead grey mullet	*Mugil cephalus*	33	16 999	16 721	13 593	10 583	10 023	10 657	9 846
Lebranche mullet	*Mugil liza*	33	3 249	3 287	1 550	2 876	2 855	3 186	2 537
Bobo mullet	*Joturus pichardi*	33	520	572	283	323	322	451	400
Mullets nei	*Mugilidae*	33	16 891	11 315	16 618	15 846	10 628	14 779	13 363
Common snook	*Centropomus undecimalis*	33	5 908	5 132	4 867	5 012	5 119	5 034	5 903
Snooks(=Robalos) nei	*Centropomus spp*	33	1 181	1 444	1 169	1 081	3 170	3 104	3 345
Black grouper	*Mycteroperca bonaci*	33	...	...	...	...	...	9	7
Gag	*Mycteroperca microlepis*	33	...	...	...	...	...	15	10
Scamp	*Mycteroperca phenax*	33	...	...	...	1	14	27	17
Yellowfin grouper	*Mycteroperca venenosa*	33	...	...	...	...	...	3	1
Nassau grouper	*Epinephelus striatus*	33	464	459	596	554	429	276	331
Yellowedge grouper	*Epinephelus flavolimbatus*	33	...	...	15	18	...	73	36
Red hind	*Epinephelus guttatus*	33	81	60	42	52	72	67	88
Red grouper	*Epinephelus morio*	33	974	305	290	97	140	153	157
Snowy grouper	*Epinephelus niveatus*	33	...	...	...	...	...	6	4
Warsaw grouper	*Epinephelus nigritus*	33	37	16	23	23	-	45	44
Groupers nei	*Epinephelus spp*	33	24 930	20 170	19 352	18 855	16 757	21 589	18 727
Black seabass	*Centropristis striata*	33	393	295	375	341	305	327	391
Groupers, seabasses nei	*Serranidae*	33	350	361	350	437	815	710	859
Southern red snapper	*Lutjanus purpureus*	33	1 597	1 296	1 361	1 163	941	1 646	1 829
Northern red snapper	*Lutjanus campechanus*	33	6 126	6 346	6 264	5 166	5 412	4 841	4 869
Lane snapper	*Lutjanus synagris*	33	3 136	3 034	3 148	3 212	2 353	2 150	2 413
Snappers nei	*Lutjanus spp*	33	377	425	788	827	1 469	1 571	1 780
Yellowtail snapper	*Ocyurus chrysurus*	33	2 176	3 165	2 436	2 828	2 426	2 297	2 514
Vermilion snapper	*Rhomboplites aurorubens*	33	...	...	678	480	731	911	1 057
Snappers, jobfishes nei	*Lutjanidae*	33	11 989	12 317	13 457	15 444	10 017	14 451	9 635
Grunts, sweetlips nei	*Haemulidae (=Pomadasyidae)*	33	14 261	13 881	18 081	15 565	11 338	10 002	12 752
Spotted weakfish	*Cynoscion nebulosus*	33	3 744	3 610	3 797	6 786	6 742	6 367	1 901
Squeteague(=Gray weakfish)	*Cynoscion regalis*	33	517	532	620	609	410	109	154
Weakfishes nei	*Cynoscion spp*	33	20 219	14 011	10 577	13 551	6 394	12 800	10 797
Whitemouth croaker	*Micropogonias furnieri*	33	7 043	6 065	3 694	4 871	1 900	3 262	6 848
Atlantic croaker	*Micropogonias undulatus*	33	551	799	1 396	1 084	729	506	608
Gulf kingcroaker	*Menticirrhus littoralis*	33	968	1 070	1 443	2 061	1 822	961	690
Black drum	*Pogonias cromis*	33	166	134	1 520	2 318	810	2 467	2 489
Spot croaker	*Leiostomus xanthurus*	33	1 539	816	1 098	1 005	884	982	1 214
Croakers, drums nei	*Sciaenidae*	33	0	1	2	183	997	1 128	137
Porgies	*Calamus spp*	33	1 266	1 052	1 541	553	1 266	512	520
Sheepshead	*Archosargus probatocephalus*	33	1 806	1 519	1 712	1 346	1 096	1 501	1 195
Porgies, seabreams nei	*Sparidae*	33	535	267	495	771	205	272	323
Goatfishes, red mullets nei	*Mullidae*	33	380	276	180	170	108	189	213
Mojarras(=Silver-biddies) nei	*Gerres spp*	33	-	-	-	-	-	159	180
Mojarras, etc. nei	*Gerreidae*	33	9 698	9 756	5 911	5 487	4 105	3 735	3 833
Sea chubs nei	*Kyphosus spp*	33	...	...	...	...	...	...	8
Wrasses, hogfishes, etc. nei	*Labridae*	33	821	529	1 456	52	82	116	68
Parrotfishes nei	*Scaridae*	33	156	99	100	118	109	101	307
Surgeonfishes nei	*Acanthuridae*	33	5	11	4	9	7	2	161
Spadefishes nei	*Ephippidae*	33	-	-	-	-	-	15	1
Boxfishes nei	*Ostraciidae*	33	...	...	1	1	55	68	118
Triggerfishes, durgons nei	*Balistidae*	33	1 457	810	551	717	555	319	295
Bearded brotula	*Brotula barbata*	34	...	...	...	5	1	1	0
Cusk-eels, brotulas nei	*Ophidiidae*	34	198	118	97	37	13	10	10
Alfonsinos nei	*Beryx spp*	34	278	22	-	-	-	-	-
Wreckfish	*Polyprion americanus*	34	112	82	14	6	1	-	-
Atlantic goldeneye tilefish	*Caulolatilus chrysops*	34	-	-	-	-	0	4	16
Great Northern tilefish	*Lopholatilus chamaeleonticeps*	34	611	114	397	80	88	173	168
Tilefishes nei	*Branchiostegidae*	34	...	...	28	347	386	528	342
Tripletail	*Lobotes surinamensis*	34	-	-	-	-	-	1	1
Escolar	*Lepidocybium flavobrunneum*	34	...	...	...	51	33	70	40
Oilfish	*Ruvettus pretiosus*	34	...	...	...	10	11	38	30
Largehead hairtail	*Trichiurus lepturus*	34	4 965	4 615	5 060	5 413	4 019	3 752	8 799
Black scabbardfish	*Aphanopus carbo*	34	-	17	-	-	-	-	-
Hairtails, scabbardfishes nei	*Trichiuridae*	34	...	...	...	6 349	6 747	7 278	5 091
Black driftfish	*Hyperoglyphe bythites*	34	...	...	...	...	2	7	5
American angler	*Lophius americanus*	34	-	-	25	12	19	9	12
Demersal percomorphs nei	*Perciformes*	34	4 560	7 448	8 787	7 176	6 185	3 325	7 581

C-31 (a)

Fish, crustaceans, molluscs, etc — Capture production by species items — Atlantic, Western Central
Poissons, crustacés, mollusques, etc — Captures par catégories d'espèces — Atlantique, centre-ouest
Peces, crustáceos, moluscos, etc — Capturas por categorías de especies — Atlántico, centro-occidental

English name Nom anglais Nombre inglés	Scientific name Nom scientifique Nombre científico	Species group Groupe d'espèces Grupo de especies	1995 mt	1996 mt	1997 mt	1998 mt	1999 mt	2000 mt	2001 mt
Round sardinella	*Sardinella aurita*	35	154 871	155 426	143 116	190 895	128 048	75 572	72 164
Brazilian sardinella	*Sardinella brasiliensis*	35	0	1	0	0	0	0	0
Atlantic menhaden	*Brevoortia tyrannus*	35	27 314	24 169	...	27 779	18 815	23 812	25 161
Gulf menhaden	*Brevoortia patronus*	35	472 039	491 612	597 565	497 461	694 242	591 434	528 500
Scaled sardines	*Harengula spp*	35	1 117	819	1 011	706	784	797	791
Atlantic thread herring	*Opisthonema oglinum*	35	7 737	5 634	15 191	14 592	17 066	14 802	6 762
Atlantic anchoveta	*Cetengraulis edentulus*	35	41	8	0	119	0	0	2
Broad-striped anchovy	*Anchoa hepsetus*	35	...	...	1	16	8	0	0
Anchovies, etc. nei	*Engraulidae*	35	1 564	1 464	903	1 762	1 613	1 163	1 328
Clupeoids nei	*Clupeoidei*	35	381	527	1 075	1 666	4 712	4 140	4 362
Atlantic bonito	*Sarda sarda*	36	3 916	3 539	4 205	4 511	4 673	3 835	3 927
Wahoo	*Acanthocybium solandri*	36	1 053	1 251	1 625	1 391	1 438	1 051	1 408
King mackerel	*Scomberomorus cavalla*	36	9 887	9 639	12 768	8 460	9 560	8 966	9 971
Atlantic Spanish mackerel	*Scomberomorus maculatus*	36	9 584	11 183	8 720	8 625	9 050	6 999	6 873
Cero	*Scomberomorus regalis*	36	429	307	481	441	230	190	147
Serra Spanish mackerel	*Scomberomorus brasiliensis*	36	6 541	5 388	5 940	6 407	4 158	5 655	5 598
Seerfishes nei	*Scomberomorus spp*	36	984	1 359	618	382	671	890	688
Frigate and bullet tunas	*Auxis thazard, A.rochei*	36	2 217	3 252	3 182	2 053	1 662	1 548	1 343
Little tunny(=Atl.black skipj)	*Euthynnus alletteratus*	36	1 674	1 879	2 341	3 011	2 504	2 152	2 203
Skipjack tuna	*Katsuwonus pelamis*	36	5 211	4 853	5 576	5 996	4 186	4 556	10 023
Atlantic bluefin tuna	*Thunnus thynnus*	36	43	75	27	490	456	395	117
Blackfin tuna	*Thunnus atlanticus*	36	3 379	2 895	2 461	2 842	2 915	2 451	4 234
Albacore	*Thunnus alalunga*	36	3 360	3 706	2 957	1 401	5 477	9 537	15 579
Yellowfin tuna	*Thunnus albacares*	36	25 850	24 881	26 230	27 554	24 130	24 652	29 732
Bigeye tuna	*Thunnus obesus*	36	729	1 091	1 325	1 756	2 941	4 187	4 168
Tunas nei	*Thunnini*	36	23	6	33	17	25	33	47
Atlantic sailfish	*Istiophorus albicans*	36	575	516	525	623	696	642	654
Atlantic blue marlin	*Makaira nigricans*	36	388	421	493	508	603	517	326
Atlantic white marlin	*Tetrapturus albidus*	36	354	254	171	163	160	116	106
Longbill spearfish	*Tetrapturus pfluegeri*	36	-	1	0	1	0	-	4
Marlins,sailfishes,etc. nei	*Istiophoridae*	36	27	112	98	197	105	490	536
Swordfish	*Xiphias gladius*	36	3 521	1 854	2 751	3 040	3 382	3 750	3 081
Tuna-like fishes nei	*Scombroidei*	36	1 007	15 485	1 373	23 687	989	7 009	5 576
Needlefishes, etc. nei	*Belonidae*	37	159	66	33	67	65	81	86
Ballyhoo halfbeak	*Hemiramphus brasiliensis*	37	...	...	...	...	32	47	41
Halfbeaks nei	*Hemiramphus spp*	37	529	399	295	443	92	415	355
Flyingfishes nei	*Exocoetidae*	37	1 843	2 148	1 623	2 835	2 165	2 053	2 059
Opah	*Lampris guttatus*	37	...	...	...	...	-	-	1
Bluefish	*Pomatomus saltatrix*	37	1 458	758	1 147	899	756	1 112	1 326
Cobia	*Rachycentron canadum*	37	499	392	757	717	630	399	1 489
Scads nei	*Decapterus spp*	37	94	82	61	101	59	38	52
Blue runner	*Caranx crysos*	37	736	313	408	408	184	204	225
Crevalle jack	*Caranx hippos*	37	146	136	264	393	91	316	304
Bar jack	*Caranx ruber*	37	-	-	-	-	3	5	7
Jacks, crevalles nei	*Caranx spp*	37	10 274	12 030	12 833	12 376	9 517	9 917	8 801
Atlantic moonfish	*Selene setapinnis*	37	2 544	1 766	2 110	2 338	1 529	976	3 443
Florida pompano	*Trachinotus carolinus*	37	54	-	296	319	68	242	181
Pompanos nei	*Trachinotus spp*	37	919	1 099	733	1 038	693	552	457
Greater amberjack	*Seriola dumerili*	37	...	...	...	560	149	538	452
Amberjacks nei	*Seriola spp*	37	2 789	2 620	2 820	2 203	2 261	2 801	800
Rainbow runner	*Elagatis bipinnulata*	37	32	14	20	20	36	23	15
Bigeye scad	*Selar crumenophthalmus*	37	2 884	2 453	2 622	2 726	3 873	1 876	1 899
Carangids nei	*Carangidae*	37	458	436	676	586	561	380	299
Common dolphinfish	*Coryphaena hippurus*	37	3 799	3 549	4 267	3 476	3 960	3 740	2 893
Chub mackerel	*Scomber japonicus*	37	379	418	556	771	641	409	527
Mackerels nei	*Scombridae*	37	1 400	1 550	1 700	1 430	1 430	1 600	1 600
North Atlantic harvestfish	*Peprilus alepidotus*	37	25	14	15	13	16	23	18
Gulf butterfishes, etc. nei	*Peprilus spp*	37	1 881	1 889	568	1 175	1 004	809	742
Butterfishes, pomfrets nei	*Stromateidae*	37	789	782	65	633	644	681	539
Barracudas nei	*Sphyraena spp*	37	1 742	1 596	2 130	2 072	1 923	1 312	1 289
Pelagic percomorphs nei	*Perciformes*	37	3 765	857	539	464	219	182	109
Bigeye thresher	*Alopias superciliosus*	38	-	-	1	...	...	...	...
Shortfin mako	*Isurus oxyrinchus*	38	-	-	73	33	...	5	5
Longfin mako	*Isurus paucus*	38	-	-	1	1	-	3	3
Nurse shark	*Ginglymostoma cirratum*	38	-	-	-	-	-	407	89
Blue shark	*Prionace glauca*	38	-	-	1 700	435	...	...	...
Blacktip shark	*Carcharhinus limbatus*	38	...	3	9	10	11	580	520
Requiem sharks nei	*Carcharhinidae*	38	12 209	12 145	9 785	8 588	6 294	6 491	5 907
Hammerhead sharks, etc. nei	*Sphyrnidae*	38	-	-	3	2	...	...	...
Smooth-hounds nei	*Mustelus spp*	38	46	253	27	45	3	30	30
Dogfish sharks nei	*Squalidae*	38	26	138	310	334	222	104	8
Rays, stingrays, mantas nei	*Rajiformes*	38	7 710	7 591	8 852	9 885	8 021	6 584	6 281
Sharks, rays, skates, etc. nei	*Elasmobranchii*	38	12 518	11 499	12 372	10 093	11 166	8 868	8 597
Marine fishes nei	*Osteichthyes*	39	233 163	223 781	259 985	220 341	174 182	159 083	167 559
Black stone crab	*Menippe mercenaria*	42	2 569	3 675	3 355	3 347	3 194	4 023	3 931
Portunus swimcrabs nei	*Portunus spp*	42	1	1	289	224	4 042	4 995	4 002
Blue crab	*Callinectes sapidus*	42	45 859	48 222	58 859	59 162	44 846	50 237	40 286
Harbour spidercrab	*Mithrax armatus*	42	-	-	-	114	110	54	35
Red crab	*Geryon quinquedens*	42	-	-	-	-	-	2	-
Marine crabs nei	*Brachyura*	42	6 241	4 765	6 058	4 511	373	1 145	744
Caribbean spiny lobster	*Panulirus argus*	43	29 278	29 472	29 172	27 186	30 655	33 439	27 698
Tropical spiny lobsters nei	*Panulirus spp*	43	-	-	-	5	5	5	4
Northern brown shrimp	*Penaeus aztecus*	45	57 111	54 703	44 459	50 722	60 527	62 713	68 411

C-31 (a)

Fish, crustaceans, molluscs, etc	Capture production by species items	Atlantic, Western Central
Poissons, crustacés, mollusques, etc	Captures par catégories d'espèces	Atlantique, centre-ouest
Peces, crustáceos, moluscos, etc	Capturas por categorías de especies	Atlántico, centro-occidental

English name Nom anglais Nombre inglés	Scientific name Nom scientifique Nombre científico	Species group Groupe d'espèces Grupo de especies	1995 mt	1996 mt	1997 mt	1998 mt	1999 mt	2000 mt	2001 mt
Northern pink shrimp	*Penaeus duorarum*	45	11 121	4 757	10 761	12 452	3 806	8 513	10 083
Northern white shrimp	*Penaeus setiferus*	45	39 959	27 461	31 928	39 799	44 014	52 280	40 696
Penaeus shrimps nei	*Penaeus spp*	45	53 471	50 099	53 816	50 679	40 810	48 871	53 455
Atlantic seabob	*Xiphopenaeus kroyeri*	45	11 524	17 310	21 464	16 603	18 648	24 782	32 351
Whitebelly prawn	*Nematopalaemon schmitti*	45	-	-	-	-	-	1 464	1 382
Rock shrimp	*Sicyonia brevirostris*	45	3 848	1 079	1 781	4 409	1 345	3 254	2 909
Royal red shrimp	*Pleoticus robustus*	45	252	184	205	182	208	369	281
Natantian decapods nei	*Natantia*	45	668	289	150	150	217	1 249	218
Marine crustaceans nei	*Crustacea*	47	440	363	373	422	200	347	186
Stromboid conchs nei	*Strombus spp*	52	13 761	12 346	17 703	14 386	15 347	18 219	18 071
Whelks	*Busycon spp*	52	1 491	1 193	1 017	908	832	600	480
Gastropods nei	*Gastropoda*	52	375	246	183	139	121	100	100
Mangrove cupped oyster	*Crassostrea rhizophorae*	53	5 224	4 110	3 702	4 895	3 906	3 467	5 632
American cupped oyster	*Crassostrea virginica*	53	125 652	108 971	95 608	101 957	85 150	222 866	156 485
South American rock mussel	*Perna perna*	54	155	223	295	3 802	451	316	1 081
American sea scallop	*Placopecten magellanicus*	55	805	459	154	150	27	115	236
Calico scallop	*Argopecten gibbus*	55	10 003	...	...	...	...	...	...
Atlantic bay scallop	*Argopecten irradians*	55	1 505	225	427	684	196	141	17
Ark clams nei	*Arca spp*	56	35 892	33 888	42 117	30 880	41 145	47 209	46 175
Northern quahog(=Hard clam)	*Mercenaria mercenaria*	56	...	...	...	...	...	5 506	5 113
Venus clams nei	*Veneridae*	56	1 517	1 350	1 279	1 296	1 222	1 756	2 222
Clams, etc. nei	*Bivalvia*	56	22 298	4 455	6 739	6 392	3 972	1 154	2 796
Cuttlefish,bobtail squids nei	*Sepiidae, Sepiolidae*	57	-	-	-	-	32	-	-
Longfin squid	*Loligo pealei*	57	39	-	-	-	-	-	-
Common squids nei	*Loligo spp*	57	862	921	883	942	205	604	1 404
Northern shortfin squid	*Illex illecebrosus*	57	20	9	2	2	-	1	0
Various squids nei	*Loliginidae, Ommastrephidae*	57	55	33	140	119	45	65	93
Common octopus	*Octopus vulgaris*	57	18 991	28 608	17 809	16 565	19 120	22 562	20 653
Octopuses, etc. nei	*Octopodidae*	57	724	1 548	1 917	6 507	711	1 152	1 087
Marine molluscs nei	*Mollusca*	58	3 836	673	1 872	2 921	6 706	2 716	1 913
Green turtle	*Chelonia mydas*	72	53	42	24	26	13	13	15
Hawksbill turtle	*Eretmochelys imbricata*	72	20	23	19	18	12	12	12
Loggerhead turtle	*Caretta caretta*	72	11	10	7	7	5	5	5
Marine turtles nei	*Testudinata*	72	125	153	36	90	1	2	4
Horseshoe crab	*Limulus polyphemus*	75	-	-	1	70	159	139	228
Echinoderms	*Echinodermata*	76	0	0	0	0	0	0	0
Sea urchins nei	*Strongylocentrotus spp*	76	15	10	15	15	15	10	10
Sea cucumbers nei	*Holothurioidea*	76	-	-	-	-	-	-	39
Jellyfishes	*Rhopilema spp*	77	...	...	...	...	...	137	-
Aquatic invertebrates nei	*Invertebrata*	77	11	6	2	4	0	2	-
Total			***1 767 597***	***1 693 548***	***1 804 094***	***1 773 670***	***1 763 871***	***1 819 345***	***1 689 685***

C-31 (b)

Fish, crustaceans, molluscs, etc — **Capture production by countries or areas** — **Atlantic, Western Central**
Poissons, crustacés, mollusques, etc — **Captures par pays ou zones** — **Atlantique, centre-ouest**
Peces, crustáceos, moluscos, etc — **Capturas por países o áreas** — **Atlántico, centro-occidental**

Country or area Pays ou zone País o área	1992 mt	1993 mt	1994 mt	1995 mt	1996 mt	1997 mt	1998 mt	1999 mt	2000 mt	2001 mt
Anguilla	386	330	333	150 F	200 F	250 F	250 F	250 F	250 F	250 F
Antigua Barb	1 712	642	696	1 311	1 209	1 437	1 415	1 361	1 481	1 583
Aruba	300	260	260	140	160	205	182	175	163	163
Bahamas	9 846	10 073	10 311	9 915	10 197	10 439	10 124	10 473	11 070	9 290
Barbados	3 574	3 214	2 818	3 581	3 512	2 809	3 644	3 250	3 100	2 676
Belize	2 413	1 784	2 102	1 877	2 048	2 835	2 557	1 871	2 434	2 517
Bermuda	432	404	394	449	465	461	466	453	286	315
Br Virgin Is	453 F	343	470	532	506	105	116	115	43	50 F
Cayman Is	125	125	125	125	110	125	125	125	125	125
China,Taiwan	5 908 F	6 979 F	3 627 F	3 558	4 516	3 669	2 430	5 663	6 039	10 597
Colombia	14 959	9 578	9 354	15 539	23 888	6 235	26 825	3 040	15 196	15 000 F
Costa Rica	261	199	269	422	437	420	402	666	1 088	816
Cuba	54 404	41 608	51 659	50 000	48 841	58 896	53 386	51 533	51 500 F	51 400 F
Dominica	711	794	882	950	1 030	1 079	1 212	1 200 F	1 200 F	1 150 F
Dominican Rp	11 816	10 820	19 058	15 768	12 606	13 793	9 188	7 835	10 842	12 059
Fr Guiana	7 617	6 931	7 819	8 089	7 377 F	6 602	6 709	6 271 F	5 237 F	5 194 F
Grenada	2 052	2 103	1 599	1 497	1 577	1 530	1 837	1 658	1 701	2 247
Guadeloupe	8 540	8 600	8 800	9 500	9 570	10 480	9 084	9 114	10 100	10 100 F
Guatemala	100	92	179	390	390	285	328	292	366	350 F
Guyana	40 452	43 323	45 567	47 200 F	47 783	53 373	52 215	53 241	48 087	52 605
Haiti	4 500 F	4 550 F	5 000 F	5 017 F	4 745 F	4 801 F	4 759 F	4 500 F	4 500 F	4 500 F
Honduras	4 630	4 274	4 526	6 686	2 691	6 560	2 332 F	1 865	4 359	2 539
Iceland	-	-	-	-	24	-	-	-	-	-
Jamaica	8 900 F	9 550 F	9 877 F	9 917 F	12 054 F	7 748	6 110	8 058	5 226	5 250 F
Japan	1 985	1 532	1 293	969	1 454	1 262	1 921	3 581	3 637	3 996
Korea Rep	4 627	1 462	750	653	626	143	621	1 789	3 327	237
Lithuania	61	-	-	-	-	-	-	-	-	-
Martinique	4 538	5 850	5 800 F	5 300	3 500 F	5 500	5 500	6 000	6 310	6 200
Mexico	330 207	320 824	329 521	272 178	294 231	320 829	302 157	285 833	274 532	259 156
Montserrat	23	58	62	48	38	45	46	50 F	50 F	50 F
NethAntilles	1 150 F	1 200 F	1 100 F	1 020 F	1 000 F	950 F	950 F	950 F	950 F	950 F
Nicaragua	4 212	5 001	6 738	5 957	9 685	9 451	12 011	13 127	14 838	13 444
Panama	-	-	-	-	-	-	-	20	714	...
Philippines	-	-	-	-	-	-	28	549	376	37
Portugal	-	-	-	-	-	-	17	22	-	-
Puerto Rico	1 812	1 877	2 275	3 173	2 701	3 187	3 006	3 020	4 154	3 794
Russian Fed	-	-	-	360	20	-	-	-	-	-
St Kitts Nev	300 F	250 F	212	192	352	272	533	555	492	591
St Lucia	1 073	1 336	1 252	1 188	1 274	1 308	1 589	1 718	1 855	1 983
St Vincent	1 381	1 479	1 090	944	889 F	948	1 365	1 032	7 325	9 020
Seychelles	-	-	-	-	-	-	-	-	127	-
Spain	4 020	3 027	1 624	1 955	906	3 145	2 090	1 998	2 224	4 631
Suriname	10 369	9 313	14 327	12 860 F	12 850 F	13 800 F	15 995	16 000	17 300 F	18 715
Trinidad Tob	13 000 F	9 637	14 483	12 000 F	9 435	11 283	9 175	8 826	9 786	11 408
Turks Caicos	1 191	1 458	1 419 F	1 395 F	1 297 F	1 250 F	1 318	1 300 F	1 300 F	1 300 F
USA	748 992	942 154	1 238 611	856 754	771 970	867 630	822 594	943 641	1 021 580	878 443
US Virgin Is	750 F	650 F	550 F	470 F	400 F	350 F	300 F	263	300 F	300 F
Venezuela	256 100	316 550	354 366	393 120	378 796	362 474	388 419	292 750	258 236	281 731
Other nei	3 994	2 959	4 584	4 448	6 188	6 130	8 339	7 838	5 539	2 923
Total	***1 573 876***	***1 793 193***	***2 165 782***	***1 767 597***	***1 693 548***	***1 804 094***	***1 773 670***	***1 763 871***	***1 819 345***	***1 689 685***

C-34 (a)

Fish, crustaceans, molluscs, etc	Capture production by species items	Atlantic, Eastern Central
Poissons, crustacés, mollusques, etc	Captures par catégories d'espèces	Atlantique, centre-est
Peces, crustáceos, moluscos, etc	Capturas por categorías de especies	Atlántico, centro-oriental

English name Nom anglais Nombre inglés	Scientific name Nom scientifique Nombre científico	Species group Groupe d'espèces Grupo de especies	1995 mt	1996 mt	1997 mt	1998 mt	1999 mt	2000 mt	2001 mt
Tilapias nei	*Oreochromis (=Tilapia) spp*	12	2 476	2 603	2 582	1 365	1 760	2 062	2 324
European eel	*Anguilla anguilla*	22	0	0	27	4	0	-	-
Shads nei	*Alosa spp*	24	-	-	13	40	-	22	62
West African ilisha	*Ilisha africana*	24	6 918	11 585	8 797	8 073	4 608	7 467	5 571
Lefteye flounders nei	*Bothidae*	31	83	24	2	0	0	17	12
Common sole	*Solea solea*	31	2 332	1 929	2 939	7 660	3 431	6 510	9 171
Wedge sole	*Dicologlossa cuneata*	31	59	87	8	0	9	10	9
Soles nei	*Soleidae*	31	9 182	12 606	13 552	16 050	9 132	5 127	4 242
Tonguefishes	*Cynoglossidae*	31	6 205	5 499	5 849	5 489	7 900	12 728	13 004
Megrim	*Lepidorhombus whiffiagonis*	31	-	-	-	36	-	38	52
Spottail spiny turbot	*Psettodes belcheri*	31	0	0	2	0	-	1	2
Flatfishes nei	*Pleuronectiformes*	31	17 073	15 178	16 019	16 999	21 132	26 109	23 288
Greater forkbeard	*Phycis blennoides*	32	447	466	115	231	369	269	353
Pouting(=Bib)	*Trisopterus luscus*	32	2 083	2 100	1 073	1 243	573	1 216	1 013
Blue whiting(=Poutassou)	*Micromesistius poutassou*	32	1	311	68	48	-	-	-
European hake	*Merluccius merluccius*	32	12 651	9 678	6 155	3 139	3 354	2 343	4 304
Senegalese hake	*Merluccius senegalensis*	32	14 779	16 485	16 430	17 396	18 187	13 515	17 923
Benguela hake	*Merluccius polli*	32	0	39	0	38	147	44	139
Hakes nei	*Merluccius spp*	32	117	1 705	1 093	2 064	1 674	1 940	1 468
Gadiformes nei	*Gadiformes*	32	268	663	716	991	903	146	114
West African ladyfish	*Elops lacerta*	33	1 474	379	735	1 546	1 822	755	558
Bonefish	*Albula vulpes*	33	20	78	1 542	1 640	245	239	515
Sea catfishes nei	*Ariidae*	33	24 447	25 916	31 530	26 145	28 932	32 143	34 749
Morays	*Muraenidae*	33	837	710	268	165	142	194	212
Flathead grey mullet	*Mugil cephalus*	33	1	-	1	3	-	-	-
Mullets nei	*Mugilidae*	33	12 720	11 526	18 071	17 703	23 243	23 010	22 975
Dusky grouper	*Epinephelus marginatus*	33	5	5	5	4	10	3	4
White grouper	*Epinephelus aeneus*	33	738	677	599	557	627	635	648
Dungat grouper	*Epinephelus goreensis*	33	102	196	63	76	290	277	62
Groupers nei	*Epinephelus spp*	33	1 180	1 317	1 017	2 225	943	690	697
Groupers, seabasses nei	*Serranidae*	33	7 289	5 778	6 520	5 748	6 248	5 795	6 805
Spotted seabass	*Dicentrarchus punctatus*	33	149	779	602	717	270	112	125
European seabass	*Dicentrarchus labrax*	33	137	108	19	7	26	9	20
Seabasses nei	*Dicentrarchus spp*	33	-	-	-	-	-	-	11
Bigeyes nei	*Priacanthus spp*	33	15	20	20	25	29	12	1
Cardinalfishes, etc. nei	*Apogonidae*	33	11	11	0	-	-	0	-
Snappers nei	*Lutjanus spp*	33	5 292	5 813	8 314	11 650	12 869	10 349	9 981
Rubberlip grunt	*Plectorhinchus mediterraneus*	33	1 079	1 865	899	455	639	220	257
Bastard grunt	*Pomadasys incisus*	33	1	1	5	-	-	0	0
Sompat grunt	*Pomadasys jubelini*	33	1 101	1 439	1 101	950	951	922	724
Bigeye grunt	*Brachydeuterus auritus*	33	22 939	19 887	29 046	18 591	22 218	21 748	19 986
Grunts, sweetlips nei	*Haemulidae (=Pomadasyidae)*	33	16 633	23 889	19 571	19 045	21 740	18 443	21 248
Brown meagre	*Sciaena umbra*	33	234	208	...	...	...	...	...
Shi drum	*Umbrina cirrosa*	33	63	86	52	43	83	93	57
Canary drum (=Baardman)	*Umbrina canariensis*	33	5	10	10	10	6	2	...
Meagre	*Argyrosomus regius*	33	1 821	1 965	2 702	2 216	3 436	2 728	2 531
Boe drum	*Pteroscion peli*	33	7 820	1 879	1 291	1 190	1 260	1 126	1 159
Law croaker	*Pseudotolithus brachygnathus*	33	813	715	618	354	318	811	1 083
Cassava croaker	*Pseudotolithus senegalensis*	33	2 936	2 727	2 781	4 659	4 837	3 949	4 800
Bobo croaker	*Pseudotolithus elongatus*	33	13 095	12 662	11 497	9 461	6 855	13 405	9 761
West African croakers nei	*Pseudotolithus spp*	33	19 791	25 305	22 156	25 794	23 733	23 515	22 077
Croakers, drums nei	*Sciaenidae*	33	18 544	20 755	32 395	25 995	30 306	30 027	29 989
Atlantic emperor	*Lethrinus atlanticus*	33	137	231	334	380	...	...	...
Blackspot(=red) seabream	*Pagellus bogaraveo*	33	27	179	108	141	145	85	12
Common pandora	*Pagellus erythrinus*	33	7	20	40	141	205	0	-
Axillary seabream	*Pagellus acarne*	33	8	18	11	127	1	1	1
Red pandora	*Pagellus bellottii*	33	4 627	7 621	8 047	10 278	13 546	2 980	3 253
Pandoras nei	*Pagellus spp*	33	8 295	8 121	6 301	6 020	8 030	6 936	5 781
Sargo breams nei	*Diplodus spp*	33	940	1 114	1 188	1 416	1 532	2 009	2 183
Large-eye dentex	*Dentex macrophthalmus*	33	338	1 101	572	2 307	3 121	1 954	1 086
Canary dentex	*Dentex canariensis*	33	-	-	-	-	-	20	-
Common dentex	*Dentex dentex*	33	1 322	995	239	389	255	134	336
Angolan dentex	*Dentex angolensis*	33	591	489	490	1 416	1 767	504	838
Congo dentex	*Dentex congoensis*	33	43	102	119	392	1 272	350	571
Dentex nei	*Dentex spp*	33	7 446	4 931	4 792	5 152	4 807	3 444	3 405
Black seabream	*Spondyliosoma cantharus*	33	1 353	1 775	1 780	1 079	1 129	1 370	1 421
Saddled seabream	*Oblada melanura*	33	29	20	0	1	-	0	0
Red porgy	*Pagrus pagrus*	33	1 038	974	639	887	641	607	846
Pargo breams nei	*Pagrus spp*	33	3 800	4 585	5 204	4 624	3 092	3 619	4 997
Gilthead seabream	*Sparus aurata*	33	2	4	16	28	735	599	1 288
Bogue	*Boops boops*	33	672	589	616	472	460	2 197	977
Sand steenbras	*Lithognathus mormyrus*	33	-	-	-	223	45	8	77
Porgies, seabreams nei	*Sparidae*	33	20 709	21 447	23 466	21 010	28 437	22 937	20 918
Picarels nei	*Spicara spp*	33	-	-	17	8	-	-	-
Surmullets(=Red mullets) nei	*Mullus spp*	33	1 037	5 981	2 120	1 746	1 882	2 607	3 459
West African goatfish	*Pseudupeneus prayensis*	33	1 025	1 719	1 808	1 396	1 146	833	3 371
Goatfishes, red mullets nei	*Mullidae*	33	18	-	-	-	-	-	-
African sicklefish	*Drepane africana*	33	1 706	2 774	3 790	2 268	1 545	2 856	2 790
Wrasses, hogfishes, etc. nei	*Labridae*	33	-	-	-	-	-	1	1
Parrotfish	*Sparisoma cretense*	33	-	-	-	-	-	2	1
Parrotfishes nei	*Scaridae*	33	...	...	...	...	...	13	1
Giant African threadfin	*Polydactylus quadrifilis*	33	3 598	7 359	10 705	10 023	10 770	12 237	12 447
Lesser African threadfin	*Galeoides decadactylus*	33	12 903	17 768	18 385	12 890	10 764	13 948	17 101
Royal threadfin	*Pentanemus quinquarius*	33	3 734	3 005	3 356	4 430	4 162	3 612	3 947
Threadfins, tasselfishes nei	*Polynemidae*	33	2 810	3 507	3 759	691	468	350	424
Spadefishes nei	*Ephippidae*	33	2	3	6	5	4	4	2

English name Nom anglais Nombre inglés	Scientific name Nom scientifique Nombre científico	Species group Groupe d'espèces Grupo de especies	1995 mt	1996 mt	1997 mt	1998 mt	1999 mt	2000 mt	2001 mt
Puffers nei	*Tetraodontidae*	33	125	28	100	...	...	4	35
Triggerfishes, durgons nei	*Balistidae*	33	20	93	58	255	72	95	97
European conger	*Conger conger*	34	2 114	2 669	2 031	2 155	1 641	1 213	1 166
Conger eels, etc. nei	*Congridae*	34	543	435	616	857	532	427	332
Longspine snipefish	*Macroramphosus scolopax*	34	20	-	89	443	-	-	44
Bearded brotula	*Brotula barbata*	34	504	811	1 842	2 208	3 861	1 719	1 926
Alfonsinos nei	*Beryx spp*	34	33	150	105	79	63	84	249
Slimeheads nei	*Trachichthyidae*	34	-	-	-	-	-	-	235
John dory	*Zeus faber*	34	1 269	1 706	779	781	813	1 294	708
Silvery John dory	*Zenopsis conchifer*	34	3	9	6	6	3	-	0
Wreckfish	*Polyprion americanus*	34	78	233	75	74	53	4	13
Tilefishes nei	*Branchiostegidae*	34	14	20	30	40	45	44	15
Oilfish	*Ruvettus pretiosus*	34	-	-	-	-	-	5	2
Largehead hairtail	*Trichiurus lepturus*	34	32 262	61 362	47 293	44 985	21 758	16 695	16 637
Silver scabbardfish	*Lepidopus caudatus*	34	2 571	9 575	4 731	3 357	3 030	0	4
Black scabbardfish	*Aphanopus carbo*	34	3 469	6 748	4 023	4 440	4 405	4 203	4 008
Hairtails, scabbardfishes nei	*Trichiuridae*	34	5 580	5 305	7 402	8 521	6 364	1 658	2 185
Scorpionfishes nei	*Scorpaenidae*	34	2 677	1 902	2 010	2 645	1 716	4 226	3 486
Gurnards, searobins nei	*Triglidae*	34	2 034	2 329	2 306	3 586	3 314	4 937	4 420
Blackbellied angler	*Lophius budegassa*	34	900	1 021	735	1 156	654	383	520
Shortspine African angler	*Lophius vaillanti*	34	...	...	...	...	...	7	5
Demersal percomorphs nei	*Perciformes*	34	140	310	1 840	1 405	1 379	1 344	1 337
Round sardinella	*Sardinella aurita*	35	250 337	399 024	318 076	317 627	282 951	285 048	228 201
Madeiran sardinella	*Sardinella maderensis*	35	80 281	125 701	139 827	131 970	129 066	140 943	128 945
Sardinellas nei	*Sardinella spp*	35	185 612	348 899	410 658	505 326	419 246	363 069	366 626
Bonga shad	*Ethmalosa fimbriata*	35	138 469	140 658	175 495	171 841	179 748	159 159	167 843
European pilchard(=Sardine)	*Sardina pilchardus*	35	779 123	614 634	649 764	588 637	533 114	600 132	794 600
European anchovy	*Engraulis encrasicolus*	35	81 736	121 552	132 015	151 329	145 984	198 646	178 765
Clupeoids nei	*Clupeoidei*	35	3 339	2 813	2 395	2 322	1 841	5 146	3 943
Atlantic bonito	*Sarda sarda*	36	1 226	5 115	9 854	12 988	7 066	5 949	6 273
Plain bonito	*Orcynopsis unicolor*	36	525	2 004	249	94	644	1 054	899
Wahoo	*Acanthocybium solandri*	36	423	608	680	510	667	571	668
West African Spanish mackerel	*Scomberomorus tritor*	36	2 051	1 296	1 377	1 456	703	834	963
Seerfishes nei	*Scomberomorus spp*	36	145	93	-	85	-	8	-
Frigate tuna	*Auxis thazard*	36	717	1 134	791	487	486	505	872
Frigate and bullet tunas	*Auxis thazard, A.rochei*	36	1 621	7 049	8 646	10 008	6 943	7 987	2 354
Little tunny(=Atl.black skipj)	*Euthynnus alletteratus*	36	2 477	4 573	6 445	4 392	4 282	5 678	3 212
Skipjack tuna	*Katsuwonus pelamis*	36	137 107	112 791	105 654	108 976	130 389	105 424	109 464
Atlantic bluefin tuna	*Thunnus thynnus*	36	1 393	3 035	4 030	3 644	2 580	3 408	5 096
Albacore	*Thunnus alalunga*	36	3 067	2 592	2 092	1 071	2 430	1 123	2 037
Yellowfin tuna	*Thunnus albacares*	36	110 268	112 629	98 982	101 832	102 497	90 912	114 280
Bigeye tuna	*Thunnus obesus*	36	87 453	88 617	73 844	82 705	87 979	67 945	67 275
Atlantic sailfish	*Istiophorus albicans*	36	1 061	1 377	1 092	812	701	985	344
Atlantic blue marlin	*Makaira nigricans*	36	1 915	2 224	1 928	1 733	1 800	1 545	756
Black marlin	*Makaira indica*	36	21	14	6	45	109	31	...
Atlantic white marlin	*Tetrapturus albidus*	36	235	332	310	246	287	460	41
Longbill spearfish	*Tetrapturus pfluegeri*	36	81	60	68	86	83	63	...
Marlins,sailfishes,etc. nei	*Istiophoridae*	36	130	768	163	824	636	844	813
Swordfish	*Xiphias gladius*	36	4 610	4 961	3 098	2 730	2 378	2 624	3 027
Tuna-like fishes nei	*Scombroidei*	36	1 497	3 376	5 732	8 340	3 359	2 706	7 661
Garfish	*Belone belone*	37	-	-	-	-	13	82	-
Needlefishes, etc. nei	*Belonidae*	37	655	1 161	398	330	416	646	656
Halfbeaks nei	*Hemiramphus spp*	37	402	295	130	399	278	264	279
Flyingfishes nei	*Exocoetidae*	37	498	574	1 259	1 727	1 395	1 186	1 356
Bluefish	*Pomatomus saltatrix*	37	1 641	3 805	3 829	4 596	2 223	1 439	2 315
Cobia	*Rachycentron canadum*	37	33	0	6	0	0	-	-
Atlantic horse mackerel	*Trachurus trachurus*	37	2	224	512	666	576	-	-
Cunene horse mackerel	*Trachurus trecae*	37	4 382	7 888	7 113	11 812	10 784	1 021	2 475
Jack and horse mackerels nei	*Trachurus spp*	37	220 925	160 097	133 054	221 552	210 857	278 103	236 094
Scads nei	*Decapterus spp*	37	3 839	3 814	4 257	6 350	2 817	5 448	6 735
Crevalle jack	*Caranx hippos*	37	4 422	2 884	2 249	2 946	3 966	139	2 585
False scad	*Caranx rhonchus*	37	3 483	3 301	3 344	2 864	2 710	3 935	281
Jacks, crevalles nei	*Caranx spp*	37	1 696	2 234	4 006	1 461	3 873	2 181	1 574
African moonfish	*Selene dorsalis*	37	1 073	1 015	832	755	772	1 236	1 936
Pompanos nei	*Trachinotus spp*	37	75	155	154	310	223	28	187
Amberjacks nei	*Seriola spp*	37	10	229	213	324	182	160	288
Leerfish	*Lichia amia*	37	2 516	496	1 350	1 448	1 619	1 139	604
Alexandria pompano	*Alectis alexandrinus*	37	124	160	742	1 023	563	502	862
Rainbow runner	*Elagatis bipinnulata*	37	-	-	1	-	27	4	1
Atlantic bumper	*Chloroscombrus chrysurus*	37	3 729	3 582	5 149	8 420	8 731	8 237	11 951
Carangids nei	*Carangidae*	37	11 462	7 780	8 751	11 312	13 002	13 812	7 664
Atlantic pomfret	*Brama brama*	37	2	-	26	-	103	33	3
Common dolphinfish	*Coryphaena hippurus*	37	2	4	34	124	177	234	477
Suckerfishes, remoras nei	*Echeneidae*	37	-	-	-	3	1	12	...
Chub mackerel	*Scomber japonicus*	37	170 015	224 169	246 040	212 934	142 812	228 945	135 419
Atlantic mackerel	*Scomber scombrus*	37	6	6	13	...	...	...	...
Scomber mackerels nei	*Scomber spp*	37	-	1 489	-	-	23	-	-
Mackerels nei	*Scombridae*	37	2 412	3 824	3 406	3 769	3 087	3 024	3 970
Blue butterfish	*Stromateus fiatola*	37	466	240	322	271	619	292	4 432
Barracudas nei	*Sphyraena spp*	37	8 371	10 187	12 180	12 731	17 188	16 697	19 863
Pelagic percomorphs nei	*Perciformes*	37	100	220	4 694	5 069	4 663	4 844	4 300
Thresher	*Alopias vulpinus*	38	...	...	30	45	151	146	7
Bigeye thresher	*Alopias superciliosus*	38	...	...	148	114	...	...	...
Thresher sharks nei	*Alopias spp*	38	...	...	34	55	66	...	...
Shortfin mako	*Isurus oxyrinchus*	38	...	...	...	...	7	48	43

C-34 (a)

Fish, crustaceans, molluscs, etc — **Capture production by species items** — **Atlantic, Eastern Central**
Poissons, crustacés, mollusques, etc — **Captures par catégories d'espèces** — **Atlantique, centre-est**
Peces, crustáceos, moluscos, etc — **Capturas por categorías de especies** — **Atlántico, centro-oriental**

English name Nom anglais Nombre inglés	Scientific name Nom scientifique Nombre científico	Species group Groupe d'espèces Grupo de especies	1995 mt	1996 mt	1997 mt	1998 mt	1999 mt	2000 mt	2001 mt
Mako sharks	*Isurus spp*	38	12	...	92	38	...	116	...
Porbeagle	*Lamna nasus*	38	...	...	...	...	...	10	2
Blue shark	*Prionace glauca*	38	...	...	...	...	76	421	557
Silky shark	*Carcharhinus falciformis*	38	18	...	2	...	110	99	...
Smooth hammerhead	*Sphyrna zygaena*	38	...	...	...	...	...	7	1
Scalloped hammerhead	*Sphyrna lewini*	38	12	12	10	10	10	10	10
Hammerhead sharks, etc. nei	*Sphyrnidae*	38	69	...	995	1 020	147	1 457	1 451
Smooth-hounds nei	*Mustelus spp*	38	199	392	51	57	28	503	1 929
Tope shark	*Galeorhinus galeus*	38	...	...	...	...	...	2	1
Leafscale gulper shark	*Centrophorus squamosus*	38	...	...	...	...	...	28	27
Dogfish sharks nei	*Squalidae*	38	2	5	2	1	0	0	-
Dogfish sharks, etc. nei	*Squaliformes*	38	220	180	140	100	60	12	10
Guitarfishes, etc. nei	*Rhinobatidae*	38	...	...	...	225	234	1 946	1 772
Sawfishes	*Pristidae*	38	...	...	48	...	41	42	...
Mantas	*Mobulidae*	38	...	...	...	342	802	931	106
Rays, stingrays, mantas nei	*Rajiformes*	38	11 463	12 093	14 079	13 865	14 727	11 192	12 504
Sharks, rays, skates, etc. nei	*Elasmobranchii*	38	16 160	22 753	52 302	36 392	34 647	33 953	31 890
Marine fishes nei	*Osteichthyes*	39	418 352	287 958	295 819	283 454	396 161	271 779	431 215
Red crab	*Geryon quinquedens*	42	-	-	-	-	-	-	31
Geryons nei	*Geryon spp*	42	...	...	...	...	...	6	14
Marine crabs nei	*Brachyura*	42	773	2 331	4 114	6 558	5 586	5 225	6 644
Tropical spiny lobsters nei	*Panulirus spp*	43	252	150	268	104	66	107	379
Palinurid spiny lobsters nei	*Palinurus spp*	43	1 868	1 679	8 816	3 577	1 590	2 345	2 470
Slipper lobsters nei	*Scyllaridae*	43	-	-	-	1	2	4	6
Norway lobster	*Nephrops norvegicus*	43	3	310	264	297	1 144	410	104
European lobster	*Homarus gammarus*	43	20	35	30	31	19	20	23
Caramote prawn	*Penaeus kerathurus*	45	373	159	100	40	19	8	1
Southern pink shrimp	*Penaeus notialis*	45	21 484	18 371	22 233	19 172	32 721	25 250	23 659
Penaeus shrimps nei	*Penaeus spp*	45	4 524	4 397	6 155	6 284	7 589	8 369	6 170
Deepwater rose shrimp	*Parapenaeus longirostris*	45	3 297	3 618	4 897	11 127	5 840	5 607	7 651
Scarlet shrimp	*Plesiopenaeus edwardsianus*	45	400	352	479	337	505	14	39
Natantian decapods nei	*Natantia*	45	24 853	30 913	19 342	46 545	23 687	20 108	20 219
Marine crustaceans nei	*Crustacea*	47	1 668	1 034	1 317	1 743	875	249	562
Murex	*Murex spp*	52	748	1 267	1 223	2 543	1 255	1 529	2 080
Volutes nei	*Cymbium spp*	52	7 454	6 648	5 166	4 678	5 737	4 989	5 422
Gastropods nei	*Gastropoda*	52	238	320	209	2 628	298	119	2 157
Cupped oysters nei	*Crassostrea spp*	53	223	100	109	89	125	101	151
Sea mussels nei	*Mytilidae*	54	8	10	6	7	...	401	102
Common edible cockle	*Cerastoderma edule*	56	178	54	69	139	147	117	105
Clams, etc. nei	*Bivalvia*	56	-	-	-	-	1	3	6
Common cuttlefish	*Sepia officinalis*	57	1 277	1 094	806	603	548	233	730
Cuttlefish,bobtail squids nei	*Sepiidae, Sepiolidae*	57	42 205	39 546	46 902	41 665	51 802	57 054	44 823
Common squids nei	*Loligo spp*	57	190	7	16	0	3	3	152
Various squids nei	*Loliginidae, Ommastrephidae*	57	29 582	29 860	13 457	20 824	23 018	20 175	15 873
Common octopus	*Octopus vulgaris*	57	31 247	27 119	23 158	39 169	23 538	12 753	16 302
Octopuses, etc. nei	*Octopodidae*	57	86 675	86 754	60 377	72 917	151 147	125 763	134 777
Cephalopods nei	*Cephalopoda*	57	-	6 240	5 636	8 551	238	440	263
Marine molluscs nei	*Mollusca*	58	1 806	746	1 489	12 599	2 258	1 189	384
Marine turtles nei	*Testudinata*	72	22	42	164	189	463	122	315
Total			***3 395 675***	***3 573 113***	***3 630 738***	***3 788 220***	***3 684 078***	***3 637 192***	***3 817 448***

C-34 (b)

Fish, crustaceans, molluscs, etc — Capture production by countries or areas — Atlantic, Eastern Central
Poissons, crustacés, mollusques, etc — Captures par pays ou zones — Atlantique, centre-est
Peces, crustáceos, moluscos, etc — Capturas por países o áreas — Atlántico, centro-oriental

Country or area Pays ou zone País o área	1992 mt	1993 mt	1994 mt	1995 mt	1996 mt	1997 mt	1998 mt	1999 mt	2000 mt	2001 mt
Bahamas	-	-	-	29	-	-	-	-	-	-
Belize	-	633	97	460	-	9 025	14 642	31 076	43 025	9 272
Benin	5 922	6 416	7 216	6 930	7 982	10 914	10 361	8 542	5 924	8 415
Bulgaria	2 182	-	-	-	-	-	8 189	-	-	-
Cameroon	49 975	42 257	52 000 F	64 131	63 400	62 000 F	61 800 F	60 000 F	57 109	58 531
Cape Verde	6 573	7 000	8 256	8 495	9 155	9 705	9 424	10 360	10 586	9 653
Cayman Is	700	320	...	...	...	...	...	...	...	...
Chile	-	-	25	-	-	-	-	-	-	-
China	5 385	9 809	2 414	21 464	18 302	20 810	20 403	28 433	25 408	36 228
China,Taiwan	2 556 F	6 467 F	12 929 F	13 199	17 517	13 158	10 623	8 890	14 494	13 677
Congo Dem R	3 800	4 200	3 780	3 876	3 973	3 844	3 954	3 945	3 945 F	3 945 F
Congo Rep	18 943	18 898	17 912	18 965	19 600	19 095	17 500 F	18 241	18 000 F	17 500 F
Côte dIvoire	71 635	63 495	58 374	58 854	57 606	52 137	57 071	63 709	65 270	63 026
Cuba	2 668	3 047	2 975	2 839	2 364	269	...	...	...	...
Cyprus	-	-	-	-	2 632	13 640	16 387	37 327	65 174	73 475
Egypt	-	-	-	-	-	-	-	4	-	-
Eq Guinea	3 230 F	2 907	4 369	1 856	4 140	5 240	5 035	5 900 F	2 558	2 500 F
Estonia	39 542	56 889	17 563	5 174	7 063	4 955	12 405	7 512	5	6
France	72 056	93 042	88 282	76 313	79 858	66 818	74 055	60 249	60 491	54 973
Gabon	22 000 F	28 289	26 515	32 789	36 680	34 143	44 609	41 143	37 053	30 607
Gambia	15 545	18 908	20 381	21 252	29 101	29 754	26 502	27 500	26 516	32 027
Georgia	24 775 F	13 500 F	5 000 F	1 000 F	-	-	-	-	-	-
Germany	-	-	-	39	11 222	28 748	23 099	26 678	-	2 673
Ghana	367 425	320 619	280 737	292 844	403 593	377 088	368 141	418 276	370 441	364 455
Greece	13 747	14 743	9 316	8 684	8 019	5 085	5 917	5 933	5 150	6 170
Guinea	51 000 F	56 000 F	60 000 F	64 760	60 580	58 841	65 764	83 314	87 513	86 000 F
GuineaBissau	5 000 F	5 100 F	5 750 F	6 078 F	6 750 F	7 000 F	5 800 F	4 800 F	4 800 F	4 800 F
Honduras	8 224	9 562	993	8 841	6 998	5 084	1 908	3 355	1 711	1 120
Ireland	-	-	-	-	-	-	-	-	-	52 980
Italy	50 932	43 445	43 578	8 067	3 428	7 180	5 891	5 845	4 077	7 359
Japan	14 431	22 773	20 358	24 307	24 766	14 934	20 294	13 487	16 218	10 973
Korea D P Rp	-	398	-	-	-	-	-	-	850	-
Korea Rep	9 890	15 510	23 057	19 324	19 380	21 851	18 481	25 219	23 978	26 809
Latvia	71 608	63 216	65 689	82 702	63 711	18 018	22 530	43 552	52 065	44 592
Liberia	4 891	3 782	3 721	4 829	3 408	4 580	6 830	11 472	7 726	7 286
Libya	820	1 085	500	400	976	877	911	850	887	239
Lithuania	101 111	57 469	22 880	9 572	33 330	25 680	45 804	46 910	53 445	111 100
Malta	-	-	1 236	3 465	8 197	-	-	-	-	-
Marshall Is	-	-	-	-	2 403	-	-	-	-	-
Mauritania	61 054 F	54 452 F	46 746 F	48 147 F	55 324 F	52 756 F	56 660 F	71 026 F	75 849 F	78 596 F
Morocco	509 904	591 270	717 578	807 775	601 734	758 321	680 075	705 978	856 362	1 054 147
Netherlands	-	-	-	-	47 515	100 969	124 475	123 937	157 195	176 922
NethAntilles	-	-	-	-	-	-	-	-	19 544	...
Nigeria	208 046	142 782	163 259	231 579	248 472	294 279	324 004	316 235	309 062	297 971
Norway	-	-	87	-	-	-	3 421	-	71	-
Panama	31 749	36 181	38 686	39 456	19 718	8 299	4 889	3 404	3 353	...
Philippines	-	-	-	-	-	-	1 189	1 854	801	29
Poland	-	-	2 220	-	19 766	5 268	-	-	-	13 185
Portugal	30 422	29 775	24 553	22 303	39 435	22 176	18 021	13 985	8 950	9 603
Romania	57 026	1 305	...	...	...	...	...	...	...	...
Russian Fed	368 890	198 630	167 472	277 631	384 330	312 524	341 413	286 179	211 018	129 436
St Vincent	776	466	0	61	30	5 144	32 608	16 727	20 369	34 940
Sao Tome Prn	2 094	2 334	3 391	3 565	3 980	3 338	3 477	3 756	3 500 F	3 500 F
Senegal	345 492	354 562	322 421	323 617	388 759	426 366	382 872	378 125	379 597	385 409
Sierra Leone	54 510	52 288	47 439	49 870	52 804 F	58 128	48 875	44 927	60 730	61 210
Spain	308 218	317 907 F	326 916 F	334 932 F	344 222	315 145	373 415	241 272	149 224	160 932
Togo	5 249	10 964	8 052	7 203	10 098	9 290	11 655	17 924	17 277	18 163
Ukraine	224 812	118 508	99 006	213 737	285 054	278 133	346 280	281 959	249 594	188 145
UK	-	-	-	-	-	-	-	-	-	31
Vanuatu	214	345	70	170	212	196	91	-	-	-
Westn Sahara	0	0	0	0	0	0	0	0	0	0
Other nei	48 972	37 194	44 630	164 091	55 526	49 933	50 470	74 268	50 277	64 838
Total	***3 303 994***	***2 938 742***	***2 878 429***	***3 395 675***	***3 573 113***	***3 630 738***	***3 788 220***	***3 684 078***	***3 637 192***	***3 817 448***

C-37 (a)

Fish, crustaceans, molluscs, etc — Capture production by species items — Mediterranean and Black Sea
Poissons, crustacés, mollusques, etc — Captures par catégories d'espèces — Méditerranée et mer Noire
Peces, crustáceos, moluscos, etc — Capturas por categorías de especies — Mediterráneo y Mar Negro

English name Nom anglais Nombre inglés	Scientific name Nom scientifique Nombre científico	Species group Groupe d'espèces Grupo de especies	1995 mt	1996 mt	1997 mt	1998 mt	1999 mt	2000 mt	2001 mt
Freshwater bream	*Abramis brama*	11	622	346	658	495	335	336	108
Common carp	*Cyprinus carpio*	11	2	2	3	3	1	-	2
Roach	*Rutilus rutilus*	11	8	1	0	1	1	2	7
Roaches nei	*Rutilus spp*	11	250	80	107	13	78	73	114
Silver carp	*Hypophthalmichthys molitrix*	11	2	1	-	-	-	-	-
Sichel	*Pelecus cultratus*	11	54	46	113	105	228	276	185
Cyprinids nei	*Cyprinidae*	11	-	-	1	-	-	167	159
European perch	*Perca fluviatilis*	13	-	-	-	-	-	-	1
Pike-perch	*Stizostedion lucioperca*	13	1 303	2 822	2 825	3 031	2 568	2 956	3 504
Freshwater fishes nei	*Osteichthyes*	13	-	-	-	-	-	-	17
Danube sturgeon(=Osetr)	*Acipenser gueldenstaedtii*	21	135	129	134	114	36	20	8
Sterlet sturgeon	*Acipenser ruthenus*	21	-	-	-	-	-	0	-
Starry sturgeon	*Acipenser stellatus*	21	9	18	49	13	11	5	3
Beluga	*Huso huso*	21	4	5	11	12	10	1	0
Sturgeons nei	*Acipenseridae*	21	678	432	443	290	185	59	22
European eel	*Anguilla anguilla*	22	849	919	1 254	914	682	464	504
Salmonoids nei	*Salmonoidei*	23	-	18	22	26	-	-	0
Pontic shad	*Alosa pontica*	24	126	131	82	153	48	15	21
Shads nei	*Alosa spp*	24	2 591	2 283	1 835	2 742	2 640	2 120	2 929
Azov sea sprat	*Clupeonella cultriventris*	24	7 805	1 448	1 522	3 496	10 862	12 006	27 777
Three-spined stickleback	*Gasterosteus aculeatus*	25	-	-	-	-	-	-	8
European plaice	*Pleuronectes platessa*	31	-	-	-	-	0	6	7
European flounder	*Platichthys flesus*	31	28	35	50	69	62	56	29
Common sole	*Solea solea*	31	9 198	6 949	5 917	5 044	4 179	5 131	4 941
Soles nei	*Soleidae*	31	-	-	-	-	18	14	12
Megrim	*Lepidorhombus whiffiagonis*	31	227	286	295	118	108	202	205
Brill	*Scophthalmus rhombus*	31	...	18	20	22	24	26	26
Turbot	*Psetta maxima*	31	3 240	2 341	1 333	2 178	2 125	2 967	2 802
Turbots nei	*Scophthalmidae*	31	1 923	1 377	964	528	478	643	622
Flatfishes nei	*Pleuronectiformes*	31	1 411	2 334	3 139	3 100	3 475	2 089	2 209
Blue ling	*Molva dypterygia*	32	-	1	1	1	-	-	-
Greater forkbeard	*Phycis blennoides*	32	246	204	241	346	209	428	74
Rocklings nei	*Gaidropsarus spp*	32	...	...	...	...	20	19	19
Poor cod	*Trisopterus minutus*	32	586	603	428	428	637	890	755
Pouting(=Bib)	*Trisopterus luscus*	32	5	313	388	402	669	1 947	657
Blue whiting(=Poutassou)	*Micromesistius poutassou*	32	17 929	18 854	20 965	32 278	23 530	26 317	27 933
Whiting	*Merlangius merlangus*	32	18 675	22 059	16 038	14 030	14 643	18 661	11 045
European hake	*Merluccius merluccius*	32	52 008	43 816	29 956	26 545	24 144	24 205	24 753
Gadiformes nei	*Gadiformes*	32	4 557	4 858	4 055	1 341	789	871	1 097
Brushtooth lizardfish	*Saurida undosquamis*	33	80	124	61	37	48	21	20
Lizardfishes nei	*Synodontidae*	33	1 135	1 484	1 257	1 138	1 382	1 545	1 150
Flathead grey mullet	*Mugil cephalus*	33	5 154	5 573	4 876	3 759	2 665	4 055	1 878
So-iuy mullet	*Mugil soiuy*	33	775	1 039	2 718	3 661	5 174	5 406	2 428
Leaping mullet	*Liza saliens*	33	1	3	2	2	2	0	10
Mullets nei	*Mugilidae*	33	27 279	34 198	31 718	37 205	37 607	37 890	34 704
Dusky grouper	*Epinephelus marginatus*	33	2 266	1 383	1 348	833	314	290	449
White grouper	*Epinephelus aeneus*	33	5	5	4	4	6	48	7
Groupers nei	*Epinephelus spp*	33	5 471	5 824	5 764	5 786	6 859	6 137	6 103
Groupers, seabasses nei	*Serranidae*	33	1 106	1 565	1 935	4 483	1 780	1 029	1 153
Spotted seabass	*Dicentrarchus punctatus*	33	73	424	119	135	203	186	238
European seabass	*Dicentrarchus labrax*	33	5 649	3 754	3 025	2 927	3 151	3 501	4 154
Seabasses nei	*Dicentrarchus spp*	33	2 498	2 846	3 550	4 486	4 354	2 677	1 786
Rubberlip grunt	*Plectorhinchus mediterraneus*	33	-	4	5	4	-	-	2
Brown meagre	*Sciaena umbra*	33	135	222	222	240	186	128	171
Shi drum	*Umbrina cirrosa*	33	1 163	710	493	386	386	289	393
Meagre	*Argyrosomus regius*	33	911	1 052	666	804	961	968	1 257
Croakers, drums nei	*Sciaenidae*	33	22	-	-	18	29	164	32
Blackspot(=red) seabream	*Pagellus bogaraveo*	33	12	12	20	20	38	38	38
Common pandora	*Pagellus erythrinus*	33	3 693	4 592	5 553	4 803	4 622	5 200	5 085
Axillary seabream	*Pagellus acarne*	33	445	90	477	670	339	284	206
Pandoras nei	*Pagellus spp*	33	5 013	2 897	3 379	2 131	2 756	3 252	3 339
White seabream	*Diplodus sargus*	33	539	409	640	433	640	695	529
Sargo breams nei	*Diplodus spp*	33	6 311	6 896	6 188	5 181	6 444	5 979	5 547
Large-eye dentex	*Dentex macrophthalmus*	33	710	651	1 010	535	497	378	318
Common dentex	*Dentex dentex*	33	3 061	2 287	1 458	1 237	1 172	1 241	767
Black seabream	*Spondyliosoma cantharus*	33	498	570	484	348	416	477	259
Saddled seabream	*Oblada melanura*	33	1 069	1 020	848	794	734	822	590
Red porgy	*Pagrus pagrus*	33	3 892	3 398	2 924	2 740	4 338	3 388	1 766
Pargo breams nei	*Pagrus spp*	33	812	666	602	738	751	1 010	765
Gilthead seabream	*Sparus aurata*	33	4 720	4 490	4 328	4 568	5 751	5 486	6 026
Bogue	*Boops boops*	33	28 962	29 016	25 414	27 606	24 400	24 509	22 479
Sand steenbras	*Lithognathus mormyrus*	33	919	718	1 011	1 184	1 004	1 006	1 090
Salema	*Sarpa salpa*	33	1 936	1 937	2 413	2 235	1 963	1 802	1 797
Porgies, seabreams nei	*Sparidae*	33	10 577	10 625	8 235	7 374	9 018	8 798	8 805
Blotched picarel	*Spicara maena*	33	218	285	505	395	419	255	381
Picarels nei	*Spicara spp*	33	10 168	12 312	12 254	13 147	9 324	8 192	8 928
Red mullet	*Mullus surmuletus*	33	12 022	11 689	11 818	9 077	9 272	9 642	9 598
Striped mullet	*Mullus barbatus*	33	5 115	5 331	4 531	4 878	5 290	4 170	5 405
Surmullets(=Red mullets) nei	*Mullus spp*	33	17 299	19 990	15 576	14 418	18 981	18 293	15 846
Parrotfishes nei	*Scaridae*	33	29	32	37	78	58	33	42
Sandeels(=Sandlances) nei	*Ammodytes spp*	33	...	74	164	283	368	308	106
Greater weever	*Trachinus draco*	33	8	83	154	207	24	53	80
Black goby	*Gobius niger*	33	-	-	-	-	-	1	1

C-37 (a)

Fish, crustaceans, molluscs, etc — Capture production by species items — Mediterranean and Black Sea
Poissons, crustacés, mollusques, etc — Captures par catégories d'espèces — Méditerranée et mer Noire
Peces, crustáceos, moluscos, etc — Capturas por categorías de especies — Mediterráneo y Mar Negro

English name Nom anglais Nombre inglés	Scientific name Nom scientifique Nombre científico	Species group Groupe d'espèces Grupo de especies	1995 mt	1996 mt	1997 mt	1998 mt	1999 mt	2000 mt	2001 mt
Gobies nei	*Gobiidae*	33	2 485	2 274	1 919	1 917	2 682	2 353	3 090
Spinefeet(=Rabbitfishes) nei	*Siganus spp*	33	297	763	390	425	510	646	945
Grey triggerfish	*Balistes carolinensis*	33	59	38	52	58	34	50	62
Argentines	*Argentina spp*	34	-	-	-	-	30	42	69
European conger	*Conger conger*	34	2 448	2 620	3 330	2 621	2 434	2 522	2 406
Conger eels, etc. nei	*Congridae*	34	304	416	310	300	680	200	340
Alfonsinos nei	*Beryx spp*	34	-	-	-	-	-	8	
John dory	*Zeus faber*	34	477	596	424	531	418	385	502
Wreckfish	*Polyprion americanus*	34	8	12	17	15	37	35	12
Largehead hairtail	*Trichiurus lepturus*	34	565	919	698	774	711	809	1 107
Silver scabbardfish	*Lepidopus caudatus*	34	7	1 619	1 365	1 713	1 830	707	3 421
Scorpionfishes nei	*Scorpaenidae*	34	2 210	2 240	2 309	2 453	1 842	2 067	2 293
Grey gurnard	*Eutrigla gurnardus*	34	1 070	1 264	1 158	1 067	1 159	1 078	778
Gurnards, searobins nei	*Triglidae*	34	6 482	7 035	5 051	4 672	4 168	3 570	3 464
Angler(=Monk)	*Lophius piscatorius*	34	4 132	3 674	7 956	3 982	3 825	4 859	4 269
Demersal percomorphs nei	*Perciformes*	34	493	609	453	629	549	1 276	1 233
Round sardinella	*Sardinella aurita*	35	250	242	435	557	306	88	80
Sardinellas nei	*Sardinella spp*	35	52 107	42 616	53 908	58 717	77 020	59 843	69 490
Red-eye round herring	*Etrumeus teres*	35	-	-	-	-	1 403	...	...
European pilchard(=Sardine)	*Sardina pilchardus*	35	247 581	224 546	209 851	204 094	212 244	223 255	195 961
European sprat	*Sprattus sprattus*	35	21 818	28 260	28 243	38 333	39 396	42 007	63 125
European anchovy	*Engraulis encrasicolus*	35	499 898	377 078	348 084	321 191	447 686	389 629	436 007
Clupeoids nei	*Clupeoidei*	35	2 812	1 997	1 864	2 879	3 206	3 215	3 212
Atlantic bonito	*Sarda sarda*	36	15 371	17 354	16 596	29 763	25 655	18 760	19 161
Plain bonito	*Orcynopsis unicolor*	36	115	132	227	130	217	145	154
Frigate and bullet tunas	*Auxis thazard, A.rochei*	36	5 205	5 909	3 324	2 291	2 011	2 439	3 810
Little tunny(=Atl.black skipj)	*Euthynnus alletteratus*	36	1 894	2 128	1 608	2 375	2 096	2 533	2 071
Skipjack tuna	*Katsuwonus pelamis*	36	43	9	4	176	53	90	77
Atlantic bluefin tuna	*Thunnus thynnus*	36	37 560	37 960	33 684	28 079	22 853	23 092	23 725
Albacore	*Thunnus alalunga*	36	1 587	3 125	2 541	2 698	4 851	5 578	4 743
Tunas nei	*Thunnini*	36	1	-	1	-	-	-	
Atlantic white marlin	*Tetrapturus albidus*	36	1	-	1	1	-	1	...
Marlins,sailfishes,etc. nei	*Istiophoridae*	36	1	1	29	-	-	1	25
Swordfish	*Xiphias gladius*	36	12 418	12 040	14 659	14 319	13 665	15 555	14 988
Tuna-like fishes nei	*Scombroidei*	36	1 263	2 010	2 446	1 374	3 389	3 383	2 422
Garfish	*Belone belone*	37	1 589	1 563	1 661	1 562	1 214	843	1 477
Big-scale sand smelt	*Atherina boyeri*	37	15	31	13	25	20	84	97
Silversides(=Sand smelts) nei	*Atherinidae*	37	6 306	7 575	6 652	7 699	5 807	3 928	5 631
Bluefish	*Pomatomus saltatrix*	37	6 642	6 171	5 014	4 582	4 367	6 256	15 291
Atlantic horse mackerel	*Trachurus trachurus*	37	9 677	10 177	8 367	6 076	5 155	8 189	11 602
Mediterranean horse mackerel	*Trachurus mediterraneus*	37	20 162	21 002	17 065	15 122	12 898	19 111	19 308
Jack and horse mackerels nei	*Trachurus spp*	37	28 882	23 580	26 087	22 899	27 324	27 956	27 497
Jacks, crevalles nei	*Caranx spp*	37	433	716	441	402	326	525	1 351
Greater amberjack	*Seriola dumerili*	37	713	1 057	724	755	1 320	1 445	1 414
Leerfish	*Lichia amia*	37	2 394	2 299	2 222	2 343	3 168	637	607
Carangids nei	*Carangidae*	37	634	457	354	413	412	512	801
Atlantic pomfret	*Brama brama*	37	...	26	157	87	118	33	109
Common dolphinfish	*Coryphaena hippurus*	37	768	706	833	1 451	1 360	1 038	1 157
Chub mackerel	*Scomber japonicus*	37	28 224	19 529	21 621	16 403	17 613	15 120	10 803
Atlantic mackerel	*Scomber scombrus*	37	7 424	7 158	4 711	7 000	4 827	7 290	6 208
Scomber mackerels nei	*Scomber spp*	37	10 621	11 680	11 878	10 879	8 941	8 572	9 148
Mackerels nei	*Scombridae*	37	500	500	300	300	300	353	357
Barracudas nei	*Sphyraena spp*	37	2 609	1 804	2 485	2 467	2 047	1 988	1 924
Basking shark	*Cetorhinus maximus*	38	-	2	6	6	-	-	-
Thresher	*Alopias vulpinus*	38	...	...	...	...	14	12	21
Shortfin mako	*Isurus oxyrinchus*	38	-	-	-	-	-	1	6
Porbeagle	*Lamna nasus*	38	0	1	0	0	0	0	1
Small-spotted catshark	*Scyliorhinus canicula*	38	...	...	...	...	...	30	31
Catsharks, nursehounds nei	*Scyliorhinus spp*	38	48	36	72	...	262	457	501
Blue shark	*Prionace glauca*	38	...	...	...	...	3	4	42
Smooth-hounds nei	*Mustelus spp*	38	8 272	5 896	2 980	3 249	3 658	4 100	1 950
Picked dogfish	*Squalus acanthias*	38	182	143	95	97	143	204	287
Portuguese dogfish	*Centroscymnus coelolepis*	38	-	-	-	-	-	7	23
Dogfish sharks nei	*Squalidae*	38	1 190	646	1 319	449	414	1 034	1 238
Angelshark	*Squatina squatina*	38	35	18	34	44	25	20	22
Angelsharks, sand devils nei	*Squatinidae*	38	31	95	35	171	100	90	36
Guitarfishes, etc. nei	*Rhinobatidae*	38	79	113	69	93	78	99	94
Thornback ray	*Raja clavata*	38	...	17	9	24	76	53	82
Raja rays nei	*Raja spp*	38	1 120	1 002	900	715	718	746	579
Common stingray	*Dasyatis pastinaca*	38	...	...	...	-	-	4	11
Rays, stingrays, mantas nei	*Rajiformes*	38	5 820	4 173	7 177	5 125	3 552	3 580	2 990
Sharks, rays, skates, etc. nei	*Elasmobranchii*	38	2 445	3 905	3 638	4 913	3 114	2 596	3 559
Marine fishes nei	*Osteichthyes*	39	112 010	100 777	83 640	87 647	63 217	75 454	63 933
Mediterranean shore crab	*Carcinus aestuarii*	42	16	65	44	66	44	30	37
Spinous spider crab	*Maja squinado*	42	43	31	59	55	16	21	27
Marine crabs nei	*Brachyura*	42	1 209	2 099	2 001	2 399	4 431	2 840	2 515
Pink spiny lobster	*Palinurus mauritanicus*	43	0	0	0	0	0	-	-
Common spiny lobster	*Palinurus elephas*	43	211	336	387	227	172	141	193
Palinurid spiny lobsters nei	*Palinurus spp*	43	197	184	200	297	188	168	217
Norway lobster	*Nephrops norvegicus*	43	6 569	6 658	6 165	4 155	4 068	3 522	3 558
European lobster	*Homarus gammarus*	43	575	277	451	248	195	235	254
Lobsters nei	*Reptantia*	43	-	-	-	-	0	2	2
Kuruma prawn	*Penaeus japonicus*	45	-	-	84	...	...	...	...

C-37 (a)

Fish, crustaceans, molluscs, etc — **Capture production by species items** — **Mediterranean and Black Sea**
Poissons, crustacés, mollusques, etc — **Captures par catégories d'espèces** — **Méditerranée et mer Noire**
Peces, crustáceos, moluscos, etc — **Capturas por categorías de especies** — **Mediterráneo y Mar Negro**

English name Nom anglais Nombre inglés	Scientific name Nom scientifique Nombre científico	Species group Groupe d'espèces Grupo de especies	1995 mt	1996 mt	1997 mt	1998 mt	1999 mt	2000 mt	2001 mt
Caramote prawn	*Penaeus kerathurus*	45	4 504	4 380	5 978	5 458	5 947	7 692	4 951
Speckled shrimp	*Metapenaeus monoceros*	45	-	-	-	-	-	-	757
Deepwater rose shrimp	*Parapenaeus longirostris*	45	9 816	9 101	10 119	8 127	9 261	11 727	11 093
Scarlet shrimp	*Plesiopenaeus edwardsianus*	45	-	-	-	-	100	40	-
Blue and red shrimp	*Aristeus antennatus*	45	1 333	1 428	1 359	1 672	1 332	1 688	2 359
Aristeid shrimps nei	*Aristeidae*	45	2 551	2 258	2 406	1 231	2 128	4 463	1 833
Common prawn	*Palaemon serratus*	45	37	40	35	89	9	10	10
Common shrimp	*Crangon crangon*	45	230	222	150	100	87	-	2
Natantian decapods nei	*Natantia*	45	8 879	7 955	10 148	10 848	10 543	8 911	8 186
Spottail mantis squillid	*Squilla mantis*	47	4 613	5 446	4 507	3 680	6 023	6 331	6 606
Marine crustaceans nei	*Crustacea*	47	5 750	4 358	4 320	4 193	3 904	3 869	3 276
Common periwinkle	*Littorina littorea*	52	-	132	147	117	126	139	-
Murex	*Murex spp*	52	43	37	41	41	35	51	52
Sea snails	*Rapana spp*	52	4 123	4 349	5 494	4 671	4 419	4 713	3 753
Gastropods nei	*Gastropoda*	52	-	-	-	-	965	813	870
European flat oyster	*Ostrea edulis*	53	2 955	2 187	1 849	1 190	904	271	157
Pacific cupped oyster	*Crassostrea gigas*	53	50	59	14	14	48	48	-
Mediterranean mussel	*Mytilus galloprovincialis*	54	40 764	39 046	53 310	38 899	44 453	46 143	46 092
Great Atlantic scallop	*Pecten maximus*	55	...	...	...	...	12	18	4
Great Mediterranean scallop	*Pecten jacobaeus*	55	23	52	95	50	68	570	150
Scallops nei	*Pectinidae*	55	-	41	42	15	-	-	-
Striped venus	*Chamelea gallina*	56	45 155	48 060	36 239	32 811	40 421	44 580	42 657
Pullet carpet shell	*Venerupis pullastra*	56	-	5	5	-	141	27	-
Grooved carpet shell	*Ruditapes decussatus*	56	1 379	488	135	172	143	796	569
Venus clams nei	*Veneridae*	56	-	-	-	-	18	19	22
Razor clams nei	*Solen spp*	56	...	2	4	1	5	5	5
Common edible cockle	*Cerastoderma edule*	56	-	-	-	-	1	-	-
Clams, etc. nei	*Bivalvia*	56	4 000	4 078	2 963	3 527	2 520	1 976	1 676
Common cuttlefish	*Sepia officinalis*	57	7 619	8 104	11 031	8 767	11 485	8 750	10 698
Cuttlefish,bobtail squids nei	*Sepiidae, Sepiolidae*	57	13 505	10 467	10 005	9 354	8 127	8 849	8 206
Common squids nei	*Loligo spp*	57	8 260	7 828	6 872	4 985	4 836	4 641	4 677
Broadtail shortfin squid	*Illex coindetii*	57	...	...	...	...	34	42	31
European flying squid	*Todarodes sagittatus*	57	4 808	4 899	2 850	4 203	2 202	2 856	2 466
Various squids nei	*Loliginidae, Ommastrephidae*	57	938	1 457	1 105	1 358	1 351	1 666	1 052
Common octopus	*Octopus vulgaris*	57	16 526	17 209	19 346	16 731	13 385	14 539	15 325
Horned and musky octopuses	*Eledone spp*	57	2 500	1 939	2 258	1 758	1 433	1 990	2 357
Octopuses, etc. nei	*Octopodidae*	57	7 908	6 830	6 131	6 714	6 969	9 354	7 837
Cephalopods nei	*Cephalopoda*	57	481	533	541	411	870	1 159	1 124
Marine molluscs nei	*Mollusca*	58	16 650	18 120	15 480	15 120	13 389	10 245	15 404
Grooved sea squirt	*Microcosmus sulcatus*	74	...	28	22	22	30	30	76
Echinoderms	*Echinodermata*	76	-	2	4	9	5	3	2
Stony sea urchin	*Paracentrotus lividus*	76	71	63	47	47	70	192	98
Sea cucumbers nei	*Holothurioidea*	76	-	4	4	4	1	9	4
Jellyfishes	*Rhopilema spp*	77	487	904	900	1 750	1 203	900	2 000
Aquatic invertebrates nei	*Invertebrata*	77	-	5	4	14	534	8	592
Total			***1 700 787***	***1 527 116***	***1 439 482***	***1 395 558***	***1 524 723***	***1 492 302***	***1 535 345***

C-37 (b)

Fish, crustaceans, molluscs, etc — **Capture production by countries or areas** — **Mediterranean and Black Sea**

Poissons, crustacés, mollusques, etc — **Captures par pays ou zones** — **Méditerranée et mer Noire**

Peces, crustáceos, moluscos, etc — **Capturas por países o áreas** — **Mediterráneo y Mar Negro**

Country or area Pays ou zone País o área	1992 mt	1993 mt	1994 mt	1995 mt	1996 mt	1997 mt	1998 mt	1999 mt	2000 mt	2001 mt
Albania	1 780	1 650 F	1 400 F	1 127	1 768	833	1 860	1 931	2 365	1 844
Algeria	95 270	101 895	135 402	105 872	81 989	91 580	92 344	102 396	100 000 F	100 000 F
Bosnia Herzg	0	0	0	0	0	0	0	0	0	0
Bulgaria	2 524	2 318	5 340 F	7 250	7 727	9 356	8 421	8 081	6 137	4 880
China	-	-	97	137	93	49	-	-	-	-
China,Taiwan	-	328	713	493	373	399	-	58	31	197
Croatia	26 530	26 342	17 137	15 901	17 799	16 627	21 928	18 890	20 701	18 056
Cyprus	2 676	2 696	2 762	2 505	2 550	2 309	2 408	2 241	2 230	2 258
Egypt	39 600	40 200	41 400	39 461	46 298	48 225	62 041	81 000	54 872	59 652
France	50 734	45 601	39 710	37 967	27 813	33 012	33 220	38 948	45 540	43 059
Gaza Strip	...	...	...	1 229	2 493	3 791	3 625	3 600 F	3 600 F	3 000 F
Georgia	7 785	2 191	1 397	2 470	2 447	2 582	2 997	1 396	2 178	1 822
Gibraltar	0	0	0	0	0	0	0	0	0	0
Greece	136 491	141 395	168 354	139 498	138 513	149 402	99 845	109 558	90 697	85 037
Israel	3 148	3 232	2 473	3 577	3 159	3 557	3 999	3 641	3 966	3 400 F
Italy	319 469	330 092	330 704	375 970	354 551	326 260	289 545	264 619	293 405	294 312
Japan	189	616	530	749	673	172	417	376	143	188
Korea Rep	-	-	-	484	701	686	-	-	-	-
Lebanon	1 700 F	2 000	2 205	4 065	4 115	3 635	3 500	3 540	3 646	3 650
Libya	28 000 F	30 000 F	33 000 F	34 000 F	32 000 F	31 000 F	32 000 F	32 000 F	32 500 F	33 000 F
Malta	565	823	1 115	1 156	986	1 019	1 143	1 224	1 059	882
Monaco	3 F	3 F	3 F	3 F	3 F	3 F	3 F	3 F	3 F	3 F
Morocco	39 239	31 624	34 999	39 676	39 652	31 485	28 658	37 290	38 650	28 146
Panama	484	467	1 499	1 498	2 850	236	-	-	-	-
Portugal	320	183	428	446	274	37	54	76	96	288
Romania	3 845	3 952	3 060	2 719	2 682	3 872	4 431	2 507	2 476	2 431
Russian Fed	14 044	6 384	13 888	15 540	8 745	8 917	9 760	14 340	22 294	33 444
Slovenia	3 612	1 987	2 007	1 851	2 078	2 065	1 959	1 784	1 630	1 621
Spain	145 000 F	146 000 F	148 000 F	149 000 F	150 496	133 302	123 325	122 359	140 203	138 568
Syria	1 954	2 019	1 950	1 950	2 670	2 574	2 750	2 600	2 581	2 322
Tunisia	86 397	82 514	85 367	82 915	83 028	86 002	87 179	92 378	94 718	97 622
Turkey	407 003	503 957	544 736	585 994	478 226	408 693	432 700	523 634	460 524	484 407
Ukraine	31 792	26 394	35 213	43 570	29 266	35 987	42 981	51 495	65 507	90 840
Yugoslavia	223	277	260	364	377	373	410	423	424	416
Other nei	3 201	1 000	1 700	1 350	721	1 442	2 055	2 335	126	-
Total	***1 453 578***	***1 538 140***	***1 656 849***	***1 700 787***	***1 527 116***	***1 439 482***	***1 395 558***	***1 524 723***	***1 492 302***	***1 535 345***

C-41 (a)

Fish, crustaceans, molluscs, etc — **Capture production by species items** — **Atlantic, Southwest**
Poissons, crustacés, mollusques, etc — **Captures par catégories d'espèces** — **Atlantique, sud-ouest**
Peces, crustáceos, moluscos, etc — **Capturas por categorías de especies** — **Atlántico, sudoccidental**

English name Nom anglais Nombre inglés	Scientific name Nom scientifique Nombre científico	Species group Groupe d'espèces Grupo de especies	1995 mt	1996 mt	1997 mt	1998 mt	1999 mt	2000 mt	2001 mt
Lefteye flounders nei	*Bothidae*	31	-	-	-	-	-	-	20
Tonguefishes	*Cynoglossidae*	31	36	23	-	-	19	48	-
Bastard halibuts nei	*Paralichthys spp*	31	10 343	9 236	10 546	9 247	7 081	6 746	5 680
Flatfishes nei	*Pleuronectiformes*	31	1 491	1 097	1 430	1 655	1 601	1 887	1 824
Tadpole codling	*Salilota australis*	32	9 946	7 590	6 094	11 454	15 643	14 541	5 528
Brazilian codling	*Urophycis brasiliensis*	32	3 703	3 152	2 570	6 081	2 842	3 496	2 564
Southern blue whiting	*Micromesistius australis*	32	138 360	104 669	102 450	108 558	83 695	84 499	79 722
Southern hake	*Merluccius australis*	32	3 899	4 115	3 011	3 125	3 471	7 035	4 649
Argentine hake	*Merluccius hubbsi*	32	636 385	681 999	648 301	527 229	372 167	242 727	302 798
Hakes nei	*Merluccius spp*	32	16	370	33	-	-	-	354
Patagonian grenadier	*Macruronus magellanicus*	32	36 338	55 782	52 199	119 810	137 340	142 260	135 520
Grenadiers nei	*Macrourus spp*	32	744	901	1 041	2 972	2 593	10 503	9
Gadiformes nei	*Gadiformes*	32	293	1 540	252	237	352	1 364	3 142
Ladyfish	*Elops saurus*	33	147	-	-	-	-	-	-
Tarpon	*Megalops atlanticus*	33	1 221	348	720	551	315	331	330
Sea catfishes nei	*Ariidae*	33	14 361	16 970	17 764	20 531	25 307	29 529	29 242
Mullets nei	*Mugilidae*	33	13 442	8 078	10 486	10 288	10 953	12 186	12 119
Snooks(=Robalos) nei	*Centropomus spp*	33	1 820	1 686	1 866	2 996	3 602	4 366	4 300
Brazilian groupers nei	*Mycteroperca spp*	33	1 117	423	584	518	628	536	530
Red grouper	*Epinephelus morio*	33	872	320	238	212	666	1 220	1 200
Groupers nei	*Epinephelus spp*	33	2 286	2 341	2 170	2 315	1 710	2 670	2 630
Argentine seabass	*Acanthistius brasilianus*	33	10 987	10 799	9 164	5 258	5 914	4 204	4 701
Groupers, seabasses nei	*Serranidae*	33	-	-	15	2	-	-	1
Bigeyes nei	*Priacanthus spp*	33	70	59	55	46	60	67	70
Southern red snapper	*Lutjanus purpureus*	33	5 816	5 104	6 085	5 937	9 790	6 580	6 500
Lane snapper	*Lutjanus synagris*	33	506	648	659	1 038	1 031	1 014	1 000
Yellowtail snapper	*Ocyurus chrysurus*	33	4 766	4 167	5 000	3 317	4 541	4 165	4 100
Snappers, jobfishes nei	*Lutjanidae*	33	4 550	3 690	4 488	4 390	4 549	3 859	3 800
Barred grunt	*Conodon nobilis*	33	339	172	118	114	84	39	40
Grunts, sweetlips nei	*Haemulidae (=Pomadasyidae)*	33	1 812	2 081	1 936	2 000	2 382	3 024	3 000
Striped weakfish	*Cynoscion striatus*	33	32 635	31 641	39 318	32 394	19 588	22 874	22 193
Weakfishes nei	*Cynoscion spp*	33	27 075	26 470	25 960	26 677	39 332	43 523	43 000
Whitemouth croaker	*Micropogonias furnieri*	33	81 526	72 633	73 163	59 631	43 217	58 207	60 301
Kingcroakers nei	*Menticirrhus spp*	33	1 330	1 144	1 303	1 402	1 062	1 352	1 336
Argentine croaker	*Umbrina canosai*	33	14 537	13 018	6 658	5 284	7 660	9 221	9 116
King weakfish	*Macrodon ancylodon*	33	5 345	7 266	7 625	9 229	5 595	6 642	6 886
Black drum	*Pogonias cromis*	33	652	332	163	938	282	357	336
Croakers, drums nei	*Sciaenidae*	33	89	150	52	117	746	688	300
South American silver porgy	*Diplodus argenteus*	33	8	39	84	15	5	2	...
Red porgy	*Pagrus pagrus*	33	1 298	2 475	2 743	2 026	3 535	2 825	2 292
Porgies, seabreams nei	*Sparidae*	33	-	-	-	-	2	1 054	1 221
Surmullets(=Red mullets) nei	*Mullus spp*	33	1 027	605	822	939	1 346	2 254	1 730
Patagonian blennie	*Eleginops maclovinus*	33	70	152	61	1 198	2 027	1 755	61
Antarctic rockcods, noties nei	*Nototheniidae*	33	-	-	-	-	207	-	-
Percoids nei	*Percoidei*	33	720	664	692	1 838	947	1 465	1 450
Brazilian flathead	*Percophis brasilianus*	33	9 320	9 231	11 887	10 191	7 236	7 862	7 640
Argentinian sandperch	*Pseudopercis semifasciata*	33	3 023	3 438	2 483	2 222	2 679	1 905	1 887
Spadefishes nei	*Ephippidae*	33	40	71	55	51	283	325	320
Puffers nei	*Tetraodontidae*	33	468	18	88	23	16	35	30
Argentine conger	*Conger orbignyanus*	34	203	89	138	189	180	122	100
Conger eels, etc. nei	*Congridae*	34	-	-	-	-	-	20	-
Pink cusk-eel	*Genypterus blacodes*	34	24 887	23 608	22 948	27 290	24 186	17 190	22 535
Cusk-eels nei	*Genypterus spp*	34	-	-	-	-	-	57	-
Cusk-eels, brotulas nei	*Ophidiidae*	34	106	176	258	452	544	507	518
Alfonsinos nei	*Beryx spp*	34	-	-	-	144	-	4	-
Wreckfish	*Polyprion americanus*	34	...	...	...	...	...	129	35
Tilefishes nei	*Branchiostegidae*	34	812	1 098	1 000	786	524	547	540
Castaneta	*Cheilodactylus bergi*	34	13 443	3 142	4 893	11 508	3 260	1 431	1 318
Patagonian toothfish	*Dissostichus eleginoides*	34	19 442	14 951	9 599	13 328	11 300	11 122	10 013
Patagonian rockcod	*Patagonotothen brevicauda*	34	-	-	-	13	-	-	-
White snake mackerel	*Thyrsitops lepidopoides*	34	66	1 232	309	285	136	-	-
Largehead hairtail	*Trichiurus lepturus*	34	1 197	736	938	1 405	1 230	1 665	1 650
Hairtails, scabbardfishes nei	*Trichiuridae*	34	-	-	-	51	1 205	1 888	155
Choicy ruff	*Seriolella porosa*	34	...	...	...	...	...	3 542	3 990
Blackbelly rosefish	*Helicolenus dactylopterus*	34	7 404	3 703	5 967	5 581	5 943	5 169	1 836
Scorpionfishes nei	*Scorpaenidae*	34	-	293	99	-	-	-	4
Atlantic searobins	*Prionotus spp*	34	922	792	540	858	1 149	1 839	1 800
Anglerfishes nei	*Lophiidae*	34	366	294	399	542	794	1 937	4 008
Demersal percomorphs nei	*Perciformes*	34	7 064	8 145	12 600	12 454	15 930	7 699	7 634
Brazilian sardinella	*Sardinella brasiliensis*	35	60 212	97 092	117 642	82 283	25 518	17 053	35 000
Brazilian menhaden	*Brevoortia aurea*	35	11 133	6 294	2 396	2 936	2 202	1 123	1 100
Argentine menhaden	*Brevoortia pectinata*	35	424	511	1 112	519	291	337	103
Scaled sardines	*Harengula spp*	35	-	20	12	2	162	126	120
Atlantic thread herring	*Opisthonema oglinum*	35	6 661	3 762	3 712	8 082	8 204	12 455	12 300
Falkland sprat	*Sprattus fuegensis*	35	0	0	0	29	17	97	4
Argentine anchovy	*Engraulis anchoita*	35	24 498	21 023	25 211	13 417	13 025	12 163	11 602
Anchovies, etc. nei	*Engraulidae*	35	3 806	7 672	7 102	2 654	2 682	4 039	4 000
Clupeoids nei	*Clupeoidei*	35	2 775	3 464	4 314	6 042	7 527	11 457	11 300
Atlantic bonito	*Sarda sarda*	36	275	108	130	12	38	20	23
Wahoo	*Acanthocybium solandri*	36	1	16	58	41	-	-	-
King mackerel	*Scomberomorus cavalla*	36	1 328	2 890	2 398	3 595	3 595	2 344	1 251
Serra Spanish mackerel	*Scomberomorus brasiliensis*	36	1 308	3 047	2 125	1 516	1 516	988	251
Frigate and bullet tunas	*Auxis thazard, A.rochei*	36	558	527	215	162	166	106	98
Little tunny(=Atl.black skipj)	*Euthynnus alletteratus*	36	1 059	834	507	920	930	615	615
Skipjack tuna	*Katsuwonus pelamis*	36	16 560	22 531	26 576	23 790	23 188	25 820	24 155

C-41 (a)

Fish, crustaceans, molluscs, etc	Capture production by species items	Atlantic, Southwest
Poissons, crustacés, mollusques, etc	Captures par catégories d'espèces	Atlantique, sud-ouest
Peces, crustáceos, moluscos, etc	Capturas por categorías de especies	Atlántico, sudoccidental

English name Nom anglais Nombre inglés	Scientific name Nom scientifique Nombre científico	Species group Groupe d'espèces Grupo de especies	1995 mt	1996 mt	1997 mt	1998 mt	1999 mt	2000 mt	2001 mt
Atlantic bluefin tuna	*Thunnus thynnus*	36	2	0	0	1	13	0	1
Blackfin tuna	*Thunnus atlanticus*	36	153	649	418	55	55	38	149
Albacore	*Thunnus alalunga*	36	8 615	16 141	13 944	11 668	12 189	8 630	11 610
Southern bluefin tuna	*Thunnus maccoyii*	36	7	18	5	6	3	3	-
Yellowfin tuna	*Thunnus albacares*	36	5 228	5 430	4 507	4 946	6 163	7 494	7 651
Bigeye tuna	*Thunnus obesus*	36	8 128	11 697	9 370	6 061	5 588	8 099	7 885
Atlantic sailfish	*Istiophorus albicans*	36	288	320	165	216	429	645	419
Atlantic blue marlin	*Makaira nigricans*	36	433	658	513	762	701	634	932
Atlantic white marlin	*Tetrapturus albidus*	36	461	398	407	361	293	291	192
Longbill spearfish	*Tetrapturus pfluegeri*	36	1	-	0	-	22	62	56
Marlins,sailfishes,etc. nei	*Istiophoridae*	36	140	33	30	144	0	18	10
Swordfish	*Xiphias gladius*	36	11 110	7 789	10 275	8 841	8 927	10 898	9 551
Tuna-like fishes nei	*Scombroidei*	36	67	2	498	359	774	337	29
Ballyhoo halfbeak	*Hemiramphus brasiliensis*	37	631	494	442	1 500	832	989	980
Flyingfishes nei	*Exocoetidae*	37	1 036	743	1 082	1 084	760	388	380
Silversides(=Sand smelts) nei	*Atherinidae*	37	453	637	584	102	637	57	41
Bluefish	*Pomatomus saltatrix*	37	8 209	6 219	3 043	2 603	2 368	3 778	3 393
Cobia	*Rachycentron canadum*	37	256	498	367	622	1 818	1 580	1 550
Rough scad	*Trachurus lathami*	37	732	976	601	328	495	107	40
Blue runner	*Caranx crysos*	37	338	503	515	443	590	625	620
Jacks, crevalles nei	*Caranx spp*	37	3 656	3 905	5 838	5 635	4 164	6 950	6 900
Atlantic moonfish	*Selene setapinnis*	37	2 468	1 679	1 914	1 512	1 514	1 386	1 370
Pompanos nei	*Trachinotus spp*	37	424	534	166	172	144	286	280
Yellowtail amberjack	*Seriola lalandi*	37	662	554	483	887	553	621	600
Amberjacks nei	*Seriola spp*	37	856	825	1 048	119	104	128	120
Atlantic bumper	*Chloroscombrus chrysurus*	37	1 868	952	1 035	4 065	3 519	1 573	1 550
Parona leatherjacket	*Parona signata*	37	1 869	2 364	2 236	2 209	2 070	1 850	1 478
Common dolphinfish	*Coryphaena hippurus*	37	3 186	2 500	4 028	4 117	2 848	4 359	4 300
Chub mackerel	*Scomber japonicus*	37	21 496	16 869	18 779	12 647	8 612	16 499	9 660
Mackerels nei	*Scombridae*	37	-	-	-	-	21	0	-
Butterfishes, pomfrets nei	*Stromateidae*	37	-	-	-	5	-	-	20
Barracudas nei	*Sphyraena spp*	37	20	11	12	828	217	527	520
Pelagic percomorphs nei	*Perciformes*	37	3	2	243	150	10	18	0
Thresher	*Alopias vulpinus*	38	...	...	...	...	...	...	3
Shortfin mako	*Isurus oxyrinchus*	38	...	83	190	...	100	135	146
Blue shark	*Prionace glauca*	38	...	743	1 103	...	500	636	1 558
Silky shark	*Carcharhinus falciformis*	38	...	502	279	...	70	80	80
Smooth hammerhead	*Sphyrna zygaena*	38	...	...	...	...	...	3	...
Scalloped hammerhead	*Sphyrna lewini*	38	...	25	170	...	30	38	40
Narrownose smooth-hound	*Mustelus schmitti*	38	11 343	10 456	10 130	13 422	12 274	8 156	9 933
Tope shark	*Galeorhinus galeus*	38	104	92	103	92	89	109	37
Argentine angelshark	*Squatina argentina*	38	3 802	4 281	4 410	4 311	3 368	3 123	3 339
Angelsharks, sand devils nei	*Squatinidae*	38	113	1 587	...	...	...	...	...
Chola guitarfish	*Rhinobatos percellens*	38	162	404	...	...	...	...	...
Rays, stingrays, mantas nei	*Rajiformes*	38	19 030	24 406	22 838	21 715	24 419	26 800	28 285
Elephantfishes nei	*Callorhinchus spp*	38	921	815	1 329	1 770	1 977	1 378	...
Sharks, rays, skates, etc. nei	*Elasmobranchii*	38	15 575	13 201	18 144	16 586	18 481	19 726	19 359
Marine fishes nei	*Osteichthyes*	39	172 960	158 570	165 118	157 405	181 165	194 956	201 983
Dana swimcrab	*Callinectes danae*	42	2 062	2 020	2 600	3 014	1 626	1 597	1 580
Red crab	*Geryon quinquedens*	42	783	1 513	3 432	2 743	3 382	5 259	2 685
Marine crabs nei	*Brachyura*	42	12 262	10 312	10 636	9 528	11 083	11 144	11 040
Caribbean spiny lobster	*Panulirus argus*	43	10 817	8 026	7 502	6 002	6 334	6 469	6 400
Southern king crab	*Lithodes antarcticus*	44	380	200	413	456	270	102	58
Softshell red crab	*Paralomis granulosa*	44	1	1	1	1	1	267	160
Redspotted shrimp	*Penaeus brasiliensis*	45	6 565	8 743	10 758	7 796	9 092	10 728	10 600
Sao Paulo shrimp	*Penaeus paulensis*	45	0	-	177	13	12	56	23
Penaeus shrimps nei	*Penaeus spp*	45	27 155	15 037	14 506	13 998	12 550	16 509	16 300
Atlantic seabob	*Xiphopenaeus kroyeri*	45	7 278	11 764	15 257	13 755	9 964	11 948	11 800
Argentine stiletto shrimp	*Artemesia longinaris*	45	250	263	166	146	37	37	...
Argentine red shrimp	*Pleoticus muelleri*	45	6 705	9 874	6 479	23 203	15 928	36 769	78 077
Natantian decapods nei	*Natantia*	45	-	-	-	-	1 090	556	2 816
Antarctic krill	*Euphausia superba*	46	-	-	-	74	-	4	-
Marine crustaceans nei	*Crustacea*	47	860	313	849	2 161	6 293	227	1 375
Angulate volute	*Zidona dufresnei*	52	574	558	1 322	1 010	683	1 621	1 406
Cupped oysters nei	*Crassostrea spp*	53	726	873	828	744	1 547	884	870
Chilean mussel	*Mytilus chilensis*	54	1	1	1	1	1	1	0
River Plata mussel	*Mytilus platensis*	54	687	370	354	375	474	412	492
Sea mussels nei	*Mytilidae*	54	2 217	863	1 193	1 230	906	802	800
Patagonean scallop	*Zygochlamis patagonica*	55	10 592	36 952	39 817	28 441	42 700	37 404	42 598
Triangular tivela	*Tivela mactroides*	56	...	126	196	664	1 684	273	270
Donax clams	*Donax spp*	56	0	0	0	0	0	0	0
Clams, etc. nei	*Bivalvia*	56	...	...	0	0	759	6	17
Patagonian squid	*Loligo gahi*	57	85 174	68 504	21 711	51 693	42 505	67 016	57 730
Common squids nei	*Loligo spp*	57	11 619	8 305	5 095	4 178	3 752	1 455	1 170
Argentine shortfin squid	*Illex argentinus*	57	520 938	656 481	980 300	693 542	1 144 998	930 781	743 024
Sevenstar flying squid	*Martialia hyadesi*	57	23 986	3 793	8 348	2	27	686	2
Various squids nei	*Loliginidae, Ommastrephidae*	57	2 760	10 157	415	-	-	-	-

C-41 (a)

Fish, crustaceans, molluscs, etc — Capture production by species items — Atlantic, Southwest
Poissons, crustacés, mollusques, etc — Captures par catégories d'espèces — Atlantique, sud-ouest
Peces, crustáceos, moluscos, etc — Capturas por categorías de especies — Atlántico, sudoccidental

English name Nom anglais Nombre inglés	Scientific name Nom scientifique Nombre científico	Species group Groupe d'espèces Grupo de especies	1995 mt	1996 mt	1997 mt	1998 mt	1999 mt	2000 mt	2001 mt
Octopuses, etc. nei	*Octopodidae*	57	757	635	688	571	1 456	1 084	1 002
Marine molluscs nei	*Mollusca*	58	3 678	1 175	1 851	1 408	4 937	2 474	2 839
Marine turtles nei	*Testudinata*	72	0	0	0	0	0	0	0
Total			***2 336 972***	***2 480 181***	***2 756 887***	***2 379 440***	***2 622 540***	***2 375 434***	***2 265 714***

C-41 (b)

Fish, crustaceans, molluscs, etc — Capture production by countries or areas — Atlantic, Southwest
Poissons, crustacés, mollusques, etc — Captures par pays ou zones — Atlantique, sud-ouest
Peces, crustáceos, moluscos, etc — Capturas por países o áreas — Atlántico, sudoccidental

Country or area Pays ou zone País o área	1992 mt	1993 mt	1994 mt	1995 mt	1996 mt	1997 mt	1998 mt	1999 mt	2000 mt	2001 mt
Argentina	692 198	919 503	938 590	1 152 054	1 272 057	1 377 427	1 141 632	1 044 221	883 204	899 462
Australia	-	-	-	-	-	-	3 594	3 711	-	-
Belize	-	-	-	-	-	-	-	4 500	6 729	2 581
Brazil	558 780 F	530 100 F	548 615 F	513 666	522 173	565 714	532 599	518 470	567 687	570 000 F
Bulgaria	14 929	4 200	-	-	-	-	-	-	-	-
Cambodia	-	-	-	-	-	-	-	-	2 768	1 200
Chile	2 432	846	23	302	-	3 744	14 755	5 226	2 749	8 849
China	-	-	3	3	3	3	31 806	61 664	93 156	93 973
China,Taiwan	146 676 F	145 840 F	125 529 F	138 326	132 450	216 202	177 096	279 001	247 505	156 502
Estonia	6 568	1 338	-	-	2 538	-	-	-	-	1 941
Falkland Is	1 855	1 974	5 914	27 190	31 539	17 112	43 615	39 163	62 927	59 823
France	-	-	1 946	7 332	4 580	1 545	4 179	2 379	2 053	-
Honduras	2 762	3 655	2 976	2 771	849	-	-	-	-	-
Italy	6 792	5 792	5 810	960	408	-	-	-	-	-
Japan	112 424	151 049	127 068	97 448	92 775	147 786	99 299	175 714	138 098	82 127
Korea Rep	232 692	149 776	97 665	141 635	158 911	220 586	101 165	283 455	170 429	167 817
Latvia	11 637	5 029	6 693	1 825	4 041	-	-	-	-	-
Lithuania	19 262	15 171	3 275	-	3 400	-	-	-	-	-
Namibia	-	-	-	-	-	304	677	746	-	-
Panama	4 027	1 059	598	460	706	-	-	159	1 342	...
Philippines	-	-	-	-	-	-	124	47	-	330
Poland	42 510	21 395	13 641	9 205	3 549	-	93	5 224	970	756
Portugal	1 549	1 810	3 766	6 847	9 488	1 595	2 219	80	388	2 794
Russian Fed	93 117	47 925	23 689	11 762	20 254	884	-	-	3 404	3 214
St Vincent	-	-	-	-	-	-	-	-	-	1 818
Seychelles	-	-	-	-	-	1 253	101	-	-	-
Spain	91 069	63 047	72 592	96 511	93 032	65 264	84 266	95 453	84 874	108 669
Ukraine	9 415	4 719	3 202	998	340	-	-	-	-	-
UK	-	446	1 256	2 080	4 356	2 730	3 706	3 259	5 501	7 007
Uruguay	125 428	118 194	119 766	125 597	122 732	134 738	138 514	100 068	101 650	96 851
Total	***2 176 122***	***2 192 868***	***2 102 617***	***2 336 972***	***2 480 181***	***2 756 887***	***2 379 440***	***2 622 540***	***2 375 434***	***2 265 714***

C-47 (a)

Fish, crustaceans, molluscs, etc — Capture production by species items — Atlantic, Southeast
Poissons, crustacés, mollusques, etc — Captures par catégories d'espèces — Atlantique, sud-est
Peces, crustáceos, moluscos, etc — Capturas por categorías de especies — Atlántico, sudoriental

English name Nom anglais Nombre inglés	Scientific name Nom scientifique Nombre científico	Species group Groupe d'espèces Grupo de especies	1995 mt	1996 mt	1997 mt	1998 mt	1999 mt	2000 mt	2001 mt
West coast sole	*Austroglossus microlepis*	31	463	409	517	393	463	571	589
Mud sole	*Austroglossus pectoralis*	31	769	909	837	859	768	800	844
Southeast Atlantic soles nei	*Austroglossus spp*	31	1 365	480	1 197	773	926	965	650
Tonguefishes	*Cynoglossidae*	31	4	3	-	-	-	-	10
Flatfishes nei	*Pleuronectiformes*	31	229	195	120	935	912	592	1 928
Shallow-water Cape hake	*Merluccius capensis*	32	26	832	777	906	2 060	658	1 863
Cape hakes	*Merluccius capensis,M.paradox.*	32	277 186	286 443	260 955	306 269	309 954	300 743	323 257
Gadiformes nei	*Gadiformes*	32	-	-	-	3	-	-	-
Bonefish	*Albula vulpes*	33	-	154	-	-	-	-	-
Sea catfishes nei	*Ariidae*	33	21	19	76	132	102	459	2 435
Mullets nei	*Mugilidae*	33	1 259	1 086	972	908	757	561	125
Groupers nei	*Epinephelus spp*	33	32	48	41	59	33	17	18
Groupers, seabasses nei	*Serranidae*	33	235	1 002	460	1 417	347	2 341	4 229
Bigeyes nei	*Priacanthus spp*	33	3	1	1	3	7	2	1
Snappers, jobfishes nei	*Lutjanidae*	33	-	-	-	-	-	-	1
Bigeye grunt	*Brachydeuterus auritus*	33	124	146	228	262	167	191	1 181
Grunts, sweetlips nei	*Haemulidae (=Pomadasyidae)*	33	149	1 031	557	440	412	817	1 715
Canary drum (=Baardman)	*Umbrina canariensis*	33	...	...	...	...	830	...	2 202
Southern meagre(=Mulloway)	*Argyrosomus hololepidotus*	33	1 268	1 293	932	1 114	1 203	1 009	3 734
Geelbek croaker	*Atractoscion aequidens*	33	353	346	474	605	415	380	383
West African croakers nei	*Pseudotolithus spp*	33	1 035	4 044	1 606	7 096	8 936	8 005	8 744
Croakers, drums nei	*Sciaenidae*	33	4 858	4 004	4 637	3 119	2 920	2 295	2 640
Emperors(=Scavengers) nei	*Lethrinidae*	33	-	-	-	-	-	-	3
Red pandora	*Pagellus bellottii*	33	...	...	...	...	188	...	744
Sargo breams nei	*Diplodus spp*	33	-	-	-	-	-	-	4
Large-eye dentex	*Dentex macrophthalmus*	33	410	682	1 125	736	1 859	2 249	862
Angolan dentex	*Dentex angolensis*	33	-	274	-	-	-	-	-
Dentex nei	*Dentex spp*	33	8 389	3 086	1 567	8 683	8 758	13 765	16 080
Black seabream	*Spondyliosoma cantharus*	33	...	...	...	93	129	212	494
Carpenter seabream	*Argyrozona argyrozona*	33	729	883	780	505	541	500	287
Santer seabream	*Cheimerius nufar*	33	41	40	34	0	29	25	25
Pargo breams nei	*Pagrus spp*	33	39	95	716	1 077	183	1 875	4 606
Red steenbras	*Petrus rupestris*	33	56	28	35	...	22	10	7
Panga seabream	*Pterogymnus laniarius*	33	999	1 200	1 323	1 042	1 349	900	1 058
White stumpnose	*Rhabdosargus globiceps*	33	168	237	165	296	335	300	169
Daggerhead breams nei	*Chrysoblephus spp*	33	197	176	139	154	74	70	78
White steenbras	*Lithognathus lithognathus*	33	11	7	8	9	2	...	...
Sand steenbras	*Lithognathus mormyrus*	33	279	214	176	1 072	726	999	1 735
Steenbrasses nei	*Lithognathus spp*	33	195	6	136	2	116	56	20
Salema	*Sarpa salpa*	33	1	4	3	0	1	...	1
Porgies, seabreams nei	*Sparidae*	33	3 122	2 705	2 919	1 805	4 958	9 869	6 255
Picarels nei	*Spicara spp*	33	9	7	0	55	...	13	-
Goatfishes, red mullets nei	*Mullidae*	33	0	-	0	-	-	-	-
Threadfins, tasselfishes nei	*Polynemidae*	33	284	370	267	520	1 378	872	1 834
Gobies nei	*Gobiidae*	33	5	552	287	20 998	16	3	...
Hector's lanternfish	*Lampanyctodes hectoris*	34	0	33	243	6 553	0	...	...
Conger eels, etc. nei	*Congridae*	34	5	1	3	3	3	3	6
Kingklip	*Genypterus capensis*	34	6 651	6 807	5 955	6 203	7 820	7 922	11 462
Alfonsinos nei	*Beryx spp*	34	909	2 552	4 261	1 810	126	302	318
Orange roughy	*Hoplostethus atlanticus*	34	6 377	13 379	18 538	10 957	3 473	1 542	857
John dory	*Zeus faber*	34	1 145	1 471	1 593	2 328	2 706	2 586	2 618
Boarfishes nei	*Caproidae*	34	-	-	-	5	-	-	-
Oreo dories nei	*Oreosomatidae*	34	6	17	188	6	42	10	54
Wreckfish	*Polyprion americanus*	34	-	-	6	42	20	8	-
Cape bonnetmouth	*Emmelichthys nitidus*	34	96	216	191	124	113	50	37
Bonnetmouths, rubyfishes nei	*Emmelichthyidae*	34	-	-	-	-	-	-	6
Pelagic armourhead	*Pseudopentaceros richardsoni*	34	49	281	18	-	-	-	-
Patagonian toothfish	*Dissostichus eleginoides*	34	-	-	-	-	-	320	5
Snoek	*Thyrsites atun*	34	16 155	13 037	12 033	14 836	12 402	11 166	12 027
Oilfish	*Ruvettus pretiosus*	34	-	-	5	-	-	-	-
Largehead hairtail	*Trichiurus lepturus*	34	13	...	97	119	468	1 895	1 871
Silver scabbardfish	*Lepidopus caudatus*	34	4 931	3 196	2 003	4 557	2 560	2 300	2 316
Hairtails, scabbardfishes nei	*Trichiuridae*	34	-	-	5	-	-	-	-
Ruffs, barrelfishes nei	*Centrolophidae*	34	...	36	-	-	-	-	-
Cape redfish	*Sebastes capensis*	34	1 353	1 759	1 630	1 095	1 214	1 389	1 673
Scorpionfishes nei	*Scorpaenidae*	34	-	-	53	21	4	2	-
Cape gurnard	*Chelidonichthys capensis*	34	559	497	642	839	578	686	609
Gurnards, searobins nei	*Triglidae*	34	0	0	0	-	-	-	-
Devil anglerfish	*Lophius vomerinus*	34	16 526	15 433	17 222	24 729	21 783	21 402	22 007
Demersal percomorphs nei	*Perciformes*	34	319	27	1 147	25	...	...	...
Sardinellas nei	*Sardinella spp*	35	41 452	17 713	23 753	55 020	79 748	113 856	58 339
Southern African pilchard	*Sardinops ocellatus*	35	158 002	106 381	144 680	196 581	175 969	161 448	200 100
Whitehead's round herring	*Etrumeus whiteheadi*	35	78 792	67 773	97 279	57 669	59 032	38 877	56 762
Southern African anchovy	*Engraulis capensis*	35	218 331	41 792	62 640	110 296	180 954	267 986	289 323
Atlantic bonito	*Sarda sarda*	36	9	39	32	-	2	159	1 193
Wahoo	*Acanthocybium solandri*	36	25	23	19	10	15	15	22
Frigate and bullet tunas	*Auxis thazard, A.rochei*	36	21	29	12	31	2	50	231
Little tunny(=Atl.black skipj)	*Euthynnus alletteratus*	36	117	235	75	406	118	132	231
Skipjack tuna	*Katsuwonus pelamis*	36	179	137	397	338	47	739	1 012
Atlantic bluefin tuna	*Thunnus thynnus*	36	120	16	120	3	-	-	-
Albacore	*Thunnus alalunga*	36	17 476	12 367	12 186	20 609	15 406	13 716	14 891
Southern bluefin tuna	*Thunnus maccoyii*	36	2 451	2 242	451	2 069	2 094	2 333	2 516
Yellowfin tuna	*Thunnus albacares*	36	6 321	4 535	3 047	6 029	5 535	4 885	2 829
Bigeye tuna	*Thunnus obesus*	36	24 325	19 778	15 560	15 807	22 292	23 478	19 638
Atlantic sailfish	*Istiophorus albicans*	36	43	50	37	157	75	89	16
Atlantic blue marlin	*Makaira nigricans*	36	810	832	666	617	587	370	499

C-47 (a)

Fish, crustaceans, molluscs, etc — Capture production by species items — Atlantic, Southeast
Poissons, crustacés, mollusques, etc — Captures par catégories d'espèces — Atlantique, sud-est
Peces, crustáceos, moluscos, etc — Capturas por categorías de especies — Atlántico, sudoriental

English name Nom anglais Nombre inglés	Scientific name Nom scientifique Nombre científico	Species group Groupe d'espèces Grupo de especies	1995 mt	1996 mt	1997 mt	1998 mt	1999 mt	2000 mt	2001 mt
Black marlin	*Makaira indica*	36	10	3	1	97	22	62	618
Atlantic white marlin	*Tetrapturus albidus*	36	371	180	136	248	215	307	58
Longbill spearfish	*Tetrapturus pfluegeri*	36	-	-	1	3	28	11	...
Marlins,sailfishes,etc. nei	*Istiophoridae*	36	-	13	60	32	-	56	-
Swordfish	*Xiphias gladius*	36	7 039	8 195	5 623	4 511	5 717	4 666	4 398
Tuna-like fishes nei	*Scombroidei*	36	151	46	217	86	421	270	391
Bluefish	*Pomatomus saltatrix*	37	57	38	15	4	33	20	170
Cape horse mackerel	*Trachurus capensis*	37	506 180	470 372	407 482	481 500	400 279	423 607	354 046
Cunene horse mackerel	*Trachurus trecae*	37	79 068	78 230	87 380	46 004	81 692	69 630	46 832
Jack and horse mackerels nei	*Trachurus spp*	37	...	33	45	...	...	...	...
Scads nei	*Decapterus spp*	37	181	...	...	-	-	-	-
Crevalle jack	*Caranx hippos*	37	42	39	16	89	302	1 835	674
Jacks, crevalles nei	*Caranx spp*	37	-	-	-	-	-	-	1
Yellowtail amberjack	*Seriola lalandi*	37	775	488	477	519	302	300	315
Leerfish	*Lichia amia*	37	-	-	-	-	-	-	1
Atlantic bumper	*Chloroscombrus chrysurus*	37	-	-	2	-	-	-	-
Carangids nei	*Carangidae*	37	9	42	230	83	3	39	806
Atlantic pomfret	*Brama brama*	37	1 729	1 192	420	322	432	843	1 121
Common dolphinfish	*Coryphaena hippurus*	37	-	-	-	-	-	-	1
Chub mackerel	*Scomber japonicus*	37	5 057	4 212	11 550	5 598	2 927	5 427	10 050
Mackerels nei	*Scombridae*	37	-	-	11	-	-	-	-
Blue butterfish	*Stromateus fiatola*	37	0	0	1	21	17	28	-
Barracudas nei	*Sphyraena spp*	37	...	...	...	...	198	-	1
Ocean sunfish	*Mola mola*	37	...	...	...	...	...	...	2
Pelagic percomorphs nei	*Perciformes*	37	-	-	-	-	2 427	...	-
Shortfin mako	*Isurus oxyrinchus*	38	1	...	587	308	338	764	388
Blue shark	*Prionace glauca*	38	21	...	3 560	1 471	2 251	4 711	3 734
Smooth hammerhead	*Sphyrna zygaena*	38	...	...	220	103	...	4	5
Raja rays nei	*Raja spp*	38	1 144	1 047	1 269	1 824	2 399	1 593	2 581
Rays, stingrays, mantas nei	*Rajiformes*	38	602	360	474	523	720	1 341	1 527
Cape elephantfish	*Callorhinchus capensis*	38	386	366	484	482	356	380	405
Sharks, rays, skates, etc. nei	*Elasmobranchii*	38	2 167	1 285	1 187	1 820	2 270	1 895	6 581
Marine fishes nei	*Osteichthyes*	39	54 442	90 578	106 531	60 388	56 941	57 768	90 626
Geryons nei	*Geryon spp*	42	7 163	5 329	3 415	4 145	3 939	5 835	4 485
Marine crabs nei	*Brachyura*	42	...	...	...	1 494	916	1 039	1 090
Tropical spiny lobsters nei	*Panulirus spp*	43	0	0	-	-	-	-	-
Cape rock lobster	*Jasus lalandii*	43	2 180	1 767	1 879	2 076	2 097	2 058	1 974
Tristan da Cunha rock lobster	*Jasus tristani*	43	344	327	321	376	336	316	425
Southern spiny lobster	*Palinurus gilchristi*	43	966	918	892	864	429	305	1 053
Slipper lobsters nei	*Scyllaridae*	43	0	0	0	1	1	1	0
Lobsters nei	*Reptantia*	43	9	20	25	31	68	10	-
Caramote prawn	*Penaeus kerathurus*	45	3	3	-	-	-	-	-
Deepwater rose shrimp	*Parapenaeus longirostris*	45	2 716	3 479	4 246	3 535	1 891	4 209	5 626
Striped red shrimp	*Aristeus varidens*	45	1 570	1 578	1 323	2 808	1 827	3 024	3 405
Natantian decapods nei	*Natantia*	45	-	855	-	9 887	1 615	3 188	4 298
Antarctic krill	*Euphausia superba*	46	-	-	-	254	-	-	-
Marine crustaceans nei	*Crustacea*	47	5	2	-	-	-	-	1 062
Perlemoen abalone	*Haliotis midae*	52	615	735	330	524	481	490	5 265
Donax clams	*Donax spp*	56	0	0	-	-	-	-	-
Cuttlefish,bobtail squids nei	*Sepiidae, Sepiolidae*	57	18	124	-	88	33	90	31
Cape Hope squid	*Loligo reynaudi*	57	7 047	7 549	3 696	6 670	7 169	6 000	3 373
Various squids nei	*Loliginidae, Ommastrephidae*	57	243	106	259	461	665	1 111	1 121
Octopuses, etc. nei	*Octopodidae*	57	125	391	190	582	434	305	129
Marine molluscs nei	*Mollusca*	58	17	-	-	-	-	-	-
Jellyfishes	*Rhopilema spp*	77	-	-	-	-	-	106	44
Total			***1 590 334***	***1 325 597***	***1 355 409***	***1 542 066***	***1 529 255***	***1 635 484***	***1 652 992***

C-47 (b)

Fish, crustaceans, molluscs, etc	Capture production by countries or areas	Atlantic, Southeast
Poissons, crustacés, mollusques, etc	Captures par pays ou zones	Atlantique, sud-est
Peces, crustáceos, moluscos, etc	Capturas por países o áreas	Atlántico, sudoriental

Country or area Pays ou zone País o área	1992 mt	1993 mt	1994 mt	1995 mt	1996 mt	1997 mt	1998 mt	1999 mt	2000 mt	2001 mt
Angola	106 625	119 200	125 413	116 781	131 815	140 304	157 149	169 799	232 351	246 518
Cambodia	-	-	-	-	-	-	-	56	-	-
Chile	-	-	-	-	-	-	-	-	-	5
China	445	-	24	29	24	121	48	5 743	7 209	4 959
China,Taiwan	9 050 F	14 958 F	18 655 F	19 249	11 394	8 934	18 173	18 811	18 970	16 118
Cuba	69	-	-	-	-	-	-	2 427	...	-
Estonia	33 127	31 447	31 372	28 836	-	-	-	-	-	-
France	449	564	129	82	190	38	40	13	83	16
Georgia	5 350 F	2 000 F	1 000 F	-	-	-	-	-	-	-
Honduras	...	193	...	48	10	25	9	20	...	...
Iceland	-	-	-	-	-	924	340	-	-	-
Italy	-	-	-	109	46	-	-	-	-	-
Japan	32 934	39 774	38 106	32 690	26 160	18 924	20 659	18 284	17 286	11 729
Korea Rep	10 551	7 711	11 242	9 216	6 414	9 548	8 020	7 822	7 225	5 026
Lithuania	20 167	20 784	2 823	-	-	-	-	-	-	-
Namibia	654 008	789 132	647 999	568 633	516 628	511 412	610 166	577 838	588 405	545 992
Norway	...	...	...	-	-	864	1 087	-	242	-
Panama	8 294	28	147	-	-	-	-	25	675	...
Philippines	-	-	-	-	-	-	3	48	-	-
Poland	-	-	1	3 178	1 736	1 964	797	-	-	3 100
Portugal	596	798	1 556	1 482	1 130	996	830	1 797	3 732	2 742
Russian Fed	189 058	220 550	226 712	178 288	138 583	129 014	128 280	123 453	82 283	39 850
St Helena	651	726	702	915	819	897	1 060	632	718	866
Seychelles	-	-	-	-	-	-	-	-	6	-
South Africa	690 240	560 716	522 129	574 374	439 229	511 249	555 852	585 240	640 000 F	752 208
Spain	28 852	25 145 F	20 759 F	16 750 F	12 555	14 331	20 990	12 572	17 160	19 580
Ukraine	66 356	96 420	53 356	19 759	28 405	5 701	18 345	4 675	18 096	4 283
Uruguay	-	-	-	-	-	-	-	-	320	-
Other nei	122	253	3 208	19 915	10 459	163	218	-	723	-
Total	***1 856 944***	***1 930 399***	***1 705 333***	***1 590 334***	***1 325 597***	***1 355 409***	***1 542 066***	***1 529 255***	***1 635 484***	***1 652 992***

C-48 (a)

Fish, crustaceans, molluscs, etc — Capture production by species items — Atlantic, Antarctic
Poissons, crustacés, mollusques, etc — Captures par catégories d'espèces — Atlantique, Antarctique
Peces, crustáceos, moluscos, etc — Capturas por categorías de especies — Atlántico, Antártico

English name Nom anglais Nombre inglés	Scientific name Nom scientifique Nombre científico	Species group Groupe d'espèces Grupo de especies	1995 mt	1996 mt	1997 mt	1998 mt	1999 mt	2000 mt	2001 mt
Blue antimora	*Antimora rostrata*	32	1	0	2	1	-	-	-
Whitson's grenadier	*Macrourus whitsoni*	32	-	-	-	-	-	-	1
Grenadiers nei	*Macrourus spp*	32	10	24	15	21	12	5	2
Marbled rockcod	*Notothenia rossii*	33	2	-	-	-	0	0	0
Humped rockcod	*Notothenia gibberifrons*	33	1	-	-	-	5	1	2
Yellowbelly rockcod	*Notothenia neglecta*	33	-	-	-	-	0	-	2
Grey rockcod	*Notothenia squamifrons*	33	-	-	-	-	5	5	0
Yellowfin notie	*Nototheniops nudifrons*	33	-	-	-	-	-	0	0
Antarctic rockcods nei	*Trematomus spp*	33	-	-	-	-	0	-	0
Antarctic silverfish	*Pleuragramma antarcticum*	33	-	-	-	-	-	-	0
Antarctic rockcods, noties nei	*Nototheniidae*	33	2	-	-	0	0	-	0
Lanternfishes nei	*Myctophidae*	34	-	-	-	-	5	67	-
Antarctic toothfish	*Dissostichus mawsoni*	34	-	-	-	1	0	-	0
Patagonian toothfish	*Dissostichus eleginoides*	34	3 262	3 822	2 389	3 259	4 291	4 693	3 348
Patagonian rockcod	*Patagonotothen brevicauda*	34	1	-	-	-	3	0	-
Blackfin icefish	*Chaenocephalus aceratus*	34	-	-	-	-	1	0	1
Mackerel icefish	*Champsocephalus gunnari*	34	10	-	-	6	266	4 114	960
South Georgia icefish	*Pseudochaenichthys georgianus*	34	-	-	-	-	3	0	6
Ocellated icefish	*Chionodraco rastrospinosus*	34	-	-	-	-	1	-	1
Spiny icefish	*Chaenodraco wilsoni*	34	-	-	-	-	0	-	0
Icefishes nei	*Channichthyidae*	34	-	-	-	-	0	-	0
Antarctic starry skate	*Raja georgiana*	38	-	-	-	-	-	0	0
Rays, stingrays, mantas nei	*Rajiformes*	38	90	40	30	14	15	4	13
Marine fishes nei	*Osteichthyes*	39	10	-	-	7	-	-	-
Antarctic stone crab	*Paralomis spinosissima*	44	-	497	-	-	-	0	3
Globose king crab	*Paralomis formosa*	44	-	-	-	-	-	3	11
Antarctic krill	*Euphausia superba*	46	117 448	101 708	82 508	80 874	103 318	104 259	98 209
Sevenstar flying squid	*Martialia hyadesi*	57	-	52	28	53	-	-	2
Jellyfishes	*Rhopilema spp*	77	-	-	-	-	-	5	-
Total			***120 837***	***106 143***	***84 972***	***84 236***	***107 925***	***113 156***	***102 561***

C-48 (b)

Fish, crustaceans, molluscs, etc — Capture production by countries or areas — Atlantic, Antarctic
Poissons, crustacés, mollusques, etc — Captures par pays ou zones — Atlantique, Antarctique
Peces, crustáceos, moluscos, etc — Capturas por países o áreas — Atlántico, Antártico

Country or area Pays ou zone País o área	1992 mt	1993 mt	1994 mt	1995 mt	1996 mt	1997 mt	1998 mt	1999 mt	2000 mt	2001 mt
Argentina	-	-	12	879	109	-	-	6 534	-	-
Bulgaria	115	223	70	179	-	-	-	-	-	-
Chile	8 986	5 386	3 985	1 896	3 098	1 275	1 492	1 668	2 324	896
Estonia	2 352	-	-	-	-	-	-	-	-	-
France	-	-	-	-	-	-	-	-	-	386
Japan	74 275	53 510	61 423	59 037	60 546	58 798	63 309	71 318	67 188	73 523
Korea Rep	519	-	146	423	441	487	1 852	1 487	5 825	5 601
Latvia	-	71	71	-	-	-	-	-	-	-
Panama	-	-	-	141	496	-	-	-	-	-
Poland	8 607	13 406	7 915	9 384	20 610	19 156	15 386	18 554	20 721	14 568
Russian Fed	199 046	4 532	1 231	12	103	-	-	273	3 462	89
South Africa	-	-	3	-	-	-	508	451	324	227
Spain	-	-	-	-	-	294	199	154	264	487
Ukraine	67 028	6 541	8 852	48 886	20 056	4 246	-	5 694	1 113	3 461
UK	17	-	12	-	-	716	1 228	1 255	1 247	1 225
USA	-	299	-	-	684	-	-	16	-	1 638
Uruguay	-	-	-	-	-	-	262	521	10 688	460
Total	***360 945***	***83 968***	***83 720***	***120 837***	***106 143***	***84 972***	***84 236***	***107 925***	***113 156***	***102 561***

Data in these tables refer to catches for which the split-year (1 July - 30 June) is used. Split-year data are shown under the calendar year in which the split-year ends.

Les données dans ces tableaux se réfèrent aux captures pour lesquelles on utilise l'année fractionnée (1er juillet - 30 juin). Les captures relatives à des années fractionnées figurent sous l'année civile durant laquelle se termine l'année fractionnée.

Los datos en estos cuadros se refieren a las capturas para las que se utiliza el año emergente (1°de julio - 30 de junio). Los datos correspondientes a los años emergentes se incluyen en el año civil en que termina el año emergente.

C-51 (a)

Fish, crustaceans, molluscs, etc	Capture production by species items	Indian Ocean, Western
Poissons, crustacés, mollusques, etc	Captures par catégories d'espèces	Océan Indien, ouest
Peces, crustáceos, moluscos, etc	Capturas por categorías de especies	Océano Indico, occidental

English name Nom anglais Nombre inglés	Scientific name Nom scientifique Nombre científico	Species group Groupe d'espèces Grupo de especies	1995 mt	1996 mt	1997 mt	1998 mt	1999 mt	2000 mt	2001 mt
Kelee shad	*Hilsa kelee*	24	2 552	3 837	3 634	3 077	3 076	3 896	6 458
Hilsa shad	*Tenualosa ilisha*	24	1 674	1 710	1 631	1 530	1 472	840	507
Milkfish	*Chanos chanos*	25	189	181	125	139	137	125	134
Barramundi(=Giant seaperch)	*Lates calcarifer*	25	187	214	209	196	204	-	-
Lefteye flounders nei	*Bothidae*	31	-	-	0	9	27	129	125
Tonguefishes	*Cynoglossidae*	31	2 075	2 286	2 523	2 158	2 066	2 137	1 923
Indian halibut	*Psettodes erumei*	31	1 021	880	811	945	850	802	1 025
Flatfishes nei	*Pleuronectiformes*	31	18 622	17 564	21 965	17 564	12 935	24 804	11 359
Unicorn cod	*Bregmaceros mcclellandi*	32	1 071	1 251	1 338	1 449	743	1 470	2 435
Gadiformes nei	*Gadiformes*	32	76	-	10	7	4	2	1
Bombay-duck	*Harpadon nehereus*	33	133 884	159 463	187 982	144 865	146 663	133 221	142 999
Greater lizardfish	*Saurida tumbil*	33	43	46	28	22	0	0	0
Brushtooth lizardfish	*Saurida undosquamis*	33	31	47	53	34	20	30	32
Lizardfishes nei	*Synodontidae*	33	26 493	20 846	16 562	20 221	20 871	17 918	13 357
Sea catfishes nei	*Ariidae*	33	80 770	80 113	89 654	96 464	94 676	76 749	89 526
Squirrelfishes nei	*Holocentridae*	33	7	40	50	34	53	46	40
Flathead grey mullet	*Mugil cephalus*	33	96	39	65	69	89	60	34
Mullets nei	*Mugilidae*	33	25 801	27 126	30 450	28 006	24 220	21 140	22 026
Fusiliers	*Caesio spp*	33	...	...	...	15	18	17	15
Groupers nei	*Epinephelus spp*	33	15 145	16 578	16 536	22 076	26 494	27 215	27 012
Groupers, seabasses nei	*Serranidae*	33	13 671	13 858	14 551	15 452	16 142	32 564	32 949
Therapon pearch	*Terapon spp*	33	-	-	-	1	2	0	1
Spotted seabass	*Dicentrarchus punctatus*	33	101	2	35	36	241	10	18
Bigeyes nei	*Priacanthus spp*	33	5	0	...	2	1	0	1
Sillago-whitings	*Sillaginidae*	33	423	289	266	218	201	194	204
Mangrove red snapper	*Lutjanus argentimaculatus*	33	3 145	2 002	2 394	3 192	3 195	3 003	2 900
Snappers nei	*Lutjanus spp*	33	908	1 036	903	1 025	1 384	778	1 287
Snappers, jobfishes nei	*Lutjanidae*	33	13 612	11 846	13 960	17 856	13 553	15 508	14 279
Threadfin breams nei	*Nemipterus spp*	33	1 983	2 303	1 676	3 926	9 843	13 774	15 562
Threadfin and dwarf breams nei	*Nemipteridae*	33	3 438	2 750	4 703	3 469	4 066	2 917	4 506
Ponyfishes(=Slipmouths) nei	*Leiognathidae*	33	7 915	11 074	13 122	9 899	9 757	5 369	7 868
Silver grunt	*Pomadasys argenteus*	33	...	0	0	3	...	472	...
Grunts, sweetlips nei	*Haemulidae (=Pomadasyidae)*	33	12 675	12 320	12 444	12 582	14 784	20 525	19 251
Meagre	*Argyrosomus regius*	33	45	0	0	19	82	-	-
Southern meagre(=Mulloway)	*Argyrosomus hololepidotus*	33	22	21	17	...	...	19	20
Geelbek croaker	*Atractoscion aequidens*	33	17	15	18	...	...	...	...
Tigertooth croaker	*Otolithes ruber*	33	-	-	-	13	10	...	...
Croakers, drums nei	*Sciaenidae*	33	273 893	275 086	300 376	256 965	309 854	261 700	239 478
Emperors(=Scavengers) nei	*Lethrinidae*	33	45 862	43 755	45 071	46 725	49 274	60 100	61 453
Pandoras nei	*Pagellus spp*	33	...	...	...	...	5	...	...
Dentex nei	*Dentex spp*	33	-	-	-	2	-	-	-
King soldier bream	*Argyrops spinifer*	33	2 711	2 784	3 013	2 992	3 015	3 803	4 026
Santer seabream	*Cheimerius nufar*	33	25	33	28	...	...	...	...
Daggerhead breams nei	*Chrysoblephus spp*	33	70	92	85	...	...	...	0
Yellowfin seabream	*Acanthopagrus latus*	33	273	234	249	280	464	350	271
Porgies, seabreams nei	*Sparidae*	33	17 193	15 085	14 731	11 840	17 598	19 063	16 536
Goatfishes	*Upeneus spp*	33	11 881	11 036	6 766	7 621	8 253	9 171	15 627
Goatfishes, red mullets nei	*Mullidae*	33	1 222	886	930	638	1 088	1 093	2 770
Mojarras(=Silver-biddies) nei	*Gerres spp*	33	1 772	1 864	2 164	2 208	2 209	2 592	2 614
Blue sea chub	*Kyphosus cinerascens*	33	-	-	8	-	-	-	-
Wrasses, hogfishes, etc. nei	*Labridae*	33	2 975	3 833	3 305	3 066	3 155	3 291	3 579
Parrotfishes nei	*Scaridae*	33	915	842	1 179	1 048	1 642	790	781
Angelfishes nei	*Pomacanthidae*	33	-	0	0	7	17	1	3
Fourfinger threadfin	*Eleutheronema tetradactylum*	33	812	516	1 783	969	...	63	55
Threadfins, tasselfishes nei	*Polynemidae*	33	3 837	1 823	1 535	2 000	2 829	2 248	2 827
Percoids nei	*Percoidei*	33	11 799	90 233	59 477	52 583	54 015	58 778	40 405
Surgeonfishes nei	*Acanthuridae*	33	113	238	173	253	205	149	343
Batfishes	*Platax spp*	33	-	-	0	1	7	1	0
Spadefishes nei	*Ephippidae*	33	-	-	-	17	1	1	2
Spinefeet(=Rabbitfishes) nei	*Siganus spp*	33	9 229	10 651	9 765	8 857	8 341	10 614	11 621
Flatheads nei	*Platycephalidae*	33	14	0	0	18	28	10	17
Puffers nei	*Tetraodontidae*	33	-	-	-	-	0	0	18
Triggerfishes, durgons nei	*Balistidae*	33	1	291	26	8	11	6	7
Lanternfishes nei	*Myctophidae*	34	2 002	0	0	0	0	0	335
Pike-congers nei	*Muraenesox spp*	34	10 922	11 418	12 965	11 456	13 796	11 177	10 670
Conger eels, etc. nei	*Congridae*	34	-	-	6	64	19	53	-
Alfonsinos nei	*Beryx spp*	34	2 249	3 079	1 031	859	1 964	1 668	585
Orange roughy	*Hoplostethus atlanticus*	34	-	-	-	-	-	1 265	711
John dory	*Zeus faber*	34	-	-	3	4	5	6	6
Boarfishes nei	*Caproidae*	34	-	-	-	-	-	-	7
Oreo dories nei	*Oreosomatidae*	34	-	-	-	-	-	175	180
Wreckfish	*Polyprion americanus*	34	-	-	-	-	-	-	1
Bonnetmouths, rubyfishes nei	*Emmelichthyidae*	34	144	28	7	275	181	-	86
Pelagic armourhead	*Pseudopentaceros richardsoni*	34	54	17	33	78	108	121	12
Patagonian toothfish	*Dissostichus eleginoides*	34	-	-	-	-	-	1 628	7 002
Antarctic toothfishes nei	*Dissostichus spp*	34	-	-	-	-	-	-	122
Largehead hairtail	*Trichiurus lepturus*	34	6 093	9 073	11 583	12 337	31 635	28 756	27 355
Hairtails, scabbardfishes nei	*Trichiuridae*	34	37 924	52 550	148 035	70 469	101 970	110 164	98 330
Ruffs, barrelfishes nei	*Centrolophidae*	34	485	254	440	395	753	396	299
Blackbelly rosefish	*Helicolenus dactylopterus*	34	-	-	-	-	-	-	1
Scorpionfishes nei	*Scorpaenidae*	34	-	-	3	1	1	2	2
Gurnards, searobins nei	*Triglidae*	34	-	-	2	-	...	0	0
Demersal percomorphs nei	*Perciformes*	34	7 729	14 719	15 632	22 357	16 488	13 417	16 841
Indian oil sardine	*Sardinella longiceps*	35	125 036	152 012	197 296	184 002	183 437	353 353	383 289
Sardinellas nei	*Sardinella spp*	35	16 412	31 089	20 839	19 659	29 843	27 897	26 983

C-51 (a)

Fish, crustaceans, molluscs, etc — **Capture production by species items** — **Indian Ocean, Western**
Poissons, crustacés, mollusques, etc — **Captures par catégories d'espèces** — **Océan Indien, ouest**
Peces, crustáceos, moluscos, etc — **Capturas por categorías de especies** — **Océano Indico, occidental**

English name Nom anglais Nombre inglés	Scientific name Nom scientifique Nombre científico	Species group Groupe d'espèces Grupo de especies	1995 mt	1996 mt	1997 mt	1998 mt	1999 mt	2000 mt	2001 mt
Southern African pilchard	*Sardinops ocellatus*	35	-	-	1	...	...	0	0
Red-eye round herring	*Etrumeus teres*	35	-	-	-	-	2 135	...	...
Stolephorus anchovies	*Stolephorus spp*	35	9 533	9 633	10 295	10 329	10 587	2 729	2 730
Anchovies, etc. nei	*Engraulidae*	35	68 641	83 500	86 400	83 493	70 797	76 254	81 070
Dorab wolf-herring	*Chirocentrus dorab*	35	2 289	1 580	1 931	2 059	2 273	2 785	2 613
Wolf-herrings nei	*Chirocentrus spp*	35	8 422	5 874	11 998	10 279	4 743	6 888	10 108
Clupeoids nei	*Clupeoidei*	35	77 900	89 567	104 273	105 952	103 738	76 761	74 928
Striped bonito	*Sarda orientalis*	36	788	370	498	162	134	95	287
Wahoo	*Acanthocybium solandri*	36	82	94	79	76	77	69	4
Narrow-barred Spanish mackerel	*Scomberomorus commerson*	36	80 844	72 383	75 367	79 829	74 284	82 213	86 891
Indo-Pacific king mackerel	*Scomberomorus guttatus*	36	17 510	12 350	12 674	18 360	16 009	17 785	49 739
Streaked seerfish	*Scomberomorus lineolatus*	36	59	59	558	67	58	64	...
Seerfishes nei	*Scomberomorus spp*	36	3 886	4 417	6 590	4 471	4 590	4 120	4 211
Frigate and bullet tunas	*Auxis thazard, A.rochei*	36	15 796	18 732	13 341	12 186	18 029	19 485	13 089
Kawakawa	*Euthynnus affinis*	36	32 030	30 491	39 489	33 488	38 273	39 501	22 000
Skipjack tuna	*Katsuwonus pelamis*	36	254 657	224 876	226 851	225 668	309 973	303 875	295 640
Longtail tuna	*Thunnus tonggol*	36	56 194	41 564	43 465	38 899	41 933	58 255	51 920
Albacore	*Thunnus alalunga*	36	13 036	14 692	19 196	30 154	20 490	11 401	8 750
Southern bluefin tuna	*Thunnus maccoyii*	36	1 808	2 600	3 977	3 939	2 495	2 580	3 068
Yellowfin tuna	*Thunnus albacares*	36	256 144	257 041	233 939	199 911	230 738	236 291	206 950
Bigeye tuna	*Thunnus obesus*	36	74 603	73 184	85 251	85 350	94 225	95 862	76 227
Tunas nei	*Thunnini*	36	...	...	...	70	96	90	110
Indo-Pacific sailfish	*Istiophorus platypterus*	36	5 911	4 312	3 727	2 976	4 373	4 642	6 309
Indo-Pacific blue marlin	*Makaira mazara*	36	2 589	2 825	3 338	5 114	4 841	5 730	4 526
Atlantic blue marlin	*Makaira nigricans*	36	-	-	-	-	-	-	2
Black marlin	*Makaira indica*	36	601	403	369	403	266	625	105
Striped marlin	*Tetrapturus audax*	36	3 933	3 334	2 864	2 715	1 849	2 493	1 801
Shortbill spearfish	*Tetrapturus angustirostris*	36	-	2	2	5	7	5	0
Marlins,sailfishes,etc. nei	*Istiophoridae*	36	6 466	8 441	7 470	5 961	5 331	5 610	3 146
Swordfish	*Xiphias gladius*	36	23 423	17 113	22 021	26 726	24 076	21 366	16 961
Tuna-like fishes nei	*Scombroidei*	36	11 986	8 849	10 376	17 505	10 950	14 943	19 172
Needlefishes nei	*Tylosurus spp*	37	974	359	420	276	397	250	286
Halfbeaks nei	*Hemiramphus spp*	37	5 497	2 576	3 064	2 545	3 189	2 955	2 824
Flyingfishes nei	*Exocoetidae*	37	167	133	184	202	401	166	159
Silversides(=Sand smelts) nei	*Atherinidae*	37	-	-	-	-	78	-	-
False trevally	*Lactarius lactarius*	37	4 520	6 403	7 060	8 000	5 503	4 623	5 992
Cobia	*Rachycentron canadum*	37	2 803	2 016	1 913	1 624	1 516	3 862	3 832
Jack and horse mackerels nei	*Trachurus spp*	37	4 415	1 380	1 440	1 615	1 611	1 400	1 813
Scads nei	*Decapterus spp*	37	3 710	2 819	3 158	5 464	6 673	5 205	4 961
Jacks, crevalles nei	*Caranx spp*	37	20 725	64 921	50 385	70 017	71 079	37 641	45 552
Snubnose pompano	*Trachinotus blochii*	37	-	-	-	4	0	31	-
Pompanos nei	*Trachinotus spp*	37	6 765	5 706	3 697	2 533	2 329	12	11
Yellowtail amberjack	*Seriola lalandi*	37	-	-	1	...	...	...	...
Amberjacks nei	*Seriola spp*	37	304	130	334	340	179	146	152
Black pomfret	*Parastromateus niger*	37	3 066	2 221	2 322	2 109	2 917	2 027	1 975
Rainbow runner	*Elagatis bipinnulata*	37	...	...	415	98	132	5	52
Golden trevally	*Gnathanodon speciosus*	37	434	441	471	536	489	1 125	1 144
Torpedo scad	*Megalaspis cordyla*	37	7 455	3 954	3 119	2 122	2 498	3 117	2 925
Queenfishes	*Scomberoides spp*	37	3 361	3 239	3 288	3 327	3 436	1 889	1 759
Yellowstripe scad	*Selaroides leptolepis*	37	2 878	2 908	3 108	3 118	3 196	2 635	2 630
Carangids nei	*Carangidae*	37	48 809	66 644	67 290	58 616	60 933	42 490	38 764
Common dolphinfish	*Coryphaena hippurus*	37	2 625	1 897	1 719	1 953	3 393	2 203	2 060
Chub mackerel	*Scomber japonicus*	37	1 926	2 053	2 397	813	385	3 561	2 747
Indian mackerel	*Rastrelliger kanagurta*	37	190 465	282 311	173 395	157 031	153 400	77 074	47 508
Indian mackerels nei	*Rastrelliger spp*	37	455	409	356	351	449	424	303
Mackerels nei	*Scombridae*	37	11	12	10	23	41	24	23
Silver pomfret	*Pampus argenteus*	37	1 142	893	576	515	289	243	158
Butterfishes, pomfrets nei	*Stromateidae*	37	39 168	22 558	22 008	19 400	19 672	14 362	11 547
Barracudas nei	*Sphyraena spp*	37	15 323	14 902	14 151	20 704	22 118	12 066	12 026
Pelagic percomorphs nei	*Perciformes*	37	62 276	64 966	64 007	74 296	85 318	69 631	85 271
Shortfin mako	*Isurus oxyrinchus*	38	-	-	-	18	...	58	95
Porbeagle	*Lamna nasus*	38	-	-	-	-	-	-	1
Blue shark	*Prionace glauca*	38	-	-	-	60	...	575	1 123
Dusky shark	*Carcharhinus obscurus*	38	-	-	7	0	...	...	...
Requiem sharks nei	*Carcharhinidae*	38	32 459	34 483	31 235	36 184	32 573	28 384	26 641
Guitarfishes, etc. nei	*Rhinobatidae*	38	1 209	1 422	1 481	1 564	1 643	2 185	1 944
Sawfishes	*Pristidae*	38	23	-	-	-	-	-	-
Rays, stingrays, mantas nei	*Rajiformes*	38	20 373	20 050	19 844	21 550	24 841	24 700	24 660
Sharks, rays, skates, etc. nei	*Elasmobranchii*	38	63 984	121 757	65 384	61 382	56 532	70 640	63 192
Marine fishes nei	*Osteichthyes*	39	816 105	600 508	649 586	617 389	648 207	628 354	646 074
Portunus swimcrabs nei	*Portunus spp*	42	807	1 047	1 289	1 017	2 179	2 380	2 557
Indo-Pacific swamp crab	*Scylla serrata*	42	22	23	19	23	21	25	24
Geryons nei	*Geryon spp*	42	414	564	1 144	1 002	875	886	682
Marine crabs nei	*Brachyura*	42	27 290	6 018	7 362	9 041	8 764	17 661	18 330
Tropical spiny lobsters nei	*Panulirus spp*	43	2 525	2 488	2 547	2 822	2 459	2 239	2 243
St.Paul rock lobster	*Jasus paulensis*	43	439	357	295	308	345	192	183
Natal spiny lobster	*Palinurus delagoae*	43	13	10	10	6	7	8	10
Spiny lobsters nei	*Palinuridae*	43	248	332	233	239	204	228	199
Slipper lobsters nei	*Scyllaridae*	43	158	133	74	67	20	9	26
Mozambique lobster	*Metanephrops mozambicus*	43	179	132	156	192	152	180	141
Lobsters nei	*Reptantia*	43	-	-	-	-	-	62	2
Giant tiger prawn	*Penaeus monodon*	45	140 160	112 483	106 762	168 026	169 080	145 996	136 583
Green tiger prawn	*Penaeus semisulcatus*	45	0	0	-	0	0	0	0
Penaeus shrimps nei	*Penaeus spp*	45	22 806	25 820	26 056	25 181	20 895	24 555	24 378

C-51 (a)

Fish, crustaceans, molluscs, etc — Capture production by species items — Indian Ocean, Western
Poissons, crustacés, mollusques, etc — Captures par catégories d'espèces — Océan Indien, ouest
Peces, crustáceos, moluscos, etc — Capturas por categorías de especies — Océano Indico, occidental

English name Nom anglais Nombre inglés	Scientific name Nom scientifique Nombre científico	Species group Groupe d'espèces Grupo de especies	1995 mt	1996 mt	1997 mt	1998 mt	1999 mt	2000 mt	2001 mt
Metapenaeus shrimps nei	*Metapenaeus spp*	45	6 981	7 602	6 801	6 204	6 791	7 126	7 246
Deepwater rose shrimp	*Parapenaeus longirostris*	45	-	-	-	-	-	-	9
Parapenaeopsis shrimps nei	*Parapenaeopsis spp*	45	12 919	14 047	16 722	14 689	12 889	11 945	11 576
Knife shrimp	*Haliporoides triarthrus*	45	2 036	1 771	1 510	1 882	1 611	1 766	1 738
Natantian decapods nei	*Natantia*	45	107 164	136 123	115 045	120 574	111 085	119 866	113 782
Marine crustaceans nei	*Crustacea*	47	12 078	15 600	12 367	12 797	11 460	11 251	9 416
Abalones nei	*Haliotis spp*	52	43	43	40	40	29	45	51
Cupped oysters nei	*Crassostrea spp*	53	14	32	24	28	34	32	29
Cuttlefish,bobtail squids nei	*Sepiidae, Sepiolidae*	57	8 882	12 907	23 236	14 778	23 302	19 975	22 086
Common squids nei	*Loligo spp*	57	827	30	37	37	10	22	41
Various squids nei	*Loliginidae, Ommastrephidae*	57	3 876	3 628	5 820	4 764	6 714	5 130	5 187
Octopuses, etc. nei	*Octopodidae*	57	1 372	1 199	1 454	1 277	1 326	1 169	1 290
Cephalopods nei	*Cephalopoda*	57	95 684	77 104	118 032	91 576	89 953	92 724	111 520
Marine molluscs nei	*Mollusca*	58	3 363	15 996	7 363	5 015	3 747	5 034	2 452
Sea squirts nei	*Ascidiacea*	74	-	-	-	-	-	95	-
Sea cucumbers nei	*Holothurioidea*	76	3 415	3 658	3 693	2 407	767	1 139	1 230
Total			***3 807 617***	***3 884 762***	***3 978 510***	***3 792 545***	***4 037 596***	***3 997 956***	***3 948 676***

C-51 (b)

Fish, crustaceans, molluscs, etc — Capture production by countries or areas — Indian Ocean, Western
Poissons, crustacés, mollusques, etc — Captures par pays ou zones — Océan Indien, ouest
Peces, crustáceos, moluscos, etc — Capturas por países o áreas — Océano Indico, occidental

Country or area Pays ou zone País o área	1992 mt	1993 mt	1994 mt	1995 mt	1996 mt	1997 mt	1998 mt	1999 mt	2000 mt	2001 mt
Bahrain	7 983	8 958	7 628	9 389	12 940	10 050	9 849	10 620	11 718	11 230
Br Ind Oc Tr	0	0	0	0	0	0	0	0	0	0
China	-	-	-	-	-	-	-	294	2 587	2 773
China,Taiwan	51 904 F	121 654 F	63 484 F	88 540 F	77 278	74 522	96 352	88 423	92 017	95 003
Comoros	11 825	11 645	12 976	13 000 F	12 700 F	12 500 F	12 500 F	12 000	13 200	12 180
Djibouti	275 F	300 F	320 F	350 F	350 F	350 F	350 F	350 F	350 F	350 F
Egypt	38 700	45 800	43 200	42 399	43 272	50 925	50 519	74 200	75 972	73 577
Eritrea	...	475	2 706	3 559	3 252	1 038	1 629	6 891	12 612	8 820
Ethiopia	100	-	-	-	-	-	-	-	-	-
France	95 614	93 057	99 908	95 917	82 854	68 920	48 720	80 443	84 544	69 152
Fr South Tr	464	460	524	519 F	437 F	375 F	388 F	425 F	272 F	263 F
India	1 779 782	1 744 857	1 904 799	1 847 631	1 954 910	2 006 621	1 832 799	1 911 318	1 848 678	1 896 206
Iran	243 969	246 985	218 944	252 583	242 437	238 486	226 500	243 800	260 500	262 805
Iraq	543	2 133	4 221	5 253	11 688	10 783	13 463	13 093	12 389	12 400 F
Israel	98	80	110	150	225	171	137	98	-	-
Italy	10 189	8 689	8 717	1 650	702	3 351	4 520	6 890	102	3 199
Japan	53 961	52 958	32 730	14 273	21 152	28 796	31 030	20 377	19 542	24 739
Jordan	30	45	60	75	90	100	120	160	150	170
Kenya	6 566	5 617	3 772	5 465	6 296	6 099	6 600	6 634	4 763	7 388
Korea Rep	30 405	29 879	25 322	22 765	31 573	37 976	25 692	19 943	21 435	10 878
Kuwait	7 871	8 466	7 752	8 616	8 255	7 826	7 799	6 271	6 000 F	5 846
Lithuania	-	-	11	-	-	-	-	-	-	-
Madagascar	76 905	84 261	86 431	85 653	84 475	86 391	96 395	99 630	102 093	105 583
Maldives	90 399	100 851	118 673	119 521	121 137	122 387	132 566	134 962	136 420	125 814
Mauritius	18 861	20 576	18 145	16 395	11 869	14 025	12 093	12 205	9 615	10 694
Mayotte	1 100 F	500 F	600 F	700 F	1 000	1 300 F	2 000 F	2 000	5 500	5 500
Mozambique	27 808	25 506	22 531	21 740	27 405	28 035	27 683	23 746	25 977	24 436
Norway	-	-	-	-	-	-	-	-	870	-
Oman	109 218	105 772	118 572	139 861	121 618	118 995	106 171	108 809	120 421	126 531
Pakistan	431 267	499 159	418 574	404 444	395 397	422 265	433 456	474 665	437 601	426 920
Philippines	-	-	-	-	-	-	2 513	2 190	1 985	1 453
Portugal	-	-	-	-	-	-	513	804	1 484	3 536
Qatar	7 845	6 994	5 086	4 271	4 739	5 032	5 279	4 207	7 142	8 606
Réunion	1 103	1 679	2 531	2 500	3 607	4 288	4 579	4 043	4 079	3 635
Russian Fed	17 558	10 893	19 444	-	-	-	-	-	-	221
Saudi Arabia	45 910	48 021	54 612	45 609	47 698	49 314	51 206	46 618	49 650	49 167
Seychelles	6 663	5 178	4 469	4 008	4 707	11 741	20 000	34 235	31 982	46 994
Somalia	25 500 F	27 500 F	29 600	27 700 F	25 800 F	23 900 F	22 000 F	20 000 F	20 000 F	19 800 F
South Africa	1 267	826	517	373	349	890	1 100	1 000 F	926	1 017
Spain	89 623	105 651	113 354	147 545	138 928	141 340	98 223	144 645	148 235	127 481
Sudan	2 000	2 500	4 000	4 000	4 500	5 000	5 500	5 500	5 000 F	5 000
Tanzania	56 085	36 685	40 785	42 771	61 645	50 210	48 000	50 489	52 779	52 900
Ukraine	2 629	1 491	2 633	2 970	3 480	1 570	1 626	3 356	2 015	810
Untd Arab Em	95 046	99 600	108 600	105 884	107 000	114 358	114 739	117 607	110 056	110 000 F
Uruguay	-	-	-	-	-	-	-	-	1 628	7 150
Yemen	79 547	82 356	81 885	107 970	104 955	115 600	127 620	124 385	114 751	142 198
Other nei	56 243	88 442	76 156	111 568	104 042	102 980	110 316	120 270	140 916	46 251
Total	***3 582 856***	***3 736 499***	***3 764 382***	***3 807 617***	***3 884 762***	***3 978 510***	***3 792 545***	***4 037 596***	***3 997 956***	***3 948 676***

English name Nom anglais Nombre inglés	Scientific name Nom scientifique Nombre científico	Species group Groupe d'espèces Grupo de especies	1995 mt	1996 mt	1997 mt	1998 mt	1999 mt	2000 mt	2001 mt
Short-finned eel	*Anguilla australis*	22	259	208	203	160	129	133	157
Chacunda gizzard shad	*Anodontostoma chacunda*	24	2 338	2 167	1 916	1 474	3 009	2 645	2 769
Kelee shad	*Hilsa kelee*	24	40 714	48 452	47 524	48 475	41 468	43 604	36 325
Hilsa shad	*Tenualosa ilisha*	24	129 115	135 358	131 204	124 105	140 710	140 367	140 000
Toli shad	*Tenualosa toli*	24	456	429	476	617	630	528	570
Indian pellona	*Pellona ditchela*	24	6 091	11 227	10 919	6 113	4 448	7 069	6 884
Diadromous clupeoids nei	*Clupeoidei*	24	30	107	102	22	45	13	50
Barramundi(=Giant seaperch)	*Lates calcarifer*	25	12 274	13 991	13 792	23 062	15 687	15 923	17 360
Sand flounders	*Rhombosolea spp*	31	34	26	26	26	29	1	...
Tonguefishes	*Cynoglossidae*	31	6 813	10 285	10 915	11 682	11 316	11 008	10 384
Indian halibut	*Psettodes erumei*	31	6 093	10 863	12 212	7 779	7 472	7 592	9 068
Flatfishes nei	*Pleuronectiformes*	31	7 975	9 650	10 290	8 717	8 148	8 545	6 289
Unicorn cod	*Bregmaceros mcclellandi*	32	161	36	13	1 138	1 352	159	32
Blue grenadier	*Macruronus novaezelandiae*	32	3 239	2 680	2 851	4 752	6 209	9 493	7 561
Gadiformes nei	*Gadiformes*	32	-	-	-	-	5	1	30
Indo-Pacific tarpon	*Megalops cyprinoides*	33	7	6	4	13	29	31	12
Bombay-duck	*Harpadon nehereus*	33	25 803	27 132	27 642	37 823	38 037	1 580	1 587
Lizardfishes nei	*Synodontidae*	33	21 386	21 665	21 510	51 137	22 861	58 627	58 844
Sea catfishes nei	*Ariidae*	33	54 222	60 907	59 149	53 098	64 274	66 992	67 799
Eeltail catfishes	*Plotosus spp*	33	1 104	881	712	742	888	1 444	1 231
Mullets nei	*Mugilidae*	33	18 824	23 954	23 796	23 814	24 590	26 860	26 110
Fusiliers	*Caesio spp*	33	5 390	4 198	5 329	4 862	5 569	5 107	5 184
Groupers nei	*Epinephelus spp*	33	14 160	15 399	17 005	18 211	16 307	16 311	16 898
Groupers, seabasses nei	*Serranidae*	33	4 155	4 159	3 296	4 294	2 645	2 624	2 616
Bigeyes nei	*Priacanthus spp*	33	12 508	15 911	16 371	16 721	13 213	13 032	13 189
Sillago-whitings	*Sillaginidae*	33	4 126	4 330	5 550	5 302	8 094	8 133	8 475
Ruff	*Arripis georgianus*	33	1 063	1 302	1 287	1 008	1 066	1 143	1 005
Australian salmon	*Arripis trutta*	33	4 840	3 507	3 457	3 623	3 354	4 087	3 799
Mangrove red snapper	*Lutjanus argentimaculatus*	33	659	662	658	1 227	615	789	608
Snappers nei	*Lutjanus spp*	33	11 110	13 479	11 887	13 127	14 236	13 890	14 166
Snappers, jobfishes nei	*Lutjanidae*	33	6 385	6 322	5 439	4 288	6 528	5 819	5 939
Threadfin breams nei	*Nemipterus spp*	33	41 490	41 795	40 520	54 583	40 546	39 463	37 334
Monocle breams	*Scolopsis spp*	33	2	1	1	2	96	86	88
Ponyfishes(=Slipmouths)	*Leiognathus spp*	33	34	101	78	86	84	161	71
Ponyfishes(=Slipmouths) nei	*Leiognathidae*	33	68 757	73 735	72 277	57 754	69 136	62 360	62 689
Silver grunt	*Pomadasys argenteus*	33	323	625	539	666	548	581	507
Grunts, sweetlips nei	*Haemulidae (=Pomadasyidae)*	33	3 755	4 359	4 079	4 274	4 553	4 777	5 011
Southern meagre(=Mulloway)	*Argyrosomus hololepidotus*	33	26	44	44	44	96	73	138
Croakers, drums nei	*Sciaenidae*	33	85 596	92 873	86 686	92 464	80 541	77 267	75 041
Emperors(=Scavengers) nei	*Lethrinidae*	33	4 950	5 046	5 921	6 791	7 060	6 648	6 460
Silver seabream	*Pagrus auratus*	33	3 025	3 293	3 213	2 487	2 588	2 326	...
Porgies, seabreams nei	*Sparidae*	33	555	278	116	159	991	1 992	966
Goatfishes	*Upeneus spp*	33	17 546	19 553	24 306	26 294	28 522	23 328	25 986
Spotted sicklefish	*Drepane punctata*	33	17	51	23	30	58	53	26
Wrasses, hogfishes, etc. nei	*Labridae*	33	129	84	80	80	100	137	86
Threadfins, tasselfishes nei	*Polynemidae*	33	14 643	13 576	13 197	15 136	15 040	12 482	13 067
Percoids nei	*Percoidei*	33	26 716	26 513	28 741	10 268	36 350	36 264	40 301
Spinefeet(=Rabbitfishes) nei	*Siganus spp*	33	86	81	88	237	27	109	87
Flatheads nei	*Platycephalidae*	33	1 360	1 640	1 477	989	3 102	2 779	2 423
Puffers nei	*Tetraodontidae*	33	62	158	122	115	3	2	...
Triggerfishes, durgons nei	*Balistidae*	33	540	708	543	250	248	115	193
Daggertooth pike conger	*Muraenesox cinereus*	34	2 467	2 793	4 441	4 277	3 568	3 322	3 238
Pike-congers nei	*Muraenesox spp*	34	2 466	4 097	4 192	5 189	8 182	5 748	6 273
Pink cusk-eel	*Genypterus blacodes*	34	0	1 397	1 923	1 830	1 881	2 039	1 696
Orange roughy	*Hoplostethus atlanticus*	34	227	357	350	4 857	7 553	4 974	5 145
John dory	*Zeus faber*	34	0	3	0	1	21	561	238
Mirror dory	*Zenopsis nebulosus*	34	0	4	9	37	1 033	500	528
Morwongs	*Nemadactylus spp*	34	75	77	94	95	638	630	775
Trumpeters nei	*Latridae*	34	0	0	0	0	153	166	69
Antarctic toothfishes nei	*Dissostichus spp*	34	-	-	-	-	-	-	450
Snoek	*Thyrsites atun*	34	400	400	300	200	88	120	154
Silver gemfish	*Rexea solandri*	34	3 000	2 000	...	...	4	2	...
Largehead hairtail	*Trichiurus lepturus*	34	12 668	14 236	16 405	30 261	15 504	11 336	11 003
Hairtails, scabbardfishes nei	*Trichiuridae*	34	36 220	28 219	31 620	34 760	42 058	38 193	68 704
South Pacific breams nei	*Seriolella spp*	34	2 558	3 406	4 020	3 356	3 327	3 449	4 190
Bluefin gurnard	*Chelidonichthys kumu*	34	1	0	1	1	0	-	-
Latchet(=Sharpbeak gurnard)	*Pterygotrigla polyommata*	34	0	0	60	25	55	66	78
Demersal percomorphs nei	*Perciformes*	34	10 088	11 468	9 100	9 210	10 440	14 910	12 290
Goldstripe sardinella	*Sardinella gibbosa*	35	26 450	29 836	26 335	31 715	34 785	37 358	38 310
Indian oil sardine	*Sardinella longiceps*	35	72 565	71 343	91 846	63 063	25 461	49 583	54 039
Bali sardinella	*Sardinella lemuru*	35	40 363	29 647	72 666	92 294	34 143	31 237	33 190
Sardinellas nei	*Sardinella spp*	35	54 849	53 086	51 084	56 813	54 321	53 577	53 250
Rainbow sardine	*Dussumieria acuta*	35	4 604	5 013	4 909	4 729	6 072	5 081	5 190
Stolephorus anchovies	*Stolephorus spp*	35	74 584	74 649	81 167	85 443	77 780	75 085	73 221
Anchovies, etc. nei	*Engraulidae*	35	58 237	53 524	52 432	48 490	45 730	44 488	37 946
Dorab wolf-herring	*Chirocentrus dorab*	35	6 090	7 551	7 744	7 648	8 451	8 335	8 284
Wolf-herrings nei	*Chirocentrus spp*	35	19 847	20 784	20 021	12 861	20 884	17 743	17 440
Clupeoids nei	*Clupeoidei*	35	87 064	90 378	89 619	87 706	86 057	89 150	85 200
Wahoo	*Acanthocybium solandri*	36	129	138	168	209	513	569	529
Narrow-barred Spanish mackerel	*Scomberomorus commerson*	36	19 235	27 015	24 807	34 183	40 602	45 470	40 558
Indo-Pacific king mackerel	*Scomberomorus guttatus*	36	15 840	18 608	18 202	20 416	27 477	29 482	12 567
Streaked seerfish	*Scomberomorus lineolatus*	36	28	37	343	40	77	83	381
Seerfishes nei	*Scomberomorus spp*	36	11 143	9 846	10 770	11 667	9 408	9 944	7 662
Frigate and bullet tunas	*Auxis thazard, A.rochei*	36	8 254	10 536	11 346	12 733	10 659	13 038	8 240

C-57 (a)

Fish, crustaceans, molluscs, etc — Capture production by species items — Indian Ocean, Eastern
Poissons, crustacés, mollusques, etc — Captures par catégories d'espèces — Océan Indien, est
Peces, crustáceos, moluscos, etc — Capturas por categorías de especies — Océano Indico, oriental

English name Nom anglais Nombre inglés	Scientific name Nom scientifique Nombre científico	Species group Groupe d'espèces Grupo de especies	1995 mt	1996 mt	1997 mt	1998 mt	1999 mt	2000 mt	2001 mt
Kawakawa	*Euthynnus affinis*	36	99 308	100 196	98 024	117 856	116 414	113 971	100 886
Skipjack tuna	*Katsuwonus pelamis*	36	65 165	73 311	88 476	100 765	108 990	94 358	88 007
Longtail tuna	*Thunnus tonggol*	36	21 594	26 227	24 205	22 677	19 824	19 239	6 026
Albacore	*Thunnus alalunga*	36	6 315	8 709	6 719	7 259	7 035	4 279	1 627
Southern bluefin tuna	*Thunnus maccoyii*	36	5 742	8 450	7 135	8 580	10 449	7 727	7 823
Yellowfin tuna	*Thunnus albacares*	36	22 982	30 269	36 016	46 162	45 315	36 327	37 927
Bigeye tuna	*Thunnus obesus*	36	27 111	28 068	30 766	30 482	25 224	27 910	16 552
Indo-Pacific sailfish	*Istiophorus platypterus*	36	3 462	5 396	6 585	8 223	7 078	9 214	8 123
Indo-Pacific blue marlin	*Makaira mazara*	36	799	781	871	1 009	1 356	1 896	426
Black marlin	*Makaira indica*	36	278	229	208	207	225	447	13
Striped marlin	*Tetrapturus audax*	36	1 231	1 395	1 250	1 205	1 141	1 130	277
Shortbill spearfish	*Tetrapturus angustirostris*	36	-	-	-	-	1	-	0
Marlins,sailfishes,etc. nei	*Istiophoridae*	36	5 502	7 777	9 373	10 572	10 674	9 487	8 286
Swordfish	*Xiphias gladius*	36	5 732	7 811	9 573	9 813	8 169	13 783	9 509
Tuna-like fishes nei	*Scombroidei*	36	27 376	26 881	26 175	74 730	32 665	50 145	51 474
Needlefishes nei	*Tylosurus spp*	37	6 081	6 584	8 335	8 335	7 848	9 154	9 660
Halfbeaks nei	*Hemiramphus spp*	37	5 020	5 199	4 967	3 940	3 461	5 993	4 696
Flyingfishes nei	*Exocoetidae*	37	5 255	6 458	7 200	8 117	6 219	7 224	7 269
False trevally	*Lactarius lactarius*	37	2 595	532	1 097	1 745	1 416	1 456	1 293
Bluefish	*Pomatomus saltatrix*	37	54	56	-	-	-	-	-
Cobia	*Rachycentron canadum*	37	62	76	111	87	111	136	150
Indian scad	*Decapterus russelli*	37	33 373	41 305	37 611	41 459	36 536	32 845	33 370
Scads nei	*Decapterus spp*	37	33 763	36 299	33 653	39 497	41 602	42 051	42 290
Jacks, crevalles nei	*Caranx spp*	37	32 185	31 815	31 296	30 370	28 286	33 278	32 209
Pompanos nei	*Trachinotus spp*	37	2 132	1 078	-	205	-	-	-
Amberjacks nei	*Seriola spp*	37	105	100	10	10	1	2	2
Black pomfret	*Parastromateus niger*	37	9 598	10 967	11 404	10 872	12 116	11 593	11 467
Rainbow runner	*Elagatis bipinnulata*	37	2 139	1 739	1 573	6 565	1 910	1 721	1 555
Torpedo scad	*Megalaspis cordyla*	37	27 681	29 963	29 607	32 212	27 379	25 193	25 241
Queenfishes	*Scomberoides spp*	37	3 941	5 067	2 519	2 645	2 968	2 723	2 657
Bigeye scad	*Selar crumenophthalmus*	37	-	1 984	1 904	3 830	3 442	3 395	3 374
Yellowstripe scad	*Selaroides leptolepis*	37	2 903	3 656	5 505	3 643	4 871	3 437	3 693
Blackbanded trevally	*Seriolina nigrofasciata*	37	2 882	3 666	3 564	2 954	2 700	2 663	2 647
Carangids nei	*Carangidae*	37	52 360	47 458	47 043	53 672	54 419	54 424	56 465
Common dolphinfish	*Coryphaena hippurus*	37	-	-	-	3	-	-	-
Indian mackerel	*Rastrelliger kanagurta*	37	53 594	49 635	56 231	66 039	62 265	37 481	49 528
Indian mackerels nei	*Rastrelliger spp*	37	213 311	189 159	184 121	195 400	191 519	179 218	180 989
Mackerels nei	*Scombridae*	37	18 113	18 202	20 002	20 900	22 828	22 265	19 285
Silver pomfret	*Pampus argenteus*	37	9 280	8 572	9 202	9 569	8 022	7 415	8 003
Butterfishes, pomfrets nei	*Stromateidae*	37	23 523	18 207	18 970	18 397	16 453	15 123	15 772
Barracudas nei	*Sphyraena spp*	37	17 969	21 566	21 537	23 592	23 659	23 131	23 890
Pelagic percomorphs nei	*Perciformes*	37	5 000	4 000	-	-	-	-	-
Silky shark	*Carcharhinus falciformis*	38	21 400	21 000	15 000	20 875	20 700	16 130	14 620
Smooth-hounds nei	*Mustelus spp*	38	3 911	3 878	4 169	2 858	2 462	2 198	2 579
Rays, stingrays, mantas nei	*Rajiformes*	38	17 954	19 143	21 247	25 116	22 000	21 713	22 150
Sharks, rays, skates, etc. nei	*Elasmobranchii*	38	72 844	73 185	75 750	82 348	81 091	81 320	78 566
Marine fishes nei	*Osteichthyes*	39	1 490 614	1 350 902	1 499 417	1 682 392	1 625 684	1 874 633	1 893 936
Blue swimming crab	*Portunus pelagicus*	42	8 363	8 048	8 120	13 287	11 697	11 396	12 364
Indo-Pacific swamp crab	*Scylla serrata*	42	5 859	5 124	5 135	5 180	5 611	5 436	5 594
Marine crabs nei	*Brachyura*	42	9 869	10 112	11 273	12 911	14 121	29 440	28 280
Australian spiny lobster	*Panulirus cygnus*	43	10 886	9 902	9 896	10 400	17 720	19 380	16 017
Tropical spiny lobsters nei	*Panulirus spp*	43	1 419	1 260	1 325	1 052	1 302	1 433	1 736
Green rock lobster	*Jasus verreauxi*	43	4 507	4 856	4 888	4 615	1 926	2 055	2 106
Flathead lobster	*Thenus orientalis*	43	335	299	177	533	37	36	36
Slipper lobsters nei	*Scyllaridae*	43	13	21	0	0	0	0	-
Banana prawn	*Penaeus merguiensis*	45	15 952	16 830	15 537	15 232	14 660	14 927	15 415
Giant tiger prawn	*Penaeus monodon*	45	45 249	45 069	49 815	45 690	64 050	67 970	69 608
Green tiger prawn	*Penaeus semisulcatus*	45	1 909	2 427	2 449	2 180	2 212	2 182	2 168
Western king prawn	*Penaeus latisulcatus*	45	1 475	1 674	1 724	1 501	1 518	1 497	1 488
Penaeus shrimps nei	*Penaeus spp*	45	16 265	18 033	18 323	16 336	18 481	18 062	16 928
Metapenaeus shrimps nei	*Metapenaeus spp*	45	13 415	11 160	13 922	12 424	13 978	12 616	13 818
Sergestid shrimps nei	*Sergestidae*	45	16 993	17 797	15 629	10 741	9 172	10 355	8 280
Natantian decapods nei	*Natantia*	45	101 466	104 708	114 177	78 443	110 769	119 401	111 016
Stomatopods nei	*Stomatopoda*	47	-	-	0	1	116	114	114
Marine crustaceans nei	*Crustacea*	47	35 664	40 568	45 940	52 271	54 329	36 447	36 509
Blacklip abalone	*Haliotis rubra*	52	4 896	5 081	4 905	4 920	5 297	5 204	5 301
Cupped oysters nei	*Crassostrea spp*	53	-	-	-	-	-	350	380
Australian mussel	*Mytilus planulatus*	54	243	71	-	-	-	-	-
Scallops nei	*Pectinidae*	55	8 159	6 872	6 093	5 543	7 622	8 301	4 024
Blood cockle	*Anadara granosa*	56	-	-	-	20	7	7	6
Anadara clams nei	*Anadara spp*	56	16 712	21 227	22 348	15 443	10 603	9 538	8 960
Hard clams nei	*Meretrix spp*	56	3 074	3 693	3 722	6 639	3 420	2 606	2 660
Short neck clams nei	*Paphia spp*	56	10 800	18 329	11 625	6 204	27 366	26 991	26 827
Pipi wedge clam	*Paphies australis*	56	351	454	830	1 041	976	1 085	1 249
Clams, etc. nei	*Bivalvia*	56	15 250	17 745	24 663	5 097	572	5 115	2 983
Cuttlefish,bobtail squids nei	*Sepiidae, Sepiolidae*	57	26 263	35 675	35 752	37 298	36 962	39 829	36 516
Common squids nei	*Loligo spp*	57	28 178	31 230	31 568	34 619	30 727	29 991	30 918
Various squids nei	*Loliginidae, Ommastrephidae*	57	16 101	15 498	20 686	17 956	20 953	29 787	23 631
Octopuses, etc. nei	*Octopodidae*	57	6 416	11 649	12 226	16 360	14 232	14 007	14 142

C-57 (a)

Fish, crustaceans, molluscs, etc — Capture production by species items — Indian Ocean, Eastern
Poissons, crustacés, mollusques, etc — Captures par catégories d'espèces — Océan Indien, est
Peces, crustáceos, moluscos, etc — Capturas por categorías de especies — Océano Indico, oriental

English name Nom anglais Nombre inglés	Scientific name Nom scientifique Nombre científico	Species group Groupe d'espèces Grupo de especies	1995 mt	1996 mt	1997 mt	1998 mt	1999 mt	2000 mt	2001 mt
Cephalopods nei	*Cephalopoda*	57	11 405	10 596	9 212	9 497	8 917	10 470	10 779
Marine molluscs nei	*Mollusca*	58	401	381	263	844	8 396	8 307	8 517
Marine turtles nei	*Testudinata*	72	291	353	248	337	273	372	381
Sea cucumbers nei	*Holothurioidea*	76	327	419	610	833	859	1 048	1 090
Jellyfishes	*Rhopilema spp*	77	20 386	22 807	29 547	6 660	34 980	35 154	36 864
Aquatic invertebrates nei	*Invertebrata*	77	211	720	200	417	544	530	590
Total			***4 161 740***	***4 141 894***	***4 394 809***	***4 703 942***	***4 590 510***	***4 830 059***	***4 767 860***

C-57 (b)

Fish, crustaceans, molluscs, etc — Capture production by countries or areas — Indian Ocean, Eastern
Poissons, crustacés, mollusques, etc — Captures par pays ou zones — Océan Indien, est
Peces, crustáceos, moluscos, etc — Capturas por países o áreas — Océano Indico, oriental

Country or area Pays ou zone País o área	1992 mt	1993 mt	1994 mt	1995 mt	1996 mt	1997 mt	1998 mt	1999 mt	2000 mt	2001 mt
Australia	124 698 F	127 895 F	105 408 F	107 555 F	105 022 F	102 142	98 875	127 563	118 202	109 909
Bangladesh	280 127	312 715	253 044	264 650	279 170	295 141	300 452	309 797	333 799	330 000 F
China	-	-	-	445	1 497	2 964	3 080	5 868	4 919	4 009
China,Taiwan	16 854 F	18 314 F	16 961 F	25 964 F	32 507	41 049	22 549	22 732	23 241	14 308
Christmas Is	0	0	0	0	0	0	0	0	0	0
Cocos Is	0	0	0	0	0	0	0	0	0	0
France	-	-	-	-	80	1 947	10 849	1 652	108	-
Honduras	-	-	-	-	-	-	-	-	-	637
India	691 033	741 950	799 934	809 231	859 606	875 052	848 254	864 749	937 998	891 684
Indonesia	647 452	732 103	755 652	763 171	798 203	856 545	1 029 867	918 151	913 252	947 680
Iran	-	-	-	-	2	6	449	-	-	-
Italy	-	-	-	-	-	212	1 473	-	...	-
Japan	5 028	6 790	11 519	40 437	27 076	21 221	22 301	24 091	19 898	17 676
Korea Rep	517	-	78	321	718	512	269	4 983	5 273	4 550
Malaysia	473 995	446 515	439 341	504 068	494 091	499 288	508 128	485 741	519 785	475 354
Myanmar	590 263	597 637	599 876	602 885	455 294	631 226	680 838	759 664	880 018	931 492
Philippines	-	-	-	-	-	-	657	1 017	219	182
Seychelles	-	-	-	-	-	1 049	3 785	62	244	556
Spain	-	-	-	-	237	379	13 379	451	168	...
Sri Lanka	185 168	200 400	208 900	214 171	206 695	207 849	233 430	244 630	260 710	249 770
Thailand	656 172	823 696	776 667	822 673	869 484	850 994	909 617	808 257	799 097	789 697
Timor-Leste	...	...	...	...	...	...	...	408 F	362	356
Other nei	4 680	3 896	5 512	6 169	12 212	7 233	15 690	10 694	12 766	...
Total	***3 675 987***	***4 011 911***	***3 972 892***	***4 161 740***	***4 141 894***	***4 394 809***	***4 703 942***	***4 590 510***	***4 830 059***	***4 767 860***

C-58 (a) Fish, crustaceans, molluscs, etc / Poissons, crustacés, mollusques, etc / Peces, crustáceos, moluscos, etc

Capture production by species items / Captures par catégories d'espèces / Capturas por categorías de especies

Indian Ocean, Antarctic / Océan Indien, Antarctique / Océano Indico, Antártico

English name Nom anglais Nombre inglés	Scientific name Nom scientifique Nombre científico	Species group Groupe d'espèces Grupo de especies	1995 mt	1996 mt	1997 mt	1998 mt	1999 mt	2000 mt	2001 mt
Blue antimora	*Antimora rostrata*	32	-	-	-	2	6	10	22
Grenadiers nei	*Macrourus spp*	32	-	-	1	48	46	333	244
Marbled rockcod	*Notothenia rossii*	33	-	-	1	-	1	-	-
Grey rockcod	*Notothenia squamifrons*	33	-	15	4	3	10	-	0
Antarctic rockcods, noties nei	*Nototheniidae*	33	-	-	-	-	3	-	0
Patagonian toothfish	*Dissostichus eleginoides*	34	5 649	4 919	7 983	7 909	12 987	8 996	9 352
Mackerel icefish	*Champsocephalus gunnari*	34	3 936	5	217	67	73	81	930
Ocellated icefish	*Chionodraco rastrospinosus*	34	-	-	1	-	-	-	-
Unicorn icefish	*Channichthys rhinoceratus*	34	-	1	7	6	3	2	1
Spiny icefish	*Chaenodraco wilsoni*	34	-	-	-	-	-		11
Porbeagle	*Lamna nasus*	38	-	-	2	-	-	-	-
Pacific sleeper shark	*Somniosus pacificus*	38	-	-	-	-	1	-	-
Rays, stingrays, mantas nei	*Rajiformes*	38	-	1	3	23	13	98	78
Sharks, rays, skates, etc. nei	*Elasmobranchii*	38	-	-	2	-	1	-	-
Marine fishes nei	*Osteichthyes*	39	61	-	1	0	0	0	0
King crabs, stone crabs nei	*Lithodidae*	44	-	-	-	0	-	3	0
Antarctic krill	*Euphausia superba*	46	1 266	6	-	-	-	-	-
Cephalopods nei	*Cephalopoda*	57	-	-	1	-	-	-	-
Sea squirts nei	*Ascidiacea*	74	-	7	-	-	-	-	-
Jellyfishes	*Rhopilema spp*	77	-	-	10	-	2	-	-
Aquatic invertebrates nei	*Invertebrata*	77	-	-	1	-	-	-	-
Total			***10 912***	***4 954***	***8 234***	***8 058***	***13 146***	***9 523***	***10 638***

C-58 (b) Fish, crustaceans, molluscs, etc / Poissons, crustacés, mollusques, etc / Peces, crustáceos, moluscos, etc

Capture production by countries or areas / Captures par pays ou zones / Capturas por países o áreas

Indian Ocean, Antarctic / Océan Indien, Antarctique / Océano Indico, Antártico

Country or area Pays ou zone País o área	1992 mt	1993 mt	1994 mt	1995 mt	1996 mt	1997 mt	1998 mt	1999 mt	2000 mt	2001 mt
Australia	5	-	4	-	-	1 105	2 493	5 547	2 665	2 707
France	1 589	826	4 211	4 173	3 673	3 681	3 879	6 286	5 770	6 806
India	-	-	-	-	13	-	-	-	-	-
Japan	-	5 762	899	1 266	264	335	-	-	-	-
Russian Fed	1 274	-	-	-	-	-	-	-	-	-
South Africa	-	-	-	-	-	2 106	689	553	1 088	961
Ukraine	5 958	1 890	942	5 473	1 004	1 007	997	760	-	65
Uruguay	-	-	-	-	-	-	-	-	-	99
Total	***8 826***	***8 478***	***6 056***	***10 912***	***4 954***	***8 234***	***8 058***	***13 146***	***9 523***	***10 638***

Data in these tables refer to catches for which the split-year (1 July - 30 June) is used. Split-year data are shown under the calendar year in which the split-year ends.

Les données dans ces tableaux se réfèrent aux captures pour lesquelles on utilise l'année fractionnée (1er juillet - 30 juin). Les captures relatives à des années fractionnées figurent sous l'année civile durant laquelle se termine l'année fractionnée.

Los datos en estos cuadros se refieren a las capturas para las que se utiliza el año emergente (1°de julio - 30 de junio). Los datos correspondientes a los años emergentes se incluyen en el año civil en que termina el año emergente.

C-61 (a)

Fish, crustaceans, molluscs, etc — **Capture production by species items** — **Pacific, Northwest**
Poissons, crustacés, mollusques, etc — **Captures par catégories d'espèces** — **Pacifique, nord-ouest**
Peces, crustáceos, moluscos, etc — **Capturas por categorías de especies** — **Pacífico, noroeste**

English name Nom anglais Nombre inglés	Scientific name Nom scientifique Nombre científico	Species group Groupe d'espèces Grupo de especies	1995 mt	1996 mt	1997 mt	1998 mt	1999 mt	2000 mt	2001 mt
Cyprinids nei	*Cyprinidae*	11	61	110	101	100	56	8	52
Pink(=Humpback)salmon	*Oncorhynchus gorbuscha*	23	162 895	133 838	183 967	200 628	174 535	157 506	158 810
Chum(=Keta=Dog)salmon	*Oncorhynchus keta*	23	262 614	297 268	266 736	209 290	191 207	173 067	225 700
Masu(=Cherry) salmon	*Oncorhynchus masou*	23	1 435	1 681	994	1 738	1 133	955	780
Sockeye(=Red)salmon	*Oncorhynchus nerka*	23	12 153	19 281	15 856	11 470	10 452	14 771	13 087
Chinook(=Spring=King)salmon	*Oncorhynchus tshawytscha*	23	332	370	1 016	750	580	362	447
Coho(=Silver)salmon	*Oncorhynchus kisutch*	23	536	1 100	987	1 616	1 122	1 235	1 446
Smelts nei	*Osmerus spp, Hypomesus spp*	23	130	122	321	322	390	735	612
Salmonoids nei	*Salmonoidei*	23	2 332	2 119	1 541	2 407	884	2 140	747
Chinese gizzard shad	*Clupanodon thrissa*	24	7 961	4 933	13 846	11 360	9 538	6 387	9 194
Dotted gizzard shad	*Konosirus punctatus*	24	23 707	18 647	14 850	20 787	17 770	12 335	17 210
Elongate ilisha	*Ilisha elongata*	24	47 101	51 969	77 881	84 671	110 624	107 823	101 382
Milkfish	*Chanos chanos*	25	1	4	-	-	-	-	1
Barramundi(=Giant seaperch)	*Lates calcarifer*	25	0	14	9	26	47	32	80
Kamchatka flounder	*Atheresthes evermanni*	31	7 039	9 739	8 476	9 567	10 743	23 473	19 049
Yellow striped flounder	*Pseudopleuronectes herzenst.*	31	13 683	18 066	18 079	20 135	19 569	15 423	14 503
Tonguefishes	*Cynoglossidae*	31	3 270	2 804	2 550	1 831	1 636	2 148	2 113
Bastard halibut	*Paralichthys olivaceus*	31	9 472	10 628	9 953	9 617	8 877	9 179	8 436
Flatfishes nei	*Pleuronectiformes*	31	121 877	140 157	147 859	156 720	169 910	175 459	159 817
Moras nei	*Moridae*	32	3 758	-	152	11 115	31 295	39 316	32 392
Pacific cod	*Gadus macrocephalus*	32	159 767	154 027	142 388	157 774	163 730	129 565	116 443
Saffron cod	*Eleginus gracilis*	32	25 566	21 110	27 803	40 426	47 032	35 763	33 753
Alaska pollock(=Walleye poll.)	*Theragra chalcogramma*	32	3 440 810	3 293 782	3 265 068	2 766 246	2 214 037	1 746 748	1 692 548
Polar cod	*Boreogadus saida*	32	2 400	-	-	-	-	13	-
Grenadiers, rattails nei	*Macrouridae*	32	222	310	1 044	62	52	579	2 232
Gadiformes nei	*Gadiformes*	32	3	10	-	2	22	6	1
Greater lizardfish	*Saurida tumbil*	33	15 908	14 705	10 846	10 909	10 791	11 203	8 342
Brushtooth lizardfish	*Saurida undosquamis*	33	209	162	207	215	35	795	758
Lizardfishes nei	*Synodontidae*	33	10 888	9 034	8 070	6 619	4 700	5 800	6 400
Sea catfishes nei	*Ariidae*	33	274	122	134	208	257	725	435
Korean sandlance	*Hypoptychus dybowskii*	33	9 677	6 613	8 832	4 801	4 806	-	462
Flathead grey mullet	*Mugil cephalus*	33	10 533	9 770	11 943	9 571	14 170	14 345	14 800
Mullets nei	*Mugilidae*	33	71 443	67 423	85 967	99 024	113 461	107 276	119 340
Groupers nei	*Epinephelus spp*	33	24 770	25 698	32 588	37 512	41 510	42 715	46 585
Groupers, seabasses nei	*Serranidae*	33	1 949	1 785	1 901	2 005	1 451	2 047	1 840
Japanese seabass	*Lateolabrax japonicus*	33	8 831	9 512	10 558	12 180	12 167	10 538	12 256
Red bigeye	*Priacanthus macracanthus*	33	5 486	7 386	4 824	3 670	2 756	3 986	3 079
Bigeyes nei	*Priacanthus spp*	33	6 781	5 391	4 962	4 149	3 000	3 600	4 000
Sillago-whitings	*Sillaginidae*	33	606	795	600	394	299	288	142
Moonfish	*Mene maculata*	33	859	725	902	1 010	1 747	1 496	1 582
Snappers, jobfishes nei	*Lutjanidae*	33	2 001	2 764	1 847	1 112	717	652	715
Golden threadfin bream	*Nemipterus virgatus*	33	228 238	242 261	263 399	266 383	250 591	296 319	290 920
Threadfin breams nei	*Nemipterus spp*	33	23 772	19 568	20 024	19 449	14 000	17 000	18 800
Grunts, sweetlips nei	*Haemulidae (=Pomadasyidae)*	33	5 555	5 303	4 801	5 703	5 285	5 136	4 995
Honnibe croaker	*Nibea mitsukurii*	33	2 164	1 940	1 177	1 285	1 566	1 999	2 156
Large yellow croaker	*Larimichthys croceus*	33	89 630	102 297	71 515	71 229	66 076	123 274	76 715
Yellow croaker	*Larimichthys polyactis*	33	181 553	280 036	236 960	278 807	323 564	425 047	330 455
Blackmouth croaker	*Atrobucca nibe*	33	197	297	288	237	223	450	644
Silver croaker	*Pennahia argentata*	33	7 369	8 599	6 085	5 089	4 940	4 180	3 796
Croakers, drums nei	*Sciaenidae*	33	195 394	71 182	81 180	79 062	90 920	50 313	40 589
Largeeye breams	*Gymnocranius spp*	33	196	208	170	143	100	120	130
Silver seabream	*Pagrus auratus*	33	17 139	19 039	19 205	19 105	19 988	20 753	22 268
Blackhead seabream	*Acanthopagrus schlegeli*	33	197	207	277	404	604	717	878
Porgies, seabreams nei	*Sparidae*	33	79 203	76 145	92 153	95 905	98 484	123 895	147 901
Goatfishes	*Upeneus spp*	33	992	1 226	917	728	778	827	700
Parrotfishes nei	*Scaridae*	33	8	11	17	12	5	29	27
Fourfinger threadfin	*Eleutheronema tetradactylum*	33	3 469	3 157	2 831	3 248	2 414	6 917	1 715
Threadfins, tasselfishes nei	*Polynemidae*	33	...	283 784	340 302	517 528	565 764	496 566	476 690
Pacific sandlance	*Ammodytes personatus*	33	108 124	115 766	108 666	90 688	82 918	66 129	92 967
Gobies nei	*Gobiidae*	33	16 012	16 120	17 898	21 850	31 069	33 521	28 024
Atka mackerel	*Pleurogrammus azonus*	33	202 537	213 363	249 356	292 771	213 769	223 256	214 392
Bartail flathead	*Platycephalus indicus*	33	2 520	2 900	4 196	2 857	2 248	2 310	1 699
Purple puffer	*Takifugu vermicularis*	33	10 178	9 708	7 471	3 897	4 787	-	-
Puffers nei	*Tetraodontidae*	33	8 991	8 238	7 103	8 329	9 647	13 967	11 555
Filefishes	*Cantherhines(=Navodon) spp*	33	124 038	211 059	297 227	236 189	241 209	225 403	204 270
Threadsail filefish	*Stephanolepis cirrhifer*	33	1 755	1 772	16 318	9 364	2 999	-	-
Deepsea smelt	*Glossanodon semifasciatus*	34	7 705	8 134	7 431	7 142	6 312	5 970	5 414
Daggertooth pike conger	*Muraenesox cinereus*	34	163 074	183 716	192 667	249 194	245 803	230 633	252 790
Pike-congers nei	*Muraenesox spp*	34	3 129	3 281	2 903	2 450	1 700	2 100	2 300
Whitespotted conger	*Conger myriaster*	34	19 667	17 314	19 136	11 913	10 160	8 304	7 676
Conger eels, etc. nei	*Congridae*	34	12 978	12 007	11 706	9 444	8 168	8 364	7 999
Alfonsinos nei	*Beryx spp*	34	-	6	-	4	38	18	14
Oreo dories nei	*Oreosomatidae*	34	-	-	-	-	2	-	14
Tilefishes nei	*Branchiostegidae*	34	7 356	8 061	7 807	7 535	7 596	8 090	8 092
Atlantic wolffish	*Anarhichas lupus*	34	-	-	-	-	38	36	33
Eelpouts	*Lycodes spp*	34	45	18	-	2	1	28	47
Japanese sandfish	*Arctoscopus japonicus*	34	7 571	9 220	8 403	8 285	9 064	8 223	10 039
Oilfish	*Ruvettus pretiosus*	34	3 729	2 634	2 622	3 043	2 661	2 584	3 678
Largehead hairtail	*Trichiurus lepturus*	34	1 178 697	1 184 849	1 111 655	1 328 470	1 322 474	1 396 737	1 388 045
Hairtails, scabbardfishes nei	*Trichiuridae*	34	3 470	3 326	2 522	1 690	1 200	1 500	1 650
Indian driftfish	*Ariomma indica*	34	287	120	79	49	35	40	45
Pacific rudderfish	*Psenopsis anomala*	34	14 191	13 290	11 777	13 734	10 871	10 721	10 942
Pacific ocean perch	*Sebastes alutus*	34	4 988	4 345	2 862	2 440	1 630	1 475	1 461
Scorpionfishes nei	*Scorpaenidae*	34	8 789	6 964	7 496	6 240	5 979	6 168	6 338
Bluefin gurnard	*Chelidonichthys kumu*	34	40	75	86	146	43	79	140

C-61 (a)

Fish, crustaceans, molluscs, etc — Capture production by species items — Pacific, Northwest
Poissons, crustacés, mollusques, etc — Captures par catégories d'espèces — Pacifique, nord-ouest
Peces, crustáceos, moluscos, etc — Capturas por categorías de especies — Pacífico, noroeste

English name Nom anglais Nombre inglés	Scientific name Nom scientifique Nombre científico	Species group Groupe d'espèces Grupo de especies	1995 mt	1996 mt	1997 mt	1998 mt	1999 mt	2000 mt	2001 mt
Sablefish	*Anoplopoma fimbria*	34	8	502	-	-	-	6	6
Anglerfishes nei	*Lophiidae*	34	8 173	11 720	5 045	3 970	3 406	4 930	5 813
Pacific herring	*Clupea pallasii*	35	131 607	181 021	344 317	433 262	401 155	393 962	341 337
Japanese sardinella	*Sardinella zunasi*	35	18 345	10 663	5 593	1 973	6 674	4 603	766
Sardinellas nei	*Sardinella spp*	35	24	30	12	8	5	5	6
Japanese pilchard	*Sardinops melanostictus*	35	733 427	430 837	417 939	295 788	515 477	305 767	339 377
Slender rainbow sardine	*Dussumieria elopsoides*	35	154	48	19	17	10	15	17
Red-eye round herring	*Etrumeus teres*	35	49 243	51 214	57 195	49 689	29 605	24 951	32 831
Silver-stripe round herring	*Spratelloides gracilis*	35	827	517	650	886	521	639	505
Japanese anchovy	*Engraulis japonicus*	35	972 008	1 254 487	1 666 503	2 093 888	1 820 259	1 725 685	1 836 502
Stolephorus anchovies	*Stolephorus spp*	35	89	31	20	17	10	15	17
Anchovies, etc. nei	*Engraulidae*	35	19	22	5	23	34	69	114
Dorab wolf-herring	*Chirocentrus dorab*	35	3	1	2	1	4	1	-
Clupeoids nei	*Clupeoidei*	35	61 756	62 105	63 841	56 765	83 126	78 802	62 679
Narrow-barred Spanish mackerel	*Scomberomorus commerson*	36	3 211	2 541	2 701	2 417	2 674	3 551	4 153
Indo-Pacific king mackerel	*Scomberomorus guttatus*	36	1 814	1 611	1 607	1 128	1 298	1 249	1 395
Japanese Spanish mackerel	*Scomberomorus niphonius*	36	259 301	301 356	365 585	551 780	595 103	539 094	522 756
Seerfishes nei	*Scomberomorus spp*	36	3 875	3 139	2 854	3 201	4 083	3 760	3 083
Frigate and bullet tunas	*Auxis thazard, A.rochei*	36	30 257	22 821	36 775	26 231	34 072	30 212	39 323
Kawakawa	*Euthynnus affinis*	36	500	0	0	0	0	0	0
Skipjack tuna	*Katsuwonus pelamis*	36	123 882	105 603	187 992	205 919	138 180	177 180	156 363
Pacific bluefin tuna	*Thunnus orientalis*	36	5 875	6 808	6 974	4 328	11 147	11 058	6 256
Longtail tuna	*Thunnus tonggol*	36	595	514	391	303	6 434	10 087	14 321
Albacore	*Thunnus alalunga*	36	38 126	51 026	70 005	54 857	86 968	47 162	50 748
Yellowfin tuna	*Thunnus albacares*	36	18 473	16 129	14 593	16 185	19 442	21 773	15 345
Bigeye tuna	*Thunnus obesus*	36	8 182	6 610	9 674	10 145	10 173	9 378	8 573
Indo-Pacific sailfish	*Istiophorus platypterus*	36	1 678	2 643	3 428	1 821	3 841	2 080	583
Indo-Pacific blue marlin	*Makaira mazara*	36	6 185	9 225	5 486	4 905	3 880	3 306	3 168
Black marlin	*Makaira indica*	36	483	830	659	285	681	332	193
Striped marlin	*Tetrapturus audax*	36	4 470	3 860	2 468	4 328	4 093	3 381	2 876
Marlins,sailfishes,etc. nei	*Istiophoridae*	36	-	-	-	2	3	-	146
Swordfish	*Xiphias gladius*	36	7 884	5 982	5 754	8 019	7 617	9 862	7 320
Tuna-like fishes nei	*Scombroidei*	36	40 517	26 600	31 117	29 036	31 896	43 874	13 410
Capelin	*Mallotus villosus*	37	44	180	160	405	70	291	1 468
Pacific saury	*Cololabis saira*	37	350 287	276 111	388 643	180 973	187 144	305 009	375 750
Japanese halfbeak	*Hyporhamphus sajori*	37	1 531	990	1 193	1 160	913	956	613
Japanese flyingfish	*Cypselurus agoo*	37	7 881	8 501	7 486	8 933	6 738	9 615	8 286
Flyingfishes nei	*Exocoetidae*	37	572	662	2 497	1 077	618	287	366
Cobia	*Rachycentron canadum*	37	779	692	987	815	655	1 014	486
Japanese jack mackerel	*Trachurus japonicus*	37	330 104	349 166	350 619	340 564	227 290	271 501	235 874
Japanese scad	*Decapterus maruadsi*	37	75 735	62 378	69 764	71 168	67 689	40 803	47 460
Indian scad	*Decapterus russelli*	37	299	382	687	6 158	7 703	3 927	598
Scads nei	*Decapterus spp*	37	529 720	645 640	522 877	539 120	507 526	507 380	550 298
Jacks, crevalles nei	*Caranx spp*	37	6 701	7 131	11 466	8 248	4 892	6 370	3 290
Japanese amberjack	*Seriola quinqueradiata*	37	7 564	246	...	...	...	...	...
Amberjacks nei	*Seriola spp*	37	57 846	54 426	53 275	55 104	63 571	82 275	73 400
Black pomfret	*Parastromateus niger*	37	5 444	3 427	2 534	1 744	1 307	1 671	2 129
Torpedo scad	*Megalaspis cordyla*	37	65	233	6 847	52	113	327	198
Common dolphinfish	*Coryphaena hippurus*	37	21 980	11 373	18 307	32 066	17 829	14 511	16 550
Chub mackerel	*Scomber japonicus*	37	1 096 121	1 603 739	1 465 627	1 104 873	1 007 284	871 513	984 850
Silver pomfret	*Pampus argenteus*	37	11 784	9 919	3 622	2 650	5 851	2 852	4 245
Silver pomfrets nei	*Pampus spp*	37	209 031	220 364	242 547	303 024	337 919	338 848	352 493
Butterfishes, pomfrets nei	*Stromateidae*	37	13 550	11 844	13 079	15 613	17 015	9 869	9 065
Barracudas nei	*Sphyraena spp*	37	429	547	761	771	519	362	439
Whip stingray	*Dasyatis akajei*	38	3 985	4 029	3 959	4 329	4 407	5 388	4 312
Rays, stingrays, mantas nei	*Rajiformes*	38	6 888	9 250	8 014	4 629	4 487	4 343	2 664
Sharks, rays, skates, etc. nei	*Elasmobranchii*	38	22 635	17 922	25 392	25 425	48 993	46 348	28 994
Marine fishes nei	*Osteichthyes*	39	4 058 374	4 864 097	4 442 968	4 372 457	4 368 442	4 055 185	3 641 860
Blue swimming crab	*Portunus pelagicus*	42	27 976	52 679	46 890	49 156	53 088	58 058	61 119
Gazami crab	*Portunus trituberculatus*	42	265 295	303 170	252 502	283 971	284 851	351 051	350 168
Indo-Pacific swamp crab	*Scylla serrata*	42	1 339	935	180	213	268	289	230
Tanner crabs nei	*Chionoecetes spp*	42	9 090	26 216	31 982	25 525	26 126	44 525	43 822
Hair crab	*Erimacrus isenbeckii*	42	...	789	612	409	440	198	162
Marine crabs nei	*Brachyura*	42	103 964	104 811	99 419	83 218	72 467	46 580	45 021
Longlegged spiny lobster	*Panulirus longipes*	43	1 154	1 106	1 082	1 098	1 166	1 716	1 924
Tropical spiny lobsters nei	*Panulirus spp*	43	20	0	0	0	0	0	0
Japanese fan lobster	*Ibacus ciliatus*	43	1 224	1 115	642	696	676	1 600	1 607
Red king crab	*Paralithodes camtschaticus*	44	54 986	34 300	23 262	32 486	37 072	28 554	16 064
Blue king crab	*Paralithodes platypus*	44	...	8 762	10 268	4 508	5 455	5 233	4 500
Brown king crab	*Paralithodes brevipes*	44	...	204	418	194	256	347	254
King crabs	*Paralithodes spp*	44	260	322	154	132	117	89	181
Golden king crab	*Lithodes aequispina*	44	...	4 666	4 917	3 897	2 746	1 797	2 245
Kuruma prawn	*Penaeus japonicus*	45	7 027	5 606	6 911	5 113	6 096	10 193	5 963
Giant tiger prawn	*Penaeus monodon*	45	67	49	45	38	92	57	91
Fleshy prawn	*Penaeus chinensis*	45	44 449	56 534	71 317	79 595	70 725	85 545	97 583
Redtail prawn	*Penaeus penicillatus*	45	3 564	5 020	2 473	647	316	308	312
Shiba shrimp	*Metapenaeus joyneri*	45	2 168	2 211	1 976	3 651	3 633	2 621	2 385
Metapenaeus shrimps nei	*Metapenaeus spp*	45	3 767	626	369	299	210	202	241
Southern rough shrimp	*Trachypenaeus curvirostris*	45	154 623	167 723	180 255	179 544	403 027	312 968	247 940
Coonstripe shrimp	*Pandalus hypsinotus*	45	...	...	467	388	288	275	359
Northern prawn	*Pandalus borealis*	45	-	-	1 786	3 127	5 096	8 520	8 457
Humpy shrimp	*Pandalus goniurus*	45	...	...	-	1 199	330	1 200	247

C-61 (a)

Fish, crustaceans, molluscs, etc — Capture production by species items — Pacific, Northwest
Poissons, crustacés, mollusques, etc — Captures par catégories d'espèces — Pacifique, nord-ouest
Peces, crustáceos, moluscos, etc — Capturas por categorías de especies — Pacífico, noroeste

English name Nom anglais Nombre inglés	Scientific name Nom scientifique Nombre científico	Species group Groupe d'espèces Grupo de especies	1995 mt	1996 mt	1997 mt	1998 mt	1999 mt	2000 mt	2001 mt
Hokkai shrimp	*Pandalus kessleri*	45	...	...	123	55	97	75	94
Pandalus shrimps nei	*Pandalus spp*	45	2 398	3 000	-	-	-	-	-
Morotoge shrimp	*Pandalopsis japonica*	45	...	...	12	86	35	11	36
Akiami paste shrimp	*Acetes japonicus*	45	406 495	460 871	495 680	587 376	598 602	639 219	577 497
Sculptured shrimps nei	*Sclerocrangon spp*	45	...	...	38	59	82	20	45
Natantian decapods nei	*Natantia*	45	82 162	75 282	73 290	71 003	61 480	57 906	49 750
Marine crustaceans nei	*Crustacea*	47	934 355	1 007 693	1 165 004	1 314 628	1 198 281	1 303 443	1 333 362
Japanese corbicula	*Corbicula japonica*	51	-	-	-	-	-	-	74
Giant abalone	*Haliotis gigantea*	52	1 980	1 941	2 218	2 269	2 109	2 146	1 982
Abalones nei	*Haliotis spp*	52	274	196	220	73	81	155	105
Horned turban	*Turbo cornutus*	52	18 864	17 480	19 010	21 748	18 397	17 120	16 515
Gastropods nei	*Gastropoda*	52	6 222	3 735	2 636	1 450	1 670	828	783
Pacific cupped oyster	*Crassostrea gigas*	53	18 269	18 259	17 210	9 905	11 609	15 939	10 060
Cupped oysters nei	*Crassostrea spp*	53	-	-	-	-	1	4	32
Korean mussel	*Mytilus coruscus*	54	2 942	2 191	3 211	1 469	1 414	1 133	1 085
Sea mussels nei	*Mytilidae*	54	3 713	837	4 098	6 465	6 851	5 819	3 870
Yesso scallop	*Patinopecten yessoensis*	55	279 382	276 406	266 957	294 211	305 510	310 104	293 268
Ark clams nei	*Arca spp*	56	473	782	667	1 311	534	318	153
Half-crenated ark	*Scapharca subcrenata*	56	15 426	16 328	14 133	10 120	10 413	7 308	4 899
Blood cockle	*Anadara granosa*	56	1 415	995	493	12 114	6 503	4 184	1 023
Japanese hard clam	*Meretrix lusoria*	56	5 031	4 394	4 133	6 147	4 110	2 973	2 289
Japanese carpet shell	*Ruditapes philippinarum*	56	64 523	56 095	56 514	51 392	57 051	56 540	51 026
Imperial surf clam	*Pseudocardium sybillae*	56	17 056	14 791	14 720	11 104	16 428	13 033	12 323
Cockles nei	*Cardiidae*	56	428	2 583	1 587	1 078	112	411	2 002
Clams, etc. nei	*Bivalvia*	56	81 078	85 807	51 838	63 249	50 047	48 022	46 155
Cuttlefish,bobtail squids nei	*Sepiidae, Sepiolidae*	57	239 755	185 530	255 542	241 895	232 788	269 796	324 915
Common squids nei	*Loligo spp*	57	6 143	14 853	19 221	20 074	12 712	8 210	6 689
Neon flying squid	*Ommastrephes bartrami*	57	...	...	47 852	53 638	34 551	46 604	21 678
Japanese flying squid	*Todarodes pacificus*	57	513 413	715 908	603 367	378 605	497 887	570 427	528 523
Various squids nei	*Loliginidae, Ommastrephidae*	57	172 418	176 002	147 722	250 739	262 688	260 097	190 193
Octopuses, etc. nei	*Octopodidae*	57	74 810	73 170	80 370	88 667	77 213	66 838	65 511
Cephalopods nei	*Cephalopoda*	57	-	-	-	-	-	-	3
Marine molluscs nei	*Mollusca*	58	1 302 416	938 870	1 502 215	1 488 978	1 457 426	1 410 260	1 405 909
Sea squirts nei	*Ascidiacea*	74	5 767	16 747	2 780	891	1 171	1 443	1 053
Echinoderms	*Echinodermata*	76	3 707	2 802	2 771	1 410	1 182	1 461	1 454
Sea urchins nei	*Strongylocentrotus spp*	76	16 432	15 013	15 861	15 552	15 108	14 324	13 170
Japanese sea cucumber	*Stichopus japonicus*	76	8 494	9 205	9 377	8 391	7 866	8 376	8 129
Jellyfishes	*Rhopilema spp*	77	171 905	265 325	400 483	430 784	402 206	334 869	331 439
Aquatic invertebrates nei	*Invertebrata*	77	38 376	13 410	10 118	10 660	8 131	11 574	8 708
Total			***21 751 722***	***23 482 610***	***24 606 259***	***24 753 317***	***24 112 564***	***23 180 042***	***22 532 921***

C-61 (b)

Fish, crustaceans, molluscs, etc — Capture production by countries or areas — Pacific, Northwest
Poissons, crustacés, mollusques, etc — Captures par pays ou zones — Pacifique, nord-ouest
Peces, crustáceos, moluscos, etc — Capturas por países o áreas — Pacífico, noroeste

Country or area Pays ou zone País o área	1992 mt	1993 mt	1994 mt	1995 mt	1996 mt	1997 mt	1998 mt	1999 mt	2000 mt	2001 mt
China	7 275 876	8 111 441	9 482 098	10 861 731	12 332 922	13 727 656	14 840 900	14 843 663	14 609 252	14 215 449
China,H.Kong	220 181	217 544	211 010	194 999	183 856	186 000	180 000	127 780	157 012	173 972
China, Macao	2 668	1 898	1 890	1 604	1 418	1 500 F	1 500 F	1 500 F	1 500 F	1 500 F
China,Taiwan	553 997 F	522 490 F	405 001 F	420 781 F	393 351	406 022	378 364	373 608	361 423	357 797
Japan	6 594 758	6 218 515	5 628 874	5 092 787	5 191 508	5 162 694	4 546 845	4 451 587	4 206 983	3 996 910
Korea D P Rp	865 000 F	835 000 F	351 961	307 083	233 125	216 462	200 000 F	190 000 F	180 000 F	180 000 F
Korea Rep	1 582 368	1 718 865	1 799 694	1 787 770	1 933 176	1 640 712	1 553 273	1 500 272	1 288 737	1 469 171
Poland	297 732	235 208	269 979	249 365	116 266	125 414	81 889	65 508	33 217	16 590
Russian Fed	3 226 838	2 741 954	2 256 773	2 835 602	3 096 988	3 139 799	2 970 546	2 558 646	2 341 918	2 121 532
Total	***20 619 418***	***20 602 915***	***20 407 280***	***21 751 722***	***23 482 610***	***24 606 259***	***24 753 317***	***24 112 564***	***23 180 042***	***22 532 921***

C-67 (a)

Fish, crustaceans, molluscs, etc — Capture production by species items — Pacific, Northeast
Poissons, crustacés, mollusques, etc — Captures par catégories d'espèces — Pacifique, nord-est
Peces, crustáceos, moluscos, etc — Capturas por categorías de especies — Pacífico, nordeste

English name Nom anglais Nombre inglés	Scientific name Nom scientifique Nombre científico	Species group Groupe d'espèces Grupo de especies	1995 mt	1996 mt	1997 mt	1998 mt	1999 mt	2000 mt	2001 mt
White sturgeon	*Acipenser transmontanus*	21	-	-	-	-	-	206	185
Green sturgeon	*Acipenser medirostris*	21	-	-	-	-	-	36	10
Sturgeons nei	*Acipenseridae*	21	4	1	0	0	0	-	-
Pink(=Humpback)salmon	*Oncorhynchus gorbuscha*	23	221 148	149 136	115 175	154 776	182 844	101 598	183 642
Chum(=Keta=Dog)salmon	*Oncorhynchus keta*	23	133 678	86 864	54 741	78 316	69 907	76 357	58 232
Sockeye(=Red)salmon	*Oncorhynchus nerka*	23	169 151	159 855	112 574	63 247	112 433	103 074	83 403
Chinook(=Spring=King)salmon	*Oncorhynchus tshawytscha*	23	7 590	5 548	7 180	6 443	4 336	4 807	6 731
Coho(=Silver)salmon	*Oncorhynchus kisutch*	23	26 727	21 497	10 607	15 466	12 991	14 575	16 810
Rainbow trout	*Oncorhynchus mykiss*	23	4	3	1	4	1	1	-
Pacific salmons nei	*Oncorhynchus spp*	23	-	-	-	-	-	50	204
Surf smelt	*Hypomesus pretiosus*	23	1	1	-	-	-	-	-
Eulachon	*Thaleichthys pacificus*	23	26	30	0	3	-	13	143
Smelts nei	*Osmerus spp, Hypomesus spp*	23	71	78	839	490	432	442	218
American shad	*Alosa sapidissima*	24	158	15	34	62	-	105	178
Pacific halibut	*Hippoglossus stenolepis*	31	26 296	28 106	39 289	41 606	43 618	40 848	40 161
English sole	*Pleuronectes vetulus*	31	...	...	...	198	874	963	1 198
Greenland halibut	*Reinhardtius hippoglossoides*	31	5 860	4 652	...	307	14	6 186	4 391
Arrow-tooth flounder	*Atheresthes stomias*	31	5 166	10 680	6 319	8 279	12 081	18 774	14 401
Petrale sole	*Eopsetta jordani*	31	...	...	1 454	1 197	1 221	1 632	1 554
Rex sole	*Glyptocephalus zachirus*	31	...	...	...	121	391	413	473
Flathead sole	*Hippoglossoides elassodon*	31	9 073	11 392	12 867	18 886	14 318	16 266	16 092
Yellowfin sole	*Limanda aspera*	31	96 765	101 354	149 302	80 500	56 830	69 971	54 918
Rock sole	*Lepidopsetta bilineata*	31	24 962	26 198	32 743	15 189	17 185	27 510	24 206
Dover sole	*Microstomus pacificus*	31	...	...	8 496	8 174	8 384	7 546	6 038
Pacific sand sole	*Psettichthys melanostictus*	31	...	...	...	15	582	44	85
Curlfin sole	*Pleuronichthys decurrens*	31	-	-	-	-	-	1	5
Pacific sanddab	*Citharichthys sordidus*	31	-	-	-	-	-	150	
California flounder	*Paralichthys californicus*	31	...	...	...	13	3	0	4
Flatfishes nei	*Pleuronectiformes*	31	34 676	34 731	23 040	23 886	18 691	13 504	10 207
Pacific cod	*Gadus macrocephalus*	32	270 429	275 448	301 507	253 597	238 515	241 347	214 441
Pacific tomcod	*Microgadus proximus*	32	...	...	...	3	1	-	0
Alaska pollock(=Walleye poll.)	*Theragra chalcogramma*	32	1 368 201	1 254 803	1 221 442	1 282 977	1 056 249	1 183 482	1 443 917
North Pacific hake	*Merluccius productus*	32	177 038	195 289	226 615	227 502	216 889	206 327	172 050
Grenadiers, rattails nei	*Macrouridae*	32	...	...	...	83	102	170	160
Lingcod	*Ophiodon elongatus*	33	5 103	4 402	3 225	2 464	3 061	3 186	2 664
Atka mackerel	*Pleurogrammus azonus*	33	67 156	88 030	59 538	51 579	51 436	44 592	57 096
Cabezon	*Scorpaenichthys marmoratus*	33			...	7	33	37	50
Sculpins nei	*Cottidae*	33	-	-	-		1	63	71
Widow rockfish	*Sebastes entomelas*	34	...	...	6 886	4 357	3 841	3 575	2 481
Yellowtail rockfish	*Sebastes flavidus*	34	...	...	2 512	3 349	3 427	3 138	2 073
Pacific ocean perch	*Sebastes alutus*	34	20 820	27 178	25 362	24 824	28 071	24 144	23 528
Bocaccio rockfish	*Sebastes paucispinis*	34	...	...	448	491	135	3	0
Canary rockfish	*Sebastes pinniger*	34	...	...	1 133	1 222	736	57	45
Chilipepper rockfish	*Sebastes goodei*	34	...	...	38	11	10	31	9
Black rockfish	*Sebastes melanops*	34	...	...	...	...	...	109	148
Shortspine thornyhead	*Sebastolobus alascanus*	34	...	...	...	...	...	331	268
Scorpionfishes nei	*Scorpaenidae*	34	48 976	51 931	48 369	31 128	35 551	30 850	28 787
Sablefish	*Anoplopoma fimbria*	34	32 650	28 126	26 255	23 408	25 339	24 257	21 984
Pacific herring	*Clupea pallasii*	35	75 446	71 521	84 392	73 378	67 885	60 382	63 080
California pilchard	*Sardinops caeruleus*	35	1	-	-	22	775	14 368	23 908
Californian anchovy	*Engraulis mordax*	35	130	86	59	103	98	0	70
Skipjack tuna	*Katsuwonus pelamis*	36	-	-	-	-	-	0	1
Pacific bluefin tuna	*Thunnus orientalis*	36	1	2	3	8	7	3	0
Albacore	*Thunnus alalunga*	36	6 671	10 038	14 588	11 517	4 675	11 182	12 133
Yellowfin tuna	*Thunnus albacares*	36	0	-	0	0	-	1	13
Bigeye tuna	*Thunnus obesus*	36	-	-	-	1	-	-	1
Tunas nei	*Thunnini*	36	0	-	1 263	-	1	-	0
Swordfish	*Xiphias gladius*	36	22	23	34	128	27	34	14
Pacific jack mackerel	*Trachurus symmetricus*	37	147	-	2	961	176	181	215
Chub mackerel	*Scomber japonicus*	37	1	388	204	632	75	121	322
Thresher	*Alopias vulpinus*	38	...	...	...	17	2	48	76
Shortfin mako	*Isurus oxyrinchus*	38	...	...	...	3	-	0	0
Blue shark	*Prionace glauca*	38	-	-	-	-	-	1	2
Tope shark	*Galeorhinus galeus*	38	...	...	...	1	-	3	1
Picked dogfish	*Squalus acanthias*	38	2 744	4 000	2 100	2 501	6 439	5 363	5 181
Dogfish sharks nei	*Squalidae*	38	1 300	2 053	625	555	1	-	-
Rays, stingrays, mantas nei	*Rajiformes*	38	1 352	2 637	3 757	2 045	2 817	3 210	3 017
Sharks, rays, skates, etc. nei	*Elasmobranchii*	38	10	10	14	20	81	3	36
Marine fishes nei	*Osteichthyes*	39	64 914	106 820	106 601	97 853	98 191	31 003	74 345
Dungeness crab	*Cancer magister*	42	24 702	33 606	19 658	17 002	18 290	19 169	21 370
Pacific rock crab	*Cancer productus*	42	...	...	...	3	5	4	5
Tanner crabs nei	*Chionoecetes spp*	42	36 658	30 785	53 932	114 230	83 989	15 665	12 177
Marine crabs nei	*Brachyura*	42	2 185	647	333	133	99	1	42
Blue mud shrimp	*Upogebia pugettensis*	43	...	...	...	...	...	10	4
Ghost shrimps	*Callianassa spp*	43	-	-	-	-	25	17	14
King crabs	*Paralithodes spp*	44	6 656	9 526	8 177	10 941	7 675	6 848	7 281
Pacific shrimps nei	*Pandalus spp, Pandalopsis spp*	45	13 591	15 030	19 131	6 268	14 015	16 158	18 740
Natantian decapods nei	*Natantia*	45	8 557	9 392	5 200	5 203	4 071	4 223	4 212

C-67 (a)

Fish, crustaceans, molluscs, etc — **Capture production by species items** — **Pacific, Northeast**
Poissons, crustacés, mollusques, etc — **Captures par catégories d'espèces** — **Pacifique, nord-est**
Peces, crustáceos, moluscos, etc — **Capturas por categorías de especies** — **Pacífico, nordeste**

English name Nom anglais Nombre inglés	Scientific name Nom scientifique Nombre científico	Species group Groupe d'espèces Grupo de especies	1995 mt	1996 mt	1997 mt	1998 mt	1999 mt	2000 mt	2001 mt
Brine shrimp	*Artemia salina*	47	...	...	...	...	...	49	63
Abalones nei	*Haliotis spp*	52	29	-	-	-	-	-	3
Gastropods nei	*Gastropoda*	52	4 228	6 369	...	...	...	-	1
Olympia flat oyster	*Ostrea lurida*	53	...	...	...	...	6	7	8
Pacific cupped oyster	*Crassostrea gigas*	53	...	4 989	6 187	11	...	...	...
Cupped oysters nei	*Crassostrea spp*	53	...	...	...	...	...	7	12
Sea mussels nei	*Mytilidae*	54	1 317	455	...	...	1	675	448
Weathervane scallop	*Patinopecten caurinus*	55	1 950	2 372	2	3 228	2 642	2 012	1 052
Butter clam	*Saxidomus giganteus*	56	1 728	1 431	1 293	1 726	1 019	1 113	1 159
Pacific littleneck clam	*Protothaca staminea*	56	-	-	-	-	-	-	102
Pacific horse clams nei	*Tresus spp*	56	-	-	-	-	-	1	2
Pacific razor clam	*Siliqua patula*	56	-	-	-	-	-	14	61
Sand gaper	*Mya arenaria*	56	-	-	-	-	-	-	123
Pacific geoduck	*Panopea abrupta*	56	2 056	1 768	3 719	4 027	4 139	3 821	3 894
Cockles nei	*Cardiidae*	56	-	-	22	-	18	49	67
Clams, etc. nei	*Bivalvia*	56	11 215	3 912	381	2 509	1 601	1 519	1 250
Neon flying squid	*Ommastrephes bartrami*	57	-	-	2 008	1 200	1 521	683	2 061
Various squids nei	*Loliginidae, Ommastrephidae*	57	225	488	571	832	1 551	6	852
Octopuses, etc. nei	*Octopodidae*	57	108	197	208	261	60	59	84
Marine molluscs nei	*Mollusca*	58	2 155	1 873	4 593	2 879	1 850	1 904	1 649
Sea urchins nei	*Strongylocentrotus spp*	76	8 640	7 085	9 044	8 081	7 101	6 841	6 378
Sea cucumbers nei	*Holothurioidea*	76	729	491	-	-	228	274	300
Total			***3 031 197***	***2 893 342***	***2 846 092***	***2 792 459***	***2 551 689***	***2 477 803***	***2 759 090***

C-67 (b)

Fish, crustaceans, molluscs, etc — **Capture production by countries or areas** — **Pacific, Northeast**
Poissons, crustacés, mollusques, etc — **Captures par pays ou zones** — **Pacifique, nord-est**
Peces, crustáceos, moluscos, etc — **Capturas por países o áreas** — **Pacífico, nordeste**

Country or area Pays ou zone País o área	1992 mt	1993 mt	1994 mt	1995 mt	1996 mt	1997 mt	1998 mt	1999 mt	2000 mt	2001 mt
Canada	295 024	286 283	309 848	217 121	235 074	247 383	226 013	212 576	143 072	183 886
China	40 000	40 000	40 000	60 000	60 000	80 000	50 000	...	...	...
China,Taiwan	-	-	-	26	0	0	0	55	1 165	310
Japan	24 888	700	683	...	...	2 008	1 201	1 521	683	2 061
Korea Rep	84 018	47 264	15 875	10 967	2 480	-	311	1 330	-	-
Poland	-	-	-	-	-	-	-	-	998	4
Russian Fed	-	-	-	-	540	-	466	1 624	6	-
USA	2 754 059	2 965 279	2 822 291	2 743 083	2 595 248	2 516 701	2 514 468	2 334 583	2 331 879	2 572 829
Total	***3 197 989***	***3 339 526***	***3 188 697***	***3 031 197***	***2 893 342***	***2 846 092***	***2 792 459***	***2 551 689***	***2 477 803***	***2 759 090***

C-71 (a)

Fish, crustaceans, molluscs, etc — Capture production by species items — Pacific, Western Central
Poissons, crustacés, mollusques, etc — Captures par catégories d'espèces — Pacifique, centre-ouest
Peces, crustáceos, moluscos, etc — Capturas por categorías de especies — Pacífico, centro-occidental

English name Nom anglais Nombre inglés	Scientific name Nom scientifique Nombre científico	Species group Groupe d'espèces Grupo de especies	1995 mt	1996 mt	1997 mt	1998 mt	1999 mt	2000 mt	2001 mt
Chacunda gizzard shad	*Anodontostoma chacunda*	24	4 155	4 585	2 630	2 376	2 479	2 511	2 457
Toli shad	*Tenualosa toli*	24	1 420	1 702	2 535	3 097	2 875	2 117	2 310
Indian pellona	*Pellona ditchela*	24	5 045	4 850	3 257	4 650	5 566	6 475	6 933
Diadromous clupeoids nei	*Clupeoidei*	24	770	1 278	574	559	577	652	1 056
Milkfish	*Chanos chanos*	25	3 807	3 626	679	173	342	2 438	332
Barramundi(=Giant seaperch)	*Lates calcarifer*	25	40 653	40 598	45 025	47 105	53 629	57 292	62 508
Lefteye flounders nei	*Bothidae*	31	86	38	27	30	15	8	10
Tonguefishes	*Cynoglossidae*	31	9 084	7 487	7 181	7 438	7 379	7 249	7 142
Indian halibut	*Psettodes erumei*	31	3 974	3 828	10 286	4 902	6 945	5 865	6 702
Flatfishes nei	*Pleuronectiformes*	31	3 209	3 569	7 493	4 421	5 942	3 423	3 647
Gadiformes nei	*Gadiformes*	32	34	24	21	33	765	29	33
Indo-Pacific tarpon	*Megalops cyprinoides*	33	1 105	1 090	1 154	1 045	1 404	1 253	1 330
Bombay-duck	*Harpadon nehereus*	33	10 717	9 836	10 867	11 500	9 607	8 449	8 130
Lizardfishes nei	*Synodontidae*	33	76 335	71 051	88 525	59 940	87 516	85 714	87 953
Sea catfishes nei	*Ariidae*	33	69 751	76 357	92 217	77 680	85 743	84 346	85 982
Eeltail catfishes	*Plotosus spp*	33	1 529	1 763	1 498	2 118	1 492	1 400	1 375
Mullets nei	*Mugilidae*	33	48 242	47 065	48 597	49 632	53 045	50 677	52 854
Fusiliers	*Caesio spp*	33	52 818	43 985	51 751	46 546	48 664	43 111	44 130
Groupers nei	*Epinephelus spp*	33	28 326	32 316	35 444	37 395	41 297	45 816	46 609
Groupers, seabasses nei	*Serranidae*	33	19 691	18 656	17 868	18 214	19 140	17 885	18 642
Bigeyes nei	*Priacanthus spp*	33	60 585	69 890	65 902	68 125	74 504	74 773	75 005
Cardinalfishes, etc. nei	*Apogonidae*	33	...	200	180	60	71	60	67
Sillago-whitings	*Sillaginidae*	33	14 709	13 066	14 074	14 014	13 616	12 993	12 514
Moonfish	*Mene maculata*	33	4 207	4 815	11 940	11 529	12 385	11 718	12 998
Mangrove red snapper	*Lutjanus argentimaculatus*	33	10 522	10 824	10 190	4 935	12 319	10 650	10 436
Snappers nei	*Lutjanus spp*	33	46 045	51 819	62 548	59 075	58 808	54 432	55 018
Snappers, jobfishes nei	*Lutjanidae*	33	34 957	31 849	34 178	34 908	37 835	35 870	36 629
Threadfin breams nei	*Nemipterus spp*	33	146 871	142 017	136 767	143 945	160 811	148 611	147 968
Monocle breams	*Scolopsis spp*	33	1 340	1 341	1 293	1 471	994	1 668	1 607
Ponyfishes(=Slipmouths)	*Leiognathus spp*	33	2 370	2 589	2 422	3 125	3 096	2 412	2 324
Ponyfishes(=Slipmouths) nei	*Leiognathidae*	33	112 799	113 772	133 408	118 593	136 982	119 136	119 430
Silver grunt	*Pomadasys argenteus*	33	852	1 019	617	2 178	1 687	919	1 086
Grunts, sweetlips nei	*Haemulidae (=Pomadasyidae)*	33	10 503	11 661	11 697	12 034	12 377	14 116	14 514
Southern meagre(=Mulloway)	*Argyrosomus hololepidotus*	33	45	44	45	48	49	93	142
Croakers, drums nei	*Sciaenidae*	33	48 792	54 682	56 345	60 945	77 527	71 641	79 468
Largeeye breams	*Gymnocranius spp*	33	19	20	22	25	28	25	30
Emperors(=Scavengers) nei	*Lethrinidae*	33	30 201	32 747	29 753	27 409	29 846	28 912	30 204
Silver seabream	*Pagrus auratus*	33	783	644	649	872	395	311	394
Porgies, seabreams nei	*Sparidae*	33	3 329	2 208	603	251	218	227	268
Goatfishes	*Upeneus spp*	33	17 698	17 689	20 590	17 095	17 809	18 331	20 893
Goatfishes, red mullets nei	*Mullidae*	33	25 511	23 674	15 893	14 183	14 528	14 719	14 876
Mojarras(=Silver-biddies) nei	*Gerres spp*	33	8 200	7 635	6 025	5 433	5 909	5 801	6 528
Spotted sicklefish	*Drepane punctata*	33	682	888	734	582	918	739	780
Wrasses, hogfishes, etc. nei	*Labridae*	33	19 509	17 520	17 781	16 266	15 119	14 883	15 395
Parrotfishes nei	*Scaridae*	33	87	59	45	30	73	124	133
Threadfins, tasselfishes nei	*Polynemidae*	33	33 050	32 078	31 591	35 249	33 138	37 679	39 424
Glassfishes	*Ambassidae*	33	3 554	3 075	2 594	2 186	1 733	1 575	1 664
Percoids nei	*Percoidei*	33	40 413	38 415	21 792	30 749	45 305	20 544	22 267
Gobies nei	*Gobiidae*	33	8 780	8 404	7 542	6 756	8 021	7 719	7 898
Surgeonfishes nei	*Acanthuridae*	33	5 142	4 695	7 495	7 961	6 066	5 703	5 997
Batfishes	*Platax spp*	33	1 600	1 584	2 975	2 792	2 870	2 597	2 607
Scats	*Scatophagus spp*	33	4 613	4 035	2 057	2 476	2 501	2 637	2 745
Spinefeet(=Rabbitfishes) nei	*Siganus spp*	33	19 915	18 205	17 077	19 600	21 351	21 627	22 264
Flatheads nei	*Platycephalidae*	33	...	...	...	...	72	72	57
Triggerfishes, durgons nei	*Balistidae*	33	858	902	1 406	1 553	2 592	1 323	2 002
Daggertooth pike conger	*Muraenesox cinereus*	34	4 981	3 355	4 003	4 701	4 607	4 409	4 080
Conger eels, etc. nei	*Congridae*	34	3 379	2 756	2 110	2 540	2 467	2 398	2 422
Alfonsinos nei	*Beryx spp*	34	0	0	0	0	0	0	0
Largehead hairtail	*Trichiurus lepturus*	34	8 213	7 349	12 624	12 539	18 602	16 072	14 413
Hairtails, scabbardfishes nei	*Trichiuridae*	34	31 068	29 357	28 789	28 320	29 614	30 625	32 118
Ruffs, barrelfishes nei	*Centrolophidae*	34	0	0	0	0	0	0	0
Scorpionfishes nei	*Scorpaenidae*	34	-	2	-	-	-	-	-
Bluefin gurnard	*Chelidonichthys kumu*	34	-	14	0	0	0	-	-
Demersal percomorphs nei	*Perciformes*	34	100	480	250	-	-	-	-
Goldstripe sardinella	*Sardinella gibbosa*	35	134 646	127 268	130 579	142 976	127 925	134 861	138 300
Bali sardinella	*Sardinella lemuru*	35	58 542	58 943	65 970	61 671	55 143	57 507	61 090
Sardinellas nei	*Sardinella spp*	35	406 095	419 695	454 189	431 674	408 403	425 234	420 928
Rainbow sardine	*Dussumieria acuta*	35	35 300	36 334	33 136	38 231	42 193	31 673	32 454
Silver-stripe round herring	*Spratelloides gracilis*	35	150	140	135	10	40	30	...
Bluestripe herring	*Herklotsichthys quadrimaculat.*	35	185	140	120	22	20	20	...
Stolephorus anchovies	*Stolephorus spp*	35	176 711	182 949	204 874	184 065	186 469	201 005	204 374
Anchovies, etc. nei	*Engraulidae*	35	123 095	122 423	117 229	121 443	103 445	102 029	101 406
Dorab wolf-herring	*Chirocentrus dorab*	35	9 897	2 741	2 715	4 724	6 598	6 508	6 468
Wolf-herrings nei	*Chirocentrus spp*	35	22 752	20 197	24 285	22 200	24 638	25 252	25 966
Clupeoids nei	*Clupeoidei*	35	39 303	44 122	37 340	44 721	44 412	33 627	39 277
Wahoo	*Acanthocybium solandri*	36	206	154	169	180	183	172	192
Narrow-barred Spanish mackerel	*Scomberomorus commerson*	36	62 144	64 589	73 198	68 345	72 929	77 333	81 642
Indo-Pacific king mackerel	*Scomberomorus guttatus*	36	9 272	9 314	9 098	10 925	10 457	12 781	13 000
Seerfishes nei	*Scomberomorus spp*	36	22 998	21 036	19 031	21 025	23 725	21 006	21 330
Frigate and bullet tunas	*Auxis thazard, A.rochei*	36	107 686	107 854	125 573	124 048	129 004	130 027	132 918
Kawakawa	*Euthynnus affinis*	36	200 968	239 881	258 820	264 349	284 840	305 485	300 604
Skipjack tuna	*Katsuwonus pelamis*	36	893 808	896 513	755 449	1 037 506	977 437	1 019 229	991 858
Pacific bluefin tuna	*Thunnus orientalis*	36	627	801	1 314	1 233	3 104	1 755	1 654
Longtail tuna	*Thunnus tonggol*	36	42 850	35 661	31 637	40 968	48 590	49 533	49 215
Albacore	*Thunnus alalunga*	36	27 579	10 246	14 418	20 778	16 163	23 222	27 471

C-71 (a)

Fish, crustaceans, molluscs, etc	Capture production by species items	Pacific, Western Central
Poissons, crustacés, mollusques, etc	Captures par catégories d'espèces	Pacifique, centre-ouest
Peces, crustáceos, moluscos, etc	Capturas por categorías de especies	Pacífico, centro-occidental

English name Nom anglais Nombre inglés	Scientific name Nom scientifique Nombre científico	Species group Groupe d'espèces Grupo de especies	1995 mt	1996 mt	1997 mt	1998 mt	1999 mt	2000 mt	2001 mt
Southern bluefin tuna	*Thunnus maccoyii*	36	...	...	...	...	0	3	-
Yellowfin tuna	*Thunnus albacares*	36	260 365	201 054	324 264	350 952	310 788	297 622	339 789
Bigeye tuna	*Thunnus obesus*	36	29 586	21 910	44 208	42 821	47 585	33 865	42 851
Tunas nei	*Thunnini*	36	-	168	-	-		-	
Indo-Pacific sailfish	*Istiophorus platypterus*	36	654	317	318	3 104	180	63	104
Indo-Pacific blue marlin	*Makaira mazara*	36	7 901	6 754	6 923	9 488	10 924	11 206	10 895
Black marlin	*Makaira indica*	36	331	262	334	1 332	396	44	174
Striped marlin	*Tetrapturus audax*	36	649	457	1 080	1 652	436	279	253
Marlins,sailfishes,etc. nei	*Istiophoridae*	36	5 445	5 210	6 623	8 188	9 784	8 716	8 710
Swordfish	*Xiphias gladius*	36	1 431	1 123	2 990	2 158	2 945	624	801
Tuna-like fishes nei	*Scombroidei*	36	87 377	93 796	100 090	107 993	116 747	124 480	136 662
Needlefishes nei	*Tylosurus spp*	37	34 209	29 558	31 429	30 938	31 884	33 800	35 357
Halfbeaks nei	*Hemiramphus spp*	37	3 604	3 301	1 717	1 998	2 342	2 362	2 559
Flyingfishes nei	*Exocoetidae*	37	28 774	31 001	40 492	50 058	53 064	52 239	55 262
Silversides(=Sand smelts) nei	*Atherinidae*	37	894	897	694	676	709	623	630
False trevally	*Lactarius lactarius*	37	418	410	552	576	824	657	709
Bluefish	*Pomatomus saltatrix*	37	181	103	179	128	204	155	259
Cobia	*Rachycentron canadum*	37	1 282	1 256	1 436	1 671	1 832	1 550	1 477
Japanese jack mackerel	*Trachurus japonicus*	37	-	-	-	1	-	-	-
Indian scad	*Decapterus russelli*	37	99 872	103 633	111 729	98 130	118 198	134 774	126 931
Scads nei	*Decapterus spp*	37	478 228	443 976	478 339	489 129	473 880	474 500	502 494
Jacks, crevalles nei	*Caranx spp*	37	40 987	43 238	49 990	38 632	59 151	61 317	62 341
Amberjacks nei	*Seriola spp*	37	72	34	34	-	-	-	0
Black pomfret	*Parastromateus niger*	37	26 910	28 061	32 204	29 799	31 250	32 844	34 007
Rainbow runner	*Elagatis bipinnulata*	37	11 279	10 649	11 138	12 257	13 769	13 170	14 610
Torpedo scad	*Megalaspis cordyla*	37	29 059	25 402	30 486	43 139	50 123	50 712	50 414
Queenfishes	*Scomberoides spp*	37	17 384	16 761	16 557	19 310	21 226	19 976	19 942
Bigeye scad	*Selar crumenophthalmus*	37	43 592	68 193	76 355	86 930	91 742	96 975	104 761
Yellowstripe scad	*Selaroides leptolepis*	37	20 956	24 467	26 526	29 091	36 470	40 816	36 165
Blackbanded trevally	*Seriolina nigrofasciata*	37	4 048	3 603	3 532	3 088	3 345	3 299	3 279
Carangids nei	*Carangidae*	37	175 876	170 669	171 314	165 656	169 901	169 826	173 656
Common dolphinfish	*Coryphaena hippurus*	37	4 765	3 952	429	309	280	254	304
Chub mackerel	*Scomber japonicus*	37	3 307	2 991	2 025	1 507	1 485	1 422	1 520
Short mackerel	*Rastrelliger brachysoma*	37	26 200	25 224	22 978	23 350	25 713	26 771	28 289
Indian mackerel	*Rastrelliger kanagurta*	37	97 142	67 992	74 320	71 536	81 239	82 282	86 336
Indian mackerels nei	*Rastrelliger spp*	37	266 125	235 953	242 756	262 610	285 551	287 834	291 344
Mackerels nei	*Scombridae*	37	126	204	189	195	1 953	2 661	2 593
Silver pomfret	*Pampus argenteus*	37	16 547	18 068	17 969	17 518	18 084	21 998	23 668
Butterfishes, pomfrets nei	*Stromateidae*	37	2 693	2 601	1 287	1 686	2 046	1 925	1 880
Barracudas nei	*Sphyraena spp*	37	32 108	31 880	30 661	33 574	39 808	35 294	36 164
Pelagic percomorphs nei	*Perciformes*	37	20	20	-	-	-	-	-
Rays, stingrays, mantas nei	*Rajiformes*	38	50 874	49 780	53 489	51 254	57 369	56 159	58 425
Sharks, rays, skates, etc. nei	*Elasmobranchii*	38	99 528	89 194	86 559	92 046	84 562	88 171	100 561
Marine fishes nei	*Osteichthyes*	39	2 144 552	2 206 263	2 122 217	2 120 946	2 261 522	2 305 847	2 354 058
Blue swimming crab	*Portunus pelagicus*	42	78 449	78 423	82 962	74 935	82 354	85 163	88 874
Indo-Pacific swamp crab	*Scylla serrata*	42	12 711	10 628	8 409	8 117	10 344	10 530	10 835
Marine crabs nei	*Brachyura*	42	43 304	37 755	36 185	61 370	49 324	53 114	63 499
Tropical spiny lobsters nei	*Panulirus spp*	43	3 318	3 072	4 578	3 641	4 289	4 400	5 131
Green rock lobster	*Jasus verreauxi*	43	...	...	...	...	...	269	...
Slipper lobsters nei	*Scyllaridae*	43	2 864	3 659	3 521	3 871	1 858	1 832	1 837
Lobsters nei	*Reptantia*	43	...	...	2 000	1 000	200	500	700
Banana prawn	*Penaeus merguiensis*	45	55 203	55 916	56 272	64 442	67 393	68 029	75 087
Giant tiger prawn	*Penaeus monodon*	45	21 621	19 453	21 517	26 547	30 996	37 053	43 084
Green tiger prawn	*Penaeus semisulcatus*	45	711	1 042	1 068	839	609	600	597
Western king prawn	*Penaeus latisulcatus*	45	1 597	1 676	1 905	1 885	2 627	2 602	2 567
Penaeus shrimps nei	*Penaeus spp*	45	80 822	81 786	78 933	57 040	50 941	51 144	51 305
Endeavour shrimp	*Metapenaeus endeavouri*	45	2 182	2 400	2 339	2 930	2 691	2 299	2 173
Metapenaeus shrimps nei	*Metapenaeus spp*	45	27 373	28 934	36 215	44 904	35 932	41 861	45 485
Sergestid shrimps nei	*Sergestidae*	45	43 384	43 814	32 661	32 686	34 230	27 770	28 225
Natantian decapods nei	*Natantia*	45	183 995	200 400	209 616	182 994	213 171	204 233	209 238
Squillids nei	*Squillidae*	47	982	695	2 133	2 219	2 068	2 141	2 561
Stomatopods nei	*Stomatopoda*	47	-	181	176	457	750	740	735
Marine crustaceans nei	*Crustacea*	47	1 981	1 875	5 966	4 683	1 680	2 322	3 647
Abalones nei	*Haliotis spp*	52	483	448	183	348	282	241	250
Gastropods nei	*Gastropoda*	52	-	-	-	0	0	-	-
Slipper cupped oyster	*Crassostrea iredalei*	53	324	291	152	89	95	79	83
Cupped oysters nei	*Crassostrea spp*	53	1 331	1 596	1 678	2 029	2 025	1 931	2 120
Green mussel	*Perna viridis*	54	20 413	20 216	18 017	17 454	6 554	6 462	6 422
Scallops nei	*Pectinidae*	55	6 986	7 276	3 488	6 392	5 403	5 093	6 086
Blood cockle	*Anadara granosa*	56	-	-	-	421	1 750	1 726	1 716
Anadara clams nei	*Anadara spp*	56	28 468	26 006	21 524	18 952	26 301	27 910	26 517
Hard clams nei	*Meretrix spp*	56	14 949	9 788	10 305	10 507	11 347	11 571	11 820
Short neck clams nei	*Paphia spp*	56	20 090	34 591	24 229	43 458	42 613	42 031	41 774
Clams, etc. nei	*Bivalvia*	56	1 745	1 700	1 537	1 193	1 442	1 383	1 237
Cuttlefish,bobtail squids nei	*Sepiidae, Sepiolidae*	57	60 483	65 259	68 051	66 514	66 764	66 482	65 766
Common squids nei	*Loligo spp*	57	148 154	138 046	152 941	146 132	141 421	141 218	146 067
Various squids nei	*Loliginidae, Ommastrephidae*	57	20 345	23 654	23 096	23 689	25 109	27 304	27 298
Octopuses, etc. nei	*Octopodidae*	57	23 790	24 953	23 446	26 788	21 349	20 284	21 341
Cephalopods nei	*Cephalopoda*	57	103 000	92 000	92 500	103 000	110 000	180 000	130 000

C-71 (a)

Fish, crustaceans, molluscs, etc — Capture production by species items — Pacific, Western Central
Poissons, crustacés, mollusques, etc — Captures par catégories d'espèces — Pacifique, centre-ouest
Peces, crustáceos, moluscos, etc — Capturas por categorías de especies — Pacífico, centro-occidental

English name Nom anglais Nombre inglés	Scientific name Nom scientifique Nombre científico	Species group Groupe d'espèces Grupo de especies	1995 mt	1996 mt	1997 mt	1998 mt	1999 mt	2000 mt	2001 mt
Marine molluscs nei	*Mollusca*	58	59 506	53 322	54 772	59 749	59 051	53 484	60 027
Green turtle	*Chelonia mydas*	72	6	22	4	2	2	2	3
Marine turtles nei	*Testudinata*	72	294	395	399	450	446	382	390
Sea urchins nei	*Strongylocentrotus spp*	76	525	492	391	264	243	215	223
Sea cucumbers nei	*Holothurioidea*	76	7 322	7 880	7 257	6 561	5 856	6 269	5 997
Jellyfishes	*Rhopilema spp*	77	145 983	37 814	85 808	74 205	90 434	87 820	88 198
Aquatic invertebrates nei	*Invertebrata*	77	241	475	775	224	1 314	1 534	1 725
Total			***8 902 133***	***8 730 401***	***8 966 335***	***9 282 908***	***9 597 552***	***9 740 281***	***9 939 821***

C-71 (b)

Fish, crustaceans, molluscs, etc — Capture production by countries or areas — Pacific, Western Central
Poissons, crustacés, mollusques, etc — Captures par pays ou zones — Pacifique, centre-ouest
Peces, crustáceos, moluscos, etc — Capturas por países o áreas — Pacífico, centro-occidental

Country or area Pays ou zone País o área	1992 mt	1993 mt	1994 mt	1995 mt	1996 mt	1997 mt	1998 mt	1999 mt	2000 mt	2001 mt
Australia	35 502 F	38 006 F	39 400 F	43 719 F	46 781 F	42 075	46 536	48 905	43 246	52 475
Brunei Darsm	1 667	1 703	4 441	4 712	7 390	4 504	5 014	3 160	2 464	1 476
Cambodia	33 700	33 100	30 000	30 500	31 200	29 800	32 200	38 100	36 000	36 000 F
China	2 818	7 797	14 415	11 512	6 406	3 774	3 446	9 003	8 882	7 519
China,Taiwan	213 067 F	254 607 F	291 612 F	280 936 F	271 520	254 979	365 733	283 184	309 239	321 673
Fiji Islands	20 521	24 910	26 000 F	25 250	21 995	23 430	23 047	31 091	34 300 F	37 051
Guam	345	431	435	185	121	158	247	223	275	278
Indonesia	1 942 970	2 036 789	2 214 080	2 418 184	2 419 377	2 630 437	2 646 364	2 741 141	2 851 227	2 949 590
Japan	301 927	282 242	277 415	272 490	226 315	210 396	254 848	220 024	242 406	225 024
Kiribati	28 211	27 072	27 753	30 306	31 871	30 052	35 304	52 741	25 563	32 375
Korea Rep	210 100	157 088	244 170	213 720	175 466	182 399	218 092	157 499	180 224	190 038
Malaysia	549 521	600 835	626 245	604 368	632 598	669 685	640 965	762 661	765 911	755 933
Marshall Is	326 F	471 F	393 F	381 F	369	400 F	500 F	500 F	8 155	37 098
Micronesia	16 542 F	17 884 F	22 922 F	7 624 F	9 064 F	10 122 F	16 576 F	12 833 F	23 710 F	18 057 F
Nauru	377	500	500	450 F	400 F	400 F	400 F	400 F	400 F	400 F
NewCaledonia	3 184	2 670	3 185	2 644	2 998	2 438	3 105	3 152	3 386	3 337 F
N Marianas	140	136	176	191	225	250	235	193	189	197
Palau	1 918	2 066	1 820	1 956	1 990	1 751	1 777	1 800 F	2 000 F	2 000 F
Papua N Guin	12 741 F	12 241 F	12 719 F	25 577 F	24 323 F	32 685 F	64 529 F	51 931 F	82 569 F	108 919 F
Philippines	1 655 368	1 623 548	1 641 010	1 674 695	1 606 246	1 646 453	1 682 862	1 720 879	1 737 883	1 807 341
Russian Fed	2 150	18 937	5 840	4 981	-	-	-	-	-	-
Singapore	9 177	9 279	11 278	10 102	9 943	9 250	7 733	6 489	5 371	3 342
Solomon Is	45 601 F	42 970 F	46 904 F	62 019 F	50 990 F	64 005 F	61 332 F	58 428 F	24 788 F	30 075 F
Thailand	2 084 045	1 928 853	2 042 132	2 021 736	1 938 574	1 848 233	1 820 022	1 937 211	1 910 976	1 881 642
Tuvalu	499	1 460	561	399	400 F	500 F	500 F	500 F	500 F	500 F
USA	154 413	176 253	194 218	153 840	143 348	140 391	154 610	188 040	121 797	113 658
Vanuatu	2 427	2 470 F	3 020 F	9 236 F	11 603 F	28 856 F	41 477 F	49 971 F	37 930 F	2 400 F
Viet Nam	729 953	785 304	946 322	990 250	1 058 708	1 098 736	1 155 154	1 217 193	1 280 590	1 321 123
Wallis Fut I	143	150	193	170	180	176	300	300 F	300 F	300 F
Total	***8 059 353***	***8 089 772***	***8 729 159***	***8 902 133***	***8 730 401***	***8 966 335***	***9 282 908***	***9 597 552***	***9 740 281***	***9 939 821***

C-77 (a)

Fish, crustaceans, molluscs, etc — Capture production by species items — Pacific, Eastern Central
Poissons, crustacés, mollusques, etc — Captures par catégories d'espèces — Pacifique, centre-est
Peces, crustáceos, moluscos, etc — Capturas por categorías de especies — Pacífico, centro-oriental

English name Nom anglais Nombre inglés	Scientific name Nom scientifique Nombre científico	Species group Groupe d'espèces Grupo de especies	1995 mt	1996 mt	1997 mt	1998 mt	1999 mt	2000 mt	2001 mt
Pink(=Humpback)salmon	*Oncorhynchus gorbuscha*	23	-	-	0	-	-	-	0
Chinook(=Spring=King)salmon	*Oncorhynchus tshawytscha*	23	2 993	2 078	2 737	937	1 991	2 613	1 193
Coho(=Silver)salmon	*Oncorhynchus kisutch*	23	5	6	3	0	-	4	1
Smelts nei	*Osmerus spp, Hypomesus spp*	23	-	-	7	7	5	6	12
American shad	*Aiosa sapidissima*	24	-	-	0	-	-	-	-
Milkfish	*Chanos chanos*	25	0	0	1	0	0	7	...
Lefteye flounders nei	*Bothidae*	31	-	-	-	-	9	-	-
Pacific halibut	*Hippoglossus stenolepis*	31	3	4	2	3	2	-	0
English sole	*Pleuronectes vetulus*	31	...	...	...	227	226	177	197
Arrow-tooth flounder	*Atheresthes stomias*	31	1	1	1	1	2	0	0
Petrale sole	*Eopsetta jordani*	31	...	...	483	267	259	238	268
Rex sole	*Glyptocephalus zachirus*	31	...	...	...	168	153	128	98
Rock sole	*Lepidopsetta bilineata*	31	7	4	9	10	7	7	7
Dover sole	*Microstomus pacificus*	31	...	...	3 846	1 818	2 174	1 866	1 403
Pacific sand sole	*Psettichthys melanostictus*	31	...	...	...	19	23	35	44
Tonguefishes	*Cynoglossidae*	31	-	-	-	-	105	95	-
California flounder	*Paralichthys californicus*	31	...	...	...	510	617	390	389
Flatfishes nei	*Pleuronectiformes*	31	8 466	9 220	4 596	1 830	2 893	3 293	2 350
Pacific tomcod	*Microgadus proximus*	32	...	...	...	0	-	0	-
North Pacific hake	*Merluccius productus*	32	79	117	1 090	252	111	393	1
Grenadiers, rattails nei	*Macrouridae*	32	...	...	...	417	210	285	145
Gadiformes nei	*Gadiformes*	32	10	21	183	137	58	40	...
Sea catfishes nei	*Ariidae*	33	1 328	1 208	1 400	1 869	1 704	1 549	1 964
Squirrelfishes nei	*Holocentridae*	33	0	1	1	0	1	1	0
Bobo mullet	*Joturus pichardi*	33	94	87	56	51	82	46	104
Mullets nei	*Mugilidae*	33	3 960	3 774	3 474	2 993	3 454	3 134	3 171
Snooks(=Robalos) nei	*Centropomus spp*	33	563	587	684	752	996	809	862
Groupers nei	*Epinephelus spp*	33	2 713	5 367	7 228	6 250	6 881	5 367	4 098
Groupers, seabasses nei	*Serranidae*	33	281	177	387	644	995	581	438
Giant seabass	*Stereolepis gigas*	33	0	1	1	3	2	2	3
Yellow snapper	*Lutjanus argentiventris*	33	3 810	4 917	3 123	3 390	2 994	3 392	3 388
Snappers nei	*Lutjanus spp*	33	266	235	207	177	2 295	1 831	2 568
Snappers, jobfishes nei	*Lutjanidae*	33	7 421	5 955	6 321	7 405	12 573	10 922	11 155
Grunts, sweetlips nei	*Haemulidae (=Pomadasyidae)*	33	604	484	868	1 276	375	471	480
Weakfishes nei	*Cynoscion spp*	33	2 281	3 402	4 042	4 602	5 507	2 494	5 356
Croakers nei	*Micropogonias spp*	33	3 432	5 036	4 966	5 476	4 602	3 705	3 728
White weakfish	*Atractoscion nobilis*	33	33	46	28	71	112	101	124
White croaker	*Genyonemus lineatus*	33	256	242	167	64	74	105	137
Black drum	*Pogonias cromis*	33	0	0	0	0	1	0	...
Croakers, drums nei	*Sciaenidae*	33	1 698	3 005	2 994	3 557	9 913	8 172	5 316
Emperors(=Scavengers) nei	*Lethrinidae*	33	2	5	2	0	1	4	6
Porgies, seabreams nei	*Sparidae*	33	-	-	-	-	-	2	-
Mojarras, etc. nei	*Gerreidae*	33	7 085	6 841	3 946	1 945	2 114	1 724	1 800
Opaleye	*Girella nigricans*	33	-	-	-	-	-	3	2
California sheephead	*Semicossyphus pulcher*	33	...	...	...	118	58	78	68
Parrotfishes nei	*Scaridae*	33	0	8	3	10	7	6	3
Surgeonfishes nei	*Acanthuridae*	33	...	8	-	13	16	10	3
Lingcod	*Ophiodon elongatus*	33	381	352	364	95	93	28	34
Cabezon	*Scorpaenichthys marmoratus*	33	...	...	...	161	140	111	69
Sculpins nei	*Cottidae*	33	-	-	-	-	-	-	3
Triggerfishes, durgons nei	*Balistidae*	33	906	1 266	5 787	3 259	2 338	1 645	...
Cusk-eels, brotulas nei	*Ophidiidae*	34	-	-	-	-	-	1	-
Ocean whitefish	*Caulolatilus princeps*	34	...	...	...	535	627	1 077	984
Escolar	*Lepidocybium flavobrunneum*	34	...	...	...	2	1	6	3
Hairtails, scabbardfishes nei	*Trichiuridae*	34	10	1	...	-	607	540	4
Widow rockfish	*Sebastes entomelas*	34	...	...	865	540	402	29	128
Yellowtail rockfish	*Sebastes flavidus*	34	...	...	232	253	41	32	4
Pacific ocean perch	*Sebastes alutus*	34	-	-	-	-	16	12	12
Bocaccio rockfish	*Sebastes paucispinis*	34	...	...	276	105	62	24	33
Canary rockfish	*Sebastes pinniger*	34	...	...	129	91	36	3	4
Chilipepper rockfish	*Sebastes goodei*	34	...	...	1 812	1 262	908	414	609
Scorpionfishes nei	*Scorpaenidae*	34	8 363	7 678	4 482	4 330	1 614	1 291	1 679
Sablefish	*Anoplopoma fimbria*	34	1 786	2 136	1 751	824	1 132	1 088	958
Demersal percomorphs nei	*Perciformes*	34	0	1	0	0	0	0	0
Pacific herring	*Clupea pallasii*	35	4 623	5 328	9 188	1 987	2 267	2 853	2 521
California pilchard	*Sardinops caeruleus*	35	363 463	450 596	498 653	381 876	404 627	531 711	661 589
Red-eye round herring	*Etrumeus teres*	35	-	-	72	7	58	42	2
Pacific thread herring	*Opisthonema libertate*	35	21 224	32 517	26 266	49 472	38 746	63 532	71 280
Californian anchovy	*Engraulis mordax*	35	27 020	14 017	7 866	2 232	11 039	19 460	19 607
Pacific anchoveta	*Cetengraulis mysticetus*	35	106 743	60 322	77 726	107 730	27 356	86 681	100 000
Clupeoids nei	*Clupeoidei*	35	364	492	1 343	1 143	1 893	1 715	2 222
Eastern Pacific bonito	*Sarda chiliensis*	36	6 789	848	1 165	1 517	1 862	473	152
Wahoo	*Acanthocybium solandri*	36	5	5	126	209	288	249	307
Pacific sierra	*Scomberomorus sierra*	36	5 600	6 231	6 070	4 878	5 424	7 656	6 959
Black skipjack	*Euthynnus lineatus*	36	101	444	84	251	90	20	150
Skipjack tuna	*Katsuwonus pelamis*	36	74 942	43 805	65 787	63 602	65 064	42 533	40 497
Pacific bluefin tuna	*Thunnus orientalis*	36	743	8 479	2 641	1 997	2 551	3 345	1 179
Longtail tuna	*Thunnus tonggol*	36	388	327	249	-	-	-	-
Albacore	*Thunnus alalunga*	36	15 943	19 292	22 535	26 423	28 861	26 228	30 832
Yellowfin tuna	*Thunnus albacares*	36	191 317	211 266	232 583	211 710	204 632	206 071	266 915
Bigeye tuna	*Thunnus obesus*	36	71 313	55 804	57 005	78 120	61 343	66 863	77 809
Tunas nei	*Thunnini*	36	7	8	15	10	14	4	5
Indo-Pacific sailfish	*Istiophorus platypterus*	36	460	459	1 447	666	575	459	441
Indo-Pacific blue marlin	*Makaira mazara*	36	6 926	3 099	4 248	4 088	2 806	2 650	3 453

C-77 (a)

Fish, crustaceans, molluscs, etc — Capture production by species items — Pacific, Eastern Central
Poissons, crustacés, mollusques, etc — Captures par catégories d'espèces — Pacifique, centre-est
Peces, crustáceos, moluscos, etc — Capturas por categorías de especies — Pacífico, centro-oriental

English name Nom anglais Nombre inglés	Scientific name Nom scientifique Nombre científico	Species group Groupe d'espèces Grupo de especies	1995 mt	1996 mt	1997 mt	1998 mt	1999 mt	2000 mt	2001 mt
Black marlin	*Makaira indica*	36	411	320	192	419	810	777	1 163
Striped marlin	*Tetrapturus audax*	36	3 336	2 051	3 130	2 322	2 064	1 483	1 111
Marlins,sailfishes,etc. nei	*Istiophoridae*	36	7 445	7 270	5 947	7 944	4 706	4 143	4 859
Swordfish	*Xiphias gladius*	36	6 063	6 854	9 511	11 664	8 592	11 031	10 395
Tuna-like fishes nei	*Scombroidei*	36	3 113	6 213	3 982	6 426	3 687	4 115	5 904
Pacific saury	*Cololabis saira*	37	-	-	-	-	754	1 060	423
Flyingfishes nei	*Exocoetidae*	37	38	30	85	122	30	118	112
Opah	*Lampris guttatus*	37	93	96	...	114	205	181	196
Silversides(=Sand smelts) nei	*Atherinidae*	37	...	...	...	1 062	906	793	704
Pacific jack mackerel	*Trachurus symmetricus*	37	1 728	2 176	1 158	832	950	1 135	3 624
Scads nei	*Decapterus spp*	37	156	237	130	149	226	187	270
Jacks, crevalles nei	*Caranx spp*	37	1 774	2 089	2 321	2 238	1 971	2 527	2 908
Pompanos nei	*Trachinotus spp*	37	286	206	200	329	309	274	222
Yellowtail amberjack	*Seriola lalandi*	37	-	-	-	111	30	50	39
Amberjacks nei	*Seriola spp*	37	352	477	345	475	172	445	400
Carangids nei	*Carangidae*	37	62	74	320	1 629	736	381	1
Pomfrets, ocean breams nei	*Bramidae*	37	-	-	-	-	-	-	5
Common dolphinfish	*Coryphaena hippurus*	37	216	338	563	565	1 273	11 615	14 365
Chub mackerel	*Scomber japonicus*	37	31 438	22 548	42 431	42 804	78 018	66 423	10 738
Butterfishes, pomfrets nei	*Stromateidae*	37	-	-	-	1	2	3	7
Barracudas nei	*Sphyraena spp*	37	35	70	489	621	259	180	72
Thresher	*Alopias vulpinus*	38	...	...	...	303	262	249	299
Bigeye thresher	*Alopias superciliosus*	38	...	...	...	11	5	5	2
Shortfin mako	*Isurus oxyrinchus*	38	...	...	...	93	63	106	99
Great white shark	*Carcharodon carcharias*	38	-	-	-	-	-	1	0
Blue shark	*Prionace glauca*	38	...	...	...	1	-	0	0
Requiem sharks nei	*Carcharhinidae*	38	6 334	5 849	3 482	2 988	2 949	3 320	3 059
Hammerhead sharks, etc. nei	*Sphyrnidae*	38	...	...	...	1	-	-	-
Brown smooth-hound	*Mustelus henlei*	38	...	...	...	3	5	3	4
Tope shark	*Galeorhinus galeus*	38	...	...	...	51	73	45	44
Picked dogfish	*Squalus acanthias*	38	1	-	0	5	24	8	3
Rays, stingrays, mantas nei	*Rajiformes*	38	4 495	5 527	5 132	5 114	6 149	5 707	4 612
Sharks, rays, skates, etc. nei	*Elasmobranchii*	38	26 957	26 812	26 514	29 776	29 661	34 283	28 810
Marine fishes nei	*Osteichthyes*	39	269 721	213 172	181 727	159 041	114 636	126 141	130 511
Dungeness crab	*Cancer magister*	42	1 580	897	1 593	1 485	734	772	840
Pacific rock crab	*Cancer productus*	42	...	...	...	571	354	490	532
Indo-Pacific swamp crab	*Scylla serrata*	42	-	-	2	2	2	2	2
Marine crabs nei	*Brachyura*	42	10 943	14 234	10 860	6 605	6 728	12 118	11 266
Green spiny lobster	*Panulirus gracilis*	43	14	101	152	66	131	476	652
Tropical spiny lobsters nei	*Panulirus spp*	43	1 945	2 520	2 582	2 440	2 175	3 096	2 744
Pelagic red crab	*Pleuroncodes planipes*	44	356	164	-	-	-	-	-
King crabs	*Paralithodes spp*	44	-	-	-	-	-	1	2
Yellowleg shrimp	*Penaeus californiensis*	45	283	272	126	271	230	138	202
Whiteleg shrimp	*Penaeus vannamei*	45	1 258	1 238	1 046	1 277	1 082	519	460
Crystal shrimp	*Penaeus brevirostris*	45	3 676	2 136	2 776	2 090	2 462	2 604	2 729
Penaeus shrimps nei	*Penaeus spp*	45	52 177	49 878	54 489	50 810	52 714	46 747	42 475
Pacific seabobs	*Xiphopenaeus,Trachypenaeus spp*	45	9 898	12 156	7 354	9 881	6 457	4 720	3 755
Northern nylon shrimp	*Heterocarpus vicarius*	45	-	-	-	40	37	20	131
Pacific shrimps nei	*Pandalus spp, Pandalopsis spp*	45	1 539	1 707	1 288	622	636	372	309
Crangonid shrimps nei	*Crangonidae*	45	-	-	31	41	-	42	45
Pacific rock shrimp	*Sicyonia ingentis*	45	...	...	...	185	630	756	165
Natantian decapods nei	*Natantia*	45	4 075	5 480	6 274	1 574	3 169	2 744	2 158
Brine shrimp	*Artemia salina*	47	...	...	...	513	691	512	475
Marine crustaceans nei	*Crustacea*	47	832	748	714	666	561	795	894
Abalones nei	*Haliotis spp*	52	1 377	1 202	1 026	709	582	535	482
Stromboid conchs nei	*Strombus spp*	52	4 163	3 091	2 427	1 500	1 473	2 102	2 672
Gastropods nei	*Gastropoda*	52	18	16	12	165	53	36	28
Flat oysters nei	*Ostrea spp*	53	1 801	1 354	1 903	1 276	2 489	1 828	2 234
Pacific cupped oyster	*Crassostrea gigas*	53	1 709	1 717	2 176	2 107	539	1 917	1 083
Sea mussels nei	*Mytilidae*	54	290	583	2 206	2 230	832	643	90
Pacific calico scallop	*Argopecten ventricosus*	55	1 256	17 290	2 320	2 726	1 864	6 287	3 100
Ark clams nei	*Arca spp*	56	1 033	834	1 367	1 018	880	675	600
Venus clams nei	*Veneridae*	56	6 244	5 043	3 915	3 672	4 486	4 835	3 406
Clams, etc. nei	*Bivalvia*	56	1 673	1 393	1 510	1 080	1 547	1 633	1 705
Neon flying squid	*Ommastrephes bartrami*	57	-	-	10	113	4	81	131
Jumbo flying squid	*Dosidicus gigas*	57	39 657	121 063	140 868	26 718	58 351	82 058	73 952
Various squids nei	*Loliginidae, Ommastrephidae*	57	70 688	79 080	71 206	2 768	91 019	119 394	92 049
Octopuses, etc. nei	*Octopodidae*	57	971	1 344	1 052	807	1 211	998	936
Marine molluscs nei	*Mollusca*	58	1 380	1 307	1 857	1 462	1 742	3 816	2 951
Marine turtles nei	*Testudinata*	72	0	0	0	3	1	21	...
Echinoderms	*Echinodermata*	76	2 791	3 027	2 109	1 148	2 052	2 828	2 267
Starfishes nei	*Asteroidea*	76	-	-	-	-	1	-	0
Sea urchins nei	*Strongylocentrotus spp*	76	10 016	9 131	8 212	4 712	6 395	6 074	5 891
Sea cucumbers nei	*Holothurioidea*	76	...	86	80	361	234	426	442
Marine worms	*Polychaeta*	77	...	42	67	51	31	28	450

C-77 (a)

Fish, crustaceans, molluscs, etc — Capture production by species items — Pacific, Eastern Central
Poissons, crustacés, mollusques, etc — Captures par catégories d'espèces — Pacifique, centre-est
Peces, crustáceos, moluscos, etc — Capturas por categorías de especies — Pacífico, centro-oriental

English name Nom anglais Nombre inglés	Scientific name Nom scientifique Nombre científico	Species group Groupe d'espèces Grupo de especies	1995 mt	1996 mt	1997 mt	1998 mt	1999 mt	2000 mt	2001 mt
Aquatic invertebrates nei	*Invertebrata*	77	1 346	830	284	8	0	220	-
Total			***1 555 851***	***1 619 983***	***1 705 176***	***1 407 631***	***1 445 221***	***1 713 080***	***1 830 592***

C-77 (b)

Fish, crustaceans, molluscs, etc — Capture production by countries or areas — Pacific, Eastern Central
Poissons, crustacés, mollusques, etc — Captures par pays ou zones — Pacifique, centre-est
Peces, crustáceos, moluscos, etc — Capturas por países o áreas — Pacífico, centro-oriental

Country or area Pays ou zone País o área	1992 mt	1993 mt	1994 mt	1995 mt	1996 mt	1997 mt	1998 mt	1999 mt	2000 mt	2001 mt
Amer Samoa	45	27	111	152	210	431	586	518	830	3 663
Belize	-	-	240	200	1 430	1 110	680	970	560	...
China	-	-	-	-	-	-	-	-	2 682	14 547
China,Taiwan	19 059 F	15 328 F	11 038 F	5 492	5 205	5 443	6 679	10 917	8 034	10 461
Colombia	...	...	...	...	...	...	...	...	7 630	7 530
Cook Is	967	985 F	932 F	1 080	900 F	820 F	770 F	750 F	720 F	700 F
Costa Rica	15 342	15 261	16 023	15 845	21 540	25 409	23 355	26 552	33 310	32 917
Cyprus	870	960	180	830	1 730	1 920	-	-	-	-
Ecuador	1 990	1 050	1 710	-	270	-	-	-	-	150
El Salvador	6 940	7 990	9 740	10 747	11 466	9 089	10 214	7 841	6 759	16 055
Estonia	-	-	-	-	-	-	-	10	-	-
Fr Polynesia	3 425	8 075	8 175	9 020	9 910	11 617	12 420	12 283	13 846	15 351
Guatemala	2 776	3 262	3 256	3 838	3 263	1 490	3 996	8 930	12 101	2 450 F
Honduras	785	509	692	864	2 412	4 002	2 155 F	1 702	2 973	3 044
Japan	129 529	102 740	95 131	82 190	69 159	82 917	63 507	46 161	76 384	54 890
Korea Rep	43 020	33 550	36 937	37 833	37 320	44 473	59 919	49 305	51 303	53 668
Mexico	712 900	665 867	743 251	921 820	1 027 461	1 042 150	767 021	817 557	928 727	1 039 262
Nicaragua	2 158	2 621	2 472	4 500	4 615	5 432	6 625	9 662	10 434	8 304
Niue	120	120 F	150 F	150 F	200 F	200 F	200 F	200 F	200 F	200 F
Panama	122 004	143 959	139 709	155 219	122 975	147 295	197 397	111 140	195 841	220 390 F
Samoa	1 296	673	1 108	2 519	2 727	7 041	7 547	10 204	13 003	12 965 F
Spain	10	740	160	980	1 910	1 950	21 960	37 753	27 416	15 722
Tokelau	191	200 F	200 F	200 F	200 F	200 F	200 F	200 F	200 F	200 F
Tonga	2 236	2 374	2 499	2 597	2 915	2 871	4 076	4 221	3 760	4 673
USA	196 843	143 488	174 088	233 025	220 175	237 646	152 124	222 855	256 127	212 580
US Minor Is	0	0	0	0	0	0	0	0	0	0
Vanuatu	20 320	14 270	13 090	21 460	13 390	22 240	20 590	16 090	5 590	6 960
Venezuela	37 240	32 700	26 870	45 030	58 600	49 430	45 300	49 400	53 120	69 560
Other nei	-	-	50	260	-	-	310	-	1 530	24 350
Total	***1 320 066***	***1 196 749***	***1 287 812***	***1 555 851***	***1 619 983***	***1 705 176***	***1 407 631***	***1 445 221***	***1 713 080***	***1 830 592***

C-81 (a)

Fish, crustaceans, molluscs, etc — **Capture production by species items** — **Pacific, Southwest**
Poissons, crustacés, mollusques, etc — **Captures par catégories d'espèces** — **Pacifique, sud-ouest**
Peces, crustáceos, moluscos, etc — **Capturas por categorías de especies** — **Pacífico, sudoccidental**

English name Nom anglais Nombre inglés	Scientific name Nom scientifique Nombre científico	Species group Groupe d'espèces Grupo de especies	1995 mt	1996 mt	1997 mt	1998 mt	1999 mt	2000 mt	2001 mt
River eels nei	*Anguilla spp*	22	1 183	524	588	400	416	380	313
Chinook(=Spring=King)salmon	*Oncorhynchus tshawytscha*	23	2	1	1	3	1	1	1
Sand flounders	*Rhombosolea spp*	31	135	225	252	2	28	...	37
Tonguefishes	*Cynoglossidae*	31	-	-	-	-	9	-	-
Flatfishes nei	*Pleuronectiformes*	31	5 250	4 289	7 753	4 276	3 517	2 939	3 220
Common mora	*Mora moro*	32	1 192	694	1 410	1 324	1 122	1 358	1 211
Red codling	*Pseudophycis bachus*	32	16 159	10 596	11 087	16 495	12 555	5 365	4 530
Southern blue whiting	*Micromesistius australis*	32	32 322	26 040	34 624	66 163	70 145	43 598	54 012
Southern hake	*Merluccius australis*	32	10 952	9 119	11 370	17 617	19 433	16 041	15 188
Blue grenadier	*Macruronus novaezelandiae*	32	208 745	198 337	282 729	319 998	274 896	274 195	250 221
Thorntooth grenadier	*Lepidorhynchus denticulatus*	32	745	670	2 361	4 627	3 678	3 833	4 783
Grenadiers, rattails nei	*Macrouridae*	32	909	1 579	5 197	4 350	3 670	2 394	3 094
Gadiformes nei	*Gadiformes*	32	5 894	4 319	3 173	4 640	4 442	2 867	5 872
Sea catfishes nei	*Ariidae*	33	0	21	21	21	32	...	...
Mullets nei	*Mugilidae*	33	6 560	5 155	5 178	5 171	4 302	3 846	4 294
Orange perch	*Lepidoperca pulchella*	33	...	...	...	193	32	...	46
Groupers, seabasses nei	*Serranidae*	33	95	173	145	162	21	33	23
Sillago-whitings	*Sillaginidae*	33	1 826	1 169	700	607	1 804	1 096	1 185
Australian salmon	*Arripis trutta*	33	5 505	4 658	4 093	3 224	4 952	4 274	3 654
Snappers, jobfishes nei	*Lutjanidae*	33	...	...	...	...	301	291	275
Southern meagre(=Mulloway)	*Argyrosomus hololepidotus*	33	140	95	88	91	94	78	64
Geelbek croaker	*Atractoscion aequidens*	33	27	20	27	27	29	...	...
Croakers, drums nei	*Sciaenidae*	33	-	-	-	9	305	-	-
Emperors(=Scavengers) nei	*Lethrinidae*	33	-	-	4	-	-	-	-
Silver seabream	*Pagrus auratus*	33	6 568	6 138	6 551	6 550	8 494	7 917	7 008
Porgies, seabreams nei	*Sparidae*	33	1 691	1 935	2 221	1 361	1 805	2 627	2 715
Parore	*Girella tricuspidata*	33	83	561	588	576	579	631	558
Wrasses, hogfishes, etc. nei	*Labridae*	33	0	0	24	2	4	8	6
Antarctic rockcods, noties nei	*Nototheniidae*	33	34	16	34	3	19	3	15
Percoids nei	*Percoidei*	33	130	146	131	131	-	-	-
New Zealand blue cod	*Parapercis colias*	33	11 262	2 115	2 227	2 313	2 286	2 130	2 441
Flatheads nei	*Platycephalidae*	33	2 762	2 749	3 099	2 298	1 570	1 940	1 640
Puffers nei	*Tetraodontidae*	33	211	70	97	69	-	-	-
Velvet leatherjacket	*Parika scaber*	33	329	442	1 095	312	738	1 279	1 142
Triggerfishes, durgons nei	*Balistidae*	33	-	-	154	-	-	-	-
Argentines	*Argentina spp*	34	26	9	41	68	63	42	56
Conger eels, etc. nei	*Congridae*	34	93	109	113	97	220	96	106
Pink cusk-eel	*Genypterus blacodes*	34	19 030	12 858	23 942	23 446	23 330	23 559	19 932
Alfonsinos nei	*Beryx spp*	34	2 177	2 159	2 617	3 593	2 579	2 880	3 044
Redfish	*Centroberyx affinis*	34	89	1 407	1 551	1 984	1 689	1 185	977
Orange roughy	*Hoplostethus atlanticus*	34	40 148	33 170	23 686	24 695	24 038	18 007	14 302
Slimeheads nei	*Trachichthyidae*	34	-	-	-	7	3	4	2
John dory	*Zeus faber*	34	1 545	969	1 235	983	1 584	1 114	1 124
Mirror dory	*Zenopsis nebulosus*	34	361	348	1 347	1 794	2	2	1
Dories nei	*Zeidae*	34	338	288	641	754	763	778	685
Oreo dories nei	*Oreosomatidae*	34	21 833	18 782	21 850	21 096	22 646	22 775	24 165
Hapuku wreckfish	*Polyprion oxygeneios*	34	1 536	1 155	1 657	1 571	1 547	1 497	1 579
Cape bonnetmouth	*Emmelichthys nitidus*	34	2 392	1 845	1 635	2 064	2 846	2 825	1 881
Bonnetmouths, rubyfishes nei	*Emmelichthyidae*	34	1 117	596	431	378	271	582	434
Giant boarfish	*Paristiopterus labiosus*	34	21	19	27	75	6	9	3
Pelagic armourhead	*Pseudopentaceros richardsoni*	34	3	7	2	78	13	6	7
Tarakihi	*Nemadactylus macropterus*	34	5 149	4 366	5 441	5 239	5 589	5 739	6 129
Morwongs	*Nemadactylus spp*	34	1 345	1 200	1 526	1 192	774	610	510
Trumpeters nei	*Latridae*	34	685	747	776	601	582	485	518
Patagonian toothfish	*Dissostichus eleginoides*	34	...	1 061	5	43	1	0	14
Black cardinal fish	*Epigonus telescopus*	34	3 650	3 002	4 334	2 568	2 869	4 095	1 957
Giant stargazer	*Kathetostoma giganteum*	34	9 597	2 123	3 991	2 196	3 370	3 638	4 233
Snoek	*Thyrsites atun*	34	21 214	18 571	25 957	29 856	28 599	28 097	28 410
Escolar	*Lepidocybium flavobrunneum*	34	0	0	1	1	22	47	87
Oilfish	*Ruvettus pretiosus*	34	38	53	34	57	53	61	86
Silver gemfish	*Rexea solandri*	34	4 026	2 344	2 253	1 899	1 356	1 476	1 282
Silver scabbardfish	*Lepidopus caudatus*	34	1 955	941	2 342	3 344	2 638	1 619	2 887
Hairtails, scabbardfishes nei	*Trichiuridae*	34	-	678	1 188	545	422	1 099	227
Common warehou	*Seriolella brama*	34	1 081	1 760	4 108	3 101	3 881	4 259	4 101
Silver warehou	*Seriolella punctata*	34	13 377	4 926	11 253	10 993	9 029	11 218	11 268
White warehou	*Seriolella caerulea*	34	2 348	1 467	2 432	2 296	2 366	2 407	1 962
South Pacific breams nei	*Seriolella spp*	34	150	790	1 605	358	356	1 558	666
Bluenose warehou	*Hyperoglyphe antarctica*	34	2 719	2 432	2 974	2 630	2 755	2 793	2 954
Ruffs, barrelfishes nei	*Centrolophidae*	34	169	169	214	202	742	591	420
Scorpionfishes nei	*Scorpaenidae*	34	986	1 582	2 204	2 313	2 815	2 401	2 468
Bluefin gurnard	*Chelidonichthys kumu*	34	11 956	2 306	2 974	2 825	2 199	2 663	3 670
Latchet(=Sharpbeak gurnard)	*Pterygotrigla polyommata*	34	121	58	76	67	39	53	70
Spotted gurnard	*Pterygotrigla picta*	34	107	87	71	87	74	55	65
Demersal percomorphs nei	*Perciformes*	34	6 136	4 459	3 661	3 653	3 008	2 891	2 073
Australian pilchard	*Sardinops neopilchardus*	35	209	169	385	519	894	1 253	1 399
Anchovies, etc. nei	*Engraulidae*	35	13	12	19	19	2	39	21
Clupeoids nei	*Clupeoidei*	35	432	344	425	418	12	11	11
Wahoo	*Acanthocybium solandri*	36	-	-	6	-	-	-	-
Narrow-barred Spanish mackerel	*Scomberomorus commerson*	36	17	9	23	23	-	-	-
Seerfishes nei	*Scomberomorus spp*	36	1 228	1 769	3 111	2 897	1 237	1 601	2 079
Frigate and bullet tunas	*Auxis thazard, A.rochei*	36	0	0	2	4	-	5	5
Skipjack tuna	*Katsuwonus pelamis*	36	5 749	6 585	11 754	9 706	7 344	12 284	5 300
Pacific bluefin tuna	*Thunnus orientalis*	36	5	8	20	27	35	24	54
Longtail tuna	*Thunnus tonggol*	36	32	23	15	-	3	-	-
Albacore	*Thunnus alalunga*	36	14 411	18 723	12 359	18 255	11 565	13 567	19 629

C-81 (a)

Fish, crustaceans, molluscs, etc — Capture production by species items — Pacific, Southwest
Poissons, crustacés, mollusques, etc — Captures par catégories d'espèces — Pacifique, sud-ouest
Peces, crustáceos, moluscos, etc — Capturas por categorías de especies — Pacífico, sudoccidental

English name Nom anglais Nombre inglés	Scientific name Nom scientifique Nombre científico	Species group Groupe d'espèces Grupo de especies	1995 mt	1996 mt	1997 mt	1998 mt	1999 mt	2000 mt	2001 mt
Southern bluefin tuna	*Thunnus maccoyii*	36	3 167	1 677	1 894	1 832	2 565	1 527	2 136
Yellowfin tuna	*Thunnus albacares*	36	4 150	4 626	3 297	2 698	1 715	1 782	2 193
Bigeye tuna	*Thunnus obesus*	36	833	1 341	1 483	3 186	2 374	2 321	2 754
Indo-Pacific sailfish	*Istiophorus platypterus*	36	17	9	8	14	282	52	46
Indo-Pacific blue marlin	*Makaira mazara*	36	67	35	56	107	28	26	41
Black marlin	*Makaira indica*	36	-	-	1	2	0	9	-
Striped marlin	*Tetrapturus audax*	36	846	784	500	994	428	322	302
Shortbill spearfish	*Tetrapturus angustirostris*	36	8	42	18	-	-	-	-
Marlins,sailfishes,etc. nei	*Istiophoridae*	36	-	-	-	12	-	-	-
Swordfish	*Xiphias gladius*	36	1 215	1 599	1 260	1 856	1 872	1 618	1 534
Tuna-like fishes nei	*Scombroidei*	36	54	325	364	394	180	236	316
Needlefishes, etc. nei	*Belonidae*	37	2	1	0	0	-	-	-
Halfbeaks nei	*Hemiramphus spp*	37	211	115	132	143	24	18	13
Flyingfishes nei	*Exocoetidae*	37	4	4	2	3	1	2	1
Opah	*Lampris guttatus*	37	130	76	126	254	335	283	340
Dealfishes	*Trachipterus spp*	37	127	49	60	74	65	67	103
King of herrings	*Regalecus glesne*	37	5	10	64	60	34	20	1
Bluefish	*Pomatomus saltatrix*	37	96	101	64	64	46	...	...
Greenback horse mackerel	*Trachurus declivis*	37	10 592	15 441	10 656	9 379	15 545	12 239	7 634
Jack and horse mackerels nei	*Trachurus spp*	37	36 504	30 577	36 174	38 050	36 196	22 815	29 469
White trevally	*Pseudocaranx dentex*	37	3 910	3 833	3 813	4 166	4 671	3 984	3 495
Scads nei	*Decapterus spp*	37	33	23	11	63	29	82	87
Amberjacks nei	*Seriola spp*	37	663	575	639	403	527	297	278
Atlantic pomfret	*Brama brama*	37	449	480	413	488	465	401	903
Pomfrets, ocean breams nei	*Bramidae*	37	-	-	-	-	-	2	-
Common dolphinfish	*Coryphaena hippurus*	37	1	0	0	0	1	0	15
Blue mackerel	*Scomber australasicus*	37	7 967	2 994	8 777	7 260	15 874	12 108	11 801
Mackerels nei	*Scombridae*	37	408	400	...	...	458	901	447
Butterfishes, pomfrets nei	*Stromateidae*	37	1 314	1 518	2 126	2 233	1 837	2 077	2 211
Barracudas nei	*Sphyraena spp*	37	515	1 176	479	126	1 114	432	512
Pelagic percomorphs nei	*Perciformes*	37	1 002	-	-	-	-	-	-
Broadnose sevengill shark	*Notorynchus cepedianus*	38	...	...	...	2	3	4	5
Basking shark	*Cetorhinus maximus*	38	14	2	2	49	129	95	84
Thresher	*Alopias vulpinus*	38	15	13	24	21	32	51	57
Shortfin mako	*Isurus oxyrinchus*	38	33	52	40	74	110	208	327
Porbeagle	*Lamna nasus*	38	5	16	21	164	246	188	127
Blue shark	*Prionace glauca*	38	111	246	120	540	593	1 169	1 328
Copper shark	*Carcharhinus brachyurus*	38	...	...	...	15	14	25	38
Smooth hammerhead	*Sphyrna zygaena*	38	12	10	3	6	11	13	17
Spotted estuary smooth-hound	*Mustelus lenticulatus*	38	2 787	1 350	3 464	1 707	1 662	1 643	1 563
Tope shark	*Galeorhinus galeus*	38	3 705	3 044	2 864	3 083	3 633	3 100	3 091
Picked dogfish	*Squalus acanthias*	38	2 753	2 477	7 232	3 064	4 409	3 362	4 192
Leafscale gulper shark	*Centrophorus squamosus*	38	...	...	...	4	1	0	0
Lanternsharks nei	*Etmopterus spp*	38	3	0	2	-	-	-	4
Birdbeak dogfish	*Deania calcea*	38	...	...	...	36	17	28	66
Kitefin shark	*Dalatias licha*	38	303	175	352	434	328	317	375
Dogfish sharks nei	*Squalidae*	38	413	693	1 705	701	1 010	770	705
Eagle rays	*Myliobatidae*	38	0	0	1	1	2	2	5
Rays, stingrays, mantas nei	*Rajiformes*	38	2 361	1 787	2 346	2 401	3 061	2 757	2 989
Dark ghost shark	*Hydrolagus novaezealandiae*	38	1 593	1 614	2 064	1 956	1 975	1 819	1 572
Ratfishes nei	*Hydrolagus spp*	38	...	...	0	36	453	975	2 184
Ghost shark	*Callorhinchus milii*	38	769	595	913	951	1 260	1 228	1 189
Chimaeras, etc. nei	*Chimaeriformes*	38	5	49	5	5	21	40	76
Sharks, rays, skates, etc. nei	*Elasmobranchii*	38	6 130	4 952	3 562	3 523	3 614	2 318	2 397
Marine fishes nei	*Osteichthyes*	39	21 989	49 339	48 209	27 498	10 734	12 611	42 338
Blue swimming crab	*Portunus pelagicus*	42	788	800	761	745	22	30	30
Marine crabs nei	*Brachyura*	42	9 231	294	359	393	407	355	360
Green rock lobster	*Jasus verreauxi*	43	199	161	158	124	123	152	115
Red rock lobster	*Jasus edwardsii*	43	3 568	3 121	5 009	2 707	2 818	2 789	2 551
Slipper lobsters nei	*Scyllaridae*	43	1	1	158	160	7	1	4
New Zealand lobster	*Metanephrops challengeri*	43	1 078	670	1 093	989	925	1 034	1 093
Penaeus shrimps nei	*Penaeus spp*	45	2 153	1 822	1 926	1 683	2 399	2 647	2 588
Natantian decapods nei	*Natantia*	45	1	2	23	1	11	-	5
Marine crustaceans nei	*Crustacea*	47	523	156	157	92	670	802	537
Blacklip abalone	*Haliotis rubra*	52	312	344	335	327	...	...	...
Abalones nei	*Haliotis spp*	52	1 280	1 020	1 180	1 300	1 170	1 265	1 064
New Zealand dredge oyster	*Ostrea lutaria*	53	1	2	2	2	3	2	2
Pacific cupped oyster	*Crassostrea gigas*	53	8	2	2	0	62	0	0
Australian mussel	*Mytilus planulatus*	54	4	4	1	1	1	1	1
Sea mussels nei	*Mytilidae*	54	191	450	0	664	2 977	4 467	2 270
New Zealand scallop	*Pecten novaezelandiae*	55	14 160	5 080	18 848	4 592	6 152	2 912	6 792
Delicate scallop	*Zygochlamis delicatula*	55	135	124	201	91	128	0	222
Scallops nei	*Pectinidae*	55	23	5	1	1	1	0	1
Stutchbury's venus	*Chione stutchburyi*	56	1 220	815	541	1 325	1 396	1 789	1 748
Short neck clams nei	*Paphia spp*	56	317	211	114	204	181	131	202
Pipi wedge clam	*Paphies australis*	56	254	248	465	455	...	...	...
Cuttlefish,bobtail squids nei	*Sepiidae, Sepiolidae*	57	392	378	374	251	381	258	219
Wellington flying squid	*Nototodarus sloani*	57	94 098	53 699	64 602	55 570	31 358	25 603	45 119
Various squids nei	*Loliginidae, Ommastrephidae*	57	36 027	18 995	19 080	14 304	11 735	9 519	11 795

C-81 (a)

Fish, crustaceans, molluscs, etc — Capture production by species items — Pacific, Southwest
Poissons, crustacés, mollusques, etc — Captures par catégories d'espèces — Pacifique, sud-ouest
Peces, crustáceos, moluscos, etc — Capturas por categorías de especies — Pacífico, sudoccidental

English name Nom anglais Nombre inglés	Scientific name Nom scientifique Nombre científico	Species group Groupe d'espèces Grupo de especies	1995 mt	1996 mt	1997 mt	1998 mt	1999 mt	2000 mt	2001 mt
Octopuses, etc. nei	*Octopodidae*	57	9 064	227	777	881	151	119	140
Marine molluscs nei	*Mollusca*	58	266	342	173	230	1 386	1 311	1 483
Echinoderms	*Echinodermata*	76	804	277	627	832	643	712	853
Starfishes nei	*Asteroidea*	76	9	4	0	13	2	6	9
Sea cucumbers nei	*Holothurioidea*	76	4	1	0	-	-	-	2
Total			***813 727***	***661 821***	***837 649***	***861 701***	***782 603***	***713 373***	***750 967***

C-81 (b)

Fish, crustaceans, molluscs, etc — Capture production by countries or areas — Pacific, Southwest
Poissons, crustacés, mollusques, etc — Captures par pays ou zones — Pacifique, sud-ouest
Peces, crustáceos, moluscos, etc — Capturas por países o áreas — Pacífico, sudoccidental

Country or area Pays ou zone País o área	1992 mt	1993 mt	1994 mt	1995 mt	1996 mt	1997 mt	1998 mt	1999 mt	2000 mt	2001 mt
Australia	68 905 F	59 617 F	57 088 F	52 466 F	47 741 F	49 719	49 969	26 900	28 534	25 652
Canada	235	235	235	235	136	149	167	253	351	206
China,Taiwan	43 126 F	24 494 F	14 485 F	11 429	20 599	12 902	13 176	6 983	8 200	5 901
Japan	198 973	140 989	113 146	95 100	77 078	72 426	74 231	58 178	43 416	56 855
Korea Rep	29 729	20 861	25 371	30 434	26 689	35 111	31 886	34 784	34 298	38 139
New Zealand	459 570	425 508	448 560	554 444	422 556	609 229	637 875	596 979	546 596	559 356
Norfolk Is	0	0	0	0	0	0	0	0	0	0
Norway	12 791	23 973	15 290	6 366	7 187	5 932	5 033	-	-	-
Pitcairn Is	8	8 F	8 F	8 F	8 F	8 F	8 F	8 F	8 F	8 F
Poland	961	-	-	-	-	-	-	-	-	-
Russian Fed	82 865	43 672	53 227	28 017	17 108	12 027	2 175	3 332	-	-
Ukraine	16 556	28 306	35 805	33 058	39 496	40 146	47 181	55 186	51 970	58 908
USA	2 918	1 123	605	2 170	3 223	-	-	-	-	5 942
Total	***916 637***	***768 786***	***763 820***	***813 727***	***661 821***	***837 649***	***861 701***	***782 603***	***713 373***	***750 967***

C-87 (a)

Fish, crustaceans, molluscs, etc — Capture production by species items — Pacific, Southeast
Poissons, crustacés, mollusques, etc — Captures par catégories d'espèces — Pacifique, sud-est
Peces, crustáceos, moluscos, etc — Capturas por categorías de especies — Pacífico, sudoriental

English name Nom anglais Nombre inglés	Scientific name Nom scientifique Nombre científico	Species group Groupe d'espèces Grupo de especies	1995 mt	1996 mt	1997 mt	1998 mt	1999 mt	2000 mt	2001 mt
Lefteye flounders nei	*Bothidae*	31	-	-	-	-	-	5	-
Tonguefishes	*Cynoglossidae*	31	-	-	-	-	13	38	-
Flatfishes nei	*Pleuronectiformes*	31	1 790	761	411	436	398	322	439
Tadpole codling	*Salilota australis*	32	1 826	481	647	352	245	372	641
Southern blue whiting	*Micromesistius australis*	32	20 917	25 445	29 131	26 642	31 470	24 733	22 046
Southern hake	*Merluccius australis*	32	24 612	23 788	24 666	22 458	24 656	29 402	28 805
South Pacific hake	*Merluccius gayi*	32	256 585	323 470	265 840	162 683	141 053	193 754	246 656
Patagonian grenadier	*Macruronus magellanicus*	32	206 734	379 015	71 479	353 823	309 723	91 310	160 774
Sea catfishes nei	*Ariidae*	33	114	425	350	121	97	100	100
Mullets nei	*Mugilidae*	33	16 904	14 180	13 368	29 179	21 005	26 476	27 451
Snooks(=Robalos) nei	*Centropomus spp*	33	86	341	73	96	62	70	70
Broomtail grouper	*Mycteroperca xenarcha*	33	207	180	125	155	82	80	210
Spotted grouper	*Epinephelus analogus*	33	22	43	48	32	39	30	28
Groupers nei	*Epinephelus spp*	33	398	421	185	256	233	161	144
Peruvian rock seabass	*Paralabrax humeralis*	33	5 837	4 954	2 789	2 554	3 278	4 373	2 011
Groupers, seabasses nei	*Serranidae*	33	23	30	24	10	19	7	5
Yellow snapper	*Lutjanus argentiventris*	33	120	43	40	100	82	80	...
Snappers nei	*Lutjanus spp*	33	578	577	118	660	1 088	804	857
Cabinza grunt	*Isacia conceptionis*	33	1 507	2 048	2 034	2 133	2 947	3 293	3 321
Grunts, sweetlips nei	*Haemulidae (=Pomadasyidae)*	33	890	194	379	1 345	818	183	419
Corvina	*Sciaena gilberti*	33	5 592	9 099	3 565	6 154	6 442	4 744	4 328
Peruvian weakfish	*Cynoscion analis*	33	9 406	7 887	5 797	11 187	8 929	6 339	4 438
Croakers nei	*Micropogonias spp*	33	920	1 026	886	1 233	789	1 125	1 069
Peruvian banded croaker	*Paralonchurus peruanus*	33	5 543	4 263	2 737	4 363	6 063	5 729	4 167
Croakers, drums nei	*Sciaenidae*	33	30 941	2 537	342	3 350	8 032	845	1 678
Porgies, seabreams nei	*Sparidae*	33	-	-	-	-	8	8	-
Threadfins, tasselfishes nei	*Polynemidae*	33	-	16	4	3	16	10	10
Patagonian blennie	*Eleginops maclovinus*	33	274	287	133	103	193	164	109
...A	*Normanichthys crockeri*	33	2 690	3 921	20 426	236	4 843	853	223
Conger eels, etc. nei	*Congridae*	34	-	-	-	23	-	-	-
Pink cusk-eel	*Genypterus blacodes*	34	5 438	5 780	6 410	6 836	5 721	6 269	7 522
Red cusk-eel	*Genypterus chilensis*	34	1 082	982	745	584	415	608	730
Black cusk-eel	*Genypterus maculatus*	34	1 193	1 343	1 661	2 753	1 943	3 542	3 889
Cusk-eels nei	*Genypterus spp*	34	1 631	1 121	439	425	196	557	552
Alfonsinos nei	*Beryx spp*	34	-	-	-	-	706	4 366	5 182
Slimeheads nei	*Trachichthyidae*	34	-	-	-	-	779	1 482	1 868
John dory	*Zeus faber*	34	-	-	-	-	5	-	-
Hapuku wreckfish	*Polyprion oxygeneios*	34	33	23	30	26	8	7	10
Tilefishes nei	*Branchiostegidae*	34	1 796	1 009	320	199	280	272	1 538
Peruvian morwong	*Cheilodactylus variegatus*	34	93	283	462	140	288	380	306
Patagonian toothfish	*Dissostichus eleginoides*	34	15 694	6 993	8 059	9 172	10 328	10 676	6 579
Cardinal fishes nei	*Epigonus spp*	34	232	513	1 727	5 284	2 999	5 792	4 648
Snoek	*Thyrsites atun*	34	687	821	1 337	1 022	604	851	830
Hairtails, scabbardfishes nei	*Trichiuridae*	34	...	...	...	...	85	13 666	3 382
South Pacific breams nei	*Seriolella spp*	34	10 930	6 186	3 297	3 176	4 936	3 975	6 536
Scorpionfishes nei	*Scorpaenidae*	34	227	212	532	208	258	141	177
Gurnards, searobins nei	*Triglidae*	34	...	...	...	...	22 634	27 129	1 428
Demersal percomorphs nei	*Perciformes*	34	1 058	170	220	161	294	200	200
South American pilchard	*Sardinops sagax*	35	1 503 131	1 493 936	722 807	937 269	442 790	338 131	135 712
Red-eye round herring	*Etrumeus teres*	35	4 946	34 349	1 095	8 873	3 636	4 414	28
Pacific thread herring	*Opisthonema libertate*	35	40 911	41 041	43 145	40 530	22 253	20 519	19 886
Pacific menhaden	*Ethmidium maculatum*	35	5 487	9 792	13 241	40 845	29 436	23 991	14 016
Araucanian herring	*Strangomera bentincki*	35	126 715	446 669	441 154	317 564	782 142	722 522	324 617
Anchoveta(=Peruvian anchovy)	*Engraulis ringens*	35	8 644 576	8 863 714	7 685 098	1 729 064	8 723 265	11 276 357	7 213 077
Pacific anchoveta	*Cetengraulis mysticetus*	35	55 241	52 698	117 904	72 975	43 002	34 262	98 636
Anchovies, etc. nei	*Engraulidae*	35	189 389	59 639	24 703	706 167	11 242	3 868	137 098
Clupeoids nei	*Clupeoidei*	35	120	1	429	273	0	3 596	4 420
Eastern Pacific bonito	*Sarda chiliensis*	36	28 386	23 079	17 764	5 722	1 325	499	1 319
Pacific sierra	*Scomberomorus sierra*	36	1 046	923	743	1 773	3 233	1 430	681
Frigate and bullet tunas	*Auxis thazard, A.rochei*	36	7 396	2 537	6 857	4 201	48 913	9 648	5 738
Black skipjack	*Euthynnus lineatus*	36	160	100	50	360	50	270	1 660
Skipjack tuna	*Katsuwonus pelamis*	36	70 331	79 317	130 690	102 634	222 468	168 174	112 426
Pacific bluefin tuna	*Thunnus orientalis*	36	-	-	2	-	1	-	-
Albacore	*Thunnus alalunga*	36	978	796	601	558	563	1 811	1 735
Southern bluefin tuna	*Thunnus maccoyii*	36	17	-	-	-	-	-	-
Yellowfin tuna	*Thunnus albacares*	36	88 242	82 379	108 882	102 411	127 989	106 132	177 650
Bigeye tuna	*Thunnus obesus*	36	29 442	46 005	48 433	30 063	36 247	67 876	45 959
Indo-Pacific sailfish	*Istiophorus platypterus*	36	7	8	44	54	10	11	35
Indo-Pacific blue marlin	*Makaira mazara*	36	493	256	284	544	123	73	164
Black marlin	*Makaira indica*	36	-	1	1	4	0	1	-
Striped marlin	*Tetrapturus audax*	36	621	1 084	729	1 078	200	165	666
Marlins,sailfishes,etc. nei	*Istiophoridae*	36	-	-	-	3 727	6 962	...	...
Swordfish	*Xiphias gladius*	36	4 037	4 326	6 389	6 600	4 496	5 164	5 716
Tuna-like fishes nei	*Scombroidei*	36	51	2 922	8 151	27 949	1 565	270	24 405
Atlantic saury	*Scomberesox saurus*	37	8	1	175	407	580	296	4 178
Flyingfishes nei	*Exocoetidae*	37	35 619	639	1 056	4 506	298 373	41 059	5 035
Silversides(=Sand smelts) nei	*Atherinidae*	37	2 915	4 492	5 678	605	10 106	12 572	8 361
Chilean jack mackerel	*Trachurus murphyi*	37	4 955 186	4 378 843	3 597 117	2 025 758	1 423 447	1 540 494	2 508 834
Jacks, crevalles nei	*Caranx spp*	37	129	667	520	347	880	235	149
Pompanos nei	*Trachinotus spp*	37	566	460	268	779	2 801	1 220	632
Amberjacks nei	*Seriola spp*	37	6 598	1 558	4 699	21 213	2 131	11 229	28 116
Pacific bumper	*Chloroscombrus orqueta*	37	17 999	1 706	952	565	1 409	...	1 008
Carangids nei	*Carangidae*	37	288	161	250	1 187	1 835	307	431
Atlantic pomfret	*Brama brama*	37	3 930	5 585	5 998	6 332	6 830	8 160	15 156
Common dolphinfish	*Coryphaena hippurus*	37	104	179	124	80	109	131	136

C-87 (a)

Fish, crustaceans, molluscs, etc — **Capture production by species items** — **Pacific, Southeast**
Poissons, crustacés, mollusques, etc — **Captures par catégories d'espèces** — **Pacifique, sud-est**
Peces, crustáceos, moluscos, etc — **Capturas por categorías de especies** — **Pacífico, sudoriental**

English name Nom anglais Nombre inglés	Scientific name Nom scientifique Nombre científico	Species group Groupe d'espèces Grupo de especies	1995 mt	1996 mt	1997 mt	1998 mt	1999 mt	2000 mt	2001 mt
Chub mackerel	*Scomber japonicus*	37	212 419	275 354	610 014	518 388	676 160	254 524	627 466
Butterfishes, pomfrets nei	*Stromateidae*	37	7	9	432	11	17	10	11
Pelagic percomorphs nei	*Perciformes*	37	382	858	120	33	1 073	771	371
Shortfin mako	*Isurus oxyrinchus*	38	475	320	1 218	1 757	379	592	964
Porbeagle	*Lamna nasus*	38	-	-	...	7	...	...	...
Blue shark	*Prionace glauca*	38	39	11	114	824	7	262	456
Hammerhead sharks, etc. nei	*Sphyrnidae*	38	-	-	...	5	...	...	...
Smooth-hounds nei	*Mustelus spp*	38	4 480	4 209	3 682	8 410	3 485	4 515	5 057
Angelsharks, sand devils nei	*Squatinidae*	38	289	358	189	101	262	406	510
Pacific guitarfish	*Rhinobatos planiceps*	38	121	460	333	344	95	2 624	1 060
Rays, stingrays, mantas nei	*Rajiformes*	38	4 463	3 804	4 137	3 490	6 407	8 425	5 047
Elephantfishes nei	*Callorhinchus spp*	38	920	1 450	822	1 416	632	603	1 125
Sharks, rays, skates, etc. nei	*Elasmobranchii*	38	1 365	2 363	2 735	5 502	4 166	6 521	4 268
Marine fishes nei	*Osteichthyes*	39	68 806	72 907	97 328	390 422	271 652	323 750	190 872
Marine crabs nei	*Brachyura*	42	9 093	6 372	4 629	6 427	18 509	9 352	8 958
Green spiny lobster	*Panulirus gracilis*	43	221	106	69	725	547	329	113
Juan Fernandez rock lobster	*Jasus frontalis*	43	29	36	32	21	22	17	21
Carrot squat lobster	*Pleuroncodes monodon*	44	4 938	7 726	8 939	12 602	12 710	11 129	1 754
Blue squat lobster	*Cervimunida johni*	44	5 743	6 402	10 322	9 426	7 273	5 069	2 178
Craylets, squat lobsters	*Galatheidae*	44	-	-	-	-	-	254	-
Southern king crab	*Lithodes antarcticus*	44	1 906	1 759	2 160	2 766	2 155	2 902	2 937
Softshell red crab	*Paralomis granulosa*	44	1 316	1 273	1 477	1 501	1 438	4 938	6 527
Yellowleg shrimp	*Penaeus californiensis*	45	73	73	60	50	20	20	40
Whiteleg shrimp	*Penaeus vannamei*	45	6 000	4 497	4 000	3 000	1 100	1 100	2 250
Western white shrimp	*Penaeus occidentalis*	45	686	1 158	1 505	1 261	2 706	2 020	1 259
Penaeus shrimps nei	*Penaeus spp*	45	11 221	9 589	15 998	18 052	7 380	1 977	2 339
Pacific seabob	*Xiphopenaeus riveti*	45	1 037	2 683	2 830	1 970	1 752	2 000	2 341
Chilean nylon shrimp	*Heterocarpus reedi*	45	10 620	10 535	10 239	7 301	7 951	5 448	4 863
Kolibri shrimp	*Solenocera agassizii*	45	...	...	...	...	...	...	686
Chilean knife shrimp	*Haliporoides diomedeae*	45	5	15	32	29	135	169	309
Natantian decapods nei	*Natantia*	45	369	317	559	345	592	588	591
Giant barnacle	*Megabalanus psittacus*	47	681	879	579	683	620	620	685
Marine crustaceans nei	*Crustacea*	47	-	1	6	15	67	7	825
False abalone	*Concholepas concholepas*	52	4 031	5 269	7 520	3 394	4 583	2 524	1 372
Stromboid conchs nei	*Strombus spp*	52	22	10	10	10	10	10	10
Gastropods nei	*Gastropoda*	52	10 096	8 310	12 475	7 779	11 739	9 686	10 444
Chilean flat oyster	*Ostrea chilensis*	53	-	-	5	1	6	9	202
Pacific cupped oyster	*Crassostrea gigas*	53	5	5	5	5	5	5	5
Chilean mussel	*Mytilus chilensis*	54	5 128	5 714	4 723	4 899	4 343	5 236	6 758
Choro mussel	*Choromytilus chorus*	54	307	323	266	127	155	217	166
Cholga mussel	*Aulacomya ater*	54	17 580	13 428	16 078	22 831	19 738	18 933	22 584
Sea mussels nei	*Mytilidae*	54	3	3	3	5	5	5	5
Peruvian calico scallop	*Argopecten purpuratus*	55	3 113	2 095	4 013	23 546	30 141	11 830	6 544
Scallops nei	*Pectinidae*	55	1 365	1 577	2 598	3 662	1 715	332	141
Gay's little venus	*Tawera gayi*	56	...	...	...	...	...	1	291
Taca clam	*Protothaca thaca*	56	17 162	20 016	12 475	24 254	16 429	16 303	26 483
Venus clams nei	*Veneridae*	56	0	0	0	0	0	0	0
Taquilla clams	*Mulinia spp*	56	1 852	999	2 757	2 549	1 536	1 491	1 699
Macha clam	*Mesodesma donacium*	56	8 113	7 204	7 831	7 042	1 728	1 259	1 396
Chilean semele	*Semele solida*	56	2 523	4 418	2 199	1 900	2 071	4 212	3 054
Clams, etc. nei	*Bivalvia*	56	20 650	19 145	18 131	12 068	21 158	17 527	17 744
Common squids nei	*Loligo spp*	57	7 856	10 340	3 906	387	1 453	24 648	18 838
Jumbo flying squid	*Dosidicus gigas*	57	96 631	21 123	21 636	748	76 422	100 341	149 832
Various squids nei	*Loliginidae, Ommastrephidae*	57	69	321	293	214	137	124	3 681
Octopuses, etc. nei	*Octopodidae*	57	4 596	4 237	6 260	10 005	4 766	2 506	2 648
Marine molluscs nei	*Mollusca*	58	557	980	3 447	1 702	3 730	2 513	2 273
Marine turtles nei	*Testudinata*	72	14	10	11	12	11	11	12
Red sea squirt	*Pyura chilensis*	74	3 297	4 549	3 174	2 530	2 704	2 290	1 298
Echinoderms	*Echinodermata*	76	131	461	424	90	1 204	1 626	2 114
Sea urchins nei	*Strongylocentrotus spp*	76	0	0	0	0	0	0	0
Chilean sea urchin	*Loxechinus albus*	76	54 609	51 437	45 560	44 843	55 654	54 096	46 794
Sea cucumbers nei	*Holothurioidea*	76	118	127	16	45	123	1 525	122
Total			***17 067 207***	***17 068 356***	***14 427 245***	***8 034 222***	***14 176 392***	***15 810 412***	***12 655 449***

C-87 (b)

Fish, crustaceans, molluscs, etc — Capture production by countries or areas — Pacific, Southeast
Poissons, crustacés, mollusques, etc — Captures par pays ou zones — Pacifique, sud-est
Peces, crustáceos, moluscos, etc — Capturas por países o áreas — Pacífico, sudoriental

Country or area Pays ou zone País o área	1992 mt	1993 mt	1994 mt	1995 mt	1996 mt	1997 mt	1998 mt	1999 mt	2000 mt	2001 mt
Belize	-	1 110	5 230	7 120	8 560	11 740	8 430	6 170	7 700	...
Chile	6 420 892	5 943 326	7 716 540	7 431 704	6 687 844	5 805 745	3 249 132	5 043 634	4 295 087	3 787 393
China,Taiwan	184	498	142	3	252	306	89	62	1 500	882
Colombia	92 452	83 381	53 462	81 636	83 880	121 073	84 410	86 167	81 964	77 470 F
Costa Rica	...	...	...	270	1 260	...	...	...	...	...
Cuba	3 197	0	-	-	-	-	-	4 234	...	...
Cyprus	5 790	6 360	6 480	5 920	5 550	6 880	430	-	-	-
Ecuador	224 495	285 211	328 238	505 095	702 404	548 588	309 622	497 472	592 147	586 020
El Salvador	...	...	...	...	...	...	...	6 030	...	...
Estonia	376	-	-	-	-	-	-	14	-	-
Ghana	-	-	-	-	-	-	-	-	7 129	6 332
Guatemala	-	-	-	-	-	-	-	4 820	18 210	...
Honduras	-	-	-	-	470	2 560	870	5 710	2 580	...
Japan	68 422	69 982	102 495	48 922	11 871	11 266	12 847	10 251	42 174	89 530
Korea Rep	36 163	59 225	67 092	34 724	11 784	2 629	508	25 464	23 980	10 097
Latvia	3 023	-	-	-	-	-	-	-	-	-
Liberia	-	-	-	-	900	-	-	-	-	-
Lithuania	7 888	-	-	-	-	-	-	-	-	-
Mexico	5 759	5 063	7 749	13 322	19 995	12 581	10 347	10 751	5 505	8 020
Nicaragua	-	-	-	-	-	-	-	-	1 660	...
Panama	5 230	2 890	3 860	5 800	3 570	6 370	320	6 080	20 550	14 590
Peru	7 469 458	8 966 487	11 950 380	8 886 553	9 486 158	7 837 650	4 303 110	8 392 378	10 626 323	7 950 450
Russian Fed	32 354	-	-	-	-	-	-	-	-	-
St Vincent	-	-	2 750	-	-	-	-	-	-	-
Spain	5 895	7 818	4 326	9 428	9 153	16 478	15 169	27 745	28 933	26 817
USA	3 503	-	-	-	5 315	2 979	1 568	-	3 310	18
Vanuatu	21 830	23 430	31 010	26 780	20 410	19 570	12 750	26 980	25 740	17 330
Venezuela	17 510	18 360	21 570	8 720	8 980	17 470	23 750	22 430	22 020	42 330
Other nei	570	1 910	1 880	1 210	-	3 360	870	-	3 900	38 170
Total	***14 424 991***	***15 475 051***	***20 303 204***	***17 067 207***	***17 068 356***	***14 427 245***	***8 034 222***	***14 176 392***	***15 810 412***	***12 655 449***

C-88 (a)

Fish, crustaceans, molluscs, etc — **Capture production by species items** — **Pacific, Antarctic**
Poissons, crustacés, mollusques, etc — **Captures par catégories d'espèces** — **Pacifique, Antarctique**
Peces, crustáceos, moluscos, etc — **Capturas por categorías de especies** — **Pacífico, Antártico**

English name Nom anglais Nombre inglés	Scientific name Nom scientifique Nombre científico	Species group Groupe d'espèces Grupo de especies	1995 mt	1996 mt	1997 mt	1998 mt	1999 mt	2000 mt	2001 mt
Smalleye moray cod	*Muraenolepis microps*	32	-	-	-	-	4	5	-
Moray cods nei	*Muraenolepis spp*	32	-	-	-	0	1	2	3
Blue antimora	*Antimora rostrata*	32	-	-	-	0	0	0	4
Whitson's grenadier	*Macrourus whitsoni*	32	-	-	-	-	1	5	48
Grenadiers nei	*Macrourus spp*	32	-	-	-	-	-	-	6
Grenadiers, rattails nei	*Macrouridae*	32	-	-	-	9	22	70	-
Antarctic rockcods, noties nei	*Nototheniidae*	33	-	-	-	0	0	0	1
Antarctic toothfish	*Dissostichus mawsoni*	34	-	-	-	41	296	751	626
Patagonian toothfish	*Dissostichus eleginoides*	34	...	-	0	1	1	0	34
Icefishes nei	*Channichthyidae*	34	-	-	-	0	0	0	2
Antarctic starry skate	*Raja georgiana*	38	-	-	-	...	11	36	7
Eaton's skate	*Bathyraja eatonii*	38	-	-	-	...	1	5	0
Bathyraja rays nei	*Bathyraja spp*	38	-	-	-	-	1	-	-
Rays and skates nei	*Rajidae*	38	-	-	-	...	6	-	-
Rays, stingrays, mantas nei	*Rajiformes*	38	-	-	-	5	-	0	-
Starfishes nei	*Asteroidea*	76	-	-	-	-	-	-	2
Total			***...***	***-***	***0***	***56***	***344***	***874***	***733***

C-88 (b)

Fish, crustaceans, molluscs, etc — **Capture production by countries or areas** — **Pacific, Antarctic**
Poissons, crustacés, mollusques, etc — **Captures par pays ou zones** — **Pacifique, Antarctique**
Peces, crustáceos, moluscos, etc — **Capturas por países o áreas** — **Pacífico, Antártico**

Country or area Pays ou zone País o área	1992 mt	1993 mt	1994 mt	1995 mt	1996 mt	1997 mt	1998 mt	1999 mt	2000 mt	2001 mt
Japan	50	-	-	-	-	-	-	-	-	-
New Zealand	...	...	...	...	-	0	56	344	874	678
South Africa	-	-	-	-	-	-	-	-	-	32
Uruguay	-	-	-	-	-	-	-	-	-	23
Total	***50***	***...***	***...***	***...***	***-***	***0***	***56***	***344***	***874***	***733***

Data in these tables refer to catches for which the split-year (1 July - 30 June) is used. Split-year data are shown under the calendar year in which the split-year ends.

Les données dans ces tableaux se réfèrent aux captures pour lesquelles on utilise l'année fractionnée (1er juillet - 30 juin). Les captures relatives à des années fractionnées figurent sous l'année civile durant laquelle se termine l'année fractionnée.

Los datos en estos cuadros se refieren a las capturas para las que se utiliza el año emergente (1°de julio - 30 de junio). Los datos correspondientes a los años emergentes se incluyen en el año civil en que termina el año emergente.

D - Capture production: by continents

D - Captures: par continents

D - Capturas: por continentes

	Fishing area Zone de pêche Area de pesca	1992 mt	1993 mt	1994 mt	1995 mt	1996 mt	1997 mt	1998 mt	1999 mt	2000 mt	2001 mt
Algeria	01	0	0	0	0	0	0	2	0	0	0
	37	95 270	101 895	135 402	105 872	81 989	91 580	92 344	102 396	100 000 F	100 000 F
	Country total	*95 270*	*101 895*	*135 402*	*105 872*	*81 989*	*91 580*	*92 346*	*102 396*	*100 000 F*	*100 000 F*
Angola	01	7 000	7 000	7 000	6 000	6 000	6 000	6 000	6 000 F	6 000 F	6 000 F
	47	106 625	119 200	125 413	116 781	131 815	140 304	157 149	169 799	232 351	246 518
	Country total	*113 625*	*126 200*	*132 413*	*122 781*	*137 815*	*146 304*	*163 149*	*175 799*	*238 351*	*252 518*
Benin	01	26 566	32 805	32 707	37 449	34 193	32 871	31 778	31 894	26 400	30 000
	34	5 922	6 416	7 216	6 930	7 982	10 914	10 361	8 542	5 924	8 415
	Country total	*32 488*	*39 221*	*39 923*	*44 379*	*42 175*	*43 785*	*42 139*	*40 436*	*32 324*	*38 415*
Botswana	01	800 F	600 F	400 F	200 F	81	160	191	157	166	118
	Country total	*800 F*	*600 F*	*400 F*	*200 F*	*81*	*160*	*191*	*157*	*166*	*118*
Br Ind Oc Tr	01	0	0	0	0	0	0	0	0	0	0
	51	0	0	0	0	0	0	0	0	0	0
	Country total	*0*	*0*	*0*	*0*	*0*	*0*	*0*	*0*	*0*	*0*
Burkina Faso	01	7 500	7 000	8 000	8 000	8 000	8 000	8 335	7 600	8 500	8 500 F
	Country total	*7 500*	*7 000*	*8 000*	*8 000*	*8 000*	*8 000*	*8 335*	*7 600*	*8 500*	*8 500 F*
Burundi	01	24 073	17 000 F	22 000 F	21 101	3 041	20 296	13 426	9 199	17 315	8 964
	Country total	*24 073*	*17 000 F*	*22 000 F*	*21 101*	*3 041*	*20 296*	*13 426*	*9 199*	*17 315*	*8 964*
Cameroon	01	22 000	23 000	27 000 F	30 000 F	35 000 F	40 000 F	45 000 F	50 000 F	55 000	52 500 F
	34	49 975	42 257	52 000 F	64 131	63 400	62 000 F	61 800 F	60 000 F	57 109	58 531
	Country total	*71 975*	*65 257*	*79 000 F*	*94 131 F*	*98 400 F*	*102 000 F*	*106 800 F*	*110 000 F*	*112 109*	*111 031*
Cape Verde	01	0	0	0	0	0	0	0	0	0	0
	34	6 573	7 000	8 256	8 495	9 155	9 705	9 424	10 360	10 586	9 653
	Country total	*6 573*	*7 000*	*8 256*	*8 495*	*9 155*	*9 705*	*9 424*	*10 360*	*10 586*	*9 653*
Cent Afr Rep	01	13 000	13 250 F	13 500 F	13 750 F	14 000 F	14 250 F	14 500 F	15 000	15 000 F	15 000 F
	Country total	*13 000*	*13 250 F*	*13 500 F*	*13 750 F*	*14 000 F*	*14 250 F*	*14 500 F*	*15 000*	*15 000 F*	*15 000 F*
Chad	01	80 000	87 300	80 000	90 000	100 000	85 000	84 000	84 000 F	84 000 F	84 000 F
	Country total	*80 000*	*87 300*	*80 000*	*90 000*	*100 000*	*85 000*	*84 000*	*84 000 F*	*84 000 F*	*84 000 F*
Comoros	01	0	0	0	0	0	0	0	0	0	0
	51	11 825	11 645	12 976	13 000 F	12 700 F	12 500 F	12 500 F	12 000	13 200	12 180
	Country total	*11 825*	*11 645*	*12 976*	*13 000 F*	*12 700 F*	*12 500 F*	*12 500 F*	*12 000*	*13 200*	*12 180*
Congo Dem R	01	184 040	192 589	152 117	154 751	159 037	158 367	174 087	204 503	204 503 F	204 503 F
	34	3 800	4 200	3 780	3 876	3 973	3 844	3 954	3 945	3 945 F	3 945 F
	Country total	*187 840*	*196 789*	*155 897*	*158 627*	*163 010*	*162 211*	*178 041*	*208 448*	*208 448 F*	*208 448 F*
Congo Rep	01	21 049	27 850	24 752	26 811	25 873	18 987	25 455	25 455 F	26 000 F	24 500 F
	34	18 943	18 898	17 912	18 965	19 600	19 095	17 500 F	18 241	18 000 F	17 500 F
	Country total	*39 992*	*46 748*	*42 664*	*45 776*	*45 473*	*38 082*	*42 955 F*	*43 696 F*	*44 000 F*	*42 000 F*
Côte dIvoire	01	15 404	13 477	15 604	11 335	11 562	12 032	12 501	10 656	10 502	10 530 F
	34	71 635	63 495	58 374	58 854	57 606	52 137	57 071	63 709	65 270	63 026
	Country total	*87 039*	*76 972*	*73 978*	*70 189*	*69 168*	*64 169*	*69 572*	*74 365*	*75 772*	*73 556*
Djibouti	01	0	0	0	0	0	0	0	0	0	0
	51	275 F	300 F	320 F	350 F	350 F	350 F	350 F	350 F	350 F	350 F
	Country total	*275 F*	*300 F*	*320 F*	*350 F*	*350 F*	*350 F*	*350 F*	*350 F*	*350 F*	*350 F*
Egypt	01	180 400	186 700	199 300	228 930	230 660	243 609	250 181	225 300	253 470	295 422
	34	-	-	-	-	-	-	-	4	-	-
	37	39 600	40 200	41 400	39 461	46 298	48 225	62 041	81 000	54 872	59 652
	51	38 700	45 800	43 200	42 399	43 272	50 925	50 519	74 200	75 972	73 577
	Country total	*258 700*	*272 700*	*283 900*	*310 790*	*320 230*	*342 759*	*362 741*	*380 504*	*384 314*	*428 651*
Eq Guinea	01	370 F	600	700	450	900	850	970	1 101 F	1 076	1 000 F
	34	3 230 F	2 907	4 369	1 856	4 140	5 240	5 035	5 900 F	2 558	2 500 F
	Country total	*3 600 F*	*3 507*	*5 069*	*2 306*	*5 040*	*6 090*	*6 005*	*7 001*	*3 634*	*3 500 F*
Eritrea	01	...	0	0	0	0	0	0	0	0	0
	51	...	475	2 706	3 559	3 252	1 038	1 629	6 891	12 612	8 820
	Country total	*...*	*475*	*2 706*	*3 559*	*3 252*	*1 038*	*1 629*	*6 891*	*12 612*	*8 820*
Ethiopia	01	4 485	4 175	5 285	6 325	8 770	10 370	14 000	15 858	15 681	15 390
	51	100	-	-	-	-	-	-	-	-	-
	Country total	*4 585*	*4 175*	*5 285*	*6 325*	*8 770*	*10 370*	*14 000*	*15 858*	*15 681*	*15 390*
Fr South Tr	01	0	0	0	0	0	0	0	0	0	0
	51	464	460	524	519 F	437 F	375 F	388 F	425 F	272 F	263 F
	Country total	*464*	*460*	*524*	*519 F*	*437 F*	*375 F*	*388 F*	*425 F*	*272 F*	*263 F*
Gabon	01	2 000 F	3 500 F	4 500 F	7 648 F	9 433	9 441	9 000	10 000	10 417	9 850
	34	22 000 F	28 289	26 515	32 789	36 680	34 143	44 609	41 143	37 053	30 607
	Country total	*24 000 F*	*31 789 F*	*31 015 F*	*40 437 F*	*46 113*	*43 584*	*53 609*	*51 143*	*47 470*	*40 457*
Gambia	01	2 500	2 400	2 400	2 500	2 500	2 500	2 500	2 500 F	2 500 F	2 500 F
	34	15 545	18 908	20 381	21 252	29 101	29 754	26 502	27 500	26 516	32 027
	Country total	*18 045*	*21 308*	*22 781*	*23 752*	*31 601*	*32 254*	*29 002*	*30 000*	*29 016*	*34 527*

D-1 Fish crustaceans, molluscs, etc — Capture production by countries or areas and fishing areas — Africa
Poissons, crustacés, mollusques, etc — Captures par pays ou zones et zones de pêche — Afrique
Peces, crustáceos, moluscos, etc — Capturas por países o áreas y áreas de pesca — Africa

	Fishing area Zone de pêche Area de pesca	1992 mt	1993 mt	1994 mt	1995 mt	1996 mt	1997 mt	1998 mt	1999 mt	2000 mt	2001 mt
Ghana	01	56 000	52 000	54 700	60 000	73 580	70 000	74 500	74 500	74 500	74 500
	34	367 425	320 619	280 737	292 844	403 593	377 088	368 141	418 276	370 441	364 455
	87	-	-	-	-	-	-	-	-	7 129	6 332
	Country total	*423 425*	*372 619*	*335 437*	*352 844*	*477 173*	*447 088*	*442 641*	*492 776*	*452 070*	*445 287*
Guinea	01	4 000 F	4 600	3 800	3 100	2 780	3 600	4 000	4 000	4 000	4 000 F
	34	51 000 F	56 000 F	60 000 F	64 760	60 580	58 841	65 764	83 314	87 513	86 000 F
	Country total	*55 000 F*	*60 600 F*	*63 800 F*	*67 860*	*63 360*	*62 441*	*69 764*	*87 314*	*91 513*	*90 000 F*
GuineaBissau	01	200 F	250 F	250 F	250 F	250 F	250 F	200 F	200 F	200 F	200 F
	34	5 000 F	5 100 F	5 750 F	6 078 F	6 750 F	7 000 F	5 800 F	4 800 F	4 800 F	4 800 F
	Country total	*5 200 F*	*5 350 F*	*6 000 F*	*6 328*	*7 000 F*	*7 250 F*	*6 000 F*	*5 000 F*	*5 000 F*	*5 000 F*
Kenya	01	155 644	176 435	198 805	187 241	174 692	154 955	165 992	198 653	210 343	156 763
	51	6 566	5 617	3 772	5 465	6 296	6 099	6 600	6 634	4 763	7 388
	Country total	*162 210*	*182 052*	*202 577*	*192 706*	*180 988*	*161 054*	*172 592*	*205 287*	*215 106*	*164 151*
Lesotho	01	16 F	22 F	22 F	26 F	28 F	30 F	30 F	30	32	24
	Country total	*16 F*	*22 F*	*22 F*	*26 F*	*28 F*	*30 F*	*30 F*	*30*	*32*	*24*
Liberia	01	4 000	4 000	4 000	4 000	4 000	4 000	4 000	4 000	4 000	4 000
	34	4 891	3 782	3 721	4 829	3 408	4 580	6 830	11 472	7 726	7 286
	87	-	-	-	-	900	-	-	-	-	-
	Country total	*8 891*	*7 782*	*7 721*	*8 829*	*8 308*	*8 580*	*10 830*	*15 472*	*11 726*	*11 286*
Libya	01	0	0	0	0	0	0	0	0	0	0
	34	820	1 085	500	400	976	877	911	850	887	239
	37	28 000 F	30 000 F	33 000 F	34 000 F	32 000 F	31 000 F	32 000 F	32 000 F	32 500 F	33 000 F
	Country total	*28 820 F*	*31 085 F*	*33 500 F*	*34 400 F*	*32 976 F*	*31 877 F*	*32 911 F*	*32 850 F*	*33 387 F*	*33 239 F*
Madagascar	01	27 500	30 000	30 000	30 000	30 000	30 000	30 000	30 000	30 000	30 000
	51	76 905	84 261	86 431	85 653	84 475	86 391	96 395	99 630	102 093	105 583
	Country total	*104 405*	*114 261*	*116 431*	*115 653*	*114 475*	*116 391*	*126 395*	*129 630*	*132 093*	*135 583*
Malawi	01	69 261	67 951	58 579	53 664	63 569	56 340	41 111	45 392	43 000 F	40 619
	Country total	*69 261*	*67 951*	*58 579*	*53 664*	*63 569*	*56 340*	*41 111*	*45 392*	*43 000 F*	*40 619*
Mali	01	68 467	64 300	62 850	132 900	111 910	99 550	98 000	98 536	109 870	100 000 F
	Country total	*68 467*	*64 300*	*62 850*	*132 900*	*111 910*	*99 550*	*98 000*	*98 536*	*109 870*	*100 000 F*
Mauritania	01	5 000 F	5 000 F	5 000 F	5 000 F	5 000 F	5 000 F	5 000 F	5 000 F	5 000 F	5 000 F
	34	61 054 F	54 452 F	46 746 F	48 147 F	55 324 F	52 756 F	56 660 F	71 026 F	75 849 F	78 596 F
	Country total	*66 054 F*	*59 452 F*	*51 746 F*	*53 147 F*	*60 324 F*	*57 756 F*	*61 660 F*	*76 026 F*	*80 849 F*	*83 596 F*
Mauritius	01	0	3	0	0	0	0	0	0	0	0
	51	18 861	20 576	18 145	16 395	11 869	14 025	12 093	12 205	9 615	10 694
	Country total	*18 861*	*20 579*	*18 145*	*16 395*	*11 869*	*14 025*	*12 093*	*12 205*	*9 615*	*10 694*
Mayotte	51	1 100 F	500 F	600 F	700 F	1 000	1 300 F	2 000 F	2 000	5 500	5 500
	Country total	*1 100 F*	*500 F*	*600 F*	*700 F*	*1 000*	*1 300 F*	*2 000 F*	*2 000*	*5 500*	*5 500*
Morocco	01	1 794	1 617	1 750	1 500	1 500	2 100	1 703	2 163	1 608	983
	34	509 904	591 270	717 578	807 775	601 734	758 321	680 075	705 978	856 362	1 054 147
	37	39 239	31 624	34 999	39 676	39 652	31 485	28 658	37 290	38 650	28 146
	Country total	*550 937*	*624 511*	*754 327*	*848 951*	*642 886*	*791 906*	*710 436*	*745 431*	*896 620*	*1 083 276*
Mozambique	01	3 800	4 689	4 925	5 093	7 510	11 668	8 994	10 243	13 088	8 076
	51	27 808	25 506	22 531	21 740	27 405	28 035	27 683	23 746	25 977	24 436
	Country total	*31 608*	*30 195*	*27 456*	*26 833*	*34 915*	*39 703*	*36 677*	*33 989*	*39 065*	*32 512*
Namibia	01	1 102	1 200	1 200	1 200	1 200	1 500	1 500	1 500	1 500	1 500
	41	-	-	-	-	-	304	677	746	-	-
	47	654 008	789 132	647 999	568 633	516 628	511 412	610 166	577 838	588 405	545 992
	Country total	*655 110*	*790 332*	*649 199*	*569 833*	*517 828*	*513 216*	*612 343*	*580 084*	*589 905*	*547 492*
Niger	01	2 456	2 162	2 516	3 616	4 156	6 328	7 013	11 000	16 250	20 800
	Country total	*2 456*	*2 162*	*2 516*	*3 616*	*4 156*	*6 328*	*7 013*	*11 000*	*16 250*	*20 800*
Nigeria	01	93 281	95 627	103 800	117 903	89 521	93 644	139 020	139 393	132 315	154 175
	34	208 046	142 782	163 259	231 579	248 472	294 279	324 004	316 235	309 062	297 971
	Country total	*301 327*	*238 409*	*267 059*	*349 482*	*337 993*	*387 923*	*463 024*	*455 628*	*441 377*	*452 146*
Réunion	01	0	0	0	0	0	0	0	0	0	0
	51	1 103	1 679	2 531	2 500	3 607	4 288	4 579	4 043	4 079	3 635
	Country total	*1 103*	*1 679*	*2 531*	*2 500*	*3 607*	*4 288*	*4 579*	*4 043*	*4 079*	*3 635*
Rwanda	01	3 644	3 500 F	3 400 F	3 300 F	2 952	4 428	6 641	6 433	6 726	6 828
	Country total	*3 644*	*3 500 F*	*3 400 F*	*3 300 F*	*2 952*	*4 428*	*6 641*	*6 433*	*6 726*	*6 828*
St Helena	01	0	0	0	0	0	-	-	-	-	-
	47	651	726	702	915	819	897	1 060	632	718	866
	Country total	*651*	*726*	*702*	*915*	*819*	*897*	*1 060*	*632*	*718*	*866*
Sao Tome Prn	01	0	0	0	0	0	0	0	0	0	0
	34	2 094	2 334	3 391	3 565	3 980	3 338	3 477	3 756	3 500 F	3 500 F
	Country total	*2 094*	*2 334*	*3 391*	*3 565*	*3 980*	*3 338*	*3 477*	*3 756*	*3 500 F*	*3 500 F*

D-1 Fish crustaceans, molluscs, etc — Capture production by countries or areas and fishing areas — Africa
Poissons, crustacés, mollusques, etc — Captures par pays ou zones et zones de pêche — Afrique
Peces, crustáceos, moluscos, etc — Capturas por países o áreas y áreas de pesca — Africa

	Fishing area Zone de pêche Area de pescá	1992 mt	1993 mt	1994 mt	1995 mt	1996 mt	1997 mt	1998 mt	1999 mt	2000 mt	2001 mt
Senegal	01	24 750	27 650	30 000 F	31 000 F	23 000 F	31 000 F	21 000 F	34 000 F	22 450	20 000 F
	34	345 492	354 562	322 421	323 617	388 759	426 366	382 872	378 125	379 597	385 409
	Country total	*370 242*	*382 212*	*352 421*	*354 617*	*411 759*	*457 366*	*403 872*	*412 125*	*402 047*	*405 409*
Seychelles	01	0	0	0	0	0	0	0	0	0	0
	31	-	-	-	-	-	-	-	-	127	-
	41	-	-	-	-	-	1 253	101	-	-	-
	47	-	-	-	-	-	-	-	-	6	-
	51	6 663	5 178	4 469	4 008	4 707	11 741	20 000	34 235	31 982	46 994
	57	-	-	-	-	-	1 049	3 785	62	244	556
	Country total	*6 663*	*5 178*	*4 469*	*4 008*	*4 707*	*14 043*	*23 886*	*34 297*	*32 359*	*47 550*
Sierra Leone	01	14 000	14 000	15 000	15 000	14 500 F	14 500	14 190	14 480	14 000	14 000
	34	54 510	52 288	47 439	49 870	52 804 F	58 128	48 875	44 927	60 730	61 210
	Country total	*68 510*	*66 288*	*62 439*	*64 870*	*67 304 F*	*72 628*	*63 065*	*59 407*	*74 730*	*75 210*
Somalia	01	250 F	250 F	250 F	250 F	250 F	250 F	250 F	250 F	200 F	200 F
	51	25 500 F	27 500 F	29 600	27 700 F	25 800 F	23 900 F	22 000 F	20 000 F	20 000 F	19 800 F
	Country total	*25 750 F*	*27 750 F*	*29 850 F*	*27 950 F*	*26 050 F*	*24 150 F*	*22 250 F*	*20 250 F*	*20 200 F*	*20 000 F*
South Africa	01	832	832	800	800	850	850	900	900 F	900 F	900 F
	47	690 240	560 716	522 129	574 374	439 229	511 249	555 852	585 240	640 000 F	752 208
	48	-	-	3	-	-	-	508	451	324	227
	51	1 267	826	517	373	349	890	1 100	1 000 F	926	1 017
	58	-	-	-	-	-	2 106	689	553	1 088	961
	88	-	-	-	-	-	-	-	-	-	32
	Country total	*692 339*	*562 374*	*523 449*	*575 547*	*440 428*	*515 095*	*559 049*	*588 144*	*643 238*	*755 345*
Sudan	01	33 000	37 500	40 000	40 000	40 500	42 000	44 000	44 000	48 000 F	53 000
	51	2 000	2 500	4 000	4 000	4 500	5 000	5 500	5 500	5 000 F	5 000
	Country total	*35 000*	*40 000*	*44 000*	*44 000*	*45 000*	*47 000*	*49 500*	*49 500*	*53 000 F*	*58 000*
Swaziland	01	60 F	68 F	65 F	60 F	60 F	65 F	70 F	70 F	70 F	70 F
	Country total	*60 F*	*68 F*	*65 F*	*60 F*	*60 F*	*65 F*	*70 F*	*70 F*	*70 F*	*70 F*
Tanzania	01	275 150	294 582	247 614	317 029	262 276	306 750	300 000	260 020	280 000	283 000
	51	56 085	36 685	40 785	42 771	61 645	50 210	48 000	50 489	52 779	52 900
	Country total	*331 235*	*331 267*	*288 399*	*359 800*	*323 921*	*356 960*	*348 000*	*310 509*	*332 779*	*335 900*
Togo	01	5 500	6 000	5 000	4 998	5 000	5 000	5 000	5 000	5 000	5 000
	34	5 249	10 964	8 052	7 203	10 098	9 290	11 655	17 924	17 277	18 163
	Country total	*10 749*	*16 964*	*13 052*	*12 201*	*15 098*	*14 290*	*16 655*	*22 924*	*22 277*	*23 163*
Tunisia	01	0	400	243	440	706	1 010	896	808	832	860
	37	86 397	82 514	85 367	82 915	83 028	86 002	87 179	92 378	94 718	97 622
	Country total	*86 397*	*82 914*	*85 610*	*83 355*	*83 734*	*87 012*	*88 075*	*93 186*	*95 550*	*98 482*
Uganda	01	264 900	219 814	213 129	208 789	195 088	218 026	220 628	226 097	219 356	220 726
	Country total	*264 900*	*219 814*	*213 129*	*208 789*	*195 088*	*218 026*	*220 628*	*226 097*	*219 356*	*220 726*
Westn Sahara	34	0	0	0	0	0	0	0	0	0	0
	Country total	*0*	*0*	*0*	*0*	*0*	*0*	*0*	*0*	*0*	*0*
Zambia	01	67 864	65 768	70 057	70 546	66 332	65 923	69 938	67 327	66 671	65 000 F
	Country total	*67 864*	*65 768*	*70 057*	*70 546*	*66 332*	*65 923*	*69 938*	*67 327*	*66 671*	*65 000 F*
Zimbabwe	01	21 601	21 230	20 219	16 463	16 387	18 156	16 371	12 410	13 114	13 000 F
	Country total	*21 601*	*21 230*	*20 219*	*16 463*	*16 387*	*18 156*	*16 371*	*12 410*	*13 114*	*13 000 F*
Total		***5 623 619***	***5 633 819***	***5 531 157***	***5 850 992***	***5 578 784***	***5 941 285***	***6 107 738***	***6 325 388***	***6 616 512***	***6 890 230***

	Fishing area Zone de pêche Area de pesca	1992 mt	1993 mt	1994 mt	1995 mt	1996 mt	1997 mt	1998 mt	1999 mt	2000 mt	2001 mt
Anguilla	02	0	0	0	0	0	0	0	0	0	0
	31	386	330	333	150 F	200 F	250 F	250 F	250 F	250 F	250 F
	Country total	*386*	*330*	*333*	*150 F*	*200 F*	*250 F*	*250 F*	*250 F*	*250 F*	*250 F*
Antigua Barb	02	0	0	0	0	0	0	0	0	0	0
	31	1 712	642	696	1 311	1 209	1 437	1 415	1 361	1 481	1 583
	Country total	*1 712*	*642*	*696*	*1 311*	*1 209*	*1 437*	*1 415*	*1 361*	*1 481*	*1 583*
Aruba	02	0	0	0	0	0	0	0	-	-	-
	31	300	260	260	140	160	205	182	175	163	163
	Country total	*300*	*260*	*260*	*140*	*160*	*205*	*182*	*175*	*163*	*163*
Bahamas	02	0	0	0	0	0	0	0	0	0	0
	31	9 846	10 073	10 311	9 915	10 197	10 439	10 124	10 473	11 070	9 290
	34	-	-	-	29	-	-	-	-	-	-
	Country total	*9 846*	*10 073*	*10 311*	*9 944*	*10 197*	*10 439*	*10 124*	*10 473*	*11 070*	*9 290*
Barbados	02	0	0	0	0	0	0	0	0	0	-
	31	3 574	3 214	2 818	3 581	3 512	2 809	3 644	3 250	3 100	2 676
	Country total	*3 574*	*3 214*	*2 818*	*3 581*	*3 512*	*2 809*	*3 644*	*3 250*	*3 100*	*2 676*
Belize	02	1	1	1	0	0	0	0	0	0	0
	31	2 413	1 784	2 102	1 877	2 048	2 835	2 557	1 871	2 434	2 517
	34	-	633	97	460	-	9 025	14 642	31 076	43 025	9 272
	41	-	-	-	-	-	-	-	4 500	6 729	2 581
	77	-	-	240	200	1 430	1 110	680	970	560	...
	87	-	1 110	5 230	7 120	8 560	11 740	8 430	6 170	7 700	...
	Country total	*2 414*	*3 528*	*7 670*	*9 657*	*12 038*	*24 710*	*26 309*	*44 587*	*60 448*	*14 370*
Bermuda	02	0	0	0	0	0	0	0	0	0	0
	31	432	404	394	449	465	461	466	453	286	315
	Country total	*432*	*404*	*394*	*449*	*465*	*461*	*466*	*453*	*286*	*315*
Br Virgin Is	02	0	0	0	0	0	0	0	0	0	0
	31	453 F	343	470	532	506	105	116	115	43	50 F
	Country total	*453 F*	*343*	*470*	*532*	*506*	*105*	*116*	*115*	*43*	*50 F*
Canada	02	42 633	36 327	36 333	38 756	38 295	38 798	40 744	40 587	40 667	35 120
	21	954 158	810 017	676 031	593 299	631 357	685 964	747 053	774 095	826 399	830 296
	67	295 024	286 283	309 848	217 121	235 074	247 383	226 013	212 576	143 072	183 886
	81	235	235	235	235	136	149	167	253	351	206
	Country total	*1 292 050*	*1 132 862*	*1 022 447*	*849 411*	*904 862*	*972 294*	*1 013 977*	*1 027 511*	*1 010 489*	*1 049 508*
Cayman Is	02	0	0	0	0	0	0	0	0	0	0
	31	125	125	125	125	110	125	125	125	125	125
	34	700	320	...	...	...	...	...	...	...	...
	Country total	*825*	*445*	*125*	*125*	*110*	*125*	*125*	*125*	*125*	*125*
Costa Rica	02	406	710	840	900	1 090	840	1 000 F	1 000 F	1 000 F	1 000 F
	31	261	199	269	422	437	420	402	666	1 088	816
	77	15 342	15 261	16 023	15 845	21 540	25 409	23 355	26 552	33 310	32 917
	87	...	...	...	270	1 260	...	...	...	...	...
	Country total	*16 009*	*16 170*	*17 132*	*17 437*	*24 327*	*26 669*	*24 757*	*28 218*	*35 398*	*34 733*
Cuba	02	13 820	10 178	9 823	8 893	10 324	8 525	5 954	4 624	4 600 F	4 600 F
	21	25 972	29 919	12 333	18 327	24 074	17 221	7 736	4 563	46	...
	31	54 404	41 608	51 659	50 000	48 841	58 896	53 386	51 533	51 500 F	51 400 F
	34	2 668	3 047	2 975	2 839	2 364	269	...	...	...	...
	47	69	-	-	-	-	-	-	2 427	...	-
	87	3 197	0	-	-	-	-	-	4 234	...	...
	Country total	*100 130*	*84 752*	*76 790*	*80 059*	*85 603*	*84 911*	*67 076*	*67 381*	*56 146 F*	*56 000 F*
Dominica	02	0	0	0	0	0	0	0	0	0	0
	31	711	794	882	950	1 030	1 079	1 212	1 200 F	1 200 F	1 150 F
	Country total	*711*	*794*	*882*	*950*	*1 030*	*1 079*	*1 212*	*1 200 F*	*1 200 F*	*1 150 F*
Dominican Rp	02	1 024	2 037	3 774	2 106	288	1 067	1 095	598	187	1 158
	31	11 816	10 820	19 058	15 768	12 606	13 793	9 188	7 835	10 842	12 059
	Country total	*12 840*	*12 857*	*22 832*	*17 874*	*12 894*	*14 860*	*10 283*	*8 433*	*11 029*	*13 217*
El Salvador	02	5 136	4 461	3 819	4 324	2 968	2 808	2 443	2 653	2 831	1 692
	77	6 940	7 990	9 740	10 747	11 466	9 089	10 214	7 841	6 759	16 055
	87	...	...	...	...	...	...	...	6 030	...	...
	Country total	*12 076*	*12 451*	*13 559*	*15 071*	*14 434*	*11 897*	*12 657*	*16 524*	*9 590*	*17 747*
Greenland	02	0	0	0	0	0	0	0	0	0	0
	21	101 982	94 937	101 952	108 574	94 356	94 096	97 953	117 841	121 885	124 149
	27	11 285	21 713	15 465	20 316	21 662	26 500	30 589	42 412	37 826	34 336
	Country total	*113 267*	*116 650*	*117 417*	*128 890*	*116 018*	*120 596*	*128 542*	*160 253*	*159 711*	*158 485*
Grenada	02	0	0	0	0	0	0	0	0	0	0
	31	2 052	2 103	1 599	1 497	1 577	1 530	1 837	1 658	1 701	2 247
	Country total	*2 052*	*2 103*	*1 599*	*1 497*	*1 577*	*1 530*	*1 837*	*1 658*	*1 701*	*2 247*
Guadeloupe	02	0	0	0	0	0	0	0	0	0	0
	31	8 540	8 600	8 800	9 500	9 570	10 480	9 084	9 114	10 100	10 100 F
	Country total	*8 540*	*8 600*	*8 800*	*9 500*	*9 570*	*10 480*	*9 084*	*9 114*	*10 100*	*10 100 F*

	Fishing area Zone de pêche Area de pesca	1992 mt	1993 mt	1994 mt	1995 mt	1996 mt	1997 mt	1998 mt	1999 mt	2000 mt	2001 mt
Guatemala	02	3 702	4 228	3 776	4 025	4 000	5 121	6 523	6 976	7 301	7 300 F
	31	100	92	179	390	390	285	328	292	366	350 F
	77	2 776	3 262	3 256	3 838	3 263	1 490	3 996	8 930	12 101	2 450 F
	87	-	-	-	-	-	-	-	4 820	18 210	...
	Country total	*6 578*	*7 582*	*7 211*	*8 253*	*7 653*	*6 896*	*10 847*	*21 018*	*37 978*	*10 100 F*
Haiti	02	500 F	600 F	500 F	500 F	500 F	500 F	500 F	500 F	500 F	500 F
	31	4 500 F	4 550 F	5 000 F	5 017 F	4 745 F	4 801 F	4 759 F	4 500 F	4 500 F	4 500 F
	Country total	*5 000 F*	*5 150 F*	*5 500 F*	*5 517 F*	*5 245 F*	*5 301 F*	*5 259 F*	*5 000 F*	*5 000 F*	*5 000 F*
Honduras	02	85	86	92	127	98	126	119	102	61	111
	21	-	1 298	-	-	-	-	-	-	-	-
	31	4 630	4 274	4 526	6 686	2 691	6 560	2 332 F	1 865	4 359	2 539
	34	8 224	9 562	993	8 841	6 998	5 084	1 908	3 355	1 711	1 120
	41	2 762	3 655	2 976	2 771	849	-	-	-	-	-
	47	...	193	...	48	10	25	9	20	...	...
	57	-	-	-	-	-	-	-	-	-	637
	77	785	509	692	864	2 412	4 002	2 155 F	1 702	2 973	3 044
	87	-	-	-	-	470	2 560	870	5 710	2 580	...
	Country total	*16 486*	*19 577*	*9 279*	*19 337*	*13 528*	*18 357*	*7 393 F*	*12 754*	*11 684*	*7 451*
Jamaica	02	450 F	450 F	450 F	450 F	450 F	450 F	450 F	450 F	450 F	450 F
	31	8 900 F	9 550 F	9 877 F	9 917 F	12 054 F	7 748	6 110	8 058	5 226	5 250 F
	Country total	*9 350 F*	*10 000 F*	*10 327 F*	*10 367 F*	*12 504 F*	*8 198*	*6 560*	*8 508*	*5 676*	*5 700 F*
Martinique	02	0	3	0	0	0	0	0	0	0	0
	31	4 538	5 850	5 800 F	5 300	3 500 F	5 500	5 500	6 000	6 310	6 200
	Country total	*4 538*	*5 853*	*5 800 F*	*5 300*	*3 500 F*	*5 500*	*5 500*	*6 000*	*6 310*	*6 200*
Mexico	02	108 707	111 178	111 125	122 020	122 501	113 552	100 335	91 462	106 817	92 154
	31	330 207	320 824	329 521	272 178	294 231	320 829	302 157	285 833	274 532	259 156
	77	712 900	665 867	743 251	921 820	1 027 461	1 042 150	767 021	817 557	928 727	1 039 262
	87	5 759	5 063	7 749	13 322	19 995	12 581	10 347	10 751	5 505	8 020
	Country total	*1 157 573*	*1 102 932*	*1 191 646*	*1 329 340*	*1 464 188*	*1 489 112*	*1 179 860*	*1 205 603*	*1 315 581*	*1 398 592*
Montserrat	02	0	0	0	0	0	0	0	0	0	0
	31	23	58	62	48	38	45	46	50 F	50 F	50 F
	Country total	*23*	*58*	*62*	*48*	*38*	*45*	*46*	*50 F*	*50 F*	*50 F*
NethAntilles	02	0	0	0	0	0	0	0	0	0	0
	31	1 150 F	1 200 F	1 100 F	1 020 F	1 000 F	950 F	950 F	950 F	950 F	950 F
	34	-	-	-	-	-	-	-	-	19 544	...
	Country total	*1 150 F*	*1 200 F*	*1 100 F*	*1 020 F*	*1 000 F*	*950 F*	*950 F*	*950 F*	*20 494 F*	*950 F*
Nicaragua	02	348	547	824	538	1 142	1 293	1 256	1 120	1 076	1 051
	31	4 212	5 001	6 738	5 957	9 685	9 451	12 011	13 127	14 838	13 444
	77	2 158	2 621	2 472	4 500	4 615	5 432	6 625	9 662	10 434	8 304
	87	-	-	-	-	-	-	-	-	1 660	...
	Country total	*6 718*	*8 169*	*10 034*	*10 995*	*15 442*	*16 176*	*19 892*	*23 909*	*28 008*	*22 799*
Panama	02	80	28	285	130	80	91	23	20	20	20 F
	27	382	210	363	289	369	58	58	-	136	...
	31	-	-	-	-	-	-	-	20	714	...
	34	31 749	36 181	38 686	39 456	19 718	8 299	4 889	3 404	3 353	...
	37	484	467	1 499	1 498	2 850	236	-	-	-	-
	41	4 027	1 059	598	460	706	-	-	159	1 342	...
	47	8 294	28	147	-	-	-	-	25	675	...
	48	-	-	-	141	496	-	-	-	-	-
	77	122 004	143 959	139 709	155 219	122 975	147 295	197 397	111 140	195 841	220 390 F
	87	5 230	2 890	3 860	5 800	3 570	6 370	320	6 080	20 550	14 590
	Country total	*172 250*	*184 822*	*185 147*	*202 993*	*150 764*	*162 349*	*202 687*	*120 848*	*222 631*	*235 000 F*
Puerto Rico	02	0	0	0	0	0	0	0	0	0	0
	31	1 812	1 877	2 275	3 173	2 701	3 187	3 006	3 020	4 154	3 794
	Country total	*1 812*	*1 877*	*2 275*	*3 173*	*2 701*	*3 187*	*3 006*	*3 020*	*4 154*	*3 794*
St Kitts Nev	02	0	0	0	0	0	0	0	0	0	0
	31	300 F	250 F	212	192	352	272	533	555	492	591
	Country total	*300 F*	*250 F*	*212*	*192*	*352*	*272*	*533*	*555*	*492*	*591*
St Lucia	02	0	0	0	0	0	0	0	0	0	0
	31	1 073	1 336	1 252	1 188	1 274	1 308	1 589	1 718	1 855	1 983
	Country total	*1 073*	*1 336*	*1 252*	*1 188*	*1 274*	*1 308*	*1 589*	*1 718*	*1 855*	*1 983*
St Pier Mq	02	0	0	0	0	0	0	0	0	0	0
	21	16 443	282	294	317	747	3 571	6 108	5 892	6 485	3 802
	Country total	*16 443*	*282*	*294*	*317*	*747*	*3 571*	*6 108*	*5 892*	*6 485*	*3 802*
St Vincent	02	0	0	0	0	2	1	0	0	0	0
	31	1 381	1 479	1 090	944	889 F	948	1 365	1 032	7 325	9 020
	34	776	466	0	61	30	5 144	32 608	16 727	20 369	34 940
	41	-	-	-	-	-	-	-	-	-	1 818
	87	-	-	2 750	-	-	-	-	-	-	-
	Country total	*2 157*	*1 945*	*3 840*	*1 005*	*921 F*	*6 093*	*33 973*	*17 759*	*27 694*	*45 778*
Trinidad Tob	02	0	0	0	0	0	0	0	0	0	0
	31	13 000 F	9 637	14 483	12 000 F	9 435	11 283	9 175	8 826	9 786	11 408
	Country total	*13 000 F*	*9 637*	*14 483*	*12 000 F*	*9 435*	*11 283*	*9 175*	*8 826*	*9 786*	*11 408*

	Fishing area Zone de pêche Area de pesca	1992 mt	1993 mt	1994 mt	1995 mt	1996 mt	1997 mt	1998 mt	1999 mt	2000 mt	2001 mt
Turks Caicos	02	0	0	0	0	0	0	0	0	0	0
	31	1 191	1 458	1 419 F	1 395 F	1 297 F	1 250 F	1 318	1 300 F	1 300 F	1 300 F
	Country total	*1 191*	*1 458*	*1 419 F*	*1 395 F*	*1 297 F*	*1 250 F*	*1 318*	*1 300 F*	*1 300 F*	*1 300 F*
USA	02	48 639	54 377	42 648	36 688	33 471	38 347	36 129	36 413	25 339	29 578
	21	1 281 350	1 240 243	1 062 863	1 199 006	1 228 049	1 179 746	1 027 487	1 024 098	985 289	1 129 720
	31	748 992	942 154	1 238 611	856 754	771 970	867 630	822 594	943 641	1 021 580	878 443
	48	-	299	-	-	684	-	-	16	-	1 638
	67	2 754 059	2 965 279	2 822 291	2 743 083	2 595 248	2 516 701	2 514 468	2 334 583	2 331 879	2 572 829
	71	154 413	176 253	194 218	153 840	143 348	140 391	154 610	188 040	121 797	113 658
	77	196 843	143 488	174 088	233 025	220 175	237 646	152 124	222 855	256 127	212 580
	81	2 918	1 123	605	2 170	3 223	-	-	-	-	5 942
	87	3 503	-	-	-	5 315	2 979	1 568	-	3 310	18
	Country total	*5 190 717*	*5 523 216*	*5 535 324*	*5 224 566*	*5 001 483*	*4 983 440*	*4 708 980*	*4 749 646*	*4 745 321*	*4 944 406*
US Virgin Is	02	0	0	0	0	0	0	0	0	0	-
	31	750 F	650 F	550 F	470 F	400 F	350 F	300 F	263	300 F	300 F
	Country total	*750 F*	*650 F*	*550 F*	*470 F*	*400 F*	*350 F*	*300 F*	*263*	*300 F*	*300 F*
Total		***8 184 726***	***8 292 472***	***8 290 290***	***7 984 054***	***7 891 184***	***8 008 495***	***7 516 032***	***7 574 700***	***7 823 129***	***8 077 213***

D-3

Fish crustaceans, molluscs, etc	Capture production by countries or areas and fishing areas	America, South
Poissons, crustacés, mollusques, etc	Captures par pays ou zones et zones de pêche	Amérique du Sud
Peces, crustáceos, moluscos, etc	Capturas por países o áreas y áreas de pesca	América del Sur

	Fishing area Zone de pêche Area de pesca	1992 mt	1993 mt	1994 mt	1995 mt	1996 mt	1997 mt	1998 mt	1999 mt	2000 mt	2001 mt
Argentina	03	11 227	11 800	12 785	17 191	19 189	22 735	23 197	27 558	30 418	23 860
	41	692 198	919 503	938 590	1 152 054	1 272 057	1 377 427	1 141 632	1 044 221	883 204	899 462
	48	-	-	12	879	109	-	-	6 534	-	-
	Country total	*703 425*	*931 303*	*951 387*	*1 170 124*	*1 291 355*	*1 400 162*	*1 164 829*	*1 078 313*	*913 622*	*923 322*
Bolivia	03	4 905	5 518	5 353	5 692	5 988	6 038	6 055	6 052	6 106	5 940
	Country total	*4 905*	*5 518*	*5 353*	*5 692*	*5 988*	*6 038*	*6 055*	*6 052*	*6 106*	*5 940*
Brazil	03	182 540 F	186 990 F	191 485 F	193 042	193 309	178 871	174 190	185 471	199 159	200 000 F
	41	558 780 F	530 100 F	548 615 F	513 666	522 173	565 714	532 599	518 470	567 687	570 000 F
	Country total	*741 320 F*	*717 090 F*	*740 100 F*	*706 708*	*715 482*	*744 585*	*706 789*	*703 941*	*766 846*	*770 000 F*
Chile	03	14	7	5	...	...	...	4	...	...	...
	34	-	-	25	-	-	-	-	-	-	-
	41	2 432	846	23	302	-	3 744	14 755	5 226	2 749	8 849
	47	-	-	-	-	-	-	-	-	-	5
	48	8 986	5 386	3 985	1 896	3 098	1 275	1 492	1 668	2 324	896
	87	6 420 892	5 943 326	7 716 540	7 431 704	6 687 844	5 805 745	3 249 132	5 043 634	4 295 087	3 787 393
	Country total	*6 432 324*	*5 949 565*	*7 720 578*	*7 433 902*	*6 690 942*	*5 810 764*	*3 265 383*	*5 050 528*	*4 300 160*	*3 797 143*
Colombia	03	33 759	30 538	34 983	23 524	23 061	20 610	21 673	28 788	24 854	25 000 F
	31	14 959	9 578	9 354	15 539	23 888	6 235	26 825	3 040	15 196	15 000 F
	77	...	...	...	...	...	...	...	...	7 630	7 530
	87	92 452	83 381	53 462	81 636	83 880	121 073	84 410	86 167	81 964	77 470 F
	Country total	*141 170*	*123 497*	*97 799*	*120 699*	*130 829*	*147 918*	*132 908*	*117 995*	*129 644*	*125 000 F*
Ecuador	03	332	372	300	300	300	400	400	400	400	400
	77	1 990	1 050	1 710	-	270	-	-	-	-	150
	87	224 495	285 211	328 238	505 095	702 404	548 588	309 622	497 472	592 147	586 020
	Country total	*226 817*	*286 633*	*330 248*	*505 395*	*702 974*	*548 988*	*310 022*	*497 872*	*592 547*	*586 570*
Falkland Is	03	-	-	-	-	1	1	1	1	1	1
	41	1 855	1 974	5 914	27 190	31 539	17 112	43 615	39 163	62 927	59 823
	Country total	*1 855*	*1 974*	*5 914*	*27 190*	*31 540*	*17 113*	*43 616*	*39 164*	*62 928*	*59 824*
Fr Guiana	03	0	0	0	0	0	0	0	0	0	0
	31	7 617	6 931	7 819	8 089	7 377 F	6 602	6 709	6 271 F	5 237 F	5 194 F
	Country total	*7 617*	*6 931*	*7 819*	*8 089*	*7 377 F*	*6 602*	*6 709*	*6 271 F*	*5 237 F*	*5 194 F*
Guyana	03	800	800	800	700 F	800	625	625	603	800	800
	31	40 452	43 323	45 567	47 200 F	47 783	53 373	52 215	53 241	48 087	52 605
	Country total	*41 252*	*44 123*	*46 367*	*47 900 F*	*48 583*	*53 998*	*52 840*	*53 844*	*48 887*	*53 405*
Paraguay	03	17 925	19 000 F	20 000 F	21 000 F	22 000	28 000	25 000 F	25 000 F	25 000 F	25 000 F
	Country total	*17 925*	*19 000 F*	*20 000 F*	*21 000 F*	*22 000*	*28 000*	*25 000 F*	*25 000 F*	*25 000 F*	*25 000 F*
Peru	03	32 734	38 290	48 837	50 789	28 890	32 221	35 327	36 223	32 297	35 653
	87	7 469 458	8 966 487	11 950 380	8 886 553	9 486 158	7 837 650	4 303 110	8 392 378	10 626 323	7 950 450
	Country total	*7 502 192*	*9 004 777*	*11 999 217*	*8 937 342*	*9 515 048*	*7 869 871*	*4 338 437*	*8 428 601*	*10 658 620*	*7 986 103*
Suriname	03	561	187	138	140 F	150 F	200 F	200 F	200 F	200 F	200
	31	10 369	9 313	14 327	12 860 F	12 850 F	13 800 F	15 995	16 000	17 300 F	18 715
	Country total	*10 930*	*9 500*	*14 465*	*13 000 F*	*13 000 F*	*14 000 F*	*16 195*	*16 200*	*17 500 F*	*18 915*
Uruguay	03	323	621	966	849	598	2 216	1 931	2 423	2 302	451
	41	125 428	118 194	119 766	125 597	122 732	134 738	138 514	100 068	101 650	96 851
	47	-	-	-	-	-	-	-	-	320	-
	48	-	-	-	-	-	-	262	521	10 688	460
	51	-	-	-	-	-	-	-	-	1 628	7 150
	58	-	-	-	-	-	-	-	-	-	99
	88	-	-	-	-	-	-	-	-	-	23
	Country total	*125 751*	*118 815*	*120 732*	*126 446*	*123 330*	*136 954*	*140 707*	*103 012*	*116 588*	*105 034*
Venezuela	03	20 114	28 251	35 412	54 175	49 911	40 881	46 035	35 219	23 739	24 326
	31	256 100	316 550	354 366	393 120	378 796	362 474	388 419	292 750	258 236	281 731
	77	37 240	32 700	26 870	45 030	58 600	49 430	45 300	49 400	53 120	69 560
	87	17 510	18 360	21 570	8 720	8 980	17 470	23 750	22 430	22 020	42 330
	Country total	*330 964*	*395 861*	*438 218*	*501 045*	*496 287*	*470 255*	*503 504*	*399 799*	*357 115*	*417 947*
Total		***16 288 447***	***17 614 587***	***22 498 197***	***19 624 532***	***19 794 735***	***17 255 248***	***10 712 994***	***16 526 592***	***18 000 800***	***14 879 397***

	Fishing area Zone de pêche Area de pesca	1992 mt	1993 mt	1994 mt	1995 mt	1996 mt	1997 mt	1998 mt	1999 mt	2000 mt	2001 mt
Afghanistan	04	1 200 F	1 200 F	1 300 F	1 300 F	1 300 F	1 250 F	1 200 F	1 200 F	1 000 F	800 F
	Country total	*1 200 F*	*1 200 F*	*1 300 F*	*1 300 F*	*1 300 F*	*1 250 F*	*1 200 F*	*1 200 F*	*1 000 F*	*800 F*
Armenia	04	1 885	1 850	1 033	821	580	580	698	1 144	1 133	866
	Country total	*1 885*	*1 850*	*1 033*	*821*	*580*	*580*	*698*	*1 144*	*1 133*	*866*
Azerbaijan	04	30 339	21 733	18 901	10 545	6 702	5 161	4 760	20 866	18 797	10 893
	Country total	*30 339*	*21 733*	*18 901*	*10 545*	*6 702*	*5 161*	*4 760*	*20 866*	*18 797*	*10 893*
Bahrain	04	0	0	0	0	0	0	0	0	-	-
	51	7 983	8 958	7 628	9 389	12 940	10 050	9 849	10 620	11 718	11 230
	Country total	*7 983*	*8 958*	*7 628*	*9 389*	*12 940*	*10 050*	*9 849*	*10 620*	*11 718*	*11 230*
Bangladesh	04	429 205	452 109	517 746	527 739	535 617	534 285	538 689	649 418	670 465	670 000 F
	57	280 127	312 715	253 044	264 650	279 170	295 141	300 452	309 797	333 799	330 000 F
	Country total	*709 332*	*764 824*	*770 790*	*792 389*	*814 787*	*829 426*	*839 141*	*959 215*	*1 004 264*	*1 000 000 F*
Bhutan	04	315 F	320 F	310 F	310 F	300 F	300 F	300 F	300 F	300 F	300 F
	Country total	*315 F*	*320 F*	*310 F*	*310 F*	*300 F*	*300 F*	*300 F*	*300 F*	*300 F*	*300 F*
Brunei Darsm	04	25	25	4	7	15	17	35	26	23	16
	71	1 667	1 703	4 441	4 712	7 390	4 504	5 014	3 160	2 464	1 476
	Country total	*1 692*	*1 728*	*4 445*	*4 719*	*7 405*	*4 521*	*5 049*	*3 186*	*2 487*	*1 492*
Cambodia	04	68 900	67 900	65 000	72 499	63 510	73 000	75 700	231 000	245 600	360 000 F
	41	-	-	-	-	-	-	-	-	2 768	1 200
	47	-	-	-	-	-	-	-	56	-	-
	71	33 700	33 100	30 000	30 500	31 200	29 800	32 200	38 100	36 000	36 000 F
	Country total	*102 600*	*101 000*	*95 000*	*102 999*	*94 710*	*102 800*	*107 900*	*269 156*	*284 368*	*397 200 F*
China	04	998 028	1 182 390	1 327 785	1 607 385	1 762 860	1 886 967	2 280 244	2 285 364	2 233 230	2 149 932
	34	5 385	9 809	2 414	21 464	18 302	20 810	20 403	28 433	25 408	36 228
	37	-	-	97	137	93	49	-	-	-	-
	41	-	-	3	3	3	3	31 806	61 664	93 156	93 973
	47	445	-	24	29	24	121	48	5 743	7 209	4 959
	51	-	-	-	-	-	-	-	294	2 587	2 773
	57	-	-	-	445	1 497	2 964	3 080	5 868	4 919	4 009
	61	7 275 876	8 111 441	9 482 098	10 861 731	12 332 922	13 727 656	14 840 900	14 843 663	14 609 252	14 215 449
	67	40 000	40 000	40 000	60 000	60 000	80 000	50 000	...	...	...
	71	2 818	7 797	14 415	11 512	6 406	3 774	3 446	9 003	8 882	7 519
	77	-	-	-	-	-	-	-	-	2 682	14 547
	Country total	*8 322 552*	*9 351 437*	*10 866 836*	*12 562 706*	*14 182 107*	*15 722 344*	*17 229 927*	*17 240 032*	*16 987 325*	*16 529 389*
China,H.Kong	04	0	0	0	0	0	0	0	0	0	0
	61	220 181	217 544	211 010	194 999	183 856	186 000	180 000	127 780	157 012	173 972
	Country total	*220 181*	*217 544*	*211 010*	*194 999*	*183 856*	*186 000*	*180 000*	*127 780*	*157 012*	*173 972*
China, Macao	04	0	0	0	0	0	0	0	0	0	0
	61	2 668	1 898	1 890	1 604	1 418	1 500 F	1 500 F	1 500 F	1 500 F	1 500 F
	Country total	*2 668*	*1 898*	*1 890*	*1 604*	*1 418*	*1 500 F*	*1 500 F*	*1 500 F*	*1 500 F*	*1 500 F*
China,Taiwan	04	1 566	1 216	1 059	695	407	403	449	561	549	591
	21	0	25	392	206	83	30	8	127	351	618
	27	0	478	1 562	1 125	31	30	47	640	1 131	564
	31	5 908 F	6 979 F	3 627 F	3 558	4 516	3 669	2 430	5 663	6 039	10 597
	34	2 556 F	6 467 F	12 929 F	13 199	17 517	13 158	10 623	8 890	14 494	13 677
	37	-	328	713	493	373	399	-	58	31	197
	41	146 676 F	145 840 F	125 529 F	138 326	132 450	216 202	177 096	279 001	247 505	156 502
	47	9 050 F	14 958 F	18 655 F	19 249	11 394	8 934	18 173	18 811	18 970	16 118
	51	51 904 F	121 654 F	63 484 F	88 540 F	77 278	74 522	96 352	88 423	92 017	95 003
	57	16 854 F	18 314 F	16 961 F	25 964 F	32 507	41 049	22 549	22 732	23 241	14 308
	61	553 997 F	522 490 F	405 001 F	420 781 F	393 351	406 022	378 364	373 608	361 423	357 797
	67	-	-	-	26	0	0	0	55	1 165	310
	71	213 067 F	254 607 F	291 612 F	280 936 F	271 520	254 979	365 733	283 184	309 239	321 673
	77	19 059 F	15 328 F	11 038 F	5 492	5 205	5 443	6 679	10 917	8 034	10 461
	81	43 126 F	24 494 F	14 485 F	11 429	20 599	12 902	13 176	6 983	8 200	5 901
	87	184	498	142	3	252	306	89	62	1 500	882
	Country total	*1 063 947*	*1 133 676*	*967 189*	*1 010 022*	*967 483*	*1 038 048*	*1 091 768*	*1 099 715*	*1 093 889*	*1 005 199*
Cyprus	04	...	...	5	65	64	70	70	70	78	70 F
	34	-	-	-	-	2 632	13 640	16 387	37 327	65 174	73 475
	37	2 676	2 696	2 762	2 505	2 550	2 309	2 408	2 241	2 230	2 258
	77	870	960	180	830	1 730	1 920	-	-	-	-
	87	5 790	6 360	6 480	5 920	5 550	6 880	430	-	-	-
	Country total	*9 336*	*10 016*	*9 427*	*9 320*	*12 526*	*24 819*	*19 295*	*39 638*	*67 482*	*75 803*
Gaza Strip	04	...	...	...	0	0	0	0	0	0	0
	37	...	...	...	1 229	2 493	3 791	3 625	3 600 F	3 600 F	3 000 F
	Country total	*...*	*...*	*...*	*1 229*	*2 493*	*3 791*	*3 625*	*3 600 F*	*3 600 F*	*3 000 F*
Georgia	04	190	549	16	90	6	1	4	17	22	8
	34	24 775 F	13 500 F	5 000 F	1 000 F	-	-	-	-	-	-
	37	7 785	2 191	1 397	2 470	2 447	2 582	2 997	1 396	2 178	1 822
	47	5 350 F	2 000 F	1 000 F	-	-	-	-	-	-	-
	Country total	*38 100 F*	*18 240 F*	*7 413 F*	*3 560 F*	*2 453*	*2 583*	*3 001*	*1 413*	*2 200*	*1 830*
India	04	373 287	575 905	552 874	608 378	633 425	641 775	692 439	696 083	955 620	974 710
	51	1 779 782	1 744 857	1 904 799	1 847 631	1 954 910	2 006 621	1 832 799	1 911 318	1 848 678	1 896 206

D-4

Fish crustaceans, molluscs, etc	Capture production by countries or areas and fishing areas	Asia
Poissons, crustacés, mollusques, etc	Captures par pays ou zones et zones de pêche	Asie
Peces, crustáceos, moluscos, etc	Capturas por países o áreas y áreas de pesca	Asia

	Fishing area Zone de pêche Area de pesca	1992 mt	1993 mt	1994 mt	1995 mt	1996 mt	1997 mt	1998 mt	1999 mt	2000 mt	2001 mt
	57	691 033	741 950	799 934	809 231	859 606	875 052	848 254	864 749	937 998	891 684
	58	-	-	-	-	13	-	-	-	-	-
	Country total	*2 844 102*	*3 062 712*	*3 257 607*	*3 265 240*	*3 447 954*	*3 523 448*	*3 373 492*	*3 472 150*	*3 742 296*	*3 762 600*
Indonesia	04	300 896	308 648	336 083	329 790	335 696	304 258	288 666	327 627	305 212	306 560
	57	647 452	732 103	755 652	763 171	798 203	856 545	1 029 867	918 151	913 252	947 680
	71	1 942 970	2 036 789	2 214 080	2 418 184	2 419 377	2 630 437	2 646 364	2 741 141	2 851 227	2 949 590
	Country total	*2 891 318*	*3 077 540*	*3 305 815*	*3 511 145*	*3 553 276*	*3 791 240*	*3 964 897*	*3 986 919*	*4 069 691*	*4 203 830*
Iran	04	61 737	75 021	89 157	88 800	109 286	103 795	140 263	143 400	123 500	73 645
	51	243 969	246 985	218 944	252 583	242 437	238 486	226 500	243 800	260 500	262 805
	57	-	-	-	-	2	6	449	-	-	-
	Country total	*305 706*	*322 006*	*308 101*	*341 383*	*351 725*	*342 287*	*367 212*	*387 200*	*384 000*	*336 450*
Iraq	04	17 530	17 808	20 926	22 955	19 049	20 519	9 111	9 330	8 378	8 400 F
	51	543	2 133	4 221	5 253	11 688	10 783	13 463	13 093	12 389	12 400 F
	Country total	*18 073*	*19 941*	*25 147*	*28 208*	*30 737*	*31 302*	*22 574*	*22 423*	*20 767*	*20 800 F*
Israel	04	2 214	1 813	1 478	1 214	1 845	1 476	2 164	2 145	1 852	1 600 F
	37	3 148	3 232	2 473	3 577	3 159	3 557	3 999	3 641	3 966	3 400 F
	51	98	80	110	150	225	171	137	98	-	-
	Country total	*5 460*	*5 125*	*4 061*	*4 941*	*5 229*	*5 204*	*6 300*	*5 884*	*5 818*	*5 000 F*
Japan	04	96 739	91 032	92 219	91 455	93 501	85 612	78 822	71 270	70 612	61 354
	21	14 774	9 064	6 147	5 741	5 616	3 928	4 961	5 693	4 429	5 671
	27	5 293	4 915	7 082	6 595	3 745	2 638	2 907	2 273	2 315	1 906
	31	1 985	1 532	1 293	969	1 454	1 262	1 921	3 581	3 637	3 996
	34	14 431	22 773	20 358	24 307	24 766	14 934	20 294	13 487	16 218	10 973
	37	189	616	530	749	673	172	417	376	143	188
	41	112 424	151 049	127 068	97 448	92 775	147 786	99 299	175 714	138 098	82 127
	47	32 934	39 774	38 106	32 690	26 160	18 924	20 659	18 284	17 286	11 729
	48	74 275	53 510	61 423	59 037	60 546	58 798	63 309	71 318	67 188	73 523
	51	53 961	52 958	32 730	14 273	21 152	28 796	31 030	20 377	19 542	24 739
	57	5 028	6 790	11 519	40 437	27 076	21 221	22 301	24 091	19 898	17 676
	58	-	5 762	899	1 266	264	335	-	-	-	-
	61	6 594 758	6 218 515	5 628 874	5 092 787	5 191 508	5 162 694	4 546 845	4 451 587	4 206 983	3 996 910
	67	24 888	700	683	...	...	2 008	1 201	1 521	683	2 061
	71	301 927	282 242	277 415	272 490	226 315	210 396	254 848	220 024	242 406	225 024
	77	129 529	102 740	95 131	82 190	69 159	82 917	63 507	46 161	76 384	54 890
	81	198 973	140 989	113 146	95 100	77 078	72 426	74 231	58 178	43 416	56 855
	87	68 422	69 982	102 495	48 922	11 871	11 266	12 847	10 251	42 174	89 530
	88	50	-	-	-	-	-	-	-	-	-
	Country total	*7 730 580*	*7 254 943*	*6 617 118*	*5 966 456*	*5 933 659*	*5 926 113*	*5 299 399*	*5 194 186*	*4 971 412*	*4 719 152*
Jordan	04	350	350	350	350	350	350	350	350	400	350
	51	30	45	60	75	90	100	120	160	150	170
	Country total	*380*	*395*	*410*	*425*	*440*	*450*	*470*	*510*	*550*	*520*
Kazakhstan	04	65 124	56 700	46 433	48 402	44 273	31 826	25 000 F	36 170	36 620	30 654
	Country total	*65 124*	*56 700*	*46 433*	*48 402*	*44 273*	*31 826*	*25 000 F*	*36 170*	*36 620*	*30 654*
Korea D P Rp	04	35 000 F	35 000 F	20 000	20 000	20 000	20 000	20 000 F	20 000 F	20 000 F	20 000 F
	34	-	398	-	-	-	-	-	-	850	-
	61	865 000 F	835 000 F	351 961	307 083	233 125	216 462	200 000 F	190 000 F	180 000 F	180 000 F
	Country total	*900 000 F*	*870 398 F*	*371 961*	*327 083*	*253 125*	*236 462*	*220 000 F*	*210 000 F*	*200 850 F*	*200 000 F*
Korea Rep	04	25 483	12 263	10 492	9 646	8 034	6 934	6 845	6 316	7 141	5 971
	21	20 965	3 738	-	-	-	-	-	-	-	-
	27	1	-	-	-	-	-	-	-	-	-
	31	4 627	1 462	750	653	626	143	621	1 789	3 327	237
	34	9 890	15 510	23 057	19 324	19 380	21 851	18 481	25 219	23 978	26 809
	37	-	-	-	484	701	686	-	-	-	-
	41	232 692	149 776	97 665	141 635	158 911	220 586	101 165	283 455	170 429	167 817
	47	10 551	7 711	11 242	9 216	6 414	9 548	8 020	7 822	7 225	5 026
	48	519	-	146	423	441	487	1 852	1 487	5 825	5 601
	51	30 405	29 879	25 322	22 765	31 573	37 976	25 692	19 943	21 435	10 878
	57	517	-	78	321	718	512	269	4 983	5 273	4 550
	61	1 582 368	1 718 865	1 799 694	1 787 770	1 933 176	1 640 712	1 553 273	1 500 272	1 288 737	1 469 171
	67	84 018	47 264	15 875	10 967	2 480	-	311	1 330	-	-
	71	210 100	157 088	244 170	213 720	175 466	182 399	218 092	157 499	180 224	190 038
	77	43 020	33 550	36 937	37 833	37 320	44 473	59 919	49 305	51 303	53 668
	81	29 729	20 861	25 371	30 434	26 689	35 111	31 886	34 784	34 298	38 139
	87	36 163	59 225	67 092	34 724	11 784	2 629	508	25 464	23 980	10 097
	Country total	*2 321 048*	*2 257 192*	*2 357 891*	*2 319 915*	*2 413 713*	*2 204 047*	*2 026 934*	*2 119 668*	*1 823 175*	*1 988 002*
Kuwait	04	0	0	0	0	0	0	0	0	0	0
	51	7 871	8 466	7 752	8 616	8 255	7 826	7 799	6 271	6 000 F	5 846
	Country total	*7 871*	*8 466*	*7 752*	*8 616*	*8 255*	*7 826*	*7 799*	*6 271*	*6 000 F*	*5 846*
Kyrgyzstan	04	201	127	131	185	160 F	120 F	80 F	48	52	57
	Country total	*201*	*127*	*131*	*185*	*160 F*	*120 F*	*80 F*	*48*	*52*	*57*
Laos	04	19 240	19 500 F	23 800	27 370	23 000 F	18 857	19 642	30 041	29 250	30 000 F
	Country total	*19 240*	*19 500 F*	*23 800*	*27 370*	*23 000 F*	*18 857*	*19 642*	*30 041*	*29 250*	*30 000 F*
Lebanon	04	20 F	20	20	20	20	20	20	20	20	20
	37	1 700 F	2 000	2 205	4 065	4 115	3 635	3 500	3 540	3 646	3 650
	Country total	*1 720 F*	*2 020*	*2 225*	*4 085*	*4 135*	*3 655*	*3 520*	*3 560*	*3 666*	*3 670*

D-4

Fish crustaceans, molluscs, etc — Capture production by countries or areas and fishing areas — Asia
Poissons, crustacés, mollusques, etc — Captures par pays ou zones et zones de pêche — Asie
Peces, crustáceos, moluscos, etc — Capturas por países o áreas y áreas de pesca — Asia

	Fishing area / Zone de pêche / Area de pesca	1992 mt	1993 mt	1994 mt	1995 mt	1996 mt	1997 mt	1998 mt	1999 mt	2000 mt	2001 mt
Malaysia	04	1 773	1 971	2 064	3 939	3 683	3 949	4 626	3 366	3 549	3 446
	57	473 995	446 515	439 341	504 068	494 091	499 288	508 128	485 741	519 785	475 354
	71	549 521	600 835	626 245	604 368	632 598	669 685	640 965	762 661	765 911	755 933
	Country total	*1 025 289*	*1 049 321*	*1 067 650*	*1 112 375*	*1 130 372*	*1 172 922*	*1 153 719*	*1 251 768*	*1 289 245*	*1 234 733*
Maldives	04	0	0	0	0	0	0	0	0	0	0
	51	90 399	100 851	118 673	119 521	121 137	122 387	132 566	134 962	136 420	125 814
	Country total	*90 399*	*100 851*	*118 673*	*119 521*	*121 137*	*122 387*	*132 566*	*134 962*	*136 420*	*125 814*
Mongolia	04	120 F	165	184	158	221	180	311	524	425	117
	Country total	*120 F*	*165*	*184*	*158*	*221*	*180*	*311*	*524*	*425*	*117*
Myanmar	04	141 281	142 065	146 365	148 347	146 494	149 069	149 279	159 746	189 708	235 376
	57	590 263	597 637	599 876	602 885	455 294	631 226	680 838	759 664	880 018	931 492
	Country total	*731 544*	*739 702*	*746 241*	*751 232*	*601 788*	*780 295*	*830 117*	*919 410*	*1 069 726*	*1 166 868*
Nepal	04	5 895	7 418	7 340	11 230	11 230	11 230	12 000	12 752	16 700	16 700
	Country total	*5 895*	*7 418*	*7 340*	*11 230*	*11 230*	*11 230*	*12 000*	*12 752*	*16 700*	*16 700*
Oman	04	0	0	0	0	0	0	0	0	0	0
	51	109 218	105 772	118 572	139 861	121 618	118 995	106 171	108 809	120 421	126 531
	Country total	*109 218*	*105 772*	*118 572*	*139 861*	*121 618*	*118 995*	*106 171*	*108 809*	*120 421*	*126 531*
Pakistan	04	109 087	109 185	118 703	121 405	142 092	167 530	163 524	179 865	176 468	180 100
	51	431 267	499 159	418 574	404 444	395 397	422 265	433 456	474 665	437 601	426 920
	Country total	*540 354*	*608 344*	*537 277*	*525 849*	*537 489*	*589 795*	*596 980*	*654 530*	*614 069*	*607 020*
Philippines	04	229 673	210 775	204 325	186 006	177 355	159 353	146 004	146 234	151 753	135 845
	31	-	-	-	-	-	-	28	549	376	37
	34	-	-	-	-	-	-	1 189	1 854	801	29
	41	-	-	-	-	-	-	124	47	-	330
	47	-	-	-	-	-	-	3	48	-	-
	51	-	-	-	-	-	-	2 513	2 190	1 985	1 453
	57	-	-	-	-	-	-	657	1 017	219	182
	71	1 655 368	1 623 548	1 641 010	1 674 695	1 606 246	1 646 453	1 682 862	1 720 879	1 737 883	1 807 341
	Country total	*1 885 041*	*1 834 323*	*1 845 335*	*1 860 701*	*1 783 601*	*1 805 806*	*1 833 380*	*1 872 818*	*1 893 017*	*1 945 217*
Qatar	04	0	0	0	0	0	0	0	0	0	0
	51	7 845	6 994	5 086	4 271	4 739	5 032	5 279	4 207	7 142	8 606
	Country total	*7 845*	*6 994*	*5 086*	*4 271*	*4 739*	*5 032*	*5 279*	*4 207*	*7 142*	*8 606*
Saudi Arabia	04	0	0	0	0	0	0	0	0	0	0
	51	45 910	48 021	54 612	45 609	47 698	49 314	51 206	46 618	49 650	49 167
	Country total	*45 910*	*48 021*	*54 612*	*45 609*	*47 698*	*49 314*	*51 206*	*46 618*	*49 650*	*49 167*
Singapore	04	24	25	23	0	0	0	0	0	0	0
	71	9 177	9 279	11 278	10 102	9 943	9 250	7 733	6 489	5 371	3 342
	Country total	*9 201*	*9 304*	*11 301*	*10 102*	*9 943*	*9 250*	*7 733*	*6 489*	*5 371*	*3 342*
Sri Lanka	04	17 500	15 000	9 500	15 000	22 250	27 250	29 900	31 450	36 700	29 870
	57	185 168	200 400	208 900	214 171	206 695	207 849	233 430	244 630	260 710	249 770
	Country total	*202 668*	*215 400*	*218 400*	*229 171*	*228 945*	*235 099*	*263 330*	*276 080*	*297 410*	*279 640*
Syria	04	2 584	2 535	3 570	3 832	3 103	3 557	4 347	5 338	3 991	5 969
	37	1 954	2 019	1 950	1 950	2 670	2 574	2 750	2 600	2 581	2 322
	Country total	*4 538*	*4 554*	*5 520*	*5 782*	*5 773*	*6 131*	*7 097*	*7 938*	*6 572*	*8 291*
Tajikistan	04	149	253	127	100	40 F	75 F	100 F	80 F	78 F	137
	Country total	*149*	*253*	*127*	*100*	*40 F*	*75 F*	*100 F*	*80 F*	*78 F*	*137*
Thailand	04	135 239	175 140	196 397	186 665	205 903	203 671	200 715	206 540	201 100	209 977
	57	656 172	823 696	776 667	822 673	869 484	850 994	909 617	808 257	799 097	789 697
	71	2 084 045	1 928 853	2 042 132	2 021 736	1 938 574	1 848 233	1 820 022	1 937 211	1 910 976	1 881 642
	Country total	*2 875 456*	*2 927 689*	*3 015 196*	*3 031 074*	*3 013 961*	*2 902 898*	*2 930 354*	*2 952 008*	*2 911 173*	*2 881 316*
Timor-Leste	04	...	...	...	...	...	...	...	0	0	0
	57	...	...	...	...	...	...	...	408 F	362	356
	Country total	*...*	*...*	*...*	*...*	*...*	*...*	*...*	*408 F*	*362*	*356*
Turkey	04	40 495	44 801	45 067	47 976	49 600	50 460	54 500	50 190	42 824	43 323
	37	407 003	503 957	544 736	585 994	478 226	408 693	432 700	523 634	460 524	484 407
	Country total	*447 498*	*548 758*	*589 803*	*633 970*	*527 826*	*459 153*	*487 200*	*573 824*	*503 348*	*527 730*
Turkmenistan	04	31 777	16 080	15 140	9 740	9 014	8 179	7 014	9 058	12 228	12 749
	Country total	*31 777*	*16 080*	*15 140*	*9 740*	*9 014*	*8 179*	*7 014*	*9 058*	*12 228*	*12 749*
Untd Arab Em	04	0	0	0	0	0	0	0	0	0	0
	51	95 046	99 600	108 600	105 884	107 000	114 358	114 739	117 607	110 056	110 000 F
	Country total	*95 046*	*99 600*	*108 600*	*105 884*	*107 000*	*114 358*	*114 739*	*117 607*	*110 056*	*110 000 F*
Uzbekistan	04	9 346	4 358	3 095	3 611	1 494	3 075	2 799	2 871	3 387	4 070
	Country total	*9 346*	*4 358*	*3 095*	*3 611*	*1 494*	*3 075*	*2 799*	*2 871*	*3 387*	*4 070*
Viet Nam	04	138 154	146 839	79 587	94 689	164 936	177 589	138 800	169 107	170 000 F	170 000 F
	71	729 953	785 304	946 322	990 250	1 058 708	1 098 736	1 155 154	1 217 193	1 280 590	1 321 123
	Country total	*868 107*	*932 143*	*1 025 909*	*1 084 939*	*1 223 644*	*1 276 325*	*1 293 954*	*1 386 300*	*1 450 590 F*	*1 491 123 F*

D-4 Fish crustaceans, molluscs, etc / Poissons, crustacés, mollusques, etc / Peces, crustáceos, moluscos, etc — Capture production by countries or areas and fishing areas / Captures par pays ou zones et zones de pêche / Capturas por países o áreas y áreas de pesca — Asia / Asie / Asia

	Fishing area Zone de pêche Area de pesca	1992 mt	1993 mt	1994 mt	1995 mt	1996 mt	1997 mt	1998 mt	1999 mt	2000 mt	2001 mt
Yemen	51	79 547	82 356	81 885	107 970	104 955	115 600	127 620	124 385	114 751	142 198
	Country total	*79 547*	*82 356*	*81 885*	*107 970*	*104 955*	*115 600*	*127 620*	*124 385*	*114 751*	*142 198*
Total		***36 687 901***	***37 932 895***	***38 867 570***	***40 352 516***	***41 953 229***	***43 842 109***	***44 799 001***	***45 717 841***	***45 543 363***	***45 261 780***

D-5

Fish crustaceans, molluscs, etc — **Capture production by countries or areas and fishing areas** — **Europe**
Poissons, crustacés, mollusques, etc — **Captures par pays ou zones et zones de pêche** — **Europe**
Peces, crustáceos, moluscos, etc — **Capturas por países o áreas y áreas de pesca** — **Europa**

	Fishing area Zone de pêche Area de pesca	1992 mt	1993 mt	1994 mt	1995 mt	1996 mt	1997 mt	1998 mt	1999 mt	2000 mt	2001 mt
Albania	05	916	850 F	700 F	252	357	180	823	814	955	1 466
	37	1 780	1 650 F	1 400 F	1 127	1 768	833	1 860	1 931	2 365	1 844
	Country total	*2 696*	*2 500 F*	*2 100 F*	*1 379*	*2 125*	*1 013*	*2 683*	*2 745*	*3 320*	*3 310*
Andorra	05	0	0	0	0	0	0	0	0	0	0
	Country total	*0*	*0*	*0*	*0*	*0*	*0*	*0*	*0*	*0*	*0*
Austria	05	479	420	388	404	450	465	451	432	439	362
	Country total	*479*	*420*	*388*	*404*	*450*	*465*	*451*	*432*	*439*	*362*
Belarus	05	1 507	2 993	786	715	821	499	457	514	553	943
	Country total	*1 507*	*2 993*	*786*	*715*	*821*	*499*	*457*	*514*	*553*	*943*
Belgium	05	511	511	511	511	511	511	511	536	511	511
	27	36 609	35 578	33 743	35 088	30 312	29 988	30 324	29 340	29 289	29 698
	Country total	*37 120*	*36 089*	*34 254*	*35 599*	*30 823*	*30 499*	*30 835*	*29 876*	*29 800*	*30 209*
Bosnia Herzg	05	2 000 F	2 500 F	2 500 F	2 500 F	2 500 F	2 500 F	2 500 F	2 500 F	2 500 F	2 500 F
	37	0	0	0	0	0	0	0	0	0	0
	Country total	*2 000 F*	*2 500 F*	*2 500 F*	*2 500 F*	*2 500 F*	*2 500 F*	*2 500 F*	*2 500 F*	*2 500 F*	*2 500 F*
Bulgaria	05	1 611	1 675	995 F	762	1 127	1 881	2 336	2 475	861	1 650
	27	2 625	5 272	-	-	-	-	-	-	-	-
	34	2 182	-	-	-	-	-	8 189	-	-	-
	37	2 524	2 318	5 340 F	7 250	7 727	9 356	8 421	8 081	6 137	4 880
	41	14 929	4 200	-	-	-	-	-	-	-	-
	48	115	223	70	179	-	-	-	-	-	-
	Country total	*23 986*	*13 688*	*6 405 F*	*8 191*	*8 854*	*11 237*	*18 946*	*10 556*	*6 998*	*6 530*
Channel Is	05	0	0	0	0	0	0	0	0	0	0
	27	2 835	2 854 F	2 783	2 949 F	4 346	4 238	4 117	3 601	3 589	3 927
	Country total	*2 835*	*2 854 F*	*2 783*	*2 949 F*	*4 346*	*4 238*	*4 117*	*3 601*	*3 589*	*3 927*
Croatia	05	198	284	340	364	434	408	10 F	10	17	34
	37	26 530	26 342	17 137	15 901	17 799	16 627	21 928	18 890	20 701	18 056
	Country total	*26 728*	*26 626*	*17 477*	*16 265*	*18 233*	*17 035*	*21 938*	*18 900*	*20 718*	*18 090*
Czech Rep	05	...	3 185	3 955	3 929	3 524	3 321	3 952	4 190	4 654	4 646
	Country total	*...*	*3 185*	*3 955*	*3 929*	*3 524*	*3 321*	*3 952*	*4 190*	*4 654*	*4 646*
Czechoslovak	05	4 350 F	-	-	-	-	-	-	-	-	-
	Country total	*4 350 F*	-	-	-	-	-	-	-	-	-
Denmark	05	653	337	243	264	196	232	349	206	183	77
	21	3 491	897	245	447	403	421	556	235	-	93
	27	1 949 684	1 613 055	1 872 828	1 998 322	1 680 918	1 826 199	1 556 430	1 404 564	1 533 906	1 510 269
	Country total	*1 953 828*	*1 614 289*	*1 873 316*	*1 999 033*	*1 681 517*	*1 826 852*	*1 557 335*	*1 405 005*	*1 534 089*	*1 510 439*
Estonia	05	3 509	2 411	1 909	2 366	2 361	2 439	3 878	3 108	3 190	2 461
	21	33	5 397	1 186	3 242	1 898	3 240	5 533	10 834	13 685	12 402
	27	45 840	49 703	71 650	92 410	94 586	112 979	96 898	90 315	96 266	88 357
	34	39 542	56 889	17 563	5 174	7 063	4 955	12 405	7 512	5	6
	41	6 568	1 338	-	-	2 538	-	-	-	-	1 941
	47	33 127	31 447	31 372	28 836	-	-	-	-	-	-
	48	2 352	-	-	-	-	-	-	-	-	-
	77	-	-	-	-	-	-	-	10	-	-
	87	376	-	-	-	-	-	-	14	-	-
	Country total	*131 347*	*147 185*	*123 680*	*132 028*	*108 446*	*123 613*	*118 714*	*111 793*	*113 146*	*105 167*
Faeroe Is	05	0	0	0	0	0	0	0	0	0	0
	21	5 069	15 138	10 024	6 884	10 209	8 451	10 094	10 560	8 543	11 000 F
	27	243 361	231 156	227 684	280 912	294 378	320 694	353 722	347 573	445 987	513 837
	Country total	*248 430*	*246 294*	*237 708*	*287 796*	*304 587*	*329 145*	*363 816*	*358 133*	*454 530*	*524 837*
Finland	05	51 407	51 522	47 895	48 436	47 618	47 618	36 813	36 813	34 840	34 820
	27	100 346	104 772	116 374	119 048	131 459	132 567	134 868	123 747	121 661	115 276
	Country total	*151 753*	*156 294*	*164 269*	*167 484*	*179 077*	*180 185*	*171 681*	*160 560*	*156 501*	*150 096*
France	05	4 350	4 400	4 450	4 500	4 540	4 540	4 500	2 134 F	2 131 F	2 130 F
	27	360 301	379 989	383 736	384 524	353 922	386 921	363 118	394 383	423 035	429 672
	34	72 056	93 042	88 282	76 313	79 858	66 818	74 055	60 249	60 491	54 973
	37	50 734	45 601	39 710	37 967	27 813	33 012	33 220	38 948	45 540	43 059
	41	-	-	1 946	7 332	4 580	1 545	4 179	2 379	2 053	-
	47	449	564	129	82	190	38	40	13	83	16
	48	-	-	-	-	-	-	-	-	-	386
	51	95 614	93 057	99 908	95 917	82 854	68 920	48 720	80 443	84 544	69 152
	57	-	-	-	-	80	1 947	10 849	1 652	108	-
	58	1 589	826	4 211	4 173	3 673	3 681	3 879	6 286	5 770	6 806
	Country total	*585 093*	*617 479*	*622 372*	*610 808*	*557 510*	*567 422*	*542 560*	*586 487*	*623 755*	*606 194*
Germany	05	8 770	10 837	10 908	22 987	22 987	22 916	22 916	22 868	22 868	22 818
	21	5 373	343	305	0	495	449	355	569	4 920	1 359
	27	202 747	241 830	218 948	216 864	201 707	207 239	220 252	188 810	177 901	184 432
	34	-	-	-	39	11 222	28 748	23 099	26 678	-	2 673
	Country total	*216 890*	*253 010*	*230 161*	*239 890*	*236 411*	*259 352*	*266 622*	*238 925*	*205 689*	*211 282*
Gibraltar	37	0	0	0	0	0	0	0	0	0	0
	Country total	*0*	*0*	*0*	*0*	*0*	*0*	*0*	*0*	*0*	*0*

D-5

Fish crustaceans, molluscs, etc	Capture production by countries or areas and fishing areas	Europe
Poissons, crustacés, mollusques, etc	Captures par pays ou zones et zones de pêche	Europe
Peces, crustáceos, moluscos, etc	Capturas por países o áreas y áreas de pesca	Europa

	Fishing area Zone de pêche Area de pesca	1992 mt	1993 mt	1994 mt	1995 mt	1996 mt	1997 mt	1998 mt	1999 mt	2000 mt	2001 mt
Greece	05	2 370	2 960	3 452	3 606	2 903	2 601	2 818	3 280	3 433	3 181
	34	13 747	14 743	9 316	8 684	8 019	5 085	5 917	5 933	5 150	6 170
	37	136 491	141 395	168 354	139 498	138 513	149 402	99 845	109 558	90 697	85 037
	Country total	*152 608*	*159 098*	*181 122*	*151 788*	*149 435*	*157 088*	*108 580*	*118 771*	*99 280*	*94 388*
Hungary	05	8 678	7 886	8 307	7 314	7 606	7 406	7 265	7 514	7 101	6 638
	Country total	*8 678*	*7 886*	*8 307*	*7 314*	*7 606*	*7 406*	*7 265*	*7 514*	*7 101*	*6 638*
Iceland	05	886	907	698	739	608	404	416	370	176	160
	21	...	2 196	2 462	8 232	20 682	7 197	6 572	9 147	9 340	5 079
	27	1 573 796	1 712 478	1 553 802	1 603 577	2 038 854	2 197 419	1 674 623	1 726 750	1 973 006	1 975 476
	31	-	-	-	-	24	-	-	-	-	-
	47	-	-	-	-	-	924	340	-	-	-
	Country total	*1 574 682*	*1 715 581*	*1 556 962*	*1 612 548*	*2 060 168*	*2 205 944*	*1 681 951*	*1 736 267*	*1 982 522*	*1 980 715*
Ireland	05	2 991	2 929	3 604	3 761	3 806	3 804	3 976	865	961	798
	27	246 395	275 845	290 561	380 871	329 224	288 869	320 298	280 092	280 845	302 531
	34	-	-	-	-	-	-	-	-	-	52 980
	Country total	*249 386*	*278 774*	*294 165*	*384 632*	*333 030*	*292 673*	*324 274*	*280 957*	*281 806*	*356 309*
Isle of Man	05	0	0	0	0	0	0	0	0	0	0
	27	4 481	4 850	3 571	3 734	3 537	4 289	2 214	2 609	3 552	3 112
	Country total	*4 481*	*4 850*	*3 571*	*3 734*	*3 537*	*4 289*	*2 214*	*2 609*	*3 552*	*3 112*
Italy	05	9 075	9 515	9 921	10 035	6 764	6 690	4 667	5 436	4 565	5 527
	34	50 932	43 445	43 578	8 067	3 428	7 180	5 891	5 845	4 077	7 359
	37	319 469	330 092	330 704	375 970	354 551	326 260	289 545	264 619	293 405	294 312
	41	6 792	5 792	5 810	960	408	-	-	-	-	-
	47	-	-	-	109	46	-	-	-	-	-
	51	10 189	8 689	8 717	1 650	702	3 351	4 520	6 890	102	3 199
	57	-	-	-	-	-	212	1 473	-	...	-
	Country total	*396 457*	*397 533*	*398 730*	*396 791*	*365 899*	*343 693*	*306 096*	*282 790*	*302 149*	*310 397*
Latvia	05	551	553	495	514	536	544	501	610	612	581
	21	7 472	8 874	473	1 026	1 253	997	1 191	3 080	3 397	3 330
	27	62 446	64 230	64 755	63 127	73 103	86 123	78 109	78 147	80 329	76 930
	34	71 608	63 216	65 689	82 702	63 711	18 018	22 530	43 552	52 065	44 592
	41	11 637	5 029	6 693	1 825	4 041	-	-	-	-	-
	48	-	71	71	-	-	-	-	-	-	-
	87	3 023	-	-	-	-	-	-	-	-	-
	Country total	*156 737*	*141 973*	*138 176*	*149 194*	*142 644*	*105 682*	*102 331*	*125 389*	*136 403*	*125 433*
Liechtensten	05	0	0	0	0	0	0	0	0	0	0
	Country total	*0*	*0*	*0*	*0*	*0*	*0*	*0*	*0*	*0*	*0*
Lithuania	05	1 522	1 146	1 187	1 260	1 295	1 713	1 737	1 715	1 911	1 854
	21	4 657	3 904	900	980	1 585	1 785	3 107	3 370	4 047	7 596
	27	33 799	18 697	18 086	45 556	48 904	14 824	15 930	20 967	19 584	31 381
	31	61	-	-	-	-	-	-	-	-	-
	34	101 111	57 469	22 880	9 572	33 330	25 680	45 804	46 910	53 445	111 100
	41	19 262	15 171	3 275	-	3 400	-	-	-	-	-
	47	20 167	20 784	2 823	-	-	-	-	-	-	-
	51	-	-	11	-	-	-	-	-	-	-
	87	7 888	-	-	-	-	-	-	-	-	-
	Country total	*188 467*	*117 171*	*49 162*	*57 368*	*88 514*	*44 002*	*66 578*	*72 962*	*78 987*	*151 931*
Luxembourg	05	0	0	0	0	0	0	0	0	0	0
	Country total	*0*	*0*	*0*	*0*	*0*	*0*	*0*	*0*	*0*	*0*
Macedonia	05	195	164	196	208	78	130	131	135	208	128
	Country total	*195*	*164*	*196*	*208*	*78*	*130*	*131*	*135*	*208*	*128*
Malta	05	0	0	0	0	0	0	0	0	0	0
	34	-	-	1 236	3 465	8 197	-	-	-	-	-
	37	565	823	1 115	1 156	986	1 019	1 143	1 224	1 059	882
	Country total	*565*	*823*	*2 351*	*4 621*	*9 183*	*1 019*	*1 143*	*1 224*	*1 059*	*882*
Moldova Rep	05	410	630	708	709	603	569	491	309	344	387
	Country total	*410*	*630*	*708*	*709*	*603*	*569*	*491*	*309*	*344*	*387*
Monaco	37	3 F	3 F	3 F	3 F	3 F	3 F	3 F	3 F	3 F	3 F
	Country total	*3 F*	*3 F*	*3 F*	*3 F*	*3 F*	*3 F*	*3 F*	*3 F*	*3 F*	*3 F*
Netherlands	05	2 287	1 601	2 446	4 107	2 157	2 293	1 547	2 303	2 280 F	2 200 F
	27	430 683	460 155	417 607	433 985	361 126	348 537	410 604	388 371	336 329	339 040
	34	-	-	-	-	47 515	100 969	124 475	123 937	157 195	176 922
	Country total	*432 970*	*461 756*	*420 053*	*438 092*	*410 798*	*451 799*	*536 626*	*514 611*	*495 804*	*518 162*
Norway	05	580	435	432	413	338	439	507	514	578	570 F
	21	2 558	9 955	11 755	11 772	7 437	3 709	2 686	4 313	4 296	14 715
	27	2 414 894	2 380 768	2 338 555	2 505 560	2 633 495	2 852 115	2 848 489	2 622 707	2 697 358	2 672 018
	34	-	-	87	-	-	-	3 421	-	71	-
	47	...	...	...	-	-	864	1 087	-	242	-
	51	-	-	-	-	-	-	-	-	870	-
	81	12 791	23 973	15 290	6 366	7 187	5 932	5 033	-	-	-
	Country total	*2 430 823*	*2 415 131*	*2 366 119*	*2 524 111*	*2 648 457*	*2 863 059*	*2 861 223*	*2 627 534*	*2 703 415*	*2 687 303*

D-5

Fish crustaceans, molluscs, etc	Capture production by countries or areas and fishing areas	Europe
Poissons, crustacés, mollusques, etc	Captures par pays ou zones et zones de pêche	Europe
Peces, crustáceos, moluscos, etc	Capturas por países o áreas y áreas de pesca	Europa

	Fishing area Zone de pêche Area de pesca	1992 mt	1993 mt	1994 mt	1995 mt	1996 mt	1997 mt	1998 mt	1999 mt	2000 mt	2001 mt
Poland	05	20 950	31 391	27 500	24 889	22 037	13 832	13 236	13 875	17 543	17 789
	21	-	-	-	-	-	824	148	894	1 732	760
	27	104 937	103 022	116 694	130 214	157 335	187 203	126 787	131 056	143 173	159 164
	34	-	-	2 220	-	19 766	5 268	-	-	-	13 185
	41	42 510	21 395	13 641	9 205	3 549	-	93	5 224	970	756
	47	-	-	1	3 178	1 736	1 964	797	-	-	3 100
	48	8 607	13 406	7 915	9 384	20 610	19 156	15 386	18 554	20 721	14 568
	61	297 732	235 208	269 979	249 365	116 266	125 414	81 889	65 508	33 217	16 590
	67	-	-	-	-	-	-	-	-	998	4
	81	961	-	-	-	-	-	-	-	-	-
	Country total	*475 697*	*404 422*	*437 950*	*426 235*	*341 299*	*353 661*	*238 336*	*235 111*	*218 354*	*225 916*
Portugal	05	3	3	4	2	0	0	0	0	0	1
	21	36 243	35 550	30 157	12 532	9 184	8 998	9 616	16 662	13 180	15 003
	27	224 037	219 683	202 601	216 641	200 335	188 077	192 964	175 033	159 740	157 247
	31	-	-	-	-	-	-	17	22	-	-
	34	30 422	29 775	24 553	22 303	39 435	22 176	18 021	13 985	8 950	9 603
	37	320	183	428	446	274	37	54	76	96	288
	41	1 549	1 810	3 766	6 847	9 488	1 595	2 219	80	388	2 794
	47	596	798	1 556	1 482	1 130	996	830	1 797	3 732	2 742
	51	-	-	-	-	-	-	513	804	1 484	3 536
	Country total	*293 170*	*287 802*	*263 065*	*260 253*	*259 846*	*221 879*	*224 234*	*208 459*	*187 570*	*191 214*
Romania	05	9 890	8 562	10 598	9 048	6 145	4 574	4 630	5 336	4 896	5 206
	27	-	-	8 593	37 508	9 432	-	-	-	-	-
	34	57 026	1 305	...	...	...	...	...	...	...	...
	37	3 845	3 952	3 060	2 719	2 682	3 872	4 431	2 507	2 476	2 431
	Country total	*70 761*	*13 819*	*22 251*	*49 275*	*18 259*	*8 446*	*9 061*	*7 843*	*7 372*	*7 637*
Russian Fed	05	275 125	216 866	217 950	212 874	233 272	227 091	271 311	307 823	292 368	206 430
	21	37 093	26 525	9 187	10 383	6 021	1 465	2 872	5 979	27 660	32 138
	27	970 582	833 147	709 669	736 359	769 774	830 132	727 936	839 509	989 122	1 061 969
	31	-	-	-	360	20	-	-	-	-	-
	34	368 890	198 630	167 472	277 631	384 330	312 524	341 413	286 179	211 018	129 436
	37	14 044	6 384	13 888	15 540	8 745	8 917	9 760	14 340	22 294	33 444
	41	93 117	47 925	23 689	11 762	20 254	884	-	-	3 404	3 214
	47	189 058	220 550	226 712	178 288	138 583	129 014	128 280	123 453	82 283	39 850
	48	199 046	4 532	1 231	12	103	-	-	273	3 462	89
	51	17 558	10 893	19 444	-	-	-	-	-	-	221
	58	1 274	-	-	-	-	-	-	-	-	-
	61	3 226 838	2 741 954	2 256 773	2 835 602	3 096 988	3 139 799	2 970 546	2 558 646	2 341 918	2 121 532
	67	-	-	-	-	540	-	466	1 624	6	-
	71	2 150	18 937	5 840	4 981	-	-	-	-	-	-
	81	82 865	43 672	53 227	28 017	17 108	12 027	2 175	3 332	-	-
	87	32 354	-	-	-	-	-	-	-	-	-
	Country total	*5 509 994*	*4 370 015*	*3 705 082*	*4 311 809*	*4 675 738*	*4 661 853*	*4 454 759*	*4 141 158*	*3 973 535*	*3 628 323*
Slovakia	05	...	1 185	1 627	1 950	1 414	1 364	1 361	1 396	1 368	1 531
	Country total	*...*	*1 185*	*1 627*	*1 950*	*1 414*	*1 364*	*1 361*	*1 396*	*1 368*	*1 531*
Slovenia	05	293	297	339	316	289	302	269	243	226	206
	37	3 612	1 987	2 007	1 851	2 078	2 065	1 959	1 784	1 630	1 621
	Country total	*3 905*	*2 284*	*2 346*	*2 167*	*2 367*	*2 367*	*2 228*	*2 027*	*1 856*	*1 827*
Spain	05	8 796	9 215	6 284	8 869	8 710 F	8 710 F	8 710 F	8 710 F	8 710 F	8 710 F
	21	55 177	47 139	54 117	20 069	20 864	27 938	30 973	37 239	45 095	40 235
	27	341 104	360 270	346 392	395 247	393 622	476 637	470 576	439 111	392 982	432 988
	31	4 020	3 027	1 624	1 955	906	3 145	2 090	1 998	2 224	4 631
	34	308 218	317 907 F	326 916 F	334 932 F	344 222	315 145	373 415	241 272	149 224	160 932
	37	145 000 F	146 000 F	148 000 F	149 000 F	150 496	133 302	123 325	122 359	140 203	138 568
	41	91 069	63 047	72 592	96 511	93 032	65 264	84 266	95 453	84 874	108 669
	47	28 852	25 145 F	20 759 F	16 750 F	12 555	14 331	20 990	12 572	17 160	19 580
	48	-	-	-	-	-	294	199	154	264	487
	51	89 623	105 651	113 354	147 545	138 928	141 340	98 223	144 645	148 235	127 481
	57	-	-	-	-	237	379	13 379	451	168	...
	77	10	740	160	980	1 910	1 950	21 960	37 753	27 416	15 722
	87	5 895	7 818	4 326	9 428	9 153	16 478	15 169	27 745	28 933	26 817
	Country total	*1 077 764 F*	*1 085 959 F*	*1 094 524 F*	*1 181 286 F*	*1 174 635*	*1 204 913*	*1 263 275*	*1 169 462*	*1 045 488*	*1 084 820*
Svalbard Is	27	0	0	0	0	0	0	0	0	0	0
	Country total	*0*	*0*	*0*	*0*	*0*	*0*	*0*	*0*	*0*	*0*
Sweden	05	2 308	2 273	2 254	1 934	1 810	2 011	1 559	1 478	1 459	1 234
	27	305 237	339 624	384 560	402 638	369 071	355 395	409 327	349 776	337 075	310 582
	Country total	*307 545*	*341 897*	*386 814*	*404 572*	*370 881*	*357 406*	*410 886*	*351 254*	*338 534*	*311 816*
Switzerland	05	2 715	1 822	1 481	1 588	1 841	1 859	1 809	1 840	1 659	1 715
	Country total	*2 715*	*1 822*	*1 481*	*1 588*	*1 841*	*1 859*	*1 809*	*1 840*	*1 659*	*1 715*
Ukraine	05	24 801	13 189	14 786	6 847	9 468	6 215	4 898	4 728	4 429	4 343
	21	-	-	-	-	-	-	-	-	-	405
	27	9 592	7 876	13 385	3 352	550	-	-	-	-	-
	34	224 812	118 508	99 006	213 737	285 054	278 133	346 280	281 959	249 594	188 145
	37	31 792	26 394	35 213	43 570	29 266	35 987	42 981	51 495	65 507	90 840
	41	9 415	4 719	3 202	998	340	-	-	-	-	-
	47	66 356	96 420	53 356	19 759	28 405	5 701	18 345	4 675	18 096	4 283
	48	67 028	6 541	8 852	48 886	20 056	4 246	-	5 694	1 113	3 461
	51	2 629	1 491	2 633	2 970	3 480	1 570	1 626	3 356	2 015	810

D-5

Fish crustaceans, molluscs, etc	Capture production by countries or areas and fishing areas	Europe
Poissons, crustacés, mollusques, etc	Captures par pays ou zones et zones de pêche	Europe
Peces, crustáceos, moluscos, etc	Capturas por países o áreas y áreas de pesca	Europa

	Fishing area Zone de pêche Area de pesca	1992 mt	1993 mt	1994 mt	1995 mt	1996 mt	1997 mt	1998 mt	1999 mt	2000 mt	2001 mt
	58	5 958	1 890	942	5 473	1 004	1 007	997	760	-	65
	81	16 556	28 306	35 805	33 058	39 496	40 146	47 181	55 186	51 970	58 908
	Country total	*458 939*	*305 334*	*267 180*	*378 650*	*417 119*	*373 005*	*462 308*	*407 853*	*392 724*	*351 260*
UK	05	2 035	1 909	2 191	2 146	1 930	1 491	4 569	4 835	2 743	3 142
	21	169	-	49	-	129	23	294	-	-	-
	27	810 861	857 874	874 510	905 702	858 730	881 301	913 288	831 549	737 867	729 701
	34	-	-	-	-	-	-	-	-	-	31
	41	-	446	1 256	2 080	4 356	2 730	3 706	3 259	5 501	7 007
	48	17	-	12	-	-	716	1 228	1 255	1 247	1 225
	Country total	*813 082*	*860 229*	*878 018*	*909 928*	*865 145*	*886 261*	*923 085*	*840 898*	*747 358*	*741 106*
Yugoslavia	05	5 111	3 797	3 912	3 803	3 653	3 500	2 200 F	828	672	672 F
	37	223	277	260	364	377	373	410	423	424	416
	Country total	*5 334*	*4 074*	*4 172*	*4 167*	*4 030*	*3 873*	*2 610 F*	*1 251*	*1 096*	*1 088*
Total		***18 004 410***	***16 505 621***	***15 806 289***	***17 171 963***	***17 491 753***	***17 911 619***	***17 099 465***	***16 073 844***	***16 169 828***	***15 962 573***

D-6

Fish crustaceans, molluscs, etc — Capture production by countries or areas and fishing areas — Oceania
Poissons, crustacés, mollusques, etc — Captures par pays ou zones et zones de pêche — Océanie
Peces, crustáceos, moluscos, etc — Capturas por países o áreas y áreas de pesca — Oceanía

	Fishing area Zone de pêche Area de pesca	1992 mt	1993 mt	1994 mt	1995 mt	1996 mt	1997 mt	1998 mt	1999 mt	2000 mt	2001 mt
Amer Samoa	06	0	0	0	0	0	0	0	0	0	0
	77	45	27	111	152	210	431	586	518	830	3 663
	Country total	*45*	*27*	*111*	*152*	*210*	*431*	*586*	*518*	*830*	*3 663*
Australia	06	3 757	3 211	584	1 724	1 766	1 790	1 867	2 328	1 984 F	1 939 F
	41	-	-	-	-	-	-	3 594	3 711	-	-
	57	124 698 F	127 895 F	105 408 F	107 555 F	105 022 F	102 142	98 875	127 563	118 202	109 909
	58	5	-	4	-	-	1 105	2 493	5 547	2 665	2 707
	71	35 502 F	38 006 F	39 400 F	43 719 F	46 781 F	42 075	46 536	48 905	43 246	52 475
	81	68 905 F	59 617 F	57 088 F	52 466 F	47 741 F	49 719	49 969	26 900	28 534	25 652
	Country total	*232 867*	*228 729*	*202 484*	*205 464*	*201 310*	*196 831*	*203 334*	*214 954*	*194 631*	*192 682*
Christmas Is	06	0	0	0	0	0	0	0	0	0	0
	57	0	0	0	0	0	0	0	0	0	0
	Country total	*0*	*0*	*0*	*0*	*0*	*0*	*0*	*0*	*0*	*0*
Cocos Is	06	0	0	0	0	0	0	0	0	0	0
	57	0	0	0	0	0	0	0	0	0	0
	Country total	*0*	*0*	*0*	*0*	*0*	*0*	*0*	*0*	*0*	*0*
Cook Is	06	0	0	0	10	0	0	0	0	0	0
	77	967	985 F	932 F	1 080	900 F	820 F	770 F	750 F	720 F	700 F
	Country total	*967*	*985 F*	*932 F*	*1 090*	*900 F*	*820 F*	*770 F*	*750 F*	*720 F*	*700 F*
Fiji Islands	06	4 084	2 907	2 964 F	3 586	3 034	4 325	5 111	5 622	5 700 F	5 921
	71	20 521	24 910	26 000 F	25 250	21 995	23 430	23 047	31 091	34 300 F	37 051
	Country total	*24 605*	*27 817*	*28 964 F*	*28 836*	*25 029*	*27 755*	*28 158*	*36 713*	*40 000 F*	*42 972*
Fr Polynesia	06	0	0	0	0	0	53	53	53	53	53
	77	3 425	8 075	8 175	9 020	9 910	11 617	12 420	12 283	13 846	15 351
	Country total	*3 425*	*8 075*	*8 175*	*9 020*	*9 910*	*11 670*	*12 473*	*12 336*	*13 899*	*15 404*
Guam	06	0	0	0	0	0	-	6	-	-	-
	71	345	431	435	185	121	158	247	223	275	278
	Country total	*345*	*431*	*435*	*185*	*121*	*158*	*253*	*223*	*275*	*278*
Kiribati	06	0	0	0	0	0	0	0	0	0	0
	71	28 211	27 072	27 753	30 306	31 871	30 052	35 304	52 741	25 563	32 375
	Country total	*28 211*	*27 072*	*27 753*	*30 306*	*31 871*	*30 052*	*35 304*	*52 741*	*25 563*	*32 375*
Marshall Is	06	0	0	0	0	0	0	0	0	0	0
	34	-	-	-	-	2 403	-	-	-	-	-
	71	326 F	471 F	393 F	381 F	369	400 F	500 F	500 F	8 155	37 098
	Country total	*326 F*	*471 F*	*393 F*	*381 F*	*2 772*	*400 F*	*500 F*	*500 F*	*8 155*	*37 098*
Micronesia	06	5 F	4 F	4 F	5 F	5 F	5 F	5 F	5 F	5 F	5 F
	71	16 542 F	17 884 F	22 922 F	7 624 F	9 064 F	10 122 F	16 576 F	12 833 F	23 710 F	18 057 F
	Country total	*16 547 F*	*17 888 F*	*22 926 F*	*7 629 F*	*9 069 F*	*10 127 F*	*16 581 F*	*12 838 F*	*23 715 F*	*18 062 F*
Nauru	71	377	500	500	450 F	400 F	400 F	400 F	400 F	400 F	400 F
	Country total	*377*	*500*	*500*	*450 F*	*400 F*	*400 F*	*400 F*	*400 F*	*400 F*	*400 F*
NewCaledonia	06	0	0	0	0	0	0	0	0	0	0
	71	3 184	2 670	3 185	2 644	2 998	2 438	3 105	3 152	3 386	3 337 F
	Country total	*3 184*	*2 670*	*3 185*	*2 644*	*2 998*	*2 438*	*3 105*	*3 152*	*3 386*	*3 337 F*
New Zealand	06	1 252	1 160	1 100	1 115	1 079	1 025	1 307	1 152	1 089	1 076
	81	459 570	425 508	448 560	554 444	422 556	609 229	637 875	596 979	546 596	559 356
	88	...	...	...	...	-	0	56	344	874	678
	Country total	*460 822*	*426 668*	*449 660*	*555 559*	*423 635*	*610 254*	*639 238*	*598 475*	*548 559*	*561 110*
Niue	06	0	0	0	0	0	0	0	0	0	0
	77	120	120 F	150 F	150 F	200 F	200 F	200 F	200 F	200 F	200 F
	Country total	*120*	*120 F*	*150 F*	*150 F*	*200 F*	*200 F*	*200 F*	*200 F*	*200 F*	*200 F*
Norfolk Is	06	0	0	0	0	0	0	0	0	0	0
	81	0	0	0	0	0	0	0	0	0	0
	Country total	*0*	*0*	*0*	*0*	*0*	*0*	*0*	*0*	*0*	*0*
N Marianas	06	0	0	0	1	0	0	0	0	0	0
	71	140	136	176	191	225	250	235	193	189	197
	Country total	*140*	*136*	*176*	*192*	*225*	*250*	*235*	*193*	*189*	*197*
Palau	06	0	0	0	0	0	0	0	0	0	0
	71	1 918	2 066	1 820	1 956	1 990	1 751	1 777	1 800 F	2 000 F	2 000 F
	Country total	*1 918*	*2 066*	*1 820*	*1 956*	*1 990*	*1 751*	*1 777*	*1 800 F*	*2 000 F*	*2 000 F*
Papua N Guin	06	13 500 F	13 500 F	13 500 F	13 500 F	13 500 F	13 500 F	13 500 F	13 500 F	13 500 F	13 500 F
	71	12 741 F	12 241 F	12 719 F	25 577 F	24 323 F	32 685 F	64 529 F	51 931 F	82 569 F	108 919 F
	Country total	*26 241 F*	*25 741 F*	*26 219 F*	*39 077 F*	*37 823 F*	*46 185 F*	*78 029 F*	*65 431 F*	*96 069 F*	*122 419 F*
Pitcairn Is	06	0	0	0	0	0	0	0	0	0	0
	81	8	8 F	8 F	8 F	8 F	8 F	8 F	8 F	8 F	8 F
	Country total	*8*	*8 F*	*8 F*	*8 F*	*8 F*	*8 F*	*8 F*	*8 F*	*8 F*	*8 F*
Samoa	06	0	0	0	0	0	0	0	0	1	1 F
	77	1 296	673	1 108	2 519	2 727	7 041	7 547	10 204	13 003	12 965 F
	Country total	*1 296*	*673*	*1 108*	*2 519*	*2 727*	*7 041*	*7 547*	*10 204*	*13 004*	*12 966 F*

D-6

Fish crustaceans, molluscs, etc — Capture production by countries or areas and fishing areas — Oceania
Poissons, crustacés, mollusques, etc — Captures par pays ou zones et zones de pêche — Océanie
Peces, crustáceos, moluscos, etc — Capturas por países o áreas y áreas de pesca — Oceanía

	Fishing area Zone de pêche Area de pesca	1992 mt	1993 mt	1994 mt	1995 mt	1996 mt	1997 mt	1998 mt	1999 mt	2000 mt	2001 mt
Solomon Is	06	0	0	0	0	0	0	0	0	0	0
	71	45 601 F	42 970 F	46 904 F	62 019 F	50 990 F	64 005 F	61 332 F	58 428 F	24 788 F	30 075 F
	Country total	*45 601 F*	*42 970 F*	*46 904 F*	*62 019 F*	*50 990 F*	*64 005 F*	*61 332 F*	*58 428 F*	*24 788 F*	*30 075 F*
Tokelau	06	0	0	0	0	0	0	0	0	0	0
	77	191	200 F	200 F	200 F	200 F	200 F	200 F	200 F	200 F	200 F
	Country total	*191*	*200 F*	*200 F*	*200 F*	*200 F*	*200 F*	*200 F*	*200 F*	*200 F*	*200 F*
Tonga	06	1	1	0	0	0	0	0	0	0	0
	77	2 236	2 374	2 499	2 597	2 915	2 871	4 076	4 221	3 760	4 673
	Country total	*2 237*	*2 375*	*2 499*	*2 597*	*2 915*	*2 871*	*4 076*	*4 221*	*3 760*	*4 673*
Tuvalu	06	0	0	0	0	0	0	0	0	0	0
	71	499	1 460	561	399	400 F	500 F	500 F	500 F	500 F	500 F
	Country total	*499*	*1 460*	*561*	*399*	*400 F*	*500 F*	*500 F*	*500 F*	*500 F*	*500 F*
US Minor Is	77	0	0	0	0	0	0	0	0	0	0
	Country total	*0*	*0*	*0*	*0*	*0*	*0*	*0*	*0*	*0*	*0*
Vanuatu	06	0	0	0	0	0	0	0	0	0	0
	34	214	345	70	170	212	196	91	-	-	-
	71	2 427	2 470 F	3 020 F	9 236 F	11 603 F	28 856 F	41 477 F	49 971 F	37 930 F	2 400 F
	77	20 320	14 270	13 090	21 460	13 390	22 240	20 590	16 090	5 590	6 960
	87	21 830	23 430	31 010	26 780	20 410	19 570	12 750	26 980	25 740	17 330
	Country total	*44 791*	*40 515 F*	*47 190 F*	*57 646 F*	*45 615 F*	*70 862 F*	*74 908 F*	*93 041 F*	*69 260 F*	*26 690 F*
Wallis Fut I	06	0	0	0	0	0	0	0	0	0	0
	71	143	150	193	170	180	176	300	300 F	300 F	300 F
	Country total	*143*	*150*	*193*	*170*	*180*	*176*	*300*	*300 F*	*300 F*	*300 F*
Total		***894 906***	***857 747***	***872 546***	***1 008 649***	***851 498***	***1 085 385***	***1 169 814***	***1 168 126***	***1 070 411***	***1 108 309***

D-9

Fish crustaceans, molluscs, etc	Capture production by countries or areas and fishing areas	Other nei
Poissons, crustacés, mollusques, etc	Captures par pays ou zones et zones de pêche	Autres nca
Peces, crustáceos, moluscos, etc	Capturas por países o áreas y áreas de pesca	Otros nep

	Fishing area Zone de pêche Area de pesca	1992 mt	1993 mt	1994 mt	1995 mt	1996 mt	1997 mt	1998 mt	1999 mt	2000 mt	2001 mt
Other nei	27	144	223	68	189	71	212	-	-	-	-
	31	3 994	2 959	4 584	4 448	6 188	6 130	8 339	7 838	5 539	2 923
	34	48 972	37 194	44 630	164 091	55 526	49 933	50 470	74 268	50 277	64 838
	37	3 201	1 000	1 700	1 350	721	1 442	2 055	2 335	126	-
	47	122	253	3 208	19 915	10 459	163	218	-	723	-
	51	56 243	88 442	76 156	111 568	104 042	102 980	110 316	120 270	140 916	46 251
	57	4 680	3 896	5 512	6 169	12 212	7 233	15 690	10 694	12 766	...
	77	-	-	50	260	-	-	310	-	1 530	24 350
	87	570	1 910	1 880	1 210	-	3 360	870	-	3 900	38 170
	Country total	*117 926*	*135 877*	*137 788*	*309 200*	*189 219*	*171 453*	*188 268*	*215 405*	*215 777*	*176 532*
Total		***117 926***	***135 877***	***137 788***	***309 200***	***189 219***	***171 453***	***188 268***	***215 405***	***215 777***	***176 532***

E - Capture production: by countries or areas

E - Captures: par pays ou zones

E - Capturas: por países o áreas

English name Nom anglais Nombre inglés	Scientific name Nom scientifique Nombre científico	1995 mt	1996 mt	1997 mt	1998 mt	1999 mt	2000 mt	2001 mt
Algeria								
Northern pike	*Esox lucius*	-	-	-	2	...	...	...
Freshwater fishes nei	*Osteichthyes*	0	0	0	0	0	0	0
European eel	*Anguilla anguilla*	...	...	...	10	...	...	...
Common sole	*Solea solea*	313	387	333	285	234	240 F	240 F
Greater forkbeard	*Phycis blennoides*	...	...	...	...	4	4 F	4 F
European hake	*Merluccius merluccius*	1 115	1 154	1 012	1 310	1 681	1 650 F	1 650 F
Groupers, seabasses nei	*Serranidae*	...	...	...	...	6	6 F	6 F
Common pandora	*Pagellus erythrinus*	1 204	1 633	2 661	1 503	1 145	1 100 F	1 100 F
Axillary seabream	*Pagellus acarne*	332	...	...	362	...	...	...
Pargo breams nei	*Pagrus spp*	71	81	58	255	254	250 F	250 F
Bogue	*Boops boops*	2 282	1 425	1 631	3 234	3 452	3 400 F	3 400 F
Porgies, seabreams nei	*Sparidae*	-	-	-	-	7	10 F	10 F
Surmullets(=Red mullets) nei	*Mullus spp*	1 576	1 950	1 320	1 480	1 420	1 400 F	1 400 F
Sardinellas nei	*Sardinella spp*	17 887	8 150	15 741	9 992	11 393	11 000 F	11 000 F
European pilchard(=Sardine)	*Sardina pilchardus*	58 989	49 906	49 142	49 295	56 724	55 000 F	55 000 F
European anchovy	*Engraulis encrasicolus*	1 913	1 330	1 855	3 511	3 141	3 000 F	3 000 F
Clupeoids nei	*Clupeoidei*	2 012	1 182	1 261	2 279	2 634	2 500 F	2 500 F
Atlantic bonito	*Sarda sarda*	506	277	357	511	475	405	350
Plain bonito	*Orcynopsis unicolor*	92	119	224	128	216	135	145
Frigate and bullet tunas	*Auxis thazard, A.rochei*	230	237	179	299	173	225	230
Little tunny(=Atl.black skipj)	*Euthynnus alletteratus*	552	554	448	384	562	494	407
Skipjack tuna	*Katsuwonus pelamis*	-	-	-	171	43	89	77
Atlantic bluefin tuna	*Thunnus thynnus*	156	156	157	1 947	2 142	2 330	2 012
Swordfish	*Xiphias gladius*	807	807	807	825	709	816	1 081
Jack and horse mackerels nei	*Trachurus spp*	6 552	3 644	5 491	4 827	6 212	6 000 F	6 000 F
Greater amberjack	*Seriola dumerili*	87	32	19	48	75	70 F	70 F
Atlantic mackerel	*Scomber scombrus*	1 243	590	750	1 055	1 053	1 000 F	1 000 F
Barracudas nei	*Sphyraena spp*	...	...	...	...	6	6 F	6 F
Rays, stingrays, mantas nei	*Rajiformes*	124	272	120	450	207	200 F	200 F
Sharks, rays, skates, etc. nei	*Elasmobranchii*	1 000	965	415	867	854	850 F	850 F
Marine fishes nei	*Osteichthyes*	3 890	3 964	3 799	2 363	3 403	3 700 F	3 892 F
Palinurid spiny lobsters nei	*Palinurus spp*	22	9	44	141	70	70 F	70 F
Norway lobster	*Nephrops norvegicus*	129	100	89	138	96	100 F	100 F
European lobster	*Homarus gammarus*	0	0	0	0	-	-	-
Deepwater rose shrimp	*Parapenaeus longirostris*	539	918	1 433	1 639	2 147	2 100 F	2 100 F
Blue and red shrimp	*Aristeus antennatus*	1 305	1 230	1 186	1 485	880	900 F	900 F
Common prawn	*Palaemon serratus*	37	40	35	89	9	10 F	10 F
Marine crustaceans nei	*Crustacea*	73	67	71	154	83	80 F	80 F
Common cuttlefish	*Sepia officinalis*	351	567	619	492	312	300 F	300 F
Common squids nei	*Loligo spp*	101	53	138	202	240	230 F	230 F
Octopuses, etc. nei	*Octopodidae*	382	190	185	543	305	300 F	300 F
Cephalopods nei	*Cephalopoda*	-	-	-	70	29	30 F	30 F
	Country total	*105 872*	*81 989*	*91 580*	*92 346*	*102 396*	*100 000 F*	*100 000 F*
Angola								
Freshwater fishes nei	*Osteichthyes*	6 000	6 000	6 000	6 000	6 000 F	6 000 F	6 000 F
Flatfishes nei	*Pleuronectiformes*	97	195	120	928	912	592	1 928
Shallow-water Cape hake	*Merluccius capensis*	26	832	777	906	2 060	658	1 863
Sea catfishes nei	*Ariidae*	8	1	43	90	69	407	2 424
Mullets nei	*Mugilidae*	0	0	0		-	-	-
Groupers, seabasses nei	*Serranidae*	235	1 002	460	1 326	347	2 341	4 214
Bigeye grunt	*Brachydeuterus auritus*	124	146	228	262	167	191	1 176
Grunts, sweetlips nei	*Haemulidae (=Pomadasyidae)*	149	1 031	557	438	412	799	1 688
Canary drum (=Baardman)	*Umbrina canariensis*	...	...	...	...	830	...	2 202
Southern meagre(=Mulloway)	*Argyrosomus hololepidotus*	...	...	...	...	259	...	2 540
West African croakers nei	*Pseudotolithus spp*	1 035	4 044	1 606	7 095	8 936	8 005	8 725
Croakers, drums nei	*Sciaenidae*	5	-	12	70	91	29	594
Red pandora	*Pagellus bellottii*	...	...	...	...	188	...	744
Dentex nei	*Dentex spp*	7 283	2 624	1 567	8 683	8 758	13 765	16 080
Black seabream	*Spondyliosoma cantharus*	...	...	...	93	129	212	494
Pargo breams nei	*Pagrus spp*	39	95	716	1 077	183	1 840	4 484
Sand steenbras	*Lithognathus mormyrus*	279	214	176	1 072	726	999	1 735
Picarels nei	*Spicara spp*	9	7	0	55	...	13	-
Goatfishes, red mullets nei	*Mullidae*	0	-	0	-	-	-	-
Threadfins, tasselfishes nei	*Polynemidae*	259	370	265	520	1 378	872	1 834
John dory	*Zeus faber*	72	315	307	1 339	1 668	1 582	1 303
Largehead hairtail	*Trichiurus lepturus*	...	...	...	...	28	-	-
Gurnards, searobins nei	*Triglidae*	0	0	0	-	-	-	-
Sardinellas nei	*Sardinella spp*	34 211	17 484	21 014	45 483	57 579	108 211	57 896
Atlantic bonito	*Sarda sarda*	9	39	32	-	2	118	1 157
Frigate and bullet tunas	*Auxis thazard, A.rochei*	21	29	12	31	2	38	206
Little tunny(=Atl.black skipj)	*Euthynnus alletteratus*	117	235	75	406	118	132	231
Skipjack tuna	*Katsuwonus pelamis*	3	15	52	2	32	14	687
Yellowfin tuna	*Thunnus albacares*	216	78	70	115	170	35	...
Swordfish	*Xiphias gladius*	0	0	0	0	0	0	...
Bluefish	*Pomatomus saltatrix*	-	-	-	-	-	-	167
Cunene horse mackerel	*Trachurus trecae*	25 308	19 764	35 797	39 739	47 719	53 243	40 898
Crevalle jack	*Caranx hippos*	42	39	16	89	302	1 835	674
Blue butterfish	*Stromateus fiatola*	0	0	-	-	-	-	-
Barracudas nei	*Sphyraena spp*	...	...	...	...	119	-	-
Raja rays nei	*Raja spp*	215	21	16	750	1 399	593	1 430
Sharks, rays, skates, etc. nei	*Elasmobranchii*	755	379	90	376	...	157	3 354
Marine fishes nei	*Osteichthyes*	44 922	80 633	74 578	43 038	32 185	31 674	80 740
Marine crabs nei	*Brachyura*	...	...	...	...	602	800	836
Deepwater rose shrimp	*Parapenaeus longirostris*	682	1 367	1 175	1 900	1 101	1 600	1 660
Striped red shrimp	*Aristeus varidens*	538	814	543	876	820	1 154	1 200

English name Nom anglais Nombre inglés	Scientific name Nom scientifique Nombre científico	1995 mt	1996 mt	1997 mt	1998 mt	1999 mt	2000 mt	2001 mt
Natantian decapods nei	*Natantia*	-	-	-	-	0	-	-
Marine crustaceans nei	*Crustacea*	-	-	-	-	-	-	1 062
Various squids nei	*Loliginidae, Ommastrephidae*	122	42	...	390	463	442	292
Octopuses, etc. nei	*Octopodidae*	-	-	-	-	45	-	-
	Country total	*122 781*	*137 815*	*146 304*	*163 149*	*175 799*	*238 351*	*252 518*
Benin								
Tilapias nei	*Oreochromis (=Tilapia) spp*	11 713	11 500 F	11 500 F	11 500 F	11 648	9 600 F	10 900 F
African lungfishes	*Protopterus spp*	121	110 F	105 F	100 F	100 F	100 F	110 F
Bonytongues nei	*Heterotis spp*	742	700 F	650 F	600 F	600 F	500 F	550 F
Black catfishes nei	*Chrysichthys spp*	1 289	1 300 F	1 400 F	1 500 F	1 542	1 300 F	1 500 F
Torpedo-shaped catfishes nei	*Clarias spp*	2 325	2 100 F	1 900 F	1 700 F	1 537	1 300 F	1 500 F
Upsidedown catfishes	*Synodontis spp*	125	120 F	110 F	110 F	110 F	100 F	110 F
Freshwater gobies nei	*Gobiidae*	914	900 F	850 F	800 F	800 F	700 F	800 F
Freshwater fishes nei	*Osteichthyes*	9 552	6 563 F	5 096 F	3 838 F	3 733 F	3 000 F	3 440 F
West African ilisha	*Ilisha africana*	528 F	602 F	822 F	781 F	671	450 F	923
Diadromous clupeoids nei	*Clupeoidei*	260	250 F	230 F	210 F	210 F	200 F	230 F
Flatfishes nei	*Pleuronectiformes*	23 F	30 F	37 F	45 F	52	30 F	14
West African ladyfish	*Elops lacerta*	...	...	...	...	...	...	15
Sea catfishes nei	*Ariidae*	22 F	20 F	17 F	15 F	12	8 F	2
Mullets nei	*Mugilidae*	637	1 000 F	1 500 F	2 000 F	2 201	1 802 F	2 055 F
Groupers nei	*Epinephelus spp*	22 F	25 F	25 F	23 F	20	14 F	16
Snappers nei	*Lutjanus spp*	58 F	66 F	90 F	86 F	92	60 F	32
Grunts, sweetlips nei	*Haemulidae (=Pomadasyidae)*	200 F	228 F	311 F	296 F	269	180 F	222
Boe drum	*Pteroscion peli*	...	...	...	...	61	40 F	115
Pandoras nei	*Pagellus spp*	14 F	16 F	22 F	20 F	15	10 F	...
Dentex nei	*Dentex spp*	83 F	95 F	100 F	90 F	83	60 F	20
Black seabream	*Spondyliosoma cantharus*	20 F	30 F	45 F	48 F	56	40 F	93
African sicklefish	*Drepane africana*	15 F	15 F	18 F	14 F	12	8 F	6
Threadfins, tasselfishes nei	*Polynemidae*	550 F	500 F	500 F	450 F	358	250 F	324
Spadefishes nei	*Ephippidae*	2 F	3 F	5 F	4 F	4	3 F	2
Puffers nei	*Tetraodontidae*	...	...	...	...	...	...	2
Largehead hairtail	*Trichiurus lepturus*	250 F	300 F	350 F	400 F	543	370 F	584
Gurnards, searobins nei	*Triglidae*	...	...	...	...	11	7 F	-
Sardinellas nei	*Sardinella spp*	1 150 F	1 100 F	1 200 F	1 100 F	1 072	750 F	1 434
Bonga shad	*Ethmalosa fimbriata*	1 963	1 900 F	1 850 F	1 800 F	1 806	1 505 F	1 708 F
European anchovy	*Engraulis encrasicolus*	200 F	250 F	350 F	400 F	478	330 F	852
Clupeoids nei	*Clupeoidei*	...	...	...	...	13	10 F	-
Plain bonito	*Orcynopsis unicolor*	1 F	1 F	3	1	1	-	-
West African Spanish mackerel	*Scomberomorus tritor*	188	188 F	362	511	205	205 F	203
Little tunny(=Atl.black skipj)	*Euthynnus alletteratus*	58	58 F	196	83	69	69 F	69
Skipjack tuna	*Katsuwonus pelamis*	2 F	2 F	7	3	2	2 F	...
Yellowfin tuna	*Thunnus albacares*	1 F	1 F	3	1	1	1 F	1
Bigeye tuna	*Thunnus obesus*	9 F	9 F	30	13	11	...	...
Atlantic sailfish	*Istiophorus albicans*	20	19	6	4	5	5 F	12
Atlantic blue marlin	*Makaira nigricans*	5 F	5 F	5	5	5	5 F	...
Swordfish	*Xiphias gladius*	24	24 F	10	0	3	...	41
Needlefishes, etc. nei	*Belonidae*	45 F	40 F	35 F	30 F	27	20 F	38
Flyingfishes nei	*Exocoetidae*	32 F	28 F	25 F	20 F	15	10 F	96
Bluefish	*Pomatomus saltatrix*	1 050 F	1 000 F	950 F	900 F	875	600 F	697
African moonfish	*Selene dorsalis*	24 F	27 F	37 F	35 F	57	40 F	121
Atlantic bumper	*Chloroscombrus chrysurus*	-	-	-	-	-	-	853
Carangids nei	*Carangidae*	600 F	700 F	800 F	900 F	1 027	700 F	475
Common dolphinfish	*Coryphaena hippurus*	...	...	...	...	4	3 F	48
Chub mackerel	*Scomber japonicus*	300 F	280 F	300 F	250 F	205	140 F	314
Barracudas nei	*Sphyraena spp*	328 F	330 F	400 F	350 F	297	200 F	262
Shortfin mako	*Isurus oxyrinchus*	...	...	...	...	4	3 F	1
Rays, stingrays, mantas nei	*Rajiformes*	54 F	62 F	70 F	60 F	47	30 F	38
Sharks, rays, skates, etc. nei	*Elasmobranchii*	120 F	100 F	100 F	80 F	59	40 F	87
Marine fishes nei	*Osteichthyes*	889 F	1 785 F	3 628 F	3 294 F	1 677	1 150 F	370
River prawns nei	*Macrobrachium spp*	188	150 F	130 F	120 F	120 F	100 F	110 F
Freshwater crustaceans nei	*Crustacea*	4 503	4 600 F	4 700 F	4 800 F	4 924	4 100 F	4 600 F
Marine crabs nei	*Brachyura*	40 F	40 F	50 F	45 F	32	20 F	9
Palinurid spiny lobsters nei	*Palinurus spp*	3 F	3 F	5 F	4 F	3	2 F	2
Penaeus shrimps nei	*Penaeus spp*	3 117	3 000 F	2 850 F	2 700 F	2 574	2 100 F	2 400 F
Natantian decapods nei	*Natantia*	0	0	0	0	31	20 F	3
Cuttlefish,bobtail squids nei	*Sepiidae, Sepiolidae*	...	...	...	...	18	12 F	1
Marine turtles nei	*Testudinata*	...	...	...	...	29	20 F	5
	Country total	*44 379*	*42 175*	*43 785*	*42 139*	*40 436*	*32 324*	*38 415*
Botswana								
Tilapias nei	*Oreochromis (=Tilapia) spp*	100 F	48	80	88	93	92	88
Freshwater fishes nei	*Osteichthyes*	100 F	33	80	103	64	74	30
	Country total	*200 F*	*81*	*160*	*191*	*157*	*166*	*118*
Br Ind Oc Tr								
Freshwater fishes nei	*Osteichthyes*	0	0	0	0	0	0	0
Marine fishes nei	*Osteichthyes*	0	0	0	0	0	0	0
	Country total	*0*	*0*	*0*	*0*	*0*	*0*	*0*
Burkina Faso								
Freshwater fishes nei	*Osteichthyes*	8 000	8 000	8 000	8 335	7 600	8 500	8 500 F
	Country total	*8 000*	*8 000*	*8 000*	*8 335*	*7 600*	*8 500*	*8 500 F*

E-1

Fish crustaceans, molluscs, etc	Capture production by countries or areas and species	Africa
Poissons, crustacés, mollusques, etc	Captures par pays ou zones et espèces	Afrique
Peces, crustáceos, moluscos, etc	Capturas por países o áreas y especies	Africa

English name Nom anglais Nombre inglés	Scientific name Nom scientifique Nombre científico	1995 mt	1996 mt	1997 mt	1998 mt	1999 mt	2000 mt	2001 mt
Burundi								
Nile tilapia	*Oreochromis niloticus*	50	50	50	50	50	120	120
Dagaas	*Stolothrissa, Limnothrissa spp*	18 163	1 508	17 868	8 646	7 030	10 839	6 138
Freshwater perches nei	*Lates spp*	2 888	1 205	2 378	3 925	1 523	3 070	1 657
Freshwater fishes nei	*Osteichthyes*	...	278	...	805	596	3 286	1 049
	Country total	*21 101*	*3 041*	*20 296*	*13 426*	*9 199*	*17 315*	*8 964*
Cameroon								
Freshwater fishes nei	*Osteichthyes*	30 000 F	35 000 F	40 000 F	45 000 F	50 000 F	55 000	52 500 F
Lefteye flounders nei	*Bothidae*	0	0	0	0	0	0	-
Tonguefishes	*Cynoglossidae*	753	748	700 F	594	455	488	439
Sea catfishes nei	*Ariidae*	976	886	640 F	370 F	520 F	455	410
Groupers nei	*Epinephelus spp*	2	0	0	1	0	1	1
Snappers nei	*Lutjanus spp*	510	511	400 F	320 F	337 F	337	293
Sompat grunt	*Pomadasys jubelini*	534	533	400 F	337 F	345 F	304	48
Cassava croaker	*Pseudotolithus senegalensis*	2 686	2 540	2 600 F	2 900 F	3 200 F	1 972	2 679
Bobo croaker	*Pseudotolithus elongatus*	4 436	4 371	4 400 F	4 400 F	1 900 F	3 400	3 218
Croakers, drums nei	*Sciaenidae*	3 679	3 539	3 500 F	3 500 F	3 500 F	3 260	2 399
Porgies, seabreams nei	*Sparidae*	0	2	2 F	5 F	5 F	4	34
African sicklefish	*Drepane africana*	111	72	75 F	78	94	73	70
Giant African threadfin	*Polydactylus quadrifilis*	76	54	40 F	12	34	43	170
Lesser African threadfin	*Galeoides decadactylus*	949	810	500 F	13	150	477	462
Royal threadfin	*Pentanemus quinquarius*	610	584	1 000 F	2 368	1 778	1 363	3 276
Conger eels, etc. nei	*Congridae*	5	6	6 F	5 F	5 F	8	7
Largehead hairtail	*Trichiurus lepturus*	1	4	10 F	16	1	6	59
Sardinellas nei	*Sardinella spp*	24 002	24 002	23 500 F	23 000 F	23 500 F	21 611	21 645
Bonga shad	*Ethmalosa fimbriata*	24 000	24 000	23 500 F	23 000 F	23 500 F	21 609	21 645
Tuna-like fishes nei	*Scombroidei*	1	0	0	0	0	1	0
Carangids nei	*Carangidae*	6	5	5 F	2	2	4	5
Barracudas nei	*Sphyraena spp*	27	37	30 F	5	28	23	293
Rays, stingrays, mantas nei	*Rajiformes*	160	186	170 F	161	211	150	130
Sharks, rays, skates, etc. nei	*Elasmobranchii*	59	48	50 F	55	86	67	146
Marine fishes nei	*Osteichthyes*	-	-	-	-	-	954	907
Marine crabs nei	*Brachyura*	32	18	20 F	20 F	20 F	27	0
Tropical spiny lobsters nei	*Panulirus spp*	0	0	0	1	1	0	0
Penaeus shrimps nei	*Penaeus spp*	514	442	450 F	635	326	471	168
Marine crustaceans nei	*Crustacea*	-	-	-	-	-	-	26
Cuttlefish,bobtail squids nei	*Sepiidae, Sepiolidae*	2	2	2 F	2 F	2 F	1	1
Marine molluscs nei	*Mollusca*	0	0	0	0	0	0	0
	Country total	*94 131 F*	*98 400 F*	*102 000 F*	*106 800 F*	*110 000 F*	*112 109*	*111 031*
Cape Verde								
Freshwater fishes nei	*Osteichthyes*	0	0	0	0	0	0	0
Demersal percomorphs nei	*Perciformes*	...	...	1 450	1 150	1 079	1 314	1 307
Wahoo	*Acanthocybium solandri*	408	503	603	429	587	487	578
Frigate and bullet tunas	*Auxis thazard, A.rochei*	13	6	22	191	154	81	171
Little tunny(=Atl.black skipj)	*Euthynnus alletteratus*	72	63	86	110	776	491	178
Skipjack tuna	*Katsuwonus pelamis*	1 314	470	591	684	962	789	794
Yellowfin tuna	*Thunnus albacares*	1 781	1 448	1 721	1 418	1 663	1 851	1 684
Bigeye tuna	*Thunnus obesus*	66	16	10	1	1	2	-
Tuna-like fishes nei	*Scombroidei*	-	-	245	-	-	-	-
Chub mackerel	*Scomber japonicus*	...	...	...	182	...	...	...
Pelagic percomorphs nei	*Perciformes*	...	...	4 414	4 899	4 463	4 823	4 288
Sharks, rays, skates, etc. nei	*Elasmobranchii*	1	...	...	...	...	...	...
Marine fishes nei	*Osteichthyes*	4 780	6 612	538	333	640	719	633
Tropical spiny lobsters nei	*Panulirus spp*	20	12	8	9	12	10	7
Palinurid spiny lobsters nei	*Palinurus spp*	40	25	17	18	23	19	13
Marine molluscs nei	*Mollusca*	0	0	0	0	0	0	-
Marine turtles nei	*Testudinata*	0	0	0	0	-	-	-
	Country total	*8 495*	*9 155*	*9 705*	*9 424*	*10 360*	*10 586*	*9 653*
Cent Afr Rep								
Freshwater fishes nei	*Osteichthyes*	13 750 F	14 000 F	14 250 F	14 500 F	15 000 F	15 000 F	15 000 F
	Country total	*13 750 F*	*14 000 F*	*14 250 F*	*14 500 F*	*15 000*	*15 000 F*	*15 000 F*
Chad								
Freshwater fishes nei	*Osteichthyes*	90 000	100 000	85 000	84 000	84 000 F	84 000 F	84 000 F
	Country total	*90 000*	*100 000*	*85 000*	*84 000*	*84 000 F*	*84 000 F*	*84 000 F*
Comoros								
Freshwater fishes nei	*Osteichthyes*	0	0	0	0	0	0	0
Sardinellas nei	*Sardinella spp*	1 000 F	1 000 F	1 000 F	1 000 F	950 F	1 050 F	1 000
Anchovies, etc. nei	*Engraulidae*	900 F	900 F	900 F	900 F	850 F	950 F	850
Seerfishes nei	*Scomberomorus spp*	271	270	260	260	250	270 F	250
Kawakawa	*Euthynnus affinis*	180	170	160	160	150	170 F	160
Skipjack tuna	*Katsuwonus pelamis*	2 185	2 150	2 070	2 070	2 000	2 200 F	2 000
Yellowfin tuna	*Thunnus albacares*	5 609	5 520	5 310	5 310	5 200	5 600 F	5 200
Bigeye tuna	*Thunnus obesus*	32	30	30	30	30	30 F	30
Indo-Pacific sailfish	*Istiophorus platypterus*	250	250	240	240	200	250 F	200
Marlins,sailfishes,etc. nei	*Istiophoridae*	136	130	120	120	100	130 F	120
Tuna-like fishes nei	*Scombroidei*	513	510	490	490	450	520 F	450
Carangids nei	*Carangidae*	500 F	490 F	470 F	470 F	450 F	500 F	500
Indian mackerels nei	*Rastrelliger spp*	250 F	250 F	240 F	240 F	230 F	250 F	220
Marine fishes nei	*Osteichthyes*	1 159 F	1 010 F	1 190 F	1 190 F	1 120 F	1 260 F	1 180

English name Nom anglais Nombre inglés	Scientific name Nom scientifique Nombre científico	1995 mt	1996 mt	1997 mt	1998 mt	1999 mt	2000 mt	2001 mt
Natantian decapods nei	*Natantia*	15 F	20 F	20 F	20 F	20 F	20 F	20
Marine molluscs nei	*Mollusca*	0	0	0	0	0	0	-
	Country total	*13 000 F*	*12 700 F*	*12 500 F*	*12 500 F*	*12 000*	*13 200*	*12 180*
Congo Dem R								
Freshwater fishes nei	*Osteichthyes*	154 751	159 037	158 367	174 087	204 503	204 503 F	204 503 F
Tonguefishes	*Cynoglossidae*	82 F	84 F	81 F	83 F	80 F	80 F	80 F
Bigeye grunt	*Brachydeuterus auritus*	388 F	398 F	385 F	396 F	400 F	400 F	400 F
Conger eels, etc. nei	*Congridae*	71 F	73 F	71 F	73 F	70 F	70 F	70 F
Sardinellas nei	*Sardinella spp*	1 551 F	1 590 F	1 538 F	1 582 F	1 595 F	1 595 F	1 595 F
Jack and horse mackerels nei	*Trachurus spp*	1 397 F	1 432 F	1 386 F	1 426 F	1 400 F	1 400 F	1 400 F
Marine fishes nei	*Osteichthyes*	387 F	396 F	383 F	394 F	400 F	400 F	400 F
	Country total	*158 627*	*163 010*	*162 211*	*178 041*	*208 448*	*208 448 F*	*208 448 F*
Congo Rep								
Freshwater fishes nei	*Osteichthyes*	26 811	25 873	18 987	25 455	25 455 F	26 000 F	24 500 F
Common sole	*Solea solea*	230	200 F	200 F	100 F	100 F	...	50 F
Tonguefishes	*Cynoglossidae*	300 F	270 F	240 F	200 F	180 F	158	150 F
Sea catfishes nei	*Ariidae*	230 F	200 F	170 F	130 F	110 F	73	70 F
Groupers nei	*Epinephelus spp*	5	5 F	4 F	3 F	3 F	2	2 F
Snappers nei	*Lutjanus spp*	50 F	40 F	30 F	20 F	10 F	1	1 F
Sompat grunt	*Pomadasys jubelini*	5 F	5 F	5 F	5 F	5 F	...	4 F
Bigeye grunt	*Brachydeuterus auritus*	126	125 F	115 F	100 F	100 F	95	90 F
Grunts, sweetlips nei	*Haemulidae (=Pomadasyidae)*	15	12 F	9 F	6 F	3 F	...	-
Boe drum	*Pteroscion peli*	637	630 F	620 F	600 F	600 F	602	600 F
Bobo croaker	*Pseudotolithus elongatus*	3 F	3 F	3 F	3 F	3 F	...	2 F
West African croakers nei	*Pseudotolithus spp*	581	500 F	450 F	400 F	350 F	298	300 F
Croakers, drums nei	*Sciaenidae*	15 F	10 F	10 F	5 F	5 F	0	0
Red pandora	*Pagellus bellottii*	7	5 F	4 F	3 F	2 F	...	2 F
Dentex nei	*Dentex spp*	220 F	180 F	140 F	100 F	60 F	22	20 F
African sicklefish	*Drepane africana*	15	15 F	15 F	15 F	20 F	24	20 F
Lesser African threadfin	*Galeoides decadactylus*	350 F	300 F	250 F	190 F	150 F	63	60 F
Royal threadfin	*Pentanemus quinquarius*	200 F	200 F	200 F	180 F	190 F	189	180 F
Conger eels, etc. nei	*Congridae*	250 F	210 F	170 F	120 F	90 F	52	50 F
Bearded brotula	*Brotula barbata*	20 F	25 F	30 F	30 F	40 F	46	40 F
Hairtails, scabbardfishes nei	*Trichiuridae*	150 F	120 F	90 F	60 F	30 F	-	-
Shortspine African angler	*Lophius vaillanti*	...	...	...	...	...	7	5 F
Sardinellas nei	*Sardinella spp*	11 600 F	12 000 F	11 600 F	11 000 F	11 200 F	12 000 F	11 500 F
Bonga shad	*Ethmalosa fimbriata*	20 F	20 F	15 F	10 F	10 F	...	10 F
Clupeoids nei	*Clupeoidei*	1 684	1 200 F	900 F	600 F	300 F	1	10 F
Skipjack tuna	*Katsuwonus pelamis*	7	6	6	6	6	6	...
Yellowfin tuna	*Thunnus albacares*	13	12	12	12	12	12	12
Bigeye tuna	*Thunnus obesus*	9	8	8	8	8	8	8
Jack and horse mackerels nei	*Trachurus spp*	64	60 F	50 F	40 F	30 F	23	20 F
Jacks, crevalles nei	*Caranx spp*	15 F	12 F	9 F	5 F	3 F	0	0
African moonfish	*Selene dorsalis*	50 F	55 F	60 F	60 F	70 F	72	70 F
Blue butterfish	*Stromateus fiatola*	...	...	...	...	...	1	1 F
Barracudas nei	*Sphyraena spp*	0	0	0	0	0	1	1 F
Dogfish sharks, etc. nei	*Squaliformes*	220 F	180 F	140 F	100 F	60 F	12	10 F
Rays, stingrays, mantas nei	*Rajiformes*	160 F	135 F	110 F	85 F	60 F	33	30 F
Marine fishes nei	*Osteichthyes*	1 383 F	2 256 F	3 100 F	2 850 F	4 009 F	3 606 F	3 765 F
Marine crabs nei	*Brachyura*	5 F	10 F	5 F	20 F	30 F	44	10 F
Palinurid spiny lobsters nei	*Palinurus spp*	3 F	4 F	3 F	7 F	9 F	10	4 F
Natantian decapods nei	*Natantia*	322	584	320 F	420 F	374	529	400 F
Cuttlefish,bobtail squids nei	*Sepiidae, Sepiolidae*	1 F	3 F	2 F	7 F	9 F	10	3 F
	Country total	*45 776*	*45 473*	*38 082*	*42 955 F*	*43 696 F*	*44 000 F*	*42 000 F*
Côte dIvoire								
Freshwater fishes nei	*Osteichthyes*	10 835	11 162 F	11 732 F	12 301 F	10 556 F	10 475	10 500
West African ilisha	*Ilisha africana*	288	491	534	452	314	152	50
Tonguefishes	*Cynoglossidae*	208	177	139	217	217	211	197
Bonefish	*Albula vulpes*	20 F	43	1 430	1 617	29	21	13
Sea catfishes nei	*Ariidae*	19	16	21	25	16	20	13
White grouper	*Epinephelus aeneus*	1 F	1 F	1 F	1 F	...	2	3
Bigeyes nei	*Priacanthus spp*	15 F	20 F	20 F	25 F	29	12	1
Sompat grunt	*Pomadasys jubelini*	235 F	247	143	227	236	231	148
Bigeye grunt	*Brachydeuterus auritus*	4 848	4 632	4 732	1 617	3 964	5 175	1 689
Meagre	*Argyrosomus regius*	15 F	10 F	10 F	5 F	3	1	1
Cassava croaker	*Pseudotolithus senegalensis*	...	5 F	10 F	20 F	30	39	39
Bobo croaker	*Pseudotolithus elongatus*	0	5	3	4	5	...	4
West African croakers nei	*Pseudotolithus spp*	653	551	466	545	378	356	407
Pandoras nei	*Pagellus spp*	361	328	346	1 093	1 137	913	697
Porgies, seabreams nei	*Sparidae*	1	1 F	1 F	1 F	1	3	...
West African goatfish	*Pseudupeneus prayensis*	53 F	68	44	38	61	44	40
African sicklefish	*Drepane africana*	35	23	18	28	21	35	20
Lesser African threadfin	*Galeoides decadactylus*	374	268	173	278	196	224	168
Conger eels, etc. nei	*Congridae*	20 F	22	26	32	12	12	17
Bearded brotula	*Brotula barbata*	298	192	158	342	513	280	156
Tilefishes nei	*Branchiostegidae*	14 F	20 F	30 F	40 F	45	44	15
Largehead hairtail	*Trichiurus lepturus*	231 F	180	265	492	518	321	200
Round sardinella	*Sardinella aurita*	12 098	18 491	10 002	10 884	12 206	19 358	18 864
Madeiran sardinella	*Sardinella maderensis*	2 273	1 500 F	1 000 F	500 F	175	426	716
Bonga shad	*Ethmalosa fimbriata*	10 000 F	9 500 F	9 500 F	11 000 F	11 006 F	11 009 F	10 512 F
Frigate and bullet tunas	*Auxis thazard, A.rochei*	...	5 269	4 458	4 502	5 772	6 768	...
Little tunny(=Atl.black skipj)	*Euthynnus alletteratus*	253	2 337	1 880	1 818	2 352	2 789	...
Yellowfin tuna	*Thunnus albacares*	-	...	2	...	...	...	...

English name Nom anglais Nombre inglés	Scientific name Nom scientifique Nombre científico	1995 mt	1996 mt	1997 mt	1998 mt	1999 mt	2000 mt	2001 mt
Atlantic sailfish	*Istiophorus albicans*	66	91	65	35	80	45	47
Atlantic blue marlin	*Makaira nigricans*	177	157	222	182	275	206	196
Atlantic white marlin	*Tetrapturus albidus*	-	1	2	1	5	1	2
Swordfish	*Xiphias gladius*	19	26	18	25	26	20	19
Needlefishes, etc. nei	*Belonidae*	1 F	1 F	1 F	1 F	...	...	3
Jacks, crevalles nei	*Caranx spp*	128	69	51	282	290	144	119
Leerfish	*Lichia amia*	-	-	-	-	20	16	3
Atlantic bumper	*Chloroscombrus chrysurus*	1 188	1 116	1 302	1 107	1 120	1 374	2 364
Carangids nei	*Carangidae*	200	169	1 400	1 267	1 025	509	500 F
Chub mackerel	*Scomber japonicus*	679	313	1 355	1 340	2 494	259	110
Blue butterfish	*Stromateus fiatola*	6 F	10 F	15 F	20 F	22	45	27
Barracudas nei	*Sphyraena spp*	287 F	366	151	195	184	160	75
Mako sharks	*Isurus spp*	12	...	92	38	...	...	...
Silky shark	*Carcharhinus falciformis*	18	...	2	...	...	...	...
Hammerhead sharks, etc. nei	*Sphyrnidae*	69	...	190	125	...	...	...
Rays, stingrays, mantas nei	*Rajiformes*	...	218	146	202	227	241	168
Sharks, rays, skates, etc. nei	*Elasmobranchii*	159 F	70	71	42	38	521	19
Marine fishes nei	*Osteichthyes*	23 132	10 290 F	11 378 F	16 101 F	18 239 F	12 053	25 100
Freshwater crustaceans nei	*Crustacea*	500	400 F	300 F	200 F	100 F	27	30 F
Marine crabs nei	*Brachyura*	0	0	0	0	1	2	2 F
Palinurid spiny lobsters nei	*Palinurus spp*	0	2 F	4 F	5 F	6	10	2
Penaeus shrimps nei	*Penaeus spp*	400	310 F	260 F	300 F	340 F	851	1
Cuttlefish,bobtail squids nei	*Sepiidae, Sepiolidae*	-	-	-	-	81	291	140
Octopuses, etc. nei	*Octopodidae*	-	-	-	-	-	-	88
Marine molluscs nei	*Mollusca*	-	-	-	-	-	26	-
Marine turtles nei	*Testudinata*	...	...	...	...	...	50	71
	Country total	*70 189*	*69 168*	*64 169*	*69 572*	*74 365*	*75 772*	*73 556*
Djibouti								
Freshwater fishes nei	*Osteichthyes*	0	0	0	0	0	0	0
Mullets nei	*Mugilidae*	0	0	0	0	0	0	0
Groupers nei	*Epinephelus spp*	100 F	100 F	100 F	100 F	100 F	100 F	100 F
Snappers, jobfishes nei	*Lutjanidae*	80 F	80 F	80 F	80 F	80 F	80 F	80 F
Porgies, seabreams nei	*Sparidae*	40 F	40 F	40 F	40 F	40 F	40 F	40 F
Seerfishes nei	*Scomberomorus spp*	67	65	61	60	60 F	60 F	60 F
Tuna-like fishes nei	*Scombroidei*	15	15	14	15	15 F	15 F	15 F
Carangids nei	*Carangidae*	20 F	20 F	20 F	20 F	20 F	20 F	20 F
Barracudas nei	*Sphyraena spp*	18 F	20 F	20 F	20 F	20 F	20 F	20 F
Marine fishes nei	*Osteichthyes*	10 F	10 F	15 F	15 F	15 F	15 F	15 F
Tropical spiny lobsters nei	*Panulirus spp*	0	0	0	0	0	0	0
Various squids nei	*Loliginidae, Ommastrephidae*	0	0	0	0	0	0	0
	Country total	*350 F*	*350 F*	*350 F*	*350 F*	*350 F*	*350 F*	*350 F*
Egypt								
Grass carp(=White amur)	*Ctenopharyngodon idellus*	10 000	15 343	16 553	4 233	707	12 826	17 918
Cyprinids nei	*Cyprinidae*	1 059	1 261	1 386	2 117	4 515	1 232	5 040
Nile tilapia	*Oreochromis niloticus*	122 207	125 307	130 992	128 446	112 811	131 276	145 291
Mudfish	*Clarias anguillaris*	19 608	20 370	22 433	21 618	21 511	31 506	39 535
Upsidedown catfishes	*Synodontis spp*	...	...	...	...	...	2 678	3 937
Nile perch	*Lates niloticus*	2 495	2 730	2 551	3 255	2 572	3 278	5 328
Freshwater fishes nei	*Osteichthyes*	47 525	45 498	45 989	64 964	58 697	34 196	37 282
European eel	*Anguilla anguilla*	798	537	585	501	709	2 064	1 979
Diadromous clupeoids nei	*Clupeoidei*	106	40	23	148	60	84	28
Common sole	*Solea solea*	473	751	1 309	1 034	1 728	3 502	3 175
Lizardfishes nei	*Synodontidae*	5 689	5 458	6 192	8 935	8 375	11 808	8 978
Mullets nei	*Mugilidae*	18 174	14 799	16 609	18 012	20 847	20 458	26 533
Groupers nei	*Epinephelus spp*	1 108	1 657	1 467	1 627	3 158	4 747	4 699
Spotted seabass	*Dicentrarchus punctatus*	174	426	154	171	444	196	256
European seabass	*Dicentrarchus labrax*	429	727	453	559	662	626	800
Seabasses nei	*Dicentrarchus spp*	2 706	2 397	2 233	2 738	815	4 075	3 789
Snappers, jobfishes nei	*Lutjanidae*	5 053	4 044	5 165	8 784	4 265	8 830	5 014
Threadfin breams nei	*Nemipterus spp*	...	...	...	...	...	2 081	4 974
Meagre	*Argyrosomus regius*	713	1 076	743	893	1 056	776	1 038
Emperors(=Scavengers) nei	*Lethrinidae*	429	425	574	577	1 040	1 147	2 696
Sargo breams nei	*Diplodus spp*	346	590	382	390	841	812	793
Saddled seabream	*Oblada melanura*	...	...	33	67	47	158	70
Red porgy	*Pagrus pagrus*	1 021	1 825	1 425	1 230	2 984	1 847	575
Gilthead seabream	*Sparus aurata*	1 359	1 228	1 087	1 225	1 955	2 478	2 312
Bogue	*Boops boops*	2 173	2 609	2 499	1 956	...	1 450	1 222
Salema	*Sarpa salpa*	0	0	0	-	-	-	-
Porgies, seabreams nei	*Sparidae*	4 012	2 428	2 175	1 282	182	4 731	3 545
Surmullets(=Red mullets) nei	*Mullus spp*	1 512	2 314	2 533	1 681	2 727	1 611	1 717
Goatfishes, red mullets nei	*Mullidae*	1 077	716	744	439	876	914	2 590
Parrotfishes nei	*Scaridae*	...	...	227	283	918	...	...
Spinefeet(=Rabbitfishes) nei	*Siganus spp*	431	878	501	554	680	666	2 306
Largehead hairtail	*Trichiurus lepturus*	564	914	679	774	723	811	1 107
Grey gurnard	*Eutrigla gurnardus*	1 070	1 264	1 158	1 067	1 159	1 078	778
Sardinellas nei	*Sardinella spp*	11 757	15 100	16 260	28 893	44 946	25 394	24 380
Red-eye round herring	*Etrumeus teres*	-	-	-	-	3 538	...	...
Atlantic bonito	*Sarda sarda*	697	985	725	724	1 442	1 128	1 072
Narrow-barred Spanish mackerel	*Scomberomorus commerson*	1 297	8 880	8 503	9 933	7 442	14 879	16 352
Kawakawa	*Euthynnus affinis*	138	318	755	841	326	344	209
Tuna-like fishes nei	*Scombroidei*	530	1 071	594	576	2 004	1 676	778
Needlefishes nei	*Tylosurus spp*	123	48	32	17	28	11	28
Silversides(=Sand smelts) nei	*Atherinidae*	3 837	6 642	5 295	5 708	4 275	4 914	8 644
Bluefish	*Pomatomus saltatrix*	174	307	147	56	153	326	468
Jacks, crevalles nei	*Caranx spp*	433	716	441	402	326	525	1 351

English name Nom anglais Nombre inglés	Scientific name Nom scientifique Nombre científico	1995 mt	1996 mt	1997 mt	1998 mt	1999 mt	2000 mt	2001 mt
Chub mackerel	*Scomber japonicus*	1 926	2 042	2 392	810	378	3 561	2 747
Indian mackerel	*Rastrelliger kanagurta*	1 363	1 151	1 914	652	1 004	430	1 442
Barracudas nei	*Sphyraena spp*	1 743	1 101	1 722	1 764	1 162	2 390	1 570
Sharks, rays, skates, etc. nei	*Elasmobranchii*	1 309	1 242	1 809	1 346	1 565	1 441	2 406
Marine fishes nei	*Osteichthyes*	10 939	12 856	21 140	18 538	33 680	12 340	11 544
Freshwater prawns, shrimps nei	*Palaemonidae*	3 169	2 286	2 387	2 886	1 388	2 411	2 450
Freshwater crustaceans nei	*Crustacea*	950	557	838	918	1 237	2 473	2 192
Marine crabs nei	*Brachyura*	11 075	981	1 067	1 398	4 701	3 866	1 904
Natantian decapods nei	*Natantia*	4 734	3 808	5 490	5 507	8 273	7 063	5 372
Freshwater molluscs nei	*Mollusca*	488	530	610	625	598	916	845
Cuttlefish,bobtail squids nei	*Sepiidae, Sepiolidae*	1 657	1 764	1 784	1 389	4 974	2 546	2 919
Marine molluscs nei	*Mollusca*	140	233	0	198	0	1 718	4 534
Sea cucumbers nei	*Holothurioidea*	-	-	-	-	-	20	139
	Country total	*310 790*	*320 230*	*342 759*	*362 741*	*380 504*	*384 314*	*428 651*
Eq Guinea								
Freshwater fishes nei	*Osteichthyes*	450	900	850	970	1 101 F	1 076	1 000 F
Flatfishes nei	*Pleuronectiformes*	190 F	420 F	530 F	325	380 F	40 F	40 F
Gadiformes nei	*Gadiformes*	200 F	450 F	570 F	603	710 F	80 F	80 F
Demersal percomorphs nei	*Perciformes*	140 F	310 F	390 F	255	300 F	30 F	30 F
Clupeoids nei	*Clupeoidei*	100 F	220 F	280 F	850	1 000 F	2 000	1 900 F
Tuna-like fishes nei	*Scombroidei*	340	216	570 F	392	460 F	50 F	50 F
Chub mackerel	*Scomber japonicus*	150 F	330 F	420 F	400	470 F	50 F	50 F
Pelagic percomorphs nei	*Perciformes*	100 F	220 F	280 F	170	200 F	20 F	10 F
Sharks, rays, skates, etc. nei	*Elasmobranchii*	220 F	490 F	620 F	779	910 F	100 F	100 F
Marine fishes nei	*Osteichthyes*	96 F	789 F	695 F	700	820 F	117 F	169 F
Marine crustaceans nei	*Crustacea*	230 F	510 F	650 F	430	500 F	50 F	50 F
Marine molluscs nei	*Mollusca*	80 F	180 F	230 F	122	140 F	20 F	20 F
Marine turtles nei	*Testudinata*	10 F	5 F	5 F	9	10 F	1 F	1 F
	Country total	*2 306*	*5 040*	*6 090*	*6 005*	*7 001*	*3 634*	*3 500 F*
Eritrea								
Freshwater fishes nei	*Osteichthyes*	0	0	0	0	0	0	0
Milkfish	*Chanos chanos*	2	0	2	3	1	3	1
Lefteye flounders nei	*Bothidae*	-	-	0	1	23	119	125
Indian halibut	*Psettodes erumei*	4	8	1	0	-	-	-
Lizardfishes nei	*Synodontidae*	166	3	5	0	1 905	3 177	2 574
Sea catfishes nei	*Ariidae*	60	9	0	149	205	851	436
Mullets nei	*Mugilidae*	...	3	4	2	4	3	3
Groupers nei	*Epinephelus spp*	...	1	56	50	...	48	...
Groupers, seabasses nei	*Serranidae*	83	140	39	67	257	378	270
Bigeyes nei	*Priacanthus spp*	5	0	-	-	0	-	-
Snappers nei	*Lutjanus spp*	205	281	219	294	365	149	280
Snappers, jobfishes nei	*Lutjanidae*	0	43	59	69	95	392	29
Threadfin breams nei	*Nemipterus spp*	206	4	15	0	1 674	1 757	1 115
Ponyfishes(=Slipmouths) nei	*Leiognathidae*	-	-	-	-	19	38	137
Silver grunt	*Pomadasys argenteus*	...	0	0	3	...	472	...
Grunts, sweetlips nei	*Haemulidae (=Pomadasyidae)*	469	367	13	45	594	536	435
Emperors(=Scavengers) nei	*Lethrinidae*	1 010	767	90	104	371	443	346
Porgies, seabreams nei	*Sparidae*	22	12	0	0	15	48	11
Goatfishes	*Upeneus spp*	7	2	1	1	21	19	9
Parrotfishes nei	*Scaridae*	15	9	0	0	3	2	2
Angelfishes nei	*Pomacanthidae*	-	0	0	0	3	0	0
Batfishes	*Platax spp*	-	-	0	1	7	1	0
Spinefeet(=Rabbitfishes) nei	*Siganus spp*	-	0	0	1	1	1	0
Flatheads nei	*Platycephalidae*	14	0	0	0	21	5	11
Triggerfishes, durgons nei	*Balistidae*	1	0	0	0	3	1	0
Sardinellas nei	*Sardinella spp*	...	...	...	0	...	2	-
Narrow-barred Spanish mackerel	*Scomberomorus commerson*	49	191	200	210	250	217	280
Kawakawa	*Euthynnus affinis*	...	4	6	...	...	0	36
Longtail tuna	*Thunnus tonggol*	...	...	6	22	...	0	-
Indo-Pacific sailfish	*Istiophorus platypterus*	-	-	1	0	0	1	-
Tuna-like fishes nei	*Scombroidei*	1	30	44	111	64	42	0
Cobia	*Rachycentron canadum*	10	38	2	6	8	31	31
Queenfishes	*Scomberoides spp*	...	8	4	...	44	243	...
Carangids nei	*Carangidae*	697	818	13	184	386	2 026	1 172
Indian mackerel	*Rastrelliger kanagurta*	75	58	2	0	20	197	...
Barracudas nei	*Sphyraena spp*	109	185	21	57	150	684	497
Requiem sharks nei	*Carcharhinidae*	6	15	13	17	38	130	109
Guitarfishes, etc. nei	*Rhinobatidae*	1	0	0	0	-	0	-
Sharks, rays, skates, etc. nei	*Elasmobranchii*	...	...	6	7	6	...	...
Marine fishes nei	*Osteichthyes*	314	202	215	211	246	8	0
Portunus swimcrabs nei	*Portunus spp*	-	-	-	-	-	-	1
Spiny lobsters nei	*Palinuridae*	...	1	1	2	1	0	0
Penaeus shrimps nei	*Penaeus spp*	13	2	0	9	75	519	790
Cuttlefish,bobtail squids nei	*Sepiidae, Sepiolidae*	...	51	0	3	11	31	35
Various squids nei	*Loliginidae, Ommastrephidae*	15	0	0	0	5	38	85
	Country total	*3 559*	*3 252*	*1 038*	*1 629*	*6 891*	*12 612*	*8 820*
Ethiopia								
Common carp	*Cyprinus carpio*	74	94	27	62	71	75	74
Crucian carp	*Carassius carassius*	47	61	191	88	101	110	108
Rhinofishes nei	*Labeo spp*	...	1 994	2 007	3 168	3 621	3 451	3 387
Cyprinids nei	*Cyprinidae*	362	387	639	768	878	860	843
Tilapias nei	*Oreochromis (=Tilapia) spp*	3 703	5 066	6 076	7 952	9 088	7 000	6 870
Naked catfishes	*Bagrus spp*	...	...	...	94	107	120	119

English name Nom anglais Nombre inglés	Scientific name Nom scientifique Nombre científico	1995 mt	1996 mt	1997 mt	1998 mt	1999 mt	2000 mt	2001 mt
North African catfish	*Clarias gariepinus*	643	812	1 200	1 677	1 917	4 000	3 926
Nile perch	*Lates niloticus*	908	270	230	191	75	65	63
Freshwater fishes nei	*Osteichthyes*	588	86	-	-	-	-	-
	Country total	*6 325*	*8 770*	*10 370*	*14 000*	*15 858*	*15 681*	*15 390*
Fr South Tr								
Freshwater fishes nei	*Osteichthyes*	0	0	0	0	0	0	0
Marine fishes nei	*Osteichthyes*	75 F	75 F	75 F	75 F	80 F	80 F	80 F
St.Paul rock lobster	*Jasus paulensis*	439	357	295	308	345	192	183
Octopuses, etc. nei	*Octopodidae*	5 F	5 F	5 F	5 F	...	...	...
	Country total	*519 F*	*437 F*	*375 F*	*388 F*	*425 F*	*272 F*	*263 F*
Gabon								
Tilapias nei	*Oreochromis (=Tilapia) spp*	2 600 F	3 620	4 150	3 519	4 178	4 221	3 423
Black catfishes nei	*Chrysichthys spp*	2 000 F	2 124	1 131	1 950	380	887	797
Freshwater fishes nei	*Osteichthyes*	3 018	4 251	4 636	3 500	5 814	5 726	5 922
Tonguefishes	*Cynoglossidae*	209	310	411	492	386	620	538
Flatfishes nei	*Pleuronectiformes*	24	75	93	65	84	242	...
Bonefish	*Albula vulpes*	...	25	21	2	6	10	...
Sea catfishes nei	*Ariidae*	415	2 780	1 565	1 290	1 351	921	1 140
Mullets nei	*Mugilidae*	51	271	254	233	382	365	1 061
Groupers nei	*Epinephelus spp*	264	250	94	190	136	105	40
Snappers nei	*Lutjanus spp*	463	611	480	907	941	795	765
Bigeye grunt	*Brachydeuterus auritus*	200	...	...	...	...	...	...
Grunts, sweetlips nei	*Haemulidae (=Pomadasyidae)*	897	780	853	1 728	1 062	819	1 050
Bobo croaker	*Pseudotolithus elongatus*	986	946	769	1 584	2 079	1 784	1 555
West African croakers nei	*Pseudotolithus spp*	2 095	4 020	3 068	4 653	3 367	2 344	...
Pandoras nei	*Pagellus spp*	538	200	191	301	127	376	274
Dentex nei	*Dentex spp*	776	820	522	1 047	423	371	498
Surmullets(=Red mullets) nei	*Mullus spp*	224	100	573	76	57	448	311
African sicklefish	*Drepane africana*	330	270	222	281	290	86	71
Lesser African threadfin	*Galeoides decadactylus*	1 876	3 805	3 174	3 484	2 808	2 516	3 201
Royal threadfin	*Pentanemus quinquarius*	...	...	...	0	292	207	389
Sardinellas nei	*Sardinella spp*	1 174	1 897	878	746	128	1 414	1 083
Bonga shad	*Ethmalosa fimbriata*	11 787	13 046	14 695	19 284	17 408	14 788	12 733
Seerfishes nei	*Scomberomorus spp*	145	79	-	85	-	-	-
Little tunny(=Atl.black skipj)	*Euthynnus alletteratus*	-	182	-	18	159	301	213
Skipjack tuna	*Katsuwonus pelamis*	51	26	...	59	76	21	101
Yellowfin tuna	*Thunnus albacares*	218	225	225	295	225	162	270
Bigeye tuna	*Thunnus obesus*	10	-	-	-	184	150	121
Marlins,sailfishes,etc. nei	*Istiophoridae*	0	523	7	-	-	-	1
Scads nei	*Decapterus spp*	106	33	20	18	76	21	...
Jacks, crevalles nei	*Caranx spp*	47	594	404	91	4	29	31
Mackerels nei	*Scombridae*	145	64	65	114	158	304	201
Barracudas nei	*Sphyraena spp*	420	1 055	883	1 238	975	976	1 636
Rays, stingrays, mantas nei	*Rajiformes*	33	172	173	90	197	141	88
Sharks, rays, skates, etc. nei	*Elasmobranchii*	22	1 267	626	1 933	1 338	659	375
Marine fishes nei	*Osteichthyes*	8 256	278	1 325	1 154	3 949	2 896	...
Freshwater crustaceans nei	*Crustacea*	30	36	44	50	6	4	9
Marine crabs nei	*Brachyura*	65	120	145	158	283	289	142
Palinurid spiny lobsters nei	*Palinurus spp*	42	103	33	85	50	32	57
Southern pink shrimp	*Penaeus notialis*	867	950	828	2 272	93	62	...
Penaeus shrimps nei	*Penaeus spp*	-	-	556	190	987	1 682	1 892
Deepwater rose shrimp	*Parapenaeus longirostris*	5	6	257	56	76	356	55
Cuttlefish,bobtail squids nei	*Sepiidae, Sepiolidae*	33	157	52	186	92	281	166
Various squids nei	*Loliginidae, Ommastrephidae*	3	5	2	0	1	8	10
Octopuses, etc. nei	*Octopodidae*	-	-	-	5	91	0	0
Marine turtles nei	*Testudinata*	12	37	159	180	424	51	238
	Country total	*40 437 F*	*46 113*	*43 584*	*53 609*	*51 143*	*47 470*	*40 457*
Gambia								
Tilapias nei	*Oreochromis (=Tilapia) spp*	1 144	1 248	1 135	1 073	1 070 F	1 062 F	1 077 F
Freshwater fishes nei	*Osteichthyes*	1 450	1 450	1 450	1 450	1 450 F	1 450 F	1 450 F
European eel	*Anguilla anguilla*	-	-	26	0	0	-	-
Tonguefishes	*Cynoglossidae*	859	541	307	441	450	725	2 262
West African ladyfish	*Elops lacerta*	...	...	...	...	...	...	12
Sea catfishes nei	*Ariidae*	846	158	1 234	517	540	749	950
Mullets nei	*Mugilidae*	279	475	278	66	70	123	69
Groupers nei	*Epinephelus spp*	118	62	53	30	30	49	63
Snappers nei	*Lutjanus spp*	127	27	150	86	90	90	126
Rubberlip grunt	*Plectorhinchus mediterraneus*	102	156	101	160	170	107	124
Sompat grunt	*Pomadasys jubelini*	307	498	350	220	230	276	423
Meagre	*Argyrosomus regius*	50	125	72	8	10	22	33
Law croaker	*Pseudotolithus brachygnathus*	449	355	225	264	270	454	856
Cassava croaker	*Pseudotolithus senegalensis*	230	182	160	109	120	58	400
Bobo croaker	*Pseudotolithus elongatus*	328	242	214	163	170	138	120
West African croakers nei	*Pseudotolithus spp*	414	473	412	327	340	629	504
Pandoras nei	*Pagellus spp*	123	76	9	12	20	12	...
Porgies, seabreams nei	*Sparidae*	446	129	...	...	-	-	-
African sicklefish	*Drepane africana*	118	151	104	23	30	60	85
Parrotfishes nei	*Scaridae*	...	...	...	...	...	13	1
Giant African threadfin	*Polydactylus quadrifilis*	94	179	110	83	90	169	141
Lesser African threadfin	*Galeoides decadactylus*	88	140	146	57	60	87	116
Puffers nei	*Tetraodontidae*	125	28	100	-	-	4	33
Triggerfishes, durgons nei	*Balistidae*	3	51	9	2	2	0	3
Hairtails, scabbardfishes nei	*Trichiuridae*	0	10	12	5	5	-	28

English name Nom anglais Nombre inglés	Scientific name Nom scientifique Nombre científico	1995 mt	1996 mt	1997 mt	1998 mt	1999 mt	2000 mt	2001 mt
Sardinellas nei	*Sardinella spp*	1	11	86	64	70	10	81
Bonga shad	*Ethmalosa fimbriata*	13 897	22 648	21 523	21 952	22 750	20 508	18 516
Clupeoids nei	*Clupeoidei*	53	27	8	12	10		-
Yellowfin tuna	*Thunnus albacares*	14	...	...	1	1	5	1
Bluefish	*Pomatomus saltatrix*	23	31	8	75	80	35	70
Cobia	*Rachycentron canadum*	33	0	6	0	0	-	-
Jack and horse mackerels nei	*Trachurus spp*	336	312	133	119	130	175	246
Jacks, crevalles nei	*Caranx spp*	73	174	124	147	160	137	288
Pompanos nei	*Trachinotus spp*	8	20	7	1	2	1	22
Barracudas nei	*Sphyraena spp*	170	355	144	116	120	284	631
Sharks, rays, skates, etc. nei	*Elasmobranchii*	498	415	3 223	606	630	720	3 982
Marine fishes nei	*Osteichthyes*	27	18	33	-	-	-	6
Marine crabs nei	*Brachyura*	...	...	...	...	...	6	2
Palinurid spiny lobsters nei	*Palinurus spp*	26	84	37	86	80	130	75
Southern pink shrimp	*Penaeus notialis*	367	339	...	399	400	301	211
Gastropods nei	*Gastropoda*	194	227	128	230	250	5	20
Cuttlefish,bobtail squids nei	*Sepiidae, Sepiolidae*	325	184	137	98	100	422	1 499
Various squids nei	*Loliginidae, Ommastrephidae*	1	...	...	...	...	...	...
Octopuses, etc. nei	*Octopodidae*	6	...	...	...	...	...	1
	Country total	*23 752*	*31 601*	*32 254*	*29 002*	*30 000*	*29 016*	*34 527*
Ghana								
Freshwater fishes nei	*Osteichthyes*	60 000	73 580	70 000	74 500	74 500	74 500	74 500
West African ilisha	*Ilisha africana*	3 051	7 343	4 305	3 632	3 262	3 341	3 600
Tonguefishes	*Cynoglossidae*	407	295	339	347	284	394	227
Benguela hake	*Merluccius polli*	0	1	0	34	3	0	-
Sea catfishes nei	*Ariidae*	0	2	24	6	0	1	0
Groupers nei	*Epinephelus spp*	306	426	478	1 361	181	94	138
Snappers nei	*Lutjanus spp*	626	328	181	294	163	491	447
Bigeye grunt	*Brachydeuterus auritus*	14 807	13 552	19 816	12 059	12 724	10 032	13 845
Grunts, sweetlips nei	*Haemulidae (=Pomadasyidae)*	523	1 191	477	411	476	150	847
West African croakers nei	*Pseudotolithus spp*	698	1 128	1 995	962	937	739	1 070
Red pandora	*Pagellus bellottii*	4 505	7 541	7 933	10 029	13 265	2 916	3 206
Angolan dentex	*Dentex angolensis*	591	489	490	1 416	1 767	504	838
Congo dentex	*Dentex congoensis*	43	102	119	392	1 272	350	571
Dentex nei	*Dentex spp*	1 372	1 584	1 084	1 278	1 874	892	675
Pargo breams nei	*Pagrus spp*	1 546	1 448	1 347	1 682	...	...	2 189
Porgies, seabreams nei	*Sparidae*	157	220	129	316	3 986	1 238	532
West African goatfish	*Pseudupeneus prayensis*	65	586	1 035	553	247	39	285
African sicklefish	*Drepane africana*	34	6	46	4	24	2	8
Lesser African threadfin	*Galeoides decadactylus*	1 969	3 146	1 477	774	586	1 947	2 892
Triggerfishes, durgons nei	*Balistidae*	2	17	-	1	-	2	2
Largehead hairtail	*Trichiurus lepturus*	1 823	2 543	2 866	2 047	1 267	1 664	1 849
Gurnards, searobins nei	*Triglidae*	8	0	76	...	53	0	37
Demersal percomorphs nei	*Perciformes*	0	0	0	0	-	-	-
Round sardinella	*Sardinella aurita*	67 835	118 408	49 394	55 965	57 170	102 043	67 321
Madeiran sardinella	*Sardinella maderensis*	13 211	13 619	14 183	15 468	12 105	14 970	15 906
Sardinellas nei	*Sardinella spp*	14 074	20 070	35 985	33 264	19 664	18 358	8 345
Bonga shad	*Ethmalosa fimbriata*	1 073	1 197	9 762	158	766	963	282
European anchovy	*Engraulis encrasicolus*	65 497	98 341	82 724	44 644	32 107	83 501	68 175
Clupeoids nei	*Clupeoidei*	236	477	1 060	162	132	3 789	6 144
Little tunny(=Atl.black skipj)	*Euthynnus alletteratus*	513	113	2 025	359	306	707	730
Skipjack tuna	*Katsuwonus pelamis*	18 607	19 602	27 667	34 150	43 460	29 950	43 340
Yellowfin tuna	*Thunnus albacares*	9 268	12 160	16 504	17 807	28 328	17 010	30 642
Bigeye tuna	*Thunnus obesus*	5 517	5 805	7 431	13 252	11 460	5 586	14 095
Atlantic sailfish	*Istiophorus albicans*	353	303	196	351	305	275	...
Atlantic blue marlin	*Makaira nigricans*	472	422	491	447	624	639	...
Atlantic white marlin	*Tetrapturus albidus*	2	1	3	7	6	8	21
Swordfish	*Xiphias gladius*	103	140	44	106	121	117	...
Chilean jack mackerel	*Trachurus murphyi*	-	-	-	-	-	2 472	1 157
Cunene horse mackerel	*Trachurus trecae*	4 215	7 714	6 962	11 690	9 964	572	1 540
Jack and horse mackerels nei	*Trachurus spp*	5 289	3 201	10 512	4 892	1 904	2 183	2 109
Scads nei	*Decapterus spp*	1 760	1 462	1 654	2 989	3	263	2 466
Crevalle jack	*Caranx hippos*	4 422	2 884	2 207	2 891	3 906	79	2 525
False scad	*Caranx rhonchus*	3 483	3 301	3 337	2 753	2 275	3 800	147
Jacks, crevalles nei	*Caranx spp*	57	13	11	1	-	1	-
African moonfish	*Selene dorsalis*	976	907	712	381	469	738	1 100
Atlantic bumper	*Chloroscombrus chrysurus*	2 482	2 466	3 847	7 264	7 611	6 861	6 418
Chub mackerel	*Scomber japonicus*	12 473	15 590	19 883	30 160	15 482	29 613	15 193
Barracudas nei	*Sphyraena spp*	1 372	1 763	1 684	872	2 591	759	1 156
Rays, stingrays, mantas nei	*Rajiformes*	338	261	185	172	869	231	814
Sharks, rays, skates, etc. nei	*Elasmobranchii*	1 115	1 106	709	1 764	3 998	1 670	2 092
Marine fishes nei	*Osteichthyes*	19 946	25 128	27 896	43 399	115 933	22 248	40 993
Marine crabs nei	*Brachyura*	218	399	576	271	145	74	155
Tropical spiny lobsters nei	*Panulirus spp*	230	134	203	65	...	39	342
Natantian decapods nei	*Natantia*	2 228	1 554	1 602	1 448	87	1 446	1 361
Cuttlefish,bobtail squids nei	*Sepiidae, Sepiolidae*	2 891	2 967	3 355	3 288	4 095	1 805	2 866
Octopuses, etc. nei	*Octopodidae*	55	137	67	103	19	4	94
	Country total	*352 844*	*477 173*	*447 088*	*442 641*	*492 776*	*452 070*	*445 287*
Guinea								
Freshwater fishes nei	*Osteichthyes*	3 100	2 780	3 600	4 000	4 000	4 000	4 000 F
Flatfishes nei	*Pleuronectiformes*	350	254	256	179	148	1 032	1 000 F
Hakes nei	*Merluccius spp*	-	-	-	-	-	-	3
Sea catfishes nei	*Ariidae*	4 381	3 589	2 462	2 610	2 378	4 593	4 500 F
Mullets nei	*Mugilidae*	1 800	1 901	1 244	1 534	1 600	1 894	1 850 F
Grunts, sweetlips nei	*Haemulidae (=Pomadasyidae)*	341	193	198	185	1 598	430	420 F

English name Nom anglais Nombre inglés	Scientific name Nom scientifique Nombre científico	1995 mt	1996 mt	1997 mt	1998 mt	1999 mt	2000 mt	2001 mt
Bobo croaker	*Pseudotolithus elongatus*	3 685	3 386	2 751	2 781	2 142	4 015	4 000 F
West African croakers nei	*Pseudotolithus spp*	4 115	2 856	2 206	2 570	1 796	3 423	3 400 F
Atlantic emperor	*Lethrinus atlanticus*	137	231	334	380	...	...	...
Porgies, seabreams nei	*Sparidae*	4 709	3 814	3 649	3 019	3 265	1 838	1 800 F
African sicklefish	*Drepane africana*	89	324	149	...	...	...	...
Lesser African threadfin	*Galeoides decadactylus*	431	338	204	43	88	123	120 F
Royal threadfin	*Pentanemus quinquarius*	1 020	435	233	...	...	...	...
Largehead hairtail	*Trichiurus lepturus*	324	401	409	400	400	418	400 F
Sardinellas nei	*Sardinella spp*	5 281	6 107	4 176	6 292	8 909	13 288	13 000 F
Bonga shad	*Ethmalosa fimbriata*	23 596	26 051	29 529	27 852	33 780	29 015	28 500 F
Jack and horse mackerels nei	*Trachurus spp*	4 781	3 576	1 546	7 109	508	6 084	6 000 F
Carangids nei	*Carangidae*	253	311	257	392	326	764	750 F
Chub mackerel	*Scomber japonicus*	...	2 043	1 006	849	...	...	...
Barracudas nei	*Sphyraena spp*	452	309	431	400	300	226	220 F
Sharks, rays, skates, etc. nei	*Elasmobranchii*	726	506	505	700	800	969	950 F
Marine fishes nei	*Osteichthyes*	8 001	3 695	7 089	7 631	23 580	18 646	18 043 F
Penaeus shrimps nei	*Penaeus spp*	97	196	98	216	396	526	701
Cuttlefish,bobtail squids nei	*Sepiidae, Sepiolidae*	61	41	95	386	181	124	236
Various squids nei	*Loliginidae, Ommastrephidae*	16	1	2	53	27	9	10 F
Octopuses, etc. nei	*Octopodidae*	114	22	12	183	1 092	96	97
	Country total	*67 860*	*63 360*	*62 441*	*69 764*	*87 314*	*91 513*	*90 000 F*
GuineaBissau								
Freshwater fishes nei	*Osteichthyes*	250 F	250 F	250 F	200 F	200 F	200 F	200 F
Flatfishes nei	*Pleuronectiformes*	87	64	70 F	60 F	50 F	50 F	50 F
Sea catfishes nei	*Ariidae*	211	195	200 F	170 F	140 F	140 F	140 F
Mullets nei	*Mugilidae*	2 930 F	3 050 F	3 200 F	2 650 F	2 200 F	2 200 F	2 200 F
Sompat grunt	*Pomadasys jubelini*	20	153	160 F	130	100 F	100 F	100 F
Bigeye grunt	*Brachydeuterus auritus*	12 F	14 F	15 F	10 F	10 F	10 F	10 F
Meagre	*Argyrosomus regius*	222	482	500 F	430 F	350 F	350 F	350 F
Croakers, drums nei	*Sciaenidae*	1 000 F	1 020 F	1 040 F	850 F	650 F	650 F	650 F
Porgies, seabreams nei	*Sparidae*	28	14 F	15 F	10 F	10 F	10 F	10 F
African sicklefish	*Drepane africana*	71	172	180 F	150 F	120 F	120 F	120 F
Giant African threadfin	*Polydactylus quadrifilis*	22	55	60 F	50 F	40 F	40 F	40 F
Lesser African threadfin	*Galeoides decadactylus*	101	108	110 F	100 F	90 F	90 F	90 F
Largehead hairtail	*Trichiurus lepturus*	17	17	20 F	15 F	10 F	10 F	10 F
Tuna-like fishes nei	*Scombroidei*	6 F	6 F	6 F	5 F	4 F	4 F	4 F
Bluefish	*Pomatomus saltatrix*	2 F	3 F	3 F	3 F	3 F	3 F	3 F
Jacks, crevalles nei	*Caranx spp*	100 F	100 F	100 F	80 F	70 F	70 F	70 F
Mackerels nei	*Scombridae*	12 F	14 F	15 F	14 F	10 F	10 F	10 F
Scalloped hammerhead	*Sphyrna lewini*	12	12	10 F	10 F	10 F	10 F	10 F
Marine fishes nei	*Osteichthyes*	17	-	-	-	-	-	-
Marine crabs nei	*Brachyura*	27 F	30 F	30 F	25 F	20 F	20 F	20 F
Deepwater rose shrimp	*Parapenaeus longirostris*	40	124	148	120 F	100 F	100 F	100 F
Natantian decapods nei	*Natantia*	1 070 F	1 100 F	1 100 F	900 F	800 F	800 F	800 F
Cuttlefish,bobtail squids nei	*Sepiidae, Sepiolidae*	17	2	2 F	2 F	2 F	2 F	2 F
Various squids nei	*Loliginidae, Ommastrephidae*	8	-	-	-	-	-	-
Common octopus	*Octopus vulgaris*	34	1	1 F	1 F	1 F	1 F	1 F
Marine molluscs nei	*Mollusca*	12 F	14 F	15 F	15 F	10 F	10 F	10 F
	Country total	*6 328*	*7 000 F*	*7 250 F*	*6 000 F*	*5 000 F*	*5 000 F*	*5 000 F*
Kenya								
Common carp	*Cyprinus carpio*	360	334	216	48	52	47	49
Rhinofishes nei	*Labeo spp*	605	1 462	355	207	219	221	222
Silver cyprinid	*Rastrineobola argentea*	56 827	49 670	40 315	42 336	48 816	49 618	41 384
Cyprinids nei	*Cyprinidae*	118	0	0	0	0	0	0
Nile tilapia	*Oreochromis niloticus*	11 827	10 765	13 953	14 652	17 524	19 347	7 292
Tilapias nei	*Oreochromis (=Tilapia) spp*	7 568	7 635	19 329	19 732	18 203	19 853	19 121
Mouthbrooding cichlids	*Haplochromis spp*	4 822	3 914	2 453	2 577	2 731	2 699	1 195
African lungfishes	*Protopterus spp*	408	164	1 704	1 717	1 600	1 608	1 822
Naked catfishes	*Bagrus spp*	127	157	158	107	147	161	178
Torpedo-shaped catfishes nei	*Clarias spp*	574	339	1 724	1 532	1 570	1 592	1 611
Upsidedown catfishes	*Synodontis spp*	28	24	408	418	506	443	48
Nile perch	*Lates niloticus*	102 546	97 145	73 555	76 663	103 014	109 815	78 534
Freshwater fishes nei	*Osteichthyes*	1 215	3 059	745	5 960	4 221	4 886	5 237
Salmonoids nei	*Salmonoidei*	196	11	16	24	29	31	67
Mullets nei	*Mugilidae*	127	153	120	107	146	181	199
Snappers, jobfishes nei	*Lutjanidae*	112	147	144	106	151	120	177
Grunts, sweetlips nei	*Haemulidae (=Pomadasyidae)*	67	65	72	55	78	63	85
Emperors(=Scavengers) nei	*Lethrinidae*	396	433	361	412	358	334	466
Spinefeet(=Rabbitfishes) nei	*Siganus spp*	387	404	347	356	304	299	403
Demersal percomorphs nei	*Perciformes*	211	1 247	1 188	1 196	1 366	1 224	2 060
Clupeoids nei	*Clupeoidei*	112	217	189	155	167	119	164
Narrow-barred Spanish mackerel	*Scomberomorus commerson*	74	93	69	139	122	94	136
Skipjack tuna	*Katsuwonus pelamis*	116	108	114	98	109	86	183
Marlins,sailfishes,etc. nei	*Istiophoridae*	-	73	53	38	82	80	78
Amberjacks nei	*Seriola spp*	89	79	63	60	78	71	92
Carangids nei	*Carangidae*	76	82	111	86	101	85	119
Barracudas nei	*Sphyraena spp*	65	54	54	71	108	83	99
Pelagic percomorphs nei	*Perciformes*	286	358	351	611	528	378	1 003
Sharks, rays, skates, etc. nei	*Elasmobranchii*	176	191	140	134	131	115	175
Marine fishes nei	*Osteichthyes*	1 842	1 187	1 006	1 675	1 774	474	667
Red swamp crawfish	*Procambarus clarkii*	20	13	24	19	21	22	3
Marine crabs nei	*Brachyura*	70	112	100	117	135	166	134
Tropical spiny lobsters nei	*Panulirus spp*	119	177	136	39	52	52	76
Natantian decapods nei	*Natantia*	207	378	491	774	513	458	690
Marine crustaceans nei	*Crustacea*	59	185	223	139	104	101	136

English name Nom anglais Nombre inglés	Scientific name Nom scientifique Nombre científico	1995 mt	1996 mt	1997 mt	1998 mt	1999 mt	2000 mt	2001 mt
Cupped oysters nei	*Crassostrea spp*	14	32	16	9	8	2	1
Various squids nei	*Loliginidae, Ommastrephidae*	345	389	317	30	35	42	78
Octopuses, etc. nei	*Octopodidae*	460	117	393	155	169	106	154
Sea cucumbers nei	*Holothurioidea*	55	15	41	38	15	30	13
	Country total	*192 706*	*180 988*	*161 054*	*172 592*	*205 287*	*215 106*	*164 151*
Lesotho								
Common carp	*Cyprinus carpio*	16 F	16 F	18 F	18 F	18 F	18 F	8
North African catfish	*Clarias gariepinus*	2 F	2 F	2 F	2 F	2 F	2 F	2
Freshwater fishes nei	*Osteichthyes*	8 F	10 F	10 F	10 F	10 F	12 F	14
	Country total	*26 F*	*28 F*	*30 F*	*30 F*	*30*	*32*	*24*
Liberia								
Freshwater fishes nei	*Osteichthyes*	4 000	4 000	4 000	4 000	4 000	4 000	4 000
West African ilisha	*Ilisha africana*	6	124	26	63	242	110	198
Common sole	*Solea solea*	158	48	150	149	217	129	206
Benguela hake	*Merluccius polli*	-	38	-	-	-	-	-
Bonefish	*Albula vulpes*	-	-	-	21	104	27	6
Sea catfishes nei	*Ariidae*	7	77	-	4	31	21	210
Mullets nei	*Mugilidae*	-	-	22	18	63	85	...
Groupers nei	*Epinephelus spp*	22	-	5	110	71	25	44
Snappers nei	*Lutjanus spp*	1	9	17	27	339	132	201
Grunts, sweetlips nei	*Haemulidae (=Pomadasyidae)*	196	105	99	85	216	102	180
Cassava croaker	*Pseudotolithus senegalensis*	20	...	11	...	...	...	...
Bobo croaker	*Pseudotolithus elongatus*	7	109	27	...	...	...	...
West African croakers nei	*Pseudotolithus spp*	1 008	364	510	433	1 025	327	210
Croakers, drums nei	*Sciaenidae*	87	-	-	-	-	-	-
Large-eye dentex	*Dentex macrophthalmus*	...	9	39	...	...	...	...
Dentex nei	*Dentex spp*	313	327	346	974	936	588	671
Black seabream	*Spondyliosoma cantharus*	-	-	-	-	-	-	10
Porgies, seabreams nei	*Sparidae*	42	-	-	-	-	-	-
Surmullets(=Red mullets) nei	*Mullus spp*	-	-	46	-	-	-	-
African sicklefish	*Drepane africana*	30	15	-	70	256	94	104
Lesser African threadfin	*Galeoides decadactylus*	26	-	-	-	-	-	-
Royal threadfin	*Pentanemus quinquarius*	134	36	155	138	118	92	102
Triggerfishes, durgons nei	*Balistidae*	-	-	5	2	-	-	-
Conger eels, etc. nei	*Congridae*	70	41	117	85	128	49	76
Bearded brotula	*Brotula barbata*	46	4	10	66	48	52	...
Largehead hairtail	*Trichiurus lepturus*	12	7	10	9	33	34	169
Sardinellas nei	*Sardinella spp*	876	199	485	620	1 112	887	1 358
Bonga shad	*Ethmalosa fimbriata*	9	70	17	33	123	37	123
Clupeoids nei	*Clupeoidei*	75	189	147	383	386	318	208
Skipjack tuna	*Katsuwonus pelamis*	-	410	-	-	-	-	-
Albacore	*Thunnus alalunga*	...	41	-	-	-	-	-
Yellowfin tuna	*Thunnus albacares*	...	180	185	310	369	227	166
Bigeye tuna	*Thunnus obesus*	57	415	340	108	112	201	175
Atlantic blue marlin	*Makaira nigricans*	...	114	122	59	37	187	131
Marlins,sailfishes,etc. nei	*Istiophoridae*	120	145	71	781	513	683	163
Swordfish	*Xiphias gladius*	28	112	543	21	39	42	34
Tuna-like fishes nei	*Scombroidei*	-	-	-	3	2	14	8
Needlefishes, etc. nei	*Belonidae*	-	-	-	54	90	30	31
Halfbeaks nei	*Hemiramphus spp*	-	-	-	77	85	96	97
Flyingfishes nei	*Exocoetidae*	...	69	8	10	-	38	19
Cunene horse mackerel	*Trachurus trecae*	12	-	-	-	-	-	-
False scad	*Caranx rhonchus*	-	-	7	111	435	135	134
Jacks, crevalles nei	*Caranx spp*	76	62	30	178	394	235	271
Leerfish	*Lichia amia*	-	2	0	-	-	-	-
Common dolphinfish	*Coryphaena hippurus*	...	...	31	20	48	45	22
Suckerfishes, remoras nei	*Echeneidae*	-	-	-	3	1	12	-
Chub mackerel	*Scomber japonicus*	134	24	8	45	218	238	34
Blue butterfish	*Stromateus fiatola*	382	143	236	248	573	181	183
Barracudas nei	*Sphyraena spp*	202	35	-	67	343	174	196
Thresher	*Alopias vulpinus*	...	...	...	...	151	146	...
Mako sharks	*Isurus spp*	...	...	...	...	...	116	...
Blue shark	*Prionace glauca*	...	...	...	...	76	70	...
Silky shark	*Carcharhinus falciformis*	...	...	...	...	110	99	...
Hammerhead sharks, etc. nei	*Sphyrnidae*	...	...	...	...	127	152	...
Guitarfishes, etc. nei	*Rhinobatidae*	...	...	...	54	175	16	...
Sawfishes	*Pristidae*	...	...	48	...	39	42	...
Mantas	*Mobulidae*	...	...	...	342	802	931	106
Rays, stingrays, mantas nei	*Rajiformes*	33	12	38	50	119	103	29
Sharks, rays, skates, etc. nei	*Elasmobranchii*	358	207	386	210	-	-	512
Marine fishes nei	*Osteichthyes*	134	362	157	539	797	281	640
Marine crabs nei	*Brachyura*	28	-	35	38	32	27	122
Palinurid spiny lobsters nei	*Palinurus spp*	10	-	8	26	4	41	36
Deepwater rose shrimp	*Parapenaeus longirostris*	...	...	8	...	...	...	...
Natantian decapods nei	*Natantia*	110	28	73	113	302	25	31
Cuttlefish,bobtail squids nei	*Sepiidae, Sepiolidae*	-	-	-	12	8	14	29
Octopuses, etc. nei	*Octopodidae*	-	176	2	61	23	16	41
	Country total	*8 829*	*8 308*	*8 580*	*10 830*	*15 472*	*11 726*	*11 286*
Libya								
Freshwater fishes nei	*Osteichthyes*	0	0	0	0	0	0	0
Groupers nei	*Epinephelus spp*	4 100 F	4 000 F	4 000 F	4 000 F	4 000 F	4 000 F	4 000
Bogue	*Boops boops*	2 550 F	2 500 F	2 500 F	2 500 F	2 500 F	2 500 F	2 500
Porgies, seabreams nei	*Sparidae*	4 100 F	4 000 F	4 000 F	4 000 F	4 000 F	4 000 F	4 000

E-1

Fish crustaceans, molluscs, etc	Capture production by countries or areas and species	Africa
Poissons, crustacés, mollusques, etc	Captures par pays ou zones et espèces	Afrique
Peces, crustáceos, moluscos, etc	Capturas por países o áreas y especies	Africa

English name Nom anglais Nombre inglés	Scientific name Nom scientifique Nombre científico	1995 mt	1996 mt	1997 mt	1998 mt	1999 mt	2000 mt	2001 mt
Red mullet	*Mullus surmuletus*	4 100 F	4 000 F	4 000 F	4 000 F	4 000 F	4 000 F	4 000 F
Sardinellas nei	*Sardinella spp*	7 100 F	7 000 F	7 000 F	7 000 F	7 000 F	7 000 F	7 000 F
Little tunny(=Atl.black skipj)	*Euthynnus alletteratus*	-	-	45	52	-	5	4
Atlantic bluefin tuna	*Thunnus thynnus*	1 540	1 388	1 029	1 331	1 195	1 550	1 940
Yellowfin tuna	*Thunnus albacares*	-	-	-	-	-	-	208
Bigeye tuna	*Thunnus obesus*	400	400	400	400	400	400	31
Swordfish	*Xiphias gladius*	-	-	-	11	...	8	6
Jack and horse mackerels nei	*Trachurus spp*	3 050	3 000 F	3 000 F	3 000 F	3 000 F	3 000 F	3 000 F
Scomber mackerels nei	*Scomber spp*	3 050 F	3 000 F	3 000 F	3 000 F	3 000 F	3 000 F	3 000 F
Marine fishes nei	*Osteichthyes*	4 410 F	3 688 F	2 903 F	3 617 F	3 755 F	3 924 F	3 550 F
	Country total	*34 400 F*	*32 976 F*	*31 877 F*	*32 911 F*	*32 850 F*	*33 387 F*	*33 239 F*
Madagascar								
Cyprinids nei	*Cyprinidae*	4 123	4 000	4 000	4 000	4 000	4 000	4 000
Cichlids nei	*Cichlidae*	21 600	21 500	21 500	21 500	21 500	21 500	21 500
Freshwater fishes nei	*Osteichthyes*	4 277	4 500	4 500	4 500	4 500	4 500	4 500
Narrow-barred Spanish mackerel	*Scomberomorus commerson*	10 000	10 000	10 000	12 000	12 000	12 000	12 000
Marine fishes nei	*Osteichthyes*	61 544	59 965	61 596	69 652	74 417	75 107	78 601
Marine crabs nei	*Brachyura*	1 300	1 000	1 000	1 500	868	1 030	1 347
Tropical spiny lobsters nei	*Panulirus spp*	390	390	390	341	338	329	359
Natantian decapods nei	*Natantia*	9 919	10 470	10 755	11 470	10 507	12 127	11 776
Cephalopods nei	*Cephalopoda*	350	500	500	550	600	600	600
Marine molluscs nei	*Mollusca*	350	350	350	400	400	400	400
Sea cucumbers nei	*Holothurioidea*	1 800	1 800	1 800	482	500	500	500
	Country total	*115 653*	*114 475*	*116 391*	*126 395*	*129 630*	*132 093*	*135 583*
Malawi								
Cyprinids nei	*Cyprinidae*	538	505	522	299	8 302	7 500 F	6 291
Tilapias nei	*Oreochromis (=Tilapia) spp*	3 863	5 080	4 472	5 104	6 808	6 200 F	5 154
Cichlids nei	*Cichlidae*	...	...	...	...	18 841	19 000 F	20 533
Torpedo-shaped catfishes nei	*Clarias spp*	...	4 556	...	3 935	8 471	7 600 F	6 367
Freshwater fishes nei	*Osteichthyes*	49 263	53 428	51 346	31 773	2 970	2 700 F	2 274
	Country total	*53 664*	*63 569*	*56 340*	*41 111*	*45 392*	*43 000 F*	*40 619*
Mali								
Nile tilapia	*Oreochromis niloticus*	39 842	30 450	1 719	26 675	26 821	32 961	30 000 F
Bonytongues nei	*Heterotis spp*	1 328	453	329	397	399	1 099	1 000 F
Bottlenose fishes nei	*Mormyrus spp*	...	...	...	...	...	7 691	7 000 F
Characins nei	*Characidae*	6 650	9 504	6 969	8 316	8 362	5 493	5 000 F
Black catfishes nei	*Chrysichthys spp*	5 320	4 172	3 982	3 650	3 670	4 395	4 000 F
North African catfish	*Clarias gariepinus*	33 210	17 129	14 933	15 009	15 091	27 468	25 000 F
Upsidedown catfishes	*Synodontis spp*	3 990	4 657	5 973	4 075	4 097	3 296	3 000 F
Nile perch	*Lates niloticus*	7 980	5 712	4 978	5 024	5 052	6 592	6 000 F
Freshwater fishes nei	*Osteichthyes*	34 580	39 833	60 667	34 854	35 044	20 875	19 000 F
	Country total	*132 900*	*111 910*	*99 550*	*98 000*	*98 536*	*109 870*	*100 000 F*
Mauritania								
Freshwater fishes nei	*Osteichthyes*	5 000 F	5 000 F	5 000 F	5 000 F	5 000 F	5 000 F	5 000 F
Soles nei	*Soleidae*	680 F	760 F	1 120 F	830 F	449 F	444 F	...
Flatfishes nei	*Pleuronectiformes*	570 F	580 F	700 F	600 F	1 751 F	1 756 F	2 200 F
Hakes nei	*Merluccius spp*	40 F	150 F	110 F	818 F	940	1 558	1 270
Sea catfishes nei	*Ariidae*	...	...	...	...	750 F	750 F	750 F
Mullets nei	*Mugilidae*	...	...	...	...	2 000 F	2 000 F	2 000 F
White grouper	*Epinephelus aeneus*	420 F	340 F	450 F	390 F	450 F	450 F	450 F
Groupers nei	*Epinephelus spp*	240 F	180 F	240 F	220 F	300 F	300 F	300 F
Spotted seabass	*Dicentrarchus punctatus*	140 F	160 F	130 F	190 F	100 F	45 F	...
Rubberlip grunt	*Plectorhinchus mediterraneus*	20 F	20 F	20 F	10 F	...	9 F	...
Grunts, sweetlips nei	*Haemulidae (=Pomadasyidae)*	10 F	10 F	10 F	10 F	...	2 F	...
Canary drum (=Baardman)	*Umbrina canariensis*	5 F	10 F	10 F	10 F	6 F	2 F	...
Meagre	*Argyrosomus regius*	500 F	480 F	580 F	510 F	600 F	600 F	600 F
West African croakers nei	*Pseudotolithus spp*	10 F	10 F	10 F	96 F	...	9 F	...
Croakers, drums nei	*Sciaenidae*	200 F	200 F	200 F	200 F	300 F	300 F	300 F
Sargo breams nei	*Diplodus spp*	20 F	20 F	20 F	70 F	99	145	277
Porgies, seabreams nei	*Sparidae*	350 F	490 F	520 F	870 F	900 F	900 F	900 F
Surmullets(=Red mullets) nei	*Mullus spp*	30 F	30 F	30 F	141 F	150 F	150 F	150 F
John dory	*Zeus faber*	10 F	5 F	10 F	20 F	...	...	...
Largehead hairtail	*Trichiurus lepturus*	100 F	160 F	110 F	-	...	...	...
Scorpionfishes nei	*Scorpaenidae*	5 F	5 F	5 F	0	1 F	0	...
Gurnards, searobins nei	*Triglidae*	10 F	10 F	0	70 F	...	...	...
Blackbellied angler	*Lophius budegassa*	20 F	20 F	20 F	20 F	20	45	23
Sardinellas nei	*Sardinella spp*	3 020 F	5 440 F	4 870 F	1 060 F	4 000 F	4 000 F	4 000 F
Bonga shad	*Ethmalosa fimbriata*	...	...	...	...	...	2 F	...
Clupeoids nei	*Clupeoidei*	-	-	-	-	-	-	1 F
West African Spanish mackerel	*Scomberomorus tritor*	...	...	...	...	...	12 F	...
Tuna-like fishes nei	*Scombroidei*	263	2 479	2 170	1 304	...		...
Bluefish	*Pomatomus saltatrix*	...	...	...	...	...	1 F	...
Jack and horse mackerels nei	*Trachurus spp*	1 510 F	1 220 F	1 060 F	940 F	...	...	58
Carangids nei	*Carangidae*	100 F	100 F	100 F	100 F	100 F	112 F	...
Chub mackerel	*Scomber japonicus*	-	110 F	80 F	150 F	...	...	
Rays, stingrays, mantas nei	*Rajiformes*	10 F	10 F	20 F	180 F	350 F	350 F	350 F
Sharks, rays, skates, etc. nei	*Elasmobranchii*	80 F	10 F	10 F	350 F	500 F	500 F	500 F
Marine fishes nei	*Osteichthyes*	15 404 F	18 595 F	19 711 F	22 179 F	37 674 F	39 997 F	42 659 F
Geryons nei	*Geryon spp*	...	...	...	...	...	6	14
Palinurid spiny lobsters nei	*Palinurus spp*	30 F	50 F	110 F	100 F	80 F	82 F	83 F
Natantian decapods nei	*Natantia*	130 F	160 F	190 F	270 F	222	779	1 378

E-1 Fish crustaceans, molluscs, etc — Capture production by countries or areas and species — Africa
Poissons, crustacés, mollusques, etc — Captures par pays ou zones et espèces — Afrique
Peces, crustáceos, moluscos, etc — Capturas por países o áreas y especies — Africa

English name Nom anglais Nombre inglés	Scientific name Nom scientifique Nombre científico	1995 mt	1996 mt	1997 mt	1998 mt	1999 mt	2000 mt	2001 mt
Marine crustaceans nei	*Crustacea*	-	-	-	30 F	9	6	20
Cuttlefish,bobtail squids nei	*Sepiidae, Sepiolidae*	5 710 F	4 510 F	5 240 F	5 000 F	4 129 F	4 344 F	4 744 F
Various squids nei	*Loliginidae, Ommastrephidae*	260 F	230 F	280 F	1 472 F	2 359 F	2 354 F	2 125 F
Octopuses, etc. nei	*Octopodidae*	18 250 F	18 770 F	14 620 F	18 420 F	12 758 F	13 709 F	13 305 F
Cephalopods nei	*Cephalopoda*	-	-	-	10 F	16	129	134
Marine molluscs nei	*Mollusca*	-	-	-	20 F	13	1	5
	Country total	*53 147 F*	*60 324 F*	*57 756 F*	*61 660 F*	*76 026 F*	*80 849 F*	*83 596 F*
Mauritius								
Tilapias nei	*Oreochromis (=Tilapia) spp*	0	0	0	0	0	0	0
Unicorn cod	*Bregmaceros mcclellandi*	301	312	367	336	285	347	340
Mullets nei	*Mugilidae*	115	121	103	111	120	76	92
Groupers, seabasses nei	*Serranidae*	1 022	905	933	931	826	863	938
Snappers, jobfishes nei	*Lutjanidae*	...	...	...	...	...	...	2 184
Emperors(=Scavengers) nei	*Lethrinidae*	6 216	5 137	5 018	4 981	4 598	4 698	4 008
Goatfishes	*Upeneus spp*	457	512	479	436	509	541	556
Percoids nei	*Percoidei*	186	183	158	153	165	181	152
Spinefeet(=Rabbitfishes) nei	*Siganus spp*	465	514	430	494	461	448	450
Clupeoids nei	*Clupeoidei*	0	0	-	-	-	-	-
Skipjack tuna	*Katsuwonus pelamis*	3 848	1 898	3 055	1 685	2 361	305	8
Albacore	*Thunnus alalunga*	2	2	7	15	12	...	18
Yellowfin tuna	*Thunnus albacares*	1 725	713	1 095	1 443	742	226	125
Bigeye tuna	*Thunnus obesus*	570	271	546	260	250	37	5
Marlins,sailfishes,etc. nei	*Istiophoridae*	196	190	639	295	287	287	2
Swordfish	*Xiphias gladius*	...	...	...	...	...	...	34
Tuna-like fishes nei	*Scombroidei*	189	70	199	44	681	726	745
Carangids nei	*Carangidae*	165	43	58	46	33	53	51
Mackerels nei	*Scombridae*	11	12	10	11	11	13	13
Rays, stingrays, mantas nei	*Rajiformes*	2	2	2	2	2	2	2
Sharks, rays, skates, etc. nei	*Elasmobranchii*	15	17	58	9	9	25	12
Marine fishes nei	*Osteichthyes*	550	589	525	502	515	441	568
Indo-Pacific swamp crab	*Scylla serrata*	22	23	19	23	21	25	24
Tropical spiny lobsters nei	*Panulirus spp*	12	14	17	17	17	17	19
Natantian decapods nei	*Natantia*	1	0	1	0	1	1	1
Octopuses, etc. nei	*Octopodidae*	325	341	306	299	299	303	347
	Country total	*16 395*	*11 869*	*14 025*	*12 093*	*12 205*	*9 615*	*10 694*
Mayotte								
Marine fishes nei	*Osteichthyes*	700 F	1 000	1 300 F	2 000 F	2 000	5 500	5 500
	Country total	*700 F*	*1 000*	*1 300 F*	*2 000 F*	*2 000*	*5 500*	*5 500*
Morocco								
Cyprinids nei	*Cyprinidae*	900	800	800	900	1 200	1 000	600
Freshwater fishes nei	*Osteichthyes*	500	600	900	500	700	500	300
European eel	*Anguilla anguilla*	100	100	401	304	250	100	73
Shads nei	*Alosa spp*	7	7	10	5	-	47	62
Common sole	*Solea solea*	1 545	1 404	1 675	1 540	2 614	5 951	3 788
Flatfishes nei	*Pleuronectiformes*	3 746	3 750	2 287	4 915	5 773	5 038	3 898
Greater forkbeard	*Phycis blennoides*	428	440	145	257	345	634	376
Pouting(=Bib)	*Trisopterus luscus*	2 088	2 105	1 078	1 251	594	2 217	1 017
European hake	*Merluccius merluccius*	6 133	3 170	2 547	1 257	2 350	2 792	4 009
Mullets nei	*Mugilidae*	1 636	1 630	2 628	2 907	3 750	2 758	4 347
White grouper	*Epinephelus aeneus*	26	30	27	47	35	94	40
Grunts, sweetlips nei	*Haemulidae (=Pomadasyidae)*	1 665	1 670	2 590	3 049	2 228	2 715	3 158
Meagre	*Argyrosomus regius*	904	869	1 542	1 263	2 473	1 785	1 538
Sargo breams nei	*Diplodus spp*	929	940	1 049	982	1 073	1 457	1 288
Black seabream	*Spondyliosoma cantharus*	156	229	185	126	146	435	300
Pargo breams nei	*Pagrus spp*	21	25	23	136	92	424	268
Gilthead seabream	*Sparus aurata*	0	0	0	4	0	206	25
Bogue	*Boops boops*	3 381	3 400	2 510	3 201	3 336	4 888	3 256
Porgies, seabreams nei	*Sparidae*	9 732	10 988	10 380	7 451	10 150	11 020	10 724
Surmullets(=Red mullets) nei	*Mullus spp*	774	682	902	806	822	1 489	1 424
European conger	*Conger conger*	1 927	2 005	1 635	1 386	1 351	1 321	1 094
John dory	*Zeus faber*	561	461	590	567	622	942	512
Largehead hairtail	*Trichiurus lepturus*	5 451	5 505	8 359	8 064	6 595	6 176	5 124
Scorpionfishes nei	*Scorpaenidae*	219	230	226	199	263	573	392
Gurnards, searobins nei	*Triglidae*	2 011	2 286	2 216	2 834	3 152	4 874	4 169
Angler(=Monk)	*Lophius piscatorius*	107	110	113	83	66	785	310
Blackbellied angler	*Lophius budegassa*	348	350	320	579	498	85	126
Sardinellas nei	*Sardinella spp*	3 024	2 900	2 186	1 245	742	803	3 301
European pilchard(=Sardine)	*Sardina pilchardus*	570 914	393 362	497 821	435 799	429 732	539 785	763 223
European anchovy	*Engraulis encrasicolus*	11 180	12 447	24 955	40 954	40 213	22 096	47 448
Atlantic bonito	*Sarda sarda*	736	961	1 304	1 596	1 510	2 278	1 705
Plain bonito	*Orcynopsis unicolor*	547	2 016	249	30	627	1 058	839
Frigate and bullet tunas	*Auxis thazard, A.rochei*	1 266	2 216	3 176	3 277	1 176	1 345	674
Little tunny(=Atl.black skipj)	*Euthynnus alletteratus*	231	588	196	203	75	101	87
Skipjack tuna	*Katsuwonus pelamis*	5 024	684	4 513	2 486	858	1 199	268
Atlantic bluefin tuna	*Thunnus thynnus*	1 713	1 621	2 603	2 430	2 227	2 923	3 008
Bigeye tuna	*Thunnus obesus*	-	-	-	-	700	770	857
Swordfish	*Xiphias gladius*	1 775	3 196	5 167	3 419	3 357	2 822	3 549
Tuna-like fishes nei	*Scombroidei*	-	-	-	-	153	-	-
Bluefish	*Pomatomus saltatrix*	12	15	74	163	47	101	198
Jack and horse mackerels nei	*Trachurus spp*	30 474	15 471	12 512	9 816	12 787	22 095	12 167
Chub mackerel	*Scomber japonicus*	30 106	16 988	28 775	11 621	17 383	63 381	26 021
Rays, stingrays, mantas nei	*Rajiformes*	1 670	1 680	1 380	1 206	1 463	2 139	1 510
Sharks, rays, skates, etc. nei	*Elasmobranchii*	1 636	1 625	1 255	2 243	2 004	3 460	2 192

English name / Nom anglais / Nombre inglés	Scientific name / Nom scientifique / Nombre científico	1995 mt	1996 mt	1997 mt	1998 mt	1999 mt	2000 mt	2001 mt
Marine fishes nei	*Osteichthyes*	43 794	42 273	88 681	67 777	54 828	7 600	14 461
Freshwater prawns, shrimps nei	*Palaemonidae*	0	0	0	2	5	3	4
Marine crabs nei	*Brachyura*	46	50	286	236	400	370	365
Palinurid spiny lobsters nei	*Palinurus spp*	14	26	61	33	38	42	456
Norway lobster	*Nephrops norvegicus*	3	7	4	5	3	5	15
European lobster	*Homarus gammarus*	8	20	14	15	8	11	11
Natantian decapods nei	*Natantia*	9 143	9 338	7 276	10 770	9 840	12 652	7 585
Marine crustaceans nei	*Crustacea*	18	29	36	17	-	34	29
Freshwater molluscs nei	*Mollusca*	-	-	-	-	8	5	6
Mediterranean mussel	*Mytilus galloprovincialis*	0	0	0	0	14	60	32
Sea mussels nei	*Mytilidae*	8	10	6	7	...	401	102
Clams, etc. nei	*Bivalvia*	0	0	0	0	-	-	-
Cuttlefish,bobtail squids nei	*Sepiidae, Sepiolidae*	12 396	12 678	16 787	14 668	20 168	32 904	17 562
Various squids nei	*Loliginidae, Ommastrephidae*	19 904	20 010	8 989	11 352	8 082	15 432	10 633
Octopuses, etc. nei	*Octopodidae*	57 834	58 636	38 187	42 533	84 579	99 391	112 634
Marine molluscs nei	*Mollusca*	180	253	275	1 920	1 926	1 048	116
	Country total	*848 951*	*642 886*	*791 906*	*710 436*	*745 431*	*896 620*	*1 083 276*
Mozambique								
Dagaas	*Stolothrissa, Limnothrissa spp*	3 093	5 574	9 921	7 313	9 052	11 813	7 076
Freshwater fishes nei	*Osteichthyes*	2 000	1 936	1 747	1 681	1 191	1 275	1 000
Swordfish	*Xiphias gladius*	312	358	524	1 039	447	...	...
Tuna-like fishes nei	*Scombroidei*	3 347	2 461	3 602	7 140	2 635	5 081	3 096
Requiem sharks nei	*Carcharhinidae*	165	21	...	...	...	...	...
Marine fishes nei	*Osteichthyes*	6 209	13 058	10 374	6 990	8 363	7 816	8 766
Geryons nei	*Geryon spp*	414	564	1 144	911	795	832	629
Spiny lobsters nei	*Palinuridae*	248	331	232	237	203	228	199
Mozambique lobster	*Metanephrops mozambicus*	179	132	156	147	92	105	69
Penaeus shrimps nei	*Penaeus spp*	8 615	8 183	9 825	8 559	8 846	9 460	9 479
Knife shrimp	*Haliporoides triarthrus*	2 036	1 771	1 510	1 882	1 611	1 766	1 738
Cupped oysters nei	*Crassostrea spp*	-	-	8	19	26	30	28
Various squids nei	*Loliginidae, Ommastrephidae*	142	366	602	677	616	538	385
Octopuses, etc. nei	*Octopodidae*	29	49	51	80	104	109	35
Marine molluscs nei	*Mollusca*	38	57	-	-	-	-	-
Sea cucumbers nei	*Holothurioidea*	6	54	7	2	8	12	12
	Country total	*26 833*	*34 915*	*39 703*	*36 677*	*33 989*	*39 065*	*32 512*
Namibia								
Freshwater fishes nei	*Osteichthyes*	1 200	1 200	1 500	1 500	1 500	1 500	1 500
West coast sole	*Austroglossus microlepis*	462	339	514	393	463	571	589
Tadpole codling	*Salilota australis*	-	-	20	99	128	-	-
Southern blue whiting	*Micromesistius australis*	-	-	83	282	29	-	-
Argentine hake	*Merluccius hubbsi*	-	-	12	15	37	-	-
Cape hakes	*Merluccius capensis,M.paradox.*	130 374	129 462	117 683	154 422	166 562	162 803	173 461
Patagonian grenadier	*Macruronus magellanicus*	-	-	98	205	308	-	-
Sea catfishes nei	*Ariidae*	13	16	31	42	32	52	11
Mullets nei	*Mugilidae*	112	...	...	...	...	161	117
Southern meagre(=Mulloway)	*Argyrosomus hololepidotus*	496	464	184	424	273	409	596
Panga seabream	*Pterogymnus laniarius*	291	470	416	199	383	...	...
Steenbrasses nei	*Lithognathus spp*	195	6	136	2	116	56	20
Porgies, seabreams nei	*Sparidae*	2 178	2 279	2 608	1 364	4 434	8 890	6 100
Gobies nei	*Gobiidae*	5	552	287	20 998	16	3	...
Pink cusk-eel	*Genypterus blacodes*	-	-	5	24	45	-	-
Kingklip	*Genypterus capensis*	3 853	3 667	2 506	2 820	3 706	3 922	6 609
Alfonsinos nei	*Beryx spp*	909	1 805	369	147	123	59	300
Orange roughy	*Hoplostethus atlanticus*	6 377	13 379	18 516	10 945	3 473	1 542	857
John dory	*Zeus faber*	3	0	5	25	14	4	138
Oreo dories nei	*Oreosomatidae*	6	17	188	...	42	10	54
Patagonian toothfish	*Dissostichus eleginoides*	-	-	2	21	28	-	-
Snoek	*Thyrsites atun*	856	622	895	701	1 212	966	1 699
Largehead hairtail	*Trichiurus lepturus*	13	...	...	...	...	346	691
Cape redfish	*Sebastes capensis*	271	576	673	343	448	639	891
Cape gurnard	*Chelidonichthys capensis*	57	51	214	216	71	236	144
Devil anglerfish	*Lophius vomerinus*	10 130	9 236	10 259	16 701	14 802	14 358	12 401
Southern African pilchard	*Sardinops ocellatus*	42 797	1 171	27 685	68 562	44 653	25 388	7 940
Whitehead's round herring	*Etrumeus whiteheadi*	1 934	20 656	5 070	5 193	176	1 127	1 432
Southern African anchovy	*Engraulis capensis*	48 023	1 080	2 545	2 748	412	146	2 133
Skipjack tuna	*Katsuwonus pelamis*	27	1	1	...	...	...	8
Albacore	*Thunnus alalunga*	1 861	1 521	1 199	1 422	1 162	2 418	3 419
Yellowfin tuna	*Thunnus albacares*	19	3	69	3	147	59	165
Bigeye tuna	*Thunnus obesus*	352	63	45	16	423	589	640
Swordfish	*Xiphias gladius*	-	-	-	-	730	469	751
Cape horse mackerel	*Trachurus capensis*	310 836	321 322	301 847	311 836	322 075	344 314	309 381
Atlantic pomfret	*Brama brama*	20	19	9	15	99	493	684
Chub mackerel	*Scomber japonicus*	...	...	...	...	...	1 641	6 329
Ocean sunfish	*Mola mola*	...	...	...	...	...	...	2
Rays, stingrays, mantas nei	*Rajiformes*	62	133	194	94	389	966	1 204
Sharks, rays, skates, etc. nei	*Elasmobranchii*	7	5	4	6	1	769	1 875
Marine fishes nei	*Osteichthyes*	3 846	5 727	15 556	7 884	9 112	11 801	1 502
Geryons nei	*Geryon spp*	2 008	1 709	1 478	2 283	2 074	2 699	2 667
Cape rock lobster	*Jasus lalandii*	224	251	199	350	304	365	363
Patagonian squid	*Loligo gahi*	-	-	74	1	0	-	-
Argentine shortfin squid	*Illex argentinus*	-	-	3	0	63	-	-
Various squids nei	*Loliginidae, Ommastrephidae*	16	26	34	42	19	28	775
Jellyfishes	*Rhopilema spp*	-	-	-	-	-	106	44
	Country total	*569 833*	*517 828*	*513 216*	*612 343*	*580 084*	*589 905*	*547 492*

E-1

Fish crustaceans, molluscs, etc — Capture production by countries or areas and species — Africa
Poissons, crustacés, mollusques, etc — Captures par pays ou zones et espèces — Afrique
Peces, crustáceos, moluscos, etc — Capturas por países o áreas y especies — Africa

English name / Nom anglais / Nombre inglés	Scientific name / Nom scientifique / Nombre científico	1995 mt	1996 mt	1997 mt	1998 mt	1999 mt	2000 mt	2001 mt
Niger								
Freshwater fishes nei	*Osteichthyes*	3 616	4 156	6 328	7 013	11 000	16 250	20 800
	Country total	*3 616*	*4 156*	*6 328*	*7 013*	*11 000*	*16 250*	*20 800*
Nigeria								
Cyprinids nei	*Cyprinidae*	4 476	4 460	6 199	5 006	6 296	4 823	5 901
Tilapias nei	*Oreochromis (=Tilapia) spp*	9 060	10 074	12 614	16 300	19 662	13 402	18 332
African lungfishes	*Protopterus spp*	678	1 284	3 950	1 679	536	1 364	2 183
Bonytongues nei	*Heterotis spp*	4 832	3 298	4 737	7 388	8 873	7 915	8 577
Knifefishes	*Notopterus spp*	383	10	1	144	33	39	22
Characins nei	*Characidae*	8 802	7 141	7 646	19 835	11 883	12 500	10 274
Bagrid catfish	*Chrysichthys nigrodigitatus*	4 000	8 505	5 814	15 538	10 679	7 311	12 413
Naked catfishes	*Bagrus spp*	6 531	3 594	4 726	5 463	5 421	4 539	4 902
North African catfish	*Clarias gariepinus*	-	2 658	8 568	5 268	9 994	4 474	10 419
Torpedo-shaped catfishes nei	*Clarias spp*	20 745	8 480	10 202	17 220	10 963	14 012	10 925
Upsidedown catfishes	*Synodontis spp*	9 709	4 466	5 069	8 781	8 143	6 054	6 642
Nile perch	*Lates niloticus*	4 993	3 746	4 224	5 250	6 366	4 447	6 139
Snakeheads(=Murrels) nei	*Channa spp*	368	921	86	2 589	2 038	2 990	2 951
Freshwater fishes nei	*Osteichthyes*	43 326	30 884	19 808	28 559	38 506	48 445	54 495
Soles nei	*Soleidae*	3 807	4 640	6 084	7 633	6 583	3 301	3 171
Tonguefishes	*Cynoglossidae*	1 771	1 281	1 311	1 781	3 812	7 057	6 655
Benguela hake	*Merluccius polli*	-	-	-	-	64	-	-
West African ladyfish	*Elops lacerta*	1 474	379	735	1 546	1 822	755	531
Sea catfishes nei	*Ariidae*	12 570	12 676	14 062	10 862	17 014	14 885	16 537
Mullets nei	*Mugilidae*	3 421	1 364	7 073	8 450	10 226	8 535	8 728
Groupers, seabasses nei	*Serranidae*	3 102	1 285	2 001	1 741	2 486	2 117	2 487
Snappers nei	*Lutjanus spp*	2 999	3 779	6 473	8 685	8 345	7 235	6 619
Bigeye grunt	*Brachydeuterus auritus*	...	556	2 537	2 887	2 141	3 345	275
Grunts, sweetlips nei	*Haemulidae (=Pomadasyidae)*	4 152	2 585	2 611	5 000	4 561	5 152	4 804
West African croakers nei	*Pseudotolithus spp*	9 264	14 544	11 762	15 035	14 591	12 604	15 084
Croakers, drums nei	*Sciaenidae*	4 605	2 165	6 335	4 056	7 846	5 826	5 900
Porgies, seabreams nei	*Sparidae*	395	-	-	-	-	-	-
Surmullets(=Red mullets) nei	*Mullus spp*	...	4 575	...	...	...	-	-
African sicklefish	*Drepane africana*	597	954	2 047	603	...	1 316	1 429
Giant African threadfin	*Polydactylus quadrifilis*	2 755	6 415	9 815	9 803	10 544	11 287	11 645
Lesser African threadfin	*Galeoides decadactylus*	983	3 194	5 974	2 343	2 884	4 279	5 131
Threadfins, tasselfishes nei	*Polynemidae*	2 156	2 930	3 184	-	-	-	-
Conger eels, etc. nei	*Congridae*	...	...	...	301	...	0	-
Hairtails, scabbardfishes nei	*Trichiuridae*	5 175	4 826	6 399	7 750	5 366	754	1 156
Scorpionfishes nei	*Scorpaenidae*	1 893	1 032	1 278	2 071	1 149	3 170	2 568
Madeiran sardinella	*Sardinella maderensis*	...	2 230	9 387	12 226	11 041	10 158	11 931
Sardinellas nei	*Sardinella spp*	76 585	102 230	97 286	95 495	83 593	80 892	61 927
Bonga shad	*Ethmalosa fimbriata*	15 072	4 643	28 000	30 216	18 529	17 570	19 049
Swordfish	*Xiphias gladius*	-	9	-	-	-	-	-
Tuna-like fishes nei	*Scombroidei*	119	200	55	73	7	51	5
Needlefishes, etc. nei	*Belonidae*	590	1 082	272	99	166	535	431
Flyingfishes nei	*Exocoetidae*	7	17	...	87	69	8	55
Jacks, crevalles nei	*Caranx spp*	636	591	2 525	414	1 099	825	317
Carangids nei	*Carangidae*	5 199	2 923	1 035	700	4 451	3 521	833
Mackerels nei	*Scombridae*	2 254	3 745	3 326	3 640	2 919	2 710	3 759
Blue butterfish	*Stromateus fiatola*	-	-	-	-	-	-	4 106
Barracudas nei	*Sphyraena spp*	3 238	3 556	5 212	7 671	10 490	10 799	10 947
Rays, stingrays, mantas nei	*Rajiformes*	3 670	4 103	5 004	7 382	7 622	5 753	7 352
Sharks, rays, skates, etc. nei	*Elasmobranchii*	2 801	4 285	3 817	6 587	7 751	7 485	7 274
Marine fishes nei	*Osteichthyes*	39 520	32 995	21 516	40 295	33 770	51 541	49 205
Marine crabs nei	*Brachyura*	48	57	593	3 384	4 018	3 211	4 374
Palinurid spiny lobsters nei	*Palinurus spp*	1 496	1 133	8 315	3 071	1 245	1 939	1 699
Southern pink shrimp	*Penaeus notialis*	14 742	12 073	14 799	12 396	27 341	18 882	18 805
Natantian decapods nei	*Natantia*	4 474	3 417	3 456	7 364	2 690	1 564	909
Marine crustaceans nei	*Crustacea*	9	0	0	0	0	0	-
Gastropods nei	*Gastropoda*	...	...	...	2 357	...	...	2 084
Cuttlefish,bobtail squids nei	*Sepiidae, Sepiolidae*	...	3	...	...	...	...	189
	Country total	*349 482*	*337 993*	*387 923*	*463 024*	*455 628*	*441 377*	*452 146*
Réunion								
Freshwater fishes nei	*Osteichthyes*	0	0	0	0	0	0	0
Groupers, seabasses nei	*Serranidae*	...	...	...	...	47	20	11
Snappers nei	*Lutjanus spp*	...	...	...	...	112	38	45
Threadfins, tasselfishes nei	*Polynemidae*	...	...	...	...	49	5	10
Clupeoids nei	*Clupeoidei*	5	5	6	7	10	12	4
Wahoo	*Acanthocybium solandri*	67	80	66	59	57	50	4
Kawakawa	*Euthynnus affinis*	28	26	24	28	22	21	13
Skipjack tuna	*Katsuwonus pelamis*	105	91	77	92	89	84	79
Albacore	*Thunnus alalunga*	163	347	306	318	357	579	648
Yellowfin tuna	*Thunnus albacares*	402	628	636	609	534	656	584
Bigeye tuna	*Thunnus obesus*	15	98	91	112	213	167	64
Indo-Pacific sailfish	*Istiophorus platypterus*	7	10	11	17	18	30	17
Striped marlin	*Tetrapturus audax*		...		...	...	120	117
Shortbill spearfish	*Tetrapturus angustirostris*	-	2	1	3	5	5	-
Marlins,sailfishes,etc. nei	*Istiophoridae*	87	120	110	136	109	3	1
Swordfish	*Xiphias gladius*	769	1 336	1 586	2 080	1 930	1 744	1 514
Tuna-like fishes nei	*Scombroidei*	-	38	-	-	-	3	79
Carangids nei	*Carangidae*	...	...	...	...	107	130	131
Common dolphinfish	*Coryphaena hippurus*	...	...	...	...	221	194	141
Sharks, rays, skates, etc. nei	*Elasmobranchii*	37	46	89	111	81	78	60
Marine fishes nei	*Osteichthyes*	810	775	1 279	1 000	67	127	98

English name Nom anglais Nombre inglés	Scientific name Nom scientifique Nombre científico	1995 mt	1996 mt	1997 mt	1998 mt	1999 mt	2000 mt	2001 mt
Marine crabs nei	*Brachyura*	5	5	6	7	5	7	11
Marine crustaceans nei	*Crustacea*	...	...	...	...	2	1	1
Octopuses, etc. nei	*Octopodidae*	...	...	...	...	8	5	3
	Country total	*2 500*	*3 607*	*4 288*	*4 579*	*4 043*	*4 079*	*3 635*
Rwanda								
Nile tilapia	*Oreochromis niloticus*	...	...	...	2 233	2 640	2 646	2 650
Freshwater fishes nei	*Osteichthyes*	3 300 F	2 952	4 428	4 408	3 793	4 080	4 178
	Country total	*3 300 F*	*2 952*	*4 428*	*6 641*	*6 433*	*6 726*	*6 828*
St Helena								
Freshwater fishes nei	*Osteichthyes*	0	0	-	-	-	-	-
Groupers nei	*Epinephelus spp*	32	48	41	59	33	17	18
Bigeyes nei	*Priacanthus spp*	3	1	1	3	7	2	1
Conger eels, etc. nei	*Congridae*	5	1	3	3	3	3	5
Wahoo	*Acanthocybium solandri*	25	23	19	10	15	15	22
Skipjack tuna	*Katsuwonus pelamis*	115	86	294	298	13	64	205
Albacore	*Thunnus alalunga*	82	47	18	1	1	58	12
Yellowfin tuna	*Thunnus albacares*	181	151	109	181	116	136	70
Bigeye tuna	*Thunnus obesus*	10	10	12	17	6	8	4
Carangids nei	*Carangidae*	2	2	3	2	3	1	1
Chub mackerel	*Scomber japonicus*	7	7	12	2	8	4	11
Sharks, rays, skates, etc. nei	*Elasmobranchii*	...	...	...	...	...	...	6
Marine fishes nei	*Osteichthyes*	82	82	39	59	68	69	70
Tropical spiny lobsters nei	*Panulirus spp*	0	0	-	-	-	-	-
Tristan da Cunha rock lobster	*Jasus tristani*	344	327	321	376	336	316	425
Slipper lobsters nei	*Scyllaridae*	0	0	0	1	1	1	0
Octopuses, etc. nei	*Octopodidae*	27	34	25	48	22	24	16
	Country total	*915*	*819*	*897*	*1 060*	*632*	*718*	*866*
Sao Tome Prn								
Freshwater fishes nei	*Osteichthyes*	0	0	0	0	0	0	0
Groupers nei	*Epinephelus spp*	44 F	33	31	40 F	45 F	40 F	40 F
Grunts, sweetlips nei	*Haemulidae (=Pomadasyidae)*	1 F	1	0	0	0	0	0
Croakers, drums nei	*Sciaenidae*	101 F	71	30	40 F	45 F	40 F	40 F
Pandoras nei	*Pagellus spp*	121 F	88	86	100 F	110 F	100 F	100 F
Bogue	*Boops boops*	...	...	17	20 F	25 F	20 F	20 F
Porgies, seabreams nei	*Sparidae*	50 F	37	49	60 F	70 F	60 F	60 F
African sicklefish	*Drepane africana*	...	...	15	20 F	25 F	20 F	20 F
Threadfins, tasselfishes nei	*Polynemidae*	102 F	75	75	100 F	110 F	100 F	100 F
Triggerfishes, durgons nei	*Balistidae*	...	...	34	40 F	45 F	40 F	40 F
European pilchard(=Sardine)	*Sardina pilchardus*	26 F	20	...	20 F	30 F	30 F	30 F
Wahoo	*Acanthocybium solandri*	-	80	52	52 F	52 F	52 F	52 F
West African Spanish mackerel	*Scomberomorus tritor*	-	8	-	-	-	-	-
Frigate and bullet tunas	*Auxis thazard, A.rochei*	-	79	323	...	...	...	...
Little tunny(=Atl.black skipj)	*Euthynnus alletteratus*	-	40	159	...	...	...	...
Skipjack tuna	*Katsuwonus pelamis*	...	...	7	-	-	-	-
Yellowfin tuna	*Thunnus albacares*	...	1	4	4 F	4 F	4 F	-
Bigeye tuna	*Thunnus obesus*	-	-	5	-	-	-	-
Atlantic sailfish	*Istiophorus albicans*	...	...	139	...	...	...	...
Atlantic blue marlin	*Makaira nigricans*	-	-	35	-	-	-	-
Atlantic white marlin	*Tetrapturus albidus*	-	-	45	-	-	-	-
Swordfish	*Xiphias gladius*	-	-	14	14 F	14 F	-	-
Tuna-like fishes nei	*Scombroidei*	-	-	9	-	-	-	-
Halfbeaks nei	*Hemiramphus spp*	401 F	293	126	150 F	160 F	150 F	150 F
Flyingfishes nei	*Exocoetidae*	350 F	256	939	1 000 F	1 100 F	1 000 F	1 000 F
Scads nei	*Decapterus spp*	...	...	43	55 F	60 F	60 F	60 F
Crevalle jack	*Caranx hippos*	...	...	42	55 F	60 F	60 F	60 F
Jacks, crevalles nei	*Caranx spp*	...	...	129	150 F	160 F	150 F	150 F
Leerfish	*Lichia amia*	34 F	25	36	45 F	50 F	50 F	50 F
Barracudas nei	*Sphyraena spp*	30 F	22	56	70 F	80 F	70 F	70 F
Sharks, rays, skates, etc. nei	*Elasmobranchii*	337 F	247	130	175 F	190 F	180 F	180 F
Marine fishes nei	*Osteichthyes*	1 933 F	2 579	683	1 167 F	1 284 F	1 234 F	1 238 F
Marine crustaceans nei	*Crustacea*	6 F	4	5	75	7 F	10 F	10 F
Marine molluscs nei	*Mollusca*	29 F	21	20	25 F	30 F	30 F	30 F
	Country total	*3 565*	*3 980*	*3 338*	*3 477*	*3 756*	*3 500 F*	*3 500 F*
Senegal								
Tilapias nei	*Oreochromis (=Tilapia) spp*	2 382	1 807	1 977	1 323	1 362	10 762	10 096 F
Black catfishes nei	*Chrysichthys spp*	...	...	...	...	...	1 218	1 100 F
Nile perch	*Lates niloticus*	...	...	...	...	...	1 451	1 300 F
Freshwater fishes nei	*Osteichthyes*	31 000 F	23 000 F	31 000 F	21 000 F	34 000 F	10 648	9 500 F
Flatfishes nei	*Pleuronectiformes*	10 510	8 113	8 002	7 132	7 335	8 113	9 059
Senegalese hake	*Merluccius senegalensis*	1	7	162	22	335	113	98
Sea catfishes nei	*Ariidae*	3 912	4 457	10 162	10 094	6 006	7 432	9 279
Morays	*Muraenidae*	837	710	268	165	142	189	209
Mullets nei	*Mugilidae*	1 947	2 109	2 780	1 690	2 512	4 226	2 546
Groupers, seabasses nei	*Serranidae*	4 175	4 411	4 317	3 488	3 162	3 067	3 685
Spotted seabass	*Dicentrarchus punctatus*	9	619	472	527	170	67	125
Snappers nei	*Lutjanus spp*	368	236	296	237	221	366	499
Bigeye grunt	*Brachydeuterus auritus*	1 255	610	1 445	1 481	2 414	1 949	2 701
Grunts, sweetlips nei	*Haemulidae (=Pomadasyidae)*	2 121	11 187	11 283	7 828	10 902	7 403	8 048
Boe drum	*Pteroscion peli*	6 833	667	527	489	581	472	444
Cassava croaker	*Pseudotolithus senegalensis*	...	...	...	1 630	1 487	1 880	1 682
Bobo croaker	*Pseudotolithus elongatus*	...	...	...	137	146	33	114

E-1

Fish crustaceans, molluscs, etc — **Capture production by countries or areas and species** — **Africa**
Poissons, crustacés, mollusques, etc — **Captures par pays ou zones et espèces** — **Afrique**
Peces, crustáceos, moluscos, etc — **Capturas por países o áreas y especies** — **Africa**

English name Nom anglais Nombre inglés	Scientific name Nom scientifique Nombre científico	1995 mt	1996 mt	1997 mt	1998 mt	1999 mt	2000 mt	2001 mt
Croakers, drums nei	*Sciaenidae*	2 115	8 047	7 620	5 572	3 295	3 199	3 700
Pandoras nei	*Pagellus spp*	4 679	4 380	5 086	3 719	4 319	4 728	3 932
Sargo breams nei	*Diplodus spp*	-	-	4	221	194	364	453
Large-eye dentex	*Dentex macrophthalmus*	...	447	...	451	531	684	499
Dentex nei	*Dentex spp*	4 318	1 608	2 301	1 132	788	1 033	833
Black seabream	*Spondyliosoma cantharus*	862	1 196	1 251	597	922	1 067	1 035
Pargo breams nei	*Pagrus spp*	1 775	2 570	2 763	1 881	1 931	2 206	2 488
Bogue	*Boops boops*	139	10	20	38	7	17	14
Sand steenbras	*Lithognathus mormyrus*	-	-	-	223	45	8	77
Porgies, seabreams nei	*Sparidae*	-	-	-	-	-	-	295
Surmullets(=Red mullets) nei	*Mullus spp*	352	731	896	1 087	1 223	1 110	1 908
West African goatfish	*Pseudupeneus prayensis*	-	-	-	-	-	-	1 846
African sicklefish	*Drepane africana*	248	741	606	325	439	356	421
Lesser African threadfin	*Galeoides decadactylus*	5 357	5 255	5 453	4 849	3 016	2 844	3 849
Triggerfishes, durgons nei	*Balistidae*	-	-	-	10	7	16	52
Conger eels, etc. nei	*Congridae*	127	83	226	241	227	236	112
Bearded brotula	*Brotula barbata*	140	590	1 644	1 770	3 260	1 325	1 669
John dory	*Zeus faber*	661	1 157	142	53	139	282	161
Largehead hairtail	*Trichiurus lepturus*	345	432	1 114	948	838	781	1 083
Scorpionfishes nei	*Scorpaenidae*	539	581	491	343	267	471	471
Gurnards, searobins nei	*Triglidae*	...	...	...	31	81	41	94
Round sardinella	*Sardinella aurita*	116 766	142 457	152 713	129 429	93 512	112 970	112 120
Madeiran sardinella	*Sardinella maderensis*	64 711	108 186	114 824	103 363	105 120	114 749	99 944
Bonga shad	*Ethmalosa fimbriata*	17 277	17 783	17 147	16 496	29 468	22 032	31 675
European anchovy	*Engraulis encrasicolus*	73	34	31	307	1 209	964	1 015
Atlantic bonito	*Sarda sarda*	354	570	564	1 723	349	179	120
Plain bonito	*Orcynopsis unicolor*	-	-	-	65	17	6	69
West African Spanish mackerel	*Scomberomorus tritor*	1 863	1 056	1 015	931	479	589	756
Little tunny(=Atl.black skipj)	*Euthynnus alletteratus*	1 133	1 066	1 662	1 604	460	1 146	1 613
Skipjack tuna	*Katsuwonus pelamis*	293	265	430	1 836	1 422	1 009	3 768
Yellowfin tuna	*Thunnus albacares*	108	68	152	222	358	218	1 118
Bigeye tuna	*Thunnus obesus*	177	135	218	791	2 007	860	2 169
Atlantic sailfish	*Istiophorus albicans*	204	206	509	192	79	447	266
Swordfish	*Xiphias gladius*	-	-	-	169	127	39	35
Tuna-like fishes nei	*Scombroidei*	52	-	-	-	-	-	6
Needlefishes, etc. nei	*Belonidae*	...	...	...	5	37	8	0
Halfbeaks nei	*Hemiramphus spp*				70	30	14	14
Bluefish	*Pomatomus saltatrix*	461	691	2 075	3 118	327	277	1 149
Jack and horse mackerels nei	*Trachurus spp*	1 027	521	1 078	1 061	2 108	1 020	1 877
Scads nei	*Decapterus spp*	1 973	2 265	2 428	3 249	2 401	4 963	4 209
African moonfish	*Selene dorsalis*	-	-	9	278	176	251	326
Pompanos nei	*Trachinotus spp*	64	132	83	28	207	27	158
Amberjacks nei	*Seriola spp*	-	-	-	62	60	64	165
Leerfish	*Lichia amia*	2 450	228	257	540	489	490	341
Alexandria pompano	*Alectis alexandrinus*	124	160	742	1 023	563	502	862
Atlantic bumper	*Chloroscombrus chrysurus*	-	-	-	-	-	-	2 045
Carangids nei	*Carangidae*	4 772	3 299	4 806	7 491	4 456	5 725	2 616
Common dolphinfish	*Coryphaena hippurus*	-	-	-	103	115	150	258
Chub mackerel	*Scomber japonicus*	1 748	947	1 670	2 117	988	2 656	2 710
Blue butterfish	*Stromateus fiatola*	-	-	-	-	20	43	57
Barracudas nei	*Sphyraena spp*	1 117	1 530	2 339	1 456	1 474	1 680	1 519
Hammerhead sharks, etc. nei	*Sphyrnidae*	-	-	-	156	20	1 305	1 451
Smooth-hounds nei	*Mustelus spp*	...	...	...	...	...	464	1 908
Guitarfishes, etc. nei	*Rhinobatidae*	...	...	...	171	59	1 930	1 772
Sawfishes	*Pristidae*	...	...	...	...	2	-	-
Rays, stingrays, mantas nei	*Rajiformes*	3 308	2 962	4 515	4 051	3 332	1 585	1 667
Sharks, rays, skates, etc. nei	*Elasmobranchii*	4 169	3 803	4 470	4 887	4 808	5 473	3 260
Marine fishes nei	*Osteichthyes*	18 088	13 522	13 782	7 590	6 679	12 146	14 708
Marine crabs nei	*Brachyura*	158	217	727	356	186	343	389
Palinurid spiny lobsters nei	*Palinurus spp*	121	130	196	144	39	37	53
Slipper lobsters nei	*Scyllaridae*	-	-	-	1	2	4	6
Southern pink shrimp	*Penaeus notialis*	4 828	4 302	6 606	4 105	4 887	6 005	4 643
Deepwater rose shrimp	*Parapenaeus longirostris*	699	954	2 998	3 888	1 167	2 451	2 199
Natantian decapods nei	*Natantia*	-	-	-	-	-	-	19
Murex	*Murex spp*	748	1 267	1 223	2 543	1 255	1 529	2 080
Volutes nei	*Cymbium spp*	7 454	6 648	5 166	4 678	5 737	4 989	5 422
Gastropods nei	*Gastropoda*	24	45	40	...	...	65	11
Cupped oysters nei	*Crassostrea spp*	223	100	109	89	125	101	151
Common edible cockle	*Cerastoderma edule*	178	54	69	139	147	117	105
Cuttlefish,bobtail squids nei	*Sepiidae, Sepiolidae*	6 399	5 919	6 864	6 365	5 709	3 953	3 941
Various squids nei	*Loliginidae, Ommastrephidae*	7	49	92	17	140	152	107
Octopuses, etc. nei	*Octopodidae*	4 191	4 180	2 701	5 530	37 257	6 058	2 985
Marine molluscs nei	*Mollusca*	268	247	748	699	212	105	75
	Country total	*354 617*	*411 759*	*457 366*	*403 872*	*412 125*	*402 047*	*405 409*
Seychelles								
Freshwater fishes nei	*Osteichthyes*	0	0	0	0	0	0	0
Tadpole codling	*Salilota australis*	-	-	56	-	-	-	-
Argentine hake	*Merluccius hubbsi*	-	-	27	-	-	-	-
Patagonian grenadier	*Macruronus magellanicus*	-	-	35	5	-	-	-
Groupers, seabasses nei	*Serranidae*	71	66	124	45	143	56	107
Snappers nei	*Lutjanus spp*	358	381	242	424	489	213	602
Snappers, jobfishes nei	*Lutjanidae*	572	466	464	602	825	552	703
Emperors(=Scavengers) nei	*Lethrinidae*	294	295	220	280	285	423	483
Pink cusk-eel	*Genypterus blacodes*	-	-	10	-	-	-	-
Patagonian toothfish	*Dissostichus eleginoides*	-	-	1	-	-	-	-
Demersal percomorphs nei	*Perciformes*	293	344	163	145	247	372	244
Wahoo	*Acanthocybium solandri*	10	-	-	3	4	6	-

English name Nom anglais Nombre inglés	Scientific name Nom scientifique Nombre científico	1995 mt	1996 mt	1997 mt	1998 mt	1999 mt	2000 mt	2001 mt
Kawakawa	*Euthynnus affinis*	125	93	97	41	158	81	52
Skipjack tuna	*Katsuwonus pelamis*	-	-	4 940	10 705	15 846	11 604	26 147
Albacore	*Thunnus alalunga*	-	-	-	183	74	247	218
Yellowfin tuna	*Thunnus albacares*	5	67	2 878	7 452	9 966	11 785	13 158
Bigeye tuna	*Thunnus obesus*	5	75	936	2 085	3 154	2 271	3 165
Indo-Pacific sailfish	*Istiophorus platypterus*	-	1	8	2	17	-	7
Indo-Pacific blue marlin	*Makaira mazara*	-	-	-	1	15	28	11
Black marlin	*Makaira indica*	-	-	-	1	1	3	1
Striped marlin	*Tetrapturus audax*	-	-	-	-	6	16	7
Marlins,sailfishes,etc. nei	*Istiophoridae*	2	80	32	6	9	6	3
Swordfish	*Xiphias gladius*	22	141	317	216	333	417	320
Tuna-like fishes nei	*Scombroidei*	7	1	6	12	28	1 379	109
Carangids nei	*Carangidae*	1 284	1 568	1 562	1 008	1 474	1 764	1 285
Indian mackerel	*Rastrelliger kanagurta*	322	425	298	25	101	295	182
Indian mackerels nei	*Rastrelliger spp*	205	159	116	111	219	174	83
Barracudas nei	*Sphyraena spp*	163	190	183	150	323	217	253
Rays, stingrays, mantas nei	*Rajiformes*	-	-	4	1	-	-	-
Sharks, rays, skates, etc. nei	*Elasmobranchii*	116	84	57	102	68	151	97
Marine fishes nei	*Osteichthyes*	123	221	114	129	416	245	228
Marine crabs nei	*Brachyura*	9	17	20	24	11	11	26
Tropical spiny lobsters nei	*Panulirus spp*	2	1	-	0	7	14	5
Patagonian squid	*Loligo gahi*	-	-	1 114	88	-	-	-
Octopuses, etc. nei	*Octopodidae*	20	32	19	40	78	29	54
	Country total	*4 008*	*4 707*	*14 043*	*23 886*	*34 297*	*32 359*	*47 550*
Sierra Leone								
Freshwater fishes nei	*Osteichthyes*	15 000	14 500 F	14 500	14 190	14 480	14 000	14 000
West African ilisha	*Ilisha africana*	3 013	3 010 F	3 009	3 110	5	2 996	4
Tonguefishes	*Cynoglossidae*	377	380 F	585	206	714	2 326	500
Benguela hake	*Merluccius polli*	-	-	0	-	-	-	-
Bonefish	*Albula vulpes*	...	...	91	...	106	181	496
Sea catfishes nei	*Ariidae*	858	860 F	973	52	62	2 092	748
Mullets nei	*Mugilidae*	598	600 F	597	150	233	595	...
Dungat grouper	*Epinephelus goreensis*	102	196	63	76	290	277	62
Snappers nei	*Lutjanus spp*	3	3 F	155	...	1 728	294	381
Bigeye grunt	*Brachydeuterus auritus*	...	...	...	...	...	...	132
Grunts, sweetlips nei	*Haemulidae (=Pomadasyidae)*	800	800 F	1 086	433	413	1 357	1 975
Boe drum	*Pteroscion peli*	...	...	8	0	...	10	-
Law croaker	*Pseudotolithus brachygnathus*	364	360 F	393	90	48	357	227
Bobo croaker	*Pseudotolithus elongatus*	3 650	3 600 F	3 330	389	410	4 035	748
West African croakers nei	*Pseudotolithus spp*	677	680 F	1 210	687	869	2 014	1 011
Pargo breams nei	*Pagrus spp*	239	240 F	933	713	752	1 106	...
Porgies, seabreams nei	*Sparidae*	885	2 565	1 967	522	500	...	1 189
West African goatfish	*Pseudupeneus prayensis*	...	...	37	...	...	35	119
African sicklefish	*Drepane africana*	12	10 F	293	657	214	662	416
Giant African threadfin	*Polydactylus quadrifilis*	635	640 F	679	75	61	690	445
Lesser African threadfin	*Galeoides decadactylus*	392	390 F	921	758	716	1 149	914
Royal threadfin	*Pentanemus quinquarius*	1 770	1 750 F	1 768	1 744	1 784	1 761	...
Threadfins, tasselfishes nei	*Polynemidae*	2	2 F	-	-	-	-	-
Triggerfishes, durgons nei	*Balistidae*	1	1 F	3	195	...	5	-
European conger	*Conger conger*	...	...	...	348	...	...	...
Largehead hairtail	*Trichiurus lepturus*	...	...	...	204	75	35	48
Sardinellas nei	*Sardinella spp*	7 669	7 670 F	7 661	7 450	7 850	7 628	9 845
Bonga shad	*Ethmalosa fimbriata*	21 738	21 700 F	21 807	21 840	22 400	21 621	24 790
European anchovy	*Engraulis encrasicolus*	182	180 F	182	150	43	181	...
Clupeoids nei	*Clupeoidei*	713	700 F	...	315	-	6	-
Atlantic bonito	*Sarda sarda*	0	0	0	4	...	11	245
Atlantic bluefin tuna	*Thunnus thynnus*	...	...	...	...	...	93	118
Albacore	*Thunnus alalunga*	...	...	...	...	...	...	91
Bigeye tuna	*Thunnus obesus*	...	...	...	...	...	6	2
Swordfish	*Xiphias gladius*	...	...	...	...	...	2	2
Tuna-like fishes nei	*Scombroidei*	598	254	2 618	4 980	2 166	623	6 639
Cunene horse mackerel	*Trachurus trecae*	4	11	44	40	5	-	434
Scads nei	*Decapterus spp*	...	...	112	39	277	141	...
Jacks, crevalles nei	*Caranx spp*	467	470 F	474	18	7	587	327
African moonfish	*Selene dorsalis*	1	1 F	...	...	...	135	319
Pompanos nei	*Trachinotus spp*	3	3 F	63	279	7	-	-
Atlantic bumper	*Chloroscombrus chrysurus*	0	0	-	-	-	-	227
Carangids nei	*Carangidae*	...	...	52	158	-	39	-
Barracudas nei	*Sphyraena spp*	681	680 F	772	237	124	1 095	2 757
Rays, stingrays, mantas nei	*Rajiformes*	1 401	1 400 F	1 404	15	10	18	...
Sharks, rays, skates, etc. nei	*Elasmobranchii*	2	2 F	1	68	41	1 672	164
Marine fishes nei	*Osteichthyes*	0	0	0	0	4	2 732	3 608
Marine crabs nei	*Brachyura*	14	15 F	140	83	107	142	269
Tropical spiny lobsters nei	*Panulirus spp*	2	4 F	57	29	51	56	30
Penaeus shrimps nei	*Penaeus spp*	868	973	2 147	1 690	2 155	1 663	1 280
Cuttlefish,bobtail squids nei	*Sepiidae, Sepiolidae*	602	1 621	1 626	600	512	294	574
Various squids nei	*Loliginidae, Ommastrephidae*	166	121	90	143	...	...	...
Octopuses, etc. nei	*Octopodidae*	381	912	777	328	188	8	11
Marine molluscs nei	*Mollusca*	-	-	-	-	-	-	63
	Country total	*64 870*	*67 304 F*	*72 628*	*63 065*	*59 407*	*74 730*	*75 210*
Somalia								
Freshwater fishes nei	*Osteichthyes*	250 F	250 F	250 F	250 F	250 F	200 F	200 F
Marine fishes nei	*Osteichthyes*	26 800 F	24 900 F	23 000 F	21 100 F	19 100 F	19 100 F	18 900 F
Tropical spiny lobsters nei	*Panulirus spp*	400 F	400 F	400 F	400 F	400 F	400 F	400 F
Cephalopods nei	*Cephalopoda*	500 F	500 F	500 F	500 F	500 F	500 F	500 F

English name Nom anglais Nombre inglés	Scientific name Nom scientifique Nombre científico	1995 mt	1996 mt	1997 mt	1998 mt	1999 mt	2000 mt	2001 mt
	Country total	*27 950 F*	*26 050 F*	*24 150 F*	*22 250 F*	*20 250 F*	*20 200 F*	*20 000 F*
South Africa								
Freshwater fishes nei	*Osteichthyes*	800	850	850	900	900 F	900 F	900 F
West coast sole	*Austroglossus microlepis*	1	16	3	...	...	...	...
Mud sole	*Austroglossus pectoralis*	769	909	837	859	768	800 F	844
Tonguefishes	*Cynoglossidae*	-	-	3	7	5 F	0	17
Blue antimora	*Antimora rostrata*	-	-	-	3	6	10	23
Cape hakes	*Merluccius capensis,M.paradox.*	137 742	155 155	141 076	151 317	141 165	135 000 F	146 393
Grenadiers nei	*Macrourus spp*	-	-	-	36	46	157	147
Gadiformes nei	*Gadiformes*	-	-	10	7	4 F	2	1
Sea catfishes nei	*Ariidae*	0	2	2	0	1	-	-
Mullets nei	*Mugilidae*	1 147	1 086	972	908	757	400 F	8
Groupers, seabasses nei	*Serranidae*	25	30	-	-	...	...	15
Snappers, jobfishes nei	*Lutjanidae*	-	2	-	-	...	...	1
Threadfin and dwarf breams nei	*Nemipteridae*	3	-	-	-	...	...	...
Grunts, sweetlips nei	*Haemulidae (=Pomadasyidae)*	2	-	6	11	5 F	1	4
Southern meagre(=Mulloway)	*Argyrosomus hololepidotus*	794	850	765	690	671	619 F	617
Geelbek croaker	*Atractoscion aequidens*	370	361	492	605	415	380 F	383
Tigertooth croaker	*Otolithes ruber*	-	-	-	13	10 F	...	...
Croakers, drums nei	*Sciaenidae*	9	8	7	...	...	1	5
Emperors(=Scavengers) nei	*Lethrinidae*	-	-	-	-	-	-	3
Carpenter seabream	*Argyrozona argyrozona*	729	883	780	505	541	500 F	287
Santer seabream	*Cheimerius nufar*	66	73	62	0	29	25 F	25
Red steenbras	*Petrus rupestris*	56	28	35	...	22	10 F	7
Panga seabream	*Pterogymnus laniarius*	708	730	907	843	966	900 F	1 058
White stumpnose	*Rhabdosargus globiceps*	168	237	165	296	335	300 F	169
Daggerhead breams nei	*Chrysoblephus spp*	267	268	224	154	74	70 F	78
White steenbras	*Lithognathus lithognathus*	11	7	8	9	2	...	...
Salema	*Sarpa salpa*	1	4	3	0	1	...	...
Porgies, seabreams nei	*Sparidae*	280	324	326	366	279 F	250 F	147
Hector's lanternfish	*Lampanyctodes hectoris*	0	33	243	6 553	0	...	...
Pike-congers nei	*Muraenesox spp*	-	-	1	9	5 F	4	5
Kingklip	*Genypterus capensis*	2 798	3 140	3 418	3 381	4 114	4 000 F	4 848
Alfonsinos nei	*Beryx spp*	-	-	-	-	-	-	10
John dory	*Zeus faber*	1 070	1 156	1 277	968	1 027 F	1 006 F	1 183
Cape bonnetmouth	*Emmelichthys nitidus*	96	216	121	117	113	50 F	37
Antarctic toothfish	*Dissostichus mawsoni*	-	-	-	-	-	-	21
Patagonian toothfish	*Dissostichus eleginoides*	-	-	2 106	1 150	948	1 238	1 020
Snoek	*Thyrsites atun*	15 299	12 400	11 126	14 135	11 188	10 200 F	10 328
Silver scabbardfish	*Lepidopus caudatus*	4 931	3 196	2 001	4 557	2 560	2 300 F	2 316
Hairtails, scabbardfishes nei	*Trichiuridae*	-	-	6	0	...	...	...
Cape redfish	*Sebastes capensis*	1 082	1 183	957	752	766	750 F	782
Scorpionfishes nei	*Scorpaenidae*	-	-	3	1	1 F	2	2
Cape gurnard	*Chelidonichthys capensis*	502	446	428	623	507	450 F	465
Gurnards, searobins nei	*Triglidae*	-	-	2	-	...	0	0
Devil anglerfish	*Lophius vomerinus*	6 246	6 139	6 958	7 903	6 949	7 000 F	9 554
Southern African pilchard	*Sardinops ocellatus*	115 205	105 210	116 996	128 019	131 316	136 060	192 160
Whitehead's round herring	*Etrumeus whiteheadi*	76 858	47 117	92 209	52 476	58 856	37 750	55 330
Southern African anchovy	*Engraulis capensis*	170 308	40 712	60 095	107 548	180 542	267 840	287 190
Atlantic bonito	*Sarda sarda*	0	0	-	-	-	-	-
Narrow-barred Spanish mackerel	*Scomberomorus commerson*	4	13	...	...	...	...	...
Indo-Pacific king mackerel	*Scomberomorus guttatus*	-	-	20	46	16	22	49
Seerfishes nei	*Scomberomorus spp*	-	-	4	5	3	4	4
Kawakawa	*Euthynnus affinis*	1	-	1	2	2	0	0
Skipjack tuna	*Katsuwonus pelamis*	5	1	7	3	3	1	1
Albacore	*Thunnus alalunga*	5 216	5 634	6 708	8 419	5 102	2 098	7 257
Southern bluefin tuna	*Thunnus maccoyii*	-	-	-	1	-	4	0
Yellowfin tuna	*Thunnus albacares*	209	157	144	335	467	552	412
Bigeye tuna	*Thunnus obesus*	27	7	10	49	54	254	192
Indo-Pacific sailfish	*Istiophorus platypterus*	-	-	0	0	0	0	0
Marlins,sailfishes,etc. nei	*Istiophoridae*	-	-	2	3	5	4	4
Swordfish	*Xiphias gladius*	4	1	1	405	124	250	498
Tuna-like fishes nei	*Scombroidei*	11	-	8	72	25	12	7
Bluefish	*Pomatomus saltatrix*	57	38	15	4	33	20 F	3
Cape horse mackerel	*Trachurus capensis*	10 262	31 995	31 206	46 384	17 970	15 000 F	9 659
Yellowtail amberjack	*Seriola lalandi*	769	488	478	519	302	300 F	315
Atlantic pomfret	*Brama brama*	1 709	1 070	381	307	324	350 F	437
Common dolphinfish	*Coryphaena hippurus*	-	-	1	-	-	-	1
Chub mackerel	*Scomber japonicus*	4 677	4 150	8 044	3 204	1 768 F	1 600 F	1 620
Mackerels nei	*Scombridae*	-	-	-	-	-	0	0
Shortfin mako	*Isurus oxyrinchus*	...	...	...	...	...	...	2
Blue shark	*Prionace glauca*	...	...	...	...	...	...	1
Dusky shark	*Carcharhinus obscurus*	-	-	7	0	...	...	...
Raja rays nei	*Raja spp*	929	1 026	1 239	1 074	1 000	1 000 F	1 151
Rays, stingrays, mantas nei	*Rajiformes*	-	-	-	8	3	4	9
Cape elephantfish	*Callorhinchus capensis*	386	366	484	482	356	380 F	405
Sharks, rays, skates, etc. nei	*Elasmobranchii*	518	327	663	699	671	579 F	426
Marine fishes nei	*Osteichthyes*	1 773	1 634	13 212	1 096	3 686 F	2 870 F	4 641
Geryons nei	*Geryon spp*	-	-	-	91	80 F	54	53
Marine crabs nei	*Brachyura*	-	-	113	4	5 F	...	2
Cape rock lobster	*Jasus lalandii*	1 956	1 516	1 680	1 726	1 793	1 693	1 611
Natal spiny lobster	*Palinurus delagoae*	13	10	10	6	7 F	8	10
Southern spiny lobster	*Palinurus gilchristi*	966	918	892	864	429	305	1 053
Slipper lobsters nei	*Scyllaridae*	-	-	2	1	1 F	2	4
Mozambique lobster	*Metanephrops mozambicus*	-	-	-	45	60 F	75	72
King crabs, stone crabs nei	*Lithodidae*	-	-	-	0	-	3	0
Penaeus shrimps nei	*Penaeus spp*	-	-	127	178	180 F	168	249

English name Nom anglais Nombre inglés	Scientific name Nom scientifique Nombre científico	1995 mt	1996 mt	1997 mt	1998 mt	1999 mt	2000 mt	2001 mt
Perlemoen abalone	*Haliotis midae*	615	735	330	524	481	490	5 265
Donax clams	*Donax spp*	0	0	-	-	-	-	-
Cuttlefish,bobtail squids nei	*Sepiidae, Sepiolidae*	-	-	9	5	10 F	16	16
Cape Hope squid	*Loligo reynaudi*	7 047	7 491	3 696	6 670	7 169	6 000 F	3 373
Various squids nei	*Loliginidae, Ommastrephidae*	17	34	47	29	24	40 F	53
Octopuses, etc. nei	*Octopodidae*	33	46	74	68	97 F	105 F	107
	Country total	*575 547*	*440 428*	*515 095*	*559 049*	*588 144*	*643 238*	*755 345*
Sudan								
Nile tilapia	*Oreochromis niloticus*	10 000	10 500	11 000	16 000	16 000	18 000 F	20 000
Freshwater fishes nei	*Osteichthyes*	30 000	30 000	31 000	28 000	28 000	30 000 F	33 000
Narrow-barred Spanish mackerel	*Scomberomorus commerson*	...	...	...	19	24	19	34
Sharks, rays, skates, etc. nei	*Elasmobranchii*	...	...	...	45	56	44	79
Marine fishes nei	*Osteichthyes*	4 000	4 500	5 000	5 436	5 420	4 937 F	4 887
	Country total	*44 000*	*45 000*	*47 000*	*49 500*	*49 500*	*53 000 F*	*58 000*
Swaziland								
Freshwater fishes nei	*Osteichthyes*	60 F	60 F	65 F	70 F	70 F	70 F	70 F
	Country total	*60 F*	*60 F*	*65 F*	*70 F*	*70 F*	*70 F*	*70 F*
Tanzania								
Tilapias nei	*Oreochromis (=Tilapia) spp*	25 869	35 160	25 100	24 000	38 000	40 000	45 000
Mouthbrooding cichlids	*Haplochromis spp*	11 960	6 558	10 600	10 500	8 000	9 000	8 500
Dagaas	*Stolothrissa, Limnothrissa spp*	50 360	40 179	48 000	46 800	42 000	40 000	43 000
Naked catfishes	*Bagrus spp*	2 900	1 136	2 800	2 500	2 000	2 300	2 000
Torpedo-shaped catfishes nei	*Clarias spp*	7 920	2 107	7 750	7 500	2 500	2 000	2 500
Nile perch	*Lates niloticus*	155 860	121 161	152 000	150 000	100 000	90 000	96 000
Freshwater fishes nei	*Osteichthyes*	62 160	55 975	60 500	58 700	67 520	96 700	86 000
Indian halibut	*Psettodes erumei*	43	148	50	45	50	52	75
Sea catfishes nei	*Ariidae*	780	913	850	810	850	880	850
Mullets nei	*Mugilidae*	470	618	550	300	350	400	250
Groupers nei	*Epinephelus spp*	290	652	400	350	500	450	500
Threadfin breams nei	*Nemipterus spp*	206	937	250	195	500	400	400
Emperors(=Scavengers) nei	*Lethrinidae*	6 490	7 304	7 350	7 025	7 500	8 000	7 850
Wrasses, hogfishes, etc. nei	*Labridae*	2 810	3 725	3 200	3 000	3 000	3 200	3 500
Percoids nei	*Percoidei*	340	333	350	200	250	300	200
Spinefeet(=Rabbitfishes) nei	*Siganus spp*	3 470	3 816	4 000	3 550	3 500	3 525	3 000
Sardinellas nei	*Sardinella spp*	3 750	14 323	5 000	4 450	14 000	15 000	15 500
Seerfishes nei	*Scomberomorus spp*	680	766	750	750	650	400	500
Yellowfin tuna	*Thunnus albacares*	-	-	-	-	350	700	800
Marlins,sailfishes,etc. nei	*Istiophoridae*	580	656	700	800	780	800	850
Tuna-like fishes nei	*Scombroidei*	670	757	850	650	500	250	250
Halfbeaks nei	*Hemiramphus spp*	1 640	1 483	1 850	1 175	1 200	1 250	1 300
Amberjacks nei	*Seriola spp*	215	51	250	275	100	75	60
Carangids nei	*Carangidae*	1 380	3 669	1 800	2 000	2 000	2 500	2 000
Indian mackerel	*Rastrelliger kanagurta*	3 490	4 618	4 000	4 050	4 500	4 500	5 000
Barracudas nei	*Sphyraena spp*	26	29	30	15	20	25	25
Rays, stingrays, mantas nei	*Rajiformes*	3 200	4 006	3 500	3 350	3 500	3 600	3 500
Sharks, rays, skates, etc. nei	*Elasmobranchii*	1 310	1 594	1 500	1 325	1 375	1 400	1 500
Marine fishes nei	*Osteichthyes*	6 721	6 334	8 300	8 395	2 125	2 000	2 000
Natantian decapods nei	*Natantia*	2 260	2 664	2 500	2 800	2 100	2 100	2 000
Octopuses, etc. nei	*Octopodidae*	490	605	653	690	600	600	650
Sea cucumbers nei	*Holothurioidea*	1 460	1 644	1 527	1 800	189	372	340
	Country total	*359 800*	*323 921*	*356 960*	*348 000*	*310 509*	*332 779*	*335 900*
Togo								
Tilapias nei	*Oreochromis (=Tilapia) spp*	3 500	3 500	3 500	3 500	3 500	3 500	3 500
Freshwater siluroids nei	*Siluroidei*	500	500	500	500	500	500	500
Freshwater fishes nei	*Osteichthyes*	998	1 000	1 000	1 000	1 000	1 000	1 000
West African ilisha	*Ilisha africana*	32	15	101	35	113	418	796
Soles nei	*Soleidae*	2	10	16	8	6	48	19
Spottail spiny turbot	*Psettodes belcheri*	0	0	2	0	-	1	2
Mullets nei	*Mugilidae*	0	0	0	1	-	3	0
Groupers nei	*Epinephelus spp*	41	14	19	215	120	16	12
Snappers nei	*Lutjanus spp*	87	203	42	433	129	24	16
Sompat grunt	*Pomadasys jubelini*	0	3	43	31	35	11	1
Bigeye grunt	*Brachydeuterus auritus*	21	0	1	5	465	742	844
Shi drum	*Umbrina cirrosa*	0	0	0	0	-	-	-
West African croakers nei	*Pseudotolithus spp*	15	22	16	11	6	37	43
Red pandora	*Pagellus bellottii*	115	74	67	202	237	64	45
Pandoras nei	*Pagellus spp*	208	91	65	365	1 055	120	75
Large-eye dentex	*Dentex macrophthalmus*	0	0	0	1	-	-	-
Black seabream	*Spondyliosoma cantharus*	10	6	5	2	6	25	12
Bogue	*Boops boops*	3	40	18	109	198	71	99
West African goatfish	*Pseudupeneus prayensis*	0	0	0	0	-	-	-
African sicklefish	*Drepane africana*	1	6	2	0	0	-	-
Giant African threadfin	*Polydactylus quadrifilis*	16	16	1	0	1	8	6
Lesser African threadfin	*Galeoides decadactylus*	7	14	3	1	20	149	98
Spadefishes nei	*Ephippidae*	-	-	1	1	0	1	-
Triggerfishes, durgons nei	*Balistidae*	0	0	3	4	0	32	0
European conger	*Conger conger*	-	-	0	0	1	0	0
Largehead hairtail	*Trichiurus lepturus*	0	0	1	0	6	2	4
Round sardinella	*Sardinella aurita*	335	739	1 300	1 045	1 921	1 912	3 123
Madeiran sardinella	*Sardinella maderensis*	86	166	432	408	614	638	445
European anchovy	*Engraulis encrasicolus*	4 779	7 072	4 758	6 325	6 678	7 164	6 660

English name Nom anglais Nombre inglés	Scientific name Nom scientifique Nombre científico	1995 mt	1996 mt	1997 mt	1998 mt	1999 mt	2000 mt	2001 mt
Clupeoids nei	*Clupeoidei*	453	0	0	0	0	0	0
Atlantic bonito	*Sarda sarda*	145	197	338	294	426	423	663
Bigeye tuna	*Thunnus obesus*	6	33	17	6	66	32	26
Marlins,sailfishes,etc. nei	*Istiophoridae*	...	...	32	...	110	77	205
Swordfish	*Xiphias gladius*	14	64	2	23	37	7	17
Needlefishes, etc. nei	*Belonidae*	18	37	90	122	96	53	153
Halfbeaks nei	*Hemiramphus spp*	1	2	4	102	3	4	18
Flyingfishes nei	*Exocoetidae*	109	204	287	610	211	130	186
Cunene horse mackerel	*Trachurus trecae*	149	163	107	82	815	449	501
African moonfish	*Selene dorsalis*	2	10	2	0	-	0	-
Rainbow runner	*Elagatis bipinnulata*	-	-	1	-	27	4	1
Carangids nei	*Carangidae*	332	273	260	291	1 613	2 433	2 482
Common dolphinfish	*Coryphaena hippurus*	1	3	3	0	10	36	148
Chub mackerel	*Scomber japonicus*	158	112	252	183	454	308	204
Barracudas nei	*Sphyraena spp*	33	96	38	25	144	226	54
Dogfish sharks, etc. nei	*Squaliformes*	-	-	-	-	-	0	0
Rays, stingrays, mantas nei	*Rajiformes*	5	11	2	0	2	16	5
Sharks, rays, skates, etc. nei	*Elasmobranchii*	15	202	57	67	230	132	130
Marine fishes nei	*Osteichthyes*	-	176	823	594	2 059	1 424	1 027
Marine crabs nei	*Brachyura*	0	0	0	1	3	33	23
Tropical spiny lobsters nei	*Panulirus spp*	0	0	0	0	2	2	0
Penaeus shrimps nei	*Penaeus spp*	1	12	2	2	0	2	0
Marine crustaceans nei	*Crustacea*	1	0	0	0	0	0	0
Cuttlefish,bobtail squids nei	*Sepiidae, Sepiolidae*	1	12	77	51	5	0	20
Marine molluscs nei	*Mollusca*	1	0	0	0	0	0	0
	Country total	*12 201*	*15 098*	*14 290*	*16 655*	*22 924*	*22 277*	*23 163*
Tunisia								
Freshwater fishes nei	*Osteichthyes*	440	706	1 010	896	808	832	860
European eel	*Anguilla anguilla*	130	192	202	150	206	108	135
Shads nei	*Alosa spp*	0	0	0	-	-	-	-
Common sole	*Solea solea*	327	569	642	480	425	402	456
Turbot	*Psetta maxima*	0	0	0	-	-	-	-
Flatfishes nei	*Pleuronectiformes*	96	52	110	178	217	190	205
European hake	*Merluccius merluccius*	744	685	922	537	1 069	974	1 284
Flathead grey mullet	*Mugil cephalus*	2 062	1 345	1 309	1 525	238	2 048	229
Mullets nei	*Mugilidae*	375	932	1 472	2 272	1 383	2 306	3 016
Groupers nei	*Epinephelus spp*	236	265	416	334	382	103	539
Groupers, seabasses nei	*Serranidae*	75	84	130	3 161	748	98	98
Seabasses nei	*Dicentrarchus spp*	379	430	96	529	540	573	317
Brown meagre	*Sciaena umbra*	90	164	180	192	114	104	138
Shi drum	*Umbrina cirrosa*	29	25	51	32	40	28	33
Common pandora	*Pagellus erythrinus*	2 292	2 746	2 506	2 967	3 128	3 105	3 219
Sargo breams nei	*Diplodus spp*	2 292	2 706	3 177	2 055	3 036	3 084	3 241
Common dentex	*Dentex dentex*	167	227	346	375	291	250	219
Saddled seabream	*Oblada melanura*	24	43	36	29	146	84	51
Red porgy	*Pagrus pagrus*	272	254	327	362	388	460	416
Gilthead seabream	*Sparus aurata*	125	107	265	333	409	757	399
Bogue	*Boops boops*	1 636	1 928	2 205	2 466	3 890	3 052	2 852
Sand steenbras	*Lithognathus mormyrus*	917	694	661	781	683	656	781
Salema	*Sarpa salpa*	916	750	1 127	1 151	1 124	951	1 024
Porgies, seabreams nei	*Sparidae*	2 693	2 381	2 714	1 841	2 973	3 084	2 825
Blotched picarel	*Spicara maena*	218	285	505	395	419	255	381
Picarels nei	*Spicara spp*	382	335	485	3 258	1 005	646	572
Red mullet	*Mullus surmuletus*	1 140	1 161	2 406	1 459	1 504	1 527	2 163
Striped mullet	*Mullus barbatus*	1 108	1 287	1 394	1 228	1 250	1 600	2 632
Greater weever	*Trachinus draco*	5	19	46	103	16	19	20
Grey triggerfish	*Balistes carolinensis*	59	38	52	58	34	50	62
European conger	*Conger conger*	6	0	14	14	5	42	7
John dory	*Zeus faber*	19	28	32	107	59	56	43
Silver scabbardfish	*Lepidopus caudatus*	-	26	7	57	423	411	375
Scorpionfishes nei	*Scorpaenidae*	311	339	472	755	442	400	444
Gurnards, searobins nei	*Triglidae*	79	107	87	115	141	168	198
Angler(=Monk)	*Lophius piscatorius*	7	6	33	30	44	39	35
Sardinellas nei	*Sardinella spp*	8 591	7 738	8 084	6 327	7 491	11 800	12 942
European pilchard(=Sardine)	*Sardina pilchardus*	11 940	10 389	10 767	8 907	13 394	15 001	13 988
European anchovy	*Engraulis encrasicolus*	119	19	55	114	199	2	269
Atlantic bonito	*Sarda sarda*	413	560	611	855	1 350	1 528	1 183
Frigate and bullet tunas	*Auxis thazard, A.rochei*	14	13	26	87	38	7	2 292
Little tunny(=Atl.black skipj)	*Euthynnus alletteratus*	696	824	333	1 113	752	1 453	1 036
Atlantic bluefin tuna	*Thunnus thynnus*	1 897	2 393	2 200	1 745	2 352	2 184	2 493
Swordfish	*Xiphias gladius*	378	352	346	414	468	483	567
Tuna-like fishes nei	*Scombroidei*	115	215	657	6	814	905	989
Garfish	*Belone belone*	48	141	273	272	382	98	307
Silversides(=Sand smelts) nei	*Atherinidae*	594	141	116	338	6	57	309
Bluefish	*Pomatomus saltatrix*	653	1 317	1 426	787	748	1 096	1 037
Jack and horse mackerels nei	*Trachurus spp*	4 116	4 755	6 326	3 899	5 635	4 917	4 204
Greater amberjack	*Seriola dumerili*	59	75	113	75	88	400	122
Leerfish	*Lichia amia*	101	93	145	160	112	90	134
Common dolphinfish	*Coryphaena hippurus*	434	399	538	1 088	919	667	784
Chub mackerel	*Scomber japonicus*	2 562	1 439	4 118	3 233	3 515	2 200	3 727
Barracudas nei	*Sphyraena spp*	36	49	69	100	103	95	96
Catsharks, nursehounds nei	*Scyliorhinus spp*	48	36	72	...	49	126	122
Smooth-hounds nei	*Mustelus spp*	128	640	132	826	997	121	192
Dogfish sharks nei	*Squalidae*	596	19	806	44	25	680	883
Angelshark	*Squatina squatina*	35	18	34	44	25	20	22
Rays, stingrays, mantas nei	*Rajiformes*	460	489	803	836	922	974	1 113
Marine fishes nei	*Osteichthyes*	18 828	18 813	5 112	12 965	9 617	5 125	4 269

E-1

Fish crustaceans, molluscs, etc	Capture production by countries or areas and species	Africa
Poissons, crustacés, mollusques, etc	Captures par pays ou zones et espèces	Afrique
Peces, crustáceos, moluscos, etc	Capturas por países o áreas y especies	Africa

English name Nom anglais Nombre inglés	Scientific name Nom scientifique Nombre científico	1995 mt	1996 mt	1997 mt	1998 mt	1999 mt	2000 mt	2001 mt
Mediterranean shore crab	*Carcinus aestuarii*	-	-	-	-	-	-	11
Palinurid spiny lobsters nei	*Palinurus spp*	44	40	71	68	61	49	41
Norway lobster	*Nephrops norvegicus*	2	3	1	4	2	4	4
European lobster	*Homarus gammarus*	0	1	1	-	1	1	0
Caramote prawn	*Penaeus kerathurus*	3 587	3 057	4 367	3 785	4 729	6 165	3 356
Speckled shrimp	*Metapenaeus monoceros*	-	-	-	-	-	-	757
Deepwater rose shrimp	*Parapenaeus longirostris*	213	161	641	838	1 014	1 283	1 454
Blue and red shrimp	*Aristeus antennatus*	28	195	173	187	38	16	41
Marine crustaceans nei	*Crustacea*	1	8	1	72	1	1	0
European flat oyster	*Ostrea edulis*	-	-	-	-	-	1	1
Grooved carpet shell	*Ruditapes decussatus*	1 343	396	77	57	48	755	544
Common cuttlefish	*Sepia officinalis*	3 517	4 340	6 479	4 935	6 622	6 002	7 148
Common squids nei	*Loligo spp*	240	230	253	288	348	310	327
Common octopus	*Octopus vulgaris*	1 868	3 460	5 681	3 129	2 349	2 103	1 752
Horned and musky octopuses	*Eledone spp*	...	...	668	252	392	369	696
Marine molluscs nei	*Mollusca*	-	-	-	-	-	-	1
	Country total	*83 355*	*83 734*	*87 012*	*88 075*	*93 186*	*95 550*	*98 482*
Uganda								
Cyprinids nei	*Cyprinidae*	2 948	1 539	13 476	15 710	17 600	12 181	12 182
Tilapias nei	*Oreochromis (=Tilapia) spp*	83 223	75 027	81 379	78 500	84 540	96 468	96 172
African lungfishes	*Protopterus spp*	7 724	6 095	6 554	7 595	9 192	5 755	5 796
Characins nei	*Characidae*	10 400	12 160	11 718	11 744	13 165	10 331	10 331
Naked catfishes	*Bagrus spp*	4 879	4 978	4 774	5 222	8 564	4 375	4 375
Torpedo-shaped catfishes nei	*Clarias spp*	2 284	1 916	2 130	2 170	2 833	2 987	2 987
Freshwater siluroids nei	*Siluroidei*	1 600	120	868	887	1 000	...	...
Nile perch	*Lates niloticus*	92 722	81 253	91 706	98 800	89 203	87 257	88 881
Freshwater fishes nei	*Osteichthyes*	3 009	12 000	5 421	-	-	2	2
	Country total	*208 789*	*195 088*	*218 026*	*220 628*	*226 097*	*219 356*	*220 726*
Westn Sahara								
Marine fishes nei	*Osteichthyes*	0	0	0	0	0	0	0
Marine crustaceans nei	*Crustacea*	0	0	0	0	0	0	0
Marine molluscs nei	*Mollusca*	0	0	0	0	0	0	0
	Country total	*0*	*0*	*0*	*0*	*0*	*0*	*0*
Zambia								
Dagaas	*Stolothrissa, Limnothrissa spp*	8 674	7 593	7 813	9 822	8 955	8 863	8 500 F
Freshwater fishes nei	*Osteichthyes*	61 872	58 739	58 110	60 116	58 372	57 808	56 500 F
	Country total	*70 546*	*66 332*	*65 923*	*69 938*	*67 327*	*66 671*	*65 000 F*
Zimbabwe								
Tilapias nei	*Oreochromis (=Tilapia) spp*	300	320	523	412	420	830	800 F
Dagaas	*Stolothrissa, Limnothrissa spp*	15 280	15 423	17 034	15 288	11 208	10 477	10 400 F
Freshwater fishes nei	*Osteichthyes*	883	644	599	671	782	1 807	1 800 F
	Country total	*16 463*	*16 387*	*18 156*	*16 371*	*12 410*	*13 114*	*13 000 F*
Total		***5 850 992***	***5 578 784***	***5 941 285***	***6 107 738***	***6 325 388***	***6 616 512***	***6 890 230***

English name Nom anglais Nombre inglés	Scientific name Nom scientifique Nombre científico	1995 mt	1996 mt	1997 mt	1998 mt	1999 mt	2000 mt	2001 mt
Anguilla								
Freshwater fishes nei	*Osteichthyes*	0	0	0	0	0	0	0
Marine fishes nei	*Osteichthyes*	105 F	140 F	180 F	180 F	180 F	180 F	180 F
Caribbean spiny lobster	*Panulirus argus*	40 F	50 F	60 F	60 F	60 F	60 F	60 F
Stromboid conchs nei	*Strombus spp*	5 F	10 F	10 F	10 F	10 F	10 F	10 F
	Country total	*150 F*	*200 F*	*250 F*	*250 F*	*250 F*	*250 F*	*250 F*
Antigua Barb								
Freshwater fishes nei	*Osteichthyes*	0	0	0	0	0	0	0
Squirrelfishes nei	*Holocentridae*	...	...	...	...	...	...	29
Groupers, seabasses nei	*Serranidae*	...	...	...	...	...	...	217
Snappers, jobfishes nei	*Lutjanidae*	...	...	...	...	...	...	284
Grunts, sweetlips nei	*Haemulidae (=Pomadasyidae)*	...	...	...	...	...	...	167
Porgies, seabreams nei	*Sparidae*	...	...	...	...	...	...	9
Sea chubs nei	*Kyphosus spp*	...	...	...	...	...	...	8
Parrotfishes nei	*Scaridae*	...	...	...	...	...	...	173
Surgeonfishes nei	*Acanthuridae*	...	...	...	...	...	...	158
Boxfishes nei	*Ostraciidae*	...	...	...	...	...	...	66
Triggerfishes, durgons nei	*Balistidae*	...	...	...	...	...	...	18
Tuna-like fishes nei	*Scombroidei*	...	...	...	...	...	...	28
Carangids nei	*Carangidae*	...	...	...	...	...	...	33
Common dolphinfish	*Coryphaena hippurus*	...	...	...	...	...	...	4
Barracudas nei	*Sphyraena spp*	...	...	...	...	...	...	6
Sharks, rays, skates, etc. nei	*Elasmobranchii*	...	...	...	...	...	...	8
Marine fishes nei	*Osteichthyes*	1 116	1 045	1 242	1 013	1 041	1 164	66
Caribbean spiny lobster	*Panulirus argus*	149	125	160	357	274	275	272
Stromboid conchs nei	*Strombus spp*	46	39	35	45	46	42	37
	Country total	*1 311*	*1 209*	*1 437*	*1 415*	*1 361*	*1 481*	*1 583*
Aruba								
Freshwater fishes nei	*Osteichthyes*	0	0	0	0	-	-	-
Groupers nei	*Epinephelus spp*	20	20	25	22	25	18	15
Snappers, jobfishes nei	*Lutjanidae*	50	60	60	50	50	45	45
Wahoo	*Acanthocybium solandri*	40	50	65	70	60	60	60
Atlantic sailfish	*Istiophorus albicans*	10	10	...	...	...	...	...
Marine fishes nei	*Osteichthyes*	20	20	55	40	40	40	43
	Country total	*140*	*160*	*205*	*182*	*175*	*163*	*163*
Bahamas								
Freshwater fishes nei	*Osteichthyes*	0	0	0	0	0	0	0
Nassau grouper	*Epinephelus striatus*	358	331	514	511	381	226	281
Groupers nei	*Epinephelus spp*	513	475	246	227	193	132	177
Snappers nei	*Lutjanus spp*	297	341	751	781	866	721	777
Grunts, sweetlips nei	*Haemulidae (=Pomadasyidae)*	8	14	67	90	66	62	67
Croakers, drums nei	*Sciaenidae*	25	-	-	-	-	-	-
Carangids nei	*Carangidae*	72	91	103	92	79	81	101
Sharks, rays, skates, etc. nei	*Elasmobranchii*	0	5	3	2	1	0	0
Marine fishes nei	*Osteichthyes*	381	385	264	156	139	110	133
Black stone crab	*Menippe mercenaria*	40	25	42	39	50	46	47
Caribbean spiny lobster	*Panulirus argus*	7 750	7 938	7 798	7 553	8 225	9 023	7 042
Stromboid conchs nei	*Strombus spp*	494	589	648	670	472	667	661
Octopuses, etc. nei	*Octopodidae*	4	-	-	-	-	-	-
Marine turtles nei	*Testudinata*	2	3	3	3	1	2	4
	Country total	*9 944*	*10 197*	*10 439*	*10 124*	*10 473*	*11 070*	*9 290*
Barbados								
Freshwater fishes nei	*Osteichthyes*	0	0	0	0	0	0	-
Snappers, jobfishes nei	*Lutjanidae*	41	40	26	32	24	25	20
Wahoo	*Acanthocybium solandri*	42	35	52	52	41	41	24
Seerfishes nei	*Scomberomorus spp*	42	49	...	...	...	...	...
Skipjack tuna	*Katsuwonus pelamis*	6	5	5	10	3	3	1
Albacore	*Thunnus alalunga*	-	-	1	1	1	...	2
Yellowfin tuna	*Thunnus albacares*	255	160	149	150	155	155	142
Bigeye tuna	*Thunnus obesus*	-	-	24	17	18	18	6
Atlantic sailfish	*Istiophorus albicans*	74	25	71	58	44	44	...
Atlantic blue marlin	*Makaira nigricans*	31	25	30	25	19	19	...
Atlantic white marlin	*Tetrapturus albidus*	43	15	41	33	25	25	...
Marlins,sailfishes,etc. nei	*Istiophoridae*	-	-	-	-	-	-	85
Swordfish	*Xiphias gladius*	-	33	16	16	12	13	19
Tuna-like fishes nei	*Scombroidei*	255	68	-	-	-	11	-
Flyingfishes nei	*Exocoetidae*	1 766	2 042	1 566	2 680	2 075	1 916	1 673
Carangids nei	*Carangidae*	24	28	19	14	6	28	11
Common dolphinfish	*Coryphaena hippurus*	758	849	721	482	745	728	574
Sharks, rays, skates, etc. nei	*Elasmobranchii*	24	25	14	12	10	14	10
Marine fishes nei	*Osteichthyes*	220	113	74	62	72	60	109
Marine crustaceans nei	*Crustacea*	0	0	0	0	0	0	-
Marine molluscs nei	*Mollusca*	0	0	0	0	0	0	-
	Country total	*3 581*	*3 512*	*2 809*	*3 644*	*3 250*	*3 100*	*2 676*
Belize								
Freshwater fishes nei	*Osteichthyes*	0	0	0	0	0	0	0
Tadpole codling	*Salilota australis*	-	-	-	-	28	237	42
Southern blue whiting	*Micromesistius australis*	-	-	-	-	-	257	206

English name Nom anglais Nombre inglés	Scientific name Nom scientifique Nombre científico	1995 mt	1996 mt	1997 mt	1998 mt	1999 mt	2000 mt	2001 mt
Argentine hake	*Merluccius hubbsi*	-	-	-	-	35	63	4
Benguela hake	*Merluccius polli*	-	-	-	-	54	...	...
Hakes nei	*Merluccius spp*	-	-	30	29	60	165	8
Patagonian grenadier	*Macruronus magellanicus*		-	-	-	84	1 720	374
Croakers, drums nei	*Sciaenidae*	67	-	1	65	299	580	100
Porgies, seabreams nei	*Sparidae*	88	-	183	1 152	2 815	2 162	30
Pink cusk-eel	*Genypterus blacodes*	-	-	-	-	15	87	8
Patagonian toothfish	*Dissostichus eleginoides*	-	-	-	-	16	27	11
Largehead hairtail	*Trichiurus lepturus*	-	-	300	317	644	265	7
Sardinellas nei	*Sardinella spp*	-	-	2 400	2 389	5 445	5 535	1 271
European pilchard(=Sardine)	*Sardina pilchardus*	-	-	2 126	2 229	595	903	127
European anchovy	*Engraulis encrasicolus*	-	-	300	278	2 118	6 686	2 245
Black skipjack	*Euthynnus lineatus*	-	40	-	20	-	-	-
Skipjack tuna	*Katsuwonus pelamis*	5 160	5 140	6 260	4 440	4 330	5 880	...
Albacore	*Thunnus alalunga*	2	-	-	-	8	2	...
Yellowfin tuna	*Thunnus albacares*	890	2 210	1 870	3 490	2 490	1 820	...
Bigeye tuna	*Thunnus obesus*	1 270	2 600	4 720	1 160	320	560	...
Black marlin	*Makaira indica*	-	-	-	-	10	4	...
Atlantic white marlin	*Tetrapturus albidus*	-	-	1	-	1	0	...
Swordfish	*Xiphias gladius*	1	-	-	-	17	8	...
Tuna-like fishes nei	*Scombroidei*	-	-	-	-	107	143	115
Jack and horse mackerels nei	*Trachurus spp*	-	-	1 600	1 626	4 788	7 619	3 492
Chub mackerel	*Scomber japonicus*	-	-	1 100	1 051	1 522	4 850	995
Rays, stingrays, mantas nei	*Rajiformes*	-	-	-	-	519	48	201
Sharks, rays, skates, etc. nei	*Elasmobranchii*	...	...	1	0	2	6	...
Marine fishes nei	*Osteichthyes*	193	400	493	1 744	3 335	7 169	919
Black stone crab	*Menippe mercenaria*	28	53	210	25	29	16	8
Caribbean spiny lobster	*Panulirus argus*	608	448	534	468	552	503	394
Penaeus shrimps nei	*Penaeus spp*	49	38	43	589	1 318	1 236	110
Natantian decapods nei	*Natantia*	-	-	-	2	6	6	11
Stromboid conchs nei	*Strombus spp*	1 026	1 105	1 926	1 891	1 051	1 745	1 980
Cuttlefish,bobtail squids nei	*Sepiidae, Sepiolidae*	141	-	406	2 445	4 803	4 862	19
Patagonian squid	*Loligo gahi*	-	-	-	-	-	2	-
Argentine shortfin squid	*Illex argentinus*	-	-	-	-	3 796	4 066	1 692
Various squids nei	*Loliginidae, Ommastrephidae*	48	4	21	114	128	31	1
Octopuses, etc. nei	*Octopodidae*	86	-	185	785	3 247	1 185	...
	Country total	*9 657*	*12 038*	*24 710*	*26 309*	*44 587*	*60 448*	*14 370*
Bermuda								
Freshwater fishes nei	*Osteichthyes*	0	0	0	0	0	0	0
Groupers nei	*Epinephelus spp*	42	42	42	43	48	27	45
Snappers, jobfishes nei	*Lutjanidae*	37	38	29	29	28	23	32
Wahoo	*Acanthocybium solandri*	93	99	105	108	104	61	56
Little tunny(=Atl.black skipj)	*Euthynnus alletteratus*	6	7	6	5	4	2	4
Skipjack tuna	*Katsuwonus pelamis*	0	0	0	0	0	0	1
Atlantic bluefin tuna	*Thunnus thynnus*	-	1	2	2	1	1	1
Blackfin tuna	*Thunnus atlanticus*	4	5	4	6	6	5	4
Albacore	*Thunnus alalunga*	-	-	1	...	2	2	2
Yellowfin tuna	*Thunnus albacares*	44	67	55	53	59	31	37
Atlantic blue marlin	*Makaira nigricans*	15	15	3	5	1	2	2
Atlantic white marlin	*Tetrapturus albidus*	1	1	1	1	1	0	1
Marlins,sailfishes,etc. nei	*Istiophoridae*	-	-	3	1	-	-	-
Swordfish	*Xiphias gladius*	1	1	5	5	3	3	2
Carangids nei	*Carangidae*	86	70	50	48	51	30	41
Sharks, rays, skates, etc. nei	*Elasmobranchii*	17	13	9	12	24	10	5
Marine fishes nei	*Osteichthyes*	93	96	108	113	80	55	57
Caribbean spiny lobster	*Panulirus argus*	10	10	38	30	36	29	21
Tropical spiny lobsters nei	*Panulirus spp*	-	-	-	5	5	5	4
	Country total	*449*	*465*	*461*	*466*	*453*	*286*	*315*
Br Virgin Is								
Freshwater fishes nei	*Osteichthyes*	0	0	0	0	0	0	0
Groupers nei	*Epinephelus spp*	12	69	1	3	4	1	1 F
Snappers nei	*Lutjanus spp*	20	6	1	4	3	1	1 F
Yellowtail snapper	*Ocyurus chrysurus*	...	...	5	9	9	0	0
Boxfishes nei	*Ostraciidae*	...	...	1	1	1	0	0
Clupeoids nei	*Clupeoidei*	...	...	5	5	5	0	0
Atlantic bonito	*Sarda sarda*	0	6	8	6	9	0	0
Seerfishes nei	*Scomberomorus spp*	...	...	6	6	5	0	0
Yellowfin tuna	*Thunnus albacares*	...	...	3	2	3	1	1 F
Swordfish	*Xiphias gladius*	19	54	6	5	5	2	2 F
Tuna-like fishes nei	*Scombroidei*	15	19			-		-
Jacks, crevalles nei	*Caranx spp*	...	...	13	15	14	1	1 F
Common dolphinfish	*Coryphaena hippurus*	2	6	3	2	3	1	1 F
Sharks, rays, skates, etc. nei	*Elasmobranchii*	...	...	1	1	1	0	0
Marine fishes nei	*Osteichthyes*	389	265	39	42	41	27	34 F
Caribbean spiny lobster	*Panulirus argus*	32	27	5	6	4	3	3 F
Stromboid conchs nei	*Strombus spp*	43	54	8	9	8	6	6 F
	Country total	*532*	*506*	*105*	*116*	*115*	*43*	*50 F*
Canada								
Common carp	*Cyprinus carpio*	619	687	543	354	741	516	506
Northern pike	*Esox lucius*	2 496	2 406	2 334	2 568	2 615	2 808	2 464
Catfishes nei	*Ictalurus spp*	37	37	29	27	35	19	19
American yellow perch	*Perca flavescens*	1 282	1 440	2 073	2 083	1 884	1 847	1 965
Walleye	*Stizostedion vitreum*	7 832	7 718	7 751	8 206	8 782	9 160	6 949

English name Nom anglais Nombre inglés	Scientific name Nom scientifique Nombre científico	1995 mt	1996 mt	1997 mt	1998 mt	1999 mt	2000 mt	2001 mt
Freshwater fishes nei	*Osteichthyes*	8 445	9 649	8 089	8 914	9 317	10 756	8 070
Sturgeons nei	*Acipenseridae*	447	299	421	418	291	284	10
American eel	*Anguilla rostrata*	950	840	766	774	631	681	249
Atlantic salmon	*Salmo salar*	103	82	77	7	1		-
Trouts nei	*Salmo spp*	20	4	1	0	0	0	-
Pink(=Humpback)salmon	*Oncorhynchus gorbuscha*	19 767	8 597	12 241	3 920	9 529	7 158	10 575
Chum(=Keta=Dog)salmon	*Oncorhynchus keta*	12 115	6 524	8 685	19 903	4 937	2 783	5 549
Sockeye(=Red)salmon	*Oncorhynchus nerka*	10 533	15 525	25 353	5 041	1 653	8 665	6 231
Chinook(=Spring=King)salmon	*Oncorhynchus tshawytscha*	1 510	455	1 662	1 386	742	507	636
Coho(=Silver)salmon	*Oncorhynchus kisutch*	4 866	3 871	751	16	14	31	46
Rainbow trout	*Oncorhynchus mykiss*	4	3	1	4	1	1	-
Lake trout(=Char)	*Salvelinus namaycush*	631	689	617	554	491	556	679
Chars nei	*Salvelinus spp*	32	11	31	35	46	46	34
Rainbow smelt	*Osmerus mordax*	1 132	1 078	1 227	1 255	356	970	308
Pond smelt	*Hypomesus olidus*	5 517	3 981	5 964	6 471	5 691	3 254	4 241
Surf smelt	*Hypomesus pretiosus*	1	1	-	-	-	-	-
Eulachon	*Thaleichthys pacificus*	26	30	0	0	-	-	1
Lake(=Common)whitefish	*Coregonus clupeaformis*	9 196	9 121	8 376	8 599	8 330	9 028	9 624
Lake cisco	*Coregonus artedi*	1 101	919	1 008	764	565	581	567
Salmonoids nei	*Salmonoidei*	139	112	-	2	1	-	-
American shad	*Alosa sapidissima*	143	138	140	64	64	79	22
Alewife	*Alosa pseudoharengus*	7 023	6 289	6 991	8 468	7 673	6 783	2 790
Striped bass	*Morone saxatilis*	18	15	0	-	-	-	-
Atlantic halibut	*Hippoglossus hippoglossus*	895	1 097	1 362	1 300	1 188	1 219	1 647
Pacific halibut	*Hippoglossus stenolepis*	5 745	5 430	7 448	7 714	7 103	6 095	4 766
Greenland halibut	*Reinhardtius hippoglossoides*	7 752	10 700	13 153	12 269	12 442	16 331	13 814
Arrow-tooth flounder	*Atheresthes stomias*	75	24	-	33	44	39	59
Witch flounder	*Glyptocephalus cynoglossus*	1 244	1 600	1 675	2 094	1 806	2 168	2 128
Amer. plaice(=Long rough dab)	*Hippoglossoides platessoides*	3 095	2 525	3 583	3 546	4 359	3 904	4 466
Yellowtail flounder	*Limanda ferruginea*	1 483	1 178	1 728	5 242	8 222	12 790	15 648
Winter flounder	*Pseudopleuronectes americanus*	2 804	2 468	2 548	1 804	1 763	2 342	2 257
Flatfishes nei	*Pleuronectiformes*	10 177	6 741	5 583	6 025	6 635	6 773	6 076
Tusk(=Cusk)	*Brosme brosme*	2 010	1 405	1 801	1 641	1 065	1 097	1 498
Atlantic cod	*Gadus morhua*	12 438	15 541	29 899	37 741	55 410	46 046	40 325
Pacific cod	*Gadus macrocephalus*	2 172	700	1 537	1 400	836	712	474
Greenland cod	*Gadus ogac*	0	-	-	-	-	-	-
Red hake	*Urophycis chuss*	135	372	248	120	156	81	130
White hake	*Urophycis tenuis*	6 492	4 827	4 309	3 102	3 318	4 246	4 140
Haddock	*Melanogrammus aeglefinus*	7 933	10 311	9 776	11 771	10 550	12 683	15 594
Atlantic tomcod	*Microgadus tomcod*	131	140	90	119	42	55	57
Saithe(=Pollock)	*Pollachius virens*	10 222	9 739	12 604	15 092	8 552	6 528	7 190
Alaska pollock(=Walleye poll.)	*Theragra chalcogramma*	3 295	2 150	1 800	800	1 233	1 044	1 747
Silver hake	*Merluccius bilinearis*	283	3 485	5 333	10 489	9 676	10 378	17 929
Roundnose grenadier	*Coryphaenoides rupestris*	695	236	75	2	-	1	2
Sandeels(=Sandlances) nei	*Ammodytes spp*	-	2	1	0	-	2	-
Lingcod	*Ophiodon elongatus*	3 793	2 500	1 700	1 900	2 523	3 042	2 511
Argentines	*Argentina spp*	229	259	591	51	12	8	17
Wolffishes(=Catfishes) nei	*Anarhichas spp*	303	422	856	526	694	678	578
Pacific ocean perch	*Sebastes alutus*	5 207	6 174	5 782	6 184	5 849	6 177	5 832
Atlantic redfishes nei	*Sebastes spp*	17 954	21 571	18 751	25 763	19 682	19 852	19 710
Scorpionfishes nei	*Scorpaenidae*	16 462	16 626	14 118	15 356	18 086	17 188	16 351
Sablefish	*Anoplopoma fimbria*	4 542	3 069	4 000	4 500	4 583	2 811	2 967
Lumpfish(=Lumpsucker)	*Cyclopterus lumpus*	25	18	175	8	1	8	-
American angler	*Lophius americanus*	1 722	1 623	2 073	1 466	1 274	1 265	1 900
Atlantic herring	*Clupea harengus*	193 690	188 843	186 550	190 861	202 046	203 261	199 295
Pacific herring	*Clupea pallasii*	26 780	22 221	31 500	33 500	28 847	29 290	24 189
Skipjack tuna	*Katsuwonus pelamis*	-	-	0	-	-	-	-
Atlantic bluefin tuna	*Thunnus thynnus*	578	599	507	596	452	550	512
Albacore	*Thunnus alalunga*	1 038	617	260	328	600	3 008	3 318
Yellowfin tuna	*Thunnus albacares*	174	155	100	57	22	105	125
Bigeye tuna	*Thunnus obesus*	148	144	166	129	263	327	241
Swordfish	*Xiphias gladius*	1 568	739	1 089	1 115	1 118	968	1 079
Tuna-like fishes nei	*Scombroidei*	5	-	-	0	-	-	-
Capelin	*Mallotus villosus*	293	32 628	21 945	38 249	23 495	21 352	19 747
Atlantic silverside	*Menidia menidia*	223	151	238	231	558	304	646
Atlantic mackerel	*Scomber scombrus*	18 300	21 027	21 305	19 557	16 317	17 637	24 488
Porbeagle	*Lamna nasus*	1 388	1 029	1 343	1 006	958	904	499
Picked dogfish	*Squalus acanthias*	3 701	4 431	2 546	3 581	8 353	7 524	8 303
Dogfish sharks, etc. nei	*Squaliformes*	293	106	169	115	143	120	92
Raja rays nei	*Raja spp*	6 263	3 900	4 373	3 044	3 119	2 066	2 625
Rays, stingrays, mantas nei	*Rajiformes*	982	1 293	1 584	900	1 406	1 661	1 618
Groundfishes nei	*Osteichthyes*	872	597	475	365	770	1 003	831
Pelagic fishes nei	*Osteichthyes*	-	36	128	25	31	29	208
Finfishes nei	*Osteichthyes*	125	1	-	-	1	-	-
Marine fishes nei	*Osteichthyes*	62 475	105 185	102 100	96 130	97 253	26 979	69 582
Atlantic rock crab	*Cancer irroratus*	5 415	4 286	6 410	6 338	5 732	7 672	7 968
Dungeness crab	*Cancer magister*	4 586	5 025	3 923	2 968	2 944	2 832	5 685
Queen crab	*Chionoecetes opilio*	65 372	65 825	71 416	75 216	95 148	93 505	95 299
Marine crabs nei	*Brachyura*	2 489	1 588	1 948	2 267	1 835	3 052	4 306
American lobster	*Homarus americanus*	40 508	39 369	40 079	41 030	43 428	45 331	51 412
Northern prawn	*Pandalus borealis*	30 213	31 340	48 310	78 867	85 331	100 091	94 328
Pandalus shrimps nei	*Pandalus spp*	24 369	24 976	28 601	29 042	30 603	35 180	31 234
Natantian decapods nei	*Natantia*	8 557	9 392	5 200	5 203	4 071	4 223	4 212
Periwinkles nei	*Littorina spp*	166	200	274	198	149	98	92
Whelks	*Busycon spp*	935	1 291	1 275	1 062	1 557	1 923	2 041
American cupped oyster	*Crassostrea virginica*	2 312	2 132	1 726	3 124	3 225	4 133	3 009
Blue mussel	*Mytilus edulis*	4 505	8 147	9 163	10 273	11 565	15 089	11 829
American sea scallop	*Placopecten magellanicus*	58 770	47 807	53 881	56 399	54 756	83 852	89 196
Iceland scallop	*Chlamys islandica*	9 900	12 180	12 145	6 632	3 160	2 764	1 347

English name Nom anglais Nombre inglés	Scientific name Nom scientifique Nombre científico	1995 mt	1996 mt	1997 mt	1998 mt	1999 mt	2000 mt	2001 mt
Ocean quahog	*Arctica islandica*	124	114	-	-	66	-	164
Butter clam	*Saxidomus giganteus*	1 728	1 431	1 206	1 600	945	949	1 111
Northern quahog(=Hard clam)	*Mercenaria mercenaria*	830	1 109	1 656	1 234	2 536	1 252	1 657
Atlantic surf clam	*Spisula solidissima*	674	1 204	800	444	303	496	602
Stimpson's surf clam	*Spisula polynyma*	24 008	25 597	27 322	25 976	26 699	22 979	20 257
Sand gaper	*Mya arenaria*	3 841	1 469	2 323	2 638	2 680	3 026	3 152
Pacific geoduck	*Panopea abrupta*	2 056	1 768	1 757	1 784	1 688	1 562	1 413
Clams, etc. nei	*Bivalvia*	263	773	259	49	791	345	1 078
Northern shortfin squid	*Illex illecebrosus*	999	8 822	15 775	1 944	314	368	65
Octopuses, etc. nei	*Octopodidae*	108	197	208	139	54	56	61
Marine molluscs nei	*Mollusca*	2 656	1 999	3 308	2 003	1 633	2 535	2 181
Sea urchins nei	*Strongylocentrotus spp*	9 833	9 665	9 221	9 867	9 052	8 012	7 063
	Country total	*849 411*	*904 862*	*972 294*	*1 013 977*	*1 027 511*	*1 010 489*	*1 049 508*
Cayman Is								
Freshwater fishes nei	*Osteichthyes*	0	0	0	0	0	0	0
Marine fishes nei	*Osteichthyes*	125	110	125	125	125	125	125
	Country total	*125*	*110*	*125*	*125*	*125*	*125*	*125*
Costa Rica								
Freshwater fishes nei	*Osteichthyes*	900	1 090	840	1 000 F	1 000 F	1 000 F	1 000 F
Groupers, seabasses nei	*Serranidae*	281	177	387	644	396	204	103
Snappers, jobfishes nei	*Lutjanidae*	550	352	244	478	526	417	347
Clupeoids nei	*Clupeoidei*	311	438	1 175	906	1 788	1 628	2 207
Skipjack tuna	*Katsuwonus pelamis*	120	620	...	...	...	...	...
Yellowfin tuna	*Thunnus albacares*	10	50	...	...	...	...	...
Bigeye tuna	*Thunnus obesus*	150	840	...	...	...	...	...
Atlantic white marlin	*Tetrapturus albidus*	...	...	...	...	3	14	-
Marlins,sailfishes,etc. nei	*Istiophoridae*	1 764	1 892	1 876	1 647	2 143	2 035	2 206
Swordfish	*Xiphias gladius*	29	433	1 072	419	100	409	653
Tuna-like fishes nei	*Scombroidei*	275	2 572	990	1 213	1 063	1 136	1 163
Common dolphinfish	*Coryphaena hippurus*	...	...	...	...	...	8 370	11 221
Sharks, rays, skates, etc. nei	*Elasmobranchii*	2 941	3 497	5 549	7 724	7 897	12 901	9 659
Marine fishes nei	*Osteichthyes*	5 421	7 616	10 727	7 868	9 979	4 541	4 127
Marine crabs nei	*Brachyura*	1	3	50	9	8	4	3
Caribbean spiny lobster	*Panulirus argus*	93	196	196	40	163	271	39
Tropical spiny lobsters nei	*Panulirus spp*	5	7	7	3	4	14	17
Northern brown shrimp	*Penaeus aztecus*	4	8	0	0	0	0	-
Yellowleg shrimp	*Penaeus californiensis*	8	6	2	12	12	6	2
Crystal shrimp	*Penaeus brevirostris*	526	408	392	276	548	350	456
Penaeus shrimps nei	*Penaeus spp*	1 172	1 923	1 615	1 105	1 141	771	717
Pacific seabobs	*Xiphopenaeus,Trachypenaeus spp*	544	576	314	586	310	198	228
Natantian decapods nei	*Natantia*	2 118	1 362	1 022	672	1 004	1 009	508
Various squids nei	*Loliginidae, Ommastrephidae*	13	20	11	1	2	7	11
Octopuses, etc. nei	*Octopodidae*	26	27	58	6	76	71	66
Marine molluscs nei	*Mollusca*	74	65	109	62	55	42	-
Marine turtles nei	*Testudinata*	101	149	33	86	0	0	-
	Country total	*17 437*	*24 327*	*26 669*	*24 757*	*28 218*	*35 398*	*34 733*
Cuba								
Blue tilapia	*Oreochromis aureus*	8 386	9 672	7 553	5 122	4 564	4 500 F	4 500 F
Freshwater fishes nei	*Osteichthyes*	161	230	...	...	4	40 F	40 F
Atlantic halibut	*Hippoglossus hippoglossus*	1	-	0	0	0	-	-
Witch flounder	*Glyptocephalus cynoglossus*	-	-	2	-	-	-	-
Amer. plaice(=Long rough dab)	*Hippoglossoides platessoides*	26	-	22	51	19	0	...
Flatfishes nei	*Pleuronectiformes*	13	-	-	-	1	0	...
Atlantic cod	*Gadus morhua*	1	2	0	1	-	-	-
Red hake	*Urophycis chuss*	170	430	259	118	86	0	...
Haddock	*Melanogrammus aeglefinus*	32	40	27	12	4	-	-
Saithe(=Pollock)	*Pollachius virens*	61	124	57	8	6	-	-
Silver hake	*Merluccius bilinearis*	16 785	22 279	12 697	6 281	3 852	-	-
Mullets nei	*Mugilidae*	108	93	159	122	21	20 F	20 F
Nassau grouper	*Epinephelus striatus*	81	117	77	40	48	50 F	50 F
Red grouper	*Epinephelus morio*	211	276	198	86	119	120 F	120 F
Groupers nei	*Epinephelus spp*	85	130	89	89	44	40 F	40 F
Southern red snapper	*Lutjanus purpureus*	809	963	865	716	698	700 F	700 F
Lane snapper	*Lutjanus synagris*	1 943	1 848	2 472	2 609	1 712	1 700 F	1 700 F
Yellowtail snapper	*Ocyurus chrysurus*	592	1 176	727	457	409	400 F	400 F
Snappers, jobfishes nei	*Lutjanidae*	416	601	403	509	95	100 F	100 F
Grunts, sweetlips nei	*Haemulidae (=Pomadasyidae)*	2 128	1 723	1 451	1 207	1 303	1 300 F	1 300 F
Porgies	*Calamus spp*	378	385	333	270	259	260 F	260 F
Porgies, seabreams nei	*Sparidae*	104	354	97	126	72	70 F	70 F
Mojarras, etc. nei	*Gerreidae*	2 221	1 719	1 346	1 199	1 013	1 000 F	1 000 F
Argentines	*Argentina spp*	113	-	553	4	5	-	-
Cusk-eels, brotulas nei	*Ophidiidae*	198	118	97	37	13	10 F	10 F
Atlantic redfishes nei	*Sebastes spp*	70	-	88	13	39	0	...
American angler	*Lophius americanus*	5	-	-	-	-	-	-
Atlantic herring	*Clupea harengus*	6	213	253	134	121	-	-
Round sardinella	*Sardinella aurita*	...	...	5	93	-	-	-
Scaled sardines	*Harengula spp*	1 045	707	947	649	766	770 F	770 F
Atlantic thread herring	*Opisthonema oglinum*	2 005	2 361	1 900	1 286	1 756	1 750 F	1 750 F
Seerfishes nei	*Scomberomorus spp*	548	613	466	236	247	247 F	247 F
Little tunny(=Atl.black skipj)	*Euthynnus alletteratus*	27	23	17	9	-	-	-
Skipjack tuna	*Katsuwonus pelamis*	886	1 000	1 282	1 302	750	750 F	750 F
Blackfin tuna	*Thunnus atlanticus*	156	287	301	226	309	309 F	309 F
Albacore	*Thunnus alalunga*	0	-	-	-	-	-	-

English name Nom anglais Nombre inglés	Scientific name Nom scientifique Nombre científico	1995 mt	1996 mt	1997 mt	1998 mt	1999 mt	2000 mt	2001 mt
Yellowfin tuna	*Thunnus albacares*	266	297	275	14	34	34 F	34 F
Bigeye tuna	*Thunnus obesus*	7	5	-	-	-	-	-
Atlantic sailfish	*Istiophorus albicans*	109	570	41	28	199	199 F	199 F
Atlantic blue marlin	*Makaira nigricans*	85	43	53	12	30	30 F	30 F
Swordfish	*Xiphias gladius*	864	67	9	10	5	5 F	5 F
Tuna-like fishes nei	*Scombroidei*	3	21	-	-	-	-	-
Jacks, crevalles nei	*Caranx spp*	344	348	234	211	167	170 F	170 F
Atlantic mackerel	*Scomber scombrus*	61	78	115	7	13	-	-
Pelagic percomorphs nei	*Perciformes*	-	-	-	-	2 427	...	-
Picked dogfish	*Squalus acanthias*	-	-	6	-	-	-	-
Raja rays nei	*Raja spp*	5	-	0	1	-	-	-
Rays, stingrays, mantas nei	*Rajiformes*	-	955	1 359	1 335	1 352	1 350 F	1 350 F
Sharks, rays, skates, etc. nei	*Elasmobranchii*	3 056	2 460	1 932	1 737	1 495	1 500 F	1 500 F
Groundfishes nei	*Osteichthyes*	19	415	164	2	5	-	-
Pelagic fishes nei	*Osteichthyes*	-	-	-	19	13	-	-
Marine fishes nei	*Osteichthyes*	18 589	15 615	24 015	20 732	23 355	19 150 F	19 050 F
Freshwater crustaceans nei	*Crustacea*	284	350	922	802	30	30 F	30 F
Black stone crab	*Menippe mercenaria*	5	28	10	35	883	880 F	880 F
Blue crab	*Callinectes sapidus*	745	904	704	1 065	567	570 F	570 F
Marine crabs nei	*Brachyura*	0	-	-	-	-	-	-
Caribbean spiny lobster	*Panulirus argus*	9 405	9 375	8 996	9 417	9 879	9 850 F	9 850 F
Northern pink shrimp	*Penaeus duorarum*	1 851	1 710	2 001	1 734	2 943	2 940 F	2 940 F
Penaeus shrimps nei	*Penaeus spp*	...	...	-	...	1	1 F	1 F
Northern prawn	*Pandalus borealis*	-	-	-	-	119	46	...
Marine crustaceans nei	*Crustacea*	79	22	31	71	3	-	-
Stromboid conchs nei	*Strombus spp*	32	717	1 234	487	831	830 F	830 F
Gastropods nei	*Gastropoda*	46	134	14	4	-	-	-
Mangrove cupped oyster	*Crassostrea rhizophorae*	1 408	1 409	1 991	2 281	1 865	1 870 F	1 870 F
Ark clams nei	*Arca spp*	1 905	1 963	2 989	2 899	2 499	2 500 F	2 500 F
Northern shortfin squid	*Illex illecebrosus*	959	493	2 978	1 085	280	-	-
Various squids nei	*Loliginidae, Ommastrephidae*	...	...	1	0	-	-	-
Frogs	*Rana spp*	62	69	46	28	26	30 F	30 F
Green turtle	*Chelonia mydas*	46	34	18	20	8	8 F	8 F
Hawksbill turtle	*Eretmochelys imbricata*	20	23	19	18	12	12 F	12 F
Loggerhead turtle	*Caretta caretta*	11	10	7	7	5	5 F	5 F
River and lake turtles nei	*Testudinata*	0	3	4	2	-	-	-
Marine turtles nei	*Testudinata*	22	1	0	1	-	-	-
	Country total	*80 059*	*85 603*	*84 911*	*67 076*	*67 381*	*56 146 F*	*56 000 F*
Dominica								
Freshwater fishes nei	*Osteichthyes*	0	0	0	0	0	0	0
Wahoo	*Acanthocybium solandri*	58	58	58	58	50	46	46
King mackerel	*Scomberomorus cavalla*	-	-	-	-	36	35	35
Skipjack tuna	*Katsuwonus pelamis*	33	33	33	33	85	86	...
Blackfin tuna	*Thunnus atlanticus*	30	...	...	...	79	83	83
Yellowfin tuna	*Thunnus albacares*	9	...	...	...	80	78	78
Marine fishes nei	*Osteichthyes*	820	939	988	1 121	870 F	872 F	908 F
	Country total	*950*	*1 030*	*1 079*	*1 212*	*1 200 F*	*1 200 F*	*1 150 F*
Dominican Rp								
Common carp	*Cyprinus carpio*	597	62	109	27	180	...	397
Tilapias nei	*Oreochromis (=Tilapia) spp*	1 188	100	207	77	320	...	529
Freshwater fishes nei	*Osteichthyes*	315	105	751	990	96	179	183
American eel	*Anguilla rostrata*	6	...	...	1	2	7	1
Tarpon	*Megalops atlanticus*	268	31	27	47	12	14	14
Bonefish	*Albula vulpes*	143	151	16	34	125	10	11
Squirrelfishes nei	*Holocentridae*	81	70	35	56	16	24	22
Mullets nei	*Mugilidae*	320	43	59	41	22	21	28
Snooks(=Robalos) nei	*Centropomus spp*	35	20	24	64	12	32	17
Red grouper	*Epinephelus morio*	763	29	92	11	21	33	37
Groupers nei	*Epinephelus spp*	2 316	455	338	379	662	873	942
Southern red snapper	*Lutjanus purpureus*	484	316	496	157	71	126	355
Yellowtail snapper	*Ocyurus chrysurus*	248	793	529	190	234	249	356
Snappers, jobfishes nei	*Lutjanidae*	84	130	235	875	49	753	231
Grunts, sweetlips nei	*Haemulidae (=Pomadasyidae)*	278	249	325	205	423	348	385
Weakfishes nei	*Cynoscion spp*	83	56	14	14	7	10	10
Porgies	*Calamus spp*	631	428	1 112	17	430	17	10
Goatfishes, red mullets nei	*Mullidae*	376	265	178	168	90	171	161
Mojarras, etc. nei	*Gerreidae*	60	45	12	20	9	14	15
Wrasses, hogfishes, etc. nei	*Labridae*	821	529	1 456	52	52	69	22
Parrotfishes nei	*Scaridae*	149	80	94	109	7	18	19
Triggerfishes, durgons nei	*Balistidae*	607	31	26	67	7	10	21
Scaled sardines	*Harengula spp*	72	111	64	55	18	27	21
Atlantic thread herring	*Opisthonema oglinum*	369	127	188	111	99	130	106
Wahoo	*Acanthocybium solandri*	...	...	325	112	31	42	37
King mackerel	*Scomberomorus cavalla*	2 042	1 648	589	288	...	271	261
Cero	*Scomberomorus regalis*	29	57	231	191	230	190	147
Skipjack tuna	*Katsuwonus pelamis*	146	123	231	158	72	23	32
Blackfin tuna	*Thunnus atlanticus*	892	518	323	89	73	114	517
Yellowfin tuna	*Thunnus albacares*	-	-	-	-	-	272	263
Atlantic sailfish	*Istiophorus albicans*	40	25	49	89	56	81	81
Atlantic blue marlin	*Makaira nigricans*	-	-	-	-	-	23	23
Tuna-like fishes nei	*Scombroidei*	138	85	624	196	394	249	-
Needlefishes, etc. nei	*Belonidae*	146	39	5	4	7	15	45
Blue runner	*Caranx crysos*	205	194	256	132	53	74	68
Jacks, crevalles nei	*Caranx spp*	144	424	453	75	23	43	43
Pompanos nei	*Trachinotus spp*	76	57	19	47	5	10	7

English name Nom anglais Nombre inglés	Scientific name Nom scientifique Nombre científico	1995 mt	1996 mt	1997 mt	1998 mt	1999 mt	2000 mt	2001 mt
Amberjacks nei	*Seriola spp*	59	50	46	50	12	16	18
Carangids nei	*Carangidae*	80	30	10	27	6	22	28
Common dolphinfish	*Coryphaena hippurus*	89	237	113	151	175	255	232
Barracudas nei	*Sphyraena spp*	14	3	-	8	4	5	7
Nurse shark	*Ginglymostoma cirratum*	-	-	-	-	-	407	89
Rays, stingrays, mantas nei	*Rajiformes*	90	39	96	62	134	111	123
Marine fishes nei	*Osteichthyes*	-	2 650	2 292	994	1 998	2 382	4 391
Freshwater crustaceans nei	*Crustacea*	-	21	-	-	-	1	48
Marine crabs nei	*Brachyura*	75	32	14	37	7	14	52
Caribbean spiny lobster	*Panulirus argus*	619	420	1 061	863	828	1 286	1 209
Penaeus shrimps nei	*Penaeus spp*	375	47	79	77	49	...	32
Stromboid conchs nei	*Strombus spp*	2 210	1 889	1 594	2 683	1 257	1 778	1 437
Mangrove cupped oyster	*Crassostrea rhizophorae*	41	6	6	12	4	7	8
Common squids nei	*Loligo spp*	37	38	24	84	12	56	51
Common octopus	*Octopus vulgaris*	33	36	33	87	39	99	57
Marine molluscs nei	*Mollusca*	...	...	...	...	...	48	18
	Country total	*17 874*	*12 894*	*14 860*	*10 283*	*8 433*	*11 029*	*13 217*
El Salvador								
Nile tilapia	*Oreochromis niloticus*	2 494	1 265	1 297	1 017	1 216	1 171	560
Jaguar guapote	*Cichlasoma managuense*	608	566	367	344	351	324	410
Catfishes nei	*Ictalurus spp*	250	240	230	213	208	142	169
Freshwater fishes nei	*Osteichthyes*	967	891	912	863	866	1 177	535
Sea catfishes nei	*Ariidae*	94	107	62	114	101	172	192
Snappers, jobfishes nei	*Lutjanidae*	334	197	117	252	230	282	481
Croakers, drums nei	*Sciaenidae*	233	313	337	252	334	279	575
Skipjack tuna	*Katsuwonus pelamis*	...	...	...	750	3 130	...	4 476
Yellowfin tuna	*Thunnus albacares*	...	...	...	920	3 250	...	2 165
Bigeye tuna	*Thunnus obesus*	-	-	-	10	230	-	2 059
Scads nei	*Decapterus spp*	156	237	130	149	226	187	270
Sharks, rays, skates, etc. nei	*Elasmobranchii*	759	347	1 186	266	176	364	759
Marine fishes nei	*Osteichthyes*	2 553	2 267	1 791	1 595	1 842	1 925	1 678
Freshwater crustaceans nei	*Crustacea*	5	5	0	3	11	9	16
Tropical spiny lobsters nei	*Panulirus spp*	-	-	-	-	-	3	2
Pelagic red crab	*Pleuroncodes planipes*	356	164	-	-	-	-	-
Whiteleg shrimp	*Penaeus vannamei*	1 258	1 238	1 046	1 277	1 082	519	460
Crystal shrimp	*Penaeus brevirostris*	860	397	318	483	177	80	153
Penaeus shrimps nei	*Penaeus spp*	167	65	96	156	24	8	19
Pacific seabobs	*Xiphopenaeus,Trachypenaeus spp*	2 812	5 038	3 079	3 015	1 656	1 472	1 525
Marine crustaceans nei	*Crustacea*	564	498	392	355	294	337	409
Freshwater molluscs nei	*Mollusca*	-	1	2	3	1	8	2
Various squids nei	*Loliginidae, Ommastrephidae*	-	-	-	-	-	16	9
Marine molluscs nei	*Mollusca*	601	598	535	620	1 119	1 115	823
	Country total	*15 071*	*14 434*	*11 897*	*12 657*	*16 524*	*9 590*	*17 747*
Greenland								
Freshwater fishes nei	*Osteichthyes*	0	0	0	0	0	0	0
Atlantic salmon	*Salmo salar*	70	82	44	...	...	24	42
Arctic char	*Salvelinus alpinus*	55	43	78	76	24	29	20
Atlantic halibut	*Hippoglossus hippoglossus*	60	51	26	34	46	19	19
European plaice	*Pleuronectes platessa*	-	-	-	2	-	-	-
Greenland halibut	*Reinhardtius hippoglossoides*	18 417	20 438	24 275	22 851	39 451	24 975	20 684
Amer. plaice(=Long rough dab)	*Hippoglossoides platessoides*	0	0	0	5	3	0	4
Tusk(=Cusk)	*Brosme brosme*	20	-	-		-	-	-
Atlantic cod	*Gadus morhua*	9 203	7 486	7 681	5 598	4 117	2 998	5 614
Greenland cod	*Gadus ogac*	2 528	2 121	1 729	1 697	1 903	931	1 133
Blue ling	*Molva dypterygia*	2	-	-		-	-	-
Haddock	*Melanogrammus aeglefinus*	1 351	1 524	1 876	762	...	176	547
Saithe(=Pollock)	*Pollachius virens*	53	165	318	437	...	601	1 526
Polar cod	*Boreogadus saida*	-	4	-		-	-	-
Roundnose grenadier	*Coryphaenoides rupestris*	60	119	172	22	34	28	27
Atlantic wolffish	*Anarhichas lupus*	6	15	6	42	7	12	16
Wolffishes(=Catfishes) nei	*Anarhichas spp*	50	47	67	30	26	47	58
Golden redfish	*Sebastes marinus*	...	...	...	...	...	41	13
Beaked redfish	*Sebastes mentella*	...	...	...	...	...	4 180	3 382
Atlantic redfishes nei	*Sebastes spp*	5 721	1 144	1 162	2 405	5 216	860	399
Lumpfish(=Lumpsucker)	*Cyclopterus lumpus*	448	426	1 157	2 143	3 057	1 211	3 216
Capelin	*Mallotus villosus*	1 865	7 182	12 162	16 935	24 295	24 645	18 641
Dogfish sharks nei	*Squalidae*	67	136	6	...	...	-	-
Groundfishes nei	*Osteichthyes*	623	691	1 273	588	-	-	-
Finfishes nei	*Osteichthyes*	-	-	-	-	-	769	589
Marine fishes nei	*Osteichthyes*	80	168	189	1 187	...	200	264
Queen crab	*Chionoecetes opilio*	998	817	2 557	1 947	2 896	10 236	14 247
Northern prawn	*Pandalus borealis*	81 926	71 986	63 932	69 570	79 178	85 402	85 842
Aesop shrimp	*Pandalus montagui*	...	...	...	...	...	697	609
Iceland scallop	*Chlamys islandica*	5 287	1 373	1 886	2 211	...	1 630	1 593
	Country total	*128 890*	*116 018*	*120 596*	*128 542*	*160 253*	*159 711*	*158 485*
Grenada								
Freshwater fishes nei	*Osteichthyes*	0	0	0	0	0	0	0
Squirrelfishes nei	*Holocentridae*	1	1	1	3	2	3	3
Snooks(=Robalos) nei	*Centropomus spp*	0	0	0	0	1	0	0
Red hind	*Epinephelus guttatus*	81	60	42	52	72	67	88
Groupers, seabasses nei	*Serranidae*	4	10	11	16	20	22	38
Snappers, jobfishes nei	*Lutjanidae*	22	32	25	28	36	48	56
Grunts, sweetlips nei	*Haemulidae (=Pomadasyidae)*	2	1	1	1	3	1	1

English name Nom anglais Nombre inglés	Scientific name Nom scientifique Nombre científico	1995 mt	1996 mt	1997 mt	1998 mt	1999 mt	2000 mt	2001 mt
Goatfishes, red mullets nei	*Mullidae*	1	1	1	0	0	0	0
Parrotfishes nei	*Scaridae*	0	0	1	1	42	18	41
Surgeonfishes nei	*Acanthuridae*	0	0	0	0	1	1	2
Triggerfishes, durgons nei	*Balistidae*	0	0	0	0	0	0	0
Brazilian sardinella	*Sardinella brasiliensis*	0	1	0	0	0	0	0
Scaled sardines	*Harengula spp*	0	1	0	2	0	0	0
Atlantic thread herring	*Opisthonema oglinum*	0	0	0	0	0	0	0
Broad-striped anchovy	*Anchoa hepsetus*	...	...	1	16	8	0	0
Atlantic bonito	*Sarda sarda*	-	24	6	14	16	7	10
Wahoo	*Acanthocybium solandri*	49	56	56	59	82	51	71
King mackerel	*Scomberomorus cavalla*	-	2	4	28	14	9	4
Serra Spanish mackerel	*Scomberomorus brasiliensis*	0	0	0	1	1	1	0
Frigate and bullet tunas	*Auxis thazard, A.rochei*	-	0	1	0	0	0	1
Skipjack tuna	*Katsuwonus pelamis*	12	11	15	23	23	23	15
Blackfin tuna	*Thunnus atlanticus*	123	164	126	233	94	164	222
Albacore	*Thunnus alalunga*	2	1	6	7	6	12	21
Yellowfin tuna	*Thunnus albacares*	410	523	411	484	430	403	759
Bigeye tuna	*Thunnus obesus*	10	-	1	-	-	-	-
Atlantic sailfish	*Istiophorus albicans*	119	56	83	151	148	164	186
Atlantic blue marlin	*Makaira nigricans*	50	26	47	60	100	87	103
Atlantic white marlin	*Tetrapturus albidus*	-	-	-	-	-	1	15
Swordfish	*Xiphias gladius*	1	4	47	33	42	84	73
Needlefishes, etc. nei	*Belonidae*	1	0	2	3	0	0	0
Halfbeaks nei	*Hemiramphus spp*	1	1	2	2	0	1	1
Flyingfishes nei	*Exocoetidae*	6	12	0	5	1	14	10
Scads nei	*Decapterus spp*	94	82	61	101	59	38	52
Atlantic moonfish	*Selene setapinnis*	0	0	0	0	0	0	0
Rainbow runner	*Elagatis bipinnulata*	32	14	20	20	36	23	15
Bigeye scad	*Selar crumenophthalmus*	48	100	181	53	72	137	97
Carangids nei	*Carangidae*	7	10	17	10	16	12	14
Common dolphinfish	*Coryphaena hippurus*	182	130	171	153	162	167	221
Barracudas nei	*Sphyraena spp*	24	40	30	41	35	57	43
Sharks, rays, skates, etc. nei	*Elasmobranchii*	14	4	9	18	24	29	29
Marine fishes nei	*Osteichthyes*	135	173	130	158	29	5	10
Caribbean spiny lobster	*Panulirus argus*	57	23	14	31	72	47	32
Stromboid conchs nei	*Strombus spp*	2	6	1	24	6	0	2
Various squids nei	*Loliginidae, Ommastrephidae*	0	0	0	0	0	0	5
Green turtle	*Chelonia mydas*	7	8	6	6	5	5	7
Sea urchins nei	*Strongylocentrotus spp*	0	0	0	0	0	0	-
	Country total	*1 497*	*1 577*	*1 530*	*1 837*	*1 658*	*1 701*	*2 247*
Guadeloupe								
Freshwater fishes nei	*Osteichthyes*	0	0	0	0	0	0	0
Blackfin tuna	*Thunnus atlanticus*	480	500	500	500	500	500	500
Common dolphinfish	*Coryphaena hippurus*	700 F	730	800 F	670 F	670 F	700 F	700 F
Mackerels nei	*Scombridae*	1 400 F	1 550	1 700 F	1 430 F	1 430 F	1 600 F	1 600 F
Marine fishes nei	*Osteichthyes*	6 270 F	6 220	6 860 F	5 800 F	5 800 F	6 600 F	6 600 F
Marine crustaceans nei	*Crustacea*	150 F	140	150	134	134	150	150
Stromboid conchs nei	*Strombus spp*	500 F	430	470	550	580	550	550
Marine turtles nei	*Testudinata*	0	0	0	0	-	-	-
	Country total	*9 500*	*9 570*	*10 480*	*9 084*	*9 114*	*10 100*	*10 100 F*
Guatemala								
Cichlids nei	*Cichlidae*	5	0	21	157	173	192	200 F
Freshwater fishes nei	*Osteichthyes*	4 020	4 000	5 100	6 366	6 803	7 109	7 100 F
Skipjack tuna	*Katsuwonus pelamis*	-	-	-	-	6 750	12 870	...
Yellowfin tuna	*Thunnus albacares*	-	-	-	-	1 660	4 650	...
Bigeye tuna	*Thunnus obesus*	-	-	-	-	1 580	10 690	...
Sharks, rays, skates, etc. nei	*Elasmobranchii*	207	81	146	237	203	151	150 F
Marine fishes nei	*Osteichthyes*	645	645	500	701	733	513	660 F
Tropical spiny lobsters nei	*Panulirus spp*	3	5	7	16	18	2	5 F
Yellowleg shrimp	*Penaeus californiensis*	275	266	124	259	218	132	200 F
Crystal shrimp	*Penaeus brevirostris*	58	33	18	21	19	16	20 F
Penaeus shrimps nei	*Penaeus spp*	677	404	596	1 712	1 199	1 043	1 150 F
Pacific seabobs	*Xiphopenaeus,Trachypenaeus spp*	2 352	2 203	355	1 320	1 619	365	400 F
Marine crustaceans nei	*Crustacea*	8	0	8	19	20	14	15 F
Various squids nei	*Loliginidae, Ommastrephidae*	3	16	21	39	23	231	200 F
	Country total	*8 253*	*7 653*	*6 896*	*10 847*	*21 018*	*37 978*	*10 100 F*
Haiti								
Freshwater fishes nei	*Osteichthyes*	500 F	500 F	500 F	500 F	500 F	500 F	500 F
Marine fishes nei	*Osteichthyes*	3 600 F	4 000 F	4 000 F	4 000 F	3 800 F	3 800 F	3 800 F
Marine crabs nei	*Brachyura*	17	5	71	59	50 F	50 F	50 F
Caribbean spiny lobster	*Panulirus argus*	900 F	190 F	200 F	200 F	200 F	200 F	200 F
Natantian decapods nei	*Natantia*	150 F	150 F	150 F	150 F	150 F	150 F	150 F
Stromboid conchs nei	*Strombus spp*	350 F	400 F	380 F	350 F	300 F	300 F	300 F
	Country total	*5 517 F*	*5 245 F*	*5 301 F*	*5 259 F*	*5 000 F*	*5 000 F*	*5 000 F*
Honduras								
Freshwater fishes nei	*Osteichthyes*	127	98	126	119	102	61	111
Tadpole codling	*Salilota australis*	106	189	-	-	-	-	-
Southern blue whiting	*Micromesistius australis*	-	3	-	-	-	-	-
Argentine hake	*Merluccius hubbsi*	78	19	-	-	-	-	-
Hakes nei	*Merluccius spp*	2	-	-	-	-	-	-
Patagonian grenadier	*Macruronus magellanicus*	1 053	92	-	-	-	-	-

English name Nom anglais Nombre inglés	Scientific name Nom scientifique Nombre científico	1995 mt	1996 mt	1997 mt	1998 mt	1999 mt	2000 mt	2001 mt
Croakers, drums nei	*Sciaenidae*	192	91	130	19	73	19	62
Porgies, seabreams nei	*Sparidae*	1 773	736	565	233	705	120	438
Pink cusk-eel	*Genypterus blacodes*	61	59	-	-	-	-	-
Patagonian toothfish	*Dissostichus eleginoides*	26	7	-	-	-	-	-
European pilchard(=Sardine)	*Sardina pilchardus*	1	-	-	-	-	-	-
Skipjack tuna	*Katsuwonus pelamis*	-	780	3 110	630	4 280	2 010	...
Yellowfin tuna	*Thunnus albacares*	-	40	230	870	1 640	620	...
Bigeye tuna	*Thunnus obesus*	-	1 230	1 230	140	420	20	...
Indo-Pacific sailfish	*Istiophorus platypterus*	-	-	-	-	-	-	205
Marlins,sailfishes,etc. nei	*Istiophoridae*	-	-	-	-	-	-	182
Swordfish	*Xiphias gladius*	-	-	-	-	-	-	165
Tuna-like fishes nei	*Scombroidei*	13	10	15	5	20	-	-
Cunene horse mackerel	*Trachurus trecae*	2	-	-	-	-	-	-
Jack and horse mackerels nei	*Trachurus spp*	1	-	-	-	-	0	-
Rays, stingrays, mantas nei	*Rajiformes*	615	460	-	-	-	-	-
Sharks, rays, skates, etc. nei	*Elasmobranchii*	...	...	10	4	...	...	85
Marine fishes nei	*Osteichthyes*	3 029	2 817	3 347	1 569	2 176	1 585	1 805
Marine crabs nei	*Brachyura*	70	117	110	35	6	60	...
Caribbean spiny lobster	*Panulirus argus*	1 520	470	1 006	306	570	670	850
Tropical spiny lobsters nei	*Panulirus spp*	2	2	9	2	1	-	-
Penaeus shrimps nei	*Penaeus spp*	3 944	1 328	2 590	1 790	951	4 227	3 548
Natantian decapods nei	*Natantia*	-	-	0	-	-	-	-
Marine crustaceans nei	*Crustacea*	150	117	208	100 F	47	232	...
Stromboid conchs nei	*Strombus spp*	410	490	2 987	500 F	44	832	...
Venus clams nei	*Veneridae*	5	5	3	0	0	...	...
Cuttlefish,bobtail squids nei	*Sepiidae, Sepiolidae*	2 877	3 108	2 076	889	1 077	687	-
Patagonian squid	*Loligo gahi*	31	8	-	-	-	-	-
Argentine shortfin squid	*Illex argentinus*	679	-	-	-	-	-	-
Sevenstar flying squid	*Martialia hyadesi*	118	-	-	-	-	-	-
Various squids nei	*Loliginidae, Ommastrephidae*	644	212	129	30	24	1	...
Octopuses, etc. nei	*Octopodidae*	1 662	1 040	468	129	590	54	-
Marine molluscs nei	*Mollusca*	146	...	8	20	27	465	...
Marine turtles nei	*Testudinata*	-	-	-	3 F	1	21	...
	Country total	*19 337*	*13 528*	*18 357*	*7 393 F*	*12 754*	*11 684*	*7 451*
Jamaica								
Nile tilapia	*Oreochromis niloticus*	150 F	150 F	150 F	150 F	150 F	150 F	150 F
Freshwater fishes nei	*Osteichthyes*	300 F	300 F	300 F	300 F	300 F	300 F	300 F
Tuna-like fishes nei	*Scombroidei*	...	239	275	...	...	78	86
Marine fishes nei	*Osteichthyes*	7 300 F	8 500 F	5 305	4 161	6 283	4 586	4 616 F
Marine crabs nei	*Brachyura*	34	11	9	9	9	8	8 F
Caribbean spiny lobster	*Panulirus argus*	350 F	394	271	170	330	517	500 F
Penaeus shrimps nei	*Penaeus spp*	100 F	60	67	70	70	37	40 F
Stromboid conchs nei	*Strombus spp*	2 133	2 850	1 821	1 700	1 366	0	...
	Country total	*10 367 F*	*12 504 F*	*8 198*	*6 560*	*8 508*	*5 676*	*5 700 F*
Martinique								
Freshwater fishes nei	*Osteichthyes*	0	0	0	0	0	0	0
Clupeoids nei	*Clupeoidei*	50 F	50 F	100 F	500 F	3 500	3 700	4 000
Atlantic bonito	*Sarda sarda*	990 F	610	610 F	610 F	610 F	610 F	530 F
Cero	*Scomberomorus regalis*	400 F	250	250 F	250 F	...	...	...
Blackfin tuna	*Thunnus atlanticus*	890 F	540 F	540 F	540 F	540 F	540 F	470 F
Flyingfishes nei	*Exocoetidae*	0	0	0	0	0	0	0
Common dolphinfish	*Coryphaena hippurus*	350 F	250 F	350 F	320 F	300 F	250 F	220 F
Rays, stingrays, mantas nei	*Rajiformes*	5 F	3 F	5 F	5 F	5 F	5 F	5 F
Sharks, rays, skates, etc. nei	*Elasmobranchii*	100 F	70 F	90 F	80 F	70 F	50 F	40 F
Marine fishes nei	*Osteichthyes*	2 390 F	1 647 F	3 430 F	3 060 F	810 F	45 F	35 F
Caribbean spiny lobster	*Panulirus argus*	110 F	70 F	110 F	120 F	150 F	200	190
Clams, etc. nei	*Bivalvia*	...	...	...	...	...	900	700
Marine turtles nei	*Testudinata*	0	0	0	0	-	-	-
Sea urchins nei	*Strongylocentrotus spp*	15 F	10 F	15 F	15 F	15 F	10	10
	Country total	*5 300*	*3 500 F*	*5 500*	*5 500*	*6 000*	*6 310*	*6 200*
Mexico								
Common carp	*Cyprinus carpio*	22 677	21 237	16 787	14 345	10 945	15 300	14 700
Cyprinids nei	*Cyprinidae*	1 739	3 732	4 838	3 982	6 805	3 988	3 806
Tilapias nei	*Oreochromis (=Tilapia) spp*	74 646	74 354	74 814	65 178	59 343	68 772	60 336
Catfishes nei	*Ictalurus spp*	4 423	5 358	4 691	4 027	4 232	3 845	3 135
Largemouth black bass	*Micropterus salmoides*	1 212	1 014	1 058	684	784	903	693
Freshwater fishes nei	*Osteichthyes*	10 724	9 766	5 185	6 482	3 284	8 716	5 574
American eel	*Anguilla rostrata*	43	35	19	9	2	1	...
Rainbow trout	*Oncorhynchus mykiss*	1 349	1 653	102	95	91	232	223
Milkfish	*Chanos chanos*	0	0	1	0	0	7	...
Flatfishes nei	*Pleuronectiformes*	2 146	2 524	2 741	1 390	2 268	2 681	1 839
North Pacific hake	*Merluccius productus*	78	116	1 089	250	111	392	...
Gadiformes nei	*Gadiformes*	10	21	183	137	32	40	...
Tarpon	*Megalops atlanticus*	2	1	14	4	4	4	...
Sea catfishes nei	*Ariidae*	6 272	6 186	7 240	8 025	6 936	7 385	8 080
Flathead grey mullet	*Mugil cephalus*	7 636	7 882	7 128	4 299	4 872	5 190	4 838
Bobo mullet	*Joturus pichardi*	614	659	339	374	404	497	504
Mullets nei	*Mugilidae*	10 615	9 787	11 481	10 884	11 760	12 186	10 262
Common snook	*Centropomus undecimalis*	2 885	2 955	3 307	2 990	3 415	3 184	4 000
Snooks(=Robalos) nei	*Centropomus spp*	1 687	1 900	1 824	1 769	2 121	1 788	2 222
Groupers nei	*Epinephelus spp*	17 781	17 643	18 909	17 850	19 831	18 595	14 609
Yellow snapper	*Lutjanus argentiventris*	3 810	4 917	3 123	3 390	2 994	3 392	3 388
Northern red snapper	*Lutjanus campechanus*	4 714	4 555	4 219	3 392	3 445	2 726	2 717

English name Nom anglais Nombre inglés	Scientific name Nom scientifique Nombre científico	1995 mt	1996 mt	1997 mt	1998 mt	1999 mt	2000 mt	2001 mt
Lane snapper	*Lutjanus synagris*	968	950	641	591	630	430	693
Yellowtail snapper	*Ocyurus chrysurus*	825	858	840	1 900	1 554	1 357	1 600
Snappers, jobfishes nei	*Lutjanidae*	2 180	2 290	2 649	4 853	5 045	4 996	3 124
Grunts, sweetlips nei	*Haemulidae (=Pomadasyidae)*	5 636	6 385	10 513	7 352	5 414	4 065	1 641
Spotted weakfish	*Cynoscion nebulosus*	3 107	3 213	3 448	6 600	6 549	6 227	1 785
Weakfishes nei	*Cynoscion spp*	2 281	3 402	4 042	4 602	5 507	2 494	5 356
Croakers nei	*Micropogonias spp*	3 432	5 036	4 966	5 476	4 602	3 705	3 728
Gulf kingcroaker	*Menticirrhus littoralis*	948	1 070	1 220	1 824	1 576	606	484
Black drum	*Pogonias cromis*	166	134	362	392	180	207	...
Croakers, drums nei	*Sciaenidae*	326	711	692	946	900	602	-
Porgies	*Calamus spp*	257	239	96	266	577	235	250
Mojarras, etc. nei	*Gerreidae*	14 464	14 708	8 493	6 190	5 183	4 425	4 600
Triggerfishes, durgons nei	*Balistidae*	1 398	1 712	6 236	3 660	2 686	1 718	...
Ocean whitefish	*Caulolatilus princeps*	...	...	...	535	622	1 073	979
Tilefishes nei	*Branchiostegidae*	...	...	...	53	68	45	28
Hairtails, scabbardfishes nei	*Trichiuridae*	...	...	...	6 349	6 724	7 265	5 089
Demersal percomorphs nei	*Perciformes*	198	170	190	180	149	171	3 046
Round sardinella	*Sardinella aurita*	1 643	1 073	2 028	4 253	1 384	1 424	373
California pilchard	*Sardinops caeruleus*	320 999	418 093	455 837	339 317	345 458	478 191	609 777
Californian anchovy	*Engraulis mordax*	24 071	9 598	2 147	782	5 814	7 973	418
Anchovies, etc. nei	*Engraulidae*	1 564	1 464	903	1 762	1 613	1 163	1 328
Clupeoids nei	*Clupeoidei*	304	276	797	569	371	275	21
Atlantic bonito	*Sarda sarda*	1 143	1 279	2 040	2 194	2 314	1 721	1 506
Eastern Pacific bonito	*Sarda chiliensis*	6 718	399	875	423	1 775	429	146
King mackerel	*Scomberomorus cavalla*	3 050	4 377	5 370	4 598	5 002	4 576	5 199
Atlantic Spanish mackerel	*Scomberomorus maculatus*	7 673	11 049	7 389	7 381	8 382	5 717	5 320
Pacific sierra	*Scomberomorus sierra*	5 137	5 742	5 405	3 896	5 265	6 261	5 959
Skipjack tuna	*Katsuwonus pelamis*	31 760	18 330	25 171	17 697	19 160	14 850	8 221
Atlantic bluefin tuna	*Thunnus thynnus*	5	14	7	14	16	35	10
Pacific bluefin tuna	*Thunnus orientalis*	83	3 700	370	34	2 370	3 030	860
Albacore	*Thunnus alalunga*	5	21	53	8	32	159	40
Yellowfin tuna	*Thunnus albacares*	107 158	127 815	140 286	117 762	121 884	102 340	135 494
Bigeye tuna	*Thunnus obesus*	299	495	108	13	97	8	92
Atlantic sailfish	*Istiophorus albicans*	6	10	7	21	33	37	36
Marlins,sailfishes,etc. nei	*Istiophoridae*	226	626	1 468	2 539	387	375	347
Swordfish	*Xiphias gladius*	437	439	2 365	3 603	1 136	2 216	2 327
Silversides(=Sand smelts) nei	*Atherinidae*	...	...	...	1 434	1 393	1 185	1 262
Cobia	*Rachycentron canadum*	347	347	630	588	565	303	1 405
Jacks, crevalles nei	*Caranx spp*	7 160	7 960	10 679	10 933	8 296	8 266	8 870
Pompanos nei	*Trachinotus spp*	885	941	629	885	851	684	555
Amberjacks nei	*Seriola spp*	1 159	1 116	1 602	1 580	1 628	1 832	...
Carangids nei	*Carangidae*	225	274	772	1 992	1 077	490	-
Common dolphinfish	*Coryphaena hippurus*	114	314	984	1 193	480	509	...
Chub mackerel	*Scomber japonicus*	22 834	12 961	24 246	22 990	69 375	45 205	3 813
Barracudas nei	*Sphyraena spp*	814	722	1 587	1 460	1 151	549	...
Requiem sharks nei	*Carcharhinidae*	11 075	11 015	7 267	6 979	6 070	6 318	6 055
Rays, stingrays, mantas nei	*Rajiformes*	9 576	9 932	10 074	10 954	9 076	7 817	7 023
Sharks, rays, skates, etc. nei	*Elasmobranchii*	22 819	24 258	18 324	18 599	20 093	21 125	19 640
Marine fishes nei	*Osteichthyes*	316 911	262 207	239 935	205 232	137 455	143 601	147 902
River prawns nei	*Macrobrachium spp*	4 202	4 261	3 580	3 244	4 161	3 478	3 112
Euro-American crayfishes nei	*Astacidae, Cambaridae*	170	198	101	99	163	108	17
Black stone crab	*Menippe mercenaria*	120	353	610	681	355	120	...
Blue crab	*Callinectes sapidus*	10 657	13 736	14 412	12 920	12 714	8 602	7 291
Marine crabs nei	*Brachyura*	10 395	13 602	10 073	6 503	6 670	12 036	11 204
Caribbean spiny lobster	*Panulirus argus*	896	756	844	613	645	747	782
Tropical spiny lobsters nei	*Panulirus spp*	1 421	1 799	1 708	1 599	1 328	2 052	1 727
Penaeus shrimps nei	*Penaeus spp*	70 034	65 564	71 067	66 586	66 491	61 597	57 509
Marine crustaceans nei	*Crustacea*	0	0	2	3	1	1	-
Freshwater molluscs nei	*Mollusca*	198	240	120	177	239	557	...
Abalones nei	*Haliotis spp*	1 227	1 075	924	709	574	535	482
Stromboid conchs nei	*Strombus spp*	9 126	5 657	7 639	4 266	8 591	10 392	11 397
Flat oysters nei	*Ostrea spp*	1 801	1 354	1 903	1 276	2 489	1 828	2 234
American cupped oyster	*Crassostrea virginica*	27 609	34 726	38 515	30 715	39 268	48 101	48 570
Sea mussels nei	*Mytilidae*	118	412	2 036	2 056	737	547	...
Pacific calico scallop	*Argopecten ventricosus*	1 256	17 290	2 320	2 726	1 864	6 287	3 100
Ark clams nei	*Arca spp*	1 033	834	1 367	1 018	880	675	600
Venus clams nei	*Veneridae*	7 239	5 852	4 658	4 673	5 330	6 238	5 240
Common squids nei	*Loligo spp*	69	112	140	71	91	85	92
Jumbo flying squid	*Dosidicus gigas*	39 657	107 967	120 877	26 611	57 985	56 153	73 741
Common octopus	*Octopus vulgaris*	18 958	28 572	17 776	16 478	19 081	22 463	20 596
Octopuses, etc. nei	*Octopodidae*	877	1 257	944	755	1 094	883	837
Frogs	*Rana spp*	497	351	1 979	1 167	311	295	...
Echinoderms	*Echinodermata*	2 791	3 027	2 099	1 138	2 042	2 813	2 252
Sea cucumbers nei	*Holothurioidea*	...	...	...	271	234	426	481
Marine worms	*Polychaeta*	59	58	57	58	48	27	...
Aquatic invertebrates nei	*Invertebrata*	1 481	1 115	526	437	569	426	...
	Country total	*1 329 340*	*1 464 188*	*1 489 112*	*1 179 860*	*1 205 603*	*1 315 581*	*1 398 592*
Montserrat								
Freshwater fishes nei	*Osteichthyes*	0	0	0	0	0	0	0
Marine fishes nei	*Osteichthyes*	48	38	45	46	50 F	50 F	50 F
	Country total	*48*	*38*	*45*	*46*	*50 F*	*50 F*	*50 F*
NethAntilles								
Freshwater fishes nei	*Osteichthyes*	0	0	0	0	0	0	0
Atlantic bonito	*Sarda sarda*	-	-	-	-	-	2	...
Wahoo	*Acanthocybium solandri*	230	230 F	230 F	230 F	230 F	230 F	230 F

English name Nom anglais Nombre inglés	Scientific name Nom scientifique Nombre científico	1995 mt	1996 mt	1997 mt	1998 mt	1999 mt	2000 mt	2001 mt
Frigate tuna	*Auxis thazard*	-	-	-	-	-	215	...
Skipjack tuna	*Katsuwonus pelamis*	35	30	30 F	30 F	30 F	11 104 F	30 F
Blackfin tuna	*Thunnus atlanticus*	50	45	45 F	45 F	45 F	45 F	45 F
Yellowfin tuna	*Thunnus albacares*	140	130	130 F	130 F	130 F	5 756 F	130 F
Bigeye tuna	*Thunnus obesus*	-	-	-	-	-	2 627	...
Atlantic sailfish	*Istiophorus albicans*	15 F	15 F	15 F	15 F	15 F	15 F	15 F
Atlantic blue marlin	*Makaira nigricans*	40 F	40 F	40 F	40 F	40 F	40 F	40 F
Marine fishes nei	*Osteichthyes*	505 F	505 F	455 F	455 F	455 F	455 F	455 F
Stromboid conchs nei	*Strombus spp*	5 F	5 F	5 F	5 F	5 F	5 F	5 F
	Country total	*1 020 F*	*1 000 F*	*950 F*	*950 F*	*950 F*	*20 494 F*	*950 F*
Nicaragua								
Tilapias nei	*Oreochromis (=Tilapia) spp*	...	...	...	...	750	680	660
Freshwater fishes nei	*Osteichthyes*	538	1 142	1 293	1 256	363	396	388
Snooks(=Robalos) nei	*Centropomus spp*	...	...	...	...	2 000	2 060	1 960
Groupers, seabasses nei	*Serranidae*	...	...	...	...	1 040	840	770
Snappers nei	*Lutjanus spp*	...	...	...	...	2 640	2 380	3 320
Seerfishes nei	*Scomberomorus spp*	...	...	...	...	240	250	260
Skipjack tuna	*Katsuwonus pelamis*	-	-	-	-	250	430	...
Yellowfin tuna	*Thunnus albacares*	-	-	-	-	3 060	4 950	...
Bigeye tuna	*Thunnus obesus*	-	-	-	-	30	10	...
Common dolphinfish	*Coryphaena hippurus*	...	...	...	...	710	2 540	2 470
Requiem sharks nei	*Carcharhinidae*	...	...	...	...	200	150	375
Marine fishes nei	*Osteichthyes*	5 261	5 859	6 886	9 530	1 667	1 388	1 875
River prawns nei	*Macrobrachium spp*	-	-	-	-	7	0	3
Blue crab	*Callinectes sapidus*	-	-	-	131	106	71	69
Harbour spidercrab	*Mithrax armatus*	-	-	-	114	110	54	35
Caribbean spiny lobster	*Panulirus argus*	2 260	4 357	4 012	3 729	5 141	6 327	3 909
Green spiny lobster	*Panulirus gracilis*	14	101	152	66	131	476	652
Penaeus shrimps nei	*Penaeus spp*	2 922	3 983	3 833	4 864	5 218	4 431	4 966
Northern nylon shrimp	*Heterocarpus vicarius*	-	-	-	40	37	20	131
Stromboid conchs nei	*Strombus spp*	-	-	-	162	209	555	956
	Country total	*10 995*	*15 442*	*16 176*	*19 892*	*23 909*	*28 008*	*22 799*
Panama								
Tilapias nei	*Oreochromis (=Tilapia) spp*	10	15	56	11	13	16	16 F
Peacock cichlid	*Cichla ocellaris*	120	65	35	12	7	4	4 F
Tadpole codling	*Salilota australis*	90	93	-	-	-	-	-
Argentine hake	*Merluccius hubbsi*	15	14	-	-	-	-	-
Hakes nei	*Merluccius spp*	-	40	29	-	-	-	-
Patagonian grenadier	*Macruronus magellanicus*	223	339	-	-	-	-	-
Snappers, jobfishes nei	*Lutjanidae*	5 170	3 857	4 659	5 358	10 433	8 480	8 500 F
Croakers, drums nei	*Sciaenidae*	400	666	667	1 138	1 533	4 352	4 300 F
Porgies, seabreams nei	*Sparidae*	396	402	46	319	66	-	-
Pink cusk-eel	*Genypterus blacodes*	33	46	-	-	-	-	-
Patagonian toothfish	*Dissostichus eleginoides*	9	-	-	-	-	-	-
Pacific thread herring	*Opisthonema libertate*	21 224	32 517	26 266	49 472	38 746	63 532	71 280 F
Pacific anchoveta	*Cetengraulis mysticetus*	106 743	60 322	77 726	107 730	27 356	86 681	100 000 F
Pacific sierra	*Scomberomorus sierra*	463	489	665	982	159	1 395	1 000 F
Frigate tuna	*Auxis thazard*	327	240	91	-	-	-	-
Black skipjack	*Euthynnus lineatus*	-	-	-	10	-	10	10
Skipjack tuna	*Katsuwonus pelamis*	18 803	9 295	5 290	1 542	6 388	15 735	6 650
Atlantic bluefin tuna	*Thunnus thynnus*	1 517	3 400	491	-	13	-	-
Albacore	*Thunnus alalunga*	301	391	58	61	14	-	-
Yellowfin tuna	*Thunnus albacares*	15 502	11 353	9 873	5 631	7 842	7 924	12 980
Bigeye tuna	*Thunnus obesus*	10 557	5 847	3 418	1 262	1 108	4 926	1 720
Atlantic blue marlin	*Makaira nigricans*	-	-	-	-	-	41	...
Swordfish	*Xiphias gladius*	-	-	-	-	122	-	-
Tuna-like fishes nei	*Scombroidei*	-	20	-	-	77	45	...
Amberjacks nei	*Seriola spp*	341	470	333	469	159	418	400 F
Chub mackerel	*Scomber japonicus*	201	-	-	397	-	-	-
Rays, stingrays, mantas nei	*Rajiformes*	85	170	-	-	-	-	-
Sharks, rays, skates, etc. nei	*Elasmobranchii*	-	-	-	-	202	...	-
Marine fishes nei	*Osteichthyes*	6 769	5 969	17 570	15 360	15 624	18 263	18 000 F
Tropical spiny lobsters nei	*Panulirus spp*	197	291	309	417	486	612	610 F
Crystal shrimp	*Penaeus brevirostris*	2 232	1 298	2 048	1 310	1 718	2 158	2 100 F
Penaeus shrimps nei	*Penaeus spp*	3 110	2 566	2 053	3 200	2 512	1 952	2 000 F
Pacific seabobs	*Xiphopenaeus,Trachypenaeus spp*	4 190	4 339	3 606	4 960	2 872	2 685	1 602
Natantian decapods nei	*Natantia*	1 945	4 122	5 254	902	1 674	1 661	1 650 F
Antarctic krill	*Euphausia superba*	141	496	-	-	-	-	-
Marine crustaceans nei	*Crustacea*	...	3	3	2	1	1	-
Stromboid conchs nei	*Strombus spp*	...	...	6	527	125	5	5 F
Gastropods nei	*Gastropoda*	0	5	0	116	17	1	3 F
Clams, etc. nei	*Bivalvia*	1 669	1 391	1 507	1 072	1 518	1 630	1 700 F
Cuttlefish,bobtail squids nei	*Sepiidae, Sepiolidae*	78	0	53	82	-	-	-
Patagonian squid	*Loligo gahi*	1	1	-	-	-	-	-
Various squids nei	*Loliginidae, Ommastrephidae*	4	53	58	20	16	...	...
Octopuses, etc. nei	*Octopodidae*	82	89	41	102	-	-	-
Marine molluscs nei	*Mollusca*	45	48	71	172	16	76	20
Marine turtles nei	*Testudinata*	0	0	0	0	0	0	...
Marine worms	*Polychaeta*	...	42	67	51	31	28	450 F
	Country total	*202 993*	*150 764*	*162 349*	*202 687*	*120 848*	*222 631*	*235 000 F*
Puerto Rico								
Freshwater fishes nei	*Osteichthyes*	0	0	0	0	0	0	0
Squirrelfishes nei	*Holocentridae*	...	...	...	...	9	13	12

English name Nom anglais Nombre inglés	Scientific name Nom scientifique Nombre científico	1995 mt	1996 mt	1997 mt	1998 mt	1999 mt	2000 mt	2001 mt
Mullets nei	*Mugilidae*	...	...	...	...	39	44	41
Snooks(=Robalos) nei	*Centropomus spp*	...	...	...	...	32	33	8
Groupers nei	*Epinephelus spp*	...	...	...	...	89	104	111
Snappers, jobfishes nei	*Lutjanidae*	...	...	...	...	566	762	744
Grunts, sweetlips nei	*Haemulidae (=Pomadasyidae)*	...	...	...	...	76	97	104
Porgies, seabreams nei	*Sparidae*	...	...	...	...	22	24	25
Goatfishes, red mullets nei	*Mullidae*	...	...	...	...	17	17	15
Mojarras, etc. nei	*Gerreidae*	...	...	...	...	14	15	13
Wrasses, hogfishes, etc. nei	*Labridae*	...	...	...	...	30	47	46
Parrotfishes nei	*Scaridae*	...	...	...	...	52	63	66
Boxfishes nei	*Ostraciidae*	...	...	...	...	54	68	52
Triggerfishes, durgons nei	*Balistidae*	...	...	...	...	32	34	41
Demersal percomorphs nei	*Perciformes*	741	1 059	1 249	1 375	-	-	-
Clupeoids nei	*Clupeoidei*	20	18	19	14	20	20	17
Wahoo	*Acanthocybium solandri*	...	...	...	...	...	...	6
Seerfishes nei	*Scomberomorus spp*	134	106	119	109	123	145	124
Little tunny(=Atl.black skipj)	*Euthynnus alletteratus*	...	...	...	...	...	...	14
Skipjack tuna	*Katsuwonus pelamis*	...	...	...	...	...	...	26
Blackfin tuna	*Thunnus atlanticus*	...	...	...	...	...	...	17
Yellowfin tuna	*Thunnus albacares*	...	...	...	...	...	...	24
Bigeye tuna	*Thunnus obesus*	-	-	54	-	-	-	-
Tuna-like fishes nei	*Scombroidei*	82	88	72	121	99	111	17
Ballyhoo halfbeak	*Hemiramphus brasiliensis*	...	...	...	...	32	47	41
Carangids nei	*Carangidae*	...	...	...	...	50	70	66
Common dolphinfish	*Coryphaena hippurus*	...	...	...	...	83	111	74
Barracudas nei	*Sphyraena spp*	...	...	...	...	16	21	13
Pelagic percomorphs nei	*Perciformes*	186	98	97	83	-	-	-
Sharks, rays, skates, etc. nei	*Elasmobranchii*	...	...	...	...	28	35	32
Marine fishes nei	*Osteichthyes*	983	569	691	50	264	298	174
Marine crabs nei	*Brachyura*	...	...	...	...	2	2	6
Caribbean spiny lobster	*Panulirus argus*	...	...	...	...	209	212	190
Marine crustaceans nei	*Crustacea*	199	189	171	184	-	-	-
Stromboid conchs nei	*Strombus spp*	758	450	638	1 025	1 025	1 710	1 643
Octopuses, etc. nei	*Octopodidae*	...	...	...	...	28	39	23
Marine molluscs nei	*Mollusca*	70	124	77	45	9	12	9
Marine turtles nei	*Testudinata*	0	0	0	0	-	-	-
	Country total	*3 173*	*2 701*	*3 187*	*3 006*	*3 020*	*4 154*	*3 794*
St Kitts Nev								
Freshwater fishes nei	*Osteichthyes*	0	0	0	0	0	0	0
Squirrelfishes nei	*Holocentridae*	3	9	5	8	9	1	1
Groupers nei	*Epinephelus spp*	8	18	10	11	11	7	4
Snappers nei	*Lutjanus spp*	4	9	5	8	15	21	10
Grunts, sweetlips nei	*Haemulidae (=Pomadasyidae)*	1	10	1	3	3	1	1
Goatfishes, red mullets nei	*Mullidae*	3	10	1	2	1	1	1
Parrotfishes nei	*Scaridae*	7	19	5	8	8	2	8
Surgeonfishes nei	*Acanthuridae*	5	11	4	9	6	1	1
Triggerfishes, durgons nei	*Balistidae*	3	7	2	5	2	2	2
Tuna-like fishes nei	*Scombroidei*	1	3	7	16	24	24	24
Needlefishes, etc. nei	*Belonidae*	12	27	26	60	58	66	41
Flyingfishes nei	*Exocoetidae*	21	54	23	38	22	24	53
Bigeye scad	*Selar crumenophthalmus*	-	-	16	20	36	35	28
Common dolphinfish	*Coryphaena hippurus*	3	13	20	34	13	29	29
Marine fishes nei	*Osteichthyes*	80	96	97	142	222	175	278
Caribbean spiny lobster	*Panulirus argus*	12	17	12	29	34	27	35
Stromboid conchs nei	*Strombus spp*	29	49	38	140	91	76	75
	Country total	*192*	*352*	*272*	*533*	*555*	*492*	*591*
St Lucia								
Freshwater fishes nei	*Osteichthyes*	0	0	0	0	0	0	0
Snappers nei	*Lutjanus spp*	56	69	31	34	45	68	82
Gulf kingcroaker	*Menticirrhus littoralis*	20	...	...	...	...	...	...
Atlantic bonito	*Sarda sarda*	1	1	0	0	0	0	0
Wahoo	*Acanthocybium solandri*	80	221	224	223	310	243	213
Seerfishes nei	*Scomberomorus spp*	80	51	4	0	...	60	7
Little tunny(=Atl.black skipj)	*Euthynnus alletteratus*	-	-	2	2	2	-	1
Skipjack tuna	*Katsuwonus pelamis*	72	38	100	263	153	216	151
Atlantic bluefin tuna	*Thunnus thynnus*	9	3	-	-	-	-	-
Blackfin tuna	*Thunnus atlanticus*	47	35	40	100	41	45	108
Albacore	*Thunnus alalunga*	1	1	0	0	0	1	3
Yellowfin tuna	*Thunnus albacares*	144	110	109	276	123	134	145
Bigeye tuna	*Thunnus obesus*	0	0	0	0	0	-	1
Marlins,sailfishes,etc. nei	*Istiophoridae*	-	-	4	1	-	14	5
Swordfish	*Xiphias gladius*	0	-	-	-	-	-	-
Tuna-like fishes nei	*Scombroidei*	10	8	1	3	3	1	-
Flyingfishes nei	*Exocoetidae*	50	40	34	112	67	99	323
Common dolphinfish	*Coryphaena hippurus*	200	351	455	264	588	552	427
Sharks, rays, skates, etc. nei	*Elasmobranchii*	6	11	3	8	6	5	5
Marine fishes nei	*Osteichthyes*	385	308	264	243	325	352	435
Marine crustaceans nei	*Crustacea*	12	12	12	32	30	25	36
Stromboid conchs nei	*Strombus spp*	15	15	25	28	25	40	41
	Country total	*1 188*	*1 274*	*1 308*	*1 589*	*1 718*	*1 855*	*1 983*
St Pier Mq								
Freshwater fishes nei	*Osteichthyes*	0	0	0	0	0	0	0
Atlantic salmon	*Salmo salar*	0	1	1	1	1	1	1

English name Nom anglais Nombre inglés	Scientific name Nom scientifique Nombre científico	1995 mt	1996 mt	1997 mt	1998 mt	1999 mt	2000 mt	2001 mt
Atlantic halibut	*Hippoglossus hippoglossus*	-	-	1	1	1	0	1
Greenland halibut	*Reinhardtius hippoglossoides*	-	-	439	1 431	1 132	2	7
Witch flounder	*Glyptocephalus cynoglossus*	-	-	8	57	35	7	120
Amer. plaice(=Long rough dab)	*Hippoglossoides platessoides*	0	0	23	27	24	41	112
Yellowtail flounder	*Limanda ferruginea*	-	-	18	59	33	60	152
Atlantic cod	*Gadus morhua*	60	44	1 547	3 123	3 171	4 682	2 350
White hake	*Urophycis tenuis*	-	-	0	1	9	122	10
Haddock	*Melanogrammus aeglefinus*	-	-	9	27	16	10	78
Saithe(=Pollock)	*Pollachius virens*	-	-	14	13	6	38	13
Roundnose grenadier	*Coryphaenoides rupestris*	-	-	4	-	-	-	-
Wolffishes(=Catfishes) nei	*Anarhichas spp*	-	-	3	3	2	0	1
Atlantic redfishes nei	*Sebastes spp*	-	-	430	654	423	196	129
Lumpfish(=Lumpsucker)	*Cyclopterus lumpus*	226	218	363	249	422	536	146
American angler	*Lophius americanus*	-	-	0	1	0	0	1
Atlantic herring	*Clupea harengus*	-	-	-	-	2	0	0
Atlantic bluefin tuna	*Thunnus thynnus*	-	-	-	-	1	0	0
Capelin	*Mallotus villosus*	1	0	1	-	2	0	1
Atlantic mackerel	*Scomber scombrus*	1	1	1	3	1	26	7
Porbeagle	*Lamna nasus*	7	40	13	20	0	23	2
Picked dogfish	*Squalus acanthias*	0	0	0	-	-	0	0
Raja rays nei	*Raja spp*	4	3	3	9	4	21	38
Marine fishes nei	*Osteichthyes*	-	-	18	18	-	-	-
Queen crab	*Chionoecetes opilio*	2	189	368	354	589	511	498
American lobster	*Homarus americanus*	1	1	1	-	1	1	2
Whelk	*Buccinum undatum*	...	...	32	45	...	0	-
Blue mussel	*Mytilus edulis*	8	4	2	4	0	0	0
American sea scallop	*Placopecten magellanicus*	4	23	27	3	0	0	2
Iceland scallop	*Chlamys islandica*	2	221	239	5	0	3	16
Northern shortfin squid	*Illex illecebrosus*	0	0	6	-	-	-	-
Marine molluscs nei	*Mollusca*	-	1	0	0	17	205	115
Sea urchins nei	*Strongylocentrotus spp*	1	1	-	-	-	0	0
	Country total	*317*	*747*	*3 571*	*6 108*	*5 892*	*6 485*	*3 802*
St Vincent								
Freshwater fishes nei	*Osteichthyes*	0	2	1	0	0	0	0
Tadpole codling	*Salilota australis*	-	-	-	-	-	-	14
Argentine hake	*Merluccius hubbsi*	-	-	-	-	-	-	5
Hakes nei	*Merluccius spp*	-	-	10	106	31	28	51
Croakers, drums nei	*Sciaenidae*	33	-	-	-	-	-	-
Porgies, seabreams nei	*Sparidae*	16	30	10	59	39	5	30
Largehead hairtail	*Trichiurus lepturus*	-	-	130	1 393	35	24	700
Sardinellas nei	*Sardinella spp*	-	-	2 000	10 038	7 962	3 369	6 971
European pilchard(=Sardine)	*Sardina pilchardus*	-	-	46	26	96	340	41
European anchovy	*Engraulis encrasicolus*	-	-	1 100	7 182	3 702	4 841	5 653
Wahoo	*Acanthocybium solandri*	16	23	10	65	52	46	311
Cero	*Scomberomorus regalis*	0	-	-	-	-	-	-
Seerfishes nei	*Scomberomorus spp*	-	1	1	1	1	138	0
Little tunny(=Atl.black skipj)	*Euthynnus alletteratus*	0	-	-	-	-	-	-
Skipjack tuna	*Katsuwonus pelamis*	53	37	42	57	37	68	97
Blackfin tuna	*Thunnus atlanticus*	20	18	22	17	15	23	24
Albacore	*Thunnus alalunga*	0	-	-	-	1	2 820	5 662
Yellowfin tuna	*Thunnus albacares*	43	37	35	48	38	1 989	1 365
Bigeye tuna	*Thunnus obesus*	0	4	2	2	1	1 216	506
Atlantic sailfish	*Istiophorus albicans*	2	1	3	-	1	-	2
Atlantic blue marlin	*Makaira nigricans*	2	0	1	-	-	-	-
Atlantic white marlin	*Tetrapturus albidus*	0	0	-	-	-	-	-
Marlins,sailfishes,etc. nei	*Istiophoridae*	-	1	0	2	1	343	339
Swordfish	*Xiphias gladius*	4	3	1	0	1	0	22
Tuna-like fishes nei	*Scombroidei*	-	-	50	387	167	...	...
Jack and horse mackerels nei	*Trachurus spp*	-	-	1 300	9 765	3 546	8 275	16 369
Chub mackerel	*Scomber japonicus*	-	-	350	2 644	988	2 951	4 475
Sharks, rays, skates, etc. nei	*Elasmobranchii*	...	2	...	...	3	...	2
Marine fishes nei	*Osteichthyes*	783	737	969	2 160	1 035	1 211	1 297
Penaeus shrimps nei	*Penaeus spp*	1	-	-	-	-	-	-
Stromboid conchs nei	*Strombus spp*	30	25 F	10	21	7	7	37
Cuttlefish,bobtail squids nei	*Sepiidae, Sepiolidae*	2	-	-	-	-	-	2
Patagonian squid	*Loligo gahi*	-	-	-	-	-	-	1 795
Argentine shortfin squid	*Illex argentinus*	-	-	-	-	-	-	4
Various squids nei	*Loliginidae, Ommastrephidae*	-	-	-	-	-	-	4
	Country total	*1 005*	*921 F*	*6 093*	*33 973*	*17 759*	*27 694*	*45 778*
Trinidad Tob								
Freshwater fishes nei	*Osteichthyes*	0	0	0	0	0	0	0
Demersal percomorphs nei	*Perciformes*	2 400 F	2 656	2 572	2 001	1 918	1 812	2 828
Clupeoids nei	*Clupeoidei*	60 F	58	196	619	859	82	189
Atlantic bonito	*Sarda sarda*	169	266	220	30	117	117	56
Wahoo	*Acanthocybium solandri*	-	0	1	1	1	2	1
King mackerel	*Scomberomorus cavalla*	471	1 029	875	746	447	432	410
Serra Spanish mackerel	*Scomberomorus brasiliensis*	1 816	1 568	1 699	2 130	1 328	1 722	2 207
Frigate and bullet tunas	*Auxis thazard, A.rochei*	56	199	368	127	138	138	...
Skipjack tuna	*Katsuwonus pelamis*	3	...	0	-	-	-	-
Albacore	*Thunnus alalunga*	...	...	2	1	1	2	11
Yellowfin tuna	*Thunnus albacares*	79	183	223	213	163	112	122
Bigeye tuna	*Thunnus obesus*	27	37	36	24	19	5	11
Atlantic sailfish	*Istiophorus albicans*	101	104	10	...	4	3	7
Atlantic blue marlin	*Makaira nigricans*	46	21	81	70	33	55	17
Marlins,sailfishes,etc. nei	*Istiophoridae*	-	-	-	-	-	-	2

English name Nom anglais Nombre inglés	Scientific name Nom scientifique Nombre científico	1995 mt	1996 mt	1997 mt	1998 mt	1999 mt	2000 mt	2001 mt
Swordfish	*Xiphias gladius*	150	158	110	130	138	41	75
Tuna-like fishes nei	*Scombroidei*	25	134	206	92	81	1 380	405
Jacks, crevalles nei	*Caranx spp*	400 F	504	562	203	189	202	210
Blacktip shark	*Carcharhinus limbatus*	...	3	8	10	11	...	...
Sharks, rays, skates, etc. nei	*Elasmobranchii*	550 F	621	545	635	701	755	756
Marine fishes nei	*Osteichthyes*	4 947 F	1 609	2 818	1 424	1 928	2 074	3 151
Portunus swimcrabs nei	*Portunus spp*	-	-	-	0	1	1	1
Caribbean spiny lobster	*Panulirus argus*	-	-	-	2	0	1	0
Penaeus shrimps nei	*Penaeus spp*	700 F	285	751	648	658	755	856
Atlantic seabob	*Xiphopenaeus kroyeri*	...	...	...	69	89	94	79
Various squids nei	*Loliginidae, Ommastrephidae*	-	-	-	0	2	1	14
	Country total	*12 000 F*	*9 435*	*11 283*	*9 175*	*8 826*	*9 786*	*11 408*
Turks Caicos								
Freshwater fishes nei	*Osteichthyes*	0	0	0	0	0	0	0
Marine fishes nei	*Osteichthyes*	300 F	300 F	300 F	300	300 F	300 F	300 F
Caribbean spiny lobster	*Panulirus argus*	400 F	350 F	300 F	230	230 F	230 F	230 F
Stromboid conchs nei	*Strombus spp*	695	647	650 F	788	770 F	770 F	770 F
	Country total	*1 395 F*	*1 297 F*	*1 250 F*	*1 318*	*1 300 F*	*1 300 F*	*1 300 F*
USA								
Buffalofishes nei	*Ictiobus spp*	480	793	991	959	697	1 280	1 569
Suckers nei	*Catostomidae*	662	542	550	464	465	231	187
Common carp	*Cyprinus carpio*	852	1 034	1 076	1 072	1 103	724	704
Goldfish	*Carassius auratus*	5	5	10	10	7	9	10
Grass carp(=White amur)	*Ctenopharyngodon idellus*	19	10	15	13	11	17	31
Cyprinids nei	*Cyprinidae*	4	3	5	15	17	29	33
Tilapias nei	*Oreochromis (=Tilapia) spp*	2 717	4	0	0	2 651	1 040	1 277
Catfishes nei	*Ictalurus spp*	5 264	4 110	6 872	5 257	9 235	7 561	7 478
Burbot	*Lota lota*	15	14	9	28	13	20	9
American yellow perch	*Perca flavescens*	1 246	711	622	556	537	567	640
Walleye	*Stizostedion vitreum*	17	18	4	6	2	4	10
Sauger	*Stizostedion canadense*	4	1	-	0	-	-	-
Freshwater fishes nei	*Osteichthyes*	2 298	328	690	467	447	756	782
White sturgeon	*Acipenser transmontanus*	-	-	-	-	-	206	185
Green sturgeon	*Acipenser medirostris*	-	-	-	-	-	36	10
Sturgeons nei	*Acipenseridae*	147	314	285	176	138	0	-
American eel	*Anguilla rostrata*	285	441	485	460	490	649	393
Pink(=Humpback)salmon	*Oncorhynchus gorbuscha*	201 700	140 542	102 965	150 856	173 316	94 440	173 067
Chum(=Keta=Dog)salmon	*Oncorhynchus keta*	122 081	81 907	46 794	59 394	65 295	73 633	52 687
Sockeye(=Red)salmon	*Oncorhynchus nerka*	158 618	144 445	87 299	58 396	110 836	94 422	77 172
Chinook(=Spring=King)salmon	*Oncorhynchus tshawytscha*	11 219	9 282	9 876	7 361	6 929	7 303	7 524
Coho(=Silver)salmon	*Oncorhynchus kisutch*	22 307	21 653	10 555	16 305	13 259	15 350	17 424
Rainbow trout	*Oncorhynchus mykiss*	161	168	137	358	82	145	221
Pacific salmons nei	*Oncorhynchus spp*	-	-	-	-	-	50	204
Lake trout(=Char)	*Salvelinus namaycush*	406	286	492	500	494	573	451
Rainbow smelt	*Osmerus mordax*	968	711	522	321	328	398	209
Eulachon	*Thaleichthys pacificus*	200	4	27	6	8	13	142
Smelts nei	*Osmerus spp, Hypomesus spp*	71	78	846	497	437	448	230
Lake(=Common)whitefish	*Coregonus clupeaformis*	5 310	5 272	5 842	5 678	5 353	5 199	4 484
Lake cisco	*Coregonus artedi*	316	226	215	284	239	255	303
Whitefishes nei	*Coregonus spp*	2 246	1 415	1 506	2 178	1 586	1 030	768
American shad	*Alosa sapidissima*	1 183	1 731	1 387	1 984	1 195	1 201	1 622
Alewife	*Alosa pseudoharengus*	387	442	526	594	682	291	712
Hickory shad	*Alosa mediocris*	36	88	75	47	62	51	90
American gizzard shad	*Dorosoma cepedianum*	906	1 272	1 989	1 291	3 007	977	1 781
Lampreys nei	*Petromyzontidae*	13	-	-	0	4	0	-
Striped bass	*Morone saxatilis*	1 644	2 138	2 802	3 046	3 004	3 135	2 949
White perch	*Morone americana*	802	995	1 164	966	1 042	1 221	1 226
Atlantic halibut	*Hippoglossus hippoglossus*	16	13	14	8	12	11	11
Pacific halibut	*Hippoglossus stenolepis*	20 554	22 680	31 843	33 895	36 517	34 753	35 391
English sole	*Pleuronectes vetulus*	...	...	...	425	1 100	1 140	1 395
Greenland halibut	*Reinhardtius hippoglossoides*	5 860	4 652	...	307	14	6 186	4 391
Arrow-tooth flounder	*Atheresthes stomias*	5 092	10 657	6 320	8 247	12 039	18 735	14 342
Petrale sole	*Eopsetta jordani*	...	...	1 937	1 464	1 480	1 870	1 822
Witch flounder	*Glyptocephalus cynoglossus*	2 197	2 082	1 775	1 855	2 123	2 439	3 020
Rex sole	*Glyptocephalus zachirus*	...	...	...	289	544	541	571
Amer. plaice(=Long rough dab)	*Hippoglossoides platessoides*	4 612	4 397	3 937	3 662	3 134	4 213	4 425
Flathead sole	*Hippoglossoides elassodon*	9 073	11 392	12 867	18 886	14 318	16 266	16 092
Yellowfin sole	*Limanda aspera*	96 765	101 354	149 302	80 500	56 830	69 971	54 918
Yellowtail flounder	*Limanda ferruginea*	1 882	2 403	2 864	3 656	4 431	6 928	7 304
Rock sole	*Lepidopsetta bilineata*	24 969	26 202	32 752	15 199	17 192	27 517	24 213
Dover sole	*Microstomus pacificus*	...	...	12 342	9 992	10 558	9 412	7 441
Pacific sand sole	*Psettichthys melanostictus*	...	...	...	34	605	79	129
Winter flounder	*Pseudopleuronectes americanus*	4 002	5 687	5 765	5 100	4 654	5 818	6 930
Curlfin sole	*Pleuronichthys decurrens*	-	-	-	-	-	1	5
Windowpane flounder	*Scophthalmus aquosus*	759	46	48	521	166	268	177
Pacific sanddab	*Citharichthys sordidus*	-	-	-	-	-	150	-
California flounder	*Paralichthys californicus*	...	...	...	523	620	390	393
Summer flounder	*Paralichthys dentatus*	6 943	5 425	4 775	6 348	5 795	6 063	6 554
Bastard halibuts nei	*Paralichthys spp*	1 926	2 192	1 059	533	391	503	-
Flatfishes nei	*Pleuronectiformes*	33 852	36 603	21 152	19 893	14 246	8 736	5 735
Tusk(=Cusk)	*Brosme brosme*	772	468	443	354	230	188	180
Atlantic cod	*Gadus morhua*	13 440	14 253	12 982	11 119	9 727	11 367	15 064
Pacific cod	*Gadus macrocephalus*	268 257	274 569	299 970	252 197	237 679	240 635	213 967
Red hake	*Urophycis chuss*	1 607	1 087	1 329	1 343	1 556	1 571	1 732
White hake	*Urophycis tenuis*	4 279	3 289	2 217	2 365	2 625	2 985	3 483

English name Nom anglais Nombre inglés	Scientific name Nom scientifique Nombre científico	1995 mt	1996 mt	1997 mt	1998 mt	1999 mt	2000 mt	2001 mt
Haddock	*Melanogrammus aeglefinus*	398	570	1 504	2 836	3 146	4 002	5 826
Pacific tomcod	*Microgadus proximus*	...	...	...	3	1	0	0
Saithe(=Pollock)	*Pollachius virens*	3 244	2 962	4 251	5 583	4 595	4 043	4 109
Alaska pollock(=Walleye poll.)	*Theragra chalcogramma*	1 293 939	1 189 844	1 139 642	1 232 177	1 055 016	1 182 438	1 442 170
Silver hake	*Merluccius bilinearis*	15 169	16 025	15 535	14 959	14 039	12 176	13 007
North Pacific hake	*Merluccius productus*	177 039	195 290	226 616	227 504	216 889	205 351	172 051
Offshore silver hake	*Merluccius albidus*	...	32	...	5	12	5	2
Grenadiers, rattails nei	*Macrouridae*	...	...	...	500	312	315	305
Gadiformes nei	*Gadiformes*	48	1	4	20	9	21	5
Ladyfish	*Elops saurus*	-	-	726	976	1 967	121	460
Sea catfishes nei	*Ariidae*	-	-	-	-	-	0	3
Mullets nei	*Mugilidae*	10 091	7 723	8 907	8 130	4 542	6 153	6 560
Black grouper	*Mycteroperca bonaci*	...	...	...	...	...	9	7
Gag	*Mycteroperca microlepis*	...	...	...	...	...	15	10
Scamp	*Mycteroperca phenax*	...	...	...	1	14	27	17
Yellowfin grouper	*Mycteroperca venenosa*	...	...	...	...	...	3	1
Yellowedge grouper	*Epinephelus flavolimbatus*	...	...	15	18	...	73	36
Snowy grouper	*Epinephelus niveatus*	...	...	...	...	...	6	4
Warsaw grouper	*Epinephelus nigritus*	37	16	23	23	-	45	44
Groupers nei	*Epinephelus spp*	4 701	4 387	4 583	4 399	1 038	5 665	5 984
Black seabass	*Centropristis striata*	1 297	1 978	1 590	1 531	1 717	1 516	1 667
Groupers, seabasses nei	*Serranidae*	...	...	...	23	4 702	73	34
Giant seabass	*Stereolepis gigas*	0	1	1	3	2	2	3
Northern red snapper	*Lutjanus campechanus*	1 412	1 791	2 045	1 774	1 967	2 115	2 152
Snappers nei	*Lutjanus spp*	242	215	187	157	175	191	138
Vermilion snapper	*Rhomboplites aurorubens*	...	...	678	480	731	911	1 057
Snappers, jobfishes nei	*Lutjanidae*	2 699	2 407	1 916	1 648	1 826	1 624	1 654
Pigfish	*Orthopristis chrysoptera*	1	-	1	-	-	0	0
Grunts, sweetlips nei	*Haemulidae (=Pomadasyidae)*	387	293	243	292	316	403	429
Spotted weakfish	*Cynoscion nebulosus*	843	483	427	265	395	259	152
Squeteague(=Gray weakfish)	*Cynoscion regalis*	3 095	3 261	3 317	3 821	3 146	2 438	2 273
Weakfishes nei	*Cynoscion spp*	91	76	74	56	95	74	53
Atlantic croaker	*Micropogonias undulatus*	7 013	9 049	12 434	11 522	12 180	12 138	13 018
Northern kingfish	*Menticirrhus saxatilis*	34	49	21	17	13	28	39
Gulf kingcroaker	*Menticirrhus littoralis*	-	-	223	237	246	355	206
White weakfish	*Atractoscion nobilis*	33	46	28	71	112	101	124
White croaker	*Genyonemus lineatus*	256	242	167	64	74	105	137
Black drum	*Pogonias cromis*	32	39	1 230	1 967	719	2 312	2 528
Spot croaker	*Leiostomus xanthurus*	3 521	2 554	3 073	3 359	2 599	3 141	3 091
Red drum	*Sciaenops ocellatus*	2	1	2	3	6	6	3
Sheepshead	*Archosargus probatocephalus*	2 305	2 024	2 292	1 939	1 653	2 105	1 720
Scup	*Stenotomus chrysops*	2 845	2 952	2 196	1 893	1 676	1 206	1 845
Porgies, seabreams nei	*Sparidae*	431	176	398	336	93	168	219
Goatfishes, red mullets nei	*Mullidae*	-	-	-	-	-	-	36
Mojarras(=Silver-biddies) nei	*Gerres spp*	-	-	-	-	-	159	180
Opaleye	*Girella nigricans*	-	-	-	-	-	3	2
Tautog	*Tautoga onitis*	148	119	116	116	95	111	138
Cunner	*Tautogolabrus adspersus*	0	1	1	3	4	3	9
California sheephead	*Semicossyphus pulcher*	...	...	...	118	58	78	68
Marbled rockcod	*Notothenia rossii*	-	-	-	-	0	-	0
Humped rockcod	*Notothenia gibberifrons*	-	-	-	-	5	-	2
Yellowbelly rockcod	*Notothenia neglecta*	-	-	-	-	0	-	2
Grey rockcod	*Notothenia squamifrons*	-	-	-	-	5	-	0
Yellowfin notie	*Nototheniops nudifrons*	-	-	-	-	-	-	0
Antarctic rockcods nei	*Trematomus spp*	-	-	-	-	0	-	0
Antarctic silverfish	*Pleuragramma antarcticum*	-	-	-	-	-	-	0
Antarctic rockcods, noties nei	*Nototheniidae*	-	-	-	-	0	-	0
Ocean pout	*Macrozoarces americanus*	24	41	15	17	18	19	18
Sandeels(=Sandlances) nei	*Ammodytes spp*	1	-	1	1	1	0	0
Spadefishes nei	*Ephippidae*	5	6	-	14	21	29	21
Lingcod	*Ophiodon elongatus*	1 691	2 254	1 889	659	631	172	187
Atka mackerel	*Pleurogrammus azonus*	67 156	88 030	59 538	51 579	51 436	44 592	57 096
Cabezon	*Scorpaenichthys marmoratus*	...	...	...	168	173	148	119
Sculpins nei	*Cottidae*	1	-	2	5	3	64	75
Northern puffer	*Sphoeroides maculatus*	19	11	18	17	37	36	35
Triggerfishes, durgons nei	*Balistidae*	355	326	80	249	196	168	180
Toadfishes, etc. nei	*Batrachoididae*	...	...	1	6	4	21	27
American conger	*Conger oceanicus*	32	29	17	48	43	49	40
Bearded brotula	*Brotula barbata*	...	...	...	5	1	1	0
Silvery John dory	*Zenopsis conchifer*	34	27	6	49	19	28	62
Wreckfish	*Polyprion americanus*	112	82	14	6	1	-	-
Ocean whitefish	*Caulolatilus princeps*	-	-	-	-	5	4	5
Atlantic goldeneye tilefish	*Caulolatilus chrysops*	-	-	-	-	0	6	806
Great Northern tilefish	*Lopholatilus chamaeleonticeps*	1 284	1 463	1 890	1 419	616	688	285
Tilefishes nei	*Branchiostegidae*	-	-	28	294	318	483	314
Tripletail	*Lobotes surinamensis*	-	-	-	-	-	1	1
Antarctic toothfish	*Dissostichus mawsoni*	-	-	-	-	0	-	0
Patagonian toothfish	*Dissostichus eleginoides*	-	187	-	-	-	-	-
Blackfin icefish	*Chaenocephalus aceratus*	-	-	-	-	1	-	1
Mackerel icefish	*Champsocephalus gunnari*	-	-	-	-	1	-	1
South Georgia icefish	*Pseudochaenichthys georgianus*	-	-	-	-	3	-	0
Ocellated icefish	*Chionodraco rastrospinosus*	-	-	-	-	1	-	1
Spiny icefish	*Chaenodraco wilsoni*	-	-	-	-	0	-	0
Icefishes nei	*Channichthyidae*	-	-	-	-	0	-	0
Wolffishes(=Catfishes) nei	*Anarhichas spp*	464	363	309	296	258	200	250
Escolar	*Lepidocybium flavobrunneum*	...	...	...	54	35	78	47
Oilfish	*Ruvettus pretiosus*	...	...	...	10	11	38	30
Largehead hairtail	*Trichiurus lepturus*	32	1	10	5	6	41	7
Black driftfish	*Hyperoglyphe bythites*	...	...	...	...	2	7	5

English name Nom anglais Nombre inglés	Scientific name Nom scientifique Nombre científico	1995 mt	1996 mt	1997 mt	1998 mt	1999 mt	2000 mt	2001 mt
Widow rockfish	*Sebastes entomelas*	...	...	7 751	4 897	4 243	3 604	2 609
Yellowtail rockfish	*Sebastes flavidus*	...	...	2 744	3 602	3 468	3 170	2 077
Pacific ocean perch	*Sebastes alutus*	15 613	21 004	19 580	18 445	20 616	17 952	17 708
Bocaccio rockfish	*Sebastes paucispinis*	...	...	724	596	197	27	33
Canary rockfish	*Sebastes pinniger*	...	...	1 262	1 313	772	60	49
Chilipepper rockfish	*Sebastes goodei*	...	...	1 850	1 273	918	445	618
Black rockfish	*Sebastes melanops*	...	...	...	...	...	109	148
Atlantic redfishes nei	*Sebastes spp*	436	322	251	320	353	322	363
Blackbelly rosefish	*Helicolenus dactylopterus*	-	-	-	-	-	3	0
Shortspine thornyhead	*Sebastolobus alascanus*	...	...	...	...	...	331	268
Scorpionfishes nei	*Scorpaenidae*	40 877	42 983	38 733	20 098	19 079	14 953	14 115
Atlantic searobins	*Prionotus spp*	111	20	10	31	38	24	39
Sablefish	*Anoplopoma fimbria*	29 894	27 193	24 006	19 732	21 888	22 534	19 975
American angler	*Lophius americanus*	25 142	24 203	27 506	26 357	25 046	20 807	23 268
Atlantic herring	*Clupea harengus*	76 980	104 002	97 706	81 813	79 597	74 037	106 561
Pacific herring	*Clupea pallasii*	53 289	54 628	62 080	41 865	41 305	33 945	41 412
Round sardinella	*Sardinella aurita*	191	571	512	489	536	614	623
California pilchard	*Sardinops caeruleus*	42 465	32 503	42 816	42 581	59 944	67 888	75 720
Atlantic menhaden	*Brevoortia tyrannus*	365 736	304 665	322 239	276 230	208 000	207 122	261 407
Gulf menhaden	*Brevoortia patronus*	472 039	491 612	597 565	497 461	694 242	591 434	528 500
Red-eye round herring	*Etrumeus teres*	-	-	72	7	58	42	2
Atlantic thread herring	*Opisthonema oglinum*	5 065	4 538	7 548	2 576	1 570	3 056	1 256
Californian anchovy	*Engraulis mordax*	3 079	4 505	5 778	1 553	5 323	11 487	19 259
Atlantic bonito	*Sarda sarda*	212	158	161	84	83	48	48
Eastern Pacific bonito	*Sarda chiliensis*	71	449	290	1 094	87	44	6
Wahoo	*Acanthocybium solandri*	...	...	2	74	33	81	58
King mackerel	*Scomberomorus cavalla*	2 049	2 069	2 515	2 361	2 410	2 246	2 194
Atlantic Spanish mackerel	*Scomberomorus maculatus*	2 007	1 545	1 695	1 459	1 362	1 650	1 916
Frigate and bullet tunas	*Auxis thazard, A.rochei*	0	-	-	1	17	9	3
Little tunny(=Atl.black skipj)	*Euthynnus alletteratus*	150	90	451	300	514	220	357
Black skipjack	*Euthynnus lineatus*	101	84	44	231	90	0	0
Skipjack tuna	*Katsuwonus pelamis*	156 289	128 548	112 154	123 667	151 340	97 429	89 038
Atlantic bluefin tuna	*Thunnus thynnus*	906	741	1 024	1 059	1 042	1 121	1 232
Pacific bluefin tuna	*Thunnus orientalis*	643	4 769	2 272	1 962	177	316	317
Blackfin tuna	*Thunnus atlanticus*	65	53	67	53	41	50	35
Albacore	*Thunnus alalunga*	11 855	18 669	18 907	20 273	14 093	13 504	18 647
Yellowfin tuna	*Thunnus albacares*	44 594	47 187	64 225	61 911	43 623	34 276	34 610
Bigeye tuna	*Thunnus obesus*	9 399	6 523	6 659	7 977	6 909	5 715	5 899
Tunas nei	*Thunnini*	75	297	1 329	42	69	72	65
Marlins,sailfishes,etc. nei	*Istiophoridae*	1 530	1 368	100	134	214	123	250
Swordfish	*Xiphias gladius*	5 916	5 842	6 163	6 846	7 277	8 076	4 268
Halfbeaks nei	*Hemiramphus spp*	528	398	293	441	350	414	354
Opah	*Lampris guttatus*	...	...	...	115	68	58	49
Atlantic silverside	*Menidia menidia*	13	19	21	24	25	15	15
Bluefish	*Pomatomus saltatrix*	3 801	4 244	4 222	3 764	3 359	3 661	3 993
Cobia	*Rachycentron canadum*	170	187	133	142	137	112	97
Pacific jack mackerel	*Trachurus symmetricus*	1 875	2 176	1 160	1 563	1 116	1 316	3 839
Blue runner	*Caranx crysos*	531	119	152	276	180	131	157
Crevalle jack	*Caranx hippos*	146	136	264	397	323	318	305
Bar jack	*Caranx ruber*	-	-	-	-	3	5	7
Jacks, crevalles nei	*Caranx spp*	-	-	-	-	-	0	-
Florida pompano	*Trachinotus carolinus*	54	44	297	321	207	242	181
Greater amberjack	*Seriola dumerili*	...	...	...	560	517	538	452
Yellowtail amberjack	*Seriola lalandi*	-	-	-	111	30	50	39
Amberjacks nei	*Seriola spp*	1 294	1 123	829	247	189	378	386
Pomfrets, ocean breams nei	*Bramidae*	-	-	-	-	-	-	4
Common dolphinfish	*Coryphaena hippurus*	980	739	768	325	572	541	418
Chub mackerel	*Scomber japonicus*	8 607	9 977	18 396	20 423	8 724	21 349	7 260
Atlantic mackerel	*Scomber scombrus*	8 495	16 022	15 395	14 429	12 041	5 649	12 339
North Atlantic harvestfish	*Peprilus alepidotus*	25	14	15	13	16	23	18
Butterfishes, pomfrets nei	*Stromateidae*	2 916	4 393	3 422	2 578	2 762	2 122	4 962
Barracudas nei	*Sphyraena spp*	...	...	0	60	95	129	127
Sand tiger shark	*Carcharias taurus*	-	-	-	-	-	1	-
Thresher	*Alopias vulpinus*	1	...	...	320	264	305	386
Bigeye thresher	*Alopias superciliosus*	...	...	...	11	5	5	2
Shortfin mako	*Isurus oxyrinchus*	5	...	...	96	61	103	70
Longfin mako	*Isurus paucus*	0	-	1	1	-	4	3
Mako sharks	*Isurus spp*	-	-	-	-	-	-	47
Porbeagle	*Lamna nasus*	0	-	-	-	-	3	1
Great white shark	*Carcharodon carcharias*	-	-	-	-	-	2	0
Nurse shark	*Ginglymostoma cirratum*	214	-	-	-	-	0	-
Blue shark	*Prionace glauca*	...	...	...	1	-	1	2
Sandbar shark	*Carcharhinus plumbeus*	1	-	-	-	-	41	24
Blacktip shark	*Carcharhinus limbatus*	-	-	1	0	-	601	521
Dusky shark	*Carcharhinus obscurus*	0	-	-	-	-	80	0
Tiger shark	*Galeocerdo cuvier*	-	-	-	-	-	-	1
Hammerhead sharks, etc. nei	*Sphyrnidae*	...	...	...	1	-	-	-
Dusky smooth-hound	*Mustelus canis*	0	...	...	...	...	334	321
Brown smooth-hound	*Mustelus henlei*	...	...	...	3	5	3	4
Tope shark	*Galeorhinus galeus*	...	...	...	52	73	48	45
Picked dogfish	*Squalus acanthias*	129	...	0	6	566	8 548	2 875
Dogfish sharks nei	*Squalidae*	23 903	29 638	21 021	22 268	16 081	1 857	295
Dogfish sharks, etc. nei	*Squaliformes*	1 470	3 317	1 083	772	1 530	590	784
Raja rays nei	*Raja spp*	6 454	13 891	10 142	13 932	12 619	13 335	13 122
Rays, stingrays, mantas nei	*Rajiformes*	430	1 554	2 488	1 340	1 616	1 716	1 500
Sharks, rays, skates etc nei	*Elasmobranchii*	4 947	3 643	5 689	5 757	4 739	3 358	2 069
Groundfishes nei	*Osteichthyes*	0	-	-	-	-	-	0
Finfishes nei	*Osteichthyes*	15 595	12 935	12 153	4 431	6 398	5 907	2 566
Marine fish	*Osteichthyes*	19 345	12 585	13 351	12 233	9 842	12 539	11 436

English name Nom anglais Nombre inglés	Scientific name Nom scientifique Nombre científico	1995 mt	1996 mt	1997 mt	1998 mt	1999 mt	2000 mt	2001 mt
River prawns nei	*Macrobrachium spp*	19	-	-	-	-	0	-
Euro-American crayfishes nei	*Astacidae, Cambaridae*	7 073	5 689	10 496	10 082	5 322	217	4 676
Atlantic rock crab	*Cancer irroratus*	844	55	1 120	1 358	1 301	1 844	326
Dungeness crab	*Cancer magister*	21 696	29 478	17 328	15 519	16 080	17 109	16 525
Jonah crab	*Cancer borealis*	332	334	745	1 255	1 549	1 114	1 245
Pacific rock crab	*Cancer productus*	...	...	...	574	359	494	537
Black stone crab	*Menippe mercenaria*	2 376	3 216	2 483	2 567	1 877	2 961	2 996
Blue crab	*Callinectes sapidus*	92 197	100 919	106 527	101 880	91 851	83 402	71 867
Green crab	*Carcinus maenas*	0	-	54	86	14	16	67
Tanner crabs nei	*Chionoecetes spp*	36 658	30 785	53 932	114 230	83 989	15 665	12 177
Red crab	*Geryon quinquedens*	521	465	96	-	-	3 132	4 004
Marine crabs nei	*Brachyura*	3 667	2 940	4 566	2 297	3 195	3 041	6 462
Caribbean spiny lobster	*Panulirus argus*	2 934	3 373	2 783	2 343	2 749	2 571	1 527
Tropical spiny lobsters nei	*Panulirus spp*	297	395	501	350	287	361	324
American lobster	*Homarus americanus*	30 122	32 496	38 066	36 125	39 676	37 730	32 389
Blue mud shrimp	*Upogebia pugettensis*	...	...	...	...	...	10	4
Ghost shrimps	*Callianassa spp*	-	-	-	-	25	17	14
King crabs	*Paralithodes spp*	6 656	9 526	8 177	10 941	7 675	6 849	7 283
Antarctic stone crab	*Paralomis spinosissima*	-	497	-	-	-	-	-
Northern brown shrimp	*Penaeus aztecus*	57 126	55 361	47 836	50 722	61 206	63 817	68 869
Northern pink shrimp	*Penaeus duorarum*	9 270	13 800	8 874	10 718	5 925	5 589	7 144
Northern white shrimp	*Penaeus setiferus*	39 959	28 808	32 841	39 799	44 633	52 593	40 724
Penaeus shrimps nei	*Penaeus spp*	4 498	3 793	6 675	5 708	2 358	1 800	2 663
Atlantic seabob	*Xiphopenaeus kroyeri*	1 724	4 558	5 744	3 397	3 626	3 415	3 951
Northern prawn	*Pandalus borealis*	7 416	9 932	7 239	3 926	2 832	2 625	1 309
Pacific shrimps nei	*Pandalus spp, Pandalopsis spp*	15 130	16 737	20 419	6 890	14 651	16 530	19 049
Crangonid shrimps nei	*Crangonidae*	-	-	31	41	-	42	45
Rock shrimp	*Sicyonia brevirostris*	3 848	10 549	1 789	4 409	1 826	3 254	2 909
Pacific rock shrimp	*Sicyonia ingentis*	...	...	...	185	630	756	165
Royal red shrimp	*Pleoticus robustus*	252	198	209	195	286	391	305
Antarctic krill	*Euphausia superba*	-	-	-	-	-	-	1 631
Brine shrimp	*Artemia salina*	...	...	...	513	691	561	538
Stomatopods nei	*Stomatopoda*	1	-	-	1	5	6	1
Freshwater molluscs nei	*Mollusca*	-	-	-	5	-	-	-
Periwinkles nei	*Littorina spp*	32	19	-	171	223	216	649
Abalones nei	*Haliotis spp*	179	127	102	-	8	-	3
Whelks	*Busycon spp*	3 402	4 196	3 104	2 581	4 624	2 930	4 037
Gastropods nei	*Gastropoda*	4 246	6 380	12	49	36	35	26
Olympia flat oyster	*Ostrea lurida*	...	...	...	...	6	7	8
Pacific cupped oyster	*Crassostrea gigas*	1 709	6 706	8 363	2 118	539	1 917	1 083
American cupped oyster	*Crassostrea virginica*	128 613	111 682	104 466	101 383	89 714	197 072	123 463
Cupped oysters nei	*Crassostrea spp*	...	...	...	64	49	42	26
Blue mussel	*Mytilus edulis*	11 210	9 434	8 063	4 774	4 086	6 009	6 524
Sea mussels nei	*Mytilidae*	1 489	626	170	174	96	771	538
American sea scallop	*Placopecten magellanicus*	62 456	61 552	48 120	43 030	77 206	113 141	164 998
Calico scallop	*Argopecten gibbus*	10 003	...	...	...	...	...	...
Atlantic bay scallop	*Argopecten irradians*	1 593	230	452	690	216	154	30
Iceland scallop	*Chlamys islandica*	0	-	-	-	-	-	-
Weathervane scallop	*Patinopecten caurinus*	1 950	2 372	2	3 228	2 642	2 012	1 052
Ocean quahog	*Arctica islandica*	183 777	173 685	163 799	148 506	144 366	122 547	142 662
Butter clam	*Saxidomus giganteus*	...	...	87	126	74	164	48
Pacific littleneck clam	*Protothaca staminea*	-	-	-	-	-	-	102
Northern quahog(=Hard clam)	*Mercenaria mercenaria*	16 058	21 855	4 631	...	...	13 723	10 136
Pacific horse clams nei	*Tresus spp*	-	-	-	-	-	1	2
Atlantic surf clam	*Spisula solidissima*	155 382	154 380	140 621	131 256	142 067	165 269	165 708
Stimpson's surf clam	*Spisula polynyma*	9	15	43	32	23	6	16
Atl.jackknife(=Atl.razor clam)	*Ensis directus*	-	-	14	49	64	99	36
Pacific razor clam	*Siliqua patula*	-	-	-	-	-	14	61
Sand gaper	*Mya arenaria*	4 813	4 355	4 454	5 493	5 113	5 222	6 917
Pacific geoduck	*Panopea abrupta*	-	-	1 962	2 243	2 451	2 259	2 481
Cockles nei	*Cardiidae*	-	-	22	-	18	49	67
Clams, etc. nei	*Bivalvia*	33 516	8 369	7 125	8 926	5 947	1 780	3 402
Longfin squid	*Loligo pealei*	18 926	12 490	16 161	18 879	18 749	16 942	14 211
Northern shortfin squid	*Illex illecebrosus*	14 222	17 050	13 632	22 717	7 334	9 011	4 009
Jumbo flying squid	*Dosidicus gigas*	-	-	3	107	18	1	0
Various squids nei	*Loliginidae, Ommastrephidae*	70 966	79 338	71 710	3 409	91 015	117 849	86 879
Octopuses, etc. nei	*Octopodidae*	5	-	0	128	8	5	26
Marine molluscs nei	*Mollusca*	4 488	2 787	4 026	3 347	3 377	1 133	1 893
Frogs	*Rana spp*	5	2	9	6	-	0	1
Diamond back terrapins	*Malaclemys spp*	0	-	-	0	-	0	-
River and lake turtles nei	*Testudinata*	6	15	13	24	15	53	51
Horseshoe crab	*Limulus polyphemus*	926	1 598	2 607	3 252	2 397	1 696	1 299
Starfishes nei	*Asteroidea*	0	-	-	-	1	0	0
Sea urchins nei	*Strongylocentrotus spp*	26 523	20 381	20 216	13 626	15 218	14 014	12 460
Sea cucumbers nei	*Holothurioidea*	729	1 779	-	2 406	3 732	4 583	1 804
Jellyfishes	*Rhopilema spp*	...	...	...	...	578	137	-
Marine worms	*Polychaeta*	0	-	223	299	343	202	481
	Country total	*5 224 566*	*5 001 483*	*4 983 440*	*4 708 980*	*4 749 646*	*4 745 321*	*4 944 406*

US Virgin Is

English name	Scientific name	1995	1996	1997	1998	1999	2000	2001
Freshwater fishes nei	*Osteichthyes*	0	0	0	0	0	0	-
Groupers nei	*Epinephelus spp*	...	...	...	...	21	25 F	25 F
Snappers, jobfishes nei	*Lutjanidae*	...	...	...	...	57	65 F	65 F
Triggerfishes, durgons nei	*Balistidae*	...	...	...	...	31	35 F	35 F
Marine fishes nei	*Osteichthyes*	400 F	340 F	300 F	255 F	119	139 F	139 F
Caribbean spiny lobster	*Panulirus argus*	55 F	50 F	45 F	40 F	34	35 F	35 F
Stromboid conchs nei	*Strombus spp*	15 F	10 F	5 F	5 F	1	1 F	1 F
Marine molluscs nei	*Mollusca*	0	0	0	0	0	0	0

E-2 Fish crustaceans, molluscs, etc / Poissons, crustacés, mollusques, etc / Peces, crustáceos, moluscos, etc — Capture production by countries or areas and species / Captures par pays ou zones et espèces / Capturas por países o áreas y especies — America, North / Amérique du Nord / América del Norte

English name Nom anglais Nombre inglés	Scientific name Nom scientifique Nombre científico	1995 mt	1996 mt	1997 mt	1998 mt	1999 mt	2000 mt	2001 mt
	Country total	*470 F*	*400 F*	*350 F*	*300 F*	*263*	*300 F*	*300 F*
Total		*7 984 054*	*7 891 184*	*8 008 495*	*7 516 032*	*7 574 700*	*7 823 129*	*8 077 213*

English name Nom anglais Nombre inglés	Scientific name Nom scientifique Nombre científico	1995 mt	1996 mt	1997 mt	1998 mt	1999 mt	2000 mt	2001 mt
Argentina								
Characins nei	*Characidae*	11 700 F	13 000 F	15 500 F	16 000 F	2 800 F	3 000 F	2 300 F
Prochilods nei	*Prochilodus spp*	...	...	...	...	16 000 F	17 700 F	14 000 F
Freshwater siluroids nei	*Siluroidei*	2 500 F	2 800 F	3 000 F	3 000 F	3 500 F	3 500 F	3 000 F
Freshwater fishes nei	*Osteichthyes*	2 908 F	3 297 F	4 135 F	4 097 F	5 138 F	6 098 F	4 460 F
Rainbow trout	*Oncorhynchus mykiss*	3	2	0	0	0	-	-
Atlantic sabretooth anchovy	*Lycengraulis grossidens*	80 F	90 F	100 F	100 F	120 F	120 F	100 F
Bastard halibuts nei	*Paralichthys spp*	10 213	8 753	10 044	8 751	6 668	6 490	5 373
Blue antimora	*Antimora rostrata*	1	-	-	-	-	-	-
Tadpole codling	*Salilota australis*	2 241	1 742	2 632	3 604	6 607	8 433	1 834
Brazilian codling	*Urophycis brasiliensis*	416	618	182	3 329	754	1 036	...
Southern blue whiting	*Micromesistius australis*	104 208	85 040	79 945	71 643	55 097	61 313	54 005
Southern hake	*Merluccius australis*	3 899	4 115	3 011	3 125	3 471	7 035	4 648
Argentine hake	*Merluccius hubbsi*	574 317	597 557	584 048	458 433	311 953	191 440	248 804
Patagonian grenadier	*Macruronus magellanicus*	22 796	44 065	41 835	96 157	117 571	123 480	111 349
Grenadiers nei	*Macrourus spp*	672	829	1 041	2 972	2 580	10 503	...
Sea catfishes nei	*Ariidae*	16	6	13	18	5	5	...
Mullets nei	*Mugilidae*	18	10	10	10	10	5	...
Argentine seabass	*Acanthistius brasilianus*	10 926	10 714	9 113	5 216	5 895	4 116	4 692
Striped weakfish	*Cynoscion striatus*	19 218	18 987	24 132	17 108	11 107	9 434	11 303
Whitemouth croaker	*Micropogonias furnieri*	29 989	23 514	26 108	9 451	6 641	5 264	4 479
Argentine croaker	*Umbrina canosai*	3 561	6 825	2 753	1 956	1 410	421	...
King weakfish	*Macrodon ancylodon*	149	380	224	167	515	44	...
Black drum	*Pogonias cromis*	40	159	81	260	182	13	...
South American silver porgy	*Diplodus argenteus*	8	39	84	15	5	2	...
Red porgy	*Pagrus pagrus*	1 203	1 590	1 159	574	2 159	1 301	805
Surmullets(=Red mullets) nei	*Mullus spp*	63	75	73	99	229	498	...
Patagonian blennie	*Eleginops maclovinus*	66	148	57	1 194	2 023	1 745	...
Marbled rockcod	*Notothenia rossii*	2	-	-	-	-	-	-
Humped rockcod	*Notothenia gibberifrons*	1	-	-	-	-	-	-
Antarctic rockcods, noties nei	*Nototheniidae*	2	-	-	-	-	-	-
Brazilian flathead	*Percophis brasilianus*	8 680	8 771	11 475	9 677	6 526	7 255	7 040
Argentinian sandperch	*Pseudopercis semifasciata*	3 023	3 438	2 483	2 222	2 679	1 905	1 887
Argentine conger	*Conger orbignyanus*	21	22	28	81	18	14	...
Pink cusk-eel	*Genypterus blacodes*	23 265	21 933	21 917	25 086	21 503	15 019	19 591
Wreckfish	*Polyprion americanus*	...	...	...	...	...	129	...
Castaneta	*Cheilodactylus bergi*	10 409	204	744	1 827	155	80	49
Patagonian toothfish	*Dissostichus eleginoides*	19 996	14 912	8 793	9 950	7 702	7 771	6 408
Patagonian rockcod	*Patagonotothen brevicauda*	1	-	-	-	-	-	-
Mackerel icefish	*Champsocephalus gunnari*	10	-	-	-	-	-	-
White snake mackerel	*Thyrsitops lepidopoides*	66	1 232	309	285	136	-	-
Choicy ruff	*Seriolella porosa*	...	...	...	...	...	3 542	3 990
Blackbelly rosefish	*Helicolenus dactylopterus*	4 670	1 499	2 563	1 192	3 354	3 050	...
Demersal percomorphs nei	*Perciformes*	1 763	2 420	2 714	2 537	3 540	-	-
Argentine menhaden	*Brevoortia pectinata*	294	427	893	104	205	271	...
Argentine anchovy	*Engraulis anchoita*	24 457	21 001	25 198	13 350	9 832	12 157	11 284
Atlantic bonito	*Sarda sarda*	138	108	130	12	38	19	0
Skipjack tuna	*Katsuwonus pelamis*	0	1	-	-	-	-	-
Albacore	*Thunnus alalunga*	0	0	120	-	-	-	-
Swordfish	*Xiphias gladius*	0	-	-	-	-	-	5
Tuna-like fishes nei	*Scombroidei*	0	0	-	-	-	-	-
Silversides(=Sand smelts) nei	*Atherinidae*	411	583	520	48	618	14	...
Bluefish	*Pomatomus saltatrix*	565	342	416	15	286	416	...
Rough scad	*Trachurus lathami*	196	587	288	247	470	67	...
Yellowtail amberjack	*Seriola lalandi*	9	16	17	7	9	10	...
Parona leatherjacket	*Parona signata*	1 478	1 925	1 805	1 744	1 710	1 479	860
Chub mackerel	*Scomber japonicus*	13 442	11 195	10 468	3 224	7 012	10 122	3 360
Narrownose smooth-hound	*Mustelus schmitti*	11 057	10 252	9 956	11 266	9 062	7 119	8 780
Tope shark	*Galeorhinus galeus*	104	92	103	92	89	109	...
Argentine angelshark	*Squatina argentina*	3 802	4 281	4 410	4 311	3 368	3 123	3 339
Rays, stingrays, mantas nei	*Rajiformes*	7 218	12 478	12 119	14 855	12 116	13 265	15 179
Elephantfishes nei	*Callorhinchus spp*	921	815	1 329	1 770	1 977	1 378	...
Sharks, rays, skates, etc. nei	*Elasmobranchii*	2 230	2 251	1 070	1 220	905	719	740
Marine fishes nei	*Osteichthyes*	10 955	4 960	7 855	6 702	5 998	7 115	21 767
Marine crabs nei	*Brachyura*	345	390	403	326	170	9	-
Southern king crab	*Lithodes antarcticus*	380	200	413	456	270	102	58
Softshell red crab	*Paralomis granulosa*	...	...	...	...	...	266	155
Argentine stiletto shrimp	*Artemesia longinaris*	250	263	166	146	37	37	...
Argentine red shrimp	*Pleoticus muelleri*	6 705	9 874	6 479	23 203	15 888	36 769	78 077
Antarctic krill	*Euphausia superba*	-	-	-	-	6 524	-	-
Marine crustaceans nei	*Crustacea*	80	2	288	2	6 027	-	-
Angulate volute	*Zidona dufresnei*	574	558	1 322	1 010	683	631	581
River Plata mussel	*Mytilus platensis*	388	164	180	149	332	236	186
Patagonean scallop	*Zygochlamis patagonica*	10 592	36 952	39 817	28 441	42 700	36 514	38 960
Common squids nei	*Loligo spp*	914	184	1 999	802	209	268	...
Argentine shortfin squid	*Illex argentinus*	199 048	292 628	411 994	291 174	342 691	278 988	229 874
Sevenstar flying squid	*Martialia hyadesi*	404	-	-	-	-	653	...
Octopuses, etc. nei	*Octopodidae*	34	39	48	17	34	5	-
Marine molluscs nei	*Mollusca*	13	1	50	-	1 000	-	-
	Country total	*1 170 124*	*1 291 355*	*1 400 162*	*1 164 829*	*1 078 313*	*913 622*	*923 322*
Bolivia								
Freshwater fishes nei	*Osteichthyes*	4 726	4 800	4 850	4 865	4 860	4 911	4 900
Rainbow trout	*Oncorhynchus mykiss*	116	338	338	340	342	345	280
Silversides(=Sand smelts) nei	*Atherinidae*	850	850	850	850	850	850	760
	Country total	*5 692*	*5 988*	*6 038*	*6 055*	*6 052*	*6 106*	*5 940*

E-3

English name Nom anglais Nombre inglés	Scientific name Nom scientifique Nombre científico	1995 mt	1996 mt	1997 mt	1998 mt	1999 mt	2000 mt	2001 mt
Brazil								
Cyprinids nei	*Cyprinidae*	119	284	212	258	302	355	360 F
Tilapias nei	*Oreochromis (=Tilapia) spp*	6 234	7 300	6 927	7 444	6 616	7 893	7 900 F
Cichlids nei	*Cichlidae*	12 045	8 565	8 356	7 785	7 375	8 145	8 200 F
Cachama	*Colossoma macropomum*	11 379	4 298	2 838	2 760	2 905	4 965	5 000 F
Characins nei	*Characidae*	81 433	80 598	65 273	60 983	65 149	67 297	67 500 F
Freshwater siluroids nei	*Siluroidei*	40 048	47 302	52 601	49 866	57 014	58 474	58 600 F
Freshwater fishes nei	*Osteichthyes*	40 372	42 285	40 507	43 588	42 875	49 593	50 000 F
Rainbow trout	*Oncorhynchus mykiss*	0	0	0	0	0	0	0
Flatfishes nei	*Pleuronectiformes*	1 491	1 091	1 430	1 655	1 590	1 844	1 820 F
Brazilian codling	*Urophycis brasiliensis*	2 924	2 311	2 116	2 408	1 807	2 225	2 200 F
Argentine hake	*Merluccius hubbsi*	255	0	-		128	226	220 F
Patagonian grenadier	*Macruronus magellanicus*	30	18	20	0	0	0	0
Ladyfish	*Elops saurus*	147	-	-	-	-	-	-
Tarpon	*Megalops atlanticus*	1 221	348	720	551	315	331	330 F
Sea catfishes nei	*Ariidae*	14 226	16 946	17 736	20 489	25 224	29 473	29 200 F
Mullets nei	*Mugilidae*	13 324	7 722	10 262	10 006	10 886	11 987	11 900 F
Snooks(=Robalos) nei	*Centropomus spp*	1 820	1 686	1 866	2 996	3 602	4 366	4 300 F
Brazilian groupers nei	*Mycteroperca spp*	1 117	423	584	518	628	536	530 F
Red grouper	*Epinephelus morio*	872	320	238	212	666	1 220	1 200 F
Groupers nei	*Epinephelus spp*	2 286	2 341	2 170	2 315	1 710	2 670	2 630 F
Bigeyes nei	*Priacanthus spp*	70	59	55	46	60	67	70 F
Southern red snapper	*Lutjanus purpureus*	5 816	5 104	6 085	5 937	9 790	6 580	6 500 F
Lane snapper	*Lutjanus synagris*	506	648	659	1 038	1 031	1 014	1 000 F
Yellowtail snapper	*Ocyurus chrysurus*	4 766	4 167	5 000	3 317	4 541	4 165	4 100 F
Snappers, jobfishes nei	*Lutjanidae*	4 550	3 690	4 488	4 390	4 549	3 859	3 800 F
Barred grunt	*Conodon nobilis*	339	172	118	114	84	39	40 F
Grunts, sweetlips nei	*Haemulidae (=Pomadasyidae)*	1 812	2 081	1 936	2 000	2 382	3 024	3 000 F
Weakfishes nei	*Cynoscion spp*	27 075	26 470	25 960	26 677	39 332	43 523	43 000 F
Whitemouth croaker	*Micropogonias furnieri*	22 024	23 374	23 311	27 927	21 926	28 797	28 500 F
Kingcroakers nei	*Menticirrhus spp*	1 330	1 144	1 303	1 395	1 058	1 348	1 330 F
Argentine croaker	*Umbrina canosai*	9 268	4 983	3 358	2 216	4 849	7 729	7 700 F
King weakfish	*Macrodon ancylodon*	3 677	5 143	6 006	6 708	4 278	5 475	5 400 F
Black drum	*Pogonias cromis*	38	1	-	-	-	-	-
Croakers, drums nei	*Sciaenidae*	69	150	52	57	36	34	30 F
Red porgy	*Pagrus pagrus*	83	884	1 572	1 448	1 362	1 497	1 480 F
Surmullets(=Red mullets) nei	*Mullus spp*	964	530	749	840	1 117	1 756	1 730 F
Percoids nei	*Percoidei*	720	664	692	1 838	947	1 465	1 450 F
Brazilian flathead	*Percophis brasilianus*	640	460	412	514	709	607	600 F
Spadefishes nei	*Ephippidae*	40	71	55	51	283	325	320 F
Puffers nei	*Tetraodontidae*	468	18	88	23	16	35	30 F
Argentine conger	*Conger orbignyanus*	182	67	110	108	162	108	100 F
Cusk-eels, brotulas nei	*Ophidiidae*	106	176	258	452	544	507	500 F
Tilefishes nei	*Branchiostegidae*	812	1 098	1 000	786	524	547	540 F
Largehead hairtail	*Trichiurus lepturus*	1 197	736	938	1 405	1 230	1 665	1 650 F
Atlantic searobins	*Prionotus spp*	922	792	540	858	1 149	1 839	1 800 F
Anglerfishes nei	*Lophiidae*	366	294	399	542	794	1 937	1 900 F
Demersal percomorphs nei	*Perciformes*	5 217	5 697	9 825	9 828	12 050	7 644	7 600 F
Brazilian sardinella	*Sardinella brasiliensis*	60 212	97 092	117 642	82 283	25 518	17 053	35 000
Brazilian menhaden	*Brevoortia aurea*	11 133	6 294	2 396	2 936	2 202	1 123	1 100 F
Scaled sardines	*Harengula spp*	-	20	12	2	162	126	120 F
Atlantic thread herring	*Opisthonema oglinum*	6 661	3 762	3 712	8 082	8 204	12 455	12 300 F
Anchovies, etc. nei	*Engraulidae*	3 806	7 672	7 102	2 654	2 682	4 039	4 000 F
Clupeoids nei	*Clupeoidei*	2 775	3 464	4 314	6 042	7 527	11 457	11 300 F
Atlantic bonito	*Sarda sarda*	137	-	-		-	-	-
Wahoo	*Acanthocybium solandri*	1	16	58	41	-	-	-
King mackerel	*Scomberomorus cavalla*	1 328	2 890	2 398	3 595	3 595	2 344	1 251
Serra Spanish mackerel	*Scomberomorus brasiliensis*	1 308	3 047	2 125	1 516	1 516	988	251
Frigate and bullet tunas	*Auxis thazard, A.rochei*	558	527	215	162	166	106	98
Little tunny(=Atl.black skipj)	*Euthynnus alletteratus*	1 059	834	507	920	930	615	615
Skipjack tuna	*Katsuwonus pelamis*	16 560	22 528	26 564	23 789	23 188	25 164	24 146
Atlantic bluefin tuna	*Thunnus thynnus*	0	0	0	0	13	0	0
Blackfin tuna	*Thunnus atlanticus*	153	649	418	55	55	38	149
Albacore	*Thunnus alalunga*	923	819	652	3 418	1 872	4 414	6 862
Yellowfin tuna	*Thunnus albacares*	4 021	2 767	2 705	2 514	4 127	6 145	6 239
Bigeye tuna	*Thunnus obesus*	1 935	1 707	1 237	644	2 024	2 768	2 659
Atlantic sailfish	*Istiophorus albicans*	245	310	137	184	356	598	412
Atlantic blue marlin	*Makaira nigricans*	180	331	193	486	509	467	780
Atlantic white marlin	*Tetrapturus albidus*	105	75	105	217	158	106	172
Longbill spearfish	*Tetrapturus pfluegeri*	-	-	-	-	-	12	56
Marlins,sailfishes,etc. nei	*Istiophoridae*	-	-	-	-	-	18	2
Swordfish	*Xiphias gladius*	1 975	1 892	4 100	3 847	4 721	4 697	4 082
Tuna-like fishes nei	*Scombroidei*	58	-	446	258	570	151	5
Ballyhoo halfbeak	*Hemiramphus brasiliensis*	631	494	442	1 500	832	989	980 F
Flyingfishes nei	*Exocoetidae*	1 036	743	1 082	1 084	760	388	380 F
Silversides(=Sand smelts) nei	*Atherinidae*	40	52	62	52	17	39	40 F
Bluefish	*Pomatomus saltatrix*	7 588	5 866	2 616	2 504	2 064	3 314	3 300 F
Cobia	*Rachycentron canadum*	256	498	367	622	1 818	1 580	1 550 F
Rough scad	*Trachurus lathami*	536	389	313	81	25	40	40 F
Blue runner	*Caranx crysos*	338	503	515	443	590	625	620 F
Jacks, crevalles nei	*Caranx spp*	3 656	3 905	5 838	5 635	4 164	6 950	6 900 F
Atlantic moonfish	*Selene setapinnis*	2 468	1 679	1 914	1 512	1 514	1 386	1 370 F
Pompanos nei	*Trachinotus spp*	424	534	166	172	144	286	280 F
Yellowtail amberjack	*Seriola lalandi*	653	538	466	880	544	611	600 F
Amberjacks nei	*Seriola spp*	856	825	1 048	119	104	128	120 F
Atlantic bumper	*Chloroscombrus chrysurus*	1 868	952	1 035	4 065	3 519	1 573	1 550 F
Common dolphinfish	*Coryphaena hippurus*	3 186	2 500	4 028	4 117	2 848	4 359	4 300 F
Chub mackerel	*Scomber japonicus*	8 049	5 670	8 306	9 422	1 595	6 377	6 300 F

English name Nom anglais Nombre inglés	Scientific name Nom scientifique Nombre científico	1995 mt	1996 mt	1997 mt	1998 mt	1999 mt	2000 mt	2001 mt
Barracudas nei	*Sphyraena spp*	20	11	12	828	217	527	520 F
Shortfin mako	*Isurus oxyrinchus*	...	83	190	...	100	120	120 F
Blue shark	*Prionace glauca*	...	743	1 103	...	500	580	570 F
Silky shark	*Carcharhinus falciformis*	...	502	279	...	70	80	80 F
Scalloped hammerhead	*Sphyrna lewini*	...	25	170	...	30	38	40 F
Angelsharks, sand devils nei	*Squatinidae*	113	1 587	...	...	...	...	...
Chola guitarfish	*Rhinobatos percellens*	162	404	...	...	...	...	...
Rays, stingrays, mantas nei	*Rajiformes*	3 948	3 104	3 010	4 673	4 277	4 867	4 800 F
Sharks, rays, skates, etc. nei	*Elasmobranchii*	10 658	8 446	10 189	12 596	13 576	15 900	15 700 F
Marine fishes nei	*Osteichthyes*	145 188	143 419	146 661	138 925	169 938	176 695	165 661 F
River prawns nei	*Macrobrachium spp*	1 412	2 677	2 157	1 506	3 235	2 437	2 440 F
Freshwater crustaceans nei	*Crustacea*	0	0	0	0	0	0	0
Dana swimcrab	*Callinectes danae*	2 062	2 020	2 600	3 014	1 626	1 597	1 580 F
Marine crabs nei	*Brachyura*	11 917	9 922	10 233	9 202	10 913	11 135	11 000 F
Caribbean spiny lobster	*Panulirus argus*	10 817	8 026	7 502	6 002	6 334	6 469	6 400 F
Redspotted shrimp	*Penaeus brasiliensis*	6 565	8 743	10 758	7 796	9 092	10 728	10 600 F
Penaeus shrimps nei	*Penaeus spp*	27 132	15 007	14 506	13 998	12 550	16 509	16 300 F
Atlantic seabob	*Xiphopenaeus kroyeri*	7 278	11 764	15 257	13 755	9 964	11 948	11 800 F
Marine crustaceans nei	*Crustacea*	287	288	310	244	266	227	220 F
Cupped oysters nei	*Crassostrea spp*	726	873	828	744	1 547	884	870 F
Sea mussels nei	*Mytilidae*	2 217	863	1 193	1 230	906	802	800 F
Triangular tivela	*Tivela mactroides*	...	126	196	664	1 684	273	270 F
Common squids nei	*Loligo spp*	1 317	950	1 486	758	1 696	1 187	1 170 F
Octopuses, etc. nei	*Octopodidae*	577	459	640	554	906	1 032	1 000 F
Marine molluscs nei	*Mollusca*	3 194	1 085	1 244	98	286	2 096	2 070 F
River and lake turtles nei	*Testudinata*	0	0	0	0	0	0	0
Marine turtles nei	*Testudinata*	0	0	0	0	0	0	0
	Country total	*706 708*	*715 482*	*744 585*	*706 789*	*703 941*	*766 846*	*770 000 F*
Chile								
Common carp	*Cyprinus carpio*	-	-	-	4	-	-	-
Flatfishes nei	*Pleuronectiformes*	220	203	154	75	84	95	76
Tadpole codling	*Salilota australis*	1 826	481	647	352	245	372	641
Southern blue whiting	*Micromesistius australis*	20 917	25 445	32 875	40 857	36 506	27 459	28 755
Southern hake	*Merluccius australis*	24 612	23 788	24 666	22 458	24 656	29 402	28 806
South Pacific hake	*Merluccius gayi*	75 403	88 555	87 620	80 151	103 789	110 143	121 200
Patagonian grenadier	*Macruronus magellanicus*	206 734	379 015	71 479	354 184	309 904	91 333	162 082
Grenadiers nei	*Macrourus spp*	0	12	-	0	-	-	-
Mullets nei	*Mugilidae*	278	244	78	68	134	132	93
Groupers, seabasses nei	*Serranidae*	23	30	24	10	19	7	5
Cabinza grunt	*Isacia conceptionis*	165	93	142	54	156	42	28
Corvina	*Sciaena gilberti*	1 239	1 179	1 350	1 069	747	1 052	1 033
Peruvian weakfish	*Cynoscion analis*	249	11	12	18	19	14	14
Croakers nei	*Micropogonias spp*	111	128	101	9	0	0	7
Croakers, drums nei	*Sciaenidae*	31	27	25	50	39	25	15
Patagonian blennie	*Eleginops maclovinus*	274	287	133	103	179	164	109
...A	*Normanichthys crockeri*	2 690	3 921	20 426	236	4 843	853	223
Pink cusk-eel	*Genypterus blacodes*	5 438	5 780	6 410	6 836	5 721	6 269	7 522
Red cusk-eel	*Genypterus chilensis*	1 082	982	745	584	415	608	730
Black cusk-eel	*Genypterus maculatus*	1 193	1 343	1 661	2 753	1 943	3 542	3 889
Alfonsinos nei	*Beryx spp*	-	-	-	144	706	4 366	5 182
Slimeheads nei	*Trachichthyidae*	-	-	-	-	779	1 482	1 868
Hapuku wreckfish	*Polyprion oxygeneios*	33	23	30	26	8	7	10
Tilefishes nei	*Branchiostegidae*	252	117	28	80	134	155	53
Peruvian morwong	*Cheilodactylus variegatus*	-	-	51	50	52	45	46
Antarctic toothfish	*Dissostichus mawsoni*	-	-	-	-	-	-	-
Patagonian toothfish	*Dissostichus eleginoides*	17 570	10 058	9 334	10 650	11 996	12 285	7 935
Mackerel icefish	*Champsocephalus gunnari*	-	-	-	6	-	715	365
Cardinal fishes nei	*Epigonus spp*	232	513	1 727	5 284	2 999	5 792	4 648
Snoek	*Thyrsites atun*	687	821	1 337	1 022	604	851	830
South Pacific breams nei	*Seriolella spp*	3 232	2 482	2 909	2 671	3 347	2 502	3 336
Scorpionfishes nei	*Scorpaenidae*	227	212	532	208	258	141	177
South American pilchard	*Sardinops sagax*	161 557	81 043	40 473	27 966	246 045	60 189	33 271
Pacific menhaden	*Ethmidium maculatum*	2 347	4 023	6 106	1 534	3 588	4 977	4 931
Araucanian herring	*Strangomera bentincki*	126 715	446 669	441 154	317 564	782 142	722 522	324 617
Anchoveta(=Peruvian anchovy)	*Engraulis ringens*	2 086 468	1 400 567	1 757 499	522 742	1 983 040	1 700 640	852 789
Eastern Pacific bonito	*Sarda chiliensis*	52	14	28	584	368	55	107
Skipjack tuna	*Katsuwonus pelamis*	-	-	53	47	9	0	57
Albacore	*Thunnus alalunga*	15	21	-	-	-	3	5
Yellowfin tuna	*Thunnus albacares*	43	32	57	78	48	77	66
Bigeye tuna	*Thunnus obesus*	15	16	6	29	6	20	5
Swordfish	*Xiphias gladius*	2 594	3 145	4 040	4 492	2 925	2 973	3 262
Atlantic saury	*Scomberesox saurus*	8	1	175	407	580	296	4 178
Silversides(=Sand smelts) nei	*Atherinidae*	558	690	494	560	3 414	1 357	833
Chilean jack mackerel	*Trachurus murphyi*	4 404 193	3 883 326	2 917 064	1 612 912	1 219 689	1 234 299	1 649 933
Carangids nei	*Carangidae*	89	96	205	348	203	307	363
Atlantic pomfret	*Brama brama*	3 930	5 585	5 998	6 332	6 830	8 160	15 156
Common dolphinfish	*Coryphaena hippurus*	104	179	124	80	109	131	136
Chub mackerel	*Scomber japonicus*	110 210	146 649	211 649	71 769	120 123	95 789	365 031
Butterfishes, pomfrets nei	*Stromateidae*	7	9	6	11	12	10	11
Shortfin mako	*Isurus oxyrinchus*	475	320	888	830	379	592	964
Blue shark	*Prionace glauca*	39	11	114	10	7	262	445
Smooth-hounds nei	*Mustelus spp*	193	225	108	56	208	143	128
Rays, stingrays, mantas nei	*Rajiformes*	2 642	2 696	2 958	2 015	3 369	4 151	2 974
Elephantfishes nei	*Callorhinchus spp*	920	1 450	822	1 416	632	603	1 125
Marine fishes nei	*Osteichthyes*	2 088	2 253	1 014	935	2 469	6 414	5 861
Marine crabs nei	*Brachyura*	5 733	3 989	3 541	5 063	6 501	6 758	6 770
Juan Fernandez rock lobster	*Jasus frontalis*	29	36	32	21	22	17	21

English name Nom anglais Nombre inglés	Scientific name Nom scientifique Nombre científico	1995 mt	1996 mt	1997 mt	1998 mt	1999 mt	2000 mt	2001 mt
Carrot squat lobster	*Pleuroncodes monodon*	4 938	7 726	8 939	12 602	12 710	11 129	1 754
Blue squat lobster	*Cervimunida johni*	5 743	6 402	10 322	9 426	7 273	5 069	2 178
Craylets, squat lobsters	*Galatheidae*	-	-	-	-	-	254	-
Southern king crab	*Lithodes antarcticus*	1 906	1 759	2 160	2 766	2 155	2 902	2 937
Softshell red crab	*Paralomis granulosa*	1 316	1 273	1 477	1 501	1 438	4 938	6 527
Chilean nylon shrimp	*Heterocarpus reedi*	10 620	10 535	10 239	7 301	7 951	5 448	4 863
Chilean knife shrimp	*Haliporoides diomedeae*	5	15	32	29	135	169	309
Giant barnacle	*Megabalanus psittacus*	681	879	579	683	620	620	685
Marine crustaceans nei	*Crustacea*	-	1	6	15	67	7	65
False abalone	*Concholepas concholepas*	2 670	2 541	3 154	2 564	2 294	1 274	828
Gastropods nei	*Gastropoda*	6 403	6 078	5 348	4 649	7 204	6 898	5 429
Chilean flat oyster	*Ostrea chilensis*	-	-	5	1	6	9	202
Chilean mussel	*Mytilus chilensis*	5 128	5 714	4 723	4 899	4 343	5 236	6 758
Choro mussel	*Choromytilus chorus*	307	323	266	127	155	217	166
Cholga mussel	*Aulacomya ater*	6 376	7 405	6 409	7 725	5 126	5 563	7 884
Peruvian calico scallop	*Argopecten purpuratus*	-	9	4	21	0	20	272
Scallops nei	*Pectinidae*	1 365	1 577	2 598	3 662	1 715	332	141
Gay's little venus	*Tawera gayi*	...	...	...	...	...	1	291
Taca clam	*Protothaca thaca*	17 162	20 016	12 475	24 254	16 429	16 303	26 483
Taquilla clams	*Mulinia spp*	1 852	999	2 757	2 549	1 536	1 491	1 699
Macha clam	*Mesodesma donacium*	6 913	6 144	6 770	6 464	1 728	1 249	1 396
Chilean semele	*Semele solida*	2 523	4 418	2 199	1 900	2 071	4 212	3 054
Clams, etc. nei	*Bivalvia*	20 075	18 720	17 880	11 900	20 804	16 551	16 775
Argentine shortfin squid	*Illex argentinus*	302	-	-	-	-	-	-
Various squids nei	*Loliginidae, Ommastrephidae*	55	26	110	179	99	64	3 594
Octopuses, etc. nei	*Octopodidae*	3 796	3 477	4 404	4 877	3 168	1 682	2 008
Marine molluscs nei	*Mollusca*	12	9	69	72	35	28	17
Red sea squirt	*Pyura chilensis*	3 297	4 549	3 174	2 530	2 704	2 290	1 298
Chilean sea urchin	*Loxechinus albus*	54 609	51 437	45 560	44 843	55 654	54 096	46 794
Sea cucumbers nei	*Holothurioidea*	106	115	1	30	108	1 510	107
	Country total	*7 433 902*	*6 690 942*	*5 810 764*	*3 265 383*	*5 050 528*	*4 300 160*	*3 797 143*
Colombia								
Characins nei	*Characidae*	5 698	5 504	3 730	6 427	8 667	6 600 F	6 600 F
Freshwater siluroids nei	*Siluroidei*	8 718	11 392	2 648	9 215	10 544	10 454 F	10 600 F
Freshwater fishes nei	*Osteichthyes*	9 108	16 165	14 232	6 031	9 577	7 800 F	7 800 F
Flatfishes nei	*Pleuronectiformes*	11	30	45	131	48	50 F	50 F
South Pacific hake	*Merluccius gayi*	...	...	267	165	143	250 F	391
Ladyfish	*Elops saurus*	30	143	20	4	11	20 F	20 F
Tarpon	*Megalops atlanticus*	13	135	2	2	-	-	-
Sea catfishes nei	*Ariidae*	158	449	370	138	106	120 F	120 F
Lebranche mullet	*Mugil liza*	21	39	-	2	-	-	-
Mullets nei	*Mugilidae*	35	103	60	63	44	60 F	58 F
Snooks(=Robalos) nei	*Centropomus spp*	108	452	78	96	62	70 F	70 F
Broomtail grouper	*Mycteroperca xenarcha*	207	180	125	155	82	80 F	210
Nassau grouper	*Epinephelus striatus*	25	11	5	3	0	...	...
Spotted grouper	*Epinephelus analogus*	22	43	48	32	39	30 F	28
Groupers nei	*Epinephelus spp*	207	180	125	155	82	70 F	57
Groupers, seabasses nei	*Serranidae*	67	86	17	37	26	30 F	30 F
Yellow snapper	*Lutjanus argentiventris*	120	43	40	100	82	80 F	...
Southern red snapper	*Lutjanus purpureus*	304	17	...	290	172	250 F	250 F
Lane snapper	*Lutjanus synagris*	225	236	35	12	11	20 F	20 F
Snappers nei	*Lutjanus spp*	278	460	43	452	515	600 F	802
Snappers, jobfishes nei	*Lutjanidae*	75	55	35	0	-	-	-
Grunts, sweetlips nei	*Haemulidae (=Pomadasyidae)*	115	224	12	6	1	5 F	5 F
Peruvian weakfish	*Cynoscion analis*	255	401	284	374	352	330	317
Croakers nei	*Micropogonias spp*	105	165	179	156	173	170 F	237
Croakers, drums nei	*Sciaenidae*	-	24	1	0	-	-	-
Mojarras, etc. nei	*Gerreidae*	38	125	6	23	0	5 F	5 F
Threadfins, tasselfishes nei	*Polynemidae*	-	16	4	3	16	10 F	10 F
South Pacific breams nei	*Seriolella spp*	...	...	...	...	...	...	8
Demersal percomorphs nei	*Perciformes*	1 069	202	236	165	211	220 F	220 F
Pacific anchoveta	*Cetengraulis mysticetus*	31 823	26 344	28 747	28 501	15 781	20 500 F	25 099
Clupeoids nei	*Clupeoidei*	120	180	555	469	61	250 F	250 F
Eastern Pacific bonito	*Sarda chiliensis*	3	6	5	8	9	10 F	13
Pacific sierra	*Scomberomorus sierra*	360	484	645	912	521	500 F	500 F
Seerfishes nei	*Scomberomorus spp*	180	539	22	30	26	50 F	50 F
Skipjack tuna	*Katsuwonus pelamis*	8 974	15 676	24 341	3 526	27 861	6 160	2 520
Yellowfin tuna	*Thunnus albacares*	41 943	18 521	42 390	14 547	29 374	15 600 F	24 920 F
Bigeye tuna	*Thunnus obesus*	...	7 270	3 090	560	1 420	1 030	150
Swordfish	*Xiphias gladius*	-	-	-	6	-	-	1
Tuna-like fishes nei	*Scombroidei*	267	17 568	8 299	49 408	1 881	5 220 F	5 090 F
Jacks, crevalles nei	*Caranx spp*	639	1 649	854	517	390	460 F	398 F
Amberjacks nei	*Seriola spp*	...	...	51	109	47	70 F	91
Smooth-hounds nei	*Mustelus spp*	208	1 007	435	361	388	360 F	311 F
Rays, stingrays, mantas nei	*Rajiformes*	-	3	2	2	1	1 F	1
Marine fishes nei	*Osteichthyes*	5 083	8 613	7 712	5 144	3 093	43 781 F	29 118 F
Marine crabs nei	*Brachyura*	238	49	166	44	11	120 F	120 F
Caribbean spiny lobster	*Panulirus argus*	449	185	108	319	175	250 F	250 F
Green spiny lobster	*Panulirus gracilis*	3	4	7	6	1	1 F	1 F
Western white shrimp	*Penaeus occidentalis*	619	1 091	1 445	1 211	2 686	2 000 F	1 219
Penaeus shrimps nei	*Penaeus spp*	391	710	1 745	377	775	3 039 F	3 000 F
Pacific seabob	*Xiphopenaeus riveti*	1 037	2 683	2 830	1 970	1 752	2 000 F	2 341
Kolibri shrimp	*Solenocera agassizii*	...	...	...	...	...	...	686
Natantian decapods nei	*Natantia*	887	456	559	345	555	570 F	591
Gastropods nei	*Gastropoda*	336	129	198	155	131	120 F	120 F
Mangrove cupped oyster	*Crassostrea rhizophorae*	-	-	-	8	-	-	-
Clams, etc. nei	*Bivalvia*	1	4	5	6	6	10 F	10 F

English name Nom anglais Nombre inglés	Scientific name Nom scientifique Nombre científico	1995 mt	1996 mt	1997 mt	1998 mt	1999 mt	2000 mt	2001 mt
Various squids nei	*Loliginidae, Ommastrephidae*	41	324	203	43	38	60 F	87
Marine molluscs nei	*Mollusca*	85	454	857	87	19	158 F	155 F
	Country total	*120 699*	*130 829*	*147 918*	*132 908*	*117 995*	*129 644*	*125 000 F*
Ecuador								
Freshwater fishes nei	*Osteichthyes*	300	300	400	400	400	400	400
Grunts, sweetlips nei	*Haemulidae (=Pomadasyidae)*	...	...	...	1 091	500	...	...
Croakers, drums nei	*Sciaenidae*	30 910	2 486	316	3 300	7 320	...	1 558
Hairtails, scabbardfishes nei	*Trichiuridae*	...	...	...	...	...	13 196	3 382
Gurnards, searobins nei	*Triglidae*	...	...	...	...	22 634	27 129	1 428
South American pilchard	*Sardinops sagax*	75 916	356 480	57 191	1 012	8 821	51 648	42 143
Red-eye round herring	*Etrumeus teres*	4 946	34 349	1 095	8 873	3 636	4 414	28
Pacific thread herring	*Opisthonema libertate*	40 911	41 041	43 145	40 530	22 253	20 519	19 886
Anchoveta(=Peruvian anchovy)	*Engraulis ringens*	-	-	-	-	-	-	2 071
Pacific anchoveta	*Cetengraulis mysticetus*	23 418	26 354	89 157	44 474	27 221	13 762	73 537
Frigate and bullet tunas	*Auxis thazard, A.rochei*	7 396	2 537	6 857	4 201	48 913	9 648	5 738
Black skipjack	*Euthynnus lineatus*	160	370	50	260	10	270	1 800
Skipjack tuna	*Katsuwonus pelamis*	31 599	37 468	67 400	67 453	126 992	105 146	68 217
Yellowfin tuna	*Thunnus albacares*	15 921	19 314	19 603	31 052	50 200	36 955	57 563
Bigeye tuna	*Thunnus obesus*	10 193	17 892	26 148	17 909	22 278	29 398	23 440
Marlins,sailfishes,etc. nei	*Istiophoridae*	-	-	-	3 727	6 962	...	...
Swordfish	*Xiphias gladius*	222	...	...	...	...	...	...
Tuna-like fishes nei	*Scombroidei*	-	-	-	-	-	-	23 255
Chilean jack mackerel	*Trachurus murphyi*	174 393	56 781	30 302	25 900	19 072	7 144	134 011
Pacific bumper	*Chloroscombrus orqueta*	17 999	1 706	952	565	1 409	...	1 008
Carangids nei	*Carangidae*	199	65	45	839	1 632	...	68
Chub mackerel	*Scomber japonicus*	57 950	79 484	192 182	44 716	28 307	84 324	85 378
Butterfishes, pomfrets nei	*Stromateidae*	...	...	426	...	...	...	...
Marine fishes nei	*Osteichthyes*	5 530	20 426	8 296	9 510	97 237	186 519	38 255
Marine crabs nei	*Brachyura*	750	750	750	600	600	600	600
Green spiny lobster	*Panulirus gracilis*	50	50	50	50	50	50	50
Yellowleg shrimp	*Penaeus californiensis*	73	73	60	50	20	20	40
Whiteleg shrimp	*Penaeus vannamei*	6 000	4 497	4 000	3 000	1 100	1 100	2 250
Western white shrimp	*Penaeus occidentalis*	67	67	60	50	20	20	40
Penaeus shrimps nei	*Penaeus spp*	344	344	350	300	125	125	264
Stromboid conchs nei	*Strombus spp*	22	10	10	10	10	10	10
Pacific cupped oyster	*Crassostrea gigas*	5	5	5	5	5	5	5
Sea mussels nei	*Mytilidae*	3	3	3	5	5	5	5
Venus clams nei	*Veneridae*	0	0	0	0	0	0	0
Clams, etc. nei	*Bivalvia*	6	10	10	10	10	10	10
Common squids nei	*Loligo spp*	90	90	100	100	100	100	100
Octopuses, etc. nei	*Octopodidae*	0	0	0	5	5	5	5
Marine turtles nei	*Testudinata*	10	10	10	10	10	10	10
Sea urchins nei	*Strongylocentrotus spp*	0	0	0	0	0	0	0
Sea cucumbers nei	*Holothurioidea*	12	12	15	15	15	15	15
	Country total	*505 395*	*702 974*	*548 988*	*310 022*	*497 872*	*592 547*	*586 570*
Falkland Is								
Sea trout	*Salmo trutta*	-	1	1	1	1	1	1
Tadpole codling	*Salilota australis*	1 530	2 033	817	1 491	2 692	1 886	1 371
Southern blue whiting	*Micromesistius australis*	1 616	1 083	727	1 977	2 127	2 704	4 581
Argentine hake	*Merluccius hubbsi*	194	383	267	959	1 031	1 000	562
Patagonian grenadier	*Macruronus magellanicus*	864	2 569	1 829	4 246	5 109	3 404	5 452
Patagonian blennie	*Eleginops maclovinus*	4	4	4	4	4	10	61
Pink cusk-eel	*Genypterus blacodes*	116	297	154	253	451	304	347
Patagonian toothfish	*Dissostichus eleginoides*	34	50	178	570	1 113	927	1 460
Falkland sprat	*Sprattus fuegensis*	0	0	0	29	17	97	4
Silversides(=Sand smelts) nei	*Atherinidae*	2	2	2	2	2	4	1
Rays, stingrays, mantas nei	*Rajiformes*	117	184	204	216	314	353	417
Marine fishes nei	*Osteichthyes*	50	370	181	1 033	1 217	1 250	774
Softshell red crab	*Paralomis granulosa*	1	1	1	1	1	1	5
Chilean mussel	*Mytilus chilensis*	1	1	1	1	1	1	0
Patagonian squid	*Loligo gahi*	22 310	24 366	12 710	32 029	22 502	50 270	42 909
Argentine shortfin squid	*Illex argentinus*	351	196	37	804	2 582	716	1 879
Sevenstar flying squid	*Martialia hyadesi*	0	0	0	-	0	0	-
	Country total	*27 190*	*31 540*	*17 113*	*43 616*	*39 164*	*62 928*	*59 824*
Fr Guiana								
Freshwater fishes nei	*Osteichthyes*	0	0	0	0	0	0	0
Marine fishes nei	*Osteichthyes*	3 634	3 000 F	2 500	2 500	2 500 F	2 500 F	2 500 F
Penaeus shrimps nei	*Penaeus spp*	4 455	4 377	4 102	4 209	3 771	2 737	2 694
	Country total	*8 089*	*7 377 F*	*6 602*	*6 709*	*6 271 F*	*5 237 F*	*5 194 F*
Guyana								
Freshwater fishes nei	*Osteichthyes*	700 F	800	625	625	603	800	800
Southern red snapper	*Lutjanus purpureus*	...	...	...	...	...	570	524
King mackerel	*Scomberomorus cavalla*	-	-	270	440	398	214	239
Serra Spanish mackerel	*Scomberomorus brasiliensis*	-	211	571	625	1 143	308	329
Sharks, rays, skates, etc. nei	*Elasmobranchii*	...	765	1 892	...	2 175	...	...
Marine fishes nei	*Osteichthyes*	37 000 F	33 971	34 841	38 124	37 534	27 666	24 662
Penaeus shrimps nei	*Penaeus spp*	400 F	84	79	1 935	1 595	1 132	1 698
Atlantic seabob	*Xiphopenaeus kroyeri*	9 800 F	12 752	15 720	11 091	10 396	16 733	23 771
Whitebelly prawn	*Nematopalaemon schmitti*	-	-	-	-	-	1 464	1 382
	Country total	*47 900 F*	*48 583*	*53 998*	*52 840*	*53 844*	*48 887*	*53 405*

English name Nom anglais Nombre inglés	Scientific name Nom scientifique Nombre científico	1995 mt	1996 mt	1997 mt	1998 mt	1999 mt	2000 mt	2001 mt
Paraguay								
Characins nei	*Characidae*	7 700 F	8 000 F	10 000 F	9 000 F	9 000 F	9 000 F	9 000 F
Freshwater siluroids nei	*Siluroidei*	9 500 F	10 000 F	13 000 F	12 000 F	12 000 F	12 000 F	12 000 F
Freshwater fishes nei	*Osteichthyes*	3 800 F	4 000 F	5 000 F	4 000 F	4 000 F	4 000 F	4 000 F
	Country total	*21 000 F*	*22 000*	*28 000*	*25 000 F*	*25 000 F*	*25 000 F*	*25 000 F*
Peru								
Velvety cichlids	*Astronotus spp*	320	308	177	141	251	188	183
Arapaima	*Arapaima gigas*	420	457	465	210	338	273	204
Cachama	*Colossoma macropomum*	...	693	762	254	1 016	963	904
Pirapatinga	*Piaractus brachypomus*	...	397	...	166	648	324	487
Netted prochilod	*Prochilodus reticulatus*	5 088	7 448	2 007	6 117	9 768	10 942	11 220
Freshwater fishes nei	*Osteichthyes*	44 452	19 524	27 941	28 047	24 018	19 387	22 464
Rainbow trout	*Oncorhynchus mykiss*	509	63	869	392	184	220	191
Flatfishes nei	*Pleuronectiformes*	1 559	528	212	230	263	177	313
South Pacific hake	*Merluccius gayi*	181 182	234 915	177 953	82 367	37 121	83 361	125 065
Mullets nei	*Mugilidae*	16 601	13 916	13 264	29 075	20 843	26 314	27 330
Groupers nei	*Epinephelus spp*	191	241	60	101	151	91	87
Peruvian rock seabass	*Paralabrax humeralis*	5 837	4 954	2 789	2 554	3 278	4 373	2 011
Snappers nei	*Lutjanus spp*	300	117	75	208	573	204	55
Cabinza grunt	*Isacia conceptionis*	1 342	1 955	1 892	2 079	2 791	3 251	3 293
Grunts, sweetlips nei	*Haemulidae (=Pomadasyidae)*	890	194	379	254	318	183	419
Corvina	*Sciaena gilberti*	4 353	7 920	2 215	5 085	5 695	3 692	3 295
Peruvian weakfish	*Cynoscion analis*	8 902	7 475	5 501	10 795	8 558	5 995	4 107
Croakers nei	*Micropogonias spp*	704	733	606	1 068	616	955	825
Peruvian banded croaker	*Paralonchurus peruanus*	5 543	4 263	2 737	4 363	6 063	5 729	4 167
Cusk-eels nei	*Genypterus spp*	1 631	1 121	439	425	196	557	552
Tilefishes nei	*Branchiostegidae*	1 544	892	292	119	146	117	1 485
Peruvian morwong	*Cheilodactylus variegatus*	93	283	411	90	236	335	260
South Pacific breams nei	*Seriolella spp*	7 698	3 704	388	505	1 589	1 473	3 192
South American pilchard	*Sardinops sagax*	1 265 658	1 056 413	625 143	908 291	187 924	226 294	60 298
Pacific menhaden	*Ethmidium maculatum*	3 140	5 769	7 135	39 311	25 848	19 014	9 085
Anchoveta(=Peruvian anchovy)	*Engraulis ringens*	6 558 108	7 463 147	5 927 599	1 206 322	6 740 225	9 575 717	6 358 217
Anchovies, etc. nei	*Engraulidae*	189 389	59 639	24 703	706 167	11 242	3 868	137 098
Eastern Pacific bonito	*Sarda chiliensis*	28 331	23 059	17 731	5 130	948	434	1 287
Pacific sierra	*Scomberomorus sierra*	686	439	98	861	2 712	930	181
Skipjack tuna	*Katsuwonus pelamis*	151	85	823	9 373	802	711	81
Yellowfin tuna	*Thunnus albacares*	914	953	908	12 747	2 784	2 548	4 175
Swordfish	*Xiphias gladius*	-	1	-	57	42	20	356
Flyingfishes nei	*Exocoetidae*	35 619	639	1 056	4 506	298 373	41 059	5 035
Silversides(=Sand smelts) nei	*Atherinidae*	2 357	3 802	5 184	45	6 692	11 215	7 528
Chilean jack mackerel	*Trachurus murphyi*	376 600	438 736	649 751	386 946	184 679	296 579	723 733
Jacks, crevalles nei	*Caranx spp*	-	129	16	69	607	35	11
Pompanos nei	*Trachinotus spp*	566	460	268	779	2 801	1 220	632
Amberjacks nei	*Seriola spp*	6 598	1 558	4 648	21 104	2 084	11 159	28 025
Chub mackerel	*Scomber japonicus*	44 259	49 221	206 183	401 903	527 729	73 263	176 202
Pelagic percomorphs nei	*Perciformes*	382	858	120	33	1 073	771	371
Smooth-hounds nei	*Mustelus spp*	4 125	3 230	3 166	8 038	2 892	4 042	4 648
Angelsharks, sand devils nei	*Squatinidae*	289	358	189	101	262	406	510
Pacific guitarfish	*Rhinobatos planiceps*	121	460	333	344	95	2 624	1 060
Rays, stingrays, mantas nei	*Rajiformes*	1 841	1 126	1 177	1 477	2 789	4 026	2 034
Sharks, rays, skates, etc. nei	*Elasmobranchii*	694	1 506	1 915	4 335	2 951	4 307	3 618
Marine fishes nei	*Osteichthyes*	58 779	45 788	83 203	375 784	164 855	91 905	122 447
Marine crabs nei	*Brachyura*	2 553	1 605	303	752	11 397	1 974	1 568
Green spiny lobster	*Panulirus gracilis*	168	52	12	669	496	278	62
Penaeus shrimps nei	*Penaeus spp*	10 877	9 245	15 648	17 752	7 255	1 852	2 075
Marine crustaceans nei	*Crustacea*	-	-	-	-	-	-	758
False abalone	*Concholepas concholepas*	1 361	2 728	4 366	830	2 289	1 250	544
Gastropods nei	*Gastropoda*	3 686	2 215	7 098	3 110	4 525	2 768	4 995
Cholga mussel	*Aulacomya ater*	11 204	6 023	9 669	15 106	14 612	13 370	14 700
Peruvian calico scallop	*Argopecten purpuratus*	3 113	2 086	4 009	23 525	30 141	11 810	6 272
Macha clam	*Mesodesma donacium*	1 200	1 060	1 061	578	-	10	0
Clams, etc. nei	*Bivalvia*	569	411	236	152	338	956	949
Common squids nei	*Loligo spp*	7 766	10 250	3 806	287	1 353	24 548	18 738
Jumbo flying squid	*Dosidicus gigas*	25 676	8 138	16 061	547	54 652	53 795	71 834
Octopuses, etc. nei	*Octopodidae*	800	760	1 856	5 123	1 593	819	635
Marine molluscs nei	*Mollusca*	468	537	2 538	1 546	3 676	2 312	2 116
Marine turtles nei	*Testudinata*	4	0	1	2	1	1	2
Echinoderms	*Echinodermata*	131	461	424	90	1 204	1 626	2 114
	Country total	*8 937 342*	*9 515 048*	*7 869 871*	*4 338 437*	*8 428 601*	*10 658 620*	*7 986 103*
Suriname								
Freshwater fishes nei	*Osteichthyes*	140 F	150 F	200 F	200 F	200 F	200 F	200
Marine fishes nei	*Osteichthyes*	11 855 F	10 400 F	11 777 F	11 845	9 800	10 500 F	11 300
Marine crabs nei	*Brachyura*	5 F	6 F	10 F	10 F	10 F	20	25
Penaeus shrimps nei	*Penaeus spp*	1 000 F	2 444	2 013	2 094	1 653	2 240 F	2 840
Atlantic seabob	*Xiphopenaeus kroyeri*	-	-	-	2 046	4 537	4 540 F	4 550
	Country total	*13 000 F*	*13 000 F*	*14 000 F*	*16 195*	*16 200*	*17 500 F*	*18 915*
Uruguay								
Characins nei	*Characidae*	763	338	1 924	1 687	1 762	1 690	107
Freshwater siluroids nei	*Siluroidei*	36	72	252	203	319	293	81
Freshwater fishes nei	*Osteichthyes*	45	188	40	41	342	312	254
Bastard halibuts nei	*Paralichthys spp*	130	483	502	496	413	256	307
Brazilian codling	*Urophycis brasiliensis*	363	223	272	344	281	235	361

English name Nom anglais Nombre inglés	Scientific name Nom scientifique Nombre científico	1995 mt	1996 mt	1997 mt	1998 mt	1999 mt	2000 mt	2001 mt
Southern blue whiting	*Micromesistius australis*	-	-	-	-	-	9	0
Argentine hake	*Merluccius hubbsi*	57 874	57 937	48 367	49 111	32 045	27 710	27 618
Patagonian grenadier	*Macruronus magellanicus*	...	...	150	1 824	1 461	800	635
Grenadiers nei	*Macrourus spp*	-	-	-	0	3	-	8
Sea catfishes nei	*Ariidae*	119	18	15	24	78	51	42
Mullets nei	*Mugilidae*	100	346	214	272	57	194	219
Argentine seabass	*Acanthistius brasilianus*	61	85	51	42	19	88	9
Striped weakfish	*Cynoscion striatus*	13 417	12 654	15 186	15 286	8 481	13 440	10 890
Whitemouth croaker	*Micropogonias furnieri*	29 513	25 745	23 744	22 253	14 650	24 146	27 322
Kingcroakers nei	*Menticirrhus spp*	-	-	-	7	4	4	6
Argentine croaker	*Umbrina canosai*	1 708	1 210	547	1 112	1 401	1 071	1 416
King weakfish	*Macrodon ancylodon*	1 519	1 743	1 395	2 354	802	1 123	1 486
Black drum	*Pogonias cromis*	574	172	82	678	100	344	336
Red porgy	*Pagrus pagrus*	12	1	12	4	14	27	7
Brazilian flathead	*Percophis brasilianus*	-	-	-	-	1	-	0
Pink cusk-eel	*Genypterus blacodes*	105	43	41	86	206	368	756
Castaneta	*Cheilodactylus bergi*	3 034	2 938	4 149	9 681	3 105	1 351	1 269
Antarctic toothfish	*Dissostichus mawsoni*	-	-	-	-	-	-	23
Patagonian toothfish	*Dissostichus eleginoides*	-	-	-	1 607	1 405	3 273	7 897
Blackbelly rosefish	*Helicolenus dactylopterus*	2 734	2 204	3 404	4 389	2 581	2 111	1 830
Demersal percomorphs nei	*Perciformes*	22	3	57	-	334	51	34
Argentine menhaden	*Brevoortia pectinata*	130	84	219	415	86	66	103
Argentine anchovy	*Engraulis anchoita*	41	22	13	67	3 193	6	318
Atlantic bonito	*Sarda sarda*	-	-	-	-	-	1	23
Atlantic bluefin tuna	*Thunnus thynnus*	2	-	-	-	-	-	1
Albacore	*Thunnus alalunga*	49	75	56	110	69	90	40
Yellowfin tuna	*Thunnus albacares*	53	171	53	88	52	54	112
Bigeye tuna	*Thunnus obesus*	80	124	69	59	27	29	63
Atlantic blue marlin	*Makaira nigricans*	-	-	-	23	-	-	2
Atlantic white marlin	*Tetrapturus albidus*	1	2	50	22	-	-	0
Swordfish	*Xiphias gladius*	499	644	760	889	661	713	716
Tuna-like fishes nei	*Scombroidei*	0	2	50	94	136	95	34
Bluefish	*Pomatomus saltatrix*	56	11	11	84	18	48	93
Parona leatherjacket	*Parona signata*	391	439	431	465	360	371	618
Chub mackerel	*Scomber japonicus*	5	4	5	1	5	-	0
Pelagic percomorphs nei	*Perciformes*	3	2	243	150	10	18	0
Narrownose smooth-hound	*Mustelus schmitti*	286	204	174	2 156	3 212	1 037	1 153
Rays, stingrays, mantas nei	*Rajiformes*	1 469	2 614	2 342	398	1 576	1 004	989
Sharks, rays, skates, etc. nei	*Elasmobranchii*	1 577	1 760	2 367	444	1 901	991	890
Marine fishes nei	*Osteichthyes*	4 083	3 289	4 510	6 335	936	3 669	2 686
Red crab	*Geryon quinquedens*	783	1 513	3 432	2 682	3 382	5 259	2 089
Southern king crab	*Lithodes antarcticus*	0	0	0	0	0	0	0
Sao Paulo shrimp	*Penaeus paulensis*	0	-	177	13	12	56	23
Argentine red shrimp	*Pleoticus muelleri*	-	-	-	-	40	0	0
Antarctic krill	*Euphausia superba*	-	-	-	-	-	9 921	-
Marine crustaceans nei	*Crustacea*	0	3	-	-	-	-	-
Angulate volute	*Zidona dufresnei*	...	...	...	...	...	990	825
River Plata mussel	*Mytilus platensis*	299	206	174	226	142	176	306
Patagonean scallop	*Zygochlamis patagonica*	...	...	...	...	...	890	3 638
Donax clams	*Donax spp*	0	0	0	0	0	0	0
Clams, etc. nei	*Bivalvia*	...	...	0	0	0	6	17
Argentine shortfin squid	*Illex argentinus*	4 182	5 669	20 857	13 175	13 679	12 144	7 373
Marine molluscs nei	*Mollusca*	323	89	557	1 310	3 651	0	-
Frogs	*Rana spp*	5	0	0	0	0	7	9
	Country total	*126 446*	*123 330*	*136 954*	*140 707*	*103 012*	*116 588*	*105 034*
Venezuela								
Common carp	*Cyprinus carpio*	544	218	1	3	5	1 390	0
Characins nei	*Characidae*	8 778	8 050	6 029	6 921	4 817	5 773	2 978
Prochilods nei	*Prochilodus spp*	20 430	18 740	10 927	15 670	9 320	9 613	6 743
Freshwater siluroids nei	*Siluroidei*	19 497	18 143	15 248	14 833	7 470	6 226	8 967
Freshwater fishes nei	*Osteichthyes*	12 229	11 452	11 977	11 011	17 334	7 222	12 189
Rainbow trout	*Oncorhynchus mykiss*	57	59	...	...	...	...	186
Diadromous clupeoids nei	*Clupeoidei*	345	552	355	467	403	287	200
Flatfishes nei	*Pleuronectiformes*	...	...	...	...	51	11	...
Sea catfishes nei	*Ariidae*	21 548	17 041	8 963	10 098	11 100	14 279	11 929
Flathead grey mullet	*Mugil cephalus*	9 363	8 839	6 465	6 284	5 151	5 467	5 008
Lebranche mullet	*Mugil liza*	3 228	3 248	1 550	2 874	2 855	3 186	2 537
Common snook	*Centropomus undecimalis*	3 023	2 177	1 560	2 022	1 704	1 850	1 903
Groupers nei	*Epinephelus spp*	2 040	2 185	2 235	1 990	1 591	1 408	811
Groupers, seabasses nei	*Serranidae*	279	265	322	384	328	140	123
Yellowtail snapper	*Ocyurus chrysurus*	511	338	335	272	220	291	158
Snappers, jobfishes nei	*Lutjanidae*	7 745	8 205	9 273	8 652	3 511	7 621	4 986
Grunts, sweetlips nei	*Haemulidae (=Pomadasyidae)*	6 310	5 466	6 336	7 685	4 108	4 191	9 132
Weakfishes nei	*Cynoscion spp*	20 045	13 879	10 503	13 481	6 303	12 716	10 734
Whitemouth croaker	*Micropogonias furnieri*	7 043	6 065	3 694	4 871	1 900	3 262	6 848
Porgies, seabreams nei	*Sparidae*	...	...	...	...	...	6	...
Mojarras, etc. nei	*Gerreidae*	0	0	0	0	0	0	0
Largehead hairtail	*Trichiurus lepturus*	4 933	4 609	5 050	5 408	4 017	3 716	8 793
Demersal percomorphs nei	*Perciformes*	1 210	3 532	4 760	3 616	4 118	1 322	1 687
Round sardinella	*Sardinella aurita*	153 037	153 782	140 571	186 060	126 468	73 534	71 168
Atlantic thread herring	*Opisthonema oglinum*	307	294	5 564	10 619	14 789	9 866	3 650
Atlantic anchoveta	*Cetengraulis edentulus*	41	8	0	119	0	0	2
Atlantic bonito	*Sarda sarda*	1 503	1 348	1 294	1 647	1 597	1 376	1 815
Wahoo	*Acanthocybium solandri*	445	479	498	349	448	150	297
King mackerel	*Scomberomorus cavalla*	2 555	2 139	3 530	340	2 424	1 498	1 861
Serra Spanish mackerel	*Scomberomorus brasiliensis*	4 725	3 609	3 670	3 651	1 686	3 624	3 062
Frigate and bullet tunas	*Auxis thazard, A.rochei*	2 161	3 053	2 813	1 926	1 524	1 410	1 342

English name Nom anglais Nombre inglés	Scientific name Nom scientifique Nombre científico	1995 mt	1996 mt	1997 mt	1998 mt	1999 mt	2000 mt	2001 mt
Little tunny(=Atl.black skipj)	*Euthynnus alletteratus*	1 627	1 840	2 064	2 815	2 389	2 040	1 948
Black skipjack	*Euthynnus lineatus*	-	50	40	80	40	10	-
Skipjack tuna	*Katsuwonus pelamis*	6 997	6 994	11 114	10 224	17 031	8 123	9 050
Blackfin tuna	*Thunnus atlanticus*	624	758	498	1 034	1 192	589	1 902
Albacore	*Thunnus alalunga*	279	315	49	107	91	1 374	349
Yellowfin tuna	*Thunnus albacares*	58 384	77 452	74 001	76 605	68 917	80 349	128 362
Bigeye tuna	*Thunnus obesus*	705	619	524	462	150	436	708
Atlantic sailfish	*Istiophorus albicans*	103	165	185	258	179	93	126
Atlantic blue marlin	*Makaira nigricans*	106	137	130	205	220	28	72
Atlantic white marlin	*Tetrapturus albidus*	171	164	90	80	61	13	72
Longbill spearfish	*Tetrapturus pfluegeri*	-	1	0	1	0	-	4
Marlins,sailfishes,etc. nei	*Istiophoridae*	-	-	-	-	-	-	8
Swordfish	*Xiphias gladius*	54	85	20	37	30	30	21
Tuna-like fishes nei	*Scombroidei*	-	-	-	10	-	-	13
Bluefish	*Pomatomus saltatrix*	1 024	651	825	581	542	950	1 158
Jacks, crevalles nei	*Caranx spp*	3 439	3 732	2 823	2 898	2 642	3 462	2 115
Atlantic moonfish	*Selene setapinnis*	2 544	1 766	2 110	2 338	1 529	976	3 443
Pompanos nei	*Trachinotus spp*	244	307	285	435	146	132	117
Amberjacks nei	*Seriola spp*	288	338	355	336	450	610	399
Bigeye scad	*Selar crumenophthalmus*	2 836	2 353	2 425	2 653	3 765	1 704	1 774
Carangids nei	*Carangidae*	25	6	23	31	11	26	5
Common dolphinfish	*Coryphaena hippurus*	447	...	...	...	290	141	...
Chub mackerel	*Scomber japonicus*	377	416	549	753	631	399	514
Gulf butterfishes, etc. nei	*Peprilus spp*	1 881	1 889	568	1 175	1 004	809	742
Barracudas nei	*Sphyraena spp*	923	899	998	1 123	881	731	1 165
Pelagic percomorphs nei	*Perciformes*	3 579	759	442	381	219	182	109
Requiem sharks nei	*Carcharhinidae*	7 468	6 979	6 000	4 597	2 973	3 343	2 536
Rays, stingrays, mantas nei	*Rajiformes*	2 450	1 812	1 896	2 111	2 287	2 148	2 182
Marine fishes nei	*Osteichthyes*	32 559	32 927	36 624	16 376	-	464	11 303
Portunus swimcrabs nei	*Portunus spp*	1	1	289	224	4 041	4 994	4 001
Marine crabs nei	*Brachyura*	5 551	4 591	5 340	4 180	196	539	155
Caribbean spiny lobster	*Panulirus argus*	629	648	619	260	95	105	78
Penaeus shrimps nei	*Penaeus spp*	10 786	11 735	10 949	6 910	4 607	9 882	12 128
Mangrove cupped oyster	*Crassostrea rhizophorae*	3 775	2 695	1 705	2 594	2 037	1 590	3 754
South American rock mussel	*Perna perna*	155	223	295	3 802	451	316	1 081
Ark clams nei	*Arca spp*	33 987	31 925	39 128	27 981	38 646	44 709	43 675
Venus clams nei	*Veneridae*	517	536	533	295	378	353	388
Common squids nei	*Loligo spp*	756	771	719	787	102	463	1 261
Octopuses, etc. nei	*Octopodidae*	724	1 548	1 917	6 507	526	1 110	1 064
Marine molluscs nei	*Mollusca*	1 098	425	600	1 005	5 808	2 457	1 056
Green turtle	*Chelonia mydas*	0	0	0	0	-	-	-
Hawksbill turtle	*Eretmochelys imbricata*	0	0	0	0	-	-	-
Loggerhead turtle	*Caretta caretta*	0	0	0	0	-	-	-
River and lake turtles nei	*Testudinata*	0	0	0	0	-	-	-
Marine turtles nei	*Testudinata*	0	0	0	0	-	-	-
	Country total	*501 045*	*496 287*	*470 255*	*503 504*	*399 799*	*357 115*	*417 947*
Total		***19 624 532***	***19 794 735***	***17 255 248***	***10 712 994***	***16 526 592***	***18 000 800***	***14 879 397***

English name / Nom anglais / Nombre inglés	Scientific name / Nom scientifique / Nombre científico	1995 mt	1996 mt	1997 mt	1998 mt	1999 mt	2000 mt	2001 mt
Afghanistan								
Freshwater fishes nei	*Osteichthyes*	1 300 F	1 300 F	1 250 F	1 200 F	1 200 F	1 000 F	800 F
	Country total	*1 300 F*	*1 300 F*	*1 250 F*	*1 200 F*	*1 200 F*	*1 000 F*	*800 F*
Armenia								
Common carp	*Cyprinus carpio*	1	91	32	14	12	9	7
Crucian carp	*Carassius carassius*	2	28	23	42	26	38	32
Cyprinids nei	*Cyprinidae*	42	33	7	37	21	19	15
Freshwater fishes nei	*Osteichthyes*	0	0	0	0	0	0	-
Trouts nei	*Salmo spp*	0	0	6	0	163	186	180
Whitefishes nei	*Coregonus spp*	776	428	512	605	922	881	632
	Country total	*821*	*580*	*580*	*698*	*1 144*	*1 133*	*866*
Azerbaijan								
Freshwater bream	*Abramis brama*	219	402	346	314	52	55	127
Common carp	*Cyprinus carpio*	370	84	49	87	92	93	51
Tench	*Tinca tinca*	0	2	0	0	0	0	-
Crucian carp	*Carassius carassius*	...	...	...	...	6	4	17
Roach	*Rutilus rutilus*	19	74	89	62	81	8	64
Asp	*Aspius aspius*	5	9	6	4	2	1	5
Northern pike	*Esox lucius*	15	7	23	18	21	28	25
Wels(=Som)catfish	*Silurus glanis*	6	11	9	8	8	9	8
Pike-perch	*Stizostedion lucioperca*	19	22	11	7	5	5	19
Freshwater fishes nei	*Osteichthyes*	25	16	21	13	8	-	-
Sturgeons nei	*Acipenseridae*	76	69	63	61	69	70	76
Trouts nei	*Salmo spp*	-	-	-	-	-	-	5
Caspian shads	*Caspialosa spp*	67	75	42	82	60	1	52
Azov sea sprat	*Clupeonella cultriventris*	9 651	5 828	4 420	4 043	20 460	18 520	10 389
Mullets nei	*Mugilidae*	73	103	82	61	2	3	55
	Country total	*10 545*	*6 702*	*5 161*	*4 760*	*20 866*	*18 797*	*10 893*
Bahrain								
Freshwater fishes nei	*Osteichthyes*	0	0	0	0	0	-	-
Mullets nei	*Mugilidae*	201	99	60	73	10	59	39
Groupers nei	*Epinephelus spp*	459	532	300	331	525	670	794
Snappers nei	*Lutjanus spp*	117	133	207	100	294	157	103
Grunts, sweetlips nei	*Haemulidae (=Pomadasyidae)*	282	253	223	355	325	297	236
Emperors(=Scavengers) nei	*Lethrinidae*	1 486	1 506	892	944	1 227	1 403	1 377
Porgies, seabreams nei	*Sparidae*	439	496	493	757	582	591	401
Goatfishes	*Upeneus spp*	36	139	65	16	414	82	192
Mojarras(=Silver-biddies) nei	*Gerres spp*	160	219	350	347	261	334	372
Parrotfishes nei	*Scaridae*	63	56	46	63	66	32	21
Spinefeet(=Rabbitfishes) nei	*Siganus spp*	1 543	2 185	1 612	1 523	1 241	2 114	1 899
Narrow-barred Spanish mackerel	*Scomberomorus commerson*	109	158	47	85	44	66	109
Flyingfishes nei	*Exocoetidae*	85	92	85	145	299	52	112
Cobia	*Rachycentron canadum*	32	38	19	6	9	9	20
Carangids nei	*Carangidae*	495	668	498	359	524	414	450
Barracudas nei	*Sphyraena spp*	107	159	82	156	6	8	7
Marine fishes nei	*Osteichthyes*	1 043	1 185	1 029	852	938	859	1 077
Portunus swimcrabs nei	*Portunus spp*	807	1 047	1 289	1 017	2 179	2 380	2 556
Slipper lobsters nei	*Scyllaridae*	150	125	56	51	6	2	2
Penaeus shrimps nei	*Penaeus spp*	1 662	3 565	2 571	2 530	1 622	2 104	1 359
Cuttlefish,bobtail squids nei	*Sepiidae, Sepiolidae*	113	285	126	139	48	85	104
	Country total	*9 389*	*12 940*	*10 050*	*9 849*	*10 620*	*11 718*	*11 230*
Bangladesh								
Freshwater fishes nei	*Osteichthyes*	443 319	445 377	451 055	457 055	575 609	591 300	590 000 F
Hilsa shad	*Tenualosa ilisha*	213 535	225 598	214 434	205 739	214 519	219 532	220 000 F
Seerfishes nei	*Scomberomorus spp*	50	40	50	60	60	60	60 F
Marine fishes nei	*Osteichthyes*	115 122	119 484	135 860	145 260	137 285	161 977	158 640 F
Marine crustaceans nei	*Crustacea*	20 363	24 288	28 027	31 027	31 742	31 395	31 300 F
	Country total	*792 389*	*814 787*	*829 426*	*839 141*	*959 215*	*1 004 264*	*1 000 000 F*
Bhutan								
Freshwater fishes nei	*Osteichthyes*	310 F	300 F	300 F	300 F	300 F	300 F	300 F
	Country total	*310 F*	*300 F*	*300 F*	*300 F*	*300 F*	*300 F*	*300 F*
Brunei Darsm								
Freshwater fishes nei	*Osteichthyes*	2	1	0	0	0	0	0
Marine fishes nei	*Osteichthyes*	4 389	6 818	4 405	4 930	3 081	2 356	1 186
Giant river prawn	*Macrobrachium rosenbergii*	5	14	17	35	26	23	16
Natantian decapods nei	*Natantia*	272	275	0	0	0	0	0
Marine crustaceans nei	*Crustacea*	18	281	78	65	44	78	266
Marine molluscs nei	*Mollusca*	33	16	21	19	35	30	24
	Country total	*4 719*	*7 405*	*4 521*	*5 049*	*3 186*	*2 487*	*1 492*
Cambodia								
Freshwater fishes nei	*Osteichthyes*	72 420	63 440	72 900	75 600	230 700	245 300	359 600 F
Tuna-like fishes nei	*Scombroidei*	-	-	-	-	56	-	-
Marine fishes nei	*Osteichthyes*	22 500	23 000	21 970	23 740	28 100	26 606	26 600 F
Freshwater crustaceans nei	*Crustacea*	79	70	100	100	300	300	400 F

English name Nom anglais Nombre inglés	Scientific name Nom scientifique Nombre científico	1995 mt	1996 mt	1997 mt	1998 mt	1999 mt	2000 mt	2001 mt
Marine crabs nei	*Brachyura*	2 300	2 350	2 240	2 420	2 850	3 593	3 600 F
Natantian decapods nei	*Natantia*	4 200	4 300	4 110	4 440	5 250	5 000	5 000 F
Argentine shortfin squid	*Illex argentinus*	-	-	-	-	-	2 768	1 200
Marine molluscs nei	*Mollusca*	1 500	1 550	1 480	1 600	1 900	801	800 F
	Country total	*102 999*	*94 710*	*102 800*	*107 900*	*269 156*	*284 368*	*397 200 F*
China								
Freshwater fishes nei	*Osteichthyes*	894 855	994 971	1 032 861	1 218 152	1 394 610	1 222 955	1 033 302
Elongate ilisha	*Ilisha elongata*	46 635	51 339	77 532	84 290	110 359	107 689	101 342
Alaska pollock(=Walleye poll.)	*Theragra chalcogramma*	249 459	226 900	338 478	191 433	64 520	60 338	39 665
Hakes nei	*Merluccius spp*	10	-	-	0	-	-	-
Mullets nei	*Mugilidae*	71 426	67 423	85 967	99 023	113 454	107 243	119 246
Groupers nei	*Epinephelus spp*	22 999	23 241	30 201	36 098	40 245	41 513	44 975
Golden threadfin bream	*Nemipterus virgatus*	224 574	238 000	258 998	262 702	246 601	291 495	287 384
Large yellow croaker	*Larimichthys croceus*	67 031	80 072	69 950	70 935	65 806	122 920	76 289
Yellow croaker	*Larimichthys polyactis*	153 048	253 482	212 631	262 786	308 907	404 637	321 614
Croakers, drums nei	*Sciaenidae*	108 241	530	490	1 568	1 027	2 365	2 061
Porgies, seabreams nei	*Sparidae*	59 633	57 078	74 054	77 347	80 932	108 453	129 330
Threadfins, tasselfishes nei	*Polynemidae*	...	283 784	340 302	517 528	565 764	496 566	476 690
Filefishes	*Cantherhines(=Navodon) spp*	122 358	210 188	296 781	235 603	240 214	221 683	201 733
Daggertooth pike conger	*Muraenesox cinereus*	154 867	177 470	184 843	239 874	234 314	220 497	243 888
Orange roughy	*Hoplostethus atlanticus*	-	-	-	-	-	623	710
Oreo dories nei	*Oreosomatidae*	-	-	-	-	-	-	180
Pelagic armourhead	*Pseudopentaceros richardsoni*	-	-	-	-	-	44	-
Largehead hairtail	*Trichiurus lepturus*	1 039 684	1 071 914	1 014 598	1 223 360	1 222 454	1 285 469	1 282 698
Pacific herring	*Clupea pallasii*	2 325	1 665	15 780	21 796	17 936	15 258	51 950
Japanese pilchard	*Sardinops melanostictus*	58 434	92 918	124 844	121 120	147 125	153 944	160 825
European pilchard(=Sardine)	*Sardina pilchardus*	-	-	-	-	-	-	2
Japanese anchovy	*Engraulis japonicus*	489 066	671 376	1 201 964	1 373 328	1 096 916	1 142 884	1 260 712
Japanese Spanish mackerel	*Scomberomorus niphonius*	226 520	283 784	340 302	517 528	565 764	496 566	476 690
Skipjack tuna	*Katsuwonus pelamis*	-	-	-	4	-	1 050	3 880
Atlantic bluefin tuna	*Thunnus thynnus*	137	93	49	85	103	80	68
Albacore	*Thunnus alalunga*	13	28	2	1	3 722	3 403	4 489
Yellowfin tuna	*Thunnus albacares*	6 175	3 375	2 253	2 536	6 762	6 822	6 916
Bigeye tuna	*Thunnus obesus*	5 360	4 247	4 322	5 504	11 334	11 993	16 112
Atlantic sailfish	*Istiophorus albicans*	6	6	6	8	18	8	8
Atlantic blue marlin	*Makaira nigricans*	73	62	78	120	201	24	92
Atlantic white marlin	*Tetrapturus albidus*	11	9	11	15	30	2	20
Longbill spearfish	*Tetrapturus pfluegeri*	-	-	-	2	-	-	-
Marlins,sailfishes,etc. nei	*Istiophoridae*	-	18	-	-	401	1 000	799
Swordfish	*Xiphias gladius*	175	370	295	483	1 504	879	1 083
Tuna-like fishes nei	*Scombroidei*	23 649	15 935	14 600	14 087	20 524	30 543	2 828
Jack and horse mackerels nei	*Trachurus spp*	-	-	-	1	-	0	3
Scads nei	*Decapterus spp*	515 298	607 686	505 991	532 986	502 590	502 289	544 728
Chub mackerel	*Scomber japonicus*	372 038	374 400	408 935	385 183	402 548	350 809	381 597
Silver pomfrets nei	*Pampus spp*	209 031	220 364	242 547	303 024	337 919	338 848	352 493
Shortfin mako	*Isurus oxyrinchus*	-	-	-	...	...	153	...
Sharks, rays, skates, etc. nei	*Elasmobranchii*	-	-	2	5	378	99	...
Marine fishes nei	*Osteichthyes*	3 326 683	4 115 477	3 749 397	3 706 067	3 773 149	3 376 638	2 983 732
Freshwater crustaceans nei	*Crustacea*	270 536	364 441	479 645	601 966	455 761	530 026	586 985
Blue swimming crab	*Portunus pelagicus*	25 168	51 288	45 749	48 190	52 577	56 092	59 416
Gazami crab	*Portunus trituberculatus*	243 485	283 394	237 960	266 630	270 280	335 078	333 556
Fleshy prawn	*Penaeus chinensis*	43 043	55 292	69 406	78 350	69 911	84 334	97 002
Penaeus shrimps nei	*Penaeus spp*	2 372	2 113	2 260	2 504	1 997	1 873	2 087
Southern rough shrimp	*Trachypenaeus curvirostris*	151 746	163 060	174 967	175 618	400 786	312 436	244 202
Akiami paste shrimp	*Acetes japonicus*	390 000	442 460	480 056	571 383	579 213	625 234	565 792
Marine crustaceans nei	*Crustacea*	852 257	914 672	1 086 501	1 231 473	1 131 643	1 213 725	1 264 976
Freshwater molluscs nei	*Mollusca*	441 994	403 448	374 461	460 126	434 993	480 249	529 645
Cuttlefish,bobtail squids nei	*Sepiidae, Sepiolidae*	221 883	173 142	243 431	227 520	216 303	260 906	318 367
Argentine shortfin squid	*Illex argentinus*	-	-	-	30 000	61 000	93 130	93 500
Various squids nei	*Loliginidae, Ommastrephidae*	57	229	66	117 200	132 237	124 188	87 290
Octopuses, etc. nei	*Octopodidae*	3 481	2 464	3 193	4 336	7 491	4 578	5 130
Marine molluscs nei	*Mollusca*	1 294 815	932 374	1 494 902	1 479 065	1 445 303	1 402 625	1 399 810
Sea urchins nei	*Strongylocentrotus spp*	150	200	200	200	200	200	200
Jellyfishes	*Rhopilema spp*	171 905	265 325	400 483	430 784	402 206	334 869	331 297
	Country total	*12 562 706*	*14 182 107*	*15 722 344*	*17 229 927*	*17 240 032*	*16 987 325*	*16 529 389*
China,H.Kong								
Cyprinids nei	*Cyprinidae*	0	0	0	0	0	0	0
Tonguefishes	*Cynoglossidae*	1 326	1 160	1 261	1 177	800 F	1 000 F	1 100 F
Lizardfishes nei	*Synodontidae*	10 888	9 034	8 070	6 619	4 700 F	5 800 F	6 400 F
Groupers, seabasses nei	*Serranidae*	1 392	1 349	1 318	1 240	900 F	1 100 F	1 200 F
Bigeyes nei	*Priacanthus spp*	6 781	5 391	4 962	4 149	3 000 F	3 600 F	4 000 F
Snappers, jobfishes nei	*Lutjanidae*	406	430	439	304	250 F	300 F	330 F
Threadfin breams nei	*Nemipterus spp*	23 772	19 568	20 024	19 449	14 000 F	17 000 F	18 800 F
Croakers, drums nei	*Sciaenidae*	5 547	3 776	4 881	3 992	2 800 F	3 500 F	3 900 F
Largeeye breams	*Gymnocranius spp*	196	208	170	143	100 F	120 F	130 F
Porgies, seabreams nei	*Sparidae*	1 215	925	883	852	600 F	700 F	780 F
Goatfishes	*Upeneus spp*	593	718	477	393	300 F	350 F	400 F
Pike-congers nei	*Muraenesox spp*	3 129	3 281	2 903	2 450	1 700 F	2 100 F	2 300 F
Tilefishes nei	*Branchiostegidae*	2 583	3 186	4 187	4 879	3 500 F	4 300 F	4 750 F
Hairtails, scabbardfishes nei	*Trichiuridae*	3 470	3 326	2 522	1 690	1 200 F	1 500 F	1 650 F
Indian driftfish	*Ariomma indica*	287	120	79	49	35 F	40 F	45 F
Pacific rudderfish	*Psenopsis anomala*	1 494	940	1 460	1 160	800 F	1 000 F	1 100 F
Sardinellas nei	*Sardinella spp*	24	30	12	8	5 F	5 F	6 F
Slender rainbow sardine	*Dussumieria elopsoides*	154	48	19	17	10 F	15 F	17 F
Stolephorus anchovies	*Stolephorus spp*	89	31	20	17	10 F	15 F	17 F

English name Nom anglais Nombre inglés	Scientific name Nom scientifique Nombre científico	1995 mt	1996 mt	1997 mt	1998 mt	1999 mt	2000 mt	2001 mt
Seerfishes nei	*Scomberomorus spp*	2 790	2 036	1 711	1 723	1 200 F	1 500 F	1 650 F
Tuna-like fishes nei	*Scombroidei*	18	18	1	4	0	0	0
Scads nei	*Decapterus spp*	5 254	5 368	4 143	2 962	2 100 F	2 600 F	2 900 F
Jacks, crevalles nei	*Caranx spp*	106	66	33	20	10 F	15 F	17 F
Butterfishes, pomfrets nei	*Stromateidae*	2 428	2 114	1 973	2 172	1 500 F	1 900 F	2 100 F
Sharks, rays, skates, etc. nei	*Elasmobranchii*	485	456	420	382	300 F	330 F	370 F
Marine fishes nei	*Osteichthyes*	98 215	101 269	101 062	108 206	76 460 F	94 272 F	104 550 F
Marine crabs nei	*Brachyura*	1 601	1 408	1 105	1 108	800 F	1 000 F	1 100 F
Tropical spiny lobsters nei	*Panulirus spp*	20	0	0	0	0	0	0
Natantian decapods nei	*Natantia*	7 155	6 589	6 225	5 285	3 800 F	4 600 F	5 100 F
Cuttlefish,bobtail squids nei	*Sepiidae, Sepiolidae*	1 689	991	841	510	400 F	450 F	500 F
Various squids nei	*Loliginidae, Ommastrephidae*	10 043	8 559	13 586	8 340	6 000 F	7 300 F	8 100 F
Marine molluscs nei	*Mollusca*	1 849	1 461	1 213	700	500 F	600 F	660 F
	Country total	*194 999*	*183 856*	*186 000*	*180 000*	*127 780*	*157 012*	*173 972*
China, Macao								
Freshwater fishes nei	*Osteichthyes*	0	0	0	0	0	0	0
Marine fishes nei	*Osteichthyes*	1 151	970	1 020 F	1 020 F	1 020 F	1 020 F	1 020 F
Natantian decapods nei	*Natantia*	181	218	230 F	230 F	230 F	230 F	230 F
Marine crustaceans nei	*Crustacea*	207	200	210 F	210 F	210 F	210 F	210 F
Marine molluscs nei	*Mollusca*	65	30	40 F	40 F	40 F	40 F	40 F
	Country total	*1 604*	*1 418*	*1 500 F*	*1 500 F*	*1 500 F*	*1 500 F*	*1 500 F*
China,Taiwan								
Common carp	*Cyprinus carpio*	44	26	19	22	45	49	50
Crucian carp	*Carassius carassius*	14	5	5	5	28	35	37
Mud carp	*Cirrhinus molitorella*	5	3	4	7	7	6	6
Grass carp(=White amur)	*Ctenopharyngodon idellus*	90	79	83	81	95	104	132
Silver carp	*Hypophthalmichthys molitrix*	37	29	35	34	11	9	11
Bighead carp	*Hypophthalmichthys nobilis*	105	67	72	99	241	204	202
Black carp	*Mylopharyngodon piceus*	20	15	16	22	27	34	36
Tilapias nei	*Oreochromis (=Tilapia) spp*	319	145	146	152	86	79	98
Torpedo-shaped catfishes nei	*Clarias spp*	-	-	-	6	-	-	-
Freshwater fishes nei	*Osteichthyes*	21	16	11	18	18	17	19
Ayu sweetfish	*Plecoglossus altivelis*	2	-	-	-	-	-	-
Chinese gizzard shad	*Clupanodon thrissa*	30	25	10	11	27	21	74
Milkfish	*Chanos chanos*	1	4	-	-	-	-	1
Barramundi(=Giant seaperch)	*Lates calcarifer*	0	14	9	26	47	32	80
Flatfishes nei	*Pleuronectiformes*	145	345	346	142	146	137	198
Alaska pollock(=Walleye poll.)	*Theragra chalcogramma*	37	12	7	9	9	9	-
Greater lizardfish	*Saurida tumbil*	6 744	6 561	3 208	3 049	3 075	3 612	2 248
Sea catfishes nei	*Ariidae*	274	122	134	208	257	725	435
Flathead grey mullet	*Mugil cephalus*	1 638	1 302	2 446	606	863	1 890	1 559
Groupers nei	*Epinephelus spp*	1 771	2 457	2 387	1 414	1 265	1 202	1 610
Red bigeye	*Priacanthus macracanthus*	5 486	7 386	4 824	3 670	2 756	3 986	3 079
Sillago-whitings	*Sillaginidae*	407	547	401	346	190	188	132
Moonfish	*Mene maculata*	859	725	902	1 010	1 747	1 496	1 582
Snappers, jobfishes nei	*Lutjanidae*	1 595	2 334	1 408	808	467	352	385
Golden threadfin bream	*Nemipterus virgatus*	3 664	4 261	4 401	3 681	3 990	4 824	3 536
Yellow croaker	*Larimichthys polyactis*	2 786	2 859	2 228	1 010	1 167	780	903
Blackmouth croaker	*Atrobucca nibe*	197	297	288	237	223	450	644
Silver croaker	*Pennahia argentata*	7 369	8 599	6 085	5 089	4 940	4 180	3 796
Croakers, drums nei	*Sciaenidae*	3 269	3 070	2 497	1 422	1 580	1 708	3 841
Silver seabream	*Pagrus auratus*	1 580	1 809	2 445	2 073	3 333	4 726	6 722
Blackhead seabream	*Acanthopagrus schlegeli*	197	207	277	404	604	717	878
Porgies, seabreams nei	*Sparidae*	5 626	5 401	6 334	6 440	7 312	6 288	9 953
Goatfishes	*Upeneus spp*	399	508	440	335	478	477	300
Parrotfishes nei	*Scaridae*	8	11	17	12	5	29	27
Fourfinger threadfin	*Eleutheronema tetradactylum*	3 469	3 157	2 831	3 248	2 414	6 917	1 715
Filefishes	*Cantherhines(=Navodon) spp*	1 680	871	446	586	995	829	959
Daggertooth pike conger	*Muraenesox cinereus*	3 548	2 846	3 246	5 733	9 001	5 874	5 084
Tilefishes nei	*Branchiostegidae*	579	1 227	626	372	496	448	512
Oilfish	*Ruvettus pretiosus*	3 729	2 634	2 622	3 043	2 661	2 584	3 678
Largehead hairtail	*Trichiurus lepturus*	16 210	11 830	8 955	7 991	9 375	7 271	8 834
Pacific rudderfish	*Psenopsis anomala*	9 212	8 834	6 001	8 258	5 075	4 506	5 646
Red-eye round herring	*Etrumeus teres*	1 653	1 462	2 152	1 248	893	1 118	1 306
Silver-stripe round herring	*Spratelloides gracilis*	827	517	650	886	521	639	505
Japanese anchovy	*Engraulis japonicus*	305	466	515	425	650	589	695
Dorab wolf-herring	*Chirocentrus dorab*	3	1	2	1	4	1	-
Clupeoids nei	*Clupeoidei*	6 348	3 873	4 222	4 338	3 721	4 221	4 393
Narrow-barred Spanish mackerel	*Scomberomorus commerson*	3 211	2 541	2 701	2 417	2 674	3 551	4 153
Indo-Pacific king mackerel	*Scomberomorus guttatus*	1 814	1 611	1 607	1 128	1 298	1 249	1 395
Japanese Spanish mackerel	*Scomberomorus niphonius*	8 971	7 546	11 761	8 579	4 516	5 955	11 497
Seerfishes nei	*Scomberomorus spp*	1 085	1 103	1 143	1 478	2 883	2 260	1 433
Frigate and bullet tunas	*Auxis thazard, A.rochei*	2 881	2 482	4 280	4 634	4 558	3 073	2 089
Kawakawa	*Euthynnus affinis*	3 454 F	2 616	2 577	2 307	2 065	1 977	2 136
Skipjack tuna	*Katsuwonus pelamis*	158 837	173 489	119 789	197 313	163 903	197 536	198 839
Atlantic bluefin tuna	*Thunnus thynnus*	501	472	504	456	250	407	397
Pacific bluefin tuna	*Thunnus orientalis*	314	956	1 813	1 910	3 089	2 690	2 084
Longtail tuna	*Thunnus tonggol*	11 166	8 900	6 762	6 466	9 206	14 429	18 621
Albacore	*Thunnus alalunga*	56 467	58 408	55 205	59 868	64 302	51 668	45 458
Southern bluefin tuna	*Thunnus maccoyii*	1 474	1 610	640	1 439	1 751	1 882	1 655
Yellowfin tuna	*Thunnus albacares*	85 128	82 891	101 379	122 484	95 004	99 279	109 574
Bigeye tuna	*Thunnus obesus*	60 117	64 498	74 770	76 250	76 761	82 484	81 244
Indo-Pacific sailfish	*Istiophorus platypterus*	2 033	2 296	3 350	3 975	3 610	2 527	2 216
Atlantic sailfish	*Istiophorus albicans*	193	102	46	153	99	135	-
Indo-Pacific blue marlin	*Makaira mazara*	10 519	13 507	10 782	12 557	13 418	14 236	13 723

English name Nom anglais Nombre inglés	Scientific name Nom scientifique Nombre científico	1995 mt	1996 mt	1997 mt	1998 mt	1999 mt	2000 mt	2001 mt
Atlantic blue marlin	*Makaira nigricans*	467	660	530	578	486	89	678
Black marlin	*Makaira indica*	1 340	1 469	1 357	2 063	1 432	1 296	964
Striped marlin	*Tetrapturus audax*	4 126	3 416	3 539	4 116	2 696	2 995	2 017
Atlantic white marlin	*Tetrapturus albidus*	908	567	441	507	464	723	35
Marlins,sailfishes,etc. nei	*Istiophoridae*	80	24	5	24	0	0	807
Swordfish	*Xiphias gladius*	23 030	14 315	21 863	20 940	19 216	20 330	19 661
Pacific saury	*Cololabis saira*	13 772	8 236	21 887	12 794	12 541	27 868	39 764
Flyingfishes nei	*Exocoetidae*	572	662	2 497	1 077	618	287	366
Cobia	*Rachycentron canadum*	779	692	987	815	655	1 014	486
Japanese jack mackerel	*Trachurus japonicus*	4 841	4 218	4 711	7 121	2 661	6 003	3 903
Japanese scad	*Decapterus maruadsi*	3 626	5 059	19 667	12 090	20 532	4 387	6 224
Indian scad	*Decapterus russelli*	299	382	687	6 158	7 703	3 927	598
Scads nei	*Decapterus spp*	9 168	32 586	12 743	3 172	2 836	2 491	2 670
Jacks, crevalles nei	*Caranx spp*	6 595	7 065	11 433	8 228	4 882	6 355	3 273
Black pomfret	*Parastromateus niger*	5 444	3 427	2 534	1 744	1 307	1 671	2 129
Torpedo scad	*Megalaspis cordyla*	65	233	6 847	52	113	327	198
Common dolphinfish	*Coryphaena hippurus*	12 203	3 209	8 089	17 157	8 560	5 558	7 427
Chub mackerel	*Scomber japonicus*	54 155	53 906	47 279	35 527	45 339	28 635	24 298
Silver pomfret	*Pampus argenteus*	6 978	5 113	2 703	1 621	2 477	2 852	4 245
Butterfishes, pomfrets nei	*Stromateidae*	226	246	336	231	280	131	146
Barracudas nei	*Sphyraena spp*	429	547	761	771	519	362	439
Rays, stingrays, mantas nei	*Rajiformes*	647	2 457	1 367	246	235	351	851
Sharks, rays, skates, etc. nei	*Elasmobranchii*	43 417	38 701	38 722	39 779	42 698	45 572	41 504
Marine fishes nei	*Osteichthyes*	79 743	50 983	62 078	75 195	73 823	81 581	75 968
Blue swimming crab	*Portunus pelagicus*	2 808	1 391	1 141	966	511	1 966	1 703
Indo-Pacific swamp crab	*Scylla serrata*	1 339	935	180	215	269	299	230
Marine crabs nei	*Brachyura*	4 747	4 830	4 630	4 553	5 433	5 898	7 373
Longlegged spiny lobster	*Panulirus longipes*	18	14	14	14	12	11	23
Japanese fan lobster	*Ibacus ciliatus*	1 224	1 115	642	696	676	1 600	1 607
Kuruma prawn	*Penaeus japonicus*	1 960	1 561	2 665	1 906	4 071	8 168	4 179
Giant tiger prawn	*Penaeus monodon*	67	49	45	38	92	57	91
Redtail prawn	*Penaeus penicillatus*	3 564	5 020	2 473	647	316	308	312
Metapenaeus shrimps nei	*Metapenaeus spp*	3 767	626	369	299	210	202	241
Southern rough shrimp	*Trachypenaeus curvirostris*	2 877	4 663	5 288	3 926	2 241	532	593
Natantian decapods nei	*Natantia*	24 146	23 770	22 112	15 702	14 943	11 336	11 987
Freshwater molluscs nei	*Mollusca*	-	-	-	1	2	2	-
Abalones nei	*Haliotis spp*	14	8	6	2	2	42	1
Gastropods nei	*Gastropoda*	998	760	1 019	125	0	10	53
Pacific cupped oyster	*Crassostrea gigas*	7	-	-	-	-	-	4
Yesso scallop	*Patinopecten yessoensis*	22	76	63	114	112	90	47
Japanese hard clam	*Meretrix lusoria*	256	135	161	805	526	0	-
Japanese carpet shell	*Ruditapes philippinarum*	16	-	-	-	-	-	-
Cuttlefish,bobtail squids nei	*Sepiidae, Sepiolidae*	12 084	7 191	8 621	7 198	5 019	3 880	4 546
Common squids nei	*Loligo spp*	20 455 F	24 721	28 402	26 927	15 898	9 573	9 099
Argentine shortfin squid	*Illex argentinus*	100 371	101 328	185 775	163 180	264 089	238 334	146 783
Japanese flying squid	*Todarodes pacificus*	22 243	19 101	12 430	34 840	11 261	6 833	4 716
Wellington flying squid	*Nototodarus sloani*	8 284	14 747	6 620	3 974	761	-	-
Sevenstar flying squid	*Martialia hyadesi*	23 464	3 792	8 348	-	-	-	-
Octopuses, etc. nei	*Octopodidae*	919	967	615	941	500	287	374
Cephalopods nei	*Cephalopoda*	-	-	-	-	-	-	3
Marine molluscs nei	*Mollusca*	932	464	443	663	433	914	118
Sea urchins nei	*Strongylocentrotus spp*	63	59	61	39	33	41	50
	Country total	*1 010 022*	*967 483*	*1 038 048*	*1 091 768*	*1 099 715*	*1 093 889*	*1 005 199*
Cyprus								
Freshwater fishes nei	*Osteichthyes*	65	64	70	70	70	78	70 F
European hake	*Merluccius merluccius*	7	3	4	2	5	6	8
Hakes nei	*Merluccius spp*	-	...	...	40	164	189	110
Flathead grey mullet	*Mugil cephalus*	2	5	5	2	6	7	7
Groupers nei	*Epinephelus spp*	25	22	22	29	23	27	24
Groupers, seabasses nei	*Serranidae*	13	16	15	11	17	3	5
Seabasses nei	*Dicentrarchus spp*	...	...	...	...	...	...	3
Brown meagre	*Sciaena umbra*	0	2	5	3	7	4	3
Common pandora	*Pagellus erythrinus*	32	32	25	19	37	42	34
Axillary seabream	*Pagellus acarne*	11	22	25	23	50	25	20
Common dentex	*Dentex dentex*	23	38	18	18	22	18	19
Saddled seabream	*Oblada melanura*	10	4	4	3	7	8	11
Red porgy	*Pagrus pagrus*	45	35	27	25	31	23	21
Gilthead seabream	*Sparus aurata*	1	1	0	0	0	-	37
Bogue	*Boops boops*	290	285	230	233	259	354	216
Salema	*Sarpa salpa*	4	5	3	4	1	4	5
Porgies, seabreams nei	*Sparidae*	61	65	83	102	154	156	237
Picarels nei	*Spicara spp*	711	764	650	709	546	533	671
Red mullet	*Mullus surmuletus*	215	240	228	176	184	159	132
Striped mullet	*Mullus barbatus*	88	108	119	115	145	103	91
Parrotfishes nei	*Scaridae*	29	32	37	78	58	33	42
Spinefeet(=Rabbitfishes) nei	*Siganus spp*	1	11	7	36	19	12	31
Largehead hairtail	*Trichiurus lepturus*	-	...	...	5 061	130	230	3 949
Scorpionfishes nei	*Scorpaenidae*	11	16	19	10	3	8	11
Sardinellas nei	*Sardinella spp*	-	...	...	2 172	4 492	15 043	8 418
European pilchard(=Sardine)	*Sardina pilchardus*	-	...	...	-	12	564	30
European anchovy	*Engraulis encrasicolus*	-	...	...	35	11 879	14 705	13 710
Atlantic bonito	*Sarda sarda*	-	-	-	-	-	14	13
Little tunny(=Atl.black skipj)	*Euthynnus alletteratus*	10	19	30	10	16	14	13
Skipjack tuna	*Katsuwonus pelamis*	4 200	3 210	5 200	300	-	-	-
Atlantic bluefin tuna	*Thunnus thynnus*	10	10	10	21	31	61	90
Albacore	*Thunnus alalunga*	-	-	-	-	-	6	-
Yellowfin tuna	*Thunnus albacares*	950	2 890	2 150	100	-	-	-

English name Nom anglais Nombre inglés	Scientific name Nom scientifique Nombre científico	1995 mt	1996 mt	1997 mt	1998 mt	1999 mt	2000 mt	2001 mt
Bigeye tuna	*Thunnus obesus*	1 600	1 180	1 450	30	-	-	-
Swordfish	*Xiphias gladius*	89	40	51	61	92	82	135
Tuna-like fishes nei	*Scombroidei*	-	...	...	39	72	180	262
Jack and horse mackerels nei	*Trachurus spp*	-	...	...	5 968	17 892	27 326	40 612
Greater amberjack	*Seriola dumerili*	16	28	21	8	17	25	14
Chub mackerel	*Scomber japonicus*	...	...	...	2 758	1 906	6 207	5 503
Sharks, rays, skates, etc. nei	*Elasmobranchii*	21	14	17	10	12	14	28
Marine fishes nei	*Osteichthyes*	322	3 057	14 003	635	944	921	928
Common spiny lobster	*Palinurus elephas*	2	5	2	4	2	4	5
Palinurid spiny lobsters nei	*Palinurus spp*	-	-	-	-	-	0	1
Deepwater rose shrimp	*Parapenaeus longirostris*	3	2	1	1	5	4	2
Common cuttlefish	*Sepia officinalis*	153	111	89	146	140	123	106
Various squids nei	*Loliginidae, Ommastrephidae*	-	-	-	-	9	1	-
Octopuses, etc. nei	*Octopodidae*	300	190	199	228	179	166	176
	Country total	*9 320*	*12 526*	*24 819*	*19 295*	*39 638*	*67 482*	*75 803*
Gaza Strip								
Freshwater fishes nei	*Osteichthyes*	0	0	0	0	0	0	0
Flatfishes nei	*Pleuronectiformes*	...	7	15	25	25 F	25 F	20 F
Lizardfishes nei	*Synodontidae*	...	16	32	32	30 F	30 F	30 F
Mullets nei	*Mugilidae*	...	32	19	35	35 F	35 F	30 F
Groupers nei	*Epinephelus spp*	...	46	45	44	45 F	45 F	40 F
Meagre	*Argyrosomus regius*	...	39	7	4	5 F	5 F	5 F
Sargo breams nei	*Diplodus spp*	...	4	17	24	20 F	20 F	20 F
Red porgy	*Pagrus pagrus*	...	35	47	35	35 F	35 F	30 F
Bogue	*Boops boops*	...	9	81	162	160 F	160 F	130 F
Surmullets(=Red mullets) nei	*Mullus spp*	...	56	57	85	85 F	85 F	70 F
Spinefeet(=Rabbitfishes) nei	*Siganus spp*	...	2	11	10	10 F	10 F	10 F
Sardinellas nei	*Sardinella spp*	...	978	2 483	1 780	1 750 F	1 750 F	1 450 F
Little tunny(=Atl.black skipj)	*Euthynnus alletteratus*	...	90	59	61	60 F	60 F	50 F
Tuna-like fishes nei	*Scombroidei*	...	50	102	92	100 F	100 F	80 F
Bluefish	*Pomatomus saltatrix*	...	36	8	31	30 F	30 F	30 F
Jack and horse mackerels nei	*Trachurus spp*	...	90	100	115	115 F	115 F	100 F
Chub mackerel	*Scomber japonicus*	...	130	120	337	340 F	340 F	280 F
Barracudas nei	*Sphyraena spp*	...	49	101	90	90 F	90 F	80 F
Smooth-hounds nei	*Mustelus spp*	...	24	11	9	10 F	10 F	10 F
Guitarfishes, etc. nei	*Rhinobatidae*	...	...	6	6	5 F	5 F	5 F
Rays, stingrays, mantas nei	*Rajiformes*	...	29	16	23	20 F	20 F	20 F
Marine fishes nei	*Osteichthyes*	1 057	559	140	194	200 F	200 F	160 F
Marine crabs nei	*Brachyura*	66	72	25	65	65 F	65 F	50 F
Natantian decapods nei	*Natantia*	50	55	161	161	160 F	160 F	130 F
Cuttlefish,bobtail squids nei	*Sepiidae, Sepiolidae*	41	62	98	144	145 F	145 F	120 F
Common squids nei	*Loligo spp*	15	23	30	61	60 F	60 F	50 F
	Country total	*1 229*	*2 493*	*3 791*	*3 625*	*3 600 F*	*3 600 F*	*3 000 F*
Georgia								
Freshwater bream	*Abramis brama*	27	3	1	2	1	3	1
Common carp	*Cyprinus carpio*	49	0	0	2	11	12	5
Wels(=Som)catfish	*Silurus glanis*	2	-	-	-	-	-	-
Freshwater fishes nei	*Osteichthyes*	0	0	0	0	5	7	2
Sturgeons nei	*Acipenseridae*	...	...	...	...	3	4	3
Rainbow trout	*Oncorhynchus mykiss*	12	3	...	...	...	...	...
Flatfishes nei	*Pleuronectiformes*	-	-	-	-	5	9	11
Whiting	*Merlangius merlangus*	146	223	58	53	41	...	32
Mullets nei	*Mugilidae*	-	-	-	-	9	19	28
Surmullets(=Red mullets) nei	*Mullus spp*	-	-	14	11	8	3	22
Gobies nei	*Gobiidae*	...	...	...	...	2	3	5
European pilchard(=Sardine)	*Sardina pilchardus*	900 F	-	-	-	-	-	-
European sprat	*Sprattus sprattus*	292	185	85	24	45	...	30
European anchovy	*Engraulis encrasicolus*	1 401 F	1 232	2 288	2 346	1 264	2 080	1 652
Jack and horse mackerels nei	*Trachurus spp*	-	-	18	13	...	35	7
Sharks, rays, skates, etc. nei	*Elasmobranchii*	31	71	1	550	18	21	27
Marine fishes nei	*Osteichthyes*	-	25	-	-	1	4	5
Sea snails	*Rapana spp*	700	711	118	-	-	-	-
	Country total	*3 560 F*	*2 453*	*2 583*	*3 001*	*1 413*	*2 200*	*1 830*
India								
Cyprinids nei	*Cyprinidae*	339 828	220 994	194 608	167 186	177 394	152 350	148 622
Freshwater siluroids nei	*Siluroidei*	...	...	...	63 724	71 152	114 099	76 809
Snakeheads(=Murrels) nei	*Channa spp*	93 933	94 944	105 496	159 065	50 349	58 669	58 856
Freshwater fishes nei	*Osteichthyes*	92 964	224 990	245 595	225 100	330 448	416 490	591 550
Kelee shad	*Hilsa kelee*	92 707	100 591	95 677	105 281	89 354	221 899	97 337
Flatfishes nei	*Pleuronectiformes*	24 681	24 197	29 252	23 662	18 024	30 071	13 615
Unicorn cod	*Bregmaceros mcclellandi*	931	975	984	2 251	1 810	1 282	2 127
Bombay-duck	*Harpadon nehereus*	159 719	185 112	213 116	179 915	181 820	134 197	144 499
Lizardfishes nei	*Synodontidae*	23 989	19 670	13 528	12 765	12 914	39 573	37 746
Sea catfishes nei	*Ariidae*	74 301	73 125	78 150	72 891	86 096	80 817	95 466
Mullets nei	*Mugilidae*	15 365	24 265	26 871	19 457	19 872	28 837	29 880
Ponyfishes(=Slipmouths) nei	*Leiognathidae*	63 394	68 582	67 369	46 068	55 462	49 280	51 790
Croakers, drums nei	*Sciaenidae*	287 577	299 558	313 214	275 240	317 752	268 073	246 058
Goatfishes	*Upeneus spp*	18 869	18 129	15 870	17 260	17 892	10 536	16 736
Threadfins, tasselfishes nei	*Polynemidae*	9 647	7 052	7 095	7 764	10 058	5 776	6 182
Percoids nei	*Percoidei*	37 983	115 879	87 710	62 498	89 950	94 561	80 354
Pike-congers nei	*Muraenesox spp*	9 545	11 101	11 966	11 279	14 090	11 386	11 729
Hairtails, scabbardfishes nei	*Trichiuridae*	50 582	53 941	146 118	73 396	115 206	120 596	144 474
Indian oil sardine	*Sardinella longiceps*	105 916	136 713	212 060	173 351	147 715	327 425	341 667

English name Nom anglais Nombre inglés	Scientific name Nom scientifique Nombre científico	1995 mt	1996 mt	1997 mt	1998 mt	1999 mt	2000 mt	2001 mt
Anchovies, etc. nei	*Engraulidae*	60 979	80 694	80 518	81 206	68 743	68 608	71 517
Wolf-herrings nei	*Chirocentrus spp*	20 472	19 510	25 153	15 127	17 511	16 074	18 726
Clupeoids nei	*Clupeoidei*	60 127	78 121	93 990	97 910	90 305	64 800	59 789
Wahoo	*Acanthocybium solandri*	5	23	21	23	29	31	...
Narrow-barred Spanish mackerel	*Scomberomorus commerson*	28 587	24 613	23 360	32 181	40 318	44 170	45 458
Indo-Pacific king mackerel	*Scomberomorus guttatus*	17 174	12 662	13 255	22 560	28 263	30 964	43 819
Streaked seerfish	*Scomberomorus lineolatus*	87	96	901	107	135	147	381
Seerfishes nei	*Scomberomorus spp*	851	1 216	3 335	-	-	-	-
Frigate and bullet tunas	*Auxis thazard, A.rochei*	5 917	11 119	10 564	7 680	17 183	18 911	7 031
Kawakawa	*Euthynnus affinis*	18 781	14 778	23 425	15 451	21 641	23 618	5 983
Skipjack tuna	*Katsuwonus pelamis*	6 577	6 850	6 096	1 037	5 707	5 986	21 789
Longtail tuna	*Thunnus tonggol*	7 036	4 263	5 322	4 750	2 568	2 723	541
Yellowfin tuna	*Thunnus albacares*	194	7 327	4 106	3 618	2 073	2 158	7 324
Bigeye tuna	*Thunnus obesus*	1 076	-	-	4	-	-	-
Indo-Pacific sailfish	*Istiophorus platypterus*	-	13	12	14	9	9	...
Marlins,sailfishes,etc. nei	*Istiophoridae*	1 543	3 940	4 502	3 381	4 202	4 582	1 208
Swordfish	*Xiphias gladius*	-	62	58	25	15	-	-
Tuna-like fishes nei	*Scombroidei*	5 198	22	62	6 156	-	1	...
Halfbeaks nei	*Hemiramphus spp*	8 674	6 118	5 574	4 672	5 389	7 643	6 170
Flyingfishes nei	*Exocoetidae*	2 345	1 886	2 094	612	736	2 711	2 446
False trevally	*Lactarius lactarius*	7 087	6 930	7 942	9 528	6 827	6 061	7 283
Jacks, crevalles nei	*Caranx spp*	21 439	67 879	50 833	66 910	65 499	39 571	46 840
Pompanos nei	*Trachinotus spp*	8 897	6 784	3 697	2 738	2 329	12	11
Carangids nei	*Carangidae*	22 489	34 225	32 752	26 271	29 916	11 129	13 091
Indian mackerel	*Rastrelliger kanagurta*	197 233	293 526	190 372	184 419	179 378	78 821	58 906
Butterfishes, pomfrets nei	*Stromateidae*	57 978	37 496	36 731	33 319	31 185	25 391	23 756
Barracudas nei	*Sphyraena spp*	9 070	8 873	7 827	14 965	14 936	4 162	4 864
Sharks, rays, skates, etc. nei	*Elasmobranchii*	77 078	132 160	71 991	74 704	76 802	76 057	72 568
Marine fishes nei	*Osteichthyes*	661 209	455 702	503 234	499 863	473 445	561 368	559 785
Freshwater crustaceans nei	*Crustacea*	26	803	71	114	85	112	77
Marine crabs nei	*Brachyura*	16 705	1 547	2 237	1 789	1 693	23 650	25 044
Giant tiger prawn	*Penaeus monodon*	176 289	150 593	146 111	204 565	223 703	204 588	195 356
Penaeus shrimps nei	*Penaeus spp*	18 564	14 645	20 558	8 783	7 830	10 873	11 775
Natantian decapods nei	*Natantia*	113 746	155 181	134 352	121 044	120 664	128 399	127 911
Antarctic krill	*Euphausia superba*	-	6	-	-	-	-	-
Marine crustaceans nei	*Crustacea*	23 685	28 047	25 339	31 261	29 747	14 750	11 504
Cephalopods nei	*Cephalopoda*	103 739	85 120	117 624	95 834	93 709	96 408	114 681
Marine molluscs nei	*Mollusca*	2 452	15 299	4 750	2 718	2 217	1 796	1 469
Sea squirts nei	*Ascidiacea*	-	7	-	-	-	95	-
	Country total	*3 265 240*	*3 447 954*	*3 523 448*	*3 373 492*	*3 472 150*	*3 742 296*	*3 762 600*
Indonesia								
Common carp	*Cyprinus carpio*	5 613	7 081	6 644	7 082	7 127	6 826	6 770
Hoven's carp	*Leptobarbus hoeveni*	5 454	6 892	5 836	3 241	4 608	3 149	3 260
Java barb	*Puntius javanicus*	18 102	19 601	19 469	20 189	17 939	17 124	17 080
Asian barbs nei	*Puntius spp*	12 344	13 254	11 976	12 131	11 263	11 745	11 420
Mozambique tilapia	*Oreochromis mossambicus*	13 293	16 943	17 715	17 865	21 172	19 119	19 550
Knifefishes	*Notopterus spp*	4 790	3 881	3 455	3 599	3 845	4 007	4 050
Glass catfishes	*Kryptopterus spp*	15 283	17 560	15 938	13 943	15 927	12 404	13 490
Torpedo-shaped catfishes nei	*Clarias spp*	9 911	8 388	7 257	7 535	6 413	5 343	5 790
Freshwater siluroids nei	*Siluroidei*	13 215	13 508	10 117	12 466	13 854	13 167	11 810
Gudgeons, sleepers nei	*Eleotridae*	1 514	1 111	1 290	1 078	1 243	985	970
Snakeskin gourami	*Trichogaster pectoralis*	24 904	30 408	21 375	20 936	23 265	20 875	21 260
Kissing gourami	*Helostoma temminckii*	19 166	12 614	18 376	16 598	23 320	17 927	18 320
Indonesian snakehead	*Channa micropeltes*	9 021	11 615	10 117	8 253	8 787	7 771	7 060
Snakeheads(=Murrels) nei	*Channa spp*	31 940	32 356	31 326	26 108	36 309	31 381	31 820
Freshwater fishes nei	*Osteichthyes*	122 559	115 327	103 812	96 597	110 347	108 387	107 180
River eels nei	*Anguilla spp*	926	4 340	657	787	1 212	4 446	4 790
Toli shad	*Tenualosa toli*	1 876	2 131	3 011	3 714	3 505	2 645	2 880
Barramundi(=Giant seaperch)	*Lates calcarifer*	47 627	48 312	55 942	65 193	65 173	68 788	75 300
Indian halibut	*Psettodes erumei*	5 669	7 397	15 075	9 939	12 071	11 143	13 470
Flatfishes nei	*Pleuronectiformes*	2 214	3 414	7 407	3 720	5 074	4 236	5 160
Bombay-duck	*Harpadon nehereus*	11 035	11 218	13 280	14 182	12 415	8 988	8 650
Lizardfishes nei	*Synodontidae*	7 062	7 226	15 158	11 998	12 944	13 383	16 730
Sea catfishes nei	*Ariidae*	58 695	62 412	78 578	66 298	69 646	70 266	73 110
Mullets nei	*Mugilidae*	31 928	35 450	35 478	35 582	35 437	36 077	36 240
Fusiliers	*Caesio spp*	42 693	32 714	38 358	34 142	37 944	33 712	34 240
Groupers nei	*Epinephelus spp*	34 004	38 286	42 164	43 766	43 472	48 422	51 400
Bigeyes nei	*Priacanthus spp*	3 914	3 679	4 448	4 612	4 982	6 203	7 090
Snappers nei	*Lutjanus spp*	52 827	60 338	69 585	66 280	66 492	62 306	63 020
Threadfin breams nei	*Nemipterus spp*	27 460	31 593	29 340	30 937	39 197	34 218	35 270
Ponyfishes(=Slipmouths) nei	*Leiognathidae*	66 220	71 401	89 403	79 532	91 219	69 512	70 390
Grunts, sweetlips nei	*Haemulidae (=Pomadasyidae)*	13 209	14 854	14 657	15 203	15 348	17 368	18 090
Croakers, drums nei	*Sciaenidae*	39 798	45 233	44 837	50 114	56 991	52 254	54 380
Emperors(=Scavengers) nei	*Lethrinidae*	31 802	34 559	31 908	31 706	32 557	30 657	29 780
Goatfishes	*Upeneus spp*	17 612	20 724	24 203	25 207	26 252	27 948	30 150
Threadfins, tasselfishes nei	*Polynemidae*	32 279	30 711	31 709	34 220	33 335	38 282	40 520
Hairtails, scabbardfishes nei	*Trichiuridae*	26 589	28 371	32 981	37 154	36 658	38 077	41 070
Goldstripe sardinella	*Sardinella gibbosa*	161 096	157 104	156 914	174 691	162 710	172 219	176 610
Bali sardinella	*Sardinella lemuru*	98 905	88 590	138 636	153 965	89 286	88 744	94 280
Rainbow sardine	*Dussumieria acuta*	20 649	21 906	22 829	22 321	24 185	23 761	24 270
Stolephorus anchovies	*Stolephorus spp*	157 216	161 781	183 591	166 808	163 117	173 944	177 760
Wolf-herrings nei	*Chirocentrus spp*	26 213	23 059	26 527	25 598	27 513	29 154	30 310
Narrow-barred Spanish mackerel	*Scomberomorus commerson*	63 432	68 454	74 392	75 201	77 711	85 430	90 350
Indo-Pacific king mackerel	*Scomberomorus guttatus*	19 873	23 098	22 250	22 746	21 674	24 449	24 860
Kawakawa	*Euthynnus affinis*	184 400	208 507	212 511	236 673	236 111	250 522	249 400
Skipjack tuna	*Katsuwonus pelamis*	159 667	182 149	187 206	227 068	244 847	236 275	253 050
Tuna-like fishes nei	*Scombroidei*	101 688	115 550	116 214	168 122	136 474	163 241	166 630

English name Nom anglais Nombre inglés	Scientific name Nom scientifique Nombre científico	1995 mt	1996 mt	1997 mt	1998 mt	1999 mt	2000 mt	2001 mt
Needlefishes nei	*Tylosurus spp*	30 119	26 688	29 797	29 741	29 409	32 870	34 690
Flyingfishes nei	*Exocoetidae*	13 180	16 534	16 126	18 961	17 822	19 980	21 040
Scads nei	*Decapterus spp*	247 305	251 289	276 924	277 593	261 138	255 375	256 850
Jacks, crevalles nei	*Caranx spp*	29 025	30 045	32 097	39 443	34 220	36 321	38 370
Black pomfret	*Parastromateus niger*	26 034	27 995	32 638	32 751	31 860	34 093	35 900
Rainbow runner	*Elagatis bipinnulata*	7 715	7 493	7 055	14 133	10 314	9 983	11 590
Torpedo scad	*Megalaspis cordyla*	14 103	14 915	14 937	17 561	19 457	20 485	22 220
Queenfishes	*Scomberoides spp*	14 907	15 623	13 573	15 872	15 979	14 552	14 390
Carangids nei	*Carangidae*	116 769	116 094	125 504	128 459	128 795	129 913	133 650
Indian mackerels nei	*Rastrelliger spp*	193 890	188 910	201 404	204 763	201 466	207 037	211 920
Silver pomfret	*Pampus argenteus*	18 748	19 932	20 461	21 308	21 340	25 492	27 170
Barracudas nei	*Sphyraena spp*	14 228	16 552	18 206	20 440	20 578	19 558	20 440
Rays, stingrays, mantas nei	*Rajiformes*	34 817	36 457	43 324	48 291	45 327	45 260	47 870
Sharks, rays, skates, etc. nei	*Elasmobranchii*	63 281	57 939	52 674	62 497	63 066	68 366	71 590
Marine fishes nei	*Osteichthyes*	381 065	433 477	413 080	456 546	470 576	508 966	510 660
Giant river prawn	*Macrobrachium rosenbergii*	5 524	6 127	5 208	5 362	5 937	5 199	5 520
Freshwater prawns, shrimps nei	*Palaemonidae*	4 024	3 354	3 242	3 871	3 900	4 116	4 350
Freshwater crustaceans nei	*Crustacea*	86	223	49	93	132	132	150
Blue swimming crab	*Portunus pelagicus*	10 884	11 558	15 388	12 370	14 276	14 053	17 530
Indo-Pacific swamp crab	*Scylla serrata*	7 980	7 342	8 298	8 161	8 707	8 774	9 190
Tropical spiny lobsters nei	*Panulirus spp*	2 852	2 464	4 021	2 394	3 244	3 596	4 220
Banana prawn	*Penaeus merguiensis*	50 477	53 914	53 924	62 192	64 179	66 644	70 370
Giant tiger prawn	*Penaeus monodon*	24 501	19 396	25 929	30 047	34 223	40 987	48 430
Metapenaeus shrimps nei	*Metapenaeus spp*	22 863	22 288	32 588	40 717	33 847	38 925	43 610
Natantian decapods nei	*Natantia*	88 029	95 780	102 120	94 481	111 277	106 113	109 440
Marine crustaceans nei	*Crustacea*	2 593	907	2 583	808	503	928	1 000
Freshwater molluscs nei	*Mollusca*	932	1 343	909	806	597	734	650
Cupped oysters nei	*Crassostrea spp*	1 331	1 596	1 678	2 029	2 025	2 281	2 500
Scallops nei	*Pectinidae*	1 103	1 142	424	837	839	578	580
Anadara clams nei	*Anadara spp*	43 661	46 183	41 919	31 591	33 910	34 695	32 590
Hard clams nei	*Meretrix spp*	18 023	13 481	14 027	17 146	14 767	14 177	14 480
Clams, etc. nei	*Bivalvia*	835	844	956	914	304	303	260
Cuttlefish,bobtail squids nei	*Sepiidae, Sepiolidae*	5 939	8 354	8 201	11 473	12 331	11 945	13 210
Common squids nei	*Loligo spp*	27 575	27 169	41 755	31 850	36 707	39 838	44 140
Octopuses, etc. nei	*Octopodidae*	664	645	1 269	3 519	2 337	1 985	3 100
Marine molluscs nei	*Mollusca*	474	1 148	382	674	617	524	590
Frogs	*Rana spp*	2 194	1 795	1 390	1 667	1 317	1 880	1 970
River and lake turtles nei	*Testudinata*	-	36	48	27	1	3	-
Marine turtles nei	*Testudinata*	575	721	638	781	706	744	760
Sea cucumbers nei	*Holothurioidea*	2 562	2 443	3 138	3 058	2 617	3 041	3 250
Jellyfishes	*Rhopilema spp*	123 076	6 740	17 719	3 861	32 652	29 516	30 850
Aquatic invertebrates nei	*Invertebrata*	1 515	1 240	1 685	709	2 649	2 849	3 660
	Country total	*3 511 145*	*3 553 276*	*3 791 240*	*3 964 897*	*3 986 919*	*4 069 691*	*4 203 830*
Iran								
Freshwater breams nei	*Abramis spp*	5	3	7	20	9	20	10
Kutum	*Rutilus frisii*	8 435	9 210	2 320	6 624	6 905	10 120	7 199
Roaches nei	*Rutilus spp*	1 178	878	203	607	626	1 515	1 316
Grass carp(=White amur)	*Ctenopharyngodon idellus*	4 392	4 205	4 517	5 039	4 600	2 500	580
Silver carp	*Hypophthalmichthys molitrix*	16 965	17 510	15 825	11 629	14 400	8 750	3 680
Bighead carp	*Hypophthalmichthys nobilis*	1 414	1 751	1 565	2 522	3 600	4 250	245
Cyprinids nei	*Cyprinidae*	7 802	12 364	12 870	19 573	10 800	10 000	2 680
Freshwater siluroids nei	*Siluroidei*	5	22	6	32	24	25	1
Pike-perch	*Stizostedion lucioperca*	10	6	4	105	19	20	26
Freshwater fishes nei	*Osteichthyes*	157	1 845	1 131	2 529	1 637	1 575	5 970
Sturgeons nei	*Acipenseridae*	1 500	1 600	1 300	1 200	1 000	1 000	870
Salmonoids nei	*Salmonoidei*	433	8	7	8	3	5	2
Caspian shads	*Caspialosa spp*	41 000	57 000	60 400	85 000	95 000	78 000	45 180
Diadromous clupeoids nei	*Clupeoidei*	490	330	566	817	697	595	623
Mullets nei	*Mugilidae*	5 014	2 554	3 074	4 558	4 080	5 125	5 263
Lanternfishes nei	*Myctophidae*	2 000	0	0		0	0	335
Clupeoids nei	*Clupeoidei*	8 000	10 000	10 000	9 708	13 030	14 955	17 453
Narrow-barred Spanish mackerel	*Scomberomorus commerson*	11 067	3 560	4 290	4 034	4 609	7 075	2 474
Indo-Pacific king mackerel	*Scomberomorus guttatus*	5 418	4 340	4 129	3 883	3 476	4 100	6 071
Frigate and bullet tunas	*Auxis thazard, A.rochei*	4 438	755	544	509	590	785	562
Kawakawa	*Euthynnus affinis*	3 911	5 665	8 451	7 947	10 858	13 500	12 474
Skipjack tuna	*Katsuwonus pelamis*	1 133	3 241	5 973	6 671	16 583	20 091	26 058
Longtail tuna	*Thunnus tonggol*	27 188	17 147	20 112	19 694	23 465	41 407	34 896
Albacore	*Thunnus alalunga*	-	10	16	9	-	-	-
Yellowfin tuna	*Thunnus albacares*	27 175	30 233	21 250	21 530	26 871	15 743	20 153
Bigeye tuna	*Thunnus obesus*		153	262	405	592	347	-
Indo-Pacific sailfish	*Istiophorus platypterus*	3 619	2 306	1 928	1 813	3 193	3 080	3 160
Black marlin	*Makaira indica*	-	3	3	5	1	1	-
Swordfish	*Xiphias gladius*	-	9	9	13	2	1	-
Tuna-like fishes nei	*Scombroidei*	-	4	5	9	-	-	9 152
Sharks, rays, skates, etc. nei	*Elasmobranchii*	...	...		1	...	...	...
Marine fishes nei	*Osteichthyes*	149 090	157 397	143 212	139 107	130 225	122 620	115 396
Tropical spiny lobsters nei	*Panulirus spp*	30	49	76	65	35	25	20
Natantian decapods nei	*Natantia*	6 864	5 837	7 620	5 774	4 570	9 850	6 940
Cephalopods nei	*Cephalopoda*	2 500	1 580	8 620	4 189	4 060	5 685	6 517
Marine molluscs nei	*Mollusca*	150	150	1 992	1 583	1 640	1 235	1 144
	Country total	*341 383*	*351 725*	*342 287*	*367 212*	*387 200*	*384 000*	*336 450*
Iraq								
Common carp	*Cyprinus carpio*	5 691	5 538	4 163	2 336	...	...	...
Cyprinids nei	*Cyprinidae*	9 981	7 694	4 782	1 703	5 723	4 013	4 000
Freshwater siluroids nei	*Siluroidei*	1 416	1 436	1 738	711	1 402	953	1 000

English name Nom anglais Nombre inglés	Scientific name Nom scientifique Nombre científico	1995 mt	1996 mt	1997 mt	1998 mt	1999 mt	2000 mt	2001 mt
Freshwater fishes nei	*Osteichthyes*	5 867	4 381	9 836	4 361	2 205	3 412	3 400 F
Marine fishes nei	*Osteichthyes*	5 253	11 688	10 783	13 463	13 093	12 389	12 400 F
	Country total	*28 208*	*30 737*	*31 302*	*22 574*	*22 423*	*20 767*	*20 800 F*
Israel								
Common carp	*Cyprinus carpio*	89	35	38	150	165	189	160 F
Silver carp	*Hypophthalmichthys molitrix*	40	40	17	11	9	32	30 F
Cyprinids nei	*Cyprinidae*	483	1 167	641	1 235	1 128	1 112	960 F
Blue tilapia	*Oreochromis aureus*	69	32	101	88	66	67	60 F
Mango tilapia	*Sarotherodon galilaeus*	316	350	462	391	405	262	230 F
Cichlids nei	*Cichlidae*	64	40	48	58	57	52	40 F
Common sole	*Solea solea*	100	...	...	...	...	...	...
European hake	*Merluccius merluccius*	120	131	86	134	60	62	50 F
Brushtooth lizardfish	*Saurida undosquamis*	80	124	61	37	48	21	20 F
Flathead grey mullet	*Mugil cephalus*	252	439	550	514	599	310	270 F
Groupers nei	*Epinephelus spp*	298	406	384	394	381	263	220 F
Meagre	*Argyrosomus regius*	80	33	2	2	9	288	200 F
Common pandora	*Pagellus erythrinus*	150	100	65	47	67	517	400 F
Bogue	*Boops boops*	102	34	34	47	73	91	80 F
Porgies, seabreams nei	*Sparidae*	100	100	...	...	...	...	...
Surmullets(=Red mullets) nei	*Mullus spp*	162	227	185	276	350	340	300 F
Spinefeet(=Rabbitfishes) nei	*Siganus spp*	50	...	...	...	...	...	...
Round sardinella	*Sardinella aurita*	250	242	435	557	306	88	80 F
Little tunny(=Atl.black skipj)	*Euthynnus alletteratus*	215	119	103	73	90	113	100 F
Atlantic bluefin tuna	*Thunnus thynnus*	-	14	-	-	-	-	-
Atlantic horse mackerel	*Trachurus trachurus*	340	65	226	172	178	175	160 F
Greater amberjack	*Seriola dumerili*	79	33	80	185	98	307	250 F
Carangids nei	*Carangidae*	171	...	...	...	...	...	...
Chub mackerel	*Scomber japonicus*	12	71	47	54	44	8	10 F
Barracudas nei	*Sphyraena spp*	84	76	104	105	160	77	70 F
Sharks, rays, skates, etc. nei	*Elasmobranchii*	48	330	49	59	58	...	40 F
Marine fishes nei	*Osteichthyes*	877	721	1 083	1 410	1 227	1 143	1 000 F
Kuruma prawn	*Penaeus japonicus*	-	-	84	...	...	...	...
Deepwater rose shrimp	*Parapenaeus longirostris*	60	50	...	...	...	...	...
Natantian decapods nei	*Natantia*	200	200	221	225	186	184	170 F
Cephalopods nei	*Cephalopoda*	50	50	98	76	120	117	100 F
	Country total	*4 941*	*5 229*	*5 204*	*6 300*	*5 884*	*5 818*	*5 000 F*
Japan								
Common carp	*Cyprinus carpio*	4 896	4 771	4 607	4 477	4 259	4 079	3 558
Crucian carp	*Carassius carassius*	4 286	4 205	4 008	3 881	3 493	3 423	2 948
Freshwater fishes nei	*Osteichthyes*	12 943	13 344	13 163	12 073	11 020	10 341	9 610
Japanese eel	*Anguilla japonica*	899	901	860	860	817	765	677
Trouts nei	*Salmo spp*	1 080	1 222	1 207	1 187	1 122	1 136	1 053
Pink(=Humpback)salmon	*Oncorhynchus gorbuscha*	25 037	32 595	15 844	25 337	16 902	26 602	9 765
Chum(=Keta=Dog)salmon	*Oncorhynchus keta*	267 718	299 881	269 183	206 622	182 866	163 449	217 359
Masu(=Cherry) salmon	*Oncorhynchus masou*	2 273	2 507	1 800	2 570	1 979	1 811	1 605
Sockeye(=Red)salmon	*Oncorhynchus nerka*	6 234	5 723	9 246	2 768	2 750	2 147	2 740
Chinook(=Spring=King)salmon	*Oncorhynchus tshawytscha*	195	250	825	534	270	147	111
Coho(=Silver)salmon	*Oncorhynchus kisutch*	270	701	575	746	508	376	502
Rainbow trout	*Oncorhynchus mykiss*	597	728	672	618	562	536	484
Ayu sweetfish	*Plecoglossus altivelis*	13 700	12 732	12 619	11 386	11 380	11 172	11 148
Dotted gizzard shad	*Konosirus punctatus*	23 707	18 647	14 850	20 787	17 770	12 335	17 210
Greenland halibut	*Reinhardtius hippoglossoides*	-	-	-	-	-	-	2 814
Witch flounder	*Glyptocephalus cynoglossus*	-	-	-	-	-	-	3
Amer. plaice(=Long rough dab)	*Hippoglossoides platessoides*	-	-	-	-	-	-	6
Bastard halibut	*Paralichthys olivaceus*	7 558	8 311	8 361	7 615	7 198	7 572	6 729
Flatfishes nei	*Pleuronectiformes*	75 294	82 994	78 578	74 944	71 291	71 066	61 016
Red codling	*Pseudophycis bachus*	105	24	14	15	27	70	26
Pacific cod	*Gadus macrocephalus*	56 561	57 576	58 477	57 243	55 292	51 052	43 550
Alaska pollock(=Walleye poll.)	*Theragra chalcogramma*	338 507	331 163	338 785	315 987	382 385	300 001	241 881
Blue whiting(=Poutassou)	*Micromesistius poutassou*	1 127	-	-	-	-	-	-
Southern blue whiting	*Micromesistius australis*	30 205	31 690	36 513	40 309	40 855	31 217	32 874
Argentine hake	*Merluccius hubbsi*	72	83	53	30	27	59	3
Cape hakes	*Merluccius capensis,M.paradox.*	55	-	-	-	-	-	-
Patagonian grenadier	*Macruronus magellanicus*	568	544	644	844	400	1 889	866
Blue grenadier	*Macruronus novaezelandiae*	31 463	26 031	25 349	26 802	17 269	12 139	15 734
Roundnose grenadier	*Coryphaenoides rupestris*	106	35	40	37	40	27	146
Gadiformes nei	*Gadiformes*	5 596	4 131	3 186	4 098	3 415	2 519	2 356
Greater lizardfish	*Saurida tumbil*	9 164	8 144	7 638	7 860	7 716	7 591	6 094
Flathead grey mullet	*Mugil cephalus*	4 579	4 268	3 933	4 003	3 629	3 725	3 457
Mullets nei	*Mugilidae*	832	850	1 053	887	1 000	1 140	1 084
Groupers, seabasses nei	*Serranidae*	10	3	13	21	10	11	13
Japanese seabass	*Lateolabrax japonicus*	7 713	8 334	9 057	9 223	9 234	9 337	10 690
Grunts, sweetlips nei	*Haemulidae (=Pomadasyidae)*	5 555	5 303	4 801	5 703	5 282	5 136	4 995
Croakers, drums nei	*Sciaenidae*	9 008	7 062	5 998	5 430	4 850	4 791	4 362
Silver seabream	*Pagrus auratus*	15 011	16 468	15 611	15 375	15 731	15 041	14 633
Porgies, seabreams nei	*Sparidae*	11 502	11 605	11 256	11 284	10 675	9 066	9 545
Pacific sandlance	*Ammodytes personatus*	108 124	115 766	108 666	90 688	82 918	49 819	88 164
Atka mackerel	*Pleurogrammus azonus*	176 603	181 513	206 763	240 971	169 481	165 118	161 160
Puffers nei	*Tetraodontidae*	8 991	8 238	7 103	8 329	9 647	10 989	7 820
Argentines	*Argentina spp*	-	1	-	-	-	-	-
Deepsea smelt	*Glossanodon semifasciatus*	7 705	8 134	7 431	7 142	6 312	5 970	5 414
Daggertooth pike conger	*Muraenesox cinereus*	3 055	1 989	2 060	2 081	2 298	2 400	2 738
Conger eels, etc. nei	*Congridae*	12 978	12 007	11 706	9 444	8 168	8 364	7 999
John dory	*Zeus faber*	38	16	19	17	13	2	46
Tilefishes nei	*Branchiostegidae*	4 194	3 648	2 994	2 284	1 949	1 678	1 781

E-4 Fish crustaceans, molluscs, etc / Poissons, crustacés, mollusques, etc / Peces, crustáceos, moluscos, etc — Capture production by countries or areas and species / Captures par pays ou zones et espèces / Capturas por países o áreas y especies — Asia / Asie / Asia

English name Nom anglais Nombre inglés	Scientific name Nom scientifique Nombre científico	1995 mt	1996 mt	1997 mt	1998 mt	1999 mt	2000 mt	2001 mt
Tarakihi	*Nemadactylus macropterus*	41	-	-	-	-	-	-
Patagonian toothfish	*Dissostichus eleginoides*	-	264	335	76	-	-	-
Atlantic wolffish	*Anarhichas lupus*	5	-	-	-	-	-	-
Wolffishes(=Catfishes) nei	*Anarhichas spp*	33	20	17	26	21	15	53
Japanese sandfish	*Arctoscopus japonicus*	5 506	6 719	6 209	6 795	6 615	6 652	8 753
Snoek	*Thyrsites atun*	588	28	12	3	23	59	189
Largehead hairtail	*Trichiurus lepturus*	28 207	26 644	20 932	22 268	26 200	22 947	16 615
Pacific rudderfish	*Psenopsis anomala*	3 485	3 516	4 316	4 316	4 996	5 215	4 196
Pacific ocean perch	*Sebastes alutus*	1 449	1 730	767	896	766	458	668
Atlantic redfishes nei	*Sebastes spp*	2 016	1 585	596	922	460	140	187
Scorpionfishes nei	*Scorpaenidae*	2 888	2 317	2 082	1 638	1 314	1 244	1 132
Bluefin gurnard	*Chelidonichthys kumu*	4	...	...	...	...	...	...
Demersal percomorphs nei	*Perciformes*	5 517	4 511	4 812	3 767	3 097	2 895	6 489
Pacific herring	*Clupea pallasii*	3 873	2 021	1 926	2 531	2 579	2 260	2 385
Japanese pilchard	*Sardinops melanostictus*	661 391	319 354	284 054	167 073	351 207	149 616	178 423
Red-eye round herring	*Etrumeus teres*	47 590	49 752	55 043	48 441	28 712	23 833	31 525
Japanese anchovy	*Engraulis japonicus*	251 958	345 517	233 113	470 616	484 230	381 020	301 168
Clupeoids nei	*Clupeoidei*	55 408	58 232	59 619	52 427	79 405	74 581	58 286
Japanese Spanish mackerel	*Scomberomorus niphonius*	6 381	3 607	2 349	2 864	5 321	10 932	9 056
Seerfishes nei	*Scomberomorus spp*	28	7	-	-	-	-	-
Frigate and bullet tunas	*Auxis thazard, A.rochei*	27 386	20 374	32 574	21 613	29 517	27 139	37 247
Skipjack tuna	*Katsuwonus pelamis*	308 966	286 671	311 542	385 462	287 317	341 414	276 680
Atlantic bluefin tuna	*Thunnus thynnus*	5 171	4 561	3 496	4 262	3 822	3 140	2 815
Pacific bluefin tuna	*Thunnus orientalis*	6 209	6 668	6 487	3 667	11 188	10 128	5 832
Albacore	*Thunnus alalunga*	62 543	61 059	83 592	73 756	99 027	61 249	69 196
Southern bluefin tuna	*Thunnus maccoyii*	5 672	6 667	5 100	7 716	7 728	6 092	7 240
Yellowfin tuna	*Thunnus albacares*	113 222	80 135	114 156	98 430	96 434	103 368	98 466
Bigeye tuna	*Thunnus obesus*	126 616	101 591	105 580	115 571	92 673	96 067	104 247
Tunas nei	*Thunnini*	1	-	1	-	-	-	-
Indo-Pacific sailfish	*Istiophorus platypterus*	1 143	1 095	844	1 583	1 360	1 109	896
Atlantic sailfish	*Istiophorus albicans*	84	80	60	91	71	73	33
Indo-Pacific blue marlin	*Makaira mazara*	12 421	7 135	7 824	8 495	6 458	6 413	6 220
Atlantic blue marlin	*Makaira nigricans*	1 427	1 685	1 175	1 098	840	859	331
Black marlin	*Makaira indica*	89	-	-	-	-	-	-
Striped marlin	*Tetrapturus audax*	9 283	7 613	6 537	8 136	5 951	4 545	4 783
Atlantic white marlin	*Tetrapturus albidus*	59	115	55	59	50	89	60
Swordfish	*Xiphias gladius*	17 018	15 652	12 595	16 898	13 585	13 181	12 818
Tuna-like fishes nei	*Scombroidei*	18 451	9 481	14 384	12 131	9 924	10 826	8 953
Pacific saury	*Cololabis saira*	273 510	229 227	290 812	144 983	141 011	216 471	269 797
Japanese flyingfish	*Cypselurus agoo*	7 881	8 501	7 486	8 933	6 738	9 615	8 286
Japanese jack mackerel	*Trachurus japonicus*	312 994	330 406	323 142	311 311	211 077	245 988	214 434
Chilean jack mackerel	*Trachurus murphyi*	-	-	-	-	7	-	-
Cape horse mackerel	*Trachurus capensis*	18	-	-	-	-	-	-
Jack and horse mackerels nei	*Trachurus spp*	4 111	335	29	8	11	12	655
Japanese scad	*Decapterus maruadsi*	72 109	57 319	50 097	59 078	47 157	36 416	41 236
Japanese amberjack	*Seriola quinqueradiata*	7 564	246	...	...	...	...	...
Amberjacks nei	*Seriola spp*	54 101	50 333	47 211	45 484	54 918	77 461	66 925
Common dolphinfish	*Coryphaena hippurus*	10 212	8 527	10 229	14 944	9 280	8 959	9 142
Chub mackerel	*Scomber japonicus*	469 447	760 430	848 967	511 238	381 866	346 220	375 273
Blue mackerel	*Scomber australasicus*	358	-	-	-	-	-	-
Butterfishes, pomfrets nei	*Stromateidae*	1 314	1 518	2 126	2 194	1 842	2 077	2 211
Pelagic percomorphs nei	*Perciformes*	2	-	-	-	-	-	-
Whip stingray	*Dasyatis akajei*	3 985	4 029	3 959	4 329	4 407	5 388	4 312
Sharks, rays, skates, etc. nei	*Elasmobranchii*	27 161	20 177	25 438	29 336	28 703	26 500	23 609
Groundfishes nei	*Osteichthyes*	130	57	32	35	31	11	6
Marine fishes nei	*Osteichthyes*	317 912	292 617	292 089	269 472	261 613	275 442	251 517
Freshwater prawns, shrimps nei	*Palaemonidae*	2 717	2 222	2 413	1 872	1 942	1 676	1 158
Gazami crab	*Portunus trituberculatus*	4 159	4 022	3 112	3 528	2 752	3 131	3 596
Tanner crabs nei	*Chionoecetes spp*	9 090	3 447	4 870	4 677	4 892	5 640	5 355
Geryons nei	*Geryon spp*	4 855	3 250	1 937	1 562	1 730	2 980	1 650
Marine crabs nei	*Brachyura*	38 812	37 265	34 894	33 678	30 859	30 310	27 353
Longlegged spiny lobster	*Panulirus longipes*	1 136	1 092	1 068	1 084	1 154	1 244	1 486
King crabs	*Paralithodes spp*	260	322	154	132	117	89	181
Kuruma prawn	*Penaeus japonicus*	2 668	2 262	2 144	2 069	1 523	1 447	1 271
Penaeus shrimps nei	*Penaeus spp*	349	192	41	-	-	-	-
Natantian decapods nei	*Natantia*	31 764	28 641	27 114	25 283	25 630	25 898	24 411
Antarctic krill	*Euphausia superba*	60 303	60 546	58 798	63 233	71 318	67 192	73 523
Marine crustaceans nei	*Crustacea*	60 785	58 892	63 028	67 945	49 783	74 072	49 619
Japanese corbicula	*Corbicula japonica*	26 938	26 714	21 822	19 932	20 009	19 295	17 295
Freshwater molluscs nei	*Mollusca*	847	1 305	1 361	1 132	900	628	583
Giant abalone	*Haliotis gigantea*	1 980	1 941	2 218	2 269	2 109	2 146	1 982
Horned turban	*Turbo cornutus*	9 943	10 119	12 132	12 556	11 000	9 839	10 241
Yesso scallop	*Patinopecten yessoensis*	274 879	271 124	261 164	287 802	299 628	304 286	290 974
Half-crenated ark	*Scapharca subcrenata*	15 426	16 328	14 133	10 120	10 413	7 308	4 899
Japanese hard clam	*Meretrix lusoria*	2 060	1 944	1 897	1 870	1 785	1 543	1 245
Japanese carpet shell	*Ruditapes philippinarum*	49 466	43 703	39 660	36 807	43 088	35 558	31 022
Imperial surf clam	*Pseudocardium sybillae*	8 018	8 738	7 541	8 227	8 808	8 883	8 914
Clams, etc. nei	*Bivalvia*	50 015	52 361	42 987	47 584	35 319	35 529	30 133
Cuttlefish,bobtail squids nei	*Sepiidae, Sepiolidae*	9 643	9 545	8 052	9 790	9 942	8 241	8 297
Common squids nei	*Loligo spp*	5 561	3 187	1 562	2 618	1 847	-	-
Neon flying squid	*Ommastrephes bartrami*	...	...	49 870	54 951	36 076	46 963	23 770
Northern shortfin squid	*Illex illecebrosus*	-	1	-	-	-	-	-
Argentine shortfin squid	*Illex argentinus*	75 691	73 896	126 504	77 452	154 129	119 966	70 652
Jumbo flying squid	*Dosidicus gigas*	36 515	14 297	23 179	-	3 305	56 825	72 412
Japanese flying squid	*Todarodes pacificus*	290 273	444 189	365 978	180 749	237 346	337 285	298 191
Wellington flying squid	*Nototodarus sloani*	19 687	11 342	5 182	3 734	1 853	1 853	1 396
Sevenstar flying squid	*Martialia hyadesi*	-	-	-	2	27	33	2
Various squids nei	*Loliginidae, Ommastrephidae*	110 441	121 718	63 116	56 371	56 575	49 607	46 451
Octopuses, etc. nei	*Octopodidae*	51 874	50 584	56 593	61 260	57 427	47 374	45 200

E-4

Fish crustaceans, molluscs, etc — **Capture production by countries or areas and species** — **Asia**
Poissons, crustacés, mollusques, etc — **Captures par pays ou zones et espèces** — **Asie**
Peces, crustáceos, moluscos, etc — **Capturas por países o áreas y especies** — **Asia**

English name Nom anglais Nombre inglés	Scientific name Nom scientifique Nombre científico	1995 mt	1996 mt	1997 mt	1998 mt	1999 mt	2000 mt	2001 mt
Sea urchins nei	*Strongylocentrotus spp*	13 735	12 996	14 297	13 653	13 530	12 455	11 208
Japanese sea cucumber	*Stichopus japonicus*	6 602	7 226	7 160	6 952	6 662	6 957	7 229
Aquatic invertebrates nei	*Invertebrata*	38 912	13 982	10 770	11 021	8 346	10 932	8 339
	Country total	*5 966 456*	*5 933 659*	*5 926 113*	*5 299 399*	*5 194 186*	*4 971 412*	*4 719 152*
Jordan								
Freshwater fishes nei	*Osteichthyes*	350	350	350	350	350	400	350
Fusiliers	*Caesio spp*	...	...	...	15	18	17	15
Emperors(=Scavengers) nei	*Lethrinidae*	...	...	...	2	2	1	2
Spinefeet(=Rabbitfishes) nei	*Siganus spp*	...	...	...	10	12	10	11
Tunas nei	*Thunnini*	...	...	...	70	96	90	110
Scads nei	*Decapterus spp*	...	...	...	20	25	25	26
Marine fishes nei	*Osteichthyes*	75	90	100	3	7	7	6
	Country total	*425*	*440*	*450*	*470*	*510*	*550*	*520*
Kazakhstan								
Freshwater bream	*Abramis brama*	20 520	18 770 F	11 000 F	9 800 F	11 000 F	12 000 F	12 630
Common carp	*Cyprinus carpio*	477	440 F	320 F	230 F	500 F	650 F	710
Crucian carp	*Carassius carassius*	4 317	3 950 F	2 840 F	2 060 F	1 900 F	1 800 F	1 707
Roaches nei	*Rutilus spp*	2 650	2 420 F	1 740 F	1 260 F	1 000 F	850 F	579
Grass carp(=White amur)	*Ctenopharyngodon idellus*	2	2 F	1 F	1 F	1 F	-	-
Silver carp	*Hypophthalmichthys molitrix*	267	240 F	170 F	120 F	70 F	30 F	-
Sichel	*Pelecus cultratus*	27	25 F	20 F	15 F	10 F	10 F	8
Asp	*Aspius aspius*	415	380 F	270 F	195 F	250 F	350 F	476
Northern pike	*Esox lucius*	637	580 F	420 F	300 F	550 F	700 F	819
Wels(=Som)catfish	*Silurus glanis*	1 513	1 380 F	990 F	700 F	750 F	750 F	780
European perch	*Perca fluviatilis*	736	670 F	480 F	350 F	450 F	500 F	547
Pike-perch	*Stizostedion lucioperca*	6 089	5 570 F	4 000 F	2 900 F	2 500 F	2 000 F	1 628
Snakeheads(=Murrels) nei	*Channa spp*	10	10 F	7 F	5 F	10 F	10 F	12
Freshwater fishes nei	*Osteichthyes*	6	269 F	358 F	364 F	10 809 F	13 715 F	10 526
Sturgeons nei	*Acipenseridae*	563	510 F	370 F	270 F	240 F	215 F	190
Whitefishes nei	*Coregonus spp*	29	27 F	20 F	15 F	20 F	30 F	35
Caspian shads	*Caspialosa spp*	0	0	0	0	0	0	0
Azov sea sprat	*Clupeonella cultriventris*	10 113	9 000 F	8 800	6 400	6 100	3 000	-
Mullets nei	*Mugilidae*	31	30 F	20 F	15 F	10 F	10 F	7
	Country total	*48 402*	*44 273*	*31 826*	*25 000 F*	*36 170*	*36 620*	*30 654*
Korea D P Rp								
Freshwater fishes nei	*Osteichthyes*	20 000	20 000	20 000	20 000 F	20 000 F	20 000 F	20 000 F
Flatfishes nei	*Pleuronectiformes*	2 953	1 966	6 972	4 000 F	4 000 F	3 800 F	3 800 F
Alaska pollock(=Walleye poll.)	*Theragra chalcogramma*	120 219	15 369	66 578	60 000 F	55 000 F	52 000 F	52 000 F
Croakers, drums nei	*Sciaenidae*	-	-	-	-	-	224	-
Porgies, seabreams nei	*Sparidae*	-	-	-	-	-	273	-
Atka mackerel	*Pleurogrammus azonus*	3 480	6 487	3 535	3 000 F	3 000 F	2 800 F	2 800 F
Japanese pilchard	*Sardinops melanostictus*	63	5	0	...	...	...	...
Marine fishes nei	*Osteichthyes*	155 973	166 325	109 770	107 900 F	102 900 F	97 654 F	97 500 F
Natantian decapods nei	*Natantia*	-	-	-	-	-	52	-
Marine crustaceans nei	*Crustacea*	21 091	33 931	15 265	15 000 F	15 000 F	14 300 F	14 300 F
Cuttlefish,bobtail squids nei	*Sepiidae, Sepiolidae*	-	-	-	-	-	61	-
Various squids nei	*Loliginidae, Ommastrephidae*	3 164	8 892	14 192	10 000 F	10 000 F	9 500 F	9 500 F
Octopuses, etc. nei	*Octopodidae*	-	-	-	-	-	86	-
Sea urchins nei	*Strongylocentrotus spp*	140	150	150	100 F	100 F	100 F	100 F
	Country total	*327 083*	*253 125*	*236 462*	*220 000 F*	*210 000 F*	*200 850 F*	*200 000 F*
Korea Rep								
Common carp	*Cyprinus carpio*	1 684	1 979	842	874	438	...	...
Cyprinids nei	*Cyprinidae*	3 311	1 010	2 221	1 977	2 294	...	...
Freshwater siluroids nei	*Siluroidei*	-	279	251	...	136	...	...
Snakehead	*Channa argus*	-	169	120	...	28	...	...
Freshwater fishes nei	*Osteichthyes*	3 309	3 418	2 633	2 755	1 681	6 402	5 254
Japanese eel	*Anguilla japonica*	124	113	56	44	13	...	...
Rainbow trout	*Oncorhynchus mykiss*	13	...	...	...	...	...	...
Ayu sweetfish	*Plecoglossus altivelis*	-	-	5	-	-	-	-
Salmonoids nei	*Salmonoidei*	1 516	1 463	652	1 122	276	28	19
Chinese gizzard shad	*Clupanodon thrissa*	7 931	4 908	13 836	11 349	9 511	6 366	9 120
Elongate ilisha	*Ilisha elongata*	466	630	349	381	265	134	40
Lefteye flounders nei	*Bothidae*	151	40	27	38	28	40	30
Yellow striped flounder	*Pseudopleuronectes herzenst.*	13 683	18 066	18 079	20 135	19 569	15 423	14 503
Southeast Atlantic soles nei	*Austroglossus spp*	1 365	480	1 197	773	926	965	650
Soles nei	*Soleidae*	29	102	-	-	256	1 016	-
Tonguefishes	*Cynoglossidae*	3 217	3 119	3 192	1 867	2 443	2 108	3 069
Bastard halibut	*Paralichthys olivaceus*	1 914	2 317	1 592	2 002	1 679	1 607	1 707
Flatfishes nei	*Pleuronectiformes*	195	377	323	97	415	218	3
Blue antimora	*Antimora rostrata*	-	0	1	0	-	-	-
Pacific cod	*Gadus macrocephalus*	2 476	2 740	3 984	6 249	6 509	10 098	13 110
Alaska pollock(=Walleye poll.)	*Theragra chalcogramma*	345 888	226 910	223 065	236 278	146 165	86 143	197 396
Southern hake	*Merluccius australis*	...	454	1 178	1 976	2 512	2 142	2 224
Benguela hake	*Merluccius polli*	-	-	-	4	-	-	-
Hakes nei	*Merluccius spp*	16	370	33	-	-	-	15
Patagonian grenadier	*Macruronus magellanicus*	76	33	-	31	282	1 076	1 553
Blue grenadier	*Macruronus novaezelandiae*	340	1 725	6 091	4 883	5 834	8 694	9 484
Grenadiers nei	*Macrourus spp*	-	4	9	3	-	-	-
Grenadiers, rattails nei	*Macrouridae*	-	-	-	-	-	140	-
Gadiformes nei	*Gadiformes*	530	1 694	278	834	2 243	1 718	3 079
Brushtooth lizardfish	*Saurida undosquamis*	209	162	207	215	35	795	758

English name Nom anglais Nombre inglés	Scientific name Nom scientifique Nombre científico	1995 mt	1996 mt	1997 mt	1998 mt	1999 mt	2000 mt	2001 mt
Korean sandlance	*Hypoptychus dybowskii*	9 677	6 613	8 832	4 801	4 806	-	462
Flathead grey mullet	*Mugil cephalus*	4 316	4 200	5 564	4 962	9 678	8 730	9 784
Mullets nei	*Mugilidae*	-	-	-	-	133	-	-
Groupers nei	*Epinephelus spp*	681	389	174	120	1 054	353	259
Groupers, seabasses nei	*Serranidae*	557	436	760	1 121	964	1 483	991
Japanese seabass	*Lateolabrax japonicus*	1 118	1 178	1 501	2 957	2 933	1 201	1 566
Sillago-whitings	*Sillaginidae*	199	248	199	48	109	100	10
Snappers nei	*Lutjanus spp*	-	-	-	555	474	524	601
Snappers, jobfishes nei	*Lutjanidae*	-	-	-	14	-	-	-
Grunts, sweetlips nei	*Haemulidae (=Pomadasyidae)*	5 620	5 532	-	130	20	25	540
Brown meagre	*Sciaena umbra*	234	208	...	...	...	...	...
Honnibe croaker	*Nibea mitsukurii*	2 164	1 940	1 177	1 285	1 566	1 999	2 156
Large yellow croaker	*Larimichthys croceus*	22 599	22 225	1 565	294	270	354	426
Yellow croaker	*Larimichthys polyactis*	25 719	23 695	22 101	15 011	13 490	19 630	7 938
Croakers, drums nei	*Sciaenidae*	89 956	73 653	95 281	89 021	114 160	67 563	50 789
Emperors(=Scavengers) nei	*Lethrinidae*	-	-	-	93	165	98	459
Angolan dentex	*Dentex angolensis*	-	274	-	-	-	-	-
Dentex nei	*Dentex spp*	4	4	-	-	-	-	-
Silver seabream	*Pagrus auratus*	552	762	1 149	1 657	924	986	913
Pargo breams nei	*Pagrus spp*	30	10	-	-	-	-	-
Porgies, seabreams nei	*Sparidae*	10 035	8 009	7 648	7 152	8 316	8 505	6 767
Goatfishes, red mullets nei	*Mullidae*	18	-	-	-	-	-	-
Threadfins, tasselfishes nei	*Polynemidae*	25	239	-	1 241	-	-	-
Pacific sandlance	*Ammodytes personatus*	-	-	-	-	-	16 293	4 803
Gobies nei	*Gobiidae*	1 941	1 598	1 722	1 299	897	788	703
Atka mackerel	*Pleurogrammus azonus*	3 342	4 103	2 983	7 911	1 005	2 554	1 261
Bartail flathead	*Platycephalus indicus*	2 520	2 900	4 196	2 857	2 248	2 310	1 699
Purple puffer	*Takifugu vermicularis*	10 178	9 708	7 471	3 897	4 787	-	-
Puffers nei	*Tetraodontidae*	-	-	-	-	-	2 978	3 735
Filefishes	*Cantherhines(=Navodon) spp*	-	-	-	-	-	2 891	1 578
Threadsail filefish	*Stephanolepis cirrhifer*	1 755	1 772	16 318	9 364	2 999	-	-
Triggerfishes, durgons nei	*Balistidae*	10	306	-	-	16	-	-
Daggertooth pike conger	*Muraenesox cinereus*	1 604	1 411	2 518	1 506	190	1 862	1 080
Whitespotted conger	*Conger myriaster*	19 667	17 314	19 136	11 913	10 160	8 304	7 676
Conger eels, etc. nei	*Congridae*	318	69	63	87	159	122	78
Pink cusk-eel	*Genypterus blacodes*	549	920	1 378	1 242	1 894	1 684	1 604
Cusk-eels nei	*Genypterus spp*	-	-	-	-	-	57	-
Bearded brotula	*Brotula barbata*	-	-	-	-	-	16	-
Cusk-eels, brotulas nei	*Ophidiidae*	-	-	-	-	-	1	-
Alfonsinos nei	*Beryx spp*	-	-	-	77	-	4	-
Orange roughy	*Hoplostethus atlanticus*	-	-	-	-	230	-	47
John dory	*Zeus faber*	265	105	298	36	540	636	225
Tilefishes nei	*Branchiostegidae*	-	-	-	-	1 651	1 664	1 049
Patagonian toothfish	*Dissostichus eleginoides*	381	366	939	1 221	1 188	1 672	1 153
Antarctic toothfishes nei	*Dissostichus spp*	-	-	-	-	-	-	572
Japanese sandfish	*Arctoscopus japonicus*	2 065	2 501	2 194	1 490	2 449	1 571	1 286
Largehead hairtail	*Trichiurus lepturus*	94 596	74 461	67 170	74 851	64 445	81 050	79 898
Hairtails, scabbardfishes nei	*Trichiuridae*	7 570	8 392	11 572	9 585	12 866	12 135	3 503
Ruffs, barrelfishes nei	*Centrolophidae*	-	-	-	-	493	317	209
Scorpionfishes nei	*Scorpaenidae*	5 901	5 070	5 725	4 705	5 036	5 098	5 354
Bluefin gurnard	*Chelidonichthys kumu*	40	75	86	146	43	79	140
Anglerfishes nei	*Lophiidae*	8 173	11 720	5 045	3 970	3 406	4 930	5 813
Pacific herring	*Clupea pallasii*	8 622	5 525	13 214	13 340	21 446	15 203	8 491
Japanese sardinella	*Sardinella zunasi*	18 345	10 663	5 593	1 973	6 674	4 603	766
Japanese pilchard	*Sardinops melanostictus*	13 539	18 560	9 041	7 595	17 142	2 207	129
Japanese anchovy	*Engraulis japonicus*	230 679	237 128	230 911	249 519	238 463	201 192	273 927
Clupeoids nei	*Clupeoidei*	25	-	-	-	1	-	-
Japanese Spanish mackerel	*Scomberomorus niphonius*	17 429	6 419	11 173	22 809	19 502	25 641	25 513
Seerfishes nei	*Scomberomorus spp*	1 253	1 830	3 112	2 935	1 242	1 609	2 097
Frigate and bullet tunas	*Auxis thazard, A.rochei*	7	-	-	-	-	-	-
Skipjack tuna	*Katsuwonus pelamis*	137 848	129 888	115 927	143 390	109 780	137 015	137 569
Atlantic bluefin tuna	*Thunnus thynnus*	663	683	613	-	-	-	-
Albacore	*Thunnus alalunga*	74	716	1 944	3 998	1 179	684	1 920
Southern bluefin tuna	*Thunnus maccoyii*	315	911	1 228	1 562	1 264	958	729
Yellowfin tuna	*Thunnus albacares*	51 951	35 267	61 244	73 354	43 255	49 461	58 957
Bigeye tuna	*Thunnus obesus*	26 819	28 418	31 191	33 042	26 558	30 079	31 335
Indo-Pacific sailfish	*Istiophorus platypterus*	224	250	1 302	389	496	185	33
Indo-Pacific blue marlin	*Makaira mazara*	8	24	398	805	300	301	350
Atlantic blue marlin	*Makaira nigricans*	-	-	7	-	-	-	-
Black marlin	*Makaira indica*	492	317	235	427	813	776	1 291
Striped marlin	*Tetrapturus audax*	412	451	1 172	795	589	637	349
Marlins,sailfishes,etc. nei	*Istiophoridae*	4 731	5 356	3 234	3 849	1 413	1 327	1 657
Swordfish	*Xiphias gladius*	98	129	967	573	474	665	1 031
Tuna-like fishes nei	*Scombroidei*	5 354	6 869	8 193	8 473	6 469	6 952	7 512
Pacific saury	*Cololabis saira*	37 865	28 368	68 853	18 531	29 538	44 340	26 205
Japanese halfbeak	*Hyporhamphus sajori*	1 531	990	1 193	1 160	913	956	613
Japanese jack mackerel	*Trachurus japonicus*	12 269	14 542	22 766	22 133	13 552	19 510	17 537
Jack and horse mackerels nei	*Trachurus spp*	486	1 202	2 133	1 983	2 182	259	313
Jacks, crevalles nei	*Caranx spp*	-	46	-	11	-	-	-
Amberjacks nei	*Seriola spp*	3 817	4 127	6 119	9 625	8 654	4 814	6 475
Chub mackerel	*Scomber japonicus*	200 490	415 003	160 479	172 925	177 609	145 945	203 743
Blue mackerel	*Scomber australasicus*	-	1	-	5	-	-	-
Silver pomfret	*Pampus argenteus*	4 806	4 806	919	1 029	3 374	-	-
Butterfishes, pomfrets nei	*Stromateidae*	10 896	9 484	10 770	13 249	15 235	7 838	6 819
Barracudas nei	*Sphyraena spp*	-	9	80	1	349	432	512
Rays, stingrays, mantas nei	*Rajiformes*	16 130	14 682	14 405	8 183	13 978	13 034	9 121
Sharks, rays, skates, etc. nei	*Elasmobranchii*	1 808	911	1 495	2 127	2 419	2 361	2 010
Marine fishes nei	*Osteichthyes*	133 631	183 016	165 430	140 162	109 914	106 683	129 001
Freshwater crustaceans nei	*Crustacea*	216	248	-	2	111	93	48

E-4

Fish crustaceans, molluscs, etc	Capture production by countries or areas and species	Asia
Poissons, crustacés, mollusques, etc	Captures par pays ou zones et espèces	Asie
Peces, crustáceos, moluscos, etc	Capturas por países o áreas y especies	Asia

English name Nom anglais Nombre inglés	Scientific name Nom scientifique Nombre científico	1995 mt	1996 mt	1997 mt	1998 mt	1999 mt	2000 mt	2001 mt
Gazami crab	*Portunus trituberculatus*	17 651	15 754	11 430	13 813	11 819	12 842	13 016
Tanner crabs nei	*Chionoecetes spp*	-	-	-	-	-	17 037	13 974
Marine crabs nei	*Brachyura*	58 804	61 513	58 855	43 999	35 439	9 564	9 031
Longlegged spiny lobster	*Panulirus longipes*	-	-	-	-	-	461	415
Kuruma prawn	*Penaeus japonicus*	2 399	1 783	2 102	1 138	502	578	513
Fleshy prawn	*Penaeus chinensis*	1 406	1 242	1 911	1 245	814	1 211	581
Penaeus shrimps nei	*Penaeus spp*	653	560	143	136	502	-	-
Shiba shrimp	*Metapenaeus joyneri*	2 168	2 211	1 976	3 651	3 633	2 621	2 385
Southern rough shrimp	*Trachypenaeus curvirostris*	-	-	-	-	-	-	3 145
Akiami paste shrimp	*Acetes japonicus*	16 495	18 411	15 624	15 993	19 389	13 985	11 705
Natantian decapods nei	*Natantia*	18 968	16 296	17 768	24 588	18 698	17 640	12 471
Antarctic krill	*Euphausia superba*	-	-	-	1 621	1 228	5 444	5 125
Marine crustaceans nei	*Crustacea*	721	53	265	1 916	1 658	1 144	5 958
Japanese corbicula	*Corbicula japonica*	659	600	387	677	1 006	-	-
Freshwater molluscs nei	*Mollusca*	245	200	290	409	255	645	669
Abalones nei	*Haliotis spp*	260	188	214	71	79	113	104
Horned turban	*Turbo cornutus*	8 921	7 361	6 878	9 192	7 397	7 281	6 274
Gastropods nei	*Gastropoda*	5 224	2 975	1 617	1 325	1 670	818	730
Pacific cupped oyster	*Crassostrea gigas*	18 262	18 259	17 210	9 905	11 609	15 939	10 056
Korean mussel	*Mytilus coruscus*	2 942	2 191	3 211	1 469	1 414	1 133	1 085
Sea mussels nei	*Mytilidae*	3 685	837	4 098	6 456	6 771	5 795	3 828
Yesso scallop	*Patinopecten yessoensis*	74	122	196	42	6	0	11
Ark clams nei	*Arca spp*	473	782	667	1 311	534	318	153
Blood cockle	*Anadara granosa*	1 415	995	493	12 114	6 503	4 184	923
Japanese hard clam	*Meretrix lusoria*	2 715	2 315	2 075	3 472	1 799	1 430	1 044
Japanese carpet shell	*Ruditapes philippinarum*	15 041	12 392	16 854	14 585	13 963	20 982	20 004
Imperial surf clam	*Pseudocardium sybillae*	9 038	6 053	7 179	2 877	7 620	4 150	3 409
Cockles nei	*Cardiidae*	428	2 583	1 587	1 078	112	411	2 002
Clams, etc. nei	*Bivalvia*	31 063	33 446	8 851	15 665	14 728	12 277	15 705
Cuttlefish,bobtail squids nei	*Sepiidae, Sepiolidae*	3 691	2 832	4 946	4 457	8 309	2 107	2 940
Common squids nei	*Loligo spp*	-	-	-	-	1 629	885	898
Argentine shortfin squid	*Illex argentinus*	124 005	144 750	208 160	92 397	271 716	150 149	142 585
Jumbo flying squid	*Dosidicus gigas*	34 440	11 784	2 384	201	18 813	15 625	5 797
Japanese flying squid	*Todarodes pacificus*	200 897	252 618	224 959	163 016	249 280	226 309	225 616
Sevenstar flying squid	*Martialia hyadesi*	-	52	28	53	-	-	2
Various squids nei	*Loliginidae, Ommastrephidae*	36 317	15 609	18 399	23 333	18 666	11 999	14 710
Octopuses, etc. nei	*Octopodidae*	22 142	21 707	23 084	27 174	22 674	19 605	19 965
Marine molluscs nei	*Mollusca*	652	1 031	1 017	2 308	683	847	1 270
Sea squirts nei	*Ascidiacea*	5 767	16 747	2 780	891	1 171	1 443	1 053
Echinoderms	*Echinodermata*	3 707	2 802	2 771	1 410	1 182	1 461	1 454
Japanese sea cucumber	*Stichopus japonicus*	1 892	1 979	2 217	1 439	1 204	1 419	900
Aquatic invertebrates nei	*Invertebrata*	1 040	873	1 094	1 047	1 003	1 883	1 302
	Country total	*2 319 915*	*2 413 713*	*2 204 047*	*2 026 934*	*2 119 668*	*1 823 175*	*1 988 002*
Kuwait								
Freshwater fishes nei	*Osteichthyes*	0	0	0	0	0	0	0
Hilsa shad	*Tenualosa ilisha*	1 198	1 148	1 034	919	970	650 F	337
Brushtooth lizardfish	*Saurida undosquamis*	31	47	53	34	20	30 F	32
Flathead grey mullet	*Mugil cephalus*	96	39	65	69	89	60 F	34
Mullets nei	*Mugilidae*	579	540	628	1 068	965	700 F	422
Groupers nei	*Epinephelus spp*	341	287	241	264	237	250 F	268
Snappers nei	*Lutjanus spp*	117	84	78	64	51	40 F	27
Threadfin breams nei	*Nemipterus spp*	6	10	17	49	19	20 F	24
Grunts, sweetlips nei	*Haemulidae (=Pomadasyidae)*	195	180	163	174	245	210 F	191
Croakers, drums nei	*Sciaenidae*	1 572	974	1 127	1 211	1 385	1 100 F	853
Emperors(=Scavengers) nei	*Lethrinidae*	15	15	20	38	35	50 F	78
Yellowfin seabream	*Acanthopagrus latus*	273	234	249	280	464	350 F	271
Narrow-barred Spanish mackerel	*Scomberomorus commerson*	68	85	73	124	127	130 F	135
Indo-Pacific king mackerel	*Scomberomorus guttatus*	156	172	206	166	211	210 F	204
Cobia	*Rachycentron canadum*	...	...	...	...	...	...	38
Carangids nei	*Carangidae*	111	97	138	74	73	150 F	242
Silver pomfret	*Pampus argenteus*	1 101	862	560	501	259	200 F	133
Butterfishes, pomfrets nei	*Stromateidae*	289	230	186	111	170	-	-
Marine fishes nei	*Osteichthyes*	729	893	922	1 055	231	550 F	580
Natantian decapods nei	*Natantia*	1 739	2 358	2 066	1 598	720	1 300 F	1 977
	Country total	*8 616*	*8 255*	*7 826*	*7 799*	*6 271*	*6 000 F*	*5 846*
Kyrgyzstan								
Freshwater bream	*Abramis brama*	11	9 F	7 F	5 F	3 F	3 F	4
Common carp	*Cyprinus carpio*	36	31 F	23 F	15 F	9 F	10 F	11
Goldfish	*Carassius auratus*	2	2 F	1 F	1 F	1 F	1 F	2
Silver carp	*Hypophthalmichthys molitrix*	28	24 F	18 F	12 F	7 F	8 F	19
Cyprinids nei	*Cyprinidae*	39	35 F	26 F	18 F	11 F	12 F	8
Pike-perch	*Stizostedion lucioperca*	7	6 F	5 F	3 F	2 F	2 F	1
Whitefishes nei	*Coregonus spp*	62	53 F	40 F	26 F	15 F	16 F	12
	Country total	*185*	*160 F*	*120 F*	*80 F*	*48*	*52*	*57*
Laos								
Cyprinids nei	*Cyprinidae*	4 000 F	3 500 F	2 800 F	3 000 F	4 500 F	4 400 F	4 500 F
Freshwater fishes nei	*Osteichthyes*	23 370 F	19 500 F	16 057 F	16 642 F	25 541 F	24 850 F	25 500 F
	Country total	*27 370*	*23 000 F*	*18 857*	*19 642*	*30 041*	*29 250*	*30 000 F*
Lebanon								
Cyprinids nei	*Cyprinidae*	10	10	10	10	10	10	10
Freshwater fishes nei	*Osteichthyes*	10	10	10	10	10	10	10

English name Nom anglais Nombre inglés	Scientific name Nom scientifique Nombre científico	1995 mt	1996 mt	1997 mt	1998 mt	1999 mt	2000 mt	2001 mt
Flatfishes nei	*Pleuronectiformes*	5	5	5	10	5	11	15
Gadiformes nei	*Gadiformes*	30	30	25	30	30	32	30
Mullets nei	*Mugilidae*	400	400	300	300	300	...	400
Groupers, seabasses nei	*Serranidae*	250	250	150	250	250	230	240
Porgies, seabreams nei	*Sparidae*	450	450	350	400	450	450	400
Picarels nei	*Spicara spp*	50	100	100	100	100	100	95
Surmullets(=Red mullets) nei	*Mullus spp*	200	200	150	200	200	250	200
European conger	*Conger conger*	5	5	5	10	5	8	10
Scorpionfishes nei	*Scorpaenidae*	100	100	100	100	150	150	100
Clupeoids nei	*Clupeoidei*	800	800	600	600	500	700	500
Tuna-like fishes nei	*Scombroidei*	500	500	700	400	400	500	450
Silversides(=Sand smelts) nei	*Atherinidae*	25	25	50	50	50	50	50
Carangids nei	*Carangidae*	450	450	350	400	350	450	400
Mackerels nei	*Scombridae*	500	500	300	300	300	350	350
Barracudas nei	*Sphyraena spp*	200	200	200	200	200	200	250
Sharks, rays, skates, etc. nei	*Elasmobranchii*	50	50	50	50	50	60	55
Marine crustaceans nei	*Crustacea*	25	25	150	50	125	55	55
Cuttlefish,bobtail squids nei	*Sepiidae, Sepiolidae*	25	25	50	25	50	25	25
Common octopus	*Octopus vulgaris*	...	...	...	25	25	25	25
	Country total	*4 085*	*4 135*	*3 655*	*3 520*	*3 560*	*3 666*	*3 670*
Malaysia								
Freshwater fishes nei	*Osteichthyes*	3 939	3 683	3 949	4 626	3 366	3 549	3 446
Chacunda gizzard shad	*Anodontostoma chacunda*	3 538	3 991	3 286	2 659	4 373	4 016	4 037
Indian pellona	*Pellona ditchela*	9 959	15 285	13 374	9 921	9 188	12 705	12 664
Diadromous clupeoids nei	*Clupeoidei*	800	1 385	676	581	622	665	1 106
Barramundi(=Giant seaperch)	*Lates calcarifer*	1 319	1 322	874	1 880	1 397	1 701	1 518
Tonguefishes	*Cynoglossidae*	3 094	2 740	3 029	2 940	3 306	3 076	2 417
Flatfishes nei	*Pleuronectiformes*	2 015	2 233	2 193	2 557	3 026	1 641	1 712
Indo-Pacific tarpon	*Megalops cyprinoides*	133	99	514	338	226	133	104
Lizardfishes nei	*Synodontidae*	8 853	12 049	14 490	14 424	15 050	17 567	17 624
Sea catfishes nei	*Ariidae*	13 822	15 191	13 599	15 182	15 341	14 805	14 076
Eeltail catfishes	*Plotosus spp*	2 010	1 803	1 447	1 659	1 749	2 222	1 987
Mullets nei	*Mugilidae*	5 211	5 330	4 550	5 627	4 862	5 349	4 003
Fusiliers	*Caesio spp*	1 600	1 681	1 959	1 317	1 324	980	987
Groupers nei	*Epinephelus spp*	7 036	8 274	9 124	10 601	12 229	12 174	10 476
Sillago-whitings	*Sillaginidae*	1 948	1 895	1 934	1 809	2 014	1 812	2 057
Mangrove red snapper	*Lutjanus argentimaculatus*	11 181	11 486	10 848	6 162	12 934	11 439	11 044
Snappers nei	*Lutjanus spp*	2 863	3 247	3 210	4 486	4 666	4 270	4 332
Snappers, jobfishes nei	*Lutjanidae*	3 736	5 338	4 694	6 085	7 787	5 229	4 771
Threadfin breams nei	*Nemipterus spp*	31 323	29 534	30 102	40 327	39 694	32 510	28 910
Monocle breams	*Scolopsis spp*	1 265	1 244	950	829	1 036	1 701	1 642
Ponyfishes(=Slipmouths)	*Leiognathus spp*	2 284	2 539	2 362	3 090	3 049	2 461	2 285
Silver grunt	*Pomadasys argenteus*	1 175	1 644	1 156	2 844	2 235	1 500	1 593
Grunts, sweetlips nei	*Haemulidae (=Pomadasyidae)*	1 016	1 113	1 074	1 078	1 557	1 507	1 417
Croakers, drums nei	*Sciaenidae*	17 392	20 350	20 368	22 470	22 188	23 439	28 760
Goatfishes	*Upeneus spp*	9 601	8 107	10 302	7 650	8 606	10 845	12 960
Spotted sicklefish	*Drepane punctata*	636	875	583	516	843	678	689
Wrasses, hogfishes, etc. nei	*Labridae*	200	226	2 302	2 697	2 402	2 139	1 948
Threadfins, tasselfishes nei	*Polynemidae*	4 207	5 098	3 701	4 943	3 911	4 315	4 328
Spinefeet(=Rabbitfishes) nei	*Siganus spp*	1 144	1 152	1 335	1 464	1 208	1 393	1 494
Triggerfishes, durgons nei	*Balistidae*	1 392	1 606	1 941	1 793	2 826	1 428	2 195
Daggertooth pike conger	*Muraenesox cinereus*	3 395	4 533	6 844	7 335	5 830	5 418	5 019
Largehead hairtail	*Trichiurus lepturus*	6 247	6 368	11 273	23 951	18 009	11 574	9 711
Stolephorus anchovies	*Stolephorus spp*	22 563	24 361	23 772	25 651	23 045	22 516	17 723
Wolf-herrings nei	*Chirocentrus spp*	4 049	4 013	4 223	4 143	4 733	4 184	3 989
Clupeoids nei	*Clupeoidei*	38 993	44 525	38 110	46 315	45 517	33 613	40 750
Seerfishes nei	*Scomberomorus spp*	14 901	14 400	13 734	16 278	17 247	15 323	14 970
Kawakawa	*Euthynnus affinis*	27 442	33 944	48 498	49 409	57 281	58 132	56 111
Longtail tuna	*Thunnus tonggol*	1 233	2 045	2 125	3 106	2 824	3 586	-
Marlins,sailfishes,etc. nei	*Istiophoridae*	453	274	362	324	2 046	161	155
False trevally	*Lactarius lactarius*	418	387	561	575	681	422	439
Cobia	*Rachycentron canadum*	341	368	385	523	756	640	557
Indian scad	*Decapterus russelli*	55 623	59 733	71 184	53 426	70 160	84 203	77 394
Jacks, crevalles nei	*Caranx spp*	26 530	26 369	29 880	11 553	36 644	39 614	37 125
Black pomfret	*Parastromateus niger*	4 811	4 650	4 717	5 002	5 627	4 546	3 811
Rainbow runner	*Elagatis bipinnulata*	737	390	1 581	390	863	565	195
Torpedo scad	*Megalaspis cordyla*	19 470	14 912	13 662	18 785	19 895	17 332	13 952
Queenfishes	*Scomberoides spp*	3 312	3 240	2 872	3 253	3 811	3 482	3 527
Yellowstripe scad	*Selaroides leptolepis*	23 859	28 123	32 031	32 734	41 341	44 253	39 858
Indian mackerels nei	*Rastrelliger spp*	126 170	95 364	86 801	102 072	111 365	98 055	99 469
Silver pomfret	*Pampus argenteus*	5 193	5 020	5 093	5 401	4 371	3 531	4 114
Butterfishes, pomfrets nei	*Stromateidae*	476	460	467	495	570	461	377
Barracudas nei	*Sphyraena spp*	4 696	4 933	6 654	7 154	8 103	7 447	6 611
Rays, stingrays, mantas nei	*Rajiformes*	15 707	15 928	17 282	16 104	17 033	16 573	16 532
Sharks, rays, skates, etc. nei	*Elasmobranchii*	8 437	8 079	7 483	7 839	8 092	7 948	8 663
Marine fishes nei	*Osteichthyes*	367 526	352 837	331 727	382 806	374 189	411 228	413 249
Marine crabs nei	*Brachyura*	10 197	9 002	9 731	14 243	14 430	12 639	12 298
Tropical spiny lobsters nei	*Panulirus spp*	702	814	836	1 037	1 094	1 103	1 612
Sergestid shrimps nei	*Sergestidae*	17 654	20 827	15 240	11 336	16 480	10 448	8 367
Natantian decapods nei	*Natantia*	75 132	79 410	76 205	35 895	73 994	85 528	69 101
Clams, etc. nei	*Bivalvia*	15 982	18 606	25 500	6 001	1 677	6 231	3 949
Cuttlefish,bobtail squids nei	*Sepiidae, Sepiolidae*	15 476	18 776	20 591	26 182	22 068	26 314	21 467
Various squids nei	*Loliginidae, Ommastrephidae*	31 254	36 270	38 491	38 697	40 283	54 339	45 277
Octopuses, etc. nei	*Octopodidae*	1 012	989	1 270	1 347	1 512	1 551	1 480
Jellyfishes	*Rhopilema spp*	7 692	19 902	53 811	11 802	7 182	9 036	10 299
	Country total	*1 112 375*	*1 130 372*	*1 172 922*	*1 153 719*	*1 251 768*	*1 289 245*	*1 234 733*

English name Nom anglais Nombre inglés	Scientific name Nom scientifique Nombre científico	1995 mt	1996 mt	1997 mt	1998 mt	1999 mt	2000 mt	2001 mt
Maldives								
Freshwater fishes nei	*Osteichthyes*	0	0	0	0	0	0	0
Frigate and bullet tunas	*Auxis thazard, A.rochei*	3 938	6 484	2 489	4 218	3 401	3 991	3 982
Kawakawa	*Euthynnus affinis*	2 694	3 789	2 089	3 624	1 692	1 898	2 149
Skipjack tuna	*Katsuwonus pelamis*	70 372	66 502	69 015	78 410	92 888	79 683	88 044
Yellowfin tuna	*Thunnus albacares*	12 504	12 440	18 619	17 164	15 079	15 706	15 247
Bigeye tuna	*Thunnus obesus*	473	630	540	606	604	472	-
Tuna-like fishes nei	*Scombroidei*	438	625	489	470	426	451	647
Sharks, rays, skates, etc. nei	*Elasmobranchii*	11 245	11 856	10 643	10 887	6 883	13 523	11 935
Marine fishes nei	*Osteichthyes*	17 620	18 526	17 914	16 788	13 450	19 618	3 571
Tropical spiny lobsters nei	*Panulirus spp*	...	...	...	...	...	7	13
Marine molluscs nei	*Mollusca*	143	140	271	314	485	866	-
Sea cucumbers nei	*Holothurioidea*	94	145	318	85	54	205	226
	Country total	*119 521*	*121 137*	*122 387*	*132 566*	*134 962*	*136 420*	*125 814*
Mongolia								
Freshwater fishes nei	*Osteichthyes*	158	221	180	311	524	425	117
	Country total	*158*	*221*	*180*	*311*	*524*	*425*	*117*
Myanmar								
Freshwater fishes nei	*Osteichthyes*	148 347	146 494	149 069	149 279	159 746	189 708	235 376
Marine fishes nei	*Osteichthyes*	581 073	435 868	607 814	656 406	731 664	849 018	900 492
Natantian decapods nei	*Natantia*	20 000 F	16 000 F	22 000 F	24 000 F	27 000 F	30 000 F	30 000 F
Jellyfishes	*Rhopilema spp*	1 812	3 426	1 412	432	1 000 F	1 000 F	1 000 F
	Country total	*751 232*	*601 788*	*780 295*	*830 117*	*919 410*	*1 069 726*	*1 166 868*
Nepal								
Freshwater fishes nei	*Osteichthyes*	11 230	11 230	11 230	12 000	12 752	16 700	16 700
	Country total	*11 230*	*11 230*	*11 230*	*12 000*	*12 752*	*16 700*	*16 700*
Oman								
Freshwater fishes nei	*Osteichthyes*	0	0	0	0	0	0	0
Sea catfishes nei	*Ariidae*	499	372	679	1 024	1 161	1 306	681
Mullets nei	*Mugilidae*	98	79	139	176	158	123	543
Groupers nei	*Epinephelus spp*	4 031	3 409	3 366	5 345	4 829	5 013	3 799
Snappers, jobfishes nei	*Lutjanidae*	689	426	657	474	597	669	380
Threadfin and dwarf breams nei	*Nemipteridae*	260	233	2 068	466	1 166	317	1 206
Grunts, sweetlips nei	*Haemulidae (=Pomadasyidae)*	1 122	803	1 111	713	627	1 747	541
Croakers, drums nei	*Sciaenidae*	4 070	3 709	5 468	2 218	2 121	1 926	1 873
Emperors(=Scavengers) nei	*Lethrinidae*	6 013	4 087	5 767	6 630	6 954	7 664	6 526
Porgies, seabreams nei	*Sparidae*	5 090	4 692	4 322	4 016	6 098	4 419	3 976
Spinefeet(=Rabbitfishes) nei	*Siganus spp*	142	59	259	122	97	131	363
Hairtails, scabbardfishes nei	*Trichiuridae*	8 163	8 132	10 384	4 767	1 776	4 367	2 617
Demersal percomorphs nei	*Perciformes*	5 175	7 529	4 554	6 938	8 760	5 735	2 421
Indian oil sardine	*Sardinella longiceps*	33 053	26 741	16 765	20 650	21 710	40 044	58 960
Anchovies, etc. nei	*Engraulidae*	2 073	1 017	1 189	941	485	5 126	970
Striped bonito	*Sarda orientalis*	788	370	498	162	134	95	287
Narrow-barred Spanish mackerel	*Scomberomorus commerson*	6 185	5 243	5 944	3 145	3 390	2 559	2 785
Frigate and bullet tunas	*Auxis thazard, A.rochei*	786	613	846	611	583	488	638
Kawakawa	*Euthynnus affinis*	2 064	2 335	2 388	1 731	1 522	1 550	1 961
Skipjack tuna	*Katsuwonus pelamis*	775	408	730	227	320	293	1
Longtail tuna	*Thunnus tonggol*	3 967	5 316	5 020	4 379	4 798	5 318	6 011
Yellowfin tuna	*Thunnus albacares*	28 477	20 718	15 905	14 897	7 377	8 377	7 945
Indo-Pacific sailfish	*Istiophorus platypterus*	664	581	1 261	591	399	448	218
Tuna-like fishes nei	*Scombroidei*	240	184	366	124	99	521	188
Needlefishes nei	*Tylosurus spp*	508	201	252	116	209	131	130
Cobia	*Rachycentron canadum*	234	103	180	115	124	100	54
Jacks, crevalles nei	*Caranx spp*	3 392	2 927	3 896	2 957	2 552	1 806	1 218
Queenfishes	*Scomberoides spp*	1 019	823	703	528	693	408	528
Carangids nei	*Carangidae*	3 607	2 708	3 216	2 323	2 556	2 485	1 302
Indian mackerel	*Rastrelliger kanagurta*	2 122	1 712	2 207	1 994	2 024	2 426	3 223
Barracudas nei	*Sphyraena spp*	2 244	1 948	2 023	1 789	2 347	1 431	1 781
Pelagic percomorphs nei	*Perciformes*	1 269	2 045	2 949	6 055	10 790	5 353	3 391
Rays, stingrays, mantas nei	*Rajiformes*	538	372	359	189	289	240	198
Sharks, rays, skates, etc. nei	*Elasmobranchii*	6 566	5 870	6 342	4 805	4 020	3 651	3 632
Marine fishes nei	*Osteichthyes*	2	101	355	432	1	384	1 273
Tropical spiny lobsters nei	*Panulirus spp*	608	397	263	336	180	402	379
Natantian decapods nei	*Natantia*	340	276	376	65	356	432	627
Abalones nei	*Haliotis spp*	43	43	40	40	29	45	51
Cuttlefish,bobtail squids nei	*Sepiidae, Sepiolidae*	2 945	5 036	6 148	4 080	7 478	2 891	3 854
	Country total	*139 861*	*121 618*	*118 995*	*106 171*	*108 809*	*120 421*	*126 531*
Pakistan								
Freshwater fishes nei	*Osteichthyes*	121 405	142 092	167 530	163 524	179 865	176 468	180 100
Hilsa shad	*Tenualosa ilisha*	476	562	597	611	502	190	170
Barramundi(=Giant seaperch)	*Lates calcarifer*	187	214	209	196	204	-	-
Tonguefishes	*Cynoglossidae*	1 982	2 205	2 390	2 149	2 037	2 124	1 915
Bombay-duck	*Harpadon nehereus*	98	101	95	91	72	65	55
Greater lizardfish	*Saurida tumbil*	43	45	28	22	-	-	-
Sea catfishes nei	*Ariidae*	45 444	49 428	54 437	55 934	51 665	39 168	38 215
Mullets nei	*Mugilidae*	17 280	17 631	18 935	17 580	12 336	9 618	9 723
Groupers nei	*Epinephelus spp*	8 600	9 793	10 474	13 991	17 355	16 012	15 928
Sillago-whitings	*Sillaginidae*	423	289	266	218	201	194	204

English name Nom anglais Nombre inglés	Scientific name Nom scientifique Nombre científico	1995 mt	1996 mt	1997 mt	1998 mt	1999 mt	2000 mt	2001 mt
Mangrove red snapper	*Lutjanus argentimaculatus*	3 145	2 002	2 394	3 192	3 195	3 003	2 900
Threadfin breams nei	*Nemipterus spp*	952	825	884	3 192	7 166	8 940	8 466
Grunts, sweetlips nei	*Haemulidae (=Pomadasyidae)*	5 537	5 268	6 010	6 221	8 147	9 961	9 752
Croakers, drums nei	*Sciaenidae*	25 201	19 934	20 428	19 625	24 665	21 976	21 725
Emperors(=Scavengers) nei	*Lethrinidae*	1 643	1 549	1 911	2 334	3 323	5 173	5 044
Porgies, seabreams nei	*Sparidae*	3 358	3 097	3 058	1 255	4 220	4 510	4 411
Fourfinger threadfin	*Eleutheronema tetradactylum*	812	516	1 783	969	...	63	55
Pike-congers nei	*Muraenesox spp*	4 692	4 904	5 637	5 627	8 377	5 937	5 834
Largehead hairtail	*Trichiurus lepturus*	6 093	9 073	11 583	12 337	31 623	28 754	27 355
Indian oil sardine	*Sardinella longiceps*	55 177	52 290	51 930	44 079	30 629	31 167	31 201
Anchovies, etc. nei	*Engraulidae*	17 564	14 091	16 113	13 165	15 154	15 191	15 001
Dorab wolf-herring	*Chirocentrus dorab*	2 289	1 580	1 931	2 051	2 266	2 775	2 604
Clupeoids nei	*Clupeoidei*	31 426	27 576	26 650	25 487	26 934	24 810	24 306
Narrow-barred Spanish mackerel	*Scomberomorus commerson*	8 618	10 108	12 009	12 232	11 734	9 366	8 405
Frigate and bullet tunas	*Auxis thazard, A.rochei*	36	49	54	56	59	42	150
Kawakawa	*Euthynnus affinis*	1 449	2 351	2 571	2 684	2 715	2 340	1 755
Skipjack tuna	*Katsuwonus pelamis*	7 089	4 140	4 480	4 372	4 505	4 308	3 968
Longtail tuna	*Thunnus tonggol*	5 006	4 121	5 360	5 220	5 600	5 315	6 935
Yellowfin tuna	*Thunnus albacares*	5 140	5 250	3 838	3 795	8 884	4 946	5 921
Indo-Pacific sailfish	*Istiophorus platypterus*	910	980	41	45	46	...	998
Marlins,sailfishes,etc. nei	*Istiophoridae*	2 684	2 834	2 198	2 264	2 340	2 215	1 177
Tuna-like fishes nei	*Scombroidei*	-	1 990	1 500	1 592	4 610	4 240	2 877
False trevally	*Lactarius lactarius*	3	2	4	5	-	-	-
Cobia	*Rachycentron canadum*	2 306	1 574	1 449	1 254	1 136	2 896	2 797
Scads nei	*Decapterus spp*	1 920	1 010	1 225	3 505	4 661	4 600	4 355
Jacks, crevalles nei	*Caranx spp*	4 631	3 972	5 391	6 523	8 407	9 111	8 928
Black pomfret	*Parastromateus niger*	3 066	2 221	2 322	2 109	2 917	2 027	1 975
Torpedo scad	*Megalaspis cordyla*	6 511	3 000	2 100	1 100	1 450	2 017	1 825
Carangids nei	*Carangidae*	16 495	15 957	19 002	18 689	17 779	16 545	15 988
Common dolphinfish	*Coryphaena hippurus*	2 570	1 841	1 658	1 892	3 109	1 954	1 869
Butterfishes, pomfrets nei	*Stromateidae*	4 156	2 799	3 786	4 089	4 605	3 945	3 454
Barracudas nei	*Sphyraena spp*	2 324	2 878	2 683	2 664	3 520	3 981	3 889
Requiem sharks nei	*Carcharhinidae*	32 288	34 447	31 179	35 357	32 535	28 245	26 524
Guitarfishes, etc. nei	*Rhinobatidae*	1 208	1 422	1 481	1 564	1 643	2 185	1 944
Sawfishes	*Pristidae*	23	-	-	-	-	-	-
Rays, stingrays, mantas nei	*Rajiformes*	16 445	15 563	15 769	17 576	20 780	20 740	20 801
Marine fishes nei	*Osteichthyes*	14 742	16 311	21 042	35 352	39 473	36 451	35 450
Marine crabs nei	*Brachyura*	877	3 200	3 989	5 680	5 109	5 187	5 099
Tropical spiny lobsters nei	*Panulirus spp*	615	724	765	782	1 077	807	756
Giant tiger prawn	*Penaeus monodon*	132	141	140	122	138	139	140
Penaeus shrimps nei	*Penaeus spp*	5 591	5 982	5 975	5 189	5 874	5 920	5 974
Metapenaeus shrimps nei	*Metapenaeus spp*	6 981	7 602	6 801	6 204	6 791	7 126	7 246
Parapenaeopsis shrimps nei	*Parapenaeopsis spp*	12 919	14 047	16 722	14 689	12 889	11 945	11 576
Cuttlefish,bobtail squids nei	*Sepiidae, Sepiolidae*	2 455	3 308	4 528	3 225	5 146	5 307	5 256
Various squids nei	*Loliginidae, Ommastrephidae*	2 832	2 600	4 460	3 300	5 062	4 070	4 024
	Country total	*525 849*	*537 489*	*589 795*	*596 980*	*654 530*	*614 069*	*607 020*
Philippines								
Cyprinids nei	*Cyprinidae*	8 880	6 497	5 717	6 453	4 677	5 032	5 562
Tilapias nei	*Oreochromis (=Tilapia) spp*	21 244	17 663	20 935	23 477	25 278	28 874	28 881
Torpedo-shaped catfishes nei	*Clarias spp*	2 669	2 696	2 396	1 628	2 058	2 200	2 366
Freshwater gobies nei	*Gobiidae*	3 431	3 585	4 300	3 803	4 027	4 563	4 280
Striped snakehead	*Channa striata*	6 018	5 457	4 547	4 856	5 789	6 386	6 698
Freshwater fishes nei	*Osteichthyes*	9 425	7 143	10 413	6 632	8 309	9 800	9 384
River eels nei	*Anguilla spp*	128	100	116	73	114	193	201
Chacunda gizzard shad	*Anodontostoma chacunda*	2 955	2 761	1 260	1 191	1 115	1 140	1 189
Indian pellona	*Pellona ditchela*	1 177	792	802	842	826	839	1 153
Milkfish	*Chanos chanos*	7 499	3 671	397	3 152	315	396	443
Barramundi(=Giant seaperch)	*Lates calcarifer*	2 821	3 049	553	655	784	642	678
Flatfishes nei	*Pleuronectiformes*	805	829	627	659	722	729	720
Indo-Pacific tarpon	*Megalops cyprinoides*	979	997	644	720	1 207	1 151	1 238
Lizardfishes nei	*Synodontidae*	8 630	8 435	6 671	7 345	7 649	5 539	5 499
Sea catfishes nei	*Ariidae*	10 160	10 497	9 025	7 797	8 312	8 396	8 422
Mullets nei	*Mugilidae*	16 322	15 409	15 039	14 204	15 660	14 951	15 617
Fusiliers	*Caesio spp*	13 914	13 788	16 754	15 931	14 952	13 516	14 086
Groupers, seabasses nei	*Serranidae*	14 226	12 776	12 197	13 160	13 675	12 492	13 285
Sillago-whitings	*Sillaginidae*	7 417	6 800	8 662	8 918	9 529	9 472	9 481
Moonfish	*Mene maculata*	4 201	4 810	11 933	11 516	12 372	11 703	12 991
Snappers, jobfishes nei	*Lutjanidae*	16 332	15 005	18 378	15 442	19 192	19 154	19 603
Threadfin breams nei	*Nemipterus spp*	35 538	32 884	29 839	30 511	29 301	29 487	29 870
Ponyfishes(=Slipmouths) nei	*Leiognathidae*	59 134	57 867	61 254	59 862	68 371	67 255	66 890
Goatfishes, red mullets nei	*Mullidae*	25 510	23 662	15 884	14 182	14 527	14 718	14 875
Mojarras(=Silver-biddies) nei	*Gerres spp*	6 491	5 909	4 285	4 854	5 008	5 117	5 062
Spotted sicklefish	*Drepane punctata*	63	64	174	96	133	114	117
Wrasses, hogfishes, etc. nei	*Labridae*	19 099	17 097	15 107	13 391	12 318	12 384	13 045
Threadfins, tasselfishes nei	*Polynemidae*	2 548	2 383	2 933	3 017	2 524	2 371	2 455
Glassfishes	*Ambassidae*	3 554	3 075	2 594	2 186	1 733	1 575	1 664
Percoids nei	*Percoidei*	38 843	36 825	20 212	20 036	20 717	19 912	21 198
Gobies nei	*Gobiidae*	8 780	8 404	7 542	6 756	8 021	7 719	7 898
Surgeonfishes nei	*Acanthuridae*	4 752	4 474	7 367	7 770	5 797	5 392	5 663
Batfishes	*Platax spp*	1 600	1 584	2 975	2 792	2 870	2 597	2 607
Scats	*Scatophagus spp*	4 782	4 204	2 152	2 496	2 541	2 698	2 800
Spinefeet(=Rabbitfishes) nei	*Siganus spp*	18 766	17 012	15 720	18 246	19 977	20 108	20 614
Conger eels, etc. nei	*Congridae*	3 061	2 687	2 053	2 540	2 459	2 349	2 344
Hairtails, scabbardfishes nei	*Trichiuridae*	12 522	12 266	9 412	9 877	10 363	8 641	8 797
Sardinellas nei	*Sardinella spp*	264 675	257 804	302 341	302 599	279 864	298 466	294 968
Rainbow sardine	*Dussumieria acuta*	19 255	19 441	15 216	20 639	24 080	12 993	13 374
Stolephorus anchovies	*Stolephorus spp*	71 516	71 456	78 678	77 049	78 087	79 630	82 112

English name Nom anglais Nombre inglés	Scientific name Nom scientifique Nombre científico	1995 mt	1996 mt	1997 mt	1998 mt	1999 mt	2000 mt	2001 mt
Wolf-herrings nei	*Chirocentrus spp*	135	114	245	317	377	354	389
Clupeoids nei	*Clupeoidei*	286	266	325	663	634	602	650
Narrow-barred Spanish mackerel	*Scomberomorus commerson*	10 593	10 557	11 237	10 772	9 137	8 889	9 252
Frigate and bullet tunas	*Auxis thazard, A.rochei*	88 426	88 969	108 494	106 433	111 301	112 227	115 905
Kawakawa	*Euthynnus affinis*	27 308	24 345	26 573	24 424	25 406	27 963	27 890
Skipjack tuna	*Katsuwonus pelamis*	110 111	110 004	110 097	116 673	108 778	113 011	112 238
Albacore	*Thunnus alalunga*	-	-		506	198	101	68
Yellowfin tuna	*Thunnus albacares*	60 957	61 280	67 342	80 000	91 251	90 832	96 786
Bigeye tuna	*Thunnus obesus*	-	-	-	2 705	4 004	2 436	1 388
Indo-Pacific blue marlin	*Makaira mazara*	1 178	1 191	1 096	1 539	2 322	2 297	2 345
Atlantic blue marlin	*Makaira nigricans*	-	-	-	7	71	38	-
Striped marlin	*Tetrapturus audax*	-	-	-	121	54	32	13
Atlantic white marlin	*Tetrapturus albidus*	-	-	-	1	12	-	-
Marlins,sailfishes,etc. nei	*Istiophoridae*	4 648	4 317	4 799	5 338	5 082	4 969	5 247
Swordfish	*Xiphias gladius*	-	-	-	272	325	193	156
Tuna-like fishes nei	*Scombroidei*	4 202	4 002	5 554	4 804	3 914	3 621	3 809
Needlefishes nei	*Tylosurus spp*	10 171	9 454	9 967	9 532	10 323	10 084	10 327
Halfbeaks nei	*Hemiramphus spp*	3 555	3 231	1 655	1 949	2 281	2 309	2 503
Flyingfishes nei	*Exocoetidae*	16 826	17 300	27 801	36 134	38 280	36 050	37 498
Silversides(=Sand smelts) nei	*Atherinidae*	864	866	669	596	618	543	544
False trevally	*Lactarius lactarius*	25	26	202	213	235	253	272
Cobia	*Rachycentron canadum*	1 003	964	1 162	1 235	1 187	1 046	1 070
Scads nei	*Decapterus spp*	264 472	228 757	234 849	250 809	254 178	260 999	287 810
Rainbow runner	*Elagatis bipinnulata*	4 966	4 505	4 075	4 298	4 500	4 342	4 378
Torpedo scad	*Megalaspis cordyla*	6 097	5 864	11 429	14 817	16 033	16 274	17 802
Queenfishes	*Scomberoides spp*	3 106	2 965	2 631	2 830	4 404	4 665	4 682
Bigeye scad	*Selar crumenophthalmus*	43 592	43 660	54 167	61 999	65 776	71 365	79 307
Carangids nei	*Carangidae*	39 682	37 456	32 175	31 472	35 204	34 713	35 889
Common dolphinfish	*Coryphaena hippurus*	4 253	3 544	359	189	223	211	226
Chub mackerel	*Scomber japonicus*	3 307	2 991	2 025	1 507	1 485	1 422	1 520
Short mackerel	*Rastrelliger brachysoma*	26 200	25 224	22 978	23 350	25 713	26 771	28 289
Indian mackerel	*Rastrelliger kanagurta*	51 352	46 264	54 732	51 919	53 606	55 088	59 232
Butterfishes, pomfrets nei	*Stromateidae*	2 382	2 281	1 001	1 366	1 547	1 522	1 564
Barracudas nei	*Sphyraena spp*	10 654	10 003	6 707	7 722	8 976	7 778	8 363
Shortfin mako	*Isurus oxyrinchus*	-	-	-	-	3	-	-
Rays, stingrays, mantas nei	*Rajiformes*	4 980	4 756	2 125	2 174	2 299	2 248	2 405
Sharks, rays, skates, etc. nei	*Elasmobranchii*	4 079	3 839	1 690	2 119	2 188	2 080	2 252
Marine fishes nei	*Osteichthyes*	19 052	37 385	10 176	13 104	12 783	12 043	13 428
Freshwater prawns, shrimps nei	*Palaemonidae*	4 581	5 214	4 282	2 014	3 915	4 020	4 507
Freshwater crustaceans nei	*Crustacea*	73	55	34	32	1	5	67
Blue swimming crab	*Portunus pelagicus*	29 536	27 660	30 358	23 919	34 076	36 303	36 973
Indo-Pacific swamp crab	*Scylla serrata*	4 835	4 258	1 133	1 124	1 211	1 247	1 412
Tropical spiny lobsters nei	*Panulirus spp*	559	522	421	214	249	250	269
Slipper lobsters nei	*Scyllaridae*	350	334	8	65	89	90	105
Giant tiger prawn	*Penaeus monodon*	431	360	292	422	169	232	328
Penaeus shrimps nei	*Penaeus spp*	11 309	10 462	11 508	10 796	11 255	11 063	11 869
Endeavour shrimp	*Metapenaeus endeavouri*	237	225	94	45	116	136	174
Metapenaeus shrimps nei	*Metapenaeus spp*	6 054	5 431	5 420	6 545	6 419	5 987	6 167
Sergestid shrimps nei	*Sergestidae*	18 997	18 657	15 562	16 719	19 262	20 122	20 629
Squillids nei	*Squillidae*	982	695	2 133	2 219	2 068	2 141	2 561
Freshwater molluscs nei	*Mollusca*	120 547	122 636	101 841	90 154	87 259	85 575	68 958
Abalones nei	*Haliotis spp*	483	448	183	347	282	241	250
Slipper cupped oyster	*Crassostrea iredalei*	324	291	152	89	95	79	83
Green mussel	*Perna viridis*	470	334	47	22	20	17	17
Scallops nei	*Pectinidae*	156	139	61	40	62	53	54
Anadara clams nei	*Anadara spp*	6	6	3	4	4	3	3
Short neck clams nei	*Paphia spp*	30	31	2	1	1	2	2
Clams, etc. nei	*Bivalvia*	921	725	375	227	277	222	223
Cuttlefish,bobtail squids nei	*Sepiidae, Sepiolidae*	2 836	2 702	2 803	2 204	2 093	2 016	2 066
Common squids nei	*Loligo spp*	56 415	52 458	54 155	48 678	47 115	46 778	47 854
Octopuses, etc. nei	*Octopodidae*	9 729	9 025	7 991	5 235	5 813	5 502	6 088
Marine molluscs nei	*Mollusca*	66	62	-	-	-	-	-
Marine turtles nei	*Testudinata*	1	1	-	2	2	1	1
Sea urchins nei	*Strongylocentrotus spp*	466	452	296	161	143	125	127
Sea cucumbers nei	*Holothurioidea*	2 062	2 123	1 191	830	849	730	791
Jellyfishes	*Rhopilema spp*	61	57	20	10	12	12	12
	Country total	*1 860 701*	*1 783 601*	*1 805 806*	*1 833 380*	*1 872 818*	*1 893 017*	*1 945 217*
Qatar								
Freshwater fishes nei	*Osteichthyes*	0	0	0	0	0	0	0
Greater lizardfish	*Saurida tumbil*	0	1	0	0	0	0	0
Sea catfishes nei	*Ariidae*	0	0	0	0	0	0	0
Mullets nei	*Mugilidae*	8	8	7	9	5	4	4
Groupers nei	*Epinephelus spp*	728	768	736	804	913	1 215	1 820
Snappers nei	*Lutjanus spp*	111	157	157	143	73	181	230
Threadfin and dwarf breams nei	*Nemipteridae*	5	13	15	3	0	0	0
Grunts, sweetlips nei	*Haemulidae (=Pomadasyidae)*	606	653	542	583	433	789	900
Emperors(=Scavengers) nei	*Lethrinidae*	722	1 031	1 172	1 326	798	1 442	1 820
King soldier bream	*Argyrops spinifer*	85	130	177	146	98	199	426
Porgies, seabreams nei	*Sparidae*	124	126	221	188	182	248	288
Goatfishes, red mullets nei	*Mullidae*	15	16	14	14	20	1	0
Mojarras(=Silver-biddies) nei	*Gerres spp*	53	50	56	92	77	78	82
Spinefeet(=Rabbitfishes) nei	*Siganus spp*	161	225	237	285	240	387	451
Sardinellas nei	*Sardinella spp*	0	0	-	0	0	0	0
Dorab wolf-herring	*Chirocentrus dorab*	0	0	-	0	0	0	0
Narrow-barred Spanish mackerel	*Scomberomorus commerson*	255	307	411	552	496	768	1 019
Tuna-like fishes nei	*Scombroidei*	0	0	-	0	0	0	0
Needlefishes nei	*Tylosurus spp*	19	14	25	12	20	1	2

English name Nom anglais Nombre inglés	Scientific name Nom scientifique Nombre científico	1995 mt	1996 mt	1997 mt	1998 mt	1999 mt	2000 mt	2001 mt
Cobia	*Rachycentron canadum*	47	56	52	56	44	56	95
Scads nei	*Decapterus spp*	0	0	0	0	0	0	0
Jacks, crevalles nei	*Caranx spp*	175	222	247	308	289	409	388
Golden trevally	*Gnathanodon speciosus*	98	101	108	172	116	185	204
Queenfishes	*Scomberoides spp*	49	57	56	80	42	57	78
Carangids nei	*Carangidae*	291	308	308	326	287	420	552
Barracudas nei	*Sphyraena spp*	48	54	59	21	2	62	86
Requiem sharks nei	*Carcharhinidae*	0	0	-	0	0	0	0
Marine fishes nei	*Osteichthyes*	593	357	332	38	10	587	79
Marine crabs nei	*Brachyura*	39	47	47	69	39	26	21
Slipper lobsters nei	*Scyllaridae*	8	8	16	15	13	5	20
Green tiger prawn	*Penaeus semisulcatus*	0	0	-	0	0	0	0
Common squids nei	*Loligo spp*	31	30	37	37	10	22	41
	Country total	*4 271*	*4 739*	*5 032*	*5 279*	*4 207*	*7 142*	*8 606*
Saudi Arabia								
Freshwater fishes nei	*Osteichthyes*	0	0	0	0	0	0	0
Milkfish	*Chanos chanos*	137	130	70	82	81	64	73
Flatfishes nei	*Pleuronectiformes*	51	58	75	90	84	84	85
Lizardfishes nei	*Synodontidae*	195	172	188	215	214	169	199
Sea catfishes nei	*Ariidae*	303	302	366	315	309	564	534
Squirrelfishes nei	*Holocentridae*	7	40	50	34	53	46	40
Mullets nei	*Mugilidae*	400	320	369	510	317	326	397
Groupers, seabasses nei	*Serranidae*	3 514	4 207	4 403	5 053	5 430	5 402	5 273
Therapon pearch	*Terapon spp*	-	-	-	1	2	0	1
Bigeyes nei	*Priacanthus spp*	-	-	-	2	1	0	1
Snappers, jobfishes nei	*Lutjanidae*	1 662	1 704	2 148	2 302	2 022	1 647	2 092
Threadfin breams nei	*Nemipterus spp*	221	131	87	66	49	144	153
Grunts, sweetlips nei	*Haemulidae (=Pomadasyidae)*	814	1 063	1 014	779	846	1 350	1 206
Emperors(=Scavengers) nei	*Lethrinidae*	6 598	7 314	6 904	6 796	7 233	7 078	7 448
Porgies, seabreams nei	*Sparidae*	1 469	1 722	2 481	2 822	2 771	2 488	2 646
Goatfishes	*Upeneus spp*	65	83	73	107	123	72	122
Mojarras(=Silver-biddies) nei	*Gerres spp*	337	360	438	445	520	529	510
Blue sea chub	*Kyphosus cinerascens*	-	-	8	-	-	-	-
Wrasses, hogfishes, etc. nei	*Labridae*	165	108	105	66	155	91	79
Parrotfishes nei	*Scaridae*	837	777	906	702	655	756	758
Angelfishes nei	*Pomacanthidae*	-	-	-	7	14	1	3
Surgeonfishes nei	*Acanthuridae*	113	238	173	253	205	149	343
Spadefishes nei	*Ephippidae*	-	-	-	17	1	1	2
Spinefeet(=Rabbitfishes) nei	*Siganus spp*	2 341	2 779	2 173	1 761	1 691	1 832	1 822
Flatheads nei	*Platycephalidae*	-	-	-	18	7	5	6
Puffers nei	*Tetraodontidae*	-	-	-	-	0	0	18
Triggerfishes, durgons nei	*Balistidae*	-	-	26	8	8	5	7
Dorab wolf-herring	*Chirocentrus dorab*	-	-	-	8	7	10	9
Clupeoids nei	*Clupeoidei*	259	201	560	108	162	187	178
Narrow-barred Spanish mackerel	*Scomberomorus commerson*	6 342	5 276	5 511	6 722	6 032	6 057	5 292
Indo-Pacific king mackerel	*Scomberomorus guttatus*	-	-	114	300	303	303	303
Kawakawa	*Euthynnus affinis*	121	162	304	332	256	264	260
Longtail tuna	*Thunnus tonggol*	234	115	101	181	136	143	180
Indo-Pacific sailfish	*Istiophorus platypterus*	-	-	1	2	1	1	8
Tuna-like fishes nei	*Scombroidei*	690	804	861	773	797	783	854
Needlefishes nei	*Tylosurus spp*	324	96	111	131	140	107	126
Cobia	*Rachycentron canadum*	124	155	155	130	137	138	167
Snubnose pompano	*Trachinotus blochii*	-	-	-	4	0	31	-
Rainbow runner	*Elagatis bipinnulata*	-	-	415	98	132	5	52
Queenfishes	*Scomberoides spp*	312	349	385	572	456	572	543
Carangids nei	*Carangidae*	4 547	5 524	5 125	4 795	5 534	6 612	5 943
Indian mackerel	*Rastrelliger kanagurta*	3 069	1 549	1 990	2 072	1 979	1 525	1 803
Mackerels nei	*Scombridae*	...	...	...	12	16	11	10
Silver pomfret	*Pampus argenteus*	41	31	16	14	30	43	25
Barracudas nei	*Sphyraena spp*	1 065	1 251	1 246	1 489	1 267	1 563	1 562
Rays, stingrays, mantas nei	*Rajiformes*	-	-	-	4	8	4	6
Sharks, rays, skates, etc. nei	*Elasmobranchii*	467	398	543	697	497	653	651
Marine fishes nei	*Osteichthyes*	1 480	1 205	820	768	587	574	804
Marine crabs nei	*Brachyura*	1 035	1 060	1 371	1 062	959	1 021	1 002
Tropical spiny lobsters nei	*Panulirus spp*	21	13	18	13	19	8	14
Penaeus shrimps nei	*Penaeus spp*	5 941	7 423	7 011	7 812	3 612	5 639	4 761
Cuttlefish,bobtail squids nei	*Sepiidae, Sepiolidae*	308	578	598	656	760	593	796
Octopuses, etc. nei	*Octopodidae*	-	-	1	-	-	0	-
	Country total	*45 609*	*47 698*	*49 314*	*51 206*	*46 618*	*49 650*	*49 167*
Singapore								
Freshwater fishes nei	*Osteichthyes*	0	0	0	0	0	0	0
Barramundi(=Giant seaperch)	*Lates calcarifer*	51	39	58	39	29	41	52
Lizardfishes nei	*Synodontidae*	186	172	125	90	51	28	6
Sea catfishes nei	*Ariidae*	359	333	385	358	241	141	76
Mullets nei	*Mugilidae*	173	54	55	34	32	42	27
Fusiliers	*Caesio spp*	1	-	9	18	13	10	1
Groupers nei	*Epinephelus spp*	109	120	94	72	56	43	38
Sillago-whitings	*Sillaginidae*	32	48	45	41	24	10	14
Moonfish	*Mene maculata*	6	5	7	13	13	15	7
Snappers nei	*Lutjanus spp*	310	321	289	238	154	122	80
Snappers, jobfishes nei	*Lutjanidae*	52	61	66	50	31	32	16
Threadfin breams nei	*Nemipterus spp*	255	209	239	158	128	96	48
Ponyfishes(=Slipmouths)	*Leiognathus spp*	83	75	63	52	47	32	23
Grunts, sweetlips nei	*Haemulidae (=Pomadasyidae)*	33	53	45	27	25	18	18
Croakers, drums nei	*Sciaenidae*	136	123	180	160	114	68	45

E-4

Fish crustaceans, molluscs, etc	Capture production by countries or areas and species	Asia
Poissons, crustacés, mollusques, etc	Captures par pays ou zones et espèces	Asie
Peces, crustáceos, moluscos, etc	Capturas por países o áreas y especies	Asia

English name Nom anglais Nombre inglés	Scientific name Nom scientifique Nombre científico	1995 mt	1996 mt	1997 mt	1998 mt	1999 mt	2000 mt	2001 mt
Goatfishes	*Upeneus spp*	29	24	29	29	25	8	15
Threadfins, tasselfishes nei	*Polynemidae*	5	5	6	7	13	25	11
Spinefeet(=Rabbitfishes) nei	*Siganus spp*	-	14	21	24	28	5	8
Largehead hairtail	*Trichiurus lepturus*	137	144	170	140	127	83	50
Wolf-herrings nei	*Chirocentrus spp*	81	87	79	77	51	42	30
Clupeoids nei	*Clupeoidei*	240	379	351	329	206	78	73
Seerfishes nei	*Scomberomorus spp*	76	76	71	70	79	78	46
Skipjack tuna	*Katsuwonus pelamis*	5	5	47	12	23	2	10
Tuna-like fishes nei	*Scombroidei*	0	0	-	-	-	-	-
Scads nei	*Decapterus spp*	209	227	212	222	156	163	106
Jacks, crevalles nei	*Caranx spp*	297	312	313	234	175	139	66
Carangids nei	*Carangidae*	22	37	33	36	30	21	12
Indian mackerels nei	*Rastrelliger spp*	151	12	51	165	129	97	68
Butterfishes, pomfrets nei	*Stromateidae*	103	100	94	103	94	75	48
Barracudas nei	*Sphyraena spp*	195	170	134	149	105	79	86
Rays, stingrays, mantas nei	*Rajiformes*	320	327	308	336	250	261	187
Sharks, rays, skates, etc. nei	*Elasmobranchii*	104	94	93	80	59	43	32
Marine fishes nei	*Osteichthyes*	4 465	4 308	3 933	2 812	2 740	2 339	1 325
Indo-Pacific swamp crab	*Scylla serrata*	27	19	15	9	9	28	9
Marine crabs nei	*Brachyura*	196	176	203	264	175	189	203
Tropical spiny lobsters nei	*Panulirus spp*	1	-	5	11	8	8	7
Slipper lobsters nei	*Scyllaridae*	11	9	11	12	9	6	7
Natantian decapods nei	*Natantia*	767	857	706	621	522	422	250
Green mussel	*Perna viridis*	-	188	-	-	-	-	-
Cuttlefish,bobtail squids nei	*Sepiidae, Sepiolidae*	196	214	235	179	142	134	56
Common squids nei	*Loligo spp*	679	546	470	462	376	348	186
	Country total	*10 102*	*9 943*	*9 250*	*7 733*	*6 489*	*5 371*	*3 342*
Sri Lanka								
Tilapias nei	*Oreochromis (=Tilapia) spp*	15 000	22 250	27 250	29 900	28 520	33 220	27 230
Freshwater fishes nei	*Osteichthyes*	...	...	...	...	2 930	3 480	2 640
Demersal percomorphs nei	*Perciformes*	7 088	8 968	9 100	9 210	10 440	14 910	12 290
Clupeoids nei	*Clupeoidei*	49 785	48 221	47 200	50 800	51 370	56 250	53 230
Wahoo	*Acanthocybium solandri*	129	128	156	196	488	545	520
Narrow-barred Spanish mackerel	*Scomberomorus commerson*	199	817	999	1 246	856	766	2 180
Indo-Pacific king mackerel	*Scomberomorus guttatus*	1	0	-	-	-	-	-
Streaked seerfish	*Scomberomorus lineolatus*	0	0	-	-	-	-	-
Seerfishes nei	*Scomberomorus spp*	-	-	-	-	168	20	-
Frigate and bullet tunas	*Auxis thazard, A.rochei*	4 006	5 334	6 521	8 133	3 515	4 583	8 240
Kawakawa	*Euthynnus affinis*	2 086	2 262	2 765	3 449	2 167	2 167	2 130
Skipjack tuna	*Katsuwonus pelamis*	18 288	22 754	27 815	34 691	51 940	51 940	45 640
Longtail tuna	*Thunnus tonggol*	0	0	-	-	-	-	-
Southern bluefin tuna	*Thunnus maccoyii*	-	68	83	104	121	120	...
Yellowfin tuna	*Thunnus albacares*	8 696	12 889	15 756	19 651	27 538	22 091	27 910
Bigeye tuna	*Thunnus obesus*	2 108	491	600	749	462	348	470
Indo-Pacific sailfish	*Istiophorus platypterus*	3 335	5 360	6 552	8 172	6 979	8 878	7 680
Indo-Pacific blue marlin	*Makaira mazara*	37	0	-	-	-	-	-
Black marlin	*Makaira indica*	0	39	48	59	69	68	...
Striped marlin	*Tetrapturus audax*	-	0	0	0	-	-	-
Marlins,sailfishes,etc. nei	*Istiophoridae*	5 196	5 675	6 937	8 653	7 253	5 728	6 340
Swordfish	*Xiphias gladius*	2 558	2 591	3 167	3 950	2 132	5 545	2 860
Tuna-like fishes nei	*Scombroidei*	2	0	0	0	11	17	90
Carangids nei	*Carangidae*	6 910	6 088	6 900	8 500	8 680	10 450	9 950
Mackerels nei	*Scombridae*	17 642	17 700	20 000	20 900	21 350	21 480	18 760
Silky shark	*Carcharhinus falciformis*	21 400	21 000	15 000	20 875	20 700	16 130	14 620
Sharks, rays, skates, etc. nei	*Elasmobranchii*	7 077	6 954	11 920	7 625	8 660	11 884	8 240
Marine fishes nei	*Osteichthyes*	55 428	36 404	23 398	25 084	16 106	26 090	26 740
Marine crustaceans nei	*Crustacea*	1 800	2 502	2 360	880	3 080	230	1 450
Cuttlefish,bobtail squids nei	*Sepiidae, Sepiolidae*	300	300	300	300	365	310	290
Marine molluscs nei	*Mollusca*	0	0	-	-	10	15	20
Sea cucumbers nei	*Holothurioidea*	100	150	272	203	170	145	120
	Country total	*229 171*	*228 945*	*235 099*	*263 330*	*276 080*	*297 410*	*279 640*
Syria								
Freshwater fishes nei	*Osteichthyes*	3 832	3 103	3 557	4 347	5 338	3 991	5 969
European hake	*Merluccius merluccius*	128 F	250	300	125	110	87	52
Pandoras nei	*Pagellus spp*	77 F	134	98	125	100	65	85
Pargo breams nei	*Pagrus spp*	39 F	50	80	90	62	74	77
Surmullets(=Red mullets) nei	*Mullus spp*	116 F	232	250	116	130	122	125
Gurnards, searobins nei	*Triglidae*	128 F	0	60	46	35	32	32
Demersal percomorphs nei	*Perciformes*	492 F	606	450	626	530	392	449
Sardinellas nei	*Sardinella spp*	336 F	550	300	292	338	251	197
Little tunny(=Atl.black skipj)	*Euthynnus alletteratus*	155	270	350	417	390	370	370
Atlantic horse mackerel	*Trachurus trachurus*	39 F	60	77	40	30	36	56
Greater amberjack	*Seriola dumerili*	52 F	90	75	108	88	52	96
Atlantic mackerel	*Scomber scombrus*	116 F	116	210	274	245	384	187
Barracudas nei	*Sphyraena spp*	52 F	82	90	88	130	76	77
Smooth-hounds nei	*Mustelus spp*	39 F	50	-	-	-	-	-
Marine fishes nei	*Osteichthyes*	91 F	90	150	328	342	580	462
Marine crustaceans nei	*Crustacea*	90 F	90	84	75	70	60	57
	Country total	*5 782*	*5 773*	*6 131*	*7 097*	*7 938*	*6 572*	*8 291*
Tajikistan								
Freshwater bream	*Abramis brama*	58	...	...	...	...	...	37
Common carp	*Cyprinus carpio*	...	...	...	...	48	59	24
Sichel	*Pelecus cultratus*	0	...	...	...	...	...	8

English name Nom anglais Nombre inglés	Scientific name Nom scientifique Nombre científico	1995 mt	1996 mt	1997 mt	1998 mt	1999 mt	2000 mt	2001 mt
Asp	*Aspius aspius*	6	...	...	...	...	...	6
Cyprinids nei	*Cyprinidae*	-	-	-	-	-	-	31
Wels(=Som)catfish	*Silurus glanis*	12	...	...	...	...	...	10
Pike-perch	*Stizostedion lucioperca*	20	...	...	...	...	...	21
Freshwater fishes nei	*Osteichthyes*	4	40 F	75 F	100 F	32 F	19 F	-
	Country total	*100*	*40 F*	*75 F*	*100 F*	*80 F*	*78 F*	*137*
Thailand								
Common carp	*Cyprinus carpio*	10 144	7 420	7 418	11 508	13 689	7 000	7 310
Asian barbs nei	*Puntius spp*	22 468	25 750	25 296	44 349	45 511	41 000	42 800
Nile tilapia	*Oreochromis niloticus*	55 746	29 253	28 727	40 173	49 840	40 000	41 740
Torpedo-shaped catfishes nei	*Clarias spp*	8 080	5 800	3 410	10 934	12 111	19 600	20 470
Pangas catfishes nei	*Pangasius spp*	1 000	541	522	917	1 061	1 300	1 358
Climbing perch	*Anabas testudineus*	6 651	3 905	3 754	4 637	6 340	6 700	6 999
Snakeskin gourami	*Trichogaster pectoralis*	186	385	353	1 486	511	700	730
Striped snakehead	*Channa striata*	21 810	25 509	24 099	16 664	17 995	20 500	21 400
Freshwater fishes nei	*Osteichthyes*	60 272	105 726	108 551	70 011	59 376	64 300	67 170
Barramundi(=Giant seaperch)	*Lates calcarifer*	60	589	169	1 049	85	84	83
Tonguefishes	*Cynoglossidae*	12 801	15 011	15 030	16 097	15 333	15 123	15 030
Indian halibut	*Psettodes erumei*	4 398	7 294	7 423	2 742	2 346	2 314	2 300
Lizardfishes nei	*Synodontidae*	70 437	61 684	71 315	76 467	73 308	72 280	71 863
Sea catfishes nei	*Ariidae*	623	5 684	5 704	7 541	12 598	12 426	12 349
Eeltail catfishes	*Plotosus spp*	623	841	763	1 201	631	622	619
Mullets nei	*Mugilidae*	4 471	5 087	5 397	5 549	5 498	5 423	5 390
Groupers, seabasses nei	*Serranidae*	9 209	9 349	8 944	9 310	8 050	7 940	7 891
Bigeyes nei	*Priacanthus spp*	69 179	82 122	77 825	80 234	82 735	81 602	81 104
Sillago-whitings	*Sillaginidae*	6 566	5 433	5 738	5 501	7 986	7 877	7 829
Snappers, jobfishes nei	*Lutjanidae*	18 211	14 653	12 828	14 747	13 241	13 060	12 980
Threadfin breams nei	*Nemipterus spp*	93 785	89 592	87 767	96 595	93 037	91 763	91 204
Monocle breams	*Scolopsis spp*	77	98	344	644	54	53	53
Croakers, drums nei	*Sciaenidae*	23 945	29 761	29 961	33 646	36 591	36 090	35 870
Threadfins, tasselfishes nei	*Polynemidae*	2 204	1 550	1 079	1 305	432	426	423
Daggertooth pike conger	*Muraenesox cinereus*	4 053	1 615	1 600	1 643	2 345	2 313	2 299
Largehead hairtail	*Trichiurus lepturus*	14 497	15 073	17 586	18 709	15 970	15 751	15 655
Sardinellas nei	*Sardinella spp*	196 029	214 857	202 792	185 858	182 813	180 310	179 210
Anchovies, etc. nei	*Engraulidae*	167 987	161 970	157 341	157 214	134 740	132 896	132 084
Dorab wolf-herring	*Chirocentrus dorab*	15 987	10 292	10 459	12 372	15 049	14 843	14 752
Seerfishes nei	*Scomberomorus spp*	18 440	15 242	14 500	16 230	15 533	15 443	13 856
Frigate and bullet tunas	*Auxis thazard, A.rochei*	23 487	21 839	19 723	20 671	20 383	20 445	17 000
Kawakawa	*Euthynnus affinis*	54 864	65 462	59 257	62 868	73 340	72 332	58 691
Skipjack tuna	*Katsuwonus pelamis*	-	-	-	-	-	1 110	-
Longtail tuna	*Thunnus tonggol*	57 533	54 383	49 162	53 431	62 494	60 455	50 600
Albacore	*Thunnus alalunga*	-	-	-	-	-	12	-
Yellowfin tuna	*Thunnus albacares*	-	-	-	-	-	478	...
Bigeye tuna	*Thunnus obesus*	-	-	-	-	-	280	...
Marlins,sailfishes,etc. nei	*Istiophoridae*	-	-	-	-	-	16	...
Swordfish	*Xiphias gladius*	-	-	-	-	-	19	...
Indian scad	*Decapterus russelli*	77 622	85 205	78 156	86 163	84 574	83 416	82 907
Black pomfret	*Parastromateus niger*	5 663	6 383	6 253	2 918	5 879	5 798	5 763
Torpedo scad	*Megalaspis cordyla*	17 070	19 674	20 065	24 188	22 117	21 814	21 681
Bigeye scad	*Selar crumenophthalmus*	-	26 517	24 092	28 761	29 408	29 005	28 828
Blackbanded trevally	*Seriolina nigrofasciata*	6 930	7 269	7 096	6 042	6 045	5 962	5 926
Carangids nei	*Carangidae*	55 682	53 028	49 747	45 994	44 250	43 645	43 378
Indian mackerel	*Rastrelliger kanagurta*	70 456	45 129	42 676	43 682	47 885	47 230	46 942
Indian mackerels nei	*Rastrelliger spp*	159 225	140 826	138 621	151 010	164 110	161 863	160 876
Silver pomfret	*Pampus argenteus*	1 886	1 688	1 617	378	395	390	387
Butterfishes, pomfrets nei	*Stromateidae*	-	-	-	-	-	16	-
Barracudas nei	*Sphyraena spp*	12 061	14 389	14 245	14 025	16 860	16 629	16 527
Rays, stingrays, mantas nei	*Rajiformes*	9 968	9 978	10 353	8 289	12 279	12 111	12 037
Sharks, rays, skates, etc. nei	*Elasmobranchii*	5 313	7 775	7 616	7 737	10 118	9 980	9 919
Marine fishes nei	*Osteichthyes*	1 127 308	1 036 562	988 317	952 477	945 303	933 410	934 909
Giant river prawn	*Macrobrachium rosenbergii*	308	1 614	1 541	36	106	-	-
Blue swimming crab	*Portunus pelagicus*	41 195	41 915	40 089	46 678	41 250	40 685	40 437
Indo-Pacific swamp crab	*Scylla serrata*	5 776	4 243	4 031	3 732	5 736	5 657	5 623
Marine crabs nei	*Brachyura*	5 335	6 601	6 874	7 535	8 457	8 341	8 290
Flathead lobster	*Thenus orientalis*	335	299	177	533	37	36	36
Slipper lobsters nei	*Scyllaridae*	1 730	2 716	2 785	3 005	1 760	1 736	1 725
Banana prawn	*Penaeus merguiensis*	15 325	13 645	12 633	12 498	12 780	12 605	12 528
Giant tiger prawn	*Penaeus monodon*	1 172	2 565	2 468	1 212	2 252	2 221	2 207
Green tiger prawn	*Penaeus semisulcatus*	2 620	3 469	3 517	3 019	2 821	2 782	2 765
Western king prawn	*Penaeus latisulcatus*	2 961	3 274	3 562	3 276	4 064	4 008	3 984
Penaeus shrimps nei	*Penaeus spp*	72 258	74 145	71 525	47 393	45 112	44 494	44 223
Metapenaeus shrimps nei	*Metapenaeus spp*	11 651	12 156	11 970	9 879	9 365	9 237	9 180
Sergestid shrimps nei	*Sergestidae*	23 726	22 127	17 488	15 372	7 660	7 555	7 509
Stomatopods nei	*Stomatopoda*	-	181	176	458	866	854	849
Green mussel	*Perna viridis*	20 079	19 698	17 970	17 432	6 534	6 445	6 405
Scallops nei	*Pectinidae*	606	552	508	1 185	551	543	540
Blood cockle	*Anadara granosa*	-	-	-	441	1 757	1 733	1 722
Short neck clams nei	*Paphia spp*	30 860	52 889	35 852	49 661	69 978	69 020	68 599
Cuttlefish,bobtail squids nei	*Sepiidae, Sepiolidae*	61 919	70 525	71 588	63 340	66 247	65 340	64 941
Common squids nei	*Loligo spp*	78 109	79 235	78 948	92 908	83 135	81 997	81 497
Octopuses, etc. nei	*Octopodidae*	16 369	23 423	23 112	31 908	25 000	24 653	24 507
Marine molluscs nei	*Mollusca*	5	-	-	126	1 689	1 666	1 656
Jellyfishes	*Rhopilema spp*	33 728	30 496	42 393	64 760	84 568	83 410	82 901
Aquatic invertebrates nei	*Invertebrata*	-	-	-	-	34	-	-
	Country total	*3 031 074*	*3 013 961*	*2 902 898*	*2 930 354*	*2 952 008*	*2 911 173*	*2 881 316*

English name Nom anglais Nombre inglés	Scientific name Nom scientifique Nombre científico	1995 mt	1996 mt	1997 mt	1998 mt	1999 mt	2000 mt	2001 mt
Timor-Leste								
Freshwater fishes nei	*Osteichthyes*	...	...	...	...	0	0	0
Yellowfin tuna	*Thunnus albacares*	...	...	...	...	1	3	...
Tuna-like fishes nei	*Scombroidei*	...	...	...	...	1	3	...
Marine fishes nei	*Osteichthyes*	...	...	...	...	400 F	350	350
Marine crabs nei	*Brachyura*	...	...	...	...	1 F	1	1
Tropical spiny lobsters nei	*Panulirus spp*	...	...	...	...	2 F	2	2
Natantian decapods nei	*Natantia*	...	...	...	...	1 F	1	1
Cephalopods nei	*Cephalopoda*	...	...	...	...	1 F	1	1
Marine turtles nei	*Testudinata*	...	...	...	...	1 F	1	1
	Country total	...	...	...	...	*408 F*	*362*	*356*
Turkey								
Freshwater bream	*Abramis brama*	...	...	...	...	259	200	151
Common carp	*Cyprinus carpio*	17 081	15 631	16 000	20 000	17 797	14 137	12 265
Tench	*Tinca tinca*	...	...	...	...	...	690	778
Common dace	*Leuciscus leuciscus*	223	215	250	300	176	104	91
Cyprinids nei	*Cyprinidae*	1 535	1 380	1 900	1 800	406	699	626
Northern pike	*Esox lucius*	453	225	350	200	276	224	192
Wels(=Som)catfish	*Silurus glanis*	896	705	1 000	1 000	958	1 019	813
North African catfish	*Clarias gariepinus*	...	...	...	...	216	576	520
Pike-perch	*Stizostedion lucioperca*	5 877	8 042	1 500	3 000	1 906	1 633	1 644
Freshwater gobies nei	*Gobiidae*	262	185	200	200	118	107	116
Freshwater fishes nei	*Osteichthyes*	3 424	4 621	1 800	1 700	2 434	1 697	3 290
European eel	*Anguilla anguilla*	390	342	400	300	99	176	122
Sea trout	*Salmo trutta*	594	395	200	200	263	277	364
Shads nei	*Alosa spp*	1 590	1 166	505	880	680	720	690
Turbot	*Psetta maxima*	2 955	2 035	980	1 860	1 870	2 700	2 455
Flatfishes nei	*Pleuronectiformes*	1 092	1 947	2 300	2 000	2 400	1 000	1 250
Greater forkbeard	*Phycis blennoides*	15	24	50	150	50	50	35
Blue whiting(=Poutassou)	*Micromesistius poutassou*	9 716	11 518	15 000	27 200	16 975	18 180	20 810
Whiting	*Merlangius merlangus*	18 094	21 450	15 500	13 150	14 110	18 000	10 000
European hake	*Merluccius merluccius*	1	150	25	15	5	10	-
Lizardfishes nei	*Synodontidae*	142	161	150	165	190	250	55
Mullets nei	*Mugilidae*	31 477	37 515	42 500	45 350	46 752	43 352	38 558
Dusky grouper	*Epinephelus marginatus*	620	700	600	640	135	85	80
Groupers, seabasses nei	*Serranidae*	339	790	1 000	700	350	400	410
Seabasses nei	*Dicentrarchus spp*	2 116	2 411	3 450	3 950	3 650	1 900	1 200
Brown meagre	*Sciaena umbra*	43	50	35	45	65	20	20
Shi drum	*Umbrina cirrosa*	91	116	65	150	155	45	55
Meagre	*Argyrosomus regius*	290	71	40	30	65	70	50
Sargo breams nei	*Diplodus spp*	2 398	2 345	1 750	2 210	2 110	1 420	655
Common dentex	*Dentex dentex*	191	442	330	370	220	100	60
Black seabream	*Spondyliosoma cantharus*	72	73	35	40	50	45	45
Saddled seabream	*Oblada melanura*	196	184	190	180	115	80	90
Red porgy	*Pagrus pagrus*	2 012	750	560	675	480	540	320
Gilthead seabream	*Sparus aurata*	1 432	1 340	1 200	1 400	1 665	830	1 070
Bogue	*Boops boops*	3 196	3 736	2 450	4 100	1 620	1 500	1 000
Salema	*Sarpa salpa*	256	331	340	300	155	200	160
Porgies, seabreams nei	*Sparidae*	334	412	220	180	240	110	135
Picarels nei	*Spicara spp*	1 210	1 525	1 650	3 700	1 680	1 500	2 250
Red mullet	*Mullus surmuletus*	3 602	3 962	2 950	2 050	2 100	2 300	1 570
Striped mullet	*Mullus barbatus*	3 906	3 936	3 000	3 500	3 865	2 450	2 455
Gobies nei	*Gobiidae*	233	390	305	250	325	300	335
Conger eels, etc. nei	*Congridae*	304	416	310	300	680	200	340
John dory	*Zeus faber*	35	73	50	120	135	100	130
Scorpionfishes nei	*Scorpaenidae*	539	585	435	515	315	360	640
Gurnards, searobins nei	*Triglidae*	1 500	2 319	750	700	710	260	200
European pilchard(=Sardine)	*Sardina pilchardus*	33 812	18 972	20 500	23 600	22 000	16 500	10 000
European anchovy	*Engraulis encrasicolus*	387 574	290 680	241 000	228 000	350 000	280 000	320 000
Atlantic bonito	*Sarda sarda*	8 944	10 284	7 810	24 000	17 900	12 000	13 460
Atlantic bluefin tuna	*Thunnus thynnus*	4 220	4 616	5 093	5 899	1 200	1 070	2 100
Swordfish	*Xiphias gladius*	306	320	350	450	230	373	360
Garfish	*Belone belone*	581	395	470	450	500	300	640
Silversides(=Sand smelts) nei	*Atherinidae*	2 080	974	2 240	2 300	2 755	2 083	2 260
Bluefish	*Pomatomus saltatrix*	5 456	4 117	3 050	3 350	2 995	4 250	13 060
Atlantic horse mackerel	*Trachurus trachurus*	7 431	7 559	5 100	4 500	4 000	7 200	10 635
Mediterranean horse mackerel	*Trachurus mediterraneus*	11 260	12 500	9 500	10 500	9 220	15 000	15 545
Leerfish	*Lichia amia*	1 345	1 544	1 650	1 950	2 780	320	255
Chub mackerel	*Scomber japonicus*	17 410	10 444	10 850	10 120	10 200	9 000	4 500
Atlantic mackerel	*Scomber scombrus*	296	592	300	650	850	900	550
Barracudas nei	*Sphyraena spp*	506	250	200	120	170	150	130
Smooth-hounds nei	*Mustelus spp*	1 783	2 158	1 720	1 450	1 625	2 880	1 000
Angelsharks, sand devils nei	*Squatinidae*	31	42	15	140	70	60	20
Rays, stingrays, mantas nei	*Rajiformes*	337	524	340	385	420	1 100	555
Marine fishes nei	*Osteichthyes*	1 829	1 855	1 095	1 861	1 375	7 365	1 230
Euro-American crayfishes nei	*Astacidae, Cambaridae*	551	850	1 100	1 500	1 372	1 681	1 634
Marine crabs nei	*Brachyura*	117	305	318	243	179	187	273
Common spiny lobster	*Palinurus elephas*	4	10	45	40	6	11	18
European lobster	*Homarus gammarus*	33	34	40	60	10	15	10
Natantian decapods nei	*Natantia*	1 976	1 100	1 380	1 400	890	2 000	3 000
Freshwater molluscs nei	*Mollusca*	1 150	1 500	2 000	1 500	1 585	1 592	1 601
European flat oyster	*Ostrea edulis*	1 836	1 140	1 495	1 050	840	150	10
Mediterranean mussel	*Mytilus galloprovincialis*	6 042	3 500	6 450	3 880	1 800	1 200	1 500
Great Mediterranean scallop	*Pecten jacobaeus*	23	52	95	50	68	570	150
Striped venus	*Chamelea gallina*	11 864	10 925	7 150	3 550	3 585	10 000	7 500
Common cuttlefish	*Sepia officinalis*	933	644	900	750	537	550	465
Common squids nei	*Loligo spp*	331	364	420	500	360	400	230

English name Nom anglais Nombre inglés	Scientific name Nom scientifique Nombre científico	1995 mt	1996 mt	1997 mt	1998 mt	1999 mt	2000 mt	2001 mt
Common octopus	*Octopus vulgaris*	602	802	1 000	1 450	510	680	1 400
Marine molluscs nei	*Mollusca*	1 224	2 466	2 092	4 077	3 646	2 168	2 671
Frogs	*Rana spp*	864	740	160	100	118	77	873
Jellyfishes	*Rhopilema spp*	487	904	900	1 750	1 203	900	2 000
	Country total	*633 970*	*527 826*	*459 153*	*487 200*	*573 824*	*503 348*	*527 730*
Turkmenistan								
Freshwater bream	*Abramis brama*	137	75	100	142	147	153	126
Common carp	*Cyprinus carpio*	484	115	154	140	150	144	93
Crucian carp	*Carassius carassius*	63	85	1	225	233	228	154
Roach	*Rutilus rutilus*	88	71	69	48	39	41	1
Sichel	*Pelecus cultratus*	16	13	1	0	0	0	-
Asp	*Aspius aspius*	6	10	6	12	8	11	16
Wels(=Som)catfish	*Silurus glanis*	33	42	7	8	9	7	3
Pike-perch	*Stizostedion lucioperca*	47	30	28	109	87	96	42
Snakeheads(=Murrels) nei	*Channa spp*	0	0	1	0	0	0	1
Freshwater fishes nei	*Osteichthyes*	6	27	2	2	1	2	9
Sturgeons nei	*Acipenseridae*	...	...	...	...	11	3	3
Azov sea sprat	*Clupeonella cultriventris*	8 860	8 546	7 800	6 324	8 370	11 540	12 300
Mullets nei	*Mugilidae*	...	...	10	4	3	3	1
	Country total	*9 740*	*9 014*	*8 179*	*7 014*	*9 058*	*12 228*	*12 749*
Untd Arab Em								
Freshwater fishes nei	*Osteichthyes*	0	0	0	0	0	0	0
Milkfish	*Chanos chanos*	50	51	53	54	55	58	60 F
Sea catfishes nei	*Ariidae*	139	140	150	150	154	763	760 F
Mullets nei	*Mugilidae*	846	1 360	1 453	1 458	1 494	86	90 F
Groupers, seabasses nei	*Serranidae*	6 696	6 767	7 232	7 256	7 437	24 045	24 050 F
Snappers, jobfishes nei	*Lutjanidae*	3 438	3 474	3 713	3 725	3 818	1 718	1 720 F
Threadfin breams nei	*Nemipterus spp*	392	396	423	424	435	432	430 F
Ponyfishes(=Slipmouths) nei	*Leiognathidae*	723	731	781	784	804	780	780 F
Grunts, sweetlips nei	*Haemulidae (=Pomadasyidae)*	1 768	1 787	1 910	1 916	1 964	4 196	4 200 F
Emperors(=Scavengers) nei	*Lethrinidae*	11 336	11 455	12 242	12 283	12 590	19 647	19 650 F
King soldier bream	*Argyrops spinifer*	2 626	2 654	2 836	2 846	2 917	3 604	3 600 F
Goatfishes, red mullets nei	*Mullidae*	130	154	172	185	192	178	180 F
Mojarras(=Silver-biddies) nei	*Gerres spp*	1 222	1 235	1 320	1 324	1 351	1 651	1 650 F
Spinefeet(=Rabbitfishes) nei	*Siganus spp*	535	541	578	580	595	1 825	1 820 F
Hairtails, scabbardfishes nei	*Trichiuridae*	51	52	65	72	78	80	80 F
Indian oil sardine	*Sardinella longiceps*	3 455	3 491	4 077	4 085	4 144	...	...
Sardinellas nei	*Sardinella spp*	8 840	8 933	9 200	9 236	9 510	6 140	6 140 F
Stolephorus anchovies	*Stolephorus spp*	9 533	9 633	10 295	10 329	10 587	2 729	2 730 F
Wolf-herrings nei	*Chirocentrus spp*	71	72	77	78	80	75	70 F
Narrow-barred Spanish mackerel	*Scomberomorus commerson*	6 584	6 653	7 110	7 133	7 311	6 644	6 640 F
Seerfishes nei	*Scomberomorus spp*	1 267	1 594	2 500	2 886	3 117	2 876	2 870 F
Frigate and bullet tunas	*Auxis thazard, A.rochei*	572	578	618	620	636	376	380 F
Kawakawa	*Euthynnus affinis*	2 394	2 418	2 500	2 512	2 605	868	870 F
Longtail tuna	*Thunnus tonggol*	5 715	5 775	3 671	3 678	3 739	1 725	1 720 F
Marlins,sailfishes,etc. nei	*Istiophoridae*	232	239	230	233	233	250	250 F
Halfbeaks nei	*Hemiramphus spp*	29	36	42	58	61	55	50 F
Cobia	*Rachycentron canadum*	50	52	56	57	58	632	630 F
Scads nei	*Decapterus spp*	1 790	1 809	1 933	1 939	1 987	580	580 F
Jacks, crevalles nei	*Caranx spp*	6 833	6 905	7 379	7 403	7 588	3 107	3 100 F
Golden trevally	*Gnathanodon speciosus*	336	340	363	364	373	940	940 F
Torpedo scad	*Megalaspis cordyla*	944	954	1 019	1 022	1 048	1 100	1 100 F
Queenfishes	*Scomberoides spp*	1 981	2 002	2 140	2 147	2 201	609	610 F
Yellowstripe scad	*Selaroides leptolepis*	2 878	2 908	3 108	3 118	3 196	2 635	2 630 F
Carangids nei	*Carangidae*	5 800	5 861	6 205	6 810	7 020	3 107	3 100 F
Common dolphinfish	*Coryphaena hippurus*	55	56	60	61	63	55	50 F
Indian mackerel	*Rastrelliger kanagurta*	4 309	4 354	4 653	4 669	4 786	4 775	4 770 F
Barracudas nei	*Sphyraena spp*	2 094	2 116	2 261	2 269	2 326	543	540 F
Sharks, rays, skates, etc. nei	*Elasmobranchii*	1 553	1 902	1 832	1 881	1 945	1 530	1 530 F
Marine fishes nei	*Osteichthyes*	8 548	7 441	10 015	9 007	9 019	7 441	7 440 F
Marine crabs nei	*Brachyura*	47	56	60	60	62	1 710	1 700 F
Cuttlefish,bobtail squids nei	*Sepiidae, Sepiolidae*	22	25	26	27	28	491	490 F
	Country total	*105 884*	*107 000*	*114 358*	*114 739*	*117 607*	*110 056*	*110 000 F*
Uzbekistan								
Freshwater bream	*Abramis brama*	474	220 F	289	387	353	335	540
Common carp	*Cyprinus carpio*	864	193 F	843	804	826	617	906
Crucian carp	*Carassius carassius*	...	...	...	...	...	...	331
Goldfish	*Carassius auratus*	361	170 F	340	300	300	298	110
Roach	*Rutilus rutilus*	200	90 F	379	392	613	1 035	1 300
Grass carp(=White amur)	*Ctenopharyngodon idellus*	19	10 F	30	92	7	17	13
Silver carp	*Hypophthalmichthys molitrix*	1 003	481 F	893	249	322	586	544
Asp	*Aspius aspius*	44	20 F	38	32	37	56	22
Cyprinids nei	*Cyprinidae*	0	-	-	-	-	87	-
Northern pike	*Esox lucius*	36	20 F	7	10	60	23	24
Wels(=Som)catfish	*Silurus glanis*	20	10 F	9	14	16	8	16
Pike-perch	*Stizostedion lucioperca*	282	130 F	117	175	118	127	136
Snakeheads(=Murrels) nei	*Channa spp*	242	110 F	...	275	209	198	127
Freshwater fishes nei	*Osteichthyes*	66	40 F	130	69	10	-	1
	Country total	*3 611*	*1 494*	*3 075*	*2 799*	*2 871*	*3 387*	*4 070*
Viet Nam								
Freshwater fishes nei	*Osteichthyes*	94 189	163 936	176 589	137 800	168 107	169 000 F	169 000 F

English name Nom anglais Nombre inglés	Scientific name Nom scientifique Nombre científico	1995 mt	1996 mt	1997 mt	1998 mt	1999 mt	2000 mt	2001 mt
Tuna-like fishes nei	*Scombroidei*	...	...	3 200 F	7 400 F	7 000 F	6 500 F	15 800 F
Marine fishes nei	*Osteichthyes*	722 055	808 226	832 118	849 310	922 690	927 205	984 623
Freshwater crustaceans nei	*Crustacea*	500 F	1 000 F	1 000 F	1 000 F	1 000 F	1 000 F	1 000 F
Marine crabs nei	*Brachyura*	32 000 F	28 300 F	26 400 F	48 000 F	35 800 F	40 000 F	50 000 F
Lobsters nei	*Reptantia*	...	...	2 000 F	1 000 F	200 F	500 F	700 F
Natantian decapods nei	*Natantia*	82 758	86 166	98 401	93 541	91 500	81 700 F	90 000 F
Cephalopods nei	*Cephalopoda*	103 000 F	92 000 F	92 500 F	103 000 F	110 000 F	180 000 F	130 000 F
Marine molluscs nei	*Mollusca*	50 437 F	44 016 F	44 117 F	52 903 F	50 003 F	44 685 F	50 000 F
	Country total	*1 084 939*	*1 223 644*	*1 276 325*	*1 293 954*	*1 386 300*	*1 450 590 F*	*1 491 123 F*
Yemen								
Indian halibut	*Psettodes erumei*	974	724	760 F	900 F	800 F	750 F	950 F
Sea catfishes nei	*Ariidae*	1 697	1 700	1 780 F	2 000 F	1 900 F	1 750 F	2 200 F
Mullets nei	*Mugilidae*	391	380	400 F	500 F	400 F	400 F	500 F
Groupers, seabasses nei	*Serranidae*	2 260	1 743	1 820 F	2 100 F	2 000 F	1 800 F	2 300 F
Snappers, jobfishes nei	*Lutjanidae*	2 006	1 460	1 530 F	1 700 F	1 700 F	1 500 F	1 900 F
Threadfin and dwarf breams nei	*Nemipteridae*	3 170	2 504	2 620 F	3 000 F	2 900 F	2 600 F	3 300 F
Grunts, sweetlips nei	*Haemulidae (=Pomadasyidae)*	1 813	1 318	1 380 F	1 600 F	1 500 F	1 350 F	1 700 F
Emperors(=Scavengers) nei	*Lethrinidae*	3 214	2 437	2 550 F	2 900 F	2 800 F	2 500 F	3 200 F
Demersal percomorphs nei	*Perciformes*	2 050	5 599	9 727 F	14 078 F	6 115 F	6 086 F	7 700 F
Indian oil sardine	*Sardinella longiceps*	...	4 120	4 310 F	4 900 F	4 700 F	4 300 F	5 500 F
Narrow-barred Spanish mackerel	*Scomberomorus commerson*	3 047	3 521	3 680	3 580	3 580 F	3 580 F	3 580 F
Seerfishes nei	*Scomberomorus spp*	500	500	520	510	510 F	510 F	510 F
Frigate and bullet tunas	*Auxis thazard, A.rochei*	20	20	20	20	20 F	20 F	20 F
Kawakawa	*Euthynnus affinis*	1 226	1 183	1 240	1 210	1 210 F	1 210 F	1 210 F
Skipjack tuna	*Katsuwonus pelamis*	15	88	90	90	90 F	90 F	90 F
Longtail tuna	*Thunnus tonggol*	2 204	1 887	1 970	1 920	1 920 F	1 920 F	1 920 F
Yellowfin tuna	*Thunnus albacares*	800	800	840	820	820 F	820 F	820 F
Tuna-like fishes nei	*Scombroidei*	300	300	310	300	300 F	300 F	300 F
Jack and horse mackerels nei	*Trachurus spp*	4 412	1 380	1 440 F	1 600 F	1 600 F	1 400 F	1 800 F
Jacks, crevalles nei	*Caranx spp*	431	413	430 F	500 F	500 F	400 F	500 F
Indian mackerel	*Rastrelliger kanagurta*	6 958	752	790	900 F	900 F	900 F	1 150 F
Barracudas nei	*Sphyraena spp*	2 356	1 813	1 900 F	2 100 F	2 100 F	1 900 F	2 400 F
Pelagic percomorphs nei	*Perciformes*	60 721	62 563	60 707 F	67 630 F	74 000 F	63 900 F	80 877 F
Rays, stingrays, mantas nei	*Rajiformes*	156	...	100 F	100 F	100 F	100 F	130 F
Sharks, rays, skates, etc. nei	*Elasmobranchii*	4 480	4 878	5 000 F	5 800 F	5 600 F	5 000 F	6 300 F
Marine crabs nei	*Brachyura*	...	...	...	...	...	...	37
Tropical spiny lobsters nei	*Panulirus spp*	328	323	482	828	334	178	202
Penaeus shrimps nei	*Penaeus spp*	984	665	547	904	686	526	1 600
Natantian decapods nei	*Natantia*	...	...	...	38	7	44	151
Cuttlefish,bobtail squids nei	*Sepiidae, Sepiolidae*	1 457	1 884	8 657	5 092	5 292	8 917	9 330
Octopuses, etc. nei	*Octopodidae*	...	...	...	...	...	...	21
Sea cucumbers nei	*Holothurioidea*	-	-	-	-	1	-	-
	Country total	*107 970*	*104 955*	*115 600*	*127 620*	*124 385*	*114 751*	*142 198*
Total		***40 352 516***	***41 953 229***	***43 842 109***	***44 799 001***	***45 717 841***	***45 543 363***	***45 261 780***

E-5 Fish crustaceans, molluscs, etc — Capture production by countries or areas and species — Europe
Poissons, crustacés, mollusques, etc — Captures par pays ou zones et espèces — Europe
Peces, crustáceos, moluscos, etc — Capturas por países o áreas y especies — Europa

English name Nom anglais Nombre inglés	Scientific name Nom scientifique Nombre científico	1995 mt	1996 mt	1997 mt	1998 mt	1999 mt	2000 mt	2001 mt
Albania								
Common carp	*Cyprinus carpio*	34	45	38	230	216	230	300
Bleak	*Alburnus alburnus*	151	162	68	149	160	190	478
Crucian carp	*Carassius carassius*	...	...	...	77	62	65	326
Grass carp(=White amur)	*Ctenopharyngodon idellus*	0	1	0	3	5	45	10
Silver carp	*Hypophthalmichthys molitrix*	11	52	33	104	130	140	101
Wuchang bream	*Megalobrama amblycephala*	0	0	0	-	-	-	-
Pike-perch	*Stizostedion lucioperca*	5	10	4	-	-	-	45
European eel	*Anguilla anguilla*	39	50	21	58	63	70	98
Salmonoids nei	*Salmonoidei*	11	35	15	102	104	110	56
Shads nei	*Alosa spp*	1	2	1	-	-	-	2
European flounder	*Platichthys flesus*	0	3	9	42	41	45	15
Common sole	*Solea solea*	25	27	21	35	31	41	14
Megrim	*Lepidorhombus whiffiagonis*	1	1	0	-	-	-	1
Blue whiting(=Poutassou)	*Micromesistius poutassou*	0	2	0	-	-	-	-
European hake	*Merluccius merluccius*	227	293	185	340	341	330	380
Mullets nei	*Mugilidae*	52	105	42	136	140	150	180
Groupers nei	*Epinephelus spp*	0	2	1	-	-	-	-
European seabass	*Dicentrarchus labrax*	14	32	14	30	30	50	70
Brown meagre	*Sciaena umbra*	2	6	2	-	-	-	10
Pandoras nei	*Pagellus spp*	12	27	25	33	35	34	15
Common dentex	*Dentex dentex*	...	...	...	...	...	...	26
Gilthead seabream	*Sparus aurata*	17	27	11	20	20	23	90
Bogue	*Boops boops*	52	104	65	220	220	220	120
Porgies, seabreams nei	*Sparidae*	3	9	1	-	-	-	1
Picarels nei	*Spicara spp*	...	11	0	7	7	10	5
Surmullets(=Red mullets) nei	*Mullus spp*	76	64	61	143	145	140	170
Gobies nei	*Gobiidae*	6	10	4	-	-	-	5
European conger	*Conger conger*	0	1	2	-	-	-	-
John dory	*Zeus faber*	2	0	0	-	-	-	-
Wreckfish	*Polyprion americanus*	0	1	0	-	-	-	-
Silver scabbardfish	*Lepidopus caudatus*	7	0	0	18	19	18	0
Gurnards, searobins nei	*Triglidae*	2	15	0	-	-	-	34
Angler(=Monk)	*Lophius piscatorius*	0	46	0	42	48	44	-
European pilchard(=Sardine)	*Sardina pilchardus*	235	196	28	28	40	45	123
European anchovy	*Engraulis encrasicolus*	0	2	0	-	-	-	4
Atlantic bonito	*Sarda sarda*	1	2	0	12	30	25	30
Swordfish	*Xiphias gladius*	0	13	0	-	-	-	2
Silversides(=Sand smelts) nei	*Atherinidae*	8	20	8	11	15	20	10
Jack and horse mackerels nei	*Trachurus spp*	50	68	18	85	92	90	21
Greater amberjack	*Seriola dumerili*	0	2	1	-	-	-	2
Scomber mackerels nei	*Scomber spp*	0	10	5	4	4	4	11
Smooth-hounds nei	*Mustelus spp*	20	12	3	12	12	32	5
Dogfish sharks nei	*Squalidae*	1	64	13	-	-	-	10
Angelsharks, sand devils nei	*Squatinidae*	0	53	20	31	30	30	16
Guitarfishes, etc. nei	*Rhinobatidae*	0	1	0	-	-	-	-
Rays, stingrays, mantas nei	*Rajiformes*	67	23	24	86	78	85	14
Marine fishes nei	*Osteichthyes*	105	327	108	370	375	789	313
Freshwater crustaceans nei	*Crustacea*	0	0	-	-	-	-	-
Common spiny lobster	*Palinurus elephas*	...	...	...	...	...	...	1
Norway lobster	*Nephrops norvegicus*	0	3	0	-	-	-	10
Caramote prawn	*Penaeus kerathurus*	30	8	4	18	18	20	23
Deepwater rose shrimp	*Parapenaeus longirostris*	0	20	8	-	-	-	52
Blue and red shrimp	*Aristeus antennatus*	0	3	0	-	-	-	-
Mediterranean mussel	*Mytilus galloprovincialis*	...	...	24	-	-	-	-
Striped venus	*Chamelea gallina*	0	0	-	-	-	-	-
Common cuttlefish	*Sepia officinalis*	39	33	33	51	51	50	22
Common squids nei	*Loligo spp*	7	47	34	93	93	90	65
Common octopus	*Octopus vulgaris*	66	75	59	93	90	85	24
	Country total	*1 379*	*2 125*	*1 013*	*2 683*	*2 745*	*3 320*	*3 310*
Andorra								
Freshwater fishes nei	*Osteichthyes*	0	0	0	0	0	0	0
	Country total	*0*	*0*	*0*	*0*	*0*	*0*	*0*
Austria								
Freshwater fishes nei	*Osteichthyes*	404	450	465	451	432	439	362
	Country total	*404*	*450*	*465*	*451*	*432*	*439*	*362*
Belarus								
Freshwater bream	*Abramis brama*	145	244	182	130	98	27	198
Common carp	*Cyprinus carpio*	39	39	12	8	5	17	15
Tench	*Tinca tinca*	13	8	4	3	90	41	7
Crucian carp	*Carassius carassius*	64	69	35	106	138	154	188
Roaches nei	*Rutilus spp*	83	95	42	24	19	9	33
Orfe(=Ide)	*Leuciscus idus*	18	19	11	14	11	5	18
Cyprinids nei	*Cyprinidae*	105	92	61	47	27	19	77
Northern pike	*Esox lucius*	29	27	11	9	12	162	171
Burbot	*Lota lota*	14	17	13	18	20	7	29
European perch	*Perca fluviatilis*	93	101	75	49	38	34	86
Pike-perch	*Stizostedion lucioperca*	9	6	3	3	10	17	24
European eel	*Anguilla anguilla*	15	20	15	18	16	14	25
Whitefishes nei	*Coregonus spp*	3	4	1	2	17	18	31
Three-spined stickleback	*Gasterosteus aculeatus*	85	80	34	26	13	29	41
	Country total	*715*	*821*	*499*	*457*	*514*	*553*	*943*

E-5

Fish crustaceans, molluscs, etc	Capture production by countries or areas and species	Europe
Poissons, crustacés, mollusques, etc	Captures par pays ou zones et espèces	Europe
Peces, crustáceos, moluscos, etc	Capturas por países o áreas y especies	Europa

English name Nom anglais Nombre inglés	Scientific name Nom scientifique Nombre científico	1995 mt	1996 mt	1997 mt	1998 mt	1999 mt	2000 mt	2001 mt
Belgium								
Freshwater bream	*Abramis brama*	60	60	60	60	60	60	60
Common carp	*Cyprinus carpio*	30	30	30	30	30	30	30
Tench	*Tinca tinca*	15	15	15	15	15	15	15
Roach	*Rutilus rutilus*	160	160	160	160	150	160	160
Cyprinids nei	*Cyprinidae*	50	50	50	50	50	50	50
Northern pike	*Esox lucius*	20	20	20	20	20	20	20
European perch	*Perca fluviatilis*	25	25	25	25	15	25	25
Pike-perch	*Stizostedion lucioperca*	15	15	15	15	10	15	15
European eel	*Anguilla anguilla*	30	30	30	30	30	30	30
Sea trout	*Salmo trutta*	100	100	100	100	150	100	100
Grayling	*Thymallus thymallus*	5	5	5	5	5	5	5
Whitefishes nei	*Coregonus spp*	1	1	1	1	1	1	1
Atlantic halibut	*Hippoglossus hippoglossus*	15	5	4	3	2	2	2
European plaice	*Pleuronectes platessa*	9 290	7 684	7 469	7 369	8 186	9 148	8 487
Common dab	*Limanda limanda*	557	690	789	961	980	865	850
Lemon sole	*Microstomus kitt*	1 006	1 094	975	1 256	1 020	1 057	1 076
European flounder	*Platichthys flesus*	348	278	146	307	363	322	316
Common sole	*Solea solea*	5 457	5 150	4 514	4 102	4 492	4 479	4 975
Megrim	*Lepidorhombus whiffiagonis*	225	208	188	142	143	132	87
Brill	*Scophthalmus rhombus*	378	422	349	341	365	423	491
Turbot	*Psetta maxima*	499	382	337	327	368	464	506
Atlantic cod	*Gadus morhua*	5 938	4 491	5 677	6 893	4 540	3 693	3 207
Ling	*Molva molva*	181	153	124	124	103	120	88
Haddock	*Melanogrammus aeglefinus*	648	394	746	976	569	512	840
Saithe(=Pollock)	*Pollachius virens*	236	161	264	256	208	126	30
Pollack	*Pollachius pollachius*	158	115	119	113	108	116	137
Pouting(=Bib)	*Trisopterus luscus*	305	377	336	323	364	468	561
Whiting	*Merlangius merlangus*	1 250	1 281	989	856	1 072	826	732
European hake	*Merluccius merluccius*	76	42	54	76	92	117	124
European conger	*Conger conger*	58	75	73	79	53	49	56
Atlantic wolffish	*Anarhichas lupus*	206	99	125	208	201	290	175
Atlantic redfishes nei	*Sebastes spp*	16	19	16	2	3	5	6
Red gurnard	*Chelidonichthys cuculus*	116	145	128	157	262	418	493
Grey gurnard	*Eutrigla gurnardus*	59	119	61	47	49	44	33
Gurnards, searobins nei	*Triglidae*	255	235	245	200	160	161	177
Angler(=Monk)	*Lophius piscatorius*	1 772	1 369	1 113	961	818	1 047	1 354
Atlantic herring	*Clupea harengus*	12	2	1	1	1	1	11
European sprat	*Sprattus sprattus*	2	3	7	4	2	2	2
Atlantic horse mackerel	*Trachurus trachurus*	51	28	19	19	21	19	20
Atlantic mackerel	*Scomber scombrus*	108	64	106	125	178	151	98
Picked dogfish	*Squalus acanthias*	14	16	15	17	10	11	13
Dogfishes and hounds nei	*Squalidae, Scyliorhinidae*	377	415	430	365	345	390	396
Raja rays nei	*Raja spp*	1 275	1 363	1 259	1 232	1 351	1 231	1 527
Various sharks nei	*Selachimorpha(Pleurotremata)*	20	19	18	11	14	15	18
Groundfishes nei	*Osteichthyes*	1 027	1 290	1 389	979	774	643	501
Pelagic fishes nei	*Osteichthyes*	7	4	2	1	1	2	4
Edible crab	*Cancer pagurus*	268	175	222	180	81	106	108
Norway lobster	*Nephrops norvegicus*	413	188	316	240	349	254	284
European lobster	*Homarus gammarus*	3	-	-	-	1	1	6
Common shrimp	*Crangon crangon*	1 567	903	743	378	1 053	616	988
Whelk	*Buccinum undatum*	164	226	166	117	83	101	119
Great Atlantic scallop	*Pecten maximus*	137	163	208	224	247	292	340
Common cuttlefish	*Sepia officinalis*	467	386	153	252	222	463	370
Common squids nei	*Loligo spp*	45	21	16	25	23	59	51
Octopuses, etc. nei	*Octopodidae*	62	28	46	49	41	44	29
Marine molluscs nei	*Mollusca*	20	30	31	26	22	4	10
	Country total	*35 599*	*30 823*	*30 499*	*30 835*	*29 876*	*29 800*	*30 209*
Bosnia Herzg								
Freshwater fishes nei	*Osteichthyes*	2 500 F	2 500 F	2 500 F	2 500 F	2 500 F	2 500 F	2 500 F
Marine fishes nei	*Osteichthyes*	0	0	0	0	0	0	0
	Country total	*2 500 F*	*2 500 F*	*2 500 F*	*2 500 F*	*2 500 F*	*2 500 F*	*2 500 F*
Bulgaria								
Freshwater bream	*Abramis brama*	74	91	90	82	71	25	18
Common carp	*Cyprinus carpio*	19	16	281	251	302	143	880
Bleak	*Alburnus alburnus*	2	6	62	67	92	24	4
Barbel	*Barbus barbus*	92	120	142	113	93	43	42
Common nase	*Chondrostoma nasus*	...	...	3	3	4	8	14
Crucian carp	*Carassius carassius*	...	...	328	392	360	179	138
Roach	*Rutilus rutilus*	...	...	...	10	40	-	-
Rudd	*Scardinius erythrophthalmus*	...	...	...	71	90	39	3
Orfe(=Ide)	*Leuciscus idus*	7	8	10	12	14	-	0
Chub	*Leuciscus cephalus*	...	...	92	121	150	43	4
Grass carp(=White amur)	*Ctenopharyngodon idellus*	3	8	8	17	20	12	18
Silver carp	*Hypophthalmichthys molitrix*	415	488	471	553	488	42	403
Vimba bream	*Vimba vimba*	...	...	84	15	50	4	1
Asp	*Aspius aspius*	3	4	5	7	8	9	7
Cyprinids nei	*Cyprinidae*	...	...	1	251	250	102	2
Northern pike	*Esox lucius*	0	0	19	41	74	14	2
Wels(=Som)catfish	*Silurus glanis*	30	27	34	44	106	59	25
Burbot	*Lota lota*	-	-	-	-	-	1	2
European perch	*Perca fluviatilis*	...	...	63	81	102	26	6
Pike-perch	*Stizostedion lucioperca*	22	26	45	54	68	27	7

English name Nom anglais Nombre inglés	Scientific name Nom scientifique Nombre científico	1995 mt	1996 mt	1997 mt	1998 mt	1999 mt	2000 mt	2001 mt
Freshwater fishes nei	*Osteichthyes*	39	182	0	1	-	-	26
Danube sturgeon(=Osetr)	*Acipenser gueldenstaedtii*	5	2	6	7	6	1	1
Sterlet sturgeon	*Acipenser ruthenus*	0	1	1	1	2	2	1
Starry sturgeon	*Acipenser stellatus*	0	0	0	4	6	1	1
Beluga	*Huso huso*	25	29	42	43	37	19	7
Sea trout	*Salmo trutta*	-	-	8	11	11	4	1
Rainbow trout	*Oncorhynchus mykiss*	-	-	12	10	10	5	17
Brook trout	*Salvelinus fontinalis*	-	-	1	1	1	1	3
European whitefish	*Coregonus lavaretus*	-	-	0	0	0	-	-
Houting	*Coregonus oxyrinchus*	-	-	0	0	0	-	-
Pontic shad	*Alosa pontica*	143	233	165	171	73	39	32
Turbot	*Psetta maxima*	60	62	60	64	54	55	57
Whiting	*Merlangius merlangus*	-	-	-	-	-	9	8
Senegalese hake	*Merluccius senegalensis*	-	-	-	10	-	-	-
Flathead grey mullet	*Mugil cephalus*	23	26	28	11	14	15	47
Leaping mullet	*Liza saliens*	1	3	2	2	2	0	10
Sargo breams nei	*Diplodus spp*	-	-	-	-	-	-	90
Porgies, seabreams nei	*Sparidae*	-	-	-	35	-	-	-
Striped mullet	*Mullus barbatus*	-	-	-	-	-	5	26
Gobies nei	*Gobiidae*	580	477	424	381	437	145	142
Patagonian toothfish	*Dissostichus eleginoides*	179	-	-	-	-	-	-
Largehead hairtail	*Trichiurus lepturus*	-	-	-	1 383	-	-	-
Sardinellas nei	*Sardinella spp*	-	-	-	3 216	-	-	-
European pilchard(=Sardine)	*Sardina pilchardus*	-	-	-	48	-	-	-
European sprat	*Sprattus sprattus*	2 874	3 535	3 646	3 275	3 595	1 737	695
European anchovy	*Engraulis encrasicolus*	35	23	44	896	36	64	102
Atlantic bonito	*Sarda sarda*	25	33	16	51	20	35	49
Tuna-like fishes nei	*Scombroidei*	-	-	-	225	-	-	-
Garfish	*Belone belone*	-	2	2	4	4	9	16
Big-scale sand smelt	*Atherina boyeri*	-	-	-	0	1	21	2
Bluefish	*Pomatomus saltatrix*	12	10	12	10	8	18	2
Mediterranean horse mackerel	*Trachurus mediterraneus*	70	68	36	40	30	111	130
Jack and horse mackerels nei	*Trachurus spp*	-	-	-	1 669	-	-	-
Chub mackerel	*Scomber japonicus*	-	-	-	486	-	-	-
Picked dogfish	*Squalus acanthias*	80	64	40	28	25	102	126
Marine fishes nei	*Osteichthyes*	70	49	51	376	-	-	2
Euro-American crayfishes nei	*Astacidae, Cambaridae*	-	-	-	-	-	-	1
Natantian decapods nei	*Natantia*	1	1	3	2	2	-	-
Sea snails	*Rapana spp*	3 120	3 260	4 900	4 300	3 800	3 800	3 353
Mediterranean mussel	*Mytilus galloprovincialis*	-	-	-	-	-	-	7
Striped venus	*Chamelea gallina*	182	-	-	-	-	-	-
	Country total	*8 191*	*8 854*	*11 237*	*18 946*	*10 556*	*6 998*	*6 530*

Channel Is

English name Nom anglais Nombre inglés	Scientific name Nom scientifique Nombre científico	1995 mt	1996 mt	1997 mt	1998 mt	1999 mt	2000 mt	2001 mt
Freshwater fishes nei	*Osteichthyes*	0	0	0	0	0	0	0
European plaice	*Pleuronectes platessa*	3 F	26	19	17	23	18	13
Lemon sole	*Microstomus kitt*	0	0	0	1	1	3	1
Common sole	*Solea solea*	2 F	9	26	21	21	26	25
Sand sole	*Solea lascaris*	1 F	2	3	3	1	1	1
Megrim	*Lepidorhombus whiffiagonis*	1 F	-	-	-	-	-	1
Brill	*Scophthalmus rhombus*	17 F	10	10	10	6	17	17
Turbot	*Psetta maxima*	6 F	5	5	3	4	6	9
Atlantic cod	*Gadus morhua*	3 F	6	14	19	21	11	6
Ling	*Molva molva*	26 F	20	37	25	19	13	3
Haddock	*Melanogrammus aeglefinus*	...	44	0	0	-	-	-
Saithe(=Pollock)	*Pollachius virens*	0	2	4	0	2	-	-
Pollack	*Pollachius pollachius*	22 F	27	35	52	75	97	57
Pouting(=Bib)	*Trisopterus luscus*	0	1	1	0	1	5	6
Blue whiting(=Poutassou)	*Micromesistius poutassou*	-	-	1	1	1	-	-
Whiting	*Merlangius merlangus*	1 F	1	0	3	2	3	3
European hake	*Merluccius merluccius*	0	0	0	0	0	-	-
Gadiformes nei	*Gadiformes*	-	-	-	-	1	-	-
Mullets nei	*Mugilidae*	8 F	23	14	11	13	12	9
European seabass	*Dicentrarchus labrax*	77 F	56	74	79	107	129	80
Gilthead seabream	*Sparus aurata*	...	...	...	...	15	-	-
Porgies, seabreams nei	*Sparidae*	13 F	9	48	126	132	105	108
Red mullet	*Mullus surmuletus*	5 F	4	11	17	22	23	19
Ballan wrasse	*Labrus bergylta*	-	-	-	-	2	1	-
Sandeels(=Sandlances) nei	*Ammodytes spp*	-	2	63	61	41	41	41
European conger	*Conger conger*	75 F	57	17	32	30	24	25
John dory	*Zeus faber*	1 F	0	0	1	1	2	1
Wreckfish	*Polyprion americanus*	-	-	2	1	0	-	-
Red gurnard	*Chelidonichthys cuculus*	3 F	10	8	6	10	7	-
Gurnards, searobins nei	*Triglidae*	...	...	10	31	22	10	41
Angler(=Monk)	*Lophius piscatorius*	0	2	3	3	2	3	5
Garfish	*Belone belone*	0	0	1	2	0	2	2
Atlantic horse mackerel	*Trachurus trachurus*	0	5	7	11	10	8	8
Atlantic mackerel	*Scomber scombrus*	1 F	9	9	23	18	16	14
Porbeagle	*Lamna nasus*	...	...	...	1	0	2	2
Blue shark	*Prionace glauca*	...	...	...	1	0	-	-
Dogfishes and hounds nei	*Squalidae, Scyliorhinidae*	3 F	48	30	22	15	33	51
Raja rays nei	*Raja spp*	144 F	180	33	223	262	181	241
Various sharks nei	*Selachimorpha(Pleurotremata)*	30 F	2	3	3	7	1	-
Marine fishes nei	*Osteichthyes*	-	22	203	27	40	-	-
Edible crab	*Cancer pagurus*	1 349 F	2 043	2 268	2 214	1 655	1 440	1 560
Spinous spider crab	*Maja squinado*	530 F	868	445	460	553	522	428
Marine crabs nei	*Brachyura*	-	-	-	2	2	2	-
Palinurid spiny lobsters nei	*Palinurus spp*	0	1	0	1	1	1	3

English name Nom anglais Nombre inglés	Scientific name Nom scientifique Nombre científico	1995 mt	1996 mt	1997 mt	1998 mt	1999 mt	2000 mt	2001 mt
European lobster	*Homarus gammarus*	114 F	260	221	214	212	153	178
Tuberculate abalone	*Haliotis tuberculata*	-	-	-	-	-	3	2
Whelk	*Buccinum undatum*	375 F	550	438	135	8	338	519
Queen scallop	*Aequipecten opercularis*	-	-	-	-	7	-	-
Great Atlantic scallop	*Pecten maximus*	65 F	29	109	155	203	295	439
Scallops nei	*Pectinidae*	-	-	50	80	-	-	-
Clams, etc. nei	*Bivalvia*	70 F	-	-	-	-	-	-
Common cuttlefish	*Sepia officinalis*	2 F	12	10	15	22	26	8
Common squids nei	*Loligo spp*	2 F	1	6	5	11	9	1
	Country total	*2 949 F*	*4 346*	*4 238*	*4 117*	*3 601*	*3 589*	*3 927*
Croatia								
Common carp	*Cyprinus carpio*	96	143	126	1 F	1	3	1
Tench	*Tinca tinca*	2	3	0	-	-	-	-
Bighead carp	*Hypophthalmichthys nobilis*	1	-	-	-	-	-	-
Northern pike	*Esox lucius*	30	30	30	1 F	1	1	1
Wels(=Som)catfish	*Silurus glanis*	19	24	20	1 F	1	1	1
Pike-perch	*Stizostedion lucioperca*	12	12	14	1 F	1	1	1
Freshwater fishes nei	*Osteichthyes*	202	217	213	5 F	5	8	12
European eel	*Anguilla anguilla*	7	6	7	-	-	-	-
Rainbow trout	*Oncorhynchus mykiss*	2	5	5	1 F	1	3	18
Flatfishes nei	*Pleuronectiformes*	124	130	133	150	65	113	111
European hake	*Merluccius merluccius*	1 129	929	828	935	650	583	557
Flathead grey mullet	*Mugil cephalus*	123	105	105	195	120	68	14
European seabass	*Dicentrarchus labrax*	...	...	...	31	20	22	13
Common dentex	*Dentex dentex*	53	56	76	70	50	55	10
Black seabream	*Spondyliosoma cantharus*	52	51	56	60	23	19	5
Saddled seabream	*Oblada melanura*	180	183	169	185	130	120	55
Gilthead seabream	*Sparus aurata*	17	13	44	84	27	25	11
Bogue	*Boops boops*	329	303	342	335	140	147	109
Salema	*Sarpa salpa*	149	127	132	125	42	24	28
Picarels nei	*Spicara spp*	312	290	340	255	245	231	189
Red mullet	*Mullus surmuletus*	420	311	280	285	200	277	480
European conger	*Conger conger*	145	122	114	130	49	38	25
Scorpionfishes nei	*Scorpaenidae*	172	137	133	185	46	40	39
Sardinellas nei	*Sardinella spp*	243	163	344	170	25	35	37
European pilchard(=Sardine)	*Sardina pilchardus*	6 377	9 199	6 996	12 500	10 500	11 226	9 097
European anchovy	*Engraulis encrasicolus*	359	220	545	990	3 000	3 735	2 850
Atlantic bonito	*Sarda sarda*	182	159	171	158	120	120	54
Frigate and bullet tunas	*Auxis thazard, A.rochei*	28	26	16	12	0	-	-
Little tunny(=Atl.black skipj)	*Euthynnus alletteratus*	7	9	9	16	0	-	-
Atlantic bluefin tuna	*Thunnus thynnus*	1 220	1 360	1 105	906	970	930	903
Swordfish	*Xiphias gladius*	...	...	...	10	20	...	...
Garfish	*Belone belone*	81	63	85	60	10	5	11
Silversides(=Sand smelts) nei	*Atherinidae*	175	148	171	260	50	44	20
Mediterranean horse mackerel	*Trachurus mediterraneus*	453	361	336	200	90	75	192
Greater amberjack	*Seriola dumerili*	78	80	64	62	30	25	19
Scomber mackerels nei	*Scomber spp*	615	650	998	585	175	33	91
Dogfish sharks nei	*Squalidae*	315	260	239	105	53	50	74
Rays, stingrays, mantas nei	*Rajiformes*	190	141	119	120	68	57	42
Marine fishes nei	*Osteichthyes*	738	619	613	590	619	1 080	863
Spinous spider crab	*Maja squinado*	43	25	54	50	15	14	21
Norway lobster	*Nephrops norvegicus*	524	486	473	500	240	250	274
European lobster	*Homarus gammarus*	30	31	43	40	18	18	13
European flat oyster	*Ostrea edulis*	...	...	...	35	14	12	24
Mediterranean mussel	*Mytilus galloprovincialis*	...	...	...	200	60	57	32
Cuttlefish,bobtail squids nei	*Sepiidae, Sepiolidae*	118	102	151	215	145	138	85
Common squids nei	*Loligo spp*	273	233	287	290	170	127	162
Common octopus	*Octopus vulgaris*	193	180	293	435	...	...	...
Cephalopods nei	*Cephalopoda*	431	483	443	265	646	873	885
Marine molluscs nei	*Mollusca*	16	38	313	125	40	27	69
Aquatic invertebrates nei	*Invertebrata*	-	-	-	4	5	8	592
	Country total	*16 265*	*18 233*	*17 035*	*21 938*	*18 900*	*20 718*	*. 18 090*
Czech Rep								
Freshwater bream	*Abramis brama*	286	247	232	253	297	261	247
Common carp	*Cyprinus carpio*	2 919	2 522	2 312	2 899	3 006	3 558	3 560
Tench	*Tinca tinca*	30	23	21	29	30	27	24
Goldfish	*Carassius auratus*	39	49	40	40	43	35	37
Orfe(=Ide)	*Leuciscus idus*	31	27	0	0	0	0	0
Grass carp(=White amur)	*Ctenopharyngodon idellus*	51	47	49	53	70	60	60
Bighead carp	*Hypophthalmichthys nobilis*	13	3	5	6	8	10	12
Asp	*Aspius aspius*	-	10	15	16	16	13	17
Northern pike	*Esox lucius*	82	163	159	168	183	180	176
Wels(=Som)catfish	*Silurus glanis*	42	36	44	49	52	53	57
European perch	*Perca fluviatilis*	31	33	34	36	37	34	34
Pike-perch	*Stizostedion lucioperca*	86	130	157	125	130	134	139
Freshwater fishes nei	*Osteichthyes*	192	102	121	134	168	151	131
European eel	*Anguilla anguilla*	31	28	27	28	28	24	29
Trouts nei	*Salmo spp*	54	53	57	70	64	55	56
Rainbow trout	*Oncorhynchus mykiss*	23	28	29	30	38	39	48
Brook trout	*Salvelinus fontinalis*	3	2	2	2	3	3	3
Grayling	*Thymallus thymallus*	16	19	15	13	16	16	15
European whitefish	*Coregonus lavaretus*	-	2	2	1	1	1	1
	Country total	*3 929*	*3 524*	*3 321*	*3 952*	*4 190*	*4 654*	*4 646*

English name Nom anglais Nombre inglés	Scientific name Nom scientifique Nombre científico	1995 mt	1996 mt	1997 mt	1998 mt	1999 mt	2000 mt	2001 mt
Denmark								
Freshwater bream	*Abramis brama*	87	66	94	144	80	48	4
Common carp	*Cyprinus carpio*	2	1	1	1	0	0	0
Tench	*Tinca tinca*	0	0	0	0	0	1	0
Crucian carp	*Carassius carassius*	0	0	0	-	0	0	0
Roach	*Rutilus rutilus*	70	43	38	86	39	43	8
Rudd	*Scardinius erythrophthalmus*	-	-	-	-	0	2	-
Northern pike	*Esox lucius*	11	11	14	16	11	9	11
Burbot	*Lota lota*	0	0	0	0	0	0	-
European perch	*Perca fluviatilis*	90	63	83	100	113	100	74
Pike-perch	*Stizostedion lucioperca*	14	12	51	53	15	28	17
Ruffe	*Gymnocephalus cernuus*	15	13	5	9	3	2	-
Freshwater fishes nei	*Osteichthyes*	0	3	-	-	0	0	-
Sturgeons nei	*Acipenseridae*	-	-	-	0	-	-	-
European eel	*Anguilla anguilla*	904	734	796	600	716	620	671
Atlantic salmon	*Salmo salar*	560	528	493	487	389	413	434
Sea trout	*Salmo trutta*	86	77	52	59	101	70	52
Rainbow trout	*Oncorhynchus mykiss*	1	1	1	-	6	-	0
European smelt	*Osmerus eperlanus*	76	53	34	19	20	29	25
European whitefish	*Coregonus lavaretus*	38	30	31	29	25	7	77
Houting	*Coregonus oxyrinchus*	-	-	-	1	-	22	0
Three-spined stickleback	*Gasterosteus aculeatus*	-	-	4	-	-	-	-
Atlantic halibut	*Hippoglossus hippoglossus*	61	71	70	62	51	53	53
European plaice	*Pleuronectes platessa*	24 092	22 571	24 621	18 927	23 123	23 902	26 849
Greenland halibut	*Reinhardtius hippoglossoides*	1	3	2	1	1	1	1
Witch flounder	*Glyptocephalus cynoglossus*	897	999	1 563	1 906	2 116	2 338	2 049
Amer. plaice(=Long rough dab)	*Hippoglossoides platessoides*	1	0	5	8	31	5	31
Common dab	*Limanda limanda*	4 484	3 952	3 211	2 646	2 514	2 113	2 300
Lemon sole	*Microstomus kitt*	1 208	1 108	1 172	1 591	1 812	2 037	1 820
European flounder	*Platichthys flesus*	4 235	5 469	5 378	4 725	3 528	4 604	6 066
Common sole	*Solea solea*	3 039	2 083	1 478	1 050	1 433	1 804	1 324
Megrim	*Lepidorhombus whiffiagonis*	2	7	6	26	21	30	55
Brill	*Scophthalmus rhombus*	281	220	148	189	244	237	163
Turbot	*Psetta maxima*	1 396	1 117	908	770	727	809	872
Tusk(=Cusk)	*Brosme brosme*	89	130	146	105	177	225	274
Atlantic cod	*Gadus morhua*	78 332	90 741	80 491	69 025	70 547	57 018	46 185
Ling	*Molva molva*	790	868	969	823	837	741	910
Blue ling	*Molva dypterygia*	16	8	14	4	7	15	27
Haddock	*Melanogrammus aeglefinus*	4 479	5 050	5 227	5 786	3 130	2 707	4 001
Saithe(=Pollock)	*Pollachius virens*	4 395	4 708	4 517	3 973	4 501	3 536	3 592
Pollack	*Pollachius pollachius*	1 036	1 049	637	564	480	490	358
Norway pout	*Trisopterus esmarkii*	262 515	162 943	153 047	63 678	57 441	150 040	62 913
Pouting(=Bib)	*Trisopterus luscus*	1	1	2	2	1	1	3
Blue whiting(=Poutassou)	*Micromesistius poutassou*	46 182	52 699	33 486	69 305	79 810	62 074	65 058
Whiting	*Merlangius merlangus*	789	391	196	144	175	326	326
European hake	*Merluccius merluccius*	1 487	868	670	591	846	811	1 043
Roundnose grenadier	*Coryphaenoides rupestris*	2 227	1 174	2 124	4 429	2 521	1 981	2 229
Mullets nei	*Mugilidae*	31	22	29	24	26	22	29
European seabass	*Dicentrarchus labrax*	1	1	1	2	1	5	2
Red mullet	*Mullus surmuletus*	0	1	1	1	3	2	5
Eelpout	*Zoarces viviparus*	2	3	7	2	7	3	5
Sandeels(=Sandlances) nei	*Ammodytes spp*	844 512	669 035	840 774	646 905	528 551	567 350	666 295
Greater weever	*Trachinus draco*	76	45	50	36	52	39	48
Argentines	*Argentina spp*	1 069	1 446	1 455	748	1 420	1 039	907
Atlantic wolffish	*Anarhichas lupus*	248	195	220	273	298	294	223
Golden redfish	*Sebastes marinus*	1	4	4	7	5	8	11
Beaked redfish	*Sebastes mentella*	14	17	19	20	48	35	89
Tub gurnard	*Chelidonichthys lucerna*	-	5	4	5	19	15	12
Grey gurnard	*Eutrigla gurnardus*	84	70	36	56	85	96	289
Lumpfish(=Lumpsucker)	*Cyclopterus lumpus*	1 069	1 973	1 553	188	835	425	422
Angler(=Monk)	*Lophius piscatorius*	1 350	1 835	2 040	1 873	1 858	1 724	1 917
Atlantic herring	*Clupea harengus*	191 415	153 009	125 300	139 710	137 578	153 899	141 263
European pilchard(=Sardine)	*Sardina pilchardus*	36 196	13 704	1 740	17 337	17 676	7 893	1 399
European sprat	*Sprattus sprattus*	258 179	226 135	284 489	270 439	282 299	276 878	256 517
European anchovy	*Engraulis encrasicolus*	759	-	-	-	-	-	0
Atlantic bluefin tuna	*Thunnus thynnus*	0	0	-	1	-	-	-
Capelin	*Mallotus villosus*	-	60 898	48 524	40 349	3 837	20 807	17 588
Garfish	*Belone belone*	622	708	994	648	571	714	558
Atlantic horse mackerel	*Trachurus trachurus*	56 167	63 929	63 430	32 597	32 047	25 083	23 631
Atlantic pomfret	*Brama brama*	-	-	-	-	-	-	0
Atlantic mackerel	*Scomber scombrus*	36 758	26 238	24 054	27 415	29 705	31 642	31 370
Porbeagle	*Lamna nasus*	86	71	69	85	107	73	76
Blue shark	*Prionace glauca*	-	3	1	1	1	2	1
Tope shark	*Galeorhinus galeus*	-	2	2	3	4	7	4
Picked dogfish	*Squalus acanthias*	146	142	196	126	131	146	156
Blue skate	*Raja batis*	-	32	9	7	11	47	0
Raja rays nei	*Raja spp*	57	44	40	20	46	87	122
Rabbit fish	*Chimaera monstrosa*	-	-	-	-	-	-	1
Various sharks nei	*Selachimorpha(Pleurotremata)*	4	-	-	-	-	-	-
Groundfishes nei	*Osteichthyes*	0	0	0	0	-	-	-
Finfishes nei	*Osteichthyes*	545	118	103	116	78	142	107
Euro-American crayfishes nei	*Astacidae, Cambaridae*	0	0	0	0	0	0	0
Edible crab	*Cancer pagurus*	17	15	14	21	25	33	53
Spinous spider crab	*Maja squinado*	1	0	1	1	1	1	2
Marine crabs nei	*Brachyura*	188	171	193	220	193	205	189
Norway lobster	*Nephrops norvegicus*	3 608	4 176	4 282	4 982	5 455	5 083	4 813
European lobster	*Homarus gammarus*	27	44	39	18	11	11	11
Northern prawn	*Pandalus borealis*	8 460	9 071	8 552	8 100	4 880	5 721	5 331
Common prawn	*Palaemon serratus*	195	144	182	122	141	151	141

E-5 Fish crustaceans, molluscs, etc — Capture production by countries or areas and species — Europe
Poissons, crustacés, mollusques, etc — Captures par pays ou zones et espèces — Europe
Peces, crustáceos, moluscos, etc — Capturas por países o áreas y especies — Europa

English name / Nom anglais / Nombre inglés	Scientific name / Nom scientifique / Nombre científico	1995 mt	1996 mt	1997 mt	1998 mt	1999 mt	2000 mt	2001 mt
Common shrimp	*Crangon crangon*	2 065	2 207	3 250	2 509	2 911	2 322	1 826
Norwegian krill	*Meganyctiphanes norvegica*	-	-	-	88	-	-	36
Marine crustaceans nei	*Crustacea*	563	117	0	-	-	-	-
Periwinkles nei	*Littorina spp*	0	4	1	2	-	2	0
Whelk	*Buccinum undatum*	-	-	-	-	1	-	1
European flat oyster	*Ostrea edulis*	12	9	24	6	8	9	23
Blue mussel	*Mytilus edulis*	107 377	86 002	90 765	108 329	96 215	110 618	122 480
Scallops nei	*Pectinidae*	0	0	0	0	0	0	0
Solid surf clam	*Spisula solida*	-	-	-	-	-	55	214
Common edible cockle	*Cerastoderma edule*	-	5	2 603	1 993	246	2 089	2 392
Clams, etc. nei	*Bivalvia*	3 136	-	-	-	-	-	-
Cuttlefish,bobtail squids nei	*Sepiidae, Sepiolidae*	3	2	0	-	26	20	5
Northern shortfin squid	*Illex illecebrosus*	-	-	17	16	-	-	-
Starfishes nei	*Asteroidea*	0	-	-	-	0	0	0
European edible sea urchin	*Echinus esculentus*	0	0	0	0	1	0	0
Aquatic invertebrates nei	*Invertebrata*	1	0	-	-	-	-	-
	Country total	*1 999 033*	*1 681 517*	*1 826 852*	*1 557 335*	*1 405 005*	*1 534 089*	*1 510 439*
Estonia								
Freshwater bream	*Abramis brama*	164	175	230	247	194	204	328
Roach	*Rutilus rutilus*	332	502	492	449	324	478	503
Orfe(=Ide)	*Leuciscus idus*	99	132	90	70	52	65	40
Vimba bream	*Vimba vimba*	188	165	185	165	122	101	83
Northern pike	*Esox lucius*	72	139	105	131	152	177	199
Burbot	*Lota lota*	31	36	31	21	53	40	38
European perch	*Perca fluviatilis*	1 007	1 030	1 202	1 052	976	841	686
Pike-perch	*Stizostedion lucioperca*	657	726	460	891	771	677	517
Freshwater fishes nei	*Osteichthyes*	218	123	160	134	168	170	209
European eel	*Anguilla anguilla*	38	55	56	44	60	67	67
Atlantic salmon	*Salmo salar*	9	9	10	7	14	21	14
Sea trout	*Salmo trutta*	6	15	11	8	10	13	13
European smelt	*Osmerus eperlanus*	731	484	415	1 431	1 008	1 194	762
Vendace	*Coregonus albula*	45	127	153	159	47	1	-
European whitefish	*Coregonus lavaretus*	6	21	20	20	28	33	33
Houting	*Coregonus oxyrinchus*	25	63	30	60	35	9	9
Sea lamprey	*Petromyzon marinus*	0	18	3	4	7	8	3
River lamprey	*Lampetra fluviatilis*	1	0	7	16	9	26	25
Atlantic halibut	*Hippoglossus hippoglossus*	-	-	-	-	-	-	0
Greenland halibut	*Reinhardtius hippoglossoides*	-	-	-	-	-	181	1 100
Witch flounder	*Glyptocephalus cynoglossus*	-	-	-	-	-	5	3
Amer. plaice(=Long rough dab)	*Hippoglossoides platessoides*	-	-	-	-	-	94	54
Yellowtail flounder	*Limanda ferruginea*	-	-	-	-	-	53	47
European flounder	*Platichthys flesus*	102	297	333	355	416	419	482
Winter flounder	*Pseudopleuronectes americanus*	-	-	-	-	-	25	-
Flatfishes nei	*Pleuronectiformes*	-	-	-	-	-	1	-
Atlantic cod	*Gadus morhua*	1 049	1 392	1 174	1 070	1 059	520	799
Blue ling	*Molva dypterygia*	-	-	-	-	-	-	85
White hake	*Urophycis tenuis*	-	-	-	-	-	3	2
Saithe(=Pollock)	*Pollachius virens*	-	-	16	-	-	-	-
Blue whiting(=Poutassou)	*Micromesistius poutassou*	13 715	10 982	5 678	6 321	0	-	-
Patagonian grenadier	*Macruronus magellanicus*	-	113	-	-	-	-	108
Roughhead grenadier	*Macrourus berglax*	-	-	-	-	-	1	-
Roundnose grenadier	*Coryphaenoides rupestris*	-	-	-	-	-	20	701
Large-eye dentex	*Dentex macrophthalmus*	181	-	-	-	-	-	-
Patagonian blennie	*Eleginops maclovinus*	-	-	-	-	14	-	-
Eelpout	*Zoarces viviparus*	2	2	8	9	2	1	1
Baird's slickhead	*Alepocephalus bairdii*	-	-	-	-	-	-	153
Wolffishes(=Catfishes) nei	*Anarhichas spp*	-	-	-	-	-	6	5
Black scabbardfish	*Aphanopus carbo*	-	-	-	-	-	-	224
Pacific ocean perch	*Sebastes alutus*	-	-	-	-	8	-	-
Atlantic redfishes nei	*Sebastes spp*	17 717	7 091	3 720	3 968	2 108	8 652	785
Atlantic herring	*Clupea harengus*	43 866	45 296	52 435	42 721	44 038	41 735	41 738
Sardinellas nei	*Sardinella spp*	-	252	415	3 171	2 474	-	-
European pilchard(=Sardine)	*Sardina pilchardus*	163	274	480	-	-	-	-
European sprat	*Sprattus sprattus*	13 051	22 493	39 693	32 165	36 407	41 394	40 777
Garfish	*Belone belone*	193	405	400	167	122	135	111
Atlantic horse mackerel	*Trachurus trachurus*	-	80	203	34	0	-	-
Cape horse mackerel	*Trachurus capensis*	28 655	-	-	-	-	-	-
Jack and horse mackerels nei	*Trachurus spp*	1 989	1 903	2 080	1 550	-	-	-
Jacks, crevalles nei	*Caranx spp*	-	-	-	-	1 622	-	-
Chub mackerel	*Scomber japonicus*	2 991	4 486	1 980	7 215	3 416	-	-
Atlantic mackerel	*Scomber scombrus*	2 286	3 741	6 324	7 356	3 595	2 673	218
Shortfin mako	*Isurus oxyrinchus*	-	-	-	-	2	-	-
Raja rays nei	*Raja spp*	-	-	-	-	-	240	1 079
Finfishes nei	*Osteichthyes*	29	26	23	28	32	41	91
Marine fishes nei	*Osteichthyes*	31	148	-	469	-	0	-
Euro-American crayfishes nei	*Astacidae, Cambaridae*	...	0	...	...	...	1	1
Deepwater rose shrimp	*Parapenaeus longirostris*	-	-	-	-	-	3	6
Northern prawn	*Pandalus borealis*	2 379	3 220	4 991	7 206	12 448	12 816	11 235
Argentine shortfin squid	*Illex argentinus*	-	-	-	-	-	-	1 833
Various squids nei	*Loliginidae, Ommastrephidae*	-	2 425	-	-	-	-	-
Marine molluscs nei	*Mollusca*	-	-	-	-	-	2	-
	Country total	*132 028*	*108 446*	*123 613*	*118 714*	*111 793*	*113 146*	*105 167*
Faeroe Is								
Freshwater fishes nei	*Osteichthyes*	0	0	0	0	0	0	0
Atlantic salmon	*Salmo salar*	3	-	-	5	-	0	0

English name Nom anglais Nombre inglés	Scientific name Nom scientifique Nombre científico	1995 mt	1996 mt	1997 mt	1998 mt	1999 mt	2000 mt	2001 mt
Atlantic halibut	*Hippoglossus hippoglossus*	654	459	479	391	432	318	205
European plaice	*Pleuronectes platessa*	425	443	506	443	325	259	250
Greenland halibut	*Reinhardtius hippoglossoides*	4 494	6 241	5 124	3 787	4 111	5 584	4 128 F
Witch flounder	*Glyptocephalus cynoglossus*	11	13	2	3	7	...	...
Amer. plaice(=Long rough dab)	*Hippoglossoides platessoides*	-	8	-	-	3	2	...
Common dab	*Limanda limanda*	27	37	39	39	29	...	...
Lemon sole	*Microstomus kitt*	265	236	332	464	433	389	694
Turbot	*Psetta maxima*	-	2	-	-	-	-	-
Flatfishes nei	*Pleuronectiformes*	341	2	757	1 059	292	...	...
Tusk(=Cusk)	*Brosme brosme*	4 490	2 562	2 593	2 389	2 714	2 585	2 922
Atlantic cod	*Gadus morhua*	45 348	60 504	57 921	39 689	33 725	32 601	38 706
Ling	*Molva molva*	3 686	3 132	4 056	3 547	2 998	2 358	2 558
Blue ling	*Molva dypterygia*	2 398	1 624	1 172	1 274	2 136	1 756	2 454
Haddock	*Melanogrammus aeglefinus*	8 494	13 896	21 409	22 598	19 697	16 212	16 061
Saithe(=Pollock)	*Pollachius virens*	31 979	20 398	22 598	26 751	34 423	35 997	45 792
Pollack	*Pollachius pollachius*	2	-	1	-	-	-	-
Norway pout	*Trisopterus esmarkii*	8 960	9 133	11 215	6 222	4 045	1 754	2 429
Blue whiting(=Poutassou)	*Micromesistius poutassou*	25 936	21 483	28 773	71 217	105 106	152 687	259 761
Whiting	*Merlangius merlangus*	966	1 042	1 018	1 724	1 756	1 593	1 289
European hake	*Merluccius merluccius*	14	1	6	5	5	...	...
Roundnose grenadier	*Coryphaenoides rupestris*	766	234	199	84	108	53	91
Gadiformes nei	*Gadiformes*	229	449	899	57	348	...	...
Sandeels(=Sandlances) nei	*Ammodytes spp*	7 485	5 023	11 221	11 071	7 487	8 513	6 030
...A	*Normanichthys crockeri*	2	3	-	-	-	-	-
Argentines	*Argentina spp*	7 131	9 496	8 433	17 167	8 186	6 388	9 952
Alfonsinos nei	*Beryx spp*	3	-	5	-	-	-	-
Orange roughy	*Hoplostethus atlanticus*	732	950	854	747	349	155	1
Black cardinal fish	*Epigonus telescopus*	38	31	129	94	4	...	...
Atlantic wolffish	*Anarhichas lupus*	141	146	196	264	291	...	...
Wolffishes(=Catfishes) nei	*Anarhichas spp*	38	31	64	60	118	154	155
Black scabbardfish	*Aphanopus carbo*	550	256	126	89	45	116	412
Golden redfish	*Sebastes marinus*	12 071	11 028	11 402	7 951	7 129	13 324	13 572
Atlantic redfishes nei	*Sebastes spp*	15	1	-	-	-	-	-
Gurnards, searobins nei	*Triglidae*	-	-	-	-	1	...	...
Angler(=Monk)	*Lophius piscatorius*	1 433	1 610	1 765	1 885	2 578	2 216	2 006
Atlantic herring	*Clupea harengus*	66 023	57 853	65 949	70 214	56 476	65 270	35 172
European sprat	*Sprattus sprattus*	598	100	-	753	1 719	...	...
Atlantic bluefin tuna	*Thunnus thynnus*	-	-	-	67	104	118	...
Yellowfin tuna	*Thunnus albacares*	-	-	-	-	-	1	...
Bigeye tuna	*Thunnus obesus*	-	-	-	-	11	8	...
Swordfish	*Xiphias gladius*	-	-	-	-	5	4	...
Capelin	*Mallotus villosus*	3 306	39 777	43 466	41 966	24 275	59 855	32 110
Opah	*Lampris guttatus*	-	-	-	-	1	...	...
Atlantic horse mackerel	*Trachurus trachurus*	950	863	1 005	216	3 643	2 014	180
Atlantic mackerel	*Scomber scombrus*	34 924	19 530	8 401	10 654	11 334	21 022	24 005
Porbeagle	*Lamna nasus*	44	7	9	7	10	13	8
Picked dogfish	*Squalus acanthias*	308	51	212	356	484	354	613
Leafscale gulper shark	*Centrophorus squamosus*	51	53	58	129	11	...	...
Portuguese dogfish	*Centroscymnus coelolepis*	60	282	228	80	35	...	...
Dogfish sharks nei	*Squalidae*	12	7	8	8	3	...	...
Raja rays nei	*Raja spp*	230	169	187	151	183	125	90
Groundfishes nei	*Osteichthyes*	-	-	474	133	-	-	-
Marine fishes nei	*Osteichthyes*	...	1 201	1 362	177	18	4 053	2 911
Marine crabs nei	*Brachyura*	1	3	1	4	1	6	1
Norway lobster	*Nephrops norvegicus*	91	66	40	57	80	73	51
Northern prawn	*Pandalus borealis*	9 290	10 592	10 868	12 985	14 843	12 611	16 175 F
Queen scallop	*Aequipecten opercularis*	2 781	3 559	3 581	4 751	5 993	3 989	4 053
Northern shortfin squid	*Illex illecebrosus*	-	-	2	32	23	...	...
	Country total	*287 796*	*304 587*	*329 145*	*363 816*	*358 133*	*454 530*	*524 837*
Finland								
Freshwater bream	*Abramis brama*	4 225	3 143	3 275	2 854	2 845	2 546	2 647
Roach	*Rutilus rutilus*	1 439	2 172	2 271	1 465	1 465	1 412	1 459
Roaches nei	*Rutilus spp*	6 915	6 775	6 775	5 362	5 362	3 999	3 999
Orfe(=Ide)	*Leuciscus idus*	751	492	491	400	396	468	471
Cyprinids nei	*Cyprinidae*	277	23	23	-	-	-	-
Northern pike	*Esox lucius*	11 831	12 345	12 377	11 649	11 663	10 449	10 428
Burbot	*Lota lota*	1 849	1 647	1 663	1 147	1 154	1 176	1 168
European perch	*Perca fluviatilis*	16 876	17 647	17 860	14 721	14 694	13 374	13 395
Pike-perch	*Stizostedion lucioperca*	2 346	2 476	2 630	3 241	3 188	1 824	1 786
Freshwater fishes nei	*Osteichthyes*	1 003	957	928	673	736	572	602
European eel	*Anguilla anguilla*	0	22	22	...	...	...	...
Atlantic salmon	*Salmo salar*	1 641	1 714	1 790	1 020	912	964	817
Sea trout	*Salmo trutta*	615	672	661	460	441	437	416
Trouts nei	*Salmo spp*	1 022	1 286	1 286	879	879	610	610
Rainbow trout	*Oncorhynchus mykiss*	1 034	917	918	1 185	1 146	737	788
European smelt	*Osmerus eperlanus*	2 237	2 223	1 898	1 209	1 330	779	899
Vendace	*Coregonus albula*	4 309	5 988	5 975	5 115	5 125	5 022	5 024
European whitefish	*Coregonus lavaretus*	6 154	5 126	5 003	4 882	4 703	4 541	4 247
Salmonoids nei	*Salmonoidei*	332	472	472	329	329	288	282
European flounder	*Platichthys flesus*	575	715	702	555	558	449	500
Turbot	*Psetta maxima*	...	...	...	...	...	6	4
Atlantic cod	*Gadus morhua*	1 861	3 139	1 543	1 037	1 572	1 824	1 723
Atlantic herring	*Clupea harengus*	95 897	94 548	91 544	86 350	83 042	81 648	82 867
European sprat	*Sprattus sprattus*	4 104	14 351	19 851	27 014	18 886	23 242	15 850
Euro-American crayfishes nei	*Astacidae, Cambaridae*	191	227	227	134	134	134	114
	Country total	*167 484*	*179 077*	*180 185*	*171 681*	*160 560*	*156 501*	*150 096*

E-5 Fish crustaceans, molluscs, etc / Poissons, crustacés, mollusques, etc / Peces, crustáceos, moluscos, etc

Capture production by countries or areas and species — Europe
Captures par pays ou zones et espèces — Europe
Capturas por países o áreas y especies — Europa

English name Nom anglais Nombre inglés	Scientific name Nom scientifique Nombre científico	1995 mt	1996 mt	1997 mt	1998 mt	1999 mt	2000 mt	2001 mt
France								
Freshwater fishes nei	*Osteichthyes*	4 500	4 500	4 500	4 460	2 000 F	2 000 F	2 000 F
Sturgeons nei	*Acipenseridae*	0	0	0	0	1	-	-
European eel	*Anguilla anguilla*	320	403	1 782	449	289	399	415 F
Atlantic salmon	*Salmo salar*	0	0	0	0	2	10	12
Sea trout	*Salmo trutta*	0	0	0	0	0	-	1
European smelt	*Osmerus eperlanus*	84	95	106	69	75	103	96
Salmonoids nei	*Salmonoidei*	-	-	-	-	30	16	1
Allis shad	*Alosa alosa*	57	59	56	2	...	38	45
Twaite shad	*Alosa fallax*	1	2	1	...	...	11	11
Shads nei	*Alosa spp*	1	2	3	4	49	25	14
Sea lamprey	*Petromyzon marinus*	13	13	15	11	27	22	23
Atlantic halibut	*Hippoglossus hippoglossus*	27	34	16	16	21	29	44
European plaice	*Pleuronectes platessa*	3 982	3 843	4 326	4 667	5 642	4 709	4 254
Greenland halibut	*Reinhardtius hippoglossoides*	542	603	647	282	268	158	230
Witch flounder	*Glyptocephalus cynoglossus*	734	719	816	602	484	587	582
Amer. plaice(=Long rough dab)	*Hippoglossoides platessoides*	13	13	14	11	14	10	11
Common dab	*Limanda limanda*	1 062	1 120	1 446	1 576	1 194	1 106	950
Lemon sole	*Microstomus kitt*	1 944	2 396	1 782	1 522	1 349	1 308	1 366
European flounder	*Platichthys flesus*	209	238	223	163	204	221	243
Common sole	*Solea solea*	8 876	7 338	7 281	7 073	8 402	8 447	7 828
Sand sole	*Solea lascaris*	160	127	150	137	118	131	153
Wedge sole	*Dicologlossa cuneata*	941	798	647	488	602	686	579
Thickback soles	*Microchirus spp*	48	50	57	77	71	71	80
Soles nei	*Soleidae*	-	-	-	-	31	65	26
Megrim	*Lepidorhombus whiffiagonis*	4 642	4 267	3 987	3 601	3 529	4 154	4 125
Brill	*Scophthalmus rhombus*	452	417	357	351	397	512	490
Turbot	*Psetta maxima*	822	810	646	629	553	650	639
Flatfishes nei	*Pleuronectiformes*	14	14	17	24	20	3 075	1 855
Tadpole codling	*Salilota australis*	21	31	25	11	5	29	-
Moras nei	*Moridae*	-	-	-	75	67	60	73
Tusk(=Cusk)	*Brosme brosme*	433	439	439	453	428	335	282
Atlantic cod	*Gadus morhua*	17 669	21 500	25 123	20 640	14 431	11 886	11 336
Ling	*Molva molva*	5 644	5 738	5 463	5 510	5 112	3 099	2 987
Blue ling	*Molva dypterygia*	3 590	4 127	4 736	5 955	4 794	5 571	3 666
Greater forkbeard	*Phycis blennoides*	499	562	605	476	528	733	746
Haddock	*Melanogrammus aeglefinus*	4 167	6 027	8 060	4 983	3 582	4 379	5 970
Saithe(=Pollock)	*Pollachius virens*	19 882	19 598	17 802	18 218	24 638	26 941	28 533
Pollack	*Pollachius pollachius*	3 843	3 881	3 626	3 359	2 935	3 775	3 649
Rocklings nei	*Gaidropsarus spp*	...	...	...	...	24	43	20
Poor cod	*Trisopterus minutus*	586	603	428	428	637	888	754
Pouting(=Bib)	*Trisopterus luscus*	5 678	6 119	6 283	6 108	6 333	7 619	7 293
Blue whiting(=Poutassou)	*Micromesistius poutassou*	6	6 463	12 467	8 013	6 372	16 076	19 078
Southern blue whiting	*Micromesistius australis*	0	67	-	-	-	-	-
Whiting	*Merlangius merlangus*	27 821	21 315	21 590	20 074	21 572	19 028	19 416
European hake	*Merluccius merluccius*	16 305	10 322	10 015	6 588	9 160	11 635	10 029
Argentine hake	*Merluccius hubbsi*	4	17	4	3	3	0	-
Hakes nei	*Merluccius spp*	-	-	...	44	5	-	-
Patagonian grenadier	*Macruronus magellanicus*	15	29	-	-	2	0	-
Roughhead grenadier	*Macrourus berglax*	1	21	25	29	116	4	4
Grenadiers nei	*Macrourus spp*	-	-	-	25	-	173	103
Roundnose grenadier	*Coryphaenoides rupestris*	8 414	7 650	7 389	6 820	7 851	9 968	8 494
Gadiformes nei	*Gadiformes*	27	24	442	861	10	2 436	3 093
Mullets nei	*Mugilidae*	1 144	1 086	1 292	1 299	1 458	1 272	1 485
Dusky grouper	*Epinephelus marginatus*	-	-	-	-	1	-	-
Spotted seabass	*Dicentrarchus punctatus*	76	68	42	75	68	71	79
European seabass	*Dicentrarchus labrax*	2 530	3 330	3 012	2 793	3 503	4 152	4 208
Cardinalfishes, etc. nei	*Apogonidae*	-	-	-	-	294	-	-
Canary drum (=Baardman)	*Umbrina canariensis*	1	1	-	-	3	7	7
Meagre	*Argyrosomus regius*	40	250	409	457	349	189	162
Blackspot(=red) seabream	*Pagellus bogaraveo*	20	27	39	40	69	60	51
Common pandora	*Pagellus erythrinus*	13	78	77	78	167	126	97
Axillary seabream	*Pagellus acarne*	118	80	39	36	318	282	198
White seabream	*Diplodus sargus*	45	54	41	35	55	64	51
Sargo breams nei	*Diplodus spp*	60	80	109	109	82	79	83
Common dentex	*Dentex dentex*	6	10	9	9	5	2	1
Black seabream	*Spondyliosoma cantharus*	2 219	2 803	2 885	3 116	3 031	3 029	2 799
Saddled seabream	*Oblada melanura*	4	13	14	14	10	14	8
Red porgy	*Pagrus pagrus*	2	3	1	1	2	1	1
Gilthead seabream	*Sparus aurata*	229	287	221	213	378	376	369
Bogue	*Boops boops*	98	189	203	219	217	299	329
Sand steenbras	*Lithognathus mormyrus*	80	60	121	110	112	118	97
Salema	*Sarpa salpa*	18	30	82	79	64	66	70
Porgies, seabreams nei	*Sparidae*	-	-	...	77	55	-	3
Picarels nei	*Spicara spp*	...	8	9	9	10	9	5
Red mullet	*Mullus surmuletus*	2 886	2 936	1 861	3 666	2 592	4 054	2 969
Surmullets(=Red mullets) nei	*Mullus spp*	189	149	149	149	266	270	206
Wrasses, hogfishes, etc. nei	*Labridae*	...	...	...	...	222	250	250
Marbled rockcod	*Notothenia rossii*	-	-	1	-	1	-	-
Humped rockcod	*Notothenia gibberifrons*	-	-	-	-	-	-	0
Grey rockcod	*Notothenia squamifrons*	-	15	-	-	-	-	-
Sandeels(=Sandlances) nei	*Ammodytes spp*	114	95	159	70	93	90	51
Greater weever	*Trachinus draco*	171	129	169	184	218	224	206
Gobies nei	*Gobiidae*	0	4	2	0	4	3	1
Grey triggerfish	*Balistes carolinensis*	-	-	1	1	2	-	4
Argentines	*Argentina spp*	-	-	-	-	121	59	45
European conger	*Conger conger*	4 646	4 657	4 110	4 574	4 612	5 470	5 225
Pink cusk-eel	*Genypterus blacodes*	3	2	1	0	0	-	-

E-5

Fish crustaceans, molluscs, etc	Capture production by countries or areas and species	Europe
Poissons, crustacés, mollusques, etc	Captures par pays ou zones et espèces	Europe
Peces, crustáceos, moluscos, etc	Capturas por países o áreas y especies	Europa

English name Nom anglais Nombre inglés	Scientific name Nom scientifique Nombre científico	1995 mt	1996 mt	1997 mt	1998 mt	1999 mt	2000 mt	2001 mt
Alfonsinos nei	*Beryx spp*	0	0	3	27	75	48	52
Orange roughy	*Hoplostethus atlanticus*	998	1 067	1 012	1 110	1 330	1 048	1 254
John dory	*Zeus faber*	720	712	690	787	906	1 275	1 364
Wreckfish	*Polyprion americanus*	2	4	10	13	42	52	22
Patagonian toothfish	*Dissostichus eleginoides*	4 090	3 655	3 675	3 835	6 281	5 503	6 634
Mackerel icefish	*Champsocephalus gunnari*	84	5	0	-	-	-	386
South Georgia icefish	*Pseudochaenichthys georgianus*	-	-	-	-	-	-	0
Unicorn icefish	*Channichthys rhinoceratus*	-	1	5	1	1	-	-
Icefishes nei	*Channichthyidae*	-	-	-	-	-	-	0
Black cardinal fish	*Epigonus telescopus*	95	52	52	232	...	197	153
Atlantic wolffish	*Anarhichas lupus*	4	3	1	2	14	9	7
Largehead hairtail	*Trichiurus lepturus*	-	-	...	3 400	509	-	-
Silver scabbardfish	*Lepidopus caudatus*	-	41	11	11	255	16	31
Black scabbardfish	*Aphanopus carbo*	2 448	2 868	2 118	1 710	1 833	3 707	5 070
Atlantic redfishes nei	*Sebastes spp*	2 520	2 244	2 567	1 604	1 254	943	1 076
Blackbelly rosefish	*Helicolenus dactylopterus*	91	104	115	140	129	123	125
Scorpionfishes nei	*Scorpaenidae*	41	56	143	81	38	43	42
Gurnards, searobins nei	*Triglidae*	6 228	5 895	5 683	5 444	5 851	6 218	6 536
Angler(=Monk)	*Lophius piscatorius*	18 269	18 701	17 370	14 278	10 838	13 055	13 543
Demersal percomorphs nei	*Perciformes*	1	3	3	3	19	43	26
Atlantic herring	*Clupea harengus*	33 240	11 742	21 222	23 398	25 531	25 398	27 577
Sardinellas nei	*Sardinella spp*	-	-	...	7 228	624	5 517	1 220
European pilchard(=Sardine)	*Sardina pilchardus*	24 649	16 552	20 973	18 589	38 861	29 209	29 075
European sprat	*Sprattus sprattus*	36	597	68	0	83	97	9
European anchovy	*Engraulis encrasicolus*	15 785	19 499	16 223	25 477	24 677	26 479	23 699
Clupeoids nei	*Clupeoidei*	-	-	-	-	837	180	1 103
Atlantic bonito	*Sarda sarda*	-	-	-	-	24	32	70
Frigate and bullet tunas	*Auxis thazard, A.rochei*	127	161	147	146	0	91	15
Little tunny(=Atl.black skipj)	*Euthynnus alletteratus*	59	22	215	21	86	21	-
Skipjack tuna	*Katsuwonus pelamis*	73 840	63 164	48 299	48 721	62 969	58 118	48 617
Atlantic bluefin tuna	*Thunnus thynnus*	10 329	9 690	8 470	7 713	6 741	7 322	6 749
Albacore	*Thunnus alalunga*	5 739	5 275	5 200	4 216	7 356	6 392	7 007
Yellowfin tuna	*Thunnus albacares*	69 202	69 397	61 194	53 120	62 046	67 483	63 587
Bigeye tuna	*Thunnus obesus*	15 643	16 079	13 804	12 013	14 045	12 622	10 066
Atlantic sailfish	*Istiophorus albicans*	128	97	110	138	131	98	-
Atlantic blue marlin	*Makaira nigricans*	126	96	82	80	83	79	-
Atlantic white marlin	*Tetrapturus albidus*	9	7	7	9	8	7	-
Longbill spearfish	*Tetrapturus pfluegeri*	78	59	68	86	81	60	-
Marlins,sailfishes,etc. nei	*Istiophoridae*	-	-	-	-	-	66	-
Swordfish	*Xiphias gladius*	84	97	164	110	104	126	113
Capelin	*Mallotus villosus*	-	-	-	-	-	1	1
Garfish	*Belone belone*	38	59	70	43	55	60	55
Silversides(=Sand smelts) nei	*Atherinidae*	79	99	127	120	135	83	96
Atlantic horse mackerel	*Trachurus trachurus*	16 000	25 396	26 862	28 103	27 087	21 628	19 852
Jack and horse mackerels nei	*Trachurus spp*	311	422	2 148	6 771	1 887	915	1 038
Atlantic pomfret	*Brama brama*	10	1	0	2	7	8	5
Chub mackerel	*Scomber japonicus*	374	184	98	1 663	362	470	187
Atlantic mackerel	*Scomber scombrus*	23 596	13 908	16 567	20 963	18 256	22 640	22 300
Mackerels nei	*Scombridae*	-	-	120	25	13	4	2 649
Basking shark	*Cetorhinus maximus*	0	0	1	0	3	-	-
Thresher	*Alopias vulpinus*	13	7	13	7	35	128	132
Porbeagle	*Lamna nasus*	565	267	315	219	318	410	368
Small-spotted catshark	*Scyliorhinus canicula*	5 022	4 871	5 338	5 480	5 707	5 944	6 284
Nursehound	*Scyliorhinus stellaris*	174	197	292	181	135	159	180
Catsharks, nursehounds nei	*Scyliorhinus spp*	8	18	6	51	13	68	7
Blue shark	*Prionace glauca*	266	278	213	163	233	396	207
Smooth-hounds nei	*Mustelus spp*	414	578	624	749	824	1 050	1 249
Tope shark	*Galeorhinus galeus*	317	403	454	369	386	450	469
Picked dogfish	*Squalus acanthias*	1 349	1 726	1 715	1 417	1 197	1 100	1 335
Black dogfish	*Centroscyllium fabricii*	-	-	-	-	-	269	271
Dogfish sharks nei	*Squalidae*	3 125	3 135	2 811	2 288	3 077	4 236	4 211
Angelsharks, sand devils nei	*Squatinidae*	2	1	0	0	1	1	1
Blue skate	*Raja batis*	285	295	314	296	467	653	667
Thornback ray	*Raja clavata*	1 749	1 756	1 579	1 343	1 335	1 251	1 231
Spotted ray	*Raja montagui*	925	977	1 163	1 179	1 260	1 341	1 563
Sandy ray	*Raja circularis*	431	438	438	410	435	369	330
Shagreen ray	*Raja fullonica*	51	46	39	38	65	38	68
Small-eyed ray	*Raja microocellata*	-	-	-	1	11	-	-
Cuckoo ray	*Raja naevus*	3 762	4 077	4 721	4 015	3 638	3 064	2 885
Longnosed skate	*Raja oxyrinchus*	359	346	311	327	194	140	89
Raja rays nei	*Raja spp*	2 752	2 902	3 196	2 870	3 408	3 111	3 195
Stingrays nei	*Dasyatis spp*	-	1	2	5	6	10	7
Eagle rays	*Myliobatidae*	2	0	0	0	2	2	2
Rays, stingrays, mantas nei	*Rajiformes*	20	84	78	97	53	164	133
Torpedo rays	*Torpedo spp*	20	16	18	19	34	32	43
Ratfishes nei	*Hydrolagus spp*	-	-	-	-	38	573	822
Sharks, rays, skates, etc. nei	*Elasmobranchii*	2	28	...	...	43	-	-
Groundfishes nei	*Osteichthyes*	277	228	-	-	-	-	-
Finfishes nei	*Osteichthyes*	5 167	3 771	3 732	2 344	2 179	8 759	12 661
Marine fishes nei	*Osteichthyes*	13 034	14 034	13 920	2 583	931	751	1 749
Edible crab	*Cancer pagurus*	6 185	5 927	6 554	5 598	6 503	6 549	6 604
Portunus swimcrabs nei	*Portunus spp*	254	225	261	155	163	242	267
Green crab	*Carcinus maenas*	292	350	359	204	465	558	541
Mediterranean shore crab	*Carcinus aestuarii*	...	16	16	16	7	14	9
Spinous spider crab	*Maja squinado*	3 467	3 171	3 435	3 126	3 380	4 428	5 440
Marine crabs nei	*Brachyura*	14	160	5	5	-	13	10
Pink spiny lobster	*Palinurus mauritanicus*	3	1	0	25	11	9	5
Common spiny lobster	*Palinurus elephas*	120	113	99	53	61	62	68
Palinurid spiny lobsters nei	*Palinurus spp*	-	-	-	-	1	-	-

English name Nom anglais Nombre inglés	Scientific name Nom scientifique Nombre científico	1995 mt	1996 mt	1997 mt	1998 mt	1999 mt	2000 mt	2001 mt
Norway lobster	*Nephrops norvegicus*	9 782	8 623	7 125	6 611	5 843	6 639	6 992
European lobster	*Homarus gammarus*	266	267	327	219	308	332	329
Lobsters nei	*Reptantia*	-	-	-	-	0	2	2
Craylets, squat lobsters	*Galatheidae*	85	90	94	81	102	98	84
Caramote prawn	*Penaeus kerathurus*	...	...	...	...	...	4	6
Deepwater rose shrimp	*Parapenaeus longirostris*	3	4	17	17	1	1	1
Common prawn	*Palaemon serratus*	381	311	288	213	296	307	326
Common shrimp	*Crangon crangon*	247	309	237	272	399	472	391
Natantian decapods nei	*Natantia*	4	3	0	0	15	21	26
Goose barnacles	*Lepas spp*	0	2	-	-	-	-	-
Spottail mantis squillid	*Squilla mantis*	2	15	10	10	34	44	33
Marine crustaceans nei	*Crustacea*	979	887	1 156	678	701	686	690
Periwinkles nei	*Littorina spp*	-	2	-	-	-	-	-
Murex	*Murex spp*	43	37	41	41	35	51	52
Tuberculate abalone	*Haliotis tuberculata*	49	62	75	36	37	61	60
Whelk	*Buccinum undatum*	3 679	4 406	12 887	6 266	7 691	12 724	11 030
Gastropods nei	*Gastropoda*	-	-	-	-	11	15	19
European flat oyster	*Ostrea edulis*	40	49	18	14	12	9	132
Pacific cupped oyster	*Crassostrea gigas*	50	54	14	14	50	108	63
Blue mussel	*Mytilus edulis*	8 910	197	6 972	1 355	9 564	8 660	8 015
Mediterranean mussel	*Mytilus galloprovincialis*	1 912	500	1 078	1 078	23	24	12
Sea mussels nei	*Mytilidae*	2 031	1 077	0	0	0	1 792	1 110
Queen scallop	*Aequipecten opercularis*	926	311	595	637	2 574	3 475	5 989
Great Atlantic scallop	*Pecten maximus*	12 288	12 169	14 451	12 866	13 719	13 745	16 458
Scallops nei	*Pectinidae*	213	268	185	149	328	313	430
Striped venus	*Chamelea gallina*	651	706	810	715	790	696	627
Pullet carpet shell	*Venerupis pullastra*	638	934	1 493	2 216	2 306	1 311	1 595
Grooved carpet shell	*Ruditapes decussatus*	39	106	71	510	490	101	304
Venus clams nei	*Veneridae*	-	1	99	-	207	253	311
Solid surf clam	*Spisula solida*	...	...	...	...	...	127	101
Donax clams	*Donax spp*	197	-	-	-	34	94	88
Razor clams nei	*Solen spp*	...	...	...	...	5	5	5
Common edible cockle	*Cerastoderma edule*	379	664	731	418	481	81	11
Clams, etc. nei	*Bivalvia*	2 637	3 746	2 893	5 822	4 638	4 279	4 696
Common cuttlefish	*Sepia officinalis*	100	85	85	85	88	106	83
Cuttlefish,bobtail squids nei	*Sepiidae, Sepiolidae*	14 642	14 532	11 982	13 015	14 465	18 939	13 814
Patagonian squid	*Loligo gahi*	7 245	4 394	1 512	4 146	2 309	2 024	-
Common squids nei	*Loligo spp*	150	135	465	465	241	264	177
Broadtail shortfin squid	*Illex coindetii*	571	395	411	216	338	402	250
Argentine shortfin squid	*Illex argentinus*	23	28	0	0	56	0	-
European flying squid	*Todarodes sagittatus*	19	87	87	87	-	-	-
Various squids nei	*Loliginidae, Ommastrephidae*	6 017	3 895	4 456	4 332	5 946	5 399	4 690
Common octopus	*Octopus vulgaris*	1 146	706	439	439	1 493	1 643	1 406
Octopuses, etc. nei	*Octopodidae*	126	67	75	90	246	484	388
Cephalopods nei	*Cephalopoda*	-	-	-	-	55	49	48
Grooved sea squirt	*Microcosmus sulcatus*	...	28	22	22	30	30	76
Stony sea urchin	*Paracentrotus lividus*	78	63	48	59	84	198	101
	Country total	*610 808*	*557 510*	*567 422*	*542 560*	*586 487*	*623 755*	*606 194*
Germany								
Freshwater breams nei	*Abramis spp*	181	162	275	144	249	124	174
Common carp	*Cyprinus carpio*	386	386	386	386	387	386	387
Tench	*Tinca tinca*	36	36	36	36	37	36	37
Roach	*Rutilus rutilus*	337	176	242	347	462	269	320
Grass carp(=White amur)	*Ctenopharyngodon idellus*	10	10	10	10	5	5	5
Silver carp	*Hypophthalmichthys molitrix*	76	76	76	76	39	39	39
Bighead carp	*Hypophthalmichthys nobilis*	12	12	12	12	6	6	6
Cyprinids nei	*Cyprinidae*	1 653	1 653	1 653	1 653	1 653	1 653	1 653
Northern pike	*Esox lucius*	317	288	306	304	336	322	309
Wels(=Som)catfish	*Silurus glanis*	12	12	12	12	12	12	12
Burbot	*Lota lota*	2	2	3	3	3	3	4
European perch	*Perca fluviatilis*	948	764	782	822	714	501	541
Pike-perch	*Stizostedion lucioperca*	570	546	560	491	524	508	494
Freshwater fishes nei	*Osteichthyes*	19 079	19 079	19 011	19 009	19 010	19 011	19 009
Sturgeons nei	*Acipenseridae*	-	0	0	0	0	0	0
European eel	*Anguilla anguilla*	585	696	746	717	747	686	638
Atlantic salmon	*Salmo salar*	13	27	35	42	30	45	39
Trouts nei	*Salmo spp*	1	7	8	6	9	12	11
European smelt	*Osmerus eperlanus*	12	29	87	32	46	4	6
European whitefish	*Coregonus lavaretus*	-	-	5	8	21	47	63
Whitefishes nei	*Coregonus spp*	459	459	459	459	459	459	459
Salmonoids nei	*Salmonoidei*	29	29	29	29	29	29	29
Atlantic halibut	*Hippoglossus hippoglossus*	30	43	23	28	42	23	20
European plaice	*Pleuronectes platessa*	6 533	4 935	4 304	3 050	3 462	4 501	4 842
Greenland halibut	*Reinhardtius hippoglossoides*	848	3 924	3 846	3 870	3 532	3 742	3 402
Witch flounder	*Glyptocephalus cynoglossus*	10	7	9	13	8	13	8
Amer. plaice(=Long rough dab)	*Hippoglossoides platessoides*	-	0	-	-	-	-	-
Common dab	*Limanda limanda*	2 074	1 880	1 384	1 129	1 104	1 124	1 074
Lemon sole	*Microstomus kitt*	71	67	78	151	68	74	77
European flounder	*Platichthys flesus*	4 488	1 637	2 449	2 159	2 347	2 782	2 349
Common sole	*Solea solea*	1 569	685	513	786	1 462	1 291	959
Megrim	*Lepidorhombus whiffiagonis*	2	1	2	3	1	3	1
Megrims nei	*Lepidorhombus spp*	...	...	...	...	...	...	289
Brill	*Scophthalmus rhombus*	72	47	48	60	54	80	65
Turbot	*Psetta maxima*	399	256	330	267	309	454	364
Flatfishes nei	*Pleuronectiformes*	-	404	508	279	293	370	-
Tusk(=Cusk)	*Brosme brosme*	29	59	26	19	16	13	10
Atlantic cod	*Gadus morhua*	31 892	37 629	26 491	23 075	21 990	18 414	19 222

English name Nom anglais Nombre inglés	Scientific name Nom scientifique Nombre científico	1995 mt	1996 mt	1997 mt	1998 mt	1999 mt	2000 mt	2001 mt
Ling	*Molva molva*	877	1 409	965	308	247	215	110
Blue ling	*Molva dypterygia*	202	119	11	16	15	110	26
Greater forkbeard	*Phycis blennoides*	-	-	-	-	1	8	12
Haddock	*Melanogrammus aeglefinus*	3 978	2 718	2 441	1 712	1 039	1 225	1 368
Saithe(=Pollock)	*Pollachius virens*	13 393	15 197	15 993	13 562	13 307	12 385	13 320
Pollack	*Pollachius pollachius*	87	102	117	43	63	39	41
Norway pout	*Trisopterus esmarkii*	38	-	-	-	-	2	-
Blue whiting(=Poutassou)	*Micromesistius poutassou*	6 314	6 865	4 722	17 970	3 170	12 654	19 059
Whiting	*Merlangius merlangus*	1 186	711	276	191	371	754	680
European hake	*Merluccius merluccius*	110	83	76	69	68	46	73
Silver hake	*Merluccius bilinearis*	0	-	-	-	-	-	-
Hakes nei	*Merluccius spp*	16	-	-	-	-	-	-
Roundnose grenadier	*Coryphaenoides rupestris*	22	14	38	121	101	42	23
Mullets nei	*Mugilidae*	-	1	3	6	21	13	23
Groupers, seabasses nei	*Serranidae*	-	-	-	-	-	-	1
Cardinalfishes, etc. nei	*Apogonidae*	-	-	-	-	-	-	10
Porgies, seabreams nei	*Sparidae*	23	-	-	-	-	-	-
Surmullets(=Red mullets) nei	*Mullus spp*	-	-	-	-	-	13	10
Eelpout	*Zoarces viviparus*	5	3	2	2	2	1	5
Argentines	*Argentina spp*	357	1 394	1 498	633	24	483	189
Baird's slickhead	*Alepocephalus bairdii*	-	-	-	-	-	12	1
Black cardinal fish	*Epigonus telescopus*	-	-	-	-	-	50	-
Atlantic wolffish	*Anarhichas lupus*	176	94	44	88	67	86	66
Silver scabbardfish	*Lepidopus caudatus*	-	-	-	-	64	4	-
Black scabbardfish	*Aphanopus carbo*	3	2	-	-	-	-	-
Atlantic redfishes nei	*Sebastes spp*	20 472	22 512	21 096	20 342	18 417	14 278	12 790
Gurnards, searobins nei	*Triglidae*	133	94	136	141	187	180	150
Lumpfish(=Lumpsucker)	*Cyclopterus lumpus*	1	2	2	0	4	1	3
Angler(=Monk)	*Lophius piscatorius*	893	542	1 137	1 446	847	568	364
Atlantic herring	*Clupea harengus*	55 916	42 153	42 749	47 028	50 857	47 048	50 680
Sardinellas nei	*Sardinella spp*	-	10 115	23 866	17 448	24 150	-	1 683
European pilchard(=Sardine)	*Sardina pilchardus*	35	50	3 295	4 781	1 446	307	500
European sprat	*Sprattus sprattus*	231	161	427	4 551	183	22	791
European anchovy	*Engraulis encrasicolus*	-	0	-	16	-	-	42
Atlantic bonito	*Sarda sarda*	-	714	417	42	143	-	51
Skipjack tuna	*Katsuwonus pelamis*	-	3	-	-	-	-	-
Atlantic bluefin tuna	*Thunnus thynnus*	1	-	-	-	-	-	-
Capelin	*Mallotus villosus*	-	-	-	5 001	-	-	-
Garfish	*Belone belone*	152	168	130	58	125	82	73
Atlantic horse mackerel	*Trachurus trachurus*	20 407	22 028	38 077	34 376	24 377	16 778	12 464
Jack and horse mackerels nei	*Trachurus spp*	-	-	-	-	-	-	708
Chub mackerel	*Scomber japonicus*	-	87	645	315	334	-	20
Atlantic mackerel	*Scomber scombrus*	24 417	16 229	15 864	21 490	19 960	22 980	25 325
Porbeagle	*Lamna nasus*	-	-	-	-	0	17	1
Picked dogfish	*Squalus acanthias*	-	-	-	-	45	188	303
Dogfish sharks nei	*Squalidae*	-	19	12	16	235	271	433
Raja rays nei	*Raja spp*	35	65	74	81	102	130	27
Various sharks nei	*Selachimorpha(Pleurotremata)*	292	309	139	110	-	-	-
Finfishes nei	*Osteichthyes*	316	270	291	164	150	66	37
Marine fishes nei	*Osteichthyes*	-	40	45	53	110	-	132
Freshwater crustaceans nei	*Crustacea*	12	12	12	12	12	12	12
Edible crab	*Cancer pagurus*	-	-	38	44	57	64	44
Marine crabs nei	*Brachyura*	-	13	6	4	-	-	-
Norway lobster	*Nephrops norvegicus*	17	77	70	70	110	86	141
European lobster	*Homarus gammarus*	9	17	0	0	0	-	0
Northern prawn	*Pandalus borealis*	-	-	-	-	1 585		-
Common shrimp	*Crangon crangon*	11 608	15 994	19 890	14 814	17 457	17 423	12 571
Clams, etc. nei	*Bivalvia*	5 410	-	-	-	-	-	-
Northern shortfin squid	*Illex illecebrosus*	11	-	-	-	-	-	-
Various squids nei	*Loliginidae, Ommastrephidae*	-	2	4	11	6	5	3
	Country total	*239 890*	*236 411*	*259 352*	*266 622*	*238 925*	*205 689*	*211 282*
Gibraltar								
Marine fishes nei	*Osteichthyes*	0	0	0	0	0	0	0
	Country total	*0*	*0*	*0*	*0*	*0*	*0*	*0*
Greece								
Common carp	*Cyprinus carpio*	...	208	256	279	247	220	198
Goldfish	*Carassius auratus*	...	530	296	235	548	448	415
Roaches nei	*Rutilus spp*	...	164	81	252	309	345	324
Cyprinids nei	*Cyprinidae*	...	176	107	107	74	90	80
Northern pike	*Esox lucius*	...	12	9	5	8	10	9
Wels(=Som)catfish	*Silurus glanis*	...	11	13	20	14	15	18
European perch	*Perca fluviatilis*	...	24	15	19	24	15	23
Freshwater fishes nei	*Osteichthyes*	3 606	1 730	1 624	1 410	1 592	1 803	1 663
European eel	*Anguilla anguilla*	31	31	31	43	27	34	32
Salmonoids nei	*Salmonoidei*	...	1	6	38	11	50	42
Shads nei	*Alosa spp*	856	1 010	1 284	1 784	1 898	1 283	2 217
Common sole	*Solea solea*	1 453	1 235	1 169	745	619	687	562
Turbot	*Psetta maxima*	102	60	60	47	65	63	77
Blue whiting(=Poutassou)	*Micromesistius poutassou*	1 945	1 227	1 558	846	630	566	471
European hake	*Merluccius merluccius*	5 454	4 649	4 257	3 052	3 128	2 969	2 753
Flathead grey mullet	*Mugil cephalus*	2 817	3 824	3 077	1 760	2 023	1 748	1 403
Dusky grouper	*Epinephelus marginatus*	195	112	94	57	64	90	99
White grouper	*Epinephelus aeneus*	293	306	125	123	148	137	161
Groupers nei	*Epinephelus spp*	76	54	91	65	85	63	142
Groupers, seabasses nei	*Serranidae*	429	410	619	330	336	286	463

English name Nom anglais Nombre inglés	Scientific name Nom scientifique Nombre científico	1995 mt	1996 mt	1997 mt	1998 mt	1999 mt	2000 mt	2001 mt
European seabass	*Dicentrarchus labrax*	529	455	380	258	289	345	300
Shi drum	*Umbrina cirrosa*	150	160	78	109	136	151	103
Pandoras nei	*Pagellus spp*	1 938	1 443	1 544	839	1 000	1 050	937
White seabream	*Diplodus sargus*	539	409	509	327	503	581	347
Sargo breams nei	*Diplodus spp*	31	19	7	22	18	117	88
Large-eye dentex	*Dentex macrophthalmus*	711	653	1 026	544	516	398	330
Common dentex	*Dentex dentex*	914	744	448	522	474	364	480
Black seabream	*Spondyliosoma cantharus*	344	391	318	219	330	176	138
Saddled seabream	*Oblada melanura*	680	607	399	332	239	443	287
Red porgy	*Pagrus pagrus*	1 427	1 280	933	853	887	913	1 139
Pargo breams nei	*Pagrus spp*	697	531	465	371	435	386	326
Gilthead seabream	*Sparus aurata*	201	199	138	125	142	248	176
Bogue	*Boops boops*	6 845	6 744	5 973	4 228	4 658	4 096	3 674
Salema	*Sarpa salpa*	592	687	485	369	404	408	346
Porgies, seabreams nei	*Sparidae*	0	-	-	-	-	-	-
Picarels nei	*Spicara spp*	6 354	8 307	8 371	4 549	4 663	4 029	4 031
Red mullet	*Mullus surmuletus*	2 535	2 004	1 944	1 095	1 274	1 368	1 238
Surmullets(=Red mullets) nei	*Mullus spp*	2 550	2 495	2 426	1 744	1 735	1 810	1 541
West African goatfish	*Pseudupeneus prayensis*	907	1 065	692	805	838	715	1 081
European conger	*Conger conger*	1 036	1 160	1 922	1 293	1 211	1 019	1 062
John dory	*Zeus faber*	422	487	309	274	201	196	296
Wreckfish	*Polyprion americanus*	0	-	-	-	-	-	-
Scorpionfishes nei	*Scorpaenidae*	1 067	841	783	518	634	640	726
Gurnards, searobins nei	*Triglidae*	330	376	380	237	229	207	169
Angler(=Monk)	*Lophius piscatorius*	1 508	942	912	757	739	882	694
Blackbellied angler	*Lophius budegassa*	-	-	4	7	4	27	-
European pilchard(=Sardine)	*Sardina pilchardus*	20 413	18 896	20 561	17 734	15 214	16 026	14 395
European sprat	*Sprattus sprattus*	178	262	279	216	110	266	474
European anchovy	*Engraulis encrasicolus*	13 876	15 073	14 583	17 099	16 456	9 863	10 770
Atlantic bonito	*Sarda sarda*	2 116	1 752	1 559	945	2 135	1 914	1 550
Frigate and bullet tunas	*Auxis thazard, A.rochei*	1 400	1 426	1 426	...	...	196	125
Atlantic bluefin tuna	*Thunnus thynnus*	1 007	878	1 218	287	248	622	361
Albacore	*Thunnus alalunga*	0	952	741	1 152	2 005	1 786	1 840
Swordfish	*Xiphias gladius*	976	1 238	750	1 650	1 520	1 960	1 730
Tuna-like fishes nei	*Scombroidei*	116	116	145	300	...	195	128
Garfish	*Belone belone*	323	331	253	188	83	184	126
Big-scale sand smelt	*Atherina boyeri*	-	8	110	350	350	350	326
Bluefish	*Pomatomus saltatrix*	346	128	111	144	259	265	511
Atlantic horse mackerel	*Trachurus trachurus*	1 860	2 484	2 956	1 360	942	774	747
Mediterranean horse mackerel	*Trachurus mediterraneus*	8 331	8 039	7 169	4 350	3 534	3 902	3 408
Jack and horse mackerels nei	*Trachurus spp*	-	-	-	-	-	1	-
Greater amberjack	*Seriola dumerili*	322	697	336	252	205	176	396
Carangids nei	*Carangidae*	-	-	-	11	-	4	1
Chub mackerel	*Scomber japonicus*	6 469	6 520	5 640	2 173	1 850	2 286	1 829
Atlantic mackerel	*Scomber scombrus*	686	886	305	141	200	443	155
Smooth-hounds nei	*Mustelus spp*	462	440	517	359	576	617	372
Dogfish sharks nei	*Squalidae*	268	290	243	290	258	270	224
Guitarfishes, etc. nei	*Rhinobatidae*	79	112	63	87	73	94	89
Raja rays nei	*Raja spp*	1 120	1 002	900	715	718	746	579
Marine fishes nei	*Osteichthyes*	13 395	11 086	12 433	9 083	9 606	12 595	11 353
Euro-American crayfishes nei	*Astacidae, Cambaridae*	...	...	15	23	28	23	27
Mediterranean shore crab	*Carcinus aestuarii*	16	49	28	50	37	16	17
Marine crabs nei	*Brachyura*	71	60	84	90	74	59	54
Norway lobster	*Nephrops norvegicus*	1 104	487	351	455	243	268	242
European lobster	*Homarus gammarus*	523	212	374	155	170	201	233
Caramote prawn	*Penaeus kerathurus*	867	1 261	1 654	1 499	1 116	1 448	1 503
Natantian decapods nei	*Natantia*	1 864	2 278	2 470	1 808	1 812	1 638	1 355
European flat oyster	*Ostrea edulis*	1 096	1 003	344	95	47	105	120
Mediterranean mussel	*Mytilus galloprovincialis*	10 806	12 625	24 139	6 389	15 860	469	254
Common cuttlefish	*Sepia officinalis*	3 416	2 473	2 969	2 279	3 123	1 766	2 238
Common squids nei	*Loligo spp*	945	856	623	426	397	560	405
Various squids nei	*Loliginidae, Ommastrephidae*	961	843	513	567	686	886	750
Common octopus	*Octopus vulgaris*	2 856	2 919	2 952	1 673	1 910	2 193	2 058
Octopuses, etc. nei	*Octopodidae*	1 563	904	815	782	1 781	934	813
Marine molluscs nei	*Mollusca*	2 393	3 076	2 241	1 789	1 736	1 688	2 169
	Country total	*151 788*	*149 435*	*157 088*	*108 580*	*118 771*	*99 280*	*94 388*
Hungary								
Common carp	*Cyprinus carpio*	2 856	2 717	2 255	3 373	3 279	3 212	2 470
Tench	*Tinca tinca*	-	-	5	-	-	-	-
Barbel	*Barbus barbus*	32	37	64	46	50	30	52
Grass carp(=White amur)	*Ctenopharyngodon idellus*	366	346	305	301	318	356	309
Silver carp	*Hypophthalmichthys molitrix*	521	862	1 483	731	676	365	997
Bighead carp	*Hypophthalmichthys nobilis*	127	74	83	-	-	-	-
Asp	*Aspius aspius*	20	22	44	38	42	38	21
Cyprinids nei	*Cyprinidae*	1 744	1 731	1 730	1 510	1 666	1 710	2 155
Northern pike	*Esox lucius*	37	46	203	158	241	280	191
Wels(=Som)catfish	*Silurus glanis*	205	201	121	113	145	104	120
Pike-perch	*Stizostedion lucioperca*	226	224	199	156	169	200	196
Freshwater fishes nei	*Osteichthyes*	733	733	776	648	714	718	89
Sterlet sturgeon	*Acipenser ruthenus*	36	34	14	9	35	12	11
European eel	*Anguilla anguilla*	411	579	124	182	179	76	27
	Country total	*7 314*	*7 606*	*7 406*	*7 265*	*7 514*	*7 101*	*6 638*
Iceland								
Freshwater fishes nei	*Osteichthyes*	0	0	0	0	0	0	0
Atlantic salmon	*Salmo salar*	778	597	202	202	142	87	88

E-5

Fish crustaceans, molluscs, etc	Capture production by countries or areas and species	Europe
Poissons, crustacés, mollusques, etc	Captures par pays ou zones et espèces	Europe
Peces, crustáceos, moluscos, etc	Capturas por países o áreas y especies	Europa

English name Nom anglais Nombre inglés	Scientific name Nom scientifique Nombre científico	1995 mt	1996 mt	1997 mt	1998 mt	1999 mt	2000 mt	2001 mt
Trouts nei	*Salmo spp*	250	250	250	250	250	91	72
Atlantic halibut	*Hippoglossus hippoglossus*	888	837	677	501	567	493	589
European plaice	*Pleuronectes platessa*	10 649	11 070	10 557	7 111	7 064	5 218	4 905
Greenland halibut	*Reinhardtius hippoglossoides*	27 408	22 125	18 631	10 751	11 187	15 060	16 642
Witch flounder	*Glyptocephalus cynoglossus*	1 755	1 486	1 272	947	1 408	1 098	1 132
Amer. plaice(=Long rough dab)	*Hippoglossoides platessoides*	5 418	7 027	6 468	3 329	3 833	3 176	3 473
Common dab	*Limanda limanda*	5 558	7 954	7 891	5 061	3 981	3 015	4 373
Lemon sole	*Microstomus kitt*	741	984	1 135	1 432	1 860	1 438	1 371
Megrim	*Lepidorhombus whiffiagonis*	405	419	281	221	124	97	96
Turbot	*Psetta maxima*	1	0	0	0	0	0	0
Flatfishes nei	*Pleuronectiformes*	10	11	13	3	5	-	2
Blue antimora	*Antimora rostrata*	-	2	-	-	-	-	-
Tusk(=Cusk)	*Brosme brosme*	5 245	5 226	4 847	4 118	5 796	4 741	3 425
Atlantic cod	*Gadus morhua*	202 900	204 058	208 636	242 968	260 643	238 324	240 002
Ling	*Molva molva*	3 729	3 670	3 634	3 603	3 976	3 223	2 864
Blue ling	*Molva dypterygia*	1 636	1 284	1 320	1 208	2 321	1 623	765
Greater forkbeard	*Phycis blennoides*	0	-	-	-	-	-	-
Haddock	*Melanogrammus aeglefinus*	60 125	56 223	43 256	40 712	44 729	41 698	39 825
Saithe(=Pollock)	*Pollachius virens*	47 466	39 297	36 548	30 532	30 729	32 947	31 941
Norway pout	*Trisopterus esmarkii*	0	0	-	-	-	-	160
Blue whiting(=Poutassou)	*Micromesistius poutassou*	369	513	10 480	68 514	160 424	259 157	365 101
Whiting	*Merlangius merlangus*	560	430	443	531	931	1 349	1 179
Cape hakes	*Merluccius capensis,M.paradox.*	-	-	352	206	-	-	-
Roughhead grenadier	*Macrourus berglax*	6	15	4	1	-	5	3
Roundnose grenadier	*Coryphaenoides rupestris*	398	216	207	120	146	70	57
Sandeels(=Sandlances) nei	*Ammodytes spp*	-	-	-	-	-	-	8
Argentines	*Argentina spp*	492	808	3 367	13 387	5 495	4 595	2 478
Baird's slickhead	*Alepocephalus bairdii*	1	0	0	0	0	0	2
Kingklip	*Genypterus capensis*	-	-	31	2	-	-	-
Alfonsinos nei	*Beryx spp*	-	7	466	126	-	-	-
Orange roughy	*Hoplostethus atlanticus*	64	43	79	28	14	68	18
John dory	*Zeus faber*	-	-	7	-	-	-	-
Atlantic wolffish	*Anarhichas lupus*	12 574	14 638	11 685	11 844	13 769	15 043	17 953
Spotted wolffish	*Anarhichas minor*	700	1 109	1 180	1 599	1 545	1 896	2 126
Snoek	*Thyrsites atun*	-	-	1	-	-	-	-
Black scabbardfish	*Aphanopus carbo*	0	17	1	0	9	18	8
Atlantic redfishes nei	*Sebastes spp*	118 750	120 751	111 652	116 132	110 345	116 302	92 527
Scorpionfishes nei	*Scorpaenidae*	-	-	5	-	-	-	-
Grey gurnard	*Eutrigla gurnardus*	-	1	0	0	0	0	-
Lumpfish(=Lumpsucker)	*Cyclopterus lumpus*	4 563	4 201	6 520	3 165	3 373	2 458	412
Angler(=Monk)	*Lophius piscatorius*	552	669	787	850	977	1 570	1 353
Devil anglerfish	*Lophius vomerinus*	-	-	5	2	-	-	-
Atlantic herring	*Clupea harengus*	284 473	265 413	291 117	277 461	298 435	287 663	178 950
Atlantic bluefin tuna	*Thunnus thynnus*	-	-	1	2	33	29	-
Bigeye tuna	*Thunnus obesus*	-	-	-	-	-	5	-
Swordfish	*Xiphias gladius*	-	-	-	-	2	2	-
Capelin	*Mallotus villosus*	715 551	1 179 051	1 319 191	750 065	703 694	892 405	918 417
Atlantic horse mackerel	*Trachurus trachurus*	0	-	-	-	-	-	-
Chub mackerel	*Scomber japonicus*	-	-	28	-	-	-	-
Atlantic mackerel	*Scomber scombrus*	0	92	927	357	144	0	1
Mackerels nei	*Scombridae*	-	-	11	-	-	-	-
Porbeagle	*Lamna nasus*	6	5	3	4	2	2	3
Greenland shark	*Somniosus microcephalus*	44	61	73	87	51	45	57
Picked dogfish	*Squalus acanthias*	166	157	106	78	57	109	136
Portuguese dogfish	*Centroscymnus coelolepis*	-	-	-	5	0	0	-
Black dogfish	*Centroscyllium fabricii*	1	4	0	-	-	2	-
Blue skate	*Raja batis*	245	181	118	108	80	94	85
Starry ray	*Raja radiata*	1 749	1 493	1 431	1 252	996	1 076	1 211
Shagreen ray	*Raja fullonica*	24	19	16	12	21	27	37
Raja rays nei	*Raja spp*	-	-	14	-	-	-	-
Rabbit fish	*Chimaera monstrosa*	106	21	15	29	11	5	1
Chimaeras, etc. nei	*Chimaeriformes*	2	1	0	-	-	-	-
Groundfishes nei	*Osteichthyes*	337	237	179	2	81	...	45
Marine fishes nei	*Osteichthyes*	-	-	-	4	-	-	-
Marine crabs nei	*Brachyura*	0	0	0	-	-	-	1
Norway lobster	*Nephrops norvegicus*	1 027	1 623	1 215	1 411	1 389	1 230	1 420
Northern prawn	*Pandalus borealis*	83 529	89 633	82 627	62 727	42 958	33 539	30 790
Whelk	*Buccinum undatum*	0	520	1 199	13	298	770	678
Sea mussels nei	*Mytilidae*	0	-	-	-	-	-	-
Iceland scallop	*Chlamys islandica*	8 381	8 978	10 403	10 098	8 858	9 074	6 499
Ocean quahog	*Arctica islandica*	1 980	6 315	4 351	8 776	3 501	1 584	7 434
European flying squid	*Todarodes sagittatus*	11	3	5	4	3	1	-
Various squids nei	*Loliginidae, Ommastrephidae*	-	-	4	0	-	-	-
European edible sea urchin	*Echinus esculentus*	923	423	20	-	10	-	-
Sea cucumbers nei	*Holothurioidea*	2	-	-	-	-	-	-
	Country total	*1 612 548*	*2 060 168*	*2 205 944*	*1 681 951*	*1 736 267*	*1 982 522*	*1 980 715*
Ireland								
Northern pike	*Esox lucius*	2 000	2 000	2 000	2 000	...	...	...
European perch	*Perca fluviatilis*	...	...	...	200	...	73	...
Freshwater fishes nei	*Osteichthyes*	0	0	0	0	...	...	...
European eel	*Anguilla anguilla*	600	550	550	650	500	515	110
Atlantic salmon	*Salmo salar*	876	818	675	725	1 026	611	688
Sea trout	*Salmo trutta*	1 200	1 200	1 200	1 200	110	100	...
Rainbow trout	*Oncorhynchus mykiss*	75	75	99	75	...	...	...
Atlantic halibut	*Hippoglossus hippoglossus*	6	8	4	9	11	1	17
European plaice	*Pleuronectes platessa*	1 590	1 679	1 699	1 731	1 424	932	841
Greenland halibut	*Reinhardtius hippoglossoides*	2	2	2	21	78	22	71

E-5

Fish crustaceans, molluscs, etc	Capture production by countries or areas and species	Europe
Poissons, crustacés, mollusques, etc	Captures par pays ou zones et espèces	Europe
Peces, crustáceos, moluscos, etc	Capturas por países o áreas y especies	Europa

English name Nom anglais Nombre inglés	Scientific name Nom scientifique Nombre científico	1995 mt	1996 mt	1997 mt	1998 mt	1999 mt	2000 mt	2001 mt
Witch flounder	*Glyptocephalus cynoglossus*	601	615	605	657	713	555	915
Amer. plaice(=Long rough dab)	*Hippoglossoides platessoides*	-	-	-	-	-	99	-
Common dab	*Limanda limanda*	95	76	113	109	66	39	34
Lemon sole	*Microstomus kitt*	724	581	667	527	531	468	440
European flounder	*Platichthys flesus*	24	13	13	13	13	12	18
Common sole	*Solea solea*	561	463	483	526	492	383	375
Sand sole	*Solea lascaris*	25	13	12	15	1	2	1
Megrim	*Lepidorhombus whiffiagonis*	3 839	3 507	3 063	3 383	3 162	3 364	3 713
Brill	*Scophthalmus rhombus*	128	126	181	141	126	119	95
Turbot	*Psetta maxima*	233	232	257	234	261	236	185
Flatfishes nei	*Pleuronectiformes*	68	57	74	184	37	172	15
Tusk(=Cusk)	*Brosme brosme*	76	64	45	43	43	113	122
Atlantic cod	*Gadus morhua*	5 650	7 992	5 702	5 294	3 860	2 923	2 647
Ling	*Molva molva*	1 542	1 379	1 305	1 272	1 138	1 089	1 463
Blue ling	*Molva dypterygia*	14	-	1	22	43	91	827
Greater forkbeard	*Phycis blennoides*	163	154	228	318	379	399	679
Haddock	*Melanogrammus aeglefinus*	3 417	4 462	6 234	6 572	4 898	5 812	5 404
Saithe(=Pollock)	*Pollachius virens*	2 929	2 579	1 841	1 687	1 704	2 848	2 048
Pollack	*Pollachius pollachius*	1 190	1 288	1 052	946	1 049	24	1 382
Norway pout	*Trisopterus esmarkii*	0	-	-	-	-	1	-
Pouting(=Bib)	*Trisopterus luscus*	5	2	12	1	21	10	28
Blue whiting(=Poutassou)	*Micromesistius poutassou*	222	1 709	25 987	45 538	33 687	26 067	29 910
Whiting	*Merlangius merlangus*	11 262	10 340	9 394	7 762	7 643	6 505	6 621
European hake	*Merluccius merluccius*	2 186	1 741	2 270	1 971	2 090	2 037	1 124
Roundnose grenadier	*Coryphaenoides rupestris*	59	1	4	-	1	45	-
Gadiformes nei	*Gadiformes*	55	105	190	-	279	26	87
Mullets nei	*Mugilidae*	22	40	33	15	29	11	3
European seabass	*Dicentrarchus labrax*	0	-	-	-	-	-	-
Cardinalfishes, etc. nei	*Apogonidae*	-	-	-	-	-	-	5
Grunts, sweetlips nei	*Haemulidae (=Pomadasyidae)*	-	-	-	-	-	-	5
Blackspot(=red) seabream	*Pagellus bogaraveo*	3	8	8	6	1	...	11
Red mullet	*Mullus surmuletus*	-	-	-	38	-	-	-
Sandeels(=Sandlances) nei	*Ammodytes spp*	-	-	-	-	389	-	-
Argentines	*Argentina spp*	6	295	1 089	405	396	4 709	7 505
European conger	*Conger conger*	144	142	202	374	295	14	253
Orange roughy	*Hoplostethus atlanticus*	-	-	-	-	-	-	2 759
John dory	*Zeus faber*	147	125	112	98	145	174	169
Wreckfish	*Polyprion americanus*	-	-	-	5	-	1	1
Atlantic wolffish	*Anarhichas lupus*	42	39	22	39	35	66	27
Largehead hairtail	*Trichiurus lepturus*	-	-	-	-	-	-	776
Black scabbardfish	*Aphanopus carbo*	-	0	1	-	1	12	299
Atlantic redfishes nei	*Sebastes spp*	18	15	48	71	171	186	433
Red gurnard	*Chelidonichthys cuculus*	-	-	-	25	47	...	...
Grey gurnard	*Eutrigla gurnardus*	-	-	-	38	71	...	...
Gurnards, searobins nei	*Triglidae*	85	77	82	-	-	79	97
Angler(=Monk)	*Lophius piscatorius*	2 929	3 348	3 880	4 251	4 298	3 839	3 112
Atlantic herring	*Clupea harengus*	46 643	71 953	57 155	58 248	45 334	42 114	40 640
Sardinellas nei	*Sardinella spp*	-	-	-	-	-	-	52 980
European pilchard(=Sardine)	*Sardina pilchardus*	-	-	-	-	3 195	2 592	7 855
European sprat	*Sprattus sprattus*	799	4 214	2 085	1 578	5 826	6 032	455
Atlantic bluefin tuna	*Thunnus thynnus*	-	-	14	21	52	24	10
Albacore	*Thunnus alalunga*	918	874	1 913	3 750	4 858	3 464	2 085
Yellowfin tuna	*Thunnus albacares*	-	-	-	-	-	-	3
Bigeye tuna	*Thunnus obesus*	-	-	-	-	-	-	8
Swordfish	*Xiphias gladius*	-	15	15	132	81	36	14
Capelin	*Mallotus villosus*	-	-	0	1	-	-	-
Atlantic horse mackerel	*Trachurus trachurus*	178 355	127 876	75 002	74 253	58 201	55 438	63 497
Atlantic pomfret	*Brama brama*	-	-	-	-	-	1	184
Atlantic mackerel	*Scomber scombrus*	78 534	49 966	53 094	67 310	59 609	74 871	76 586
Porbeagle	*Lamna nasus*	-	-	-	-	8	1	6
Small-spotted catshark	*Scyliorhinus canicula*	...	...	...	...	...	...	633
Blue shark	*Prionace glauca*	-	-	-	-	67	23	66
Tope shark	*Galeorhinus galeus*	...	...	...	...	...	...	4
Picked dogfish	*Squalus acanthias*	2 435	2 095	1 407	1 259	962	880	1 301
Birdbeak dogfish	*Deania calcea*	...	...	...	...	...	...	216
Dogfish sharks nei	*Squalidae*	1 676	1 170	917	1 144	683	441	30
Raja rays nei	*Raja spp*	2 098	2 212	2 715	2 120	2 283	2 096	2 140
Rabbit fish	*Chimaera monstrosa*	...	...	...	...	...	5	15
Ratfishes nei	*Hydrolagus spp*	...	...	...	...	...	...	5
Various sharks nei	*Selachimorpha(Pleurotremata)*	40	23	32	-	-	233	455
Groundfishes nei	*Osteichthyes*	0	65	-	-	-	2	1 520
Pelagic fishes nei	*Osteichthyes*	-	-	-	-	-	1	13
Edible crab	*Cancer pagurus*	7 049	5 649	7 572	7 392	7 772	9 865	9 738
Portunus swimcrabs nei	*Portunus spp*	...	...	...	314	309	...	214
Green crab	*Carcinus maenas*	...	...	...	79	169	...	68
Spinous spider crab	*Maja squinado*	153	192	153	185	299	163	264
Marine crabs nei	*Brachyura*	487	312	272	-	1	268	-
Palinurid spiny lobsters nei	*Palinurus spp*	84	62	48	46	35	41	35
Norway lobster	*Nephrops norvegicus*	7 241	2 769	7 020	6 950	8 492	2 945	7 074
European lobster	*Homarus gammarus*	564	567	513	611	597	533	781
Northern prawn	*Pandalus borealis*	...	2 409	-	-	-	1 116	-
Palaemonid shrimps nei	*Palaemonidae*	312	399	358	505	551	4 047	268
Marine crustaceans nei	*Crustacea*	-	-	-	-	-	49	401
Common periwinkle	*Littorina littorea*	301	2 836	3 152	2 636	3 014	2 172	2 781
Whelk	*Buccinum undatum*	5 952	6 575	3 852	3 667	4 561	4 942	6 364
European flat oyster	*Ostrea edulis*	815	415	773	...	...	...	...
Blue mussel	*Mytilus edulis*	4 556	1 372	1 963	955	503	-	-
Queen scallop	*Aequipecten opercularis*	11	3	7	5	29	3	13
Great Atlantic scallop	*Pecten maximus*	423	560	633	693	1 497	1 579	1 413

English name Nom anglais Nombre inglés	Scientific name Nom scientifique Nombre científico	1995 mt	1996 mt	1997 mt	1998 mt	1999 mt	2000 mt	2001 mt
Grooved carpet shell	*Ruditapes decussatus*	20	21	23	3	3	3	130
Razor clams nei	*Solen spp*	-	-	28	316	407	334	201
Common edible cockle	*Cerastoderma edule*	20	10	64	296	1	8	6
Clams, etc. nei	*Bivalvia*	-	-	-	-	-	301	126
Northern shortfin squid	*Illex illecebrosus*	...	...	...	...	...	...	121
European flying squid	*Todarodes sagittatus*	...	...	...	...	...	...	14
Various squids nei	*Loliginidae, Ommastrephidae*	298	481	442	610	282	322	242
Octopuses, etc. nei	*Octopodidae*	25	13	7	3	10	10	14
Marine molluscs nei	*Mollusca*	-	-	-	-	-	66	-
European edible sea urchin	*Echinus esculentus*	10	2	5	1	2	1	5
	Country total	*384 632*	*333 030*	*292 673*	*324 274*	*280 957*	*281 806*	*356 309*
Isle of Man								
Freshwater fishes nei	*Osteichthyes*	0	0	0	0	0	0	0
Atlantic salmon	*Salmo salar*	0	-	-	-	-	-	-
Sea trout	*Salmo trutta*	0	-	-	-	-	-	-
Whitefishes nei	*Coregonus spp*	0	-	-	-	-	-	-
European plaice	*Pleuronectes platessa*	20	16	11	14	5	6	1
Witch flounder	*Glyptocephalus cynoglossus*	2	0	1	0	0	0	-
Common dab	*Limanda limanda*	0	0	0	-	-	-	-
Lemon sole	*Microstomus kitt*	2	4	0	4	3	3	1
Common sole	*Solea solea*	12	4	5	3	1	1	1
Megrim	*Lepidorhombus whiffiagonis*	-	-	3	2	...	...	-
Brill	*Scophthalmus rhombus*	1	1	0	0	1	1	0
Turbot	*Psetta maxima*	0	1	1	0	0	-	0
Atlantic cod	*Gadus morhua*	22	27	19	34	9	11	1
Ling	*Molva molva*	1	3	2	1	1	1	0
Haddock	*Melanogrammus aeglefinus*	27	38	9	13	7	19	1
Saithe(=Pollock)	*Pollachius virens*	11	11	9	7	2	1	0
Pollack	*Pollachius pollachius*	15	16	11	11	2	1	-
Whiting	*Merlangius merlangus*	41	28	24	33	5	2	1
European hake	*Merluccius merluccius*	23	18	28	30	3	3	1
Red mullet	*Mullus surmuletus*	-	-	-	-	-	-	4
European conger	*Conger conger*	0	0	0	...	-	-	-
Gurnards, searobins nei	*Triglidae*	5	2	1	1	1	...	1
Angler(=Monk)	*Lophius piscatorius*	27	34	27	28	9	5	2
Atlantic herring	*Clupea harengus*	615	693	821	0	1	...	35
Atlantic mackerel	*Scomber scombrus*	1	0	0	0	4	0	8
Dogfish sharks nei	*Squalidae*	24	25	25	12	19	11	3
Raja rays nei	*Raja spp*	9	10	6	6	3	5	1
Finfishes nei	*Osteichthyes*	0	-	-	0	-	-	-
Edible crab	*Cancer pagurus*	282	94	478	274	231	142	170
Norway lobster	*Nephrops norvegicus*	29	20	24	17	10	3	2
European lobster	*Homarus gammarus*	14	0	26	25	14	8	12
Whelk	*Buccinum undatum*	148	296	193	...	227	89	2
Queen scallop	*Aequipecten opercularis*	1 465	1 129	1 630	991	1 255	2 275	1 749
Great Atlantic scallop	*Pecten maximus*	931	1 064	933	706	794	965	1 115
Common squids nei	*Loligo spp*	7	3	2	2	2	-	1
	Country total	*3 734*	*3 537*	*4 289*	*2 214*	*2 609*	*3 552*	*3 112*
Italy								
Cyprinids nei	*Cyprinidae*	2 540	1 146	2 378	1 155	1 900	725	1 821
Freshwater fishes nei	*Osteichthyes*	3 900	3 647	2 895	2 346	2 316	2 819	2 643
European eel	*Anguilla anguilla*	886	883	1 010	682	645	549	446
Salmonoids nei	*Salmonoidei*	3 325	1 625	1 091	897	937	692	846
Common sole	*Solea solea*	6 065	3 597	3 085	2 638	2 252	2 165	2 966
Tonguefishes	*Cynoglossidae*	101	66	-	-	-	-	-
Turbots nei	*Scophthalmidae*	1 923	1 377	964	528	478	643	622
Flatfishes nei	*Pleuronectiformes*	-	-	261	244	96	107	117
Blue whiting(=Poutassou)	*Micromesistius poutassou*	1 769	1 546	1 300	1 449	1 451	1 261	1 167
European hake	*Merluccius merluccius*	38 051	30 707	17 971	13 166	9 754	9 219	9 248
Argentine hake	*Merluccius hubbsi*	44	-	-	-	-	-	-
Cape hakes	*Merluccius capensis,M.paradox.*	5	-	-	-	-	-	-
Gadiformes nei	*Gadiformes*	76	-	-	-	-	-	-
Mullets nei	*Mugilidae*	5 655	5 172	5 281	5 344	4 799	4 095	5 023
Dusky grouper	*Epinephelus marginatus*	1 454	558	640	124	89	97	252
European seabass	*Dicentrarchus labrax*	4 633	2 481	2 030	1 889	1 881	2 195	2 735
Shi drum	*Umbrina cirrosa*	956	495	351	138	138	158	259
Pandoras nei	*Pagellus spp*	2 937	1 152	1 445	836	751	1 171	949
Sargo breams nei	*Diplodus spp*	1 123	1 069	706	382	340	321	462
Common dentex	*Dentex dentex*	2 270	1 253	389	190	205	309	201
Gilthead seabream	*Sparus aurata*	2 179	1 743	1 859	1 717	1 754	1 939	2 675
Bogue	*Boops boops*	5 659	5 281	4 178	4 074	3 105	3 541	3 537
Picarels nei	*Spicara spp*	1 138	953	647	545	547	385	313
Surmullets(=Red mullets) nei	*Mullus spp*	9 441	11 325	7 499	7 491	8 771	9 044	7 121
Gobies nei	*Gobiidae*	1 452	1 311	1 085	991	800	712	665
Scorpionfishes nei	*Scorpaenidae*	-	-	-	-	-	-	14
Gurnards, searobins nei	*Triglidae*	4 116	3 908	3 473	3 293	2 668	2 168	2 264
Angler(=Monk)	*Lophius piscatorius*	2 409	2 345	6 672	2 845	1 705	1 269	1 244
European pilchard(=Sardine)	*Sardina pilchardus*	36 825	42 129	38 174	36 387	28 876	25 805	23 980
European anchovy	*Engraulis encrasicolus*	42 746	40 541	53 439	44 429	39 783	50 728	53 047
Atlantic bonito	*Sarda sarda*	1 512	2 233	4 580	2 121	1 614	1 116	1 006
Frigate and bullet tunas	*Auxis thazard, A.rochei*	1 435	229	499	254	439	215	375
Skipjack tuna	*Katsuwonus pelamis*	-	-	1 754	3 024	3 416	...	1 681
Atlantic bluefin tuna	*Thunnus thynnus*	7 062	10 006	9 548	4 059	3 279	3 845	4 377
Albacore	*Thunnus alalunga*	1 109	1 769	1 426	1 472	2 561	3 630	2 882
Yellowfin tuna	*Thunnus albacares*	-	-	1 340	2 299	2 626	...	1 332

E-5 Fish crustaceans, molluscs, etc — Capture production by countries or areas and species — Europe
Poissons, crustacés, mollusques, etc — Captures par pays ou zones et espèces — Europe
Peces, crustáceos, moluscos, etc — Capturas por países o áreas y especies — Europa

English name Nom anglais Nombre inglés	Scientific name Nom scientifique Nombre científico	1995 mt	1996 mt	1997 mt	1998 mt	1999 mt	2000 mt	2001 mt
Bigeye tuna	*Thunnus obesus*	-	-	457	612	848	...	57
Swordfish	*Xiphias gladius*	6 725	5 286	6 104	6 104	6 312	7 515	6 388
Tuna-like fishes nei	*Scombroidei*	-	-	1	-	-	-	8
Garfish	*Belone belone*	554	243	216	238	209	134	139
Silversides(=Sand smelts) nei	*Atherinidae*	1 530	1 112	1 101	883	851	725	736
Jack and horse mackerels nei	*Trachurus spp*	7 458	6 790	5 168	6 314	4 315	3 428	3 927
Leerfish	*Lichia amia*	948	643	400	197	249	185	183
Scomber mackerels nei	*Scomber spp*	6 949	8 012	7 866	7 277	5 748	5 522	6 033
Smooth-hounds nei	*Mustelus spp*	5 942	2 659	621	636	440	462	369
Rays, stingrays, mantas nei	*Rajiformes*	4 586	2 309	5 325	2 807	1 117	507	555
Marine fishes nei	*Osteichthyes*	39 476	33 212	28 258	27 773	23 355	24 132	27 185
Marine crabs nei	*Brachyura*	-	-	-	-	-	8	-
Common spiny lobster	*Palinurus elephas*	197	312	331	174	161	123	166
Norway lobster	*Nephrops norvegicus*	4 312	5 101	4 834	2 582	3 033	2 485	2 287
Caramote prawn	*Penaeus kerathurus*	196	3	-	-	-	-	-
Penaeus shrimps nei	*Penaeus spp*	23	30	-	-	-	-	-
Deepwater rose shrimp	*Parapenaeus longirostris*	7 998	7 065	7 019	4 410	4 631	7 500	6 980
Aristeid shrimps nei	*Aristeidae*	2 551	2 258	2 406	1 231	2 128	4 463	1 833
Natantian decapods nei	*Natantia*	40	306	345	456	493	370	686
Spottail mantis squillid	*Squilla mantis*	4 611	5 431	4 497	3 670	4 767	5 244	5 570
Marine crustaceans nei	*Crustacea*	4 461	3 657	3 681	3 245	2 294	2 130	1 849
Mediterranean mussel	*Mytilus galloprovincialis*	21 425	22 174	21 430	27 270 F	26 510 F	44 200 F	44 160 F
Striped venus	*Chamelea gallina*	32 609	36 707	28 604	28 830	36 462	34 191	34 916
Cuttlefish,bobtail squids nei	*Sepiidae, Sepiolidae*	12 389	8 937	9 398	8 330	7 080	6 325	7 370
Common squids nei	*Loligo spp*	5 731	5 368	4 152	2 237	1 909	1 890	2 352
European flying squid	*Todarodes sagittatus*	4 789	4 672	2 614	3 995	2 056	2 516	2 346
Various squids nei	*Loliginidae, Ommastrephidae*	507	411	36	54	101	95	11
Common octopus	*Octopus vulgaris*	9 970	9 301	9 519	10 411	8 844	9 173	10 036
Horned and musky octopuses	*Eledone spp*	2 500	1 939	1 590	1 506	1 041	1 621	1 661
Octopuses, etc. nei	*Octopodidae*	192	176	-	-	-	-	-
Marine molluscs nei	*Mollusca*	13 326	11 238	9 750	7 845	7 860	6 282	6 324
	Country total	*396 791*	*365 899*	*343 693*	*306 096*	*282 790*	*302 149*	*310 397*
Latvia								
Freshwater bream	*Abramis brama*	237	242	241	191	235	218	246
Common carp	*Cyprinus carpio*	5	3	3	6	5	3	5
Tench	*Tinca tinca*	15	15	24	22	41	29	35
Crucian carp	*Carassius carassius*	7	7	11	10	14	15	24
Roach	*Rutilus rutilus*	54	46	64	56	67	70	60
Orfe(=Ide)	*Leuciscus idus*	6	2	2	2	3	2	3
Vimba bream	*Vimba vimba*	55	58	57	92	119	92	109
Cyprinids nei	*Cyprinidae*	36	34	42	26	27	-	-
Northern pike	*Esox lucius*	55	56	47	55	72	72	73
Wels(=Som)catfish	*Silurus glanis*	1	1	2	1	2		-
European perch	*Perca fluviatilis*	64	56	56	55	87	58	83
Pike-perch	*Stizostedion lucioperca*	57	73	40	38	58	48	59
Ruffe	*Gymnocephalus cernuus*	4	5	5	-	1		-
Freshwater fishes nei	*Osteichthyes*	9	20	13	13	20	63	50
European eel	*Anguilla anguilla*	28	26	29	27	17	15	19
Atlantic salmon	*Salmo salar*	139	151	169	125	166	151	138
Sea trout	*Salmo trutta*	14	10	7	7	10	14	11
European smelt	*Osmerus eperlanus*	354	386	335	218	180	261	128
Vendace	*Coregonus albula*	5	5	5	6	7	4	5
Whitefishes nei	*Coregonus spp*	3	2	1	1	-		-
Salmonoids nei	*Salmonoidei*	1	-	-		-		-
River lamprey	*Lampetra fluviatilis*	95	140	80	79	120	135	88
Three-spined stickleback	*Gasterosteus aculeatus*	2	34	110	82	-	-	-
Greenland halibut	*Reinhardtius hippoglossoides*	-	-	-	-	-	215	291
European flounder	*Platichthys flesus*	362	294	367	364	509	418	613
Turbot	*Psetta maxima*	49	42	46	36	54	16	6
Atlantic cod	*Gadus morhua*	6 471	8 741	6 187	7 778	6 914	6 280	6 298
Southern blue whiting	*Micromesistius australis*	-	2	-	-	-	-	-
Senegalese hake	*Merluccius senegalensis*	8	68	27	16	320	280	125
Patagonian grenadier	*Macruronus magellanicus*	32	15	-	-	-	-	-
Mullets nei	*Mugilidae*	58	21	1	20	44	19	...
Grunts, sweetlips nei	*Haemulidae (=Pomadasyidae)*	-	-	2	-	-	-	-
Sargo breams nei	*Diplodus spp*	18	19	20	13	81	90	176
Large-eye dentex	*Dentex macrophthalmus*	-	-	24	57	91	190	71
Common dentex	*Dentex dentex*	694	436	8	...	...	...	19
Porgies, seabreams nei	*Sparidae*	-	-	-	48	80	53	-
Eelpout	*Zoarces viviparus*	143	139	80	41	32	23	26
Largehead hairtail	*Trichiurus lepturus*	-	12	-	1 232	1 502	544	13
Silver scabbardfish	*Lepidopus caudatus*	8	-	-	-	-	-	-
Atlantic redfishes nei	*Sebastes spp*	5 307	1 084	-	-	-	13	11
Atlantic herring	*Clupea harengus*	24 972	27 523	29 330	24 417	27 163	26 768	26 652
Sardinellas nei	*Sardinella spp*	17 032	24 209	6 497	6 064	15 031	7 886	7 306
European pilchard(=Sardine)	*Sardina pilchardus*	-	474	-	7	23	633	54
European sprat	*Sprattus sprattus*	24 383	34 211	49 314	44 858	42 834	46 186	42 769
European anchovy	*Engraulis encrasicolus*	-	-	-	1 978	4 876	10 142	8 623
Atlantic bonito	*Sarda sarda*	19	301	887	318	510	416	396
Yellowfin tuna	*Thunnus albacares*	55	151	223	97	25	36	72
Tuna-like fishes nei	*Scombroidei*	-	-	-	147	27	-	-
Garfish	*Belone belone*	-	-	-	-	-	-	11
Bluefish	*Pomatomus saltatrix*	7	155	116	31	116	144	17
Atlantic horse mackerel	*Trachurus trachurus*	6	-	-	-	-	-	-
Jack and horse mackerels nei	*Trachurus spp*	38 824	14 818	4 881	8 710	14 284	22 591	17 617
Jacks, crevalles nei	*Caranx spp*	-	-	11	-	-	-	-
Leerfish	*Lichia amia*	18	152	236	127	172	274	96

English name Nom anglais Nombre inglés	Scientific name Nom scientifique Nombre científico	1995 mt	1996 mt	1997 mt	1998 mt	1999 mt	2000 mt	2001 mt
Carangids nei	*Carangidae*	-	-	36	-	-	-	-
Chub mackerel	*Scomber japonicus*	3 494	3 765	1 931	2 562	3 123	7 151	9 227
Atlantic mackerel	*Scomber scombrus*	534	233	-	-	-	-	-
Groundfishes nei	*Osteichthyes*	43	-	-	-	-	-	-
Marine fishes nei	*Osteichthyes*	23 015	19 198	3 118	1 107	3 247	1 616	759
Northern prawn	*Pandalus borealis*	679	1 253	997	1 191	3 080	3 169	3 028
Natantian decapods nei	*Natantia*	-	-	-	-	-	-	19
Various squids nei	*Loliginidae, Ommastrephidae*	1 717	3 956	-	-	-	-	-
Octopuses, etc. nei	*Octopodidae*	-	-	-	-	-	-	2
	Country total	*149 194*	*142 644*	*105 682*	*102 331*	*125 389*	*136 403*	*125 433*
Liechtensten								
Freshwater fishes nei	*Osteichthyes*	0	0	0	0	0	0	0
	Country total	*0*	*0*	*0*	*0*	*0*	*0*	*0*
Lithuania								
Freshwater bream	*Abramis brama*	420	397	448	454	466	467	470
Freshwater breams nei	*Abramis spp*	4	7	10	12	11	13	26
Common carp	*Cyprinus carpio*	27	31	14	13	12	16	16
Tench	*Tinca tinca*	10	9	12	9	11	13	15
Bleak	*Alburnus alburnus*	-	-	-	-	-	3	4
Goldfish	*Carassius auratus*	41	30	30	38	32	45	38
Roach	*Rutilus rutilus*	314	431	593	645	647	635	643
Rudd	*Scardinius erythrophthalmus*	17	14	18	14	13	17	20
Orfe(=Ide)	*Leuciscus idus*	3	1	1	-	-	1	1
Chub	*Leuciscus cephalus*	-	-	-	-	-	3	3
Vimba bream	*Vimba vimba*	3	2	3	3	11	48	40
Sichel	*Pelecus cultratus*	-	-	-	4	3	8	12
Asp	*Aspius aspius*	3	4	6	9	6	6	5
Cyprinids nei	*Cyprinidae*	3	1	2	-	-	-	19
Northern pike	*Esox lucius*	78	66	71	56	62	71	68
Wels(=Som)catfish	*Silurus glanis*	0	0	-	-	-	0	0
Burbot	*Lota lota*	5	5	14	10	13	13	9
European perch	*Perca fluviatilis*	114	83	114	104	116	115	115
Pike-perch	*Stizostedion lucioperca*	44	56	65	51	60	78	115
Ruffe	*Gymnocephalus cernuus*	-	-	-	-	52	97	64
Freshwater fishes nei	*Osteichthyes*	50	45	90	88	10	6	2
European eel	*Anguilla anguilla*	10	12	11	17	18	11	12
Atlantic salmon	*Salmo salar*	2	10	4	5	6	6	4
Sea trout	*Salmo trutta*	3	-	2	3	4	5	3
European smelt	*Osmerus eperlanus*	105	81	190	334	365	214	360
Whitefishes nei	*Coregonus spp*	8	18	20	10	11	7	10
River lamprey	*Lampetra fluviatilis*	-	1	-	-	-	-	3
Three-spined stickleback	*Gasterosteus aculeatus*	-	-	-	-	13	22	18
European plaice	*Pleuronectes platessa*	194	330	624	736	571	618	1 137
Greenland halibut	*Reinhardtius hippoglossoides*	-	-	-	-	-	21	395
Witch flounder	*Glyptocephalus cynoglossus*	-	-	-	-	-	-	3
Amer. plaice(=Long rough dab)	*Hippoglossoides platessoides*	-	-	-	-	-	-	3
Yellowtail flounder	*Limanda ferruginea*	-	-	-	-	-	-	1
Turbot	*Psetta maxima*	-	-	-	62	58	23	18
Atlantic cod	*Gadus morhua*	3 629	5 520	4 694	3 296	4 371	4 721	3 852
Blue ling	*Molva dypterygia*	-	-	-	-	-	-	16
Blue whiting(=Poutassou)	*Micromesistius poutassou*	400	651	-	-	1 231	-	-
Senegalese hake	*Merluccius senegalensis*	...	...	...	180	307	180	42
Roughhead grenadier	*Macrourus berglax*	-	-	-	-	-	1	28
Roundnose grenadier	*Coryphaenoides rupestris*	-	-	-	-	-	-	137
Dentex nei	*Dentex spp*	8	-	-	-	-	-	-
Porgies, seabreams nei	*Sparidae*	...	...	...	192	157	155	32
Baird's slickhead	*Alepocephalus bairdii*	-	-	-	-	-	-	460
Largehead hairtail	*Trichiurus lepturus*	...	...	...	9 708	13	32	69
Black scabbardfish	*Aphanopus carbo*	-	-	-	-	-	-	3
Atlantic redfishes nei	*Sebastes spp*	22 893	10 649	-	1 769	3 884	6 687	20 182
Atlantic herring	*Clupea harengus*	7 058	4 257	3 330	2 368	1 313	1 198	1 639
Round sardinella	*Sardinella aurita*	...	...	...	10 575	8 680	6 324	4 167
European pilchard(=Sardine)	*Sardina pilchardus*	...	...	...	20	6	292	22
European sprat	*Sprattus sprattus*	4 799	10 165	6 018	4 460	3 117	1 682	3 135
European anchovy	*Engraulis encrasicolus*	...	...	...	3 612	13 774	16 137	8 441
Clupeoids nei	*Clupeoidei*	35	2 400	-	-	-	-	-
Tuna-like fishes nei	*Scombroidei*	...	...	...	467	110	80	153
Atlantic horse mackerel	*Trachurus trachurus*	232	7 400	-	-	421	5	344
Jack and horse mackerels nei	*Trachurus spp*	...	...	...	11 902	20 657	25 464	15 226
Leerfish	*Lichia amia*	1	-	-	-	-	-	-
Chub mackerel	*Scomber japonicus*	...	...	...	5 420	2 105	3 871	2 798
Atlantic mackerel	*Scomber scombrus*	6 236	7 334	-	2 823	4 936	2 085	1 949
Dogfish sharks nei	*Squalidae*	-	-	-	-	-	-	14
Raja rays nei	*Raja spp*	-	-	-	-	-	-	4
Finfishes nei	*Osteichthyes*	66	188	152	41	40	204	3
Marine fishes nei	*Osteichthyes*	9 572	33 330	25 680	3 495	1 099	899	80 150
Euro-American crayfishes nei	*Astacidae, Cambaridae*	1	1	1	0	1	1	0
Northern prawn	*Pandalus borealis*	980	1 585	1 785	3 340	4 167	6 376	5 413
Natantian decapods nei	*Natantia*	-	-	-	-	-	11	-
Various squids nei	*Loliginidae, Ommastrephidae*	...	3 400	...	233	2	-	-
	Country total	*57 368*	*88 514*	*44 002*	*66 578*	*72 962*	*78 987*	*151 931*
Luxembourg								
Freshwater fishes nei	*Osteichthyes*	0	0	0	0	0	0	0

English name Nom anglais Nombre inglés	Scientific name Nom scientifique Nombre científico	1995 mt	1996 mt	1997 mt	1998 mt	1999 mt	2000 mt	2001 mt
	Country total	*0*	*0*	*0*	*0*	*0*	*0*	*0*
Macedonia								
Common carp	*Cyprinus carpio*	32	10	9	...	...	25	6
Freshwater fishes nei	*Osteichthyes*	146	31	68	107	113	52	7
Sturgeons nei	*Acipenseridae*	0	0	2	6	-	-	-
Trouts nei	*Salmo spp*	30	37	51	18	22	131	115
	Country total	*208*	*78*	*130*	*131*	*135*	*208*	*128*
Malta								
Freshwater fishes nei	*Osteichthyes*	0	0	0	0	0	0	0
Common sole	*Solea solea*	0	0	0	0	0	0	0
Greater forkbeard	*Phycis blennoides*	-	-	2	3	4	5	0
European hake	*Merluccius merluccius*	1	2	4	5	6	6	0
Mullets nei	*Mugilidae*	0	0	0	0	0	0	0
Groupers nei	*Epinephelus spp*	15	19	27	15	0	15	15
Groupers, seabasses nei	*Serranidae*	1	1	1	1	2	2	2
European seabass	*Dicentrarchus labrax*	0	0	0	0	15	0	0
Common pandora	*Pagellus erythrinus*	3	5	6	6	6	5	2
Axillary seabream	*Pagellus acarne*	7	5	4	3	3	2	-
Sargo breams nei	*Diplodus spp*	0	2	0	4	4	2	2
Common dentex	*Dentex dentex*	0	0	1	0	1	1	1
Saddled seabream	*Oblada melanura*	1	1	2	1	2	2	1
Red porgy	*Pagrus pagrus*	6	8	9	8	6	6	4
Bogue	*Boops boops*	19	17	16	15	12	21	27
Salema	*Sarpa salpa*	0	0	0	1	0	0	0
Picarels nei	*Spicara spp*	3	7	7	7	8	9	6
Surmullets(=Red mullets) nei	*Mullus spp*	4	7	7	8	12	7	5
Greater weever	*Trachinus draco*	1	2	2	0	3	0	0
Gobies nei	*Gobiidae*	0	0	0	0	0	0	-
European conger	*Conger conger*	2	2	4	3	2	2	3
John dory	*Zeus faber*	0	0	1	1	2	1	0
Wreckfish	*Polyprion americanus*	8	9	14	8	8	8	8
Scorpionfishes nei	*Scorpaenidae*	6	8	3	8	12	11	0
Gurnards, searobins nei	*Triglidae*	1	2	0	2	4	2	1
Angler(=Monk)	*Lophius piscatorius*	0	0	1	0	2	0	-
Clupeoids nei	*Clupeoidei*	0	0	0	0	0	-	-
Atlantic bonito	*Sarda sarda*	0	2	7	2	2	1	1
Frigate and bullet tunas	*Auxis thazard, A.rochei*	2	3	6	6	3	1	1
Little tunny(=Atl.black skipj)	*Euthynnus alletteratus*	8	3	3	0	0	0	5
Atlantic bluefin tuna	*Thunnus thynnus*	587	399	393	407	447	376	219
Albacore	*Thunnus alalunga*	-	-	1	1	1	4	...
Marlins,sailfishes,etc. nei	*Istiophoridae*	1	1	1	-	-	-	-
Swordfish	*Xiphias gladius*	58	58	83	116	167	160	89
Tuna-like fishes nei	*Scombroidei*	0	0	0	0	-	-	-
Mediterranean horse mackerel	*Trachurus mediterraneus*	8	5	4	2	4	0	0
Greater amberjack	*Seriola dumerili*	4	9	6	6	6	3	2
Carangids nei	*Carangidae*	13	7	4	13	23	28	8
Common dolphinfish	*Coryphaena hippurus*	334	307	295	363	349	234	303
Chub mackerel	*Scomber japonicus*	13	23	29	40	19	34	32
Porbeagle	*Lamna nasus*	0	1	0	0	0	0	0
Picked dogfish	*Squalus acanthias*	24	28	28	23	18	19	17
Dogfish sharks nei	*Squalidae*	5	4	5	3	1	2	3
Angelsharks, sand devils nei	*Squatinidae*	0	0	0	0	0	0	0
Rays, stingrays, mantas nei	*Rajiformes*	5	7	8	5	6	7	0
Sharks, rays, skates, etc. nei	*Elasmobranchii*	4	3	2	11	4	13	0
Marine fishes nei	*Osteichthyes*	3 466	8 199	1	12	16	29	82
Natantian decapods nei	*Natantia*	5	9	16	18	24	23	36
Cuttlefish,bobtail squids nei	*Sepiidae, Sepiolidae*	0	5	3	3	5	4	0
Common squids nei	*Loligo spp*	0	2	2	2	2	3	2
European flying squid	*Todarodes sagittatus*	-	-	-	2	2	2	0
Octopuses, etc. nei	*Octopodidae*	6	11	11	9	11	9	5
	Country total	*4 621*	*9 183*	*1 019*	*1 143*	*1 224*	*1 059*	*882*
Moldova Rep								
Common carp	*Cyprinus carpio*	472	408	349	280	178	192	212
Crucian carp	*Carassius carassius*	164	128	166	159	104	132	127
Northern pike	*Esox lucius*	56	53	39	38	25	19	36
Pike-perch	*Stizostedion lucioperca*	6	5	4	1	2	1	12
Freshwater fishes nei	*Osteichthyes*	11	9	11	13	-	-	-
	Country total	*709*	*603*	*569*	*491*	*309*	*344*	*387*
Monaco								
Marine fishes nei	*Osteichthyes*	3 F	3 F	3 F	3 F	3 F	3 F	3 F
	Country total	*3 F*	*3 F*	*3 F*	*3 F*	*3 F*	*3 F*	*3 F*
Netherlands								
Freshwater bream	*Abramis brama*	...	75	65	399	355	350 F	350 F
Roaches nei	*Rutilus spp*	54	100	123	107	100	100 F	100 F
European perch	*Perca fluviatilis*	219	376	336	155	177	170 F	150 F
Pike-perch	*Stizostedion lucioperca*	79	100	89	61	104	101 F	104 F
Freshwater fishes nei	*Osteichthyes*	410	350	362	176	154	150 F	150 F
European eel	*Anguilla anguilla*	432	336	315	345	372	351 F	334 F
Atlantic salmon	*Salmo salar*	1	2	1	1	1	-	0

English name Nom anglais Nombre inglés	Scientific name Nom scientifique Nombre científico	1995 mt	1996 mt	1997 mt	1998 mt	1999 mt	2000 mt	2001 mt
Trouts nei	*Salmo spp*	0	-	-	-	-	-	-
European smelt	*Osmerus eperlanus*	2 952	856	1 033	327	1 097	1 154 F	1 161 F
Twaite shad	*Alosa fallax*	-	-	-	-	-	1	5
Atlantic halibut	*Hippoglossus hippoglossus*	5	3	5	4	2	1	-
European plaice	*Pleuronectes platessa*	44 262	35 539	34 272	30 592	37 543	35 079	33 835
Witch flounder	*Glyptocephalus cynoglossus*	7	0	1	4	9	7	1
Common dab	*Limanda limanda*	...	...	...	7 983	8 656	6 544	5 969
Lemon sole	*Microstomus kitt*	...	...	...	839	681	492	456
European flounder	*Platichthys flesus*	...	...	...	4 942	3 159	2 658	2 621
Common sole	*Solea solea*	20 927	15 563	10 370	15 308	16 329	15 343	13 737
Megrim	*Lepidorhombus whiffiagonis*	26	11	23	31	28	20	11
Brill	*Scophthalmus rhombus*	943	736	598	811	809	1 005	1 093
Turbot	*Psetta maxima*	2 476	1 780	1 866	1 700	1 812	2 287	2 277
Atlantic cod	*Gadus morhua*	11 189	9 307	11 838	14 724	9 075	6 000	3 656
Ling	*Molva molva*	-	-	-	-	-	5	4
Haddock	*Melanogrammus aeglefinus*	146	111	494	289	115	121	295
Saithe(=Pollock)	*Pollachius virens*	9	19	42	8	7	11	19
Pollack	*Pollachius pollachius*	17	19	15	7	5	5	1
Norway pout	*Trisopterus esmarkii*	138	13	85	3	1	3	-
Pouting(=Bib)	*Trisopterus luscus*	-	-	-	-	-	612	645
Blue whiting(=Poutassou)	*Micromesistius poutassou*	22 685	16 407	24 132	27 693	32 889	43 145	63 625
Whiting	*Merlangius merlangus*	3 640	3 411	2 554	1 981	1 806	1 899	2 619
European hake	*Merluccius merluccius*	78	111	62	75	98	43	72
Mullets nei	*Mugilidae*	4	-	0	-	17	36	184
Groupers, seabasses nei	*Serranidae*	-	-	-	17	14	-	-
European seabass	*Dicentrarchus labrax*	-	8	1	49	32	60	79
Blackspot(=red) seabream	*Pagellus bogaraveo*	-	38	-	-	28	71	2
Canary dentex	*Dentex canariensis*	-	-	-	-	-	20	-
Red mullet	*Mullus surmuletus*	-	1	0	-	-	235	560
Greater weever	*Trachinus draco*	-	-	-	-	-	-	6
Argentines	*Argentina spp*	4 136	3 953	4 696	4 964	8 033	3 636	3 659
European conger	*Conger conger*	-	-	-	-	-	-	1
Wolffishes(=Catfishes) nei	*Anarhichas spp*	50	6	16	36	21	10	2
Largehead hairtail	*Trichiurus lepturus*	-	33	-	103	401	115	697
Black scabbardfish	*Aphanopus carbo*	-	-	-	-	11	7	-
Atlantic redfishes nei	*Sebastes spp*	29	41	53	20	16	19	8
Tub gurnard	*Chelidonichthys lucerna*	-	-	-	-	-	1 164	-
Red gurnard	*Chelidonichthys cuculus*	-	-	-	-	-	46	1 724
Grey gurnard	*Eutrigla gurnardus*	-	-	-	-	-	459	295
Angler(=Monk)	*Lophius piscatorius*	362	227	319	259	169	170	168
Atlantic herring	*Clupea harengus*	99 447	77 605	65 448	77 090	78 741	75 221	66 357
Sardinellas nei	*Sardinella spp*	-	41 488	86 635	110 091	115 753	122 783	134 490
European pilchard(=Sardine)	*Sardina pilchardus*	116	1 242	6 488	8 998	7 624	17 862	11 786
European sprat	*Sprattus sprattus*	402	293	806	54	264	307	136
European anchovy	*Engraulis encrasicolus*	20	6	1	16	3	-	3
Atlantic bonito	*Sarda sarda*	-	1 694	1 625	2 171	966	1 507	1 791
Garfish	*Belone belone*	-	-	-	-	-	-	2
Atlantic horse mackerel	*Trachurus trachurus*	113 828	135 965	122 683	103 248	84 891	65 994	84 011
Jack and horse mackerels nei	*Trachurus spp*	-	1 938	3 245	3 163	2 847	9 053	14 476
Chub mackerel	*Scomber japonicus*	-	857	3 202	1 836	1 561	12 005	12 708
Atlantic mackerel	*Scomber scombrus*	35 787	24 246	23 702	30 163	27 816	32 403	33 109
Picked dogfish	*Squalus acanthias*	-	-	-	-	-	28	39
Raja rays nei	*Raja spp*	-	-	-	550	480	631	748
Various sharks nei	*Selachimorpha(Pleurotremata)*	-	-	-	-	-	-	3
Marine fishes nei	*Osteichthyes*	19 457	16 914	19 406	4 127	3 897	1 201	1 781
Edible crab	*Cancer pagurus*	-	-	-	-	-	146	300
Norway lobster	*Nephrops norvegicus*	253	423	627	694	662	572	853
European lobster	*Homarus gammarus*	-	-	-	-	13	12	33
Penaeus shrimps nei	*Penaeus spp*	-	-	-	-	1	1	3
Common shrimp	*Crangon crangon*	13 912	12 067	13 054	11 871	13 772	11 496	14 081
Whelk	*Buccinum undatum*	-	-	-	-	-	121	163
Great Atlantic scallop	*Pecten maximus*	-	228	188	408	306	249	274
Common edible cockle	*Cerastoderma edule*	39 594	6 300	10 923	68 133	50 888	19 633	-
Cuttlefish,bobtail squids nei	*Sepiidae, Sepiolidae*	-	-	-	-	-	101	162
Various squids nei	*Loliginidae, Ommastrephidae*	-	-	-	-	-	773	171
Octopuses, etc. nei	*Octopodidae*	-	-	-	-	-	-	7
	Country total	*438 092*	*410 798*	*451 799*	*536 626*	*514 611*	*495 804*	*518 162*
Norway								
Northern pike	*Esox lucius*	...	...	13	7	...	...	...
Burbot	*Lota lota*	...	...	1	1	...	...	...
European perch	*Perca fluviatilis*	...	...	9	3	...	...	...
Pike-perch	*Stizostedion lucioperca*	...	...	5	3	...	...	...
European eel	*Anguilla anguilla*	454	352	497	363	475	281	304
Atlantic salmon	*Salmo salar*	845	793	638	753	827	1 054	1 125 F
Chars nei	*Salvelinus spp*	104	81	78	94	93	119	111 F
Grayling	*Thymallus thymallus*	...	...	1	...	...	...	...
Vendace	*Coregonus albula*	...	...	10	4	10	6	5 F
European whitefish	*Coregonus lavaretus*	...	...	57	52	54	47	45 F
Atlantic halibut	*Hippoglossus hippoglossus*	551	678	879	672	696	1 039	868
European plaice	*Pleuronectes platessa*	1 166	1 731	2 857	1 872	1 816	1 943	2 773
Greenland halibut	*Reinhardtius hippoglossoides*	14 073	17 073	12 343	11 948	19 704	13 028	15 152
Witch flounder	*Glyptocephalus cynoglossus*	100	80	86	140	135	97	88
Amer. plaice(=Long rough dab)	*Hippoglossoides platessoides*	-	-	119	24	15	0	15
Common dab	*Limanda limanda*	-	-	-	-	-	-	54
Lemon sole	*Microstomus kitt*	31	47	63	59	59	60	53
European flounder	*Platichthys flesus*	-	-	-	-	-	-	3
Soles nei	*Soleidae*	...	...	...	...	...	...	95

English name Nom anglais Nombre inglés	Scientific name Nom scientifique Nombre científico	1995 mt	1996 mt	1997 mt	1998 mt	1999 mt	2000 mt	2001 mt
Brill	*Scophthalmus rhombus*	20	21	26	26	30	27	26
Turbot	*Psetta maxima*	53	54	57	45	48	68	94
Flatfishes nei	*Pleuronectiformes*	337	376	477	389	475	346	153
Moras nei	*Moridae*	-	-	-	-	-	-	277
Tusk(=Cusk)	*Brosme brosme*	18 682	19 483	13 797	21 032	23 274	21 912	18 778
Atlantic cod	*Gadus morhua*	365 333	358 395	401 277	321 428	256 555	220 120	208 856
Ling	*Molva molva*	18 172	18 931	15 295	22 719	19 217	16 899	13 562
Blue ling	*Molva dypterygia*	734	530	497	420	544	834	1 020
Greater forkbeard	*Phycis blennoides*	-	-	-	-	-	-	1 340
Haddock	*Melanogrammus aeglefinus*	79 834	97 115	106 155	79 008	53 243	45 920	51 648
Saithe(=Pollock)	*Pollachius virens*	218 853	221 638	183 451	194 452	198 387	169 653	169 505
Pollack	*Pollachius pollachius*	3 071	2 318	2 230	2 247	2 928	3 385	2 888
Norway pout	*Trisopterus esmarkii*	118 081	103 126	47 032	27 575	51 124	52 912	27 123
Blue whiting(=Poutassou)	*Micromesistius poutassou*	261 362	356 054	348 268	570 665	534 570	553 478	573 686
Whiting	*Merlangius merlangus*	334	210	140	116	143	145	237
European hake	*Merluccius merluccius*	783	938	981	825	609	696	635
Southern hake	*Merluccius australis*	57	210	117	16	-	-	-
Blue grenadier	*Macruronus novaezelandiae*	6 100	6 614	5 576	4 633	-	-	-
Roughhead grenadier	*Macrourus berglax*	-	-	-	-	-	-	148
Roundnose grenadier	*Coryphaenoides rupestris*	-	-	-	-	-	-	78
Gadiformes nei	*Gadiformes*	263	217	262	940	438	868	-
Wrasses, hogfishes, etc. nei	*Labridae*	-	-	-	-	-	-	6
Sandeels(=Sandlances) nei	*Ammodytes spp*	263 490	160 702	350 672	343 625	187 589	119 015	187 459
Argentines	*Argentina spp*	6 419	6 817	5 167	8 654	7 823	6 107	14 668
European conger	*Conger conger*	0	0	0	0	1	0	0
Alfonsinos nei	*Beryx spp*	-	-	836	1 066	-	324	-
Orange roughy	*Hoplostethus atlanticus*	1	5	34	15	-	642	-
Oreo dories nei	*Oreosomatidae*	-	1	-	7	-	175	-
Atlantic wolffish	*Anarhichas lupus*	...	...	...	...	...	...	1 111
Spotted wolffish	*Anarhichas minor*	...	...	...	...	...	...	1 111
Wolffishes(=Catfishes) nei	*Anarhichas spp*	7 589	6 819	12 769	16 332	6 398	6 378	12 205
Giant stargazer	*Kathetostoma giganteum*	-	1	1	1	-	-	-
Oilfish	*Ruvettus pretiosus*	-	-	5	-	-	-	-
South Pacific breams nei	*Seriolella spp*	-	121	70	4	-	-	-
Atlantic redfishes nei	*Sebastes spp*	23 282	28 679	22 687	28 560	30 856	25 653	28 657
Lumpfish(=Lumpsucker)	*Cyclopterus lumpus*	4 015	4 355	5 652	1 365	2 059	2 374	5 184
Angler(=Monk)	*Lophius piscatorius*	1 731	2 071	1 447	2 646	3 239	4 357	4 974
Atlantic herring	*Clupea harengus*	686 705	763 073	923 165	831 844	829 008	799 731	581 161
European pilchard(=Sardine)	*Sardina pilchardus*	-	-	-	3 421	-	-	-
European sprat	*Sprattus sprattus*	40 969	59 115	7 051	35 166	22 214	6 425	12 465
Atlantic bluefin tuna	*Thunnus thynnus*	-	-	-	-	5	0	-
Swordfish	*Xiphias gladius*	-	-	1	-	-	-	-
Capelin	*Mallotus villosus*	27 740	207 706	157 889	88 226	91 813	374 580	482 835
Garfish	*Belone belone*	3	1	2	1	1	0	0
Atlantic horse mackerel	*Trachurus trachurus*	96 132	15 556	46 491	13 366	46 657	2 084	7 988
Jack and horse mackerels nei	*Trachurus spp*	-	0	1	-	-	-	-
Atlantic mackerel	*Scomber scombrus*	202 209	136 699	137 256	158 340	161 046	174 173	180 603
Basking shark	*Cetorhinus maximus*	108	1 979	1 159	137	77	293	200
Porbeagle	*Lamna nasus*	27	28	17	28	33	22	17
Picked dogfish	*Squalus acanthias*	3 939	2 749	1 567	1 293	1 461	1 643	1 424
Leafscale gulper shark	*Centrophorus squamosus*	-	-	-	-	-	-	1
Portuguese dogfish	*Centroscymnus coelolepis*	-	-	-	-	-	-	13
Dogfish sharks nei	*Squalidae*	-	-	-	-	-	-	313
Blue skate	*Raja batis*	...	...	...	...	...	...	65
Raja rays nei	*Raja spp*	951	798	591	752	791	778	725
Rabbit fish	*Chimaera monstrosa*	-	-	-	-	-	-	70
Ratfishes nei	*Hydrolagus spp*	-	-	-	-	-	-	6
Various sharks nei	*Selachimorpha(Pleurotremata)*	0	0	1	0	13	119	72
Finfishes nei	*Osteichthyes*	128	158	322	2 791	3 389	3 201	1 909
Marine fishes nei	*Osteichthyes*	208	235	155	187	-	42	-
Noble crayfish	*Astacus astacus*	...	...	10	10	10	...	...
Edible crab	*Cancer pagurus*	1 807	1 889	2 204	2 984	2 837	2 897	3 476
Marine crabs nei	*Brachyura*	-	-	-	-	-	-	2
Norway lobster	*Nephrops norvegicus*	166	188	187	293	383	346	281
European lobster	*Homarus gammarus*	34	30	35	45	59	52	40
Red king crab	*Paralithodes camtschaticus*	32	70	71	124	202	203	434
Northern prawn	*Pandalus borealis*	39 250	41 505	41 961	57 141	63 538	66 208	66 336
Flat oysters nei	*Ostrea spp*	0	-	-	-	-	-	5
Blue mussel	*Mytilus edulis*	8	4	0	-	1	10	-
Horse mussels nei	*Modiolus spp*	7	20	30	20	7	2	2
Great Atlantic scallop	*Pecten maximus*	66	14	39	114	425	570	670
Iceland scallop	*Chlamys islandica*	-	-	-	-	-	-	14
Scallops nei	*Pectinidae*	7 310	3	16	21	12	14	13
Common edible cockle	*Cerastoderma edule*	-	-	-	-	-	-	33
European flying squid	*Todarodes sagittatus*	352	0	190	2	0	0	-
Marine molluscs nei	*Mollusca*	40	1	14	111	118	60	9
	Country total	*2 524 111*	*2 648 457*	*2 863 059*	*2 861 223*	*2 627 534*	*2 703 415*	*2 687 303*
Poland								
Freshwater bream	*Abramis brama*	1 505	3 622	2 367	2 057	2 321	2 945	2 576
Freshwater breams nei	*Abramis spp*	267	373	...	...	235	490	400
Common carp	*Cyprinus carpio*	-	77	82	78	37	45	50
Tench	*Tinca tinca*	70	97	101	91	102	160	113
Crucian carp	*Carassius carassius*	60	46	50	47	49	95	87
Roach	*Rutilus rutilus*	1 987	3 095	1 834	1 662	1 962	2 188	2 175
Orfe(=Ide)	*Leuciscus idus*	0	0	-	-	-	9	7
Grass carp(=White amur)	*Ctenopharyngodon idellus*	3	4	11	4	2	4	4
Silver carp	*Hypophthalmichthys molitrix*	198	211	180	106	136	185	120

English name Nom anglais Nombre inglés	Scientific name Nom scientifique Nombre científico	1995 mt	1996 mt	1997 mt	1998 mt	1999 mt	2000 mt	2001 mt
Vimba bream	*Vimba vimba*	...	...	...	...	...	5	2
Asp	*Aspius aspius*	...	...	...	...	...	6	5
Cyprinids nei	*Cyprinidae*	0	0	395	330	-	2	2
Northern pike	*Esox lucius*	245	1 076	280	226	262	363	311
Wels(=Som)catfish	*Silurus glanis*	0	0	0	1	1	7	4
Burbot	*Lota lota*	0	0	9	10	12	18	26
European perch	*Perca fluviatilis*	952	1 169	1 584	1 172	1 124	922	1 119
Pike-perch	*Stizostedion lucioperca*	440	512	480	381	537	546	478
Freshwater fishes nei	*Osteichthyes*	21 255	13 845	10 074	10 327	10 417	12 508	13 276
European eel	*Anguilla anguilla*	627	639	489	454	474	429	426
Atlantic salmon	*Salmo salar*	133	125	110	114	118	145	161
Trouts nei	*Salmo spp*	187	190	200	329	385	718	577
Rainbow trout	*Oncorhynchus mykiss*	...	...	35	27	14	12	12
European smelt	*Osmerus eperlanus*	-	-	38	53	212	19	20
Vendace	*Coregonus albula*	232	234	297	217	286	275	227
European whitefish	*Coregonus lavaretus*	28	33	24	14	21	29	13
Lampreys nei	*Petromyzontidae*	0	0	2	2	2	6	5
Atlantic halibut	*Hippoglossus hippoglossus*	-	-	-	-	-	-	488
Pacific halibut	*Hippoglossus stenolepis*	-	-	-	-	-	-	4
Greenland halibut	*Reinhardtius hippoglossoides*	-	-	12	31	8	4	4
Amer. plaice(=Long rough dab)	*Hippoglossoides platessoides*	-	-	-	-	-	-	1
Flatfishes nei	*Pleuronectiformes*	8 964	8 836	6 168	5 835	5 779	5 601	6 725
Atlantic cod	*Gadus morhua*	25 001	35 968	34 295	27 705	28 056	23 340	23 310
Ling	*Molva molva*	-	-	-	-	-	-	19
Haddock	*Melanogrammus aeglefinus*	-	18	35	27	24	16	96
Saithe(=Pollock)	*Pollachius virens*	592	365	822	813	862	747	727
Alaska pollock(=Walleye poll.)	*Theragra chalcogramma*	249 257	116 257	125 413	81 889	65 508	33 192	16 590
Southern blue whiting	*Micromesistius australis*	8 923	3 402	-	-	-	-	-
Whiting	*Merlangius merlangus*	-	-	-	1	-	-	-
Senegalese hake	*Merluccius senegalensis*	-	64	-	-	-	-	87
Argentine hake	*Merluccius hubbsi*	-	-	-	-	35	-	-
North Pacific hake	*Merluccius productus*	-	-	-	-	-	977	-
Cape hakes	*Merluccius capensis,M.paradox.*	-	3	-	-	-	-	-
Patagonian grenadier	*Macruronus magellanicus*	-	146	-	-	86	-	73
Grenadiers nei	*Macrourus spp*	-	-	-	-	13	-	-
Roundnose grenadier	*Coryphaenoides rupestris*	-	-	5 867	6 769	546	-	179
Porgies, seabreams nei	*Sparidae*	-	7	5	-	-	-	-
Antarctic rockcods, noties nei	*Nototheniidae*	-	-	-	-	207	-	-
Alfonsinos nei	*Beryx spp*	-	-	1 964	-	-	-	-
Atlantic wolffish	*Anarhichas lupus*	-	-	19	40	6	18	8
Pacific ocean perch	*Sebastes alutus*	-	-	-	-	-	21	-
Atlantic redfishes nei	*Sebastes spp*	-	-	777	12	6	2	9
Atlantic herring	*Clupea harengus*	45 676	31 246	28 939	21 873	19 229	24 516	37 611
Sardinellas nei	*Sardinella spp*	-	7 166	2 553	-	-	-	3 463
European pilchard(=Sardine)	*Sardina pilchardus*	-	2 439	1 269	-	-	-	-
European sprat	*Sprattus sprattus*	46 182	77 472	105 298	59 090	71 706	84 324	85 757
Atlantic bonito	*Sarda sarda*	-	225	39	-	-	-	521
Cape horse mackerel	*Trachurus capensis*	3 058	1 700	-	-	-	-	3 098
Jack and horse mackerels nei	*Trachurus spp*	-	3 583	281	-	-	-	1 449
Scads nei	*Decapterus spp*	-	54	-	-	-	-	-
Chub mackerel	*Scomber japonicus*	-	4 086	480	-	-	-	1 666
Atlantic mackerel	*Scomber scombrus*	-	-	22	-	-	-	-
Raja rays nei	*Raja spp*	-	-	-	-	-	-	2
Sharks, rays, skates, etc. nei	*Elasmobranchii*	-	-	-	-	-	-	11
Groundfishes nei	*Osteichthyes*	165	36	39	41	-	-	-
Pelagic fishes nei	*Osteichthyes*	281	60	66	61	-	-	-
Finfishes nei	*Osteichthyes*	53	23	34	23	-	17	307
Marine fishes nei	*Osteichthyes*	228	2 184	642	-	8	25	6 001
Natantian decapods nei	*Natantia*	-	-	824	691	894	1 732	263
Antarctic krill	*Euphausia superba*	9 384	20 610	19 156	15 714	18 554	20 721	14 568
Patagonian squid	*Loligo gahi*	-	-	-	19	4 875	-	-
Argentine shortfin squid	*Illex argentinus*	282	1	-	-	-	970	683
	Country total	*426 235*	*341 299*	*353 661*	*238 336*	*235 111*	*218 354*	*225 916*

Portugal

English name	Scientific name	1995	1996	1997	1998	1999	2000	2001
Common carp	*Cyprinus carpio*	0	0	0	0	0	0	0
Crucian carp	*Carassius carassius*	2	0	0	0	0	0	0
Northern pike	*Esox lucius*	0	0	0	-	-	0	0
Freshwater fishes nei	*Osteichthyes*	0	0	0	0	0	0	1
Sturgeons nei	*Acipenseridae*	0	0	-	-	-	-	-
European eel	*Anguilla anguilla*	...	...	...	...	30	29	37
Atlantic salmon	*Salmo salar*	-	-	-	-	0	0	-
Sea trout	*Salmo trutta*	1	0	-	-	1	1	1
Allis and twaite shads	*Alosa alosa, A.fallax*	39	18	17	21	17	20	22
West African ilisha	*Ilisha africana*	-	-	-	-	1	-	-
Diadromous clupeoids nei	*Clupeoidei*	35	35	36	30	-	-	-
Sea lamprey	*Petromyzon marinus*	4	3	2	2	4	6	6
Lefteye flounders nei	*Bothidae*	18	22	2	...	89	119	119
Atlantic halibut	*Hippoglossus hippoglossus*	17	14	17	31	51	30	45
European plaice	*Pleuronectes platessa*	147	137	89	115	95	124	145
Greenland halibut	*Reinhardtius hippoglossoides*	2 020	3 395	3 393	3 341	4 044	4 725	5 066
Witch flounder	*Glyptocephalus cynoglossus*	422	270	380	403	535	254	599
Amer. plaice(=Long rough dab)	*Hippoglossoides platessoides*	219	376	446	645	789	510	704
Yellowtail flounder	*Limanda ferruginea*	-	-	-	85	426	153	351
European flounder	*Platichthys flesus*	0	0	-	-	-	-	-
Common sole	*Solea solea*	235	167	151	113	121	152	201
Sand sole	*Solea lascaris*	155	77	95	118	90	116	142
Wedge sole	*Dicologlossa cuneata*	59	87	8	0	111	119	121

E-5 Fish crustaceans, molluscs, etc — Capture production by countries or areas and species — Europe
Poissons, crustacés, mollusques, etc — Captures par pays ou zones et espèces — Europe
Peces, crustáceos, moluscos, etc — Capturas por países o áreas y especies — Europa

English name Nom anglais Nombre inglés	Scientific name Nom scientifique Nombre científico	1995 mt	1996 mt	1997 mt	1998 mt	1999 mt	2000 mt	2001 mt
Soles nei	*Soleidae*	164	248	124	60	854	1 124	904
Megrim	*Lepidorhombus whiffiagonis*	53	58	26	47	54	47	22
Brill	*Scophthalmus rhombus*	57	48	39	33	39	46	57
Turbot	*Psetta maxima*	57	40	28	27	34	63	83
Flatfishes nei	*Pleuronectiformes*	1 662	1 304	1 083	964	11	11	6
Tadpole codling	*Salilota australis*	-	-	-	-	-	12	-
Tusk(=Cusk)	*Brosme brosme*	0	0	-	-	-	-	-
Atlantic cod	*Gadus morhua*	7 323	8 083	9 079	6 042	4 212	3 778	4 384
Blue ling	*Molva dypterygia*	29	25	21	14	10	14	9
Greater forkbeard	*Phycis blennoides*	129	113	46	45	54	98	92
Brazilian codling	*Urophycis brasiliensis*	-	-	-	-	-	-	3
Red hake	*Urophycis chuss*	230	125	56	18	77	42	273
Haddock	*Melanogrammus aeglefinus*	607	208	207	55	48	144	128
Saithe(=Pollock)	*Pollachius virens*	5	24	13	49	37	64	86
Pollack	*Pollachius pollachius*	2	2	2	1	1	15	41
Pouting(=Bib)	*Trisopterus luscus*	3 070	2 491	2 051	2 254	2 792	3 299	4 511
Blue whiting(=Poutassou)	*Micromesistius poutassou*	2 346	3 565	2 448	1 900	2 676	2 169	1 763
Southern blue whiting	*Micromesistius australis*	-	-	-	-	-	1	-
Whiting	*Merlangius merlangus*	169	184	139	115	76	77	38
European hake	*Merluccius merluccius*	3 466	3 622	2 578	2 563	3 217	3 061	3 032
Senegalese hake	*Merluccius senegalensis*	38	223	102	42	17	...	...
Argentine hake	*Merluccius hubbsi*	2 371	4 253	603	310	-	3	-
Hakes nei	*Merluccius spp*	49	1 515	914	1 027	474	-	365
Cape hakes	*Merluccius capensis,M.paradox.*	-	-	-	-	-	-	1
Patagonian grenadier	*Macruronus magellanicus*	-	-	-	-	-	32	-
Roughhead grenadier	*Macrourus berglax*	1 377	787	762	1 090	1 299	395	610
Grenadiers nei	*Macrourus spp*	80	80	-	-	-	-	1
Roundnose grenadier	*Coryphaenoides rupestris*	0	0	-	-	-	-	-
Gadiformes nei	*Gadiformes*	667	724	724	726	596	514	492
Morays	*Muraenidae*	...	...	...	...	193	169	145
Mullets nei	*Mugilidae*	398	297	303	336	324	336	376
Dusky grouper	*Epinephelus marginatus*	23	123	13	27	8	2	2
White grouper	*Epinephelus aeneus*	3	5	0	-	-	-	1
Groupers nei	*Epinephelus spp*	116	203	36	11	9	2	5
Groupers, seabasses nei	*Serranidae*	383	461	298	338	336	292	180
European seabass	*Dicentrarchus labrax*	68	57	40	38	37	49	43
Seabasses nei	*Dicentrarchus spp*	-	-	5	6	336	405	378
Cardinalfishes, etc. nei	*Apogonidae*	11	11	0	-	-	0	-
Rubberlip grunt	*Plectorhinchus mediterraneus*	712	1 436	544	115	267	87	12
Bastard grunt	*Pomadasys incisus*	1	1	0	-	-	0	0
Grunts, sweetlips nei	*Haemulidae (=Pomadasyidae)*	111	186	87	69	47	35	27
Meagre	*Argyrosomus regius*	-	0	0	3	3	4	30
Southern meagre(=Mulloway)	*Argyrosomus hololepidotus*	-	-	-	-	-	-	1
Boe drum	*Pteroscion peli*	350	582	136	101	18	2	0
West African croakers nei	*Pseudotolithus spp*	115	157	36	15	49	635	25
Croakers, drums nei	*Sciaenidae*	60	110	52	30	239	138	175
Blackspot(=red) seabream	*Pagellus bogaraveo*	1 391	1 446	1 350	1 511	1 490	1 033	1 128
Common pandora	*Pagellus erythrinus*	94	122	158	134	105	151	128
Axillary seabream	*Pagellus acarne*	1 052	1 198	970	883	997	1 298	1 202
Red pandora	*Pagellus bellottii*	-	1	-	-	-	-	-
Sargo breams nei	*Diplodus spp*	34	65	18	4	1 051	979	799
Large-eye dentex	*Dentex macrophthalmus*	20	71	74	78	66	0	0
Common dentex	*Dentex dentex*	18	20	21	38	25	13	16
Dentex nei	*Dentex spp*	11	15	2	63	28	6	2
Black seabream	*Spondyliosoma cantharus*	13	23	5	4	165	177	158
Saddled seabream	*Oblada melanura*	-	1	0	1	-	0	0
Red porgy	*Pagrus pagrus*	118	171	274	635	594	476	269
Pargo breams nei	*Pagrus spp*	124	196	31	11	4	41	145
Gilthead seabream	*Sparus aurata*	202	213	189	173	151	183	213
Bogue	*Boops boops*	451	417	420	358	426	670	958
Sand steenbras	*Lithognathus mormyrus*	...	...	...	...	178	158	109
Salema	*Sarpa salpa*	...	...	...	...	336	246	320
Porgies, seabreams nei	*Sparidae*	2 615	2 477	2 044	1 859	6	21	3
Picarels nei	*Spicara spp*	...	...	...	...	43	22	25
Red mullet	*Mullus surmuletus*	64	52	69	66	180	155	191
Surmullets(=Red mullets) nei	*Mullus spp*	0	3	0	1	0	0	-
Wrasses, hogfishes, etc. nei	*Labridae*	...	...	...	...	44	37	40
Parrotfish	*Sparisoma cretense*	-	-	-	-	56	89	162
Sandeels(=Sandlances) nei	*Ammodytes spp*	64	41	18	9	13	29	40
Greater weever	*Trachinus draco*	0	0	0	2	7	12	11
Gobies nei	*Gobiidae*	0	0	0	-	0	-	-
Triggerfishes, durgons nei	*Balistidae*	4	9	4	1	44	38	21
Toadfishes, etc. nei	*Batrachoididae*	-	-	-	-	70	91	75
European conger	*Conger conger*	3 439	3 355	2 867	2 861	2 473	2 002	1 896
Conger eels, etc. nei	*Congridae*	-	-	-	-	-	-	1
Longspine snipefish	*Macroramphosus scolopax*	0	-	-	-	-	-	-
Pink cusk-eel	*Genypterus blacodes*	-	-	-	-	-	13	89
Alfonsinos nei	*Beryx spp*	33	126	58	51	132	89	67
Orange roughy	*Hoplostethus atlanticus*	-	-	-	-	117	157	161
Slimeheads nei	*Trachichthyidae*	-	-	-	-	-	-	470
John dory	*Zeus faber*	166	166	179	317	366	431	457
Silvery John dory	*Zenopsis conchifer*	3	9	6	6	3	-	0
Wreckfish	*Polyprion americanus*	697	644	416	389	321	348	310
Patagonian toothfish	*Dissostichus eleginoides*	-	-	-	-	-	3	-
Spotted wolffish	*Anarhichas minor*	...	...	...	...	...	...	20
Wolffishes(=Catfishes) nei	*Anarhichas spp*	1 860	292	410	617	645	229	304
Stargazer	*Uranoscopus scaber*	...	...	...	...	15	50	46
Oilfish	*Ruvettus pretiosus*	-	-	-	-	14	14	11
Largehead hairtail	*Trichiurus lepturus*	0	0	0	0	3	0	0

E-5 Fish crustaceans, molluscs, etc — Capture production by countries or areas and species — Europe
Poissons, crustacés, mollusques, etc — Captures par pays ou zones et espèces — Europe
Peces, crustáceos, moluscos, etc — Capturas por países o áreas y especies — Europa

English name Nom anglais Nombre inglés	Scientific name Nom scientifique Nombre científico	1995 mt	1996 mt	1997 mt	1998 mt	1999 mt	2000 mt	2001 mt
Silver scabbardfish	*Lepidopus caudatus*	9 050	10 996	7 584	5 512	3 285	66	86
Black scabbardfish	*Aphanopus carbo*	7 744	10 434	7 576	7 583	7 181	7 070	6 753
Golden redfish	*Sebastes marinus*	1	2	7	10	2	269	2
Atlantic redfishes nei	*Sebastes spp*	9 354	5 057	5 336	6 628	10 451	9 699	8 319
Blackbelly rosefish	*Helicolenus dactylopterus*	...	...	...	...	334	436	313
Scorpionfishes nei	*Scorpaenidae*	905	730	658	685	298	273	372
Tub gurnard	*Chelidonichthys lucerna*	-	-	-	-	3	5	3
Gurnards, searobins nei	*Triglidae*	621	641	615	504	503	650	616
Angler(=Monk)	*Lophius piscatorius*	104	224	323	179	1 471	880	617
Blackbellied angler	*Lophius budegassa*	82	127	55	26	16	18	2
American angler	*Lophius americanus*	2	-	-	-	-	-	-
Atlantic herring	*Clupea harengus*	0	2	0	0	1	0	2
Madeiran sardinella	*Sardinella maderensis*	0	0	1	5	11	2	3
European pilchard(=Sardine)	*Sardina pilchardus*	87 711	86 855	81 477	82 992	71 972	66 319	71 947
European sprat	*Sprattus sprattus*	1	0	-	-	-	-	-
European anchovy	*Engraulis encrasicolus*	2 530	2 775	633	1 657	1 408	310	855
Clupeoids nei	*Clupeoidei*	1	-	-	-	-	-	-
Atlantic bonito	*Sarda sarda*	78	83	49	98	98	161	47
Frigate and bullet tunas	*Auxis thazard, A.rochei*	0	-	1	59	268	503	236
Little tunny(=Atl.black skipj)	*Euthynnus alletteratus*	72	218	320	171	14	50	-
Skipjack tuna	*Katsuwonus pelamis*	4 996	8 297	4 399	4 544	1 810	1 307	2 168
Atlantic bluefin tuna	*Thunnus thynnus*	481	473	749	377	487	502	468
Albacore	*Thunnus alalunga*	7 125	2 128	651	215	556	764	1 217
Yellowfin tuna	*Thunnus albacares*	231	288	176	267	177	205	26
Bigeye tuna	*Thunnus obesus*	9 662	5 810	5 437	6 334	3 313	1 498	1 606
Atlantic sailfish	*Istiophorus albicans*	0	0	-	-	53	18	3
Atlantic blue marlin	*Makaira nigricans*	10	7	3	47	8	17	18
Marlins,sailfishes,etc. nei	*Istiophoridae*	-	0	-	20	13	75	471
Swordfish	*Xiphias gladius*	1 997	2 092	1 344	1 318	1 443	1 400	1 996
Tuna-like fishes nei	*Scombroidei*	489	252	164	296	138	221	184
Garfish	*Belone belone*	72	41	35	43	54	55	57
Needlefishes, etc. nei	*Belonidae*	1	1	-	19	-	-	-
Atlantic saury	*Scomberesox saurus*	0	0	0	54	0	1	1
Dealfishes	*Trachipterus spp*	...	...	...	...	29	20	25
Silversides(=Sand smelts) nei	*Atherinidae*	1 002	3	6	22	6	3	0
Bluefish	*Pomatomus saltatrix*	70	62	48	37	22	20	15
Atlantic horse mackerel	*Trachurus trachurus*	17 703	14 065	18 739	21 404	15 535	15 471	15 305
Jack and horse mackerels nei	*Trachurus spp*	206	599	764	660	495	562	386
Jacks, crevalles nei	*Caranx spp*	44	134	114	95	32	1	1
Pompanos nei	*Trachinotus spp*	0	0	1	2	3	-	-
Greater amberjack	*Seriola dumerili*	...	...	...	...	10	21	20
Amberjacks nei	*Seriola spp*	10	33	45	101	32	8	7
Leerfish	*Lichia amia*	0	0	0	-	0	0	0
Carangids nei	*Carangidae*	...	...	...	...	45	29	30
Atlantic pomfret	*Brama brama*	18	34	91	9	9	8	10
Common dolphinfish	*Coryphaena hippurus*	1	1	0	1	1	1	4
Chub mackerel	*Scomber japonicus*	5 210	7 658	7 781	7 771	14 962	11 593	4 938
Atlantic mackerel	*Scomber scombrus*	3 073	3 009	2 083	2 898	2 035	2 254	3 121
Mackerels nei	*Scombridae*	107	237	340	247	35	0	-
Blue butterfish	*Stromateus fiatola*	63	87	72	24	17	29	-
Butterfishes, pomfrets nei	*Stromateidae*	-	-	-	5	94	35	94
Barracudas nei	*Sphyraena spp*	14	20	9	10	43	43	42
Basking shark	*Cetorhinus maximus*	1	1	1	-	1	1	3
Thresher	*Alopias vulpinus*	...	...	...	...	13	20	39
Shortfin mako	*Isurus oxyrinchus*	1	...	...	22	163	659	513
Porbeagle	*Lamna nasus*	0	0	0	0	0	16	4
Blue shark	*Prionace glauca*	21	...	...	85	905	3 083	4 663
Smooth hammerhead	*Sphyrna zygaena*	...	...	...	...	8	22	10
Smooth-hounds nei	*Mustelus spp*	97	187	27	14	81	41	43
Tope shark	*Galeorhinus galeus*	...	...	...	...	...	2	1
Greenland shark	*Somniosus microcephalus*	11	0	0	-	0	0	1
Picked dogfish	*Squalus acanthias*	5	2	2	2	21	2	3
Gulper shark	*Centrophorus granulosus*	...	...	...	...	73	54	93
Leafscale gulper shark	*Centrophorus squamosus*	...	...	...	...	440	506	537
Birdbeak dogfish	*Deania calcea*	...	...	...	...	...	18	50
Portuguese dogfish	*Centroscymnus coelolepis*	...	...	...	...	607	640	643
Kitefin shark	*Dalatias licha*	...	...	...	...	45	311	189
Dogfish sharks nei	*Squalidae*	1 137	977	999	905	-	-	-
Dogfishes and hounds nei	*Squalidae, Scyliorhinidae*	1 594	1 341	1 376	1 266	754	803	810
Angular roughshark	*Oxynotus centrina*	...	...	...	...	81	33	63
Raja rays nei	*Raja spp*	3 659	2 459	2 615	2 807	3 728	2 325	2 563
Eagle rays	*Myliobatidae*	...	...	...	...	11	8	9
Rays, stingrays, mantas nei	*Rajiformes*	137	261	257	189	74	0	82
Various sharks nei	*Selachimorpha(Pleurotremata)*	1 853	1 659	1 874	1 882	352	297	217
Sharks, rays, skates, etc. nei	*Elasmobranchii*	871	2 366	1 241	1 214	1 933	570	1 070
Finfishes nei	*Osteichthyes*	12 823	9 352	9 407	10 767	5 084	2 514	1 572
Marine fishes nei	*Osteichthyes*	332	741	486	410	299	72	155
Edible crab	*Cancer pagurus*	5	4	15	13	21	14	13
Portunus swimcrabs nei	*Portunus spp*	32	29	16	19	0	59	67
Green crab	*Carcinus maenas*	351	200	125	156	77	111	125
Spinous spider crab	*Maja squinado*	49	40	47	58	60	59	51
Marine crabs nei	*Brachyura*	21	33	61	36	36	120	243
Palinurid spiny lobsters nei	*Palinurus spp*	97	139	45	27	20	20	12
Norway lobster	*Nephrops norvegicus*	282	185	162	187	258	289	370
European lobster	*Homarus gammarus*	2	3	3	2	1	2	2
Lobsters nei	*Reptantia*	9	9	25	17	89	83	8
Penaeus shrimps nei	*Penaeus spp*	277	354	497	808	-	-	-
Deepwater rose shrimp	*Parapenaeus longirostris*	53	534	340	646	2 345	1 756	1 504
Blue and red shrimp	*Aristeus antennatus*	...	...	...	...	194	269	182

E-5 Fish crustaceans, molluscs, etc — Capture production by countries or areas and species — Europe
Poissons, crustacés, mollusques, etc — Captures par pays ou zones et espèces — Europe
Peces, crustáceos, moluscos, etc — Capturas por países o áreas y especies — Europa

English name Nom anglais Nombre inglés	Scientific name Nom scientifique Nombre científico	1995 mt	1996 mt	1997 mt	1998 mt	1999 mt	2000 mt	2001 mt
Northern prawn	*Pandalus borealis*	20	1	241	374	1 062	555	640
Palaemonid shrimps nei	*Palaemonidae*	1	1	15	5	38	63	13
Common shrimp	*Crangon crangon*	1	0	0		1	3	0
Natantian decapods nei	*Natantia*	884	1 273	679	803	211	379	1 074
Marine crustaceans nei	*Crustacea*	97	66	182	123	189	15	28
Periwinkles nei	*Littorina spp*	0	0	-	-	-	-	-
Gastropods nei	*Gastropoda*	20	48	41	41	189	200	188
Pacific cupped oyster	*Crassostrea gigas*	8	-	-	-	-	-	-
Blue mussel	*Mytilus edulis*	45	35	46	24	87	48	74
Queen scallop	*Aequipecten opercularis*	0	0	-	-	-	-	-
Great Atlantic scallop	*Pecten maximus*	-	1	2	0	0	0	1
Striped venus	*Chamelea gallina*	4	7	17	185	129	156	48
Pullet carpet shell	*Venerupis pullastra*	225	153	419	240	212	87	81
Grooved carpet shell	*Ruditapes decussatus*	169	185	27	33	75	16	21
Solid surf clam	*Spisula solida*	-	-	-	-	765	1 153	1 125
Donax clams	*Donax spp*	...	...	...	...	456	401	540
Razor clams nei	*Solen spp*	2 603	1 729	124	15	4	12	214
Common edible cockle	*Cerastoderma edule*	591	3 522	1 285	1 264	1 409	1 292	683
Clams, etc. nei	*Bivalvia*	1 489	1 482	741	1 007	493	395	398
Common cuttlefish	*Sepia officinalis*	1 364	2 076	1 640	1 865	1 319	1 432	1 463
Common squids nei	*Loligo spp*	5 074	4 776	1 567	1 499	704	1 006	1 274
Northern shortfin squid	*Illex illecebrosus*	-	4	-	1	-	-	-
Argentine shortfin squid	*Illex argentinus*	353	640	712	1 531	-	-	1 049
Various squids nei	*Loliginidae, Ommastrephidae*	0	0	0	2	0	10	5
Common octopus	*Octopus vulgaris*	-	-	-	-	12	624	823
Octopuses, etc. nei	*Octopodidae*	9 956	11 785	9 333	6 515	9 509	9 220	7 332
Marine molluscs nei	*Mollusca*	640	626	687	799	32	68	121
Echinoderms	*Echinodermata*	-	-	-	-	-	-	15
	Country total	*260 253*	*259 846*	*221 879*	*224 234*	*208 459*	*187 570*	*191 214*
Romania								
Freshwater bream	*Abramis brama*	1 604	827	328	951	1 052	936	800
Freshwater breams nei	*Abramis spp*	-	-	-	3	1	-	-
Common carp	*Cyprinus carpio*	577	441	173	147	310	458	566
Tench	*Tinca tinca*	-	-	-	2	7	4	31
Bleak	*Alburnus alburnus*	-	-	-	24	16	30	60
Barbel	*Barbus barbus*	-	-	-	32	11	17	34
Crucian carp	*Carassius carassius*	-	-	33	2	23	128	97
Goldfish	*Carassius auratus*	2 320	1 954	817	1 113	1 199	1 212	1 149
Roaches nei	*Rutilus spp*	693	301	125	292	234	174	278
Orfe(=Ide)	*Leuciscus idus*	-	-	1	5	1	-	-
Chub	*Leuciscus cephalus*	-	-	-	45	17	24	28
Grass carp(=White amur)	*Ctenopharyngodon idellus*	170	124	12	46	21	83	73
Silver carp	*Hypophthalmichthys molitrix*	1 950	1 272	1 940	428	1 308	634	644
Bighead carp	*Hypophthalmichthys nobilis*	700	449	299	369	396	356	313
Vimba bream	*Vimba vimba*	-	-	26	10	31	3	2
Asp	*Aspius aspius*	-	-	-	11	7	17	4
Cyprinids nei	*Cyprinidae*	153	112	16	396	126	39	298
Northern pike	*Esox lucius*	11	8	4	40	47	47	95
Wels(=Som)catfish	*Silurus glanis*	72	22	9	58	87	73	116
European perch	*Perca fluviatilis*	39	38	8	40	60	28	42
Pike-perch	*Stizostedion lucioperca*	353	94	30	78	154	155	92
Freshwater fishes nei	*Osteichthyes*	27	67	5	-	-	169	188
Danube sturgeon(=Osetr)	*Acipenser gueldenstaedtii*	-	-	-	5	10	19	17
Starry sturgeon	*Acipenser stellatus*	-	-	5	2	11	22	20
Beluga	*Huso huso*	-	-	2	7	7	32	20
Sturgeons nei	*Acipenseridae*	14	7	7	7	4	1	2
European eel	*Anguilla anguilla*	-	-	1	1	0	26	-
Sea trout	*Salmo trutta*	39	39	28	10	63	18	48
Rainbow trout	*Oncorhynchus mykiss*	...	...	3	25	25	32	65
Grayling	*Thymallus thymallus*	-	-	-	-	2	-	-
Pontic shad	*Alosa pontica*	331	400	487	446	20	106	128
Shads nei	*Alosa spp*	106	101	43	114	60	77	22
Azov sea sprat	*Clupeonella cultriventris*	42	4	2	52	4	5	11
Common sole	*Solea solea*	-	-	-	4	5	6	9
Turbot	*Psetta maxima*	4	6	1	2	2	2	13
Whiting	*Merlangius merlangus*	327	372	441	640	272	275	306
Surmullets(=Red mullets) nei	*Mullus spp*	6	1	3	3	1	2	3
Gobies nei	*Gobiidae*	13	8	2	6	30	42	24
Atlantic herring	*Clupea harengus*	6 588	1 794	-	-	-	-	-
European pilchard(=Sardine)	*Sardina pilchardus*	-	2	-	-	-	-	-
European sprat	*Sprattus sprattus*	1 982	2 014	3 318	3 293	1 933	1 803	1 792
European anchovy	*Engraulis encrasicolus*	189	138	45	146	155	204	186
Silversides(=Sand smelts) nei	*Atherinidae*	3	3	10	73	33	42	29
Bluefish	*Pomatomus saltatrix*	-	-	-	12	3	4	10
Atlantic horse mackerel	*Trachurus trachurus*	-	360	-	-	-	-	-
Mediterranean horse mackerel	*Trachurus mediterraneus*	23	13	1	15	3	8	17
Atlantic mackerel	*Scomber scombrus*	30 844	7 265	...	...	...	...	...
Picked dogfish	*Squalus acanthias*	7	-	-	-	-	-	-
Marine fishes nei	*Osteichthyes*	88	23	2	2	1	1	5
Euro-American crayfishes nei	*Astacidae, Cambaridae*	-	-	181	65	56	32	-
Frogs	*Rana spp*	...	...	38	41	35	26	-
	Country total	*49 275*	*18 259*	*8 446*	*9 061*	*7 843*	*7 372*	*7 637*
Russian Fed								
Freshwater bream	*Abramis brama*	29 760	29 374	30 143	32 189	25 687	25 473	28 754
Freshwater breams nei	*Abramis spp*	1 720	1 664	1 046	2 058	1 951	1 593	1 751

English name Nom anglais Nombre inglés	Scientific name Nom scientifique Nombre científico	1995 mt	1996 mt	1997 mt	1998 mt	1999 mt	2000 mt	2001 mt
Common carp	*Cyprinus carpio*	754	3 935	2 716	3 215	3 508	4 007	2 699
Tench	*Tinca tinca*	1 300	936	1 647	990	1 071	1 307	1 409
Roaches nei	*Rutilus spp*	19 880	21 968	15 971	15 963	10 433	14 547	15 312
Orfe(=Ide)	*Leuciscus idus*	1 977	2 914	3 037	3 280	2 604	2 378	2 257
Sichel	*Pelecus cultratus*	489	451	342	1 724	1 786	1 201	1 548
Asp	*Aspius aspius*	55	84	36	45	53	57	65
Cyprinids nei	*Cyprinidae*	11 948	9 962	17 169	23 453	20 387	20 148	18 307
Northern pike	*Esox lucius*	4 938	5 688	4 646	5 567	6 118	8 864	8 444
Wels(=Som)catfish	*Silurus glanis*	4 590	5 864	7 026	9 305	7 031	7 199	6 540
Burbot	*Lota lota*	993	1 531	1 533	2 121	1 763	1 778	1 644
European perch	*Perca fluviatilis*	2 234	2 628	3 527	4 459	3 455	3 726	3 986
Pike-perch	*Stizostedion lucioperca*	4 221	6 215	6 104	5 390	5 320	6 128	5 711
Freshwater fishes nei	*Osteichthyes*	7 671	9 737	6 172	4 071	4 171	15 857	4 468
Sturgeons nei	*Acipenseridae*	2 949	1 609	1 705	1 433	931	648	571
European eel	*Anguilla anguilla*	41	46	47	49	23	46	56
Atlantic salmon	*Salmo salar*	192	173	148	164	127	154	161
Sea trout	*Salmo trutta*	-	-	-	-	1	1	4
Pink(=Humpback)salmon	*Oncorhynchus gorbuscha*	148 231	113 181	187 667	191 439	187 181	157 138	167 566
Chum(=Keta=Dog)salmon	*Oncorhynchus keta*	22 681	23 083	22 898	26 046	28 162	36 490	32 067
Masu(=Cherry) salmon	*Oncorhynchus masou*	17	45	8	7	10	3	5
Sockeye(=Red)salmon	*Oncorhynchus nerka*	14 227	22 891	10 177	12 767	14 889	19 548	22 475
Chinook(=Spring=King)salmon	*Oncorhynchus tshawytscha*	875	521	636	556	793	479	499
Coho(=Silver)salmon	*Oncorhynchus kisutch*	1 479	1 976	1 310	2 319	1 668	2 278	2 034
European smelt	*Osmerus eperlanus*	805	1 022	760	835	409	844	976
Rainbow smelt	*Osmerus mordax*	1 006	640	1 113	1 758	1 340	609	725
Smelts nei	*Osmerus spp, Hypomesus spp*	3 248	2 049	4 416	3 424	3 130	4 560	4 180
Whitefishes nei	*Coregonus spp*	5 786	9 313	9 355	8 786	8 871	9 161	9 061
Salmonoids nei	*Salmonoidei*	1 217	1 338	1 565	2 103	1 446	2 728	1 511
Pontic shad	*Alosa pontica*	2	9	2	1	-	-	-
Caspian shads	*Caspialosa spp*	1 478	1 882	2 304	3 284	4 654	1 273	176
Azov sea sprat	*Clupeonella cultriventris*	81 025	92 190	81 911	116 414	152 934	122 811	55 717
Lampreys nei	*Petromyzontidae*	40	76	31	37	67	39	49
Three-spined stickleback	*Gasterosteus aculeatus*	337	271	314	884	1 021	1 630	1 507
Atlantic halibut	*Hippoglossus hippoglossus*	-	-	-	-	6	5	2
European plaice	*Pleuronectes platessa*	4 431	2 161	3 531	3 729	3 911	3 114	1 250
Greenland halibut	*Reinhardtius hippoglossoides*	1 481	2 565	1 075	5 144	7 630	8 879	9 568
Kamchatka flounder	*Atheresthes evermanni*	7 039	9 739	8 476	9 567	10 743	23 473	19 049
Witch flounder	*Glyptocephalus cynoglossus*	-	-	-	52	110	114	65
Amer. plaice(=Long rough dab)	*Hippoglossoides platessoides*	1 157	778	1 482	6 569	7 878	3 207	3 754
Yellowtail flounder	*Limanda ferruginea*	-	-	-	-	96	212	148
European flounder	*Platichthys flesus*	-	-	-	-	-	1 392	1 351
Turbot	*Psetta maxima*	-	-	-	-	-	53	69
Flatfishes nei	*Pleuronectiformes*	46 812	57 223	63 812	79 723	97 050	103 108	95 131
Red codling	*Pseudophycis bachus*	138	-	-	-	-	-	-
Moras nei	*Moridae*	3 758	-	152	11 115	31 295	39 316	32 392
Tusk(=Cusk)	*Brosme brosme*	-	-	-	-	-	46	83
Atlantic cod	*Gadus morhua*	297 770	309 391	316 147	248 719	215 616	171 018	188 884
Pacific cod	*Gadus macrocephalus*	100 730	93 890	79 927	94 282	101 929	68 415	59 783
Ling	*Molva molva*	-	-	-	-	-	8	2
Greater forkbeard	*Phycis blennoides*	-	-	-	-	-	2	11
Red hake	*Urophycis chuss*	-	-	-	4	2	120	118
Haddock	*Melanogrammus aeglefinus*	54 516	73 857	41 228	20 560	30 978	24 894	34 970
Navaga(=Wachna cod)	*Eleginus navaga*	1 014	659	1 152	1 185	480	674	1 166
Saffron cod	*Eleginus gracilis*	25 566	21 110	27 803	40 426	47 032	35 763	33 753
Saithe(=Pollock)	*Pollachius virens*	1 148	1 177	1 802	3 837	3 932	4 564	4 953
Alaska pollock(=Walleye poll.)	*Theragra chalcogramma*	2 208 410	2 439 980	2 252 742	1 930 650	1 500 450	1 215 065	1 145 016
Polar cod	*Boreogadus saida*	26 430	20 784	6 826	3 592	22 005	40 743	39 445
Blue whiting(=Poutassou)	*Micromesistius poutassou*	93 824	87 310	118 656	130 042	182 637	241 905	315 586
Southern blue whiting	*Micromesistius australis*	22	377	610	-	-	-	30
Whiting	*Merlangius merlangus*	91	11	10	119	184	341	642
Senegalese hake	*Merluccius senegalensis*	132	1 112	1 081	1 171	1 230	604	207
Southern hake	*Merluccius australis*	414	27	202	-	-	-	-
Silver hake	*Merluccius bilinearis*	-	639	-	163	-	1 679	2 055
Argentine hake	*Merluccius hubbsi*	-	-	-	-	-	-	197
Cape hakes	*Merluccius capensis,M.paradox.*	10	98	252	321	67	41	49
Patagonian grenadier	*Macruronus magellanicus*	-	-	-	-	-	-	173
Blue grenadier	*Macruronus novaezelandiae*	7 907	3 905	3 191	-	-	-	-
Grenadiers nei	*Macrourus spp*	2	-	-	-	-	-	0
Roundnose grenadier	*Coryphaenoides rupestris*	191	267	1 041	921	641	2 627	1 992
Grenadiers, rattails nei	*Macrouridae*	222	310	1 044	62	52	579	2 232
Gadiformes nei	*Gadiformes*	3	10	-	2	6	6	1
Bonefish	*Albula vulpes*	-	164	-	-	-	-	-
Sea catfishes nei	*Ariidae*	-	-	-	-	2	2	-
Mullets nei	*Mugilidae*	325	735	1 270	1 796	2 618	2 693	1 677
Bigeye grunt	*Brachydeuterus auritus*	1 282	-	-	36	-	-	5
Grunts, sweetlips nei	*Haemulidae (=Pomadasyidae)*	9		-	2	12	1	8
Meagre	*Argyrosomus regius*	-	-	-	-	-	5	-
West African croakers nei	*Pseudotolithus spp*	146	-	15	33	13	13	37
Croakers, drums nei	*Sciaenidae*	127	58	-	-	-	-	59
Large-eye dentex	*Dentex macrophthalmus*	558	1 210	1 566	2 242	4 003	2 385	1 181
Salema	*Sarpa salpa*	-	-	-	-	-	-	1
Porgies, seabreams nei	*Sparidae*	364	902	623	1 341	1 492	1 361	868
Surmullets(=Red mullets) nei	*Mullus spp*	324	76	68	119	92	127	119
Pacific sandlance	*Ammodytes personatus*	-	-	-	-	-	17	-
Gobies nei	*Gobiidae*	14 071	14 524	16 177	20 554	30 177	32 745	27 346
Atka mackerel	*Pleurogrammus azonus*	19 112	21 260	36 075	40 889	40 283	52 784	49 171
Triggerfishes, durgons nei	*Balistidae*	-	-	154	-	-	-	-
Argentines	*Argentina spp*	-	-	-	-	-	1 219	496
Lanternfishes nei	*Myctophidae*	-	-	-	-	5	67	-

E-5

Fish crustaceans, molluscs, etc	Capture production by countries or areas and species	Europe
Poissons, crustacés, mollusques, etc	Captures par pays ou zones et espèces	Europe
Peces, crustáceos, moluscos, etc	Capturas por países o áreas y especies	Europa

English name Nom anglais Nombre inglés	Scientific name Nom scientifique Nombre científico	1995 mt	1996 mt	1997 mt	1998 mt	1999 mt	2000 mt	2001 mt
Longspine snipefish	*Macroramphosus scolopax*	20	-	89	443	-	-	44
Kingklip	*Genypterus capensis*	-	-	-	-	-	-	5
Cusk-eels, brotulas nei	*Ophidiidae*	-	-	-	-	-	-	18
Alfonsinos nei	*Beryx spp*	819	162	48	83	59	35	231
Orange roughy	*Hoplostethus atlanticus*	-	-	-	-	-	14	-
John dory	*Zeus faber*	-	-	-	-	-	2	-
Boarfishes nei	*Caproidae*	-	-	-	5	-	-	-
Oreo dories nei	*Oreosomatidae*	-	5	-	-	2	-	14
Cape bonnetmouth	*Emmelichthys nitidus*	-	-	70	7	-	-	-
Bonnetmouths, rubyfishes nei	*Emmelichthyidae*	-	-	-	-	-	-	6
Patagonian toothfish	*Dissostichus eleginoides*	10	103	-	-	-	-	90
Patagonian rockcod	*Patagonotothen brevicauda*	-	-	-	-	3	0	-
Mackerel icefish	*Champsocephalus gunnari*	-	-	-	-	265	3 395	-
Atlantic wolffish	*Anarhichas lupus*	6 280	10 523	18 247	23 730	23 794	22 792	19 120
Wolffishes(=Catfishes) nei	*Anarhichas spp*	57	-	-	38	-	7	23
Eelpouts	*Lycodes spp*	45	18	-	2	1	28	48
Snoek	*Thyrsites atun*	-	15	11	-	-	-	-
Largehead hairtail	*Trichiurus lepturus*	23 709	51 778	33 465	7 422	7 217	4 846	2 326
Silver scabbardfish	*Lepidopus caudatus*		-	-		-	-	2
South Pacific breams nei	*Seriolella spp*	150	185	352	34	28	-	-
Pacific ocean perch	*Sebastes alutus*	3 539	2 615	2 095	1 739	2 478	1 023	793
Atlantic redfishes nei	*Sebastes spp*	56 772	47 660	41 988	31 215	29 615	39 137	47 338
Scorpionfishes nei	*Scorpaenidae*	-	5	-	-	-	46	56
Gurnards, searobins nei	*Triglidae*	-	-	-	628	2 426	26 081	3 155
Sablefish	*Anoplopoma fimbria*	8	502	-	-	-	6	6
Lumpfish(=Lumpsucker)	*Cyclopterus lumpus*	-	-	-	-	-	20	28
Angler(=Monk)	*Lophius piscatorius*	-	-	-	-	-	-	1
Atlantic herring	*Clupea harengus*	121 561	134 380	181 179	139 566	170 601	174 204	127 420
Pacific herring	*Clupea pallasii*	116 787	171 810	313 397	395 595	359 194	361 241	278 511
Round sardinella	*Sardinella aurita*	53 303	116 705	101 457	106 318	109 445	42 400	22 557
Sardinellas nei	*Sardinella spp*	7 241	229	2 739	9 537	22 169	5 645	443
Japanese pilchard	*Sardinops melanostictus*	-	-	-	-	3	-	-
European pilchard(=Sardine)	*Sardina pilchardus*	47 526	56 168	24 864	5 100	5 504	11 200	1 829
European sprat	*Sprattus sprattus*	16 197	19 824	22 900	22 321	36 100	35 912	43 081
European anchovy	*Engraulis encrasicolus*	10 487	3 503	21 581	47 358	33 407	34 140	23 325
Anchovies, etc. nei	*Engraulidae*	19	22	5	23	34	69	114
Atlantic bonito	*Sarda sarda*	6	175	1 937	4 960	2 156	878	574
West African Spanish mackerel	*Scomberomorus tritor*	-	44	-	14	19	7	4
Frigate and bullet tunas	*Auxis thazard, A.rochei*	459	46	500	2 433	460	420	1 053
Little tunny(=Atl.black skipj)	*Euthynnus alletteratus*	96	49	-	88	-	-	-
Skipjack tuna	*Katsuwonus pelamis*	1 466	381	1 146	2 086	1 426	374	33
Yellowfin tuna	*Thunnus albacares*	2 936	2 696	4 275	4 931	4 359	737	-
Bigeye tuna	*Thunnus obesus*	-	13	38	4	8	91	-
Tuna-like fishes nei	*Scombroidei*	4 994	-	-	6	11	22	18
Capelin	*Mallotus villosus*	44	180	160	405	32 555	94 984	181 566
Garfish	*Belone belone*	-	-	-	-	15	83	1
Pacific saury	*Cololabis saira*	25 140	10 280	7 091	4 665	4 808	17 390	40 407
Big-scale sand smelt	*Atherina boyeri*	15	31	13	25	30	63	95
Bluefish	*Pomatomus saltatrix*	74	96	406	283	536	226	157
Atlantic horse mackerel	*Trachurus trachurus*	1 709	804	554	345	121	86	16
Pacific jack mackerel	*Trachurus symmetricus*	-	-	-	230	10	-	-
Cape horse mackerel	*Trachurus capensis*	116 695	78 429	69 248	104 935	55 642	50 456	28 215
Cunene horse mackerel	*Trachurus trecae*	53 760	58 466	51 583	6 265	33 973	16 387	5 934
Greenback horse mackerel	*Trachurus declivis*	1 602	2 280	886	52	223	-	-
Jack and horse mackerels nei	*Trachurus spp*	86 124	73 446	56 008	84 736	71 262	70 949	56 235
Jacks, crevalles nei	*Caranx spp*	53	15	24	-	32	2	1
African moonfish	*Selene dorsalis*	20	15	12	1	-	-	-
Yellowtail amberjack	*Seriola lalandi*	6	-	-	-	-	-	-
Amberjacks nei	*Seriola spp*	-	82	232	-	213	-	-
Leerfish	*Lichia amia*	14	84	393	736	622	200	84
Atlantic bumper	*Chloroscombrus chrysurus*	59	-	2	49	-	-	-
Carangids nei	*Carangidae*	7	40	227	81	-	38	805
Atlantic pomfret	*Brama brama*	2	103	56	-	112	21	-
Chub mackerel	*Scomber japonicus*	52 631	72 940	64 748	68 368	48 365	45 836	31 710
Atlantic mackerel	*Scomber scombrus*	46 249	43 046	53 732	67 838	51 348	50 772	41 571
Blue mackerel	*Scomber australasicus*	75	-	-	-	-	-	-
Blue butterfish	*Stromateus fiatola*	15	-	-	-	4	-	-
Barracudas nei	*Sphyraena spp*	515	1 176	399	125	844	-	1
Raja rays nei	*Raja spp*	6	7	476	932	975	4 485	2 914
Rays, stingrays, mantas nei	*Rajiformes*	20	28	25	34	344	1 440	1 835
Sharks, rays, skates, etc. nei	*Elasmobranchii*	90	19	9	107	30	12	127
Pelagic fishes nei	*Osteichthyes*	102	8	-	-	-	-	-
Finfishes nei	*Osteichthyes*	1 088	801	520	603	845	22 659	698
Marine fishes nei	*Osteichthyes*	6 801	7 439	8 150	11 363	8 997	76 914	45 431
Freshwater crustaceans nei	*Crustacea*	12	28	34	33	41	957	1 733
Tanner crabs nei	*Chionoecetes spp*	...	22 769	27 112	20 848	21 234	21 848	24 493
Hair crab	*Erimacrus isenbeckii*	...	789	612	409	440	198	162
Marine crabs nei	*Brachyura*	-	-	49	-	-	-	194
Red king crab	*Paralithodes camtschaticus*	54 986	34 300	23 262	32 557	37 072	28 632	16 316
Blue king crab	*Paralithodes platypus*	...	8 762	10 268	4 508	5 455	5 233	4 500
Brown king crab	*Paralithodes brevipes*	...	204	418	194	256	347	254
Golden king crab	*Lithodes aequispina*	...	4 666	4 917	3 897	2 746	1 797	2 245
Coonstripe shrimp	*Pandalus hypsinotus*	...	...	467	388	288	275	359
Northern prawn	*Pandalus borealis*	6 961	10 191	4 369	8 022	16 964	35 253	20 057
Humpy shrimp	*Pandalus goniurus*	...	...	-	1 199	330	1 200	247
Hokkai shrimp	*Pandalus kessleri*	...	...	123	55	97	75	94
Pandalus shrimps nei	*Pandalus spp*	2 398	3 000	-	-	-	-	-
Morotoge shrimp	*Pandalopsis japonica*	...	...	12	86	35	11	36
Sculptured shrimps nei	*Sclerocrangon spp*	...	...	38	59	82	20	45

English name Nom anglais Nombre inglés	Scientific name Nom scientifique Nombre científico	1995 mt	1996 mt	1997 mt	1998 mt	1999 mt	2000 mt	2001 mt
Natantian decapods nei	*Natantia*	-	-	2	-	16	92	83
Japanese corbicula	*Corbicula japonica*	-	-	-	-	-	-	74
Cupped oysters nei	*Crassostrea spp*	-	-	-	-	1	4	32
Mediterranean mussel	*Mytilus galloprovincialis*	-	-	-	-	4	-	-
Sea mussels nei	*Mytilidae*	83	-	-	9	80	63	68
Yesso scallop	*Patinopecten yessoensis*	4 407	5 084	5 534	6 253	5 764	5 728	2 236
Scallops nei	*Pectinidae*	8 372	7 634	13 878	12 887	11 948	12 717	13 598
Blood cockle	*Anadara granosa*	-	-	-	-	-	-	100
Clams, etc. nei	*Bivalvia*	-	-	-	-	-	216	317
Cuttlefish,bobtail squids nei	*Sepiidae, Sepiolidae*	-	-	-	-	-	-	5
Neon flying squid	*Ommastrephes bartrami*	-	-	-	-	-	405	100
Northern shortfin squid	*Illex illecebrosus*	-	-	-	29	-	12	-
Argentine shortfin squid	*Illex argentinus*	11 762	20 254	884	-	-	3 404	2 578
Various squids nei	*Loliginidae, Ommastrephidae*	50 940	41 678	60 722	52 464	56 108	69 835	44 249
Octopuses, etc. nei	*Octopodidae*	317	235	210	34	24	29	64
Marine molluscs nei	*Mollusca*	4 122	3 530	4 780	6 301	10 586	5 643	4 780
Sea urchins nei	*Strongylocentrotus spp*	2 344	1 608	1 153	1 560	1 245	1 677	1 763
Jellyfishes	*Rhopilema spp*	-	-	-	-	-	-	142
	Country total	*4 311 809*	*4 675 738*	*4 661 853*	*4 454 759*	*4 141 158*	*3 973 535*	*3 628 323*
Slovakia								
Freshwater breams nei	*Abramis spp*	141	111	102	99	98	94	95
Common carp	*Cyprinus carpio*	1 063	778	746	778	822	854	967
Tench	*Tinca tinca*	8	8	9	8	8	8	5
Barbel	*Barbus barbus*	24	19	21	15	10	17	19
Common nase	*Chondrostoma nasus*	31	24	23	25	24	24	20
Crucian carp	*Carassius carassius*	-	-	-	2	1	0	...
Goldfish	*Carassius auratus*	90	76	53	54	62	0	73
Chubs nei	*Leuciscus spp*	63	47	39	32	38	33	31
Grass carp(=White amur)	*Ctenopharyngodon idellus*	23	15	21	16	17	15	9
Silver carp	*Hypophthalmichthys molitrix*	-	-	-	4	5	7	8
Vimba bream	*Vimba vimba*	40	20	25	27	14	11	10
Asp	*Aspius aspius*	13	12	9	9	9	8	8
Cyprinids nei	*Cyprinidae*	64	8	44	37	38	53	12
Northern pike	*Esox lucius*	134	103	85	68	69	76	73
Wels(=Som)catfish	*Silurus glanis*	26	21	21	20	20	22	28
Burbot	*Lota lota*	3	1	3	4	4	2	4
European perch	*Perca fluviatilis*	32	18	11	13	13	13	14
Pike-perch	*Stizostedion lucioperca*	94	76	70	65	64	56	62
Sterlet sturgeon	*Acipenser ruthenus*	1	1	0	0	0	1	-
European eel	*Anguilla anguilla*	13	7	8	8	8	4	6
Sea trout	*Salmo trutta*	46	34	40	42	41	37	38
Rainbow trout	*Oncorhynchus mykiss*	15	16	16	17	16	19	30
Brook trout	*Salvelinus fontinalis*	1	0	1	1	0	0	1
Huchen	*Hucho hucho*	1	1	1	1	1	1	1
Grayling	*Thymallus thymallus*	24	18	16	16	14	13	17
	Country total	*1 950*	*1 414*	*1 364*	*1 361*	*1 396*	*1 368*	*1 531*
Slovenia								
Freshwater bream	*Abramis brama*	16	10	11	11	10	11	10
Freshwater breams nei	*Abramis spp*	5	5	5	5	4	0	1
Common carp	*Cyprinus carpio*	89	86	90	94	78	71	75
Tench	*Tinca tinca*	2	1	2	2	3	2	1
Barbel	*Barbus barbus*	13	10	12	9	12	11	11
Common nase	*Chondrostoma nasus*	-	-	-	-	35	35	25
Crucian carp	*Carassius carassius*	-	-	-	0	1	0	0
Goldfish	*Carassius auratus*	-	-	-	0	1	1	1
Roach	*Rutilus rutilus*	2	2	2	2	2	3	4
Roaches nei	*Rutilus spp*	8	6	8	9	10	11	7
Rudd	*Scardinius erythrophthalmus*	7	7	6	4	4	1	0
Orfe(=Ide)	*Leuciscus idus*	19	17	17	16	14	15	14
Common dace	*Leuciscus leuciscus*	5	5	5	5	4	0	0
Grass carp(=White amur)	*Ctenopharyngodon idellus*	-	-	-	2	2	3	2
Silver carp	*Hypophthalmichthys molitrix*	7	7	6	4	4	-	0
Vimba bream	*Vimba vimba*	-	-	-	1	1	2	2
Asp	*Aspius aspius*	-	-	-	0	-	0	0
Cyprinids nei	*Cyprinidae*	59	54	40	42	-	-	-
Northern pike	*Esox lucius*	11	11	12	9	10	11	9
Wels(=Som)catfish	*Silurus glanis*	6	4	7	7	5	7	6
European perch	*Perca fluviatilis*	2	1	1	1	0	0	1
Pike-perch	*Stizostedion lucioperca*	5	5	5	5	4	5	4
Freshwater fishes nei	*Osteichthyes*	0	0	17	0	3	-	-
Beluga	*Huso huso*	0	1	1	1	1	0	1
Sea trout	*Salmo trutta*	14	14	13	11	10	9	7
Trouts nei	*Salmo spp*	2	3	2	2	2	2	1
Rainbow trout	*Oncorhynchus mykiss*	36	32	33	23	19	21	21
Brook trout	*Salvelinus fontinalis*	-	1	0	0	0	1	0
Arctic char	*Salvelinus alpinus*	1	0	1	0	0	1	1
Grayling	*Thymallus thymallus*	7	7	6	4	4	3	2
European flounder	*Platichthys flesus*	-	-	-	-	-	1	3
Common sole	*Solea solea*	2	1	1	1	1	2	3
Poor cod	*Trisopterus minutus*	-	-	-	-	-	2	1
Whiting	*Merlangius merlangus*	-	-	-	13	16	14	37
European hake	*Merluccius merluccius*	4	4	4	2	1	0	2
Flathead grey mullet	*Mugil cephalus*	3	12	2	27	13	22	49
European seabass	*Dicentrarchus labrax*	2	-	-	-	1	1	5
Common pandora	*Pagellus erythrinus*	-	-	-	1	1	3	7

English name / Nom anglais / Nombre inglés	Scientific name / Nom scientifique / Nombre científico	1995 mt	1996 mt	1997 mt	1998 mt	1999 mt	2000 mt	2001 mt
Sargo breams nei	*Diplodus spp*	-	-	-	-	-	1	2
Gilthead seabream	*Sparus aurata*	2	-	-	-	1	1	4
Bogue	*Boops boops*	-	-	-	2	1	3	2
Sand steenbras	*Lithognathus mormyrus*	-	-	-	-	-	1	2
Picarels nei	*Spicara spp*	-	-	-	3	3	4	5
Red mullet	*Mullus surmuletus*	-	-	-	1	1	1	4
Black goby	*Gobius niger*	-	-	-	-	-	1	1
European conger	*Conger conger*	-	-	-	-	-	0	1
Gurnards, searobins nei	*Triglidae*	-	-	-	-	-	1	1
European pilchard(=Sardine)	*Sardina pilchardus*	1 719	1 982	1 973	1 788	1 614	1 415	1 219
European sprat	*Sprattus sprattus*	11	7	1	-	2	3	8
European anchovy	*Engraulis encrasicolus*	29	24	33	51	75	96	97
Garfish	*Belone belone*	-	-	-	-	-	1	1
Silversides(=Sand smelts) nei	*Atherinidae*	-	-	-	-	-	1	2
Atlantic horse mackerel	*Trachurus trachurus*	7	9	8	4	5	4	4
Chub mackerel	*Scomber japonicus*	3	10	4	17	3	4	3
Atlantic mackerel	*Scomber scombrus*	-	5	5	10	13	14	16
Smooth-hounds nei	*Mustelus spp*	-	-	-	-	-	2	4
Picked dogfish	*Squalus acanthias*	4	0	0	1	1	-	-
Marine fishes nei	*Osteichthyes*	12	8	7	5	4	-	-
Spottail mantis squillid	*Squilla mantis*	-	-	-	-	-	0	3
European flat oyster	*Ostrea edulis*	0	0	0	-	-	0	0
Mediterranean mussel	*Mytilus galloprovincialis*	1	1	1	-	-	-	0
Striped venus	*Chamelea gallina*	-	-	-	-	-	0	1
Common cuttlefish	*Sepia officinalis*	10	6	5	18	18	11	72
Common squids nei	*Loligo spp*	2	2	4	3	2	3	8
Common octopus	*Octopus vulgaris*	34	2	7	-	-	-	-
Marine molluscs nei	*Mollusca*	6	5	10	12	8	18	54
	Country total	*2 167*	*2 367*	*2 367*	*2 228*	*2 027*	*1 856*	*1 827*
Spain								
Freshwater fishes nei	*Osteichthyes*	3 944	4 000 F	4 000 F	4 000 F	4 000 F	4 000 F	4 000 F
Sturgeons nei	*Acipenseridae*	-	-	-	1	1	-	-
European eel	*Anguilla anguilla*	68 F	68	72	23	39	70	62
Atlantic salmon	*Salmo salar*	9	10 F	10 F	10 F	10 F	10 F	10 F
Sea trout	*Salmo trutta*	2 163	2 200 F	2 200 F	2 200 F	2 200 F	2 200 F	2 200 F
Salmonoids nei	*Salmonoidei*	-	18	24	26	8	33	13
Alewife	*Alosa pseudoharengus*	...	...	...	...	...	61	-
Atlantic halibut	*Hippoglossus hippoglossus*	68	58	46	69	103	112	152
European plaice	*Pleuronectes platessa*	12	14	3	6	3	39	41
Greenland halibut	*Reinhardtius hippoglossoides*	10 323	7 525	8 200	7 484	9 611	10 401	13 337
Witch flounder	*Glyptocephalus cynoglossus*	3 425	3 387	5 527	6 154	5 328	4 443	4 225
Amer. plaice(=Long rough dab)	*Hippoglossoides platessoides*	669	878	1 118	1 346	1 522	1 622	1 412
Yellowtail flounder	*Limanda ferruginea*	65	259	656	562	752	775	622
Common dab	*Limanda limanda*	-	-	0	130	129	29	24
Lemon sole	*Microstomus kitt*	-	-	235	1 197	1 282	2 207	4 040
European flounder	*Platichthys flesus*	-	74	319	873	323	88	139
Common sole	*Solea solea*	689 F	574	553	5 783	248	1 093	4 805
West coast sole	*Austroglossus microlepis*	-	54	-	-	-	-	-
Soles nei	*Soleidae*	4 500 F	6 846	6 208	7 519	1 781	227	979
Megrim	*Lepidorhombus whiffiagonis*	175 F	175	185	44	-	101	131
Megrims nei	*Lepidorhombus spp*	6 275	6 405	7 006	7 446	6 969	8 361	8 161
Brill	*Scophthalmus rhombus*	-	478	429	60	36	29	23
Turbot	*Psetta maxima*	257 F	282	339	231	252	124	122
Flatfishes nei	*Pleuronectiformes*	5 851 F	10 253	18 234	15 696	15 850	11 066	8 123
Blue antimora	*Antimora rostrata*	-	16	-	26	24	21	-
Tadpole codling	*Salilota australis*	5 942	3 484	2 505	6 140	5 935	3 914	2 250
Tusk(=Cusk)	*Brosme brosme*	-	62	98	106	150	249	72
Atlantic cod	*Gadus morhua*	16 143	16 356	17 228	14 392	10 159	8 770	20 283
Ling	*Molva molva*	5 026	5 301	6 153	9 256	8 907	6 259	4 276
Blue ling	*Molva dypterygia*	-	299	1 166	1 405	1 710	3 113	4 472
Greater forkbeard	*Phycis blennoides*	200 F	1 741	5 209	4 671	3 675	2 369	1 181
Red hake	*Urophycis chuss*	112	893	958	1 200	1 350	1 551	1 755
White hake	*Urophycis tenuis*	39	187	304	491	426	802	689
Haddock	*Melanogrammus aeglefinus*	62	718	505	541	780	669	2 217
Saithe(=Pollock)	*Pollachius virens*	13	33	83	397	82	158	152
Pollack	*Pollachius pollachius*	216	185	213	218	175	175	436
Pouting(=Bib)	*Trisopterus luscus*	4 656	2 823	3 637	2 269	2 736	3 635	3 240
Blue whiting(=Poutassou)	*Micromesistius poutassou*	37 897 F	35 113	41 054	33 359	35 371	34 276	34 283
Southern blue whiting	*Micromesistius australis*	10 711	2 471	1 591	3 435	3 128	3 346	5 243
Whiting	*Merlangius merlangus*	6	44	72	187	233	353	299
European hake	*Merluccius merluccius*	36 568 F	27 012	26 304	27 200	29 354	31 183	19 159
Senegalese hake	*Merluccius senegalensis*	14 600 F	15 001	14 478	13 594	15 678	11 211	16 716
Silver hake	*Merluccius bilinearis*	-	-	-	-	-	4	9
Argentine hake	*Merluccius hubbsi*	1 161	21 697	14 816	18 298	26 810	22 196	25 302
Benguela hake	*Merluccius polli*	-	-	-	-	18	-	-
Cape hakes	*Merluccius capensis,M.paradox.*	9 000 F	1 724	1 592	3	2 154	2 877	3 349
Patagonian grenadier	*Macruronus magellanicus*	10 601	7 733	7 422	16 104	11 132	9 794	13 599
Roughhead grenadier	*Macrourus berglax*	-	257	3 740	6 050	5 705	8 097	1 103
Grenadiers nei	*Macrourus spp*	-	-	1	1	-	-	-
Roundnose grenadier	*Coryphaenoides rupestris*	2 678	4 066	2 476	3 935	6 224	15 465	38 225
Gadiformes nei	*Gadiformes*	34 666 F	35 932	22 256	12 029	14 338	3 724	29 664
Mullets nei	*Mugilidae*	130	489	456	660	210	117	75
Dusky grouper	*Epinephelus marginatus*	305	222	204	256	123	43	71
Groupers nei	*Epinephelus spp*	-	119	32	21	28	42	36
Groupers, seabasses nei	*Serranidae*	-	214	467	719	693	663	624
European seabass	*Dicentrarchus labrax*	446	534	474	457	383	473	326
Seabasses nei	*Dicentrarchus spp*	-	-	-	-	158	197	259

E-5

Fish crustaceans, molluscs, etc — Capture production by countries or areas and species — Europe
Poissons, crustacés, mollusques, etc — Captures par pays ou zones et espèces — Europe
Peces, crustáceos, moluscos, etc — Capturas por países o áreas y especies — Europa

English name Nom anglais Nombre inglés	Scientific name Nom scientifique Nombre científico	1995 mt	1996 mt	1997 mt	1998 mt	1999 mt	2000 mt	2001 mt
Rubberlip grunt	*Plectorhinchus mediterraneus*	245 F	257	239	174	453	94	135
Bastard grunt	*Pomadasys incisus*	-	0	5	-	-	-	-
Grunts, sweetlips nei	*Haemulidae (=Pomadasyidae)*	304	476	205	261	100	198	163
Canary drum (=Baardman)	*Umbrina canariensis*	-	-	-		-	-	4
Meagre	*Argyrosomus regius*	130 F	-	-		-		-
Croakers, drums nei	*Sciaenidae*	-	117	-	184	275	174	58
Blackspot(=red) seabream	*Pagellus bogaraveo*	1 206	739	3 075	1 395	947	699	94
Common pandora	*Pagellus erythrinus*	-	-	247	322	277	305	229
Axillary seabream	*Pagellus acarne*	-	-	424	373	-	-	-
Red pandora	*Pagellus bellottii*	-	-	43	44	42	-	-
Pandoras nei	*Pagellus spp*	2 300 F	3 083	763	708	2 122	1 609	2 056
White seabream	*Diplodus sargus*	-	-	131	106	137	114	182
Sargo breams nei	*Diplodus spp*	-	151	117	111	77	74	98
Common dentex	*Dentex dentex*	55 F	60	62	61	149	265	56
Dentex nei	*Dentex spp*	372 F	335	338	494	649	859	869
Black seabream	*Spondyliosoma cantharus*	320 F	340	360	325	7	18	23
Saddled seabream	*Oblada melanura*	-	-	30	44	78	66	81
Red porgy	*Pagrus pagrus*	120 F	142	425	841	357	222	181
Pargo breams nei	*Pagrus spp*	70 F	100	106	223	313	177	141
Gilthead seabream	*Sparus aurata*	595 F	681	546	508	956	1 229	2 164
Bogue	*Boops boops*	1 544 F	1 356	1 642	1 919	3 898	1 943	1 819
Sand steenbras	*Lithognathus mormyrus*	-	-	270	323	255	290	246
Salema	*Sarpa salpa*	-	-	242	199	167	147	164
Porgies, seabreams nei	*Sparidae*	3 572 F	3 144	2 770	3 085	2 237	3 341	3 566
Picarels nei	*Spicara spp*	-	1	-	-	494	712	759
Red mullet	*Mullus surmuletus*	-	292	300	132	136	278	352
Surmullets(=Red mullets) nei	*Mullus spp*	800 F	754	527	547	2 669	2 492	2 513
Threadfins, tasselfishes nei	*Polynemidae*	-	-	2	-	-	-	-
Antarctic rockcods, noties nei	*Nototheniidae*	-	-	-	0	-	-	-
Sandeels(=Sandlances) nei	*Ammodytes spp*	-	616	656	466	757	1 117	689
Greater weever	*Trachinus draco*	-	60	69	67	-	30	57
Gobies nei	*Gobiidae*	-	-	-	-	478	311	378
Argentines	*Argentina spp*	-	-	-		23	72	99
European conger	*Conger conger*	4 422 F	5 219	5 349	5 193	4 750	3 668	3 311
Pink cusk-eel	*Genypterus blacodes*	1 299	706	779	1 800	1 901	1 392	1 408
Bearded brotula	*Brotula barbata*	-	-	-		-	-	61
Alfonsinos nei	*Beryx spp*	-	24	233	420	-	79	247
Orange roughy	*Hoplostethus atlanticus*	-	22	26	26	38	20	16
Slimeheads nei	*Trachichthyidae*	-	833	1 052	33	25	3	6
John dory	*Zeus faber*	25 F	284	510	621	615	541	949
Boarfishes nei	*Caproidae*	-	-	-	-	-	-	7
Wreckfish	*Polyprion americanus*	42	166	280	162	186	79	133
Patagonian toothfish	*Dissostichus eleginoides*	191	79	400	552	726	802	775
Atlantic wolffish	*Anarhichas lupus*	290	7	43	44	20	68	105
Wolffishes(=Catfishes) nei	*Anarhichas spp*	195	695	535	473	435	518	785
Largehead hairtail	*Trichiurus lepturus*	-	-			-	-	3
Silver scabbardfish	*Lepidopus caudatus*	-	2 197	2 000	3 004	2 747	284	3 287
Black scabbardfish	*Aphanopus carbo*	-	41	106	137	117	1 029	1 323
Hairtails, scabbardfishes nei	*Trichiuridae*	-	-	-		-	-	13
Ruffs, barrelfishes nei	*Centrolophidae*	-	-	-		-	36	-
Beaked redfish	*Sebastes mentella*	4 554	4 307	4 438	4 587	3 526	2 530	5 575
Atlantic redfishes nei	*Sebastes spp*	695	968	5 965	5 218	8 235	5 741	4 652
Blackbelly rosefish	*Helicolenus dactylopterus*	-	-	-	-	8	8	7
Scorpionfishes nei	*Scorpaenidae*	736	1 122	1 881	1 147	1 243	1 646	985
Gurnards, searobins nei	*Triglidae*	-	1 356	1 525	868	865	1 114	852
Angler(=Monk)	*Lophius piscatorius*	3 518	3 534	5 161	6 301	7 935	6 890	6 737
Blackbellied angler	*Lophius budegassa*	450 F	524	336	524	116	208	369
American angler	*Lophius americanus*	-	-	1	2	0	3	9
Devil anglerfish	*Lophius vomerinus*	150 F	58	...	123	32	44	52
Anglerfishes nei	*Lophiidae*	-	-	-		-		3
Demersal percomorphs nei	*Perciformes*	...	...	...	...	...	841	758
Atlantic herring	*Clupea harengus*	-	-	-		-	232	232
Round sardinella	*Sardinella aurita*	-	2 224	3 210	3 411	17	41	49
Sardinellas nei	*Sardinella spp*	6 030 F	6 879	7 156	7 996	9 785	7 579	13 556
European pilchard(=Sardine)	*Sardina pilchardus*	213 590 F	221 956	192 443	200 420	128 231	81 028	71 144
European sprat	*Sprattus sprattus*	-	5	17	-	6	17	17
European anchovy	*Engraulis encrasicolus*	40 407 F	29 823	26 220	24 878	31 070	28 109	37 062
Clupeoids nei	*Clupeoidei*	433	7 873	6 542	15 572	19 461	10 945	20 242
Atlantic bonito	*Sarda sarda*	634	692	629	333	445	354	360
Wahoo	*Acanthocybium solandri*	15	25	25	29	32	32	38
Frigate and bullet tunas	*Auxis thazard, A.rochei*	1 858	4 470	1 393	1 057	692	1 356	1 436
Little tunny(=Atl.black skipj)	*Euthynnus alletteratus*	25	73	36	125	6	10	392
Skipjack tuna	*Katsuwonus pelamis*	126 641	108 604	111 076	114 784	159 685	137 211	124 098
Atlantic bluefin tuna	*Thunnus thynnus*	8 426	8 762	8 047	5 800	5 363	6 246	5 744
Albacore	*Thunnus alalunga*	21 690	17 737	18 961	13 877	16 950	16 932	9 826
Yellowfin tuna	*Thunnus albacares*	105 455	98 306	88 490	75 711	82 532	85 876	90 982
Bigeye tuna	*Thunnus obesus*	32 302	29 307	31 622	23 724	41 840	43 079	25 416
Indo-Pacific sailfish	*Istiophorus platypterus*	-	-	3	9	7	1	1
Atlantic sailfish	*Istiophorus albicans*	49	28	41	38	25	23	...
Indo-Pacific blue marlin	*Makaira mazara*	-	-	-		16	1	3
Atlantic blue marlin	*Makaira nigricans*	38	110	76	88	112	141	...
Black marlin	*Makaira indica*	1	0	-		1	0	0
Striped marlin	*Tetrapturus audax*	-	-	-		1	0	0
Atlantic white marlin	*Tetrapturus albidus*	28	107	74	69	133	186	...
Shortbill spearfish	*Tetrapturus angustirostris*	-	-	1	2	2	0	0
Longbill spearfish	*Tetrapturus pfluegeri*	4	1	1	1	52	64	...
Marlins,sailfishes,etc. nei	*Istiophoridae*	13	17	95	127	-		-
Swordfish	*Xiphias gladius*	20 339	17 156	17 412	14 125	13 790	15 209	14 068
Tuna-like fishes nei	*Scombroidei*	1	11	27	2 890	150	817	10

E-5

Fish crustaceans, molluscs, etc — **Capture production by countries or areas and species** — **Europe**
Poissons, crustacés, mollusques, etc — **Captures par pays ou zones et espèces** — **Europe**
Peces, crustáceos, moluscos, etc — **Capturas por países o áreas y especies** — **Europa**

English name Nom anglais Nombre inglés	Scientific name Nom scientifique Nombre científico	1995 mt	1996 mt	1997 mt	1998 mt	1999 mt	2000 mt	2001 mt
Garfish	*Belone belone*	295	455	509	998	600	270	380
Atlantic saury	*Scomberesox saurus*	526	483	639	907	574	184	992
Silversides(=Sand smelts) nei	*Atherinidae*	-	297	313	242	244	254	212
Bluefish	*Pomatomus saltatrix*	674	3 885	2 114	1 169	958	261	206
Cape horse mackerel	*Trachurus capensis*	-	-	2	-	-	-	-
Jack and horse mackerels nei	*Trachurus spp*	31 863	32 628	39 970	36 946	43 750	44 468	46 996
Pompanos nei	*Trachinotus spp*	-	-	-	-	4	-	7
Greater amberjack	*Seriola dumerili*	-	-	-	-	704	375	430
Amberjacks nei	*Seriola spp*	-	114	142	161	71	88	116
Leerfish	*Lichia amia*	4	24	32	38	28	43	45
Carangids nei	*Carangidae*	-	-	-	-	39	34	393
Atlantic pomfret	*Brama brama*	5 592	5 339	3 491	2 908	2 346	586	348
Common dolphinfish	*Coryphaena hippurus*	-	-	-	-	92	137	70
Chub mackerel	*Scomber japonicus*	-	894	100	378	302	3 697	1 138
Atlantic mackerel	*Scomber scombrus*	14 895 F	17 982	21 256	28 212	25 636	28 190	27 340
Scomber mackerels nei	*Scomber spp*	-	1 489	-	-	23	-	-
Mackerels nei	*Scombridae*	21 899	26 414	34 628	36 547	19 200	13 609	22 949
Blue butterfish	*Stromateus fiatola*	...	...	...	...	...	20	58
Butterfishes, pomfrets nei	*Stromateidae*	-	-	-	-	-	-	7
Barracudas nei	*Sphyraena spp*	-	24	31	19	57	125	167
Pelagic percomorphs nei	*Perciformes*	...	...	...	...	...	1	2
Basking shark	*Cetorhinus maximus*	-	2	6	6	-	-	-
Thresher	*Alopias vulpinus*	...	...	30	45	...	...	...
Bigeye thresher	*Alopias superciliosus*	...	...	149	114	...	...	...
Thresher sharks nei	*Alopias spp*	...	...	34	55	66	...	...
Shortfin mako	*Isurus oxyrinchus*	-	-	990	1 264	335	264	228
Porbeagle	*Lamna nasus*	-	31	124	686	1 001	1 184	1 009
Catsharks, nursehounds nei	*Scyliorhinus spp*	-	-	-	-	213	331	379
Blue shark	*Prionace glauca*	-	-	5 260	2 695	2 233	2 803	2 795
Requiem sharks nei	*Carcharhinidae*	-	-	43	810	-	9	8
Smooth hammerhead	*Sphyrna zygaena*	-	-	220	103	-	-	-
Hammerhead sharks, etc. nei	*Sphyrnidae*	...	...	808	746	...	...	...
Smooth-hounds nei	*Mustelus spp*	-	118	-	-	21	15	19
Tope shark	*Galeorhinus galeus*	-	-	-	-	-	-	37
Picked dogfish	*Squalus acanthias*	...	63	0	27	94	381	372
Lanternsharks nei	*Etmopterus spp*	...	...	...	...	573	...	...
Dogfish sharks nei	*Squalidae*	-	676	1 442	1 235	939	923	1 077
Raja rays nei	*Raja spp*	5 824	8 550	19 355	20 018	24 387	25 233	19 745
Rays, stingrays, mantas nei	*Rajiformes*	-	451	290	308	511	536	375
Various sharks nei	*Selachimorpha(Pleurotremata)*	18 101	3 642	30 377	20 450	23 662	31 793	30 815
Sharks, rays, skates, etc. nei	*Elasmobranchii*	455 F	5 479	40 192	18 757	13 191	18 877	11 753
Groundfishes nei	*Osteichthyes*	13 806	12 354	2 506	5 411	3 506	7 687	828
Pelagic fishes nei	*Osteichthyes*	18 450	25 609	8 169	15 685	4 426	2 947	100
Finfishes nei	*Osteichthyes*	10 103	10 721	19 532	33 711	22 091	17 492	12 646
Marine fishes nei	*Osteichthyes*	47 800 F	32 897	35 699	37 203	21 290	17 454	16 842
White-clawed crayfish	*Austropotamobius pallipes*	0	0	0	0	0	0	0
Red swamp crawfish	*Procambarus clarkii*	2 753	2 500 F	2 500 F	2 500 F	2 500 F	2 500 F	2 500 F
Edible crab	*Cancer pagurus*	149	120	51	23	49	48	35
Portunus swimcrabs nei	*Portunus spp*	169	0	6	20	18	-	-
Green crab	*Carcinus maenas*	-	1	2	8	8	10	12
Spinous spider crab	*Maja squinado*	151	185	209	196	210	199	113
Geryons nei	*Geryon spp*	300 F	370	-	300	135	156	168
Marine crabs nei	*Brachyura*	575	3 883	11 608	5 842	1 933	1 674	2 063
Palinurid spiny lobsters nei	*Palinurus spp*	137 F	149	104	103	96	46	111
Norway lobster	*Nephrops norvegicus*	1 971 F	1 735	1 928	1 647	2 542	1 579	1 540
European lobster	*Homarus gammarus*	18	30	17	15	13	10	11
Lobsters nei	*Reptantia*	-	11	-	14	1	-	-
Caramote prawn	*Penaeus kerathurus*	200 F	213	53	196	103	63	64
Southern pink shrimp	*Penaeus notialis*	680 F	707	-	-	-	-	-
Penaeus shrimps nei	*Penaeus spp*	-	373	368	498	336	318	433
Deepwater rose shrimp	*Parapenaeus longirostris*	5 534 F	4 993	5 217	9 274	6 489	5 749	9 346
Scarlet shrimp	*Plesiopenaeus edwardsianus*	400 F	352	479	337	605	54	39
Blue and red shrimp	*Aristeus antennatus*	-	-	-	-	414	772	1 418
Striped red shrimp	*Aristeus varidens*	1 032	764	780	1 932	1 007	1 870	2 205
Northern prawn	*Pandalus borealis*	279	940	1 429	1 366	1 296	2 294	1 683
Common prawn	*Palaemon serratus*	-	80	91	79	66	47	41
Common shrimp	*Crangon crangon*	230 F	222	150	100	87	-	1
Natantian decapods nei	*Natantia*	8 641 F	16 437	19 065	41 445	14 598	10 341	11 875
Spottail mantis squillid	*Squilla mantis*	-	-	-	-	1 222	1 043	1 000
Marine crustaceans nei	*Crustacea*	3 200 F	1 600	1 404	2 376	2 146	2 153	2 271
Common periwinkle	*Littorina littorea*	-	132	147	117	126	139	-
Periwinkles nei	*Littorina spp*	5	8	8	15	8	3	5
Gastropods nei	*Gastropoda*	-	-	-	-	955	799	851
European flat oyster	*Ostrea edulis*	12 F	253	388	316	443	152	37
Pacific cupped oyster	*Crassostrea gigas*	10	9	-	-	1	9	9
Blue mussel	*Mytilus edulis*	-	18	176	84	27	-	33
Mediterranean mussel	*Mytilus galloprovincialis*	-	0	29	0	19	18	24
Great Atlantic scallop	*Pecten maximus*	185	675	391	299	86	508	84
Scallops nei	*Pectinidae*	81	94	78	50	141	63	35
Striped venus	*Chamelea gallina*	510 F	2 156	2 213	3 518	3 639	4 779	5 121
Pullet carpet shell	*Venerupis pullastra*	442	1 739	2 146	2 710	2 451	990	744
Grooved carpet shell	*Ruditapes decussatus*	448	761	1 088	1 340	1 283	612	289
Razor clams nei	*Solen spp*	21	88	209	167	171	50	129
Common edible cockle	*Cerastoderma edule*	627	2 358	2 964	2 472	3 105	3 740	1 486
Clams, etc. nei	*Bivalvia*	7 210 F	5 836	5 422	6 494	5 676	3 182	2 841
Common cuttlefish	*Sepia officinalis*	-	675	776	1 561	2 396	1 895	1 712
Cuttlefish,bobtail squids nei	*Sepiidae, Sepiolidae*	3 888 F	2 532	3 059	2 386	5 529	2 901	3 115
Patagonian squid	*Loligo gahi*	53 671	35 692	3 967	8 933	8 185	9 392	9 011
Cape Hope squid	*Loligo reynaudi*	-	58	-	-	0	-	-

English name Nom anglais Nombre inglés	Scientific name Nom scientifique Nombre científico	1995 mt	1996 mt	1997 mt	1998 mt	1999 mt	2000 mt	2001 mt
Common squids nei	*Loligo spp*	1 835 F	1 520	2 982	2 595	2 378	2 062	2 279
Northern shortfin squid	*Illex illecebrosus*	2 359	2 601	2 427	765	1 932	1 831	1 503
Argentine shortfin squid	*Illex argentinus*	3 889	17 091	25 374	23 829	30 694	26 140	41 318
European flying squid	*Todarodes sagittatus*	-	1 113	2 657	2 256	2 625	2 204	1 708
Sevenstar flying squid	*Martialia hyadesi*	-	1	0	0	0	-	-
Various squids nei	*Loliginidae, Ommastrephidae*	8 000 F	9 206	4 317	9 268	14 724	3 449	3 158
Common octopus	*Octopus vulgaris*	31 000 F	26 874	22 545	38 235	21 692	11 379	14 913
Octopuses, etc. nei	*Octopodidae*	13 018 F	11 526	11 190	11 902	10 065	14 275	9 549
Cephalopods nei	*Cephalopoda*	-	6 240	5 636	8 541	242	401	190
Marine molluscs nei	*Mollusca*	1 519 F	1 301	1 543	11 230	404	294	287
Echinoderms	*Echinodermata*	-	487	590	560	621	309	306
Sea cucumbers nei	*Holothurioidea*	-	4	4	4	1	9	4
Aquatic invertebrates nei	*Invertebrata*	-	5	4	10	529	-	-
	Country total	*1 181 286 F*	*1 174 635*	*1 204 913*	*1 263 275*	*1 169 462*	*1 045 488*	*1 084 820*
Svalbard Is								
Marine fishes nei	*Osteichthyes*	0	0	0	0	0	0	0
	Country total	*0*	*0*	*0*	*0*	*0*	*0*	*0*
Sweden								
Freshwater bream	*Abramis brama*	14	25	18	22	20	6	4
Tench	*Tinca tinca*	0	0	0	1	1	-	-
Roach	*Rutilus rutilus*	1	0	1	-	-	-	-
Roaches nei	*Rutilus spp*	4	0	1	-	-	-	-
Cyprinids nei	*Cyprinidae*	0	0	0	-	-	-	-
Northern pike	*Esox lucius*	394	360	338	302	322	290	242
Burbot	*Lota lota*	104	75	66	68	65	4	3
European perch	*Perca fluviatilis*	343	267	386	371	301	231	201
Pike-perch	*Stizostedion lucioperca*	351	312	366	334	346	327	267
Freshwater fishes nei	*Osteichthyes*	206	131	230	134	157	197	155
European eel	*Anguilla anguilla*	1 099	1 042	1 073	645	734	560	580
Atlantic salmon	*Salmo salar*	697	805	715	657	433	514	412
Trouts nei	*Salmo spp*	92	131	118	138	94	88	63
Rainbow trout	*Oncorhynchus mykiss*	4	-	-	-	0	-	-
Arctic char	*Salvelinus alpinus*	47	25	24	28	21	24	18
Grayling	*Thymallus thymallus*	1	3	0	0	0	-	-
European smelt	*Osmerus eperlanus*	11	9	10	8	9	-	-
Vendace	*Coregonus albula*	1 118	1 321	1 208	676	573	830	820
European whitefish	*Coregonus lavaretus*	667	530	497	489	460	420	336
Salmonoids nei	*Salmonoidei*	0	0	1	0	0	-	-
Atlantic halibut	*Hippoglossus hippoglossus*	13	8	10	8	8	10	8
European plaice	*Pleuronectes platessa*	511	531	558	431	405	449	416
Witch flounder	*Glyptocephalus cynoglossus*	357	299	355	448	501	578	576
Amer. plaice(=Long rough dab)	*Hippoglossoides platessoides*	0	22	6	0	-	-	-
Common dab	*Limanda limanda*	59	37	46	33	16	10	14
Lemon sole	*Microstomus kitt*	96	117	121	105	94	71	61
European flounder	*Platichthys flesus*	459	1 262	1 073	526	274	341	467
Common sole	*Solea solea*	89	61	52	41	43	30	20
Megrim	*Lepidorhombus whiffiagonis*	1	-	-	-	-	-	-
Brill	*Scophthalmus rhombus*	15	7	11	13	18	17	13
Turbot	*Psetta maxima*	195	296	294	188	159	106	64
Flatfishes nei	*Pleuronectiformes*	1	0	1	0	0	-	-
Tusk(=Cusk)	*Brosme brosme*	5	7	3	3	3	8	6
Atlantic cod	*Gadus morhua*	33 186	41 827	34 797	22 475	22 597	23 174	24 111
Ling	*Molva molva*	94	73	61	44	44	46	47
Blue ling	*Molva dypterygia*	0	0	2	0	-	-	-
Haddock	*Melanogrammus aeglefinus*	1 265	1 226	1 519	1 013	895	964	1 087
Saithe(=Pollock)	*Pollachius virens*	1 998	1 773	1 649	1 857	1 929	1 468	1 628
Pollack	*Pollachius pollachius*	510	355	261	180	160	124	108
Norway pout	*Trisopterus esmarkii*	68	237	2	-	-	133	780
Blue whiting(=Poutassou)	*Micromesistius poutassou*	13 000	4 038	4 568	6 034	15 511	3 362	2 058
Whiting	*Merlangius merlangus*	670	374	101	90	128	177	153
European hake	*Merluccius merluccius*	69	45	33	26	27	34	63
Roundnose grenadier	*Coryphaenoides rupestris*	1	0	42	0	-	-	258
Gadiformes nei	*Gadiformes*	0	0	-	-	-	57	160
Sandeels(=Sandlances) nei	*Ammodytes spp*	40	-	1	8 585	23 225	28 165	50 559
Greater weever	*Trachinus draco*	25	10	1	3	12	7	26
Argentines	*Argentina spp*	...	...	541	428	0	273	1 010
Atlantic wolffish	*Anarhichas lupus*	146	117	174	157	163	154	95
Atlantic redfishes nei	*Sebastes spp*	1	0	-	0	1	-	-
Grey gurnard	*Eutrigla gurnardus*	7	4	5	8	133	5	5
Lumpfish(=Lumpsucker)	*Cyclopterus lumpus*	70	147	129	20	72	47	83
Angler(=Monk)	*Lophius piscatorius*	55	38	44	44	48	107	92
Atlantic herring	*Clupea harengus*	157 503	132 153	166 311	201 738	157 541	174 081	125 749
European pilchard(=Sardine)	*Sardina pilchardus*	-	-	-	-	-	-	1 031
European sprat	*Sprattus sprattus*	165 604	168 582	126 361	149 664	112 452	91 164	88 562
Atlantic bluefin tuna	*Thunnus thynnus*	0	-	-	-	-	-	-
Garfish	*Belone belone*	33	14	12	27	30	7	9
Atlantic horse mackerel	*Trachurus trachurus*	447	166	1 761	3 418	2 004	1 162	119
Atlantic mackerel	*Scomber scombrus*	6 268	5 387	4 390	5 161	5 003	4 500	5 098
Porbeagle	*Lamna nasus*	2	1	1	1	1	1	1
Picked dogfish	*Squalus acanthias*	104	154	197	140	114	124	238
Raja rays nei	*Raja spp*	17	9	8	2	3	3	12
Sharks, rays, skates, etc. nei	*Elasmobranchii*	0	0	0	0	0	0	-
Finfishes nei	*Osteichthyes*	12 650	3 065	3 003	338	382	540	508
Euro-American crayfishes nei	*Astacidae, Cambaridae*	5	6	9	6	8	17	37
Edible crab	*Cancer pagurus*	105	87	79	93	122	128	133

E-5 Fish crustaceans, molluscs, etc — Capture production by countries or areas and species — Europe
Poissons, crustacés, mollusques, etc — Captures par pays ou zones et espèces — Europe
Peces, crustáceos, moluscos, etc — Capturas por países o áreas y especies — Europa

English name Nom anglais Nombre inglés	Scientific name Nom scientifique Nombre científico	1995 mt	1996 mt	1997 mt	1998 mt	1999 mt	2000 mt	2001 mt
Norway lobster	*Nephrops norvegicus*	914	1 105	1 130	1 317	1 263	1 232	1 067
European lobster	*Homarus gammarus*	29	26	27	26	25	20	18
Northern prawn	*Pandalus borealis*	2 678	2 176	2 598	2 283	2 297	2 073	2 113
Marine crustaceans nei	*Crustacea*	-	-	-	-	2	1	-
European flat oyster	*Ostrea edulis*	2	3	3	2	4	2	1
Blue mussel	*Mytilus edulis*	52	-	3	36	0	70	51
Various squids nei	*Loliginidae, Ommastrephidae*	0	0	1	1	1	-	0
Marine molluscs nei	*Mollusca*	-	-	-	-	-	1	-
	Country total	*404 572*	*370 881*	*357 406*	*410 886*	*351 254*	*338 534*	*311 816*
Switzerland								
Freshwater bream	*Abramis brama*	22	24	20	16	16	13	9
Common carp	*Cyprinus carpio*	1	1	2	1	1	1	1
Tench	*Tinca tinca*	3	4	3	3	3	3	3
Roach	*Rutilus rutilus*	169	146	164	168	154	136	137
Cyprinids nei	*Cyprinidae*	34	42	20	10	6	8	13
Northern pike	*Esox lucius*	60	36	34	33	34	48	48
Wels(=Som)catfish	*Silurus glanis*	1	1	1	0	1	1	1
Burbot	*Lota lota*	11	6	9	6	7	8	8
European perch	*Perca fluviatilis*	373	475	364	422	491	395	262
Pike-perch	*Stizostedion lucioperca*	13	10	6	7	6	5	9
Freshwater fishes nei	*Osteichthyes*	8	0	0	3	0	4	1
European eel	*Anguilla anguilla*	5	3	2	3	3	2	2
Sea trout	*Salmo trutta*	10	8	13	12	11	13	13
Rainbow trout	*Oncorhynchus mykiss*	40	0	0	0	0	0	0
Arctic char	*Salvelinus alpinus*	22	28	27	21	22	22	16
Whitefishes nei	*Coregonus spp*	816	1 050	1 182	1 098	1 074	980	1 174
Shads nei	*Alosa spp*	...	7	12	6	11	20	18
	Country total	*1 588*	*1 841*	*1 859*	*1 809*	*1 840*	*1 659*	*1 715*
Ukraine								
Freshwater bream	*Abramis brama*	912	1 008	832	742	840	881	942
Freshwater breams nei	*Abramis spp*	53	101	62	86	122	80	137
Common carp	*Cyprinus carpio*	153	598	27	33	41	44	40
Goldfish	*Carassius auratus*	704	638	461	452	638	628	747
Roach	*Rutilus rutilus*	725	799	795	1 141	667	451	526
Rudd	*Scardinius erythrophthalmus*	5	15	19	34	38	38	33
Grass carp(=White amur)	*Ctenopharyngodon idellus*	1	2	-	-	-	0	-
Silver carp	*Hypophthalmichthys molitrix*	2 450	3 188	1 380	1 008	442	444	267
Vimba bream	*Vimba vimba*	24	25	10	18	13	13	20
Sichel	*Pelecus cultratus*	37	9	40	37	27	16	14
Asp	*Aspius aspius*	18	25	19	7	15	26	7
Northern pike	*Esox lucius*	8	26	23	15	18	15	14
Wels(=Som)catfish	*Silurus glanis*	1	2	1	1	1	3	3
European perch	*Perca fluviatilis*	319	126	80	107	79	155	74
Pike-perch	*Stizostedion lucioperca*	512	811	955	965	986	1 171	1 722
Freshwater gobies nei	*Gobiidae*	...	...	...	60	16	...	...
Freshwater fishes nei	*Osteichthyes*	261	213	15	1	2	...	7
Danube sturgeon(=Osetr)	*Acipenser gueldenstaedtii*	134	130	134	113	34	22	8
Starry sturgeon	*Acipenser stellatus*	9	18	49	13	11	5	3
Pontic shad	*Alosa pontica*	236	283	329	213	24	83	148
Shads nei	*Alosa spp*	59	23	57	29	5	0	-
Azov sea sprat	*Clupeonella cultriventris*	7 724	3 289	3 218	3 324	9 992	8 446	18 989
European flounder	*Platichthys flesus*	25	30	30	16	15	8	10
Turbot	*Psetta maxima*	96	120	82	63	110	118	171
Common mora	*Mora moro*	-	-	-	-	-	3	2
Red codling	*Pseudophycis bachus*	...	...	...	51	...	63	50
Southern blue whiting	*Micromesistius australis*	3 610	3 713	3 607	7 730	8 306	3 502	267
Whiting	*Merlangius merlangus*	17	3	29	55	18	20	18
Senegalese hake	*Merluccius senegalensis*	...	10	580	2 361	300	1 127	648
Southern hake	*Merluccius australis*	774	111	181	578	1 422	1100	8
Argentine hake	*Merluccius hubbsi*	-	1	-	-	-	-	-
Cape hakes	*Merluccius capensis,M.paradox.*	-	-	-	-	6	22	4
Blue grenadier	*Macruronus novaezelandiae*	10 774	14 754	12 632	16 064	15 141	19 333	1 300
Gadiformes nei	*Gadiformes*	-	-	-	-	-	-	997
Sea catfishes nei	*Ariidae*	-	-	-	-	-	1	-
So-iuy mullet	*Mugil soiuy*	775	1 039	2 718	3 674	5 364	5 575	2 810
Mullets nei	*Mugilidae*	4	3	0	5	90	121	78
Grunts, sweetlips nei	*Haemulidae (=Pomadasyidae)*	-	-	-	-	-	132	6
West African croakers nei	*Pseudotolithus spp*	...	...	...	28	12	87	5
Large-eye dentex	*Dentex macrophthalmus*	...	52	4	207	272	924	185
Dentex nei	*Dentex spp*	-	-	-	2	-	-	-
Porgies, seabreams nei	*Sparidae*	155	153	1 245	757	1 264	1 966	363
Picarels nei	*Spicara spp*		-	-		-	3	-
Striped mullet	*Mullus barbatus*	13	...	18	35	30	12	201
Grey rockcod	*Notothenia squamifrons*	-	0	-	-	-	-	-
Gobies nei	*Gobiidae*	201	72	96	286	601	825	1 510
Lanternfishes nei	*Myctophidae*	2	-	-	-	-	-	-
Pink cusk-eel	*Genypterus blacodes*	...	...	...	21	35	258	35
Alfonsinos nei	*Beryx spp*	2 249	3 826	1 423	859	1 964	1 578	371
Orange roughy	*Hoplostethus atlanticus*	-	-	-	-	-	102	195
John dory	*Zeus faber*	...	...	...	99	7	59	9
Bonnetmouths, rubyfishes nei	*Emmelichthyidae*	645	28	7	275	181	-	86
Pelagic armourhead	*Pseudopentaceros richardsoni*	103	298	51	78	108	77	12
Patagonian toothfish	*Dissostichus eleginoides*	1 560	1 003	1 007	997	760	128	164
Mackerel icefish	*Champsocephalus gunnari*	3 852	-	-	-	-	-	-
Snoek	*Thyrsites atun*	1 748	2 244	3 898	3 881	7 922	6 113	2 970

English name Nom anglais Nombre inglés	Scientific name Nom scientifique Nombre científico	1995 mt	1996 mt	1997 mt	1998 mt	1999 mt	2000 mt	2001 mt
Largehead hairtail	*Trichiurus lepturus*	-	-	-	2 490	1 314	1 423	-
Silver scabbardfish	*Lepidopus caudatus*	-	-	-	-	-	83	11
South Pacific breams nei	*Seriolella spp*	...	484	1 183	320	328	1 558	666
Ruffs, barrelfishes nei	*Centrolophidae*	485	290	440	395	753	360	299
Beaked redfish	*Sebastes mentella*	3 185	518	-	-	-	-	-
Sardinellas nei	*Sardinella spp*	21 578	83 333	97 009	161 821	74 729	38 120	10 613
European pilchard(=Sardine)	*Sardina pilchardus*	49 354	44 221	10 126	12 828	49 136	40 902	27 939
European sprat	*Sprattus sprattus*	15 218	20 720	20 208	30 282	29 238	32 655	49 004
European anchovy	*Engraulis encrasicolus*	18 505	4 398	9 444	3 914	5 527	16 390	16 668
Atlantic bonito	*Sarda sarda*	...	342	2 786	1 918	1 114	399	231
West African Spanish mackerel	*Scomberomorus tritor*	-	-	-	-	-	21	...
Frigate and bullet tunas	*Auxis thazard, A.rochei*	-	-	-	-	36	48	...
Tuna-like fishes nei	*Scombroidei*	...	...	...	303	-	28	213
Garfish	*Belone belone*	1	0	0	1	1	0	-
Silversides(=Sand smelts) nei	*Atherinidae*	195	326	396	632	388	653	332
Bluefish	*Pomatomus saltatrix*	0	-	209	13	238	97	29
Mediterranean horse mackerel	*Trachurus mediterraneus*	2	...	5	-	-	1	1
Cape horse mackerel	*Trachurus capensis*	18 060	27 121	5 179	18 345	4 592	13 837	3 693
Greenback horse mackerel	*Trachurus declivis*	8 990	13 093	9 740	9 309	15 306	12 213	7 577
Jack and horse mackerels nei	*Trachurus spp*	56 239	42 471	38 311	64 237	53 333	66 266	37 240
Scads nei	*Decapterus spp*	181	...	...	-	-	-	-
Jacks, crevalles nei	*Caranx spp*	...	...	...	-	2	-	-
Amberjacks nei	*Seriola spp*	-	-	-	-	15	...	...
Leerfish	*Lichia amia*	...	...	428	-	266	109	31
Atlantic bumper	*Chloroscombrus chrysurus*	-	-	-	-	-	2	44
Atlantic pomfret	*Brama brama*	-	-	-	-	-	9	-
Pomfrets, ocean breams nei	*Bramidae*	-	-	-	-	-	2	-
Chub mackerel	*Scomber japonicus*	66 210	98 944	120 272	72 959	42 283	44 961	19 988
Blue mackerel	*Scomber australasicus*	...	156	9	214	3 457	1 677	2 040
Blue butterfish	*Stromateus fiatola*	-	-	-	-	-	1	-
Picked dogfish	*Squalus acanthias*	67	44	20	38	94	71	134
Thornback ray	*Raja clavata*	...	17	9	24	31	24	65
Common stingray	*Dasyatis pastinaca*	...	...	...	-	-	4	11
Rays, stingrays, mantas nei	*Rajiformes*	15	1	1	-	-	-	-
Sharks, rays, skates, etc. nei	*Elasmobranchii*	...	...	...	...	...	...	1
Finfishes nei	*Osteichthyes*	167	32	-	-	-	-	-
Marine fishes nei	*Osteichthyes*	21 855	16 663	8 279	29 968	59 791	60 146	125 388
Freshwater crustaceans nei	*Crustacea*	...	...	...	...	...	0	2
Tropical spiny lobsters nei	*Panulirus spp*	-	-	-	1	-	-	-
Northern prawn	*Pandalus borealis*	-	-	-	-	-	-	405
Natantian decapods nei	*Natantia*	...	1	1	1	1	1	...
Antarctic krill	*Euphausia superba*	48 886	20 056	4 246	-	5 694	985	3 362
Sea snails	*Rapana spp*	303	378	476	371	619	913	400
Mediterranean mussel	*Mytilus galloprovincialis*	578	246	159	82	163	115	71
Wellington flying squid	*Nototodarus sloani*	6 630	4 136	7 955	5 321	1 462	2 872	8 623
Various squids nei	*Loliginidae, Ommastrephidae*	998	339	-	-	3	-	-
Marine molluscs nei	*Mollusca*	0	-	-	-	-	-	3
	Country total	*378 650*	*417 119*	*373 005*	*462 308*	*407 853*	*392 724*	*351 260*
UK								
Freshwater bream	*Abramis brama*	2	1	9	8	8	2	0
Roach	*Rutilus rutilus*	-	-	4	5	5	-	-
Northern pike	*Esox lucius*	4	5	2	4	4	6	5
Pike-perch	*Stizostedion lucioperca*	1	1	0	0	2	1	1
European eel	*Anguilla anguilla*	808	895	812	741	697	796	595
Atlantic salmon	*Salmo salar*	823	568	457	419	403	480	519
Sea trout	*Salmo trutta*	260	198	141	572	489	150	571
Trouts nei	*Salmo spp*	-	-	-	-	-	-	254
Rainbow trout	*Oncorhynchus mykiss*	300	300	100	2 831	3 224	1 267	1 174
Chars nei	*Salvelinus spp*	-	-	-	-	-	-	1
Grayling	*Thymallus thymallus*	-	-	-	-	-	0	0
European smelt	*Osmerus eperlanus*	7	149	15	1 217	888	5 316	21
Allis and twaite shads	*Alosa alosa, A.fallax*	0	6	0	1	1	3	8
River lamprey	*Lampetra fluviatilis*	-	-	-	-	-	2	2
Atlantic halibut	*Hippoglossus hippoglossus*	405	392	367	202	254	208	159
European plaice	*Pleuronectes platessa*	29 194	25 319	26 599	23 510	20 394	23 701	21 936
Greenland halibut	*Reinhardtius hippoglossoides*	2 890	2 358	1 515	1 777	1 611	1 746	1 690
Witch flounder	*Glyptocephalus cynoglossus*	3 187	3 479	3 748	3 889	3 661	4 102	4 535
Amer. plaice(=Long rough dab)	*Hippoglossoides platessoides*	-	1	0	0	0	0	15
Common dab	*Limanda limanda*	2 225	2 221	2 590	2 467	2 248	1 705	1 530
Lemon sole	*Microstomus kitt*	4 974	5 444	5 723	5 230	5 010	4 367	4 005
European flounder	*Platichthys flesus*	388	357	379	293	149	201	148
Common sole	*Solea solea*	3 530	3 022	2 671	2 561	2 808	2 443	2 720
Sand sole	*Solea lascaris*	28	27	28	32	12	47	42
Wedge sole	*Dicologlossa cuneata*	-	-	1	0	0	-	-
Megrim	*Lepidorhombus whiffiagonis*	6 563	7 212	7 479	6 520	5 606	5 513	4 672
Brill	*Scophthalmus rhombus*	618	688	568	513	410	470	535
Turbot	*Psetta maxima*	1 281	1 270	1 148	974	851	877	1 001
Flatfishes nei	*Pleuronectiformes*	132	172	172	-	-	158	171
Blue antimora	*Antimora rostrata*	-	-	1	0	-	-	-
Tadpole codling	*Salilota australis*	16	18	39	24	188	30	17
Moras nei	*Moridae*	-	-	415	0	-	2	-
Tusk(=Cusk)	*Brosme brosme*	341	391	507	723	852	1 016	872
Atlantic cod	*Gadus morhua*	78 650	78 364	74 637	77 182	51 695	41 750	32 840
Ling	*Molva molva*	14 695	13 942	12 924	13 594	11 350	9 244	8 095
Blue ling	*Molva dypterygia*	1 121	1 772	1 030	2 450	4 119	3 019	5 980
Greater forkbeard	*Phycis blennoides*	1 044	2 022	2 054	1 887	1 495	1 563	1 204
Haddock	*Melanogrammus aeglefinus*	86 315	87 420	83 388	83 436	72 001	50 644	42 865

E-5

Fish crustaceans, molluscs, etc	Capture production by countries or areas and species	Europe
Poissons, crustacés, mollusques, etc	Captures par pays ou zones et espèces	Europe
Peces, crustáceos, moluscos, etc	Capturas por países o áreas y especies	Europa

English name Nom anglais Nombre inglés	Scientific name Nom scientifique Nombre científico	1995 mt	1996 mt	1997 mt	1998 mt	1999 mt	2000 mt	2001 mt
Saithe(=Pollock)	*Pollachius virens*	15 220	15 413	14 609	12 261	12 442	10 924	10 585
Pollack	*Pollachius pollachius*	4 011	3 631	3 845	3 211	2 398	2 706	2 839
Norway pout	*Trisopterus esmarkii*	0	215	13	-	2	0	0
Pouting(=Bib)	*Trisopterus luscus*	919	962	813	554	458	885	899
Blue whiting(=Poutassou)	*Micromesistius poutassou*	5 495	14 326	33 701	98 936	106 491	45 048	51 889
Southern blue whiting	*Micromesistius australis*	30	108	20	48	85	22	30
Whiting	*Merlangius merlangus*	40 383	36 788	35 189	27 243	25 561	23 458	15 287
European hake	*Merluccius merluccius*	5 614	5 380	5 716	5 168	5 537	4 519	2 775
Argentine hake	*Merluccius hubbsi*	0	38	104	67	53	30	83
Patagonian grenadier	*Macruronus magellanicus*	80	86	166	2	347	42	30
Roughhead grenadier	*Macrourus berglax*	-	-	14	-	1 038	8	16
Whitson's grenadier	*Macrourus whitsoni*	-	-	-	-	-	-	1
Grenadiers nei	*Macrourus spp*	-	-	5	4	8	5	2
Roundnose grenadier	*Coryphaenoides rupestris*	91	184	228	-	-	587	1 030
Gadiformes nei	*Gadiformes*	-	-	-	-	-	-	3
Mullets nei	*Mugilidae*	203	63	126	89	115	67	65
Dusky grouper	*Epinephelus marginatus*	-	-	12	1	-	-	-
European seabass	*Dicentrarchus labrax*	722	582	572	501	687	406	457
Blackspot(=red) seabream	*Pagellus bogaraveo*	766	481	606	501	149	159	37
Black seabream	*Spondyliosoma cantharus*	-	-	-	5	260	240	311
Red mullet	*Mullus surmuletus*	228	192	173	180	147	220	310
Wrasses, hogfishes, etc. nei	*Labridae*	30	17	16	20	20	22	25
Marbled rockcod	*Notothenia rossii*	-	-	-	-	-	0	-
Humped rockcod	*Notothenia gibberifrons*	-	-	-	-	-	1	0
Grey rockcod	*Notothenia squamifrons*	-	-	-	-	-	5	0
Yellowfin notie	*Nototheniops nudifrons*	-	-	-	-	-	0	0
Antarctic rockcods, noties nei	*Nototheniidae*	-	-	-	-	-	-	0
Sandeels(=Sandlances) nei	*Ammodytes spp*	17 528	22 936	39 271	29 080	14 102	16 530	1 264
Argentines	*Argentina spp*	485	-	-	-	28	-	7 955
European conger	*Conger conger*	1 024	980	957	1 044	1 081	1 157	1 255
Pink cusk-eel	*Genypterus blacodes*	5	6	11	7	32	7	9
Alfonsinos nei	*Beryx spp*	-	-	4	-	-	7	16
Orange roughy	*Hoplostethus atlanticus*	-	0	0	0	12	2	35
John dory	*Zeus faber*	287	220	159	136	181	296	267
Wreckfish	*Polyprion americanus*	2	8	0	0	0	0	1
Patagonian toothfish	*Dissostichus eleginoides*	1	1	400	607	1 268	1 227	991
Patagonian rockcod	*Patagonotothen brevicauda*	-	-	-	13	-	0	-
Blackfin icefish	*Chaenocephalus aceratus*	-	-	-	-	-	0	-
Mackerel icefish	*Champsocephalus gunnari*	-	-	-	-	-	4	208
South Georgia icefish	*Pseudochaenichthys georgianus*	-	-	-	-	-	0	6
Black cardinal fish	*Epigonus telescopus*	-	-	-	-	-	1	22
Atlantic wolffish	*Anarhichas lupus*	277	1 128	199	893	928	917	140
Wolffishes(=Catfishes) nei	*Anarhichas spp*	833	0	...	...	...	...	600
Silver scabbardfish	*Lepidopus caudatus*	-	-	-	-	-	12	5
Black scabbardfish	*Aphanopus carbo*	20	40	2	159	201	428	742
Atlantic redfishes nei	*Sebastes spp*	1 346	1 776	1 507	1 554	2 672	2 495	2 875
Scorpionfishes nei	*Scorpaenidae*	0	15	...	116	253	185	188
Red gurnard	*Chelidonichthys cuculus*	81	59	17	...	...	4	207
Grey gurnard	*Eutrigla gurnardus*	21	56	59	...	...	...	46
Gurnards, searobins nei	*Triglidae*	599	583	484	629	631	953	1 131
Lumpfish(=Lumpsucker)	*Cyclopterus lumpus*	-	0	0	0	0	1	-
Angler(=Monk)	*Lophius piscatorius*	24 456	31 451	29 783	21 395	16 989	15 955	16 249
Anglerfishes nei	*Lophiidae*	-	-	-	-	-	-	2 105
Atlantic herring	*Clupea harengus*	114 571	120 935	103 405	104 627	104 752	82 658	81 363
European pilchard(=Sardine)	*Sardina pilchardus*	7 133	7 304	7 400	6 873	4 815	4 358	10 427
European sprat	*Sprattus sprattus*	5 906	7 172	8 317	7 021	15 168	8 336	5 091
European anchovy	*Engraulis encrasicolus*	...	...	...	79	3	0	274
Clupeoids nei	*Clupeoidei*	4	4	17	4	-	-	-
Atlantic bluefin tuna	*Thunnus thynnus*	1	0	1	2	12	0	-
Albacore	*Thunnus alalunga*	196	49	33	117	343	15	2
Marlins,sailfishes,etc. nei	*Istiophoridae*	-	-	-	82	-	-	-
Swordfish	*Xiphias gladius*	1	5	11	0	2	1	-
Capelin	*Mallotus villosus*	-	-	-	1 115	79	-	-
Garfish	*Belone belone*	0	0	0	1	4	3	2
Silversides(=Sand smelts) nei	*Atherinidae*	-	-	-	-	-	-	1
Atlantic horse mackerel	*Trachurus trachurus*	48 203	49 927	51 909	32 832	21 025	17 100	19 581
Atlantic mackerel	*Scomber scombrus*	218 417	144 964	149 448	179 711	166 658	193 638	198 953
Shortfin mako	*Isurus oxyrinchus*	-	-	-	0	2	3	2
Porbeagle	*Lamna nasus*	-	-	-	-	6	6	10
Small-spotted catshark	*Scyliorhinus canicula*	341	273	275	260	111	238	155
Nursehound	*Scyliorhinus stellaris*	69	109	86	77	139	115	84
Blue shark	*Prionace glauca*	-	-	-	-	-	12	9
Smooth-hound	*Mustelus mustelus*	-	-	-	-	-	15	-
Tope shark	*Galeorhinus galeus*	63	53	55	55	74	110	82
Picked dogfish	*Squalus acanthias*	10 815	9 423	8 691	8 926	7 527	7 138	7 306
Birdbeak dogfish	*Deania calcea*	-	-	-	-	-	-	1
Portuguese dogfish	*Centroscymnus coelolepis*	-	54	52	147	75	514	1 663
Longnose velvet dogfish	*Centroscymnus crepidater*	-	-	-	3	-	-	-
Dogfish sharks nei	*Squalidae*	-	-	-	4	-	-	477
Dogfishes and hounds nei	*Squalidae, Scyliorhinidae*	143	218	247	358	999	1 806	1 443
Angelsharks, sand devils nei	*Squatinidae*	0	0	47	-	-	-	-
Antarctic starry skate	*Raja georgiana*	-	-	-	-	-	0	0
Raja rays nei	*Raja spp*	8 373	9 157	8 088	7 537	6 233	6 457	6 392
Rays, stingrays, mantas nei	*Rajiformes*	12	8	21	14	49	20	38
Rabbit fish	*Chimaera monstrosa*	-	-	0	1	1	4	36
Ratfishes nei	*Hydrolagus spp*	-	-	-	-	-	-	2
Straightnose rabbitfish	*Rhinochimaera atlantica*	-	-	-	-	-	-	2
Various sharks nei	*Selachimorpha(Pleurotremata)*	2 339	2 040	3 865	2 669	2 342	954	1 640
Sharks, rays, skates, etc. nei	*Elasmobranchii*	-	-	16	31	-	-	4

English name Nom anglais Nombre inglés	Scientific name Nom scientifique Nombre científico	1995 mt	1996 mt	1997 mt	1998 mt	1999 mt	2000 mt	2001 mt
Groundfishes nei	*Osteichthyes*	5 095	4 131	3 689	3 684	2 498	911	977
Pelagic fishes nei	*Osteichthyes*	-	-	10	-	12	-	-
Marine fishes nei	*Osteichthyes*	20	48	21	60	0	13	19
Signal crayfish	*Pacifastacus leniusculus*	-	-	-	-	10	81	80
Edible crab	*Cancer pagurus*	16 166	13 182	18 516	21 799	19 988	21 607	21 909
Portunus swimcrabs nei	*Portunus spp*	3 835	1 459	2 929	2 513	2 282	1 672	2 418
Green crab	*Carcinus maenas*	470	286	411	462	333	194	331
Spinous spider crab	*Maja squinado*	2 178	1 224	1 849	1 843	1 310	990	1 199
Red crab	*Geryon quinquedens*	-	-	-	61	-	-	627
Geryons nei	*Geryon spp*	10	1 477	325	587	1 015	763	382
Marine crabs nei	*Brachyura*	1 469	639	205	57	90	17	100
Palinurid spiny lobsters nei	*Palinurus spp*	67	7	42	19	15	15	15
Norway lobster	*Nephrops norvegicus*	31 893	29 218	31 713	29 212	31 312	28 422	28 536
European lobster	*Homarus gammarus*	1 306	1 043	1 534	1 482	1 819	1 141	1 086
Craylets, squat lobsters	*Galatheidae*	7	15	12	-	28	-	-
Red king crab	*Paralithodes camtschaticus*	-	-	-	37	139	89	44
Antarctic stone crab	*Paralomis spinosissima*	-	-	-	-	-	0	3
Globose king crab	*Paralomis formosa*	-	-	-	-	-	3	11
Northern prawn	*Pandalus borealis*	1 541	1 996	417	595	1 815	539	996
Pandalus shrimps nei	*Pandalus spp*	46	39	48	1 421	46	460	74
Common prawn	*Palaemon serratus*	16	8	15	6	22	15	24
Common shrimp	*Crangon crangon*	1 131	742	598	739	1 453	1 129	2 290
Natantian decapods nei	*Natantia*	2	2	-	-	-	-	-
Antarctic krill	*Euphausia superba*	-	-	308	634	-	-	-
Marine crustaceans nei	*Crustacea*	2	2	-	-	-	-	12
Periwinkles nei	*Littorina spp*	2 331	1 756	2 870	1 925	1 336	1 059	760
Whelk	*Buccinum undatum*	5 534	12 150	8 892	3 525	4 925	10 733	11 270
European flat oyster	*Ostrea edulis*	527	584	553	1 047	407	439	611
Flat oysters nei	*Ostrea spp*	2	2	5	-	-	-	-
Pacific cupped oyster	*Crassostrea gigas*	82	125	98	59	6	5	20
Blue mussel	*Mytilus edulis*	9 534	12 337	18 994	11 434	7 972	7 468	14 905
Queen scallop	*Aequipecten opercularis*	2 857	2 181	5 630	8 102	5 888	5 149	8 660
Great Atlantic scallop	*Pecten maximus*	9 376	16 618	18 703	20 078	19 108	19 507	9 796
Scallops nei	*Pectinidae*	6 297	-	-	-	-	-	9 722
Grooved carpet shell	*Ruditapes decussatus*	-	-	-	-	-	-	29
Northern quahog(=Hard clam)	*Mercenaria mercenaria*	-	-	0	0	-	175	-
Razor clams nei	*Solen spp*	46	56	220	134	114	67	59
Common edible cockle	*Cerastoderma edule*	21 796	24 176	19 493	12 035	14 123	20 306	19 048
Clams, etc. nei	*Bivalvia*	326	49	144	13	78	15	91
Cuttlefish,bobtail squids nei	*Sepiidae, Sepiolidae*	3 631	4 607	2 202	2 760	2 260	3 076	2 705
Patagonian squid	*Loligo gahi*	1 916	4 043	2 334	3 279	2 148	5 328	4 015
Common squids nei	*Loligo spp*	2 272	3 264	3 004	3 039	2 611	1 757	850
Argentine shortfin squid	*Illex argentinus*	-	-	-	-	336	6	21
European flying squid	*Todarodes sagittatus*	-	-	18	293	204	186	193
Various squids nei	*Loliginidae, Ommastrephidae*	666	13	8	8	4	3	815
Octopuses, etc. nei	*Octopodidae*	333	229	148	111	63	135	164
Marine molluscs nei	*Mollusca*	2	2	15	-	98	13	1
Sea urchins nei	*Strongylocentrotus spp*	-	1	0	-	-	-	-
Jellyfishes	*Rhopilema spp*	-	-	-	-	-	5	-
Aquatic invertebrates nei	*Invertebrata*	-	-	-	-	-	-	24
	Country total	*909 928*	*865 145*	*886 261*	*923 085*	*840 898*	*747 358*	*741 106*
Yugoslavia								
Northern pike	*Esox lucius*	...	...	...	...	22	...	...
Wels(=Som)catfish	*Silurus glanis*	...	...	...	...	47	...	...
Freshwater fishes nei	*Osteichthyes*	3 803	3 653	3 500	2 200 F	759	672	672 F
Flatfishes nei	*Pleuronectiformes*	13	10	8	10	9	9	11
European hake	*Merluccius merluccius*	13	22	20	21	19	17	18
Flathead grey mullet	*Mugil cephalus*	26	25	24	26	25	26	25
Seabasses nei	*Dicentrarchus spp*	3	5	4	7	6	7	7
Common dentex	*Dentex dentex*	5	7	7	9	9	11	11
Black seabream	*Spondyliosoma cantharus*	2	5	3	5	3	4	4
Saddled seabream	*Oblada melanura*	3	4	4	6	7	5	6
Gilthead seabream	*Sparus aurata*	2	4	4	4	6	6	7
Bogue	*Boops boops*	18	21	20	20	22	24	25
Salema	*Sarpa salpa*	4	7	7	9	10	10	10
Picarels nei	*Spicara spp*	8	11	12	13	16	21	27
Red mullet	*Mullus surmuletus*	10	11	10	11	9	10	11
European conger	*Conger conger*	11	20	19	19	17	16	14
Scorpionfishes nei	*Scorpaenidae*	6	7	7	7	7	8	7
Sardinellas nei	*Sardinella spp*	10	11	11	13	17	13	16
European pilchard(=Sardine)	*Sardina pilchardus*	57	42	45	49	49	36	28
European anchovy	*Engraulis encrasicolus*	1	0	1	2	4	5	10
Atlantic bonito	*Sarda sarda*	6	10	12	12	14	17	17
Frigate and bullet tunas	*Auxis thazard, A.rochei*	2	6	6	6	7	8	9
Little tunny(=Atl.black skipj)	*Euthynnus alletteratus*	35	22	18	20	18	16	16
Atlantic bluefin tuna	*Thunnus thynnus*	2	4	4	6	7	4	5
Garfish	*Belone belone*	1	1	1	2	5	5	6
Silversides(=Sand smelts) nei	*Atherinidae*	6	6	7	9	9	13	14
Mediterranean horse mackerel	*Trachurus mediterraneus*	15	16	14	15	17	14	15
Greater amberjack	*Seriola dumerili*	16	11	9	11	9	12	13
Scomber mackerels nei	*Scomber spp*	7	8	9	13	14	13	13
Dogfish sharks nei	*Squalidae*	7	10	10	8	9	9	7
Rays, stingrays, mantas nei	*Rajiformes*	14	12	12	12	12	11	11
Marine fishes nei	*Osteichthyes*	21	19	25	20	25	27	-
Spinous spider crab	*Maja squinado*	0	0	0	0	0	0	0
Norway lobster	*Nephrops norvegicus*	13	5	5	7	5	7	9
European lobster	*Homarus gammarus*	1	5	5	6	5	6	7

E-5 **Fish crustaceans, molluscs, etc** — **Capture production by countries or areas and species** — **Europe**
Poissons, crustacés, mollusques, etc — **Captures par pays ou zones et espèces** — **Europe**
Peces, crustáceos, moluscos, etc — **Capturas por países o áreas y especies** — **Europa**

English name Nom anglais Nombre inglés	Scientific name Nom scientifique Nombre científico	1995 mt	1996 mt	1997 mt	1998 mt	1999 mt	2000 mt	2001 mt
European flat oyster	*Ostrea edulis*	0	0	0	0	1	1	0
Cuttlefish,bobtail squids nei	*Sepiidae, Sepiolidae*	9	10	9	10	10	10	10
Common squids nei	*Loligo spp*	13	12	13	13	12	13	14
Common octopus	*Octopus vulgaris*	4	8	8	9	9	10	12
Marine molluscs nei	*Mollusca*	0	0	0	0	0	-	1
	Country total	*4 167*	*4 030*	*3 873*	*2 610 F*	*1 251*	*1 096*	*1 088*
Total		***17 171 963***	***17 491 753***	***17 911 619***	***17 099 465***	***16 073 844***	***16 169 828***	***15 962 573***

English name Nom anglais Nombre inglés	Scientific name Nom scientifique Nombre científico	1995 mt	1996 mt	1997 mt	1998 mt	1999 mt	2000 mt	2001 mt
Amer Samoa								
Freshwater fishes nei	*Osteichthyes*	0	0	0	0	0	0	0
Squirrelfishes nei	*Holocentridae*	0	1	1	0	1	1	0
Groupers nei	*Epinephelus spp*	2	3	2	2	1	1	1
Snappers, jobfishes nei	*Lutjanidae*	7	8	7	4	5	5	9
Emperors(=Scavengers) nei	*Lethrinidae*	2	5	2	0	1	4	6
Parrotfishes nei	*Scaridae*	0	8	3	10	7	6	3
Surgeonfishes nei	*Acanthuridae*	...	8	-	13	16	10	3
Wahoo	*Acanthocybium solandri*	5	5	7	12	17	20	47
Skipjack tuna	*Katsuwonus pelamis*	80	32	16	8	20	14	56
Albacore	*Thunnus alalunga*	27	86	309	446	338	624	3 253
Yellowfin tuna	*Thunnus albacares*	2	12	22	42	64	86	183
Bigeye tuna	*Thunnus obesus*	1	4	4	10	9	21	74
Indo-Pacific sailfish	*Istiophorus platypterus*	3	2	3	2	3	1	1
Indo-Pacific blue marlin	*Makaira mazara*	11	14	18	16	13	17	5
Swordfish	*Xiphias gladius*	-	-	-	2	0	1	1
Tuna-like fishes nei	*Scombroidei*	1	3	0	0	0	0	1
Opah	*Lampris guttatus*	-	-	-	-	1	1	1
Carangids nei	*Carangidae*	1	1	2	1	1	2	1
Pomfrets, ocean breams nei	*Bramidae*	-	-	-	-	-	-	1
Common dolphinfish	*Coryphaena hippurus*	5	5	17	10	13	15	16
Barracudas nei	*Sphyraena spp*	2	2	4	1	0	0	0
Sharks, rays, skates, etc. nei	*Elasmobranchii*	0	-	4	-	-	-	-
Marine fishes nei	*Osteichthyes*	3	10	9	5	6	0	0
Tropical spiny lobsters nei	*Panulirus spp*	-	1	1	2	1	1	1
Octopuses, etc. nei	*Octopodidae*	-	-	-	0	1	0	0
	Country total	*152*	*210*	*431*	*586*	*518*	*830*	*3 663*
Australia								
Freshwater fishes nei	*Osteichthyes*	301	252	194	200	154	134	135
Short-finned eel	*Anguilla australis*	739	621	523	453	425	433 F	457 F
River eels nei	*Anguilla spp*	40	40	98	100	178	176	182
Barramundi(=Giant seaperch)	*Lates calcarifer*	1 943	2 333	2 356	2 580	3 324	3 236 F	3 388 F
Sand flounders	*Rhombosolea spp*	66	27	27	28	57	1	...
Flatfishes nei	*Pleuronectiformes*	5	7	6	6	12	...	...
Tadpole codling	*Salilota australis*	-	-	-	85	60	-	-
Southern blue whiting	*Micromesistius australis*	-	-	-	23	165	-	-
Argentine hake	*Merluccius hubbsi*	-	-	-	3	10	-	-
Patagonian grenadier	*Macruronus magellanicus*	-	-	-	31	377	-	-
Blue grenadier	*Macruronus novaezelandiae*	3 239	2 680	2 851	4 752	6 209	9 493	7 561
Grenadiers nei	*Macrourus spp*	-	-	1	-	1	3	-
Gadiformes nei	*Gadiformes*	34	24	21	33	26	23	33
Sea catfishes nei	*Ariidae*	31	256	39	21	32	...	...
Mullets nei	*Mugilidae*	7 775	6 492	6 610	6 614	6 396	5 232	6 105
Groupers, seabasses nei	*Serranidae*	496	860	155	179	71	99	92
Sillago-whitings	*Sillaginidae*	4 698	4 389	3 945	3 654	3 961	3 051	2 793
Ruff	*Arripis georgianus*	1 063	1 302	1 287	1 008	1 066	1 143	1 005
Australian salmon	*Arripis trutta*	5 926	4 669	4 773	3 919	3 526	4 536	4 519
Snappers, jobfishes nei	*Lutjanidae*	1 029	1 120	1 686	1 790	2 857	2 326	2 636
Southern meagre(=Mulloway)	*Argyrosomus hololepidotus*	211	183	177	183	239	244	344
Geelbek croaker	*Atractoscion aequidens*	27	20	27	27	29	...	...
Silver seabream	*Pagrus auratus*	4 211	4 261	4 180	3 631	4 601	3 702	1 193
Porgies, seabreams nei	*Sparidae*	1 407	940	780	755	859	821	775
Parore	*Girella tricuspidata*	...	501	507	496	503	537	487
Wrasses, hogfishes, etc. nei	*Labridae*	121	77	77	77	99	141	91
Threadfins, tasselfishes nei	*Polynemidae*	640	906	216	272	914	1 275	1 455
Grey rockcod	*Notothenia squamifrons*	-	-	4	3	10	-	0
Antarctic rockcods, noties nei	*Nototheniidae*	-	-	-	-	3	-	0
Percoids nei	*Percoidei*	136	497	131	131	-	-	-
Flatheads nei	*Platycephalidae*	4 120	4 388	4 575	3 285	4 738	4 784	4 101
Puffers nei	*Tetraodontidae*	273	228	219	184	3	2	...
Pink cusk-eel	*Genypterus blacodes*	85	1 397	1 923	1 832	1 891	2 039	1 696
Redfish	*Centroberyx affinis*	...	1 284	1 357	1 812	1 555	1 009	775
Orange roughy	*Hoplostethus atlanticus*	7 297	4 883	3 479	8 064	7 581	5 000	5 161
John dory	*Zeus faber*	401	122	118	103	175	196	177
Mirror dory	*Zenopsis nebulosus*	361	352	1 356	1 831	1 035	502	529
Morwongs	*Nemadactylus spp*	1 300	1 185	1 513	1 178	1 334	1 141	1 193
Trumpeters nei	*Latridae*	16	20	15	34	161	175	82
Patagonian toothfish	*Dissostichus eleginoides*	-	-	860	2 432	5 475	2 579	1 765
Mackerel icefish	*Champsocephalus gunnari*	-	-	217	67	73	81	930
Ocellated icefish	*Chionodraco rastrospinosus*	-	-	1	-	-	-	-
Unicorn icefish	*Channichthys rhinoceratus*	-	-	2	5	2	2	1
Spiny icefish	*Chaenodraco wilsoni*	-	-	-	-	-	-	11
Snoek	*Thyrsites atun*	850 F	850 F	300	200	102	140	183
Silver gemfish	*Rexea solandri*	5 000 F	3 000 F	339	598	458	449	455
South Pacific breams nei	*Seriolella spp*	2 558	3 406	4 020	3 356	3 327	3 449	4 190
Scorpionfishes nei	*Scorpaenidae*	...	316	349	371	356	363	373
Bluefin gurnard	*Chelidonichthys kumu*	124	60	350	372	0	...	...
Latchet(=Sharpbeak gurnard)	*Pterygotrigla polyommata*	121	58	136	92	94	119	148
Demersal percomorphs nei	*Perciformes*	4 100 F	2 600 F	-	-	-	-	-
Anchovies, etc. nei	*Engraulidae*	483	787	19	19	2	39	21
Clupeoids nei	*Clupeoidei*	13 252	13 030	13 445	8 218	4 377	2 304	1 646
Wahoo	*Acanthocybium solandri*	-	1	4	4	8	6	9
Narrow-barred Spanish mackerel	*Scomberomorus commerson*	1 246	1 160	1 475	1 617	336	307	490
Seerfishes nei	*Scomberomorus spp*	428	626	171	-	-	22	55
Kawakawa	*Euthynnus affinis*	0	0	1	-	-	-	0
Skipjack tuna	*Katsuwonus pelamis*	4 763	2 979	5 598	3 098	5 416	4 720	3 170

E-6

Fish crustaceans, molluscs, etc	Capture production by countries or areas and species	Oceania
Poissons, crustacés, mollusques, etc	Captures par pays ou zones et espèces	Océanie
Peces, crustáceos, moluscos, etc	Capturas por países o áreas y especies	Oceanía

English name Nom anglais Nombre inglés	Scientific name Nom scientifique Nombre científico	1995 mt	1996 mt	1997 mt	1998 mt	1999 mt	2000 mt	2001 mt
Longtail tuna	*Thunnus tonggol*	20	13	0	0	34	8	58
Albacore	*Thunnus alalunga*	428	472	340	444	435	387	532
Southern bluefin tuna	*Thunnus maccoyii*	5 268	5 355	5 940	4 791	5 799	4 706	5 561
Yellowfin tuna	*Thunnus albacares*	1 522	1 850	2 041	2 111	2 662	1 695	2 943
Bigeye tuna	*Thunnus obesus*	239	318	1 039	2 056	1 379	1 140	1 548
Indo-Pacific sailfish	*Istiophorus platypterus*	-	-	-	3	1	-	-
Indo-Pacific blue marlin	*Makaira mazara*	4	0	2	-	-	-	-
Black marlin	*Makaira indica*	-	-	4	-	-	-	-
Striped marlin	*Tetrapturus audax*	7	3	17	12	62	3	...
Shortbill spearfish	*Tetrapturus angustirostris*	-	-	-	-	1	-	0
Marlins,sailfishes,etc. nei	*Istiophoridae*	...	199	1 084	1 929	2 185	2 604	2 573
Swordfish	*Xiphias gladius*	62	22	43	337	1 360	1 798	2 900
Tuna-like fishes nei	*Scombroidei*	24	305	290	444	34	8	12
Needlefishes, etc. nei	*Belonidae*	2	1	0	0	-	-	-
Halfbeaks nei	*Hemiramphus spp*	367	237	680	695	0	3	-
Bluefish	*Pomatomus saltatrix*	331	260	243	192	250	155	259
Greenback horse mackerel	*Trachurus declivis*	...	68	30	18	16	26	57
White trevally	*Pseudocaranx dentex*	...	964	872	558	479	381	379
Amberjacks nei	*Seriola spp*	453	294	94	86	2	3	2
Mackerels nei	*Scombridae*	879	987	89	91	3 728	4 186	3 404
Pelagic percomorphs nei	*Perciformes*	6 020 F	4 020 F	-	-	-	-	-
Porbeagle	*Lamna nasus*	-	-	2	-	-	-	-
Smooth-hounds nei	*Mustelus spp*	3 911	3 878	4 169	2 858	2 462	2 198	2 579
Pacific sleeper shark	*Somniosus pacificus*	-	-	-	-	1	-	-
Rays, stingrays, mantas nei	*Rajiformes*	90	125	130	126	78	51	55
Sharks, rays, skates, etc. nei	*Elasmobranchii*	4 957	4 715	8 336	7 478	10 551	7 441	8 336
Marine fishes nei	*Osteichthyes*	25 728 F	29 654 F	25 438	30 495	22 199	16 828	17 613
Australian crayfish	*Euastacus armatus*	-	-	-	-	66	24	22
Blue swimming crab	*Portunus pelagicus*	5 985	6 138	6 008	6 000	4 471	5 548	6 328
Australian spiny lobster	*Panulirus cygnus*	10 886	9 902	9 896	10 400	17 720	19 380	16 017
Tropical spiny lobsters nei	*Panulirus spp*	183	205	233	551	520	432	358
Green rock lobster	*Jasus verreauxi*	4 658	4 953	4 992	4 723	2 036	2 441	2 211
Slipper lobsters nei	*Scyllaridae*	787	622	875	945	7	0	0
Banana prawn	*Penaeus merguiensis*	4 490	4 347	4 546	3 711	4 125	2 551	6 567
Giant tiger prawn	*Penaeus monodon*	4 368	3 841	3 055	3 686	3 477	2 726	2 697
Western king prawn	*Penaeus latisulcatus*	111	76	67	110	81	91	71
Penaeus shrimps nei	*Penaeus spp*	15 673	17 034	16 149	16 870	15 454	16 296	14 729
Endeavour shrimp	*Metapenaeus endeavouri*	1 945	2 175	2 245	2 885	2 575	2 163	1 999
Marine crustaceans nei	*Crustacea*	782	1 352	5 474	5 328	2 492	2 774	3 086
Blacklip abalone	*Haliotis rubra*	5 208	5 425	5 240	5 247	5 297	5 204	5 301
Australian mussel	*Mytilus planulatus*	247	75	1	1	1	1	1
Scallops nei	*Pectinidae*	13 303	12 320	8 589	9 874	11 574	12 220	8 937
Pipi wedge clam	*Paphies australis*	605	702	1 295	1 496	976	1 085	1 249
Cuttlefish,bobtail squids nei	*Sepiidae, Sepiolidae*	447	438	436	313	451	299	274
Patagonian squid	*Loligo gahi*	-	-	-	3 198	2 486	-	-
Argentine shortfin squid	*Illex argentinus*	-	-	-	-	167	-	-
Various squids nei	*Loliginidae, Ommastrephidae*	3 821	1 884	3 351	1 617	3 641	2 683	4 252
Octopuses, etc. nei	*Octopodidae*	111	74	590	783	121	147	112
Cephalopods nei	*Cephalopoda*	-	-	1	-	-	-	-
Marine molluscs nei	*Mollusca*	626	818	381	1 046	8 833	8 911	9 248
Jellyfishes	*Rhopilema spp*	-	-	10	-	2	-	-
Aquatic invertebrates nei	*Invertebrata*	-	-	1	-	-	-	-
	Country total	*205 464*	*201 310*	*196 831*	*203 334*	*214 954*	*194 631*	*192 682*
Christmas Is								
Freshwater fishes nei	*Osteichthyes*	0	0	0	0	0	0	0
Marine fishes nei	*Osteichthyes*	0	0	0	0	0	0	0
	Country total	*0*	*0*	*0*	*0*	*0*	*0*	*0*
Cocos Is								
Freshwater fishes nei	*Osteichthyes*	0	0	0	0	0	0	0
Marine fishes nei	*Osteichthyes*	0	0	0	0	0	0	0
	Country total	*0*	*0*	*0*	*0*	*0*	*0*	*0*
Cook Is								
Freshwater fishes nei	*Osteichthyes*	0	0	0	0	0	0	0
Mullets nei	*Mugilidae*	5 F	5 F	5 F	5 F	5 F	5 F	5 F
Groupers nei	*Epinephelus spp*	123 F	110 F	100 F	90 F	80 F	60 F	60 F
Snappers nei	*Lutjanus spp*	24 F	20 F	20 F	20 F	20 F	20 F	20 F
Albacore	*Thunnus alalunga*	32	14	10 F	10 F	10 F	10 F	10 F
Yellowfin tuna	*Thunnus albacares*	23	7	5 F	5 F	5 F	5 F	5 F
Bigeye tuna	*Thunnus obesus*	16	4	5 F	5 F	5 F	5 F	5 F
Marlins,sailfishes,etc. nei	*Istiophoridae*	40	30 F	30 F	30 F	30 F	20 F	20 F
Swordfish	*Xiphias gladius*	44	40 F	40 F	30 F	30 F	30 F	30 F
Flyingfishes nei	*Exocoetidae*	38 F	30 F	30 F	30 F	30 F	30 F	30 F
Jacks, crevalles nei	*Caranx spp*	51 F	40 F	40 F	40 F	40 F	40 F	40 F
Sharks, rays, skates, etc. nei	*Elasmobranchii*	30 F	20 F	20 F	20 F	20 F	20 F	20 F
Marine fishes nei	*Osteichthyes*	360 F	335 F	300 F	300 F	300 F	300 F	280 F
Oceanian crayfishes nei	*Parastacidae*	10	...	...	...	...	...	...
Marine crabs nei	*Brachyura*	5 F	5 F	5 F	5 F	5 F	5 F	5 F
Octopuses, etc. nei	*Octopodidae*	68 F	60 F	50 F	40 F	30 F	30 F	30 F
Marine molluscs nei	*Mollusca*	196 F	160 F	140 F	120 F	120 F	120 F	120 F
Sea urchins nei	*Strongylocentrotus spp*	25 F	20 F	20 F	20 F	20 F	20 F	20 F
	Country total	*1 090*	*900 F*	*820 F*	*770 F*	*750 F*	*720 F*	*700 F*

English name Nom anglais Nombre inglés	Scientific name Nom scientifique Nombre científico	1995 mt	1996 mt	1997 mt	1998 mt	1999 mt	2000 mt	2001 mt
Fiji Islands								
Nile tilapia	*Oreochromis niloticus*	6	37	40	288	290	280 F	278
Milkfish	*Chanos chanos*	20	34	26	21	30	30 F	32
Mullets nei	*Mugilidae*	1 795	759	860	1 195	3 067	2 800 F	2 915
Groupers nei	*Epinephelus spp*	854	1 017	1 060	1 160	1 600	1 450 F	1 544
Cardinalfishes, etc. nei	*Apogonidae*	...	200	180	60	71	60 F	67
Snappers nei	*Lutjanus spp*	1 117	1 347	1 315	1 155	1 710	1 600 F	1 728
Ponyfishes(=Slipmouths)	*Leiognathus spp*	37	76	75	69	84	80 F	87
Largeeye breams	*Gymnocranius spp*	19	20	22	25	28	25 F	30
Emperors(=Scavengers) nei	*Lethrinidae*	1 359	1 230	1 780	1 731	2 990	2 700 F	2 883
Goatfishes	*Upeneus spp*	9	108	190	155	160	140 F	157
Mojarras(=Silver-biddies) nei	*Gerres spp*	9	6	10	13	11	10 F	11
Wrasses, hogfishes, etc. nei	*Labridae*	218	204	375	180	399	360 F	400
Surgeonfishes nei	*Acanthuridae*	278	125	105	180	206	180 F	196
Spinefeet(=Rabbitfishes) nei	*Siganus spp*	62	73	80	93	100	90 F	96
Triggerfishes, durgons nei	*Balistidae*	6	4	8	10	14	10 F	...
Demersal percomorphs nei	*Perciformes*	...	380	250	-	-	-	...
Sardinellas nei	*Sardinella spp*	240	120	140	30	47	35 F	...
Silver-stripe round herring	*Spratelloides gracilis*	150	140	135	10	40	30 F	...
Bluestripe herring	*Herklotsichthys quadrimaculat.*	185	140	120	22	20	20 F	...
Wahoo	*Acanthocybium solandri*	179	130	145	148	160	150 F	167
Narrow-barred Spanish mackerel	*Scomberomorus commerson*	1 424	1 247	1 025	1 455	2 296	2 000 F	2 120
Skipjack tuna	*Katsuwonus pelamis*	4 319	3 124	987	459	507	343	431
Albacore	*Thunnus alalunga*	702	1 446	1 842	2 121	2 279	6 065	7 971
Yellowfin tuna	*Thunnus albacares*	1 507	1 540	1 016	869	725	2 467	2 126
Bigeye tuna	*Thunnus obesus*	378	593	409	460	462	687	662
Halfbeaks nei	*Hemiramphus spp*	49	70	62	49	61	50 F	56
Silversides(=Sand smelts) nei	*Atherinidae*	30	31	25	80	91	80 F	86
Jacks, crevalles nei	*Caranx spp*	644	384	695	647	730	650 F	709
Indian mackerel	*Rastrelliger kanagurta*	452	400	312	224	721	650 F	722
Barracudas nei	*Sphyraena spp*	1 735	567	1 562	1 626	2 979	2 700 F	2 809
Marine fishes nei	*Osteichthyes*	1 414	1 800	1 035	1 358	1 218	1 200 F	1 113
Freshwater crustaceans nei	*Crustacea*	1 011	327	315	323	332	340 F	343
Indo-Pacific swamp crab	*Scylla serrata*	234	208	290	270	281	250 F	268
Marine crabs nei	*Brachyura*	288	200	280	255	300	270 F	290
Tropical spiny lobsters nei	*Panulirus spp*	184	105	130	211	220	200 F	224
Marine crustaceans nei	*Crustacea*	180	85	78	85	91	80 F	87
Freshwater molluscs nei	*Mollusca*	2 569	2 670	3 970	4 500	5 000	5 080 F	5 300
Anadara clams nei	*Anadara spp*	1 513	1 044	1 950	2 800	2 990	2 750 F	2 884
Clams, etc. nei	*Bivalvia*	92	54	45	52	50	40 F	43
Octopuses, etc. nei	*Octopodidae*	91	48	35	62	61	50 F	57
Marine molluscs nei	*Mollusca*	2 570	2 000	3 880	3 200	3 302	3 100 F	3 150
Green turtle	*Chelonia mydas*	6	22	4	2	2	2 F	3
Marine turtles nei	*Testudinata*	7	24	7	2	8	6 F	7
Sea urchins nei	*Strongylocentrotus spp*	59	40	95	103	100	90 F	96
Sea cucumbers nei	*Holothurioidea*	835	850	790	400	880	800 F	824
	Country total	*28 836*	*25 029*	*27 755*	*28 158*	*36 713*	*40 000 F*	*42 972*
Fr Polynesia								
Freshwater fishes nei	*Osteichthyes*	0	0	50	50	50	50	50
Snappers, jobfishes nei	*Lutjanidae*	...	...	100	81	109	127	112
Wahoo	*Acanthocybium solandri*	...	...	119	188	269	229	259
Skipjack tuna	*Katsuwonus pelamis*	1 400	1 400	1 126	1 560	1 386	1 189	1 557
Albacore	*Thunnus alalunga*	1 100	1 750	2 717	3 235	2 642	3 580	4 432
Yellowfin tuna	*Thunnus albacares*	820	811	860	843	1 225	1 762	1 514
Bigeye tuna	*Thunnus obesus*	184	186	310	403	278	712	746
Marlins,sailfishes,etc. nei	*Istiophoridae*	598	587	598	518	703	566	551
Swordfish	*Xiphias gladius*	62	84	56	58	66	47	79
Tuna-like fishes nei	*Scombroidei*	...	...	...	...	2	0	0
Flyingfishes nei	*Exocoetidae*	...	...	55	92	...	88	82
Opah	*Lampris guttatus*	93	96	...	...	137	124	148
Common dolphinfish	*Coryphaena hippurus*	178	257	427	437	429	446	651
Shortfin mako	*Isurus oxyrinchus*	...	...	...	...	...	27	53
Sharks, rays, skates, etc. nei	*Elasmobranchii*	365	387	367	347	427	582	705
Marine fishes nei	*Osteichthyes*	4 200	4 332	4 820	4 585	4 538	4 274	4 362
Giant river prawn	*Macrobrachium rosenbergii*	0	0	3	3	3	3	3
Indo-Pacific swamp crab	*Scylla serrata*	-	-	2	2	2	2	2
Tropical spiny lobsters nei	*Panulirus spp*	20	20	40	51	50	51	58
Marine crustaceans nei	*Crustacea*	0	0	0	0	0	0	-
Marine molluscs nei	*Mollusca*	0	0	10	10	10	25	25
Echinoderms	*Echinodermata*	0	0	10	10	10	15	15
	Country total	*9 020*	*9 910*	*11 670*	*12 473*	*12 336*	*13 899*	*15 404*
Guam								
Tilapias nei	*Oreochromis (=Tilapia) spp*	-	-	-	6	-	-	-
Freshwater fishes nei	*Osteichthyes*	0	0	-	-	-	-	-
Groupers nei	*Epinephelus spp*	1	0	1	0	1	1	1
Snappers, jobfishes nei	*Lutjanidae*	3	1	2	3	10	6	5
Emperors(=Scavengers) nei	*Lethrinidae*	1	1	1	1	1	1	3
Wrasses, hogfishes, etc. nei	*Labridae*	-	-	-	1	2	0	1
Parrotfishes nei	*Scaridae*	0	1	1	4	1	0	0
Surgeonfishes nei	*Acanthuridae*	0	1	...	3	0	-	0
Spinefeet(=Rabbitfishes) nei	*Siganus spp*	0	0	...	1	2	0	0
Wahoo	*Acanthocybium solandri*	24	19	20	30	19	20	23
Skipjack tuna	*Katsuwonus pelamis*	23	17	24	28	19	61	60
Yellowfin tuna	*Thunnus albacares*	23	15	17	25	15	18	10
Indo-Pacific sailfish	*Istiophorus platypterus*	1	0	0	1	0	1	1

English name Nom anglais Nombre inglés	Scientific name Nom scientifique Nombre científico	1995 mt	1996 mt	1997 mt	1998 mt	1999 mt	2000 mt	2001 mt
Indo-Pacific blue marlin	*Makaira mazara*	31	15	24	13	14	30	15
Tuna-like fishes nei	*Scombroidei*	1	1	2	1	1	0	2
Scads nei	*Decapterus spp*	0	0	3	2	5	4	5
Rainbow runner	*Elagatis bipinnulata*	-	-	-	1	2	1	2
Carangids nei	*Carangidae*	1	1	1	1	2	1	1
Common dolphinfish	*Coryphaena hippurus*	72	35	43	79	40	34	53
Barracudas nei	*Sphyraena spp*	0	1	1	1	1	2	4
Sharks, rays, skates, etc. nei	*Elasmobranchii*	0	0	0	-	0	0	0
Marine fishes nei	*Osteichthyes*	4	13	18	49	85	92	89
Tropical spiny lobsters nei	*Panulirus spp*	0	0	0	1	1	2	1
Natantian decapods nei	*Natantia*	-	-	-	1	0	0	0
Octopuses, etc. nei	*Octopodidae*	0	0	-	1	2	1	2
	Country total	*185*	*121*	*158*	*253*	*223*	*275*	*278*
Kiribati								
Freshwater fishes nei	*Osteichthyes*	0	0	0	0	0	0	0
Milkfish	*Chanos chanos*	285	290	290	...	80	2 175	58
Mullets nei	*Mugilidae*	450	450	500	149	994	611	1 300
Snappers, jobfishes nei	*Lutjanidae*	1 930	1 950	1 940	1 047	1 505	2 141	2 794
Emperors(=Scavengers) nei	*Lethrinidae*	1 950	1 970	1 960	675	1 299	2 137	3 930
Goatfishes	*Upeneus spp*	440	450	450	149	582	639	1 609
Mojarras(=Silver-biddies) nei	*Gerres spp*	1 700	1 720	1 730	566	890	674	1 455
Percoids nei	*Percoidei*	1 570	1 590	1 580	10 713	24 588	632	1 069
Clupeoids nei	*Clupeoidei*	3 300	3 340	3 320	613	2 638	2 725	2 079
Skipjack tuna	*Katsuwonus pelamis*	2 520	4 111	2 855	5 544	4 493	3 701	3 286
Yellowfin tuna	*Thunnus albacares*	1 025	651	2 223	2 076	1 423	1 209	1 220
Bigeye tuna	*Thunnus obesus*	66	69	130	99	157	63	113
Flyingfishes nei	*Exocoetidae*	1 760	1 780	1 770	2 525	2 547	836	1 594
Jacks, crevalles nei	*Caranx spp*	500	500	510	2 530	1 910	1 108	2 858
Barracudas nei	*Sphyraena spp*	2 130	2 160	450	987	855	420	732
Sharks, rays, skates, etc. nei	*Elasmobranchii*	1 820	1 840	1 830	2 381	3 012	1 581	1 273
Marine fishes nei	*Osteichthyes*	2 350	2 400	2 380	3 875	4 216	2 396	3 198
Marine crustaceans nei	*Crustacea*	210	220	4	...	...	131	418
Octopuses, etc. nei	*Octopodidae*	2 200	2 230	1 874	650	687	373	69
Marine molluscs nei	*Mollusca*	4 100	4 150	4 120	571	776	1 947	3 260
Sea cucumbers nei	*Holothurioidea*	-	-	136	154	89	64	60
	Country total	*30 306*	*31 871*	*30 052*	*35 304*	*52 741*	*25 563*	*32 375*
Marshall Is								
Freshwater fishes nei	*Osteichthyes*	0	0	0	0	0	0	0
Skipjack tuna	*Katsuwonus pelamis*	-	-	-	-	-	6 715	33 468
Yellowfin tuna	*Thunnus albacares*	18	-	-	-	-	905	2 993
Bigeye tuna	*Thunnus obesus*	13	-	-	-	-	35	137
Marine fishes nei	*Osteichthyes*	350 F	2 772	400 F	500 F	500 F	500 F	500 F
Indo-Pacific swamp crab	*Scylla serrata*	0	0	0	0	0	0	0
Tropical spiny lobsters nei	*Panulirus spp*	0	0	0	0	0	-	-
Natantian decapods nei	*Natantia*	0	0	0	0	0	-	-
	Country total	*381 F*	*2 772*	*400 F*	*500 F*	*500 F*	*8 155*	*37 098*
Micronesia								
Freshwater fishes nei	*Osteichthyes*	5 F	5 F	5 F	5 F	5 F	5 F	5 F
Skipjack tuna	*Katsuwonus pelamis*	4 216	6 745	5 501	11 314	6 972	15 843	10 310
Albacore	*Thunnus alalunga*	-	-	1	-	2	3	4
Yellowfin tuna	*Thunnus albacares*	2 062	891	2 845	3 212	3 403	5 394	5 316
Bigeye tuna	*Thunnus obesus*	211	183	430	705	1 011	1 025	882
Marine fishes nei	*Osteichthyes*	1 100 F	1 200 F	1 300 F	1 300 F	1 400 F	1 400 F	1 500 F
Indo-Pacific swamp crab	*Scylla serrata*	5 F	5 F	5 F	5 F	5 F	5 F	5 F
Tropical spiny lobsters nei	*Panulirus spp*	15 F	20 F	20 F	20 F	20 F	20 F	20 F
Natantian decapods nei	*Natantia*	0	0	0	0	0	0	0
Octopuses, etc. nei	*Octopodidae*	15 F	20 F	20 F	20 F	20 F	20 F	20 F
Marine turtles nei	*Testudinata*	0	0	0	0	0	0	0
	Country total	*7 629 F*	*9 069 F*	*10 127 F*	*16 581 F*	*12 838 F*	*23 715 F*	*18 062 F*
Nauru								
Marine fishes nei	*Osteichthyes*	450 F	400 F	400 F	400 F	400 F	400 F	400 F
	Country total	*450 F*	*400 F*	*400 F*	*400 F*	*400 F*	*400 F*	*400 F*
NewCaledonia								
Freshwater fishes nei	*Osteichthyes*	0	0	0	0	0	0	0
Mullets nei	*Mugilidae*	33	20	61	64	63	75	75 F
Snappers nei	*Lutjanus spp*	38	45	36	43	22	24	24 F
Alfonsinos nei	*Beryx spp*	0	0	0	0	0	0	0
Ruffs, barrelfishes nei	*Centrolophidae*	0	0	0	0	0	0	0
Seerfishes nei	*Scomberomorus spp*	12	19	3	16	41	4	4 F
Skipjack tuna	*Katsuwonus pelamis*	2	0	1	1	0	0	0
Albacore	*Thunnus alalunga*	332	414	277	860	690	895	1 020
Yellowfin tuna	*Thunnus albacares*	839	554	466	185	373	250	570
Bigeye tuna	*Thunnus obesus*	103	233	234	498	553	517	128
Tuna-like fishes nei	*Scombroidei*	255	244	169	285	236	294	310
Mackerels nei	*Scombridae*	126	119	102	104	161	161	161 F
Marine fishes nei	*Osteichthyes*	363	415	385	458	415	509	514 F
Marine crabs nei	*Brachyura*	8	9	29	20	58	22	22 F
Tropical spiny lobsters nei	*Panulirus spp*	16	17	14	17	17	13	13 F
Penaeus shrimps nei	*Penaeus spp*	-	-	-	-	0	0	0

English name Nom anglais Nombre inglés	Scientific name Nom scientifique Nombre científico	1995 mt	1996 mt	1997 mt	1998 mt	1999 mt	2000 mt	2001 mt
Marine crustaceans nei	*Crustacea*	0	0	0	-	0	0	0
Cuttlefish,bobtail squids nei	*Sepiidae, Sepiolidae*	3	3	1	2	3	1	1 F
Marine molluscs nei	*Mollusca*	34	130	95	150	27	6	6 F
Sea cucumbers nei	*Holothurioidea*	480	776	565	402	493	615	489 F
	Country total	*2 644*	*2 998*	*2 438*	*3 105*	*3 152*	*3 386*	*3 337 F*
New Zealand								
Common carp	*Cyprinus carpio*	0	0	28	2	3	4	6
Brown bullhead	*Ameiurus nebulosus*	1	1	5	12	14	10	8
Freshwater fishes nei	*Osteichthyes*	500	600	600	400	400	400	300
River eels nei	*Anguilla spp*	1 797	1 002	980	1 293	1 151	1 055	1 075
Chinook(=Spring=King)salmon	*Oncorhynchus tshawytscha*	2	1	1	3	1	1	1
Sand flounders	*Rhombosolea spp*	103	224	251	...	...	...	37
Flatfishes nei	*Pleuronectiformes*	5 247	4 284	7 747	4 270	3 505	2 939	3 220
Smalleye moray cod	*Muraenolepis microps*	-	-	-	-	4	5	-
Moray cods nei	*Muraenolepis spp*	-	-	-	0	1	2	3
Common mora	*Mora moro*	1 192	694	1 410	1 324	1 122	1 355	1 209
Red codling	*Pseudophycis bachus*	15 916	10 572	11 073	16 429	12 528	5 232	4 454
Blue antimora	*Antimora rostrata*	-	-	-	0	0	0	3
Southern blue whiting	*Micromesistius australis*	11 357	2 753	10 234	35 059	39 012	23 000	29 789
Southern hake	*Merluccius australis*	9 707	8 317	9 692	15 047	15 499	12 799	12 956
Blue grenadier	*Macruronus novaezelandiae*	152 161	145 308	229 890	267 616	236 652	234 029	223 703
Whitson's grenadier	*Macrourus whitsoni*	-	-	-	-	1	5	48
Thorntooth grenadier	*Lepidorhynchus denticulatus*	745	670	2 361	4 627	3 678	3 833	4 783
Grenadiers, rattails nei	*Macrouridae*	909	1 579	5 197	4 359	3 692	2 464	3 094
Gadiformes nei	*Gadiformes*	73	36	22	34	11	14	8
Mullets nei	*Mugilidae*	886	897	872	720	846	748	910
Orange perch	*Lepidoperca pulchella*	...	...	...	193	32	...	46
Australian salmon	*Arripis trutta*	4 419	3 496	2 777	2 928	4 780	3 825	2 934
Emperors(=Scavengers) nei	*Lethrinidae*	-	-	4	-	-	-	-
Silver seabream	*Pagrus auratus*	6 161	5 814	6 233	6 278	6 876	6 852	6 209
Parore	*Girella tricuspidata*	83	60	81	80	76	94	71
Wrasses, hogfishes, etc. nei	*Labridae*	0	0	24	2	3	4	2
Antarctic rockcods, noties nei	*Nototheniidae*	34	16	34	3	19	3	16
New Zealand blue cod	*Parapercis colias*	11 262	2 115	2 227	2 313	2 286	2 130	2 441
Flatheads nei	*Platycephalidae*	2	1	1	2	6	7	19
Velvet leatherjacket	*Parika scaber*	329	442	1 095	312	738	1 279	1 142
Argentines	*Argentina spp*	26	8	41	68	63	42	56
Conger eels, etc. nei	*Congridae*	93	109	113	97	88	96	106
Pink cusk-eel	*Genypterus blacodes*	18 396	12 454	22 594	22 215	21 424	21 617	18 620
Alfonsinos nei	*Beryx spp*	2 177	2 159	2 617	3 516	2 579	2 880	3 044
Redfish	*Centroberyx affinis*	89	123	194	172	134	176	202
Orange roughy	*Hoplostethus atlanticus*	33 077	28 639	20 545	21 485	23 780	17 879	14 044
Slimeheads nei	*Trachichthyidae*	-	-	-	7	3	4	2
John dory	*Zeus faber*	841	729	800	828	882	841	914
Dories nei	*Zeidae*	338	288	641	754	763	778	685
Oreo dories nei	*Oreosomatidae*	21 833	18 776	21 850	21 095	22 646	22 775	24 165
Hapuku wreckfish	*Polyprion oxygeneios*	1 536	1 155	1 657	1 571	1 547	1 497	1 579
Cape bonnetmouth	*Emmelichthys nitidus*	2 392	1 845	1 635	2 064	2 846	2 825	1 881
Bonnetmouths, rubyfishes nei	*Emmelichthyidae*	616	596	431	378	271	582	434
Giant boarfish	*Paristiopterus labiosus*	21	19	27	75	6	9	3
Pelagic armourhead	*Pseudopentaceros richardsoni*	3	7	2	78	13	6	7
Tarakihi	*Nemadactylus macropterus*	5 108	4 366	5 441	5 239	5 589	5 739	6 129
Morwongs	*Nemadactylus spp*	120	92	107	109	78	99	92
Trumpeters nei	*Latridae*	669	727	761	567	574	476	505
Antarctic toothfish	*Dissostichus mawsoni*	-	-	-	41	296	751	582
Patagonian toothfish	*Dissostichus eleginoides*	...	1 061	5	44	2	0	44
Icefishes nei	*Channichthyidae*	-	-	-	0	0	0	2
Black cardinal fish	*Epigonus telescopus*	3 650	3 002	4 334	2 568	2 869	4 095	1 957
Giant stargazer	*Kathetostoma giganteum*	9 597	2 122	3 990	2 195	3 370	3 638	4 233
Snoek	*Thyrsites atun*	18 428	15 849	22 047	25 972	20 642	21 905	25 222
Escolar	*Lepidocybium flavobrunneum*	0	0	1	1	22	47	87
Oilfish	*Ruvettus pretiosus*	38	53	34	57	53	61	86
Silver gemfish	*Rexea solandri*	2 026	1 344	1 914	1 301	902	1 029	827
Silver scabbardfish	*Lepidopus caudatus*	1 955	941	2 342	3 344	2 638	1 536	2 876
Common warehou	*Seriolella brama*	1 081	1 760	4 108	3 101	3 881	4 259	4 101
Silver warehou	*Seriolella punctata*	13 377	4 926	11 253	10 993	9 029	11 218	11 268
White warehou	*Seriolella caerulea*	2 348	1 467	2 432	2 296	2 366	2 407	1 962
Bluenose warehou	*Hyperoglyphe antarctica*	2 719	2 432	2 974	2 630	2 755	2 793	2 954
Ruffs, barrelfishes nei	*Centrolophidae*	169	169	214	202	249	274	211
Scorpionfishes nei	*Scorpaenidae*	986	1 133	1 643	1 843	2 088	1 819	1 897
Bluefin gurnard	*Chelidonichthys kumu*	11 829	2 260	2 625	2 454	2 199	2 663	3 670
Spotted gurnard	*Pterygotrigla picta*	107	87	71	87	74	55	65
Australian pilchard	*Sardinops neopilchardus*	209	169	385	519	894	1 253	1 399
Clupeoids nei	*Clupeoidei*	...	11	8	1	12	11	11
Wahoo	*Acanthocybium solandri*	-	-	6	-	-	-	-
Frigate and bullet tunas	*Auxis thazard, A.rochei*	0	0	2	4	-	5	5
Skipjack tuna	*Katsuwonus pelamis*	1 428	3 631	4 792	8 156	5 688	9 699	3 691
Pacific bluefin tuna	*Thunnus orientalis*	2	5	12	20	21	21	50
Albacore	*Thunnus alalunga*	6 423	7 150	3 220	6 525	3 903	4 500	5 353
Southern bluefin tuna	*Thunnus maccoyii*	181	81	138	337	460	380	358
Yellowfin tuna	*Thunnus albacares*	138	181	118	127	153	107	137
Bigeye tuna	*Thunnus obesus*	60	86	140	388	420	421	480
Striped marlin	*Tetrapturus audax*	0	0	1	-	-	-	-
Shortbill spearfish	*Tetrapturus angustirostris*	8	42	18	-	-	-	-
Swordfish	*Xiphias gladius*	93	152	170	564	1 004	975	1 029
Tuna-like fishes nei	*Scombroidei*	41	36	83	66	101	69	88
Halfbeaks nei	*Hemiramphus spp*	18	16	17	28	24	18	13

E-6

Fish crustaceans, molluscs, etc — **Capture production by countries or areas and species** — **Oceania**
Poissons, crustacés, mollusques, etc — **Captures par pays ou zones et espèces** — **Océanie**
Peces, crustáceos, moluscos, etc — **Capturas por países o áreas y especies** — **Oceanía**

English name Nom anglais Nombre inglés	Scientific name Nom scientifique Nombre científico	1995 mt	1996 mt	1997 mt	1998 mt	1999 mt	2000 mt	2001 mt
Flyingfishes nei	*Exocoetidae*	4	4	2	3	1	2	1
Opah	*Lampris guttatus*	130	76	126	254	335	283	340
Dealfishes	*Trachipterus spp*	127	49	60	74	65	67	103
King of herrings	*Regalecus glesne*	5	10	64	60	34	20	1
Jack and horse mackerels nei	*Trachurus spp*	31 919	29 085	34 057	36 059	34 003	22 544	28 507
White trevally	*Pseudocaranx dentex*	3 910	2 869	2 941	3 608	4 192	3 603	3 116
Scads nei	*Decapterus spp*	33	23	11	63	29	82	87
Amberjacks nei	*Seriola spp*	315	381	349	327	317	296	278
Atlantic pomfret	*Brama brama*	449	480	413	488	465	401	903
Common dolphinfish	*Coryphaena hippurus*	1	0	0	0	1	0	15
Blue mackerel	*Scomber australasicus*	7 534	2 837	8 768	7 041	12 417	10 431	9 761
Broadnose sevengill shark	*Notorynchus cepedianus*	...	...	...	2	3	4	5
Basking shark	*Cetorhinus maximus*	14	2	2	49	129	95	84
Thresher	*Alopias vulpinus*	15	13	24	21	32	51	57
Shortfin mako	*Isurus oxyrinchus*	33	52	40	74	110	208	327
Porbeagle	*Lamna nasus*	5	16	21	164	246	188	127
Blue shark	*Prionace glauca*	111	246	120	540	593	1 169	1 328
Copper shark	*Carcharhinus brachyurus*	...	...	...	15	14	25	38
Smooth hammerhead	*Sphyrna zygaena*	12	10	3	6	11	13	17
Spotted estuary smooth-hound	*Mustelus lenticulatus*	2 787	1 350	3 464	1 707	1 662	1 643	1 563
Tope shark	*Galeorhinus galeus*	3 705	3 044	2 864	3 083	3 633	3 100	3 091
Picked dogfish	*Squalus acanthias*	2 753	2 477	7 232	3 064	4 409	3 362	4 192
Leafscale gulper shark	*Centrophorus squamosus*	...	...	...	4	1	0	0
Lanternsharks nei	*Etmopterus spp*	3	0	2	-	-	-	4
Birdbeak dogfish	*Deania calcea*	...	...	...	36	17	28	66
Kitefin shark	*Dalatias licha*	303	175	352	434	328	317	375
Dogfish sharks nei	*Squalidae*	413	693	1 705	701	1 010	770	705
Antarctic starry skate	*Raja georgiana*	-	-	-	...	11	36	7
Eaton's skate	*Bathyraja eatonii*	-	-	-	...	1	5	0
Bathyraja rays nei	*Bathyraja spp*	-	-	-	-	1		-
Rays and skates nei	*Rajidae*	-	-	-	...	6	-	-
Eagle rays	*Myliobatidae*	0	0	1	1	2	2	5
Rays, stingrays, mantas nei	*Rajiformes*	2 116	1 582	2 227	2 318	2 821	2 634	2 784
Dark ghost shark	*Hydrolagus novaezealandiae*	1 593	1 614	2 064	1 956	1 975	1 819	1 572
Ratfishes nei	*Hydrolagus spp*	...	...	0	36	453	975	2 184
Ghost shark	*Callorhinchus milii*	769	595	913	951	1 260	1 228	1 189
Chimaeras, etc. nei	*Chimaeriformes*	5	49	5	5	21	40	76
Sharks, rays, skates, etc. nei	*Elasmobranchii*	3 129	2 375	1 580	673	1 062	6	-
Marine fishes nei	*Osteichthyes*	5 598	26 527	26 266	65	113	122	119
Marine crabs nei	*Brachyura*	9 231	294	359	393	407	355	360
Green rock lobster	*Jasus verreauxi*	48	64	54	16	13	35	10
Red rock lobster	*Jasus edwardsii*	3 568	3 121	5 009	2 707	2 818	2 789	2 551
Slipper lobsters nei	*Scyllaridae*	0	0	0	4	0	1	4
New Zealand lobster	*Metanephrops challengeri*	1 078	670	1 093	989	925	1 034	1 093
Natantian decapods nei	*Natantia*	1	2	23	1	-	-	0
Abalones nei	*Haliotis spp*	1 280	1 020	1 180	1 300	1 170	1 265	1 064
New Zealand dredge oyster	*Ostrea lutaria*	1	2	2	2	3	2	2
Pacific cupped oyster	*Crassostrea gigas*	8	2	2	0	62	0	0
Sea mussels nei	*Mytilidae*	191	450	0	664	2 977	4 467	2 270
New Zealand scallop	*Pecten novaezelandiae*	14 160	5 080	18 848	4 592	6 152	2 912	6 792
Delicate scallop	*Zygochlamis delicatula*	135	124	201	91	128	0	222
Stutchbury's venus	*Chione stutchburyi*	1 220	815	541	1 325	1 396	1 789	1 748
Short neck clams nei	*Paphia spp*	317	211	114	204	181	131	202
Wellington flying squid	*Nototodarus sloani*	59 497	23 474	44 845	42 541	27 282	20 878	35 100
Various squids nei	*Loliginidae, Ommastrephidae*	10	7	17	27	48	74	45
Octopuses, etc. nei	*Octopodidae*	9 064	227	267	167	148	119	140
Marine molluscs nei	*Mollusca*	10	16	4	-	-	2	1
Echinoderms	*Echinodermata*	804	277	627	832	643	712	853
Starfishes nei	*Asteroidea*	9	4	0	13	2	6	11
Sea cucumbers nei	*Holothurioidea*	4	1	0	-	-	-	2
	Country total	*555 559*	*423 635*	*610 254*	*639 238*	*598 475*	*548 559*	*561 110*
Niue								
Freshwater fishes nei	*Osteichthyes*	0	0	0	0	0	0	0
Marine fishes nei	*Osteichthyes*	150 F	200 F	200 F	200 F	200 F	200 F	200 F
	Country total	*150 F*	*200 F*	*200 F*	*200 F*	*200 F*	*200 F*	*200 F*
Norfolk Is								
Freshwater fishes nei	*Osteichthyes*	0	0	0	0	0	0	0
Marine fishes nei	*Osteichthyes*	0	0	0	0	0	0	0
	Country total	*0*	*0*	*0*	*0*	*0*	*0*	*0*
N Marianas								
Freshwater fishes nei	*Osteichthyes*	1	0	0	0	0	0	0
Groupers nei	*Epinephelus spp*	1	3	5	3	2	2	3
Snappers, jobfishes nei	*Lutjanidae*	9	10	15	7	16	5	11
Emperors(=Scavengers) nei	*Lethrinidae*	2	5	13	50	4	4	8
Goatfishes, red mullets nei	*Mullidae*	1	12	9	1	1	1	1
Parrotfishes nei	*Scaridae*	1	3	7	1	2	4	13
Surgeonfishes nei	*Acanthuridae*	1	3	-	0	3	3	10
Spinefeet(=Rabbitfishes) nei	*Siganus spp*	0	...	...	2	3	5	4
Wahoo	*Acanthocybium solandri*	3	5	4	2	4	2	2
Skipjack tuna	*Katsuwonus pelamis*	60	75	64	61	48	56	61
Yellowfin tuna	*Thunnus albacares*	9	17	11	5	11	7	7
Indo-Pacific blue marlin	*Makaira mazara*	3	3	4	2	2	2	1
Tuna-like fishes nei	*Scombroidei*	4	7	7	8	8	7	3

English name Nom anglais Nombre inglés	Scientific name Nom scientifique Nombre científico	1995 mt	1996 mt	1997 mt	1998 mt	1999 mt	2000 mt	2001 mt
Scads nei	*Decapterus spp*	5	2	4	0	5	10	13
Carangids nei	*Carangidae*	0	1	2	2	2	1	3
Common dolphinfish	*Coryphaena hippurus*	11	16	17	9	6	3	6
Marine fishes nei	*Osteichthyes*	79	61	88	80	75	75	49
Tropical spiny lobsters nei	*Panulirus spp*	1	2	0	2	1	2	2
Natantian decapods nei	*Natantia*	1	-	-	-	-	-	-
Octopuses, etc. nei	*Octopodidae*	0	0	-	0	0	-	0
Marine turtles nei	*Testudinata*	0	0	-	-	-	-	-
	Country total	*192*	*225*	*250*	*235*	*193*	*189*	*197*
Palau								
Freshwater fishes nei	*Osteichthyes*	0	0	0	0	0	0	0
Groupers nei	*Epinephelus spp*	9	15	1	3	10 F	13 F	13 F
Snappers, jobfishes nei	*Lutjanidae*	40	33	8	25	25 F	27 F	27 F
Emperors(=Scavengers) nei	*Lethrinidae*	37	28	12	37	50 F	60 F	60 F
Parrotfishes nei	*Scaridae*	86	55	37	25	70 F	120 F	120 F
Surgeonfishes nei	*Acanthuridae*	111	92	23	8	60 F	128 F	128 F
Spinefeet(=Rabbitfishes) nei	*Siganus spp*	29	35	9	7	60 F	135 F	135 F
Skipjack tuna	*Katsuwonus pelamis*	100	100	100	100	100	100	100
Marlins,sailfishes,etc. nei	*Istiophoridae*	1	2	...	...	4 F	5 F	5 F
Tuna-like fishes nei	*Scombroidei*	21	27	7	18	30 F	51 F	51 F
Carangids nei	*Carangidae*	22	28	7	19	30 F	56 F	56 F
Common dolphinfish	*Coryphaena hippurus*	1	0	0	0	0	0	-
Marine fishes nei	*Osteichthyes*	1 457	1 532	1 503	1 489	1 300 F	1 230 F	1 230 F
Indo-Pacific swamp crab	*Scylla serrata*	14	15	15	15	30 F	48 F	48 F
Marine crabs nei	*Brachyura*	2	2	2	4	10 F	16 F	16 F
Tropical spiny lobsters nei	*Panulirus spp*	16	16	16	16	10 F	6 F	6 F
Sea cucumbers nei	*Holothurioidea*	6	6	7	7	6 F	...	...
Aquatic invertebrates nei	*Invertebrata*	4	4	4	4	5	5 F	5 F
	Country total	*1 956*	*1 990*	*1 751*	*1 777*	*1 800 F*	*2 000 F*	*2 000 F*
Papua N Guin								
Mozambique tilapia	*Oreochromis mossambicus*	2 310	2 310 F	2 310 F	2 310 F	2 310 F	2 310 F	2 310 F
Gudgeons, sleepers nei	*Eleotridae*	1 850	1 850 F	1 850 F	1 850 F	1 850 F	1 850 F	1 850 F
Freshwater fishes nei	*Osteichthyes*	6 645 F	6 649 F	6 649 F	6 641 F	6 649 F	6 655 F	6 654 F
Diadromous clupeoids nei	*Clupeoidei*	480 F	480 F	480 F	480 F	480 F	480 F	480 F
Barramundi(=Giant seaperch)	*Lates calcarifer*	359 F	356 F	393 F	395 F	508 F	423 F	499 F
Sea catfishes nei	*Ariidae*	1 850 F	1 850 F	1 850 F	1 850 F	1 850 F	1 850 F	1 850 F
Kawakawa	*Euthynnus affinis*	0	0	-	-	-	-	-
Skipjack tuna	*Katsuwonus pelamis*	9 811	9 512	11 270	37 214	29 949	52 289	64 355
Longtail tuna	*Thunnus tonggol*	0	0	-	-	-	-	-
Albacore	*Thunnus alalunga*	6	38	101	40	85	102	49
Yellowfin tuna	*Thunnus albacares*	2 722	971	6 968	11 726	7 610	14 446	25 812
Bigeye tuna	*Thunnus obesus*	161	50	1 060	1 461	998	1 471	4 981
Marlins,sailfishes,etc. nei	*Istiophoridae*	0	16	6	87	230	418	368
Marine fishes nei	*Osteichthyes*	10 000 F	10 500 F	10 500 F	10 000 F	10 000 F	10 000 F	10 000 F
River prawns nei	*Macrobrachium spp*	5 F	5 F	5 F	5 F	5 F	5 F	5 F
Oceanian crayfishes nei	*Parastacidae*	10	6	6	14	6	0	1
Indo-Pacific swamp crab	*Scylla serrata*	28 F	25 F	24 F	23 F	22 F	24 F	24 F
Tropical spiny lobsters nei	*Panulirus spp*	206	165	205	217	203	197	131
Banana prawn	*Penaeus merguiensis*	858	820	676	1 233	949	1 136	1 017
Giant tiger prawn	*Penaeus monodon*	137	109	99	209	164	126	117
Metapenaeus shrimps nei	*Metapenaeus spp*	220	219	159	187	279	328	346
Natantian decapods nei	*Natantia*	84	104	59	50	99	135	117
Sea cucumbers nei	*Holothurioidea*	1 335	1 788	1 515	2 037	1 185	1 824	1 453
	Country total	*39 077 F*	*37 823 F*	*46 185 F*	*78 029 F*	*65 431 F*	*96 069 F*	*122 419 F*
Pitcairn Is								
Freshwater fishes nei	*Osteichthyes*	0	0	0	0	0	0	0
Marine fishes nei	*Osteichthyes*	8 F	8 F	8 F	8 F	8 F	8 F	8 F
	Country total	*8 F*	*8 F*	*8 F*	*8 F*	*8 F*	*8 F*	*8 F*
Samoa								
Freshwater fishes nei	*Osteichthyes*	0	0	0	0	0	1	1 F
Albacore	*Thunnus alalunga*	1 883	1 775	4 108	4 742	4 027	4 067	4 820
Yellowfin tuna	*Thunnus albacares*	216	573	1 327	801	681	1 120	470
Bigeye tuna	*Thunnus obesus*	40	27	63	334	283	177	185
Tuna-like fishes nei	*Scombroidei*	-	-	-	-	-	479	470 F
Sharks, rays, skates, etc. nei	*Elasmobranchii*	...	...	...	...	...	20	20 F
Marine fishes nei	*Osteichthyes*	350	322	1 513	1 640	5 163	5 289	5 200 F
Marine crustaceans nei	*Crustacea*	10	10	10	10	30	207	200 F
Marine molluscs nei	*Mollusca*	20	20	20	20	20	1 644	1 600 F
	Country total	*2 519*	*2 727*	*7 041*	*7 547*	*10 204*	*13 004*	*12 966 F*
Solomon Is								
Freshwater fishes nei	*Osteichthyes*	0	0	0	0	0	0	0
Skipjack tuna	*Katsuwonus pelamis*	40 136	26 485	36 311	38 662	35 613	8 368	12 530
Albacore	*Thunnus alalunga*	24	100	109	370	136	224	54
Yellowfin tuna	*Thunnus albacares*	8 114	11 003	9 588	8 114	8 843	3 196	4 490
Bigeye tuna	*Thunnus obesus*	1 391	1 109	1 434	1 232	1 070	577	576
Marlins,sailfishes,etc. nei	*Istiophoridae*	50 F	50 F	50 F	50 F	50 F	50 F	50 F
Sharks, rays, skates, etc. nei	*Elasmobranchii*	80 F	50 F	4 000 F	600 F	310 F	300 F	300 F
Marine fishes nei	*Osteichthyes*	12 000 F	12 000 F	12 000 F	12 000 F	12 000 F	12 000 F	12 000 F
Banana prawn	*Penaeus merguiensis*	5 F	20 F	30 F	40 F	20 F	20 F	20 F

English name Nom anglais Nombre inglés	Scientific name Nom scientifique Nombre científico	1995 mt	1996 mt	1997 mt	1998 mt	1999 mt	2000 mt	2001 mt
Abalones nei	*Haliotis spp*	-	-	-	1	-	-	-
Gastropods nei	*Gastropoda*	-	-	-	0	0	-	-
Clams, etc. nei	*Bivalvia*	...	60	280 F	10 F	10 F	5 F	5 F
Sea cucumbers nei	*Holothurioidea*	219	113	203	253	376	48	50 F
	Country total	*62 019 F*	*50 990 F*	*64 005 F*	*61 332 F*	*58 428 F*	*24 788 F*	*30 075 F*
Tokelau								
Freshwater fishes nei	*Osteichthyes*	0	0	0	0	0	0	0
Marine fishes nei	*Osteichthyes*	200 F	200 F	200 F	200 F	200 F	200 F	200 F
	Country total	*200 F*	*200 F*	*200 F*	*200 F*	*200 F*	*200 F*	*200 F*
Tonga								
Freshwater fishes nei	*Osteichthyes*	0	0	0	0	0	0	0
Skipjack tuna	*Katsuwonus pelamis*	3	2	4	7	3	2	12
Albacore	*Thunnus alalunga*	379	431	493	616	801	862	1 268
Yellowfin tuna	*Thunnus albacares*	59	88	100	125	163	175	259
Bigeye tuna	*Thunnus obesus*	23	60	69	86	112	120	191
Black marlin	*Makaira indica*	15	20	10	14	6	13	10
Swordfish	*Xiphias gladius*	10	7	6	8	5	53	8
Tuna-like fishes nei	*Scombroidei*	8	1	8	14	34	42	85
Marine fishes nei	*Osteichthyes*	2 000	2 100	2 000	2 936	2 890	2 305	2 548
Marine crustaceans nei	*Crustacea*	100	120	100	177	200	175	270
Marine molluscs nei	*Mollusca*	...	...	1	3	7	13	22
Sea cucumbers nei	*Holothurioidea*	...	86	80	90	0	0	0
	Country total	*2 597*	*2 915*	*2 871*	*4 076*	*4 221*	*3 760*	*4 673*
Tuvalu								
Freshwater fishes nei	*Osteichthyes*	0	0	0	0	0	0	0
Skipjack tuna	*Katsuwonus pelamis*	259	260 F	300 F	300 F	300 F	300 F	300 F
Yellowfin tuna	*Thunnus albacares*	13	15 F	20 F	20 F	20 F	20 F	20 F
Tuna-like fishes nei	*Scombroidei*	15	15 F	20 F	20 F	20 F	20 F	20 F
Marine fishes nei	*Osteichthyes*	112	110 F	160 F	160 F	160 F	160 F	160 F
	Country total	*399*	*400 F*	*500 F*	*500 F*	*500 F*	*500 F*	*500 F*
US Minor Is								
Marine fishes nei	*Osteichthyes*	0	0	0	0	0	0	0
	Country total	*0*	*0*	*0*	*0*	*0*	*0*	*0*
Vanuatu								
Freshwater fishes nei	*Osteichthyes*	0	0	0	0	0	0	0
Porgies, seabreams nei	*Sparidae*	0	0	-	0	-	-	-
Black skipjack	*Euthynnus lineatus*	-	-	-	10	-	-	-
Skipjack tuna	*Katsuwonus pelamis*	20 047	19 600	30 860	40 112	57 151	44 115	7 980
Albacore	*Thunnus alalunga*	109	192	95	10	-	-	-
Yellowfin tuna	*Thunnus albacares*	25 620	13 182	31 267	28 243	28 477	16 615	10 720
Bigeye tuna	*Thunnus obesus*	9 391	10 231	6 107	4 143	5 099	6 210	3 790
Marlins,sailfishes,etc. nei	*Istiophoridae*	209	98	97	99	14	...	-
Tuna-like fishes nei	*Scombroidei*	-	-	40	-	-	20	1 800
Marine fishes nei	*Osteichthyes*	1 211 F	1 209 F	1 327 F	1 310 F	1 400 F	1 400 F	1 500 F
Penaeus shrimps nei	*Penaeus spp*	137	170	150	70	-	-	-
Natantian decapods nei	*Natantia*	-	-	-	0	-	-	-
Marine crustaceans nei	*Crustacea*	250 F	250 F	250 F	250 F	250 F	250 F	250 F
Cuttlefish,bobtail squids nei	*Sepiidae, Sepiolidae*	22	33	19	11	-	-	-
Octopuses, etc. nei	*Octopodidae*	-	-	-	0	-	-	-
Marine molluscs nei	*Mollusca*	600 F	600 F	600 F	600 F	600 F	600 F	600 F
Sea cucumbers nei	*Holothurioidea*	50 F	50 F	50 F	50 F	50 F	50 F	50 F
	Country total	*57 646 F*	*45 615 F*	*70 862 F*	*74 908 F*	*93 041 F*	*69 260 F*	*26 690 F*
Wallis Fut I								
Freshwater fishes nei	*Osteichthyes*	0	0	0	0	0	0	0
Marine fishes nei	*Osteichthyes*	164	174	170	294	294 F	294 F	294 F
Marine crabs nei	*Brachyura*	1	1	1	1	1 F	1 F	1 F
Tropical spiny lobsters nei	*Panulirus spp*	2	2	2	2	2 F	2 F	2 F
Octopuses, etc. nei	*Octopodidae*	1	1	1	1	1 F	1 F	1 F
Marine turtles nei	*Testudinata*	2	2	2	2	2 F	2 F	2 F
	Country total	*170*	*180*	*176*	*300*	*300 F*	*300 F*	*300 F*
Total		*1 008 649*	*851 498*	*1 085 385*	*1 169 814*	*1 168 126*	*1 070 411*	*1 108 309*

English name Nom anglais Nombre inglés	Scientific name Nom scientifique Nombre científico	1995 mt	1996 mt	1997 mt	1998 mt	1999 mt	2000 mt	2001 mt
Other nei								
Benguela hake	*Merluccius polli*	-	-	-	-	8	44	139
Cape hakes	*Merluccius capensis,M.paradox.*	-	1	-	-	-	-	-
Croakers, drums nei	*Sciaenidae*	-	-	-	-	-	-	12
Dentex nei	*Dentex spp*	1 106	462	-	-	-	-	-
Porgies, seabreams nei	*Sparidae*	-	-	-	-	63	49	341
Largehead hairtail	*Trichiurus lepturus*	-	-	-	-	147	948	523
Sardinellas nei	*Sardinella spp*	-	-	-	-	5 812	2 306	2 353
European pilchard(=Sardine)	*Sardina pilchardus*	-	-	-	-	226	1 876	541
European anchovy	*Engraulis encrasicolus*	-	-	-	-	3 560	4 419	3 815
Atlantic bonito	*Sarda sarda*	300	300	75	-	-	-	-
Narrow-barred Spanish mackerel	*Scomberomorus commerson*	3 060	3 060	3 060	-	-	-	-
Seerfishes nei	*Scomberomorus spp*	431	431	431	-	-	-	-
Frigate tuna	*Auxis thazard*	390	894	700	487	486	290	872
Frigate and bullet tunas	*Auxis thazard, A.rochei*	100	100	100	-	18	367	110
Little tunny(=Atl.black skipj)	*Euthynnus alletteratus*	253	200	200	203	202	3	-
Kawakawa	*Euthynnus affinis*	140	140	140	-	-	-	-
Skipjack tuna	*Katsuwonus pelamis*	58 946	56 159	50 472	53 927	65 228	66 897	59 393
Atlantic bluefin tuna	*Thunnus thynnus*	1 039	244	1 375	1 921	2 564	396	49
Longtail tuna	*Thunnus tonggol*	351	351	351	-	-	85	-
Albacore	*Thunnus alalunga*	2 413	2 932	5 911	11 324	1 705	2 533	92
Southern bluefin tuna	*Thunnus maccoyii*	282	295	333	476	483	31	...
Yellowfin tuna	*Thunnus albacares*	56 553	61 983	57 123	52 750	60 109	64 591	66 969
Bigeye tuna	*Thunnus obesus*	32 046	39 055	41 830	55 608	59 526	44 431	26 519
Atlantic sailfish	*Istiophorus albicans*	40	40	40	-	-	-	-
Indo-Pacific blue marlin	*Makaira mazara*	748	1 086	1 058	1 827	1 400	1 562	...
Atlantic blue marlin	*Makaira nigricans*	200	200	200	-	-	-	-
Black marlin	*Makaira indica*	198	214	114	225	176	167	...
Striped marlin	*Tetrapturus audax*	1 258	1 482	755	1 114	852	905	...
Atlantic white marlin	*Tetrapturus albidus*	100	100	100	-	-	-	-
Marlins,sailfishes,etc. nei	*Istiophoridae*	36	144	348	365	300	335	...
Swordfish	*Xiphias gladius*	5 762	8 782	6 169	7 746	7 014	8 339	...
Tuna-like fishes nei	*Scombroidei*	257	245	-	-	-	744	30
Cape horse mackerel	*Trachurus capensis*	18 596	9 805	-	-	-	-	-
Jack and horse mackerels nei	*Trachurus spp*	-	-	-	-	3 375	8 994	10 439
Chub mackerel	*Scomber japonicus*	-	-	-	-	781	4 002	3 338
Sharks, rays, skates, etc. nei	*Elasmobranchii*	469	501	366	295	198	250	...
Marine fishes nei	*Osteichthyes*	124 126	13	202	-	740	675	508
Penaeus shrimps nei	*Penaeus spp*	-	-	-	-	19	31	-
Natantian decapods nei	*Natantia*	-	-	-	-	-	17	-
Cuttlefish,bobtail squids nei	*Sepiidae, Sepiolidae*	-	-	-	-	138	154	148
Various squids nei	*Loliginidae, Ommastrephidae*	-	-	-	-	79	91	77
Octopuses, etc. nei	*Octopodidae*	-	-	-	-	196	245	264
	Country total	*309 200*	*189 219*	*171 453*	*188 268*	*215 405*	*215 777*	*176 532*
Total		***309 200***	***189 219***	***171 453***	***188 268***	***215 405***	***215 777***	***176 532***

Notes on individual countries or areas

Notes sur divers pays ou zones

Notas sobre los distintos países o áreas

Notes on individual countries or areas

ALGERIA

Data concerning tuna catches are reviewed in collaboration with the International Commission for the Conservation of Atlantic Tunas (ICCAT), the regional agency concerned with tuna statistics.

AMERICAN SAMOA

Data concerning tuna catches are reviewed in collaboration with the Secretariat of the Pacific Community (SPC), the regional agency concerned with tuna statistics.

ANGOLA

Capture production in Angolan waters by chartered foreign vessels have not been included pending further information on the nationality to which they should be attributed.

Data concerning tuna catches are reviewed in collaboration with ICCAT, the regional agency concerned with tuna statistics.

ANTIGUA AND BARBUDA

Catch data for 'Stromboid conchs nei' expressed on a meat-weight basis.

ARGENTINA

Data for Patagonean scallop (*Zygochlamis patagonica*) previously reported in meat weight have been revised.

AUSTRALIA

Data refer to a split-year (1 July – 30 June) shown under the calendar year in which the split-year ends.

Catch data, excluding those for tuna species, for the Southwest Atlantic area are provided by the Fisheries Department, Falkland Islands Government.

AUSTRIA

Excludes recreational fisheries.

BAHAMAS

Data for Caribbean spiny lobster and for groupers have been converted to live weight equivalents by using the conversion factors '3' and '2.5' respectively.

BANGLADESH

Data refer to a split-year (1 July – 30 June) shown under the calendar year in which the split-year ends.

Data concerning tuna catches are reviewed in collaboration with the Indian Ocean Tuna Commission (IOTC), the regional agency concerned with tuna statistics.

Notes sur divers pays ou zones

ALGÉRIE

Les données concernant les captures des thonidés sont révisées en collaboration avec la Commission internationale pour la conservation des thonidés de l'Atlantique (CICTA), l'organisation régionale chargée des statistiques des thonidés.

SAMOA AMÉRICAINES

Les données concernant les captures des thonidés les sont révisées en collaboration avec le Secrétariat général de la Communauté du Pacifique (CPS), l'organisation régionale chargée des statistiques des thonidés.

ANGOLA

Les captures dans les eaux angolaises des bateaux de pêche étrangers n'ont pas été incluses dans l'attente d'un complément d'information sur les pays auxquels elles doivent être attribuées.

Les données concernant les captures des thonidés sont révisées en collaboration avec la CICTA, l'organisation régionale chargée des statistiques des thonidés.

ANTIGUA-ET-BARBUDA

Les données relatives aux captures de 'Strombes nca' sont exprimées en poids-viande.

ARGENTINE

Les données relatives aux Peigne patagonien (*Zygochlamis patagonica*) précédemment reportées en poids-viande ont été révisées.

AUSTRALIE

Les données se réfèrent à une année fractionnée (1er juillet - 30 juin) et figurent sous l'année civile durant laquelle se termine l'année fractionnée.

Les données sur les captures, à l'exclusion de celles concernant le thon, pour l' Atlantique sud-ouest, sont fournies par le Département des pêches du Gouvernement des Îles Falkland.

AUTRICHE

Non compris la pêche récréative.

BAHAMAS

Les données relatives aux langoustes des Caraïbes et aux mérous ont été converties en équivalents poids vif en utilisant respectivement les facteurs de conversion '3' et '2,5'.

BANGLADESH

Les données se réfèrent à une année fractionnée (1er juillet - 30 juin) et figurent sous l'année civile durant laquelle se termine l'année fractionnée.

Les données concernant les captures des thonidés sont révisées en collaboration avec la Commission des thons de l'océan Indien (CTOI), l'organisation régionale chargée des statistiques des thonidés.

Notas sobre los distintos países o áreas

ARGELIA

Los datos relativos a las capturas de atún se revisan en colaboración con la Comisión Internacional para la Conservación del Atún Atlántico (CICAA), el organismo regional encargado de las estadísticas atuneras.

SAMOA AMERICANA

Los datos relativos a las capturas de atún se revisan en colaboración con la Secretaría general de la Comunidad del Pacífico (SCP), el organismo regional encargado de las estadísticas atuneras.

ANGOLA

No se han incluido capturas en aguas de Angola por buques fletados extranjeros, en espera de más información sobre la nacionalidad que se les debe atribuir.

Los datos relativos a las capturas de atún se revisan en colaboración con la CICAA, el organismo regional encargado de las estadísticas atuneras.

ANTIGUA Y BARBUDA

Los datos de capturas relativos a los 'Cobos nep' son expresados en peso carne.

ARGENTINA

Los datos relativos a Pecten patagónico (*Zygochlamis patagonica*) previamente notificados en peso carne han sido revisados.

AUSTRALIA

Los datos se refieren a un año emergente (1° de julio - 30 de junio) que se indica como el año civil en que finaliza el año emergente.

Los datos relativos a las capturas en la zona del Atlántico Sudoccidental, con exclusión de las de atunes, fueron proporcionados por el Departamento de Pesca del Gobierno de las Islas Malvinas (Falkland).

AUSTRIA

Se excluye la pesca recreativa.

BAHAMAS

Los datos de la langosta del Caribe y de los meros se han convertido en sus equivalentes de peso en vivo aplicando respectivamente los factores de conversión '3' y '2,5'.

BANGLADESH

Los datos se refieren a un año emergente (1° de julio - 30 de junio) que se indica como el año civil en que finaliza el año emergente.

Los datos relativos a las capturas de atún se revisan en colaboración con la Comisión del Atún para el Océano Indico (CAOI), el organismo regional encargado de las estadísticas atuneras.

Notes on individual countries or areas

Notes sur divers pays ou zones

Notas sobre los distintos países o áreas

BELIZE

Data concerning tuna catches are reviewed in collaboration with the Inter-American Tropical Tuna Commission (IATTC) and ICCAT, the regional agencies concerned with tuna statistics.

Catch data, excluding those for tuna species, for the Eastern Central and Southeast Atlantic areas are provided by the Las Palmas Survey (LPS) and the "Bulletin Statistique" published by IMROP, Mauritania.

Catch data, excluding those for tuna species, for the Southwest Atlantic area are provided by the Fisheries Department, Falkland Islands Government.

Data reported since 1980 as "fish fillet", "lobster tail", "crab claws", and "conch" have been converted to live weight equivalents by using the conversion factors of '2', '2', '4' and '7.5' respectively.

BELIZE

Les données concernant les captures des thonidés sont révisées en collaboration avec la Commission interaméricaine du thon tropical (CITT) et la CICTA, les organisations régionales chargées des statistiques des thonidés.

Les données sur les captures, à l'exclusion de celles concernant le thon, pour l'Atlantique du centre-est et du sud-est sont tirées de l'Enquête Las Palmas et du "Bulletin statistique" publié par le IMROP, Mauritanie.

Les données sur les captures, à l'exclusion de celles concernant le thon, pour l' Atlantique sud-ouest, sont fournies par le Département des pêches du Gouvernement des Îles Falkland.

Les données rapporté depuis 1980 comme "fish fillet", "lobster tail", "crab claws" et "conch" ont été converties en équivalents poids vif en utilisant respectivement les facteurs de conversion '2' '2', '4' et '7.5'.

BELICE

Los datos relativos a las capturas de atún se revisan en colaboración con la Comisión Interamericana del Atún Tropical (CIAT) y la CICAA, los organismos regionales encargados de las estadísticas atuneras.

Los datos relativos a las capturas en las zonas del Atlántico Centro-oriental y Sudoriental, con exclusión de las de atunes, fueron proporcionados por Las Palmas Survey (LPS) y por la publicación "Bulletin Statistique" del IMROP, Mauritania.

Los datos relativos a las capturas en la zona del Atlántico Sudoccidental, con exclusión de las de atunes, fueron proporcionados por el Departamento de Pesca del Gobierno de las Islas Malvinas (Falkland).

Los datos proporcionados desde 1980 como "fish fillet", "lobster tail", "crab claws" y "conch" se han convertido en sus equivalentes de peso en vivo aplicando respectivamente los factores de conversión '2' '2', '4' et '7.5'.

BENIN

Marine catches by Beninese canoes considerably greater than catches landed in Benin as many Beninese canoes operate from beaches in Cameroon, Congo and Gabon. Catches by these canoes included in the catches of the respective countries.

Data concerning tuna catches are reviewed in collaboration with ICCAT, the regional agency concerned with tuna statistics.

BÉNIN

Les captures maritimes des pirogues béninoises sont néanmoins nettement supérieures aux captures débarquées au Bénin, car nombre d'entre elles opèrent à partir des plages du Cameroun, du Congo et du Gabon. Leurs captures sont incluses dans les captures de ces différents pays.

Les données concernant les captures des thonidés sont révisées en collaboration avec la CICTA, l'organisation régionale chargée des statistiques des thonidés.

BENIN

Sin embargo, las capturas marinas de las canoas beninesas son considerablemente mayores que las capturas desembarcadas en Benín, ya que muchas de estas canoas faenan desde playas del Camerún, Congo y Gabón. Las capturas de estas canoas se incluyen en las capturas de los respectivos países.

Los datos relativos a las capturas de atún se revisan en colaboración con la CICAA, el organismo regional encargado de las estadísticas atuneras.

BHUTAN

Country does not submit returns. Data estimated on the basis of reports of visiting missions and other documentation.

BHOUTAN

Pays non déclarant. Données estimées à partir de rapports de missions et d'autres documents.

BHUTÁN

El país no proporciona información. Datos estimados sobre la base de informes de misiones al país y de otro tipo de documentación.

BOTSWANA

Submission of capture data resumed with 2000 statistics after twelve years of not reporting. Data before 1996 were probably overestimated.

BOTSWANA

Les données de captures fournies par le pays ont repris avec les statistiques de 2000 après douze ans de non-rapport. Les données avant 1996 ont été probablement surestimées.

BOTSWANA

El envío de datos de capturas empezó de nuevo en 2000 después de doce años de no envíos. Los datos antes de 1996 son en exceso.

BRAZIL

Since 1995, 100,000 tonnes of estimated subsistence catches have been included under 'Marine fishes nei'.

Data concerning tuna catches are reviewed in collaboration with ICCAT, the regional agency concerned with tuna statistics.

BRÉSIL

Depuis 1995 100 000 tonnes des estimations de captures de la pêche de subsistance sont incluses dans 'Poissons marins nca'.

Les données concernant les captures des thonidés sont révisées en collaboration avec la CICTA, l'organisation régionale chargée des statistiques des thonidés.

BRASIL

Desde 1995 100 000 toneladas métricas de capturas de la pesca de subsistencia se incluyen en 'Peces marinos nep'.

Los datos relativos a las capturas de atún se revisan en colaboración con la CICAA, el organismo regional encargado de las estadísticas atuneras.

BRITISH VIRGIN ISLANDS

Include Anegada, Jost Van Dyke, Tortola and Virgin Gorda.

In 1999, data reported since 1976 in thousands of pounds instead of tonnes have been revised.

ÎLES VIERGES BRITANNIQ.

Comprend Anegada, Jost Van Dyke, Tortola et Virgin Gorda.

En 1999 il a été procédé à une révision des séries de données commençant en 1976, car celles-ci étaient auparavant présentées en milliers de livres plutôt qu'en tonnes.

ISLAS VÍRGENES BRITÁN.

Incluye Anegada, Jost Van Dyke, Tortola y Virgin Gorda.

En 1999 se revisaron los datos a partir de 1976, notificados anteriormente a la FAO en miles de libras en lugar de toneladas.

BULGARIA

Inland water 1994-96 catches refer only to Danube River.

Catch data, excluding those for tuna species, for the Eastern Central Atlantic area have also been taken from the "Bulletin Statistique" published by IMROP, Mauritania.

BULGARIE

Les données de 1994-96 concernant les captures d'eau douce se rapportent uniquement au fleuve Danube.

Les données sur les captures, à l'exclusion de celles concernant le thon, pour l'Atlantique centre-est ont également été tirées du "Bulletin statistique" publié par le IMROP, Mauritanie.

BULGARIA

Los datos de capturas de 1994-96 relativos a la pesca en aguas continentales se refieren sólo al río Danubio.

Los datos sobre las capturas de la zona del Atlántico Centro-oriental, con exclusión de las de atunes, también se han extraído de la publicación "Bulletin Statistique" del IMROP, Mauritania.

Notes on individual countries or areas

BURUNDI

Decreased catch for 1996 due to limited period of fishing in the Lake Tanganyka (only two months).

CAMBODIA

Due to the introduction of a new methodology to estimate production of semi-commercial and subsistence fisheries, inland water statistics since 1999 are not comparable with those of previous years.

Catch data for the Southwest Atlantic area are provided by the Fisheries Department, Falkland Islands Government and tuna catches are reviewed in collaboration with ICCAT.

CAPE VERDE

Data concerning tuna catches are reviewed in collaboration with ICCAT, the regional agency concerned with tuna statistics.

CAYMAN IS

Data for the Eastern Central Atlantic refer to catches of vessels fishing with flag of convenience.

CHANNEL ISLANDS

The 1970-82 catch data refer to the Bailiwick of Guernsey only. The 1983-2001 catch data include also statistics relating to the Bailiwick of Jersey.

CHINA

For statistical purposes, the data for China do not include Hong Kong Special Administrative Region (Hong Kong SAR), Macao SAR and Taiwan Province of China.

Catch data, considered to be overstated since the early 1990s, under review and subject to possible downward revisions.

Data concerning tuna catches are reviewed in collaboration with ICCAT, IOTC and SPC, the regional agencies concerned with tuna statistics.

Catch data, excluding those for tuna species, for the Eastern Central and Southeast Atlantic areas are provided by the Las Palmas Survey (LPS).

Catch data, excluding those for tuna species, for the Southwest Atlantic area are provided also by the Fisheries Department, Falkland Islands Government.

Since 1970 data for "Aquatic plants nei", recorded on a dry-weight basis have been converted to wet-weight equivalents by using the conversion factor '10'.

Data for "Freshwater molluscs nei" and "Marine molluscs nei" prior to 1997 have been converted to live weight equivalents by using the conversion factor '2.13'.

Notes sur divers pays ou zones

BURUNDI

La diminution des captures enregistrées en 1996 est due à la durée limitée de pêche dans le lac Tanganyika (la pêche n'a été ouverte que pendant deux mois).

CAMBODGE

Du fait de l'introduction d'une nouvelle méthodologie d'estimation de la production des pêches semi-commerciales et de subsistance, les statistiques des pêches dans les eaux intérieures depuis 1999 ne sont plus comparables avec celles des années précédentes.

Les données sur les captures dans l'Atlantique sud-ouest sont fournies par le Département des pêches du Gouvernement des Îles Falkland et les captures des thonidés sont révisées en collaboration avec la CICTA.

CAP-VERT

Les données concernant les captures des thonidés sont révisées en collaboration avec la CICTA, l'organisation régionale chargée des statistiques des thonidés.

IS CAÏMANES

Les données dans l'Atlantique centre-est se rapportent aux captures effectuées sous pavillon de complaisance.

ÎLES ANGLO-NORMANDES

Les données relatives aux captures de 1970-82 se rapportent uniquement au bailliage de Guernsey. Les données relatives aux captures de 1983-2001 comprennent aussi les statistiques relatives au bailliage de Jersey.

CHINE

Les données statistiques relatives à la Chine ne comprennent pas celles qui concernent la Région administrative spéciale de Hong Kong (la RAS de Hong-Kong), la RAS de Macao et la province chinoise de Taïwan.

On considère les données sur les captures exagérées à partir du début des années 90. Elles sont actuellement examinées et seront éventuellement révisées à la baisse.

Les données concernant les captures des thonidés sont révisées en collaboration avec la CICTA, la CTOI et la CPS, les organisations régionales chargées des statistiques des thonidés.

Les données sur les captures, à l'exclusion de celles concernant le thon, pour l'Atlantique du centre-est et du sud-est sont tirées de l'Enquête Las Palmas.

Les données sur les captures, à l'exclusion de celles concernant le thon, pour l'Atlantique sud-ouest, sont fournies également par le Département des pêches du Gouvernement des Îles Falkland.

Depuis 1970 les données relatives aux "Plantes aquatiques nca" enregistrées en poids sec, ont été converties en leurs équivalents poids vert en utilisant le facteur de conversion '10'.

Les données relatives aux "Mollusques d'eau douce nca" et "Mollusques marins nca" avant 1997 ont été converties en équivalents poids vif en utilisant le facteur de conversion '2,13'.

Notas sobre los distintos países o áreas

BURUNDI

La diminución de captura registrada en 1996 se debe a la durata limitada de la pesca en el lago Tanganica (la pesca se permitió solamente por dos meses).

CAMBOYA

Debido a la introducción de una nueva metodología para calcular la producción de las pesquerías semicomerciales y de subsistencia, las estadísticas de la pesca continental desde el año 1999 no son comparables con las de años anteriores.

Los datos relativos a las capturas en la zona del Atlántico Sudoccidental fueron proporcionados por el Departamento de Pesca del Gobierno de las Islas Malvinas (Falkland) y las capturas de atún se revisaron en colaboración con la CICAA.

CABO VERDE

Los datos relativos a las capturas de atún se revisan en colaboración con la CICAA, el organismo regional encargado de las estadísticas atuneras.

IS CAIMÁN

Los datos en el Atlántico Centro-oriental se refieren a capturas de pabellones de conveniencia.

ISLAS ANGLONORMANDAS

Los datos de capturas de 1970-82 se refieren sólo a la bailía de Guernsey. Los datos de capturas de 1983-2001 incluyen también las estadísticas relativas a la bailía de Jersey.

CHINA

Los datos estadísticos relativos a China excluyen los datos correspondientes a la Región Administrativa Especial de Hong Kong (la RAE de Hong Kong), a la RAE de Macao y a la Provincia china de Taiwán.

Los datos relativos a las capturas se consideran exagerados desde el principio de los años noventa y están siendo examinados para una posible rectificación.

Los datos relativos a las capturas de atún se revisan en colaboración con la CICAA, la CAOI y la SCP, los organismos regionales encargados de las estadísticas atuneras.

Los datos relativos a las capturas en las zonas del Atlántico Centro-oriental y Sudoriental, con exclusión de las de atunes, fueron proporcionados por Las Palmas Survey (LPS).

Los datos relativos a las capturas en la zona del Atlántico Sudoccidental, con exclusión de las de atunes, también fueron proporcionados por el Departamento de Pesca del Gobierno de las Islas Malvinas (Falkland).

Desde 1970 los datos relativos a las "Plantas acuáticas nep" registrados en peso en seco se han convertido en sus equivalentes de peso húmedo aplicando el factor de conversión '10'.

Los datos de "Moluscos de agua dulce nep" y de "Moluscos marinos nep" antes de 1997 se han convertido en sus equivalentes de peso en vivo aplicando el factor de conversión '2,13'.

Notes on individual countries or areas

CHINA, HONG KONG SAR

Since 1999 only a total of marine capture production has been made available to FAO. Species breakdown is estimated.

CHINA, MACAO SAR

Fishery surveys have been suspended since July 1996. Data for 1997 onwards are FAO estimates.

COLOMBIA

Data concerning tuna catches are reviewed in collaboration with IATTC and ICCAT, the regional agency concerned with tuna statistics.

CONGO DEM. REP.

Formerly Zaire.

COSTA RICA

Data for shrimps have been converted to live weight equivalents by using the conversion factor '2'.

Since 1992 data for "Elasmobranchii – Sharks, rays, skates, etc. nei" include also production of shark fins converted to live weight equivalents by using the conversion factor '20'.

CÔTE D'IVOIRE

Data concerning tuna catches are reviewed in collaboration with ICCAT, the regional agency concerned with tuna statistics.

CROATIA

Catches in inland waters decreased since 1998 as statistics on recreational fishery are no longer collected.

Since 1997, "Marine shells nei" includes also significant production from aquaculture.

Data concerning tuna catches are reviewed in collaboration with ICCAT, the regional agency concerned with tuna statistics.

CUBA

In 1999, all data on catches and aquaculture for blue tilapia (*Oreochromis aureus*) were revised. Eighty per cent of the total production was considered as capture production and 20% as aquaculture production.

CYPRUS

Data refer to government-controlled area only.

Data concerning tuna catches are reviewed in collaboration with IATTC and ICCAT the regional agencies concerned with tuna statistics.

Catch data, excluding those for tuna species, for the Eastern Central Atlantic area have also been taken from the "Bulletin Statistique" published by IMROP, Mauritania.

Notes sur divers pays ou zones

CHINE, RAS DE HONG-KONG

Depuis 1999 la FAO n'a reçu que les données des captures marines totales. La répartition entre les espèces correspond à une estimation.

CHINE, RAS DE MACAO

Les prospections des pêches ont été suspendues à partir de juillet 1996. Les données à partir de 1997 sont des estimations de la FAO.

COLOMBIE

Les données concernant les captures des thonidés sont révisées en collaboration avec la CITT et la CICTA, les organisations régionales chargées des statistiques des thonidés.

REP. DÉM DU CONGO

Anciennement Zaïre.

COSTA RICA

Les données concernant les crevettes ont été converties en équivalents poids vif en utilisant le facteur de conversion '2'.

Depuis 1992 les données sous "Elasmobranchii – Requins, raies, etc. nca" comprennent également la production d'ailerons de requins, converties en équivalents poids vif en utilisant le facteur de conversion '20'.

CÔTE D'IVOIRE

Les données concernant les captures des thonidés sont révisées en collaboration avec la CICTA, l'organisation régionale chargée des statistiques des thonidés.

CROATIE

La diminution des captures dans les eaux continentales depuis 1998 est due à que les statistiques concernant la pêche récréative ne sont plus recueillies.

Depuis 1997 "Coquilles marines nca" comprend des quantités importantes provenant de l'aquaculture.

Les données concernant les captures des thonidés sont révisées en collaboration avec la CICTA, l'organisation régionale chargée des statistiques des thonidés.

CUBA

En 1999 toutes les données sur les captures et l'aquaculture pour le tilapia bleu (*Oreochromis aureus*) ont été révisées. Quatre-vingt pour cent de la production totale a été considérée comme pêches de capture et 20 pour cent comme production de l'aquaculture.

CHYPRE

Les données ne concernent que la partie du territoire sous contrôle du gouvernement.

Les données concernant les captures des thonidés sont révisées en collaboration avec la CITT et la CICTA, les organisations régionale chargées des statistiques des thonidés.

Les données sur les captures, à l'exclusion de celles concernant le thon, pour l'Atlantique centre-est ont également été tirées du "Bulletin statistique" publié par le IMROP, Mauritanie.

Notas sobre los distintos países o áreas

CHINA, RAE DE HONG KONG

Desde 1999 sólo se proporcionaron a la FAO datos de la producción total de capturas marinas. El desglose de especies constituye una estimación.

CHINA, RAE DE MACAO

Las encuestas pesqueras se suspendieron desde julio de 1996. Los datos relativos a 1997 y períodos sucesivos constituyen estimaciones de la FAO.

COLOMBIA

Los datos relativos a las capturas de atún se revisan en colaboración con la CIAT y la CICAA, los organismos regional encargados de las estadísticas atuneras.

REP. DEM. DEL CONGO

Antes Zaire.

COSTA RICA

Los datos relativos a los camarones se han convertido en sus equivalentes de peso en vivo aplicando el factor de conversión '2'.

Desde 1992 los datos incluidos en "Elasmobranchii – Tiburones, rayas, etc. nep" comprenden también la producción de aletas de tiburón, convertida en sus equivalentes de peso en vivo aplicando el factor de conversión '20'.

CÔTE D'IVOIRE

Los datos relativos a las capturas de atún se revisan en colaboración con la CICAA, el organismo regional encargado de las estadísticas atuneras.

CROACIA

La diminución de las capturas en aguas continentales desde 1998 se debe que ya no se recogen datos estadísticos sobre la pesca recreativa.

Desde 1997 "Conchas marinas nep" incluye tambien cantidades importantes de producción de acuicultura.

Los datos relativos a las capturas de atún se revisan en colaboración con la CICAA, el organismo regional encargado de las estadísticas atuneras.

CUBA

En 1999 se revisaron todos los datos relativos a las capturas y la producción en acuicultura de la tilapia *Oreochromis aureus*. Se consideró que el 80% de la producción total correspondía a las capturas y el 20% a la acuicultura.

CHIPRE

Los datos se refieren solamente a la zona controlada por el Gobierno.

Los datos relativos a las capturas de atún se revisan en colaboración con la CIAT y la CICAA, los organismo regional encargados de las estadísticas atuneras.

Los datos sobre las capturas de la zona del Atlántico Centro-oriental, con exclusión de las de atunes, también se han extraído de la publicación "Bulletin Statistique" del IMROP, Mauritania.

Notes on individual countries or areas

CZECHOSLOVAKIA

Czechoslovakia refers to the area that was formerly the Czechoslovak Socialist Republic. Since 1993, the Czech Republic and Slovakia are shown separately.

EGYPT

Catch mortality of 10% reported in the 1992-99 period not included.

Data concerning tuna catches are reviewed in collaboration with ICCAT and IOTC, the regional agencies concerned with tuna statistics.

Catch data, excluding those for tuna species, for the Eastern Central Atlantic area have also been taken from the "Bulletin Statistique" published by IMROP, Mauritania.

ERITREA

Formerly part of Ethiopia.

Data concerning tuna catches are reviewed in collaboration with IOTC, the regional agency concerned with tuna statistics.

ETHIOPIA

Up to 1992 includes Eritrea.

FIJI ISLANDS

Includes Viti Levu, Vanua Levu, and Rotuma islands.

Data concerning tuna catches are reviewed in collaboration with SPC, the regional agency concerned with tuna statistics.

Data for aquatic plants expressed in dry-weight.

FINLAND

Increase of catches in inland waters since 1980 due to changes in the statistical methods and to inclusion of recreational fishing.

FRANCE

Data concerning tuna catches are reviewed in collaboration with ICCAT and IOTC, the regional agencies concerned with tuna statistics.

Catch data, excluding those for tuna species, for the Eastern Central Atlantic area have also been taken from the "Bulletin Statistique" published by IMROP, Mauritania.

Catch data, excluding those for tuna species, for the Southwest Atlantic area are provided by the Fisheries Department, Falkland Islands Government.

Notes sur divers pays ou zones

TCHÉCOSLOVAQUIE

Par Tchécoslovaquie on entend le territoire de l'ancienne République socialiste de Tchécoslovaquie. Depuis 1993 les Républiques tchèque et Slovaque sont indiquées séparément.

EGYPTE

La mortalité des pêches de 10% reportée pour la période 1992-99 n'est pas comprise.

Les données concernant les captures des thonidés sont révisées en collaboration avec la CICTA et la CTOI, les organisations régionales chargées des statistiques des thonidés.

Les données sur les captures, à l'exclusion de celles concernant le thon, pour l'Atlantique centre-est ont également été tirées du "Bulletin statistique" publié par le IMROP, Mauritanie.

ERYTHRÉE

Anciennement partie de l'Ethiopie.

Les données concernant les captures des thonidés sont révisées en collaboration avec la CTOI, l'organisation régionale chargée des statistiques des thonidés.

ETHIOPIE

Jusqu'en 1992 comprend l'Erythrée.

ÎLES FIDJI

Comprend les îles Viti Levu, Vanua Levu et Rotuma.

Les données concernant les captures des thonidés sont révisées en collaboration avec la CPS, l'organisation régionale chargée des statistiques des thonidés.

Les données sur les plantes aquatiques sont exprimées en au poids-sec.

FINLANDE

L'augmentation des captures dans les eaux continentales depuis 1980 s'explique par une modification des méthodes statistiques et l'inclusion de la pêche récréative.

FRANCE

Les données concernant les captures des thonidés sont révisées en collaboration avec la CICTA et la CTOI, les organisations régionales chargées des statistiques des thonidés.

Les données sur les captures, à l'exclusion de celles concernant le thon, pour l'Atlantique centre-est ont également été tirées du "Bulletin statistique" publié par le IMROP, Mauritanie.

Les données sur les captures, à l'exclusion de celles concernant le thon, pour l' Atlantique sud-ouest, sont fournies par le Département des pêches du Gouvernement des Îles Falkland.

Notas sobre los distintos países o áreas

CHECOSLOVAQUIA

Para Checoslovaquia se entiende la superficie correspondiente a la antigua República Socialista de Checoslovaquia. Desde 1993 las Repúblicas Checa y Eslovaca se muestran separadas.

EGIPTO

La mortalidad por pesca de 10% notificada para el período 1992-99 no se ha incluido.

Los datos relativos a las capturas de atún se revisan en colaboración con la CICAA y la CAOI, los organismos regionales encargados de las estadísticas atuneras.

Los datos sobre las capturas de la zona del Atlántico Centro-oriental, con exclusión de las de atunes, también se han extraído de la publicación "Bulletin Statistique" del IMROP, Mauritania.

ERITREA

Antes parte de Etiopia.

Los datos relativos a las capturas de atún se revisan en colaboración con la CAOI, el organismo regional encargado de las estadísticas atuneras.

ETIOPÍA

Hasta el 1992 incluye Eritrea.

ISLAS FIJI

Incluyen las islas Viti Levu, Vanua Levu y Rotuma.

Los datos relativos a las capturas de atún se revisan en colaboración con la SCP, el organismo regional encargado de las estadísticas atuneras.

Los datos relativos a las plantas acuáticas se han expresado en peso seco.

FINLANDIA

El incremento de las capturas en aguas continentales desde 1980 se debe a cambios en los métodos estadísticos y a la inclusión de la pesca recreativa.

FRANCIA

Los datos relativos a las capturas de atún se revisan en colaboración con la CICAA y la CAOI, los organismos regionales encargados de las estadísticas atuneras.

Los datos sobre las capturas de la zona del Atlántico Centro-oriental, con exclusión de las de atunes, también se han extraído de la publicación "Bulletin Statistique" del IMROP, Mauritania.

Los datos relativos a las capturas en la zona del Atlántico Sudoccidental, con exclusión de las de atunes, fueron proporcionados por el Departamento de Pesca del Gobierno de las Islas Malvinas (Falkland).

Notes on individual countries or areas

FRENCH POLYNESIA

Up to and including 1992, data refer only to quantities sold at the principal fish markets and tuna catches were reviewed in collaboration with the SPC, the regional agency concerned with tuna statistics. Since 1993, increased catches due to complete data collection.

Since 1997, data for 'Trochus shells', 'Pearl oyster shells nei' and 'Sponges' are derived from export statistics. Data for 'Pearl oyster shells nei' include also production of pearls.

FRENCH SOUTHERN TERR.

Include Amsterdam and St. Paul Islands in area 51 (Western Indian Ocean) and Kerguelen and Crozet Islands in area 58 (Antarctic Indian Ocean). Catches reported for area 58 are included with those of France.

GERMANY

Increased catches of 'Freshwater fishes nei' since 1995 due to the inclusion of estimated catches (about 18,800 mt) from recreational fisheries.

GHANA

Catch by Ghanaian artisanal fishermen considerably greater than fish landed in Ghana, as Ghanaian fishermen operate from beaches along the West African coast, notably in Liberia, Gambia, Togo, Benin and Sierra Leone. Catches by these canoes included in totals for the respective countries.

Data concerning tuna catches are reviewed in collaboration with ICCAT, the regional agency concerned with tuna statistics.

GREECE

Since 1988, the data for catches in inland waters are those provided by the Ministry of Agriculture while data prior to 1988 were those provided by the Ministry of National Economy.

Data concerning tuna catches are reviewed in collaboration with ICCAT, the regional agency concerned with tuna statistics.

GREENLAND

Starting with 1996, data reported on marine mammals include also harvest for subsistence. Data for 2001 cover only the period between January and September.

GUATEMALA

Data concerning tuna catches are reviewed in collaboration with IATTC, the regional agency concerned with tuna statistics.

Data exclude significant quantities of unrecorded subsistence fishery.

Notes sur divers pays ou zones

POLYNÉSIE FRANÇAISE

Jusqu'à l'année 1992 comprise, les données se rapportent uniquement aux quantités vendues sur les principaux marchés de poisson et les captures des thonidés ont été révisées en collaboration avec la CPS, l'organisation régionale chargée des statistiques des thonidés. Depuis 1993 l'augmentation des captures est imputable à la collecte complète des données.

Depuis 1997 les données relatives aux 'Troques', 'Coquilles d'huîtres perl. nca' et 'Eponges' proviennent des statistiques d'exportation. Données de 'Coquilles d'huîtres perl. nca' comprennent également la production des perles.

TERRES AUSTRALES FR.

Comprennent les îles d'Amsterdam et de Saint-Paul dans la zone de pêche 51 (océan Indien, ouest) et les îles Kerguelen et Crozet dans la zone de pêche 58 (océan Indien, antarctique). Les données relatives à la zone de pêche 58 sont incluses avec celles de la France.

ALLEMAGNE

L'augmentation des captures de 'Poissons d'eau nca' douce depuis 1995 est imputable à l'inclusion de l'estimation (approximativement 18 800 tonnes) de pêche récréative.

GHANA

Les captures de pêcheurs artisanaux ghanéens sont sensiblement plus élevées que les quantités de poisson débarquées au Ghana, car les pêcheurs ghanéens opèrent à partir des plages des côtes de l'Afrique occidentale, notamment au Libéria, en Gambie, au Togo, au Bénin et en Sierra Leone. Les captures de ces piroguiers sont incluses dans les totaux de ces différents pays.

Les données concernant les captures des thonidés sont révisées en collaboration avec la CICTA, l'organisation régionale chargée des statistiques des thonidés.

GRÈCE

Depuis 1988 les données sur les captures dans les eaux continentales sont fournies par le Ministère de l'agriculture tandis que les données avant 1988 étaient fournies par le Ministère de l'économie nationale.

Les données concernant les captures des thonidés sont révisées en collaboration avec la CICTA, l'organisation régionale chargée des statistiques des thonidés.

GROENLAND

Á partir de 1996, les données reportées officiellement concernant les mammifères aquatiques comprennent également la pêche destinée à la subsistance. Les données pour 2001 se réfèrent exclusivement à la période entre janvier et septembre.

GUATEMALA

Les données concernant les captures des thonidés sont révisées en collaboration avec la CITT, l'organisation régionale chargée des statistiques des thonidés.

Non compris des quantités importantes non enregistrées concernant la pêche de subsistance.

Notas sobre los distintos países o áreas

POLINESIA FRANCESA

Hasta 1992 inclusive, los datos se refieren sólo a cantidades vendidas en los principales mercados de pescado y las capturas de atún han sido revisadas en colaboración con la SCP, el organismo regional encargado de las estadísticas atuneras. Desde 1993 el incremento de las capturas se debe a una colección completa de datos.

Desde 1997 los datos relativos a 'Tróquidos', 'Conchas de ostras perleras nep' y 'Esponjas' han sido obtenidos de las estadísticas de exportación. Datos de 'Conchas de ostras perleras nep' incluyen también la producción de perlas.

TIERRAS AUSTRALES FR.

Incluyen las islas Amsterdam y San Pablo en área de pesca 51 (Océano Indico occidental) y las islas Kerguélen y Crozet en área de pesca 58 (Océano Indico Antártico). Datos relativos a la área de pesca 58 se incluyen en los datos de la Francia.

ALEMANIA

El incremento de capturas de 'Peces de agua dulce nep' desde 1995 se debe a la inclusión de la estimación (aproximadamente 18 800 toneladas) de la pesca recreativa.

GHANA

La captura de los pescadores artesanales ghaneses es considerablemente mayor que la captura desembarcada en Ghana, ya que faenan desde playas de la costa occidental africana, sobre todo en Liberia, Gambia, Togo, Benín y Sierra Leona. Las capturas de estas canoas se incluyen en los totales de los respectivos países.

Los datos relativos a las capturas de atún se revisan en colaboración con la CICAA, el organismo regional encargado de las estadísticas atuneras.

GRECIA

Desde 1988 las cifras de las capturas en aguas continentales son aquellas proporcionadas por el Ministerio de la Agricultura mientras las cifras antecedentes fueron proporcionadas por el Ministerio de la Economía Nacional.

Los datos relativos a las capturas de atún se revisan en colaboración con la CICAA, el organismo regional encargado de las estadísticas atuneras.

GROENLANDIA

A partir de 1996, los datos proporcionados oficialmente incluyen también la recogida hecha con fin de subsistencia. Datos sobre 2001 se refieren solamente al período entre enero y septiembre.

GUATEMALA

Los datos relativos a las capturas de atún se revisan en colaboración con la CIAT, el organismo regional encargado de las estadísticas atuneras.

Se excluyen importantes cantidades no registradas relativas a la pesca de subsistencia.

Notes on individual countries or areas

Notes sur divers pays ou zones

Notas sobre los distintos países o áreas

HONDURAS

Data concerning tuna catches are reviewed in collaboration with IATTC and ICCAT, the regional agencies concerned with tuna statistics. A portion of the tunas caught in the Indian Ocean by vessels flying the Honduran flag may be included under "Other nei".

Catch data, excluding those for tuna species, for the Eastern Central and Southeast Atlantic areas are provided by the Las Palmas Survey (LPS).

HONDURAS

Les données concernant les captures des thonidés sont révisées en collaboration avec la CITT et la CICTA, les organisations régionales chargées des statistiques des thonidés. Une partie des données de thonidés attrapés dans l'océan Indien par des navires battant pavillon hondurien peut être incluse avec celles de "Autres nca".

Les données sur les captures, à l'exclusion de celles concernant le thon, pour l'Atlantique du centre-est et du sud-est sont tirées de l'Enquête Las Palmas.

HONDURAS

Los datos relativos a las capturas de atún se revisan en colaboración con la CIAT y la CICAA, los organismos regionales encargados de las estadísticas atuneras. Una porción de las capturas de atún en el Océano Indico por embarcaciones con pabellón hondureño puede ser incluida en los datos de "Otros nep".

Los datos relativos a las capturas en las zonas del Atlántico Centro-oriental y Sudoriental, con exclusión de las de atunes, fueron proporcionados por Las Palmas Survey (LPS).

INDIA

Data concerning tuna catches are reviewed in collaboration with IOTC, the regional agency concerned with tuna statistics.

Data for frogs are derived from trade statistics.

INDE

Les données concernant les captures des thonidés sont révisées en collaboration avec la CTOI, l'organisation régionale chargée des statistiques des thonidés.

Les données concernant les grenouilles proviennent des statistiques du commerce.

INDIA

Los datos relativos a las capturas de atún se revisan en colaboración con la CAOI, el organismo regional encargado de las estadísticas atuneras.

Los datos relativos a las ranas derivan de las estadísticas de comercio.

INDONESIA

Starting with the 2001 data, in accordance with the boundary change between fishing areas 57 and 71 in the Australian-Indonesian region, Indonesia has provided a first release (backwards to 1975) of its capture statistics in areas 57 and 71 revised according to the new boundary.

Data for tuna species are subject to revisions.

INDONÉSIE

Á partir des données de 2001, selon le changement de la limite entre les zones de pêche 57 et 71 dans la région australien-indonésienne, l'Indonésie a fourni un premier dégagement (à partir de 1975) de ses statistiques de captures dans les zones 57 et 71 révisées selon la nouvelle limite.

Les données concernant le thon sont sujettes à révisions.

INDONESIA

A partir de los datos de 2001, como consecuencia del cambio del límite entre las áreas de pesca 57 y 71 en la región australiana-indonesia, Indonesia ha proporcionado un primer lanzamiento (a partir de 1975) de sus estadísticas de capturas en las áreas 57 y 71 revisadas según el nuevo límite.

Los datos de atunes están sujetos a revisiones.

IRAN (ISLAMIC REP. OF)

Data concerning tuna catches are reviewed in collaboration with IOTC, the regional agency concerned with tuna statistics.

IRAN (RÉP. ISLAMIQUE D')

Les données concernant les captures des thonidés sont révisées en collaboration avec la CTOI, l'organisation régionale chargée des statistiques des thonidés.

IRÁN (REP. ISLÁMICA DEL)

Los datos relativos a las capturas de atún se revisan en colaboración con la CAOI, el organismo regional encargado de las estadísticas atuneras.

IRELAND

Since 1999 catches in inland waters are no longer collected except those for European eel, Atlantic salmon and sea trout.

Data concerning tuna catches are reviewed in collaboration with ICCAT, the regional agency concerned with tuna statistics.

IRLANDE

Depuis 1999 les données de captures dans les eaux continentales ne sont plus rassemblées à l'exception de celles relatives à anguille d'Europe, saumon de l'Atlantique et truite de mer.

Les données concernant les captures des thonidés sont révisées en collaboration avec la CICTA, l'organisation régionale chargée des statistiques des thonidés.

IRLANDA

Desde 1999 las capturas en aguas continentales ya no se recogen a excepión de datos por anguila europea, salmón del Atlántico y trucha marina.

Los datos relativos a las capturas de atún se revisan en colaboración con la CICAA, el organismo regional encargado de las estadísticas atuneras.

ISRAEL

Up to 1994 includes Gaza Strip.

ISRAËL

Jusqu'en 1994 comprend la Zone de Gaza.

ISRAEL

Hasta el 1994 incluye la Zona de Gaza.

ITALY

Data concerning tuna catches are reviewed in collaboration with ICCAT, the regional agency concerned with tuna statistics.

Catch data for 'Mediterranean mussel' are subject to revisions; increased catches since 1998 due to improved coverage.

ITALIE

Les données concernant les captures des thonidés sont révisées en collaboration avec la CICTA, l'organisation régionale chargée des statistiques des thonidés.

Les données relatives aux captures de 'Moule commune' sont sujettes à révisions; l'augmentation de captures depuis 1998 s'explique par une meilleure couverture.

ITALIA

Los datos relativos a las capturas de atún se revisan en colaboración con la CICAA, el organismo regional encargado de las estadísticas atuneras.

Los datos de capturas relativos al 'Mejillón común' están sujetos a revisiones; el incremento de capturas desde 1998 se debe a una mejor cobertura.

JAMAICA

Catch data for 'Stromboid conchs nei' expressed on a meat-weight basis.

JAMAÏQUE

Les données relatives aux captures de 'Strombes nca' sont exprimées en poids-viande.

JAMAICA

Los datos de capturas relativos a los 'Cobos nep' son expresados en peso carne.

Notes on individual countries or areas

JAPAN

Data for tuna species not comparable with those of IATTC, ICCAT, IOTC and SPC.

KENYA

Landings recording system for Lake Victoria, which accounts for 70% of the total catch, considered unreliable. Great part of the increased catch recorded since 1990 is probably due to improved coverage.

KIRIBATI

Includes Fanning Island, Washington Island and Christmas Island in the Line Islands; Ocean Island, Phoenix Islands (Birnie, Gardner, Hull, McKean, Phoenix, Sydney, Canton and Enderbury).

Data concerning tuna catches are reviewed in collaboration with the SPC, the regional agency concerned with tuna statistics.

KOREA DEM. PEOPLE'S REP.

The 1990-98 catch data have been extensively revised on the basis of the information made available by the "FAO Report of the Fisheries Development Programming Mission", November 1998.

Catch data, excluding those for tuna species, for the Eastern Central Atlantic areas are provided by the Las Palmas Survey (LPS).

KOREA REPUBLIC

Data for tuna species not comparable with those of IATTC, ICCAT, IOTC and SPC.

LATVIA

Catch data, excluding those for tuna species, for the Eastern Central Atlantic area have also been taken from the "Bulletin Statistique" published by IMROP, Mauritania.

LIBYAN ARAB JAMAHIRIYA

Data concerning tuna catches are reviewed in collaboration with ICCAT, the regional agency concerned with tuna statistics.

LITHUANIA

Catch data, excluding those for tuna species, for the Eastern Central Atlantic area have also been taken from the "Bulletin Statistique" published by IMROP, Mauritania.

MADAGASCAR

Data for sea cucumbers expressed in dry-weight.

Notes sur divers pays ou zones

JAPON

Les données concernant les thonidés ne sont pas comparables avec les renseignements fournis par la CITT, la CICTA, la CTOI et la CPS.

KENYA

Le système d'enregistrement des débarquements utilisé pour le lac Victoria, d'où proviennent 70% des captures totales, n'est pas considéré fiable. L'augmentation des captures enregistrées depuis 1990 est probablement en grande partie due à une meilleure couverture.

KIRIBATI

Comprend Fanning, Washington, l'île Christmas dans les îles de la Ligne; l'île Océan, les îles Phoenix (Birnie, Gardner, Hull, McKean, Phoenix, Sidney, Canton et Enderbury).

Les données concernant les captures des thonidés sont révisées en collaboration avec la CPS, l'organisation régionale chargée des statistiques des thonidés.

RÉP. POP. DÉM. DE CORÉE

Les données sur les captures de la période 1990-98 ont été largement révisées en fonction des informations recueillies sur "Rapport FAO de la mission de programmation de développement des pêches", novembre 1998.

Les données sur les captures, à l'exclusion de celles concernant le thon, pour l'Atlantique du centre-est sont tirées de l'Enquête Las Palmas.

RÉPUBLIC DE CORÉE

Les données concernant les thonidés ne sont pas comparables avec les renseignements fournis par la CITT, la CICTA, la CTOI et la CPS.

LETTONIE

Les données sur les captures, à l'exclusion de celles concernant le thon, pour l'Atlantique centre-est ont également été tirées du "Bulletin statistique" publié par le IMROP, Mauritanie.

JAMAHIRIYA ARABE LIBYEN.

Les données concernant les captures des thonidés sont révisées en collaboration avec la CICTA, l'organisation régionale chargée des statistiques des thonidés.

LITUANIE

Les données sur les captures, à l'exclusion de celles concernant le thon, pour l'Atlantique centre-est ont également été tirées du "Bulletin statistique" publié par le IMROP, Mauritanie.

MADAGASCAR

Les données sur les bèche-de-mer sont exprimées en poids sec.

Notas sobre los distintos países o áreas

JAPÓN

Los datos sobre las especies de atún no son comparables con la información de la CIAT, la CICAA, la CAOI y la SCP.

KENIA

No se considera fiable el sistema de registro de los desembarques del Lago Victoria, que representa el 70% de la captura total. La mayor captura registrada desde 1990 se debe probablemente en gran parte a una mejor cobertura.

KIRIBATI

Incluye las islas Fanning, Washington, y Christmas en las islas de la Línea; la isla Océano, y las islas Phoenix (Birnie, Gardner, Hull, McKean, Phoenix, Sidney, Canton y Enderbury).

Los datos relativos a las capturas de atún se revisan en colaboración con la SCP, el organismo regional encargado de las estadísticas atuneras.

REP. POP. DEM . DE COREA

Los datos de las capturas de 1990-98 han sido revisados ampliamente para tener en cuenta la información sobre "Informe FAO de la Misión de programación del desarrollo pesquero", noviembre de 1998.

Los datos relativos a las capturas en las zonas del Atlántico Centro-oriental, con exclusión de las de atunes, fueron proporcionados por Las Palmas Survey (LPS).

REPÚBLICA DE COREA

Los datos sobre las especies de atún no son comparables con la información de la CIAT, la CICAA, la CAOI y la SCP.

LETONIA

Los datos sobre las capturas de la zona del Atlántico Centro-oriental, con exclusión de las de atunes, también se han extraído de la publicación "Bulletin Statistique" del IMROP, Mauritania.

JAMAHIRIYA ARABE LIBIA

Los datos relativos a las capturas de atún se revisan en colaboración con la CICAA, el organismo regional encargado de las estadísticas atuneras.

LITUANIA

Los datos sobre las capturas de la zona del Atlántico Centro-oriental, con exclusión de las de atunes, también se han extraído de la publicación "Bulletin Statistique" del IMROP, Mauritania.

MADAGASCAR

Los datos relativos a las cohombros de mar se han expresado en peso seco.

Notes on individual countries or areas

Notes sur divers pays ou zones

Notas sobre los distintos países o áreas

MALAYSIA

Includes Peninsular Malaysia, Sabah and Sarawak.

The catch series for 1982-93 have been extensively revised by the Malaysian fisheries authorities. An improved methodology for data collection was introduced in 1987 which was based on larger sample size and full coverage of artisanal fishing villages. The new methodology showed that catches had been under-estimated by the old statistical system and revised catch estimates were prepared for the period 1982-86 during which the old system had deteriorated.

Data concerning tuna catches are reviewed in collaboration with IOTC, the regional agency concerned with tuna statistics.

MALAISIE

Comprend la Malaisie Péninsulaire, Sabah et Sarawak.

Les séries des captures pour 1982-93 ont été largement révisées par les autorités malaisiennes chargées des pêches. Une méthodologie améliorée concernant la collecte des données a été introduite en 1987; elle a été basée sur des tailles plus grandes d'échantillonnage et une couverture complète des villages pratiquant la pêche artisanale. La nouvelle méthodologie a montré que les captures avaient été sous-estimées par l'ancien système de collecte des statistiques et des estimations de captures révisées ont maintenant été préparées pour la période 1982-86 durant laquelle l'ancien système s'était détérioré.

Les données concernant les captures des thonidés sont révisées en collaboration avec la CTOI, l'organisation régionale chargée des statistiques des thonidés.

MALASIA

Incluye la Malasia Peninsular, Sabah y Sarawak.

Las series de capturas para 1982-93 han sido revisadas ampliamente por las autoridades de Malasia encargadas de la pesca. Se ha introducido en 1987 una metodología mejorada de colecta de datos basada sobre tallas de muestreo más grandes y una cobertura completa de las comunidades de pescadores artesanales. La nueva metodología ha mostrado que las capturas efectuadas bajo el viejo sistema estadístico habían sido sobrestimadas y se han preparado estimaciones de capturas revisadas para el período 1982-86 durante el cual el sistema viejo se había deteriorado.

Los datos relativos a las capturas de atún se revisan en colaboración con la CAOI, el organismo regional encargado de las estadísticas atuneras.

MALDIVES

Data concerning tuna catches are reviewed in collaboration with IOTC, the regional agency concerned with tuna statistics.

MALDIVES

Les données concernant les captures des thonidés, poissons type thon nca et poissons marins nca sont révisées en collaboration avec la CTOI, l'organisation régionale chargée des statistiques des thonidés.

MALDIVAS

Los datos relativos a la captura nominal de atunes, peces parecidos a los atunes nep y peces marinos nep se revisan en colaboración con la CAOI, el organismo regional encargado de las estadísticas atuneras.

MALTA

Data concerning tuna catches are reviewed in collaboration with ICCAT, the regional agency concerned with tuna statistics.

Catch data, excluding those for tuna species, for the Eastern Central Atlantic area have also been taken from the "Bulletin Statistique" published by IMROP, Mauritania.

MALTE

Les données concernant les captures des thonidés sont révisées en collaboration avec la CICTA, l'organisation régionale chargée des statistiques des thonidés.

Les données sur les captures, à l'exclusion de celles concernant le thon, pour l'Atlantique centre-est ont également été tirées du "Bulletin statistique" publié par le IMROP, Mauritanie.

MALTA

Los datos relativos a las capturas de atún se revisan en colaboración con la CICAA, el organismo regional encargado de las estadísticas atuneras.

Los datos sobre las capturas de la zona del Atlántico Centro-oriental, con exclusión de las de atunes, también se han extraído de la publicación "Bulletin Statistique" del IMROP, Mauritania.

MARSHALL ISLANDS

Data concerning tuna catches are reviewed in collaboration with the SPC, the regional agency concerned with tuna statistics. Increase of tuna catches in 2001 due to vessels, previously flagged in Vanuatu, re-flagged in the Marshall Islands.

Catch data, excluding those for tuna species, for the Eastern Central Atlantic area have also been taken from the "Bulletin Statistique" published by IMROP, Mauritania.

ÎLES MARSHALL

Les données concernant les captures des thonidés sont révisées en collaboration avec la CPS, l'organisation régionale chargée des statistiques des thonidés. L'augmentation en 2001 des captures pour les thonidés s'explique par les captures effectuées par bateaux, précédemment battant pavillon de Vanuatu, transférés en pavillon des Îles Marshall.

Les données sur les captures, à l'exclusion de celles concernant le thon, pour l'Atlantique centre-est ont également été tirées du "Bulletin statistique" publié par le IMROP, Mauritanie.

ISLAS MARSHALL

Los datos relativos a las capturas de atún se revisan en colaboración con la SCP, el organismo regional encargado de las estadísticas atuneras. El incremento en 2001 de las capturas de atún se debe a embarcaciones, previamente con pabellón de Vanuatu, que cambiaron de pabellón en las Islas Marshall.

Los datos sobre las capturas de la zona del Atlántico Centro-oriental, con exclusión de las de atunes, también se han extraído de la publicación "Bulletin Statistique" del IMROP, Mauritania.

MAURITANIA

Catch data revised backwards to 1991 according to new estimates of artisanal catches higher than those previously included in the "Bulletin Statistique" published by IMROP.

MAURITANIE

Les donnés sur les captures depuis 1991 ont été largement révisées selon nouvelles estimations sur les captures artisanales plus élevées que celles précédemment incluses dans le "Bulletin statistique" publié par le IMROP.

MAURITANIA

Los datos de las capturas desde 1991 han sido revisados ampliamente según nuevas estimaciones de las capturas artesanales más grandes que esas previamente incluidas en la publicación "Bulletin Statistique" del IMROP.

MAYOTTE

Includes Grand-Terre and Pamandzi.

MAYOTTE

Comprend Grande-Terre et Pamandzi.

MAYOTTE

Incluye Grande-Terre y Pamandzi.

MICRONESIA

Includes Yap, Truk, Pohnpei and Kosrae.

Data concerning tuna catches are reviewed in collaboration with the SPC, the regional agency concerned with tuna statistics.

MICRONÉSIE

Comprend Yap, Truk, Pohnpei et Kosrae.

Les données concernant les captures des thonidés sont révisées en collaboration avec la CPS, l'organisation régionale chargée des statistiques des thonidés.

MICRONESIA

Incluye Yap, Truk, Pohnpei y Kosrae.

Los datos relativos a las capturas de atún se revisan en colaboración con la SCP, el organismo regional encargado de las estadísticas atuneras.

Notes on individual countries or areas

Notes sur divers pays ou zones

Notas sobre los distintos países o áreas

MOROCCO

Data concerning tuna catches are reviewed in collaboration with ICCAT, the regional agency concerned with tuna statistics.

MYANMAR

Data refer to a fiscal year period (1 April – 31 March) shown under the calendar year in which the split-year ends.

Data for jellyfishes expressed in dry-weight.

NAMIBIA

A remarkable amount of 'Cape horse mackerel' (*Trachurus capensis*) catches has been added to Namibian capture production for the 1992-2000 period to take account of information provided by the national reporting office.

Data concerning tuna catches are reviewed in collaboration with ICCAT, the regional agency concerned with tuna statistics.

Catch data, excluding those for tuna species, for the Southwest Atlantic area are provided by the Fisheries Department, Falkland Islands Government.

Data for 'Gracilaria seaweeds' recorded on a dry-weight basis have been converted to wet-weight equivalents by using the conversion factor '8'.

NEPAL

Data refer to a split-year period (16 July - 15 July) shown under the calendar year in which the split-year ends.

NETHERLANDS ANTILLES

Include Curaçao, Bonaire, Saba, St. Eustatius and St. Martin.

Data concerning tuna catches are reviewed in collaboration with ICCAT, the regional agency concerned with tuna statistics.

NEW CALEDONIA

Data concerning tuna catches are reviewed in collaboration with the SPC, the regional agency concerned with tuna statistics.

NEW ZEALAND

For the period 1995-2000, data for "Elasmobranchii – Sharks, rays, skates, etc. nei" include also production of shark fins converted to live weight equivalents by using the conversion factor '20'.

Data for 'New Zealand scallop' since 1987 have been converted to live weight equivalents by using the conversion factor '8'

Data for 'New Zealand dredge oyster' are subject to revision.

MAROC

Les données concernant les captures des thonidés sont révisées en collaboration avec la CICTA, l'organisation régionale chargée des statistiques des thonidés.

MYANMAR

Les données se réfèrent à l'exercice financier (1er avril - 31 mars) et figurent sous l'année civile durant laquelle se termine l'année fractionnée.

Les données sur les méduses sont exprimées en poids sec.

NAMIBIE

Une quantité remarquable de captures de 'Chinchard du Cap' (*Trachurus capensis*) a été ajoutée aux captures de Namibie pour la période 1992-2000 en fonction des informations fournies par le service national déclarant.

Les données concernant les captures des thonidés sont révisées en collaboration avec la CICTA, l'organisation régionale chargée des statistiques des thonidés.

Les données sur les captures, à l'exclusion de celles concernant le thon, pour l' Atlantique sud-ouest, sont fournies par le Département des pêches du Gouvernement des Îles Falkland.

Les données relatives aux 'Algues gracilaires' enregistrées en poids sec, ont été converties en leurs équivalents poids vert en utilisant le facteur de conversion '8'.

NÉPAL

Les données se réfèrent à une année fractionnée (16 juillet - 15 juillet) et figurent sous l'année civile durant laquelle se termine l'année fractionnée.

ANTILLES NÉERLANDAISES

Comprennent Curaçao, Bonaire, Saba, Saint-Eustache et Saint-Martin.

Les données concernant les captures des thonidés sont révisées en collaboration avec la CICTA, l'organisation régionale chargée des statistiques des thonidés.

NOUVELLE-CALÉDONIE

Les données concernant les captures des thonidés sont révisées en collaboration avec la CPS, l'organisation régionale chargée des statistiques des thonidés.

NOUVELLE-ZÉLANDE

Pour la période 1995-2000 les données sous "Elasmobranchii – Requins, raies, etc. nca" comprennent également la production d'ailerons de requins, converties en équivalents poids vif en utilisant le facteur de conversion '20'.

Les données relatives aux 'Pecten de Nouvelle Zélande' depuis 1987 ont été converties en équivalents poids vif en utilisant le facteur de conversion '8'.

Les données relatives aux 'Huître plate néo-zéolandaise' sont sujettes à révision.

MARRUECOS

Los datos relativos a las capturas de atún se revisan en colaboración con la CICAA, el organismo regional encargado de las estadísticas atuneras.

MYANMAR

Los datos se refieren a un año fiscal (1º de abril - 31 de marzo) que se indica como el año civil en que finaliza el año emergente.

Los datos relativos a las medusas se han expresado en peso seco.

NAMIBIA

Una cantidad notable de 'Jurel del Cabo' (*Trachurus capensis*) se han agregado a las capturas de Namibia para el período 1992-2000 para tener en cuenta la información notificada por parte de la oficina nacional.

Los datos relativos a las capturas de atún se revisan en colaboración con la CICAA, el organismo regional encargado de las estadísticas atuneras.

Los datos relativos a las capturas en la zona del Atlántico Sudoccidental, con exclusión de las de atunes, fueron proporcionados por el Departamento de Pesca del Gobierno de las Islas Malvinas (Falkland).

Los datos relativos a las algas 'Gracilarias' registrados en peso en seco se han convertido en sus equivalentes de peso húmedo aplicando el factor de conversión '8'.

NEPAL

Los datos se refieren a un año emergente (16 de julio - 15 de julio) que se indica como el año civil en que finaliza el año emergente.

ANTILLAS NEERLANDESAS

Incluye Curaçao, Bonaire, Saba, San Eustaquio y San Martín.

Los datos relativos a las capturas de atún se revisan en colaboración con la CICAA, el organismo regional encargado de las estadísticas atuneras.

NUEVA CALEDONIA

Los datos relativos a las capturas de atún se revisan en colaboración con la SCP, el organismo regional encargado de las estadísticas atuneras.

NUEVA ZELANDIA

Para el período 1995-2000 los datos incluidos en "Elasmobranchii – Tiburones, rayas, etc. nep" comprenden también la producción de aletas de tiburón, convertida en sus equivalentes de peso en vivo aplicando el factor de conversión '20'.

Los datos de 'Vieira de Nueva Zelandia' desde 1987 se han convertido en sus equivalentes de peso en vivo aplicando el factor de conversión '8'.

Los datos de 'Ostra de Nueva Zelandia' están sujetos a revision.

Notes on individual countries or areas

NICARAGUA

Data concerning tuna catches are reviewed in collaboration with IATTC, the regional agency concerned with tuna statistics.

NORTHERN MARIANA IS.

Includes Marianas Islands except Guam.

NORWAY

Up to 1995, reported weight of the basking shark (*Cetorhinus maximus*) liver was converted to live weight equivalents by using the conversion factor '10'. Since 1996, basking shark fins have been converted to live weight equivalents by using the conversion factor '100'.

OMAN

Up to 1992 catch data for 'Pelagic Percomorphs' include significant quantities of 'Indian oil sardine'.

Data concerning tuna catches are reviewed in collaboration with IOTC, the regional agency concerned with tuna statistics.

OTHER NEI

Data refer to catches, mainly of tuna species, reported by IATTC, ICCAT, IOTC, Las Palmas Survey (LPS) and the "Bulletin Statistique" published by IMROP, Mauritania, as caught by unidentified countries.

PAKISTAN

Data concerning tuna catches are reviewed in collaboration with IOTC, the regional agency concerned with tuna statistics.

PANAMA

Data concerning tuna catches are reviewed in collaboration with IATTC and ICCAT, the regional agencies concerned with tuna statistics.

Catch data, excluding those for tuna species, for the Eastern Central and Southeast Atlantic areas are provided by the Las Palmas Survey (LPS) and the "Bulletin Statistique" published by IMROP, Mauritania.

PAPUA NEW GUINEA

Data concerning tuna catches are reviewed in collaboration with SPC, the regional agency concerned with tuna statistics.

PERU

Since 1991 the catch data of Jumbo squid (*Dosidicus gigas*) exclude the quantities caught by the foreign jigging line vessels which operate with fishing permits awarded through public tender.

Notes sur divers pays ou zones

NICARAGUA

Les données concernant les captures des thonidés sont révisées en collaboration avec la CITT, l'organisation régionale chargée des statistiques des thonidés.

ÎLES MARIANNES SEPTENTR.

Comprend les îles Mariannes sauf Guam.

NORVÈGE

Jusqu'en 1995, le poids du foie du requin pèlerin (*Cetorhinus maximus*) était converties en équivalents poids vif en utilisant le facteur de conversion '10'. Depuis 1996, les ailerons de requins pèlerins ont été converties en équivalents poids vif en utilisant le facteur de conversion '100'.

OMAN

Jusqu'à l'année 1992 comprise, les données relatives aux captures de 'Percomorphes pélagiques' comprennent des quantités importantes de 'Sardinelle indienne'.

Les données concernant les captures des thonidés sont révisées en collaboration avec la CTOI, l'organisation régionale chargée des statistiques des thonidés.

AUTRES NCA

Les données se rapportent aux captures, principalement des thonidés, indiqués par la CITT, la CICTA, la CTOI, l'Enquête Las Palmas et du "Bulletin statistique" publié par le IMROP, Mauritanie, comme captures de pays non identifiables.

PAKISTAN

Les données concernant les captures des thonidés sont révisées en collaboration avec la CTOI, l'organisation régionale chargée des statistiques des thonidés.

PANAMA

Les données concernant les captures des thonidés sont révisées en collaboration avec la CITT et la CICTA, les organisations régionales chargées des statistiques des thonidés.

Les données sur les captures, à l'exclusion de celles concernant le thon, pour l'Atlantique du centre-est et du sud-est sont tirées de l'Enquête Las Palmas et du "Bulletin statistique" publié par le IMROP, Mauritanie.

PAPOUASIE-NLLE-GUINÉE

Les données concernant les captures des thonidés sont révisées en collaboration avec la CPS, l'organisation régionale chargée des statistiques des thonidés.

PÉROU

Depuis 1991 les données relatives aux captures d'Encornet géant (*Dosidicus gigas*) ne comprennent pas les quantités capturées par les ligneurs à turluttes étrangers qui possèdent un permis de pêche délivré sur offres publiques.

Notas sobre los distintos países o áreas

NICARAGUA

Los datos relativos a las capturas de atún se revisan en colaboración con la CIAT, el organismo regional encargado de las estadísticas atuneras.

ISLAS MARIANAS SEPTENT.

Incluye las islas Marianas excepto Guam.

NORUEGA

Hasta 1995, el peso notificado de hígado de peregrino (*Cetorhinus maximus*) se convertió en sus equivalentes de peso en vivo aplicando el factor de conversión '10'. Desde 1996, las aletas de peregrinos se han convertido en sus equivalentes de peso en vivo aplicando el factor de conversión '100'.

OMÁN

Hasta 1992 inclusive, los datos relativos a las capturas de 'Percomorfos pelágicos' incluyen cantidades importantes de 'Sardinella aceitera'.

Los datos relativos a las capturas de atún se revisan en colaboración con la CAOI, el organismo regional encargado de las estadísticas atuneras.

OTROS NEP

Los datos abarcan capturas, principalmente de atún, indicado por la CIAT, la CICAA, la CAOI, Las Palmas Survey (LPS) y por la publicación "Bulletin Statistique" del IMROP, Mauritania, como capturado por países desconocidos.

PAKISTÁN

Los datos relativos a las capturas de atún se revisan en colaboración con la CAOI, el organismo regional encargado de las estadísticas atuneras.

PANAMÁ

Los datos relativos a las capturas de atún se revisan en colaboración con la CIAT y la CICAA, los organismos regionales encargados de las estadísticas atuneras.

Los datos relativos a las capturas en las zonas del Atlántico Centro-oriental y Sudoriental, con exclusión de las de atunes, fueron proporcionados por Las Palmas Survey (LPS) y por la publicación "Bulletin Statistique" del IMROP, Mauritania.

PAPUA NUEVA GUINEA

Los datos relativos a las capturas de atún se revisan en colaboración con la SCP, el organismo regional encargado de las estadísticas atuneras.

PERÚ

Desde 1991 los datos de capturas de Jibia gigante (*Dosidicus gigas*) excluyen las cantidades capturadas por las embarcaciones con calamareras extranjeras, que operan bajo "Permiso de Pesca" según concurso público.

Notes on individual countries or areas

PHILIPPINES

Data concerning tuna catches in the Atlantic and in the Indian Oceans are reviewed in collaboration with ICCAT and IOTC, the regional agencies concerned with tuna statistics.

PITCAIRN ISLANDS

Includes Henderson, Ducie and Oeno.

POLAND

Estimated catches (13,000 mt) from recreational fisheries included in 'Freshwater fishes nei'.

PORTUGAL

Data concerning tuna catches are reviewed in collaboration with ICCAT and IOTC, the regional agencies concerned with tuna statistics.

Catch data, excluding those for tuna species, for the Eastern Central Atlantic area have also been taken from the "Bulletin Statistique" published by IMROP, Mauritania.

PUERTO RICO

Data for 'Stromboid conchs nei' (1992-98 statistics taken from the "Fishery of the United States" yearly publication) have been converted to live weight equivalents by using the conversion factor '7.5'.

RÉUNION

Data concerning tuna catches are reviewed in collaboration with IOTC, the regional agency concerned with tuna statistics.

SAINT HELENA

Includes the islands of Ascensión and Tristan da Cunha.

Data for 'Tristan da Cunha rock lobster' (*Jasus tristani*), octopuses and, since 1994, for most of the quantities included under 'Marine fishes nei' refer to catches from the Tristan da Cunha area. A portion of these catches are taken by foreign fleets.

SAINT VINCENT/GRENADINES

Data concerning tuna catches are reviewed in collaboration with ICCAT, the regional agency concerned with tuna statistics.

Catch data, excluding those for tuna species, for the Eastern Central and Southeast Atlantic areas are provided by the Las Palmas Survey (LPS) and the "Bulletin Statistique" published by IMROP, Mauritania.

Catch data, excluding those for tuna species, for the Southwest Atlantic area are provided by the Fisheries Department, Falkland Islands Government.

SAMOA

Since 1993, data concerning tuna catches are reviewed in collaboration with the SPC, the regional agency concerned with tuna statistics.

Notes sur divers pays ou zones

PHILIPPINES

Les données concernant les captures des thonidés dans les océans Atlantique et Indien sont révisées en collaboration avec la CICTA et la CTOI, les organisations régionales chargées des statistiques des thonidés.

ÎLES PITCAIRN

Comprend Henderson, Ducie et Oeno.

POLOGNE

Estimation (13 000 tonnes) de pêche récréative comprises en 'Poissons d'eau douce nca'.

PORTUGAL

Les données concernant les captures des thonidés sont révisées en collaboration avec la CICTA et la CTOI, les organisations régionales chargées des statistiques des thonidés.

Les données sur les captures, à l'exclusion de celles concernant le thon, pour l'Atlantique centre-est ont également été tirées du "Bulletin statistique" publié par le IMROP, Mauritanie.

PORTO RICO

Les données relatives aux 'Strombes nca' (statistiques de la période 1992-98 été tirées de la publication annuelle "Fishery of the United States") ont été converties en équivalents poids vif en utilisant le facteur de conversion '7,5'.

RÉUNION

Les données concernant les captures des thonidés sont révisées en collaboration avec la CTOI, l'organisation régionale chargée des statistiques des thonidés.

SAINTE-HÉLÈNE

Comprend les îles de l'Ascension et Tristan da Cunha.

Les données concernant la 'Langouste de Tristan da Cunha (*Jasus tristani*), les poulpes et, depuis 1994, pour la plupart des quantités incluses sous 'Poissons marins nca' se réfèrent aux captures effectuées dans la région de Tristan da Cunha. Une partie de ces quantités sont capturées par les flottes étrangères.

SAINT-VINCENT/GRENADINES

Les données concernant les captures des thonidés sont révisées en collaboration avec la CICTA, l'organisation régionale chargée des statistiques des thonidés.

Les données sur les captures, à l'exclusion de celles concernant le thon, pour l'Atlantique du centre-est et du sud-est sont tirées de l'Enquête Las Palmas et du "Bulletin statistique" publié par le IMROP, Mauritanie.

Les données sur les captures, à l'exclusion de celles concernant le thon, pour l' Atlantique sud-ouest, sont fournies par le Département des pêches du Gouvernement des Îles Falkland.

SAMOA

Depuis 1993 les données concernant les captures des thonidés sont révisées en collaboration avec la CPS, l'organisation régionale chargée des statistiques des thonidés.

Notas sobre los distintos países o áreas

FILIPINAS

Los datos relativos a las capturas de atún en los Océanos Atlántico e Indico se revisan en colaboración con la CICAA y la CAOI, los organismos regionales encargados de las estadísticas atuneras.

ISLAS PITCAIRN

Incluye Henderson, Ducie y Oeno.

POLONIA

Estimación (13 000 toneladas) de la pesca recreativa incluida en 'Peces de agua dulce nep'.

PORTUGAL

Los datos relativos a las capturas de atún en el Atlántico se revisan en colaboración con la CICAA y la CAOI, los organismos regionales encargados de las estadísticas atuneras.

Los datos sobre las capturas de la zona del Atlántico Centro-oriental, con exclusión de las de atunes, también se han extraído de la publicación "Bulletin Statistique" del IMROP, Mauritania.

PUERTO RICO

Los datos de 'Cobos nep' (estadísticas para el período 1992-98 se han extraído de la publicación anual "Fishery of the United States") se han convertido en sus equivalentes de peso en vivo aplicando el factor de conversión '7,5'.

REUNIÓN

Los datos relativos a las capturas de atún se revisan en colaboración con la CAOI, el organismo regional encargado de las estadísticas atuneras.

SANTA ELENA

Incluye las islas de Ascensión y Tristán de Cunha.

Los datos relativos a la 'Langosta de Tristán da Cunha' (*Jasus tristani*), los pulpos y, desde 1994, para la mayoría de las cantidades incluidas en 'Peces marinos nep' se refieren a capturas desde el área de Tristan da Cunha. Una porción de estas cantidades se captura por las flotas extranjeras.

SAN VICENTE/GRENADINAS

Los datos relativos a las capturas de atún en el Atlántico se revisan en colaboración con la CICAA, el organismo regional encargado de las estadísticas atuneras.

Los datos relativos a las capturas en las zonas del Atlántico Centro-oriental y Sudoriental, con exclusión de las de atunes, fueron proporcionados por Las Palmas Survey (LPS) y por la publicación "Bulletin Statistique" del IMROP, Mauritania.

Los datos relativos a las capturas en la zona del Atlántico Sudoccidental, con exclusión de las de atunes, fueron proporcionados por el Departamento de Pesca del Gobierno de las Islas Malvinas (Falkland).

SAMOA

Desde 1993 los datos relativos a las capturas de atún se revisan en colaboración con la SCP, el organismo regional encargado de las estadísticas atuneras.

Notes on individual countries or areas

SAUDI ARABIA

Data concerning tuna catches are reviewed in collaboration with IOTC, the regional agency concerned with tuna statistics.

SENEGAL

Inland water statistics for the period 1994-99 have been revised to exclude quantities probably referring to marine species already included in the marine statistics.

SEYCHELLES

Data concerning tuna catches are reviewed in collaboration with ICCAT and IOTC, the regional agencies concerned with tuna statistics.

Increased catches of tuna species since 1997 due to fishing activities of purse seiners re-flagged under Seychelles flag.

SOLOMON ISLANDS

Data concerning tuna catches are reviewed in collaboration with SPC, the regional agency concerned with tuna statistics.

SOMALIA

Since 1991, the only data available have been estimates of the total marine capture production in 1994 provided by the FAO Representative in Somalia. Approximately half of the total capture production for 1994 was made of shark catches but this datum is included under "Marine fishes nei", as no shark statistics are available for other years.

SOUTH AFRICA

Data concerning tuna catches are reviewed in collaboration with ICCAT and IOTC, the regional agencies concerned with tuna statistics.

Data for seaweeds recorded on a dry-weight basis have been converted to wet-weight equivalents by using the conversion factor '8'.

Catch data of *Haliotis midae* (Perlemoen abalone) reported in 2001 considered to be overstated and subject to possible downward revisions.

SPAIN

Data concerning tuna catches are reviewed in collaboration with IATTC, ICCAT, IOTC and SPC, the regional agencies concerned with tuna statistics.

Catch data, excluding those for tuna species, for the Eastern Central Atlantic area have also been taken from the "Bulletin Statistique" published by IMROP, Mauritania.

Catch data, excluding those for tuna species, for the Southwest Atlantic area are provided also by the Fisheries Department, Falkland Islands Government.

Notes sur divers pays ou zones

ARABIE SAOUDITE

Les données concernant les captures des thonidés sont révisées en collaboration avec la CTOI, l'organisation régionale chargée des statistiques des thonidés.

SÉNÉGAL

Les statistiques des pêches dans les eaux intérieures pour la période 1994-99, ont été révisées afin d'exclure des quantités qui se référaient probablement aux espèces marines déjà comprises dans les statistiques de pêche maritime.

SEYCHELLES

Les données concernant les captures des thonidés sont révisées en collaboration avec la CICTA et la CTOI, les organisations régionales chargées des statistiques des thonidés.

L'augmentation de captures concernant le thon depuis 1997 s'explique par les activités de pêche de quatre senneurs à senne coulissante battant pavillon des Seychelles.

ÎLES SALOMON

Les données concernant les captures des thonidés sont révisées en collaboration avec la CPS, l'organisation régionale chargée des statistiques des thonidés.

SOMALIE

Depuis 1991, les seules données disponibles sont des estimations des captures marines totales en 1994 fournies par le Représentant de la FAO en Somalie. Environ la moitié des captures totales pour 1994 était composée de requins, mais ces données sont comprises sous "Poissons marins nca" car on ne dispose pas de statistiques sur les requins pour les autres années.

AFRIQUE DU SUD

Les données concernant les captures des thonidés sont révisées en collaboration avec la CICTA et la CTOI, les organisations régionales chargées des statistiques des thonidés.

Les données concernant les algues enregistrées en poids sec, ont été converties en leurs équivalents poids vert en utilisant le facteur de conversion '8'.

Les données des captures de *Haliotis midae* (Ormeau de Mida) en 2001 sont considère exagérées et seront éventuellement révisées à la baisse.

ESPAGNE

Les données concernant les captures des thonidés sont révisées en collaboration avec la CITT, la CICTA, la CTOI et la CPS, les organisations régionales chargées des statistiques des thonidés.

Les données sur les captures, à l'exclusion de celles concernant le thon, pour l'Atlantique centre-est ont également été tirées du "Bulletin statistique" publié par le IMROP, Mauritanie.

Les données sur les captures, à l'exclusion de celles concernant le thon, pour l'Atlantique sud-ouest, sont fournies également par le Département des pêches du Gouvernement des Îles Falkland.

Notas sobre los distintos países o áreas

ARABIA SAUDITA

Los datos relativos a las capturas de atún se revisan en colaboración con la CAOI, el organismo regional encargado de las estadísticas atuneras.

SENEGAL

Las estadísticas de la pesca continental para el período 1994-99 han sido revisadas para excluir cantidades que probablemente se referían a especies marinas ya incluidas en las estadísticas de pesca marina.

SEYCHELLES

Los datos relativos a las capturas de atún se revisan en colaboración con la CICAA y la CAOI, los organismos regionales encargados de las estadísticas atuneras.

El incremento desde 1997 de las capturas para atunes se debe a los desembarques de cuatro cerqueros con jareta con pabellón de Seychelles.

ISLAS SALOMÓN

Los datos relativos a las capturas de atún se revisan en colaboración con la SCP, el organismo regional encargado de las estadísticas atuneras.

SOMALIA

Desde 1991 los únicos datos disponibles han sido las estimaciones de la producción total de capturas marinas en 1994 proporcionadas por el Representante de la FAO en Somalia. Aproximadamente la mitad de la producción total de capturas en 1994 correspondía a tiburones, pero este dato se ha incluido en "Peces marinos nep" por no disponerse de estadísticas sobre tiburones relativas a otros años.

SUDÁFRICA

Los datos relativos a las capturas de atún se revisan en colaboración con la CICAA y la CAOI, los organismos regionales encargados de las estadísticas atuneras.

Los datos correspondientes a las algas registrados en peso en seco se han convertido en sus equivalentes de peso húmedo aplicando el factor de conversión '8'.

Los datos relativos a las capturas de *Haliotis midae* (Oreja de mar) en 2001 se consideran sobrestimados y es posible que sean revisados a la baja.

ESPAÑA

Los datos relativos a las capturas de atún se revisan en colaboración con la CIAT, la CICAA, la CAOI y la SCP, los organismos regionales encargados de las estadísticas atuneras.

Los datos sobre las capturas de la zona del Atlántico Centro-oriental, con exclusión de las de atunes, también se han extraído de la publicación "Bulletin Statistique" del IMROP, Mauritania.

Los datos relativos a las capturas en la zona del Atlántico Sudoccidental, con exclusión de las de atunes, también fueron proporcionados por el Departamento de Pesca del Gobierno de las Islas Malvinas (Falkland).

Notes on individual countries or areas

Notes sur divers pays ou zones

Notas sobre los distintos países o áreas

SRI LANKA

In 1999, fishery statistics formerly assigned to area 51 (Western Indian Ocean) have been moved to area 57 (Eastern Indian Ocean) following a modification of the boundary between the two areas.

Data concerning tuna catches are reviewed in collaboration with IOTC, the regional agency concerned with tuna statistics.

Up to 1996, data for 'Silky shark' estimated as 75% of the officially reported total catch data for 'Sharks, rays, skates, etc.'.

TANZANIA

Data concerning tuna catches are reviewed in collaboration with IOTC, the regional agency concerned with tuna statistics.

THAILAND

Data concerning tuna catches in the Indian Ocean are reviewed in collaboration with IOTC, the regional agency concerned with tuna statistics.

TIMOR-LESTE

Formerly part of Indonesia.

In 2001, fishery statistics formerly assigned to area 71 (Western Central Pacific) have been moved to area 57 (Eastern Indian Ocean) following a modification of the boundary between the two areas.

Data concerning tuna catches in the Indian Ocean are reviewed in collaboration with IOTC, the regional agency concerned with tuna statistics.

TONGA

Data concerning tuna catches are reviewed in collaboration with the SPC, the regional agency concerned with tuna statistics.

TRINIDAD AND TOBAGO

Data concerning tuna catches are reviewed in collaboration with ICCAT, the regional agency concerned with tuna statistics.

TUNISIA

Data concerning tuna catches are reviewed in collaboration with ICCAT, the regional agency concerned with tuna statistics.

TURKEY

Data concerning tuna catches are reviewed in collaboration with ICCAT, the regional agency concerned with tuna statistics.

SRI LANKA

En 1999 les statistiques des pêches ont été déplacées de la zone 51 (l'océan Indien ouest) à la zone 57 (l'océan Indien est) à la suite d'une modification des limites entre les deux zones.

Les données concernant les captures des thonidés sont révisées en collaboration avec la CTOI, l'organisation régionale chargée des statistiques des thonidés.

Jusqu'à l'année 1996 les captures de 'Requin soyeux' sont des estimations calculées comme 75% des quantités totales relatives aux 'Requins, raies, etc.' communiquées par le service national.

TANZANIE

Les données concernant les captures des thonidés sont révisées en collaboration avec la CTOI, l'organisation régionale chargée des statistiques des thonidés.

THAÏLAND

Les données concernant les captures des thonidés dans l'océan Indien sont révisées en collaboration avec la CTOI, l'organisation régionale chargée des statistiques des thonidés.

TIMOR-LESTE

Anciennement partie de l'Indonésie.

En 2001 les statistiques des pêches ont été déplacées de la zone 71 (Pacifique centre-ouest) à la zone 57 (Océan Indien est) à la suite d'une modification des limites entre les deux zones.

Les données concernant les captures des thonidés dans l'océan Indien sont révisées en collaboration avec la CTOI, l'organisation régionale chargée des statistiques des thonidés.

TONGA

Les données concernant les captures des thonidés sont révisées en collaboration avec la CPS, l'organisation régionale chargée des statistiques des thonidés.

TRINITÉ-ET-TOBAGO

Les données concernant les captures des thonidés sont révisées en collaboration avec la CICTA, l'organisation régionale chargée des statistiques des thonidés.

TUNISIE

Les données concernant les captures des thonidés sont révisées en collaboration avec la CICTA, l'organisation régionale chargée des statistiques des thonidés.

TURQUIE

Les données concernant les captures des thonidés sont révisées en collaboration avec la CICTA, l'organisation régionale chargée des statistiques des thonidés.

SRI LANKA

En 1999 las estadísticas pesqueras se desplazaron de la zona 51 (Océano Indico Occidental) a la zona 57 (Océano Indico Oriental) tras una modificación del límite entre ambas.

Los datos relativos a las capturas de atún se revisan en colaboración con la CAOI, el organismo regional encargado de las estadísticas atuneras.

Hasta el 1996 la captura de 'Tiburón jaquetón' ha sido estimada según el 75% del total de 'Tiburones, rayas, etc.' proporcionado por la oficina nacional.

TANZANÍA

Los datos relativos a las capturas de atún se revisan en colaboración con la CAOI, el organismo regional encargado de las estadísticas atuneras.

TAILANDIA

Los datos relativos a las capturas de atún en el Océano Indico oriental se revisan en colaboración con la CAOI, el organismo regional encargado de las estadísticas atuneras.

TIMOR-LESTE

Antes parte de Indonesia.

En 2001 las estadísticas pesqueras se desplazaron de la zona 71 (Pacífico centro-occidental) a la zona 57 (Océano Indico oriental) tras una modificación del límite entre ambas.

Los datos relativos a las capturas de atún en el Océano Indico oriental se revisan en colaboración con la CAOI, el organismo regional encargado de las estadísticas atuneras.

TONGA

Los datos relativos a las capturas de atún se revisan en colaboración con la SCP, el organismo regional encargado de las estadísticas atuneras.

TRINIDAD Y TABAGO

Los datos relativos a las capturas de atún en el Atlántico se revisan en colaboración con la CICAA, el organismo regional encargado de las estadísticas atuneras.

TÚNEZ

Los datos relativos a las capturas de atún en el Atlántico se revisan en colaboración con la CICAA, el organismo regional encargado de las estadísticas atuneras.

TURQUÍA

Los datos relativos a las capturas de atún en el Atlántico se revisan en colaboración con la CICAA, el organismo regional encargado de las estadísticas atuneras.

Notes on individual countries or areas

Notes sur divers pays ou zones

Notas sobre los distintos países o áreas

UK

Data for England and Wales, Scotland and Northern Ireland have been combined.

Increase of catches in inland waters since 1998 due to inclusion of significant quantities from recreational fisheries.

Catch data, excluding those for tuna species, for the Southwest Atlantic area are provided by the Fisheries Department, Falkland Islands Government.

UKRAINE

Catch data, excluding those for tuna species, for the Eastern Central Atlantic area have also been taken from the "Bulletin Statistique" published by IMROP, Mauritania.

URUGUAY

Data concerning tuna catches are reviewed in collaboration with ICCAT, the regional agency concerned with tuna statistics.

USA

Capture data for recreational fisheries are not included.

Data for tuna species not comparable with those of IATTC, ICCAT and SPC.

Catch data officially reported for selected clams, mussels and oysters, have been revised to exclude quantities derived from aquaculture.

Due to confidentiality reasons, data for calico scallop (*Argopecten gibbus*) are not available for some years.

Since 1985, data for seals refer exclusively to harvest for subsistence.

US VIRGIN ISLANDS

Data refer to a split-year period: 1 July/30 June. Includes Saint Croix, Saint John and Saint Thomas.

VANUATU

Data concerning tuna catches are reviewed in collaboration with IATTC and SPC, the regional agencies concerned with tuna statistics. Increase of tuna catches in the Western Central Pacific since 1995 due to vessels fishing with flag of convenience. In 2001, these vessels were re-flagged in the Marshall Islands.

Catch data, excluding those for tuna species, for the Eastern Central and Southeast Atlantic areas are provided by the Las Palmas Survey (LPS).

VENEZUELA

Data concerning tuna catches are reviewed in collaboration with IATTC and ICCAT, the regional agencies concerned with tuna statistics.

ROYAUME-UNI

Les chiffres concernent l'Angleterre et le pays de Galles, l'Ecosse et l'Irlande du Nord.

L'augmentation des captures dans les eaux continentales depuis 1998 s'explique par l'inclusion de quantités importantes de pêche récréative.

Les données sur les captures, à l'exclusion de celles concernant le thon, pour l' Atlantique sud-ouest, sont fournies par le Département des pêches du Gouvernement des Îles Falkland.

UKRAINE

Les données sur les captures, à l'exclusion de celles concernant le thon, pour l'Atlantique centre-est ont également été tirées du "Bulletin statistique" publié par le IMROP, Mauritanie.

URUGUAY

Les données concernant les captures des thonidés sont révisées en collaboration avec la CICTA, l'organisation régionale chargée des statistiques des thonidés.

ETATS-UNIS D'AMÉRIQUE

Les données relatives aux captures ne comprennent pas la pêche récréative.

Les données concernant les thonidés ne sont pas comparables avec les renseignements fournis par la CITT, la CICTA et la CPS.

Les données relatives aux captures de certaines clams, moules et huîtres, reportées officiellement, ont été révisées afin d'exclure les quantités provenant de l'aquaculture.

Le manque de données concernant le peigne calicot (*Argopecten gibbus*) pour certaines années s'explique en raison de leur caractère confidentiel.

Depuis 1985 les données concernant les otaries se réfèrent exclusivement à la pêche destinée à la subsistance.

ÎLES VIERGES AMÉRICAINES

Les données se réfèrent à une année fractionnée: 1er juillet/30 juin. Comprend Sainte-Croix, Saint-John et Saint-Thomas.

VANUATU

Les données concernant les captures des thonidés sont révisées en collaboration avec la CITT et la CPS, les organisations régionales chargées des statistiques des thonidés. A partir de 1995, l'augmentation des captures pour les thonidés dans le Pacifique centre-ouest s'explique par les captures effectuées sous pavillon de complaisance. En 2001 ces bateaux ont été transférés en pavillon des Îles Marshall.

Les données sur les captures, à l'exclusion de celles concernant le thon, pour l'Atlantique du centre-est et du sud-est sont tirées de l'Enquête Las Palmas.

VENEZUELA

Les données concernant les captures des thonidés sont révisées en collaboration avec la CITT et la CICTA, les organisations régionales chargées des statistiques des thonidés.

REINO UNIDO

Se han agrupado los datos sobre Inglaterra y Gales, Escocia e Irlanda del Norte.

El incremento de las capturas en aguas continentales desde 1998 se debe a la inclusión de importantes cantidades de pesca recreativa.

Los datos relativos a las capturas en la zona del Atlántico Sudoccidental, con exclusión de las de atunes, fueron proporcionados por el Departamento de Pesca del Gobierno de las Islas Malvinas (Falkland).

UCRANIA

Los datos sobre las capturas de la zona del Atlántico Centro-oriental, con exclusión de las de atunes, también se han extraído de la publicación "Bulletin Statistique" del IMROP, Mauritania.

URUGUAY

Los datos relativos a las capturas de atún en el Atlántico se revisan en colaboración con la CICAA, el organismo regional encargado de las estadísticas atuneras.

ESTADOS UNIDOS DE AMÉRICA

Los datos de capturas no incluyen la pesca recreativa.

Los datos sobre las especies de atún no son comparables con la información de la CIAT, la CICAA y la SCP.

Las capturas relativas a algunas especies de almejas, mejillones y ostras notificadas por parte de la oficina nacional, se han revisado para que excluyan cantidades procedentes de la acuicultura.

La falta de los datos para peine percal (*Argopecten gibbus*) por algunos años se debe al carácter confidencial de la información.

Desde 1985 los datos sobre los lobos se refieren exclusivamente a la recogida hecha con fin de subsistencia.

ISLAS VÍRGENES DE LOS EU

Los datos se refieren a un año emergente: 1° de julio/30 de junio. Incluye Sainte Croix, Saint John y Saint Thomas.

VANUATU

Los datos relativos a las capturas de atún se revisan en colaboración con la CIAT y la SCP, los organismos regionales encargados de las estadísticas atuneras. Desde 1995 el incremento de las capturas de atún en el Pacífico centro-occidental se debe a las capturas de pabellones de conveniencia. En 2001 estas embarcaciones cambiaron de pabellón en las Islas Marshall.

Los datos relativos a las capturas en las zonas del Atlántico Centro-oriental y Sudoriental, con exclusión de las de atunes, fueron proporcionados por Las Palmas Survey (LPS).

VENEZUELA

Los datos relativos a las capturas de atún se revisan en colaboración con la CIAT y la CICAA, los organismos regionales encargados de las estadísticas atuneras.

Notes on individual countries or areas

VIET NAM

The 1986-99 catch data have been extensively revised on the basis of the publication "Statistical data of Vietnam agriculture, forestry and fishery 1975-2000" published by the General Statistical Office, and of trade data.

YEMEN

Includes Kamaran Islands, Perim and Socotra.

The 1982-95 catch data for 'Pelagic Percomorphs' include quantities of 'Indian oil sardine'.

ZAMBIA

Data for 'Stolothrissa, Limnothrissa spp' refer only to the capture fisheries of Lake Kariba.

ZIMBABWE

Data refer only to the capture fisheries of Lake Kariba.

Notes sur divers pays ou zones

VIET NAM

Les données sur les captures de la période 1986-99 ont été largement révisées en fonction de la publication "Statistical data of Vietnam agriculture, forestry and fishery 1975-2000" publié par le Bureau Général de Statistique, et des statistiques du commerce.

YÉMEN

Comprend les îles Kamaran, Perim et Socotra

Les données de 1982-95 concernant les captures de 'Percomorphes pélagiques' comprennent des quantités de 'Sardinelle indienne'.

ZAMBIE

Les données relatives aux 'Stolothrissa, Limnothrissa spp' se réfèrent exclusivement aux pêches de capture du lac Kariba.

ZIMBABWE

Les données se réfèrent exclusivement aux pêches de capture du lac Kariba.

Notas sobre los distintos países o áreas

VIET NAM

Los datos de las capturas de 1986-99 han sido revisados ampliamente para tener en cuenta la publicación "Statistical data of Vietnam agriculture, forestry and fishery 1975-2000" publicada por la Dirección General de Estadística, y de las estadísticas de comercio.

YEMEN

Incluyen las islas Kamaran, Perim y Socotra.

Los datos de 1982-95 relativos a las capturas de de 'Percomorfos pelágicos' incluyen cantidades de 'Sardinella aceitera'.

ZAMBIA

Los datos de 'Stolothrissa, Limnothrissa spp' se refieren únicamente a las pesquerías de capturas del lago Kariba.

ZIMBABWE

Los datos se refieren únicamente a las pesquerías de capturas del lago Kariba.

Index of FAO English, French, Spanish and scientific names

Index des noms scientifiques et des noms FAO en anglais, français et espagnol

Indice de nombres científicos y de nombres FAO en inglés, francés y español

Index of FAO English, French, Spanish and scientific names

Index des noms scientifiques et des noms FAO en anglais, français et espagnol

Indice de nombres científicos y de nombres FAO en inglés, francés y español

Index of FAO English, French, Spanish and scientific names

Index des noms scientifiques et des noms FAO en anglais, français et espagnol

Indice de nombres científicos y de nombres FAO en inglés, francés y español

Index of FAO English, French, Spanish and scientific names

Index des noms scientifiques et des noms FAO en anglais, français et espagnol

Indice de nombres científicos y de nombres FAO en inglés, francés y español

Index of FAO English, French, Spanish and scientific names

Index des noms scientifiques et des noms FAO en anglais, français et espagnol

Indice de nombres científicos y de nombres FAO en inglés, francés y español

Index of FAO English, French, Spanish and scientific names

Index des noms scientifiques et des noms FAO en anglais, français et espagnol

Indice de nombres científicos y de nombres FAO en inglés, francés y español

Index of FAO English, French, Spanish and scientific names

Index des noms scientifiques et des noms FAO en anglais, français et espagnol

Indice de nombres científicos y de nombres FAO en inglés, francés y español

Index of FAO English, French, Spanish and scientific names

Index des noms scientifiques et des noms FAO en anglais, français et espagnol

Indice de nombres científicos y de nombres FAO en inglés, francés y español

Index of FAO English, French, Spanish and scientific names

Index des noms scientifiques et des noms FAO en anglais, français et espagnol

Indice de nombres científicos y de nombres FAO en inglés, francés y español

Index of FAO English, French, Spanish and scientific names

Index des noms scientifiques et des noms FAO en anglais, français et espagnol

Indice de nombres científicos y de nombres FAO en inglés, francés y español

Index of FAO English, French, Spanish and scientific names

Index des noms scientifiques et des noms FAO en anglais, français et espagnol

Indice de nombres científicos y de nombres FAO en inglés, francés y español

Index of FAO English, French, Spanish and scientific names

Index des noms scientifiques et des noms FAO en anglais, français et espagnol

Indice de nombres científicos y de nombres FAO en inglés, francés y español

Index of FAO English, French, Spanish and scientific names

Index des noms scientifiques et des noms FAO en anglais, français et espagnol

Indice de nombres científicos y de nombres FAO en inglés, francés y español

Index of FAO English, French, Spanish and scientific names

Index des noms scientifiques et des noms FAO en anglais, français et espagnol

Indice de nombres científicos y de nombres FAO en inglés, francés y español

Index of FAO English, French, Spanish and scientific names

Index des noms scientifiques et des noms FAO en anglais, français et espagnol

Indice de nombres científicos y de nombres FAO en inglés, francés y español

Index of FAO English, French, Spanish and scientific names

Index des noms scientifiques et des noms FAO en anglais, français et espagnol

Indice de nombres científicos y de nombres FAO en inglés, francés y español

Index of FAO English, French, Spanish and scientific names

Index des noms scientifiques et des noms FAO en anglais, français et espagnol

Indice de nombres científicos y de nombres FAO en inglés, francés y español

Index of FAO English, French, Spanish and scientific names

Index des noms scientifiques et des noms FAO en anglais, français et espagnol

Indice de nombres científicos y de nombres FAO en inglés, francés y español

Index of FAO English, French, Spanish and scientific names

Index des noms scientifiques et des noms FAO en anglais, français et espagnol

Indice de nombres científicos y de nombres FAO en inglés, francés y español

Index of FAO English, French, Spanish and scientific names | **Index des noms scientifiques et des noms FAO en anglais, français et espagnol** | **Indice de nombres científicos y de nombres FAO en inglés, francés y español**

Index of FAO English, French, Spanish and scientific names

Index des noms scientifiques et des noms FAO en anglais, français et espagnol

Indice de nombres científicos y de nombres FAO en inglés, francés y español

Index of FAO English, French, Spanish and scientific names

Index des noms scientifiques et des noms FAO en anglais, français et espagnol

Indice de nombres científicos y de nombres FAO en inglés, francés y español

Index of FAO English, French, Spanish and scientific names

Index des noms scientifiques et des noms FAO en anglais, français et espagnol

Indice de nombres científicos y de nombres FAO en inglés, francés y español

Index of FAO English, French, Spanish and scientific names

Index des noms scientifiques et des noms FAO en anglais, français et espagnol

Indice de nombres científicos y de nombres FAO en inglés, francés y español

Index of FAO English, French, Spanish and scientific names

Index des noms scientifiques et des noms FAO en anglais, français et espagnol

Indice de nombres científicos y de nombres FAO en inglés, francés y español

Index of FAO English, French, Spanish and scientific names

Index des noms scientifiques et des noms FAO en anglais, français et espagnol

Indice de nombres científicos y de nombres FAO en inglés, francés y español

Index of FAO English, French, Spanish and scientific names

Index des noms scientifiques et des noms FAO en anglais, français et espagnol

Indice de nombres científicos y de nombres FAO en inglés, francés y español

Index of FAO English, French, Spanish and scientific names

Index des noms scientifiques et des noms FAO en anglais, français et espagnol

Indice de nombres científicos y de nombres FAO en inglés, francés y español

List of yearbooks of fishery statistics

Liste des annuaires statistiques des pêches

Lista de los anuarios estadísticos de pesca

Volumes published in 1948-63 — Volumes publiés en 1948-63 — Volúmenes publicados en 1948-63

Production *Production* *Producción*	*Production and fishing craft* *Production et bateaux de pêche* *Producción y embarcaciones de pesca*	*International Trade* *Commerce international* *Comercio internacional*
	▸ Vol. I *(1947)*	
	▸ Vol.II *(1948-49)*	
	▸ Vol.III *(1950-51)*	
	▸ Vol.IV *(1952-53)* Part 1	▸ Vol.IV *(1952-53)* Part 2
	▸ Vol.V *(1954-55)*	
	▸ Vol.VI *(1955-56)*	
▸ Vol.VII *(1957)*		▸ Vol.VIII *(1957)*
	▸ Vol.IX *(1958)*	▸ Vol.X *(1958-59)*
▸ Vol.XI *(1959)*	▸ Vol.XII *(1960)*	▸ Vol.XIII *(1960-61)*
▸ Vol.XIV *(1961)*	▸ Vol.XV *(1962)*	

Volumes published in 1964-97 — Volumes publiés en 1964-97 — Volúmenes publicados en 1964-97

'Catches and landings' *'Captures et quantités débarquées'* *'Capturas y desembarques'*			*'Fishery Commodities'* *'Produits des pêches'* *'Productos pesqueros'*		
▸ Vol.16	('----' *1963*)	Dec. 1964	▸ Vol.17	('----' *1963*)	Jan. 1965
▸ Vol.18	('----' *1964*)	Oct. 1965	▸ Vol.19	('----' *1964*)	Dec. 1965
▸ Vol.20	('----' *1965*)	Oct. 1966	▸ Vol.21	('----' *1965*)	Dec. 1966
▸ Vol.22	('----' *1966*)	Oct. 1967	▸ Vol.23	('----' *1966*)	Dec. 1967
▸ Vol.24	('----' *1967*)	Oct. 1968	▸ Vol.25	('----' *1967*)	Dec. 1968
▸ Vol.26	('----' *1968*)	Oct. 1969	▸ Vol.27	('----' *1968*)	Dec. 1969
▸ Vol.28	('----' *1969*)	Oct. 1970	▸ Vol.29	('----' *1969*)	Dec. 1970
▸ Vol.30	('----' *1970*)	Nov. 1971	▸ Vol.31	('----' *1970*)	Dec. 1971
▸ Vol.32	('----' *1971*)	Nov. 1972	▸ Vol.33	('----' *1971*)	Dec. 1972
▸ Vol.34	('----' *1972*)	Nov. 1973	▸ Vol.35	('----' *1972*)	Dec. 1973
▸ Vol.36	('----' *1973*)	Nov. 1974	▸ Vol.37	('----' *1973*)	Dec. 1974
▸ Vol.38	('----' *1974*)	Dec. 1975	▸ Vol.39	('----' *1974*)	Dec. 1975
▸ Vol.40	('----' *1975*)	Dec. 1976	▸ Vol.41	('----' *1975*)	Dec. 1976
▸ Vol.42	('----' *1976*)	Nov. 1977	▸ Vol.43	('----' *1976*)	Dec. 1977
▸ Vol.44	('----' *1977*)	Nov. 1978	▸ Vol.45	('----' *1977*)	Dec. 1978
▸ Vol.46	('----' *1978*)	Nov. 1979	▸ Vol.47	('----' *1978*)	Dec. 1979
▸ Vol.48	('----' *1979*)	Dec. 1980	▸ Vol.49	('----' *1979*)	Dec. 1980
▸ Vol.50	('----' *1980*)	Dec. 1981	▸ Vol.51	('----' *1980*)	Dec. 1981
▸ Vol.52	('----' *1981*)	Jan. 1983	▸ Vol.53	('----' *1981*)	Feb. 1983
▸ Vol.54	('----' *1982*)	Jan. 1984	▸ Vol.55	('----' *1982*)	Jan. 1984
▸ Vol.56	('----' *1983*)	Dec. 1984	▸ Vol.57	('----' *1983*)	Dec. 1984
▸ Vol.58	('----' *1984*)	Jun. 1986	▸ Vol.59	('----' *1984*)	Jun. 1986
Vol.60	('----' *1985*)	May 1987	▸ Vol.61	('----' *1985*)	May 1987
▸ Vol.62	('----' *1986*)	Mar. 1988	▸ Vol.63	('----' *1986*)	Mar. 1988
▸ Vol.64	('----' *1987*)	Mar. 1989	▸ Vol.65	('----' *1987*)	Mar. 1989
▸ Vol.66	('----' *1988*)	Apr. 1990	▸ Vol.67	('----' *1988*)	May 1990
▸ Vol.68	('----' *1989*)	Apr. 1991	Vol.69	('----' *1989*)	May 1991
Vol.70	('----' *1990*)	Apr. 1992	Vol.71	('----' *1990*)	May 1992
Vol.72	('----' *1991*)	Apr. 1993	Vol.73	('----' *1991*)	May 1993
▸ Vol.74	('----' *1992*)	May 1994	Vol.75	('----' *1992*)	May 1994
Vol.76	('----' *1993*)	Apr. 1995	Vol.77	('----' *1993*)	May 1995
Vol.78	('----' *1994*)	Apr. 1996	Vol.79	('----' *1994*)	May 1996
Vol.80	('----' *1995*)	Apr. 1997	Vol.81	('----' *1995*)	May 1997

▸ Out of Print ▸ Epuisé ▸ Agotado

List of yearbooks of fishery statistics	**Liste des annuaires statistiques des pêches**	**Lista de los anuarios estadísticos de pesca**
Volumes published since 1998	Volumes publiés à partir de 1998	Volúmenes publicados a partir de 1998

'Capture production' 'Captures' 'Capturas'			'Aquaculture production'a/ 'Production de l'aquaculture'a/ 'Producción de acuicultura'a/			'Fishery Commodities' 'Produits des pêches' 'Productos pesqueros'		
Vol.82	('----' *1996*)	Apr. 1998				Vol.83	('----' *1996*)	Apr. 1998
Vol.84	('----' *1997*)	Apr. 1999				Vol.85	('----' *1997*)	Apr. 1999
Vol.86/1	('----' *1998*)	Apr. 2000	Vol.86/2	('----' *1998*)	Apr. 2000	Vol.87	('----' *1998*)	Apr. 2000
Vol.88/1	('----' *1999*)	Apr. 2001	Vol.88/2	('----' *1999*)	Apr. 2001	Vol.89	('----' *1999*)	Apr. 2001
Vol.90/1	('----' *2000*)	Apr. 2002	Vol.90/2	('----' 2000)	Apr. 2002	Vol.91	('----' *2000*)	Apr. 2002
Vol.92/1	('----' *2001*)	Apr. 2003	Vol.92/2	('----' 2001)	Apr. 2003	Vol.93	('----' *2001*)	Apr. 2003

a/ Aquaculture production statistics were combined with those of capture fisheries and published jointly in the "FAO yearbook. Fishery statistics. Catches and landings" until Volume 80. Since 1989, they were also published separately as "FAO Fisheries Circular No. 815: Aquaculture production statistics".

a/ Les statistiques de production de l'aquaculture étaient confondues avec celles de la production de pêche et publiées dans l'"Annuaire de la FAO. Statistiques des pêches. Captures et quantités débarquées" jusqu'au Volume 80. A partir de 1989, les données étaient aussi publiées séparément dans la "Circulaire de la FAO sur les pêches n° 815: Statistiques de la production de l'aquaculture".

a/ Las estadísticas de producción de acuicultura estaban incluidas en la producción de pesca y publicadas juntas en el "Anuario de la FAO. Estadísticas de pesca. Capturas y desembarques", hasta el Volumen 80. Desde 1989, los datos han sido publicados también por separado en la "Circular de Pesca de la FAO n° 815: Estadísticas de la producción de acuicultura".

Sales and Marketing Group, Information Division, FAO
Viale delle Terme di Caracalla, 00100 Rome, Italy
Tel.: +39 06 57051 – Fax: +39 06 5705 3360
E-mail: publications-sales@fao.org
www.fao.org/catalog/giphome.htm

أماكن بيع مطبوعات المنظمة
当地何处可以购买粮农组织出版物
WHERE TO PURCHASE FAO PUBLICATIONS LOCALLY
POINTS DE VENTE DES PUBLICATIONS DE LA FAO
PUNTOS DE VENTA DE PUBLICACIONES DE LA FAO

5/03

• *ANGOLA*
Empresa Nacional do Disco e de Publicações, ENDIPU-U.E.E.
Rua Cirilo da Conceição Silva, Nº 7
C.P. Nº 1314-C, Luanda

• *ARGENTINA*
Librería Hemisferio Sur
Pasteur 743, 1028 Buenos Aires
Correo eléctronico:
adolfop@hemisferiosur.com.ar
World Publications S.A.
Av. Córdoba 1877, 1120 Buenos Aires
Tel./Fax: (+54) 11 48158156

• *AUSTRALIA*
Hunter Publications (Tek Imaging Pty. Ltd)
PO Box 404, Abbotsford, Vic. 3067
Tel.: (+61) 3 9417 5361
Fax: (+61) 3 9419 7154
E-mail: admin@tekimaging.com.au

• *BELGIQUE*
M.J. De Lannoy
202, avenue du Roi, B-1060 Bruxelles
CCP: 000-0808993-13
Mél.: jean.de.lannoy@infoboard.be

• *BOLIVIA*
Los Amigos del Libro
Av. Heroínas 311, Casilla 450
Cochabamba;
Mercado 1315, La Paz
Correo eléctronico:
gutten@amigol.bo.net

• *BOTSWANA*
Botsalo Books (Pty) Ltd
PO Box 1532, Gaborone
Tel.: (+267) 312576
Fax: (+267) 372608
E-mail: botsalo@botsnet.bw

• *BRAZIL*
Fundação Getúlio Vargas
Praia do Botafogo 190, C.P. 9052
Rio de Janeiro
Correo eléctronico: livraria@fgv.br
Núcleo Editora da Universidade Federal Fluminense
Rua Miguel de Frias 9
Icaraí-Niterói 24
[illegible]0-000 Rio de Janeiro
[illegible]ra UFPR
[illegible]esidente Faria s/nº
[illegible]dio Histórico da UFPR
Curitiba, Paraná, CEP 80.020-300
Tel.: (+55) 41 310 2734
Web Site: www.editora.ufpr.br

• *CAMEROUN*
CADDES
Centre Africain de Diffusion et Développement Social
B.P. 7317, Douala Bassa
Tél.: (+237) 43 37 83
Télécopie: (+237) 42 77 03

• *CANADA*
Renouf Publishing
5369 chemin Canotek Road, Unit 1
Ottawa, Ontario K1J 9J3
Tel.: (+1) 613 745 2665
Fax: (+1) 613 745 7660
E-mail: order.dept@renoufbooks.com
Web site: www.renoufbooks.com

• *CHILE*
Librería - Marta Caballero
c/o FAO, Oficina Regional para América Latina y el Caribe (RLC)
Avda. Dag Hammarskjold, 3241
Vitacura, Santiago
Tel.: (+56) 2 33 72 314
Correo electrónico:
german.rojas@field.fao.org
Correo eléctronico:
caballerocastillo@hotmail.com

• *CHINA*
China National Publications Import & Export Corporation
16 Gongti East Road, Beijing 100020
Tel.: (+86) 10 6506 3070
Fax: (+86) 10 6506 3101
E-mail: serials@cnpiec.com.cn

• *COLOMBIA*
INFOENLACE LTDA
Calle 72 Nº 13-23 Piso 3
Edificio Nueva Granada
Santafé de Bogotá
Tel.: (+57) 1 6009474-6009480
Fax: (+57) 1 2480808-2176435
Correo electrónico:
servicliente@infoenlace.com.co

• *CONGO*
Office national des librairies populaires
B.P. 577, Brazzaville

• *COSTA RICA*
Librería Lehmann S.A.
Av. Central, Apartado 10011
1000 San José
Correo eléctronico:
llehmann@solracsa.co.cr

• *CÔTE D'IVOIRE*
CEDA
04 B.P. 541, Abidjan 04
Tél.: (+225) 22 20 55
Télécopie: (+225) 21 72 62

• *CUBA*
Ediciones Cubanas
Empresa de Comercio Exterior de Publicaciones
Obispo 461, Apartado 605, La Habana

• *CZECH REPUBLIC*
Myris Trade Ltd
V Stinhlach 1311/3, PO Box 2
142 01 Prague 4
Tel.: (+420) 2 34035200
Fax: (+420) 2 34035207
E-mail: myris@myris.cz
Web site: www.myris.cz

• *DENMARK*
Gad Import Booksellers
c/o Gad Direct
31-33 Fiolstraede
DK-1171 Copenhagen K
Tel.: (+45) 3313 7233
Fax: (+45) 3254 2368
E-mail: info@gaddirect.dk

• *ECUADOR*
Libri Mundi, Librería Internacional
Juan León Mera 851
Apartado Postal 3029, Quito
Correo electrónico:
librimu1@librimundi.com.ec
Web site: www.librimundi.com
Universidad Agraria del Ecuador
Centro de Información Agraria
Av. 23 de julio, Apartado 09-01-1248
Guayaquil
Librería Española
Murgeón 364 y Ulloa, Quito

• *EGYPT*
MERIC
The Middle East Readers' Information Centre
2 Baghat Aly Street, Appt. 24
El Masry Tower D
Cairo/Zamalek
Tel.: (+20) 2 3413824/34038818
Fax: (+20) 2 3419355
E-mail: info@mericonline.com

• *ESPAÑA*
Librería Agrícola
Fernando VI 2, 28004 Madrid
Librería de la Generalitat de Catalunya
Rambla dels Estudis 118 (Palau Moja)
08002 Barcelona
Tel.: (+34) 93 302 6462
Fax: (+34) 93 302 1299

Mundi Prensa Libros S.A.
Castelló 37, 28001 Madrid
Tel.: +34 91 436 37 00
Fax: +34 91 575 39 98
Sitio Web: www.mundiprensa.com
Correo electrónico:
libreria@mundiprensa.es
Mundi Prensa - Barcelona
Consejo de Ciento 391
08009 Barcelona
Tel.: (+34) 93 488 34 92
Fax: (+34) 93 487 76 59

• *FINLAND*
Akateeminen Kirjakauppa
PL 23, 00381 Helsinki
(Myymälä/Shop: Keskuskatu 1
00100 Helsinki)
Tel.: (+358) 9 121 4385
Fax: (+358) 9 121 4450
E-mail: akatilaus@akateeminen.com
Web site: www.akateeminen.com/suurasiakkaat/palvelut.htm

• *FRANCE*
Editions A. Pedone
13, rue Soufflot, 75005 Paris
Lavoisier Tec & Doc
14, rue de Provigny
94236 Cachan Cedex
Mél.: livres@lavoisier.fr
Site Web: www.lavoisier.fr
Librairie du commerce international
10, avenue d'Iéna
75783 Paris Cedex 16
Mél.: librarie@cfce.fr
Site Web: www.cfce.fr

• *GERMANY*
Alexander Horn Internationale Buchhandlung
Friedrichstrasse 34
D-65185 Wiesbaden
Tel.: +49 611 9923540/9923541
Fax: +49 611 9923543
E-mail: alexhorn1@aol.com
TRIOPS - Tropical Scientific Books
S. Toeche-Mittler Versandbuchhandlung GmbH
Hindenburstr. 33
D-64295 Darmstadt
Tel.: (+49) 6151 336 65
Fax: (+49) 6151 314 048
E-mail for orders: orders@net-library.de
E-mail for info.: info@net-library.de/
triops@triops.de
Web site: www.net-library.de/
www.triops.de
Uno Verlag
Am Hofgarten, 10
D-53113 Bonn
Tel.: (+49) 228 94 90 20
Fax: (+49) 228 94 90 222
E-mail: info@uno-verlag.de
Web site: www.uno-verlag.de

• *GHANA*
SEDCO Publishing Ltd
Sedco House, Tabon Street
Off Ring Road Central, North Ridge
PO Box 2051, Accra
Readwide Bookshop Ltd
PO Box 0600 Osu, Accra
Tel.: (+233) 21 22 1387
Fax: (+233) 21 66 3347
E-mail: readwide@africaonline.cpm.gh

• *GREECE*
Librairie Kauffmann SA
28, rue Stadiou, 10564 Athens
Tel.: (+30) 1 3236817
Fax: (+30) 1 3230320
E-mail: ord@otenet.gr

• *GUYANA*
Guyana National Trading Corporation Ltd
45-47 Water Street, PO Box 308
Georgetown

• *HONDURAS*
Escuela Agrícola Panamericana
Librería RTAC
El Zamorano, Apartado 93, Tegucigalpa
Correo electrónico:
libreriazam@zamorano.edu.hn

• *HUNGARY*
Librotrade Kft.
PO Box 126, H-1656 Budapest
Tel.: (+36) 1 256 1672
Fax: (+36) 1 256 8727

• *INDIA*
Allied Publisher Ltd
751 Mount Road
Chennai 600 002
Tel.: (+91) 44 8523938/8523984
Fax: (+91) 44 8520649
E-mail:
allied.mds@smb.sprintrpg.ems.vsnl.net.in
EWP Affiliated East-West Press PVT, Ltd
G-I/16, Ansari Road, Darya Gany
New Delhi 110 002
Tel.: (+91) 11 3264 180
Fax: (+91) 11 3260 358
E-mail: affiliat@nda.vsnl.net.in
Oxford Book and Stationery Co.
Scindia House
New Delhi 110001
Tel.: (+91) 11 3315310
Fax: (+91) 11 3713275
E-mail: oxford@vsnl.com
Periodical Expert Book Agency
G-56, 2nd Floor, Laxmi Nagar
Vikas Marg, Delhi 110092
Tel.: (+91) 11 2215045/2150534
Fax: (+91) 11 2418599
E-mail: pebe@vsnl.net.in
Bookwell
Head Office:
2/72, Nirankari Colony, New Delhi - 110009
Tel.: (+91) 11 725 1283
Fax: (+91) 11 328 13 15
Sales Office:
24/4800, Ansari Road
Darya Ganj, New Delhi - 110002
Tel.: (+91) 11 326 8786
E-mail: bkwell@nde.vsnl.net.in

• *INDONESIA*
P.F. Book
Jl. Setia Budhi No. 274, Bandung 40143
Tel.: (+62) 22 201 1149
Fax: (+62) 22 201 2840
E-mail:
pfbook@bandung.wasantara.net.id

• *IRAN*
The FAO Bureau, International and Regional Specialized Organizations Affairs
Ministry of Agriculture of the Islamic Republic of Iran
Keshavarz Bld, M.O.A., 17th floor
Teheran

• *ITALY*
FAO Bookshop
Viale delle Terme di Caracalla
00100 Roma
Tel.: (+39) 06 57052313
Fax: (+39) 06 57053360
E-mail: publications-sales@fao.org
Libreria Commissionaria Sansoni S.p.A. - Licosa
Via Duca di Calabria 1/1
50125 Firenze
Tel.: (+39) 55 64831
Fax: (+39) 55 64 2 57
E-mail: licosa@ftbcc.it
Libreria Scientifica Dott. Lucio de Biasio "Aeiou"
Via Coronelli 6, 20146 Milano

• *JAPAN*
Far Eastern Booksellers (Kyokuto Shoten Ltd)
12 Kanda-Jimbocho 2 chome
Chiyoda-ku - PO Box 72
Tokyo 101-91
Tel.: (+81) 3 3265 7531
Fax: (+81) 3 3265 4656

أماكن بيع مطبوعات المنظمة
当地何处可以购买粮农组织出版物
WHERE TO PURCHASE FAO PUBLICATIONS LOCALLY
POINTS DE VENTE DES PUBLICATIONS DE LA FAO
PUNTOS DE VENTA DE PUBLICACIONES DE LA FAO

5/03

Maruzen Company Ltd
5-7-1 Heiwajima, Ohta-Ku
Tokyo 143-0006
Tel.: (+81) 3 3763 2259
Fax: (+81) 3 3763 2830
E-mail: o_miyakawa@maruzen.co.jp

• ***KENYA***
Text Book Centre Ltd
Kijabe Street
PO Box 47540, Nairobi
Tel.: +254 2 330 342
Fax: +254 2 22 57 79
Inter Africa Book Distribution
Kencom House, Moi Avenue
PO Box 73580, Nairobi
Tel.: (+254) 2 211 184
Fax: (+254) 2 22 3 5 70
Legacy Books
Mezzanine 1, Loita House, Loita Street
Nairobi, PO Box 68077
Tel.: (+254) 2 303853
Fax: (+254) 2 330854
E-mail: info@legacybookshop.com

• ***LUXEMBOURG***
M.J. De Lannoy
202, avenue du Roi
B-1060, Bruxelles (Belgique)
Mél.: jean.de.lannoy@infoboard.be

• ***MADAGASCAR***
Centre d'Information et de Documentation Scientifique et Technique
Ministère de la recherche appliquée au développement
B.P. 6224, Tsimbazaza, Antananarivo

• ***MALAYSIA***
MDC Publishers Printers Sdn Bhd
MDC Building
2717 & 2718, Jalan Parmata Empat
Taman Permata, Ulu Kelang
53300 Kuala Lumpur
Tel.: (+60) 3 41086600
Fax: (+60) 3 41081506
E-mail: mdcpp@mdcpp.com.my
Web site: www.mdcpp.com.my

• ***MAROC***
La Librairie Internationale
70, rue T'ssoule
B.P. 302 (RP), Rabat
Tél.: (+212) 37 75 0183
Fax: (+212) 37 75 8661

• ***MÉXICO***
Librería, Universidad Autónoma de Chapingo
56230 Chapingo
Libros y Editoriales S.A.
Av. Progreso Nº 202-1º Piso A
Apartado Postal 18922
Col. Escandón, 11800 México D.F.
Correo electrónico: lyesa99@mail.com/
ventas@lyesa.com
Mundi Prensa Mexico, S.A.
Río Pánuco, 141 Col. Cuauhtémoc
C.P. 06500, México, DF
Tel.: (+52) 5 533 56 58
Fax: (+52) 5 514 67 99
Correo electrónico:
resavbp@data.net.mx

• ***NETHERLANDS***
Roodveldt Import b.v.
Brouwersgracht 288
1013 HG Amsterdam
Tel.: (+31) 20 622 80 35
Fax: (+31) 20 625 54 93
E-mail: roodboek@euronet.nl
Swets & Zeitlinger b.v.
PO Box 830, 2160 Lisse
Heereweg 347 B, 2161 CA Lisse
E-mail: infono@swets.nl
Web site: www.swets.nl

• ***NEW ZEALAND***
Legislation Direct
c/o Securacopy, PO Box 12 418
1st floor, 242 Thorndon Quay,
Wellington
Tel.: (+64) 4 496 56 94
Fax: (+64) 4 496 56 98
E-mail: Jeanette@legislationdirect.co.nz
Web site: www.gplegislation.co.nz

• ***NICARAGUA***
Librería HISPAMER
Costado Este Univ. Centroamericana
Apartado Postal A-221, Managua
Correo electrónico:
hispamer@munditel.com.ni

• ***NIGERIA***
University Bookshop (Nigeria) Ltd
University of Ibadan, Ibadan

• ***PAKISTAN***
Mirza Book Agency
65 Shahrah-e-Quaid-e-Azam
PO Box 729, Lahore 3

• ***PARAGUAY***
Librería Intercontinental
Editora e Impresora S.R.L.
Caballero 270 c/Mcal Estigarribia
Asunción

• ***PERU***
Librería de la Biblioteca Agrícola Nacional - Universidad Nacional Agraria
Av. La Universidad s/n
La Molina, Lima
Tel.: (+51) 1 3493910; Fax: (+51) 1 3493910
Correo electrónico: ban@lamolina.edu.pe
Web site: http//tumi.lamolina.edu.pe/
ban.htm

• ***PHILIPPINES***
International Booksource Center, Inc.
1127-A Antipolo St, Barangay Valenzuela
Makati City
Tel.: (+63) 2 8966501/8966505/8966507
Fax: (+63) 2 8966497
E-mail: ibcdina@pacific.net.ph

• ***POLAND***
Ars Polona Joint Stock Company
Krakowskie Przedmiescie 7
00-950 Warsaw, PO Box 1001
Tel.: (+48) 22 826 12 01
Fax: (+48) 22 826 62 40
E-mail: books119@arspolona.com.pl
Web site: www.arspolona.com.pl

• ***PORTUGAL***
Livraria Portugal, Dias e Andrade Ltda.
Rua do Carmo, 70-74
Apartado 2681, 1200 Lisboa Codex
Correo electrónico:
liv.portugal@mail.telepac.pt

• ***REPÚBLICA DOMINICANA***
CEDAF - Centro para el Desarrollo Agropecuario y Forestal, Inc.
Calle José Amado Soler, 50 - Urban.
Paraíso
Apartado Postal, 567-2, Santo Domingo
Tel.: (+001) 809 5440616/5440634/
5655603
Fax: (+001) 809 5444727/5676989
Correo electrónico: fda@Codetel.net.do
Web site: www.cedaf.org.do

• ***SERBIA AND MONTENEGRO***
Jugoslovenska Knjiga DD
Terazije 27
POB 36, 11000 Beograd
Tel.: (+381) 11 3340 025
Fax: (+381) 11 3231 079
E-mail: juknjiga@eunet.yu
or babicmius@yahoo.com

• ***SINGAPORE***
Select Books Pte Ltd
Tanglin Shopping Centre
19 Tanglin Road, #03-15,
Singapore 247909
Tel.: (+65) 732 1515
Fax: (+65) 736 0855
E-mail: info@selectbooks.com.sg
Web site: www.selectbooks.com.sg

• ***SLOVAK REPUBLIC***
Institute of Scientific and Technical Information for Agriculture
Samova 9, 950 10 Nitra
Tel.: (+421) 87 522 185
Fax: (+421) 87 525 275
E-mail: uvtip@nr.sanet.sk

• ***SOMALIA***
Samater
PO Box 936, Mogadishu

• ***SOUTH AFRICA***
Preasidium Books (Pty) Ltd
810 - 4th Street, Wynberg 2090
Tel.: (+27) 11 88 75994
Fax: (+27) 11 88 78138
E-mail: pbooks@global.co.za

• ***SUISSE***
UN Bookshop
Palais des Nations
CH-1211 Genève 1
Site Web: www.un.org
Adeco - Editions Van Diermen
Chemin du Lacuez, 41
CH-1807 Blonay
Tel.: (+41) (0) 21 943 2673
Fax: (+41) (0) 21 943 3605
E-mail: mvandier@ip-worldcom.ch
Münstergass Buchhandlung
Docudisp, PO Box 584
CH-3000 Berne 8
Tel.: (+41) 31 310 2321
Fax: (+41) 31 310 2324
E-mail: docudisp@muenstergass.ch
Web site: www.docudisp.ch

• ***SURINAME***
Vaco n.v. in Suriname
Domineestraat 26, PO Box 1841
Paramaribo

• ***SWEDEN***
Swets Blackwell AB
PO Box 1305, S-171 25 Solna
Tel.: (+46) 8 705 9750
Fax: (+46) 8 27 00 71
E-mail:
awahlquist@se.swetsblackwell.com
Web site: www.swetsblackwell.com/se/
Bokdistributören
c/o Longus Books Import
PO Box 610, S-151 27 Södertälje
Tel.: (+46) 8 55 09 49 70
Fax: (+46) 8 55 01 76 10; E-mail:
lis.ledin@hk.akademibokhandeln.se

• ***THAILAND***
Suksapan Panit
Mansion 9, Rajdamnern Avenue,
Bangkok

• ***TOGO***
Librairie du Bon Pasteur
B.P. 1164, Lomé

• ***TRINIDAD AND TOBAGO***
Systematics Studies Limited
St Augustine Shopping Centre
Eastern Main Road, St Augustine
Tel.: (+001) 868 645 8466
Fax: (+001) 868 645 8467
E-mail: tobe@trinidad.net

• ***TURKEY***
DUNYA ACTUEL A.S.
"Globus" Dunya Basinevi
100. Yil Mahallesi
34440 Bagcilar, Istanbul
Tel.: (+90) 212 629 0808
Fax: (+90) 212 629 4689
E-mail: aktuel.info@dunya.comr
Web site: www.dunyagazetesi.com.tr/

• ***UNITED ARAB EMIRATES***
Al Rawdha Bookshop
PO Box 5027, Sharjah
Tel.: (+971) 6 538 7933
Fax: (+971) 6 538 4473
E-mail: alrawdha@hotmail.com

• ***UNITED KINGDOM***
The Stationery Office
51 Nine Elms Lane
London SW8 5DR
Tel.: (+44) (0) 870 600 5522 (orders)
(+44) (0) 207 873 8372 (enquiries)
Fax: (+44) (0) 870 600 5533 (orders)
(+44) (0) 207 873 8247 (enquiries)
E-mail: ipa.enquiries@theso.co.uk
Web site: www.clicktso.com
and through The Stationery Office Bookshops
E-mail: postmaster@theso.co.uk
Web site: www.the-stationery-office.co.uk
Intermediate Technology Bookshop
103-105 Southampton Row
London WC1B 4HH
Tel.: (+44) 207 436 9761
Fax: (+44) 207 436 2013
E-mail: orders@itpubs.org.uk
Web site:
www.developmentbookshop.com

• ***UNITED STATES***
Publications:
BERNAN Associates (ex UNIPUB)
4611/F Assembly Drive
Lanham, MD 20706-4391
Toll-free: (+1) 800 274 4447
Fax: (+1) 800 865 3450
E-mail: query@bernan.com
Web site: www.bernan.com
United Nations Publications
Two UN Plaza, Room DC2-853
New York, NY 10017
Tel.: (+1) 212 963 8302/800 253 9646
Fax: (+1) 212 963 3489
E-mail: publications@un.org
Web site: www.unog.ch
UN Bookshop (direct sales)
The United Nations Bookshop
General Assembly Building Room 32
New York, NY 10017
Tel.: (+1) 212 963 7680
Fax: (+1) 212 963 4910
E-mail: bookshop@un.org
Web site: www.un.org
Periodicals:
Ebsco Subscription Services
PO Box 1943
Birmingham, AL 35201-1943
Tel.: (+1) 205 991 6600
Fax: (+1) 205 991 1449
The Faxon Company Inc.
15 Southwest Park
Westwood, MA 02090
Tel.: (+1) 617 329 3350
Telex: 95 1980
Cable: FW Faxon Wood

• ***URUGUAY***
Librería Agropecuaria S.R.L.
Buenos Aires 335, Casilla 1755
Montevideo C.P. 11000

• ***VENEZUELA***
Tecni-Ciencia Libros
CCCT Nivel C-2
Caracas
Tel.: (+58) 2 959 4747
Fax: (+58) 2 959 5636
Correo electrónico:
tclibros@attglobal.net
Fudeco, Librería
Avenida Libertador-Este
Ed. Fudeco, Apartado 254
Barquisimeto C.P. 3002, Ed. Lara
Tel.: (+58) 51 538 022
Fax: (+58) 51 544 394
Librería FAGRO
Universidad Central de Venezuela (UCV)
Maracay

• ***YUGOSLAVIA***
See Serbia and Montenegro

• ***ZIMBABWE***
Grassroots Books
The Book Café
Fife Avenue, Harare
Tel.: (+263) 4 79 31 82
Fax: (+263) 4 72 62 43